Y0-AEZ-125

Pfleger/Maurer/Weber

Mass Spectral and GC Data

of Drugs, Poisons, Pesticides, Pollutants and Their Metabolites

Second, revised and enlarged edition

Part 2
Mass Spectra (*m/z* 4 to 221 amu)

Distribution:
VCH, P. O. Box 101161, D-6940 Weinheim (Federal Republic of Germany)
Switzerland: VCH, P. O. Box, CH-4020 Basel (Switzerland)
United Kingdom and Ireland: VCH (UK) Ltd., 8 Wellington Court, Cambridge CB11HZ (UK)
USA and Canada: VCH, 220 East 23rd Street, New York, NY 10010–4606 (USA)

ISBN 3–527–26989–4 (VCH Verlagsgesellschaft)
ISBN 0–89573–855–4 (VCH Publishers)

Karl Pfleger/Hans H. Maurer/Armin Weber

Mass Spectral and GC Data

of Drugs, Poisons, Pesticides, Pollutants and Their Metabolites

Second, revised and enlarged edition

Part 2

Mass Spectra (m/z 4 to 221 amu)

Weinheim · New York · Basel · Cambridge

Univ.-Prof. Dr. Karl Pfleger
– Retired Head of the Department of Clinical Toxicology –
Univ.-Prof. Dr. Hans H. Maurer
– Head of the Department of Clinical Toxicology –
Armin Weber

Department of Clinical Toxicology
Medical Faculty of the University of Saarland
D-6650 Homburg (Saar)
Federal Republic of Germany

Second Edition 1992
First Edition 1985

Editorial Director: Dr. Christina Dyllick-Brenzinger
Poduction Manager: Myriam Nothacker

LOC Card No. applied for.

A CIP catalogue record for this book is available from the British Library

Deutsche Bibliothek Cataloguing-in-Publication Data
Pfleger, Karl:
Mass spectral and GC data of drugs, poisons, pesticides, pollutants and their metabolites / Karl Pfleger ; Hans H. Maurer ; Armin Weber. – Weinheim ; New York ; Basel ; Cambridge : VCH
Früher u. d. T.: Pfleger, Karl: Mass spectral and GC data of drugs, poisons and their metabolites
ISBN 3-527-26989-4 (Weinheim ...)
ISBN 0-89573-855-4 (New York)
NE: Maurer, Hans H.:; Weber, Armin:

Pt. 2. Mass spectra (m/z 4 to 221 amu). – 2., rev. and enl. ed. – 1992.

Printed on acid- and chlorine-free paper

Printing: Zechnersche Buchdruckerei, D-6720 Speyer
Bookbinding: Großbuchbinderei Fikentscher, D-6100 Darmstadt

Printed in the Federal Republic of Germany

Contents of Part 2 (Mass Spectra (*m/z* 4 to 221 amu))

Contents of Part 1 (Methods, Tables, Indexes)

Methods

Tables

Indexes

Contents of Part 3 (Mass Spectra (*m/z* 222 to 777 amu))

1 Explanatory Notes

1.1 Arrangement of Spectra

Parts 2 and 3 contain 4370 different mass spectra. They are arranged primarily in ascending mass of the typical fragment ions in each spectrum, the accurate masses of which are rounded off to the nearest integer. For each nominal mass value the spectra are arranged in order of ascending retention index. Spectra of compounds with the same nominal mass and the same retention index are arranged in alphabetical order.

Various criteria have been selected to aid the search for a spectrum. Because the molecular ion (M^+) normally contains the most important information the spectrum can be found under it. But if the M^+ is too low or hidden by the background the spectrum can also be found under the next highest predominant ion. In order to avoid accumulation of data these ions were chosen sparingly. Finally the spectrum can be found under the base peak. If there are two or more large fragment ions (>80 %), the spectrum can be found under both, because it is possible that their relationship could vary. Hence the reference spectra are reproduced more than once. The ion under which the spectrum is arranged on a particular page is underlined.

1.2 Lay-Out of Spectra

For easier visualization of the data the mass spectra are presented as bar graphs, in which the abscissa represents the mass to charge ratio (*m/z*) in atomic mass units (amu) and the ordinate indicates the relative intensities of the ion currents of the various fragment ions. Predominant ions are labelled with their *m/z* value. The ion under which the spectrum is arranged on a particular page is underlined.

Some spectra contain molecular ions with a relative intensity of less than 1 %. In these cases the M^+ is labelled although the fragment cannot be seen in the spectrum. In our experience the detection of these low intensity M^+ can be necessary for the identification of the compound, when the other fragment ions are not typical. In these cases the unknown spectra should be expanded by the data system. Fig. 1-1 explains the information given with each spectrum and the abbreviations used are listed in Table 2-1.

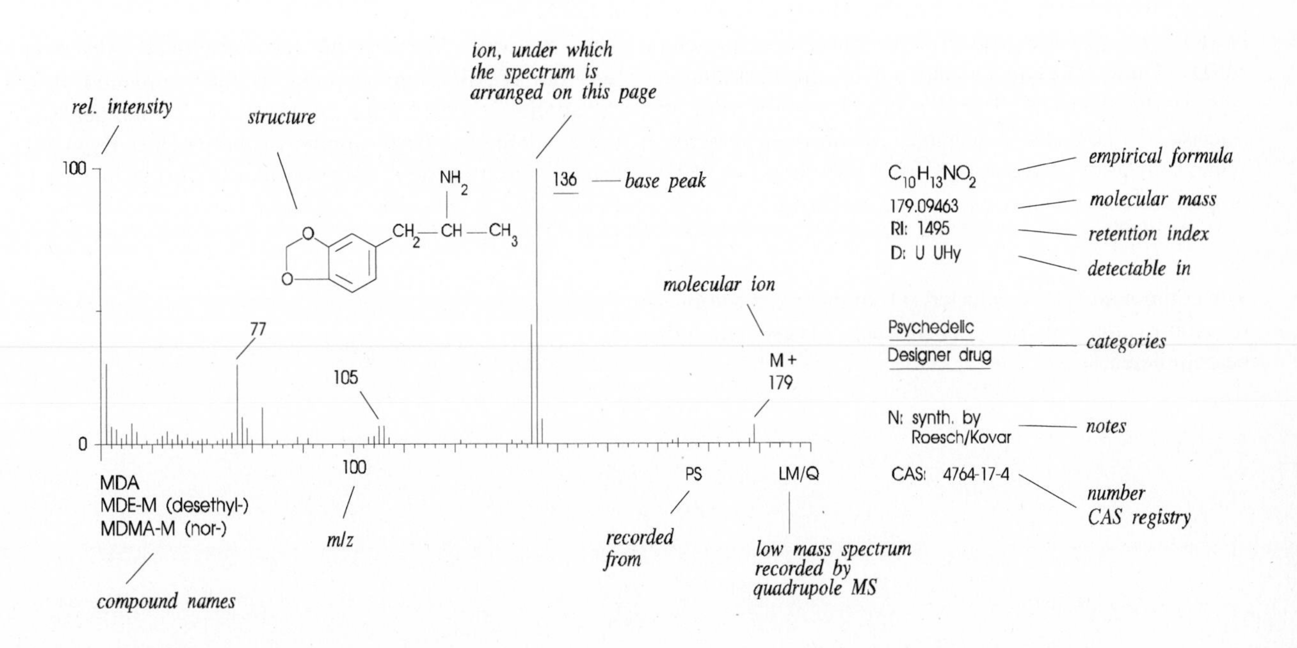

Fig. 1-1: Sample spectrum with explanations

Compound name:
The international non proprietary names for drugs (INN), the common names for pesticides and thehemical names for chemicals are used. If necessary, a synonym index (e.g. ABDA, 1988; Budavari et al., 1989; Negwer, 1987; Perkow, 1983-1988) should be consulted. Additional information from the CAS is accessible through the CAS cross indexes listed in Part 1 (Sec. 11). If the compound is a common metabolite or derivative of several parent compounds all parent compounds are given.

Structure:
The formulas are placed in the spectrum to fit the available space in the spectrum. Formulas of metabolites or artifacts are those of their probable structures (Sec. 3 in Part 1).

Empirical Formula:
The empirical formulas are given to facilitate the identification of new metabolites or derivatives. The elements are arranged according to the *Hill* convention.

Molecular mass:
The molecular masses were calculated from the atomic masses of the most abundant isotopes (Sec. 5 in Part 1).

RI:
The retention indices (RI) were measured on columns packed with OV-101 or on methyl silicone capillary columns using a temperature program (Sec. 2.3.1 in Part 1). Capillary columns were used for the determination of RI's by GC-MS (quadrupole mass spectrometer MSD). These RI's can be find out by the information "/*Q*" at the spectrum. The RI's of compounds with an asterisk (*) are not detectable by nitrogen-selective flame-ionization detection (N-FID) and therefore, FID or mass selective detection must be employed.

D:
The compound can be detected (D) in the given samples (cf. abbreviations in Sec. 2 of this part). These data will be completed.

Category:
The major pharmacological category is given.

N:
If necessary notes (N) were added (cf. abbreviations in Sec. 2 of this part).

LM, LS, LM/Q or LS/Q :
This indicates whether the low resolution mass spectrum (LM) was background-subtracted (LS). The spectra recorded by a quadrupole mass spectrometer (MSD) were marked by /*Q* (Sec. 2.3.2.1 in Part 1).
Relative ion intensities can be falsified by background-subtraction. This should be taken into account when comparing the spectra. Such variations does not impair the use of the library in our experience. With some experience it is possible to decide whether the variation is acceptable within two spectra of the same compound. If in doubt investigators should record a reference spectrum of the suspected compound on their own GC-MS.

Recorded from:
A statement of the type of sample from which the spectrum was recorded (cf. abbreviations in Sec. 2 of this part). If the spectrum was recorded from samples of biological origin, it should be remembered that fragment ions from sample impurities may be present in the spectrum. With experience it is possible to decide whether these ions can be ignored.

CAS:
The *Registry Number* of the Chemical Abstracts Services (CAS) is given here. If the compound is a metabolite or a derivative for which a CAS number was not available the CAS number of the parent compound can be found in the CAS cross indexes listed in Part 1 (Sec. 11).

2 Abbreviations

The abbreviations used in this handbook are listed in Table 2-1.

Table 2-1: Abbreviations.

Abbreviation	Meaning	see Part 1 Sec.
AC	Acetylated	2.2.3.1
(AC)	Possibly acetylated	
ALHY	Extract after alkaline hydrolysis	
Altered during HY	The altered compound can be detected in UHY	4.3-4
amu	Atomic mass unit, 1/12 of the mass of the nuclide ^{12}C	
Artifact ( )	( ) artifact	4
BP	Base peak; the most intense fragment ion in a mass spectrum	
BPH	Benzophenone	
CI	Chemical ionization	
CMP	Computer monitoring program	2.5
$-CH_3Br$	Artifact formed by elimination of methyl bromide	4.2.4
-CHNO	Artifact formed by decarbamoylation	4.3.2
$-C_2H_3NO$	Artifact formed by N-methyl decarbamoylation	4.3.2
$-C_3H_5NO$	Artifact formed by N,N-dimethyl decarbamoylation	4.3.2
$-C_5H_9NO$	Artifact formed by N-isobutyl decarbamoylation	4.3.2
$-(CH_3)_2NOH$	Artifact formed by Cope elimination of the N-oxide	4.2.2
$-(C_2H_5)_2NOH$	Artifact formed by Cope elimination of the N-oxide	4.2.2
$-C_6H_{14}N_2O_2$	Artifact formed by Cope elimination of the N-oxide	4.2.2
$-CH_2O$	Artifact formed by elimination of formaldehyde	4.2.4
$-C_2H_2O_2$	Artifact formed by decarboxylation after hydrolysis of methyl carboxylate	4.2.1
$-C_3H_4O_2$	Artifact formed by decarboxylation after hydrolysis of ethyl carboxylate	4.2.1
$-CO_2$	Artifact formed by decarboxylation	4.2.1
D:	Detectable in	
DIS	Direct insert system used for recording the spectrum	
DS	Data system	
EG	Ethylene glycol	
EI	Electron impact ionization	3.1
EIA	Enzyme immunoassay	
EMIT	Enzyme-multiplied immunoassay technique of *Syva*	
ET	Ethylated	2.2.3.3
FID	Flame ionisation detector	2.3.1.1
FPIA	Fluorescence polarization immunoassay	
G	Standard extract of gastric contents	2.2.2.1
GC	Gas chromatographic, -graph, -graphy	2.3
GC artifact	Artifact formed during GC	4.2-3
GC artifact in methanol	Artifact (of beta-adrenergic blocking agents) by reaction with methanol during GC	4.2.6
-HCl	Artifact formed by elimination of hydrogen chloride	4.2.4
-HCN	Artifact formed by elimination of hydrogen cyanide	4.2.4
HO-	Hydroxy	
$+H_2O$	Artifact formed by hydration (of an alkene)	4.4.7
$-H_2O$	Artifact formed by dehydration (of an alcohol or with rearrangement of an amino oxo compound)	4.3.1 4.4.4
HOOC-	Carboxy	

Abbreviation	Meaning	see Part 1 Sec.
HY	Acid hydrolyzed or acid hydrolysis	2.2.2.2
HY artifact	Artifact formed during acid hydrolysis	4.4
-I	Intoxication; this compound is only detectable after a toxic dosage	
I.D.	Internal diameter	
INN	International non proprietary name (WHO)	
LM	Low resolution mass spectrum	
LS	Background subtracted low resolution mass spectrum	
LM/Q	Low resolution mass spectrum recorded on a quadrupole MS	
LS/Q	Background subtracted low resolution mass spectrum recorded on a quadrupole MS	
M^+	Molecular ion	
-M	Metabolite	
-M ( )	( ) metabolite	
-M (HO-)	Hydroxy metabolite	
-M (HOOC-)	Carboxylated metabolite	
-M (nor-)	N-desmethyl metabolite	
-M (ring)	Ring compound as metabolite (e.g. of phenothiazines)	
-M artifact	Artifact of a metabolite	
-M/artifact	Metabolite or artifact	
m/z	Mass to charge ratio	3.1
ME	Methylated	2.2.3.2
(ME)	Methylated by methanol during GC	4.2.5
ME in methanol	Methylated by methanol during GC	4.2.5
MS	Mass spectrometric, -meter, -metry, mass spectrum	
N:	Notes	
N-FID	Nitrogen-sensitive flame ionisation detector	2.3.1.1
$-NH_3$	Artifact formed by elimination of ammonia	4.2.4
Not detectable after HY	Compound destroyed during acid hydrolysis	
P	Standard extract of plasma	2.2.2.1
PC	Paper chromatography	
PEG	Extract of plasma for determination of glycols	2.2.2.10
PFP	Pentafluoropropionylated	2.2.3.6
PFPA	Pentafluoropropionic acid anhydride	2.2.3.6
PFPOH	Pentafluoropropanol	2.2.3.7
PIV	Pivalylated	2.2.3.8
PS	Pure substance	
PTHCME	Extract of plasma for detection of tetrahydrocannabinol metabolites	2.2.2.7
Rat	Compound found in the urine of rats	2.1
RI	Retention index (Kovats, 1958) on OV-101 or methyl silicone capillary	2.3.1
RIA	Radio immunoassay	
$-SO_2NH$	Artifact formed by elimination of a sulfonamide group	4.2.4
SIM	Selected ion mode	
SPE	Solid phase extraction	2.2.2
STA	Systematic toxicological analysis	2.5
STED	Solvent transfer and evaporation device	2.2.1
TDx	Fluorescence polarization immunoassay of *Abbott*	
TFA	Trifluoroacetylated	2.2.3.5
THC	Tetrahydrocannabinol	
THC-COOH	11-Nor-D-9-tetrahydrocannabinol-9-carboxylic acid	
TLC	Thin layer chromatography	
TM	Trade mark	

Abbreviation	Meaning	see Part 1 Sec.
TMS	Trimethylsilylated	2.2.3.4
UA	Extract of urine for detection of amphetamines	2.2.2.4
UCO	Extract of urine for detection of cocaine	2.2.2.7
UGLUC	Extract of urine after cleavage of conjugates using glucuronidase and arylsulfatase	2.2.2.3
UHY	Extract of urine after acid hydrolysis	2.2.2.2
ULSD	Extract of urine for detection of lysergide (LSD)	2.2.2.9
UMAM	Extract of urine for detection of 6-monoacteyl morphine	2.2.2.5
UTHCME	Extract of urine for detection of tetrahydrocannabinol metabolites after methylation	2.2.2.6
*	Compound contains no nitrogen and cannot be detected by N-FID	
----	RI not determined	
9999	Compound not volatile and could not be detected by GC	

3 Compound Index

4 Mass spectra (m/z 4 to 221 amu)

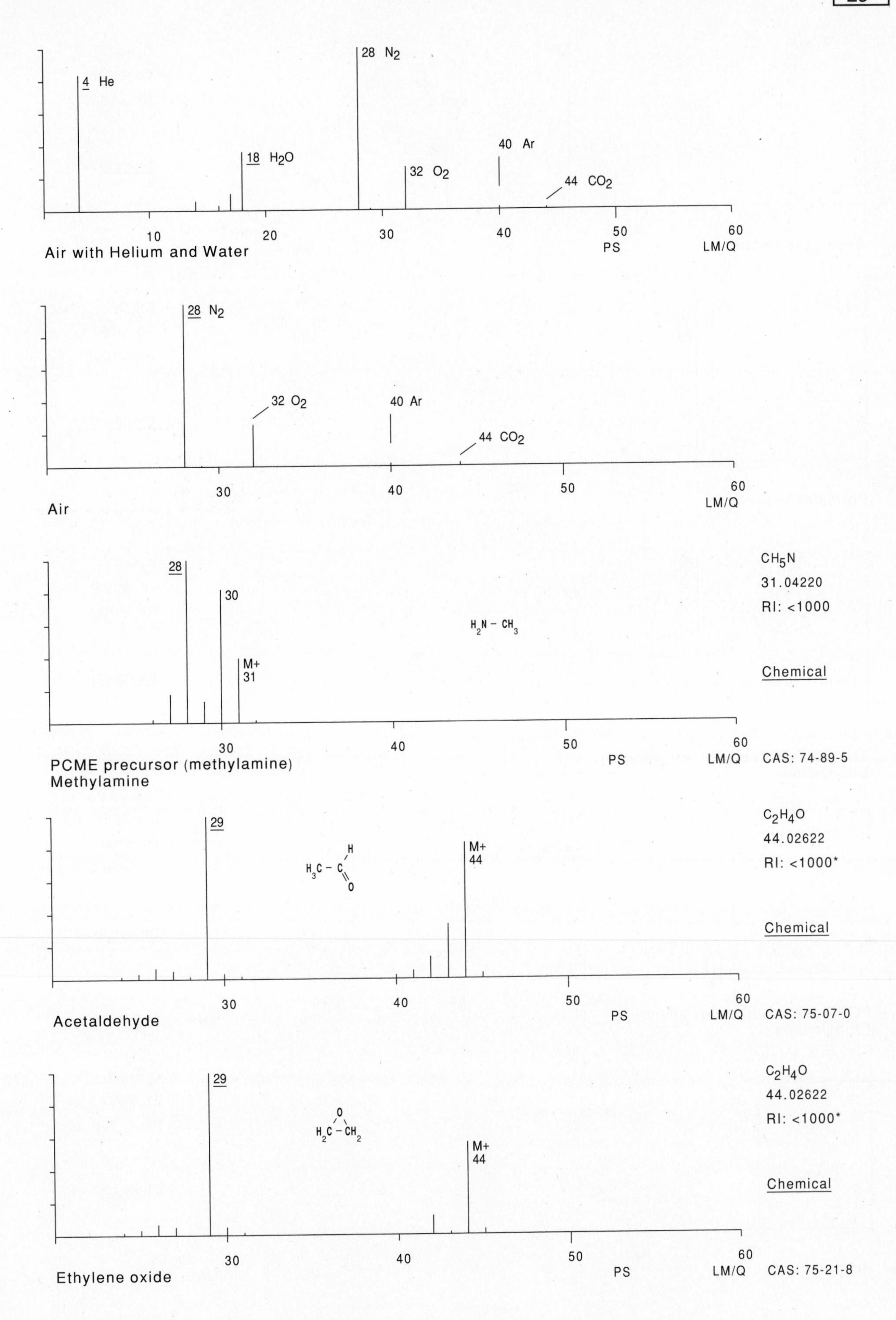

CH_5N
31.04220
RI: <1000

Chemical

CAS: 74-89-5

C_2H_4O
44.02622
RI: <1000*

Chemical

CAS: 75-07-0

C_2H_4O
44.02622
RI: <1000*

Chemical

CAS: 75-21-8

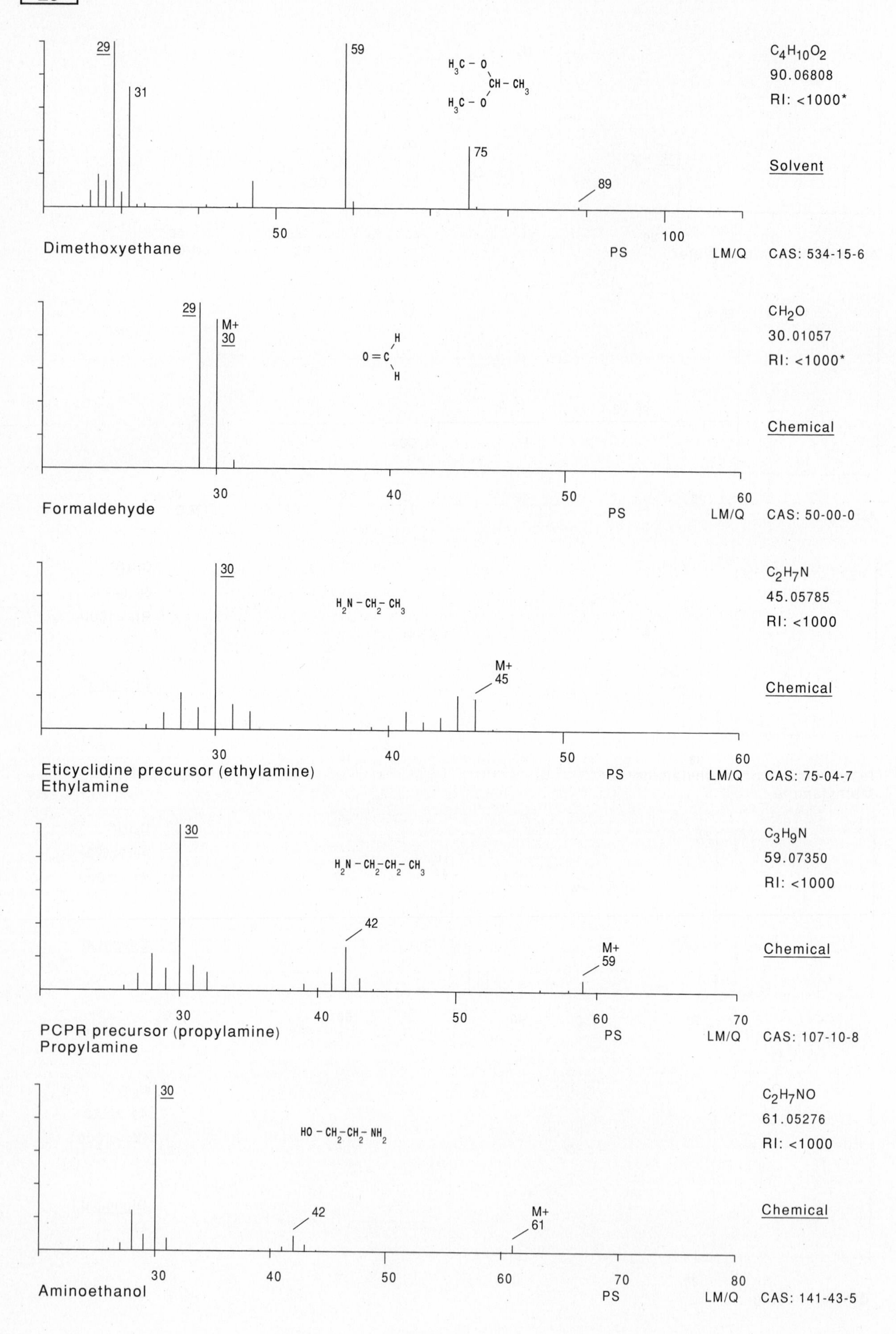

29
31
59
75
89
H_3C-O
$CH-CH_3$
H_3C-O
50
100
Dimethoxyethane
PS
LM/Q
$C_4H_{10}O_2$
90.06808
RI: <1000*
Solvent
CAS: 534-15-6
29
M+
30
O=C(H)H
30
40
50
60
Formaldehyde
PS
LM/Q
CH_2O
30.01057
RI: <1000*
Chemical
CAS: 50-00-0
30
$H_2N-CH_2-CH_3$
M+
45
30
40
50
60
Eticyclidine precursor (ethylamine)
Ethylamine
PS
LM/Q
C_2H_7N
45.05785
RI: <1000
Chemical
CAS: 75-04-7
30
$H_2N-CH_2-CH_2-CH_3$
42
M+
59
30
40
50
60
70
PCPR precursor (propylamine)
Propylamine
PS
LM/Q
C_3H_9N
59.07350
RI: <1000
Chemical
CAS: 107-10-8
30
$HO-CH_2-CH_2-NH_2$
42
M+
61
30
40
50
60
70
80
Aminoethanol
PS
LM/Q
C_2H_7NO
61.05276
RI: <1000
Chemical
CAS: 141-43-5

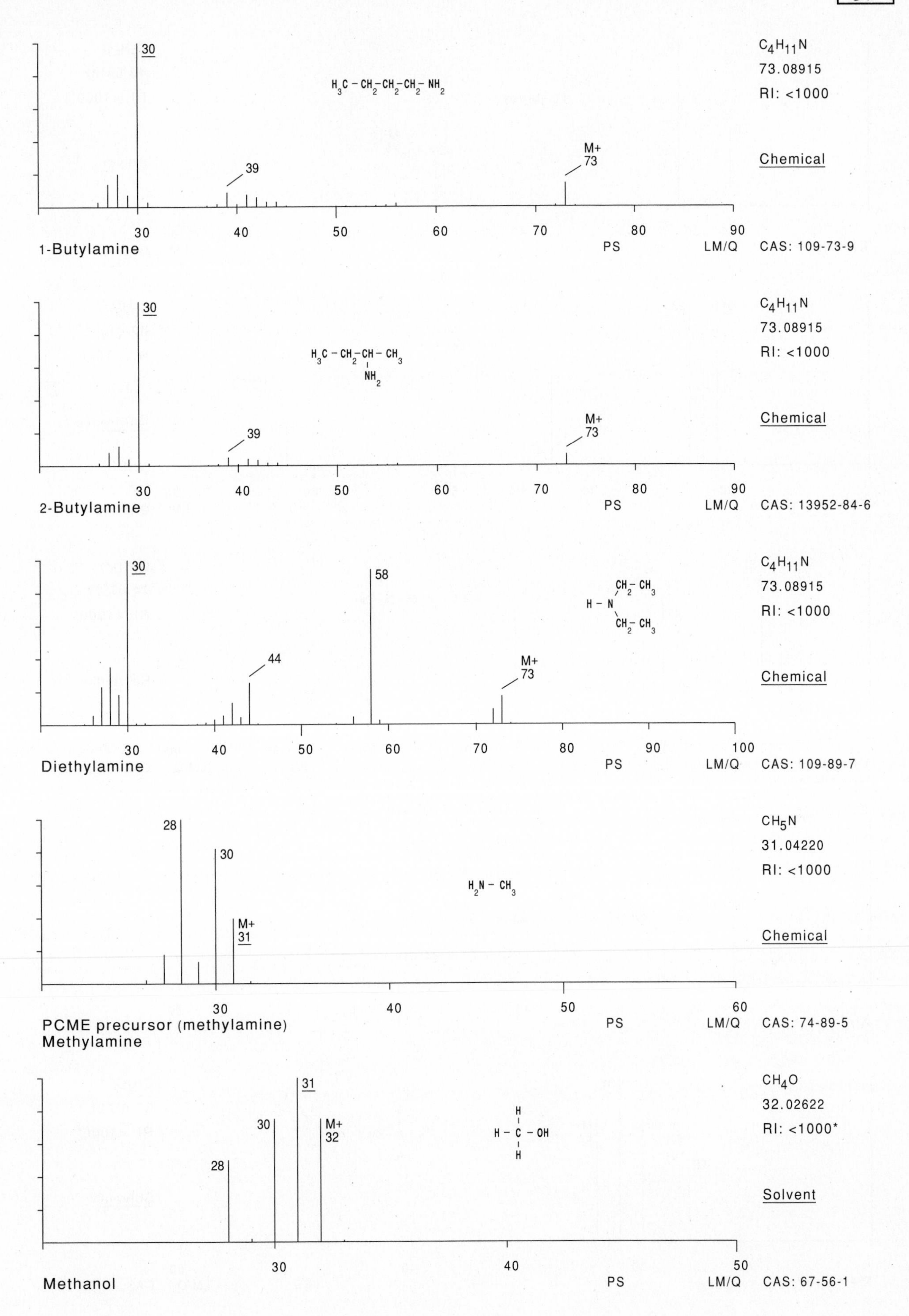
30
39
M+
73
$H_3C-CH_2-CH_2-CH_2-NH_2$
30 40 50 60 70 80 90
1-Butylamine
PS
LM/Q
$C_4H_{11}N$
73.08915
RI: <1000
Chemical
CAS: 109-73-9
30
39
M+
73
$H_3C-CH_2-CH(NH_2)-CH_3$
30 40 50 60 70 80 90
2-Butylamine
PS
LM/Q
$C_4H_{11}N$
73.08915
RI: <1000
Chemical
CAS: 13952-84-6
30
58
44
M+
73
$H-N(CH_2-CH_3)_2$
30 40 50 60 70 80 90 100
Diethylamine
PS
LM/Q
$C_4H_{11}N$
73.08915
RI: <1000
Chemical
CAS: 109-89-7
28
30
M+
31
H_2N-CH_3
30 40 50 60
PCME precursor (methylamine)
Methylamine
PS
LM/Q
CH_5N
31.04220
RI: <1000
Chemical
CAS: 74-89-5
31
30
M+
32
28
H_3C-OH
30 40 50
Methanol
PS
LM/Q
CH_4O
32.02622
RI: <1000*
Solvent
CAS: 67-56-1

31

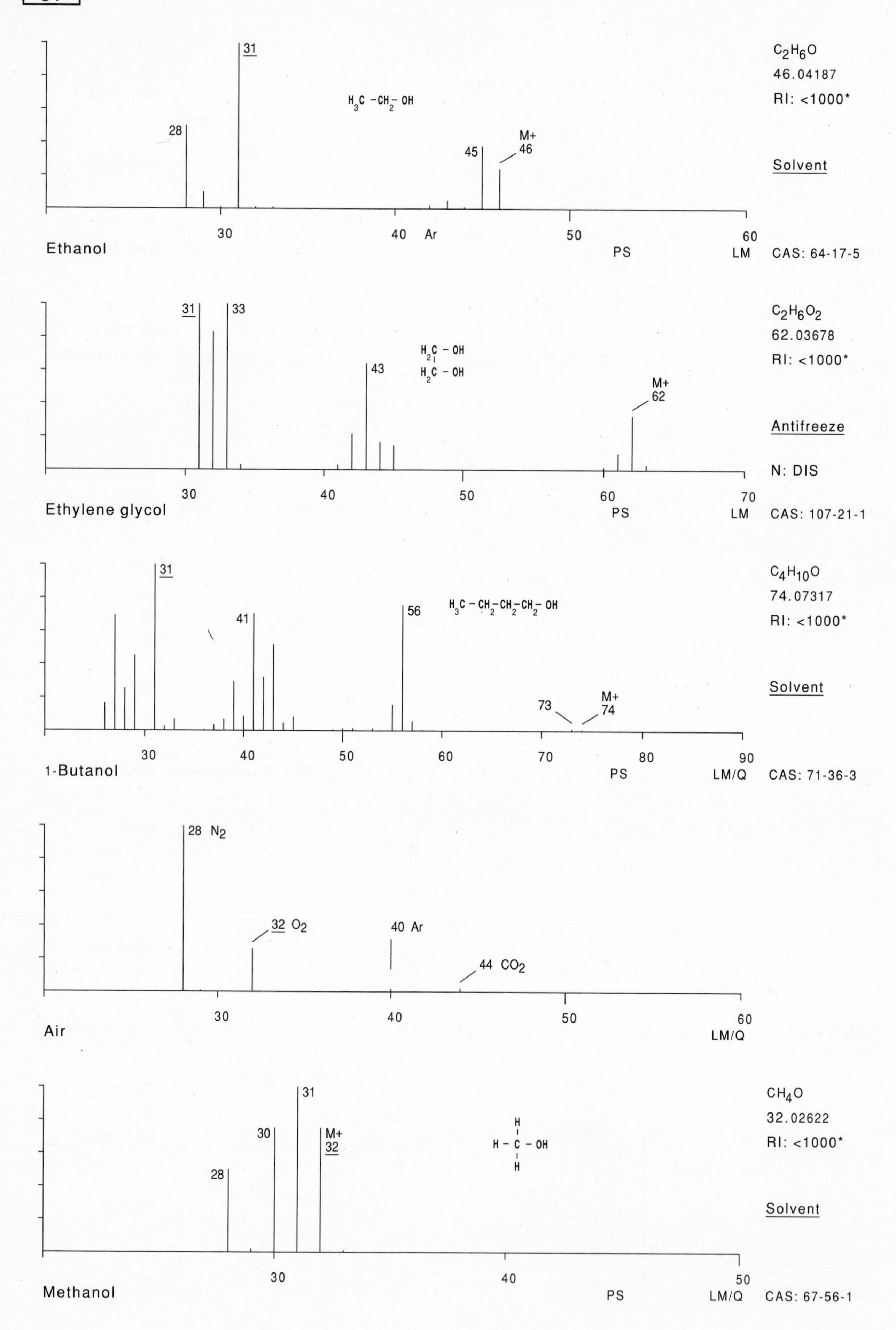
31
28
45
M+
46
H_3C-CH_2-OH
C_2H_6O
46.04187
RI: <1000*
Solvent
30
40
Ar
50
60
Ethanol
PS
LM
CAS: 64-17-5
31
33
43
M+
62
H_2C-OH
H_2C-OH
$C_2H_6O_2$
62.03678
RI: <1000*
Antifreeze
N: DIS
30
40
50
60
70
Ethylene glycol
PS
LM
CAS: 107-21-1
31
41
56
73
M+
74
$H_3C-CH_2-CH_2-CH_2-OH$
$C_4H_{10}O$
74.07317
RI: <1000*
Solvent
30
40
50
60
70
80
90
1-Butanol
PS
LM/Q
CAS: 71-36-3
28 N_2
32 O_2
40 Ar
44 CO_2
30
40
50
60
Air
LM/Q
31
30
28
M+
32
H
H - C - OH
H
CH_4O
32.02622
RI: <1000*
Solvent
30
40
50
Methanol
PS
LM/Q
CAS: 67-56-1

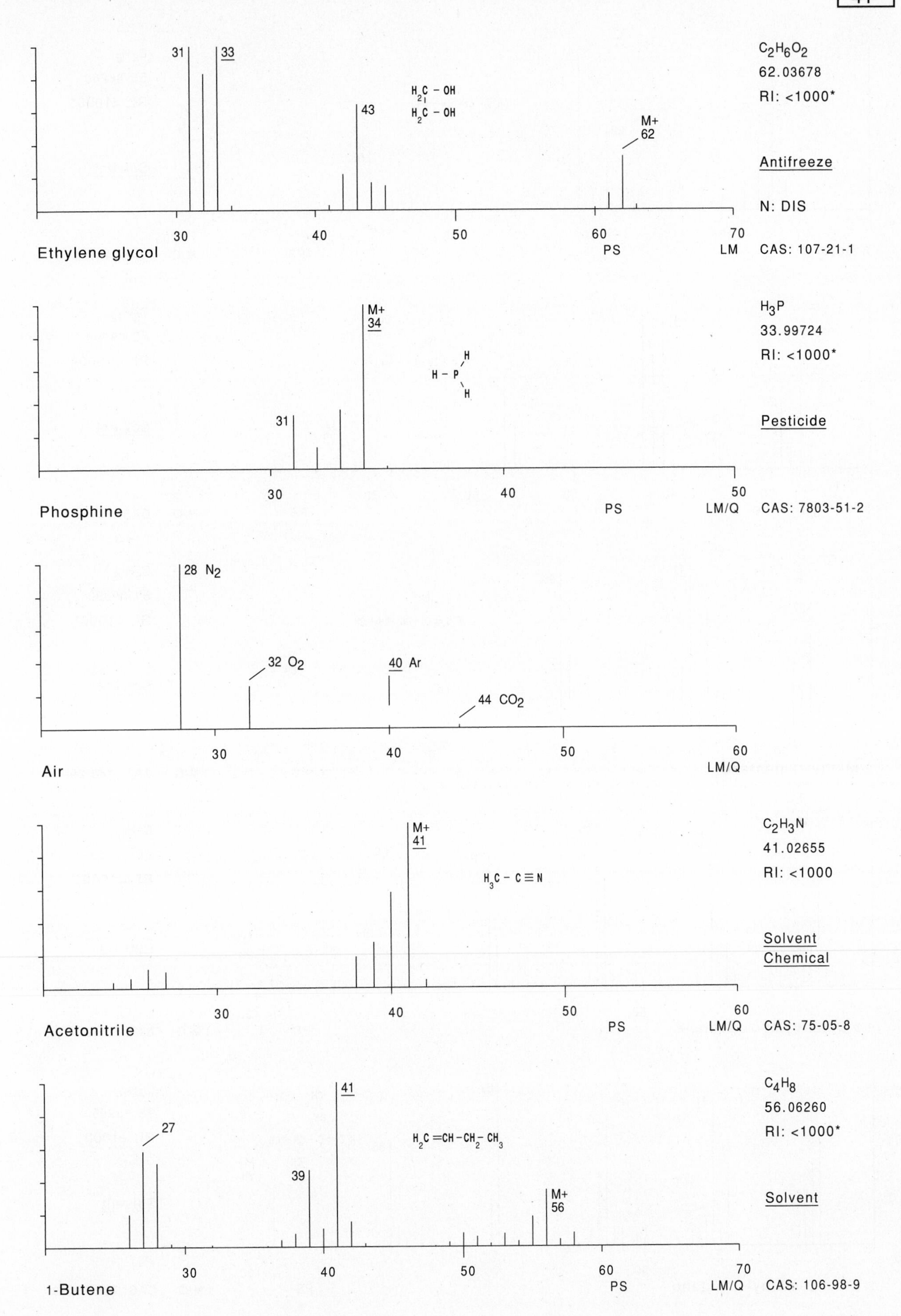

$C_2H_6O_2$
62.03678
RI: <1000*

Antifreeze

N: DIS

CAS: 107-21-1

H_3P
33.99724
RI: <1000*

Pesticide

CAS: 7803-51-2

C_2H_3N
41.02655
RI: <1000

Solvent
Chemical

CAS: 75-05-8

C_4H_8
56.06260
RI: <1000*

Solvent

CAS: 106-98-9

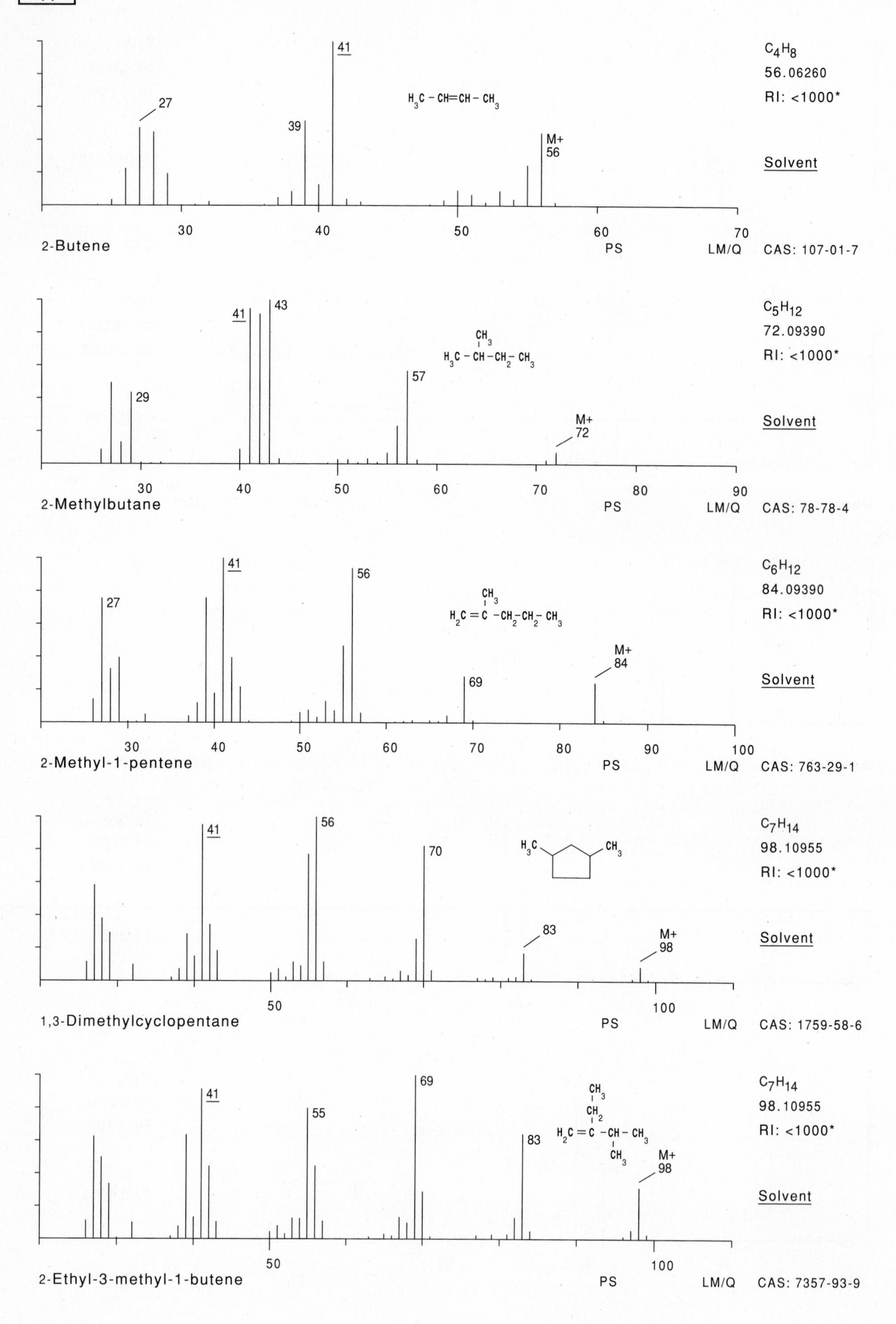

27
39
41
$H_3C-CH=CH-CH_3$
M+
56
30
40
50
60
70
2-Butene
PS
LM/Q
C_4H_8
56.06260
RI: <1000*
Solvent
CAS: 107-01-7
29
41
43
57
$H_3C-CH(CH_3)-CH_2-CH_3$
M+
72
2-Methylbutane
C_5H_{12}
72.09390
RI: <1000*
Solvent
CAS: 78-78-4
27
41
56
69
$H_2C=C(CH_3)-CH_2-CH_2-CH_3$
M+
84
2-Methyl-1-pentene
C_6H_{12}
84.09390
RI: <1000*
Solvent
CAS: 763-29-1
41
56
70
83
M+
98
1,3-Dimethylcyclopentane
C_7H_{14}
98.10955
RI: <1000*
Solvent
CAS: 1759-58-6
41
55
69
83
$H_2C=C(CH_2CH_3)-CH(CH_3)-CH_3$
M+
98
2-Ethyl-3-methyl-1-butene
C_7H_{14}
98.10955
RI: <1000*
Solvent
CAS: 7357-93-9

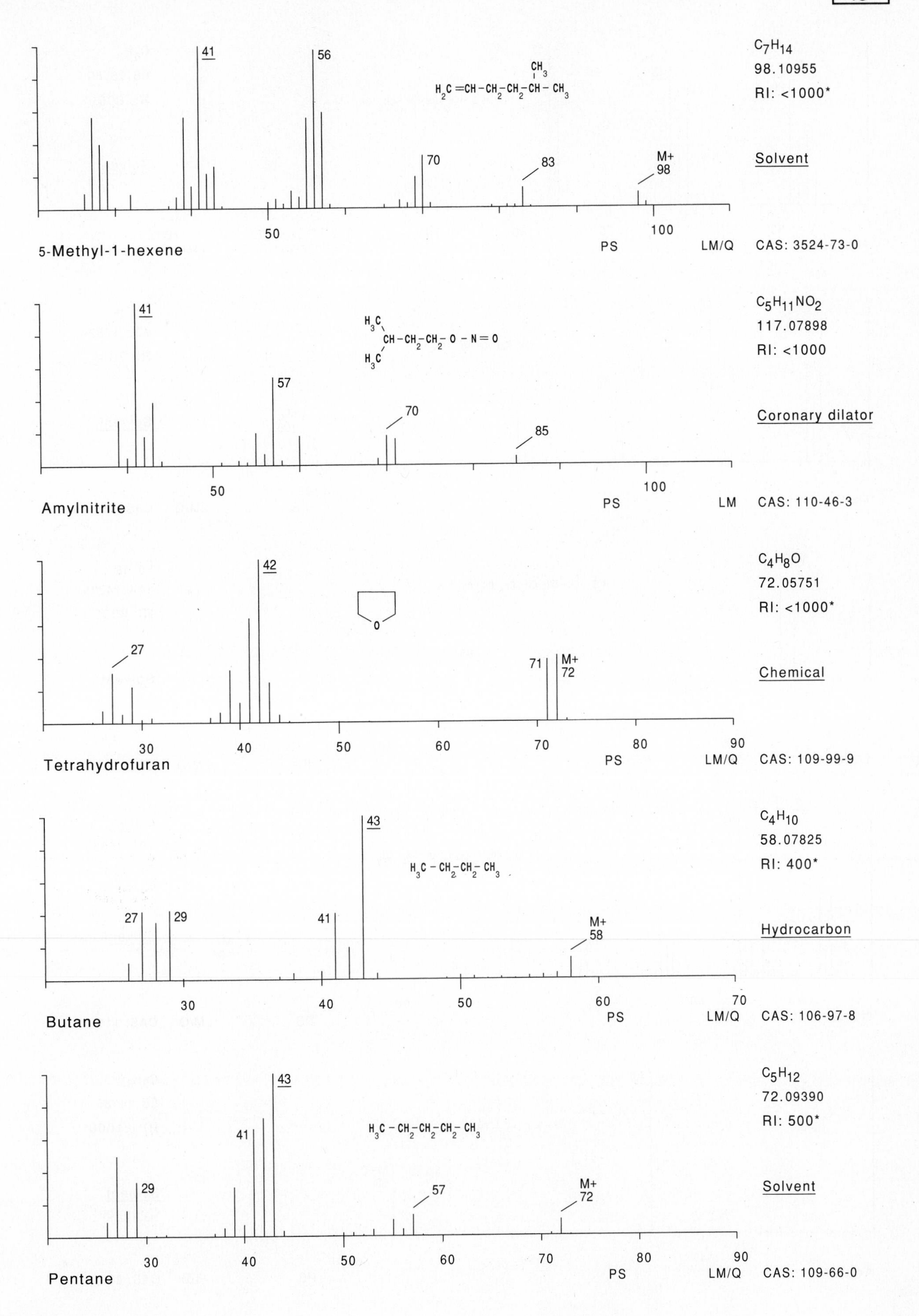

C_7H_{14}
98.10955
RI: <1000*

Solvent

CAS: 3524-73-0

$C_5H_{11}NO_2$
117.07898
RI: <1000

Coronary dilator

CAS: 110-46-3

C_4H_8O
72.05751
RI: <1000*

Chemical

CAS: 109-99-9

C_4H_{10}
58.07825
RI: 400*

Hydrocarbon

CAS: 106-97-8

C_5H_{12}
72.09390
RI: 500*

Solvent

CAS: 109-66-0

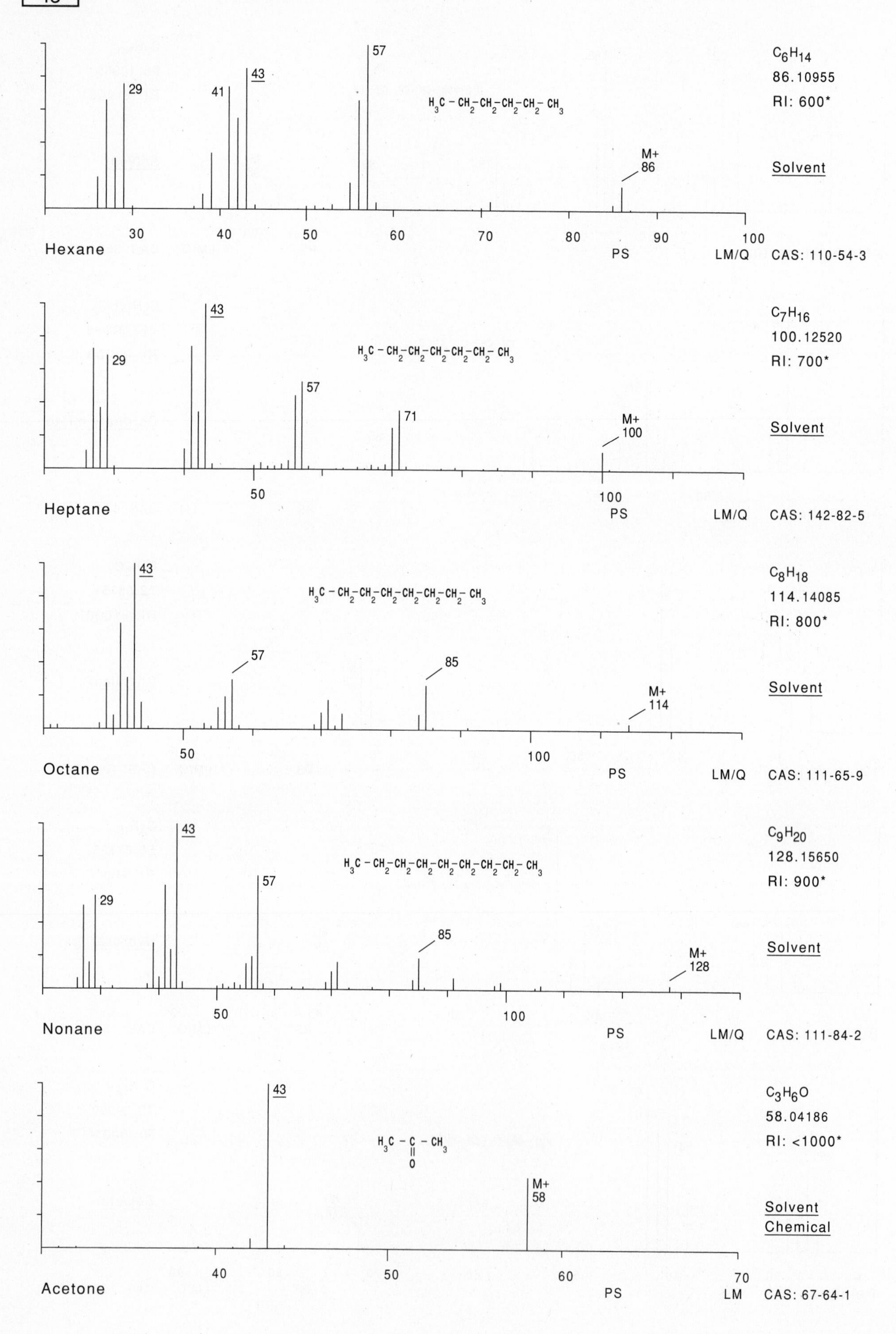

57
43
29
41
$H_3C-CH_2-CH_2-CH_2-CH_2-CH_3$
M+
86
30
40
50
60
70
80
90
100
Hexane
PS
LM/Q
C_6H_{14}
86.10955
RI: 600*
Solvent
CAS: 110-54-3
43
29
$H_3C-CH_2-CH_2-CH_2-CH_2-CH_2-CH_3$
57
71
M+
100
50
100
Heptane
PS
LM/Q
C_7H_{16}
100.12520
RI: 700*
Solvent
CAS: 142-82-5
43
$H_3C-CH_2-CH_2-CH_2-CH_2-CH_2-CH_2-CH_3$
57
85
M+
114
50
100
Octane
PS
LM/Q
C_8H_{18}
114.14085
RI: 800*
Solvent
CAS: 111-65-9
43
$H_3C-CH_2-CH_2-CH_2-CH_2-CH_2-CH_2-CH_2-CH_3$
57
29
85
M+
128
50
100
Nonane
PS
LM/Q
C_9H_{20}
128.15650
RI: 900*
Solvent
CAS: 111-84-2
43
$H_3C-C(=O)-CH_3$
M+
58
40
50
60
70
Acetone
PS
LM
C_3H_6O
58.04186
RI: <1000*
Solvent
Chemical
CAS: 67-64-1

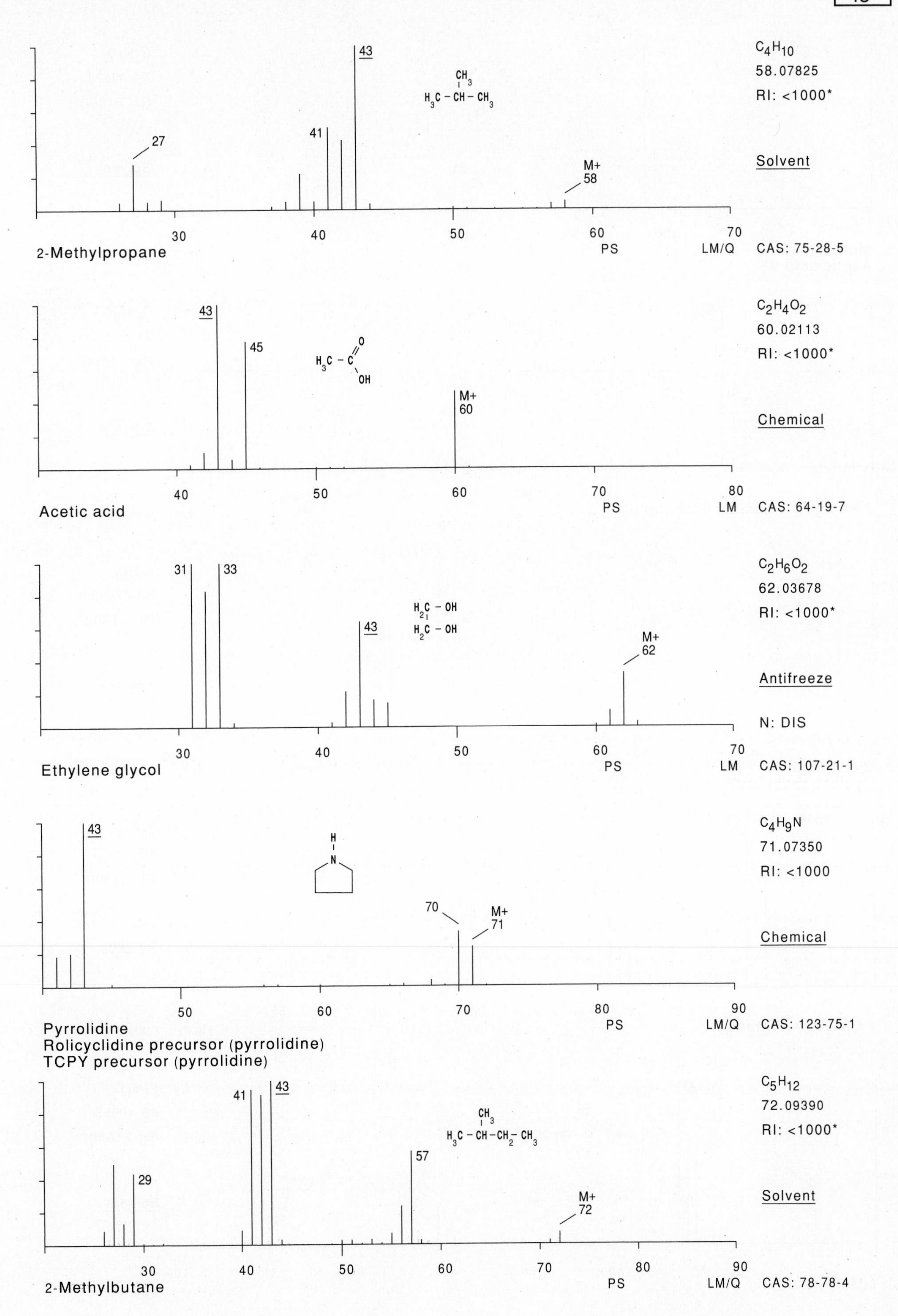

C4H10
58.07825
RI: <1000*
Solvent
2-Methylpropane
CAS: 75-28-5
C2H4O2
60.02113
RI: <1000*
Chemical
Acetic acid
CAS: 64-19-7
C2H6O2
62.03678
RI: <1000*
Antifreeze
N: DIS
Ethylene glycol
CAS: 107-21-1
C4H9N
71.07350
RI: <1000
Chemical
Pyrrolidine
Rolicyclidine precursor (pyrrolidine)
TCPY precursor (pyrrolidine)
CAS: 123-75-1
C5H12
72.09390
RI: <1000*
Solvent
2-Methylbutane
CAS: 78-78-4
PS
LM/Q
LM
M+

43

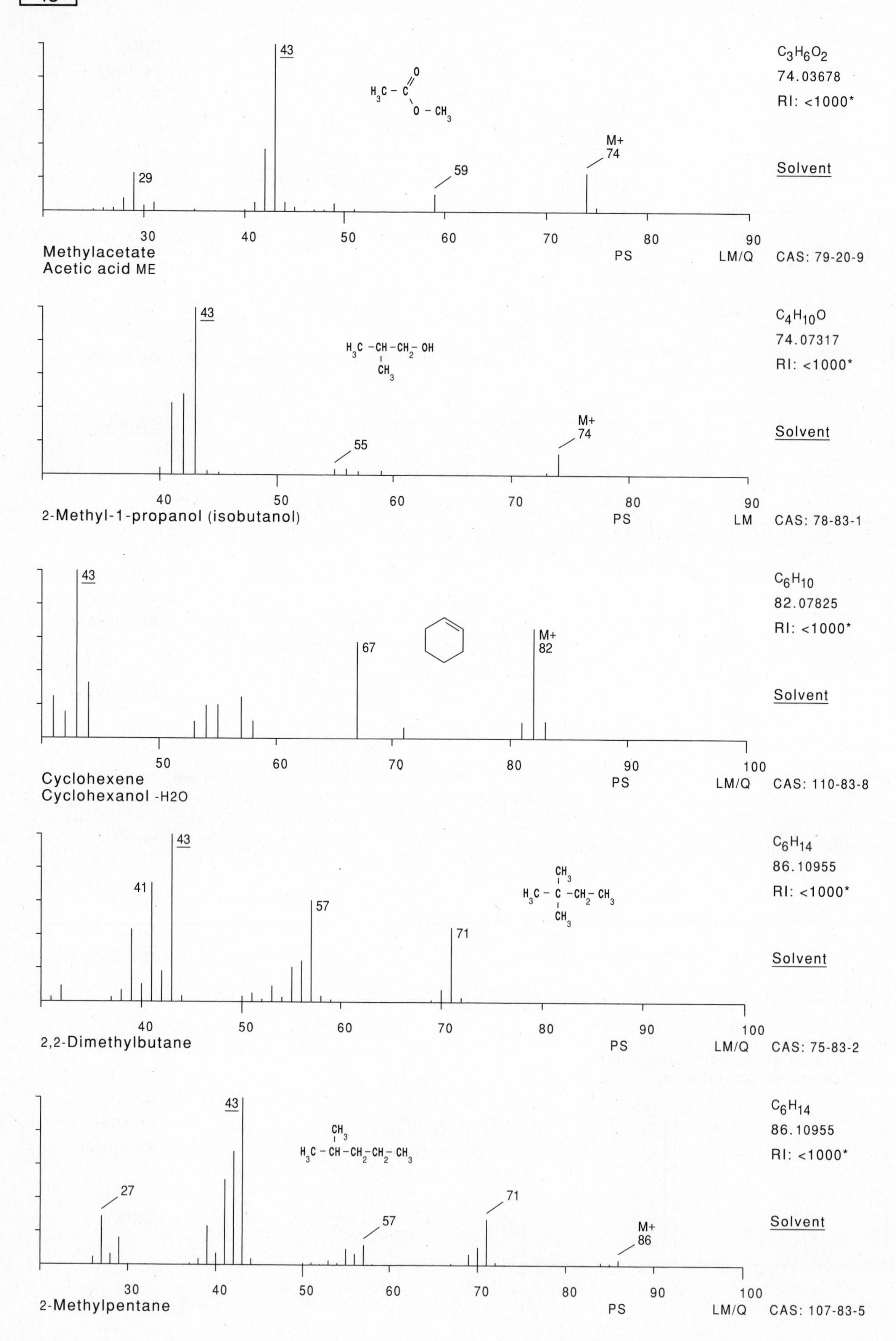

43
29
59
M+
74
30
40
50
60
70
80
90
PS
LM/Q
Methylacetate
Acetic acid ME
C3H6O2
74.03678
RI: <1000*
Solvent
CAS: 79-20-9
43
55
M+
74
2-Methyl-1-propanol (isobutanol)
PS
LM
C4H10O
74.07317
RI: <1000*
Solvent
CAS: 78-83-1
43
67
M+
82
100
Cyclohexene
Cyclohexanol -H2O
C6H10
82.07825
RI: <1000*
Solvent
CAS: 110-83-8
43
41
57
71
2,2-Dimethylbutane
C6H14
86.10955
RI: <1000*
Solvent
CAS: 75-83-2
43
27
57
71
M+
86
2-Methylpentane
C6H14
86.10955
RI: <1000*
Solvent
CAS: 107-83-5

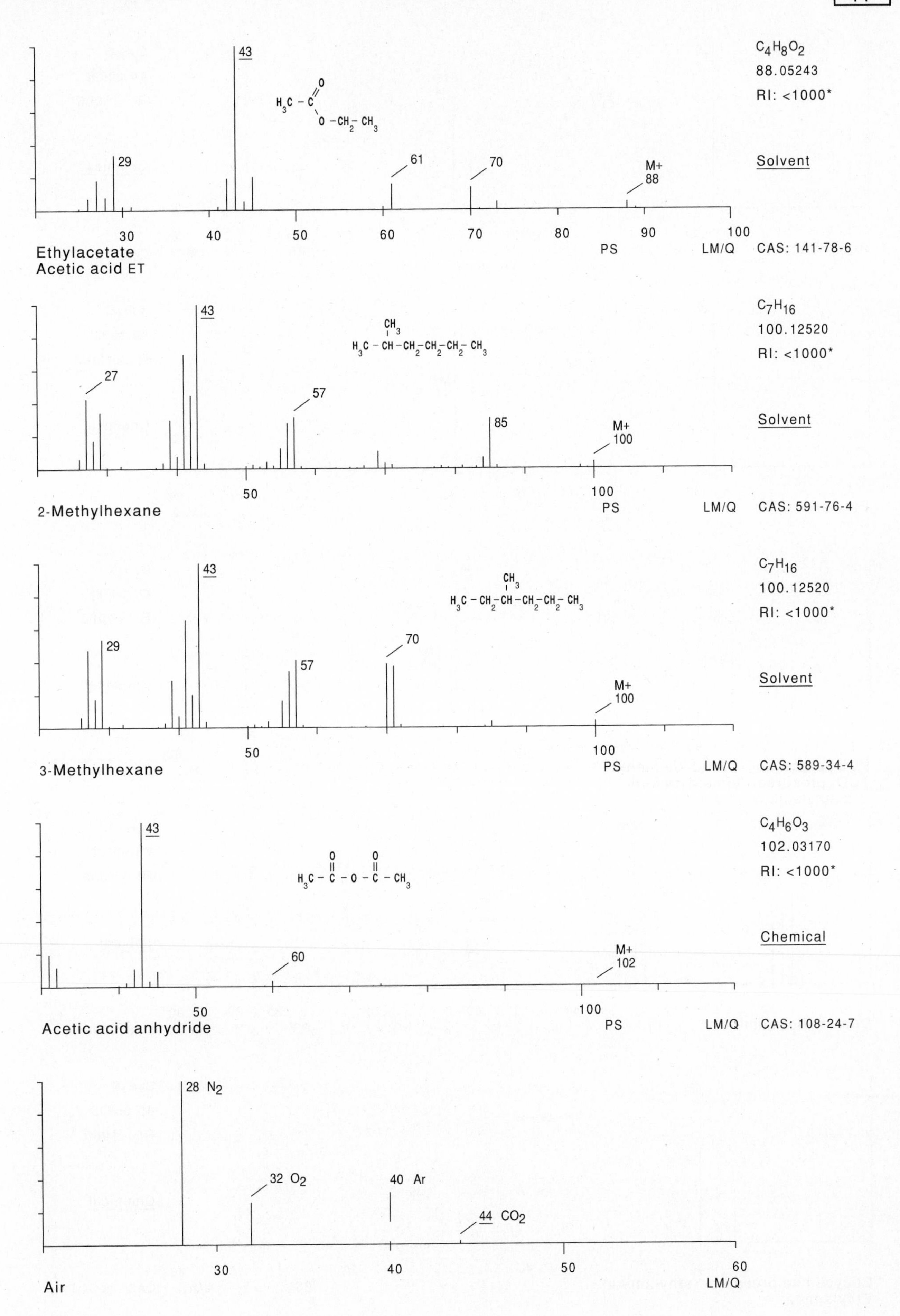

$C_4H_8O_2$
88.05243
RI: <1000*

Solvent

CAS: 141-78-6

C_7H_{16}
100.12520
RI: <1000*

Solvent

CAS: 591-76-4

C_7H_{16}
100.12520
RI: <1000*

Solvent

CAS: 589-34-4

$C_4H_6O_3$
102.03170
RI: <1000*

Chemical

CAS: 108-24-7

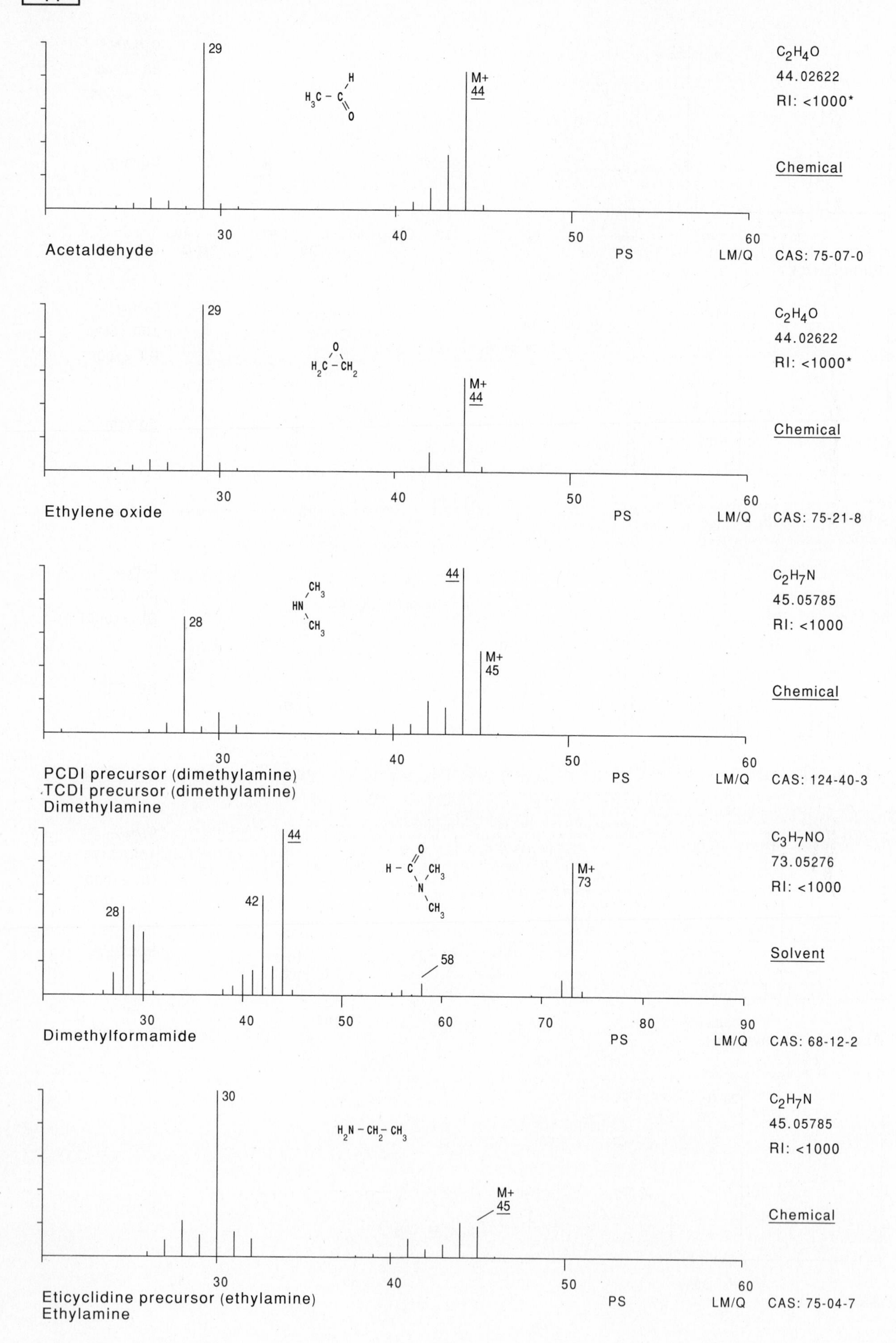
29
M+
44
C_2H_4O
44.02622
RI: <1000*
Chemical
30
40
50
60
Acetaldehyde
PS
LM/Q
CAS: 75-07-0
29
M+
44
C_2H_4O
44.02622
RI: <1000*
Chemical
30
40
50
60
Ethylene oxide
PS
LM/Q
CAS: 75-21-8
44
28
M+
45
C_2H_7N
45.05785
RI: <1000
Chemical
30
40
50
60
PCDI precursor (dimethylamine)
TCDI precursor (dimethylamine)
Dimethylamine
PS
LM/Q
CAS: 124-40-3
44
42
28
58
M+
73
C_3H_7NO
73.05276
RI: <1000
Solvent
30
40
50
60
70
80
90
Dimethylformamide
PS
LM/Q
CAS: 68-12-2
30
M+
45
C_2H_7N
45.05785
RI: <1000
Chemical
30
40
50
60
Eticyclidine precursor (ethylamine)
Ethylamine
PS
LM/Q
CAS: 75-04-7

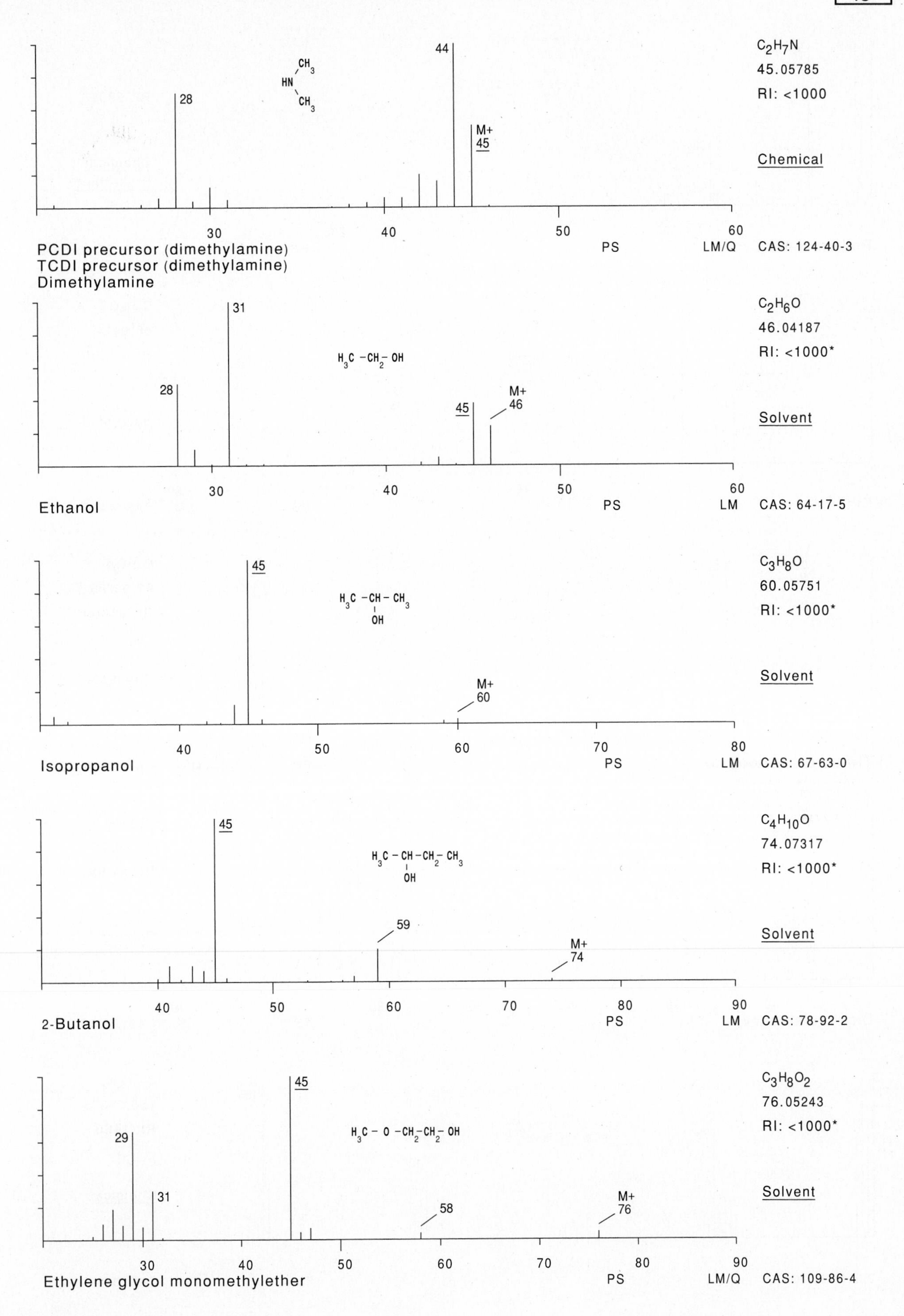

44
28
M+
45
30
40
50
60
PS
LM/Q
C_2H_7N
45.05785
RI: <1000
Chemical
CAS: 124-40-3
PCDI precursor (dimethylamine)
TCDI precursor (dimethylamine)
Dimethylamine
31
28
45
M+
46
30
40
50
60
PS
LM
C_2H_6O
46.04187
RI: <1000*
Solvent
CAS: 64-17-5
Ethanol
45
M+
60
40
50
60
70
80
PS
LM
C_3H_8O
60.05751
RI: <1000*
Solvent
CAS: 67-63-0
Isopropanol
45
59
M+
74
40
50
60
70
80
90
PS
LM
$C_4H_{10}O$
74.07317
RI: <1000*
Solvent
CAS: 78-92-2
2-Butanol
45
29
31
58
M+
76
30
40
50
60
70
80
90
PS
LM/Q
$C_3H_8O_2$
76.05243
RI: <1000*
Solvent
CAS: 109-86-4
Ethylene glycol monomethylether

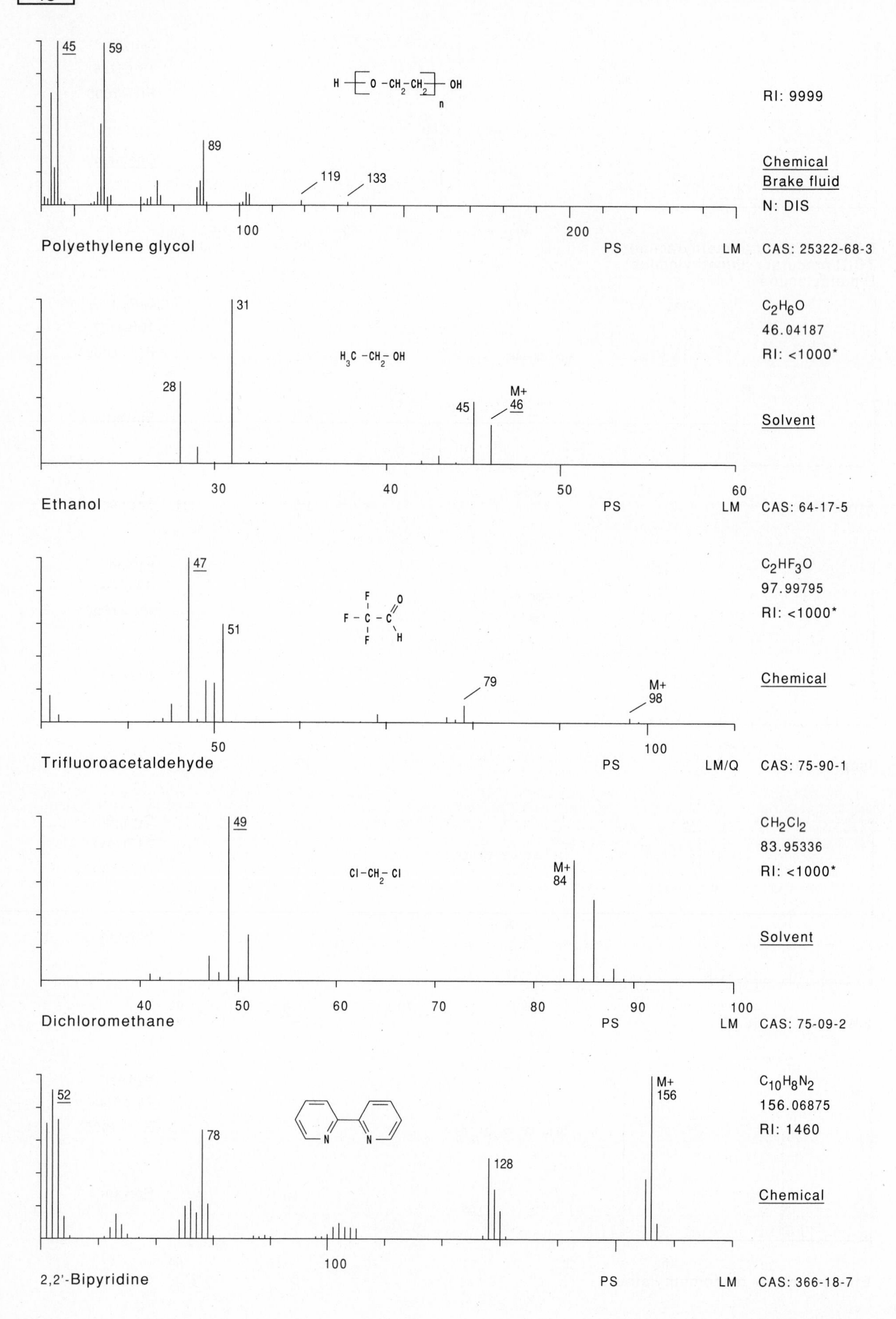

45
59
89
119
133
100
200
H-[O-CH2-CH2]n-OH
RI: 9999
Chemical
Brake fluid
N: DIS
Polyethylene glycol
PS
LM
CAS: 25322-68-3
31
28
45
M+
46
30
40
50
60
H3C-CH2-OH
C2H6O
46.04187
RI: <1000*
Solvent
Ethanol
PS
LM
CAS: 64-17-5
47
51
79
M+
98
50
100
C2HF3O
97.99795
RI: <1000*
Chemical
Trifluoroacetaldehyde
PS
LM/Q
CAS: 75-90-1
49
M+
84
40
50
60
70
80
90
100
Cl-CH2-Cl
CH2Cl2
83.95336
RI: <1000*
Solvent
Dichloromethane
PS
LM
CAS: 75-09-2
52
78
128
M+
156
100
C10H8N2
156.06875
RI: 1460
Chemical
2,2'-Bipyridine
PS
LM
CAS: 366-18-7

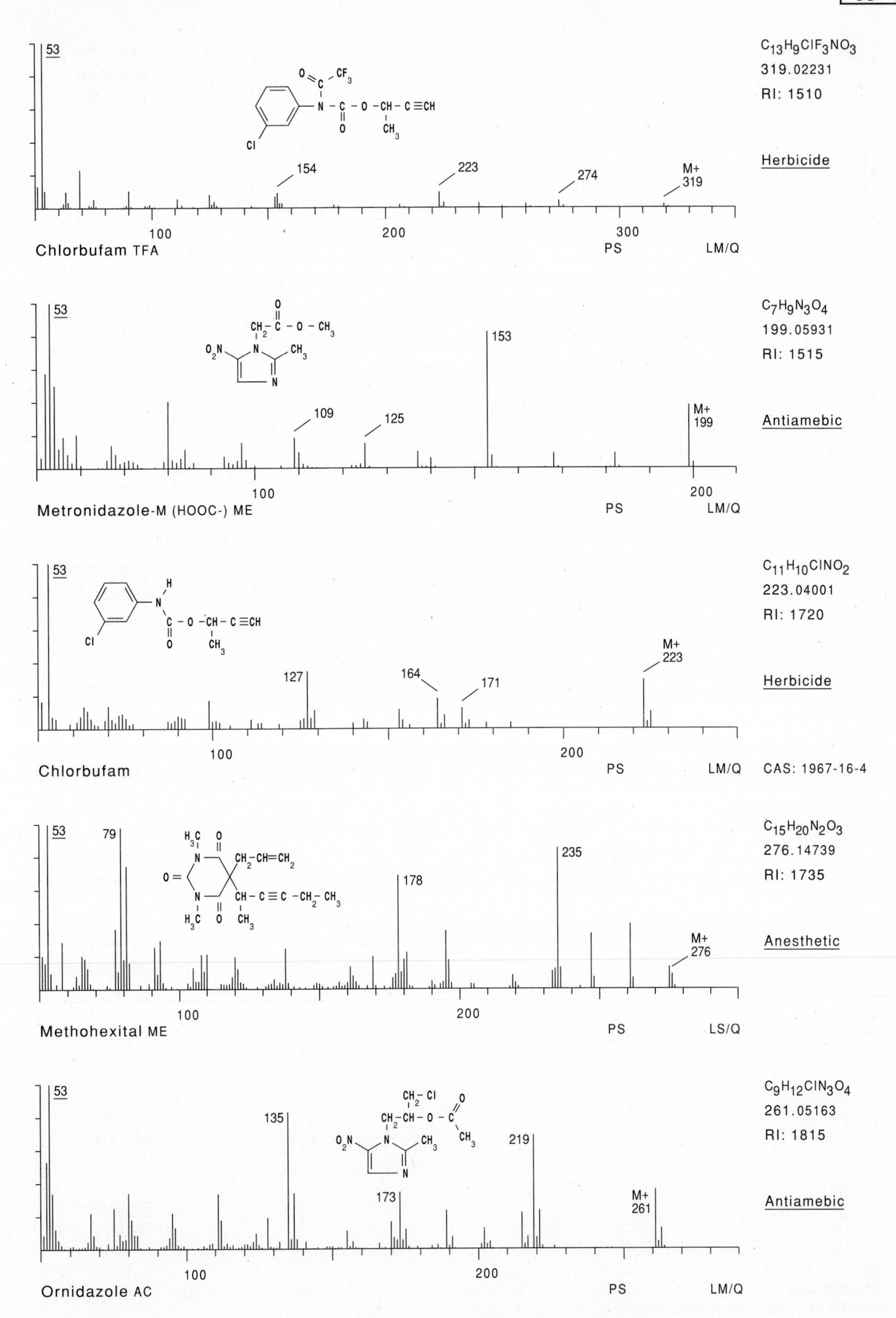
53
154
223
274
M+ 319
100
200
300
Chlorbufam TFA
PS
LM/Q
C13H9ClF3NO3
319.02231
RI: 1510
Herbicide
53
153
109
125
M+ 199
100
200
Metronidazole-M (HOOC-) ME
PS
LM/Q
C7H9N3O4
199.05931
RI: 1515
Antiamebic
53
127
164
171
M+ 223
100
200
Chlorbufam
PS
LM/Q
C11H10ClNO2
223.04001
RI: 1720
Herbicide
CAS: 1967-16-4
53
79
178
235
M+ 276
100
200
Methohexital ME
PS
LS/Q
C15H20N2O3
276.14739
RI: 1735
Anesthetic
53
135
219
173
M+ 261
100
200
Ornidazole AC
PS
LM/Q
C9H12ClN3O4
261.05163
RI: 1815
Antiamebic

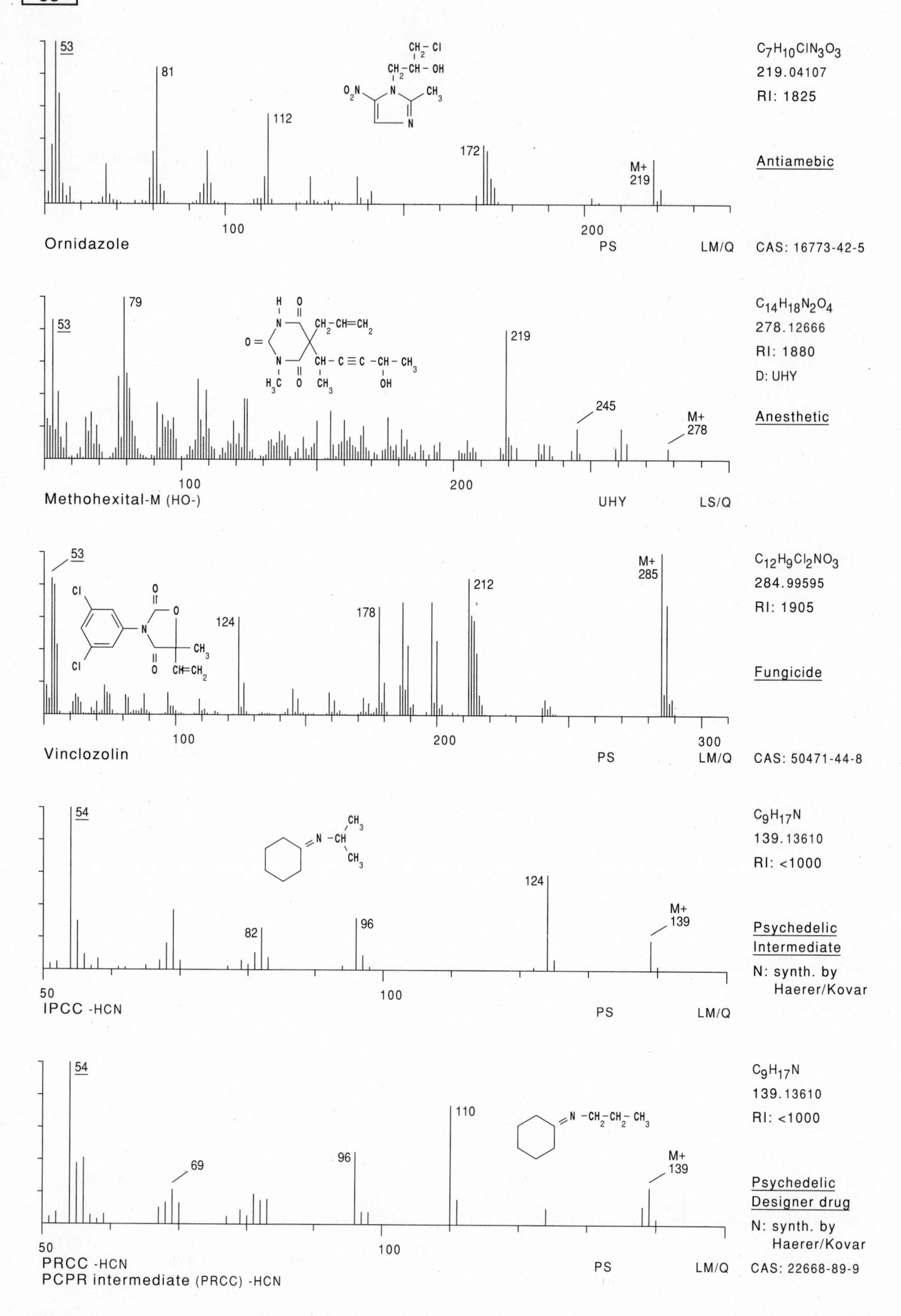

53
81
112
172
M+
219
100
200
Ornidazole
PS
LM/Q
C7H10ClN3O3
219.04107
RI: 1825
Antiamebic
CAS: 16773-42-5
79
53
219
245
M+
278
100
200
Methohexital-M (HO-)
UHY
LS/Q
C14H18N2O4
278.12666
RI: 1880
D: UHY
Anesthetic
53
124
178
212
M+
285
100
200
300
Vinclozolin
PS
LM/Q
C12H9Cl2NO3
284.99595
RI: 1905
Fungicide
CAS: 50471-44-8
54
124
82
96
M+
139
50
100
IPCC -HCN
PS
LM/Q
C9H17N
139.13610
RI: <1000
Psychedelic
Intermediate
N: synth. by
Haerer/Kovar
54
110
69
96
M+
139
50
100
PRCC -HCN
PCPR intermediate (PRCC) -HCN
PS
LM/Q
C9H17N
139.13610
RI: <1000
Psychedelic
Designer drug
N: synth. by
Haerer/Kovar
CAS: 22668-89-9

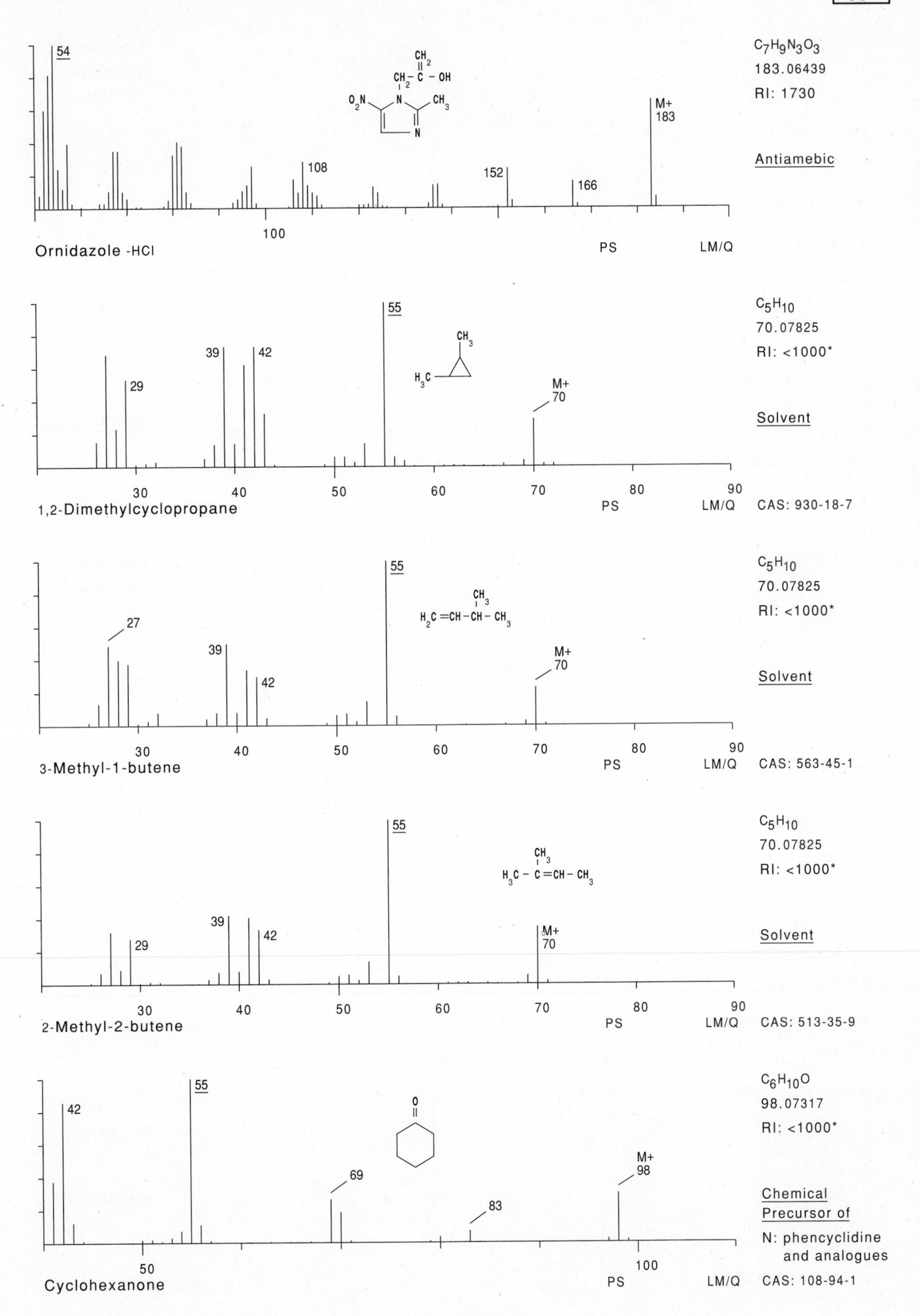

54
108
152
166
M+
183
100
Ornidazole -HCl
PS
LM/Q
C7H9N3O3
183.06439
RI: 1730
Antiamebic
55
39
42
29
M+
70
1,2-Dimethylcyclopropane
C5H10
70.07825
RI: <1000*
Solvent
CAS: 930-18-7
55
27
39
42
M+
70
3-Methyl-1-butene
C5H10
70.07825
RI: <1000*
Solvent
CAS: 563-45-1
55
39
29
42
M+
70
2-Methyl-2-butene
C5H10
70.07825
RI: <1000*
Solvent
CAS: 513-35-9
55
42
69
83
M+
98
50
100
Cyclohexanone
C6H10O
98.07317
RI: <1000*
Chemical
Precursor of
N: phencyclidine
and analogues
CAS: 108-94-1

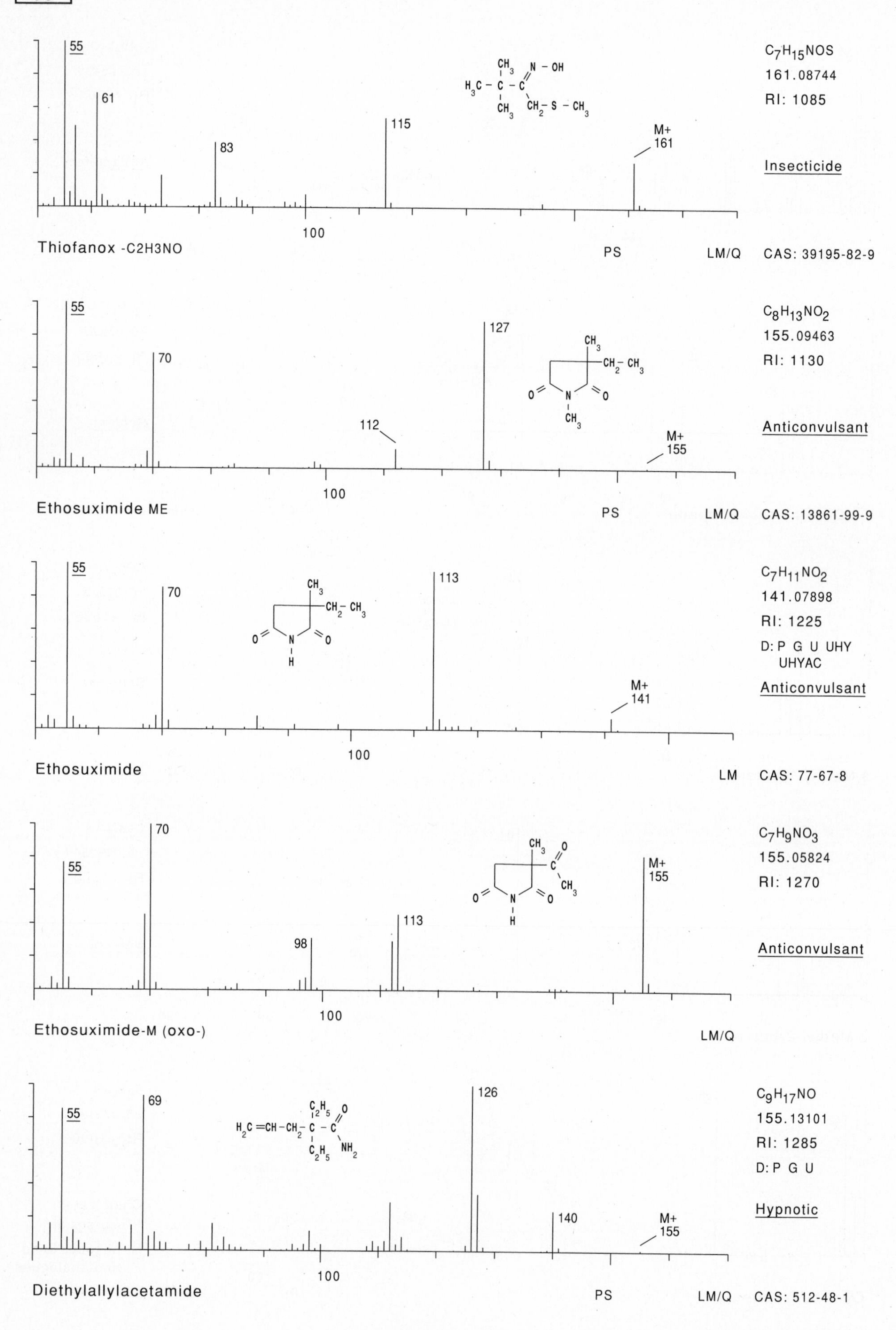

55
61
83
115
M+
161
100
Thiofanox -C2H3NO
PS
LM/Q
C7H15NOS
161.08744
RI: 1085
Insecticide
CAS: 39195-82-9
55
70
112
127
M+
155
100
Ethosuximide ME
PS
LM/Q
C8H13NO2
155.09463
RI: 1130
Anticonvulsant
CAS: 13861-99-9
55
70
113
M+
141
100
Ethosuximide
LM
C7H11NO2
141.07898
RI: 1225
D: P G U UHY
UHYAC
Anticonvulsant
CAS: 77-67-8
70
55
98
113
M+
155
100
Ethosuximide-M (oxo-)
LM/Q
C7H9NO3
155.05824
RI: 1270
Anticonvulsant
55
69
126
140
M+
155
100
Diethylallylacetamide
PS
LM/Q
C9H17NO
155.13101
RI: 1285
D: P G U
Hypnotic
CAS: 512-48-1

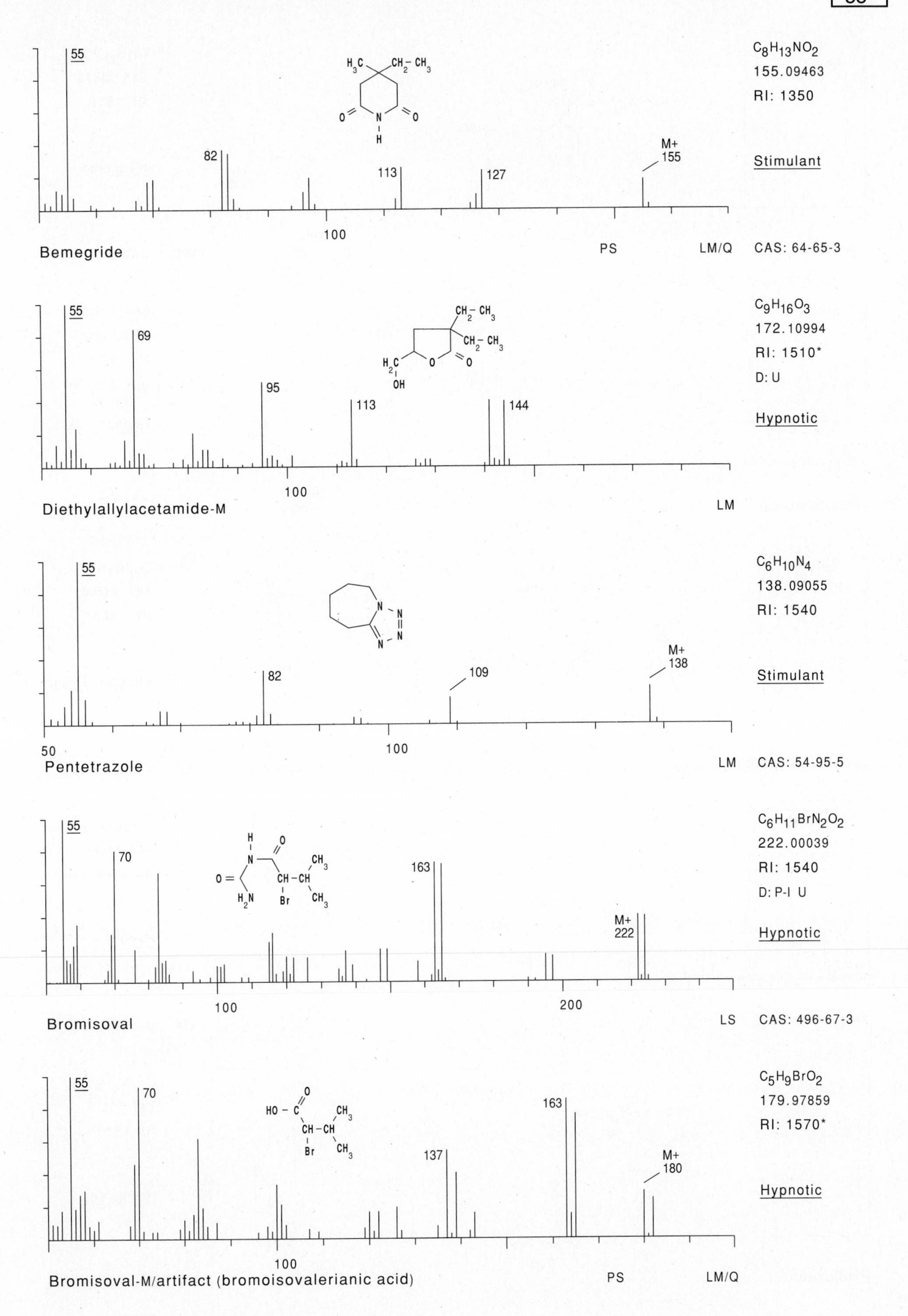

Bemegride

$C_8H_{13}NO_2$
155.09463
RI: 1350
Stimulant
CAS: 64-65-3

Diethylallylacetamide-M

$C_9H_{16}O_3$
172.10994
RI: 1510*
D: U
Hypnotic

Pentetrazole

$C_6H_{10}N_4$
138.09055
RI: 1540
Stimulant
CAS: 54-95-5

Bromisoval

$C_6H_{11}BrN_2O_2$
222.00039
RI: 1540
D: P-I U
Hypnotic
CAS: 496-67-3

Bromisoval-M/artifact (bromoisovalerianic acid)

$C_5H_9BrO_2$
179.97859
RI: 1570*
Hypnotic

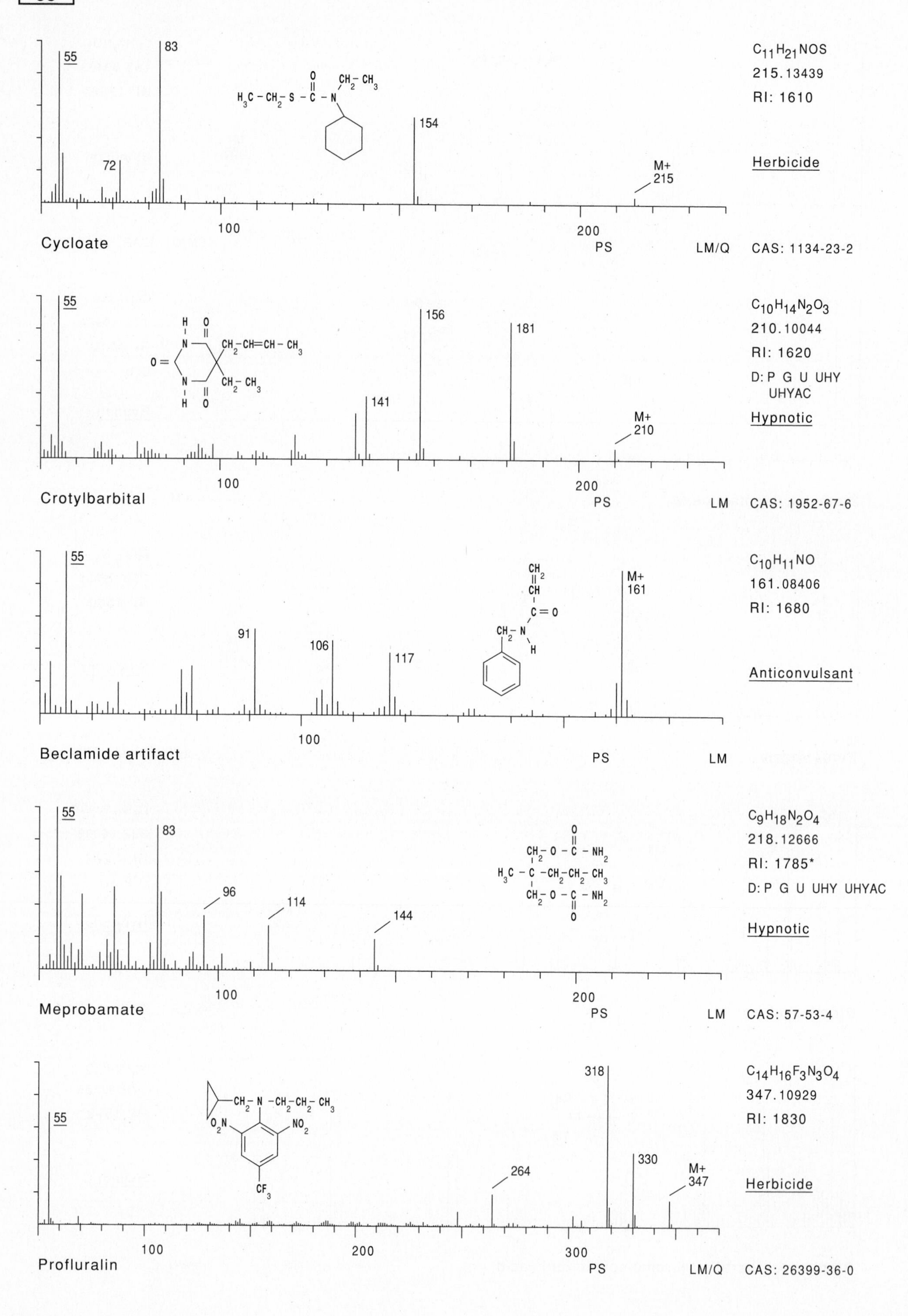
55
83
72
154
M+
215
100
200
Cycloate
PS
LM/Q
C11H21NOS
215.13439
RI: 1610
Herbicide
CAS: 1134-23-2
55
156
181
141
M+
210
Crotylbarbital
LM
C10H14N2O3
210.10044
RI: 1620
D: P G U UHY
UHYAC
Hypnotic
CAS: 1952-67-6
55
M+
161
91
106
117
Beclamide artifact
C10H11NO
161.08406
RI: 1680
Anticonvulsant
55
83
96
114
144
Meprobamate
C9H18N2O4
218.12666
RI: 1785*
D: P G U UHY UHYAC
Hypnotic
CAS: 57-53-4
318
55
264
330
M+
347
300
Profluralin
C14H16F3N3O4
347.10929
RI: 1830
Herbicide
CAS: 26399-36-0

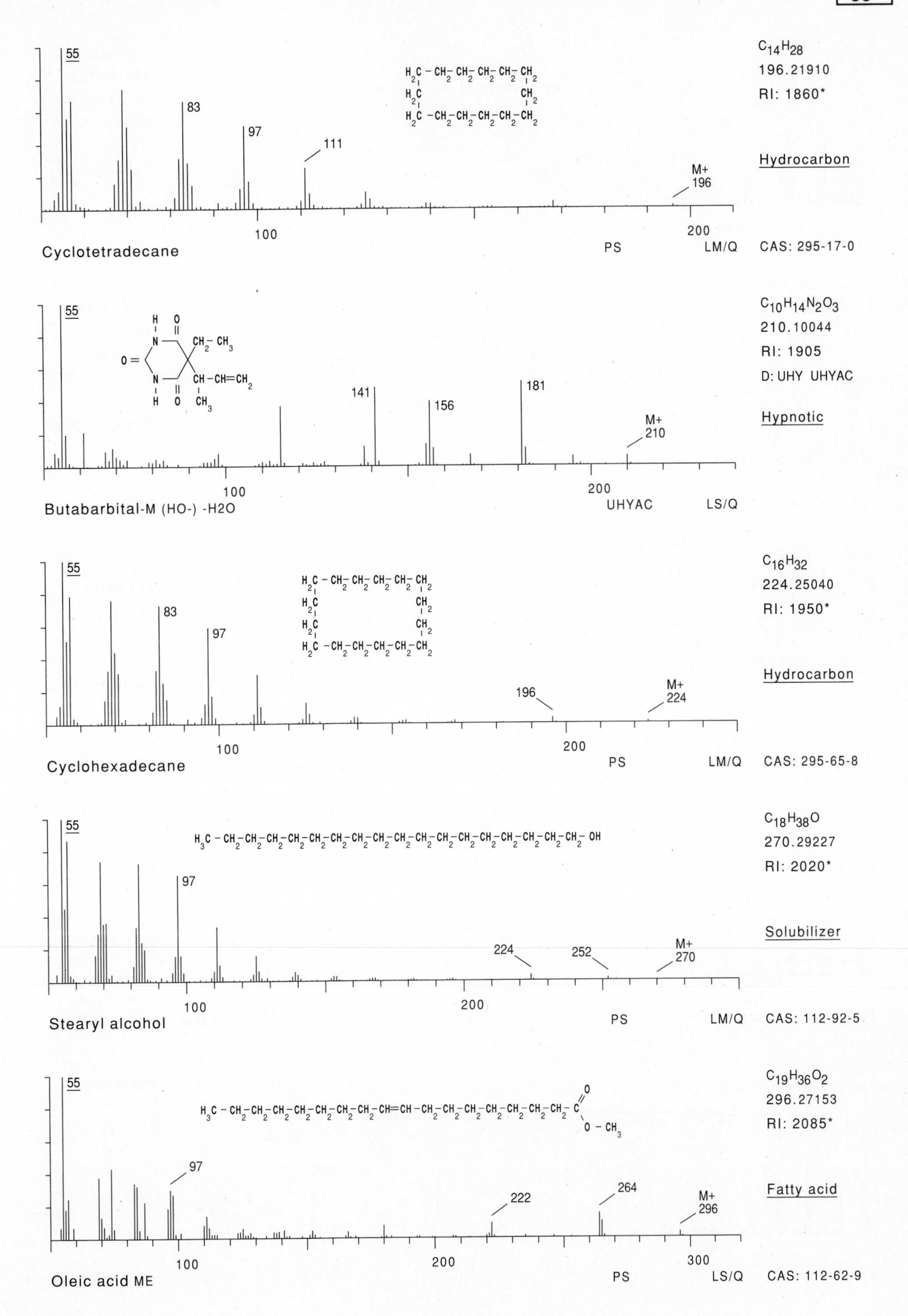

55
83
97
111
M+
196
100
200
Cyclotetradecane
PS
LM/Q
C14H28
196.21910
RI: 1860*
Hydrocarbon
CAS: 295-17-0
55
141
156
181
M+
210
100
200
Butabarbital-M (HO-) -H2O
UHYAC
LS/Q
C10H14N2O3
210.10044
RI: 1905
D: UHY UHYAC
Hypnotic
55
83
97
196
M+
224
100
200
Cyclohexadecane
PS
LM/Q
C16H32
224.25040
RI: 1950*
Hydrocarbon
CAS: 295-65-8
55
97
224
252
M+
270
100
200
Stearyl alcohol
PS
LM/Q
C18H38O
270.29227
RI: 2020*
Solubilizer
CAS: 112-92-5
55
97
222
264
M+
296
100
200
300
Oleic acid ME
PS
LS/Q
C19H36O2
296.27153
RI: 2085*
Fatty acid
CAS: 112-62-9

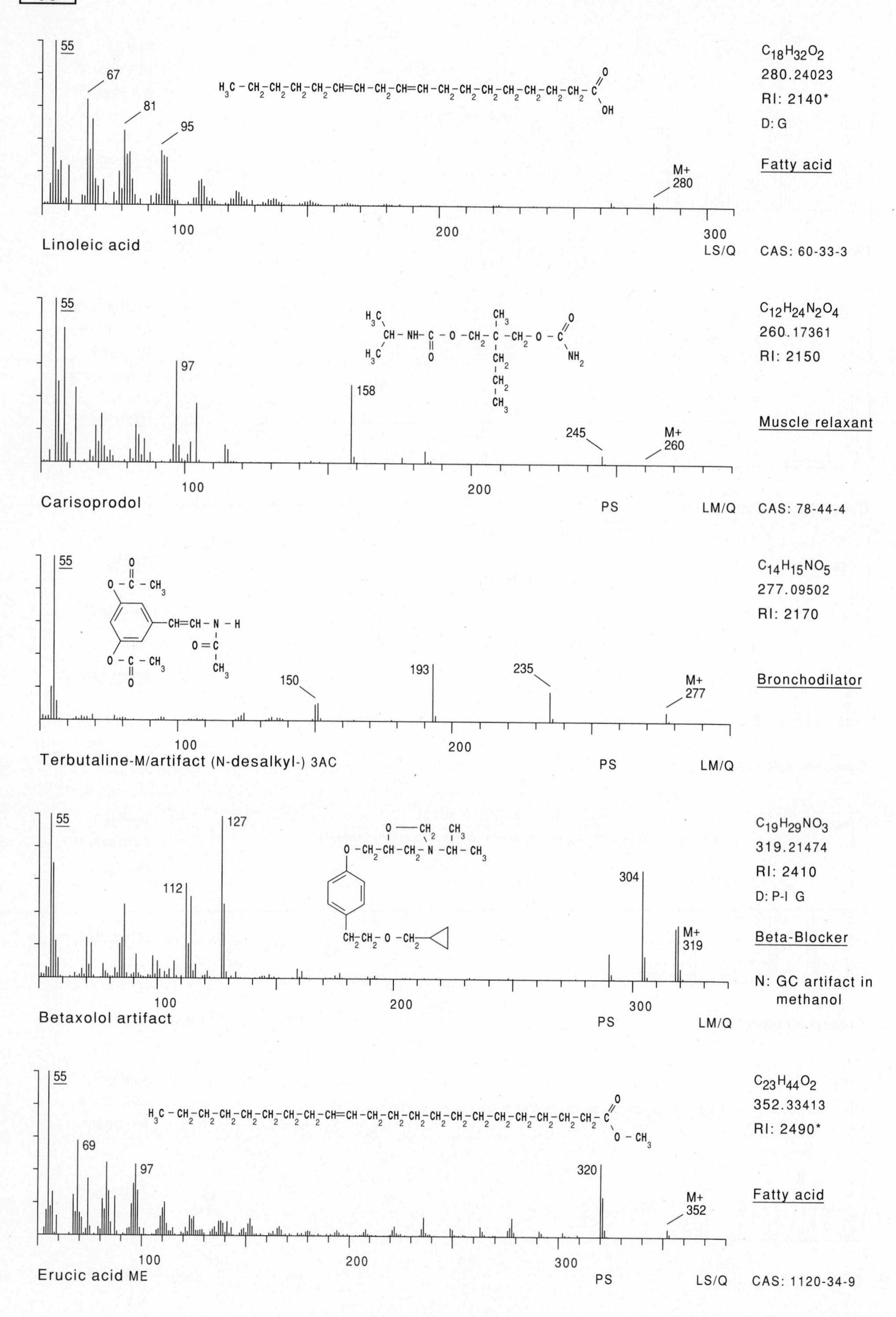

55
67
81
95
M+
280
100
200
300
Linoleic acid
LS/Q
$C_{18}H_{32}O_2$
280.24023
RI: 2140*
D: G
Fatty acid
CAS: 60-33-3
55
97
158
245
M+
260
100
200
Carisoprodol
PS
LM/Q
$C_{12}H_{24}N_2O_4$
260.17361
RI: 2150
Muscle relaxant
CAS: 78-44-4
55
150
193
235
M+
277
100
200
Terbutaline-M/artifact (N-desalkyl-) 3AC
PS
LM/Q
$C_{14}H_{15}NO_5$
277.09502
RI: 2170
Bronchodilator
55
127
112
304
M+
319
100
200
300
Betaxolol artifact
PS
LM/Q
$C_{19}H_{29}NO_3$
319.21474
RI: 2410
D: P-I G
Beta-Blocker
N: GC artifact in methanol
55
69
97
320
M+
352
100
200
300
Erucic acid ME
PS
LS/Q
$C_{23}H_{44}O_2$
352.33413
RI: 2490*
Fatty acid
CAS: 1120-34-9

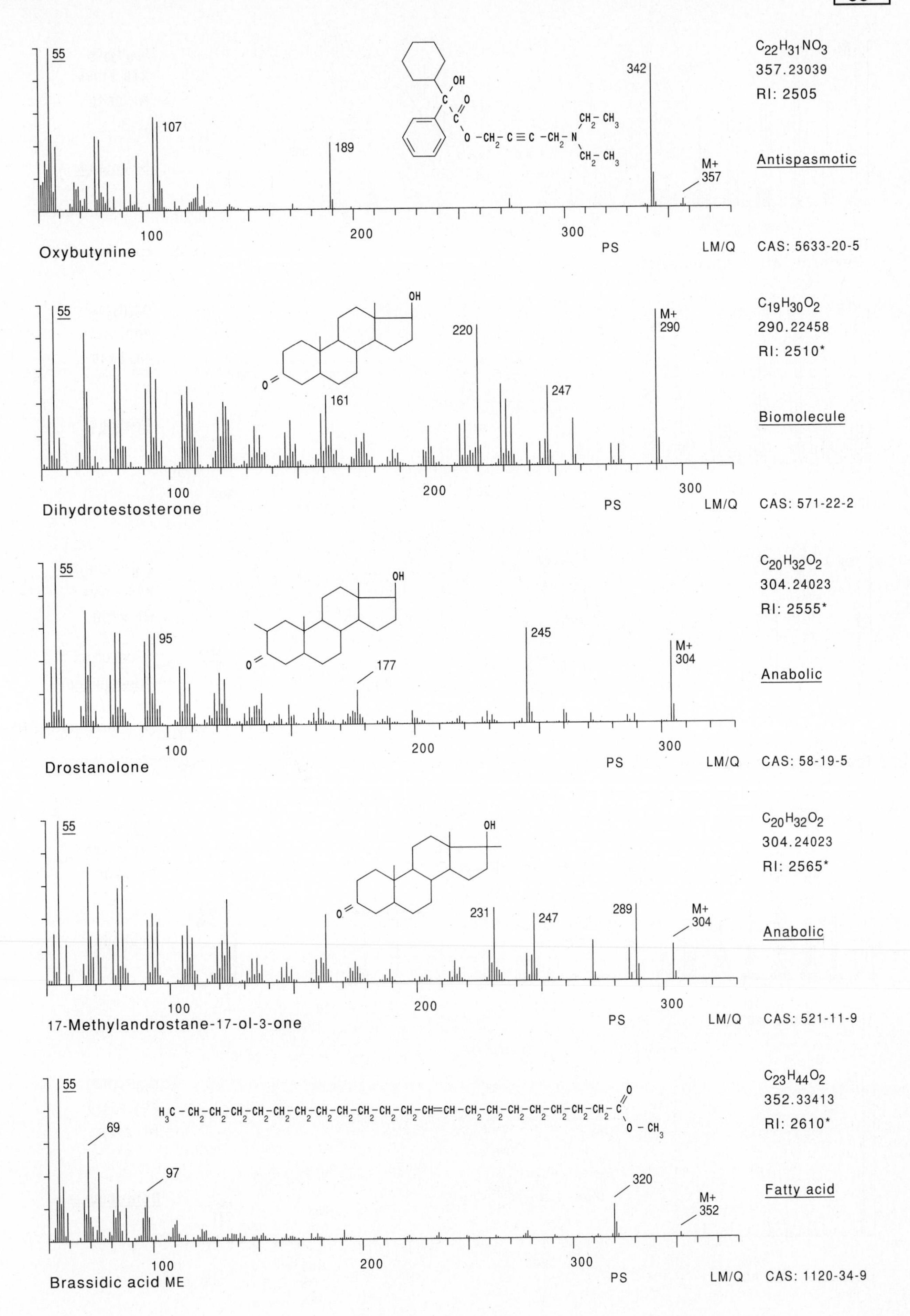

Oxybutynine

$C_{22}H_{31}NO_3$
357.23039
RI: 2505

Antispasmotic

CAS: 5633-20-5

Dihydrotestosterone

$C_{19}H_{30}O_2$
290.22458
RI: 2510*

Biomolecule

CAS: 571-22-2

Drostanolone

$C_{20}H_{32}O_2$
304.24023
RI: 2555*

Anabolic

CAS: 58-19-5

17-Methylandrostane-17-ol-3-one

$C_{20}H_{32}O_2$
304.24023
RI: 2565*

Anabolic

CAS: 521-11-9

Brassidic acid ME

$C_{23}H_{44}O_2$
352.33413
RI: 2610*

Fatty acid

CAS: 1120-34-9

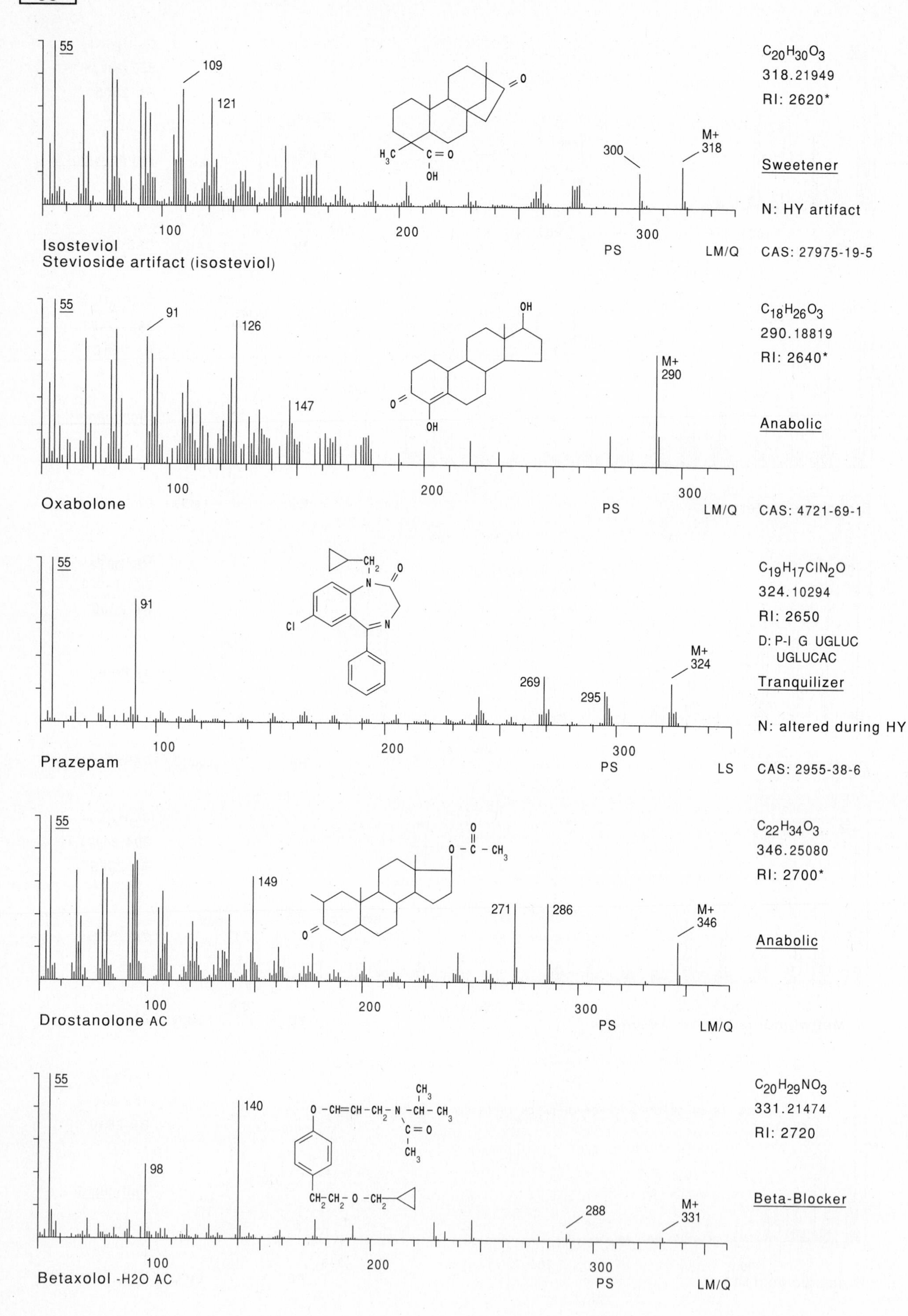
55
109
121
300
M+
318
100
200
300
PS
LM/Q
H3C
C=O
OH
=O
$C_{20}H_{30}O_3$
318.21949
RI: 2620*
Sweetener
N: HY artifact
CAS: 27975-19-5
Isosteviol
Stevioside artifact (isosteviol)
55
91
126
147
M+
290
OH
O=
OH
100
200
300
PS
LM/Q
$C_{18}H_{26}O_3$
290.18819
RI: 2640*
Anabolic
Oxabolone
CAS: 4721-69-1
55
91
269
295
M+
324
CH2
N
O
Cl
N
100
200
300
PS
LS
$C_{19}H_{17}ClN_2O$
324.10294
RI: 2650
D: P-I G UGLUC
UGLUCAC
Tranquilizer
N: altered during HY
Prazepam
CAS: 2955-38-6
55
149
271
286
M+
346
O
O - C - CH3
O=
100
200
300
PS
LM/Q
$C_{22}H_{34}O_3$
346.25080
RI: 2700*
Anabolic
Drostanolone AC
55
140
98
288
M+
331
CH3
O - CH=CH - CH2 - N - CH - CH3
C = O
CH3
CH2 - CH2 - O - CH2
100
200
300
PS
LM/Q
$C_{20}H_{29}NO_3$
331.21474
RI: 2720
Beta-Blocker
Betaxolol -H2O AC

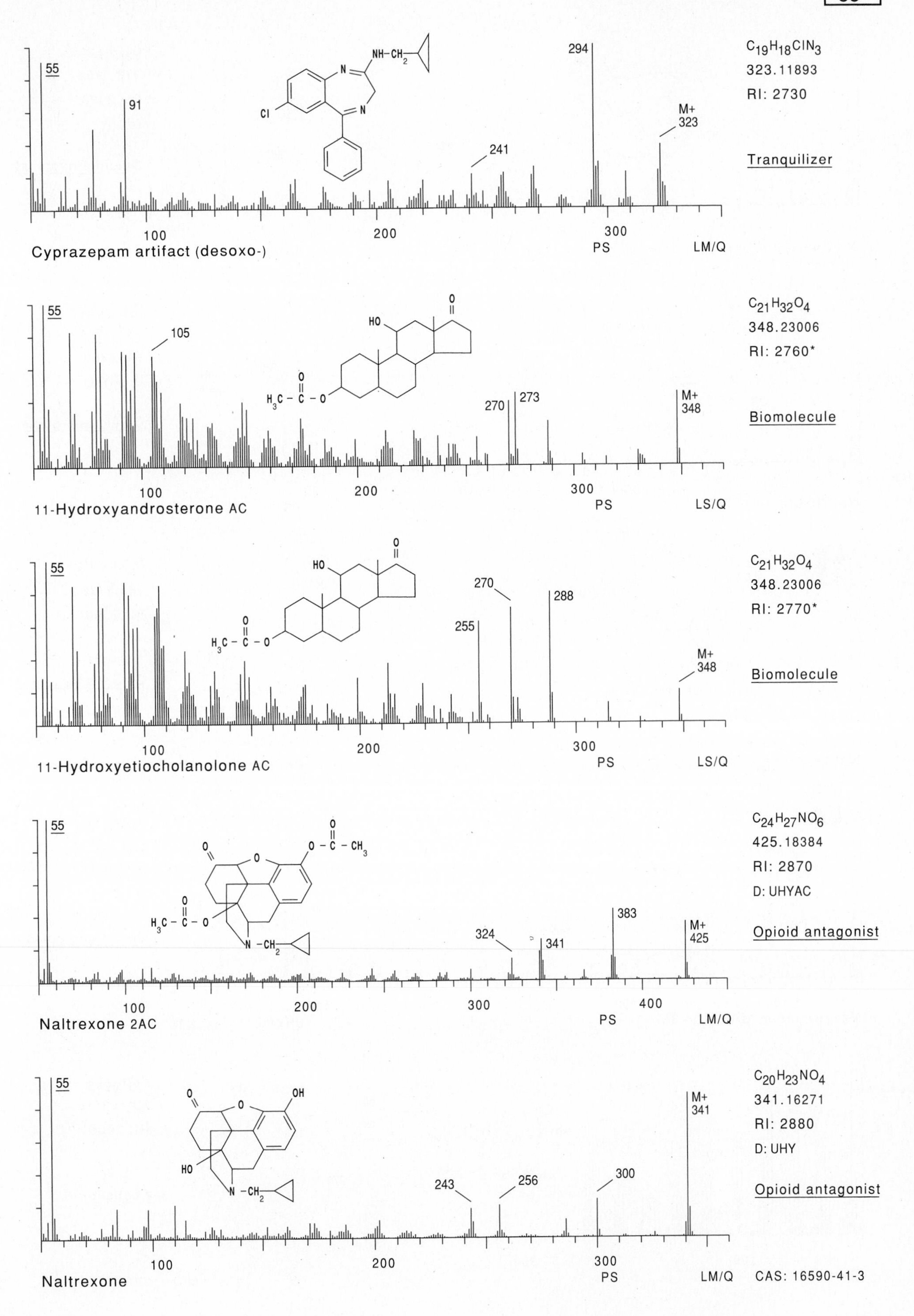

$C_{19}H_{18}ClN_3$
323.11893
RI: 2730

Tranquilizer

$C_{21}H_{32}O_4$
348.23006
RI: 2760*

Biomolecule

$C_{21}H_{32}O_4$
348.23006
RI: 2770*

Biomolecule

$C_{24}H_{27}NO_6$
425.18384
RI: 2870
D: UHYAC

Opioid antagonist

$C_{20}H_{23}NO_4$
341.16271
RI: 2880
D: UHY

Opioid antagonist

CAS: 16590-41-3

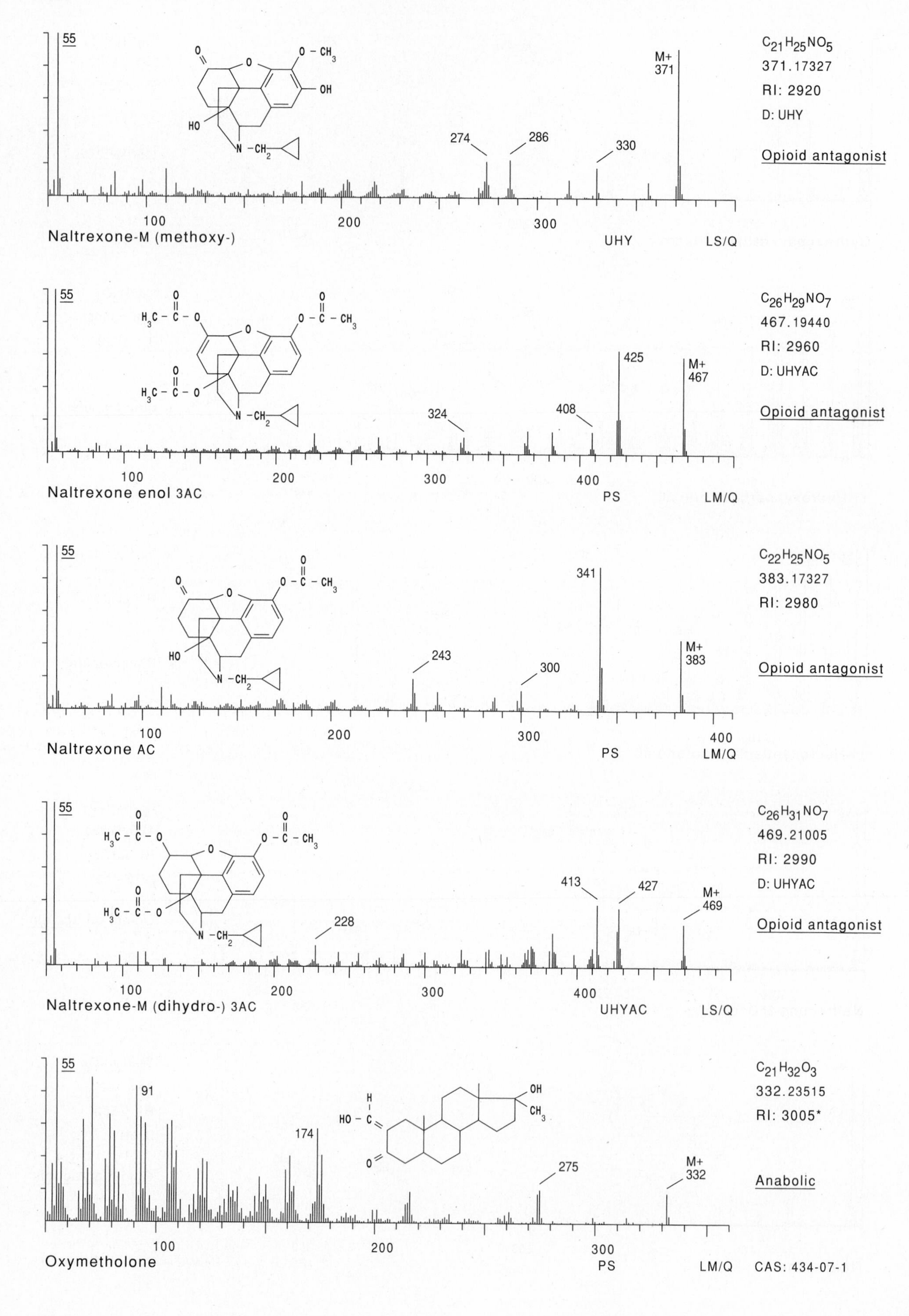
55
M+
371
274
286
330
100
200
300
Naltrexone-M (methoxy-)
UHY
LS/Q
$C_{21}H_{25}NO_5$
371.17327
RI: 2920
D: UHY
Opioid antagonist
55
425
M+
467
324
408
100
200
300
400
Naltrexone enol 3AC
PS
LM/Q
$C_{26}H_{29}NO_7$
467.19440
RI: 2960
D: UHYAC
Opioid antagonist
55
341
M+
383
243
300
100
200
300
400
Naltrexone AC
PS
LM/Q
$C_{22}H_{25}NO_5$
383.17327
RI: 2980
Opioid antagonist
55
413
427
M+
469
228
100
200
300
400
Naltrexone-M (dihydro-) 3AC
UHYAC
LS/Q
$C_{26}H_{31}NO_7$
469.21005
RI: 2990
D: UHYAC
Opioid antagonist
55
91
174
275
M+
332
100
200
300
Oxymetholone
PS
LM/Q
$C_{21}H_{32}O_3$
332.23515
RI: 3005*
Anabolic
CAS: 434-07-1

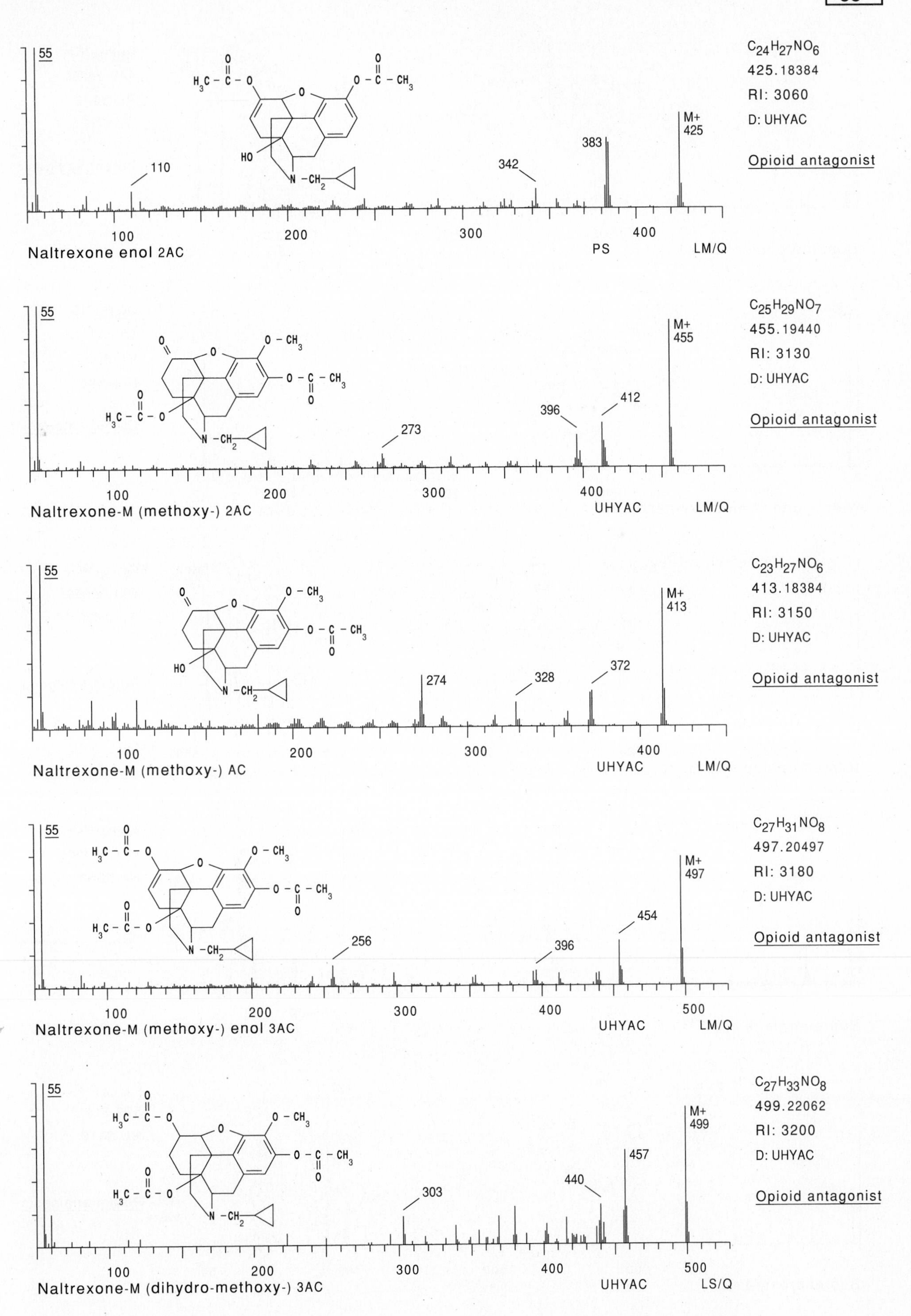

Naltrexone enol 2AC
$C_{24}H_{27}NO_6$
425.18384
RI: 3060
D: UHYAC
Opioid antagonist
M+ 425
383
342
110
55
PS
LM/Q
Naltrexone-M (methoxy-) 2AC
$C_{25}H_{29}NO_7$
455.19440
RI: 3130
D: UHYAC
Opioid antagonist
M+ 455
412
396
273
55
UHYAC
LM/Q
Naltrexone-M (methoxy-) AC
$C_{23}H_{27}NO_6$
413.18384
RI: 3150
D: UHYAC
Opioid antagonist
M+ 413
372
328
274
55
UHYAC
LM/Q
Naltrexone-M (methoxy-) enol 3AC
$C_{27}H_{31}NO_8$
497.20497
RI: 3180
D: UHYAC
Opioid antagonist
M+ 497
454
396
256
55
UHYAC
LM/Q
Naltrexone-M (dihydro-methoxy-) 3AC
$C_{27}H_{33}NO_8$
499.22062
RI: 3200
D: UHYAC
Opioid antagonist
M+ 499
457
440
303
55
UHYAC
LS/Q

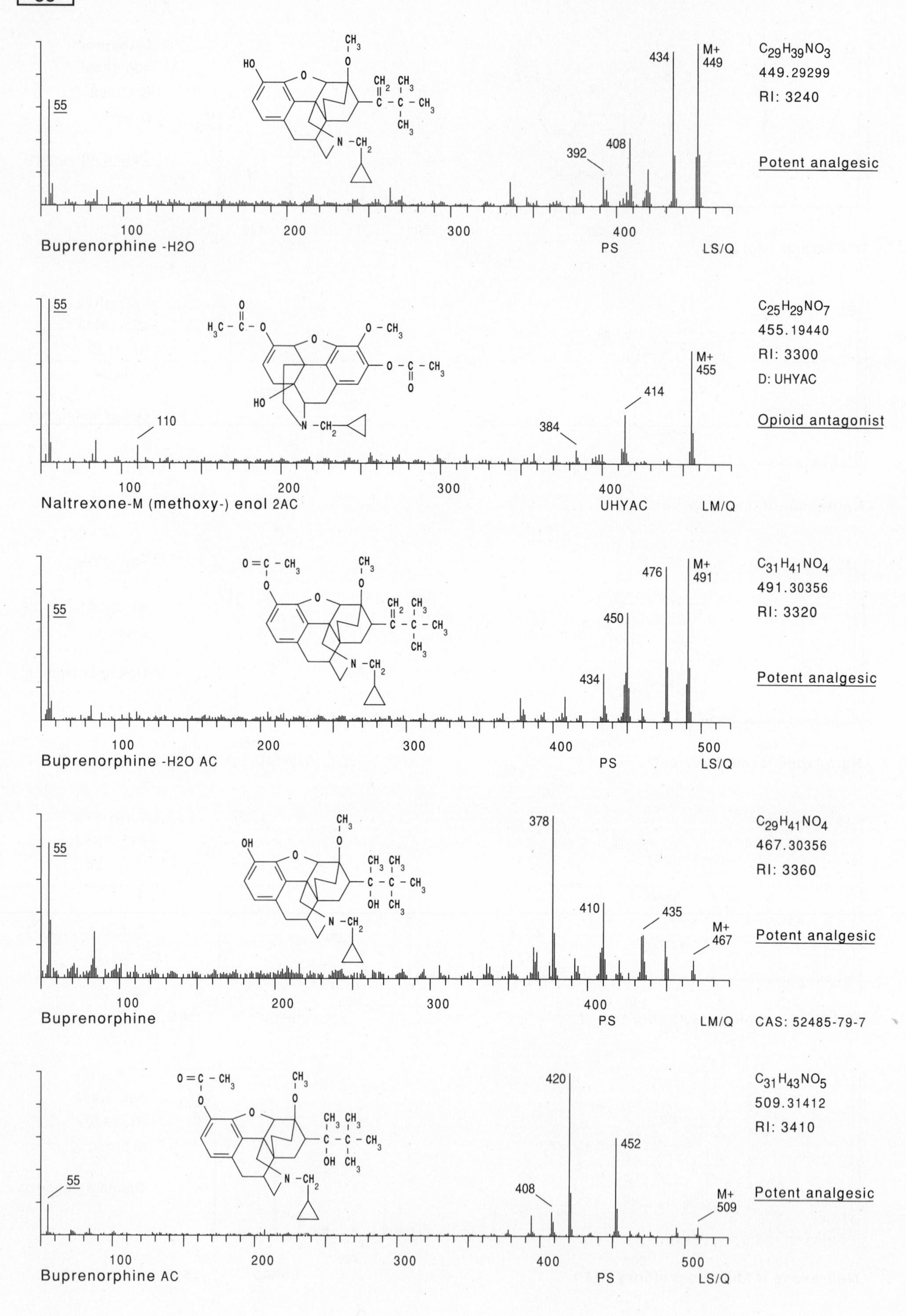
Buprenorphine -H2O
PS
LS/Q
$C_{29}H_{39}NO_3$
449.29299
RI: 3240
Potent analgesic
Naltrexone-M (methoxy-) enol 2AC
UHYAC
LM/Q
$C_{25}H_{29}NO_7$
455.19440
RI: 3300
D: UHYAC
Opioid antagonist
Buprenorphine -H2O AC
PS
LS/Q
$C_{31}H_{41}NO_4$
491.30356
RI: 3320
Potent analgesic
Buprenorphine
PS
LM/Q
$C_{29}H_{41}NO_4$
467.30356
RI: 3360
Potent analgesic
CAS: 52485-79-7
Buprenorphine AC
PS
LS/Q
$C_{31}H_{43}NO_5$
509.31412
RI: 3410
Potent analgesic

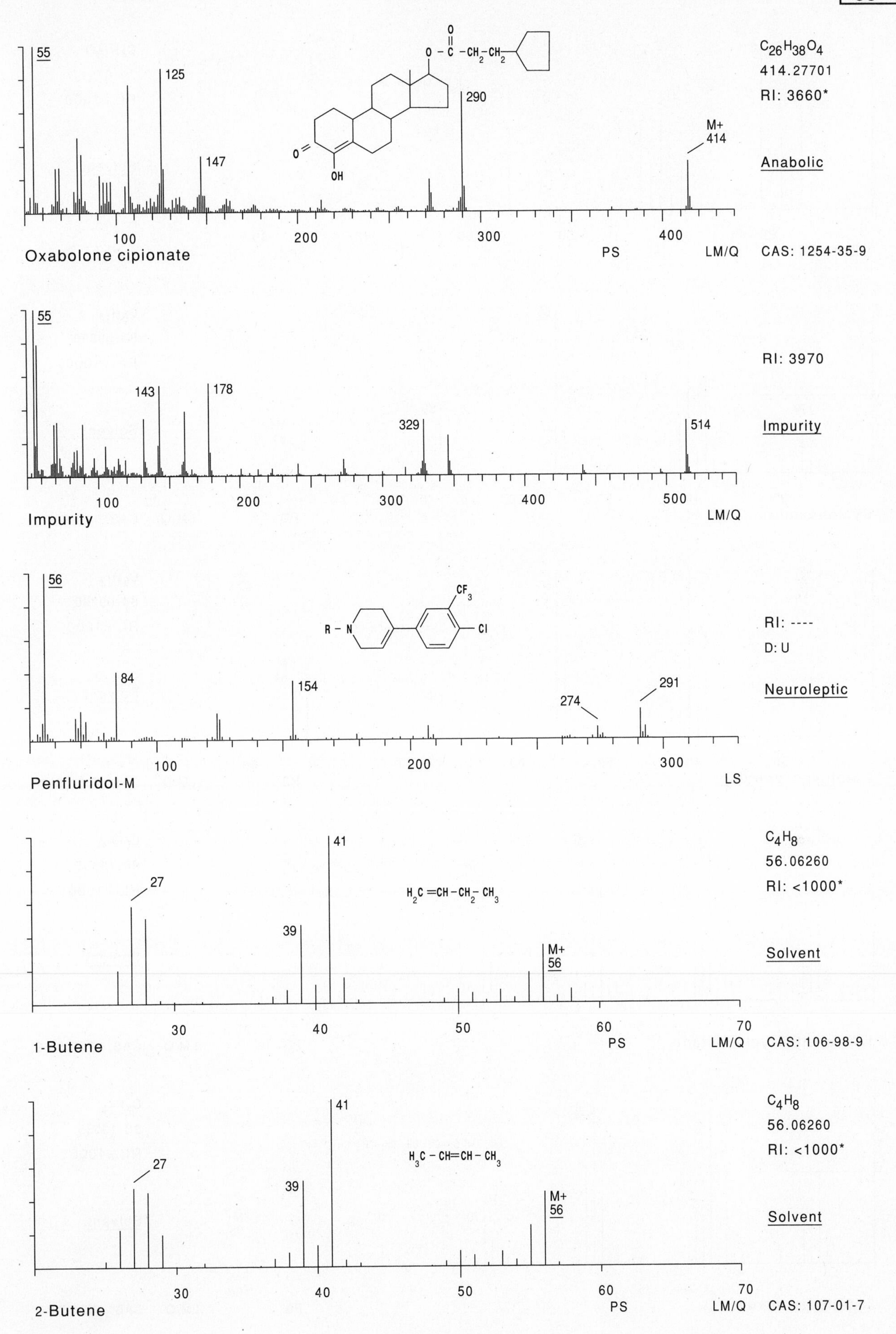

$C_{26}H_{38}O_4$
414.27701
RI: 3660*

Anabolic

CAS: 1254-35-9

RI: 3970

Impurity

RI: ----
D: U

Neuroleptic

C_4H_8
56.06260
RI: <1000*

Solvent

CAS: 106-98-9

C_4H_8
56.06260
RI: <1000*

Solvent

CAS: 107-01-7

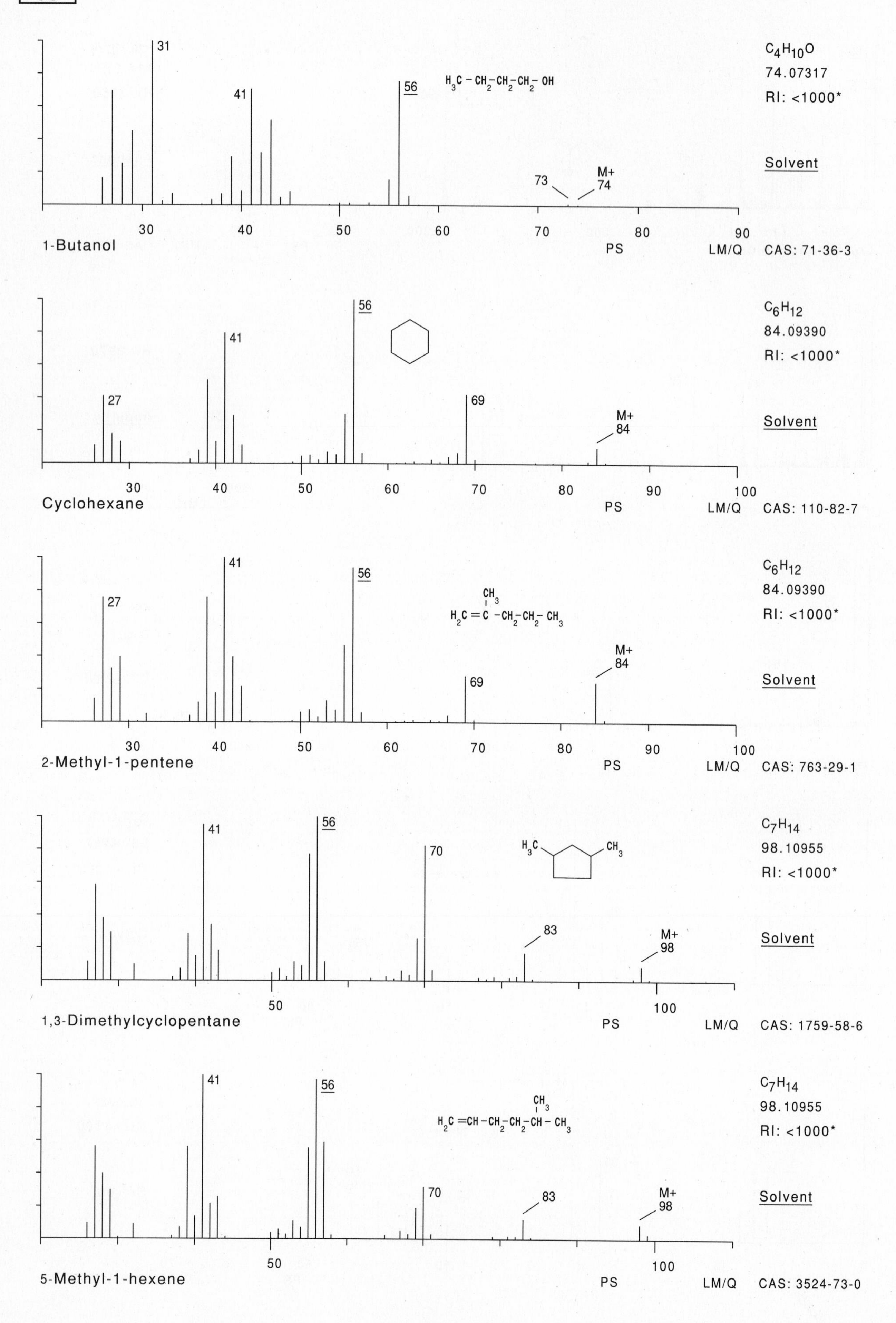
31
41
56
$H_3C-CH_2-CH_2-CH_2-OH$
73
M+
74
30
40
50
60
70
80
90
1-Butanol
PS
LM/Q
$C_4H_{10}O$
74.07317
RI: <1000*
Solvent
CAS: 71-36-3
56
41
27
69
M+
84
Cyclohexane
100
C_6H_{12}
84.09390
RI: <1000*
Solvent
CAS: 110-82-7
41
56
27
CH_3
$H_2C=C-CH_2-CH_2-CH_3$
M+
84
69
2-Methyl-1-pentene
C_6H_{12}
84.09390
RI: <1000*
Solvent
CAS: 763-29-1
41
56
70
H_3C
CH_3
83
M+
98
1,3-Dimethylcyclopentane
C_7H_{14}
98.10955
RI: <1000*
Solvent
CAS: 1759-58-6
41
56
CH_3
$H_2C=CH-CH_2-CH_2-CH-CH_3$
70
83
M+
98
5-Methyl-1-hexene
C_7H_{14}
98.10955
RI: <1000*
Solvent
CAS: 3524-73-0

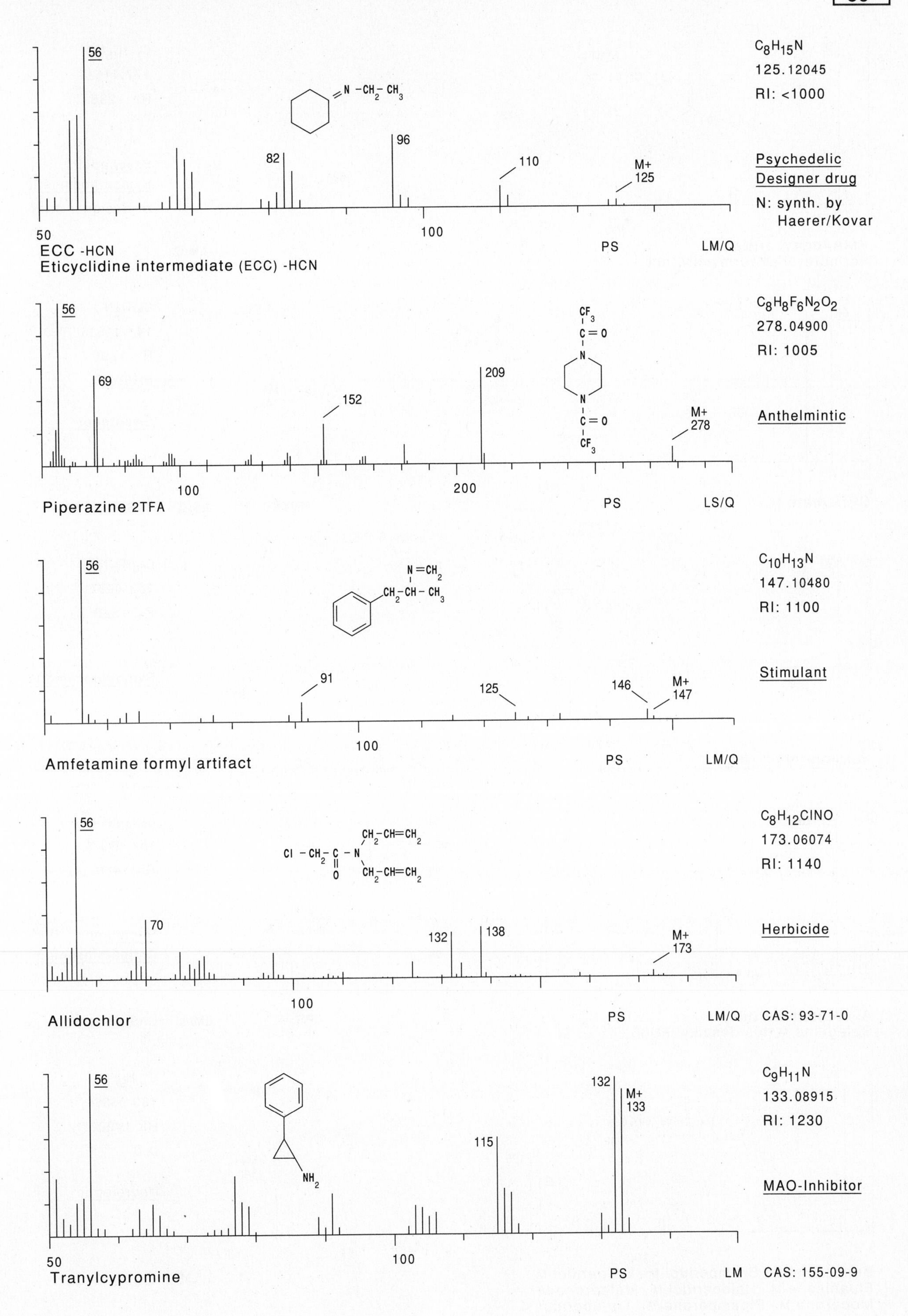
56
56
82
96
110
M+
125
50
100
ECC -HCN
Eticyclidine intermediate (ECC) -HCN
PS
LM/Q
C8H15N
125.12045
RI: <1000
Psychedelic
Designer drug
N: synth. by
Haerer/Kovar
56
69
152
209
M+
278
100
200
Piperazine 2TFA
PS
LS/Q
C8H8F6N2O2
278.04900
RI: 1005
Anthelmintic
56
91
125
146
M+
147
100
Amfetamine formyl artifact
PS
LM/Q
C10H13N
147.10480
RI: 1100
Stimulant
56
70
132
138
M+
173
100
Allidochlor
PS
LM/Q
C8H12ClNO
173.06074
RI: 1140
Herbicide
CAS: 93-71-0
56
115
132
M+
133
50
100
Tranylcypromine
PS
LM
C9H11N
133.08915
RI: 1230
MAO-Inhibitor
CAS: 155-09-9

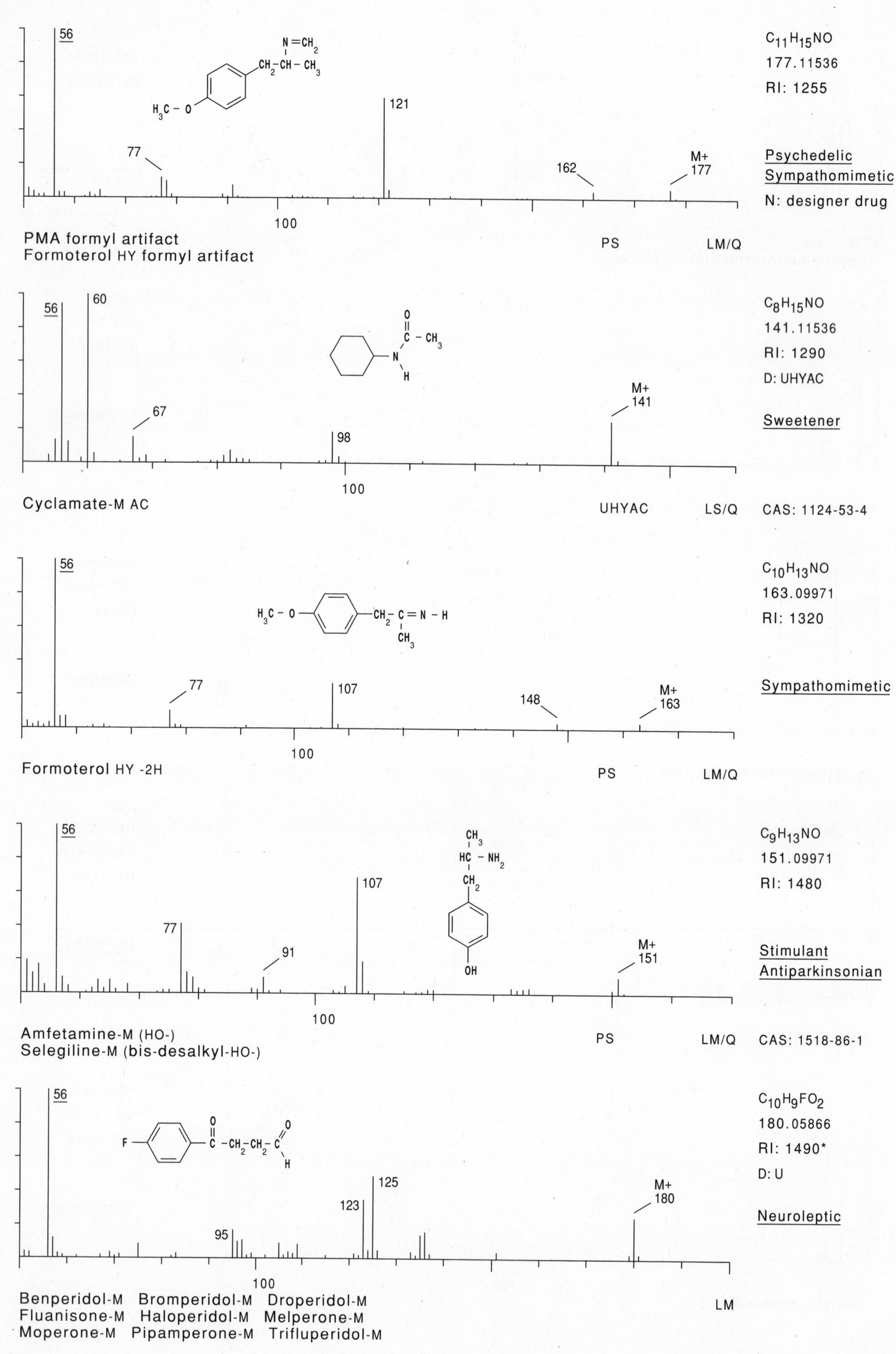
56
N=CH2
CH2-CH-CH3
H3C-O
121
77
162
M+
177
100
PS
LM/Q
C11H15NO
177.11536
RI: 1255
Psychedelic
Sympathomimetic
N: designer drug
PMA formyl artifact
Formoterol HY formyl artifact
56
60
O
C-CH3
N
H
67
98
M+
141
100
UHYAC
LS/Q
C8H15NO
141.11536
RI: 1290
D: UHYAC
Sweetener
Cyclamate-M AC
CAS: 1124-53-4
56
H3C-O
CH2-C=N-H
CH3
77
107
148
M+
163
100
PS
LM/Q
C10H13NO
163.09971
RI: 1320
Sympathomimetic
Formoterol HY -2H
56
CH3
HC-NH2
CH2
OH
107
77
91
M+
151
100
PS
LM/Q
C9H13NO
151.09971
RI: 1480
Stimulant
Antiparkinsonian
Amfetamine-M (HO-)
Selegiline-M (bis-desalkyl-HO-)
CAS: 1518-86-1
56
F
O
C-CH2-CH2-C
O
H
125
123
95
M+
180
100
LM
C10H9FO2
180.05866
RI: 1490*
D: U
Neuroleptic
Benperidol-M
Bromperidol-M
Droperidol-M
Fluanisone-M
Haloperidol-M
Melperone-M
Moperone-M
Pipamperone-M
Trifluperidol-M

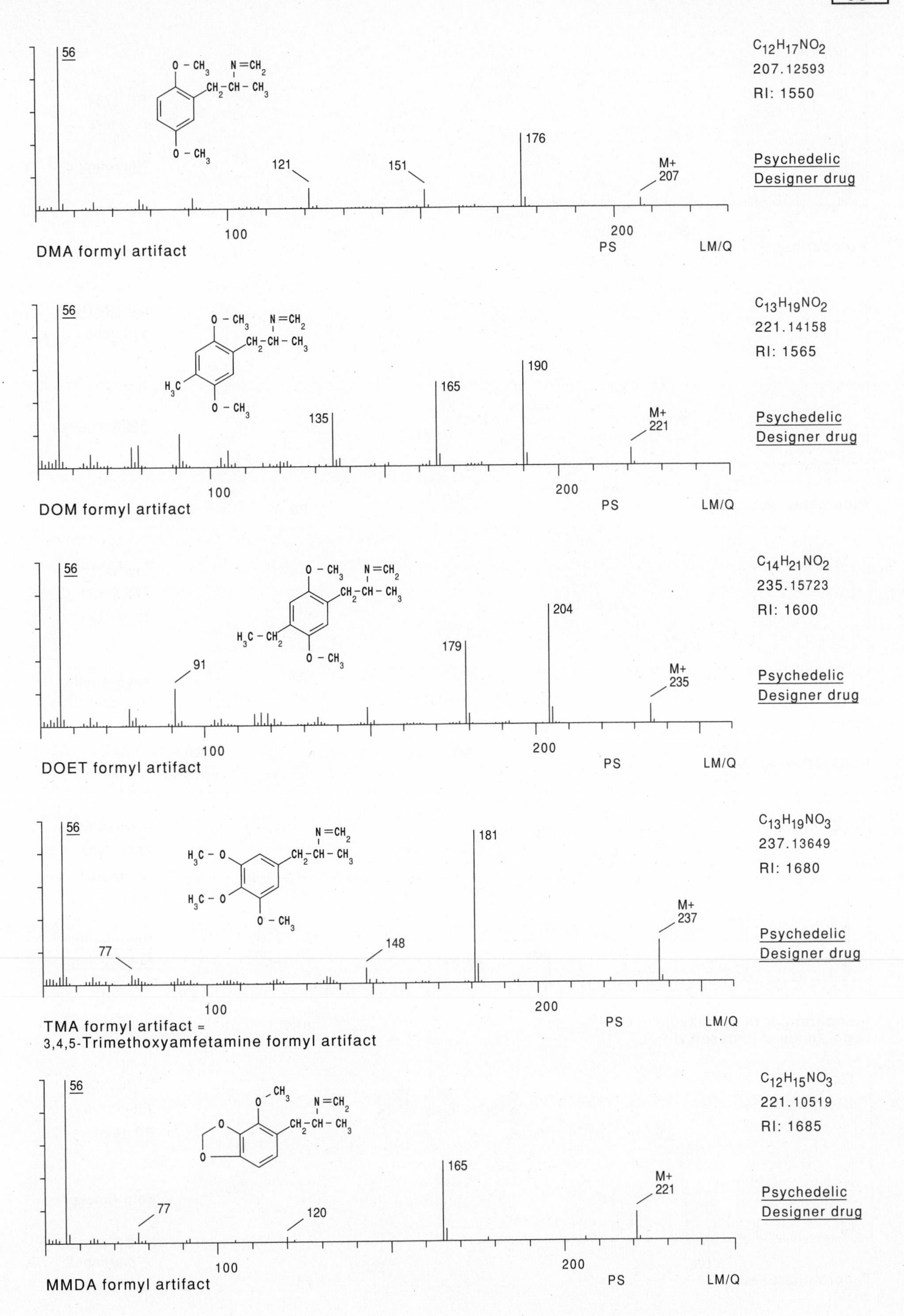

DMA formyl artifact

$C_{12}H_{17}NO_2$
207.12593
RI: 1550

Psychedelic
Designer drug

DOM formyl artifact

$C_{13}H_{19}NO_2$
221.14158
RI: 1565

Psychedelic
Designer drug

DOET formyl artifact

$C_{14}H_{21}NO_2$
235.15723
RI: 1600

Psychedelic
Designer drug

TMA formyl artifact =
3,4,5-Trimethoxyamfetamine formyl artifact

$C_{13}H_{19}NO_3$
237.13649
RI: 1680

Psychedelic
Designer drug

MMDA formyl artifact

$C_{12}H_{15}NO_3$
221.10519
RI: 1685

Psychedelic
Designer drug

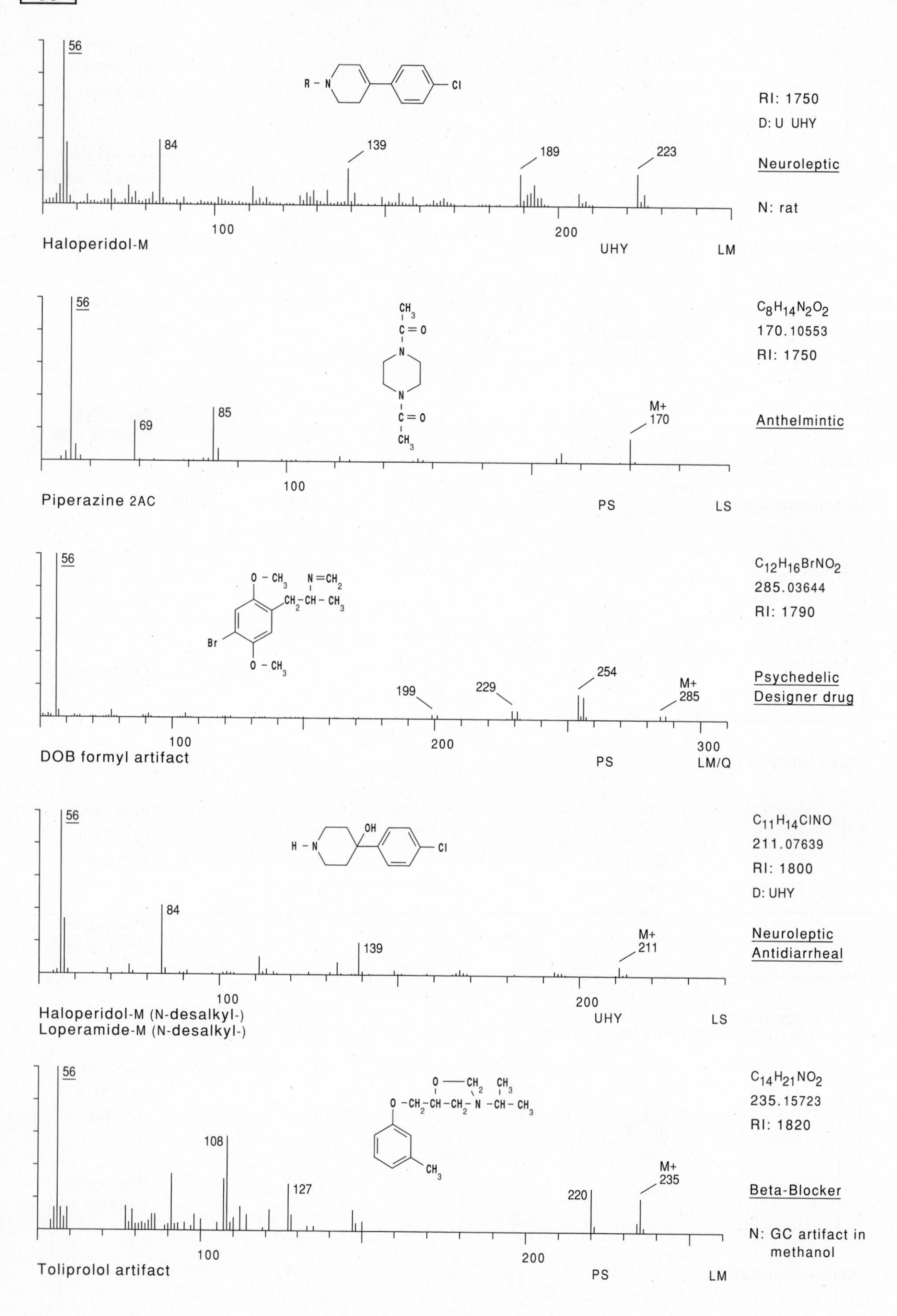
56
84
139
189
223
R - N
Cl
100
200
Haloperidol-M
UHY
LM
RI: 1750
D: U UHY
Neuroleptic
N: rat
56
69
85
M+
170
CH3
C = O
N
N
C = O
CH3
100
Piperazine 2AC
PS
LS
$C_8H_{14}N_2O_2$
170.10553
RI: 1750
Anthelmintic
56
199
229
254
M+
285
O - CH3
N = CH2
CH2 - CH - CH3
Br
O - CH3
100
200
300
DOB formyl artifact
PS
LM/Q
$C_{12}H_{16}BrNO_2$
285.03644
RI: 1790
Psychedelic
Designer drug
56
84
139
M+
211
OH
H - N
Cl
100
200
Haloperidol-M (N-desalkyl-)
Loperamide-M (N-desalkyl-)
UHY
LS
$C_{11}H_{14}ClNO$
211.07639
RI: 1800
D: UHY
Neuroleptic
Antidiarrheal
56
108
127
220
M+
235
O — CH2 CH3
O - CH2 - CH - CH2 - N - CH - CH3
CH3
100
200
Toliprolol artifact
PS
LM
$C_{14}H_{21}NO_2$
235.15723
RI: 1820
Beta-Blocker
N: GC artifact in methanol

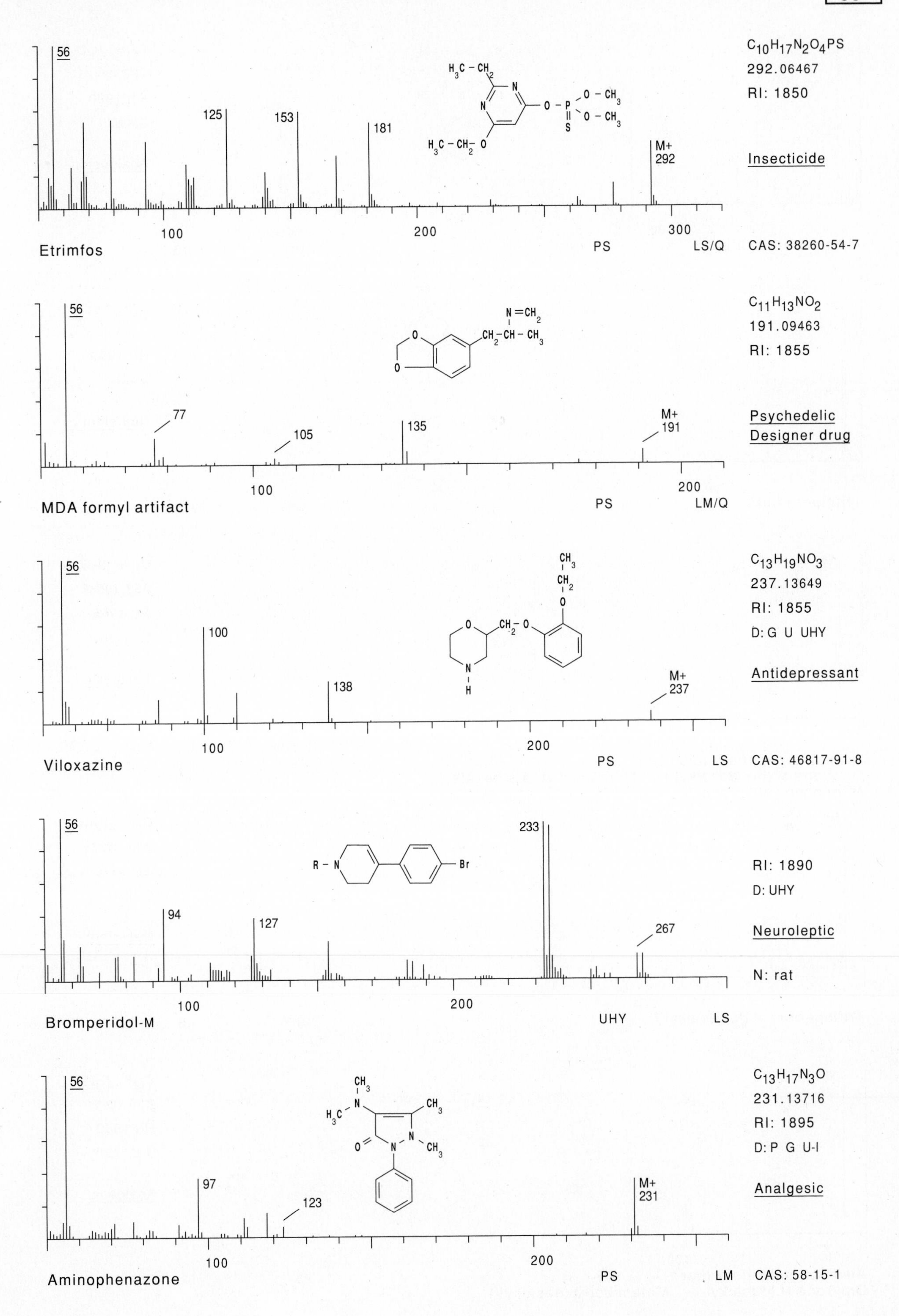

$C_{10}H_{17}N_2O_4PS$
292.06467
RI: 1850

Insecticide

CAS: 38260-54-7

$C_{11}H_{13}NO_2$
191.09463
RI: 1855

Psychedelic
Designer drug

$C_{13}H_{19}NO_3$
237.13649
RI: 1855
D: G U UHY

Antidepressant

CAS: 46817-91-8

RI: 1890
D: UHY

Neuroleptic

N: rat

$C_{13}H_{17}N_3O$
231.13716
RI: 1895
D: P G U-I

Analgesic

CAS: 58-15-1

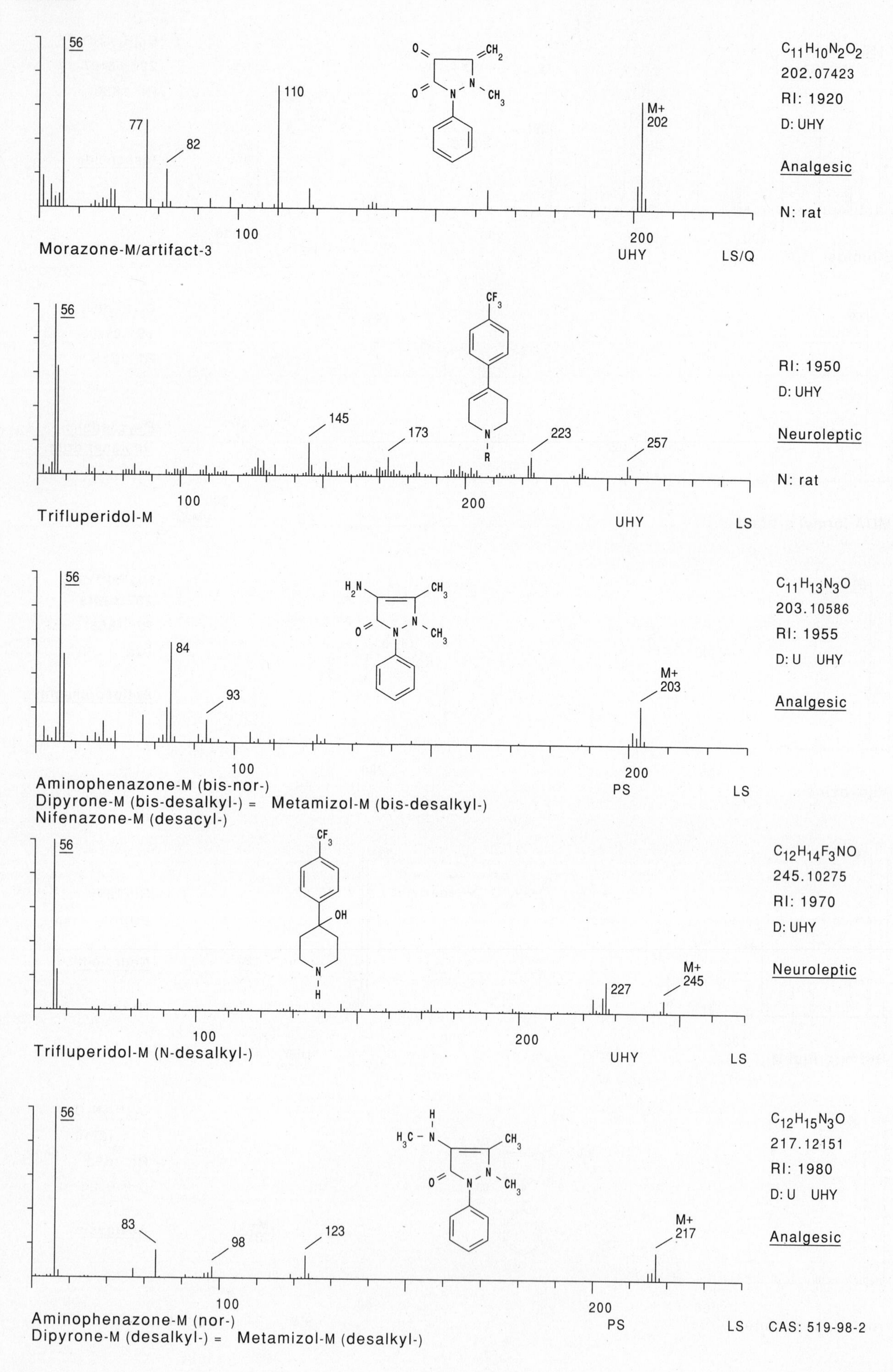
56
77
82
110
M+
202
100
200
UHY
LS/Q
$C_{11}H_{10}N_2O_2$
202.07423
RI: 1920
D: UHY
Analgesic
N: rat
Morazone-M/artifact-3
56
145
173
223
257
100
200
UHY
LS
RI: 1950
D: UHY
Neuroleptic
N: rat
Trifluperidol-M
56
84
93
M+
203
100
200
PS
LS
$C_{11}H_{13}N_3O$
203.10586
RI: 1955
D: U UHY
Analgesic
Aminophenazone-M (bis-nor-)
Dipyrone-M (bis-desalkyl-) = Metamizol-M (bis-desalkyl-)
Nifenazone-M (desacyl-)
56
227
M+
245
100
200
UHY
LS
$C_{12}H_{14}F_3NO$
245.10275
RI: 1970
D: UHY
Neuroleptic
Trifluperidol-M (N-desalkyl-)
56
83
98
123
M+
217
100
200
PS
LS
$C_{12}H_{15}N_3O$
217.12151
RI: 1980
D: U UHY
Analgesic
Aminophenazone-M (nor-)
Dipyrone-M (desalkyl-) = Metamizol-M (desalkyl-)
CAS: 519-98-2

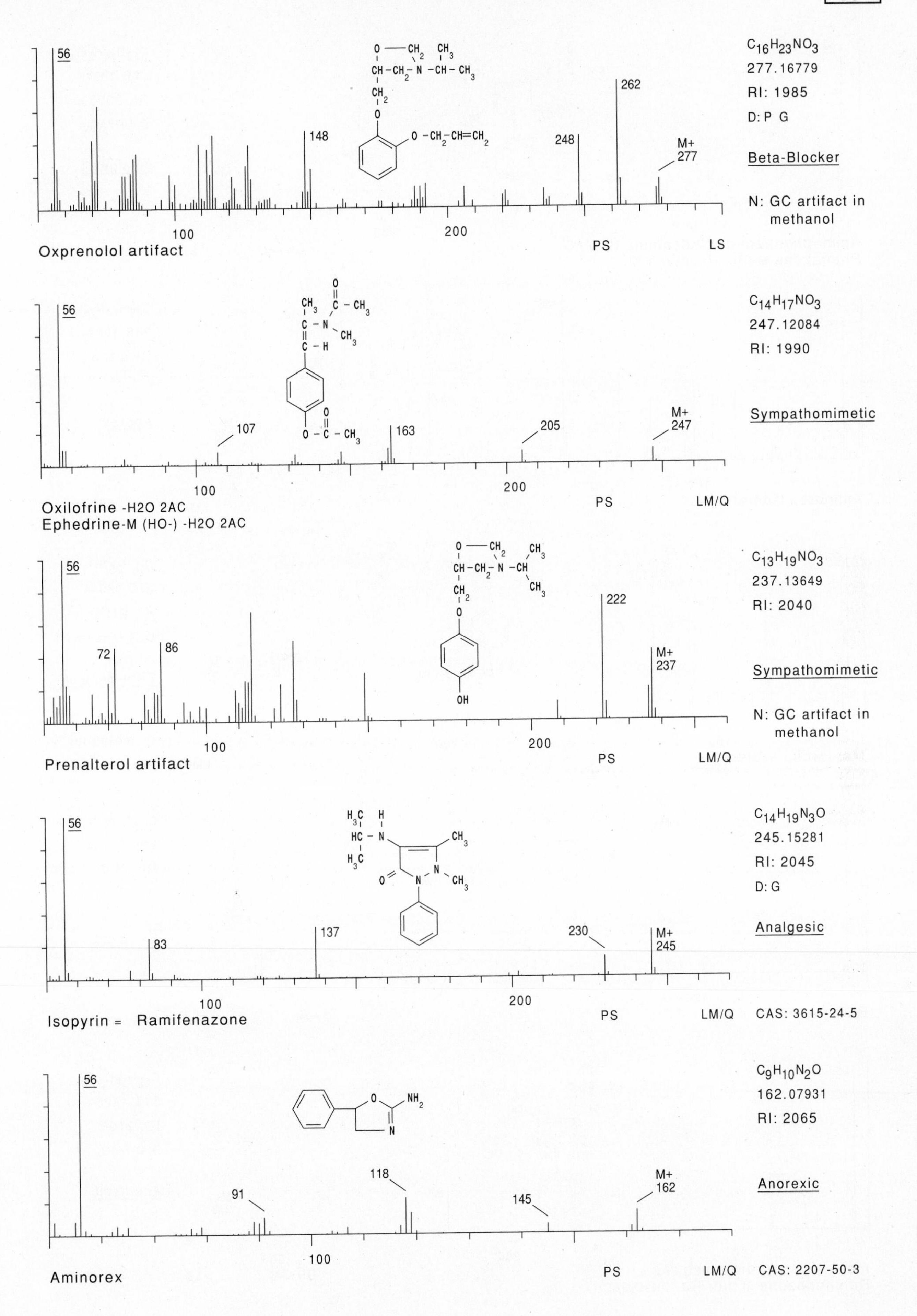

Oxprenolol artifact

$C_{16}H_{23}NO_3$
277.16779
RI: 1985
D: P G

Beta-Blocker

N: GC artifact in methanol

Oxilofrine -H2O 2AC
Ephedrine-M (HO-) -H2O 2AC

$C_{14}H_{17}NO_3$
247.12084
RI: 1990

Sympathomimetic

Prenalterol artifact

$C_{13}H_{19}NO_3$
237.13649
RI: 2040

Sympathomimetic

N: GC artifact in methanol

Isopyrin = Ramifenazone

$C_{14}H_{19}N_3O$
245.15281
RI: 2045
D: G

Analgesic

CAS: 3615-24-5

Aminorex

$C_9H_{10}N_2O$
162.07931
RI: 2065

Anorexic

CAS: 2207-50-3

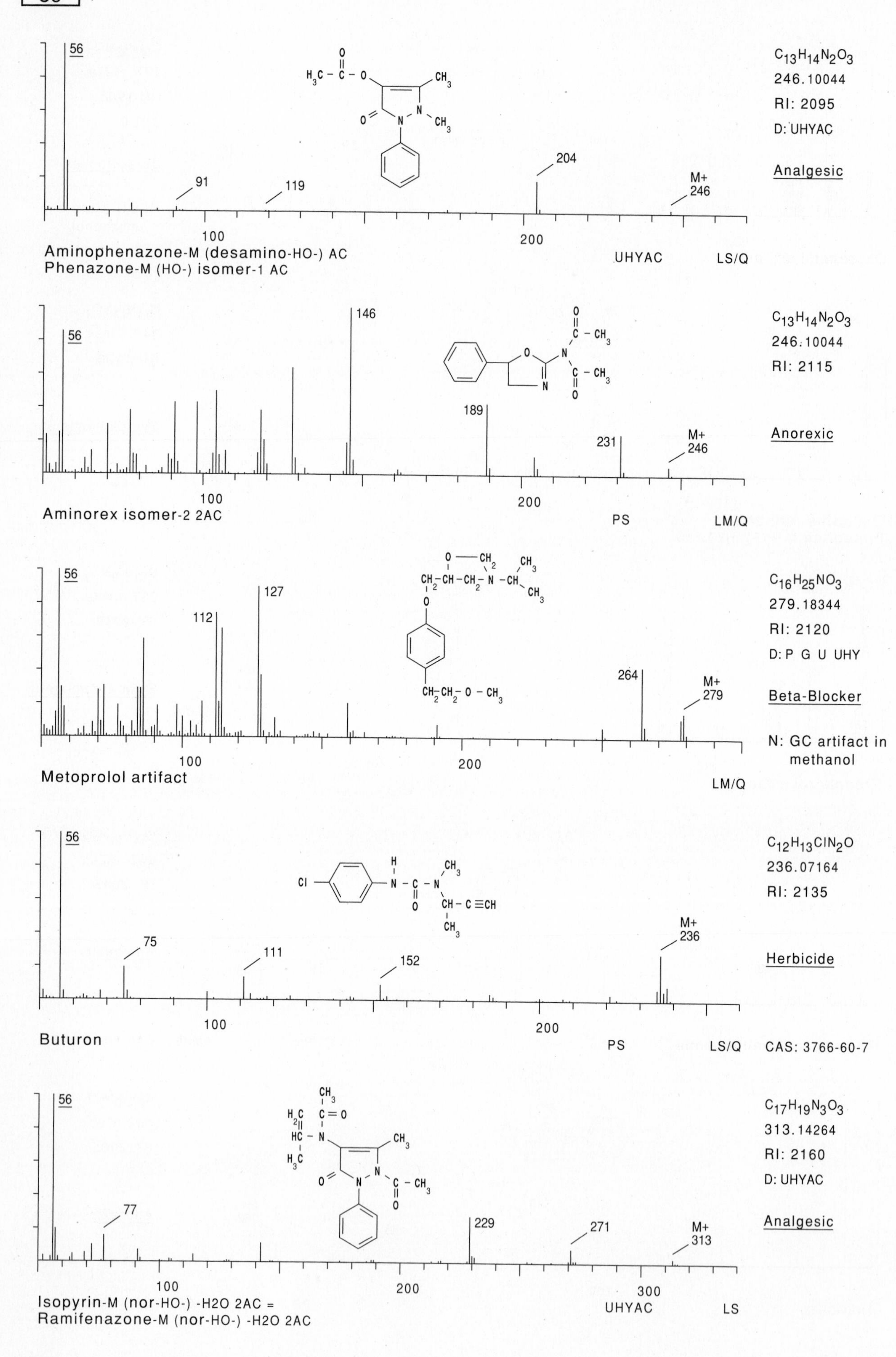
56
91
119
204
M+
246
100
200
Aminophenazone-M (desamino-HO-) AC
Phenazone-M (HO-) isomer-1 AC
UHYAC
LS/Q
C13H14N2O3
246.10044
RI: 2095
D: UHYAC
Analgesic
146
56
189
231
M+
246
100
200
Aminorex isomer-2 2AC
PS
LM/Q
C13H14N2O3
246.10044
RI: 2115
Anorexic
56
127
112
264
M+
279
100
200
Metoprolol artifact
LM/Q
C16H25NO3
279.18344
RI: 2120
D: P G U UHY
Beta-Blocker
N: GC artifact in methanol
56
75
111
152
M+
236
100
200
Buturon
PS
LS/Q
C12H13ClN2O
236.07164
RI: 2135
Herbicide
CAS: 3766-60-7
56
77
229
271
M+
313
100
200
300
Isopyrin-M (nor-HO-) -H2O 2AC =
Ramifenazone-M (nor-HO-) -H2O 2AC
UHYAC
LS
C17H19N3O3
313.14264
RI: 2160
D: UHYAC
Analgesic

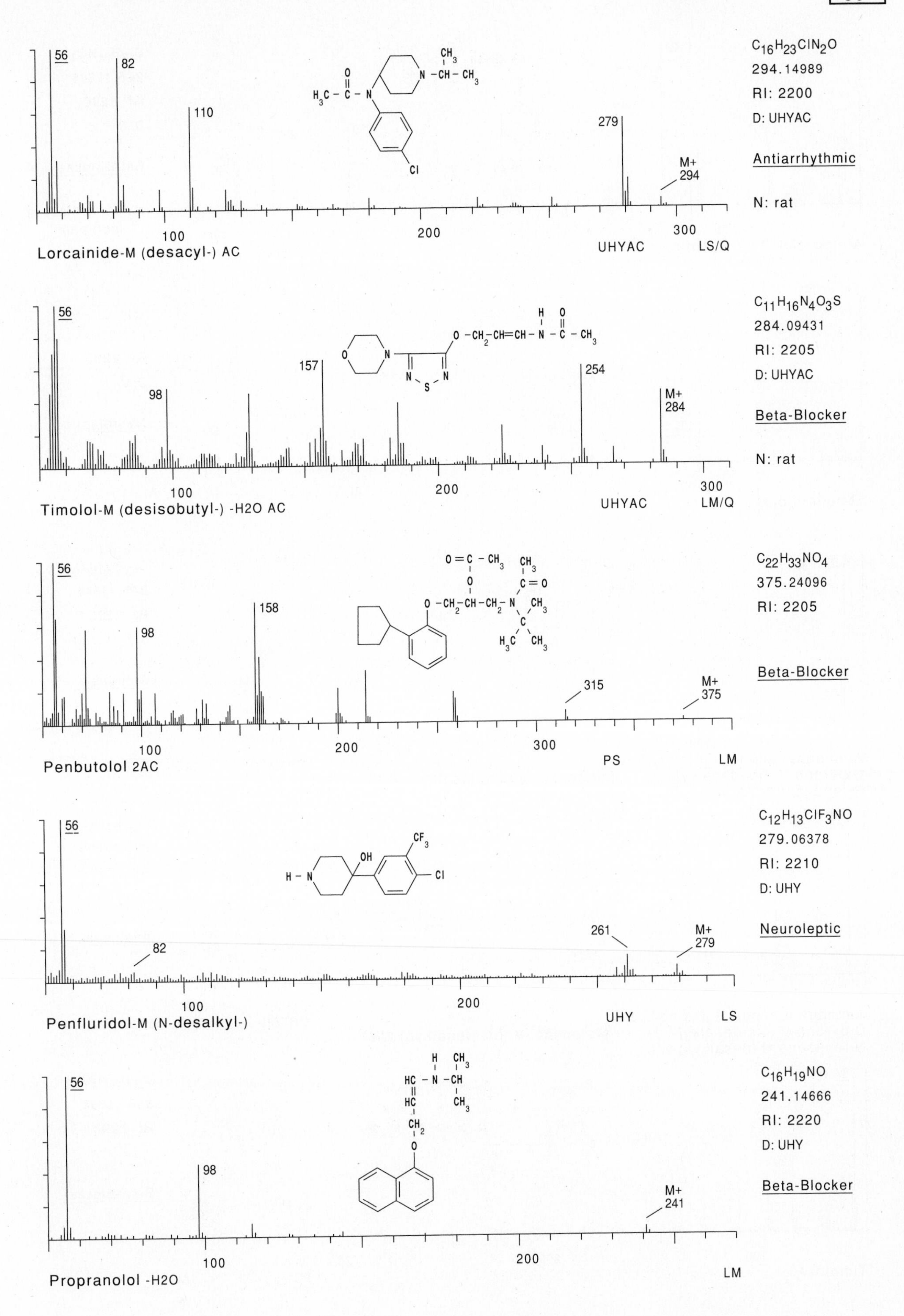

$C_{16}H_{23}ClN_2O$
294.14989
RI: 2200
D: UHYAC
Antiarrhythmic
N: rat

$C_{11}H_{16}N_4O_3S$
284.09431
RI: 2205
D: UHYAC
Beta-Blocker
N: rat

$C_{22}H_{33}NO_4$
375.24096
RI: 2205
Beta-Blocker

$C_{12}H_{13}ClF_3NO$
279.06378
RI: 2210
D: UHY
Neuroleptic

$C_{16}H_{19}NO$
241.14666
RI: 2220
D: UHY
Beta-Blocker

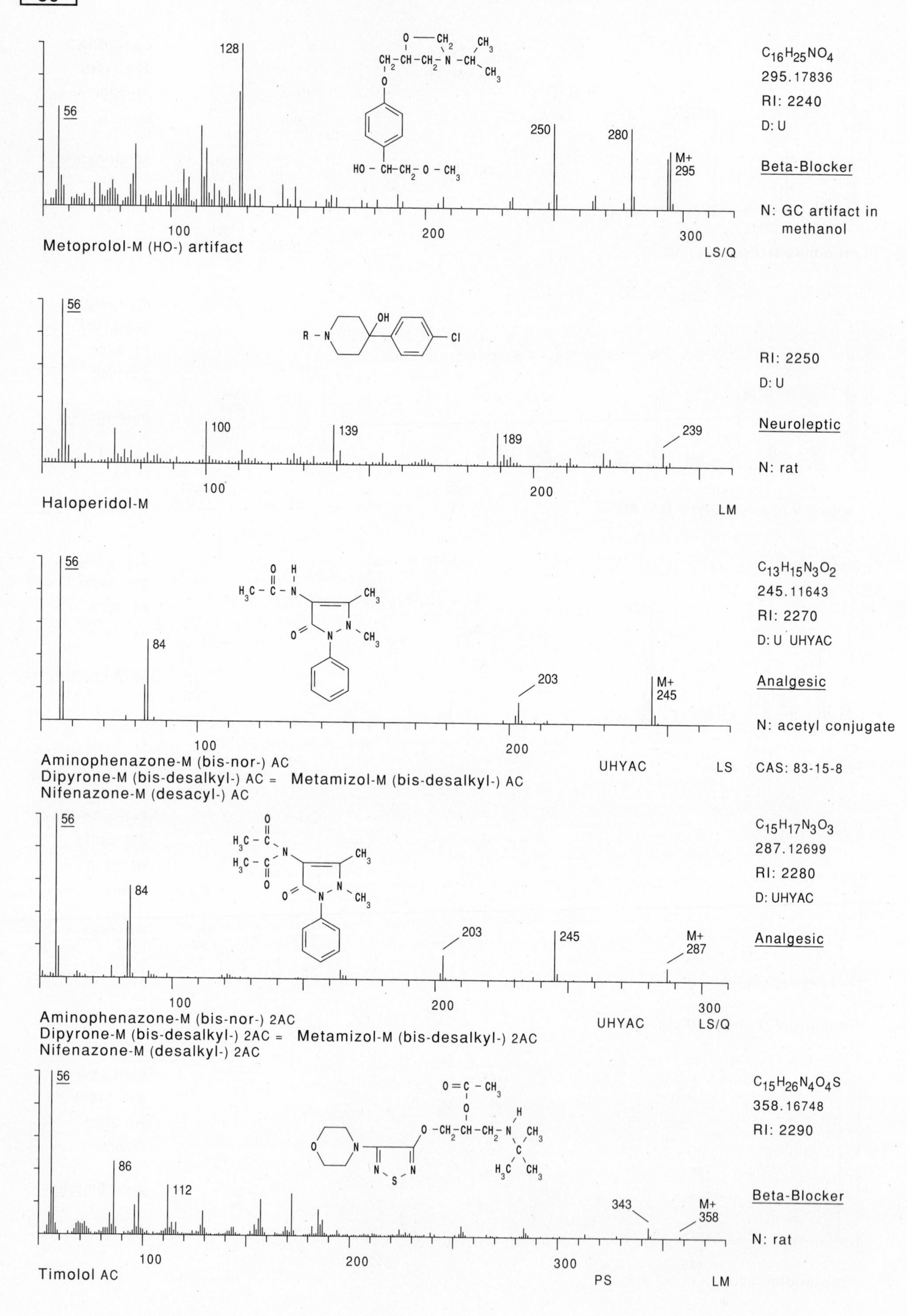
128
56
250
280
M+
295
100
200
300
LS/Q
Metoprolol-M (HO-) artifact
C16H25NO4
295.17836
RI: 2240
D: U
Beta-Blocker
N: GC artifact in methanol
56
100
139
189
239
100
200
Haloperidol-M
LM
RI: 2250
D: U
Neuroleptic
N: rat
56
84
203
M+
245
100
200
UHYAC
LS
Aminophenazone-M (bis-nor-) AC
Dipyrone-M (bis-desalkyl-) AC = Metamizol-M (bis-desalkyl-) AC
Nifenazone-M (desacyl-) AC
C13H15N3O2
245.11643
RI: 2270
D: U UHYAC
Analgesic
N: acetyl conjugate
CAS: 83-15-8
56
84
203
245
M+
287
100
200
300
UHYAC
LS/Q
Aminophenazone-M (bis-nor-) 2AC
Dipyrone-M (bis-desalkyl-) 2AC = Metamizol-M (bis-desalkyl-) 2AC
Nifenazone-M (desalkyl-) 2AC
C15H17N3O3
287.12699
RI: 2280
D: UHYAC
Analgesic
56
86
112
343
M+
358
100
200
300
PS
LM
Timolol AC
C15H26N4O4S
358.16748
RI: 2290
Beta-Blocker
N: rat

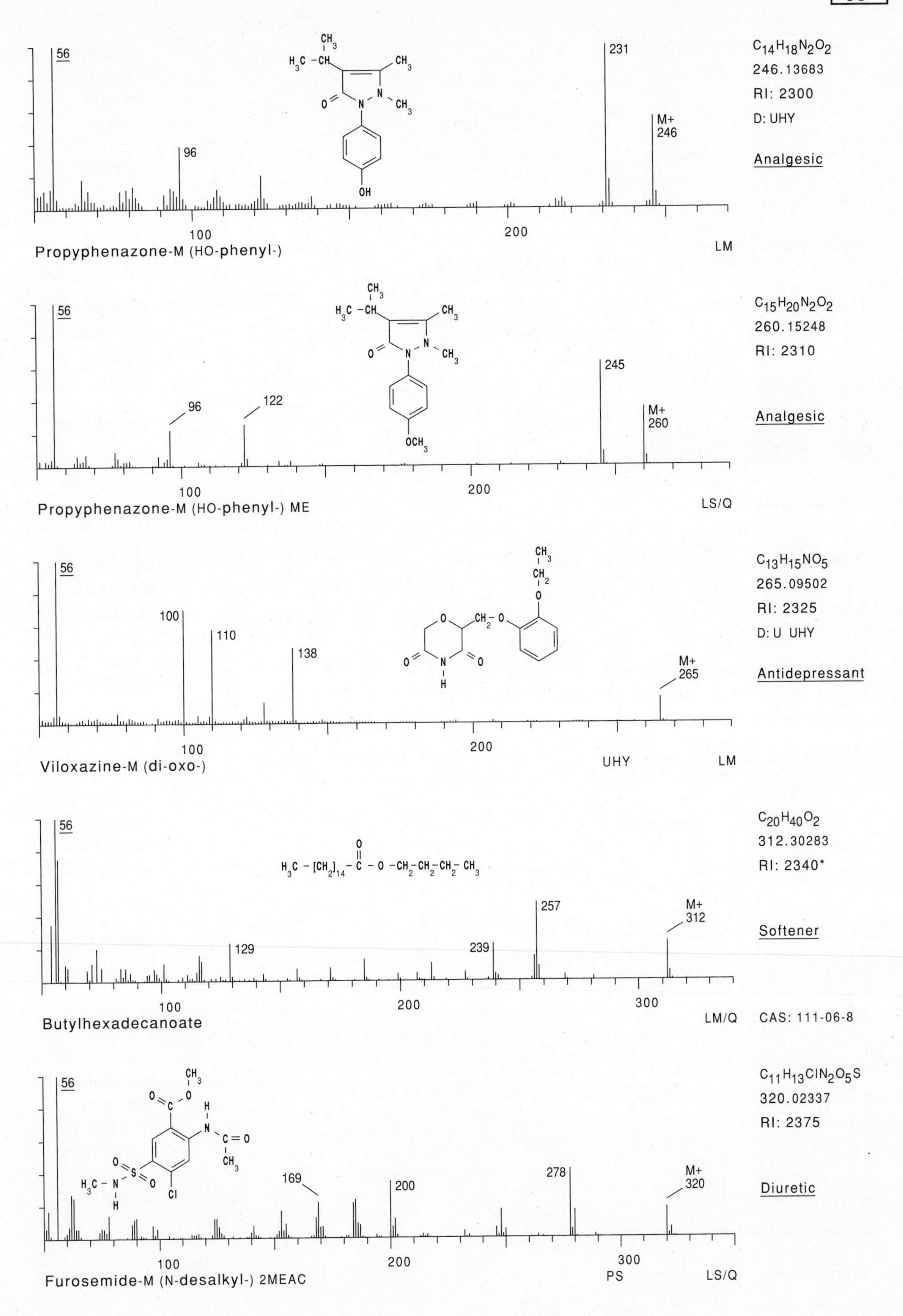
56
96
231
M+
246
100
200
LM
Propyphenazone-M (HO-phenyl-)
$C_{14}H_{18}N_2O_2$
246.13683
RI: 2300
D: UHY
Analgesic
56
96
122
245
M+
260
100
200
LS/Q
Propyphenazone-M (HO-phenyl-) ME
$C_{15}H_{20}N_2O_2$
260.15248
RI: 2310
Analgesic
56
100
110
138
M+
265
100
200
UHY
LM
Viloxazine-M (di-oxo-)
$C_{13}H_{15}NO_5$
265.09502
RI: 2325
D: U UHY
Antidepressant
56
129
239
257
M+
312
100
200
300
LM/Q
Butylhexadecanoate
$C_{20}H_{40}O_2$
312.30283
RI: 2340*
Softener
CAS: 111-06-8
56
169
200
278
M+
320
100
200
300
PS
LS/Q
Furosemide-M (N-desalkyl-) 2MEAC
$C_{11}H_{13}ClN_2O_5S$
320.02337
RI: 2375
Diuretic

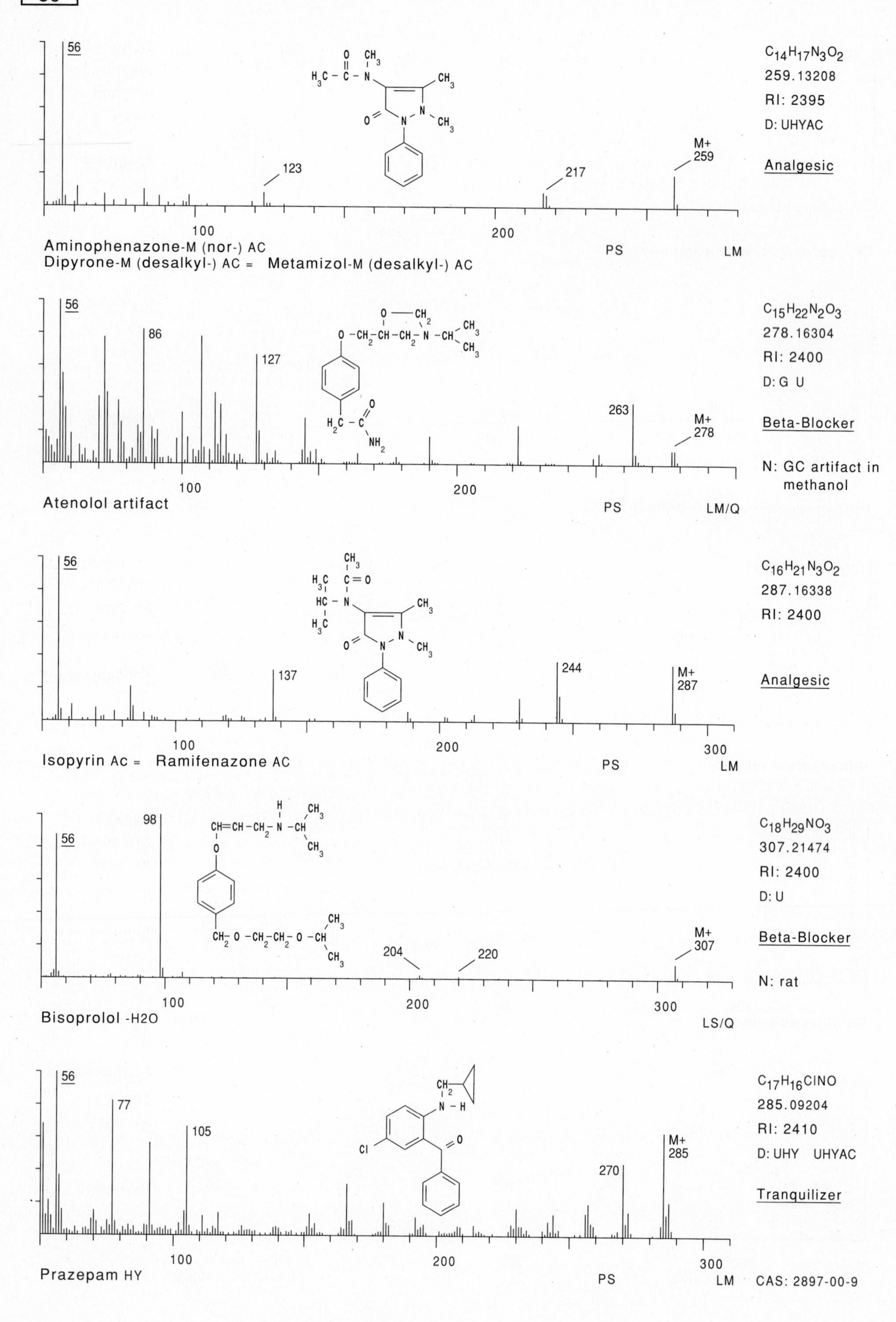

56
123
217
M+ 259
100
200
PS
LM
C14H17N3O2
259.13208
RI: 2395
D: UHYAC
Analgesic
Aminophenazone-M (nor-) AC
Dipyrone-M (desalkyl-) AC = Metamizol-M (desalkyl-) AC
56
86
127
263
M+ 278
100
200
PS
LM/Q
C15H22N2O3
278.16304
RI: 2400
D: G U
Beta-Blocker
N: GC artifact in methanol
Atenolol artifact
56
137
244
M+ 287
100
200
300
PS
LM
C16H21N3O2
287.16338
RI: 2400
Analgesic
Isopyrin AC = Ramifenazone AC
98
56
204
220
M+ 307
100
200
300
LS/Q
C18H29NO3
307.21474
RI: 2400
D: U
Beta-Blocker
N: rat
Bisoprolol -H2O
56
77
105
270
M+ 285
100
200
300
PS
LM
C17H16ClNO
285.09204
RI: 2410
D: UHY UHYAC
Tranquilizer
Prazepam HY
CAS: 2897-00-9

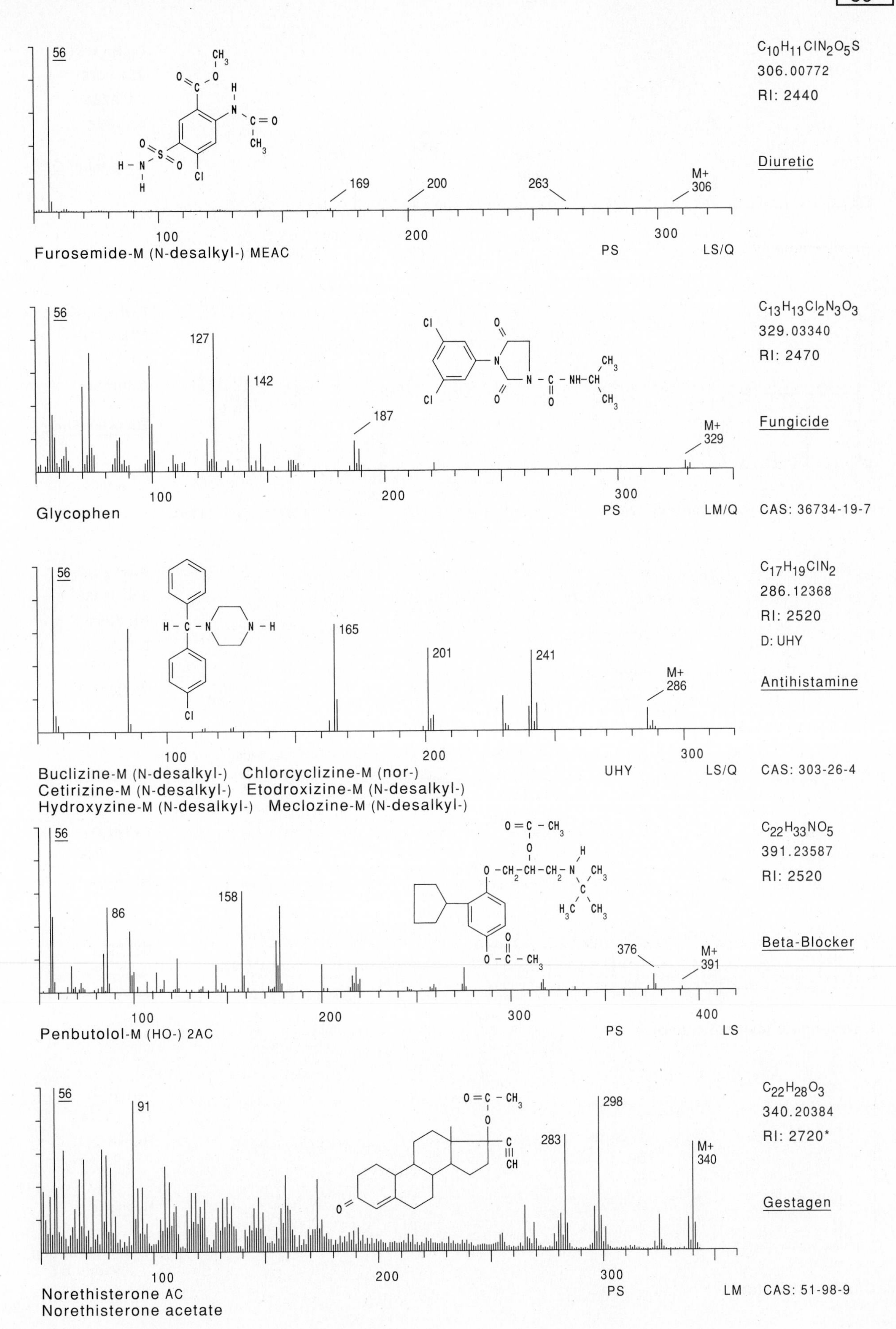

Furosemide-M (N-desalkyl-) MEAC
$C_{10}H_{11}ClN_2O_5S$
306.00772
RI: 2440
Diuretic
56
169
200
263
M+
306
100
200
300
PS
LS/Q
Glycophen
$C_{13}H_{13}Cl_2N_3O_3$
329.03340
RI: 2470
Fungicide
CAS: 36734-19-7
56
127
142
187
M+
329
100
200
300
PS
LM/Q
Buclizine-M (N-desalkyl-) Chlorcyclizine-M (nor-)
Cetirizine-M (N-desalkyl-) Etodroxizine-M (N-desalkyl-)
Hydroxyzine-M (N-desalkyl-) Meclozine-M (N-desalkyl-)
$C_{17}H_{19}ClN_2$
286.12368
RI: 2520
D: UHY
Antihistamine
CAS: 303-26-4
56
165
201
241
M+
286
100
200
300
UHY
LS/Q
Penbutolol-M (HO-) 2AC
$C_{22}H_{33}NO_5$
391.23587
RI: 2520
Beta-Blocker
56
86
158
376
M+
391
100
200
300
400
PS
LS
Norethisterone AC
Norethisterone acetate
$C_{22}H_{28}O_3$
340.20384
RI: 2720*
Gestagen
CAS: 51-98-9
56
91
283
298
M+
340
100
200
300
PS
LM

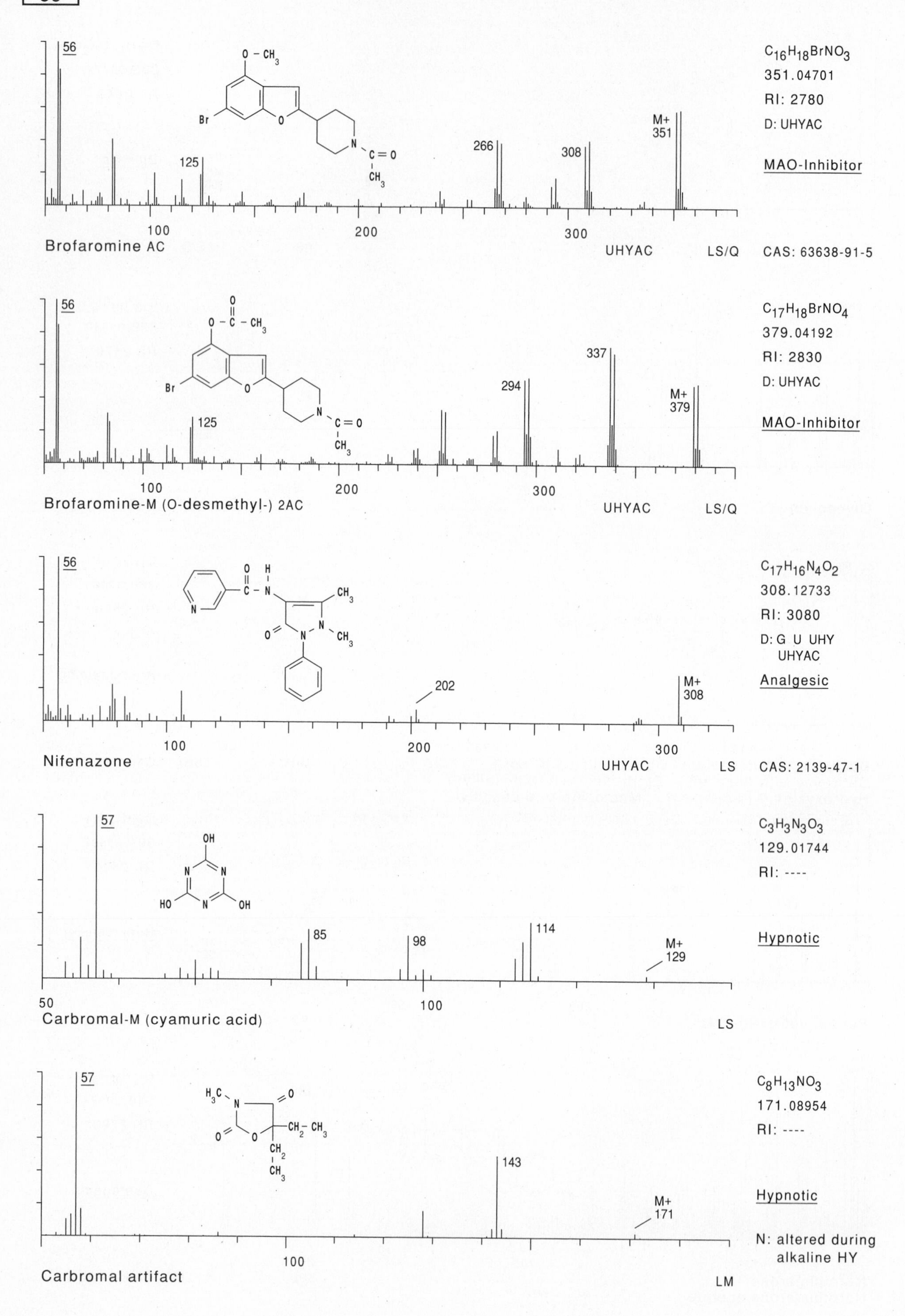

$C_{16}H_{18}BrNO_3$
351.04701
RI: 2780
D: UHYAC
MAO-Inhibitor

CAS: 63638-91-5

$C_{17}H_{18}BrNO_4$
379.04192
RI: 2830
D: UHYAC
MAO-Inhibitor

$C_{17}H_{16}N_4O_2$
308.12733
RI: 3080
D: G U UHY UHYAC
Analgesic

CAS: 2139-47-1

$C_3H_3N_3O_3$
129.01744
RI: ----

Hypnotic

$C_8H_{13}NO_3$
171.08954
RI: ----

Hypnotic

N: altered during alkaline HY

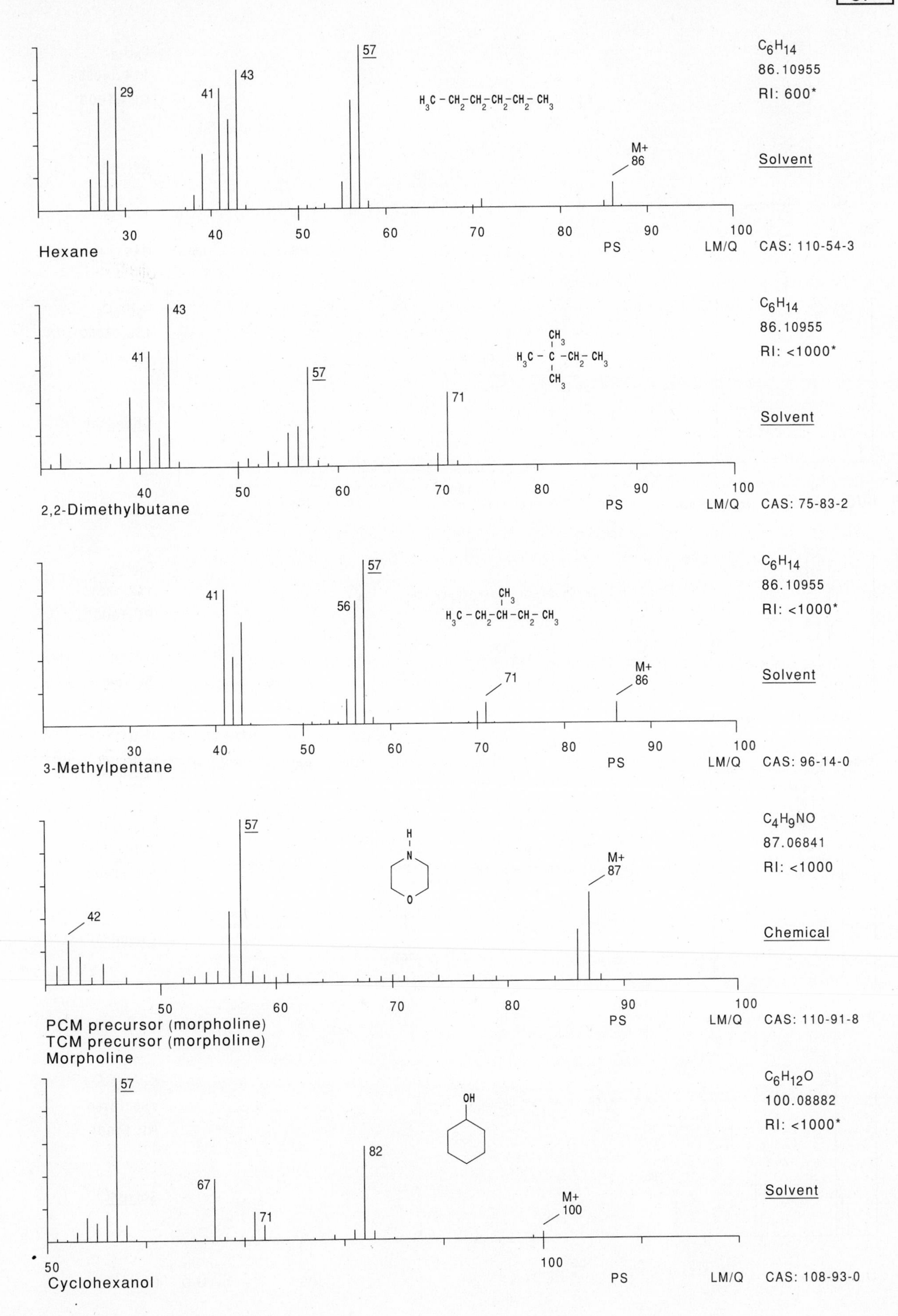

C_6H_{14}
86.10955
RI: 600*

Solvent

CAS: 110-54-3

C_6H_{14}
86.10955
RI: <1000*

Solvent

CAS: 75-83-2

C_6H_{14}
86.10955
RI: <1000*

Solvent

CAS: 96-14-0

C_4H_9NO
87.06841
RI: <1000

Chemical

CAS: 110-91-8

PCM precursor (morpholine)
TCM precursor (morpholine)
Morpholine

$C_6H_{12}O$
100.08882
RI: <1000*

Solvent

CAS: 108-93-0

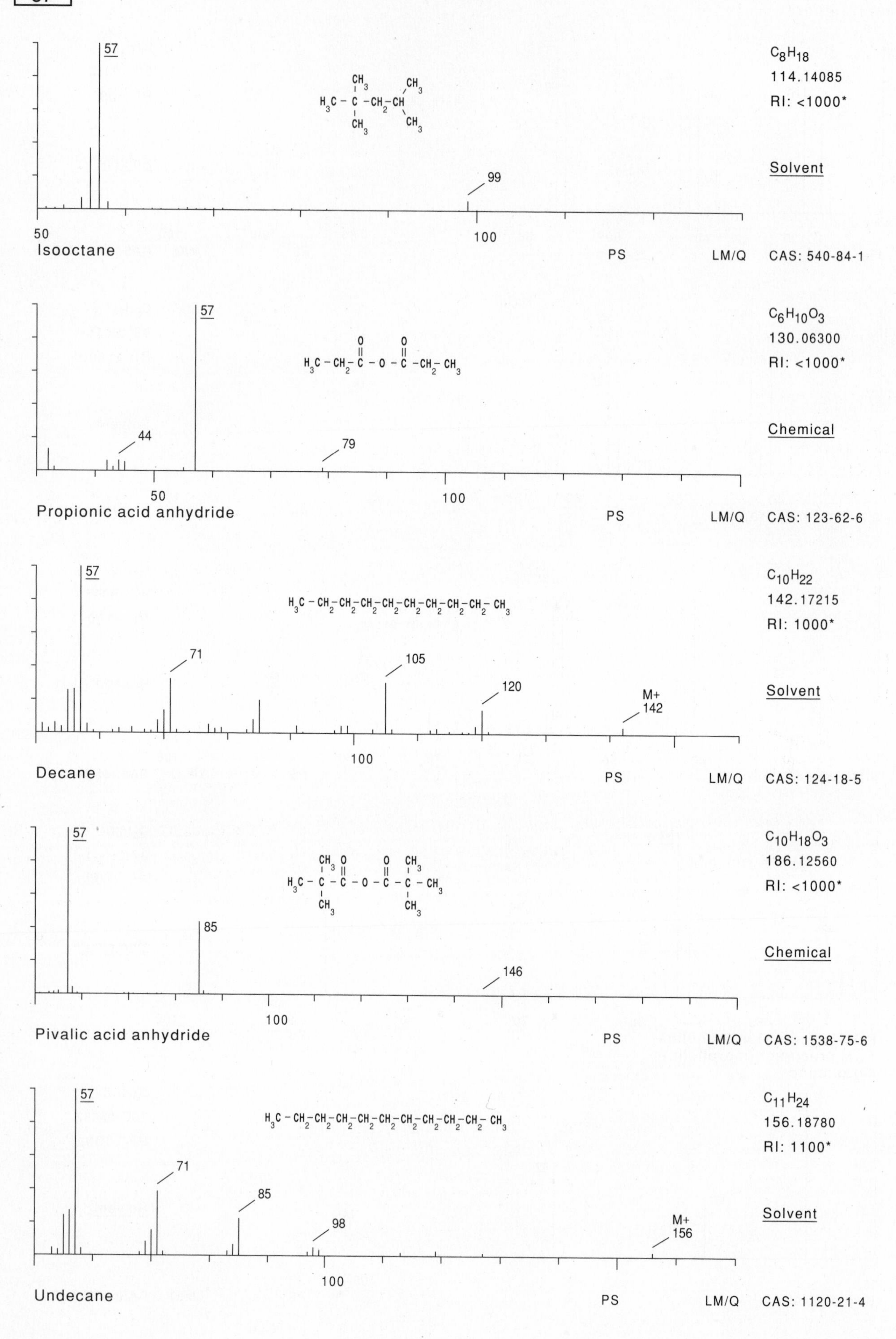
57
57
99
50
100
Isooctane
PS
LM/Q
C8H18
114.14085
RI: <1000*
Solvent
CAS: 540-84-1
57
44
79
50
100
Propionic acid anhydride
PS
LM/Q
C6H10O3
130.06300
RI: <1000*
Chemical
CAS: 123-62-6
57
71
105
120
M+
142
100
Decane
PS
LM/Q
C10H22
142.17215
RI: 1000*
Solvent
CAS: 124-18-5
57
85
146
100
Pivalic acid anhydride
PS
LM/Q
C10H18O3
186.12560
RI: <1000*
Chemical
CAS: 1538-75-6
57
71
85
98
M+
156
100
Undecane
PS
LM/Q
C11H24
156.18780
RI: 1100*
Solvent
CAS: 1120-21-4

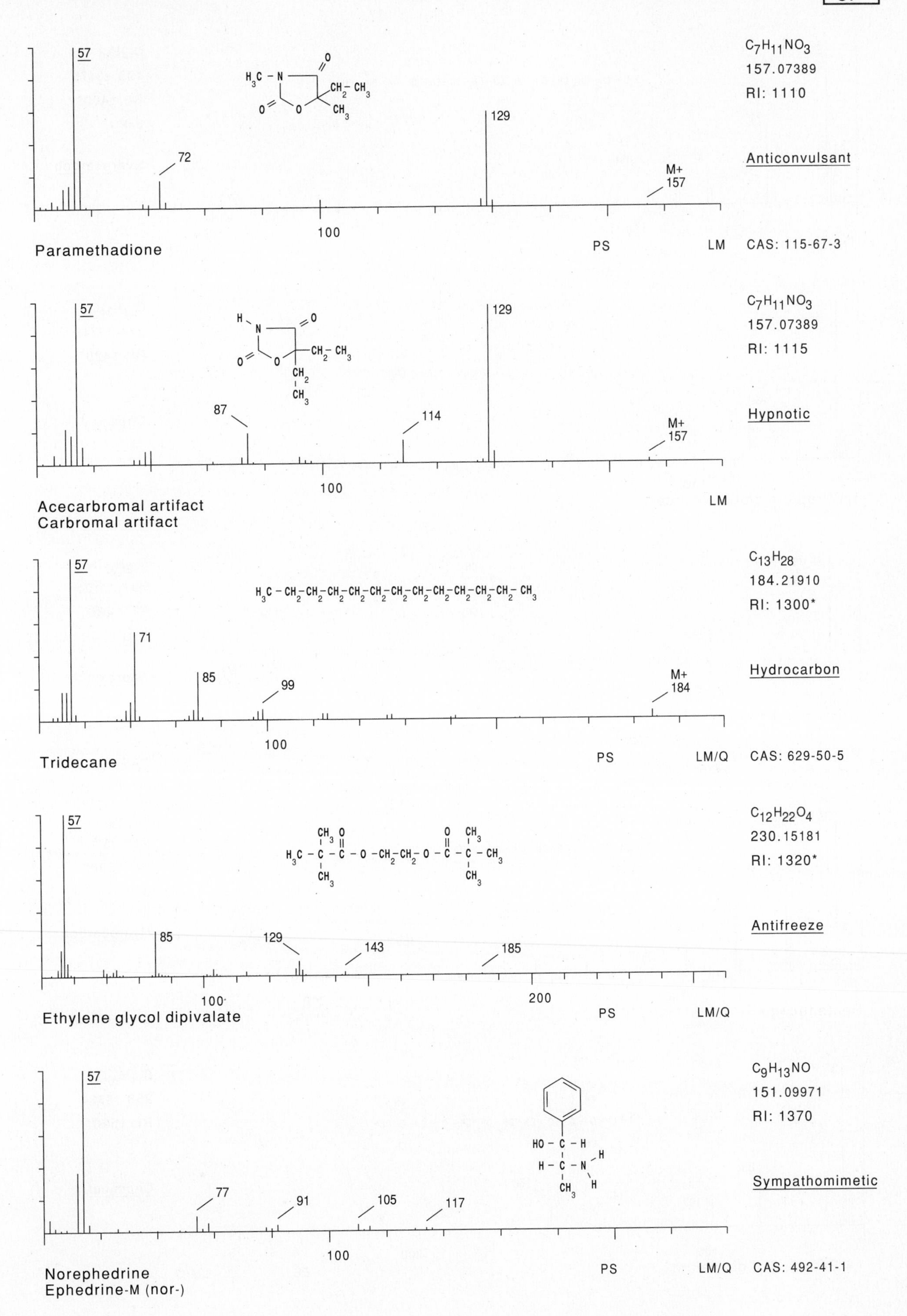

57
72
129
M+
157
100
Paramethadione
PS
LM
C7H11NO3
157.07389
RI: 1110
Anticonvulsant
CAS: 115-67-3
57
87
114
129
M+
157
100
Acecarbromal artifact
Carbromal artifact
LM
C7H11NO3
157.07389
RI: 1115
Hypnotic
57
71
85
99
M+
184
100
Tridecane
PS
LM/Q
C13H28
184.21910
RI: 1300*
Hydrocarbon
CAS: 629-50-5
57
85
129
143
185
100
200
Ethylene glycol dipivalate
PS
LM/Q
C12H22O4
230.15181
RI: 1320*
Antifreeze
57
77
91
105
117
100
Norephedrine
Ephedrine-M (nor-)
PS
LM/Q
C9H13NO
151.09971
RI: 1370
Sympathomimetic
CAS: 492-41-1

57

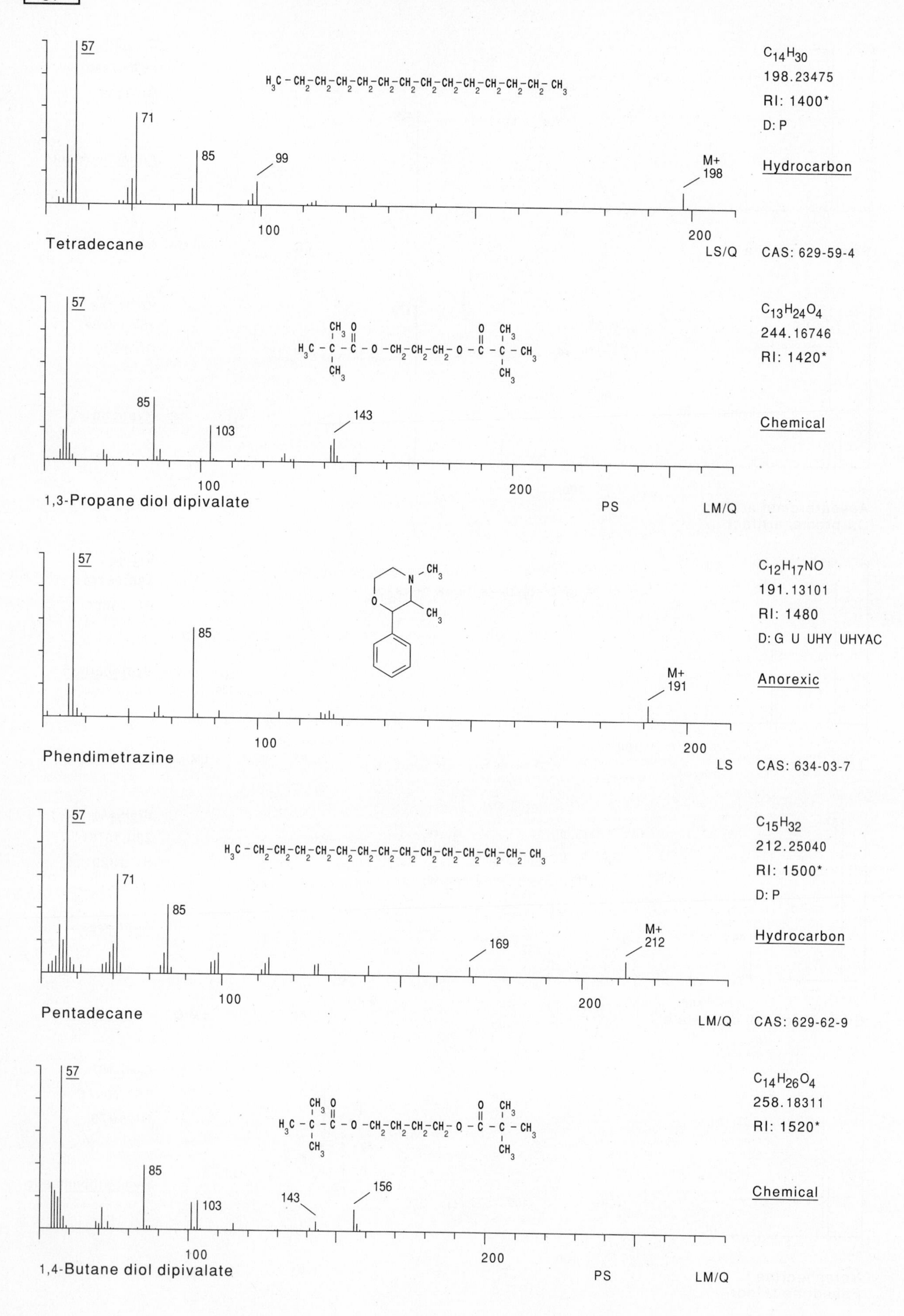
57
71
85
99
M+
198
100
200
Tetradecane
LS/Q
$C_{14}H_{30}$
198.23475
RI: 1400*
D: P
Hydrocarbon
CAS: 629-59-4
57
85
103
143
100
200
1,3-Propane diol dipivalate
PS
LM/Q
$C_{13}H_{24}O_4$
244.16746
RI: 1420*
Chemical
57
85
M+
191
100
200
Phendimetrazine
LS
$C_{12}H_{17}NO$
191.13101
RI: 1480
D: G U UHY UHYAC
Anorexic
CAS: 634-03-7
57
71
85
169
M+
212
100
200
Pentadecane
LM/Q
$C_{15}H_{32}$
212.25040
RI: 1500*
D: P
Hydrocarbon
CAS: 629-62-9
57
85
103
143
156
100
200
1,4-Butane diol dipivalate
PS
LM/Q
$C_{14}H_{26}O_4$
258.18311
RI: 1520*
Chemical

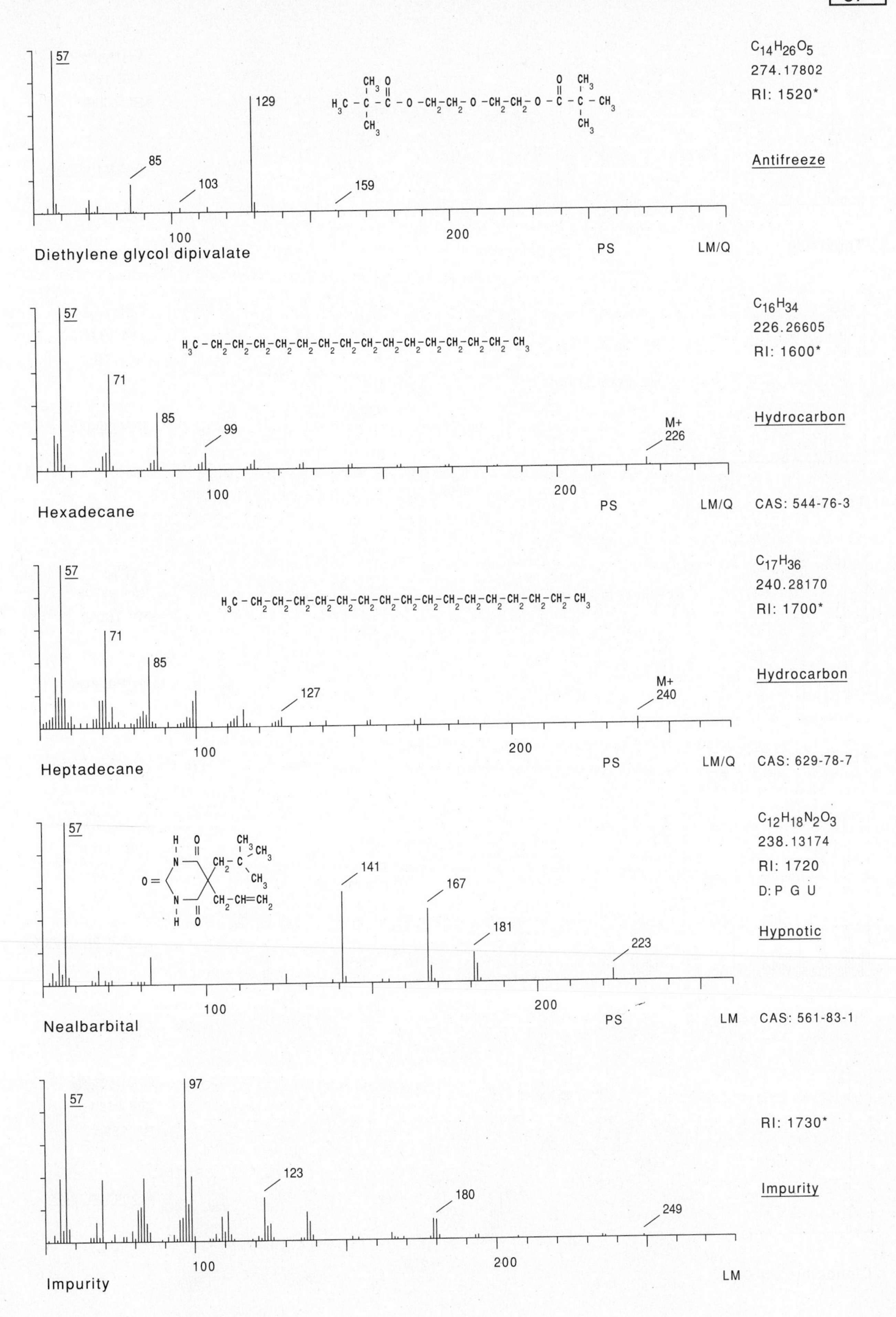
Diethylene glycol dipivalate
$C_{14}H_{26}O_5$
274.17802
RI: 1520*
Antifreeze
57
85
103
129
159
100
200
PS
LM/Q
Hexadecane
$C_{16}H_{34}$
226.26605
RI: 1600*
Hydrocarbon
CAS: 544-76-3
57
71
85
99
M+
226
100
200
PS
LM/Q
Heptadecane
$C_{17}H_{36}$
240.28170
RI: 1700*
Hydrocarbon
CAS: 629-78-7
57
71
85
127
M+
240
100
200
PS
LM/Q
Nealbarbital
$C_{12}H_{18}N_2O_3$
238.13174
RI: 1720
D: P G U
Hypnotic
CAS: 561-83-1
57
141
167
181
223
100
200
PS
LM
Impurity
RI: 1730*
Impurity
57
97
123
180
249
100
200
LM

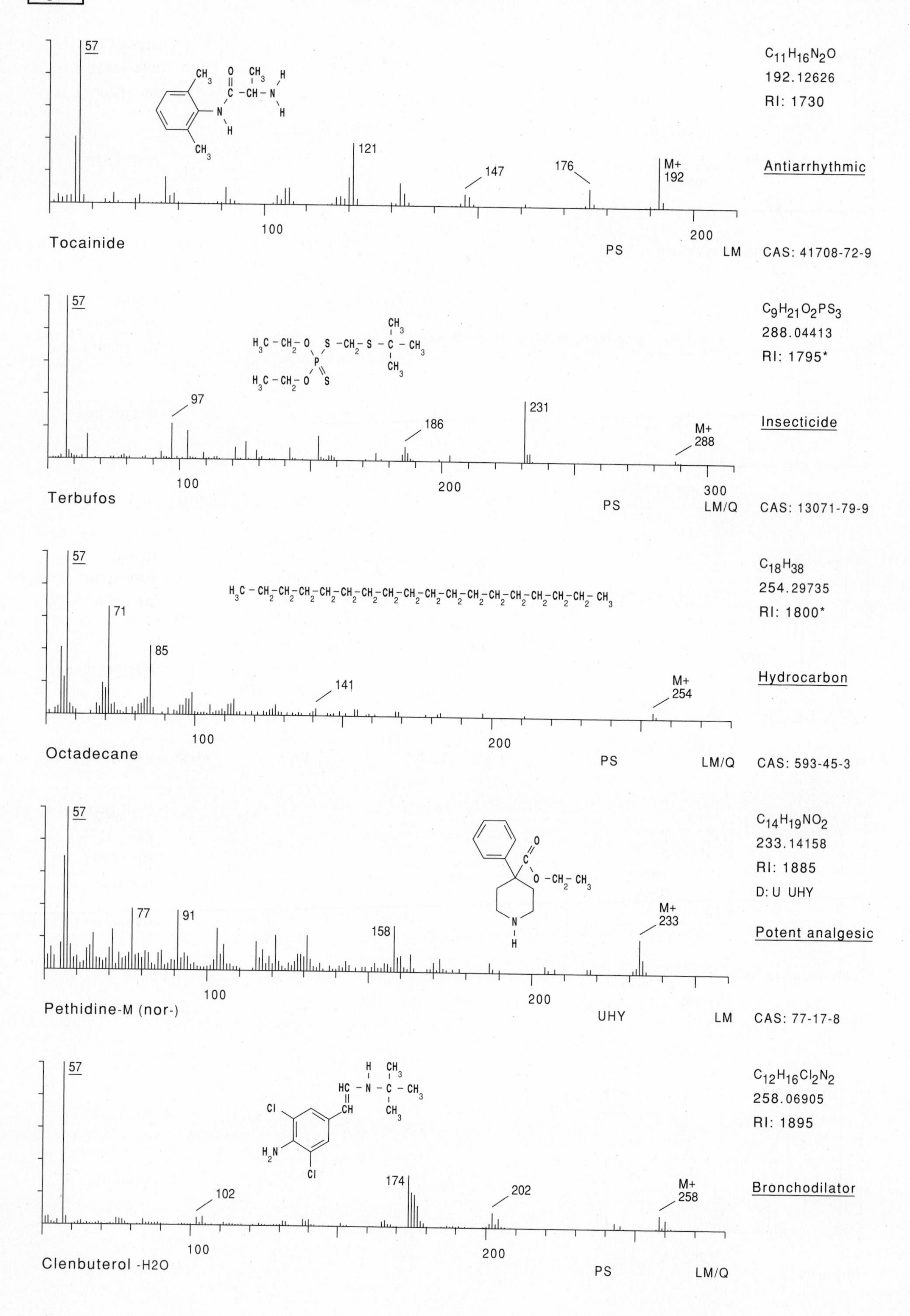
57
121
147
176
M+
192
100
200
Tocainide
PS
LM
C11H16N2O
192.12626
RI: 1730
Antiarrhythmic
CAS: 41708-72-9
57
97
186
231
M+
288
100
200
300
Terbufos
PS
LM/Q
C9H21O2PS3
288.04413
RI: 1795*
Insecticide
CAS: 13071-79-9
57
71
85
141
M+
254
100
200
Octadecane
PS
LM/Q
C18H38
254.29735
RI: 1800*
Hydrocarbon
CAS: 593-45-3
57
77
91
158
M+
233
100
200
Pethidine-M (nor-)
UHY
LM
C14H19NO2
233.14158
RI: 1885
D: U UHY
Potent analgesic
CAS: 77-17-8
57
102
174
202
M+
258
100
200
Clenbuterol -H2O
PS
LM/Q
C12H16Cl2N2
258.06905
RI: 1895
Bronchodilator

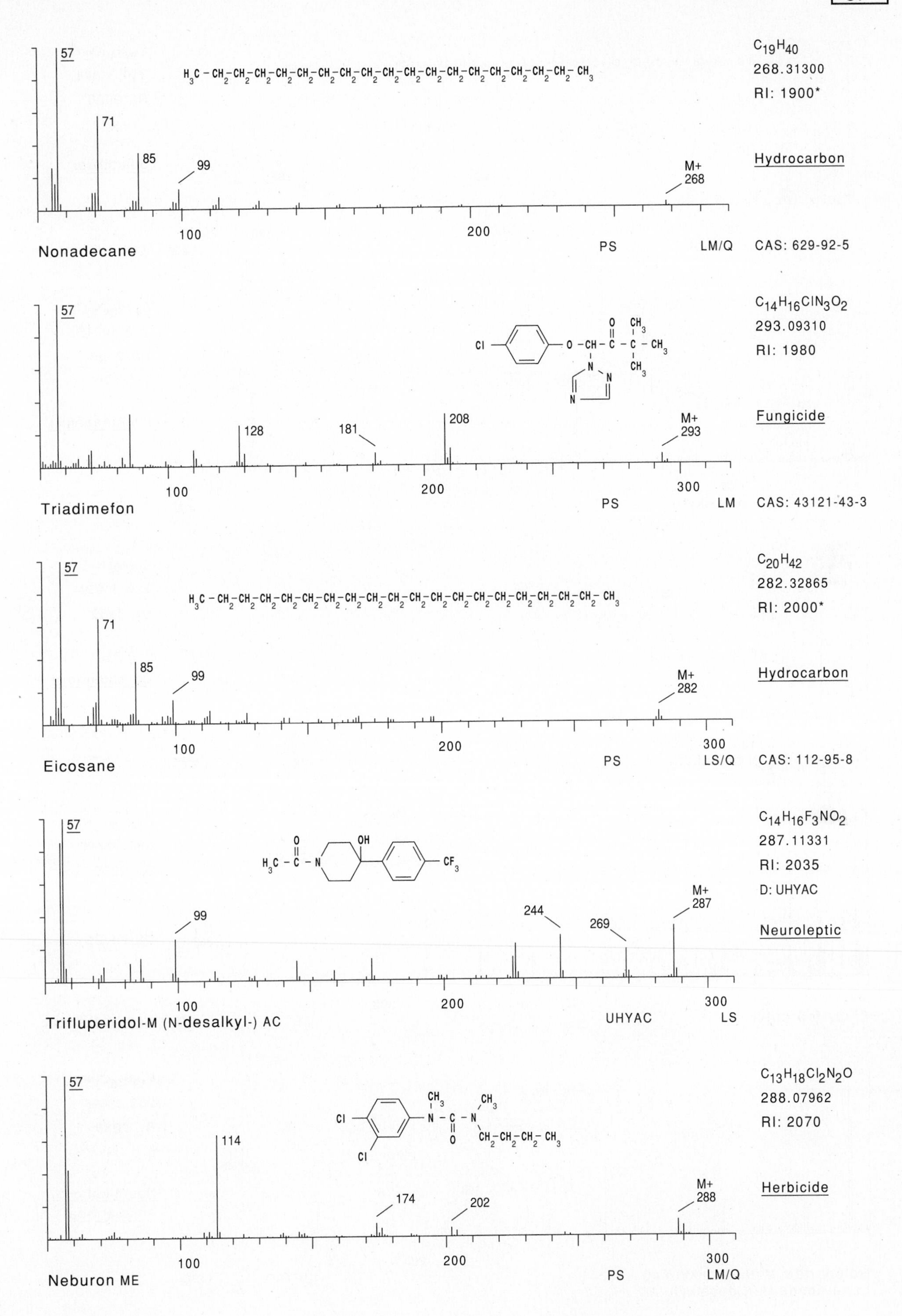

$C_{19}H_{40}$
268.31300
RI: 1900*

Hydrocarbon

CAS: 629-92-5

$C_{14}H_{16}ClN_3O_2$
293.09310
RI: 1980

Fungicide

CAS: 43121-43-3

$C_{20}H_{42}$
282.32865
RI: 2000*

Hydrocarbon

CAS: 112-95-8

$C_{14}H_{16}F_3NO_2$
287.11331
RI: 2035
D: UHYAC

Neuroleptic

$C_{13}H_{18}Cl_2N_2O$
288.07962
RI: 2070

Herbicide

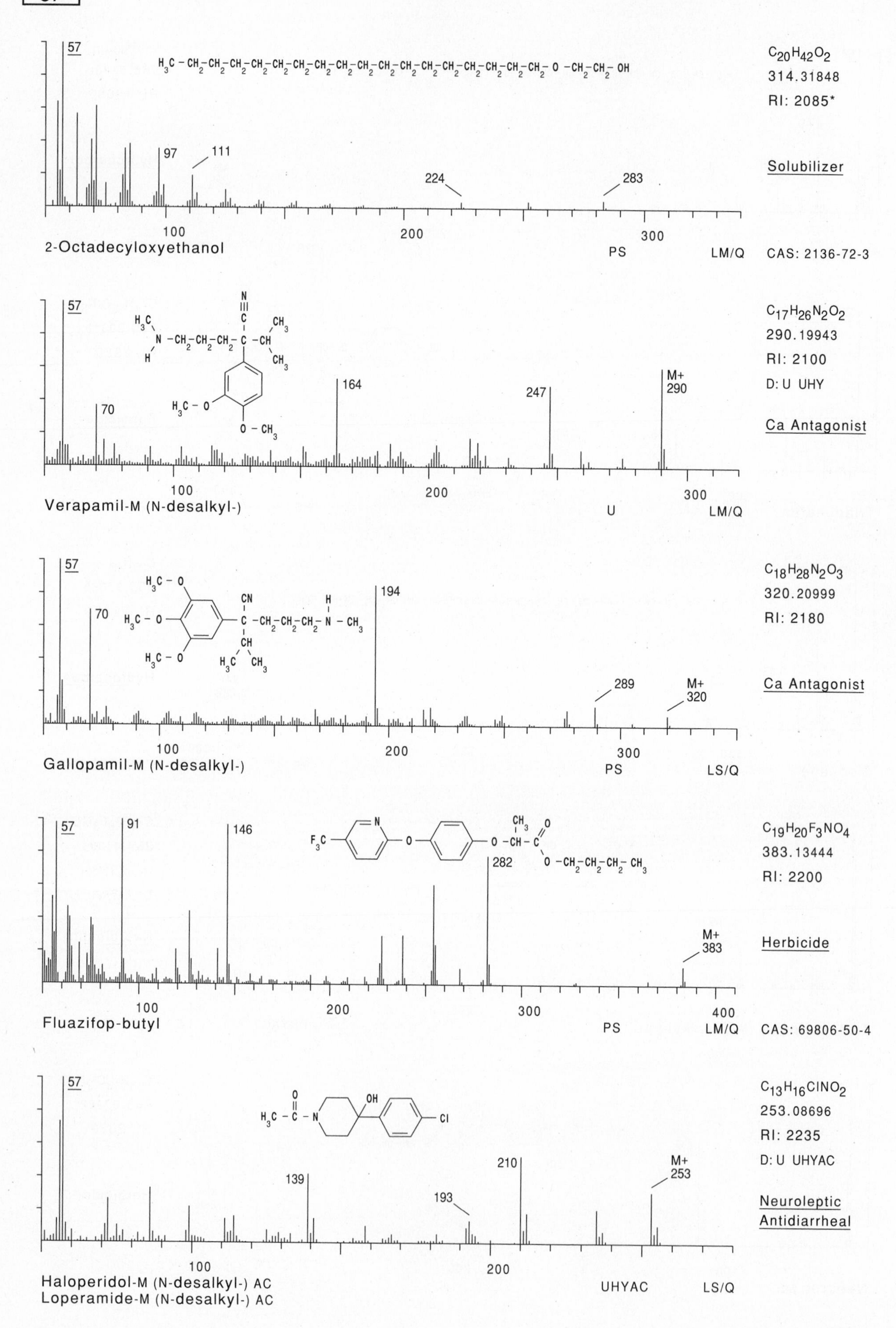

2-Octadecyloxyethanol

$C_{20}H_{42}O_2$
314.31848
RI: 2085*
Solubilizer
CAS: 2136-72-3

Verapamil-M (N-desalkyl-)

$C_{17}H_{26}N_2O_2$
290.19943
RI: 2100
D: U UHY
Ca Antagonist

Gallopamil-M (N-desalkyl-)

$C_{18}H_{28}N_2O_3$
320.20999
RI: 2180
Ca Antagonist

Fluazifop-butyl

$C_{19}H_{20}F_3NO_4$
383.13444
RI: 2200
Herbicide
CAS: 69806-50-4

Haloperidol-M (N-desalkyl-) AC
Loperamide-M (N-desalkyl-) AC

$C_{13}H_{16}ClNO_2$
253.08696
RI: 2235
D: U UHYAC
Neuroleptic
Antidiarrheal

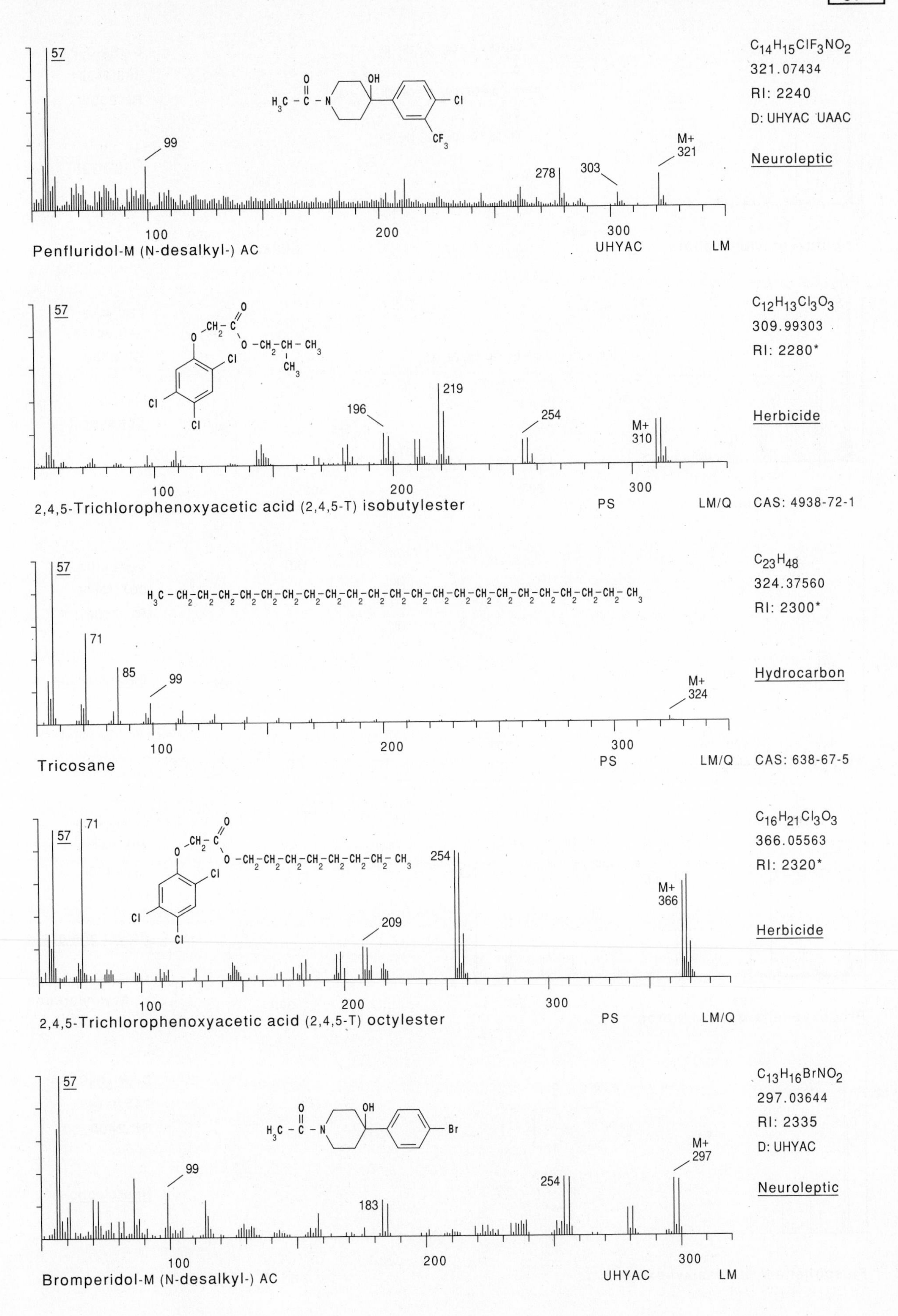

57
99
278
303
M+ 321
100
200
300
Penfluridol-M (N-desalkyl-) AC
UHYAC
LM
$C_{14}H_{15}ClF_3NO_2$
321.07434
RI: 2240
D: UHYAC UAAC
Neuroleptic
57
196
219
254
M+ 310
100
200
300
2,4,5-Trichlorophenoxyacetic acid (2,4,5-T) isobutylester
PS
LM/Q
$C_{12}H_{13}Cl_3O_3$
309.99303
RI: 2280*
Herbicide
CAS: 4938-72-1
57
71
85
99
M+ 324
100
200
300
Tricosane
PS
LM/Q
$C_{23}H_{48}$
324.37560
RI: 2300*
Hydrocarbon
CAS: 638-67-5
57
71
254
209
M+ 366
100
200
300
2,4,5-Trichlorophenoxyacetic acid (2,4,5-T) octylester
PS
LM/Q
$C_{16}H_{21}Cl_3O_3$
366.05563
RI: 2320*
Herbicide
57
99
183
254
M+ 297
100
200
300
Bromperidol-M (N-desalkyl-) AC
UHYAC
LM
$C_{13}H_{16}BrNO_2$
297.03644
RI: 2335
D: UHYAC
Neuroleptic

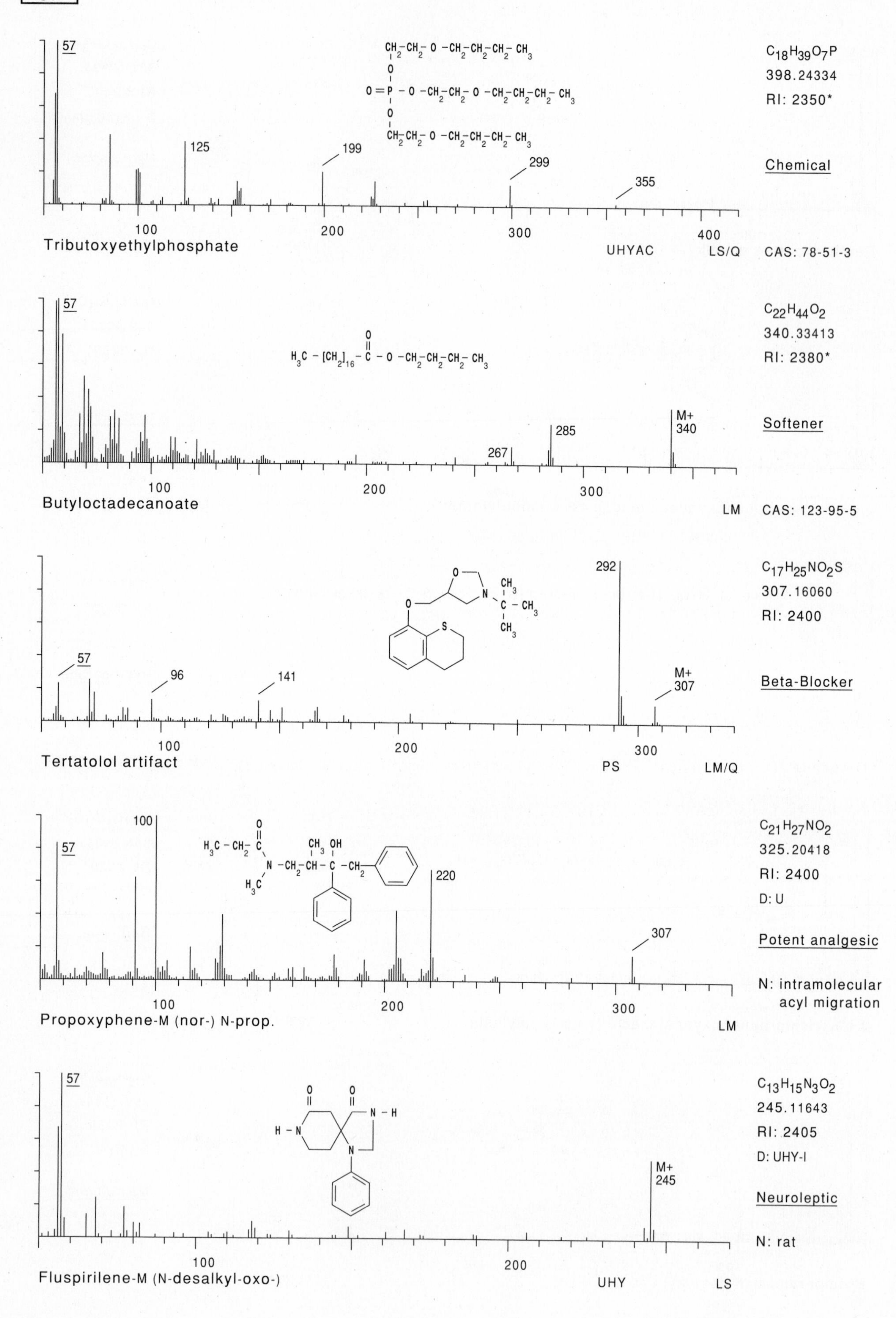
57
125
199
299
355
100
200
300
400
Tributoxyethylphosphate
UHYAC
LS/Q
$C_{18}H_{39}O_7P$
398.24334
RI: 2350*
Chemical
CAS: 78-51-3
57
285
267
M+
340
100
200
300
Butyloctadecanoate
LM
$C_{22}H_{44}O_2$
340.33413
RI: 2380*
Softener
CAS: 123-95-5
292
57
96
141
M+
307
100
200
300
Tertatolol artifact
PS
LM/Q
$C_{17}H_{25}NO_2S$
307.16060
RI: 2400
Beta-Blocker
100
57
220
307
100
200
300
Propoxyphene-M (nor-) N-prop.
LM
$C_{21}H_{27}NO_2$
325.20418
RI: 2400
D: U
Potent analgesic
N: intramolecular acyl migration
57
M+
245
100
200
Fluspirilene-M (N-desalkyl-oxo-)
UHY
LS
$C_{13}H_{15}N_3O_2$
245.11643
RI: 2405
D: UHY-I
Neuroleptic
N: rat

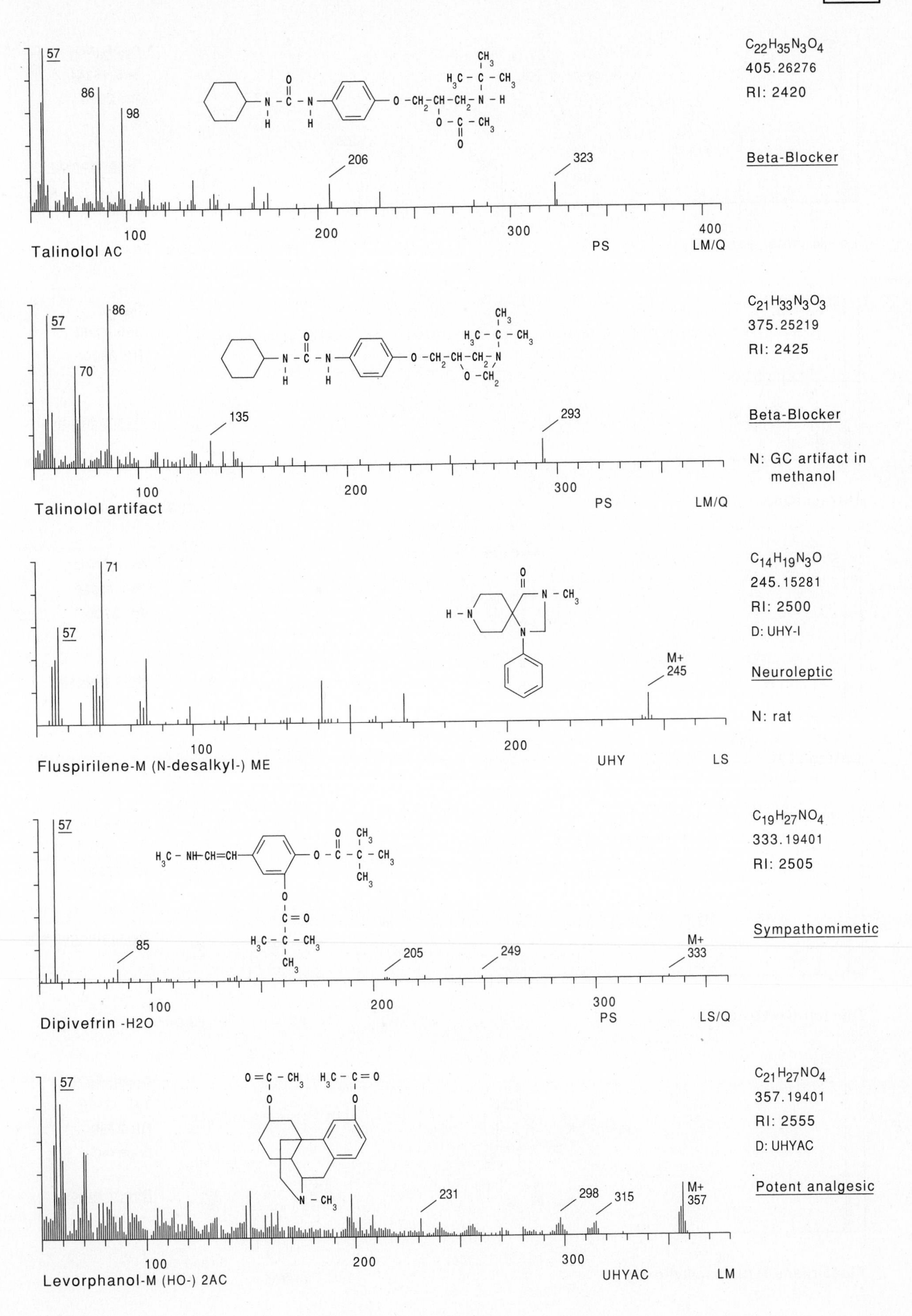
57
86
98
206
323
100
200
300
400
PS
LM/Q
Talinolol AC
C22H35N3O4
405.26276
RI: 2420
Beta-Blocker
57
86
70
135
293
100
200
300
PS
LM/Q
Talinolol artifact
C21H33N3O3
375.25219
RI: 2425
Beta-Blocker
N: GC artifact in methanol
71
57
M+
245
100
200
UHY
LS
Fluspirilene-M (N-desalkyl-) ME
C14H19N3O
245.15281
RI: 2500
D: UHY-I
Neuroleptic
N: rat
57
85
205
249
M+
333
100
200
300
PS
LS/Q
Dipivefrin -H2O
C19H27NO4
333.19401
RI: 2505
Sympathomimetic
57
231
298
315
M+
357
100
200
300
UHYAC
LM
Levorphanol-M (HO-) 2AC
C21H27NO4
357.19401
RI: 2555
D: UHYAC
Potent analgesic

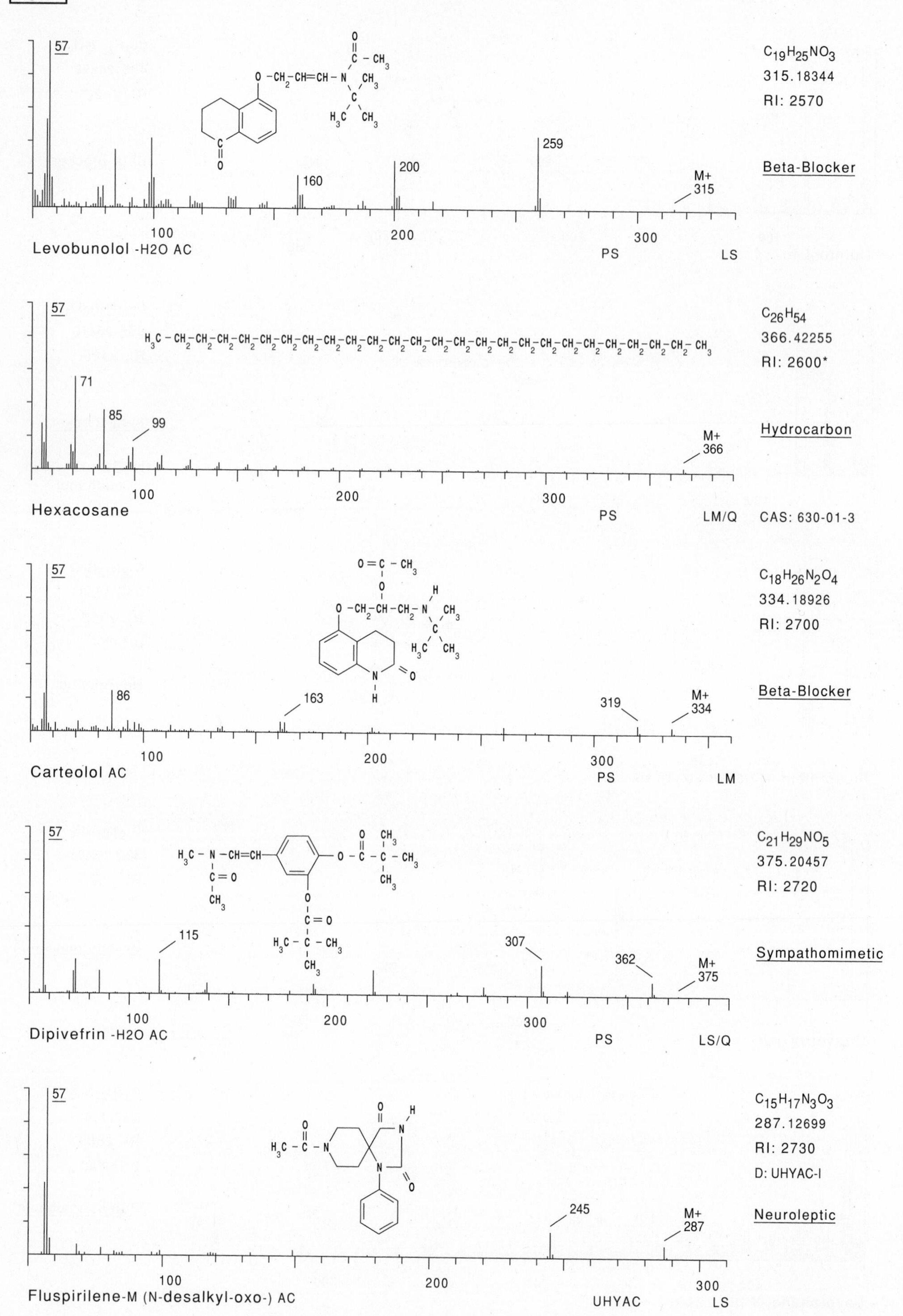

57
160
200
259
M+
315
100
200
300
Levobunolol -H2O AC
PS
LS
C19H25NO3
315.18344
RI: 2570
Beta-Blocker
57
71
85
99
M+
366
100
200
300
Hexacosane
PS
LM/Q
C26H54
366.42255
RI: 2600*
Hydrocarbon
CAS: 630-01-3
57
86
163
319
M+
334
100
200
300
Carteolol AC
PS
LM
C18H26N2O4
334.18926
RI: 2700
Beta-Blocker
57
115
307
362
M+
375
100
200
300
Dipivefrin -H2O AC
PS
LS/Q
C21H29NO5
375.20457
RI: 2720
Sympathomimetic
57
245
M+
287
100
200
300
Fluspirilene-M (N-desalkyl-oxo-) AC
UHYAC
LS
C15H17N3O3
287.12699
RI: 2730
D: UHYAC-I
Neuroleptic

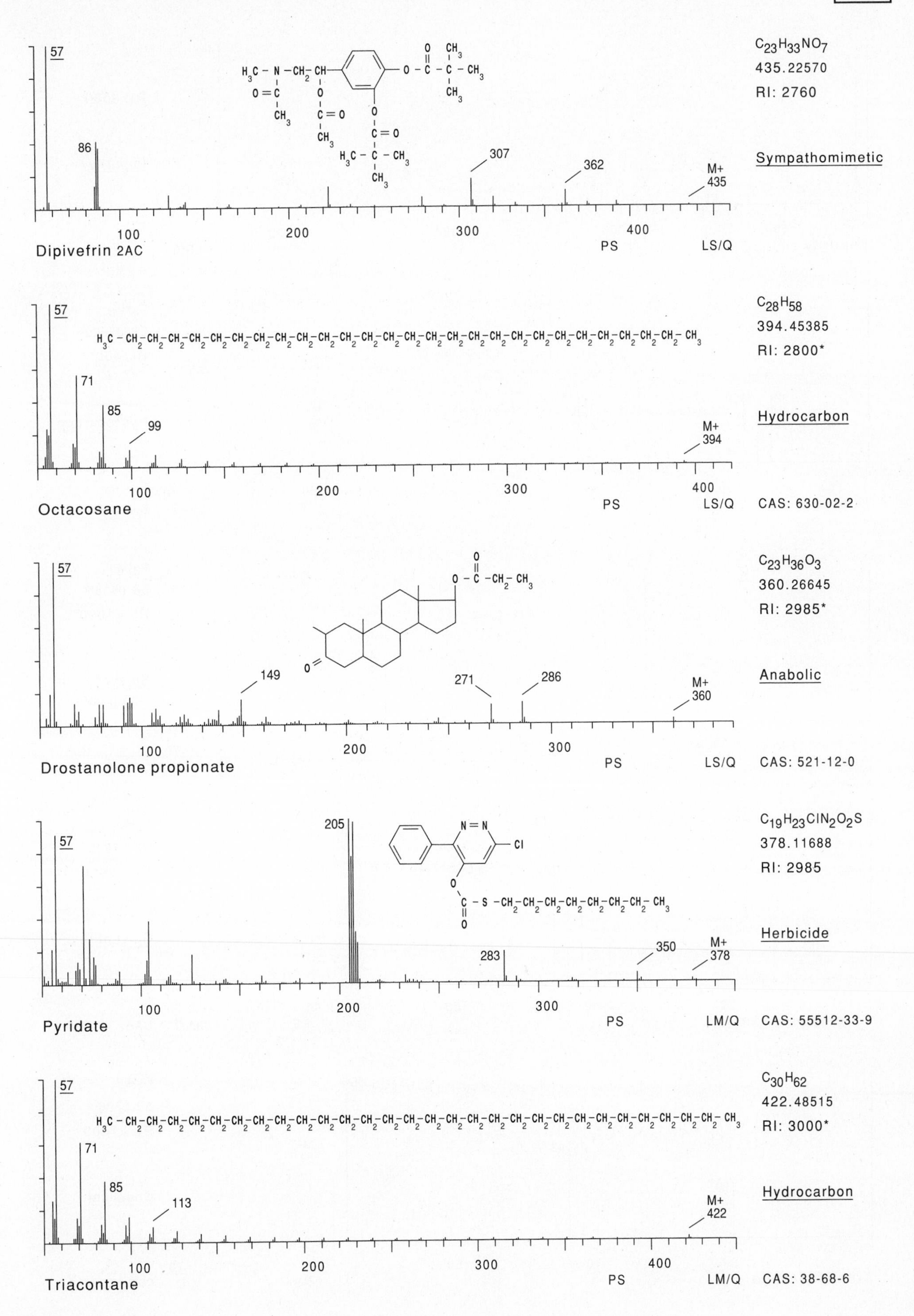
57
86
307
362
M+
435
100
200
300
400
PS
LS/Q
Dipivefrin 2AC
C23H33NO7
435.22570
RI: 2760
Sympathomimetic
57
71
85
99
M+
394
Octacosane
C28H58
394.45385
RI: 2800*
Hydrocarbon
CAS: 630-02-2
57
149
271
286
M+
360
Drostanolone propionate
C23H36O3
360.26645
RI: 2985*
Anabolic
CAS: 521-12-0
57
205
283
350
M+
378
LM/Q
Pyridate
C19H23ClN2O2S
378.11688
RI: 2985
Herbicide
CAS: 55512-33-9
57
71
85
113
M+
422
Triacontane
C30H62
422.48515
RI: 3000*
Hydrocarbon
CAS: 38-68-6

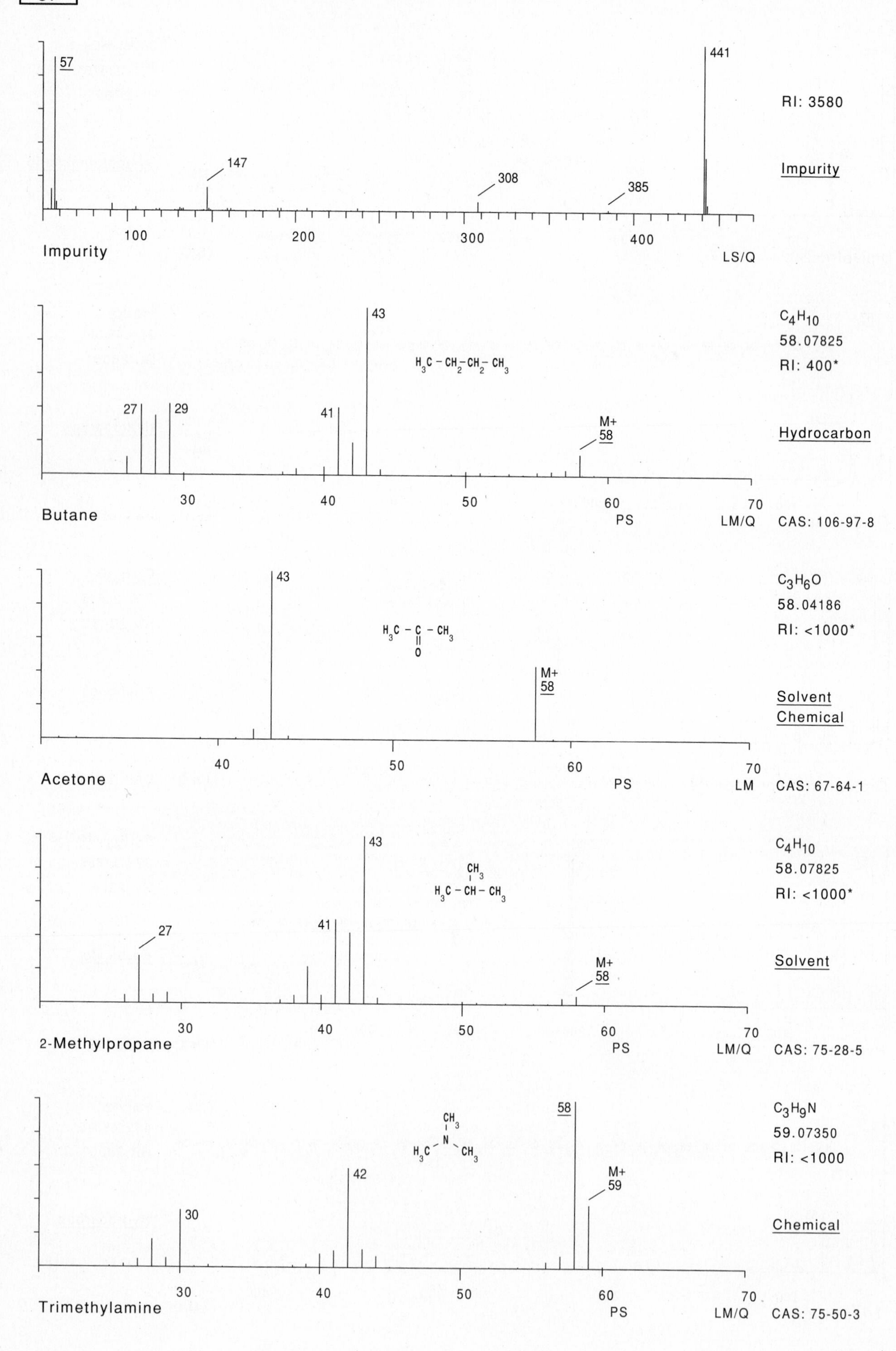
57
147
308
385
441
100
200
300
400
Impurity
LS/Q
RI: 3580
Impurity
43
27
29
41
M+
58
30
40
50
60
70
Butane
PS
LM/Q
C4H10
58.07825
RI: 400*
Hydrocarbon
CAS: 106-97-8
43
M+
58
40
50
60
70
Acetone
PS
LM
C3H6O
58.04186
RI: <1000*
Solvent
Chemical
CAS: 67-64-1
43
27
41
M+
58
30
40
50
60
70
2-Methylpropane
PS
LM/Q
C4H10
58.07825
RI: <1000*
Solvent
CAS: 75-28-5
58
42
30
M+
59
30
40
50
60
70
Trimethylamine
PS
LM/Q
C3H9N
59.07350
RI: <1000
Chemical
CAS: 75-50-3

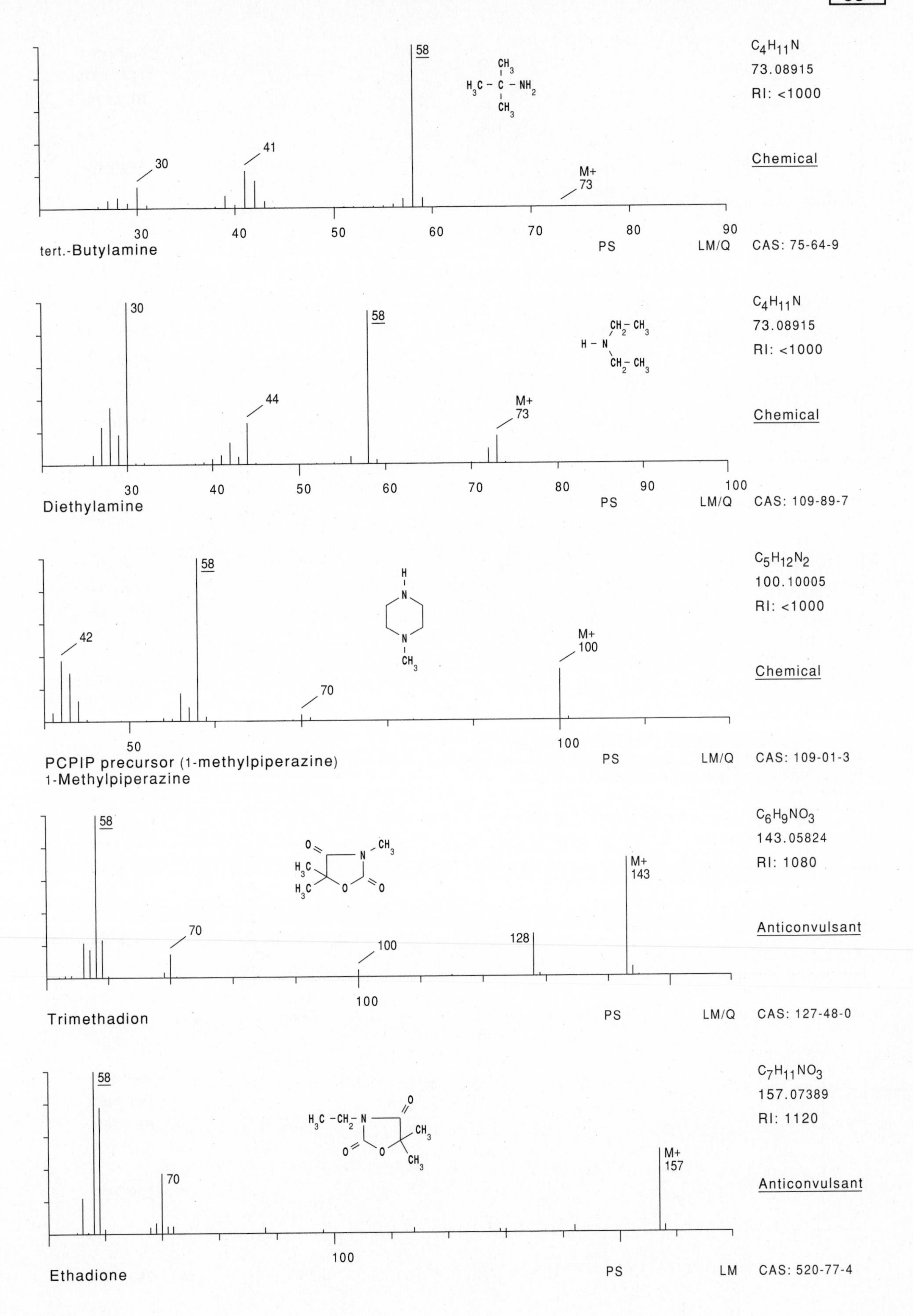

58
CH3
H3C - C - NH2
CH3
41
30
M+
73
30
40
50
60
70
80
90
tert.-Butylamine
PS
LM/Q
C4H11N
73.08915
RI: <1000
Chemical
CAS: 75-64-9
30
58
CH2- CH3
H - N
CH2- CH3
44
M+
73
30
40
50
60
70
80
90
100
Diethylamine
PS
LM/Q
C4H11N
73.08915
RI: <1000
Chemical
CAS: 109-89-7
58
H
N
N
CH3
42
M+
100
70
50
100
PCPIP precursor (1-methylpiperazine)
1-Methylpiperazine
PS
LM/Q
C5H12N2
100.10005
RI: <1000
Chemical
CAS: 109-01-3
58
O
N
CH3
H3C
H3C
O
O
M+
143
70
128
100
100
Trimethadion
PS
LM/Q
C6H9NO3
143.05824
RI: 1080
Anticonvulsant
CAS: 127-48-0
58
H3C -CH2- N
O
CH3
O
O
CH3
M+
157
70
100
Ethadione
PS
LM
C7H11NO3
157.07389
RI: 1120
Anticonvulsant
CAS: 520-77-4

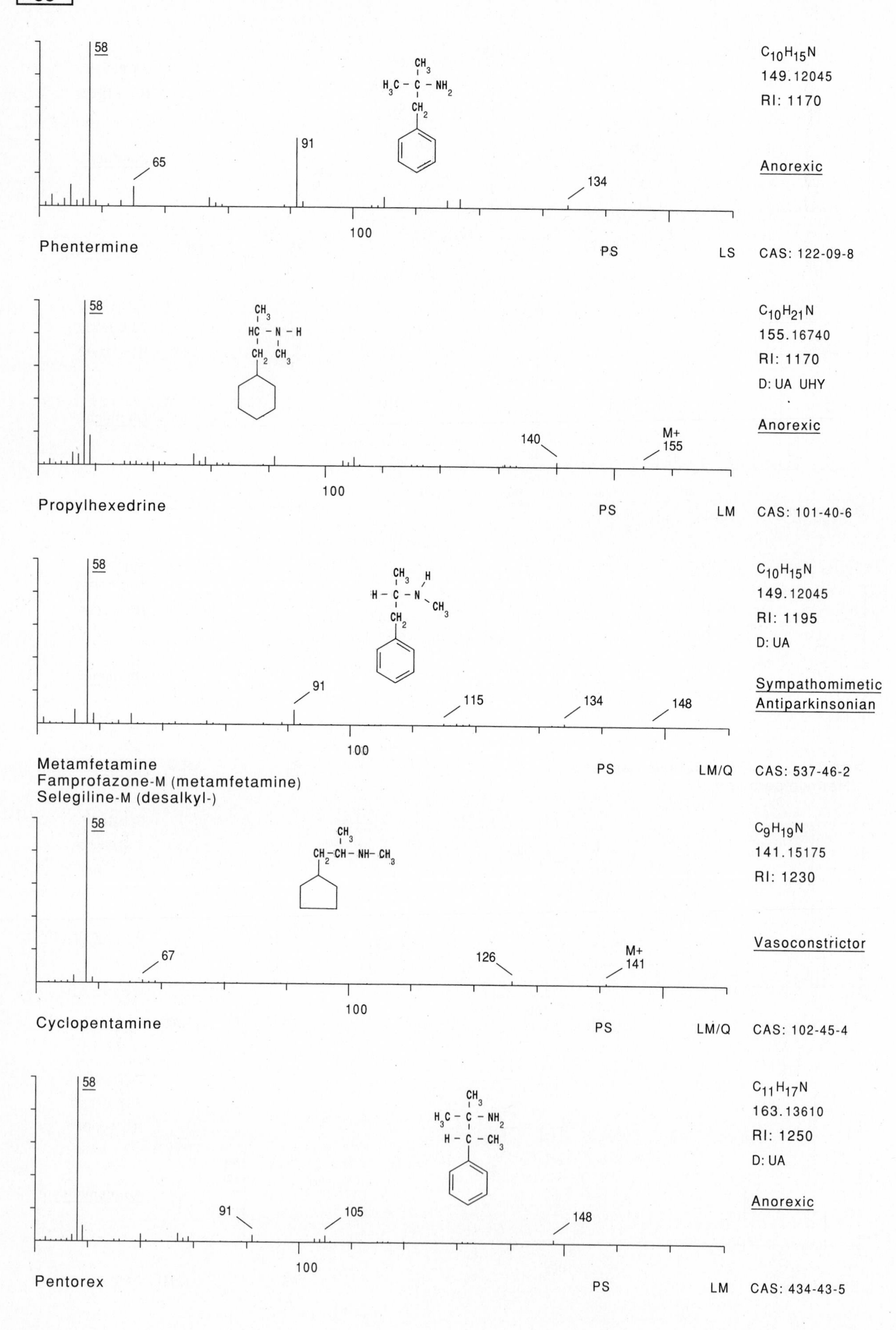

58
91
65
134
100
Phentermine
PS
LS
$C_{10}H_{15}N$
149.12045
RI: 1170
Anorexic
CAS: 122-09-8
58
140
M+
155
100
Propylhexedrine
PS
LM
$C_{10}H_{21}N$
155.16740
RI: 1170
D: UA UHY
Anorexic
CAS: 101-40-6
58
91
115
134
148
100
Metamfetamine
Famprofazone-M (metamfetamine)
Selegiline-M (desalkyl-)
PS
LM/Q
$C_{10}H_{15}N$
149.12045
RI: 1195
D: UA
Sympathomimetic
Antiparkinsonian
CAS: 537-46-2
58
67
126
M+
141
100
Cyclopentamine
PS
LM/Q
$C_9H_{19}N$
141.15175
RI: 1230
Vasoconstrictor
CAS: 102-45-4
58
91
105
148
100
Pentorex
PS
LM
$C_{11}H_{17}N$
163.13610
RI: 1250
D: UA
Anorexic
CAS: 434-43-5

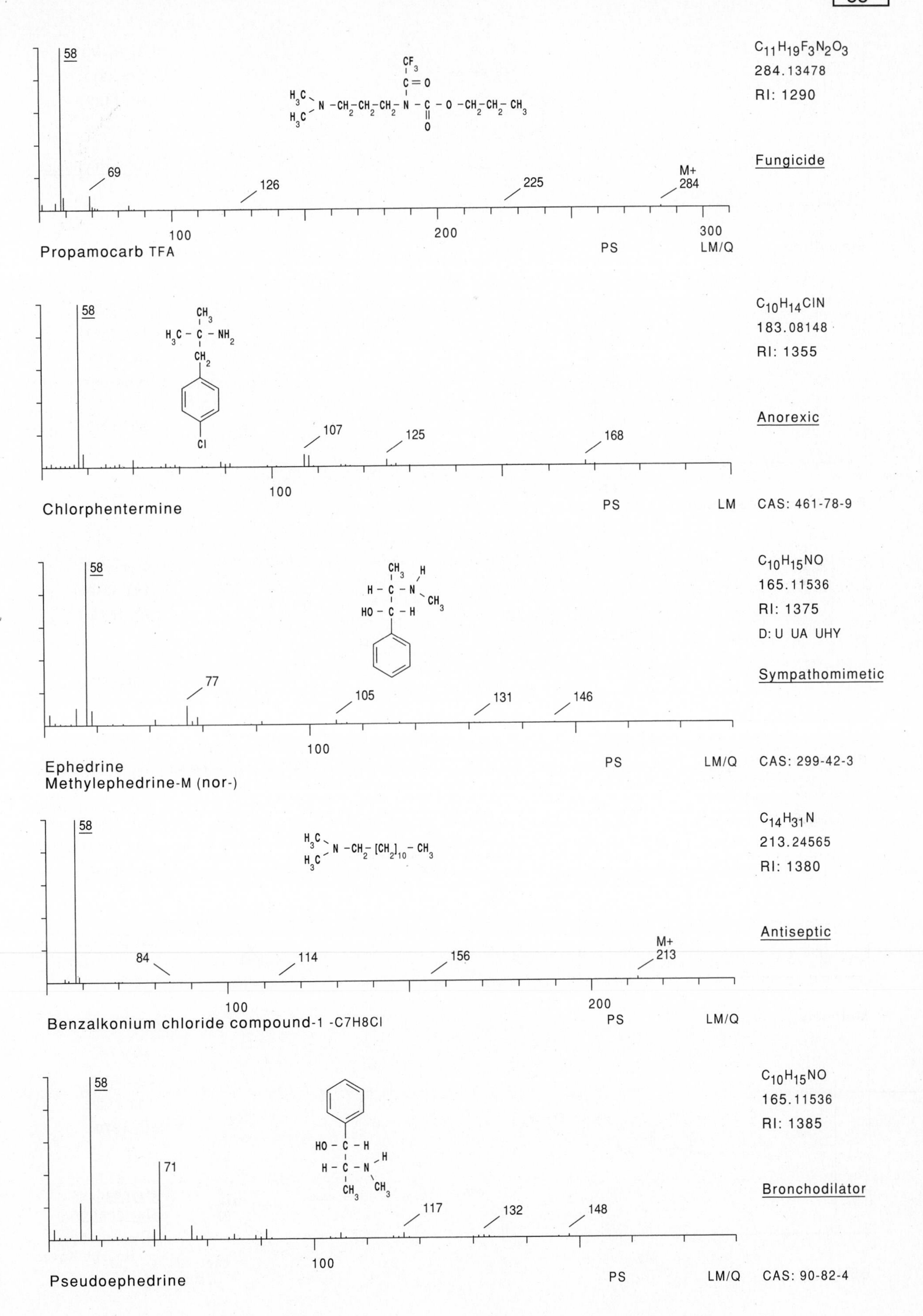

58
69
126
225
M+
284
100
200
300
Propamocarb TFA
PS
LM/Q
C11H19F3N2O3
284.13478
RI: 1290
Fungicide
58
107
125
168
100
Chlorphentermine
PS
LM
C10H14ClN
183.08148
RI: 1355
Anorexic
CAS: 461-78-9
58
77
105
131
146
100
Ephedrine
Methylephedrine-M (nor-)
PS
LM/Q
C10H15NO
165.11536
RI: 1375
D: U UA UHY
Sympathomimetic
CAS: 299-42-3
58
84
114
156
M+
213
100
200
Benzalkonium chloride compound-1 -C7H8Cl
PS
LM/Q
C14H31N
213.24565
RI: 1380
Antiseptic
58
71
117
132
148
100
Pseudoephedrine
PS
LM/Q
C10H15NO
165.11536
RI: 1385
Bronchodilator
CAS: 90-82-4

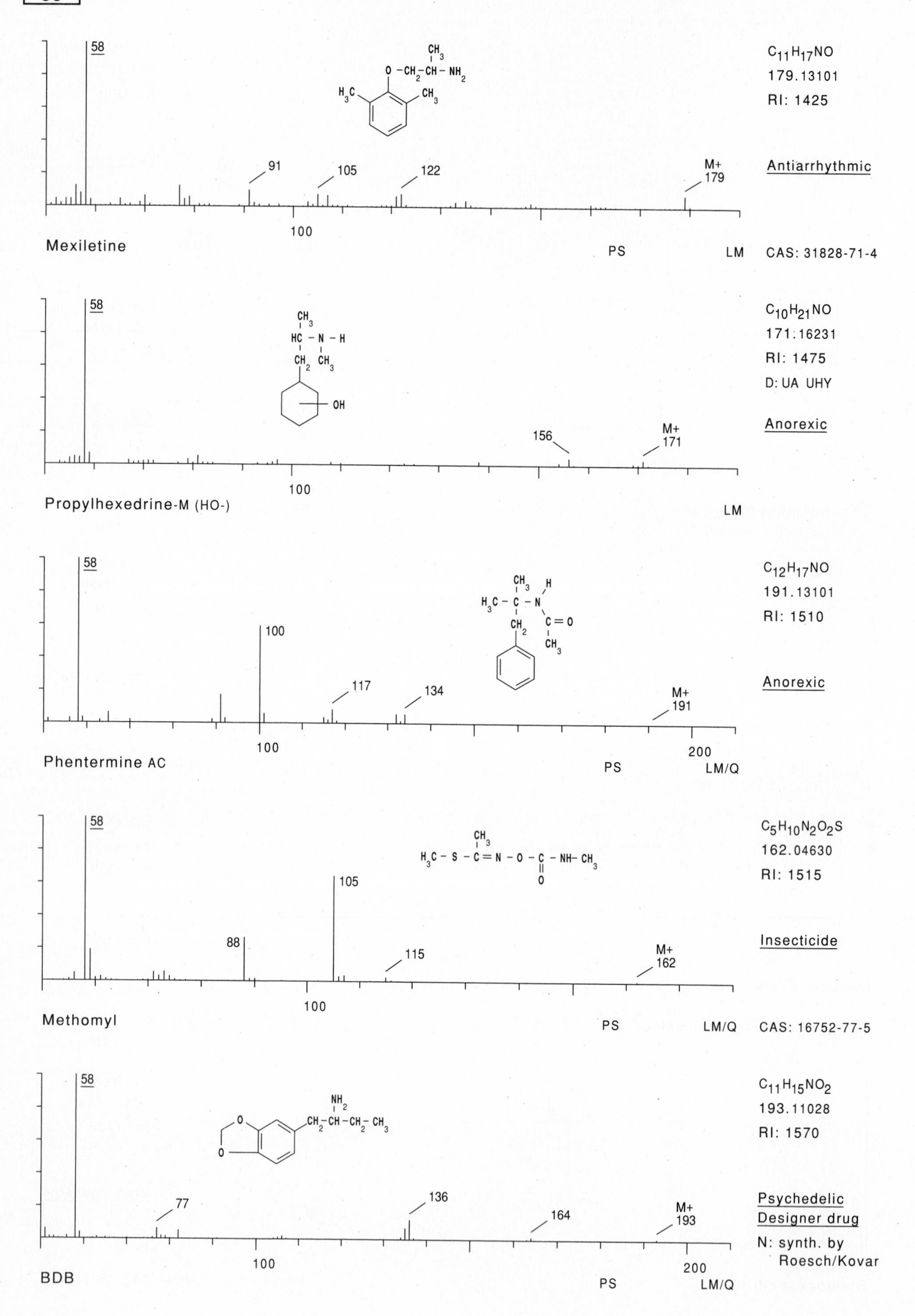
58
91
105
122
M+
179
100
Mexiletine
PS
LM
$C_{11}H_{17}NO$
179.13101
RI: 1425
Antiarrhythmic
CAS: 31828-71-4
58
156
M+
171
100
Propylhexedrine-M (HO-)
LM
$C_{10}H_{21}NO$
171.16231
RI: 1475
D: UA UHY
Anorexic
58
100
117
134
M+
191
100
200
Phentermine AC
PS
LM/Q
$C_{12}H_{17}NO$
191.13101
RI: 1510
Anorexic
58
88
105
115
M+
162
100
Methomyl
PS
LM/Q
$C_5H_{10}N_2O_2S$
162.04630
RI: 1515
Insecticide
CAS: 16752-77-5
58
77
136
164
M+
193
100
200
BDB
PS
LM/Q
$C_{11}H_{15}NO_2$
193.11028
RI: 1570
Psychedelic
Designer drug
N: synth. by Roesch/Kovar

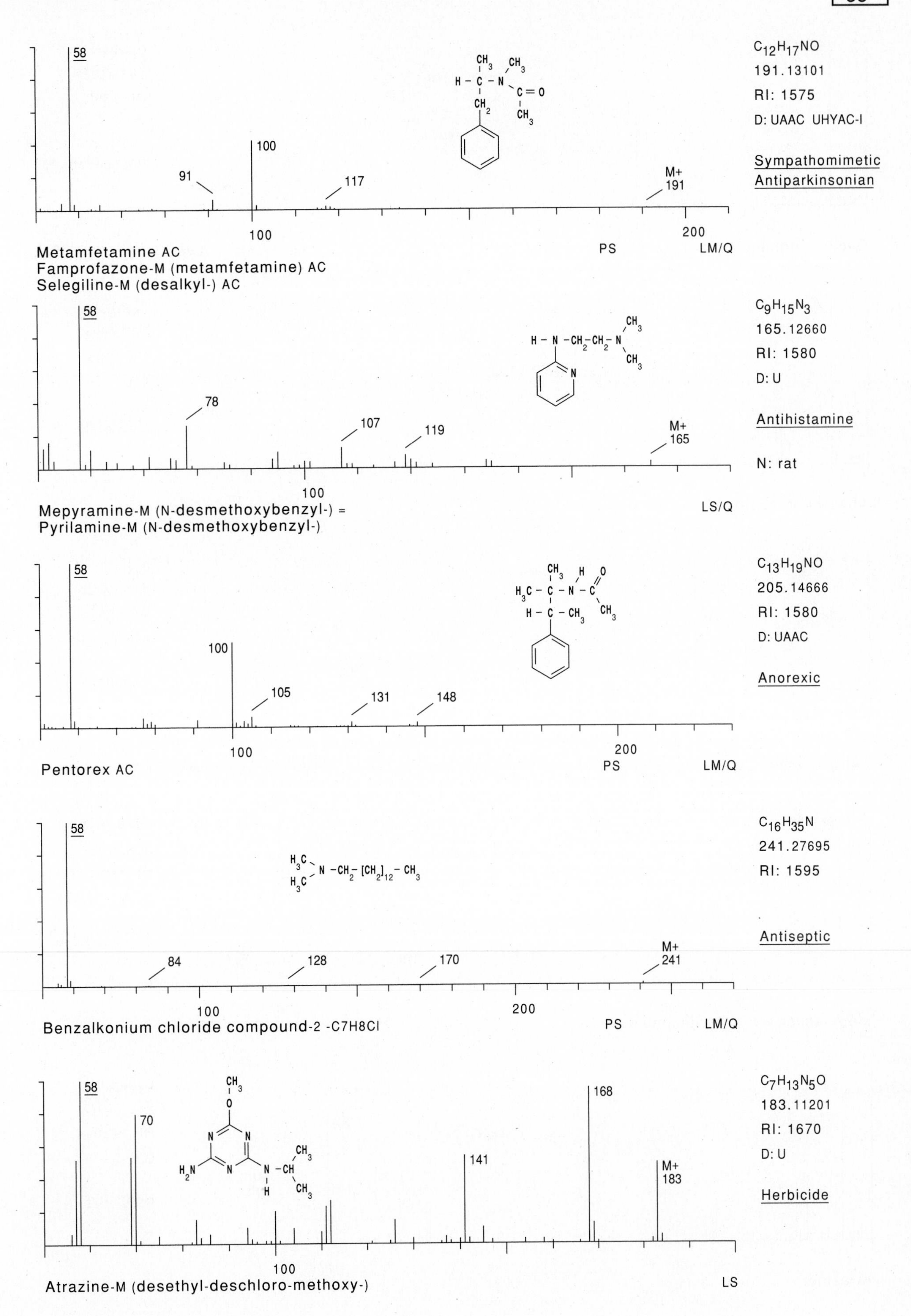

$C_{12}H_{17}NO$
191.13101
RI: 1575
D: UAAC UHYAC-I

Sympathomimetic
Antiparkinsonian

Metamfetamine AC
Famprofazone-M (metamfetamine) AC
Selegiline-M (desalkyl-) AC

$C_9H_{15}N_3$
165.12660
RI: 1580
D: U

Antihistamine

N: rat

Mepyramine-M (N-desmethoxybenzyl-) =
Pyrilamine-M (N-desmethoxybenzyl-)

$C_{13}H_{19}NO$
205.14666
RI: 1580
D: UAAC

Anorexic

Pentorex AC

$C_{16}H_{35}N$
241.27695
RI: 1595

Antiseptic

Benzalkonium chloride compound-2 -C7H8Cl

$C_7H_{13}N_5O$
183.11201
RI: 1670
D: U

Herbicide

Atrazine-M (desethyl-deschloro-methoxy-)

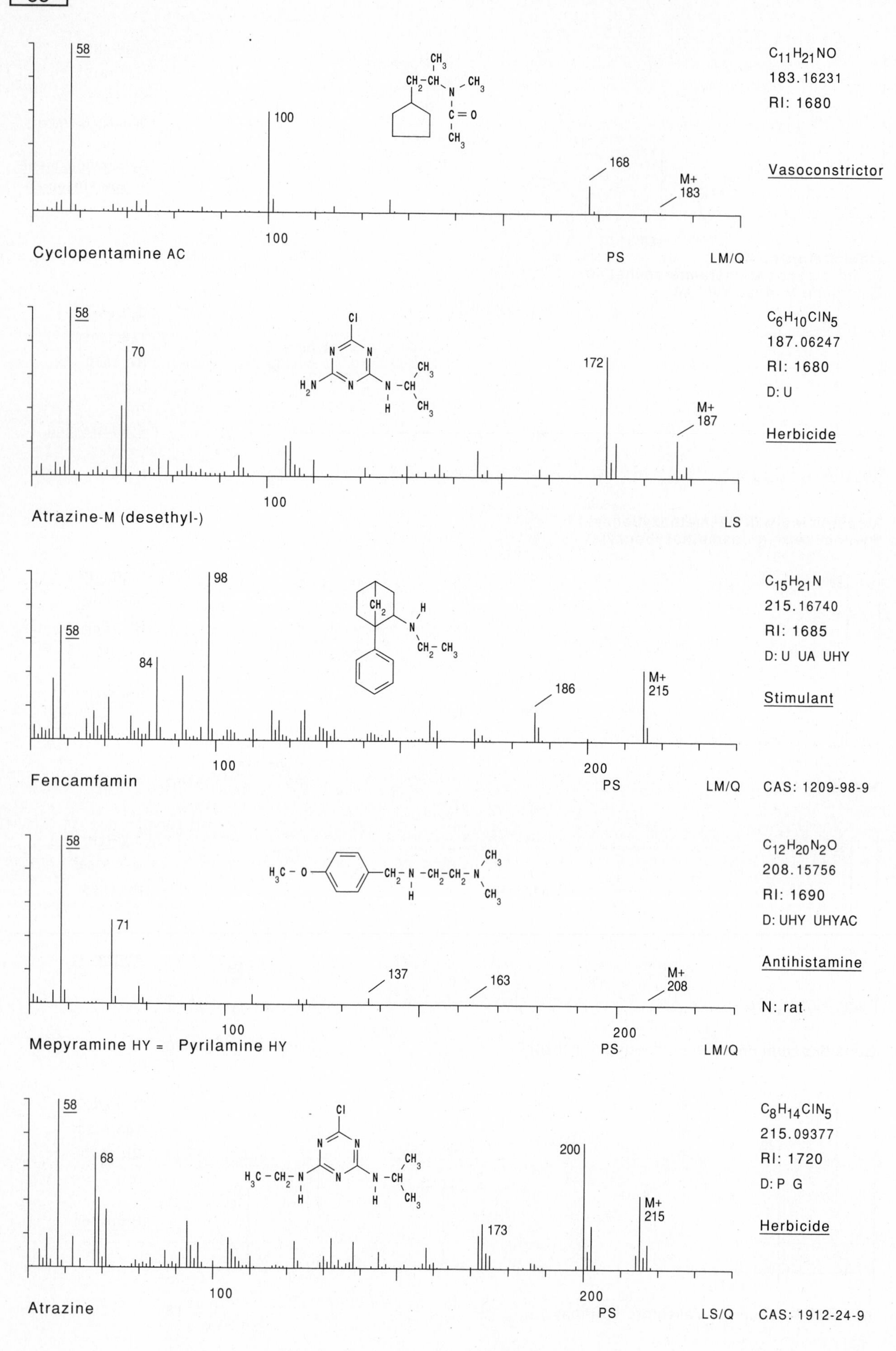
58
100
168
M+
183
100
Cyclopentamine AC
PS
LM/Q
$C_{11}H_{21}NO$
183.16231
RI: 1680
Vasoconstrictor
58
70
172
M+
187
100
Atrazine-M (desethyl-)
LS
$C_6H_{10}ClN_5$
187.06247
RI: 1680
D: U
Herbicide
98
58
84
186
M+
215
100
200
Fencamfamin
PS
LM/Q
$C_{15}H_{21}N$
215.16740
RI: 1685
D: U UA UHY
Stimulant
CAS: 1209-98-9
58
71
137
163
M+
208
100
200
Mepyramine HY = Pyrilamine HY
PS
LM/Q
$C_{12}H_{20}N_2O$
208.15756
RI: 1690
D: UHY UHYAC
Antihistamine
N: rat
58
68
200
M+
215
173
100
200
Atrazine
PS
LS/Q
$C_8H_{14}ClN_5$
215.09377
RI: 1720
D: P G
Herbicide
CAS: 1912-24-9

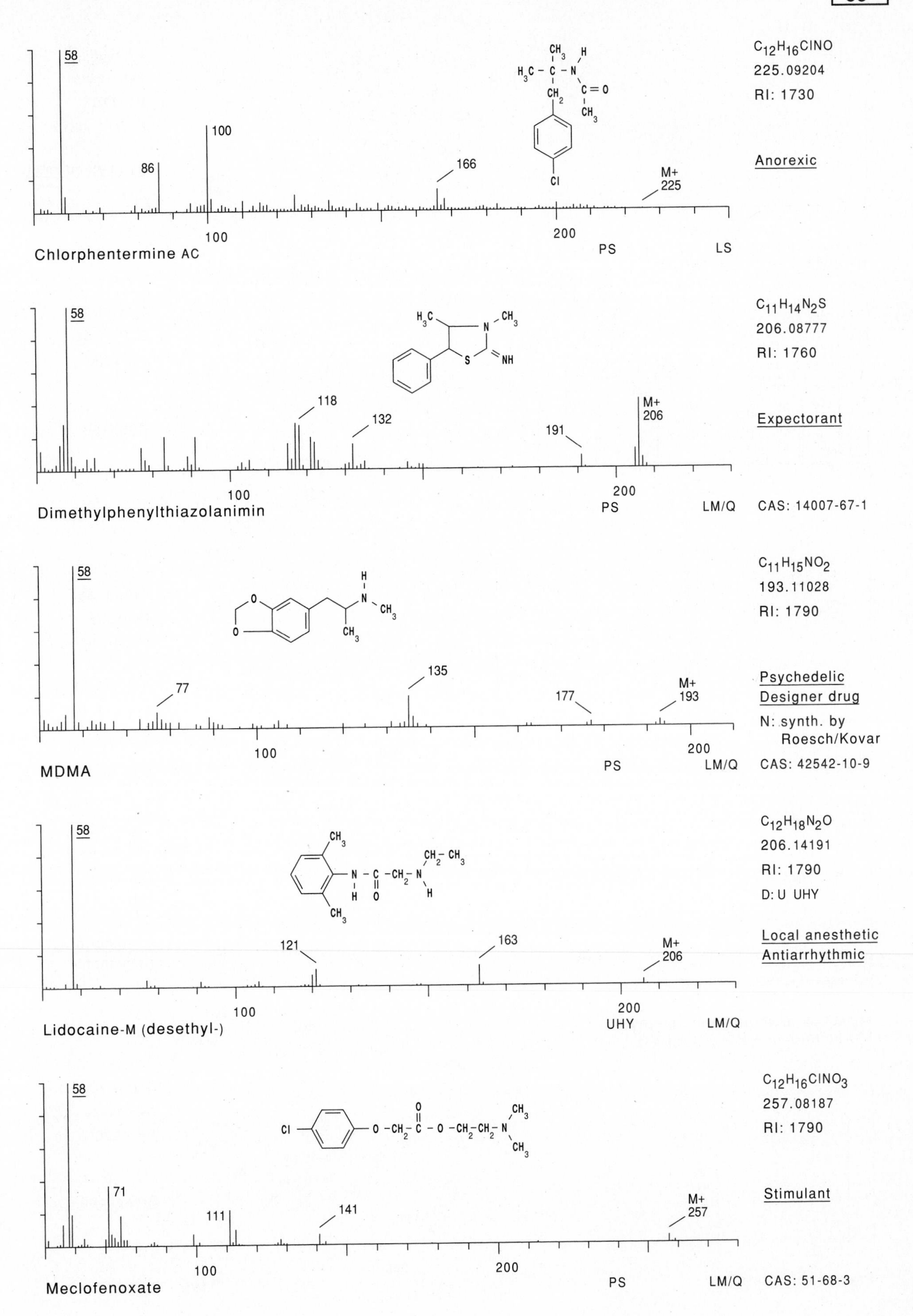

58
100
86
166
M+
225
100
200
Chlorphentermine AC
PS
LS
C12H16ClNO
225.09204
RI: 1730
Anorexic
58
118
132
191
M+
206
100
200
Dimethylphenylthiazolanimin
PS
LM/Q
C11H14N2S
206.08777
RI: 1760
Expectorant
CAS: 14007-67-1
58
77
135
177
M+
193
100
200
MDMA
PS
LM/Q
C11H15NO2
193.11028
RI: 1790
Psychedelic
Designer drug
N: synth. by
Roesch/Kovar
CAS: 42542-10-9
58
121
163
M+
206
100
200
Lidocaine-M (desethyl-)
UHY
LM/Q
C12H18N2O
206.14191
RI: 1790
D: U UHY
Local anesthetic
Antiarrhythmic
58
71
111
141
M+
257
100
200
Meclofenoxate
PS
LM/Q
C12H16ClNO3
257.08187
RI: 1790
Stimulant
CAS: 51-68-3

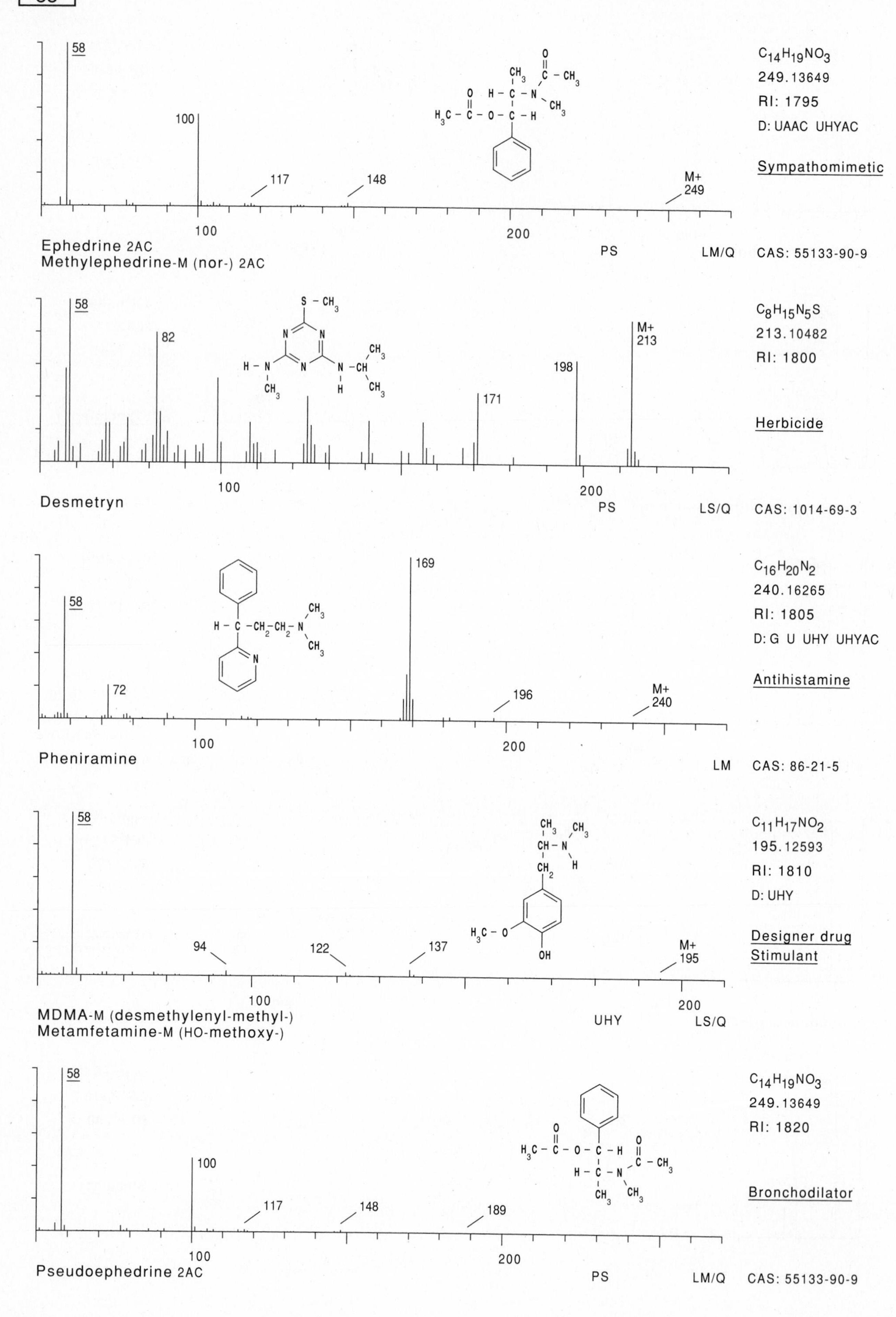
58
100
117
148
M+
249
100
200
PS
LM/Q
Ephedrine 2AC
Methylephedrine-M (nor-) 2AC
$C_{14}H_{19}NO_3$
249.13649
RI: 1795
D: UAAC UHYAC
Sympathomimetic
CAS: 55133-90-9
58
82
198
M+
213
171
100
200
PS
LS/Q
Desmetryn
$C_8H_{15}N_5S$
213.10482
RI: 1800
Herbicide
CAS: 1014-69-3
169
58
72
196
M+
240
100
200
LM
Pheniramine
$C_{16}H_{20}N_2$
240.16265
RI: 1805
D: G U UHY UHYAC
Antihistamine
CAS: 86-21-5
58
94
122
137
M+
195
100
200
UHY
LS/Q
MDMA-M (desmethylenyl-methyl-)
Metamfetamine-M (HO-methoxy-)
$C_{11}H_{17}NO_2$
195.12593
RI: 1810
D: UHY
Designer drug
Stimulant
58
100
117
148
189
100
200
PS
LM/Q
Pseudoephedrine 2AC
$C_{14}H_{19}NO_3$
249.13649
RI: 1820
Bronchodilator
CAS: 55133-90-9

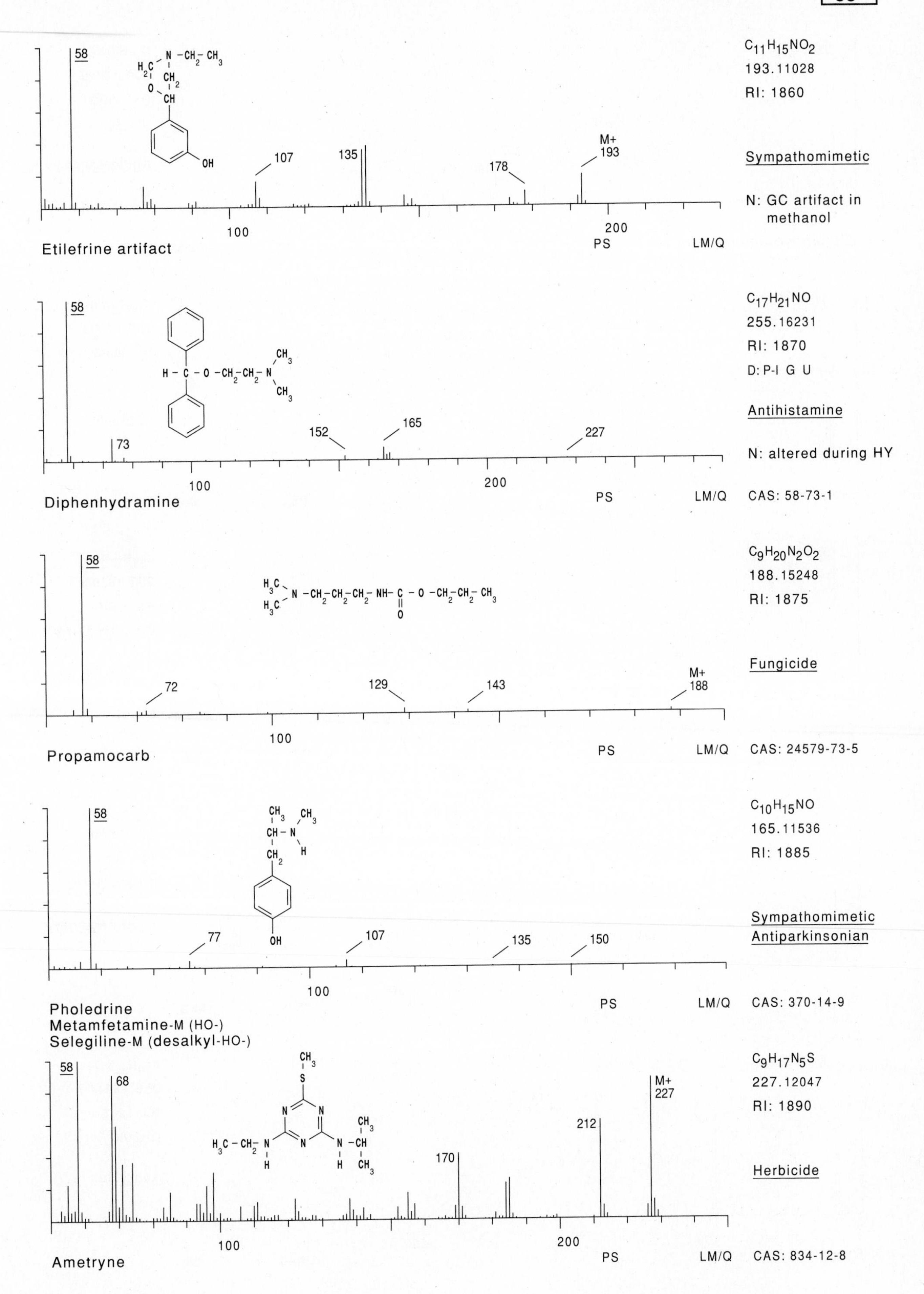
58
107
135
178
M+
193
100
200
PS
LM/Q
Etilefrine artifact
C11H15NO2
193.11028
RI: 1860
Sympathomimetic
N: GC artifact in methanol
58
73
152
165
227
100
200
PS
LM/Q
Diphenhydramine
C17H21NO
255.16231
RI: 1870
D: P-I G U
Antihistamine
N: altered during HY
CAS: 58-73-1
58
72
129
143
M+
188
100
PS
LM/Q
Propamocarb
C9H20N2O2
188.15248
RI: 1875
Fungicide
CAS: 24579-73-5
58
77
107
135
150
100
PS
LM/Q
Pholedrine
Metamfetamine-M (HO-)
Selegiline-M (desalkyl-HO-)
C10H15NO
165.11536
RI: 1885
Sympathomimetic
Antiparkinsonian
CAS: 370-14-9
58
68
170
212
M+
227
100
200
PS
LM/Q
Ametryne
C9H17N5S
227.12047
RI: 1890
Herbicide
CAS: 834-12-8

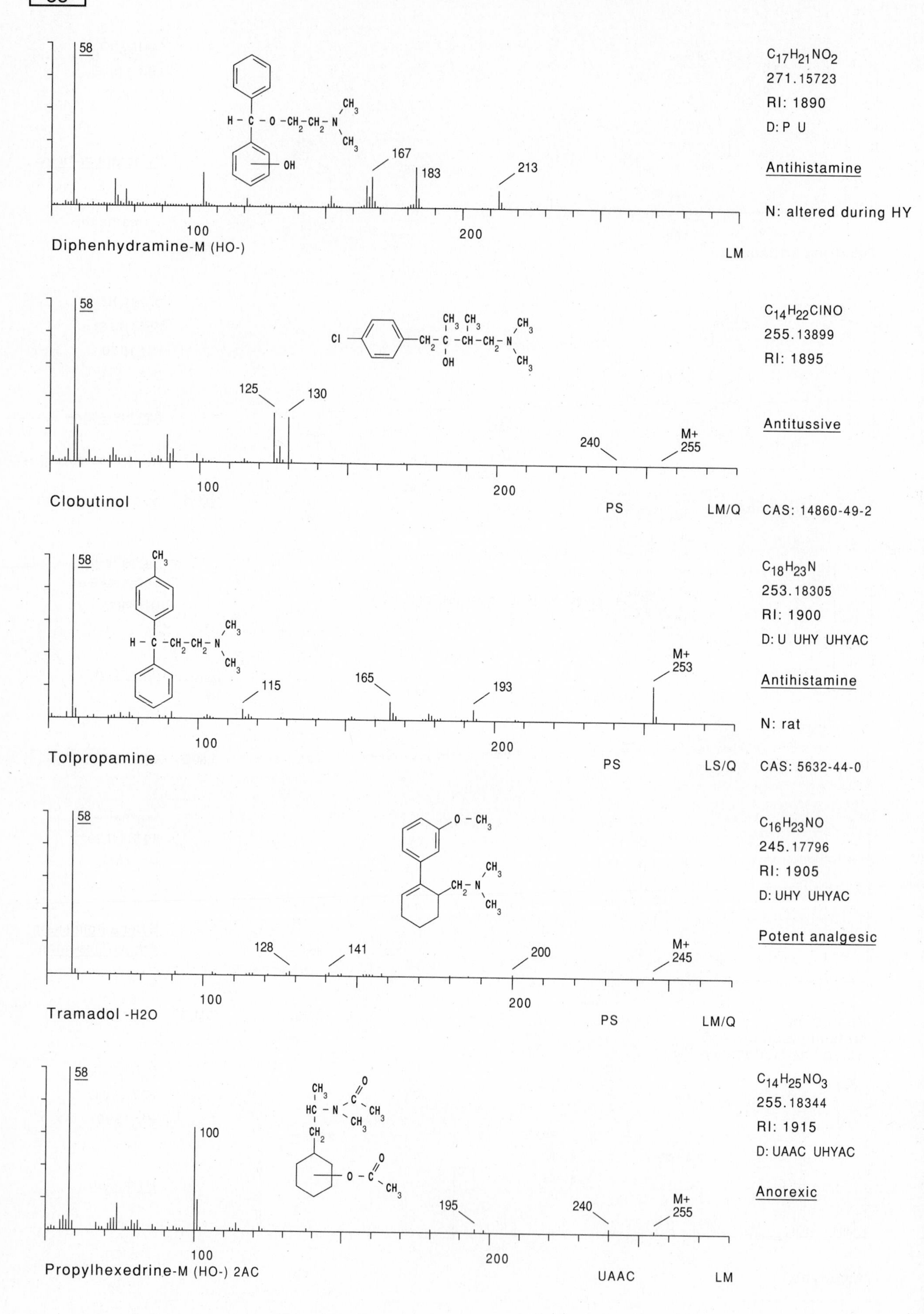
58
Diphenhydramine-M (HO-)
$C_{17}H_{21}NO_2$
271.15723
RI: 1890
D: P U
Antihistamine
N: altered during HY
LM
Clobutinol
$C_{14}H_{22}ClNO$
255.13899
RI: 1895
Antitussive
PS
LM/Q
CAS: 14860-49-2
Tolpropamine
$C_{18}H_{23}N$
253.18305
RI: 1900
D: U UHY UHYAC
Antihistamine
N: rat
PS
LS/Q
CAS: 5632-44-0
Tramadol -H2O
$C_{16}H_{23}NO$
245.17796
RI: 1905
D: UHY UHYAC
Potent analgesic
PS
LM/Q
Propylhexedrine-M (HO-) 2AC
$C_{14}H_{25}NO_3$
255.18344
RI: 1915
D: UAAC UHYAC
Anorexic
UAAC
LM

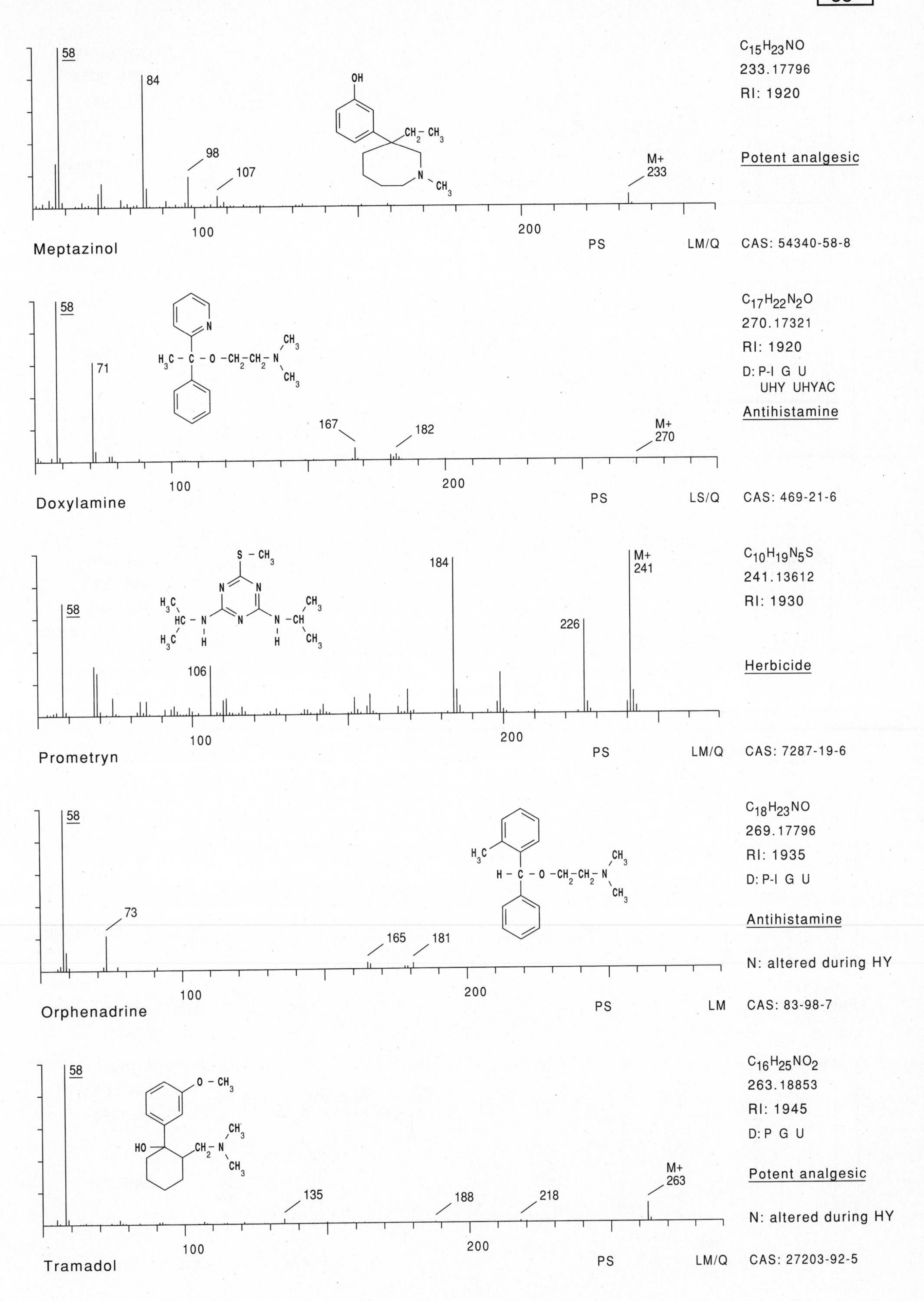

$C_{15}H_{23}NO$
233.17796
RI: 1920

Potent analgesic

CAS: 54340-58-8

$C_{17}H_{22}N_2O$
270.17321
RI: 1920
D: P-I G U
UHY UHYAC
Antihistamine

CAS: 469-21-6

$C_{10}H_{19}N_5S$
241.13612
RI: 1930

Herbicide

CAS: 7287-19-6

$C_{18}H_{23}NO$
269.17796
RI: 1935
D: P-I G U

Antihistamine

N: altered during HY

CAS: 83-98-7

$C_{16}H_{25}NO_2$
263.18853
RI: 1945
D: P G U

Potent analgesic

N: altered during HY

CAS: 27203-92-5

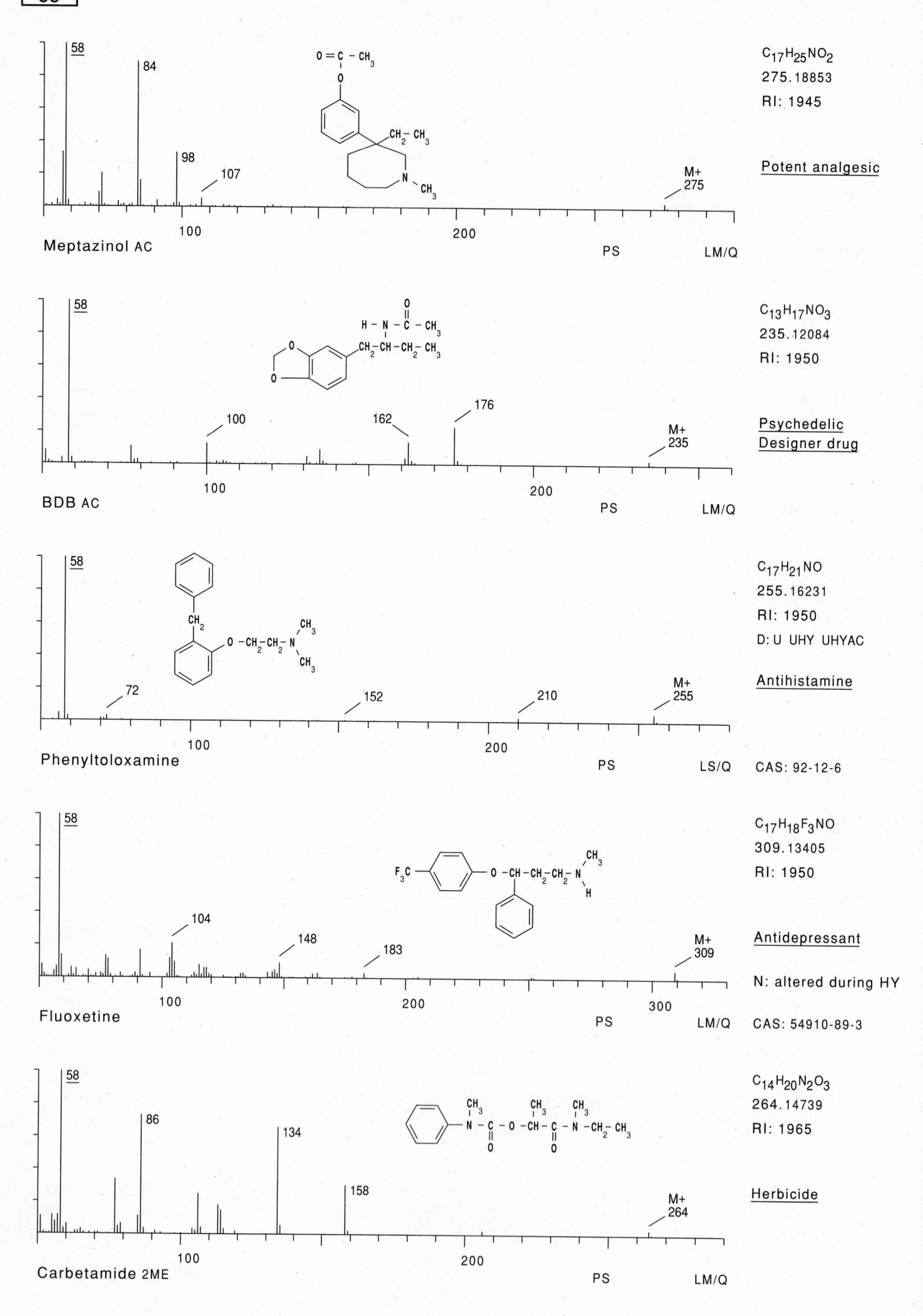

58
84
98
107
M+
275
100
200
PS
LM/Q
Meptazinol AC
C17H25NO2
275.18853
RI: 1945
Potent analgesic
58
100
162
176
M+
235
100
200
PS
LM/Q
BDB AC
C13H17NO3
235.12084
RI: 1950
Psychedelic
Designer drug
58
72
152
210
M+
255
100
200
PS
LS/Q
Phenyltoloxamine
C17H21NO
255.16231
RI: 1950
D: U UHY UHYAC
Antihistamine
CAS: 92-12-6
58
104
148
183
M+
309
100
200
300
PS
LM/Q
Fluoxetine
C17H18F3NO
309.13405
RI: 1950
Antidepressant
N: altered during HY
CAS: 54910-89-3
58
86
134
158
M+
264
100
200
PS
LM/Q
Carbetamide 2ME
C14H20N2O3
264.14739
RI: 1965
Herbicide

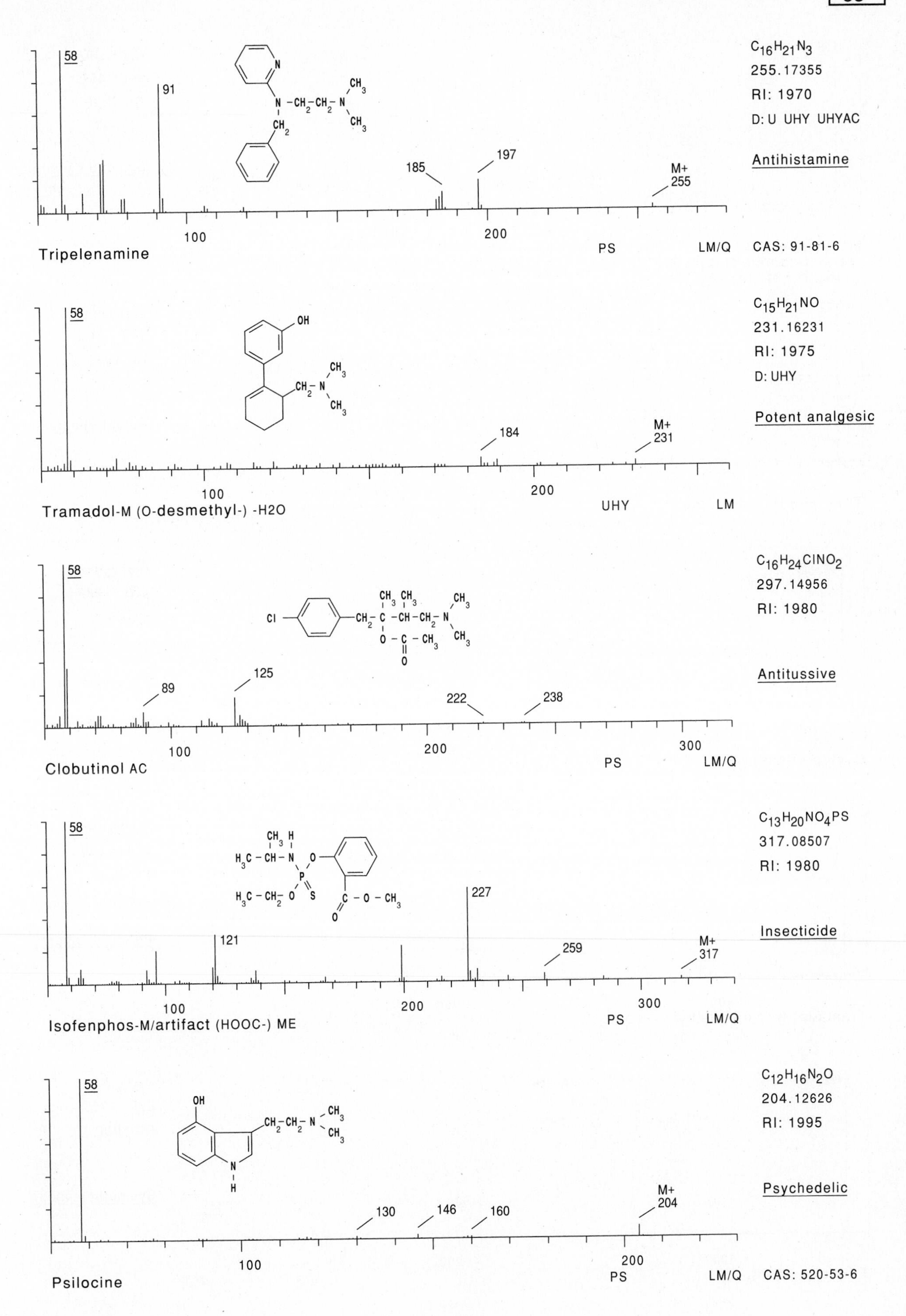
58
91
185
197
M+
255
100
200
Tripelenamine
PS
LM/Q
C16H21N3
255.17355
RI: 1970
D: U UHY UHYAC
Antihistamine
CAS: 91-81-6
58
184
M+
231
100
200
Tramadol-M (O-desmethyl-) -H2O
UHY
LM
C15H21NO
231.16231
RI: 1975
D: UHY
Potent analgesic
58
89
125
222
238
100
200
300
Clobutinol AC
PS
LM/Q
C16H24ClNO2
297.14956
RI: 1980
Antitussive
58
121
227
259
M+
317
100
200
300
Isofenphos-M/artifact (HOOC-) ME
PS
LM/Q
C13H20NO4PS
317.08507
RI: 1980
Insecticide
58
130
146
160
M+
204
100
200
Psilocine
PS
LM/Q
C12H16N2O
204.12626
RI: 1995
Psychedelic
CAS: 520-53-6

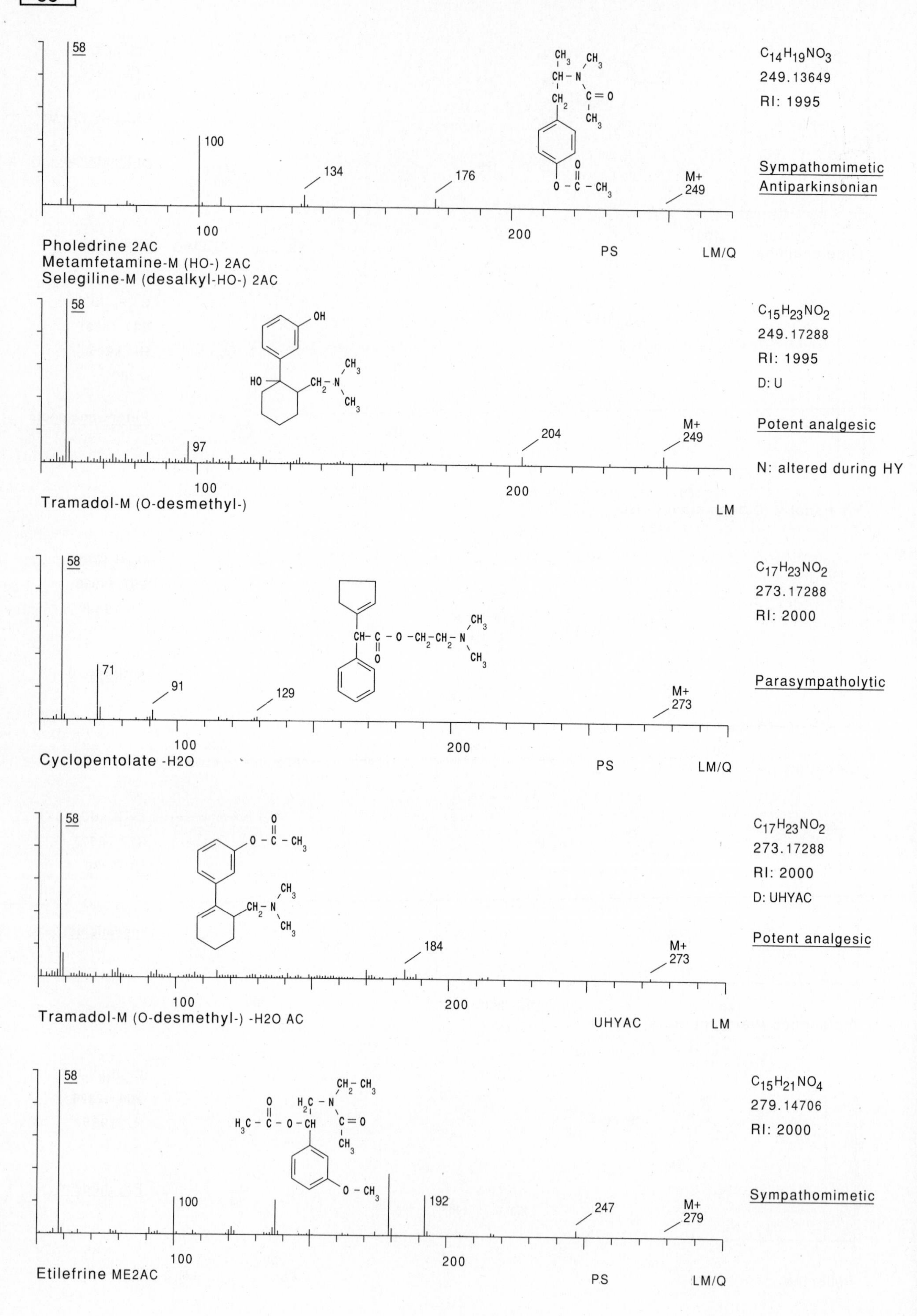

58
100
134
176
M+
249
100
200
PS
LM/Q
$C_{14}H_{19}NO_3$
249.13649
RI: 1995
Sympathomimetic
Antiparkinsonian
Pholedrine 2AC
Metamfetamine-M (HO-) 2AC
Selegiline-M (desalkyl-HO-) 2AC
58
97
204
M+
249
100
200
LM
$C_{15}H_{23}NO_2$
249.17288
RI: 1995
D: U
Potent analgesic
N: altered during HY
Tramadol-M (O-desmethyl-)
58
71
91
129
M+
273
100
200
PS
LM/Q
$C_{17}H_{23}NO_2$
273.17288
RI: 2000
Parasympatholytic
Cyclopentolate -H2O
58
184
M+
273
100
200
UHYAC
LM
$C_{17}H_{23}NO_2$
273.17288
RI: 2000
D: UHYAC
Potent analgesic
Tramadol-M (O-desmethyl-) -H2O AC
58
100
192
247
M+
279
100
200
PS
LM/Q
$C_{15}H_{21}NO_4$
279.14706
RI: 2000
Sympathomimetic
Etilefrine ME2AC

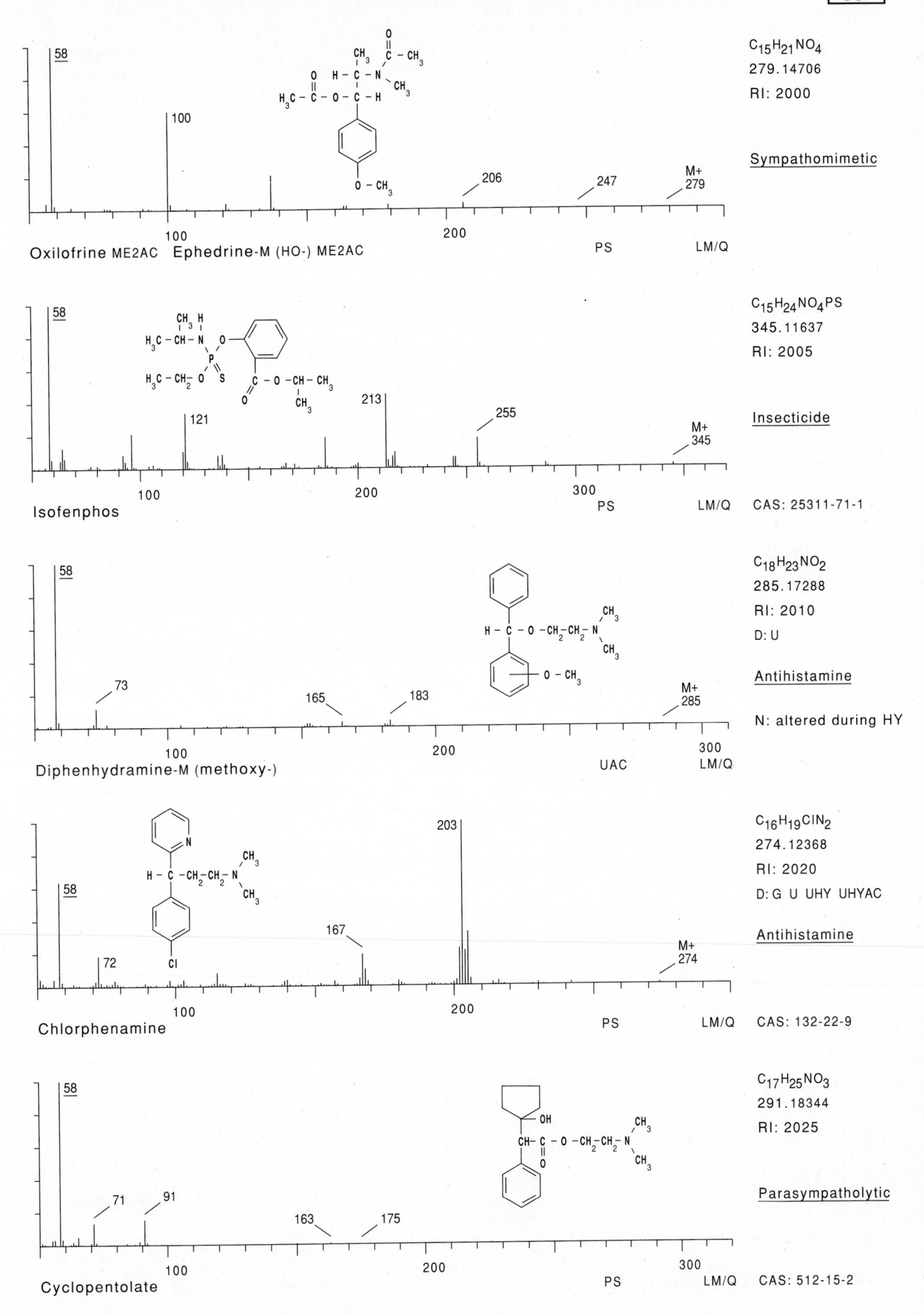

Oxilofrine ME2AC Ephedrine-M (HO-) ME2AC
C15H21NO4
279.14706
RI: 2000
Sympathomimetic
58
100
206
247
M+
279
PS
LM/Q
Isofenphos
C15H24NO4PS
345.11637
RI: 2005
Insecticide
58
121
213
255
M+
345
PS
LM/Q
CAS: 25311-71-1
Diphenhydramine-M (methoxy-)
C18H23NO2
285.17288
RI: 2010
D: U
Antihistamine
N: altered during HY
58
73
165
183
M+
285
UAC
LM/Q
Chlorphenamine
C16H19ClN2
274.12368
RI: 2020
D: G U UHY UHYAC
Antihistamine
58
72
167
203
M+
274
PS
LM/Q
CAS: 132-22-9
Cyclopentolate
C17H25NO3
291.18344
RI: 2025
Parasympatholytic
58
71
91
163
175
PS
LM/Q
CAS: 512-15-2

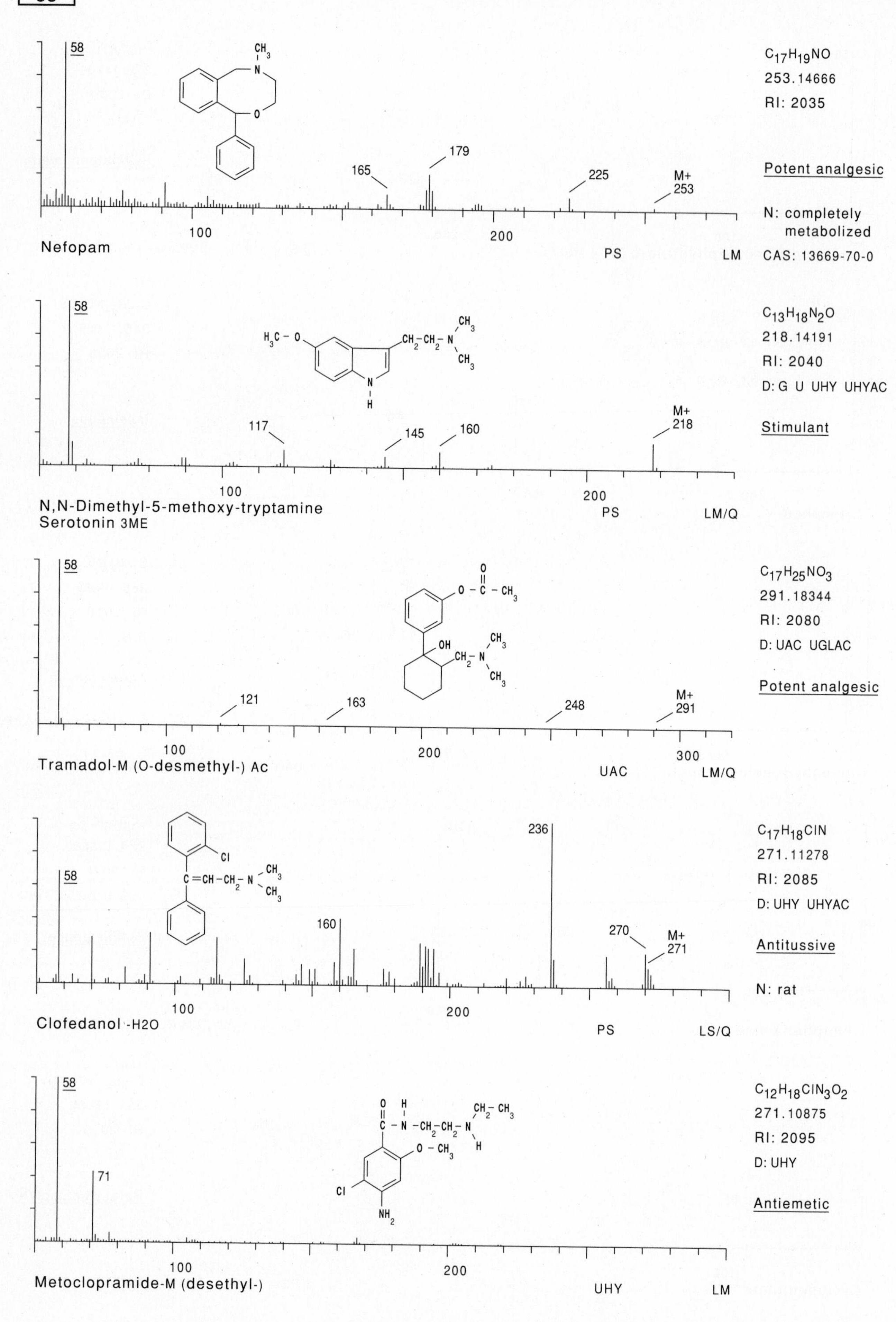
58
CH3
N
O
165
179
225
M+
253
100
200
Nefopam
PS
LM
C17H19NO
253.14666
RI: 2035
Potent analgesic
N: completely
metabolized
CAS: 13669-70-0
58
H3C – O
CH2 – CH2 – N
CH3
CH3
N
H
117
145
160
M+
218
100
200
N,N-Dimethyl-5-methoxy-tryptamine
Serotonin 3ME
PS
LM/Q
C13H18N2O
218.14191
RI: 2040
D: G U UHY UHYAC
Stimulant
58
O
O – C – CH3
OH
CH2 – N
CH3
CH3
121
163
248
M+
291
100
200
300
Tramadol-M (O-desmethyl-) Ac
UAC
LM/Q
C17H25NO3
291.18344
RI: 2080
D: UAC UGLAC
Potent analgesic
236
Cl
58
C =CH – CH2 – N
CH3
CH3
160
270
M+
271
100
200
Clofedanol -H2O
PS
LS/Q
C17H18ClN
271.11278
RI: 2085
D: UHY UHYAC
Antitussive
N: rat
58
71
O
H
C – N – CH2 – CH2 – N
CH2 – CH3
H
O – CH3
Cl
NH2
100
200
Metoclopramide-M (desethyl-)
UHY
LM
C12H18ClN3O2
271.10875
RI: 2095
D: UHY
Antiemetic

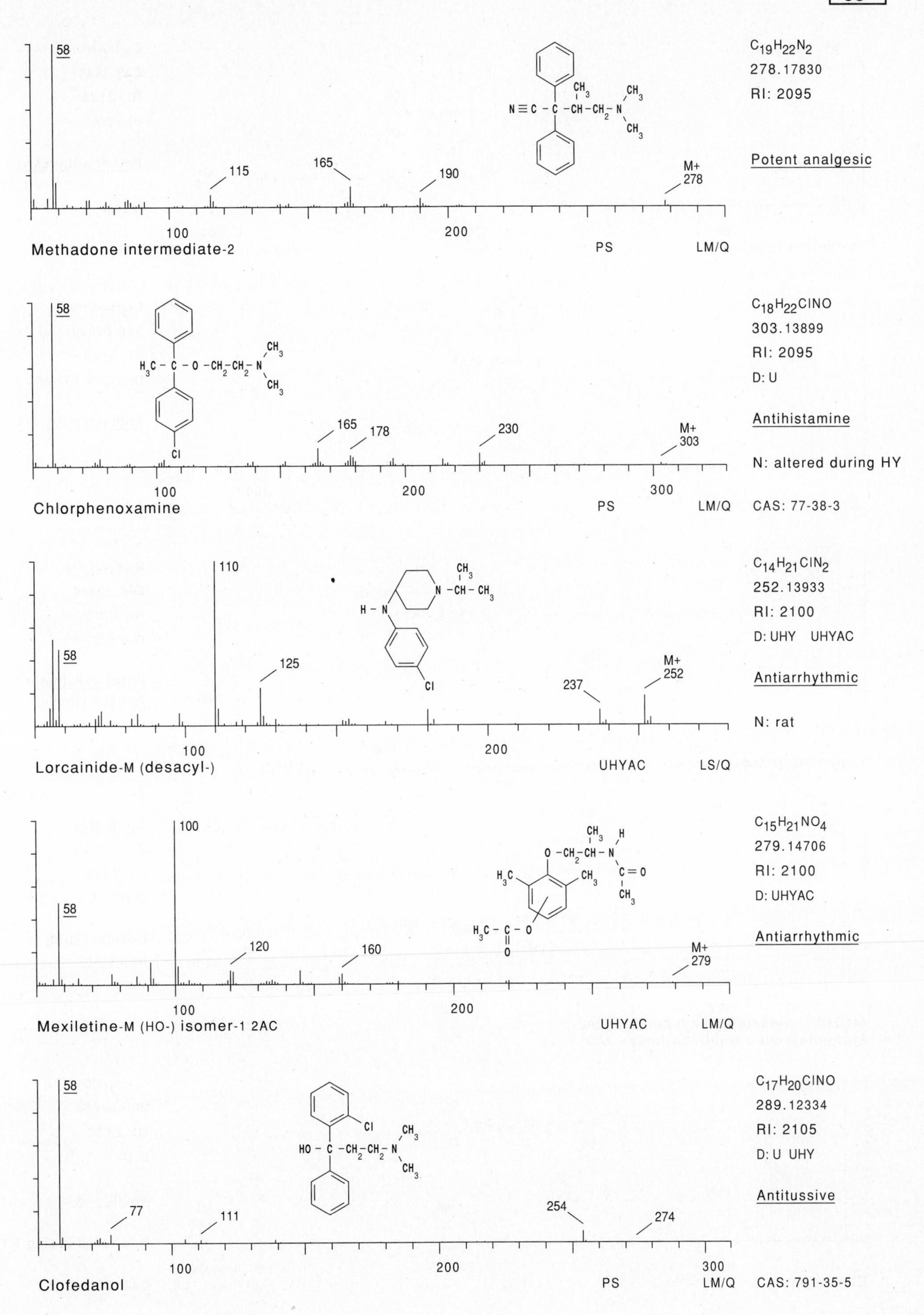

Methadone intermediate-2
PS LM/Q
$C_{19}H_{22}N_2$
278.17830
RI: 2095
Potent analgesic
M+ 278
58
115
165
190
Chlorphenoxamine
PS LM/Q
$C_{18}H_{22}ClNO$
303.13899
RI: 2095
D: U
Antihistamine
N: altered during HY
CAS: 77-38-3
M+ 303
58
165
178
230
Lorcainide-M (desacyl-)
UHYAC LS/Q
$C_{14}H_{21}ClN_2$
252.13933
RI: 2100
D: UHY UHYAC
Antiarrhythmic
N: rat
M+ 252
110
58
125
237
Mexiletine-M (HO-) isomer-1 2AC
UHYAC LM/Q
$C_{15}H_{21}NO_4$
279.14706
RI: 2100
D: UHYAC
Antiarrhythmic
M+ 279
100
58
120
160
Clofedanol
PS LM/Q
$C_{17}H_{20}ClNO$
289.12334
RI: 2105
D: U UHY
Antitussive
CAS: 791-35-5
58
77
111
254
274

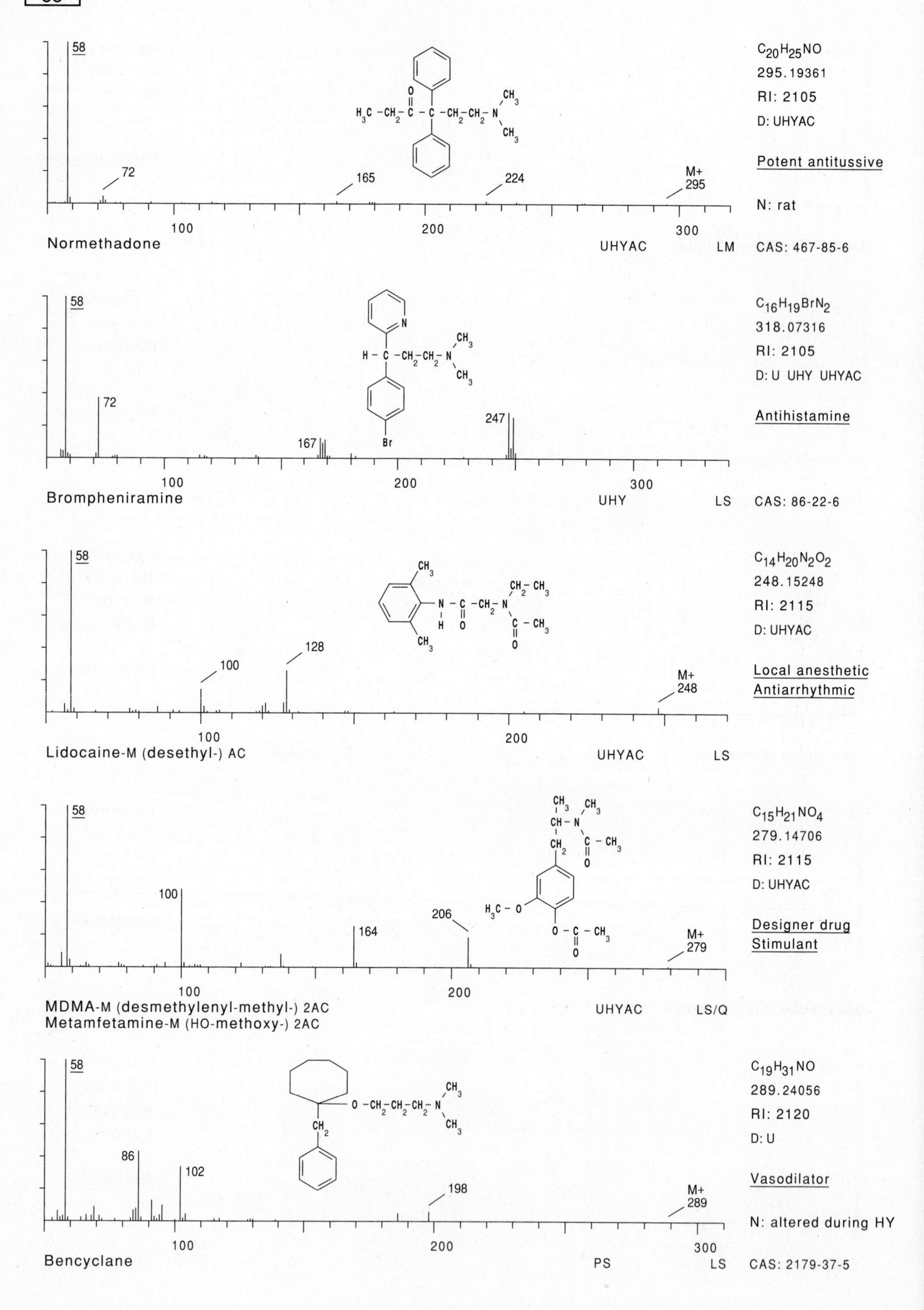
58
72
165
224
M+
295
100
200
300
Normethadone
UHYAC
LM
$C_{20}H_{25}NO$
295.19361
RI: 2105
D: UHYAC
Potent antitussive
N: rat
CAS: 467-85-6
58
72
167
247
100
200
300
Brompheniramine
UHY
LS
$C_{16}H_{19}BrN_2$
318.07316
RI: 2105
D: U UHY UHYAC
Antihistamine
CAS: 86-22-6
58
100
128
M+
248
100
200
Lidocaine-M (desethyl-) AC
UHYAC
LS
$C_{14}H_{20}N_2O_2$
248.15248
RI: 2115
D: UHYAC
Local anesthetic
Antiarrhythmic
58
100
164
206
M+
279
100
200
MDMA-M (desmethylenyl-methyl-) 2AC
Metamfetamine-M (HO-methoxy-) 2AC
UHYAC
LS/Q
$C_{15}H_{21}NO_4$
279.14706
RI: 2115
D: UHYAC
Designer drug
Stimulant
58
86
102
198
M+
289
100
200
300
Bencyclane
PS
LS
$C_{19}H_{31}NO$
289.24056
RI: 2120
D: U
Vasodilator
N: altered during HY
CAS: 2179-37-5

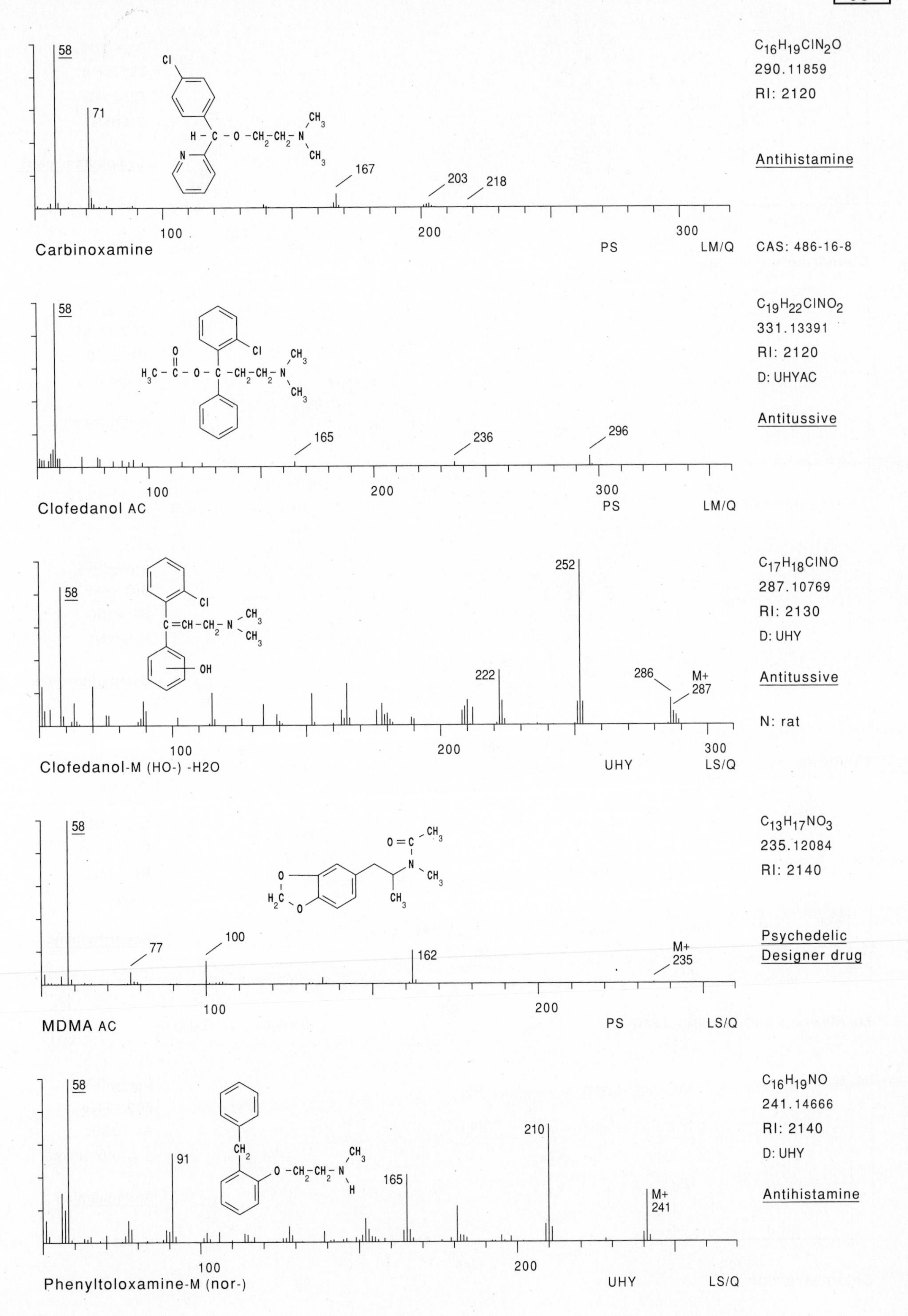

$C_{16}H_{19}ClN_2O$
290.11859
RI: 2120

Antihistamine

CAS: 486-16-8

$C_{19}H_{22}ClNO_2$
331.13391
RI: 2120
D: UHYAC

Antitussive

$C_{17}H_{18}ClNO$
287.10769
RI: 2130
D: UHY

Antitussive

N: rat

$C_{13}H_{17}NO_3$
235.12084
RI: 2140

Psychedelic
Designer drug

$C_{16}H_{19}NO$
241.14666
RI: 2140
D: UHY

Antihistamine

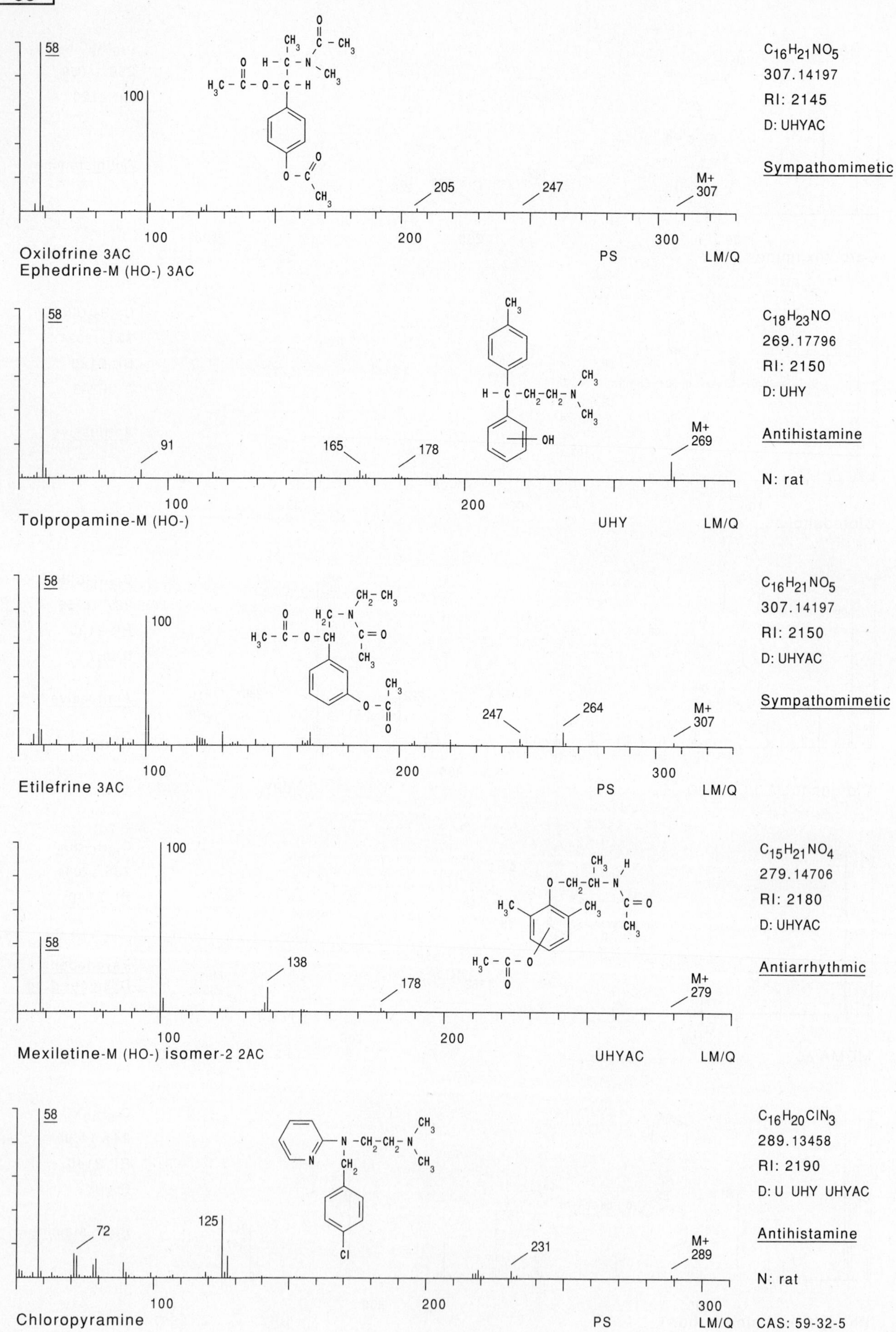

58
100
205
247
M+ 307
100
200
300
Oxilofrine 3AC
Ephedrine-M (HO-) 3AC
PS
LM/Q
$C_{16}H_{21}NO_5$
307.14197
RI: 2145
D: UHYAC
Sympathomimetic
58
91
165
178
M+ 269
100
200
Tolpropamine-M (HO-)
UHY
LM/Q
$C_{18}H_{23}NO$
269.17796
RI: 2150
D: UHY
Antihistamine
N: rat
58
100
247
264
M+ 307
100
200
300
Etilefrine 3AC
PS
LM/Q
$C_{16}H_{21}NO_5$
307.14197
RI: 2150
D: UHYAC
Sympathomimetic
100
58
138
178
M+ 279
100
200
Mexiletine-M (HO-) isomer-2 2AC
UHYAC
LM/Q
$C_{15}H_{21}NO_4$
279.14706
RI: 2180
D: UHYAC
Antiarrhythmic
58
72
125
231
M+ 289
100
200
300
Chloropyramine
PS
LM/Q
$C_{16}H_{20}ClN_3$
289.13458
RI: 2190
D: U UHY UHYAC
Antihistamine
N: rat
CAS: 59-32-5

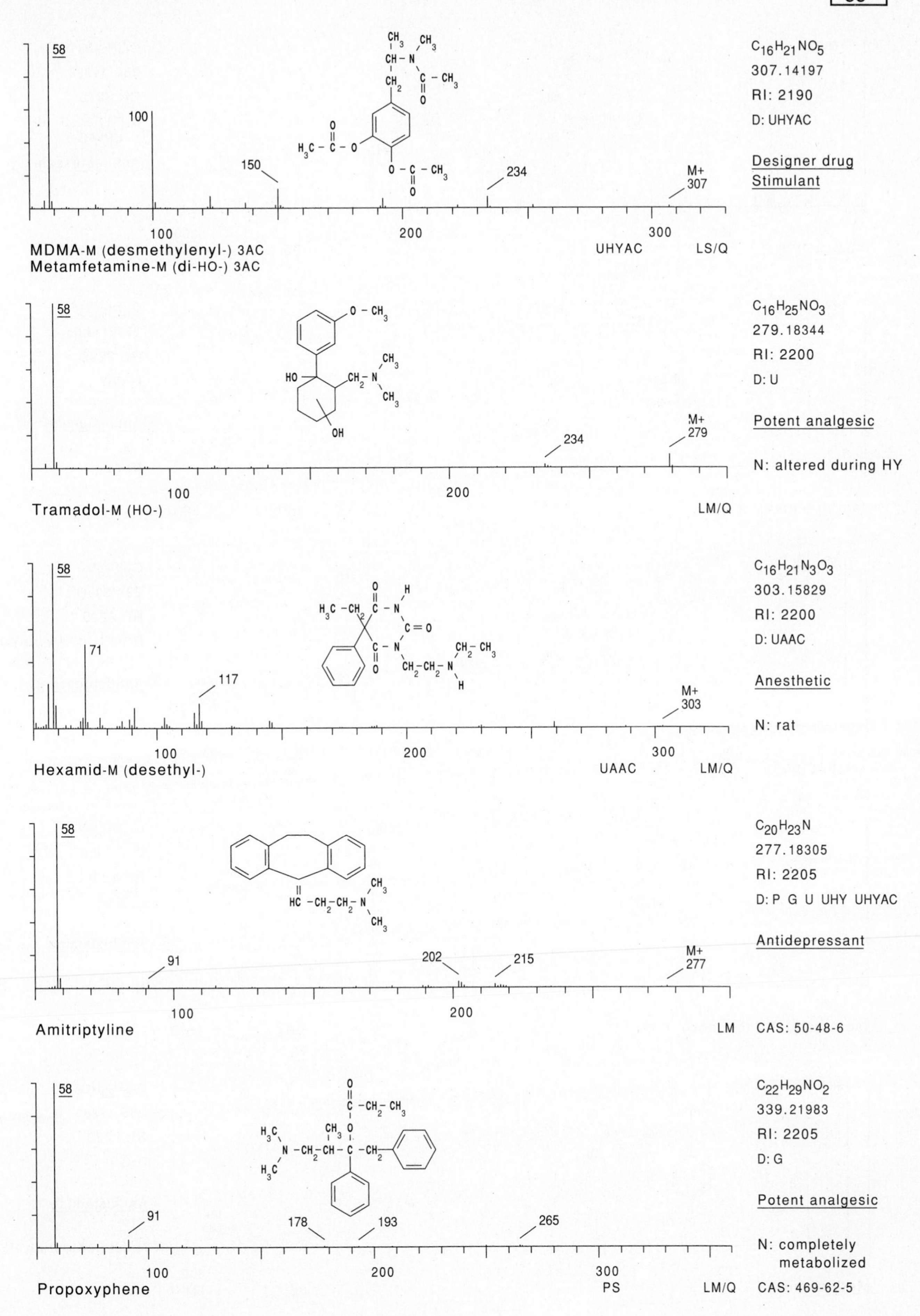

C16H21NO5
307.14197
RI: 2190
D: UHYAC
Designer drug
Stimulant
58
100
150
234
M+
307
100
200
300
UHYAC
LS/Q
MDMA-M (desmethylenyl-) 3AC
Metamfetamine-M (di-HO-) 3AC
C16H25NO3
279.18344
RI: 2200
D: U
Potent analgesic
N: altered during HY
58
234
M+
279
100
200
LM/Q
Tramadol-M (HO-)
C16H21N3O3
303.15829
RI: 2200
D: UAAC
Anesthetic
N: rat
58
71
117
M+
303
100
200
300
UAAC
LM/Q
Hexamid-M (desethyl-)
C20H23N
277.18305
RI: 2205
D: P G U UHY UHYAC
Antidepressant
58
91
202
215
M+
277
100
200
LM
Amitriptyline
CAS: 50-48-6
C22H29NO2
339.21983
RI: 2205
D: G
Potent analgesic
N: completely metabolized
58
91
178
193
265
100
200
300
PS
LM/Q
Propoxyphene
CAS: 469-62-5

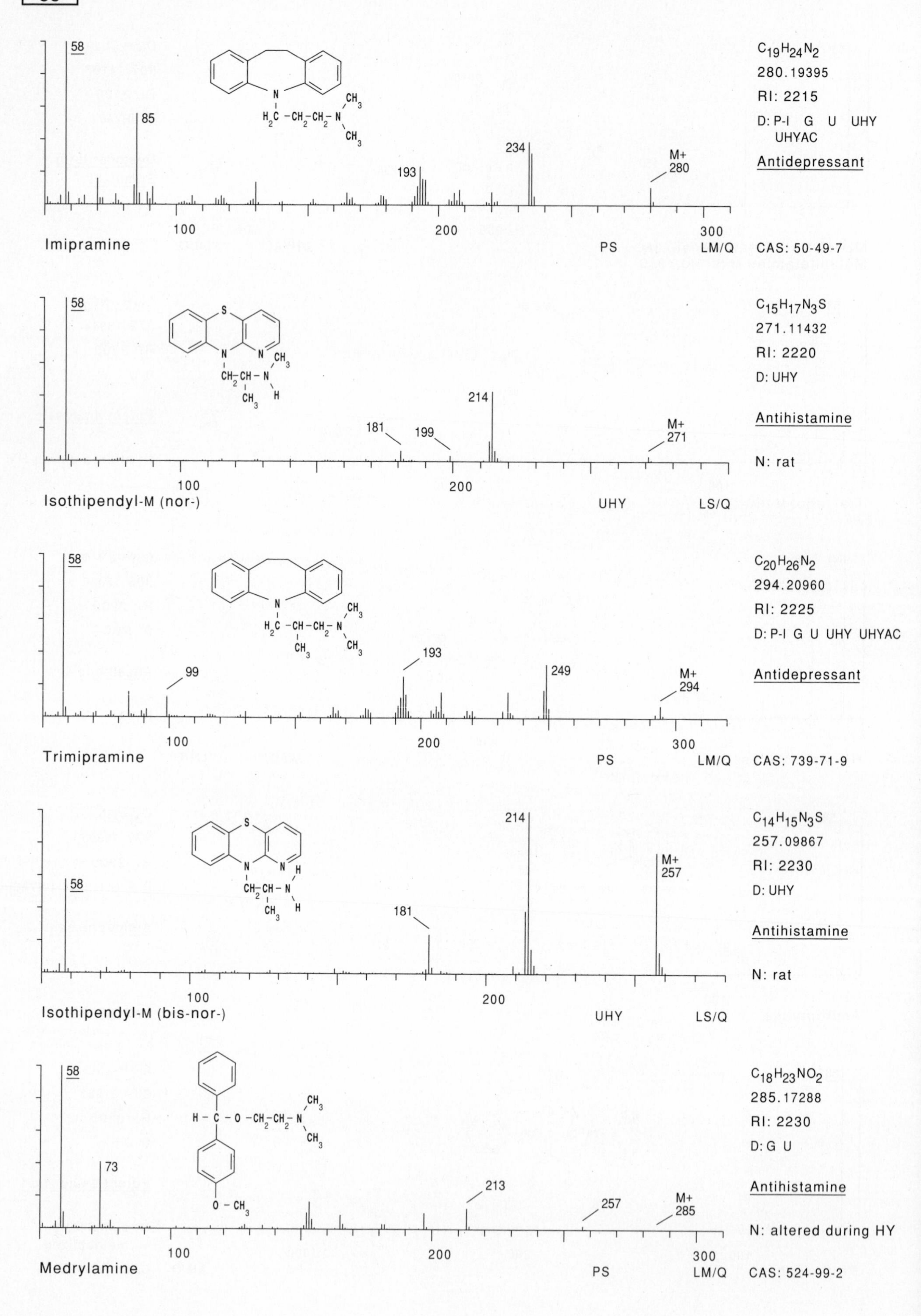

58
85
193
234
M+
280
100
200
300
Imipramine
PS
LM/Q
$C_{19}H_{24}N_2$
280.19395
RI: 2215
D: P-I G U UHY UHYAC
Antidepressant
CAS: 50-49-7
58
181
199
214
M+
271
100
200
Isothipendyl-M (nor-)
UHY
LS/Q
$C_{15}H_{17}N_3S$
271.11432
RI: 2220
D: UHY
Antihistamine
N: rat
58
99
193
249
M+
294
100
200
300
Trimipramine
PS
LM/Q
$C_{20}H_{26}N_2$
294.20960
RI: 2225
D: P-I G U UHY UHYAC
Antidepressant
CAS: 739-71-9
214
M+
257
58
181
100
200
Isothipendyl-M (bis-nor-)
UHY
LS/Q
$C_{14}H_{15}N_3S$
257.09867
RI: 2230
D: UHY
Antihistamine
N: rat
58
73
213
257
M+
285
100
200
300
Medrylamine
PS
LM/Q
$C_{18}H_{23}NO_2$
285.17288
RI: 2230
D: G U
Antihistamine
N: altered during HY
CAS: 524-99-2

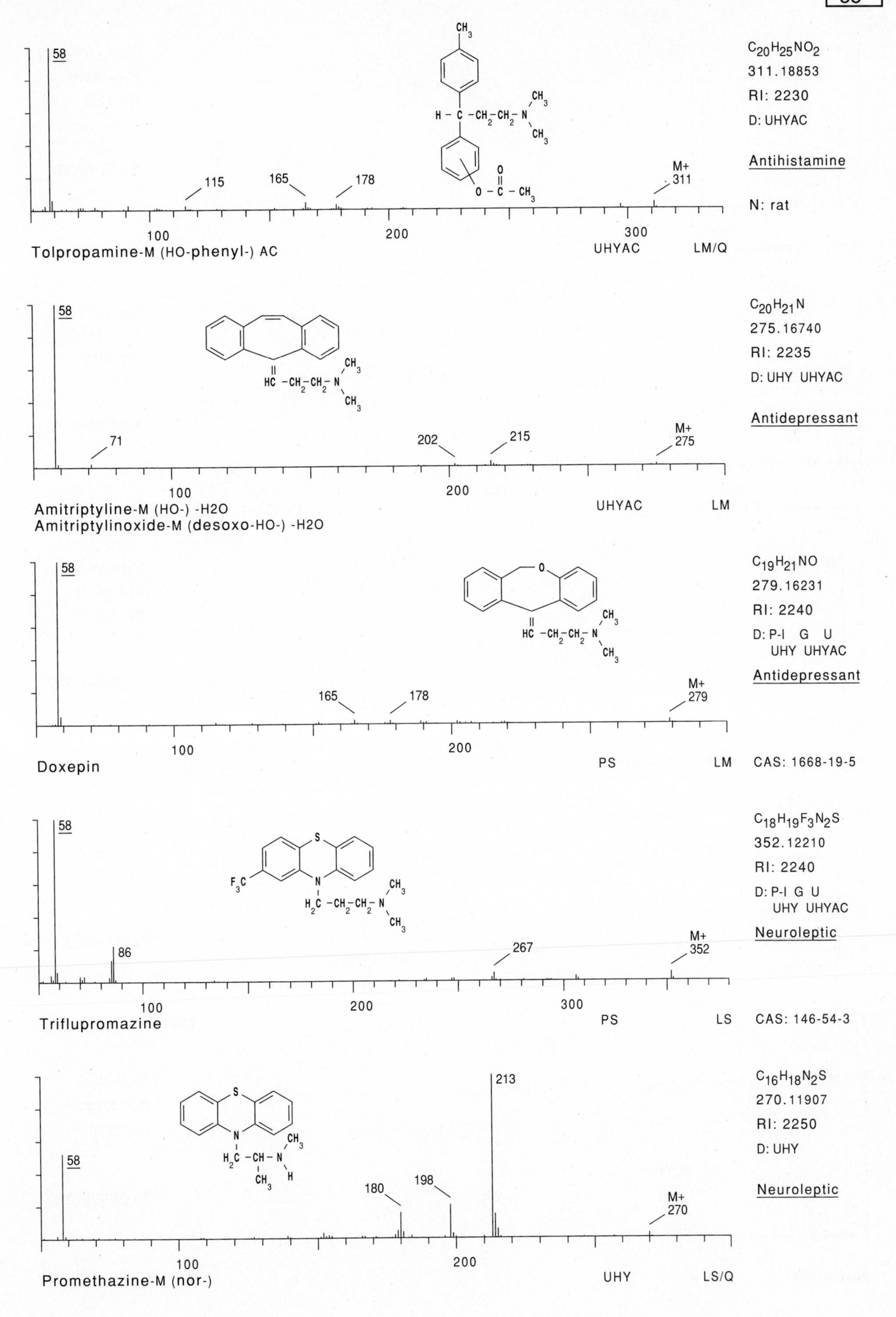

58
115
165
178
M+
311
100
200
300
Tolpropamine-M (HO-phenyl-) AC
UHYAC
LM/Q
$C_{20}H_{25}NO_2$
311.18853
RI: 2230
D: UHYAC
Antihistamine
N: rat
58
71
202
215
M+
275
100
200
Amitriptyline-M (HO-) -H2O
Amitriptylinoxide-M (desoxo-HO-) -H2O
UHYAC
LM
$C_{20}H_{21}N$
275.16740
RI: 2235
D: UHY UHYAC
Antidepressant
58
165
178
M+
279
100
200
Doxepin
PS
LM
$C_{19}H_{21}NO$
279.16231
RI: 2240
D: P-I G U
UHY UHYAC
Antidepressant
CAS: 1668-19-5
58
86
267
M+
352
100
200
300
Triflupromazine
PS
LS
$C_{18}H_{19}F_3N_2S$
352.12210
RI: 2240
D: P-I G U
UHY UHYAC
Neuroleptic
CAS: 146-54-3
213
58
180
198
M+
270
100
200
Promethazine-M (nor-)
UHY
LS/Q
$C_{16}H_{18}N_2S$
270.11907
RI: 2250
D: UHY
Neuroleptic

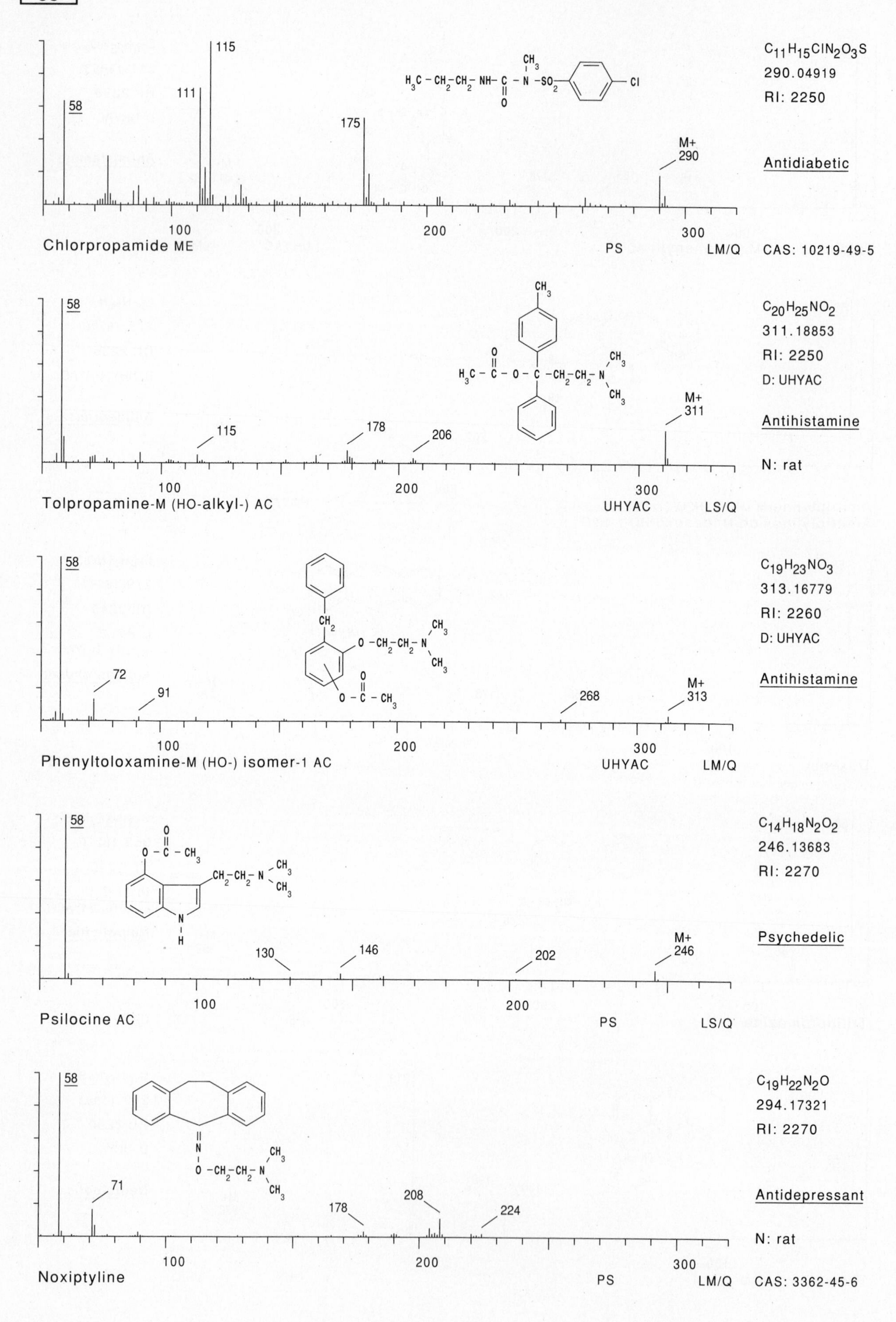
Chlorpropamide ME
PS LM/Q
C11H15ClN2O3S
290.04919
RI: 2250
Antidiabetic
CAS: 10219-49-5
Tolpropamine-M (HO-alkyl-) AC
UHYAC LS/Q
C20H25NO2
311.18853
RI: 2250
D: UHYAC
Antihistamine
N: rat
Phenyltoloxamine-M (HO-) isomer-1 AC
UHYAC LM/Q
C19H23NO3
313.16779
RI: 2260
D: UHYAC
Antihistamine
Psilocine AC
PS LS/Q
C14H18N2O2
246.13683
RI: 2270
Psychedelic
Noxiptyline
PS LM/Q
C19H22N2O
294.17321
RI: 2270
Antidepressant
N: rat
CAS: 3362-45-6

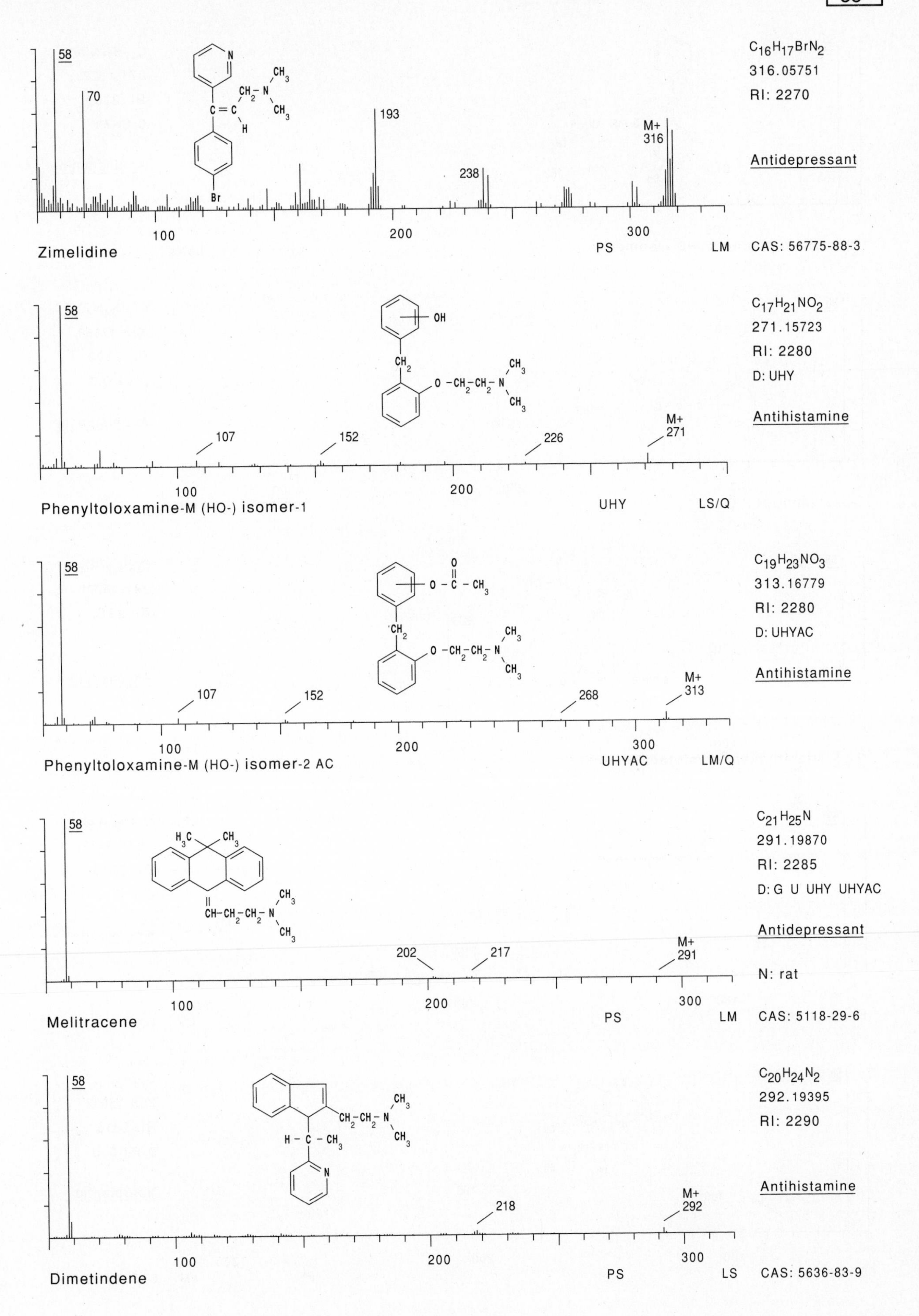

58
70
193
238
M+
316
Br
N
CH3
CH2 - N
C = C
H
100
200
300
Zimelidine
PS
LM
C16H17BrN2
316.05751
RI: 2270
Antidepressant
CAS: 56775-88-3
58
OH
CH2
O - CH2 - CH2 - N
CH3
107
152
226
M+
271
100
200
Phenyltoloxamine-M (HO-) isomer-1
UHY
LS/Q
C17H21NO2
271.15723
RI: 2280
D: UHY
Antihistamine
58
O - C - CH3
107
152
268
M+
313
100
200
300
Phenyltoloxamine-M (HO-) isomer-2 AC
UHYAC
LM/Q
C19H23NO3
313.16779
RI: 2280
D: UHYAC
Antihistamine
58
H3C
CH3
CH-CH2-CH2- N
202
217
M+
291
100
200
300
Melitracene
PS
LM
C21H25N
291.19870
RI: 2285
D: G U UHY UHYAC
Antidepressant
N: rat
CAS: 5118-29-6
58
CH2-CH2- N
H - C - CH3
218
M+
292
100
200
300
Dimetindene
PS
LS
C20H24N2
292.19395
RI: 2290
Antihistamine
CAS: 5636-83-9

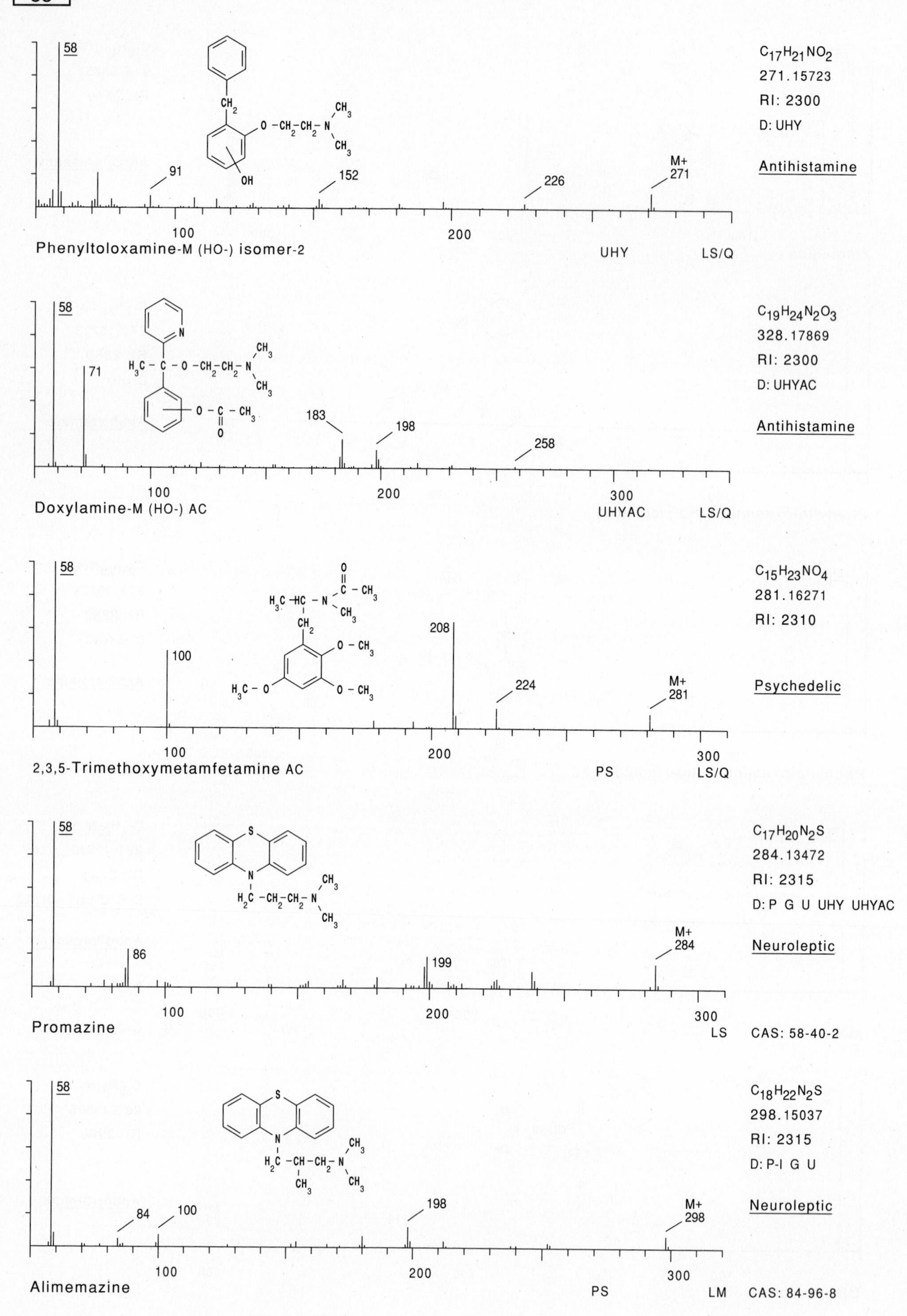
58
91
152
226
M+
271
CH2
O - CH2 - CH2 - N
CH3
CH3
OH
100
200
$C_{17}H_{21}NO_2$
271.15723
RI: 2300
D: UHY
Antihistamine
Phenyltoloxamine-M (HO-) isomer-2
UHY
LS/Q
58
71
N
H3C - C - O - CH2 - CH2 - N
CH3
CH3
O - C - CH3
O
183
198
258
100
200
300
$C_{19}H_{24}N_2O_3$
328.17869
RI: 2300
D: UHYAC
Antihistamine
Doxylamine-M (HO-) AC
UHYAC
LS/Q
58
100
O
C - CH3
H3C - HC - N
CH3
CH2
O - CH3
H3C - O
O - CH3
208
224
M+
281
100
200
300
$C_{15}H_{23}NO_4$
281.16271
RI: 2310
Psychedelic
2,3,5-Trimethoxymetamfetamine AC
PS
LS/Q
58
S
N
H2C - CH2 - CH2 - N
CH3
CH3
86
199
M+
284
100
200
300
$C_{17}H_{20}N_2S$
284.13472
RI: 2315
D: P G U UHY UHYAC
Neuroleptic
Promazine
LS
CAS: 58-40-2
58
S
N
H2C - CH - CH2 - N
CH3
CH3
CH3
84
100
198
M+
298
100
200
300
$C_{18}H_{22}N_2S$
298.15037
RI: 2315
D: P-I G U
Neuroleptic
Alimemazine
PS
LM
CAS: 84-96-8

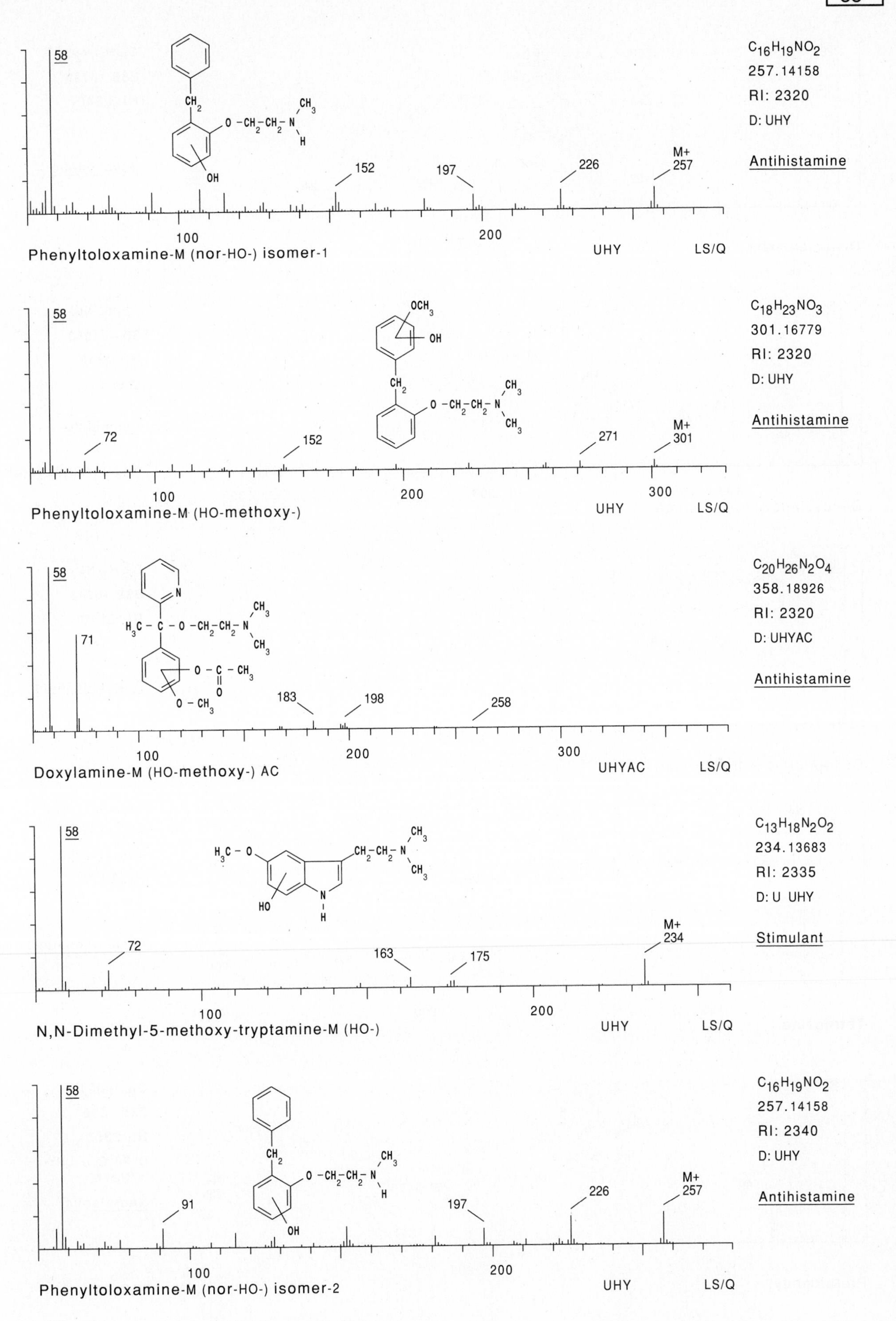

$C_{16}H_{19}NO_2$
257.14158
RI: 2320
D: UHY

Antihistamine

$C_{18}H_{23}NO_3$
301.16779
RI: 2320
D: UHY

Antihistamine

$C_{20}H_{26}N_2O_4$
358.18926
RI: 2320
D: UHYAC

Antihistamine

$C_{13}H_{18}N_2O_2$
234.13683
RI: 2335
D: U UHY

Stimulant

$C_{16}H_{19}NO_2$
257.14158
RI: 2340
D: UHY

Antihistamine

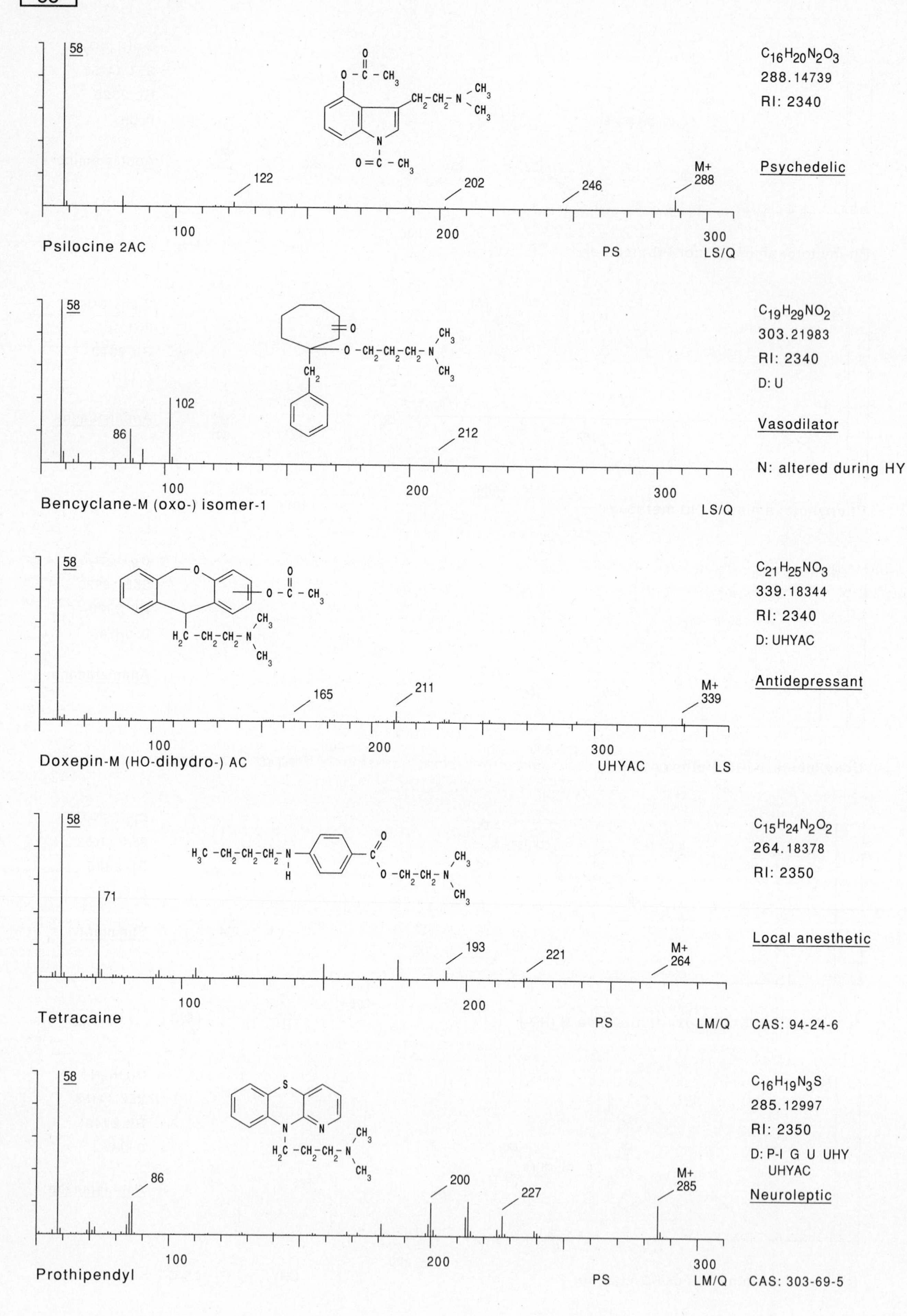
58
122
202
246
M+
288
100
200
300
Psilocine 2AC
PS
LS/Q
C16H20N2O3
288.14739
RI: 2340
Psychedelic
58
86
102
212
100
200
300
Bencyclane-M (oxo-) isomer-1
LS/Q
C19H29NO2
303.21983
RI: 2340
D: U
Vasodilator
N: altered during HY
58
165
211
M+
339
100
200
300
Doxepin-M (HO-dihydro-) AC
UHYAC
LS
C21H25NO3
339.18344
RI: 2340
D: UHYAC
Antidepressant
58
71
193
221
M+
264
100
200
Tetracaine
PS
LM/Q
C15H24N2O2
264.18378
RI: 2350
Local anesthetic
CAS: 94-24-6
58
86
200
227
M+
285
100
200
300
Prothipendyl
PS
LM/Q
C16H19N3S
285.12997
RI: 2350
D: P-I G U UHY
UHYAC
Neuroleptic
CAS: 303-69-5

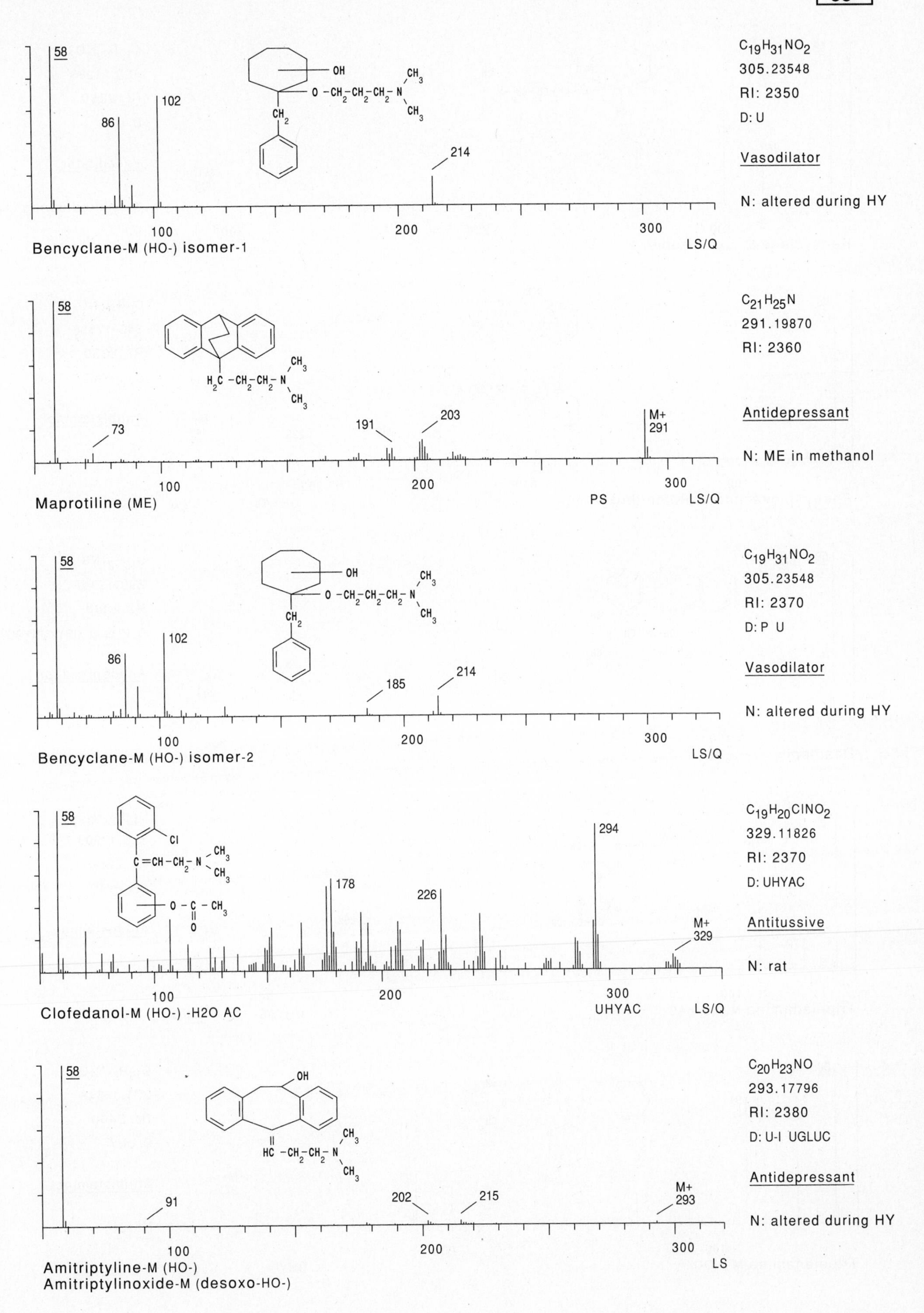
58
Bencyclane-M (HO-) isomer-1
LS/Q
$C_{19}H_{31}NO_2$
305.23548
RI: 2350
D: U
Vasodilator
N: altered during HY
Maprotiline (ME)
PS
LS/Q
$C_{21}H_{25}N$
291.19870
RI: 2360
Antidepressant
N: ME in methanol
Bencyclane-M (HO-) isomer-2
LS/Q
$C_{19}H_{31}NO_2$
305.23548
RI: 2370
D: P U
Vasodilator
N: altered during HY
Clofedanol-M (HO-) -H2O AC
UHYAC
LS/Q
$C_{19}H_{20}ClNO_2$
329.11826
RI: 2370
D: UHYAC
Antitussive
N: rat
Amitriptyline-M (HO-)
Amitriptylinoxide-M (desoxo-HO-)
LS
$C_{20}H_{23}NO$
293.17796
RI: 2380
D: U-I UGLUC
Antidepressant
N: altered during HY

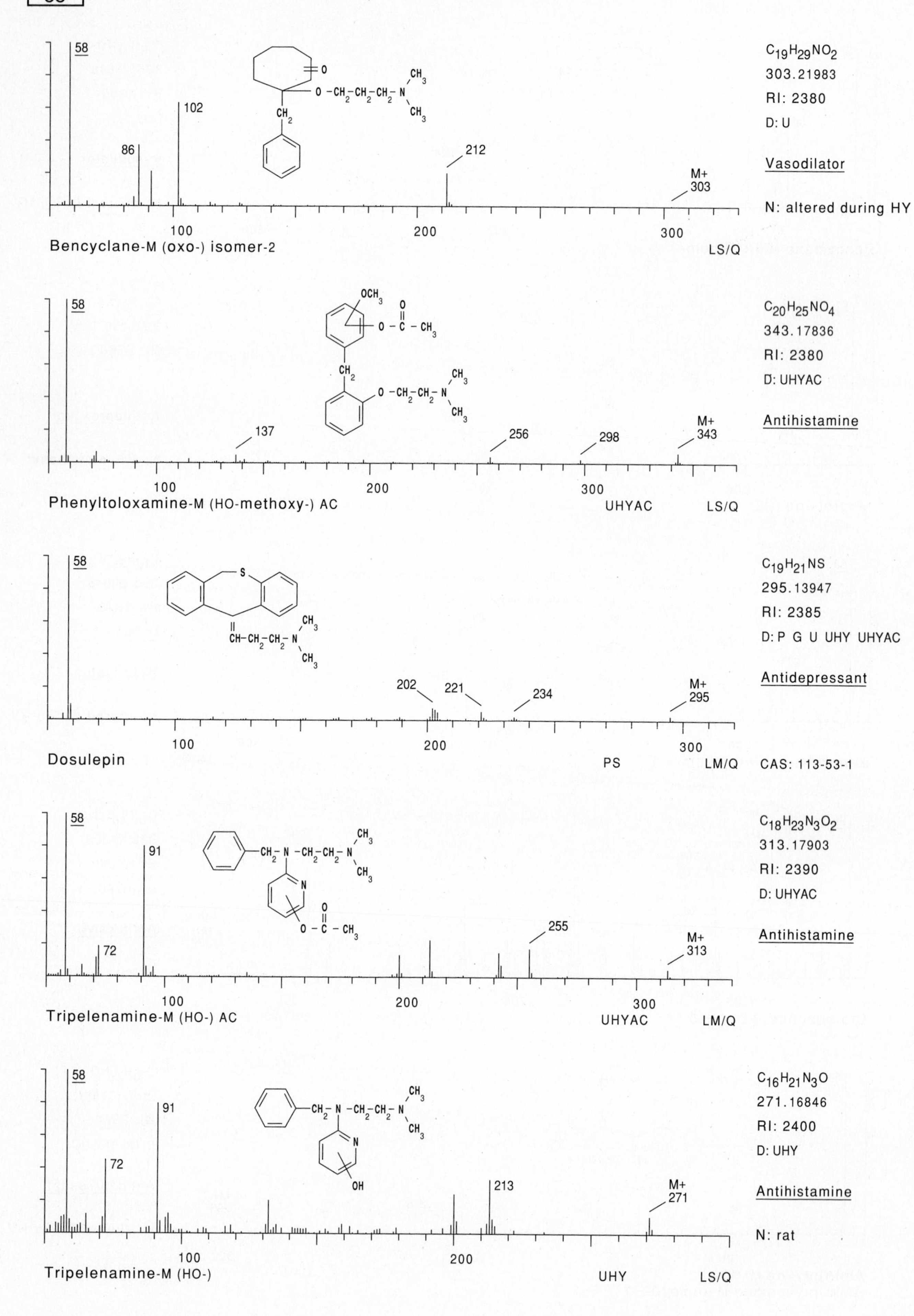

58
102
86
212
M+
303
100
200
300
Bencyclane-M (oxo-) isomer-2
LS/Q
C19H29NO2
303.21983
RI: 2380
D: U
Vasodilator
N: altered during HY
58
137
256
298
M+
343
100
200
300
Phenyltoloxamine-M (HO-methoxy-) AC
UHYAC
LS/Q
C20H25NO4
343.17836
RI: 2380
D: UHYAC
Antihistamine
58
202
221
234
M+
295
100
200
300
Dosulepin
PS
LM/Q
C19H21NS
295.13947
RI: 2385
D: P G U UHY UHYAC
Antidepressant
CAS: 113-53-1
58
91
72
255
M+
313
100
200
300
Tripelenamine-M (HO-) AC
UHYAC
LM/Q
C18H23N3O2
313.17903
RI: 2390
D: UHYAC
Antihistamine
58
91
72
213
M+
271
100
200
Tripelenamine-M (HO-)
UHY
LS/Q
C16H21N3O
271.16846
RI: 2400
D: UHY
Antihistamine
N: rat

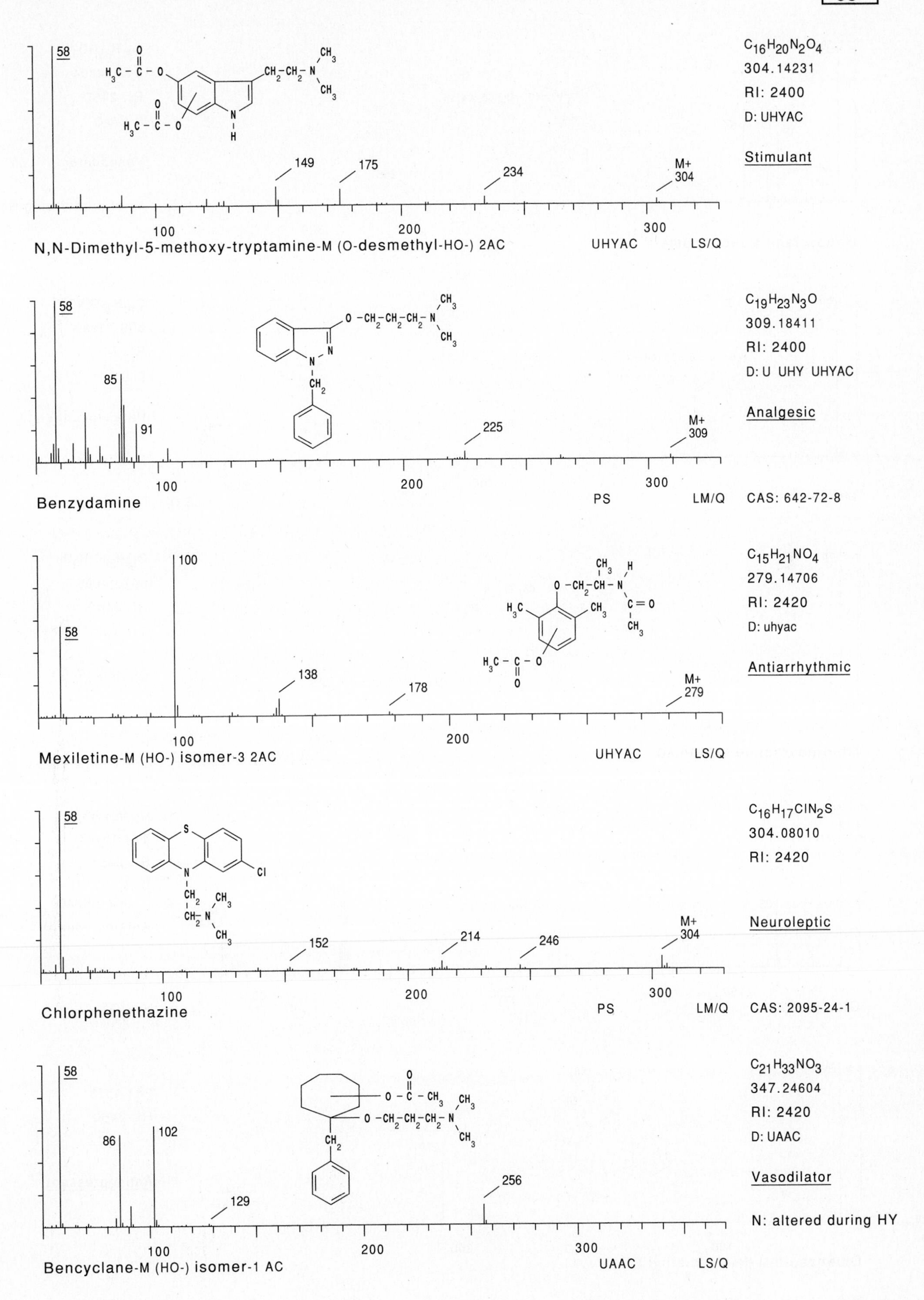
58
149
175
234
M+
304
100
200
300
N,N-Dimethyl-5-methoxy-tryptamine-M (O-desmethyl-HO-) 2AC
UHYAC
LS/Q
C16H20N2O4
304.14231
RI: 2400
D: UHYAC
Stimulant
58
85
91
225
M+
309
100
200
300
Benzydamine
PS
LM/Q
C19H23N3O
309.18411
RI: 2400
D: U UHY UHYAC
Analgesic
CAS: 642-72-8
100
58
138
178
M+
279
100
200
Mexiletine-M (HO-) isomer-3 2AC
UHYAC
LS/Q
C15H21NO4
279.14706
RI: 2420
D: uhyac
Antiarrhythmic
58
152
214
246
M+
304
100
200
300
Chlorphenethazine
PS
LM/Q
C16H17ClN2S
304.08010
RI: 2420
Neuroleptic
CAS: 2095-24-1
58
86
102
129
256
100
200
300
Bencyclane-M (HO-) isomer-1 AC
UAAC
LS/Q
C21H33NO3
347.24604
RI: 2420
D: UAAC
Vasodilator
N: altered during HY

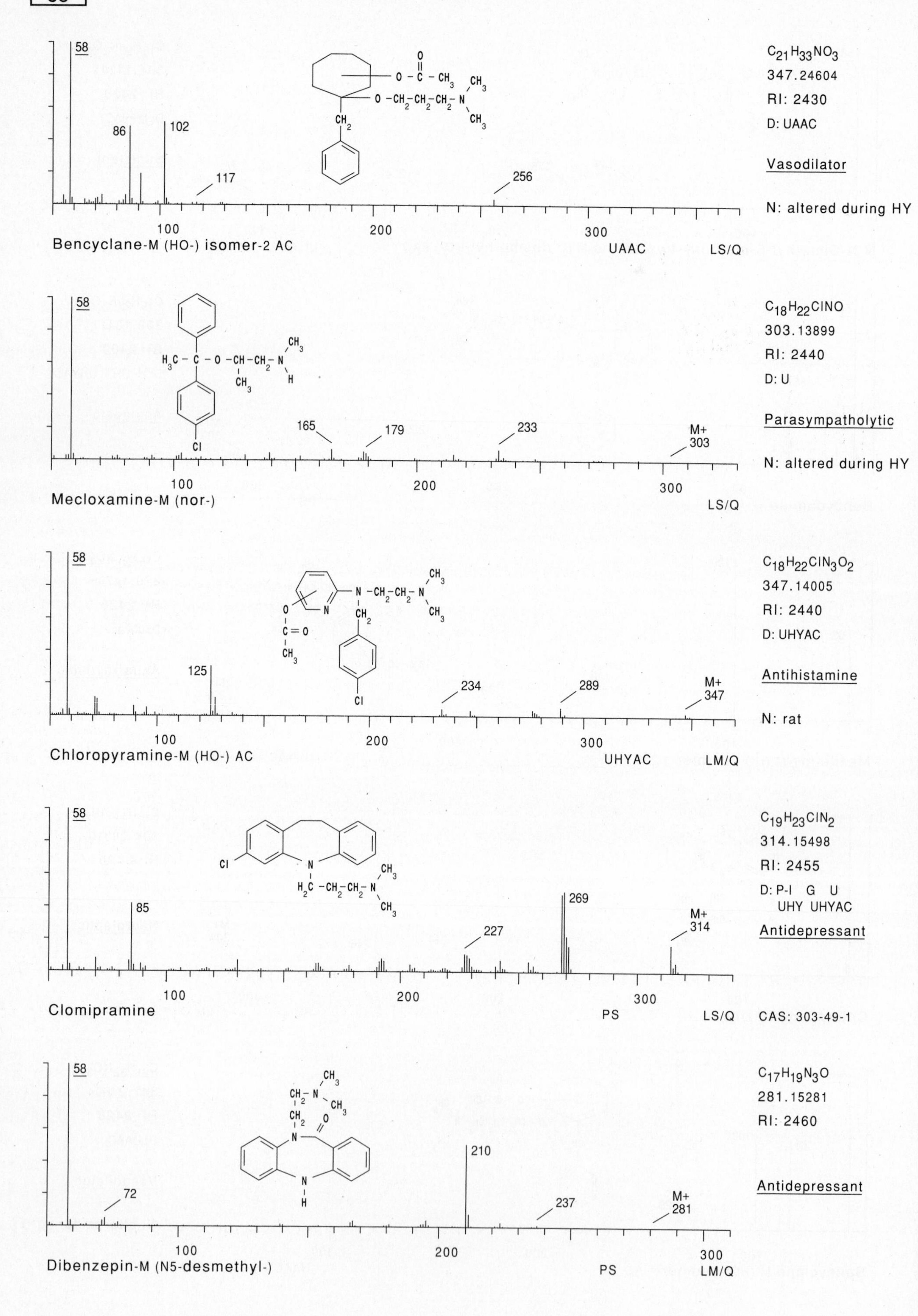
58
86
102
117
256
100
200
300
Bencyclane-M (HO-) isomer-2 AC
UAAC
LS/Q
$C_{21}H_{33}NO_3$
347.24604
RI: 2430
D: UAAC
Vasodilator
N: altered during HY
58
165
179
233
M+
303
100
200
300
Mecloxamine-M (nor-)
LS/Q
$C_{18}H_{22}ClNO$
303.13899
RI: 2440
D: U
Parasympatholytic
N: altered during HY
58
125
234
289
M+
347
100
200
300
Chloropyramine-M (HO-) AC
UHYAC
LM/Q
$C_{18}H_{22}ClN_3O_2$
347.14005
RI: 2440
D: UHYAC
Antihistamine
N: rat
58
85
227
269
M+
314
100
200
300
Clomipramine
PS
LS/Q
$C_{19}H_{23}ClN_2$
314.15498
RI: 2455
D: P-I G U
UHY UHYAC
Antidepressant
CAS: 303-49-1
58
72
210
237
M+
281
100
200
300
Dibenzepin-M (N5-desmethyl-)
PS
LM/Q
$C_{17}H_{19}N_3O$
281.15281
RI: 2460
Antidepressant

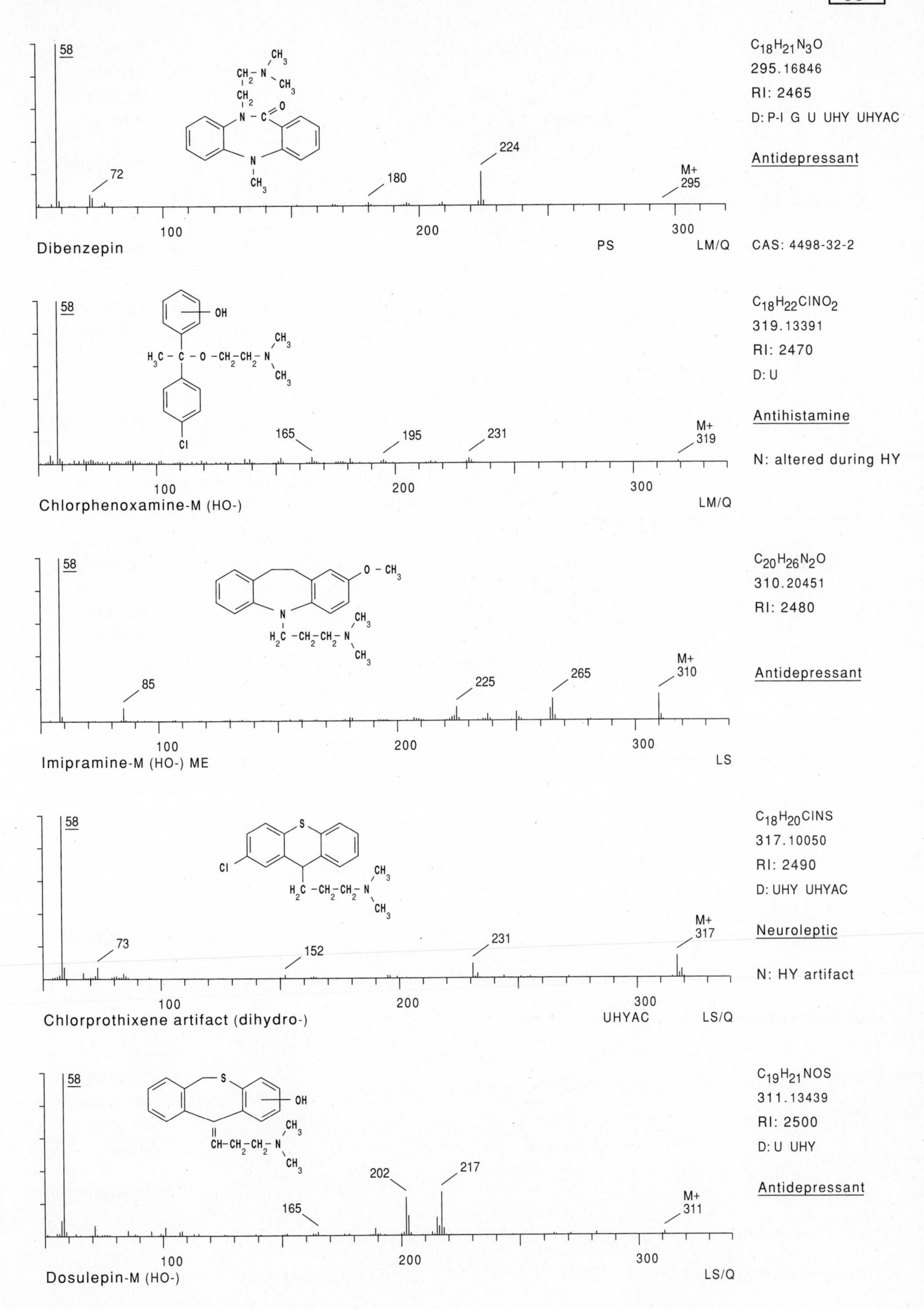

Dibenzepin

$C_{18}H_{21}N_3O$
295.16846
RI: 2465
D: P-I G U UHY UHYAC
Antidepressant
CAS: 4498-32-2

Chlorphenoxamine-M (HO-)

$C_{18}H_{22}ClNO_2$
319.13391
RI: 2470
D: U
Antihistamine
N: altered during HY

Imipramine-M (HO-) ME

$C_{20}H_{26}N_2O$
310.20451
RI: 2480
Antidepressant

Chlorprothixene artifact (dihydro-)

$C_{18}H_{20}ClNS$
317.10050
RI: 2490
D: UHY UHYAC
Neuroleptic
N: HY artifact

Dosulepin-M (HO-)

$C_{19}H_{21}NOS$
311.13439
RI: 2500
D: U UHY
Antidepressant

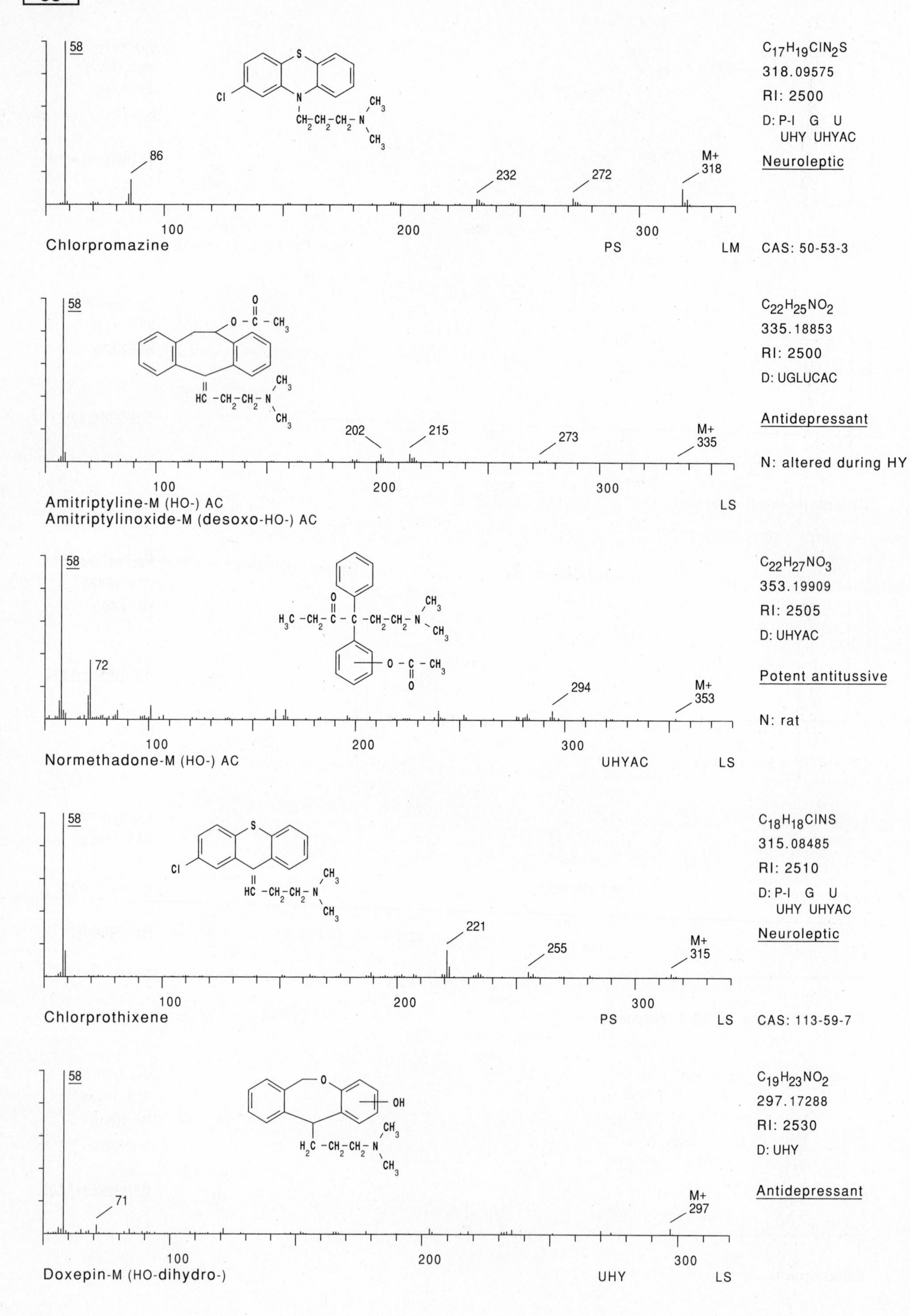

58
86
232
272
M+
318
100
200
300
Chlorpromazine
PS
LM
$C_{17}H_{19}ClN_2S$
318.09575
RI: 2500
D: P-I G U
UHY UHYAC
Neuroleptic
CAS: 50-53-3
58
202
215
273
M+
335
Amitriptyline-M (HO-) AC
Amitriptylinoxide-M (desoxo-HO-) AC
LS
$C_{22}H_{25}NO_2$
335.18853
RI: 2500
D: UGLUCAC
Antidepressant
N: altered during HY
58
72
294
M+
353
Normethadone-M (HO-) AC
UHYAC
LS
$C_{22}H_{27}NO_3$
353.19909
RI: 2505
D: UHYAC
Potent antitussive
N: rat
58
221
255
M+
315
Chlorprothixene
PS
LS
$C_{18}H_{18}ClNS$
315.08485
RI: 2510
D: P-I G U
UHY UHYAC
Neuroleptic
CAS: 113-59-7
58
71
M+
297
Doxepin-M (HO-dihydro-)
UHY
LS
$C_{19}H_{23}NO_2$
297.17288
RI: 2530
D: UHY
Antidepressant

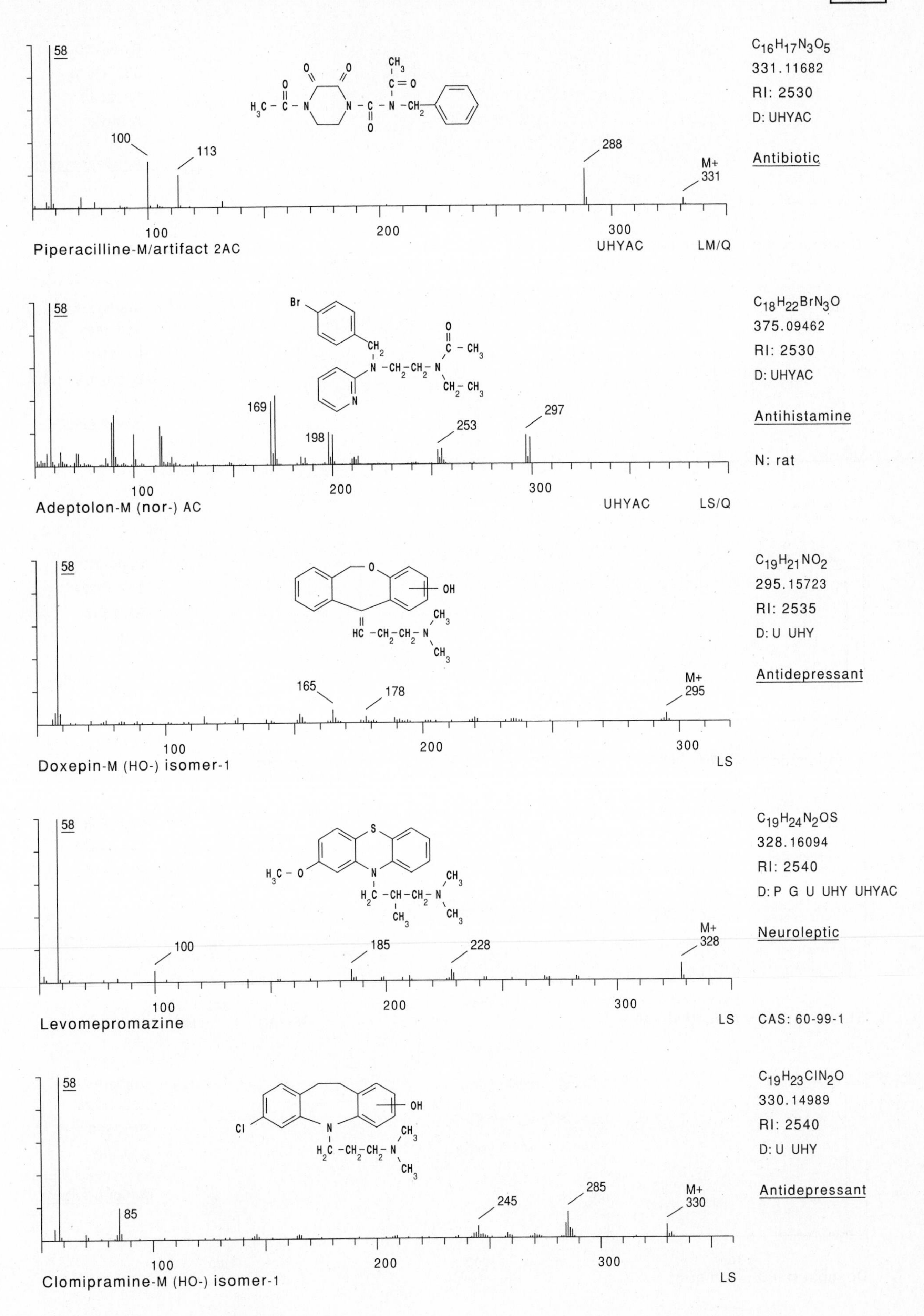
58
100
113
288
M+
331
100
200
300
Piperacilline-M/artifact 2AC
UHYAC
LM/Q
$C_{16}H_{17}N_3O_5$
331.11682
RI: 2530
D: UHYAC
Antibiotic
58
169
198
253
297
Br
N
Adeptolon-M (nor-) AC
UHYAC
LS/Q
$C_{18}H_{22}BrN_3O$
375.09462
RI: 2530
D: UHYAC
Antihistamine
N: rat
58
OH
165
178
M+
295
Doxepin-M (HO-) isomer-1
LS
$C_{19}H_{21}NO_2$
295.15723
RI: 2535
D: U UHY
Antidepressant
58
S
N
100
185
228
M+
328
Levomepromazine
LS
CAS: 60-99-1
$C_{19}H_{24}N_2OS$
328.16094
RI: 2540
D: P G U UHY UHYAC
Neuroleptic
58
OH
Cl
N
85
245
285
M+
330
Clomipramine-M (HO-) isomer-1
LS
$C_{19}H_{23}ClN_2O$
330.14989
RI: 2540
D: U UHY
Antidepressant

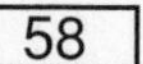

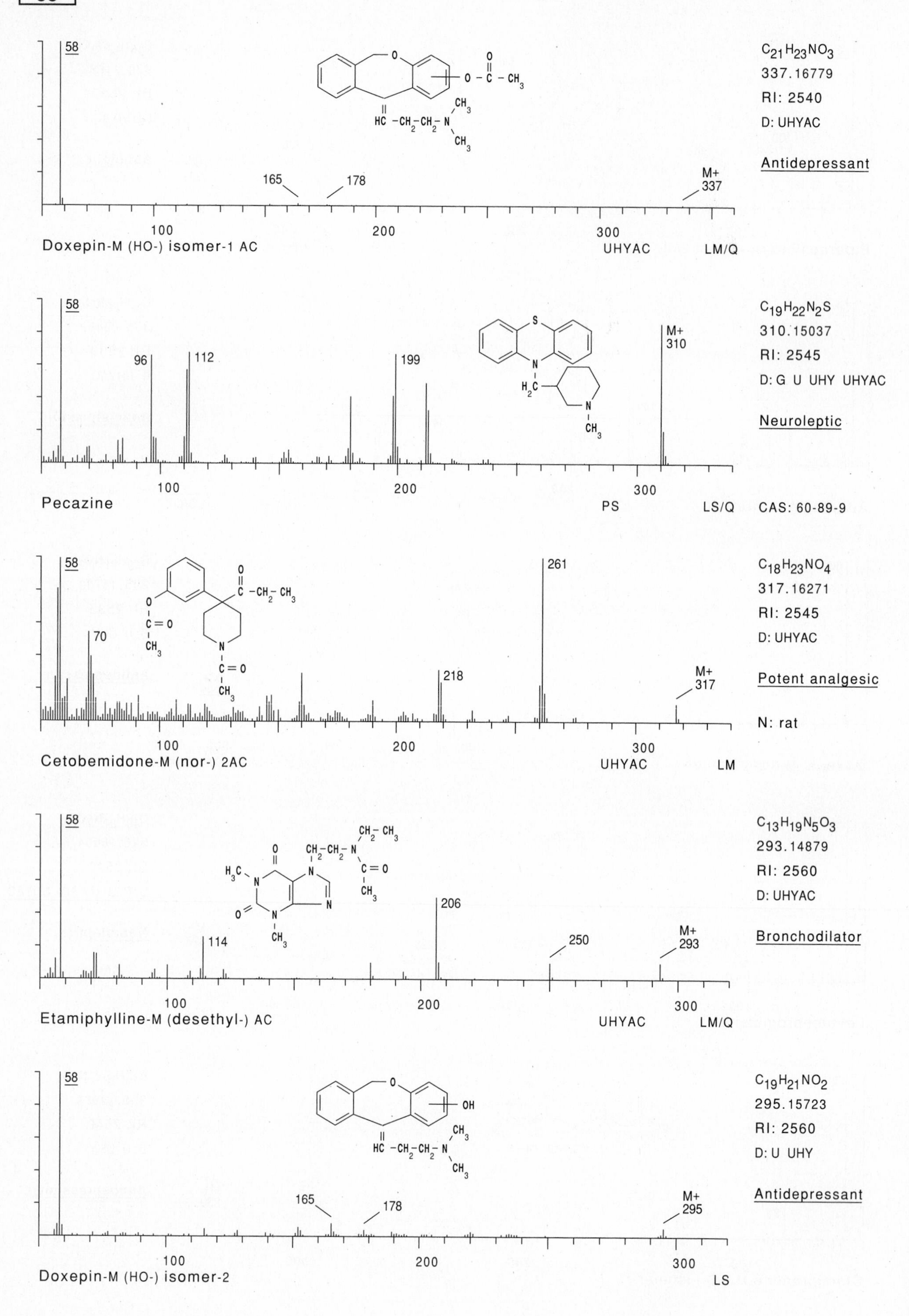

58
165
178
M+
337
100
200
300
Doxepin-M (HO-) isomer-1 AC
UHYAC
LM/Q
$C_{21}H_{23}NO_3$
337.16779
RI: 2540
D: UHYAC
Antidepressant
58
96
112
199
M+
310
100
200
300
Pecazine
PS
LS/Q
$C_{19}H_{22}N_2S$
310.15037
RI: 2545
D: G U UHY UHYAC
Neuroleptic
CAS: 60-89-9
58
70
218
261
M+
317
100
200
300
Cetobemidone-M (nor-) 2AC
UHYAC
LM
$C_{18}H_{23}NO_4$
317.16271
RI: 2545
D: UHYAC
Potent analgesic
N: rat
58
114
206
250
M+
293
100
200
300
Etamiphylline-M (desethyl-) AC
UHYAC
LM/Q
$C_{13}H_{19}N_5O_3$
293.14879
RI: 2560
D: UHYAC
Bronchodilator
58
165
178
M+
295
100
200
300
Doxepin-M (HO-) isomer-2
LS
$C_{19}H_{21}NO_2$
295.15723
RI: 2560
D: U UHY
Antidepressant

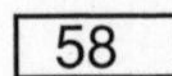
58

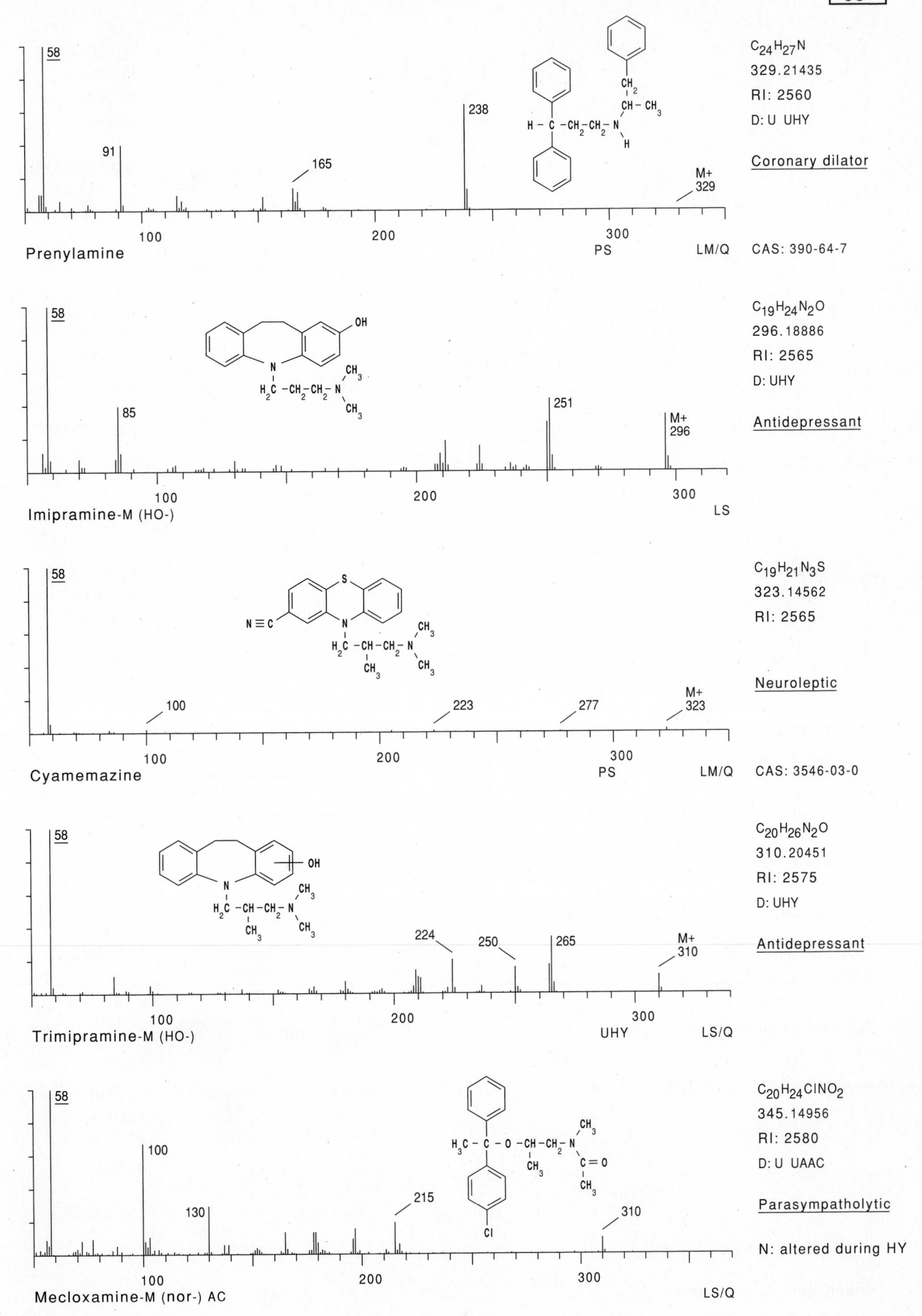
58
91
165
238
M+
329
100
200
300
PS
LM/Q
Prenylamine
C24H27N
329.21435
RI: 2560
D: U UHY
Coronary dilator
CAS: 390-64-7
58
85
251
M+
296
100
200
300
LS
Imipramine-M (HO-)
C19H24N2O
296.18886
RI: 2565
D: UHY
Antidepressant
58
100
223
277
M+
323
100
200
300
PS
LM/Q
Cyamemazine
C19H21N3S
323.14562
RI: 2565
Neuroleptic
CAS: 3546-03-0
58
224
250
265
M+
310
100
200
300
UHY
LS/Q
Trimipramine-M (HO-)
C20H26N2O
310.20451
RI: 2575
D: UHY
Antidepressant
58
100
130
215
310
100
200
300
LS/Q
Mecloxamine-M (nor-) AC
C20H24ClNO2
345.14956
RI: 2580
D: U UAAC
Parasympatholytic
N: altered during HY

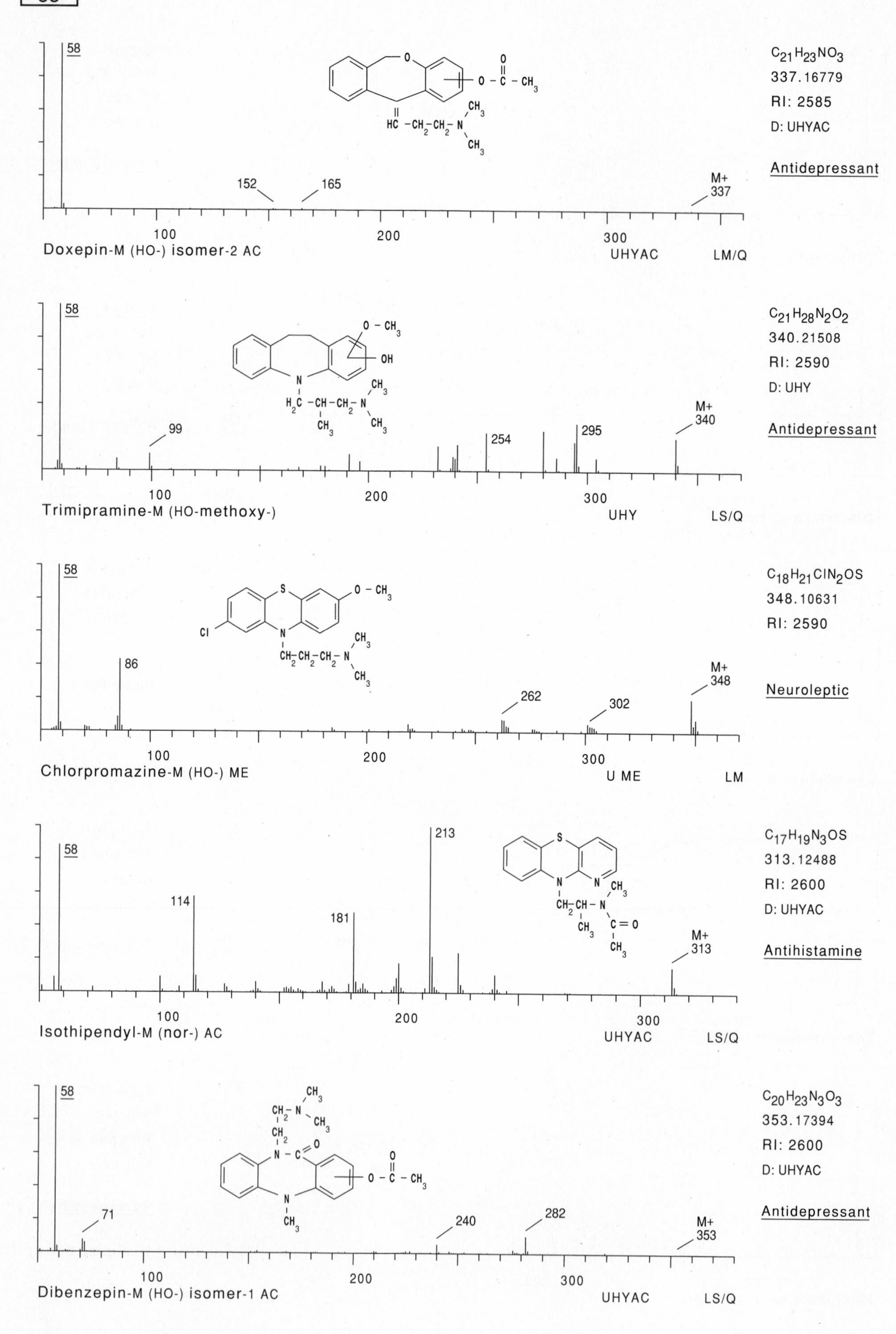
58
152
165
M+
337
100
200
300
Doxepin-M (HO-) isomer-2 AC
UHYAC
LM/Q
$C_{21}H_{23}NO_3$
337.16779
RI: 2585
D: UHYAC
Antidepressant
58
99
254
295
M+
340
Trimipramine-M (HO-methoxy-)
UHY
LS/Q
$C_{21}H_{28}N_2O_2$
340.21508
RI: 2590
D: UHY
Antidepressant
58
86
262
302
M+
348
Chlorpromazine-M (HO-) ME
U ME
LM
$C_{18}H_{21}ClN_2OS$
348.10631
RI: 2590
Neuroleptic
58
114
181
213
M+
313
Isothipendyl-M (nor-) AC
UHYAC
LS/Q
$C_{17}H_{19}N_3OS$
313.12488
RI: 2600
D: UHYAC
Antihistamine
58
71
240
282
M+
353
Dibenzepin-M (HO-) isomer-1 AC
UHYAC
LS/Q
$C_{20}H_{23}N_3O_3$
353.17394
RI: 2600
D: UHYAC
Antidepressant

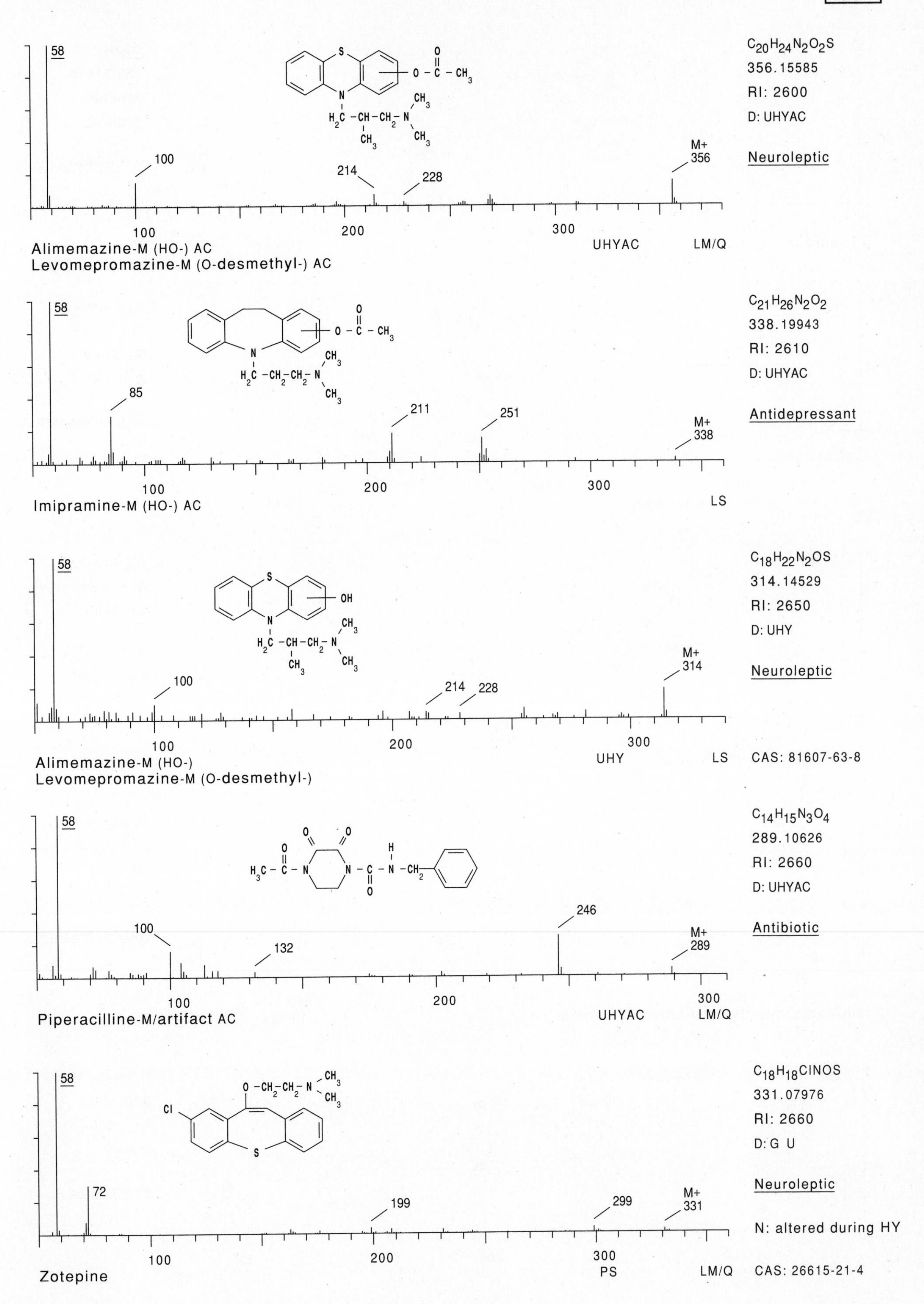

C20H24N2O2S
356.15585
RI: 2600
D: UHYAC
Neuroleptic
M+ 356
UHYAC
LM/Q
Alimemazine-M (HO-) AC
Levomepromazine-M (O-desmethyl-) AC
C21H26N2O2
338.19943
RI: 2610
D: UHYAC
Antidepressant
M+ 338
LS
Imipramine-M (HO-) AC
C18H22N2OS
314.14529
RI: 2650
D: UHY
Neuroleptic
M+ 314
UHY
LS
CAS: 81607-63-8
Alimemazine-M (HO-)
Levomepromazine-M (O-desmethyl-)
C14H15N3O4
289.10626
RI: 2660
D: UHYAC
Antibiotic
M+ 289
UHYAC
LM/Q
Piperacilline-M/artifact AC
C18H18ClNOS
331.07976
RI: 2660
D: G U
Neuroleptic
N: altered during HY
M+ 331
PS
LM/Q
CAS: 26615-21-4
Zotepine

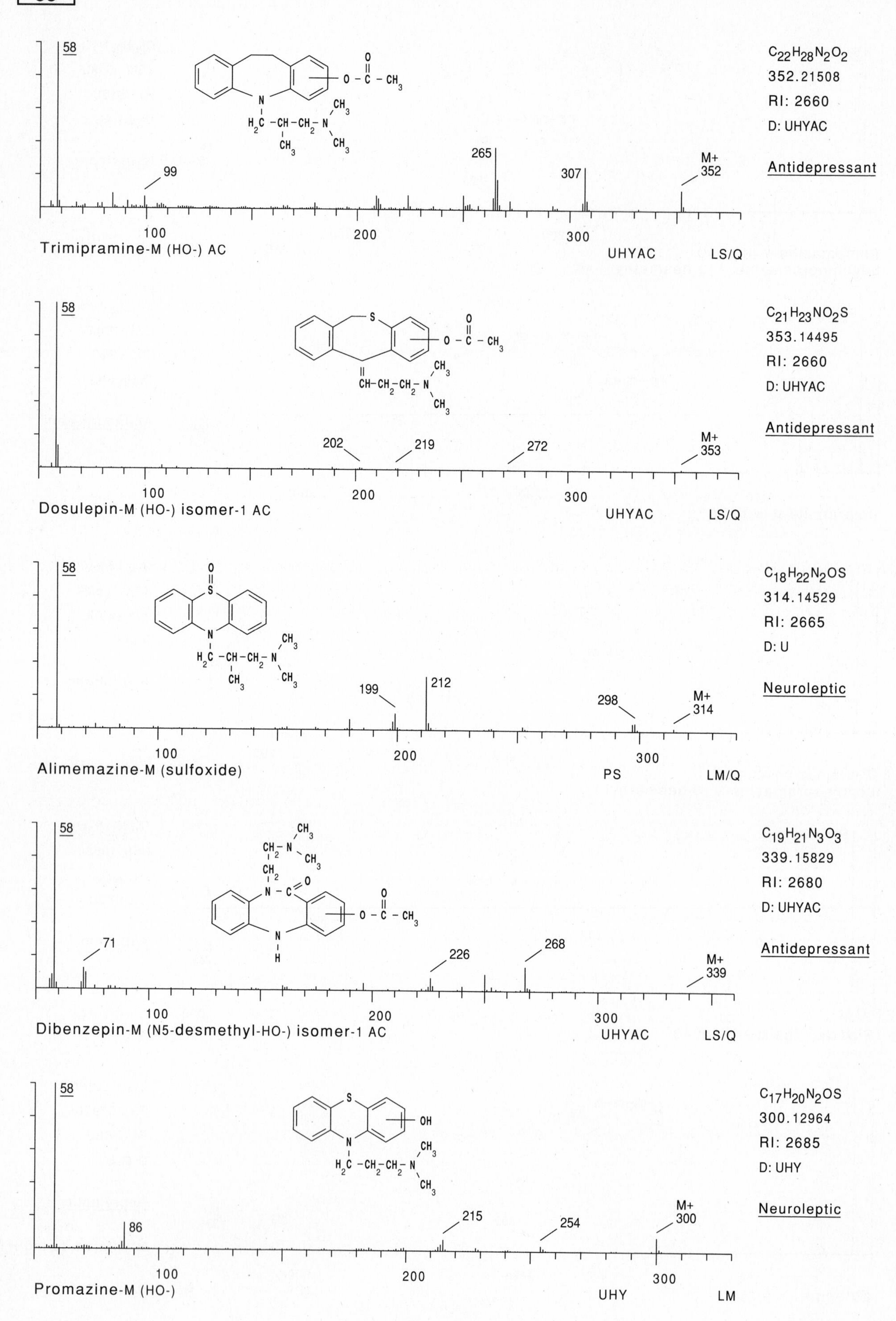

58
99
265
307
M+
352
100
200
300
Trimipramine-M (HO-) AC
UHYAC
LS/Q
C22H28N2O2
352.21508
RI: 2660
D: UHYAC
Antidepressant
58
202
219
272
M+
353
100
200
300
Dosulepin-M (HO-) isomer-1 AC
UHYAC
LS/Q
C21H23NO2S
353.14495
RI: 2660
D: UHYAC
Antidepressant
58
199
212
298
M+
314
100
200
300
Alimemazine-M (sulfoxide)
PS
LM/Q
C18H22N2OS
314.14529
RI: 2665
D: U
Neuroleptic
58
71
226
268
M+
339
100
200
300
Dibenzepin-M (N5-desmethyl-HO-) isomer-1 AC
UHYAC
LS/Q
C19H21N3O3
339.15829
RI: 2680
D: UHYAC
Antidepressant
58
86
215
254
M+
300
100
200
300
Promazine-M (HO-)
UHY
LM
C17H20N2OS
300.12964
RI: 2685
D: UHY
Neuroleptic

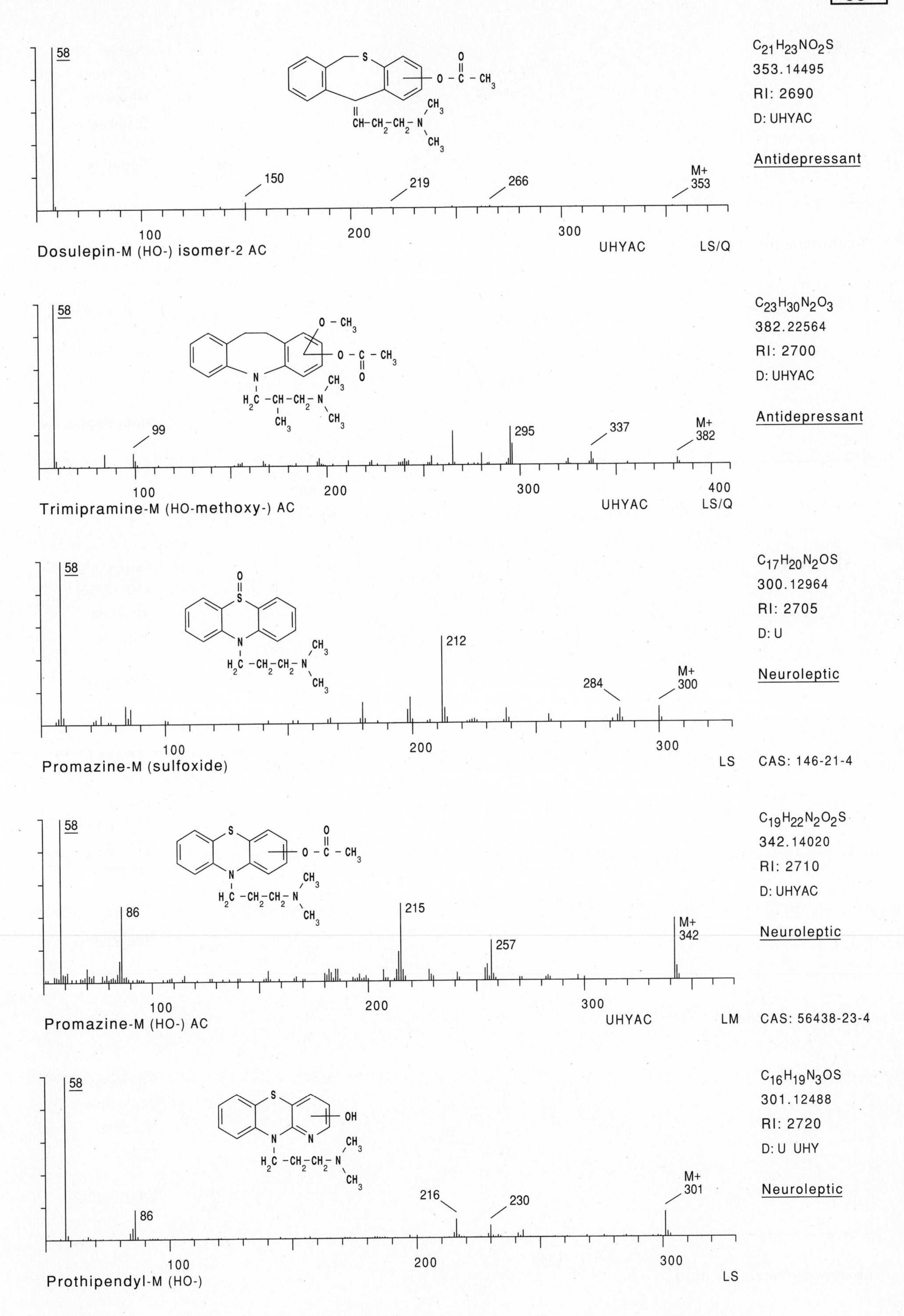
Dosulepin-M (HO-) isomer-2 AC
$C_{21}H_{23}NO_2S$
353.14495
RI: 2690
D: UHYAC
Antidepressant
58
150
219
266
M+
353
UHYAC
LS/Q
Trimipramine-M (HO-methoxy-) AC
$C_{23}H_{30}N_2O_3$
382.22564
RI: 2700
D: UHYAC
Antidepressant
58
99
295
337
M+
382
UHYAC
LS/Q
Promazine-M (sulfoxide)
$C_{17}H_{20}N_2OS$
300.12964
RI: 2705
D: U
Neuroleptic
58
212
284
M+
300
LS
CAS: 146-21-4
Promazine-M (HO-) AC
$C_{19}H_{22}N_2O_2S$
342.14020
RI: 2710
D: UHYAC
Neuroleptic
58
86
215
257
M+
342
UHYAC
LM
CAS: 56438-23-4
Prothipendyl-M (HO-)
$C_{16}H_{19}N_3OS$
301.12488
RI: 2720
D: U UHY
Neuroleptic
58
86
216
230
M+
301
LS

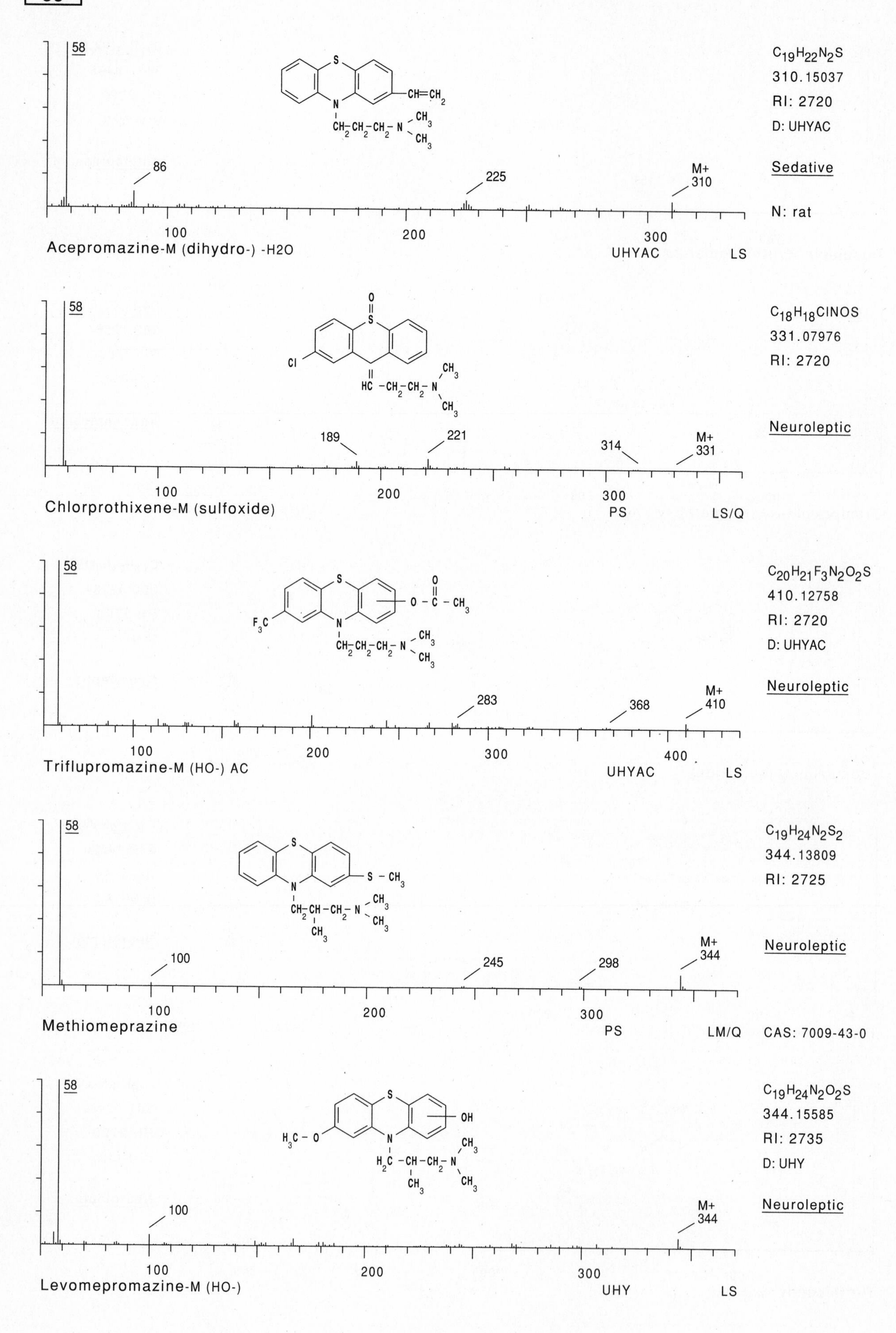
58
86
225
M+
310
100
200
300
Acepromazine-M (dihydro-) -H2O
UHYAC
LS
C19H22N2S
310.15037
RI: 2720
D: UHYAC
Sedative
N: rat
58
189
221
314
M+
331
100
200
300
Chlorprothixene-M (sulfoxide)
PS
LS/Q
C18H18ClNOS
331.07976
RI: 2720
Neuroleptic
58
283
368
M+
410
100
200
300
400
Triflupromazine-M (HO-) AC
UHYAC
LS
C20H21F3N2O2S
410.12758
RI: 2720
D: UHYAC
Neuroleptic
58
100
245
298
M+
344
100
200
300
Methiomeprazine
PS
LM/Q
CAS: 7009-43-0
C19H24N2S2
344.13809
RI: 2725
Neuroleptic
58
100
M+
344
100
200
300
Levomepromazine-M (HO-)
UHY
LS
C19H24N2O2S
344.15585
RI: 2735
D: UHY
Neuroleptic

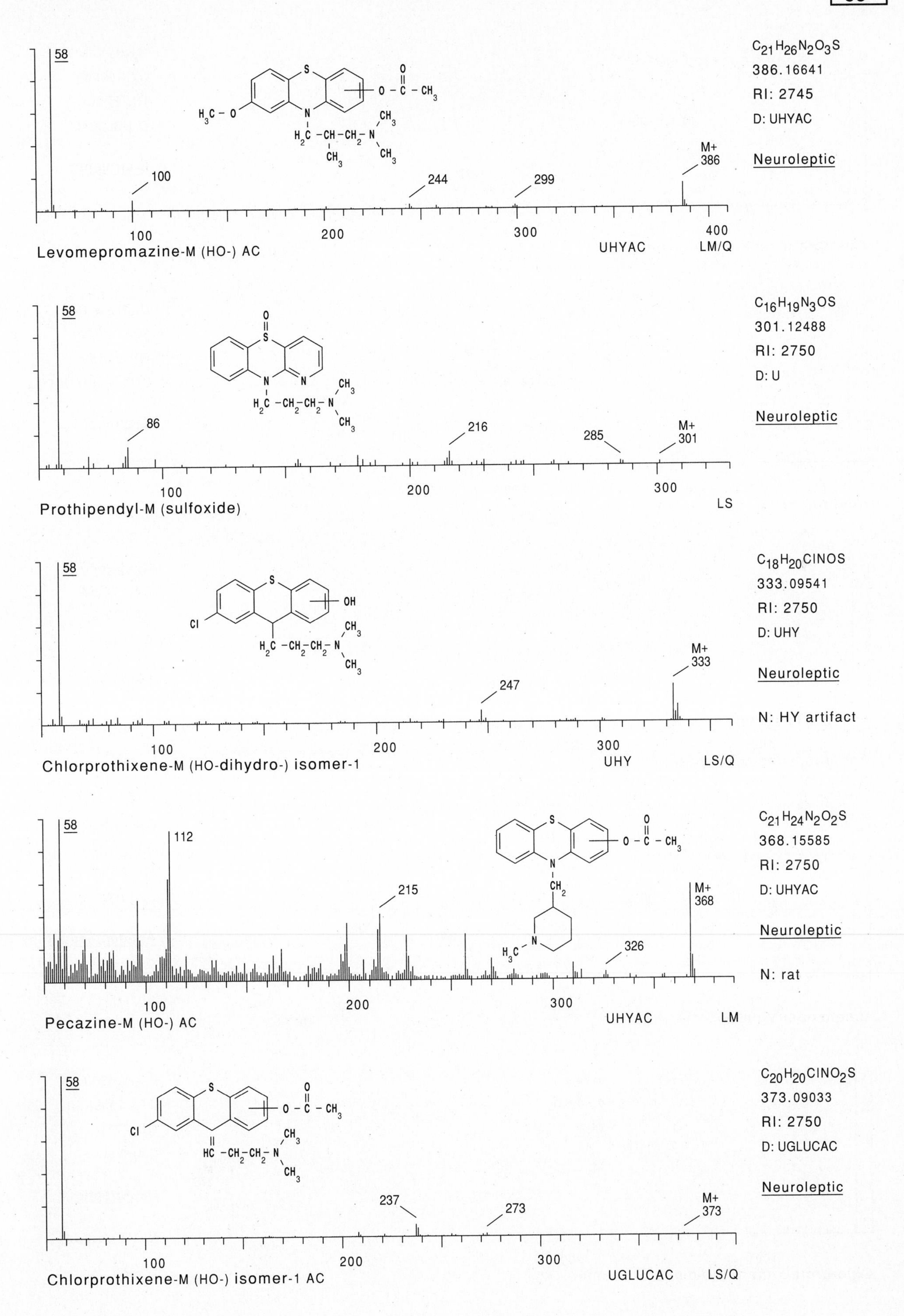

$C_{21}H_{26}N_2O_3S$
386.16641
RI: 2745
D: UHYAC

Neuroleptic

$C_{16}H_{19}N_3OS$
301.12488
RI: 2750
D: U

Neuroleptic

$C_{18}H_{20}ClNOS$
333.09541
RI: 2750
D: UHY

Neuroleptic

N: HY artifact

$C_{21}H_{24}N_2O_2S$
368.15585
RI: 2750
D: UHYAC

Neuroleptic

N: rat

$C_{20}H_{20}ClNO_2S$
373.09033
RI: 2750
D: UGLUCAC

Neuroleptic

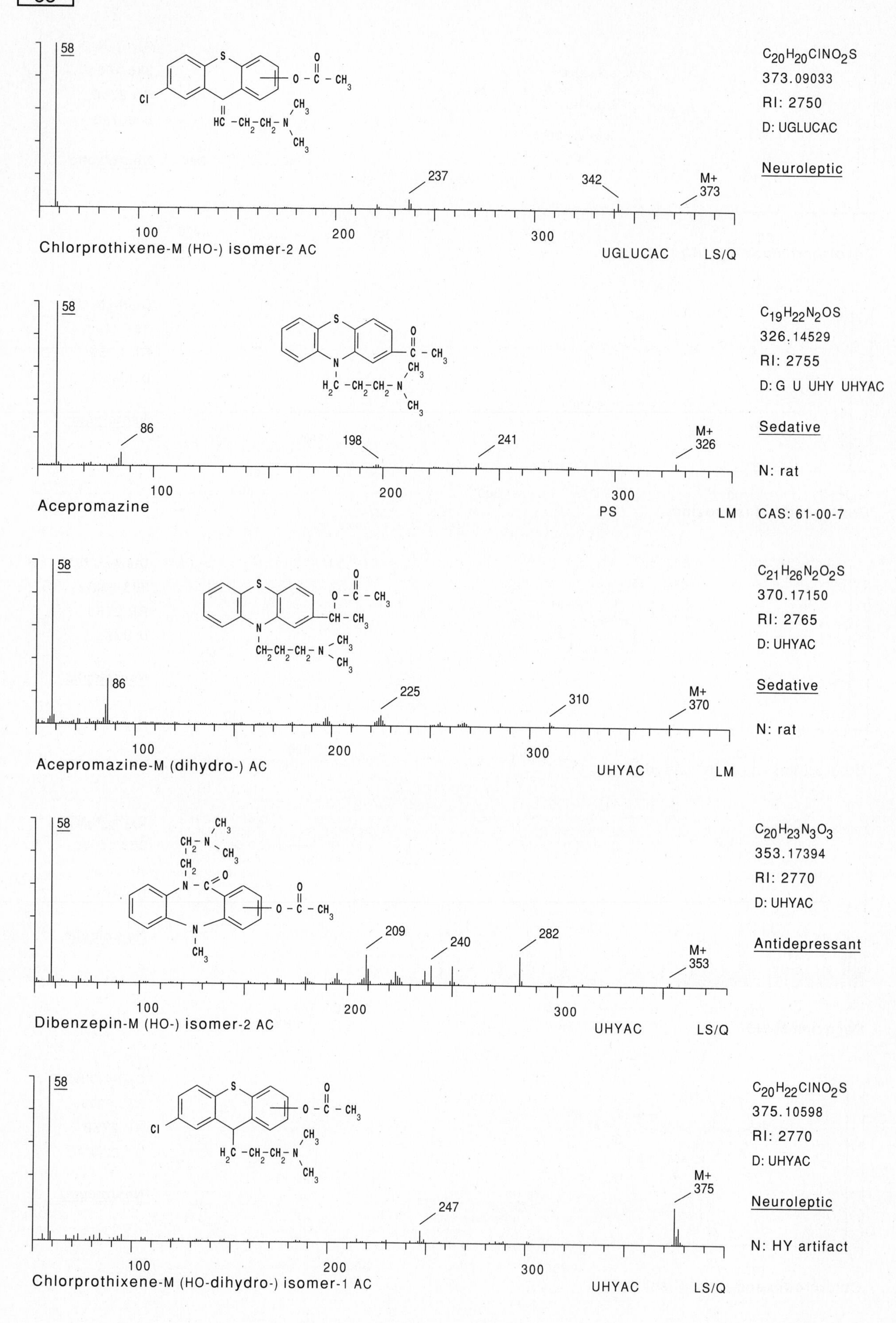

58
237
342
M+ 373
100
200
300
Chlorprothixene-M (HO-) isomer-2 AC
UGLUCAC
LS/Q
C20H20ClNO2S
373.09033
RI: 2750
D: UGLUCAC
Neuroleptic
58
86
198
241
M+ 326
Acepromazine
PS
LM
C19H22N2OS
326.14529
RI: 2755
D: G U UHY UHYAC
Sedative
N: rat
CAS: 61-00-7
58
86
225
310
M+ 370
Acepromazine-M (dihydro-) AC
UHYAC
LM
C21H26N2O2S
370.17150
RI: 2765
D: UHYAC
Sedative
N: rat
58
209
240
282
M+ 353
Dibenzepin-M (HO-) isomer-2 AC
UHYAC
LS/Q
C20H23N3O3
353.17394
RI: 2770
D: UHYAC
Antidepressant
58
247
M+ 375
Chlorprothixene-M (HO-dihydro-) isomer-1 AC
UHYAC
LS/Q
C20H22ClNO2S
375.10598
RI: 2770
D: UHYAC
Neuroleptic
N: HY artifact

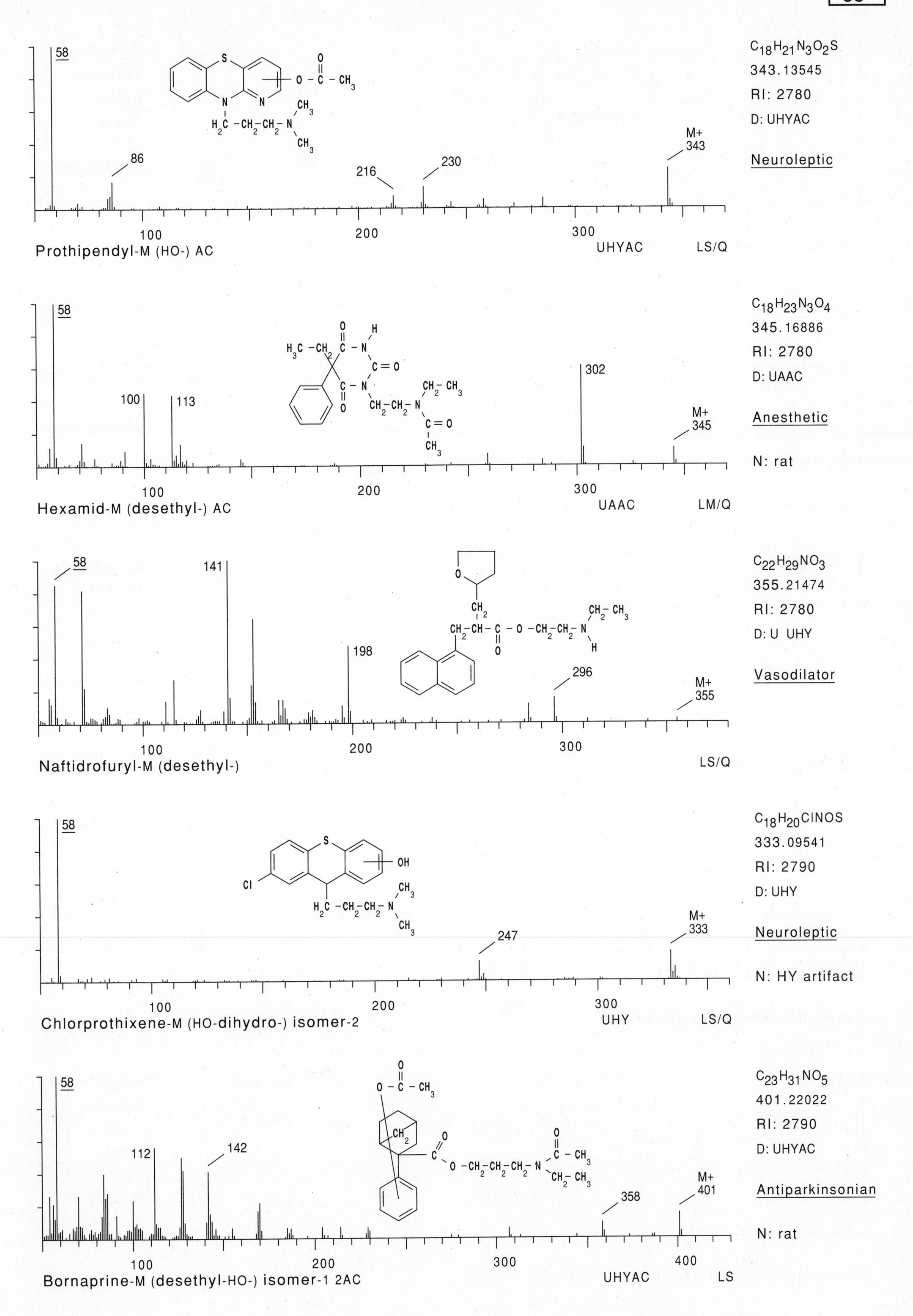
58
86
216
230
M+
343
100
200
300
UHYAC
LS/Q
Prothipendyl-M (HO-) AC
$C_{18}H_{21}N_3O_2S$
343.13545
RI: 2780
D: UHYAC
Neuroleptic
58
100
113
302
M+
345
UAAC
LM/Q
Hexamid-M (desethyl-) AC
$C_{18}H_{23}N_3O_4$
345.16886
RI: 2780
D: UAAC
Anesthetic
N: rat
58
141
198
296
M+
355
LS/Q
Naftidrofuryl-M (desethyl-)
$C_{22}H_{29}NO_3$
355.21474
RI: 2780
D: U UHY
Vasodilator
58
247
M+
333
UHY
LS/Q
Chlorprothixene-M (HO-dihydro-) isomer-2
$C_{18}H_{20}ClNOS$
333.09541
RI: 2790
D: UHY
Neuroleptic
N: HY artifact
58
112
142
358
M+
401
400
UHYAC
LS
Bornaprine-M (desethyl-HO-) isomer-1 2AC
$C_{23}H_{31}NO_5$
401.22022
RI: 2790
D: UHYAC
Antiparkinsonian
N: rat

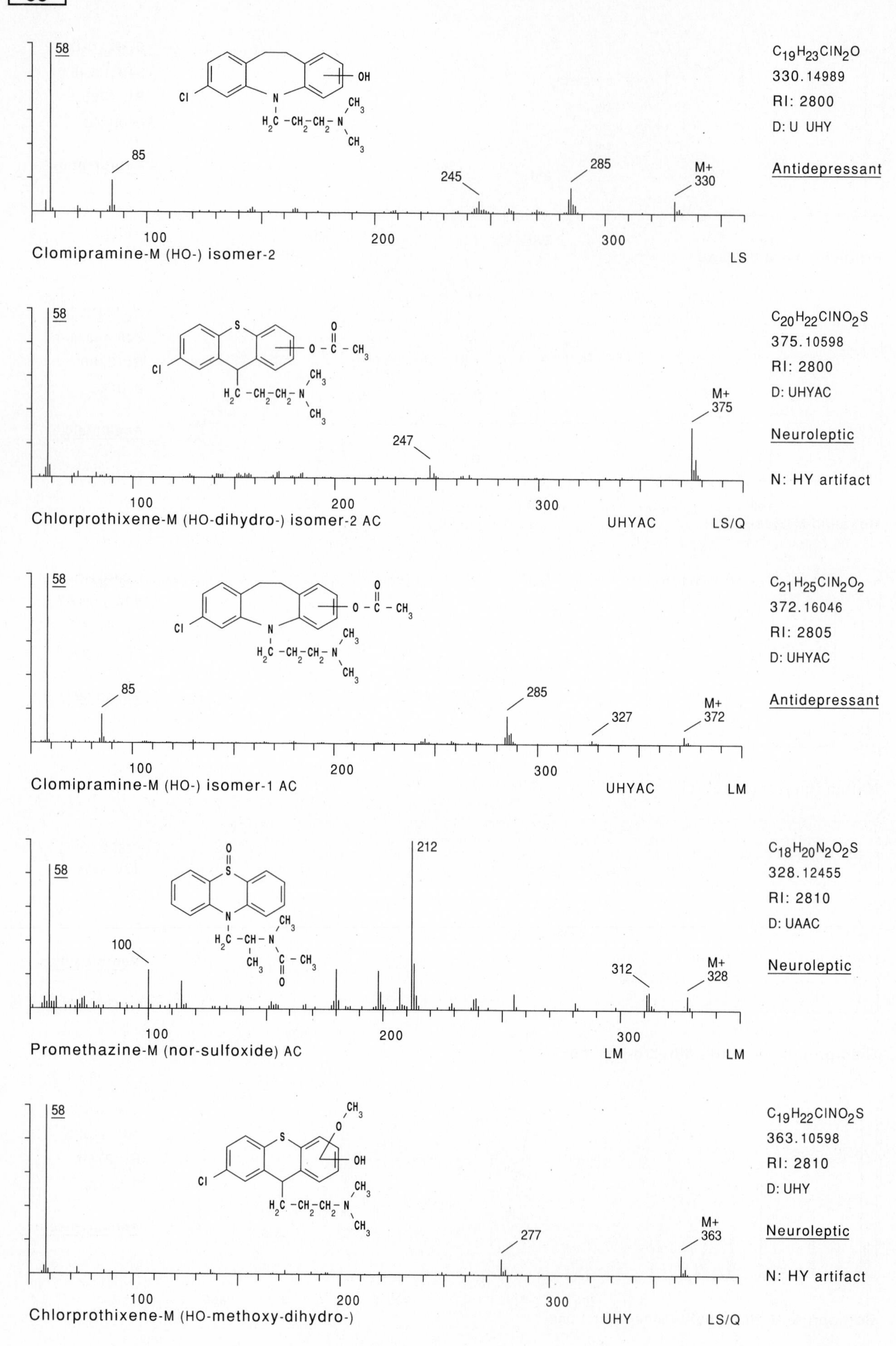

58
85
245
285
M+
330
100
200
300
Clomipramine-M (HO-) isomer-2
LS
C19H23ClN2O
330.14989
RI: 2800
D: U UHY
Antidepressant
58
247
M+
375
100
200
300
Chlorprothixene-M (HO-dihydro-) isomer-2 AC
UHYAC
LS/Q
C20H22ClNO2S
375.10598
RI: 2800
D: UHYAC
Neuroleptic
N: HY artifact
58
85
285
327
M+
372
100
200
300
Clomipramine-M (HO-) isomer-1 AC
UHYAC
LM
C21H25ClN2O2
372.16046
RI: 2805
D: UHYAC
Antidepressant
212
58
100
312
M+
328
100
200
300
Promethazine-M (nor-sulfoxide) AC
LM
LM
C18H20N2O2S
328.12455
RI: 2810
D: UAAC
Neuroleptic
58
277
M+
363
100
200
300
Chlorprothixene-M (HO-methoxy-dihydro-)
UHY
LS/Q
C19H22ClNO2S
363.10598
RI: 2810
D: UHY
Neuroleptic
N: HY artifact

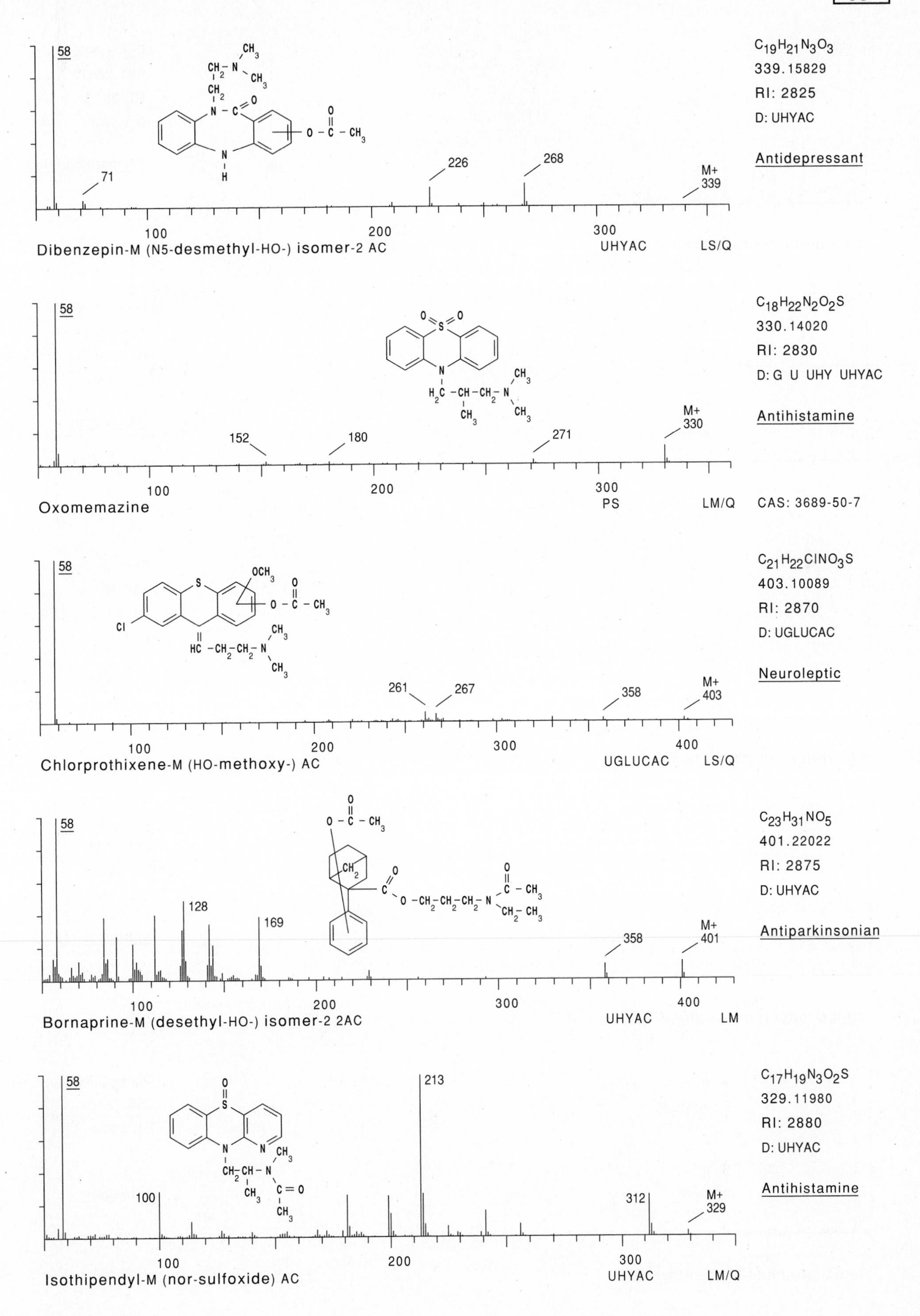

58
71
226
268
M+
339
100
200
300
Dibenzepin-M (N5-desmethyl-HO-) isomer-2 AC
UHYAC
LS/Q
C19H21N3O3
339.15829
RI: 2825
D: UHYAC
Antidepressant
58
152
180
271
M+
330
100
200
300
Oxomemazine
PS
LM/Q
C18H22N2O2S
330.14020
RI: 2830
D: G U UHY UHYAC
Antihistamine
CAS: 3689-50-7
58
261
267
358
M+
403
100
200
300
400
Chlorprothixene-M (HO-methoxy-) AC
UGLUCAC
LS/Q
C21H22ClNO3S
403.10089
RI: 2870
D: UGLUCAC
Neuroleptic
58
128
169
358
M+
401
100
200
300
400
Bornaprine-M (desethyl-HO-) isomer-2 2AC
UHYAC
LM
C23H31NO5
401.22022
RI: 2875
D: UHYAC
Antiparkinsonian
58
213
100
312
M+
329
100
200
300
Isothipendyl-M (nor-sulfoxide) AC
UHYAC
LM/Q
C17H19N3O2S
329.11980
RI: 2880
D: UHYAC
Antihistamine

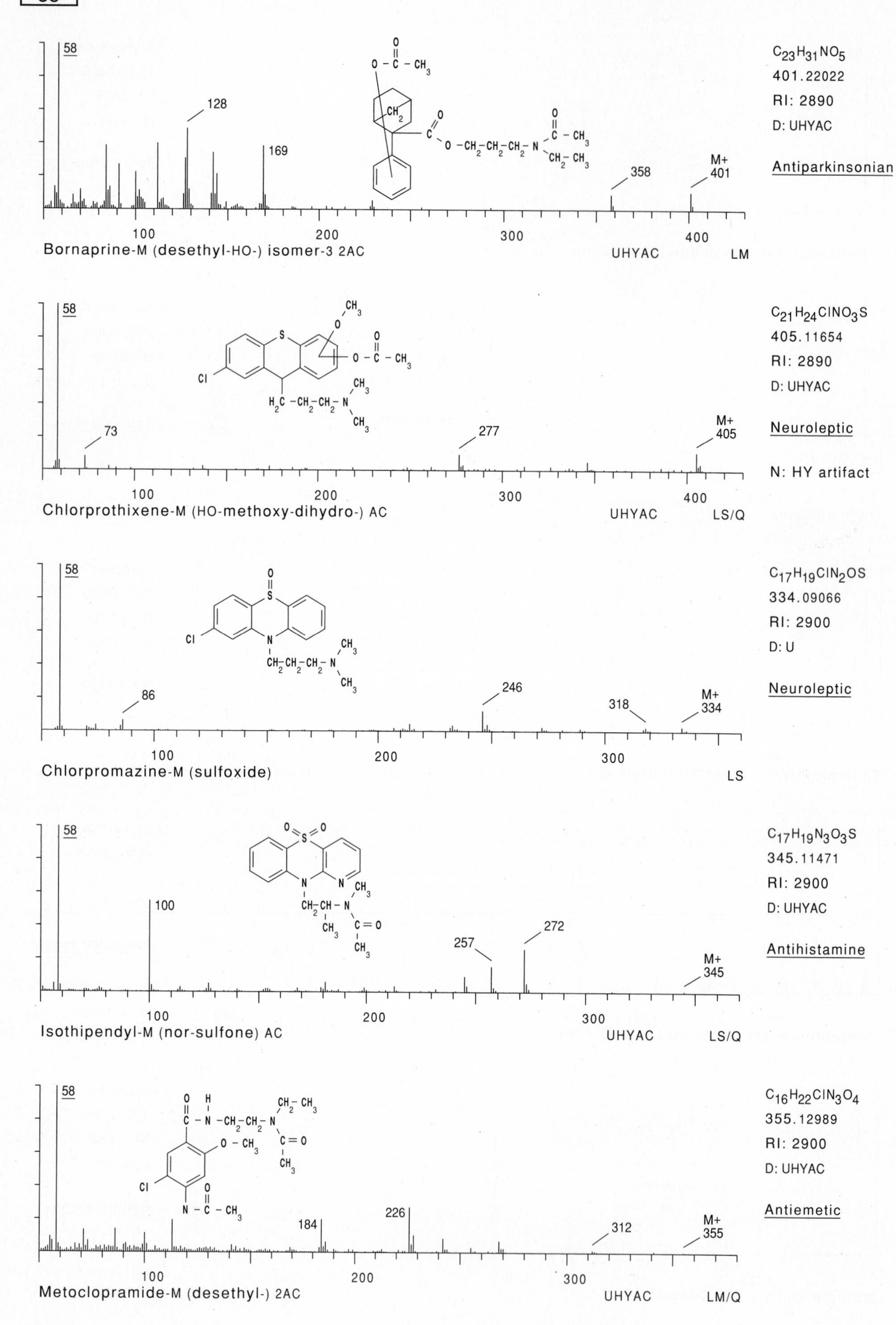
Bornaprine-M (desethyl-HO-) isomer-3 2AC
UHYAC
LM
$C_{23}H_{31}NO_5$
401.22022
RI: 2890
D: UHYAC
Antiparkinsonian
Chlorprothixene-M (HO-methoxy-dihydro-) AC
UHYAC
LS/Q
$C_{21}H_{24}ClNO_3S$
405.11654
RI: 2890
D: UHYAC
Neuroleptic
N: HY artifact
Chlorpromazine-M (sulfoxide)
LS
$C_{17}H_{19}ClN_2OS$
334.09066
RI: 2900
D: U
Neuroleptic
Isothipendyl-M (nor-sulfone) AC
UHYAC
LS/Q
$C_{17}H_{19}N_3O_3S$
345.11471
RI: 2900
D: UHYAC
Antihistamine
Metoclopramide-M (desethyl-) 2AC
UHYAC
LM/Q
$C_{16}H_{22}ClN_3O_4$
355.12989
RI: 2900
D: UHYAC
Antiemetic

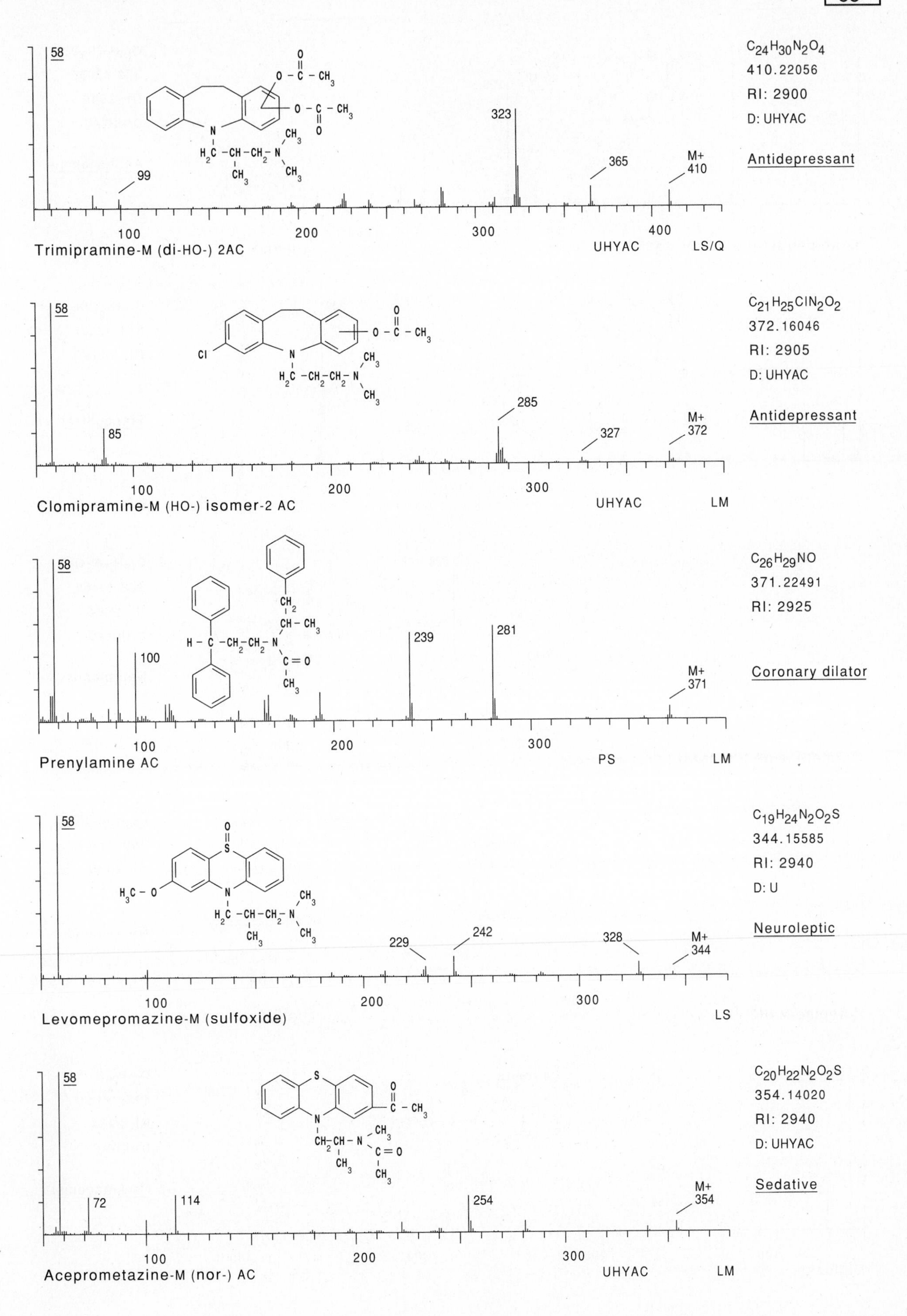

$C_{24}H_{30}N_2O_4$
410.22056
RI: 2900
D: UHYAC
Antidepressant

$C_{21}H_{25}ClN_2O_2$
372.16046
RI: 2905
D: UHYAC
Antidepressant

$C_{26}H_{29}NO$
371.22491
RI: 2925
Coronary dilator

$C_{19}H_{24}N_2O_2S$
344.15585
RI: 2940
D: U
Neuroleptic

$C_{20}H_{22}N_2O_2S$
354.14020
RI: 2940
D: UHYAC
Sedative

58

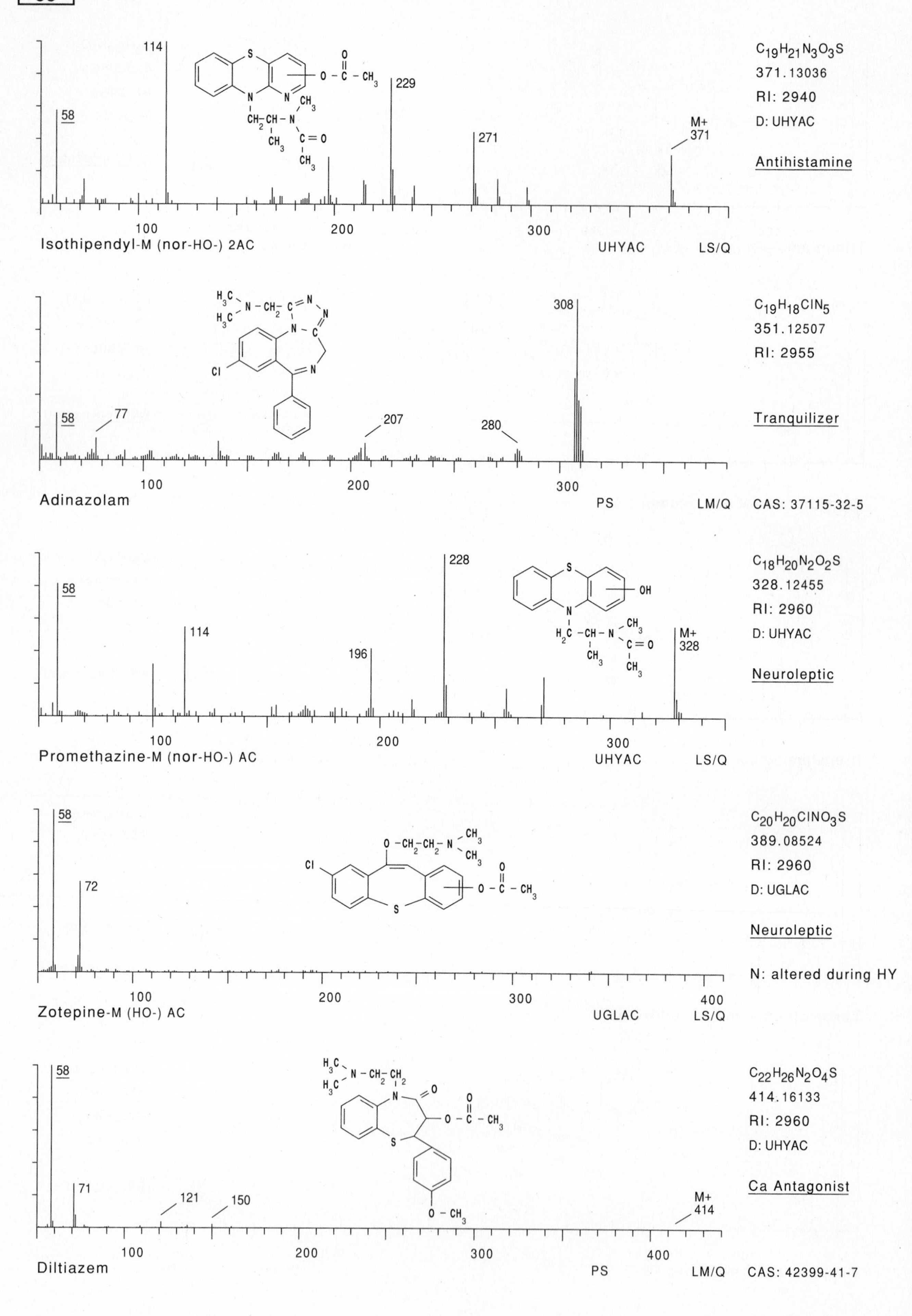

Isothipendyl-M (nor-HO-) 2AC
UHYAC
LS/Q
$C_{19}H_{21}N_3O_3S$
371.13036
RI: 2940
D: UHYAC
Antihistamine
M+ 371
Adinazolam
PS
LM/Q
$C_{19}H_{18}ClN_5$
351.12507
RI: 2955
Tranquilizer
CAS: 37115-32-5
Promethazine-M (nor-HO-) AC
UHYAC
LS/Q
$C_{18}H_{20}N_2O_2S$
328.12455
RI: 2960
D: UHYAC
Neuroleptic
M+ 328
Zotepine-M (HO-) AC
UGLAC
LS/Q
$C_{20}H_{20}ClNO_3S$
389.08524
RI: 2960
D: UGLAC
Neuroleptic
N: altered during HY
Diltiazem
PS
LM/Q
$C_{22}H_{26}N_2O_4S$
414.16133
RI: 2960
D: UHYAC
Ca Antagonist
M+ 414
CAS: 42399-41-7

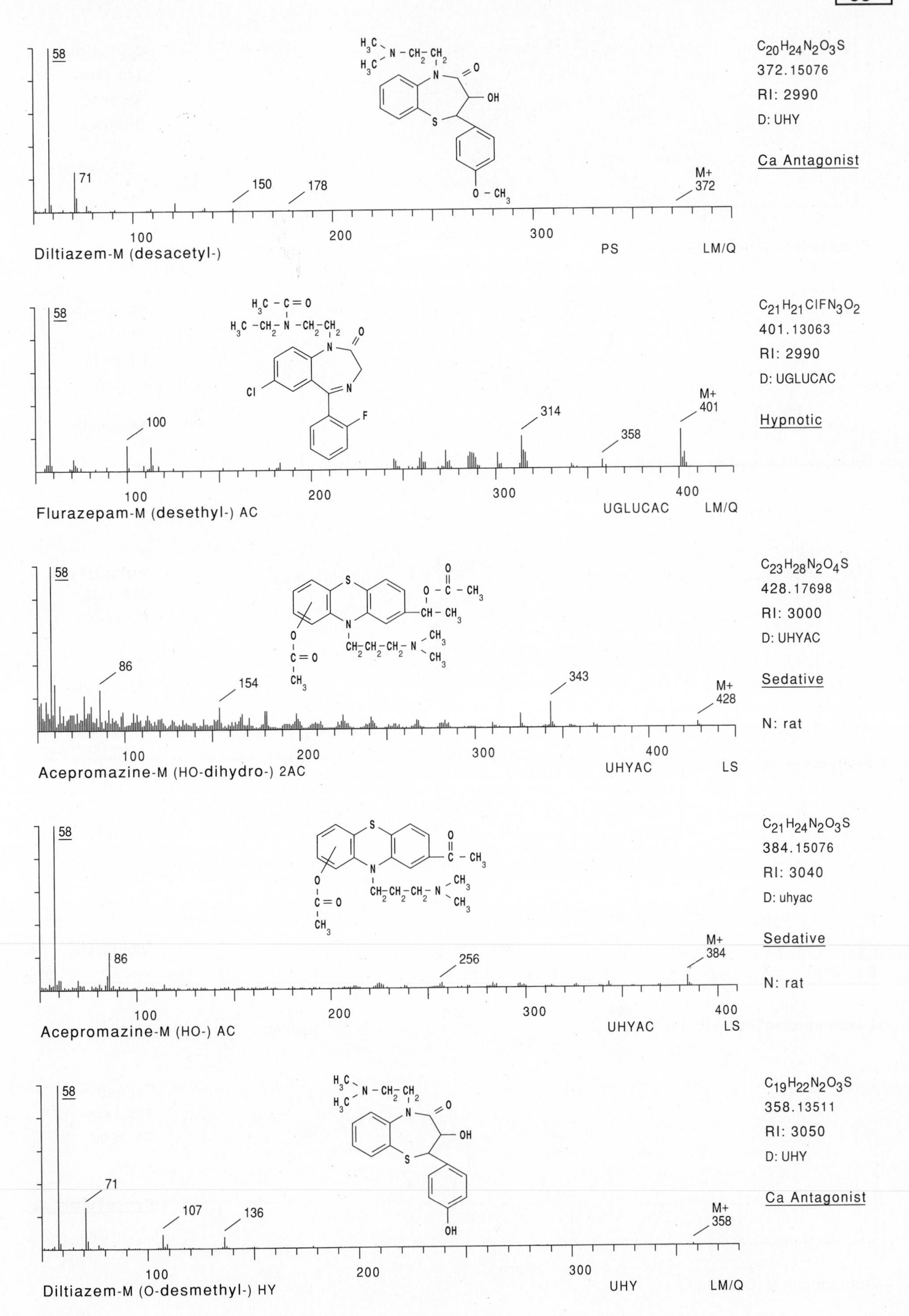
Diltiazem-M (desacetyl-)
PS
LM/Q
$C_{20}H_{24}N_2O_3S$
372.15076
RI: 2990
D: UHY
Ca Antagonist
Flurazepam-M (desethyl-) AC
UGLUCAC
LM/Q
$C_{21}H_{21}ClFN_3O_2$
401.13063
RI: 2990
D: UGLUCAC
Hypnotic
Acepromazine-M (HO-dihydro-) 2AC
UHYAC
LS
$C_{23}H_{28}N_2O_4S$
428.17698
RI: 3000
D: UHYAC
Sedative
N: rat
Acepromazine-M (HO-) AC
UHYAC
LS
$C_{21}H_{24}N_2O_3S$
384.15076
RI: 3040
D: uhyac
Sedative
N: rat
Diltiazem-M (O-desmethyl-) HY
UHY
LM/Q
$C_{19}H_{22}N_2O_3S$
358.13511
RI: 3050
D: UHY
Ca Antagonist

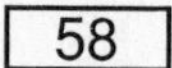

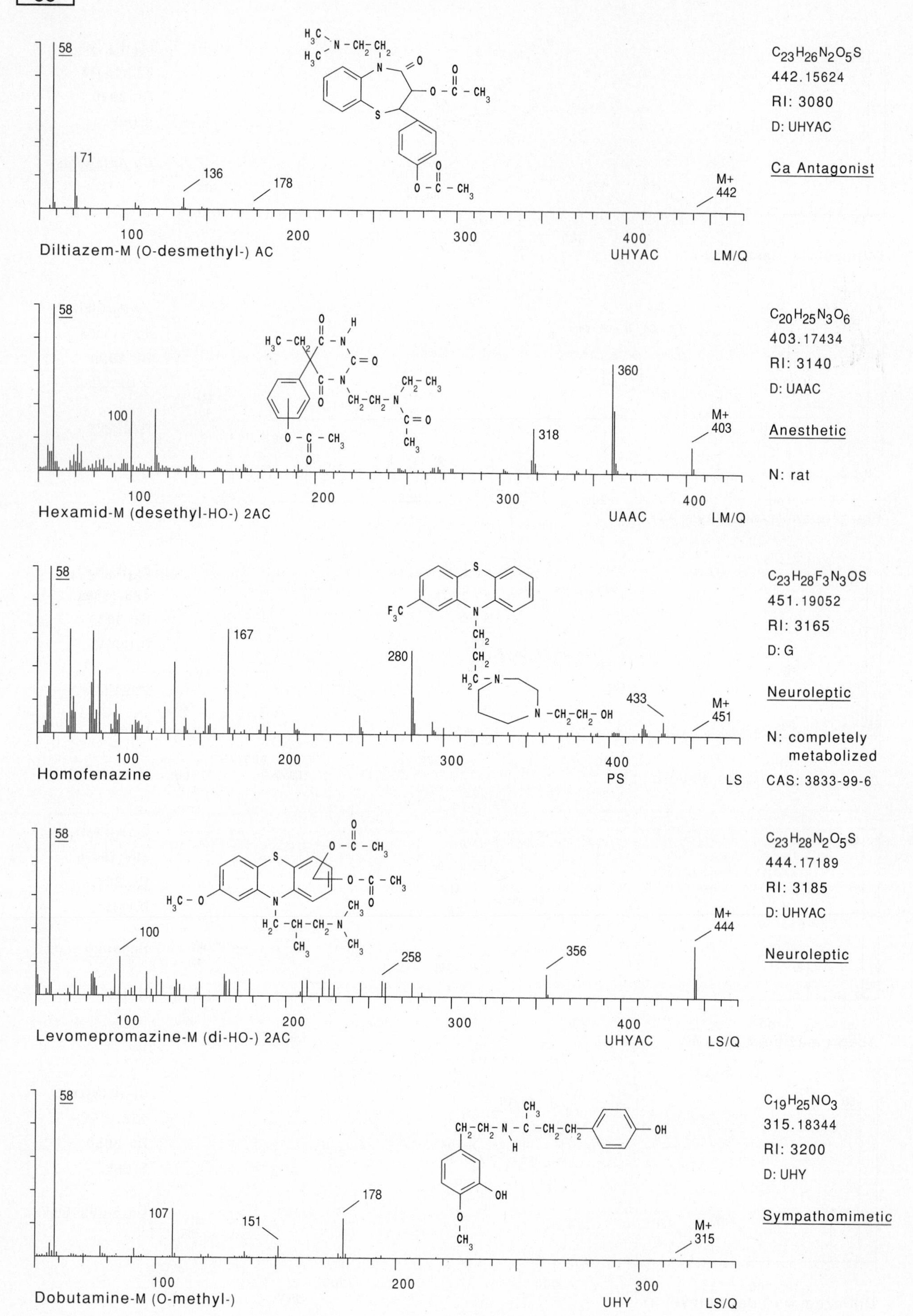

$C_{23}H_{26}N_2O_5S$
442.15624
RI: 3080
D: UHYAC
Ca Antagonist

$C_{20}H_{25}N_3O_6$
403.17434
RI: 3140
D: UAAC
Anesthetic
N: rat

$C_{23}H_{28}F_3N_3OS$
451.19052
RI: 3165
D: G
Neuroleptic
N: completely metabolized
CAS: 3833-99-6

$C_{23}H_{28}N_2O_5S$
444.17189
RI: 3185
D: UHYAC
Neuroleptic

$C_{19}H_{25}NO_3$
315.18344
RI: 3200
D: UHY
Sympathomimetic

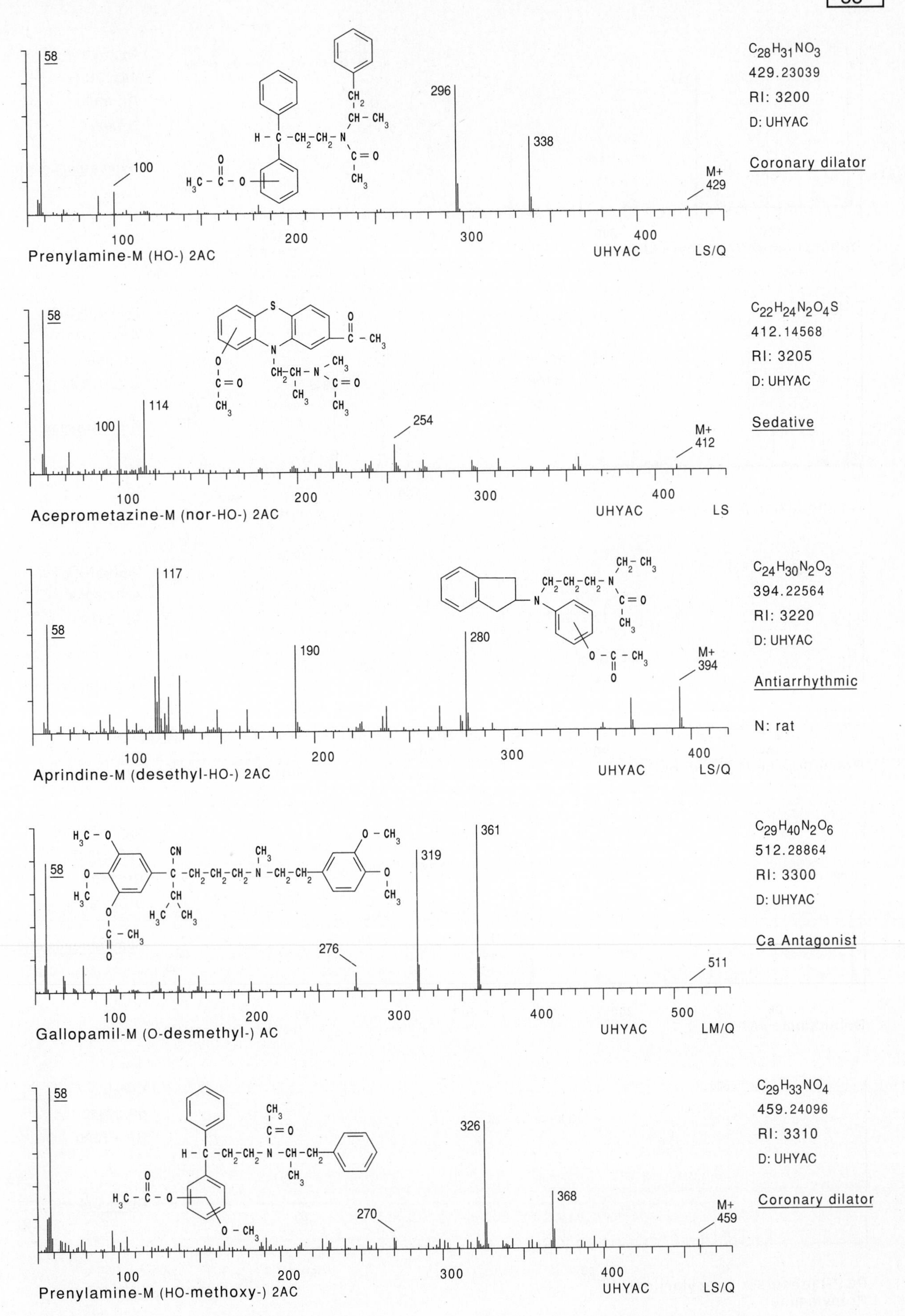

Prenylamine-M (HO-) 2AC

$C_{28}H_{31}NO_3$
429.23039
RI: 3200
D: UHYAC

Coronary dilator

Aceprometazine-M (nor-HO-) 2AC

$C_{22}H_{24}N_2O_4S$
412.14568
RI: 3205
D: UHYAC

Sedative

Aprindine-M (desethyl-HO-) 2AC

$C_{24}H_{30}N_2O_3$
394.22564
RI: 3220
D: UHYAC

Antiarrhythmic

N: rat

Gallopamil-M (O-desmethyl-) AC

$C_{29}H_{40}N_2O_6$
512.28864
RI: 3300
D: UHYAC

Ca Antagonist

Prenylamine-M (HO-methoxy-) 2AC

$C_{29}H_{33}NO_4$
459.24096
RI: 3310
D: UHYAC

Coronary dilator

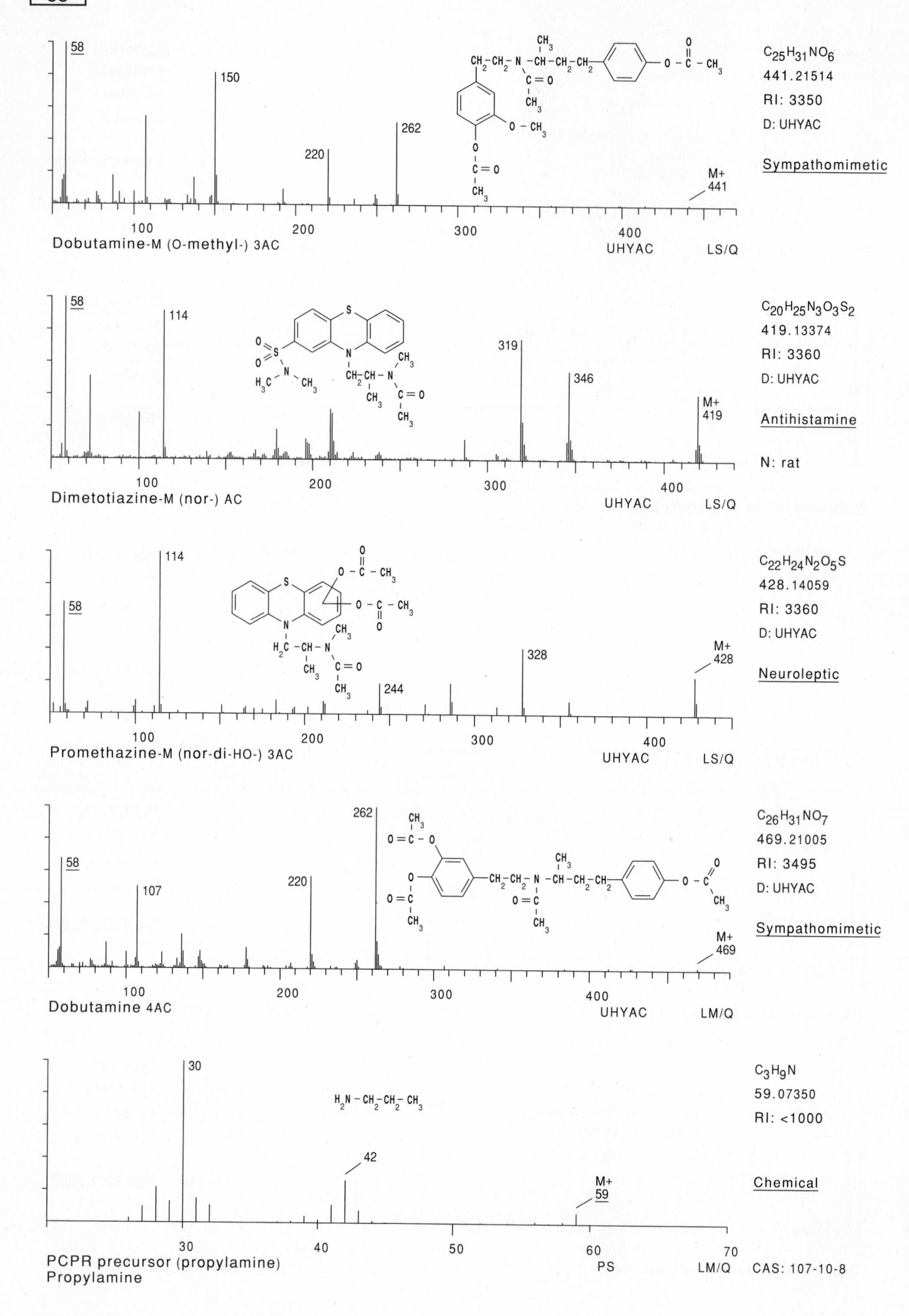

Dobutamine-M (O-methyl-) 3AC
UHYAC
LS/Q
C25H31NO6
441.21514
RI: 3350
D: UHYAC
Sympathomimetic
M+ 441
Dimetotiazine-M (nor-) AC
UHYAC
LS/Q
C20H25N3O3S2
419.13374
RI: 3360
D: UHYAC
Antihistamine
N: rat
M+ 419
Promethazine-M (nor-di-HO-) 3AC
UHYAC
LS/Q
C22H24N2O5S
428.14059
RI: 3360
D: UHYAC
Neuroleptic
M+ 428
Dobutamine 4AC
UHYAC
LM/Q
C26H31NO7
469.21005
RI: 3495
D: UHYAC
Sympathomimetic
M+ 469
PCPR precursor (propylamine)
Propylamine
PS
LM/Q
C3H9N
59.07350
RI: <1000
Chemical
CAS: 107-10-8
M+ 59

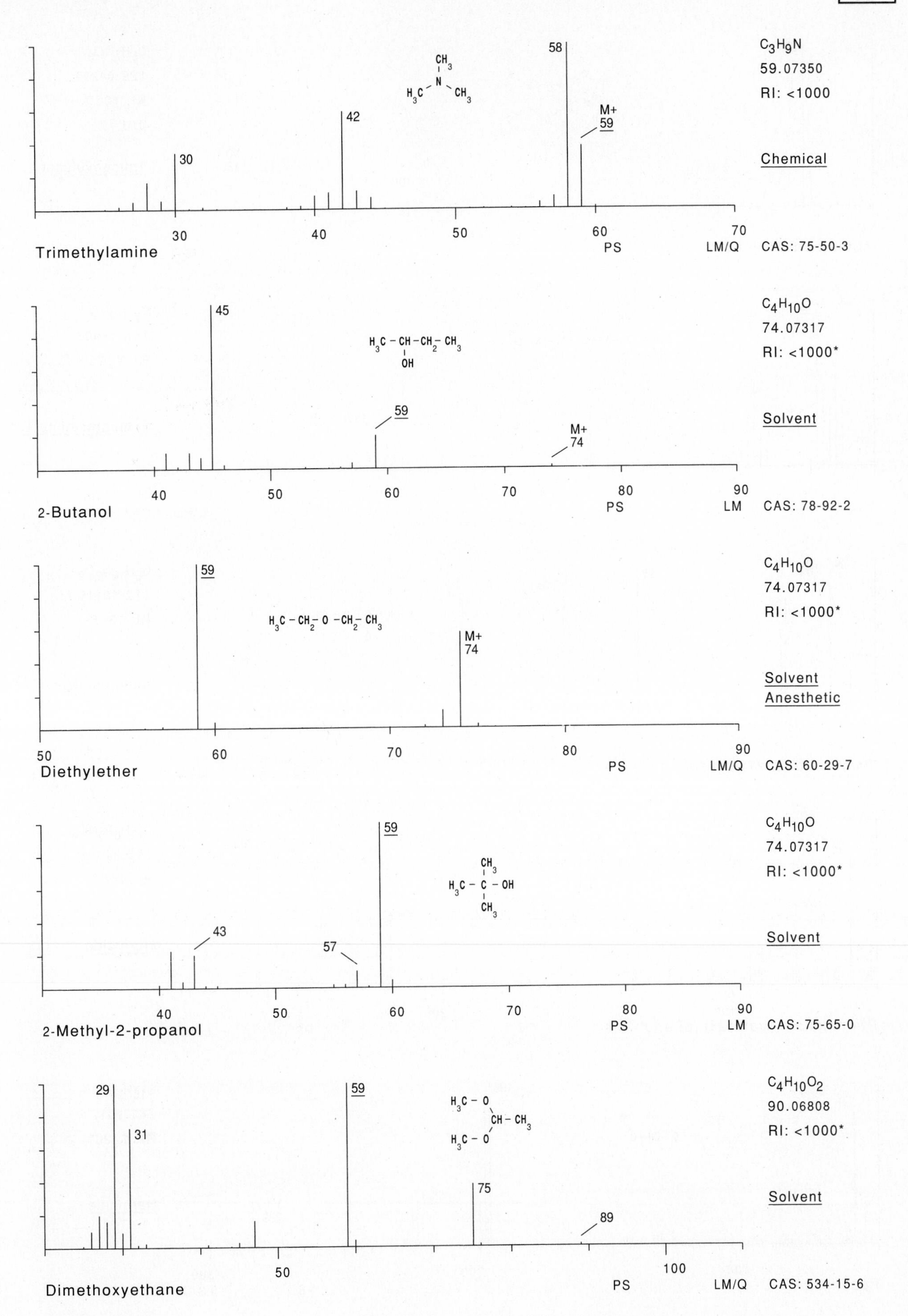

58
42
30
M+
59
30
40
50
60
70
PS
LM/Q
Trimethylamine
C3H9N
59.07350
RI: <1000
Chemical
CAS: 75-50-3
45
59
M+
74
40
50
60
70
80
90
PS
LM
2-Butanol
C4H10O
74.07317
RI: <1000*
Solvent
CAS: 78-92-2
59
M+
74
50
60
70
80
90
PS
LM/Q
Diethylether
C4H10O
74.07317
RI: <1000*
Solvent
Anesthetic
CAS: 60-29-7
59
43
57
40
50
60
70
80
90
PS
LM
2-Methyl-2-propanol
C4H10O
74.07317
RI: <1000*
Solvent
CAS: 75-65-0
29
31
59
75
89
50
100
PS
LM/Q
Dimethoxyethane
C4H10O2
90.06808
RI: <1000*
Solvent
CAS: 534-15-6

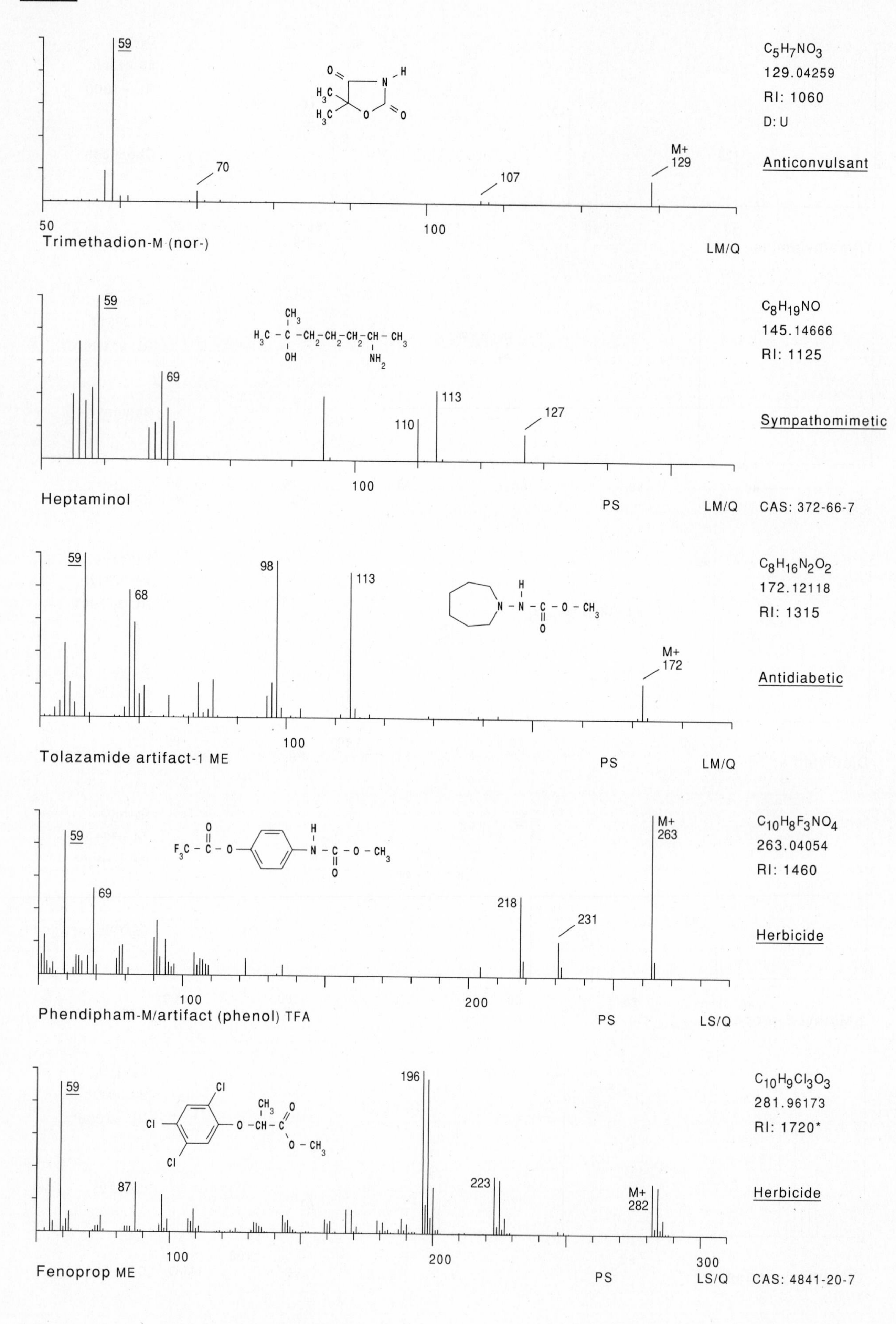
59
70
107
M+
129
50
100
Trimethadion-M (nor-)
LM/Q
C5H7NO3
129.04259
RI: 1060
D: U
Anticonvulsant
59
69
113
110
127
100
Heptaminol
PS
LM/Q
C8H19NO
145.14666
RI: 1125
Sympathomimetic
CAS: 372-66-7
59
98
113
68
M+
172
100
Tolazamide artifact-1 ME
PS
LM/Q
C8H16N2O2
172.12118
RI: 1315
Antidiabetic
59
69
218
231
M+
263
100
200
Phendipham-M/artifact (phenol) TFA
PS
LS/Q
C10H8F3NO4
263.04054
RI: 1460
Herbicide
59
196
87
223
M+
282
100
200
300
Fenoprop ME
PS
LS/Q
C10H9Cl3O3
281.96173
RI: 1720*
Herbicide
CAS: 4841-20-7

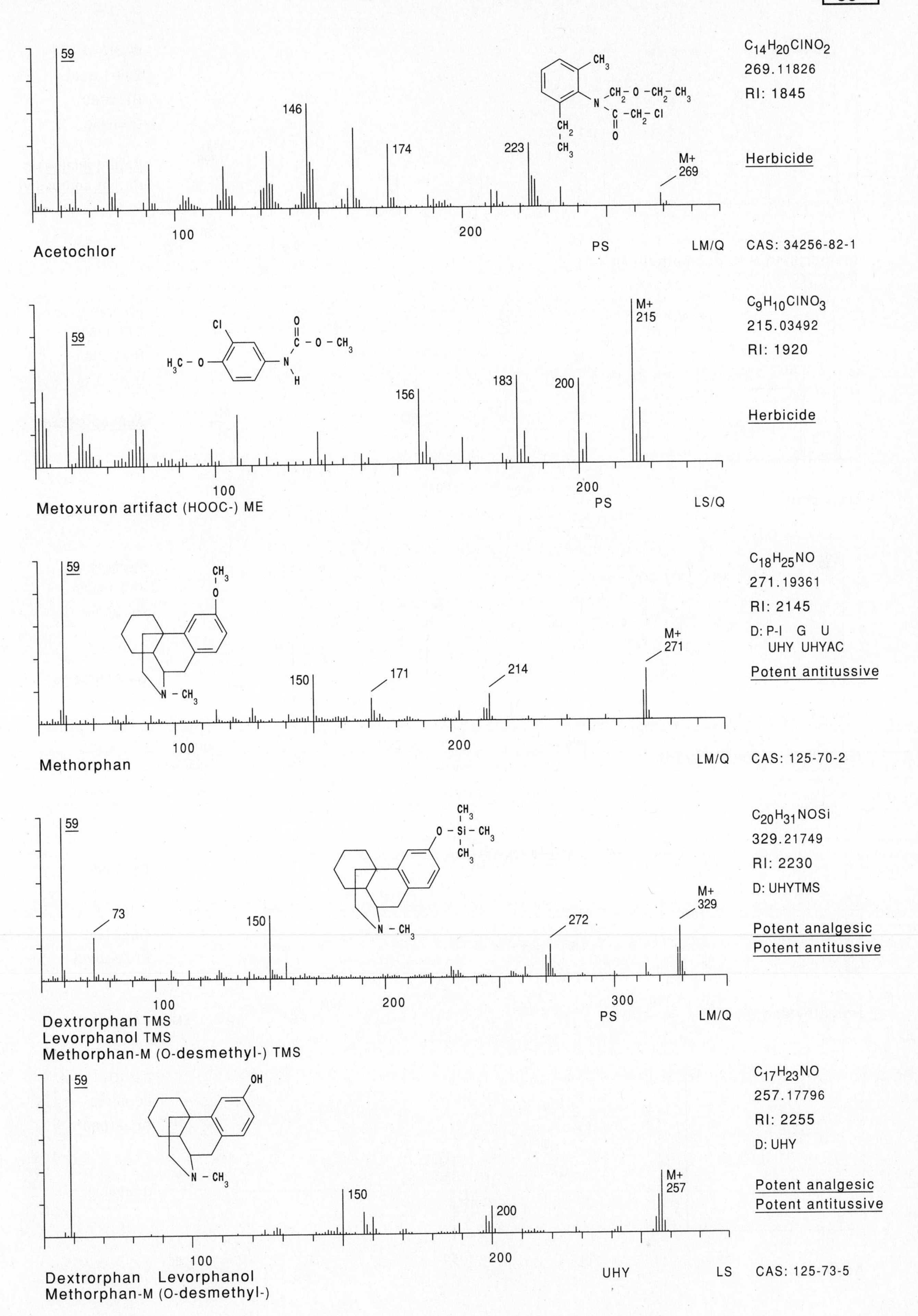
59
146
174
223
M+
269
100
200
PS
LM/Q
Acetochlor
$C_{14}H_{20}ClNO_2$
269.11826
RI: 1845
Herbicide
CAS: 34256-82-1
59
156
183
200
M+
215
100
200
PS
LS/Q
Metoxuron artifact (HOOC-) ME
$C_9H_{10}ClNO_3$
215.03492
RI: 1920
Herbicide
59
150
171
214
M+
271
100
200
LM/Q
Methorphan
$C_{18}H_{25}NO$
271.19361
RI: 2145
D: P-I G U
UHY UHYAC
Potent antitussive
CAS: 125-70-2
59
73
150
272
M+
329
100
200
300
PS
LM/Q
Dextrorphan TMS
Levorphanol TMS
Methorphan-M (O-desmethyl-) TMS
$C_{20}H_{31}NOSi$
329.21749
RI: 2230
D: UHYTMS
Potent analgesic
Potent antitussive
59
150
200
M+
257
100
200
UHY
LS
Dextrorphan Levorphanol
Methorphan-M (O-desmethyl-)
$C_{17}H_{23}NO$
257.17796
RI: 2255
D: UHY
Potent analgesic
Potent antitussive
CAS: 125-73-5

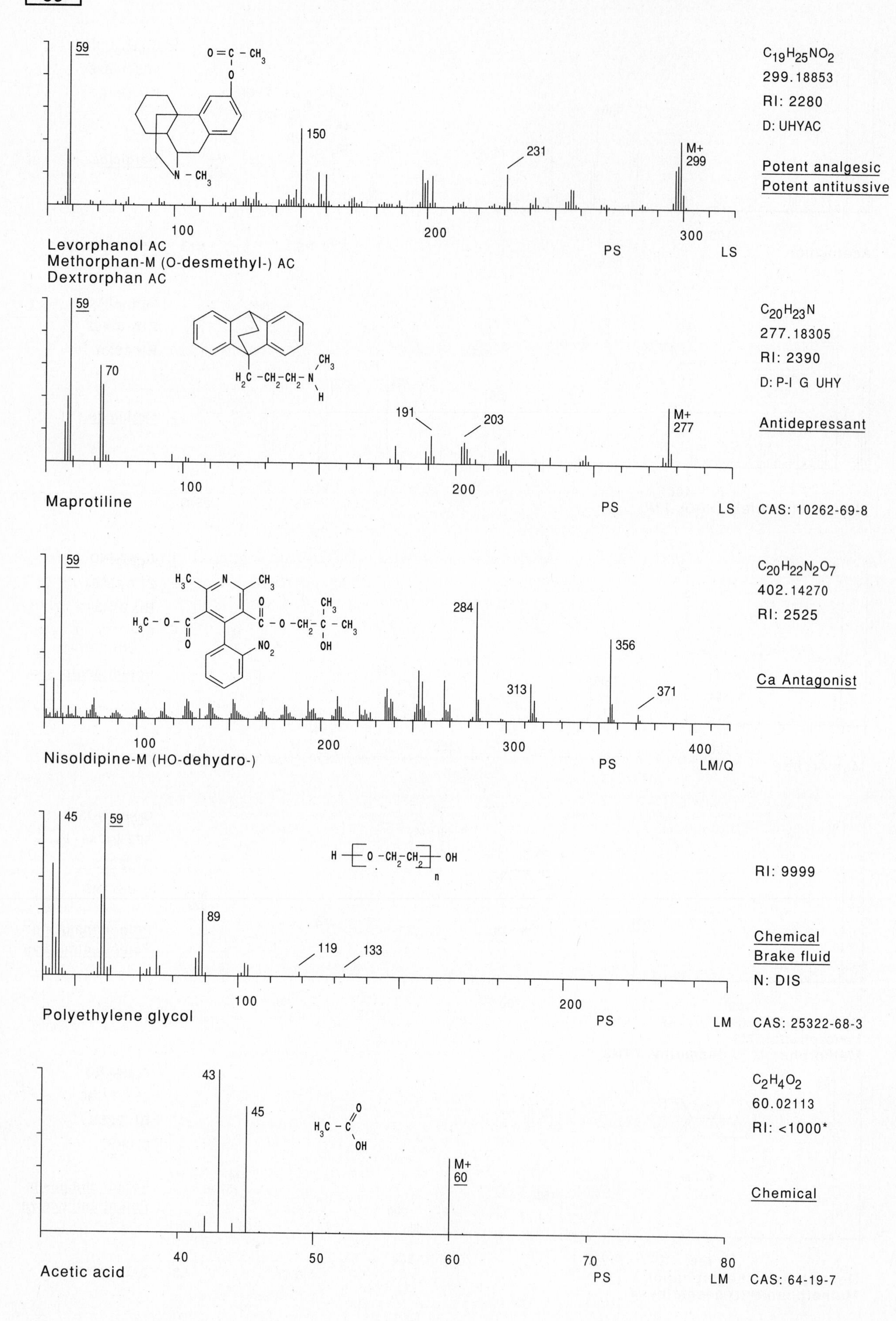

59
150
231
M+
299
100
200
300
PS
LS
C19H25NO2
299.18853
RI: 2280
D: UHYAC
Potent analgesic
Potent antitussive
Levorphanol AC
Methorphan-M (O-desmethyl-) AC
Dextrorphan AC
59
70
191
203
M+
277
100
200
PS
LS
C20H23N
277.18305
RI: 2390
D: P-I G UHY
Antidepressant
Maprotiline
CAS: 10262-69-8
59
284
356
313
371
100
200
300
400
PS
LM/Q
C20H22N2O7
402.14270
RI: 2525
Ca Antagonist
Nisoldipine-M (HO-dehydro-)
45
59
89
119
133
100
200
PS
LM
RI: 9999
Chemical
Brake fluid
N: DIS
Polyethylene glycol
CAS: 25322-68-3
43
45
M+
60
40
50
60
70
80
PS
LM
C2H4O2
60.02113
RI: <1000*
Chemical
Acetic acid
CAS: 64-19-7

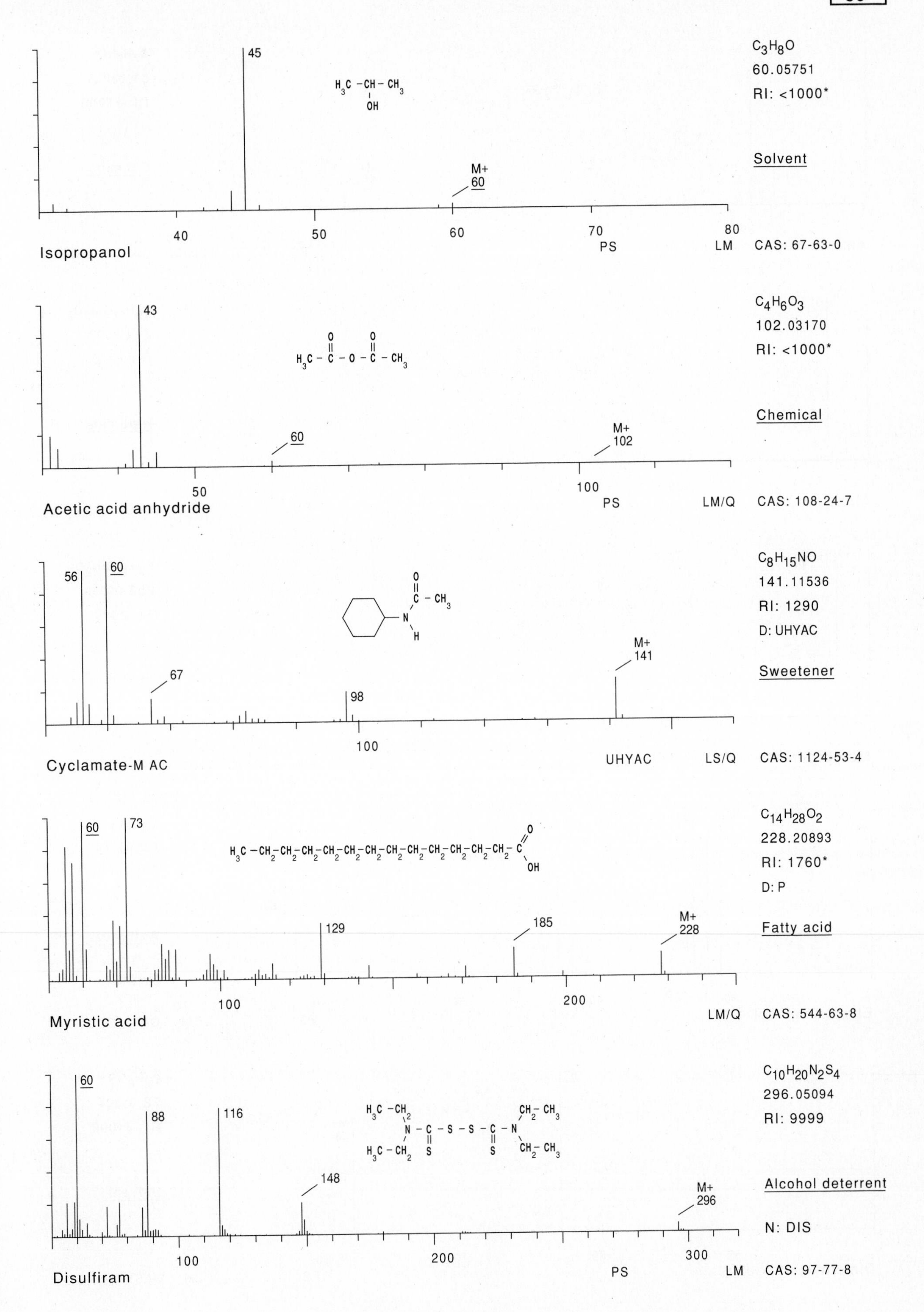

45
$H_3C-CH-CH_3$
OH
M+
60
40
50
60
70
80
PS
LM
Isopropanol
C_3H_8O
60.05751
RI: <1000*
Solvent
CAS: 67-63-0
43
$H_3C-C(=O)-O-C(=O)-CH_3$
60
M+
102
50
100
PS
LM/Q
Acetic acid anhydride
$C_4H_6O_3$
102.03170
RI: <1000*
Chemical
CAS: 108-24-7
56
60
67
98
M+
141
100
UHYAC
LS/Q
Cyclamate-M AC
$C_8H_{15}NO$
141.11536
RI: 1290
D: UHYAC
Sweetener
CAS: 1124-53-4
60
73
$H_3C-CH_2-CH_2-CH_2-CH_2-CH_2-CH_2-CH_2-CH_2-CH_2-CH_2-CH_2-CH_2-C(=O)OH$
129
185
M+
228
100
200
LM/Q
Myristic acid
$C_{14}H_{28}O_2$
228.20893
RI: 1760*
D: P
Fatty acid
CAS: 544-63-8
60
88
116
148
M+
296
100
200
300
PS
LM
Disulfiram
$C_{10}H_{20}N_2S_4$
296.05094
RI: 9999
Alcohol deterrent
N: DIS
CAS: 97-77-8

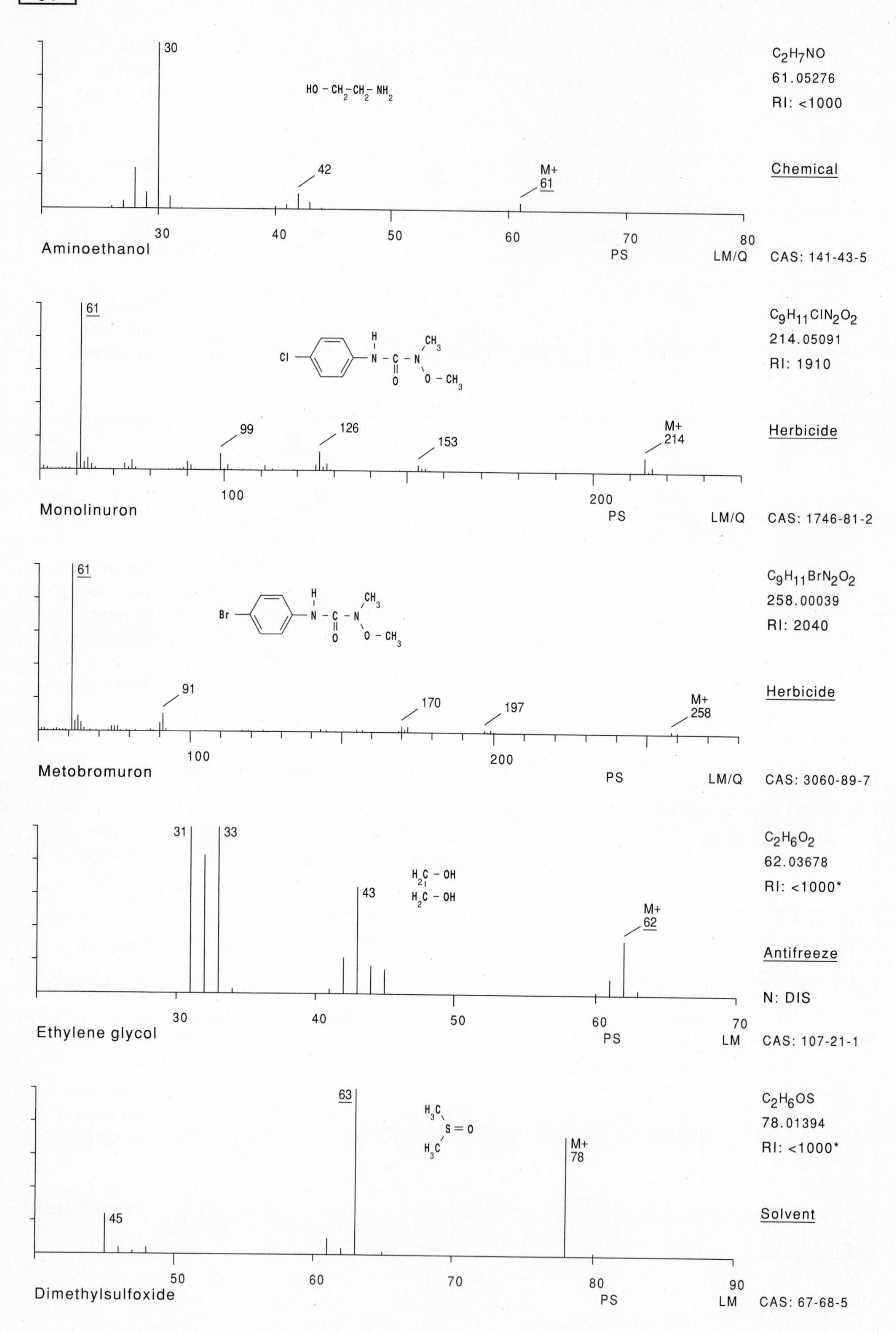
30
HO - CH2- CH2- NH2
42
M+
61
30
40
50
60
70
80
Aminoethanol
PS
LM/Q
C2H7NO
61.05276
RI: <1000
Chemical
CAS: 141-43-5
61
99
126
153
M+
214
100
200
Monolinuron
PS
LM/Q
C9H11ClN2O2
214.05091
RI: 1910
Herbicide
CAS: 1746-81-2
61
91
170
197
M+
258
100
200
Metobromuron
PS
LM/Q
C9H11BrN2O2
258.00039
RI: 2040
Herbicide
CAS: 3060-89-7
31
33
43
M+
62
30
40
50
60
70
Ethylene glycol
PS
LM
C2H6O2
62.03678
RI: <1000*
Antifreeze
N: DIS
CAS: 107-21-1
63
M+
78
45
50
60
70
80
90
Dimethylsulfoxide
PS
LM
C2H6OS
78.01394
RI: <1000*
Solvent
CAS: 67-68-5

64

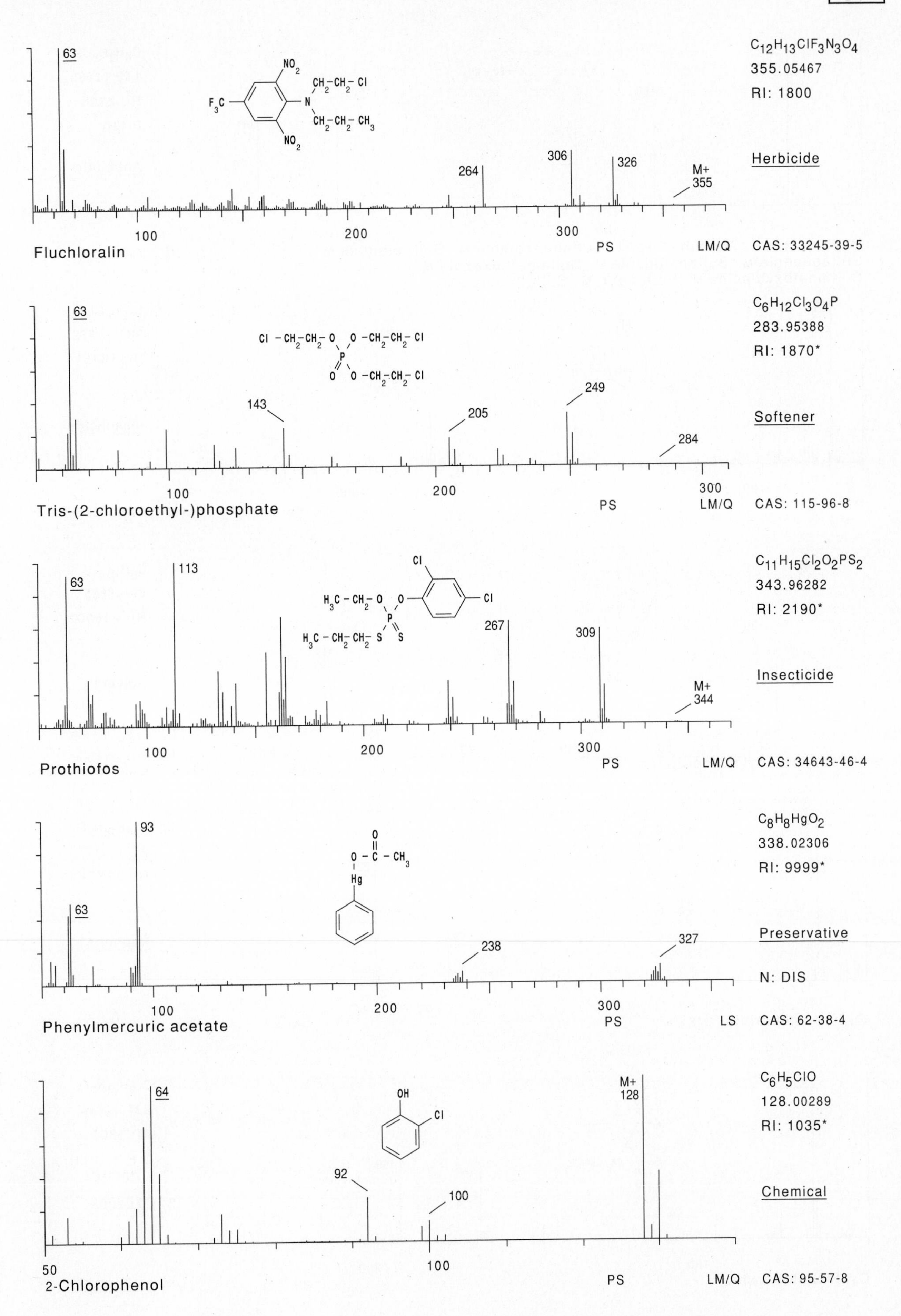

C12H13ClF3N3O4
355.05467
RI: 1800
Herbicide
63
264
306
326
M+
355
100
200
300
PS
LM/Q
Fluchloralin
CAS: 33245-39-5
C6H12Cl3O4P
283.95388
RI: 1870*
Softener
63
143
205
249
284
Tris-(2-chloroethyl-)phosphate
CAS: 115-96-8
C11H15Cl2O2PS2
343.96282
RI: 2190*
Insecticide
63
113
267
309
M+
344
Prothiofos
CAS: 34643-46-4
C8H8HgO2
338.02306
RI: 9999*
Preservative
N: DIS
63
93
238
327
LS
Phenylmercuric acetate
CAS: 62-38-4
C6H5ClO
128.00289
RI: 1035*
Chemical
64
92
100
M+
128
50
2-Chlorophenol
CAS: 95-57-8

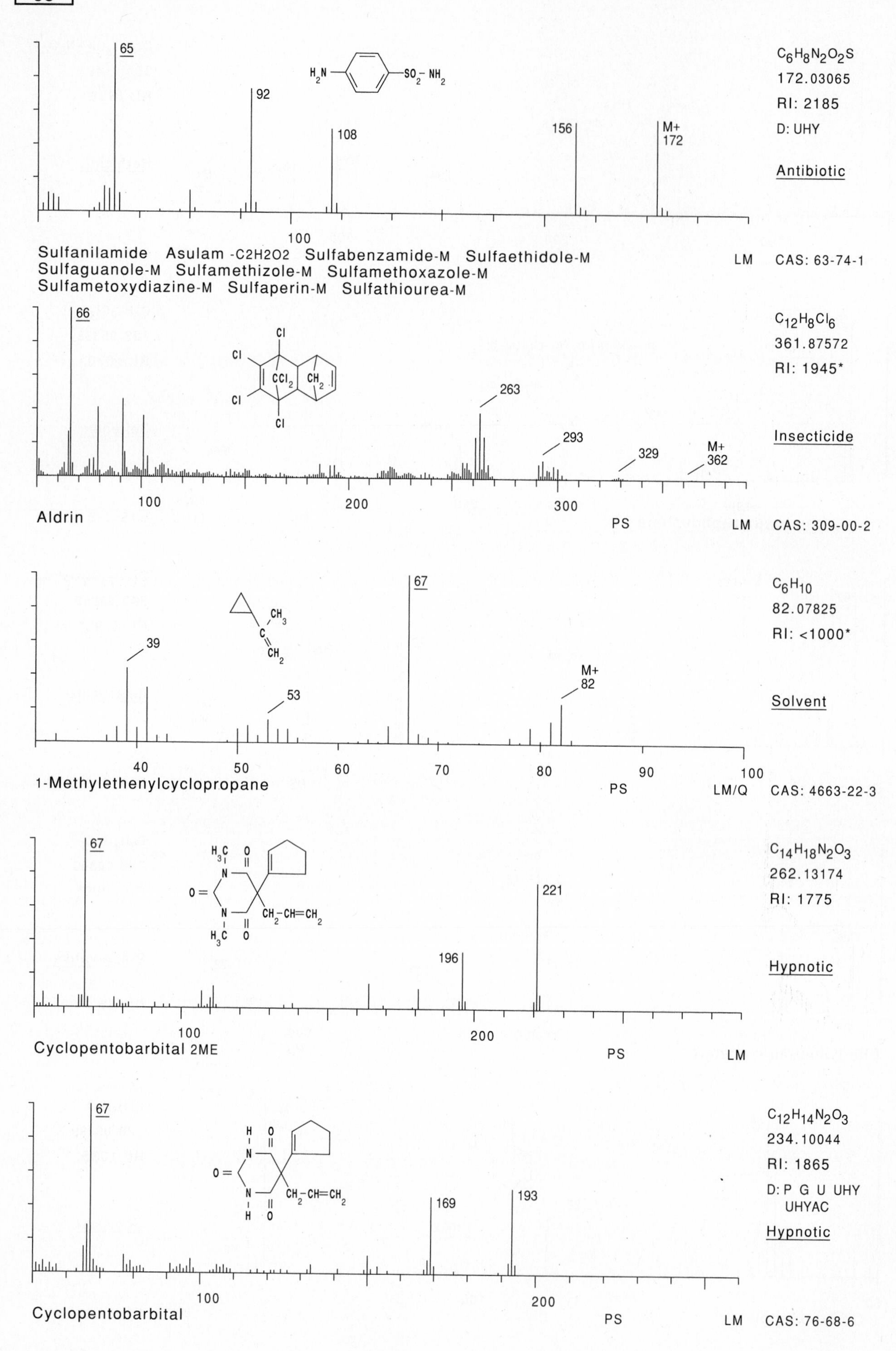
65
$C_6H_8N_2O_2S$
172.03065
RI: 2185
D: UHY
Antibiotic
Sulfanilamide Asulam -C2H2O2 Sulfabenzamide-M Sulfaethidole-M
Sulfaguanole-M Sulfamethizole-M Sulfamethoxazole-M
Sulfametoxydiazine-M Sulfaperin-M Sulfathiourea-M
LM CAS: 63-74-1
$C_{12}H_8Cl_6$
361.87572
RI: 1945*
Insecticide
Aldrin
PS LM CAS: 309-00-2
C_6H_{10}
82.07825
RI: <1000*
Solvent
1-Methylethenylcyclopropane
PS LM/Q CAS: 4663-22-3
$C_{14}H_{18}N_2O_3$
262.13174
RI: 1775
Hypnotic
Cyclopentobarbital 2ME
PS LM
$C_{12}H_{14}N_2O_3$
234.10044
RI: 1865
D: P G U UHY UHYAC
Hypnotic
Cyclopentobarbital
PS LM CAS: 76-68-6

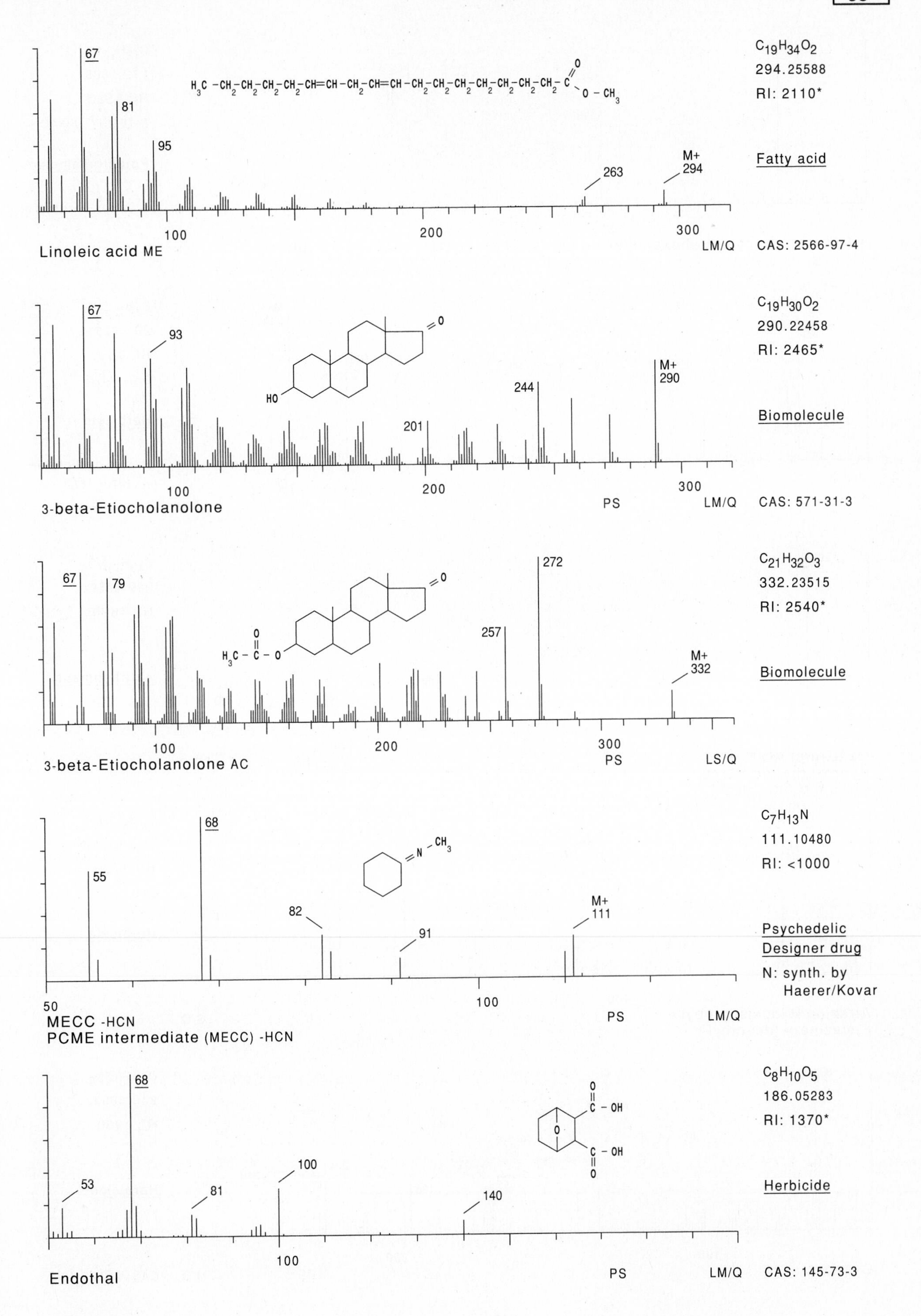

67
81
95
263
M+
294
$H_3C-CH_2-CH_2-CH_2-CH_2-CH=CH-CH_2-CH=CH-CH_2-CH_2-CH_2-CH_2-CH_2-CH_2-CH_2-C(=O)-O-CH_3$
100
200
300
LM/Q
Linoleic acid ME
$C_{19}H_{34}O_2$
294.25588
RI: 2110*
Fatty acid
CAS: 2566-97-4
67
93
201
244
M+
290
HO
100
200
300
PS
LM/Q
3-beta-Etiocholanolone
$C_{19}H_{30}O_2$
290.22458
RI: 2465*
Biomolecule
CAS: 571-31-3
67
79
257
272
M+
332
$H_3C-C(=O)-O$
100
200
300
PS
LS/Q
3-beta-Etiocholanolone AC
$C_{21}H_{32}O_3$
332.23515
RI: 2540*
Biomolecule
55
68
82
91
M+
111
$N-CH_3$
50
100
PS
LM/Q
MECC -HCN
PCME intermediate (MECC) -HCN
$C_7H_{13}N$
111.10480
RI: <1000
Psychedelic
Designer drug
N: synth. by Haerer/Kovar
53
68
81
100
140
C - OH
C - OH
O
100
PS
LM/Q
Endothal
$C_8H_{10}O_5$
186.05283
RI: 1370*
Herbicide
CAS: 145-73-3

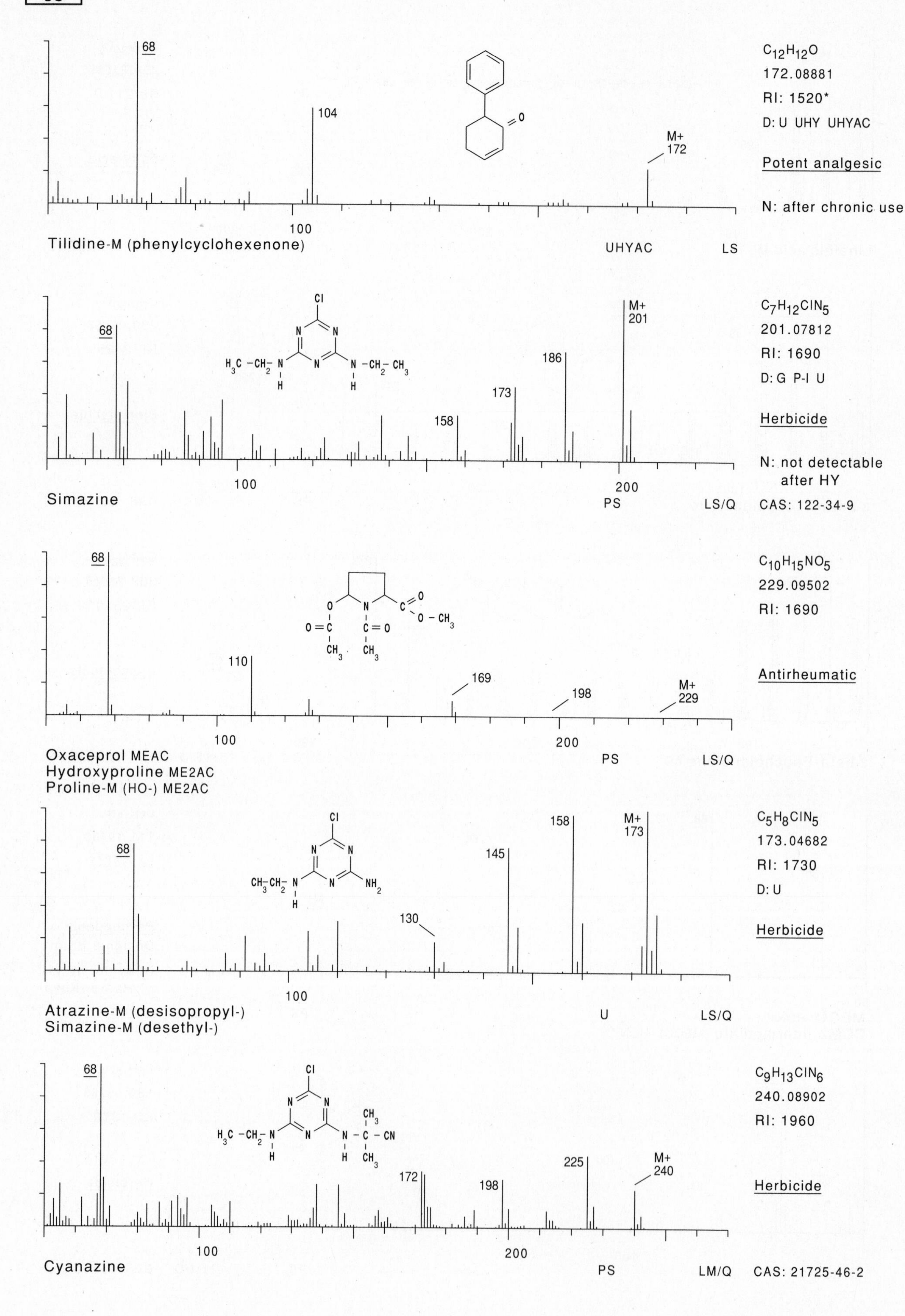

Tilidine-M (phenylcyclohexenone)
UHYAC LS
C12H12O
172.08881
RI: 1520*
D: U UHY UHYAC
Potent analgesic
N: after chronic use
Simazine
PS LS/Q
C7H12ClN5
201.07812
RI: 1690
D: G P-I U
Herbicide
N: not detectable after HY
CAS: 122-34-9
Oxaceprol MEAC
Hydroxyproline ME2AC
Proline-M (HO-) ME2AC
PS LS/Q
C10H15NO5
229.09502
RI: 1690
Antirheumatic
Atrazine-M (desisopropyl-)
Simazine-M (desethyl-)
U LS/Q
C5H8ClN5
173.04682
RI: 1730
D: U
Herbicide
Cyanazine
PS LM/Q
C9H13ClN6
240.08902
RI: 1960
Herbicide
CAS: 21725-46-2

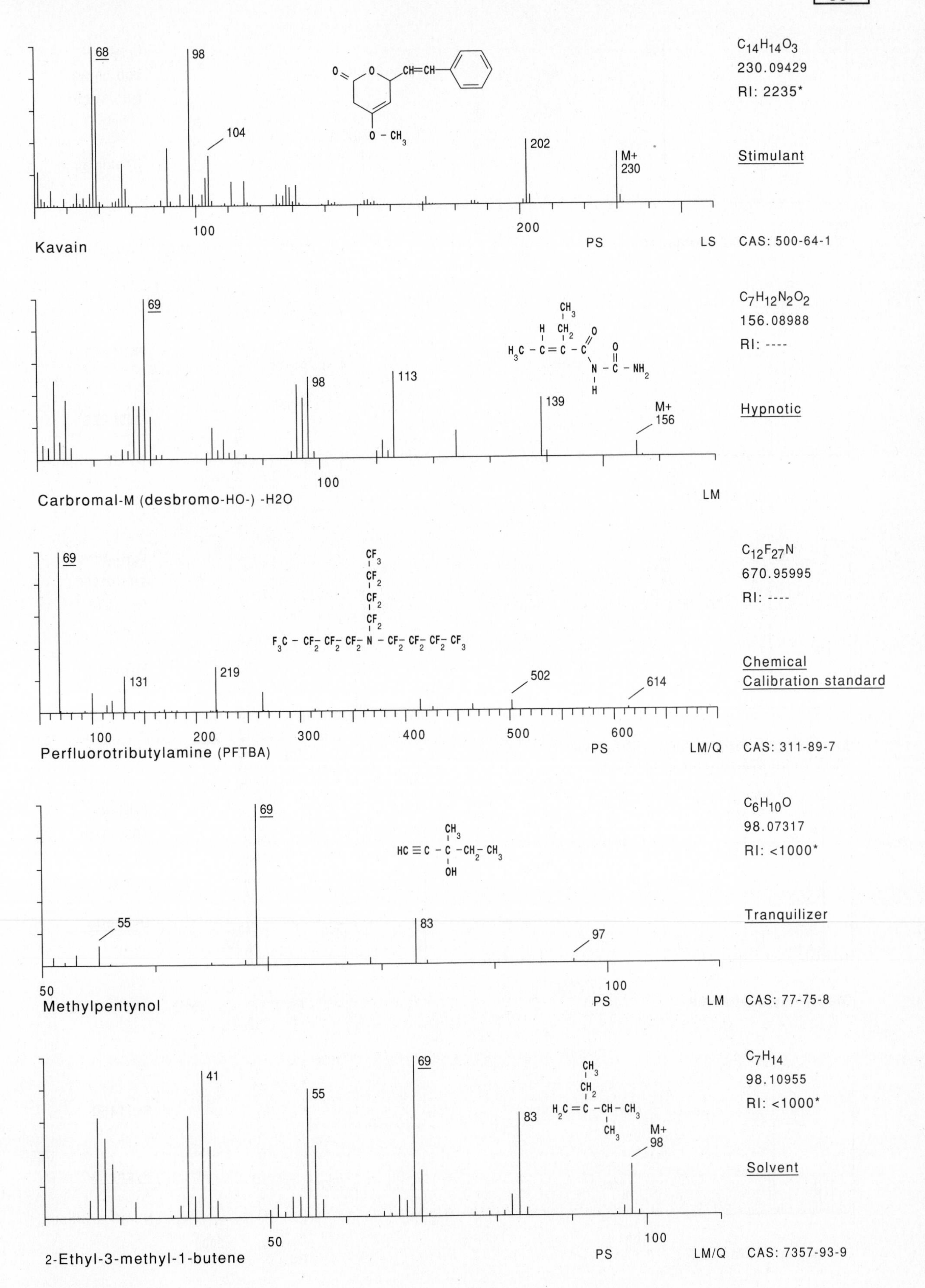
68
98
104
202
M+
230
100
200
PS
LS
Kavain
$C_{14}H_{14}O_3$
230.09429
RI: 2235*
Stimulant
CAS: 500-64-1
69
98
113
139
M+
156
100
Carbromal-M (desbromo-HO-) -H2O
LM
$C_7H_{12}N_2O_2$
156.08988
RI: ----
Hypnotic
69
131
219
502
614
100
200
300
400
500
600
Perfluorotributylamine (PFTBA)
PS
LM/Q
$C_{12}F_{27}N$
670.95995
RI: ----
Chemical
Calibration standard
CAS: 311-89-7
69
55
83
97
50
100
Methylpentynol
PS
LM
$C_6H_{10}O$
98.07317
RI: <1000*
Tranquilizer
CAS: 77-75-8
69
41
55
83
M+
98
50
100
2-Ethyl-3-methyl-1-butene
PS
LM/Q
C_7H_{14}
98.10955
RI: <1000*
Solvent
CAS: 7357-93-9

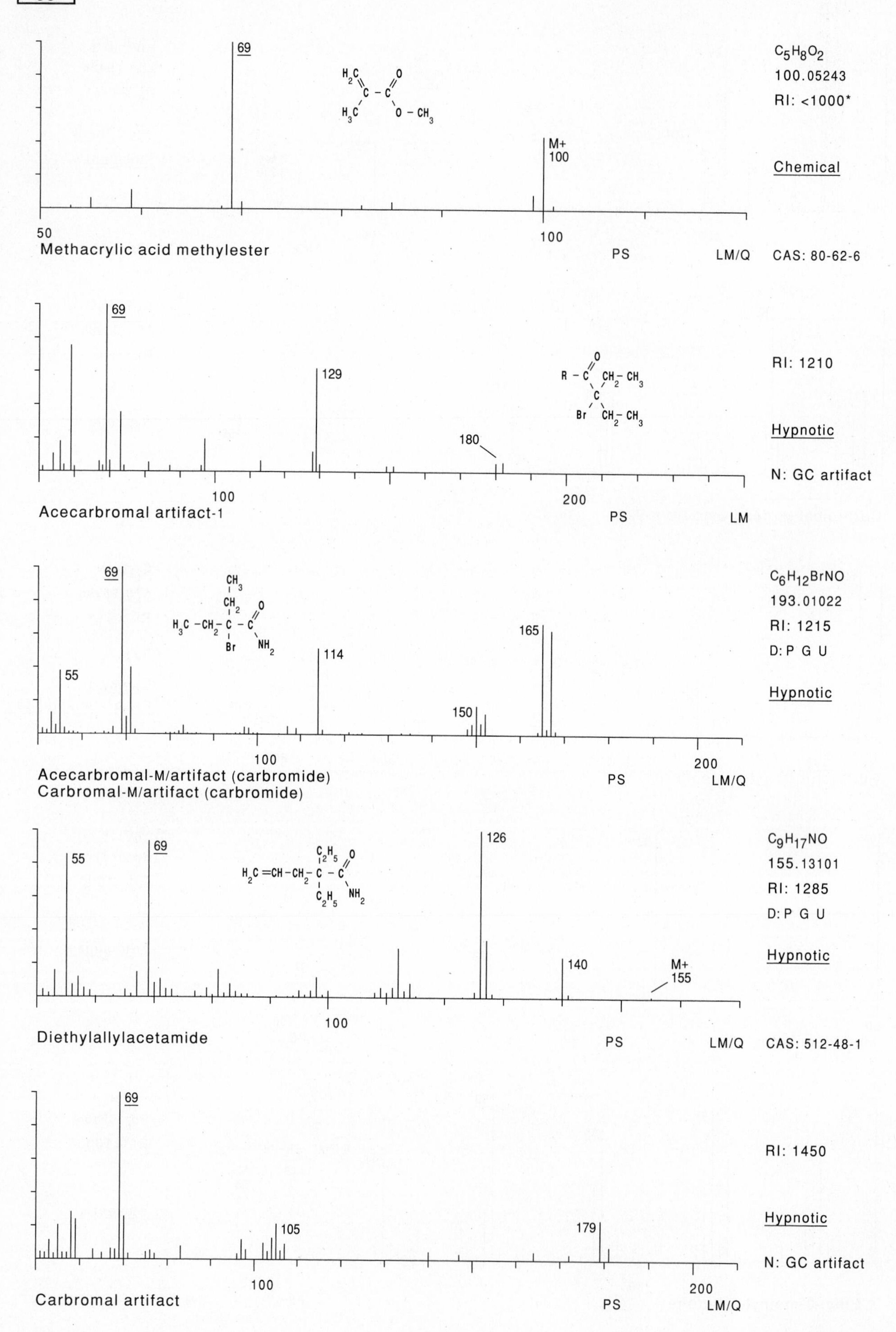
69
69
M+
100
50
100
Methacrylic acid methylester
PS
LM/Q
$C_5H_8O_2$
100.05243
RI: <1000*
Chemical
CAS: 80-62-6
69
129
180
100
200
Acecarbromal artifact-1
PS
LM
RI: 1210
Hypnotic
N: GC artifact
69
55
114
150
165
100
200
Acecarbromal-M/artifact (carbromide)
Carbromal-M/artifact (carbromide)
PS
LM/Q
$C_6H_{12}BrNO$
193.01022
RI: 1215
D: P G U
Hypnotic
55
69
126
140
M+
155
100
Diethylallylacetamide
PS
LM/Q
$C_9H_{17}NO$
155.13101
RI: 1285
D: P G U
Hypnotic
CAS: 512-48-1
69
105
179
100
200
Carbromal artifact
PS
LM/Q
RI: 1450
Hypnotic
N: GC artifact

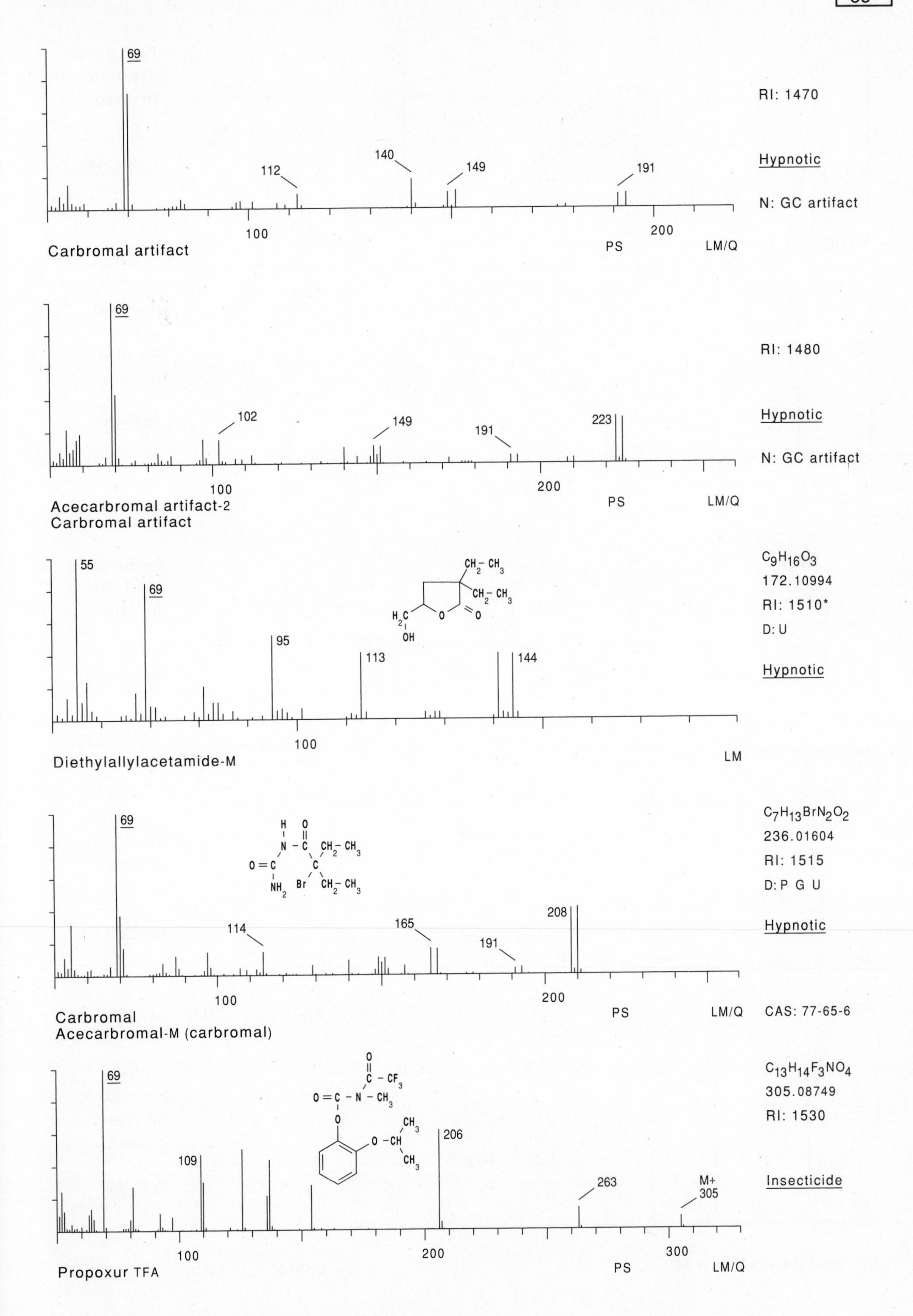

RI: 1470

Hypnotic

N: GC artifact

RI: 1480

Hypnotic

N: GC artifact

$C_9H_{16}O_3$
172.10994
RI: 1510*
D: U

Hypnotic

$C_7H_{13}BrN_2O_2$
236.01604
RI: 1515
D: P G U

Hypnotic

CAS: 77-65-6

$C_{13}H_{14}F_3NO_4$
305.08749
RI: 1530

Insecticide

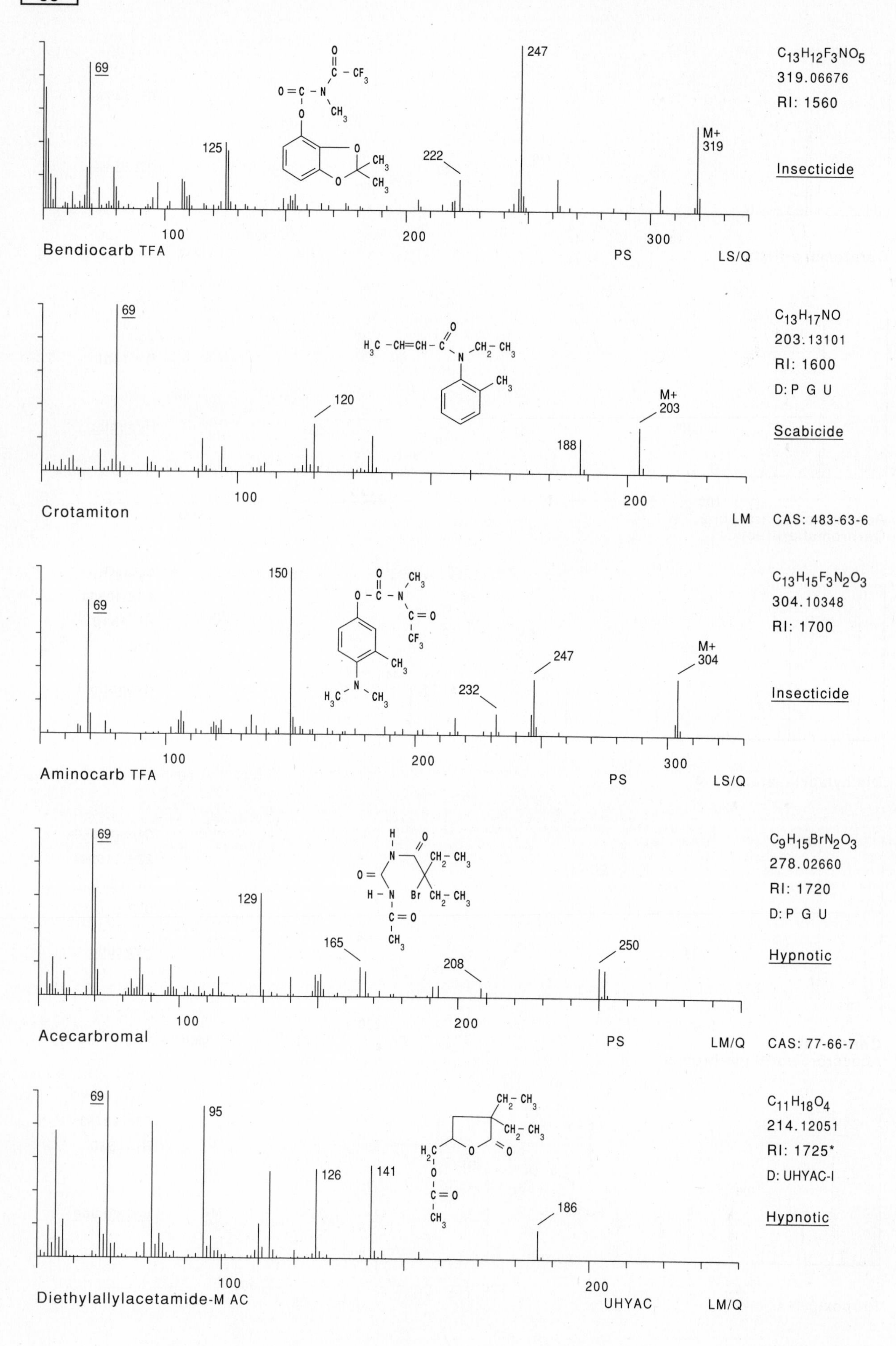

69
125
222
247
M+
319
100
200
300
Bendiocarb TFA
PS
LS/Q
$C_{13}H_{12}F_3NO_5$
319.06676
RI: 1560
Insecticide
69
120
188
M+
203
100
200
Crotamiton
LM
$C_{13}H_{17}NO$
203.13101
RI: 1600
D: P G U
Scabicide
CAS: 483-63-6
150
69
232
247
M+
304
100
200
300
Aminocarb TFA
PS
LS/Q
$C_{13}H_{15}F_3N_2O_3$
304.10348
RI: 1700
Insecticide
69
129
165
208
250
100
200
Acecarbromal
PS
LM/Q
$C_9H_{15}BrN_2O_3$
278.02660
RI: 1720
D: P G U
Hypnotic
CAS: 77-66-7
69
95
126
141
186
100
200
Diethylallylacetamide-M AC
UHYAC
LM/Q
$C_{11}H_{18}O_4$
214.12051
RI: 1725*
D: UHYAC-I
Hypnotic

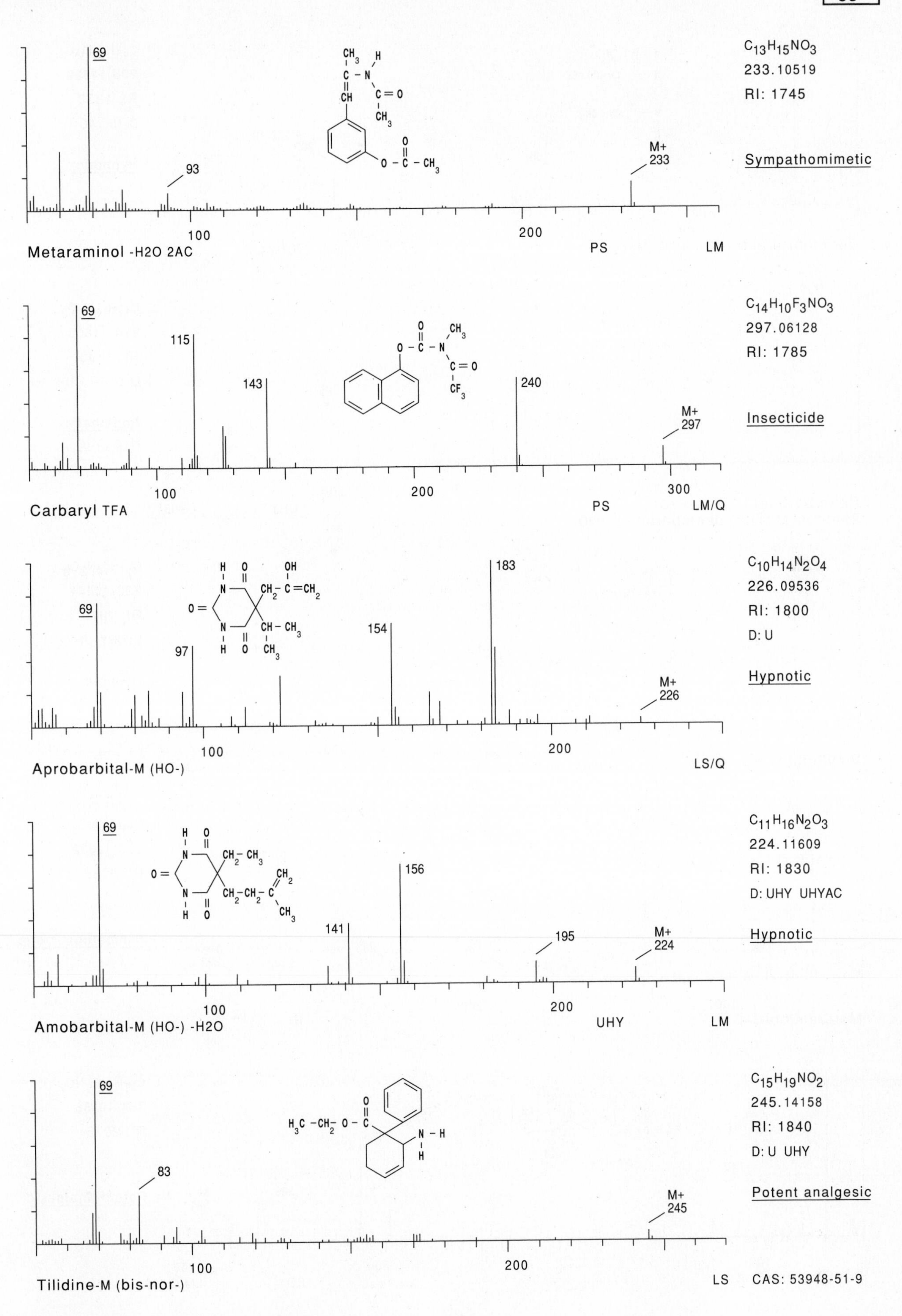

$C_{13}H_{15}NO_3$
233.10519
RI: 1745

Sympathomimetic

$C_{14}H_{10}F_3NO_3$
297.06128
RI: 1785

Insecticide

$C_{10}H_{14}N_2O_4$
226.09536
RI: 1800
D: U

Hypnotic

$C_{11}H_{16}N_2O_3$
224.11609
RI: 1830
D: UHY UHYAC

Hypnotic

$C_{15}H_{19}NO_2$
245.14158
RI: 1840
D: U UHY

Potent analgesic

CAS: 53948-51-9

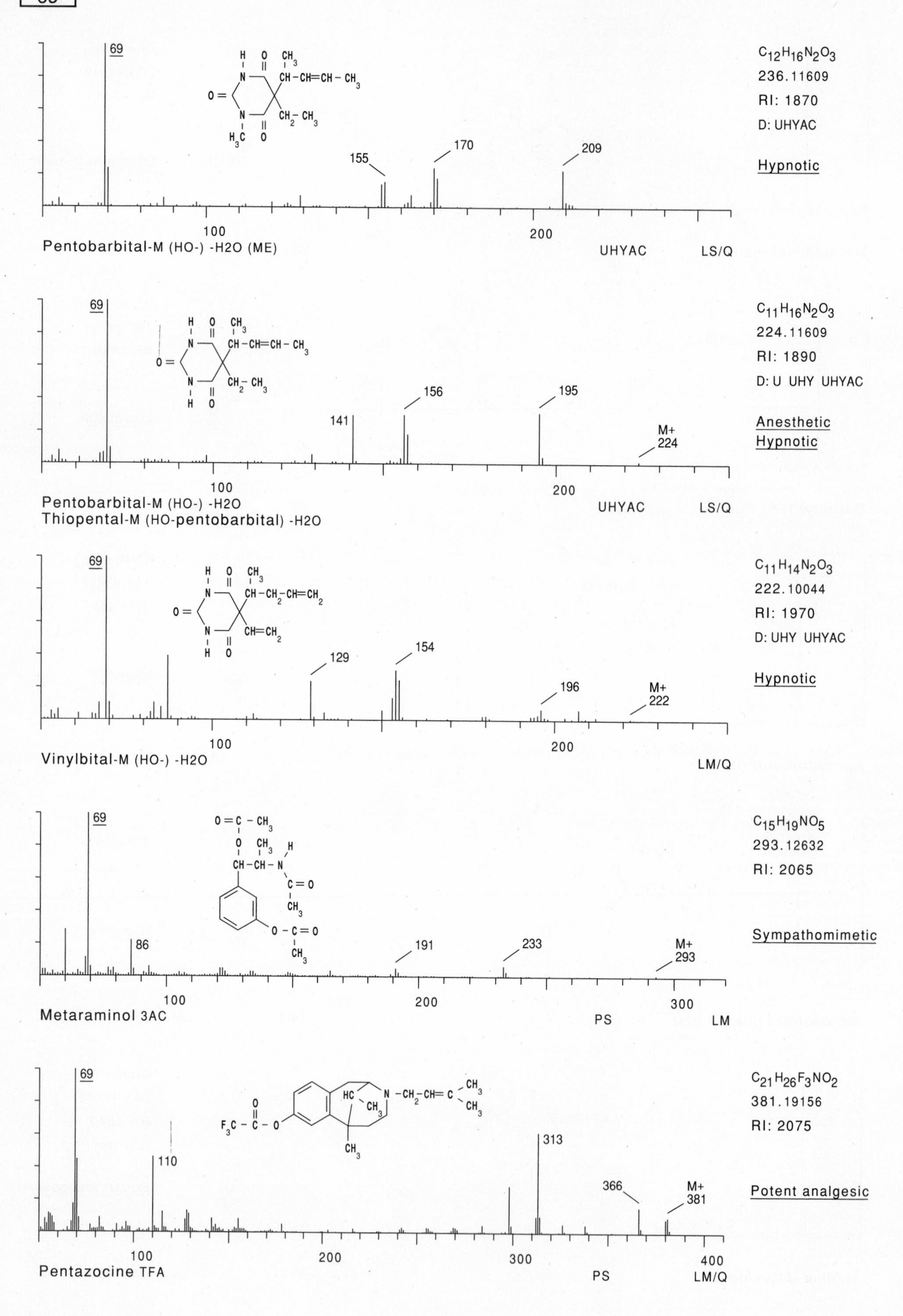

69
155
170
209
100
200
Pentobarbital-M (HO-) -H2O (ME)
UHYAC
LS/Q
C12H16N2O3
236.11609
RI: 1870
D: UHYAC
Hypnotic
69
141
156
195
M+
224
100
200
Pentobarbital-M (HO-) -H2O
Thiopental-M (HO-pentobarbital) -H2O
UHYAC
LS/Q
C11H16N2O3
224.11609
RI: 1890
D: U UHY UHYAC
Anesthetic
Hypnotic
69
129
154
196
M+
222
100
200
Vinylbital-M (HO-) -H2O
LM/Q
C11H14N2O3
222.10044
RI: 1970
D: UHY UHYAC
Hypnotic
69
86
191
233
M+
293
100
200
300
Metaraminol 3AC
PS
LM
C15H19NO5
293.12632
RI: 2065
Sympathomimetic
69
110
313
366
M+
381
100
200
300
400
Pentazocine TFA
PS
LM/Q
C21H26F3NO2
381.19156
RI: 2075
Potent analgesic

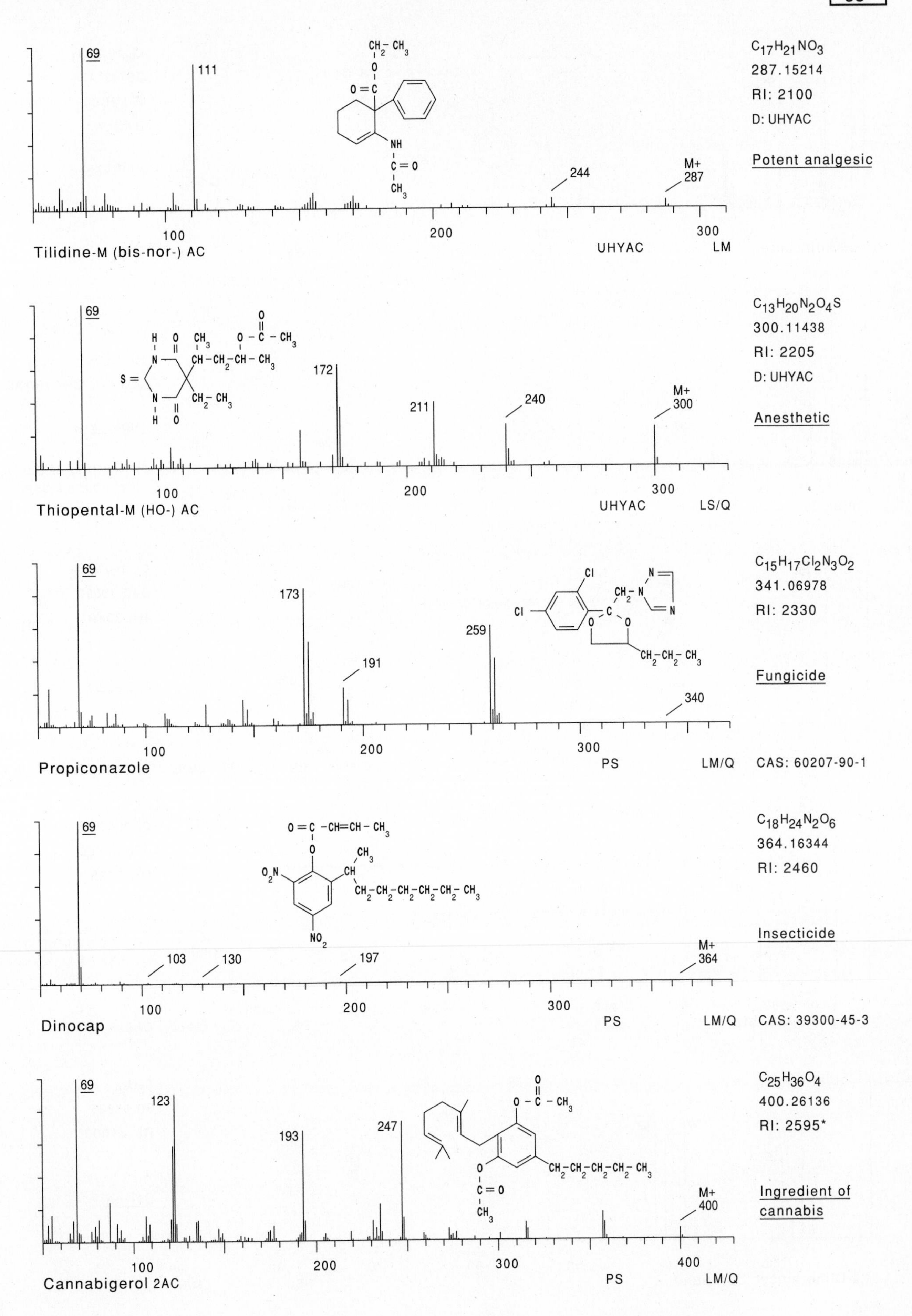
$C_{17}H_{21}NO_3$
287.15214
RI: 2100
D: UHYAC
Potent analgesic
Tilidine-M (bis-nor-) AC
UHYAC
LM
$C_{13}H_{20}N_2O_4S$
300.11438
RI: 2205
D: UHYAC
Anesthetic
Thiopental-M (HO-) AC
UHYAC
LS/Q
$C_{15}H_{17}Cl_2N_3O_2$
341.06978
RI: 2330
Fungicide
Propiconazole
PS
LM/Q
CAS: 60207-90-1
$C_{18}H_{24}N_2O_6$
364.16344
RI: 2460
Insecticide
Dinocap
PS
LM/Q
CAS: 39300-45-3
$C_{25}H_{36}O_4$
400.26136
RI: 2595*
Ingredient of cannabis
Cannabigerol 2AC
PS
LM/Q

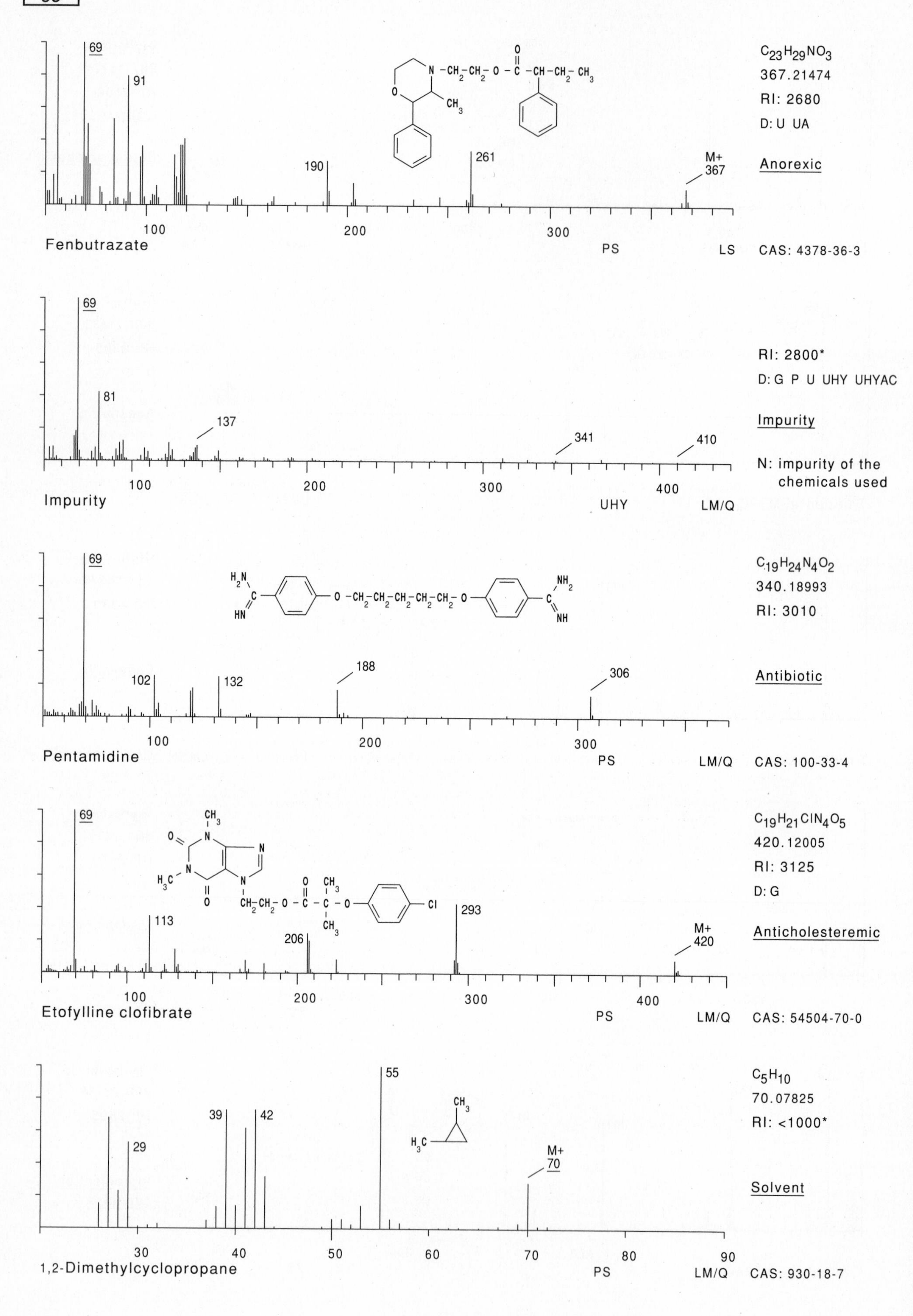
69
91
190
261
M+
367
100
200
300
Fenbutrazate
PS
LS
$C_{23}H_{29}NO_3$
367.21474
RI: 2680
D: U UA
Anorexic
CAS: 4378-36-3
69
81
137
341
410
100
200
300
400
Impurity
UHY
LM/Q
RI: 2800*
D: G P U UHY UHYAC
Impurity
N: impurity of the chemicals used
69
102
132
188
306
100
200
300
Pentamidine
PS
LM/Q
$C_{19}H_{24}N_4O_2$
340.18993
RI: 3010
Antibiotic
CAS: 100-33-4
69
113
206
293
M+
420
100
200
300
400
Etofylline clofibrate
PS
LM/Q
$C_{19}H_{21}ClN_4O_5$
420.12005
RI: 3125
D: G
Anticholesteremic
CAS: 54504-70-0
55
39
42
29
M+
70
30
40
50
60
70
80
90
1,2-Dimethylcyclopropane
PS
LM/Q
C_5H_{10}
70.07825
RI: <1000*
Solvent
CAS: 930-18-7

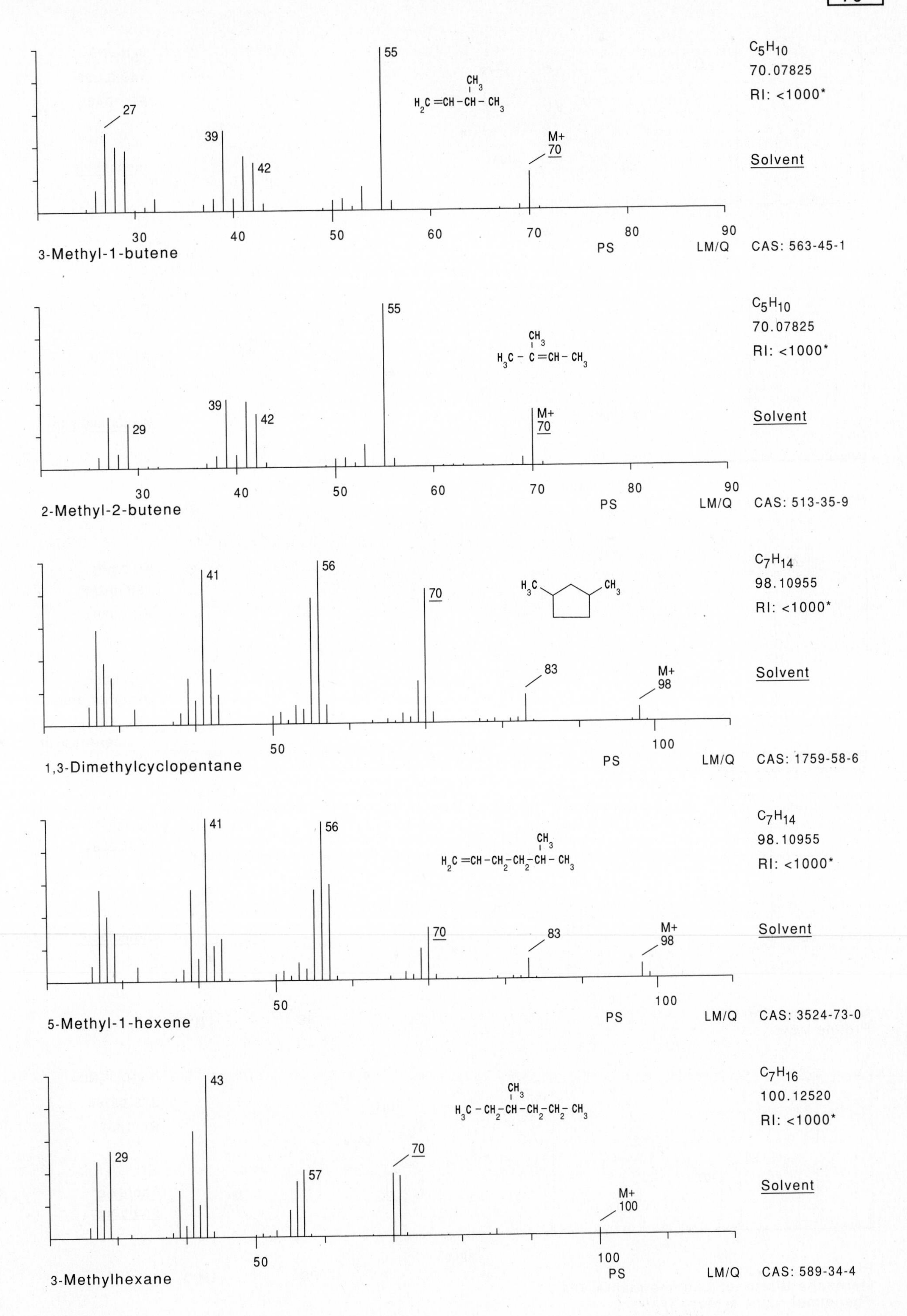

55
27
39
42
M+
70
30
40
50
60
70
80
90
PS
LM/Q
3-Methyl-1-butene
C_5H_{10}
70.07825
RI: <1000*
Solvent
CAS: 563-45-1
55
39
29
42
M+
70
2-Methyl-2-butene
C_5H_{10}
70.07825
RI: <1000*
Solvent
CAS: 513-35-9
41
56
70
83
M+
98
50
100
1,3-Dimethylcyclopentane
C_7H_{14}
98.10955
RI: <1000*
Solvent
CAS: 1759-58-6
41
56
70
83
M+
98
5-Methyl-1-hexene
C_7H_{14}
98.10955
RI: <1000*
Solvent
CAS: 3524-73-0
43
29
57
70
M+
100
3-Methylhexane
C_7H_{16}
100.12520
RI: <1000*
Solvent
CAS: 589-34-4

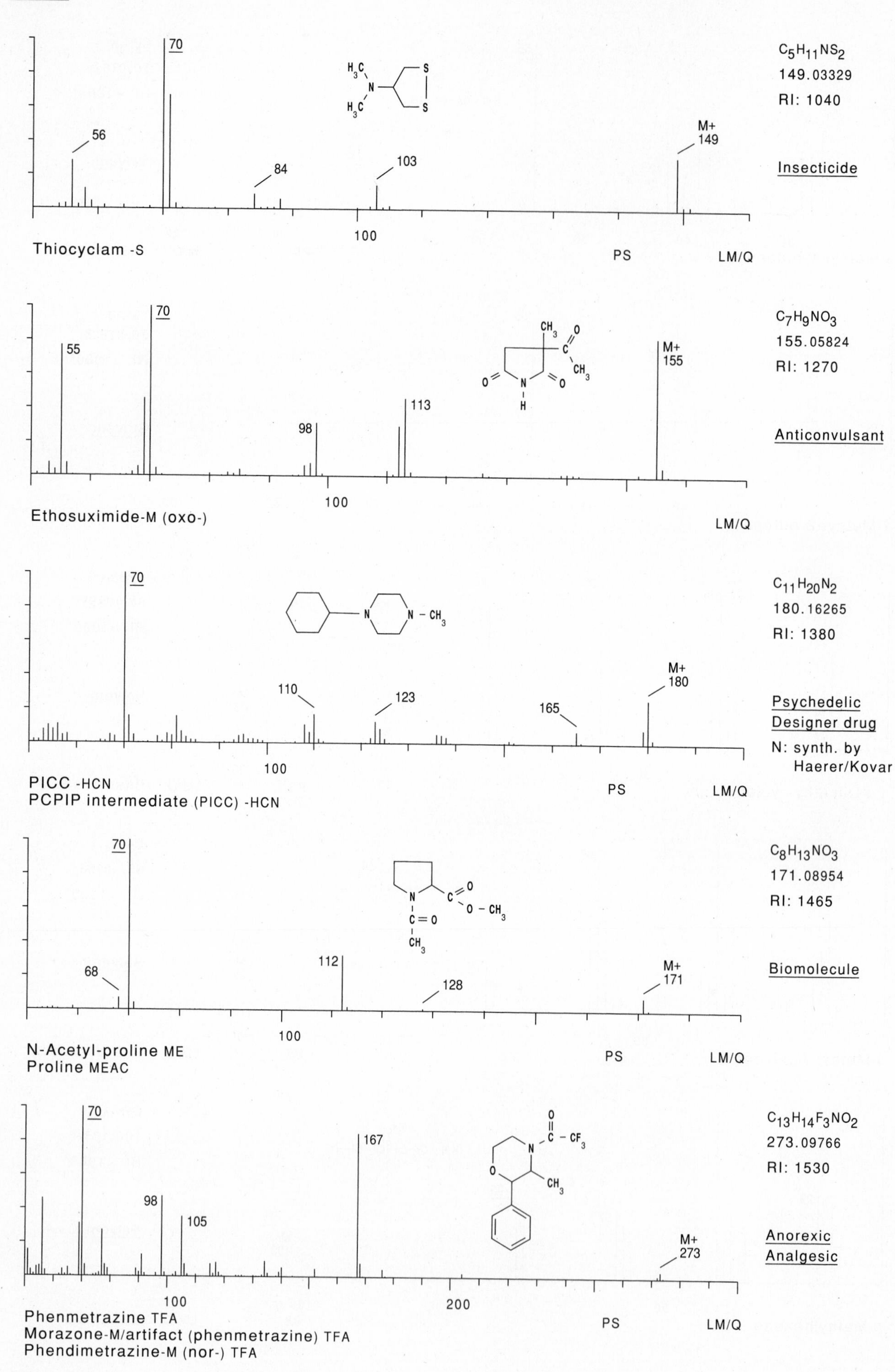
70
56
84
103
M+
149
100
PS
LM/Q
Thiocyclam -S
$C_5H_{11}NS_2$
149.03329
RI: 1040
Insecticide
70
55
98
113
M+
155
100
LM/Q
Ethosuximide-M (oxo-)
$C_7H_9NO_3$
155.05824
RI: 1270
Anticonvulsant
70
110
123
165
M+
180
100
PS
LM/Q
PICC -HCN
PCPIP intermediate (PICC) -HCN
$C_{11}H_{20}N_2$
180.16265
RI: 1380
Psychedelic
Designer drug
N: synth. by
Haerer/Kovar
70
68
112
128
M+
171
100
PS
LM/Q
N-Acetyl-proline ME
Proline MEAC
$C_8H_{13}NO_3$
171.08954
RI: 1465
Biomolecule
70
98
105
167
M+
273
100
200
PS
LM/Q
Phenmetrazine TFA
Morazone-M/artifact (phenmetrazine) TFA
Phendimetrazine-M (nor-) TFA
$C_{13}H_{14}F_3NO_2$
273.09766
RI: 1530
Anorexic
Analgesic

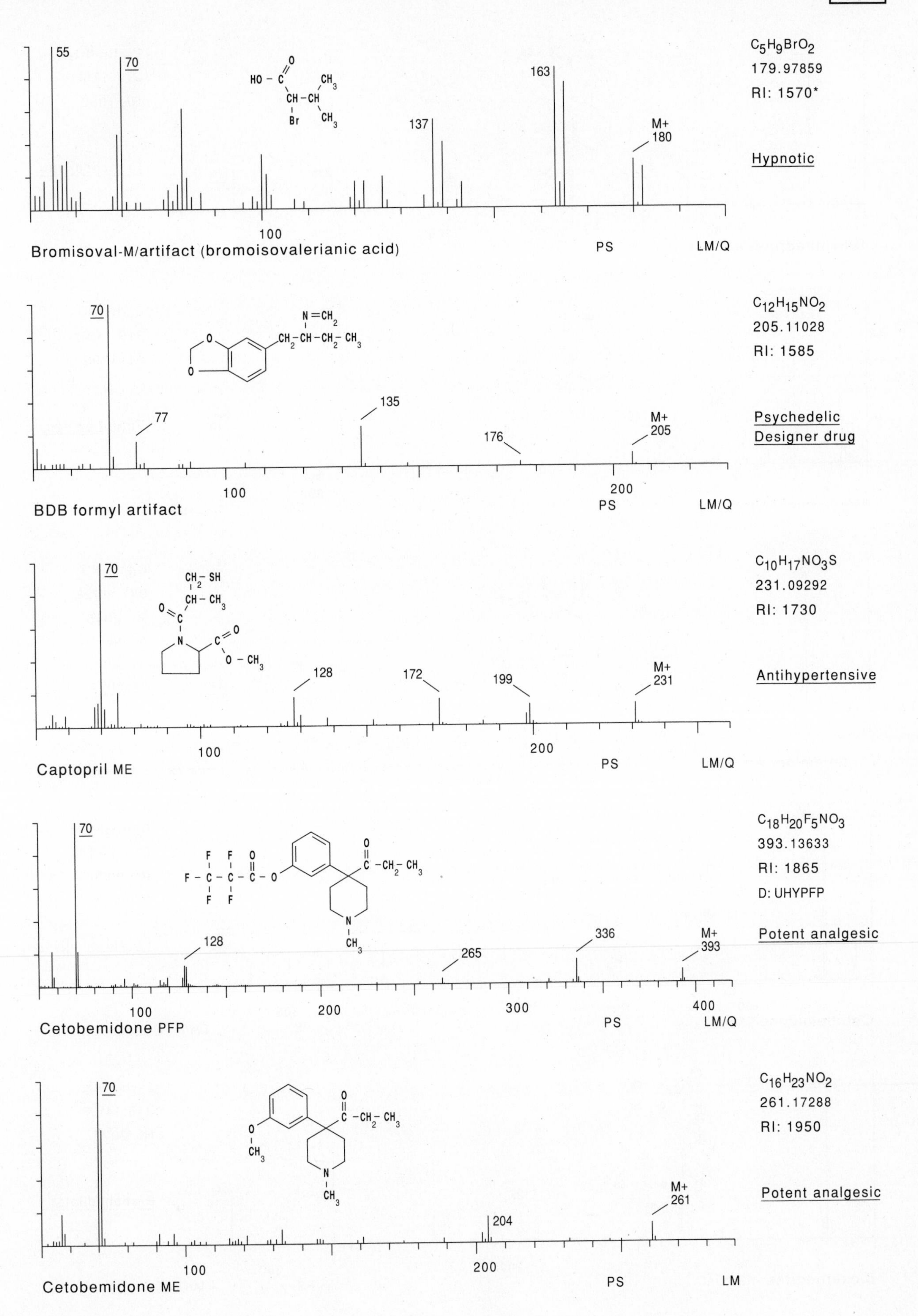

Bromisoval-M/artifact (bromoisovalerianic acid)

$C_5H_9BrO_2$
179.97859
RI: 1570*

Hypnotic

BDB formyl artifact

$C_{12}H_{15}NO_2$
205.11028
RI: 1585

Psychedelic
Designer drug

Captopril ME

$C_{10}H_{17}NO_3S$
231.09292
RI: 1730

Antihypertensive

Cetobemidone PFP

$C_{18}H_{20}F_5NO_3$
393.13633
RI: 1865
D: UHYPFP

Potent analgesic

Cetobemidone ME

$C_{16}H_{23}NO_2$
261.17288
RI: 1950

Potent analgesic

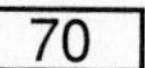

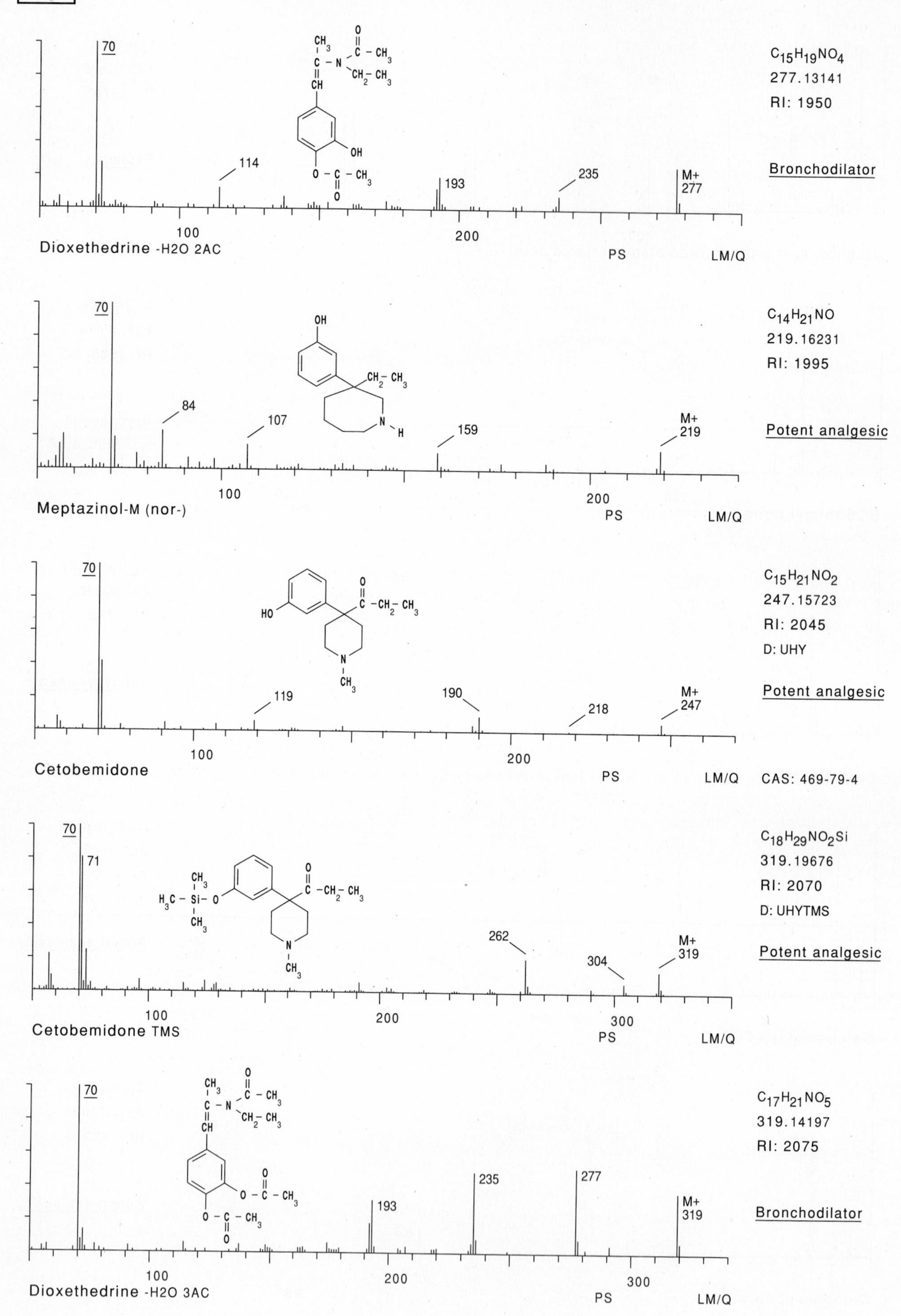

70
114
193
235
M+ 277
100
200
Dioxethedrine -H2O 2AC
PS
LM/Q
$C_{15}H_{19}NO_4$
277.13141
RI: 1950
Bronchodilator
70
84
107
159
M+ 219
100
200
Meptazinol-M (nor-)
PS
LM/Q
$C_{14}H_{21}NO$
219.16231
RI: 1995
Potent analgesic
70
119
190
218
M+ 247
100
200
Cetobemidone
PS
LM/Q
$C_{15}H_{21}NO_2$
247.15723
RI: 2045
D: UHY
Potent analgesic
CAS: 469-79-4
70
71
262
304
M+ 319
100
200
300
Cetobemidone TMS
PS
LM/Q
$C_{18}H_{29}NO_2Si$
319.19676
RI: 2070
D: UHYTMS
Potent analgesic
70
193
235
277
M+ 319
100
200
300
Dioxethedrine -H2O 3AC
PS
LM/Q
$C_{17}H_{21}NO_5$
319.14197
RI: 2075
Bronchodilator

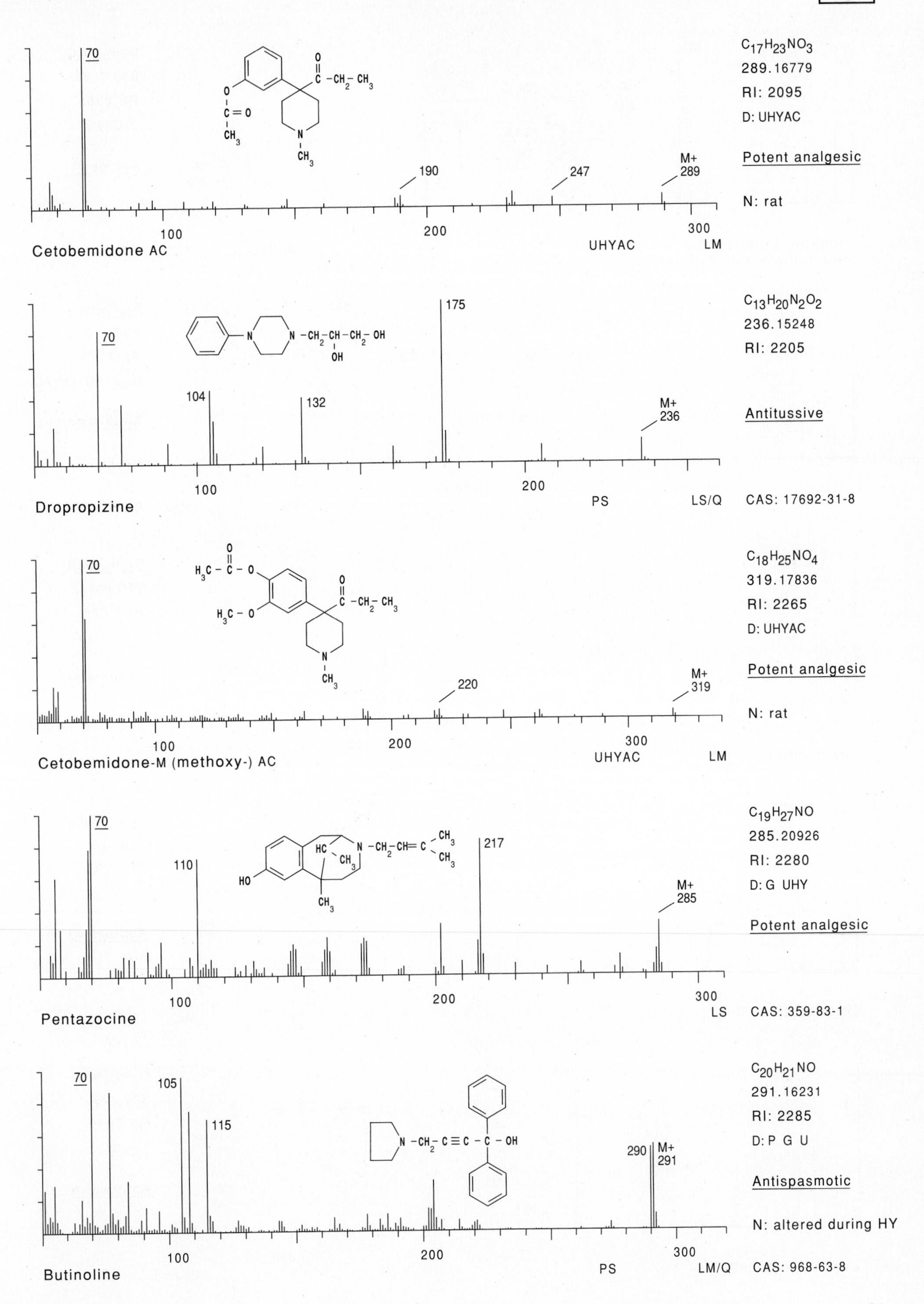
70
190
247
M+
289
100
200
300
Cetobemidone AC
UHYAC
LM
$C_{17}H_{23}NO_3$
289.16779
RI: 2095
D: UHYAC
Potent analgesic
N: rat
175
70
104
132
M+
236
Dropropizine
PS
LS/Q
$C_{13}H_{20}N_2O_2$
236.15248
RI: 2205
Antitussive
CAS: 17692-31-8
70
220
M+
319
Cetobemidone-M (methoxy-) AC
UHYAC
LM
$C_{18}H_{25}NO_4$
319.17836
RI: 2265
D: UHYAC
Potent analgesic
N: rat
70
110
217
M+
285
Pentazocine
LS
$C_{19}H_{27}NO$
285.20926
RI: 2280
D: G UHY
Potent analgesic
CAS: 359-83-1
70
105
115
290
M+
291
Butinoline
PS
LM/Q
$C_{20}H_{21}NO$
291.16231
RI: 2285
D: P G U
Antispasmotic
N: altered during HY
CAS: 968-63-8

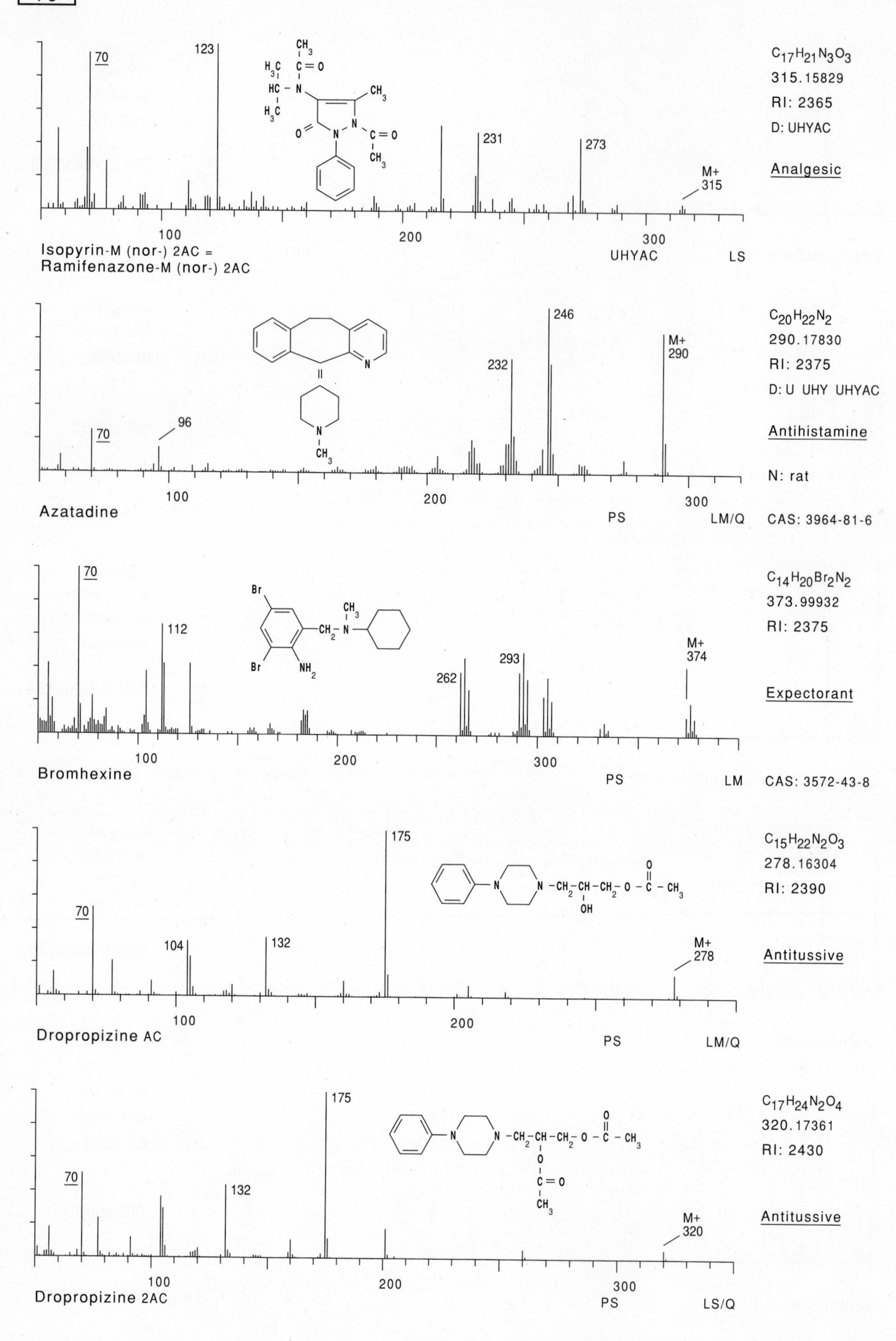

70
123
231
273
M+
315
100
200
300
Isopyrin-M (nor-) 2AC =
Ramifenazone-M (nor-) 2AC
UHYAC
LS
$C_{17}H_{21}N_3O_3$
315.15829
RI: 2365
D: UHYAC
Analgesic
246
M+
290
232
96
70
N
Azatadine
PS
LM/Q
$C_{20}H_{22}N_2$
290.17830
RI: 2375
D: U UHY UHYAC
Antihistamine
N: rat
CAS: 3964-81-6
70
112
Br
NH2
293
262
M+
374
Bromhexine
PS
LM
$C_{14}H_{20}Br_2N_2$
373.99932
RI: 2375
Expectorant
CAS: 3572-43-8
175
70
104
132
M+
278
OH
Dropropizine AC
PS
LM/Q
$C_{15}H_{22}N_2O_3$
278.16304
RI: 2390
Antitussive
175
70
132
M+
320
Dropropizine 2AC
PS
LS/Q
$C_{17}H_{24}N_2O_4$
320.17361
RI: 2430
Antitussive

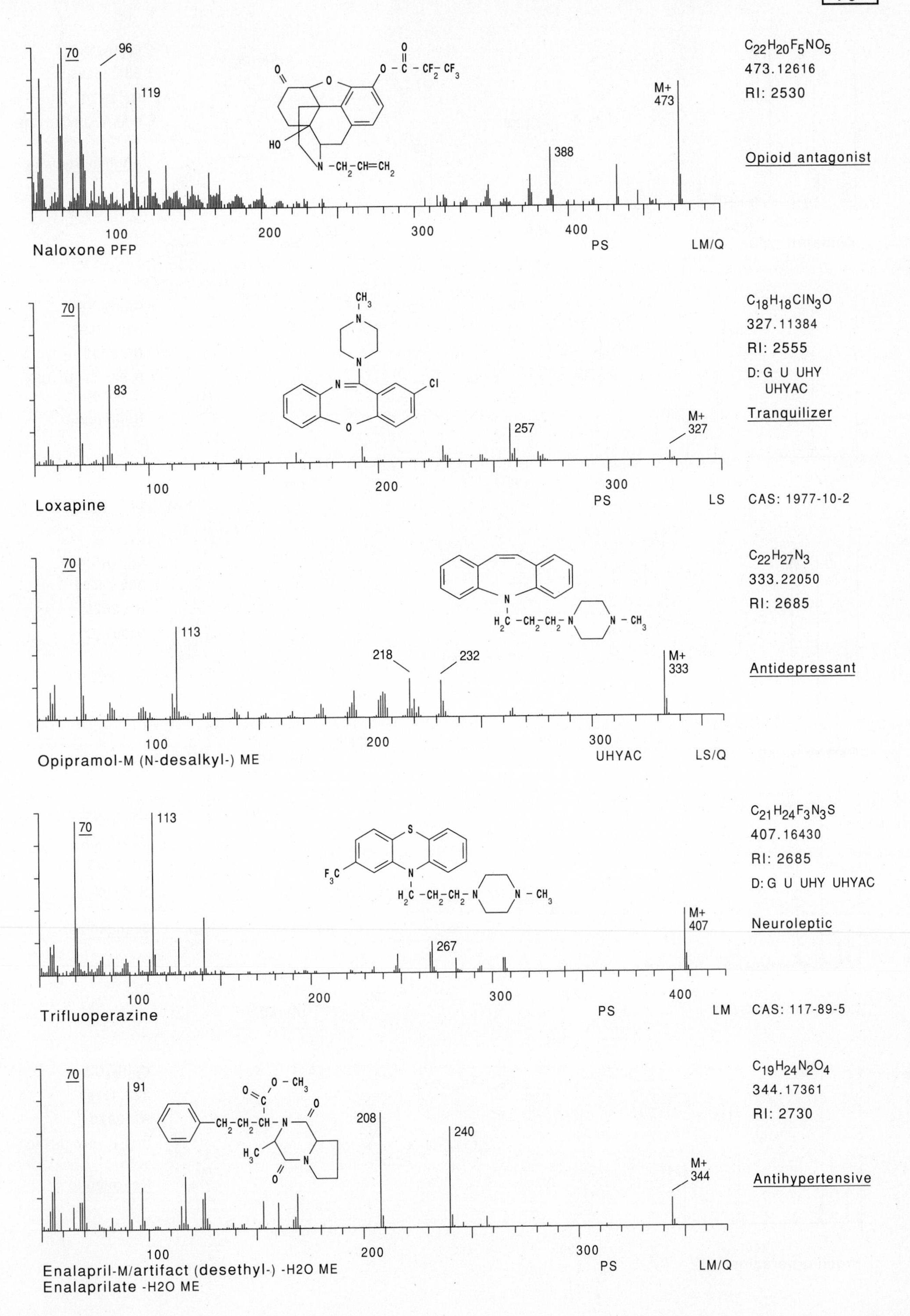

Naloxone PFP

$C_{22}H_{20}F_5NO_5$
473.12616
RI: 2530

Opioid antagonist

Loxapine

$C_{18}H_{18}ClN_3O$
327.11384
RI: 2555
D: G U UHY
UHYAC

Tranquilizer

CAS: 1977-10-2

Opipramol-M (N-desalkyl-) ME

$C_{22}H_{27}N_3$
333.22050
RI: 2685

Antidepressant

Trifluoperazine

$C_{21}H_{24}F_3N_3S$
407.16430
RI: 2685
D: G U UHY UHYAC

Neuroleptic

CAS: 117-89-5

Enalapril-M/artifact (desethyl-) -H2O ME
Enalaprilate -H2O ME

$C_{19}H_{24}N_2O_4$
344.17361
RI: 2730

Antihypertensive

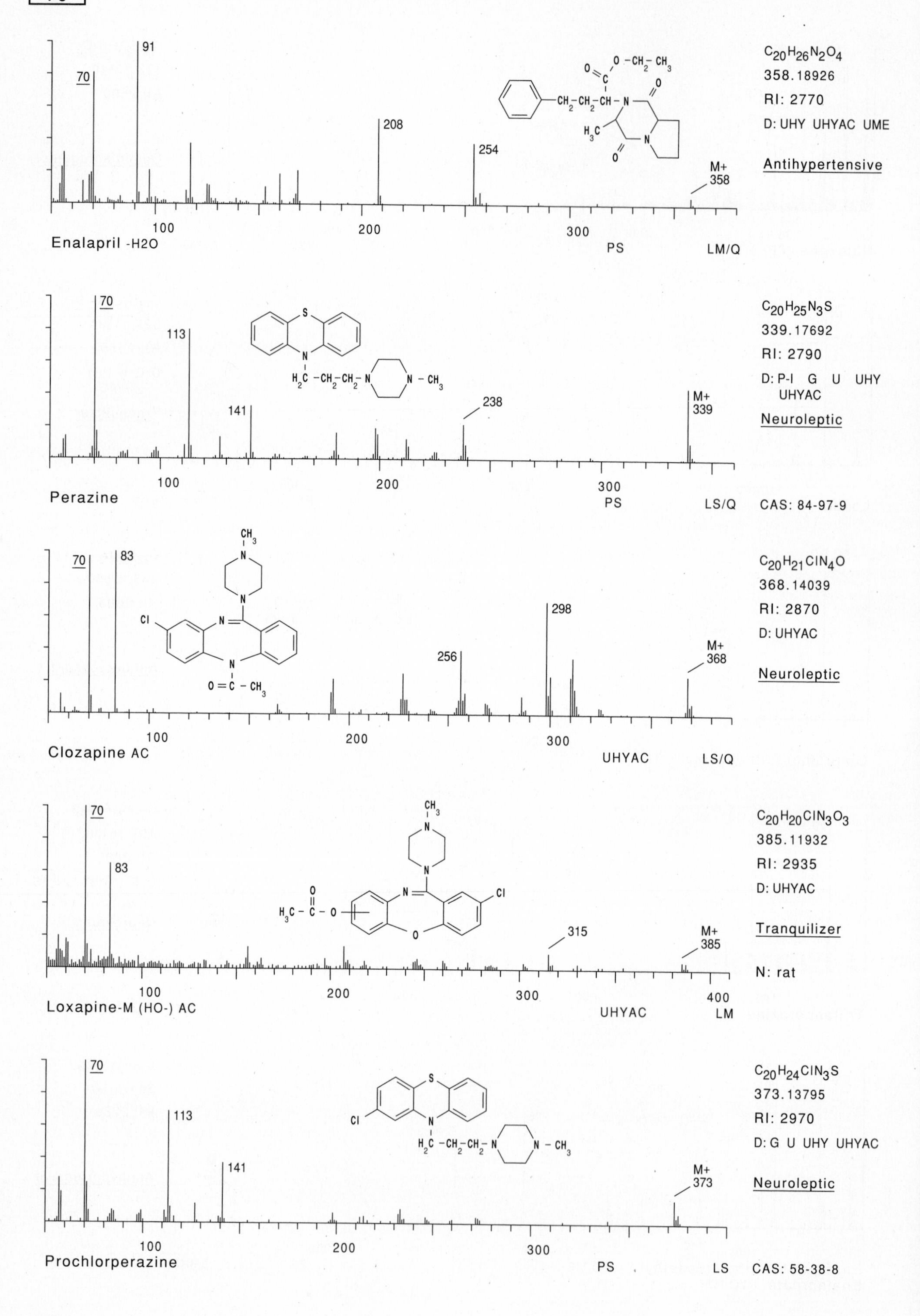
Enalapril -H2O
C20H26N2O4
358.18926
RI: 2770
D: UHY UHYAC UME
Antihypertensive
PS LM/Q
Perazine
C20H25N3S
339.17692
RI: 2790
D: P-I G U UHY UHYAC
Neuroleptic
PS LS/Q
CAS: 84-97-9
Clozapine AC
C20H21ClN4O
368.14039
RI: 2870
D: UHYAC
Neuroleptic
UHYAC LS/Q
Loxapine-M (HO-) AC
C20H20ClN3O3
385.11932
RI: 2935
D: UHYAC
Tranquilizer
N: rat
UHYAC LM
Prochlorperazine
C20H24ClN3S
373.13795
RI: 2970
D: G U UHY UHYAC
Neuroleptic
PS LS
CAS: 58-38-8

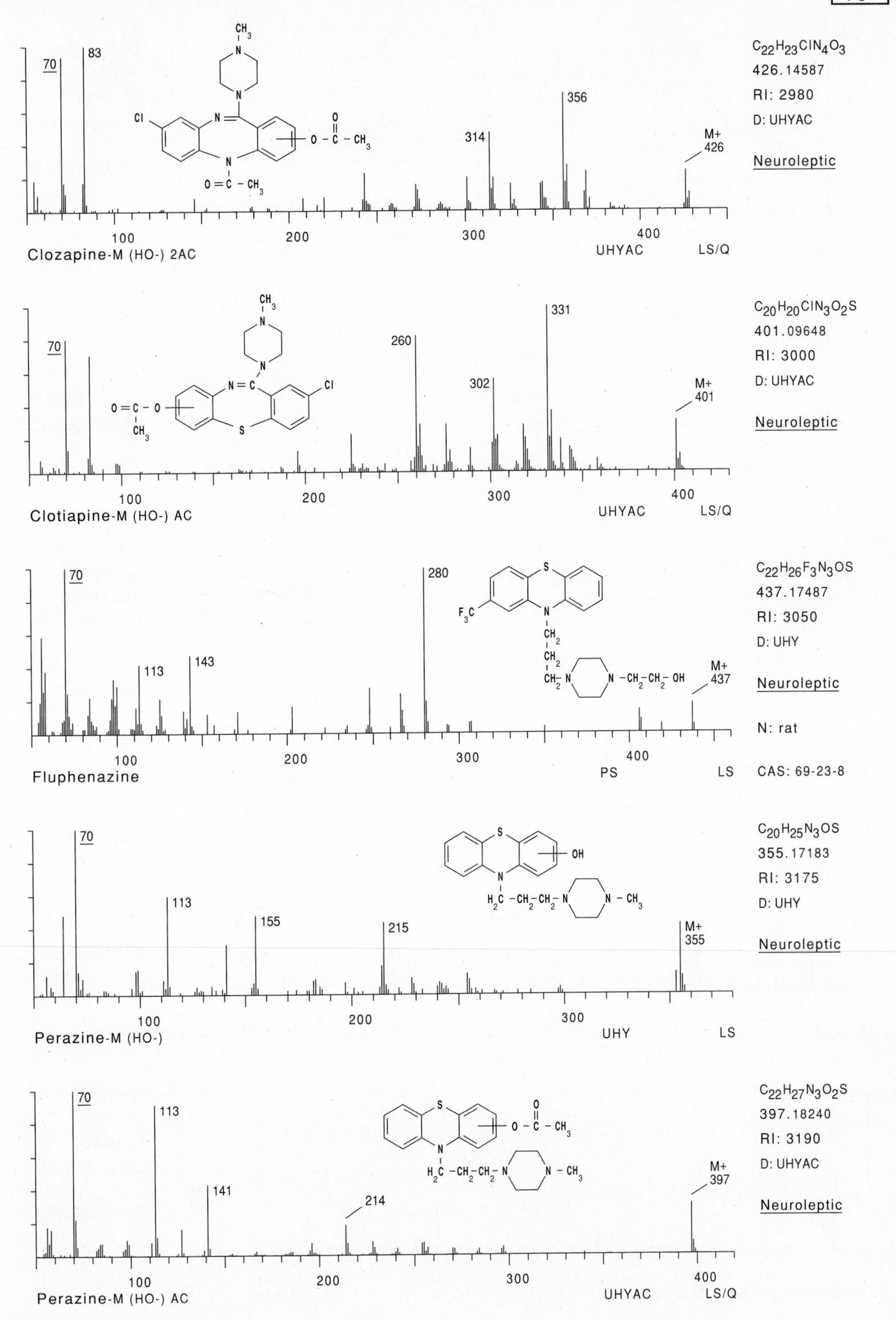

$C_{22}H_{23}ClN_4O_3$
426.14587
RI: 2980
D: UHYAC

Neuroleptic

$C_{20}H_{20}ClN_3O_2S$
401.09648
RI: 3000
D: UHYAC

Neuroleptic

$C_{22}H_{26}F_3N_3OS$
437.17487
RI: 3050
D: UHY

Neuroleptic

N: rat

CAS: 69-23-8

$C_{20}H_{25}N_3OS$
355.17183
RI: 3175
D: UHY

Neuroleptic

$C_{22}H_{27}N_3O_2S$
397.18240
RI: 3190
D: UHYAC

Neuroleptic

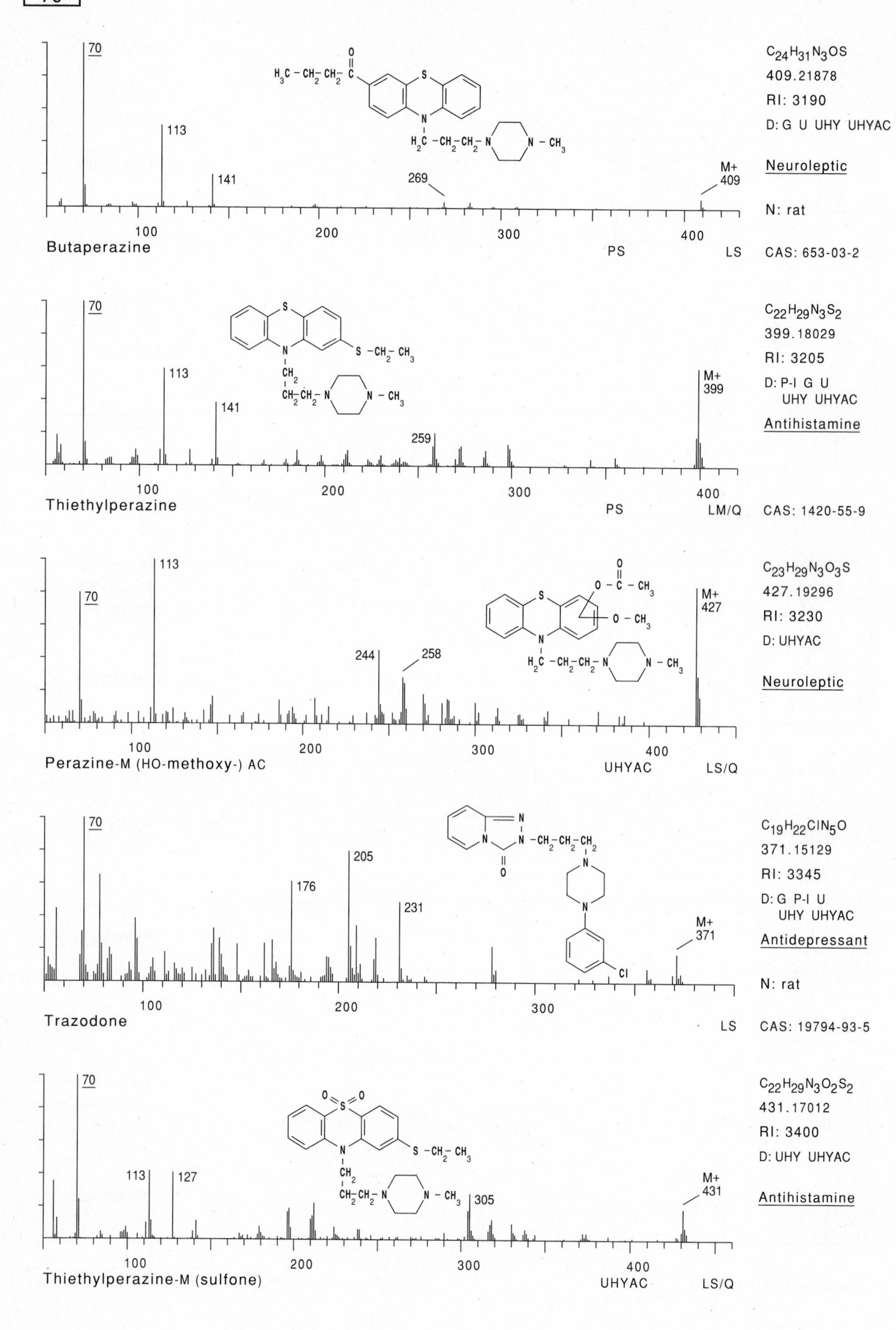
70
Butaperazine
C24H31N3OS
409.21878
RI: 3190
D: G U UHY UHYAC
Neuroleptic
N: rat
PS
LS
CAS: 653-03-2
Thiethylperazine
C22H29N3S2
399.18029
RI: 3205
D: P-I G U
UHY UHYAC
Antihistamine
PS
LM/Q
CAS: 1420-55-9
Perazine-M (HO-methoxy-) AC
C23H29N3O3S
427.19296
RI: 3230
D: UHYAC
Neuroleptic
UHYAC
LS/Q
Trazodone
C19H22ClN5O
371.15129
RI: 3345
D: G P-I U
UHY UHYAC
Antidepressant
N: rat
LS
CAS: 19794-93-5
Thiethylperazine-M (sulfone)
C22H29N3O2S2
431.17012
RI: 3400
D: UHY UHYAC
Antihistamine
UHYAC
LS/Q

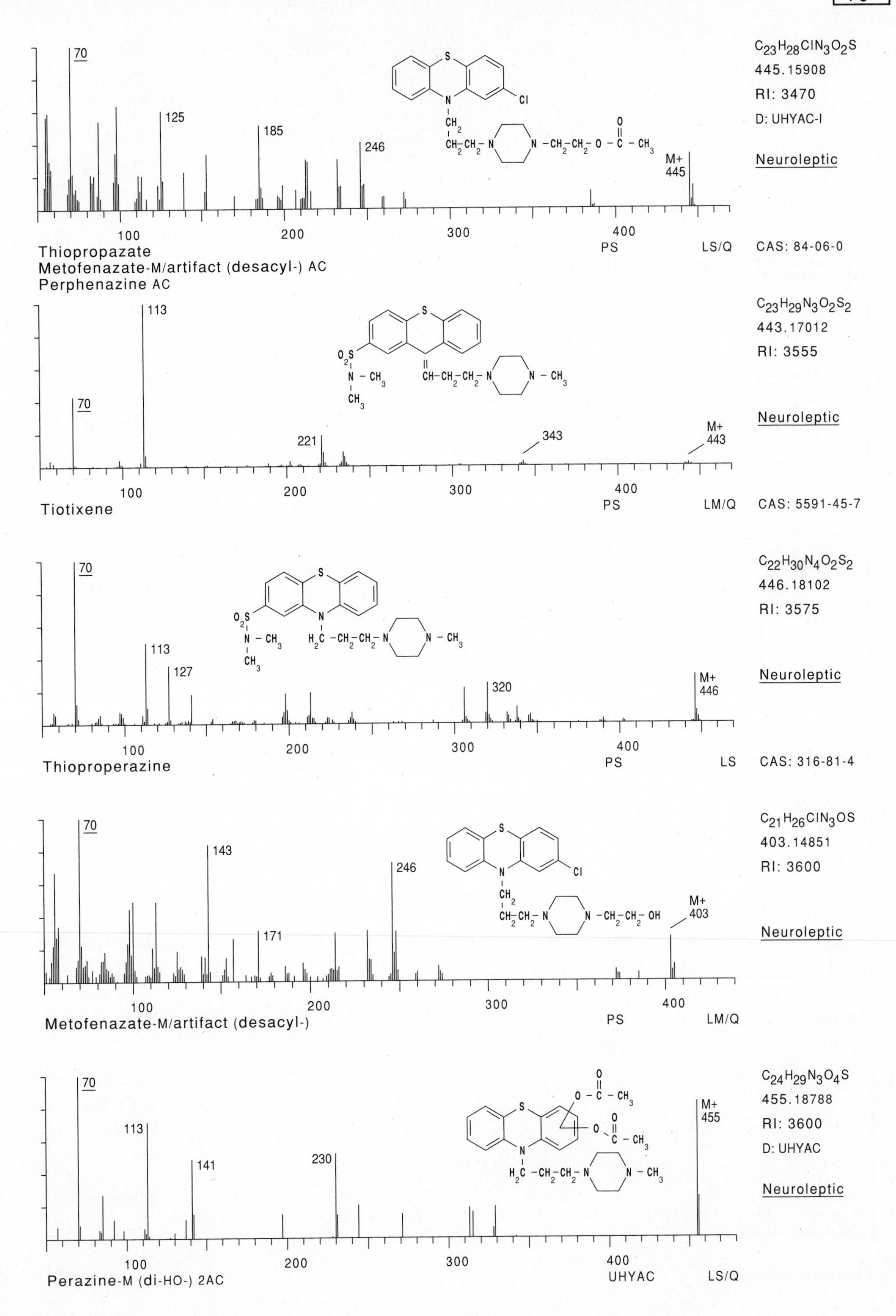

70
125
185
246
M+
445
100
200
300
400
PS
LS/Q
Thiopropazate
Metofenazate-M/artifact (desacyl-) AC
Perphenazine AC
C23H28ClN3O2S
445.15908
RI: 3470
D: UHYAC-I
Neuroleptic
CAS: 84-06-0
113
70
221
343
M+
443
LM/Q
Tiotixene
C23H29N3O2S2
443.17012
RI: 3555
Neuroleptic
CAS: 5591-45-7
70
113
127
320
M+
446
LS
Thioproperazine
C22H30N4O2S2
446.18102
RI: 3575
Neuroleptic
CAS: 316-81-4
70
143
246
171
M+
403
Metofenazate-M/artifact (desacyl-)
C21H26ClN3OS
403.14851
RI: 3600
Neuroleptic
70
113
141
230
M+
455
UHYAC
Perazine-M (di-HO-) 2AC
C24H29N3O4S
455.18788
RI: 3600
D: UHYAC
Neuroleptic

71

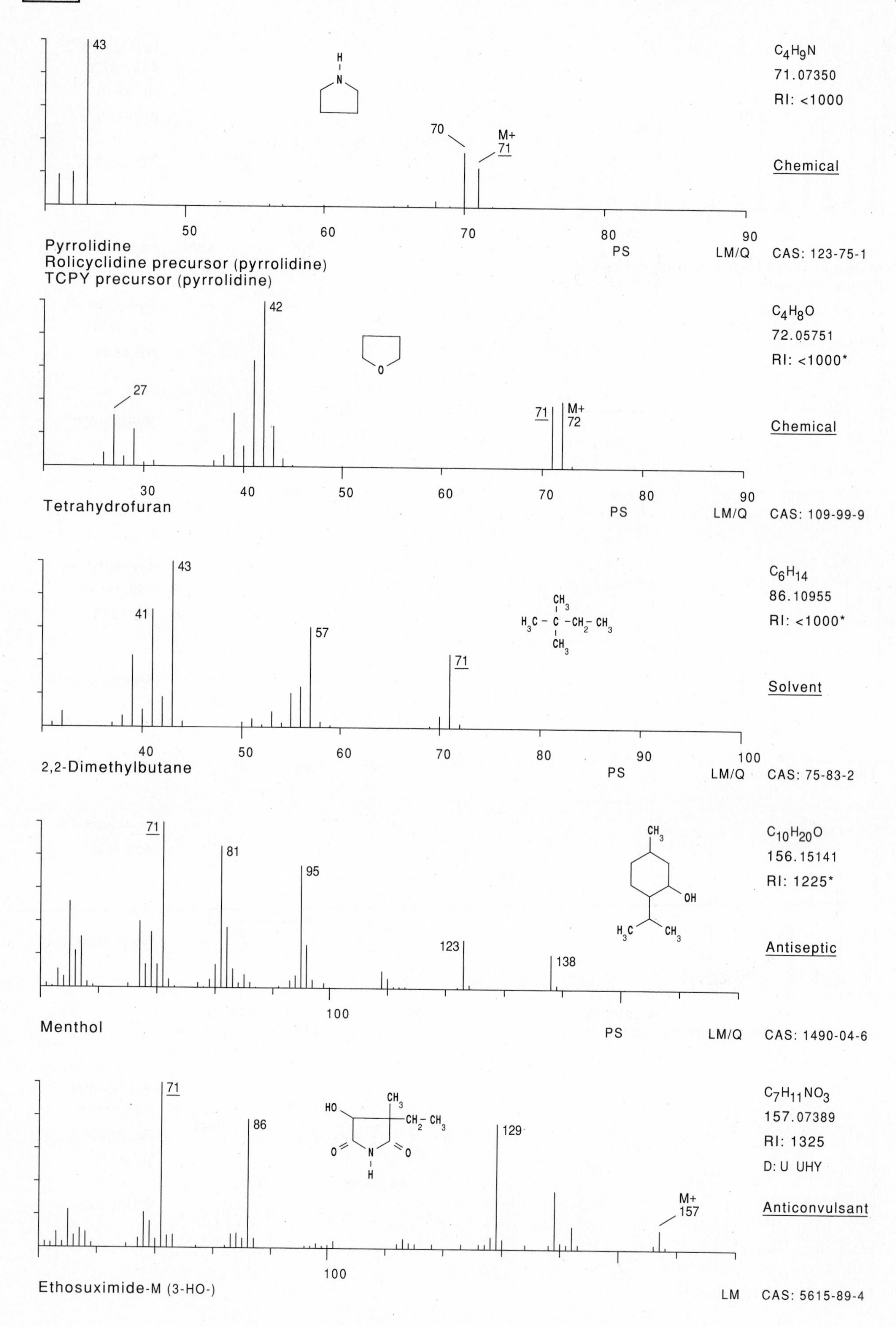

43
70
M+
71
50
60
70
80
90
PS
LM/Q
Pyrrolidine
Rolicyclidine precursor (pyrrolidine)
TCPY precursor (pyrrolidine)
C4H9N
71.07350
RI: <1000
Chemical
CAS: 123-75-1
42
27
71
M+
72
30
40
50
60
70
80
90
PS
LM/Q
Tetrahydrofuran
C4H8O
72.05751
RI: <1000*
Chemical
CAS: 109-99-9
43
41
57
71
40
50
60
70
80
90
100
PS
LM/Q
2,2-Dimethylbutane
C6H14
86.10955
RI: <1000*
Solvent
CAS: 75-83-2
71
81
95
123
138
100
PS
LM/Q
Menthol
C10H20O
156.15141
RI: 1225*
Antiseptic
CAS: 1490-04-6
71
86
129
M+
157
100
LM
Ethosuximide-M (3-HO-)
C7H11NO3
157.07389
RI: 1325
D: U UHY
Anticonvulsant
CAS: 5615-89-4

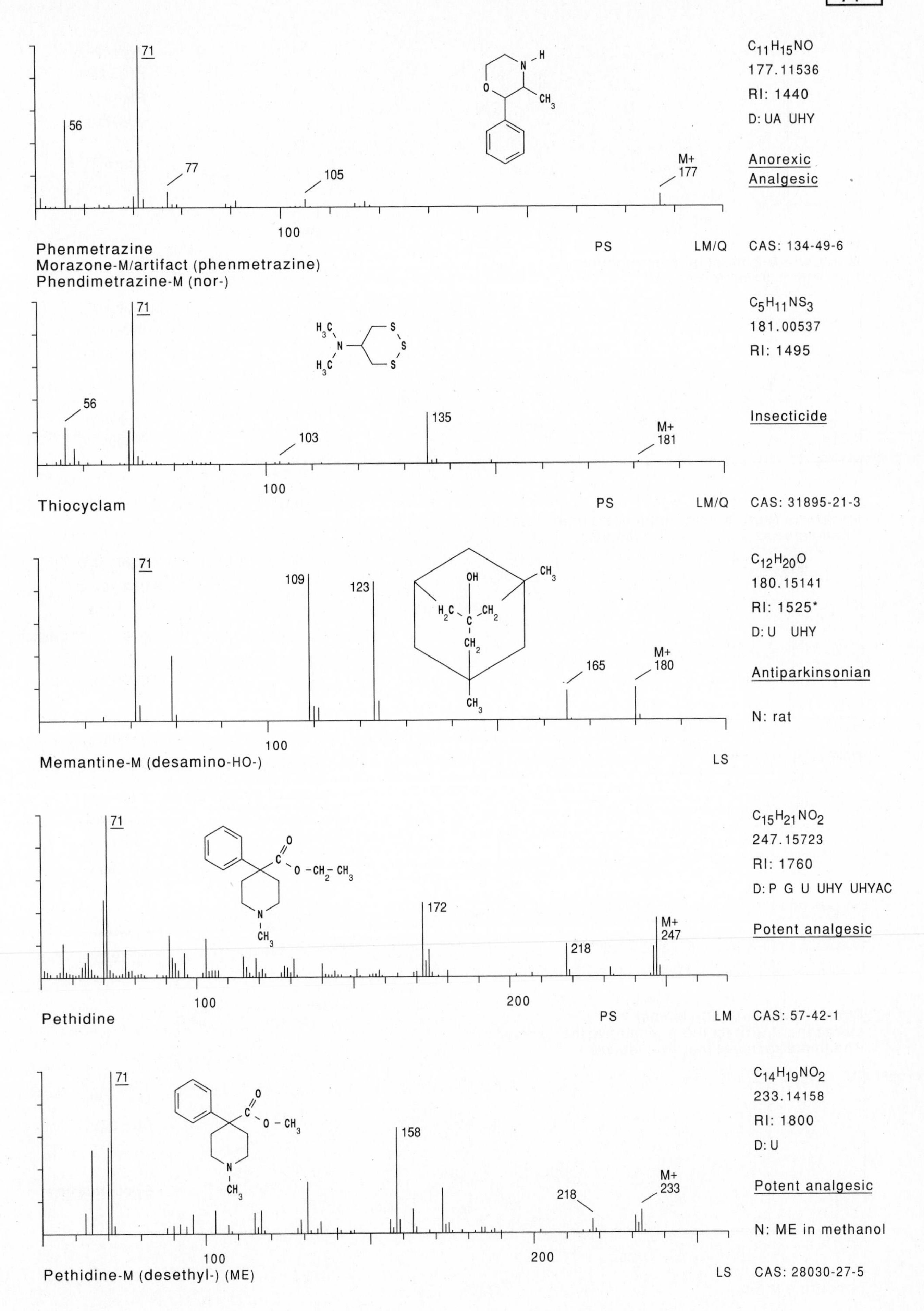

71
56
77
105
M+ 177
100
Phenmetrazine
Morazone-M/artifact (phenmetrazine)
Phendimetrazine-M (nor-)
PS
LM/Q
$C_{11}H_{15}NO$
177.11536
RI: 1440
D: UA UHY
Anorexic
Analgesic
CAS: 134-49-6
71
56
103
135
M+ 181
100
Thiocyclam
PS
LM/Q
$C_5H_{11}NS_3$
181.00537
RI: 1495
Insecticide
CAS: 31895-21-3
71
109
123
165
M+ 180
100
Memantine-M (desamino-HO-)
LS
$C_{12}H_{20}O$
180.15141
RI: 1525*
D: U UHY
Antiparkinsonian
N: rat
71
172
218
M+ 247
100
200
Pethidine
PS
LM
$C_{15}H_{21}NO_2$
247.15723
RI: 1760
D: P G U UHY UHYAC
Potent analgesic
CAS: 57-42-1
71
158
218
M+ 233
100
200
Pethidine-M (desethyl-) (ME)
LS
$C_{14}H_{19}NO_2$
233.14158
RI: 1800
D: U
Potent analgesic
N: ME in methanol
CAS: 28030-27-5

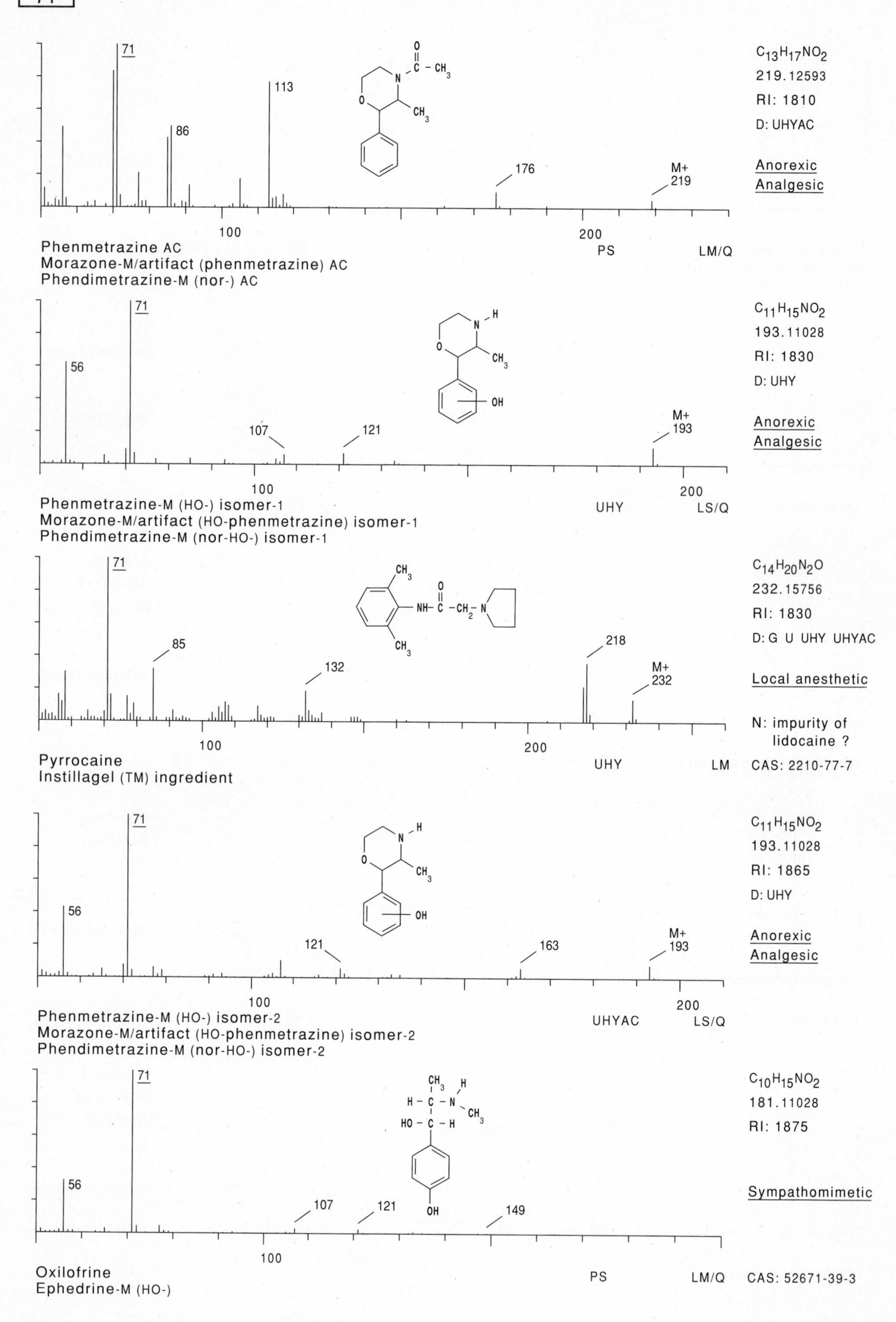
71
113
86
176
M+
219
100
200
PS
LM/Q
C13H17NO2
219.12593
RI: 1810
D: UHYAC
Anorexic
Analgesic
Phenmetrazine AC
Morazone-M/artifact (phenmetrazine) AC
Phendimetrazine-M (nor-) AC
71
56
107
121
M+
193
100
200
UHY
LS/Q
C11H15NO2
193.11028
RI: 1830
D: UHY
Anorexic
Analgesic
Phenmetrazine-M (HO-) isomer-1
Morazone-M/artifact (HO-phenmetrazine) isomer-1
Phendimetrazine-M (nor-HO-) isomer-1
71
85
132
218
M+
232
100
200
UHY
LM
C14H20N2O
232.15756
RI: 1830
D: G U UHY UHYAC
Local anesthetic
N: impurity of lidocaine ?
CAS: 2210-77-7
Pyrrocaine
Instillagel (TM) ingredient
71
56
121
163
M+
193
100
200
UHYAC
LS/Q
C11H15NO2
193.11028
RI: 1865
D: UHY
Anorexic
Analgesic
Phenmetrazine-M (HO-) isomer-2
Morazone-M/artifact (HO-phenmetrazine) isomer-2
Phendimetrazine-M (nor-HO-) isomer-2
71
56
107
121
149
100
PS
LM/Q
C10H15NO2
181.11028
RI: 1875
Sympathomimetic
CAS: 52671-39-3
Oxilofrine
Ephedrine-M (HO-)

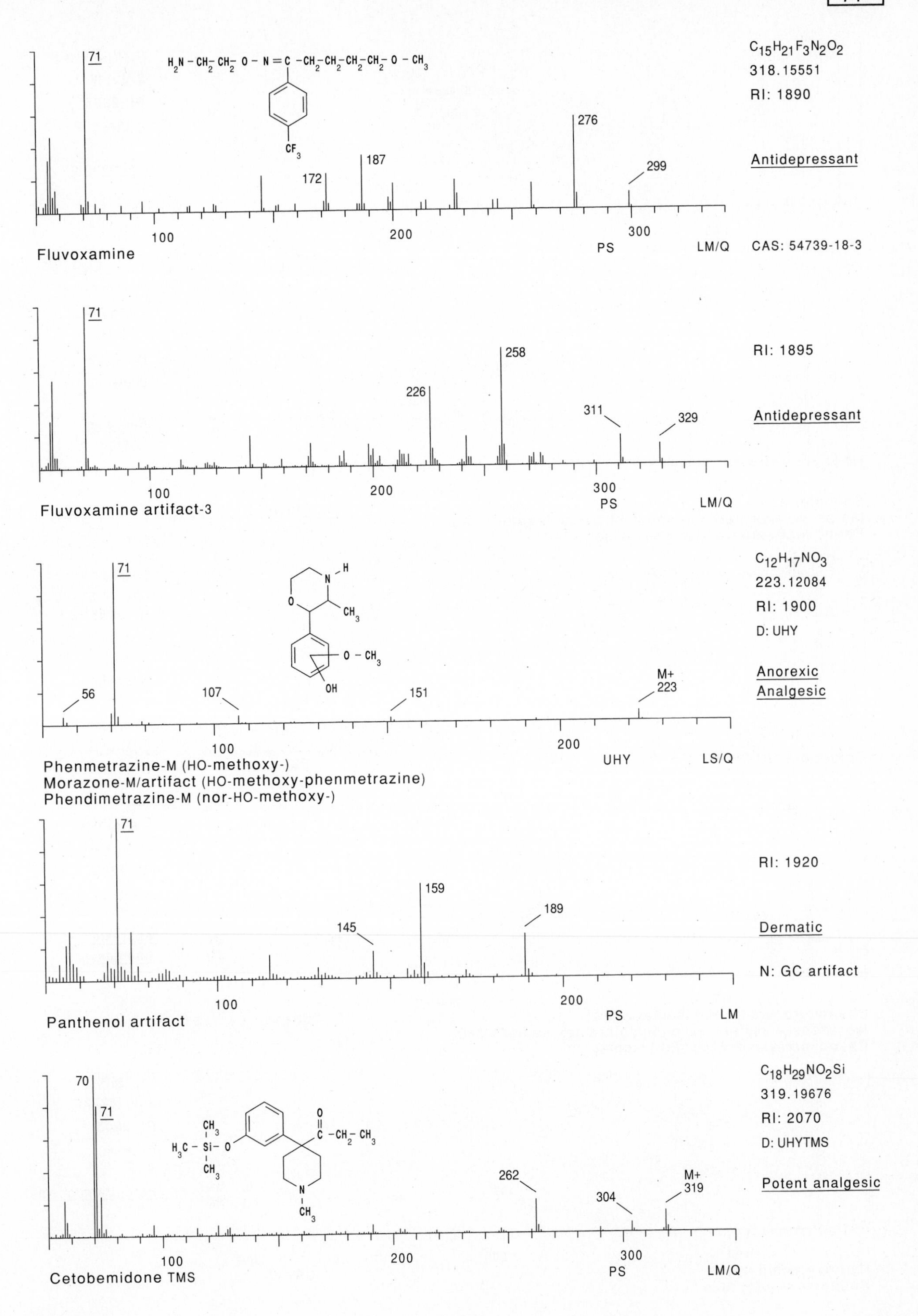
71
$H_2N-CH_2-CH_2-O-N=C-CH_2-CH_2-CH_2-CH_2-O-CH_3$
CF_3
172
187
276
299
100
200
300
PS
LM/Q
Fluvoxamine
$C_{15}H_{21}F_3N_2O_2$
318.15551
RI: 1890
Antidepressant
CAS: 54739-18-3
71
226
258
311
329
100
200
300
PS
LM/Q
Fluvoxamine artifact-3
RI: 1895
Antidepressant
71
N
H
O
CH_3
$O-CH_3$
OH
56
107
151
M+
223
100
200
UHY
LS/Q
Phenmetrazine-M (HO-methoxy-)
Morazone-M/artifact (HO-methoxy-phenmetrazine)
Phendimetrazine-M (nor-HO-methoxy-)
$C_{12}H_{17}NO_3$
223.12084
RI: 1900
D: UHY
Anorexic
Analgesic
71
145
159
189
100
200
PS
LM
Panthenol artifact
RI: 1920
Dermatic
N: GC artifact
70
71
CH_3
$H_3C-Si-O$
CH_3
O
$C-CH_2-CH_3$
N
CH_3
262
304
M+
319
100
200
300
PS
LM/Q
Cetobemidone TMS
$C_{18}H_{29}NO_2Si$
319.19676
RI: 2070
D: UHYTMS
Potent analgesic

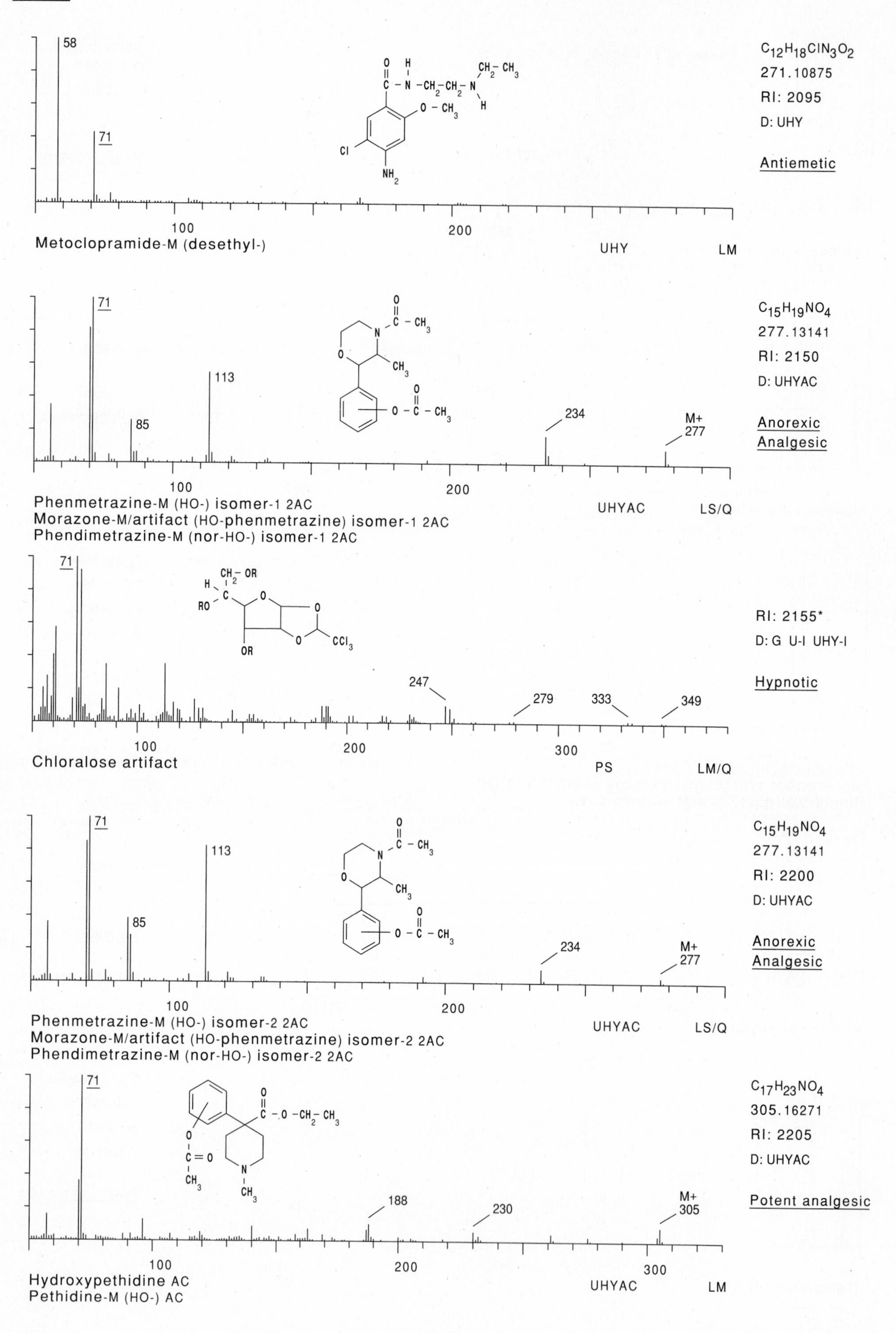

58
71
100
200
$C_{12}H_{18}ClN_3O_2$
271.10875
RI: 2095
D: UHY
Antiemetic
Metoclopramide-M (desethyl-)
UHY
LM
71
113
85
234
M+
277
$C_{15}H_{19}NO_4$
277.13141
RI: 2150
D: UHYAC
Anorexic
Analgesic
Phenmetrazine-M (HO-) isomer-1 2AC
Morazone-M/artifact (HO-phenmetrazine) isomer-1 2AC
Phendimetrazine-M (nor-HO-) isomer-1 2AC
UHYAC
LS/Q
71
247
279
333
349
300
RI: 2155*
D: G U-I UHY-I
Hypnotic
Chloralose artifact
PS
LM/Q
$C_{15}H_{19}NO_4$
277.13141
RI: 2200
D: UHYAC
Anorexic
Analgesic
Phenmetrazine-M (HO-) isomer-2 2AC
Morazone-M/artifact (HO-phenmetrazine) isomer-2 2AC
Phendimetrazine-M (nor-HO-) isomer-2 2AC
UHYAC
LS/Q
188
230
M+
305
$C_{17}H_{23}NO_4$
305.16271
RI: 2205
D: UHYAC
Potent analgesic
Hydroxypethidine AC
Pethidine-M (HO-) AC
UHYAC
LM

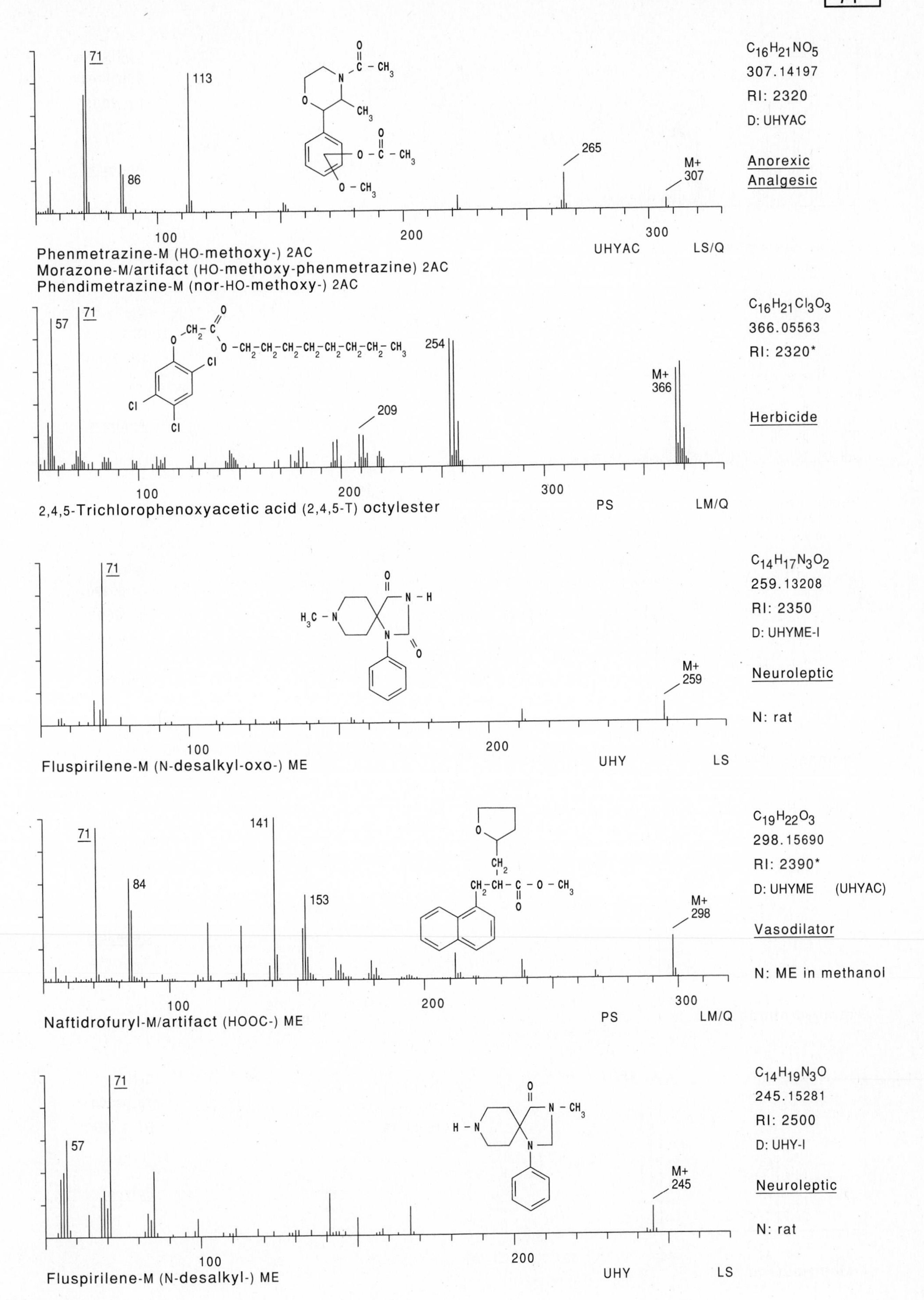
71
113
86
265
M+
307
100
200
300
UHYAC
LS/Q
$C_{16}H_{21}NO_5$
307.14197
RI: 2320
D: UHYAC
Anorexic
Analgesic
Phenmetrazine-M (HO-methoxy-) 2AC
Morazone-M/artifact (HO-methoxy-phenmetrazine) 2AC
Phendimetrazine-M (nor-HO-methoxy-) 2AC
57
71
254
M+
366
209
100
200
300
PS
LM/Q
$C_{16}H_{21}Cl_3O_3$
366.05563
RI: 2320*
Herbicide
2,4,5-Trichlorophenoxyacetic acid (2,4,5-T) octylester
71
M+
259
100
200
UHY
LS
$C_{14}H_{17}N_3O_2$
259.13208
RI: 2350
D: UHYME-I
Neuroleptic
N: rat
Fluspirilene-M (N-desalkyl-oxo-) ME
71
141
84
153
M+
298
100
200
300
PS
LM/Q
$C_{19}H_{22}O_3$
298.15690
RI: 2390*
D: UHYME (UHYAC)
Vasodilator
N: ME in methanol
Naftidrofuryl-M/artifact (HOOC-) ME
71
57
M+
245
100
200
UHY
LS
$C_{14}H_{19}N_3O$
245.15281
RI: 2500
D: UHY-I
Neuroleptic
N: rat
Fluspirilene-M (N-desalkyl-) ME

71

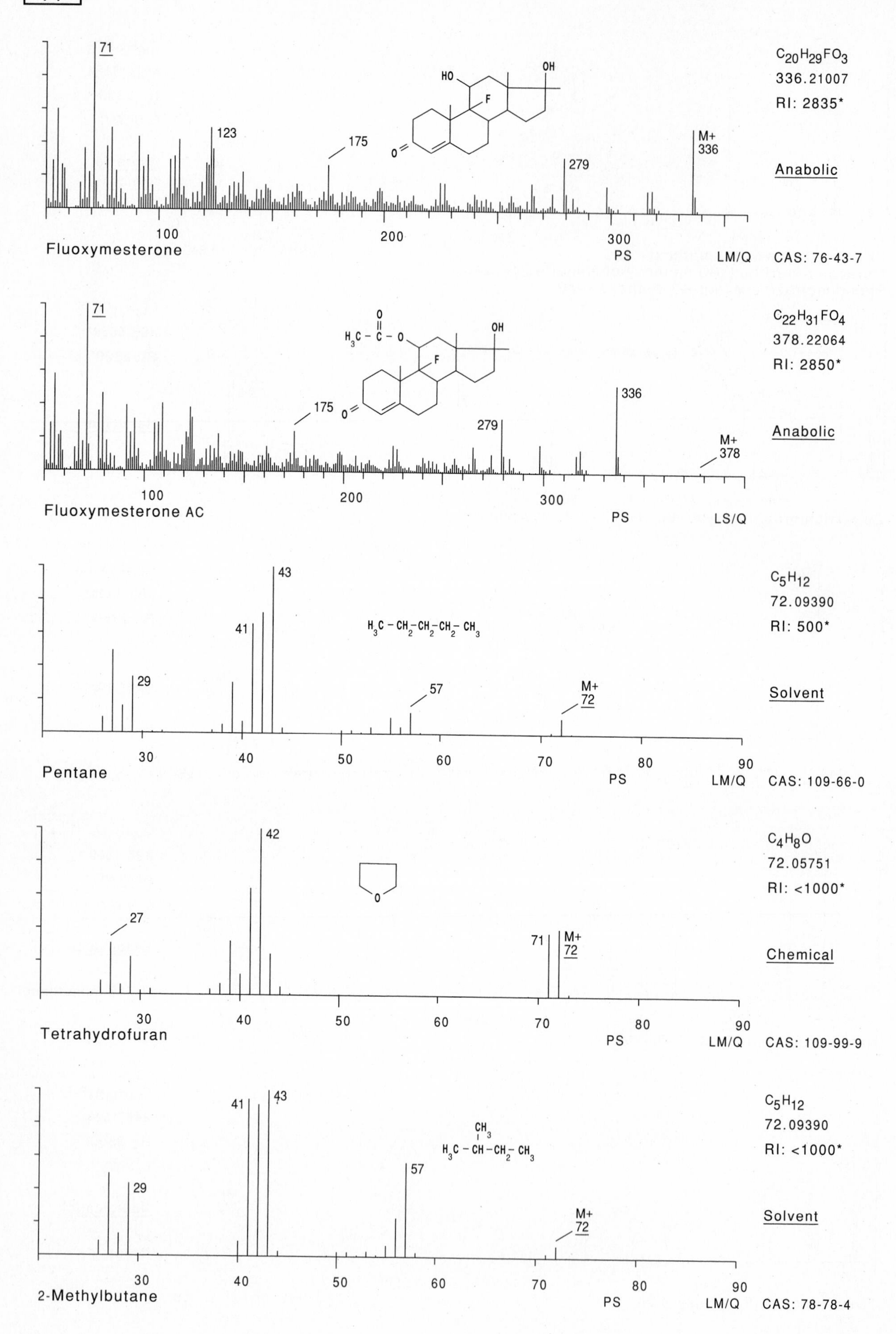

71
123
175
279
M+
336
HO
OH
F
O
100
200
300
Fluoxymesterone
PS
LM/Q
$C_{20}H_{29}FO_3$
336.21007
RI: 2835*
Anabolic
CAS: 76-43-7
71
175
279
336
M+
378
$H_3C-C(=O)-O$
OH
F
O
100
200
300
Fluoxymesterone AC
PS
LS/Q
$C_{22}H_{31}FO_4$
378.22064
RI: 2850*
Anabolic
43
41
29
57
M+
72
$H_3C-CH_2-CH_2-CH_2-CH_3$
30
40
50
60
70
80
90
Pentane
PS
LM/Q
C_5H_{12}
72.09390
RI: 500*
Solvent
CAS: 109-66-0
42
27
71
M+
72
O
30
40
50
60
70
80
90
Tetrahydrofuran
PS
LM/Q
C_4H_8O
72.05751
RI: <1000*
Chemical
CAS: 109-99-9
41
43
29
57
M+
72
CH_3
$H_3C-CH-CH_2-CH_3$
30
40
50
60
70
80
90
2-Methylbutane
PS
LM/Q
C_5H_{12}
72.09390
RI: <1000*
Solvent
CAS: 78-78-4

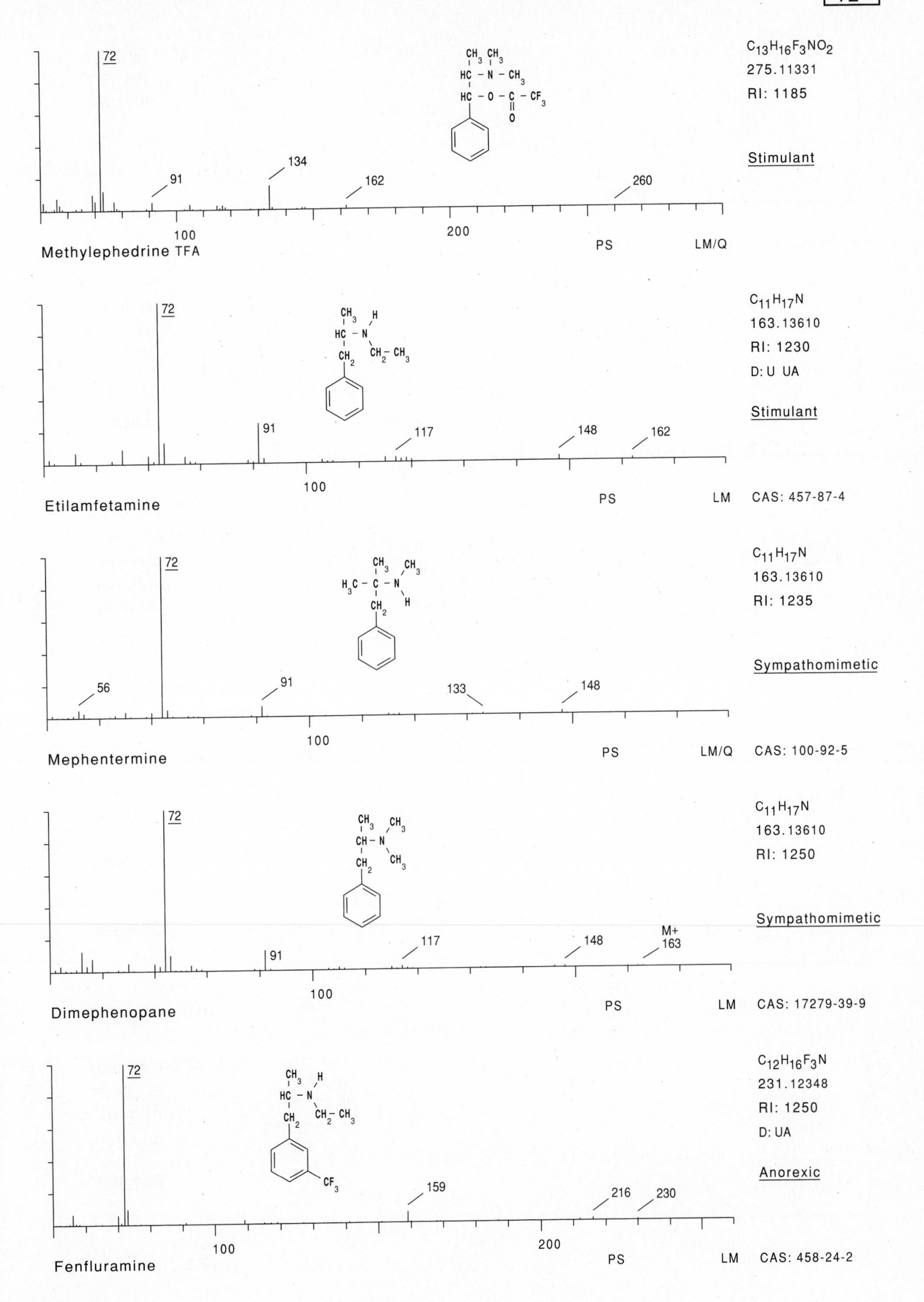

$C_{13}H_{16}F_3NO_2$
275.11331
RI: 1185

Stimulant

$C_{11}H_{17}N$
163.13610
RI: 1230
D: U UA

Stimulant

CAS: 457-87-4

$C_{11}H_{17}N$
163.13610
RI: 1235

Sympathomimetic

CAS: 100-92-5

$C_{11}H_{17}N$
163.13610
RI: 1250

Sympathomimetic

CAS: 17279-39-9

$C_{12}H_{16}F_3N$
231.12348
RI: 1250
D: UA

Anorexic

CAS: 458-24-2

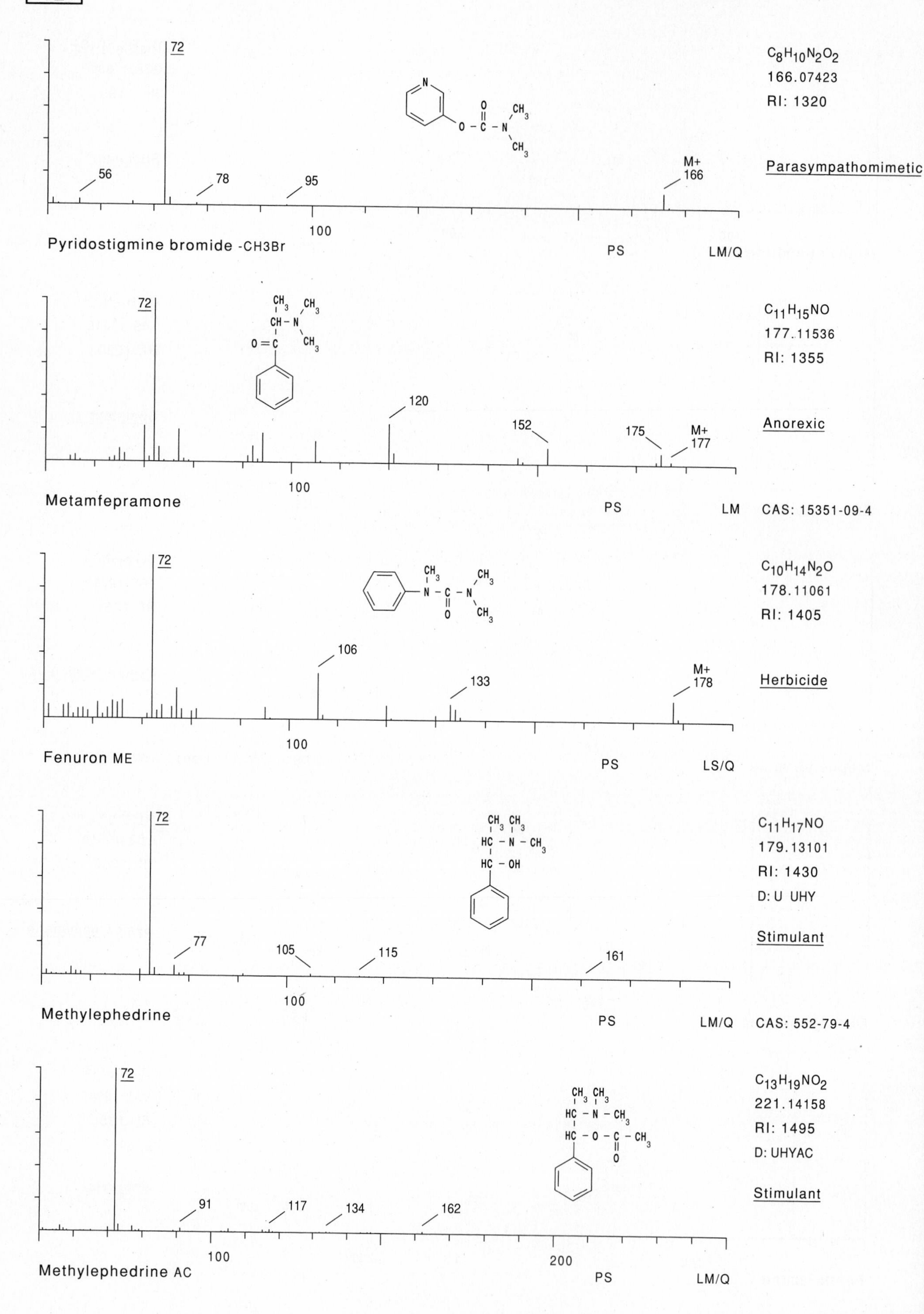
72
56
78
95
M+
166
100
N
O
CH3
O - C - N
CH3
Pyridostigmine bromide -CH3Br
PS
LM/Q
$C_8H_{10}N_2O_2$
166.07423
RI: 1320
Parasympathomimetic
72
120
152
175
M+
177
100
Metamfepramone
PS
LM
$C_{11}H_{15}NO$
177.11536
RI: 1355
Anorexic
CAS: 15351-09-4
72
106
133
M+
178
100
Fenuron ME
PS
LS/Q
$C_{10}H_{14}N_2O$
178.11061
RI: 1405
Herbicide
72
77
105
115
161
100
HC - N - CH3
HC - OH
Methylephedrine
PS
LM/Q
$C_{11}H_{17}NO$
179.13101
RI: 1430
D: U UHY
Stimulant
CAS: 552-79-4
72
91
117
134
162
100
200
HC - O - C - CH3
Methylephedrine AC
PS
LM/Q
$C_{13}H_{19}NO_2$
221.14158
RI: 1495
D: UHYAC
Stimulant

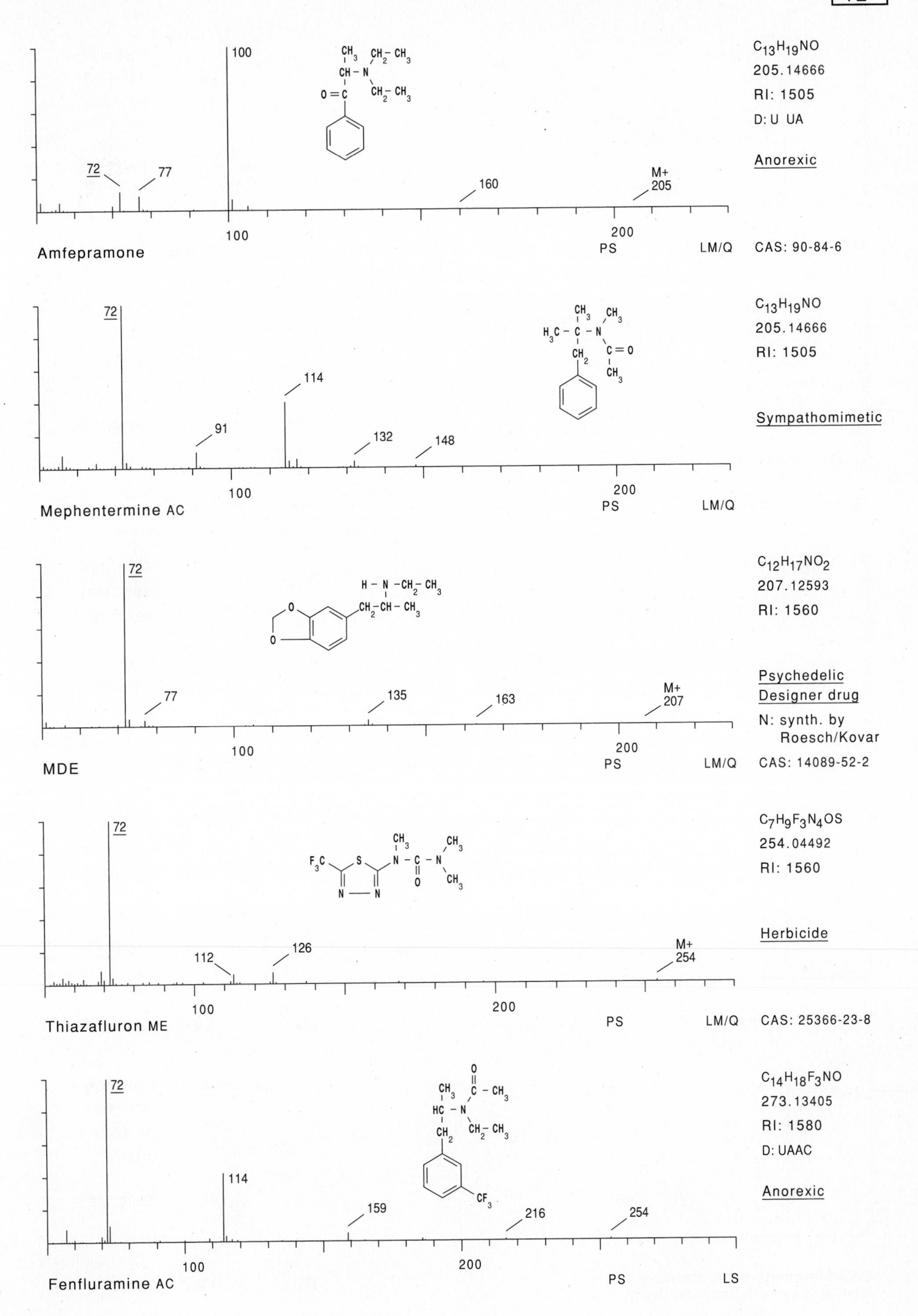

100
72
77
160
M+
205
100
200
Amfepramone
PS
LM/Q
$C_{13}H_{19}NO$
205.14666
RI: 1505
D: U UA
Anorexic
CAS: 90-84-6
72
114
91
132
148
100
200
Mephentermine AC
PS
LM/Q
$C_{13}H_{19}NO$
205.14666
RI: 1505
Sympathomimetic
72
77
135
163
M+
207
100
200
MDE
PS
LM/Q
$C_{12}H_{17}NO_2$
207.12593
RI: 1560
Psychedelic
Designer drug
N: synth. by Roesch/Kovar
CAS: 14089-52-2
72
112
126
M+
254
100
200
Thiazafluron ME
PS
LM/Q
$C_7H_9F_3N_4OS$
254.04492
RI: 1560
Herbicide
CAS: 25366-23-8
72
114
159
216
254
100
200
Fenfluramine AC
PS
LS
$C_{14}H_{18}F_3NO$
273.13405
RI: 1580
D: UAAC
Anorexic

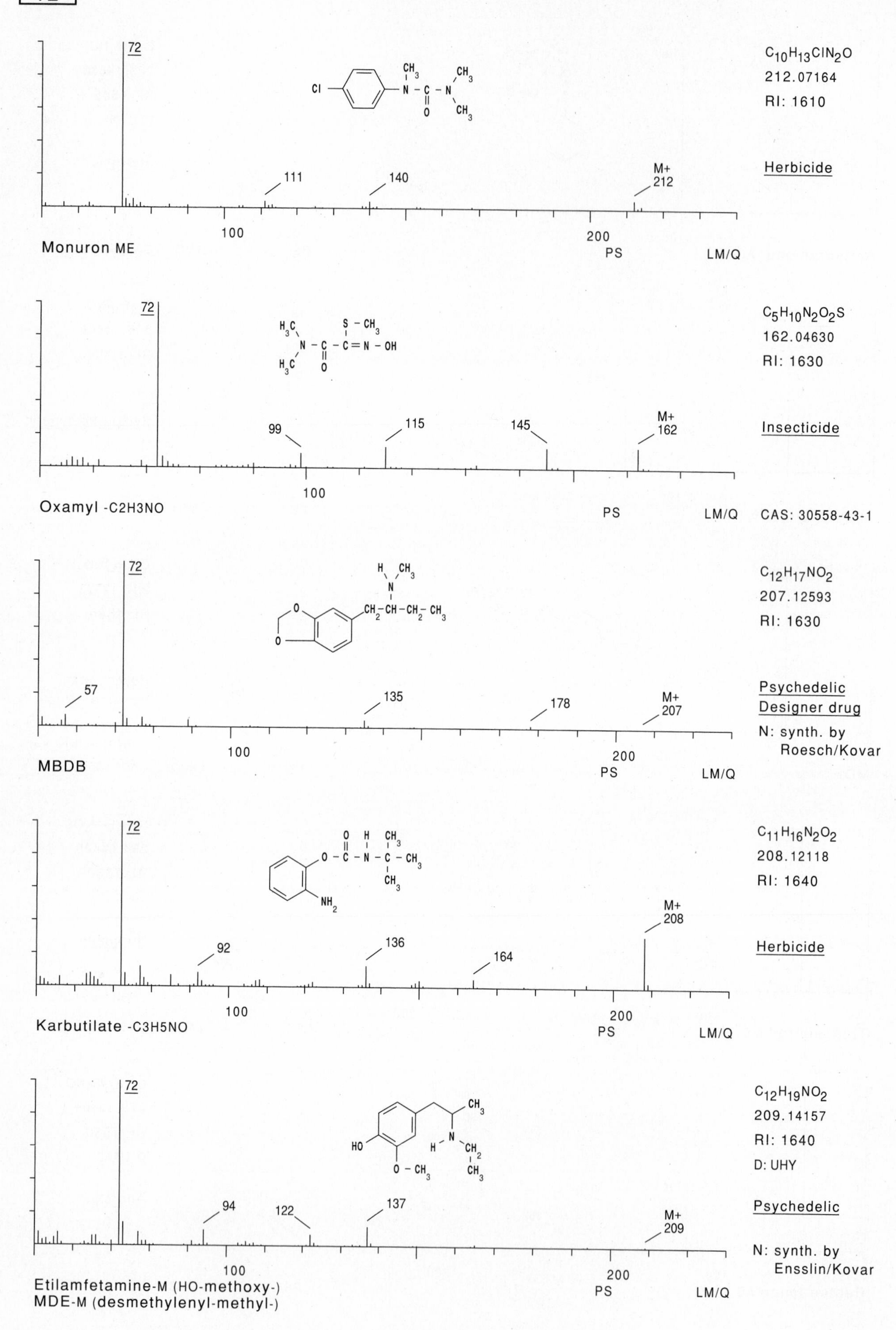

72
Cl
CH3
N - C - N
O
CH3
CH3
111
140
M+
212
100
200
Monuron ME
PS
LM/Q
C10H13ClN2O
212.07164
RI: 1610
Herbicide
72
H3C
N - C - C = N - OH
S - CH3
O
99
115
145
M+
162
Oxamyl -C2H3NO
PS
LM/Q
C5H10N2O2S
162.04630
RI: 1630
Insecticide
CAS: 30558-43-1
72
H
CH3
N
CH2 - CH - CH2 - CH3
O
57
135
178
M+
207
MBDB
PS
LM/Q
C12H17NO2
207.12593
RI: 1630
Psychedelic
Designer drug
N: synth. by
Roesch/Kovar
72
O
H
CH3
O - C - N - C - CH3
NH2
92
136
164
M+
208
Karbutilate -C3H5NO
PS
LM/Q
C11H16N2O2
208.12118
RI: 1640
Herbicide
72
HO
O - CH3
CH3
N
H
CH2
94
122
137
M+
209
Etilamfetamine-M (HO-methoxy-)
MDE-M (desmethylenyl-methyl-)
PS
LM/Q
C12H19NO2
209.14157
RI: 1640
D: UHY
Psychedelic
N: synth. by
Ensslin/Kovar

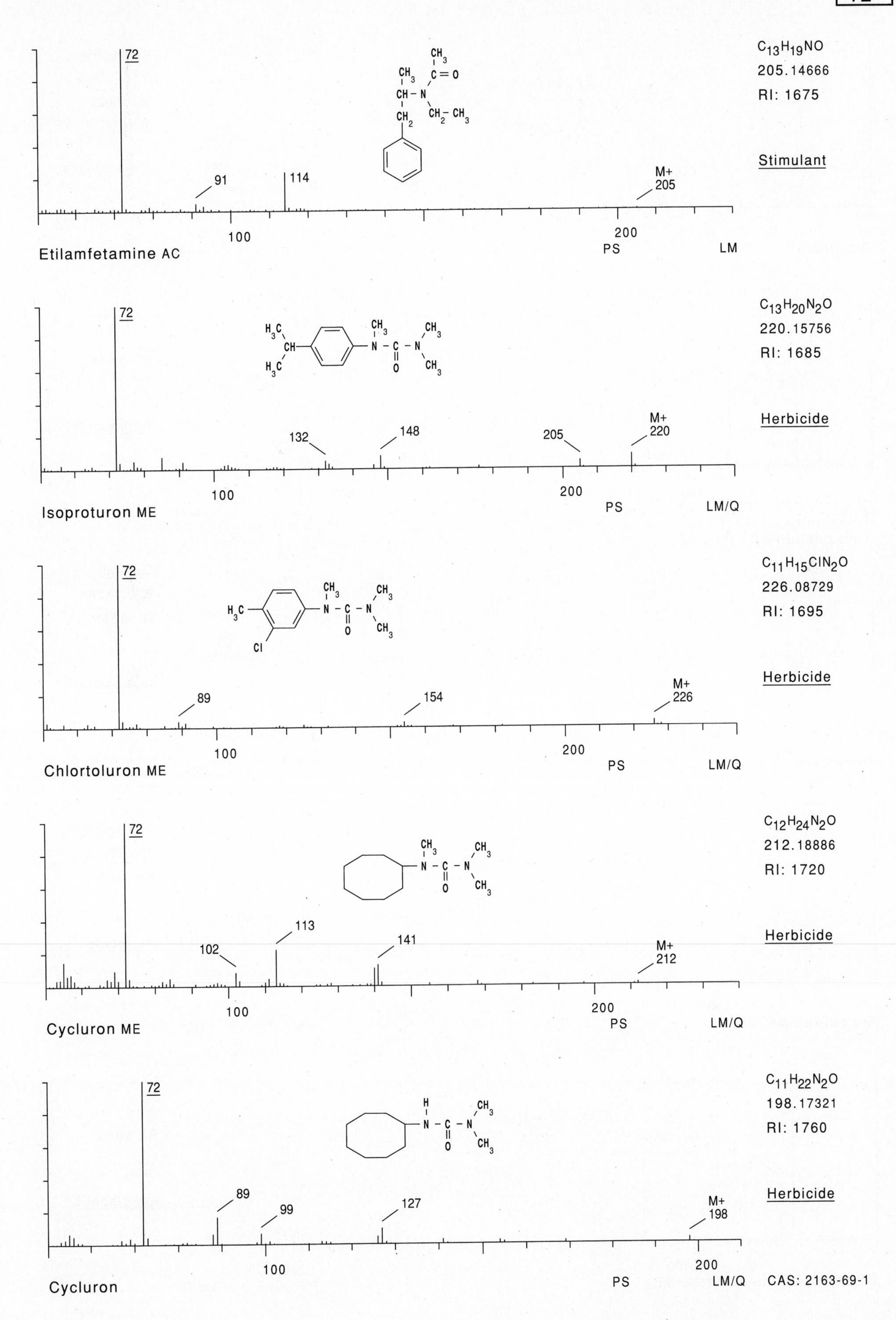

$C_{13}H_{19}NO$
205.14666
RI: 1675

Stimulant

$C_{13}H_{20}N_2O$
220.15756
RI: 1685

Herbicide

$C_{11}H_{15}ClN_2O$
226.08729
RI: 1695

Herbicide

$C_{12}H_{24}N_2O$
212.18886
RI: 1720

Herbicide

$C_{11}H_{22}N_2O$
198.17321
RI: 1760

Herbicide

CAS: 2163-69-1

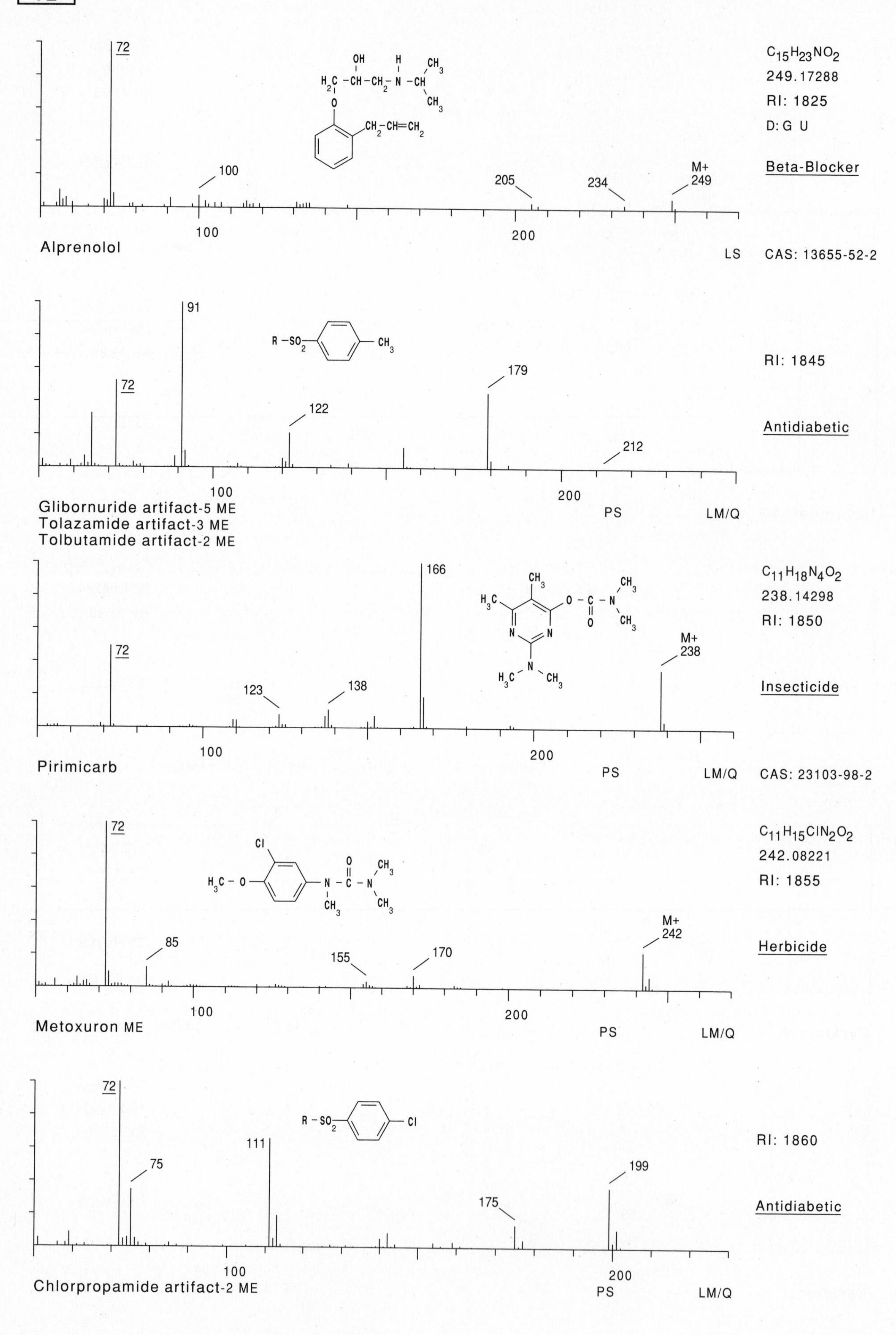

Alprenolol
$C_{15}H_{23}NO_2$
249.17288
RI: 1825
D: G U
Beta-Blocker
LS
CAS: 13655-52-2
Glibornuride artifact-5 ME
Tolazamide artifact-3 ME
Tolbutamide artifact-2 ME
RI: 1845
Antidiabetic
PS
LM/Q
Pirimicarb
$C_{11}H_{18}N_4O_2$
238.14298
RI: 1850
Insecticide
PS
LM/Q
CAS: 23103-98-2
Metoxuron ME
$C_{11}H_{15}ClN_2O_2$
242.08221
RI: 1855
Herbicide
PS
LM/Q
Chlorpropamide artifact-2 ME
RI: 1860
Antidiabetic
PS
LM/Q

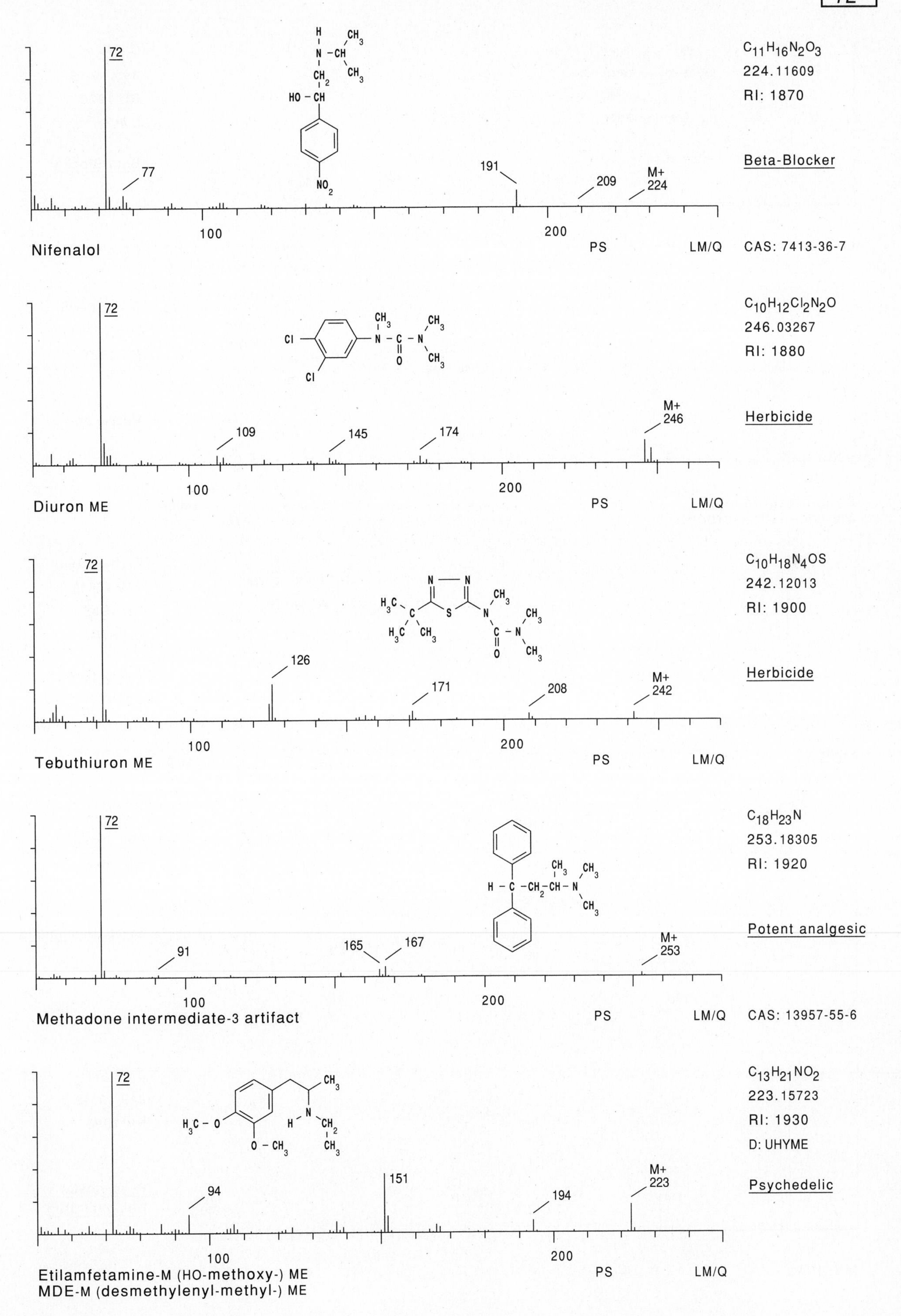

72
77
191
209
M+
224
100
200
PS
LM/Q
Nifenalol
$C_{11}H_{16}N_2O_3$
224.11609
RI: 1870
Beta-Blocker
CAS: 7413-36-7
72
109
145
174
M+
246
100
200
PS
LM/Q
Diuron ME
$C_{10}H_{12}Cl_2N_2O$
246.03267
RI: 1880
Herbicide
72
126
171
208
M+
242
100
200
PS
LM/Q
Tebuthiuron ME
$C_{10}H_{18}N_4OS$
242.12013
RI: 1900
Herbicide
72
91
165
167
M+
253
100
200
PS
LM/Q
Methadone intermediate-3 artifact
$C_{18}H_{23}N$
253.18305
RI: 1920
Potent analgesic
CAS: 13957-55-6
72
94
151
194
M+
223
100
200
PS
LM/Q
Etilamfetamine-M (HO-methoxy-) ME
MDE-M (desmethylenyl-methyl-) ME
$C_{13}H_{21}NO_2$
223.15723
RI: 1930
D: UHYME
Psychedelic

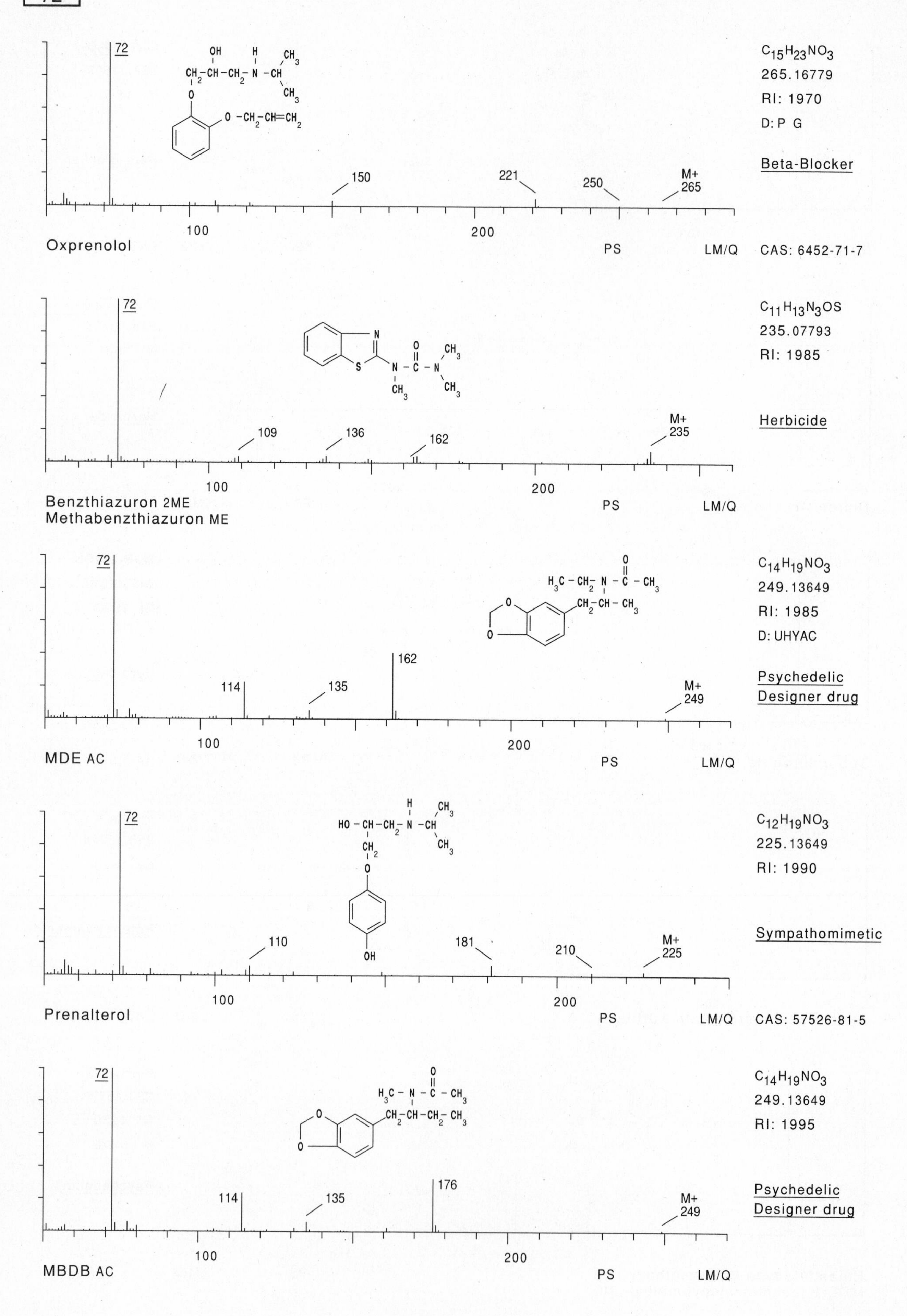
72
150
221
250
M+
265
100
200
Oxprenolol
PS
LM/Q
$C_{15}H_{23}NO_3$
265.16779
RI: 1970
D: P G
Beta-Blocker
CAS: 6452-71-7
72
109
136
162
M+
235
100
200
Benzthiazuron 2ME
Methabenzthiazuron ME
PS
LM/Q
$C_{11}H_{13}N_3OS$
235.07793
RI: 1985
Herbicide
72
114
135
162
M+
249
100
200
MDE AC
PS
LM/Q
$C_{14}H_{19}NO_3$
249.13649
RI: 1985
D: UHYAC
Psychedelic
Designer drug
72
110
181
210
M+
225
100
200
Prenalterol
PS
LM/Q
$C_{12}H_{19}NO_3$
225.13649
RI: 1990
Sympathomimetic
CAS: 57526-81-5
72
114
135
176
M+
249
100
200
MBDB AC
PS
LM/Q
$C_{14}H_{19}NO_3$
249.13649
RI: 1995
Psychedelic
Designer drug

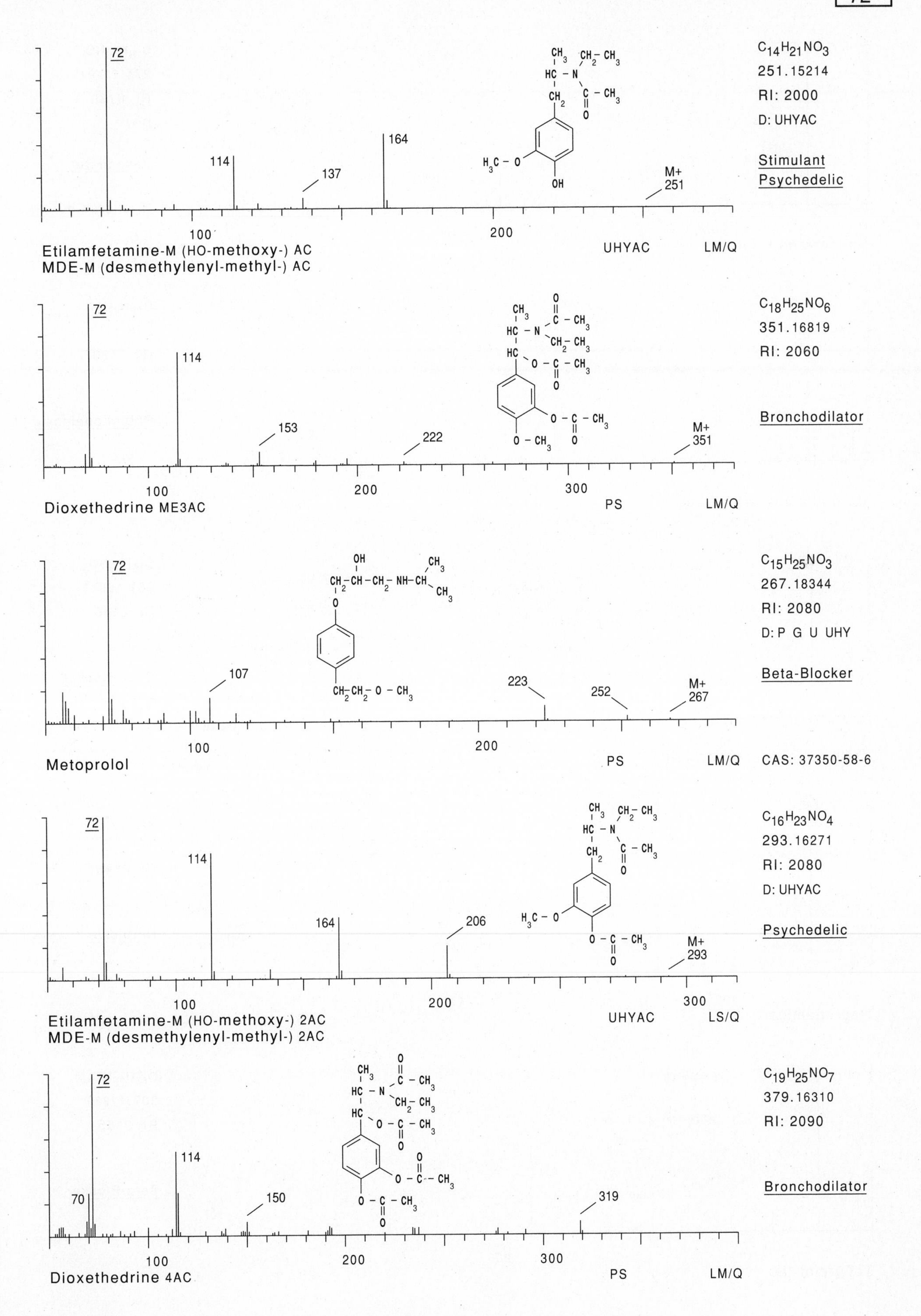

72
114
137
164
M+
251
100
200
Etilamfetamine-M (HO-methoxy-) AC
MDE-M (desmethylenyl-methyl-) AC
UHYAC
LM/Q
$C_{14}H_{21}NO_3$
251.15214
RI: 2000
D: UHYAC
Stimulant
Psychedelic
72
114
153
222
M+
351
100
200
300
Dioxethedrine ME3AC
PS
LM/Q
$C_{18}H_{25}NO_6$
351.16819
RI: 2060
Bronchodilator
72
107
223
252
M+
267
100
200
Metoprolol
PS
LM/Q
CAS: 37350-58-6
$C_{15}H_{25}NO_3$
267.18344
RI: 2080
D: P G U UHY
Beta-Blocker
72
114
164
206
M+
293
100
200
300
Etilamfetamine-M (HO-methoxy-) 2AC
MDE-M (desmethylenyl-methyl-) 2AC
UHYAC
LS/Q
$C_{16}H_{23}NO_4$
293.16271
RI: 2080
D: UHYAC
Psychedelic
72
70
114
150
319
100
200
300
Dioxethedrine 4AC
PS
LM/Q
$C_{19}H_{25}NO_7$
379.16310
RI: 2090
Bronchodilator

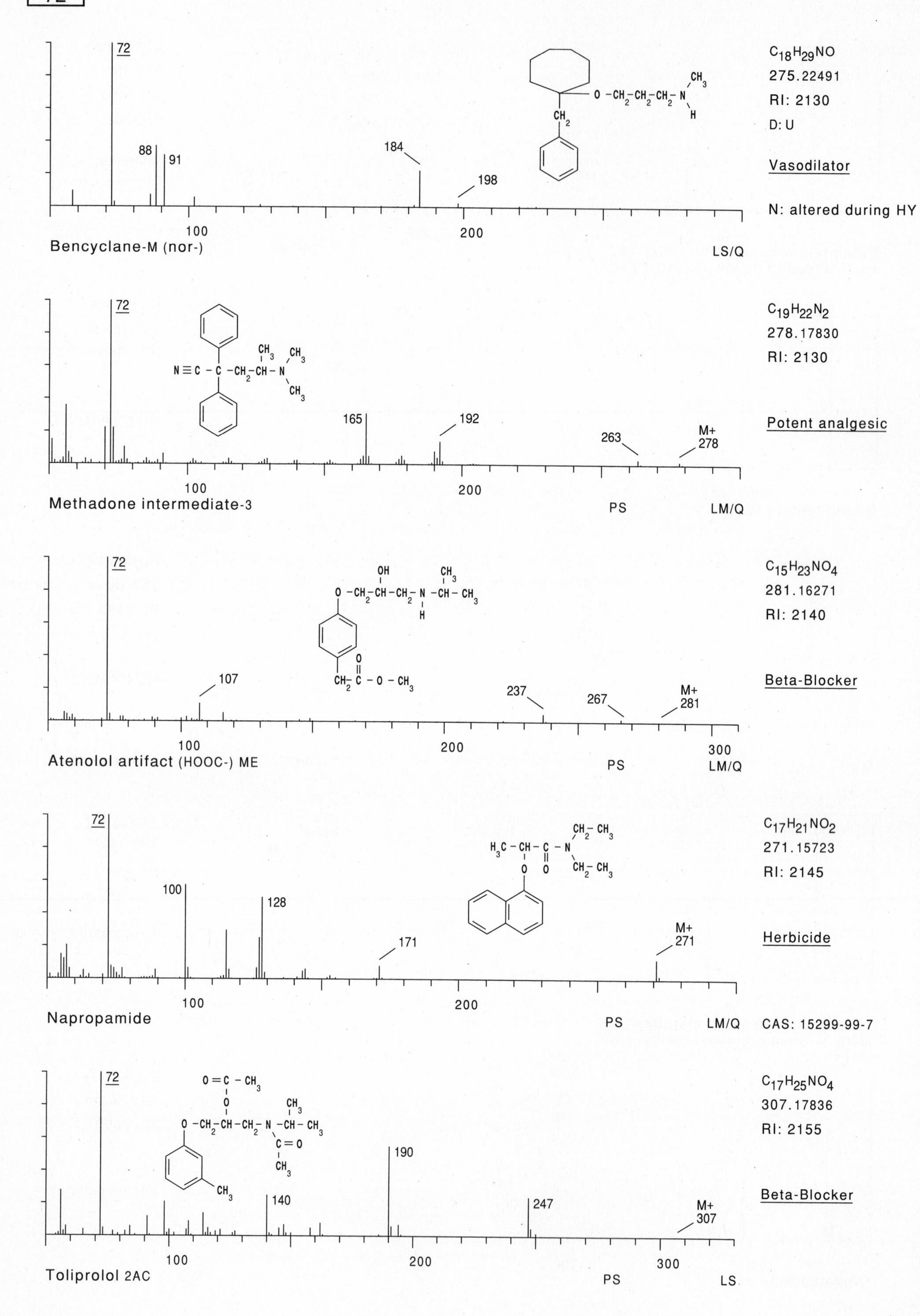
72
88
91
184
198
100
200
Bencyclane-M (nor-)
LS/Q
$C_{18}H_{29}NO$
275.22491
RI: 2130
D: U
Vasodilator
N: altered during HY
72
165
192
263
M+
278
100
200
Methadone intermediate-3
PS
LM/Q
$C_{19}H_{22}N_2$
278.17830
RI: 2130
Potent analgesic
72
107
237
267
M+
281
100
200
300
Atenolol artifact (HOOC-) ME
PS
LM/Q
$C_{15}H_{23}NO_4$
281.16271
RI: 2140
Beta-Blocker
72
100
128
171
M+
271
100
200
Napropamide
PS
LM/Q
CAS: 15299-99-7
$C_{17}H_{21}NO_2$
271.15723
RI: 2145
Herbicide
72
190
140
247
M+
307
100
200
300
Toliprolol 2AC
PS
LS
$C_{17}H_{25}NO_4$
307.17836
RI: 2155
Beta-Blocker

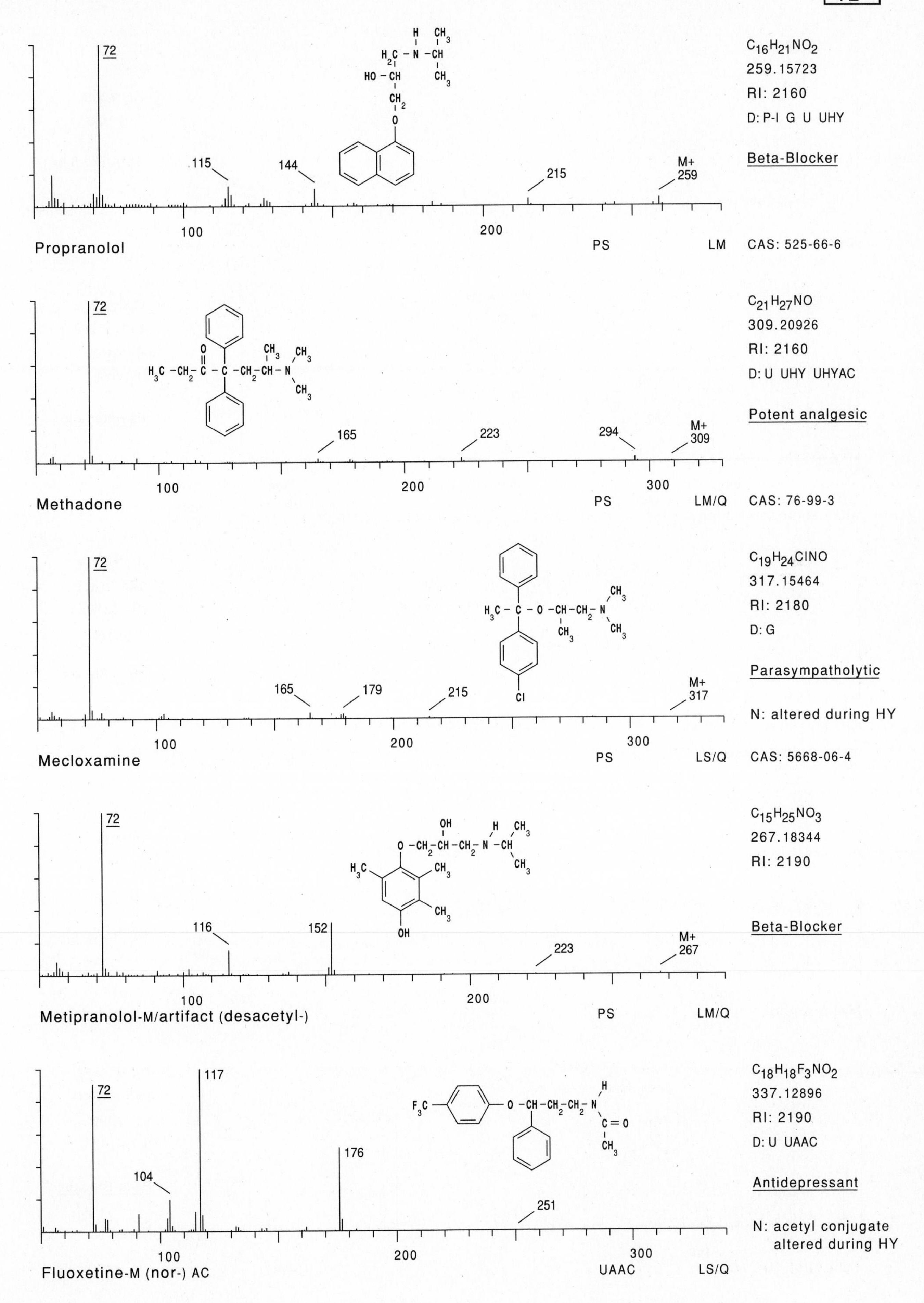

Propranolol
PS LM
C16H21NO2
259.15723
RI: 2160
D: P-I G U UHY
Beta-Blocker
CAS: 525-66-6
72
115
144
215
M+ 259
100
200
Methadone
PS LM/Q
C21H27NO
309.20926
RI: 2160
D: U UHY UHYAC
Potent analgesic
CAS: 76-99-3
72
165
223
294
M+ 309
100
200
300
Mecloxamine
PS LS/Q
C19H24ClNO
317.15464
RI: 2180
D: G
Parasympatholytic
N: altered during HY
CAS: 5668-06-4
72
165
179
215
M+ 317
100
200
300
Metipranolol-M/artifact (desacetyl-)
PS LM/Q
C15H25NO3
267.18344
RI: 2190
Beta-Blocker
72
116
152
223
M+ 267
100
200
Fluoxetine-M (nor-) AC
UAAC LS/Q
C18H18F3NO2
337.12896
RI: 2190
D: U UAAC
Antidepressant
N: acetyl conjugate altered during HY
72
104
117
176
251
100
200
300

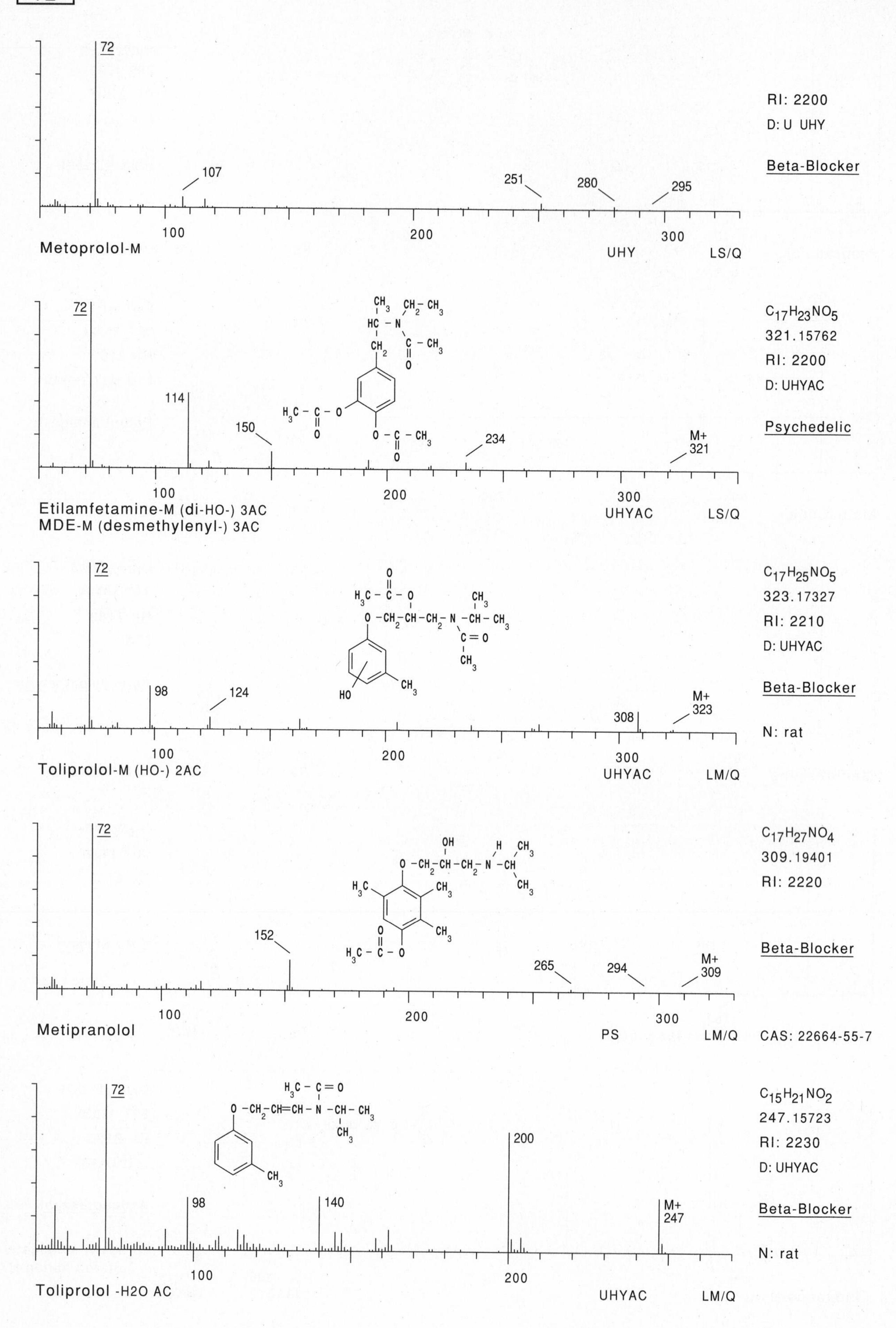
72
107
251
280
295
100
200
300
Metoprolol-M
UHY
LS/Q
RI: 2200
D: U UHY
Beta-Blocker
72
114
150
234
M+
321
Etilamfetamine-M (di-HO-) 3AC
MDE-M (desmethylenyl-) 3AC
UHYAC
C17H23NO5
321.15762
RI: 2200
D: UHYAC
Psychedelic
72
98
124
308
M+
323
Toliprolol-M (HO-) 2AC
UHYAC
LM/Q
C17H25NO5
323.17327
RI: 2210
D: UHYAC
Beta-Blocker
N: rat
72
152
265
294
M+
309
Metipranolol
PS
LM/Q
C17H27NO4
309.19401
RI: 2220
Beta-Blocker
CAS: 22664-55-7
72
98
140
200
M+
247
Toliprolol -H2O AC
UHYAC
LM/Q
C15H21NO2
247.15723
RI: 2230
D: UHYAC
Beta-Blocker
N: rat

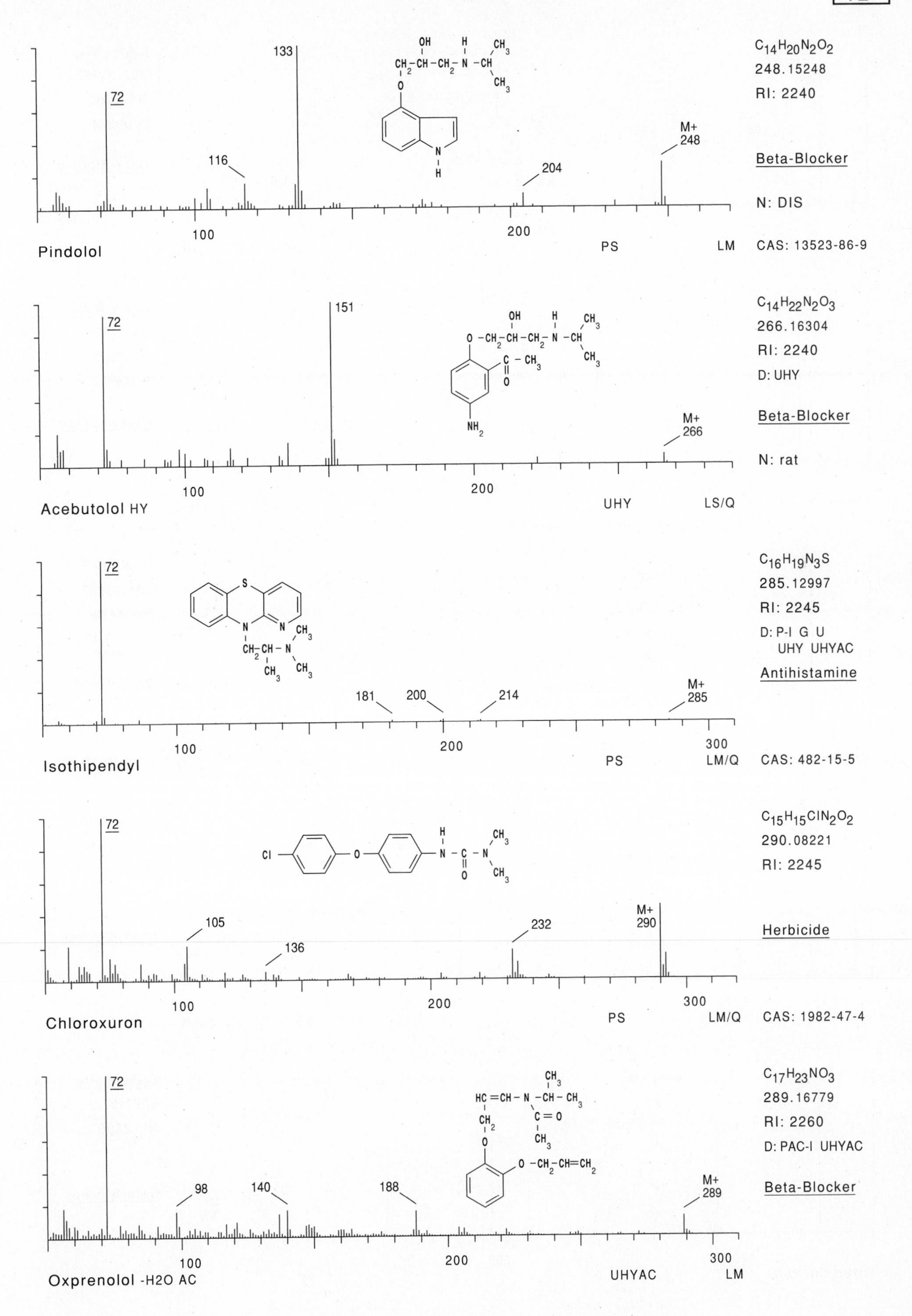

Pindolol

$C_{14}H_{20}N_2O_2$
248.15248
RI: 2240

Beta-Blocker

N: DIS

CAS: 13523-86-9

Acebutolol HY

$C_{14}H_{22}N_2O_3$
266.16304
RI: 2240
D: UHY

Beta-Blocker

N: rat

Isothipendyl

$C_{16}H_{19}N_3S$
285.12997
RI: 2245
D: P-I G U
UHY UHYAC

Antihistamine

CAS: 482-15-5

Chloroxuron

$C_{15}H_{15}ClN_2O_2$
290.08221
RI: 2245

Herbicide

CAS: 1982-47-4

Oxprenolol -H2O AC

$C_{17}H_{23}NO_3$
289.16779
RI: 2260
D: PAC-I UHYAC

Beta-Blocker

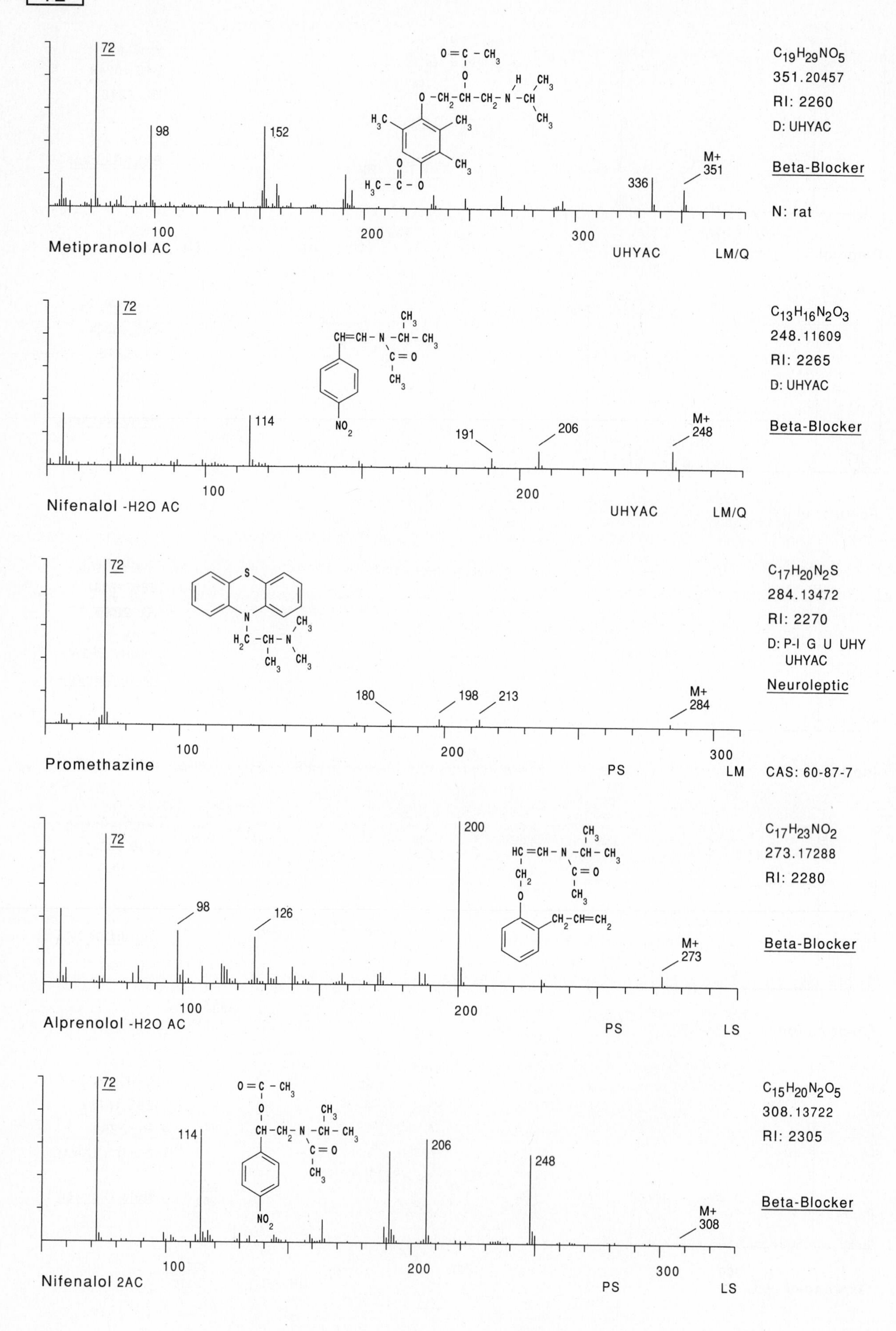

72
98
152
336
M+
351
100
200
300
Metipranolol AC
UHYAC
LM/Q
C19H29NO5
351.20457
RI: 2260
D: UHYAC
Beta-Blocker
N: rat
72
114
191
206
M+
248
100
200
Nifenalol -H2O AC
UHYAC
LM/Q
C13H16N2O3
248.11609
RI: 2265
D: UHYAC
Beta-Blocker
72
180
198
213
M+
284
100
200
300
Promethazine
PS
LM
C17H20N2S
284.13472
RI: 2270
D: P-I G U UHY
UHYAC
Neuroleptic
CAS: 60-87-7
72
98
126
200
M+
273
100
200
Alprenolol -H2O AC
PS
LS
C17H23NO2
273.17288
RI: 2280
Beta-Blocker
72
114
206
248
M+
308
100
200
300
Nifenalol 2AC
PS
LS
C15H20N2O5
308.13722
RI: 2305
Beta-Blocker

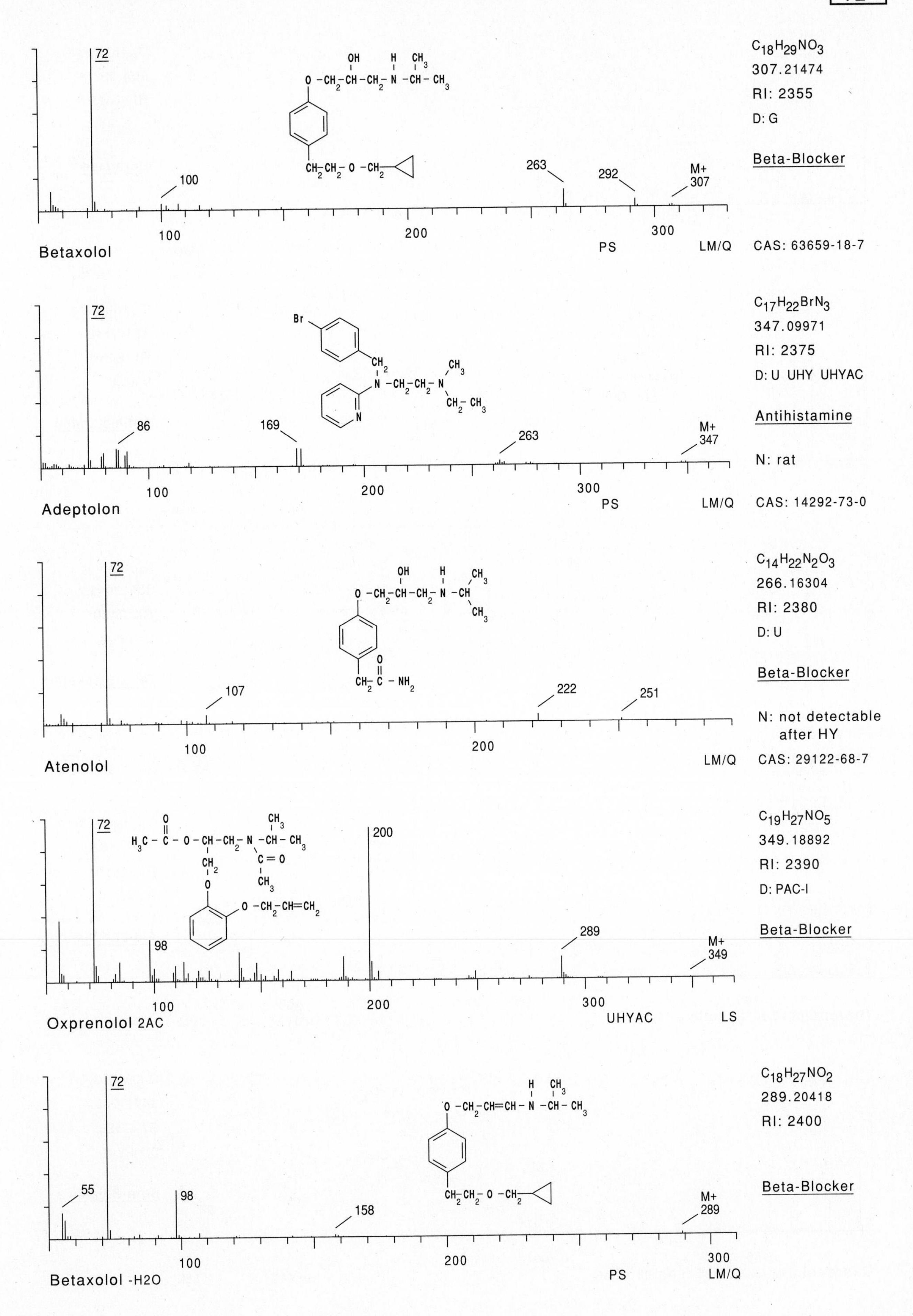

Betaxolol
72
100
263
292
M+
307
100
200
300
PS
LM/Q
$C_{18}H_{29}NO_3$
307.21474
RI: 2355
D: G
Beta-Blocker
CAS: 63659-18-7
Adeptolon
72
86
169
263
M+
347
100
200
300
PS
LM/Q
$C_{17}H_{22}BrN_3$
347.09971
RI: 2375
D: U UHY UHYAC
Antihistamine
N: rat
CAS: 14292-73-0
Atenolol
72
107
222
251
100
200
LM/Q
$C_{14}H_{22}N_2O_3$
266.16304
RI: 2380
D: U
Beta-Blocker
N: not detectable after HY
CAS: 29122-68-7
Oxprenolol 2AC
72
98
200
289
M+
349
100
200
300
UHYAC
LS
$C_{19}H_{27}NO_5$
349.18892
RI: 2390
D: PAC-I
Beta-Blocker
Betaxolol -H2O
72
55
98
158
M+
289
100
200
300
PS
LM/Q
$C_{18}H_{27}NO_2$
289.20418
RI: 2400
Beta-Blocker

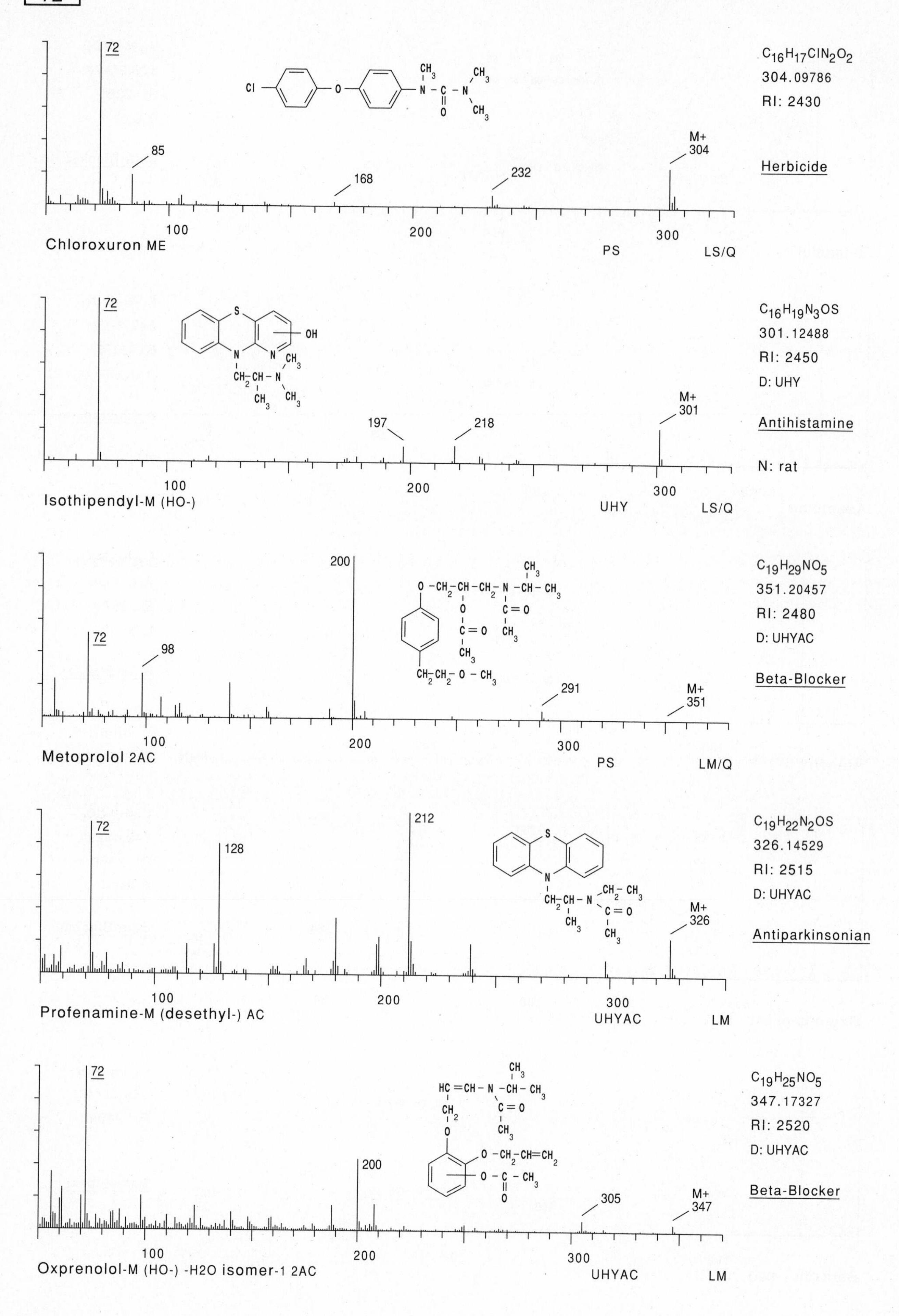

72
85
168
232
M+
304
100
200
300
Chloroxuron ME
PS
LS/Q
$C_{16}H_{17}ClN_2O_2$
304.09786
RI: 2430
Herbicide
72
197
218
M+
301
Isothipendyl-M (HO-)
UHY
LS/Q
$C_{16}H_{19}N_3OS$
301.12488
RI: 2450
D: UHY
Antihistamine
N: rat
200
72
98
291
M+
351
Metoprolol 2AC
PS
LM/Q
$C_{19}H_{29}NO_5$
351.20457
RI: 2480
D: UHYAC
Beta-Blocker
72
128
212
M+
326
Profenamine-M (desethyl-) AC
UHYAC
LM
$C_{19}H_{22}N_2OS$
326.14529
RI: 2515
D: UHYAC
Antiparkinsonian
72
200
305
M+
347
Oxprenolol-M (HO-) -H2O isomer-1 2AC
UHYAC
LM
$C_{19}H_{25}NO_5$
347.17327
RI: 2520
D: UHYAC
Beta-Blocker

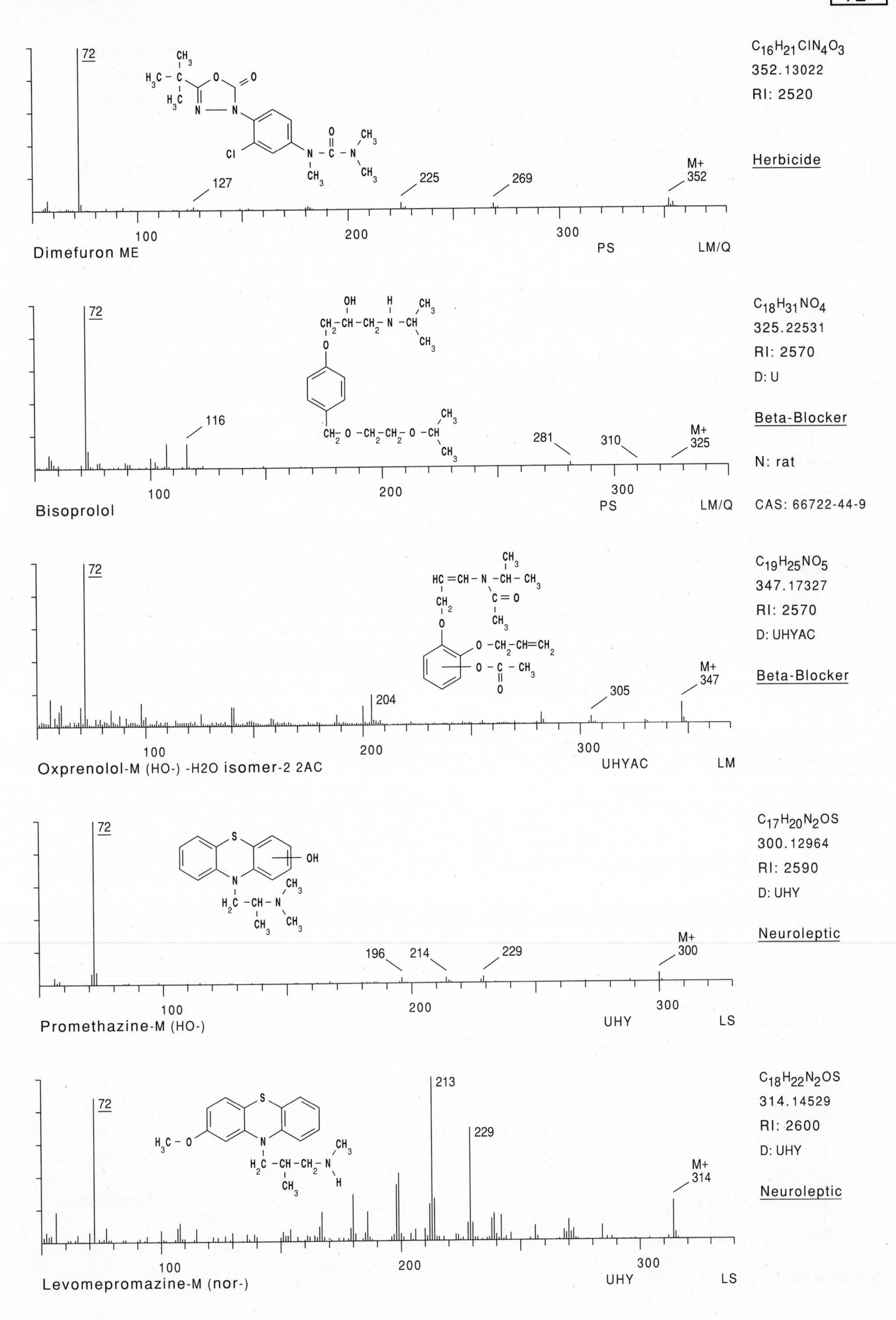

Dimefuron ME
C16H21ClN4O3
352.13022
RI: 2520
Herbicide
72
127
225
269
M+
352
PS
LM/Q
Bisoprolol
C18H31NO4
325.22531
RI: 2570
D: U
Beta-Blocker
N: rat
CAS: 66722-44-9
72
116
281
310
M+
325
PS
LM/Q
Oxprenolol-M (HO-) -H2O isomer-2 2AC
C19H25NO5
347.17327
RI: 2570
D: UHYAC
Beta-Blocker
72
204
305
M+
347
UHYAC
LM
Promethazine-M (HO-)
C17H20N2OS
300.12964
RI: 2590
D: UHY
Neuroleptic
72
196
214
229
M+
300
UHY
LS
Levomepromazine-M (nor-)
C18H22N2OS
314.14529
RI: 2600
D: UHY
Neuroleptic
72
213
229
M+
314
UHY
LS
100
200
300

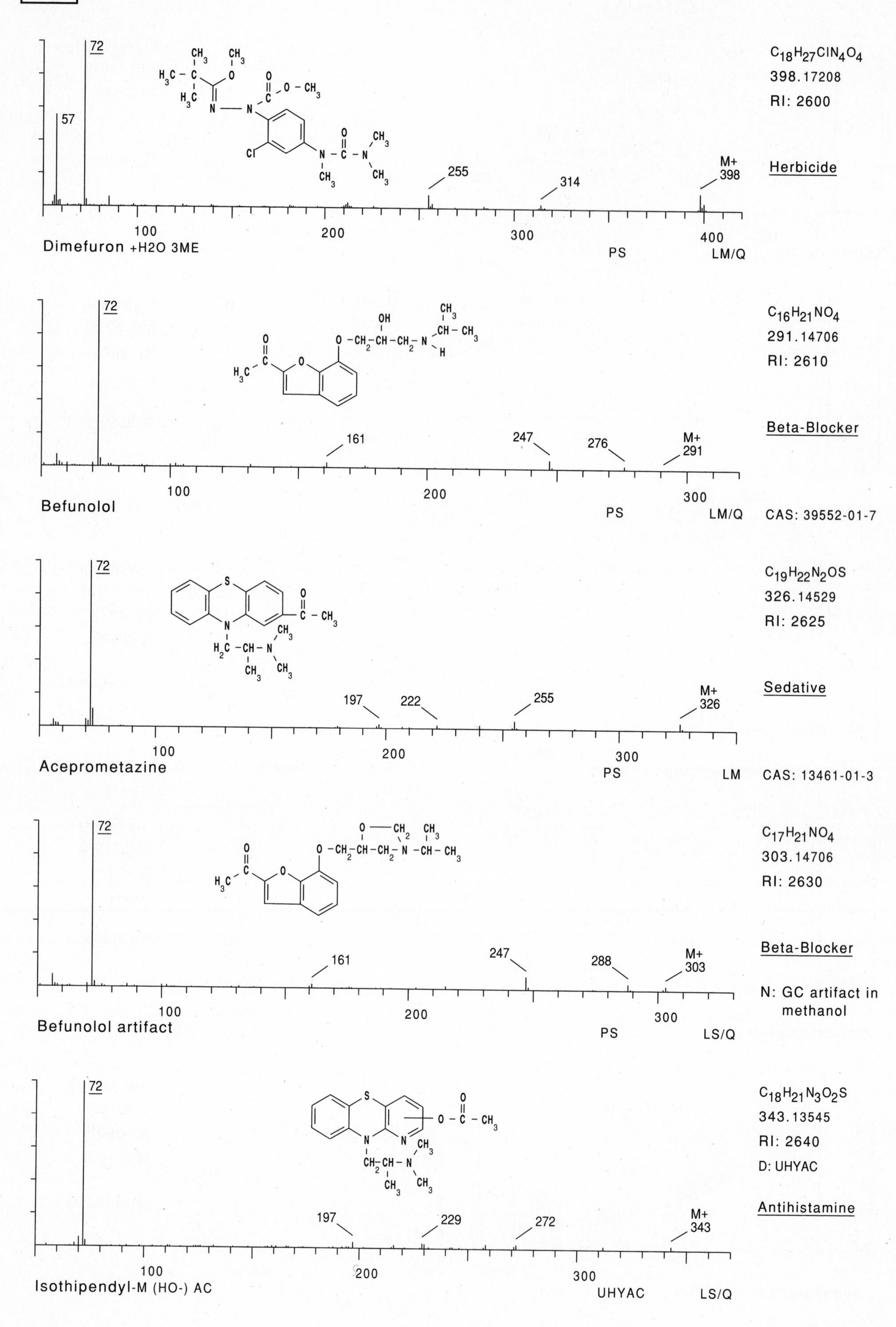
72
57
255
314
M+
398
100
200
300
400
Dimefuron +H2O 3ME
PS
LM/Q
$C_{18}H_{27}ClN_4O_4$
398.17208
RI: 2600
Herbicide
72
161
247
276
M+
291
100
200
300
Befunolol
PS
LM/Q
$C_{16}H_{21}NO_4$
291.14706
RI: 2610
Beta-Blocker
CAS: 39552-01-7
72
197
222
255
M+
326
100
200
300
Aceprometazine
PS
LM
$C_{19}H_{22}N_2OS$
326.14529
RI: 2625
Sedative
CAS: 13461-01-3
72
161
247
288
M+
303
100
200
300
Befunolol artifact
PS
LS/Q
$C_{17}H_{21}NO_4$
303.14706
RI: 2630
Beta-Blocker
N: GC artifact in methanol
72
197
229
272
M+
343
100
200
300
Isothipendyl-M (HO-) AC
UHYAC
LS/Q
$C_{18}H_{21}N_3O_2S$
343.13545
RI: 2640
D: UHYAC
Antihistamine

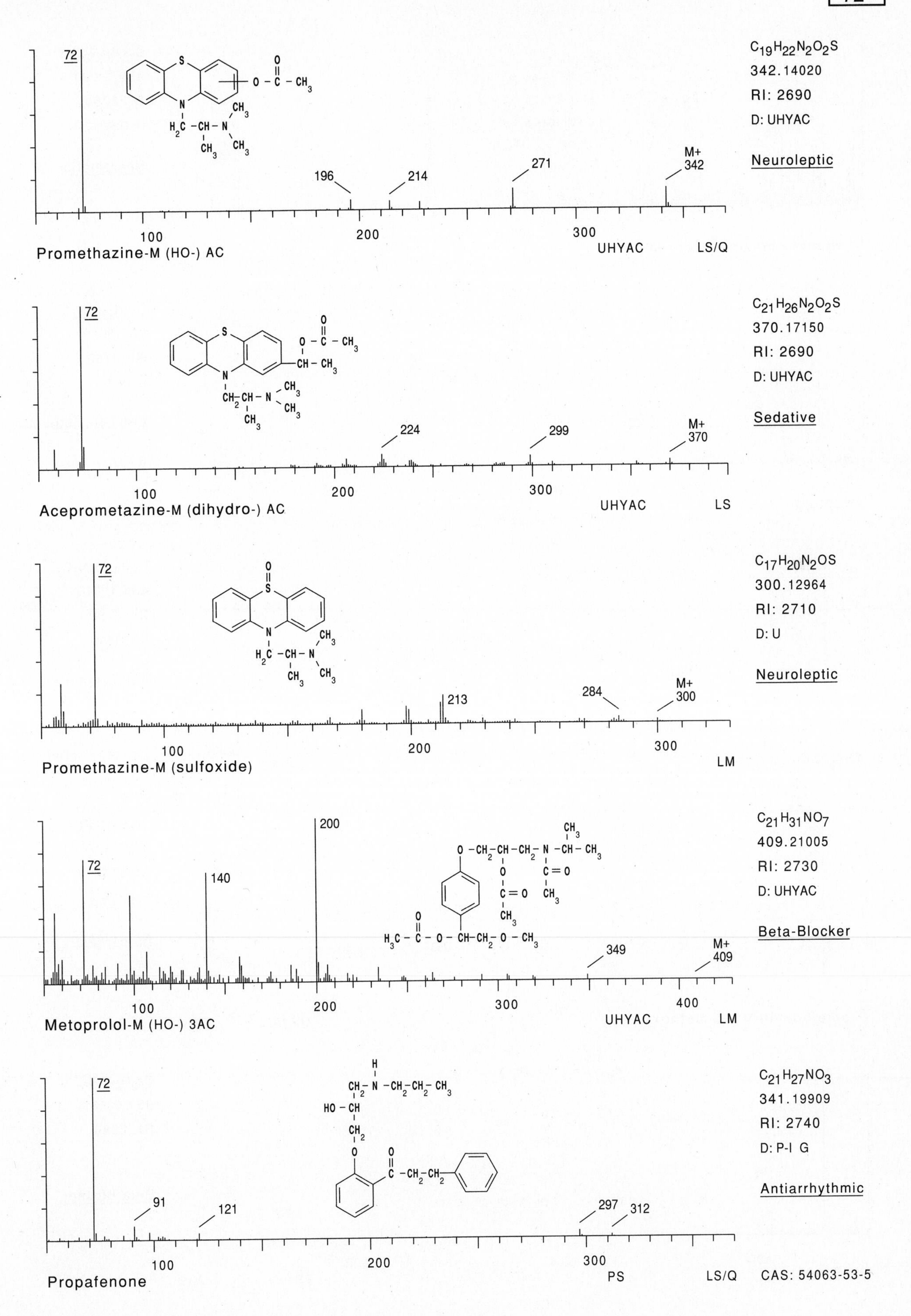

$C_{19}H_{22}N_2O_2S$
342.14020
RI: 2690
D: UHYAC
Neuroleptic

$C_{21}H_{26}N_2O_2S$
370.17150
RI: 2690
D: UHYAC
Sedative

$C_{17}H_{20}N_2OS$
300.12964
RI: 2710
D: U
Neuroleptic

$C_{21}H_{31}NO_7$
409.21005
RI: 2730
D: UHYAC
Beta-Blocker

$C_{21}H_{27}NO_3$
341.19909
RI: 2740
D: P-I G
Antiarrhythmic
CAS: 54063-53-5

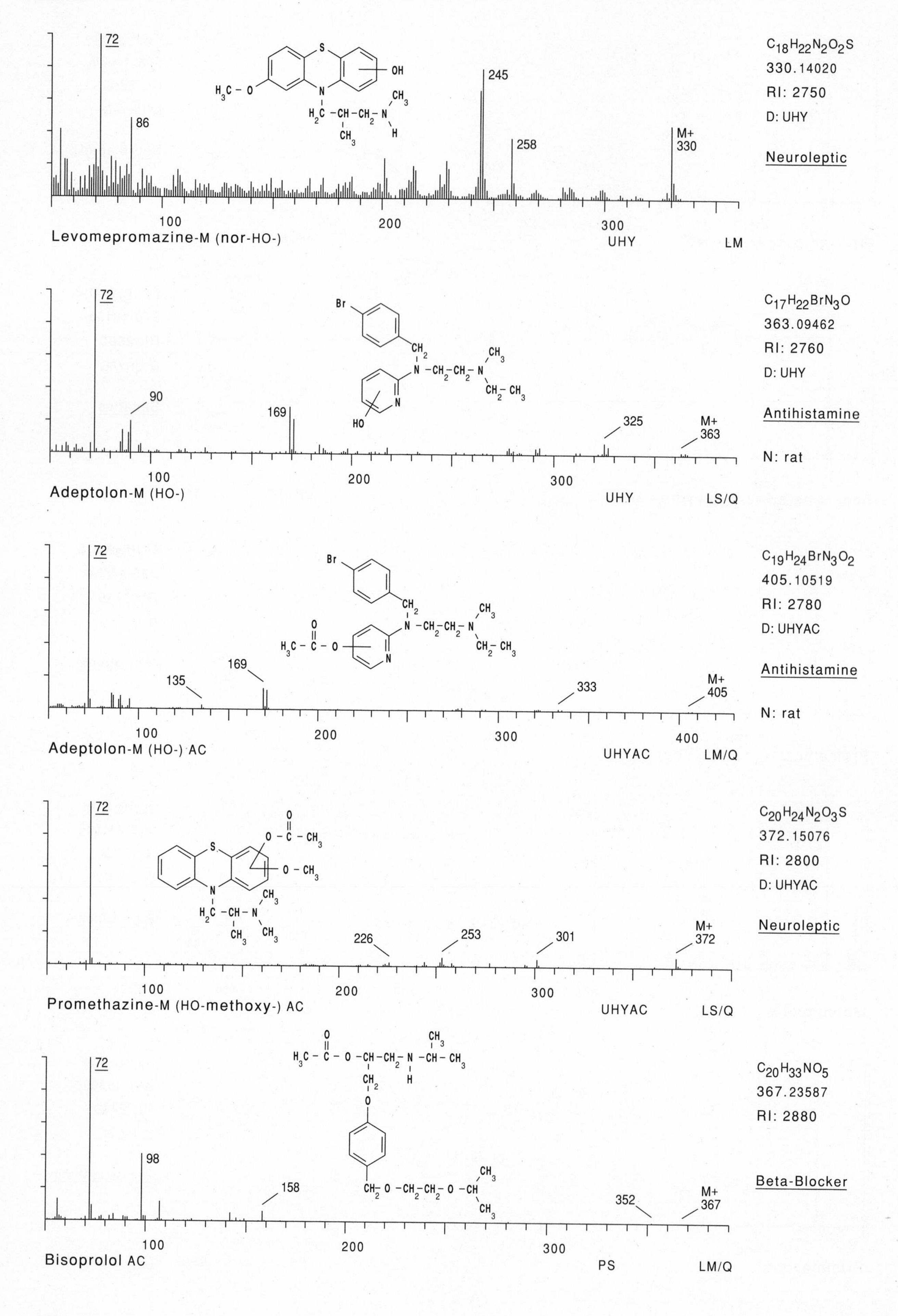

72
86
245
258
M+
330
100
200
300
Levomepromazine-M (nor-HO-)
UHY
LM
C18H22N2O2S
330.14020
RI: 2750
D: UHY
Neuroleptic
72
90
169
325
M+
363
Adeptolon-M (HO-)
UHY
LS/Q
C17H22BrN3O
363.09462
RI: 2760
D: UHY
Antihistamine
N: rat
72
135
169
333
M+
405
400
Adeptolon-M (HO-) AC
UHYAC
LM/Q
C19H24BrN3O2
405.10519
RI: 2780
D: UHYAC
Antihistamine
N: rat
72
226
253
301
M+
372
Promethazine-M (HO-methoxy-) AC
UHYAC
LS/Q
C20H24N2O3S
372.15076
RI: 2800
D: UHYAC
Neuroleptic
72
98
158
352
M+
367
Bisoprolol AC
PS
LM/Q
C20H33NO5
367.23587
RI: 2880
Beta-Blocker

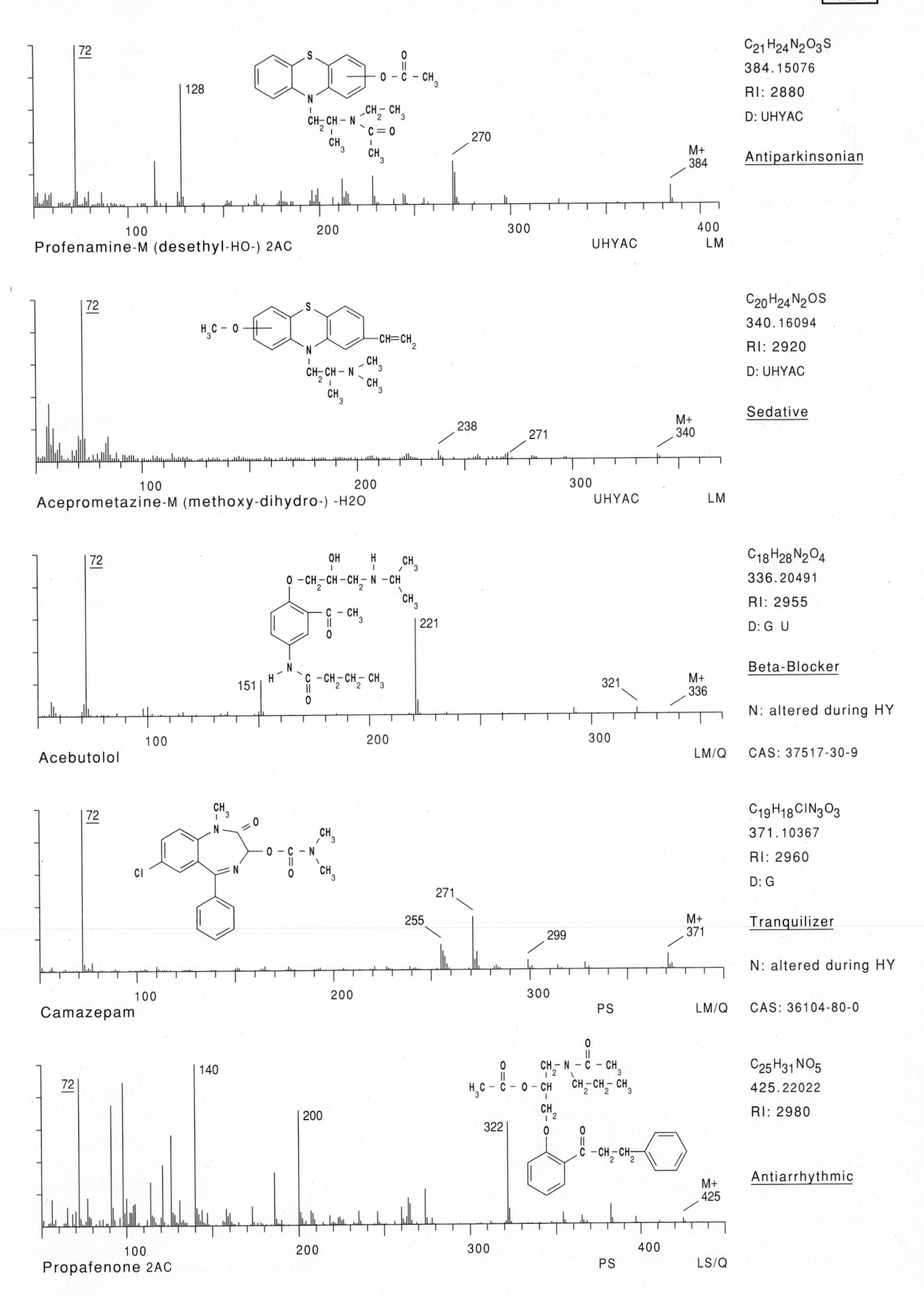

Profenamine-M (desethyl-HO-) 2AC
UHYAC
LM
72
128
270
M+ 384
$C_{21}H_{24}N_2O_3S$
384.15076
RI: 2880
D: UHYAC
Antiparkinsonian
Aceprometazine-M (methoxy-dihydro-) -H2O
UHYAC
LM
72
238
271
M+ 340
$C_{20}H_{24}N_2OS$
340.16094
RI: 2920
D: UHYAC
Sedative
Acebutolol
LM/Q
72
151
221
321
M+ 336
$C_{18}H_{28}N_2O_4$
336.20491
RI: 2955
D: G U
Beta-Blocker
N: altered during HY
CAS: 37517-30-9
Camazepam
PS
LM/Q
72
255
271
299
M+ 371
$C_{19}H_{18}ClN_3O_3$
371.10367
RI: 2960
D: G
Tranquilizer
N: altered during HY
CAS: 36104-80-0
Propafenone 2AC
PS
LS/Q
72
140
200
322
M+ 425
$C_{25}H_{31}NO_5$
425.22022
RI: 2980
Antiarrhythmic

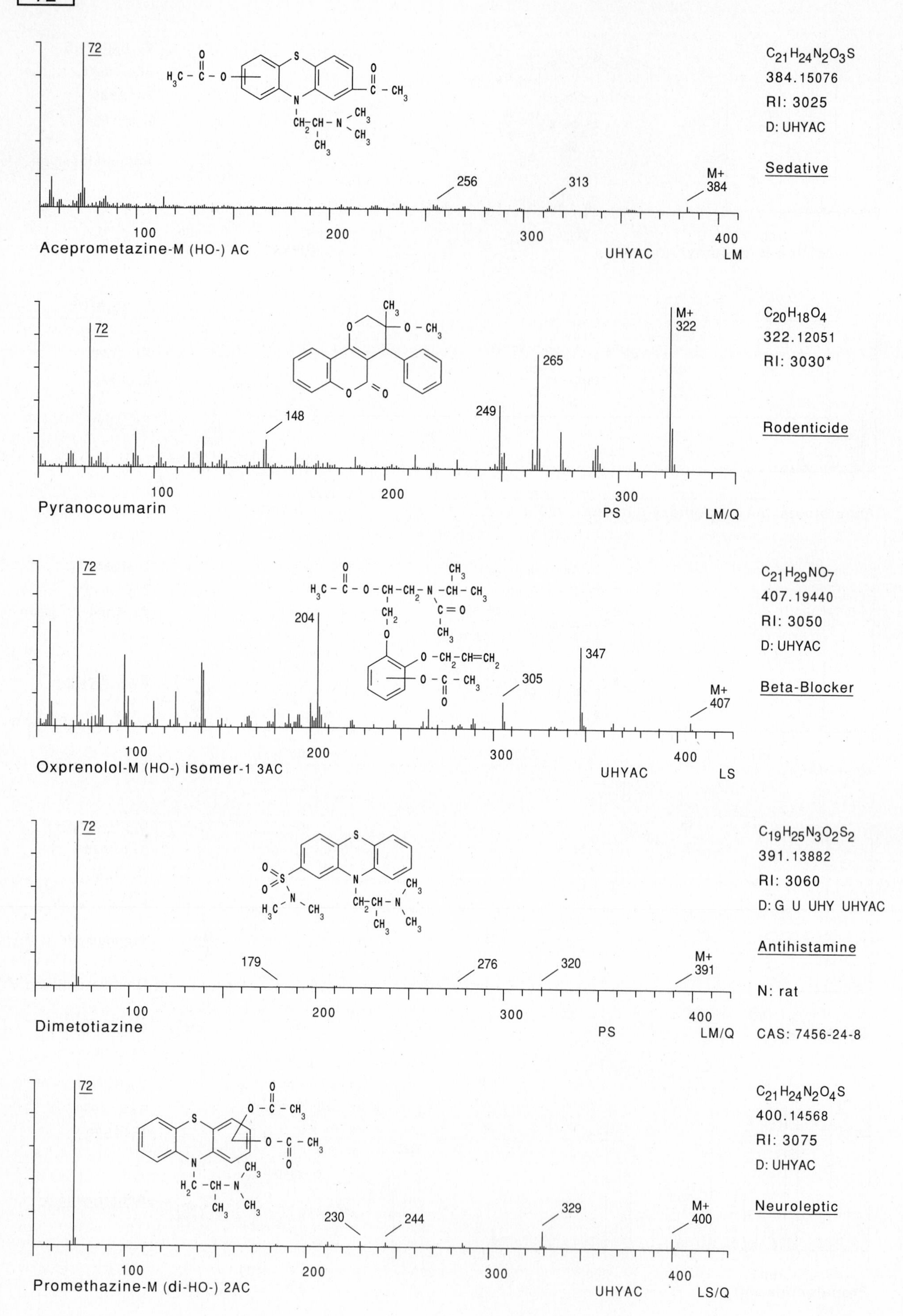
72
256
313
M+
384
100
200
300
400
Acepromethazine-M (HO-) AC
UHYAC
LM
C21H24N2O3S
384.15076
RI: 3025
D: UHYAC
Sedative
72
148
249
265
M+
322
100
200
300
Pyranocoumarin
PS
LM/Q
C20H18O4
322.12051
RI: 3030*
Rodenticide
72
204
305
347
M+
407
100
200
300
400
Oxprenolol-M (HO-) isomer-1 3AC
UHYAC
LS
C21H29NO7
407.19440
RI: 3050
D: UHYAC
Beta-Blocker
72
179
276
320
M+
391
100
200
300
400
Dimetotiazine
PS
LM/Q
C19H25N3O2S2
391.13882
RI: 3060
D: G U UHY UHYAC
Antihistamine
N: rat
CAS: 7456-24-8
72
230
244
329
M+
400
100
200
300
400
Promethazine-M (di-HO-) 2AC
UHYAC
LS/Q
C21H24N2O4S
400.14568
RI: 3075
D: UHYAC
Neuroleptic

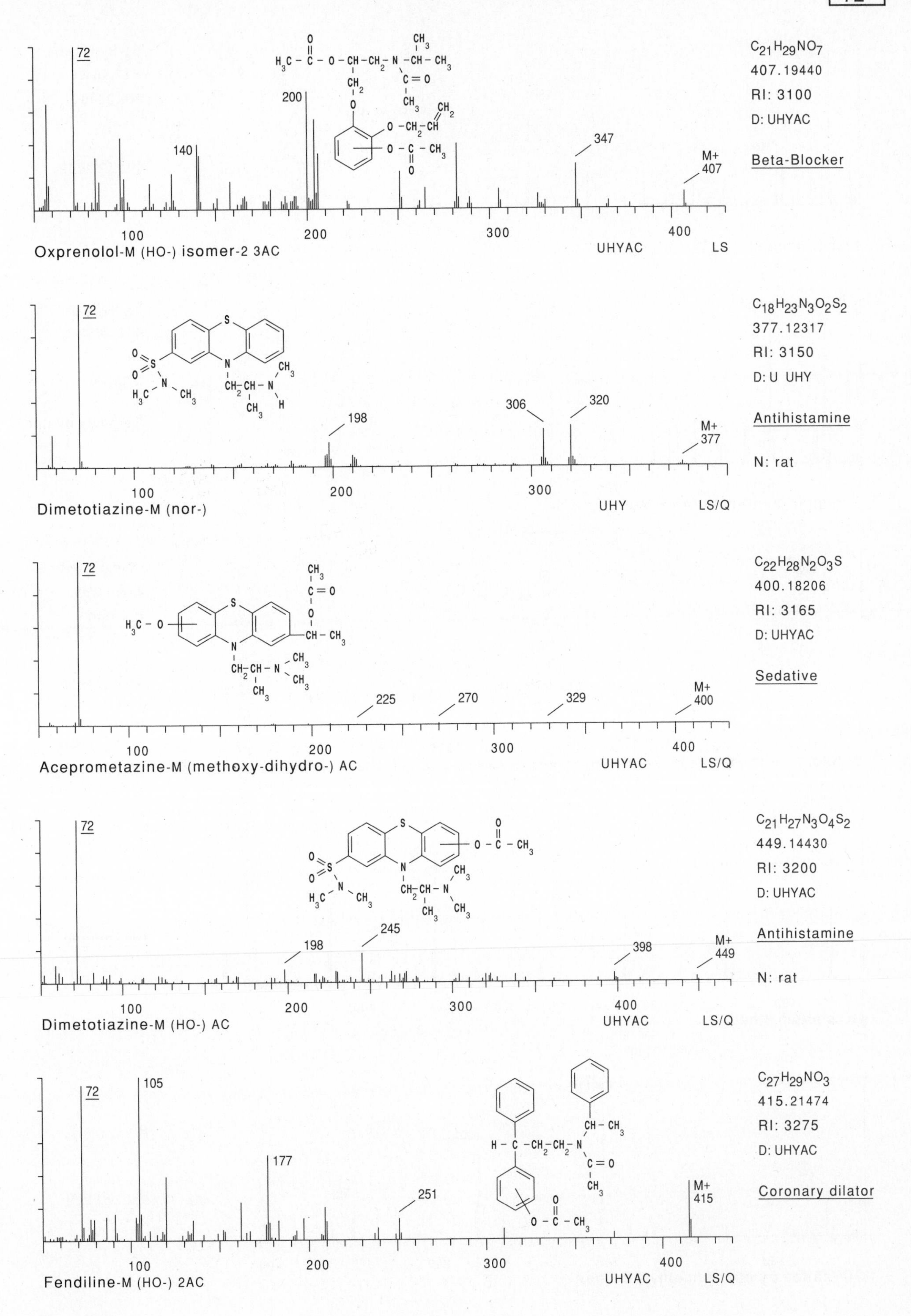

72
200
140
347
M+ 407
100
200
300
400
Oxprenolol-M (HO-) isomer-2 3AC
UHYAC
LS
C21H29NO7
407.19440
RI: 3100
D: UHYAC
Beta-Blocker
72
198
306
320
M+ 377
100
200
300
Dimetotiazine-M (nor-)
UHY
LS/Q
C18H23N3O2S2
377.12317
RI: 3150
D: U UHY
Antihistamine
N: rat
72
225
270
329
M+ 400
100
200
300
400
Aceprometazine-M (methoxy-dihydro-) AC
UHYAC
LS/Q
C22H28N2O3S
400.18206
RI: 3165
D: UHYAC
Sedative
72
198
245
398
M+ 449
100
200
300
400
Dimetotiazine-M (HO-) AC
UHYAC
LS/Q
C21H27N3O4S2
449.14430
RI: 3200
D: UHYAC
Antihistamine
N: rat
72
105
177
251
M+ 415
100
200
300
400
Fendiline-M (HO-) 2AC
UHYAC
LS/Q
C27H29NO3
415.21474
RI: 3275
D: UHYAC
Coronary dilator

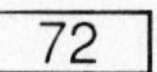

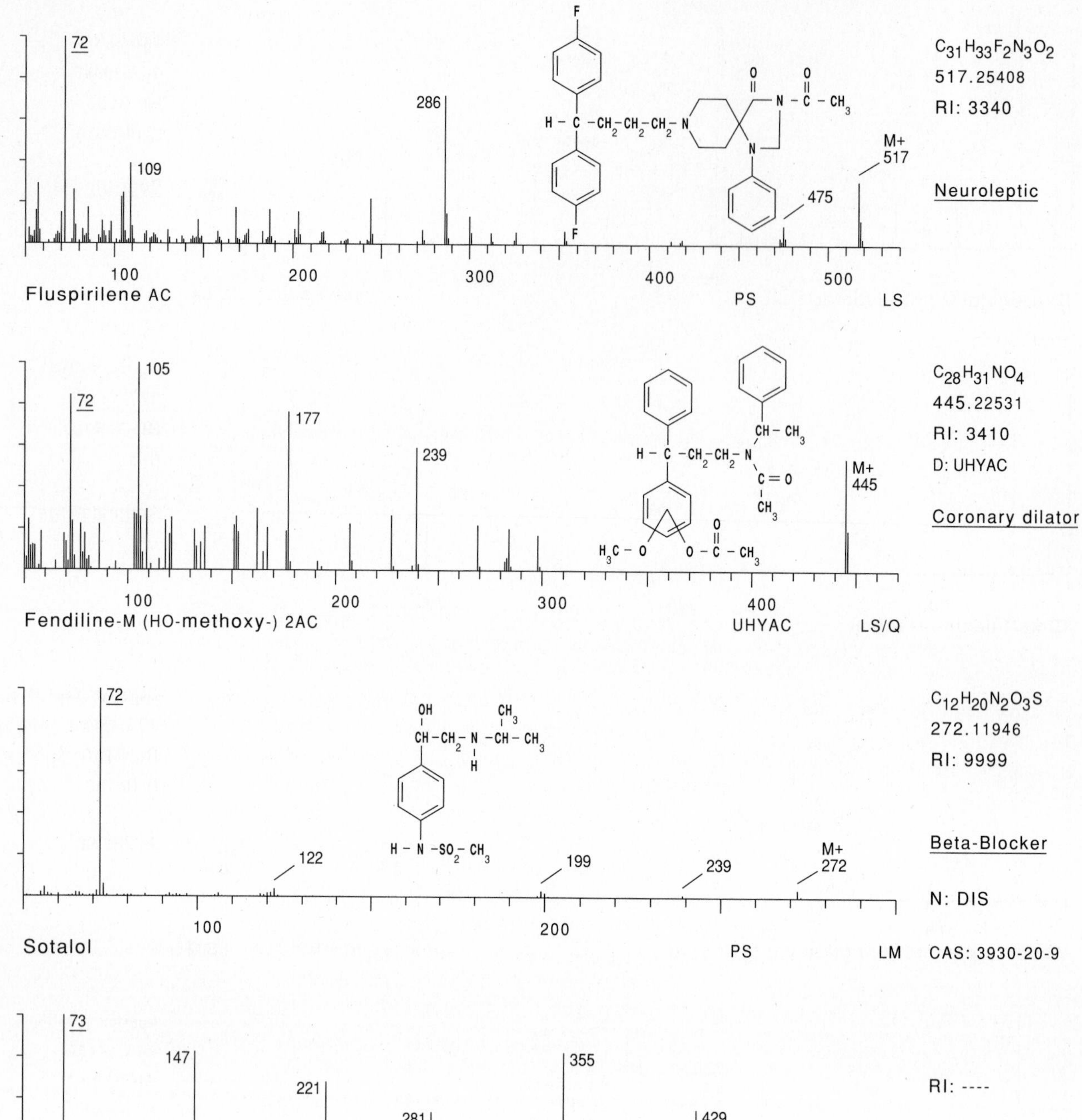

72
109
286
M+
517
475
100
200
300
400
500
Fluspirilene AC
PS
LS
$C_{31}H_{33}F_2N_3O_2$
517.25408
RI: 3340
Neuroleptic
105
72
177
239
M+
445
100
200
300
400
Fendiline-M (HO-methoxy-) 2AC
UHYAC
LS/Q
$C_{28}H_{31}NO_4$
445.22531
RI: 3410
D: UHYAC
Coronary dilator
72
122
199
239
M+
272
100
200
Sotalol
PS
LM
$C_{12}H_{20}N_2O_3S$
272.11946
RI: 9999
Beta-Blocker
N: DIS
CAS: 3930-20-9

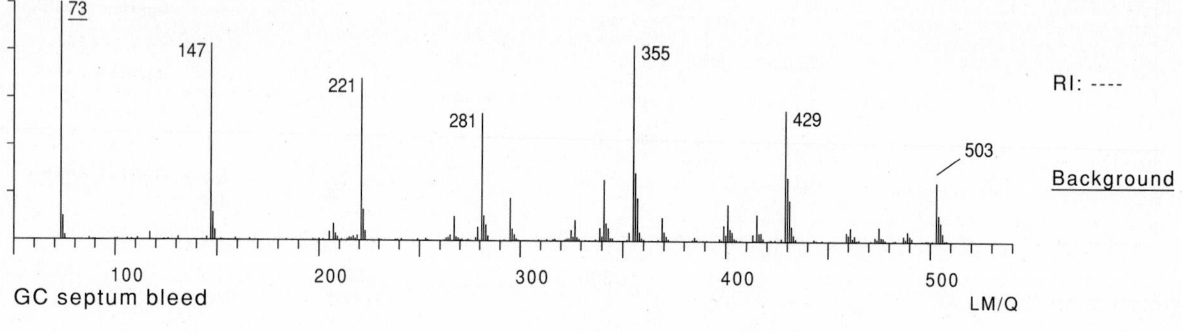

73
147
221
281
355
429
503
100
200
300
400
500
GC septum bleed
LM/Q
RI: ----
Background

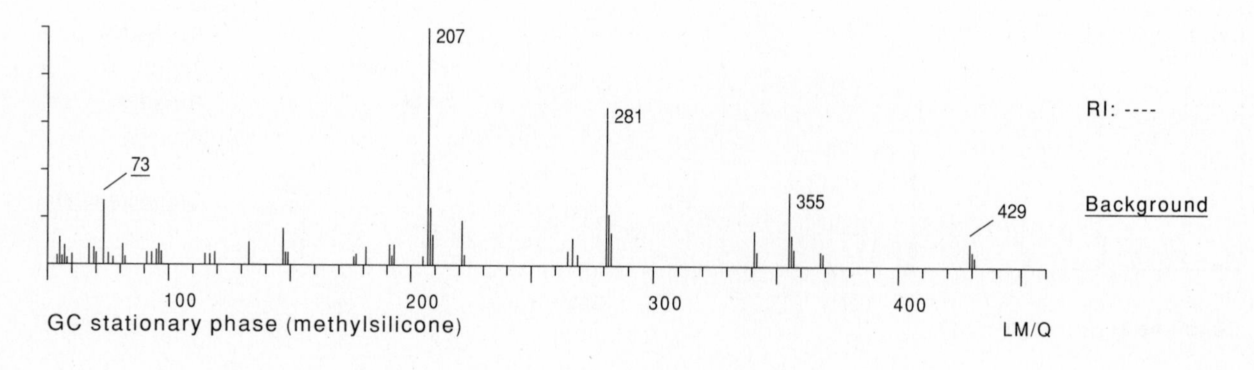

207
281
73
355
429
100
200
300
400
GC stationary phase (methylsilicone)
LM/Q
RI: ----
Background

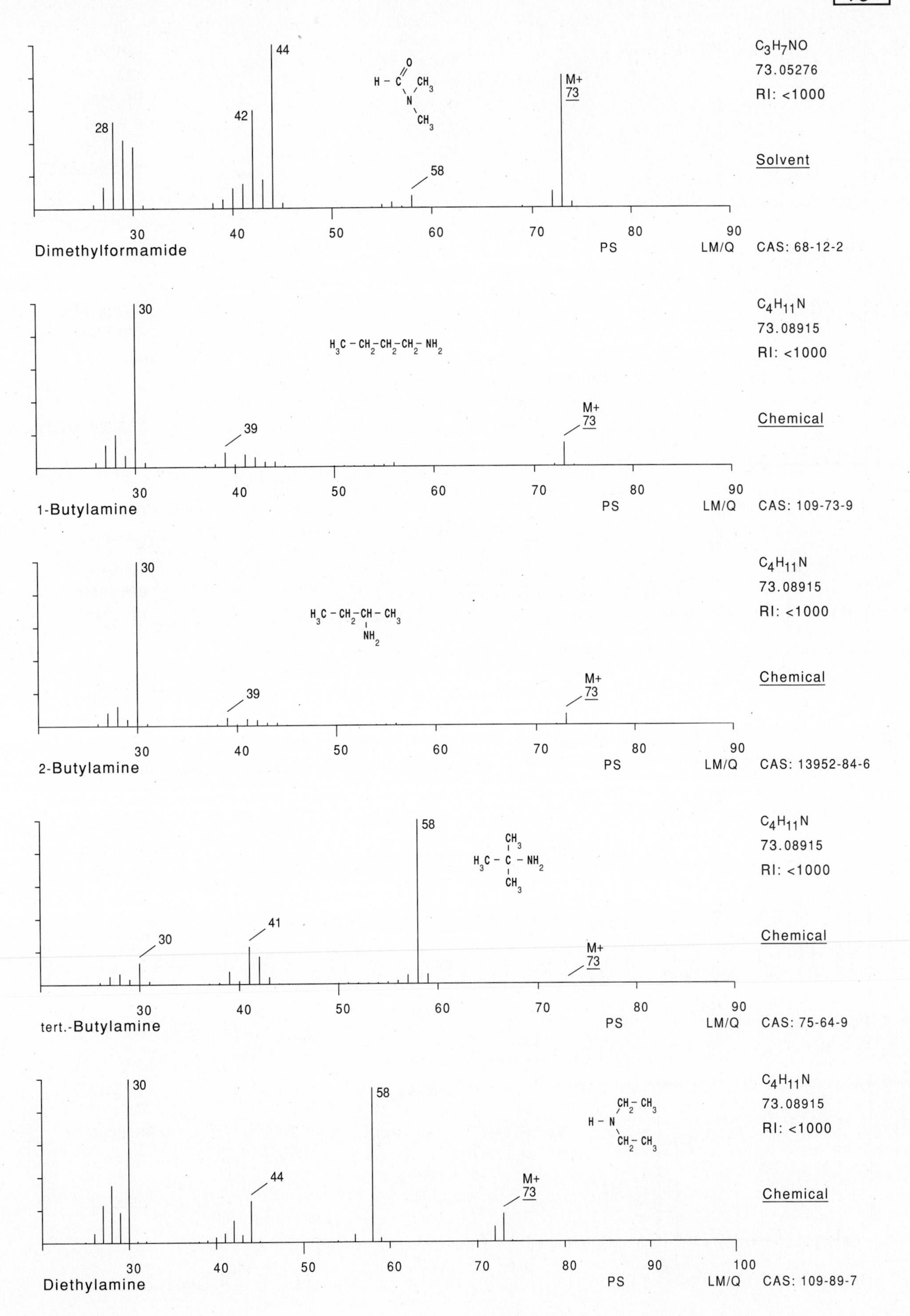

44
28
42
M+
73
58
30
40
50
60
70
80
90
Dimethylformamide
PS
LM/Q
C3H7NO
73.05276
RI: <1000
Solvent
CAS: 68-12-2
1-Butylamine
C4H11N
73.08915
RI: <1000
Chemical
CAS: 109-73-9
2-Butylamine
CAS: 13952-84-6
tert.-Butylamine
CAS: 75-64-9
Diethylamine
100
CAS: 109-89-7

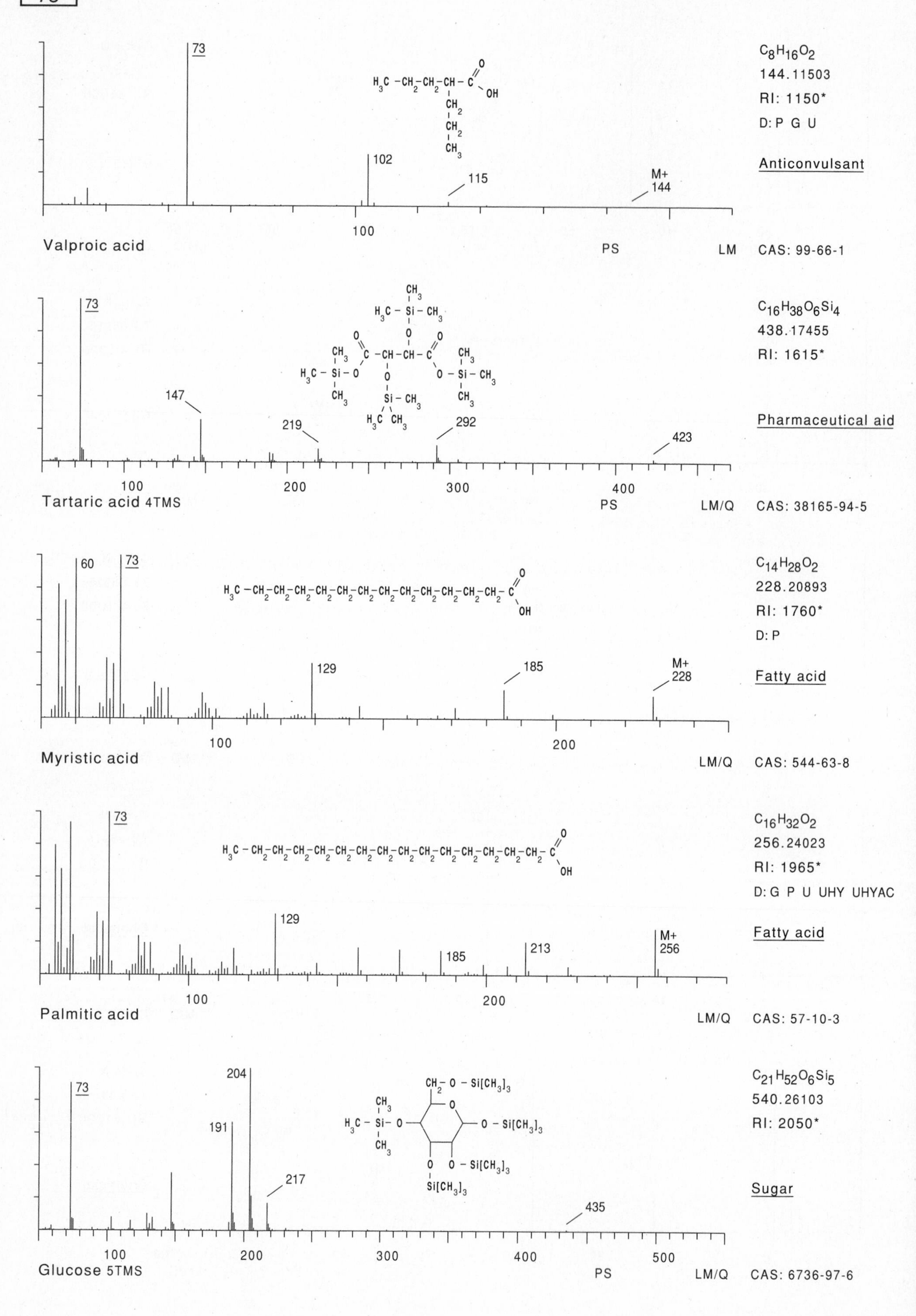
73
Valproic acid
$C_8H_{16}O_2$
144.11503
RI: 1150*
D: P G U
Anticonvulsant
CAS: 99-66-1
PS
LM
Tartaric acid 4TMS
$C_{16}H_{38}O_6Si_4$
438.17455
RI: 1615*
Pharmaceutical aid
CAS: 38165-94-5
PS
LM/Q
Myristic acid
$C_{14}H_{28}O_2$
228.20893
RI: 1760*
D: P
Fatty acid
CAS: 544-63-8
LM/Q
Palmitic acid
$C_{16}H_{32}O_2$
256.24023
RI: 1965*
D: G P U UHY UHYAC
Fatty acid
CAS: 57-10-3
LM/Q
Glucose 5TMS
$C_{21}H_{52}O_6Si_5$
540.26103
RI: 2050*
Sugar
CAS: 6736-97-6
PS
LM/Q

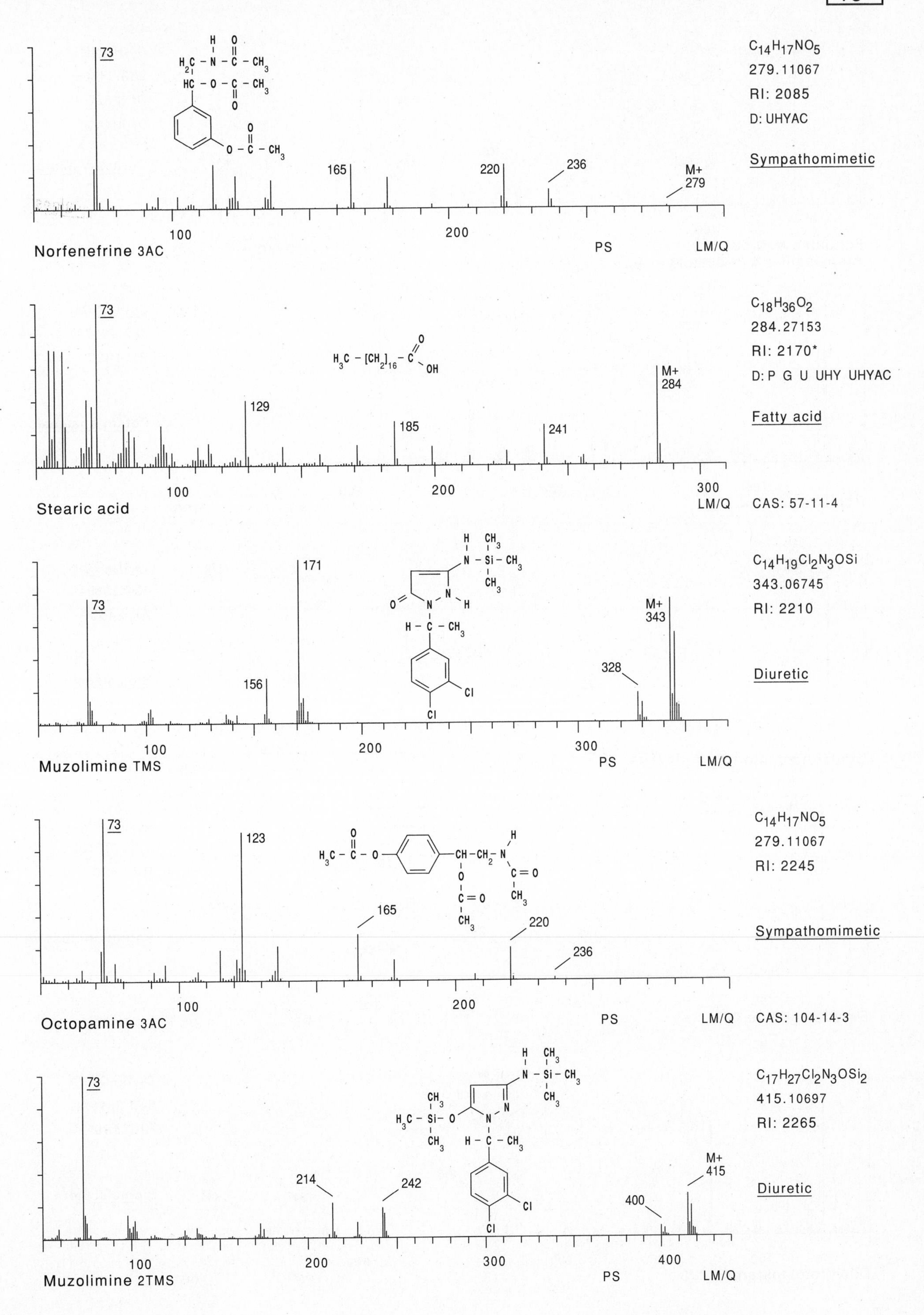

$C_{14}H_{17}NO_5$
279.11067
RI: 2085
D: UHYAC

Sympathomimetic

$C_{18}H_{36}O_2$
284.27153
RI: 2170*
D: P G U UHY UHYAC

Fatty acid

CAS: 57-11-4

$C_{14}H_{19}Cl_2N_3OSi$
343.06745
RI: 2210

Diuretic

$C_{14}H_{17}NO_5$
279.11067
RI: 2245

Sympathomimetic

CAS: 104-14-3

$C_{17}H_{27}Cl_2N_3OSi_2$
415.10697
RI: 2265

Diuretic

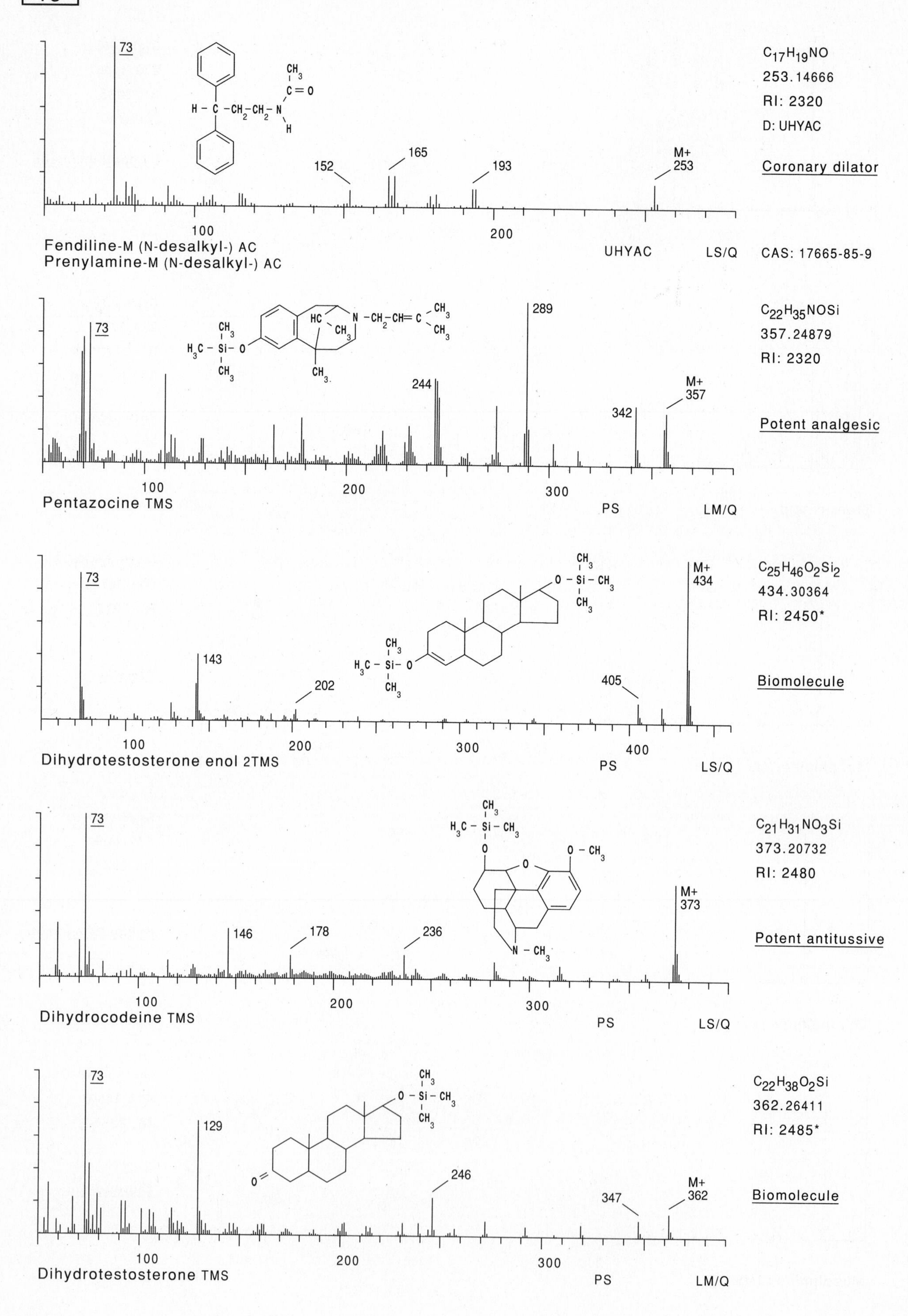
73
152
165
193
M+
253
100
200
Fendiline-M (N-desalkyl-) AC
Prenylamine-M (N-desalkyl-) AC
UHYAC
LS/Q
C17H19NO
253.14666
RI: 2320
D: UHYAC
Coronary dilator
CAS: 17665-85-9
73
289
244
342
M+
357
100
200
300
Pentazocine TMS
PS
LM/Q
C22H35NOSi
357.24879
RI: 2320
Potent analgesic
73
143
202
405
M+
434
100
200
300
400
Dihydrotestosterone enol 2TMS
PS
LS/Q
C25H46O2Si2
434.30364
RI: 2450*
Biomolecule
73
146
178
236
M+
373
100
200
300
Dihydrocodeine TMS
PS
LS/Q
C21H31NO3Si
373.20732
RI: 2480
Potent antitussive
73
129
246
347
M+
362
100
200
300
Dihydrotestosterone TMS
PS
LM/Q
C22H38O2Si
362.26411
RI: 2485*
Biomolecule

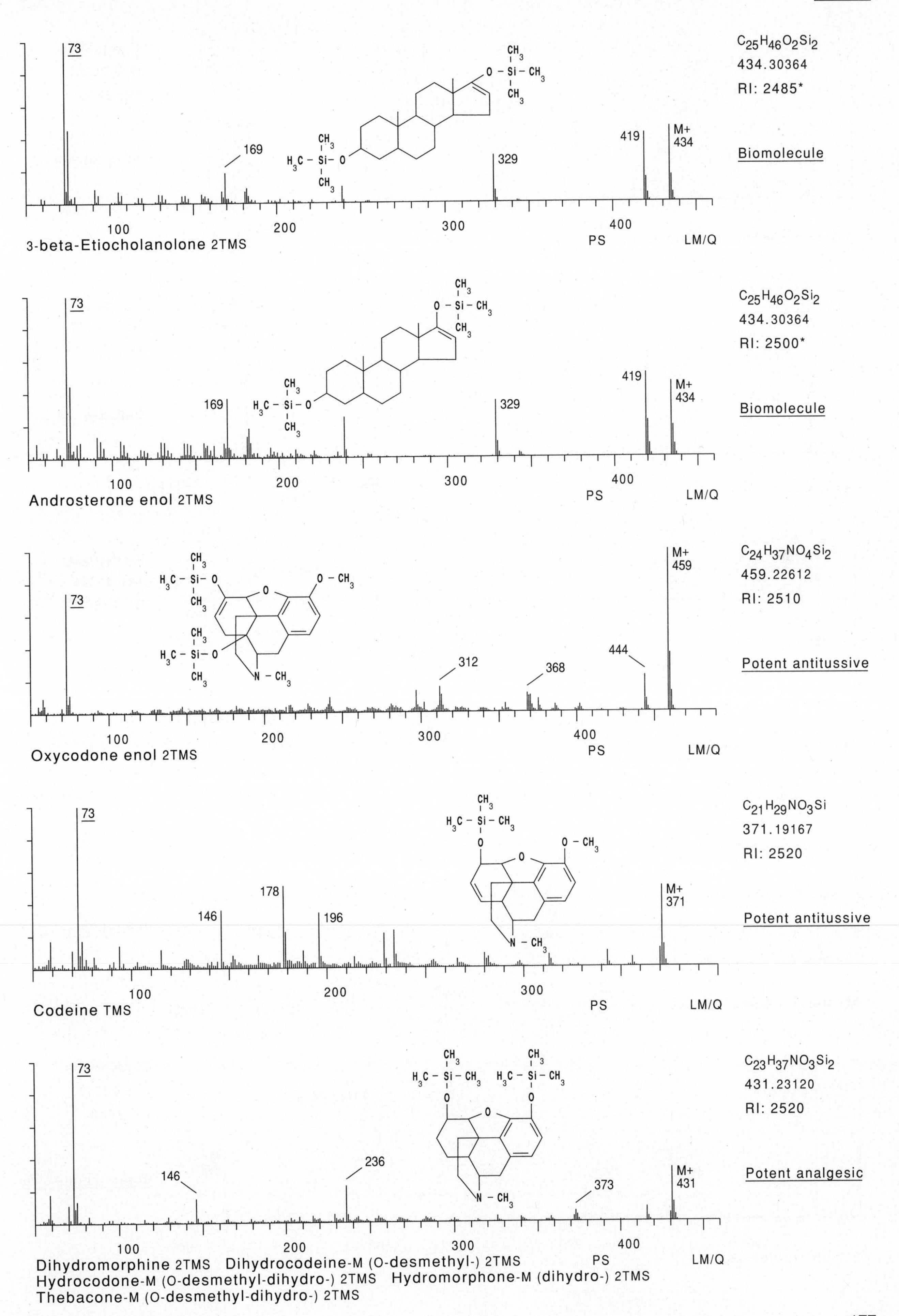

$C_{25}H_{46}O_2Si_2$
434.30364
RI: 2485*
Biomolecule

$C_{25}H_{46}O_2Si_2$
434.30364
RI: 2500*
Biomolecule

$C_{24}H_{37}NO_4Si_2$
459.22612
RI: 2510
Potent antitussive

$C_{21}H_{29}NO_3Si$
371.19167
RI: 2520
Potent antitussive

$C_{23}H_{37}NO_3Si_2$
431.23120
RI: 2520
Potent analgesic

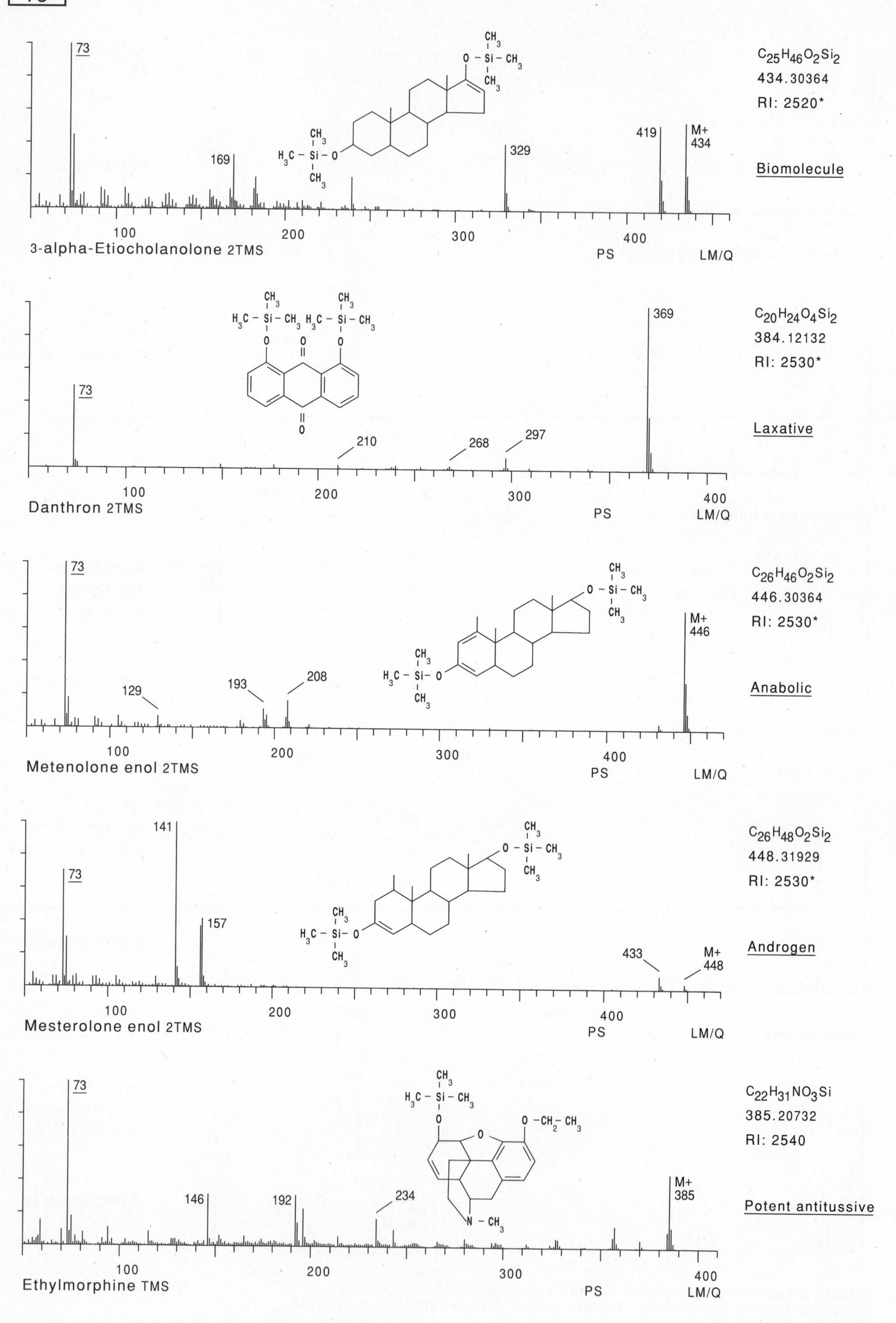
73
169
329
419
M+
434
100
200
300
400
PS
LM/Q
3-alpha-Etiocholanolone 2TMS
C25H46O2Si2
434.30364
RI: 2520*
Biomolecule
369
73
210
268
297
Danthron 2TMS
C20H24O4Si2
384.12132
RI: 2530*
Laxative
73
129
193
208
M+
446
Metenolone enol 2TMS
C26H46O2Si2
446.30364
RI: 2530*
Anabolic
141
73
157
433
M+
448
Mesterolone enol 2TMS
C26H48O2Si2
448.31929
RI: 2530*
Androgen
73
146
192
234
M+
385
Ethylmorphine TMS
C22H31NO3Si
385.20732
RI: 2540
Potent antitussive

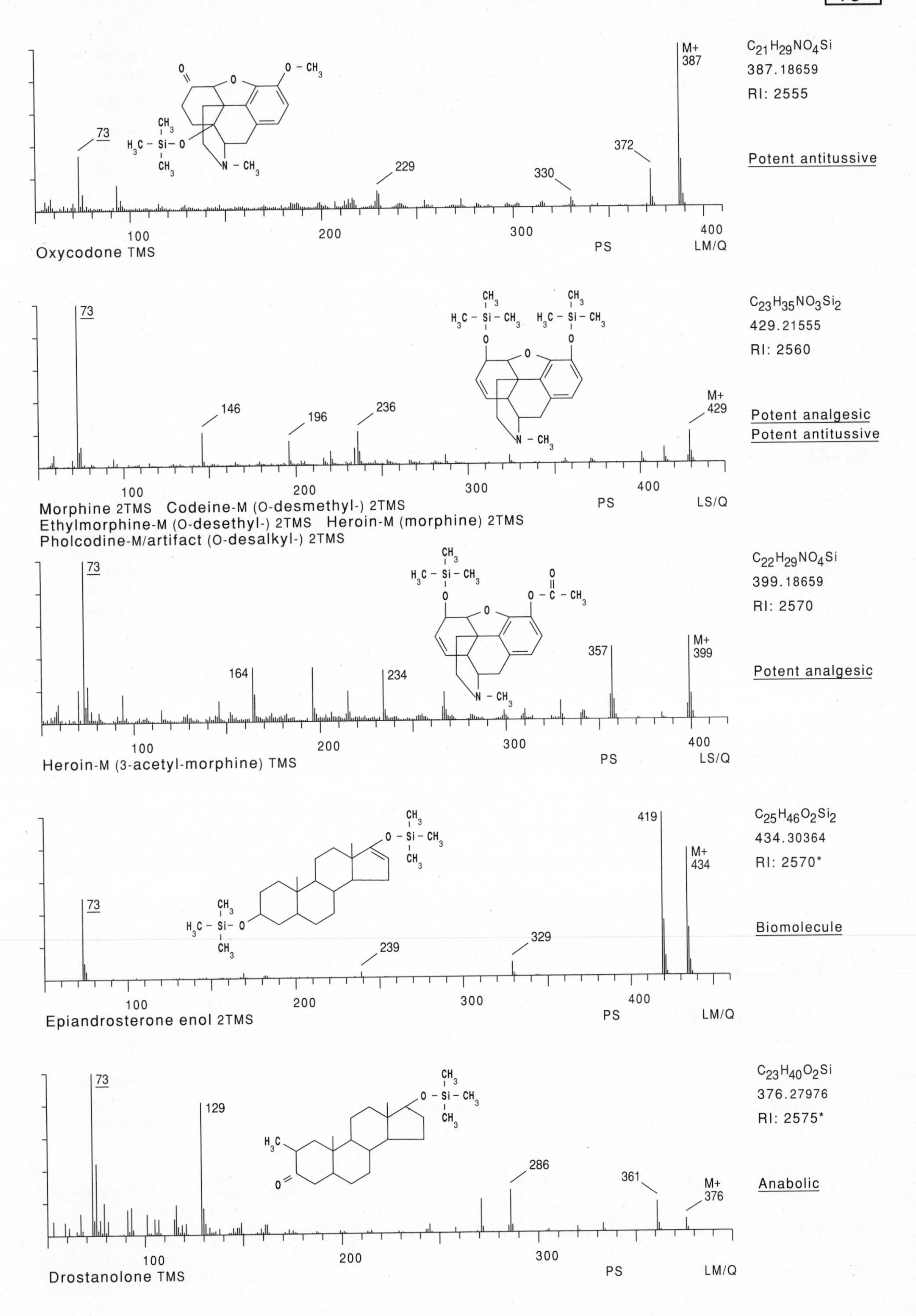

$C_{21}H_{29}NO_4Si$
387.18659
RI: 2555
Potent antitussive

$C_{23}H_{35}NO_3Si_2$
429.21555
RI: 2560
Potent analgesic
Potent antitussive

$C_{22}H_{29}NO_4Si$
399.18659
RI: 2570
Potent analgesic

$C_{25}H_{46}O_2Si_2$
434.30364
RI: 2570*
Biomolecule

$C_{23}H_{40}O_2Si$
376.27976
RI: 2575*
Anabolic

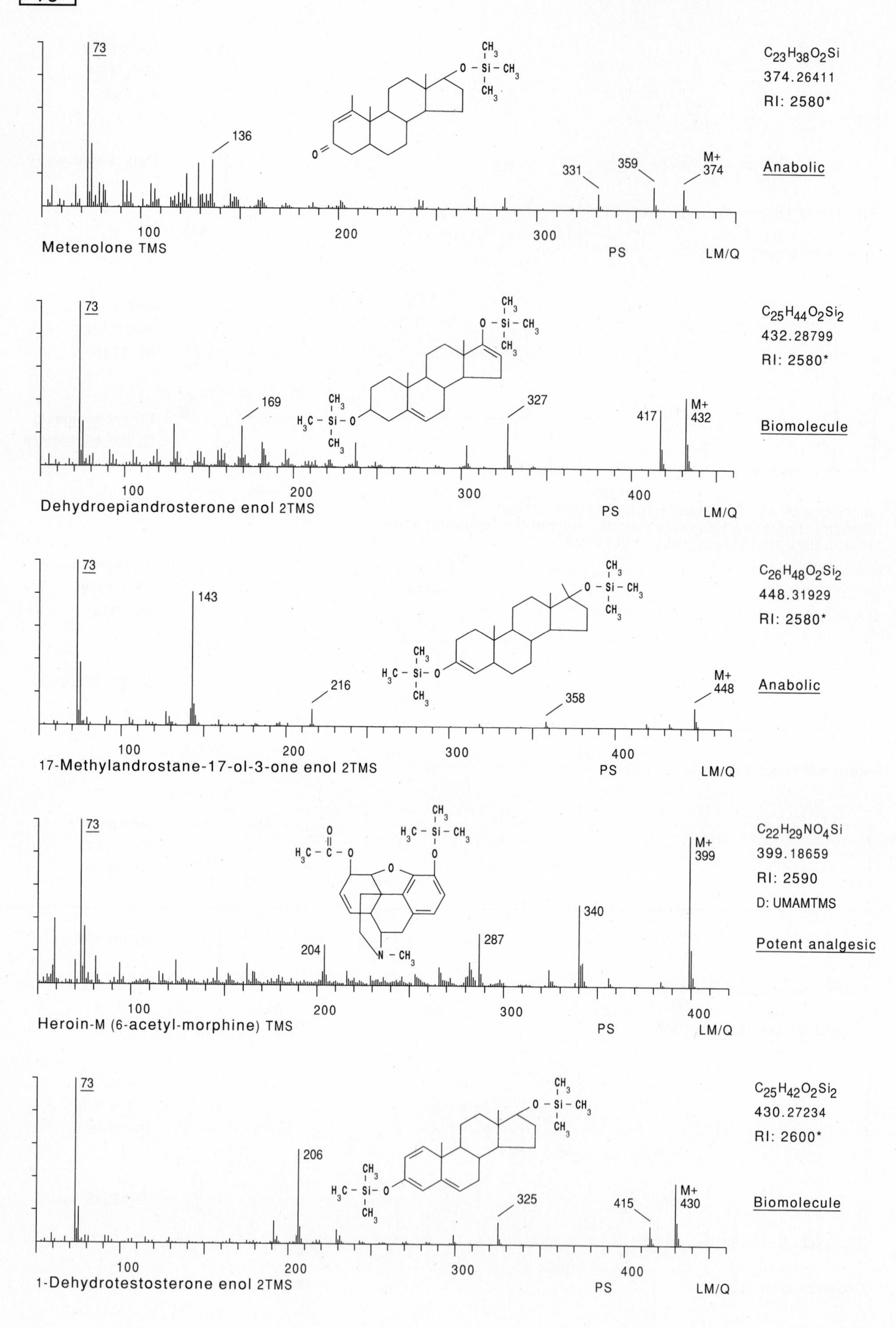

73
136
331
359
M+
374
100
200
300
Metenolone TMS
PS
LM/Q
C23H38O2Si
374.26411
RI: 2580*
Anabolic
73
169
327
417
M+
432
100
200
300
400
Dehydroepiandrosterone enol 2TMS
PS
LM/Q
C25H44O2Si2
432.28799
RI: 2580*
Biomolecule
73
143
216
358
M+
448
100
200
300
400
17-Methylandrostane-17-ol-3-one enol 2TMS
PS
LM/Q
C26H48O2Si2
448.31929
RI: 2580*
Anabolic
73
204
287
340
M+
399
100
200
300
400
Heroin-M (6-acetyl-morphine) TMS
PS
LM/Q
C22H29NO4Si
399.18659
RI: 2590
D: UMAMTMS
Potent analgesic
73
206
325
415
M+
430
100
200
300
400
1-Dehydrotestosterone enol 2TMS
PS
LM/Q
C25H42O2Si2
430.27234
RI: 2600*
Biomolecule

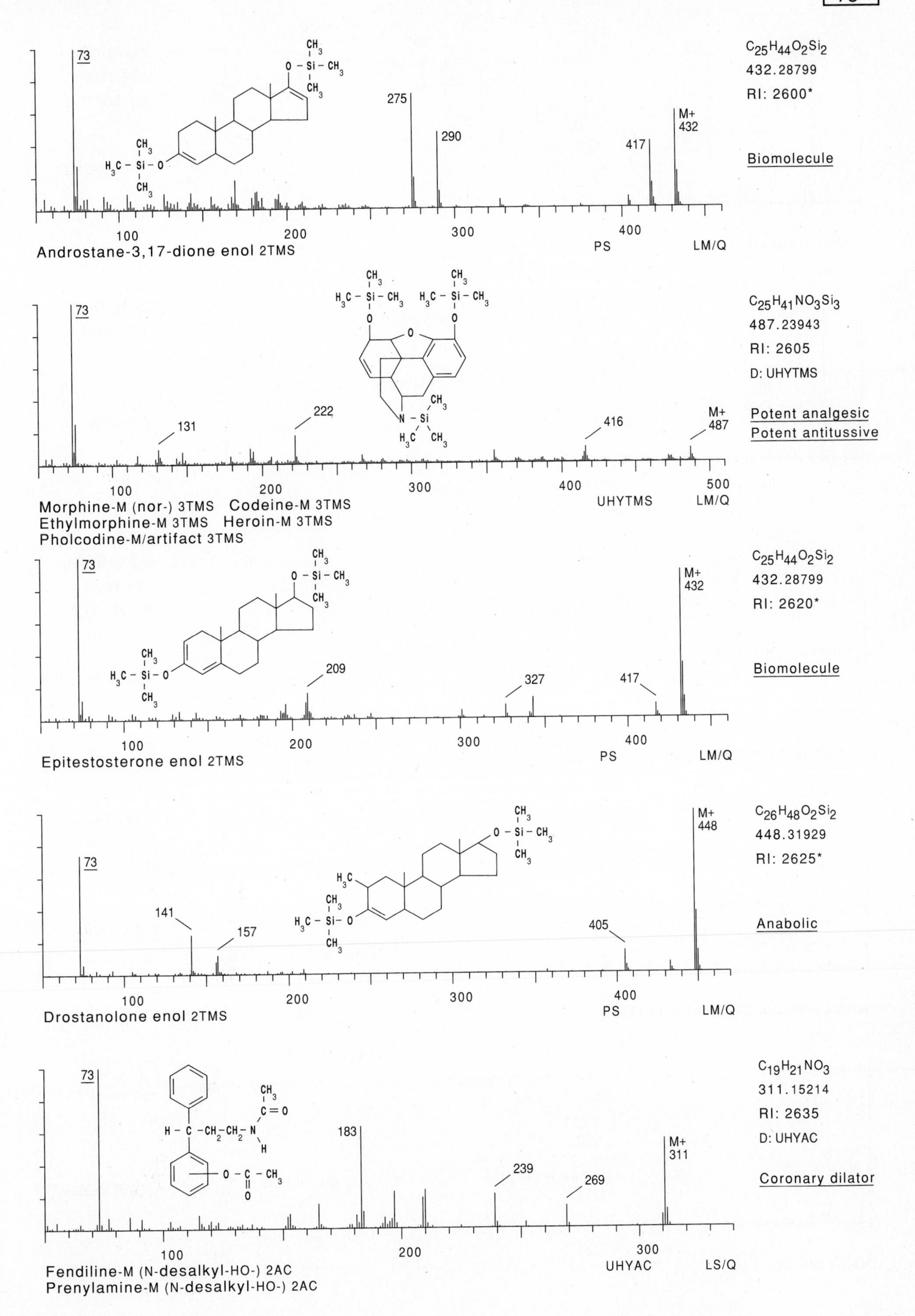

73
275
290
417
M+
432
100
200
300
400
PS
LM/Q
Androstane-3,17-dione enol 2TMS
$C_{25}H_{44}O_2Si_2$
432.28799
RI: 2600*
Biomolecule
73
131
222
416
M+
487
100
200
300
400
500
UHYTMS
LM/Q
Morphine-M (nor-) 3TMS Codeine-M 3TMS
Ethylmorphine-M 3TMS Heroin-M 3TMS
Pholcodine-M/artifact 3TMS
$C_{25}H_{41}NO_3Si_3$
487.23943
RI: 2605
D: UHYTMS
Potent analgesic
Potent antitussive
73
209
327
417
M+
432
100
200
300
400
PS
LM/Q
Epitestosterone enol 2TMS
$C_{25}H_{44}O_2Si_2$
432.28799
RI: 2620*
Biomolecule
73
141
157
405
M+
448
100
200
300
400
PS
LM/Q
Drostanolone enol 2TMS
$C_{26}H_{48}O_2Si_2$
448.31929
RI: 2625*
Anabolic
73
183
239
269
M+
311
100
200
300
UHYAC
LS/Q
Fendiline-M (N-desalkyl-HO-) 2AC
Prenylamine-M (N-desalkyl-HO-) 2AC
$C_{19}H_{21}NO_3$
311.15214
RI: 2635
D: UHYAC
Coronary dilator

73

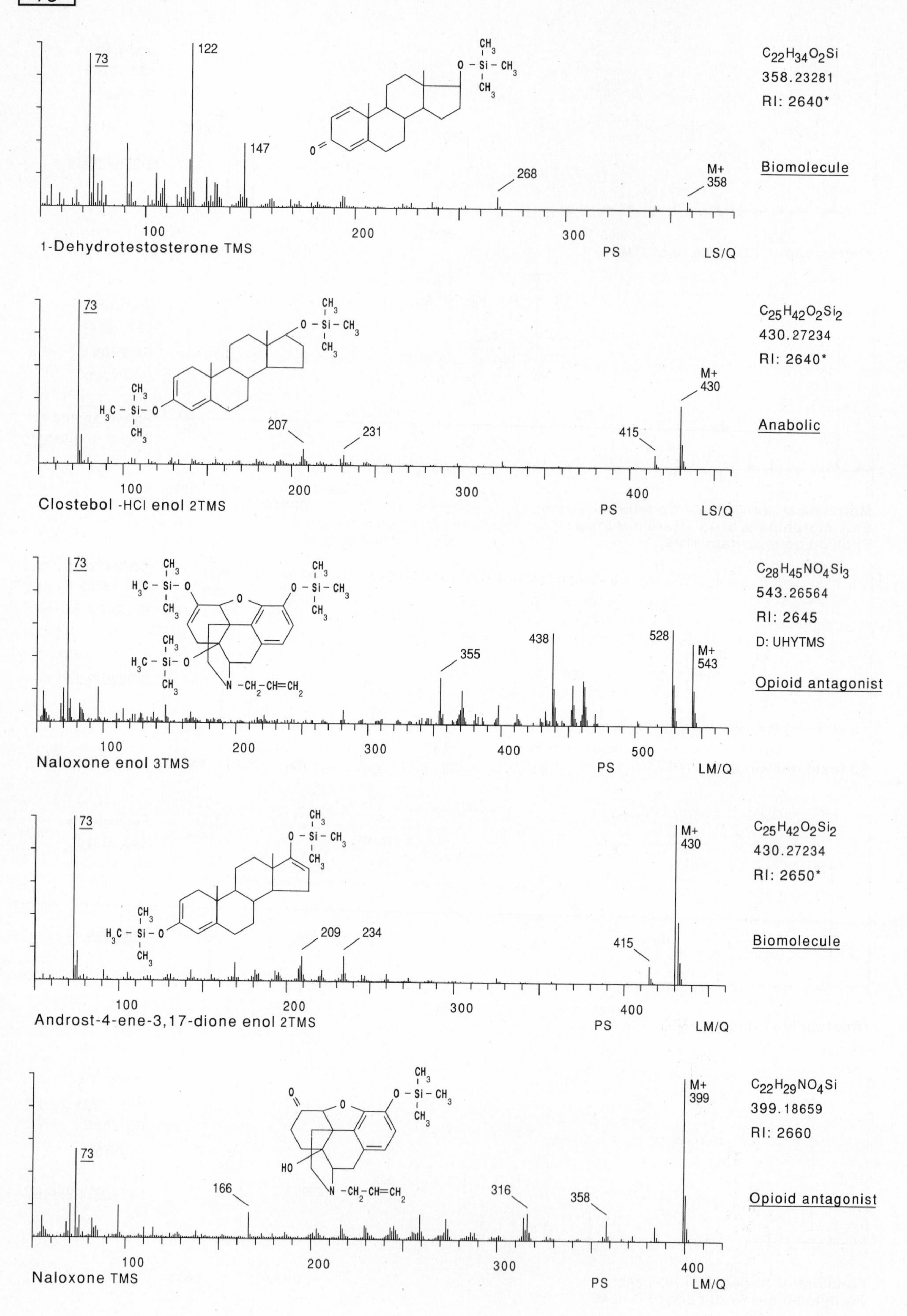

1-Dehydrotestosterone TMS
$C_{22}H_{34}O_2Si$
358.23281
RI: 2640*
Biomolecule
73
122
147
268
M+ 358
PS
LS/Q
Clostebol -HCl enol 2TMS
$C_{25}H_{42}O_2Si_2$
430.27234
RI: 2640*
Anabolic
73
207
231
415
M+ 430
PS
LS/Q
Naloxone enol 3TMS
$C_{28}H_{45}NO_4Si_3$
543.26564
RI: 2645
D: UHYTMS
Opioid antagonist
73
355
438
528
M+ 543
PS
LM/Q
Androst-4-ene-3,17-dione enol 2TMS
$C_{25}H_{42}O_2Si_2$
430.27234
RI: 2650*
Biomolecule
73
209
234
415
M+ 430
PS
LM/Q
Naloxone TMS
$C_{22}H_{29}NO_4Si$
399.18659
RI: 2660
Opioid antagonist
73
166
316
358
M+ 399
PS
LM/Q

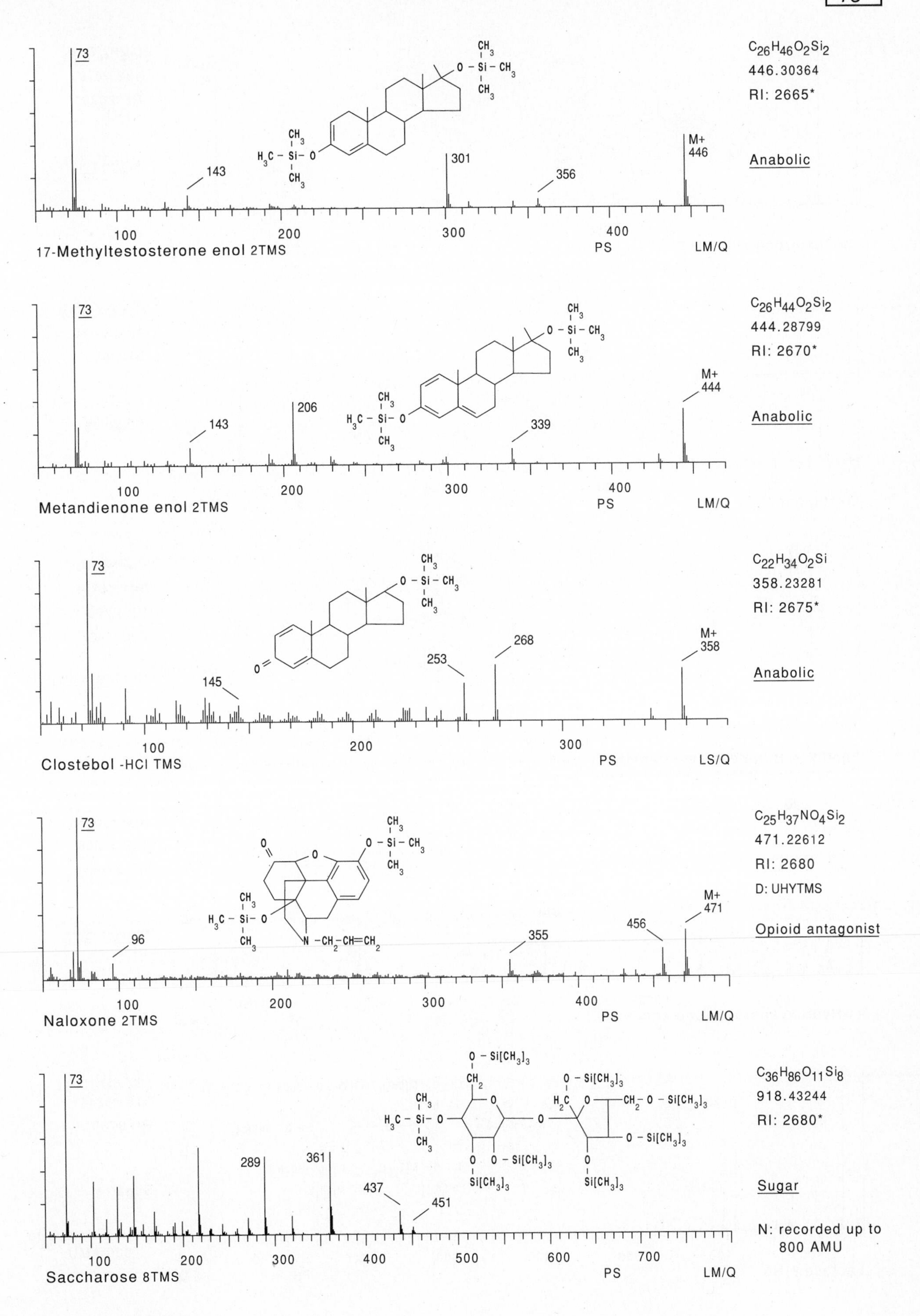

$C_{26}H_{46}O_2Si_2$
446.30364
RI: 2665*

Anabolic

$C_{26}H_{44}O_2Si_2$
444.28799
RI: 2670*

Anabolic

$C_{22}H_{34}O_2Si$
358.23281
RI: 2675*

Anabolic

$C_{25}H_{37}NO_4Si_2$
471.22612
RI: 2680
D: UHYTMS

Opioid antagonist

$C_{36}H_{86}O_{11}Si_8$
918.43244
RI: 2680*

Sugar

N: recorded up to 800 AMU

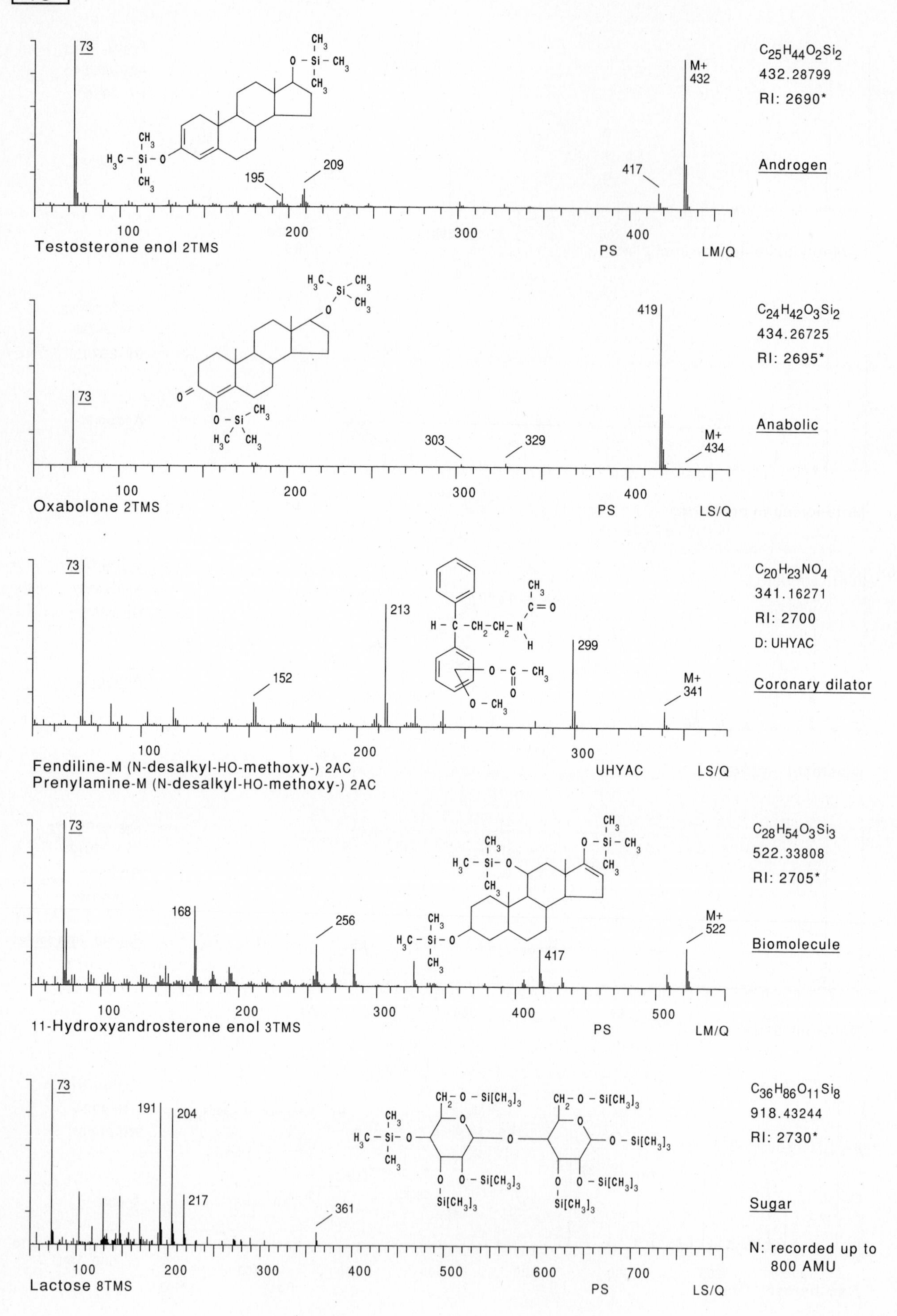

73
195
209
417
M+
432
100
200
300
400
Testosterone enol 2TMS
PS
LM/Q
$C_{25}H_{44}O_2Si_2$
432.28799
RI: 2690*
Androgen
73
303
329
419
M+
434
Oxabolone 2TMS
PS
LS/Q
$C_{24}H_{42}O_3Si_2$
434.26725
RI: 2695*
Anabolic
73
152
213
299
M+
341
Fendiline-M (N-desalkyl-HO-methoxy-) 2AC
Prenylamine-M (N-desalkyl-HO-methoxy-) 2AC
UHYAC
LS/Q
$C_{20}H_{23}NO_4$
341.16271
RI: 2700
D: UHYAC
Coronary dilator
73
168
256
417
M+
522
500
11-Hydroxyandrosterone enol 3TMS
PS
LM/Q
$C_{28}H_{54}O_3Si_3$
522.33808
RI: 2705*
Biomolecule
73
191
204
217
361
600
700
Lactose 8TMS
PS
LS/Q
$C_{36}H_{86}O_{11}Si_8$
918.43244
RI: 2730*
Sugar
N: recorded up to 800 AMU

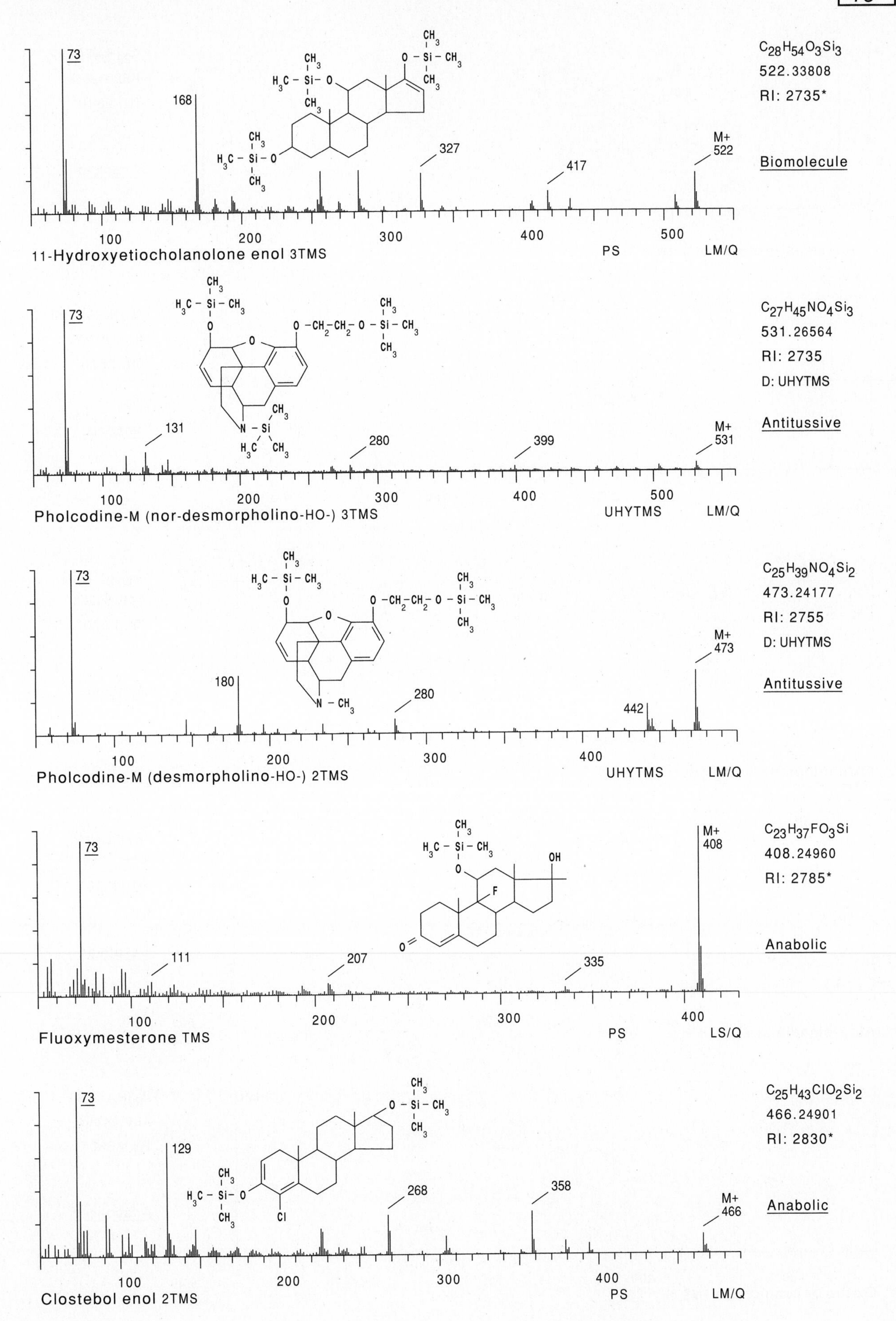

$C_{28}H_{54}O_3Si_3$
522.33808
RI: 2735*
Biomolecule

$C_{27}H_{45}NO_4Si_3$
531.26564
RI: 2735
D: UHYTMS
Antitussive

$C_{25}H_{39}NO_4Si_2$
473.24177
RI: 2755
D: UHYTMS
Antitussive

$C_{23}H_{37}FO_3Si$
408.24960
RI: 2785*
Anabolic

$C_{25}H_{43}ClO_2Si_2$
466.24901
RI: 2830*
Anabolic

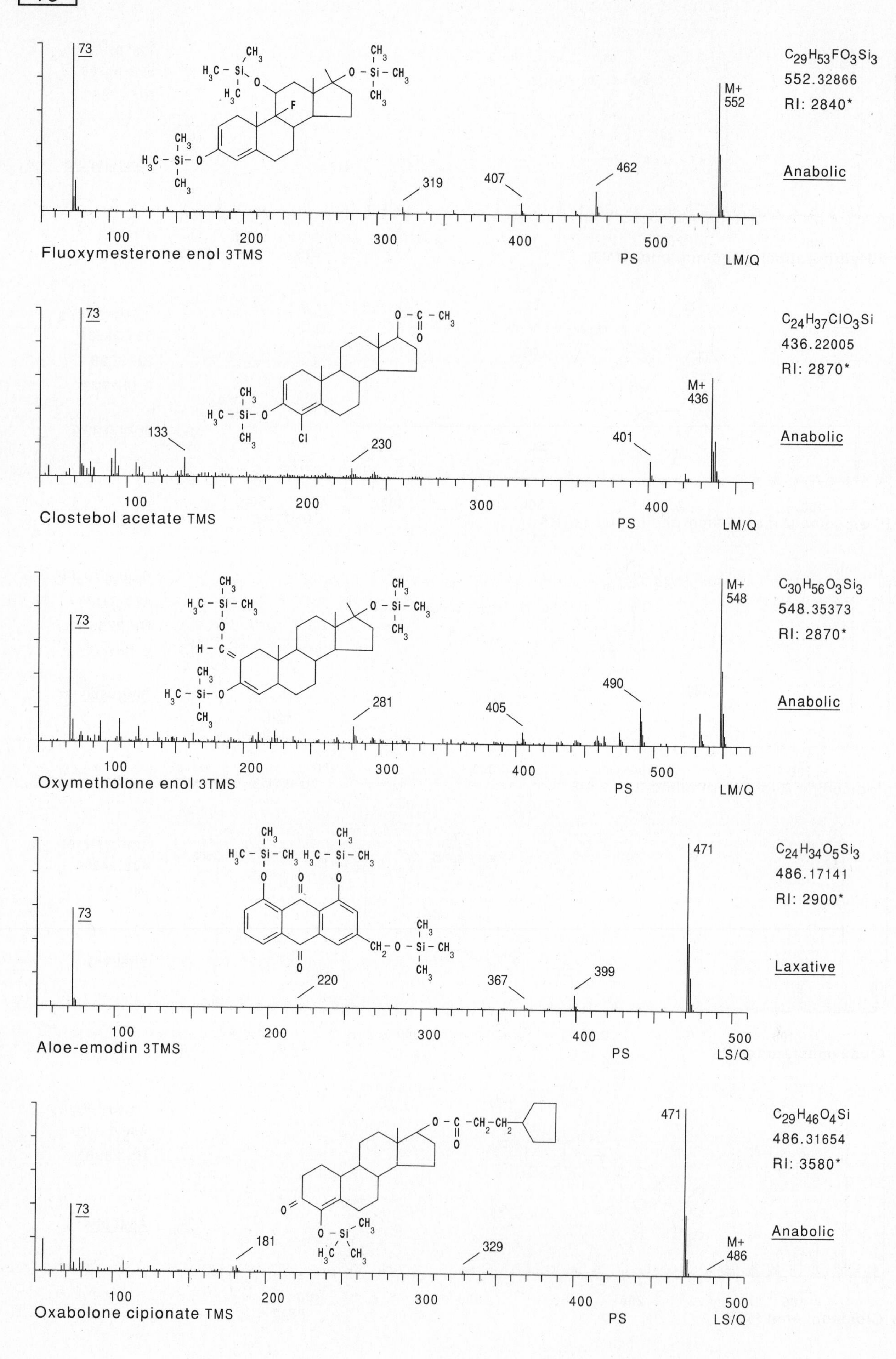
73
M+
552
319
407
462
100
200
300
400
500
Fluoxymesterone enol 3TMS
PS
LM/Q
C29H53FO3Si3
552.32866
RI: 2840*
Anabolic
73
133
230
401
M+
436
100
200
300
400
Clostebol acetate TMS
PS
LM/Q
C24H37ClO3Si
436.22005
RI: 2870*
Anabolic
73
M+
548
281
405
490
100
200
300
400
500
Oxymetholone enol 3TMS
PS
LM/Q
C30H56O3Si3
548.35373
RI: 2870*
Anabolic
73
471
220
367
399
100
200
300
400
500
Aloe-emodin 3TMS
PS
LS/Q
C24H34O5Si3
486.17141
RI: 2900*
Laxative
73
471
181
329
M+
486
100
200
300
400
500
Oxabolone cipionate TMS
PS
LS/Q
C29H46O4Si
486.31654
RI: 3580*
Anabolic

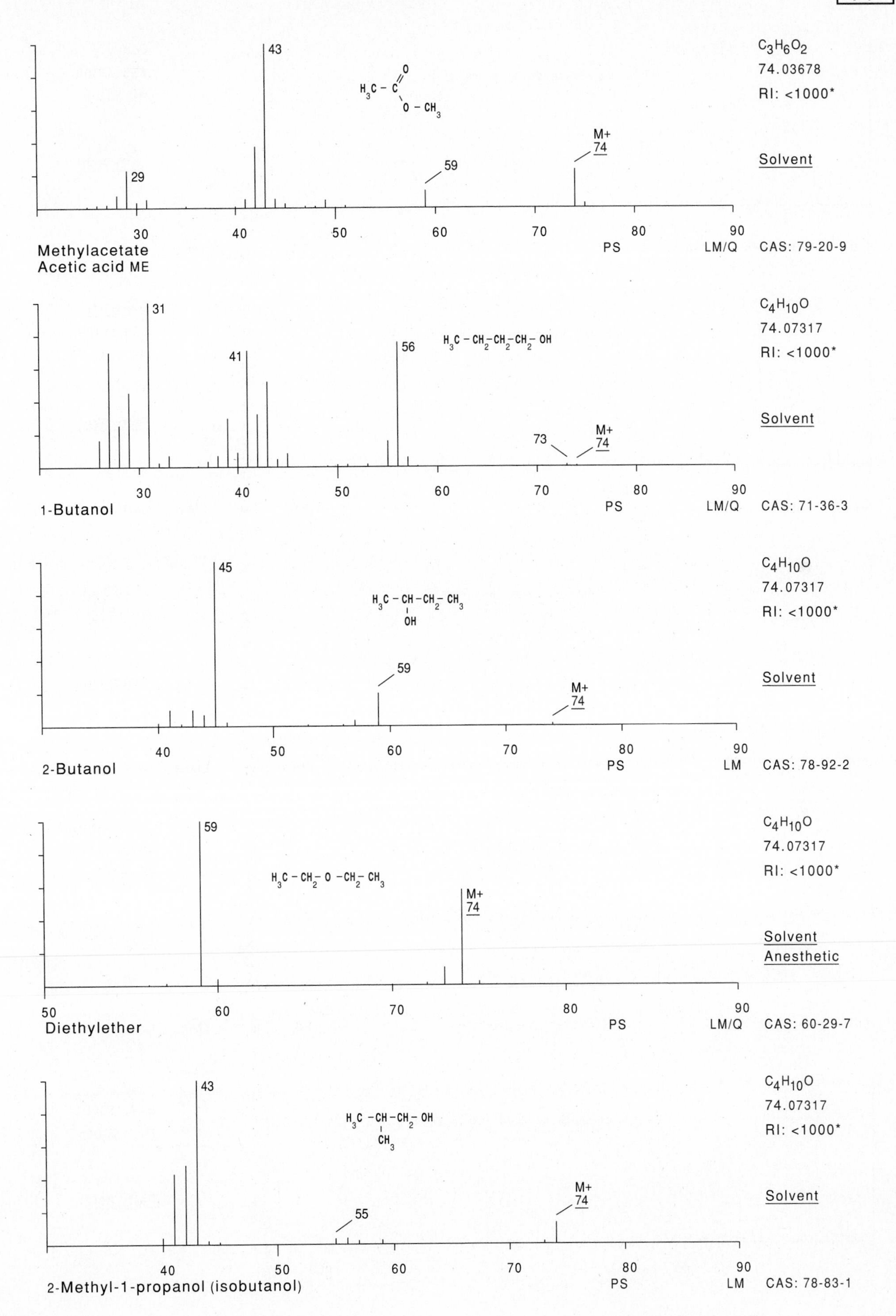

$C_3H_6O_2$
74.03678
RI: <1000*
Solvent
Methylacetate
Acetic acid ME
CAS: 79-20-9
$C_4H_{10}O$
74.07317
RI: <1000*
Solvent
1-Butanol
CAS: 71-36-3
$C_4H_{10}O$
74.07317
RI: <1000*
Solvent
2-Butanol
CAS: 78-92-2
$C_4H_{10}O$
74.07317
RI: <1000*
Solvent
Anesthetic
Diethylether
CAS: 60-29-7
$C_4H_{10}O$
74.07317
RI: <1000*
Solvent
2-Methyl-1-propanol (isobutanol)
CAS: 78-83-1

74

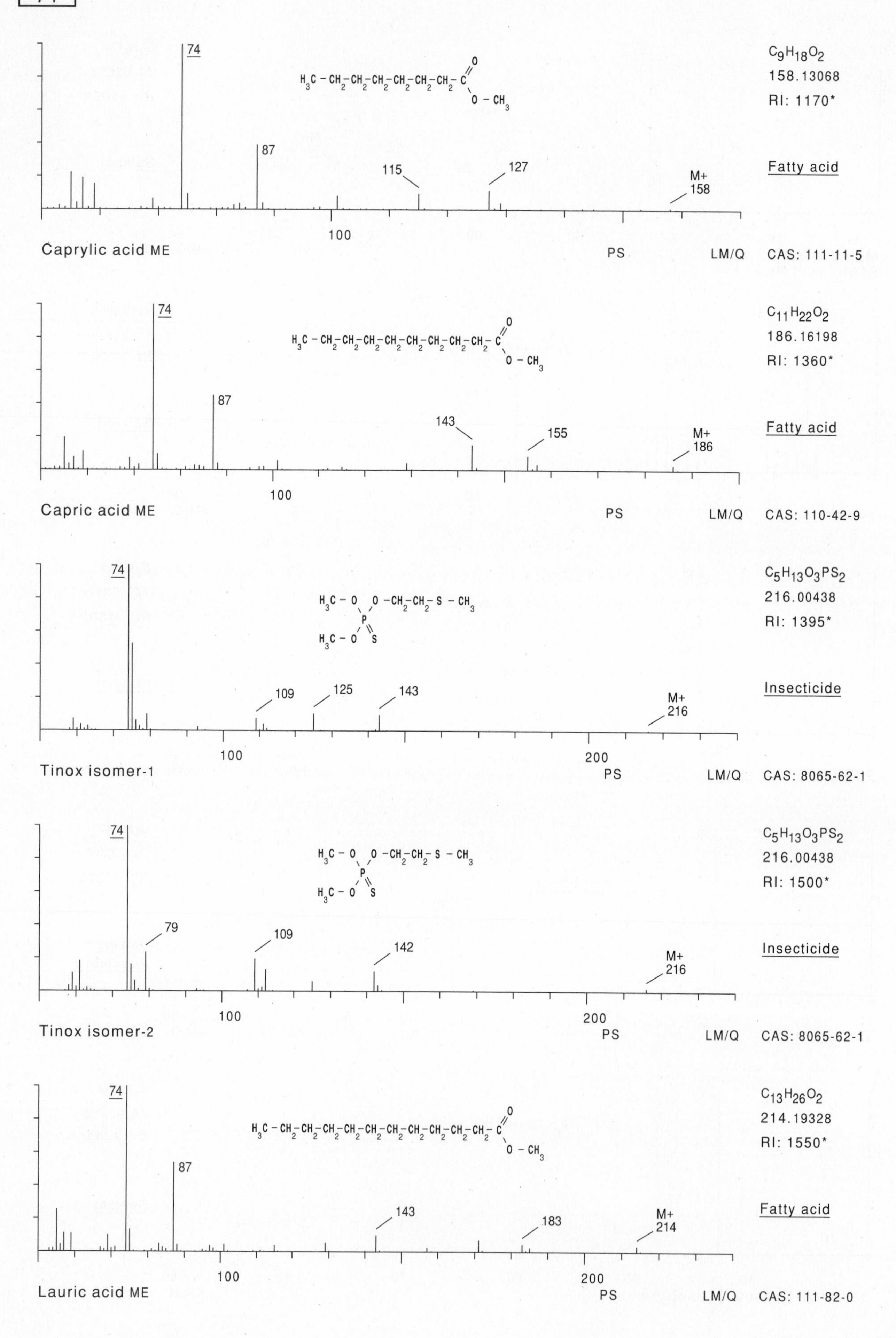

Caprylic acid ME

$C_9H_{18}O_2$
158.13068
RI: 1170*
Fatty acid
CAS: 111-11-5

Capric acid ME

$C_{11}H_{22}O_2$
186.16198
RI: 1360*
Fatty acid
CAS: 110-42-9

Tinox isomer-1

$C_5H_{13}O_3PS_2$
216.00438
RI: 1395*
Insecticide
CAS: 8065-62-1

Tinox isomer-2

$C_5H_{13}O_3PS_2$
216.00438
RI: 1500*
Insecticide
CAS: 8065-62-1

Lauric acid ME

$C_{13}H_{26}O_2$
214.19328
RI: 1550*
Fatty acid
CAS: 111-82-0

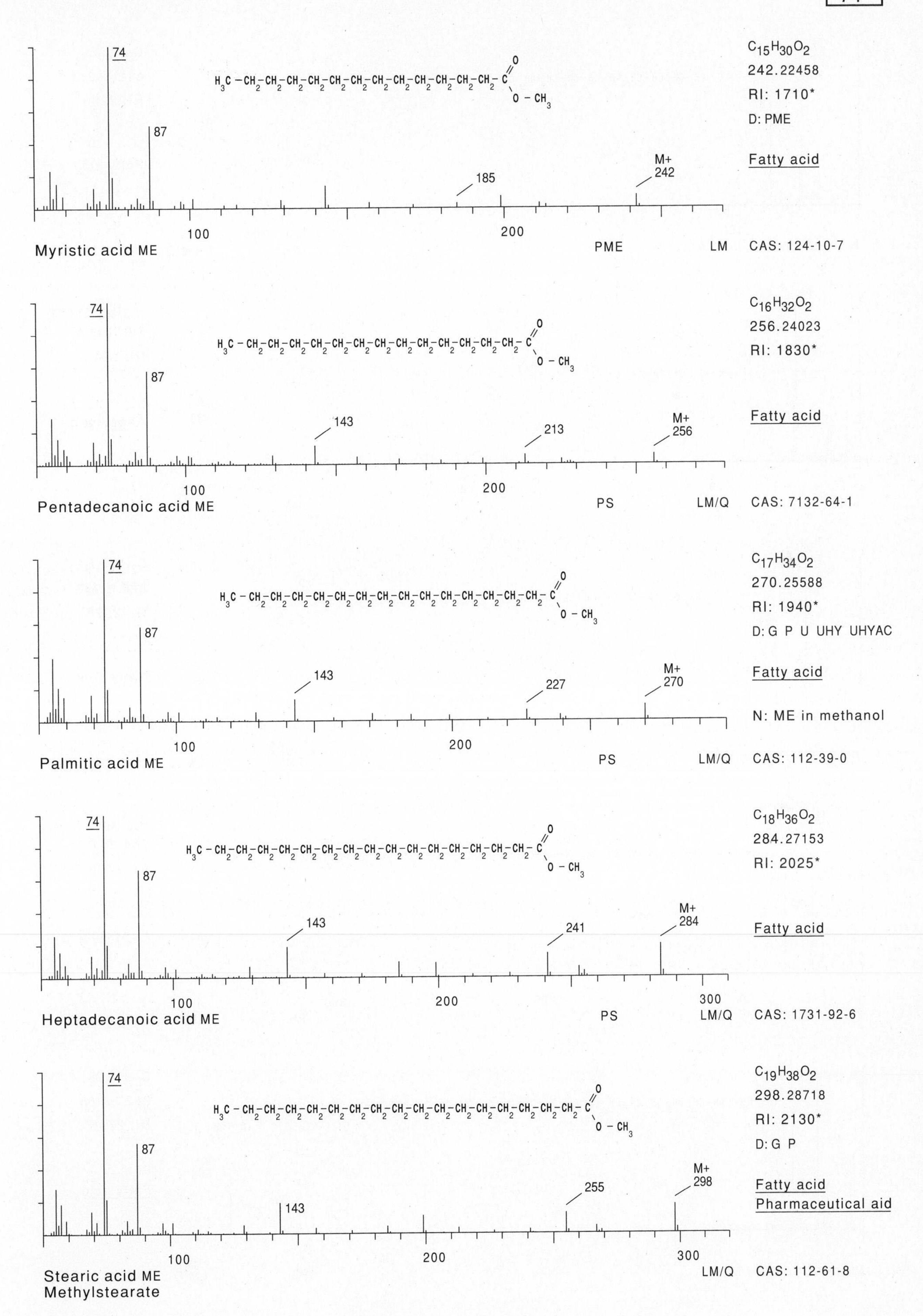

Myristic acid ME

$C_{15}H_{30}O_2$
242.22458
RI: 1710*
D: PME
Fatty acid
CAS: 124-10-7

Pentadecanoic acid ME

$C_{16}H_{32}O_2$
256.24023
RI: 1830*
Fatty acid
CAS: 7132-64-1

Palmitic acid ME

$C_{17}H_{34}O_2$
270.25588
RI: 1940*
D: G P U UHY UHYAC
Fatty acid
N: ME in methanol
CAS: 112-39-0

Heptadecanoic acid ME

$C_{18}H_{36}O_2$
284.27153
RI: 2025*
Fatty acid
CAS: 1731-92-6

Stearic acid ME
Methylstearate

$C_{19}H_{38}O_2$
298.28718
RI: 2130*
D: G P
Fatty acid
Pharmaceutical aid
CAS: 112-61-8

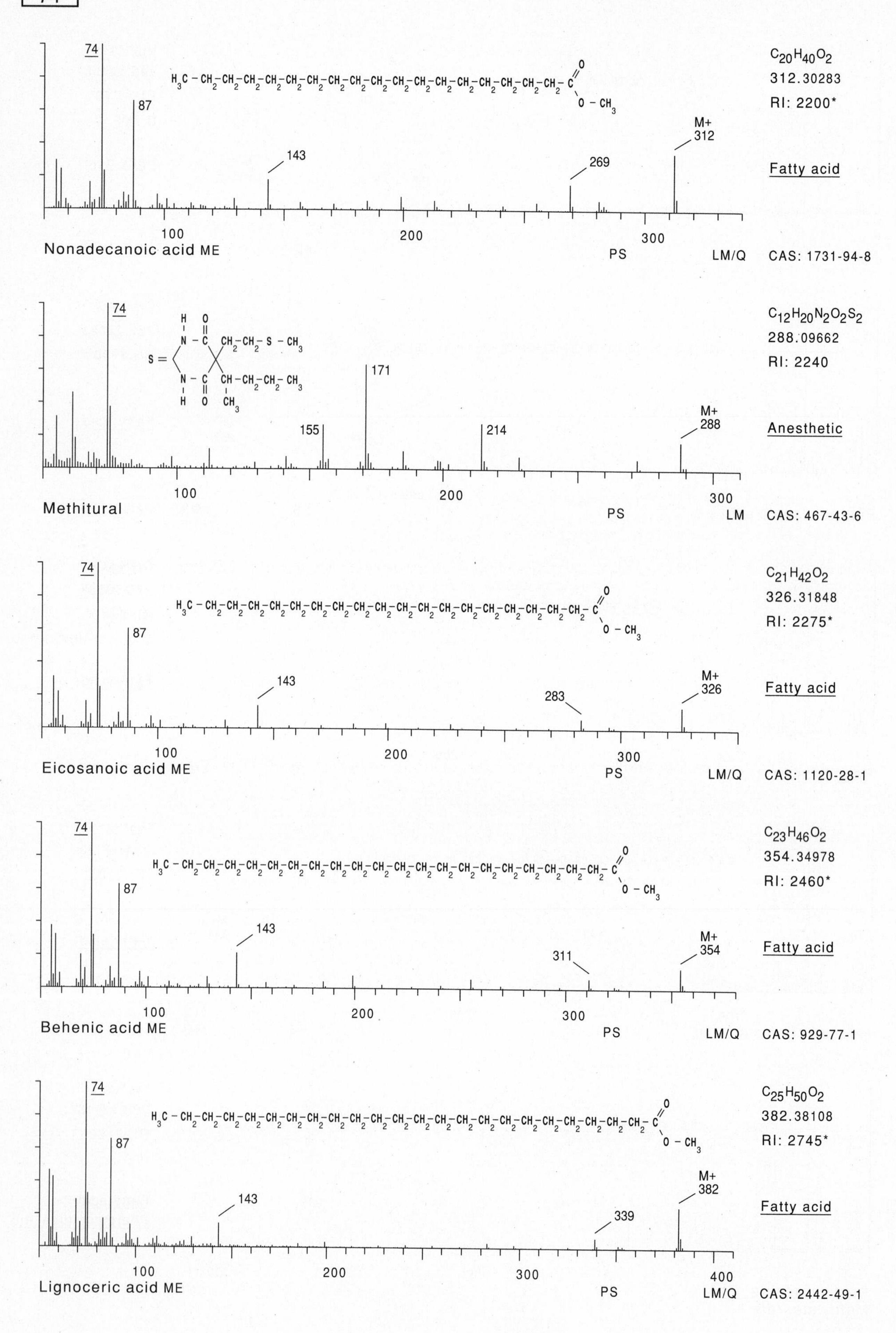
74
87
143
269
M+
312
100
200
300
PS
LM/Q
Nonadecanoic acid ME
$C_{20}H_{40}O_2$
312.30283
RI: 2200*
Fatty acid
CAS: 1731-94-8
74
155
171
214
M+
288
PS
LM
Methitural
$C_{12}H_{20}N_2O_2S_2$
288.09662
RI: 2240
Anesthetic
CAS: 467-43-6
74
87
143
283
M+
326
PS
LM/Q
Eicosanoic acid ME
$C_{21}H_{42}O_2$
326.31848
RI: 2275*
Fatty acid
CAS: 1120-28-1
74
87
143
311
M+
354
PS
LM/Q
Behenic acid ME
$C_{23}H_{46}O_2$
354.34978
RI: 2460*
Fatty acid
CAS: 929-77-1
74
87
143
339
M+
382
400
PS
LM/Q
Lignoceric acid ME
$C_{25}H_{50}O_2$
382.38108
RI: 2745*
Fatty acid
CAS: 2442-49-1

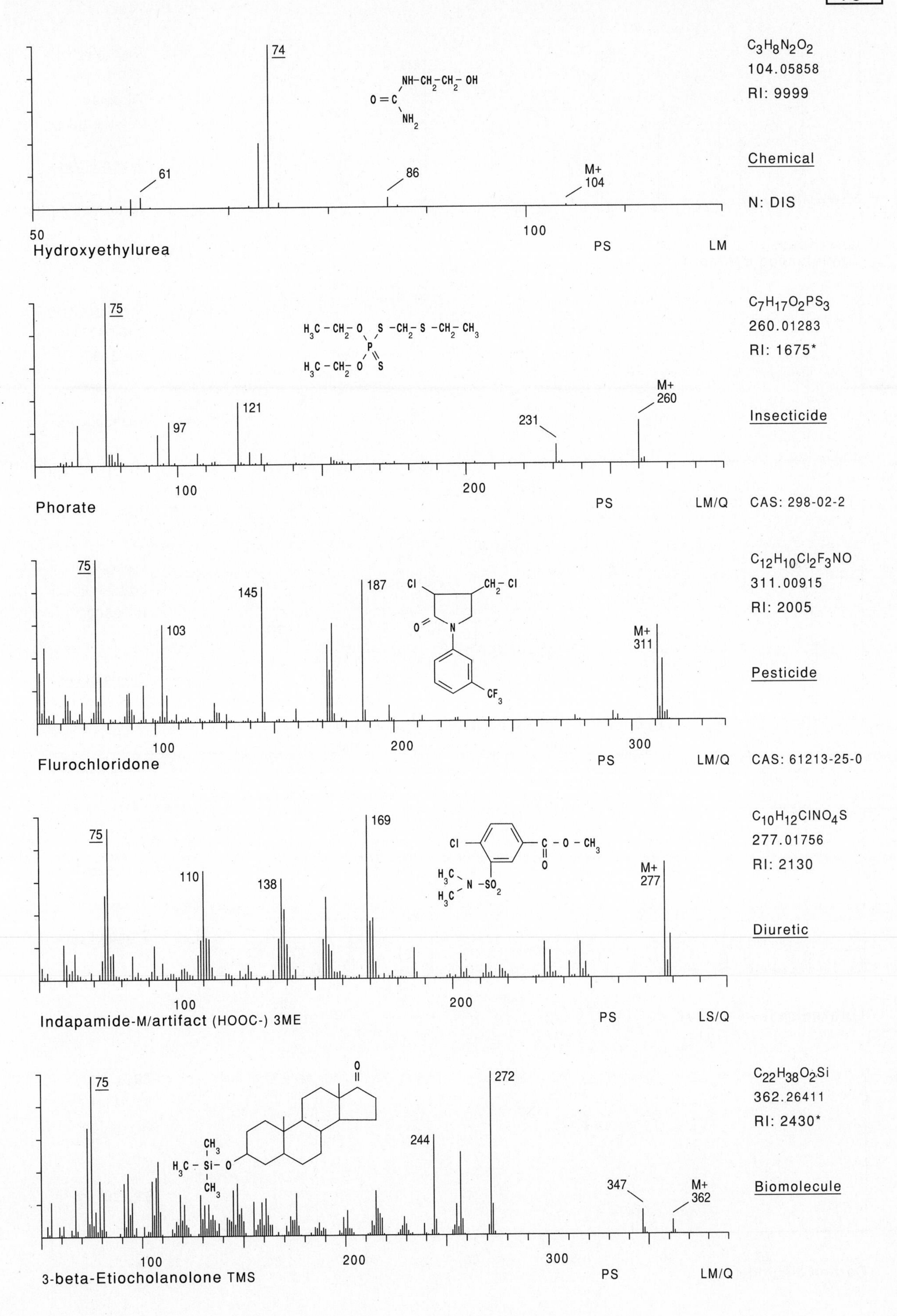
74
61
86
M+
104
50
100
Hydroxyethylurea
PS
LM
C3H8N2O2
104.05858
RI: 9999
Chemical
N: DIS
75
97
121
231
M+
260
100
200
Phorate
PS
LM/Q
C7H17O2PS3
260.01283
RI: 1675*
Insecticide
CAS: 298-02-2
75
145
187
103
M+
311
100
200
300
Flurochloridone
PS
LM/Q
C12H10Cl2F3NO
311.00915
RI: 2005
Pesticide
CAS: 61213-25-0
169
75
110
138
M+
277
100
200
Indapamide-M/artifact (HOOC-) 3ME
PS
LS/Q
C10H12ClNO4S
277.01756
RI: 2130
Diuretic
75
272
244
347
M+
362
100
200
300
3-beta-Etiocholanolone TMS
PS
LM/Q
C22H38O2Si
362.26411
RI: 2430*
Biomolecule

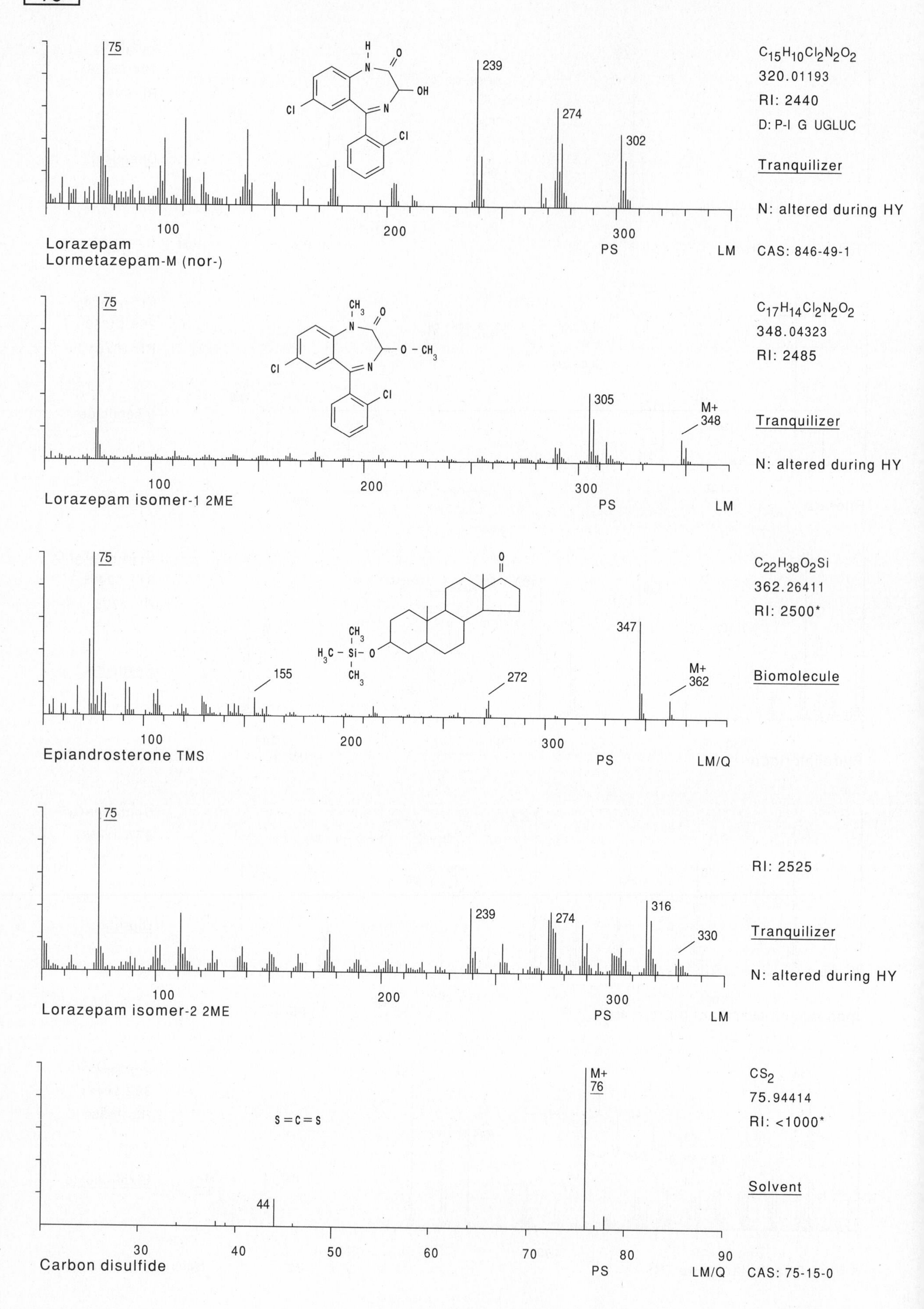

75
239
274
302
Cl
OH
100
200
300
Lorazepam
Lormetazepam-M (nor-)
PS
LM
$C_{15}H_{10}Cl_2N_2O_2$
320.01193
RI: 2440
D: P-I G UGLUC
Tranquilizer
N: altered during HY
CAS: 846-49-1
75
305
M+
348
Lorazepam isomer-1 2ME
$C_{17}H_{14}Cl_2N_2O_2$
348.04323
RI: 2485
Tranquilizer
N: altered during HY
75
155
272
347
M+
362
Epiandrosterone TMS
LM/Q
$C_{22}H_{38}O_2Si$
362.26411
RI: 2500*
Biomolecule
75
239
274
316
330
Lorazepam isomer-2 2ME
RI: 2525
Tranquilizer
N: altered during HY
M+
76
44
S=C=S
30
40
50
60
70
80
90
Carbon disulfide
CS_2
75.94414
RI: <1000*
Solvent
CAS: 75-15-0

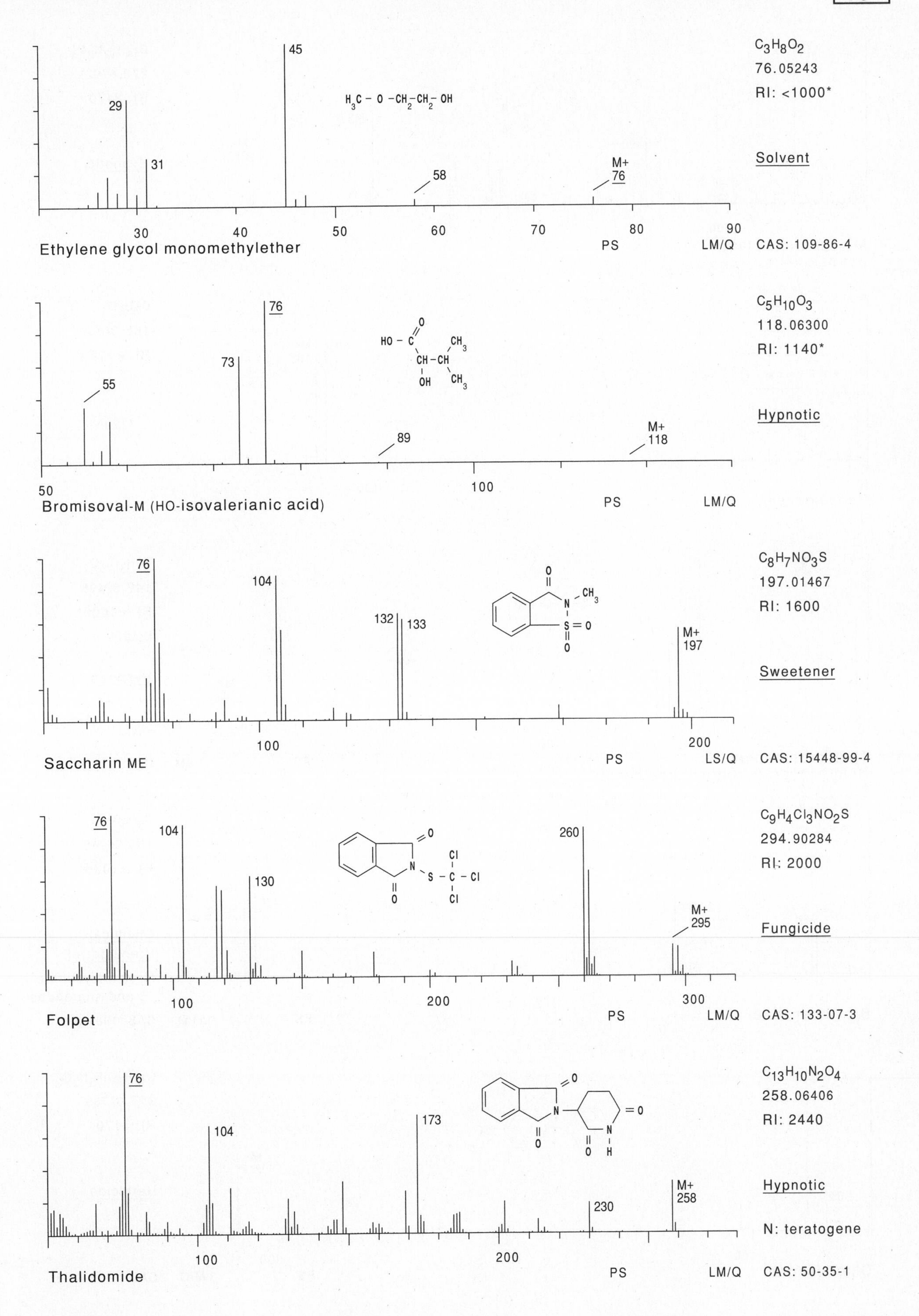
C3H8O2
76.05243
RI: <1000*
Solvent
Ethylene glycol monomethylether
CAS: 109-86-4
C5H10O3
118.06300
RI: 1140*
Hypnotic
Bromisoval-M (HO-isovalerianic acid)
C8H7NO3S
197.01467
RI: 1600
Sweetener
Saccharin ME
CAS: 15448-99-4
C9H4Cl3NO2S
294.90284
RI: 2000
Fungicide
Folpet
CAS: 133-07-3
C13H10N2O4
258.06406
RI: 2440
Hypnotic
N: teratogene
Thalidomide
CAS: 50-35-1

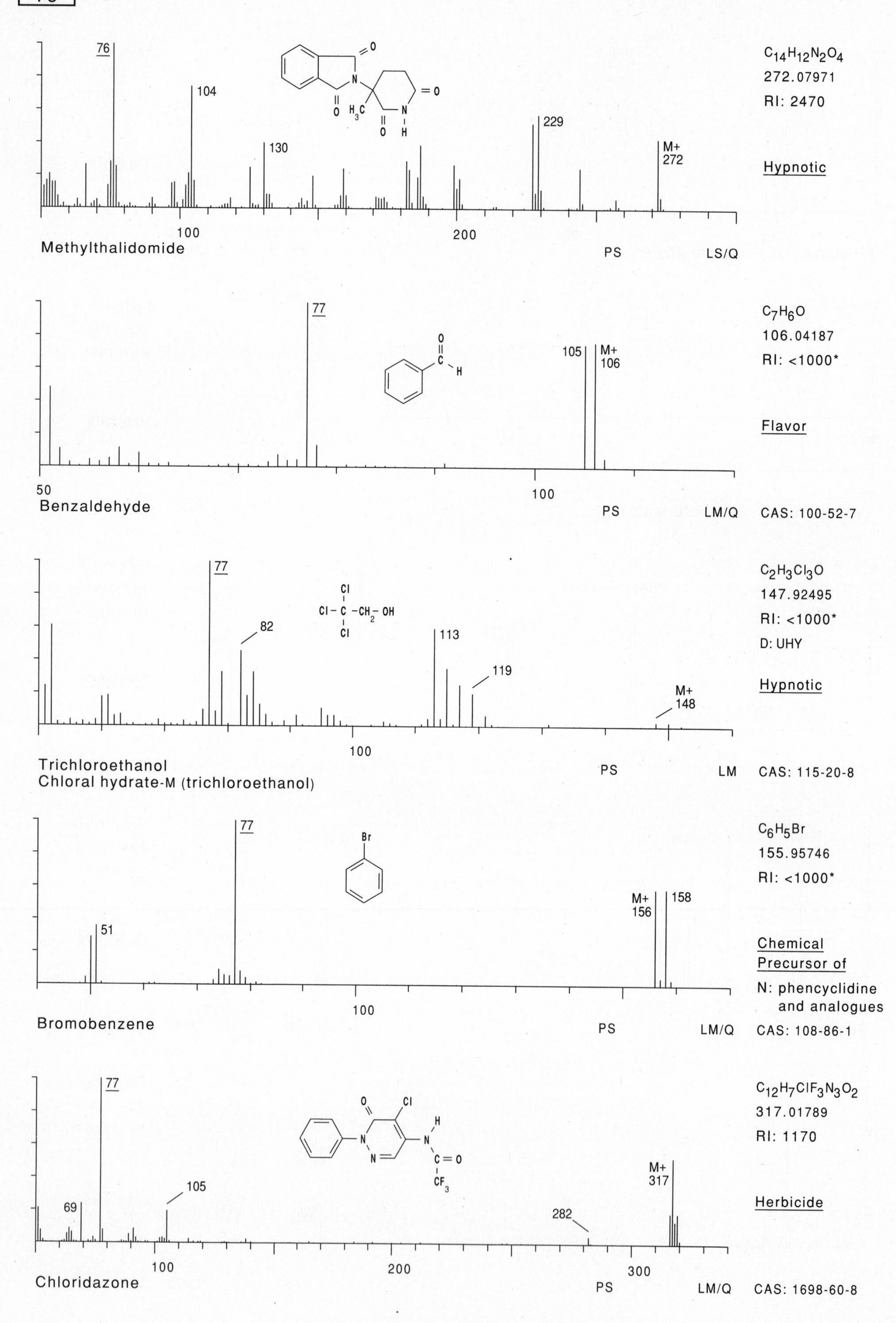

Methylthalidomide
76
104
130
229
M+
272
100
200
PS
LS/Q
C14H12N2O4
272.07971
RI: 2470
Hypnotic
Benzaldehyde
77
105
M+
106
50
100
PS
LM/Q
C7H6O
106.04187
RI: <1000*
Flavor
CAS: 100-52-7
Trichloroethanol
Chloral hydrate-M (trichloroethanol)
77
82
113
119
M+
148
100
PS
LM
C2H3Cl3O
147.92495
RI: <1000*
D: UHY
Hypnotic
CAS: 115-20-8
Bromobenzene
51
77
M+
156
158
100
PS
LM/Q
C6H5Br
155.95746
RI: <1000*
Chemical
Precursor of
N: phencyclidine
and analogues
CAS: 108-86-1
Chloridazone
69
77
105
282
M+
317
100
200
300
PS
LM/Q
C12H7ClF3N3O2
317.01789
RI: 1170
Herbicide
CAS: 1698-60-8

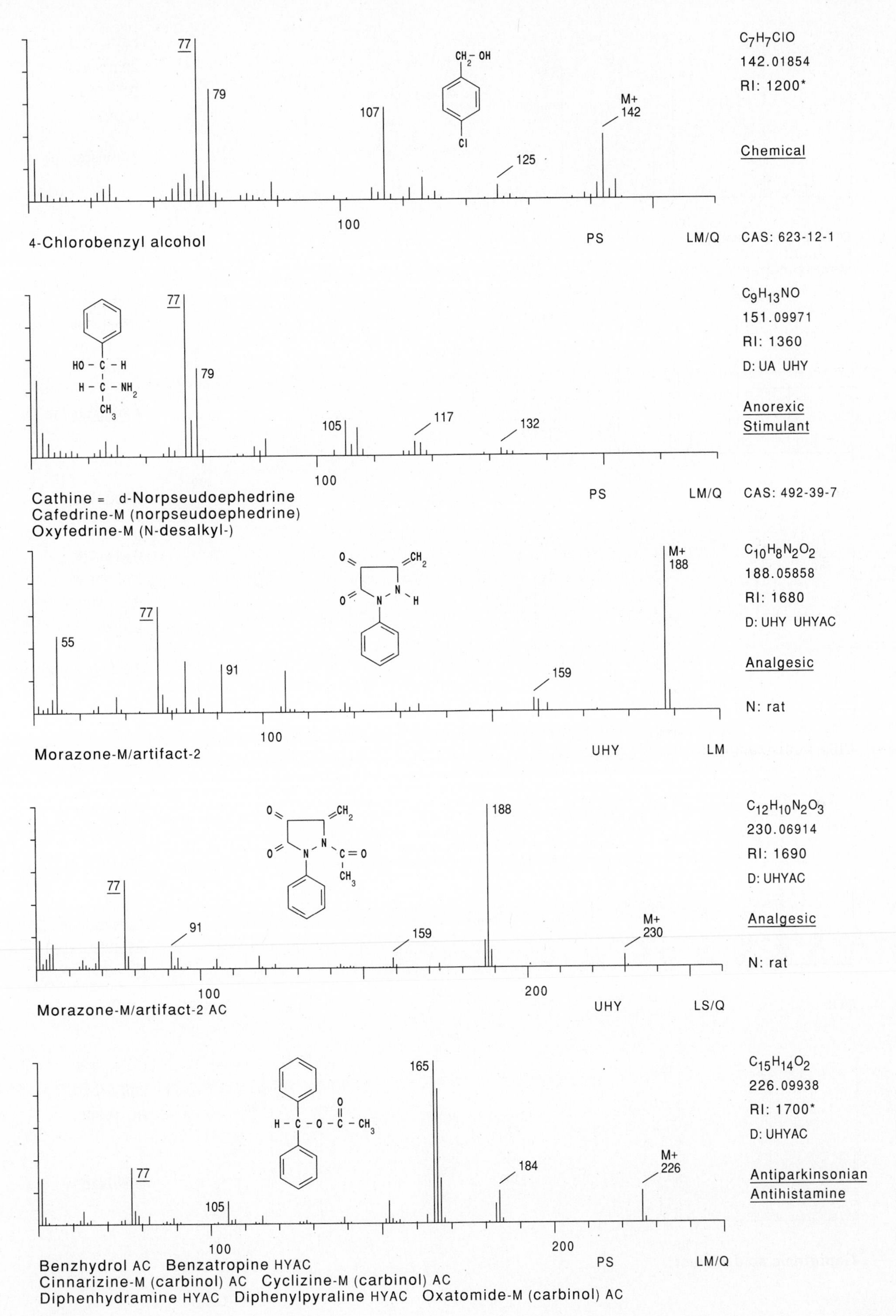
C7H7ClO
142.01854
RI: 1200*
Chemical
CAS: 623-12-1
4-Chlorobenzyl alcohol
PS
LM/Q
77
79
107
125
M+
142
100
CH2- OH
Cl
C9H13NO
151.09971
RI: 1360
D: UA UHY
Anorexic
Stimulant
CAS: 492-39-7
Cathine = d-Norpseudoephedrine
Cafedrine-M (norpseudoephedrine)
Oxyfedrine-M (N-desalkyl-)
PS
LM/Q
77
79
105
117
132
100
HO - C - H
H - C - NH2
CH3
C10H8N2O2
188.05858
RI: 1680
D: UHY UHYAC
Analgesic
N: rat
Morazone-M/artifact-2
UHY
LM
55
77
91
159
M+
188
100
C12H10N2O3
230.06914
RI: 1690
D: UHYAC
Analgesic
N: rat
Morazone-M/artifact-2 AC
UHY
LS/Q
77
91
159
188
M+
230
100
200
C15H14O2
226.09938
RI: 1700*
D: UHYAC
Antiparkinsonian
Antihistamine
Benzhydrol AC Benzatropine HYAC
Cinnarizine-M (carbinol) AC Cyclizine-M (carbinol) AC
Diphenhydramine HYAC Diphenylpyraline HYAC Oxatomide-M (carbinol) AC
PS
LM/Q
77
105
165
184
M+
226
100
200
H - C - O - C - CH3

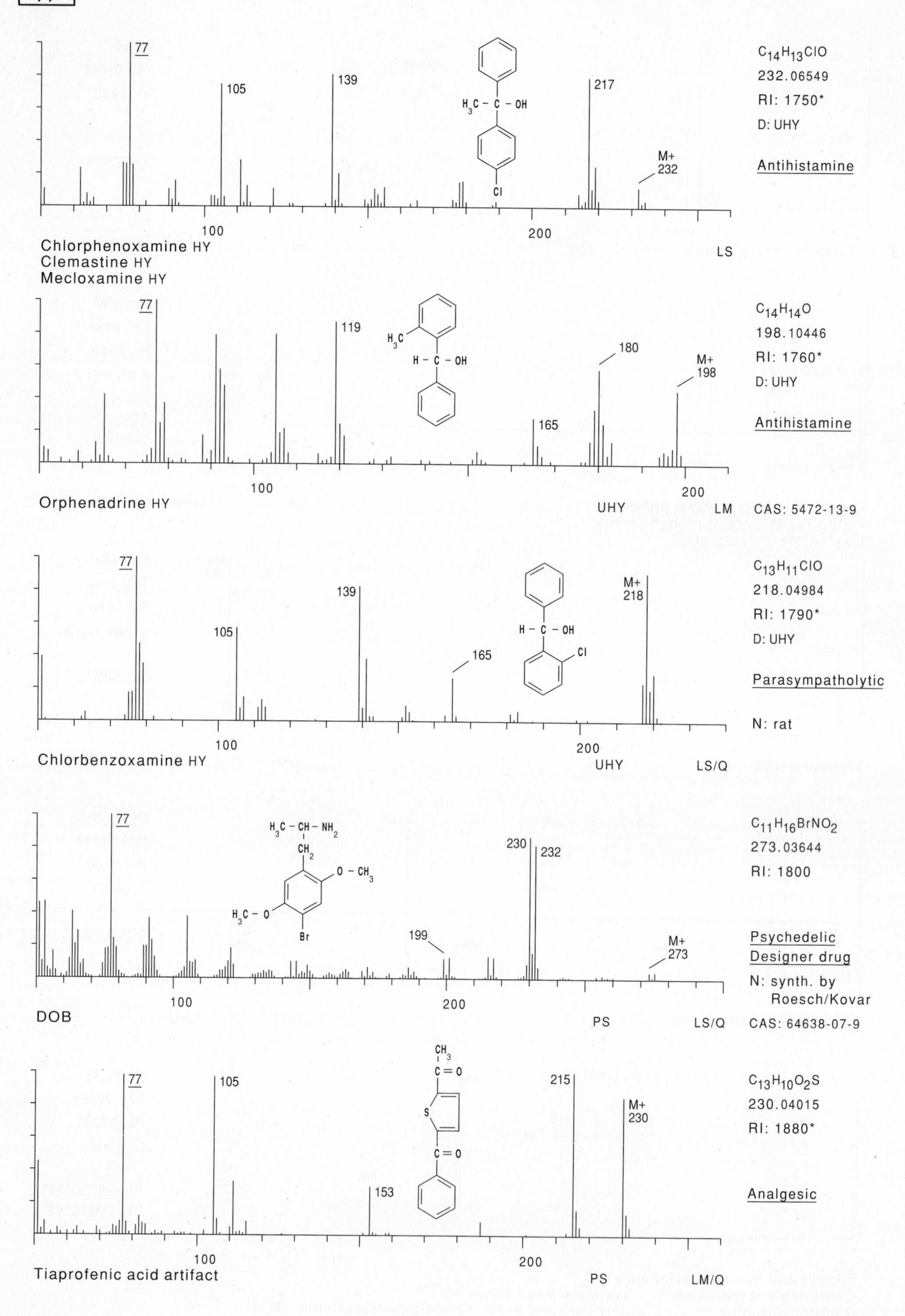
77
105
139
217
M+
232
H3C - C - OH
Cl
100
200
Chlorphenoxamine HY
Clemastine HY
Mecloxamine HY
LS
C14H13ClO
232.06549
RI: 1750*
D: UHY
Antihistamine
77
119
180
M+
198
165
H3C
H - C - OH
100
200
Orphenadrine HY
UHY
LM
C14H14O
198.10446
RI: 1760*
D: UHY
Antihistamine
CAS: 5472-13-9
77
139
105
165
M+
218
H - C - OH
Cl
100
200
Chlorbenzoxamine HY
UHY
LS/Q
C13H11ClO
218.04984
RI: 1790*
D: UHY
Parasympatholytic
N: rat
77
H3C - CH - NH2
CH2
O - CH3
H3C - O
Br
230
232
199
M+
273
100
200
DOB
PS
LS/Q
C11H16BrNO2
273.03644
RI: 1800
Psychedelic
Designer drug
N: synth. by
Roesch/Kovar
CAS: 64638-07-9
77
105
CH3
C = O
S
C = O
215
M+
230
153
100
200
Tiaprofenic acid artifact
PS
LM/Q
C13H10O2S
230.04015
RI: 1880*
Analgesic

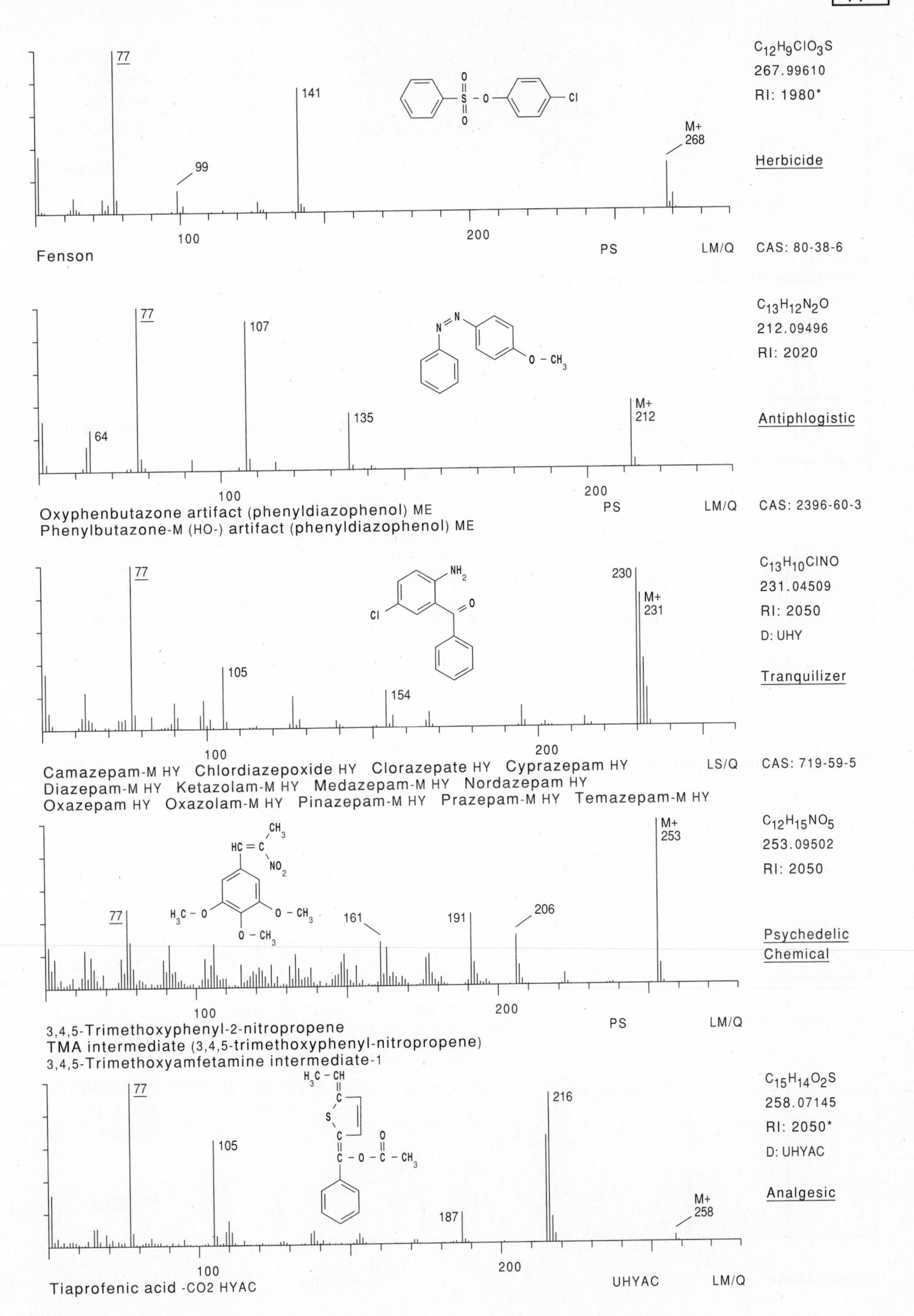

$C_{12}H_9ClO_3S$
267.99610
RI: 1980*

Herbicide

Fenson

CAS: 80-38-6

$C_{13}H_{12}N_2O$
212.09496
RI: 2020

Antiphlogistic

Oxyphenbutazone artifact (phenyldiazophenol) ME
Phenylbutazone-M (HO-) artifact (phenyldiazophenol) ME

CAS: 2396-60-3

$C_{13}H_{10}ClNO$
231.04509
RI: 2050
D: UHY

Tranquilizer

Camazepam-M HY Chlordiazepoxide HY Clorazepate HY Cyprazepam HY
Diazepam-M HY Ketazolam-M HY Medazepam-M HY Nordazepam HY
Oxazepam HY Oxazolam-M HY Pinazepam-M HY Prazepam-M HY Temazepam-M HY

CAS: 719-59-5

$C_{12}H_{15}NO_5$
253.09502
RI: 2050

Psychedelic Chemical

3,4,5-Trimethoxyphenyl-2-nitropropene
TMA intermediate (3,4,5-trimethoxyphenyl-nitropropene)
3,4,5-Trimethoxyamfetamine intermediate-1

$C_{15}H_{14}O_2S$
258.07145
RI: 2050*
D: UHYAC

Analgesic

Tiaprofenic acid -CO2 HYAC

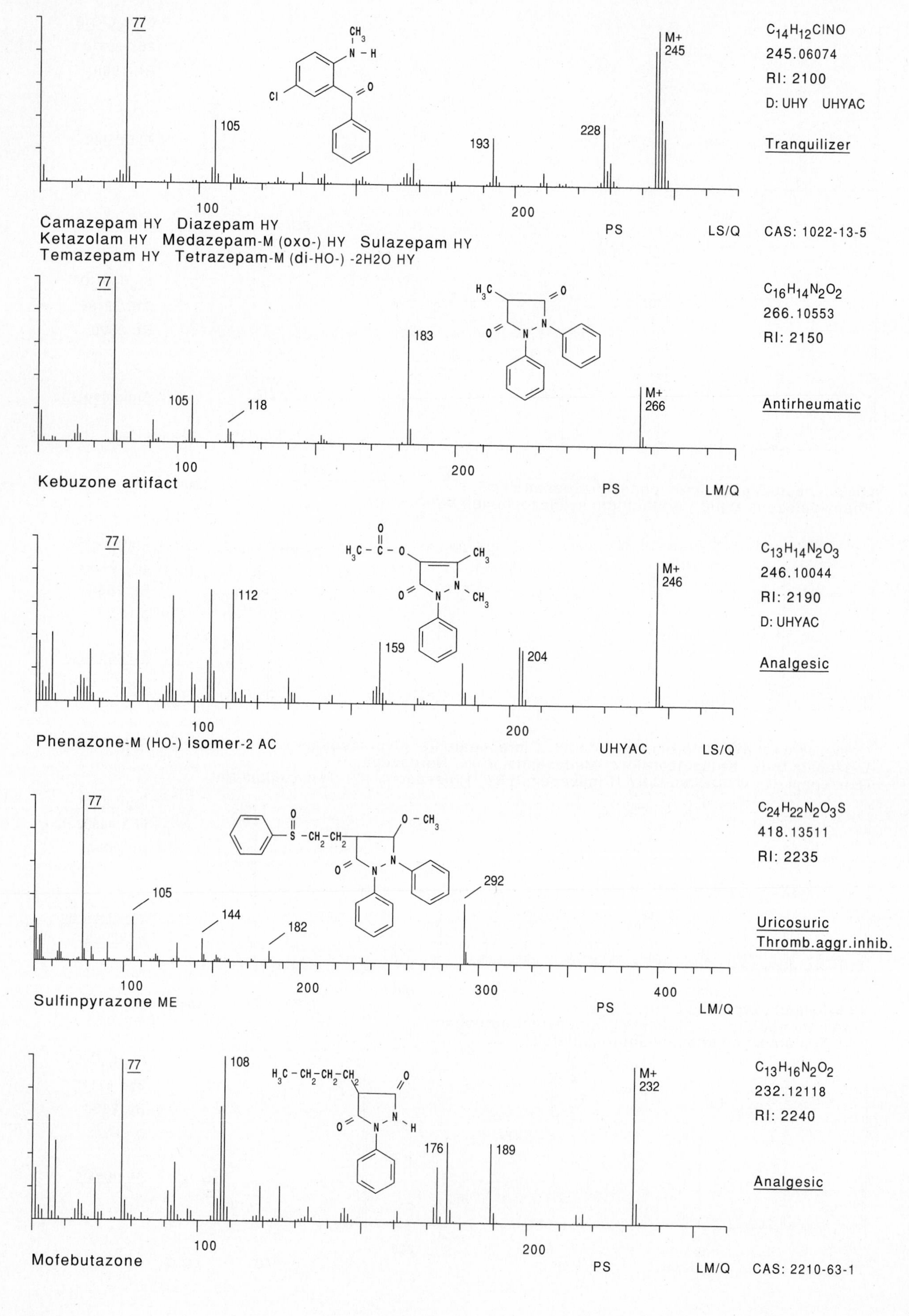

C14H12ClNO
245.06074
RI: 2100
D: UHY UHYAC
Tranquilizer
Camazepam HY Diazepam HY
Ketazolam HY Medazepam-M (oxo-) HY Sulazepam HY
Temazepam HY Tetrazepam-M (di-HO-) -2H2O HY
PS LS/Q
CAS: 1022-13-5
C16H14N2O2
266.10553
RI: 2150
Antirheumatic
Kebuzone artifact
PS LM/Q
C13H14N2O3
246.10044
RI: 2190
D: UHYAC
Analgesic
Phenazone-M (HO-) isomer-2 AC
UHYAC LS/Q
C24H22N2O3S
418.13511
RI: 2235
Uricosuric
Thromb.aggr.inhib.
Sulfinpyrazone ME
PS LM/Q
C13H16N2O2
232.12118
RI: 2240
Analgesic
Mofebutazone
PS LM/Q
CAS: 2210-63-1

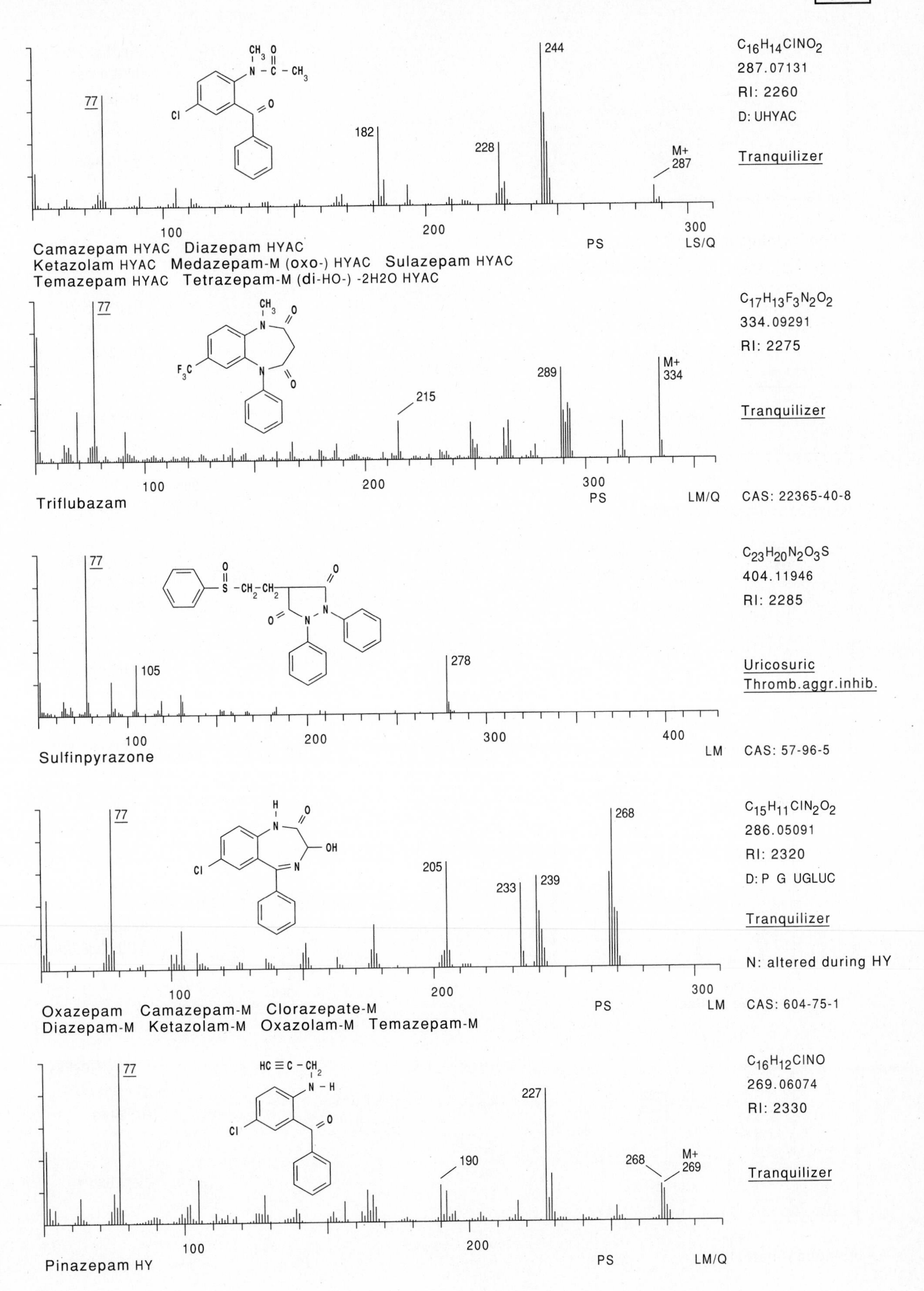

C16H14ClNO2
287.07131
RI: 2260
D: UHYAC
Tranquilizer
Camazepam HYAC Diazepam HYAC
Ketazolam HYAC Medazepam-M (oxo-) HYAC Sulazepam HYAC
Temazepam HYAC Tetrazepam-M (di-HO-) -2H2O HYAC
PS
LS/Q
C17H13F3N2O2
334.09291
RI: 2275
Tranquilizer
Triflubazam
PS
LM/Q
CAS: 22365-40-8
C23H20N2O3S
404.11946
RI: 2285
Uricosuric
Thromb.aggr.inhib.
Sulfinpyrazone
LM
CAS: 57-96-5
C15H11ClN2O2
286.05091
RI: 2320
D: P G UGLUC
Tranquilizer
N: altered during HY
Oxazepam Camazepam-M Clorazepate-M
Diazepam-M Ketazolam-M Oxazolam-M Temazepam-M
PS
LM
CAS: 604-75-1
C16H12ClNO
269.06074
RI: 2330
Tranquilizer
Pinazepam HY
PS
LM/Q

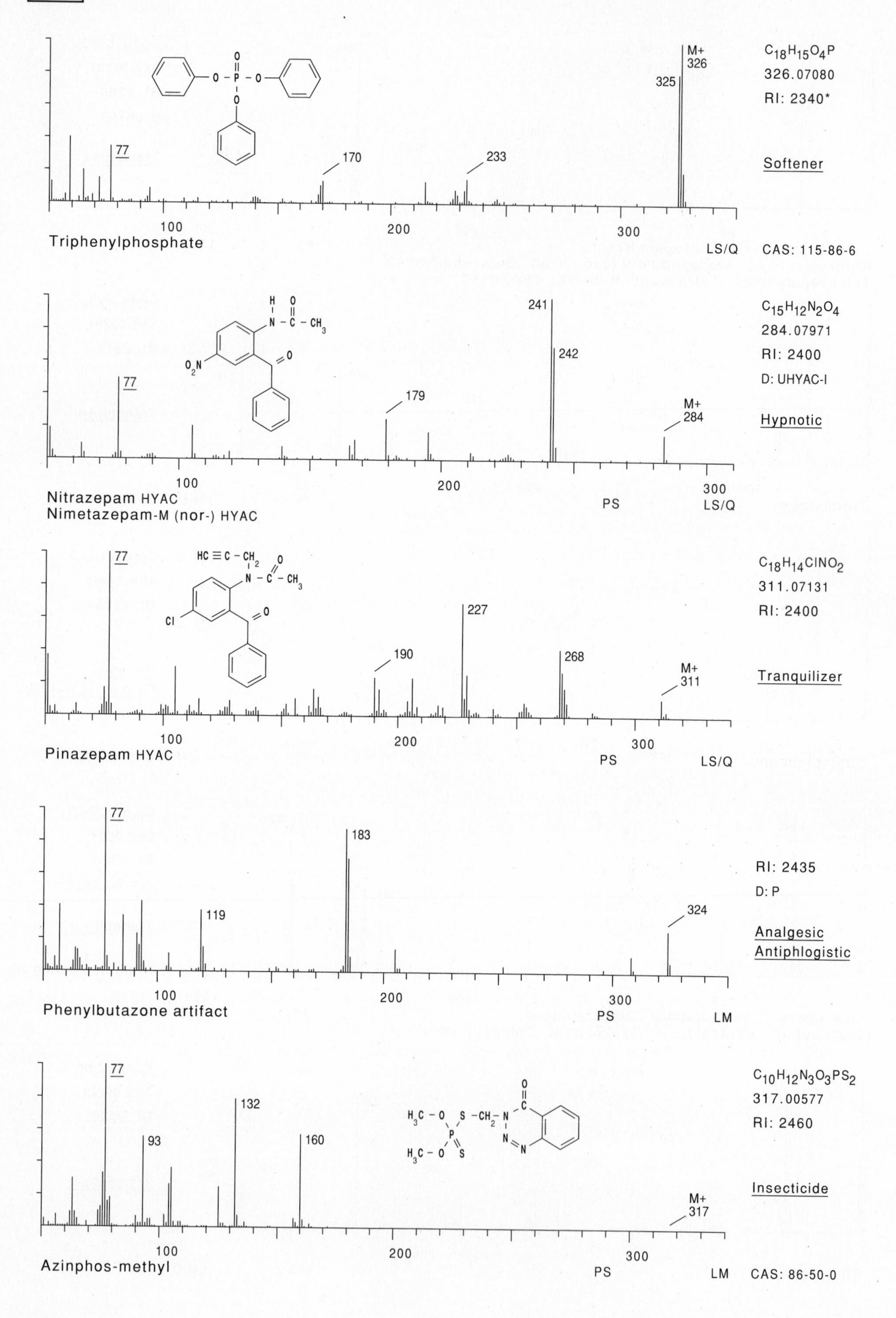

77
170
233
M+
326
325
100
200
300
Triphenylphosphate
LS/Q
C18H15O4P
326.07080
RI: 2340*
Softener
CAS: 115-86-6
77
179
241
242
M+
284
100
200
300
Nitrazepam HYAC
Nimetazepam-M (nor-) HYAC
PS
LS/Q
C15H12N2O4
284.07971
RI: 2400
D: UHYAC-I
Hypnotic
77
190
227
268
M+
311
100
200
300
Pinazepam HYAC
PS
LS/Q
C18H14ClNO2
311.07131
RI: 2400
Tranquilizer
77
119
183
324
100
200
300
Phenylbutazone artifact
PS
LM
RI: 2435
D: P
Analgesic
Antiphlogistic
77
93
132
160
M+
317
100
200
300
Azinphos-methyl
PS
LM
C10H12N3O3PS2
317.00577
RI: 2460
Insecticide
CAS: 86-50-0

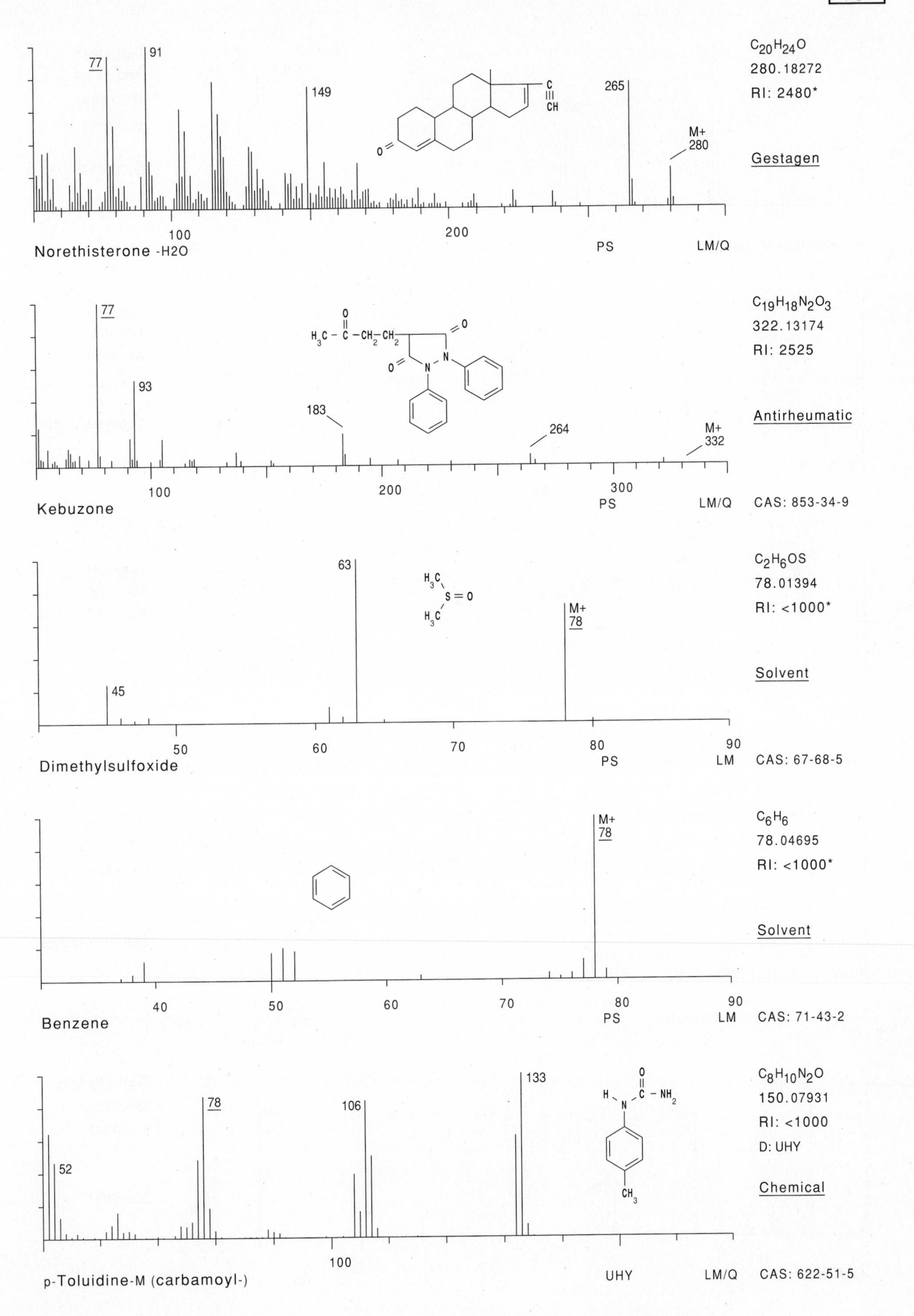

77
91
149
265
M+
280
100
200
Norethisterone -H2O
PS
LM/Q
C20H24O
280.18272
RI: 2480*
Gestagen
77
93
183
264
M+
332
100
200
300
Kebuzone
PS
LM/Q
C19H18N2O3
322.13174
RI: 2525
Antirheumatic
CAS: 853-34-9
63
45
M+
78
50
60
70
80
90
Dimethylsulfoxide
PS
LM
C2H6OS
78.01394
RI: <1000*
Solvent
CAS: 67-68-5
M+
78
40
50
60
70
80
90
Benzene
PS
LM
C6H6
78.04695
RI: <1000*
Solvent
CAS: 71-43-2
133
78
106
52
100
p-Toluidine-M (carbamoyl-)
UHY
LM/Q
C8H10N2O
150.07931
RI: <1000
D: UHY
Chemical
CAS: 622-51-5

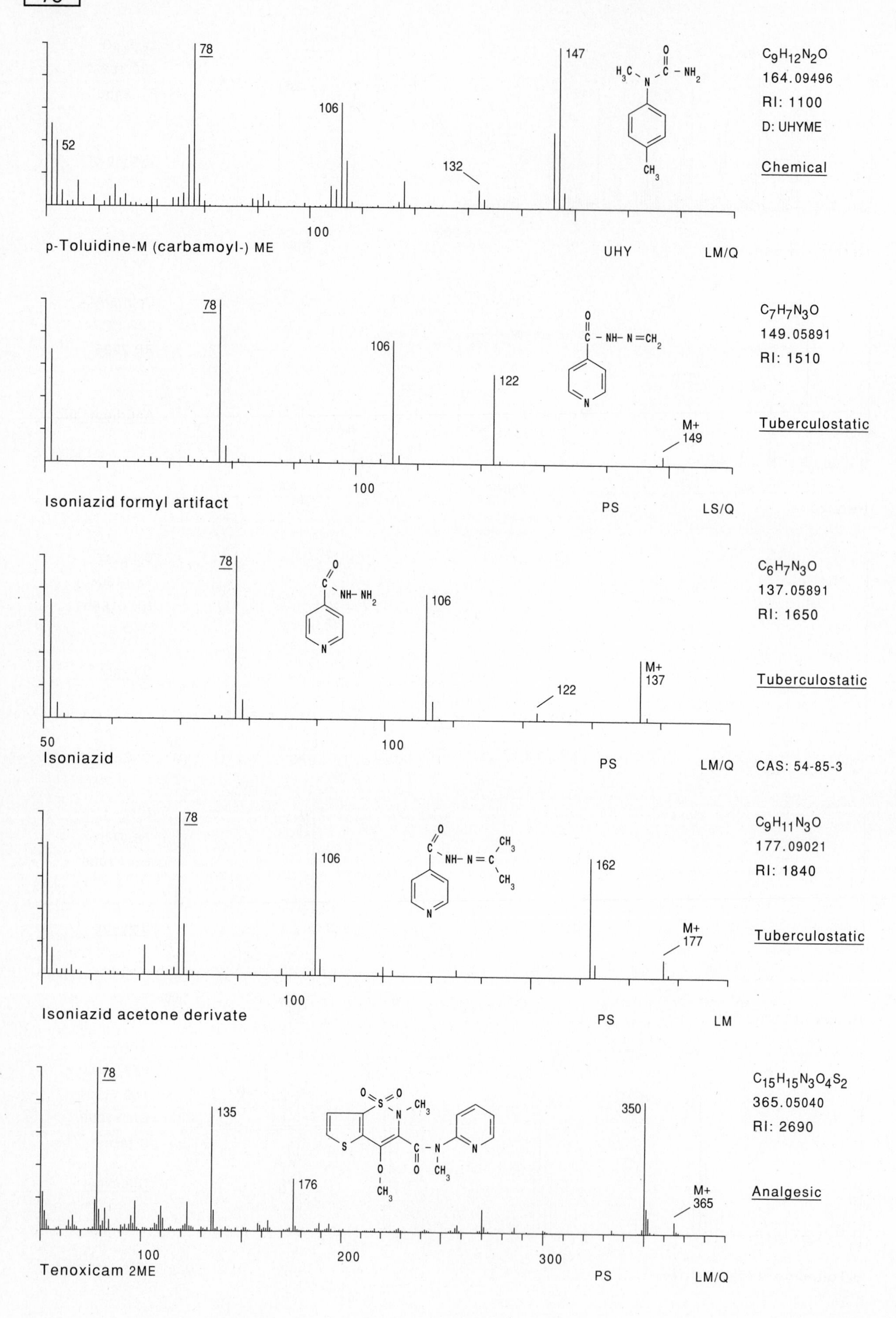
78
147
106
52
132
100
$C_9H_{12}N_2O$
164.09496
RI: 1100
D: UHYME
Chemical
p-Toluidine-M (carbamoyl-) ME
UHY
LM/Q
78
106
122
M+
149
100
$C_7H_7N_3O$
149.05891
RI: 1510
Tuberculostatic
Isoniazid formyl artifact
PS
LS/Q
78
106
122
M+
137
50
100
$C_6H_7N_3O$
137.05891
RI: 1650
Tuberculostatic
Isoniazid
PS
LM/Q
CAS: 54-85-3
78
106
162
M+
177
100
$C_9H_{11}N_3O$
177.09021
RI: 1840
Tuberculostatic
Isoniazid acetone derivate
PS
LM
78
135
176
350
M+
365
100
200
300
$C_{15}H_{15}N_3O_4S_2$
365.05040
RI: 2690
Analgesic
Tenoxicam 2ME
PS
LM/Q

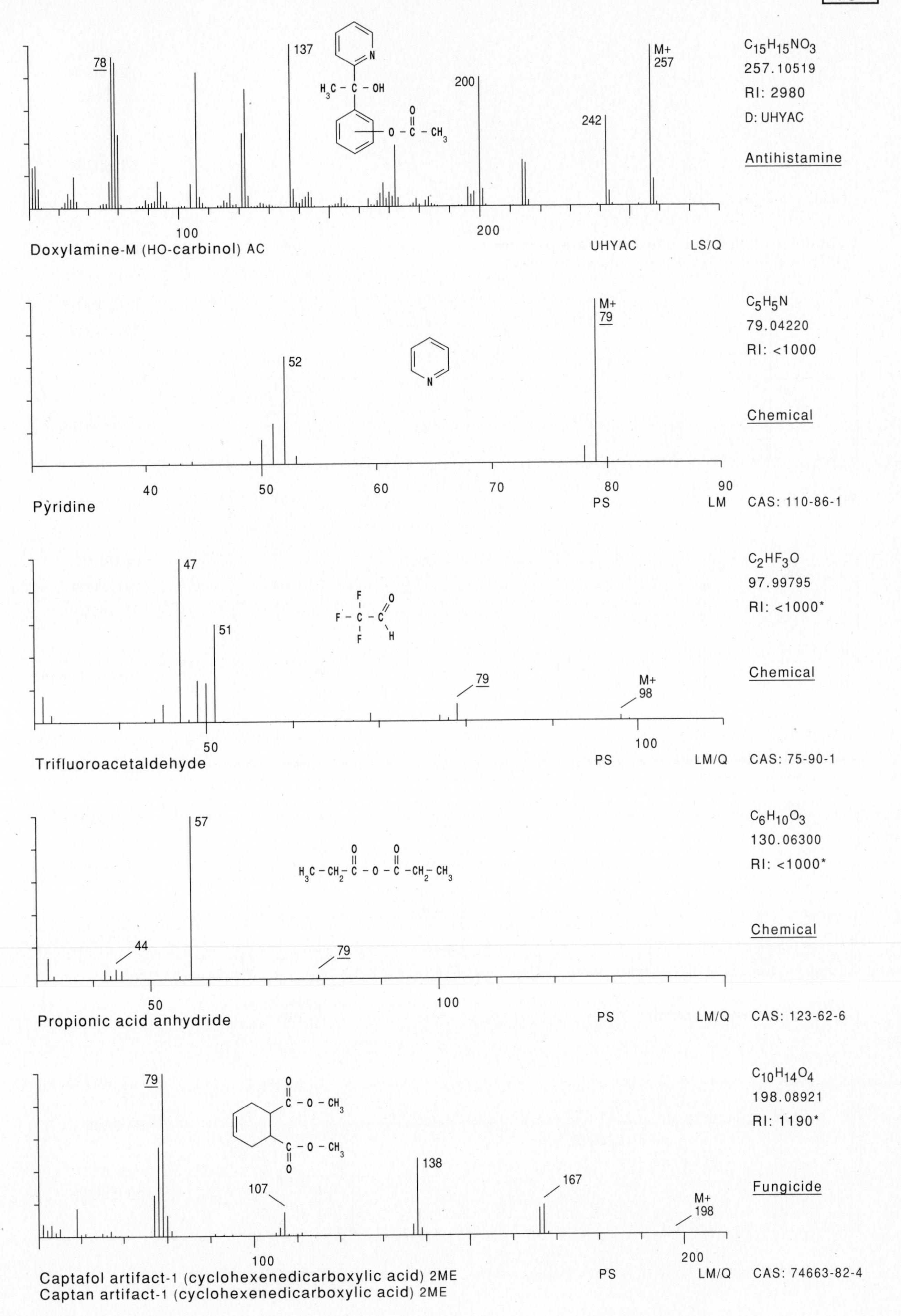

Doxylamine-M (HO-carbinol) AC

$C_{15}H_{15}NO_3$
257.10519
RI: 2980
D: UHYAC
Antihistamine

Pyridine

C_5H_5N
79.04220
RI: <1000
Chemical
CAS: 110-86-1

Trifluoroacetaldehyde

C_2HF_3O
97.99795
RI: <1000*
Chemical
CAS: 75-90-1

Propionic acid anhydride

$C_6H_{10}O_3$
130.06300
RI: <1000*
Chemical
CAS: 123-62-6

Captafol artifact-1 (cyclohexenedicarboxylic acid) 2ME
Captan artifact-1 (cyclohexenedicarboxylic acid) 2ME

$C_{10}H_{14}O_4$
198.08921
RI: 1190*
Fungicide
CAS: 74663-82-4

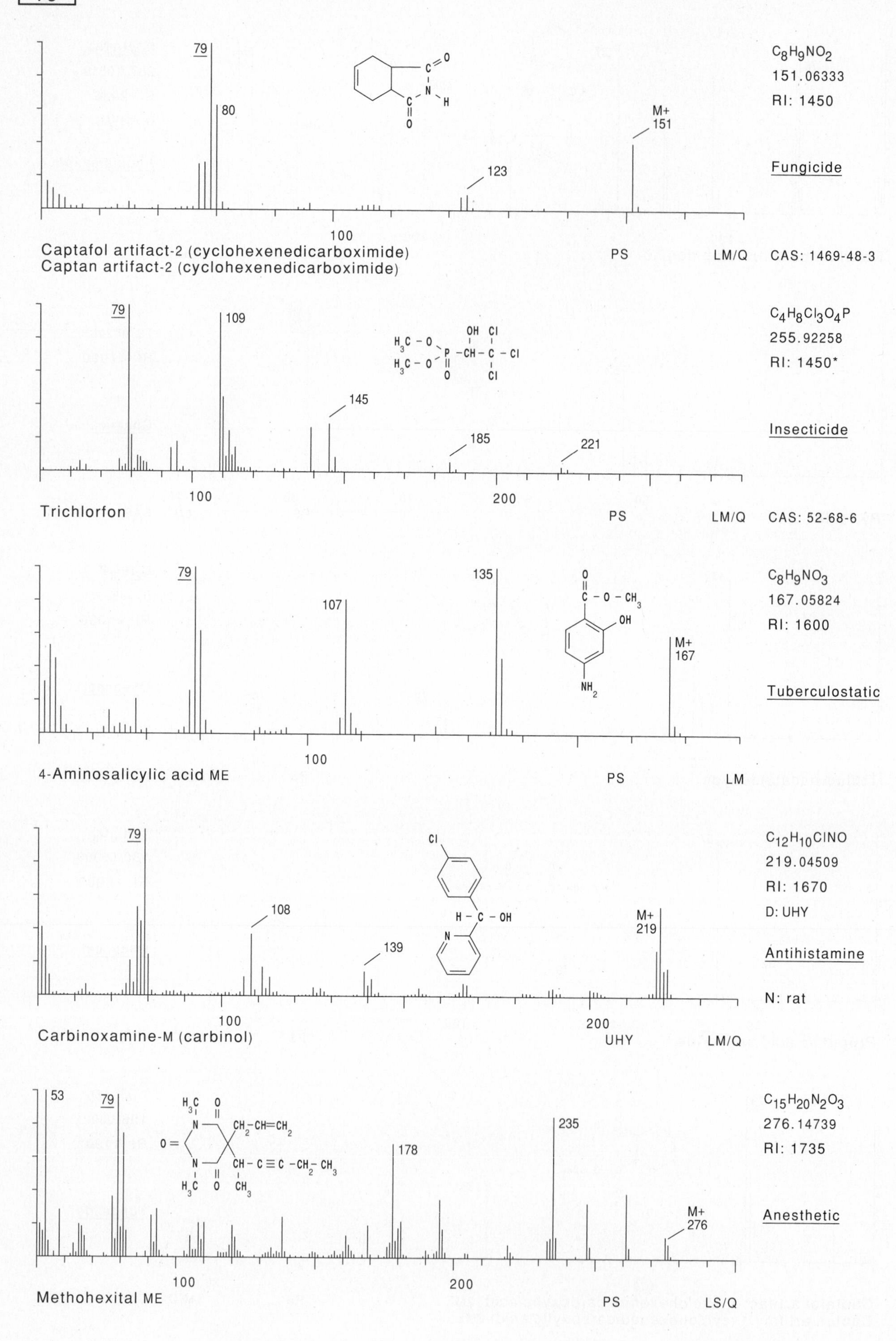

79
79
80
M+
151
123
100
C8H9NO2
151.06333
RI: 1450
Fungicide
Captafol artifact-2 (cyclohexenedicarboximide)
Captan artifact-2 (cyclohexenedicarboximide)
PS
LM/Q
CAS: 1469-48-3
79
109
145
185
221
100
200
C4H8Cl3O4P
255.92258
RI: 1450*
Insecticide
Trichlorfon
PS
LM/Q
CAS: 52-68-6
79
135
107
M+
167
100
C8H9NO3
167.05824
RI: 1600
Tuberculostatic
4-Aminosalicylic acid ME
PS
LM
79
108
139
M+
219
100
200
C12H10ClNO
219.04509
RI: 1670
D: UHY
Antihistamine
N: rat
Carbinoxamine-M (carbinol)
UHY
LM/Q
53
79
235
178
M+
276
100
200
C15H20N2O3
276.14739
RI: 1735
Anesthetic
Methohexital ME
PS
LS/Q

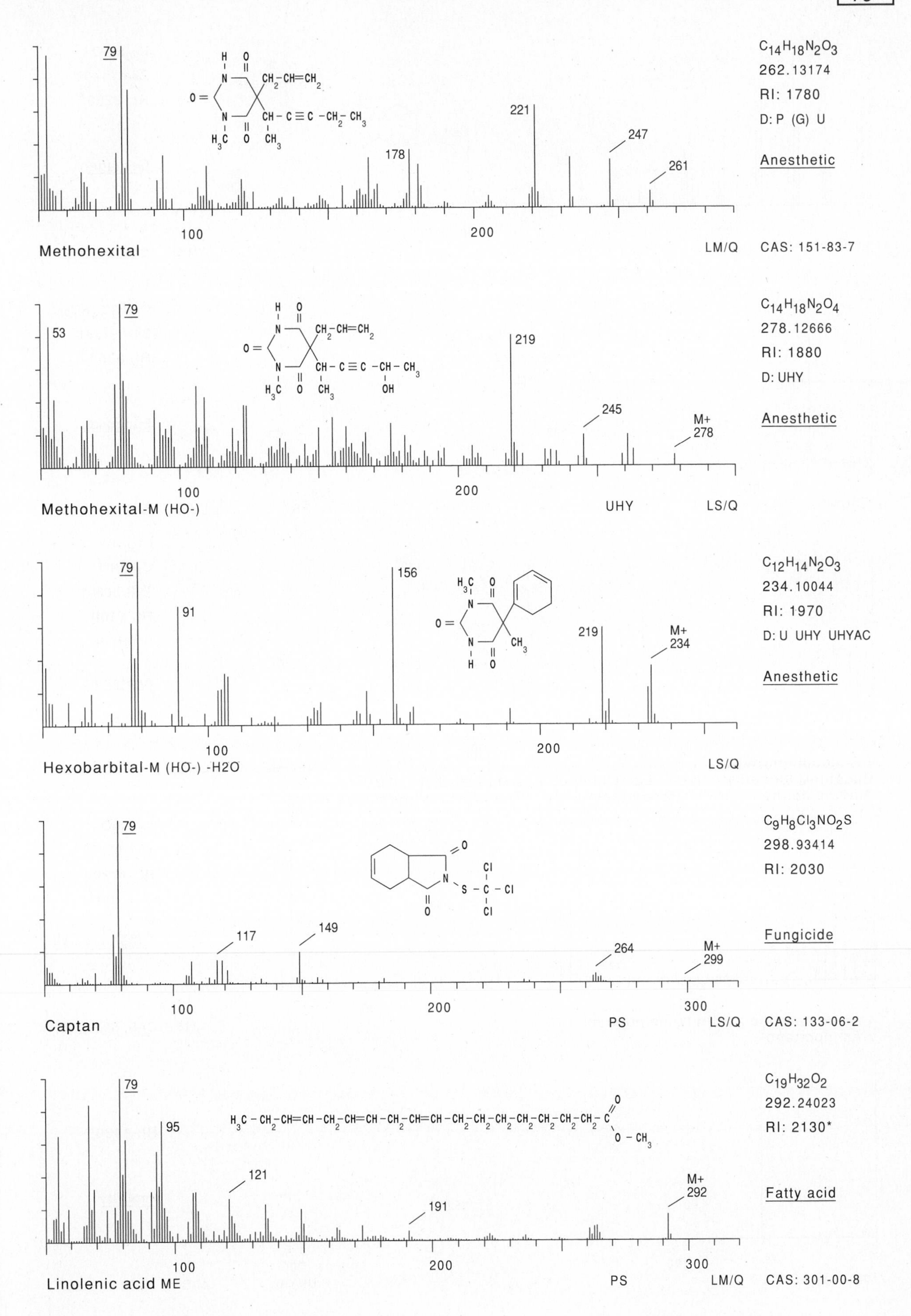

79
221
178
247
261
100
200
Methohexital
LM/Q
C14H18N2O3
262.13174
RI: 1780
D: P (G) U
Anesthetic
CAS: 151-83-7
79
53
219
245
M+
278
100
200
Methohexital-M (HO-)
UHY
LS/Q
C14H18N2O4
278.12666
RI: 1880
D: UHY
Anesthetic
79
156
91
219
M+
234
100
200
Hexobarbital-M (HO-) -H2O
LS/Q
C12H14N2O3
234.10044
RI: 1970
D: U UHY UHYAC
Anesthetic
79
117
149
264
M+
299
100
200
300
Captan
PS
LS/Q
C9H8Cl3NO2S
298.93414
RI: 2030
Fungicide
CAS: 133-06-2
79
95
121
191
M+
292
100
200
300
Linolenic acid ME
PS
LM/Q
C19H32O2
292.24023
RI: 2130*
Fatty acid
CAS: 301-00-8

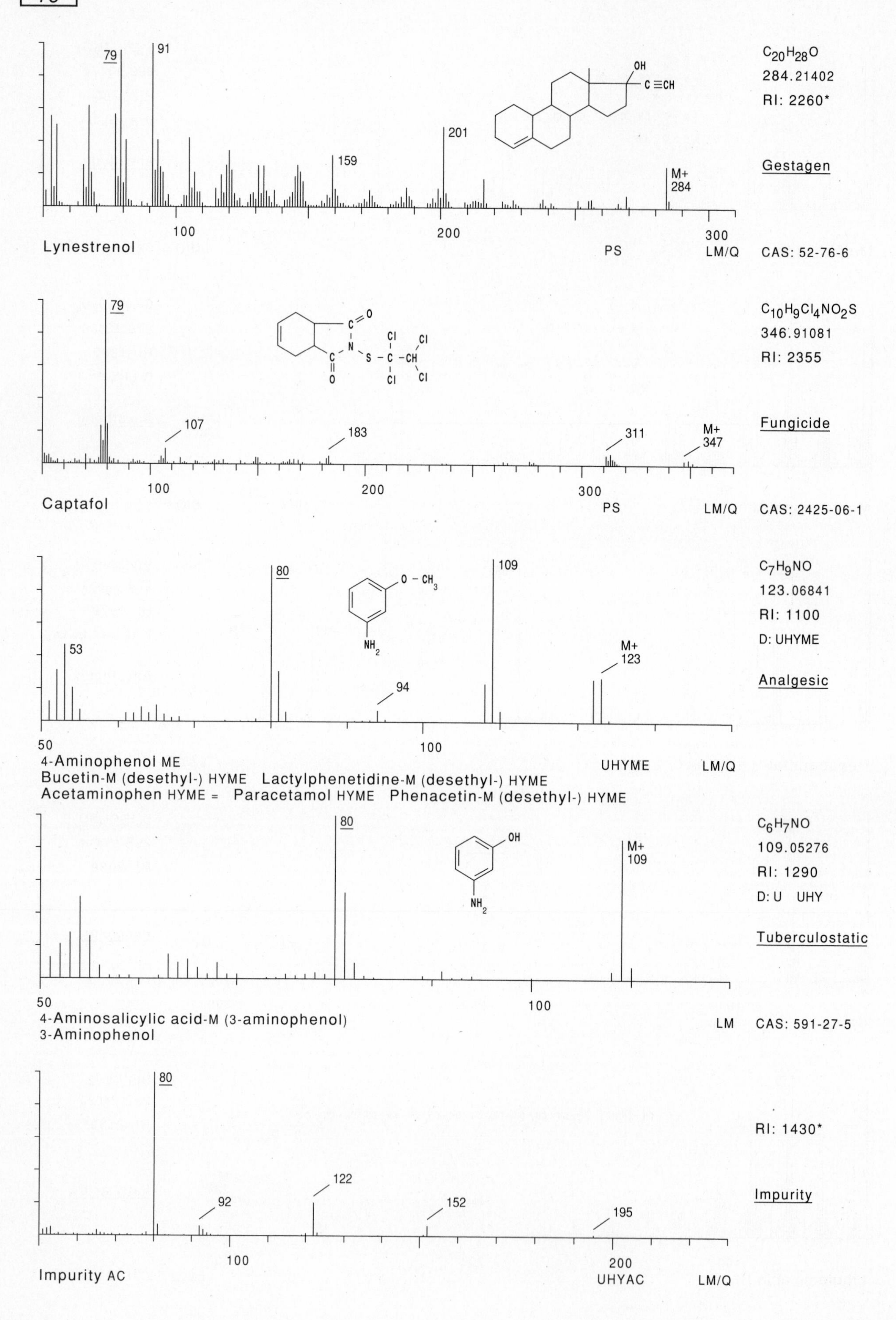
Lynestrenol
PS
LM/Q
$C_{20}H_{28}O$
284.21402
RI: 2260*
Gestagen
CAS: 52-76-6
Captafol
PS
LM/Q
$C_{10}H_9Cl_4NO_2S$
346.91081
RI: 2355
Fungicide
CAS: 2425-06-1
4-Aminophenol ME
Bucetin-M (desethyl-) HYME Lactylphenetidine-M (desethyl-) HYME
Acetaminophen HYME = Paracetamol HYME Phenacetin-M (desethyl-) HYME
UHYME
LM/Q
C_7H_9NO
123.06841
RI: 1100
D: UHYME
Analgesic
4-Aminosalicylic acid-M (3-aminophenol)
3-Aminophenol
LM
C_6H_7NO
109.05276
RI: 1290
D: U UHY
Tuberculostatic
CAS: 591-27-5
Impurity AC
UHYAC
LM/Q
RI: 1430*
Impurity

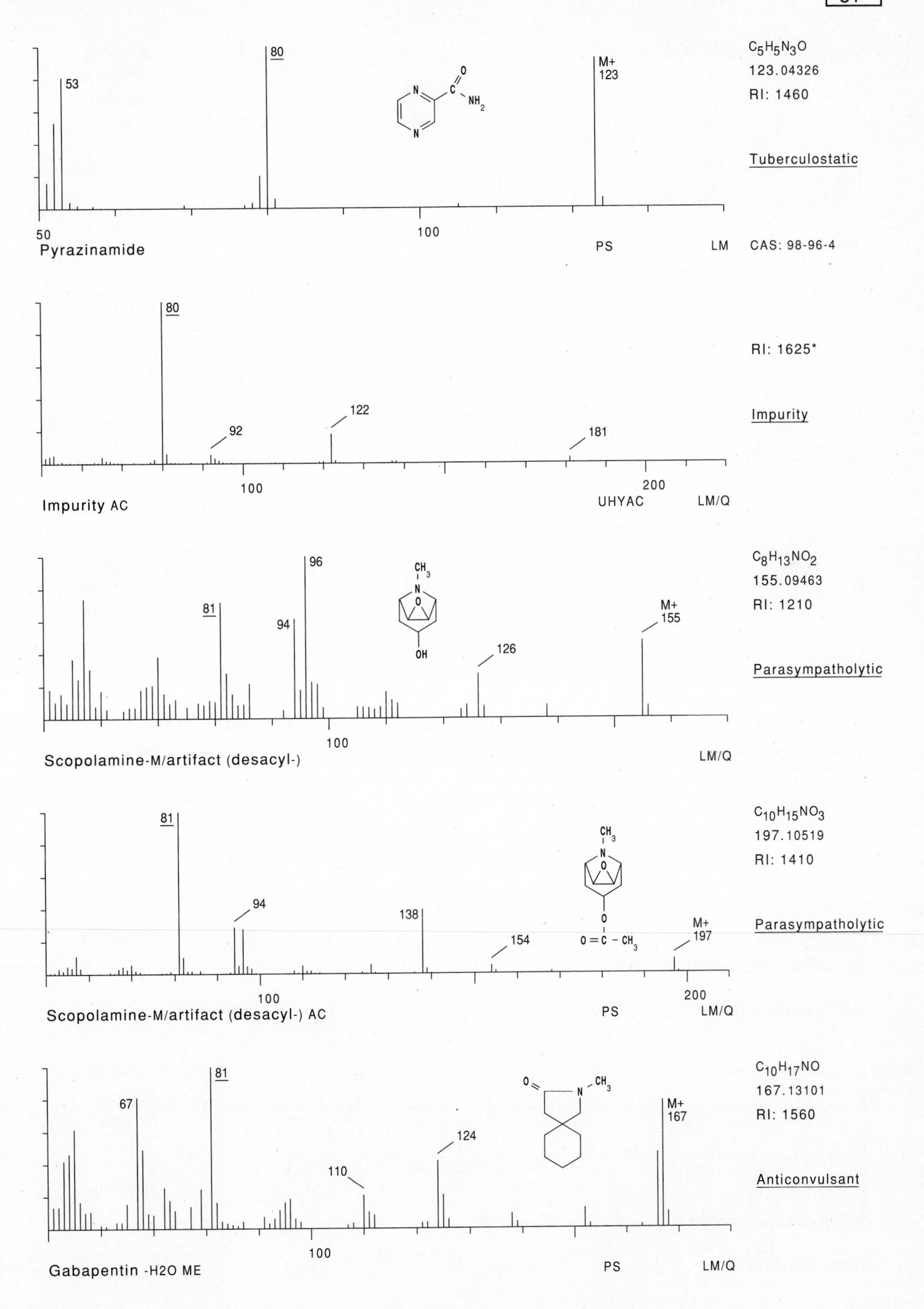

80
53
M+
123
O
N
C
NH2
N
50
100
Pyrazinamide
PS
LM
C5H5N3O
123.04326
RI: 1460
Tuberculostatic
CAS: 98-96-4
80
92
122
181
100
200
Impurity AC
UHYAC
LM/Q
RI: 1625*
Impurity
96
81
94
CH3
N
O
OH
126
M+
155
100
Scopolamine-M/artifact (desacyl-)
LM/Q
C8H13NO2
155.09463
RI: 1210
Parasympatholytic
81
94
138
154
CH3
N
O
O
O = C - CH3
M+
197
100
200
Scopolamine-M/artifact (desacyl-) AC
PS
LM/Q
C10H15NO3
197.10519
RI: 1410
Parasympatholytic
81
67
110
124
O
N
CH3
M+
167
100
Gabapentin -H2O ME
PS
LM/Q
C10H17NO
167.13101
RI: 1560
Anticonvulsant

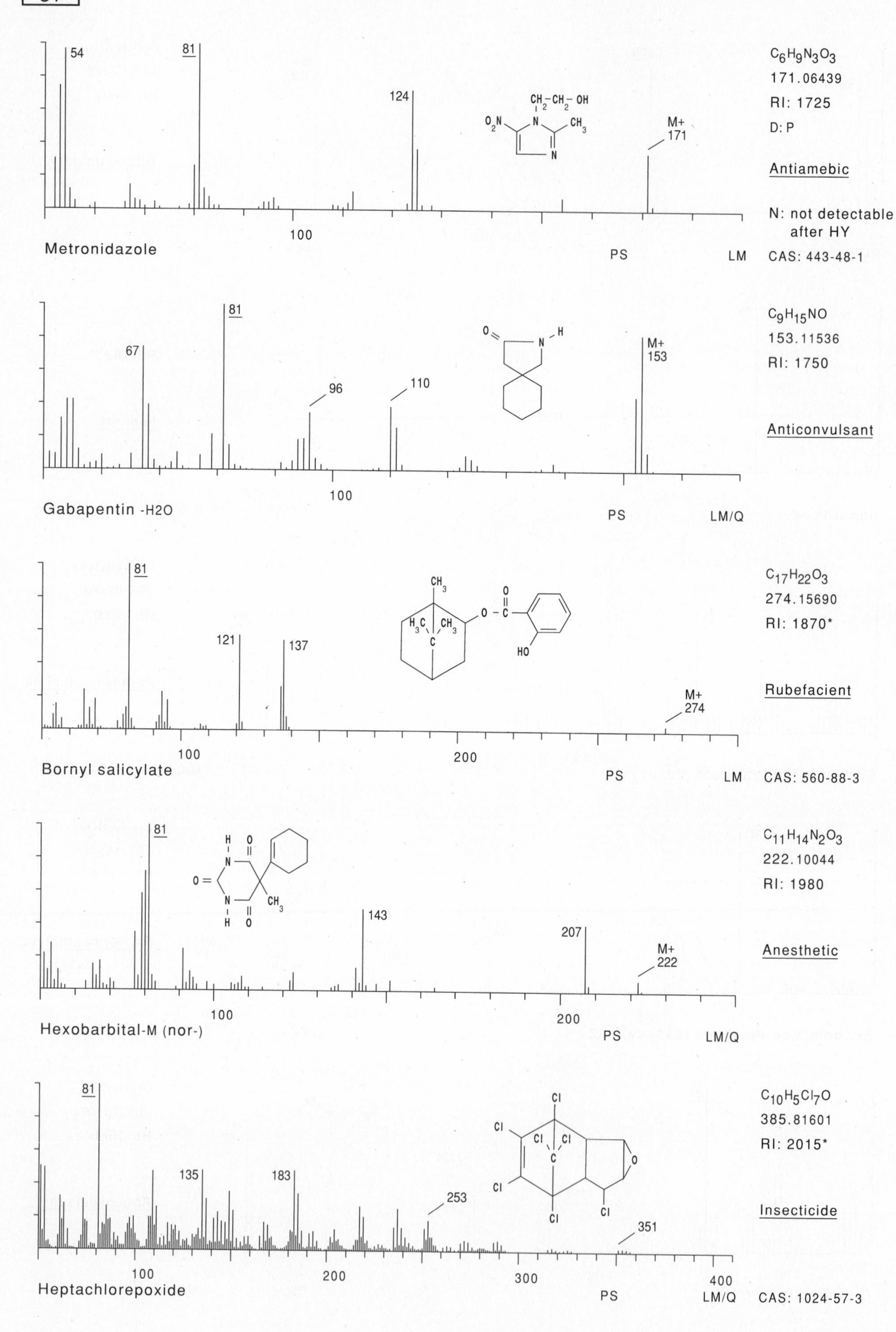

54
81
124
M+
171
100
Metronidazole
PS
LM
CH_2-CH_2-OH
O_2N
CH_3
N
$C_6H_9N_3O_3$
171.06439
RI: 1725
D: P
Antiamebic
N: not detectable after HY
CAS: 443-48-1
81
67
96
110
M+
153
100
Gabapentin -H2O
PS
LM/Q
$C_9H_{15}NO$
153.11536
RI: 1750
Anticonvulsant
81
121
137
M+
274
100
200
Bornyl salicylate
PS
LM
CH_3
H_3C
HO
$C_{17}H_{22}O_3$
274.15690
RI: 1870*
Rubefacient
CAS: 560-88-3
81
143
207
M+
222
100
200
Hexobarbital-M (nor-)
PS
LM/Q
$C_{11}H_{14}N_2O_3$
222.10044
RI: 1980
Anesthetic
81
135
183
253
351
100
200
300
400
Heptachlorepoxide
PS
LM/Q
Cl
$C_{10}H_5Cl_7O$
385.81601
RI: 2015*
Insecticide
CAS: 1024-57-3

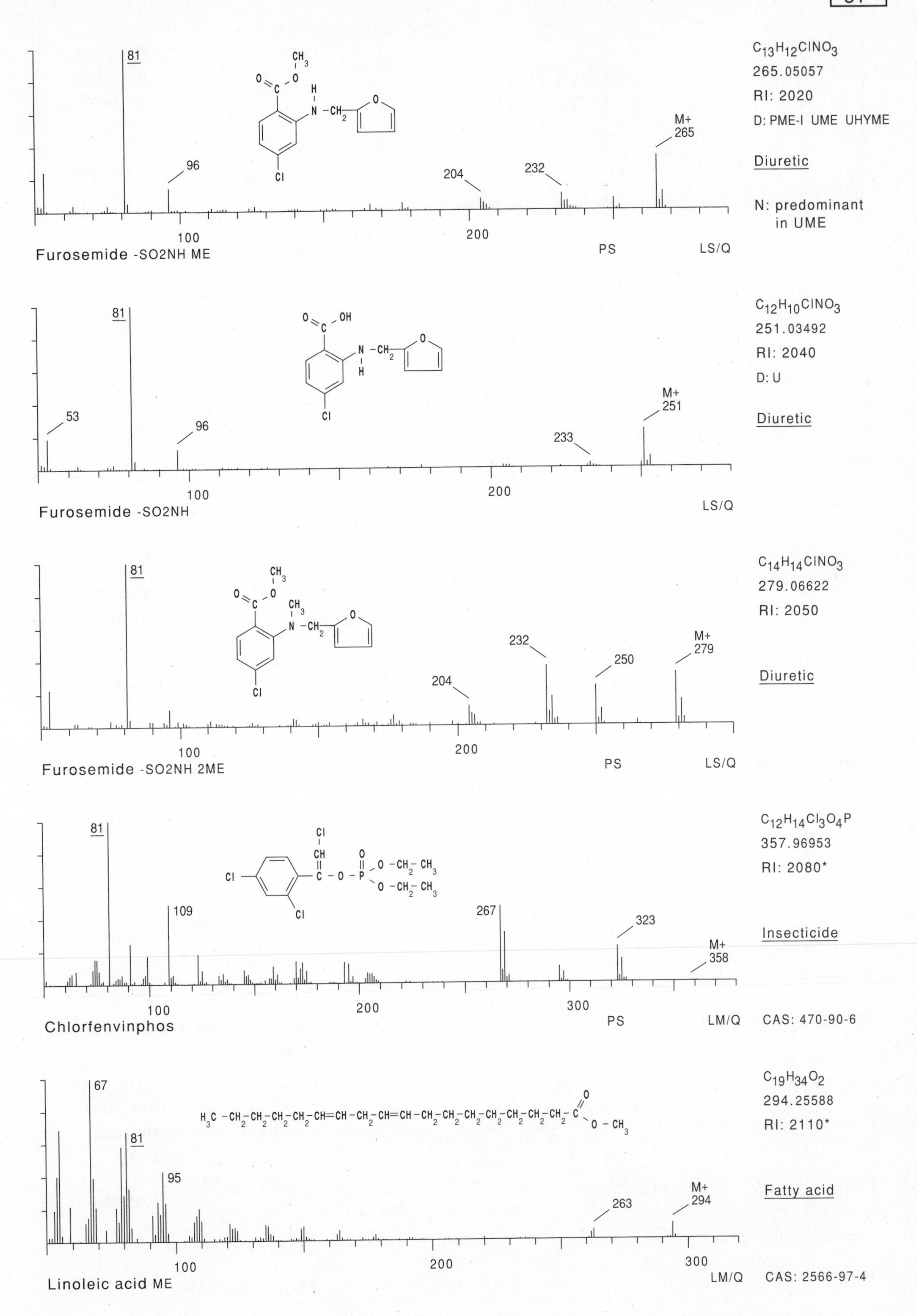

81
96
204
232
M+
265
100
200
PS
LS/Q
Furosemide -SO2NH ME
C13H12ClNO3
265.05057
RI: 2020
D: PME-I UME UHYME
Diuretic
N: predominant
in UME
81
53
96
233
M+
251
100
200
LS/Q
Furosemide -SO2NH
C12H10ClNO3
251.03492
RI: 2040
D: U
Diuretic
81
204
232
250
M+
279
100
200
PS
LS/Q
Furosemide -SO2NH 2ME
C14H14ClNO3
279.06622
RI: 2050
Diuretic
81
109
267
323
M+
358
100
200
300
PS
LM/Q
Chlorfenvinphos
C12H14Cl3O4P
357.96953
RI: 2080*
Insecticide
CAS: 470-90-6
67
81
95
263
M+
294
100
200
300
LM/Q
Linoleic acid ME
C19H34O2
294.25588
RI: 2110*
Fatty acid
CAS: 2566-97-4

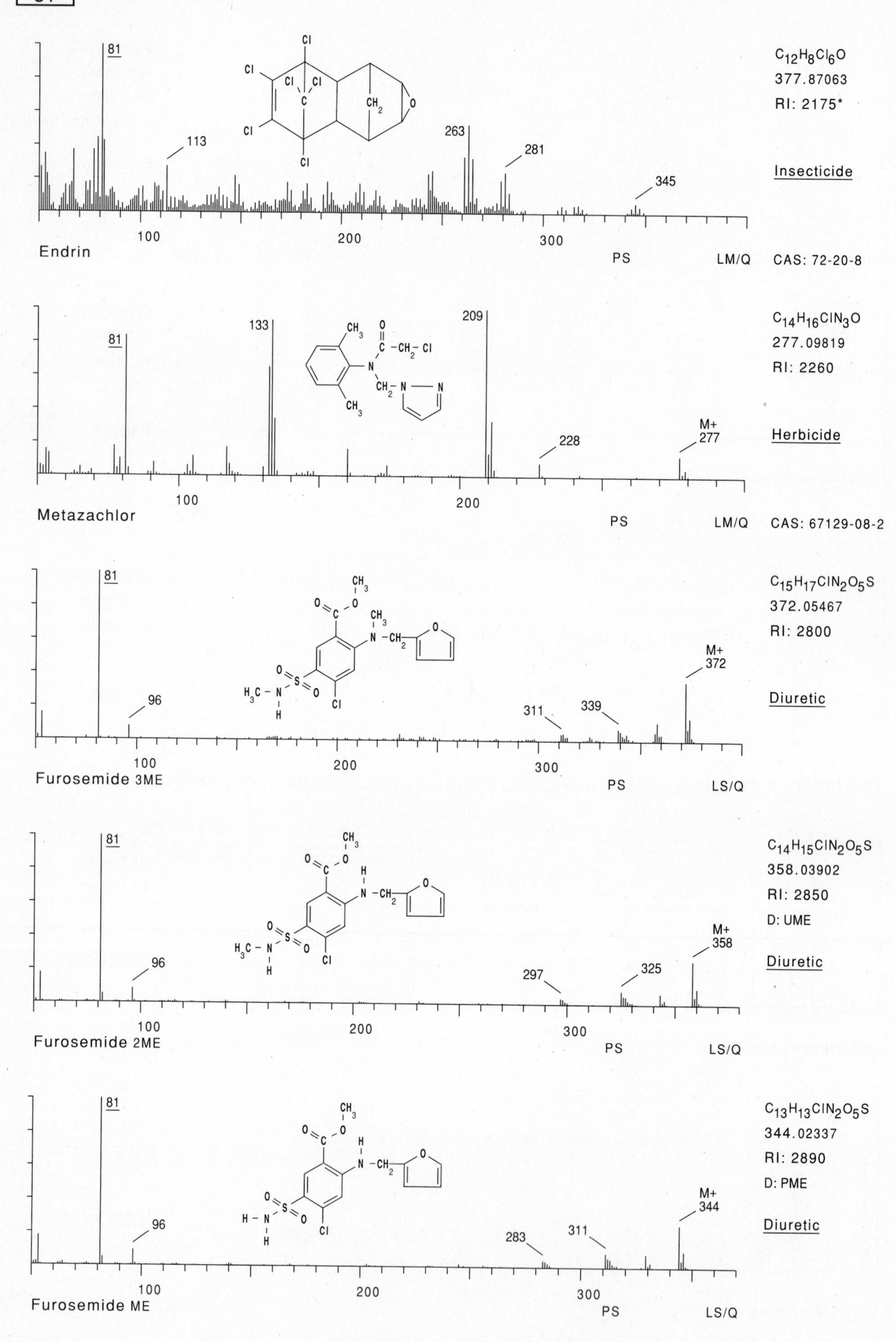

81
113
263
281
345
100
200
300
Endrin
PS
LM/Q
$C_{12}H_8Cl_6O$
377.87063
RI: 2175*
Insecticide
CAS: 72-20-8
81
133
209
228
M+
277
100
200
Metazachlor
PS
LM/Q
$C_{14}H_{16}ClN_3O$
277.09819
RI: 2260
Herbicide
CAS: 67129-08-2
81
96
311
339
M+
372
100
200
300
Furosemide 3ME
PS
LS/Q
$C_{15}H_{17}ClN_2O_5S$
372.05467
RI: 2800
Diuretic
81
96
297
325
M+
358
100
200
300
Furosemide 2ME
PS
LS/Q
$C_{14}H_{15}ClN_2O_5S$
358.03902
RI: 2850
D: UME
Diuretic
81
96
283
311
M+
344
100
200
300
Furosemide ME
PS
LS/Q
$C_{13}H_{13}ClN_2O_5S$
344.02337
RI: 2890
D: PME
Diuretic

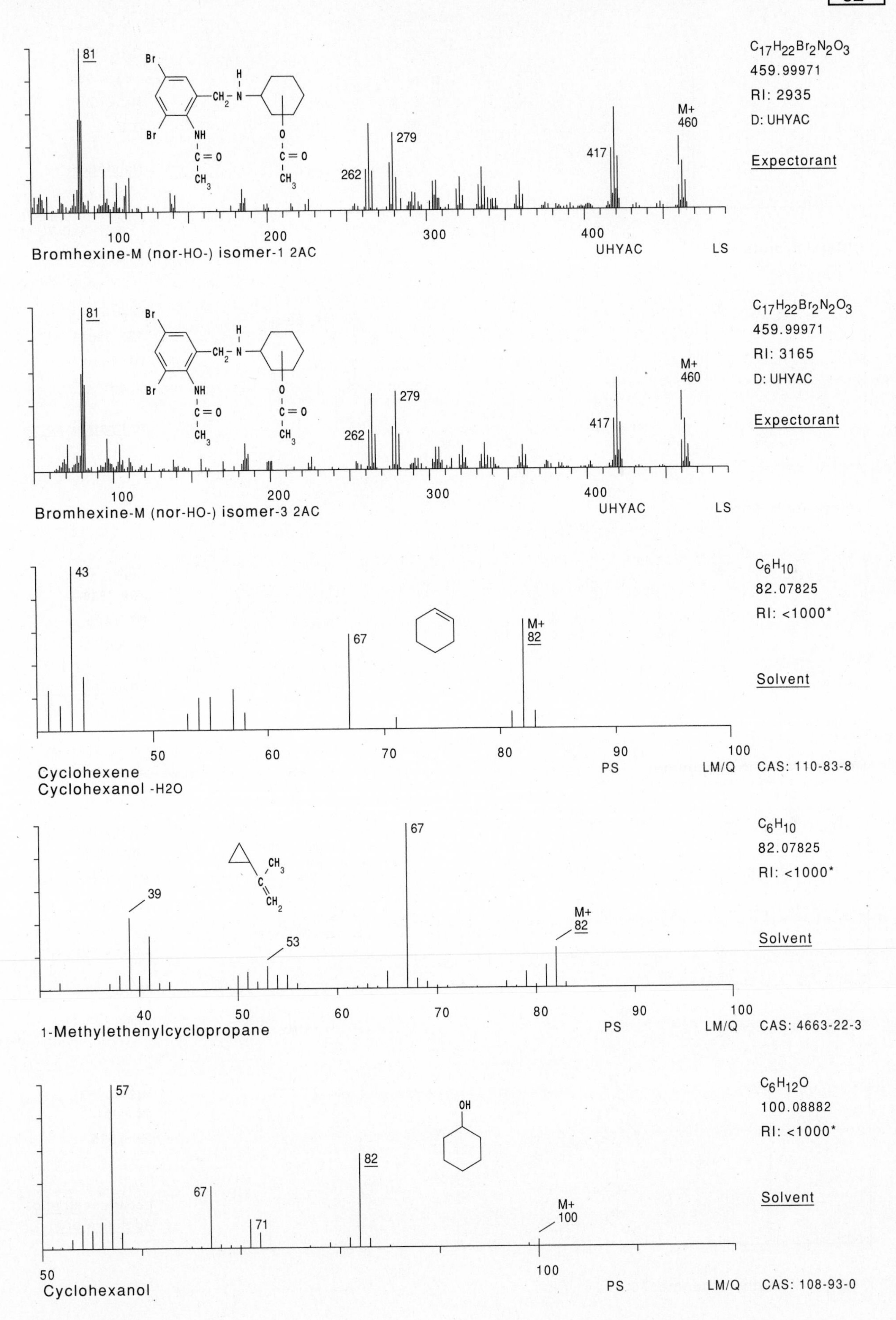

Bromhexine-M (nor-HO-) isomer-1 2AC
$C_{17}H_{22}Br_2N_2O_3$
459.99971
RI: 2935
D: UHYAC
Expectorant
UHYAC
LS
Bromhexine-M (nor-HO-) isomer-3 2AC
$C_{17}H_{22}Br_2N_2O_3$
459.99971
RI: 3165
D: UHYAC
Expectorant
UHYAC
LS
Cyclohexene
Cyclohexanol -H2O
C_6H_{10}
82.07825
RI: <1000*
Solvent
PS
LM/Q
CAS: 110-83-8
1-Methylethenylcyclopropane
C_6H_{10}
82.07825
RI: <1000*
Solvent
PS
LM/Q
CAS: 4663-22-3
Cyclohexanol
$C_6H_{12}O$
100.08882
RI: <1000*
Solvent
PS
LM/Q
CAS: 108-93-0

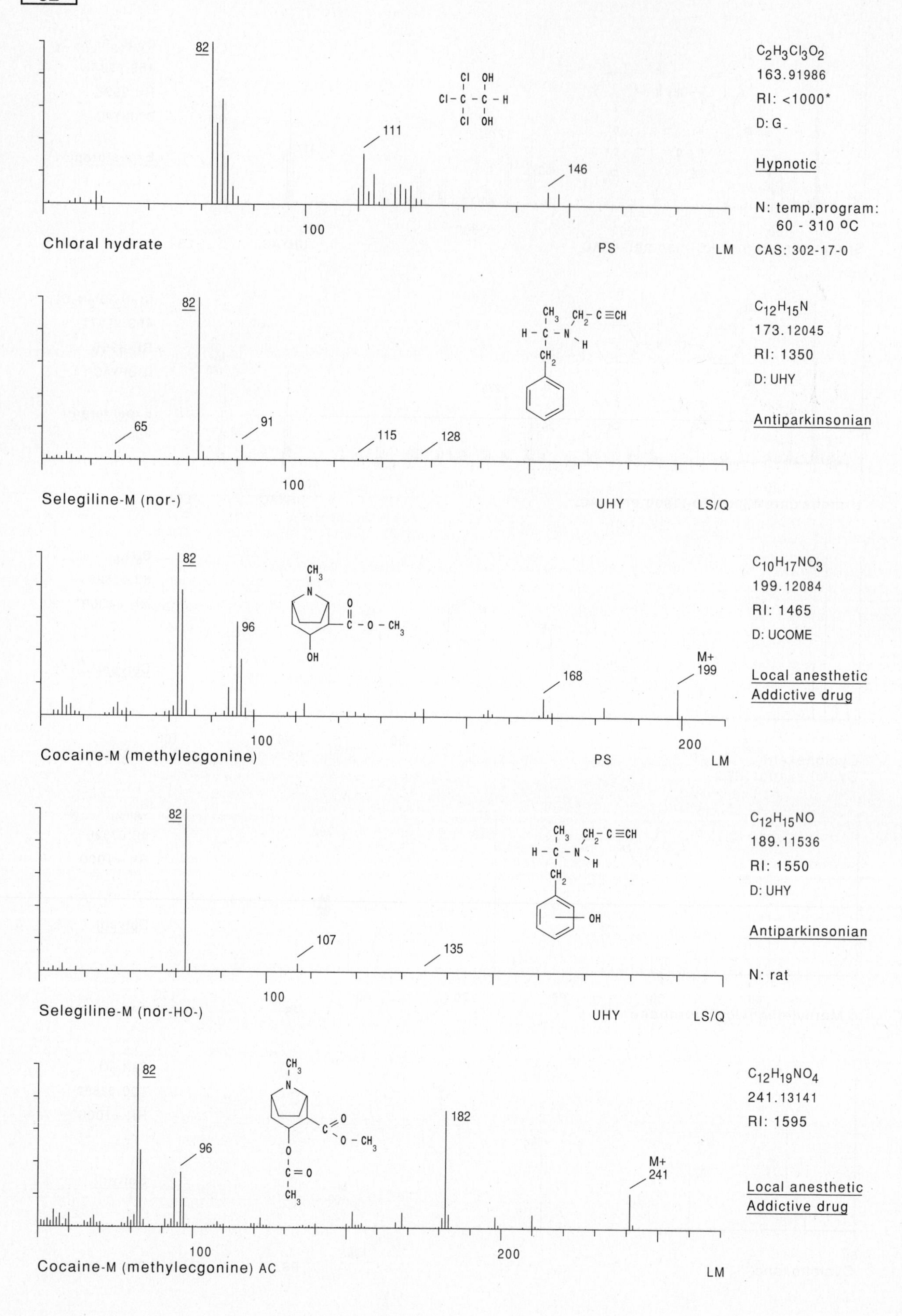
82
111
146
100
Chloral hydrate
PS
LM
$C_2H_3Cl_3O_2$
163.91986
RI: <1000*
D: G
Hypnotic
N: temp.program: 60 - 310 °C
CAS: 302-17-0
82
65
91
115
128
100
Selegiline-M (nor-)
UHY
LS/Q
$C_{12}H_{15}N$
173.12045
RI: 1350
D: UHY
Antiparkinsonian
82
96
168
M+
199
100
200
Cocaine-M (methylecgonine)
PS
LM
$C_{10}H_{17}NO_3$
199.12084
RI: 1465
D: UCOME
Local anesthetic
Addictive drug
82
107
135
100
Selegiline-M (nor-HO-)
UHY
LS/Q
$C_{12}H_{15}NO$
189.11536
RI: 1550
D: UHY
Antiparkinsonian
N: rat
82
96
182
M+
241
100
200
Cocaine-M (methylecgonine) AC
LM
$C_{12}H_{19}NO_4$
241.13141
RI: 1595
Local anesthetic
Addictive drug

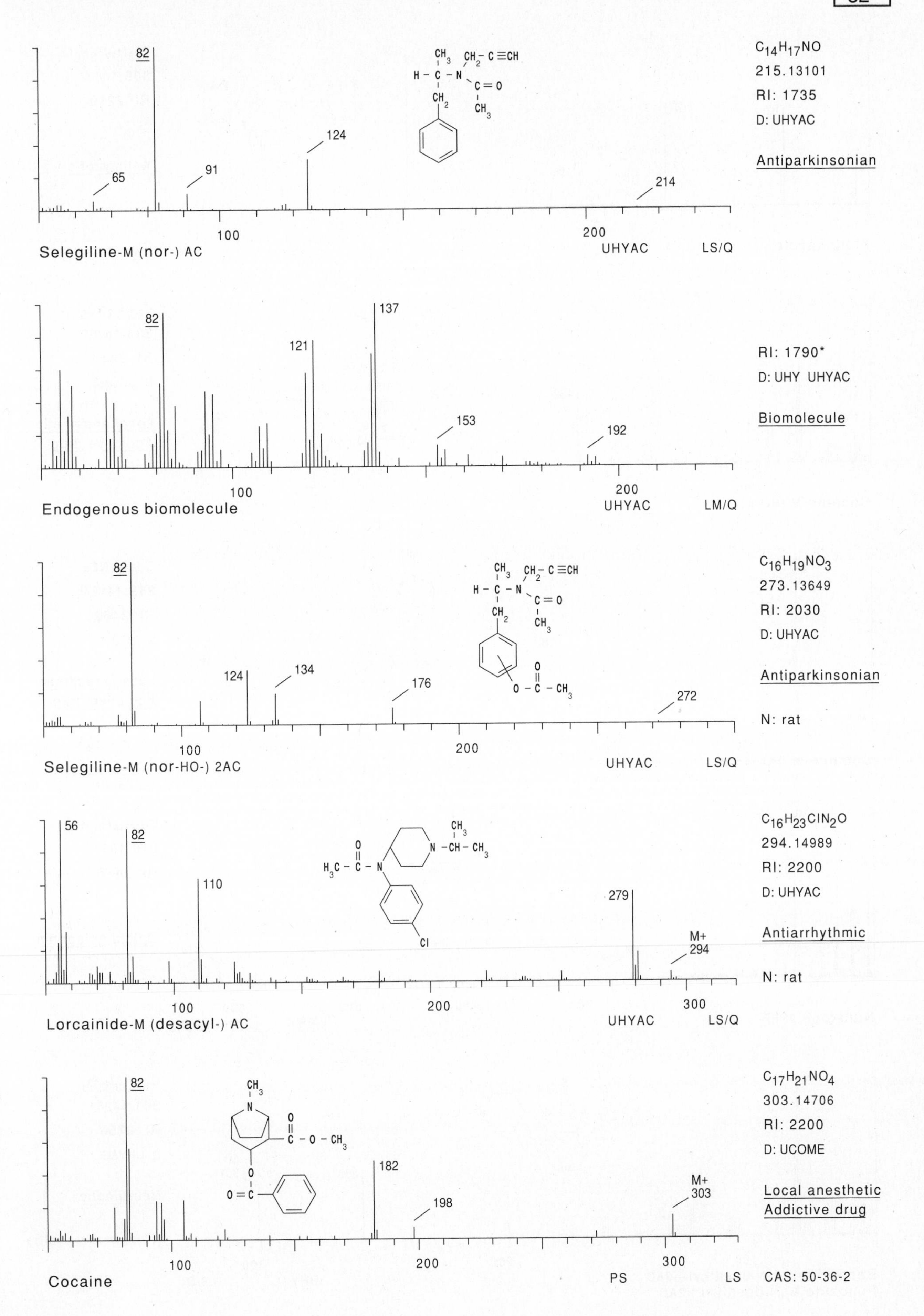

$C_{14}H_{17}NO$
215.13101
RI: 1735
D: UHYAC

Antiparkinsonian

RI: 1790*
D: UHY UHYAC

Biomolecule

$C_{16}H_{19}NO_3$
273.13649
RI: 2030
D: UHYAC

Antiparkinsonian

N: rat

$C_{16}H_{23}ClN_2O$
294.14989
RI: 2200
D: UHYAC

Antiarrhythmic

N: rat

$C_{17}H_{21}NO_4$
303.14706
RI: 2200
D: UCOME

Local anesthetic
Addictive drug

CAS: 50-36-2

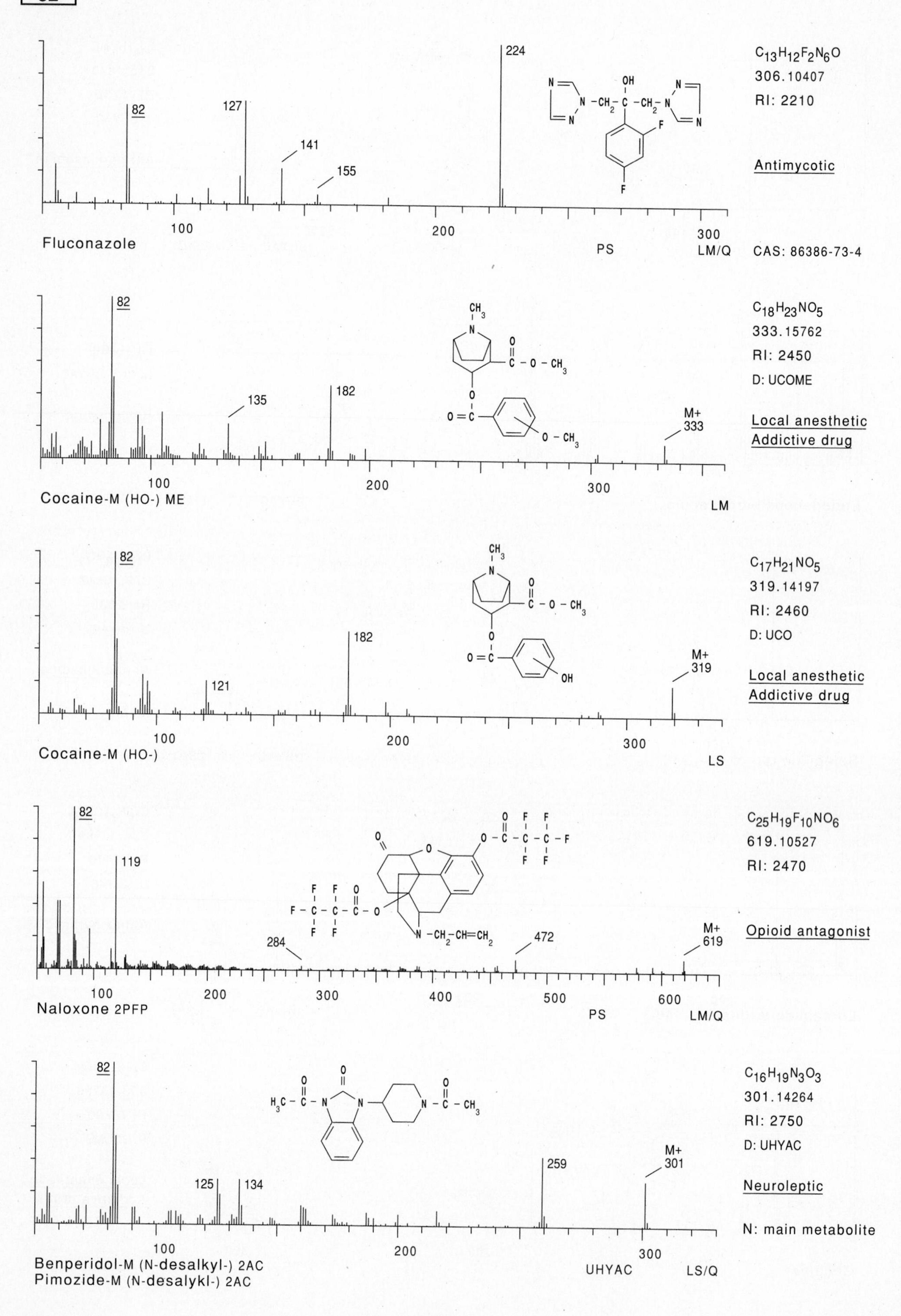

224
82
127
141
155
100
200
300
Fluconazole
PS
LM/Q
$C_{13}H_{12}F_2N_6O$
306.10407
RI: 2210
Antimycotic
CAS: 86386-73-4
82
135
182
M+
333
Cocaine-M (HO-) ME
LM
$C_{18}H_{23}NO_5$
333.15762
RI: 2450
D: UCOME
Local anesthetic
Addictive drug
82
182
121
M+
319
Cocaine-M (HO-)
LS
$C_{17}H_{21}NO_5$
319.14197
RI: 2460
D: UCO
Local anesthetic
Addictive drug
82
119
284
472
M+
619
Naloxone 2PFP
PS
LM/Q
$C_{25}H_{19}F_{10}NO_6$
619.10527
RI: 2470
Opioid antagonist
82
125
134
259
M+
301
Benperidol-M (N-desalkyl-) 2AC
Pimozide-M (N-desalykl-) 2AC
UHYAC
LS/Q
$C_{16}H_{19}N_3O_3$
301.14264
RI: 2750
D: UHYAC
Neuroleptic
N: main metabolite

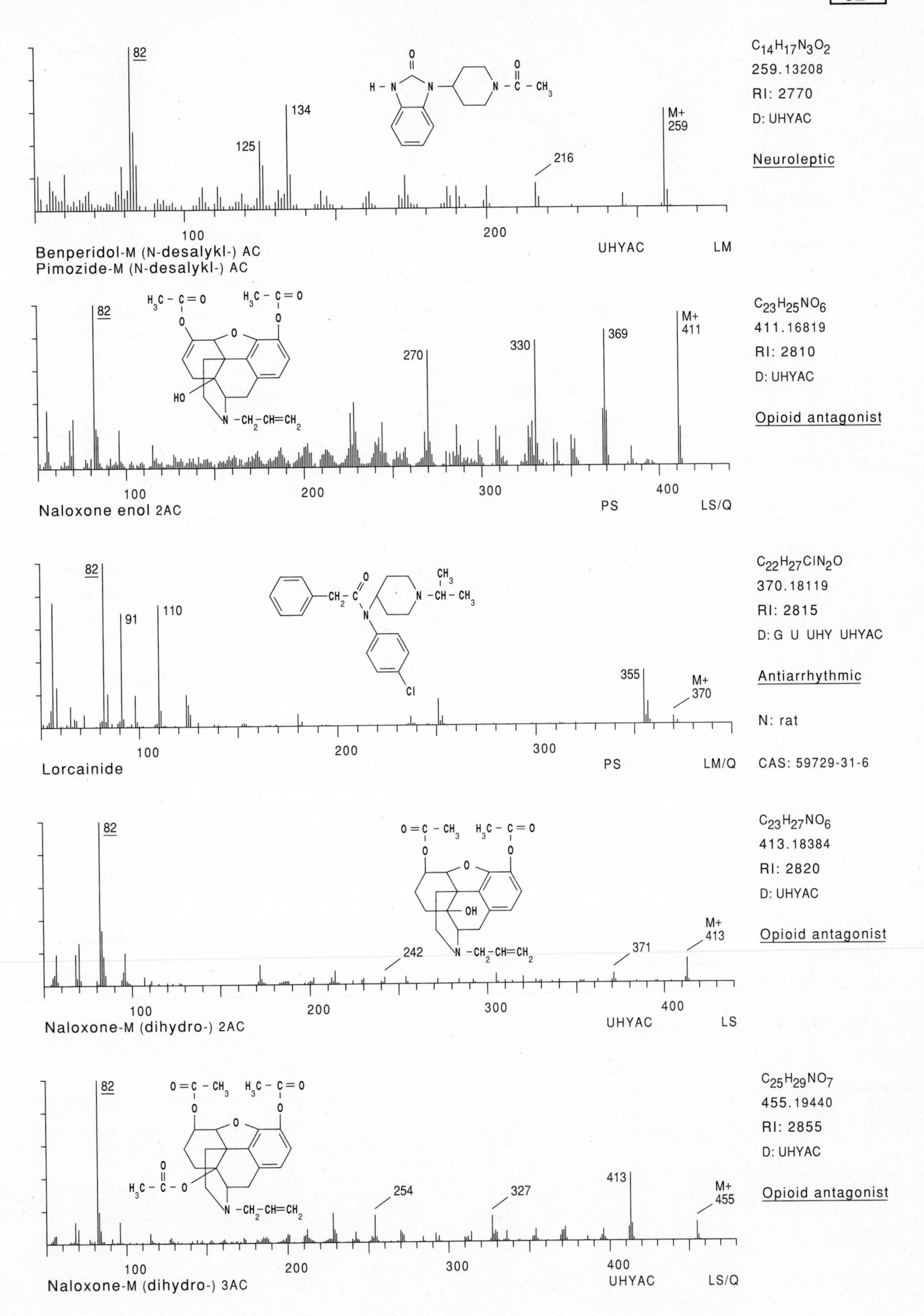

$C_{14}H_{17}N_3O_2$
259.13208
RI: 2770
D: UHYAC
Neuroleptic

$C_{23}H_{25}NO_6$
411.16819
RI: 2810
D: UHYAC
Opioid antagonist

$C_{22}H_{27}ClN_2O$
370.18119
RI: 2815
D: G U UHY UHYAC
Antiarrhythmic
N: rat
CAS: 59729-31-6

$C_{23}H_{27}NO_6$
413.18384
RI: 2820
D: UHYAC
Opioid antagonist

$C_{25}H_{29}NO_7$
455.19440
RI: 2855
D: UHYAC
Opioid antagonist

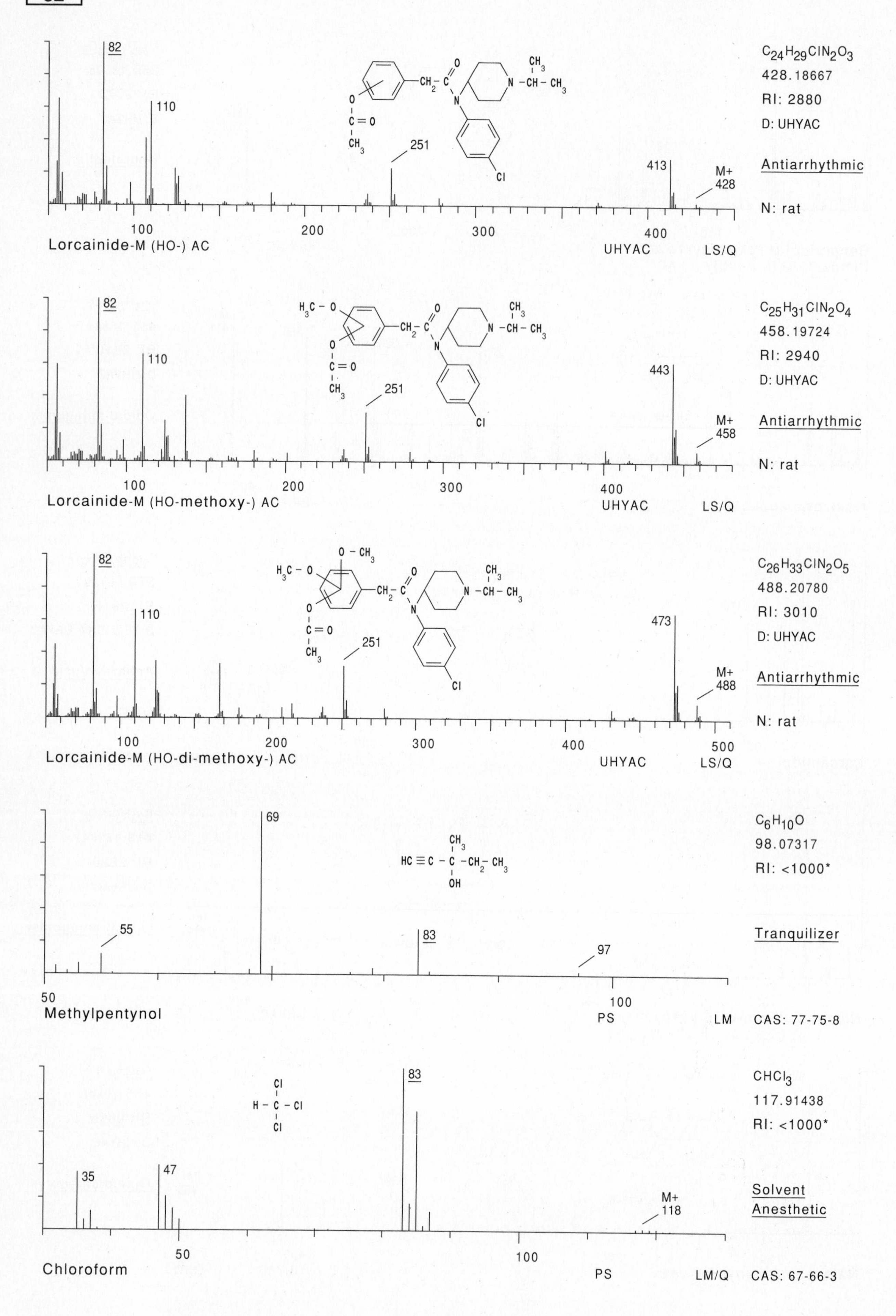
82
110
251
413
M+
428
100
200
300
400
Lorcainide-M (HO-) AC
UHYAC
LS/Q
$C_{24}H_{29}ClN_2O_3$
428.18667
RI: 2880
D: UHYAC
Antiarrhythmic
N: rat
82
110
251
443
M+
458
100
200
300
400
Lorcainide-M (HO-methoxy-) AC
UHYAC
LS/Q
$C_{25}H_{31}ClN_2O_4$
458.19724
RI: 2940
D: UHYAC
Antiarrhythmic
N: rat
82
110
251
473
M+
488
100
200
300
400
500
Lorcainide-M (HO-di-methoxy-) AC
UHYAC
LS/Q
$C_{26}H_{33}ClN_2O_5$
488.20780
RI: 3010
D: UHYAC
Antiarrhythmic
N: rat
69
55
83
97
50
100
Methylpentynol
PS
LM
$C_6H_{10}O$
98.07317
RI: <1000*
Tranquilizer
CAS: 77-75-8
83
35
47
M+
118
50
100
Chloroform
PS
LM/Q
$CHCl_3$
117.91438
RI: <1000*
Solvent
Anesthetic
CAS: 67-66-3

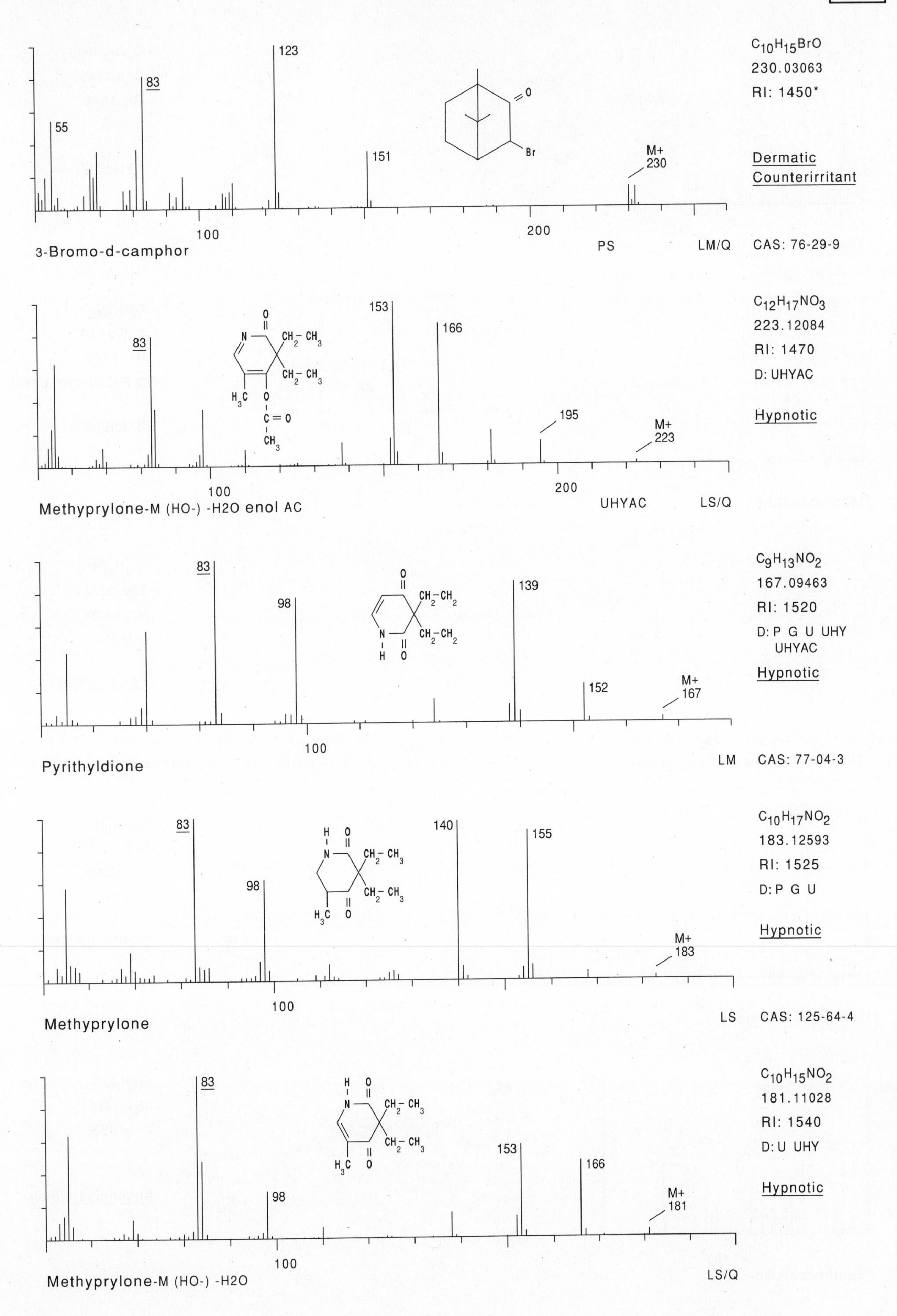

123
83
55
151
M+
230
100
200
PS
LM/Q
3-Bromo-d-camphor
C10H15BrO
230.03063
RI: 1450*
Dermatic
Counterirritant
CAS: 76-29-9
153
166
83
195
M+
223
100
200
UHYAC
LS/Q
Methyprylone-M (HO-) -H2O enol AC
C12H17NO3
223.12084
RI: 1470
D: UHYAC
Hypnotic
83
139
98
152
M+
167
100
LM
Pyrithyldione
C9H13NO2
167.09463
RI: 1520
D: P G U UHY
UHYAC
Hypnotic
CAS: 77-04-3
83
140
155
98
M+
183
100
LS
Methyprylone
C10H17NO2
183.12593
RI: 1525
D: P G U
Hypnotic
CAS: 125-64-4
83
153
166
98
M+
181
100
LS/Q
Methyprylone-M (HO-) -H2O
C10H15NO2
181.11028
RI: 1540
D: U UHY
Hypnotic

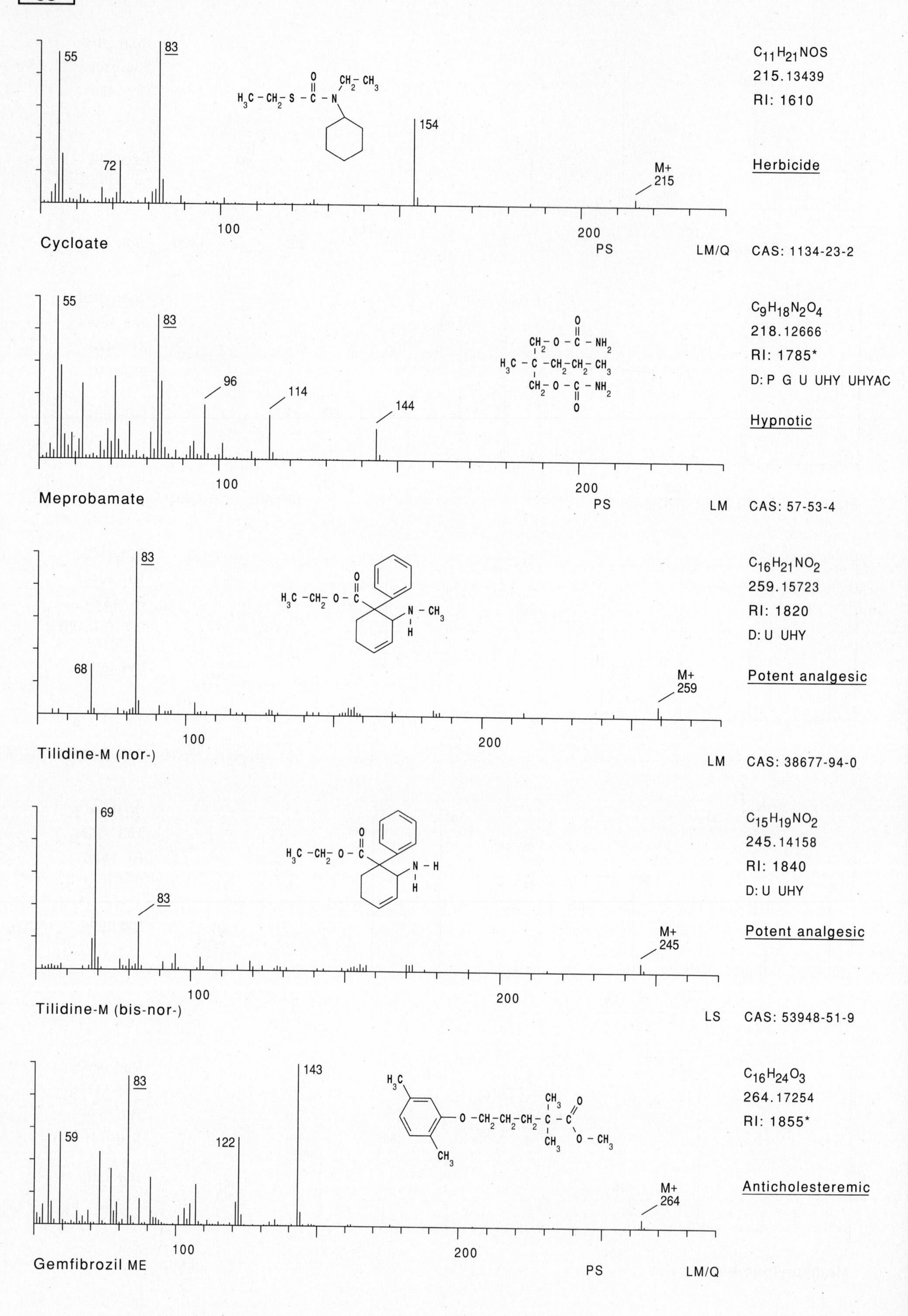
55
83
72
154
M+
215
100
200
Cycloate
PS
LM/Q
$C_{11}H_{21}NOS$
215.13439
RI: 1610
Herbicide
CAS: 1134-23-2
55
83
96
114
144
100
200
Meprobamate
PS
LM
$C_9H_{18}N_2O_4$
218.12666
RI: 1785*
D: P G U UHY UHYAC
Hypnotic
CAS: 57-53-4
83
68
M+
259
100
200
Tilidine-M (nor-)
LM
$C_{16}H_{21}NO_2$
259.15723
RI: 1820
D: U UHY
Potent analgesic
CAS: 38677-94-0
69
83
M+
245
100
200
Tilidine-M (bis-nor-)
LS
$C_{15}H_{19}NO_2$
245.14158
RI: 1840
D: U UHY
Potent analgesic
CAS: 53948-51-9
143
83
59
122
M+
264
100
200
Gemfibrozil ME
PS
LM/Q
$C_{16}H_{24}O_3$
264.17254
RI: 1855*
Anticholesteremic

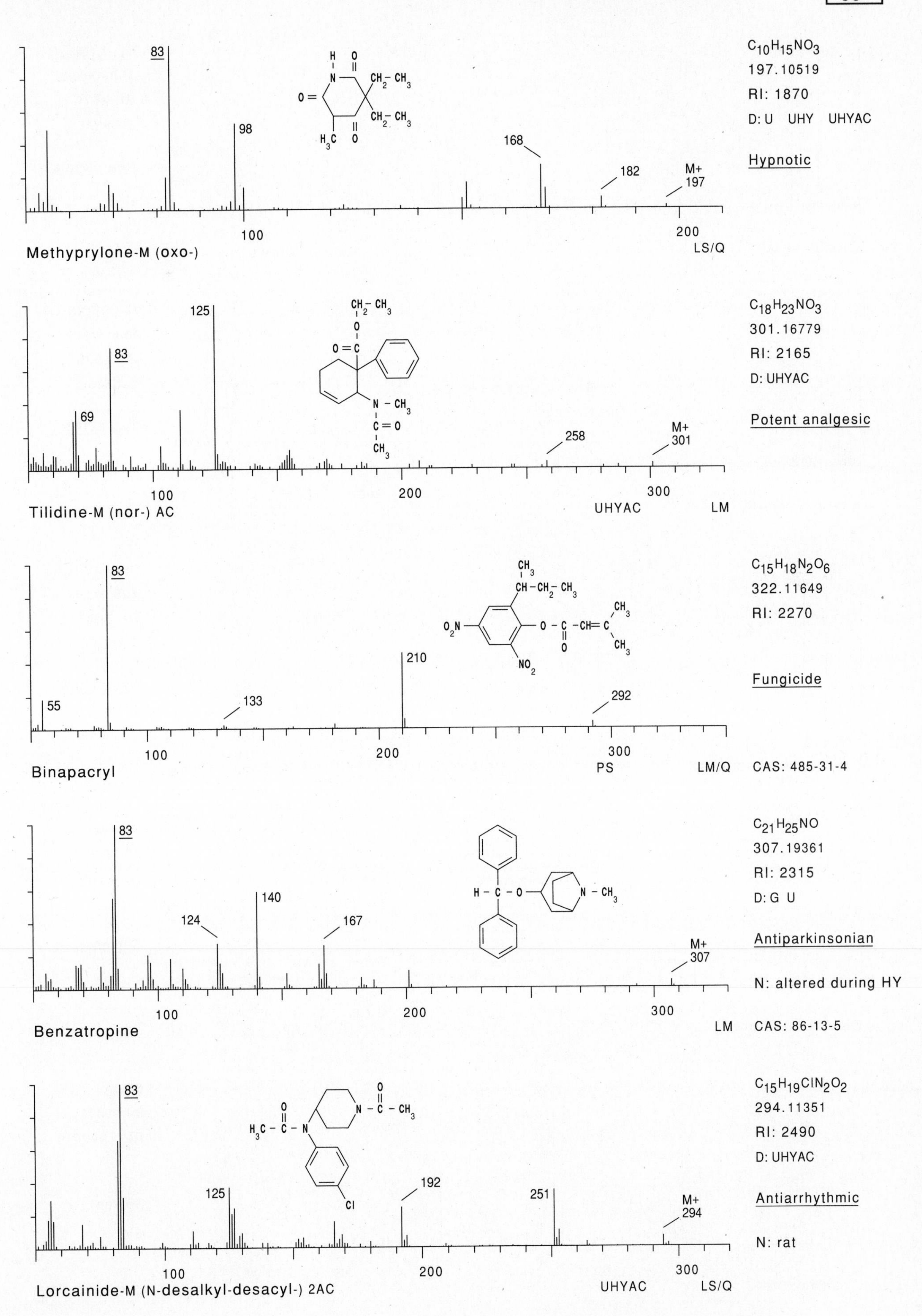

Methyprylone-M (oxo-)
LS/Q
$C_{10}H_{15}NO_3$
197.10519
RI: 1870
D: U UHY UHYAC
Hypnotic
Tilidine-M (nor-) AC
UHYAC LM
$C_{18}H_{23}NO_3$
301.16779
RI: 2165
D: UHYAC
Potent analgesic
Binapacryl
PS LM/Q
$C_{15}H_{18}N_2O_6$
322.11649
RI: 2270
Fungicide
CAS: 485-31-4
Benzatropine
LM
$C_{21}H_{25}NO$
307.19361
RI: 2315
D: G U
Antiparkinsonian
N: altered during HY
CAS: 86-13-5
Lorcainide-M (N-desalkyl-desacyl-) 2AC
UHYAC LS/Q
$C_{15}H_{19}ClN_2O_2$
294.11351
RI: 2490
D: UHYAC
Antiarrhythmic
N: rat

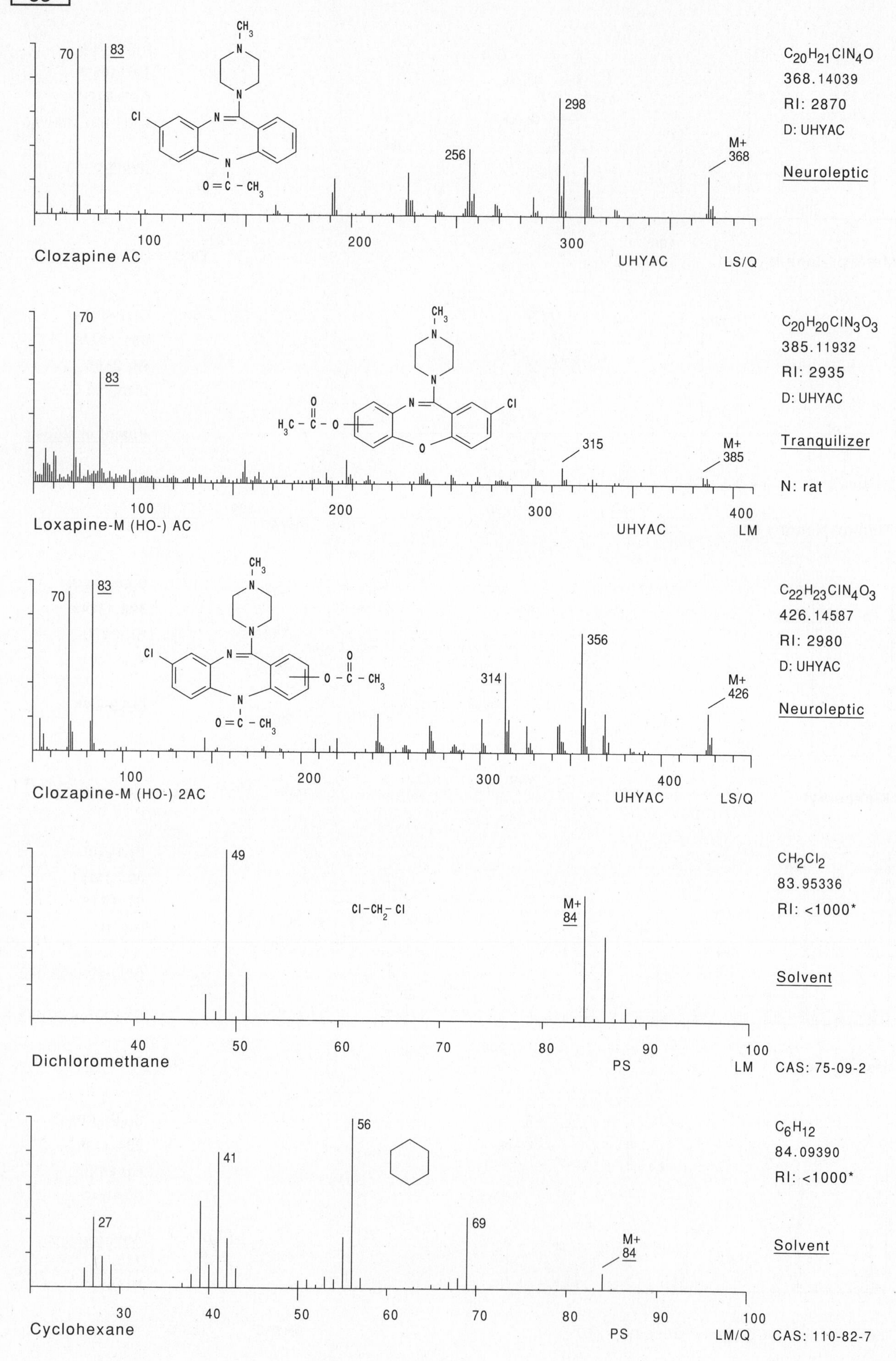

70
83
256
298
M+
368
100
200
300
Clozapine AC
UHYAC
LS/Q
$C_{20}H_{21}ClN_4O$
368.14039
RI: 2870
D: UHYAC
Neuroleptic
70
83
315
M+
385
100
200
300
400
Loxapine-M (HO-) AC
UHYAC
LM
$C_{20}H_{20}ClN_3O_3$
385.11932
RI: 2935
D: UHYAC
Tranquilizer
N: rat
70
83
314
356
M+
426
100
200
300
400
Clozapine-M (HO-) 2AC
UHYAC
LS/Q
$C_{22}H_{23}ClN_4O_3$
426.14587
RI: 2980
D: UHYAC
Neuroleptic
49
Cl-CH2-Cl
M+
84
40
50
60
70
80
90
100
Dichloromethane
PS
LM
CH_2Cl_2
83.95336
RI: <1000*
Solvent
CAS: 75-09-2
27
41
56
69
M+
84
30
40
50
60
70
80
90
100
Cyclohexane
PS
LM/Q
C_6H_{12}
84.09390
RI: <1000*
Solvent
CAS: 110-82-7

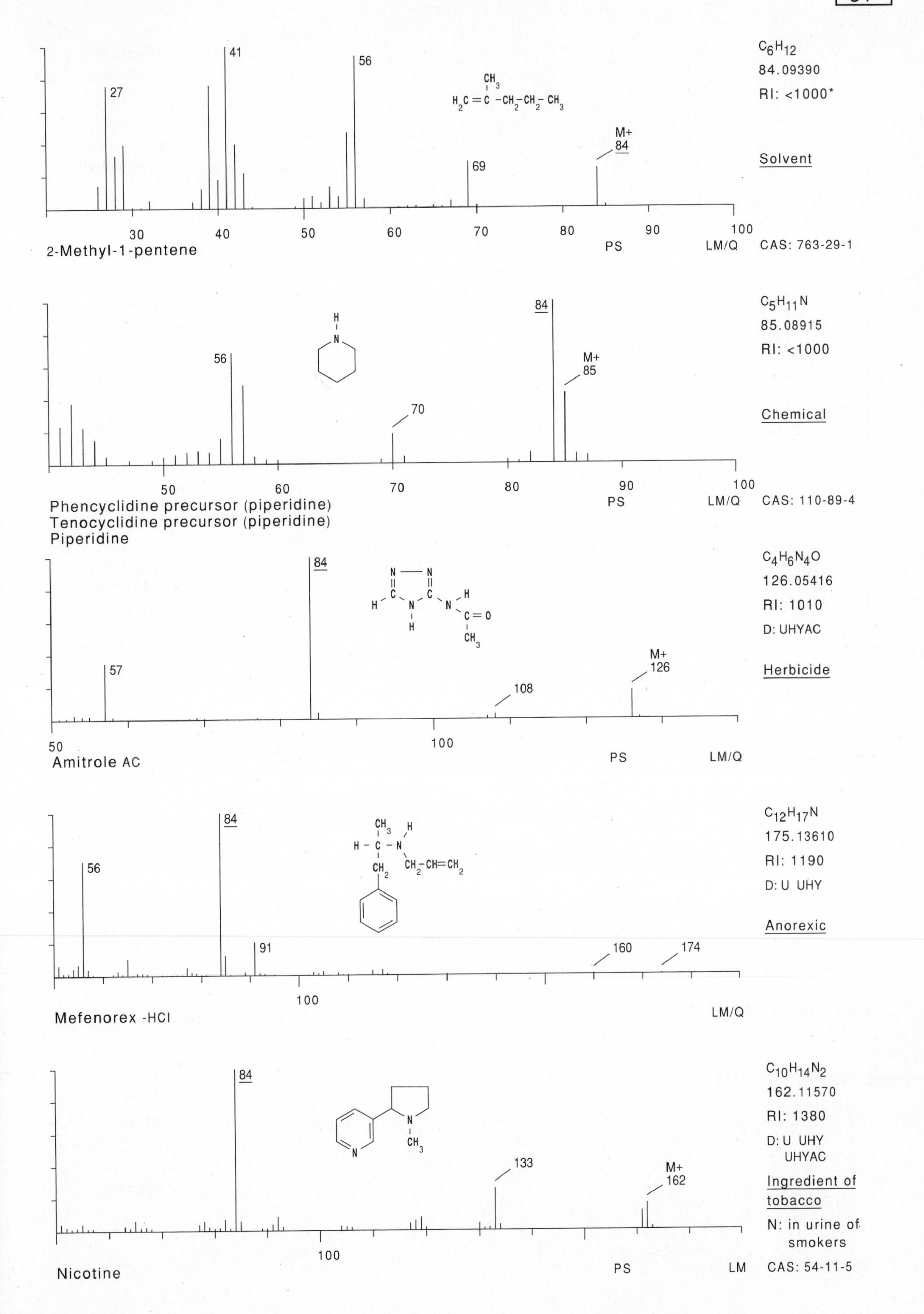

27
41
56
69
M+
84
30
40
50
60
70
80
90
100
PS
LM/Q
2-Methyl-1-pentene
C6H12
84.09390
RI: <1000*
Solvent
CAS: 763-29-1
56
70
84
M+
85
50
60
70
80
90
100
PS
LM/Q
Phencyclidine precursor (piperidine)
Tenocyclidine precursor (piperidine)
Piperidine
C5H11N
85.08915
RI: <1000
Chemical
CAS: 110-89-4
57
84
108
M+
126
50
100
PS
LM/Q
Amitrole AC
C4H6N4O
126.05416
RI: 1010
D: UHYAC
Herbicide
56
84
91
160
174
100
LM/Q
Mefenorex -HCl
C12H17N
175.13610
RI: 1190
D: U UHY
Anorexic
84
133
M+
162
100
PS
LM
Nicotine
C10H14N2
162.11570
RI: 1380
D: U UHY
UHYAC
Ingredient of tobacco
N: in urine of smokers
CAS: 54-11-5

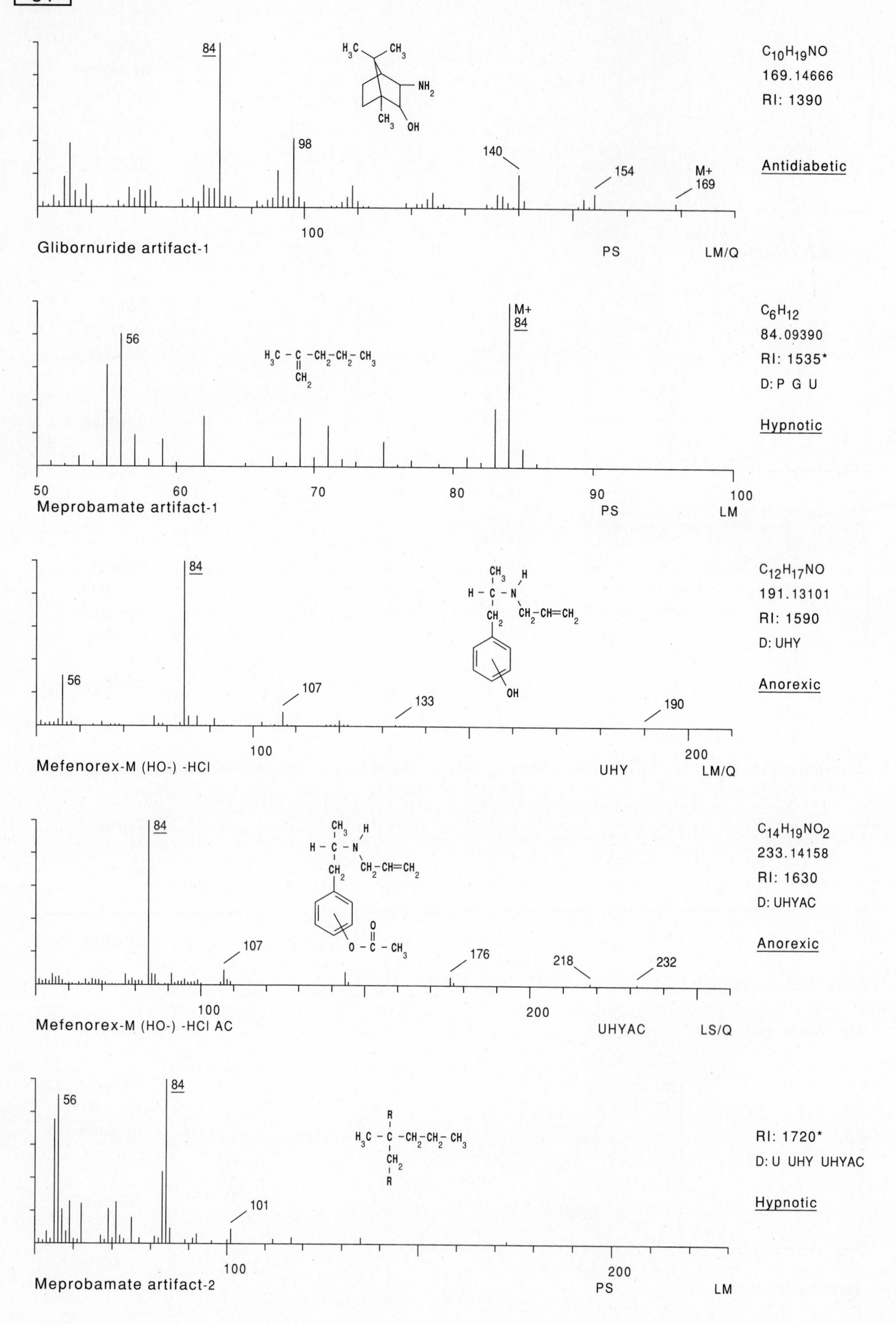

84
98
140
154
M+
169
100
Glibornuride artifact-1
PS
LM/Q
$C_{10}H_{19}NO$
169.14666
RI: 1390
Antidiabetic
M+
84
56
50
60
70
80
90
100
Meprobamate artifact-1
PS
LM
C_6H_{12}
84.09390
RI: 1535*
D: P G U
Hypnotic
84
56
107
133
190
100
200
Mefenorex-M (HO-) -HCl
UHY
LM/Q
$C_{12}H_{17}NO$
191.13101
RI: 1590
D: UHY
Anorexic
84
107
176
218
232
100
200
Mefenorex-M (HO-) -HCl AC
UHYAC
LS/Q
$C_{14}H_{19}NO_2$
233.14158
RI: 1630
D: UHYAC
Anorexic
84
56
101
100
200
Meprobamate artifact-2
PS
LM
RI: 1720*
D: U UHY UHYAC
Hypnotic

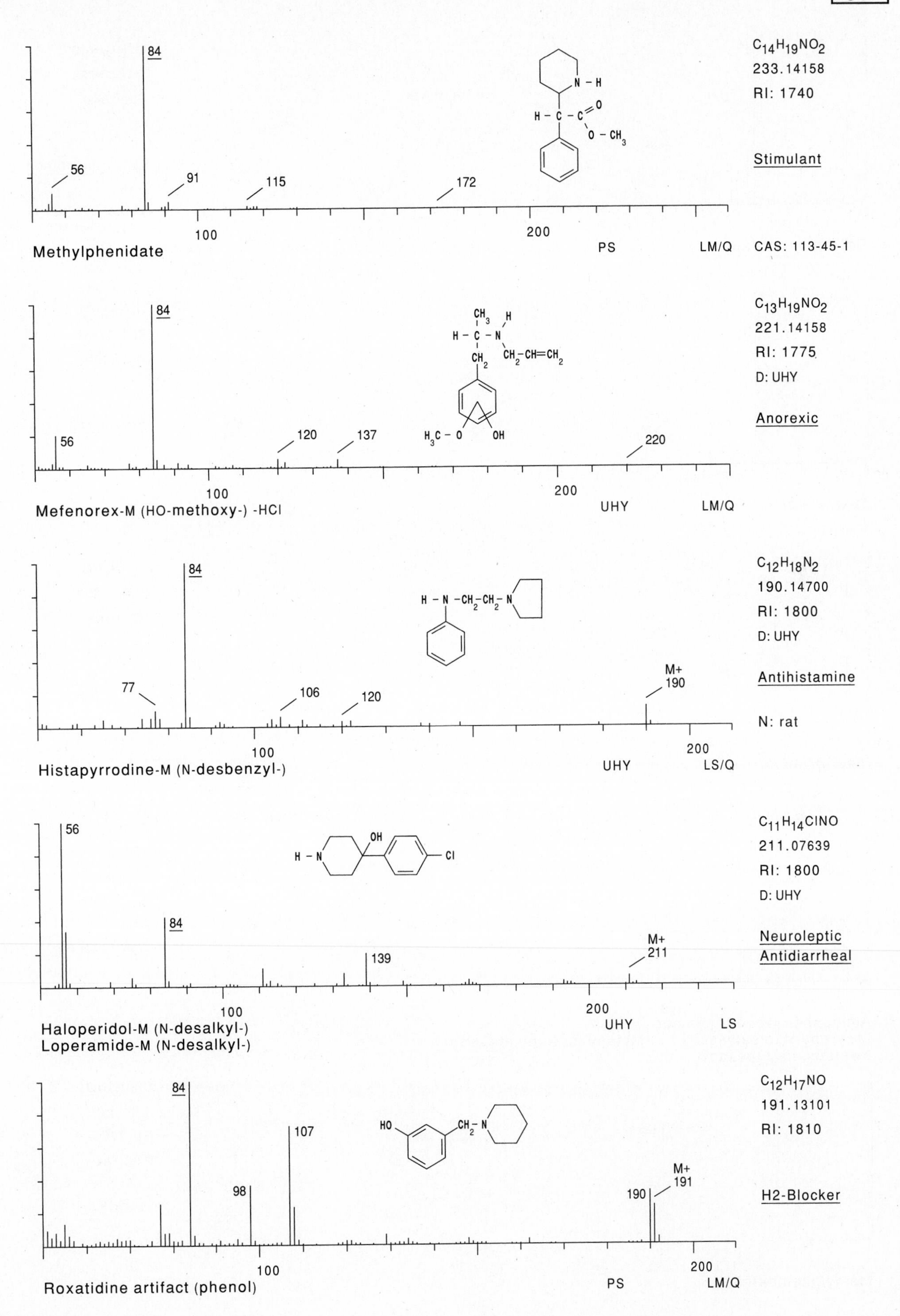
Methylphenidate
PS
LM/Q
$C_{14}H_{19}NO_2$
233.14158
RI: 1740
Stimulant
CAS: 113-45-1
Mefenorex-M (HO-methoxy-) -HCl
UHY
LM/Q
$C_{13}H_{19}NO_2$
221.14158
RI: 1775
D: UHY
Anorexic
Histapyrrodine-M (N-desbenzyl-)
UHY
LS/Q
$C_{12}H_{18}N_2$
190.14700
RI: 1800
D: UHY
Antihistamine
N: rat
Haloperidol-M (N-desalkyl-)
Loperamide-M (N-desalkyl-)
UHY
LS
$C_{11}H_{14}ClNO$
211.07639
RI: 1800
D: UHY
Neuroleptic
Antidiarrheal
Roxatidine artifact (phenol)
PS
LM/Q
$C_{12}H_{17}NO$
191.13101
RI: 1810
H2-Blocker

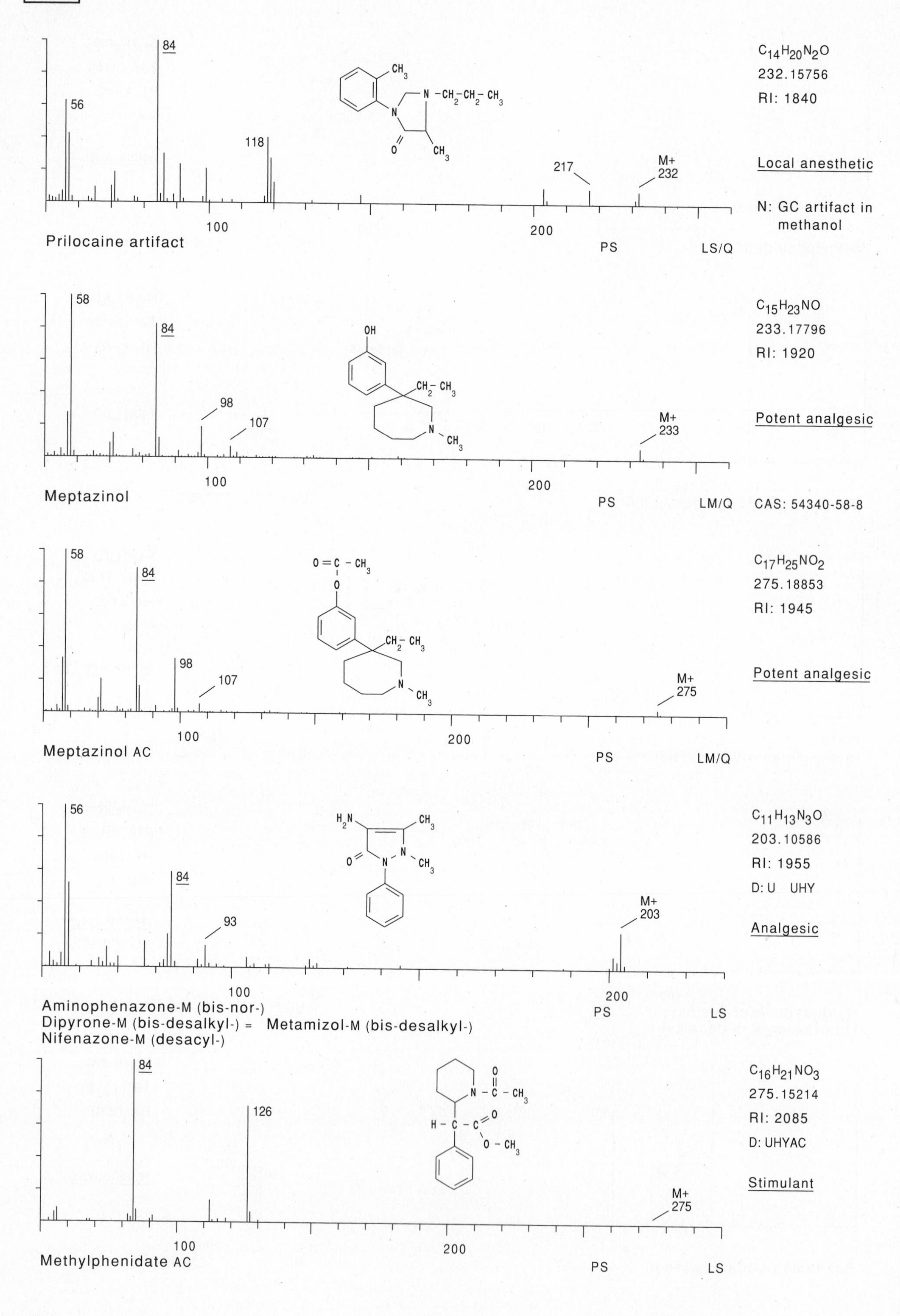

84
56
118
217
M+
232
100
200
Prilocaine artifact
PS
LS/Q
$C_{14}H_{20}N_2O$
232.15756
RI: 1840
Local anesthetic
N: GC artifact in methanol
58
84
98
107
M+
233
Meptazinol
LM/Q
$C_{15}H_{23}NO$
233.17796
RI: 1920
Potent analgesic
CAS: 54340-58-8
58
84
98
107
M+
275
Meptazinol AC
LM/Q
$C_{17}H_{25}NO_2$
275.18853
RI: 1945
Potent analgesic
56
84
93
M+
203
Aminophenazone-M (bis-nor-)
Dipyrone-M (bis-desalkyl-) = Metamizol-M (bis-desalkyl-)
Nifenazone-M (desacyl-)
LS
$C_{11}H_{13}N_3O$
203.10586
RI: 1955
D: U UHY
Analgesic
84
126
M+
275
Methylphenidate AC
LS
$C_{16}H_{21}NO_3$
275.15214
RI: 2085
D: UHYAC
Stimulant

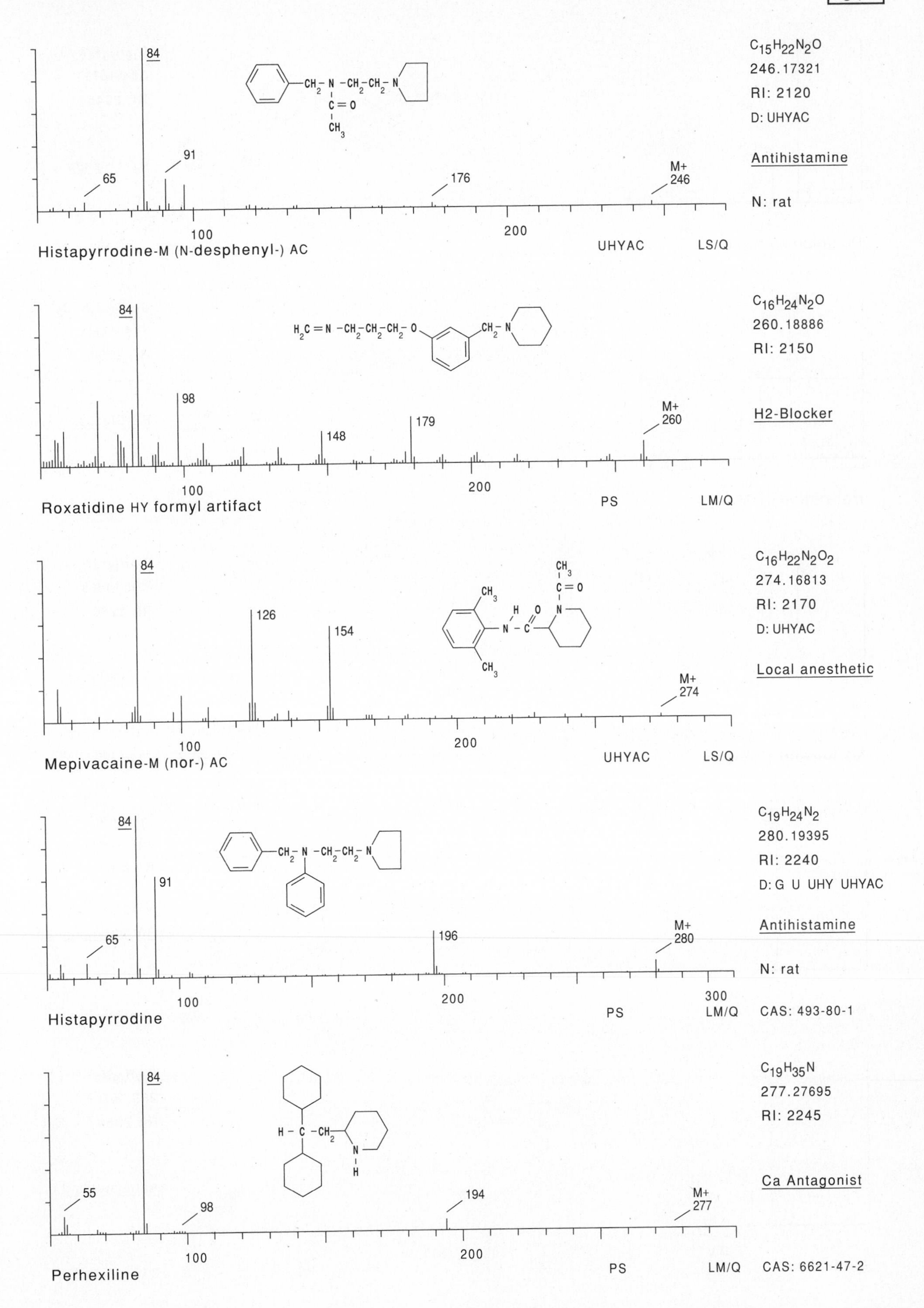

84
91
65
176
M+
246
100
200
Histapyrrodine-M (N-desphenyl-) AC
UHYAC
LS/Q
C15H22N2O
246.17321
RI: 2120
D: UHYAC
Antihistamine
N: rat
84
98
179
148
M+
260
100
200
Roxatidine HY formyl artifact
PS
LM/Q
C16H24N2O
260.18886
RI: 2150
H2-Blocker
84
126
154
M+
274
100
200
Mepivacaine-M (nor-) AC
UHYAC
LS/Q
C16H22N2O2
274.16813
RI: 2170
D: UHYAC
Local anesthetic
84
91
65
196
M+
280
100
200
300
Histapyrrodine
PS
LM/Q
C19H24N2
280.19395
RI: 2240
D: G U UHY UHYAC
Antihistamine
N: rat
CAS: 493-80-1
84
55
98
194
M+
277
100
200
Perhexiline
PS
LM/Q
C19H35N
277.27695
RI: 2245
Ca Antagonist
CAS: 6621-47-2

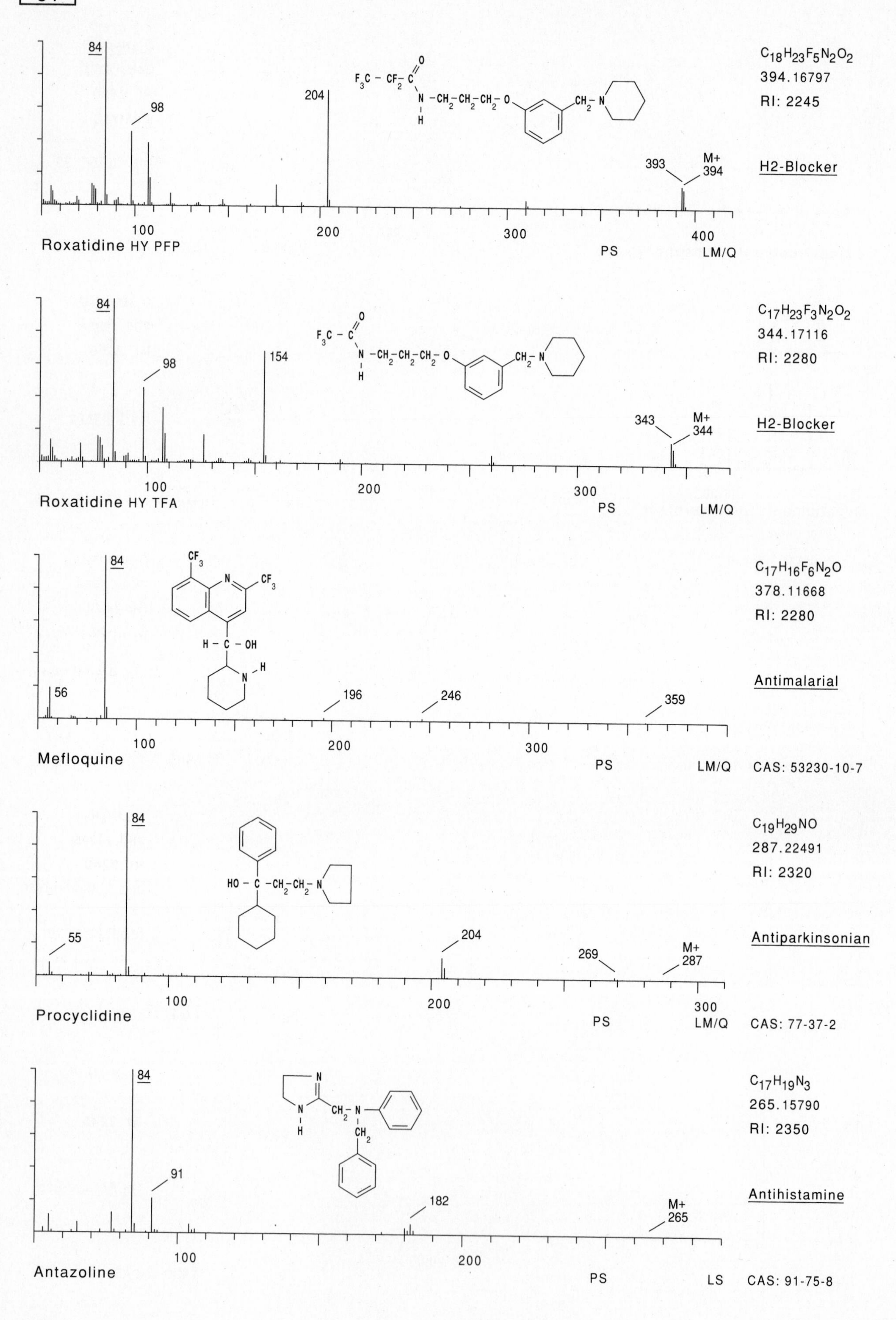

84
98
204
393
M+
394
100
200
300
400
Roxatidine HY PFP
PS
LM/Q
C18H23F5N2O2
394.16797
RI: 2245
H2-Blocker
84
98
154
343
M+
344
100
200
300
Roxatidine HY TFA
PS
LM/Q
C17H23F3N2O2
344.17116
RI: 2280
H2-Blocker
84
56
196
246
359
100
200
300
Mefloquine
PS
LM/Q
C17H16F6N2O
378.11668
RI: 2280
Antimalarial
CAS: 53230-10-7
84
55
204
269
M+
287
100
200
300
Procyclidine
PS
LM/Q
C19H29NO
287.22491
RI: 2320
Antiparkinsonian
CAS: 77-37-2
84
91
182
M+
265
100
200
Antazoline
PS
LS
C17H19N3
265.15790
RI: 2350
Antihistamine
CAS: 91-75-8

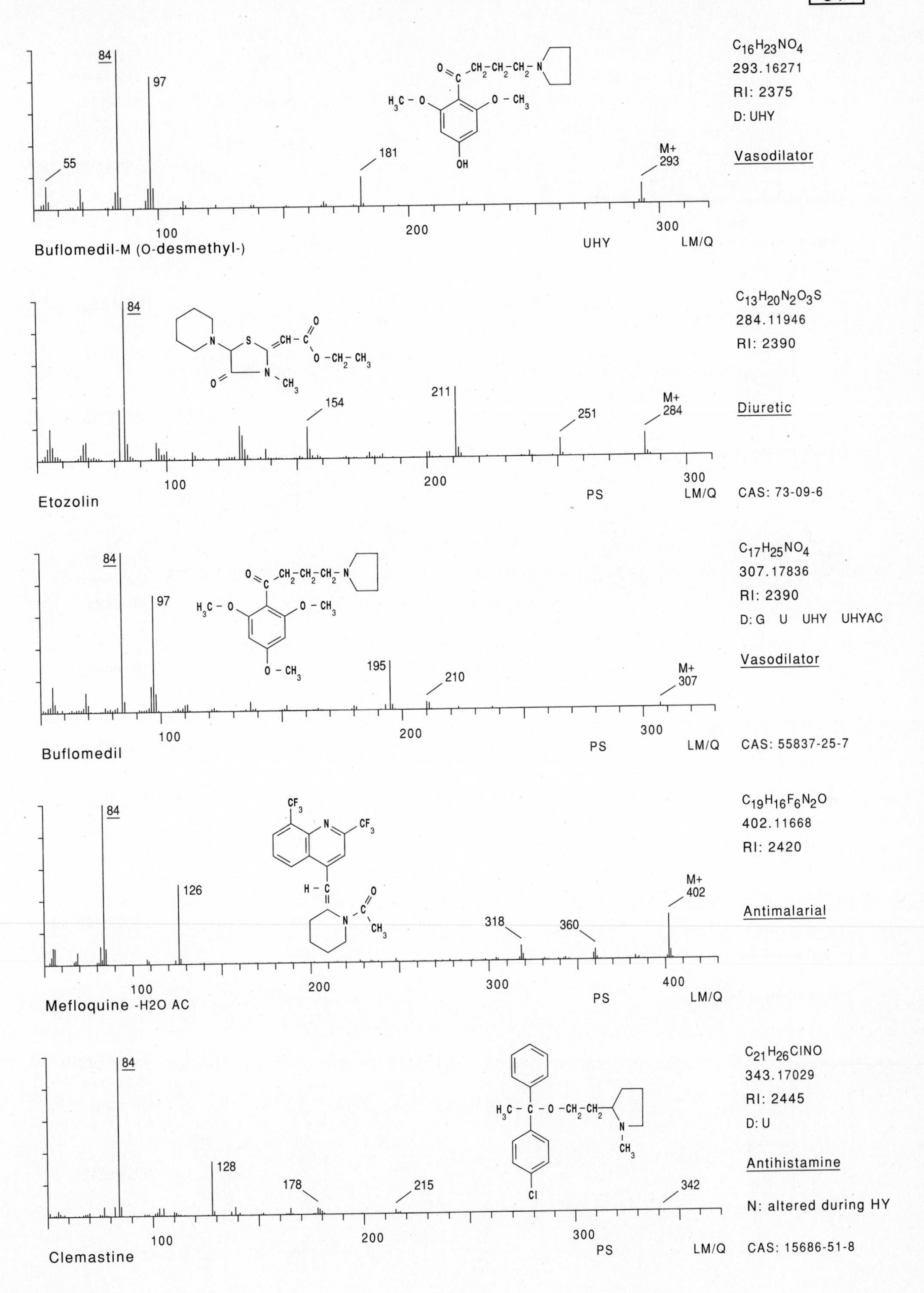
Buflomedil-M (O-desmethyl-)
$C_{16}H_{23}NO_4$
293.16271
RI: 2375
D: UHY
Vasodilator
UHY
LM/Q
Etozolin
$C_{13}H_{20}N_2O_3S$
284.11946
RI: 2390
Diuretic
PS
LM/Q
CAS: 73-09-6
Buflomedil
$C_{17}H_{25}NO_4$
307.17836
RI: 2390
D: G U UHY UHYAC
Vasodilator
PS
LM/Q
CAS: 55837-25-7
Mefloquine -H2O AC
$C_{19}H_{16}F_6N_2O$
402.11668
RI: 2420
Antimalarial
PS
LM/Q
Clemastine
$C_{21}H_{26}ClNO$
343.17029
RI: 2445
D: U
Antihistamine
N: altered during HY
PS
LM/Q
CAS: 15686-51-8

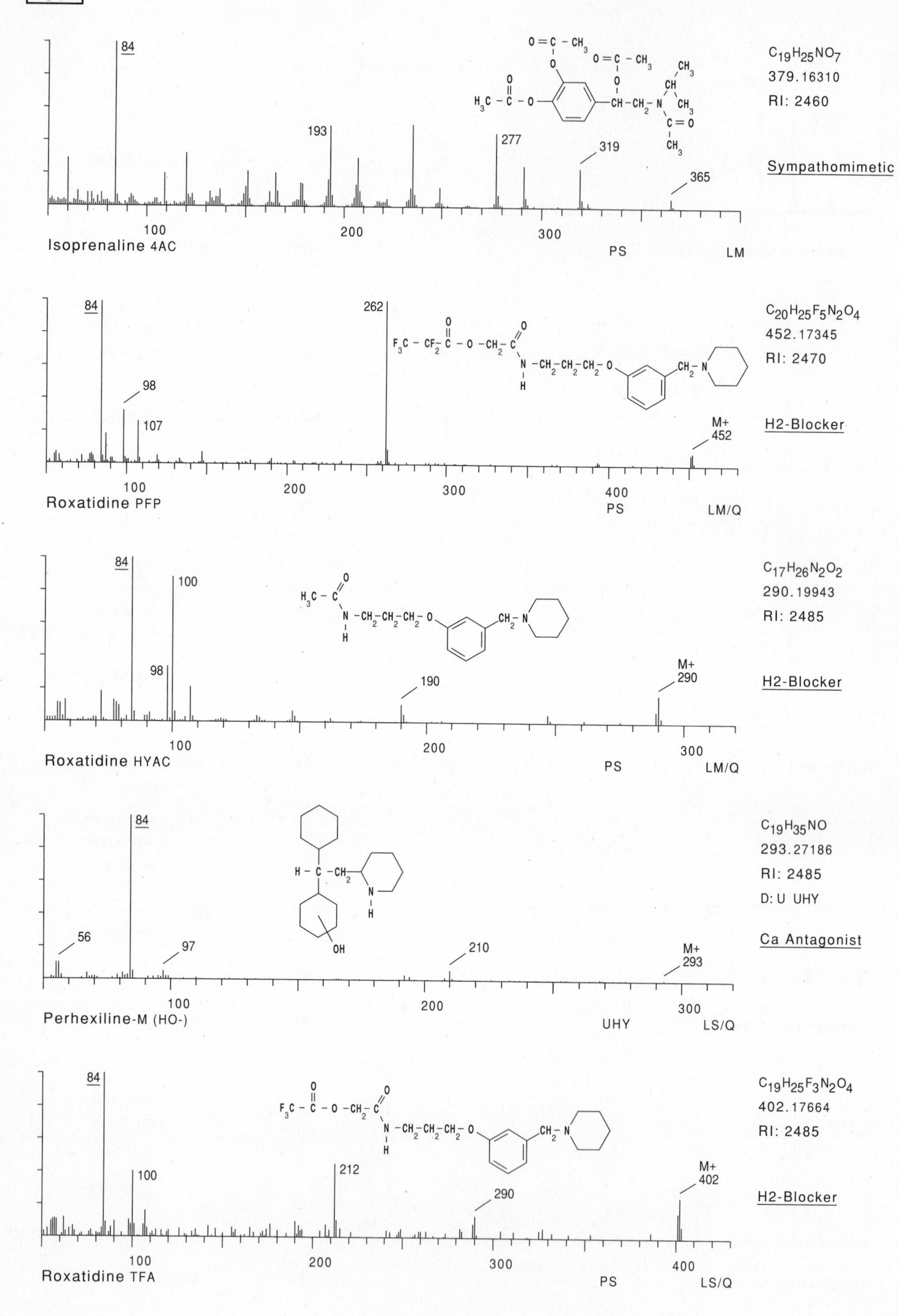
84
193
277
319
365
100
200
300
Isoprenaline 4AC
PS
LM
$C_{19}H_{25}NO_7$
379.16310
RI: 2460
Sympathomimetic
84
262
98
107
M+
452
Roxatidine PFP
PS
LM/Q
$C_{20}H_{25}F_5N_2O_4$
452.17345
RI: 2470
H2-Blocker
84
100
98
190
M+
290
Roxatidine HYAC
PS
LM/Q
$C_{17}H_{26}N_2O_2$
290.19943
RI: 2485
H2-Blocker
84
56
97
210
M+
293
Perhexiline-M (HO-)
UHY
LS/Q
$C_{19}H_{35}NO$
293.27186
RI: 2485
D: U UHY
Ca Antagonist
84
100
212
290
M+
402
Roxatidine TFA
PS
LS/Q
$C_{19}H_{25}F_3N_2O_4$
402.17664
RI: 2485
H2-Blocker

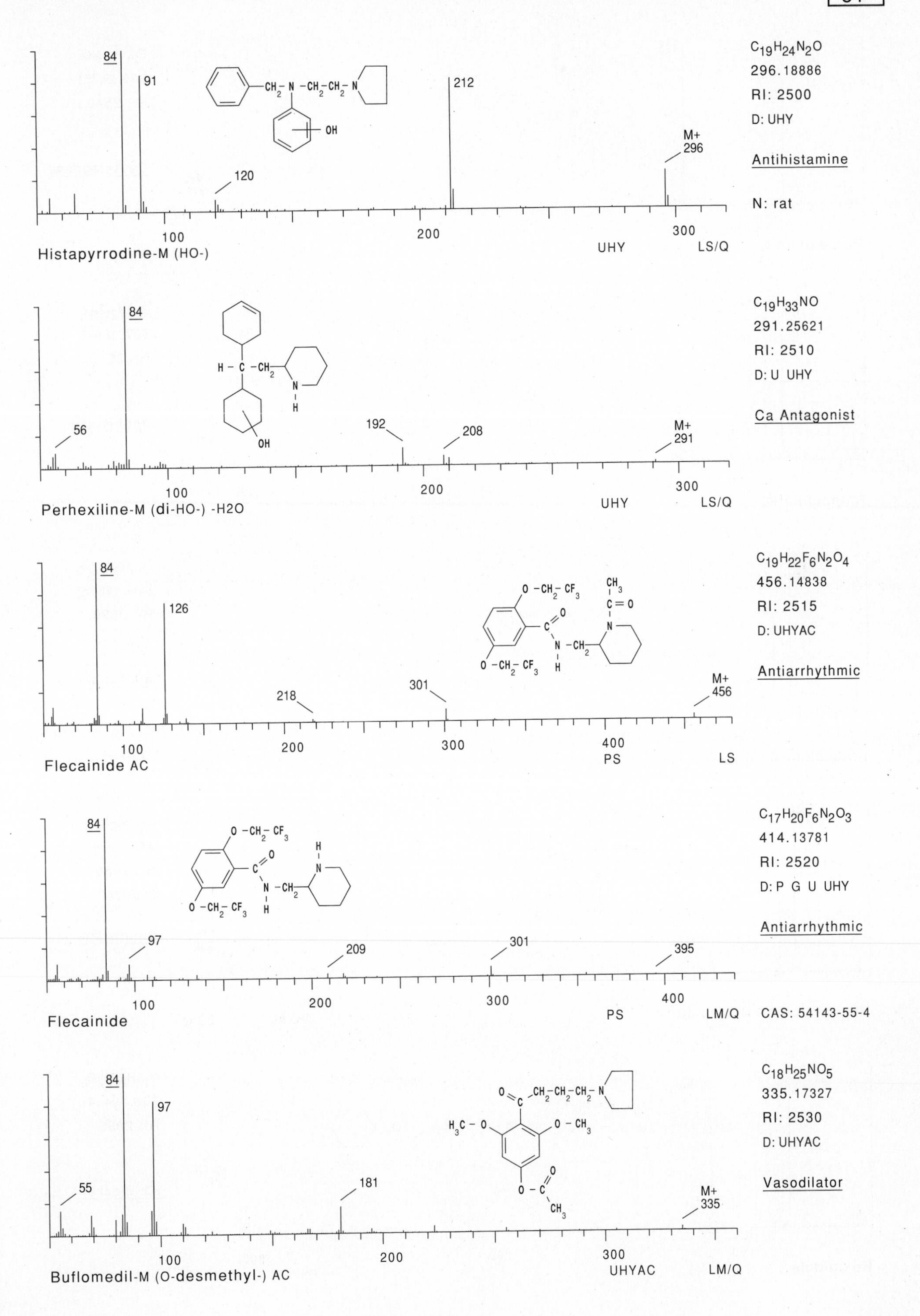

84
91
212
120
M+
296
100
200
300
UHY
LS/Q
Histapyrrodine-M (HO-)
$C_{19}H_{24}N_2O$
296.18886
RI: 2500
D: UHY
Antihistamine
N: rat
84
56
192
208
M+
291
Perhexiline-M (di-HO-) -H2O
D: U UHY
$C_{19}H_{33}NO$
291.25621
RI: 2510
Ca Antagonist
84
126
218
301
M+
456
400
PS
LS
Flecainide AC
$C_{19}H_{22}F_6N_2O_4$
456.14838
RI: 2515
D: UHYAC
Antiarrhythmic
84
97
209
301
395
LM/Q
Flecainide
$C_{17}H_{20}F_6N_2O_3$
414.13781
RI: 2520
D: P G U UHY
Antiarrhythmic
CAS: 54143-55-4
84
97
55
181
M+
335
UHYAC
Buflomedil-M (O-desmethyl-) AC
$C_{18}H_{25}NO_5$
335.17327
RI: 2530
D: UHYAC
Vasodilator

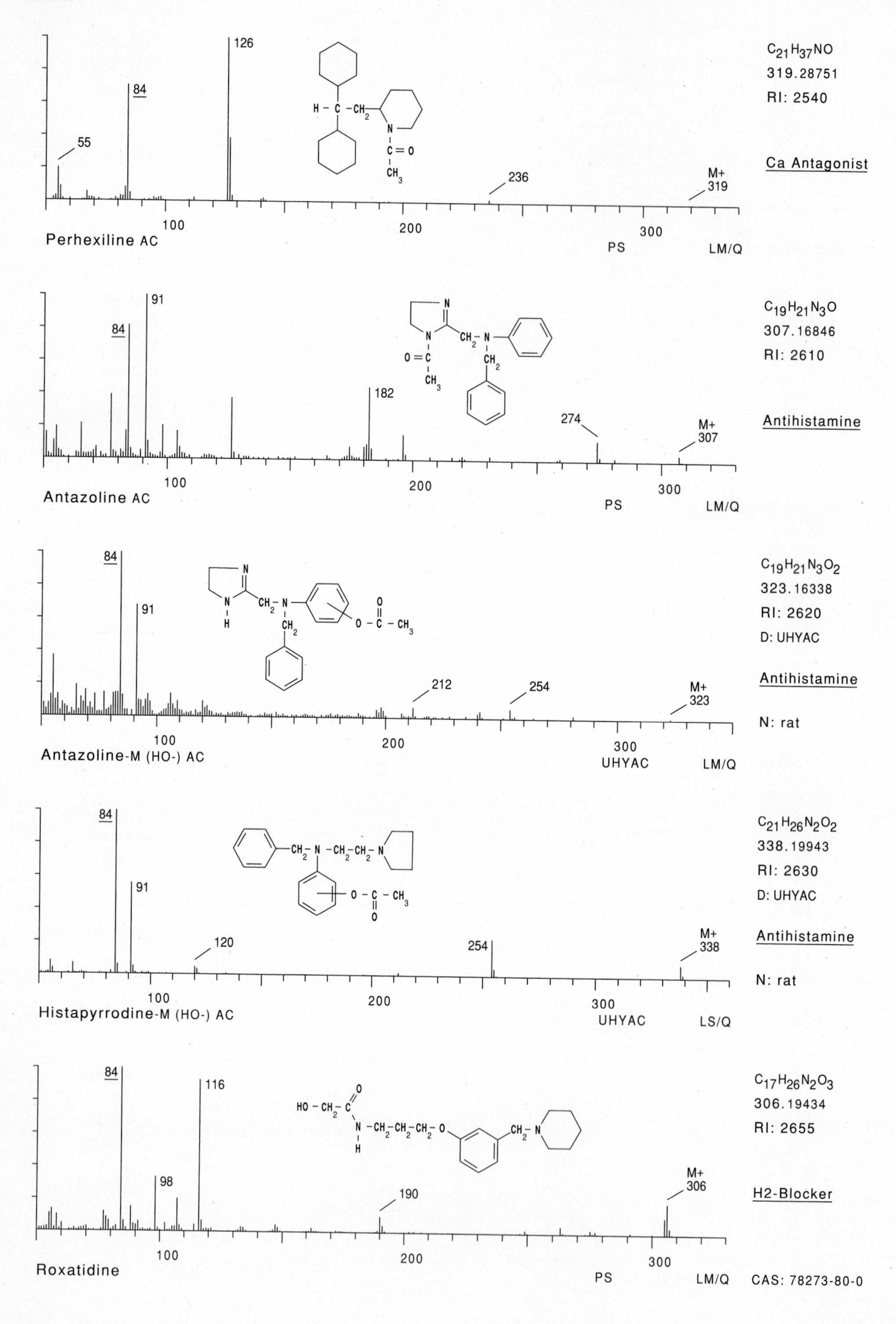

126
84
55
236
M+
319
100
200
300
Perhexiline AC
PS
LM/Q
C21H37NO
319.28751
RI: 2540
Ca Antagonist
91
84
182
274
M+
307
Antazoline AC
LM/Q
C19H21N3O
307.16846
RI: 2610
Antihistamine
84
91
212
254
M+
323
Antazoline-M (HO-) AC
UHYAC
C19H21N3O2
323.16338
RI: 2620
D: UHYAC
N: rat
84
91
120
254
M+
338
Histapyrrodine-M (HO-) AC
LS/Q
C21H26N2O2
338.19943
RI: 2630
84
116
98
190
M+
306
Roxatidine
C17H26N2O3
306.19434
RI: 2655
H2-Blocker
CAS: 78273-80-0

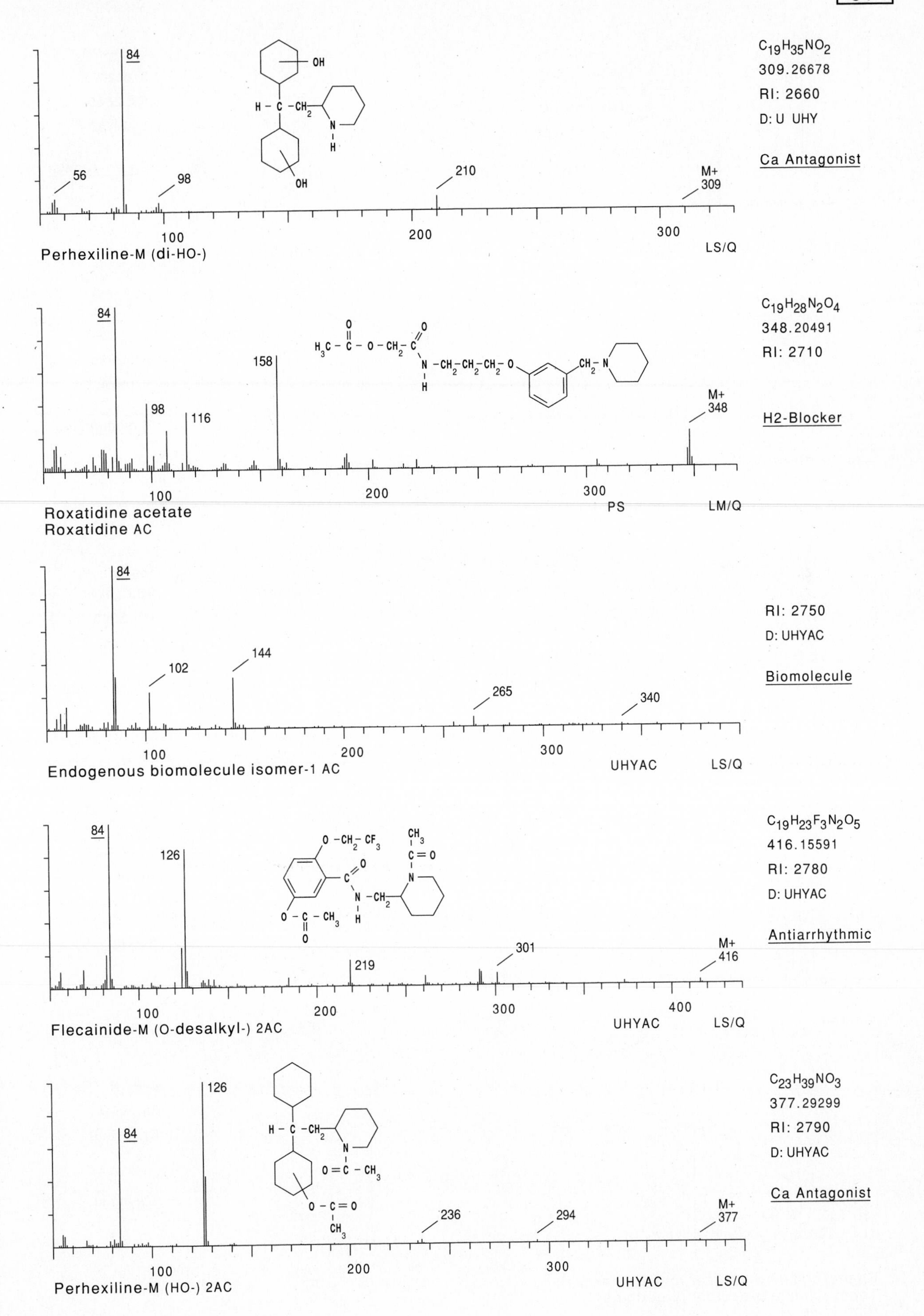

$C_{19}H_{35}NO_2$
309.26678
RI: 2660
D: U UHY
Ca Antagonist

$C_{19}H_{28}N_2O_4$
348.20491
RI: 2710
H2-Blocker

RI: 2750
D: UHYAC
Biomolecule

$C_{19}H_{23}F_3N_2O_5$
416.15591
RI: 2780
D: UHYAC
Antiarrhythmic

$C_{23}H_{39}NO_3$
377.29299
RI: 2790
D: UHYAC
Ca Antagonist

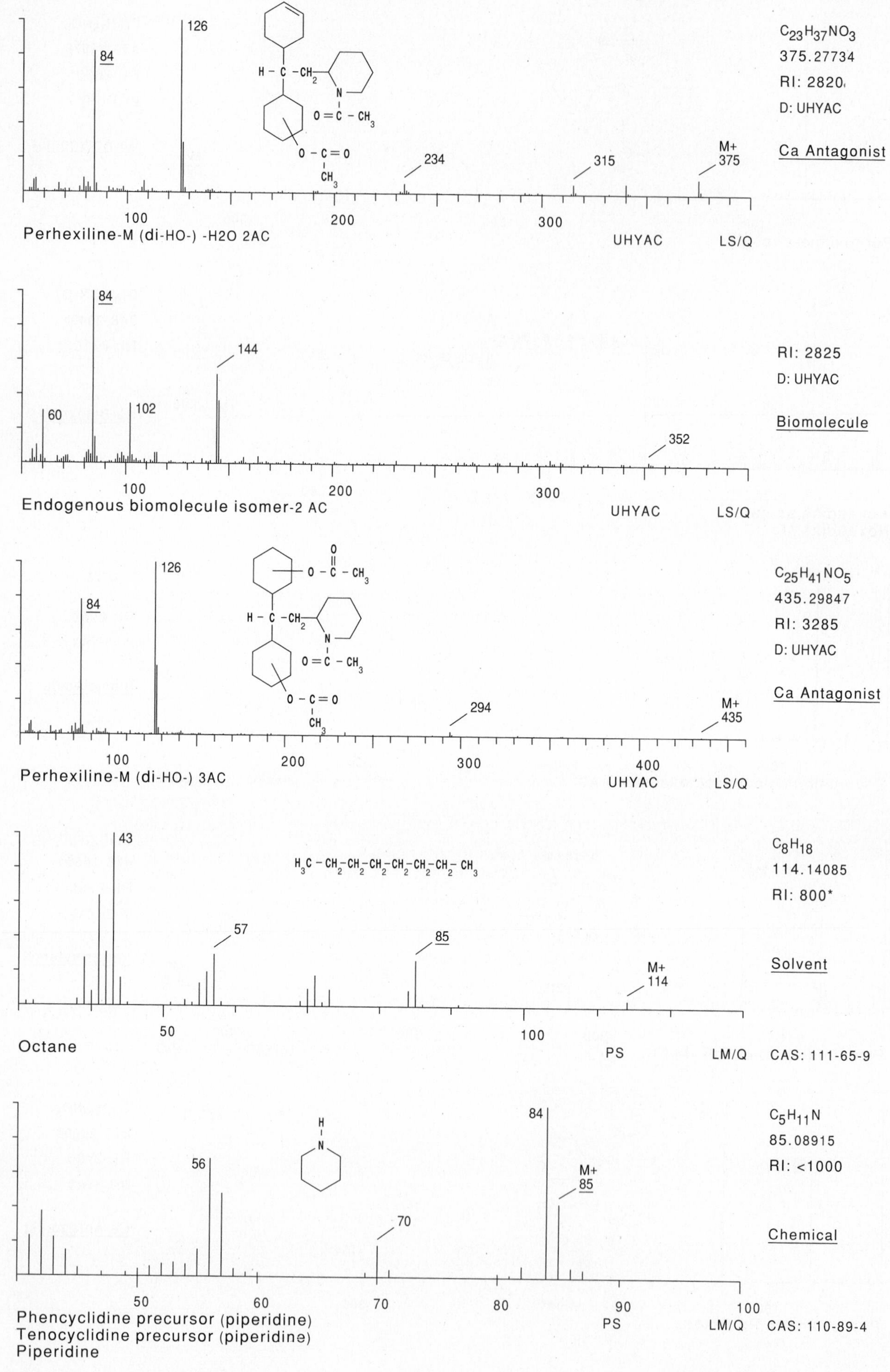

126
84
234
315
M+
375
100
200
300
Perhexiline-M (di-HO-) -H2O 2AC
UHYAC
LS/Q
$C_{23}H_{37}NO_3$
375.27734
RI: 2820
D: UHYAC
Ca Antagonist
84
144
60
102
352
100
200
300
Endogenous biomolecule isomer-2 AC
UHYAC
LS/Q
RI: 2825
D: UHYAC
Biomolecule
126
84
294
M+
435
100
200
300
400
Perhexiline-M (di-HO-) 3AC
UHYAC
LS/Q
$C_{25}H_{41}NO_5$
435.29847
RI: 3285
D: UHYAC
Ca Antagonist
43
57
85
M+
114
50
100
Octane
PS
LM/Q
CAS: 111-65-9
C_8H_{18}
114.14085
RI: 800*
Solvent
84
56
M+
85
70
50
60
70
80
90
100
Phencyclidine precursor (piperidine)
Tenocyclidine precursor (piperidine)
Piperidine
PS
LM/Q
CAS: 110-89-4
$C_5H_{11}N$
85.08915
RI: <1000
Chemical

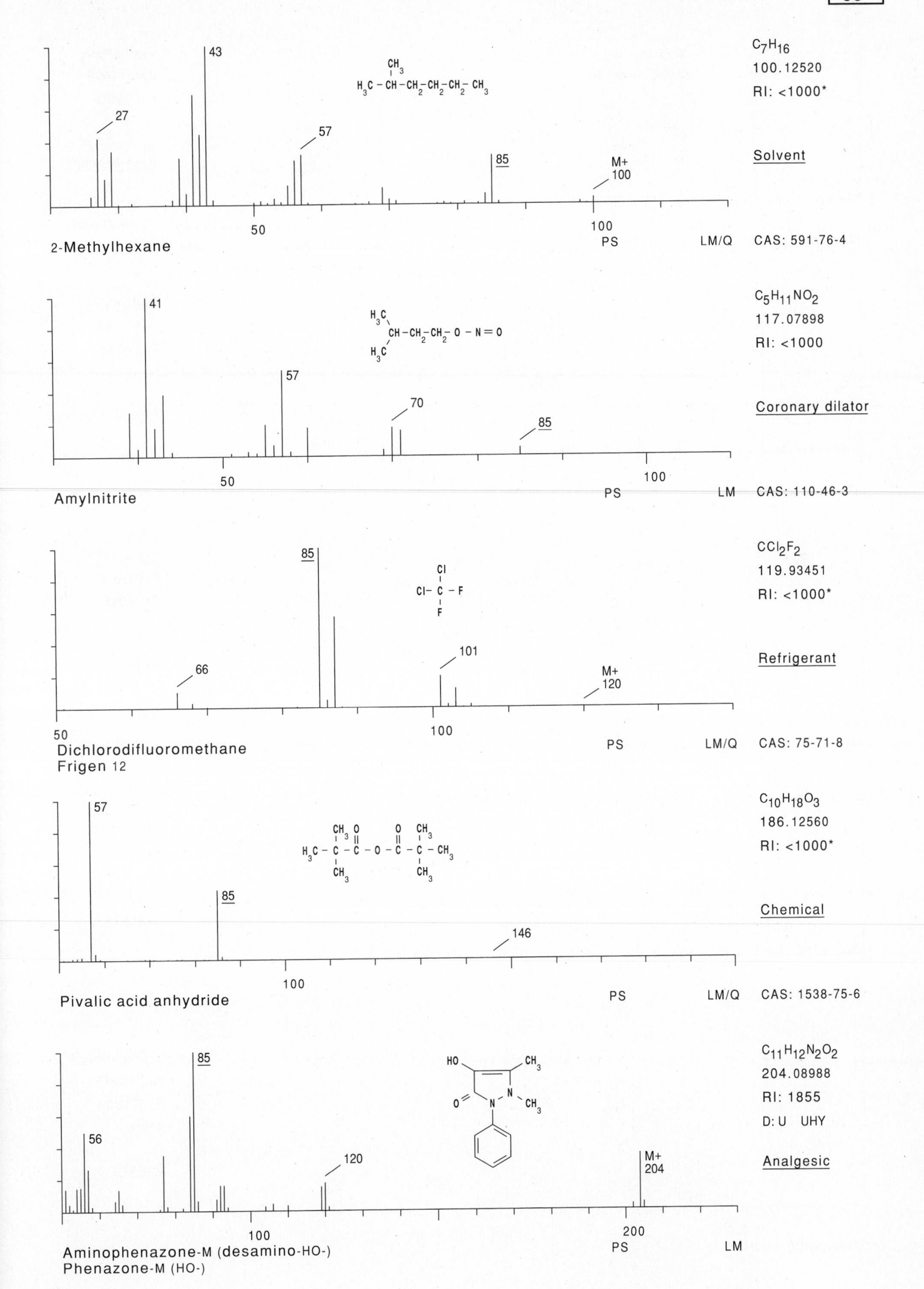

C_7H_{16}
100.12520
RI: <1000*

Solvent

2-Methylhexane

CAS: 591-76-4

$C_5H_{11}NO_2$
117.07898
RI: <1000

Coronary dilator

Amylnitrite

CAS: 110-46-3

CCl_2F_2
119.93451
RI: <1000*

Refrigerant

Dichlorodifluoromethane
Frigen 12

CAS: 75-71-8

$C_{10}H_{18}O_3$
186.12560
RI: <1000*

Chemical

Pivalic acid anhydride

CAS: 1538-75-6

$C_{11}H_{12}N_2O_2$
204.08988
RI: 1855
D: U UHY

Analgesic

Aminophenazone-M (desamino-HO-)
Phenazone-M (HO-)

85

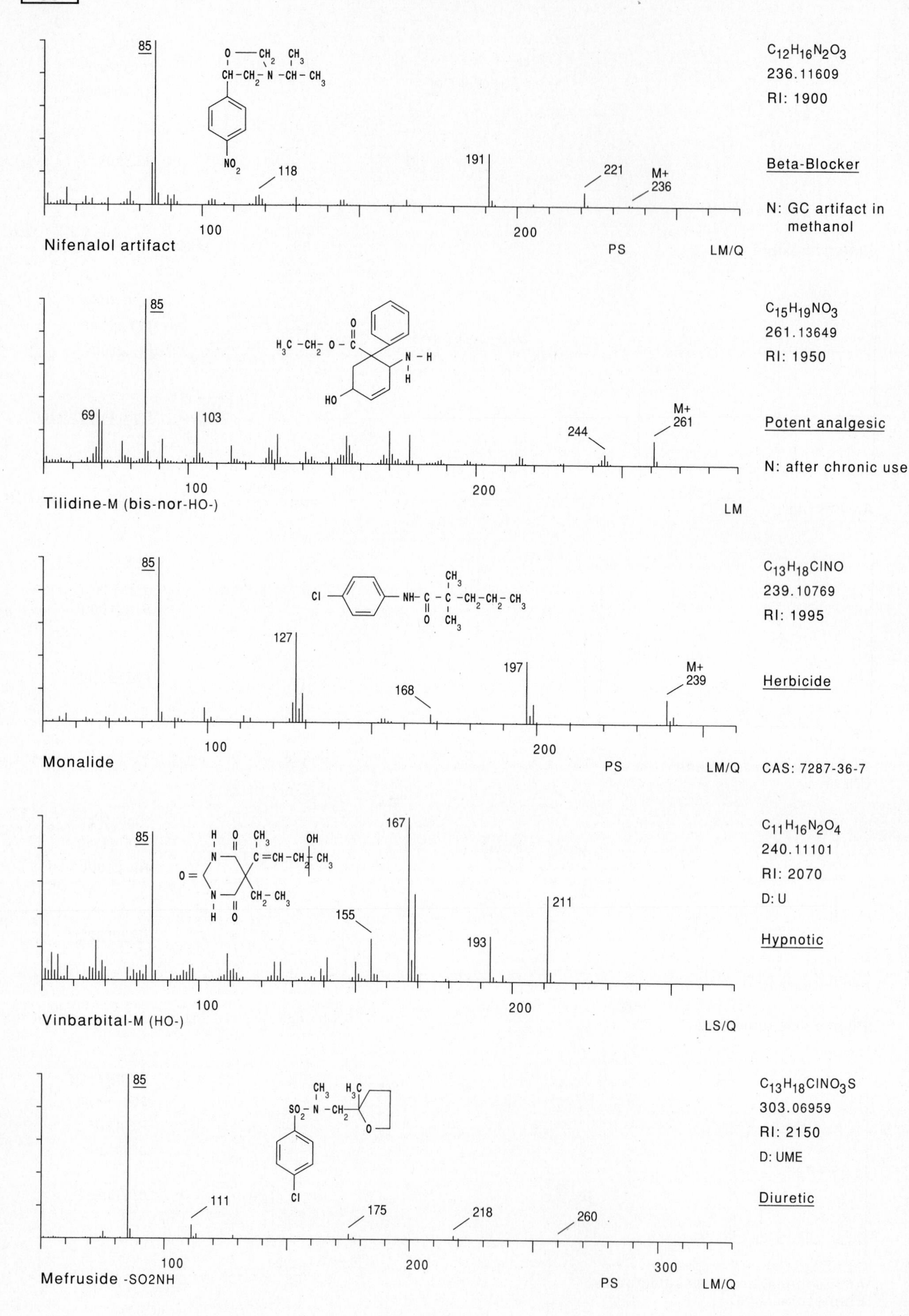

Nifenalol artifact
C12H16N2O3
236.11609
RI: 1900
Beta-Blocker
N: GC artifact in methanol
PS
LM/Q
Tilidine-M (bis-nor-HO-)
C15H19NO3
261.13649
RI: 1950
Potent analgesic
N: after chronic use
LM
Monalide
C13H18ClNO
239.10769
RI: 1995
Herbicide
PS
LM/Q
CAS: 7287-36-7
Vinbarbital-M (HO-)
C11H16N2O4
240.11101
RI: 2070
D: U
Hypnotic
LS/Q
Mefruside -SO2NH
C13H18ClNO3S
303.06959
RI: 2150
D: UME
Diuretic
PS
LM/Q

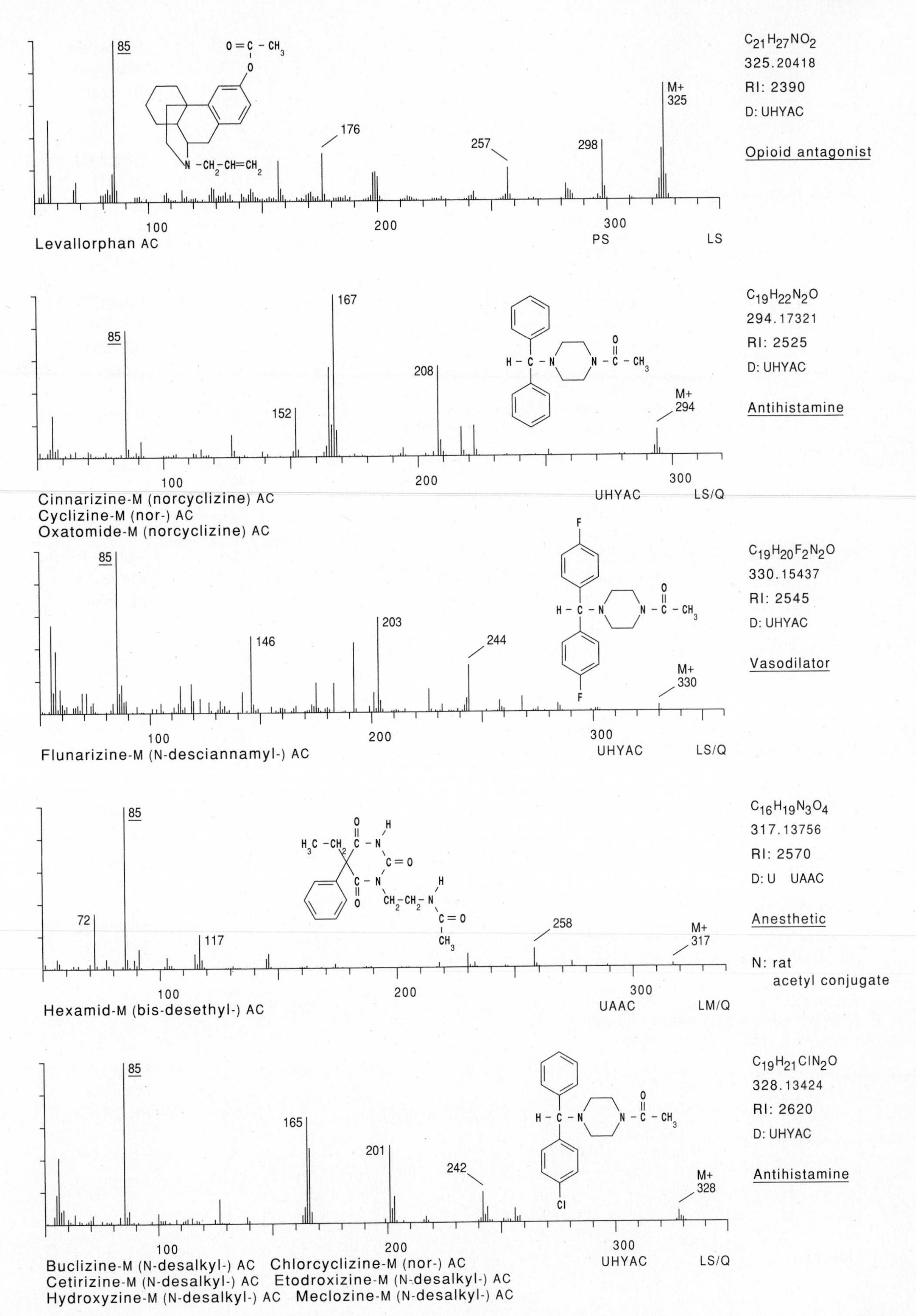
Levallorphan AC
$C_{21}H_{27}NO_2$
325.20418
RI: 2390
D: UHYAC
Opioid antagonist
85
176
257
298
M+
325
PS
LS
Cinnarizine-M (norcyclizine) AC
Cyclizine-M (nor-) AC
Oxatomide-M (norcyclizine) AC
$C_{19}H_{22}N_2O$
294.17321
RI: 2525
D: UHYAC
Antihistamine
85
152
167
208
M+
294
UHYAC
LS/Q
Flunarizine-M (N-desciannamyl-) AC
$C_{19}H_{20}F_2N_2O$
330.15437
RI: 2545
D: UHYAC
Vasodilator
85
146
203
244
M+
330
UHYAC
LS/Q
Hexamid-M (bis-desethyl-) AC
$C_{16}H_{19}N_3O_4$
317.13756
RI: 2570
D: U UAAC
Anesthetic
N: rat
acetyl conjugate
72
85
117
258
M+
317
UAAC
LM/Q
Buclizine-M (N-desalkyl-) AC
Chlorcyclizine-M (nor-) AC
Cetirizine-M (N-desalkyl-) AC
Etodroxizine-M (N-desalkyl-) AC
Hydroxyzine-M (N-desalkyl-) AC
Meclozine-M (N-desalkyl-) AC
$C_{19}H_{21}ClN_2O$
328.13424
RI: 2620
D: UHYAC
Antihistamine
85
165
201
242
M+
328
UHYAC
LS/Q

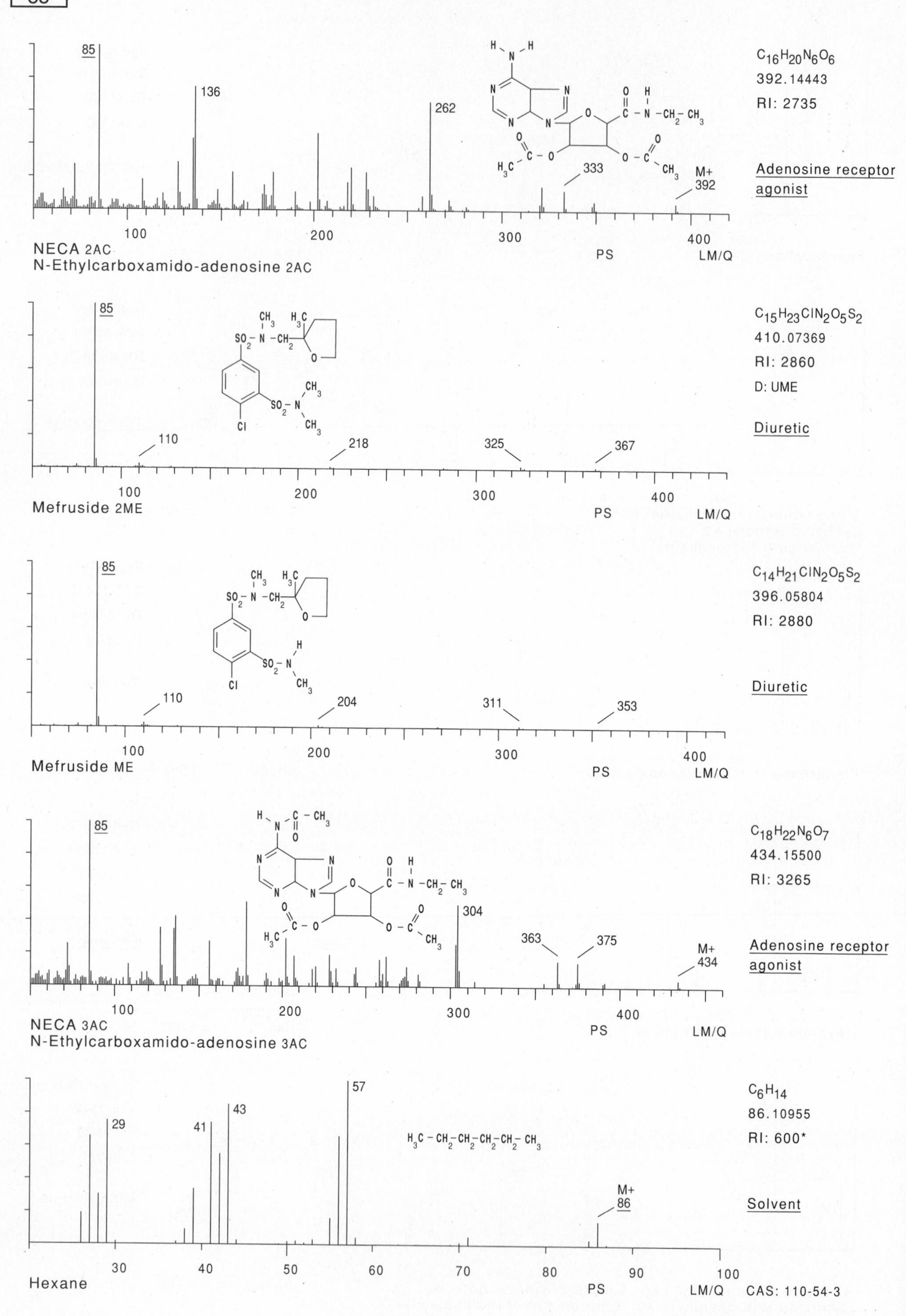
85
136
262
333
M+
392
100
200
300
400
PS
LM/Q
NECA 2AC
N-Ethylcarboxamido-adenosine 2AC
$C_{16}H_{20}N_6O_6$
392.14443
RI: 2735
Adenosine receptor agonist
85
110
218
325
367
PS
LM/Q
Mefruside 2ME
$C_{15}H_{23}ClN_2O_5S_2$
410.07369
RI: 2860
D: UME
Diuretic
85
110
204
311
353
PS
LM/Q
Mefruside ME
$C_{14}H_{21}ClN_2O_5S_2$
396.05804
RI: 2880
Diuretic
85
304
363
375
M+
434
PS
LM/Q
NECA 3AC
N-Ethylcarboxamido-adenosine 3AC
$C_{18}H_{22}N_6O_7$
434.15500
RI: 3265
Adenosine receptor agonist
57
43
29
41
M+
86
30
40
50
60
70
80
90
100
PS
LM/Q
Hexane
C_6H_{14}
86.10955
RI: 600*
Solvent
CAS: 110-54-3

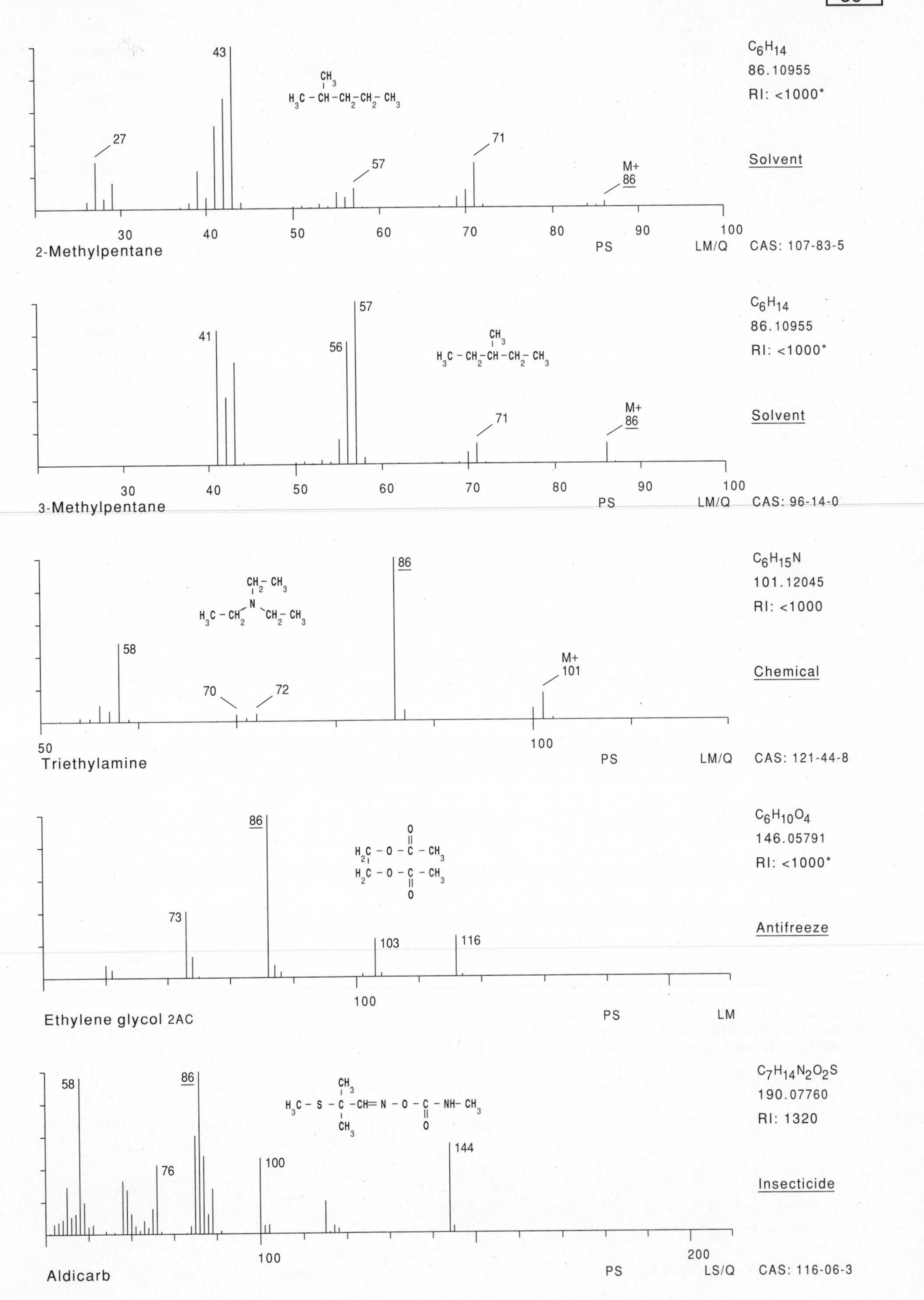

C_6H_{14}
86.10955
RI: <1000*
Solvent
CAS: 107-83-5

C_6H_{14}
86.10955
RI: <1000*
Solvent
CAS: 96-14-0

$C_6H_{15}N$
101.12045
RI: <1000
Chemical
CAS: 121-44-8

$C_6H_{10}O_4$
146.05791
RI: <1000*
Antifreeze

$C_7H_{14}N_2O_2S$
190.07760
RI: 1320
Insecticide
CAS: 116-06-3

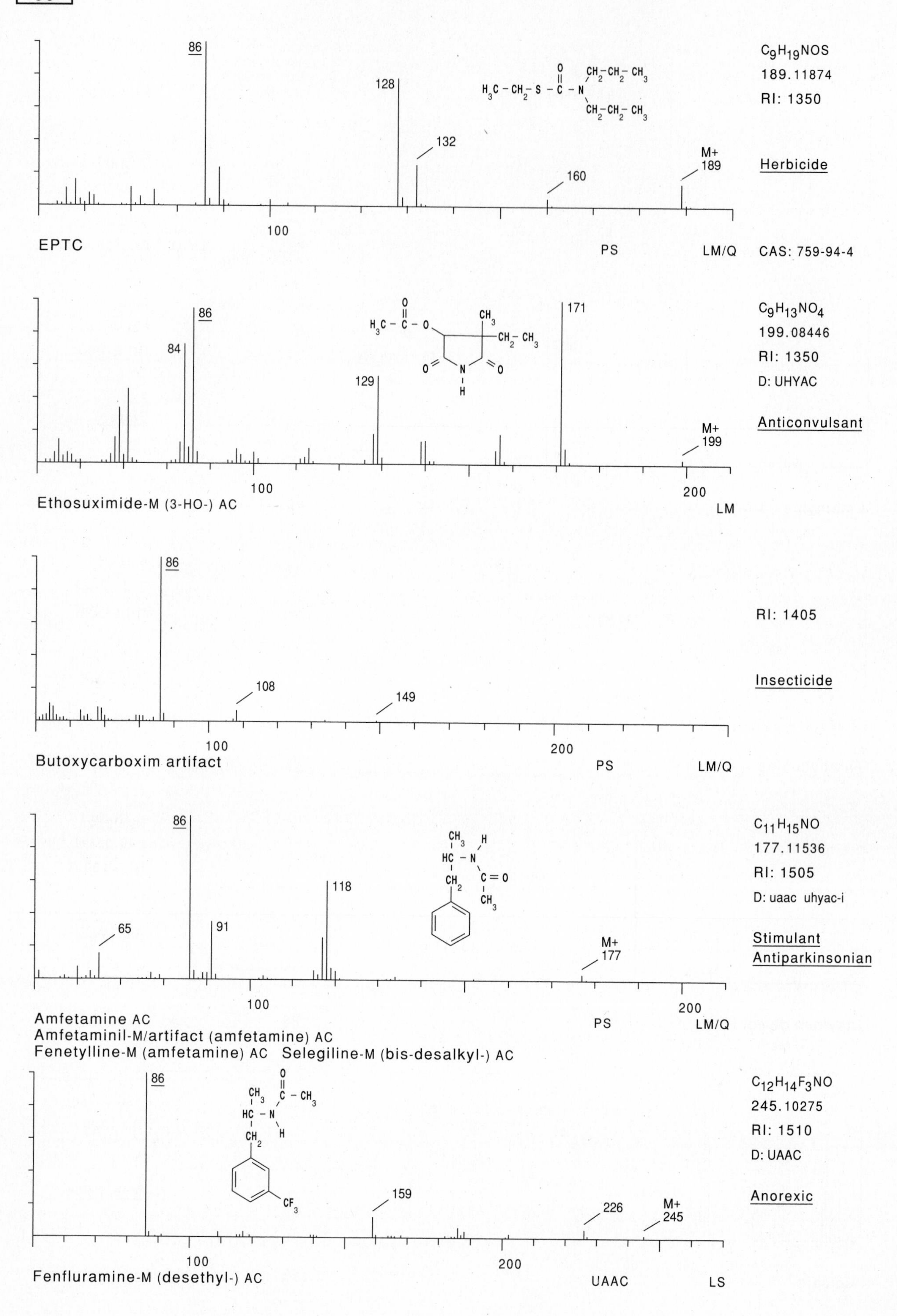

EPTC
86
128
132
160
M+ 189
100
PS
LM/Q
$C_9H_{19}NOS$
189.11874
RI: 1350
Herbicide
CAS: 759-94-4
Ethosuximide-M (3-HO-) AC
84
86
129
171
M+ 199
100
200
LM
$C_9H_{13}NO_4$
199.08446
RI: 1350
D: UHYAC
Anticonvulsant
Butoxycarboxim artifact
86
108
149
100
200
PS
LM/Q
RI: 1405
Insecticide
Amfetamine AC
Amfetaminil-M/artifact (amfetamine) AC
Fenetylline-M (amfetamine) AC Selegiline-M (bis-desalkyl-) AC
65
86
91
118
M+ 177
100
200
PS
LM/Q
$C_{11}H_{15}NO$
177.11536
RI: 1505
D: uaac uhyac-i
Stimulant
Antiparkinsonian
Fenfluramine-M (desethyl-) AC
86
159
226
M+ 245
100
200
UAAC
LS
$C_{12}H_{14}F_3NO$
245.10275
RI: 1510
D: UAAC
Anorexic

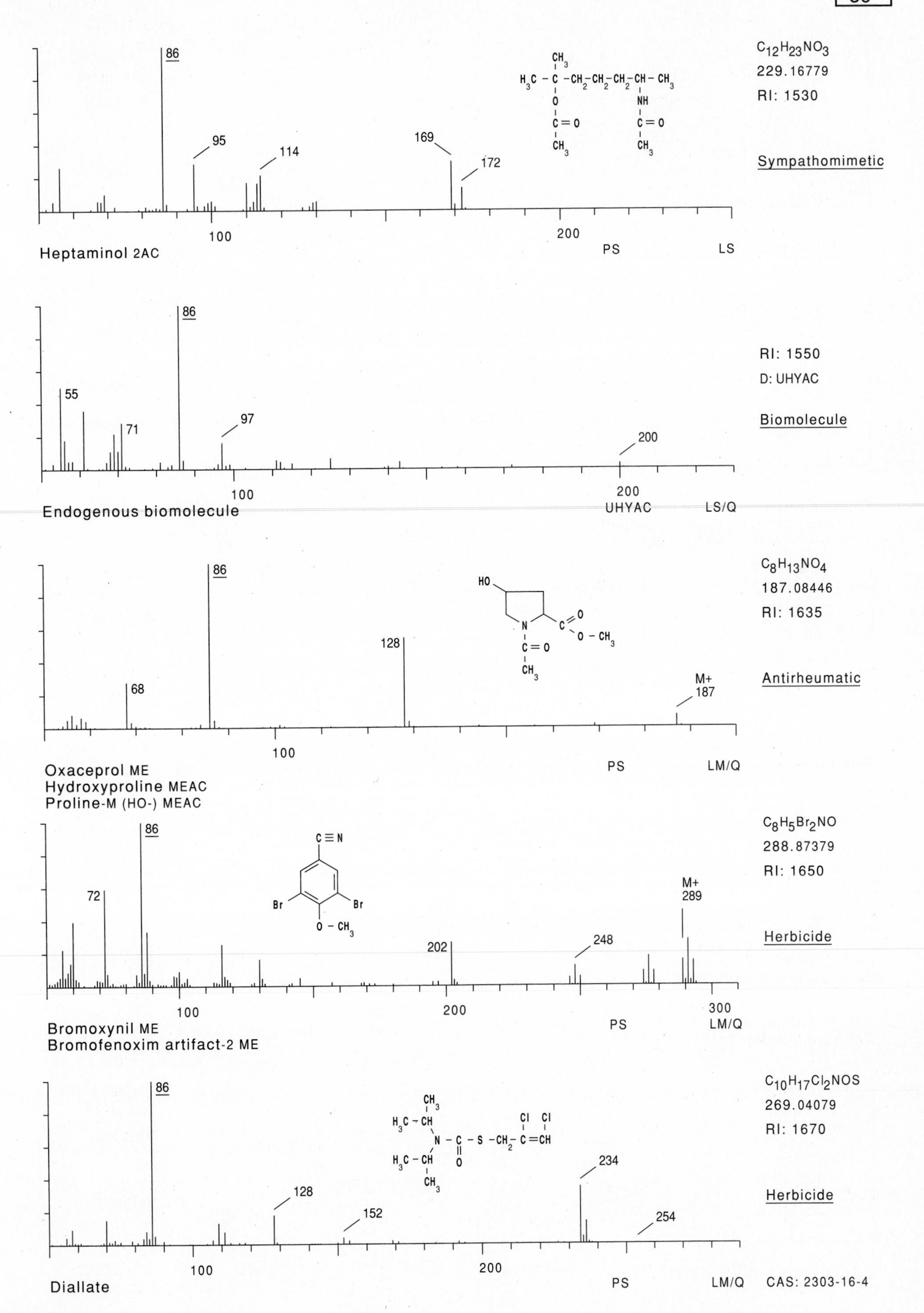
86
95
114
169
172
100
200
PS
LS
Heptaminol 2AC
$C_{12}H_{23}NO_3$
229.16779
RI: 1530
Sympathomimetic
86
55
71
97
200
100
200
UHYAC
LS/Q
Endogenous biomolecule
RI: 1550
D: UHYAC
Biomolecule
86
128
68
M+
187
100
PS
LM/Q
Oxaceprol ME
Hydroxyproline MEAC
Proline-M (HO-) MEAC
$C_8H_{13}NO_4$
187.08446
RI: 1635
Antirheumatic
86
72
202
248
M+
289
100
200
300
PS
LM/Q
Bromoxynil ME
Bromofenoxim artifact-2 ME
$C_8H_5Br_2NO$
288.87379
RI: 1650
Herbicide
86
128
152
234
254
100
200
PS
LM/Q
Diallate
$C_{10}H_{17}Cl_2NOS$
269.04079
RI: 1670
Herbicide
CAS: 2303-16-4

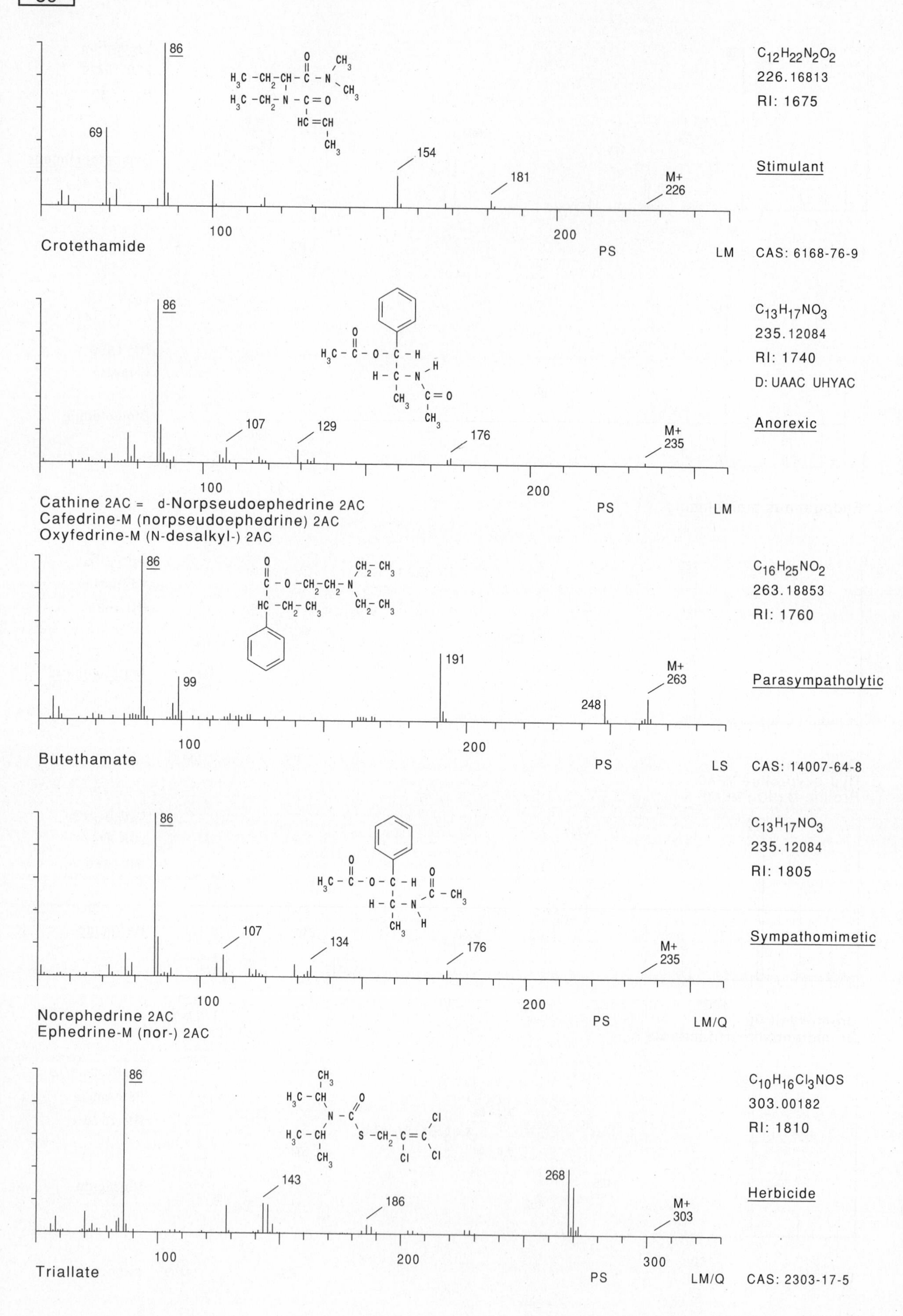
86
69
154
181
M+
226
100
200
Crotethamide
PS
LM
$C_{12}H_{22}N_2O_2$
226.16813
RI: 1675
Stimulant
CAS: 6168-76-9
86
107
129
176
M+
235
100
200
Cathine 2AC = d-Norpseudoephedrine 2AC
Cafedrine-M (norpseudoephedrine) 2AC
Oxyfedrine-M (N-desalkyl-) 2AC
PS
LM
$C_{13}H_{17}NO_3$
235.12084
RI: 1740
D: UAAC UHYAC
Anorexic
86
99
191
248
M+
263
100
200
Butethamate
PS
LS
$C_{16}H_{25}NO_2$
263.18853
RI: 1760
Parasympatholytic
CAS: 14007-64-8
86
107
134
176
M+
235
100
200
Norephedrine 2AC
Ephedrine-M (nor-) 2AC
PS
LM/Q
$C_{13}H_{17}NO_3$
235.12084
RI: 1805
Sympathomimetic
86
143
186
268
M+
303
100
200
300
Triallate
PS
LM/Q
$C_{10}H_{16}Cl_3NOS$
303.00182
RI: 1810
Herbicide
CAS: 2303-17-5

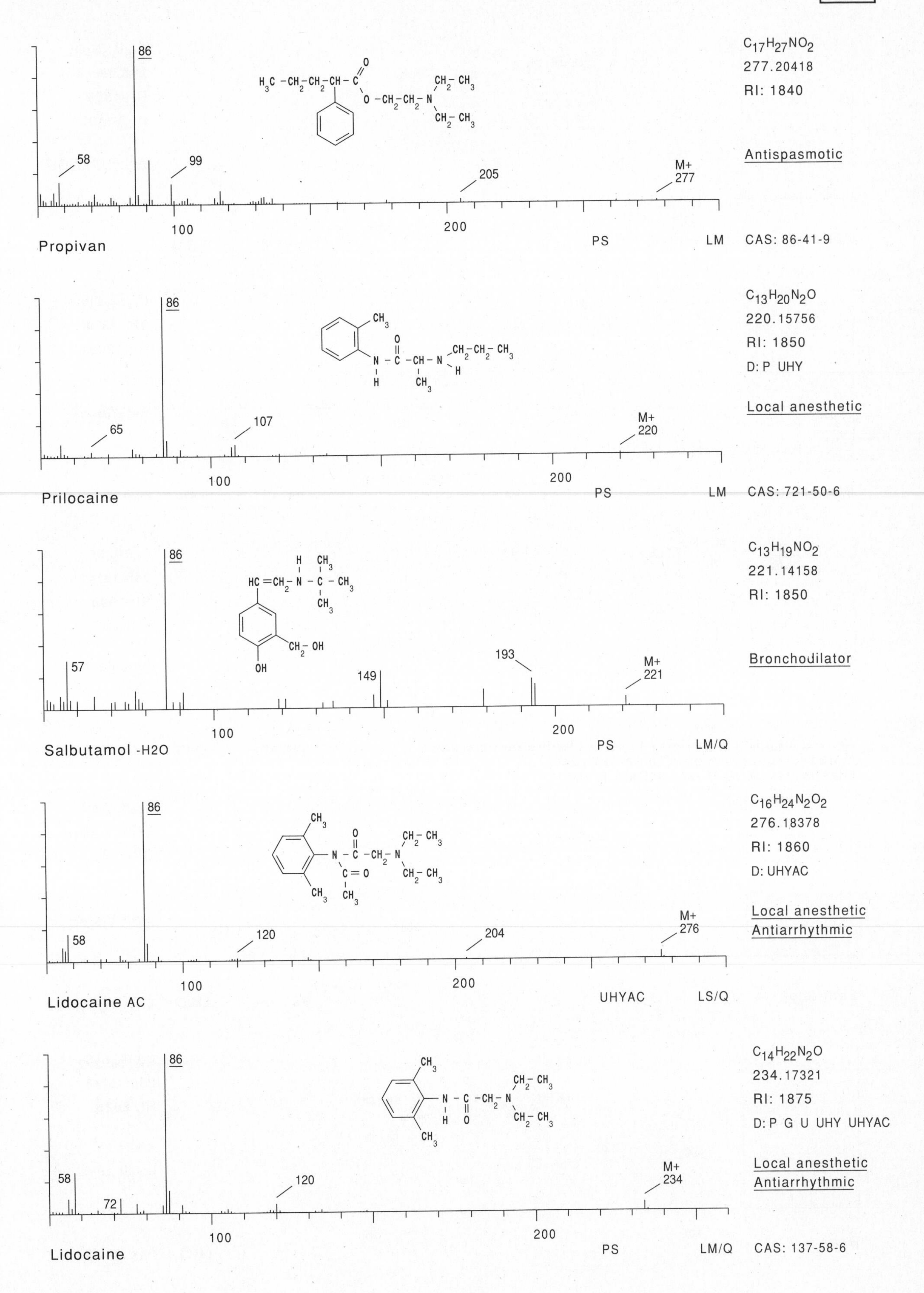
Propivan
$C_{17}H_{27}NO_2$
277.20418
RI: 1840
Antispasmotic
CAS: 86-41-9
86
58
99
205
M+ 277
100
200
PS
LM
Prilocaine
$C_{13}H_{20}N_2O$
220.15756
RI: 1850
D: P UHY
Local anesthetic
CAS: 721-50-6
86
65
107
M+ 220
100
200
PS
LM
Salbutamol -H2O
$C_{13}H_{19}NO_2$
221.14158
RI: 1850
Bronchodilator
86
57
149
193
M+ 221
100
200
PS
LM/Q
Lidocaine AC
$C_{16}H_{24}N_2O_2$
276.18378
RI: 1860
D: UHYAC
Local anesthetic
Antiarrhythmic
86
58
120
204
M+ 276
100
200
UHYAC
LS/Q
Lidocaine
$C_{14}H_{22}N_2O$
234.17321
RI: 1875
D: P G U UHY UHYAC
Local anesthetic
Antiarrhythmic
CAS: 137-58-6
86
58
72
120
M+ 234
100
200
PS
LM/Q

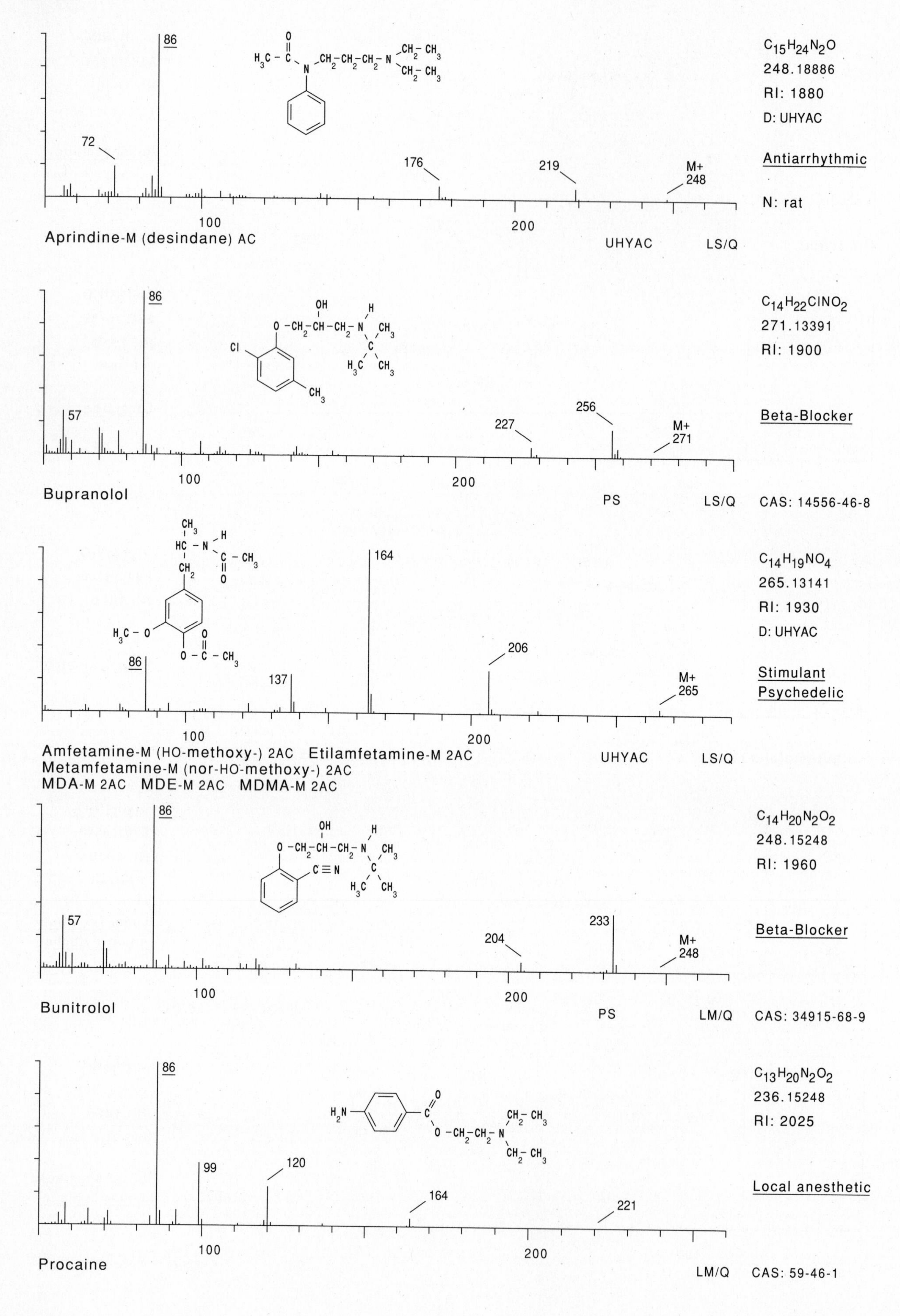
86
72
176
219
M+
248
100
200
Aprindine-M (desindane) AC
UHYAC
LS/Q
$C_{15}H_{24}N_2O$
248.18886
RI: 1880
D: UHYAC
Antiarrhythmic
N: rat
86
57
227
256
M+
271
100
200
Bupranolol
PS
LS/Q
$C_{14}H_{22}ClNO_2$
271.13391
RI: 1900
Beta-Blocker
CAS: 14556-46-8
164
86
137
206
M+
265
100
200
Amfetamine-M (HO-methoxy-) 2AC Etilamfetamine-M 2AC
Metamfetamine-M (nor-HO-methoxy-) 2AC
MDA-M 2AC MDE-M 2AC MDMA-M 2AC
UHYAC
LS/Q
$C_{14}H_{19}NO_4$
265.13141
RI: 1930
D: UHYAC
Stimulant
Psychedelic
86
57
204
233
M+
248
100
200
Bunitrolol
PS
LM/Q
$C_{14}H_{20}N_2O_2$
248.15248
RI: 1960
Beta-Blocker
CAS: 34915-68-9
86
99
120
164
221
100
200
Procaine
LM/Q
$C_{13}H_{20}N_2O_2$
236.15248
RI: 2025
Local anesthetic
CAS: 59-46-1

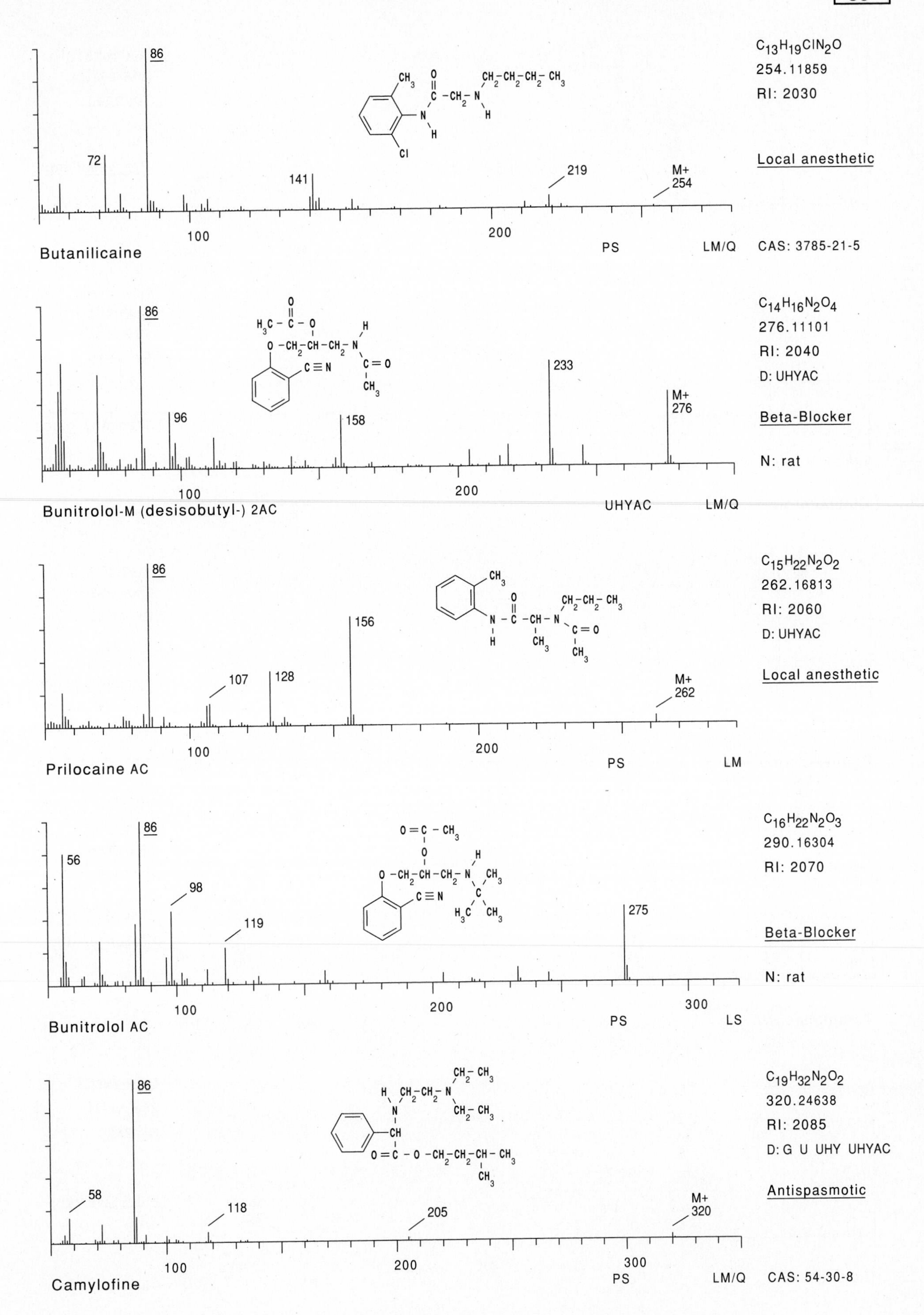

Butanilicaine

$C_{13}H_{19}ClN_2O$
254.11859
RI: 2030

Local anesthetic

CAS: 3785-21-5

Bunitrolol-M (desisobutyl-) 2AC

$C_{14}H_{16}N_2O_4$
276.11101
RI: 2040
D: UHYAC

Beta-Blocker

N: rat

Prilocaine AC

$C_{15}H_{22}N_2O_2$
262.16813
RI: 2060
D: UHYAC

Local anesthetic

Bunitrolol AC

$C_{16}H_{22}N_2O_3$
290.16304
RI: 2070

Beta-Blocker

N: rat

Camylofine

$C_{19}H_{32}N_2O_2$
320.24638
RI: 2085
D: G U UHY UHYAC

Antispasmotic

CAS: 54-30-8

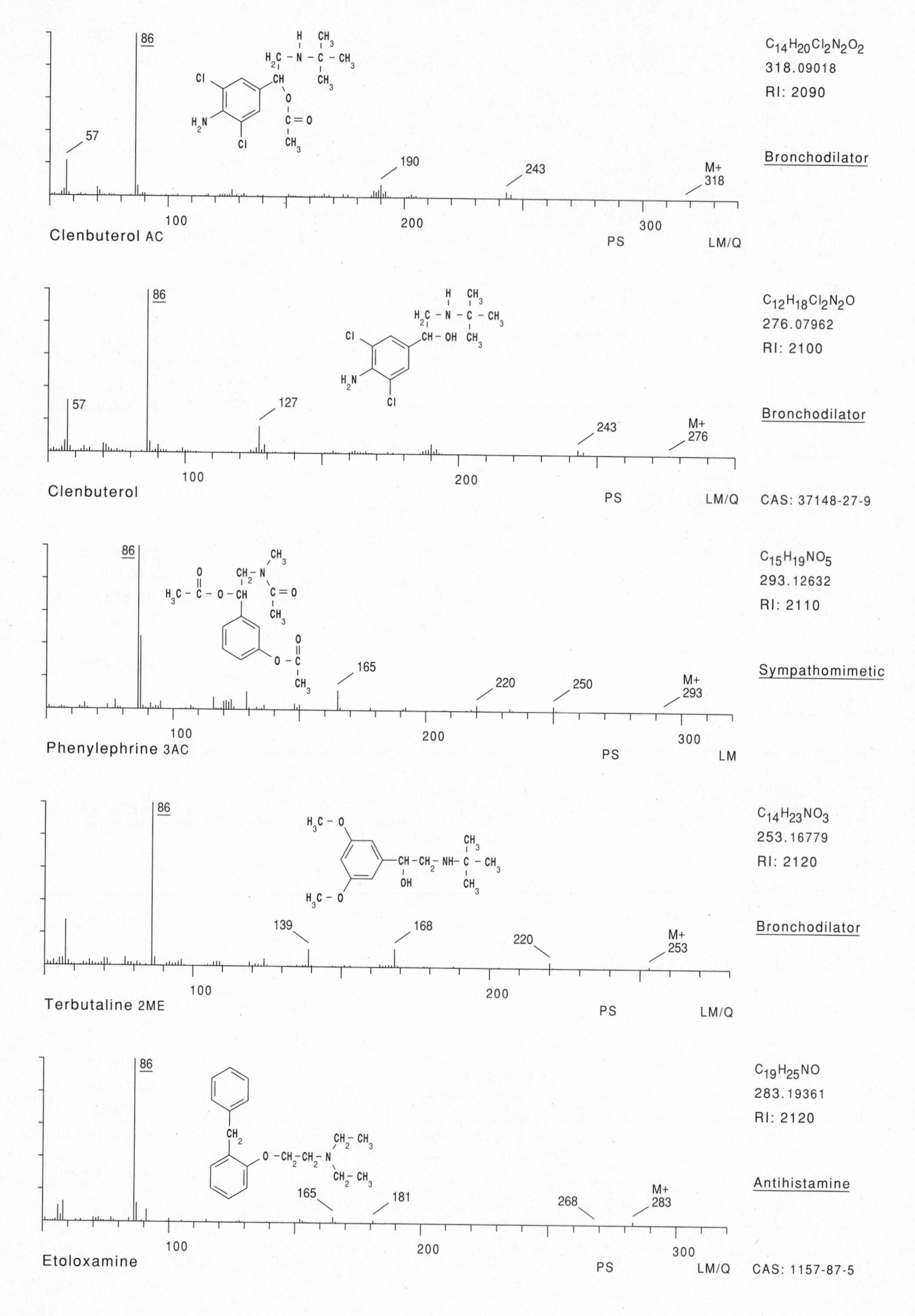
86
57
190
243
M+
318
100
200
300
Clenbuterol AC
PS
LM/Q
C14H20Cl2N2O2
318.09018
RI: 2090
Bronchodilator
86
57
127
243
M+
276
100
200
Clenbuterol
PS
LM/Q
C12H18Cl2N2O
276.07962
RI: 2100
Bronchodilator
CAS: 37148-27-9
86
165
220
250
M+
293
100
200
300
Phenylephrine 3AC
PS
LM
C15H19NO5
293.12632
RI: 2110
Sympathomimetic
86
139
168
220
M+
253
100
200
Terbutaline 2ME
PS
LM/Q
C14H23NO3
253.16779
RI: 2120
Bronchodilator
86
165
181
268
M+
283
100
200
300
Etoloxamine
PS
LM/Q
C19H25NO
283.19361
RI: 2120
Antihistamine
CAS: 1157-87-5

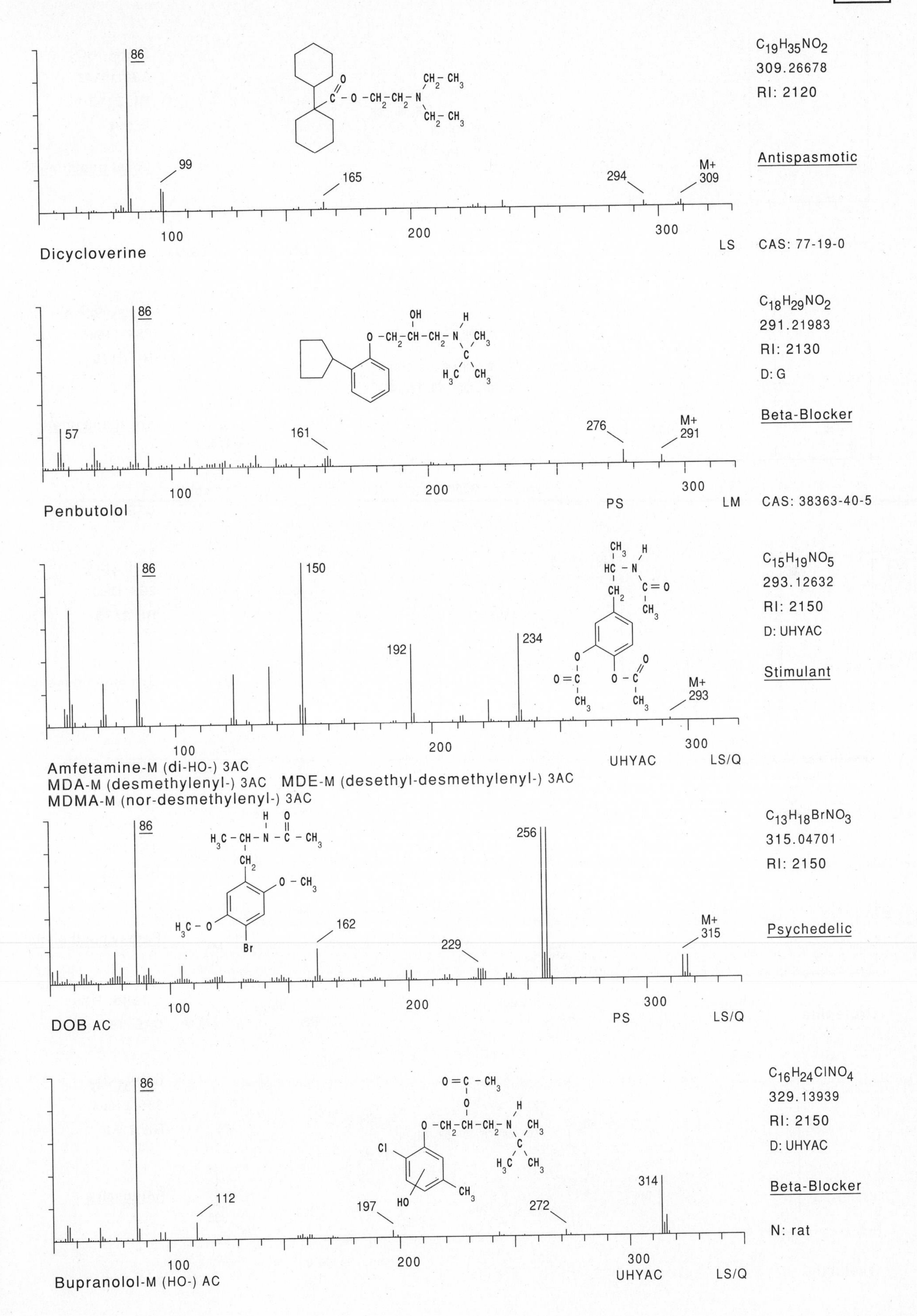

Dicycloverine

$C_{19}H_{35}NO_2$
309.26678
RI: 2120

Antispasmotic

CAS: 77-19-0

Penbutolol

$C_{18}H_{29}NO_2$
291.21983
RI: 2130
D: G

Beta-Blocker

CAS: 38363-40-5

Amfetamine-M (di-HO-) 3AC
MDA-M (desmethylenyl-) 3AC MDE-M (desethyl-desmethylenyl-) 3AC
MDMA-M (nor-desmethylenyl-) 3AC

$C_{15}H_{19}NO_5$
293.12632
RI: 2150
D: UHYAC

Stimulant

DOB AC

$C_{13}H_{18}BrNO_3$
315.04701
RI: 2150

Psychedelic

Bupranolol-M (HO-) AC

$C_{16}H_{24}ClNO_4$
329.13939
RI: 2150
D: UHYAC

Beta-Blocker

N: rat

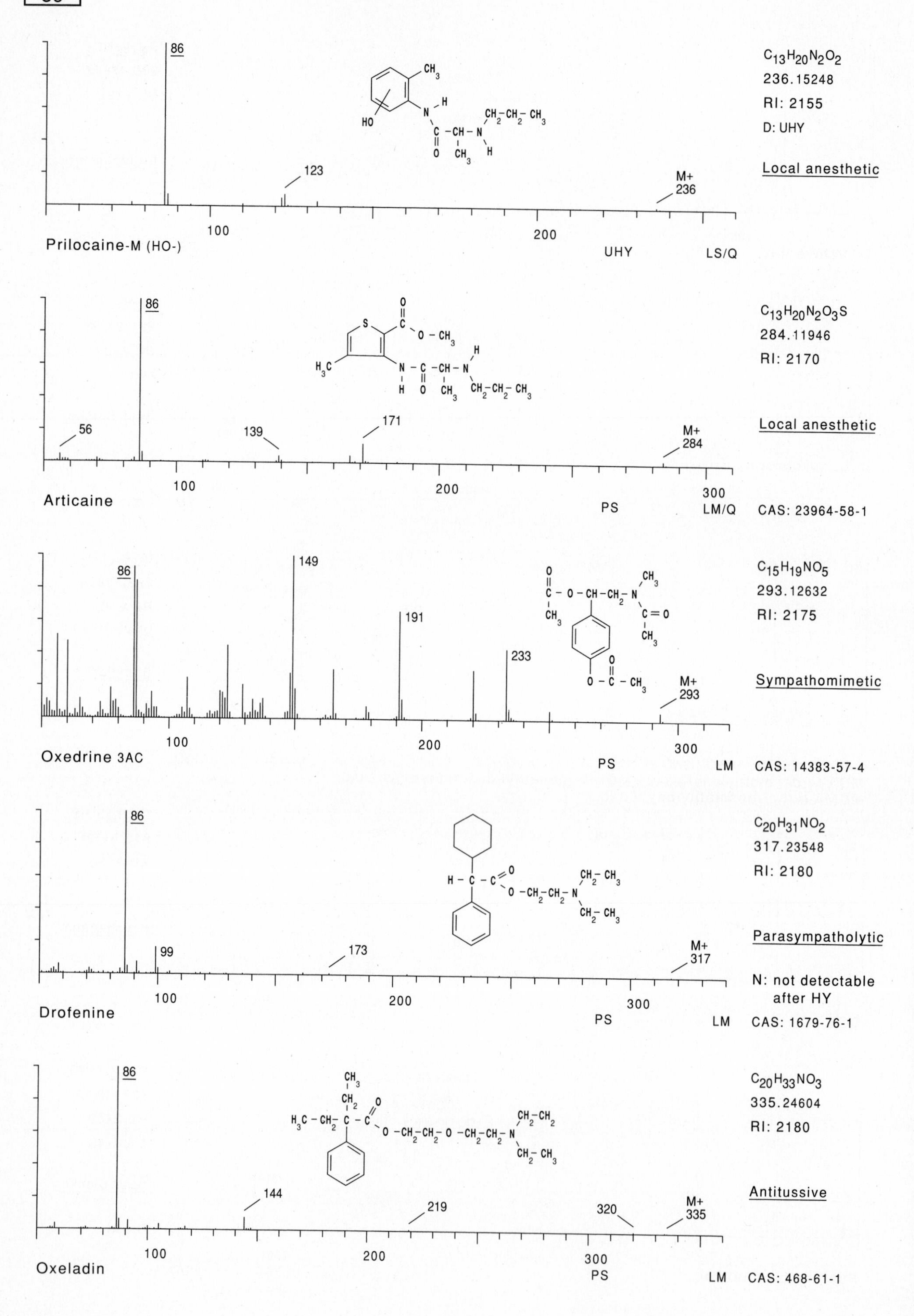

86
123
M+
236
100
200
Prilocaine-M (HO-)
UHY
LS/Q
$C_{13}H_{20}N_2O_2$
236.15248
RI: 2155
D: UHY
Local anesthetic
86
56
139
171
M+
284
100
200
300
Articaine
PS
LM/Q
$C_{13}H_{20}N_2O_3S$
284.11946
RI: 2170
Local anesthetic
CAS: 23964-58-1
86
149
191
233
M+
293
100
200
300
Oxedrine 3AC
PS
LM
$C_{15}H_{19}NO_5$
293.12632
RI: 2175
Sympathomimetic
CAS: 14383-57-4
86
99
173
M+
317
100
200
300
Drofenine
PS
LM
$C_{20}H_{31}NO_2$
317.23548
RI: 2180
Parasympatholytic
N: not detectable after HY
CAS: 1679-76-1
86
144
219
320
M+
335
100
200
300
Oxeladin
PS
LM
$C_{20}H_{33}NO_3$
335.24604
RI: 2180
Antitussive
CAS: 468-61-1

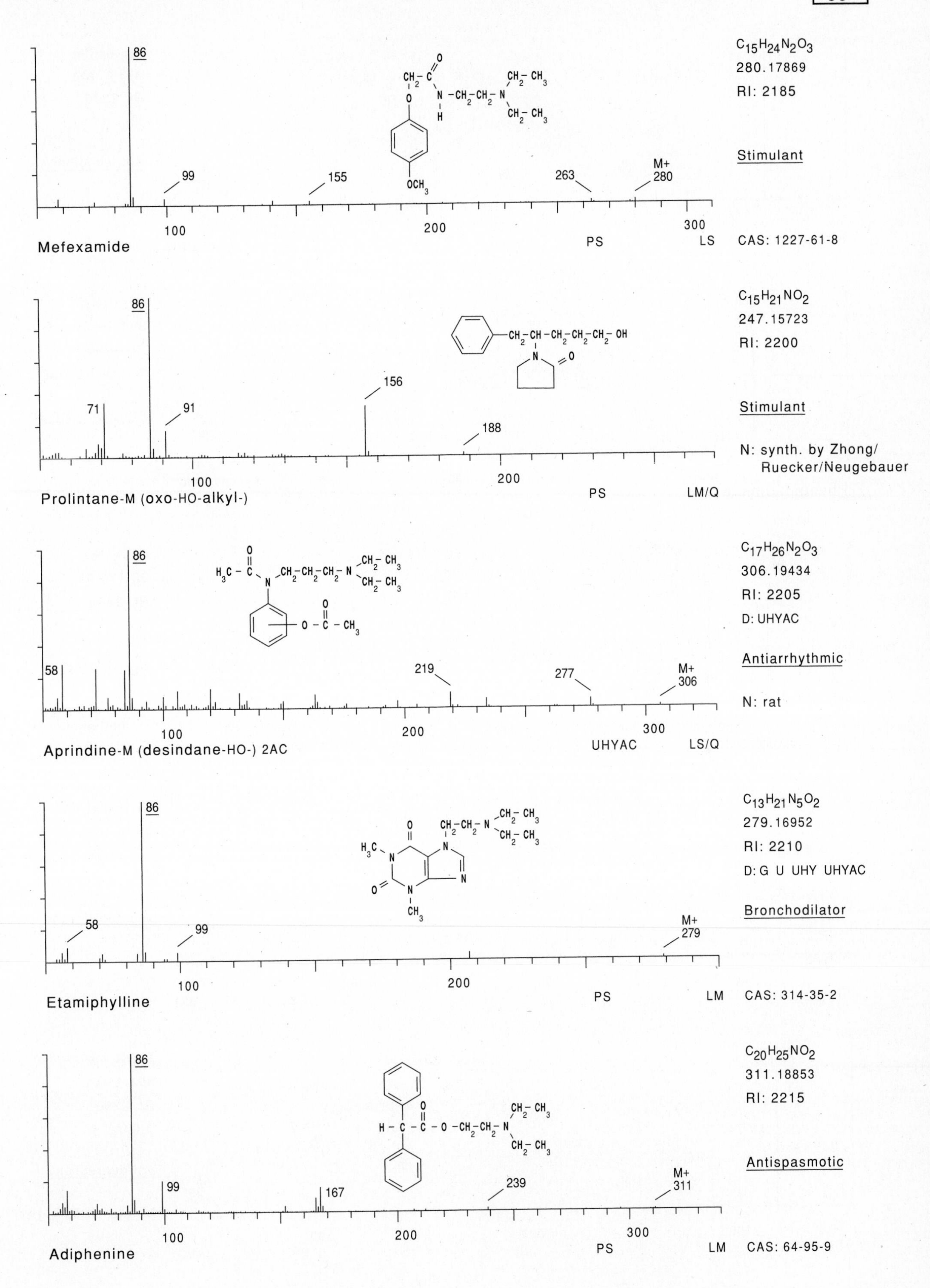

$C_{15}H_{24}N_2O_3$
280.17869
RI: 2185

Stimulant

CAS: 1227-61-8

$C_{15}H_{21}NO_2$
247.15723
RI: 2200

Stimulant

N: synth. by Zhong/Ruecker/Neugebauer

$C_{17}H_{26}N_2O_3$
306.19434
RI: 2205
D: UHYAC

Antiarrhythmic

N: rat

$C_{13}H_{21}N_5O_2$
279.16952
RI: 2210
D: G U UHY UHYAC

Bronchodilator

CAS: 314-35-2

$C_{20}H_{25}NO_2$
311.18853
RI: 2215

Antispasmodic

CAS: 64-95-9

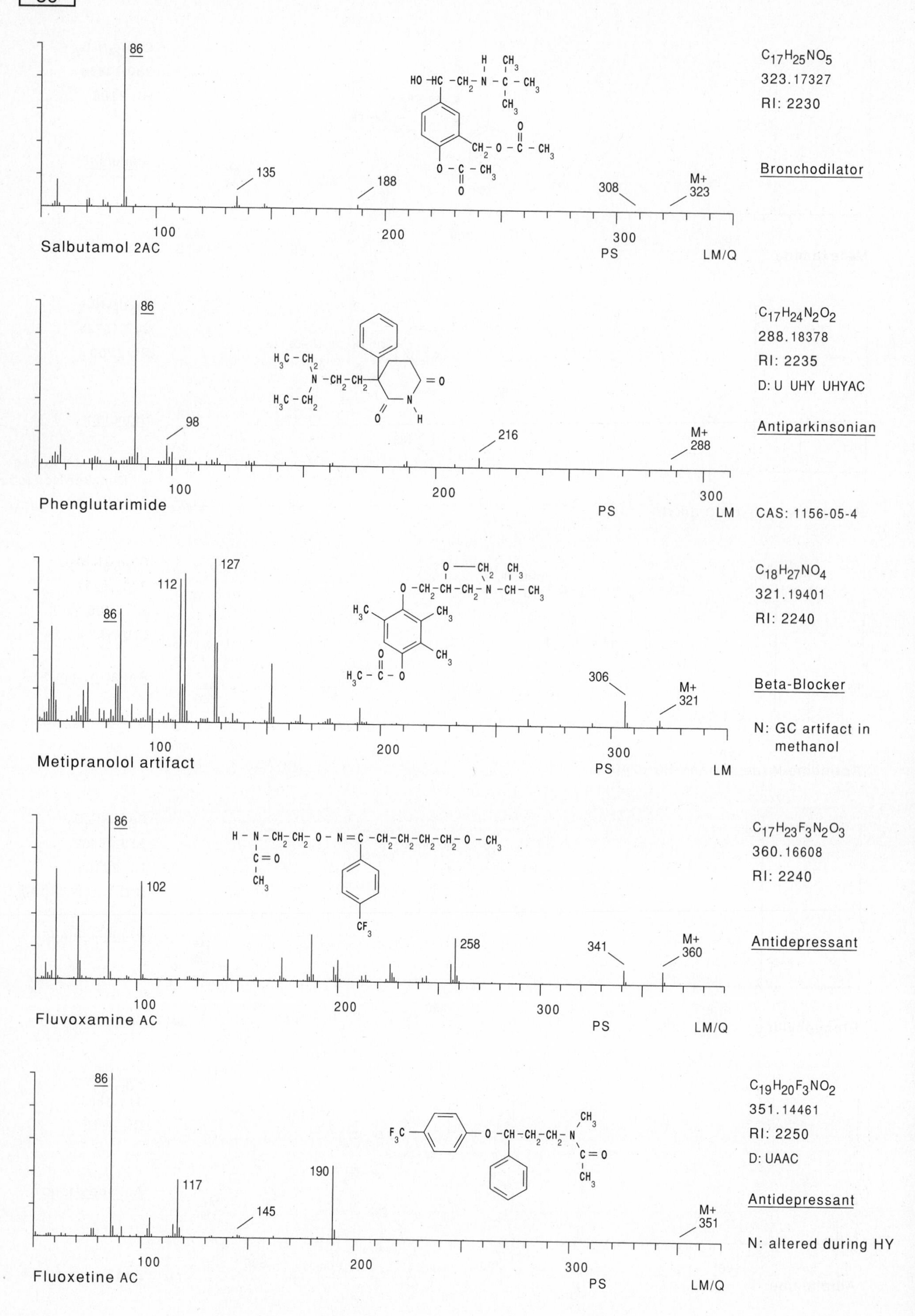

Salbutamol 2AC
86
135
188
308
M+ 323
100
200
300
PS
LM/Q
C17H25NO5
323.17327
RI: 2230
Bronchodilator
Phenglutarimide
86
98
216
M+ 288
C17H24N2O2
288.18378
RI: 2235
D: U UHY UHYAC
Antiparkinsonian
CAS: 1156-05-4
Metipranolol artifact
86
112
127
306
M+ 321
C18H27NO4
321.19401
RI: 2240
Beta-Blocker
N: GC artifact in methanol
Fluvoxamine AC
86
102
258
341
M+ 360
C17H23F3N2O3
360.16608
RI: 2240
Antidepressant
Fluoxetine AC
86
117
145
190
M+ 351
C19H20F3NO2
351.14461
RI: 2250
D: UAAC
Antidepressant
N: altered during HY

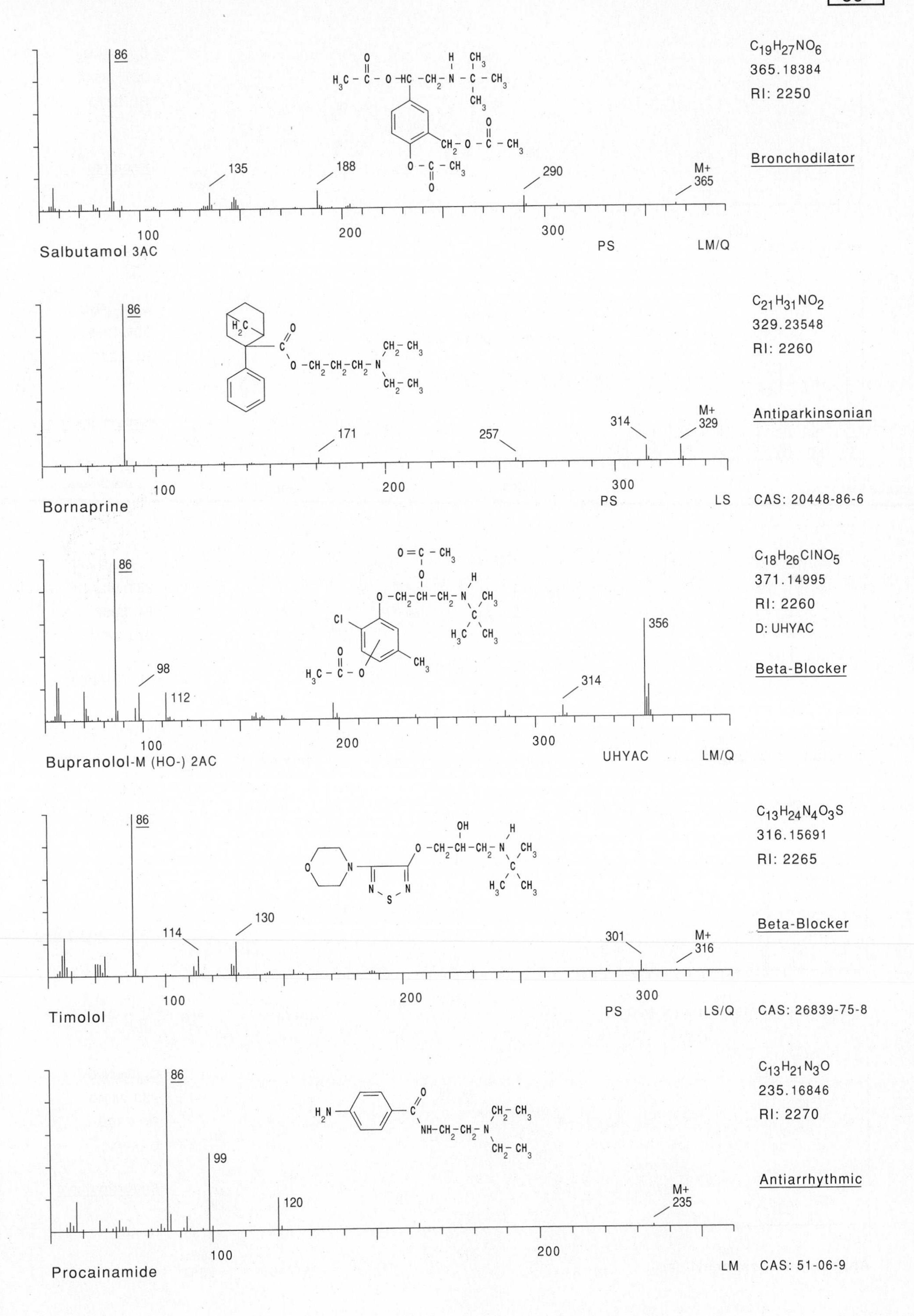

Salbutamol 3AC
C19H27NO6
365.18384
RI: 2250
Bronchodilator
86
135
188
290
M+
365
100
200
300
PS
LM/Q
Bornaprine
C21H31NO2
329.23548
RI: 2260
Antiparkinsonian
CAS: 20448-86-6
86
171
257
314
M+
329
100
200
300
PS
LS
Bupranolol-M (HO-) 2AC
C18H26ClNO5
371.14995
RI: 2260
D: UHYAC
Beta-Blocker
86
98
112
314
356
100
200
300
UHYAC
LM/Q
Timolol
C13H24N4O3S
316.15691
RI: 2265
Beta-Blocker
CAS: 26839-75-8
86
114
130
301
M+
316
100
200
300
PS
LS/Q
Procainamide
C13H21N3O
235.16846
RI: 2270
Antiarrhythmic
CAS: 51-06-9
86
99
120
M+
235
100
200
LM

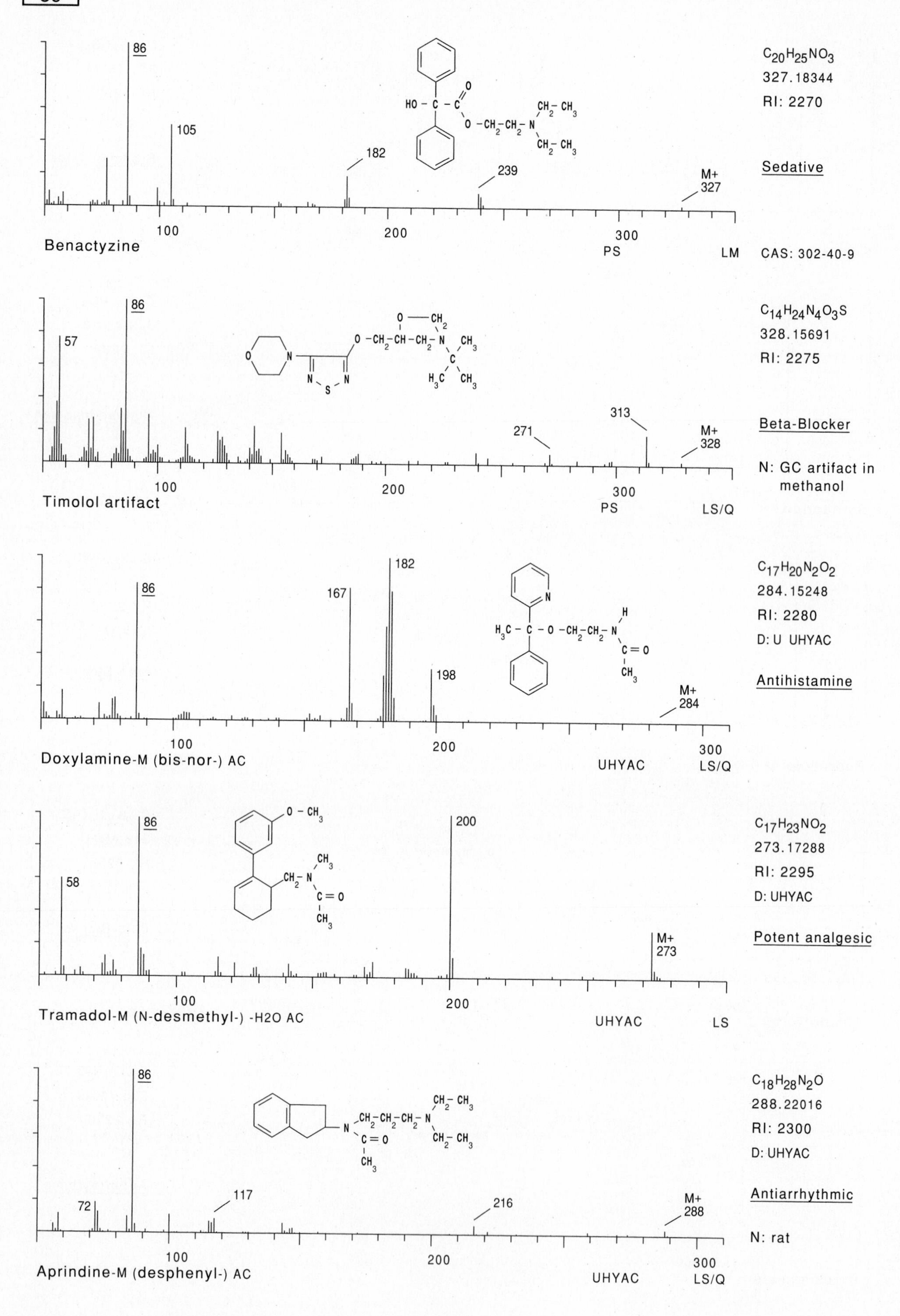

Benactyzine

$C_{20}H_{25}NO_3$
327.18344
RI: 2270

Sedative

CAS: 302-40-9

Timolol artifact

$C_{14}H_{24}N_4O_3S$
328.15691
RI: 2275

Beta-Blocker

N: GC artifact in methanol

Doxylamine-M (bis-nor-) AC

$C_{17}H_{20}N_2O_2$
284.15248
RI: 2280
D: U UHYAC

Antihistamine

Tramadol-M (N-desmethyl-) -H2O AC

$C_{17}H_{23}NO_2$
273.17288
RI: 2295
D: UHYAC

Potent analgesic

Aprindine-M (desphenyl-) AC

$C_{18}H_{28}N_2O$
288.22016
RI: 2300
D: UHYAC

Antiarrhythmic

N: rat

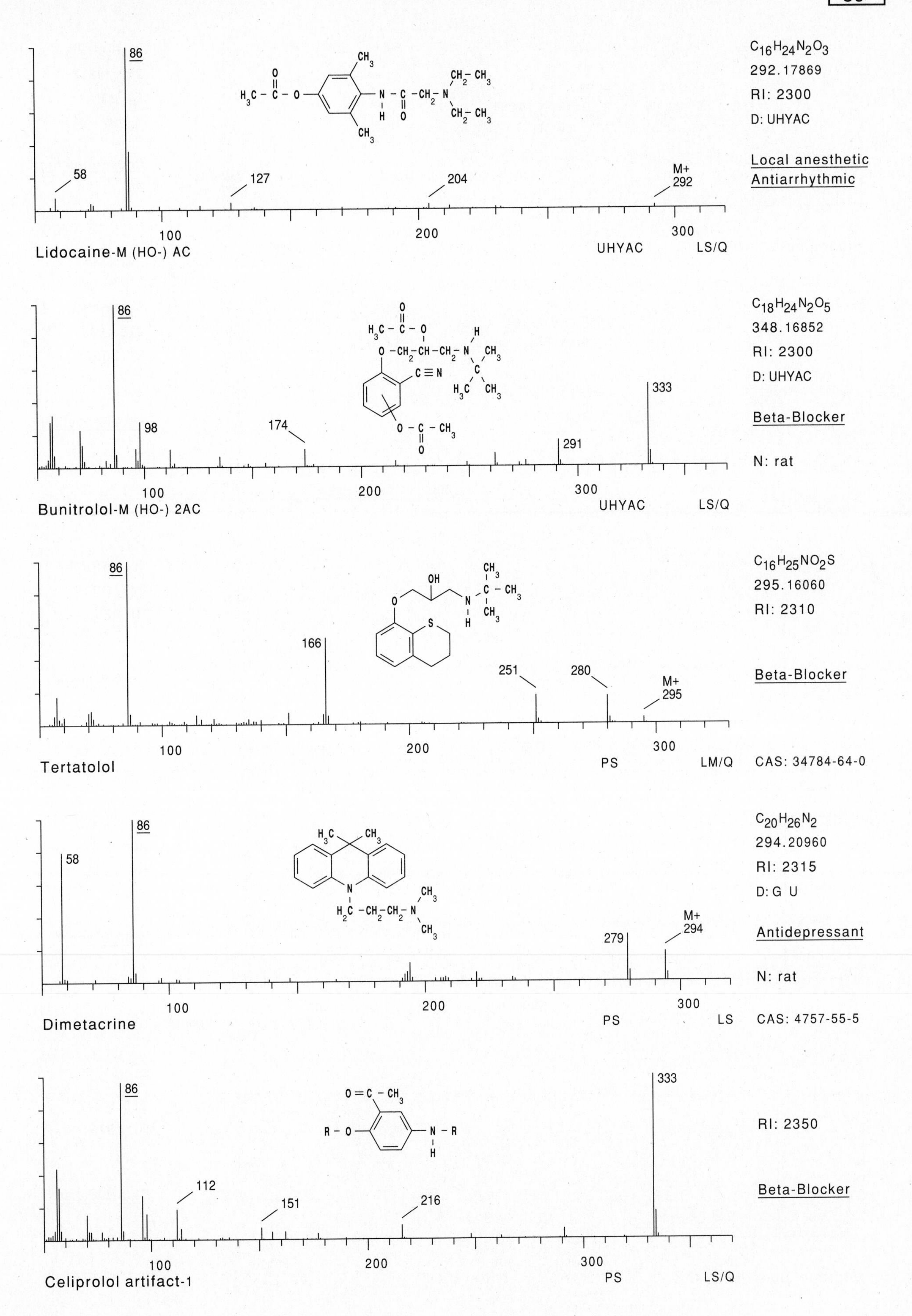

$C_{16}H_{24}N_2O_3$
292.17869
RI: 2300
D: UHYAC

Local anesthetic
Antiarrhythmic

$C_{18}H_{24}N_2O_5$
348.16852
RI: 2300
D: UHYAC

Beta-Blocker

N: rat

$C_{16}H_{25}NO_2S$
295.16060
RI: 2310

Beta-Blocker

CAS: 34784-64-0

$C_{20}H_{26}N_2$
294.20960
RI: 2315
D: G U

Antidepressant

N: rat

CAS: 4757-55-5

RI: 2350

Beta-Blocker

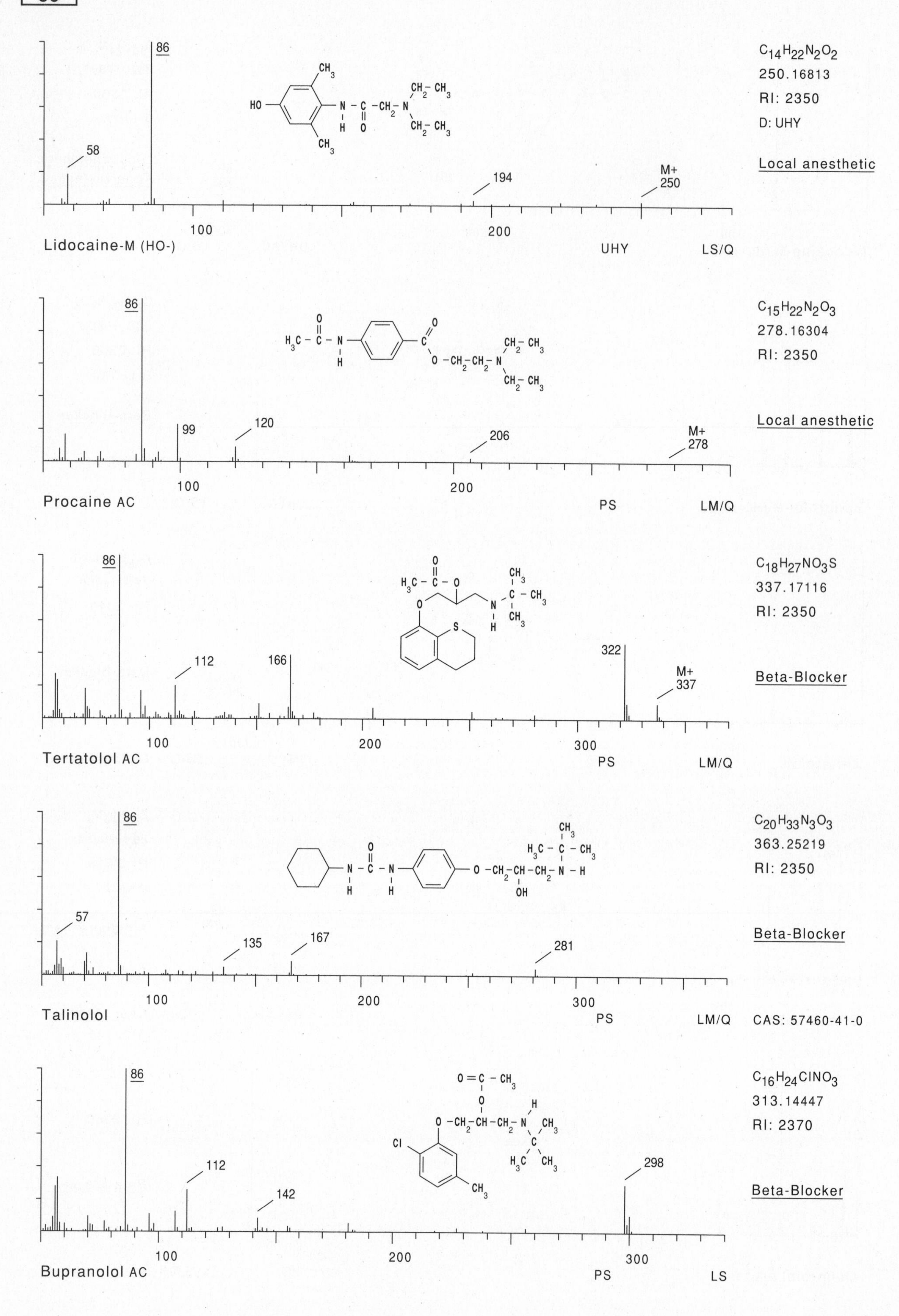

Lidocaine-M (HO-)
UHY
LS/Q
C14H22N2O2
250.16813
RI: 2350
D: UHY
Local anesthetic
86
58
194
M+
250
Procaine AC
PS
LM/Q
C15H22N2O3
278.16304
RI: 2350
Local anesthetic
86
99
120
206
M+
278
Tertatolol AC
PS
LM/Q
C18H27NO3S
337.17116
RI: 2350
Beta-Blocker
86
112
166
322
M+
337
Talinolol
PS
LM/Q
CAS: 57460-41-0
C20H33N3O3
363.25219
RI: 2350
Beta-Blocker
86
57
135
167
281
Bupranolol AC
PS
LS
C16H24ClNO3
313.14447
RI: 2370
Beta-Blocker
86
112
142
298

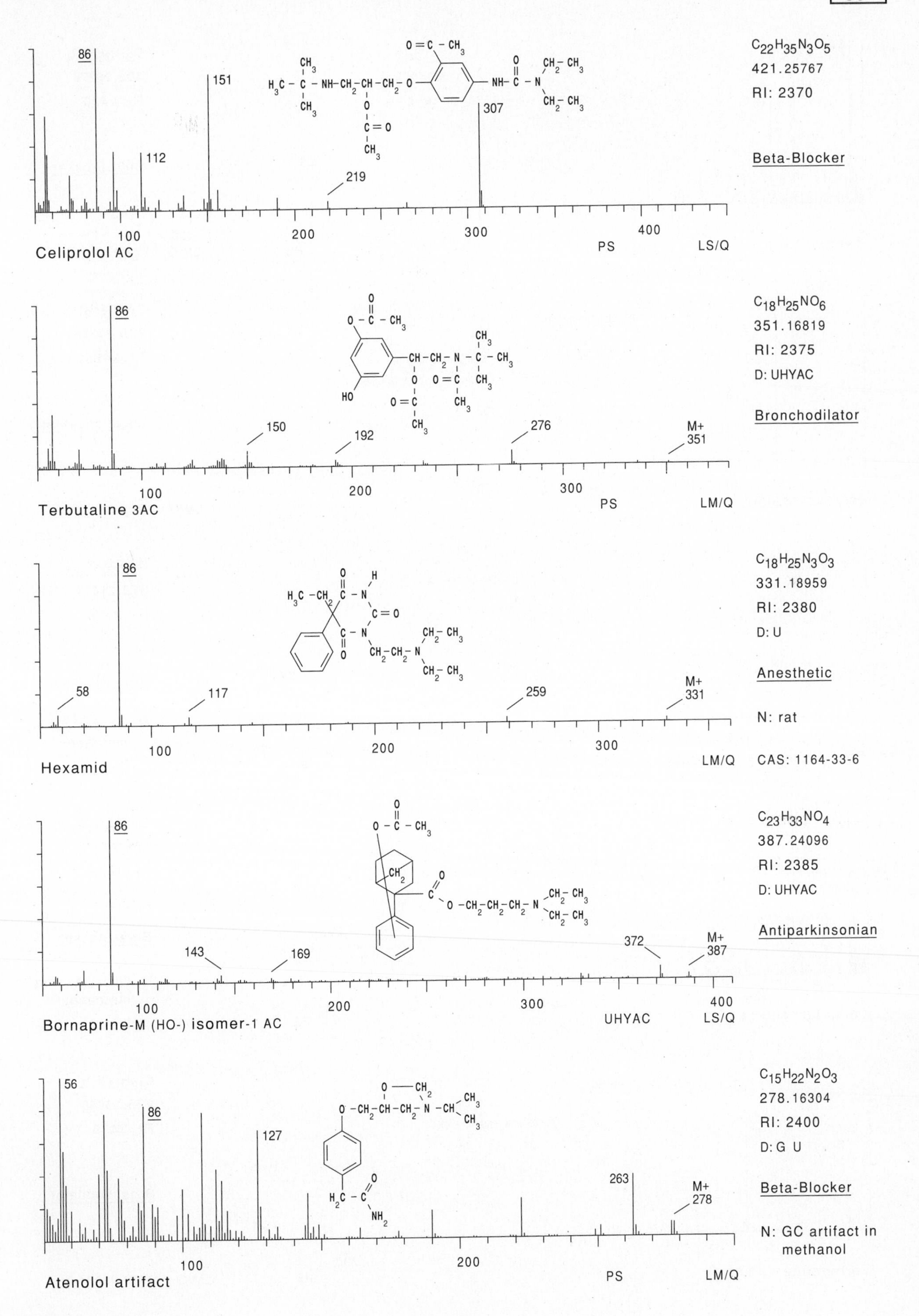
Celiprolol AC
PS
LS/Q
$C_{22}H_{35}N_3O_5$
421.25767
RI: 2370
Beta-Blocker
Terbutaline 3AC
PS
LM/Q
$C_{18}H_{25}NO_6$
351.16819
RI: 2375
D: UHYAC
Bronchodilator
Hexamid
LM/Q
$C_{18}H_{25}N_3O_3$
331.18959
RI: 2380
D: U
Anesthetic
N: rat
CAS: 1164-33-6
Bornaprine-M (HO-) isomer-1 AC
UHYAC
LS/Q
$C_{23}H_{33}NO_4$
387.24096
RI: 2385
D: UHYAC
Antiparkinsonian
Atenolol artifact
PS
LM/Q
$C_{15}H_{22}N_2O_3$
278.16304
RI: 2400
D: G U
Beta-Blocker
N: GC artifact in methanol

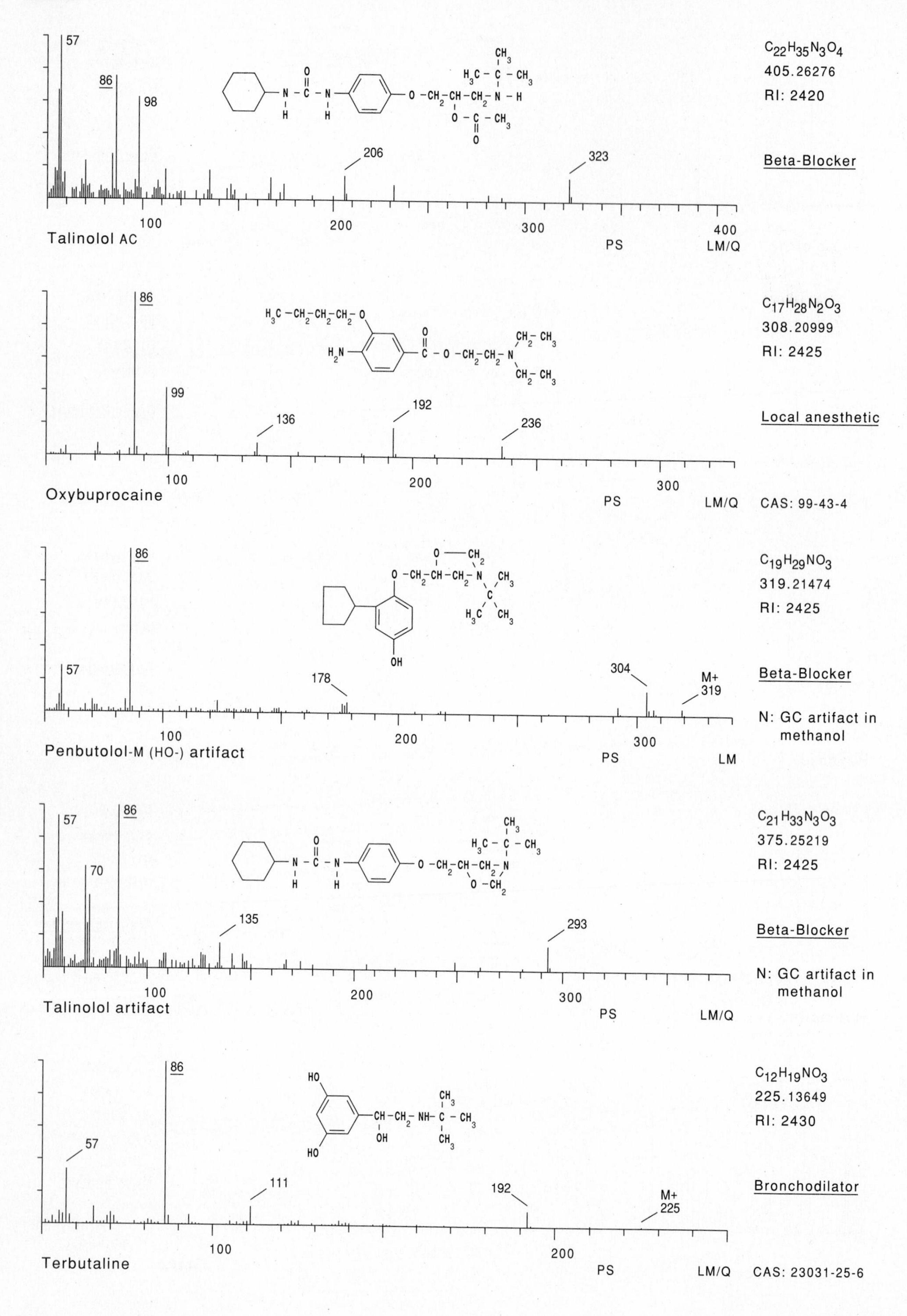
57
86
98
206
323
100
200
300
400
Talinolol AC
PS
LM/Q
$C_{22}H_{35}N_3O_4$
405.26276
RI: 2420
Beta-Blocker
86
99
136
192
236
Oxybuprocaine
LM/Q
$C_{17}H_{28}N_2O_3$
308.20999
RI: 2425
Local anesthetic
CAS: 99-43-4
86
57
178
304
M+
319
Penbutolol-M (HO-) artifact
LM
$C_{19}H_{29}NO_3$
319.21474
RI: 2425
Beta-Blocker
N: GC artifact in methanol
57
86
70
135
293
Talinolol artifact
$C_{21}H_{33}N_3O_3$
375.25219
RI: 2425
Beta-Blocker
N: GC artifact in methanol
86
57
111
192
M+
225
Terbutaline
$C_{12}H_{19}NO_3$
225.13649
RI: 2430
Bronchodilator
CAS: 23031-25-6

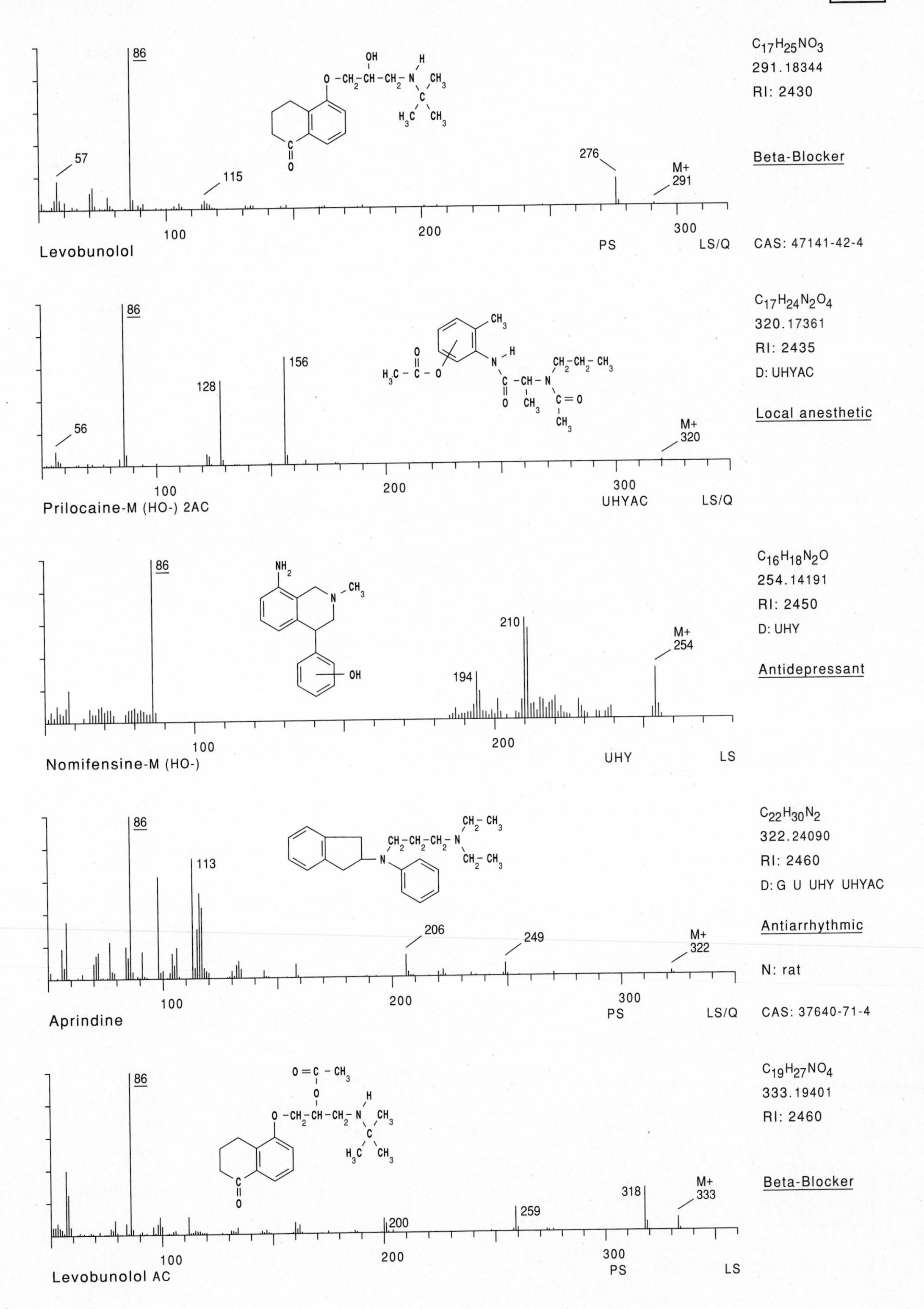
Levobunolol
$C_{17}H_{25}NO_3$
291.18344
RI: 2430
Beta-Blocker
CAS: 47141-42-4
PS LS/Q
Prilocaine-M (HO-) 2AC
$C_{17}H_{24}N_2O_4$
320.17361
RI: 2435
D: UHYAC
Local anesthetic
UHYAC LS/Q
Nomifensine-M (HO-)
$C_{16}H_{18}N_2O$
254.14191
RI: 2450
D: UHY
Antidepressant
UHY LS
Aprindine
$C_{22}H_{30}N_2$
322.24090
RI: 2460
D: G U UHY UHYAC
Antiarrhythmic
N: rat
CAS: 37640-71-4
PS LS/Q
Levobunolol AC
$C_{19}H_{27}NO_4$
333.19401
RI: 2460
Beta-Blocker
PS LS

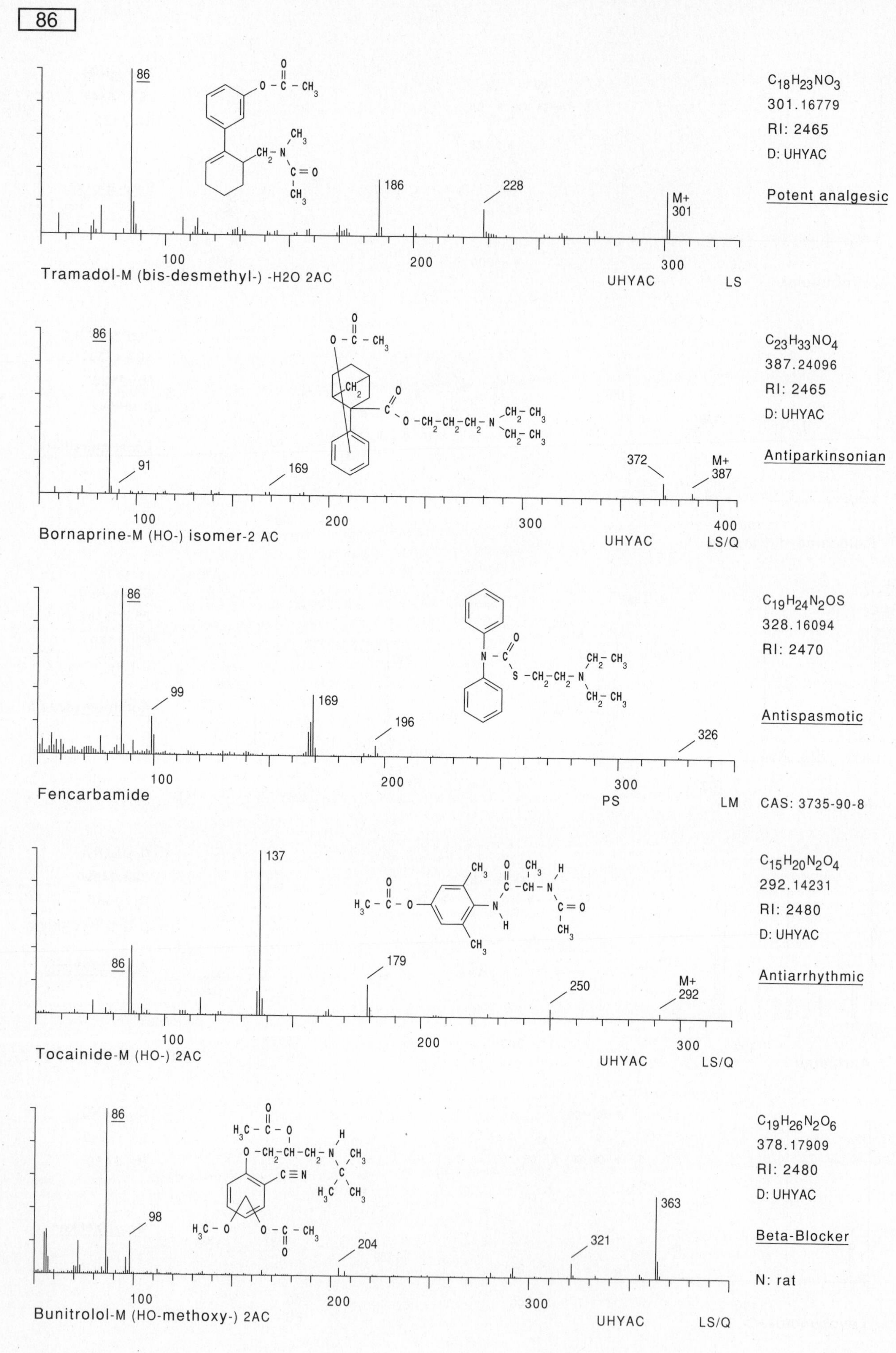
86
Tramadol-M (bis-desmethyl-) -H2O 2AC
UHYAC
LS
C18H23NO3
301.16779
RI: 2465
D: UHYAC
Potent analgesic
Bornaprine-M (HO-) isomer-2 AC
UHYAC
LS/Q
C23H33NO4
387.24096
RI: 2465
D: UHYAC
Antiparkinsonian
Fencarbamide
PS
LM
C19H24N2OS
328.16094
RI: 2470
Antispasmotic
CAS: 3735-90-8
Tocainide-M (HO-) 2AC
UHYAC
LS/Q
C15H20N2O4
292.14231
RI: 2480
D: UHYAC
Antiarrhythmic
Bunitrolol-M (HO-methoxy-) 2AC
UHYAC
LS/Q
C19H26N2O6
378.17909
RI: 2480
D: UHYAC
Beta-Blocker
N: rat

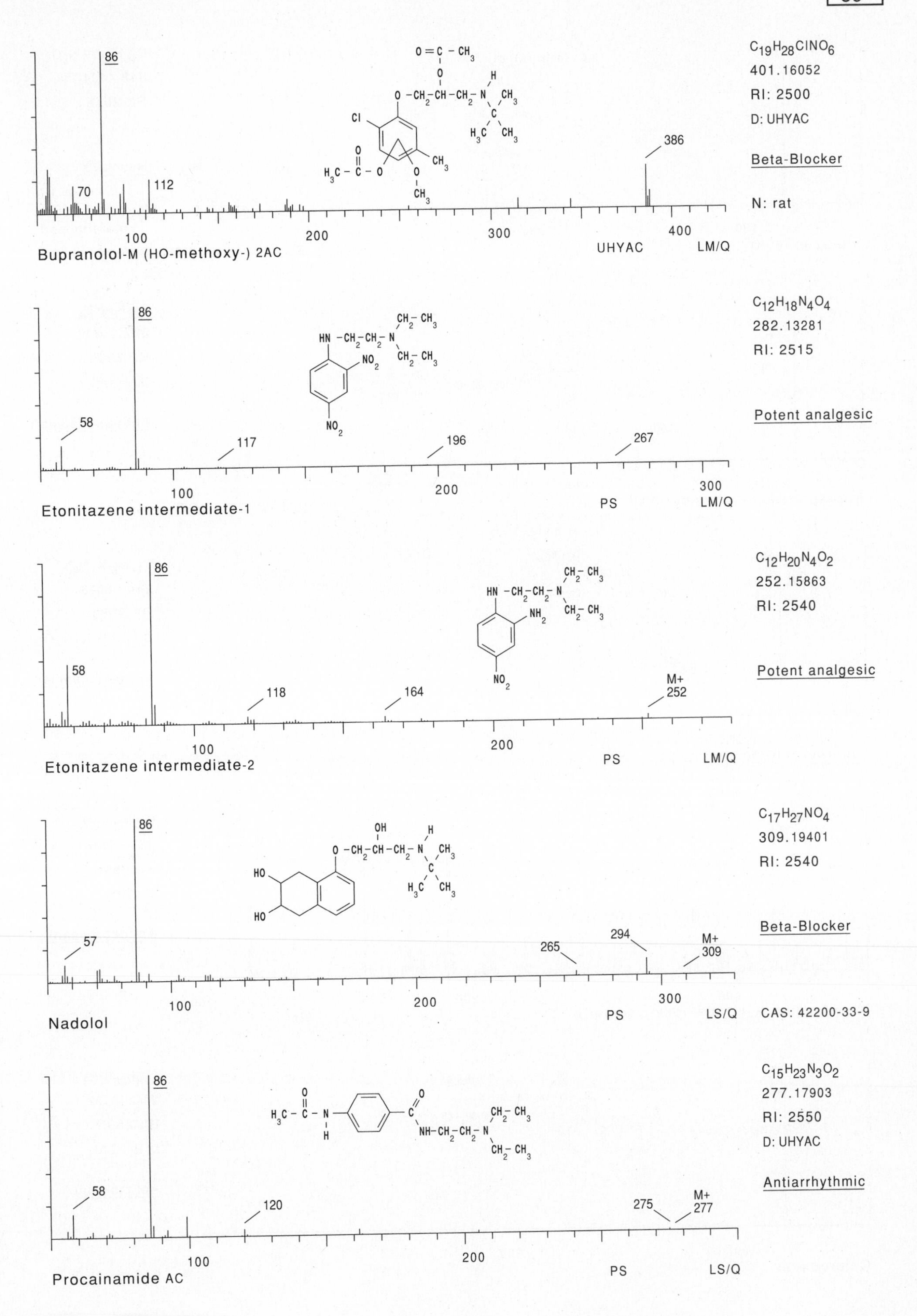

Bupranolol-M (HO-methoxy-) 2AC
UHYAC
LM/Q
C19H28ClNO6
401.16052
RI: 2500
D: UHYAC
Beta-Blocker
N: rat
Etonitazene intermediate-1
PS
LM/Q
C12H18N4O4
282.13281
RI: 2515
Potent analgesic
Etonitazene intermediate-2
PS
LM/Q
C12H20N4O2
252.15863
RI: 2540
Potent analgesic
Nadolol
PS
LS/Q
C17H27NO4
309.19401
RI: 2540
Beta-Blocker
CAS: 42200-33-9
Procainamide AC
PS
LS/Q
C15H23N3O2
277.17903
RI: 2550
D: UHYAC
Antiarrhythmic

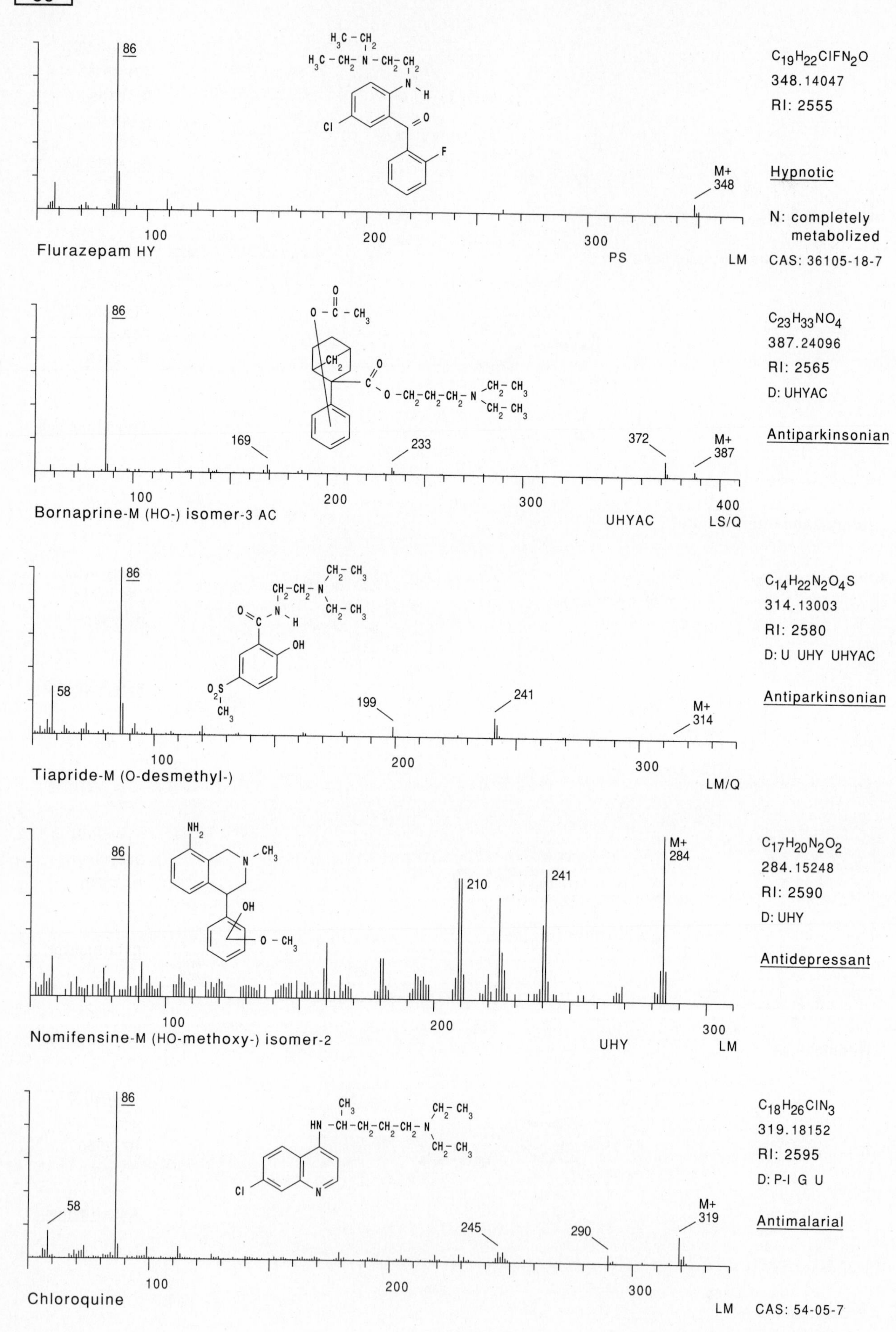

Flurazepam HY
C19H22ClFN2O
348.14047
RI: 2555
Hypnotic
N: completely metabolized
CAS: 36105-18-7
PS
LM
Bornaprine-M (HO-) isomer-3 AC
C23H33NO4
387.24096
RI: 2565
D: UHYAC
Antiparkinsonian
UHYAC
LS/Q
Tiapride-M (O-desmethyl-)
C14H22N2O4S
314.13003
RI: 2580
D: U UHY UHYAC
Antiparkinsonian
LM/Q
Nomifensine-M (HO-methoxy-) isomer-2
C17H20N2O2
284.15248
RI: 2590
D: UHY
Antidepressant
UHY
LM
Chloroquine
C18H26ClN3
319.18152
RI: 2595
D: P-I G U
Antimalarial
LM
CAS: 54-05-7

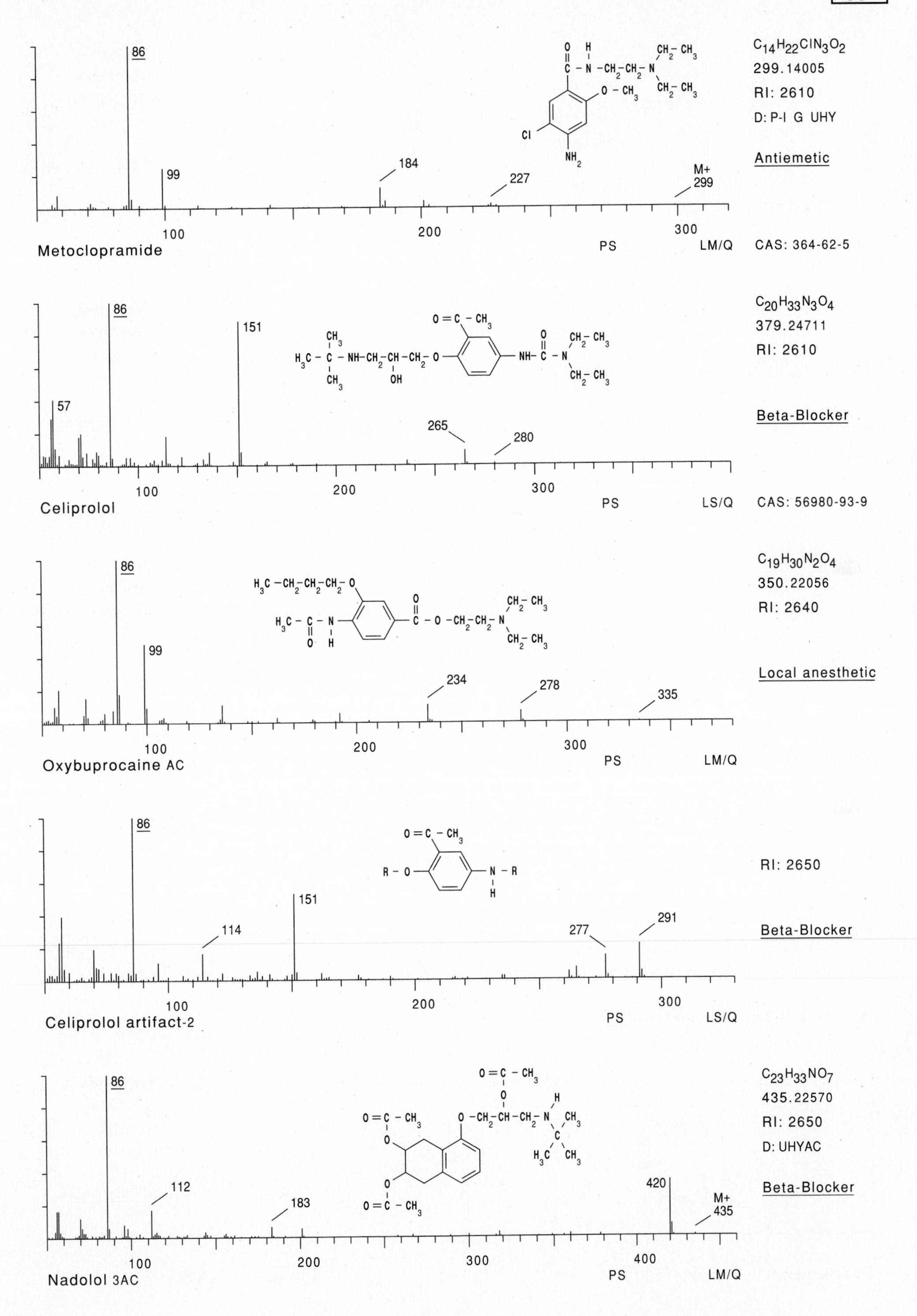

Metoclopramide
$C_{14}H_{22}ClN_3O_2$
299.14005
RI: 2610
D: P-I G UHY
Antiemetic
CAS: 364-62-5
86
99
184
227
M+ 299
PS
LM/Q
Celiprolol
$C_{20}H_{33}N_3O_4$
379.24711
RI: 2610
Beta-Blocker
CAS: 56980-93-9
86
151
57
265
280
PS
LS/Q
Oxybuprocaine AC
$C_{19}H_{30}N_2O_4$
350.22056
RI: 2640
Local anesthetic
86
99
234
278
335
PS
LM/Q
Celiprolol artifact-2
RI: 2650
Beta-Blocker
86
151
114
277
291
PS
LS/Q
Nadolol 3AC
$C_{23}H_{33}NO_7$
435.22570
RI: 2650
D: UHYAC
Beta-Blocker
86
112
183
420
M+ 435
PS
LM/Q

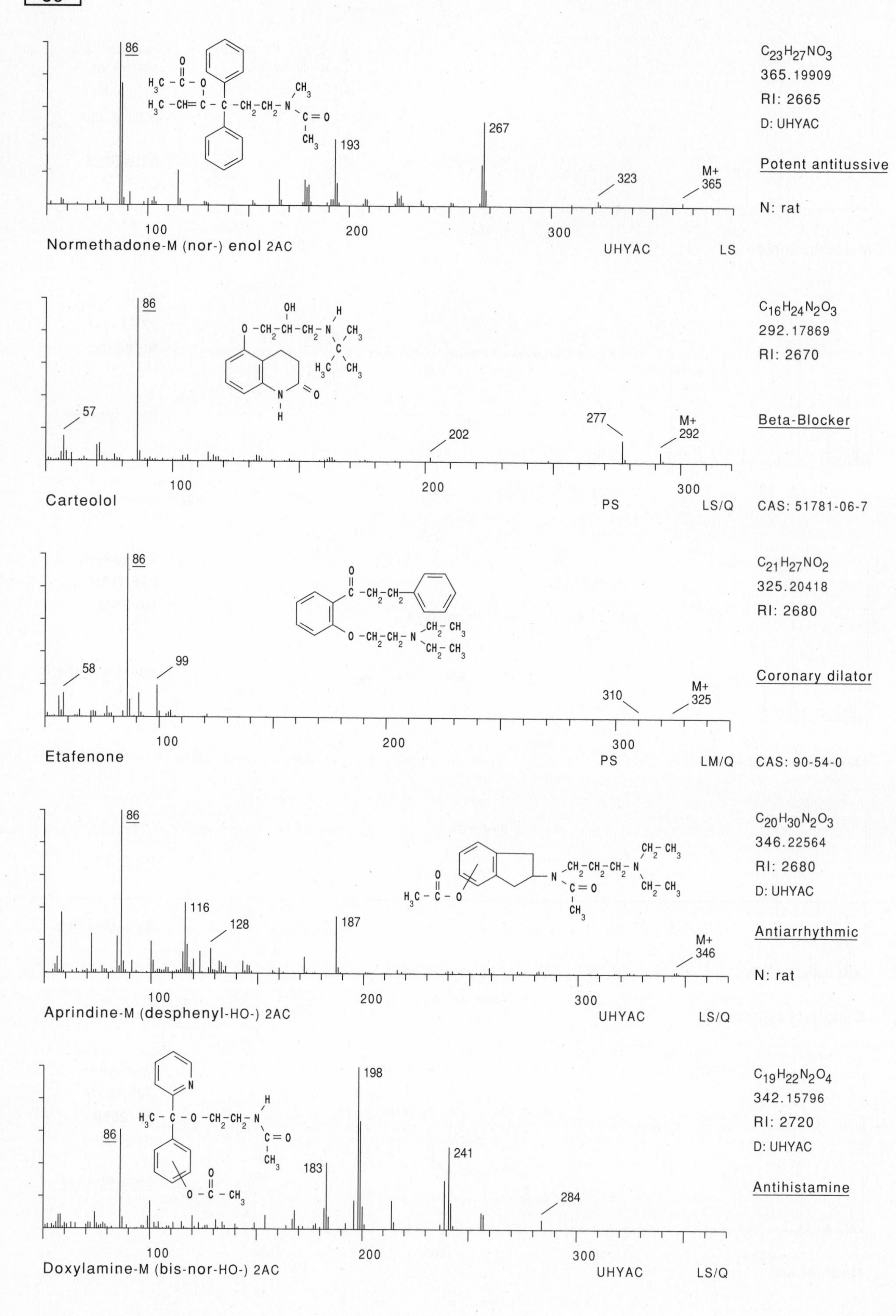

86
193
267
323
M+
365
100
200
300
Normethadone-M (nor-) enol 2AC
UHYAC
LS
$C_{23}H_{27}NO_3$
365.19909
RI: 2665
D: UHYAC
Potent antitussive
N: rat
86
57
202
277
M+
292
100
200
300
Carteolol
PS
LS/Q
$C_{16}H_{24}N_2O_3$
292.17869
RI: 2670
Beta-Blocker
CAS: 51781-06-7
86
58
99
310
M+
325
100
200
300
Etafenone
PS
LM/Q
$C_{21}H_{27}NO_2$
325.20418
RI: 2680
Coronary dilator
CAS: 90-54-0
86
116
128
187
M+
346
100
200
300
Aprindine-M (desphenyl-HO-) 2AC
UHYAC
LS/Q
$C_{20}H_{30}N_2O_3$
346.22564
RI: 2680
D: UHYAC
Antiarrhythmic
N: rat
198
86
183
241
284
100
200
300
Doxylamine-M (bis-nor-HO-) 2AC
UHYAC
LS/Q
$C_{19}H_{22}N_2O_4$
342.15796
RI: 2720
D: UHYAC
Antihistamine

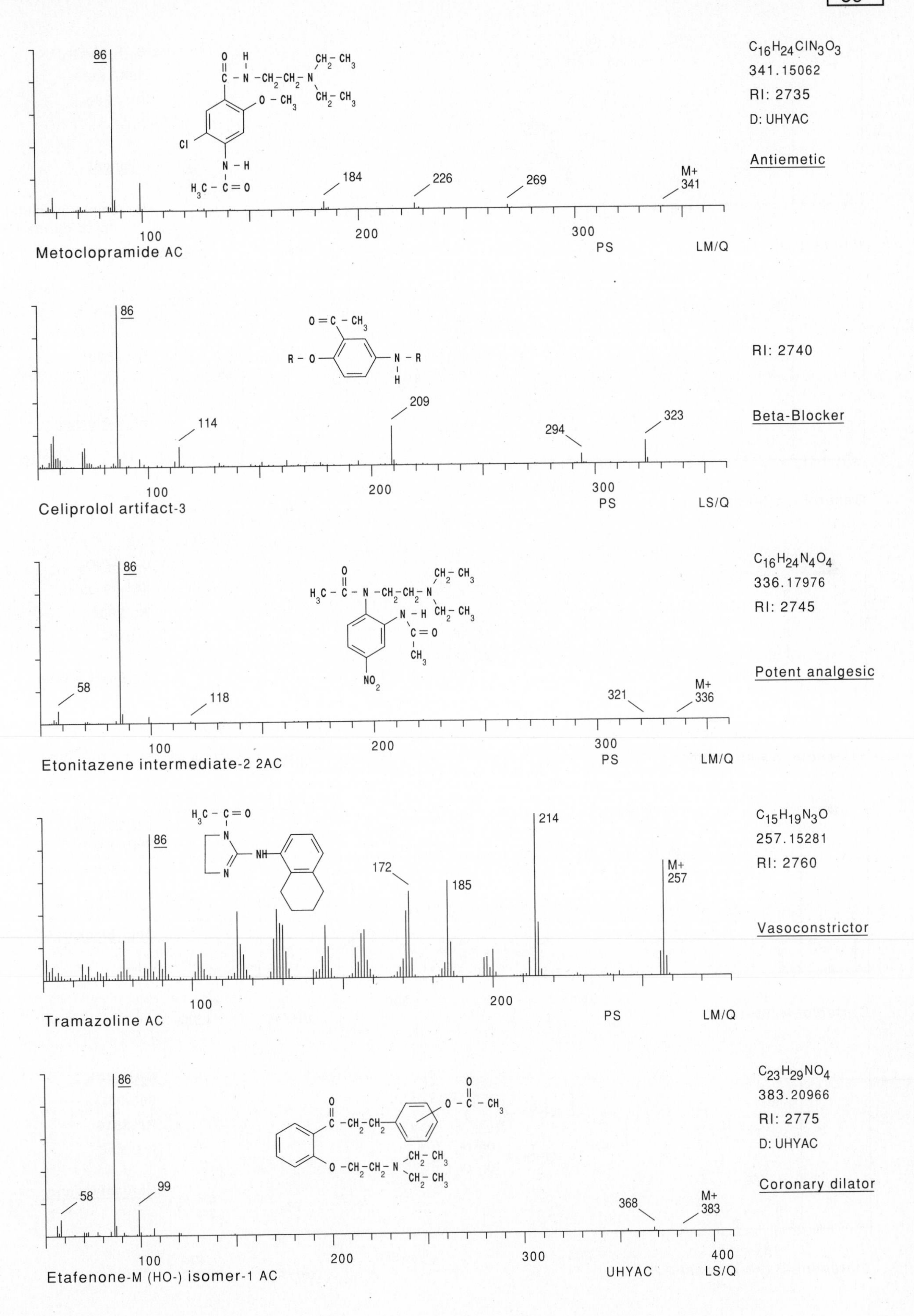
86
184
226
269
M+
341
100
200
300
PS
LM/Q
Metoclopramide AC
$C_{16}H_{24}ClN_3O_3$
341.15062
RI: 2735
D: UHYAC
Antiemetic
86
114
209
294
323
100
200
300
PS
LS/Q
Celiprolol artifact-3
RI: 2740
Beta-Blocker
86
58
118
321
M+
336
100
200
300
PS
LM/Q
Etonitazene intermediate-2 2AC
$C_{16}H_{24}N_4O_4$
336.17976
RI: 2745
Potent analgesic
86
172
185
214
M+
257
100
200
PS
LM/Q
Tramazoline AC
$C_{15}H_{19}N_3O$
257.15281
RI: 2760
Vasoconstrictor
86
58
99
368
M+
383
100
200
300
400
UHYAC
LS/Q
Etafenone-M (HO-) isomer-1 AC
$C_{23}H_{29}NO_4$
383.20966
RI: 2775
D: UHYAC
Coronary dilator

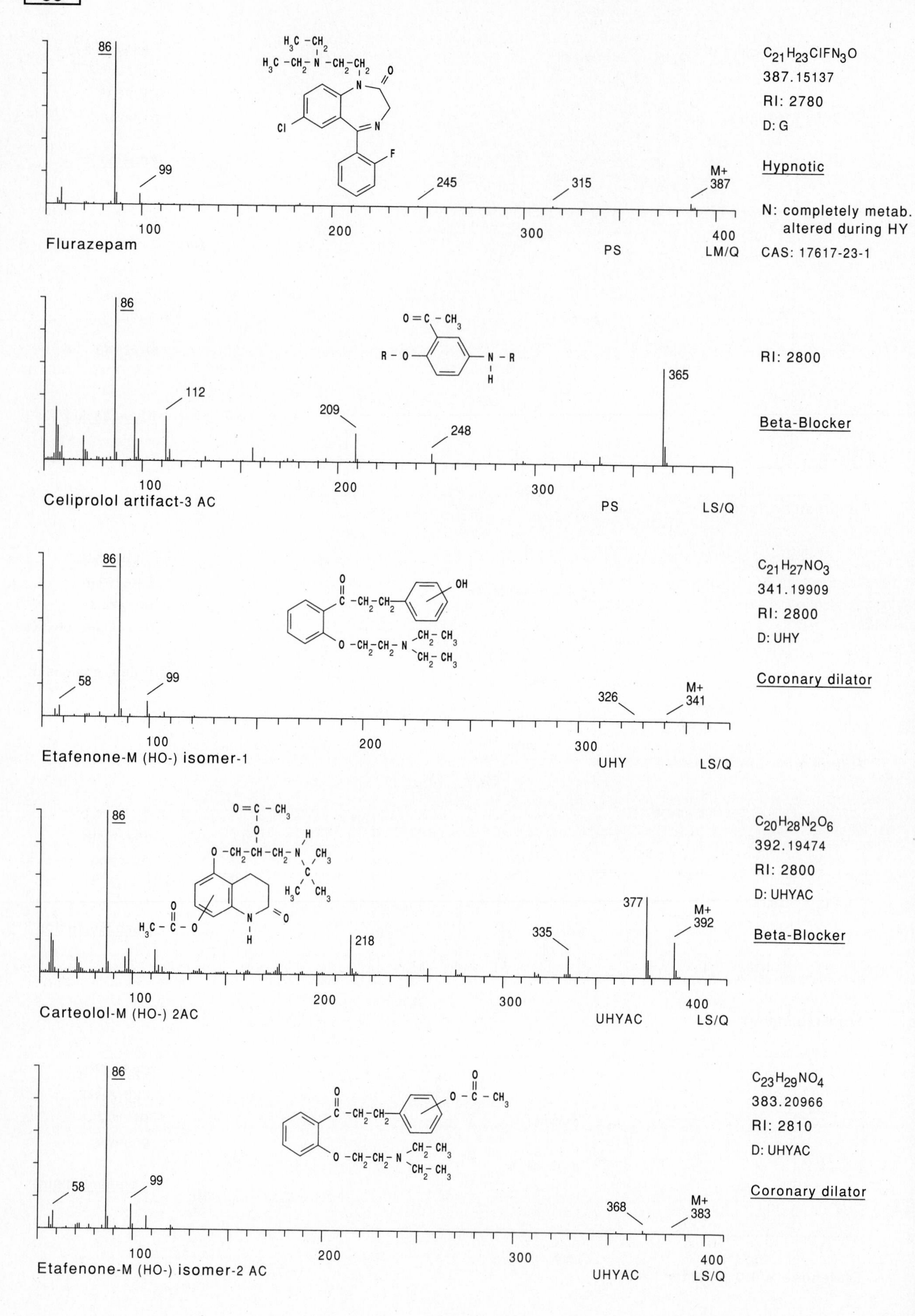

86
99
245
315
M+
387
100
200
300
400
Flurazepam
PS
LM/Q
$C_{21}H_{23}ClFN_3O$
387.15137
RI: 2780
D: G
Hypnotic
N: completely metab. altered during HY
CAS: 17617-23-1
86
112
209
248
365
100
200
300
Celiprolol artifact-3 AC
PS
LS/Q
RI: 2800
Beta-Blocker
86
58
99
326
M+
341
100
200
300
Etafenone-M (HO-) isomer-1
UHY
LS/Q
$C_{21}H_{27}NO_3$
341.19909
RI: 2800
D: UHY
Coronary dilator
86
218
335
377
M+
392
100
200
300
400
Carteolol-M (HO-) 2AC
UHYAC
LS/Q
$C_{20}H_{28}N_2O_6$
392.19474
RI: 2800
D: UHYAC
Beta-Blocker
86
58
99
368
M+
383
100
200
300
400
Etafenone-M (HO-) isomer-2 AC
UHYAC
LS/Q
$C_{23}H_{29}NO_4$
383.20966
RI: 2810
D: UHYAC
Coronary dilator

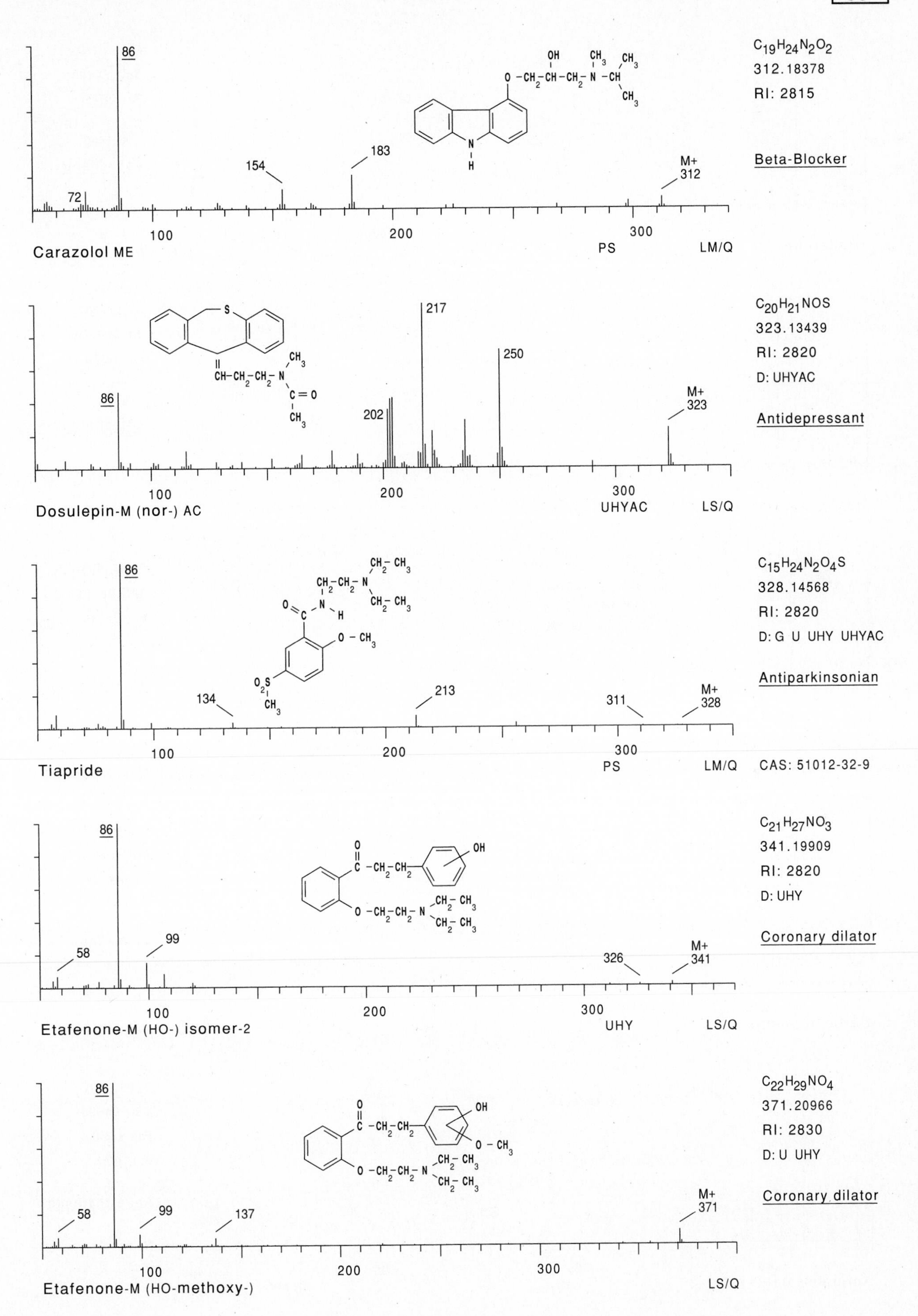

86
72
154
183
M+
312
100
200
300
Carazolol ME
PS
LM/Q
C19H24N2O2
312.18378
RI: 2815
Beta-Blocker
217
250
86
202
M+
323
100
200
300
Dosulepin-M (nor-) AC
UHYAC
LS/Q
C20H21NOS
323.13439
RI: 2820
D: UHYAC
Antidepressant
86
134
213
311
M+
328
100
200
300
Tiapride
PS
LM/Q
C15H24N2O4S
328.14568
RI: 2820
D: G U UHY UHYAC
Antiparkinsonian
CAS: 51012-32-9
86
58
99
326
M+
341
100
200
300
Etafenone-M (HO-) isomer-2
UHY
LS/Q
C21H27NO3
341.19909
RI: 2820
D: UHY
Coronary dilator
86
58
99
137
M+
371
100
200
300
Etafenone-M (HO-methoxy-)
LS/Q
C22H29NO4
371.20966
RI: 2830
D: U UHY
Coronary dilator

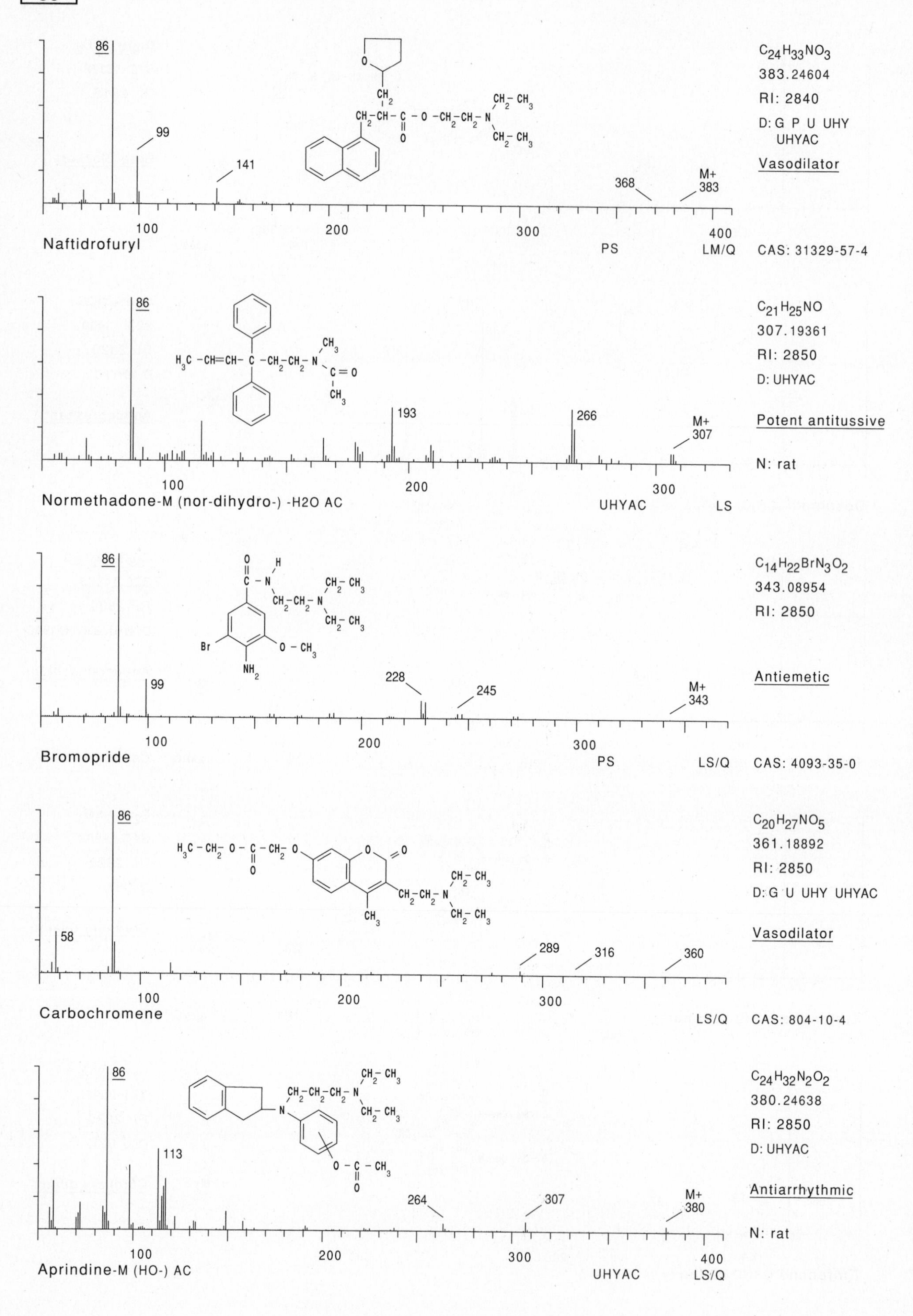
86
99
141
368
M+
383
100
200
300
400
Naftidrofuryl
PS
LM/Q
C24H33NO3
383.24604
RI: 2840
D: G P U UHY
UHYAC
Vasodilator
CAS: 31329-57-4
86
193
266
M+
307
Normethadone-M (nor-dihydro-) -H2O AC
UHYAC
LS
C21H25NO
307.19361
RI: 2850
D: UHYAC
Potent antitussive
N: rat
86
99
228
245
M+
343
Bromopride
PS
LS/Q
C14H22BrN3O2
343.08954
RI: 2850
Antiemetic
CAS: 4093-35-0
86
58
289
316
360
Carbochromene
LS/Q
C20H27NO5
361.18892
RI: 2850
D: G U UHY UHYAC
Vasodilator
CAS: 804-10-4
86
113
264
307
M+
380
Aprindine-M (HO-) AC
UHYAC
LS/Q
C24H32N2O2
380.24638
RI: 2850
D: UHYAC
Antiarrhythmic
N: rat

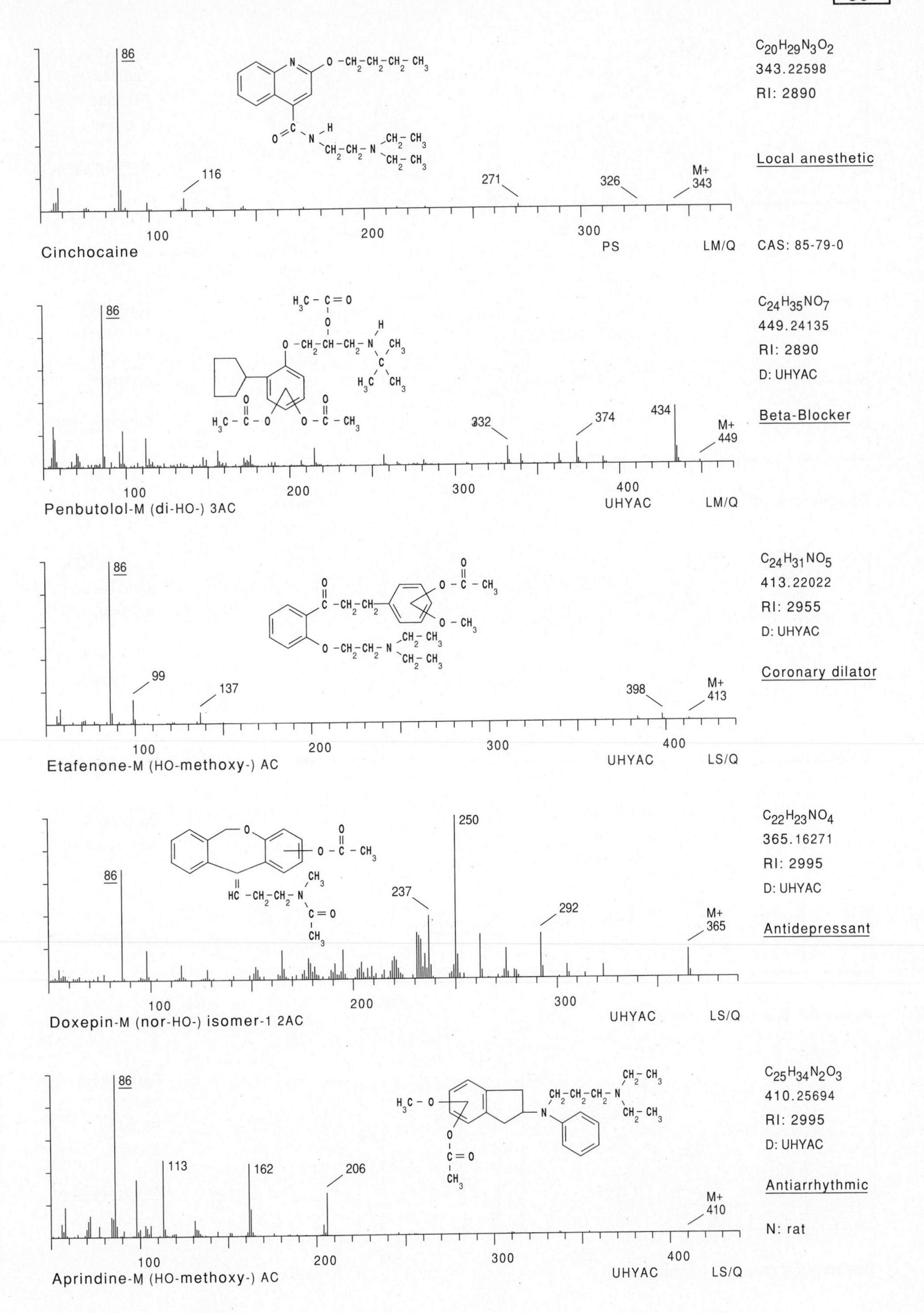
86
Cinchocaine
$C_{20}H_{29}N_3O_2$
343.22598
RI: 2890
Local anesthetic
CAS: 85-79-0
PS
LM/Q
86
116
271
326
M+ 343
Penbutolol-M (di-HO-) 3AC
$C_{24}H_{35}NO_7$
449.24135
RI: 2890
D: UHYAC
Beta-Blocker
UHYAC
LM/Q
86
332
374
434
M+ 449
Etafenone-M (HO-methoxy-) AC
$C_{24}H_{31}NO_5$
413.22022
RI: 2955
D: UHYAC
Coronary dilator
UHYAC
LS/Q
86
99
137
398
M+ 413
Doxepin-M (nor-HO-) isomer-1 2AC
$C_{22}H_{23}NO_4$
365.16271
RI: 2995
D: UHYAC
Antidepressant
UHYAC
LS/Q
86
237
250
292
M+ 365
Aprindine-M (HO-methoxy-) AC
$C_{25}H_{34}N_2O_3$
410.25694
RI: 2995
D: UHYAC
Antiarrhythmic
N: rat
UHYAC
LS/Q
86
113
162
206
M+ 410

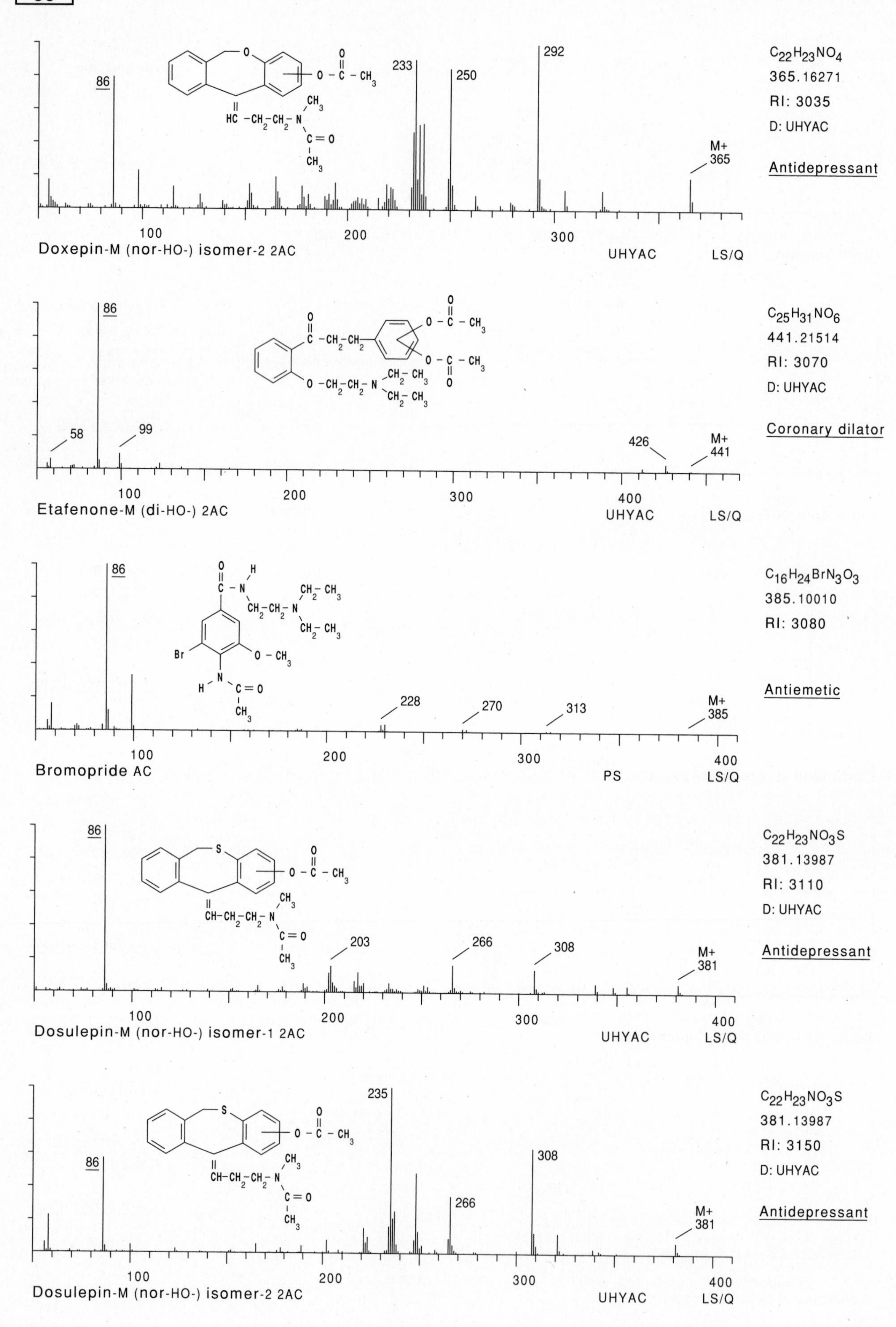

86
233
250
292
M+
365
100
200
300
Doxepin-M (nor-HO-) isomer-2 2AC
UHYAC
LS/Q
C22H23NO4
365.16271
RI: 3035
D: UHYAC
Antidepressant
86
58
99
426
M+
441
Etafenone-M (di-HO-) 2AC
400
C25H31NO6
441.21514
RI: 3070
Coronary dilator
Br
228
270
313
M+
385
Bromopride AC
PS
C16H24BrN3O3
385.10010
RI: 3080
Antiemetic
S
203
266
308
M+
381
Dosulepin-M (nor-HO-) isomer-1 2AC
C22H23NO3S
381.13987
RI: 3110
235
Dosulepin-M (nor-HO-) isomer-2 2AC
RI: 3150

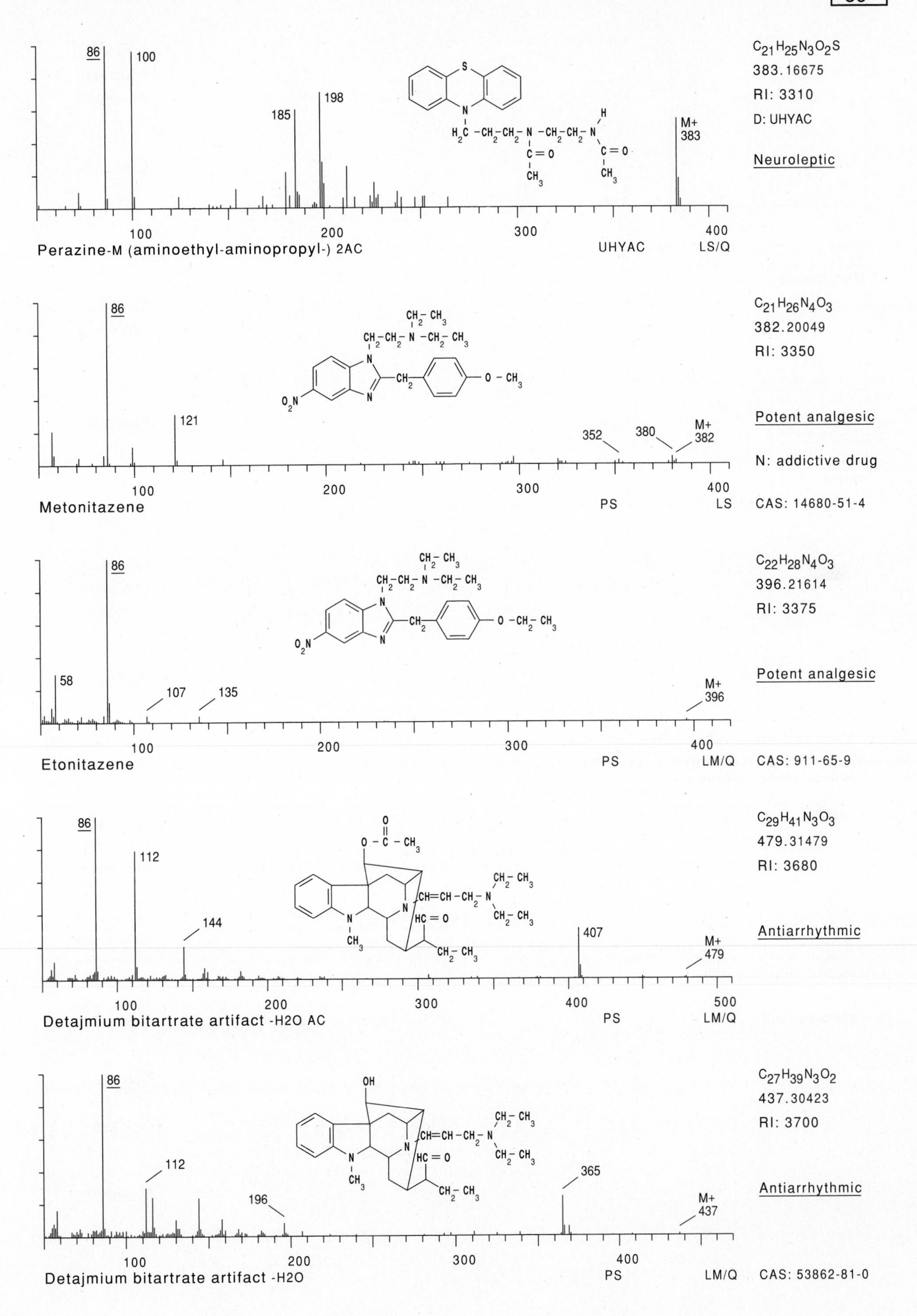

Perazine-M (aminoethyl-aminopropyl-) 2AC

$C_{21}H_{25}N_3O_2S$
383.16675
RI: 3310
D: UHYAC

Neuroleptic

Metonitazene

$C_{21}H_{26}N_4O_3$
382.20049
RI: 3350

Potent analgesic

N: addictive drug

CAS: 14680-51-4

Etonitazene

$C_{22}H_{28}N_4O_3$
396.21614
RI: 3375

Potent analgesic

CAS: 911-65-9

Detajmium bitartrate artifact -H2O AC

$C_{29}H_{41}N_3O_3$
479.31479
RI: 3680

Antiarrhythmic

Detajmium bitartrate artifact -H2O

$C_{27}H_{39}N_3O_2$
437.30423
RI: 3700

Antiarrhythmic

CAS: 53862-81-0

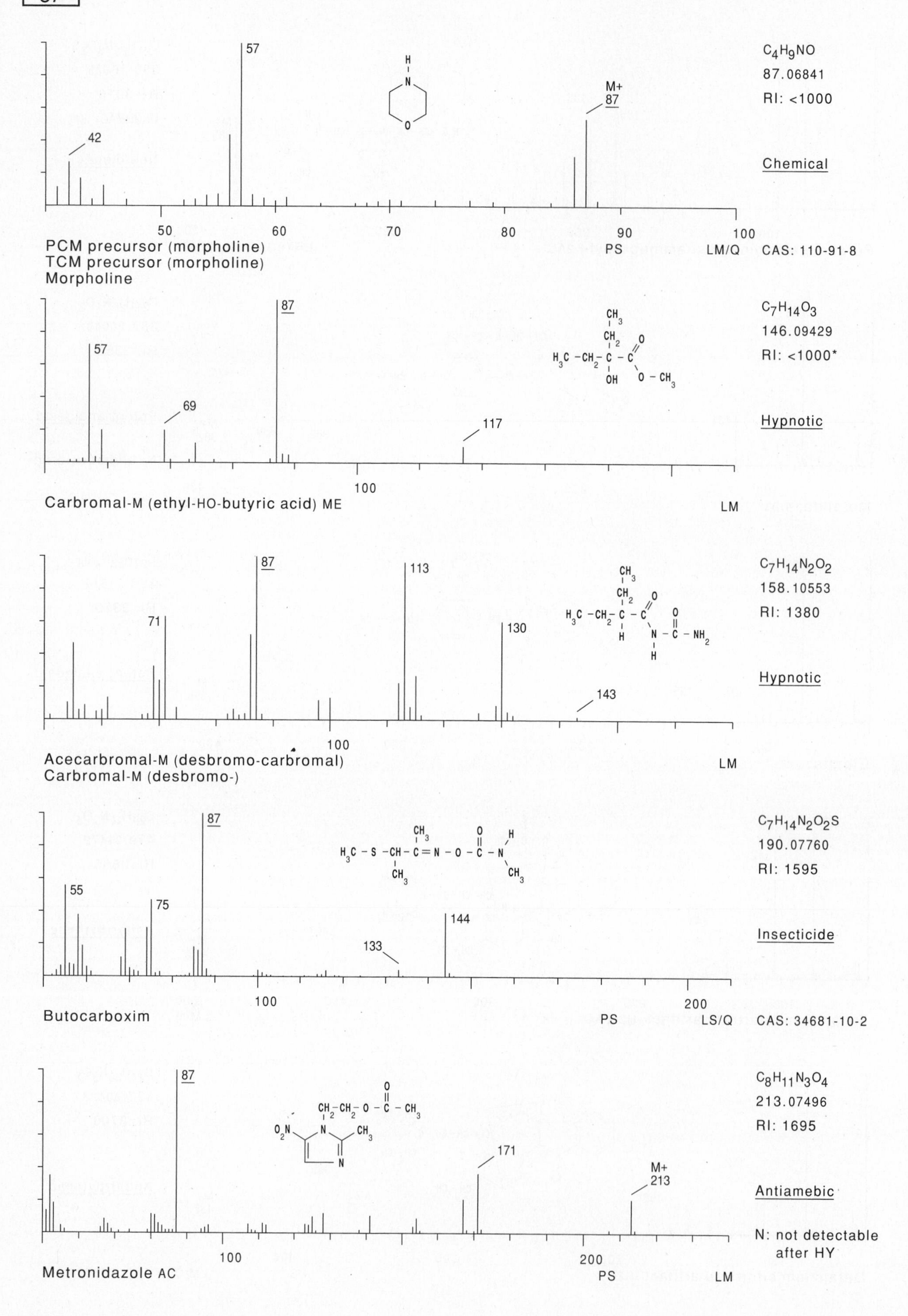

57
42
M+
87
50
60
70
80
90
100
PS
LM/Q
C4H9NO
87.06841
RI: <1000
Chemical
CAS: 110-91-8
PCM precursor (morpholine)
TCM precursor (morpholine)
Morpholine
87
57
69
117
100
LM
C7H14O3
146.09429
RI: <1000*
Hypnotic
Carbromal-M (ethyl-HO-butyric acid) ME
87
113
71
130
143
100
LM
C7H14N2O2
158.10553
RI: 1380
Hypnotic
Acecarbromal-M (desbromo-carbromal)
Carbromal-M (desbromo-)
87
55
75
144
133
100
200
PS
LS/Q
C7H14N2O2S
190.07760
RI: 1595
Insecticide
CAS: 34681-10-2
Butocarboxim
87
171
M+
213
100
200
PS
LM
C8H11N3O4
213.07496
RI: 1695
Antiamebic
N: not detectable after HY
Metronidazole AC

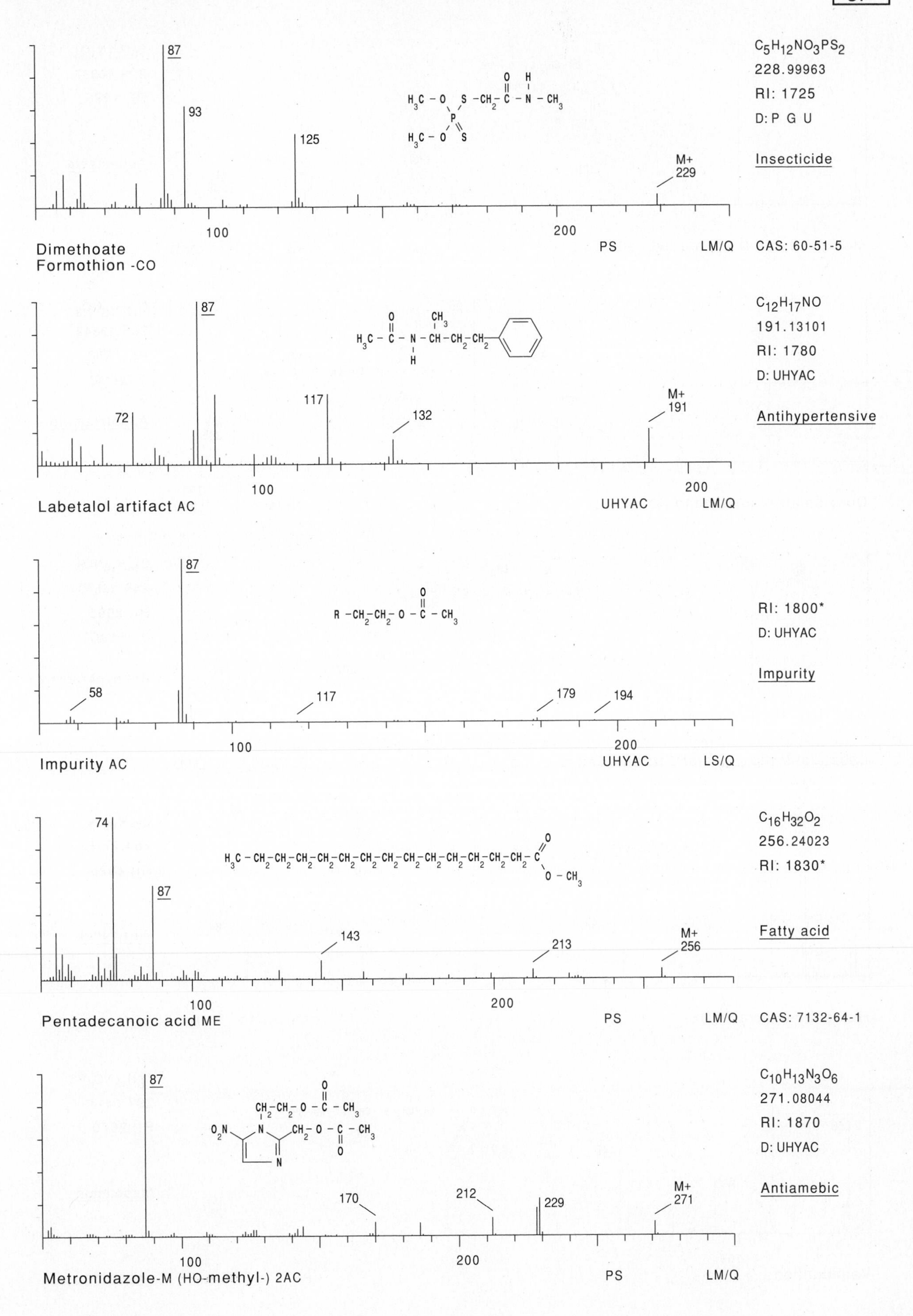

$C_5H_{12}NO_3PS_2$
228.99963
RI: 1725
D: P G U
Insecticide
CAS: 60-51-5

$C_{12}H_{17}NO$
191.13101
RI: 1780
D: UHYAC
Antihypertensive

RI: 1800*
D: UHYAC
Impurity

$C_{16}H_{32}O_2$
256.24023
RI: 1830*
Fatty acid
CAS: 7132-64-1

$C_{10}H_{13}N_3O_6$
271.08044
RI: 1870
D: UHYAC
Antiamebic

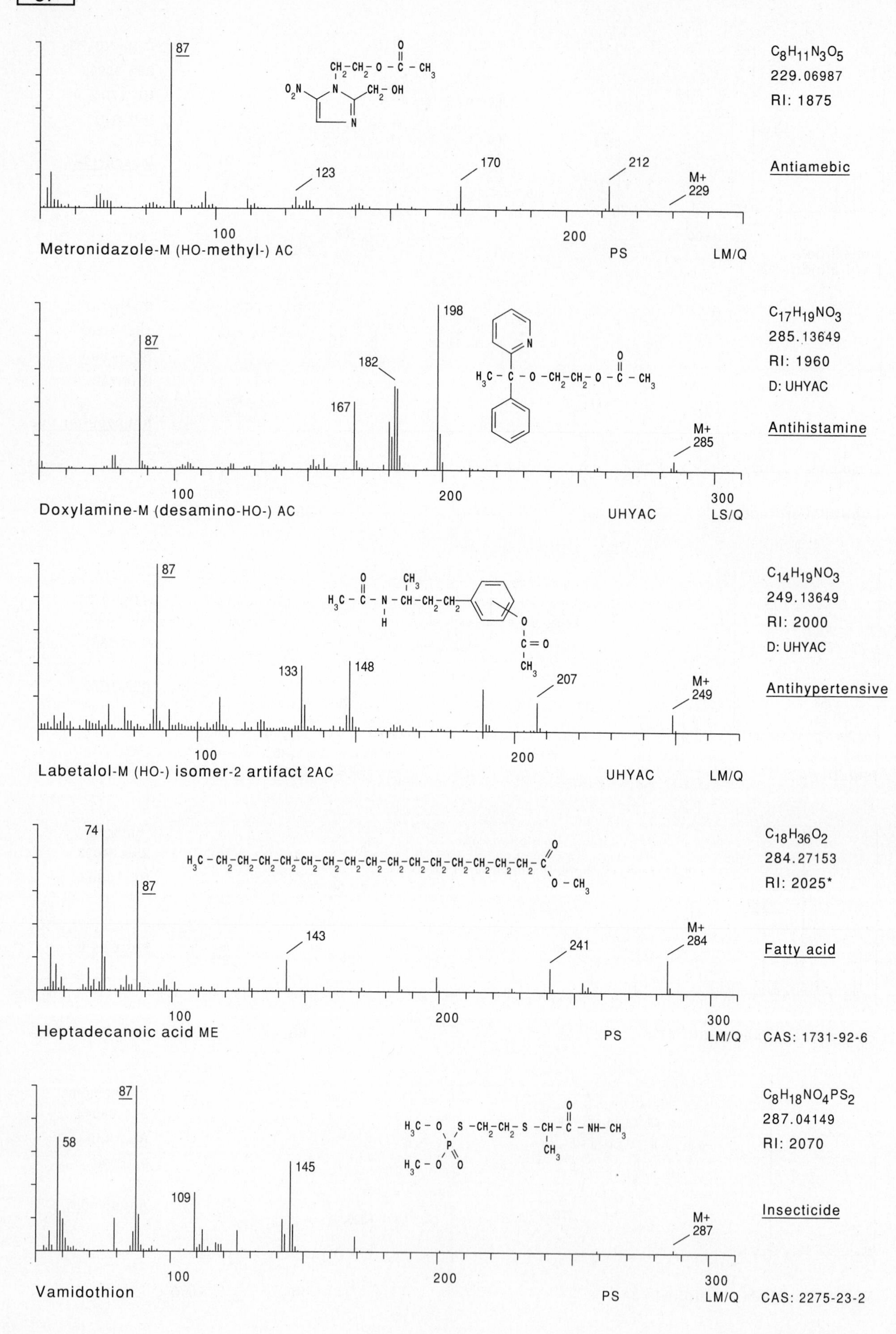

87
123
170
212
M+
229
100
200
Metronidazole-M (HO-methyl-) AC
PS
LM/Q
$C_8H_{11}N_3O_5$
229.06987
RI: 1875
Antiamebic
198
87
182
167
M+
285
100
200
300
Doxylamine-M (desamino-HO-) AC
UHYAC
LS/Q
$C_{17}H_{19}NO_3$
285.13649
RI: 1960
D: UHYAC
Antihistamine
87
133
148
207
M+
249
100
200
Labetalol-M (HO-) isomer-2 artifact 2AC
UHYAC
LM/Q
$C_{14}H_{19}NO_3$
249.13649
RI: 2000
D: UHYAC
Antihypertensive
74
87
143
241
M+
284
100
200
300
Heptadecanoic acid ME
PS
LM/Q
$C_{18}H_{36}O_2$
284.27153
RI: 2025*
Fatty acid
CAS: 1731-92-6
87
58
145
109
M+
287
100
200
300
Vamidothion
PS
LM/Q
$C_8H_{18}NO_4PS_2$
287.04149
RI: 2070
Insecticide
CAS: 2275-23-2

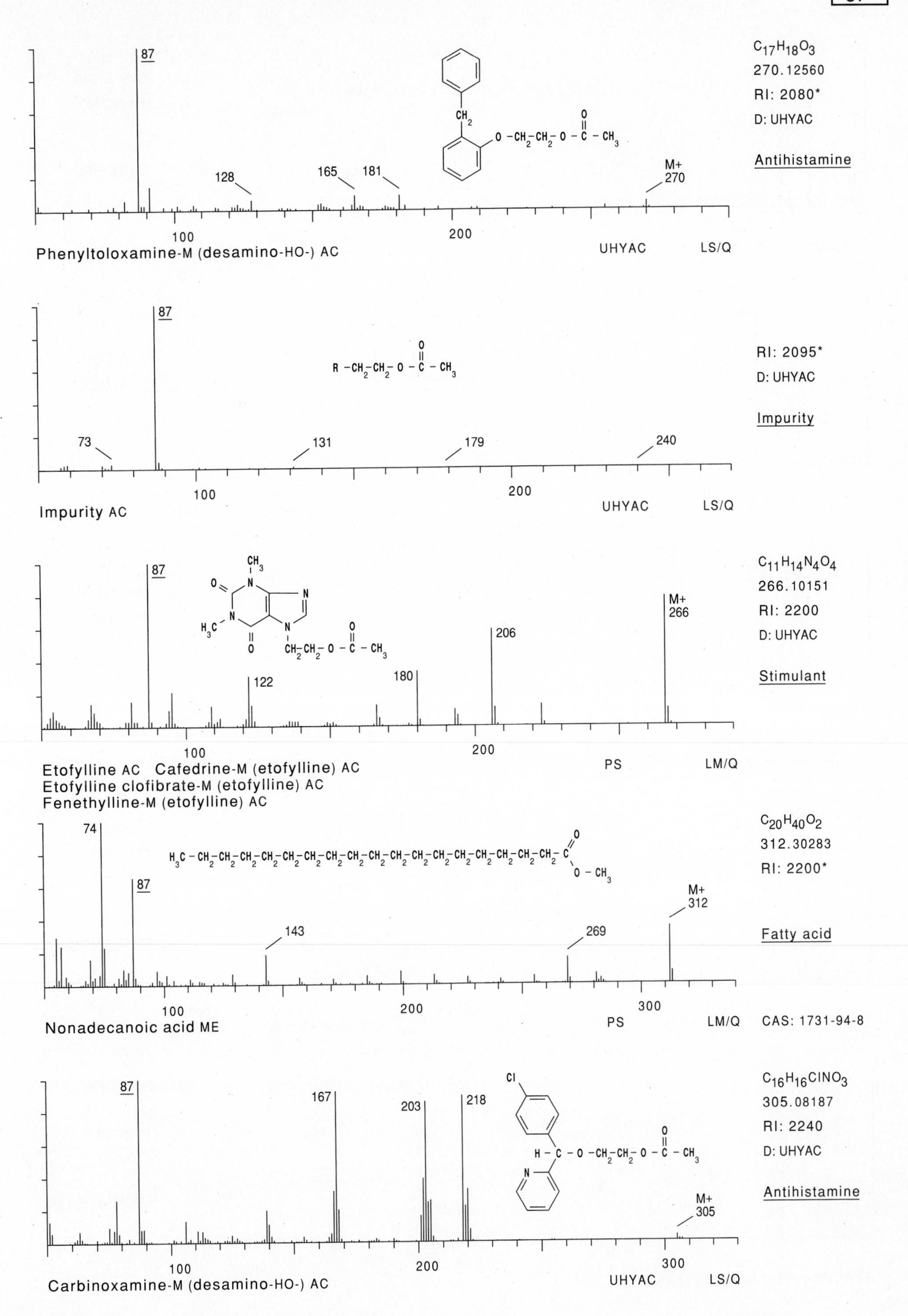

$C_{17}H_{18}O_3$
270.12560
RI: 2080*
D: UHYAC
Antihistamine

RI: 2095*
D: UHYAC
Impurity

$C_{11}H_{14}N_4O_4$
266.10151
RI: 2200
D: UHYAC
Stimulant

$C_{20}H_{40}O_2$
312.30283
RI: 2200*
Fatty acid
CAS: 1731-94-8

$C_{16}H_{16}ClNO_3$
305.08187
RI: 2240
D: UHYAC
Antihistamine

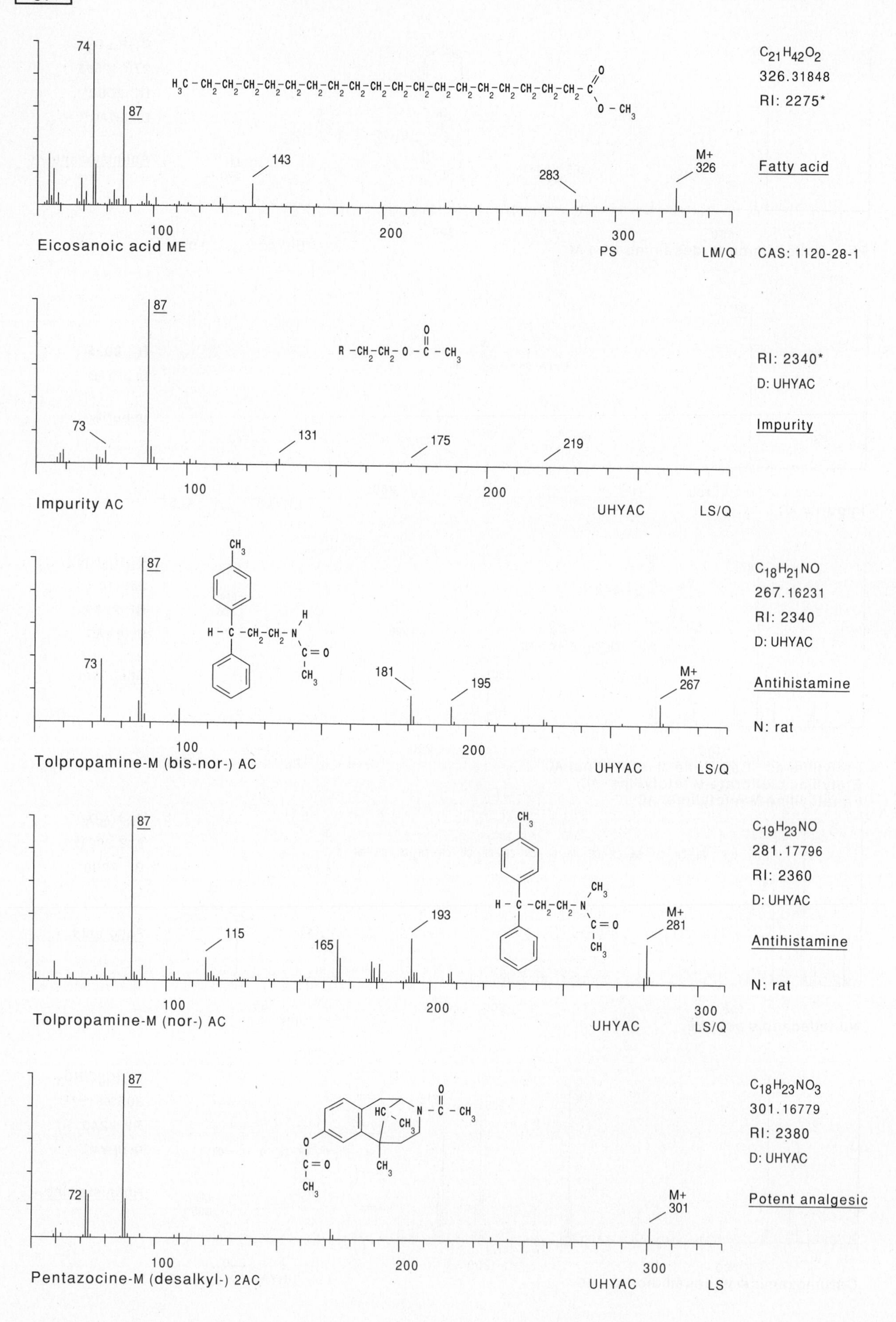
74
87
143
283
M+
326
100
200
300
Eicosanoic acid ME
PS
LM/Q
C21H42O2
326.31848
RI: 2275*
Fatty acid
CAS: 1120-28-1
87
73
131
175
219
100
200
Impurity AC
UHYAC
LS/Q
RI: 2340*
D: UHYAC
Impurity
87
73
181
195
M+
267
100
200
Tolpropamine-M (bis-nor-) AC
UHYAC
LS/Q
C18H21NO
267.16231
RI: 2340
D: UHYAC
Antihistamine
N: rat
87
115
165
193
M+
281
100
200
300
Tolpropamine-M (nor-) AC
UHYAC
LS/Q
C19H23NO
281.17796
RI: 2360
D: UHYAC
Antihistamine
N: rat
87
72
M+
301
100
200
300
Pentazocine-M (desalkyl-) 2AC
UHYAC
LS
C18H23NO3
301.16779
RI: 2380
D: UHYAC
Potent analgesic

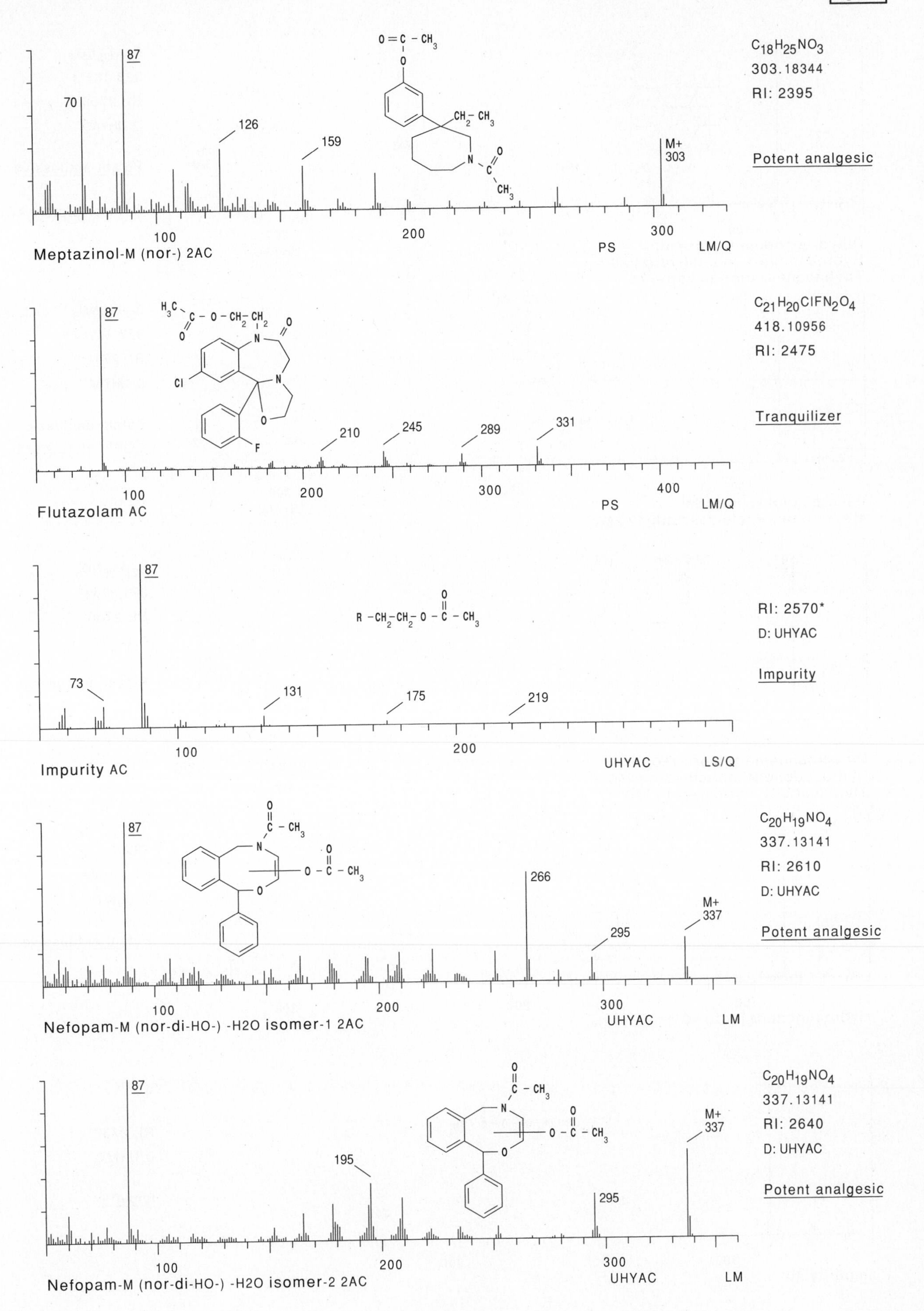

$C_{18}H_{25}NO_3$
303.18344
RI: 2395

Potent analgesic

$C_{21}H_{20}ClFN_2O_4$
418.10956
RI: 2475

Tranquilizer

RI: 2570*
D: UHYAC

Impurity

$C_{20}H_{19}NO_4$
337.13141
RI: 2610
D: UHYAC

Potent analgesic

$C_{20}H_{19}NO_4$
337.13141
RI: 2640
D: UHYAC

Potent analgesic

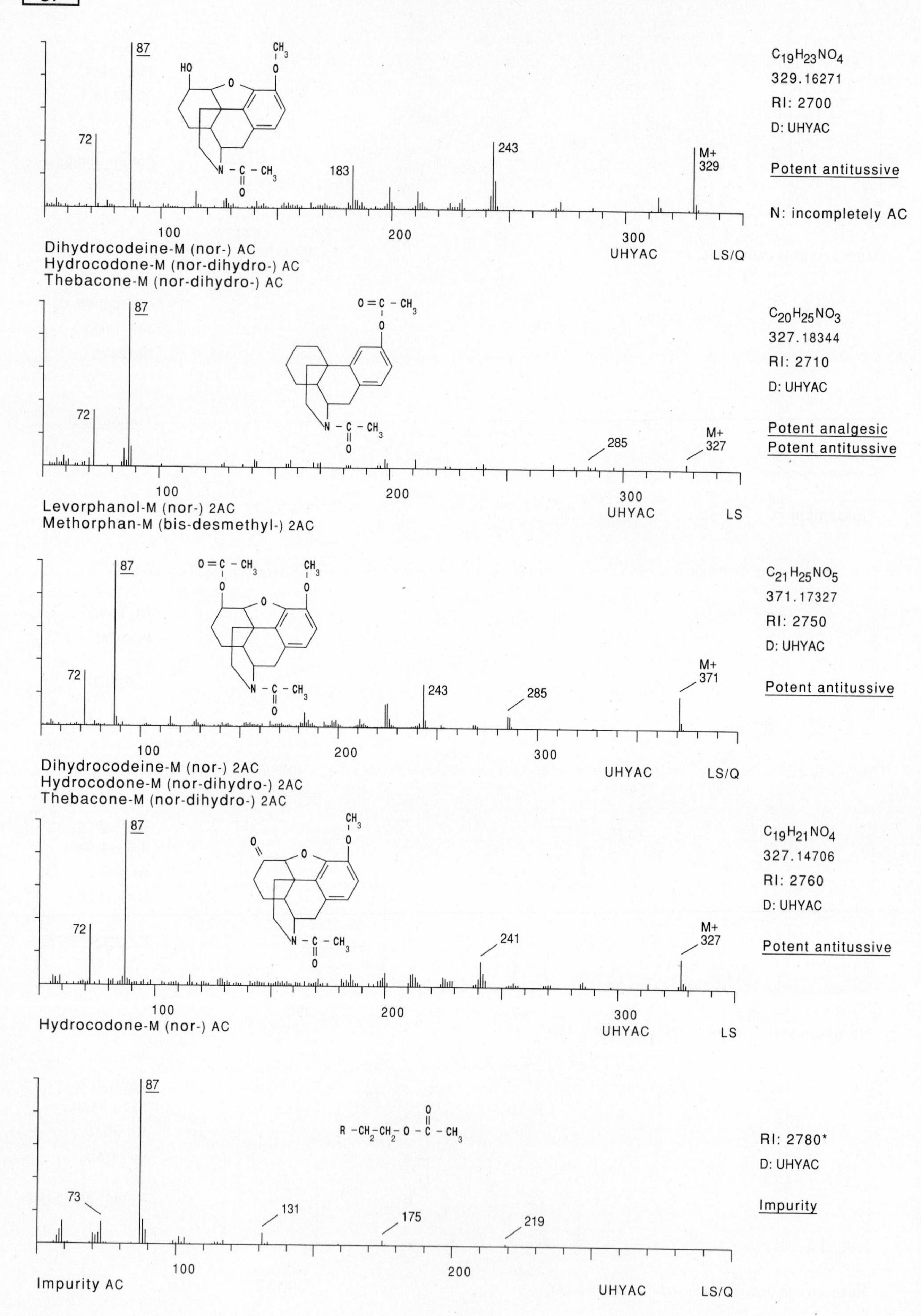

87
72
183
243
M+
329
100
200
300
UHYAC
LS/Q
Dihydrocodeine-M (nor-) AC
Hydrocodone-M (nor-dihydro-) AC
Thebacone-M (nor-dihydro-) AC
C19H23NO4
329.16271
RI: 2700
D: UHYAC
Potent antitussive
N: incompletely AC
87
72
285
M+
327
UHYAC
LS
Levorphanol-M (nor-) 2AC
Methorphan-M (bis-desmethyl-) 2AC
C20H25NO3
327.18344
RI: 2710
D: UHYAC
Potent analgesic
Potent antitussive
87
72
243
285
M+
371
UHYAC
LS/Q
Dihydrocodeine-M (nor-) 2AC
Hydrocodone-M (nor-dihydro-) 2AC
Thebacone-M (nor-dihydro-) 2AC
C21H25NO5
371.17327
RI: 2750
D: UHYAC
Potent antitussive
87
72
241
M+
327
UHYAC
LS
Hydrocodone-M (nor-) AC
C19H21NO4
327.14706
RI: 2760
D: UHYAC
Potent antitussive
87
73
131
175
219
R -CH2-CH2- O - C - CH3
UHYAC
LS/Q
Impurity AC
RI: 2780*
D: UHYAC
Impurity

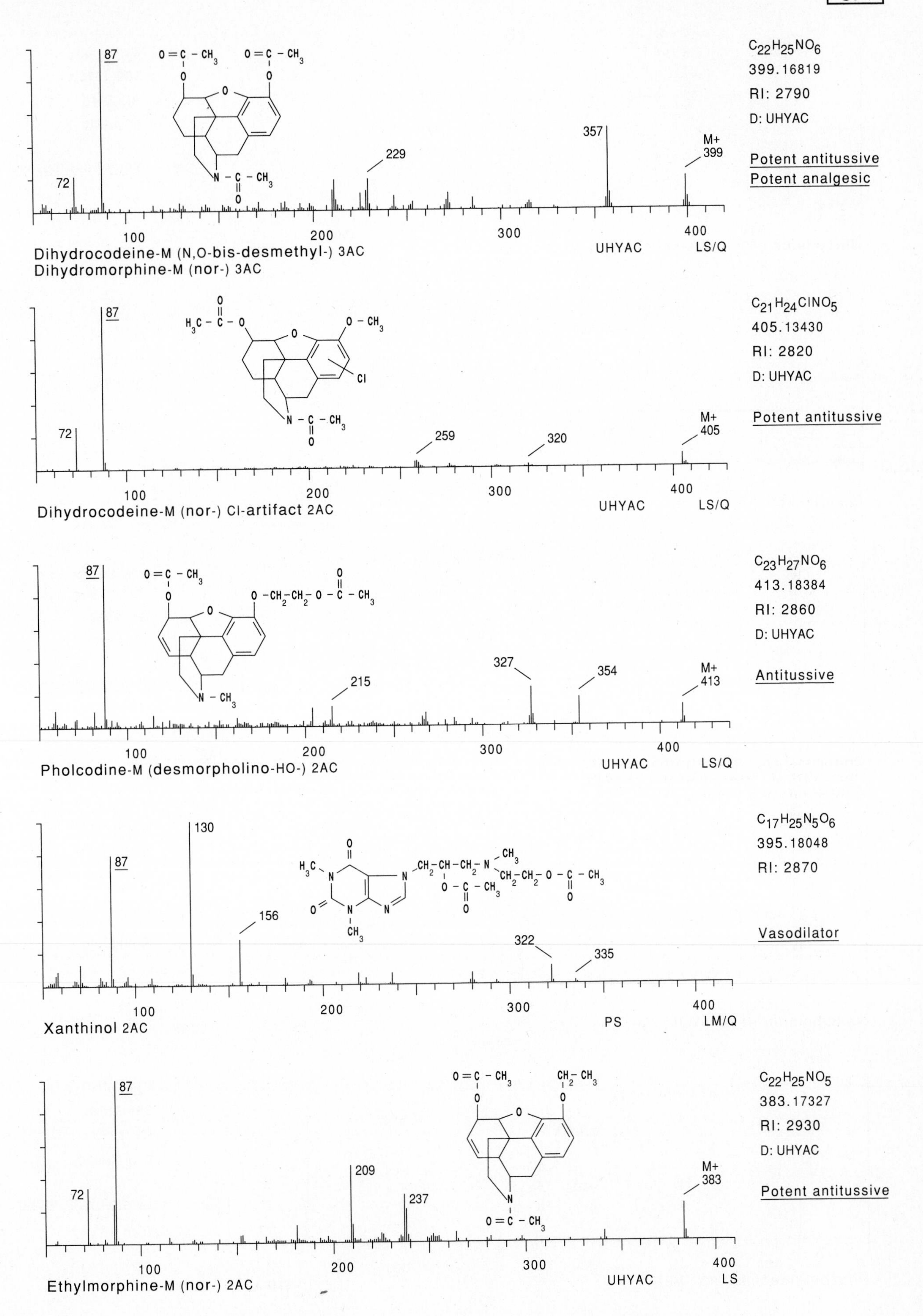

Dihydrocodeine-M (N,O-bis-desmethyl-) 3AC
Dihydromorphine-M (nor-) 3AC

$C_{22}H_{25}NO_6$
399.16819
RI: 2790
D: UHYAC

Potent antitussive
Potent analgesic

Dihydrocodeine-M (nor-) Cl-artifact 2AC

$C_{21}H_{24}ClNO_5$
405.13430
RI: 2820
D: UHYAC

Potent antitussive

Pholcodine-M (desmorpholino-HO-) 2AC

$C_{23}H_{27}NO_6$
413.18384
RI: 2860
D: UHYAC

Antitussive

Xanthinol 2AC

$C_{17}H_{25}N_5O_6$
395.18048
RI: 2870

Vasodilator

Ethylmorphine-M (nor-) 2AC

$C_{22}H_{25}NO_5$
383.17327
RI: 2930
D: UHYAC

Potent antitussive

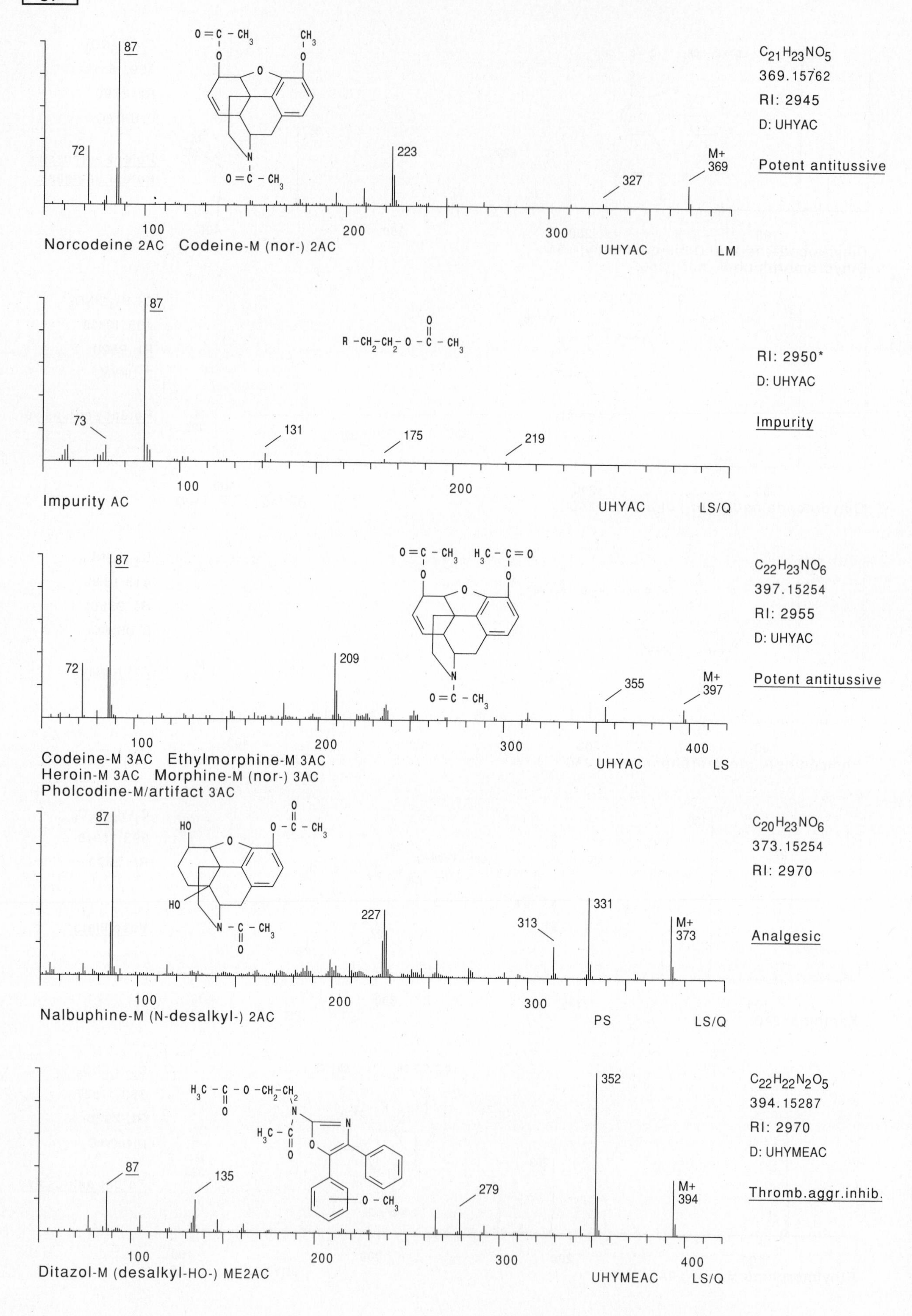

87
72
223
M+
369
327
100
200
300
Norcodeine 2AC Codeine-M (nor-) 2AC
UHYAC
LM
C21H23NO5
369.15762
RI: 2945
D: UHYAC
Potent antitussive
87
73
131
175
219
100
200
Impurity AC
UHYAC
LS/Q
RI: 2950*
D: UHYAC
Impurity
87
72
209
355
M+
397
100
200
300
400
Codeine-M 3AC Ethylmorphine-M 3AC
Heroin-M 3AC Morphine-M (nor-) 3AC
Pholcodine-M/artifact 3AC
UHYAC
LS
C22H23NO6
397.15254
RI: 2955
D: UHYAC
Potent antitussive
87
227
313
331
M+
373
100
200
300
Nalbuphine-M (N-desalkyl-) 2AC
PS
LS/Q
C20H23NO6
373.15254
RI: 2970
Analgesic
352
87
135
279
M+
394
100
200
300
400
Ditazol-M (desalkyl-HO-) ME2AC
UHYMEAC
LS/Q
C22H22N2O5
394.15287
RI: 2970
D: UHYMEAC
Thromb.aggr.inhib.

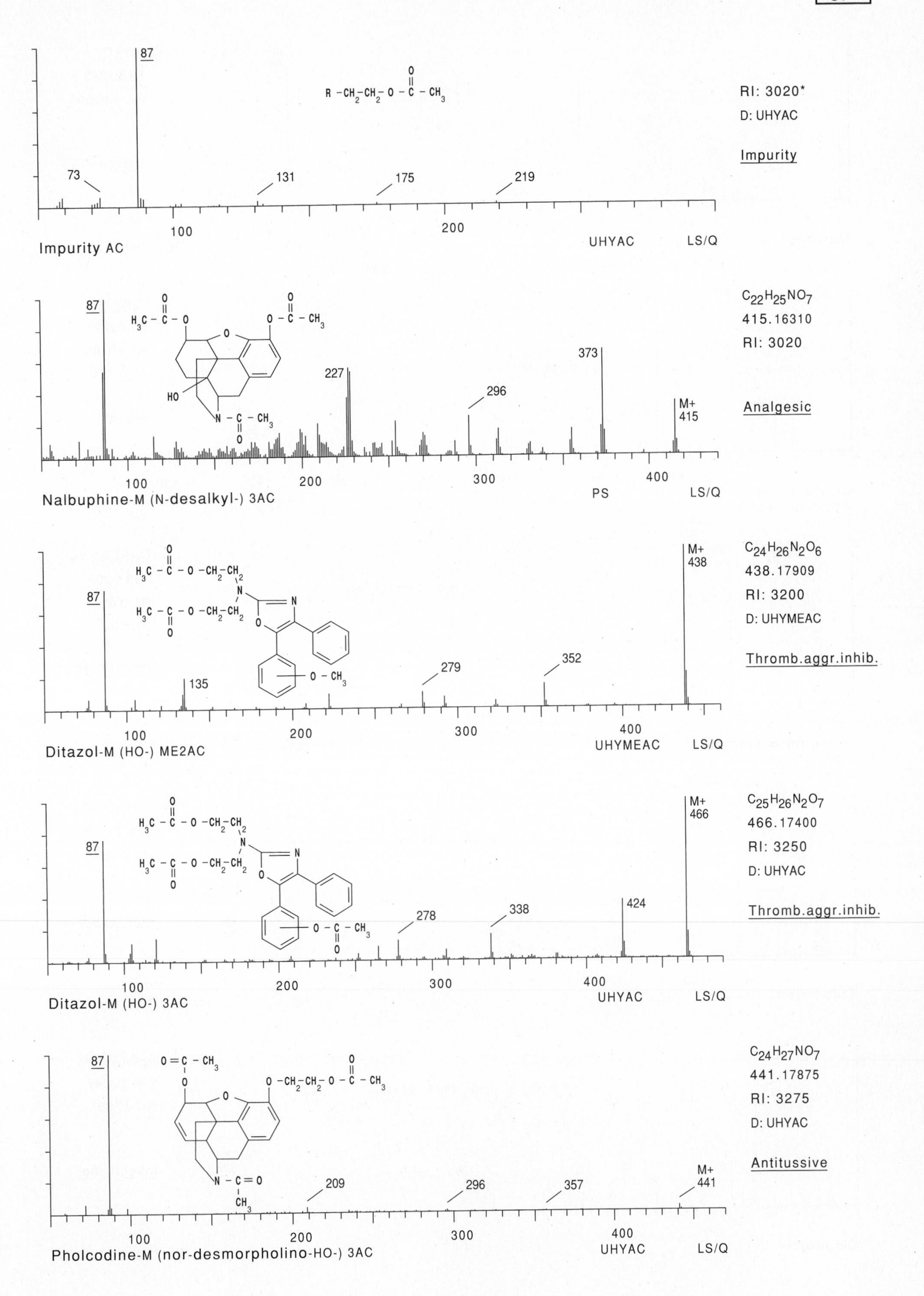

RI: 3020*
D: UHYAC
Impurity
Impurity AC
UHYAC LS/Q
$C_{22}H_{25}NO_7$
415.16310
RI: 3020
Analgesic
Nalbuphine-M (N-desalkyl-) 3AC
PS LS/Q
$C_{24}H_{26}N_2O_6$
438.17909
RI: 3200
D: UHYMEAC
Thromb.aggr.inhib.
Ditazol-M (HO-) ME2AC
UHYMEAC LS/Q
$C_{25}H_{26}N_2O_7$
466.17400
RI: 3250
D: UHYAC
Thromb.aggr.inhib.
Ditazol-M (HO-) 3AC
UHYAC LS/Q
$C_{24}H_{27}NO_7$
441.17875
RI: 3275
D: UHYAC
Antitussive
Pholcodine-M (nor-desmorpholino-HO-) 3AC
UHYAC LS/Q

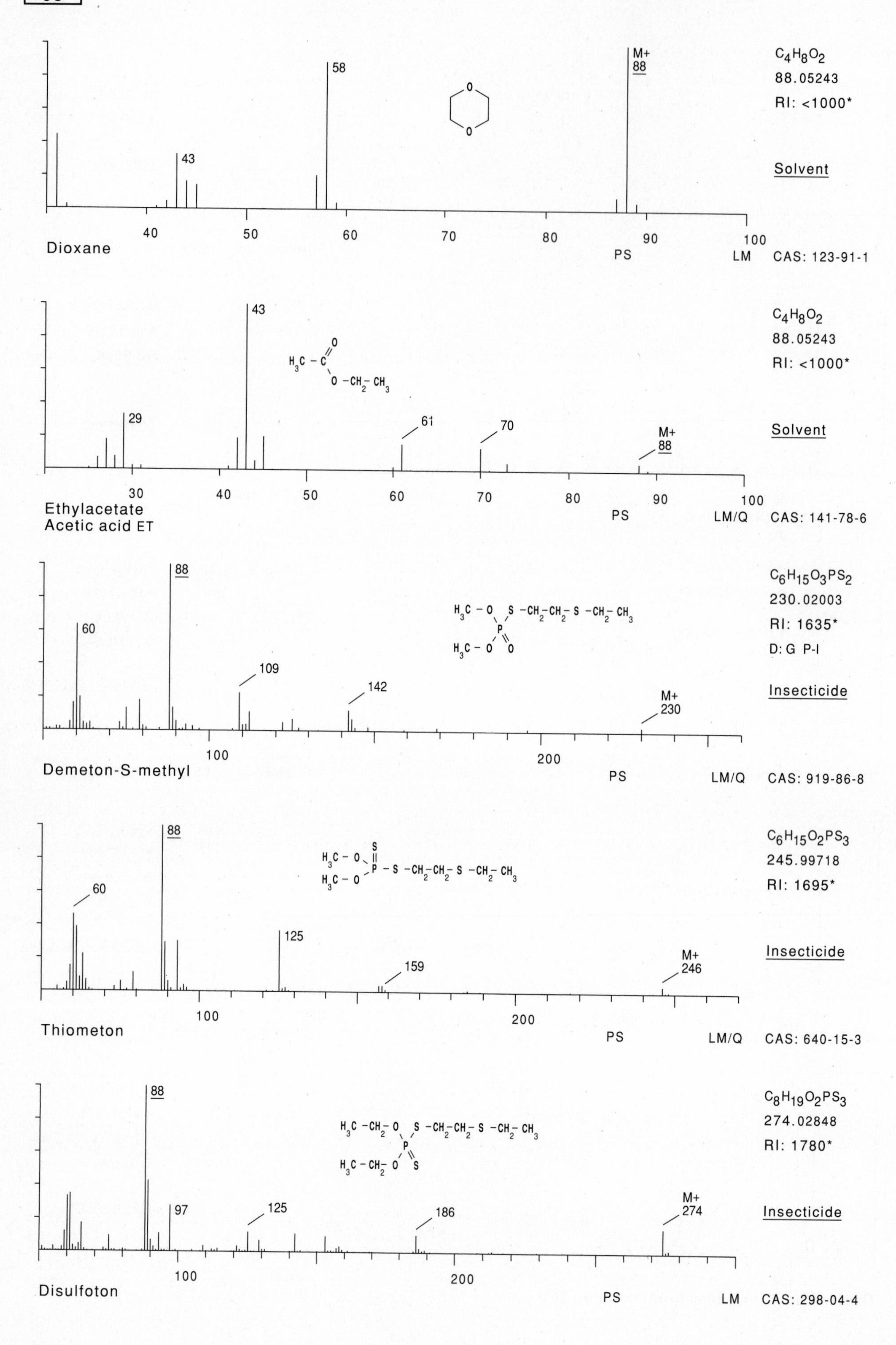

58
M+
88
43
40
50
60
70
80
90
100
Dioxane
PS
LM
$C_4H_8O_2$
88.05243
RI: <1000*
Solvent
CAS: 123-91-1
43
29
61
70
M+
88
30
Ethylacetate
Acetic acid ET
PS
LM/Q
$C_4H_8O_2$
88.05243
RI: <1000*
Solvent
CAS: 141-78-6
88
60
109
142
M+
230
100
200
Demeton-S-methyl
PS
LM/Q
$C_6H_{15}O_3PS_2$
230.02003
RI: 1635*
D: G P-I
Insecticide
CAS: 919-86-8
88
60
125
159
M+
246
Thiometon
PS
LM/Q
$C_6H_{15}O_2PS_3$
245.99718
RI: 1695*
Insecticide
CAS: 640-15-3
88
97
125
186
M+
274
Disulfoton
PS
LM
$C_8H_{19}O_2PS_3$
274.02848
RI: 1780*
Insecticide
CAS: 298-04-4

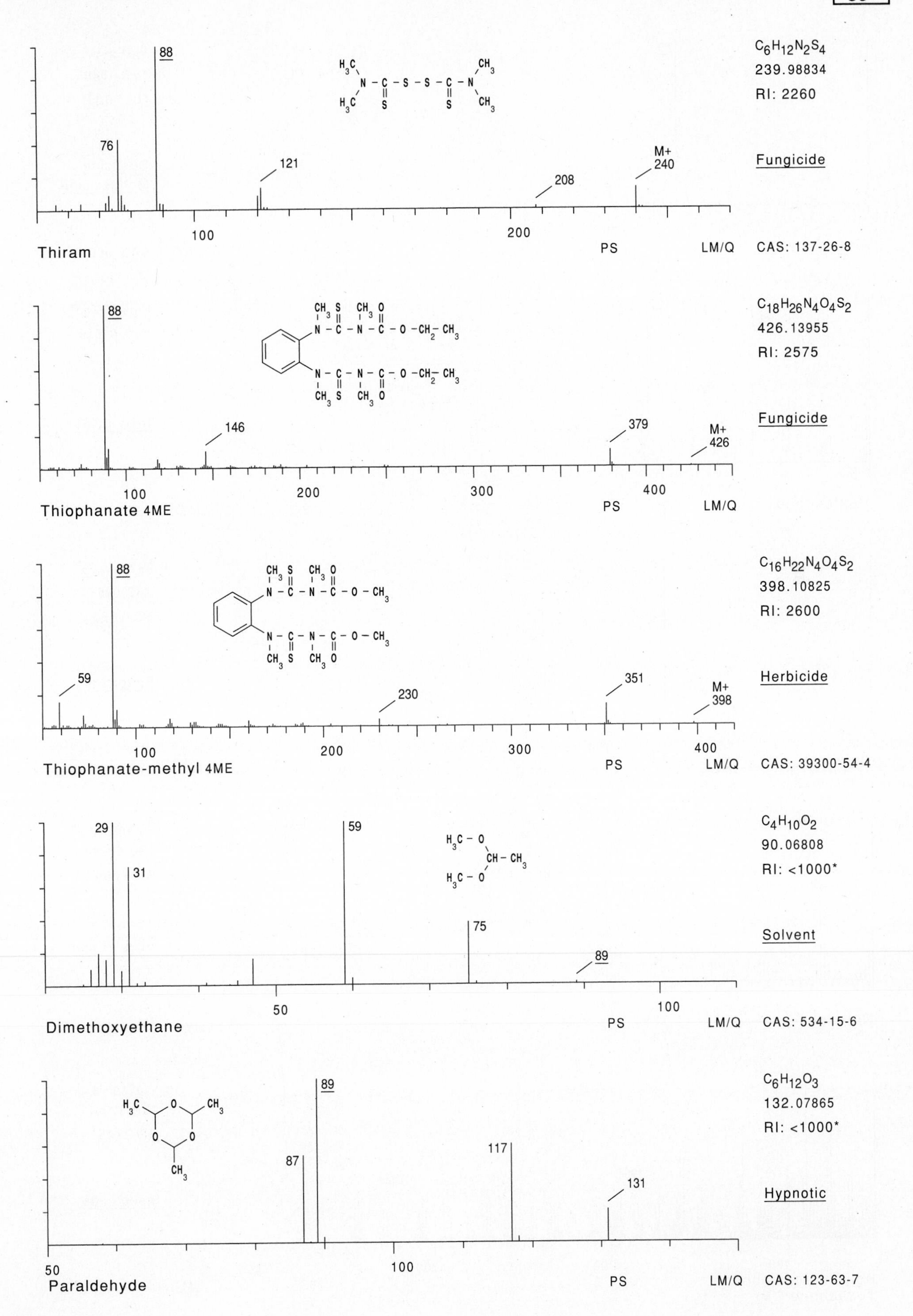

$C_6H_{12}N_2S_4$
239.98834
RI: 2260

Fungicide

CAS: 137-26-8

$C_{18}H_{26}N_4O_4S_2$
426.13955
RI: 2575

Fungicide

$C_{16}H_{22}N_4O_4S_2$
398.10825
RI: 2600

Herbicide

CAS: 39300-54-4

$C_4H_{10}O_2$
90.06808
RI: <1000*

Solvent

CAS: 534-15-6

$C_6H_{12}O_3$
132.07865
RI: <1000*

Hypnotic

CAS: 123-63-7

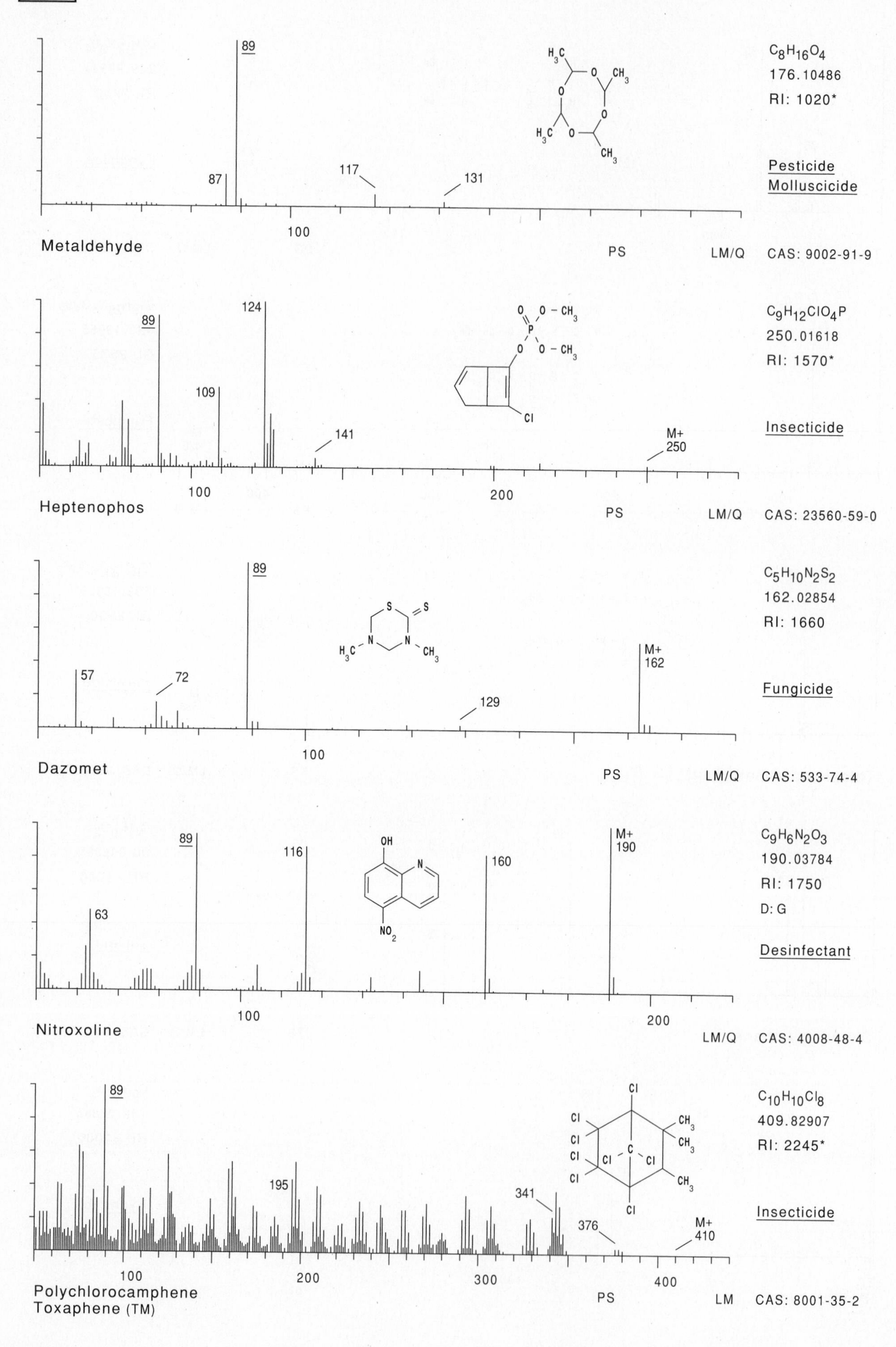

89
87
117
131
100
Metaldehyde
PS
LM/Q
$C_8H_{16}O_4$
176.10486
RI: 1020*
Pesticide
Molluscicide
CAS: 9002-91-9
124
89
109
141
M+
250
100
200
Heptenophos
PS
LM/Q
$C_9H_{12}ClO_4P$
250.01618
RI: 1570*
Insecticide
CAS: 23560-59-0
89
57
72
129
M+
162
100
Dazomet
PS
LM/Q
$C_5H_{10}N_2S_2$
162.02854
RI: 1660
Fungicide
CAS: 533-74-4
89
116
160
M+
190
63
100
200
Nitroxoline
LM/Q
$C_9H_6N_2O_3$
190.03784
RI: 1750
D: G
Desinfectant
CAS: 4008-48-4
89
195
341
376
M+
410
100
200
300
400
Polychlorocamphene
Toxaphene (TM)
PS
LM
$C_{10}H_{10}Cl_8$
409.82907
RI: 2245*
Insecticide
CAS: 8001-35-2

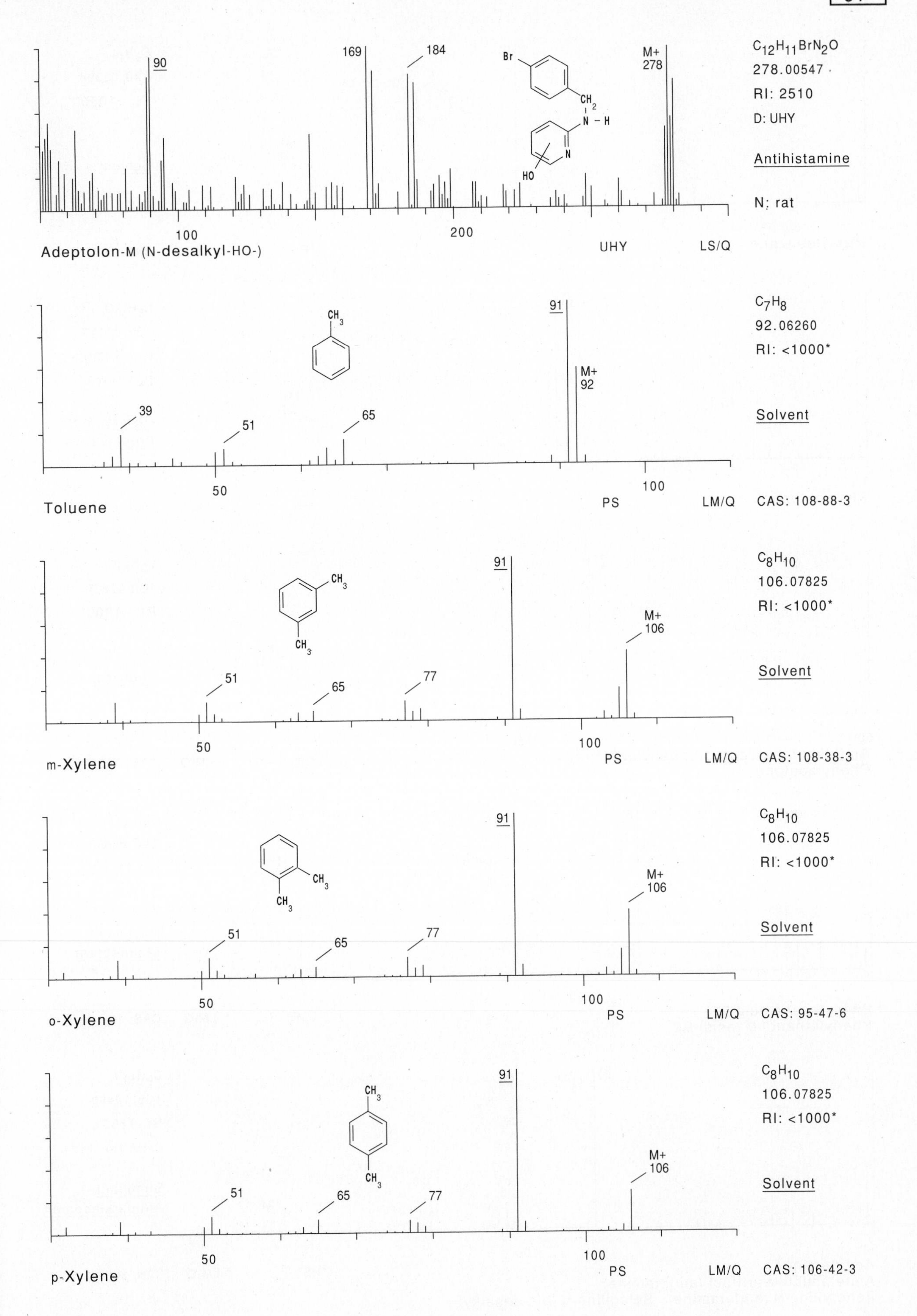

$C_{12}H_{11}BrN_2O$
278.00547
RI: 2510
D: UHY

Antihistamine

N: rat

C_7H_8
92.06260
RI: <1000*

Solvent

CAS: 108-88-3

C_8H_{10}
106.07825
RI: <1000*

Solvent

CAS: 108-38-3

C_8H_{10}
106.07825
RI: <1000*

Solvent

CAS: 95-47-6

C_8H_{10}
106.07825
RI: <1000*

Solvent

CAS: 106-42-3

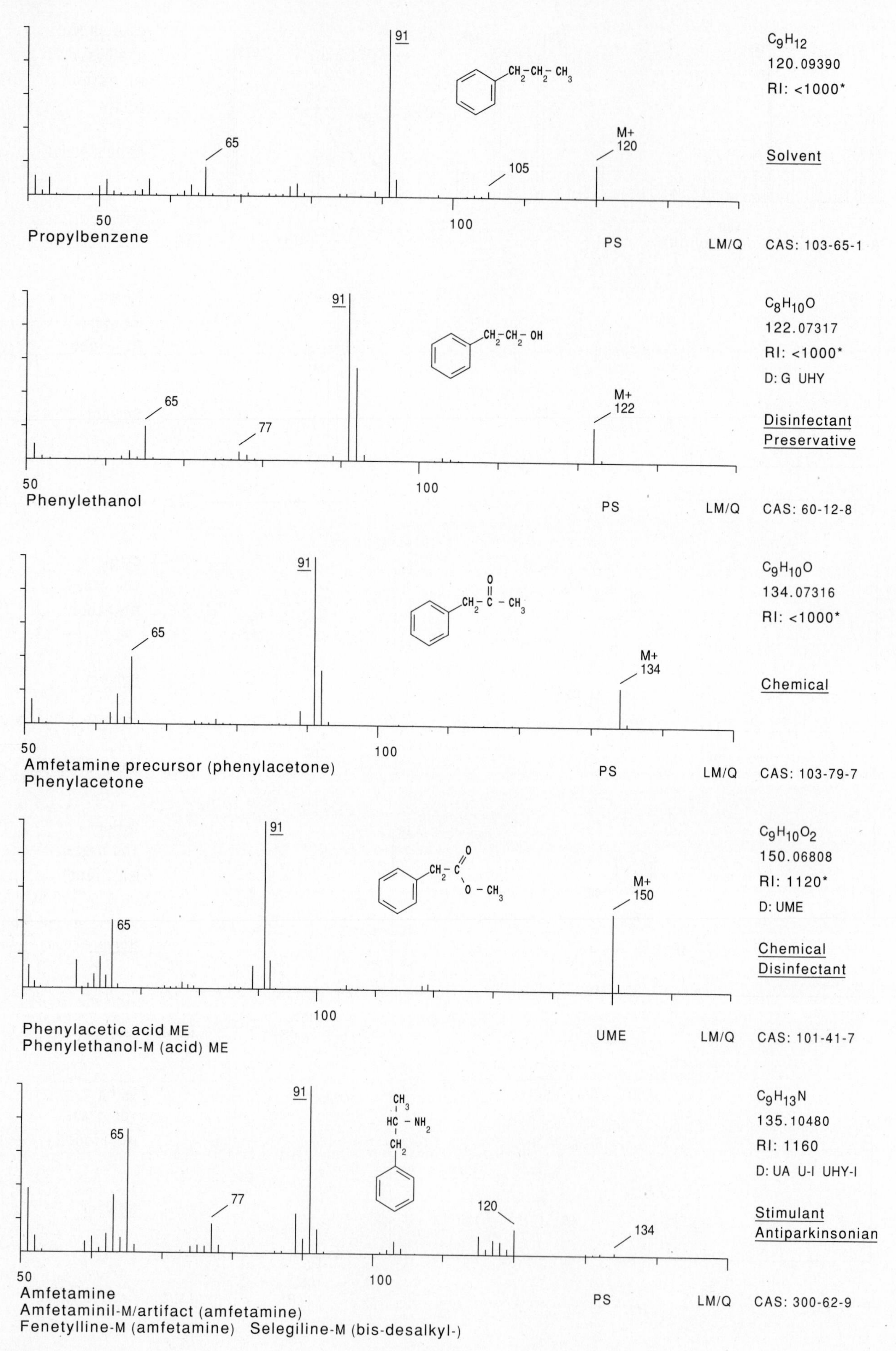
91
65
105
M+
120
50
100
Propylbenzene
PS
LM/Q
C_9H_{12}
120.09390
RI: <1000*
Solvent
CAS: 103-65-1
91
65
77
M+
122
50
100
Phenylethanol
PS
LM/Q
$C_8H_{10}O$
122.07317
RI: <1000*
D: G UHY
Disinfectant
Preservative
CAS: 60-12-8
91
65
M+
134
50
100
Amfetamine precursor (phenylacetone)
Phenylacetone
PS
LM/Q
$C_9H_{10}O$
134.07316
RI: <1000*
Chemical
CAS: 103-79-7
91
65
M+
150
100
Phenylacetic acid ME
Phenylethanol-M (acid) ME
UME
LM/Q
$C_9H_{10}O_2$
150.06808
RI: 1120*
D: UME
Chemical
Disinfectant
CAS: 101-41-7
91
65
77
120
134
50
100
Amfetamine
Amfetaminil-M/artifact (amfetamine)
Fenetylline-M (amfetamine) Selegiline-M (bis-desalkyl-)
PS
LM/Q
$C_9H_{13}N$
135.10480
RI: 1160
D: UA U-I UHY-I
Stimulant
Antiparkinsonian
CAS: 300-62-9

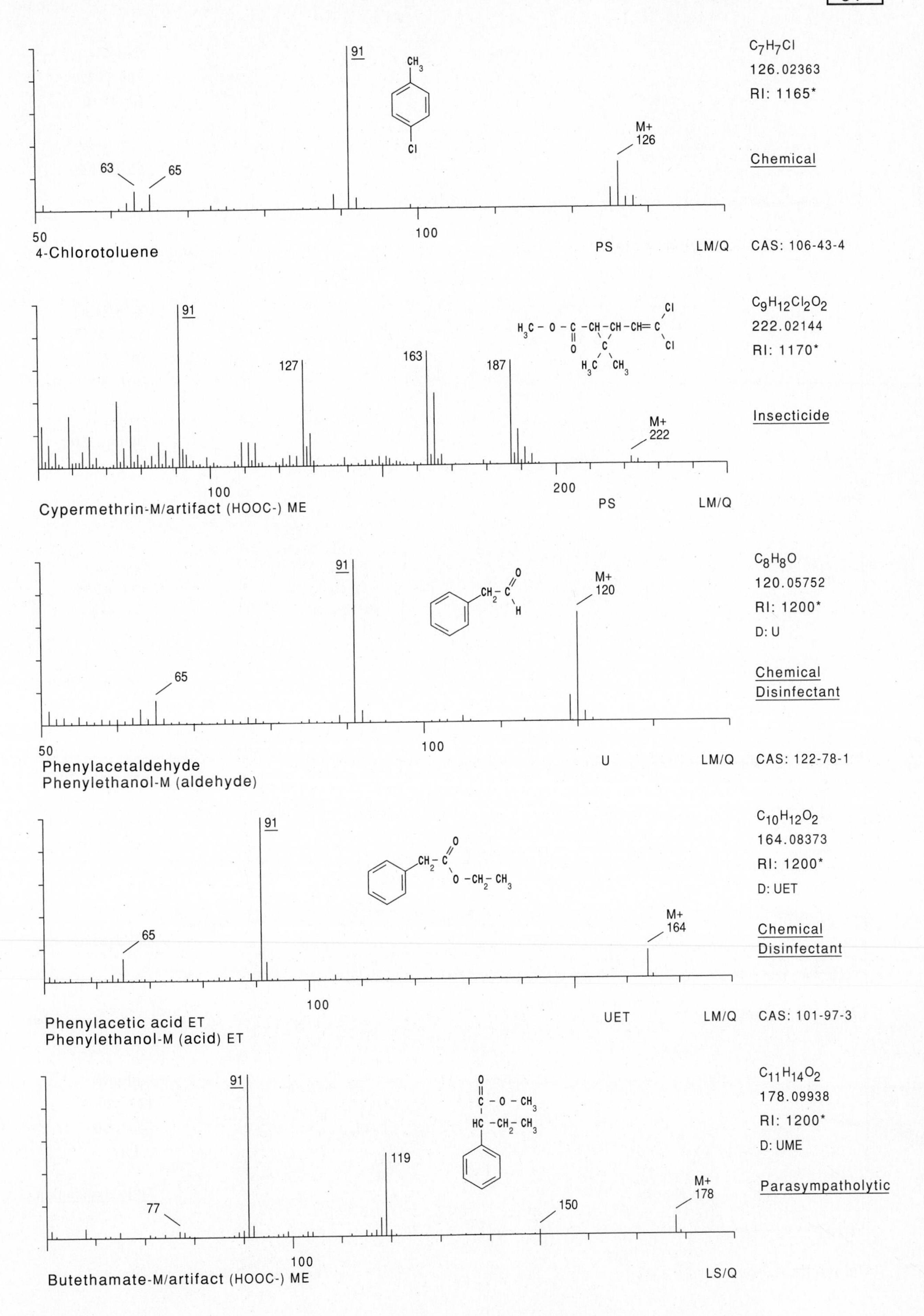

C_7H_7Cl
126.02363
RI: 1165*

Chemical

CAS: 106-43-4

$C_9H_{12}Cl_2O_2$
222.02144
RI: 1170*

Insecticide

C_8H_8O
120.05752
RI: 1200*
D: U

Chemical
Disinfectant

CAS: 122-78-1

$C_{10}H_{12}O_2$
164.08373
RI: 1200*
D: UET

Chemical
Disinfectant

CAS: 101-97-3

$C_{11}H_{14}O_2$
178.09938
RI: 1200*
D: UME

Parasympatholytic

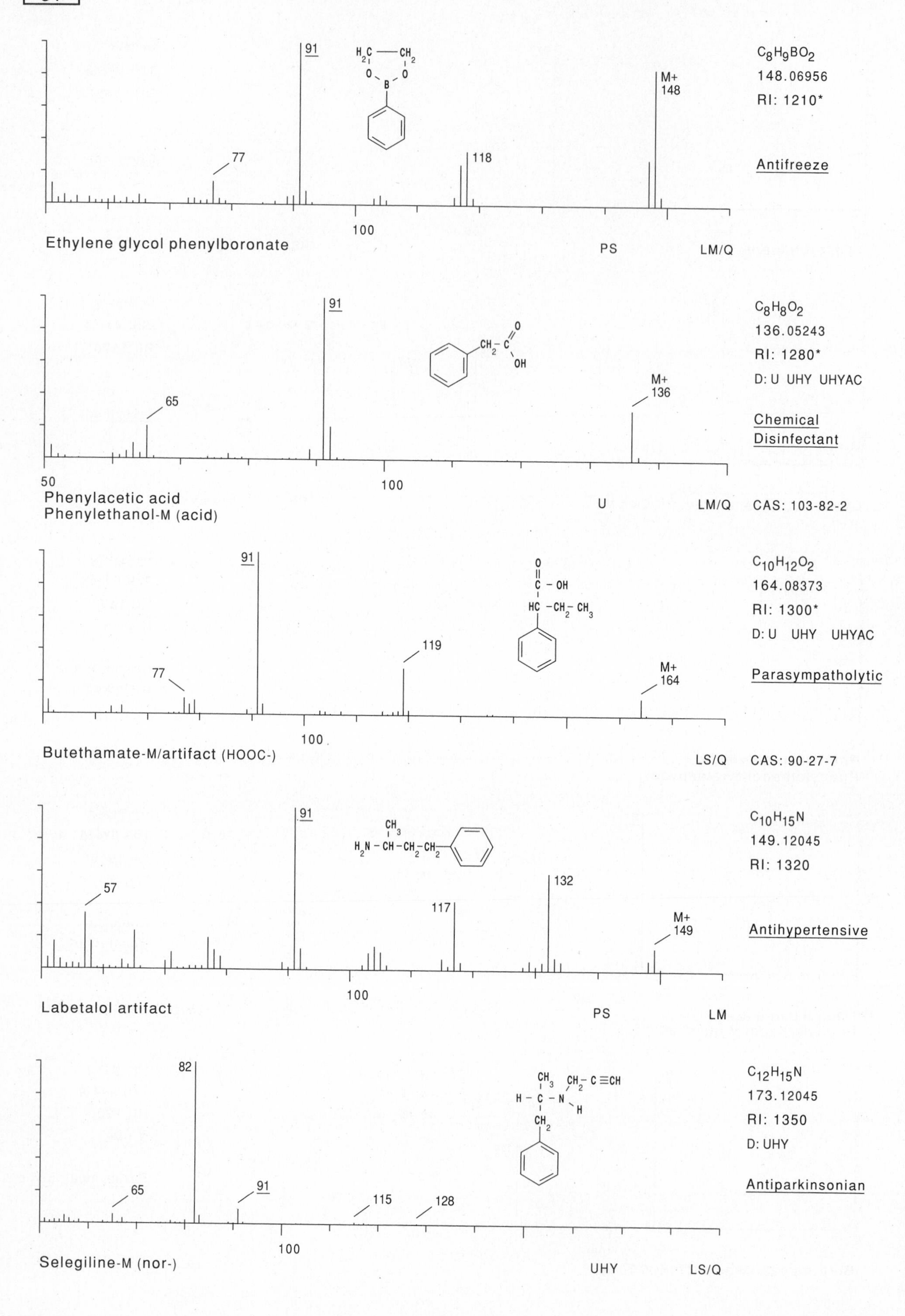

91
77
118
M+
148
100
Ethylene glycol phenylboronate
PS
LM/Q
$C_8H_9BO_2$
148.06956
RI: 1210*
Antifreeze
91
65
M+
136
50
100
Phenylacetic acid
Phenylethanol-M (acid)
U
LM/Q
$C_8H_8O_2$
136.05243
RI: 1280*
D: U UHY UHYAC
Chemical Disinfectant
CAS: 103-82-2
91
77
119
M+
164
100
Butethamate-M/artifact (HOOC-)
LS/Q
$C_{10}H_{12}O_2$
164.08373
RI: 1300*
D: U UHY UHYAC
Parasympatholytic
CAS: 90-27-7
91
57
117
132
M+
149
100
Labetalol artifact
PS
LM
$C_{10}H_{15}N$
149.12045
RI: 1320
Antihypertensive
82
65
91
115
128
100
Selegiline-M (nor-)
UHY
LS/Q
$C_{12}H_{15}N$
173.12045
RI: 1350
D: UHY
Antiparkinsonian

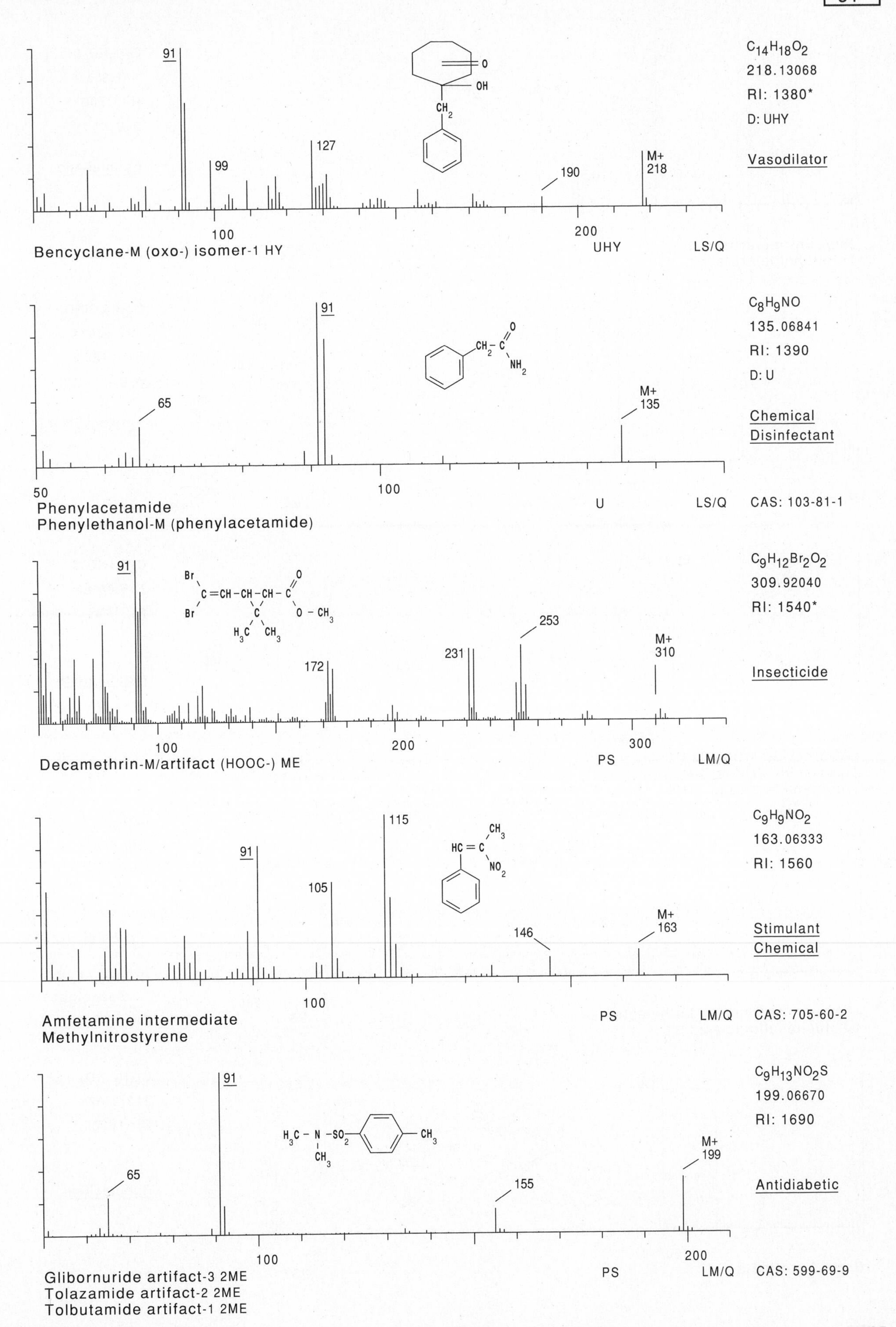

$C_{14}H_{18}O_2$
218.13068
RI: 1380*
D: UHY

Vasodilator

Bencyclane-M (oxo-) isomer-1 HY

C_8H_9NO
135.06841
RI: 1390
D: U

Chemical Disinfectant

CAS: 103-81-1

Phenylacetamide
Phenylethanol-M (phenylacetamide)

$C_9H_{12}Br_2O_2$
309.92040
RI: 1540*

Insecticide

Decamethrin-M/artifact (HOOC-) ME

$C_9H_9NO_2$
163.06333
RI: 1560

Stimulant
Chemical

CAS: 705-60-2

Amfetamine intermediate
Methylnitrostyrene

$C_9H_{13}NO_2S$
199.06670
RI: 1690

Antidiabetic

CAS: 599-69-9

Glibornuride artifact-3 2ME
Tolazamide artifact-2 2ME
Tolbutamide artifact-1 2ME

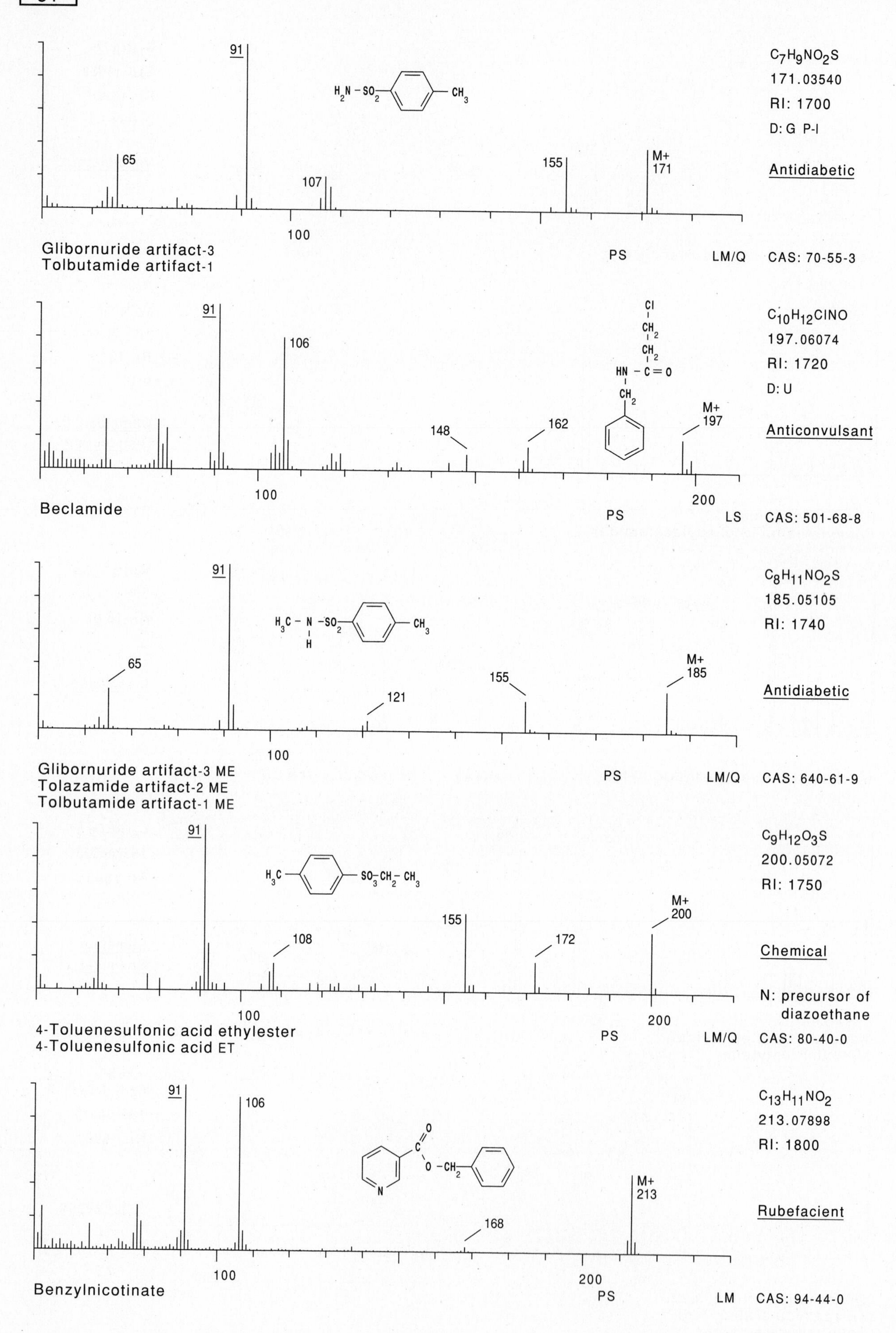
91
65
107
155
M+
171
100
$C_7H_9NO_2S$
171.03540
RI: 1700
D: G P-I
Antidiabetic
Glibornuride artifact-3
Tolbutamide artifact-1
PS
LM/Q
CAS: 70-55-3
91
106
148
162
M+
197
100
200
$C_{10}H_{12}ClNO$
197.06074
RI: 1720
D: U
Anticonvulsant
Beclamide
PS
LS
CAS: 501-68-8
91
65
121
155
M+
185
100
$C_8H_{11}NO_2S$
185.05105
RI: 1740
Antidiabetic
Glibornuride artifact-3 ME
Tolazamide artifact-2 ME
Tolbutamide artifact-1 ME
PS
LM/Q
CAS: 640-61-9
91
108
155
172
M+
200
100
200
$C_9H_{12}O_3S$
200.05072
RI: 1750
Chemical
N: precursor of
diazoethane
4-Toluenesulfonic acid ethylester
4-Toluenesulfonic acid ET
PS
LM/Q
CAS: 80-40-0
91
106
168
M+
213
100
200
$C_{13}H_{11}NO_2$
213.07898
RI: 1800
Rubefacient
Benzylnicotinate
PS
LM
CAS: 94-44-0

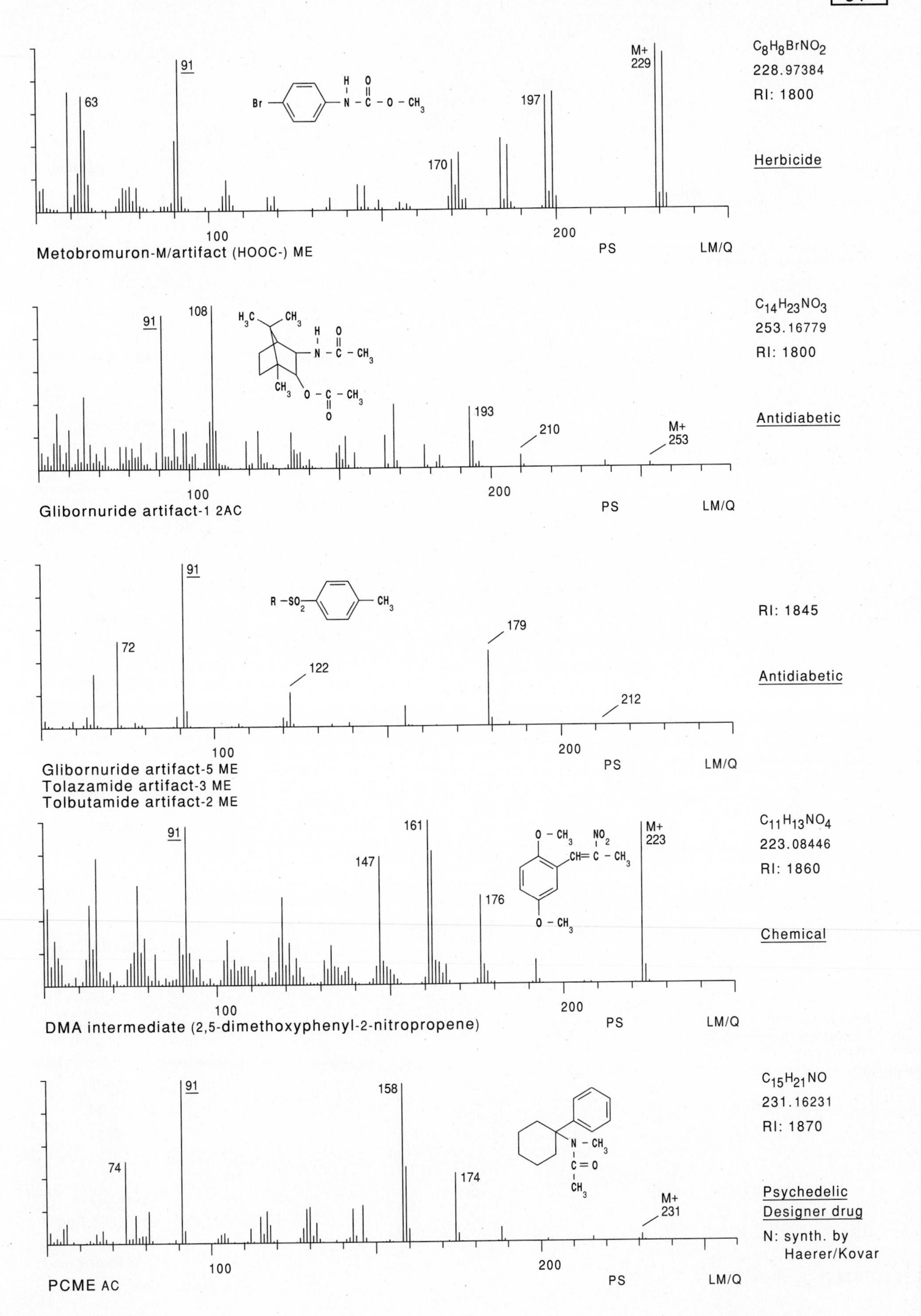

$C_8H_8BrNO_2$
228.97384
RI: 1800

Herbicide

$C_{14}H_{23}NO_3$
253.16779
RI: 1800

Antidiabetic

RI: 1845

Antidiabetic

$C_{11}H_{13}NO_4$
223.08446
RI: 1860

Chemical

$C_{15}H_{21}NO$
231.16231
RI: 1870

Psychedelic
Designer drug

N: synth. by Haerer/Kovar

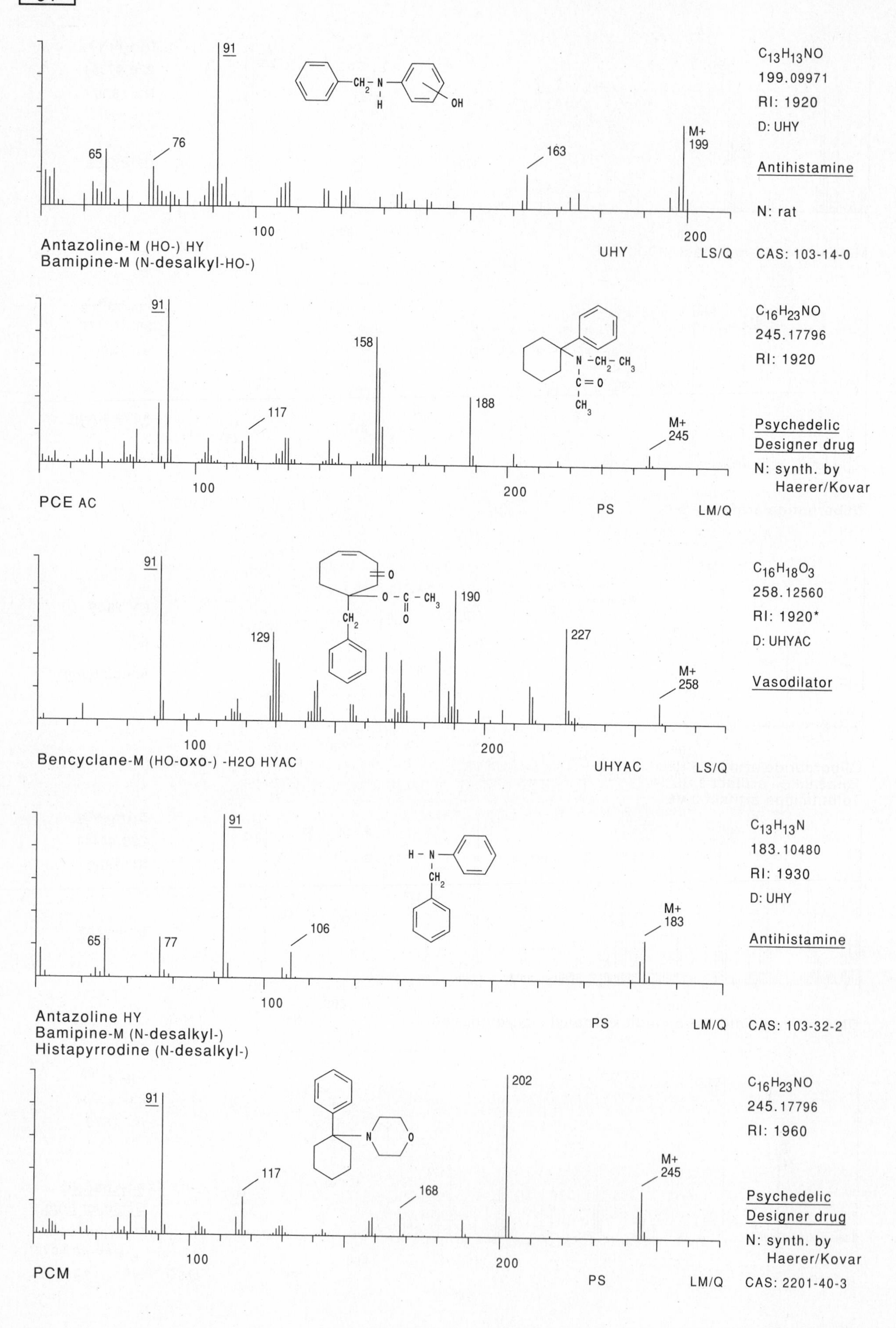

91
65
76
163
M+
199
100
200
Antazoline-M (HO-) HY
Bamipine-M (N-desalkyl-HO-)
UHY
LS/Q
C13H13NO
199.09971
RI: 1920
D: UHY
Antihistamine
N: rat
CAS: 103-14-0
91
158
117
188
M+
245
100
200
PCE AC
PS
LM/Q
C16H23NO
245.17796
RI: 1920
Psychedelic
Designer drug
N: synth. by
Haerer/Kovar
91
190
129
227
M+
258
100
200
Bencyclane-M (HO-oxo-) -H2O HYAC
UHYAC
LS/Q
C16H18O3
258.12560
RI: 1920*
D: UHYAC
Vasodilator
91
65
77
106
M+
183
100
Antazoline HY
Bamipine-M (N-desalkyl-)
Histapyrrodine (N-desalkyl-)
PS
LM/Q
C13H13N
183.10480
RI: 1930
D: UHY
Antihistamine
CAS: 103-32-2
91
202
117
168
M+
245
100
200
PCM
PS
LM/Q
C16H23NO
245.17796
RI: 1960
Psychedelic
Designer drug
N: synth. by
Haerer/Kovar
CAS: 2201-40-3

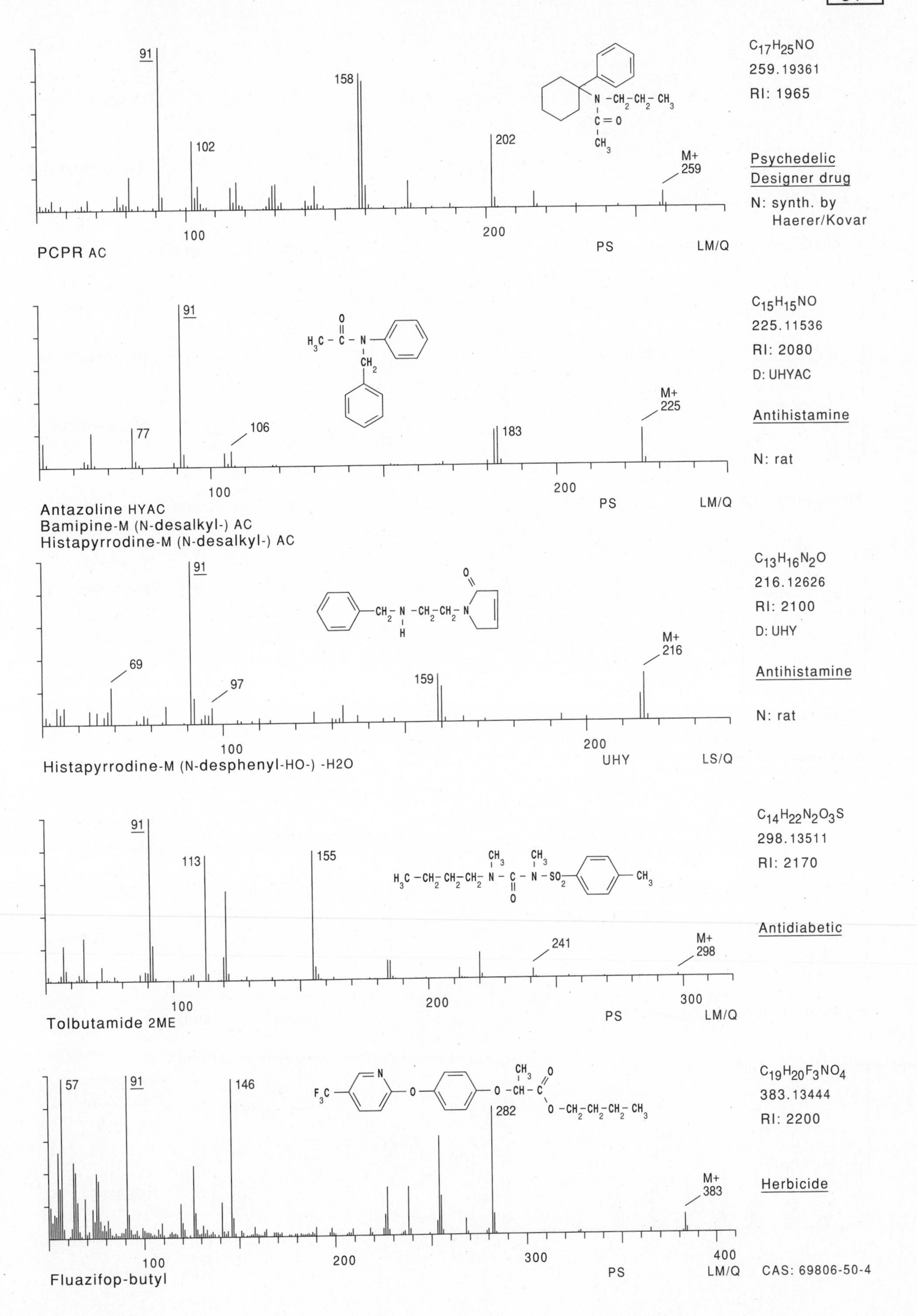

PCPR AC
C17H25NO
259.19361
RI: 1965
Psychedelic
Designer drug
N: synth. by Haerer/Kovar
91
158
102
202
M+ 259
100
200
PS
LM/Q
Antazoline HYAC
Bamipine-M (N-desalkyl-) AC
Histapyrrodine-M (N-desalkyl-) AC
C15H15NO
225.11536
RI: 2080
D: UHYAC
Antihistamine
N: rat
91
77
106
183
M+ 225
Histapyrrodine-M (N-desphenyl-HO-) -H2O
C13H16N2O
216.12626
RI: 2100
D: UHY
Antihistamine
N: rat
91
69
97
159
M+ 216
UHY
LS/Q
Tolbutamide 2ME
C14H22N2O3S
298.13511
RI: 2170
Antidiabetic
91
113
155
241
M+ 298
300
Fluazifop-butyl
C19H20F3NO4
383.13444
RI: 2200
Herbicide
CAS: 69806-50-4
57
91
146
282
M+ 383
400

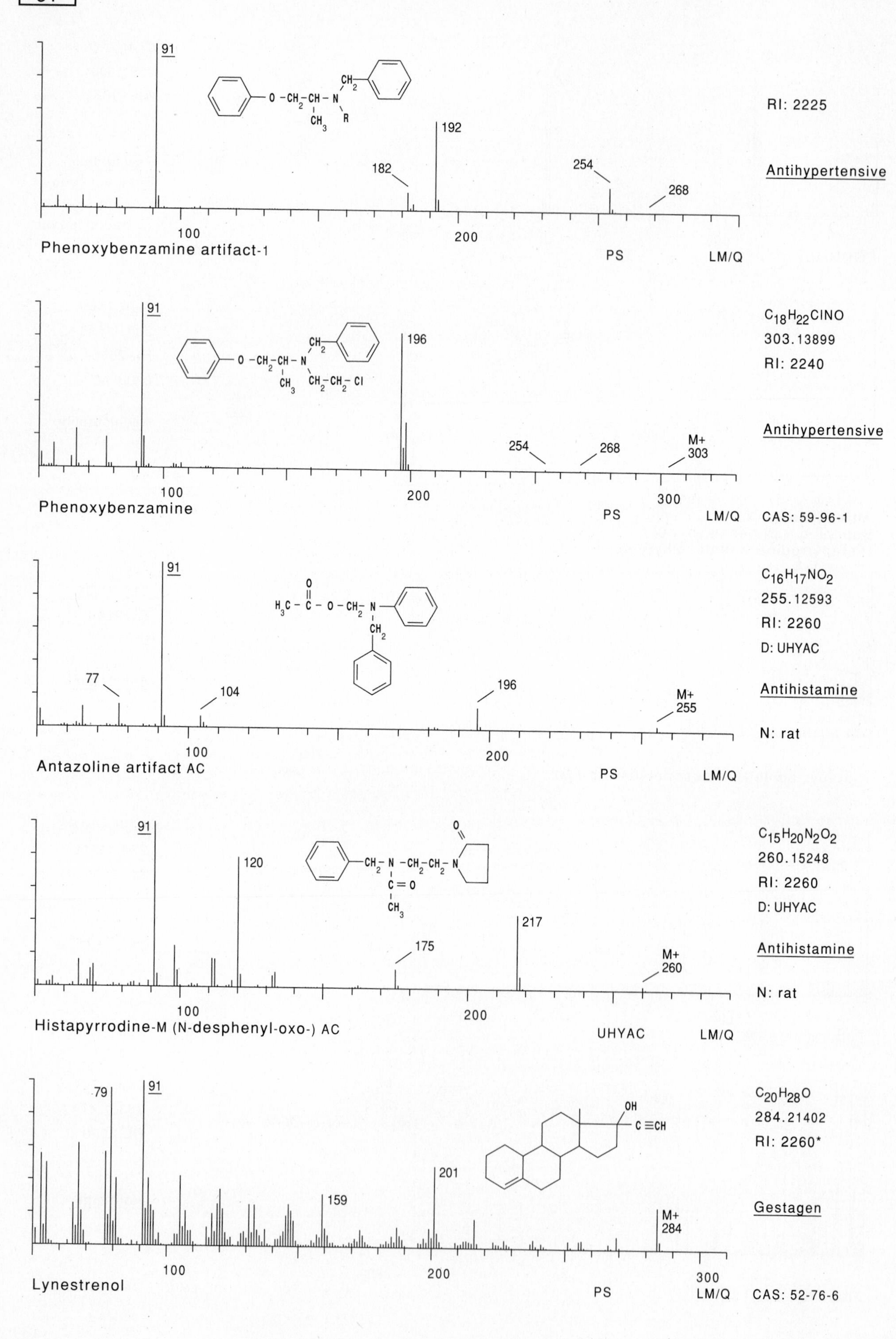
91
192
182
254
268
100
200
Phenoxybenzamine artifact-1
PS
LM/Q
RI: 2225
Antihypertensive
91
196
254
268
M+
303
100
200
300
Phenoxybenzamine
PS
LM/Q
C18H22ClNO
303.13899
RI: 2240
Antihypertensive
CAS: 59-96-1
91
77
104
196
M+
255
100
200
Antazoline artifact AC
PS
LM/Q
C16H17NO2
255.12593
RI: 2260
D: UHYAC
Antihistamine
N: rat
91
120
217
175
M+
260
100
200
Histapyrrodine-M (N-desphenyl-oxo-) AC
UHYAC
LM/Q
C15H20N2O2
260.15248
RI: 2260
D: UHYAC
Antihistamine
N: rat
79
91
201
159
M+
284
100
200
300
Lynestrenol
PS
LM/Q
C20H28O
284.21402
RI: 2260*
Gestagen
CAS: 52-76-6

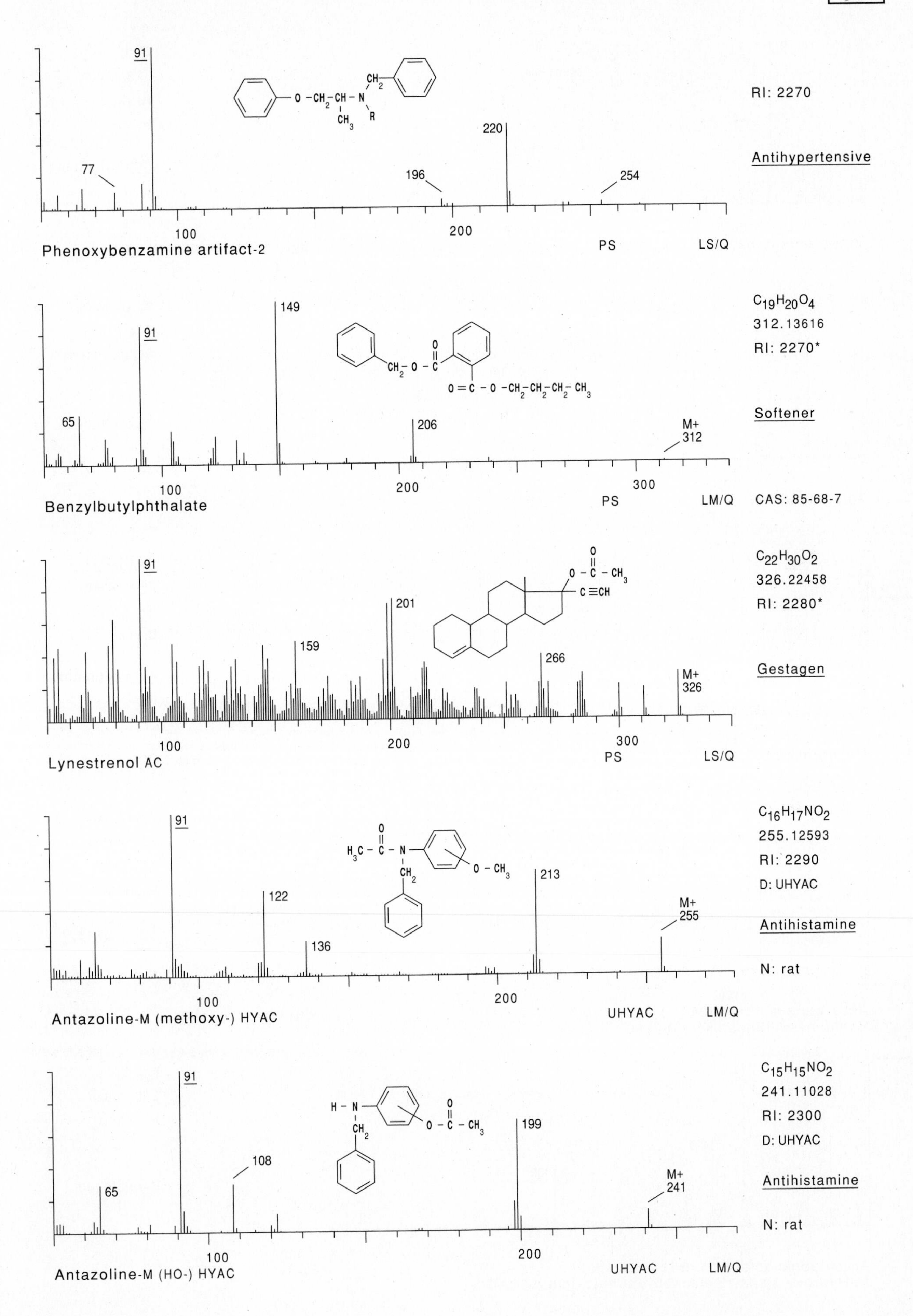
91
77
196
220
254
100
200
Phenoxybenzamine artifact-2
PS
LS/Q
RI: 2270
Antihypertensive
149
91
65
206
M+
312
100
200
300
Benzylbutylphthalate
PS
LM/Q
$C_{19}H_{20}O_4$
312.13616
RI: 2270*
Softener
CAS: 85-68-7
91
201
159
266
M+
326
100
200
300
Lynestrenol AC
PS
LS/Q
$C_{22}H_{30}O_2$
326.22458
RI: 2280*
Gestagen
91
122
136
213
M+
255
100
200
Antazoline-M (methoxy-) HYAC
UHYAC
LM/Q
$C_{16}H_{17}NO_2$
255.12593
RI: 2290
D: UHYAC
Antihistamine
N: rat
91
108
65
199
M+
241
100
200
Antazoline-M (HO-) HYAC
UHYAC
LM/Q
$C_{15}H_{15}NO_2$
241.11028
RI: 2300
D: UHYAC
Antihistamine
N: rat

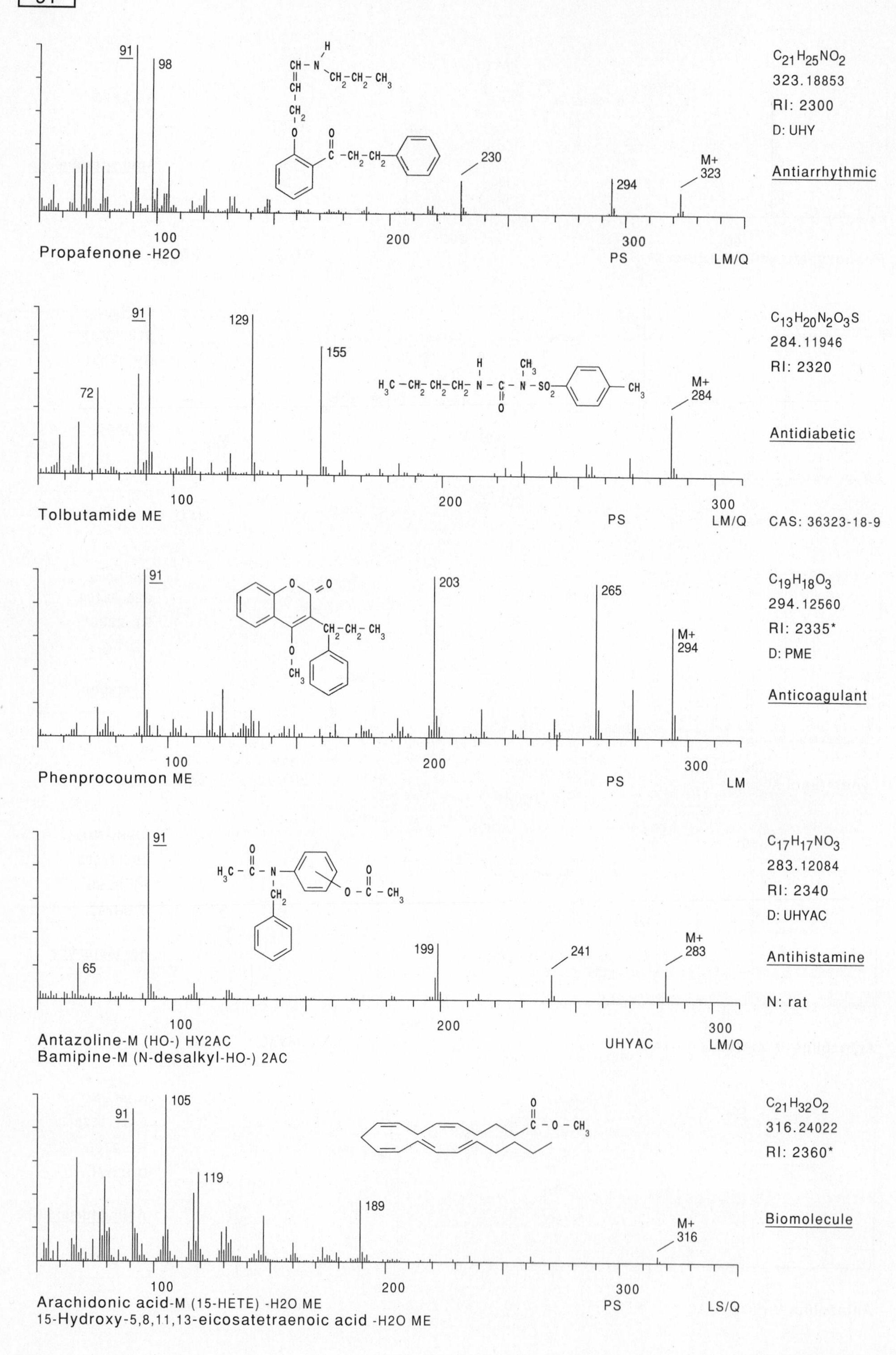

91
98
230
294
M+
323
100
200
300
Propafenone -H2O
PS
LM/Q
C21H25NO2
323.18853
RI: 2300
D: UHY
Antiarrhythmic
91
129
155
72
M+
284
Tolbutamide ME
PS
LM/Q
C13H20N2O3S
284.11946
RI: 2320
Antidiabetic
CAS: 36323-18-9
91
203
265
M+
294
Phenprocoumon ME
PS
LM
C19H18O3
294.12560
RI: 2335*
D: PME
Anticoagulant
91
65
199
241
M+
283
Antazoline-M (HO-) HY2AC
Bamipine-M (N-desalkyl-HO-) 2AC
UHYAC
LM/Q
C17H17NO3
283.12084
RI: 2340
D: UHYAC
Antihistamine
N: rat
105
91
119
189
M+
316
Arachidonic acid-M (15-HETE) -H2O ME
15-Hydroxy-5,8,11,13-eicosatetraenoic acid -H2O ME
PS
LS/Q
C21H32O2
316.24022
RI: 2360*
Biomolecule

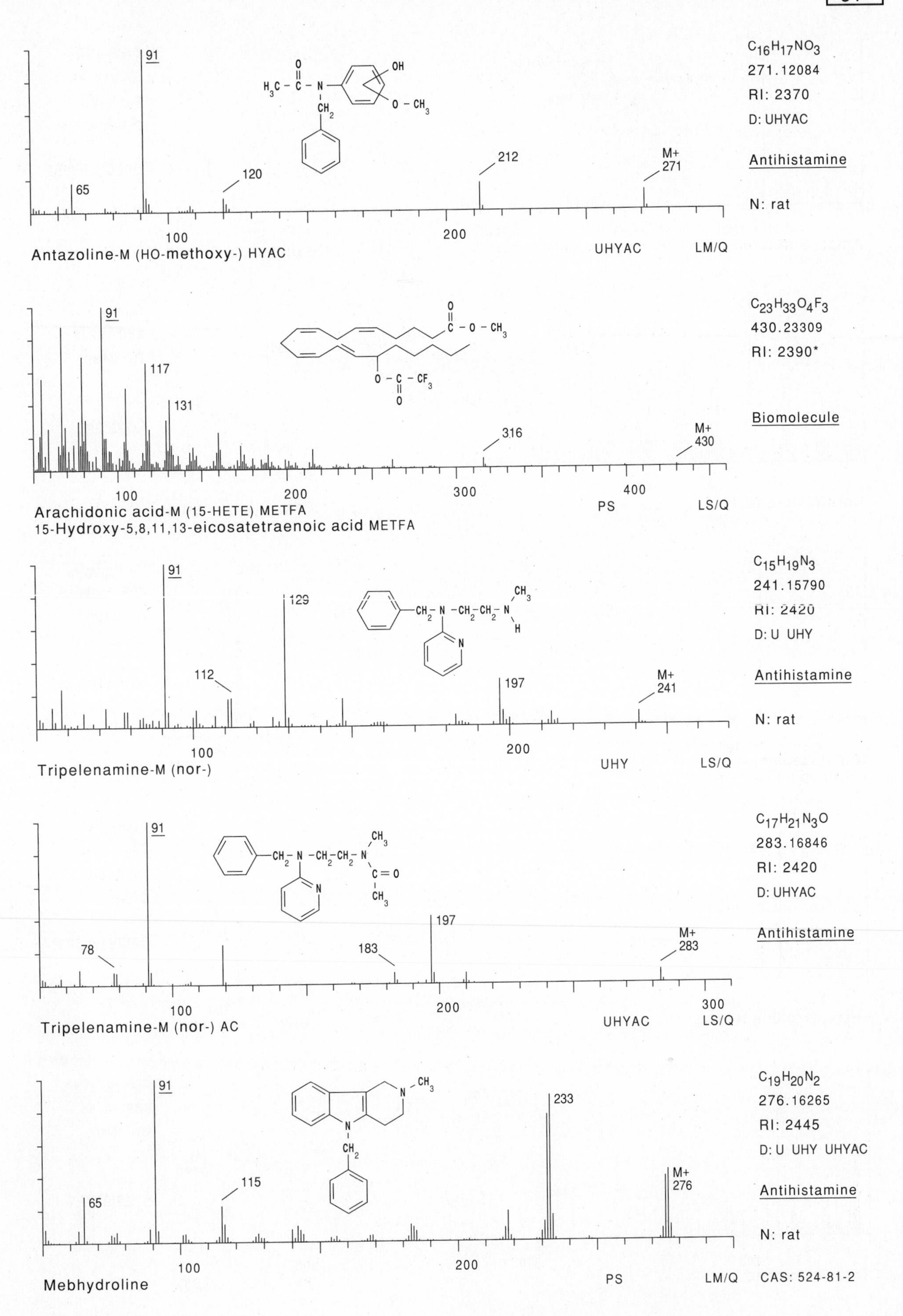

Antazoline-M (HO-methoxy-) HYAC

$C_{16}H_{17}NO_3$
271.12084
RI: 2370
D: UHYAC

Antihistamine

N: rat

Arachidonic acid-M (15-HETE) METFA
15-Hydroxy-5,8,11,13-eicosatetraenoic acid METFA

$C_{23}H_{33}O_4F_3$
430.23309
RI: 2390*

Biomolecule

Tripelenamine-M (nor-)

$C_{15}H_{19}N_3$
241.15790
RI: 2420
D: U UHY

Antihistamine

N: rat

Tripelenamine-M (nor-) AC

$C_{17}H_{21}N_3O$
283.16846
RI: 2420
D: UHYAC

Antihistamine

Mebhydroline

$C_{19}H_{20}N_2$
276.16265
RI: 2445
D: U UHY UHYAC

Antihistamine

N: rat

CAS: 524-81-2

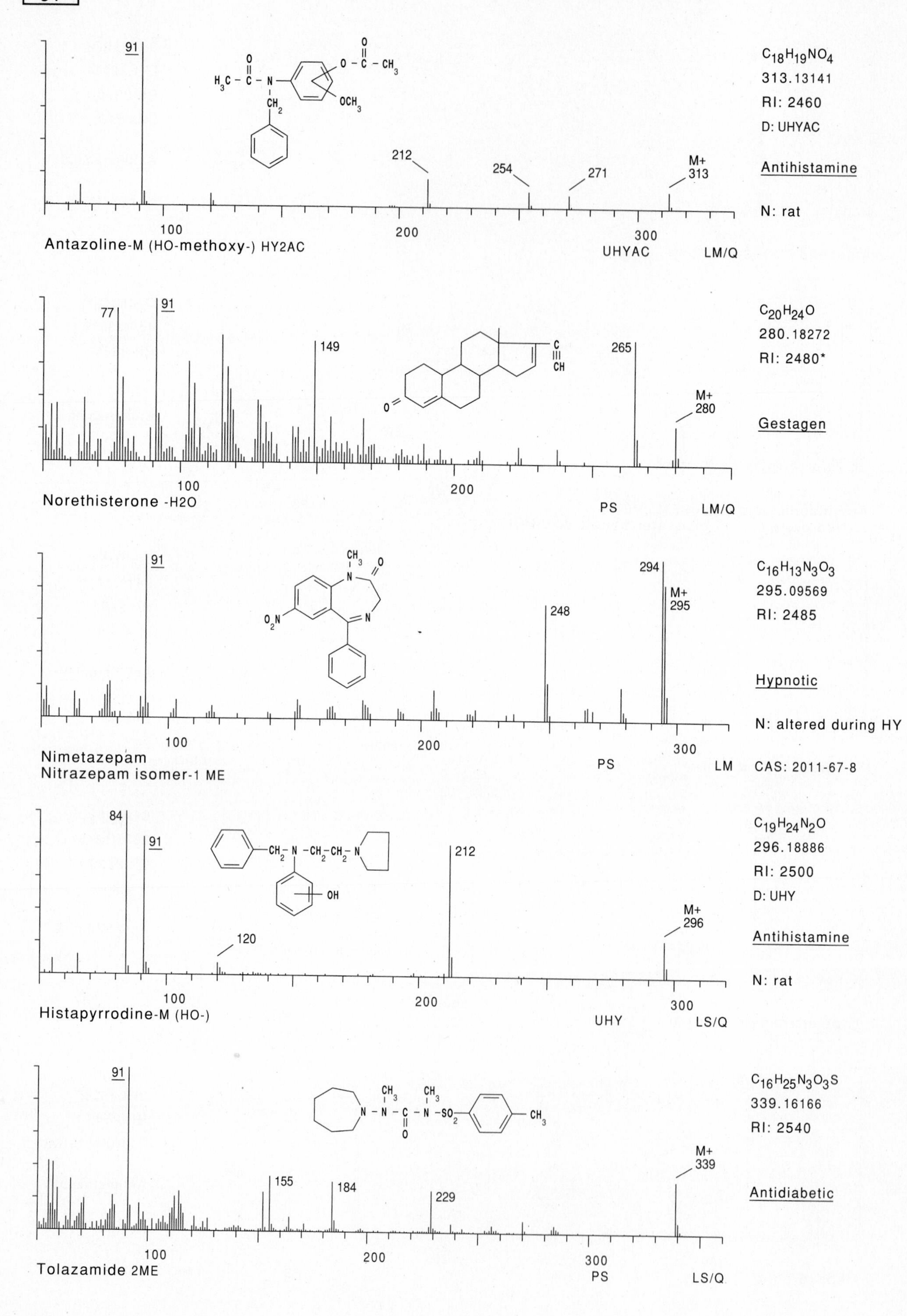

$C_{18}H_{19}NO_4$
313.13141
RI: 2460
D: UHYAC
Antihistamine
N: rat

$C_{20}H_{24}O$
280.18272
RI: 2480*
Gestagen

$C_{16}H_{13}N_3O_3$
295.09569
RI: 2485
Hypnotic
N: altered during HY
CAS: 2011-67-8

$C_{19}H_{24}N_2O$
296.18886
RI: 2500
D: UHY
Antihistamine
N: rat

$C_{16}H_{25}N_3O_3S$
339.16166
RI: 2540
Antidiabetic

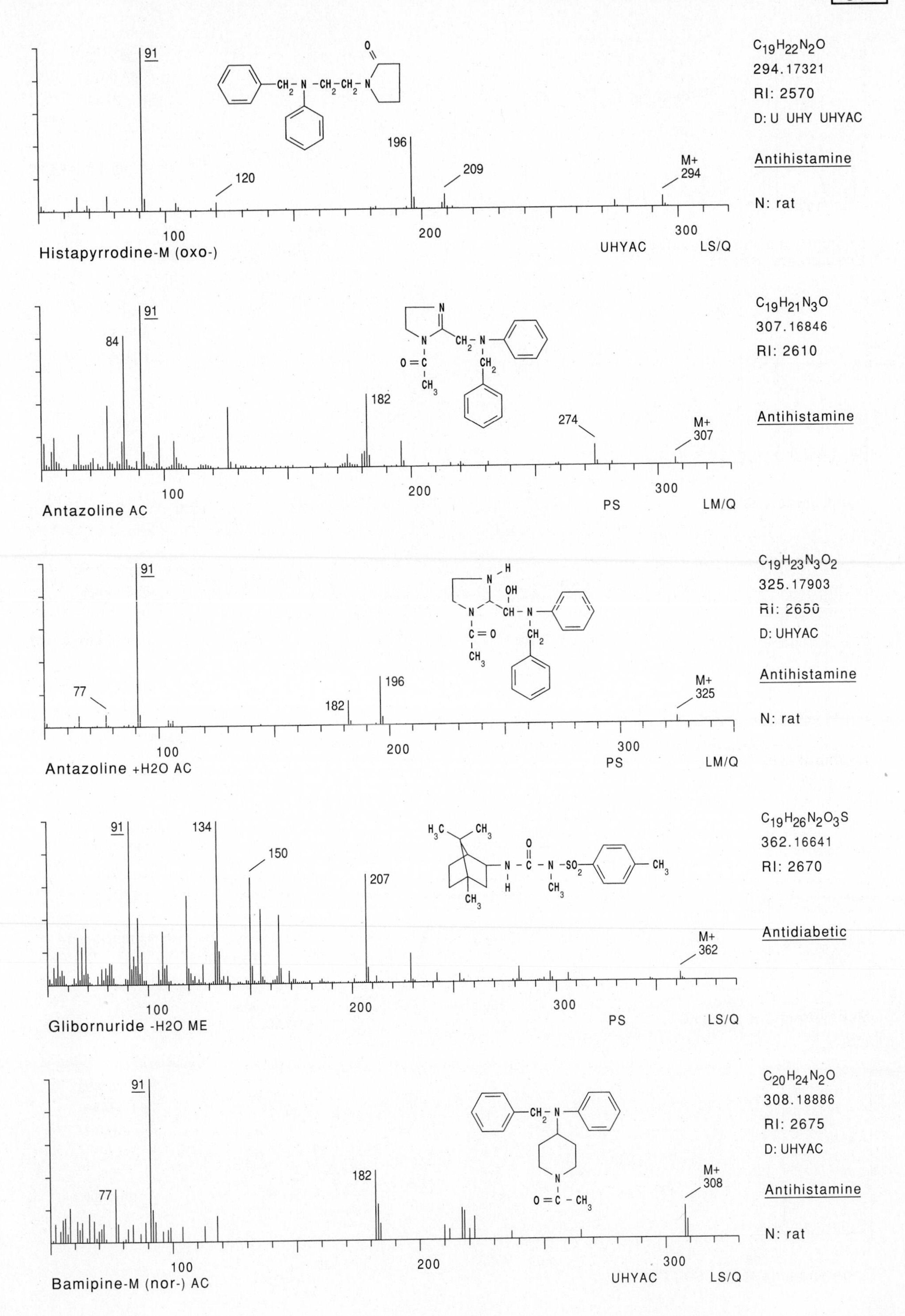
Histapyrrodine-M (oxo-)
UHYAC LS/Q
$C_{19}H_{22}N_2O$
294.17321
RI: 2570
D: U UHY UHYAC
Antihistamine
N: rat
Antazoline AC
PS LM/Q
$C_{19}H_{21}N_3O$
307.16846
RI: 2610
Antihistamine
Antazoline +H2O AC
PS LM/Q
$C_{19}H_{23}N_3O_2$
325.17903
RI: 2650
D: UHYAC
Antihistamine
N: rat
Glibornuride -H2O ME
PS LS/Q
$C_{19}H_{26}N_2O_3S$
362.16641
RI: 2670
Antidiabetic
Bamipine-M (nor-) AC
UHYAC LS/Q
$C_{20}H_{24}N_2O$
308.18886
RI: 2675
D: UHYAC
Antihistamine
N: rat

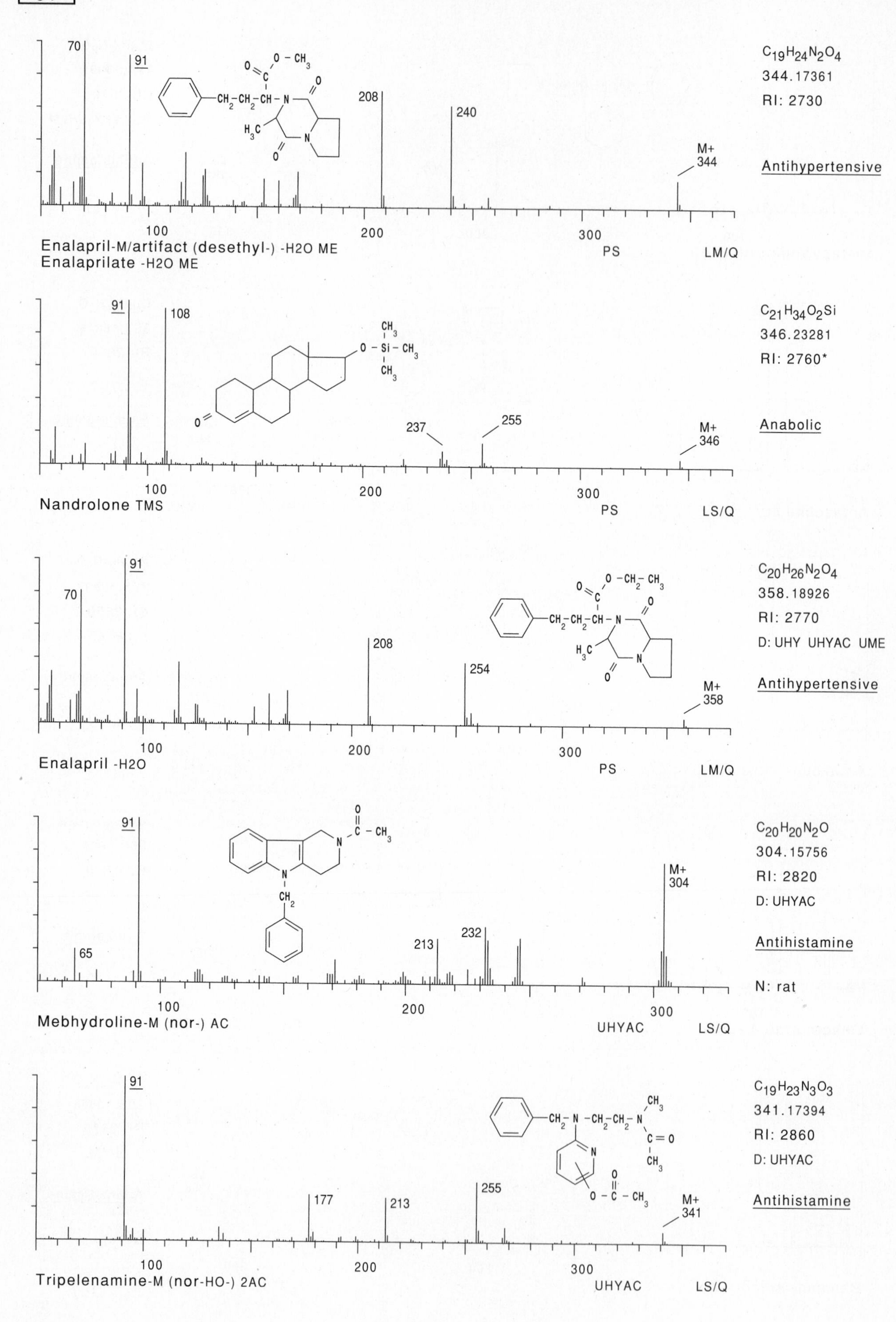
70
91
208
240
M+
344
100
200
300
Enalapril-M/artifact (desethyl-) -H2O ME
Enalaprilate -H2O ME
PS
LM/Q
C19H24N2O4
344.17361
RI: 2730
Antihypertensive
91
108
237
255
M+
346
Nandrolone TMS
PS
LS/Q
C21H34O2Si
346.23281
RI: 2760*
Anabolic
91
70
208
254
M+
358
Enalapril -H2O
PS
LM/Q
C20H26N2O4
358.18926
RI: 2770
D: UHY UHYAC UME
Antihypertensive
91
65
213
232
M+
304
Mebhydroline-M (nor-) AC
UHYAC
LS/Q
C20H20N2O
304.15756
RI: 2820
D: UHYAC
Antihistamine
N: rat
91
177
213
255
M+
341
Tripelenamine-M (nor-HO-) 2AC
UHYAC
LS/Q
C19H23N3O3
341.17394
RI: 2860
D: UHYAC
Antihistamine

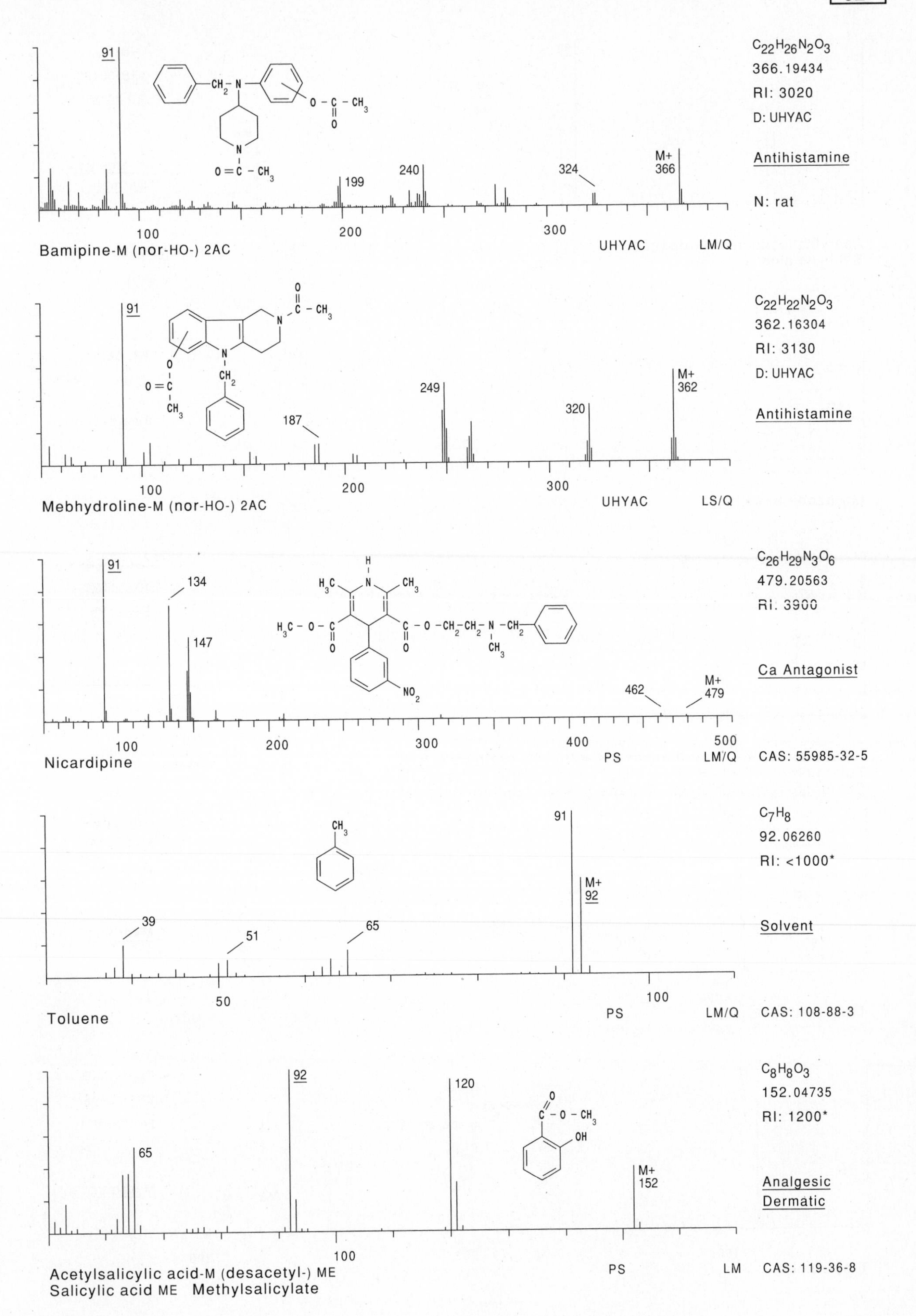
91
199
240
324
M+
366
100
200
300
UHYAC
LM/Q
Bamipine-M (nor-HO-) 2AC
C22H26N2O3
366.19434
RI: 3020
D: UHYAC
Antihistamine
N: rat
91
187
249
320
M+
362
100
200
300
UHYAC
LS/Q
Mebhydroline-M (nor-HO-) 2AC
C22H22N2O3
362.16304
RI: 3130
D: UHYAC
Antihistamine
91
134
147
462
M+
479
100
200
300
400
500
PS
LM/Q
Nicardipine
C26H29N3O6
479.20563
RI: 3900
Ca Antagonist
CAS: 55985-32-5
39
51
65
91
M+
92
50
100
PS
LM/Q
Toluene
C7H8
92.06260
RI: <1000*
Solvent
CAS: 108-88-3
65
92
120
M+
152
100
PS
LM
Acetylsalicylic acid-M (desacetyl-) ME
Salicylic acid ME Methylsalicylate
C8H8O3
152.04735
RI: 1200*
Analgesic
Dermatic
CAS: 119-36-8

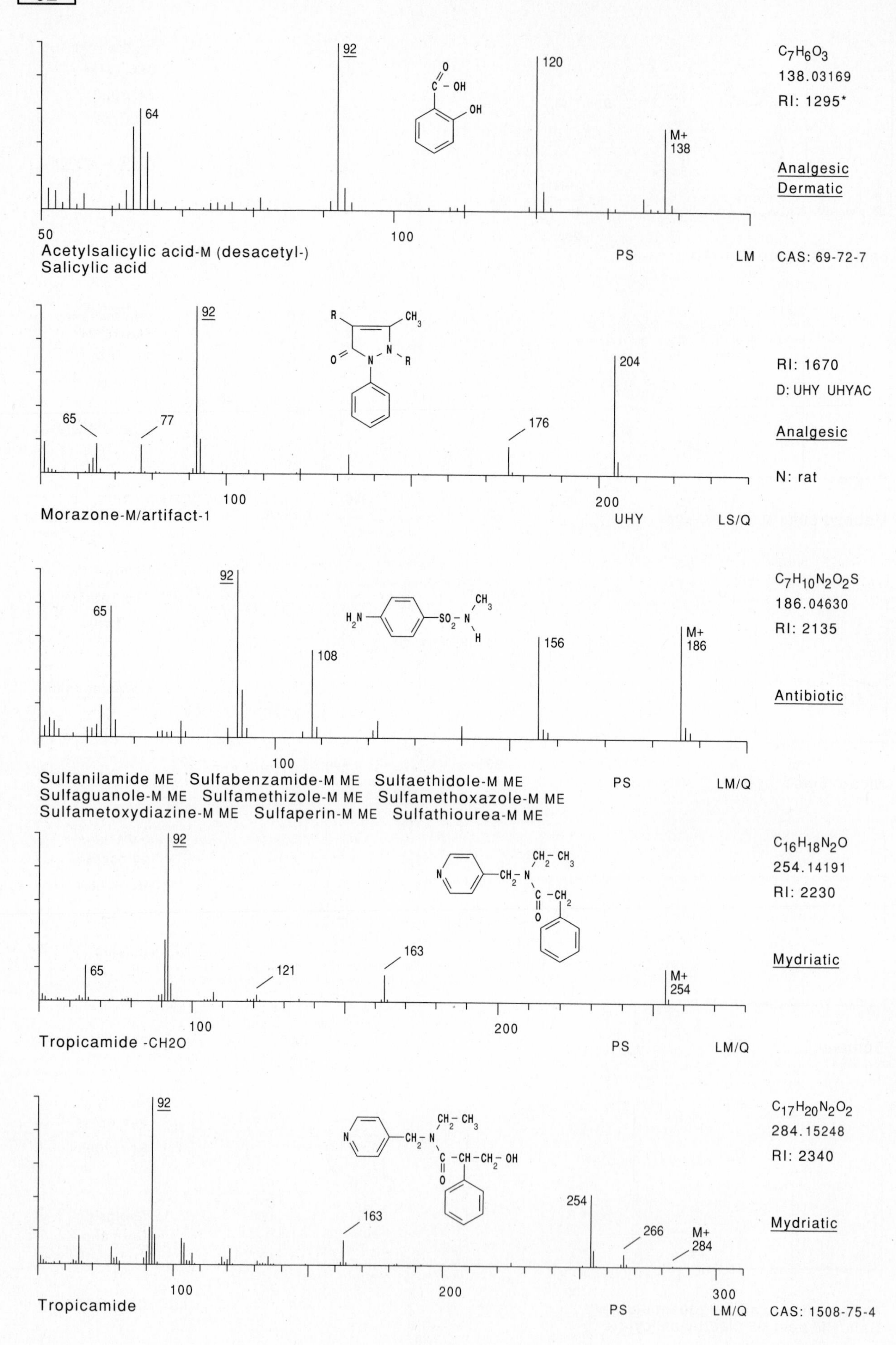

92
120
64
M+
138
50
100
C - OH
OH
$C_7H_6O_3$
138.03169
RI: 1295*
Analgesic
Dermatic
Acetylsalicylic acid-M (desacetyl-)
Salicylic acid
PS
LM
CAS: 69-72-7
92
204
65
77
176
100
200
R
CH3
N
R
RI: 1670
D: UHY UHYAC
Analgesic
N: rat
Morazone-M/artifact-1
UHY
LS/Q
92
65
108
156
M+
186
100
H2N
SO2
CH3
N
H
$C_7H_{10}N_2O_2S$
186.04630
RI: 2135
Antibiotic
Sulfanilamide ME Sulfabenzamide-M ME Sulfaethidole-M ME
Sulfaguanole-M ME Sulfamethizole-M ME Sulfamethoxazole-M ME
Sulfametoxydiazine-M ME Sulfaperin-M ME Sulfathiourea-M ME
PS
LM/Q
92
65
121
163
M+
254
100
200
N
CH2 - N
CH2 - CH3
C - CH2
O
$C_{16}H_{18}N_2O$
254.14191
RI: 2230
Mydriatic
Tropicamide -CH2O
PS
LM/Q
92
163
254
266
M+
284
100
200
300
C - CH - CH2 - OH
$C_{17}H_{20}N_2O_2$
284.15248
RI: 2340
Mydriatic
Tropicamide
PS
LM/Q
CAS: 1508-75-4

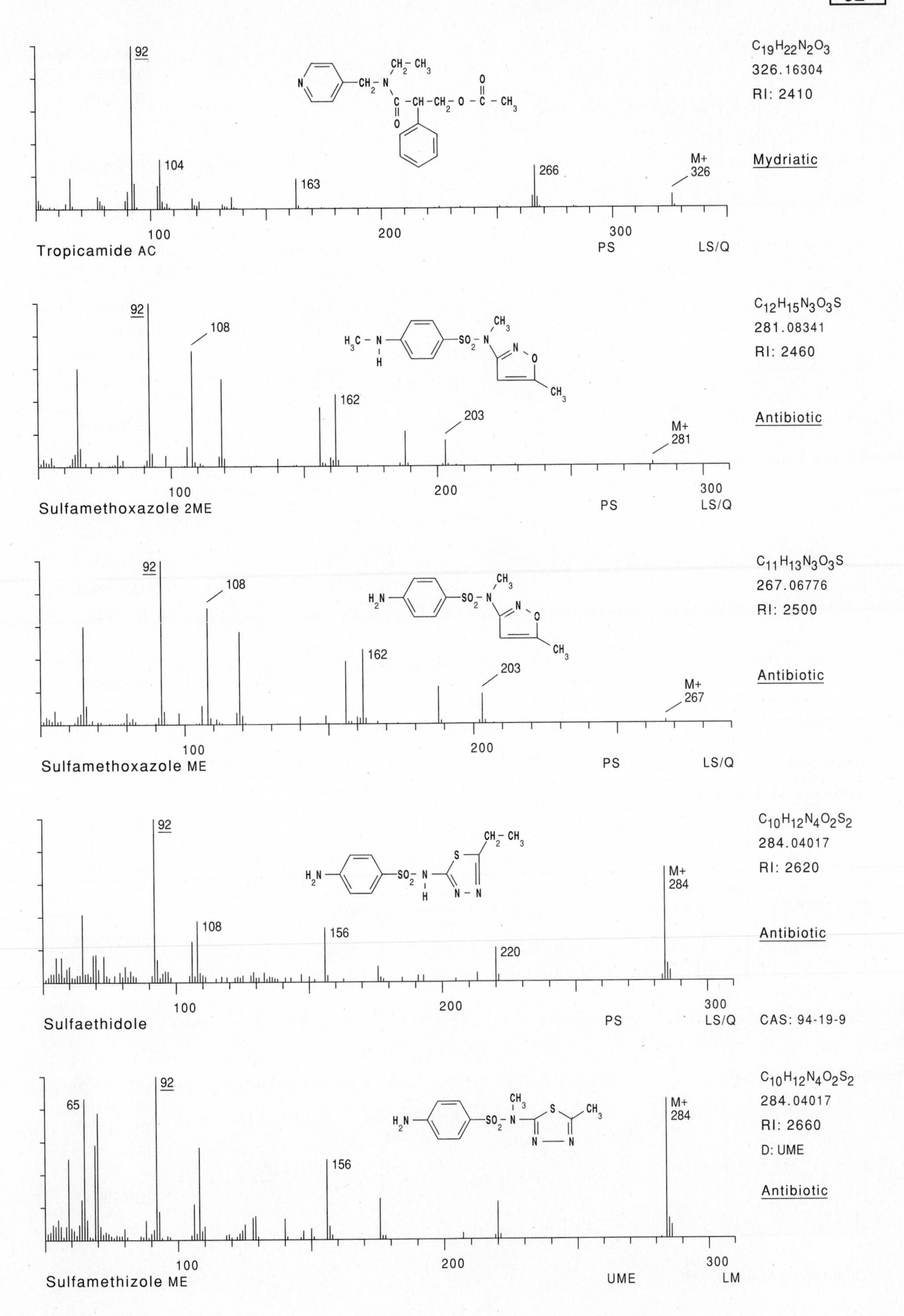

92
104
163
266
M+
326
100
200
300
PS
LS/Q
Tropicamide AC
C19H22N2O3
326.16304
RI: 2410
Mydriatic
92
108
162
203
M+
281
100
200
300
PS
LS/Q
Sulfamethoxazole 2ME
C12H15N3O3S
281.08341
RI: 2460
Antibiotic
92
108
162
203
M+
267
100
200
PS
LS/Q
Sulfamethoxazole ME
C11H13N3O3S
267.06776
RI: 2500
Antibiotic
92
108
156
220
M+
284
100
200
300
PS
LS/Q
Sulfaethidole
C10H12N4O2S2
284.04017
RI: 2620
Antibiotic
CAS: 94-19-9
65
92
156
M+
284
100
200
300
UME
LM
Sulfamethizole ME
C10H12N4O2S2
284.04017
RI: 2660
D: UME
Antibiotic

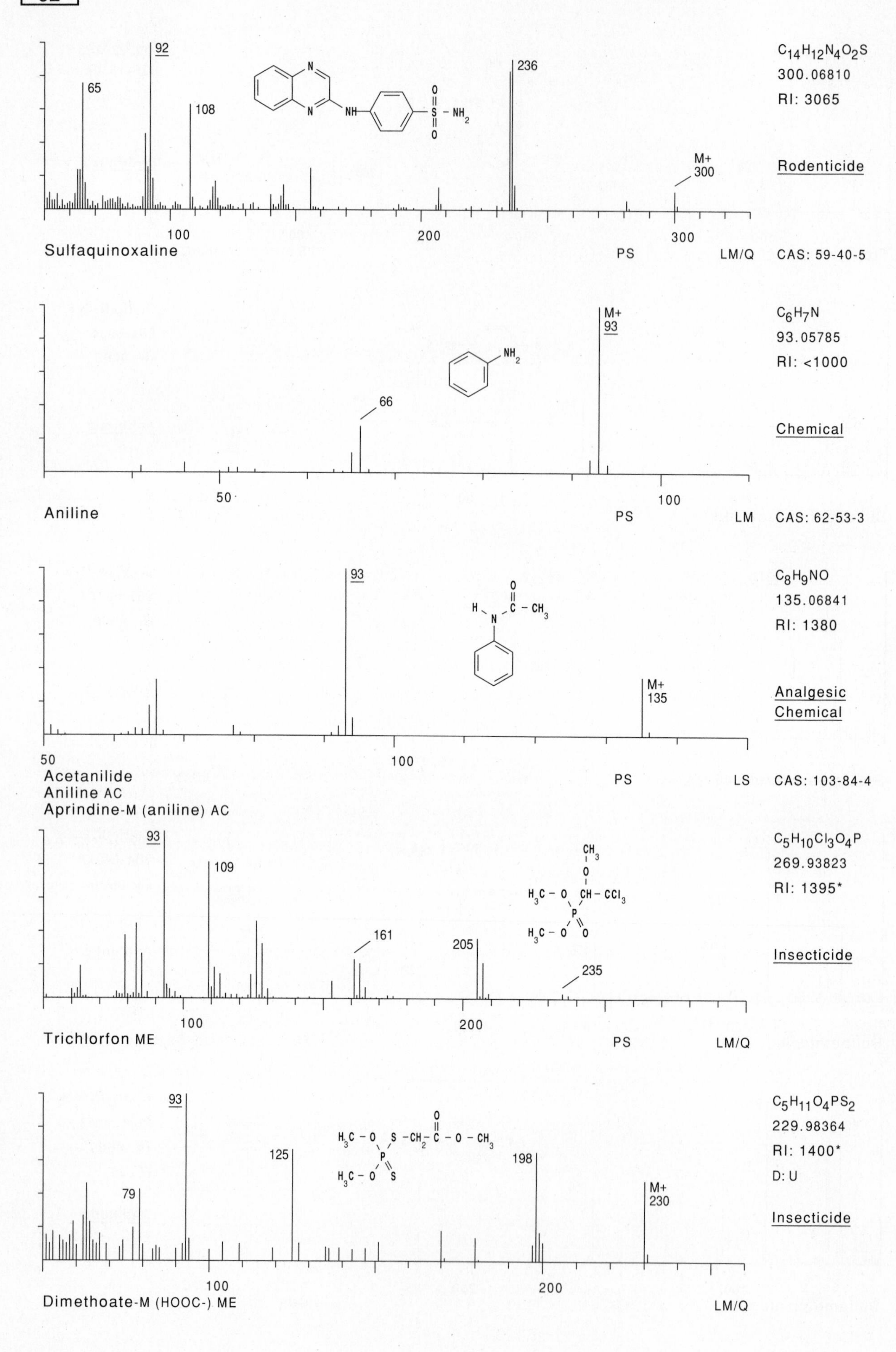

92
236
65
108
M+
300
100
200
300
Sulfaquinoxaline
PS
LM/Q
$C_{14}H_{12}N_4O_2S$
300.06810
RI: 3065
Rodenticide
CAS: 59-40-5
M+
93
66
50
100
Aniline
PS
LM
C_6H_7N
93.05785
RI: <1000
Chemical
CAS: 62-53-3
93
M+
135
50
100
Acetanilide
Aniline AC
Aprindine-M (aniline) AC
PS
LS
C_8H_9NO
135.06841
RI: 1380
Analgesic
Chemical
CAS: 103-84-4
93
109
161
205
235
100
200
Trichlorfon ME
PS
LM/Q
$C_5H_{10}Cl_3O_4P$
269.93823
RI: 1395*
Insecticide
93
125
198
79
M+
230
100
200
Dimethoate-M (HOOC-) ME
LM/Q
$C_5H_{11}O_4PS_2$
229.98364
RI: 1400*
D: U
Insecticide

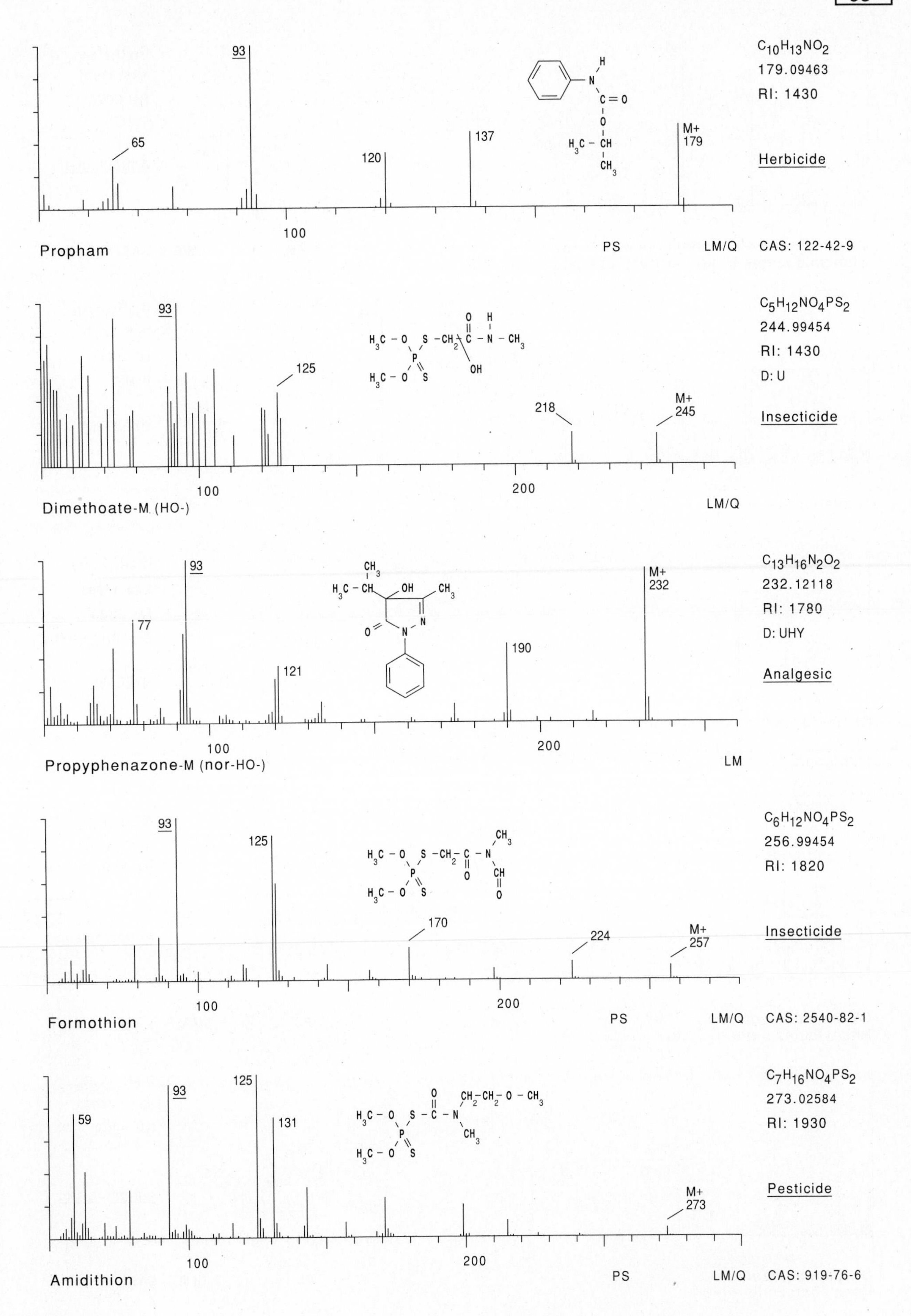

Propham
C10H13NO2
179.09463
RI: 1430
Herbicide
PS
LM/Q
CAS: 122-42-9
Dimethoate-M (HO-)
C5H12NO4PS2
244.99454
RI: 1430
D: U
Insecticide
LM/Q
Propyphenazone-M (nor-HO-)
C13H16N2O2
232.12118
RI: 1780
D: UHY
Analgesic
LM
Formothion
C6H12NO4PS2
256.99454
RI: 1820
Insecticide
PS
LM/Q
CAS: 2540-82-1
Amidithion
C7H16NO4PS2
273.02584
RI: 1930
Pesticide
PS
LM/Q
CAS: 919-76-6

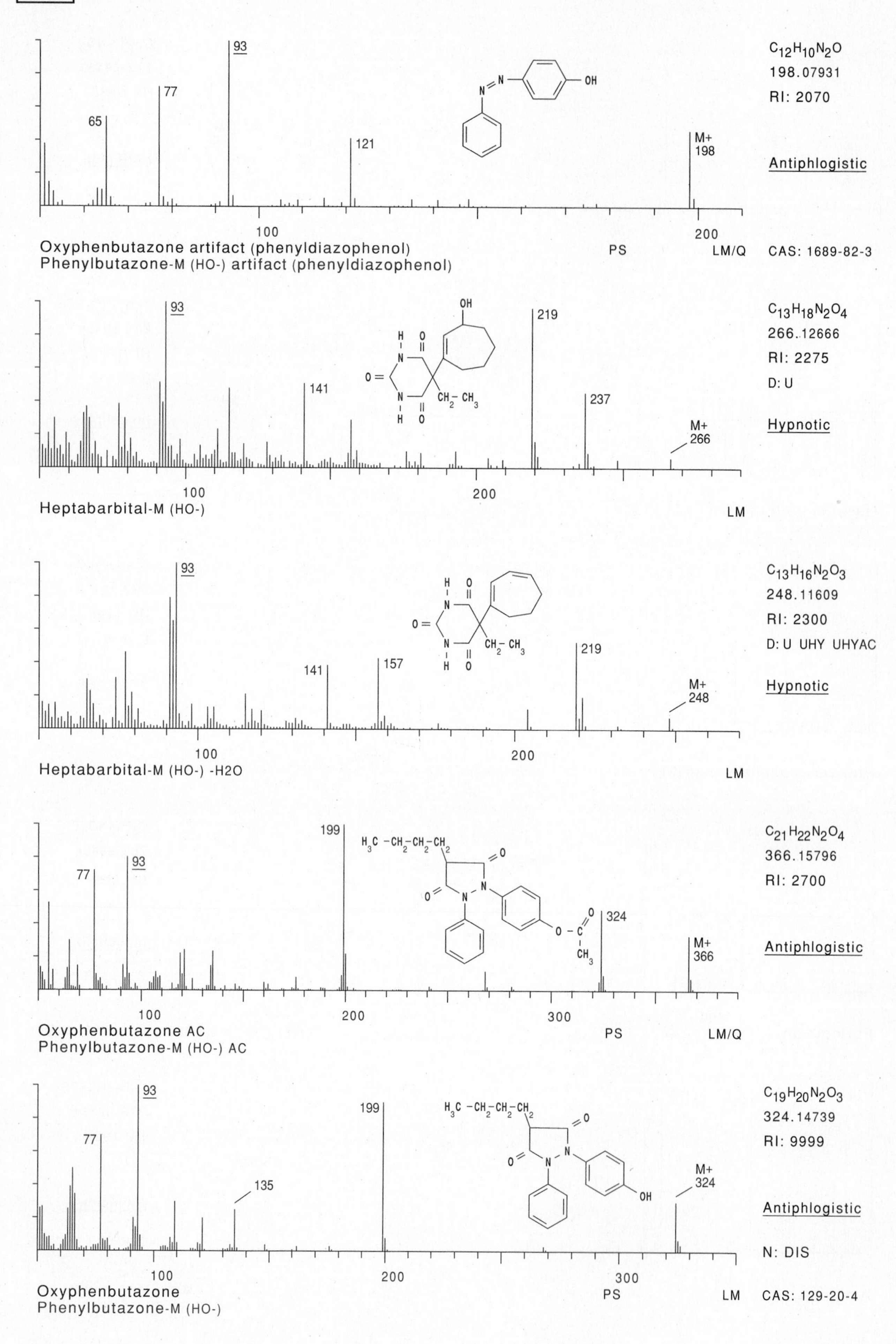

93
77
65
121
M+
198
N=N
OH
100
200
Oxyphenbutazone artifact (phenyldiazophenol)
Phenylbutazone-M (HO-) artifact (phenyldiazophenol)
PS
LM/Q
$C_{12}H_{10}N_2O$
198.07931
RI: 2070
Antiphlogistic
CAS: 1689-82-3
93
OH
219
H O
N
O=
N
H O
CH2- CH3
141
237
M+
266
100
200
Heptabarbital-M (HO-)
LM
$C_{13}H_{18}N_2O_4$
266.12666
RI: 2275
D: U
Hypnotic
93
141
157
219
M+
248
100
200
Heptabarbital-M (HO-) -H2O
LM
$C_{13}H_{16}N_2O_3$
248.11609
RI: 2300
D: U UHY UHYAC
Hypnotic
199
77
93
H3C -CH2-CH2-CH2
O
324
M+
366
O - C
CH3
100
200
300
Oxyphenbutazone AC
Phenylbutazone-M (HO-) AC
PS
LM/Q
$C_{21}H_{22}N_2O_4$
366.15796
RI: 2700
Antiphlogistic
93
199
77
135
H3C -CH2-CH2-CH2
OH
M+
324
100
200
300
Oxyphenbutazone
Phenylbutazone-M (HO-)
PS
LM
$C_{19}H_{20}N_2O_3$
324.14739
RI: 9999
Antiphlogistic
N: DIS
CAS: 129-20-4

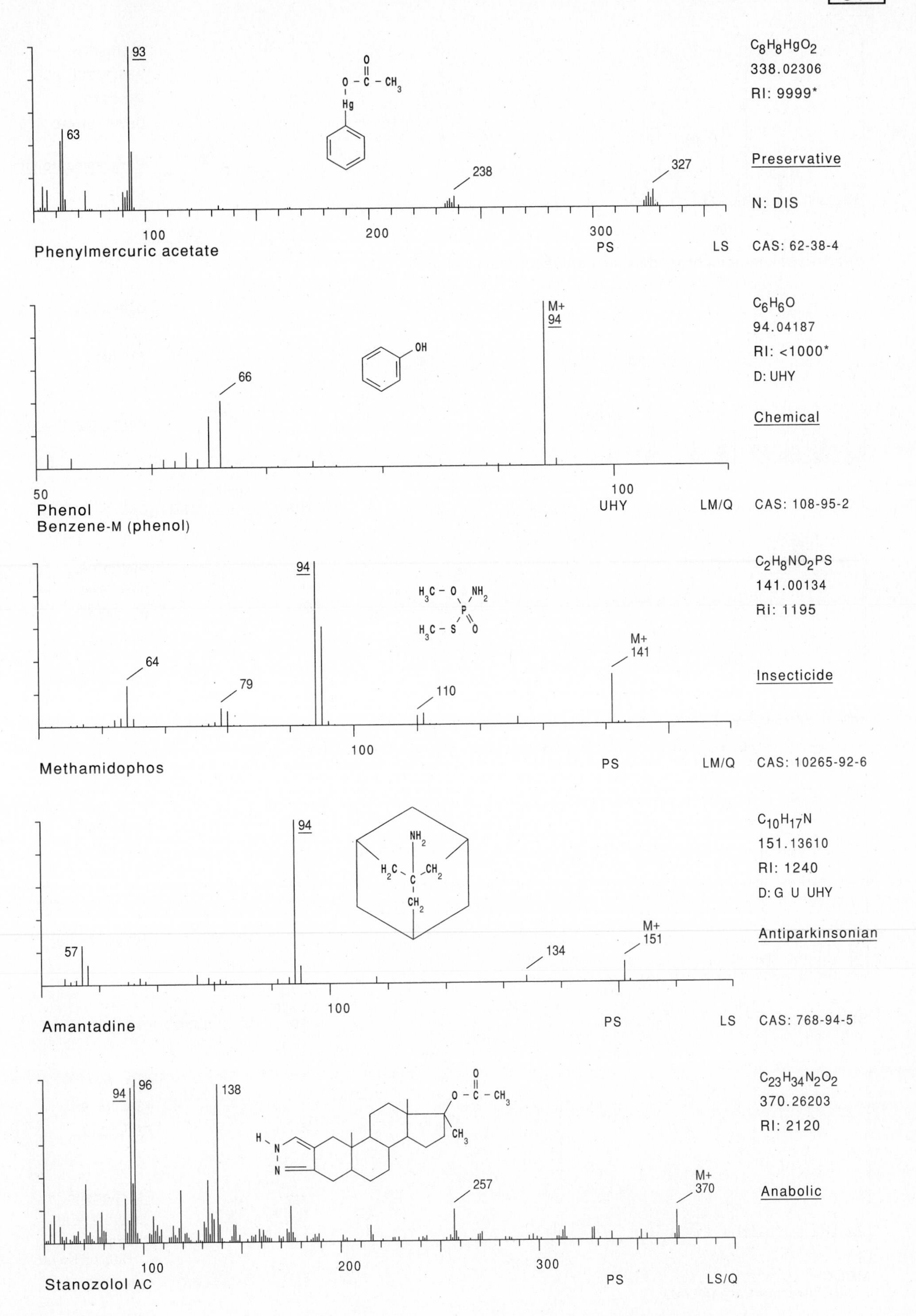

$C_8H_8HgO_2$
338.02306
RI: 9999*

Preservative

N: DIS

CAS: 62-38-4

C_6H_6O
94.04187
RI: <1000*
D: UHY

Chemical

CAS: 108-95-2

$C_2H_8NO_2PS$
141.00134
RI: 1195

Insecticide

CAS: 10265-92-6

$C_{10}H_{17}N$
151.13610
RI: 1240
D: G U UHY

Antiparkinsonian

CAS: 768-94-5

$C_{23}H_{34}N_2O_2$
370.26203
RI: 2120

Anabolic

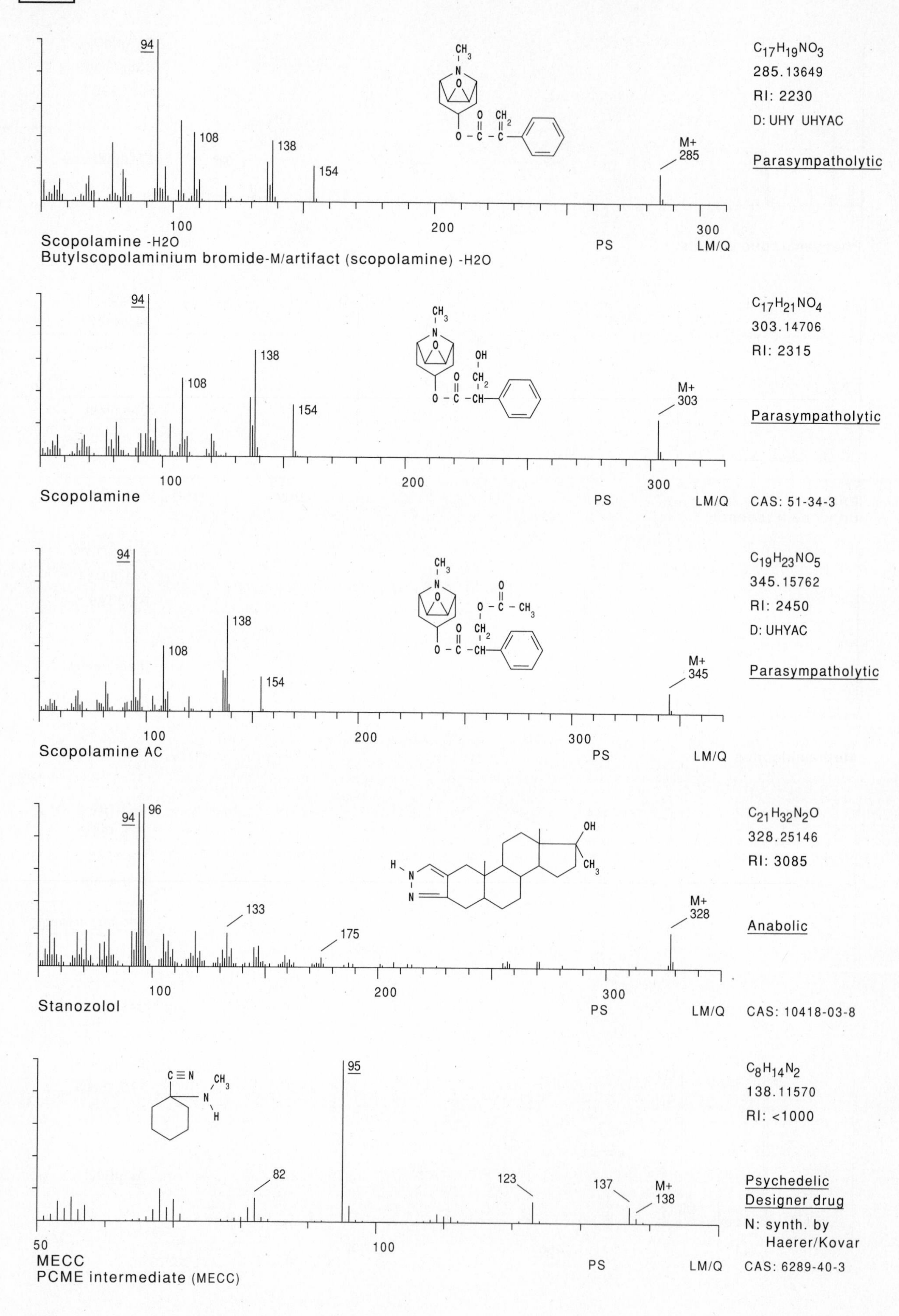

$C_{17}H_{19}NO_3$
285.13649
RI: 2230
D: UHY UHYAC
Parasympatholytic
M+ 285
Scopolamine -H2O
Butylscopolaminium bromide-M/artifact (scopolamine) -H2O
PS
LM/Q
$C_{17}H_{21}NO_4$
303.14706
RI: 2315
Parasympatholytic
M+ 303
Scopolamine
PS
LM/Q
CAS: 51-34-3
$C_{19}H_{23}NO_5$
345.15762
RI: 2450
D: UHYAC
Parasympatholytic
M+ 345
Scopolamine AC
PS
LM/Q
$C_{21}H_{32}N_2O$
328.25146
RI: 3085
Anabolic
M+ 328
Stanozolol
PS
LM/Q
CAS: 10418-03-8
$C_8H_{14}N_2$
138.11570
RI: <1000
Psychedelic
Designer drug
N: synth. by Haerer/Kovar
M+ 138
MECC
PCME intermediate (MECC)
PS
LM/Q
CAS: 6289-40-3

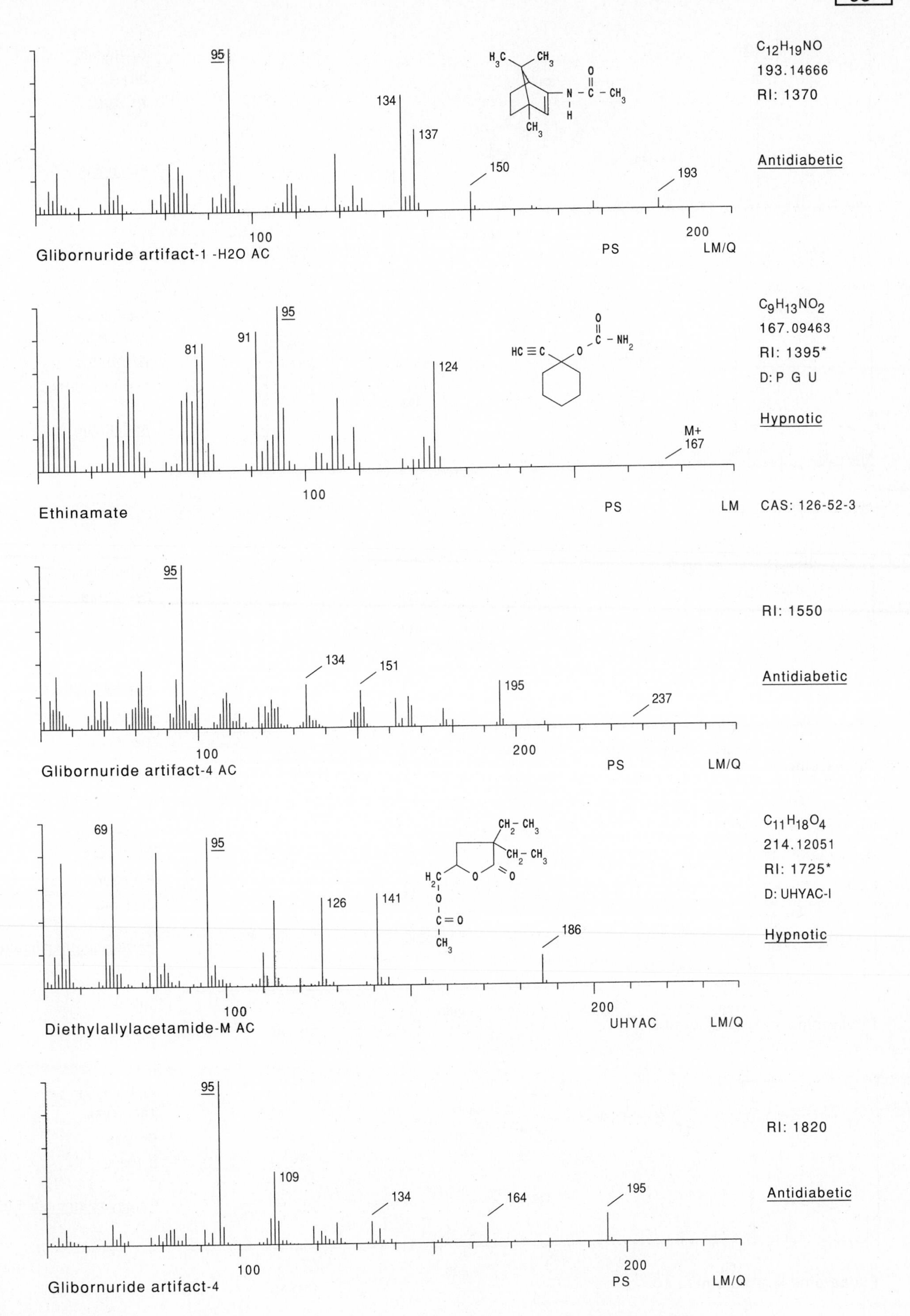
95
134
137
150
193
100
200
PS
LM/Q
Glibornuride artifact-1 -H2O AC
$C_{12}H_{19}NO$
193.14666
RI: 1370
Antidiabetic
95
91
81
124
M+
167
100
PS
LM
Ethinamate
$C_9H_{13}NO_2$
167.09463
RI: 1395*
D: P G U
Hypnotic
CAS: 126-52-3
95
134
151
195
237
100
200
PS
LM/Q
Glibornuride artifact-4 AC
RI: 1550
Antidiabetic
69
95
126
141
186
100
200
UHYAC
LM/Q
Diethylallylacetamide-M AC
$C_{11}H_{18}O_4$
214.12051
RI: 1725*
D: UHYAC-I
Hypnotic
95
109
134
164
195
100
200
PS
LM/Q
Glibornuride artifact-4
RI: 1820
Antidiabetic

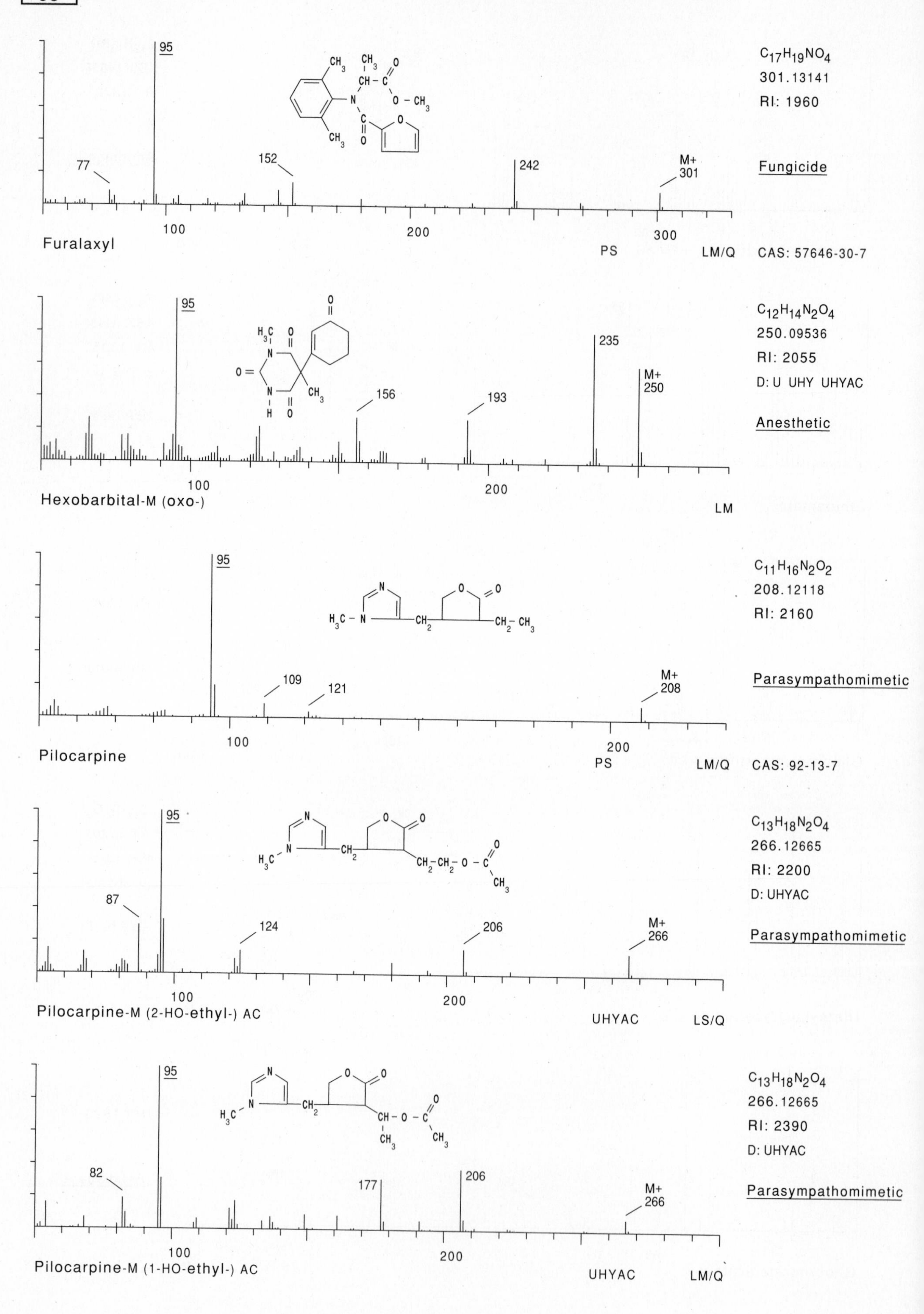

95
77
152
242
M+ 301
100
200
300
Furalaxyl
PS
LM/Q
$C_{17}H_{19}NO_4$
301.13141
RI: 1960
Fungicide
CAS: 57646-30-7
95
156
193
235
M+ 250
100
200
Hexobarbital-M (oxo-)
LM
$C_{12}H_{14}N_2O_4$
250.09536
RI: 2055
D: U UHY UHYAC
Anesthetic
95
109
121
M+ 208
100
200
Pilocarpine
PS
LM/Q
$C_{11}H_{16}N_2O_2$
208.12118
RI: 2160
Parasympathomimetic
CAS: 92-13-7
95
87
124
206
M+ 266
100
200
Pilocarpine-M (2-HO-ethyl-) AC
UHYAC
LS/Q
$C_{13}H_{18}N_2O_4$
266.12665
RI: 2200
D: UHYAC
Parasympathomimetic
95
82
177
206
M+ 266
100
200
Pilocarpine-M (1-HO-ethyl-) AC
UHYAC
LM/Q
$C_{13}H_{18}N_2O_4$
266.12665
RI: 2390
D: UHYAC
Parasympathomimetic

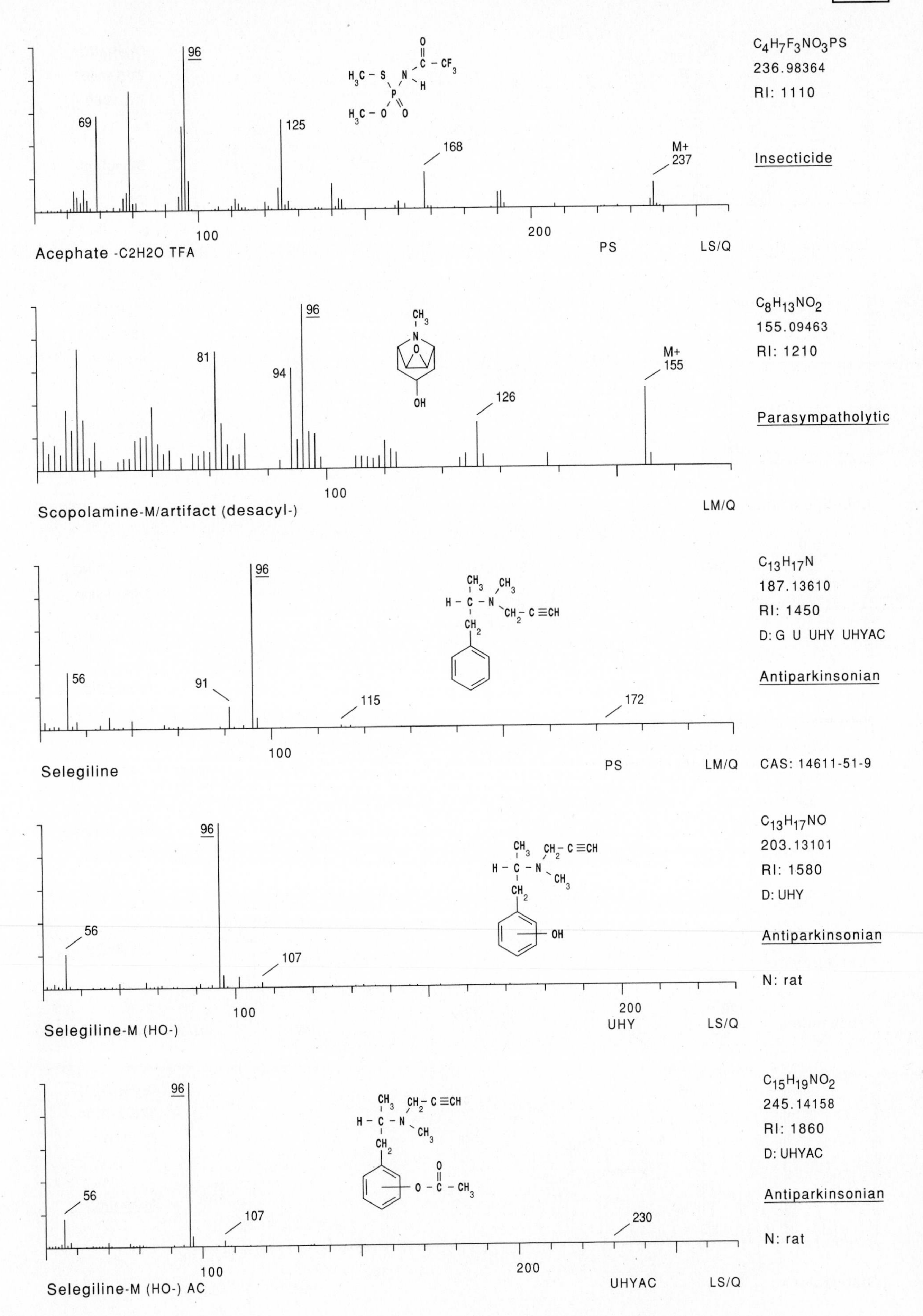

Acephate -C2H2O TFA
$C_4H_7F_3NO_3PS$
236.98364
RI: 1110
Insecticide
PS LS/Q
Scopolamine-M/artifact (desacyl-)
$C_8H_{13}NO_2$
155.09463
RI: 1210
Parasympatholytic
LM/Q
Selegiline
$C_{13}H_{17}N$
187.13610
RI: 1450
D: G U UHY UHYAC
Antiparkinsonian
PS LM/Q
CAS: 14611-51-9
Selegiline-M (HO-)
$C_{13}H_{17}NO$
203.13101
RI: 1580
D: UHY
Antiparkinsonian
N: rat
UHY LS/Q
Selegiline-M (HO-) AC
$C_{15}H_{19}NO_2$
245.14158
RI: 1860
D: UHYAC
Antiparkinsonian
N: rat
UHYAC LS/Q

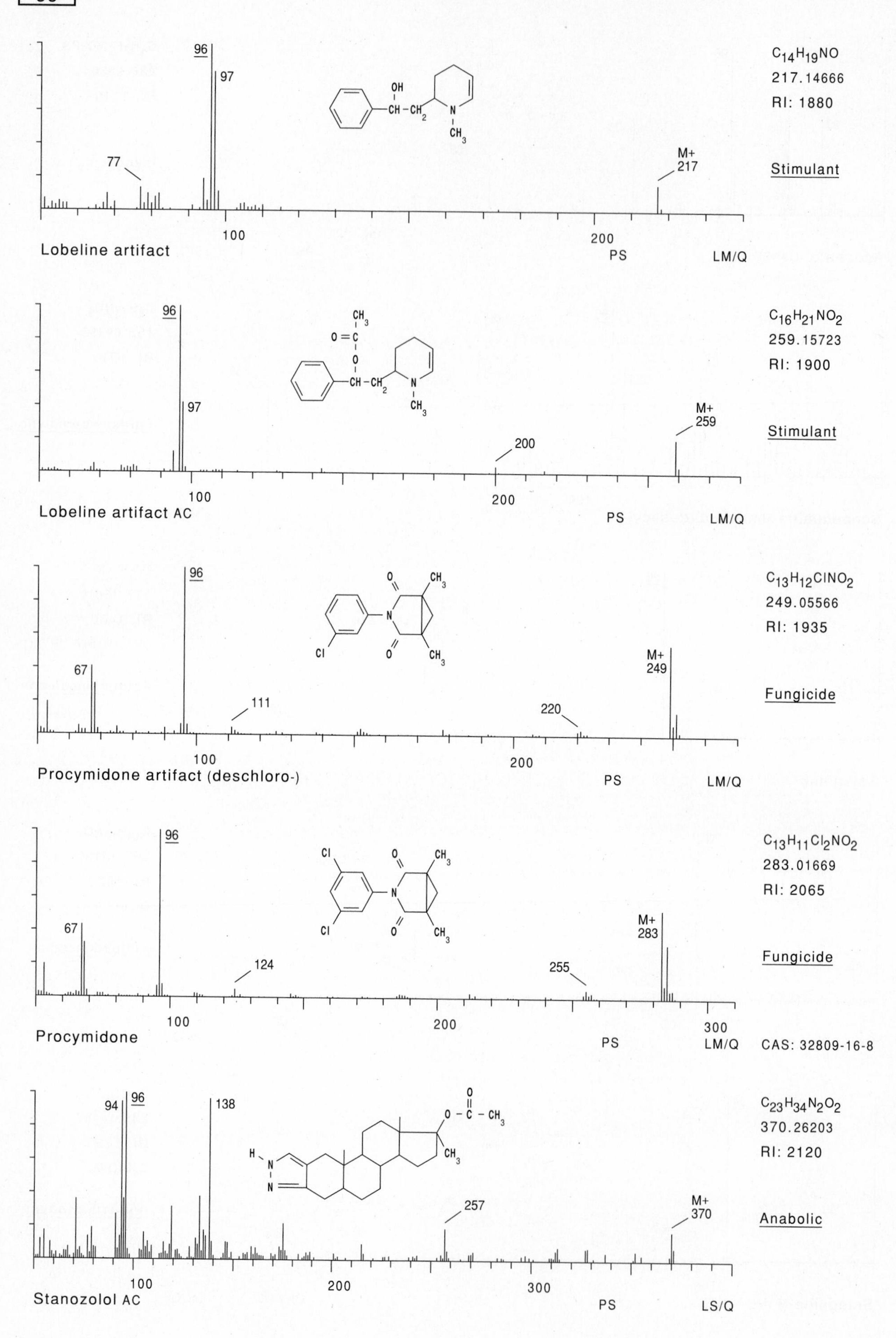
96
97
77
M+
217
100
200
PS
LM/Q
Lobeline artifact
$C_{14}H_{19}NO$
217.14666
RI: 1880
Stimulant
96
97
200
M+
259
100
200
PS
LM/Q
Lobeline artifact AC
$C_{16}H_{21}NO_2$
259.15723
RI: 1900
Stimulant
96
67
111
220
M+
249
100
200
PS
LM/Q
Procymidone artifact (deschloro-)
$C_{13}H_{12}ClNO_2$
249.05566
RI: 1935
Fungicide
96
67
124
255
M+
283
100
200
300
PS
LM/Q
Procymidone
$C_{13}H_{11}Cl_2NO_2$
283.01669
RI: 2065
Fungicide
CAS: 32809-16-8
94
96
138
257
M+
370
100
200
300
PS
LS/Q
Stanozolol AC
$C_{23}H_{34}N_2O_2$
370.26203
RI: 2120
Anabolic

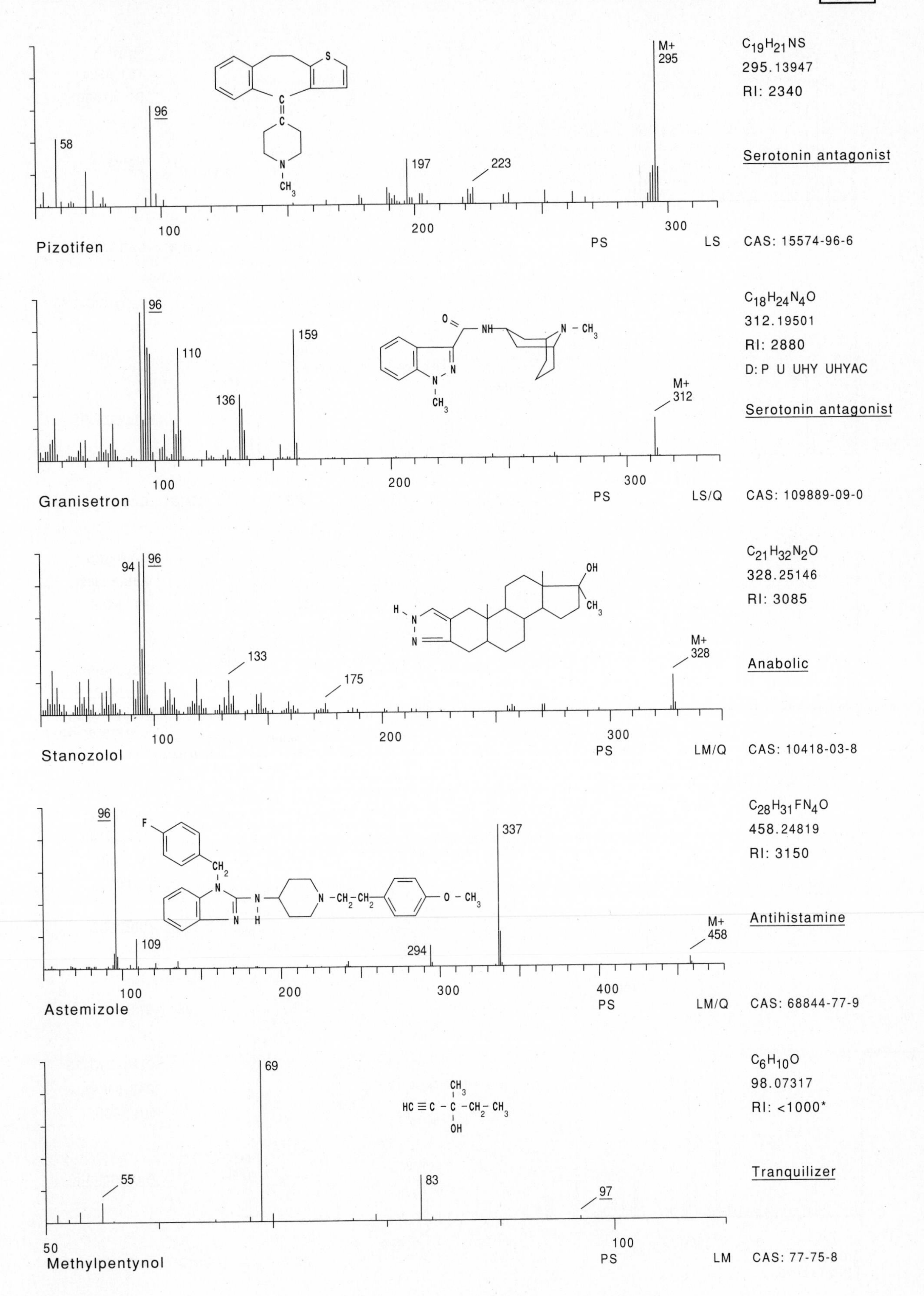

M+ 295
96
58
197
223
100
200
300
Pizotifen
PS
LS
$C_{19}H_{21}NS$
295.13947
RI: 2340
Serotonin antagonist
CAS: 15574-96-6
96
159
110
136
M+ 312
Granisetron
PS
LS/Q
$C_{18}H_{24}N_4O$
312.19501
RI: 2880
D: P U UHY UHYAC
Serotonin antagonist
CAS: 109889-09-0
94
96
133
175
M+ 328
Stanozolol
PS
LM/Q
$C_{21}H_{32}N_2O$
328.25146
RI: 3085
Anabolic
CAS: 10418-03-8
96
337
109
294
M+ 458
400
Astemizole
PS
LM/Q
$C_{28}H_{31}FN_4O$
458.24819
RI: 3150
Antihistamine
CAS: 68844-77-9
69
55
83
97
50
Methylpentynol
PS
LM
$C_6H_{10}O$
98.07317
RI: <1000*
Tranquilizer
CAS: 77-75-8

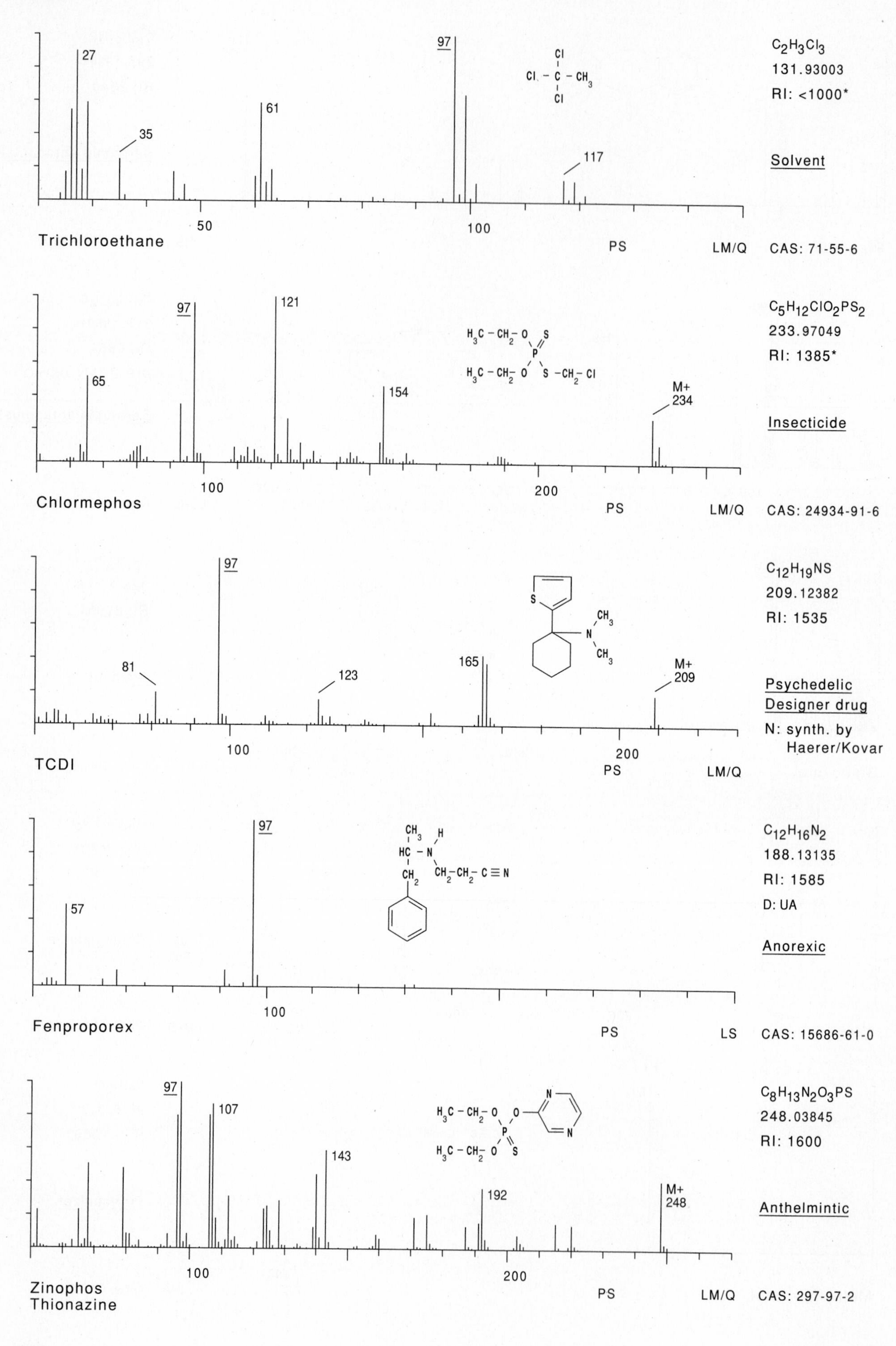

27
35
61
97
117
Cl
Cl - C - CH3
Cl
50
100
Trichloroethane
PS
LM/Q
C2H3Cl3
131.93003
RI: <1000*
Solvent
CAS: 71-55-6
65
97
121
154
M+
234
H3C - CH2 - O S
P
H3C - CH2 - O S - CH2 - Cl
100
200
Chlormephos
PS
LM/Q
C5H12ClO2PS2
233.97049
RI: 1385*
Insecticide
CAS: 24934-91-6
81
97
123
165
M+
209
S
CH3
N
CH3
100
200
TCDI
PS
LM/Q
C12H19NS
209.12382
RI: 1535
Psychedelic
Designer drug
N: synth. by
Haerer/Kovar
57
97
CH3
HC - N
H
CH2
CH2-CH2- C ≡ N
100
Fenproporex
PS
LS
C12H16N2
188.13135
RI: 1585
D: UA
Anorexic
CAS: 15686-61-0
97
107
143
192
M+
248
H3C - CH2 - O O
N
P
N
H3C - CH2 - O S
100
200
Zinophos
Thionazine
PS
LM/Q
C8H13N2O3PS
248.03845
RI: 1600
Anthelmintic
CAS: 297-97-2

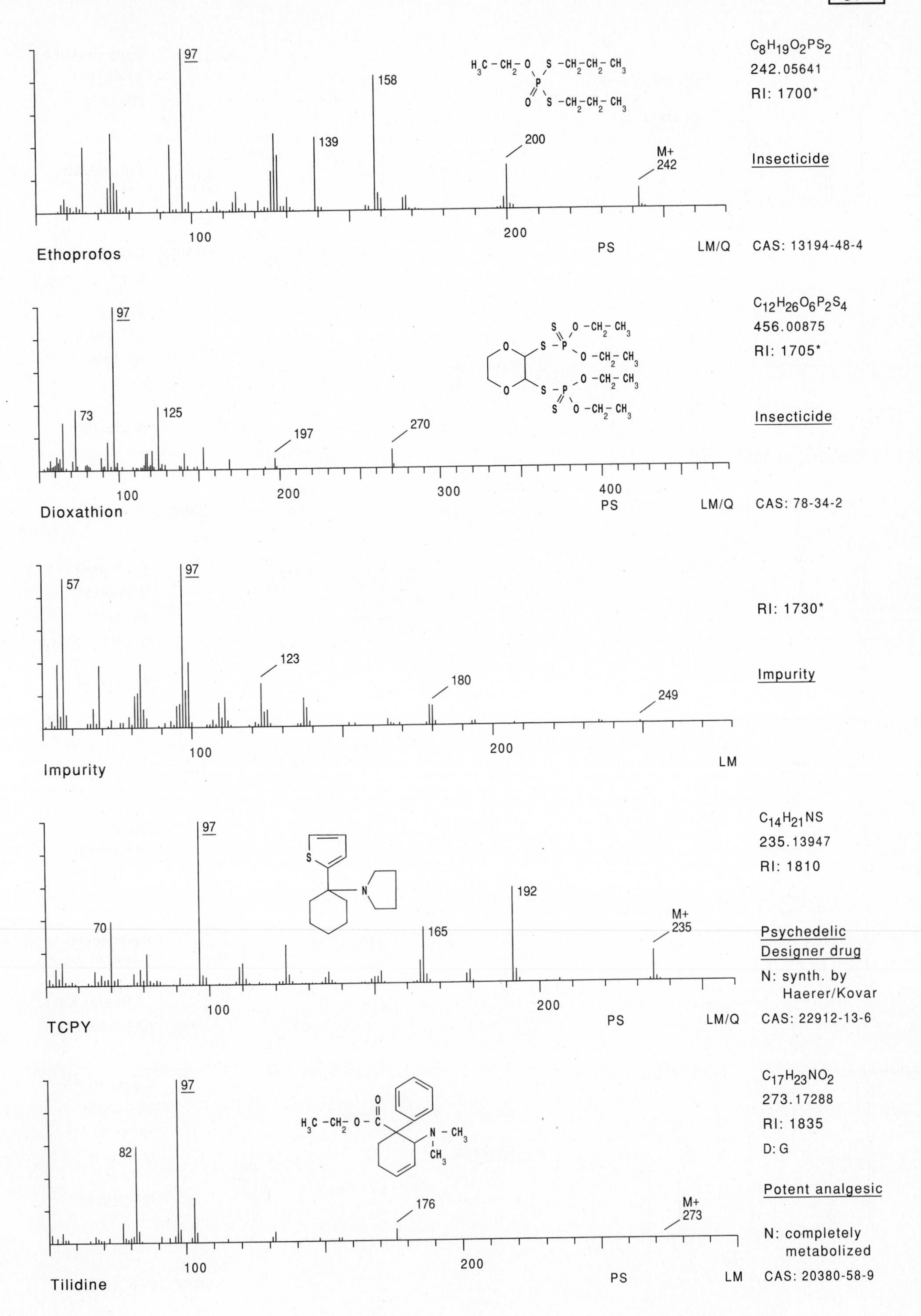
Ethoprofos
C8H19O2PS2
242.05641
RI: 1700*
Insecticide
CAS: 13194-48-4
97
158
139
200
M+
242
100
200
PS
LM/Q
Dioxathion
C12H26O6P2S4
456.00875
RI: 1705*
Insecticide
CAS: 78-34-2
97
73
125
197
270
100
200
300
400
PS
LM/Q
Impurity
RI: 1730*
Impurity
57
97
123
180
249
100
200
LM
TCPY
C14H21NS
235.13947
RI: 1810
Psychedelic
Designer drug
N: synth. by
Haerer/Kovar
CAS: 22912-13-6
97
70
165
192
M+
235
100
200
PS
LM/Q
Tilidine
C17H23NO2
273.17288
RI: 1835
D: G
Potent analgesic
N: completely
metabolized
CAS: 20380-58-9
97
82
176
M+
273
100
200
PS
LM

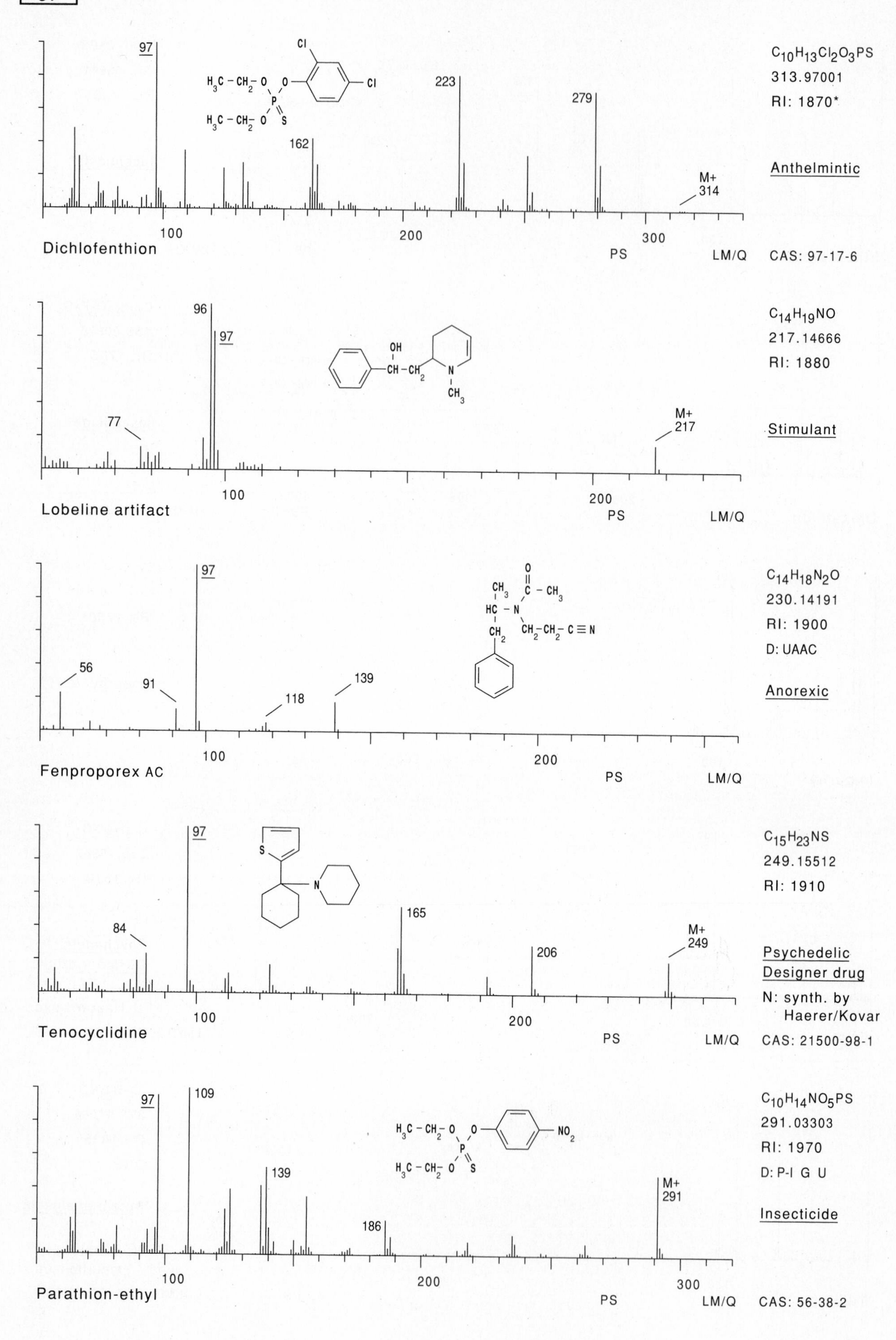
97
Dichlofenthion
$C_{10}H_{13}Cl_2O_3PS$
313.97001
RI: 1870*
Anthelmintic
CAS: 97-17-6
M+ 314
Lobeline artifact
$C_{14}H_{19}NO$
217.14666
RI: 1880
Stimulant
M+ 217
Fenproporex AC
$C_{14}H_{18}N_2O$
230.14191
RI: 1900
D: UAAC
Anorexic
Tenocyclidine
$C_{15}H_{23}NS$
249.15512
RI: 1910
Psychedelic
Designer drug
N: synth. by Haerer/Kovar
CAS: 21500-98-1
M+ 249
Parathion-ethyl
$C_{10}H_{14}NO_5PS$
291.03303
RI: 1970
D: P-I G U
Insecticide
CAS: 56-38-2
M+ 291
PS
LM/Q

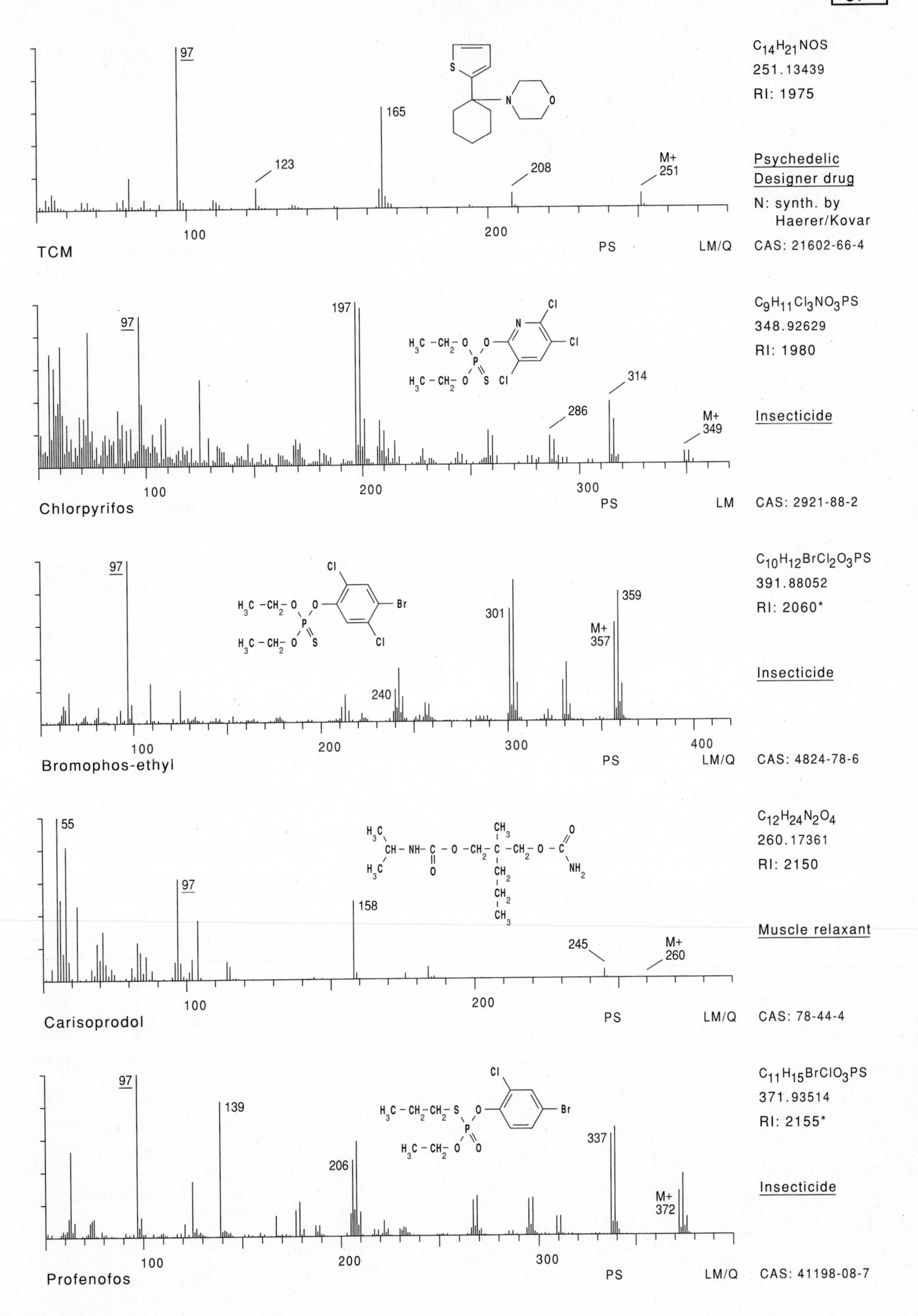
C14H21NOS
251.13439
RI: 1975
Psychedelic
Designer drug
N: synth. by
Haerer/Kovar
TCM
PS
LM/Q
CAS: 21602-66-4
C9H11Cl3NO3PS
348.92629
RI: 1980
Insecticide
Chlorpyrifos
PS
LM
CAS: 2921-88-2
C10H12BrCl2O3PS
391.88052
RI: 2060*
Insecticide
Bromophos-ethyl
PS
LM/Q
CAS: 4824-78-6
C12H24N2O4
260.17361
RI: 2150
Muscle relaxant
Carisoprodol
PS
LM/Q
CAS: 78-44-4
C11H15BrClO3PS
371.93514
RI: 2155*
Insecticide
Profenofos
PS
LM/Q
CAS: 41198-08-7

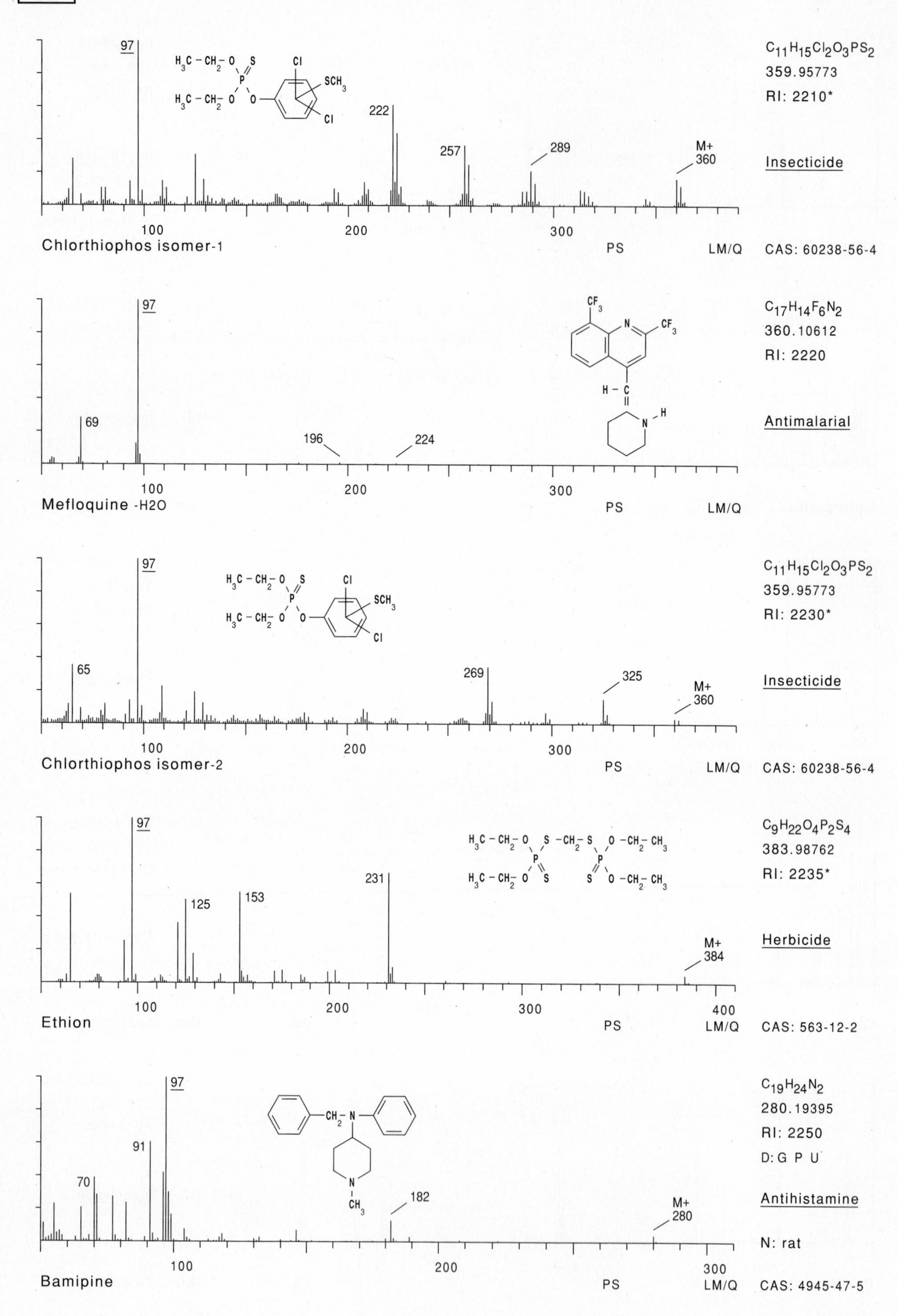
97
222
257
289
M+
360
100
200
300
Chlorthiophos isomer-1
PS
LM/Q
C11H15Cl2O3PS2
359.95773
RI: 2210*
Insecticide
CAS: 60238-56-4
97
69
196
224
100
200
300
Mefloquine -H2O
PS
LM/Q
C17H14F6N2
360.10612
RI: 2220
Antimalarial
97
65
269
325
M+
360
100
200
300
Chlorthiophos isomer-2
PS
LM/Q
C11H15Cl2O3PS2
359.95773
RI: 2230*
Insecticide
CAS: 60238-56-4
97
125
153
231
M+
384
100
200
300
400
Ethion
PS
LM/Q
C9H22O4P2S4
383.98762
RI: 2235*
Herbicide
CAS: 563-12-2
97
91
70
182
M+
280
100
200
300
Bamipine
PS
LM/Q
C19H24N2
280.19395
RI: 2250
D: G P U
Antihistamine
N: rat
CAS: 4945-47-5

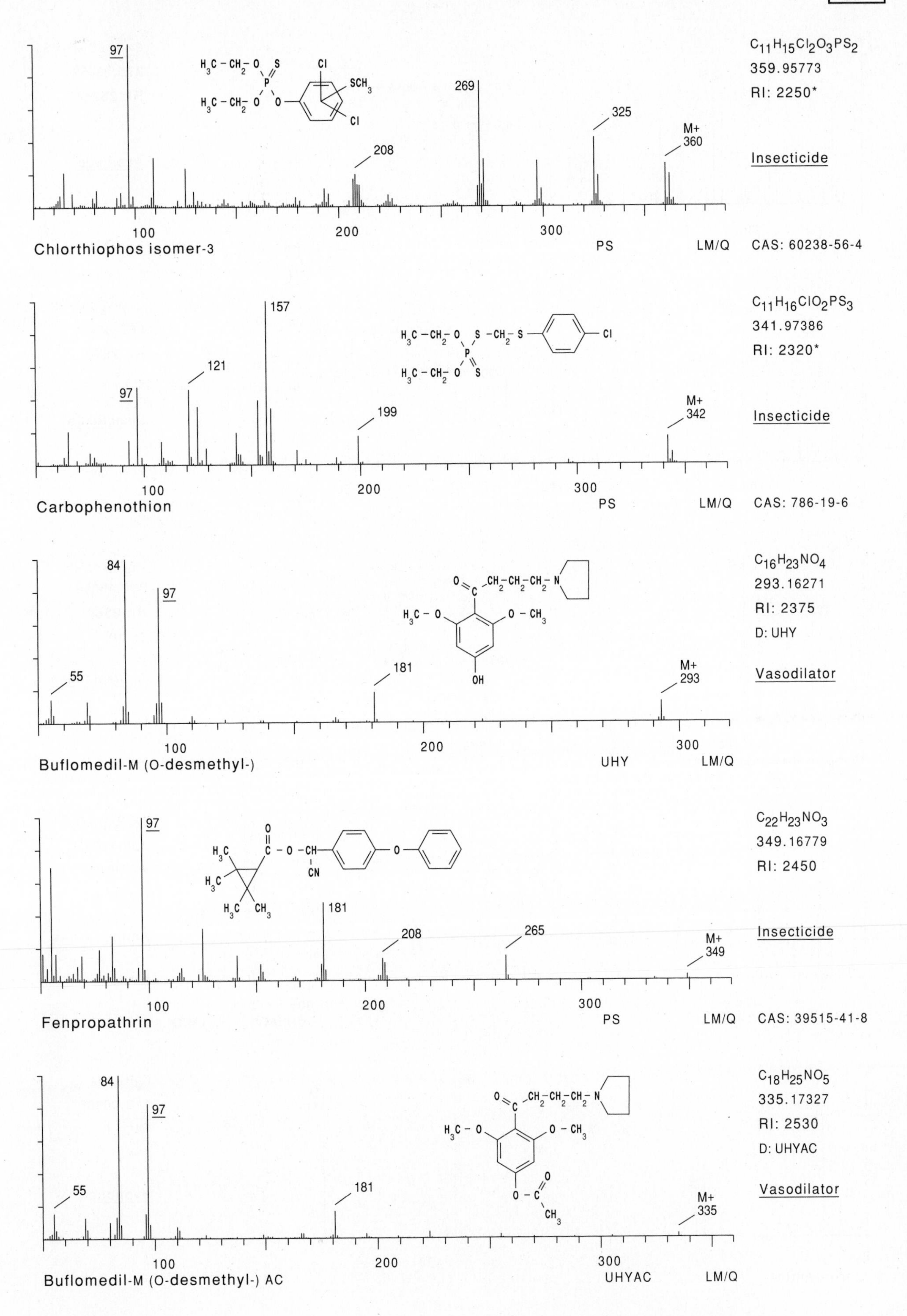

$C_{11}H_{15}Cl_2O_3PS_2$
359.95773
RI: 2250*

Insecticide

CAS: 60238-56-4

$C_{11}H_{16}ClO_2PS_3$
341.97386
RI: 2320*

Insecticide

CAS: 786-19-6

$C_{16}H_{23}NO_4$
293.16271
RI: 2375
D: UHY

Vasodilator

$C_{22}H_{23}NO_3$
349.16779
RI: 2450

Insecticide

CAS: 39515-41-8

$C_{18}H_{25}NO_5$
335.17327
RI: 2530
D: UHYAC

Vasodilator

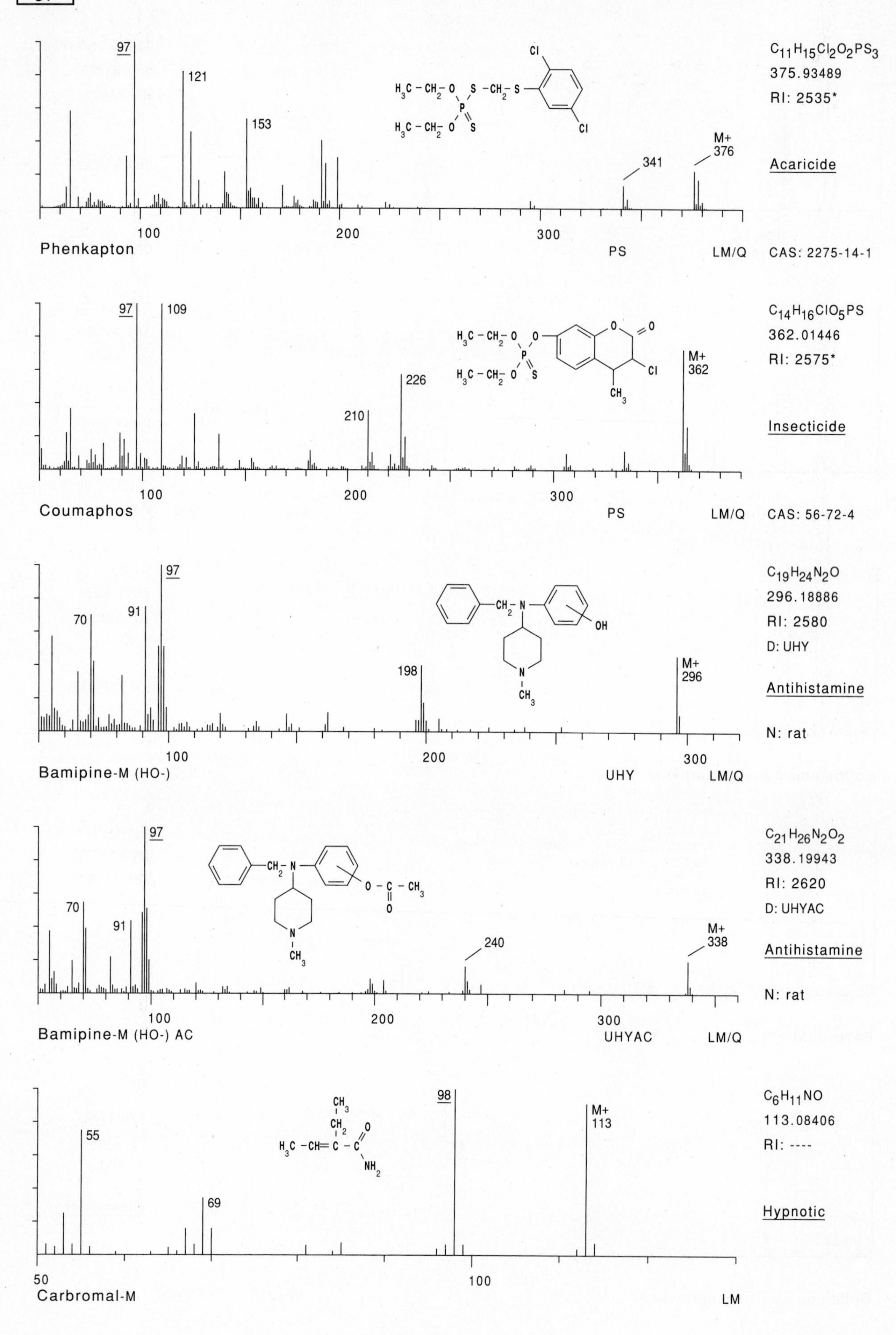
Phenkapton
PS
LM/Q
$C_{11}H_{15}Cl_2O_2PS_3$
375.93489
RI: 2535*
Acaricide
CAS: 2275-14-1
Coumaphos
PS
LM/Q
$C_{14}H_{16}ClO_5PS$
362.01446
RI: 2575*
Insecticide
CAS: 56-72-4
Bamipine-M (HO-)
UHY
LM/Q
$C_{19}H_{24}N_2O$
296.18886
RI: 2580
D: UHY
Antihistamine
N: rat
Bamipine-M (HO-) AC
UHYAC
LM/Q
$C_{21}H_{26}N_2O_2$
338.19943
RI: 2620
D: UHYAC
Antihistamine
N: rat
Carbromal-M
LM
$C_6H_{11}NO$
113.08406
RI: ----
Hypnotic

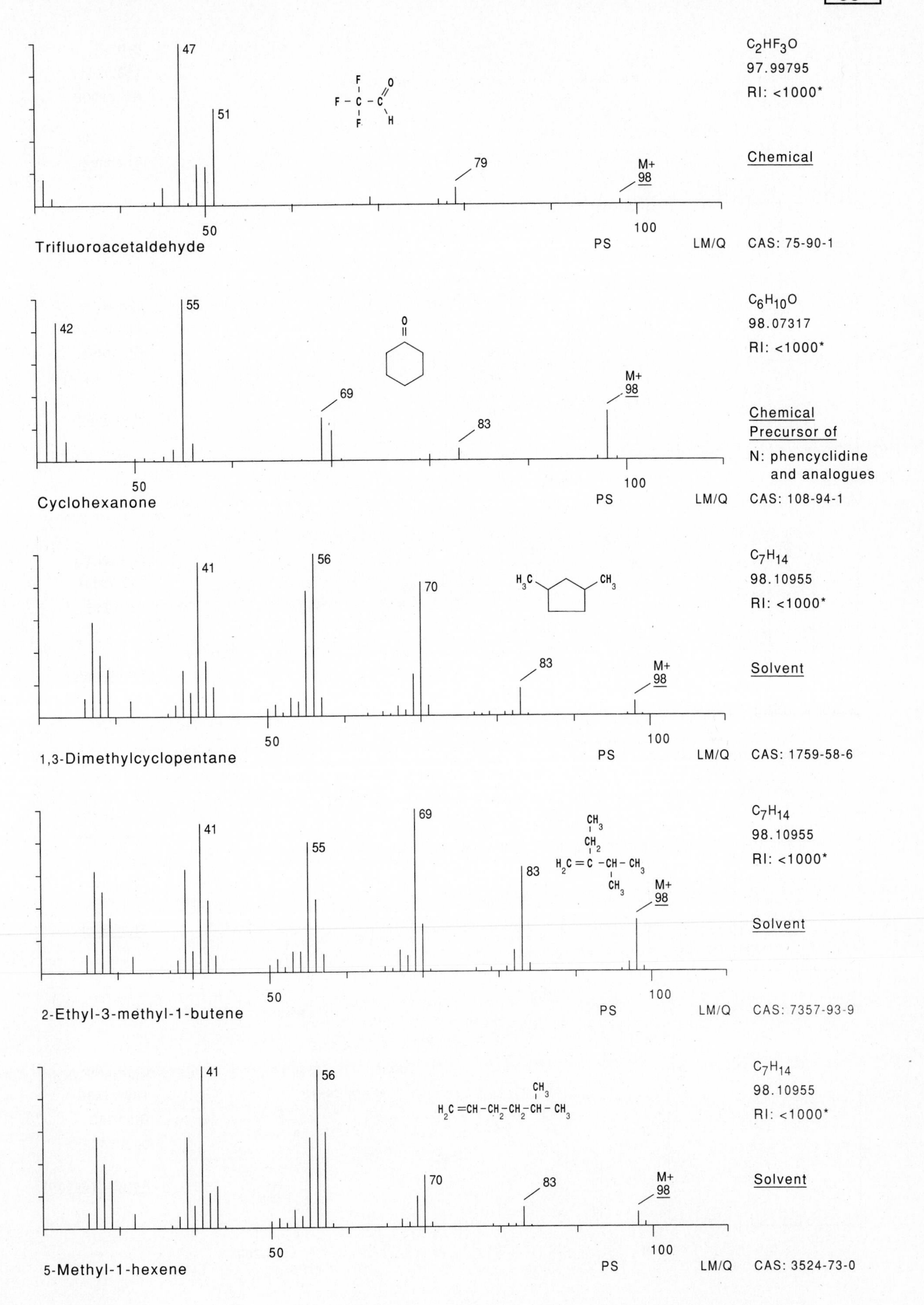
47
51
79
M+
98
50
100
Trifluoroacetaldehyde
PS
LM/Q
C_2HF_3O
97.99795
RI: <1000*
Chemical
CAS: 75-90-1
42
55
69
83
M+
98
50
100
Cyclohexanone
PS
LM/Q
$C_6H_{10}O$
98.07317
RI: <1000*
Chemical
Precursor of
N: phencyclidine
and analogues
CAS: 108-94-1
41
56
70
83
M+
98
50
100
1,3-Dimethylcyclopentane
PS
LM/Q
C_7H_{14}
98.10955
RI: <1000*
Solvent
CAS: 1759-58-6
41
55
69
83
M+
98
50
100
2-Ethyl-3-methyl-1-butene
PS
LM/Q
C_7H_{14}
98.10955
RI: <1000*
Solvent
CAS: 7357-93-9
41
56
70
83
M+
98
50
100
5-Methyl-1-hexene
PS
LM/Q
C_7H_{14}
98.10955
RI: <1000*
Solvent
CAS: 3524-73-0

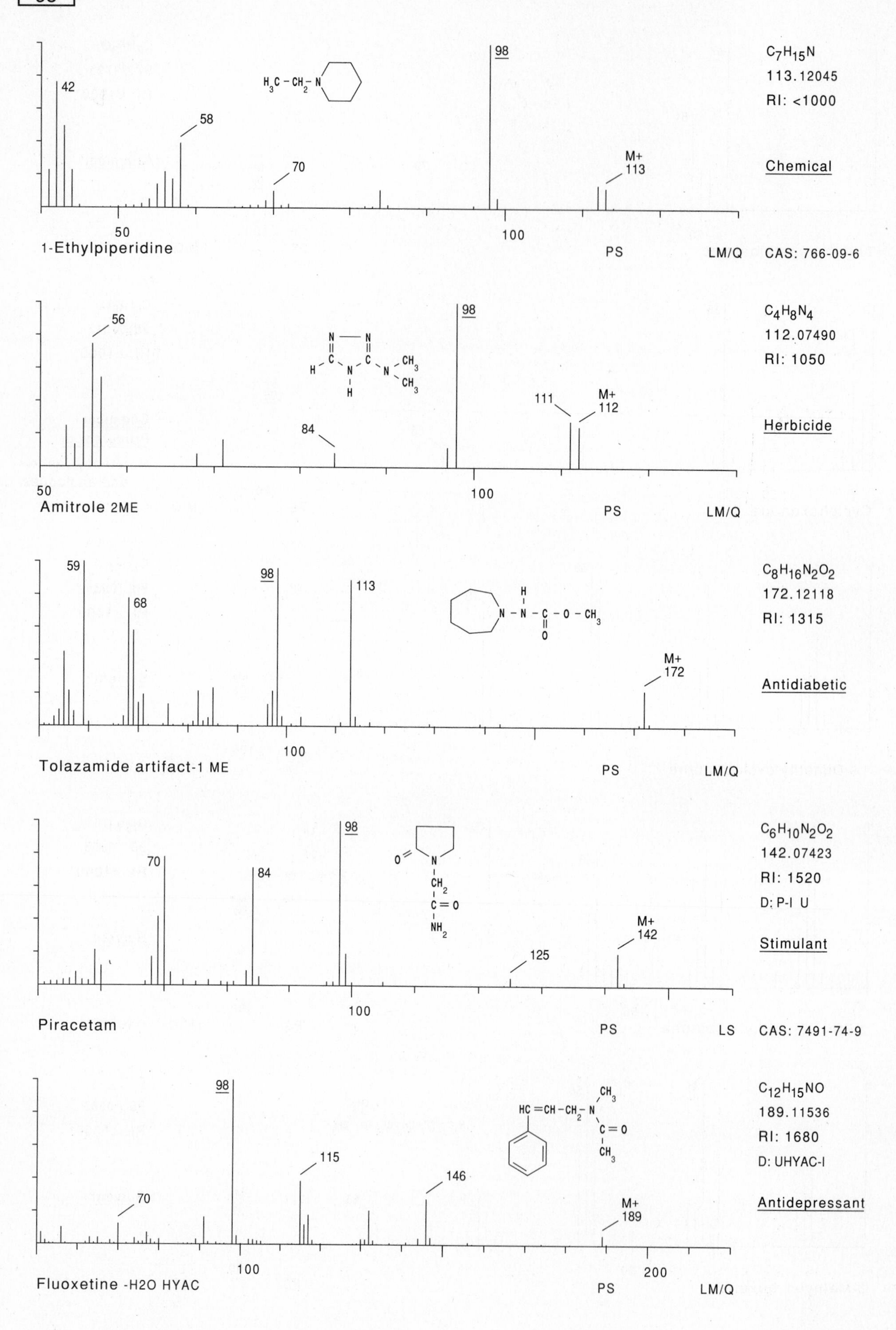

42
58
70
98
M+
113
50
100
1-Ethylpiperidine
PS
LM/Q
$C_7H_{15}N$
113.12045
RI: <1000
Chemical
CAS: 766-09-6
56
84
98
111
M+
112
50
100
Amitrole 2ME
PS
LM/Q
$C_4H_8N_4$
112.07490
RI: 1050
Herbicide
59
68
98
113
M+
172
100
Tolazamide artifact-1 ME
PS
LM/Q
$C_8H_{16}N_2O_2$
172.12118
RI: 1315
Antidiabetic
70
84
98
125
M+
142
100
Piracetam
PS
LS
$C_6H_{10}N_2O_2$
142.07423
RI: 1520
D: P-I U
Stimulant
CAS: 7491-74-9
70
98
115
146
M+
189
100
200
Fluoxetine -H2O HYAC
PS
LM/Q
$C_{12}H_{15}NO$
189.11536
RI: 1680
D: UHYAC-I
Antidepressant

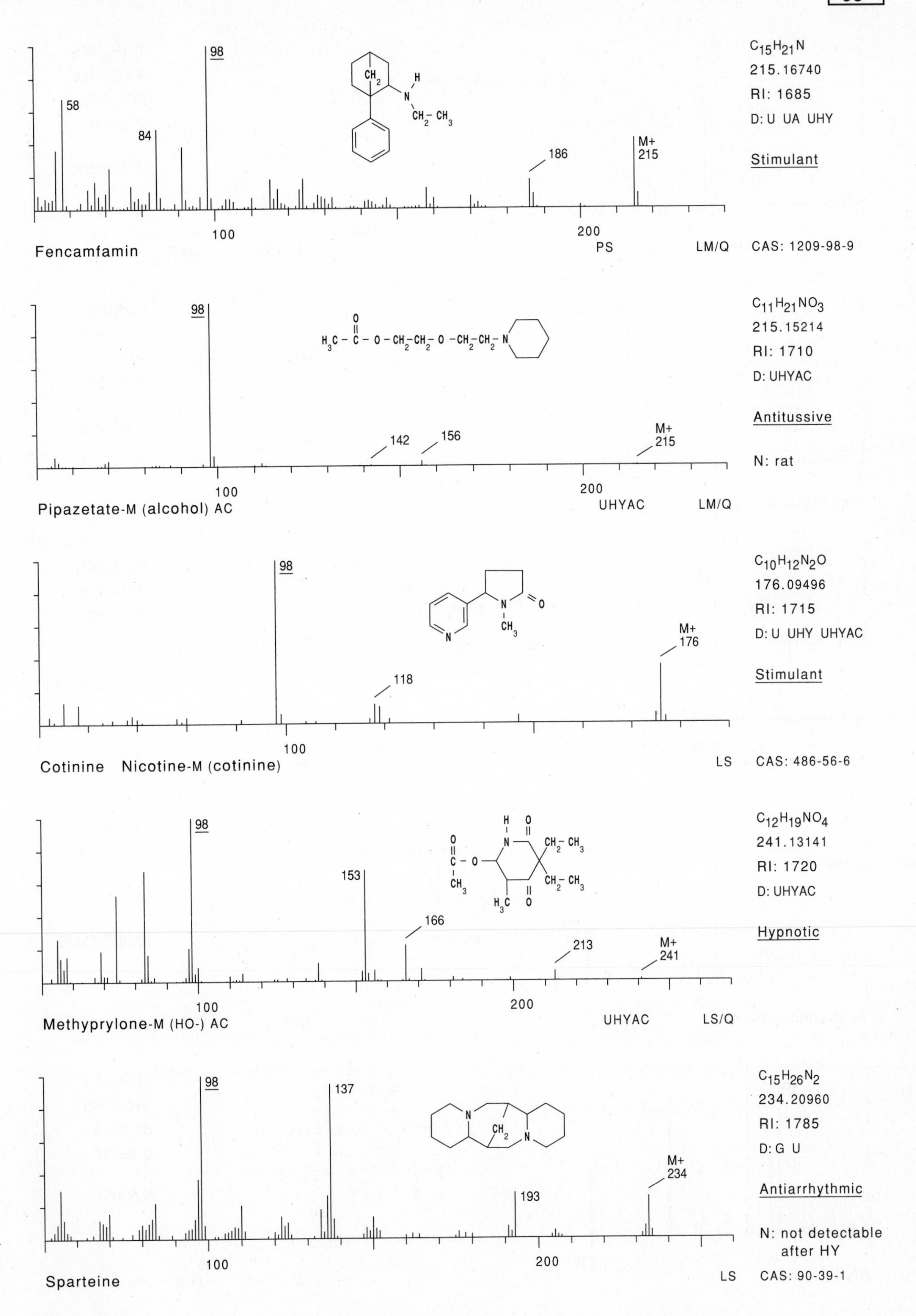

Fencamfamin

$C_{15}H_{21}N$
215.16740
RI: 1685
D: U UA UHY
Stimulant
CAS: 1209-98-9

Pipazetate-M (alcohol) AC

$C_{11}H_{21}NO_3$
215.15214
RI: 1710
D: UHYAC
Antitussive
N: rat

Cotinine Nicotine-M (cotinine)

$C_{10}H_{12}N_2O$
176.09496
RI: 1715
D: U UHY UHYAC
Stimulant
CAS: 486-56-6

Methyprylone-M (HO-) AC

$C_{12}H_{19}NO_4$
241.13141
RI: 1720
D: UHYAC
Hypnotic

Sparteine

$C_{15}H_{26}N_2$
234.20960
RI: 1785
D: G U
Antiarrhythmic
N: not detectable after HY
CAS: 90-39-1

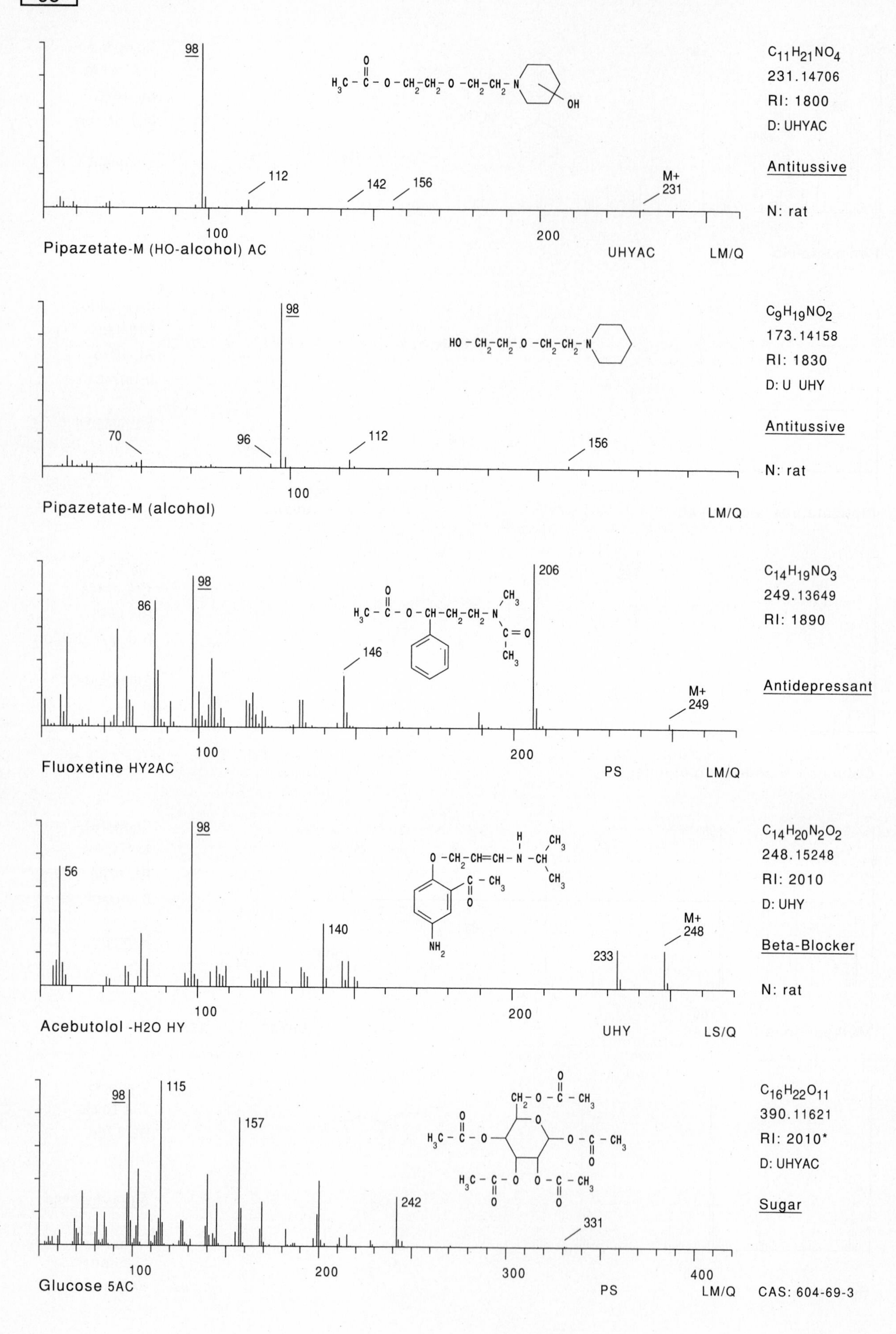
98
112
142
156
M+
231
100
200
Pipazetate-M (HO-alcohol) AC
UHYAC
LM/Q
$C_{11}H_{21}NO_4$
231.14706
RI: 1800
D: UHYAC
Antitussive
N: rat
98
70
96
112
156
100
Pipazetate-M (alcohol)
LM/Q
$C_9H_{19}NO_2$
173.14158
RI: 1830
D: U UHY
Antitussive
N: rat
206
98
86
146
M+
249
100
200
Fluoxetine HY2AC
PS
LM/Q
$C_{14}H_{19}NO_3$
249.13649
RI: 1890
Antidepressant
98
56
140
233
M+
248
100
200
Acebutolol -H2O HY
UHY
LS/Q
$C_{14}H_{20}N_2O_2$
248.15248
RI: 2010
D: UHY
Beta-Blocker
N: rat
98
115
157
242
331
100
200
300
400
Glucose 5AC
PS
LM/Q
$C_{16}H_{22}O_{11}$
390.11621
RI: 2010*
D: UHYAC
Sugar
CAS: 604-69-3

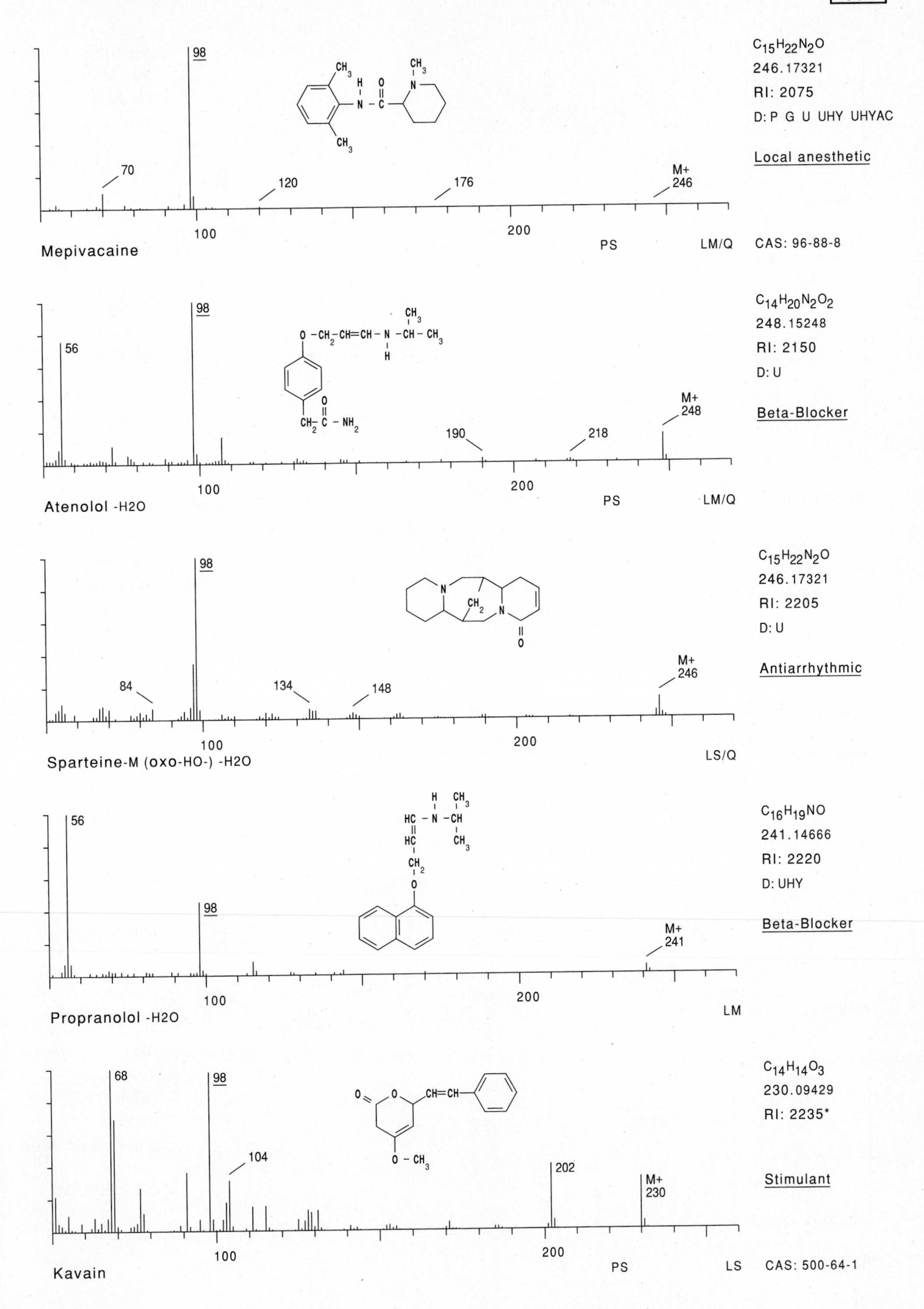
98
70
120
176
M+
246
100
200
PS
LM/Q
Mepivacaine
$C_{15}H_{22}N_2O$
246.17321
RI: 2075
D: P G U UHY UHYAC
Local anesthetic
CAS: 96-88-8
98
56
190
218
M+
248
100
200
PS
LM/Q
Atenolol -H2O
$C_{14}H_{20}N_2O_2$
248.15248
RI: 2150
D: U
Beta-Blocker
98
84
134
148
M+
246
100
200
LS/Q
Sparteine-M (oxo-HO-) -H2O
$C_{15}H_{22}N_2O$
246.17321
RI: 2205
D: U
Antiarrhythmic
56
98
M+
241
100
200
LM
Propranolol -H2O
$C_{16}H_{19}NO$
241.14666
RI: 2220
D: UHY
Beta-Blocker
68
98
104
202
M+
230
100
200
PS
LS
Kavain
$C_{14}H_{14}O_3$
230.09429
RI: 2235*
Stimulant
CAS: 500-64-1

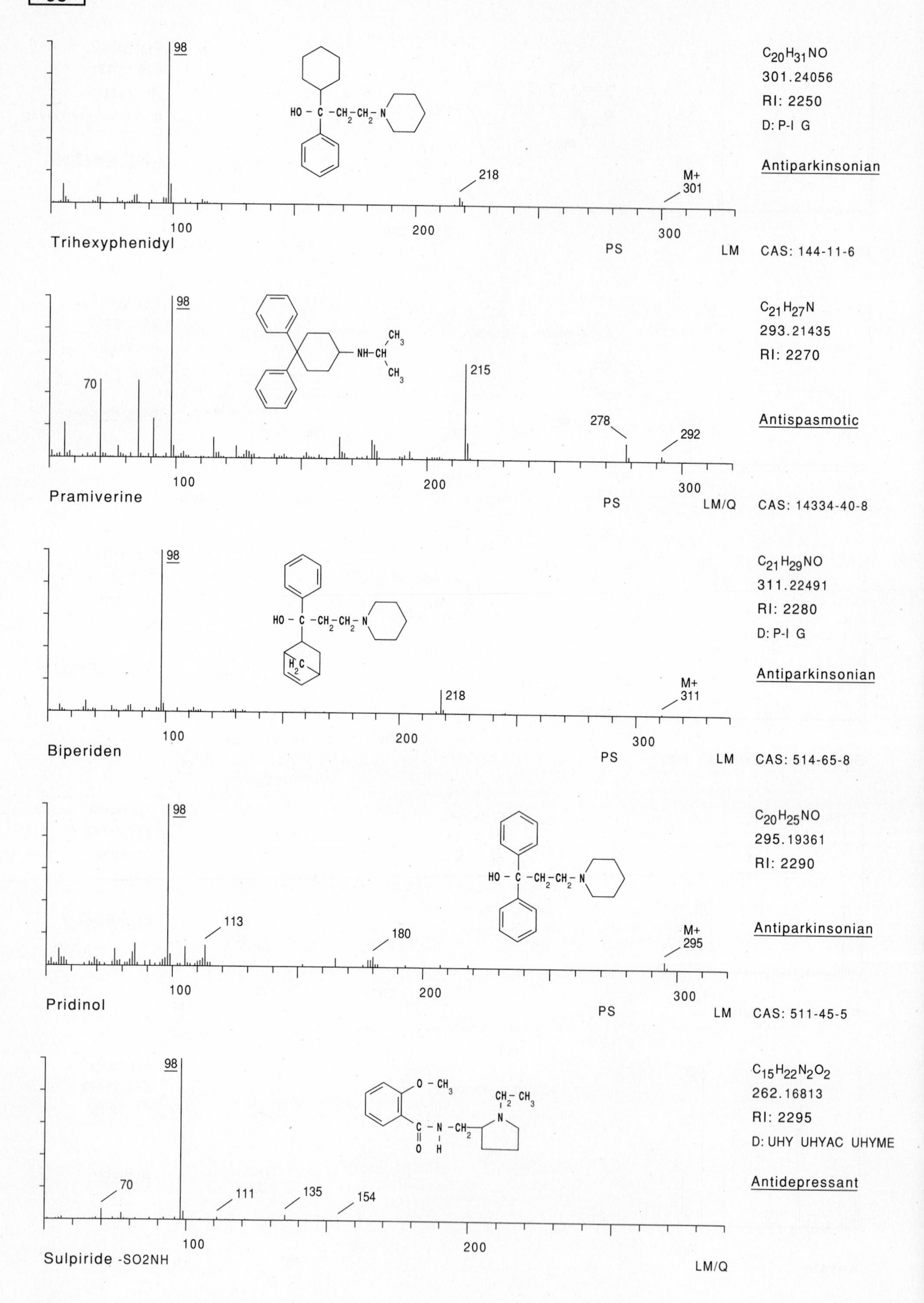
98
HO – C – CH2–CH2– N
218
M+
301
100
200
300
Trihexyphenidyl
PS
LM
C20H31NO
301.24056
RI: 2250
D: P-I G
Antiparkinsonian
CAS: 144-11-6
98
CH3
NH–CH
CH3
70
215
278
292
100
200
300
Pramiverine
PS
LM/Q
C21H27N
293.21435
RI: 2270
Antispasmotic
CAS: 14334-40-8
98
HO – C – CH2–CH2– N
H2C
218
M+
311
100
200
300
Biperiden
PS
LM
C21H29NO
311.22491
RI: 2280
D: P-I G
Antiparkinsonian
CAS: 514-65-8
98
HO – C – CH2–CH2– N
113
180
M+
295
100
200
300
Pridinol
PS
LM
C20H25NO
295.19361
RI: 2290
Antiparkinsonian
CAS: 511-45-5
98
O – CH3
CH2– CH3
N
C – N – CH2
O H
70
111
135
154
100
200
Sulpiride -SO2NH
LM/Q
C15H22N2O2
262.16813
RI: 2295
D: UHY UHYAC UHYME
Antidepressant

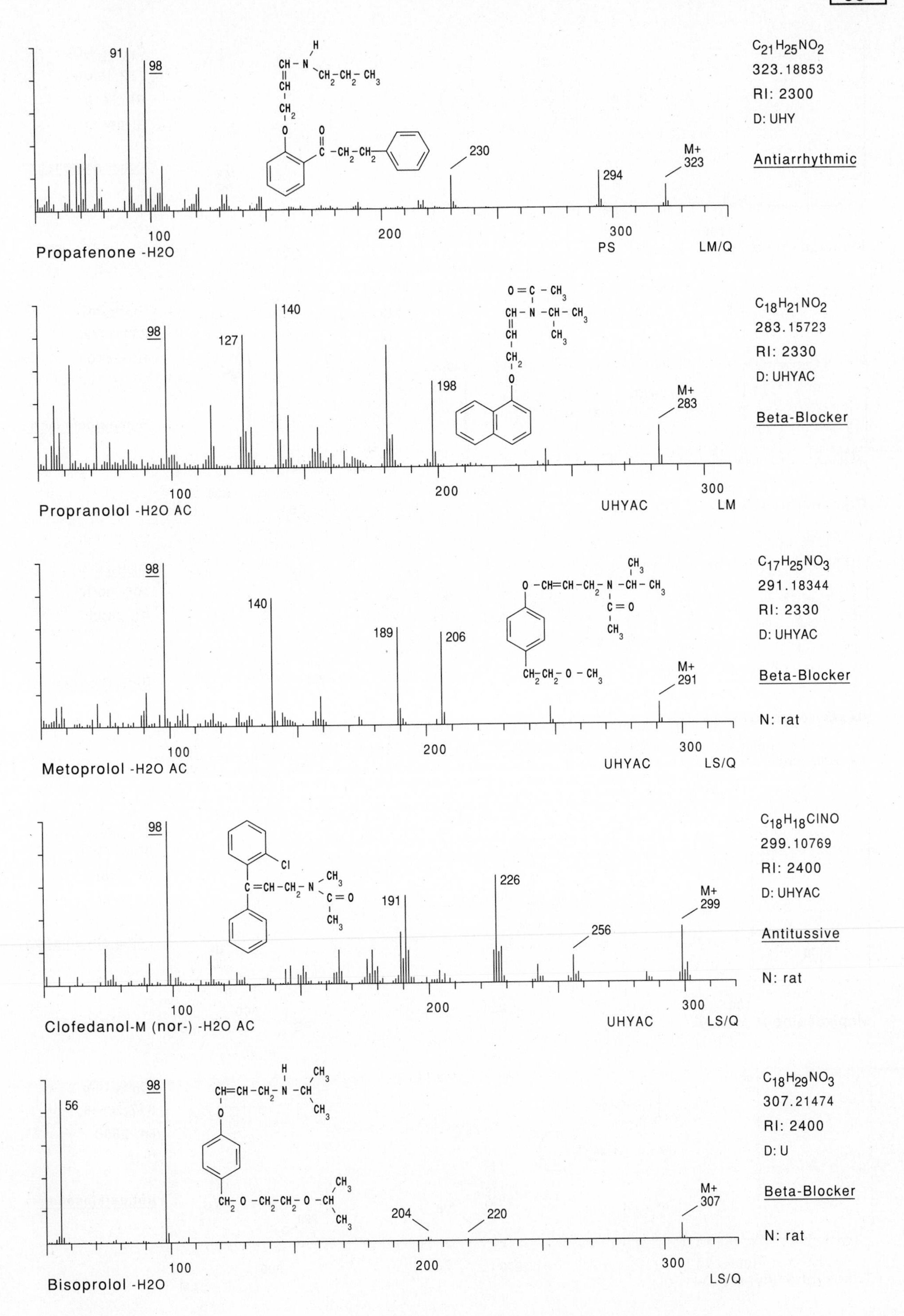

$C_{21}H_{25}NO_2$
323.18853
RI: 2300
D: UHY
Antiarrhythmic

$C_{18}H_{21}NO_2$
283.15723
RI: 2330
D: UHYAC
Beta-Blocker

$C_{17}H_{25}NO_3$
291.18344
RI: 2330
D: UHYAC
Beta-Blocker
N: rat

$C_{18}H_{18}ClNO$
299.10769
RI: 2400
D: UHYAC
Antitussive
N: rat

$C_{18}H_{29}NO_3$
307.21474
RI: 2400
D: U
Beta-Blocker
N: rat

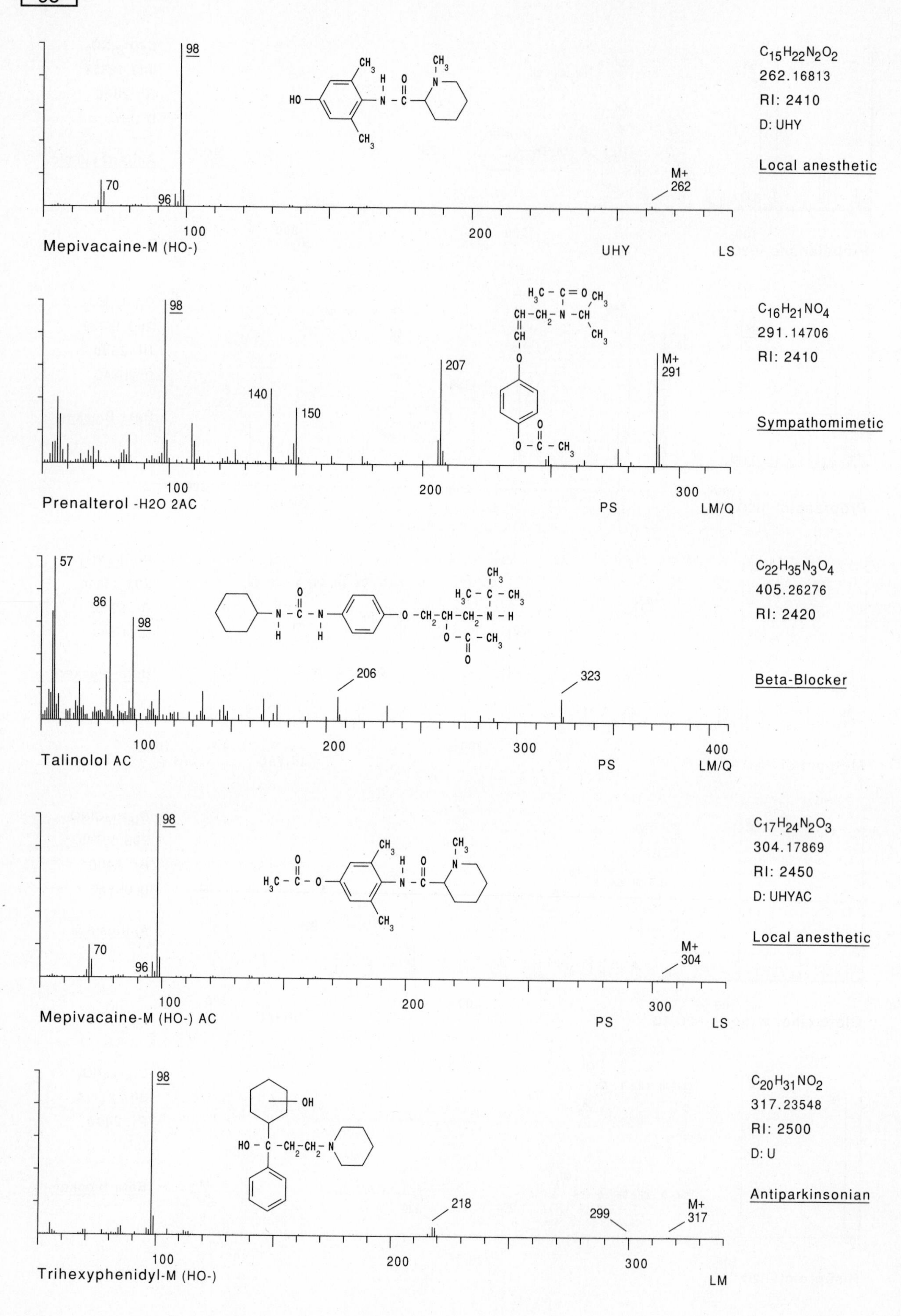
98
70
96
M+
262
100
200
Mepivacaine-M (HO-)
UHY
LS
$C_{15}H_{22}N_2O_2$
262.16813
RI: 2410
D: UHY
Local anesthetic
98
140
150
207
M+
291
100
200
300
Prenalterol -H2O 2AC
PS
LM/Q
$C_{16}H_{21}NO_4$
291.14706
RI: 2410
Sympathomimetic
57
86
98
206
323
100
200
300
400
Talinolol AC
PS
LM/Q
$C_{22}H_{35}N_3O_4$
405.26276
RI: 2420
Beta-Blocker
98
70
96
M+
304
100
200
300
Mepivacaine-M (HO-) AC
PS
LS
$C_{17}H_{24}N_2O_3$
304.17869
RI: 2450
D: UHYAC
Local anesthetic
98
218
299
M+
317
100
200
300
Trihexyphenidyl-M (HO-)
LM
$C_{20}H_{31}NO_2$
317.23548
RI: 2500
D: U
Antiparkinsonian

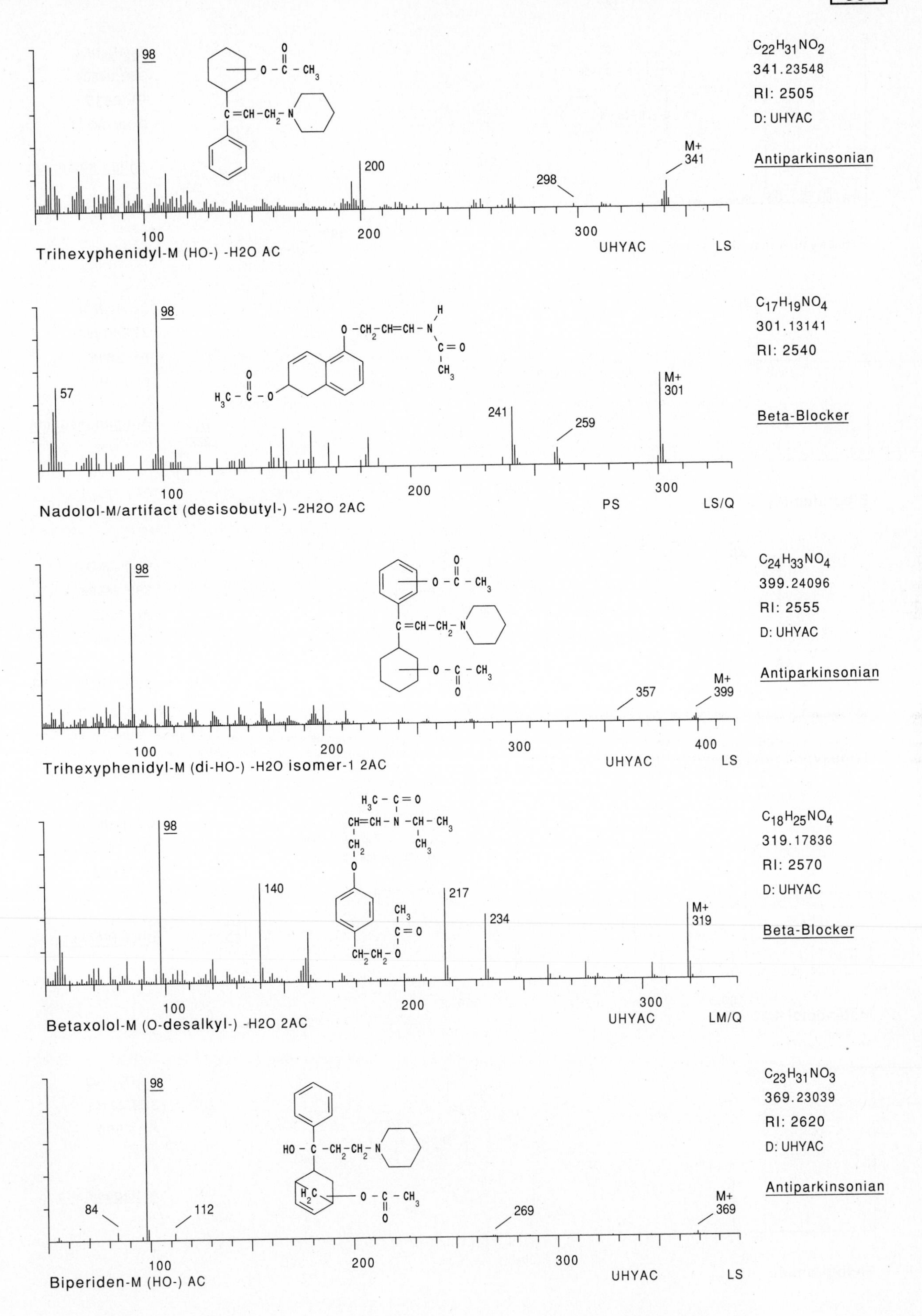

$C_{22}H_{31}NO_2$
341.23548
RI: 2505
D: UHYAC
Antiparkinsonian

$C_{17}H_{19}NO_4$
301.13141
RI: 2540
Beta-Blocker

$C_{24}H_{33}NO_4$
399.24096
RI: 2555
D: UHYAC
Antiparkinsonian

$C_{18}H_{25}NO_4$
319.17836
RI: 2570
D: UHYAC
Beta-Blocker

$C_{23}H_{31}NO_3$
369.23039
RI: 2620
D: UHYAC
Antiparkinsonian

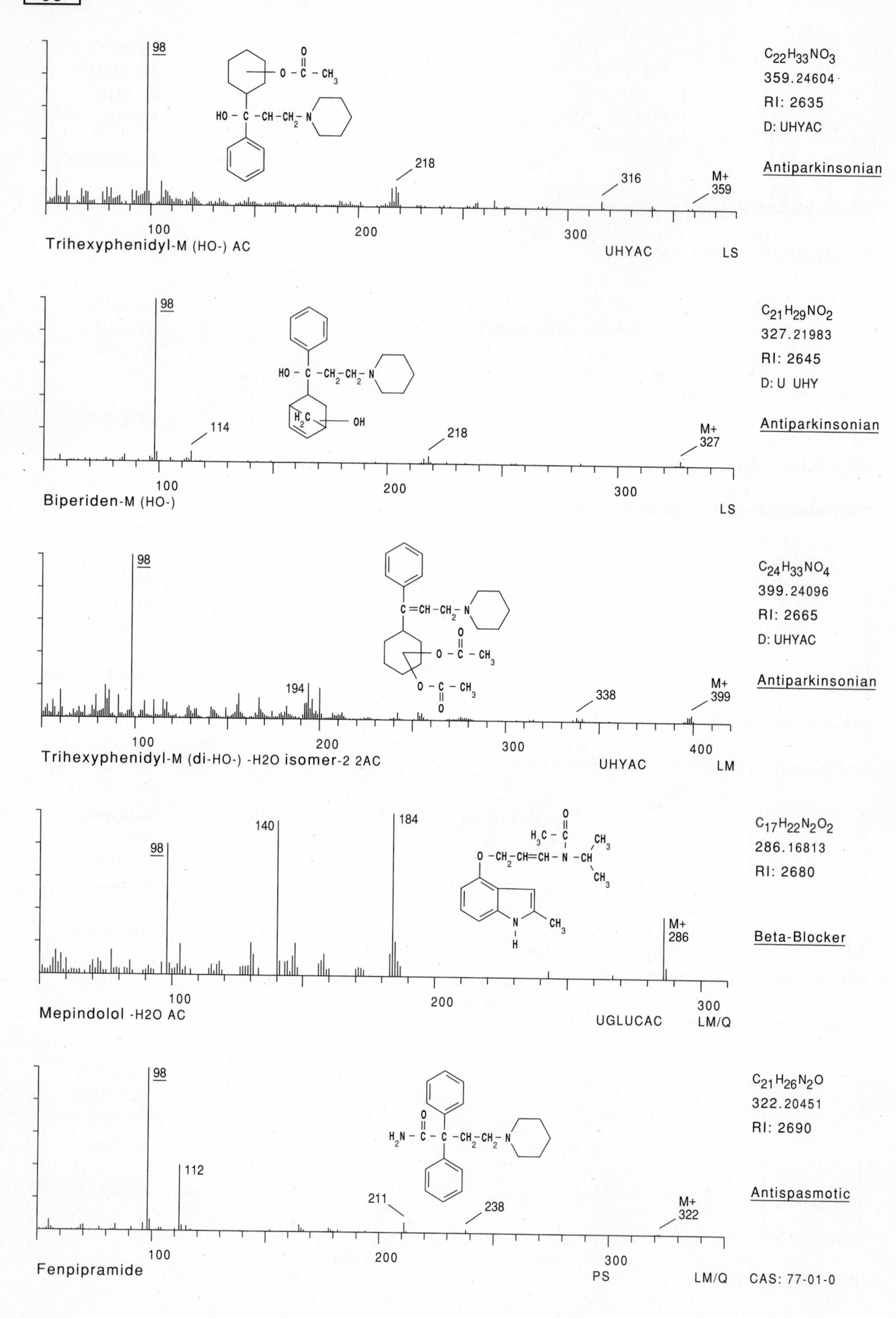

98
218
316
M+
359
100
200
300
Trihexyphenidyl-M (HO-) AC
UHYAC
LS
C22H33NO3
359.24604
RI: 2635
D: UHYAC
Antiparkinsonian
98
114
218
M+
327
Biperiden-M (HO-)
LS
C21H29NO2
327.21983
RI: 2645
D: U UHY
Antiparkinsonian
98
194
338
M+
399
400
Trihexyphenidyl-M (di-HO-) -H2O isomer-2 2AC
UHYAC
LM
C24H33NO4
399.24096
RI: 2665
D: UHYAC
Antiparkinsonian
98
140
184
M+
286
Mepindolol -H2O AC
UGLUCAC
LM/Q
C17H22N2O2
286.16813
RI: 2680
Beta-Blocker
98
112
211
238
M+
322
Fenpipramide
PS
LM/Q
C21H26N2O
322.20451
RI: 2690
Antispasmotic
CAS: 77-01-0

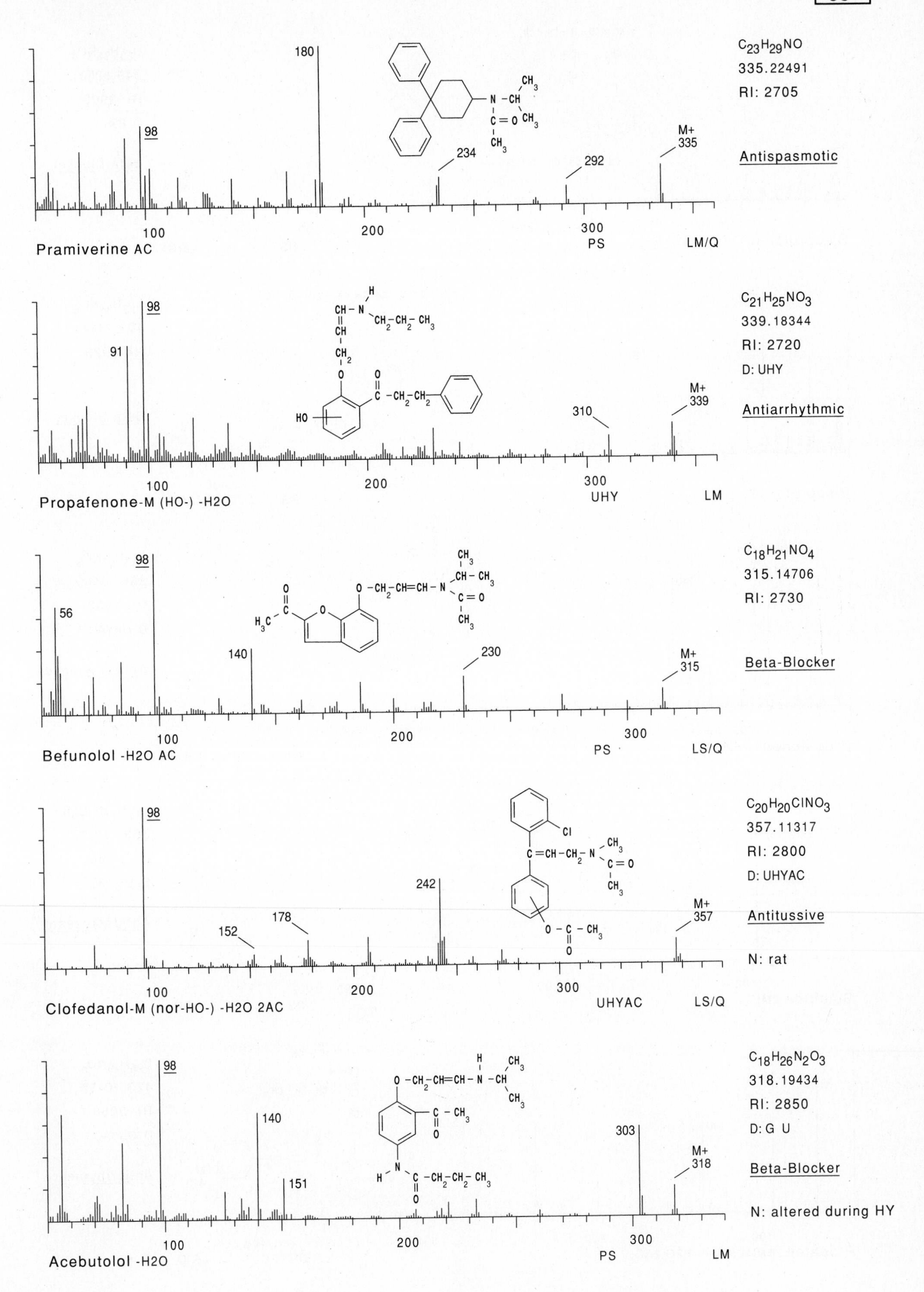

180
98
234
292
M+
335
100
200
300
PS
LM/Q
Pramiverine AC
C23H29NO
335.22491
RI: 2705
Antispasmodic
98
91
310
M+
339
100
200
300
UHY
LM
Propafenone-M (HO-) -H2O
C21H25NO3
339.18344
RI: 2720
D: UHY
Antiarrhythmic
98
56
140
230
M+
315
100
200
300
PS
LS/Q
Befunolol -H2O AC
C18H21NO4
315.14706
RI: 2730
Beta-Blocker
98
242
178
152
M+
357
100
200
300
UHYAC
LS/Q
Clofedanol-M (nor-HO-) -H2O 2AC
C20H20ClNO3
357.11317
RI: 2800
D: UHYAC
Antitussive
N: rat
98
140
151
303
M+
318
100
200
300
PS
LM
Acebutolol -H2O
C18H26N2O3
318.19434
RI: 2850
D: G U
Beta-Blocker
N: altered during HY

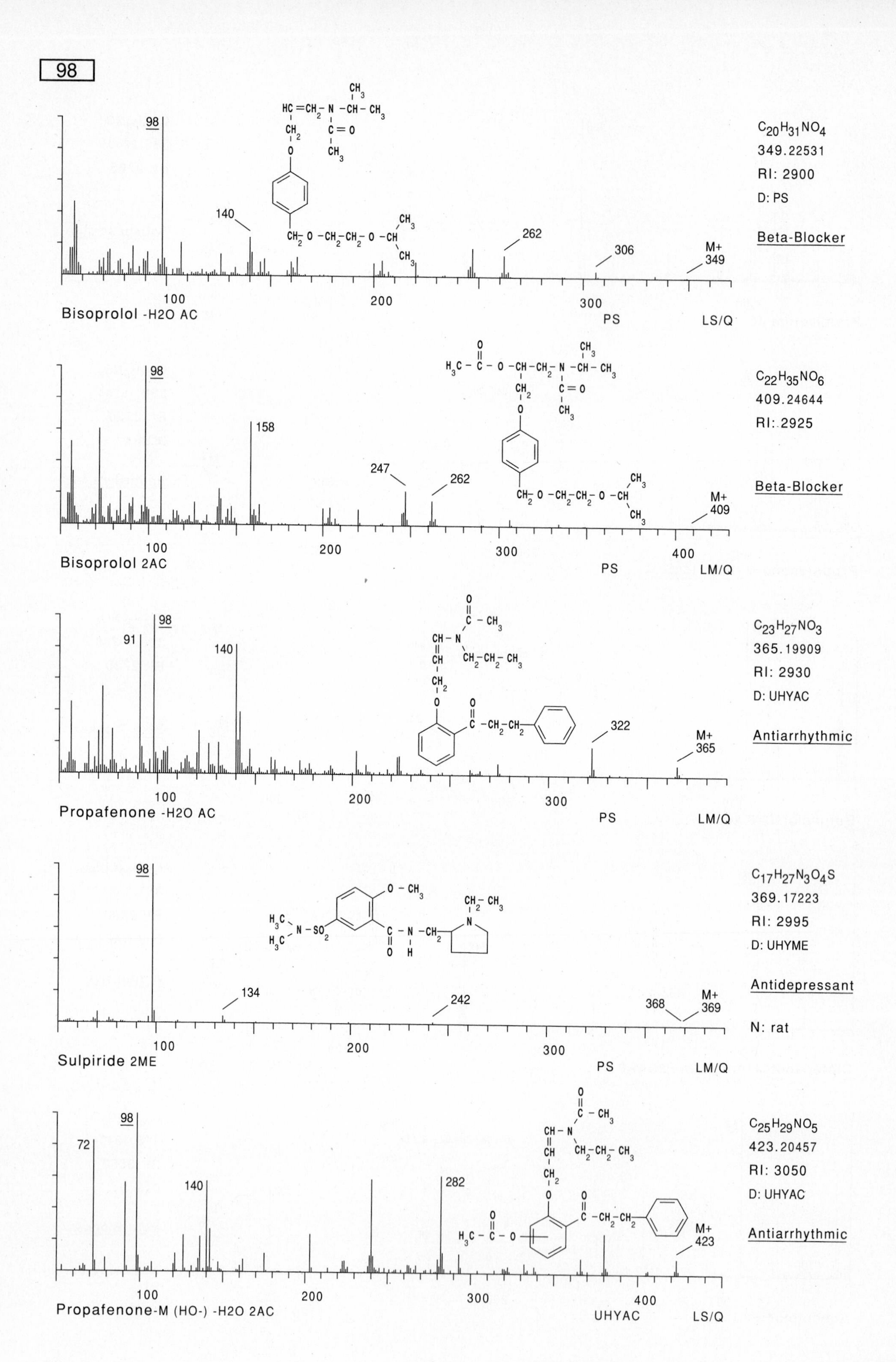
98
140
262
306
M+
349
100
200
300
Bisoprolol -H2O AC
PS
LS/Q
$C_{20}H_{31}NO_4$
349.22531
RI: 2900
D: PS
Beta-Blocker
98
158
247
262
M+
409
100
200
300
400
Bisoprolol 2AC
PS
LM/Q
$C_{22}H_{35}NO_6$
409.24644
RI: 2925
Beta-Blocker
98
91
140
322
M+
365
100
200
300
Propafenone -H2O AC
PS
LM/Q
$C_{23}H_{27}NO_3$
365.19909
RI: 2930
D: UHYAC
Antiarrhythmic
98
134
242
368
M+
369
100
200
300
Sulpiride 2ME
PS
LM/Q
$C_{17}H_{27}N_3O_4S$
369.17223
RI: 2995
D: UHYME
Antidepressant
N: rat
98
72
140
282
M+
423
100
200
300
400
Propafenone-M (HO-) -H2O 2AC
UHYAC
LS/Q
$C_{25}H_{29}NO_5$
423.20457
RI: 3050
D: UHYAC
Antiarrhythmic

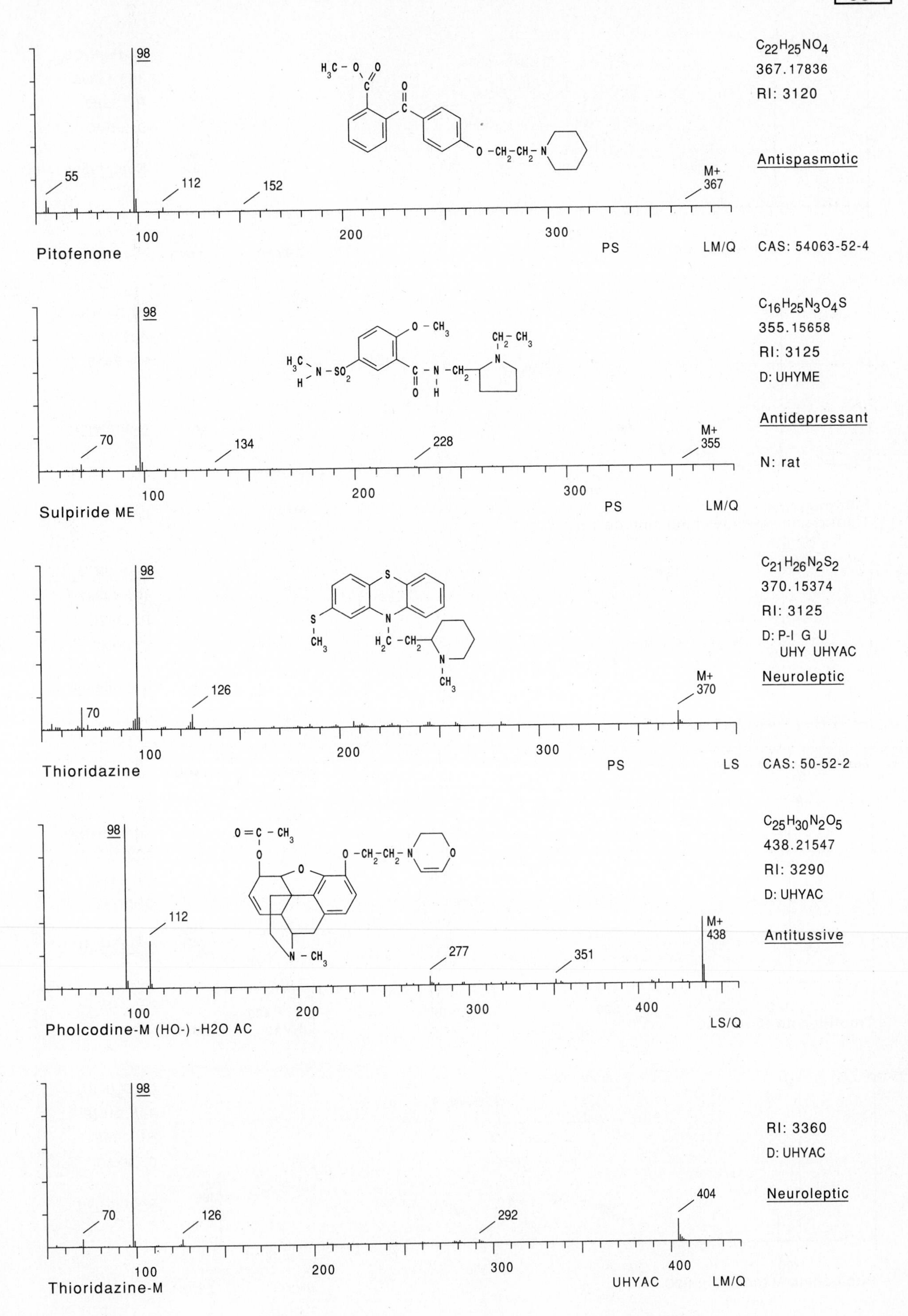

$C_{22}H_{25}NO_4$
367.17836
RI: 3120
Antispasmotic
CAS: 54063-52-4

$C_{16}H_{25}N_3O_4S$
355.15658
RI: 3125
D: UHYME
Antidepressant
N: rat

$C_{21}H_{26}N_2S_2$
370.15374
RI: 3125
D: P-I G U
UHY UHYAC
Neuroleptic
CAS: 50-52-2

$C_{25}H_{30}N_2O_5$
438.21547
RI: 3290
D: UHYAC
Antitussive

RI: 3360
D: UHYAC
Neuroleptic

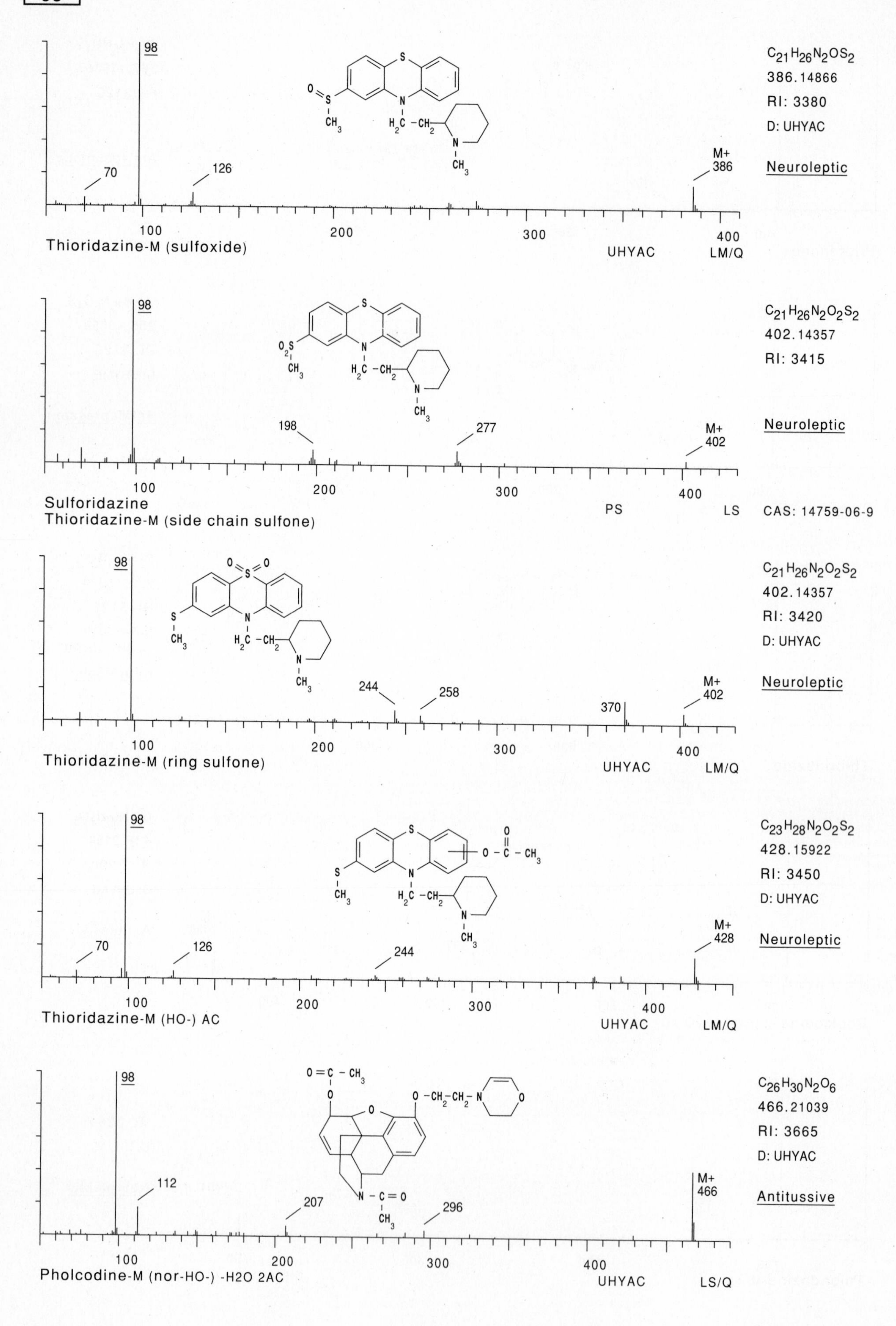
98
70
126
M+
386
100
200
300
400
Thioridazine-M (sulfoxide)
UHYAC
LM/Q
C21H26N2OS2
386.14866
RI: 3380
D: UHYAC
Neuroleptic
98
198
277
M+
402
Sulforidazine
Thioridazine-M (side chain sulfone)
PS
LS
C21H26N2O2S2
402.14357
RI: 3415
Neuroleptic
CAS: 14759-06-9
98
244
258
370
M+
402
Thioridazine-M (ring sulfone)
UHYAC
LM/Q
C21H26N2O2S2
402.14357
RI: 3420
D: UHYAC
Neuroleptic
98
70
126
244
M+
428
Thioridazine-M (HO-) AC
UHYAC
LM/Q
C23H28N2O2S2
428.15922
RI: 3450
D: UHYAC
Neuroleptic
98
112
207
296
M+
466
Pholcodine-M (nor-HO-) -H2O 2AC
UHYAC
LS/Q
C26H30N2O6
466.21039
RI: 3665
D: UHYAC
Antitussive

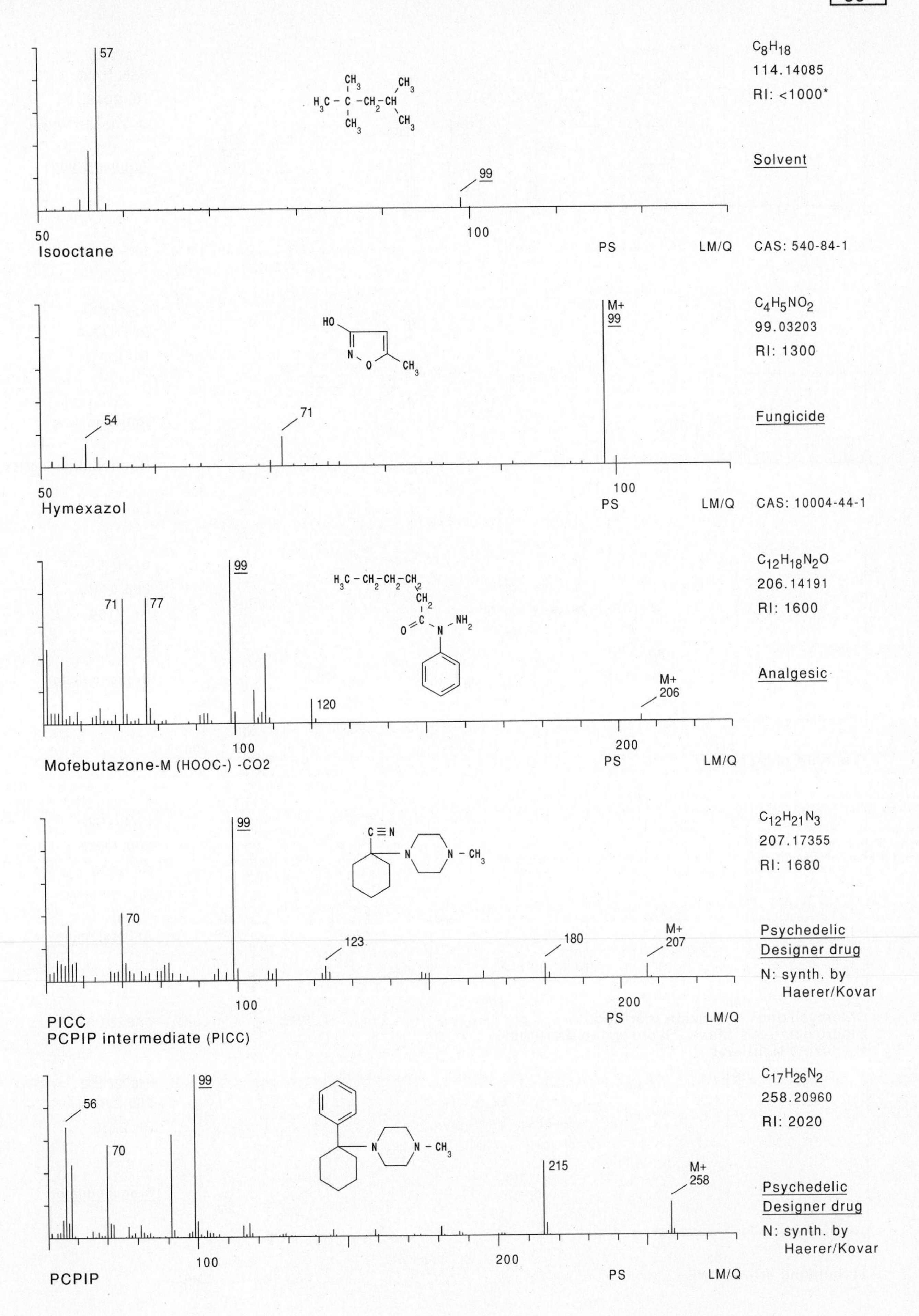

C_8H_{18}
114.14085
RI: <1000*

Solvent

CAS: 540-84-1

$C_4H_5NO_2$
99.03203
RI: 1300

Fungicide

CAS: 10004-44-1

$C_{12}H_{18}N_2O$
206.14191
RI: 1600

Analgesic

$C_{12}H_{21}N_3$
207.17355
RI: 1680

Psychedelic
Designer drug
N: synth. by Haerer/Kovar

$C_{17}H_{26}N_2$
258.20960
RI: 2020

Psychedelic
Designer drug
N: synth. by Haerer/Kovar

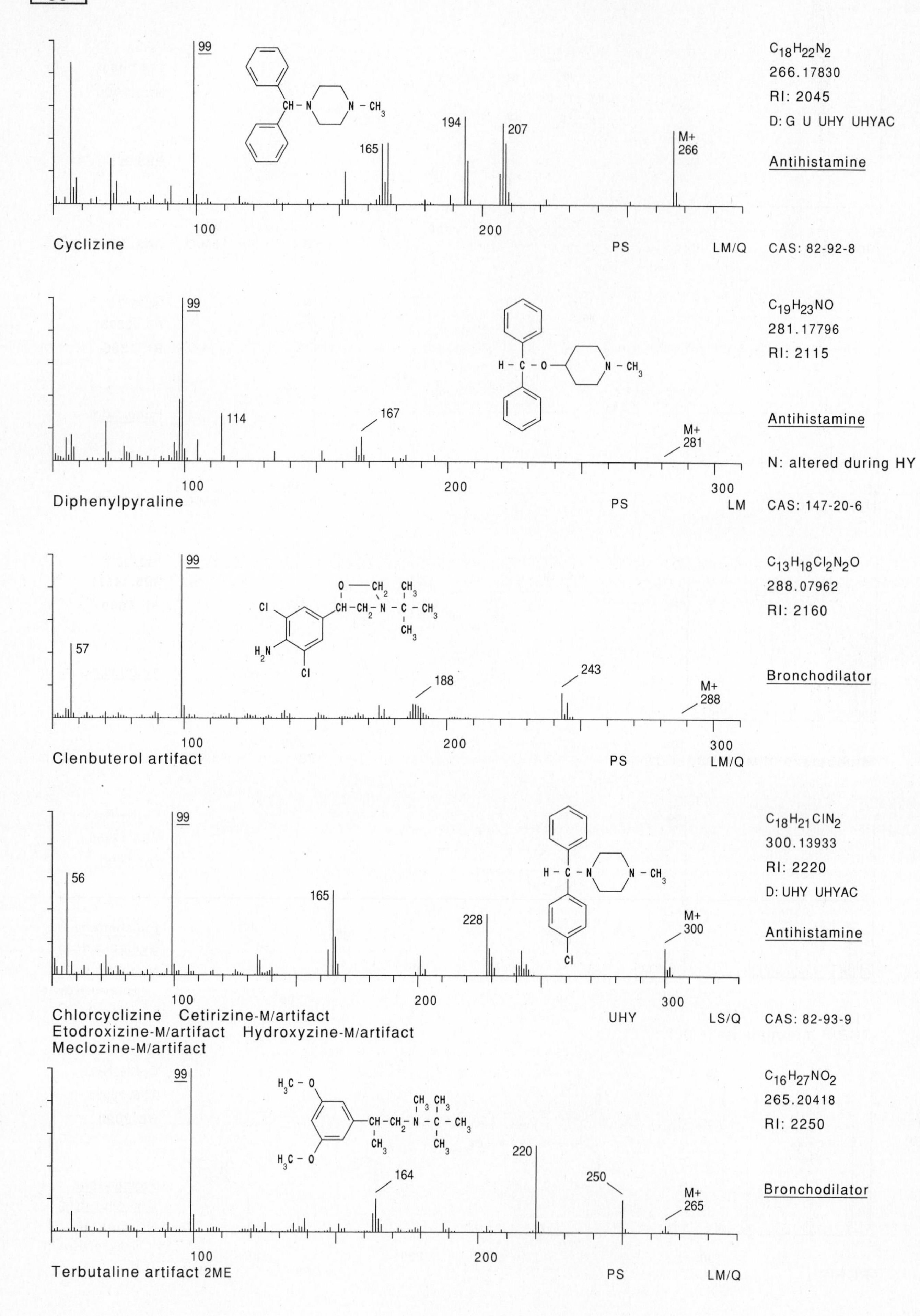

99
CH - N
N - CH3
194
207
165
M+
266
100
200
Cyclizine
PS
LM/Q
$C_{18}H_{22}N_2$
266.17830
RI: 2045
D: G U UHY UHYAC
Antihistamine
CAS: 82-92-8
99
H - C - O
N - CH3
114
167
M+
281
100
200
300
Diphenylpyraline
PS
LM
$C_{19}H_{23}NO$
281.17796
RI: 2115
Antihistamine
N: altered during HY
CAS: 147-20-6
99
57
188
243
M+
288
100
200
300
Clenbuterol artifact
PS
LM/Q
$C_{13}H_{18}Cl_2N_2O$
288.07962
RI: 2160
Bronchodilator
99
56
165
228
M+
300
100
200
300
Chlorcyclizine Cetirizine-M/artifact
Etodroxizine-M/artifact Hydroxyzine-M/artifact
Meclozine-M/artifact
UHY
LS/Q
$C_{18}H_{21}ClN_2$
300.13933
RI: 2220
D: UHY UHYAC
Antihistamine
CAS: 82-93-9
99
164
220
250
M+
265
100
200
Terbutaline artifact 2ME
PS
LM/Q
$C_{16}H_{27}NO_2$
265.20418
RI: 2250
Bronchodilator

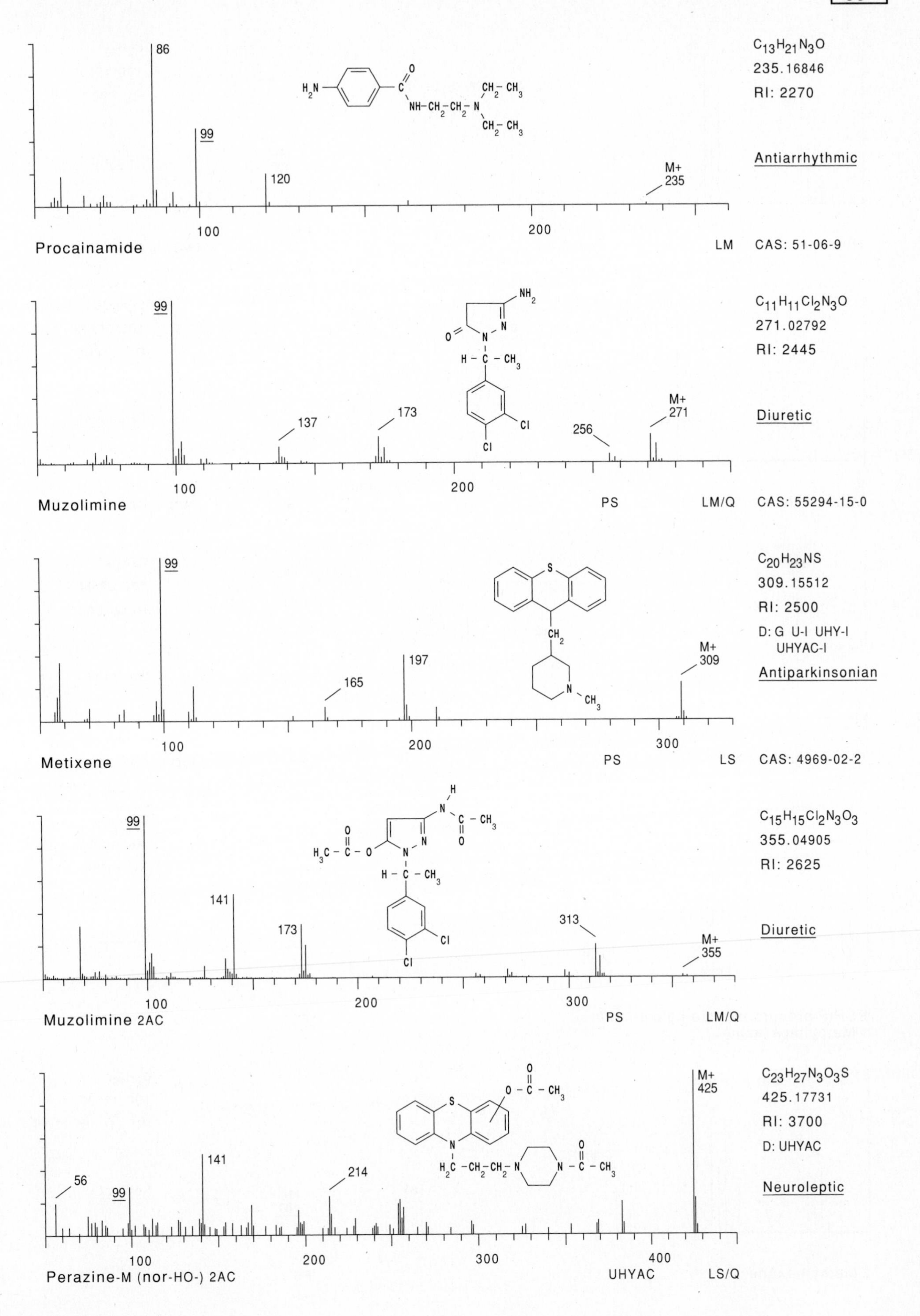
86
99
120
M+
235
100
200
Procainamide
LM
$C_{13}H_{21}N_3O$
235.16846
RI: 2270
Antiarrhythmic
CAS: 51-06-9
99
137
173
256
M+
271
Cl
Cl
100
200
Muzolimine
PS
LM/Q
$C_{11}H_{11}Cl_2N_3O$
271.02792
RI: 2445
Diuretic
CAS: 55294-15-0
99
165
197
M+
309
100
200
300
Metixene
PS
LS
$C_{20}H_{23}NS$
309.15512
RI: 2500
D: G U-I UHY-I
UHYAC-I
Antiparkinsonian
CAS: 4969-02-2
99
141
173
313
M+
355
Cl
Cl
100
200
300
Muzolimine 2AC
PS
LM/Q
$C_{15}H_{15}Cl_2N_3O_3$
355.04905
RI: 2625
Diuretic
56
99
141
214
M+
425
100
200
300
400
Perazine-M (nor-HO-) 2AC
UHYAC
LS/Q
$C_{23}H_{27}N_3O_3S$
425.17731
RI: 3700
D: UHYAC
Neuroleptic

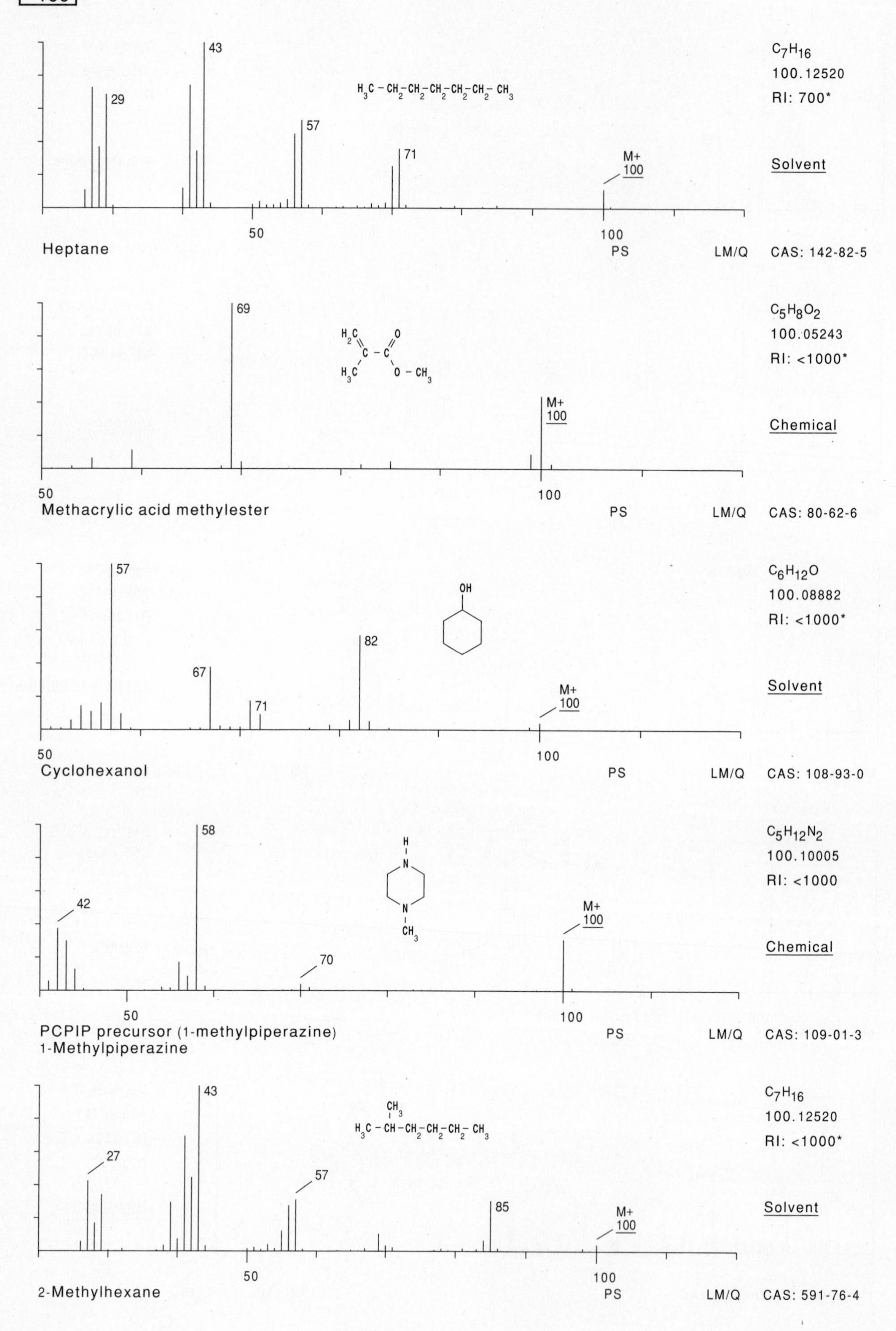

43
29
57
71
M+
100
50
100
PS
LM/Q
$H_3C-CH_2-CH_2-CH_2-CH_2-CH_3$
C7H16
100.12520
RI: 700*
Solvent
Heptane
CAS: 142-82-5
69
M+
100
50
100
PS
LM/Q
C5H8O2
100.05243
RI: <1000*
Chemical
Methacrylic acid methylester
CAS: 80-62-6
57
82
67
71
M+
100
50
100
PS
LM/Q
OH
C6H12O
100.08882
RI: <1000*
Solvent
Cyclohexanol
CAS: 108-93-0
58
42
70
M+
100
50
100
PS
LM/Q
C5H12N2
100.10005
RI: <1000
Chemical
PCPIP precursor (1-methylpiperazine)
1-Methylpiperazine
CAS: 109-01-3
43
27
57
85
M+
100
50
100
PS
LM/Q
C7H16
100.12520
RI: <1000*
Solvent
2-Methylhexane
CAS: 591-76-4

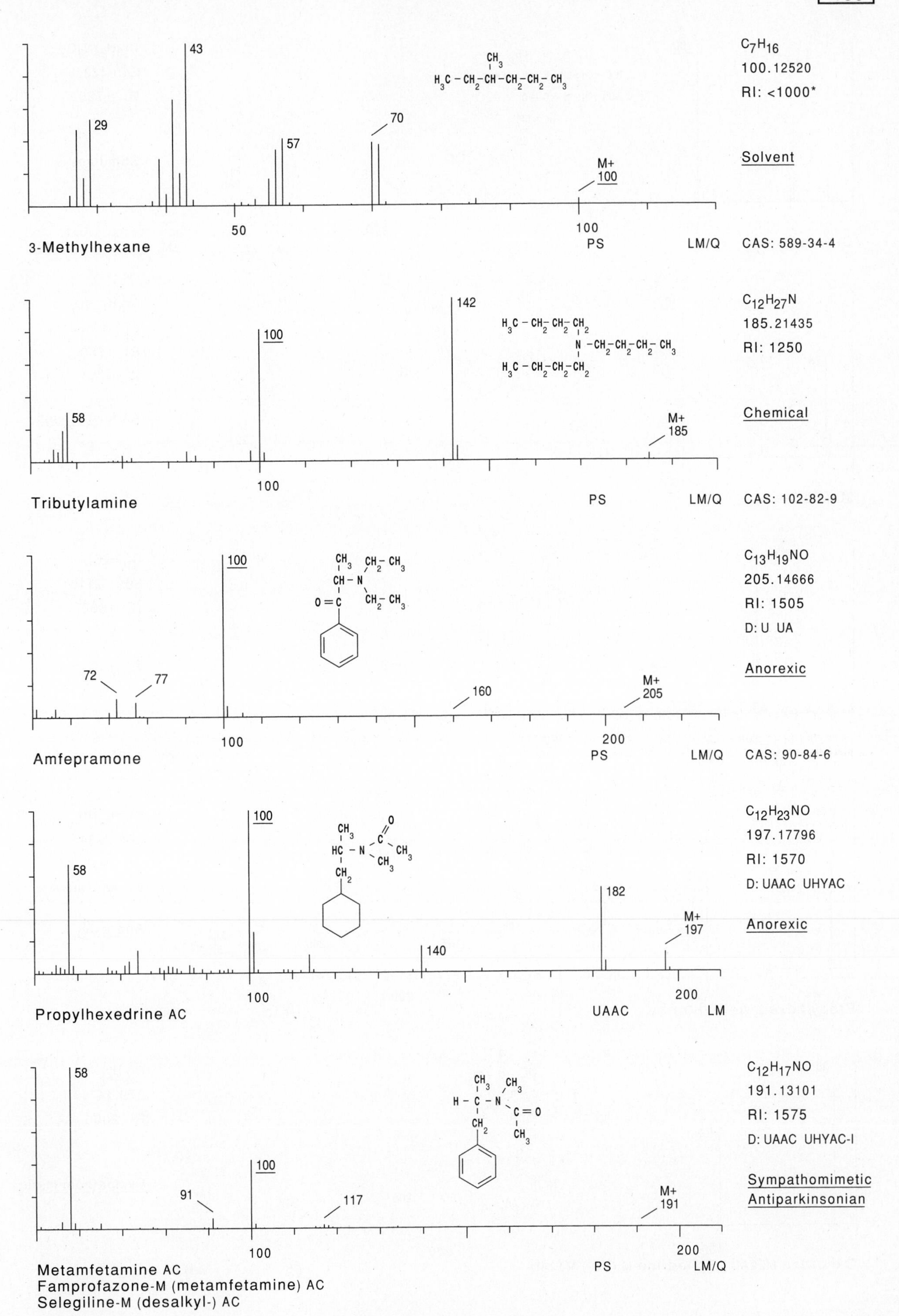
43
29
57
70
M+
100
50
100
PS
LM/Q
3-Methylhexane
C7H16
100.12520
RI: <1000*
Solvent
CAS: 589-34-4
142
100
58
M+
185
100
PS
LM/Q
Tributylamine
C12H27N
185.21435
RI: 1250
Chemical
CAS: 102-82-9
100
72
77
160
M+
205
100
200
PS
LM/Q
Amfepramone
C13H19NO
205.14666
RI: 1505
D: U UA
Anorexic
CAS: 90-84-6
100
58
182
M+
197
140
100
200
UAAC
LM
Propylhexedrine AC
C12H23NO
197.17796
RI: 1570
D: UAAC UHYAC
Anorexic
58
100
91
117
M+
191
100
200
PS
LM/Q
Metamfetamine AC
Famprofazone-M (metamfetamine) AC
Selegiline-M (desalkyl-) AC
C12H17NO
191.13101
RI: 1575
D: UAAC UHYAC-I
Sympathomimetic
Antiparkinsonian

100

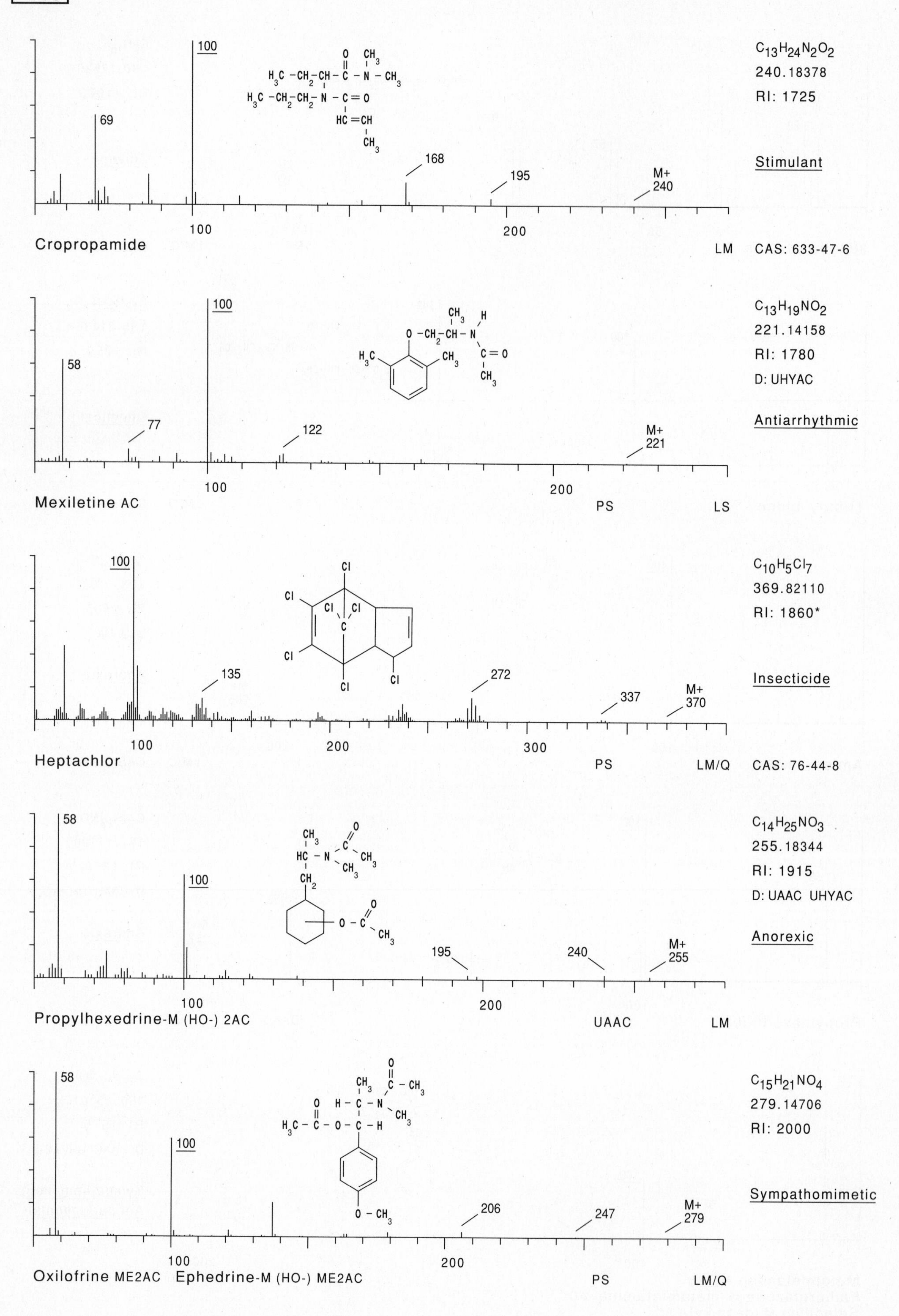

100
69
168
195
M+
240
100
200
Cropropamide
LM
$C_{13}H_{24}N_2O_2$
240.18378
RI: 1725
Stimulant
CAS: 633-47-6
100
58
77
122
M+
221
100
200
Mexiletine AC
PS
LS
$C_{13}H_{19}NO_2$
221.14158
RI: 1780
D: UHYAC
Antiarrhythmic
100
135
272
337
M+
370
100
200
300
Heptachlor
PS
LM/Q
$C_{10}H_5Cl_7$
369.82110
RI: 1860*
Insecticide
CAS: 76-44-8
58
100
195
240
M+
255
100
200
Propylhexedrine-M (HO-) 2AC
UAAC
LM
$C_{14}H_{25}NO_3$
255.18344
RI: 1915
D: UAAC UHYAC
Anorexic
58
100
206
247
M+
279
100
200
Oxilofrine ME2AC Ephedrine-M (HO-) ME2AC
PS
LM/Q
$C_{15}H_{21}NO_4$
279.14706
RI: 2000
Sympathomimetic

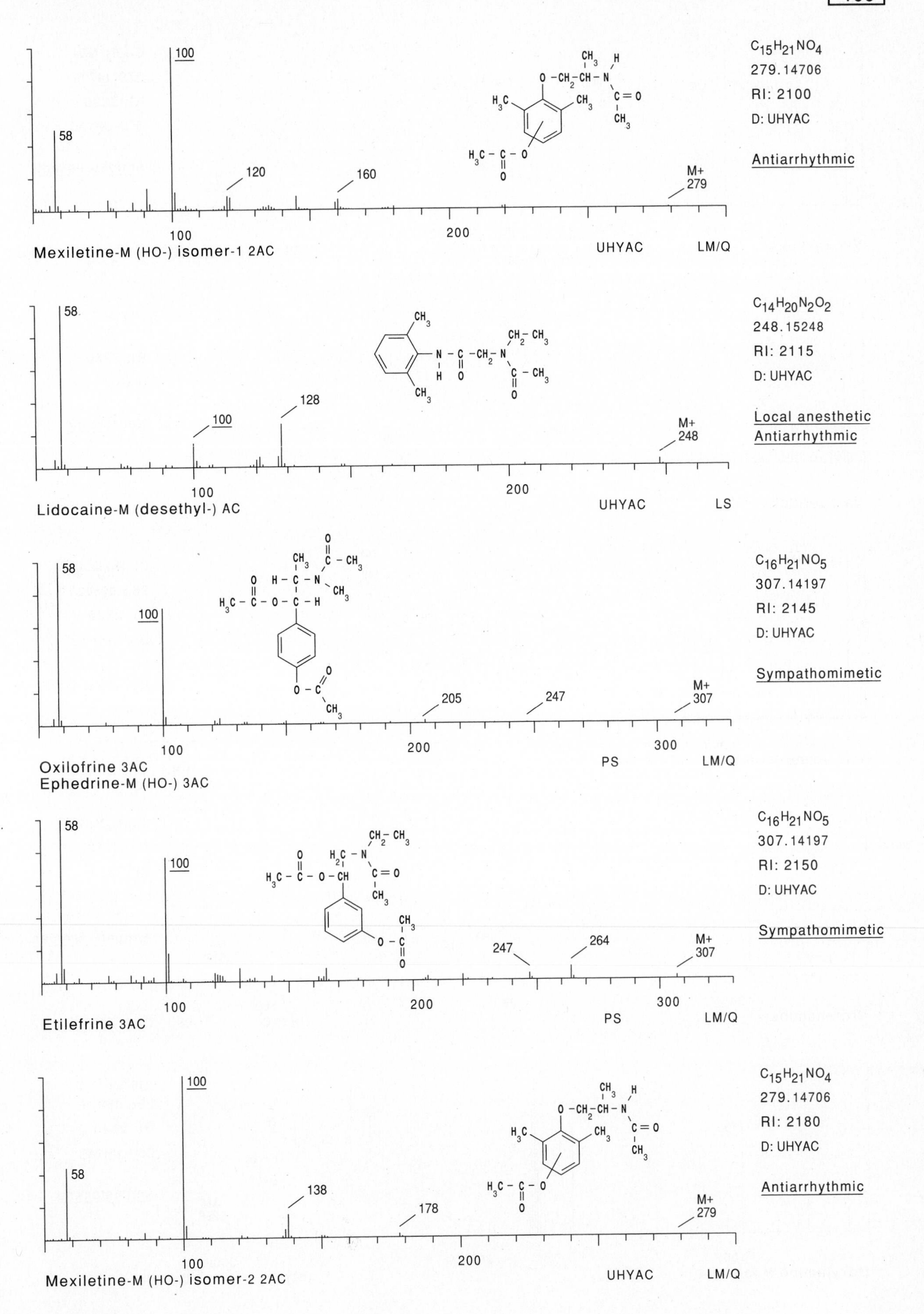

Mexiletine-M (HO-) isomer-1 2AC
UHYAC LM/Q
$C_{15}H_{21}NO_4$
279.14706
RI: 2100
D: UHYAC
Antiarrhythmic
Lidocaine-M (desethyl-) AC
UHYAC LS
$C_{14}H_{20}N_2O_2$
248.15248
RI: 2115
D: UHYAC
Local anesthetic
Antiarrhythmic
Oxilofrine 3AC
Ephedrine-M (HO-) 3AC
PS LM/Q
$C_{16}H_{21}NO_5$
307.14197
RI: 2145
D: UHYAC
Sympathomimetic
Etilefrine 3AC
PS LM/Q
$C_{16}H_{21}NO_5$
307.14197
RI: 2150
D: UHYAC
Sympathomimetic
Mexiletine-M (HO-) isomer-2 2AC
UHYAC LM/Q
$C_{15}H_{21}NO_4$
279.14706
RI: 2180
D: UHYAC
Antiarrhythmic

100

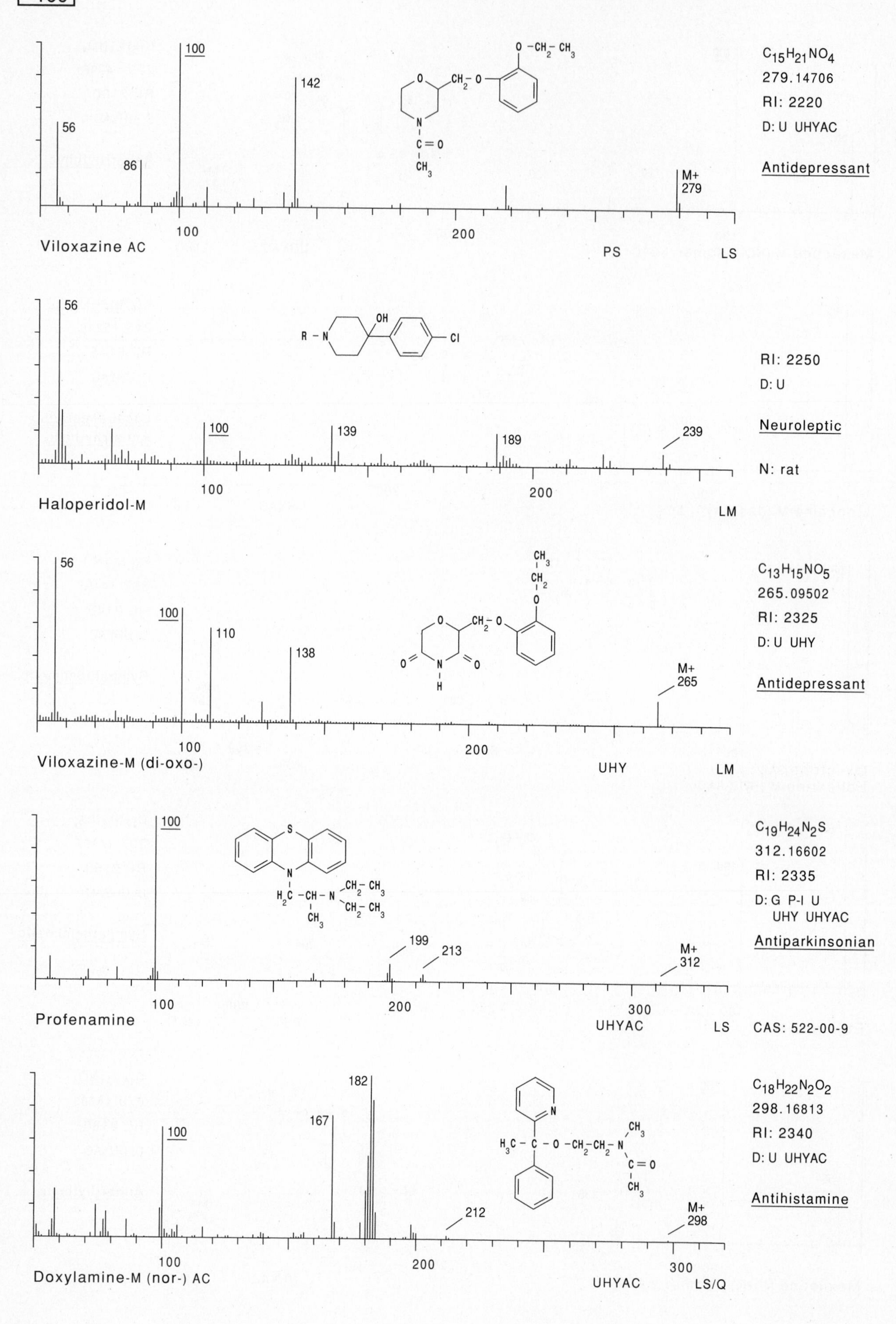

100
142
56
86
M+
279
100
200
Viloxazine AC
PS
LS
O
CH2
O
O
CH2
CH3
N
C = O
CH3
C15H21NO4
279.14706
RI: 2220
D: U UHYAC
Antidepressant
56
R
N
OH
Cl
100
139
189
239
100
200
Haloperidol-M
LM
RI: 2250
D: U
Neuroleptic
N: rat
56
100
110
138
CH3
CH2
O
CH2
O
H
M+
265
100
200
Viloxazine-M (di-oxo-)
UHY
LM
C13H15NO5
265.09502
RI: 2325
D: U UHY
Antidepressant
100
S
N
H2C
CH
N
CH2
CH3
CH3
199
213
M+
312
100
200
300
Profenamine
UHYAC
LS
C19H24N2S
312.16602
RI: 2335
D: G P-I U
UHY UHYAC
Antiparkinsonian
CAS: 522-00-9
182
167
100
N
H3C
C
O
CH2
CH2
N
CH3
C = O
CH3
212
M+
298
100
200
300
Doxylamine-M (nor-) AC
UHYAC
LS/Q
C18H22N2O2
298.16813
RI: 2340
D: U UHYAC
Antihistamine

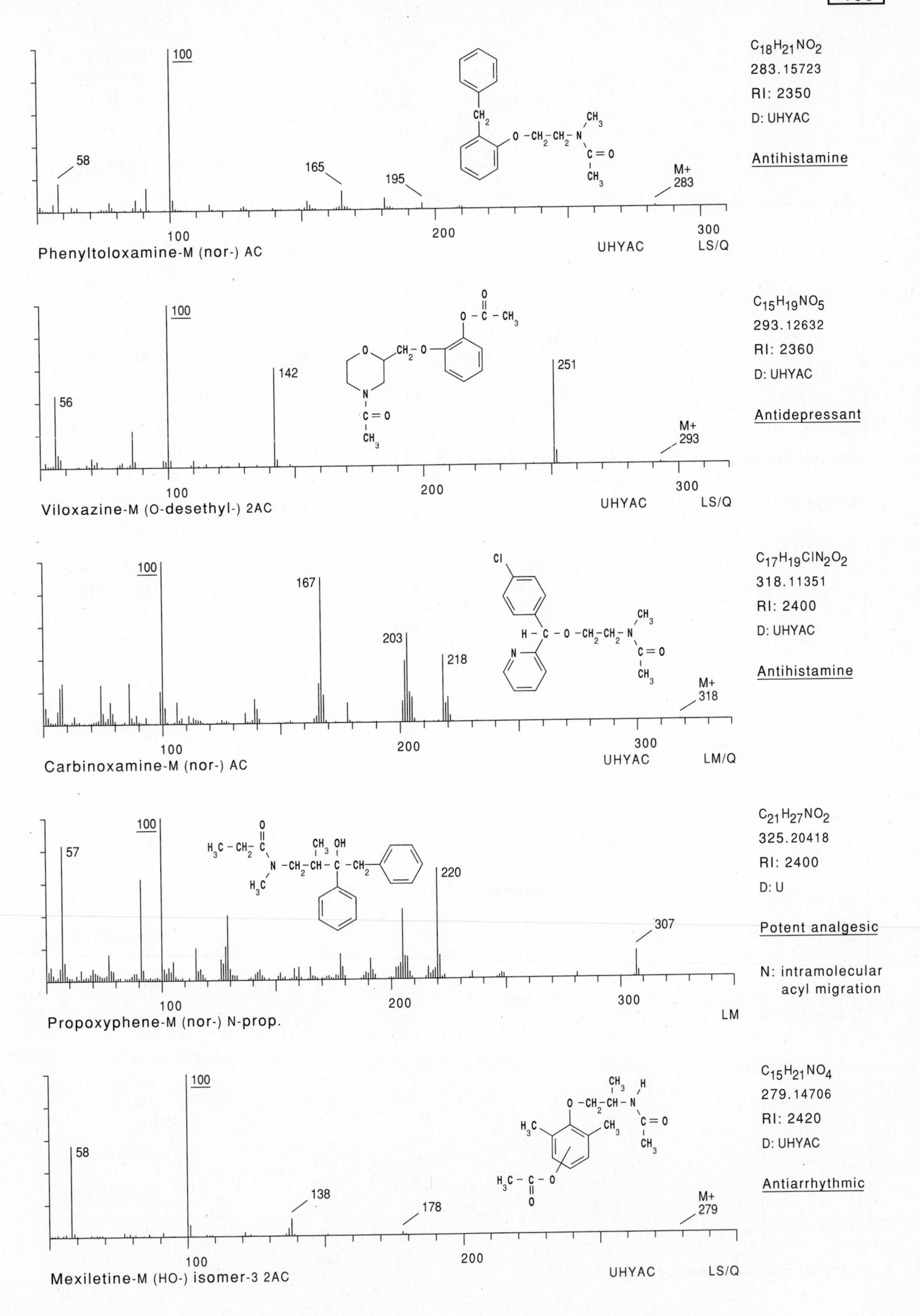
Phenyltoloxamine-M (nor-) AC
UHYAC
LS/Q
$C_{18}H_{21}NO_2$
283.15723
RI: 2350
D: UHYAC
Antihistamine
Viloxazine-M (O-desethyl-) 2AC
UHYAC
LS/Q
$C_{15}H_{19}NO_5$
293.12632
RI: 2360
D: UHYAC
Antidepressant
Carbinoxamine-M (nor-) AC
UHYAC
LM/Q
$C_{17}H_{19}ClN_2O_2$
318.11351
RI: 2400
D: UHYAC
Antihistamine
Propoxyphene-M (nor-) N-prop.
LM
$C_{21}H_{27}NO_2$
325.20418
RI: 2400
D: U
Potent analgesic
N: intramolecular acyl migration
Mexiletine-M (HO-) isomer-3 2AC
UHYAC
LS/Q
$C_{15}H_{21}NO_4$
279.14706
RI: 2420
D: UHYAC
Antiarrhythmic

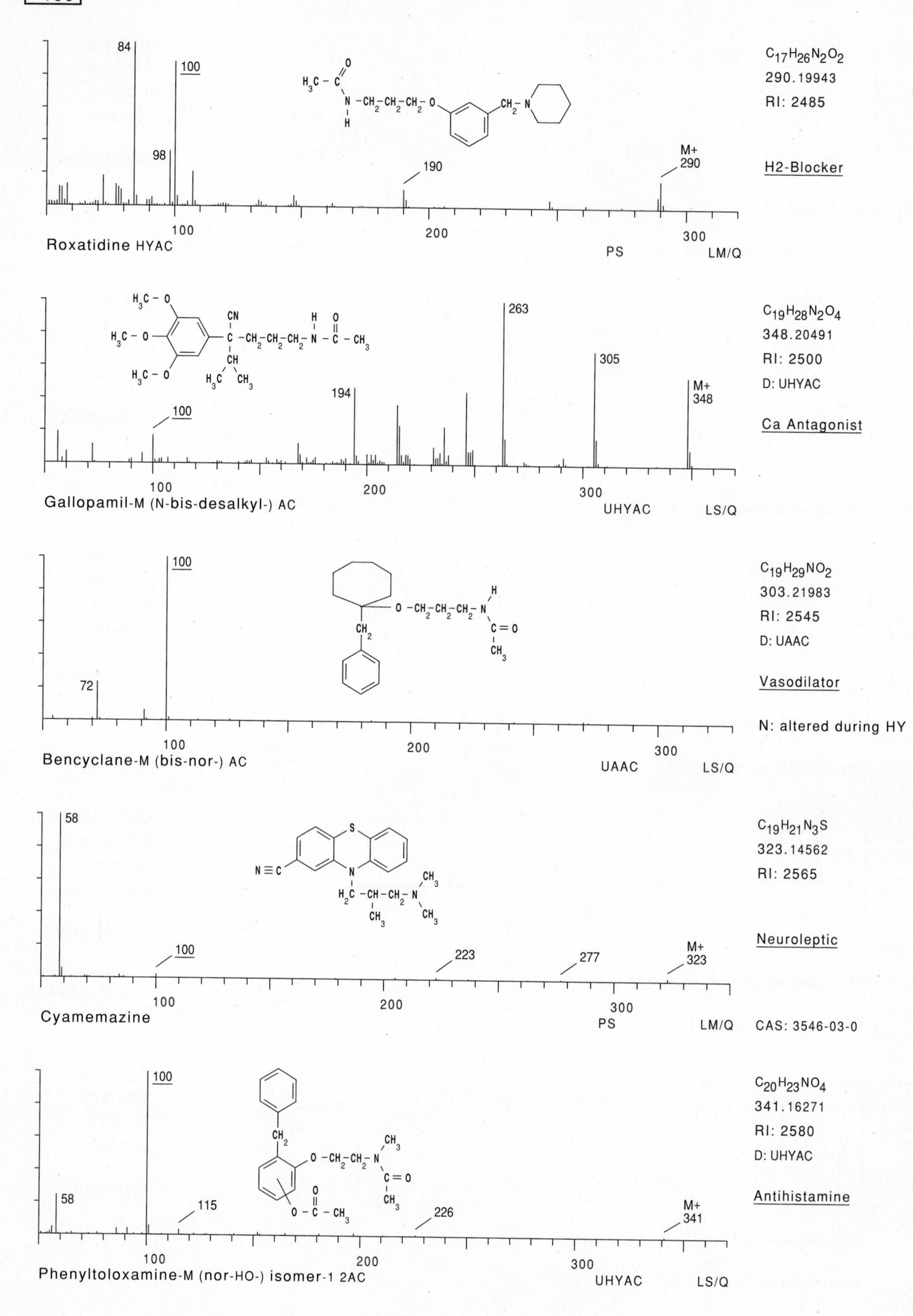
84
100
98
190
M+
290
100
200
300
Roxatidine HYAC
PS
LM/Q
C17H26N2O2
290.19943
RI: 2485
H2-Blocker
263
305
194
100
M+
348
Gallopamil-M (N-bis-desalkyl-) AC
UHYAC
LS/Q
C19H28N2O4
348.20491
RI: 2500
D: UHYAC
Ca Antagonist
100
72
Bencyclane-M (bis-nor-) AC
UAAC
LS/Q
C19H29NO2
303.21983
RI: 2545
D: UAAC
Vasodilator
N: altered during HY
58
100
223
277
M+
323
Cyamemazine
PS
LM/Q
C19H21N3S
323.14562
RI: 2565
Neuroleptic
CAS: 3546-03-0
100
58
115
226
M+
341
Phenyltoloxamine-M (nor-HO-) isomer-1 2AC
UHYAC
LS/Q
C20H23NO4
341.16271
RI: 2580
D: UHYAC
Antihistamine

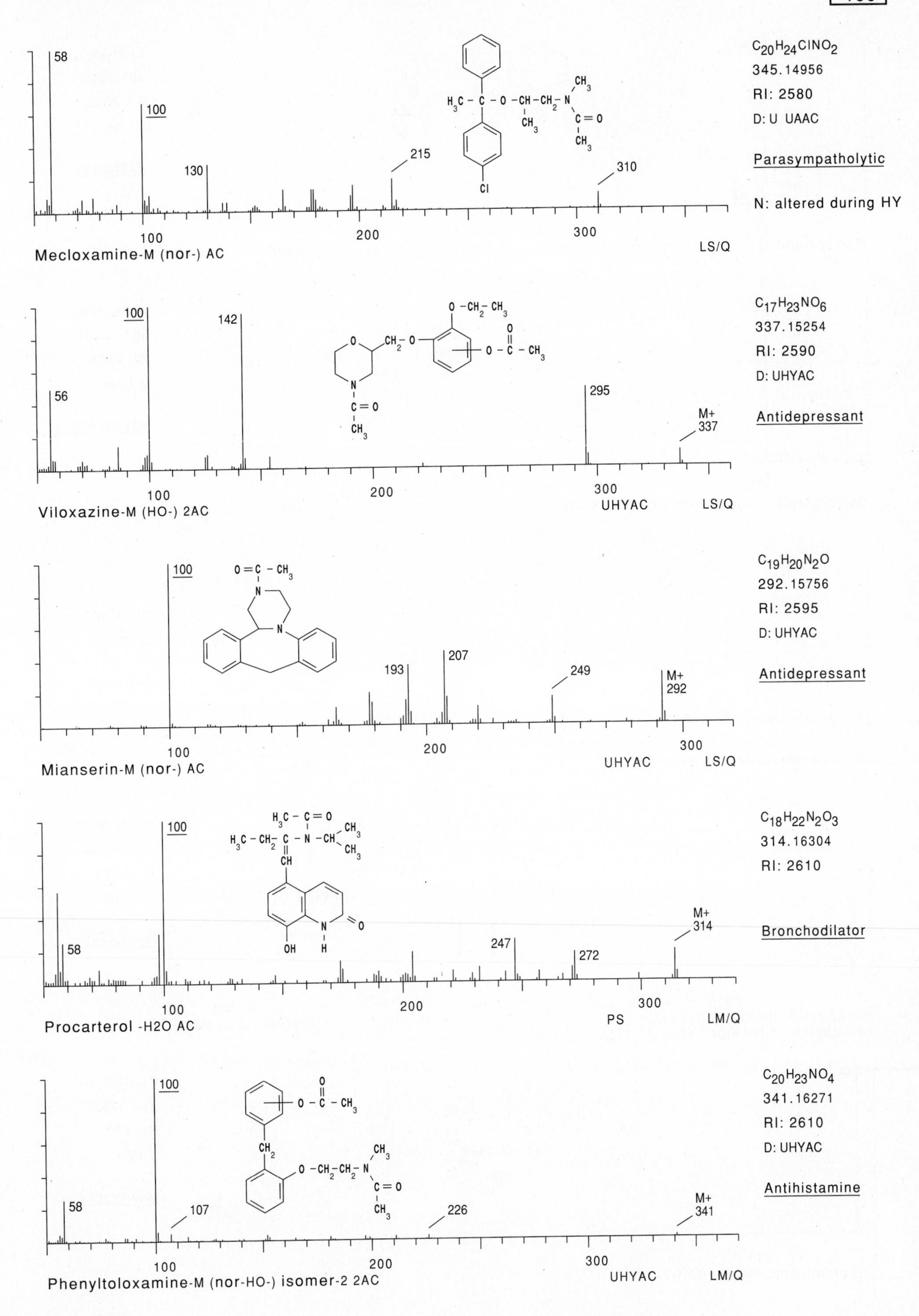
Mecloxamine-M (nor-) AC
LS/Q
C20H24ClNO2
345.14956
RI: 2580
D: U UAAC
Parasympatholytic
N: altered during HY
Viloxazine-M (HO-) 2AC
UHYAC LS/Q
C17H23NO6
337.15254
RI: 2590
D: UHYAC
Antidepressant
Mianserin-M (nor-) AC
UHYAC LS/Q
C19H20N2O
292.15756
RI: 2595
D: UHYAC
Antidepressant
Procarterol -H2O AC
PS LM/Q
C18H22N2O3
314.16304
RI: 2610
Bronchodilator
Phenyltoloxamine-M (nor-HO-) isomer-2 2AC
UHYAC LM/Q
C20H23NO4
341.16271
RI: 2610
D: UHYAC
Antihistamine

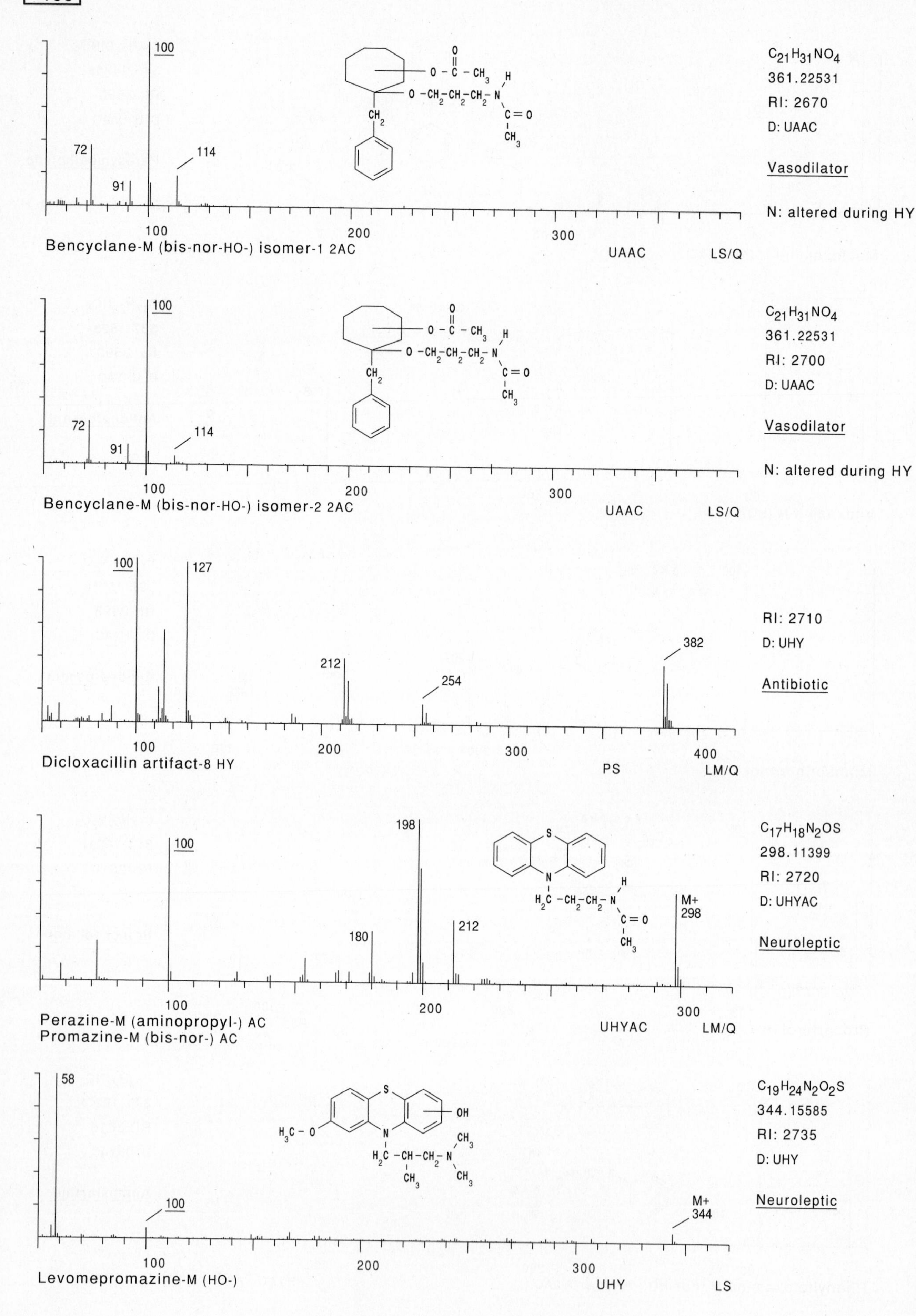
100
72
91
114
100
200
300
Bencyclane-M (bis-nor-HO-) isomer-1 2AC
UAAC
LS/Q
C21H31NO4
361.22531
RI: 2670
D: UAAC
Vasodilator
N: altered during HY
100
72
91
114
100
200
300
Bencyclane-M (bis-nor-HO-) isomer-2 2AC
UAAC
LS/Q
C21H31NO4
361.22531
RI: 2700
D: UAAC
Vasodilator
N: altered during HY
100
127
212
254
382
100
200
300
400
Dicloxacillin artifact-8 HY
PS
LM/Q
RI: 2710
D: UHY
Antibiotic
198
100
180
212
M+
298
100
200
300
Perazine-M (aminopropyl-) AC
Promazine-M (bis-nor-) AC
UHYAC
LM/Q
C17H18N2OS
298.11399
RI: 2720
D: UHYAC
Neuroleptic
58
100
M+
344
100
200
300
Levomepromazine-M (HO-)
UHY
LS
C19H24N2O2S
344.15585
RI: 2735
D: UHY
Neuroleptic

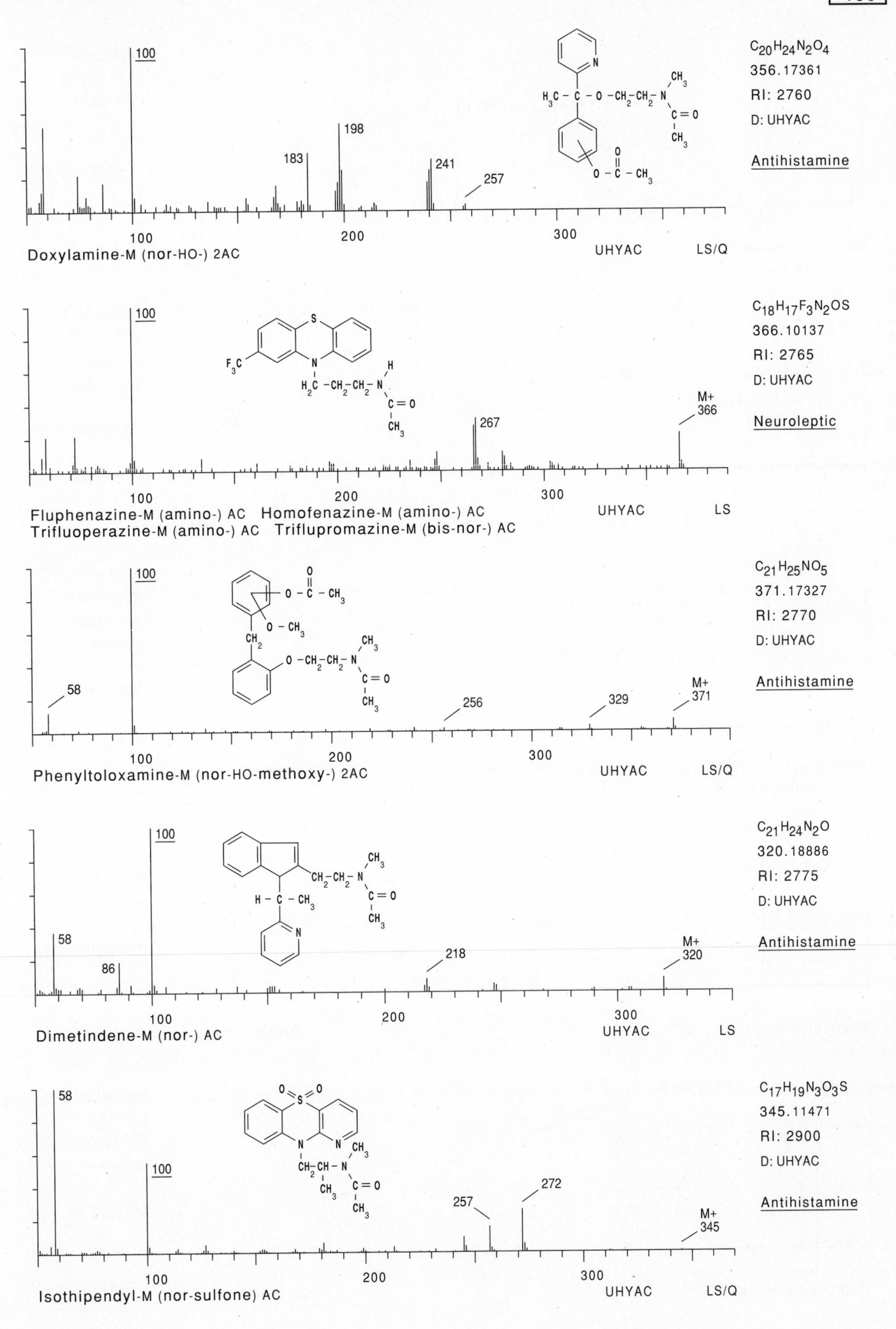

$C_{20}H_{24}N_2O_4$
356.17361
RI: 2760
D: UHYAC
Antihistamine

$C_{18}H_{17}F_3N_2OS$
366.10137
RI: 2765
D: UHYAC
Neuroleptic

$C_{21}H_{25}NO_5$
371.17327
RI: 2770
D: UHYAC
Antihistamine

$C_{21}H_{24}N_2O$
320.18886
RI: 2775
D: UHYAC
Antihistamine

$C_{17}H_{19}N_3O_3S$
345.11471
RI: 2900
D: UHYAC
Antihistamine

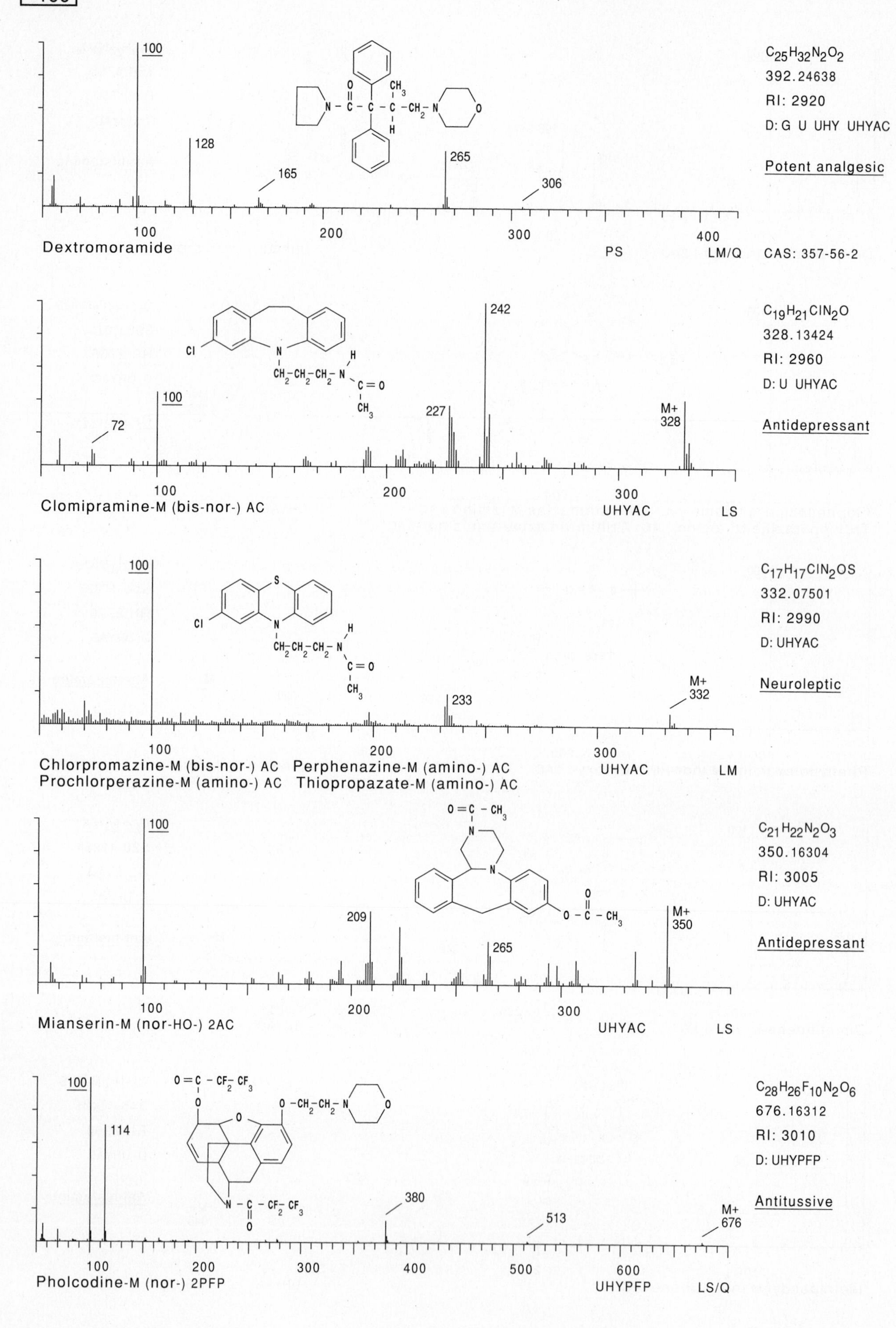
100
128
165
265
306
100
200
300
400
Dextromoramide
PS
LM/Q
C25H32N2O2
392.24638
RI: 2920
D: G U UHY UHYAC
Potent analgesic
CAS: 357-56-2
242
100
72
227
M+
328
Clomipramine-M (bis-nor-) AC
UHYAC
LS
C19H21ClN2O
328.13424
RI: 2960
D: U UHYAC
Antidepressant
233
M+
332
Chlorpromazine-M (bis-nor-) AC Perphenazine-M (amino-) AC
Prochlorperazine-M (amino-) AC Thiopropazate-M (amino-) AC
UHYAC
LM
C17H17ClN2OS
332.07501
RI: 2990
D: UHYAC
Neuroleptic
209
265
M+
350
Mianserin-M (nor-HO-) 2AC
UHYAC
LS
C21H22N2O3
350.16304
RI: 3005
D: UHYAC
Antidepressant
114
380
513
M+
676
500
600
Pholcodine-M (nor-) 2PFP
UHYPFP
LS/Q
C28H26F10N2O6
676.16312
RI: 3010
D: UHYPFP
Antitussive

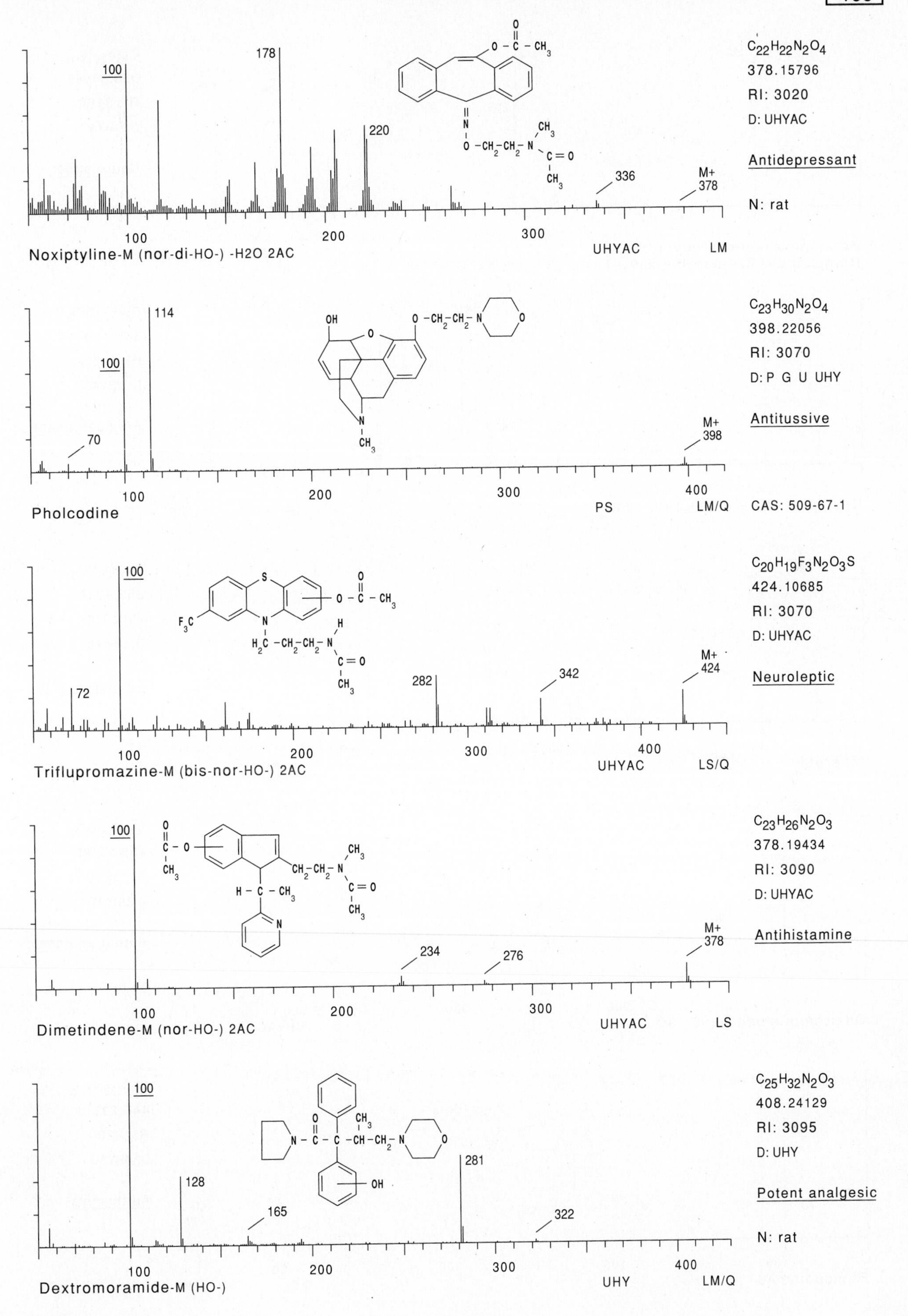

178
100
220
336
M+
378
100
200
300
Noxiptyline-M (nor-di-HO-) -H2O 2AC
UHYAC
LM
$C_{22}H_{22}N_2O_4$
378.15796
RI: 3020
D: UHYAC
Antidepressant
N: rat
114
100
70
M+
398
100
200
300
400
Pholcodine
PS
LM/Q
$C_{23}H_{30}N_2O_4$
398.22056
RI: 3070
D: P G U UHY
Antitussive
CAS: 509-67-1
100
72
282
342
M+
424
100
200
300
400
Triflupromazine-M (bis-nor-HO-) 2AC
UHYAC
LS/Q
$C_{20}H_{19}F_3N_2O_3S$
424.10685
RI: 3070
D: UHYAC
Neuroleptic
100
234
276
M+
378
100
200
300
Dimetindene-M (nor-HO-) 2AC
UHYAC
LS
$C_{23}H_{26}N_2O_3$
378.19434
RI: 3090
D: UHYAC
Antihistamine
100
128
165
281
322
100
200
300
400
Dextromoramide-M (HO-)
UHY
LM/Q
$C_{25}H_{32}N_2O_3$
408.24129
RI: 3095
D: UHY
Potent analgesic
N: rat

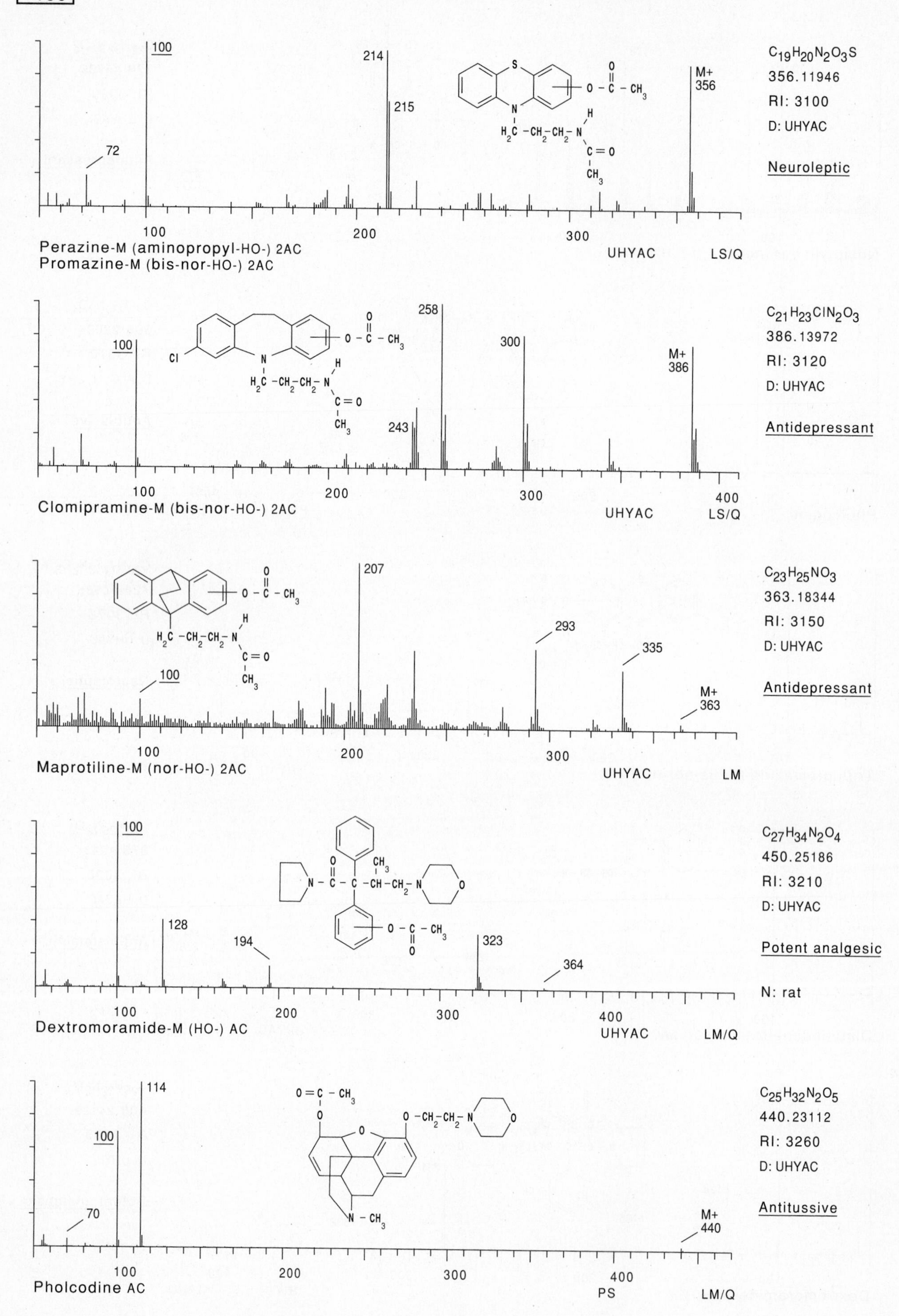

$C_{19}H_{20}N_2O_3S$
356.11946
RI: 3100
D: UHYAC

Neuroleptic

Perazine-M (aminopropyl-HO-) 2AC
Promazine-M (bis-nor-HO-) 2AC

$C_{21}H_{23}ClN_2O_3$
386.13972
RI: 3120
D: UHYAC

Antidepressant

Clomipramine-M (bis-nor-HO-) 2AC

$C_{23}H_{25}NO_3$
363.18344
RI: 3150
D: UHYAC

Antidepressant

Maprotiline-M (nor-HO-) 2AC

$C_{27}H_{34}N_2O_4$
450.25186
RI: 3210
D: UHYAC

Potent analgesic

N: rat

Dextromoramide-M (HO-) AC

$C_{25}H_{32}N_2O_5$
440.23112
RI: 3260
D: UHYAC

Antitussive

Pholcodine AC

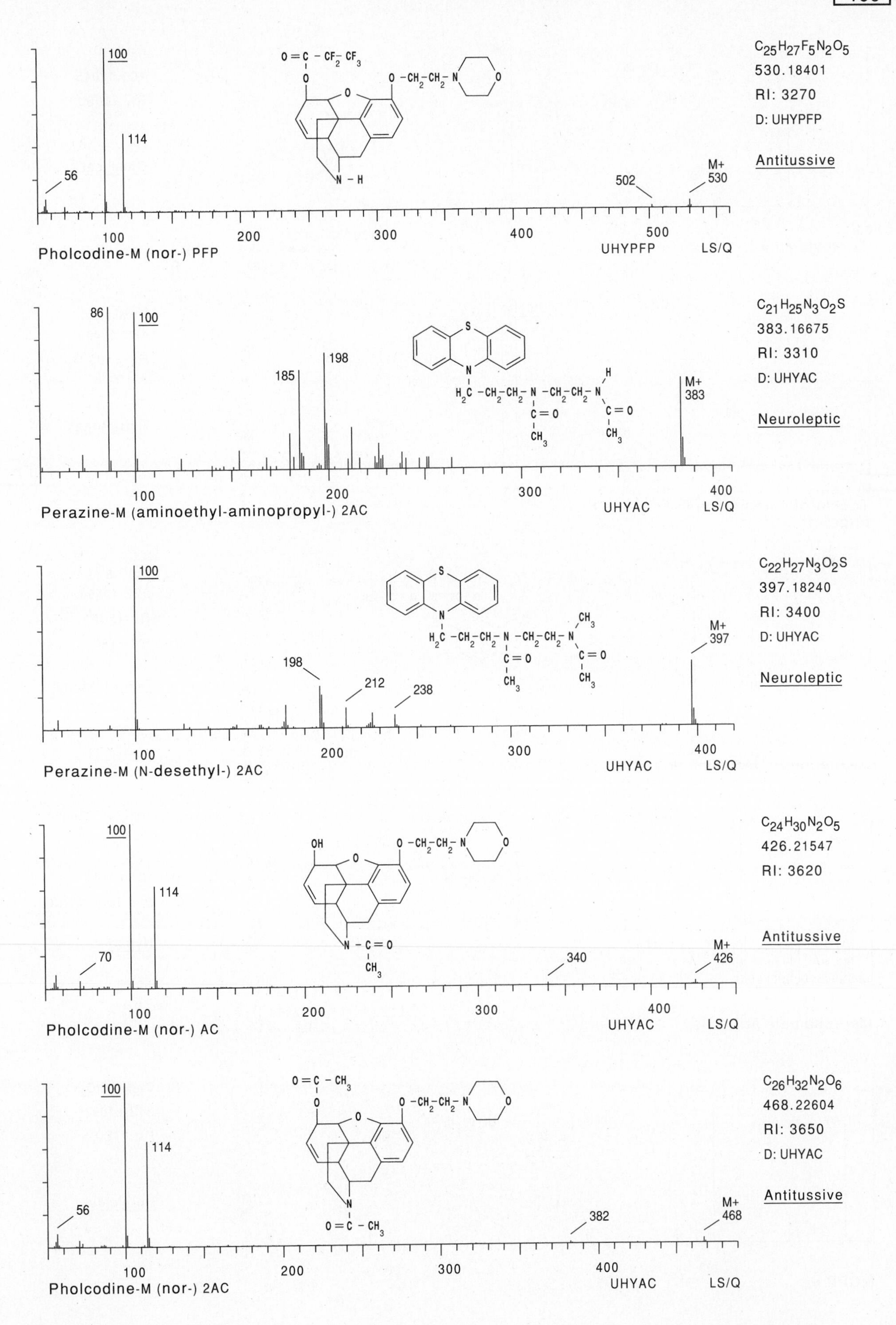

$C_{25}H_{27}F_5N_2O_5$
530.18401
RI: 3270
D: UHYPFP
Antitussive

$C_{21}H_{25}N_3O_2S$
383.16675
RI: 3310
D: UHYAC
Neuroleptic

$C_{22}H_{27}N_3O_2S$
397.18240
RI: 3400
D: UHYAC
Neuroleptic

$C_{24}H_{30}N_2O_5$
426.21547
RI: 3620
Antitussive

$C_{26}H_{32}N_2O_6$
468.22604
RI: 3650
D: UHYAC
Antitussive

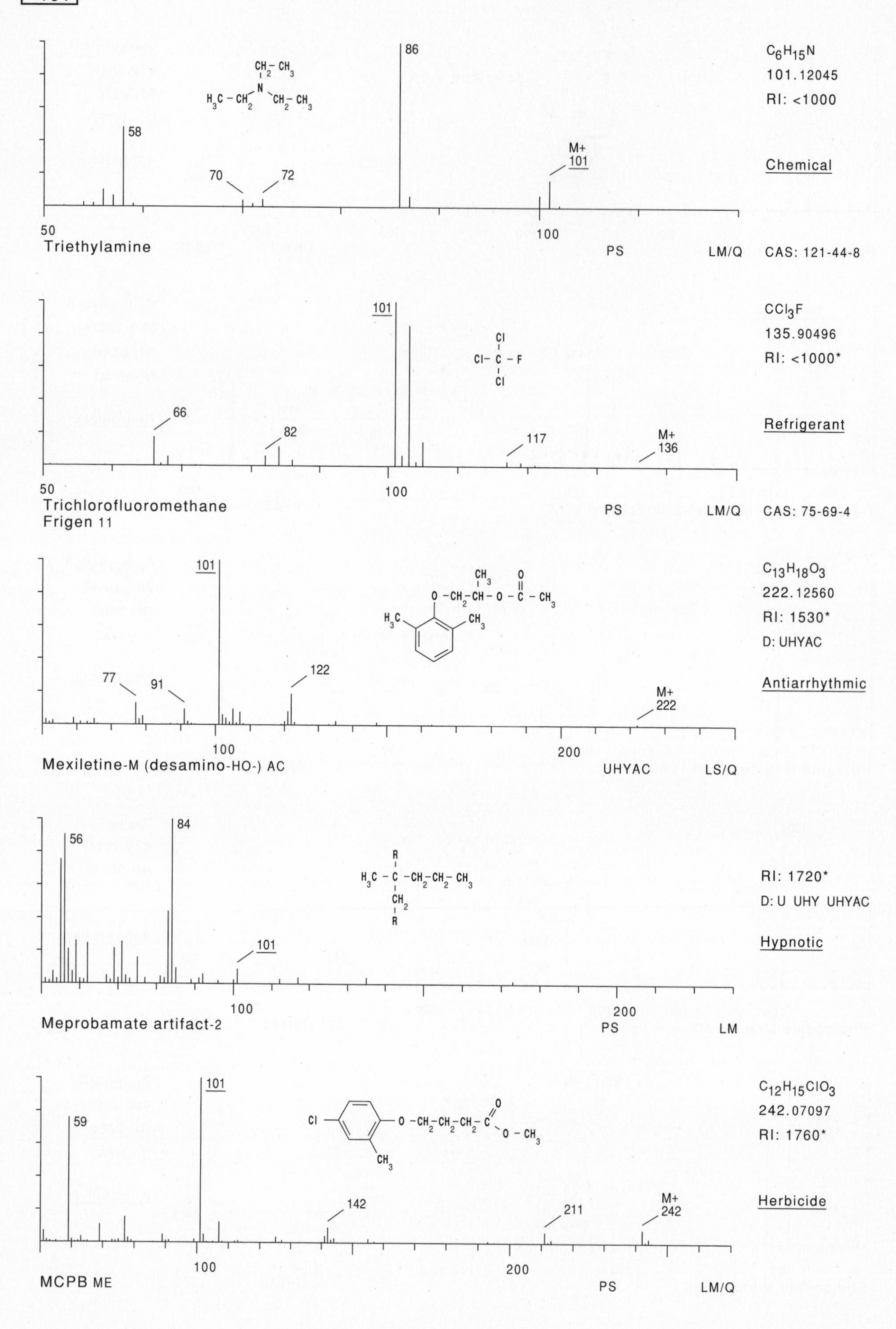
86
58
70
72
M+
101
50
100
Triethylamine
PS
LM/Q
$C_6H_{15}N$
101.12045
RI: <1000
Chemical
CAS: 121-44-8
101
66
82
117
M+
136
50
100
Trichlorofluoromethane
Frigen 11
PS
LM/Q
CCl_3F
135.90496
RI: <1000*
Refrigerant
CAS: 75-69-4
101
77
91
122
M+
222
100
200
Mexiletine-M (desamino-HO-) AC
UHYAC
LS/Q
$C_{13}H_{18}O_3$
222.12560
RI: 1530*
D: UHYAC
Antiarrhythmic
56
84
101
100
200
Meprobamate artifact-2
PS
LM
RI: 1720*
D: U UHY UHYAC
Hypnotic
101
59
142
211
M+
242
100
200
MCPB ME
PS
LM/Q
$C_{12}H_{15}ClO_3$
242.07097
RI: 1760*
Herbicide

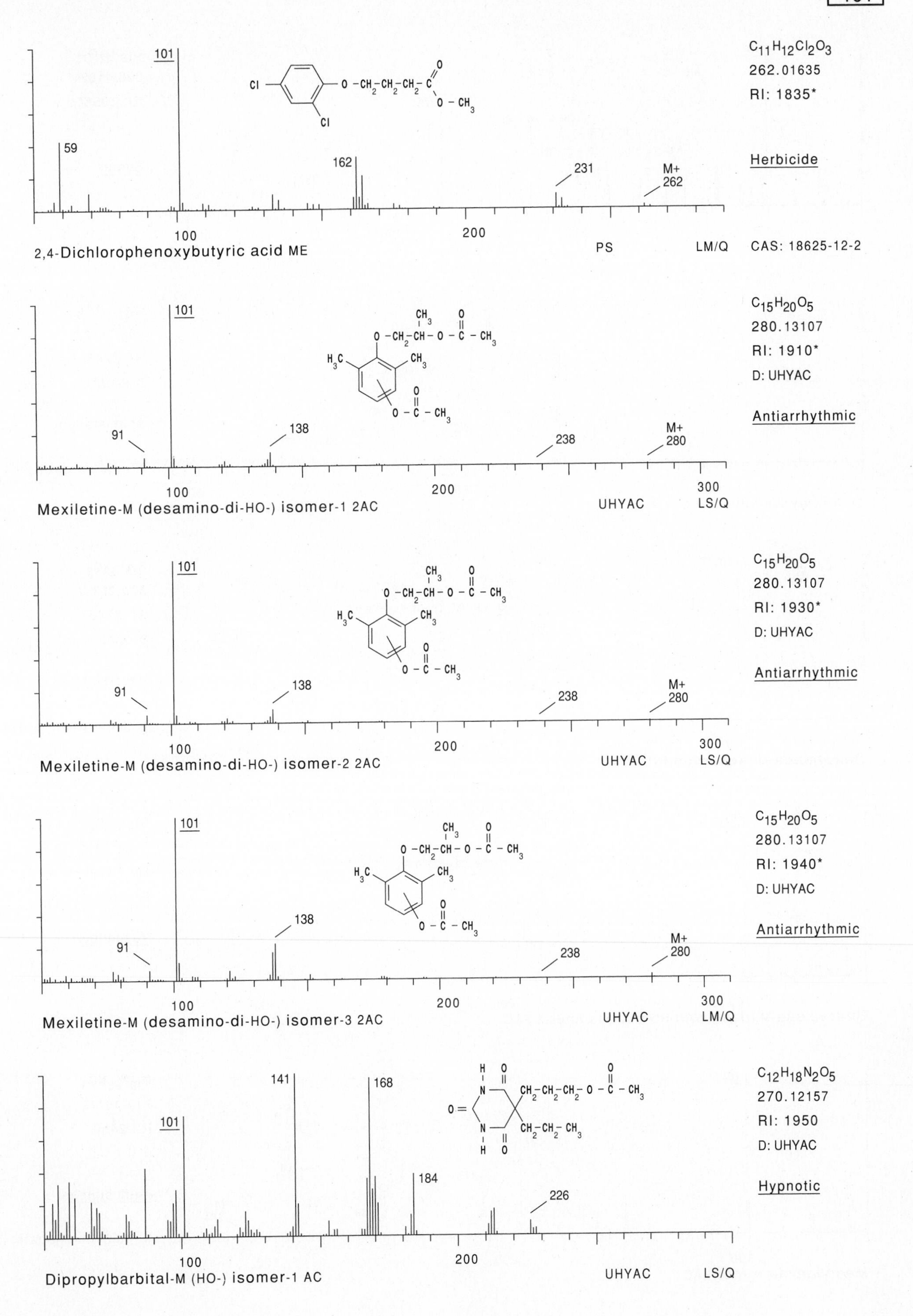

101
59
162
231
M+
262
100
200
2,4-Dichlorophenoxybutyric acid ME
PS
LM/Q
$C_{11}H_{12}Cl_2O_3$
262.01635
RI: 1835*
Herbicide
CAS: 18625-12-2
101
91
138
238
M+
280
100
200
300
Mexiletine-M (desamino-di-HO-) isomer-1 2AC
UHYAC
LS/Q
$C_{15}H_{20}O_5$
280.13107
RI: 1910*
D: UHYAC
Antiarrhythmic
101
91
138
238
M+
280
100
200
300
Mexiletine-M (desamino-di-HO-) isomer-2 2AC
UHYAC
LS/Q
$C_{15}H_{20}O_5$
280.13107
RI: 1930*
D: UHYAC
Antiarrhythmic
101
91
138
238
M+
280
100
200
300
Mexiletine-M (desamino-di-HO-) isomer-3 2AC
UHYAC
LM/Q
$C_{15}H_{20}O_5$
280.13107
RI: 1940*
D: UHYAC
Antiarrhythmic
141
168
101
184
226
100
200
Dipropylbarbital-M (HO-) isomer-1 AC
UHYAC
LS/Q
$C_{12}H_{18}N_2O_5$
270.12157
RI: 1950
D: UHYAC
Hypnotic

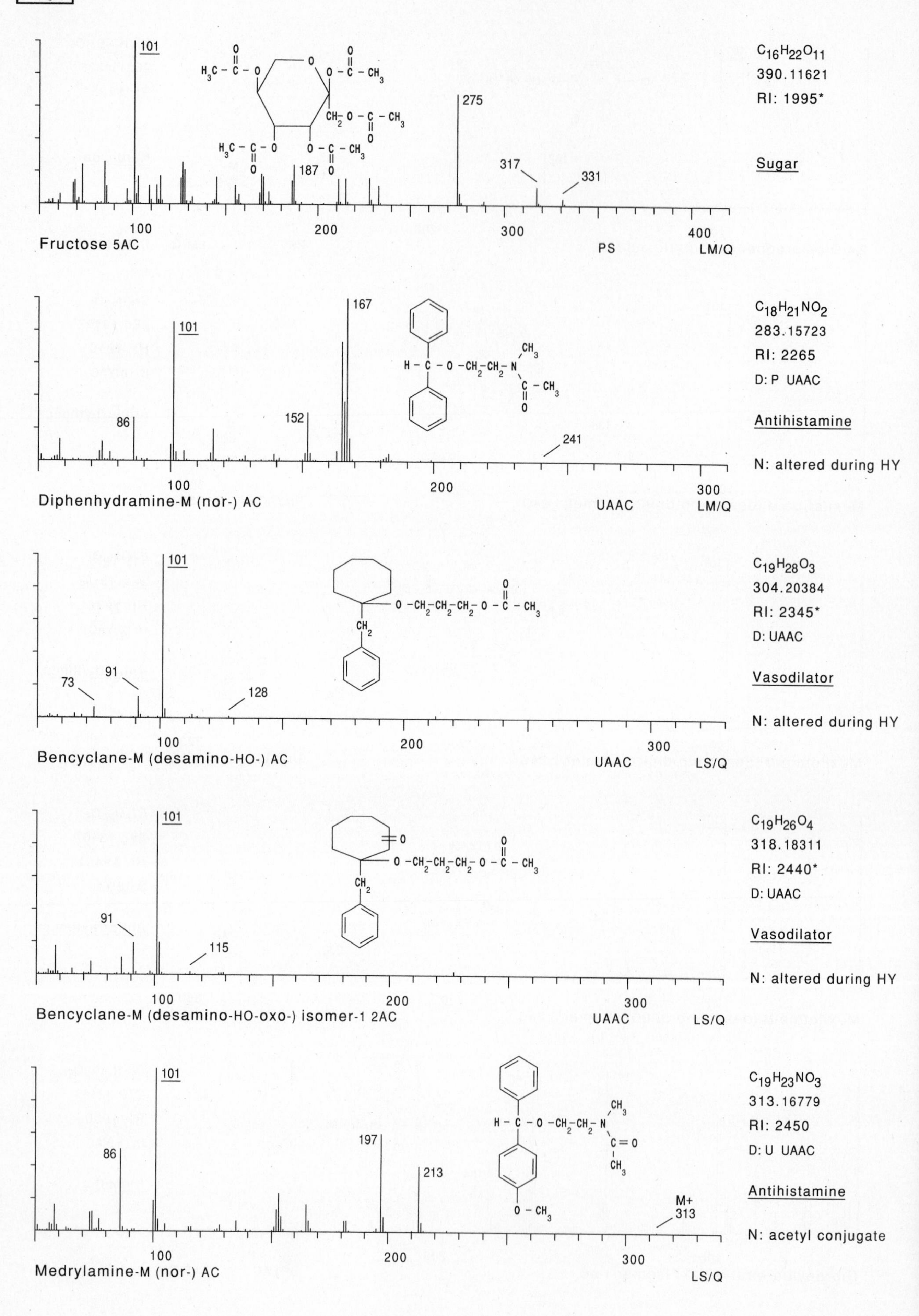

101
275
317
331
187
Fructose 5AC
PS
LM/Q
$C_{16}H_{22}O_{11}$
390.11621
RI: 1995*
Sugar
167
101
86
152
241
Diphenhydramine-M (nor-) AC
UAAC
LM/Q
$C_{18}H_{21}NO_2$
283.15723
RI: 2265
D: P UAAC
Antihistamine
N: altered during HY
101
73
91
128
Bencyclane-M (desamino-HO-) AC
UAAC
LS/Q
$C_{19}H_{28}O_3$
304.20384
RI: 2345*
D: UAAC
Vasodilator
N: altered during HY
101
91
115
Bencyclane-M (desamino-HO-oxo-) isomer-1 2AC
UAAC
LS/Q
$C_{19}H_{26}O_4$
318.18311
RI: 2440*
D: UAAC
Vasodilator
N: altered during HY
101
86
197
213
M+
313
Medrylamine-M (nor-) AC
LS/Q
$C_{19}H_{23}NO_3$
313.16779
RI: 2450
D: U UAAC
Antihistamine
N: acetyl conjugate

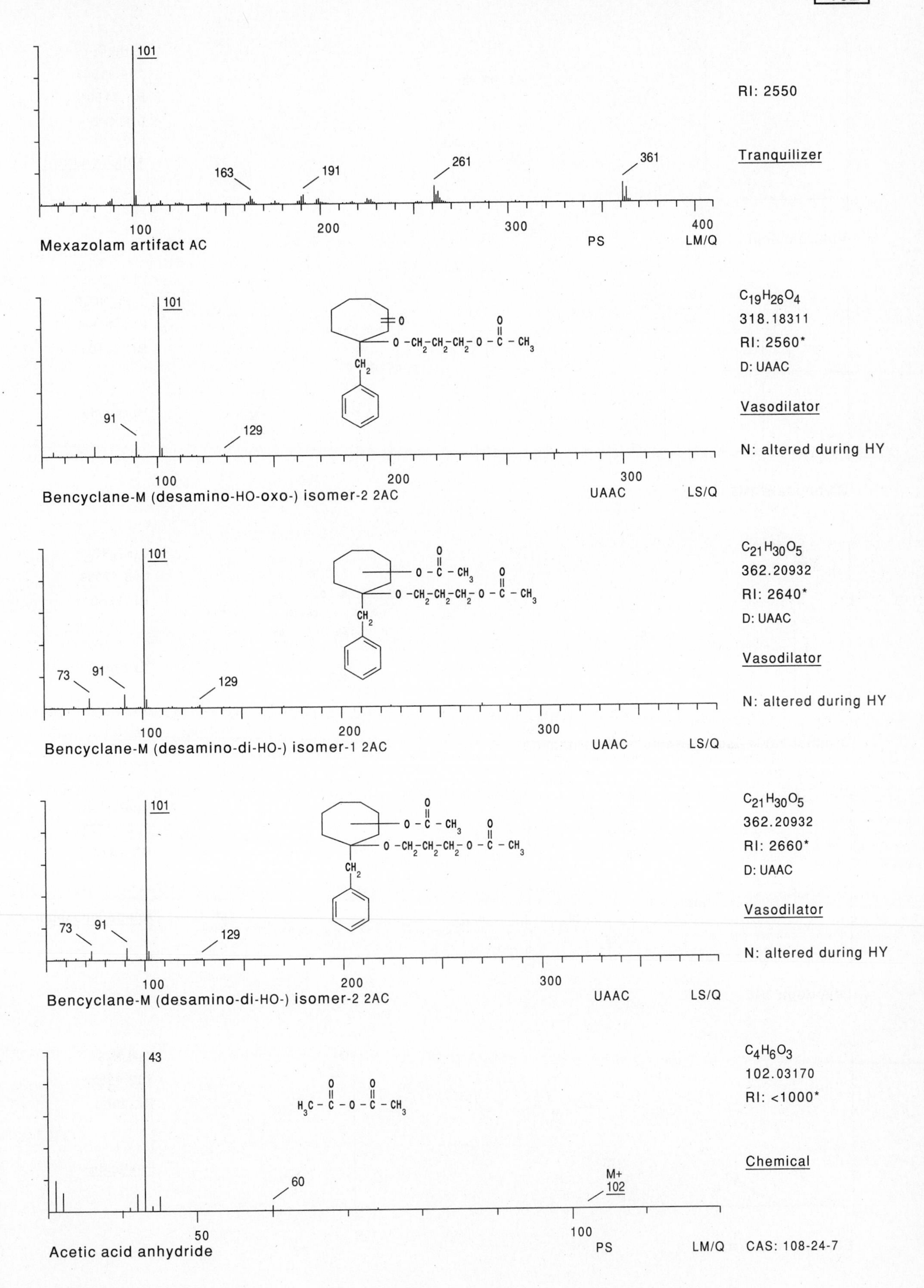
101
163
191
261
361
100
200
300
400
PS
LM/Q
Mexazolam artifact AC
RI: 2550
Tranquilizer
101
91
129
O
O -CH2-CH2-CH2- O - C - CH3
CH2
100
200
300
UAAC
LS/Q
Bencyclane-M (desamino-HO-oxo-) isomer-2 2AC
C19H26O4
318.18311
RI: 2560*
D: UAAC
Vasodilator
N: altered during HY
101
73
91
129
O - C - CH3
O -CH2-CH2-CH2- O - C - CH3
CH2
100
200
300
UAAC
LS/Q
Bencyclane-M (desamino-di-HO-) isomer-1 2AC
C21H30O5
362.20932
RI: 2640*
D: UAAC
Vasodilator
N: altered during HY
101
73
91
129
O - C - CH3
O -CH2-CH2-CH2- O - C - CH3
CH2
100
200
300
UAAC
LS/Q
Bencyclane-M (desamino-di-HO-) isomer-2 2AC
C21H30O5
362.20932
RI: 2660*
D: UAAC
Vasodilator
N: altered during HY
43
H3C - C - O - C - CH3
60
M+
102
50
100
PS
LM/Q
Acetic acid anhydride
C4H6O3
102.03170
RI: <1000*
Chemical
CAS: 108-24-7

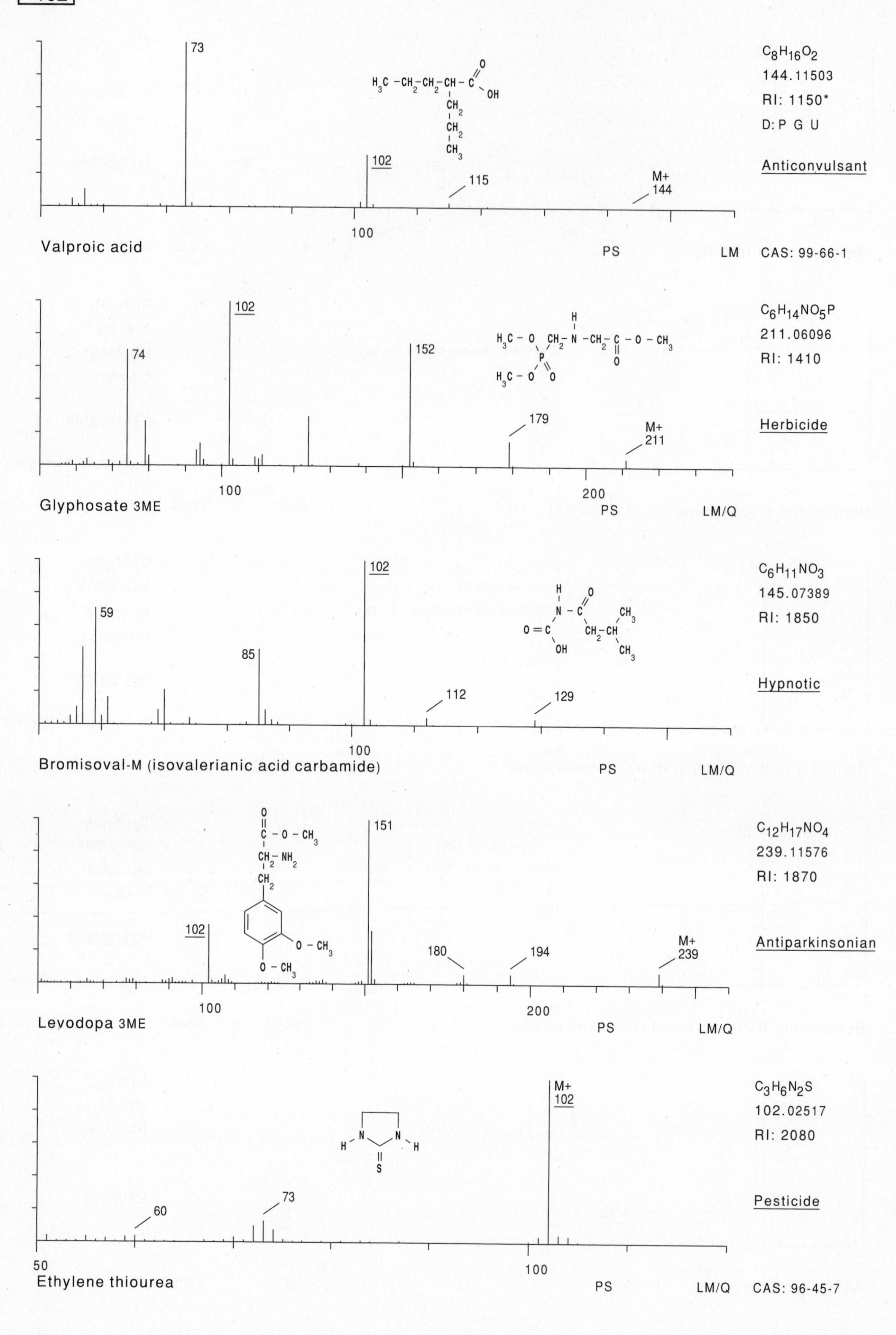

73
102
115
M+
144
100
$C_8H_{16}O_2$
144.11503
RI: 1150*
D: P G U
Anticonvulsant
Valproic acid
PS
LM
CAS: 99-66-1
74
102
152
179
M+
211
100
200
$C_6H_{14}NO_5P$
211.06096
RI: 1410
Herbicide
Glyphosate 3ME
PS
LM/Q
59
85
102
112
129
100
$C_6H_{11}NO_3$
145.07389
RI: 1850
Hypnotic
Bromisoval-M (isovalerianic acid carbamide)
PS
LM/Q
102
151
180
194
M+
239
100
200
$C_{12}H_{17}NO_4$
239.11576
RI: 1870
Antiparkinsonian
Levodopa 3ME
PS
LM/Q
60
73
M+
102
50
100
$C_3H_6N_2S$
102.02517
RI: 2080
Pesticide
Ethylene thiourea
PS
LM/Q
CAS: 96-45-7

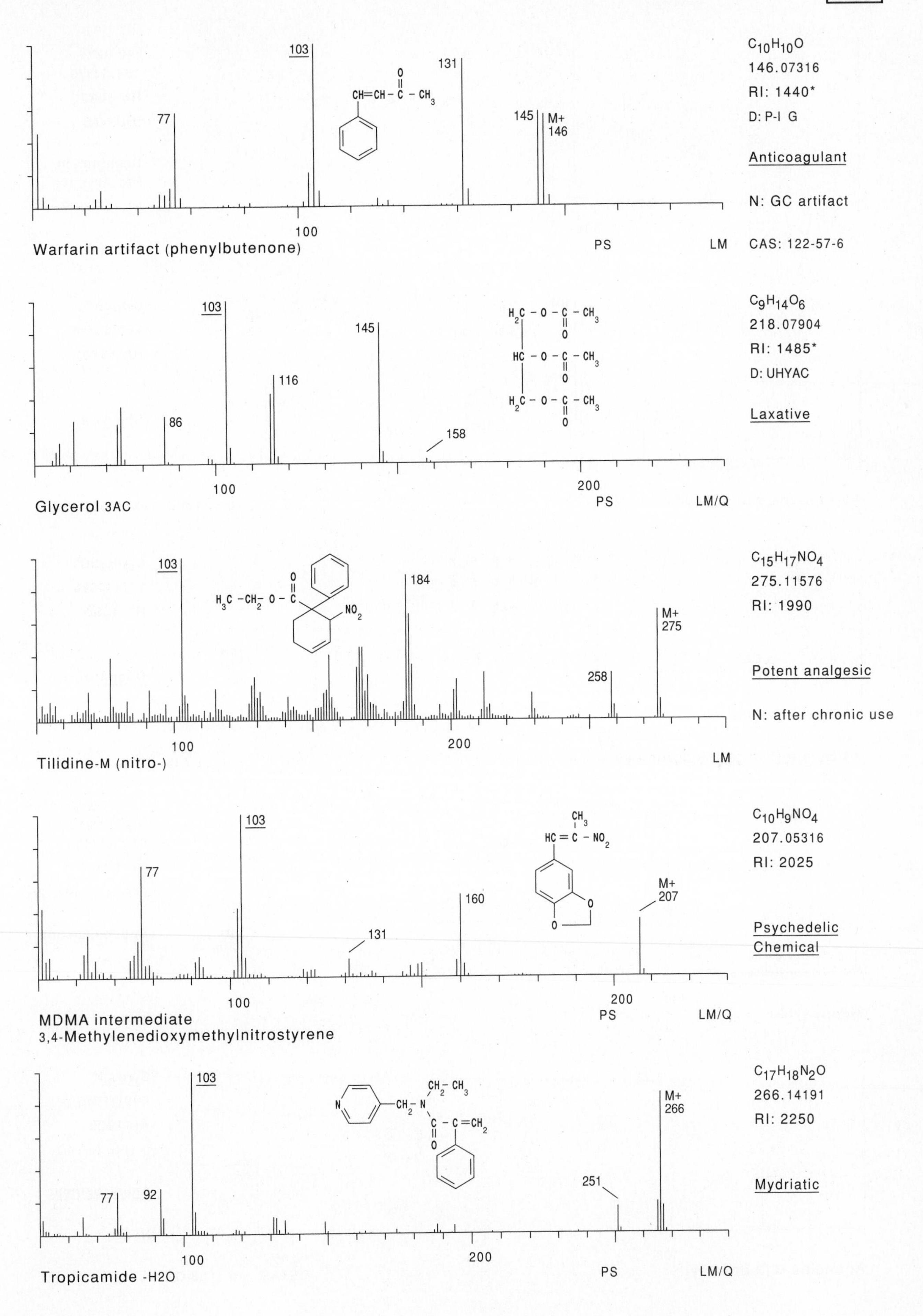

Warfarin artifact (phenylbutenone)
$C_{10}H_{10}O$
146.07316
RI: 1440*
D: P-I G
Anticoagulant
N: GC artifact
CAS: 122-57-6
PS
LM
Glycerol 3AC
$C_9H_{14}O_6$
218.07904
RI: 1485*
D: UHYAC
Laxative
PS
LM/Q
Tilidine-M (nitro-)
$C_{15}H_{17}NO_4$
275.11576
RI: 1990
Potent analgesic
N: after chronic use
LM
MDMA intermediate
3,4-Methylenedioxymethylnitrostyrene
$C_{10}H_9NO_4$
207.05316
RI: 2025
Psychedelic Chemical
PS
LM/Q
Tropicamide -H2O
$C_{17}H_{18}N_2O$
266.14191
RI: 2250
Mydriatic
PS
LM/Q

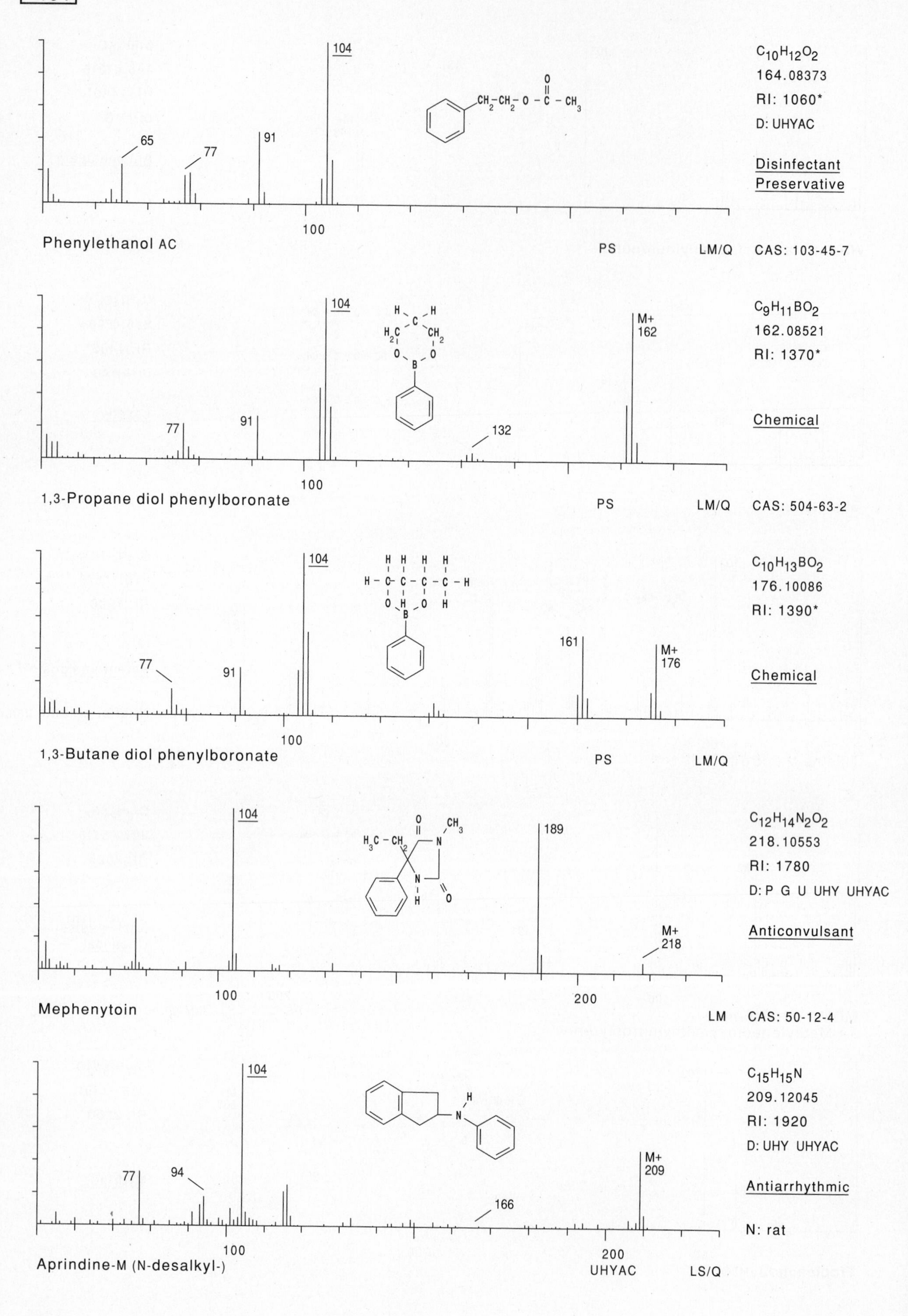

104
65
77
91
100
$CH_2-CH_2-O-C(=O)-CH_3$
Phenylethanol AC
PS
LM/Q
$C_{10}H_{12}O_2$
164.08373
RI: 1060*
D: UHYAC
Disinfectant
Preservative
CAS: 103-45-7
104
M+
162
77
91
132
100
1,3-Propane diol phenylboronate
PS
LM/Q
$C_9H_{11}BO_2$
162.08521
RI: 1370*
Chemical
CAS: 504-63-2
104
161
M+
176
77
91
100
1,3-Butane diol phenylboronate
PS
LM/Q
$C_{10}H_{13}BO_2$
176.10086
RI: 1390*
Chemical
104
189
M+
218
100
200
Mephenytoin
LM
$C_{12}H_{14}N_2O_2$
218.10553
RI: 1780
D: P G U UHY UHYAC
Anticonvulsant
CAS: 50-12-4
104
M+
209
77
94
166
100
200
Aprindine-M (N-desalkyl-)
UHYAC
LS/Q
$C_{15}H_{15}N$
209.12045
RI: 1920
D: UHY UHYAC
Antiarrhythmic
N: rat

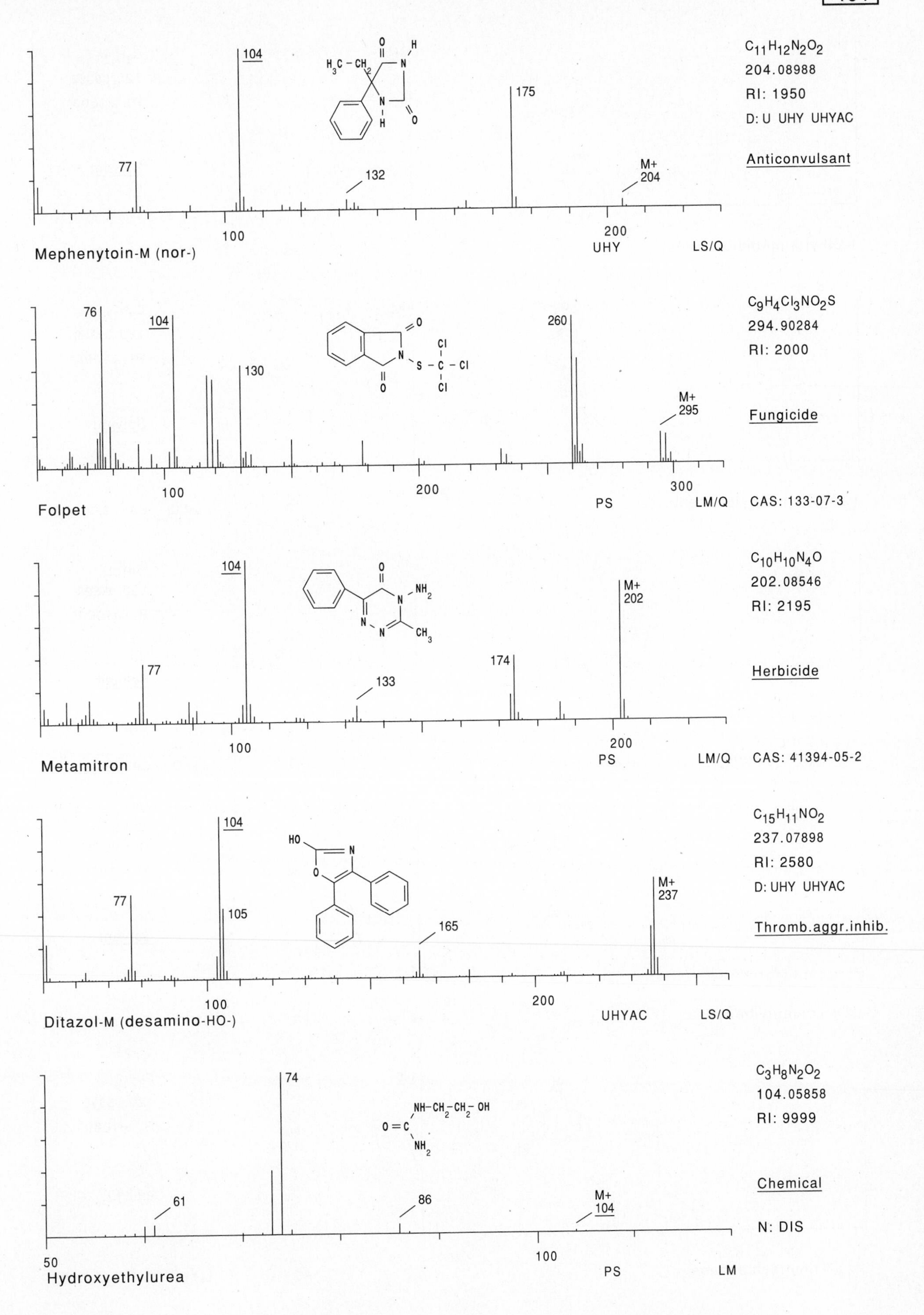

$C_{11}H_{12}N_2O_2$
204.08988
RI: 1950
D: U UHY UHYAC

Anticonvulsant

$C_9H_4Cl_3NO_2S$
294.90284
RI: 2000

Fungicide

CAS: 133-07-3

$C_{10}H_{10}N_4O$
202.08546
RI: 2195

Herbicide

CAS: 41394-05-2

$C_{15}H_{11}NO_2$
237.07898
RI: 2580
D: UHY UHYAC

Thromb.aggr.inhib.

$C_3H_8N_2O_2$
104.05858
RI: 9999

Chemical

N: DIS

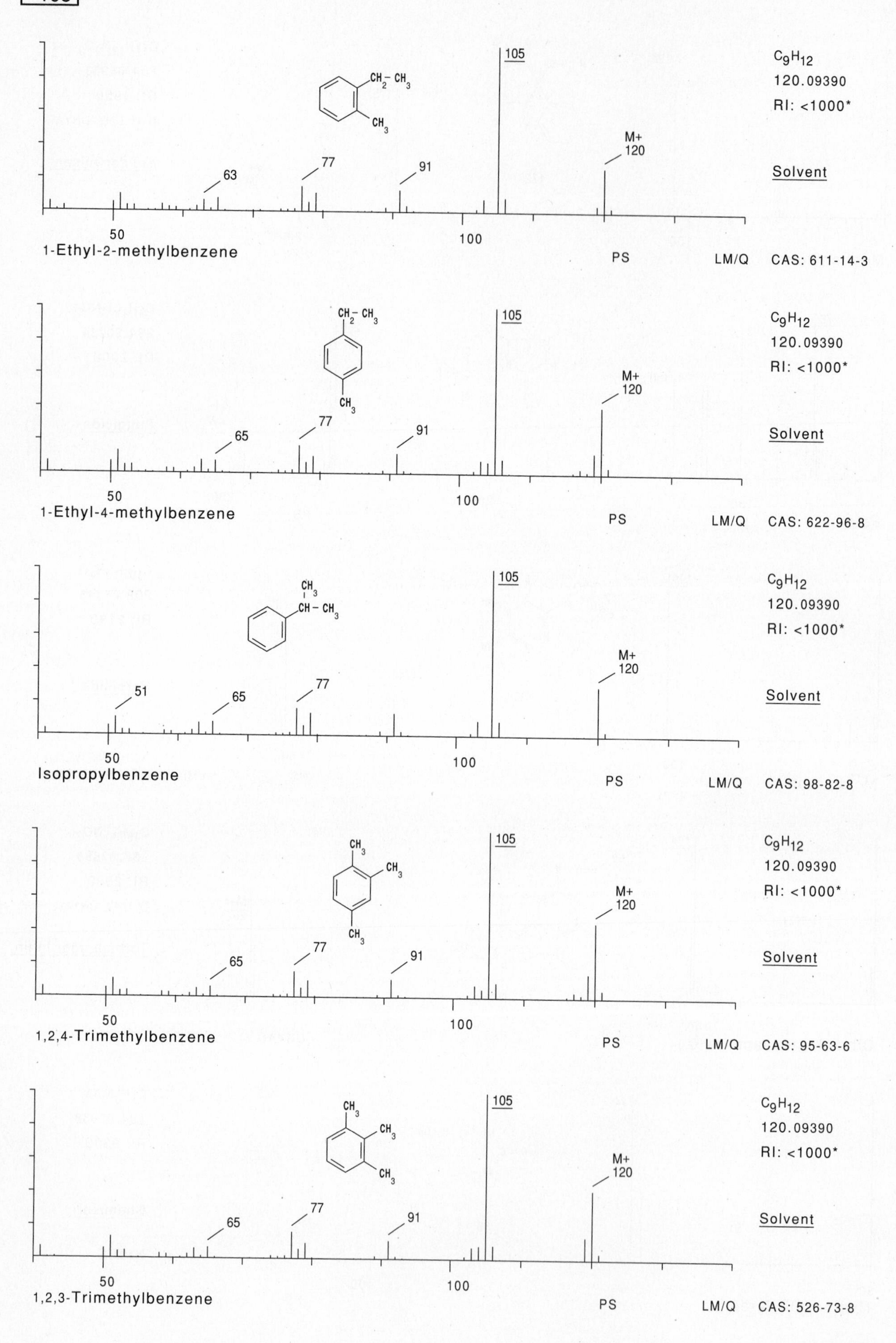
105
CH2 – CH3
CH3
M+
120
63
77
91
50
100
1-Ethyl-2-methylbenzene
C9H12
120.09390
RI: <1000*
Solvent
PS
LM/Q
CAS: 611-14-3
105
CH2 – CH3
CH3
M+
120
65
77
91
50
100
1-Ethyl-4-methylbenzene
C9H12
120.09390
RI: <1000*
Solvent
PS
LM/Q
CAS: 622-96-8
105
CH3
CH – CH3
M+
120
51
65
77
50
100
Isopropylbenzene
C9H12
120.09390
RI: <1000*
Solvent
PS
LM/Q
CAS: 98-82-8
105
CH3
CH3
CH3
M+
120
65
77
91
50
100
1,2,4-Trimethylbenzene
C9H12
120.09390
RI: <1000*
Solvent
PS
LM/Q
CAS: 95-63-6
105
CH3
CH3
CH3
M+
120
65
77
91
50
100
1,2,3-Trimethylbenzene
C9H12
120.09390
RI: <1000*
Solvent
PS
LM/Q
CAS: 526-73-8

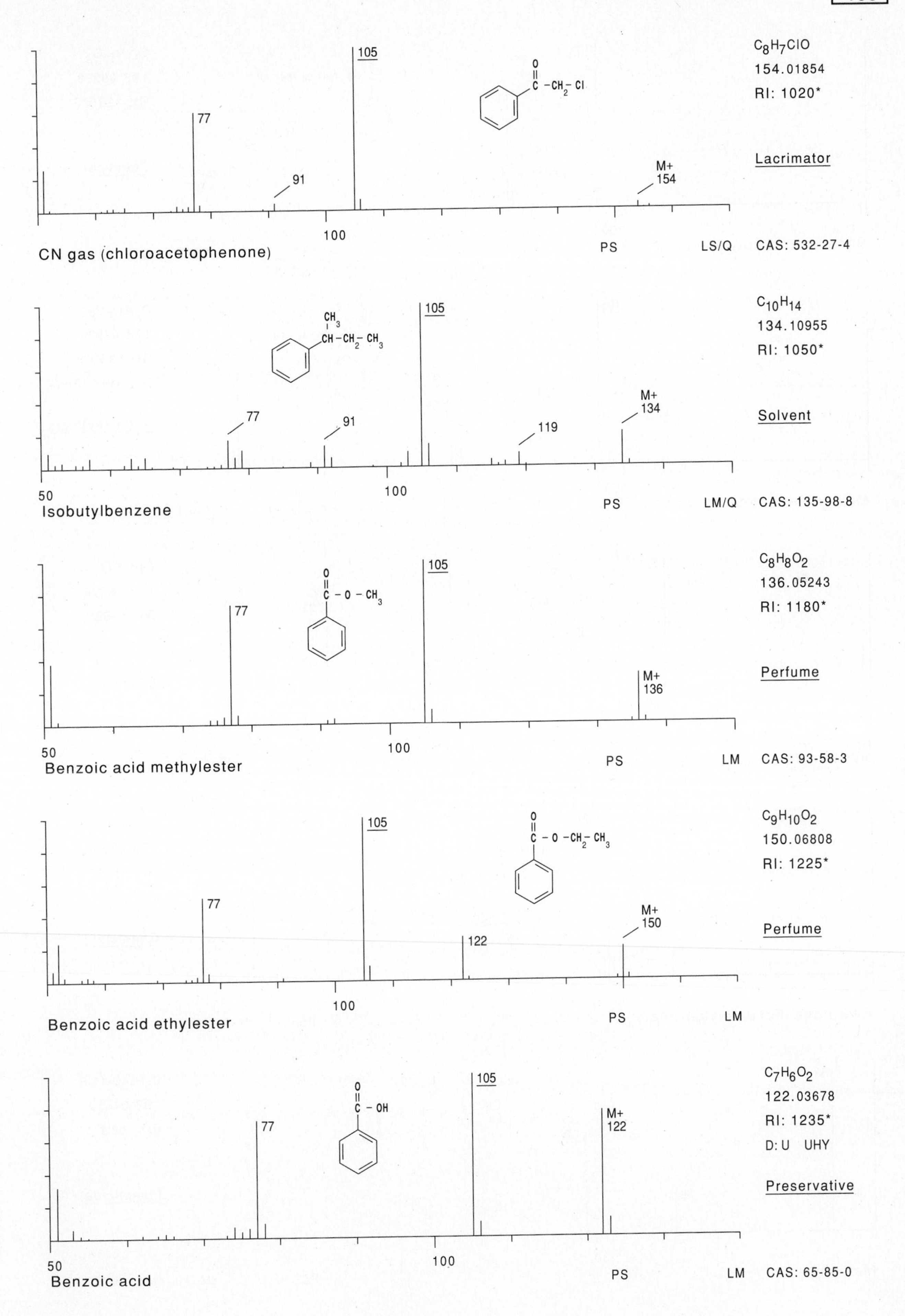
105
CN gas (chloroacetophenone)
C_8H_7ClO
154.01854
RI: 1020*
Lacrimator
LS/Q
CAS: 532-27-4
Isobutylbenzene
$C_{10}H_{14}$
134.10955
RI: 1050*
Solvent
LM/Q
CAS: 135-98-8
Benzoic acid methylester
$C_8H_8O_2$
136.05243
RI: 1180*
Perfume
LM
CAS: 93-58-3
Benzoic acid ethylester
$C_9H_{10}O_2$
150.06808
RI: 1225*
Perfume
LM
Benzoic acid
$C_7H_6O_2$
122.03678
RI: 1235*
D: U UHY
Preservative
LM
CAS: 65-85-0

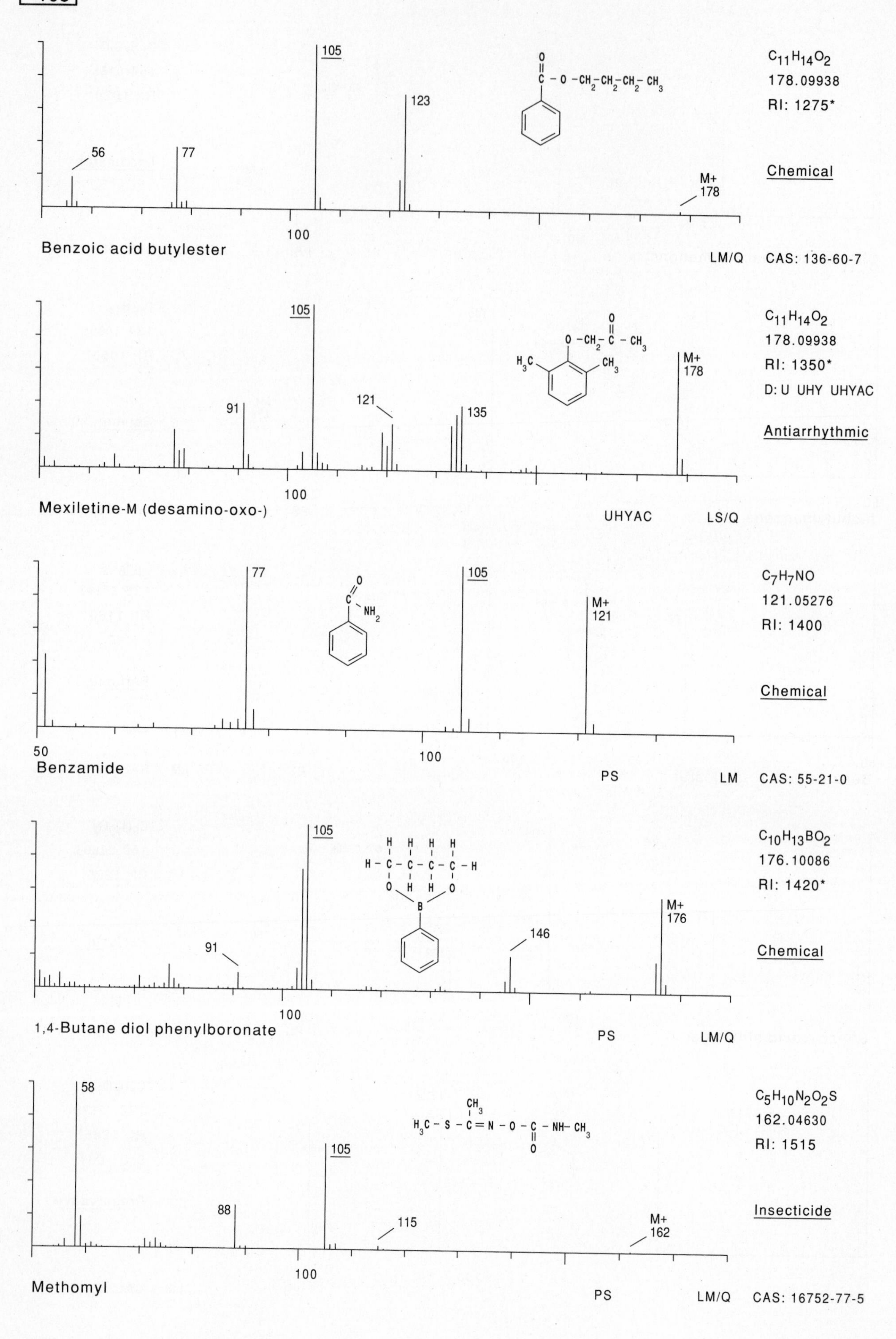
Benzoic acid butylester
$C_{11}H_{14}O_2$
178.09938
RI: 1275*
Chemical
LM/Q
CAS: 136-60-7
Mexiletine-M (desamino-oxo-)
$C_{11}H_{14}O_2$
178.09938
RI: 1350*
D: U UHY UHYAC
Antiarrhythmic
UHYAC
LS/Q
Benzamide
C_7H_7NO
121.05276
RI: 1400
Chemical
PS
LM
CAS: 55-21-0
1,4-Butane diol phenylboronate
$C_{10}H_{13}BO_2$
176.10086
RI: 1420*
Chemical
PS
LM/Q
Methomyl
$C_5H_{10}N_2O_2S$
162.04630
RI: 1515
Insecticide
PS
LM/Q
CAS: 16752-77-5

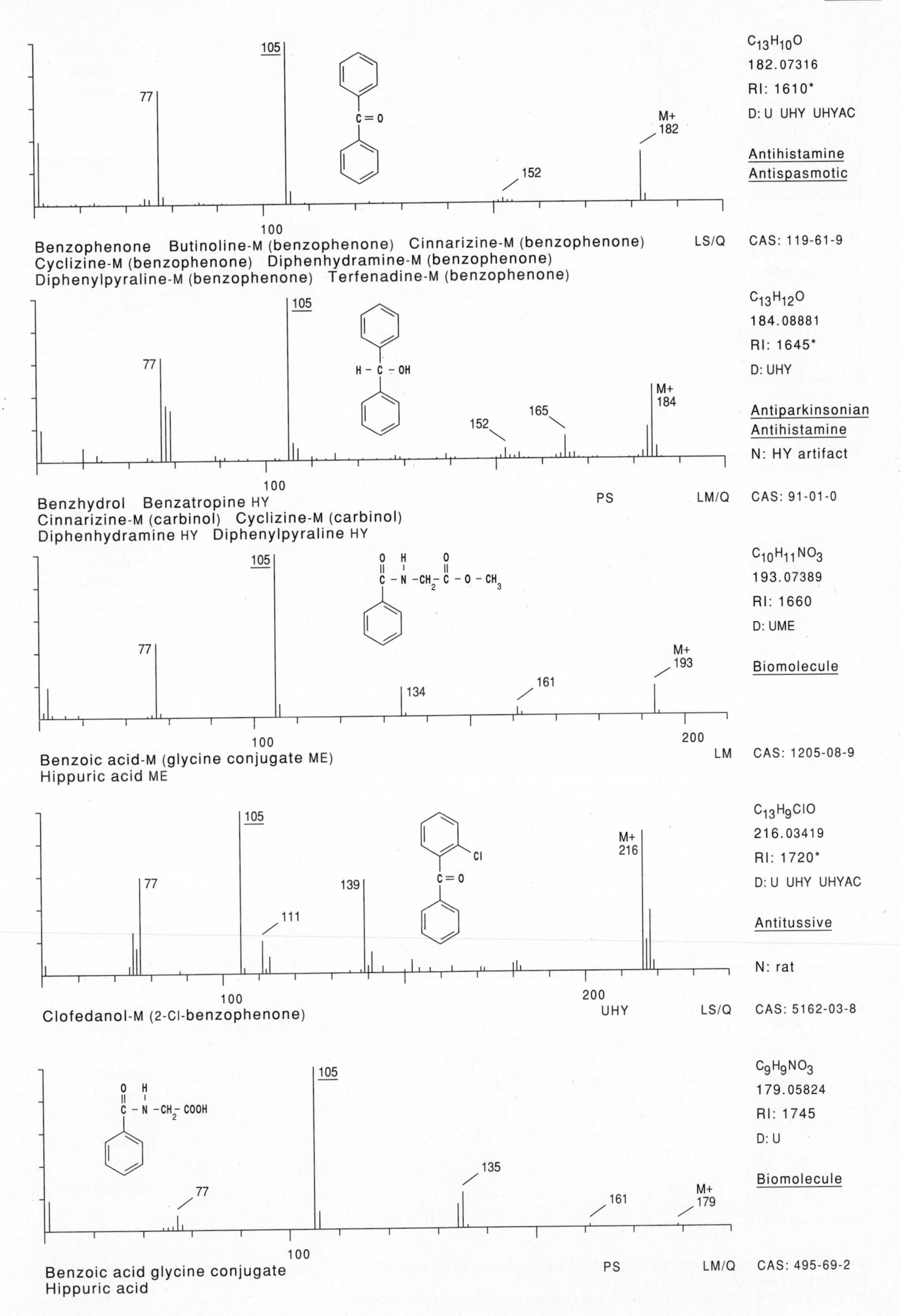

C13H10O
182.07316
RI: 1610*
D: U UHY UHYAC
Antihistamine
Antispasmotic
LS/Q
CAS: 119-61-9
Benzophenone Butinoline-M (benzophenone) Cinnarizine-M (benzophenone)
Cyclizine-M (benzophenone) Diphenhydramine-M (benzophenone)
Diphenylpyraline-M (benzophenone) Terfenadine-M (benzophenone)
C13H12O
184.08881
RI: 1645*
D: UHY
Antiparkinsonian
Antihistamine
N: HY artifact
PS
LM/Q
CAS: 91-01-0
Benzhydrol Benzatropine HY
Cinnarizine-M (carbinol) Cyclizine-M (carbinol)
Diphenhydramine HY Diphenylpyraline HY
C10H11NO3
193.07389
RI: 1660
D: UME
Biomolecule
LM
CAS: 1205-08-9
Benzoic acid-M (glycine conjugate ME)
Hippuric acid ME
C13H9ClO
216.03419
RI: 1720*
D: U UHY UHYAC
Antitussive
N: rat
UHY
LS/Q
CAS: 5162-03-8
Clofedanol-M (2-Cl-benzophenone)
C9H9NO3
179.05824
RI: 1745
D: U
Biomolecule
PS
LM/Q
CAS: 495-69-2
Benzoic acid glycine conjugate
Hippuric acid

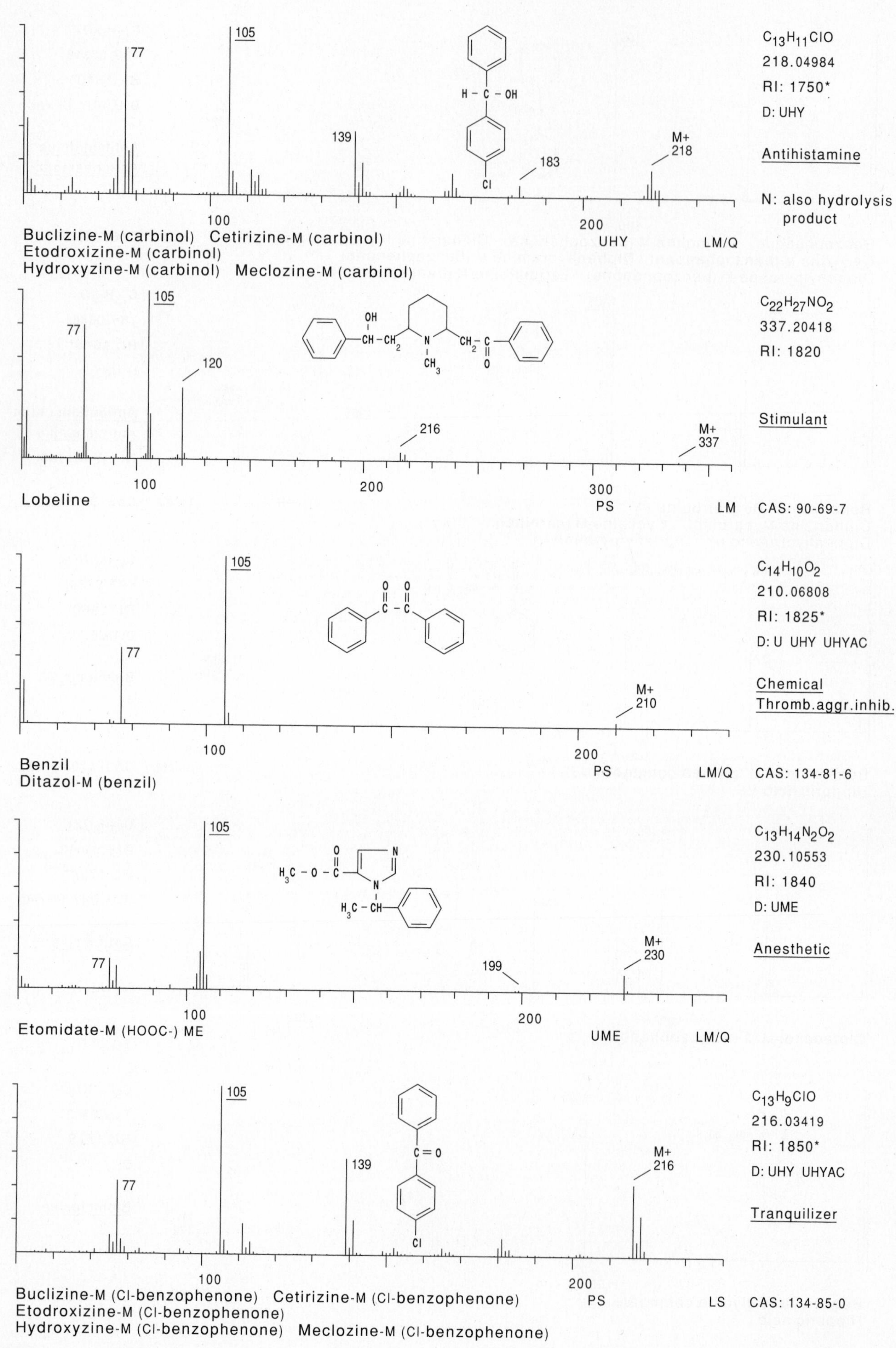

$C_{13}H_{11}ClO$
218.04984
RI: 1750*
D: UHY
Antihistamine
N: also hydrolysis product

Buclizine-M (carbinol) Cetirizine-M (carbinol)
Etodroxizine-M (carbinol)
Hydroxyzine-M (carbinol) Meclozine-M (carbinol)

$C_{22}H_{27}NO_2$
337.20418
RI: 1820
Stimulant

Lobeline
CAS: 90-69-7

$C_{14}H_{10}O_2$
210.06808
RI: 1825*
D: U UHY UHYAC
Chemical
Thromb.aggr.inhib.

Benzil
Ditazol-M (benzil)
CAS: 134-81-6

$C_{13}H_{14}N_2O_2$
230.10553
RI: 1840
D: UME
Anesthetic

Etomidate-M (HOOC-) ME

$C_{13}H_9ClO$
216.03419
RI: 1850*
D: UHY UHYAC
Tranquilizer

Buclizine-M (Cl-benzophenone) Cetirizine-M (Cl-benzophenone)
Etodroxizine-M (Cl-benzophenone)
Hydroxyzine-M (Cl-benzophenone) Meclozine-M (Cl-benzophenone)
CAS: 134-85-0

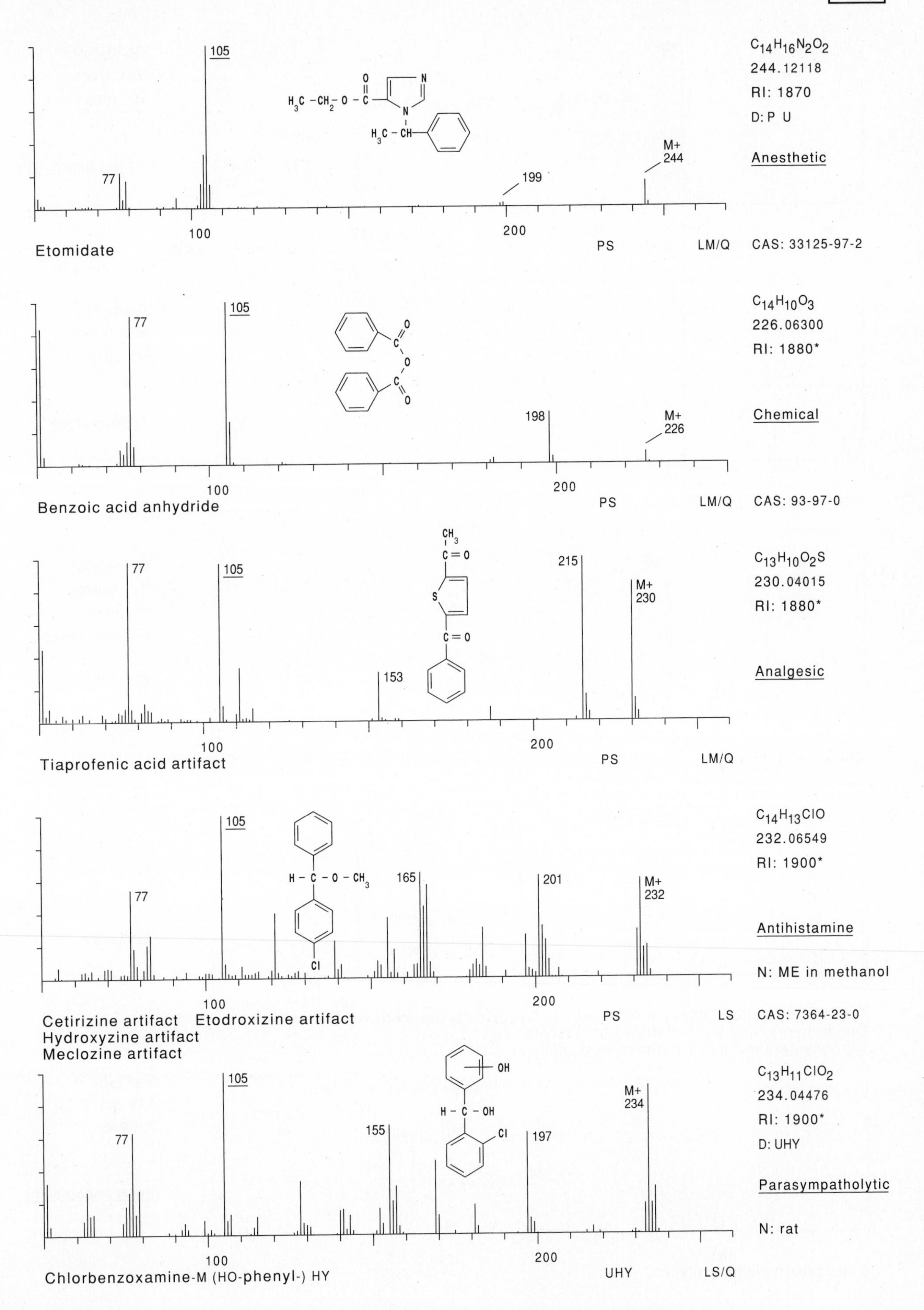

Etomidate
C14H16N2O2
244.12118
RI: 1870
D: P U
Anesthetic
PS LM/Q
CAS: 33125-97-2
Benzoic acid anhydride
C14H10O3
226.06300
RI: 1880*
Chemical
PS LM/Q
CAS: 93-97-0
Tiaprofenic acid artifact
C13H10O2S
230.04015
RI: 1880*
Analgesic
PS LM/Q
Cetirizine artifact Etodroxizine artifact
Hydroxyzine artifact
Meclozine artifact
C14H13ClO
232.06549
RI: 1900*
Antihistamine
N: ME in methanol
PS LS
CAS: 7364-23-0
Chlorbenzoxamine-M (HO-phenyl-) HY
C13H11ClO2
234.04476
RI: 1900*
D: UHY
Parasympatholytic
N: rat
UHY LS/Q

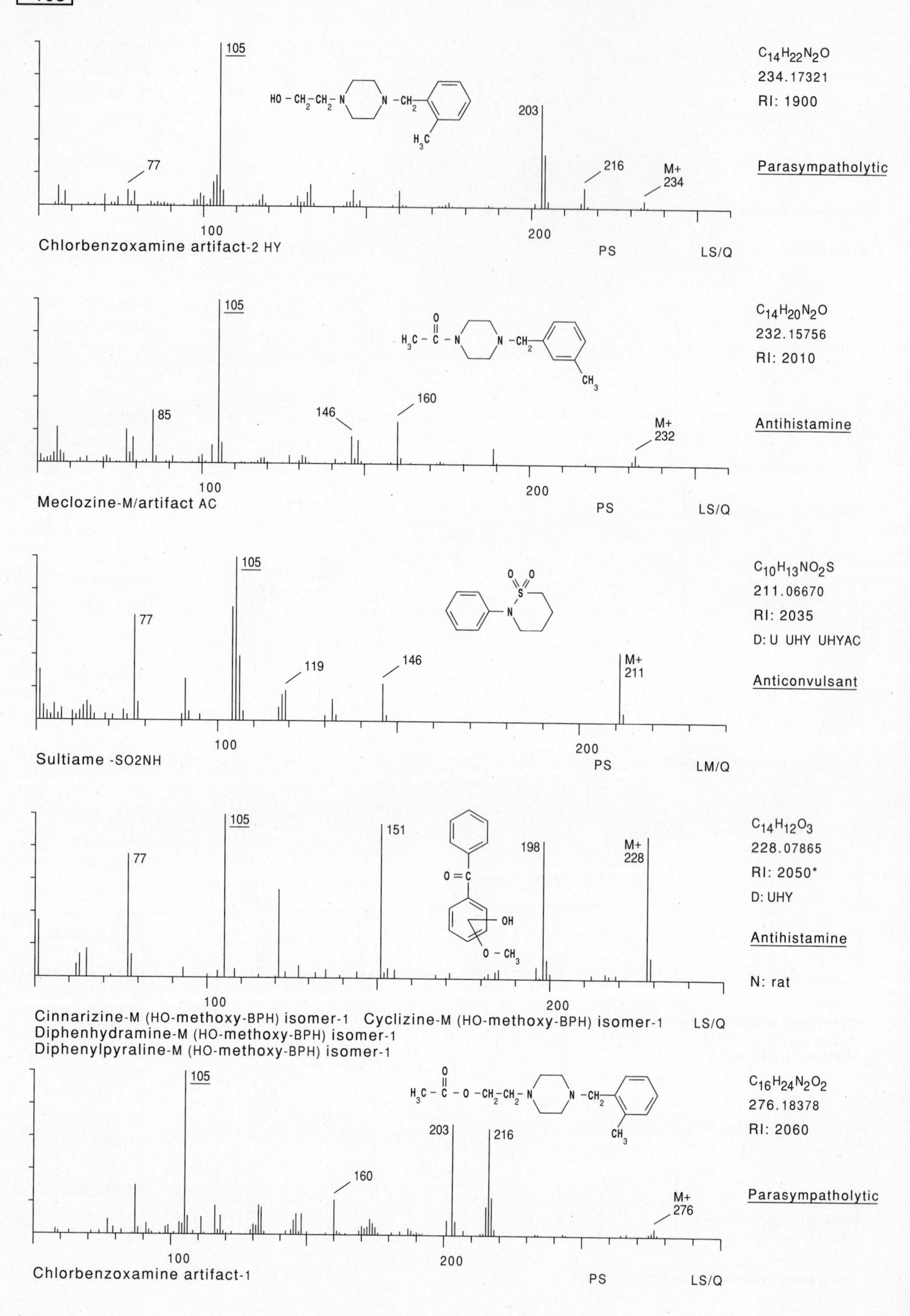

Chlorbenzoxamine artifact-2 HY

$C_{14}H_{22}N_2O$
234.17321
RI: 1900

Parasympatholytic

Meclozine-M/artifact AC

$C_{14}H_{20}N_2O$
232.15756
RI: 2010

Antihistamine

Sultiame -SO2NH

$C_{10}H_{13}NO_2S$
211.06670
RI: 2035
D: U UHY UHYAC

Anticonvulsant

Cinnarizine-M (HO-methoxy-BPH) isomer-1 Cyclizine-M (HO-methoxy-BPH) isomer-1
Diphenhydramine-M (HO-methoxy-BPH) isomer-1
Diphenylpyraline-M (HO-methoxy-BPH) isomer-1

$C_{14}H_{12}O_3$
228.07865
RI: 2050*
D: UHY

Antihistamine

N: rat

Chlorbenzoxamine artifact-1

$C_{16}H_{24}N_2O_2$
276.18378
RI: 2060

Parasympatholytic

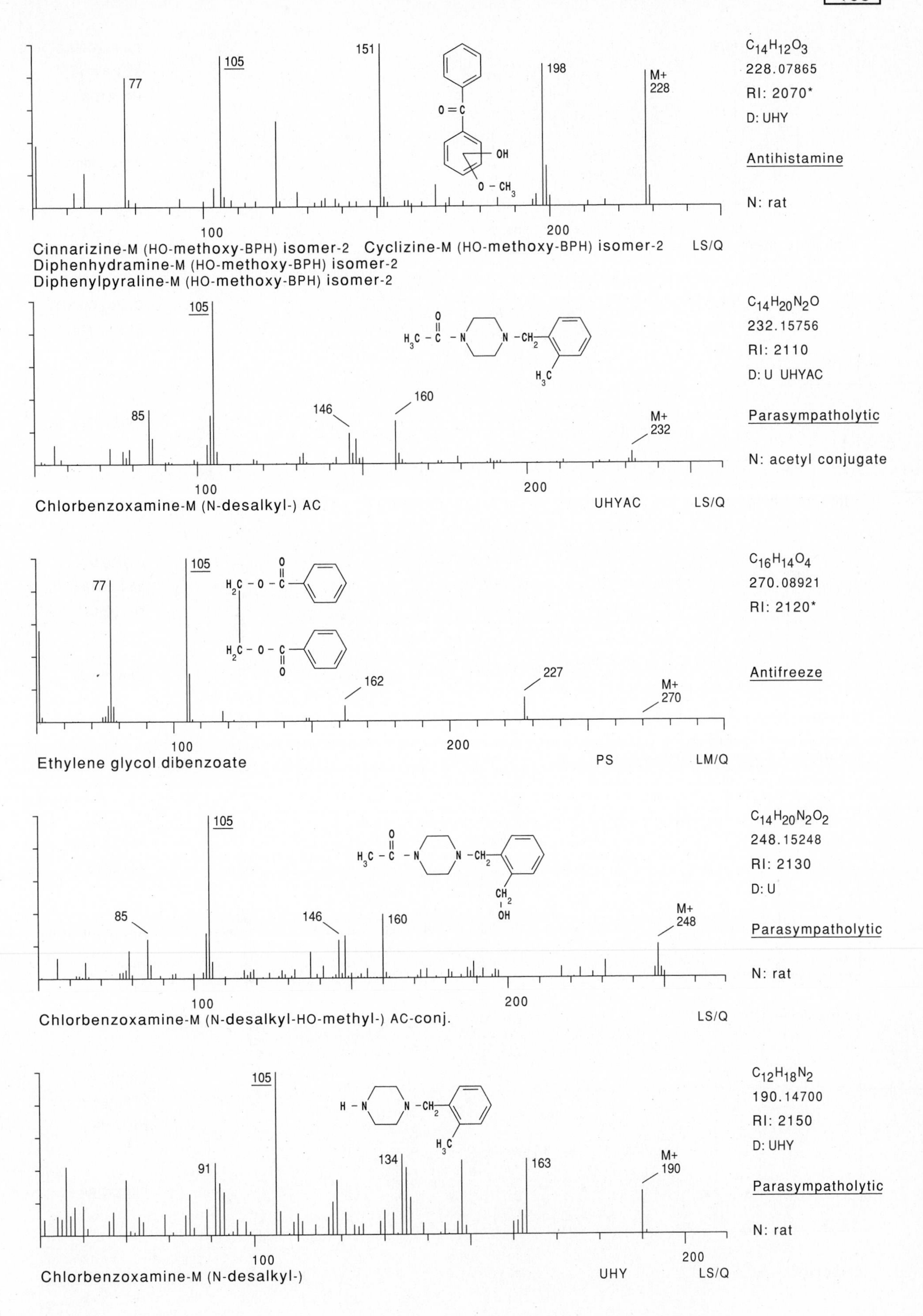

C14H12O3
228.07865
RI: 2070*
D: UHY
Antihistamine
N: rat
Cinnarizine-M (HO-methoxy-BPH) isomer-2 Cyclizine-M (HO-methoxy-BPH) isomer-2 LS/Q
Diphenhydramine-M (HO-methoxy-BPH) isomer-2
Diphenylpyraline-M (HO-methoxy-BPH) isomer-2
C14H20N2O
232.15756
RI: 2110
D: U UHYAC
Parasympatholytic
N: acetyl conjugate
Chlorbenzoxamine-M (N-desalkyl-) AC UHYAC LS/Q
C16H14O4
270.08921
RI: 2120*
Antifreeze
Ethylene glycol dibenzoate PS LM/Q
C14H20N2O2
248.15248
RI: 2130
D: U
Parasympatholytic
N: rat
Chlorbenzoxamine-M (N-desalkyl-HO-methyl-) AC-conj. LS/Q
C12H18N2
190.14700
RI: 2150
D: UHY
Parasympatholytic
N: rat
Chlorbenzoxamine-M (N-desalkyl-) UHY LS/Q

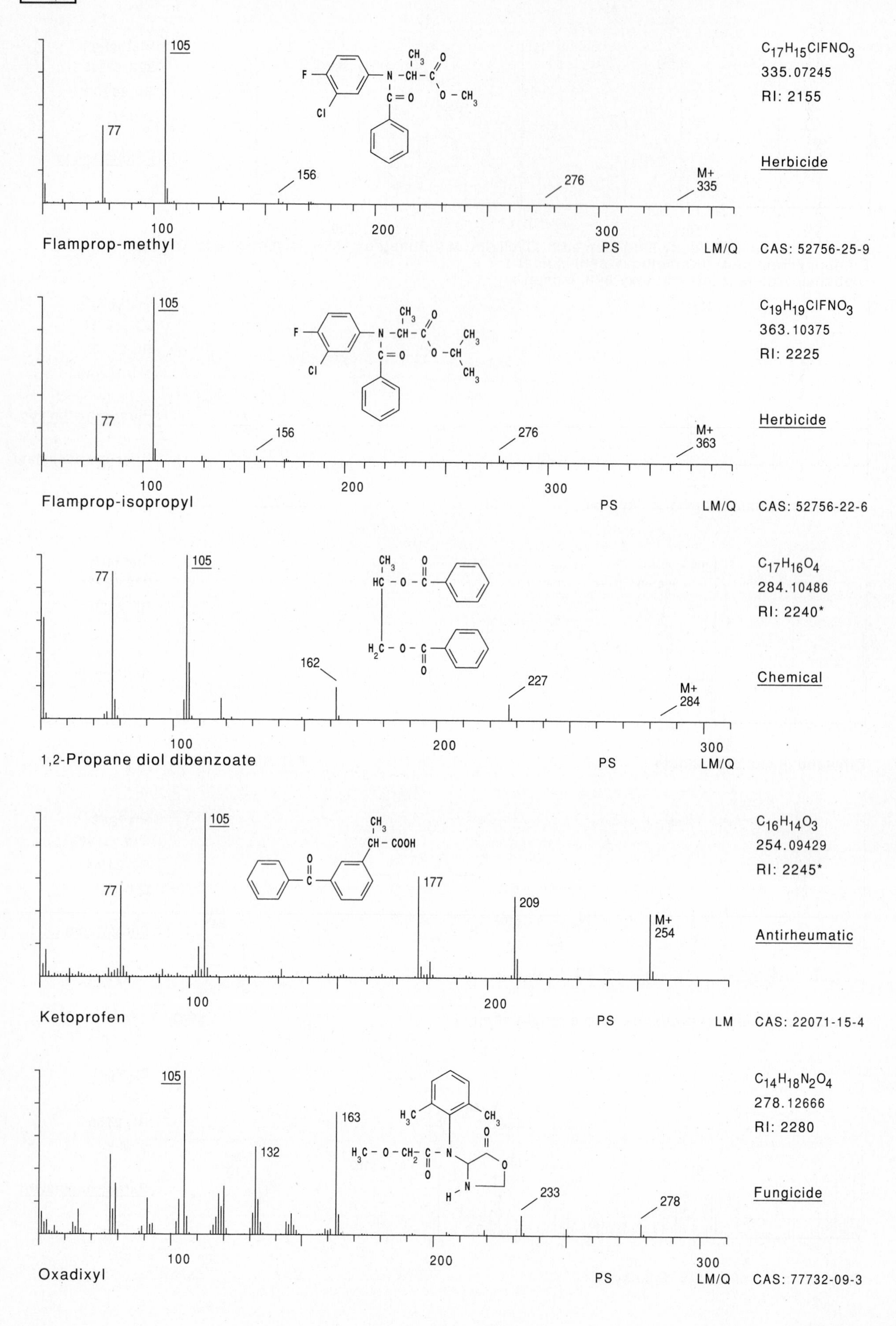
105
77
156
276
M+
335
100
200
300
Flamprop-methyl
PS
LM/Q
$C_{17}H_{15}ClFNO_3$
335.07245
RI: 2155
Herbicide
CAS: 52756-25-9
105
77
156
276
M+
363
100
200
300
Flamprop-isopropyl
PS
LM/Q
$C_{19}H_{19}ClFNO_3$
363.10375
RI: 2225
Herbicide
CAS: 52756-22-6
77
105
162
227
M+
284
100
200
300
1,2-Propane diol dibenzoate
PS
LM/Q
$C_{17}H_{16}O_4$
284.10486
RI: 2240*
Chemical
105
77
177
209
M+
254
100
200
Ketoprofen
PS
LM
$C_{16}H_{14}O_3$
254.09429
RI: 2245*
Antirheumatic
CAS: 22071-15-4
105
163
132
233
278
100
200
300
Oxadixyl
PS
LM/Q
$C_{14}H_{18}N_2O_4$
278.12666
RI: 2280
Fungicide
CAS: 77732-09-3

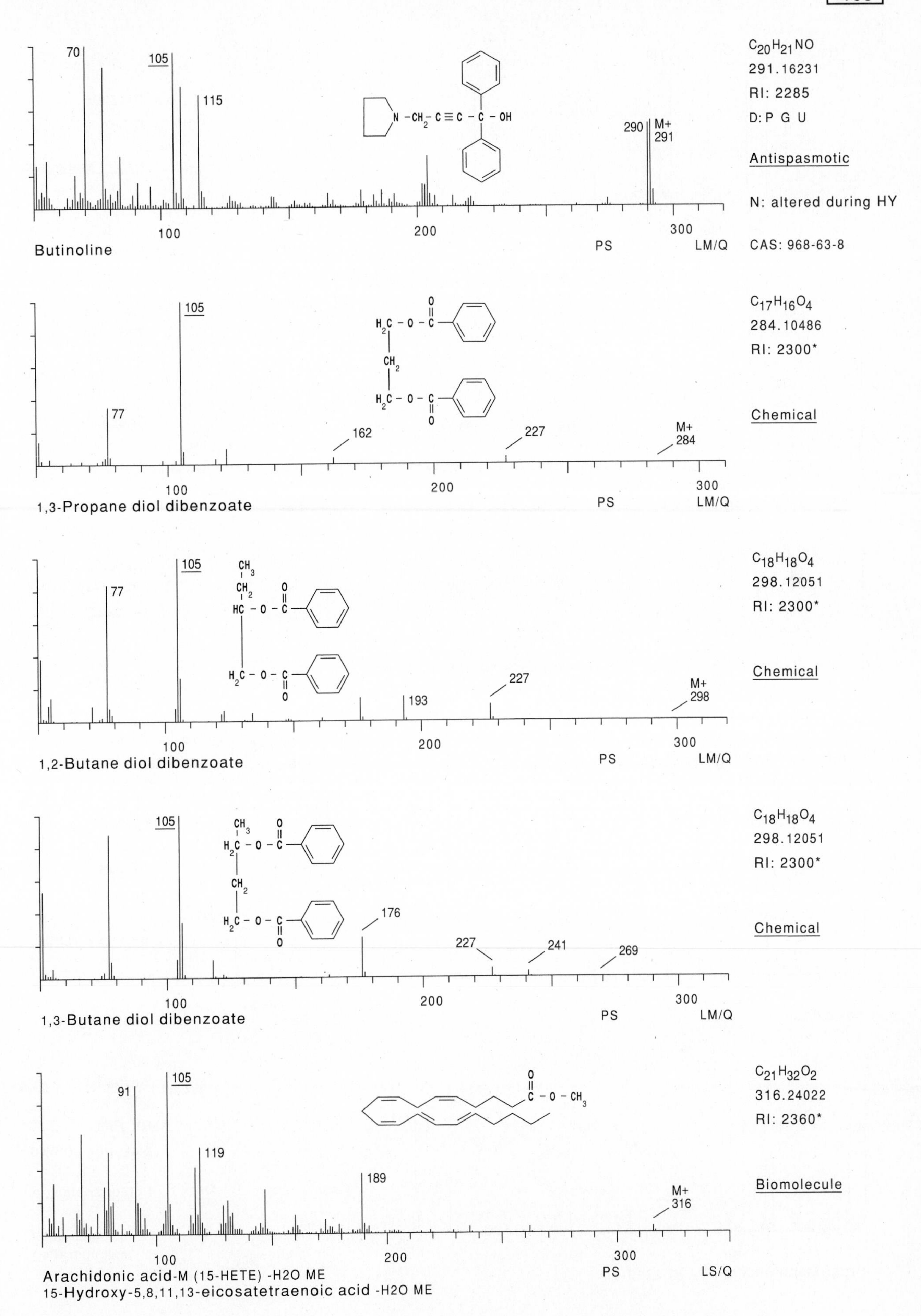

$C_{20}H_{21}NO$
291.16231
RI: 2285
D: P G U

Antispasmotic

N: altered during HY

CAS: 968-63-8

$C_{17}H_{16}O_4$
284.10486
RI: 2300*

Chemical

$C_{18}H_{18}O_4$
298.12051
RI: 2300*

Chemical

$C_{18}H_{18}O_4$
298.12051
RI: 2300*

Chemical

$C_{21}H_{32}O_2$
316.24022
RI: 2360*

Biomolecule

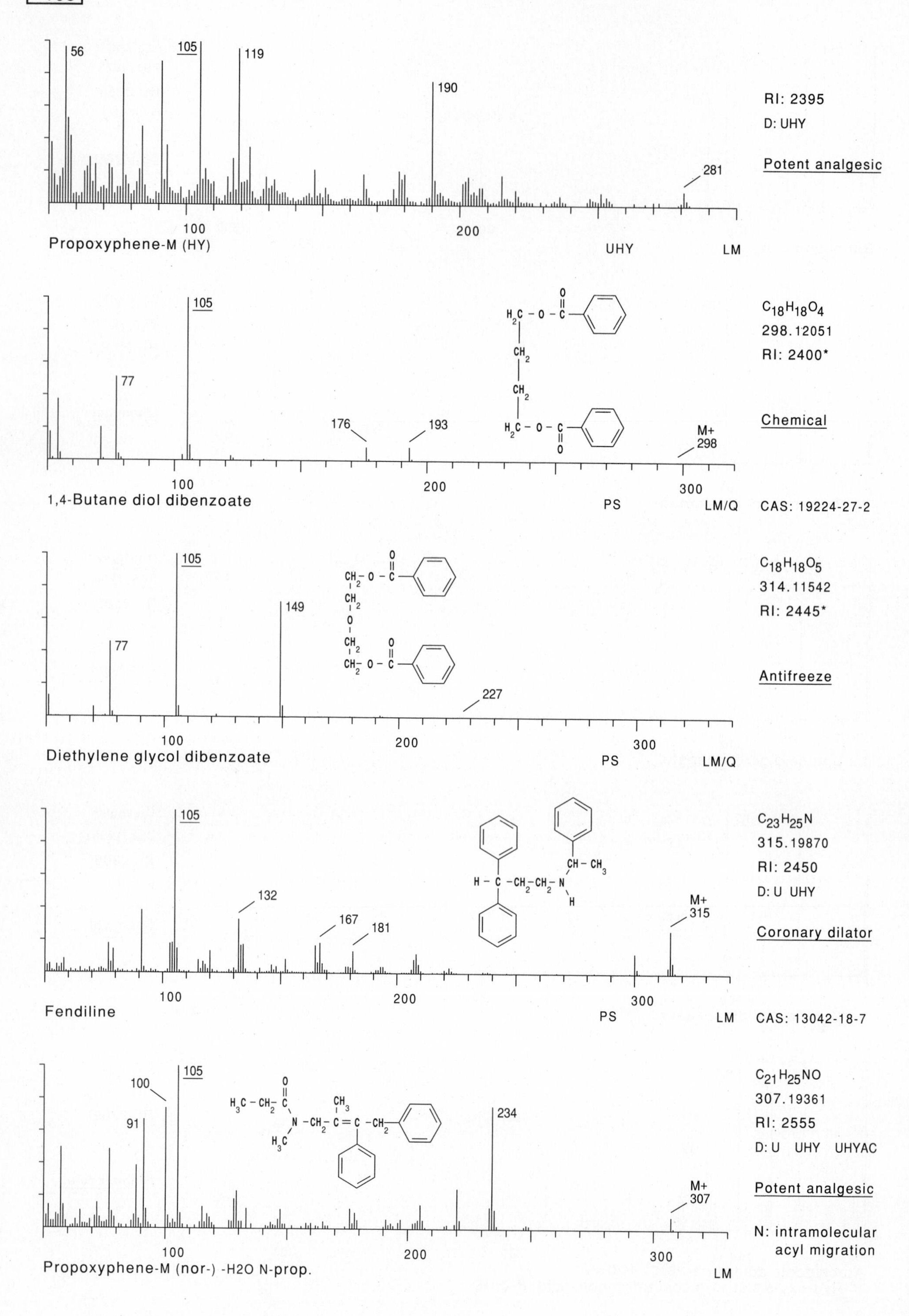
56
105
119
190
281
100
200
Propoxyphene-M (HY)
UHY
LM
RI: 2395
D: UHY
Potent analgesic
105
77
176
193
M+
298
100
200
300
1,4-Butane diol dibenzoate
PS
LM/Q
C18H18O4
298.12051
RI: 2400*
Chemical
CAS: 19224-27-2
105
149
77
227
100
200
300
Diethylene glycol dibenzoate
PS
LM/Q
C18H18O5
314.11542
RI: 2445*
Antifreeze
105
132
167
181
M+
315
100
200
300
Fendiline
PS
LM
C23H25N
315.19870
RI: 2450
D: U UHY
Coronary dilator
CAS: 13042-18-7
100
105
91
234
M+
307
100
200
300
Propoxyphene-M (nor-) -H2O N-prop.
LM
C21H25NO
307.19361
RI: 2555
D: U UHY UHYAC
Potent analgesic
N: intramolecular
acyl migration

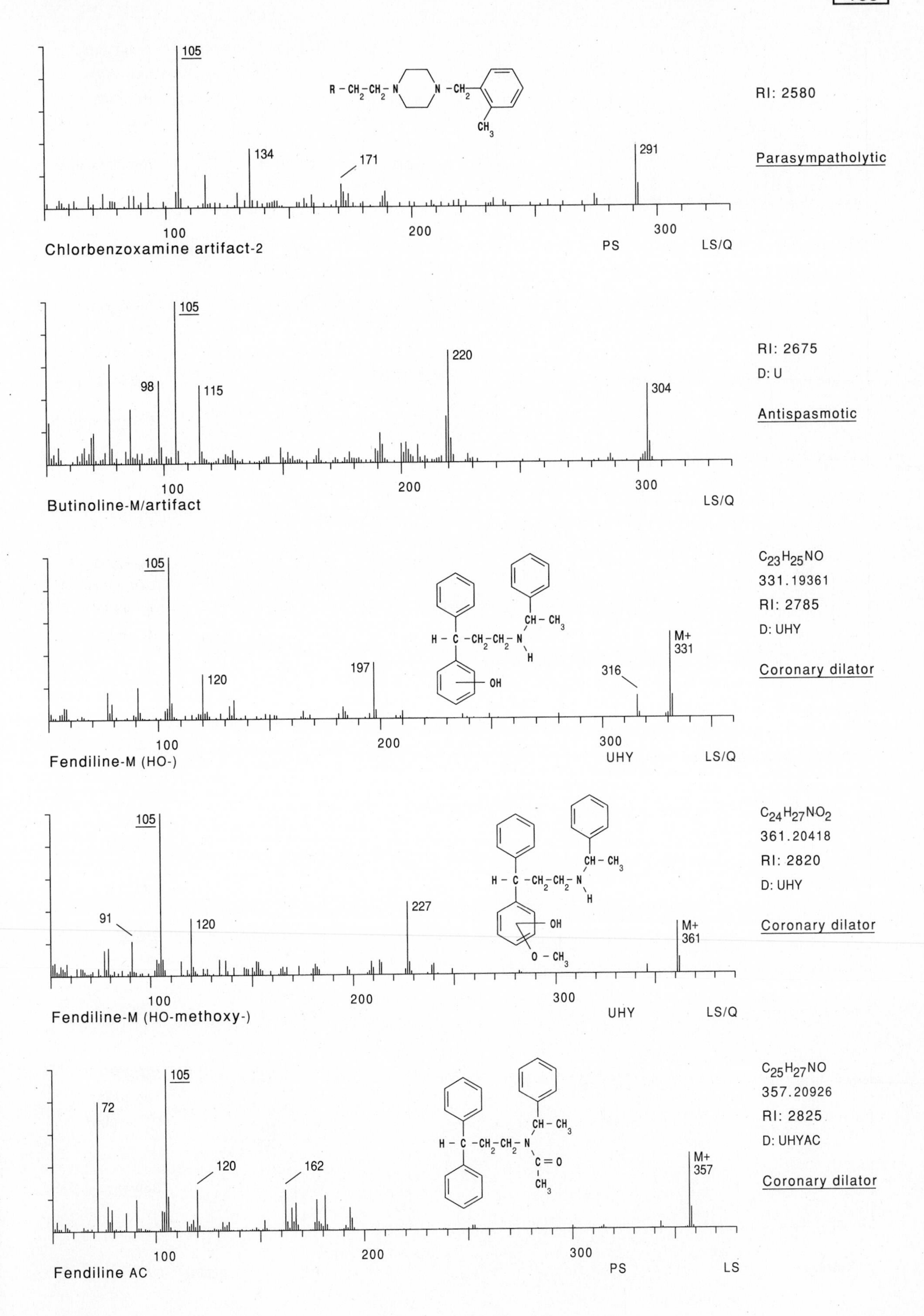

RI: 2580

Parasympatholytic

RI: 2675

D: U

Antispasmotic

$C_{23}H_{25}NO$

331.19361

RI: 2785

D: UHY

Coronary dilator

$C_{24}H_{27}NO_2$

361.20418

RI: 2820

D: UHY

Coronary dilator

$C_{25}H_{27}NO$

357.20926

RI: 2825

D: UHYAC

Coronary dilator

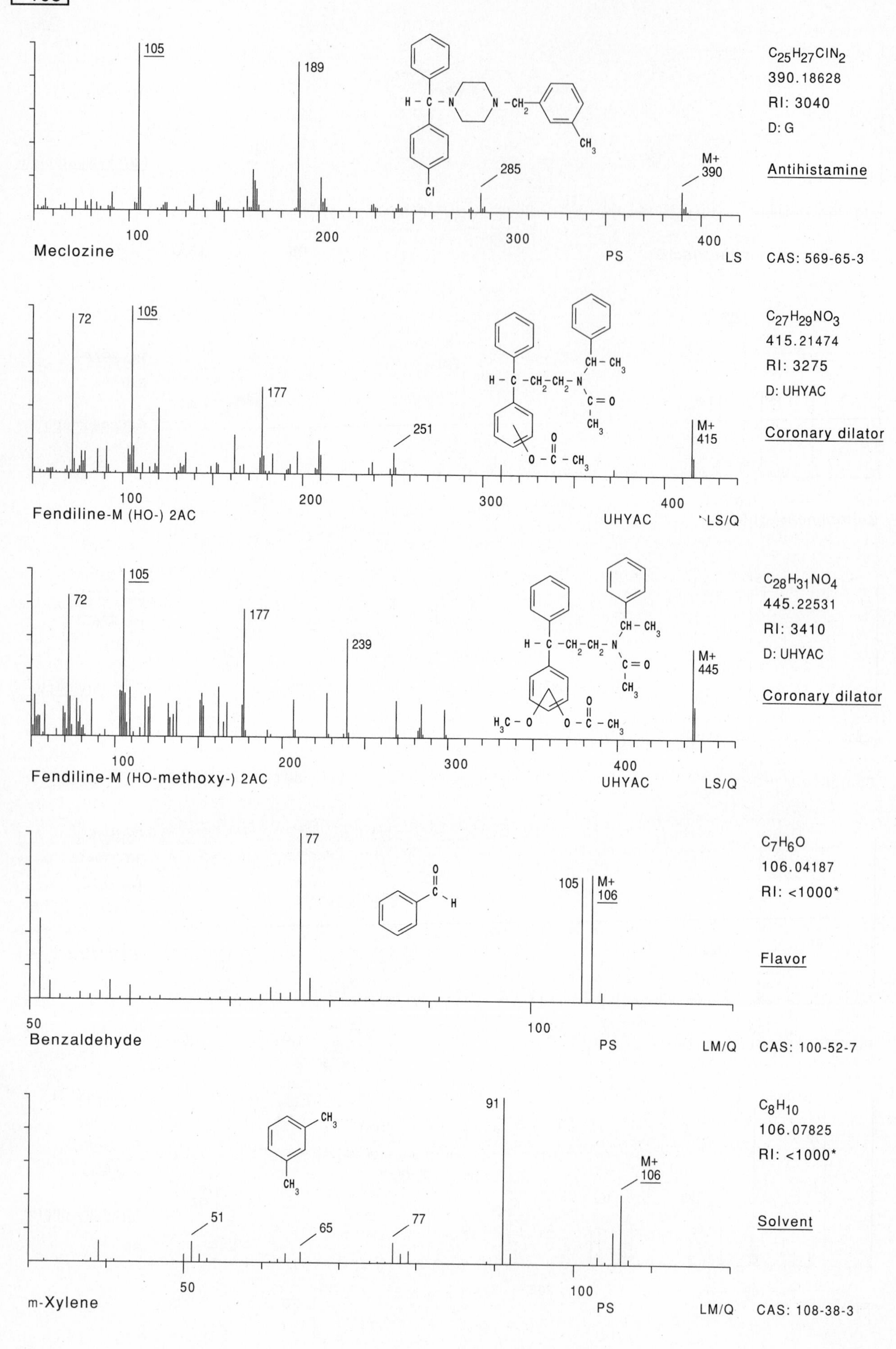
105
189
285
M+
390
100
200
300
400
Meclozine
PS
LS
$C_{25}H_{27}ClN_2$
390.18628
RI: 3040
D: G
Antihistamine
CAS: 569-65-3
72
105
177
251
M+
415
Fendiline-M (HO-) 2AC
UHYAC
LS/Q
$C_{27}H_{29}NO_3$
415.21474
RI: 3275
D: UHYAC
Coronary dilator
72
105
177
239
M+
445
Fendiline-M (HO-methoxy-) 2AC
UHYAC
LS/Q
$C_{28}H_{31}NO_4$
445.22531
RI: 3410
D: UHYAC
Coronary dilator
77
105
M+
106
50
100
Benzaldehyde
PS
LM/Q
C_7H_6O
106.04187
RI: <1000*
Flavor
CAS: 100-52-7
91
M+
106
51
65
77
50
100
m-Xylene
PS
LM/Q
C_8H_{10}
106.07825
RI: <1000*
Solvent
CAS: 108-38-3

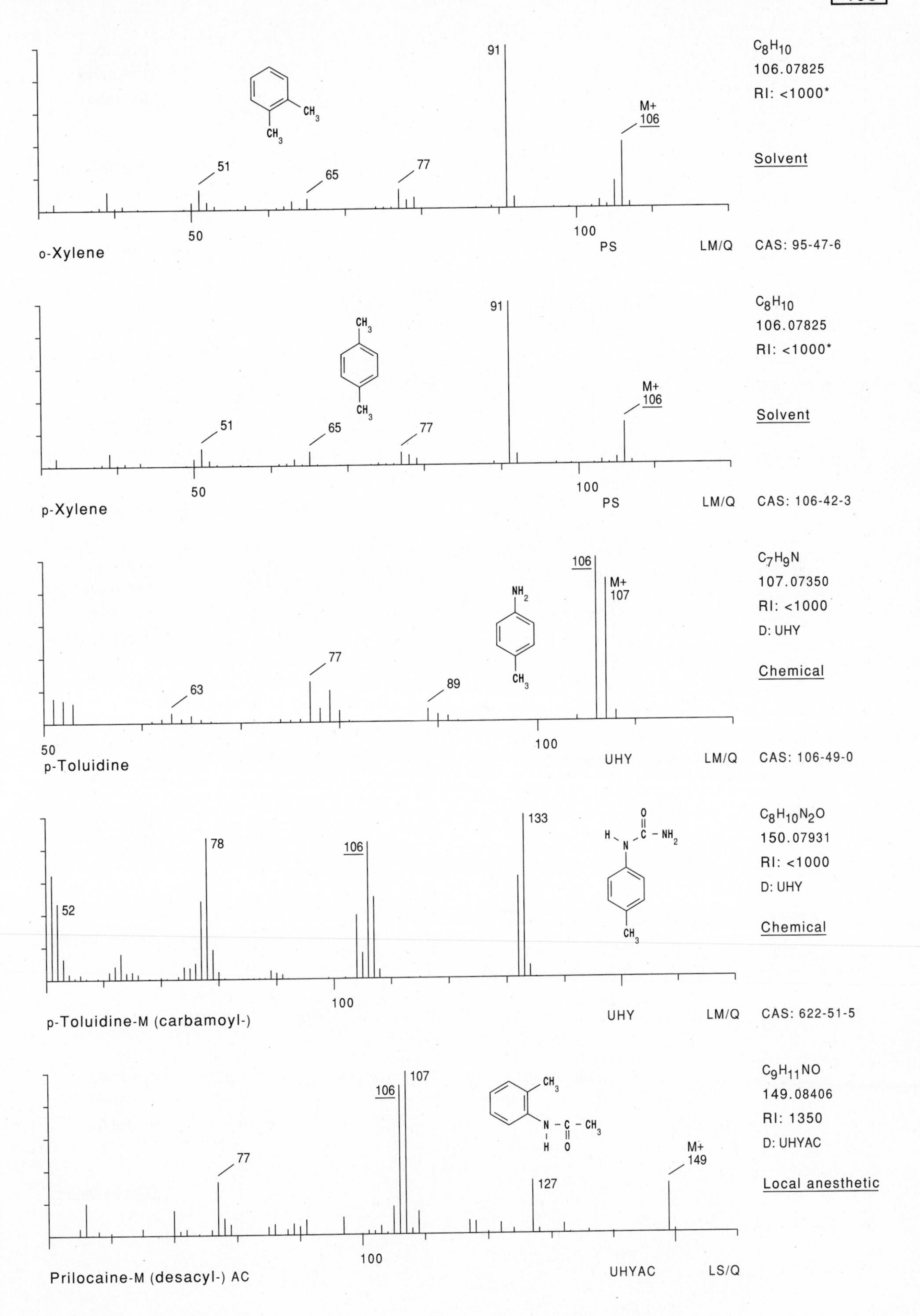
91
M+
106
51
65
77
CH3
CH3
50
100
PS
LM/Q
o-Xylene
C8H10
106.07825
RI: <1000*
Solvent
CAS: 95-47-6
91
M+
106
51
65
77
CH3
CH3
50
100
PS
LM/Q
p-Xylene
C8H10
106.07825
RI: <1000*
Solvent
CAS: 106-42-3
106
M+
107
NH2
CH3
77
63
89
50
100
UHY
LM/Q
p-Toluidine
C7H9N
107.07350
RI: <1000
D: UHY
Chemical
CAS: 106-49-0
133
78
106
52
O
H N C NH2
CH3
100
UHY
LM/Q
p-Toluidine-M (carbamoyl-)
C8H10N2O
150.07931
RI: <1000
D: UHY
Chemical
CAS: 622-51-5
107
106
CH3
N C CH3
H O
77
127
M+
149
100
UHYAC
LS/Q
Prilocaine-M (desacyl-) AC
C9H11NO
149.08406
RI: 1350
D: UHYAC
Local anesthetic

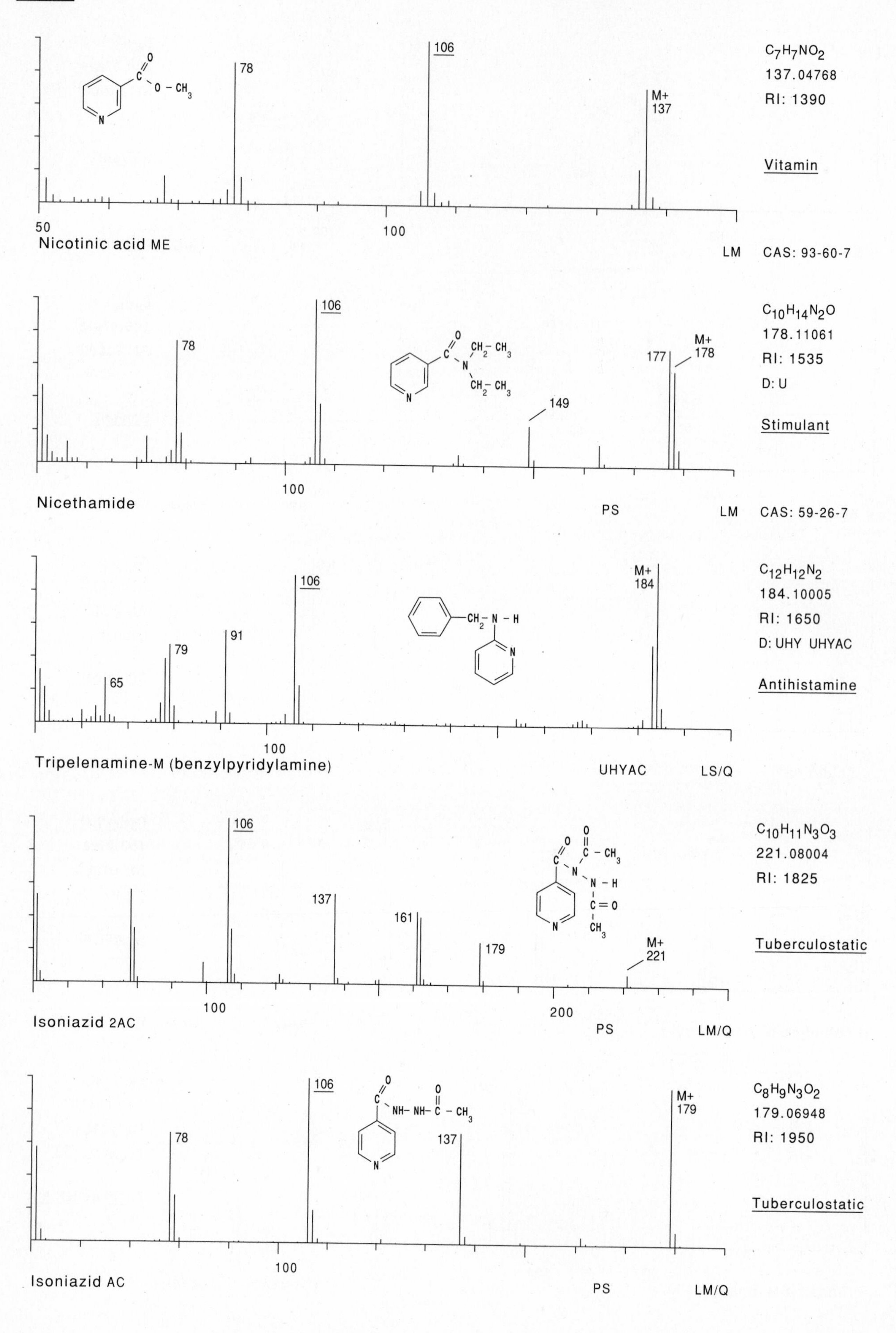

106
78
M+
137
50
100
Nicotinic acid ME
LM
$C_7H_7NO_2$
137.04768
RI: 1390
Vitamin
CAS: 93-60-7
106
78
M+
178
177
149
100
Nicethamide
PS
LM
$C_{10}H_{14}N_2O$
178.11061
RI: 1535
D: U
Stimulant
CAS: 59-26-7
M+
184
106
91
79
65
100
Tripelenamine-M (benzylpyridylamine)
UHYAC
LS/Q
$C_{12}H_{12}N_2$
184.10005
RI: 1650
D: UHY UHYAC
Antihistamine
106
137
161
179
M+
221
100
200
Isoniazid 2AC
PS
LM/Q
$C_{10}H_{11}N_3O_3$
221.08004
RI: 1825
Tuberculostatic
106
M+
179
78
137
100
Isoniazid AC
PS
LM/Q
$C_8H_9N_3O_2$
179.06948
RI: 1950
Tuberculostatic

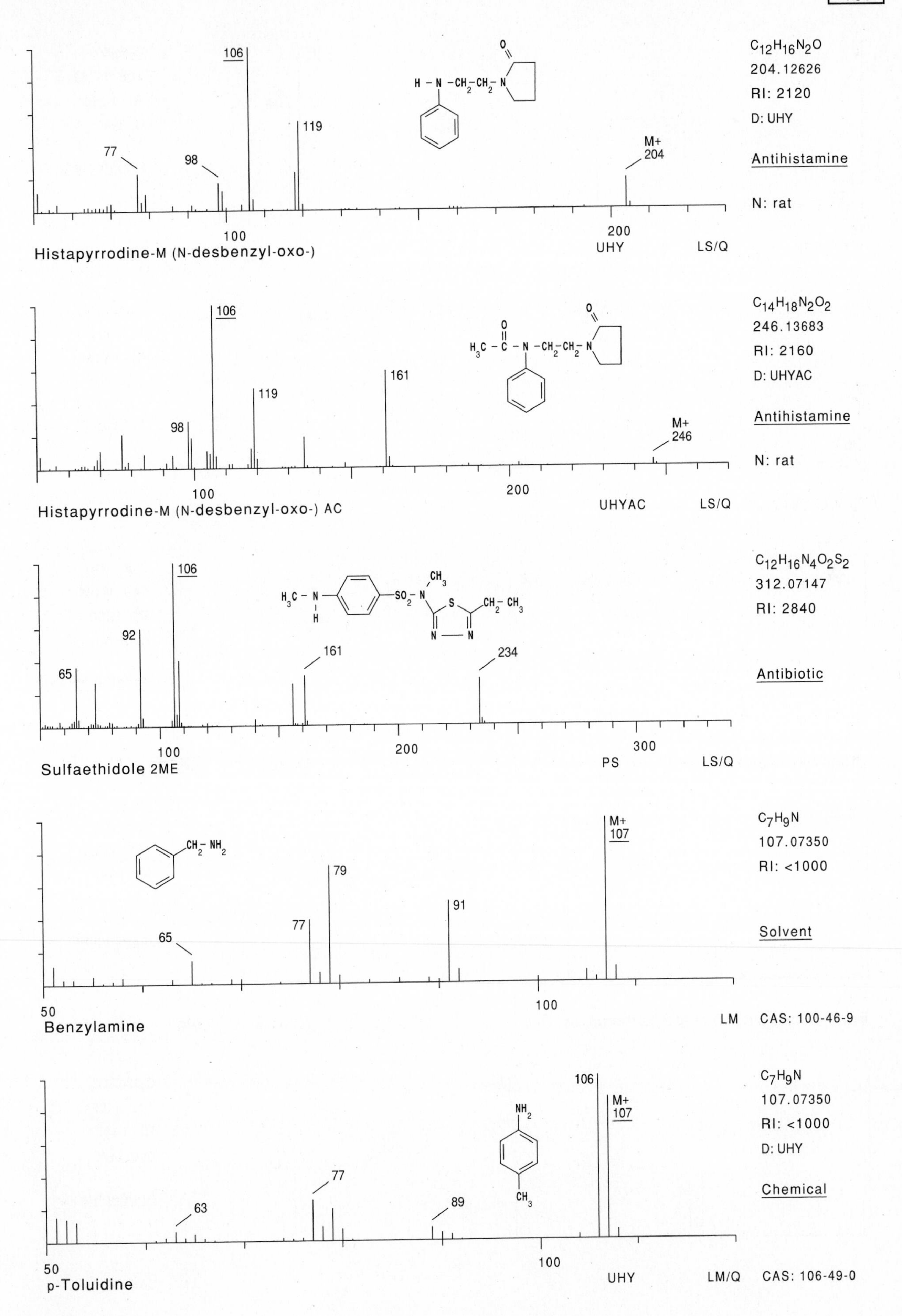

$C_{12}H_{16}N_2O$
204.12626
RI: 2120
D: UHY
Antihistamine
N: rat

$C_{14}H_{18}N_2O_2$
246.13683
RI: 2160
D: UHYAC
Antihistamine
N: rat

$C_{12}H_{16}N_4O_2S_2$
312.07147
RI: 2840
Antibiotic

C_7H_9N
107.07350
RI: <1000
Solvent
CAS: 100-46-9

C_7H_9N
107.07350
RI: <1000
D: UHY
Chemical
CAS: 106-49-0

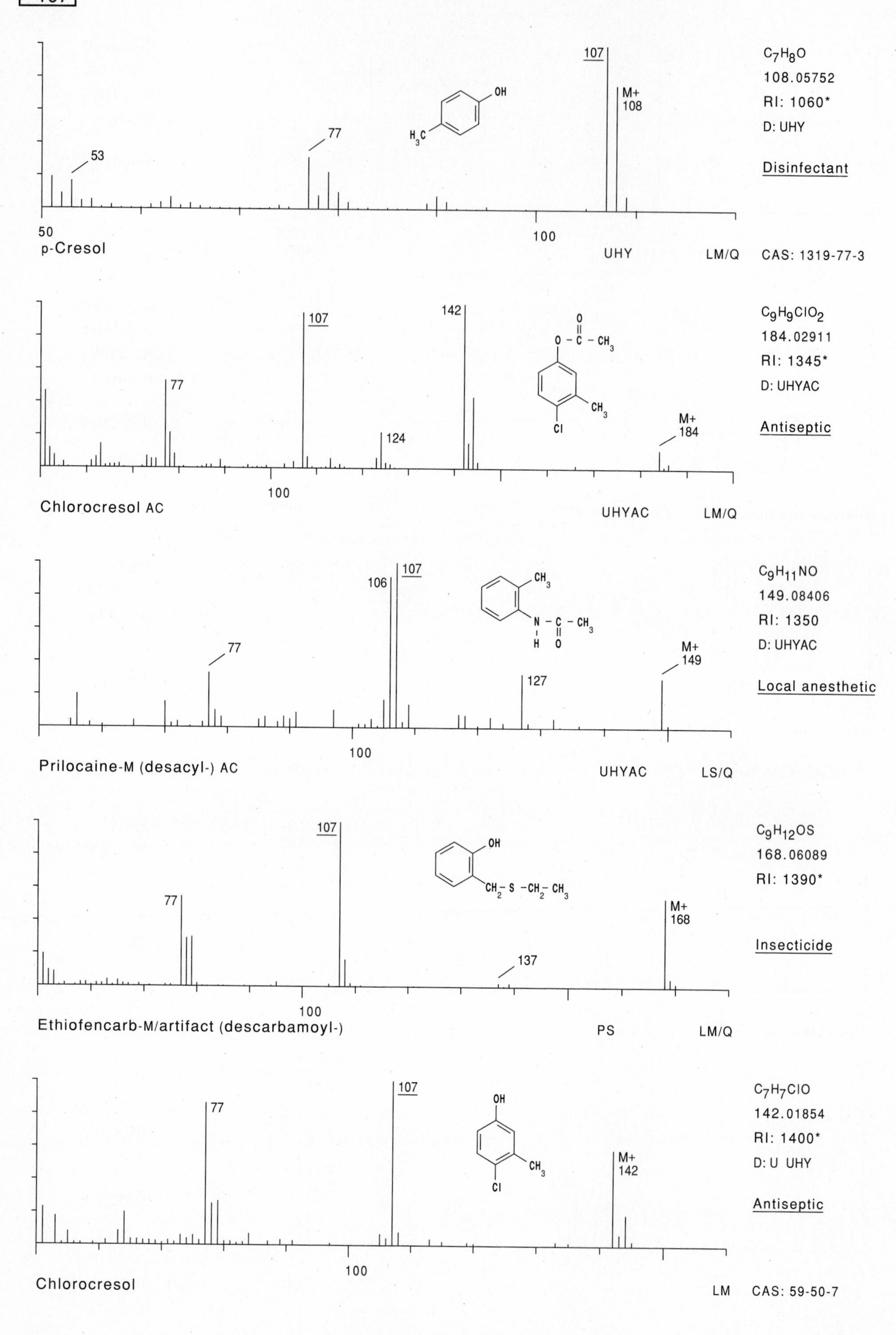
107
53
77
107
M+
108
OH
H_3C
50
100
p-Cresol
UHY
LM/Q
C_7H_8O
108.05752
RI: 1060*
D: UHY
Disinfectant
CAS: 1319-77-3
77
107
124
142
M+
184
100
Chlorocresol AC
UHYAC
LM/Q
$C_9H_9ClO_2$
184.02911
RI: 1345*
D: UHYAC
Antiseptic
77
106
107
127
M+
149
100
Prilocaine-M (desacyl-) AC
UHYAC
LS/Q
$C_9H_{11}NO$
149.08406
RI: 1350
D: UHYAC
Local anesthetic
77
107
137
M+
168
OH
100
Ethiofencarb-M/artifact (descarbamoyl-)
PS
LM/Q
$C_9H_{12}OS$
168.06089
RI: 1390*
Insecticide
77
107
M+
142
OH
Cl
100
Chlorocresol
LM
C_7H_7ClO
142.01854
RI: 1400*
D: U UHY
Antiseptic
CAS: 59-50-7

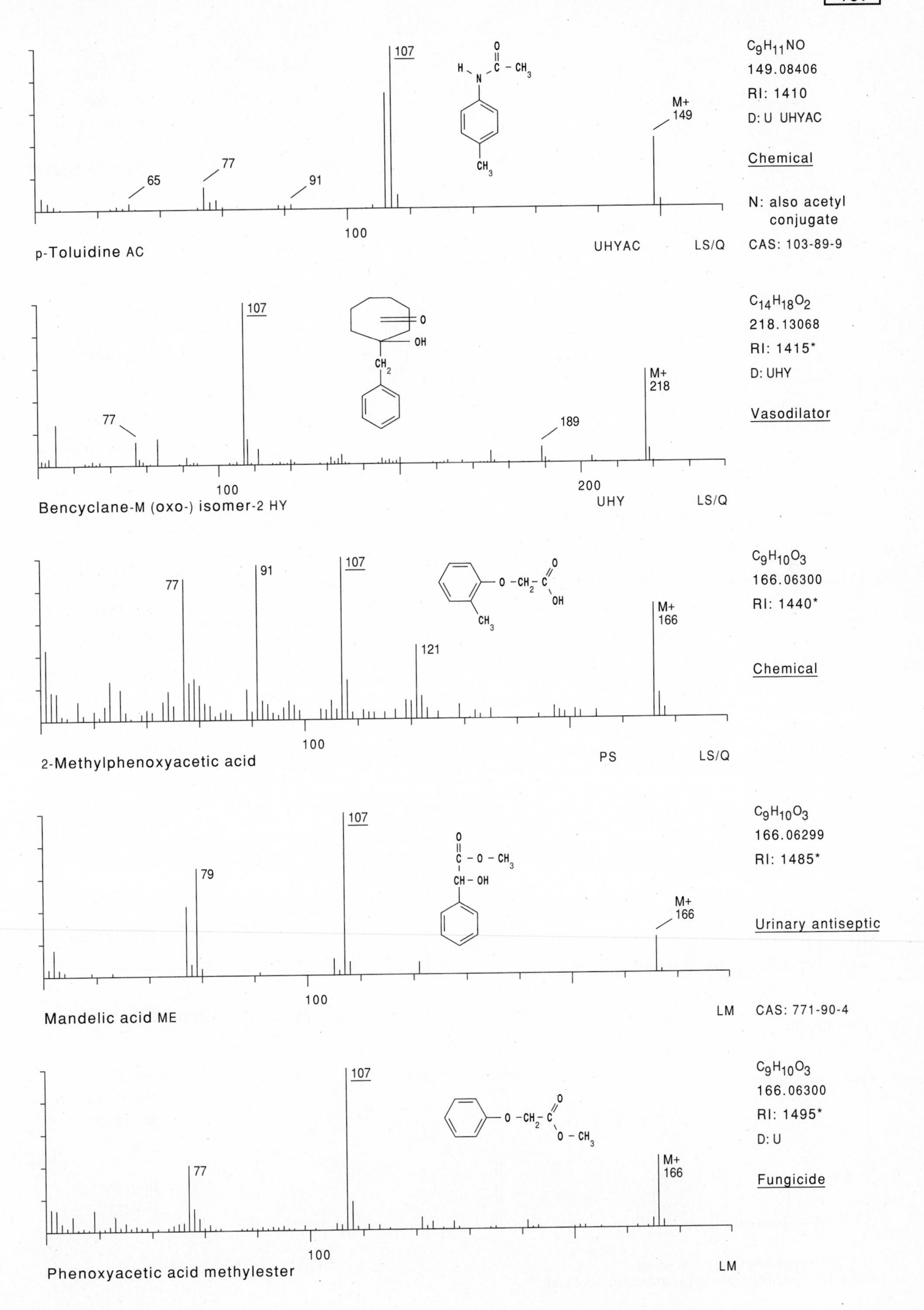
p-Toluidine AC
C9H11NO
149.08406
RI: 1410
D: U UHYAC
Chemical
N: also acetyl conjugate
CAS: 103-89-9
UHYAC LS/Q
Bencyclane-M (oxo-) isomer-2 HY
C14H18O2
218.13068
RI: 1415*
D: UHY
Vasodilator
UHY LS/Q
2-Methylphenoxyacetic acid
C9H10O3
166.06300
RI: 1440*
Chemical
PS LS/Q
Mandelic acid ME
C9H10O3
166.06299
RI: 1485*
Urinary antiseptic
LM
CAS: 771-90-4
Phenoxyacetic acid methylester
C9H10O3
166.06300
RI: 1495*
D: U
Fungicide
LM

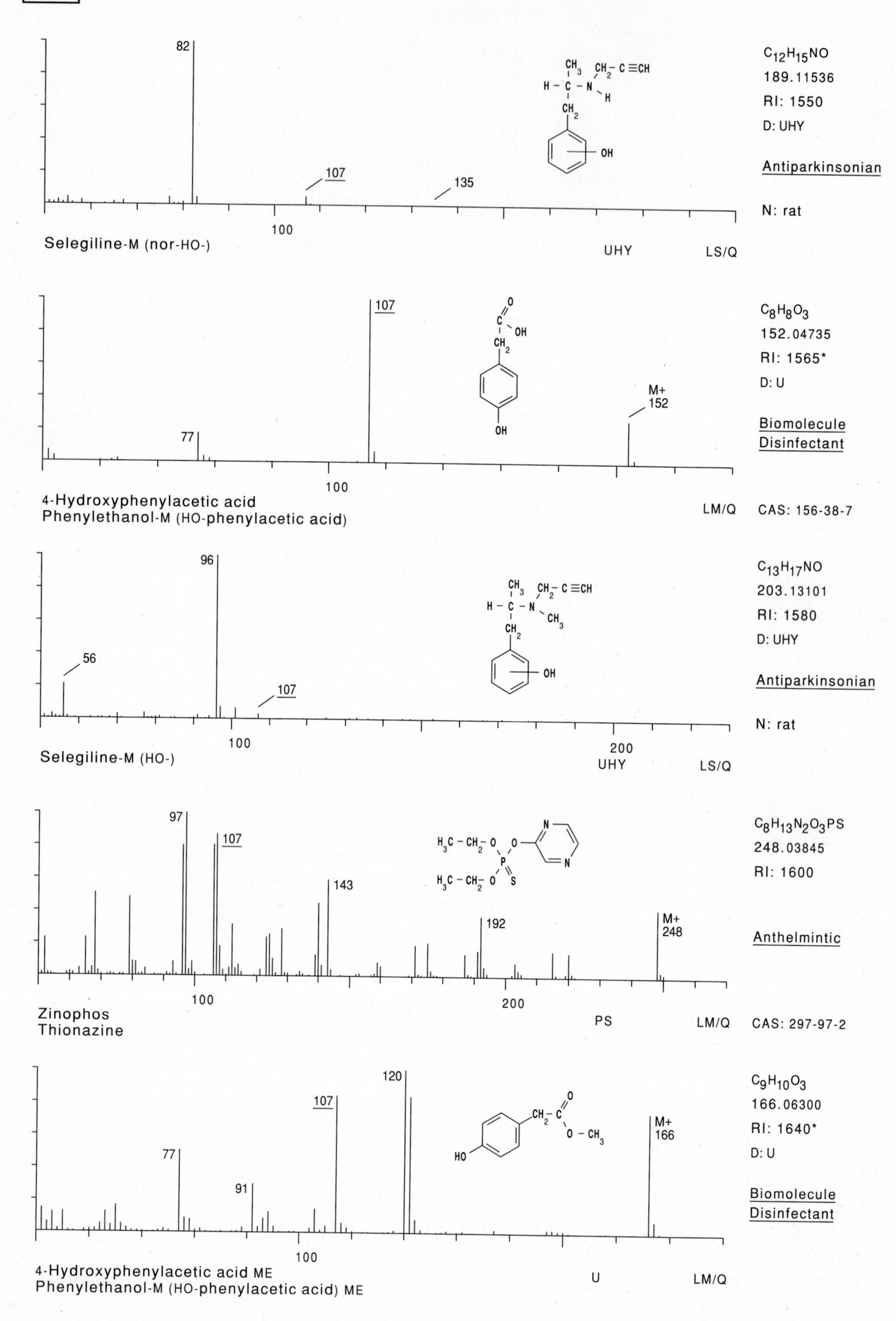

$C_{12}H_{15}NO$
189.11536
RI: 1550
D: UHY
Antiparkinsonian
N: rat
Selegiline-M (nor-HO-)
UHY
LS/Q
$C_8H_8O_3$
152.04735
RI: 1565*
D: U
Biomolecule
Disinfectant
4-Hydroxyphenylacetic acid
Phenylethanol-M (HO-phenylacetic acid)
LM/Q
CAS: 156-38-7
$C_{13}H_{17}NO$
203.13101
RI: 1580
D: UHY
Antiparkinsonian
N: rat
Selegiline-M (HO-)
UHY
LS/Q
$C_8H_{13}N_2O_3PS$
248.03845
RI: 1600
Anthelmintic
Zinophos
Thionazine
PS
LM/Q
CAS: 297-97-2
$C_9H_{10}O_3$
166.06300
RI: 1640*
D: U
Biomolecule
Disinfectant
4-Hydroxyphenylacetic acid ME
Phenylethanol-M (HO-phenylacetic acid) ME
U
LM/Q

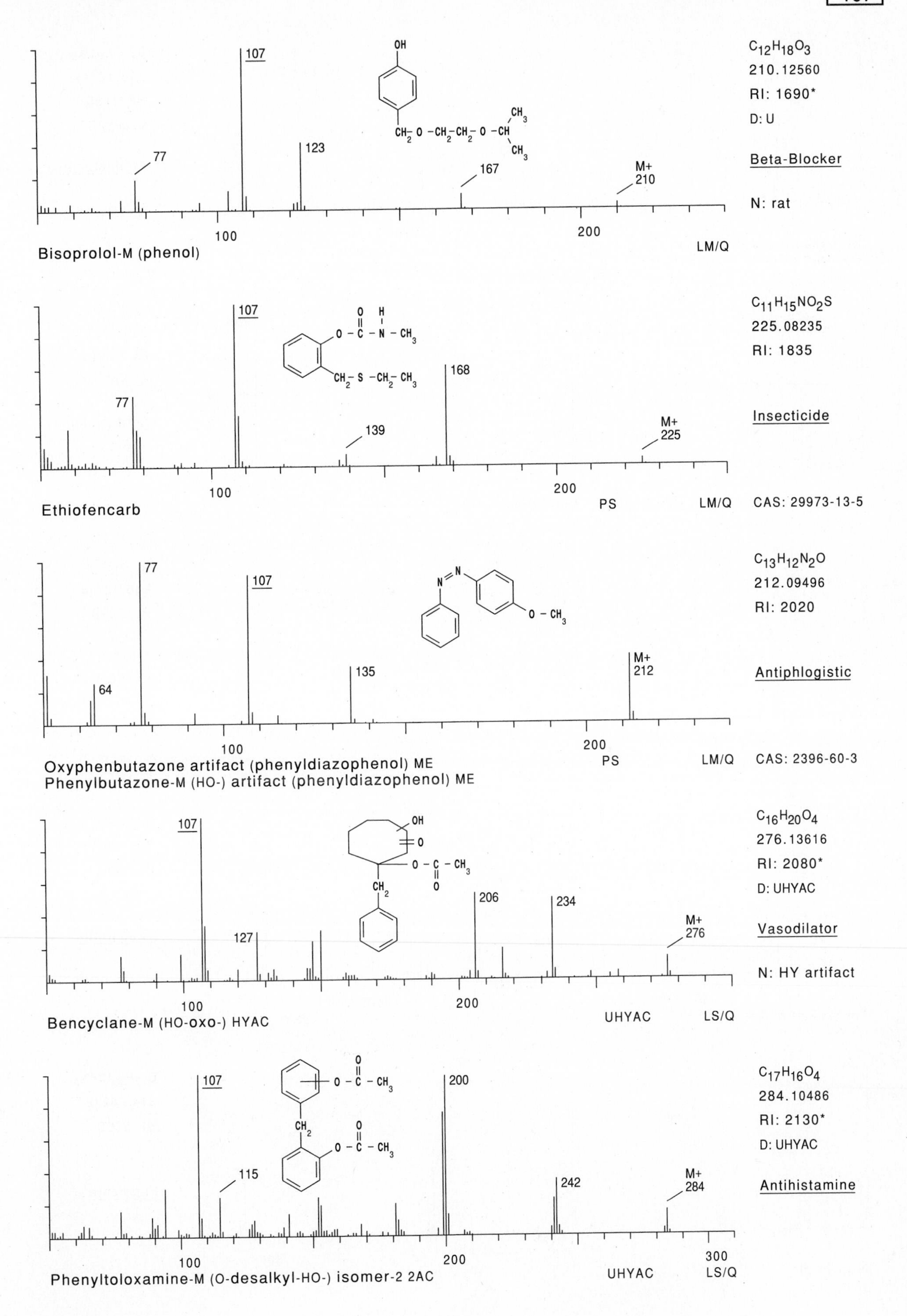
107
123
77
167
M+
210
100
200
LM/Q
Bisoprolol-M (phenol)
$C_{12}H_{18}O_3$
210.12560
RI: 1690*
D: U
Beta-Blocker
N: rat
107
168
77
139
M+
225
100
200
PS
LM/Q
Ethiofencarb
$C_{11}H_{15}NO_2S$
225.08235
RI: 1835
Insecticide
CAS: 29973-13-5
77
107
135
64
M+
212
100
200
PS
LM/Q
Oxyphenbutazone artifact (phenyldiazophenol) ME
Phenylbutazone-M (HO-) artifact (phenyldiazophenol) ME
$C_{13}H_{12}N_2O$
212.09496
RI: 2020
Antiphlogistic
CAS: 2396-60-3
107
127
206
234
M+
276
100
200
UHYAC
LS/Q
Bencyclane-M (HO-oxo-) HYAC
$C_{16}H_{20}O_4$
276.13616
RI: 2080*
D: UHYAC
Vasodilator
N: HY artifact
107
200
115
242
M+
284
100
200
300
UHYAC
LS/Q
Phenyltoloxamine-M (O-desalkyl-HO-) isomer-2 2AC
$C_{17}H_{16}O_4$
284.10486
RI: 2130*
D: UHYAC
Antihistamine

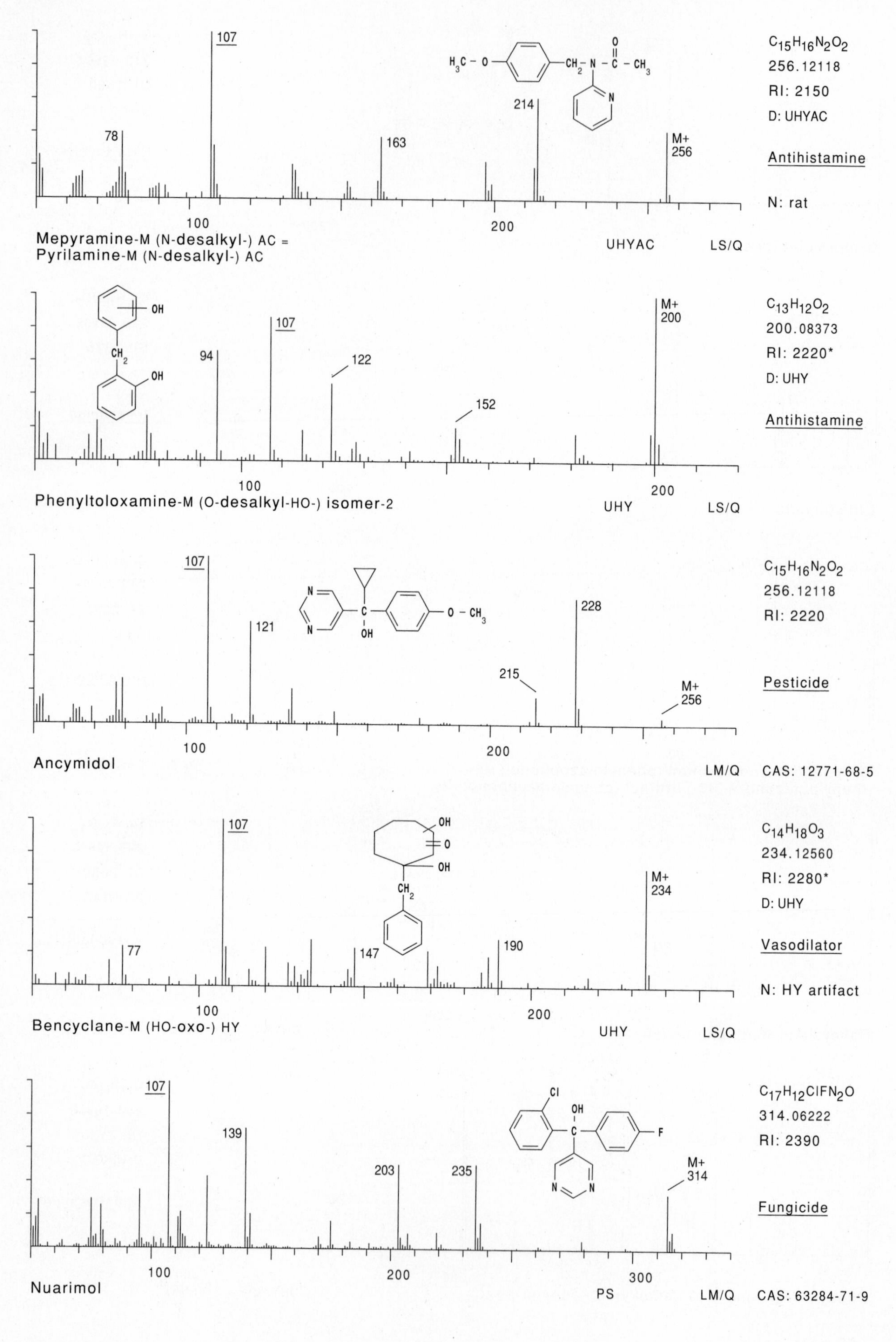
107
78
163
214
M+
256
100
200
Mepyramine-M (N-desalkyl-) AC =
Pyrilamine-M (N-desalkyl-) AC
UHYAC
LS/Q
$C_{15}H_{16}N_2O_2$
256.12118
RI: 2150
D: UHYAC
Antihistamine
N: rat
OH
CH_2
OH
94
107
122
152
M+
200
100
200
Phenyltoloxamine-M (O-desalkyl-HO-) isomer-2
UHY
LS/Q
$C_{13}H_{12}O_2$
200.08373
RI: 2220*
D: UHY
Antihistamine
107
121
228
215
M+
256
100
200
Ancymidol
LM/Q
$C_{15}H_{16}N_2O_2$
256.12118
RI: 2220
Pesticide
CAS: 12771-68-5
107
77
147
190
M+
234
100
200
Bencyclane-M (HO-oxo-) HY
UHY
LS/Q
$C_{14}H_{18}O_3$
234.12560
RI: 2280*
D: UHY
Vasodilator
N: HY artifact
107
139
203
235
M+
314
100
200
300
Nuarimol
PS
LM/Q
$C_{17}H_{12}ClFN_2O$
314.06222
RI: 2390
Fungicide
CAS: 63284-71-9

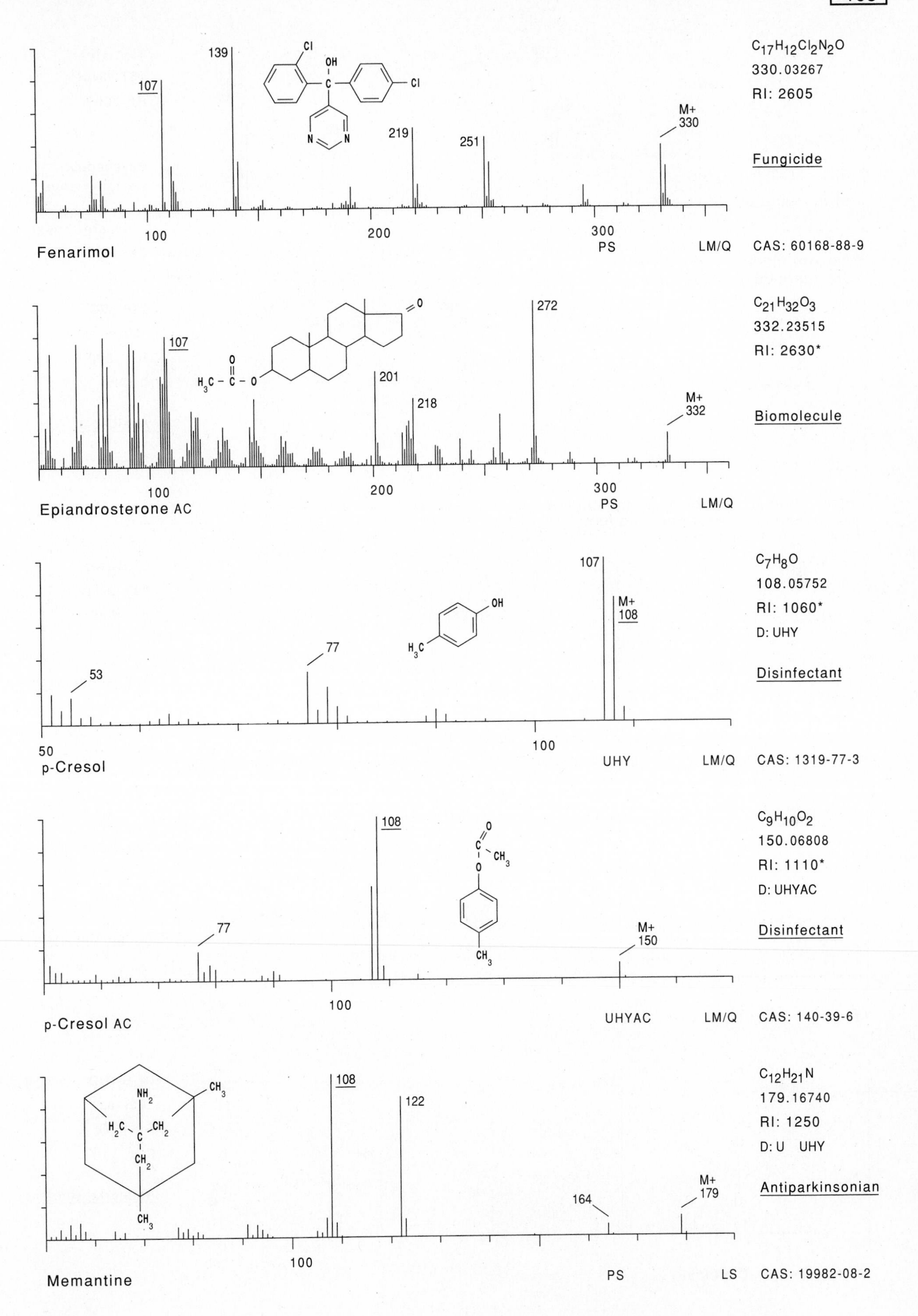

$C_{17}H_{12}Cl_2N_2O$
330.03267
RI: 2605
Fungicide
CAS: 60168-88-9

$C_{21}H_{32}O_3$
332.23515
RI: 2630*
Biomolecule

C_7H_8O
108.05752
RI: 1060*
D: UHY
Disinfectant
CAS: 1319-77-3

$C_9H_{10}O_2$
150.06808
RI: 1110*
D: UHYAC
Disinfectant
CAS: 140-39-6

$C_{12}H_{21}N$
179.16740
RI: 1250
D: U UHY
Antiparkinsonian
CAS: 19982-08-2

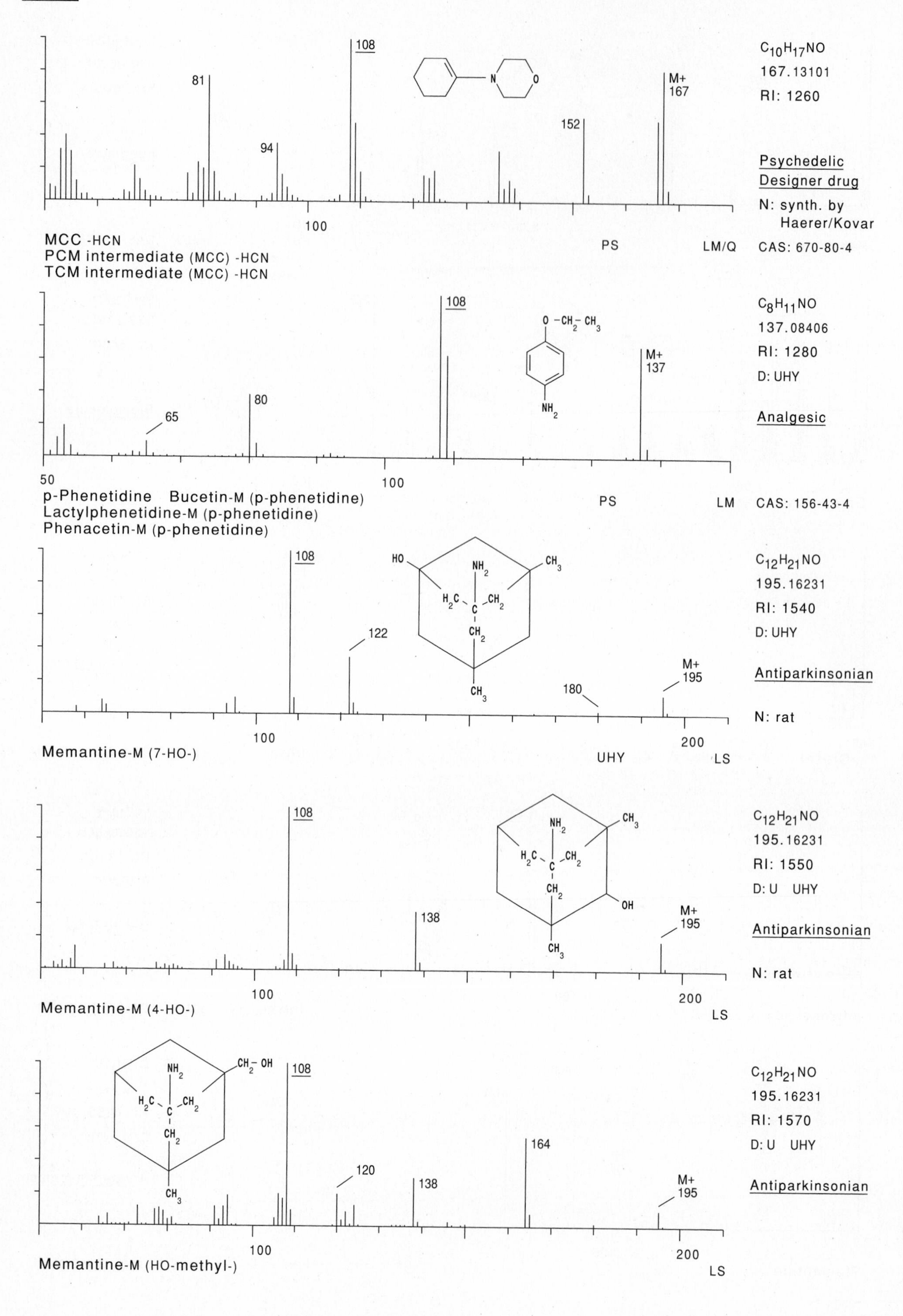

108
81
94
152
M+
167
100
$C_{10}H_{17}NO$
167.13101
RI: 1260
Psychedelic
Designer drug
N: synth. by
Haerer/Kovar
MCC -HCN
PCM intermediate (MCC) -HCN
TCM intermediate (MCC) -HCN
PS
LM/Q
CAS: 670-80-4
108
O -CH2- CH3
NH2
M+
137
80
65
50
100
$C_8H_{11}NO$
137.08406
RI: 1280
D: UHY
Analgesic
p-Phenetidine Bucetin-M (p-phenetidine)
Lactylphenetidine-M (p-phenetidine)
Phenacetin-M (p-phenetidine)
PS
LM
CAS: 156-43-4
108
HO
NH2
CH3
H2C
C
CH2
CH2
CH3
122
180
M+
195
100
200
$C_{12}H_{21}NO$
195.16231
RI: 1540
D: UHY
Antiparkinsonian
N: rat
Memantine-M (7-HO-)
UHY
LS
108
NH2
CH3
H2C
C
CH2
CH2
OH
CH3
138
M+
195
100
200
$C_{12}H_{21}NO$
195.16231
RI: 1550
D: U UHY
Antiparkinsonian
N: rat
Memantine-M (4-HO-)
LS
NH2
CH2- OH
108
H2C
C
CH2
CH2
CH3
164
120
138
M+
195
100
200
$C_{12}H_{21}NO$
195.16231
RI: 1570
D: U UHY
Antiparkinsonian
Memantine-M (HO-methyl-)
LS

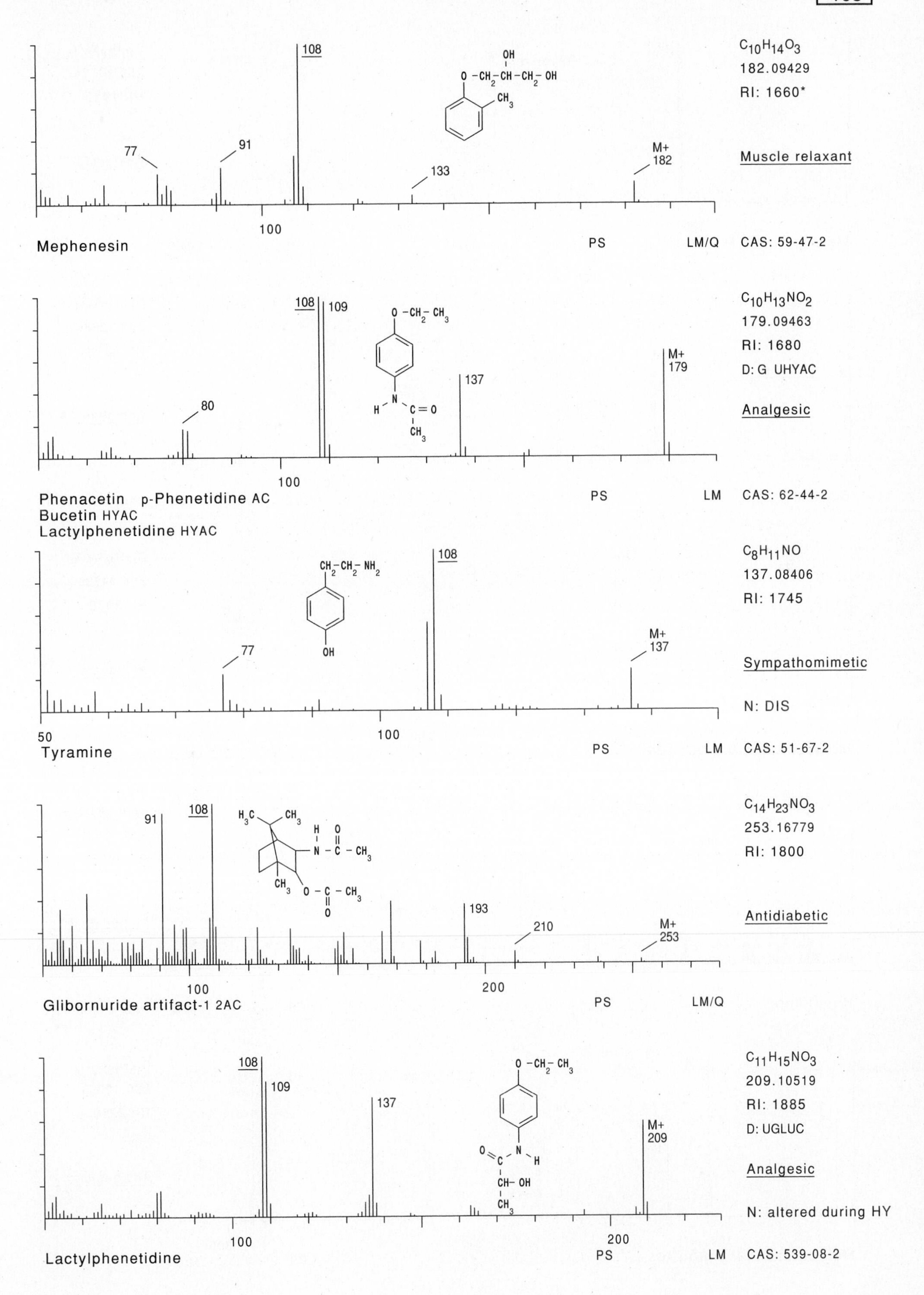

108
OH
O-CH2-CH-CH2-OH
CH3
77
91
133
M+
182
100
$C_{10}H_{14}O_3$
182.09429
RI: 1660*
Muscle relaxant
Mephenesin
PS
LM/Q
CAS: 59-47-2
108
109
O-CH2-CH3
H
N
C=O
CH3
80
137
M+
179
100
$C_{10}H_{13}NO_2$
179.09463
RI: 1680
D: G UHYAC
Analgesic
Phenacetin p-Phenetidine AC
Bucetin HYAC
Lactylphenetidine HYAC
PS
LM
CAS: 62-44-2
108
CH2-CH2-NH2
OH
77
M+
137
50
100
$C_8H_{11}NO$
137.08406
RI: 1745
Sympathomimetic
N: DIS
Tyramine
PS
LM
CAS: 51-67-2
91
108
H3C
CH3
H O
N-C-CH3
CH3
O-C-CH3
O
193
210
M+
253
100
200
$C_{14}H_{23}NO_3$
253.16779
RI: 1800
Antidiabetic
Glibornuride artifact-1 2AC
PS
LM/Q
108
109
137
O-CH2-CH3
O
C
N
H
CH-OH
CH3
M+
209
100
200
$C_{11}H_{15}NO_3$
209.10519
RI: 1885
D: UGLUC
Analgesic
N: altered during HY
Lactylphenetidine
PS
LM
CAS: 539-08-2

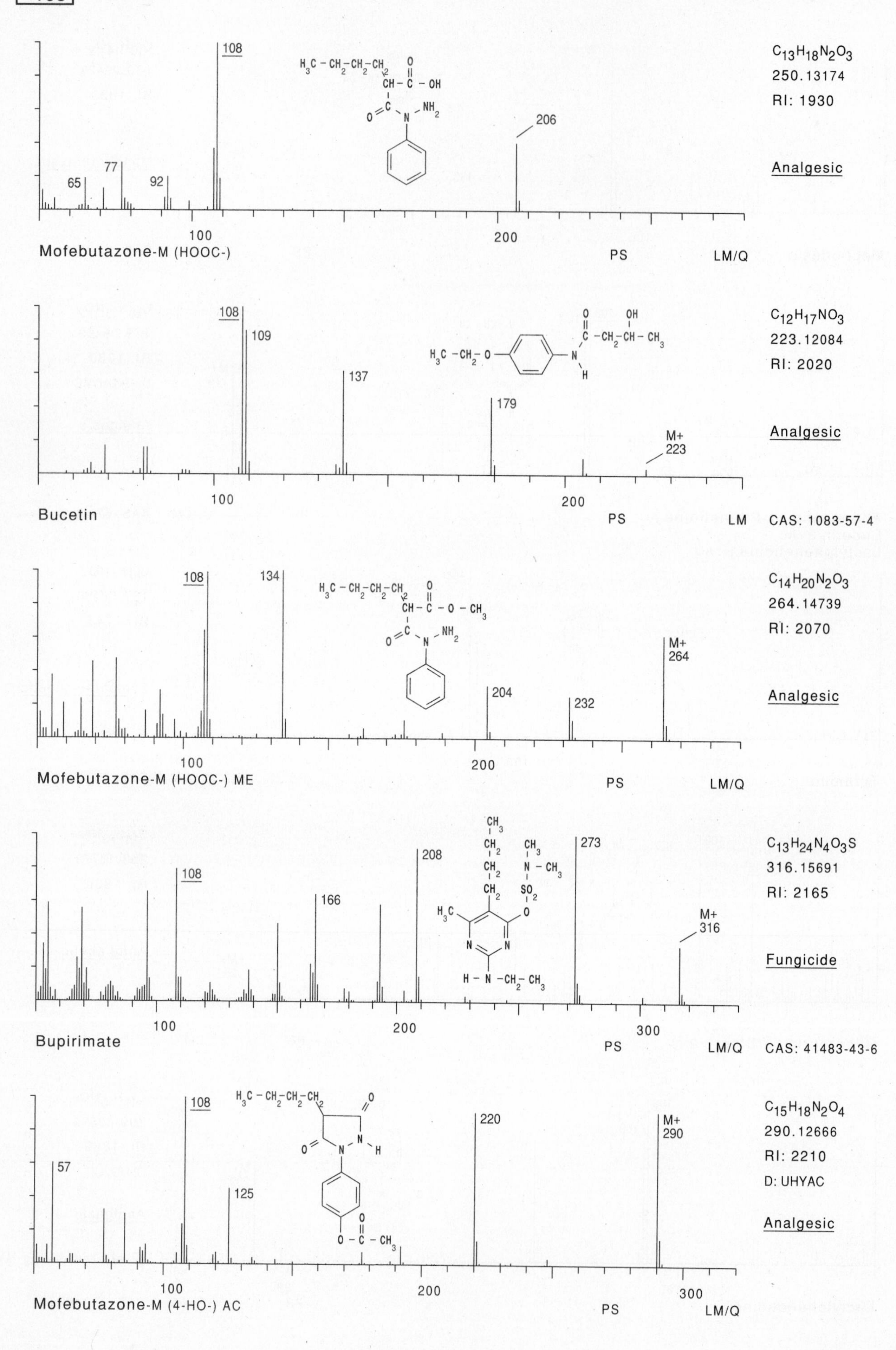

108
65
77
92
206
100
200
Mofebutazone-M (HOOC-)
PS
LM/Q
$C_{13}H_{18}N_2O_3$
250.13174
RI: 1930
Analgesic
108
109
137
179
M+
223
100
200
Bucetin
PS
LM
CAS: 1083-57-4
$C_{12}H_{17}NO_3$
223.12084
RI: 2020
Analgesic
108
134
204
232
M+
264
100
200
Mofebutazone-M (HOOC-) ME
PS
LM/Q
$C_{14}H_{20}N_2O_3$
264.14739
RI: 2070
Analgesic
108
166
208
273
M+
316
100
200
300
Bupirimate
PS
LM/Q
CAS: 41483-43-6
$C_{13}H_{24}N_4O_3S$
316.15691
RI: 2165
Fungicide
57
108
125
220
M+
290
100
200
300
Mofebutazone-M (4-HO-) AC
PS
LM/Q
$C_{15}H_{18}N_2O_4$
290.12666
RI: 2210
D: UHYAC
Analgesic

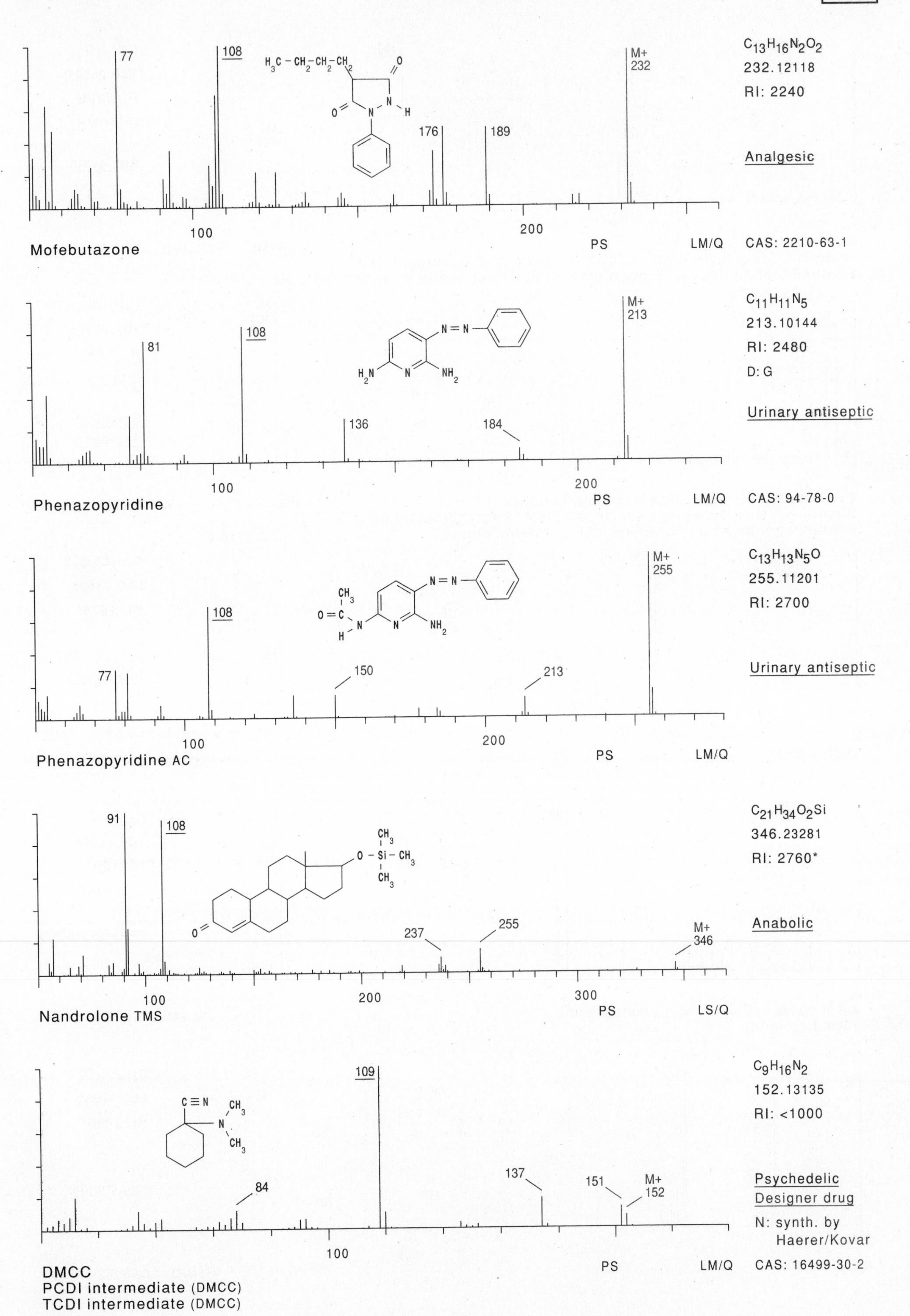

$C_{13}H_{16}N_2O_2$
232.12118
RI: 2240

Analgesic

CAS: 2210-63-1

$C_{11}H_{11}N_5$
213.10144
RI: 2480
D: G

Urinary antiseptic

CAS: 94-78-0

$C_{13}H_{13}N_5O$
255.11201
RI: 2700

Urinary antiseptic

$C_{21}H_{34}O_2Si$
346.23281
RI: 2760*

Anabolic

$C_9H_{16}N_2$
152.13135
RI: <1000

Psychedelic
Designer drug

N: synth. by Haerer/Kovar

CAS: 16499-30-2

DMCC
PCDI intermediate (DMCC)
TCDI intermediate (DMCC)

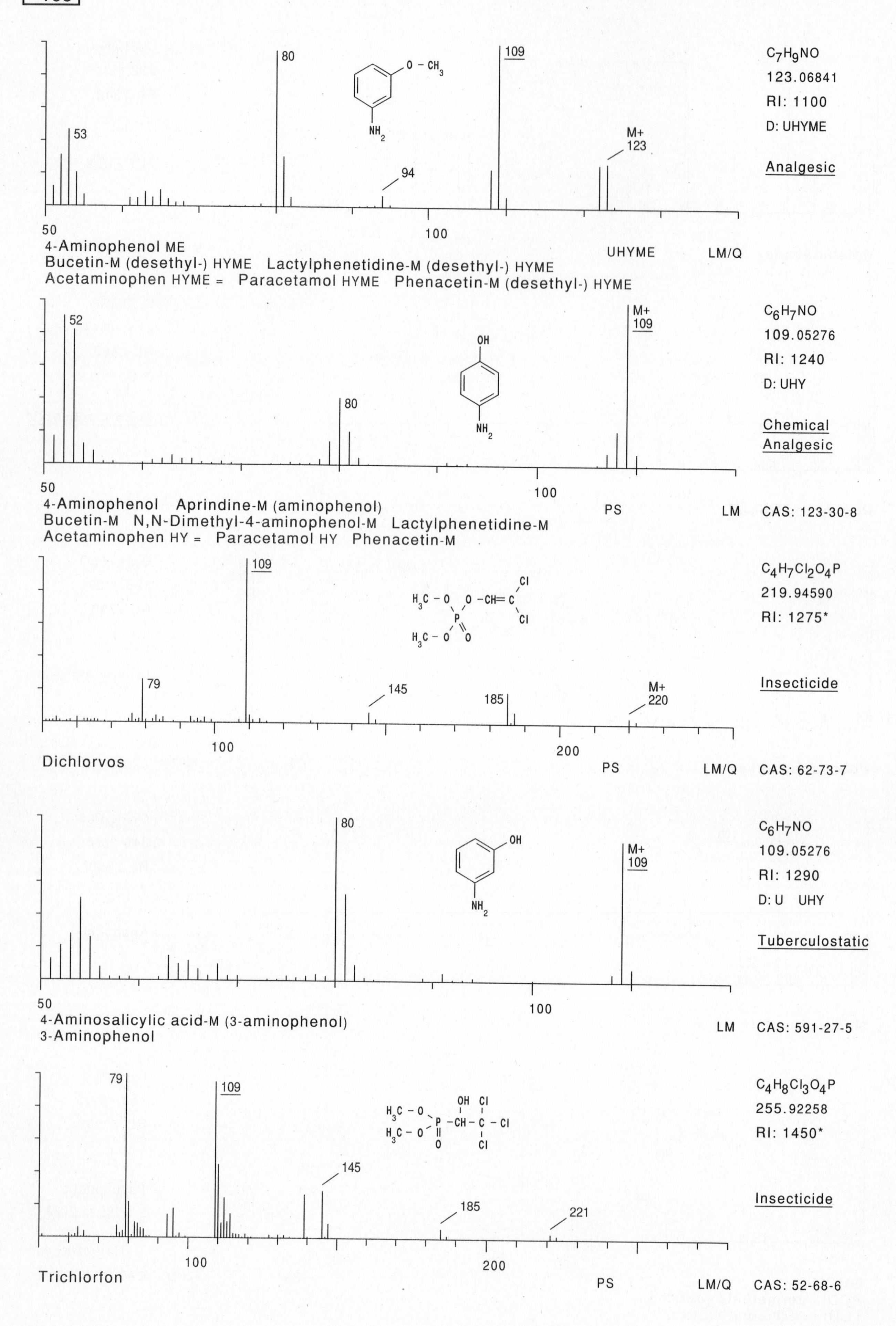
80
109
O – CH3
NH2
53
94
M+
123
50
100
C7H9NO
123.06841
RI: 1100
D: UHYME
Analgesic
4-Aminophenol ME
Bucetin-M (desethyl-) HYME Lactylphenetidine-M (desethyl-) HYME
Acetaminophen HYME = Paracetamol HYME Phenacetin-M (desethyl-) HYME
UHYME
LM/Q
52
OH
NH2
80
M+
109
50
100
C6H7NO
109.05276
RI: 1240
D: UHY
Chemical
Analgesic
4-Aminophenol Aprindine-M (aminophenol)
Bucetin-M N,N-Dimethyl-4-aminophenol-M Lactylphenetidine-M
Acetaminophen HY = Paracetamol HY Phenacetin-M
PS
LM
CAS: 123-30-8
109
H3C – O O –CH= C Cl Cl
P
H3C – O O
79
145
185
M+
220
100
200
C4H7Cl2O4P
219.94590
RI: 1275*
Insecticide
Dichlorvos
PS
LM/Q
CAS: 62-73-7
80
OH
NH2
M+
109
50
100
C6H7NO
109.05276
RI: 1290
D: U UHY
Tuberculostatic
4-Aminosalicylic acid-M (3-aminophenol)
3-Aminophenol
LM
CAS: 591-27-5
79
109
OH Cl
H3C – O
H3C – O P –CH– C – Cl
O Cl
145
185
221
100
200
C4H8Cl3O4P
255.92258
RI: 1450*
Insecticide
Trichlorfon
PS
LM/Q
CAS: 52-68-6

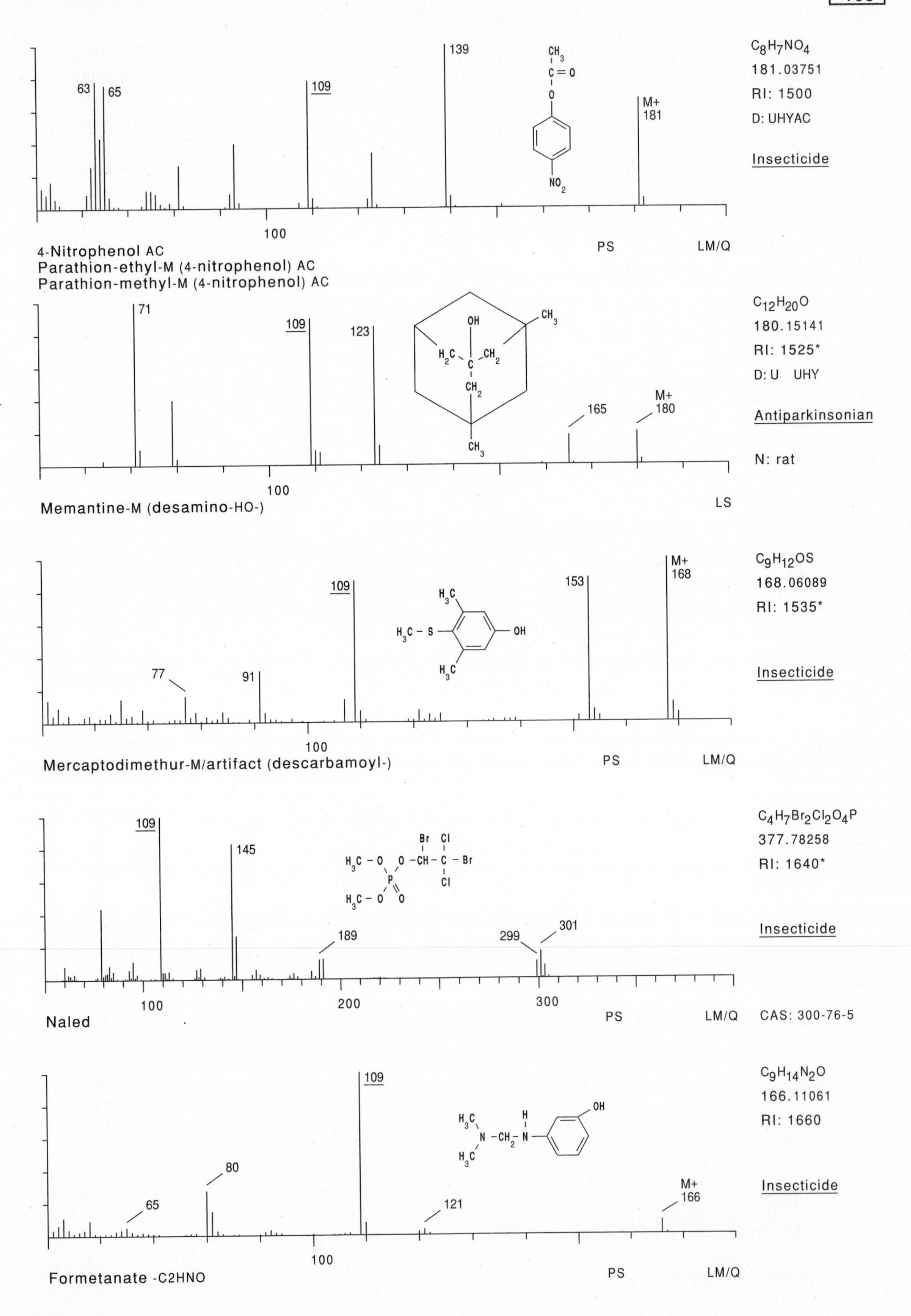

63
65
109
139
M+
181
100
PS
LM/Q
4-Nitrophenol AC
Parathion-ethyl-M (4-nitrophenol) AC
Parathion-methyl-M (4-nitrophenol) AC
C8H7NO4
181.03751
RI: 1500
D: UHYAC
Insecticide
71
109
123
165
M+
180
100
LS
Memantine-M (desamino-HO-)
C12H20O
180.15141
RI: 1525*
D: U UHY
Antiparkinsonian
N: rat
77
91
109
153
M+
168
100
PS
LM/Q
Mercaptodimethur-M/artifact (descarbamoyl-)
C9H12OS
168.06089
RI: 1535*
Insecticide
109
145
189
299
301
100
200
300
PS
LM/Q
Naled
C4H7Br2Cl2O4P
377.78258
RI: 1640*
Insecticide
CAS: 300-76-5
65
80
109
121
M+
166
100
PS
LM/Q
Formetanate -C2HNO
C9H14N2O
166.11061
RI: 1660
Insecticide

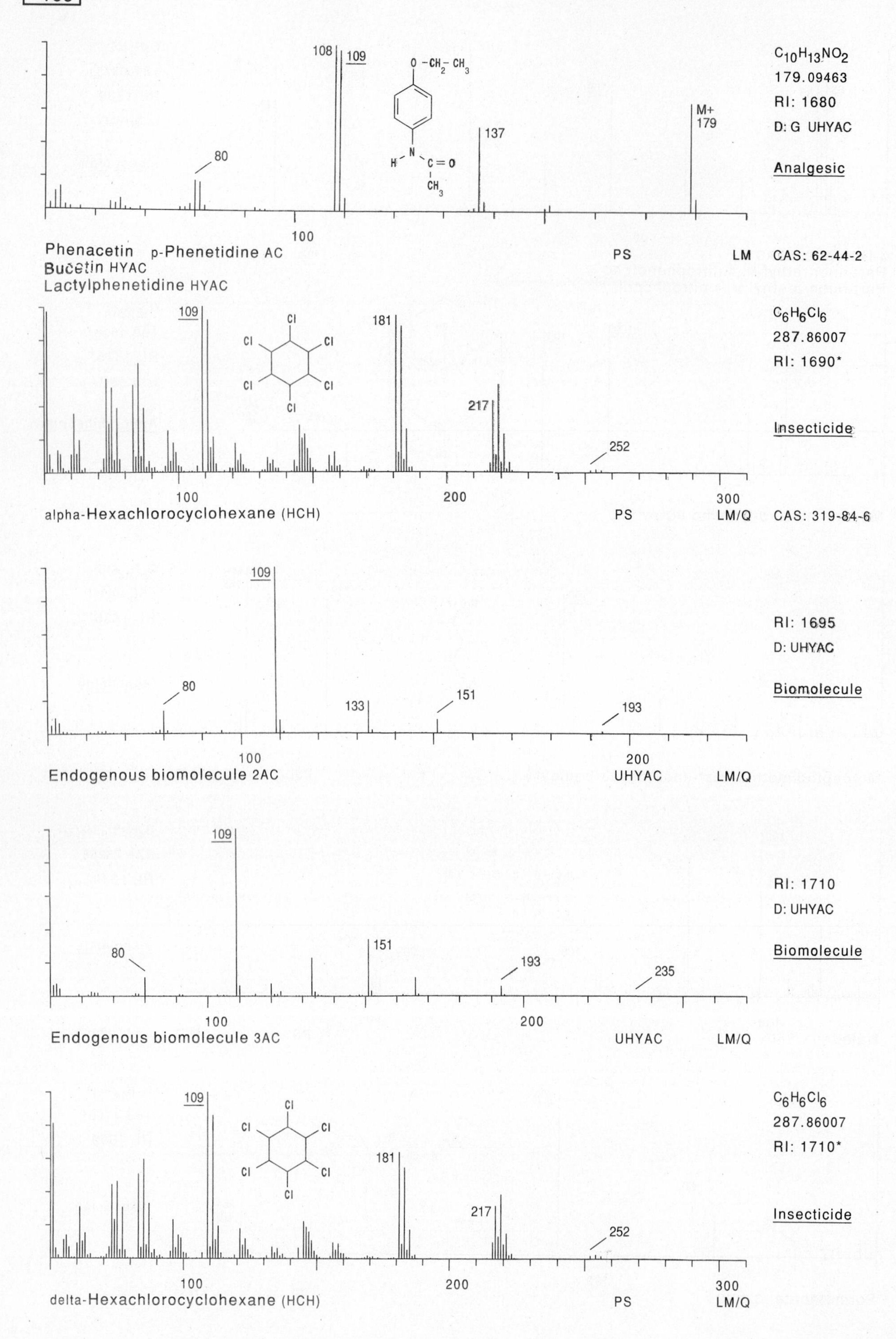

108
109
O-CH2-CH3
H-N-C=O
CH3
80
137
M+
179
100
Phenacetin p-Phenetidine AC
Bucetin HYAC
Lactylphenetidine HYAC
PS
LM
C10H13NO2
179.09463
RI: 1680
D: G UHYAC
Analgesic
CAS: 62-44-2
109
181
Cl
217
252
100
200
300
alpha-Hexachlorocyclohexane (HCH)
PS
LM/Q
C6H6Cl6
287.86007
RI: 1690*
Insecticide
CAS: 319-84-6
109
80
133
151
193
100
200
Endogenous biomolecule 2AC
UHYAC
LM/Q
RI: 1695
D: UHYAC
Biomolecule
109
80
151
193
235
100
200
Endogenous biomolecule 3AC
UHYAC
LM/Q
RI: 1710
D: UHYAC
Biomolecule
109
181
217
252
100
200
300
delta-Hexachlorocyclohexane (HCH)
PS
LM/Q
C6H6Cl6
287.86007
RI: 1710*
Insecticide

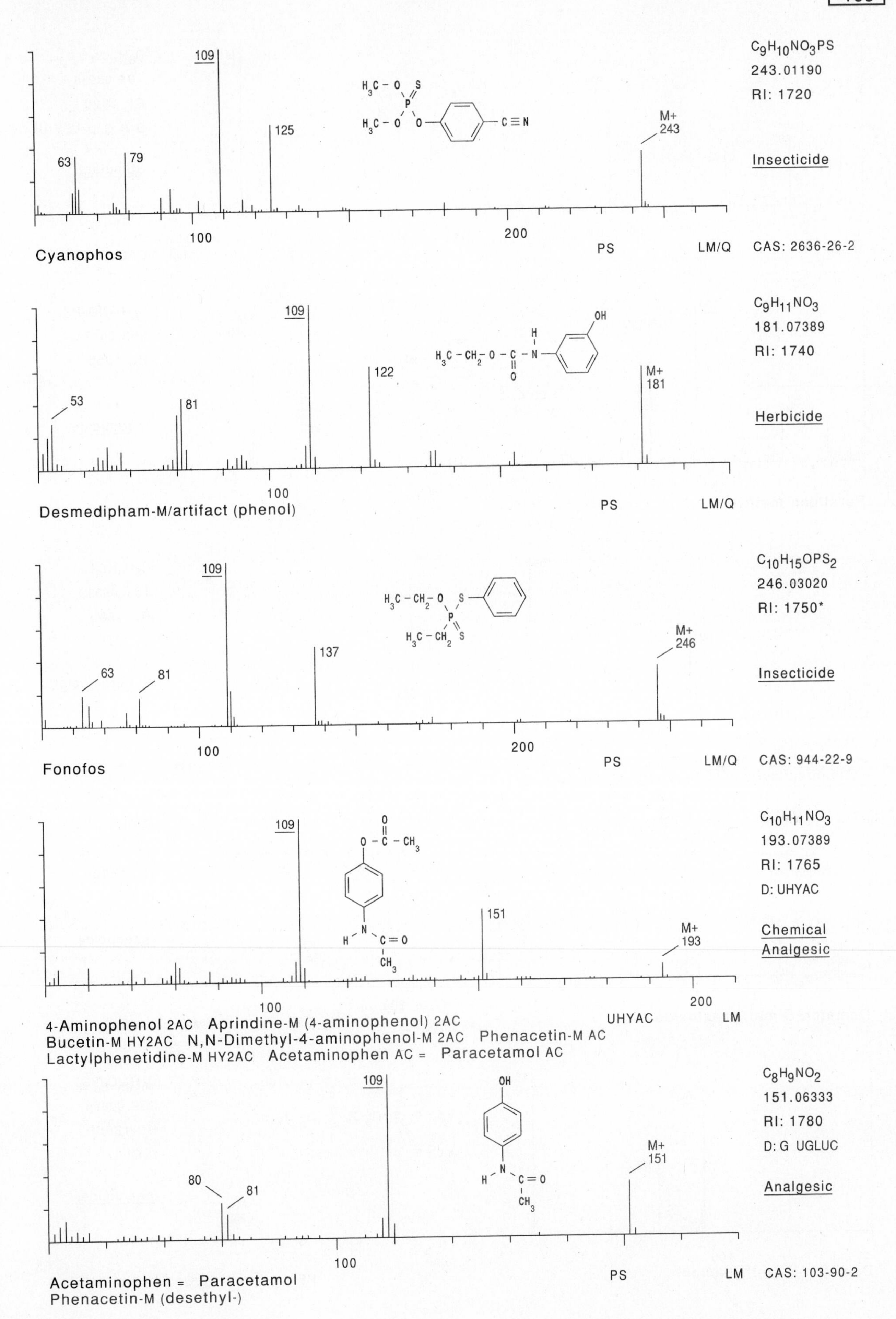
109
109
63
79
125
M+
243
100
200
Cyanophos
PS
LM/Q
C9H10NO3PS
243.01190
RI: 1720
Insecticide
CAS: 2636-26-2
109
53
81
122
M+
181
100
Desmedipham-M/artifact (phenol)
PS
LM/Q
C9H11NO3
181.07389
RI: 1740
Herbicide
109
63
81
137
M+
246
100
200
Fonofos
PS
LM/Q
C10H15OPS2
246.03020
RI: 1750*
Insecticide
CAS: 944-22-9
109
151
M+
193
100
200
UHYAC
LM
4-Aminophenol 2AC Aprindine-M (4-aminophenol) 2AC
Bucetin-M HY2AC N,N-Dimethyl-4-aminophenol-M 2AC Phenacetin-M AC
Lactylphenetidine-M HY2AC Acetaminophen AC = Paracetamol AC
C10H11NO3
193.07389
RI: 1765
D: UHYAC
Chemical
Analgesic
109
80
81
M+
151
100
PS
LM
Acetaminophen = Paracetamol
Phenacetin-M (desethyl-)
C8H9NO2
151.06333
RI: 1780
D: G UGLUC
Analgesic
CAS: 103-90-2

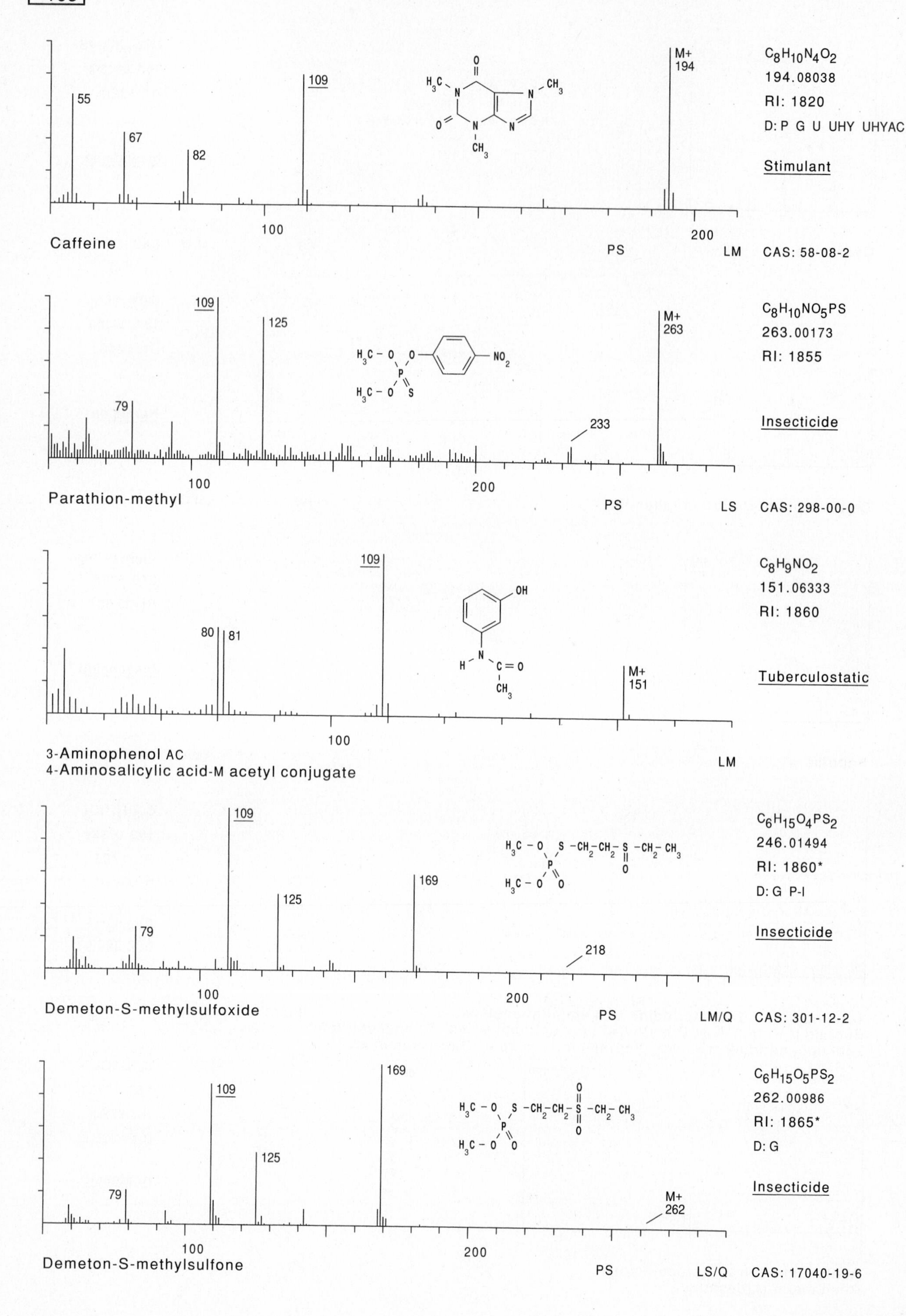

55
67
82
109
M+
194
100
200
Caffeine
PS
LM
$C_8H_{10}N_4O_2$
194.08038
RI: 1820
D: P G U UHY UHYAC
Stimulant
CAS: 58-08-2
109
125
79
233
M+
263
100
200
Parathion-methyl
PS
LS
$C_8H_{10}NO_5PS$
263.00173
RI: 1855
Insecticide
CAS: 298-00-0
109
80
81
M+
151
100
3-Aminophenol AC
4-Aminosalicylic acid-M acetyl conjugate
LM
$C_8H_9NO_2$
151.06333
RI: 1860
Tuberculostatic
109
125
169
79
218
100
200
Demeton-S-methylsulfoxide
PS
LM/Q
$C_6H_{15}O_4PS_2$
246.01494
RI: 1860*
D: G P-I
Insecticide
CAS: 301-12-2
169
109
125
79
M+
262
100
200
Demeton-S-methylsulfone
PS
LS/Q
$C_6H_{15}O_5PS_2$
262.00986
RI: 1865*
D: G
Insecticide
CAS: 17040-19-6

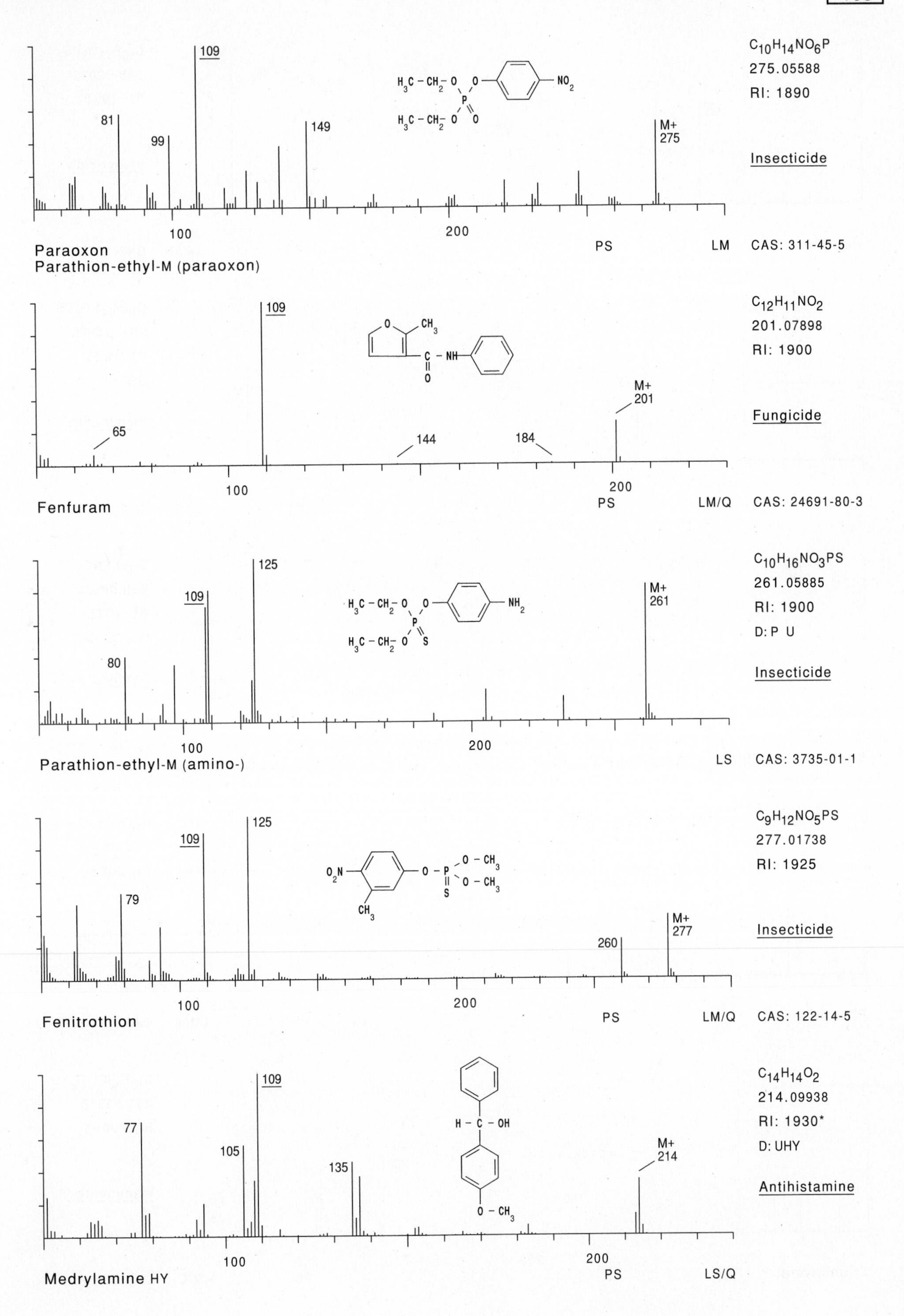

109
81
99
149
M+
275
100
200
PS
LM
Paraoxon
Parathion-ethyl-M (paraoxon)
C10H14NO6P
275.05588
RI: 1890
Insecticide
CAS: 311-45-5
109
65
144
184
M+
201
100
200
PS
LM/Q
Fenfuram
C12H11NO2
201.07898
RI: 1900
Fungicide
CAS: 24691-80-3
125
109
80
M+
261
100
200
LS
Parathion-ethyl-M (amino-)
C10H16NO3PS
261.05885
RI: 1900
D: P U
Insecticide
CAS: 3735-01-1
125
109
79
260
M+
277
100
200
PS
LM/Q
Fenitrothion
C9H12NO5PS
277.01738
RI: 1925
Insecticide
CAS: 122-14-5
109
77
105
135
M+
214
100
200
PS
LS/Q
Medrylamine HY
C14H14O2
214.09938
RI: 1930*
D: UHY
Antihistamine

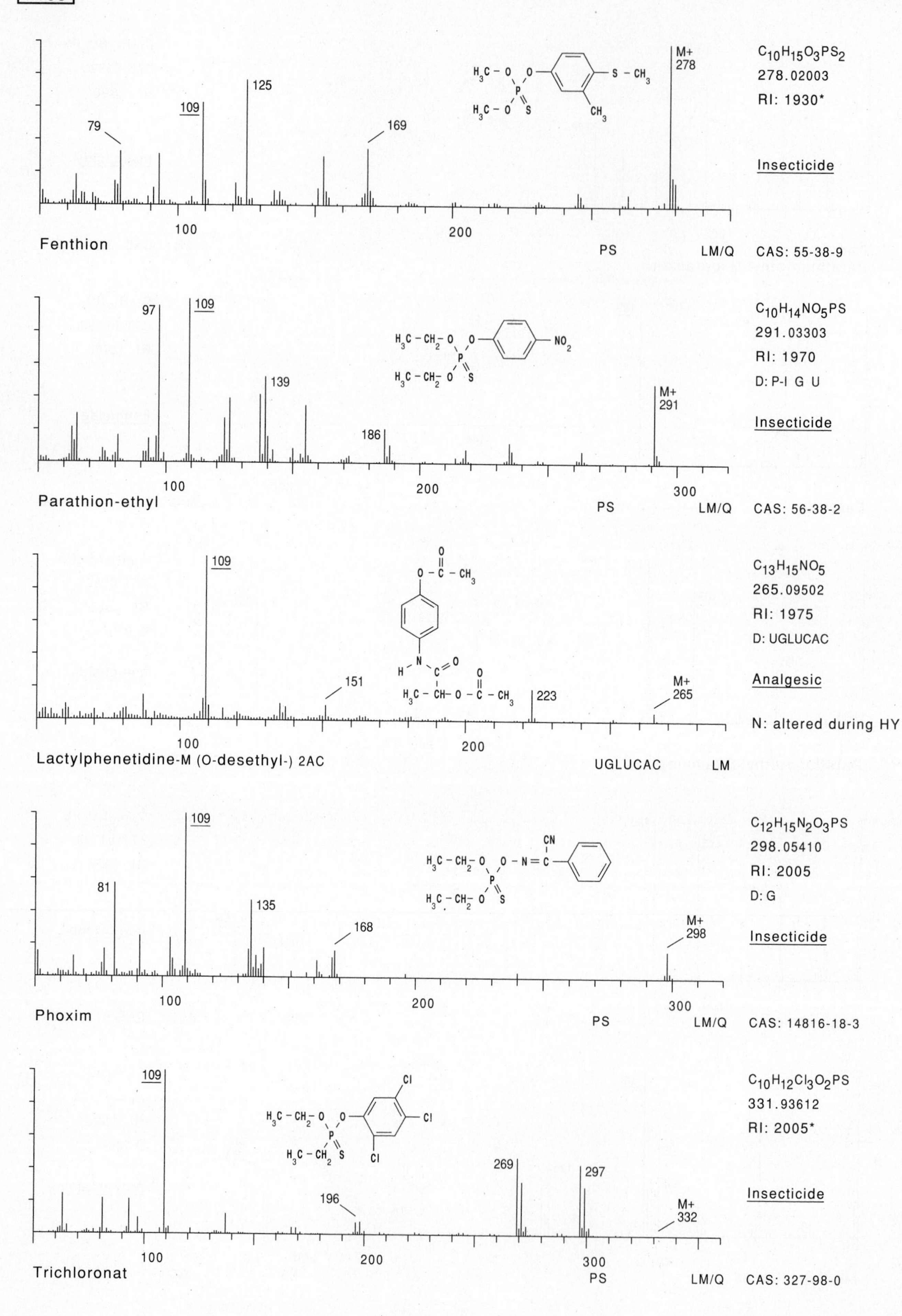

109
Fenthion
$C_{10}H_{15}O_3PS_2$
278.02003
RI: 1930*
Insecticide
PS
LM/Q
CAS: 55-38-9
Parathion-ethyl
$C_{10}H_{14}NO_5PS$
291.03303
RI: 1970
D: P-I G U
Insecticide
PS
LM/Q
CAS: 56-38-2
Lactylphenetidine-M (O-desethyl-) 2AC
$C_{13}H_{15}NO_5$
265.09502
RI: 1975
D: UGLUCAC
Analgesic
N: altered during HY
UGLUCAC
LM
Phoxim
$C_{12}H_{15}N_2O_3PS$
298.05410
RI: 2005
D: G
Insecticide
PS
LM/Q
CAS: 14816-18-3
Trichloronat
$C_{10}H_{12}Cl_3O_2PS$
331.93612
RI: 2005*
Insecticide
PS
LM/Q
CAS: 327-98-0

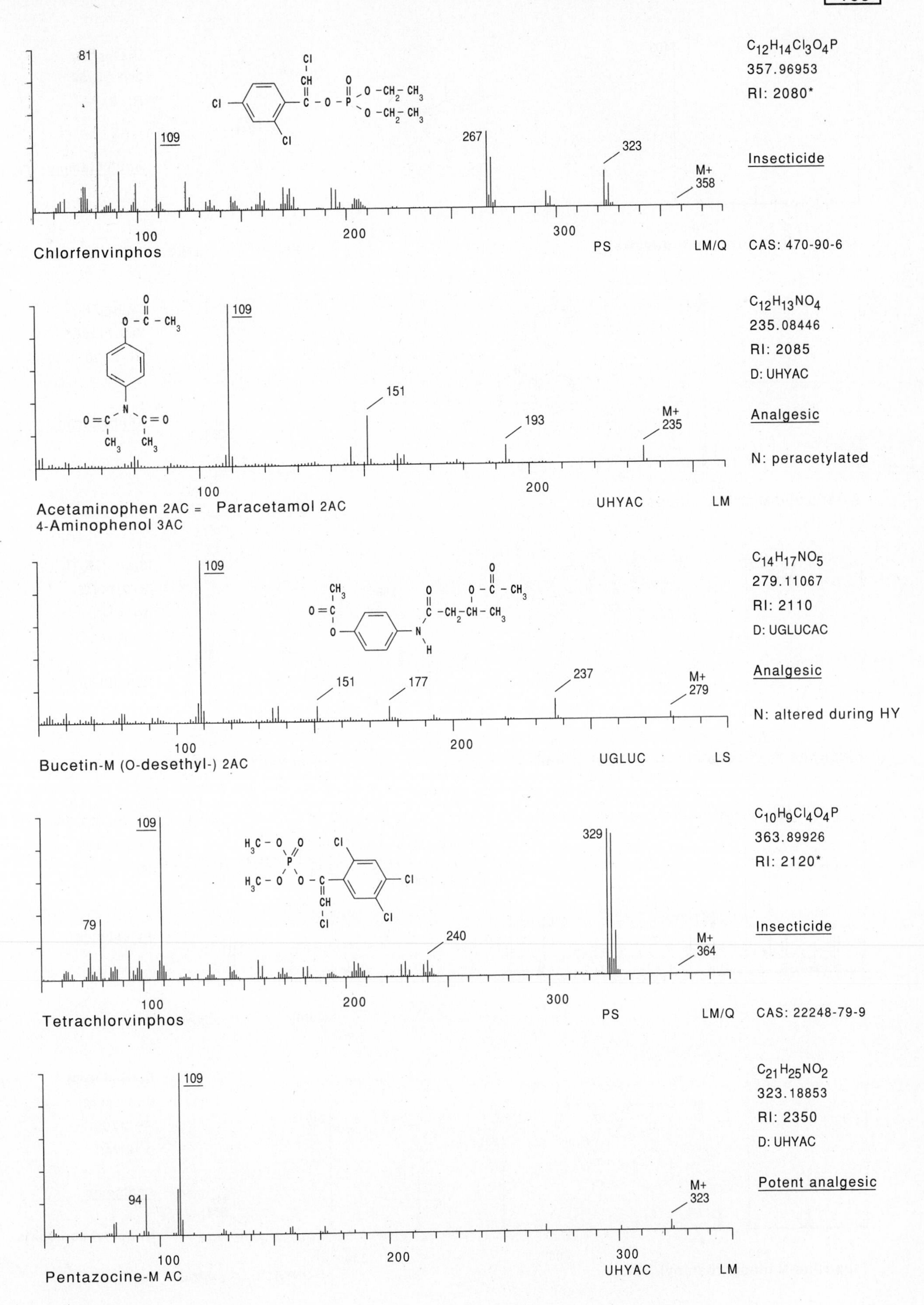

$C_{12}H_{14}Cl_3O_4P$
357.96953
RI: 2080*

Insecticide

CAS: 470-90-6

$C_{12}H_{13}NO_4$
235.08446
RI: 2085
D: UHYAC

Analgesic

N: peracetylated

$C_{14}H_{17}NO_5$
279.11067
RI: 2110
D: UGLUCAC

Analgesic

N: altered during HY

$C_{10}H_9Cl_4O_4P$
363.89926
RI: 2120*

Insecticide

CAS: 22248-79-9

$C_{21}H_{25}NO_2$
323.18853
RI: 2350
D: UHYAC

Potent analgesic

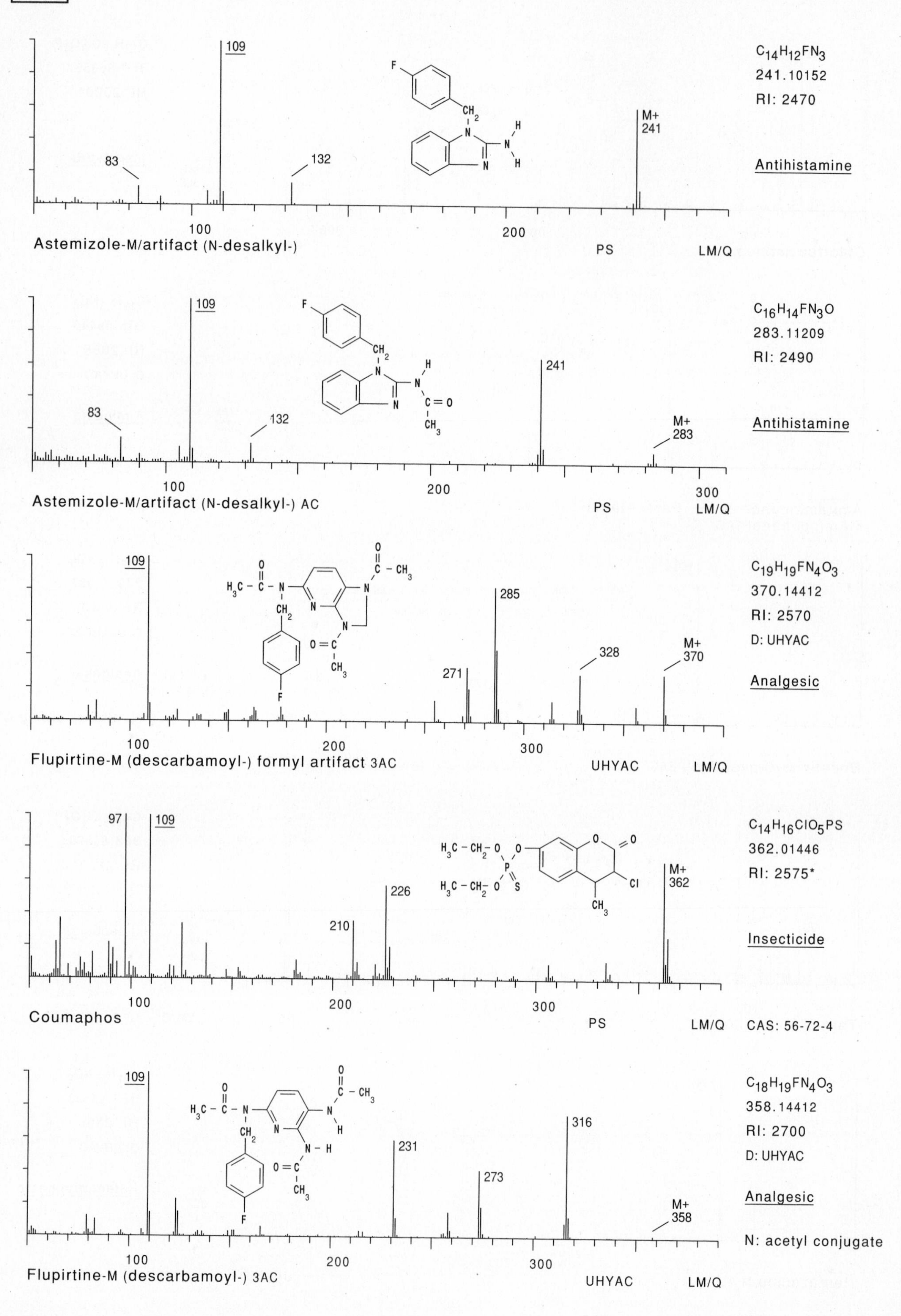
109
83
132
M+
241
F
CH2
N
H
N
N
H
100
200
Astemizole-M/artifact (N-desalkyl-)
PS
LM/Q
C14H12FN3
241.10152
RI: 2470
Antihistamine
109
83
132
241
M+
283
C = O
CH3
100
200
300
Astemizole-M/artifact (N-desalkyl-) AC
PS
LM/Q
C16H14FN3O
283.11209
RI: 2490
Antihistamine
109
285
271
328
M+
370
H3C – C – N
C – CH3
100
200
300
Flupirtine-M (descarbamoyl-) formyl artifact 3AC
UHYAC
LM/Q
C19H19FN4O3
370.14412
RI: 2570
D: UHYAC
Analgesic
97
109
226
210
M+
362
H3C – CH2– O
Cl
CH3
100
200
300
Coumaphos
PS
LM/Q
C14H16ClO5PS
362.01446
RI: 2575*
Insecticide
CAS: 56-72-4
109
316
231
273
M+
358
N – H
100
200
300
Flupirtine-M (descarbamoyl-) 3AC
UHYAC
LM/Q
C18H19FN4O3
358.14412
RI: 2700
D: UHYAC
Analgesic
N: acetyl conjugate

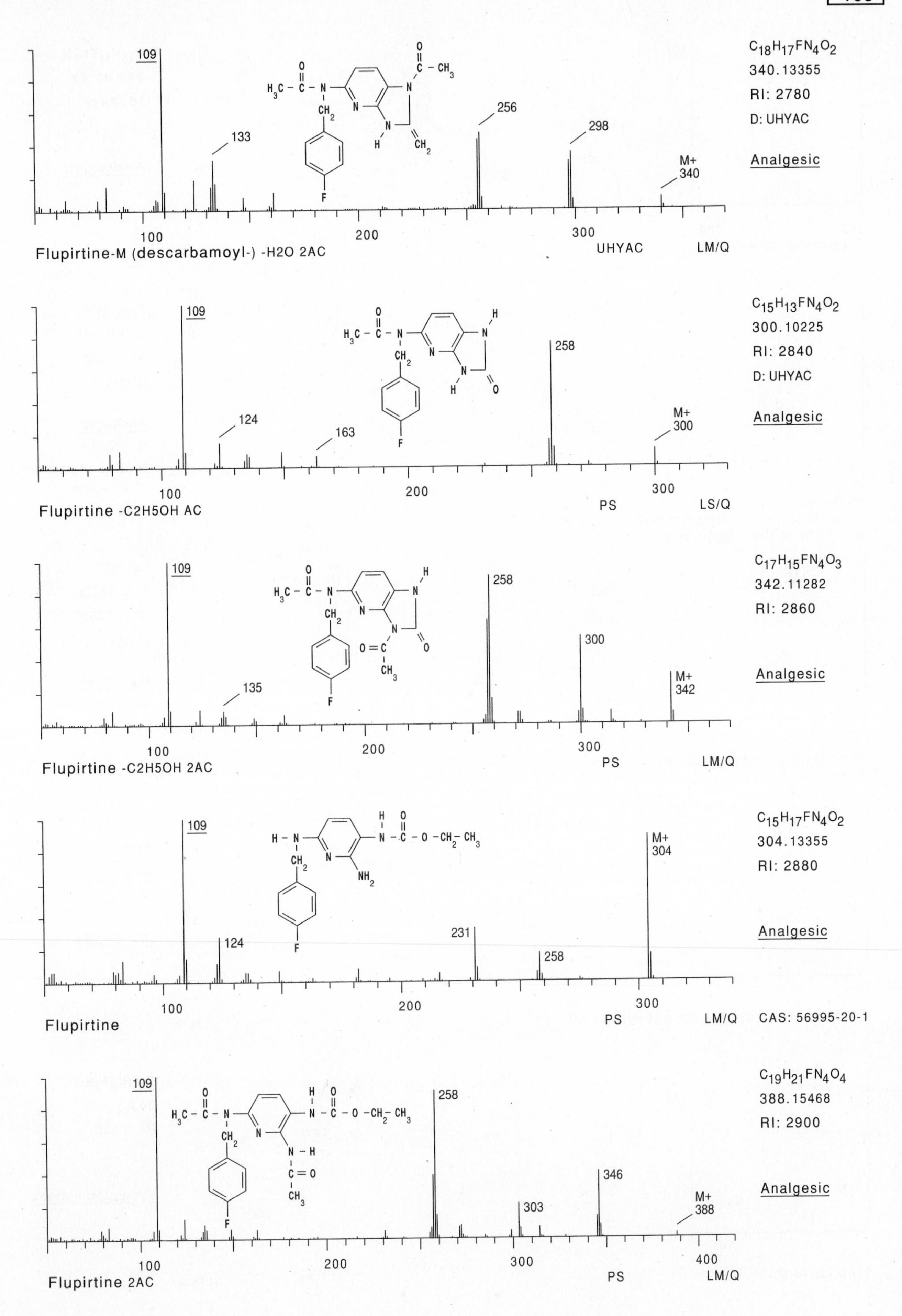
109
133
256
298
M+
340
100
200
300
UHYAC
LM/Q
Flupirtine-M (descarbamoyl-) -H2O 2AC
C18H17FN4O2
340.13355
RI: 2780
D: UHYAC
Analgesic
109
124
163
258
M+
300
100
200
300
PS
LS/Q
Flupirtine -C2H5OH AC
C15H13FN4O2
300.10225
RI: 2840
D: UHYAC
Analgesic
109
135
258
300
M+
342
100
200
300
PS
LM/Q
Flupirtine -C2H5OH 2AC
C17H15FN4O3
342.11282
RI: 2860
Analgesic
109
124
231
258
M+
304
100
200
300
PS
LM/Q
Flupirtine
C15H17FN4O2
304.13355
RI: 2880
Analgesic
CAS: 56995-20-1
109
258
303
346
M+
388
100
200
300
400
PS
LM/Q
Flupirtine 2AC
C19H21FN4O4
388.15468
RI: 2900
Analgesic

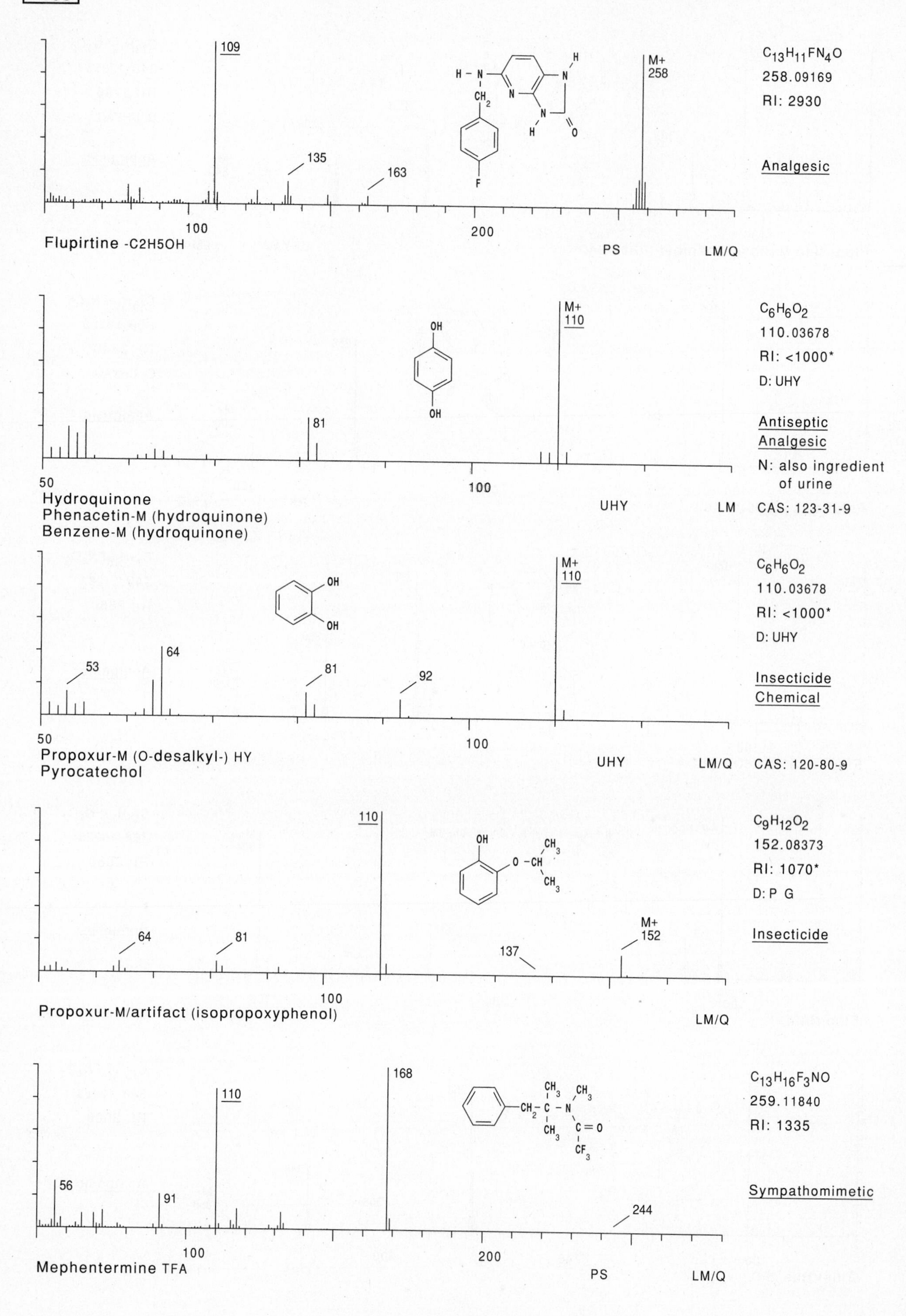

109
135
163
M+
258
100
200
Flupirtine -C2H5OH
PS
LM/Q
$C_{13}H_{11}FN_4O$
258.09169
RI: 2930
Analgesic
M+
110
81
50
100
Hydroquinone
Phenacetin-M (hydroquinone)
Benzene-M (hydroquinone)
UHY
LM
$C_6H_6O_2$
110.03678
RI: <1000*
D: UHY
Antiseptic
Analgesic
N: also ingredient of urine
CAS: 123-31-9
M+
110
53
64
81
92
50
100
Propoxur-M (O-desalkyl-) HY
Pyrocatechol
UHY
LM/Q
$C_6H_6O_2$
110.03678
RI: <1000*
D: UHY
Insecticide
Chemical
CAS: 120-80-9
110
64
81
137
M+
152
100
Propoxur-M/artifact (isopropoxyphenol)
LM/Q
$C_9H_{12}O_2$
152.08373
RI: 1070*
D: P G
Insecticide
168
110
56
91
244
100
200
Mephentermine TFA
PS
LM/Q
$C_{13}H_{16}F_3NO$
259.11840
RI: 1335
Sympathomimetic

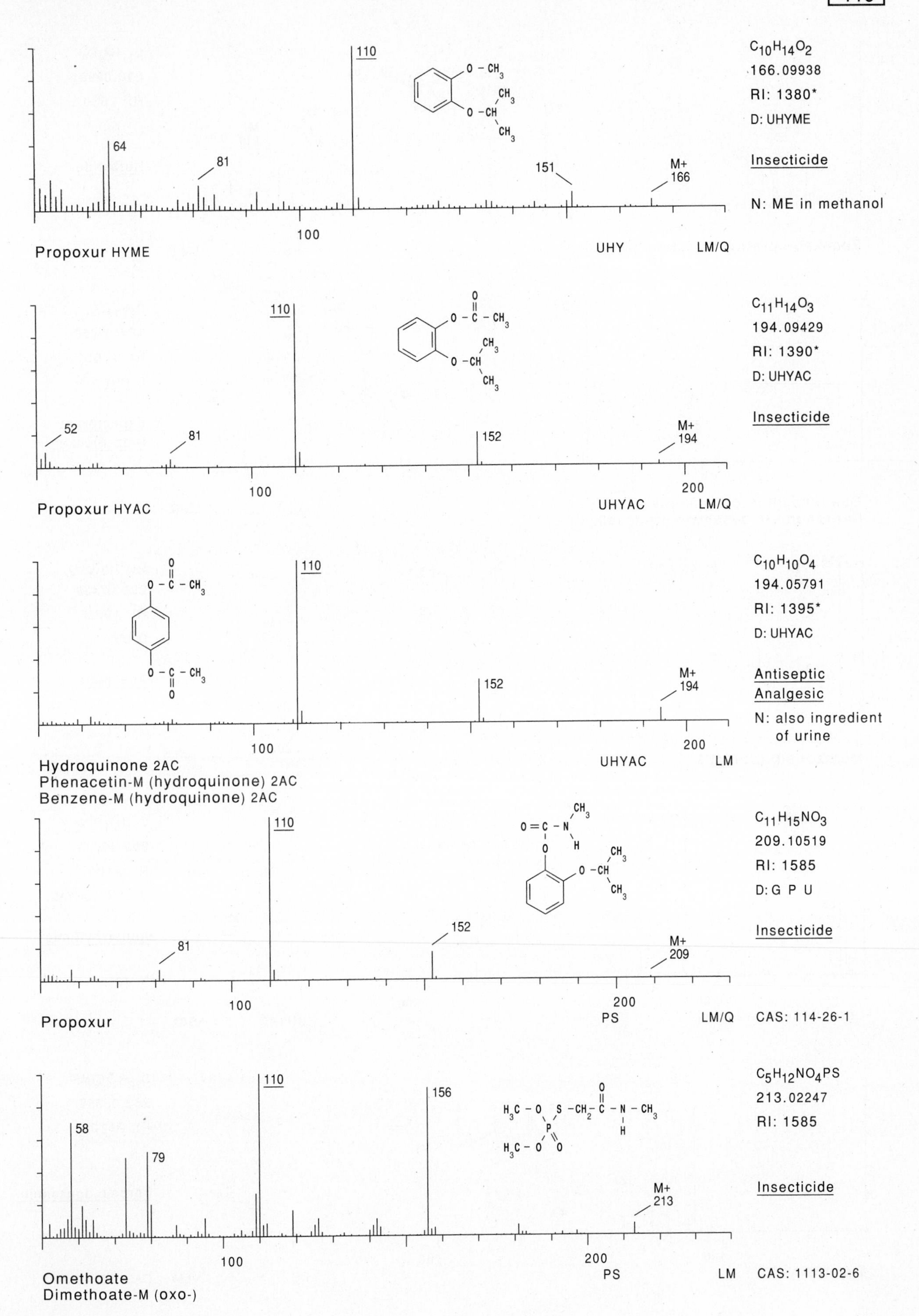

110
O - CH3
CH3
O - CH
CH3
64
81
151
M+
166
100
Propoxur HYME
UHY
LM/Q
C10H14O2
166.09938
RI: 1380*
D: UHYME
Insecticide
N: ME in methanol
110
O
O - C - CH3
CH3
O - CH
CH3
52
81
152
M+
194
100
200
Propoxur HYAC
UHYAC
LM/Q
C11H14O3
194.09429
RI: 1390*
D: UHYAC
Insecticide
O
O - C - CH3
O - C - CH3
O
110
152
M+
194
100
200
Hydroquinone 2AC
Phenacetin-M (hydroquinone) 2AC
Benzene-M (hydroquinone) 2AC
UHYAC
LM
C10H10O4
194.05791
RI: 1395*
D: UHYAC
Antiseptic
Analgesic
N: also ingredient
of urine
110
CH3
O = C - N
H
O
CH3
O - CH
CH3
81
152
M+
209
100
200
Propoxur
PS
LM/Q
C11H15NO3
209.10519
RI: 1585
D:G P U
Insecticide
CAS: 114-26-1
110
156
58
79
O
H3C - O S - CH2 - C - N - CH3
P
H
H3C - O O
M+
213
100
200
Omethoate
Dimethoate-M (oxo-)
PS
LM
C5H12NO4PS
213.02247
RI: 1585
Insecticide
CAS: 1113-02-6

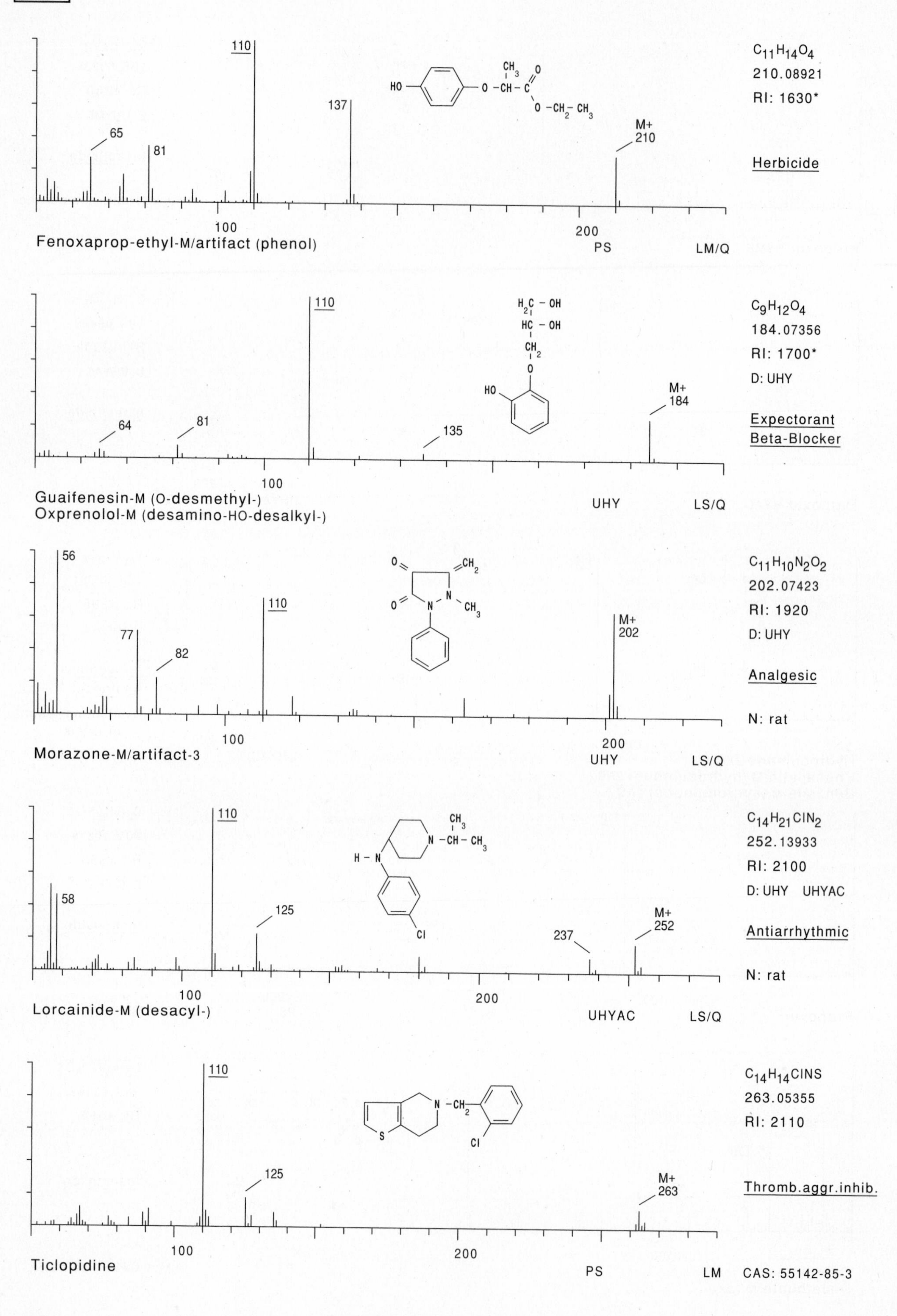

110
137
65
81
M+ 210
100
200
C11H14O4
210.08921
RI: 1630*
Herbicide
Fenoxaprop-ethyl-M/artifact (phenol)
PS
LM/Q
110
64
81
135
M+ 184
100
C9H12O4
184.07356
RI: 1700*
D: UHY
Expectorant
Beta-Blocker
Guaifenesin-M (O-desmethyl-)
Oxprenolol-M (desamino-HO-desalkyl-)
UHY
LS/Q
56
110
77
82
M+ 202
100
200
C11H10N2O2
202.07423
RI: 1920
D: UHY
Analgesic
N: rat
Morazone-M/artifact-3
UHY
LS/Q
110
58
125
237
M+ 252
100
200
C14H21ClN2
252.13933
RI: 2100
D: UHY UHYAC
Antiarrhythmic
N: rat
Lorcainide-M (desacyl-)
UHYAC
LS/Q
110
125
M+ 263
100
200
C14H14ClNS
263.05355
RI: 2110
Thromb.aggr.inhib.
Ticlopidine
PS
LM
CAS: 55142-85-3

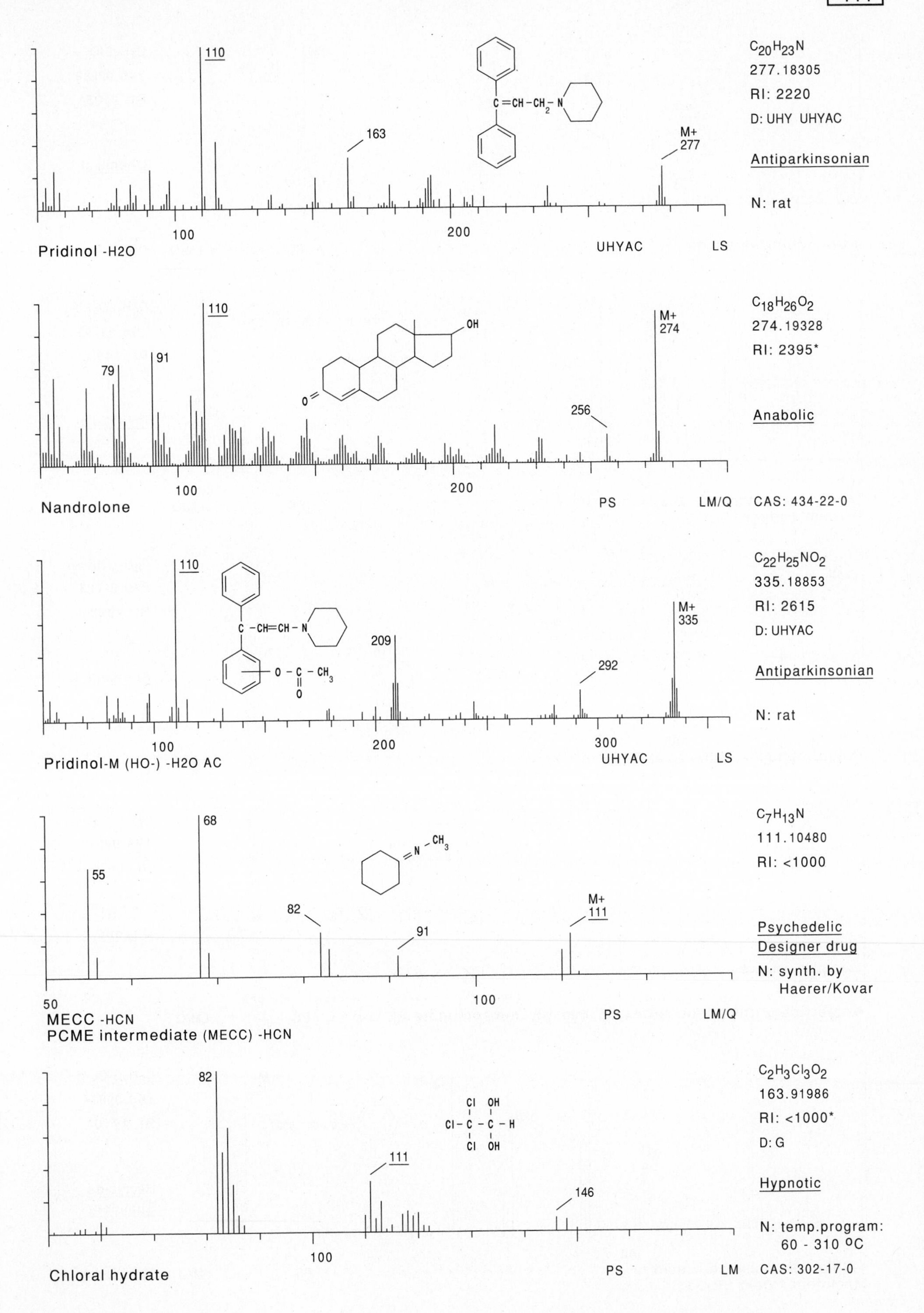

110
163
M+
277
100
200
Pridinol -H2O
UHYAC
LS
C20H23N
277.18305
RI: 2220
D: UHY UHYAC
Antiparkinsonian
N: rat
110
91
79
OH
O
256
M+
274
100
200
Nandrolone
PS
LM/Q
C18H26O2
274.19328
RI: 2395*
Anabolic
CAS: 434-22-0
110
C -CH=CH - N
O - C - CH3
O
209
292
M+
335
100
200
300
Pridinol-M (HO-) -H2O AC
UHYAC
LS
C22H25NO2
335.18853
RI: 2615
D: UHYAC
Antiparkinsonian
N: rat
68
55
N - CH3
82
91
M+
111
50
100
MECC -HCN
PCME intermediate (MECC) -HCN
PS
LM/Q
C7H13N
111.10480
RI: <1000
Psychedelic
Designer drug
N: synth. by
Haerer/Kovar
82
Cl OH
Cl - C - C - H
Cl OH
111
146
100
Chloral hydrate
PS
LM
C2H3Cl3O2
163.91986
RI: <1000*
D: G
Hypnotic
N: temp.program:
60 - 310 °C
CAS: 302-17-0

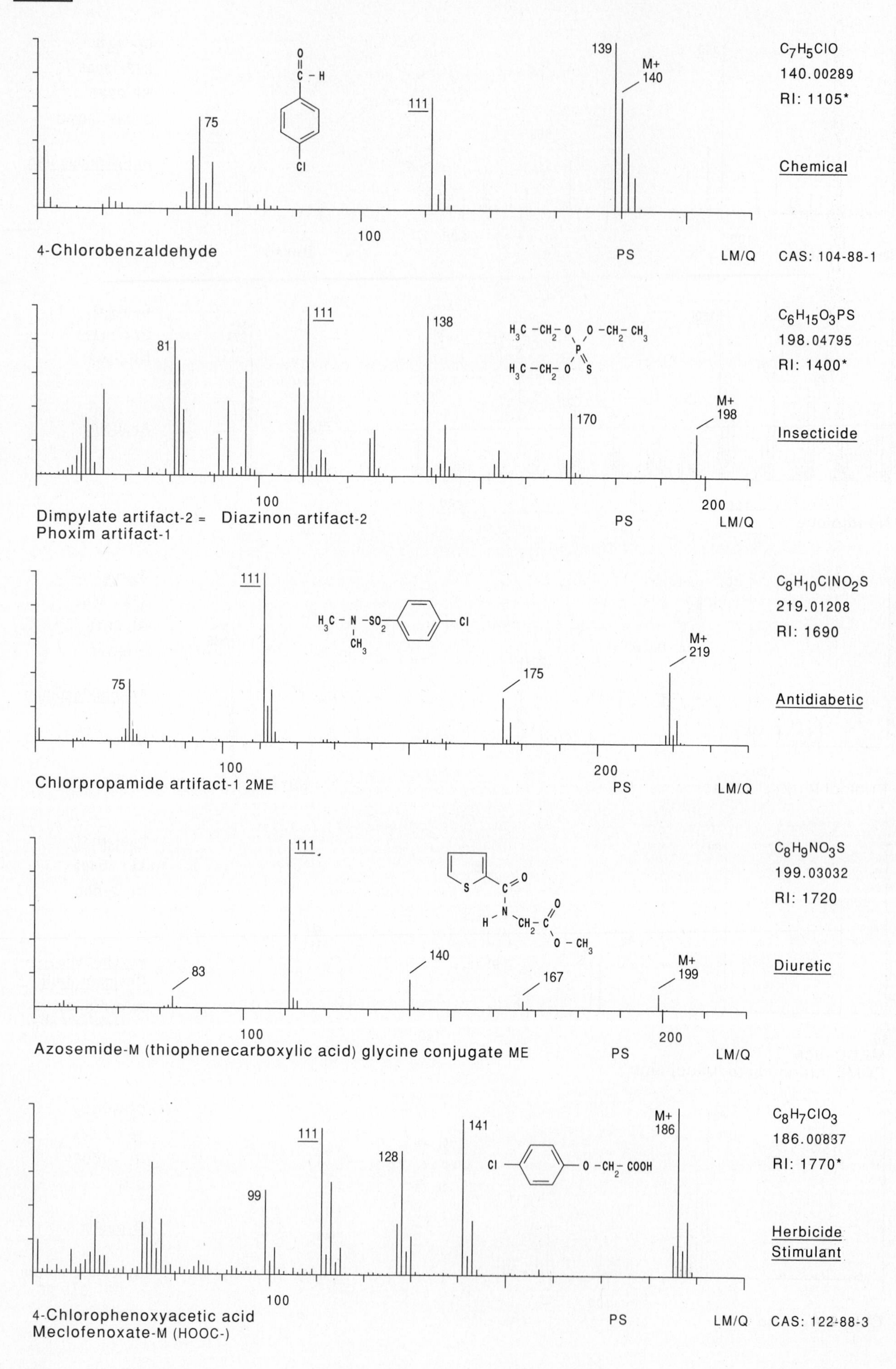
139
M+
140
111
75
C_7H_5ClO
140.00289
RI: 1105*
Chemical
100
4-Chlorobenzaldehyde
PS
LM/Q
CAS: 104-88-1
111
138
81
$C_6H_{15}O_3PS$
198.04795
RI: 1400*
M+
198
170
Insecticide
100
200
Dimpylate artifact-2 = Diazinon artifact-2
Phoxim artifact-1
PS
LM/Q
111
$C_8H_{10}ClNO_2S$
219.01208
RI: 1690
M+
219
175
75
Antidiabetic
100
200
Chlorpropamide artifact-1 2ME
PS
LM/Q
111
$C_8H_9NO_3S$
199.03032
RI: 1720
140
83
167
M+
199
Diuretic
100
200
Azosemide-M (thiophenecarboxylic acid) glycine conjugate ME
PS
LM/Q
M+
186
141
111
128
99
$C_8H_7ClO_3$
186.00837
RI: 1770*
Herbicide
Stimulant
100
4-Chlorophenoxyacetic acid
Meclofenoxate-M (HOOC-)
PS
LM/Q
CAS: 122-88-3

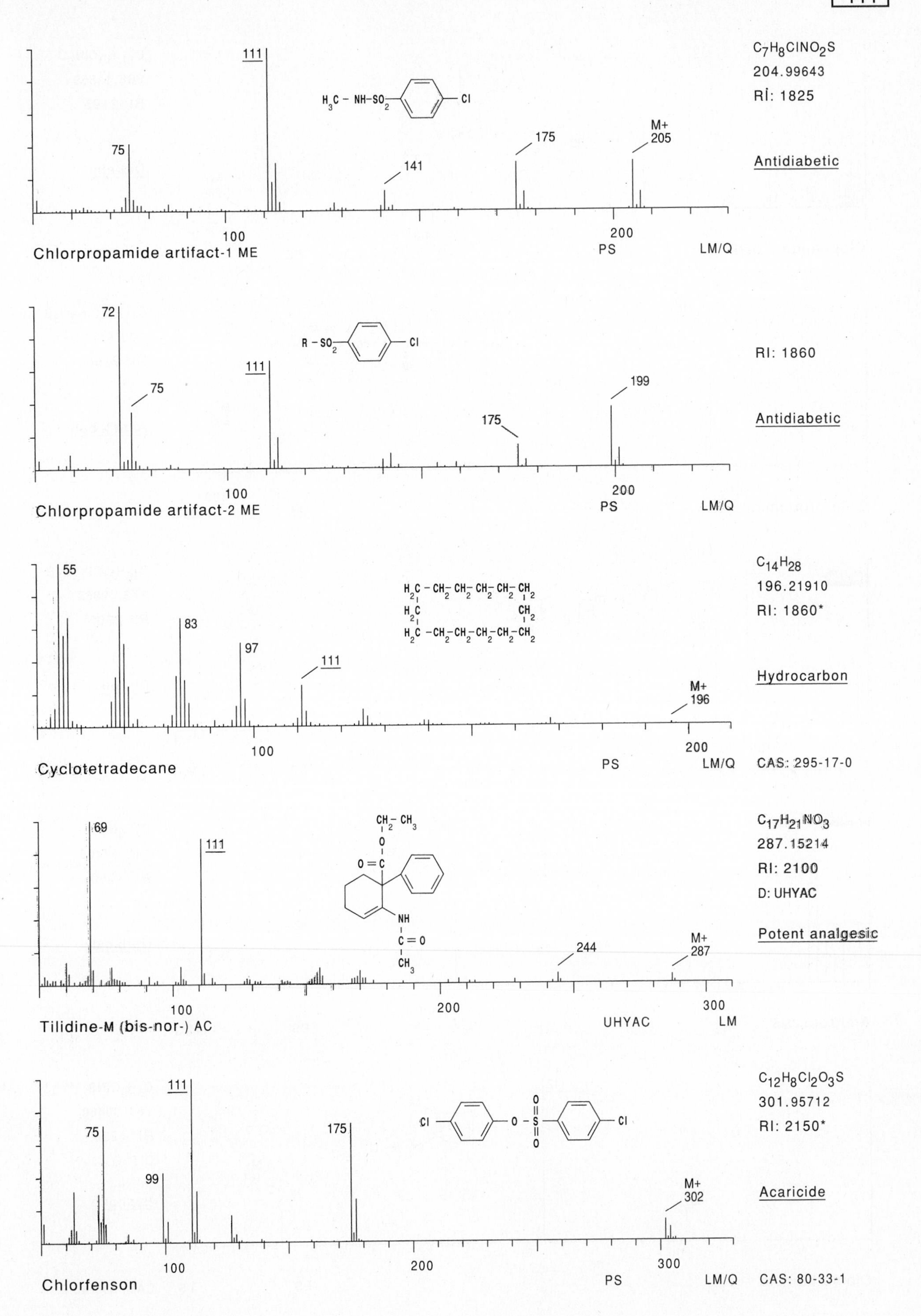

$C_7H_8ClNO_2S$
204.99643
RI: 1825

Antidiabetic

RI: 1860

Antidiabetic

$C_{14}H_{28}$
196.21910
RI: 1860*

Hydrocarbon

CAS: 295-17-0

$C_{17}H_{21}NO_3$
287.15214
RI: 2100
D: UHYAC

Potent analgesic

$C_{12}H_8Cl_2O_3S$
301.95712
RI: 2150*

Acaricide

CAS: 80-33-1

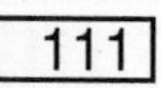
111

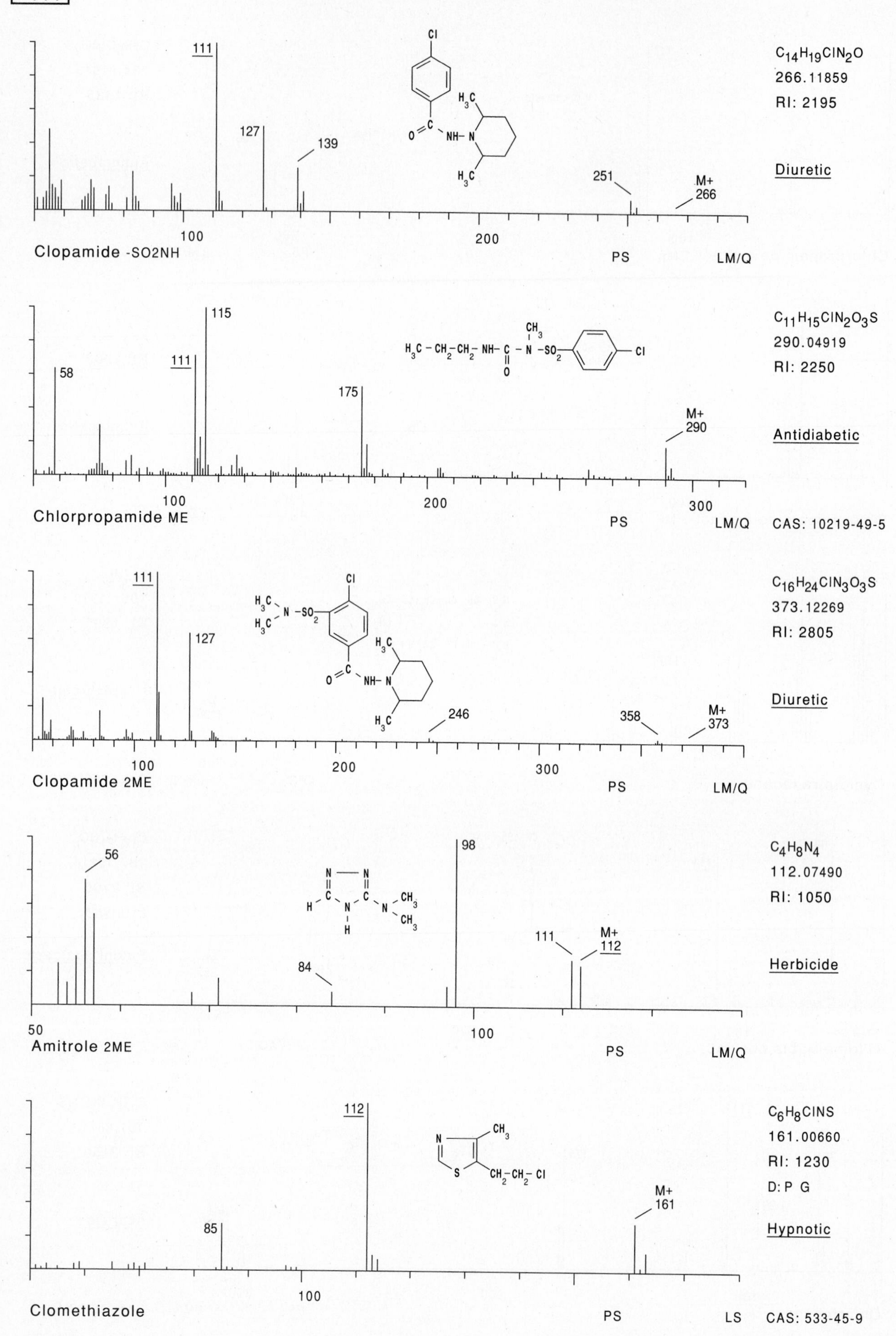
111
127
139
251
M+
266
100
200
Clopamide -SO2NH
PS
LM/Q
C14H19ClN2O
266.11859
RI: 2195
Diuretic
115
58
111
175
M+
290
100
200
300
Chlorpropamide ME
PS
LM/Q
C11H15ClN2O3S
290.04919
RI: 2250
Antidiabetic
CAS: 10219-49-5
111
127
246
358
M+
373
100
200
300
Clopamide 2ME
PS
LM/Q
C16H24ClN3O3S
373.12269
RI: 2805
Diuretic
56
98
84
111
M+
112
50
100
Amitrole 2ME
PS
LM/Q
C4H8N4
112.07490
RI: 1050
Herbicide
112
85
M+
161
100
Clomethiazole
PS
LS
C6H8ClNS
161.00660
RI: 1230
D: P G
Hypnotic
CAS: 533-45-9

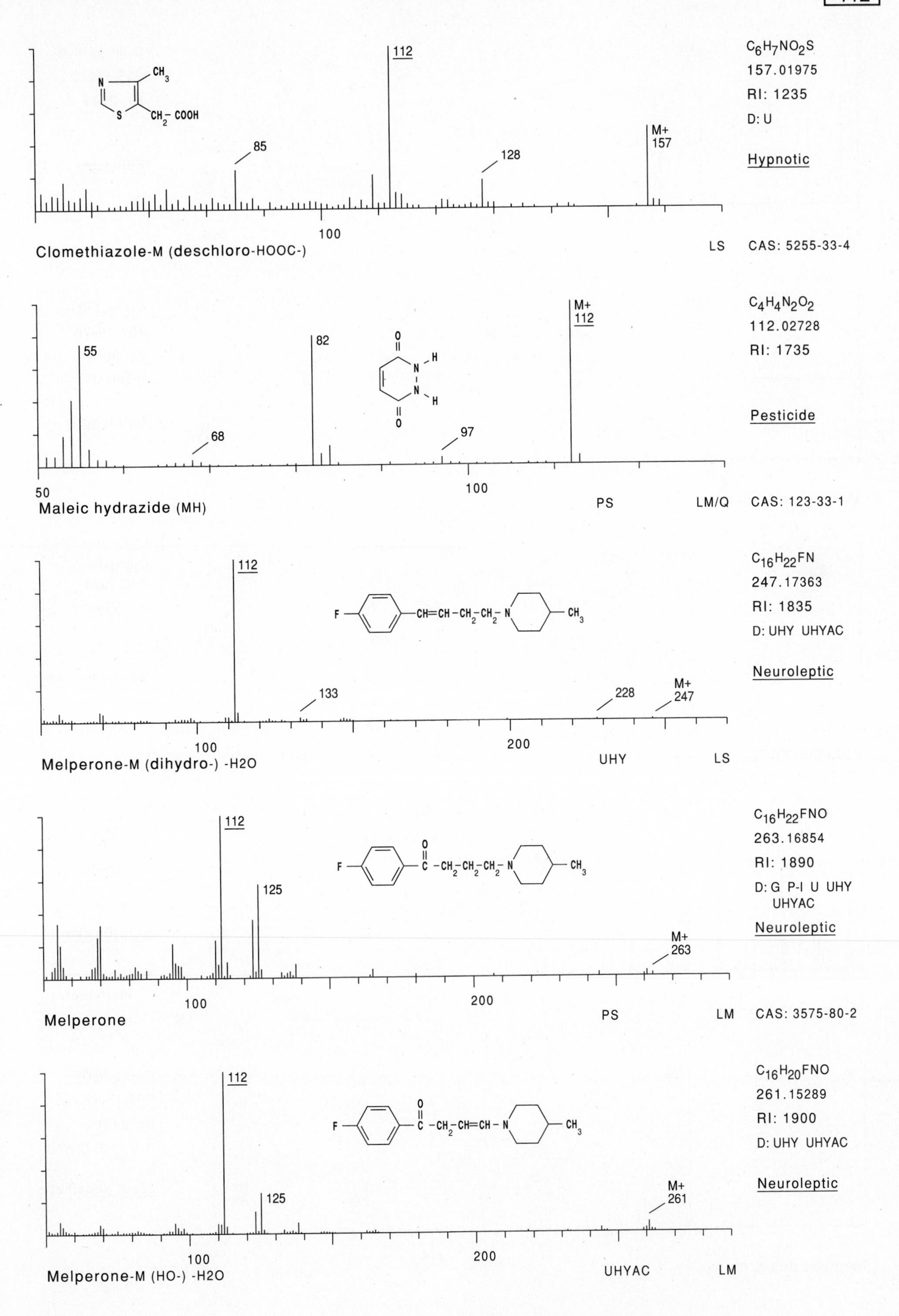

112
$C_6H_7NO_2S$
157.01975
RI: 1235
D: U
Hypnotic
Clomethiazole-M (deschloro-HOOC-)
LS
CAS: 5255-33-4
$C_4H_4N_2O_2$
112.02728
RI: 1735
Pesticide
Maleic hydrazide (MH)
PS
LM/Q
CAS: 123-33-1
$C_{16}H_{22}FN$
247.17363
RI: 1835
D: UHY UHYAC
Neuroleptic
Melperone-M (dihydro-) -H2O
UHY
LS
$C_{16}H_{22}FNO$
263.16854
RI: 1890
D: G P-I U UHY UHYAC
Neuroleptic
Melperone
PS
LM
CAS: 3575-80-2
$C_{16}H_{20}FNO$
261.15289
RI: 1900
D: UHY UHYAC
Neuroleptic
Melperone-M (HO-) -H2O
UHYAC
LM

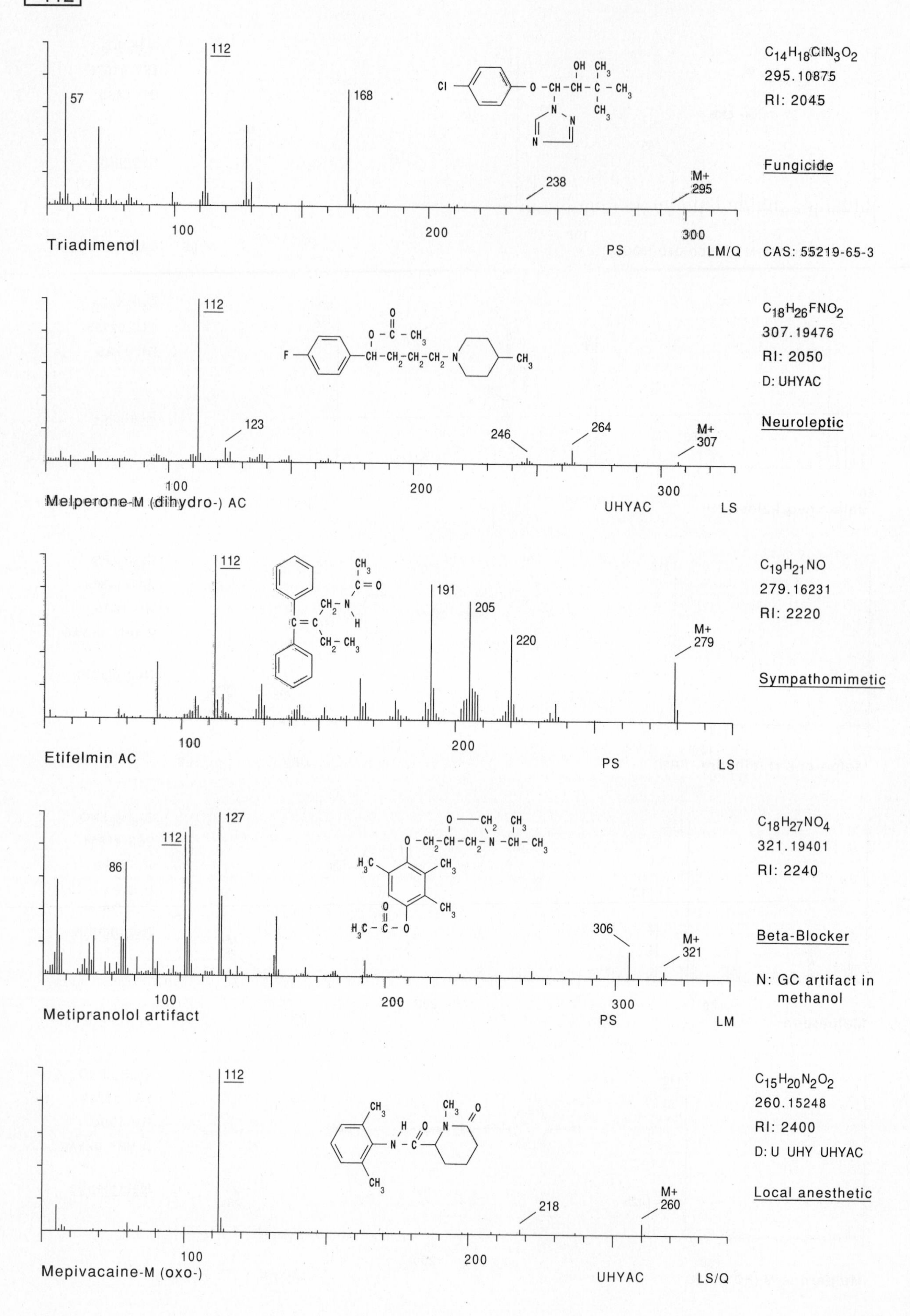
112
57
112
168
238
M+
295
100
200
300
Triadimenol
PS
LM/Q
$C_{14}H_{18}ClN_3O_2$
295.10875
RI: 2045
Fungicide
CAS: 55219-65-3
112
123
246
264
M+
307
100
200
300
Melperone-M (dihydro-) AC
UHYAC
LS
$C_{18}H_{26}FNO_2$
307.19476
RI: 2050
D: UHYAC
Neuroleptic
112
191
205
220
M+
279
100
200
Etifelmin AC
PS
LS
$C_{19}H_{21}NO$
279.16231
RI: 2220
Sympathomimetic
127
112
86
306
M+
321
100
200
300
Metipranolol artifact
PS
LM
$C_{18}H_{27}NO_4$
321.19401
RI: 2240
Beta-Blocker
N: GC artifact in methanol
112
218
M+
260
100
200
Mepivacaine-M (oxo-)
UHYAC
LS/Q
$C_{15}H_{20}N_2O_2$
260.15248
RI: 2400
D: U UHY UHYAC
Local anesthetic

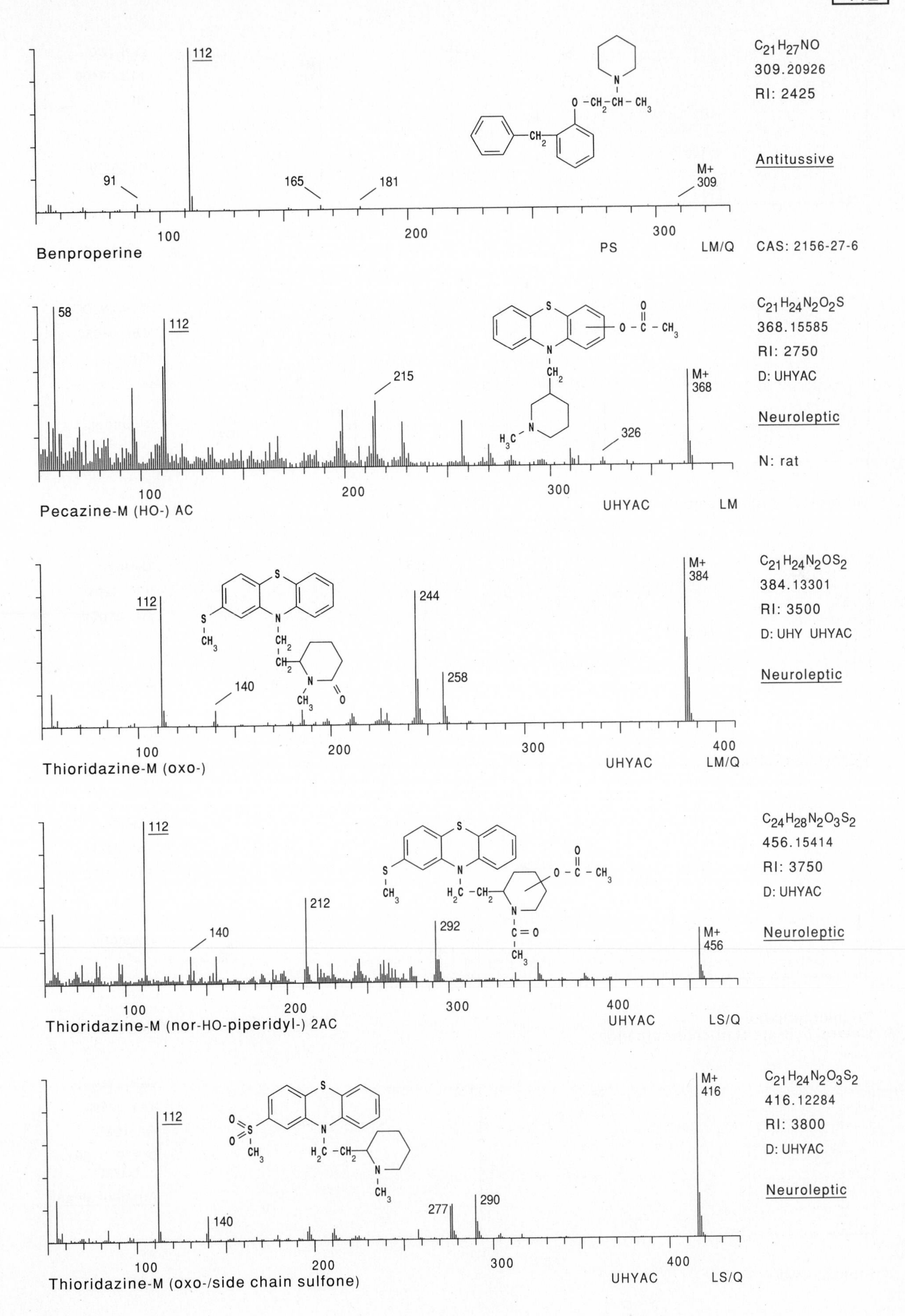
112
Benproperine
PS
LM/Q
$C_{21}H_{27}NO$
309.20926
RI: 2425
Antitussive
CAS: 2156-27-6
M+
309
Pecazine-M (HO-) AC
UHYAC
LM
$C_{21}H_{24}N_2O_2S$
368.15585
RI: 2750
D: UHYAC
Neuroleptic
N: rat
M+
368
Thioridazine-M (oxo-)
UHYAC
LM/Q
$C_{21}H_{24}N_2OS_2$
384.13301
RI: 3500
D: UHY UHYAC
Neuroleptic
M+
384
Thioridazine-M (nor-HO-piperidyl-) 2AC
UHYAC
LS/Q
$C_{24}H_{28}N_2O_3S_2$
456.15414
RI: 3750
D: UHYAC
Neuroleptic
M+
456
Thioridazine-M (oxo-/side chain sulfone)
UHYAC
LS/Q
$C_{21}H_{24}N_2O_3S_2$
416.12284
RI: 3800
D: UHYAC
Neuroleptic
M+
416

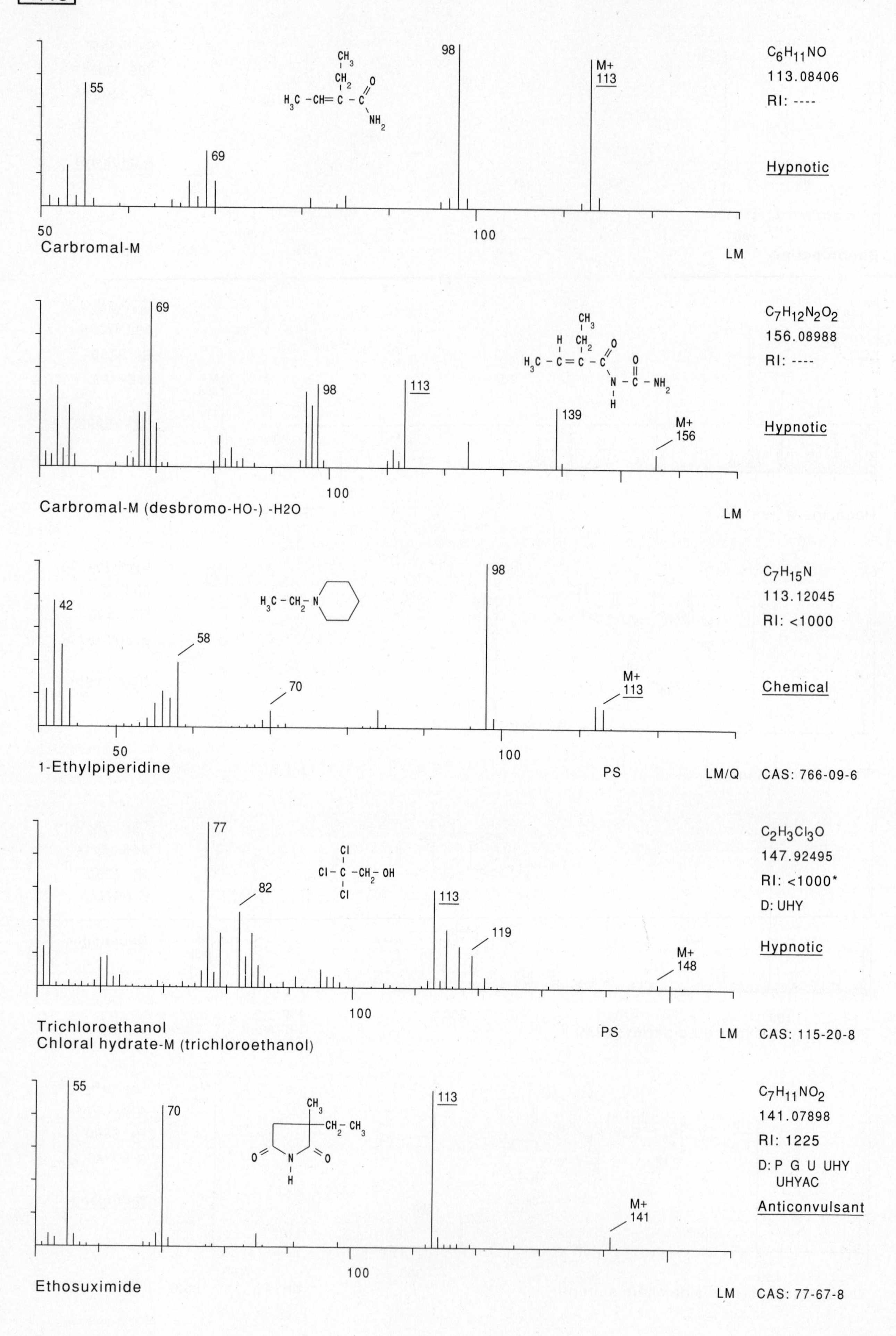

113
Carbromal-M
C6H11NO
113.08406
RI: ----
Hypnotic
LM
Carbromal-M (desbromo-HO-) -H2O
C7H12N2O2
156.08988
RI: ----
Hypnotic
LM
1-Ethylpiperidine
C7H15N
113.12045
RI: <1000
Chemical
PS
LM/Q
CAS: 766-09-6
Trichloroethanol
Chloral hydrate-M (trichloroethanol)
C2H3Cl3O
147.92495
RI: <1000*
D: UHY
Hypnotic
PS
LM
CAS: 115-20-8
Ethosuximide
C7H11NO2
141.07898
RI: 1225
D: P G U UHY
UHYAC
Anticonvulsant
LM
CAS: 77-67-8

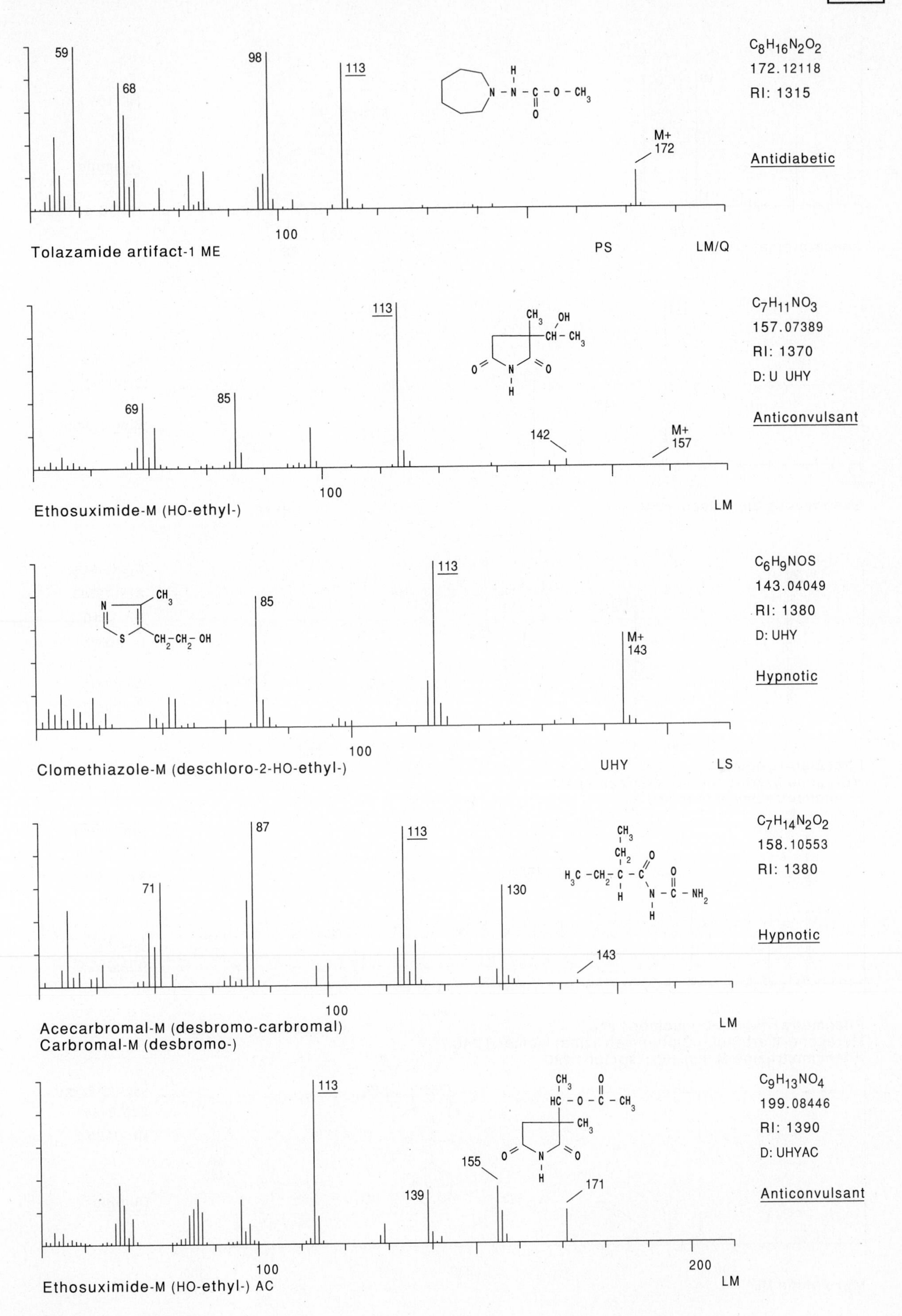

$C_8H_{16}N_2O_2$
172.12118
RI: 1315

Antidiabetic

$C_7H_{11}NO_3$
157.07389
RI: 1370
D: U UHY

Anticonvulsant

C_6H_9NOS
143.04049
RI: 1380
D: UHY

Hypnotic

$C_7H_{14}N_2O_2$
158.10553
RI: 1380

Hypnotic

$C_9H_{13}NO_4$
199.08446
RI: 1390
D: UHYAC

Anticonvulsant

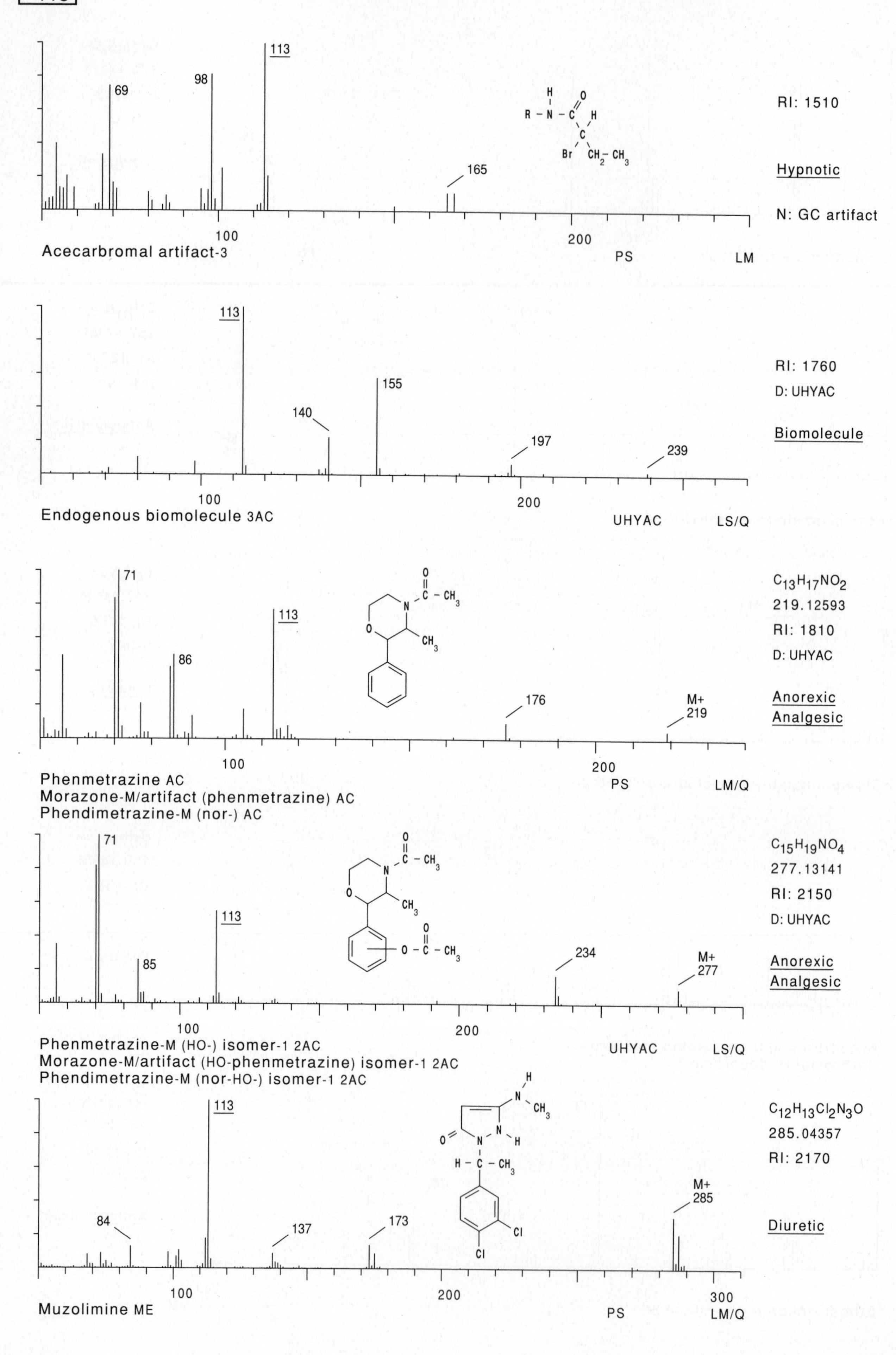

113
98
69
165
R – N – C – H
O
Br
CH2 – CH3
100
200
RI: 1510
Hypnotic
N: GC artifact
Acecarbromal artifact-3
PS
LM
113
155
140
197
239
100
200
RI: 1760
D: UHYAC
Biomolecule
Endogenous biomolecule 3AC
UHYAC
LS/Q
71
113
86
176
M+
219
C13H17NO2
219.12593
RI: 1810
D: UHYAC
Anorexic
Analgesic
Phenmetrazine AC
Morazone-M/artifact (phenmetrazine) AC
Phendimetrazine-M (nor-) AC
PS
LM/Q
71
113
85
234
M+
277
C15H19NO4
277.13141
RI: 2150
D: UHYAC
Anorexic
Analgesic
Phenmetrazine-M (HO-) isomer-1 2AC
Morazone-M/artifact (HO-phenmetrazine) isomer-1 2AC
Phendimetrazine-M (nor-HO-) isomer-1 2AC
UHYAC
LS/Q
113
84
137
173
M+
285
C12H13Cl2N3O
285.04357
RI: 2170
Diuretic
Muzolimine ME
300
PS
LM/Q

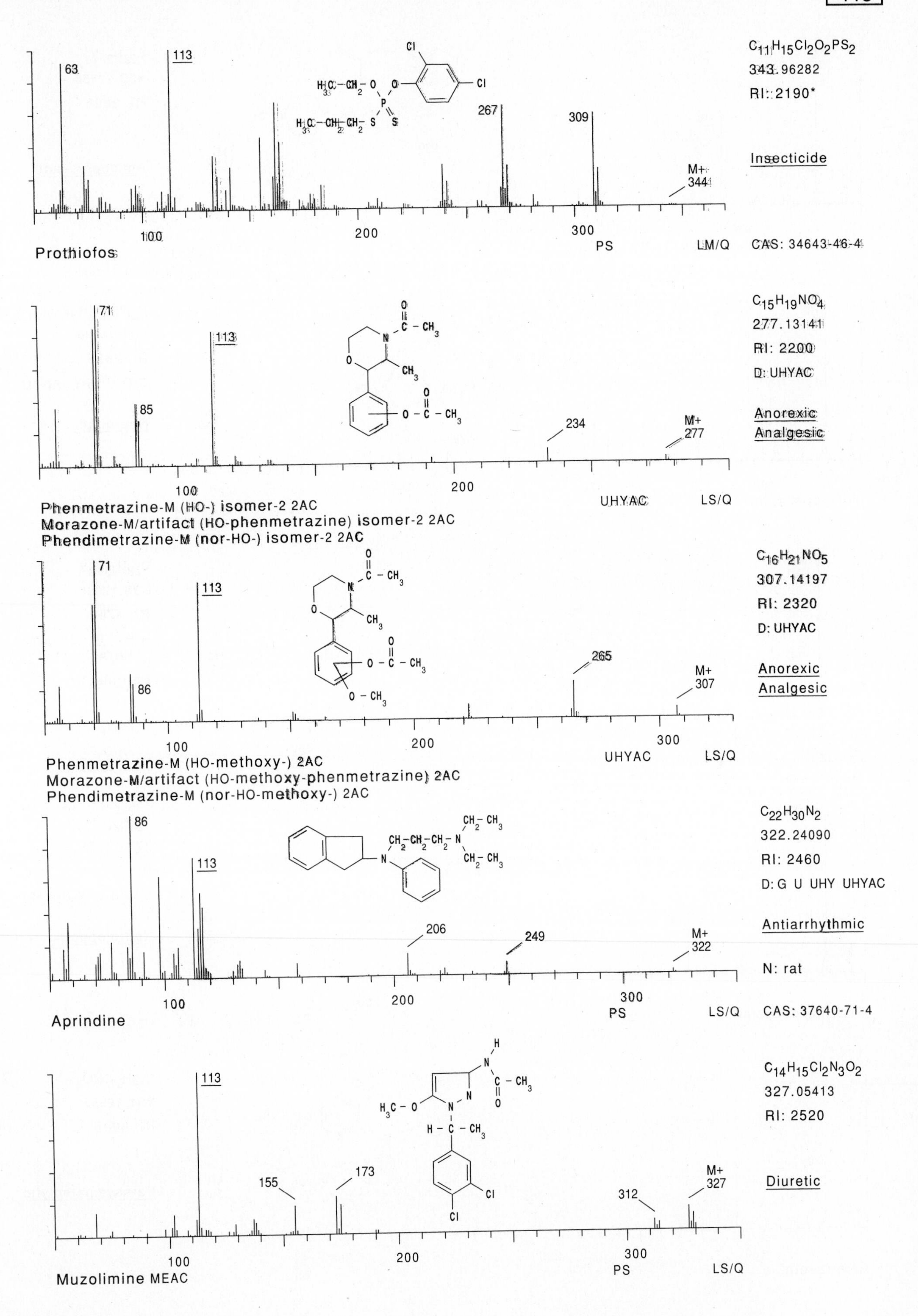

Prothiofos

$C_{11}H_{15}Cl_2O_2PS_2$
343.96282
RI: 2190*
Insecticide
CAS: 34643-46-4

Phenmetrazine-M (HO-) isomer-2 2AC
Morazone-M/artifact (HO-phenmetrazine) isomer-2 2AC
Phendimetrazine-M (nor-HO-) isomer-2 2AC

$C_{15}H_{19}NO_4$
277.13141
RI: 2200
D: UHYAC
Anorexic
Analgesic

Phenmetrazine-M (HO-methoxy-) 2AC
Morazone-M/artifact (HO-methoxy-phenmetrazine) 2AC
Phendimetrazine-M (nor-HO-methoxy-) 2AC

$C_{16}H_{21}NO_5$
307.14197
RI: 2320
D: UHYAC
Anorexic
Analgesic

Aprindine

$C_{22}H_{30}N_2$
322.24090
RI: 2460
D: G U UHY UHYAC
Antiarrhythmic
N: rat
CAS: 37640-71-4

Muzolimine MEAC

$C_{14}H_{15}Cl_2N_3O_2$
327.05413
RI: 2520
Diuretic

113

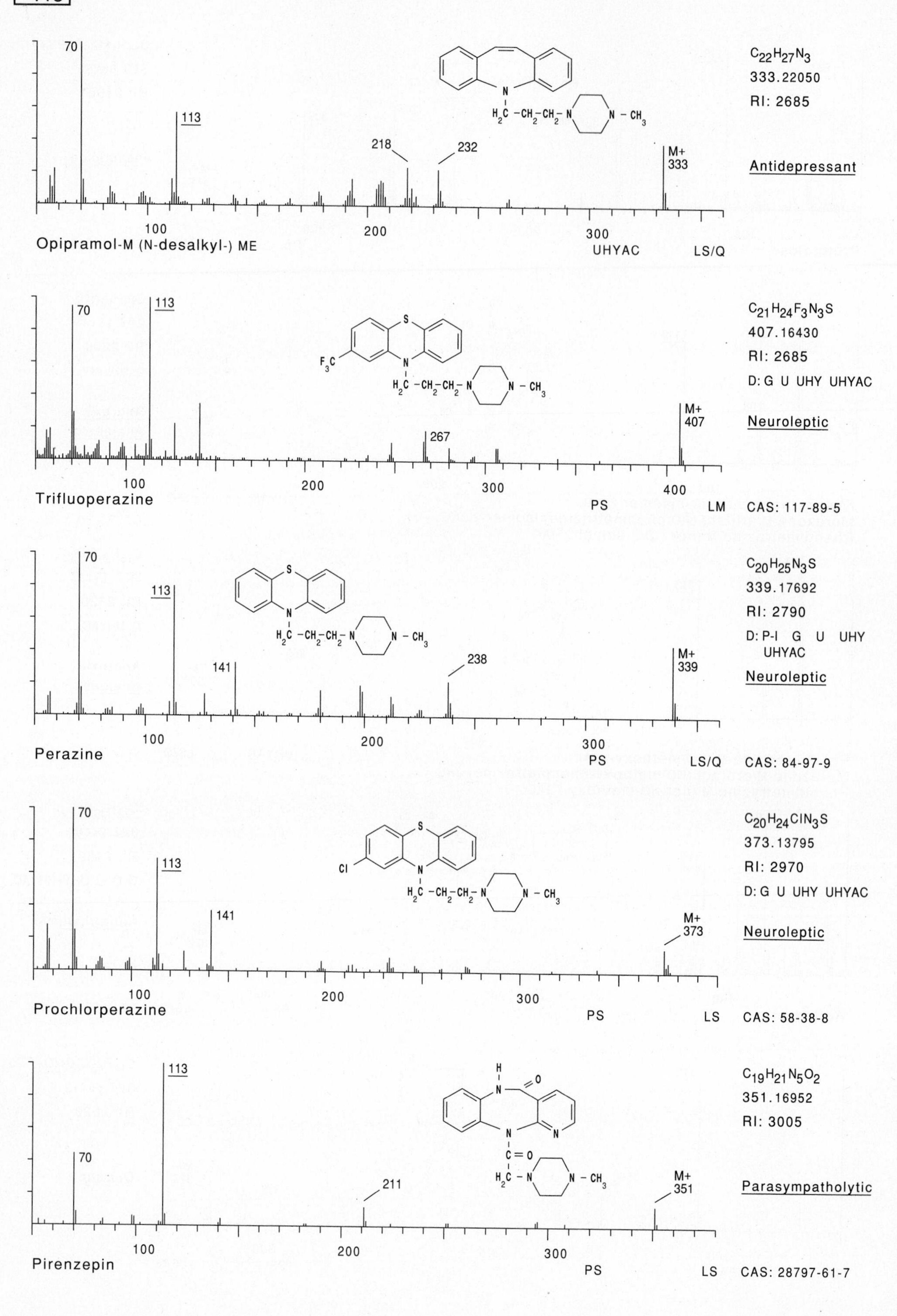
70
113
218
232
M+
333
$C_{22}H_{27}N_3$
333.22050
RI: 2685
Antidepressant
100
200
300
Opipramol-M (N-desalkyl-) ME
UHYAC
LS/Q
70
113
267
M+
407
$C_{21}H_{24}F_3N_3S$
407.16430
RI: 2685
D: G U UHY UHYAC
Neuroleptic
100
200
300
400
Trifluoperazine
PS
LM
CAS: 117-89-5
70
113
141
238
M+
339
$C_{20}H_{25}N_3S$
339.17692
RI: 2790
D: P-I G U UHY UHYAC
Neuroleptic
100
200
300
Perazine
PS
LS/Q
CAS: 84-97-9
70
113
141
M+
373
$C_{20}H_{24}ClN_3S$
373.13795
RI: 2970
D: G U UHY UHYAC
Neuroleptic
100
200
300
Prochlorperazine
PS
LS
CAS: 58-38-8
113
70
211
M+
351
$C_{19}H_{21}N_5O_2$
351.16952
RI: 3005
Parasympatholytic
100
200
300
Pirenzepin
PS
LS
CAS: 28797-61-7

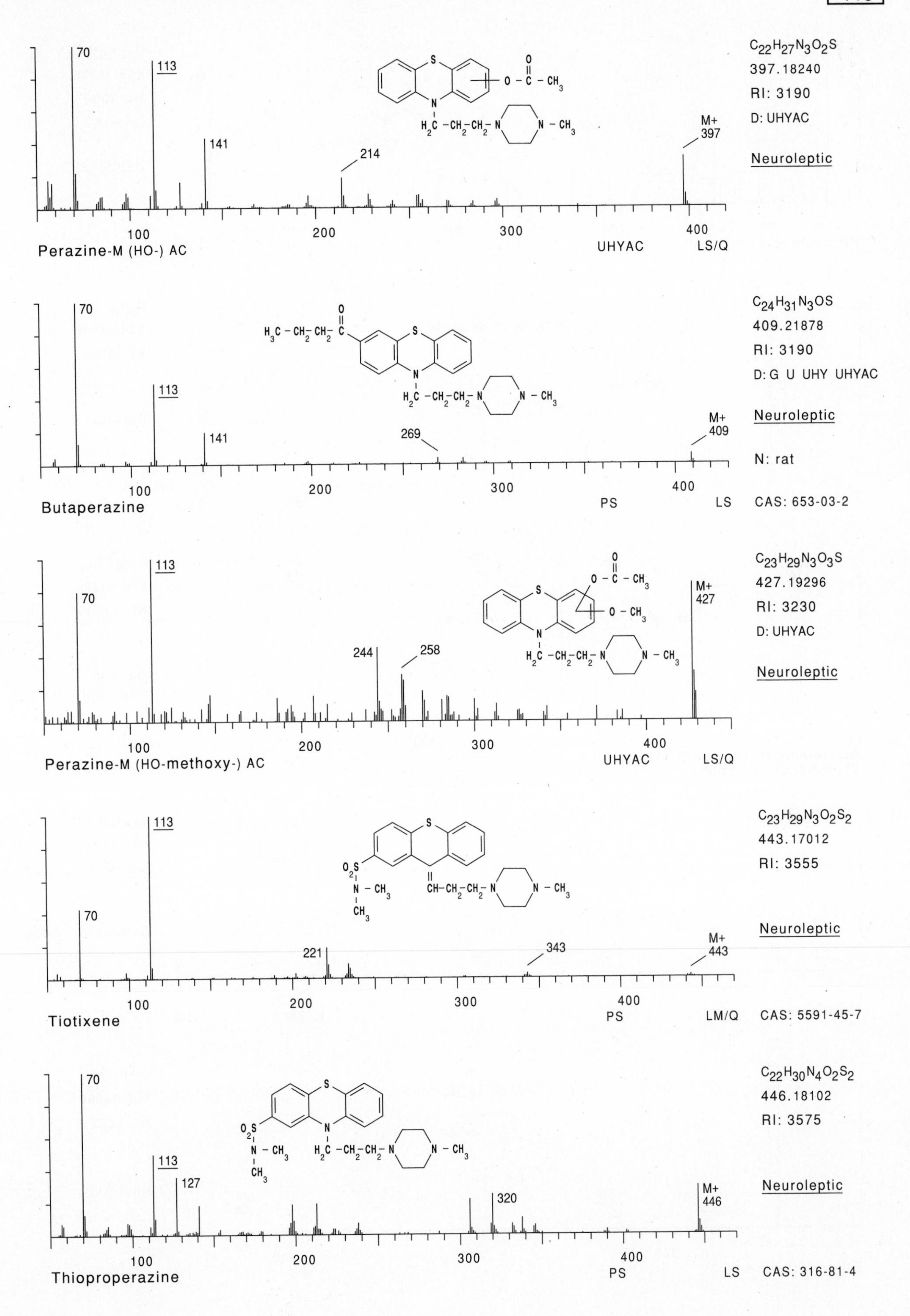

Perazine-M (HO-) AC
UHYAC LS/Q
C22H27N3O2S
397.18240
RI: 3190
D: UHYAC
Neuroleptic
Butaperazine
PS LS
C24H31N3OS
409.21878
RI: 3190
D: G U UHY UHYAC
Neuroleptic
N: rat
CAS: 653-03-2
Perazine-M (HO-methoxy-) AC
UHYAC LS/Q
C23H29N3O3S
427.19296
RI: 3230
D: UHYAC
Neuroleptic
Tiotixene
PS LM/Q
C23H29N3O2S2
443.17012
RI: 3555
Neuroleptic
CAS: 5591-45-7
Thioproperazine
PS LS
C22H30N4O2S2
446.18102
RI: 3575
Neuroleptic
CAS: 316-81-4

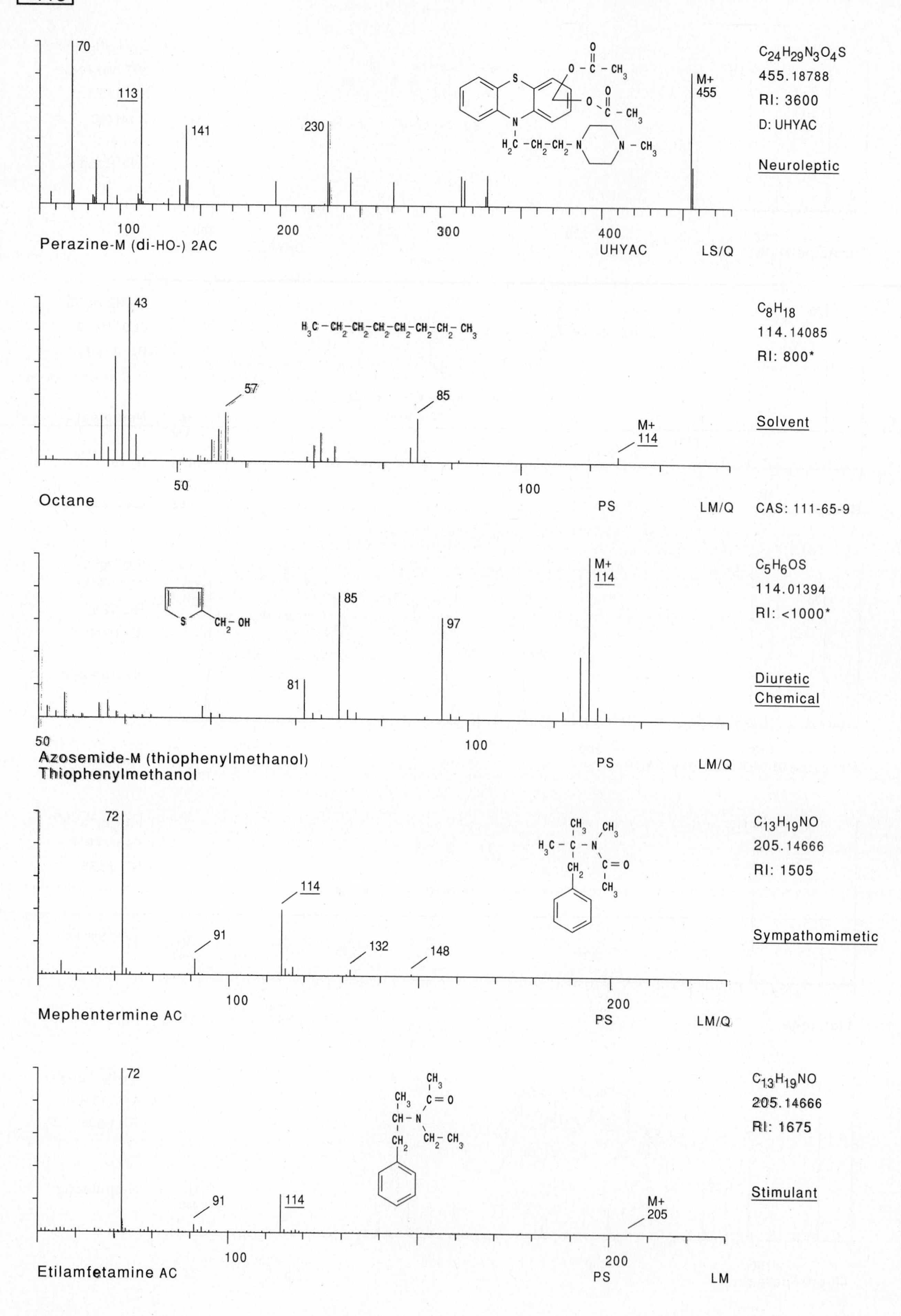
70
113
141
230
M+
455
100
200
300
400
Perazine-M (di-HO-) 2AC
UHYAC
LS/Q
$C_{24}H_{29}N_3O_4S$
455.18788
RI: 3600
D: UHYAC
Neuroleptic
43
57
85
M+
114
50
100
Octane
PS
LM/Q
C_8H_{18}
114.14085
RI: 800*
Solvent
CAS: 111-65-9
M+
114
85
97
81
50
100
Azosemide-M (thiophenylmethanol)
Thiophenylmethanol
PS
LM/Q
C_5H_6OS
114.01394
RI: <1000*
Diuretic
Chemical
72
114
91
132
148
100
200
Mephentermine AC
PS
LM/Q
$C_{13}H_{19}NO$
205.14666
RI: 1505
Sympathomimetic
72
91
114
M+
205
100
200
Etilamfetamine AC
PS
LM
$C_{13}H_{19}NO$
205.14666
RI: 1675
Stimulant

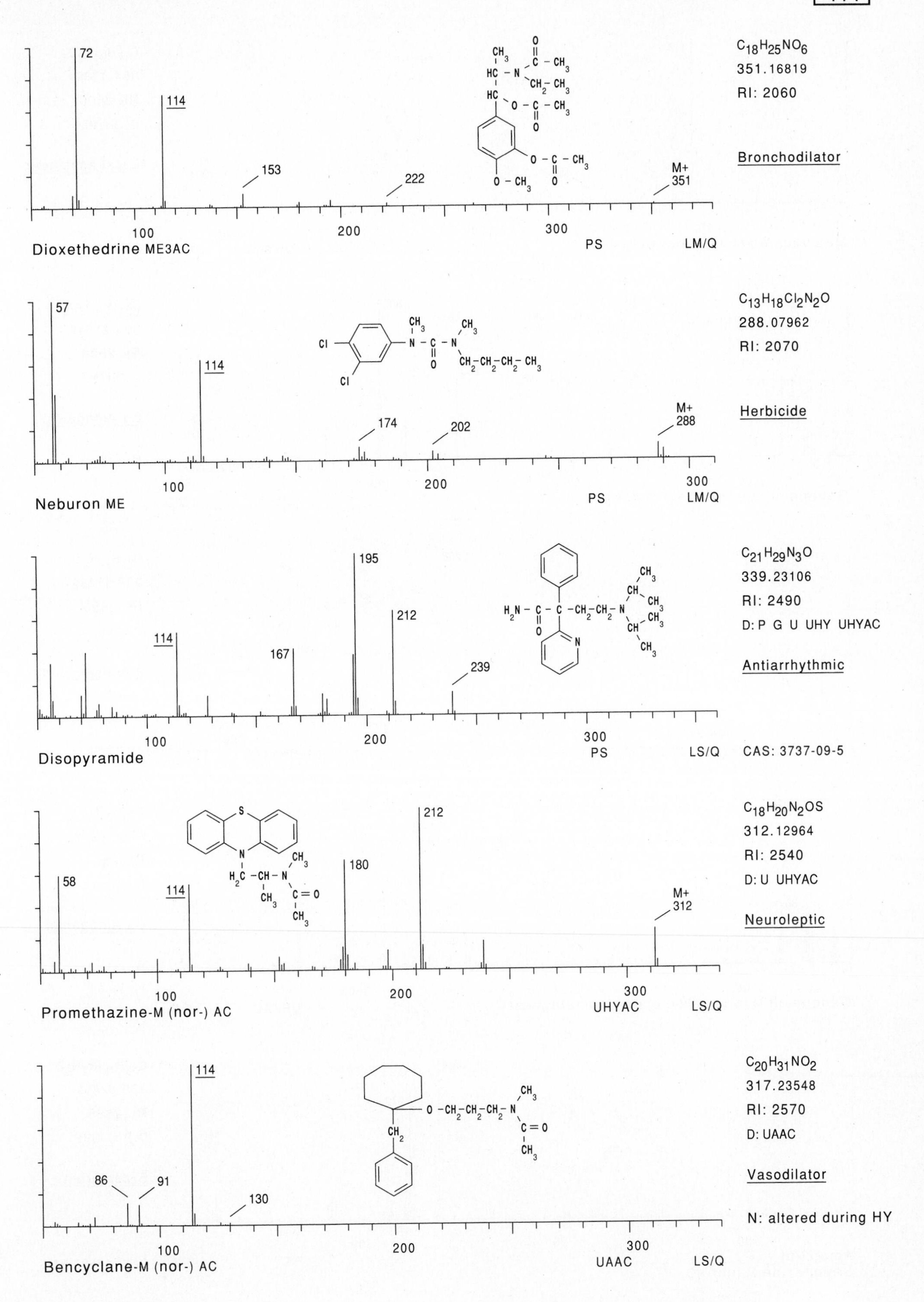

Dioxethedrine ME3AC
PS LM/Q
$C_{18}H_{25}NO_6$
351.16819
RI: 2060
Bronchodilator
Neburon ME
PS LM/Q
$C_{13}H_{18}Cl_2N_2O$
288.07962
RI: 2070
Herbicide
Disopyramide
PS LS/Q
$C_{21}H_{29}N_3O$
339.23106
RI: 2490
D: P G U UHY UHYAC
Antiarrhythmic
CAS: 3737-09-5
Promethazine-M (nor-) AC
UHYAC LS/Q
$C_{18}H_{20}N_2OS$
312.12964
RI: 2540
D: U UHYAC
Neuroleptic
Bencyclane-M (nor-) AC
UAAC LS/Q
$C_{20}H_{31}NO_2$
317.23548
RI: 2570
D: UAAC
Vasodilator
N: altered during HY

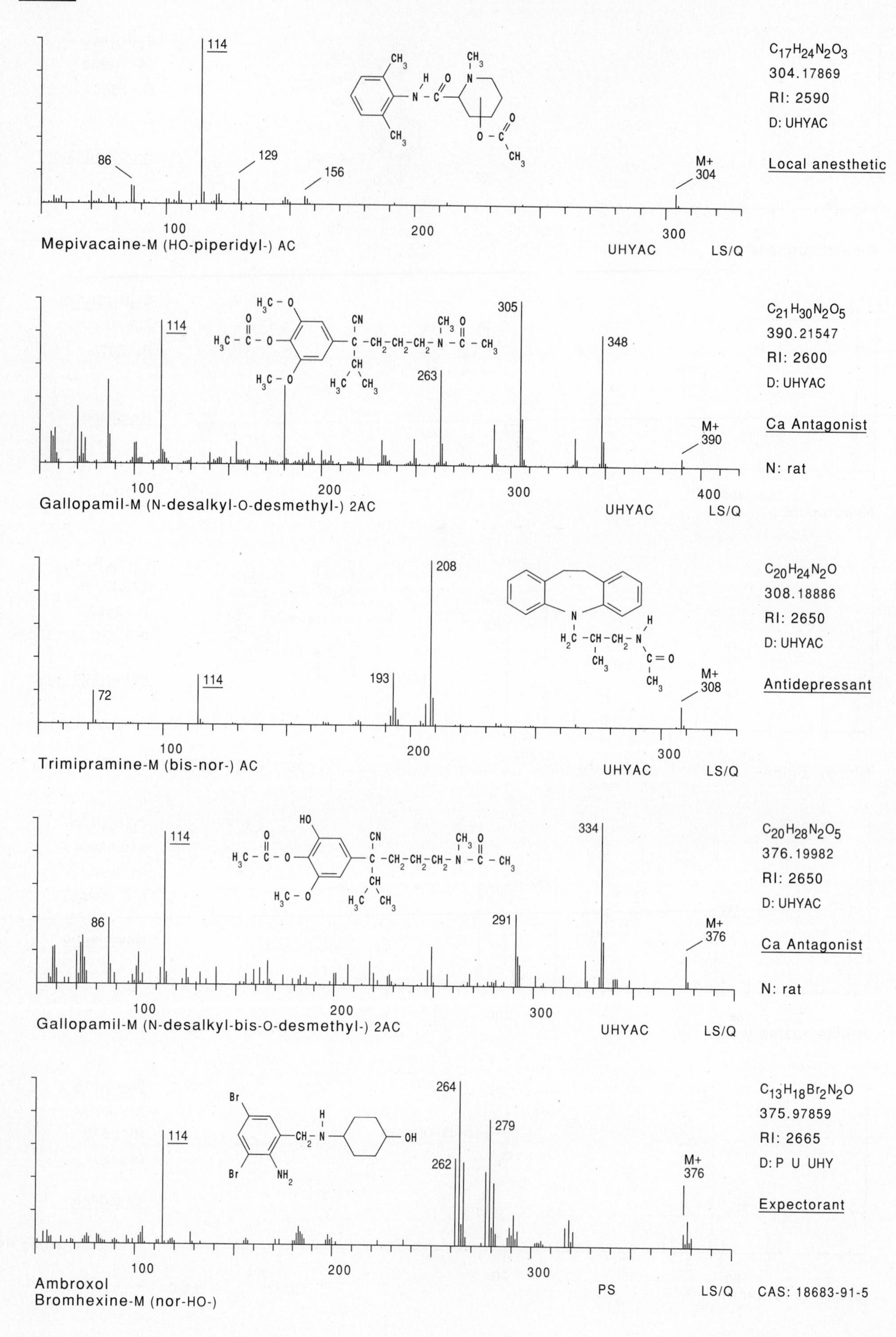
114
86
129
156
M+
304
100
200
300
Mepivacaine-M (HO-piperidyl-) AC
UHYAC
LS/Q
$C_{17}H_{24}N_2O_3$
304.17869
RI: 2590
D: UHYAC
Local anesthetic
305
114
348
263
M+
390
Gallopamil-M (N-desalkyl-O-desmethyl-) 2AC
$C_{21}H_{30}N_2O_5$
390.21547
RI: 2600
D: UHYAC
Ca Antagonist
N: rat
208
114
72
193
M+
308
Trimipramine-M (bis-nor-) AC
$C_{20}H_{24}N_2O$
308.18886
RI: 2650
D: UHYAC
Antidepressant
334
114
86
291
M+
376
Gallopamil-M (N-desalkyl-bis-O-desmethyl-) 2AC
$C_{20}H_{28}N_2O_5$
376.19982
RI: 2650
D: UHYAC
Ca Antagonist
N: rat
264
114
279
262
M+
376
Ambroxol
Bromhexine-M (nor-HO-)
PS
LS/Q
$C_{13}H_{18}Br_2N_2O$
375.97859
RI: 2665
D: P U UHY
Expectorant
CAS: 18683-91-5

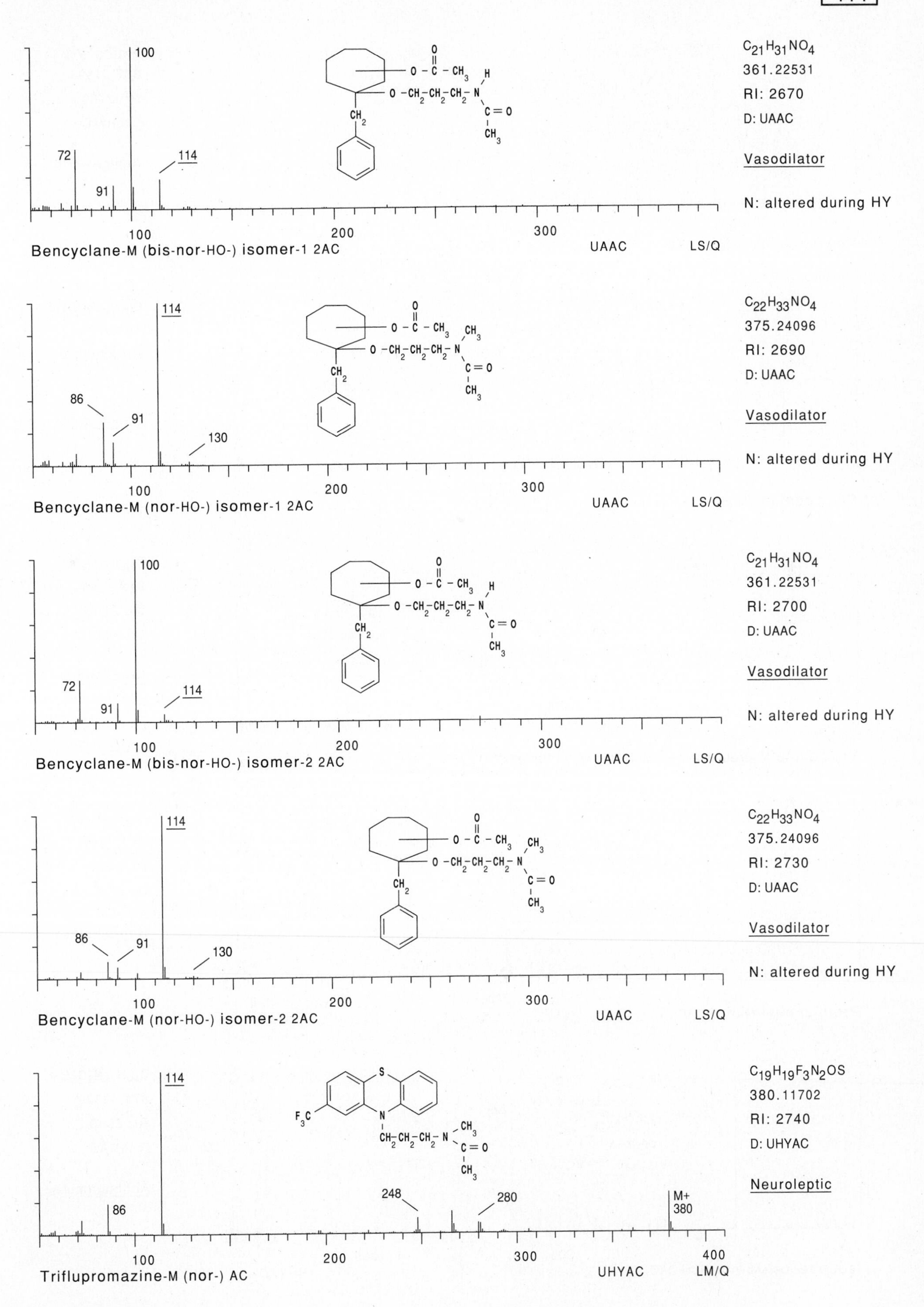
114
100
72
91
114
Bencyclane-M (bis-nor-HO-) isomer-1 2AC
UAAC
LS/Q
$C_{21}H_{31}NO_4$
361.22531
RI: 2670
D: UAAC
Vasodilator
N: altered during HY
114
86
91
130
Bencyclane-M (nor-HO-) isomer-1 2AC
UAAC
LS/Q
$C_{22}H_{33}NO_4$
375.24096
RI: 2690
D: UAAC
Vasodilator
N: altered during HY
100
72
91
114
Bencyclane-M (bis-nor-HO-) isomer-2 2AC
UAAC
LS/Q
$C_{21}H_{31}NO_4$
361.22531
RI: 2700
D: UAAC
Vasodilator
N: altered during HY
114
86
91
130
Bencyclane-M (nor-HO-) isomer-2 2AC
UAAC
LS/Q
$C_{22}H_{33}NO_4$
375.24096
RI: 2730
D: UAAC
Vasodilator
N: altered during HY
114
86
248
280
M+ 380
Triflupromazine-M (nor-) AC
UHYAC
LM/Q
$C_{19}H_{19}F_3N_2OS$
380.11702
RI: 2740
D: UHYAC
Neuroleptic

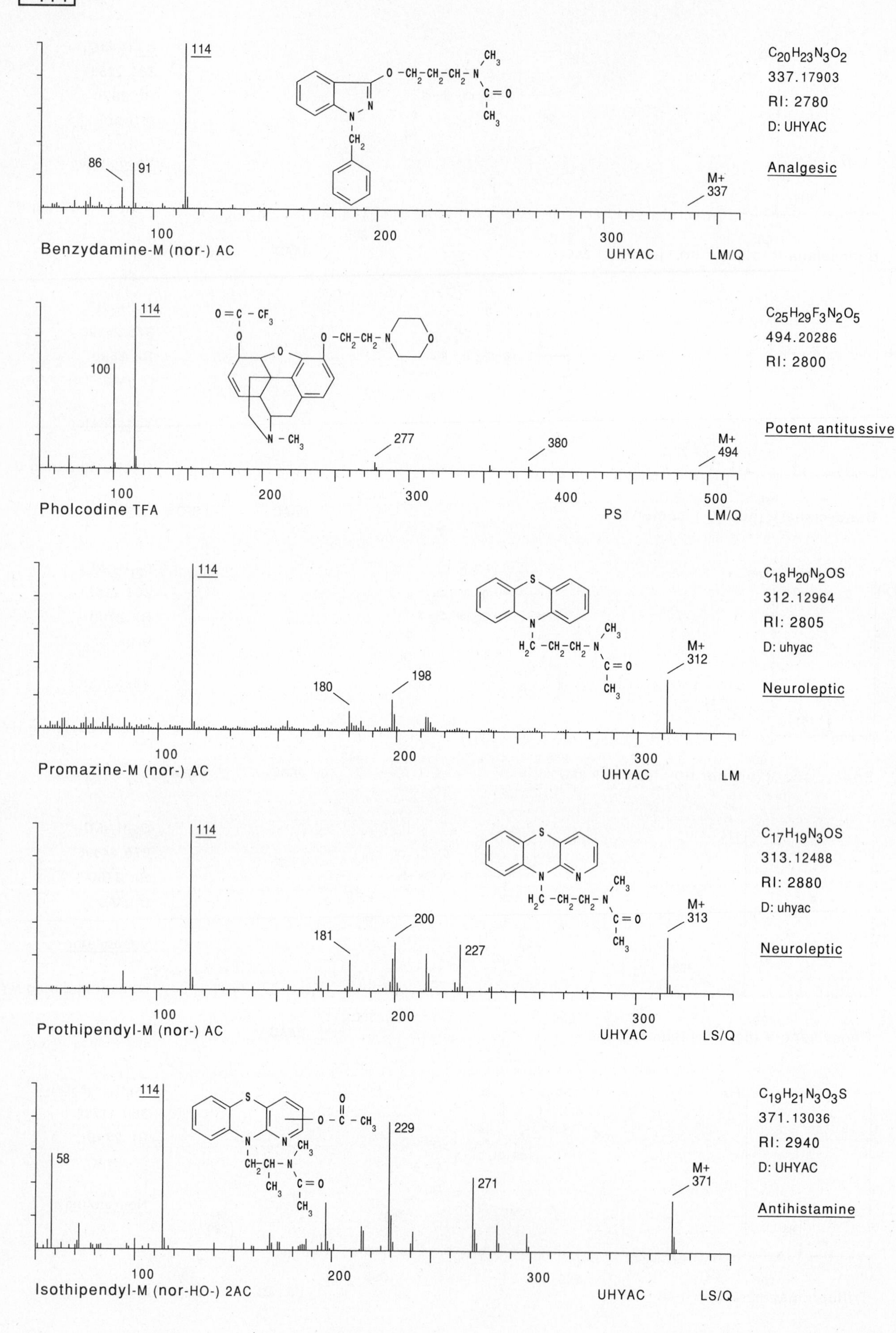

Benzydamine-M (nor-) AC
UHYAC LM/Q
$C_{20}H_{23}N_3O_2$
337.17903
RI: 2780
D: UHYAC
Analgesic
Pholcodine TFA
PS LM/Q
$C_{25}H_{29}F_3N_2O_5$
494.20286
RI: 2800
Potent antitussive
Promazine-M (nor-) AC
UHYAC LM
$C_{18}H_{20}N_2OS$
312.12964
RI: 2805
D: uhyac
Neuroleptic
Prothipendyl-M (nor-) AC
UHYAC LS/Q
$C_{17}H_{19}N_3OS$
313.12488
RI: 2880
D: uhyac
Neuroleptic
Isothipendyl-M (nor-HO-) 2AC
UHYAC LS/Q
$C_{19}H_{21}N_3O_3S$
371.13036
RI: 2940
D: UHYAC
Antihistamine

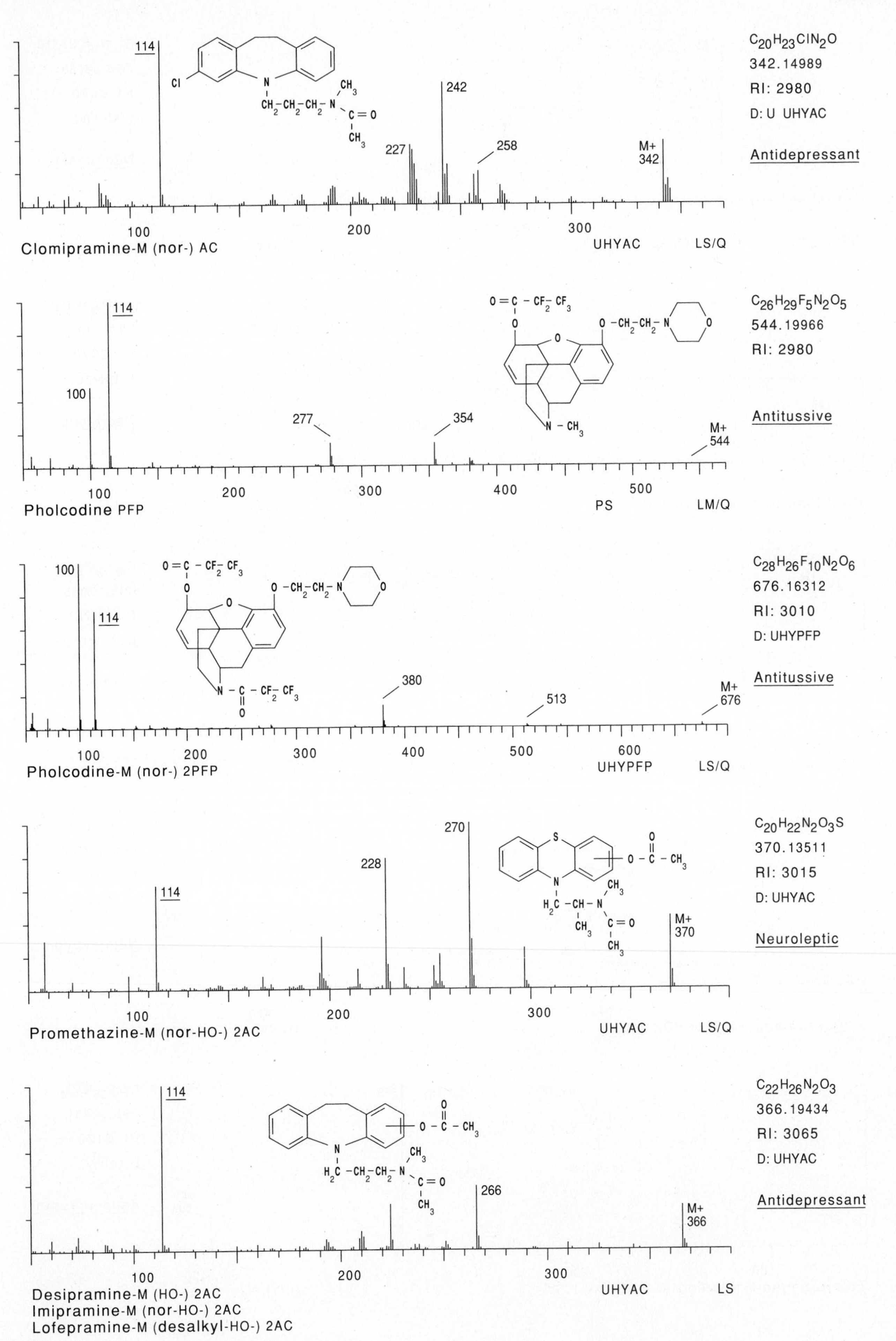

$C_{20}H_{23}ClN_2O$
342.14989
RI: 2980
D: U UHYAC

Antidepressant

$C_{26}H_{29}F_5N_2O_5$
544.19966
RI: 2980

Antitussive

$C_{28}H_{26}F_{10}N_2O_6$
676.16312
RI: 3010
D: UHYPFP

Antitussive

$C_{20}H_{22}N_2O_3S$
370.13511
RI: 3015
D: UHYAC

Neuroleptic

$C_{22}H_{26}N_2O_3$
366.19434
RI: 3065
D: UHYAC

Antidepressant

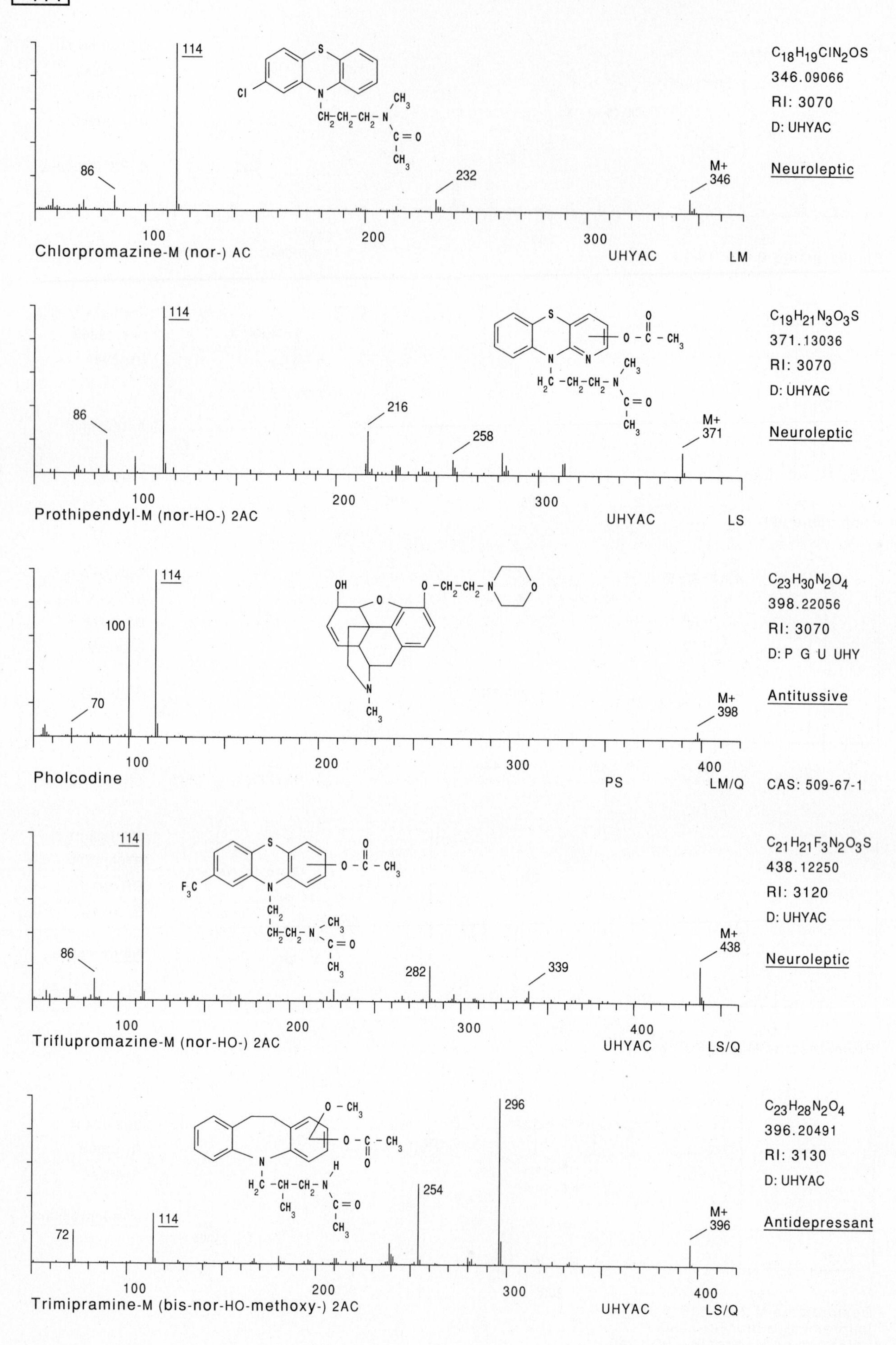

Chlorpromazine-M (nor-) AC

$C_{18}H_{19}ClN_2OS$
346.09066
RI: 3070
D: UHYAC
Neuroleptic

Prothipendyl-M (nor-HO-) 2AC

$C_{19}H_{21}N_3O_3S$
371.13036
RI: 3070
D: UHYAC
Neuroleptic

Pholcodine

$C_{23}H_{30}N_2O_4$
398.22056
RI: 3070
D: P G U UHY
Antitussive
CAS: 509-67-1

Triflupromazine-M (nor-HO-) 2AC

$C_{21}H_{21}F_3N_2O_3S$
438.12250
RI: 3120
D: UHYAC
Neuroleptic

Trimipramine-M (bis-nor-HO-methoxy-) 2AC

$C_{23}H_{28}N_2O_4$
396.20491
RI: 3130
D: UHYAC
Antidepressant

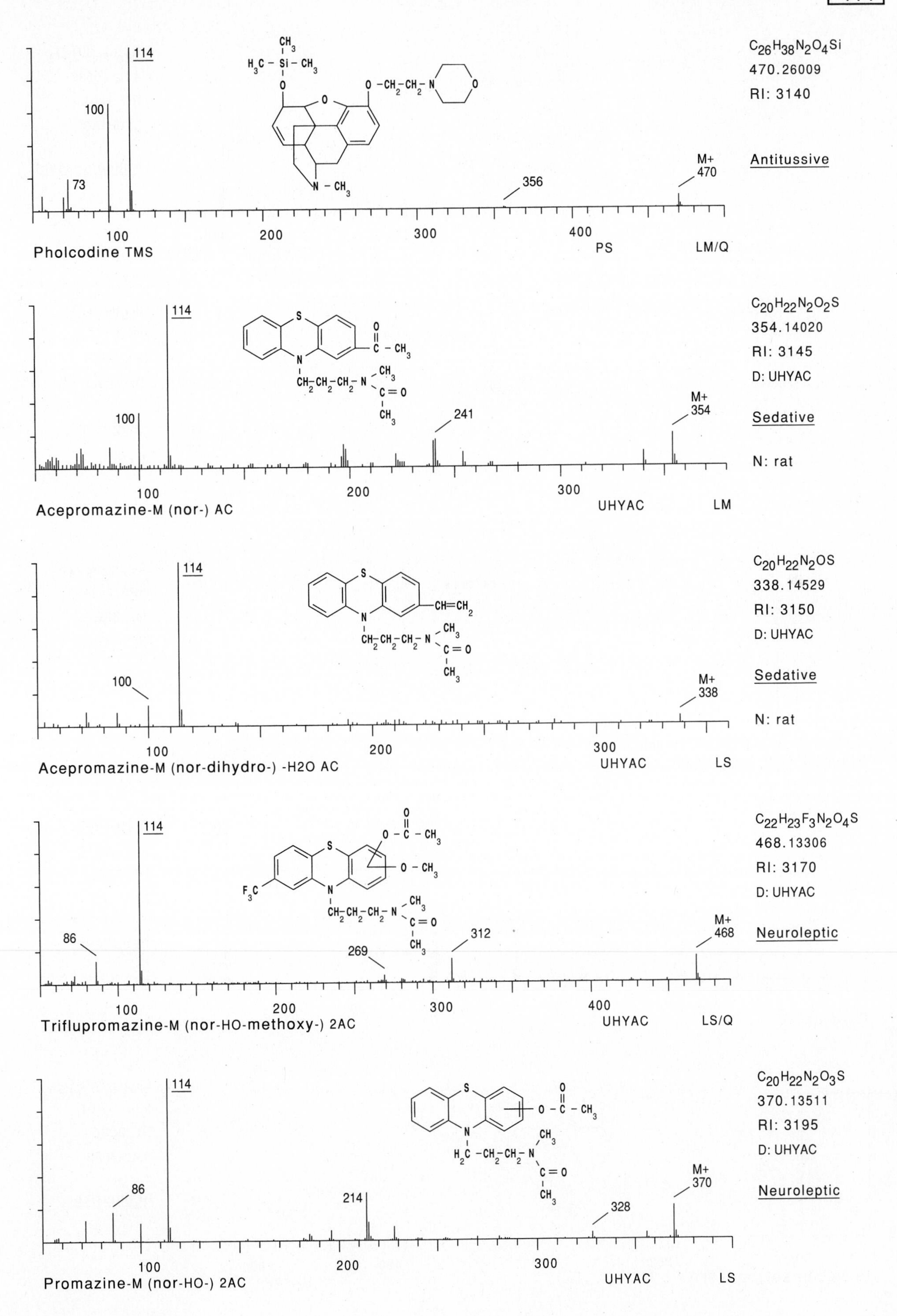

114
100
73
356
M+
470
100
200
300
400
Pholcodine TMS
PS
LM/Q
C26H38N2O4Si
470.26009
RI: 3140
Antitussive
114
100
241
M+
354
100
200
300
Acepromazine-M (nor-) AC
UHYAC
LM
C20H22N2O2S
354.14020
RI: 3145
D: UHYAC
Sedative
N: rat
114
100
M+
338
100
200
300
Acepromazine-M (nor-dihydro-) -H2O AC
UHYAC
LS
C20H22N2OS
338.14529
RI: 3150
D: UHYAC
Sedative
N: rat
114
86
269
312
M+
468
100
200
300
400
Triflupromazine-M (nor-HO-methoxy-) 2AC
UHYAC
LS/Q
C22H23F3N2O4S
468.13306
RI: 3170
D: UHYAC
Neuroleptic
114
86
214
328
M+
370
100
200
300
Promazine-M (nor-HO-) 2AC
UHYAC
LS
C20H22N2O3S
370.13511
RI: 3195
D: UHYAC
Neuroleptic

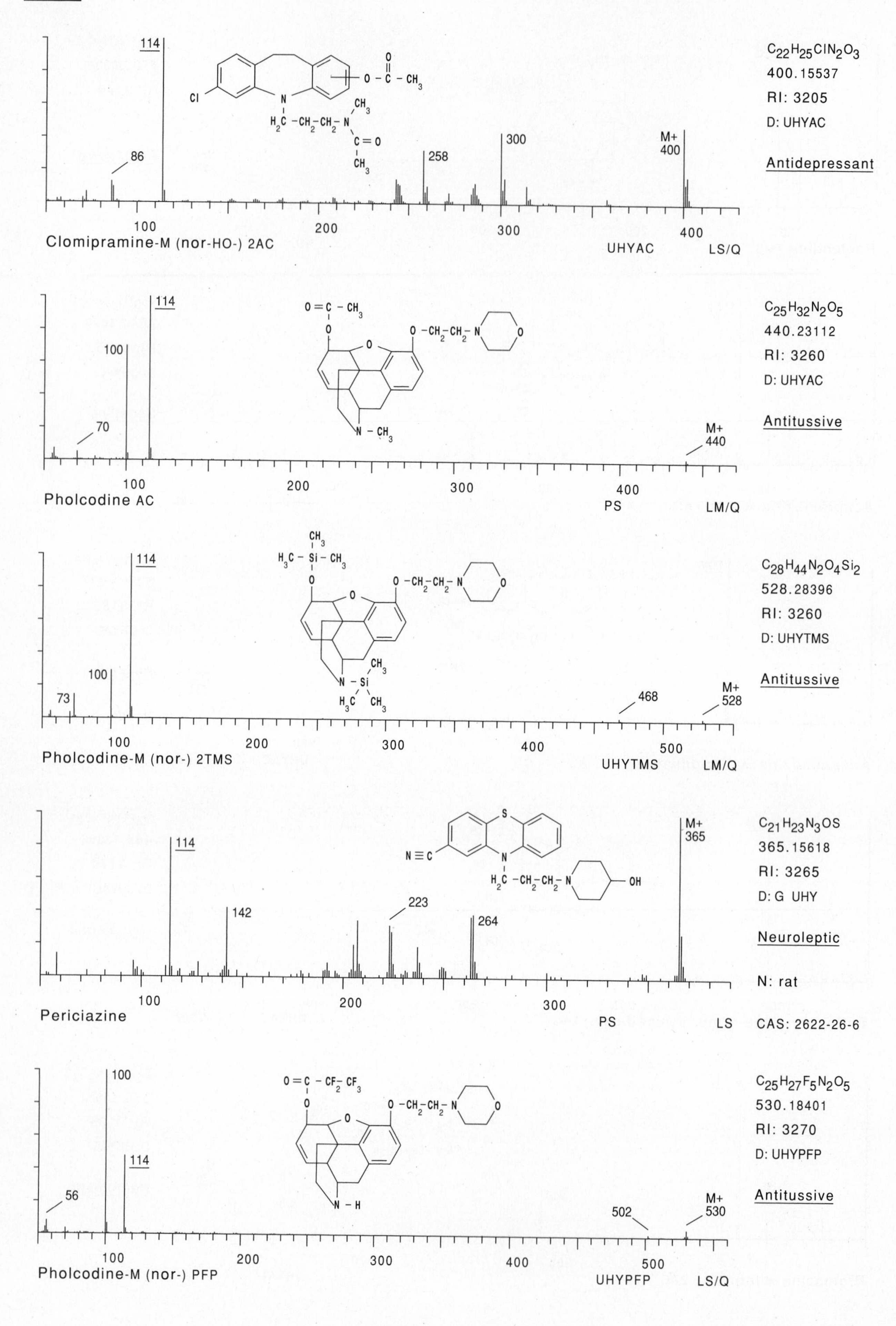

114
86
258
300
M+
400
100
200
300
400
Clomipramine-M (nor-HO-) 2AC
UHYAC
LS/Q
$C_{22}H_{25}ClN_2O_3$
400.15537
RI: 3205
D: UHYAC
Antidepressant
114
100
70
M+
440
Pholcodine AC
PS
LM/Q
$C_{25}H_{32}N_2O_5$
440.23112
RI: 3260
D: UHYAC
Antitussive
114
100
73
468
M+
528
Pholcodine-M (nor-) 2TMS
UHYTMS
LM/Q
$C_{28}H_{44}N_2O_4Si_2$
528.28396
RI: 3260
D: UHYTMS
Antitussive
M+
365
114
142
223
264
Periciazine
PS
LS
$C_{21}H_{23}N_3OS$
365.15618
RI: 3265
D: G UHY
Neuroleptic
N: rat
CAS: 2622-26-6
100
114
56
502
M+
530
Pholcodine-M (nor-) PFP
UHYPFP
LS/Q
$C_{25}H_{27}F_5N_2O_5$
530.18401
RI: 3270
D: UHYPFP
Antitussive

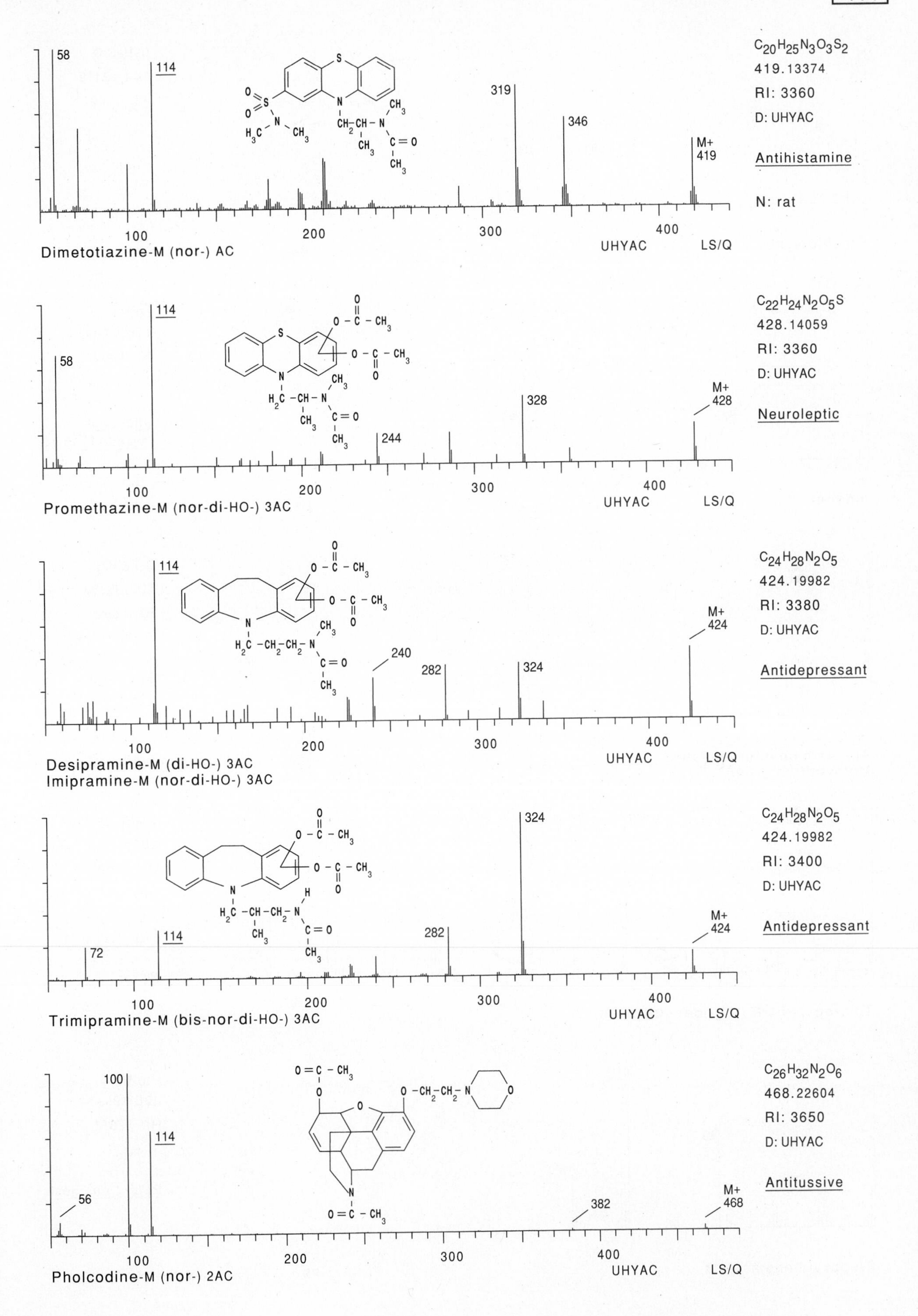

$C_{20}H_{25}N_3O_3S_2$
419.13374
RI: 3360
D: UHYAC
Antihistamine
N: rat

$C_{22}H_{24}N_2O_5S$
428.14059
RI: 3360
D: UHYAC
Neuroleptic

$C_{24}H_{28}N_2O_5$
424.19982
RI: 3380
D: UHYAC
Antidepressant

$C_{24}H_{28}N_2O_5$
424.19982
RI: 3400
D: UHYAC
Antidepressant

$C_{26}H_{32}N_2O_6$
468.22604
RI: 3650
D: UHYAC
Antitussive

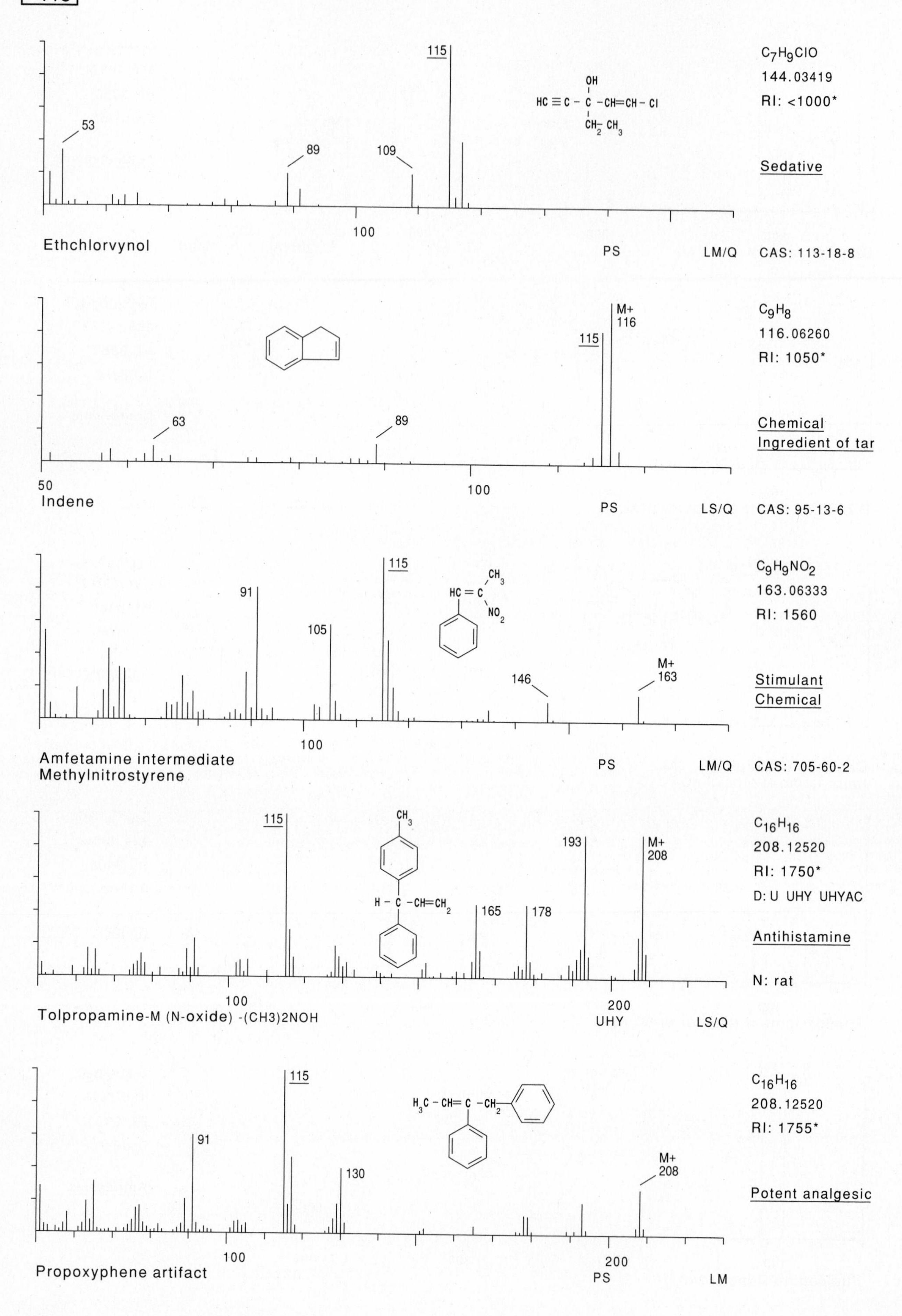
115
53
89
109
100
OH
HC ≡ C – C –CH=CH – Cl
CH2 CH3
Ethchlorvynol
PS
LM/Q
C7H9ClO
144.03419
RI: <1000*
Sedative
CAS: 113-18-8
M+
116
115
63
89
50
100
Indene
PS
LS/Q
C9H8
116.06260
RI: 1050*
Chemical
Ingredient of tar
CAS: 95-13-6
115
91
105
CH3
HC = C
NO2
146
M+
163
100
Amfetamine intermediate
Methylnitrostyrene
PS
LM/Q
C9H9NO2
163.06333
RI: 1560
Stimulant
Chemical
CAS: 705-60-2
115
CH3
H – C –CH=CH2
193
M+
208
165
178
100
200
Tolpropamine-M (N-oxide) -(CH3)2NOH
UHY
LS/Q
C16H16
208.12520
RI: 1750*
D: U UHY UHYAC
Antihistamine
N: rat
115
H3C – CH= C –CH2
91
130
M+
208
100
200
Propoxyphene artifact
PS
LM
C16H16
208.12520
RI: 1755*
Potent analgesic

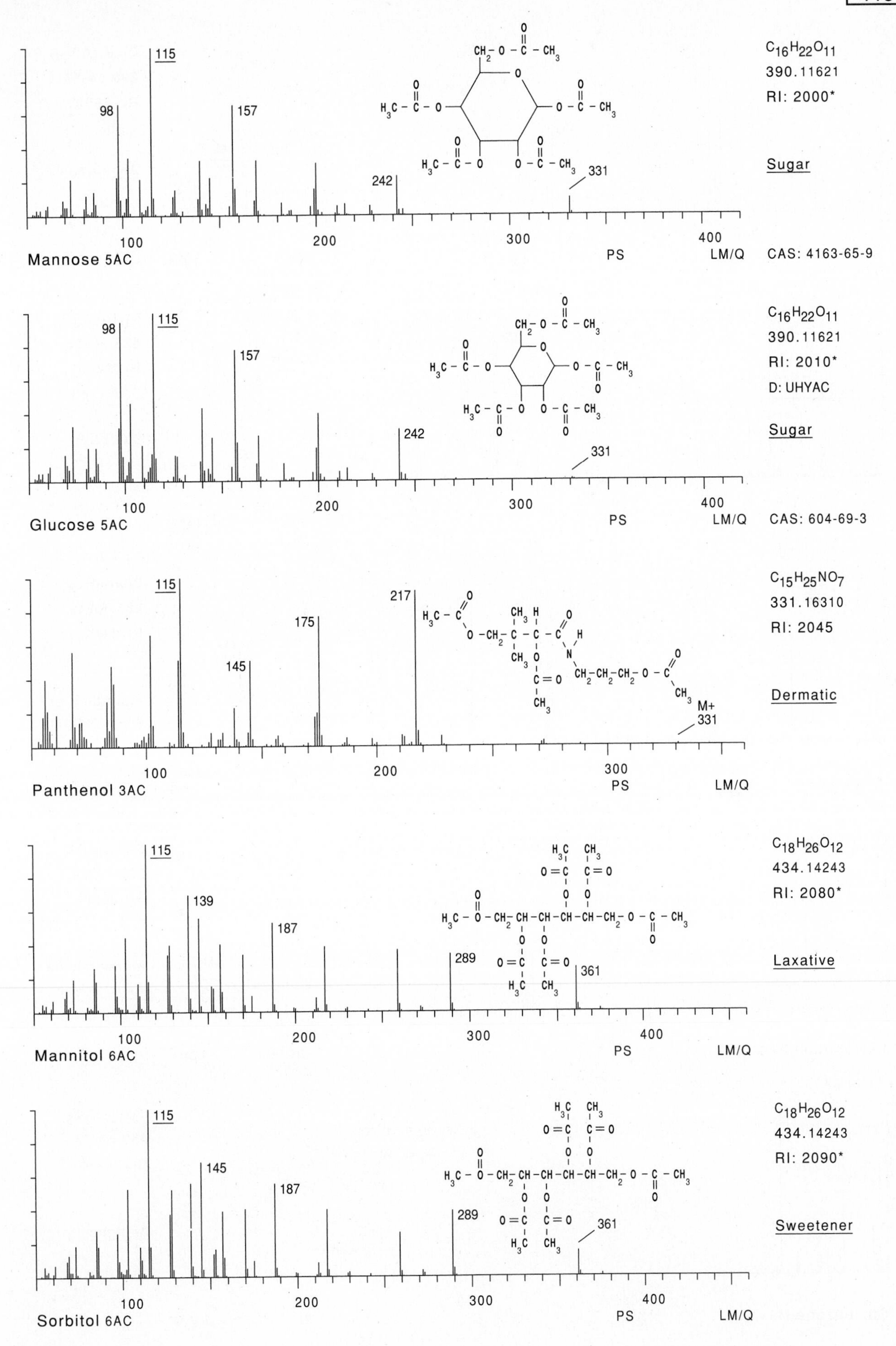
Mannose 5AC
$C_{16}H_{22}O_{11}$
390.11621
RI: 2000*
Sugar
CAS: 4163-65-9
Glucose 5AC
$C_{16}H_{22}O_{11}$
390.11621
RI: 2010*
D: UHYAC
Sugar
CAS: 604-69-3
Panthenol 3AC
$C_{15}H_{25}NO_7$
331.16310
RI: 2045
Dermatic
Mannitol 6AC
$C_{18}H_{26}O_{12}$
434.14243
RI: 2080*
Laxative
Sorbitol 6AC
$C_{18}H_{26}O_{12}$
434.14243
RI: 2090*
Sweetener
PS
LM/Q

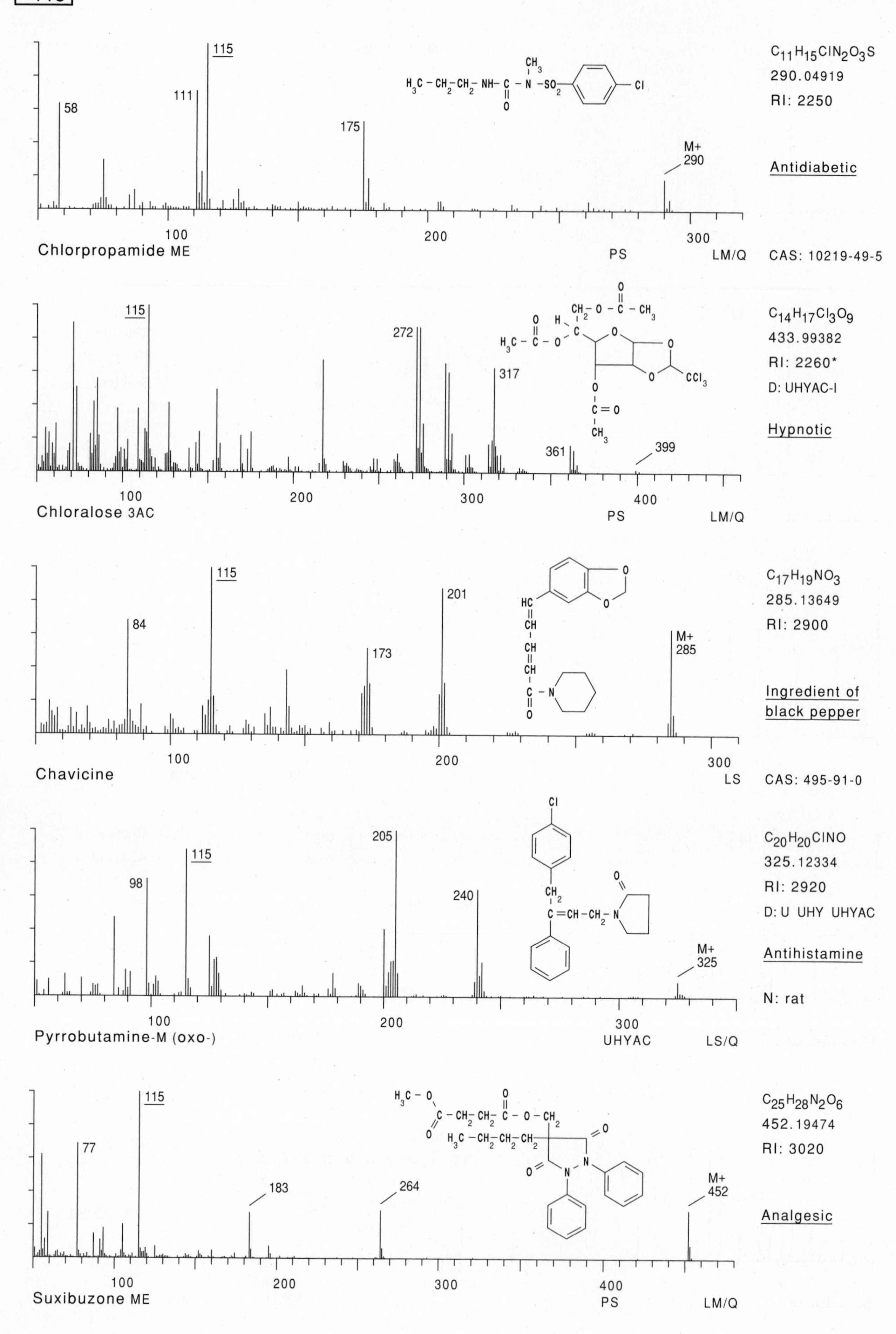

115
58
111
175
M+ 290
100
200
300
Chlorpropamide ME
PS
LM/Q
C11H15ClN2O3S
290.04919
RI: 2250
Antidiabetic
CAS: 10219-49-5
115
272
317
361
399
100
200
300
400
Chloralose 3AC
PS
LM/Q
C14H17Cl3O9
433.99382
RI: 2260*
D: UHYAC-I
Hypnotic
115
201
84
173
M+ 285
100
200
300
Chavicine
LS
C17H19NO3
285.13649
RI: 2900
Ingredient of black pepper
CAS: 495-91-0
205
115
98
240
M+ 325
100
200
300
Pyrrobutamine-M (oxo-)
UHYAC
LS/Q
C20H20ClNO
325.12334
RI: 2920
D: U UHY UHYAC
Antihistamine
N: rat
115
77
183
264
M+ 452
100
200
300
400
Suxibuzone ME
PS
LM/Q
C25H28N2O6
452.19474
RI: 3020
Analgesic

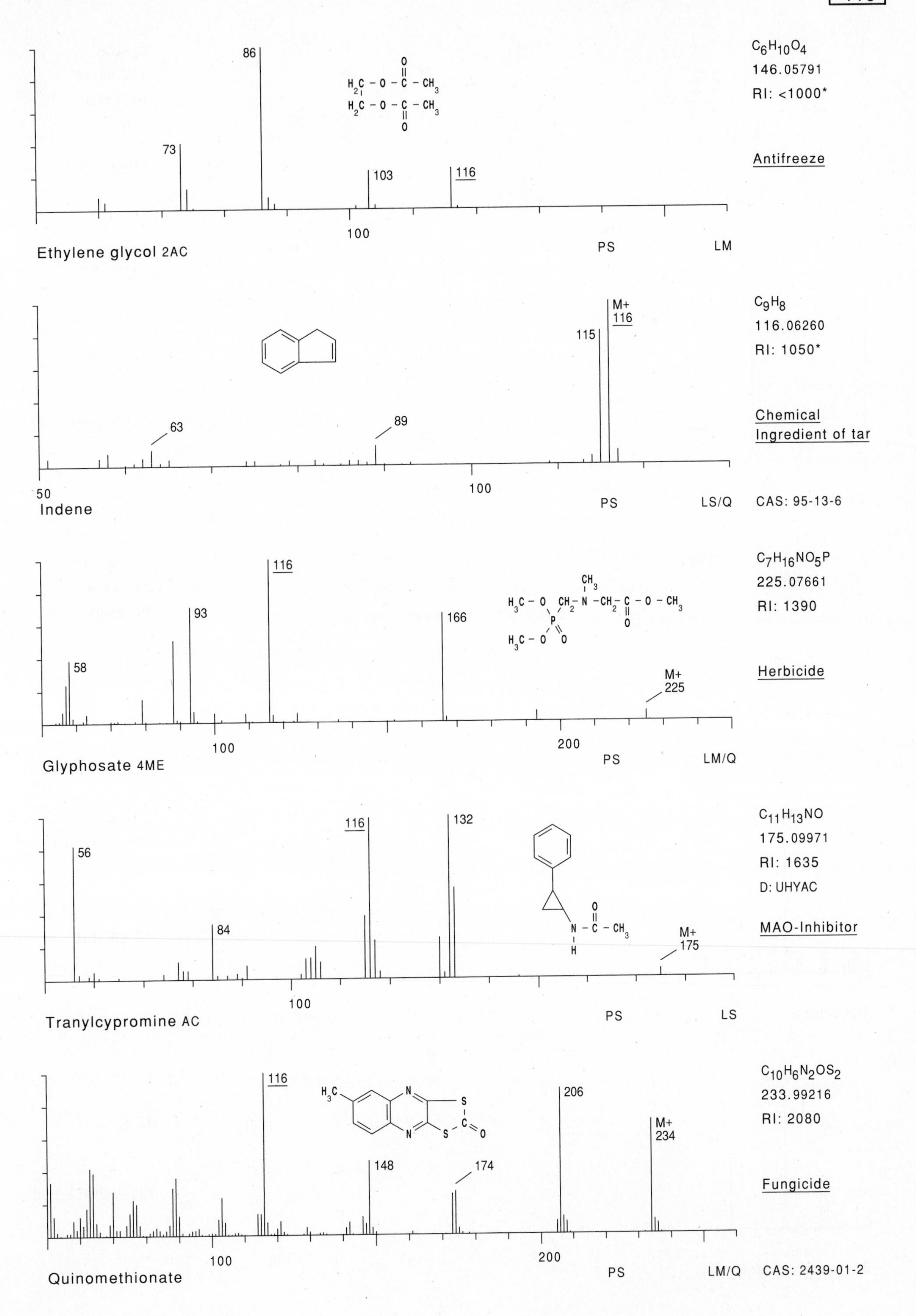

$C_6H_{10}O_4$
146.05791
RI: <1000*

Antifreeze

C_9H_8
116.06260
RI: 1050*

Chemical Ingredient of tar

CAS: 95-13-6

$C_7H_{16}NO_5P$
225.07661
RI: 1390

Herbicide

$C_{11}H_{13}NO$
175.09971
RI: 1635
D: UHYAC

MAO-Inhibitor

$C_{10}H_6N_2OS_2$
233.99216
RI: 2080

Fungicide

CAS: 2439-01-2

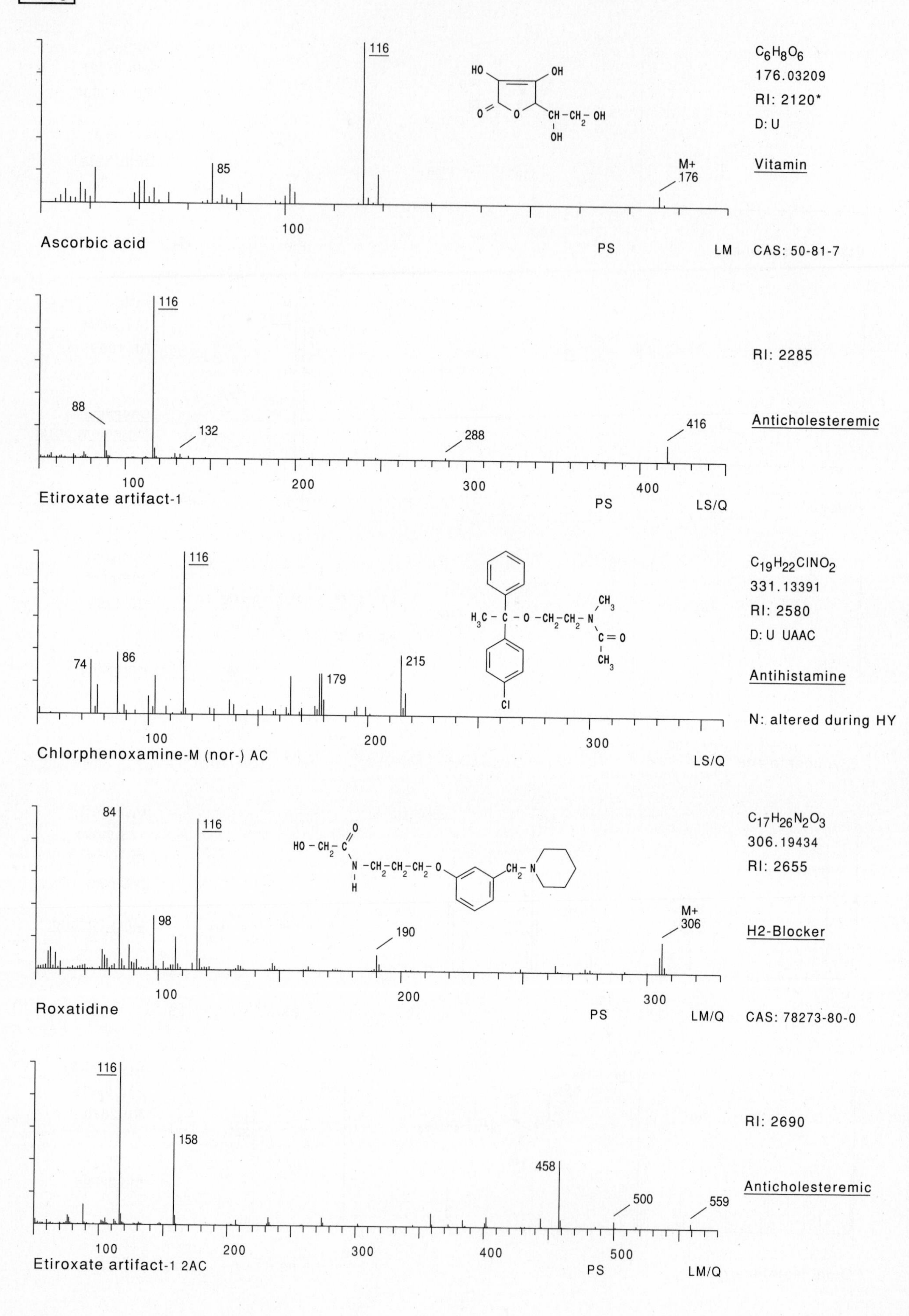

116
85
M+
176
100
Ascorbic acid
PS
LM
$C_6H_8O_6$
176.03209
RI: 2120*
D: U
Vitamin
CAS: 50-81-7
116
88
132
288
416
100
200
300
400
Etiroxate artifact-1
PS
LS/Q
RI: 2285
Anticholesteremic
116
74
86
179
215
100
200
300
Chlorphenoxamine-M (nor-) AC
LS/Q
$C_{19}H_{22}ClNO_2$
331.13391
RI: 2580
D: U UAAC
Antihistamine
N: altered during HY
84
116
98
190
M+
306
100
200
300
Roxatidine
PS
LM/Q
$C_{17}H_{26}N_2O_3$
306.19434
RI: 2655
H2-Blocker
CAS: 78273-80-0
116
158
458
500
559
100
200
300
400
500
Etiroxate artifact-1 2AC
PS
LM/Q
RI: 2690
Anticholesteremic

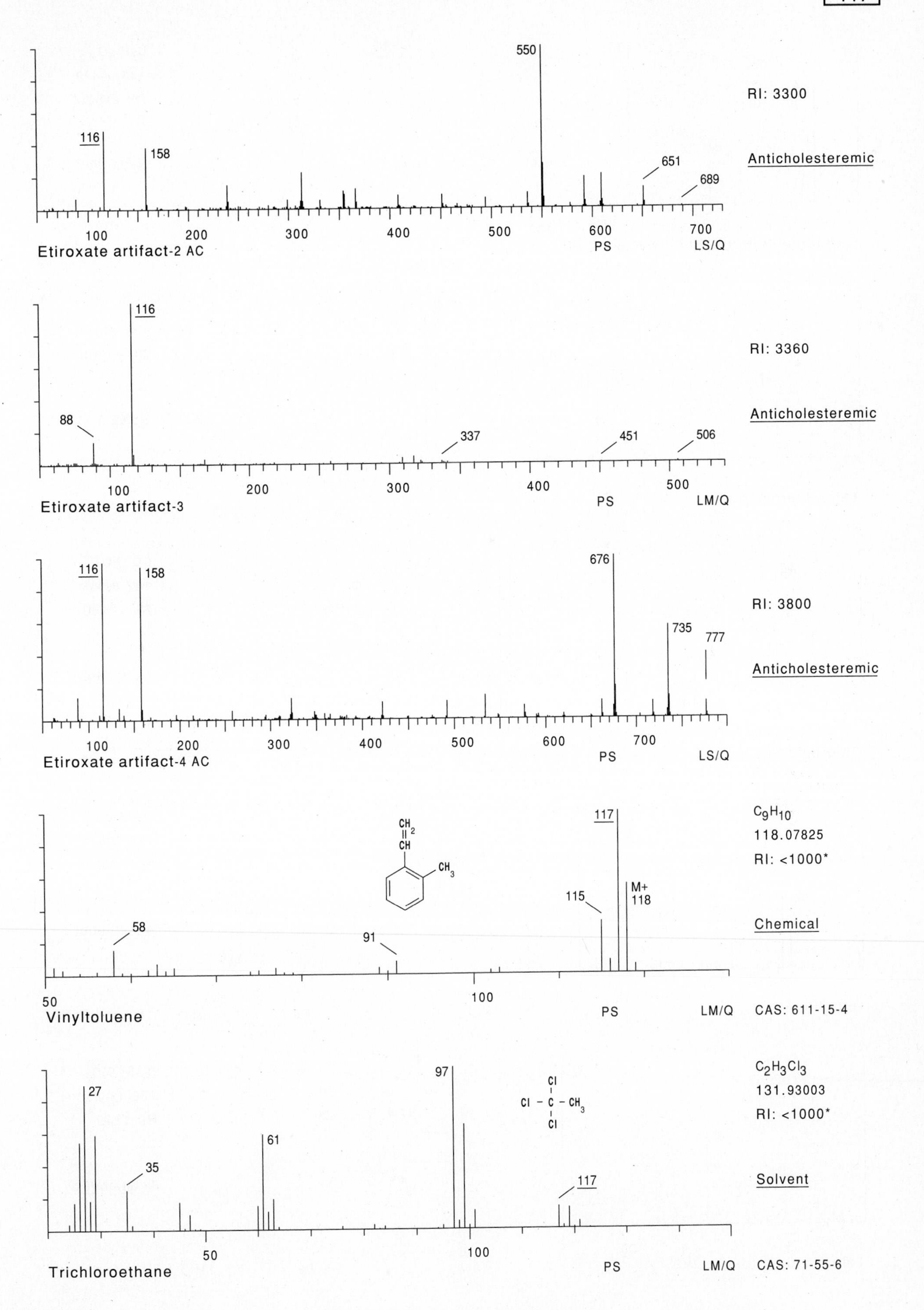
550
116
158
651
689
100
200
300
400
500
600
700
PS
LS/Q
Etiroxate artifact-2 AC
RI: 3300
Anticholesteremic
116
88
337
451
506
100
200
300
400
500
PS
LM/Q
Etiroxate artifact-3
RI: 3360
Anticholesteremic
116
158
676
735
777
100
200
300
400
500
600
700
PS
LS/Q
Etiroxate artifact-4 AC
RI: 3800
Anticholesteremic
117
M+
118
115
58
91
50
100
PS
LM/Q
Vinyltoluene
C9H10
118.07825
RI: <1000*
Chemical
CAS: 611-15-4
97
27
61
35
117
50
100
PS
LM/Q
Trichloroethane
C2H3Cl3
131.93003
RI: <1000*
Solvent
CAS: 71-55-6

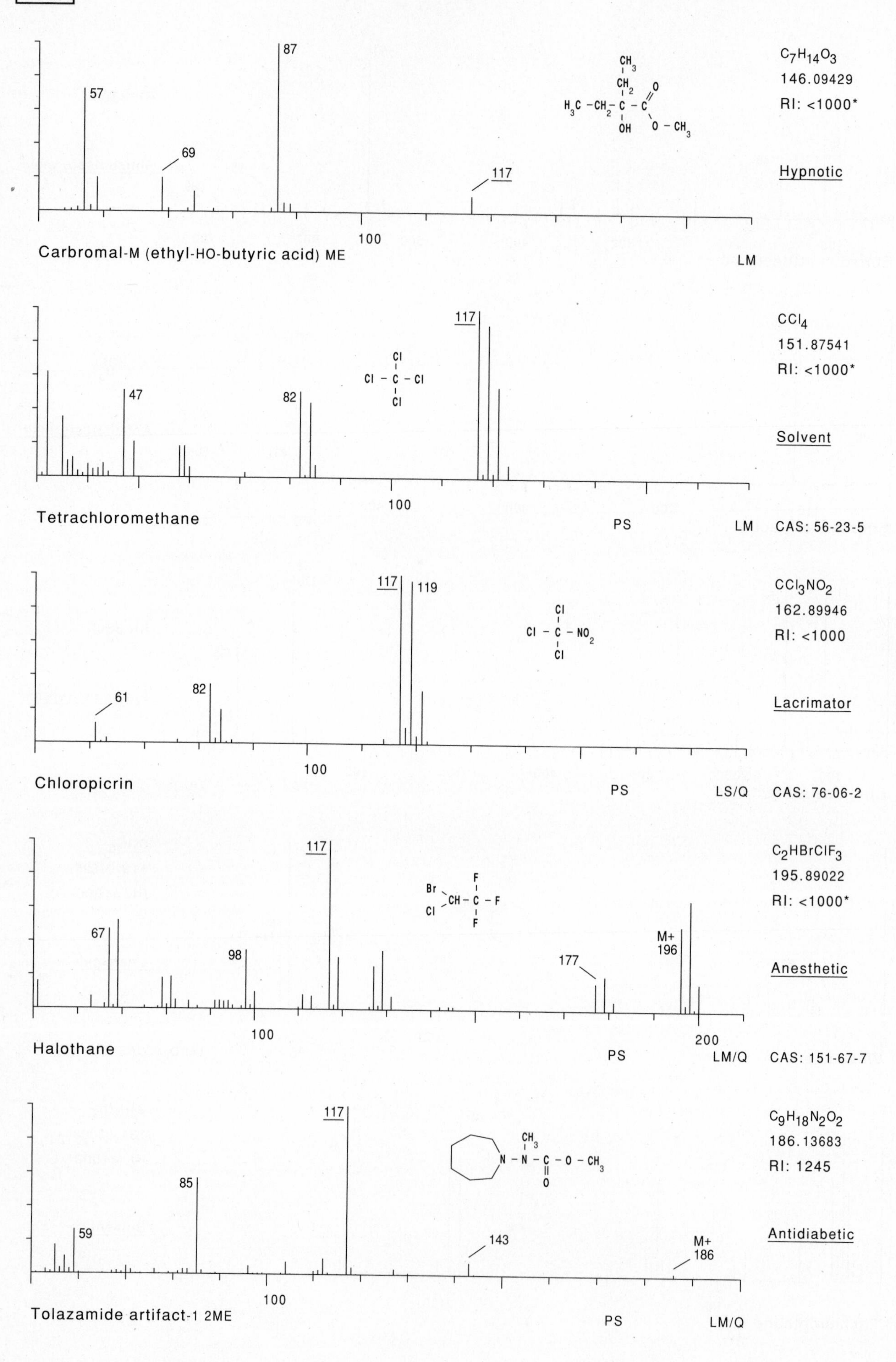

57
87
69
117
100
$C_7H_{14}O_3$
146.09429
RI: <1000*
Hypnotic
Carbromal-M (ethyl-HO-butyric acid) ME
LM
117
47
82
100
CCl_4
151.87541
RI: <1000*
Solvent
Tetrachloromethane
PS
LM
CAS: 56-23-5
117
119
61
82
100
CCl_3NO_2
162.89946
RI: <1000
Lacrimator
Chloropicrin
PS
LS/Q
CAS: 76-06-2
117
67
98
177
M+
196
100
200
$C_2HBrClF_3$
195.89022
RI: <1000*
Anesthetic
Halothane
PS
LM/Q
CAS: 151-67-7
117
85
59
143
M+
186
100
$C_9H_{18}N_2O_2$
186.13683
RI: 1245
Antidiabetic
Tolazamide artifact-1 2ME
PS
LM/Q

117

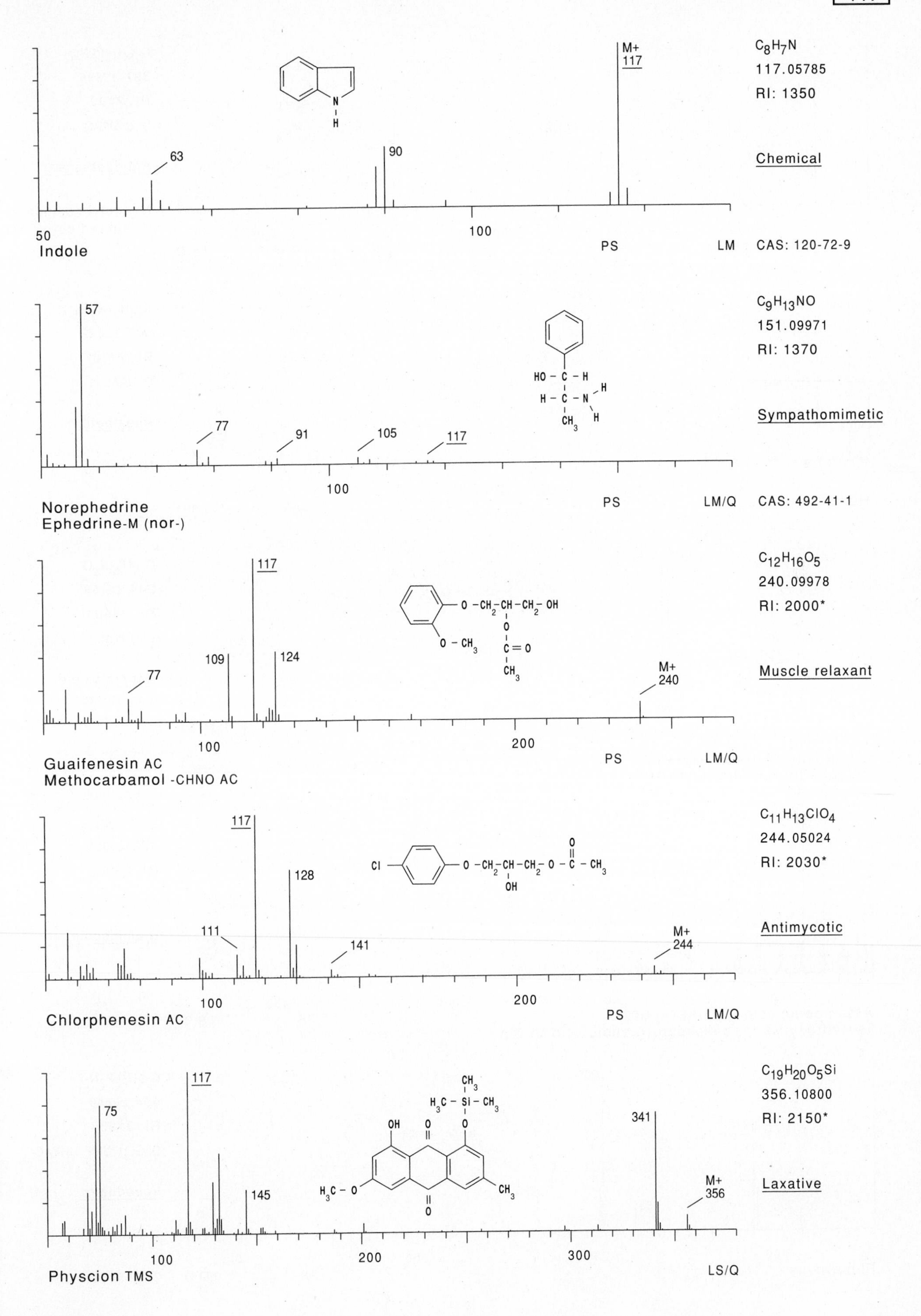
M+
117
63
90
50
100
Indole
PS
LM
C8H7N
117.05785
RI: 1350
Chemical
CAS: 120-72-9
57
77
91
105
117
100
Norephedrine
Ephedrine-M (nor-)
PS
LM/Q
C9H13NO
151.09971
RI: 1370
Sympathomimetic
CAS: 492-41-1
117
109
124
77
M+
240
100
200
Guaifenesin AC
Methocarbamol -CHNO AC
PS
LM/Q
C12H16O5
240.09978
RI: 2000*
Muscle relaxant
117
128
111
141
M+
244
100
200
Chlorphenesin AC
PS
LM/Q
C11H13ClO4
244.05024
RI: 2030*
Antimycotic
117
75
145
341
M+
356
100
200
300
Physcion TMS
LS/Q
C19H20O5Si
356.10800
RI: 2150*
Laxative

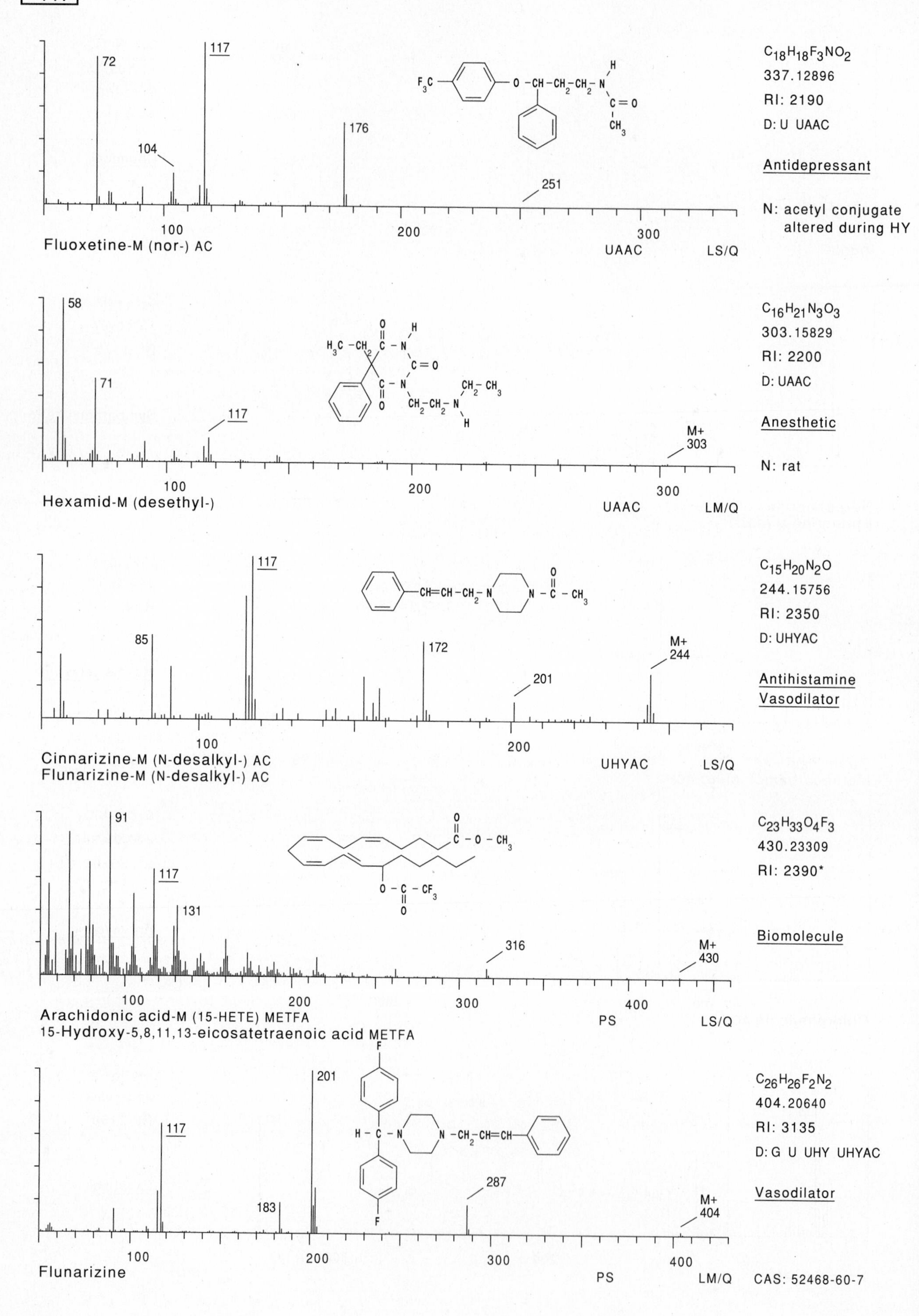

72
117
104
176
251
100
200
300
Fluoxetine-M (nor-) AC
UAAC
LS/Q
$C_{18}H_{18}F_3NO_2$
337.12896
RI: 2190
D: U UAAC
Antidepressant
N: acetyl conjugate altered during HY
58
71
117
M+ 303
Hexamid-M (desethyl-)
UAAC
LM/Q
$C_{16}H_{21}N_3O_3$
303.15829
RI: 2200
D: UAAC
Anesthetic
N: rat
117
85
172
201
M+ 244
Cinnarizine-M (N-desalkyl-) AC
Flunarizine-M (N-desalkyl-) AC
UHYAC
LS/Q
$C_{15}H_{20}N_2O$
244.15756
RI: 2350
D: UHYAC
Antihistamine
Vasodilator
91
117
131
316
M+ 430
400
Arachidonic acid-M (15-HETE) METFA
15-Hydroxy-5,8,11,13-eicosatetraenoic acid METFA
PS
LS/Q
$C_{23}H_{33}O_4F_3$
430.23309
RI: 2390*
Biomolecule
201
117
183
287
M+ 404
Flunarizine
PS
LM/Q
$C_{26}H_{26}F_2N_2$
404.20640
RI: 3135
D: G U UHY UHYAC
Vasodilator
CAS: 52468-60-7

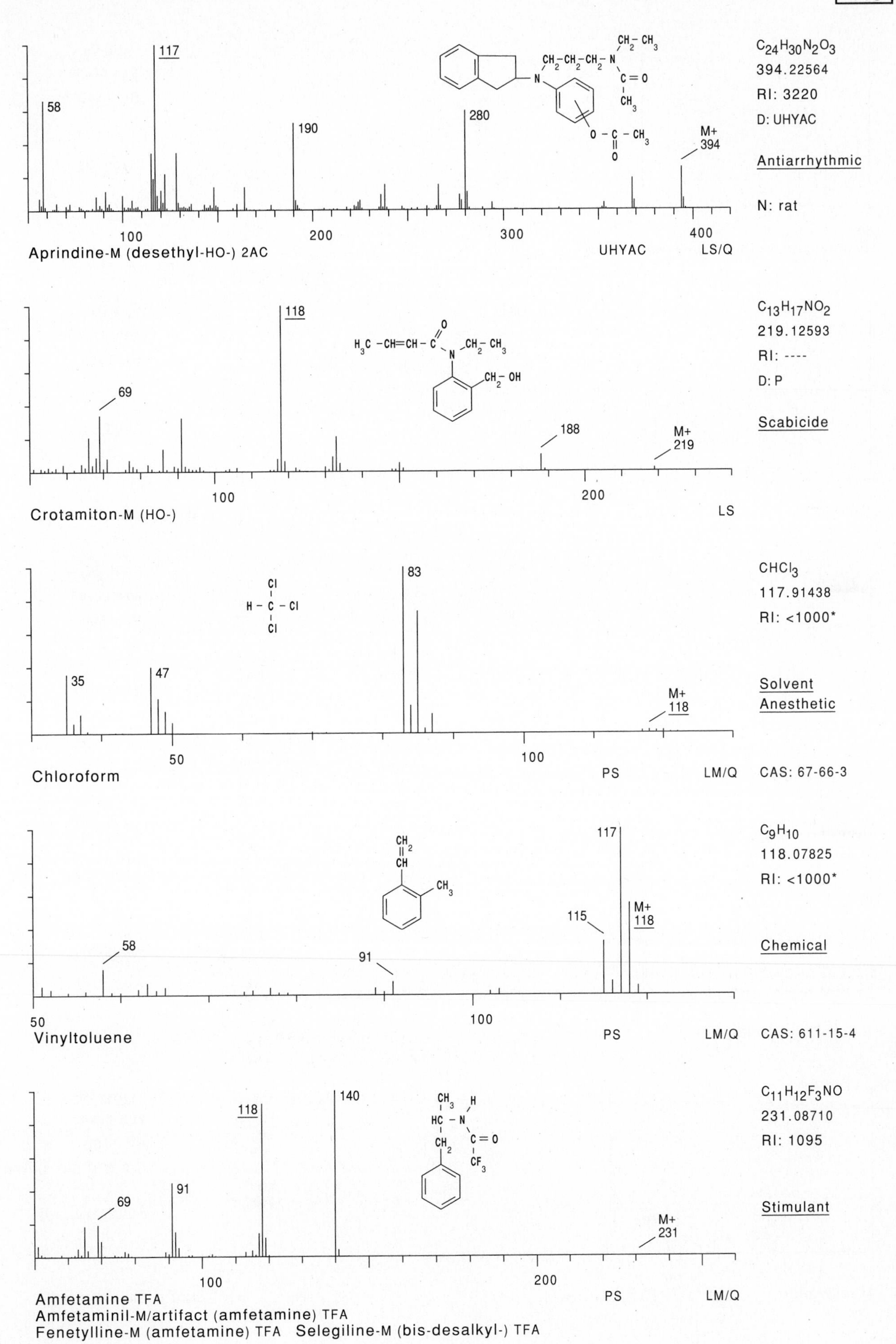

C24H30N2O3
394.22564
RI: 3220
D: UHYAC
Antiarrhythmic
N: rat
Aprindine-M (desethyl-HO-) 2AC
UHYAC
LS/Q
C13H17NO2
219.12593
RI: ----
D: P
Scabicide
Crotamiton-M (HO-)
LS
CHCl3
117.91438
RI: <1000*
Solvent
Anesthetic
Chloroform
PS
LM/Q
CAS: 67-66-3
C9H10
118.07825
RI: <1000*
Chemical
Vinyltoluene
PS
LM/Q
CAS: 611-15-4
C11H12F3NO
231.08710
RI: 1095
Stimulant
Amfetamine TFA
Amfetaminil-M/artifact (amfetamine) TFA
Fenetylline-M (amfetamine) TFA
Selegiline-M (bis-desalkyl-) TFA
PS
LM/Q

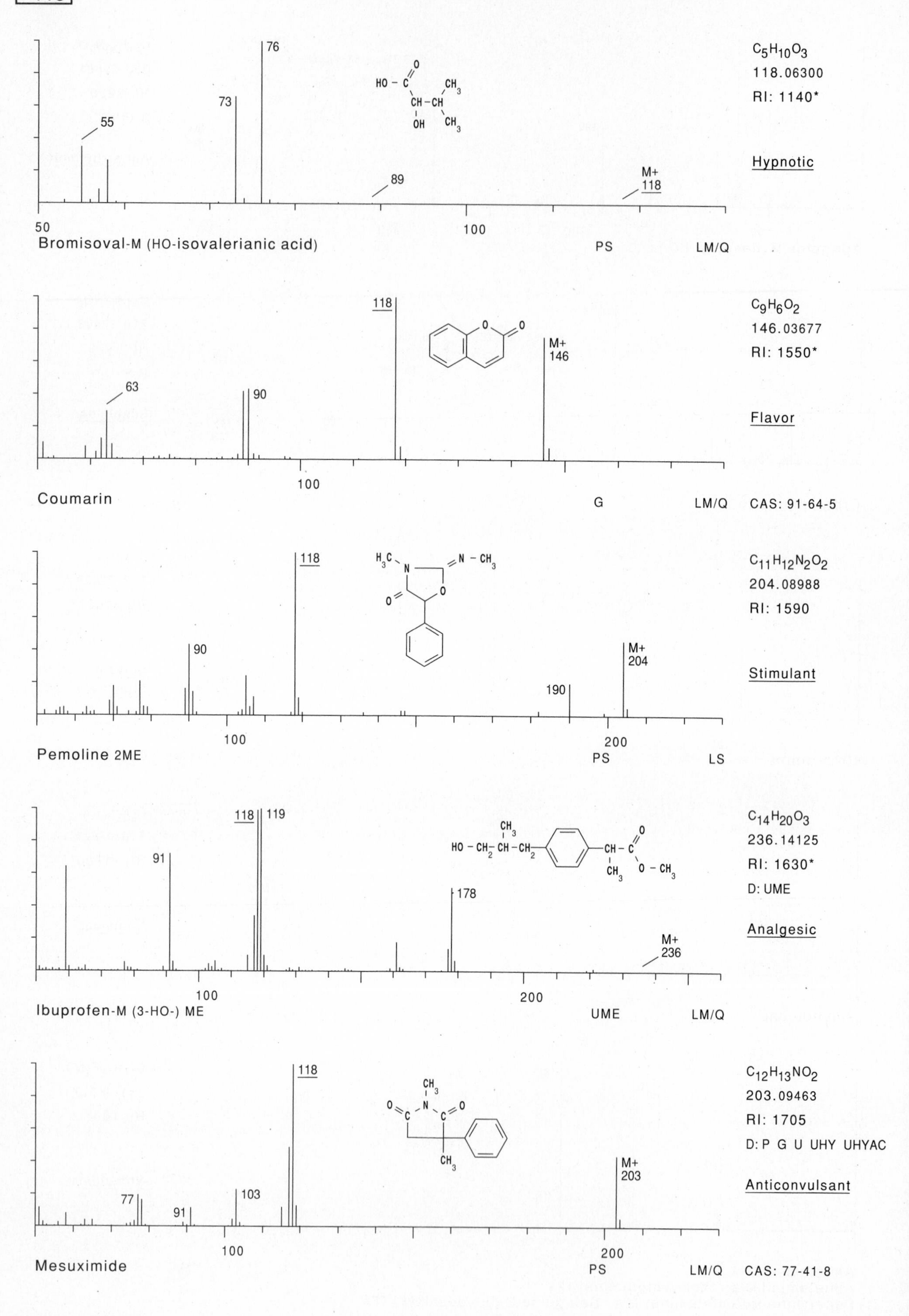

76
73
55
89
M+
118
50
100
Bromisoval-M (HO-isovalerianic acid)
PS
LM/Q
C5H10O3
118.06300
RI: 1140*
Hypnotic
118
M+
146
63
90
100
Coumarin
G
LM/Q
C9H6O2
146.03677
RI: 1550*
Flavor
CAS: 91-64-5
118
90
190
M+
204
100
200
Pemoline 2ME
PS
LS
C11H12N2O2
204.08988
RI: 1590
Stimulant
118
119
91
178
M+
236
100
200
Ibuprofen-M (3-HO-) ME
UME
LM/Q
C14H20O3
236.14125
RI: 1630*
D: UME
Analgesic
118
77
91
103
M+
203
100
200
Mesuximide
PS
LM/Q
C12H13NO2
203.09463
RI: 1705
D: P G U UHY UHYAC
Anticonvulsant
CAS: 77-41-8

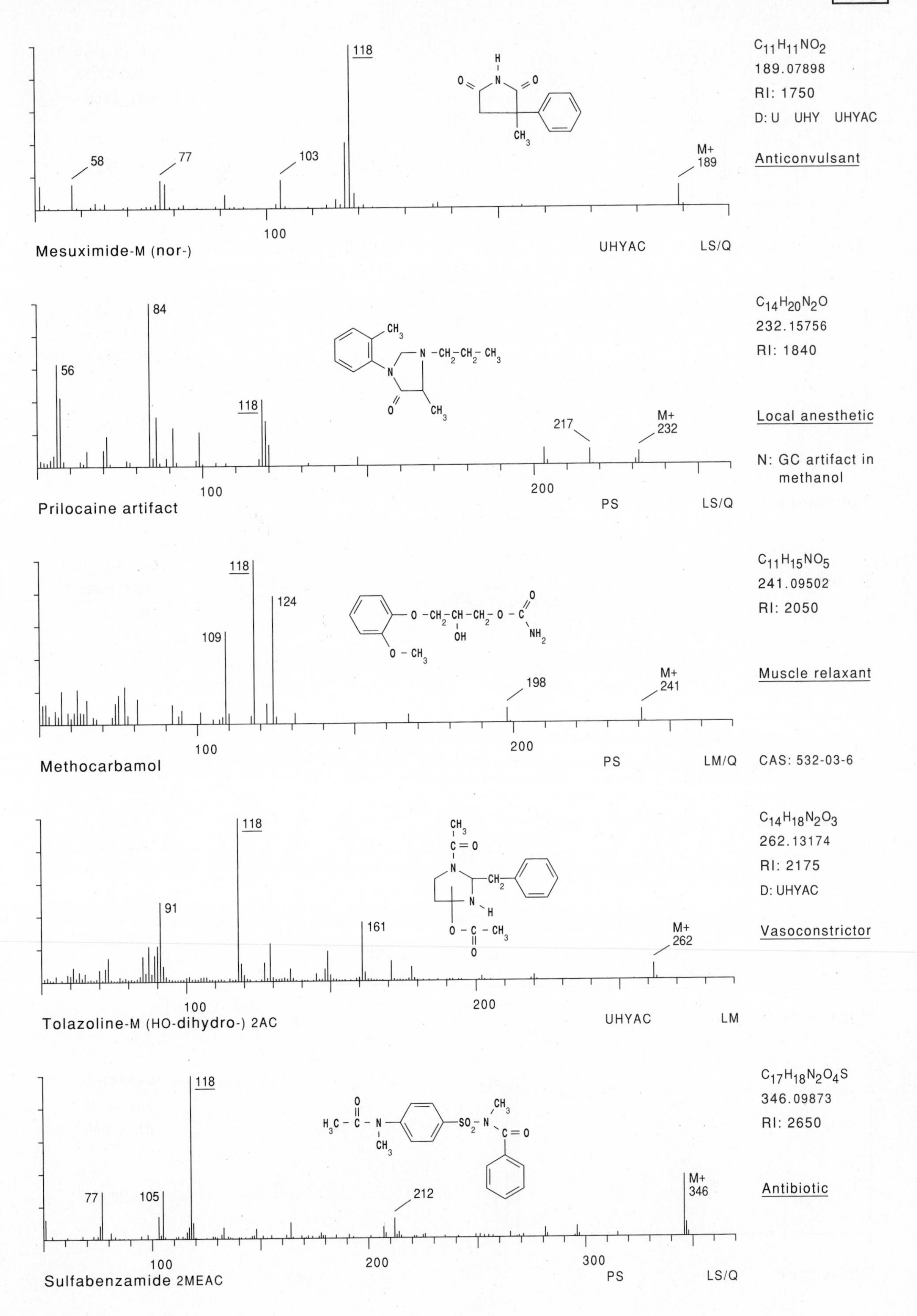

$C_{11}H_{11}NO_2$
189.07898
RI: 1750
D: U UHY UHYAC
Anticonvulsant

$C_{14}H_{20}N_2O$
232.15756
RI: 1840
Local anesthetic
N: GC artifact in methanol

$C_{11}H_{15}NO_5$
241.09502
RI: 2050
Muscle relaxant
CAS: 532-03-6

$C_{14}H_{18}N_2O_3$
262.13174
RI: 2175
D: UHYAC
Vasoconstrictor

$C_{17}H_{18}N_2O_4S$
346.09873
RI: 2650
Antibiotic

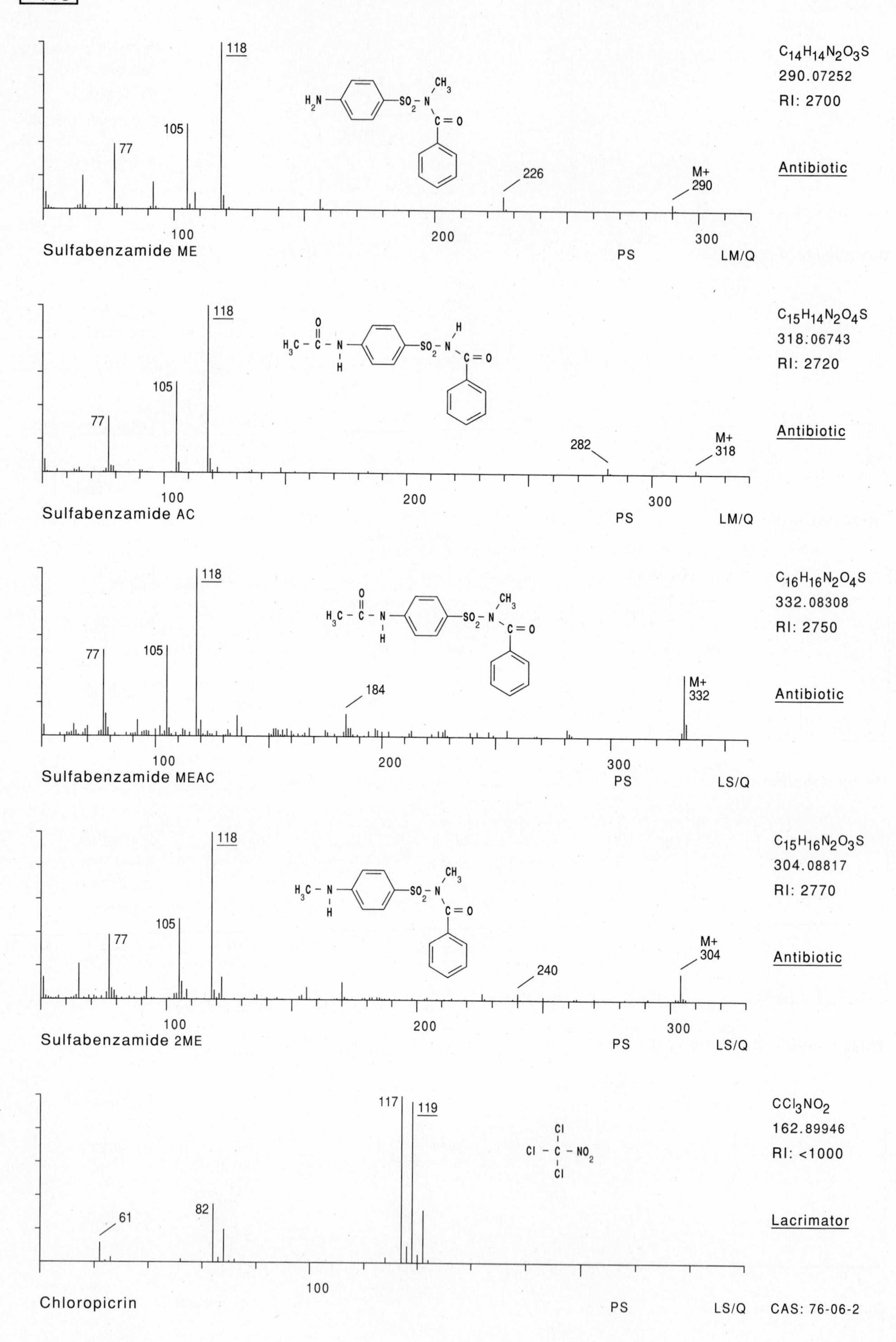
118
105
77
226
M+
290
100
200
300
Sulfabenzamide ME
PS
LM/Q
C14H14N2O3S
290.07252
RI: 2700
Antibiotic
118
105
77
282
M+
318
100
200
300
Sulfabenzamide AC
PS
LM/Q
C15H14N2O4S
318.06743
RI: 2720
Antibiotic
118
77
105
184
M+
332
100
200
300
Sulfabenzamide MEAC
PS
LS/Q
C16H16N2O4S
332.08308
RI: 2750
Antibiotic
118
105
77
240
M+
304
100
200
300
Sulfabenzamide 2ME
PS
LS/Q
C15H16N2O3S
304.08817
RI: 2770
Antibiotic
117
119
82
61
100
Chloropicrin
PS
LS/Q
CCl3NO2
162.89946
RI: <1000
Lacrimator
CAS: 76-06-2

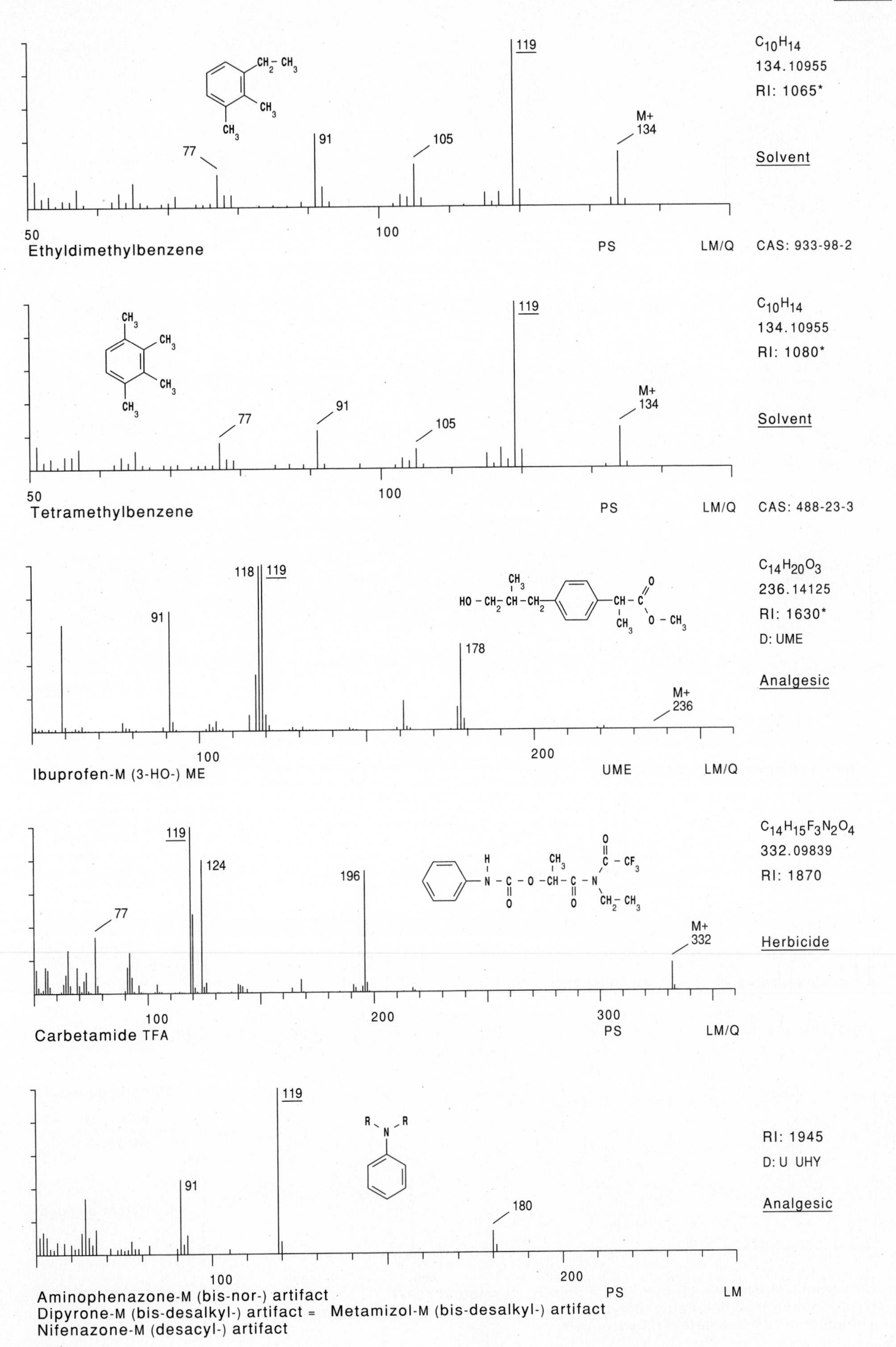

119
91
105
77
M+
134
50
100
Ethyldimethylbenzene
PS
LM/Q
$C_{10}H_{14}$
134.10955
RI: 1065*
Solvent
CAS: 933-98-2
119
91
105
77
M+
134
50
100
Tetramethylbenzene
PS
LM/Q
$C_{10}H_{14}$
134.10955
RI: 1080*
Solvent
CAS: 488-23-3
118
119
91
178
M+
236
100
200
Ibuprofen-M (3-HO-) ME
UME
LM/Q
$C_{14}H_{20}O_3$
236.14125
RI: 1630*
D: UME
Analgesic
119
124
196
77
M+
332
100
200
300
Carbetamide TFA
PS
LM/Q
$C_{14}H_{15}F_3N_2O_4$
332.09839
RI: 1870
Herbicide
119
91
180
100
200
Aminophenazone-M (bis-nor-) artifact
Dipyrone-M (bis-desalkyl-) artifact = Metamizol-M (bis-desalkyl-) artifact
Nifenazone-M (desacyl-) artifact
PS
LM
RI: 1945
D: U UHY
Analgesic

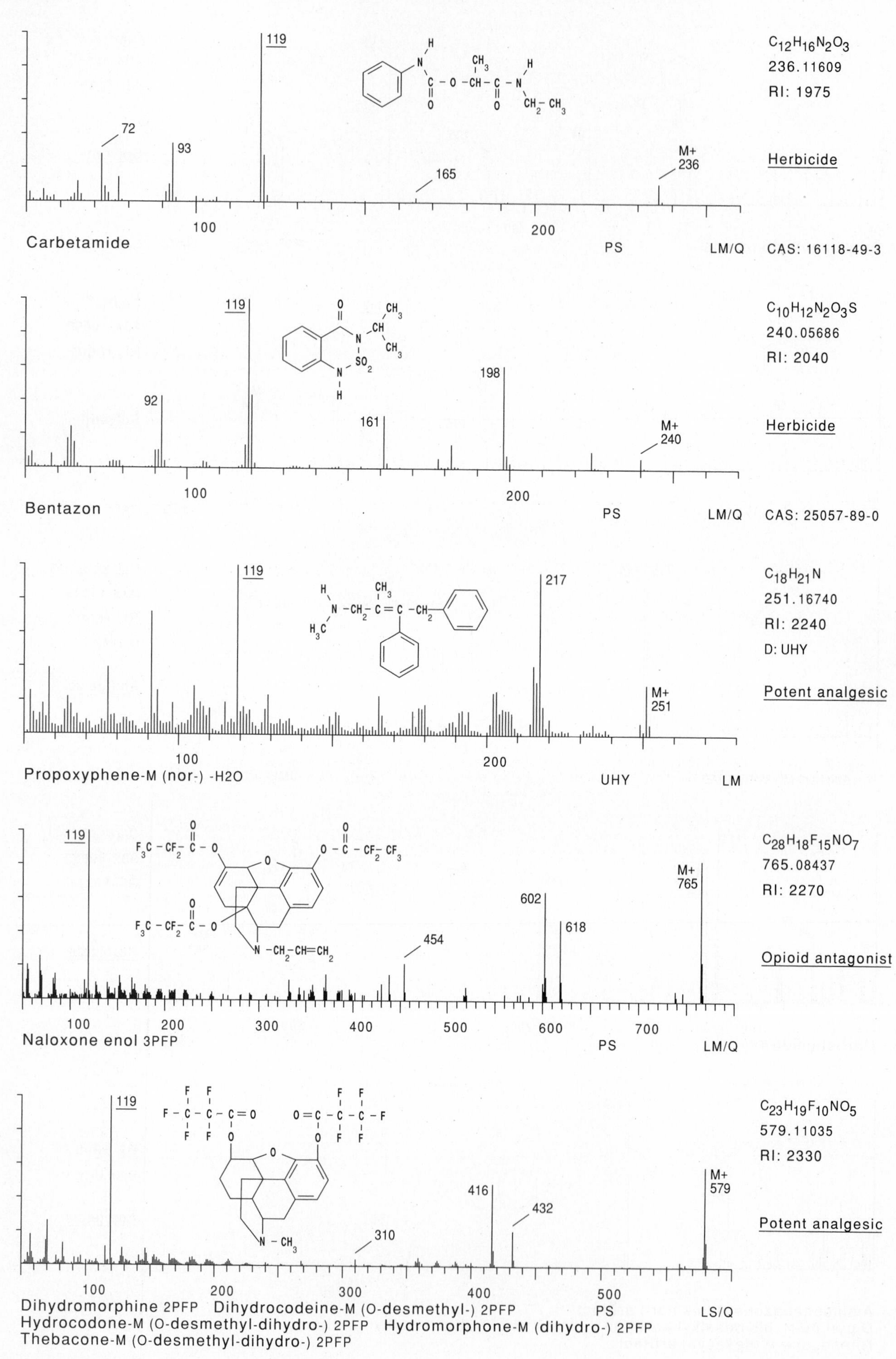
119
72
93
165
M+
236
100
200
Carbetamide
PS
LM/Q
$C_{12}H_{16}N_2O_3$
236.11609
RI: 1975
Herbicide
CAS: 16118-49-3
119
92
198
161
M+
240
100
200
Bentazon
PS
LM/Q
$C_{10}H_{12}N_2O_3S$
240.05686
RI: 2040
Herbicide
CAS: 25057-89-0
119
217
M+
251
100
200
Propoxyphene-M (nor-) -H2O
UHY
LM
$C_{18}H_{21}N$
251.16740
RI: 2240
D: UHY
Potent analgesic
119
M+
765
602
618
454
100
200
300
400
500
600
700
Naloxone enol 3PFP
PS
LM/Q
$C_{28}H_{18}F_{15}NO_7$
765.08437
RI: 2270
Opioid antagonist
119
M+
579
416
432
310
100
200
300
400
500
Dihydromorphine 2PFP Dihydrocodeine-M (O-desmethyl-) 2PFP
Hydrocodone-M (O-desmethyl-dihydro-) 2PFP Hydromorphone-M (dihydro-) 2PFP
Thebacone-M (O-desmethyl-dihydro-) 2PFP
PS
LS/Q
$C_{23}H_{19}F_{10}NO_5$
579.11035
RI: 2330
Potent analgesic

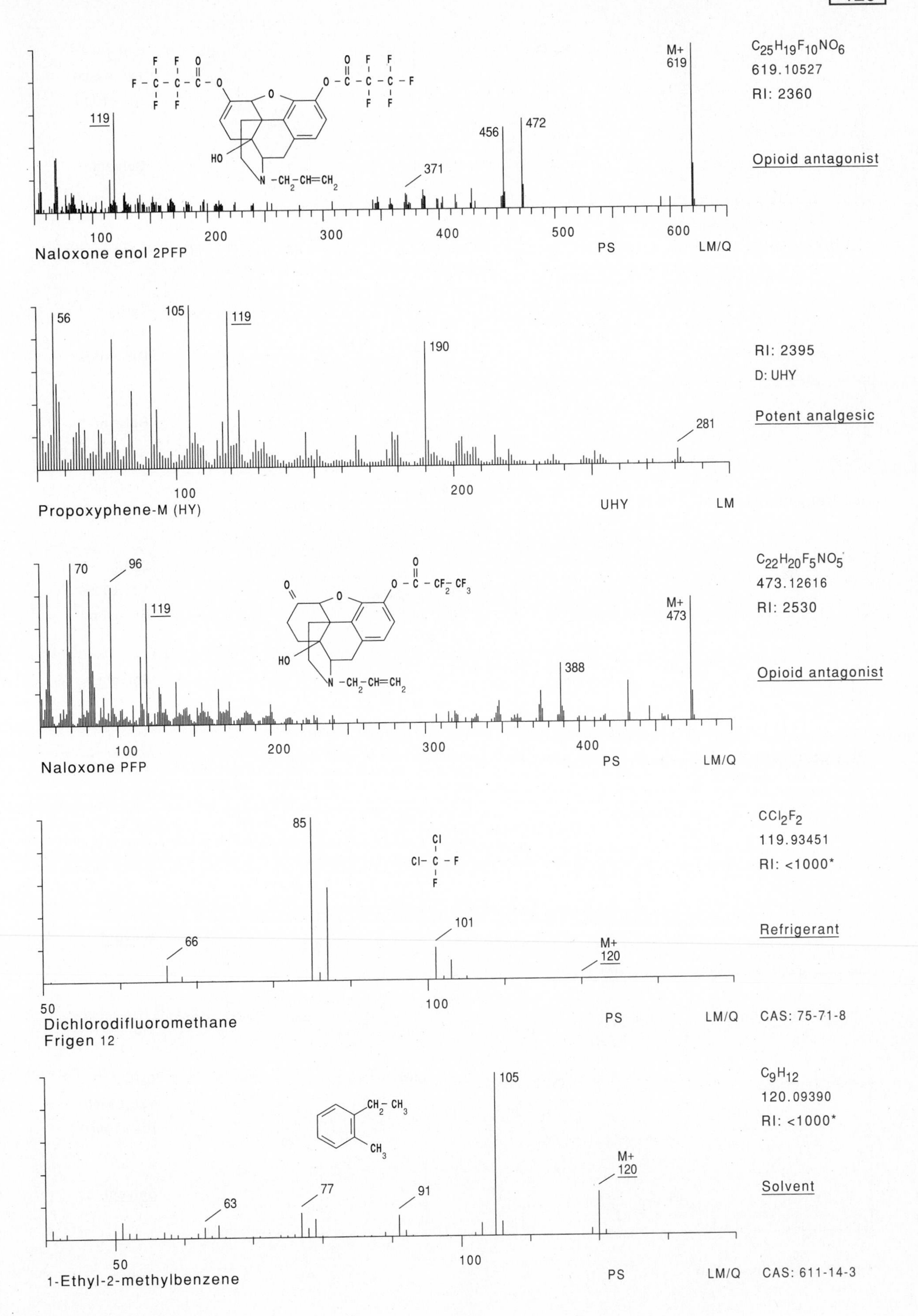

Naloxone enol 2PFP

$C_{25}H_{19}F_{10}NO_6$
619.10527
RI: 2360

Opioid antagonist

Propoxyphene-M (HY)

RI: 2395
D: UHY

Potent analgesic

Naloxone PFP

$C_{22}H_{20}F_5NO_5$
473.12616
RI: 2530

Opioid antagonist

Dichlorodifluoromethane
Frigen 12

CCl_2F_2
119.93451
RI: <1000*

Refrigerant

CAS: 75-71-8

1-Ethyl-2-methylbenzene

C_9H_{12}
120.09390
RI: <1000*

Solvent

CAS: 611-14-3

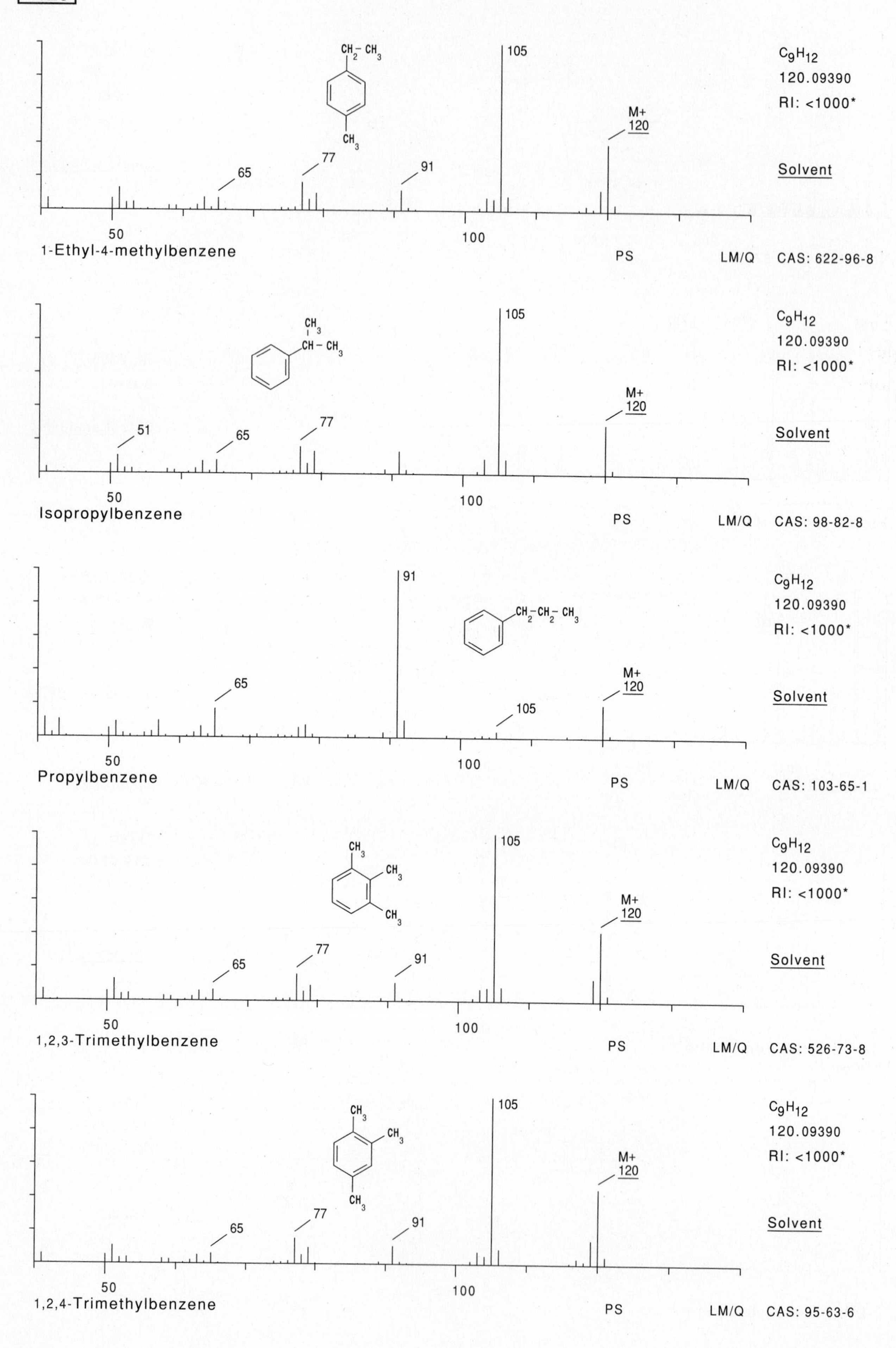

CH2- CH3
CH3
105
M+
120
65
77
91
50
100
1-Ethyl-4-methylbenzene
PS
LM/Q
C9H12
120.09390
RI: <1000*
Solvent
CAS: 622-96-8
CH3
CH - CH3
105
M+
120
51
65
77
50
100
Isopropylbenzene
PS
LM/Q
C9H12
120.09390
RI: <1000*
Solvent
CAS: 98-82-8
91
CH2- CH2- CH3
M+
120
65
105
50
100
Propylbenzene
PS
LM/Q
C9H12
120.09390
RI: <1000*
Solvent
CAS: 103-65-1
CH3
CH3
CH3
105
M+
120
77
65
91
50
100
1,2,3-Trimethylbenzene
PS
LM/Q
C9H12
120.09390
RI: <1000*
Solvent
CAS: 526-73-8
CH3
CH3
CH3
105
M+
120
65
77
91
50
100
1,2,4-Trimethylbenzene
PS
LM/Q
C9H12
120.09390
RI: <1000*
Solvent
CAS: 95-63-6

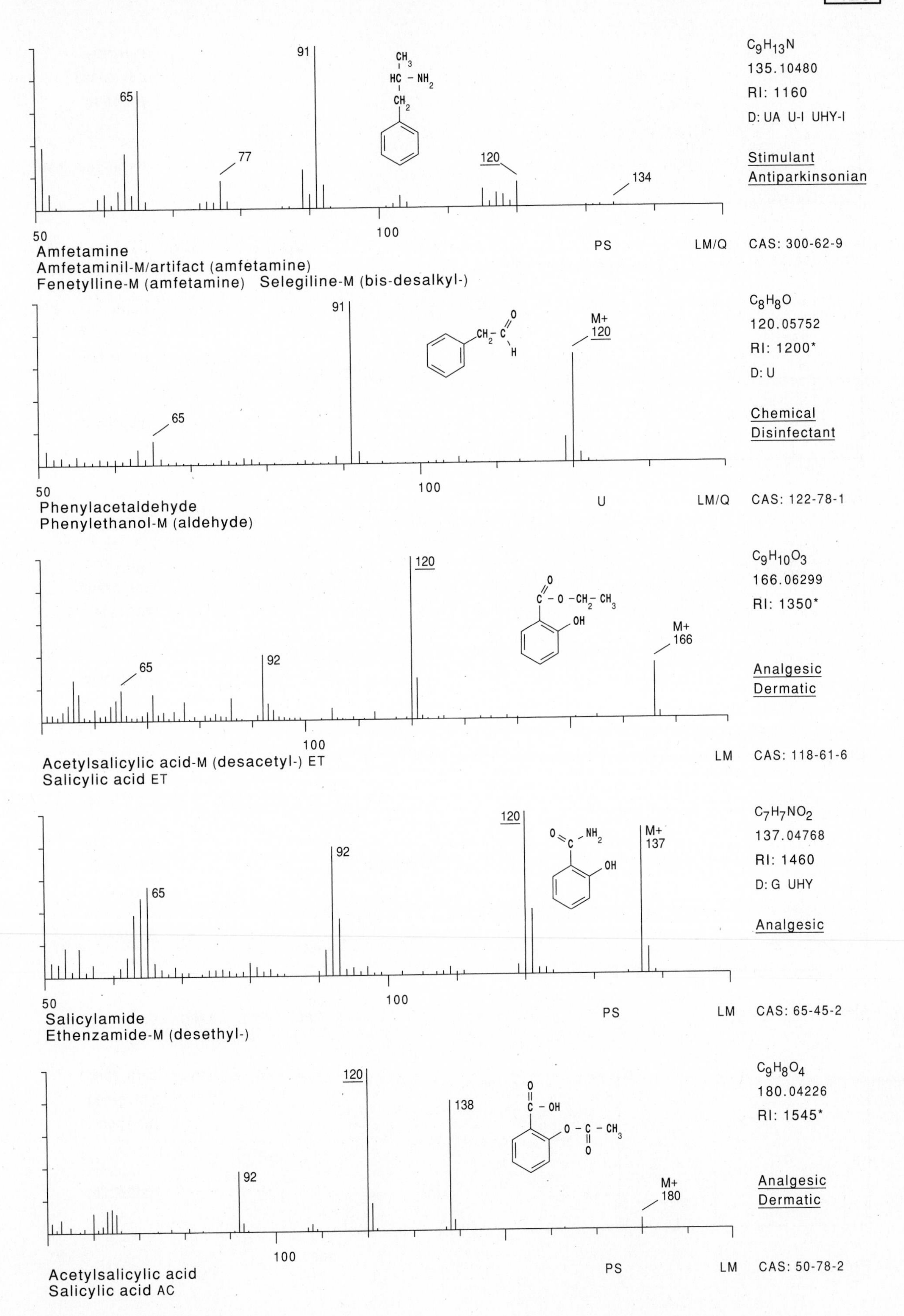

Amfetamine
Amfetaminil-M/artifact (amfetamine)
Fenetylline-M (amfetamine) Selegiline-M (bis-desalkyl-)

$C_9H_{13}N$
135.10480
RI: 1160
D: UA U-I UHY-I
Stimulant
Antiparkinsonian
CAS: 300-62-9

Phenylacetaldehyde
Phenylethanol-M (aldehyde)

C_8H_8O
120.05752
RI: 1200*
D: U
Chemical
Disinfectant
CAS: 122-78-1

Acetylsalicylic acid-M (desacetyl-) ET
Salicylic acid ET

$C_9H_{10}O_3$
166.06299
RI: 1350*
Analgesic
Dermatic
CAS: 118-61-6

Salicylamide
Ethenzamide-M (desethyl-)

$C_7H_7NO_2$
137.04768
RI: 1460
D: G UHY
Analgesic
CAS: 65-45-2

Acetylsalicylic acid
Salicylic acid AC

$C_9H_8O_4$
180.04226
RI: 1545*
Analgesic
Dermatic
CAS: 50-78-2

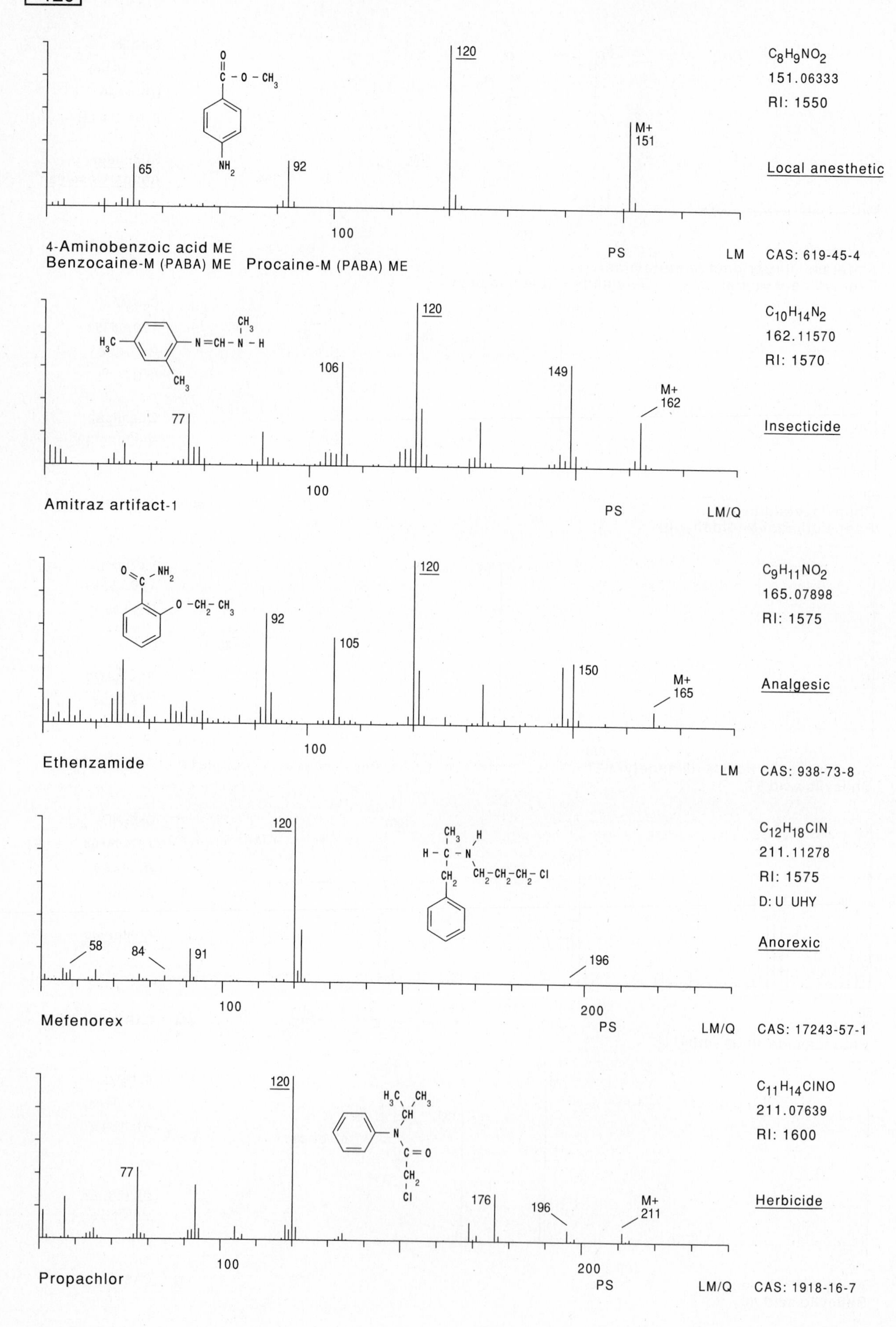

120
65
92
M+
151
100
PS
LM
$C_8H_9NO_2$
151.06333
RI: 1550
Local anesthetic
4-Aminobenzoic acid ME
Benzocaine-M (PABA) ME
Procaine-M (PABA) ME
CAS: 619-45-4
120
106
149
77
M+
162
100
PS
LM/Q
$C_{10}H_{14}N_2$
162.11570
RI: 1570
Insecticide
Amitraz artifact-1
120
92
105
150
M+
165
100
LM
$C_9H_{11}NO_2$
165.07898
RI: 1575
Analgesic
Ethenzamide
CAS: 938-73-8
120
58
84
91
196
100
200
PS
LM/Q
$C_{12}H_{18}ClN$
211.11278
RI: 1575
D: U UHY
Anorexic
Mefenorex
CAS: 17243-57-1
120
77
176
196
M+
211
100
200
PS
LM/Q
$C_{11}H_{14}ClNO$
211.07639
RI: 1600
Herbicide
Propachlor
CAS: 1918-16-7

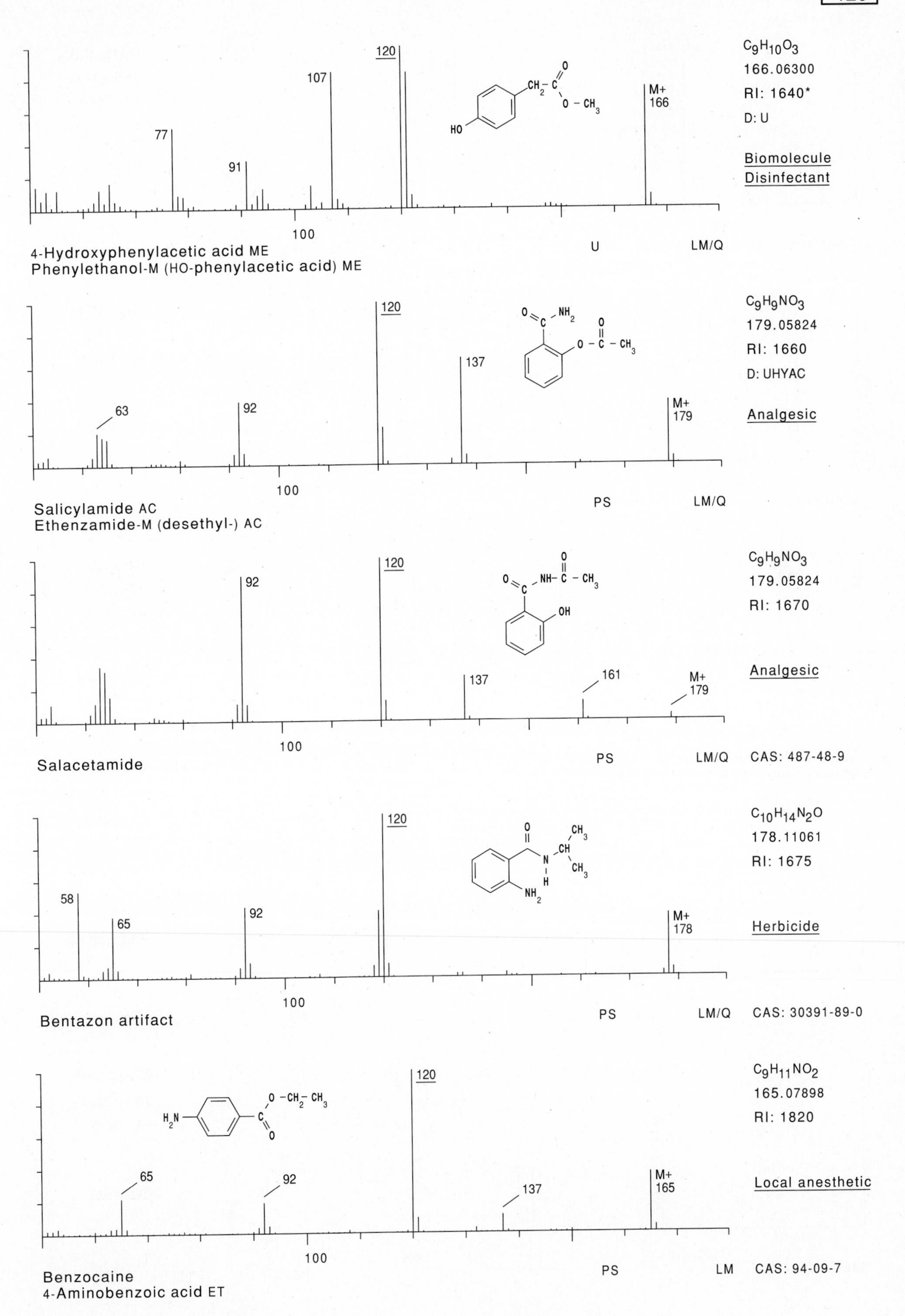
120
107
77
91
M+
166
100
U
LM/Q
C9H10O3
166.06300
RI: 1640*
D: U
Biomolecule
Disinfectant
4-Hydroxyphenylacetic acid ME
Phenylethanol-M (HO-phenylacetic acid) ME
120
137
63
92
M+
179
100
PS
LM/Q
C9H9NO3
179.05824
RI: 1660
D: UHYAC
Analgesic
Salicylamide AC
Ethenzamide-M (desethyl-) AC
120
92
137
161
M+
179
100
PS
LM/Q
C9H9NO3
179.05824
RI: 1670
Analgesic
CAS: 487-48-9
Salacetamide
120
58
65
92
M+
178
100
PS
LM/Q
C10H14N2O
178.11061
RI: 1675
Herbicide
CAS: 30391-89-0
Bentazon artifact
120
65
92
137
M+
165
100
PS
LM
C9H11NO2
165.07898
RI: 1820
Local anesthetic
CAS: 94-09-7
Benzocaine
4-Aminobenzoic acid ET

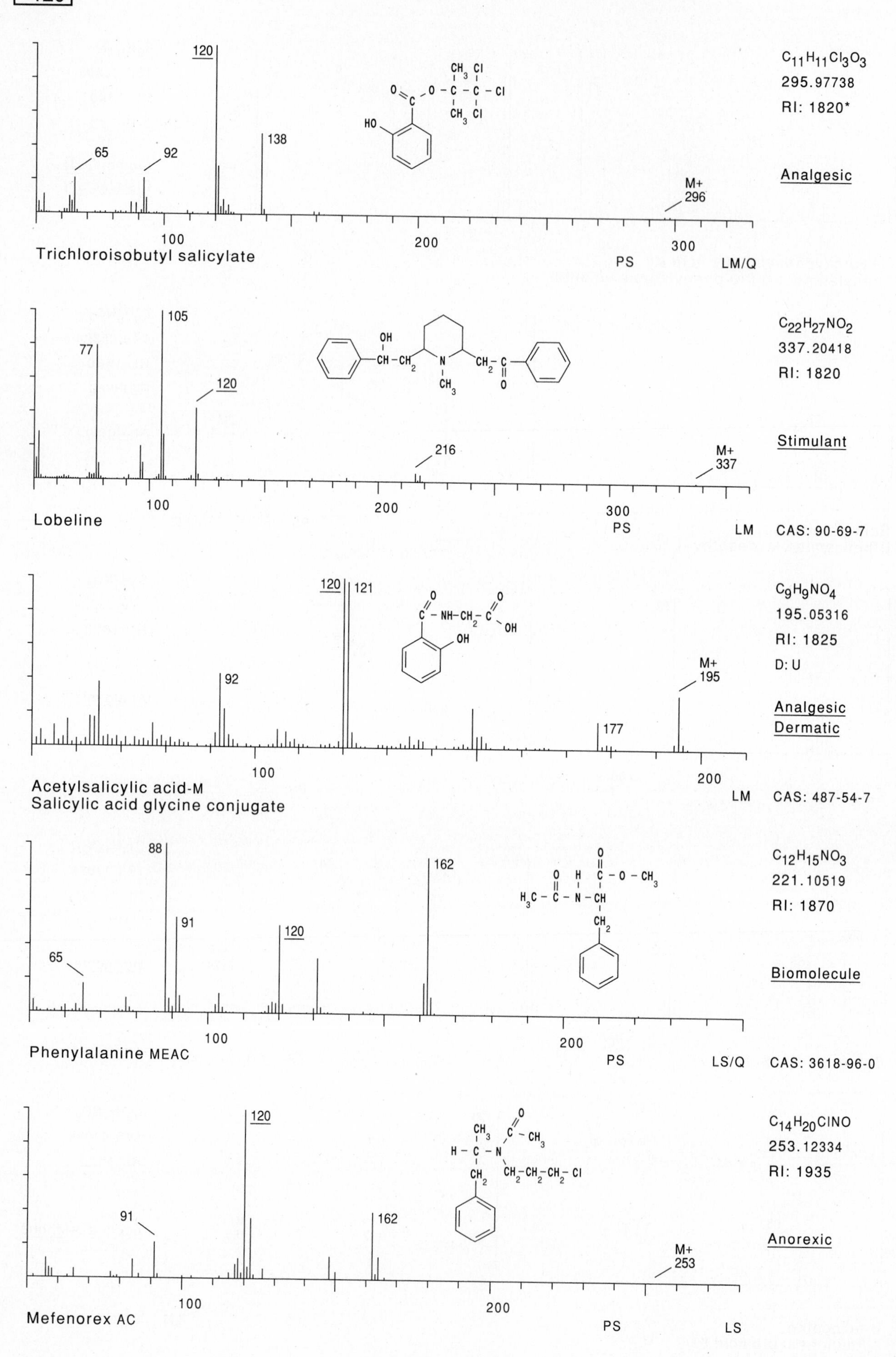
120
65
92
138
M+
296
100
200
300
PS
LM/Q
Trichloroisobutyl salicylate
$C_{11}H_{11}Cl_3O_3$
295.97738
RI: 1820*
Analgesic
105
77
120
216
M+
337
Lobeline
PS
LM
$C_{22}H_{27}NO_2$
337.20418
RI: 1820
Stimulant
CAS: 90-69-7
120
121
92
177
M+
195
Acetylsalicylic acid-M
Salicylic acid glycine conjugate
LM
$C_9H_9NO_4$
195.05316
RI: 1825
D: U
Analgesic
Dermatic
CAS: 487-54-7
88
91
65
120
162
Phenylalanine MEAC
PS
LS/Q
$C_{12}H_{15}NO_3$
221.10519
RI: 1870
Biomolecule
CAS: 3618-96-0
120
91
162
M+
253
Mefenorex AC
PS
LS
$C_{14}H_{20}ClNO$
253.12334
RI: 1935
Anorexic

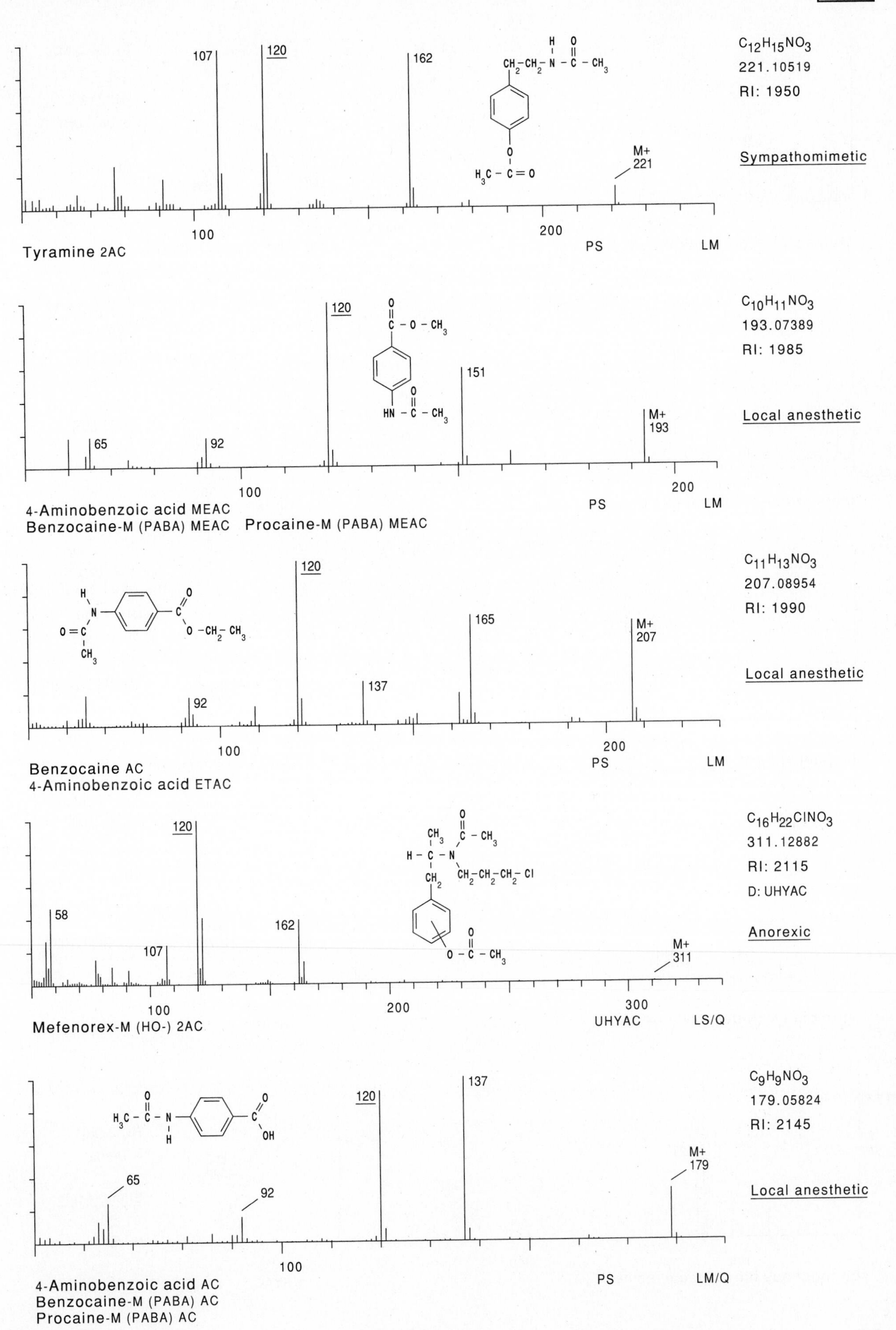

107
120
162
M+ 221
100
200
PS
LM
Tyramine 2AC
$C_{12}H_{15}NO_3$
221.10519
RI: 1950
Sympathomimetic
120
65
92
151
M+ 193
100
200
PS
LM
4-Aminobenzoic acid MEAC
Benzocaine-M (PABA) MEAC
Procaine-M (PABA) MEAC
$C_{10}H_{11}NO_3$
193.07389
RI: 1985
Local anesthetic
120
92
137
165
M+ 207
100
200
PS
LM
Benzocaine AC
4-Aminobenzoic acid ETAC
$C_{11}H_{13}NO_3$
207.08954
RI: 1990
Local anesthetic
120
58
107
162
M+ 311
100
200
300
UHYAC
LS/Q
Mefenorex-M (HO-) 2AC
$C_{16}H_{22}ClNO_3$
311.12882
RI: 2115
D: UHYAC
Anorexic
137
120
65
92
M+ 179
100
PS
LM/Q
4-Aminobenzoic acid AC
Benzocaine-M (PABA) AC
Procaine-M (PABA) AC
$C_9H_9NO_3$
179.05824
RI: 2145
Local anesthetic

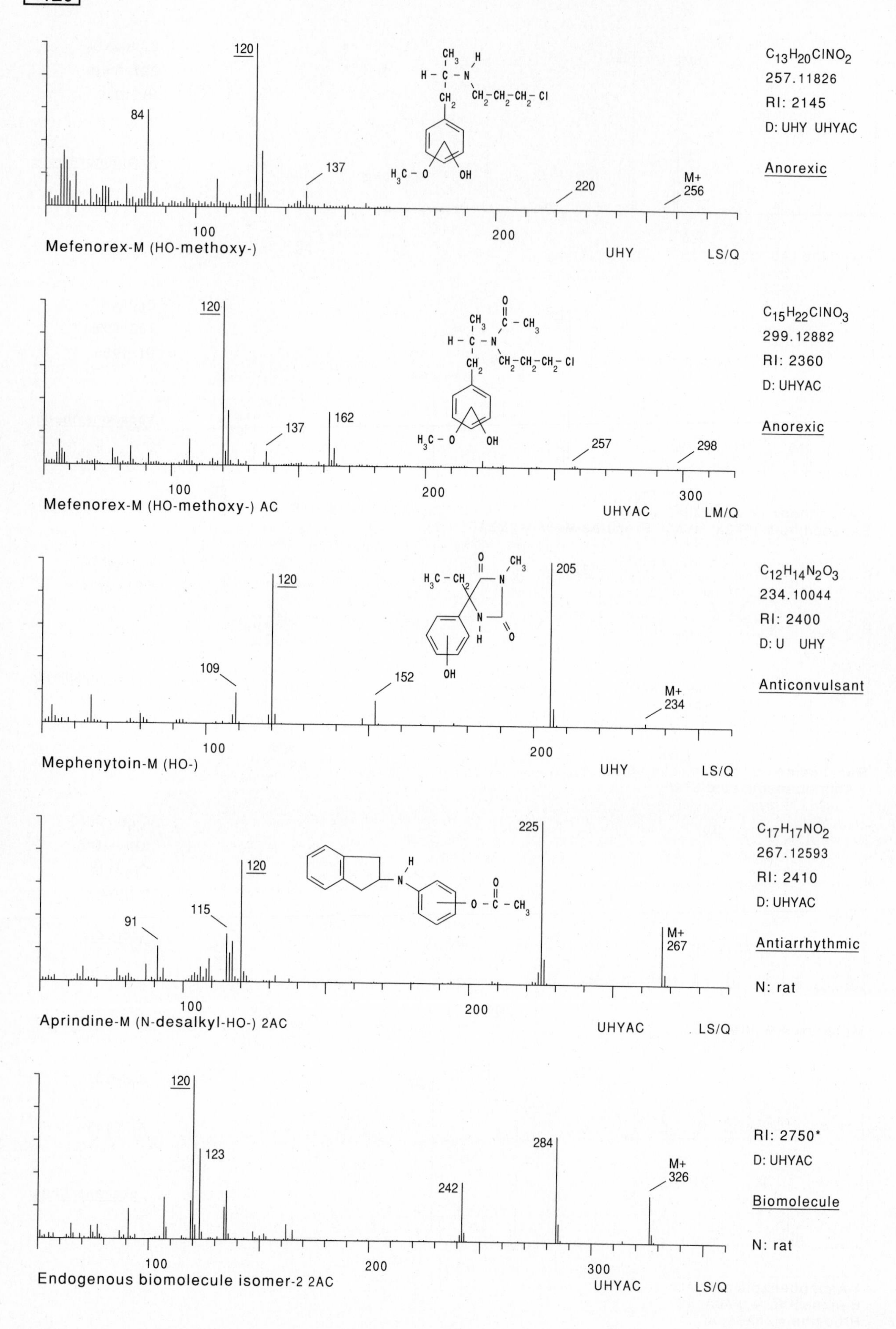

Mefenorex-M (HO-methoxy-)
UHY LS/Q
C13H20ClNO2
257.11826
RI: 2145
D: UHY UHYAC
Anorexic
Mefenorex-M (HO-methoxy-) AC
UHYAC LM/Q
C15H22ClNO3
299.12882
RI: 2360
D: UHYAC
Anorexic
Mephenytoin-M (HO-)
UHY LS/Q
C12H14N2O3
234.10044
RI: 2400
D: U UHY
Anticonvulsant
Aprindine-M (N-desalkyl-HO-) 2AC
UHYAC LS/Q
C17H17NO2
267.12593
RI: 2410
D: UHYAC
Antiarrhythmic
N: rat
Endogenous biomolecule isomer-2 2AC
UHYAC LS/Q
RI: 2750*
D: UHYAC
Biomolecule
N: rat

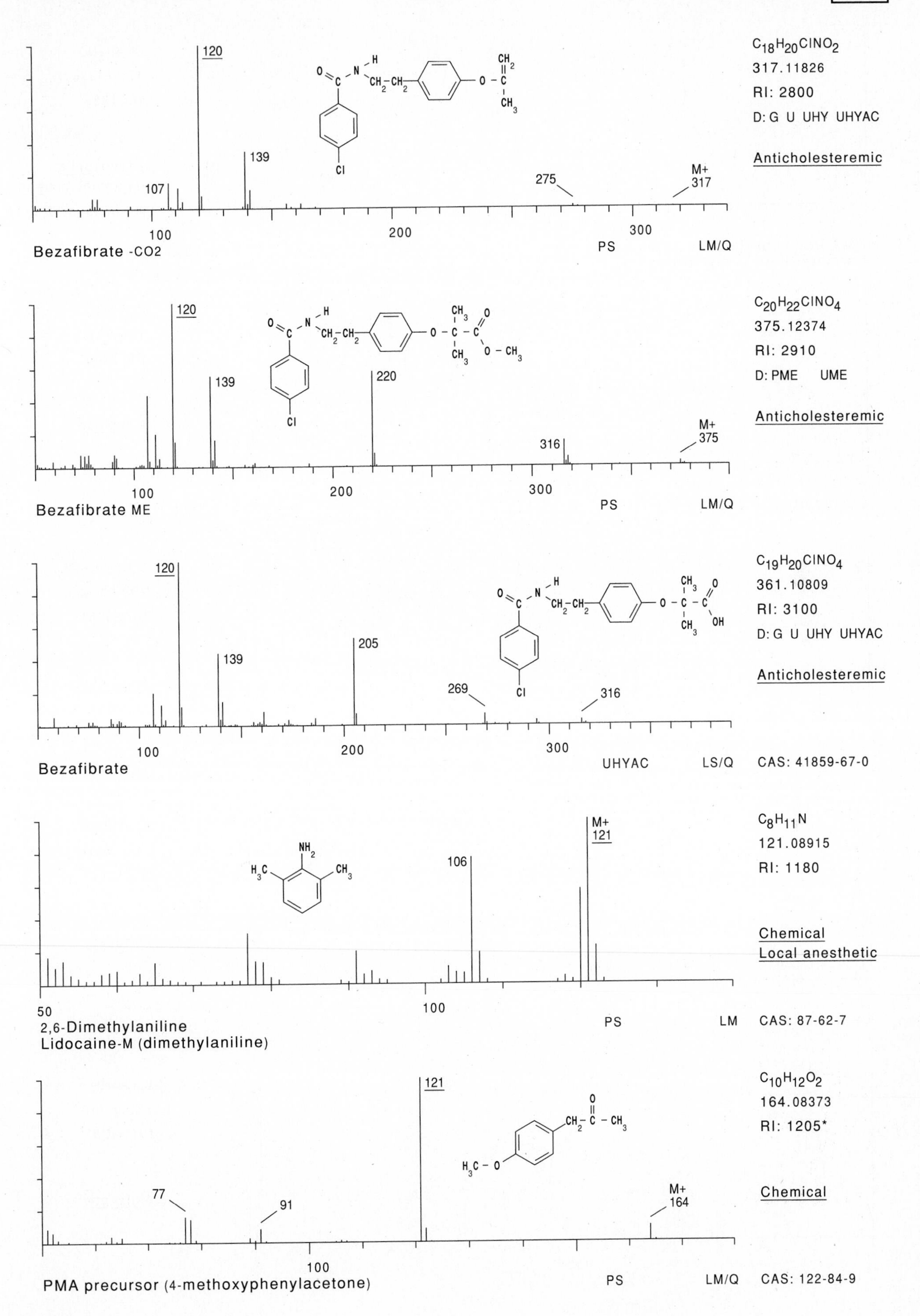
Bezafibrate -CO2
PS LM/Q
$C_{18}H_{20}ClNO_2$
317.11826
RI: 2800
D: G U UHY UHYAC
Anticholesteremic
Bezafibrate ME
PS LM/Q
$C_{20}H_{22}ClNO_4$
375.12374
RI: 2910
D: PME UME
Anticholesteremic
Bezafibrate
UHYAC LS/Q
CAS: 41859-67-0
$C_{19}H_{20}ClNO_4$
361.10809
RI: 3100
D: G U UHY UHYAC
Anticholesteremic
2,6-Dimethylaniline
Lidocaine-M (dimethylaniline)
PS LM
CAS: 87-62-7
$C_8H_{11}N$
121.08915
RI: 1180
Chemical
Local anesthetic
PMA precursor (4-methoxyphenylacetone)
PS LM/Q
CAS: 122-84-9
$C_{10}H_{12}O_2$
164.08373
RI: 1205*
Chemical

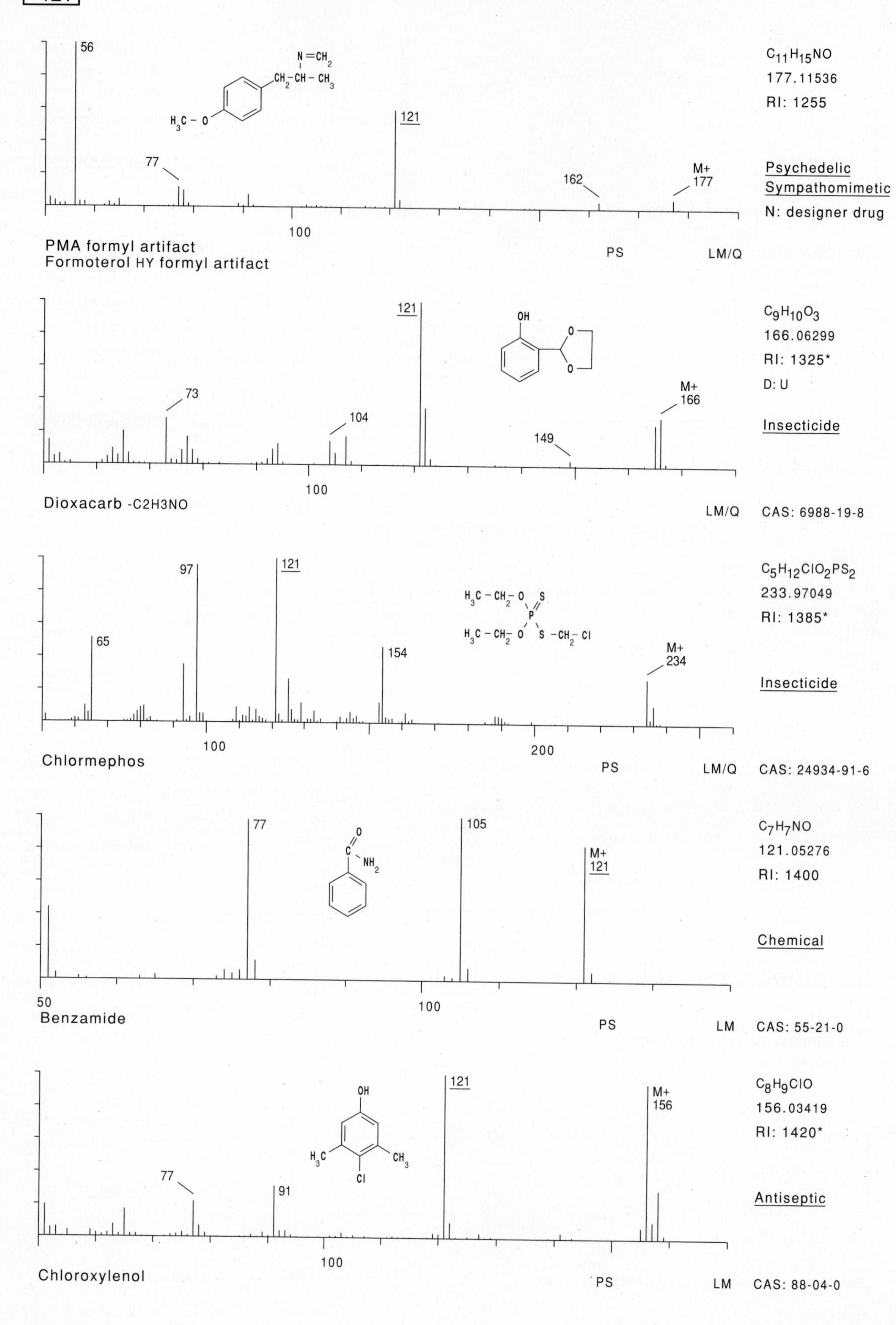

56
77
121
162
M+
177
100
$C_{11}H_{15}NO$
177.11536
RI: 1255
Psychedelic
Sympathomimetic
N: designer drug
PMA formyl artifact
Formoterol HY formyl artifact
PS
LM/Q
121
73
104
149
M+
166
100
$C_9H_{10}O_3$
166.06299
RI: 1325*
D: U
Insecticide
Dioxacarb -C2H3NO
LM/Q
CAS: 6988-19-8
97
121
65
154
M+
234
100
200
$C_5H_{12}ClO_2PS_2$
233.97049
RI: 1385*
Insecticide
Chlormephos
PS
LM/Q
CAS: 24934-91-6
77
105
M+
121
50
100
C_7H_7NO
121.05276
RI: 1400
Chemical
Benzamide
PS
LM
CAS: 55-21-0
121
M+
156
77
91
100
C_8H_9ClO
156.03419
RI: 1420*
Antiseptic
Chloroxylenol
PS
LM
CAS: 88-04-0

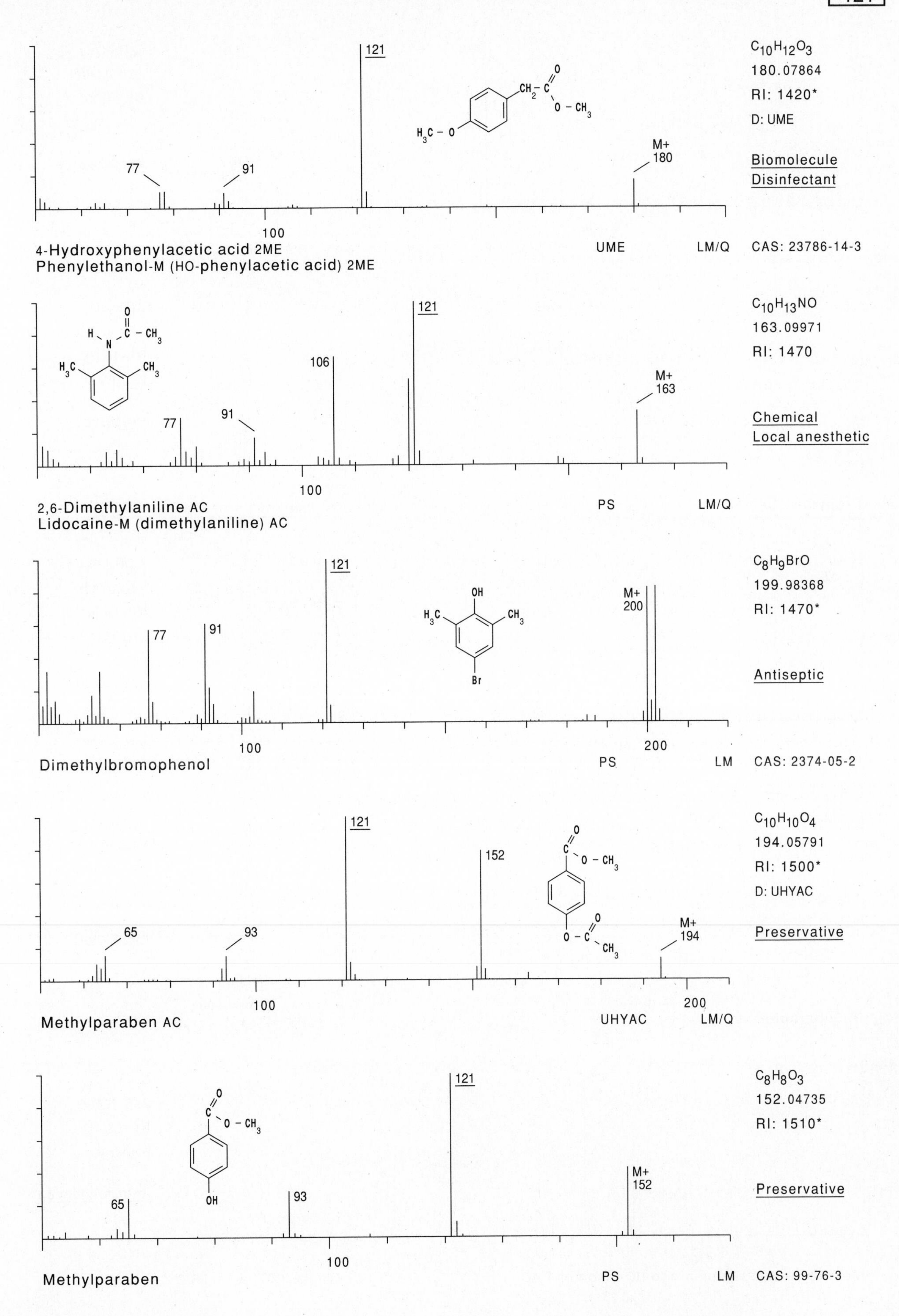

$C_{10}H_{12}O_3$
180.07864
RI: 1420*
D: UME

Biomolecule
Disinfectant

4-Hydroxyphenylacetic acid 2ME
Phenylethanol-M (HO-phenylacetic acid) 2ME
CAS: 23786-14-3

$C_{10}H_{13}NO$
163.09971
RI: 1470

Chemical
Local anesthetic

2,6-Dimethylaniline AC
Lidocaine-M (dimethylaniline) AC

C_8H_9BrO
199.98368
RI: 1470*

Antiseptic

Dimethylbromophenol
CAS: 2374-05-2

$C_{10}H_{10}O_4$
194.05791
RI: 1500*
D: UHYAC

Preservative

Methylparaben AC

$C_8H_8O_3$
152.04735
RI: 1510*

Preservative

Methylparaben
CAS: 99-76-3

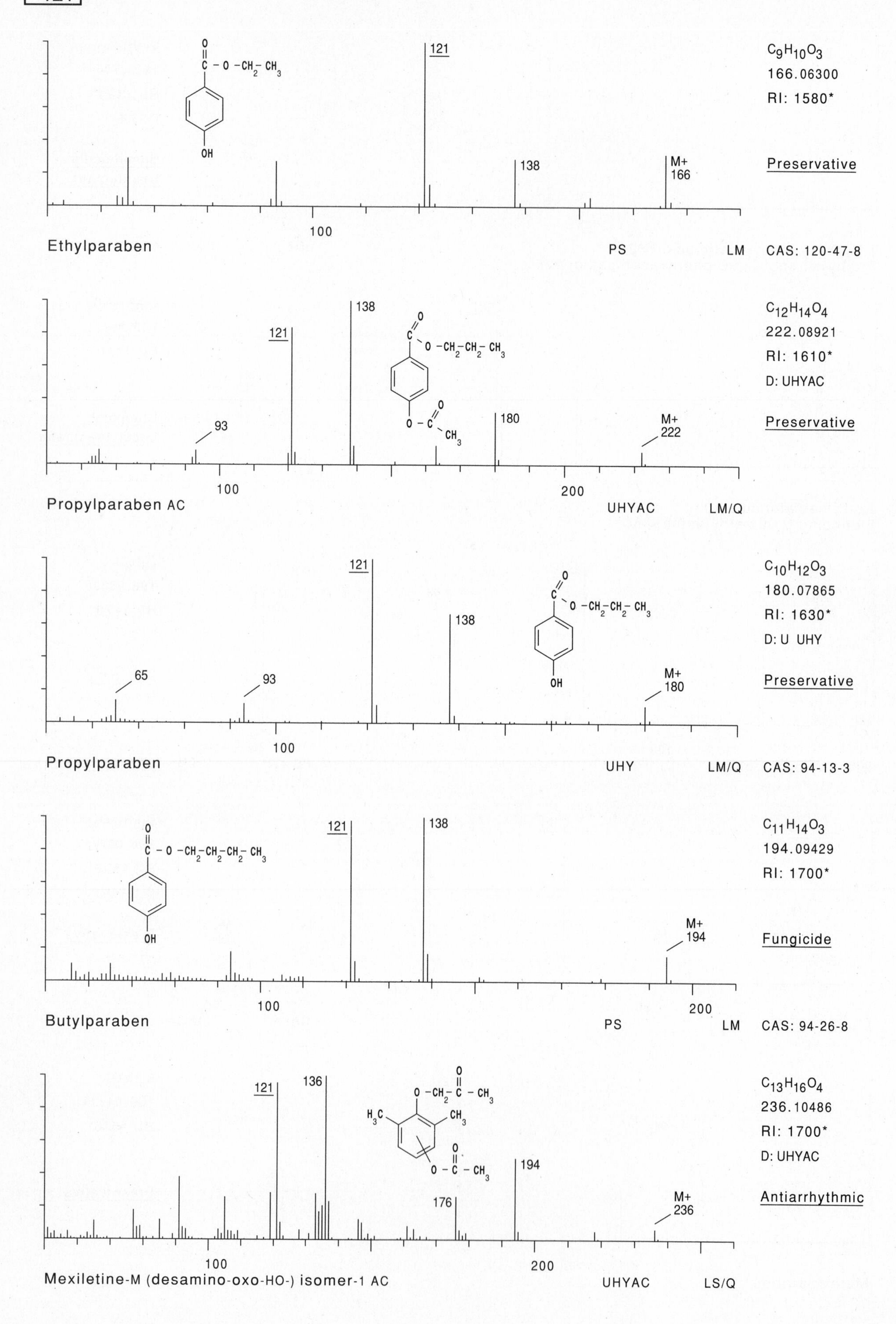
Ethylparaben
PS LM
CAS: 120-47-8
$C_9H_{10}O_3$
166.06300
RI: 1580*
Preservative
121
138
M+ 166
100
Propylparaben AC
UHYAC LM/Q
$C_{12}H_{14}O_4$
222.08921
RI: 1610*
D: UHYAC
Preservative
93
121
138
180
M+ 222
100
200
Propylparaben
UHY LM/Q
CAS: 94-13-3
$C_{10}H_{12}O_3$
180.07865
RI: 1630*
D: U UHY
Preservative
65
93
121
138
M+ 180
100
Butylparaben
PS LM
CAS: 94-26-8
$C_{11}H_{14}O_3$
194.09429
RI: 1700*
Fungicide
121
138
M+ 194
100
200
Mexiletine-M (desamino-oxo-HO-) isomer-1 AC
UHYAC LS/Q
$C_{13}H_{16}O_4$
236.10486
RI: 1700*
D: UHYAC
Antiarrhythmic
121
136
176
194
M+ 236
100
200

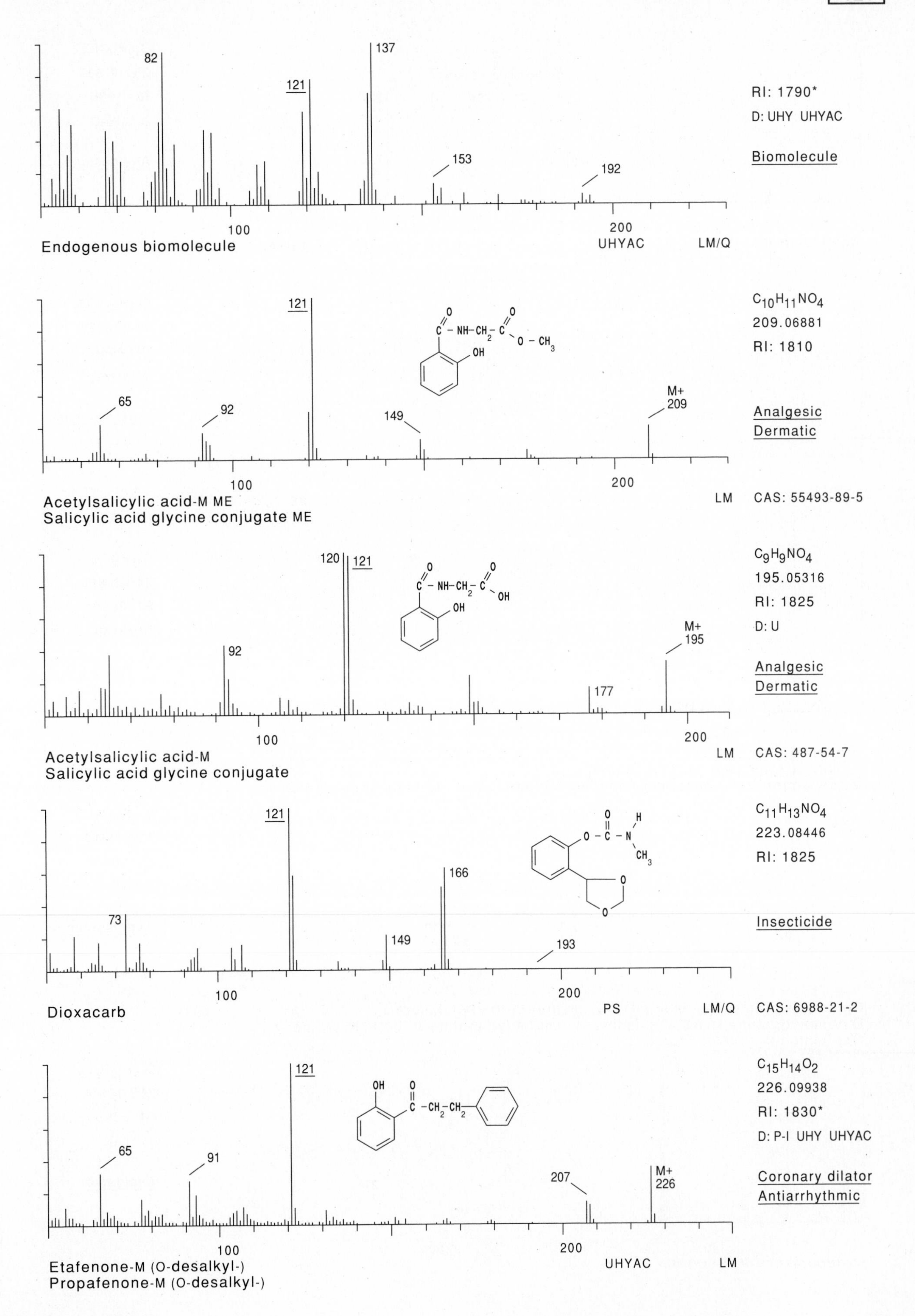

82
121
137
153
192
100
200
Endogenous biomolecule
UHYAC
LM/Q
RI: 1790*
D: UHY UHYAC
Biomolecule
121
65
92
149
M+
209
100
200
Acetylsalicylic acid-M ME
Salicylic acid glycine conjugate ME
LM
C10H11NO4
209.06881
RI: 1810
Analgesic
Dermatic
CAS: 55493-89-5
120
121
92
177
M+
195
100
200
Acetylsalicylic acid-M
Salicylic acid glycine conjugate
LM
C9H9NO4
195.05316
RI: 1825
D: U
Analgesic
Dermatic
CAS: 487-54-7
121
73
149
166
193
100
200
Dioxacarb
PS
LM/Q
C11H13NO4
223.08446
RI: 1825
Insecticide
CAS: 6988-21-2
121
65
91
207
M+
226
100
200
Etafenone-M (O-desalkyl-)
Propafenone-M (O-desalkyl-)
UHYAC
LM
C15H14O2
226.09938
RI: 1830*
D: P-I UHY UHYAC
Coronary dilator
Antiarrhythmic

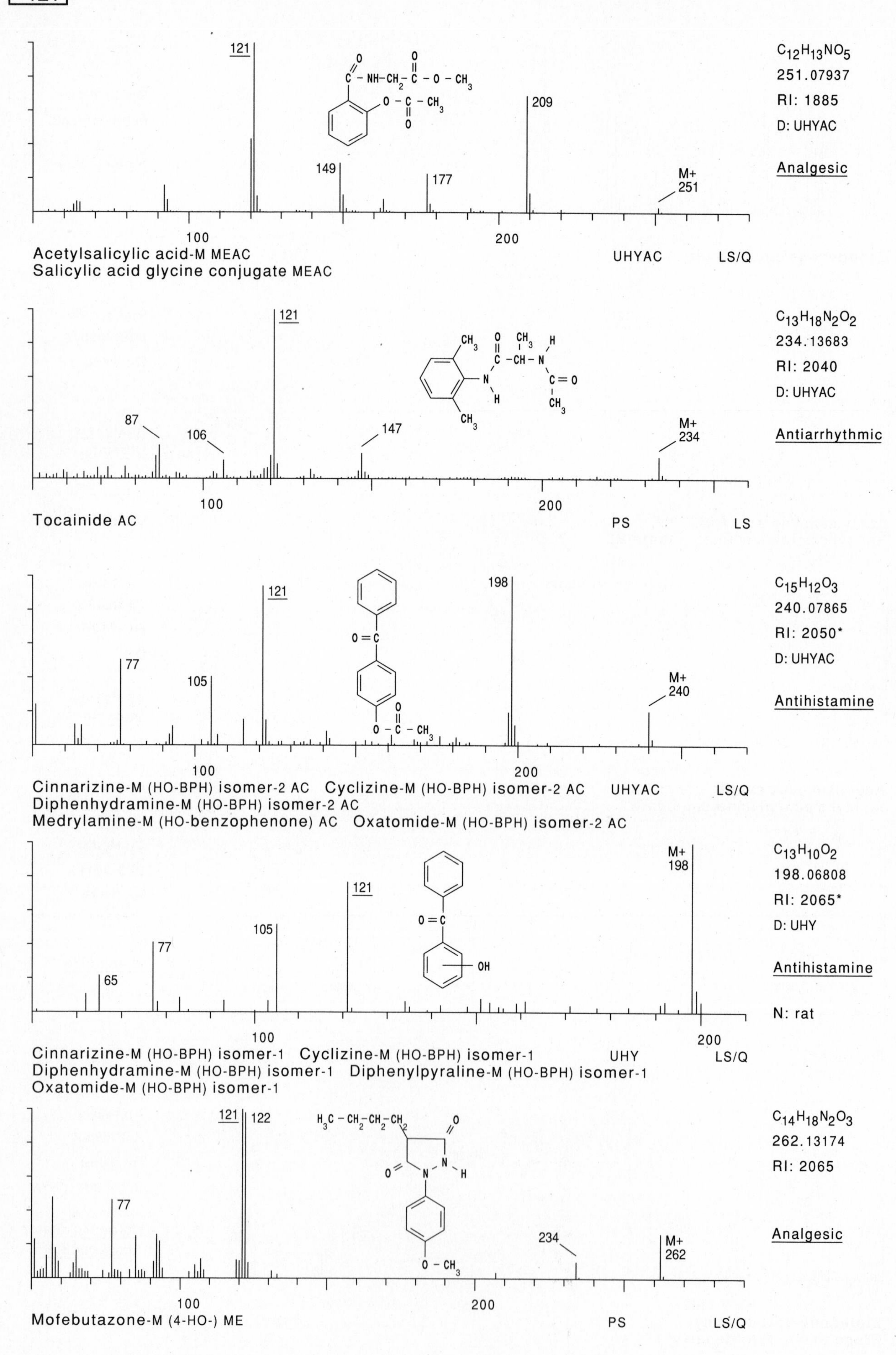

121
209
149
177
M+ 251
100
200
UHYAC
LS/Q
C12H13NO5
251.07937
RI: 1885
D: UHYAC
Analgesic
Acetylsalicylic acid-M MEAC
Salicylic acid glycine conjugate MEAC
121
87
106
147
M+ 234
PS
LS
C13H18N2O2
234.13683
RI: 2040
D: UHYAC
Antiarrhythmic
Tocainide AC
121
198
77
105
M+ 240
UHYAC
LS/Q
C15H12O3
240.07865
RI: 2050*
D: UHYAC
Antihistamine
Cinnarizine-M (HO-BPH) isomer-2 AC Cyclizine-M (HO-BPH) isomer-2 AC
Diphenhydramine-M (HO-BPH) isomer-2 AC
Medrylamine-M (HO-benzophenone) AC Oxatomide-M (HO-BPH) isomer-2 AC
M+ 198
121
105
77
65
UHY
LS/Q
C13H10O2
198.06808
RI: 2065*
D: UHY
Antihistamine
N: rat
Cinnarizine-M (HO-BPH) isomer-1 Cyclizine-M (HO-BPH) isomer-1
Diphenhydramine-M (HO-BPH) isomer-1 Diphenylpyraline-M (HO-BPH) isomer-1
Oxatomide-M (HO-BPH) isomer-1
121
122
77
234
M+ 262
PS
LS/Q
C14H18N2O3
262.13174
RI: 2065
Analgesic
Mofebutazone-M (4-HO-) ME

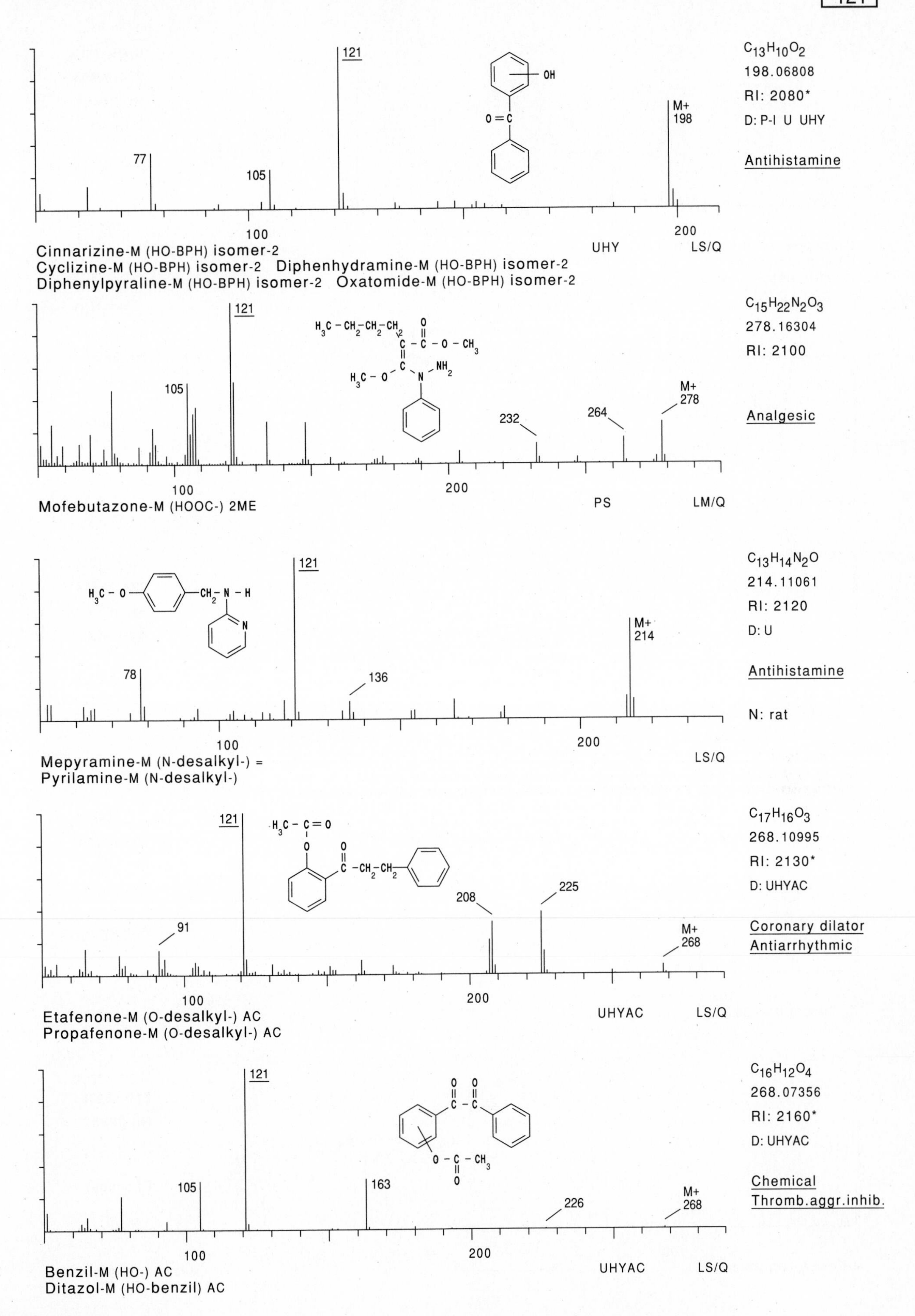

$C_{13}H_{10}O_2$
198.06808
RI: 2080*
D: P-I U UHY

Antihistamine

$C_{15}H_{22}N_2O_3$
278.16304
RI: 2100

Analgesic

$C_{13}H_{14}N_2O$
214.11061
RI: 2120
D: U

Antihistamine

N: rat

$C_{17}H_{16}O_3$
268.10995
RI: 2130*
D: UHYAC

Coronary dilator
Antiarrhythmic

$C_{16}H_{12}O_4$
268.07356
RI: 2160*
D: UHYAC

Chemical
Thromb.aggr.inhib.

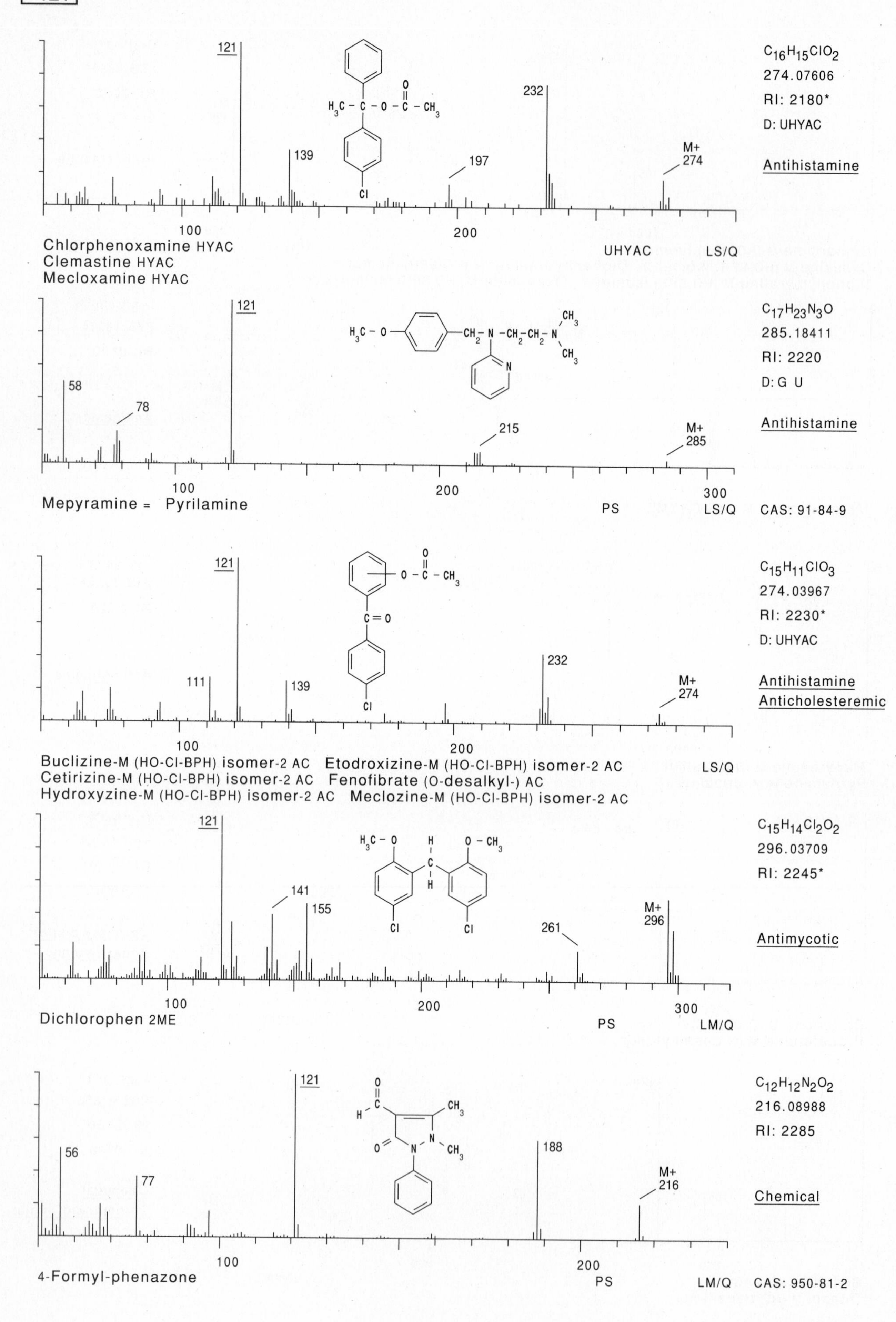
121
232
139
197
M+
274
100
200
$C_{16}H_{15}ClO_2$
274.07606
RI: 2180*
D: UHYAC
Antihistamine
Chlorphenoxamine HYAC
Clemastine HYAC
Mecloxamine HYAC
UHYAC
LS/Q
121
58
78
215
M+
285
300
$C_{17}H_{23}N_3O$
285.18411
RI: 2220
D: G U
Antihistamine
Mepyramine = Pyrilamine
PS
LS/Q
CAS: 91-84-9
121
111
139
232
M+
274
$C_{15}H_{11}ClO_3$
274.03967
RI: 2230*
D: UHYAC
Antihistamine
Anticholesteremic
Buclizine-M (HO-Cl-BPH) isomer-2 AC Etodroxizine-M (HO-Cl-BPH) isomer-2 AC
Cetirizine-M (HO-Cl-BPH) isomer-2 AC Fenofibrate (O-desalkyl-) AC
Hydroxyzine-M (HO-Cl-BPH) isomer-2 AC Meclozine-M (HO-Cl-BPH) isomer-2 AC
LS/Q
121
141
155
261
M+
296
$C_{15}H_{14}Cl_2O_2$
296.03709
RI: 2245*
Antimycotic
Dichlorophen 2ME
PS
LM/Q
121
56
77
188
M+
216
$C_{12}H_{12}N_2O_2$
216.08988
RI: 2285
Chemical
4-Formyl-phenazone
PS
LM/Q
CAS: 950-81-2

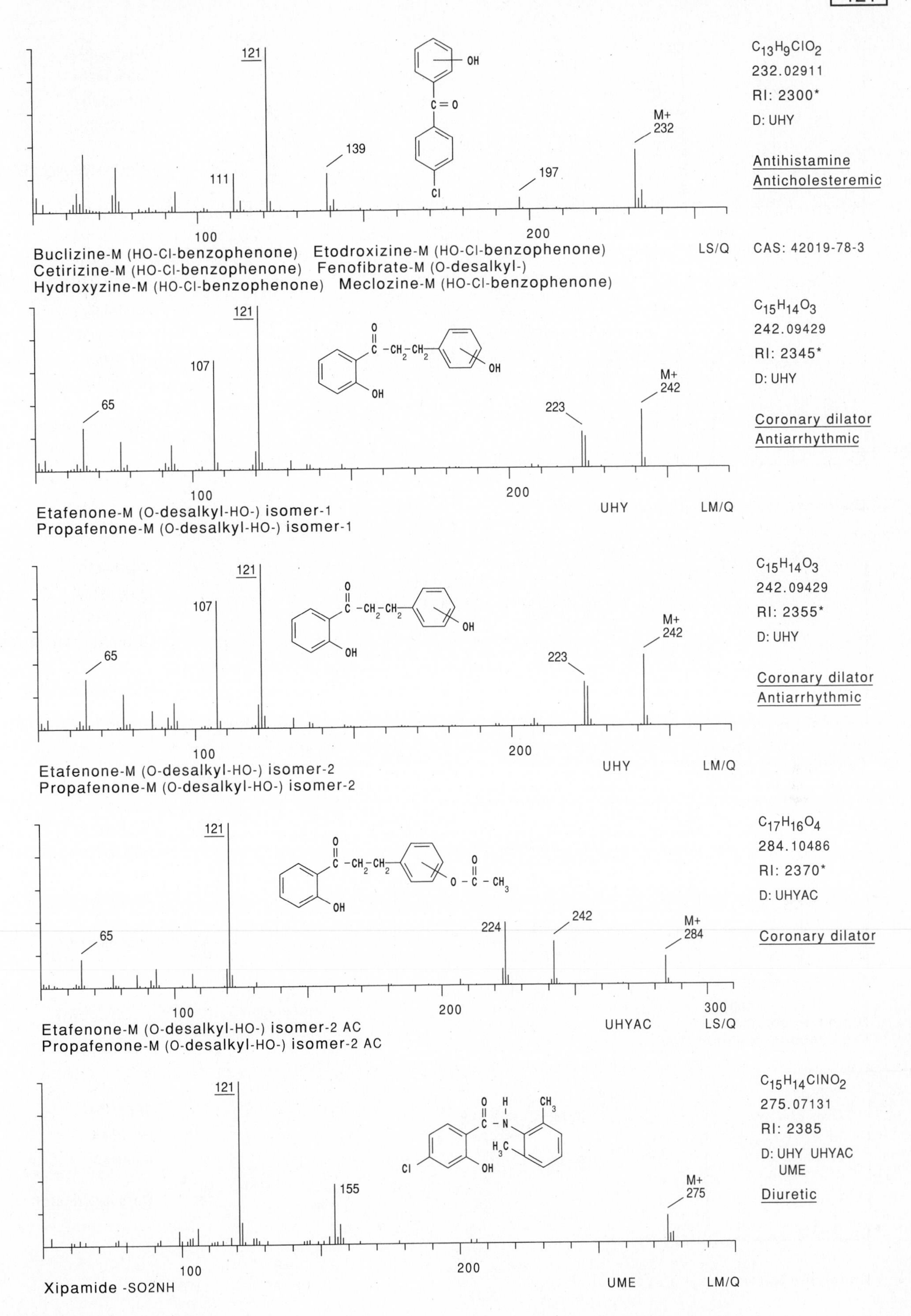

121
139
111
197
M+
232
OH
C = O
Cl
100
200
LS/Q
Buclizine-M (HO-Cl-benzophenone) Etodroxizine-M (HO-Cl-benzophenone)
Cetirizine-M (HO-Cl-benzophenone) Fenofibrate-M (O-desalkyl-)
Hydroxyzine-M (HO-Cl-benzophenone) Meclozine-M (HO-Cl-benzophenone)
C13H9ClO2
232.02911
RI: 2300*
D: UHY
Antihistamine
Anticholesteremic
CAS: 42019-78-3
121
107
65
223
M+
242
OH
100
200
UHY
LM/Q
Etafenone-M (O-desalkyl-HO-) isomer-1
Propafenone-M (O-desalkyl-HO-) isomer-1
C15H14O3
242.09429
RI: 2345*
D: UHY
Coronary dilator
Antiarrhythmic
121
107
65
223
M+
242
100
200
UHY
LM/Q
Etafenone-M (O-desalkyl-HO-) isomer-2
Propafenone-M (O-desalkyl-HO-) isomer-2
C15H14O3
242.09429
RI: 2355*
D: UHY
Coronary dilator
Antiarrhythmic
121
65
224
242
M+
284
100
200
300
UHYAC
LS/Q
Etafenone-M (O-desalkyl-HO-) isomer-2 AC
Propafenone-M (O-desalkyl-HO-) isomer-2 AC
C17H16O4
284.10486
RI: 2370*
D: UHYAC
Coronary dilator
121
155
M+
275
Cl
OH
100
200
UME
LM/Q
Xipamide -SO2NH
C15H14ClNO2
275.07131
RI: 2385
D: UHY UHYAC
UME
Diuretic

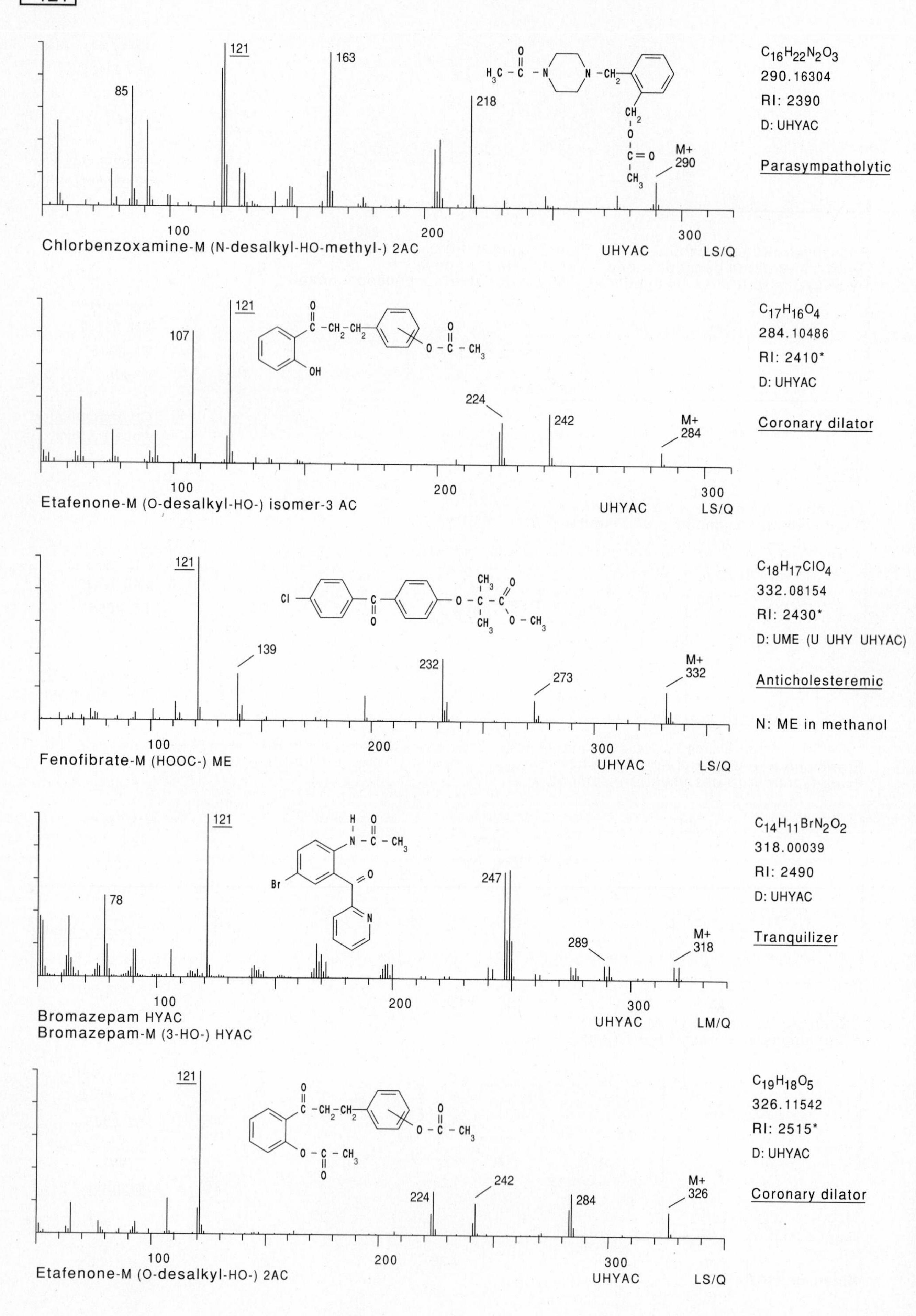
121
163
85
218
M+
290
100
200
300
Chlorbenzoxamine-M (N-desalkyl-HO-methyl-) 2AC
UHYAC
LS/Q
C16H22N2O3
290.16304
RI: 2390
D: UHYAC
Parasympatholytic
121
107
224
242
M+
284
Etafenone-M (O-desalkyl-HO-) isomer-3 AC
UHYAC
LS/Q
C17H16O4
284.10486
RI: 2410*
D: UHYAC
Coronary dilator
121
139
232
273
M+
332
Fenofibrate-M (HOOC-) ME
UHYAC
LS/Q
C18H17ClO4
332.08154
RI: 2430*
D: UME (U UHY UHYAC)
Anticholesteremic
N: ME in methanol
121
78
247
289
M+
318
Bromazepam HYAC
Bromazepam-M (3-HO-) HYAC
UHYAC
LM/Q
C14H11BrN2O2
318.00039
RI: 2490
D: UHYAC
Tranquilizer
121
224
242
284
M+
326
Etafenone-M (O-desalkyl-HO-) 2AC
UHYAC
LS/Q
C19H18O5
326.11542
RI: 2515*
D: UHYAC
Coronary dilator

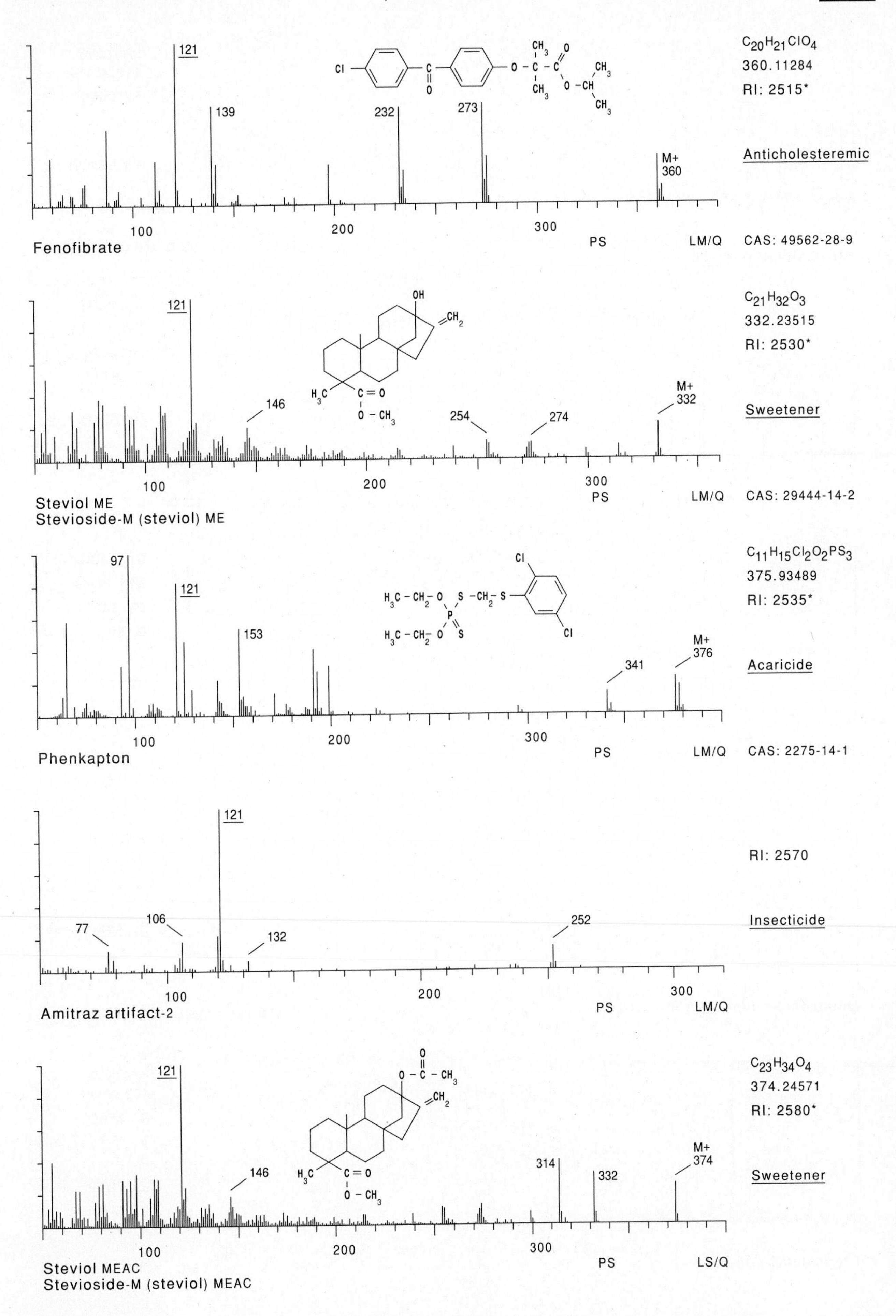

121
139
232
273
M+
360
100
200
300
PS
LM/Q
Fenofibrate
C20H21ClO4
360.11284
RI: 2515*
Anticholesteremic
CAS: 49562-28-9
121
146
254
274
M+
332
100
200
300
PS
LM/Q
Steviol ME
Stevioside-M (steviol) ME
C21H32O3
332.23515
RI: 2530*
Sweetener
CAS: 29444-14-2
97
121
153
341
M+
376
100
200
300
PS
LM/Q
Phenkapton
C11H15Cl2O2PS3
375.93489
RI: 2535*
Acaricide
CAS: 2275-14-1
121
77
106
132
252
100
200
300
PS
LM/Q
Amitraz artifact-2
RI: 2570
Insecticide
121
146
314
332
M+
374
100
200
300
PS
LS/Q
Steviol MEAC
Stevioside-M (steviol) MEAC
C23H34O4
374.24571
RI: 2580*
Sweetener

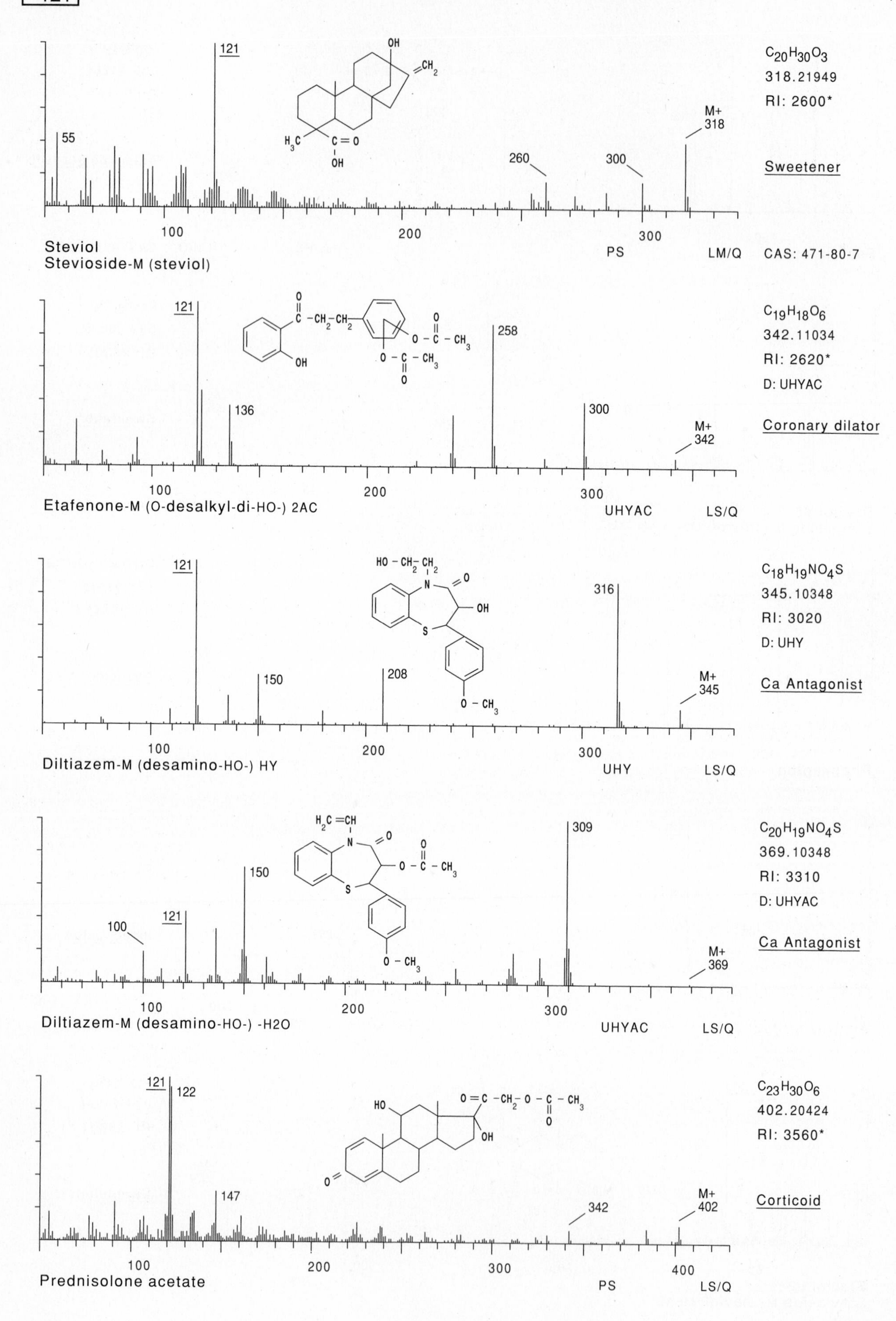
Steviol
Stevioside-M (steviol)
PS LM/Q
C20H30O3
318.21949
RI: 2600*
Sweetener
CAS: 471-80-7
Etafenone-M (O-desalkyl-di-HO-) 2AC
UHYAC LS/Q
C19H18O6
342.11034
RI: 2620*
D: UHYAC
Coronary dilator
Diltiazem-M (desamino-HO-) HY
UHY LS/Q
C18H19NO4S
345.10348
RI: 3020
D: UHY
Ca Antagonist
Diltiazem-M (desamino-HO-) -H2O
UHYAC LS/Q
C20H19NO4S
369.10348
RI: 3310
D: UHYAC
Ca Antagonist
Prednisolone acetate
PS LS/Q
C23H30O6
402.20424
RI: 3560*
Corticoid

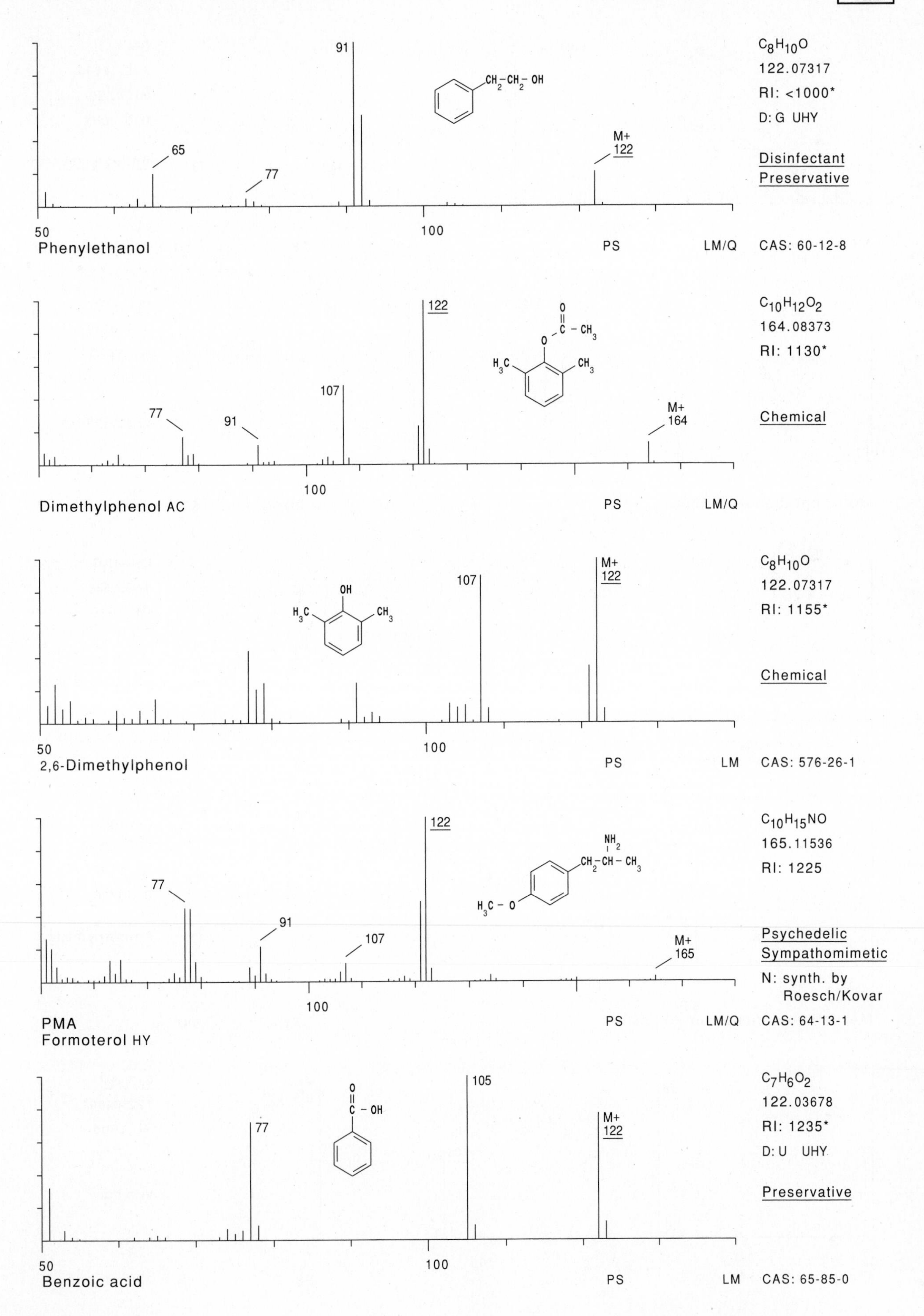

91
65
77
M+
122
50
100
Phenylethanol
PS
LM/Q
$C_8H_{10}O$
122.07317
RI: <1000*
D: G UHY
Disinfectant
Preservative
CAS: 60-12-8
122
107
77
91
M+
164
100
Dimethylphenol AC
PS
LM/Q
$C_{10}H_{12}O_2$
164.08373
RI: 1130*
Chemical
M+
122
107
50
100
2,6-Dimethylphenol
PS
LM
$C_8H_{10}O$
122.07317
RI: 1155*
Chemical
CAS: 576-26-1
122
77
91
107
M+
165
100
PMA
Formoterol HY
PS
LM/Q
$C_{10}H_{15}NO$
165.11536
RI: 1225
Psychedelic
Sympathomimetic
N: synth. by
Roesch/Kovar
CAS: 64-13-1
105
77
M+
122
50
100
Benzoic acid
PS
LM
$C_7H_6O_2$
122.03678
RI: 1235*
D: U UHY
Preservative
CAS: 65-85-0

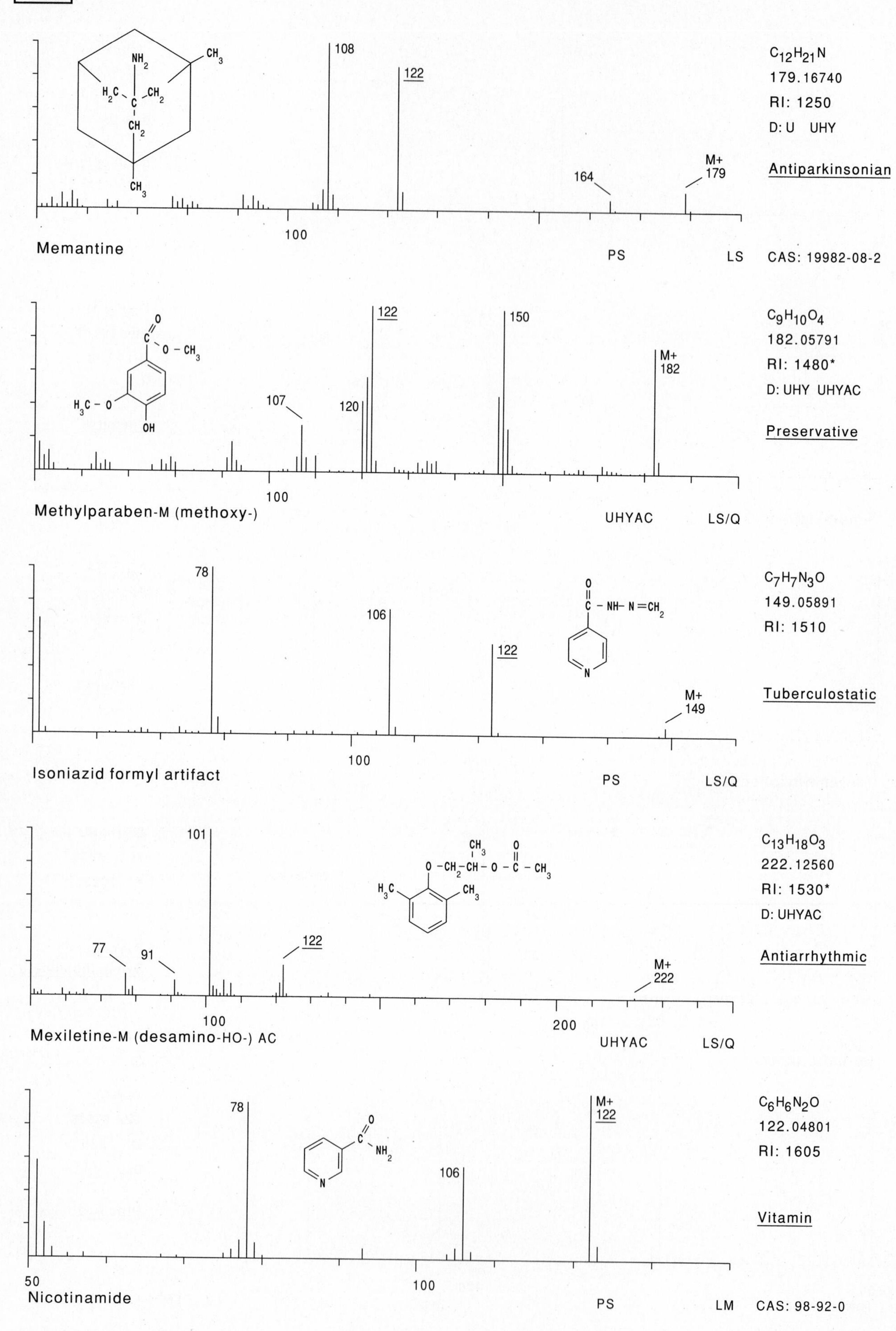

108
122
164
M+ 179
100
Memantine
PS
LS
C12H21N
179.16740
RI: 1250
D: U UHY
Antiparkinsonian
CAS: 19982-08-2
122
150
M+ 182
107
120
100
Methylparaben-M (methoxy-)
UHYAC
LS/Q
C9H10O4
182.05791
RI: 1480*
D: UHY UHYAC
Preservative
78
106
122
M+ 149
100
Isoniazid formyl artifact
PS
LS/Q
C7H7N3O
149.05891
RI: 1510
Tuberculostatic
101
77
91
122
M+ 222
100
200
Mexiletine-M (desamino-HO-) AC
UHYAC
LS/Q
C13H18O3
222.12560
RI: 1530*
D: UHYAC
Antiarrhythmic
78
106
M+ 122
50
100
Nicotinamide
PS
LM
C6H6N2O
122.04801
RI: 1605
Vitamin
CAS: 98-92-0

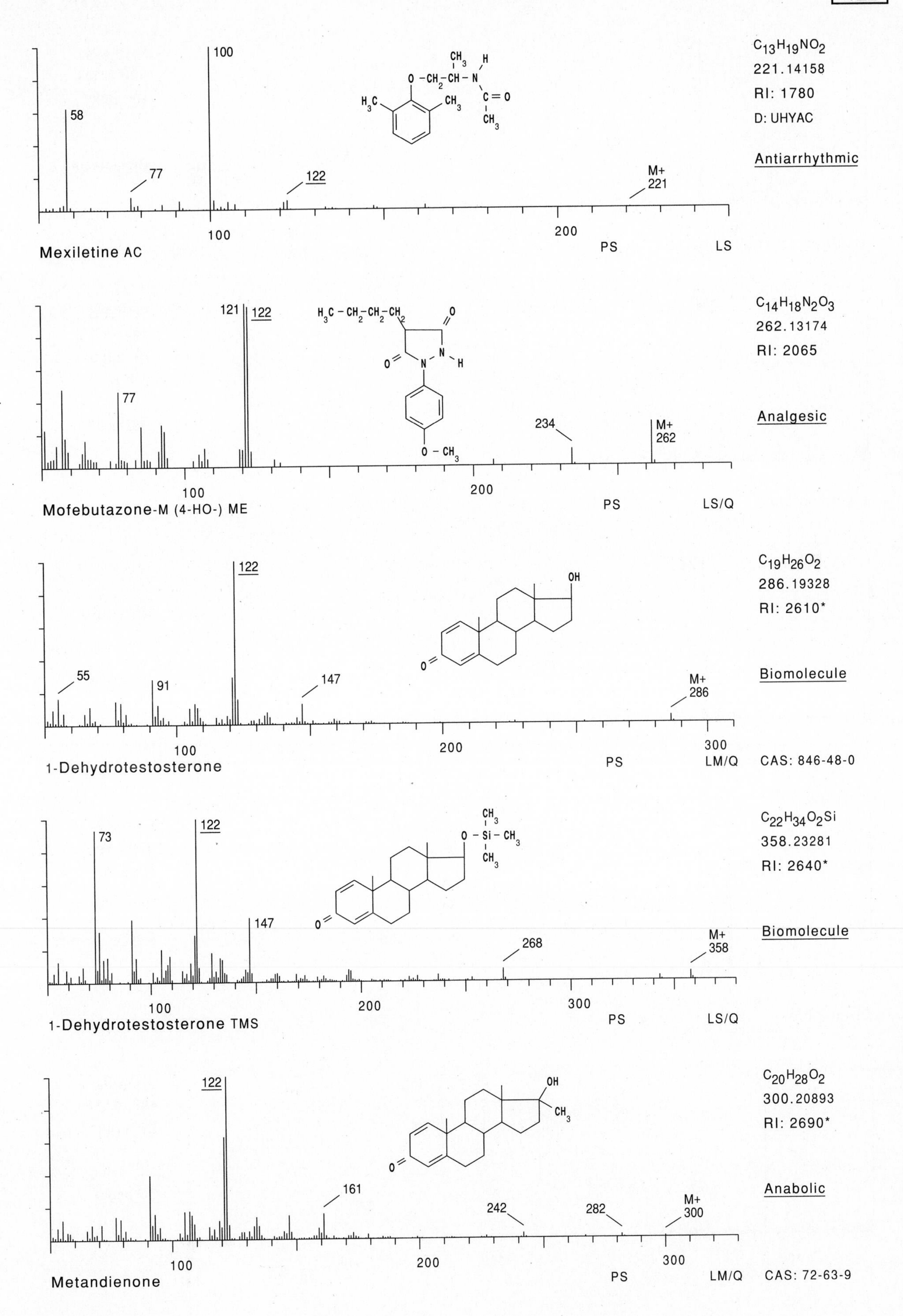

Mexiletine AC
$C_{13}H_{19}NO_2$
221.14158
RI: 1780
D: UHYAC
Antiarrhythmic
58
77
100
122
M+ 221
PS
LS
Mofebutazone-M (4-HO-) ME
$C_{14}H_{18}N_2O_3$
262.13174
RI: 2065
Analgesic
77
121
122
234
M+ 262
PS
LS/Q
1-Dehydrotestosterone
$C_{19}H_{26}O_2$
286.19328
RI: 2610*
Biomolecule
55
91
122
147
M+ 286
PS
LM/Q
CAS: 846-48-0
1-Dehydrotestosterone TMS
$C_{22}H_{34}O_2Si$
358.23281
RI: 2640*
Biomolecule
73
122
147
268
M+ 358
PS
LS/Q
Metandienone
$C_{20}H_{28}O_2$
300.20893
RI: 2690*
Anabolic
122
161
242
282
M+ 300
PS
LM/Q
CAS: 72-63-9

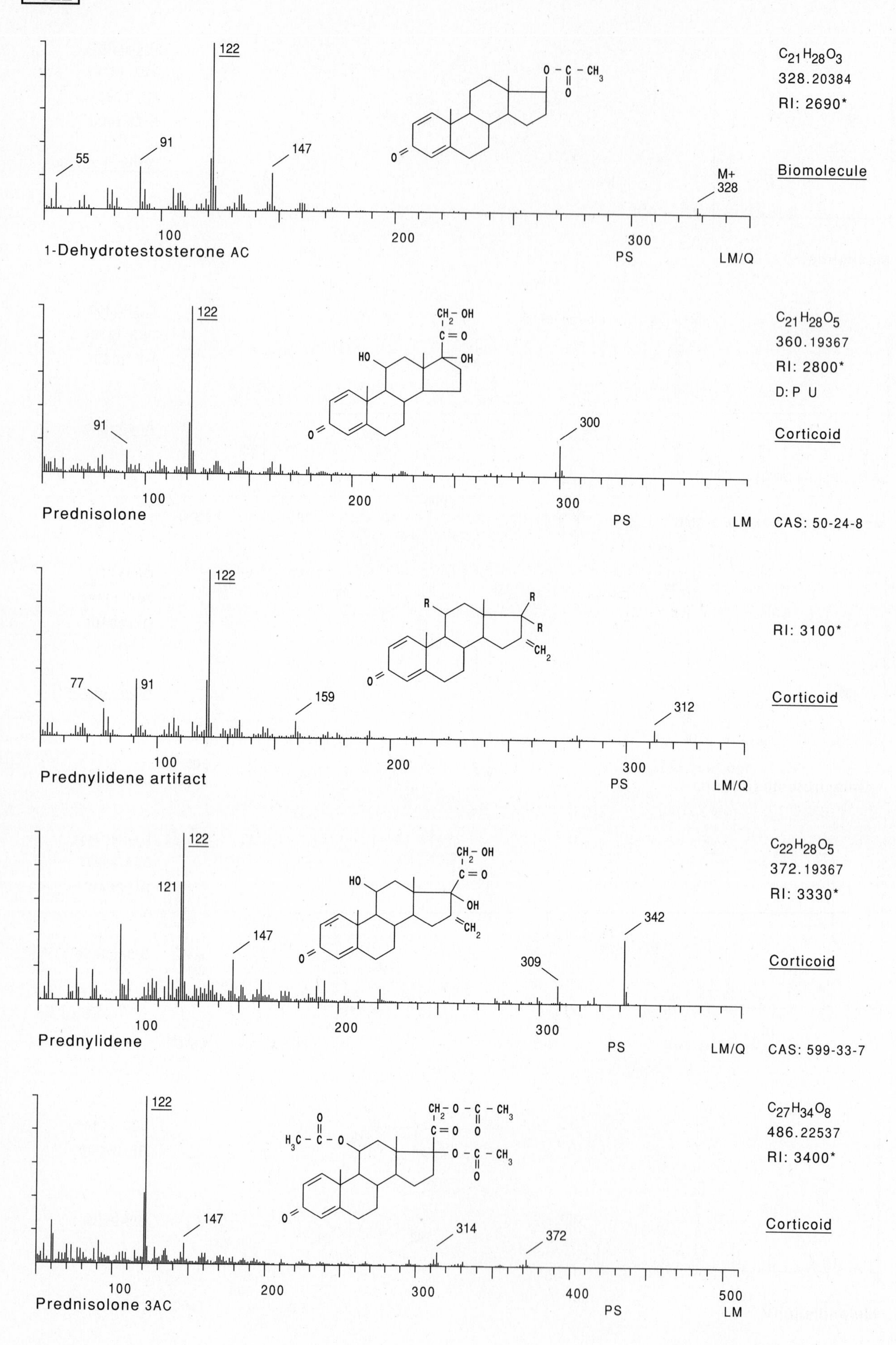
122
55
91
147
M+
328
100
200
300
1-Dehydrotestosterone AC
PS
LM/Q
$C_{21}H_{28}O_3$
328.20384
RI: 2690*
Biomolecule
122
91
300
100
200
300
Prednisolone
PS
LM
$C_{21}H_{28}O_5$
360.19367
RI: 2800*
D: P U
Corticoid
CAS: 50-24-8
122
77
91
159
312
100
200
300
Prednylidene artifact
PS
LM/Q
RI: 3100*
Corticoid
122
121
147
309
342
100
200
300
Prednylidene
PS
LM/Q
$C_{22}H_{28}O_5$
372.19367
RI: 3330*
Corticoid
CAS: 599-33-7
122
147
314
372
100
200
300
400
500
Prednisolone 3AC
PS
LM
$C_{27}H_{34}O_8$
486.22537
RI: 3400*
Corticoid

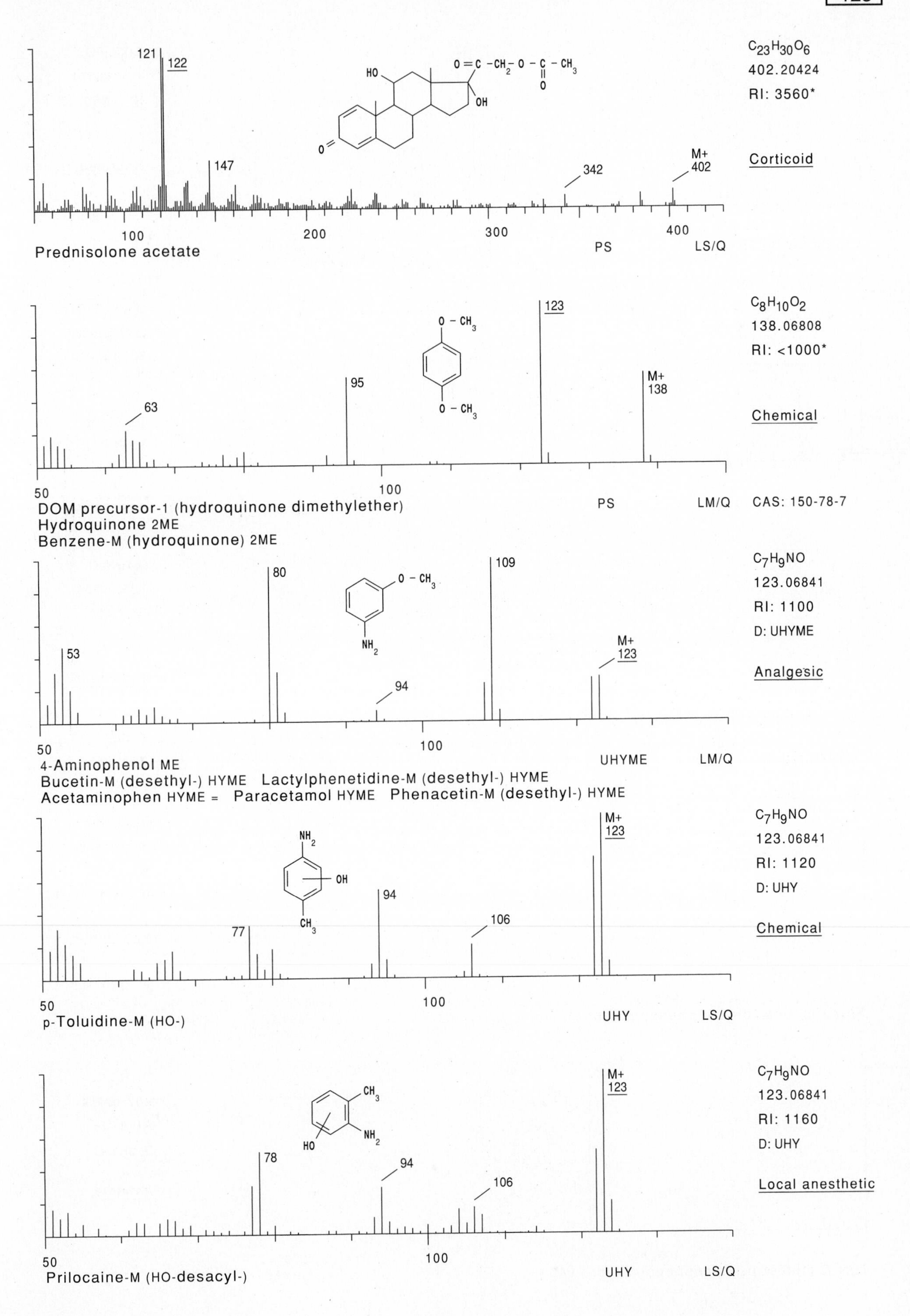
123
Prednisolone acetate
PS LS/Q
$C_{23}H_{30}O_6$
402.20424
RI: 3560*
Corticoid
DOM precursor-1 (hydroquinone dimethylether)
Hydroquinone 2ME
Benzene-M (hydroquinone) 2ME
PS LM/Q
CAS: 150-78-7
$C_8H_{10}O_2$
138.06808
RI: <1000*
Chemical
4-Aminophenol ME
Bucetin-M (desethyl-) HYME Lactylphenetidine-M (desethyl-) HYME
Acetaminophen HYME = Paracetamol HYME Phenacetin-M (desethyl-) HYME
UHYME LM/Q
C_7H_9NO
123.06841
RI: 1100
D: UHYME
Analgesic
p-Toluidine-M (HO-)
UHY LS/Q
C_7H_9NO
123.06841
RI: 1120
D: UHY
Chemical
Prilocaine-M (HO-desacyl-)
UHY LS/Q
C_7H_9NO
123.06841
RI: 1160
D: UHY
Local anesthetic

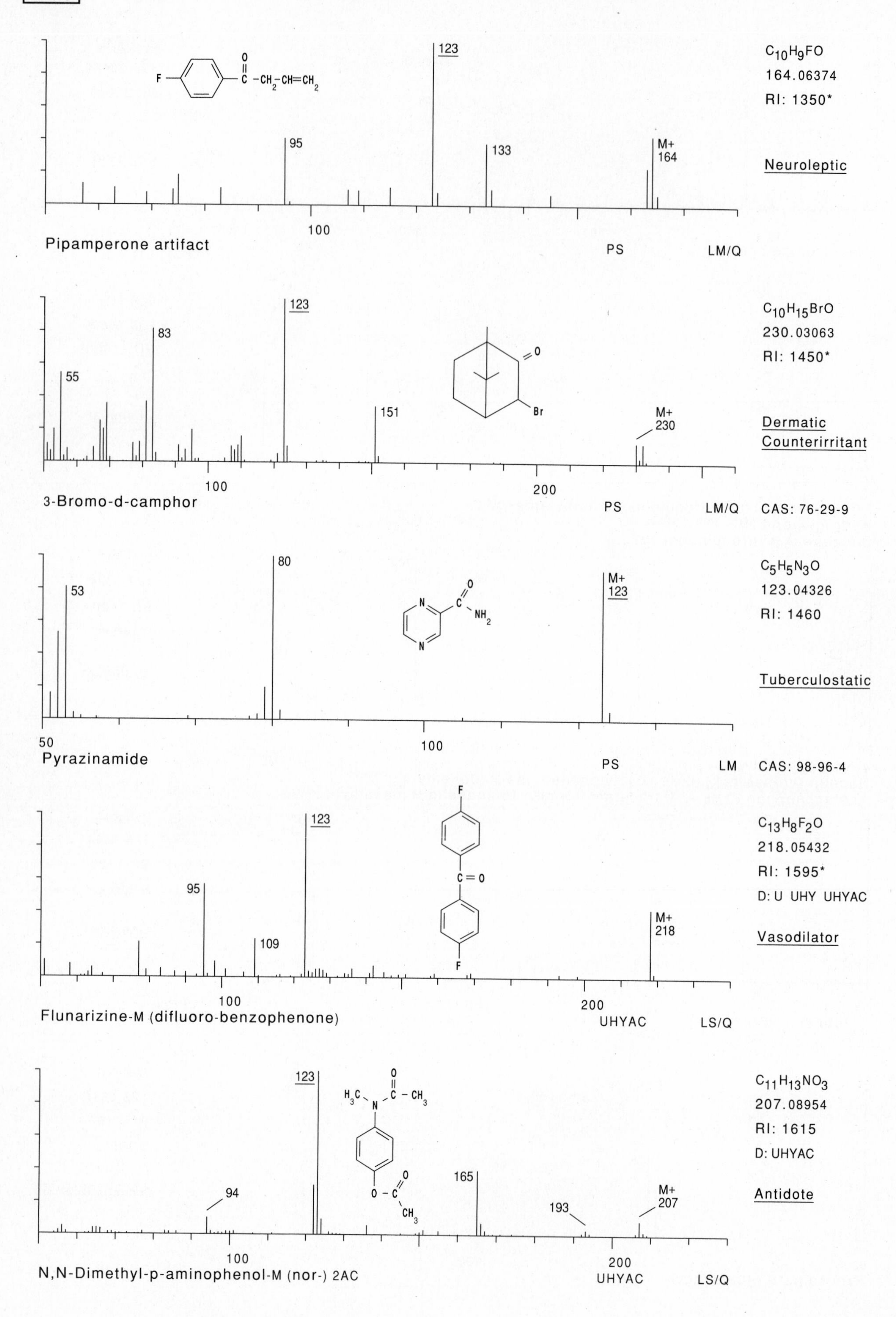

123
95
133
M+
164
100
Pipamperone artifact
PS
LM/Q
C10H9FO
164.06374
RI: 1350*
Neuroleptic
123
83
55
151
M+
230
100
200
3-Bromo-d-camphor
PS
LM/Q
C10H15BrO
230.03063
RI: 1450*
Dermatic
Counterirritant
CAS: 76-29-9
80
53
M+
123
50
100
Pyrazinamide
PS
LM
C5H5N3O
123.04326
RI: 1460
Tuberculostatic
CAS: 98-96-4
123
95
109
M+
218
100
200
Flunarizine-M (difluoro-benzophenone)
UHYAC
LS/Q
C13H8F2O
218.05432
RI: 1595*
D: U UHY UHYAC
Vasodilator
123
94
165
193
M+
207
100
200
N,N-Dimethyl-p-aminophenol-M (nor-) 2AC
UHYAC
LS/Q
C11H13NO3
207.08954
RI: 1615
D: UHYAC
Antidote

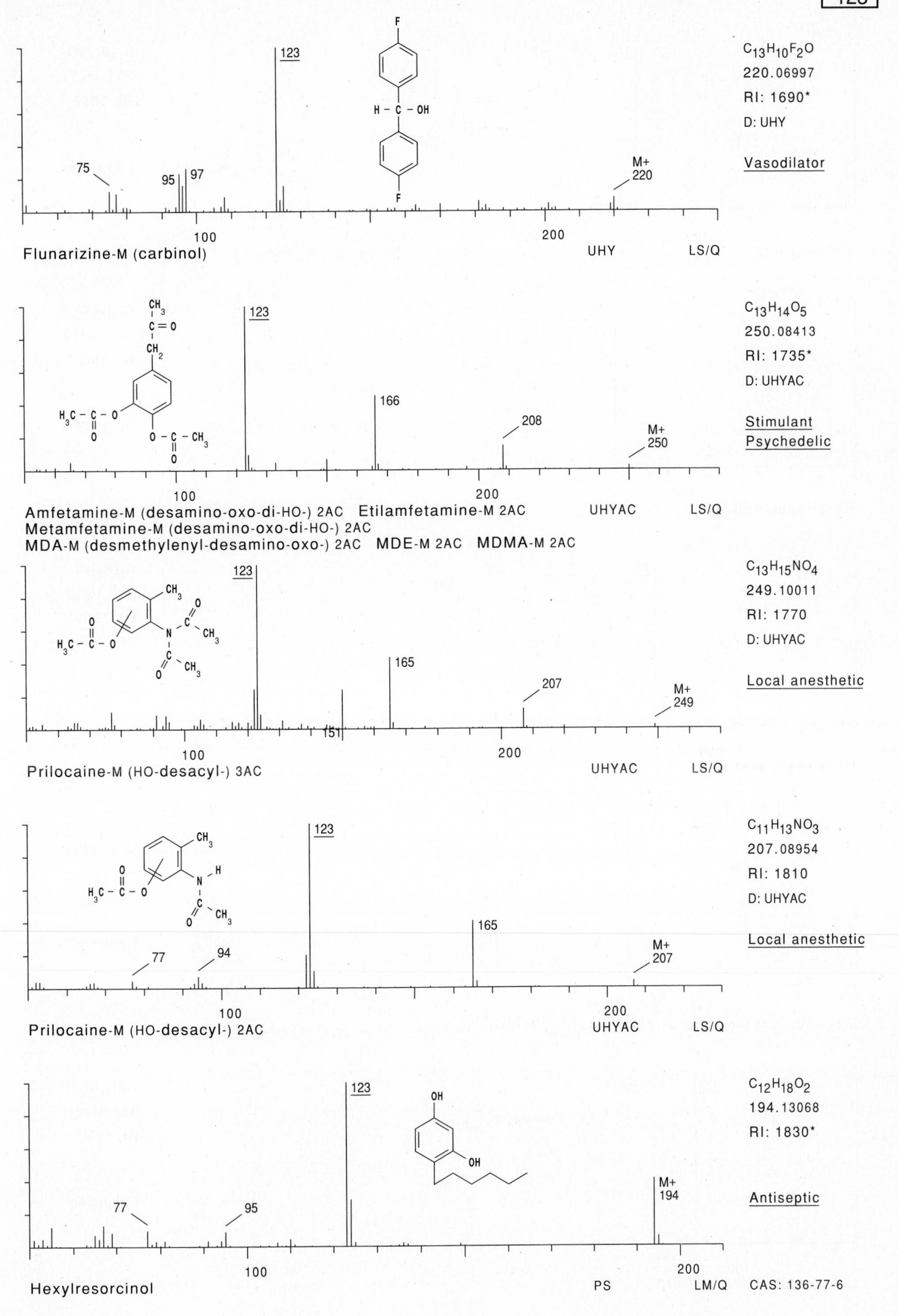

Flunarizine-M (carbinol)

$C_{13}H_{10}F_2O$
220.06997
RI: 1690*
D: UHY
Vasodilator

Amfetamine-M (desamino-oxo-di-HO-) 2AC Etilamfetamine-M 2AC
Metamfetamine-M (desamino-oxo-di-HO-) 2AC
MDA-M (desmethylenyl-desamino-oxo-) 2AC MDE-M 2AC MDMA-M 2AC

$C_{13}H_{14}O_5$
250.08413
RI: 1735*
D: UHYAC
Stimulant
Psychedelic

Prilocaine-M (HO-desacyl-) 3AC

$C_{13}H_{15}NO_4$
249.10011
RI: 1770
D: UHYAC
Local anesthetic

Prilocaine-M (HO-desacyl-) 2AC

$C_{11}H_{13}NO_3$
207.08954
RI: 1810
D: UHYAC
Local anesthetic

Hexylresorcinol

$C_{12}H_{18}O_2$
194.13068
RI: 1830*
Antiseptic
CAS: 136-77-6

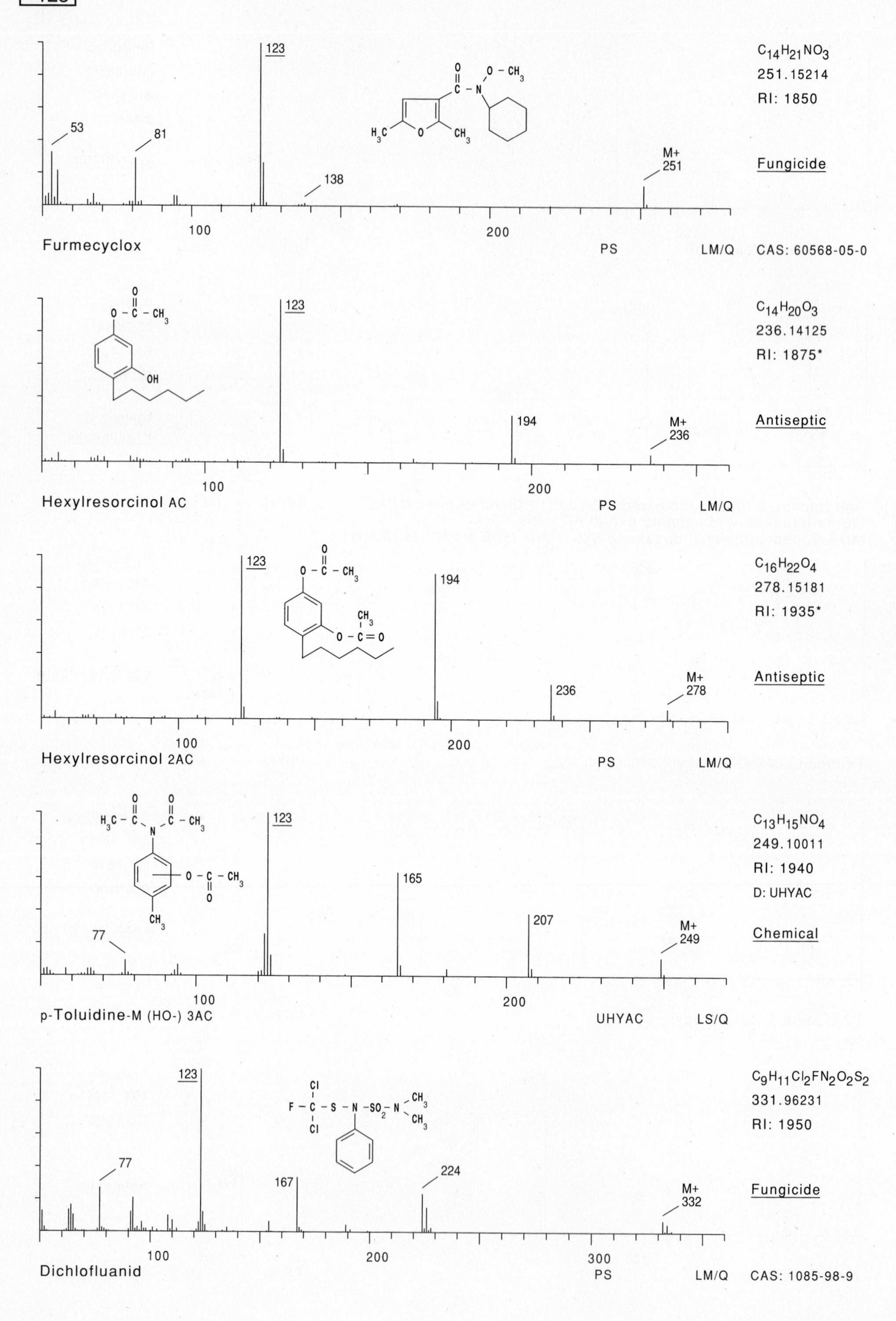

123
53
81
138
M+
251
100
200
Furmecyclox
PS
LM/Q
C14H21NO3
251.15214
RI: 1850
Fungicide
CAS: 60568-05-0
123
194
M+
236
Hexylresorcinol AC
C14H20O3
236.14125
RI: 1875*
Antiseptic
123
194
236
M+
278
Hexylresorcinol 2AC
C16H22O4
278.15181
RI: 1935*
Antiseptic
123
165
207
77
M+
249
p-Toluidine-M (HO-) 3AC
UHYAC
LS/Q
C13H15NO4
249.10011
RI: 1940
D: UHYAC
Chemical
123
77
167
224
M+
332
300
Dichlofluanid
C9H11Cl2FN2O2S2
331.96231
RI: 1950
Fungicide
CAS: 1085-98-9

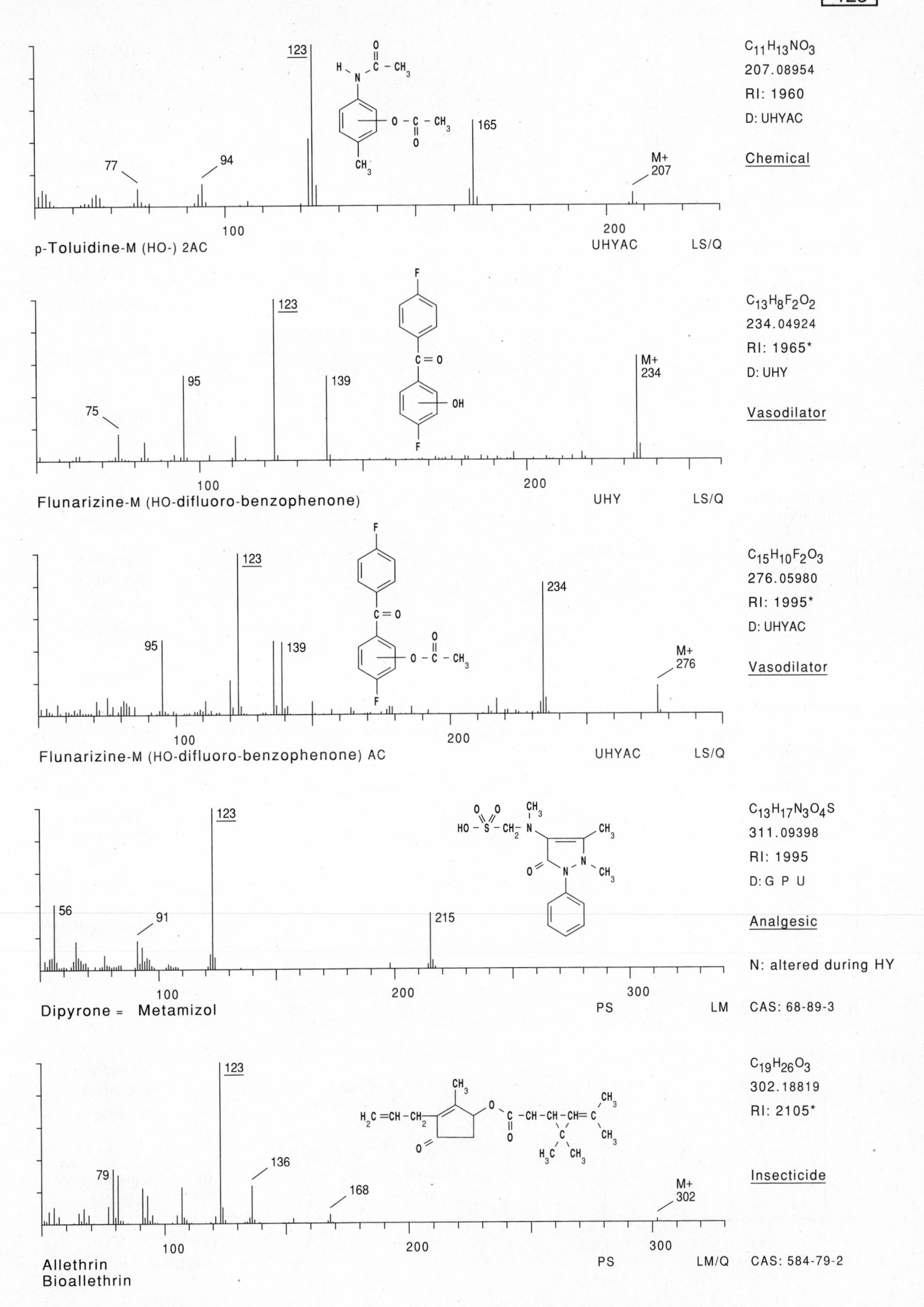

$C_{11}H_{13}NO_3$
207.08954
RI: 1960
D: UHYAC
Chemical

p-Toluidine-M (HO-) 2AC UHYAC LS/Q

$C_{13}H_8F_2O_2$
234.04924
RI: 1965*
D: UHY
Vasodilator

Flunarizine-M (HO-difluoro-benzophenone) UHY LS/Q

$C_{15}H_{10}F_2O_3$
276.05980
RI: 1995*
D: UHYAC
Vasodilator

Flunarizine-M (HO-difluoro-benzophenone) AC UHYAC LS/Q

$C_{13}H_{17}N_3O_4S$
311.09398
RI: 1995
D:G P U
Analgesic
N: altered during HY

Dipyrone = Metamizol PS LM CAS: 68-89-3

$C_{19}H_{26}O_3$
302.18819
RI: 2105*
Insecticide

Allethrin
Bioallethrin PS LM/Q CAS: 584-79-2

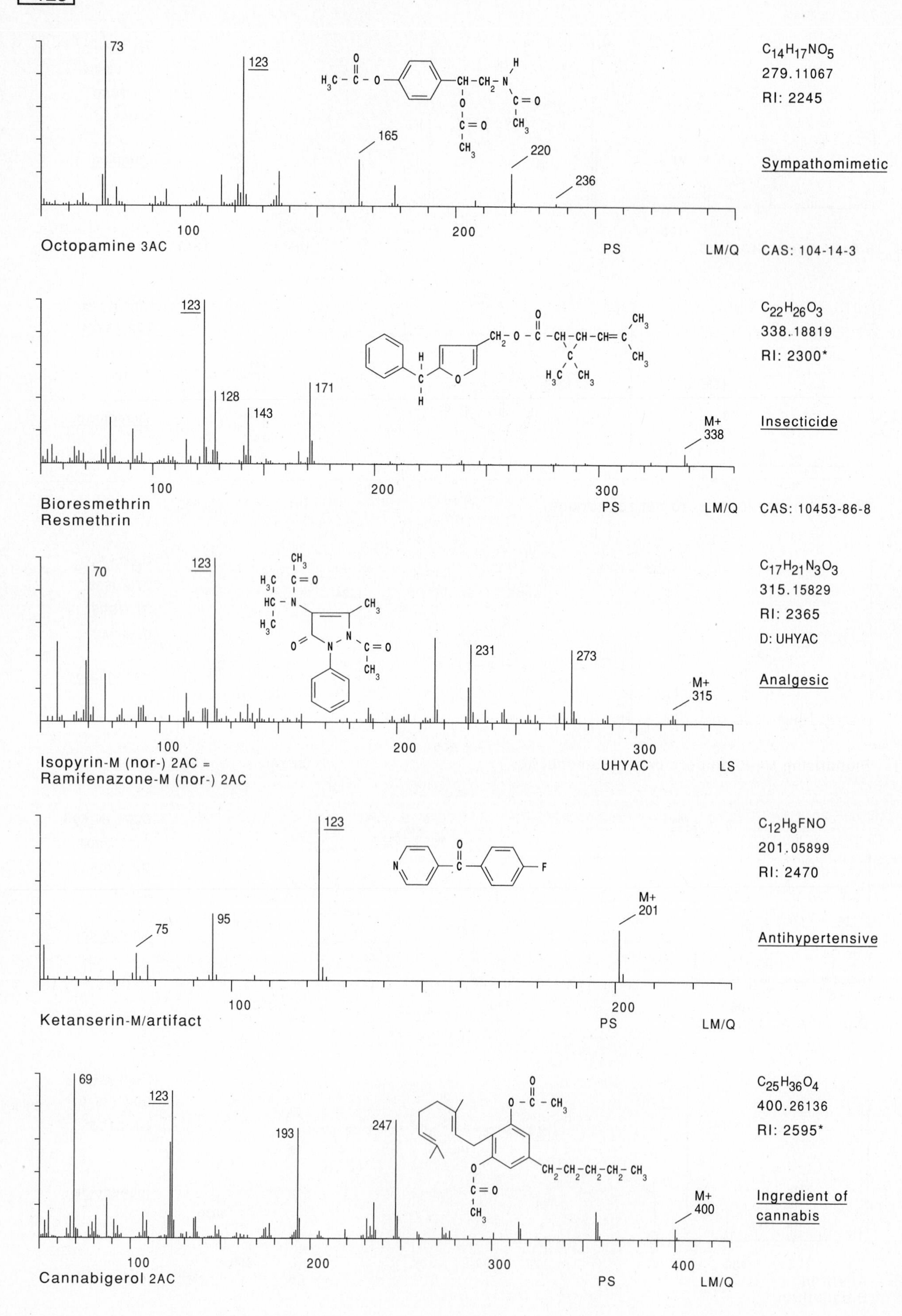
123
Octopamine 3AC
C14H17NO5
279.11067
RI: 2245
Sympathomimetic
CAS: 104-14-3
PS
LM/Q
Bioresmethrin
Resmethrin
C22H26O3
338.18819
RI: 2300*
Insecticide
CAS: 10453-86-8
PS
LM/Q
Isopyrin-M (nor-) 2AC =
Ramifenazone-M (nor-) 2AC
C17H21N3O3
315.15829
RI: 2365
D: UHYAC
Analgesic
UHYAC
LS
Ketanserin-M/artifact
C12H8FNO
201.05899
RI: 2470
Antihypertensive
PS
LM/Q
Cannabigerol 2AC
C25H36O4
400.26136
RI: 2595*
Ingredient of cannabis
PS
LM/Q

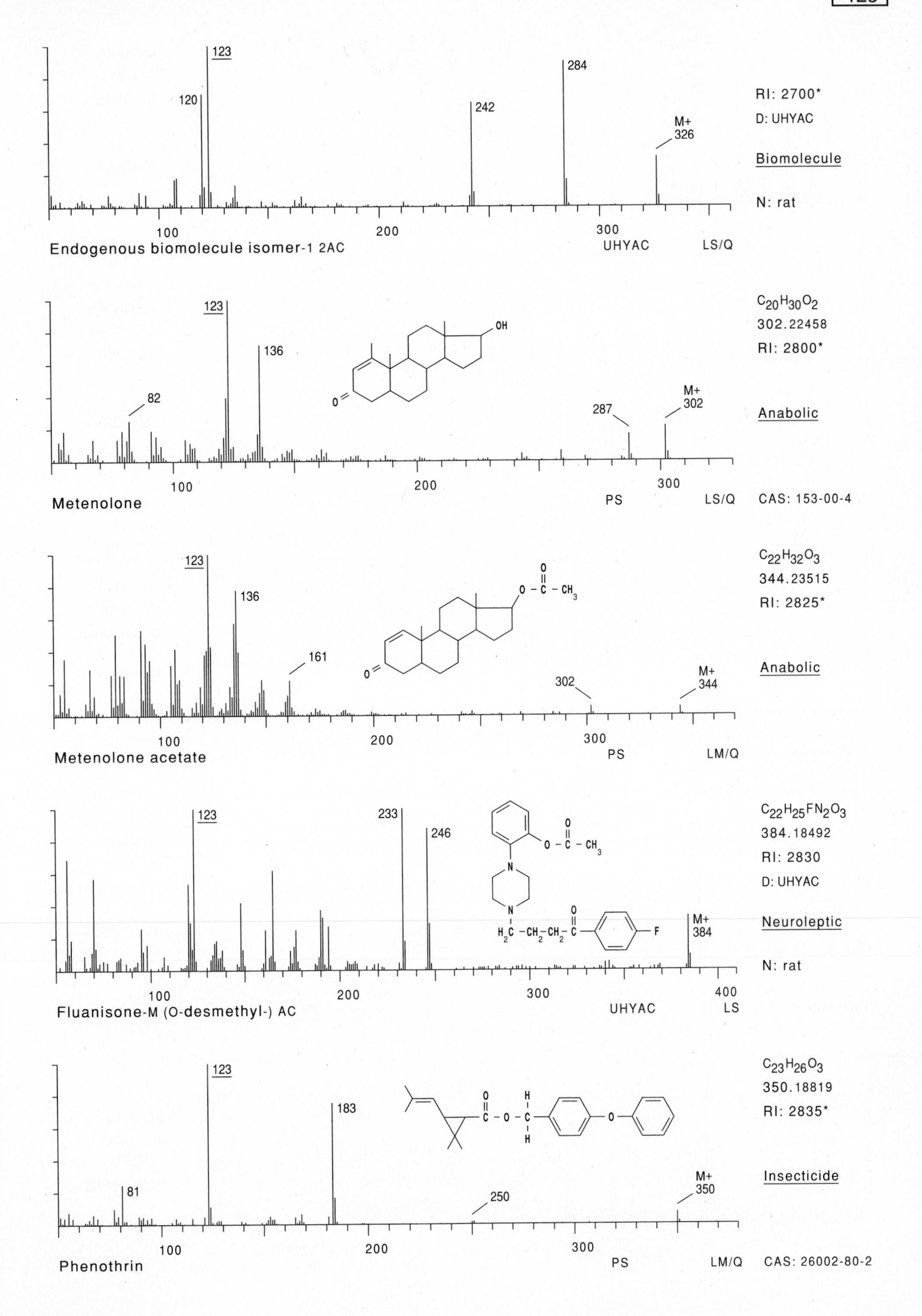

123
120
242
284
M+ 326
100
200
300
Endogenous biomolecule isomer-1 2AC
UHYAC
LS/Q
RI: 2700*
D: UHYAC
Biomolecule
N: rat
123
136
82
OH
O
287
M+ 302
100
200
300
Metenolone
PS
LS/Q
C20H30O2
302.22458
RI: 2800*
Anabolic
CAS: 153-00-4
123
136
161
O
O - C - CH3
302
M+ 344
100
200
300
Metenolone acetate
PS
LM/Q
C22H32O3
344.23515
RI: 2825*
Anabolic
123
233
246
O - C - CH3
N
H2C - CH2 - CH2 - C
F
M+ 384
100
200
300
400
Fluanisone-M (O-desmethyl-) AC
UHYAC
LS
C22H25FN2O3
384.18492
RI: 2830
D: UHYAC
Neuroleptic
N: rat
123
183
81
250
C - O - C
H
O
M+ 350
100
200
300
Phenothrin
PS
LM/Q
C23H26O3
350.18819
RI: 2835*
Insecticide
CAS: 26002-80-2

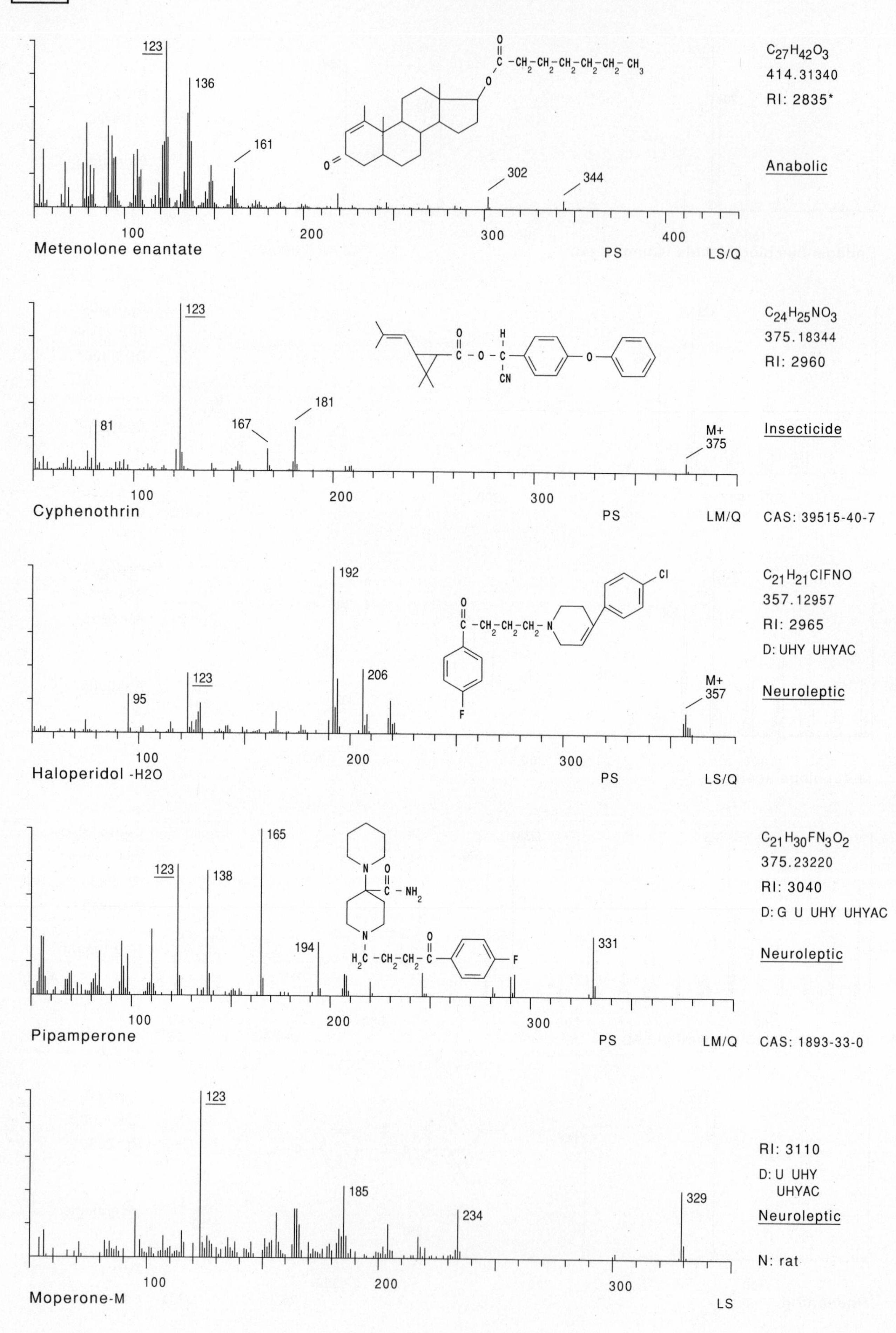

123
136
161
302
344
100
200
300
400
Metenolone enantate
PS
LS/Q
$C_{27}H_{42}O_3$
414.31340
RI: 2835*
Anabolic
123
81
167
181
M+
375
Cyphenothrin
PS
LM/Q
$C_{24}H_{25}NO_3$
375.18344
RI: 2960
Insecticide
CAS: 39515-40-7
192
123
95
206
M+
357
Haloperidol -H2O
PS
LS/Q
$C_{21}H_{21}ClFNO$
357.12957
RI: 2965
D: UHY UHYAC
Neuroleptic
165
123
138
194
331
Pipamperone
PS
LM/Q
$C_{21}H_{30}FN_3O_2$
375.23220
RI: 3040
D: G U UHY UHYAC
Neuroleptic
CAS: 1893-33-0
123
185
234
329
Moperone-M
LS
RI: 3110
D: U UHY
UHYAC
Neuroleptic
N: rat

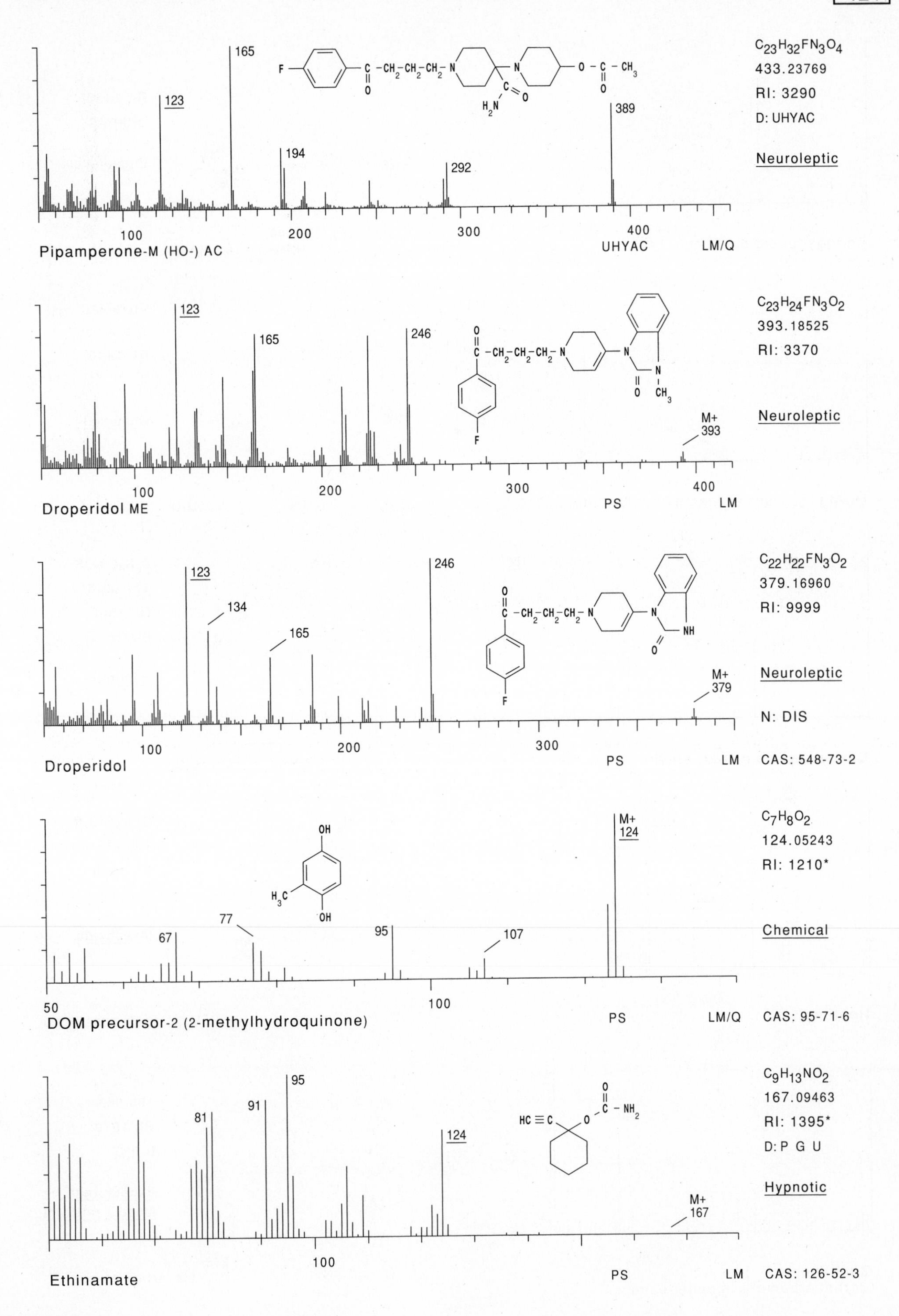
Pipamperone-M (HO-) AC
UHYAC
LM/Q
C23H32FN3O4
433.23769
RI: 3290
D: UHYAC
Neuroleptic
Droperidol ME
PS
LM
C23H24FN3O2
393.18525
RI: 3370
Neuroleptic
Droperidol
PS
LM
C22H22FN3O2
379.16960
RI: 9999
Neuroleptic
N: DIS
CAS: 548-73-2
DOM precursor-2 (2-methylhydroquinone)
PS
LM/Q
C7H8O2
124.05243
RI: 1210*
Chemical
CAS: 95-71-6
Ethinamate
PS
LM
C9H13NO2
167.09463
RI: 1395*
D: P G U
Hypnotic
CAS: 126-52-3

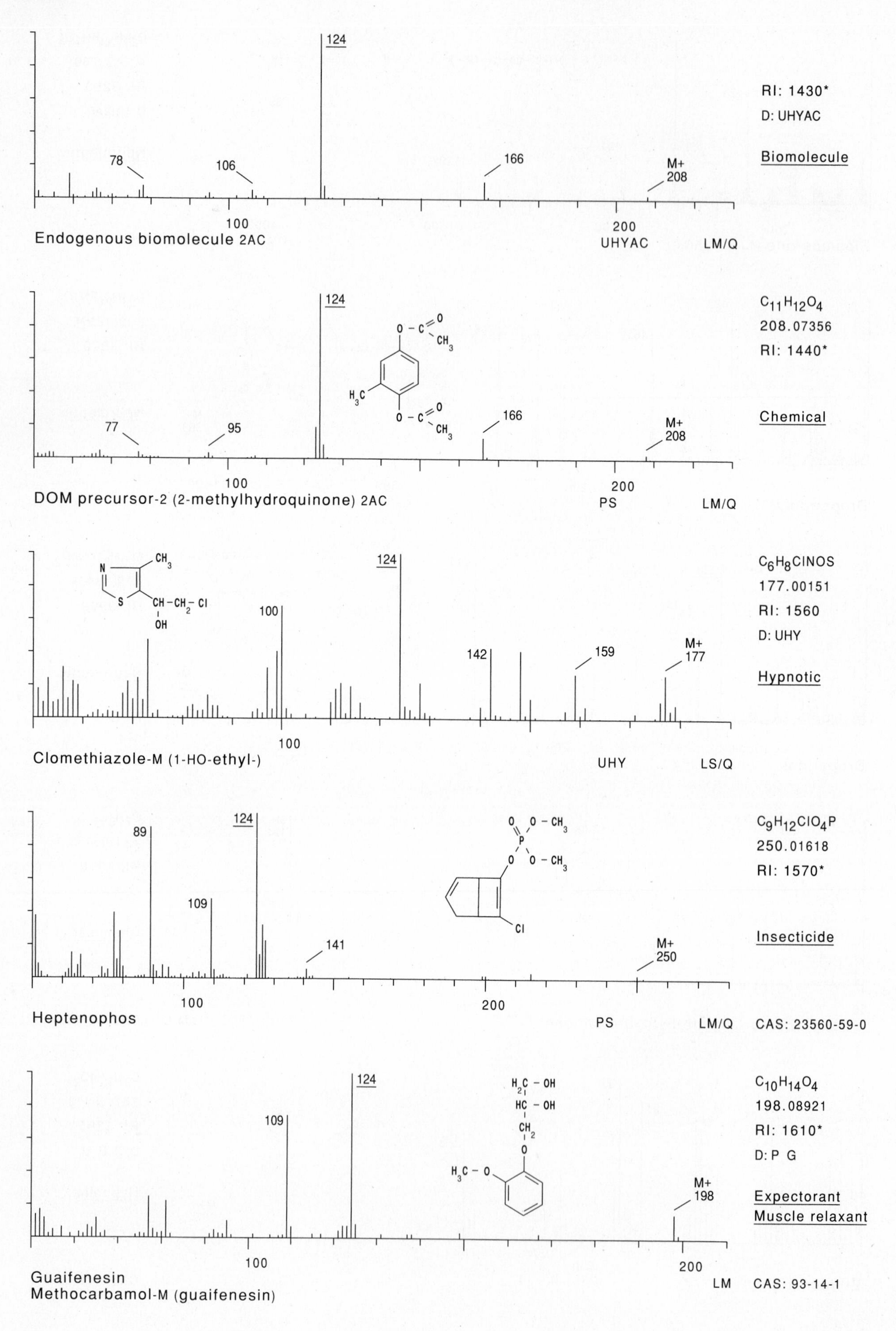

124
78
106
166
M+
208
100
200
RI: 1430*
D: UHYAC
Biomolecule
Endogenous biomolecule 2AC
UHYAC
LM/Q
124
77
95
166
M+
208
100
200
C11H12O4
208.07356
RI: 1440*
Chemical
DOM precursor-2 (2-methylhydroquinone) 2AC
PS
LM/Q
124
100
142
159
M+
177
100
C6H8ClNOS
177.00151
RI: 1560
D: UHY
Hypnotic
Clomethiazole-M (1-HO-ethyl-)
UHY
LS/Q
89
124
109
141
M+
250
100
200
C9H12ClO4P
250.01618
RI: 1570*
Insecticide
Heptenophos
PS
LM/Q
CAS: 23560-59-0
124
109
M+
198
100
200
C10H14O4
198.08921
RI: 1610*
D: P G
Expectorant
Muscle relaxant
Guaifenesin
Methocarbamol-M (guaifenesin)
LM
CAS: 93-14-1

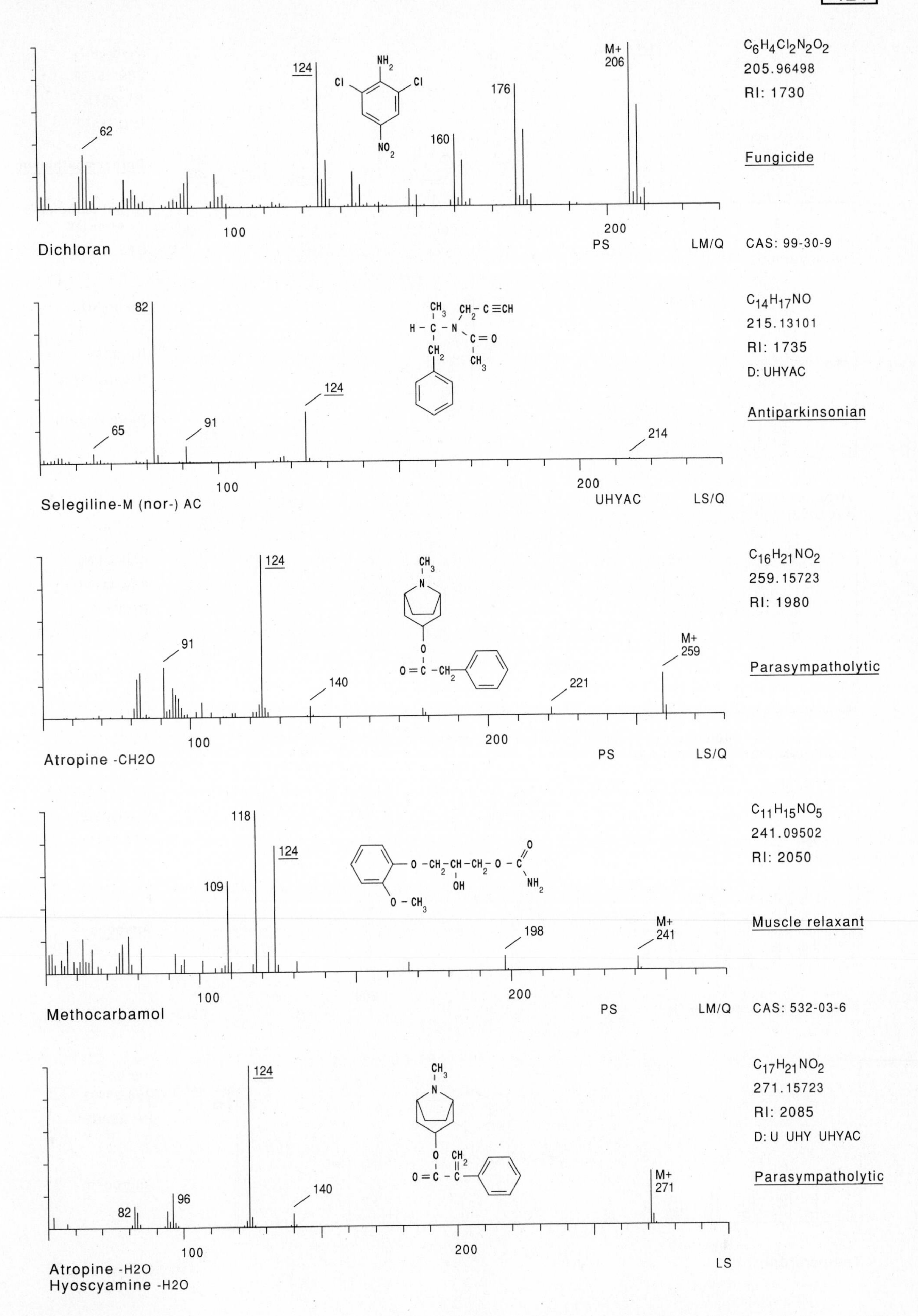

Dichloran
$C_6H_4Cl_2N_2O_2$
205.96498
RI: 1730
Fungicide
CAS: 99-30-9
PS
LM/Q
Selegiline-M (nor-) AC
$C_{14}H_{17}NO$
215.13101
RI: 1735
D: UHYAC
Antiparkinsonian
UHYAC
LS/Q
Atropine -CH2O
$C_{16}H_{21}NO_2$
259.15723
RI: 1980
Parasympatholytic
PS
LS/Q
Methocarbamol
$C_{11}H_{15}NO_5$
241.09502
RI: 2050
Muscle relaxant
PS
LM/Q
CAS: 532-03-6
Atropine -H2O
Hyoscyamine -H2O
$C_{17}H_{21}NO_2$
271.15723
RI: 2085
D: U UHY UHYAC
Parasympatholytic
LS

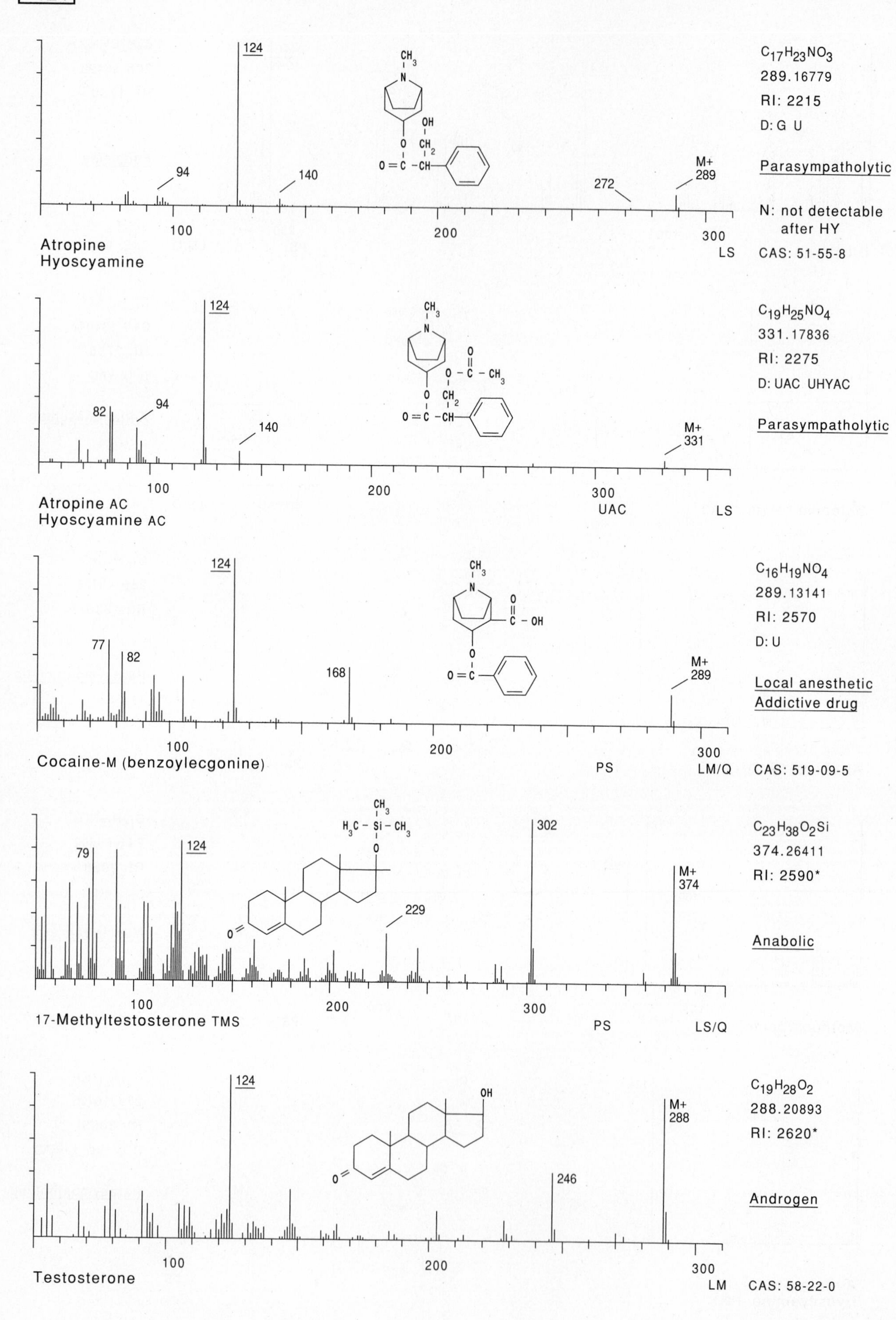
124
94
140
272
M+
289
100
200
300
LS
Atropine
Hyoscyamine
$C_{17}H_{23}NO_3$
289.16779
RI: 2215
D: G U
Parasympatholytic
N: not detectable after HY
CAS: 51-55-8
124
82
94
140
M+
331
100
200
300
UAC
LS
Atropine AC
Hyoscyamine AC
$C_{19}H_{25}NO_4$
331.17836
RI: 2275
D: UAC UHYAC
Parasympatholytic
124
77
82
168
M+
289
100
200
300
PS
LM/Q
Cocaine-M (benzoylecgonine)
$C_{16}H_{19}NO_4$
289.13141
RI: 2570
D: U
Local anesthetic
Addictive drug
CAS: 519-09-5
79
124
229
302
M+
374
100
200
300
PS
LS/Q
17-Methyltestosterone TMS
$C_{23}H_{38}O_2Si$
374.26411
RI: 2590*
Anabolic
124
246
M+
288
100
200
300
LM
Testosterone
$C_{19}H_{28}O_2$
288.20893
RI: 2620*
Androgen
CAS: 58-22-0

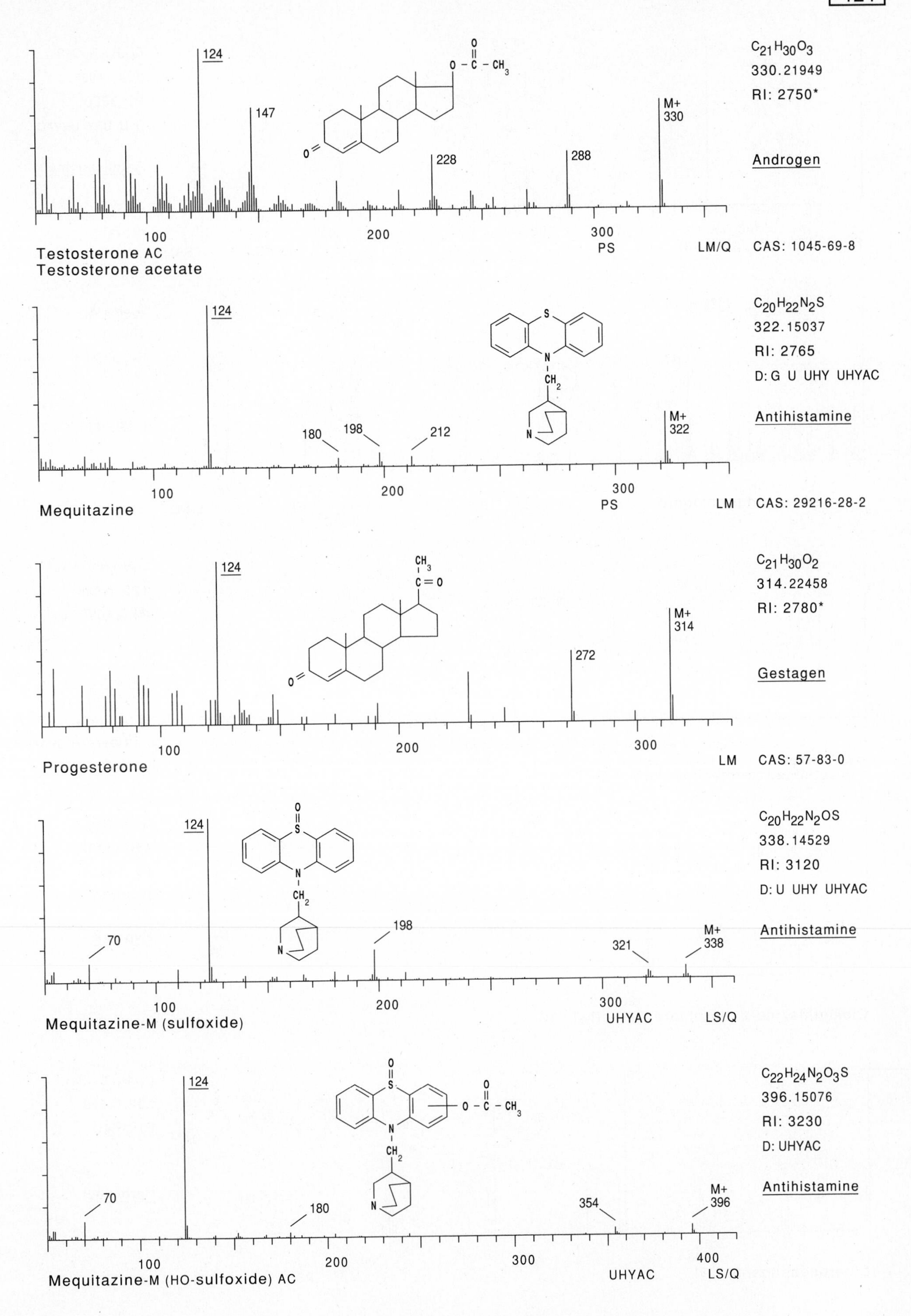

$C_{21}H_{30}O_3$
330.21949
RI: 2750*

Androgen

CAS: 1045-69-8

$C_{20}H_{22}N_2S$
322.15037
RI: 2765
D: G U UHY UHYAC

Antihistamine

CAS: 29216-28-2

$C_{21}H_{30}O_2$
314.22458
RI: 2780*

Gestagen

CAS: 57-83-0

$C_{20}H_{22}N_2OS$
338.14529
RI: 3120
D: U UHY UHYAC

Antihistamine

$C_{22}H_{24}N_2O_3S$
396.15076
RI: 3230
D: UHYAC

Antihistamine

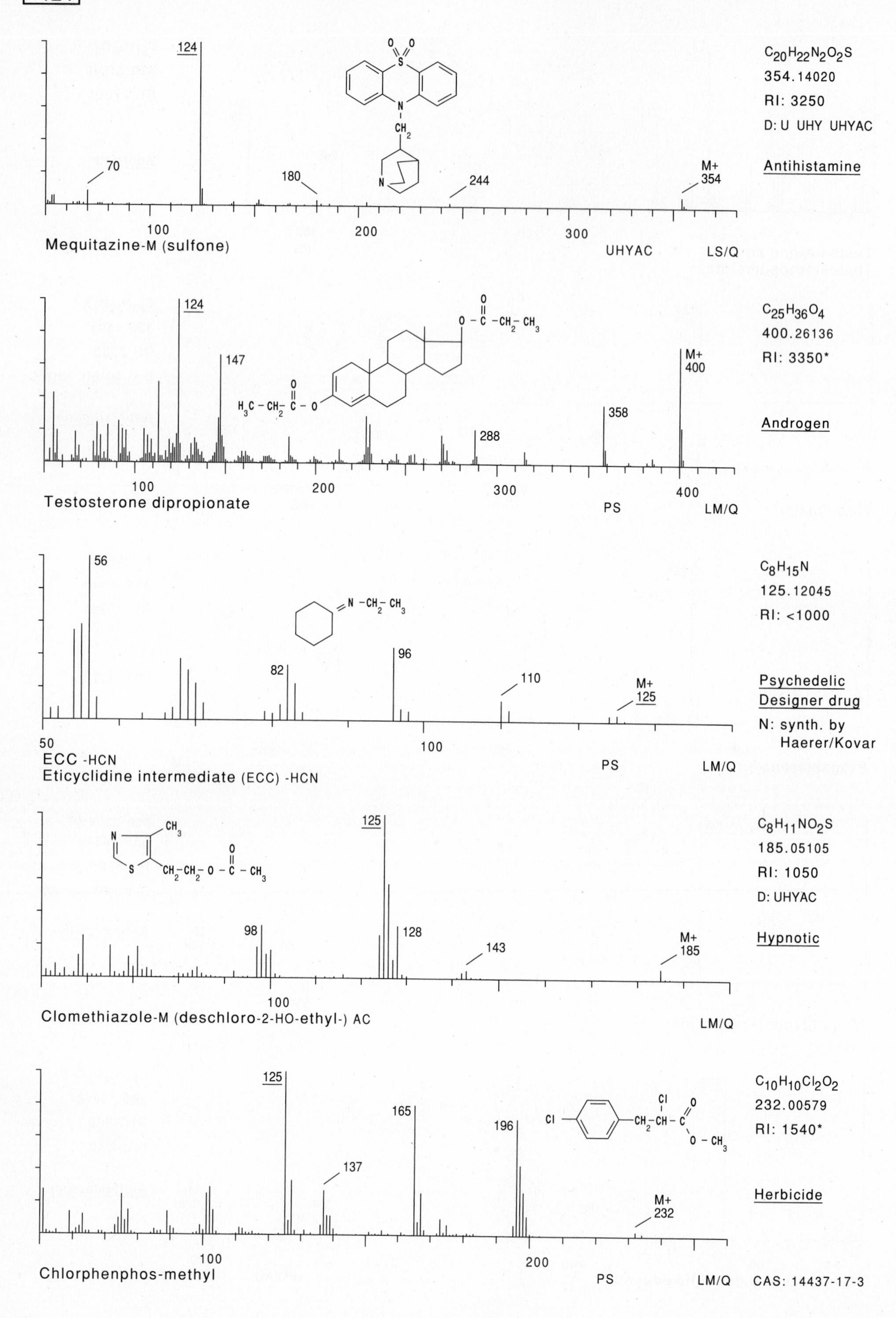

124
70
180
244
M+
354
100
200
300
Mequitazine-M (sulfone)
UHYAC
LS/Q
$C_{20}H_{22}N_2O_2S$
354.14020
RI: 3250
D: U UHY UHYAC
Antihistamine
124
147
M+
400
358
288
100
200
300
400
Testosterone dipropionate
PS
LM/Q
$C_{25}H_{36}O_4$
400.26136
RI: 3350*
Androgen
56
96
82
110
M+
125
50
100
ECC -HCN
Eticyclidine intermediate (ECC) -HCN
PS
LM/Q
$C_8H_{15}N$
125.12045
RI: <1000
Psychedelic
Designer drug
N: synth. by
Haerer/Kovar
125
98
128
143
M+
185
100
Clomethiazole-M (deschloro-2-HO-ethyl-) AC
LM/Q
$C_8H_{11}NO_2S$
185.05105
RI: 1050
D: UHYAC
Hypnotic
125
165
196
137
M+
232
100
200
Chlorphenphos-methyl
PS
LM/Q
CAS: 14437-17-3
$C_{10}H_{10}Cl_2O_2$
232.00579
RI: 1540*
Herbicide

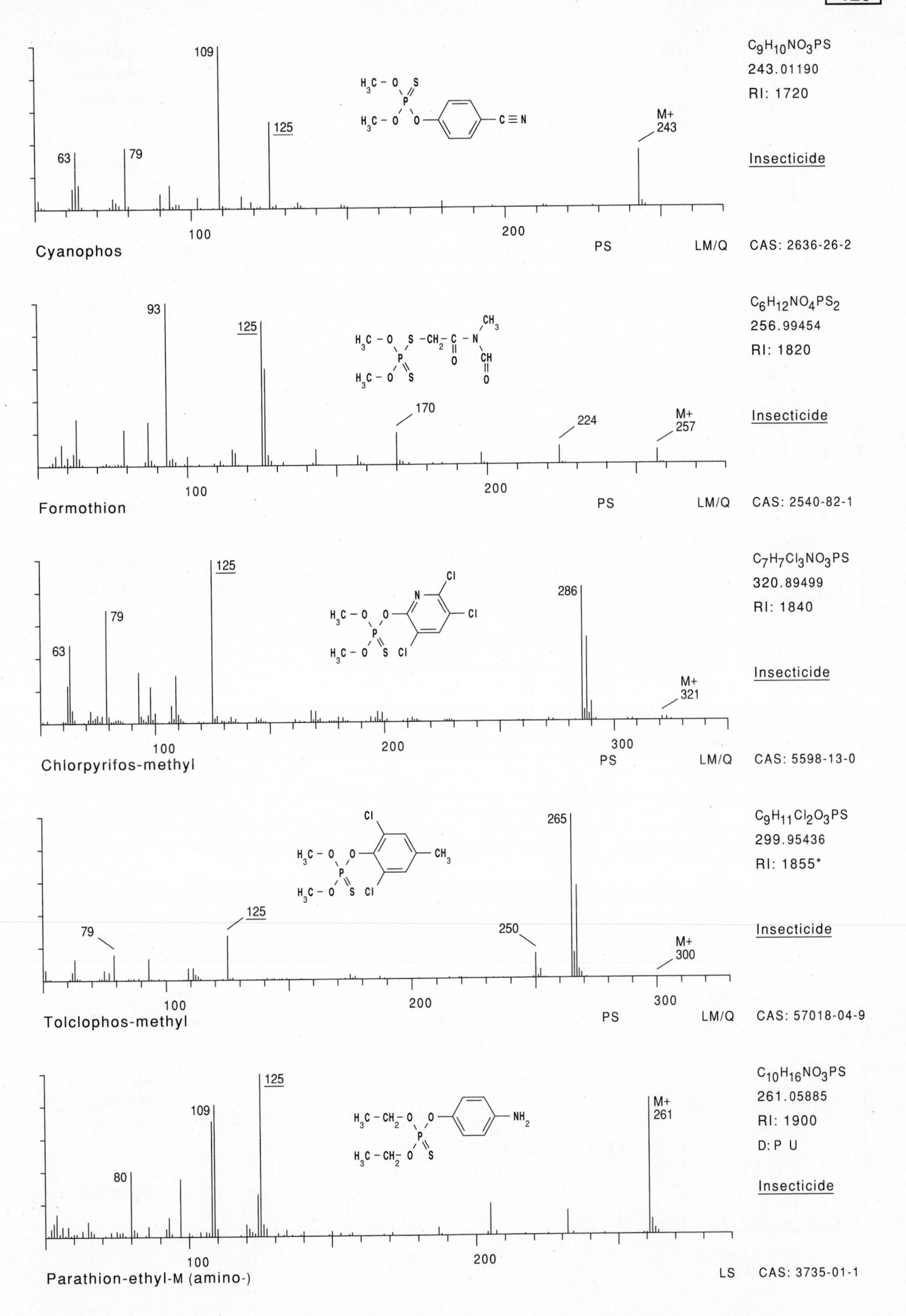

$C_9H_{10}NO_3PS$
243.01190
RI: 1720

Insecticide

CAS: 2636-26-2

$C_6H_{12}NO_4PS_2$
256.99454
RI: 1820

Insecticide

CAS: 2540-82-1

$C_7H_7Cl_3NO_3PS$
320.89499
RI: 1840

Insecticide

CAS: 5598-13-0

$C_9H_{11}Cl_2O_3PS$
299.95436
RI: 1855*

Insecticide

CAS: 57018-04-9

$C_{10}H_{16}NO_3PS$
261.05885
RI: 1900
D: P U

Insecticide

CAS: 3735-01-1

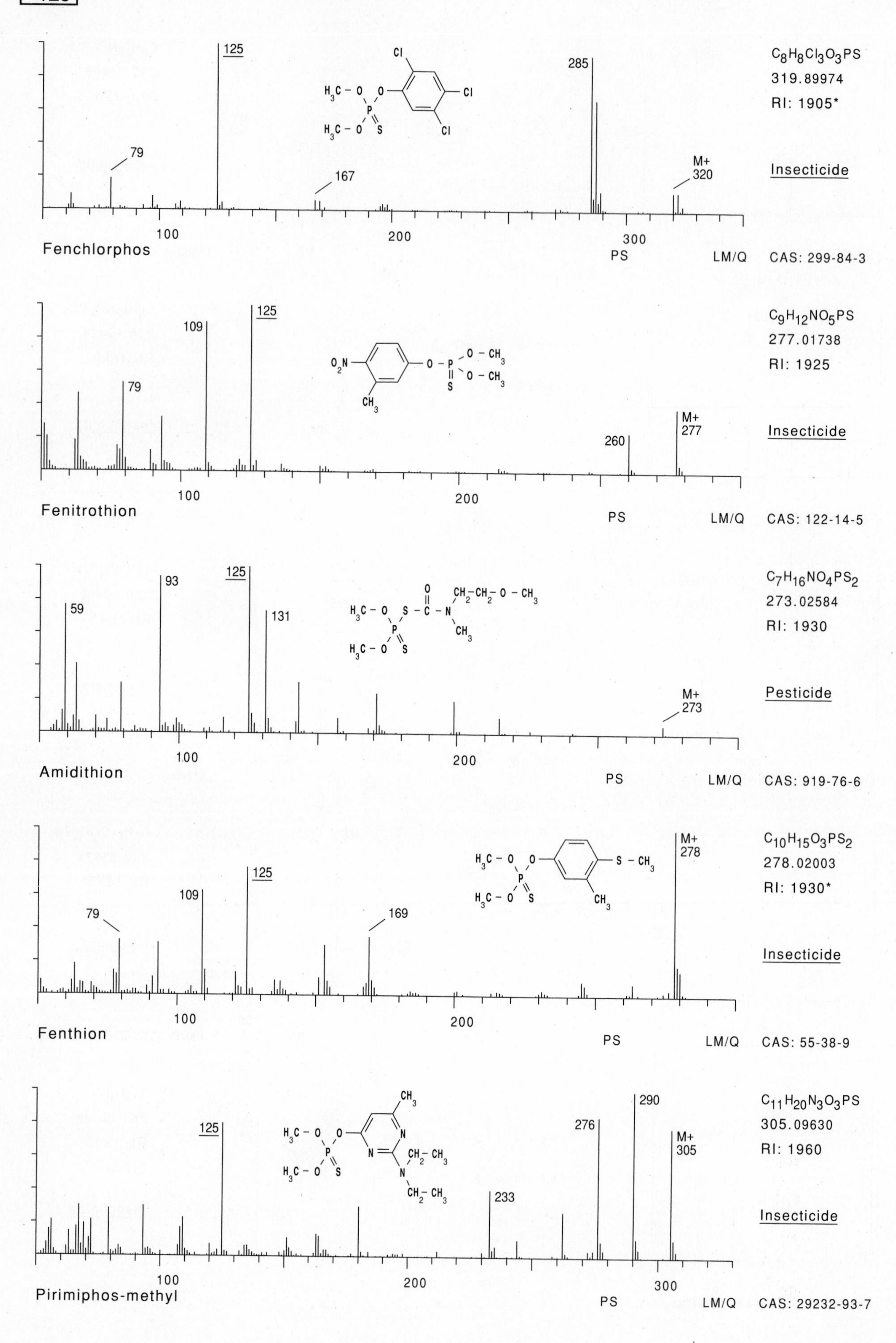
125
79
167
285
M+
320
100
200
300
Fenchlorphos
PS
LM/Q
$C_8H_8Cl_3O_3PS$
319.89974
RI: 1905*
Insecticide
CAS: 299-84-3
109
125
79
260
M+
277
100
200
Fenitrothion
PS
LM/Q
$C_9H_{12}NO_5PS$
277.01738
RI: 1925
Insecticide
CAS: 122-14-5
93
125
59
131
M+
273
100
200
Amidithion
PS
LM/Q
$C_7H_{16}NO_4PS_2$
273.02584
RI: 1930
Pesticide
CAS: 919-76-6
M+
278
125
109
79
169
100
200
Fenthion
PS
LM/Q
$C_{10}H_{15}O_3PS_2$
278.02003
RI: 1930*
Insecticide
CAS: 55-38-9
290
125
276
M+
305
233
100
200
300
Pirimiphos-methyl
PS
LM/Q
$C_{11}H_{20}N_3O_3PS$
305.09630
RI: 1960
Insecticide
CAS: 29232-93-7

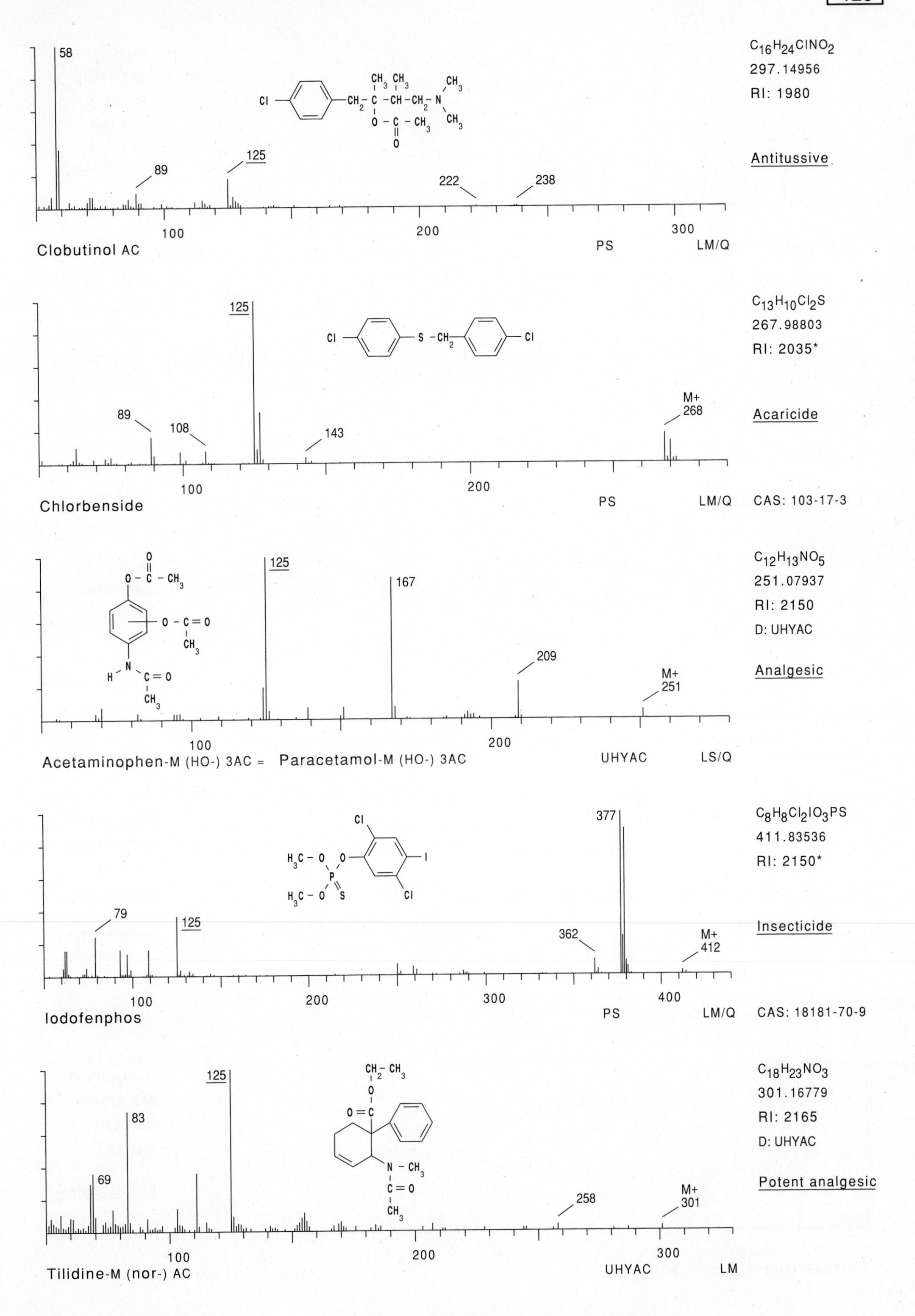

Clobutinol AC

$C_{16}H_{24}ClNO_2$
297.14956
RI: 1980

Antitussive

Chlorbenside

$C_{13}H_{10}Cl_2S$
267.98803
RI: 2035*

Acaricide

CAS: 103-17-3

Acetaminophen-M (HO-) 3AC = Paracetamol-M (HO-) 3AC

$C_{12}H_{13}NO_5$
251.07937
RI: 2150
D: UHYAC

Analgesic

Iodofenphos

$C_8H_8Cl_2IO_3PS$
411.83536
RI: 2150*

Insecticide

CAS: 18181-70-9

Tilidine-M (nor-) AC

$C_{18}H_{23}NO_3$
301.16779
RI: 2165
D: UHYAC

Potent analgesic

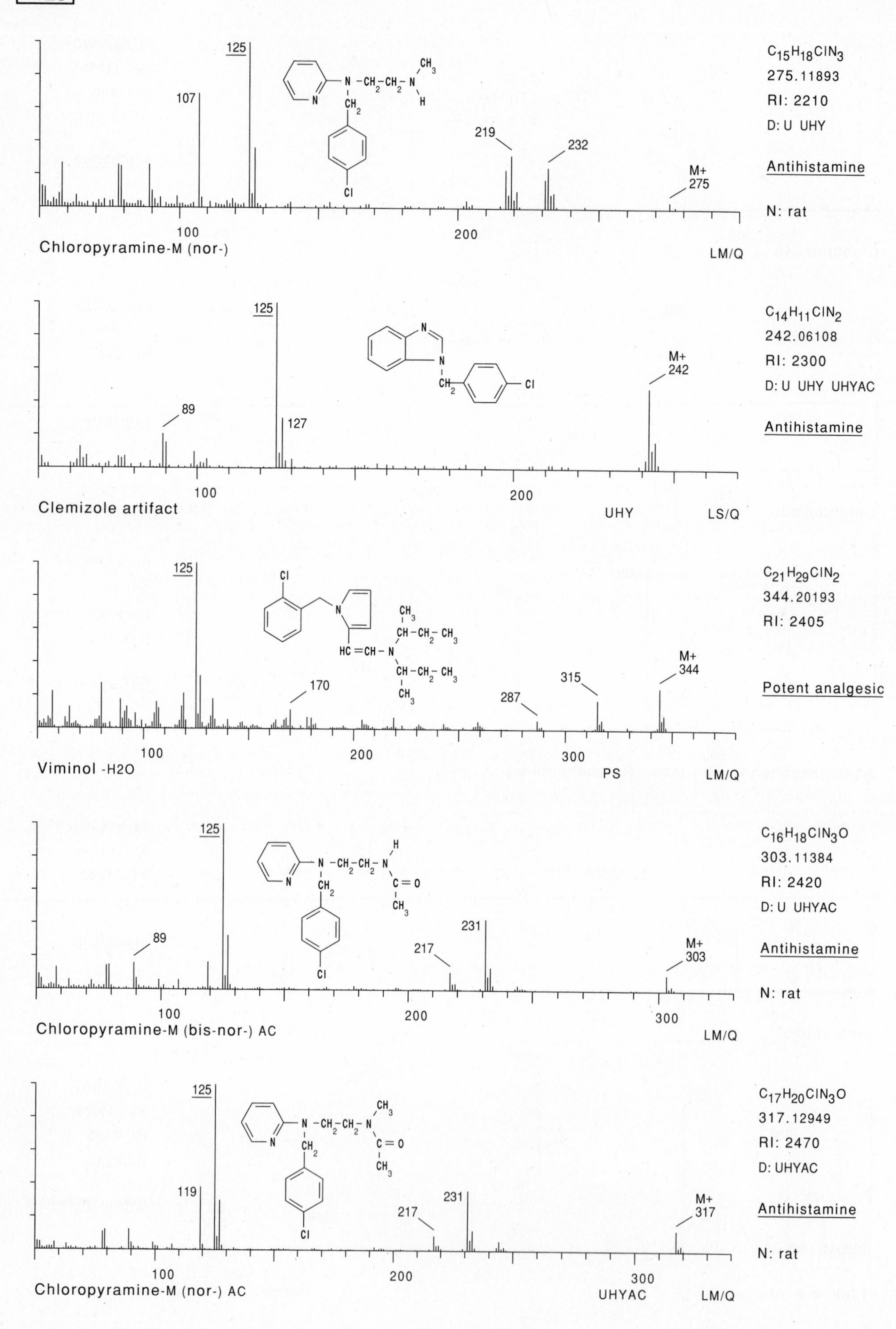

Chloropyramine-M (nor-)

$C_{15}H_{18}ClN_3$
275.11893
RI: 2210
D: U UHY
Antihistamine
N: rat

Clemizole artifact

$C_{14}H_{11}ClN_2$
242.06108
RI: 2300
D: U UHY UHYAC
Antihistamine

Viminol -H2O

$C_{21}H_{29}ClN_2$
344.20193
RI: 2405
Potent analgesic

Chloropyramine-M (bis-nor-) AC

$C_{16}H_{18}ClN_3O$
303.11384
RI: 2420
D: U UHYAC
Antihistamine
N: rat

Chloropyramine-M (nor-) AC

$C_{17}H_{20}ClN_3O$
317.12949
RI: 2470
D: UHYAC
Antihistamine
N: rat

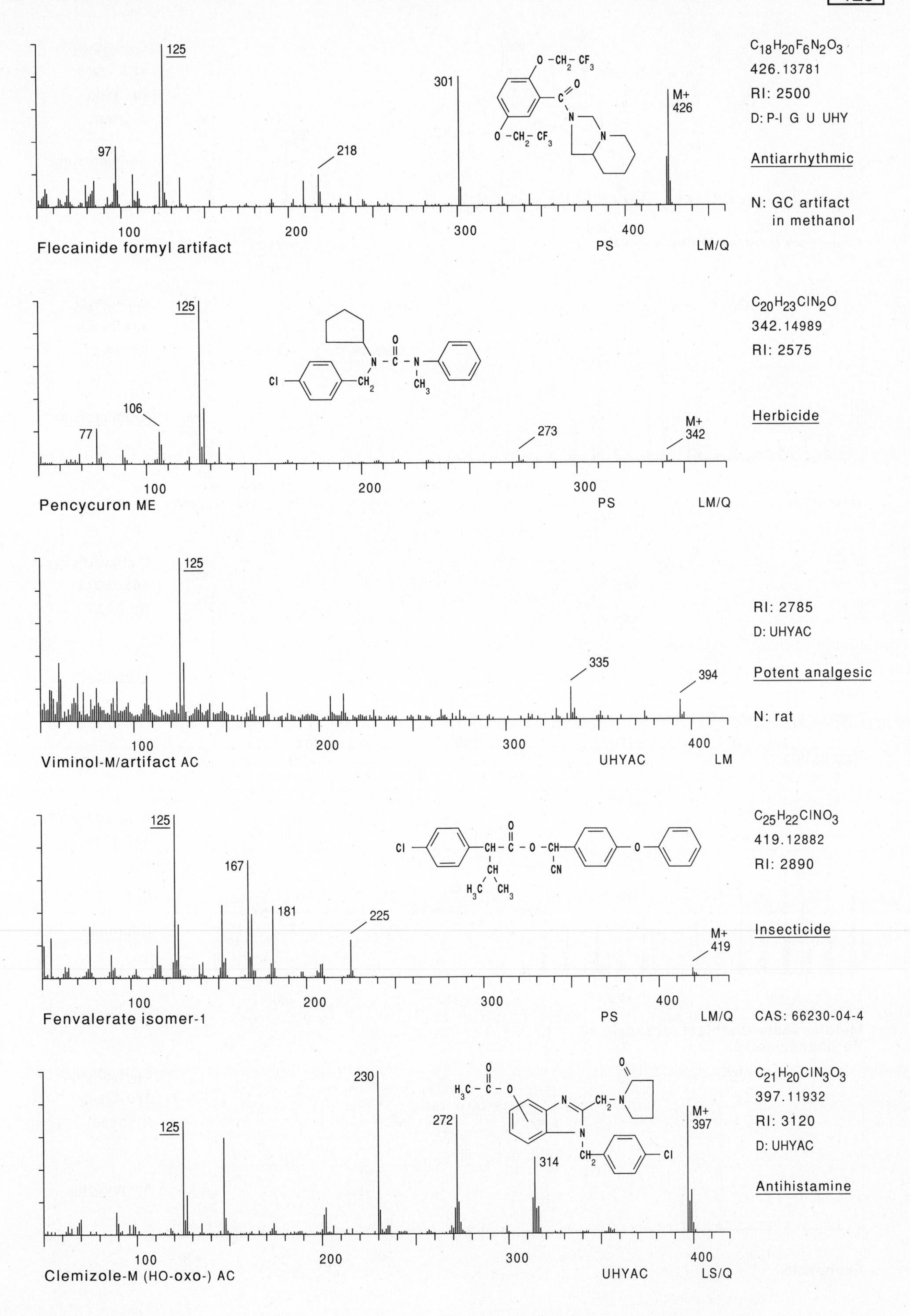

125
301
218
97
M+
426
100
200
300
400
Flecainide formyl artifact
PS
LM/Q
$C_{18}H_{20}F_6N_2O_3$
426.13781
RI: 2500
D: P-I G U UHY
Antiarrhythmic
N: GC artifact in methanol
125
106
77
273
M+
342
100
200
300
Pencycuron ME
PS
LM/Q
$C_{20}H_{23}ClN_2O$
342.14989
RI: 2575
Herbicide
125
335
394
100
200
300
400
Viminol-M/artifact AC
UHYAC
LM
RI: 2785
D: UHYAC
Potent analgesic
N: rat
125
167
181
225
M+
419
100
200
300
400
Fenvalerate isomer-1
PS
LM/Q
$C_{25}H_{22}ClNO_3$
419.12882
RI: 2890
Insecticide
CAS: 66230-04-4
230
272
125
314
M+
397
100
200
300
400
Clemizole-M (HO-oxo-) AC
UHYAC
LS/Q
$C_{21}H_{20}ClN_3O_3$
397.11932
RI: 3120
D: UHYAC
Antihistamine

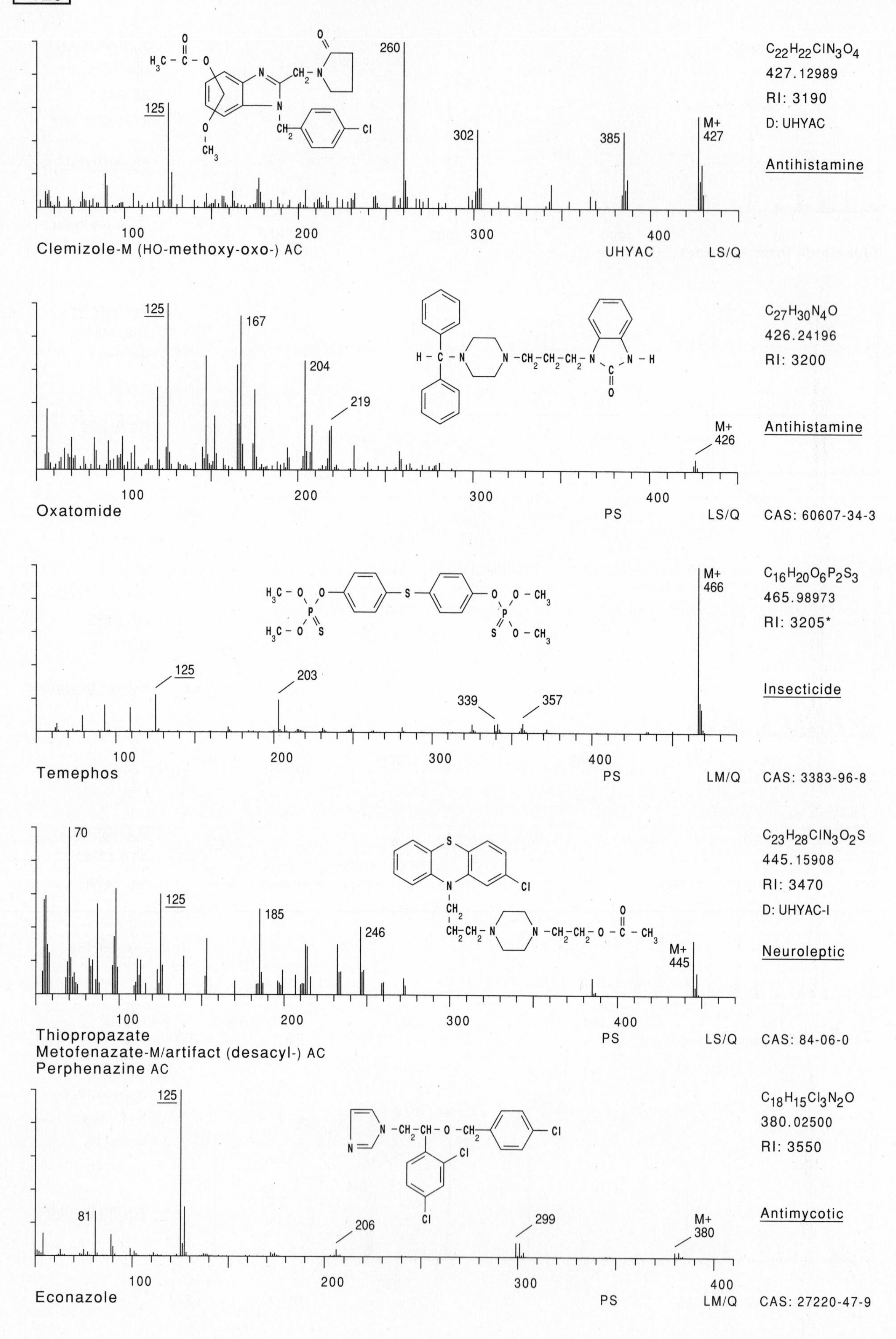
125
260
302
385
M+
427
100
200
300
400
Clemizole-M (HO-methoxy-oxo-) AC
UHYAC
LS/Q
C22H22ClN3O4
427.12989
RI: 3190
D: UHYAC
Antihistamine
125
167
204
219
M+
426
Oxatomide
PS
LS/Q
C27H30N4O
426.24196
RI: 3200
Antihistamine
CAS: 60607-34-3
M+
466
125
203
339
357
Temephos
PS
LM/Q
C16H20O6P2S3
465.98973
RI: 3205*
Insecticide
CAS: 3383-96-8
70
125
185
246
M+
445
Thiopropazate
Metofenazate-M/artifact (desacyl-) AC
Perphenazine AC
PS
LS/Q
C23H28ClN3O2S
445.15908
RI: 3470
D: UHYAC-I
Neuroleptic
CAS: 84-06-0
125
81
206
299
M+
380
Econazole
PS
LM/Q
C18H15Cl3N2O
380.02500
RI: 3550
Antimycotic
CAS: 27220-47-9

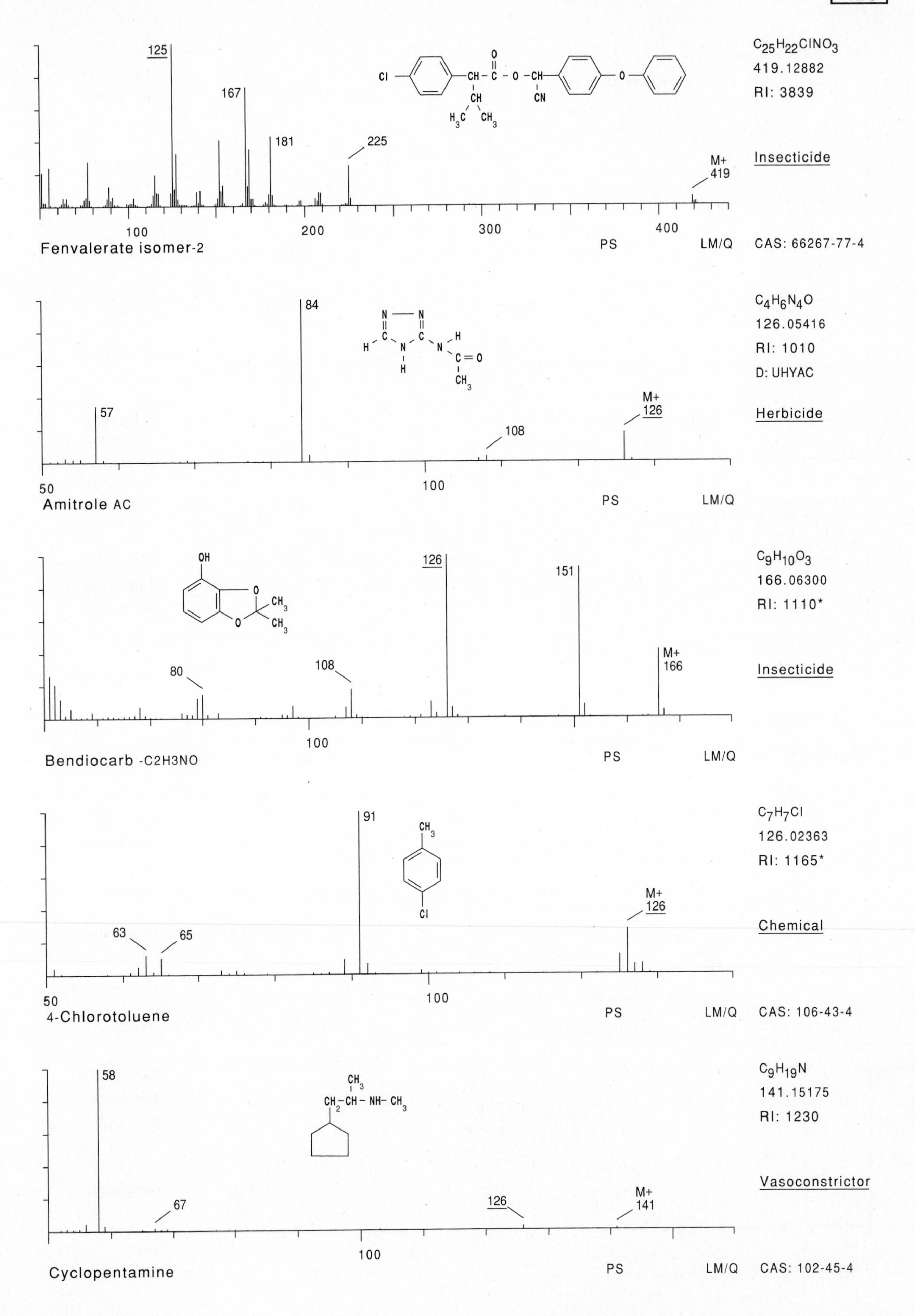

$C_{25}H_{22}ClNO_3$
419.12882
RI: 3839

Insecticide

CAS: 66267-77-4

$C_4H_6N_4O$
126.05416
RI: 1010
D: UHYAC

Herbicide

$C_9H_{10}O_3$
166.06300
RI: 1110*

Insecticide

C_7H_7Cl
126.02363
RI: 1165*

Chemical

CAS: 106-43-4

$C_9H_{19}N$
141.15175
RI: 1230

Vasoconstrictor

CAS: 102-45-4

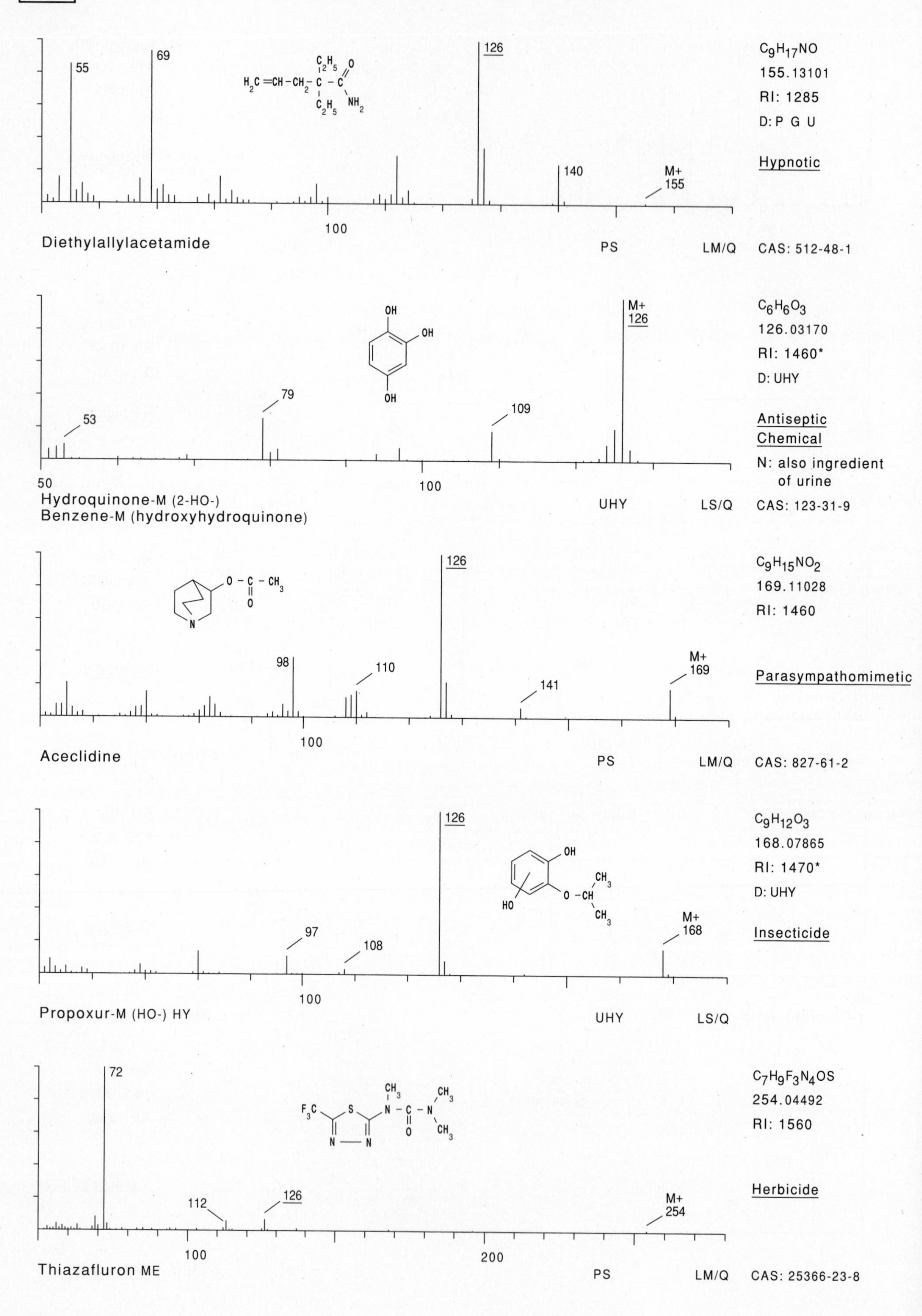
55
69
126
140
M+
155
100
Diethylallylacetamide
PS
LM/Q
C9H17NO
155.13101
RI: 1285
D: P G U
Hypnotic
CAS: 512-48-1
M+
126
53
79
109
50
100
Hydroquinone-M (2-HO-)
Benzene-M (hydroxyhydroquinone)
UHY
LS/Q
C6H6O3
126.03170
RI: 1460*
D: UHY
Antiseptic
Chemical
N: also ingredient
of urine
CAS: 123-31-9
126
98
110
141
M+
169
100
Aceclidine
PS
LM/Q
C9H15NO2
169.11028
RI: 1460
Parasympathomimetic
CAS: 827-61-2
126
97
108
M+
168
100
Propoxur-M (HO-) HY
UHY
LS/Q
C9H12O3
168.07865
RI: 1470*
D: UHY
Insecticide
72
112
126
M+
254
100
200
Thiazafluron ME
PS
LM/Q
C7H9F3N4OS
254.04492
RI: 1560
Herbicide
CAS: 25366-23-8

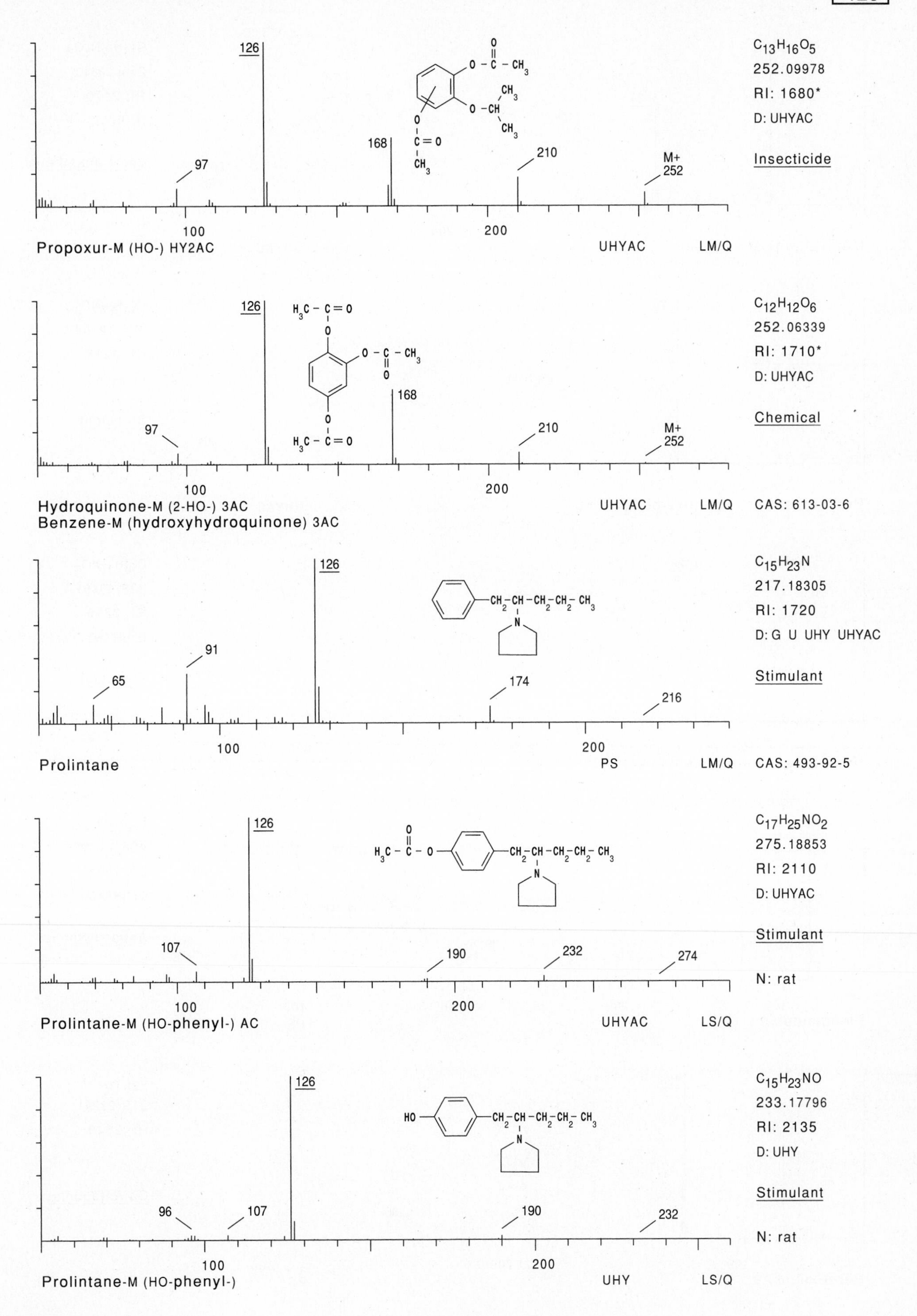

$C_{13}H_{16}O_5$
252.09978
RI: 1680*
D: UHYAC
Insecticide

$C_{12}H_{12}O_6$
252.06339
RI: 1710*
D: UHYAC
Chemical
CAS: 613-03-6

$C_{15}H_{23}N$
217.18305
RI: 1720
D: G U UHY UHYAC
Stimulant
CAS: 493-92-5

$C_{17}H_{25}NO_2$
275.18853
RI: 2110
D: UHYAC
Stimulant
N: rat

$C_{15}H_{23}NO$
233.17796
RI: 2135
D: UHY
Stimulant
N: rat

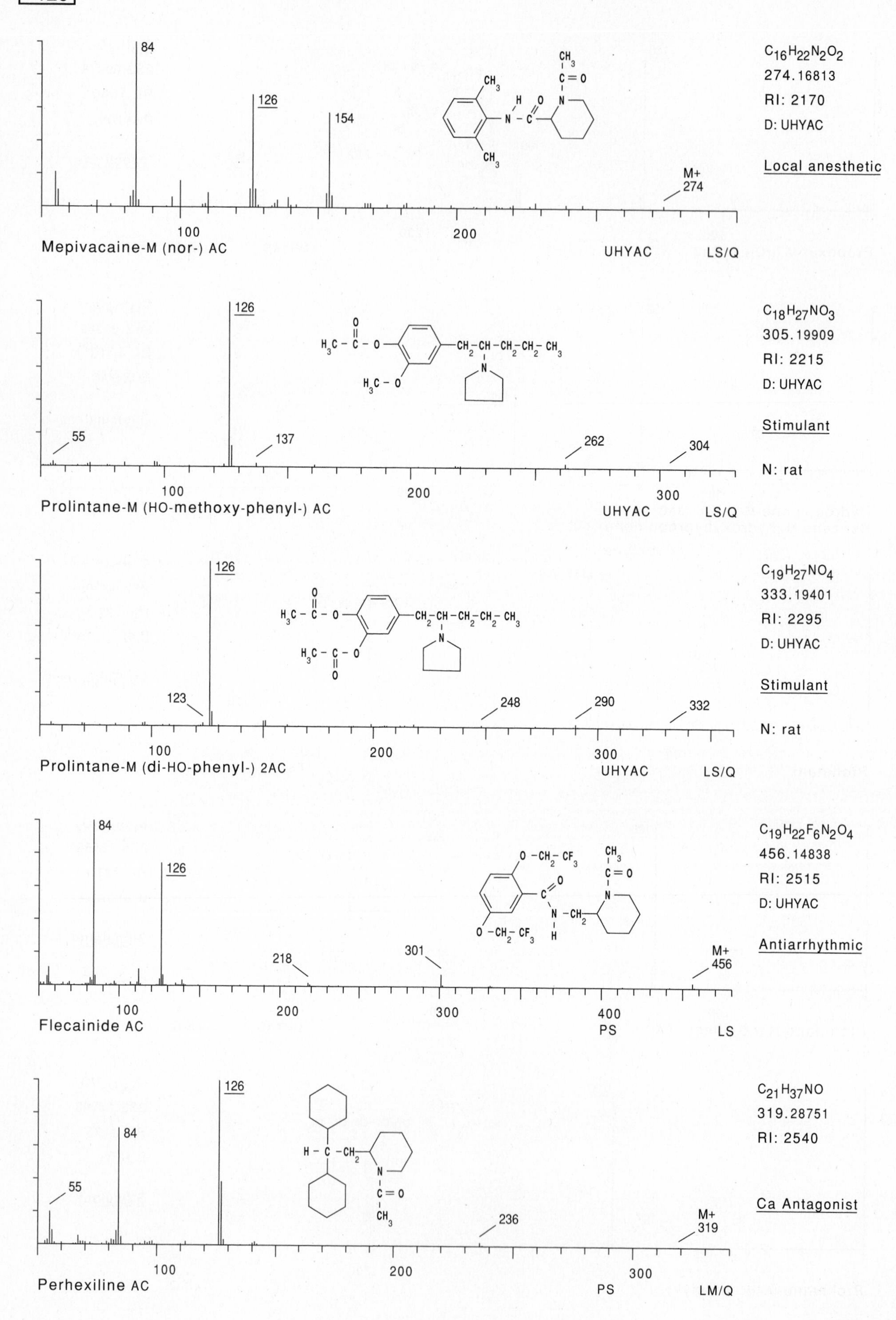
84
126
154
M+
274
100
200
Mepivacaine-M (nor-) AC
UHYAC
LS/Q
C16H22N2O2
274.16813
RI: 2170
D: UHYAC
Local anesthetic
126
55
137
262
304
100
200
300
Prolintane-M (HO-methoxy-phenyl-) AC
UHYAC
LS/Q
C18H27NO3
305.19909
RI: 2215
D: UHYAC
Stimulant
N: rat
126
123
248
290
332
100
200
300
Prolintane-M (di-HO-phenyl-) 2AC
UHYAC
LS/Q
C19H27NO4
333.19401
RI: 2295
D: UHYAC
Stimulant
N: rat
84
126
218
301
M+
456
100
200
300
400
Flecainide AC
PS
LS
C19H22F6N2O4
456.14838
RI: 2515
D: UHYAC
Antiarrhythmic
126
84
55
236
M+
319
100
200
300
Perhexiline AC
PS
LM/Q
C21H37NO
319.28751
RI: 2540
Ca Antagonist

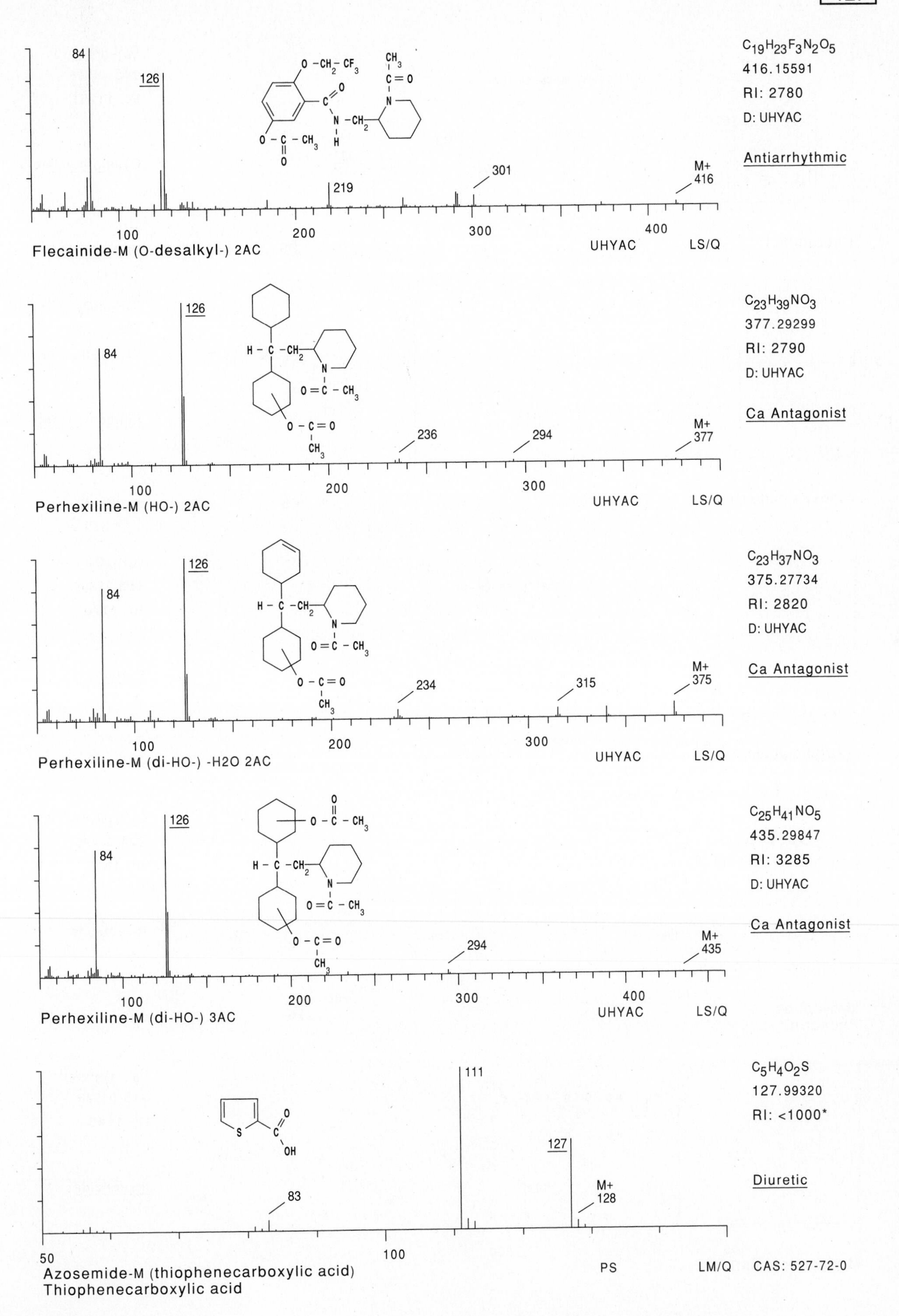

$C_{19}H_{23}F_3N_2O_5$
416.15591
RI: 2780
D: UHYAC

Antiarrhythmic

$C_{23}H_{39}NO_3$
377.29299
RI: 2790
D: UHYAC

Ca Antagonist

$C_{23}H_{37}NO_3$
375.27734
RI: 2820
D: UHYAC

Ca Antagonist

$C_{25}H_{41}NO_5$
435.29847
RI: 3285
D: UHYAC

Ca Antagonist

$C_5H_4O_2S$
127.99320
RI: <1000*

Diuretic

CAS: 527-72-0

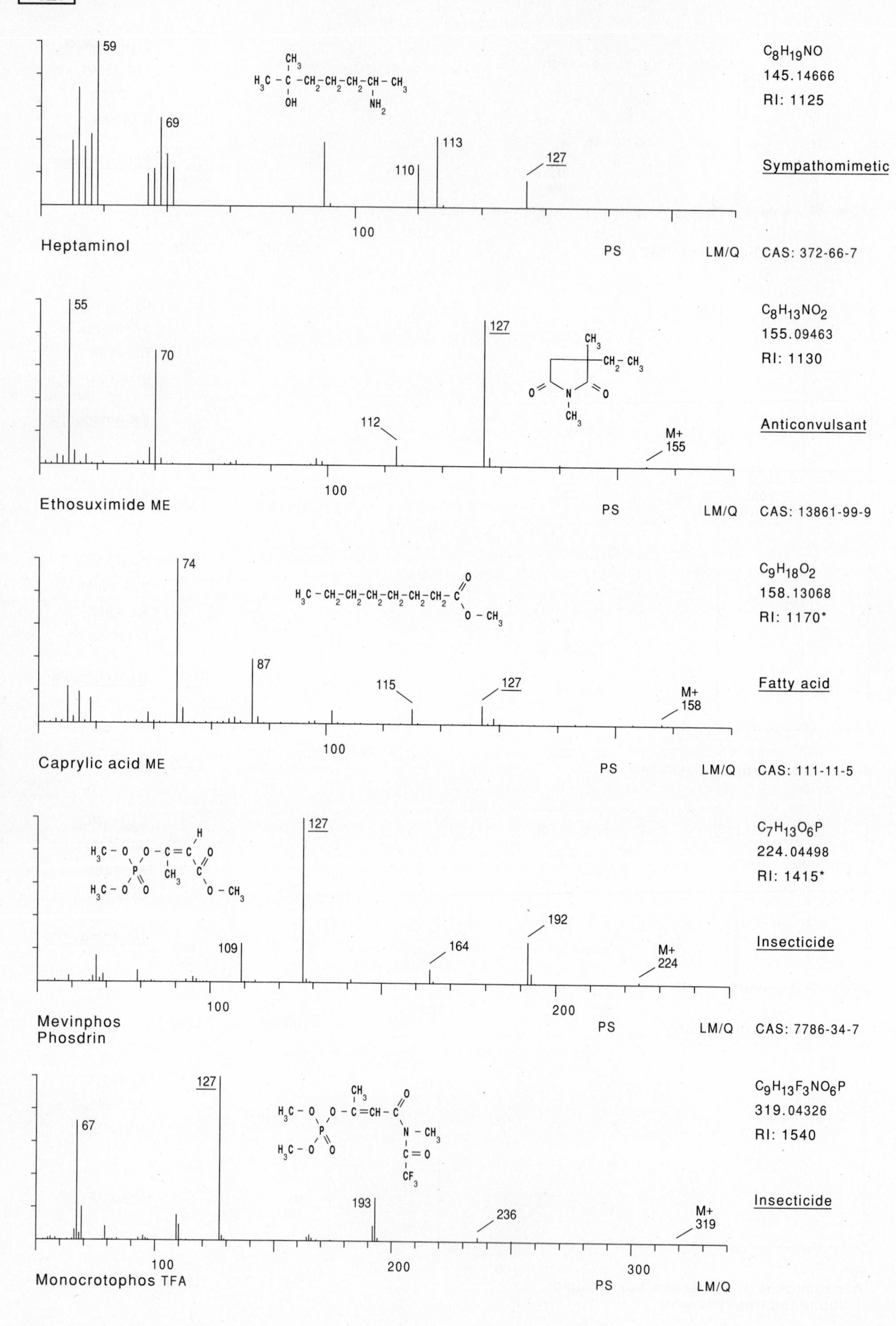

59
69
113
110
127
100
Heptaminol
PS
LM/Q
C8H19NO
145.14666
RI: 1125
Sympathomimetic
CAS: 372-66-7
55
70
127
112
M+
155
100
Ethosuximide ME
PS
LM/Q
C8H13NO2
155.09463
RI: 1130
Anticonvulsant
CAS: 13861-99-9
74
87
115
127
M+
158
100
Caprylic acid ME
PS
LM/Q
C9H18O2
158.13068
RI: 1170*
Fatty acid
CAS: 111-11-5
127
109
164
192
M+
224
100
200
Mevinphos
Phosdrin
PS
LM/Q
C7H13O6P
224.04498
RI: 1415*
Insecticide
CAS: 7786-34-7
127
67
193
236
M+
319
100
200
300
Monocrotophos TFA
PS
LM/Q
C9H13F3NO6P
319.04326
RI: 1540
Insecticide

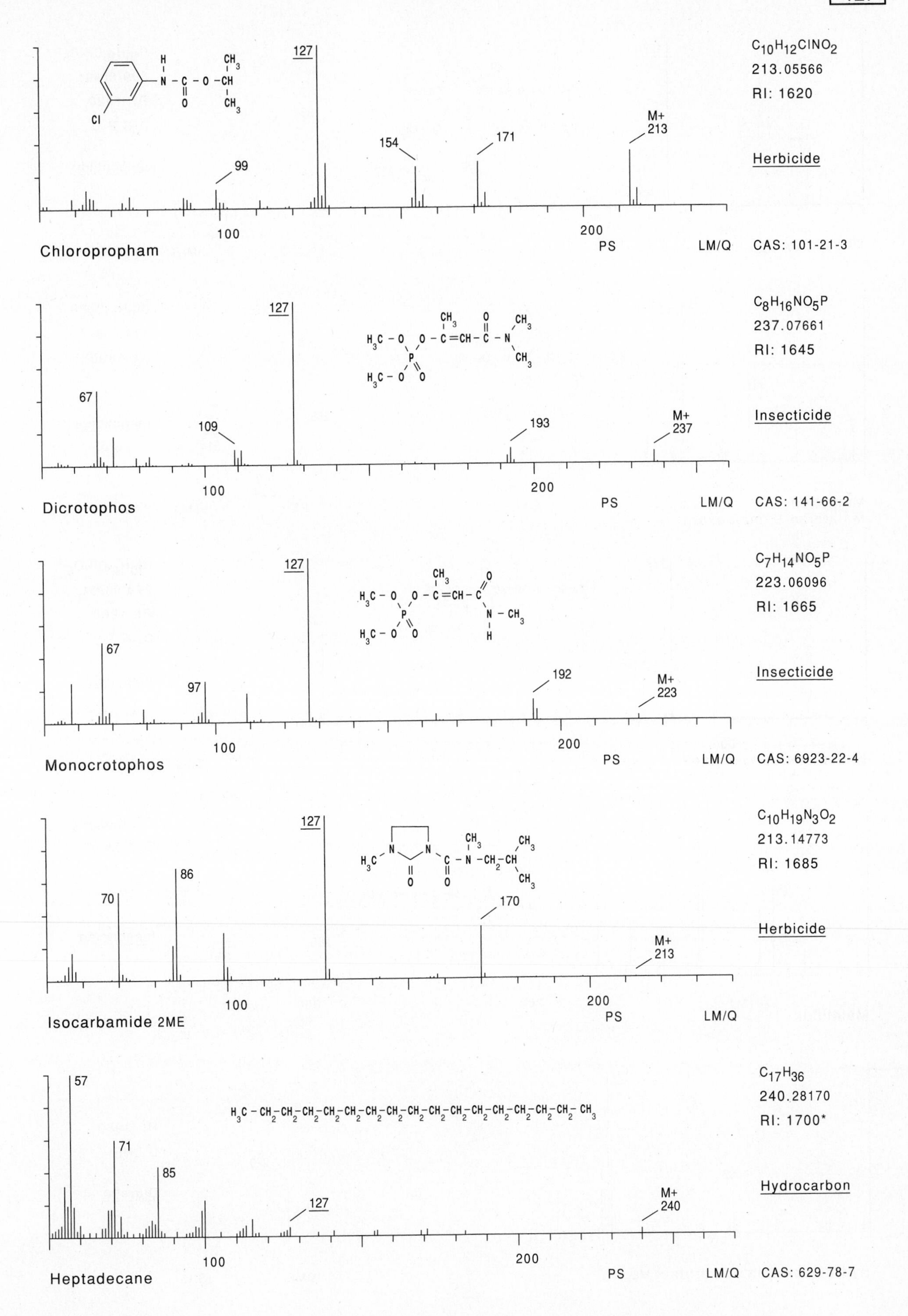

127
99
154
171
M+
213
100
200
PS
LM/Q
Chloropropham
$C_{10}H_{12}ClNO_2$
213.05566
RI: 1620
Herbicide
CAS: 101-21-3
127
67
109
193
M+
237
Dicrotophos
$C_8H_{16}NO_5P$
237.07661
RI: 1645
Insecticide
CAS: 141-66-2
127
67
97
192
M+
223
Monocrotophos
$C_7H_{14}NO_5P$
223.06096
RI: 1665
Insecticide
CAS: 6923-22-4
127
86
70
170
M+
213
Isocarbamide 2ME
$C_{10}H_{19}N_3O_2$
213.14773
RI: 1685
Herbicide
57
71
85
127
M+
240
Heptadecane
$C_{17}H_{36}$
240.28170
RI: 1700*
Hydrocarbon
CAS: 629-78-7

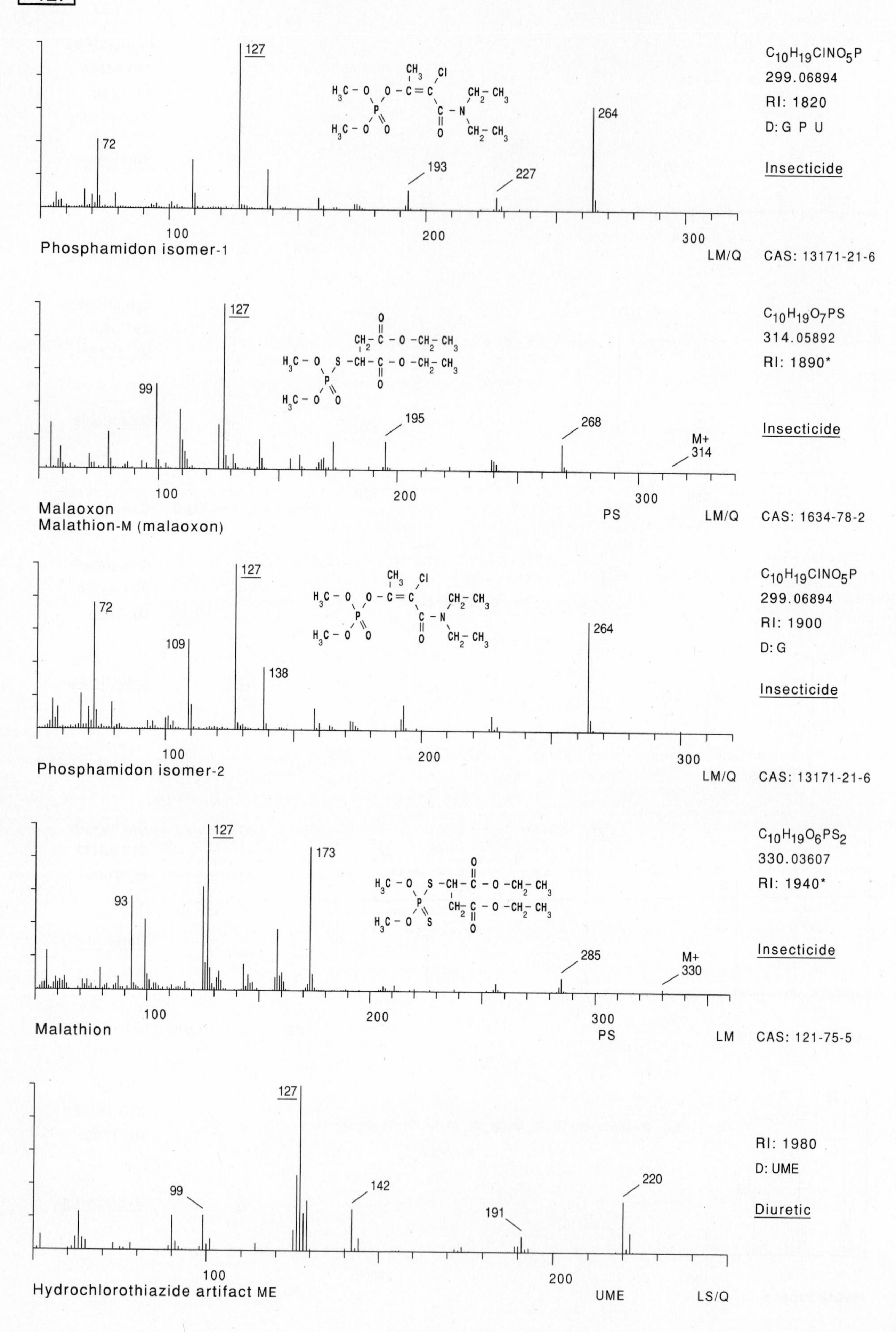

127
72
193
227
264
100
200
300
LM/Q
Phosphamidon isomer-1
C10H19ClNO5P
299.06894
RI: 1820
D: G P U
Insecticide
CAS: 13171-21-6
127
99
195
268
M+
314
100
200
300
PS
LM/Q
Malaoxon
Malathion-M (malaoxon)
C10H19O7PS
314.05892
RI: 1890*
Insecticide
CAS: 1634-78-2
127
72
109
138
264
100
200
300
LM/Q
Phosphamidon isomer-2
C10H19ClNO5P
299.06894
RI: 1900
D: G
Insecticide
CAS: 13171-21-6
127
93
173
285
M+
330
100
200
300
PS
LM
Malathion
C10H19O6PS2
330.03607
RI: 1940*
Insecticide
CAS: 121-75-5
127
99
142
191
220
100
200
UME
LS/Q
Hydrochlorothiazide artifact ME
RI: 1980
D: UME
Diuretic

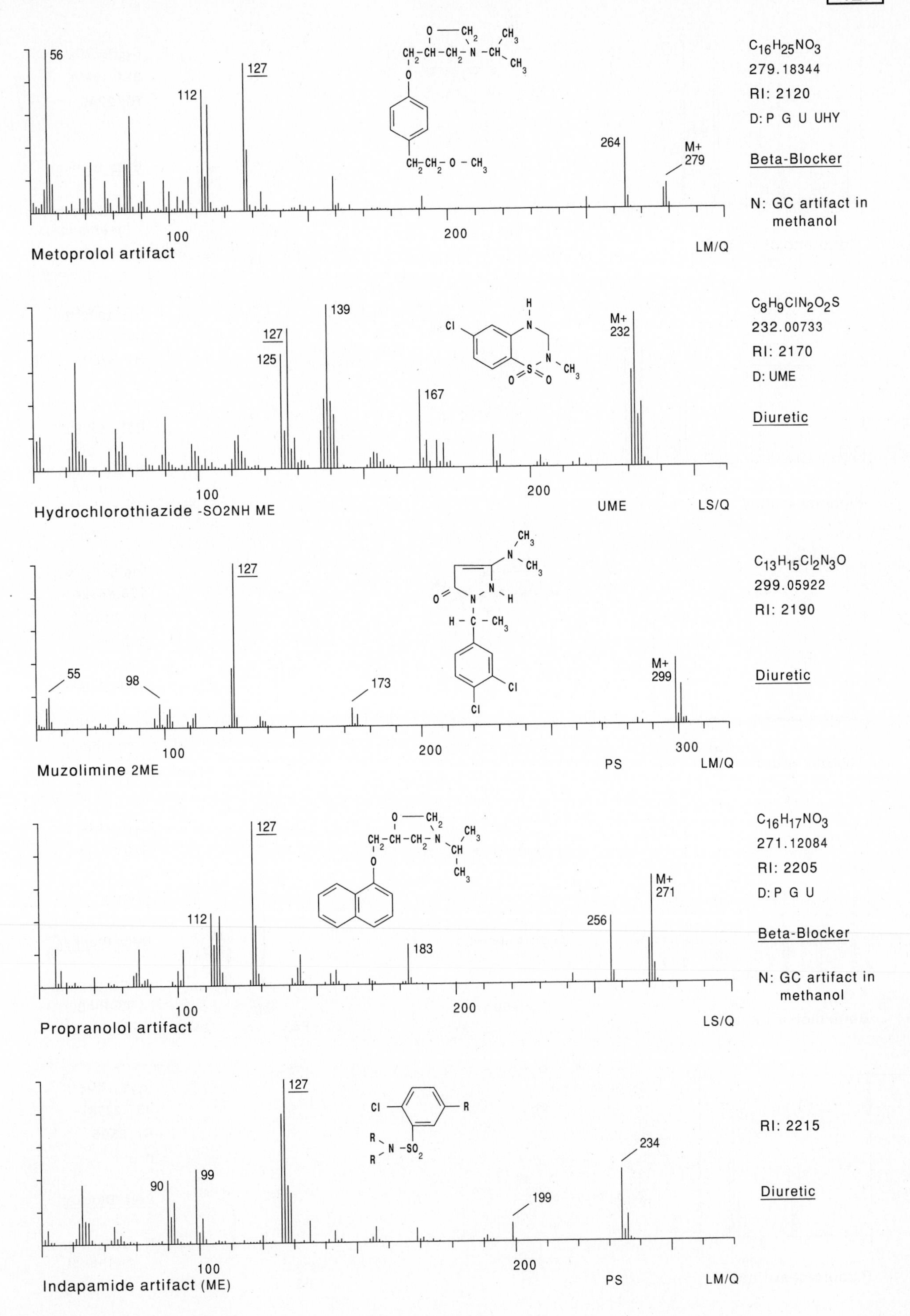

56
127
112
264
M+
279
100
200
LM/Q
Metoprolol artifact
$C_{16}H_{25}NO_3$
279.18344
RI: 2120
D: P G U UHY
Beta-Blocker
N: GC artifact in methanol
139
127
125
167
M+
232
100
200
UME
LS/Q
Hydrochlorothiazide -SO2NH ME
$C_8H_9ClN_2O_2S$
232.00733
RI: 2170
D: UME
Diuretic
127
55
98
173
M+
299
100
200
300
PS
LM/Q
Muzolimine 2ME
$C_{13}H_{15}Cl_2N_3O$
299.05922
RI: 2190
Diuretic
127
112
183
256
M+
271
100
200
LS/Q
Propranolol artifact
$C_{16}H_{17}NO_3$
271.12084
RI: 2205
D: P G U
Beta-Blocker
N: GC artifact in methanol
127
90
99
199
234
100
200
PS
LM/Q
Indapamide artifact (ME)
RI: 2215
Diuretic

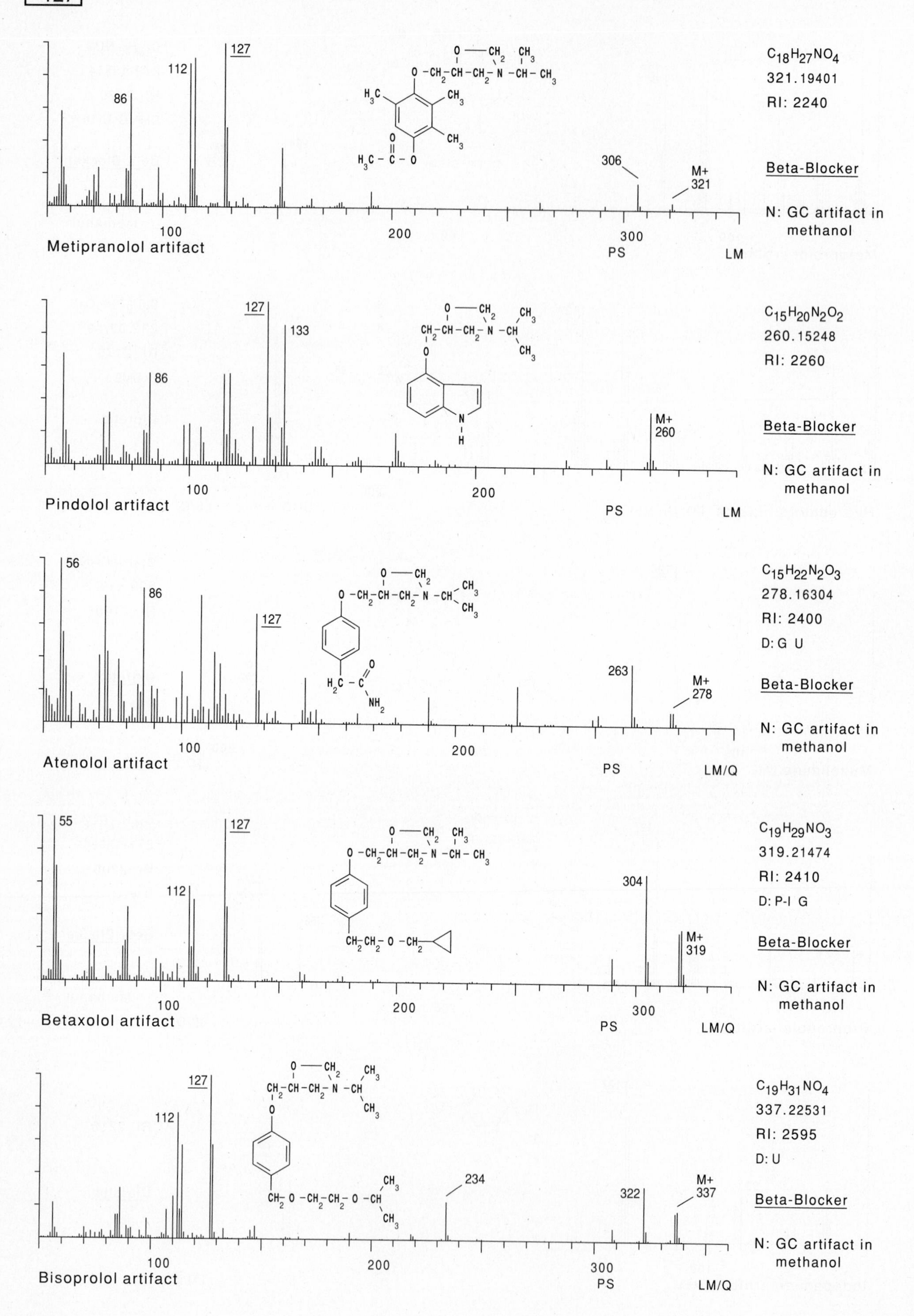

127
112
86
306
M+ 321
100
200
300
Metipranolol artifact
PS
LM
$C_{18}H_{27}NO_4$
321.19401
RI: 2240
Beta-Blocker
N: GC artifact in methanol
127
133
86
M+ 260
100
200
Pindolol artifact
PS
LM
$C_{15}H_{20}N_2O_2$
260.15248
RI: 2260
Beta-Blocker
N: GC artifact in methanol
56
86
127
263
M+ 278
100
200
Atenolol artifact
PS
LM/Q
$C_{15}H_{22}N_2O_3$
278.16304
RI: 2400
D: G U
Beta-Blocker
N: GC artifact in methanol
55
127
112
304
M+ 319
100
200
300
Betaxolol artifact
PS
LM/Q
$C_{19}H_{29}NO_3$
319.21474
RI: 2410
D: P-I G
Beta-Blocker
N: GC artifact in methanol
127
112
234
322
M+ 337
100
200
300
Bisoprolol artifact
PS
LM/Q
$C_{19}H_{31}NO_4$
337.22531
RI: 2595
D: U
Beta-Blocker
N: GC artifact in methanol

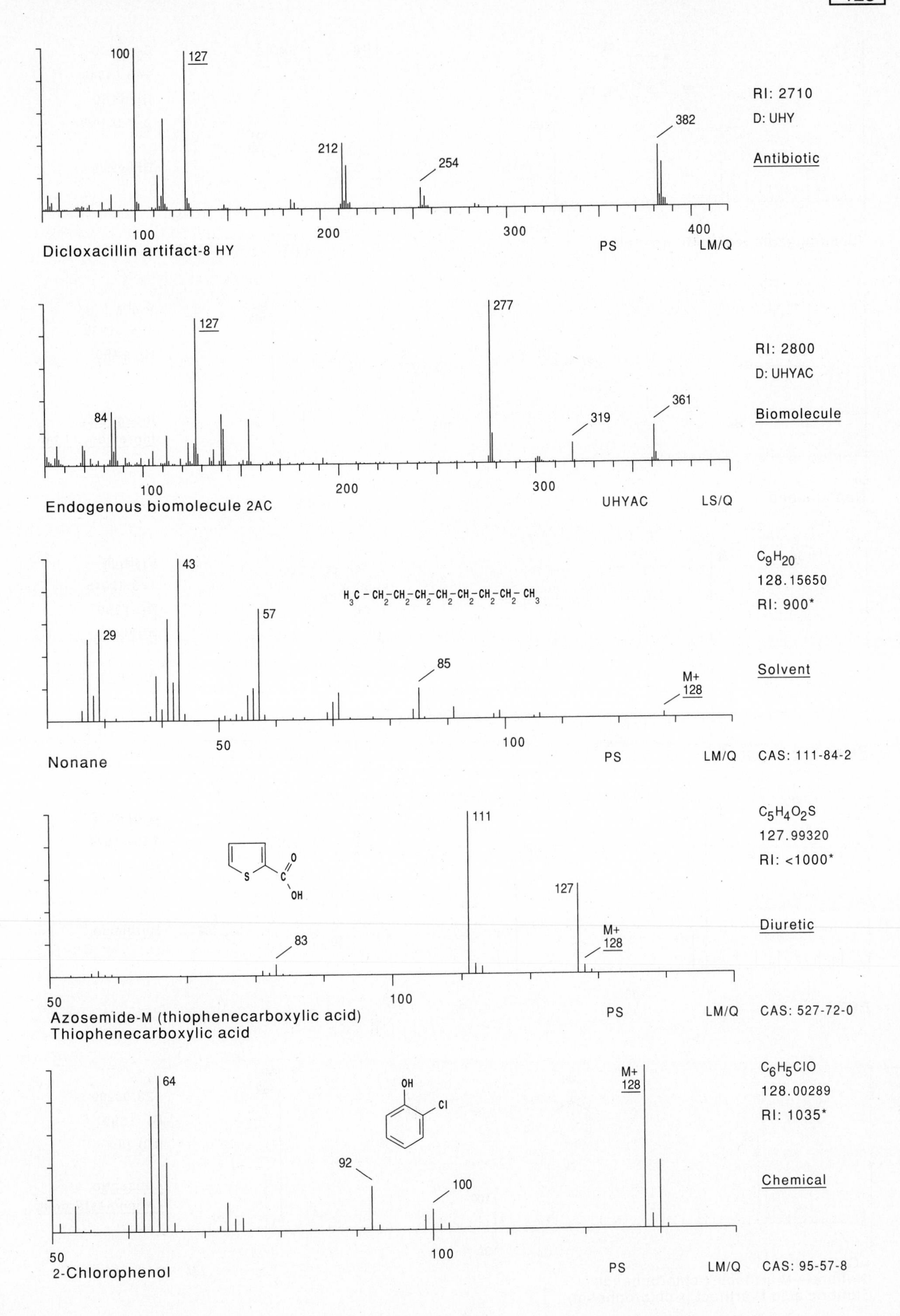

100
127
212
254
382
100
200
300
400
PS
LM/Q
RI: 2710
D: UHY
Antibiotic
Dicloxacillin artifact-8 HY
277
127
84
319
361
100
200
300
UHYAC
LS/Q
RI: 2800
D: UHYAC
Biomolecule
Endogenous biomolecule 2AC
43
29
57
85
M+
128
H3C - CH2- CH2- CH2- CH2- CH2- CH2- CH2- CH3
50
100
PS
LM/Q
C9H20
128.15650
RI: 900*
Solvent
CAS: 111-84-2
Nonane
111
127
83
M+
128
S
O
OH
50
100
PS
LM/Q
C5H4O2S
127.99320
RI: <1000*
Diuretic
CAS: 527-72-0
Azosemide-M (thiophenecarboxylic acid)
Thiophenecarboxylic acid
64
M+
128
92
100
OH
Cl
50
100
PS
LM/Q
C6H5ClO
128.00289
RI: 1035*
Chemical
CAS: 95-57-8
2-Chlorophenol

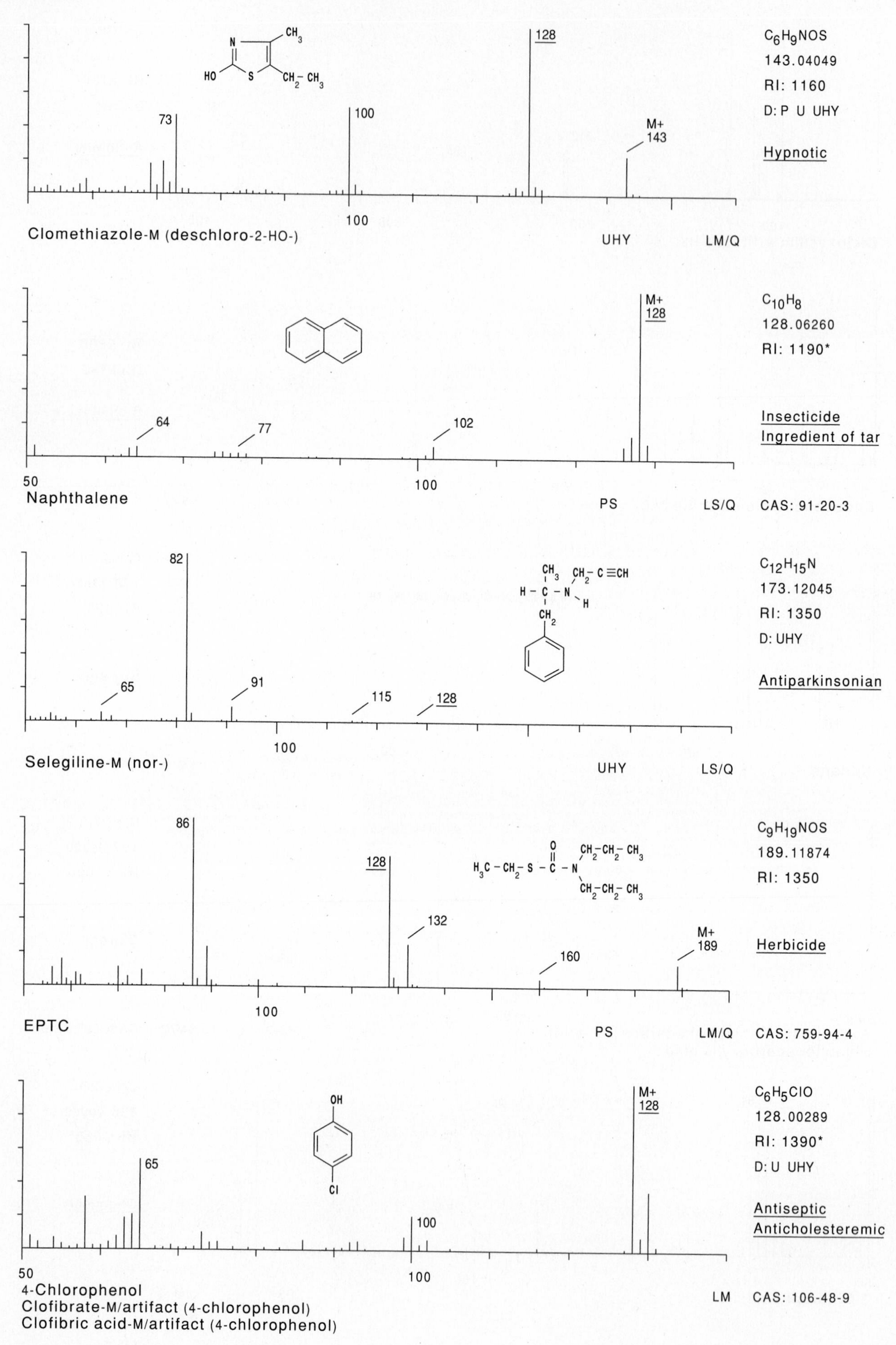

73
100
128
M+
143
100
Clomethiazole-M (deschloro-2-HO-)
UHY
LM/Q
C_6H_9NOS
143.04049
RI: 1160
D: P U UHY
Hypnotic
M+
128
64
77
102
50
100
Naphthalene
PS
LS/Q
$C_{10}H_8$
128.06260
RI: 1190*
Insecticide
Ingredient of tar
CAS: 91-20-3
82
65
91
115
128
100
Selegiline-M (nor-)
UHY
LS/Q
$C_{12}H_{15}N$
173.12045
RI: 1350
D: UHY
Antiparkinsonian
86
128
132
160
M+
189
100
EPTC
PS
LM/Q
$C_9H_{19}NOS$
189.11874
RI: 1350
Herbicide
CAS: 759-94-4
65
100
M+
128
50
100
4-Chlorophenol
Clofibrate-M/artifact (4-chlorophenol)
Clofibric acid-M/artifact (4-chlorophenol)
LM
C_6H_5ClO
128.00289
RI: 1390*
D: U UHY
Antiseptic
Anticholesteremic
CAS: 106-48-9

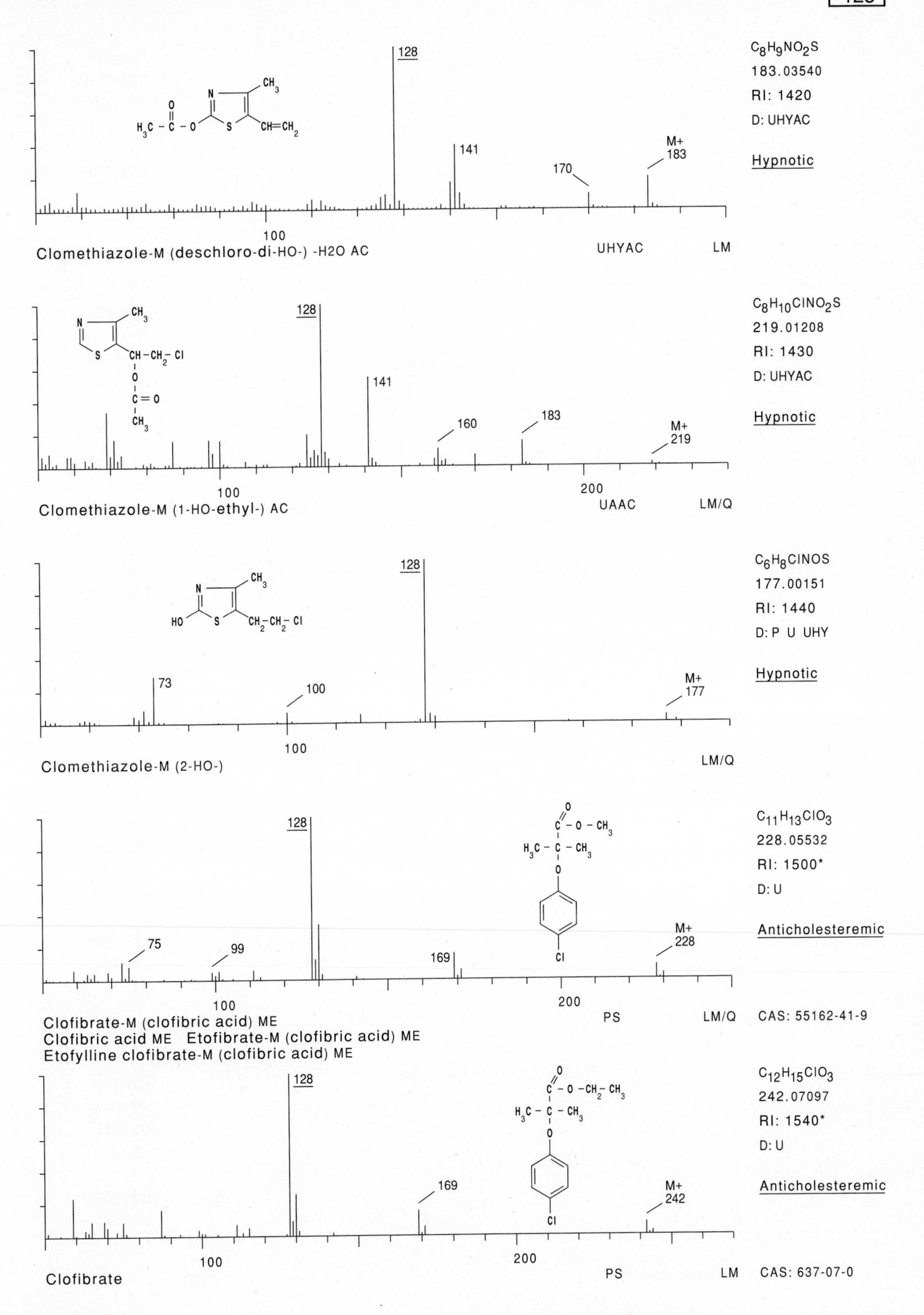

128
141
170
M+
183
100
Clomethiazole-M (deschloro-di-HO-) -H2O AC
UHYAC
LM
$C_8H_9NO_2S$
183.03540
RI: 1420
D: UHYAC
Hypnotic
128
141
160
183
M+
219
100
200
Clomethiazole-M (1-HO-ethyl-) AC
UAAC
LM/Q
$C_8H_{10}ClNO_2S$
219.01208
RI: 1430
D: UHYAC
Hypnotic
128
73
100
M+
177
100
Clomethiazole-M (2-HO-)
LM/Q
C_6H_8ClNOS
177.00151
RI: 1440
D: P U UHY
Hypnotic
128
75
99
169
M+
228
100
200
Clofibrate-M (clofibric acid) ME
Clofibric acid ME Etofibrate-M (clofibric acid) ME
Etofylline clofibrate-M (clofibric acid) ME
PS
LM/Q
$C_{11}H_{13}ClO_3$
228.05532
RI: 1500*
D: U
Anticholesteremic
CAS: 55162-41-9
128
169
M+
242
100
200
Clofibrate
PS
LM
$C_{12}H_{15}ClO_3$
242.07097
RI: 1540*
D: U
Anticholesteremic
CAS: 637-07-0

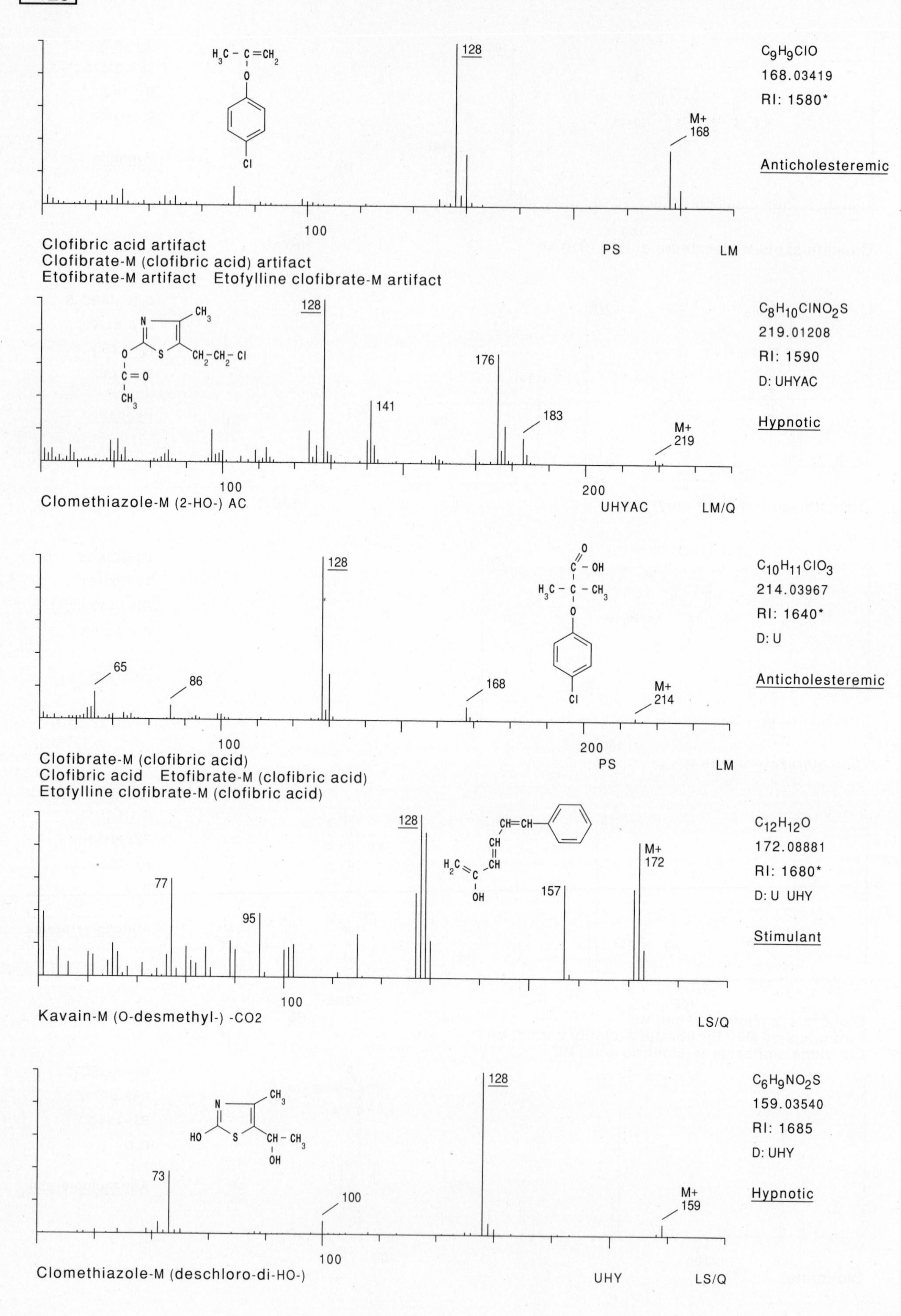

C_9H_9ClO
168.03419
RI: 1580*
Anticholesteremic
Clofibric acid artifact
Clofibrate-M (clofibric acid) artifact
Etofibrate-M artifact Etofylline clofibrate-M artifact
PS
LM
$C_8H_{10}ClNO_2S$
219.01208
RI: 1590
D: UHYAC
Hypnotic
Clomethiazole-M (2-HO-) AC
UHYAC
LM/Q
$C_{10}H_{11}ClO_3$
214.03967
RI: 1640*
D: U
Anticholesteremic
Clofibrate-M (clofibric acid)
Clofibric acid Etofibrate-M (clofibric acid)
Etofylline clofibrate-M (clofibric acid)
PS
LM
$C_{12}H_{12}O$
172.08881
RI: 1680*
D: U UHY
Stimulant
Kavain-M (O-desmethyl-) -CO2
LS/Q
$C_6H_9NO_2S$
159.03540
RI: 1685
D: UHY
Hypnotic
Clomethiazole-M (deschloro-di-HO-)
UHY
LS/Q

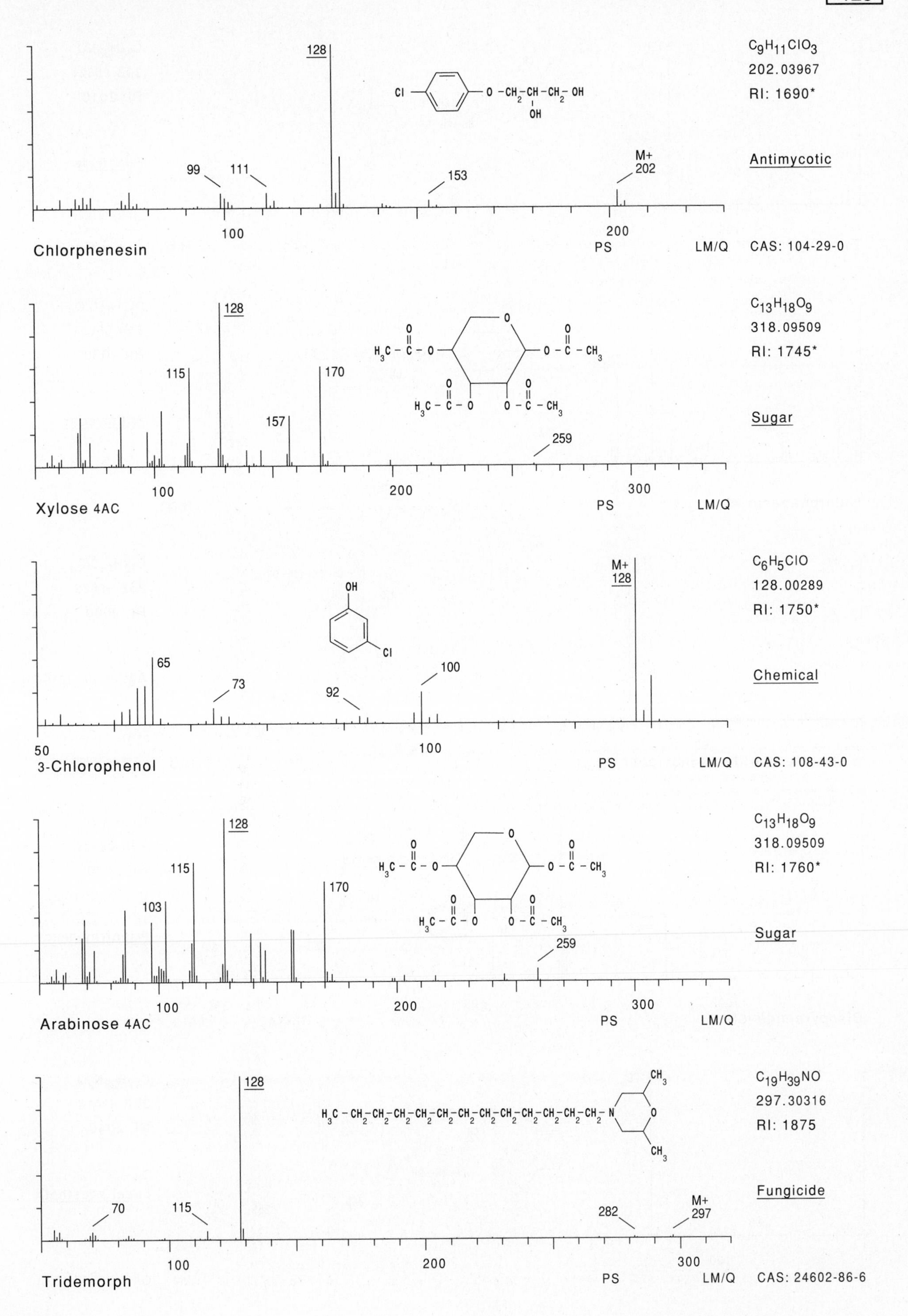

$C_9H_{11}ClO_3$
202.03967
RI: 1690*
Antimycotic
CAS: 104-29-0

$C_{13}H_{18}O_9$
318.09509
RI: 1745*
Sugar

C_6H_5ClO
128.00289
RI: 1750*
Chemical
CAS: 108-43-0

$C_{13}H_{18}O_9$
318.09509
RI: 1760*
Sugar

$C_{19}H_{39}NO$
297.30316
RI: 1875
Fungicide
CAS: 24602-86-6

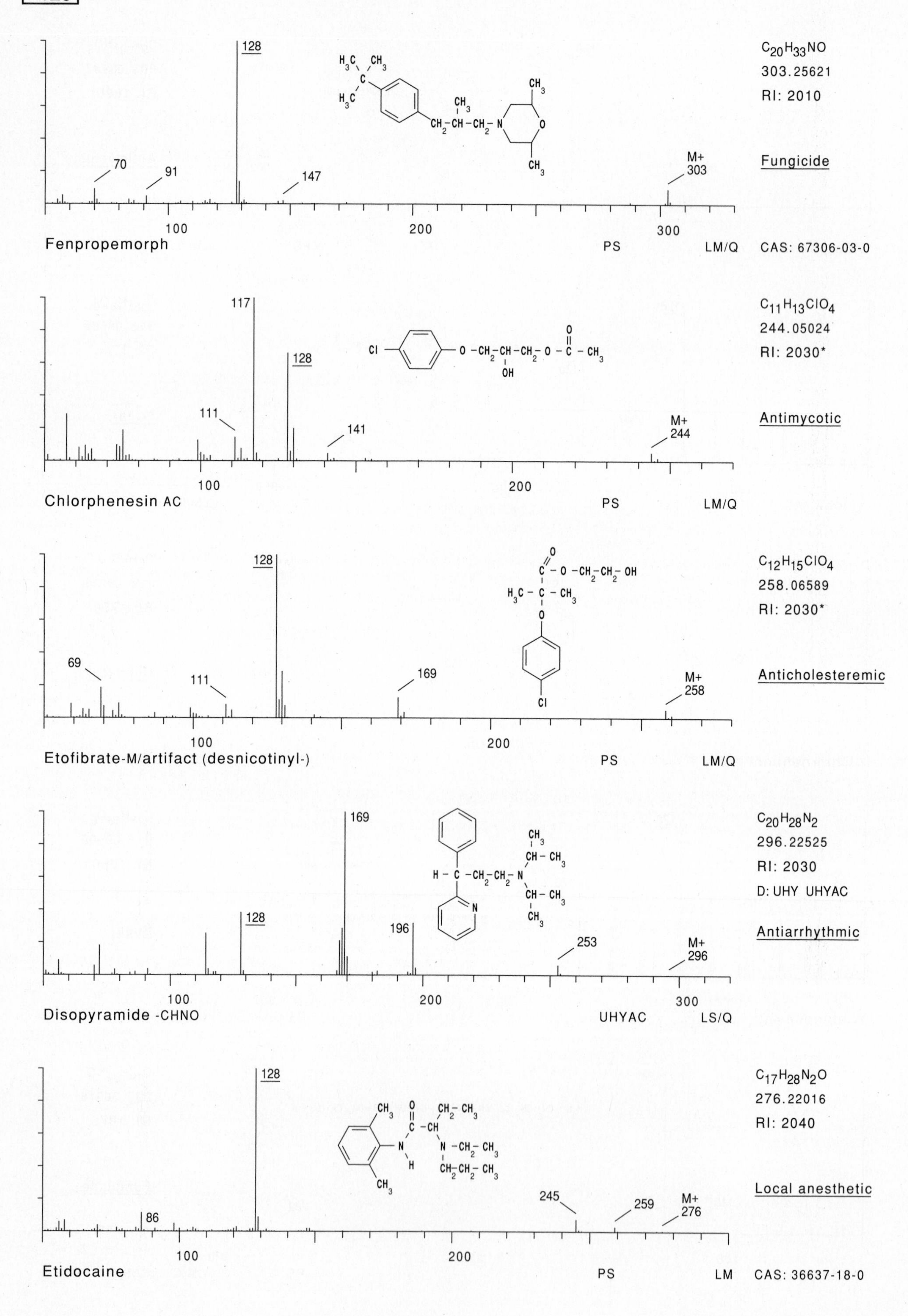
128
70
91
147
M+
303
100
200
300
Fenpropemorph
PS
LM/Q
C20H33NO
303.25621
RI: 2010
Fungicide
CAS: 67306-03-0
117
128
111
141
M+
244
100
200
Chlorphenesin AC
PS
LM/Q
C11H13ClO4
244.05024
RI: 2030*
Antimycotic
128
69
111
169
M+
258
100
200
Etofibrate-M/artifact (desnicotinyl-)
PS
LM/Q
C12H15ClO4
258.06589
RI: 2030*
Anticholesteremic
169
128
196
253
M+
296
100
200
300
Disopyramide -CHNO
UHYAC
LS/Q
C20H28N2
296.22525
RI: 2030
D: UHY UHYAC
Antiarrhythmic
128
86
245
259
M+
276
100
200
Etidocaine
PS
LM
C17H28N2O
276.22016
RI: 2040
Local anesthetic
CAS: 36637-18-0

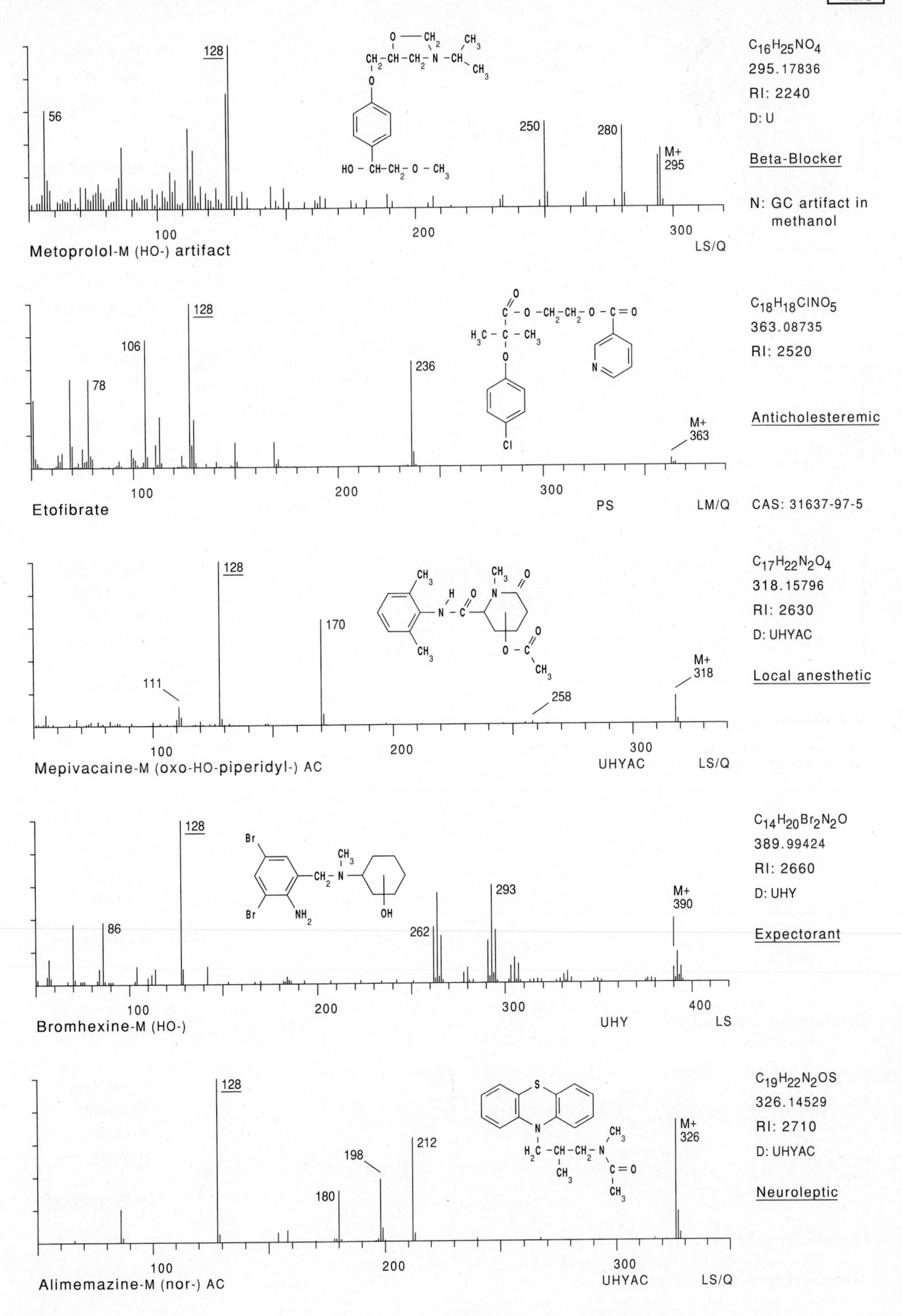
Metoprolol-M (HO-) artifact
LS/Q
M+
295
$C_{16}H_{25}NO_4$
295.17836
RI: 2240
D: U
Beta-Blocker
N: GC artifact in methanol
Etofibrate
PS
LM/Q
M+
363
$C_{18}H_{18}ClNO_5$
363.08735
RI: 2520
Anticholesteremic
CAS: 31637-97-5
Mepivacaine-M (oxo-HO-piperidyl-) AC
UHYAC
LS/Q
M+
318
$C_{17}H_{22}N_2O_4$
318.15796
RI: 2630
D: UHYAC
Local anesthetic
Bromhexine-M (HO-)
UHY
LS
M+
390
$C_{14}H_{20}Br_2N_2O$
389.99424
RI: 2660
D: UHY
Expectorant
Alimemazine-M (nor-) AC
UHYAC
LS/Q
M+
326
$C_{19}H_{22}N_2OS$
326.14529
RI: 2710
D: UHYAC
Neuroleptic

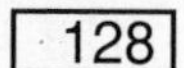
128

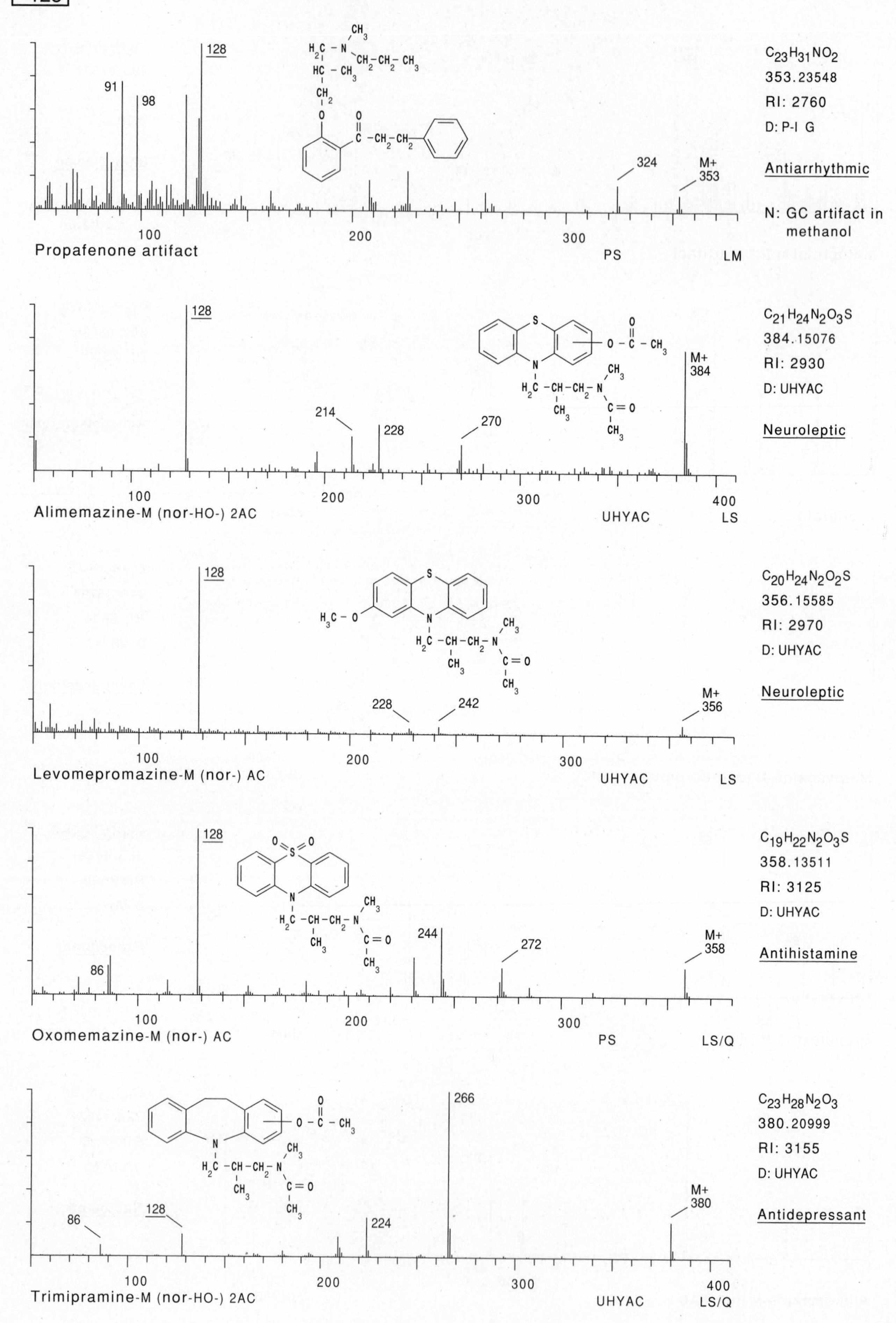
128
91
98
324
M+
353
100
200
300
Propafenone artifact
PS
LM
$C_{23}H_{31}NO_2$
353.23548
RI: 2760
D: P-I G
Antiarrhythmic
N: GC artifact in methanol
128
214
228
270
M+
384
100
200
300
400
Alimemazine-M (nor-HO-) 2AC
UHYAC
LS
$C_{21}H_{24}N_2O_3S$
384.15076
RI: 2930
D: UHYAC
Neuroleptic
128
228
242
M+
356
100
200
300
Levomepromazine-M (nor-) AC
UHYAC
LS
$C_{20}H_{24}N_2O_2S$
356.15585
RI: 2970
D: UHYAC
Neuroleptic
128
86
244
272
M+
358
100
200
300
Oxomemazine-M (nor-) AC
PS
LS/Q
$C_{19}H_{22}N_2O_3S$
358.13511
RI: 3125
D: UHYAC
Antihistamine
266
128
86
224
M+
380
100
200
300
400
Trimipramine-M (nor-HO-) 2AC
UHYAC
LS/Q
$C_{23}H_{28}N_2O_3$
380.20999
RI: 3155
D: UHYAC
Antidepressant

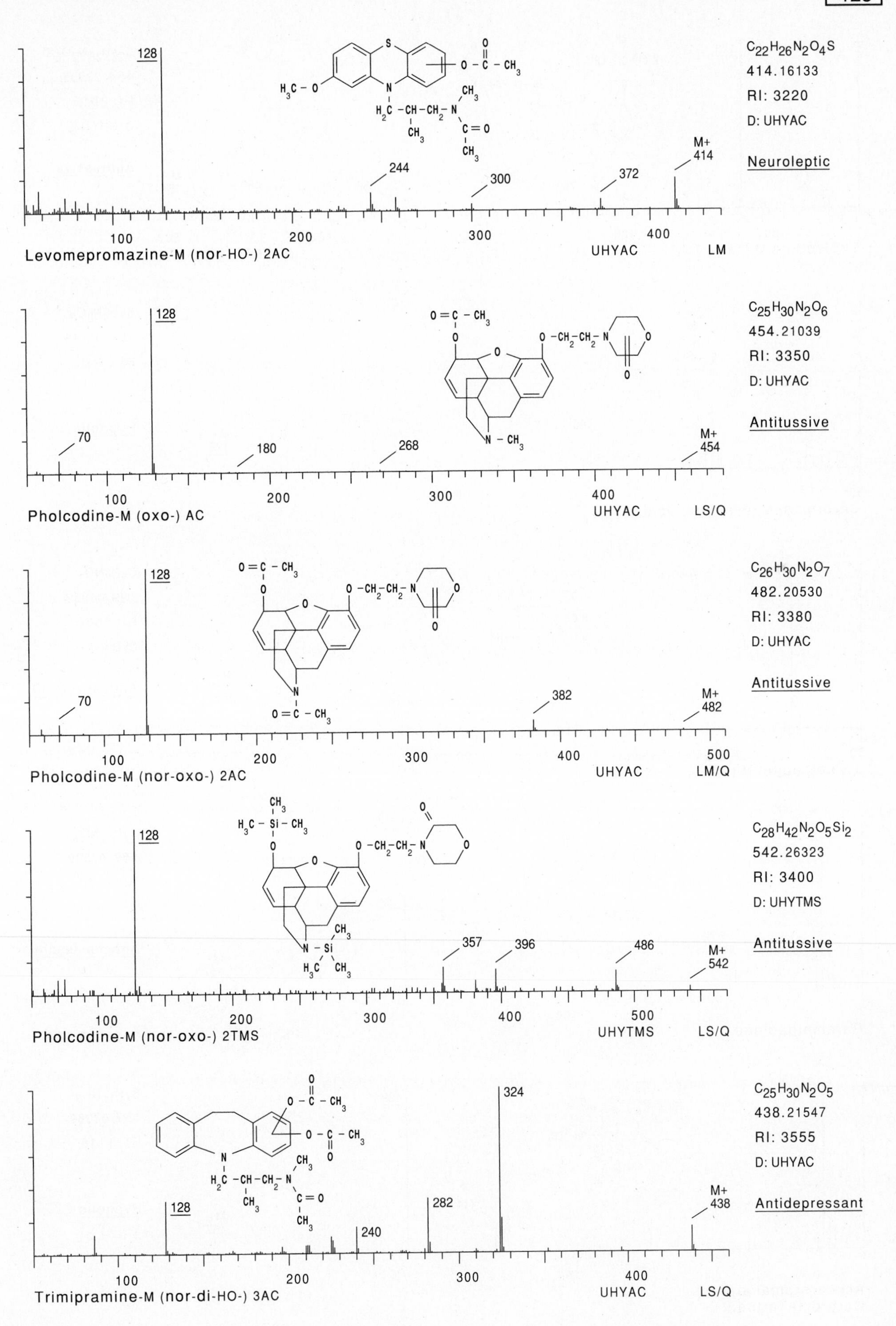

128
244
300
372
M+
414
100
200
300
400
Levomepromazine-M (nor-HO-) 2AC
UHYAC
LM
$C_{22}H_{26}N_2O_4S$
414.16133
RI: 3220
D: UHYAC
Neuroleptic
128
70
180
268
M+
454
Pholcodine-M (oxo-) AC
UHYAC
LS/Q
$C_{25}H_{30}N_2O_6$
454.21039
RI: 3350
D: UHYAC
Antitussive
128
70
382
M+
482
500
Pholcodine-M (nor-oxo-) 2AC
UHYAC
LM/Q
$C_{26}H_{30}N_2O_7$
482.20530
RI: 3380
D: UHYAC
Antitussive
128
357
396
486
M+
542
Pholcodine-M (nor-oxo-) 2TMS
UHYTMS
LS/Q
$C_{28}H_{42}N_2O_5Si_2$
542.26323
RI: 3400
D: UHYTMS
Antitussive
324
128
282
240
M+
438
Trimipramine-M (nor-di-HO-) 3AC
UHYAC
LS/Q
$C_{25}H_{30}N_2O_5$
438.21547
RI: 3555
D: UHYAC
Antidepressant

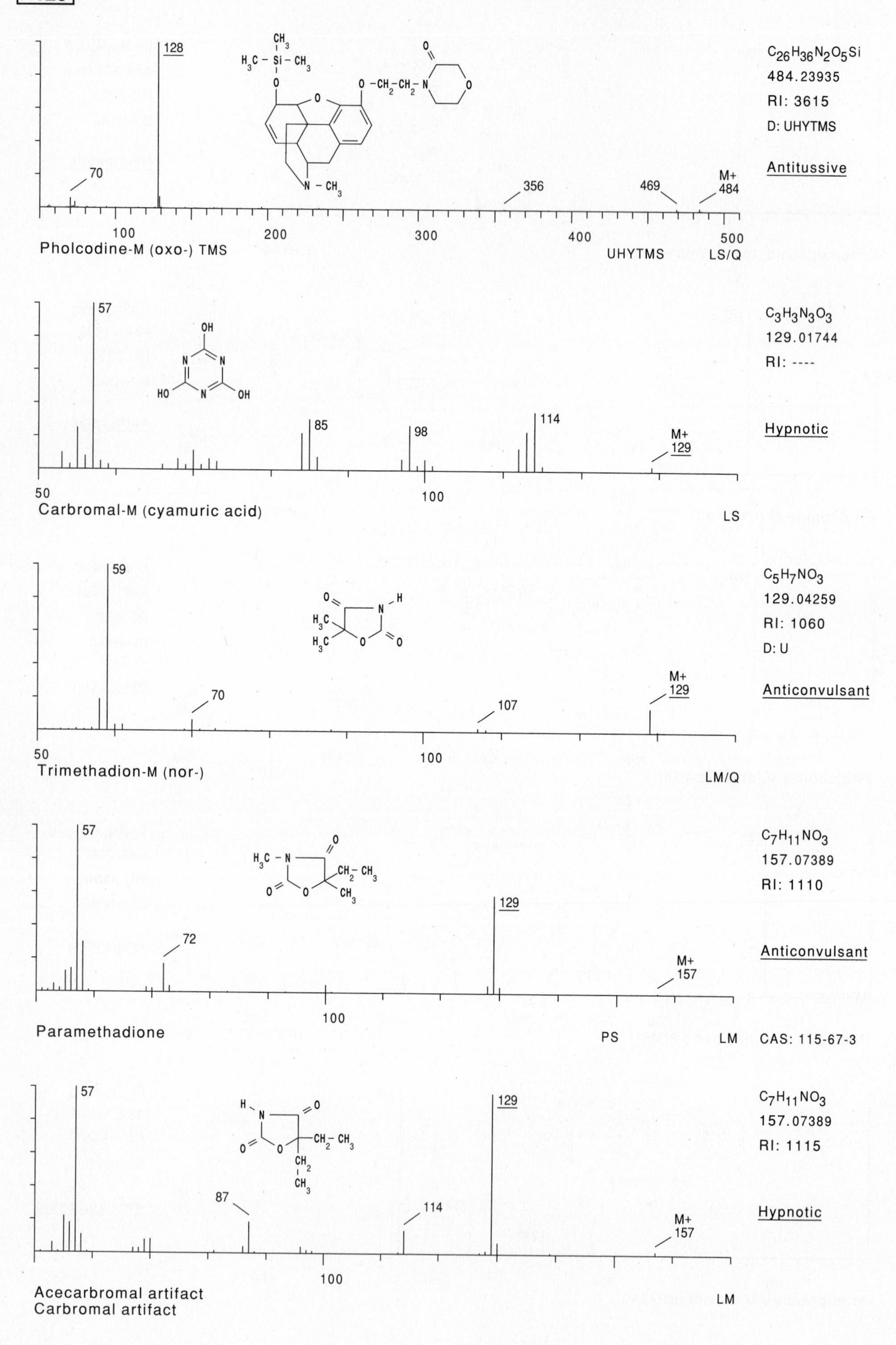

128
70
356
469
M+
484
100
200
300
400
500
Pholcodine-M (oxo-) TMS
UHYTMS
LS/Q
$C_{26}H_{36}N_2O_5Si$
484.23935
RI: 3615
D: UHYTMS
Antitussive
57
85
98
114
M+
129
50
100
Carbromal-M (cyamuric acid)
LS
$C_3H_3N_3O_3$
129.01744
RI: ----
Hypnotic
59
70
107
M+
129
50
100
Trimethadion-M (nor-)
LM/Q
$C_5H_7NO_3$
129.04259
RI: 1060
D: U
Anticonvulsant
57
72
129
M+
157
100
Paramethadione
PS
LM
CAS: 115-67-3
$C_7H_{11}NO_3$
157.07389
RI: 1110
Anticonvulsant
57
87
114
129
M+
157
100
Acecarbromal artifact
Carbromal artifact
LM
$C_7H_{11}NO_3$
157.07389
RI: 1115
Hypnotic

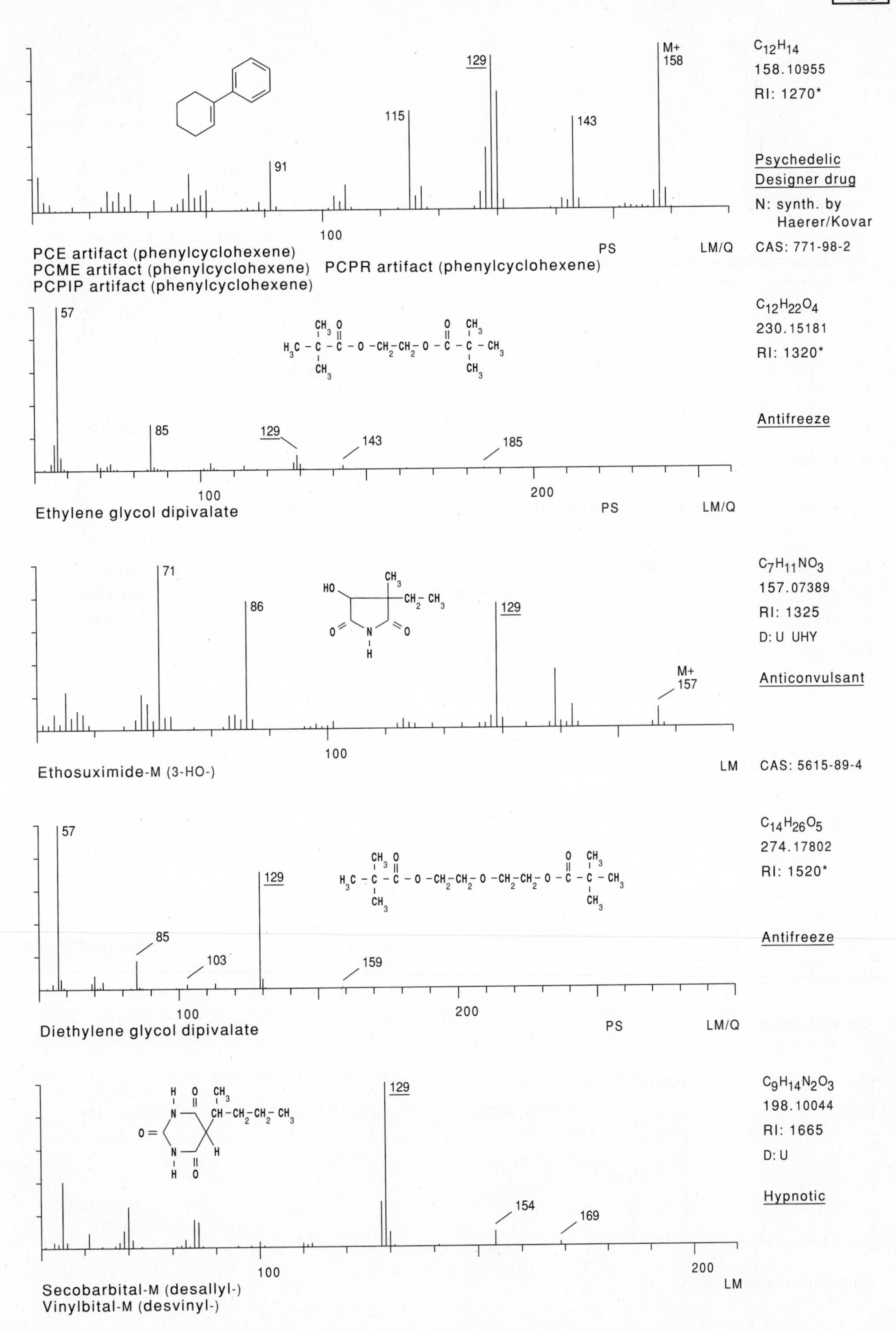
C12H14
158.10955
RI: 1270*
Psychedelic
Designer drug
N: synth. by
Haerer/Kovar
CAS: 771-98-2
PCE artifact (phenylcyclohexene)
PCME artifact (phenylcyclohexene)
PCPIP artifact (phenylcyclohexene)
PCPR artifact (phenylcyclohexene)
PS
LM/Q
C12H22O4
230.15181
RI: 1320*
Antifreeze
Ethylene glycol dipivalate
PS
LM/Q
C7H11NO3
157.07389
RI: 1325
D: U UHY
Anticonvulsant
Ethosuximide-M (3-HO-)
LM
CAS: 5615-89-4
C14H26O5
274.17802
RI: 1520*
Antifreeze
Diethylene glycol dipivalate
PS
LM/Q
C9H14N2O3
198.10044
RI: 1665
D: U
Hypnotic
Secobarbital-M (desallyl-)
Vinylbital-M (desvinyl-)
LM

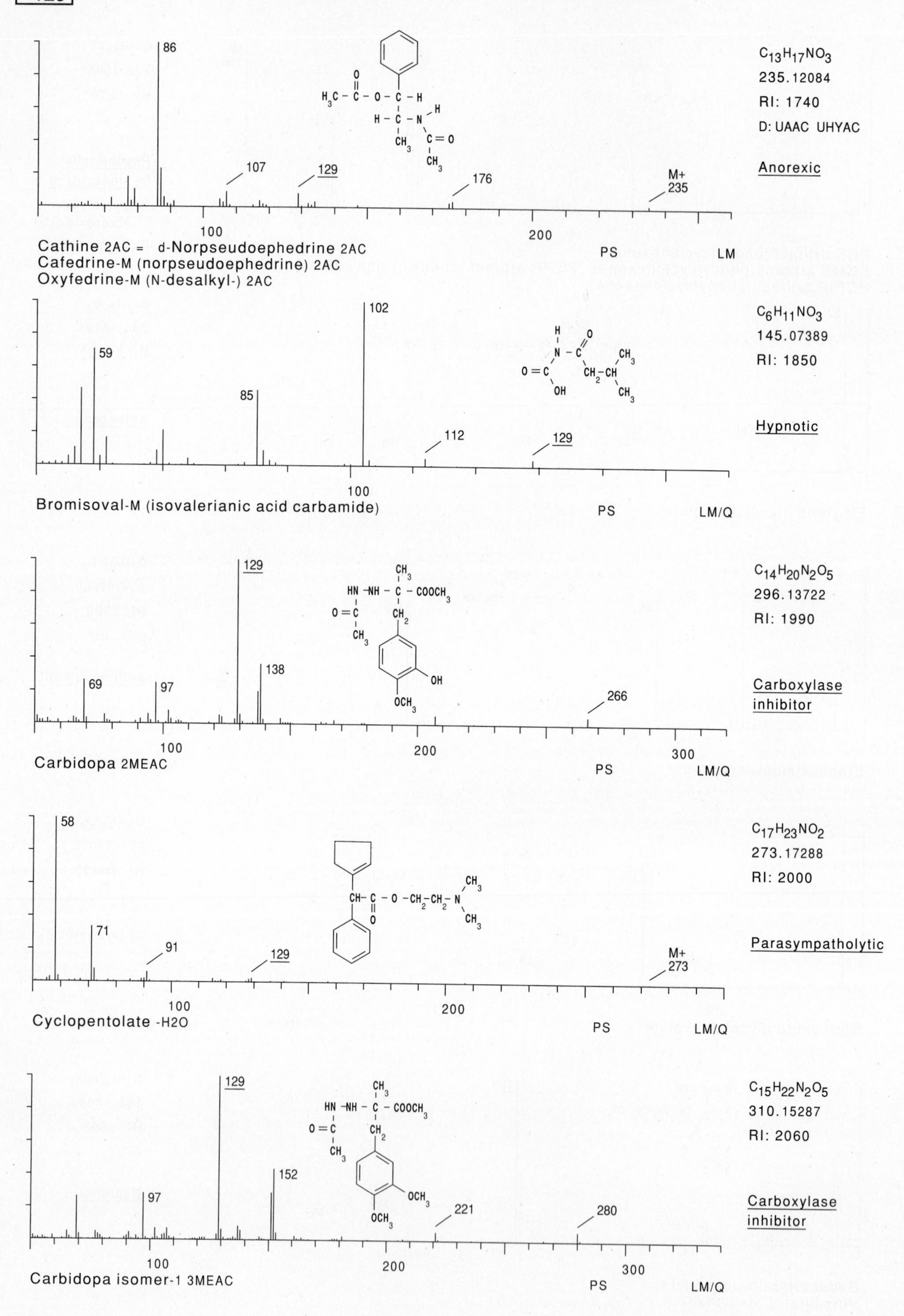
86
107
129
176
M+
235
100
200
C13H17NO3
235.12084
RI: 1740
D: UAAC UHYAC
Anorexic
PS
LM
Cathine 2AC = d-Norpseudoephedrine 2AC
Cafedrine-M (norpseudoephedrine) 2AC
Oxyfedrine-M (N-desalkyl-) 2AC
102
59
85
112
129
C6H11NO3
145.07389
RI: 1850
Hypnotic
PS
LM/Q
Bromisoval-M (isovalerianic acid carbamide)
129
138
69
97
266
300
C14H20N2O5
296.13722
RI: 1990
Carboxylase
inhibitor
Carbidopa 2MEAC
58
71
91
129
M+
273
C17H23NO2
273.17288
RI: 2000
Parasympatholytic
Cyclopentolate -H2O
129
152
97
221
280
C15H22N2O5
310.15287
RI: 2060
Carboxylase
inhibitor
Carbidopa isomer-1 3MEAC

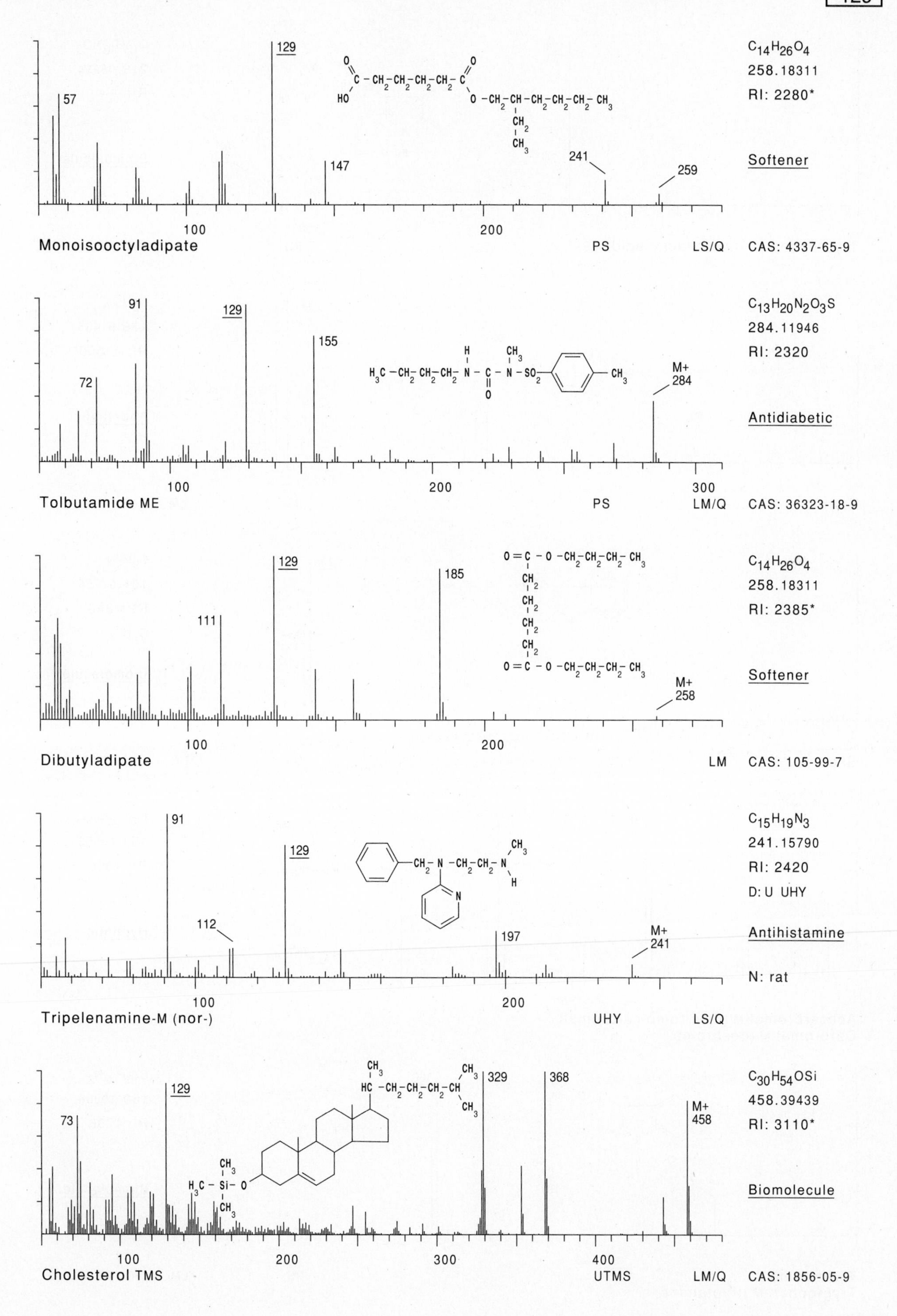

129
57
147
241
259
100
200
Monoisooctyladipate
PS
LS/Q
$C_{14}H_{26}O_4$
258.18311
RI: 2280*
Softener
CAS: 4337-65-9
91
129
155
72
M+
284
300
Tolbutamide ME
LM/Q
$C_{13}H_{20}N_2O_3S$
284.11946
RI: 2320
Antidiabetic
CAS: 36323-18-9
185
111
258
Dibutyladipate
LM
$C_{14}H_{26}O_4$
258.18311
RI: 2385*
Softener
CAS: 105-99-7
112
197
241
Tripelenamine-M (nor-)
UHY
$C_{15}H_{19}N_3$
241.15790
RI: 2420
D: U UHY
Antihistamine
N: rat
73
329
368
458
400
Cholesterol TMS
UTMS
$C_{30}H_{54}OSi$
458.39439
RI: 3110*
Biomolecule
CAS: 1856-05-9

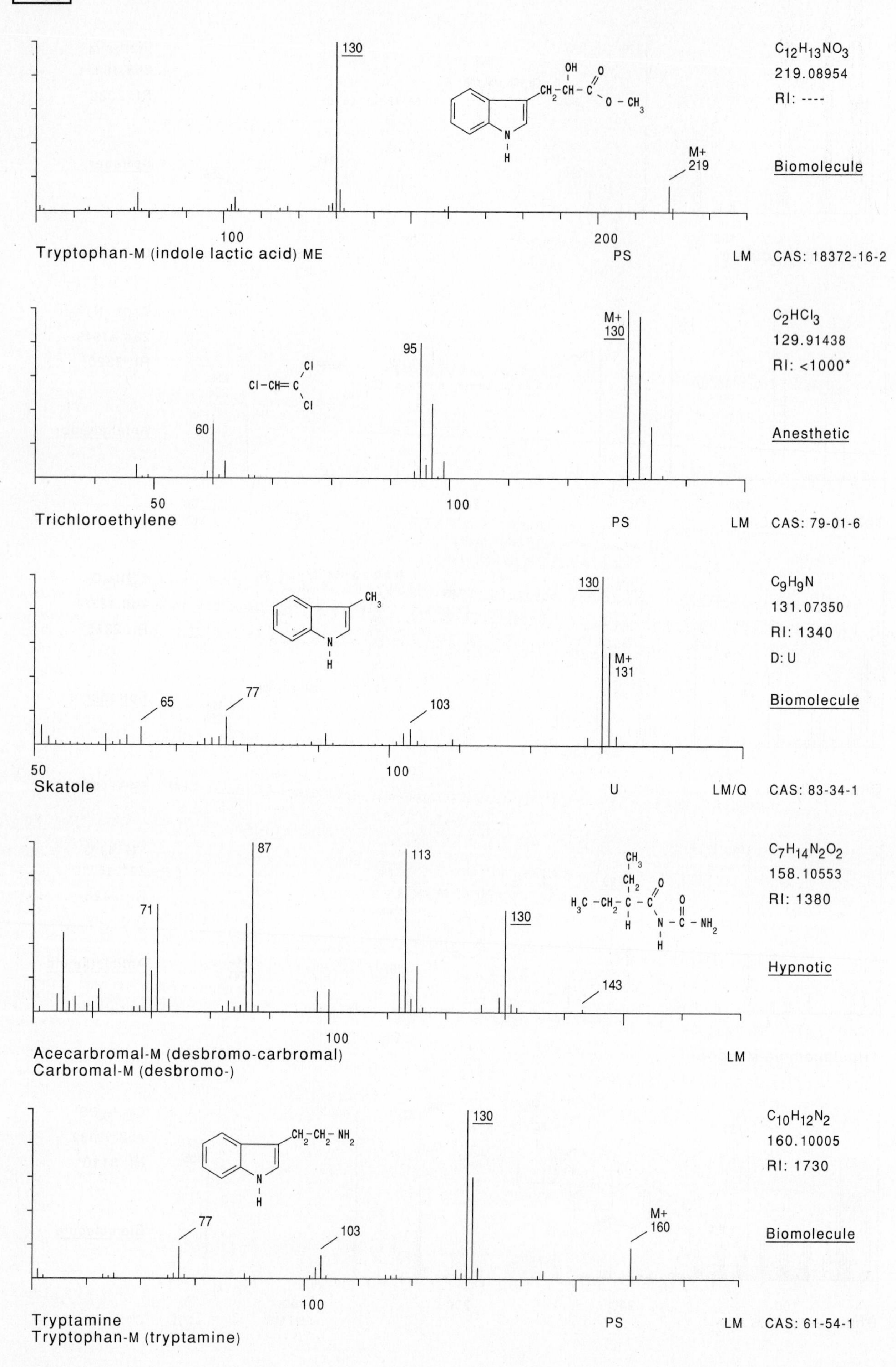
130
OH
CH2-CH-C
O
O-CH3
N
H
M+
219
100
200
Tryptophan-M (indole lactic acid) ME
PS
LM
C12H13NO3
219.08954
RI: ----
Biomolecule
CAS: 18372-16-2
M+
130
95
Cl
Cl-CH=C
Cl
60
50
100
Trichloroethylene
PS
LM
C2HCl3
129.91438
RI: <1000*
Anesthetic
CAS: 79-01-6
130
CH3
N
H
M+
131
65
77
103
50
100
Skatole
U
LM/Q
C9H9N
131.07350
RI: 1340
D: U
Biomolecule
CAS: 83-34-1
87
113
71
130
CH3
CH2
H3C-CH2-C-C
O
H
N-C-NH2
H
143
100
Acecarbromal-M (desbromo-carbromal)
Carbromal-M (desbromo-)
LM
C7H14N2O2
158.10553
RI: 1380
Hypnotic
130
CH2-CH2-NH2
N
H
77
103
M+
160
100
Tryptamine
Tryptophan-M (tryptamine)
PS
LM
C10H12N2
160.10005
RI: 1730
Biomolecule
CAS: 61-54-1

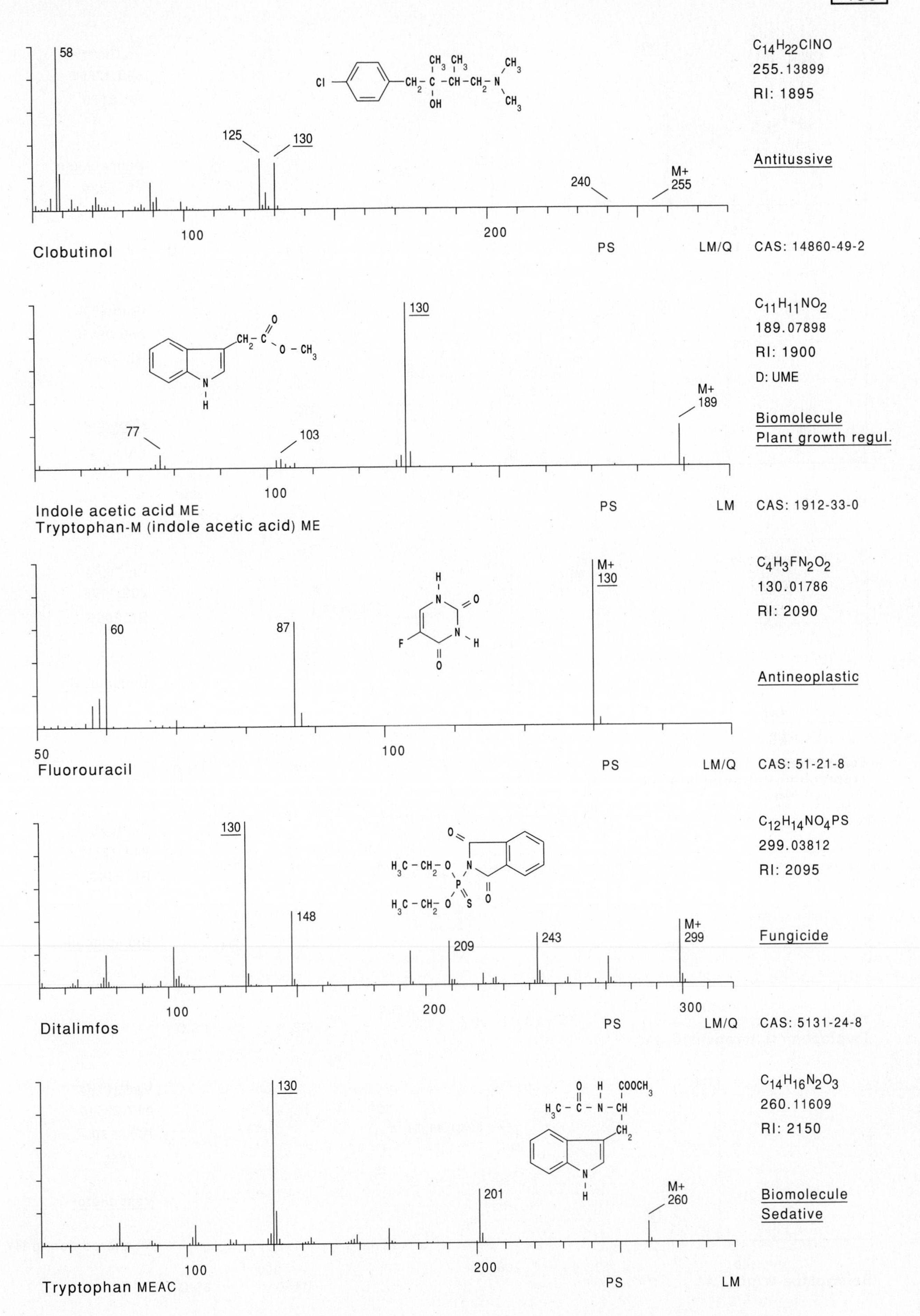

58
125
130
240
M+
255
100
200
Clobutinol
PS
LM/Q
C14H22ClNO
255.13899
RI: 1895
Antitussive
CAS: 14860-49-2
130
77
103
M+
189
Indole acetic acid ME
Tryptophan-M (indole acetic acid) ME
PS
LM
C11H11NO2
189.07898
RI: 1900
D: UME
Biomolecule
Plant growth regul.
CAS: 1912-33-0
M+
130
60
87
50
Fluorouracil
PS
LM/Q
C4H3FN2O2
130.01786
RI: 2090
Antineoplastic
CAS: 51-21-8
130
148
209
243
M+
299
300
Ditalimfos
PS
LM/Q
C12H14NO4PS
299.03812
RI: 2095
Fungicide
CAS: 5131-24-8
130
201
M+
260
Tryptophan MEAC
PS
LM
C14H16N2O3
260.11609
RI: 2150
Biomolecule
Sedative

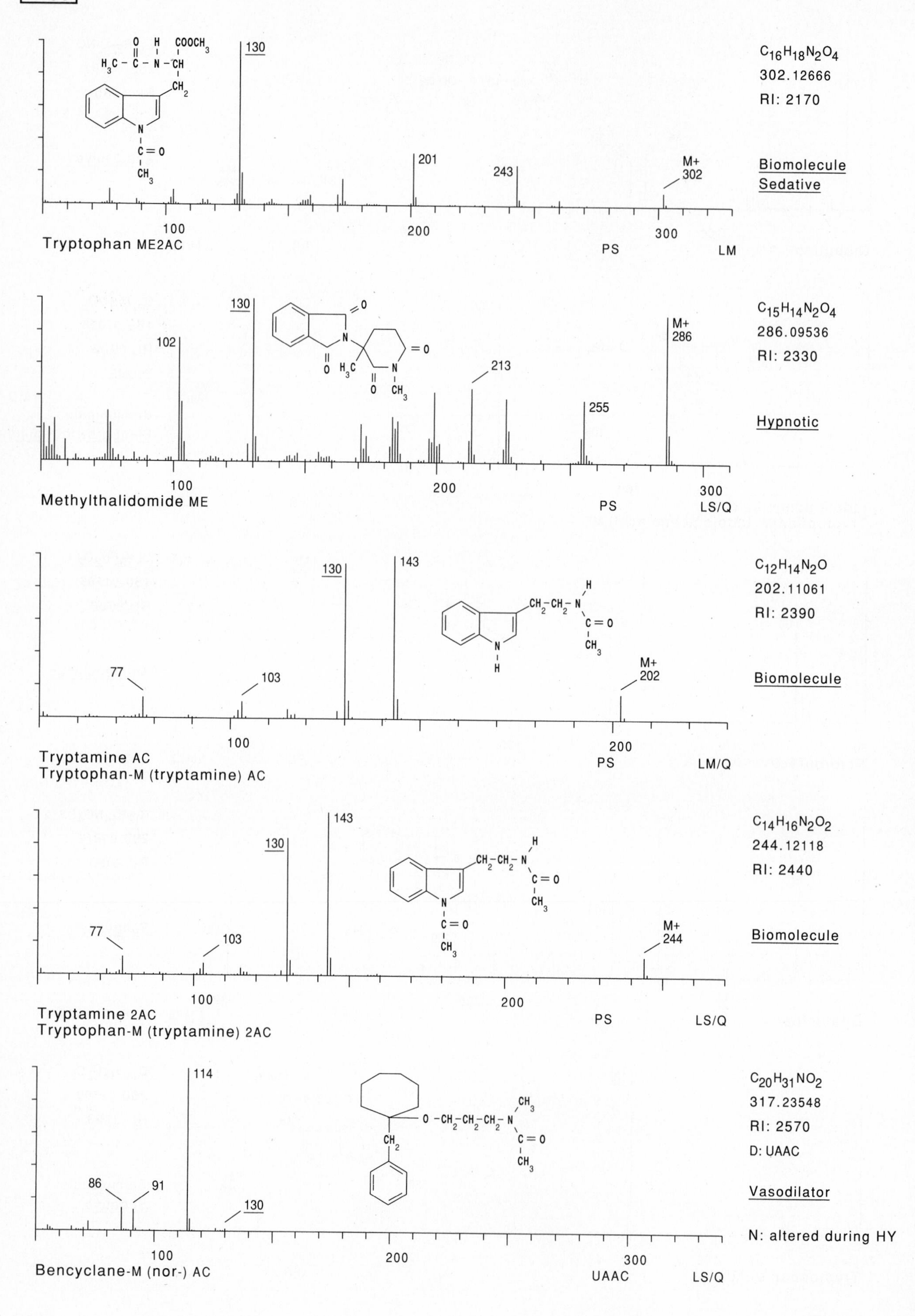

130
201
243
M+ 302
100
200
300
Tryptophan ME2AC
PS
LM
$C_{16}H_{18}N_2O_4$
302.12666
RI: 2170
Biomolecule
Sedative
130
102
213
255
M+ 286
Methylthalidomide ME
PS
LS/Q
$C_{15}H_{14}N_2O_4$
286.09536
RI: 2330
Hypnotic
130
143
77
103
M+ 202
Tryptamine AC
Tryptophan-M (tryptamine) AC
PS
LM/Q
$C_{12}H_{14}N_2O$
202.11061
RI: 2390
Biomolecule
143
130
77
103
M+ 244
Tryptamine 2AC
Tryptophan-M (tryptamine) 2AC
PS
LS/Q
$C_{14}H_{16}N_2O_2$
244.12118
RI: 2440
Biomolecule
114
86
91
130
Bencyclane-M (nor-) AC
UAAC
LS/Q
$C_{20}H_{31}NO_2$
317.23548
RI: 2570
D: UAAC
Vasodilator
N: altered during HY

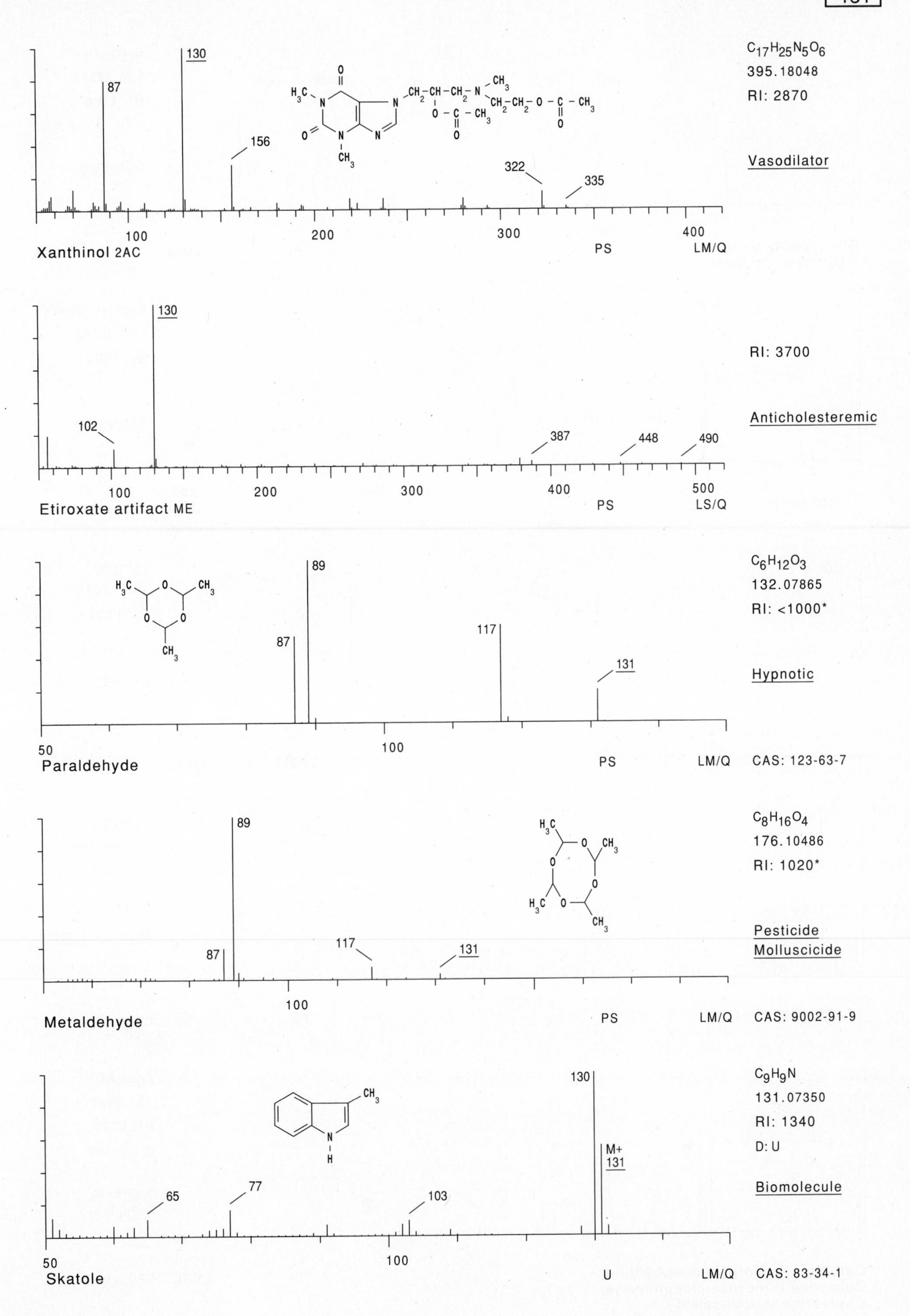

$C_{17}H_{25}N_5O_6$
395.18048
RI: 2870

Vasodilator

RI: 3700

Anticholesteremic

$C_6H_{12}O_3$
132.07865
RI: <1000*

Hypnotic

CAS: 123-63-7

$C_8H_{16}O_4$
176.10486
RI: 1020*

Pesticide
Molluscicide

CAS: 9002-91-9

C_9H_9N
131.07350
RI: 1340
D: U

Biomolecule

CAS: 83-34-1

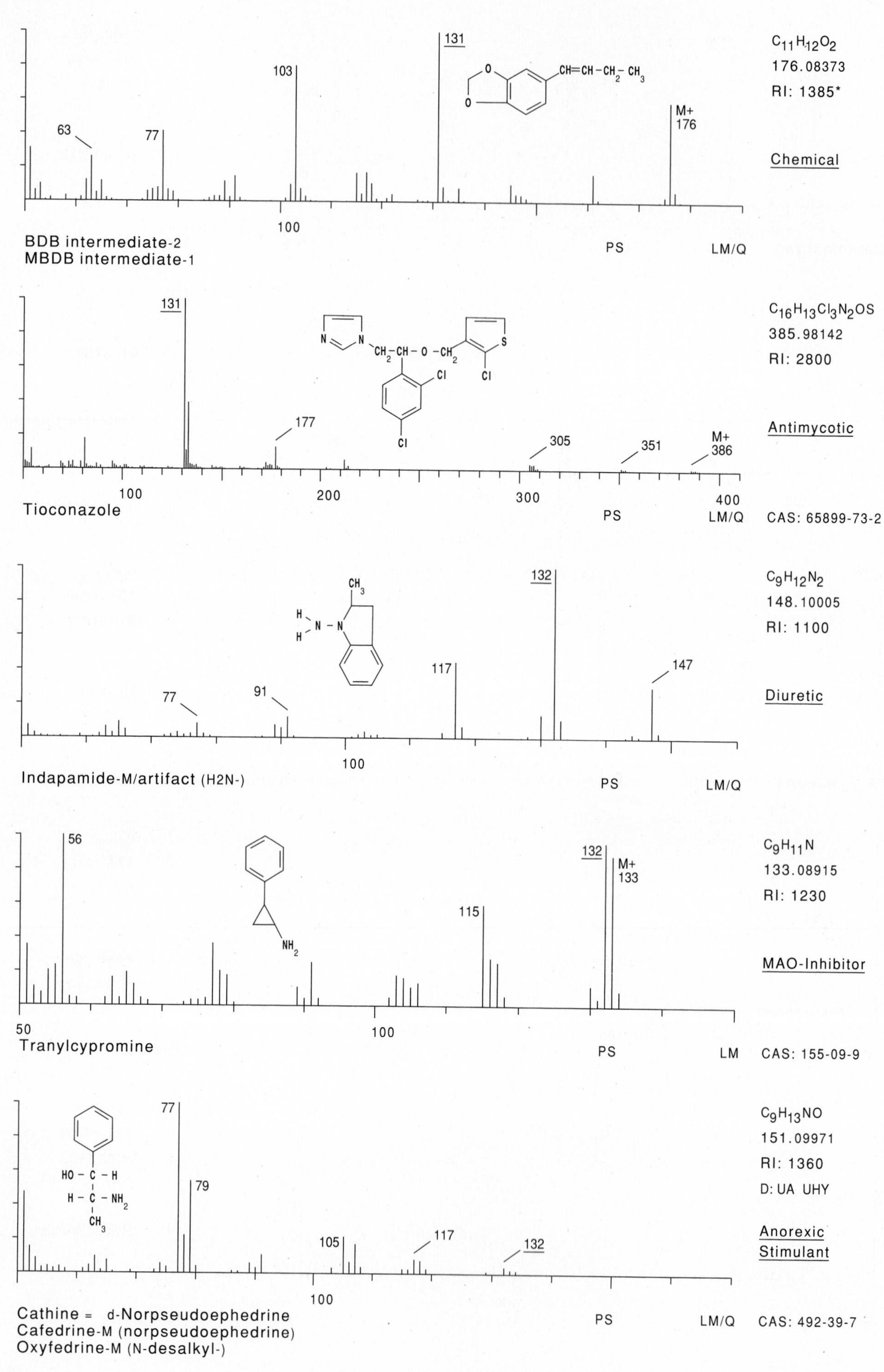
131
103
63
77
M+
176
100
CH=CH-CH2-CH3
C11H12O2
176.08373
RI: 1385*
Chemical
BDB intermediate-2
MBDB intermediate-1
PS
LM/Q
131
177
305
351
M+
386
100
200
300
400
C16H13Cl3N2OS
385.98142
RI: 2800
Antimycotic
Tioconazole
PS
LM/Q
CAS: 65899-73-2
132
117
147
77
91
100
C9H12N2
148.10005
RI: 1100
Diuretic
Indapamide-M/artifact (H2N-)
PS
LM/Q
56
132
M+
133
115
50
100
C9H11N
133.08915
RI: 1230
MAO-Inhibitor
Tranylcypromine
PS
LM
CAS: 155-09-9
77
79
105
117
132
100
HO - C - H
H - C - NH2
CH3
C9H13NO
151.09971
RI: 1360
D: UA UHY
Anorexic
Stimulant
Cathine = d-Norpseudoephedrine
Cafedrine-M (norpseudoephedrine)
Oxyfedrine-M (N-desalkyl-)
PS
LM/Q
CAS: 492-39-7

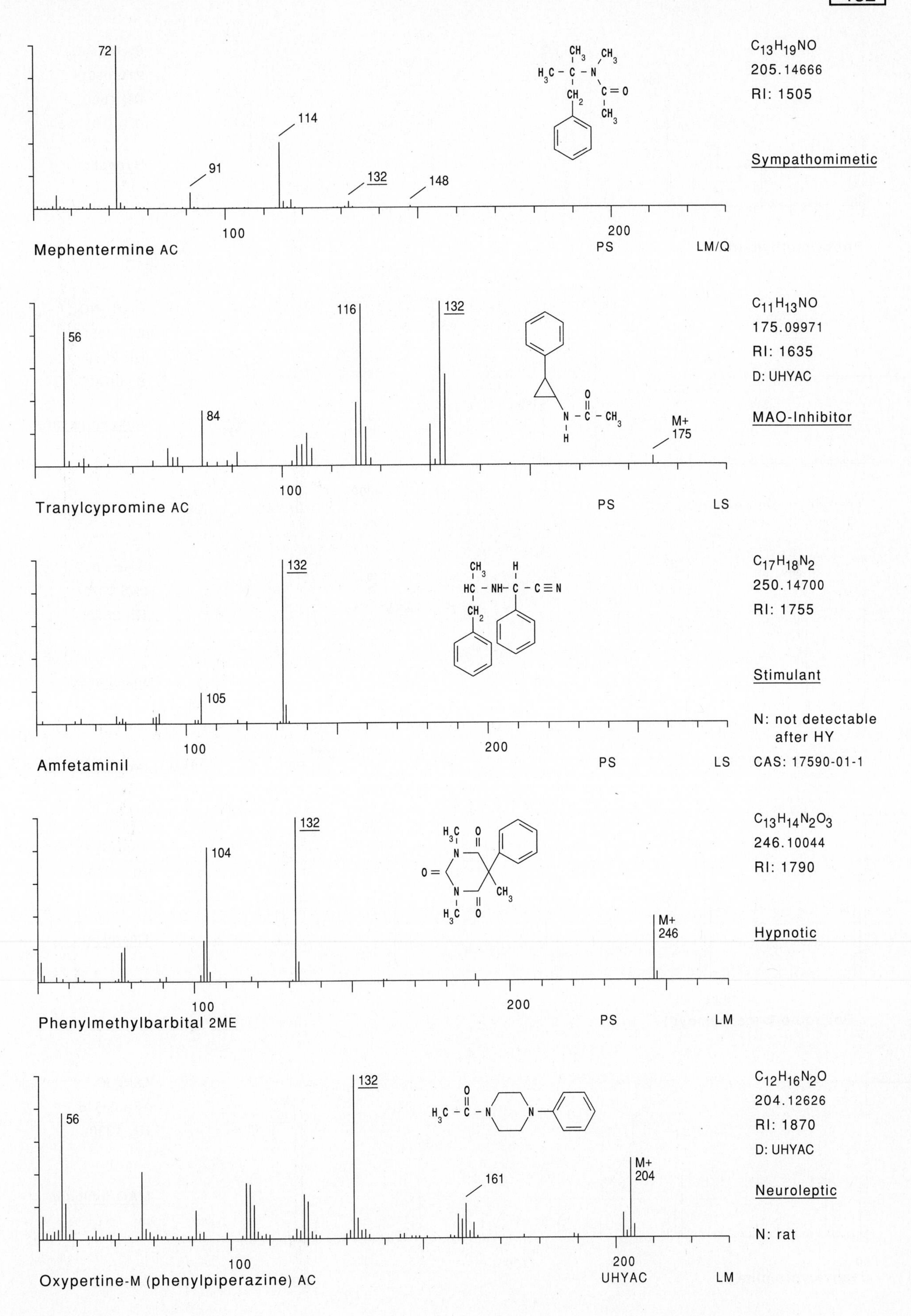
72
114
91
132
148
100
200
Mephentermine AC
PS
LM/Q
C13H19NO
205.14666
RI: 1505
Sympathomimetic
116
132
56
84
M+
175
100
Tranylcypromine AC
PS
LS
C11H13NO
175.09971
RI: 1635
D: UHYAC
MAO-Inhibitor
132
105
100
200
Amfetaminil
PS
LS
C17H18N2
250.14700
RI: 1755
Stimulant
N: not detectable
after HY
CAS: 17590-01-1
132
104
M+
246
100
200
Phenylmethylbarbital 2ME
PS
LM
C13H14N2O3
246.10044
RI: 1790
Hypnotic
132
56
161
M+
204
100
200
UHYAC
Oxypertine-M (phenylpiperazine) AC
LM
C12H16N2O
204.12626
RI: 1870
D: UHYAC
Neuroleptic
N: rat

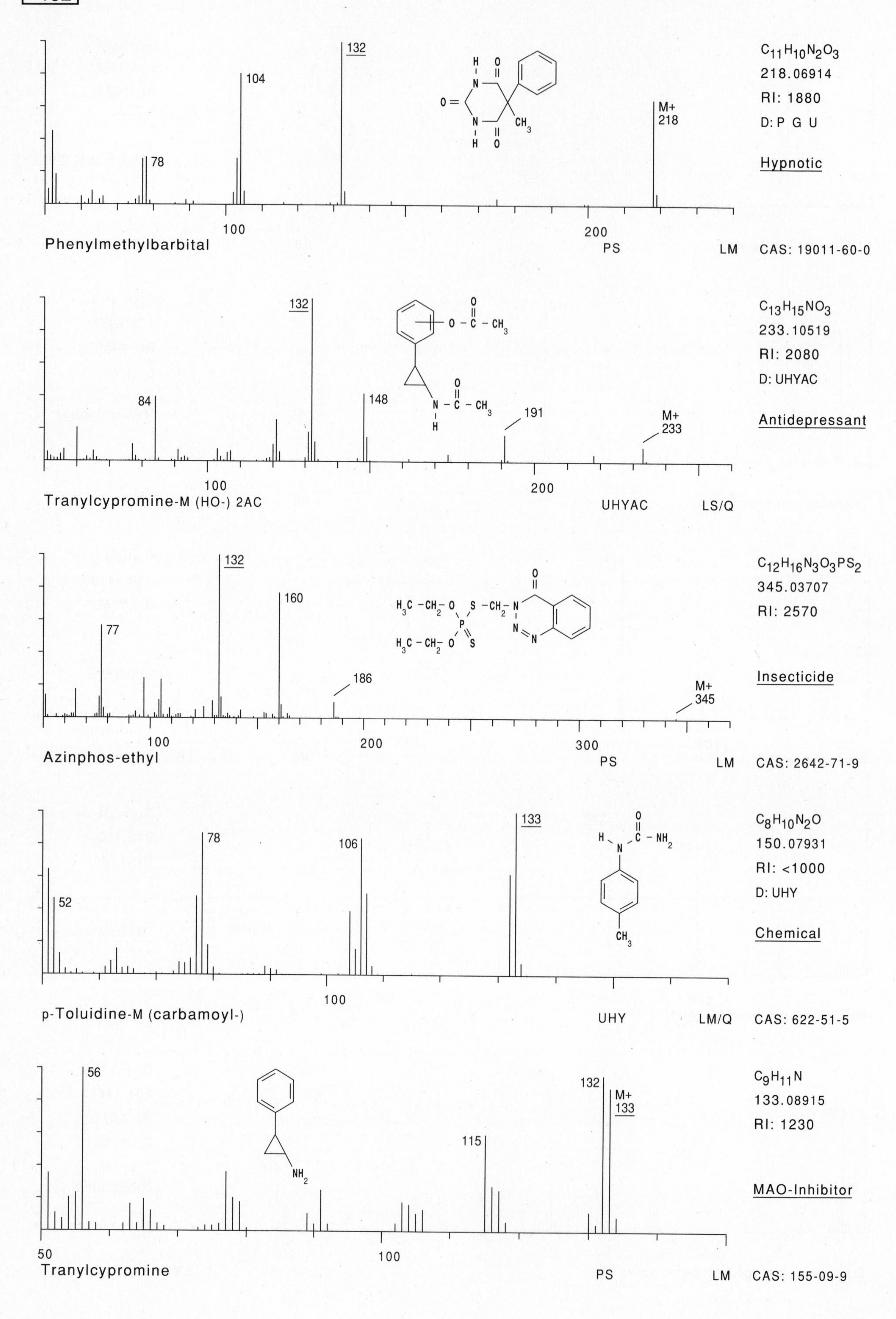
132
104
78
M+
218
100
200
Phenylmethylbarbital
PS
LM
$C_{11}H_{10}N_2O_3$
218.06914
RI: 1880
D: P G U
Hypnotic
CAS: 19011-60-0
132
84
148
191
M+
233
100
200
Tranylcypromine-M (HO-) 2AC
UHYAC
LS/Q
$C_{13}H_{15}NO_3$
233.10519
RI: 2080
D: UHYAC
Antidepressant
132
160
77
186
M+
345
100
200
300
Azinphos-ethyl
PS
LM
$C_{12}H_{16}N_3O_3PS_2$
345.03707
RI: 2570
Insecticide
CAS: 2642-71-9
133
78
106
52
100
p-Toluidine-M (carbamoyl-)
UHY
LM/Q
$C_8H_{10}N_2O$
150.07931
RI: <1000
D: UHY
Chemical
CAS: 622-51-5
56
132
M+
133
115
50
100
Tranylcypromine
PS
LM
$C_9H_{11}N$
133.08915
RI: 1230
MAO-Inhibitor
CAS: 155-09-9

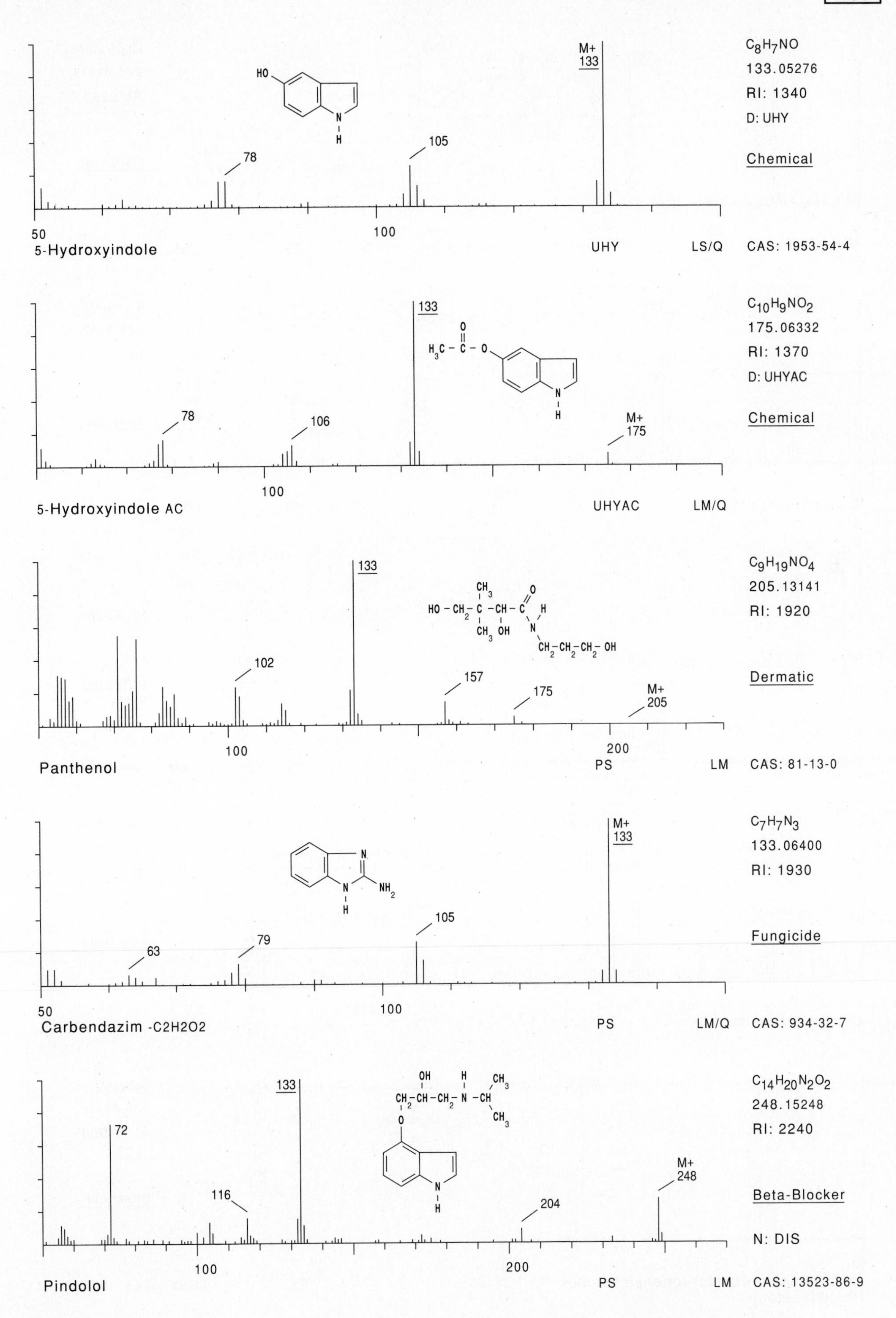

C_8H_7NO
133.05276
RI: 1340
D: UHY
Chemical
CAS: 1953-54-4

$C_{10}H_9NO_2$
175.06332
RI: 1370
D: UHYAC
Chemical

$C_9H_{19}NO_4$
205.13141
RI: 1920
Dermatic
CAS: 81-13-0

$C_7H_7N_3$
133.06400
RI: 1930
Fungicide
CAS: 934-32-7

$C_{14}H_{20}N_2O_2$
248.15248
RI: 2240
Beta-Blocker
N: DIS
CAS: 13523-86-9

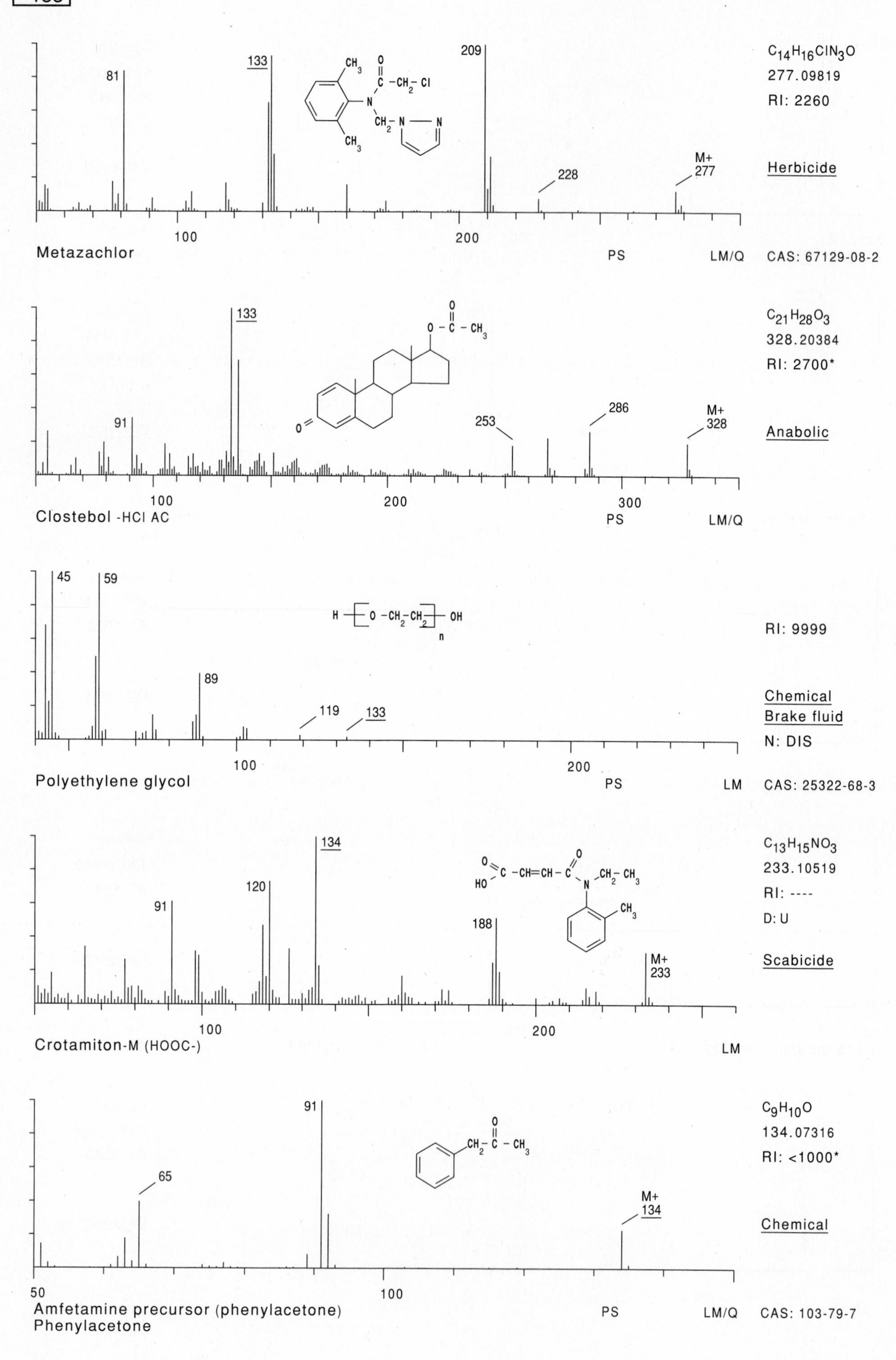
133
Metazachlor
C14H16ClN3O
277.09819
RI: 2260
Herbicide
CAS: 67129-08-2
81
133
209
228
M+
277
100
200
PS
LM/Q
Clostebol -HCl AC
C21H28O3
328.20384
RI: 2700*
Anabolic
91
133
253
286
M+
328
300
Polyethylene glycol
RI: 9999
Chemical
Brake fluid
N: DIS
CAS: 25322-68-3
45
59
89
119
LM
Crotamiton-M (HOOC-)
C13H15NO3
233.10519
RI: ----
D: U
Scabicide
91
120
134
188
M+
233
Amfetamine precursor (phenylacetone)
Phenylacetone
C9H10O
134.07316
RI: <1000*
Chemical
CAS: 103-79-7
65
M+
134
50

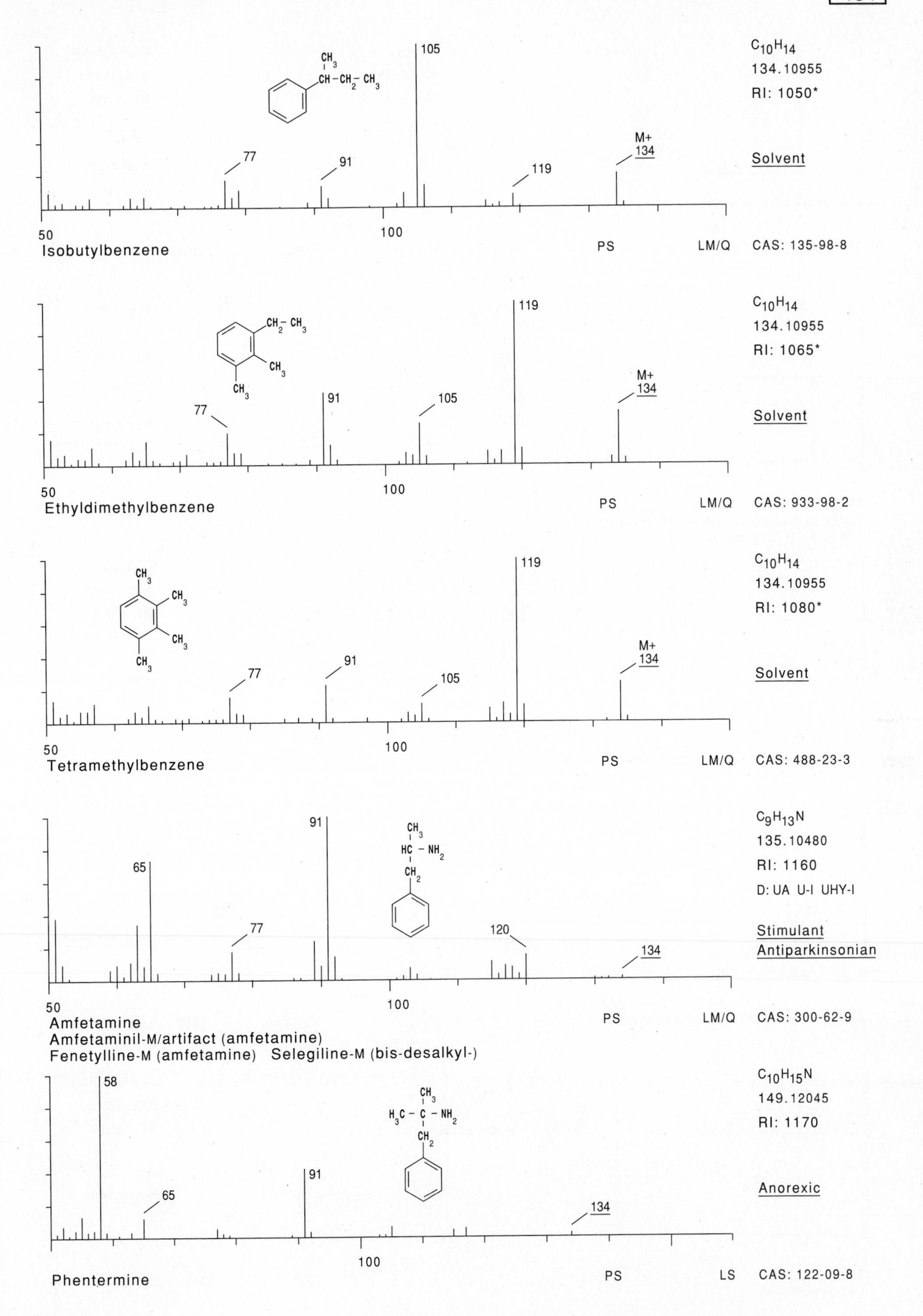

105
77
91
119
M+
134
50
100
Isobutylbenzene
PS
LM/Q
$C_{10}H_{14}$
134.10955
RI: 1050*
Solvent
CAS: 135-98-8
119
91
105
77
M+
134
50
100
Ethyldimethylbenzene
PS
LM/Q
$C_{10}H_{14}$
134.10955
RI: 1065*
Solvent
CAS: 933-98-2
119
77
91
105
M+
134
50
100
Tetramethylbenzene
PS
LM/Q
$C_{10}H_{14}$
134.10955
RI: 1080*
Solvent
CAS: 488-23-3
91
65
77
120
134
50
100
Amfetamine
Amfetaminil-M/artifact (amfetamine)
Fenetylline-M (amfetamine) Selegiline-M (bis-desalkyl-)
PS
LM/Q
$C_9H_{13}N$
135.10480
RI: 1160
D: UA U-I UHY-I
Stimulant
Antiparkinsonian
CAS: 300-62-9
58
91
65
134
100
Phentermine
PS
LS
$C_{10}H_{15}N$
149.12045
RI: 1170
Anorexic
CAS: 122-09-8

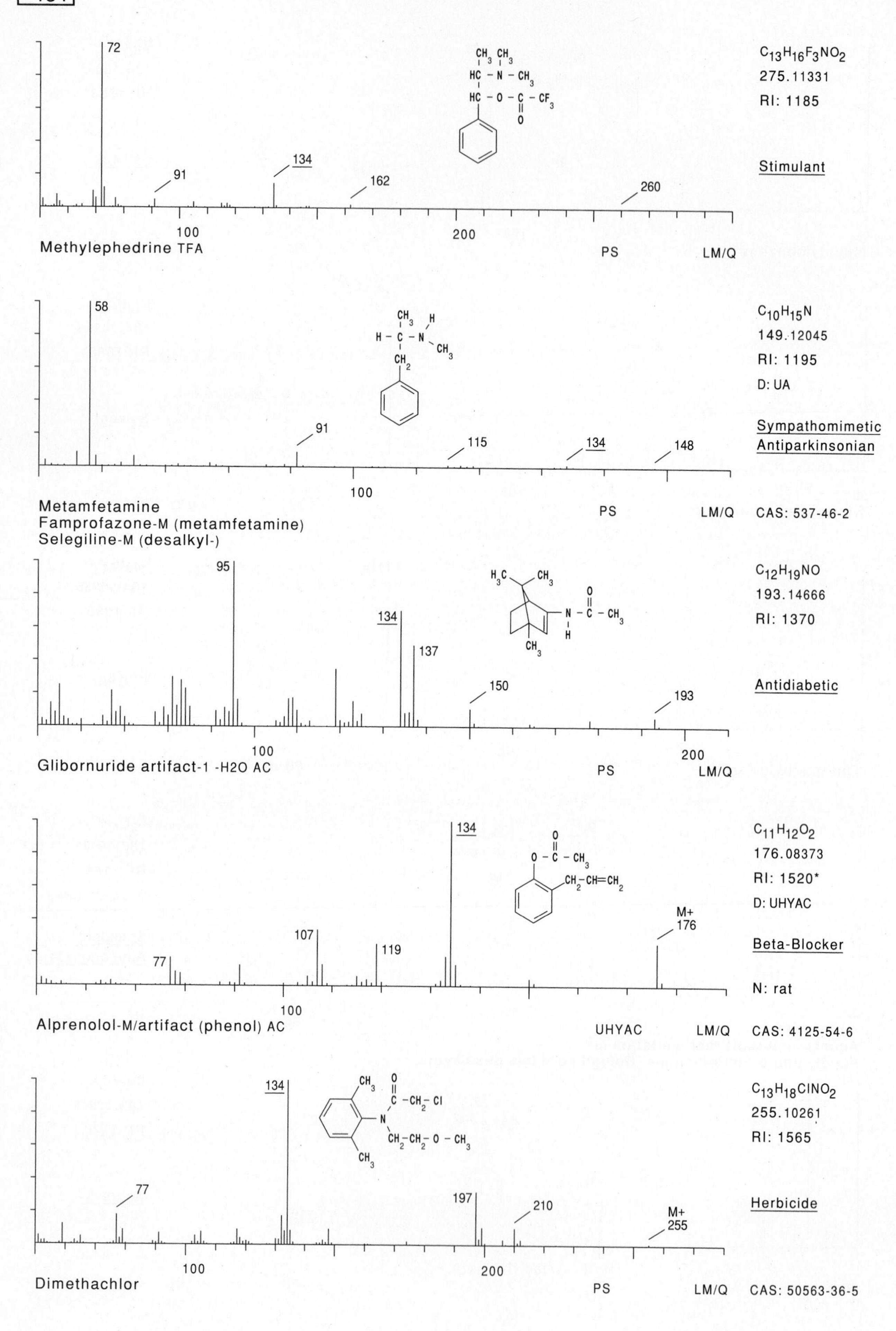

72
91
134
162
260
100
200
Methylephedrine TFA
PS
LM/Q
C13H16F3NO2
275.11331
RI: 1185
Stimulant
58
91
115
134
148
100
Metamfetamine
Famprofazone-M (metamfetamine)
Selegiline-M (desalkyl-)
PS
LM/Q
C10H15N
149.12045
RI: 1195
D: UA
Sympathomimetic
Antiparkinsonian
CAS: 537-46-2
95
134
137
150
193
100
200
Glibornuride artifact-1 -H2O AC
PS
LM/Q
C12H19NO
193.14666
RI: 1370
Antidiabetic
77
107
119
134
M+
176
100
Alprenolol-M/artifact (phenol) AC
UHYAC
LM/Q
C11H12O2
176.08373
RI: 1520*
D: UHYAC
Beta-Blocker
N: rat
CAS: 4125-54-6
77
134
197
210
M+
255
100
200
Dimethachlor
PS
LM/Q
C13H18ClNO2
255.10261
RI: 1565
Herbicide
CAS: 50563-36-5

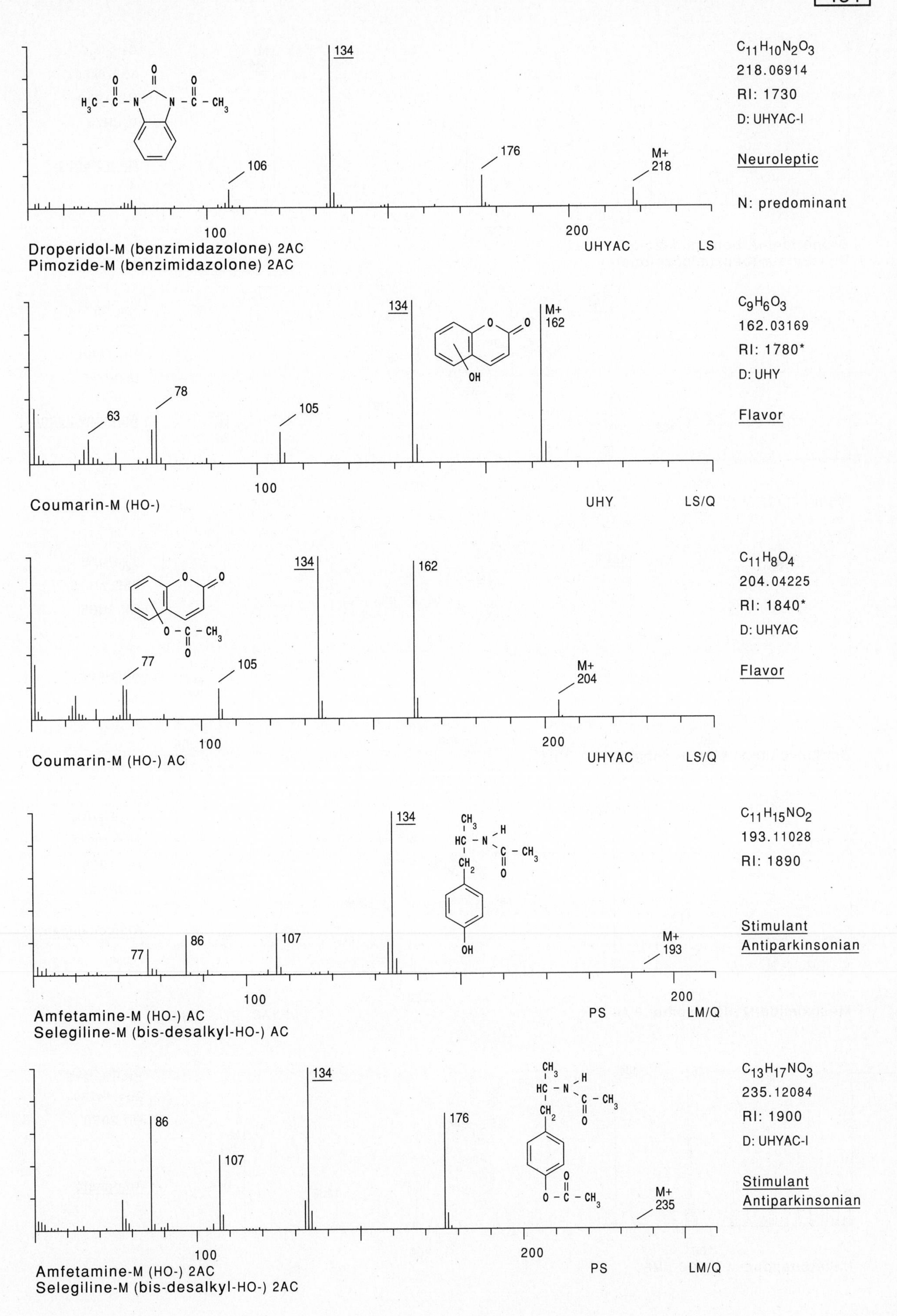

134
106
176
M+
218
100
200
UHYAC
LS
Droperidol-M (benzimidazolone) 2AC
Pimozide-M (benzimidazolone) 2AC
$C_{11}H_{10}N_2O_3$
218.06914
RI: 1730
D: UHYAC-I
Neuroleptic
N: predominant
134
M+
162
63
78
105
100
UHY
LS/Q
Coumarin-M (HO-)
$C_9H_6O_3$
162.03169
RI: 1780*
D: UHY
Flavor
134
162
77
105
M+
204
100
200
UHYAC
LS/Q
Coumarin-M (HO-) AC
$C_{11}H_8O_4$
204.04225
RI: 1840*
D: UHYAC
Flavor
134
77
86
107
M+
193
100
200
PS
LM/Q
Amfetamine-M (HO-) AC
Selegiline-M (bis-desalkyl-HO-) AC
$C_{11}H_{15}NO_2$
193.11028
RI: 1890
Stimulant
Antiparkinsonian
134
86
107
176
M+
235
100
200
PS
LM/Q
Amfetamine-M (HO-) 2AC
Selegiline-M (bis-desalkyl-HO-) 2AC
$C_{13}H_{17}NO_3$
235.12084
RI: 1900
D: UHYAC-I
Stimulant
Antiparkinsonian

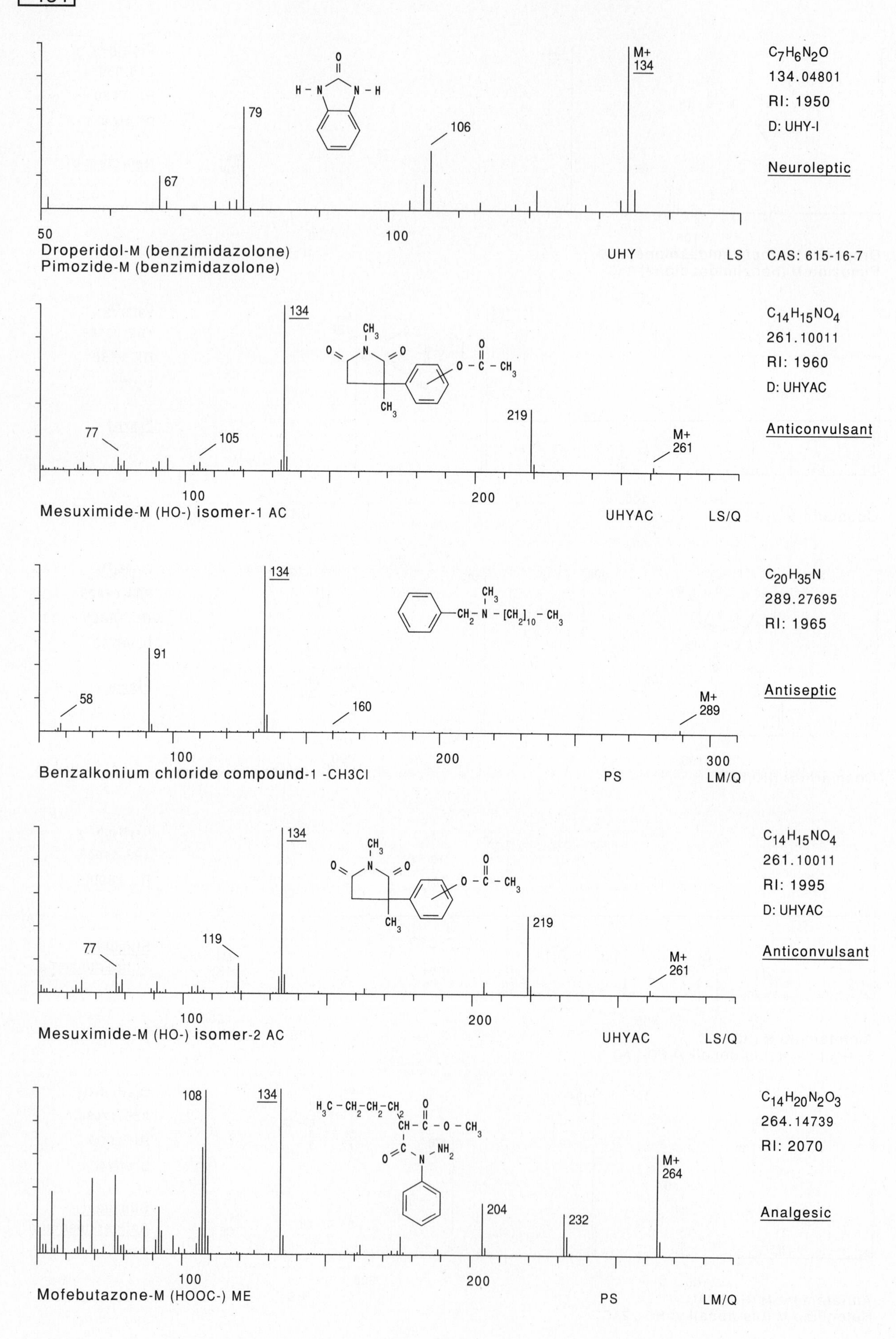

M+ 134
79
106
67
50
100
Droperidol-M (benzimidazolone)
Pimozide-M (benzimidazolone)
UHY
LS
C7H6N2O
134.04801
RI: 1950
D: UHY-I
Neuroleptic
CAS: 615-16-7
134
219
77
105
M+ 261
100
200
Mesuximide-M (HO-) isomer-1 AC
UHYAC
LS/Q
C14H15NO4
261.10011
RI: 1960
D: UHYAC
Anticonvulsant
134
91
58
160
M+ 289
100
200
300
Benzalkonium chloride compound-1 -CH3Cl
PS
LM/Q
C20H35N
289.27695
RI: 1965
Antiseptic
134
219
77
119
M+ 261
100
200
Mesuximide-M (HO-) isomer-2 AC
UHYAC
LS/Q
C14H15NO4
261.10011
RI: 1995
D: UHYAC
Anticonvulsant
108
134
M+ 264
204
232
100
200
Mofebutazone-M (HOOC-) ME
PS
LM/Q
C14H20N2O3
264.14739
RI: 2070
Analgesic

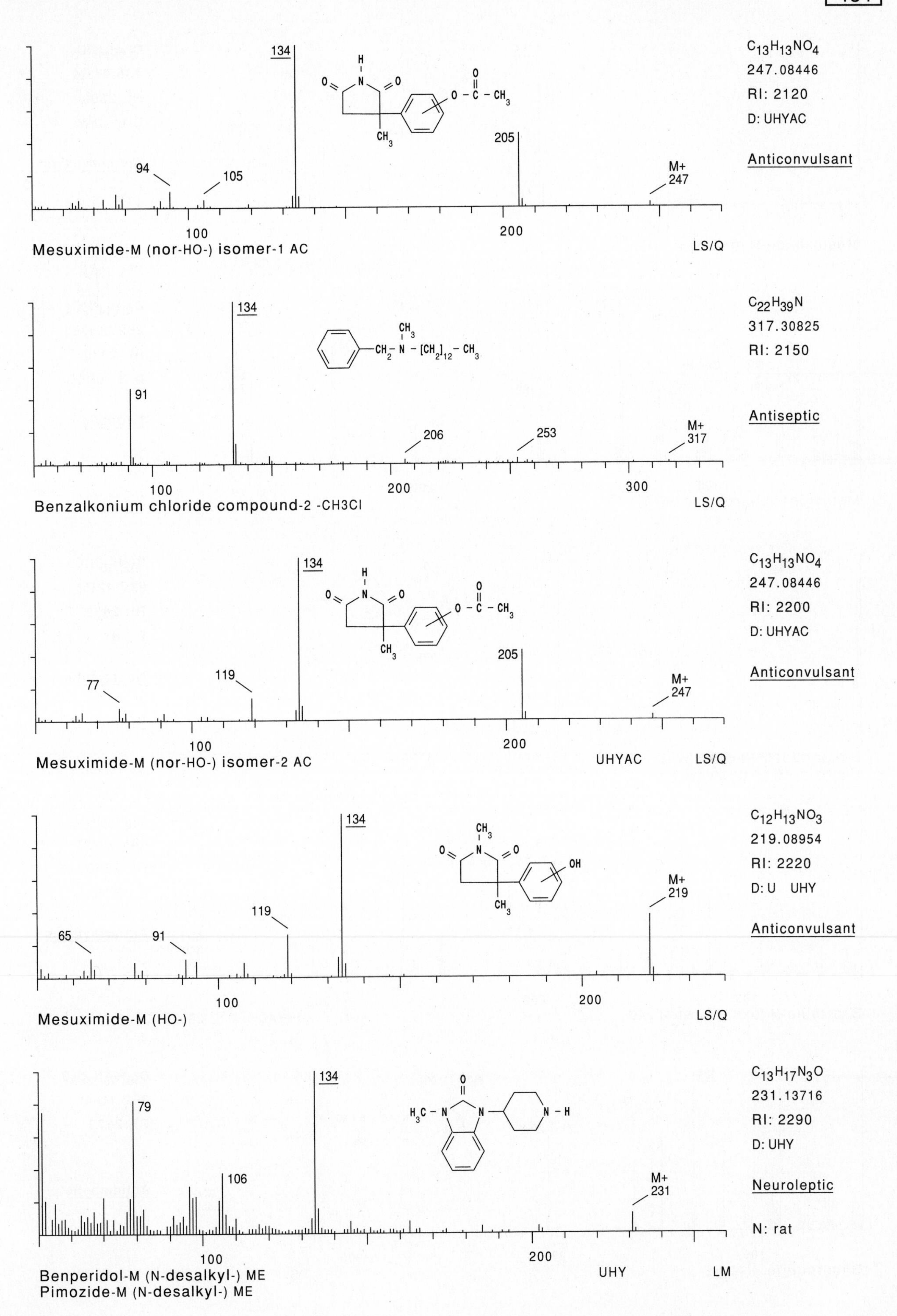

$C_{13}H_{13}NO_4$
247.08446
RI: 2120
D: UHYAC

Anticonvulsant

$C_{22}H_{39}N$
317.30825
RI: 2150

Antiseptic

$C_{13}H_{13}NO_4$
247.08446
RI: 2200
D: UHYAC

Anticonvulsant

$C_{12}H_{13}NO_3$
219.08954
RI: 2220
D: U UHY

Anticonvulsant

$C_{13}H_{17}N_3O$
231.13716
RI: 2290
D: UHY

Neuroleptic

N: rat

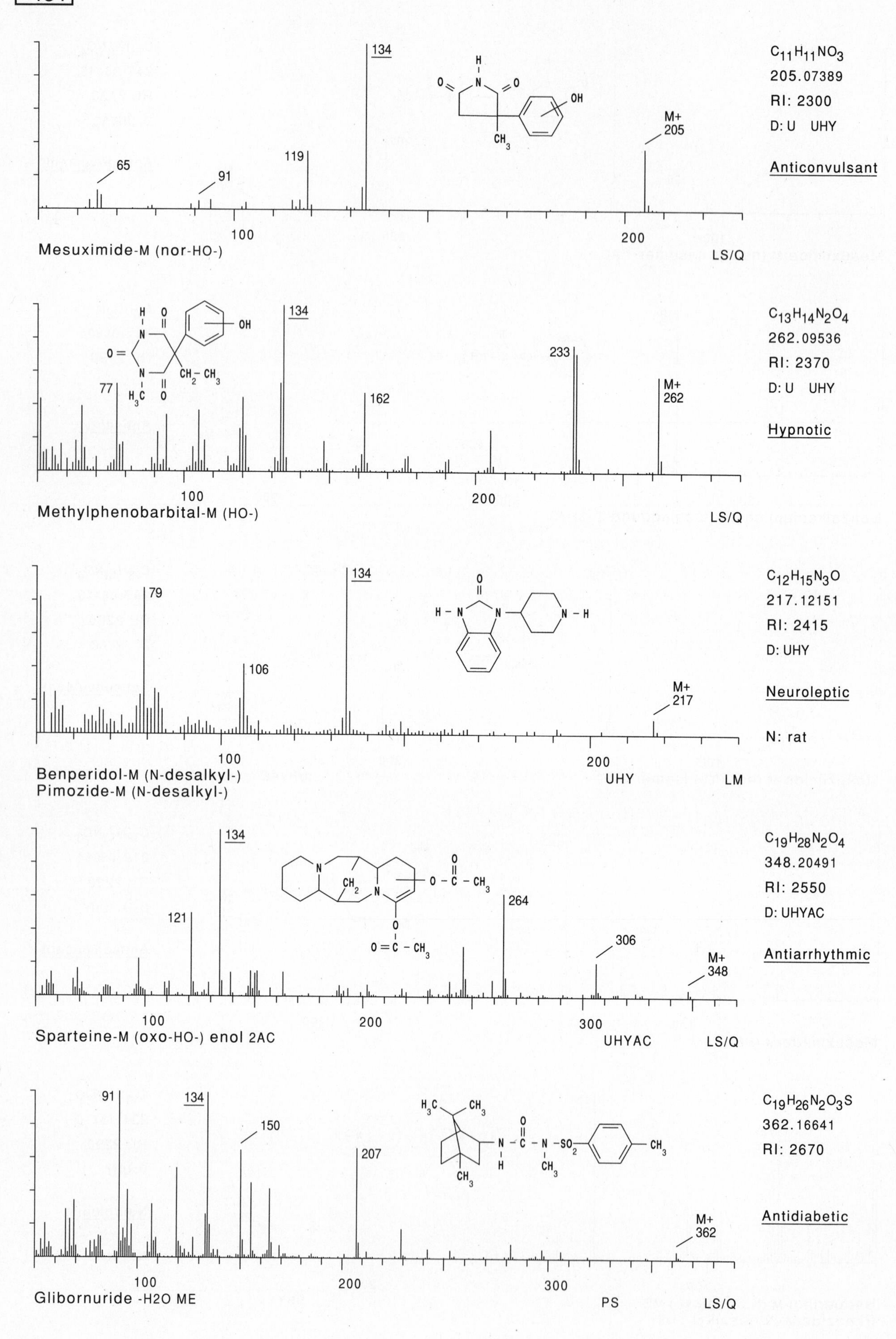

134
65
91
119
M+ 205
100
200
Mesuximide-M (nor-HO-)
LS/Q
C11H11NO3
205.07389
RI: 2300
D: U UHY
Anticonvulsant
134
233
162
77
M+ 262
100
200
Methylphenobarbital-M (HO-)
LS/Q
C13H14N2O4
262.09536
RI: 2370
D: U UHY
Hypnotic
134
79
106
M+ 217
100
200
Benperidol-M (N-desalkyl-)
Pimozide-M (N-desalkyl-)
UHY
LM
C12H15N3O
217.12151
RI: 2415
D: UHY
Neuroleptic
N: rat
134
264
121
306
M+ 348
100
200
300
Sparteine-M (oxo-HO-) enol 2AC
UHYAC
LS/Q
C19H28N2O4
348.20491
RI: 2550
D: UHYAC
Antiarrhythmic
91
134
150
207
M+ 362
100
200
300
Glibornuride -H2O ME
PS
LS/Q
C19H26N2O3S
362.16641
RI: 2670
Antidiabetic

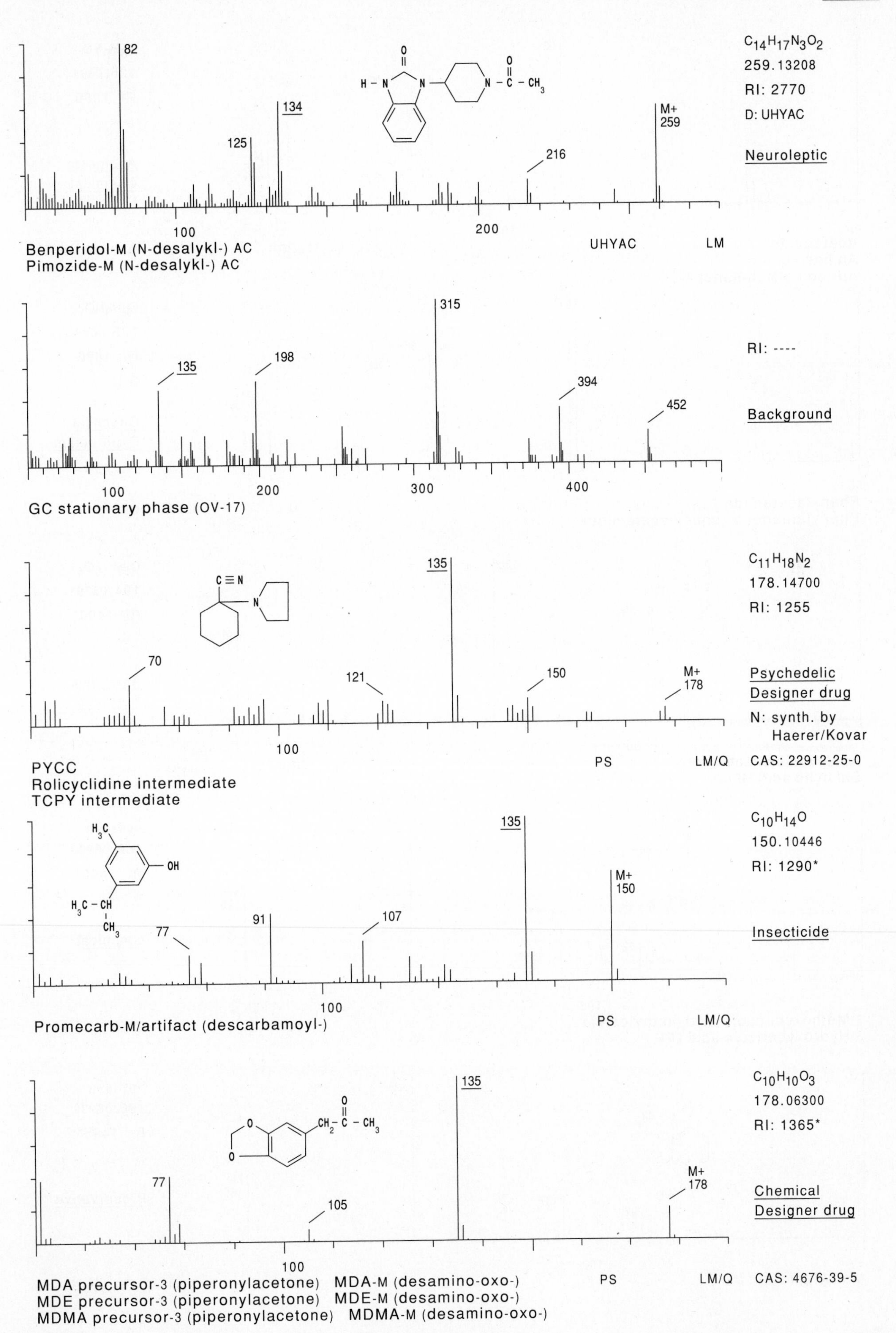

Benperidol-M (N-desalykl-) AC
Pimozide-M (N-desalykl-) AC

$C_{14}H_{17}N_3O_2$
259.13208
RI: 2770
D: UHYAC
Neuroleptic

GC stationary phase (OV-17)

RI: ----
Background

PYCC
Rolicyclidine intermediate
TCPY intermediate

$C_{11}H_{18}N_2$
178.14700
RI: 1255
Psychedelic
Designer drug
N: synth. by Haerer/Kovar
CAS: 22912-25-0

Promecarb-M/artifact (descarbamoyl-)

$C_{10}H_{14}O$
150.10446
RI: 1290*
Insecticide

MDA precursor-3 (piperonylacetone) MDA-M (desamino-oxo-)
MDE precursor-3 (piperonylacetone) MDE-M (desamino-oxo-)
MDMA precursor-3 (piperonylacetone) MDMA-M (desamino-oxo-)

$C_{10}H_{10}O_3$
178.06300
RI: 1365*
Chemical
Designer drug
CAS: 4676-39-5

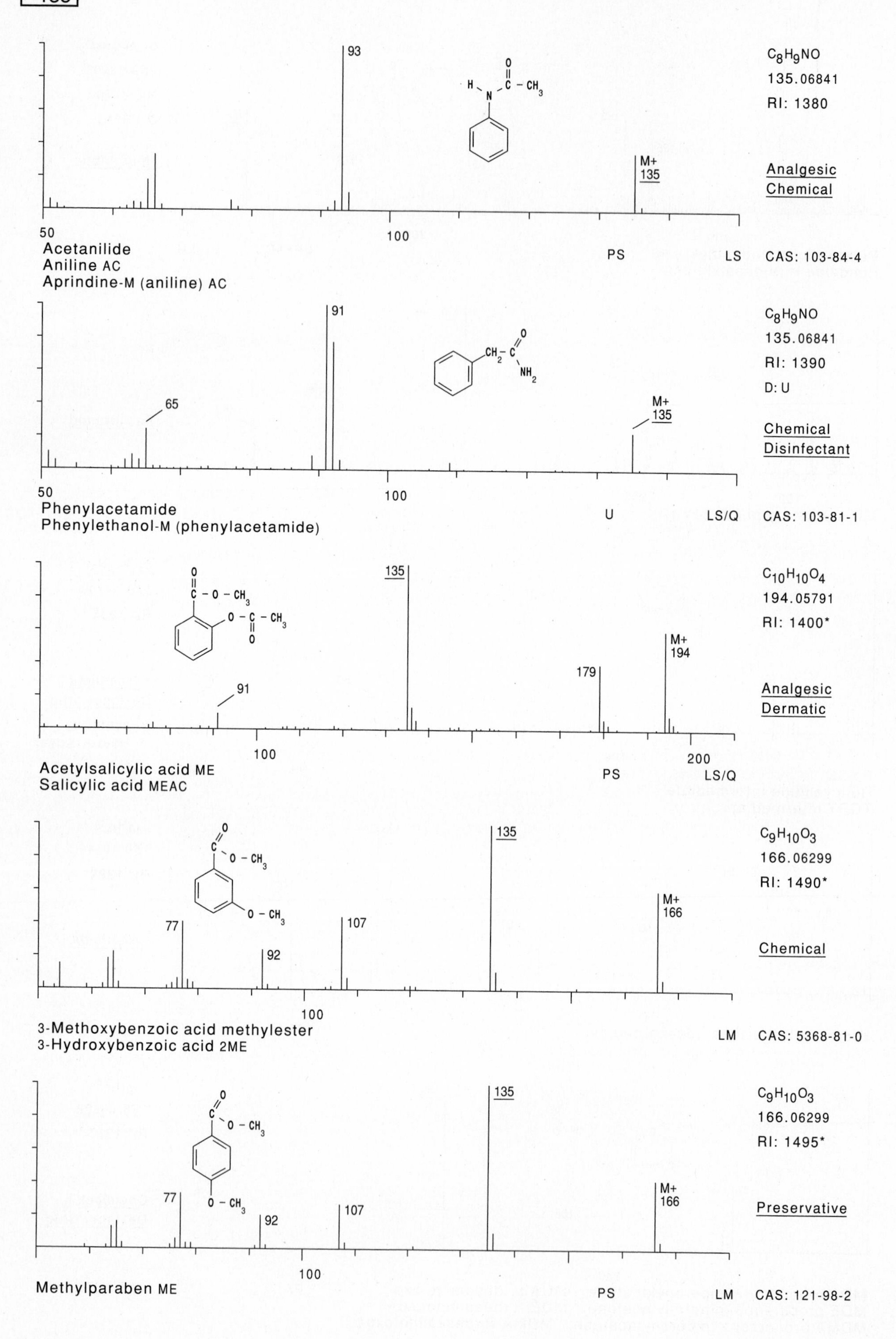
93
M+
135
50
100
C8H9NO
135.06841
RI: 1380
Analgesic
Chemical
PS
LS
CAS: 103-84-4
Acetanilide
Aniline AC
Aprindine-M (aniline) AC
91
65
M+
135
C8H9NO
135.06841
RI: 1390
D: U
Chemical
Disinfectant
U
LS/Q
CAS: 103-81-1
Phenylacetamide
Phenylethanol-M (phenylacetamide)
135
M+
194
179
91
C10H10O4
194.05791
RI: 1400*
Analgesic
Dermatic
200
PS
LS/Q
Acetylsalicylic acid ME
Salicylic acid MEAC
135
M+
166
77
107
92
C9H10O3
166.06299
RI: 1490*
Chemical
LM
CAS: 5368-81-0
3-Methoxybenzoic acid methylester
3-Hydroxybenzoic acid 2ME
C9H10O3
166.06299
RI: 1495*
Preservative
PS
LM
CAS: 121-98-2
Methylparaben ME

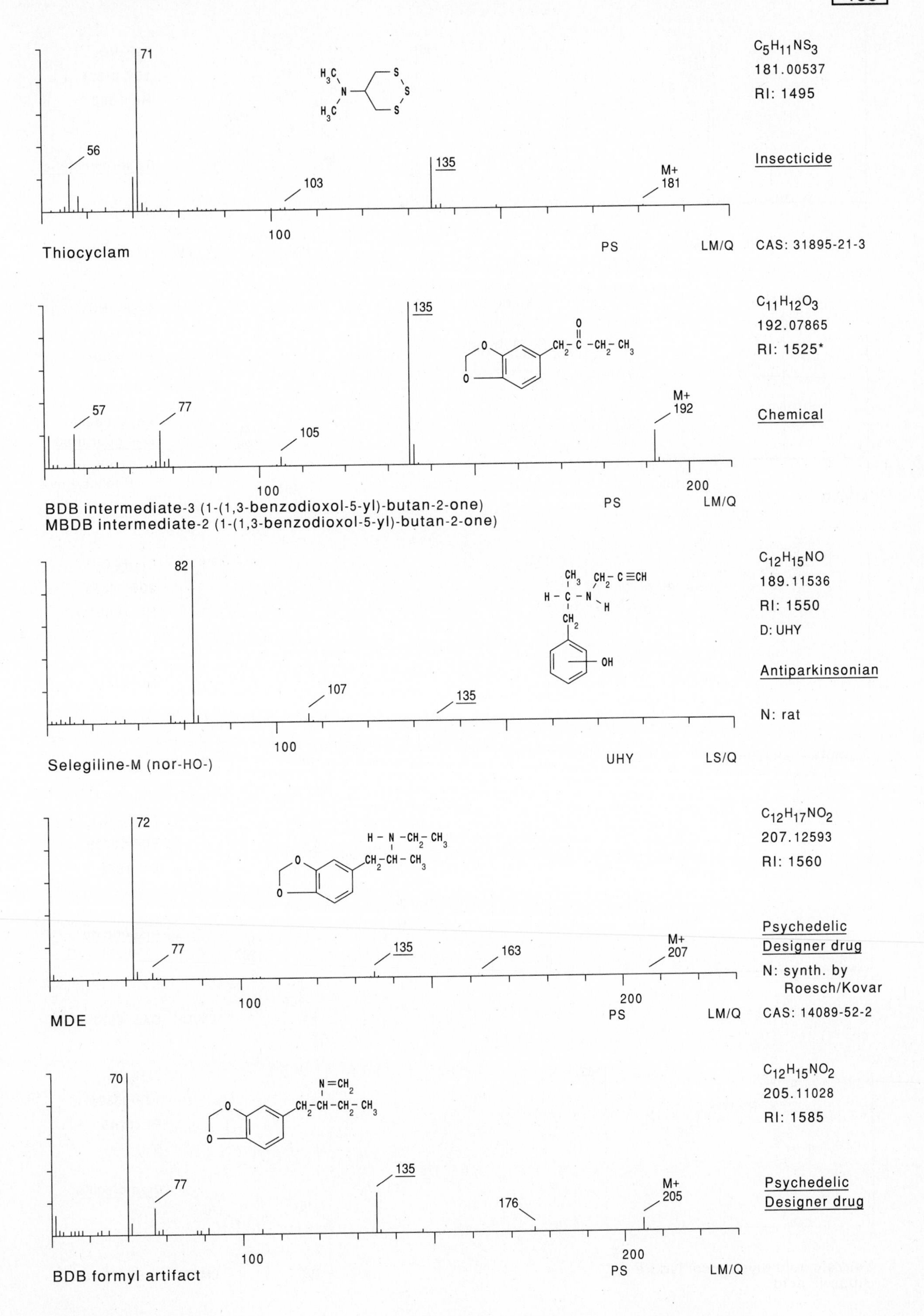

Thiocyclam
$C_5H_{11}NS_3$
181.00537
RI: 1495
Insecticide
CAS: 31895-21-3
71
56
103
135
M+ 181
100
PS
LM/Q
BDB intermediate-3 (1-(1,3-benzodioxol-5-yl)-butan-2-one)
MBDB intermediate-2 (1-(1,3-benzodioxol-5-yl)-butan-2-one)
$C_{11}H_{12}O_3$
192.07865
RI: 1525*
Chemical
135
57
77
105
M+ 192
100
200
PS
LM/Q
Selegiline-M (nor-HO-)
$C_{12}H_{15}NO$
189.11536
RI: 1550
D: UHY
Antiparkinsonian
N: rat
82
107
135
100
UHY
LS/Q
MDE
$C_{12}H_{17}NO_2$
207.12593
RI: 1560
Psychedelic
Designer drug
N: synth. by Roesch/Kovar
CAS: 14089-52-2
72
77
135
163
M+ 207
100
200
PS
LM/Q
BDB formyl artifact
$C_{12}H_{15}NO_2$
205.11028
RI: 1585
Psychedelic
Designer drug
70
77
135
176
M+ 205
100
200
PS
LM/Q

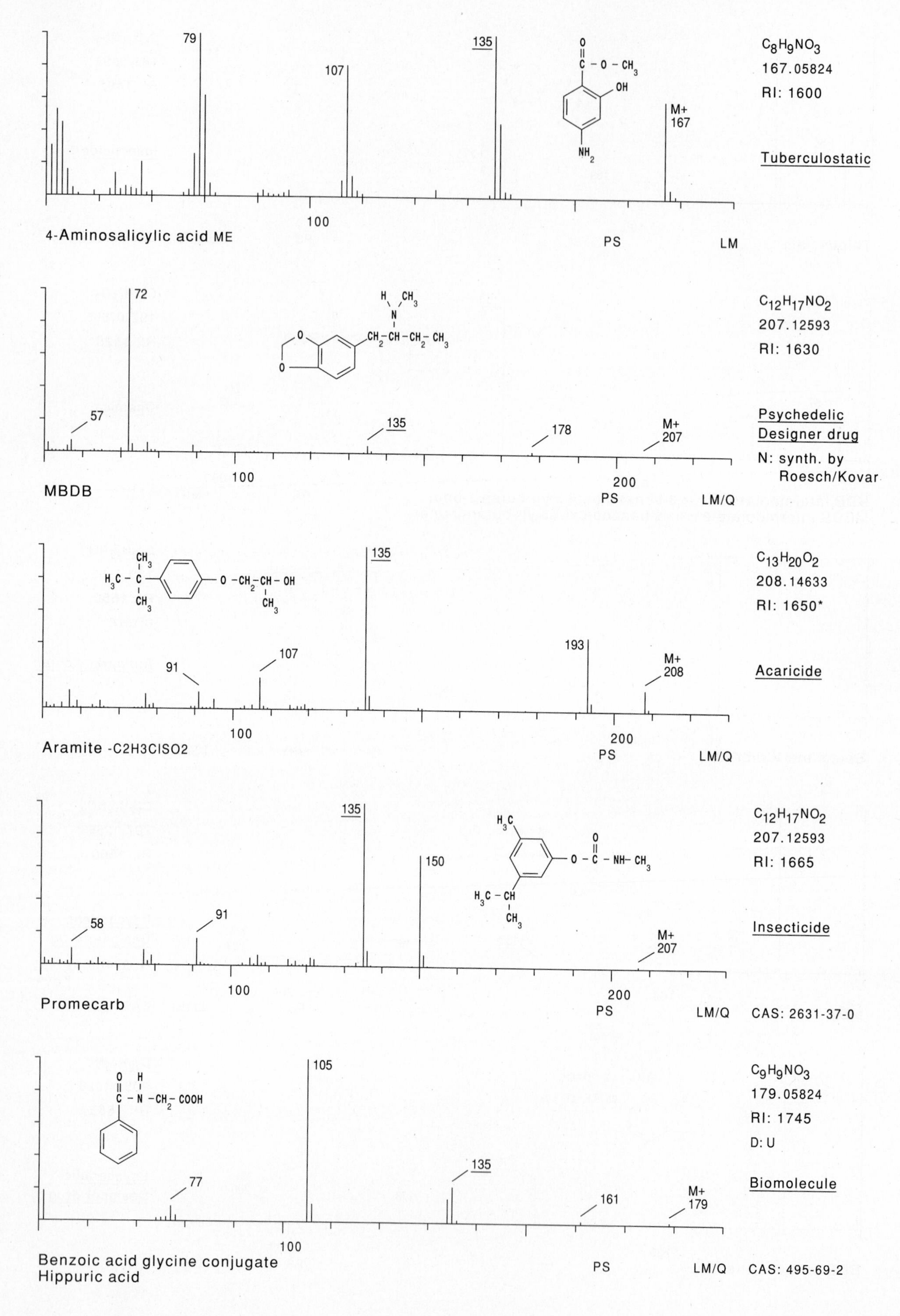
79
135
107
M+
167
100
4-Aminosalicylic acid ME
PS
LM
$C_8H_9NO_3$
167.05824
RI: 1600
Tuberculostatic
72
57
135
178
M+
207
100
200
MBDB
PS
LM/Q
$C_{12}H_{17}NO_2$
207.12593
RI: 1630
Psychedelic
Designer drug
N: synth. by
Roesch/Kovar
135
107
91
193
M+
208
100
200
Aramite -C2H3ClSO2
PS
LM/Q
$C_{13}H_{20}O_2$
208.14633
RI: 1650*
Acaricide
135
150
58
91
M+
207
100
200
Promecarb
PS
LM/Q
$C_{12}H_{17}NO_2$
207.12593
RI: 1665
Insecticide
CAS: 2631-37-0
105
77
135
161
M+
179
100
Benzoic acid glycine conjugate
Hippuric acid
PS
LM/Q
$C_9H_9NO_3$
179.05824
RI: 1745
D: U
Biomolecule
CAS: 495-69-2

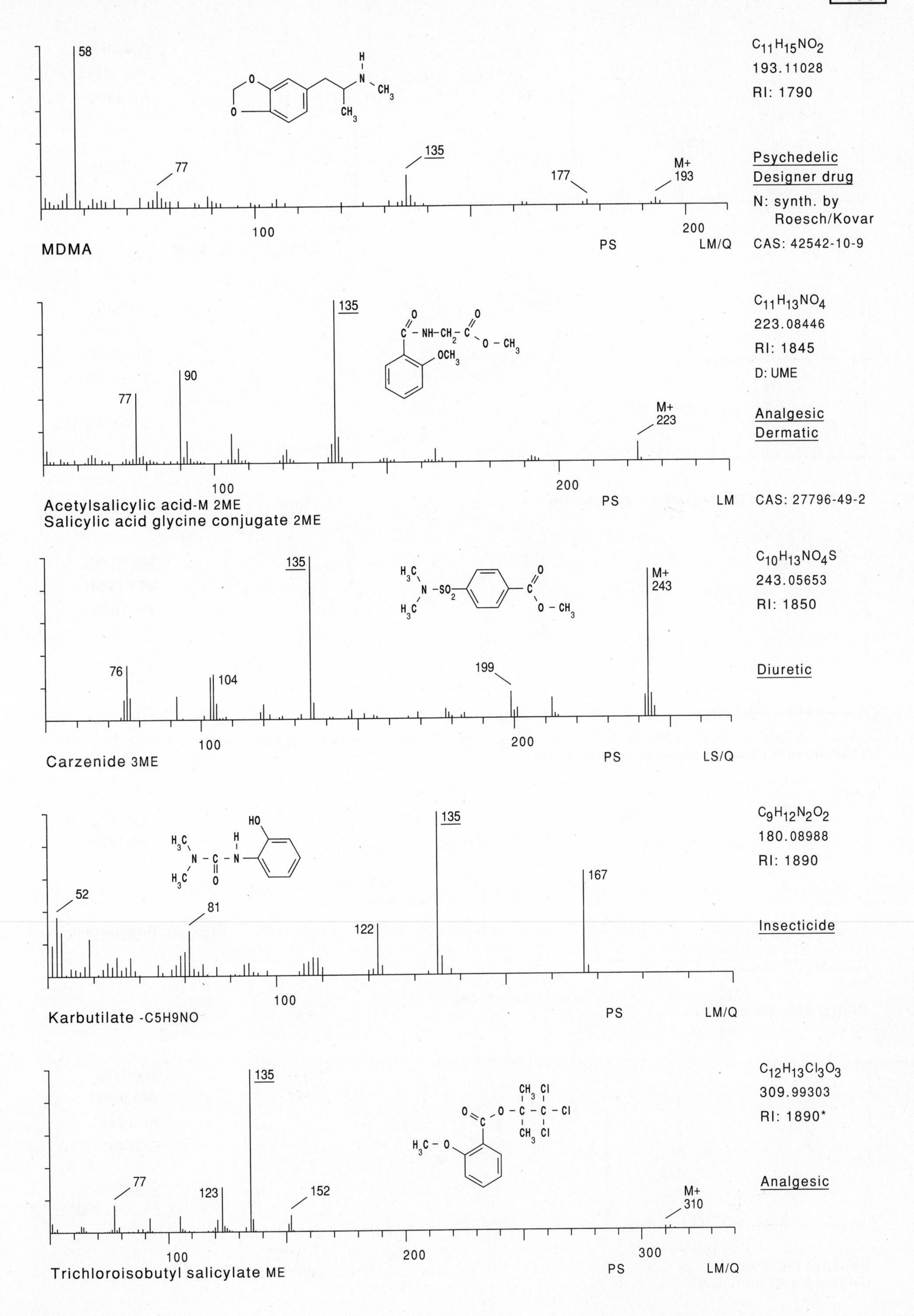
135
58
77
135
177
M+
193
100
200
PS
LM/Q
MDMA
C11H15NO2
193.11028
RI: 1790
Psychedelic
Designer drug
N: synth. by
Roesch/Kovar
CAS: 42542-10-9
135
90
77
M+
223
100
200
PS
LM
Acetylsalicylic acid-M 2ME
Salicylic acid glycine conjugate 2ME
C11H13NO4
223.08446
RI: 1845
D: UME
Analgesic
Dermatic
CAS: 27796-49-2
135
M+
243
76
104
199
100
200
PS
LS/Q
Carzenide 3ME
C10H13NO4S
243.05653
RI: 1850
Diuretic
135
167
52
81
122
100
PS
LM/Q
Karbutilate -C5H9NO
C9H12N2O2
180.08988
RI: 1890
Insecticide
135
77
123
152
M+
310
100
200
300
PS
LM/Q
Trichloroisobutyl salicylate ME
C12H13Cl3O3
309.99303
RI: 1890*
Analgesic

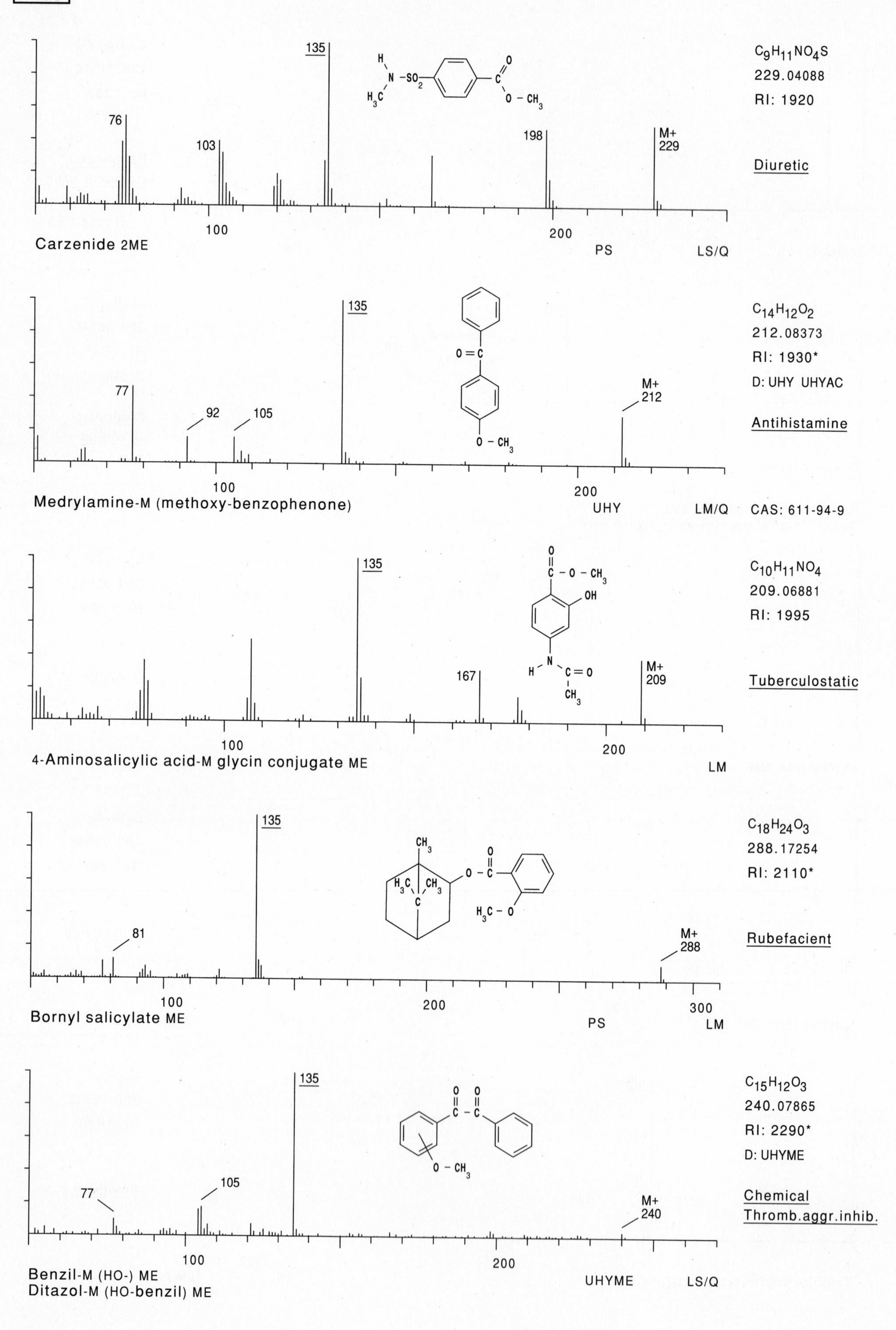
135
76
103
198
M+
229
100
200
Carzenide 2ME
PS
LS/Q
C9H11NO4S
229.04088
RI: 1920
Diuretic
135
77
92
105
M+
212
100
200
Medrylamine-M (methoxy-benzophenone)
UHY
LM/Q
C14H12O2
212.08373
RI: 1930*
D: UHY UHYAC
Antihistamine
CAS: 611-94-9
135
167
M+
209
100
200
4-Aminosalicylic acid-M glycin conjugate ME
LM
C10H11NO4
209.06881
RI: 1995
Tuberculostatic
135
81
M+
288
100
200
300
Bornyl salicylate ME
PS
LM
C18H24O3
288.17254
RI: 2110*
Rubefacient
135
77
105
M+
240
100
200
Benzil-M (HO-) ME
Ditazol-M (HO-benzil) ME
UHYME
LS/Q
C15H12O3
240.07865
RI: 2290*
D: UHYME
Chemical
Thromb.aggr.inhib.

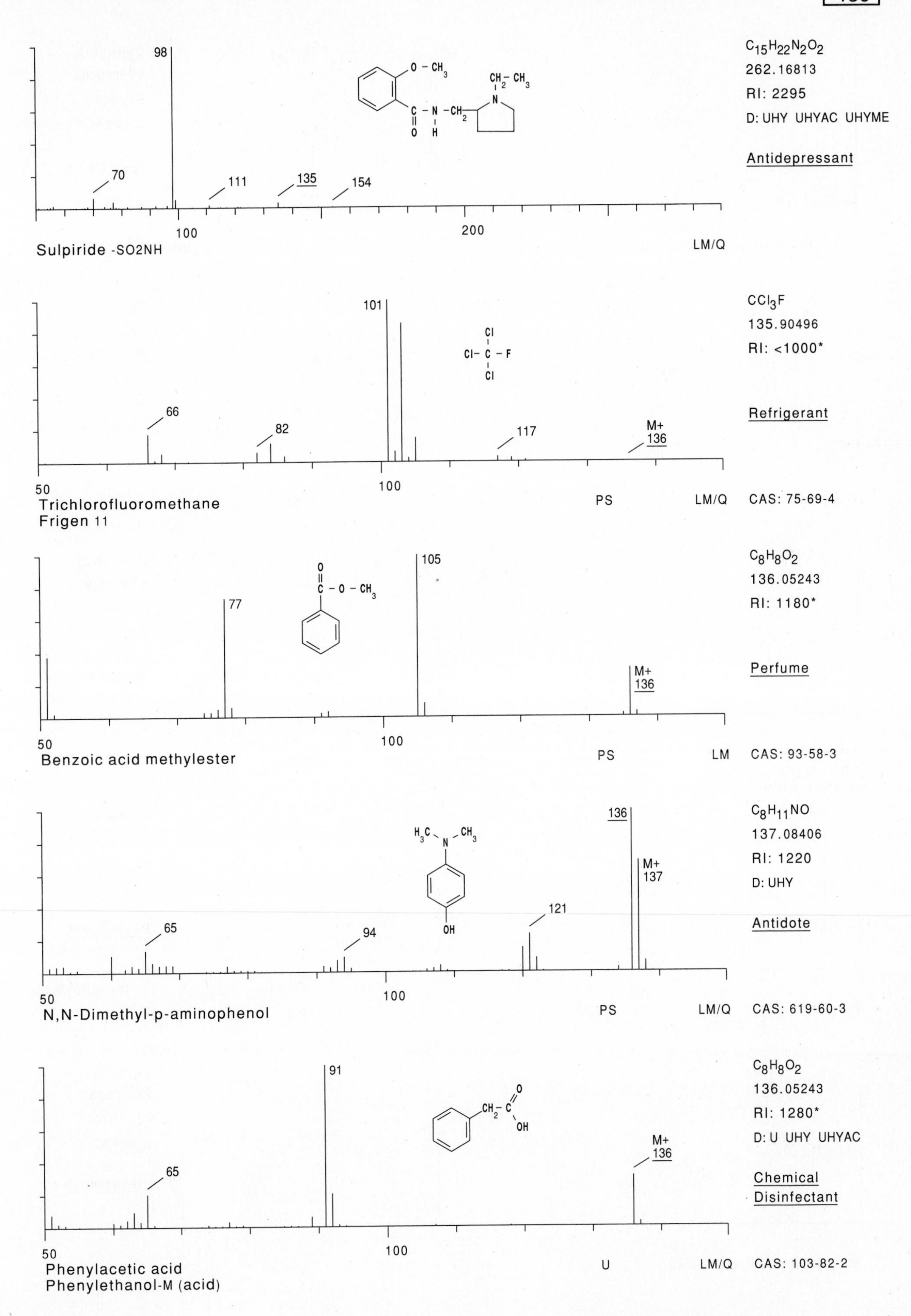
98
70
111
135
154
100
200
Sulpiride -SO2NH
LM/Q
C15H22N2O2
262.16813
RI: 2295
D: UHY UHYAC UHYME
Antidepressant
101
66
82
117
M+
136
50
100
Trichlorofluoromethane
Frigen 11
PS
LM/Q
CCl3F
135.90496
RI: <1000*
Refrigerant
CAS: 75-69-4
105
77
M+
136
50
100
Benzoic acid methylester
PS
LM
C8H8O2
136.05243
RI: 1180*
Perfume
CAS: 93-58-3
136
M+
137
121
65
94
50
100
N,N-Dimethyl-p-aminophenol
PS
LM/Q
C8H11NO
137.08406
RI: 1220
D: UHY
Antidote
CAS: 619-60-3
91
M+
136
65
50
100
Phenylacetic acid
Phenylethanol-M (acid)
U
LM/Q
C8H8O2
136.05243
RI: 1280*
D: U UHY UHYAC
Chemical
Disinfectant
CAS: 103-82-2

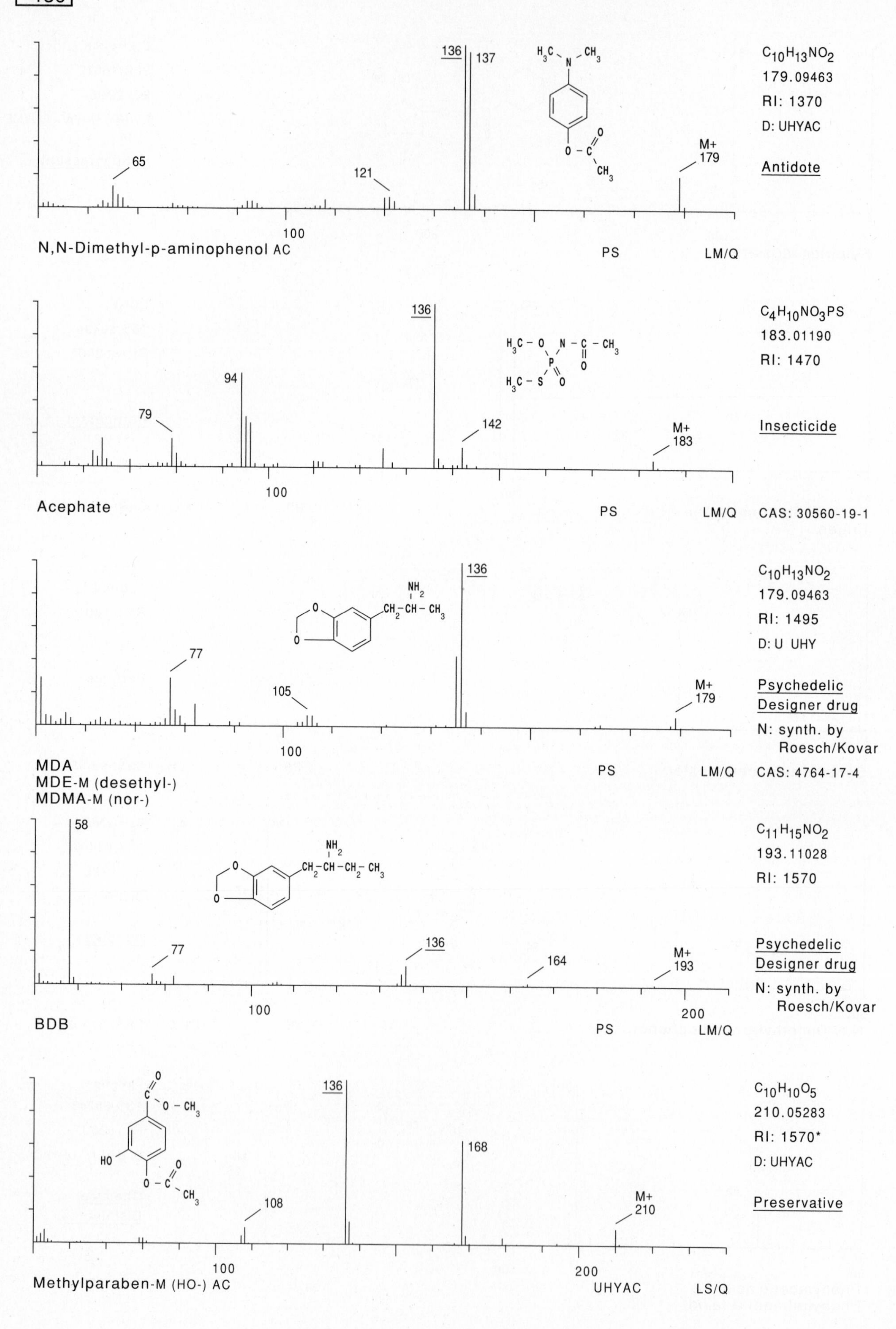
136
137
65
121
M+
179
100
$C_{10}H_{13}NO_2$
179.09463
RI: 1370
D: UHYAC
Antidote
N,N-Dimethyl-p-aminophenol AC
PS
LM/Q
136
94
79
142
M+
183
100
$C_4H_{10}NO_3PS$
183.01190
RI: 1470
Insecticide
Acephate
PS
LM/Q
CAS: 30560-19-1
136
77
105
M+
179
100
$C_{10}H_{13}NO_2$
179.09463
RI: 1495
D: U UHY
Psychedelic
Designer drug
N: synth. by
Roesch/Kovar
MDA
MDE-M (desethyl-)
MDMA-M (nor-)
PS
LM/Q
CAS: 4764-17-4
58
77
136
164
M+
193
100
200
$C_{11}H_{15}NO_2$
193.11028
RI: 1570
Psychedelic
Designer drug
N: synth. by
Roesch/Kovar
BDB
PS
LM/Q
136
168
108
M+
210
100
200
$C_{10}H_{10}O_5$
210.05283
RI: 1570*
D: UHYAC
Preservative
Methylparaben-M (HO-) AC
UHYAC
LS/Q

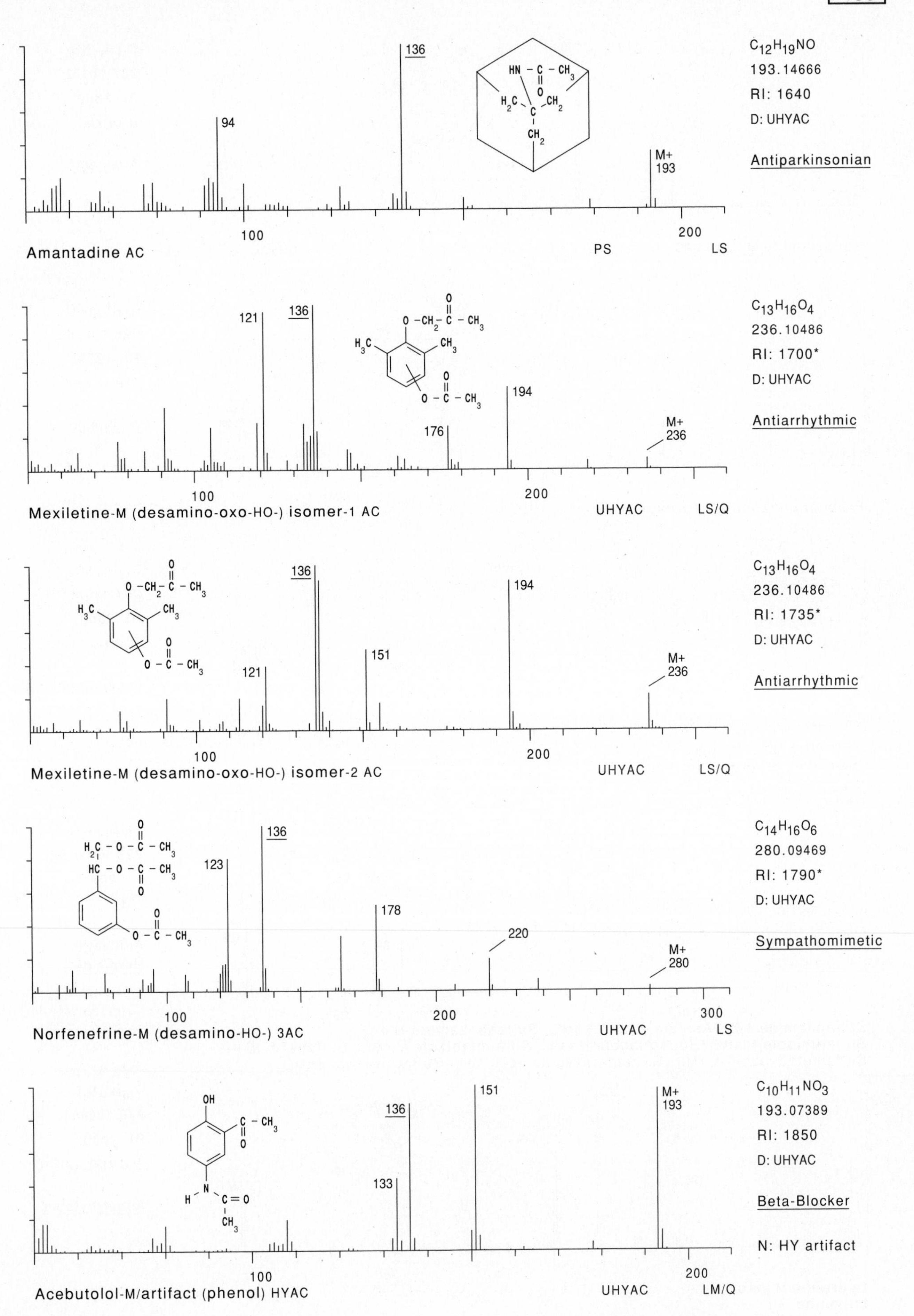
136
Amantadine AC
$C_{12}H_{19}NO$
193.14666
RI: 1640
D: UHYAC
Antiparkinsonian
PS
LS
Mexiletine-M (desamino-oxo-HO-) isomer-1 AC
$C_{13}H_{16}O_4$
236.10486
RI: 1700*
D: UHYAC
Antiarrhythmic
UHYAC
LS/Q
Mexiletine-M (desamino-oxo-HO-) isomer-2 AC
$C_{13}H_{16}O_4$
236.10486
RI: 1735*
D: UHYAC
Antiarrhythmic
UHYAC
LS/Q
Norfenefrine-M (desamino-HO-) 3AC
$C_{14}H_{16}O_6$
280.09469
RI: 1790*
D: UHYAC
Sympathomimetic
UHYAC
LS
Acebutolol-M/artifact (phenol) HYAC
$C_{10}H_{11}NO_3$
193.07389
RI: 1850
D: UHYAC
Beta-Blocker
N: HY artifact
UHYAC
LM/Q

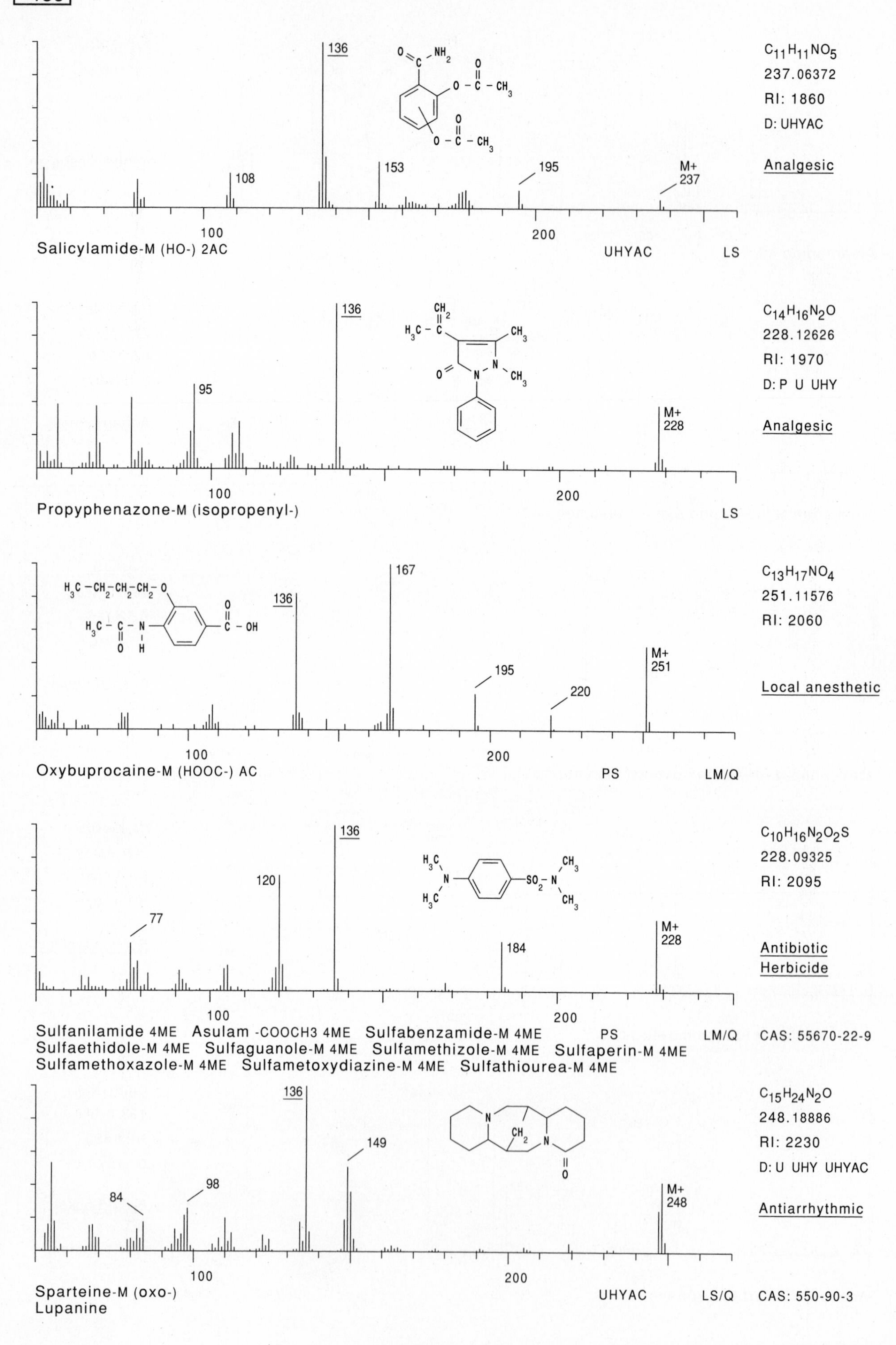

136
108
153
195
M+
237
100
200
C11H11NO5
237.06372
RI: 1860
D: UHYAC
Analgesic
Salicylamide-M (HO-) 2AC
UHYAC
LS
136
95
M+
228
100
200
C14H16N2O
228.12626
RI: 1970
D: P U UHY
Analgesic
Propyphenazone-M (isopropenyl-)
LS
167
136
195
220
M+
251
100
200
C13H17NO4
251.11576
RI: 2060
Local anesthetic
Oxybuprocaine-M (HOOC-) AC
PS
LM/Q
136
120
77
184
M+
228
100
200
C10H16N2O2S
228.09325
RI: 2095
Antibiotic
Herbicide
Sulfanilamide 4ME Asulam -COOCH3 4ME Sulfabenzamide-M 4ME
PS
LM/Q
CAS: 55670-22-9
Sulfaethidole-M 4ME Sulfaguanole-M 4ME Sulfamethizole-M 4ME Sulfaperin-M 4ME
Sulfamethoxazole-M 4ME Sulfametoxydiazine-M 4ME Sulfathiourea-M 4ME
136
149
84
98
M+
248
100
200
C15H24N2O
248.18886
RI: 2230
D: U UHY UHYAC
Antiarrhythmic
Sparteine-M (oxo-)
Lupanine
UHYAC
LS/Q
CAS: 550-90-3

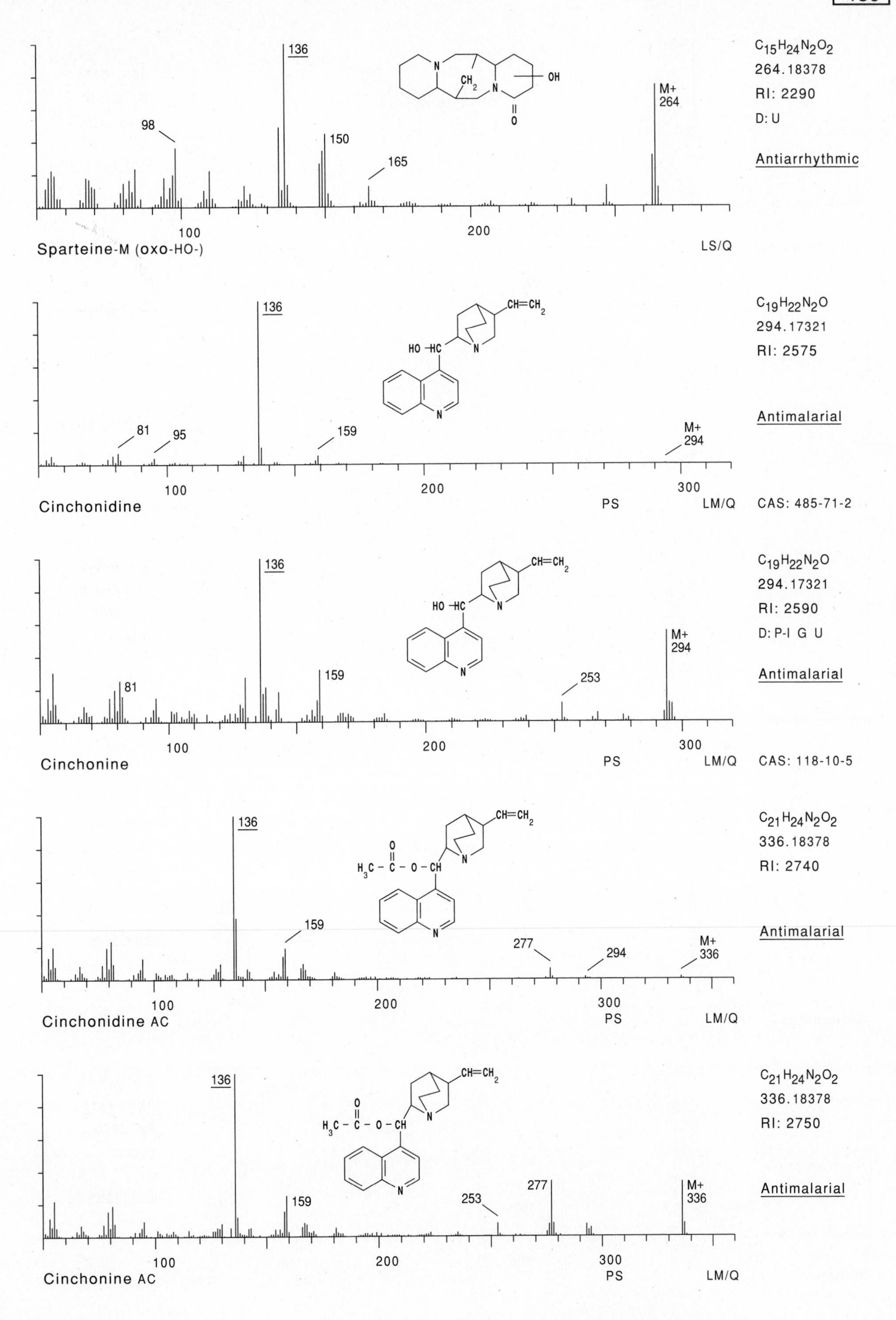

$C_{15}H_{24}N_2O_2$
264.18378
RI: 2290
D: U

Antiarrhythmic

$C_{19}H_{22}N_2O$
294.17321
RI: 2575

Antimalarial

CAS: 485-71-2

$C_{19}H_{22}N_2O$
294.17321
RI: 2590
D: P-I G U

Antimalarial

CAS: 118-10-5

$C_{21}H_{24}N_2O_2$
336.18378
RI: 2740

Antimalarial

$C_{21}H_{24}N_2O_2$
336.18378
RI: 2750

Antimalarial

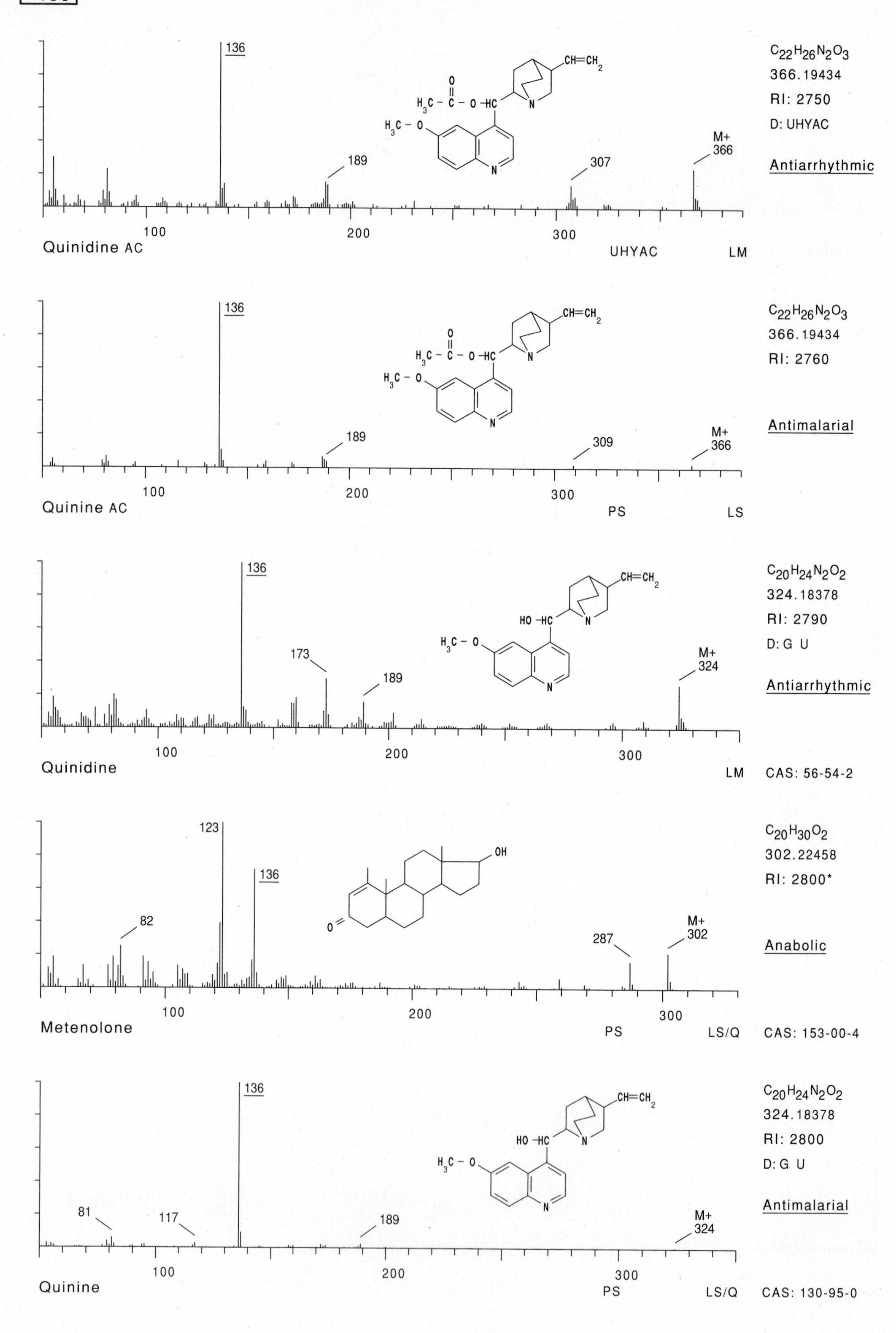
136
189
307
M+
366
CH=CH2
H3C - C - O -HC
O
N
H3C - O
N
100
200
300
Quinidine AC
UHYAC
LM
C22H26N2O3
366.19434
RI: 2750
D: UHYAC
Antiarrhythmic
136
189
309
M+
366
100
200
300
Quinine AC
PS
LS
C22H26N2O3
366.19434
RI: 2760
Antimalarial
136
173
189
M+
324
HO -HC
100
200
300
Quinidine
LM
C20H24N2O2
324.18378
RI: 2790
D: G U
Antiarrhythmic
CAS: 56-54-2
123
136
82
287
M+
302
OH
O
100
200
300
Metenolone
PS
LS/Q
C20H30O2
302.22458
RI: 2800*
Anabolic
CAS: 153-00-4
136
81
117
189
M+
324
100
200
300
Quinine
PS
LS/Q
C20H24N2O2
324.18378
RI: 2800
D: G U
Antimalarial
CAS: 130-95-0

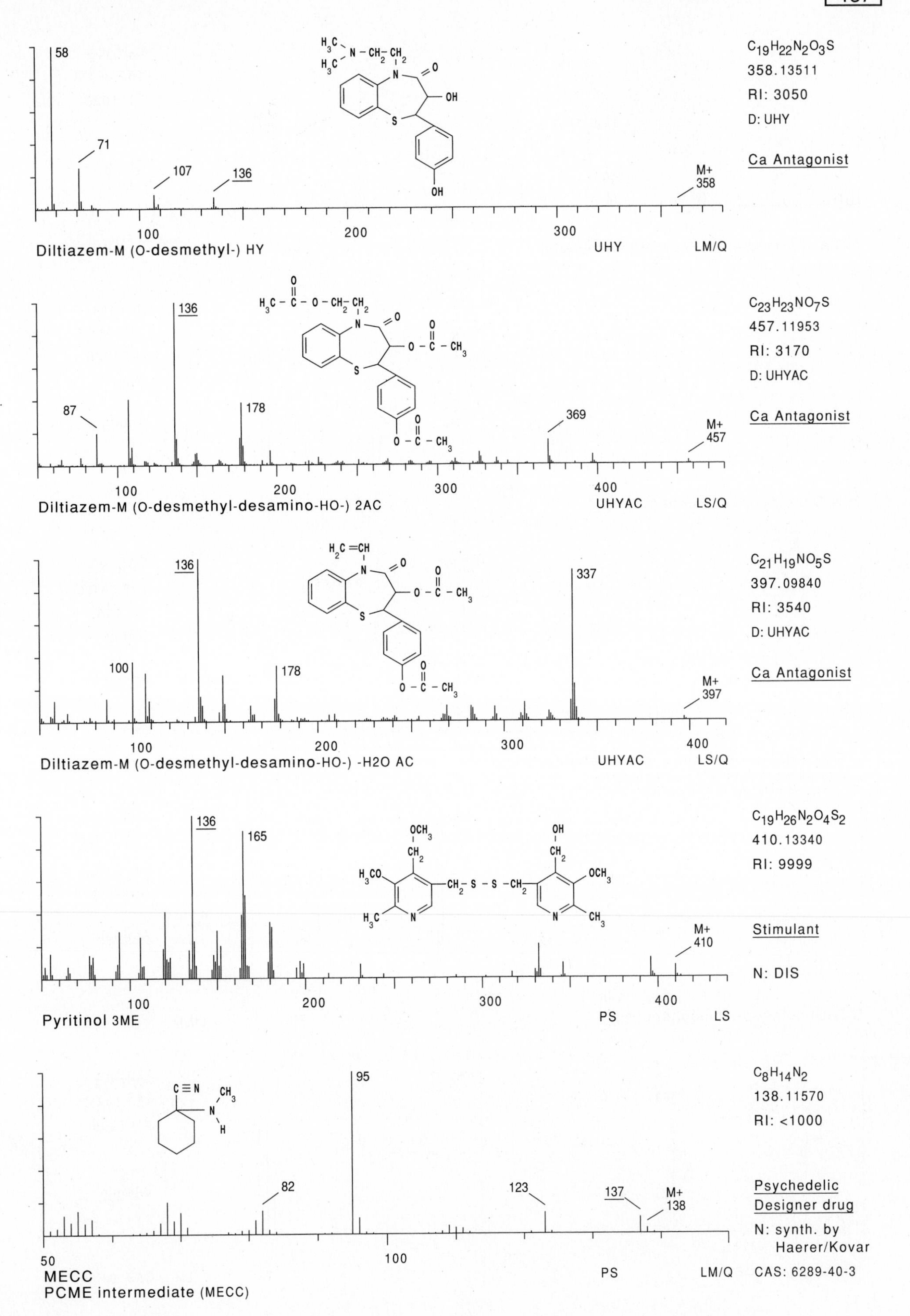
137
C19H22N2O3S
358.13511
RI: 3050
D: UHY
Ca Antagonist
58
71
107
136
M+
358
100
200
300
Diltiazem-M (O-desmethyl-) HY
UHY
LM/Q
C23H23NO7S
457.11953
RI: 3170
D: UHYAC
Ca Antagonist
136
87
178
369
M+
457
100
200
300
400
Diltiazem-M (O-desmethyl-desamino-HO-) 2AC
UHYAC
LS/Q
C21H19NO5S
397.09840
RI: 3540
D: UHYAC
Ca Antagonist
136
337
100
178
M+
397
100
200
300
400
Diltiazem-M (O-desmethyl-desamino-HO-) -H2O AC
UHYAC
LS/Q
C19H26N2O4S2
410.13340
RI: 9999
Stimulant
N: DIS
136
165
M+
410
100
200
300
400
Pyritinol 3ME
PS
LS
C8H14N2
138.11570
RI: <1000
Psychedelic
Designer drug
N: synth. by
Haerer/Kovar
CAS: 6289-40-3
95
82
123
137
M+
138
50
100
MECC
PCME intermediate (MECC)
PS
LM/Q

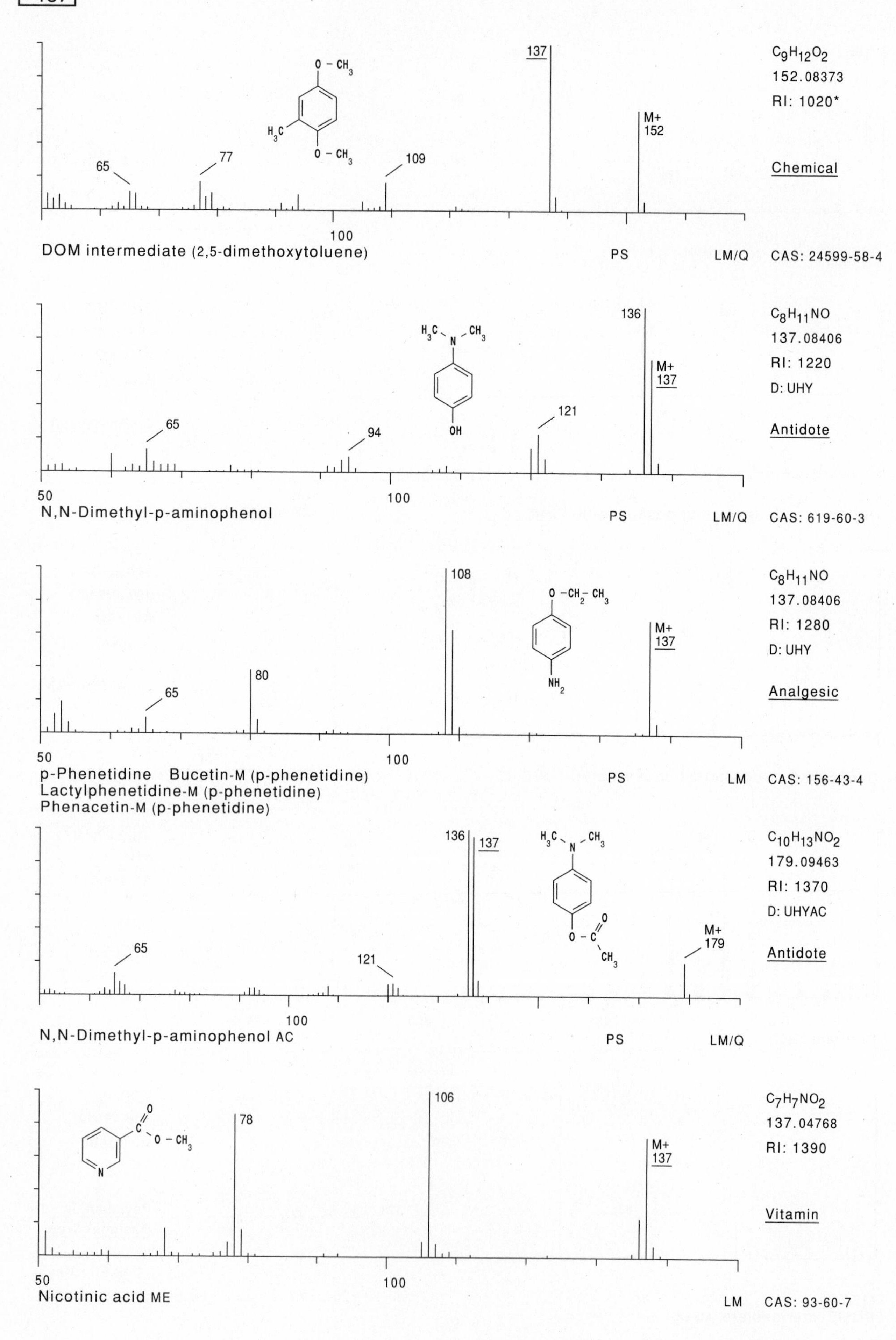

137
M+
152
65
77
109
100
$C_9H_{12}O_2$
152.08373
RI: 1020*
Chemical
DOM intermediate (2,5-dimethoxytoluene)
PS
LM/Q
CAS: 24599-58-4
136
M+
137
65
94
121
50
100
$C_8H_{11}NO$
137.08406
RI: 1220
D: UHY
Antidote
N,N-Dimethyl-p-aminophenol
PS
LM/Q
CAS: 619-60-3
108
M+
137
80
65
50
100
$C_8H_{11}NO$
137.08406
RI: 1280
D: UHY
Analgesic
p-Phenetidine Bucetin-M (p-phenetidine)
Lactylphenetidine-M (p-phenetidine)
Phenacetin-M (p-phenetidine)
PS
LM
CAS: 156-43-4
136
137
M+
179
65
121
100
$C_{10}H_{13}NO_2$
179.09463
RI: 1370
D: UHYAC
Antidote
N,N-Dimethyl-p-aminophenol AC
PS
LM/Q
106
78
M+
137
50
100
$C_7H_7NO_2$
137.04768
RI: 1390
Vitamin
Nicotinic acid ME
LM
CAS: 93-60-7

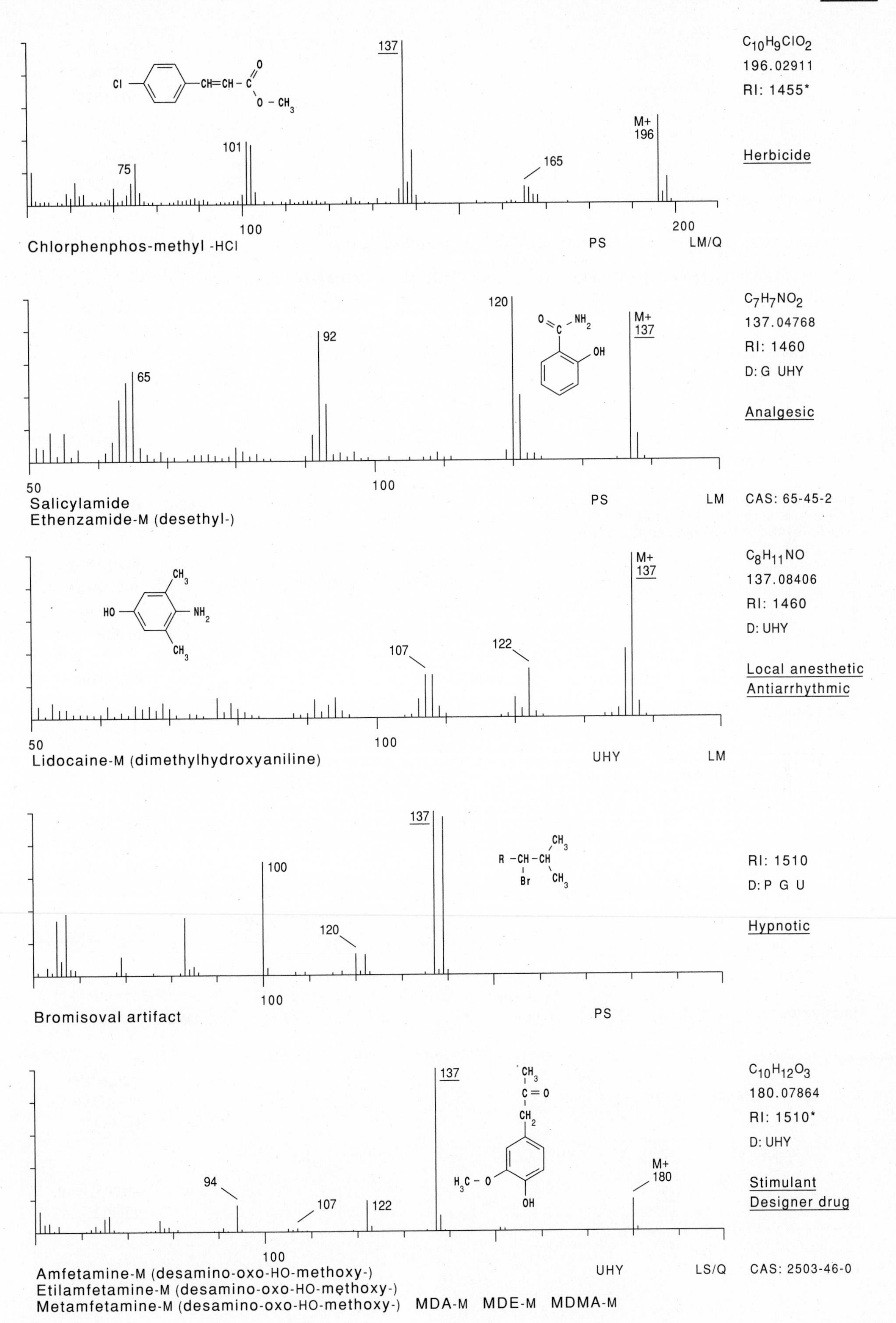

137
Cl
CH=CH-C
O
O-CH3
M+
196
101
165
75
100
200
Chlorphenphos-methyl -HCl
PS
LM/Q
C10H9ClO2
196.02911
RI: 1455*
Herbicide
120
92
65
M+
137
O
C
NH2
OH
50
100
Salicylamide
Ethenzamide-M (desethyl-)
PS
LM
C7H7NO2
137.04768
RI: 1460
D: G UHY
Analgesic
CAS: 65-45-2
M+
137
CH3
HO
NH2
CH3
107
122
50
100
Lidocaine-M (dimethylhydroxyaniline)
UHY
LM
C8H11NO
137.08406
RI: 1460
D: UHY
Local anesthetic
Antiarrhythmic
137
100
120
CH3
R-CH-CH
Br
CH3
100
Bromisoval artifact
PS
RI: 1510
D: P G U
Hypnotic
137
CH3
C=O
CH2
H3C-O
OH
94
107
122
M+
180
100
Amfetamine-M (desamino-oxo-HO-methoxy-)
Etilamfetamine-M (desamino-oxo-HO-methoxy-)
Metamfetamine-M (desamino-oxo-HO-methoxy-)
MDA-M
MDE-M
MDMA-M
UHY
LS/Q
C10H12O3
180.07864
RI: 1510*
D: UHY
Stimulant
Designer drug
CAS: 2503-46-0

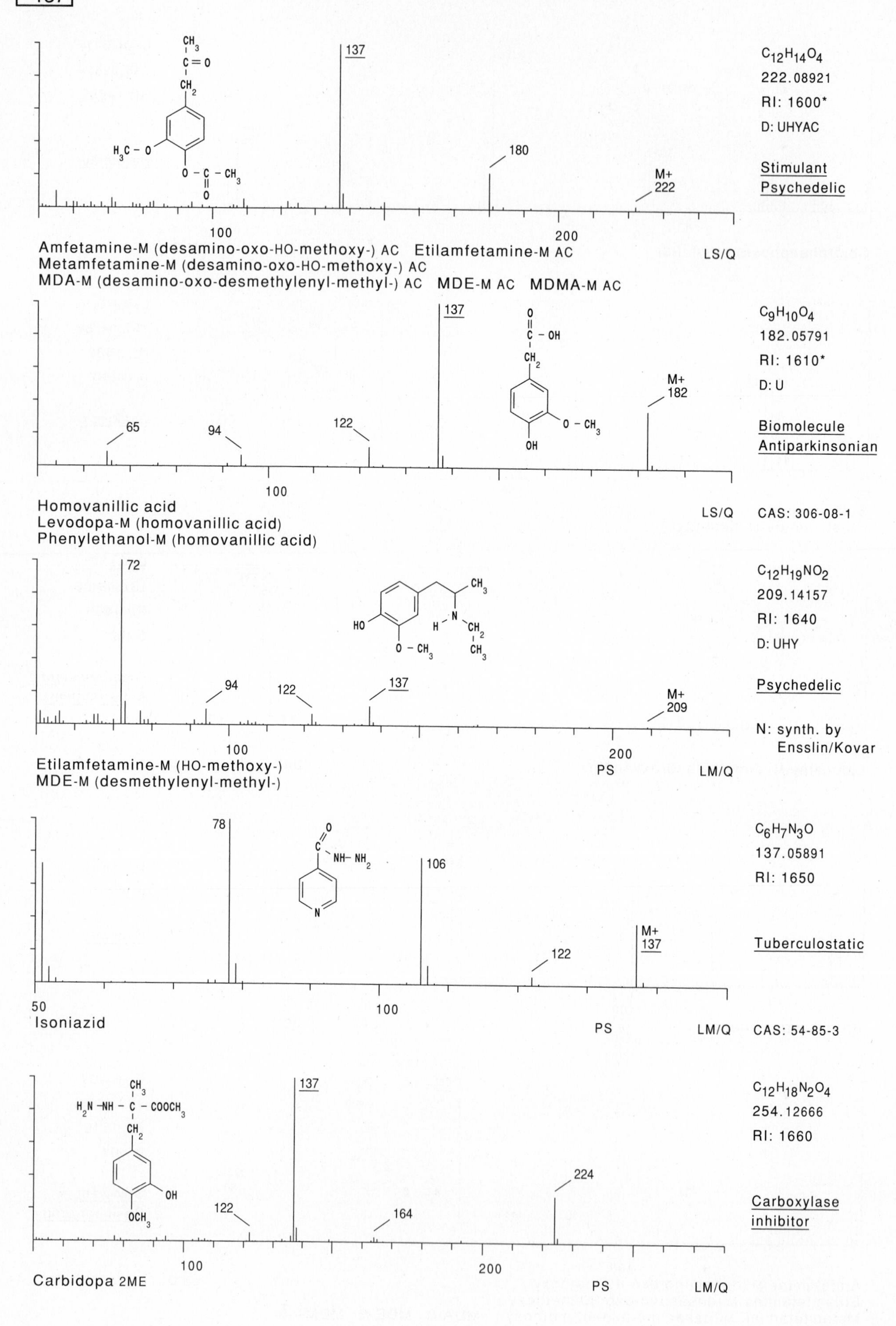

137
180
M+
222
100
200
LS/Q
C12H14O4
222.08921
RI: 1600*
D: UHYAC
Stimulant
Psychedelic
Amfetamine-M (desamino-oxo-HO-methoxy-) AC Etilamfetamine-M AC
Metamfetamine-M (desamino-oxo-HO-methoxy-) AC
MDA-M (desamino-oxo-desmethylenyl-methyl-) AC MDE-M AC MDMA-M AC
65
94
122
137
M+
182
100
LS/Q
C9H10O4
182.05791
RI: 1610*
D: U
Biomolecule
Antiparkinsonian
CAS: 306-08-1
Homovanillic acid
Levodopa-M (homovanillic acid)
Phenylethanol-M (homovanillic acid)
72
94
122
137
M+
209
100
200
PS
LM/Q
C12H19NO2
209.14157
RI: 1640
D: UHY
Psychedelic
N: synth. by Ensslin/Kovar
Etilamfetamine-M (HO-methoxy-)
MDE-M (desmethylenyl-methyl-)
78
106
122
M+
137
50
100
PS
LM/Q
C6H7N3O
137.05891
RI: 1650
Tuberculostatic
CAS: 54-85-3
Isoniazid
137
122
164
224
100
200
PS
LM/Q
C12H18N2O4
254.12666
RI: 1660
Carboxylase inhibitor
Carbidopa 2ME

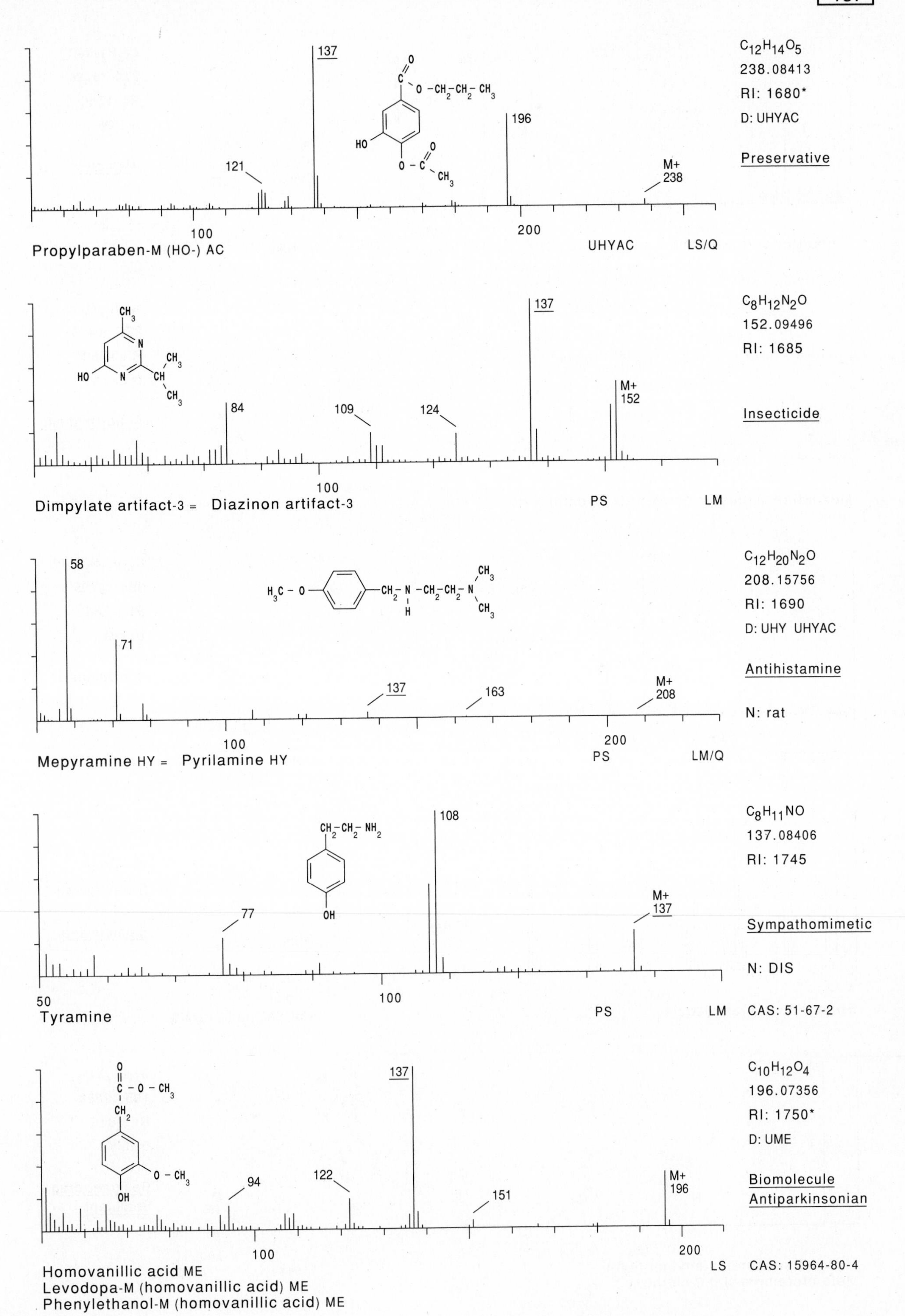
137
196
121
M+
238
100
200
Propylparaben-M (HO-) AC
UHYAC
LS/Q
C12H14O5
238.08413
RI: 1680*
D: UHYAC
Preservative
137
84
109
124
M+
152
100
Dimpylate artifact-3 = Diazinon artifact-3
PS
LM
C8H12N2O
152.09496
RI: 1685
Insecticide
58
71
137
163
M+
208
100
200
Mepyramine HY = Pyrilamine HY
PS
LM/Q
C12H20N2O
208.15756
RI: 1690
D: UHY UHYAC
Antihistamine
N: rat
108
77
M+
137
50
100
Tyramine
PS
LM
C8H11NO
137.08406
RI: 1745
Sympathomimetic
N: DIS
CAS: 51-67-2
137
94
122
151
M+
196
100
200
Homovanillic acid ME
Levodopa-M (homovanillic acid) ME
Phenylethanol-M (homovanillic acid) ME
LS
C10H12O4
196.07356
RI: 1750*
D: UME
Biomolecule
Antiparkinsonian
CAS: 15964-80-4

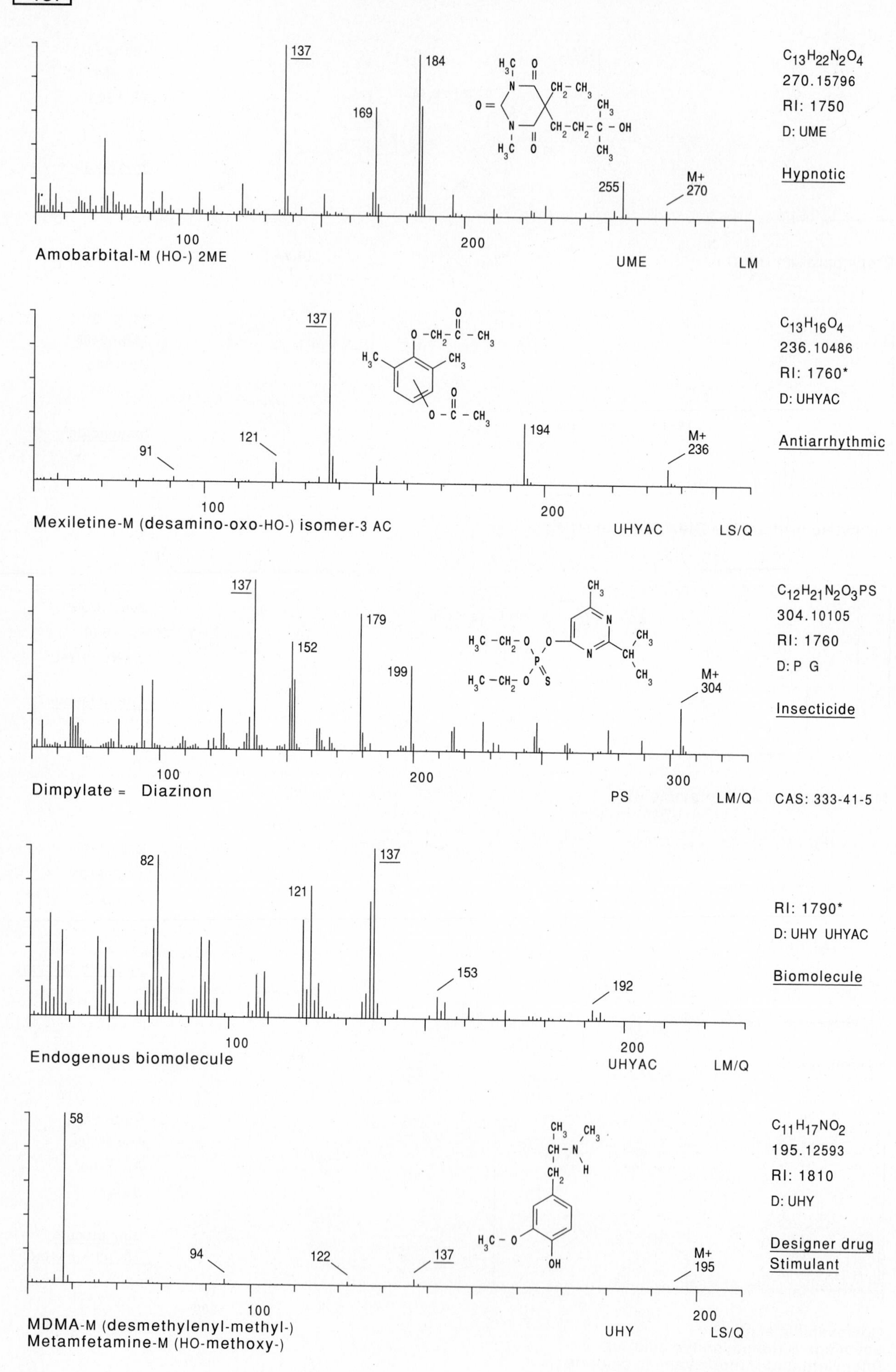
137
184
169
255
M+
270
100
200
Amobarbital-M (HO-) 2ME
UME
LM
$C_{13}H_{22}N_2O_4$
270.15796
RI: 1750
D: UME
Hypnotic
137
194
121
91
M+
236
100
200
Mexiletine-M (desamino-oxo-HO-) isomer-3 AC
UHYAC
LS/Q
$C_{13}H_{16}O_4$
236.10486
RI: 1760*
D: UHYAC
Antiarrhythmic
137
179
152
199
M+
304
100
200
300
Dimpylate = Diazinon
PS
LM/Q
$C_{12}H_{21}N_2O_3PS$
304.10105
RI: 1760
D: P G
Insecticide
CAS: 333-41-5
82
137
121
153
192
100
200
Endogenous biomolecule
UHYAC
LM/Q
RI: 1790*
D: UHY UHYAC
Biomolecule
58
94
122
137
M+
195
100
200
MDMA-M (desmethylenyl-methyl-)
Metamfetamine-M (HO-methoxy-)
UHY
LS/Q
$C_{11}H_{17}NO_2$
195.12593
RI: 1810
D: UHY
Designer drug
Stimulant

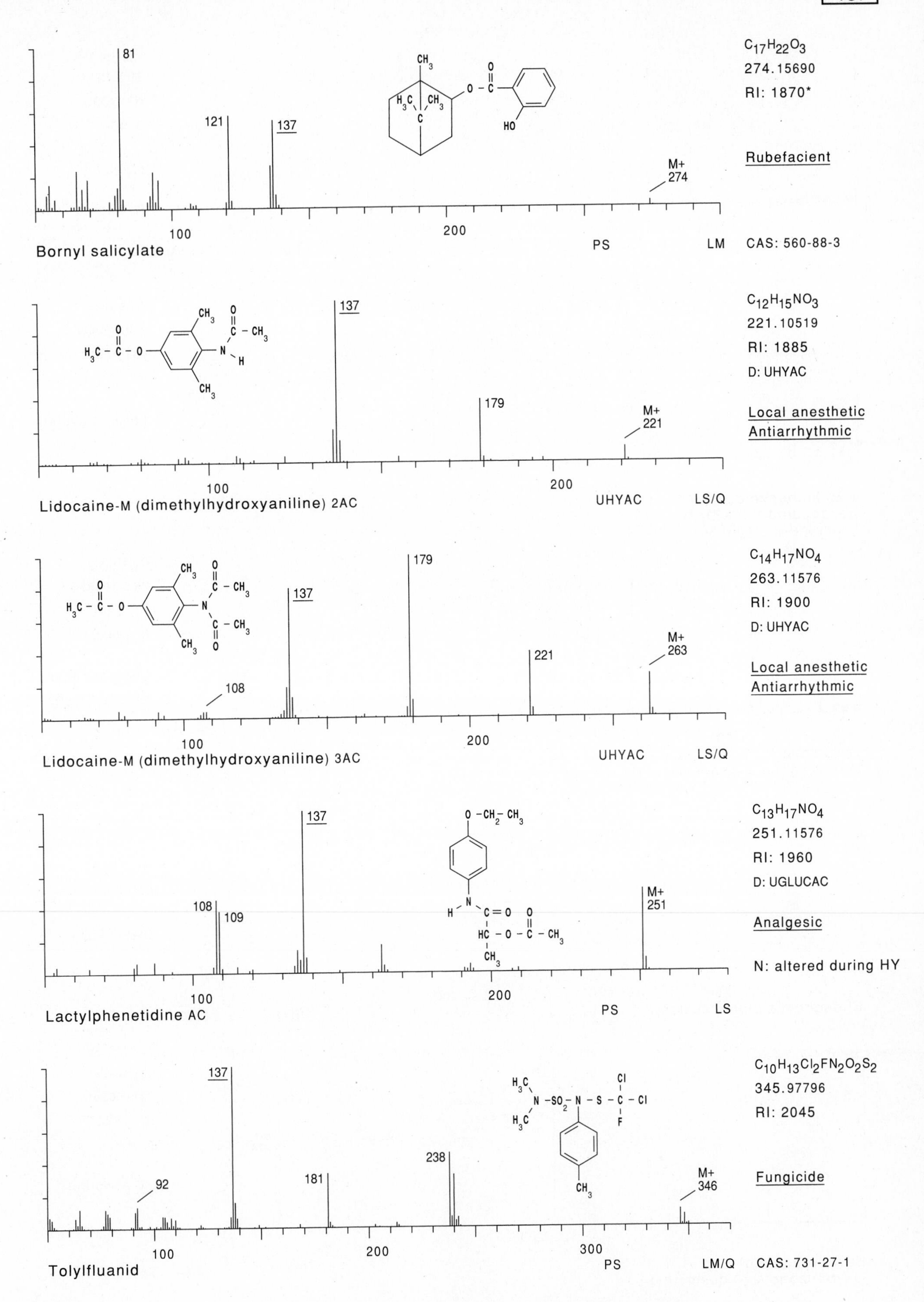

81
121
137
M+
274
100
200
PS
LM
Bornyl salicylate
C17H22O3
274.15690
RI: 1870*
Rubefacient
CAS: 560-88-3
137
179
M+
221
100
200
UHYAC
LS/Q
Lidocaine-M (dimethylhydroxyaniline) 2AC
C12H15NO3
221.10519
RI: 1885
D: UHYAC
Local anesthetic
Antiarrhythmic
179
137
221
M+
263
108
100
200
UHYAC
LS/Q
Lidocaine-M (dimethylhydroxyaniline) 3AC
C14H17NO4
263.11576
RI: 1900
D: UHYAC
Local anesthetic
Antiarrhythmic
137
108
109
M+
251
100
200
PS
LS
Lactylphenetidine AC
C13H17NO4
251.11576
RI: 1960
D: UGLUCAC
Analgesic
N: altered during HY
137
238
181
92
M+
346
100
200
300
PS
LM/Q
Tolylfluanid
C10H13Cl2FN2O2S2
345.97796
RI: 2045
Fungicide
CAS: 731-27-1

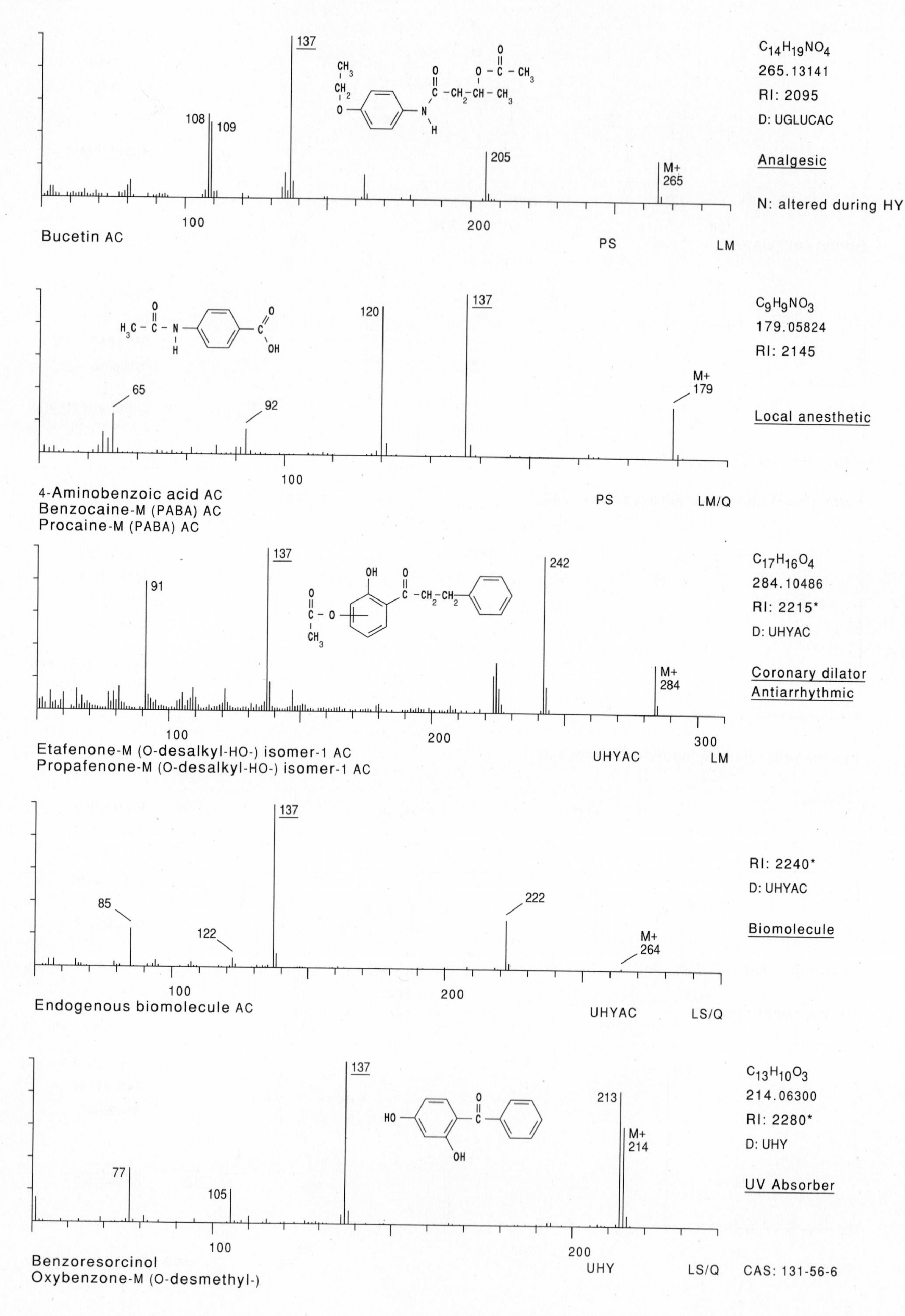

137
108
109
205
M+
265
100
200
Bucetin AC
PS
LM
C14H19NO4
265.13141
RI: 2095
D: UGLUCAC
Analgesic
N: altered during HY
120
137
65
92
M+
179
100
4-Aminobenzoic acid AC
Benzocaine-M (PABA) AC
Procaine-M (PABA) AC
PS
LM/Q
C9H9NO3
179.05824
RI: 2145
Local anesthetic
91
137
242
M+
284
100
200
300
Etafenone-M (O-desalkyl-HO-) isomer-1 AC
Propafenone-M (O-desalkyl-HO-) isomer-1 AC
UHYAC
LM
C17H16O4
284.10486
RI: 2215*
D: UHYAC
Coronary dilator
Antiarrhythmic
137
85
122
222
M+
264
100
200
Endogenous biomolecule AC
UHYAC
LS/Q
RI: 2240*
D: UHYAC
Biomolecule
137
77
105
213
M+
214
100
200
Benzoresorcinol
Oxybenzone-M (O-desmethyl-)
UHY
LS/Q
C13H10O3
214.06300
RI: 2280*
D: UHY
UV Absorber
CAS: 131-56-6

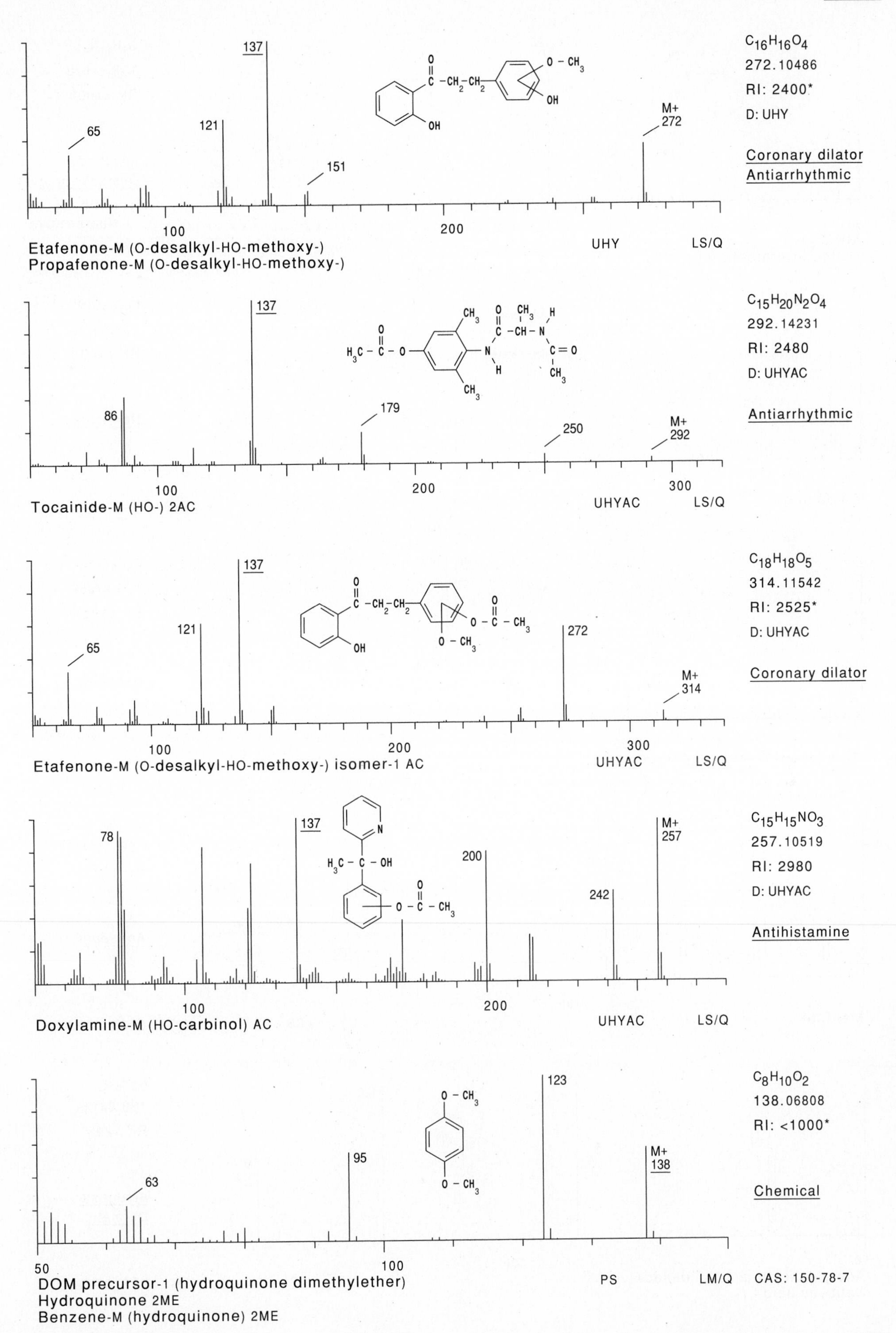

Etafenone-M (O-desalkyl-HO-methoxy-)
Propafenone-M (O-desalkyl-HO-methoxy-)

$C_{16}H_{16}O_4$
272.10486
RI: 2400*
D: UHY

Coronary dilator
Antiarrhythmic

Tocainide-M (HO-) 2AC

$C_{15}H_{20}N_2O_4$
292.14231
RI: 2480
D: UHYAC

Antiarrhythmic

Etafenone-M (O-desalkyl-HO-methoxy-) isomer-1 AC

$C_{18}H_{18}O_5$
314.11542
RI: 2525*
D: UHYAC

Coronary dilator

Doxylamine-M (HO-carbinol) AC

$C_{15}H_{15}NO_3$
257.10519
RI: 2980
D: UHYAC

Antihistamine

DOM precursor-1 (hydroquinone dimethylether)
Hydroquinone 2ME
Benzene-M (hydroquinone) 2ME

$C_8H_{10}O_2$
138.06808
RI: <1000*

Chemical

CAS: 150-78-7

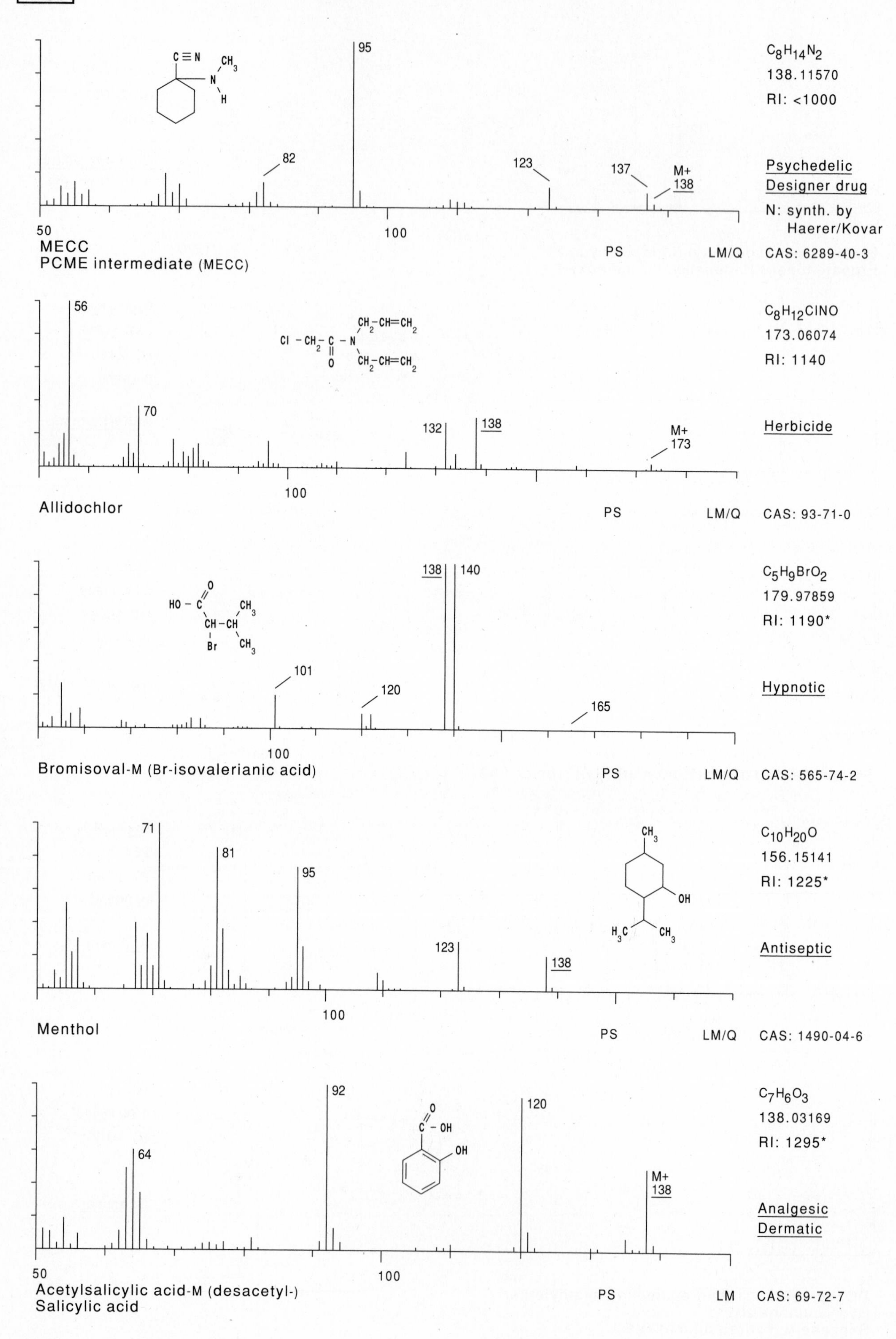

95
82
123
137
M+
138
50
100
MECC
PCME intermediate (MECC)
PS
LM/Q
$C_8H_{14}N_2$
138.11570
RI: <1000
Psychedelic
Designer drug
N: synth. by
Haerer/Kovar
CAS: 6289-40-3
56
70
132
138
M+
173
100
Allidochlor
PS
LM/Q
$C_8H_{12}ClNO$
173.06074
RI: 1140
Herbicide
CAS: 93-71-0
138
140
101
120
165
100
Bromisoval-M (Br-isovalerianic acid)
PS
LM/Q
$C_5H_9BrO_2$
179.97859
RI: 1190*
Hypnotic
CAS: 565-74-2
71
81
95
123
138
100
Menthol
PS
LM/Q
$C_{10}H_{20}O$
156.15141
RI: 1225*
Antiseptic
CAS: 1490-04-6
92
120
64
M+
138
50
100
Acetylsalicylic acid-M (desacetyl-)
Salicylic acid
PS
LM
$C_7H_6O_3$
138.03169
RI: 1295*
Analgesic
Dermatic
CAS: 69-72-7

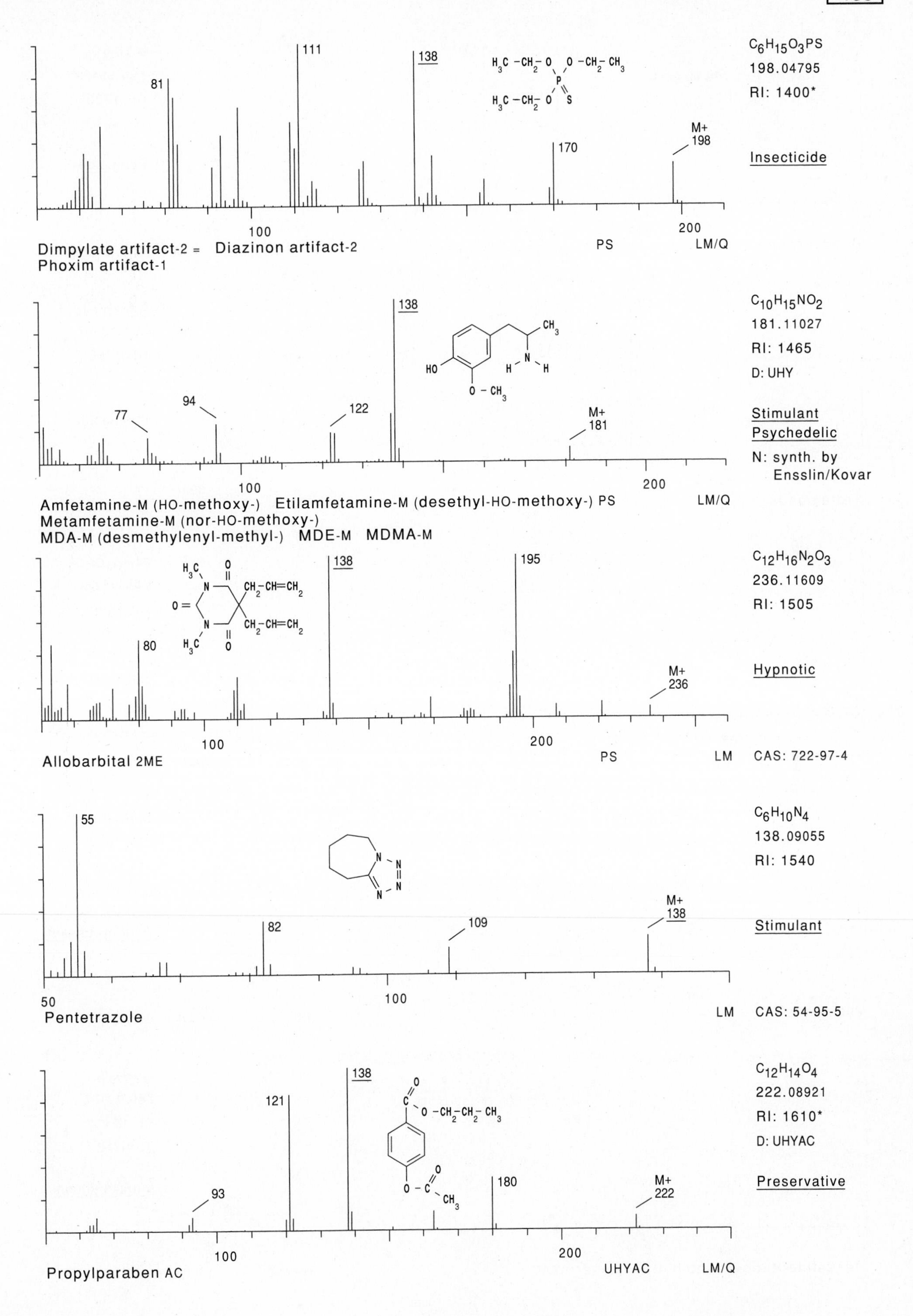

111
138
81
170
M+
198
100
200
PS
LM/Q
Dimpylate artifact-2 = Diazinon artifact-2
Phoxim artifact-1
C6H15O3PS
198.04795
RI: 1400*
Insecticide
138
94
77
122
M+
181
100
200
Amfetamine-M (HO-methoxy-) Etilamfetamine-M (desethyl-HO-methoxy-) PS
LM/Q
Metamfetamine-M (nor-HO-methoxy-)
MDA-M (desmethylenyl-methyl-) MDE-M MDMA-M
C10H15NO2
181.11027
RI: 1465
D: UHY
Stimulant
Psychedelic
N: synth. by
Ensslin/Kovar
138
195
80
M+
236
100
200
PS
LM
Allobarbital 2ME
C12H16N2O3
236.11609
RI: 1505
Hypnotic
CAS: 722-97-4
55
82
109
M+
138
50
100
LM
Pentetrazole
C6H10N4
138.09055
RI: 1540
Stimulant
CAS: 54-95-5
138
121
93
180
M+
222
100
200
UHYAC
LM/Q
Propylparaben AC
C12H14O4
222.08921
RI: 1610*
D: UHYAC
Preservative

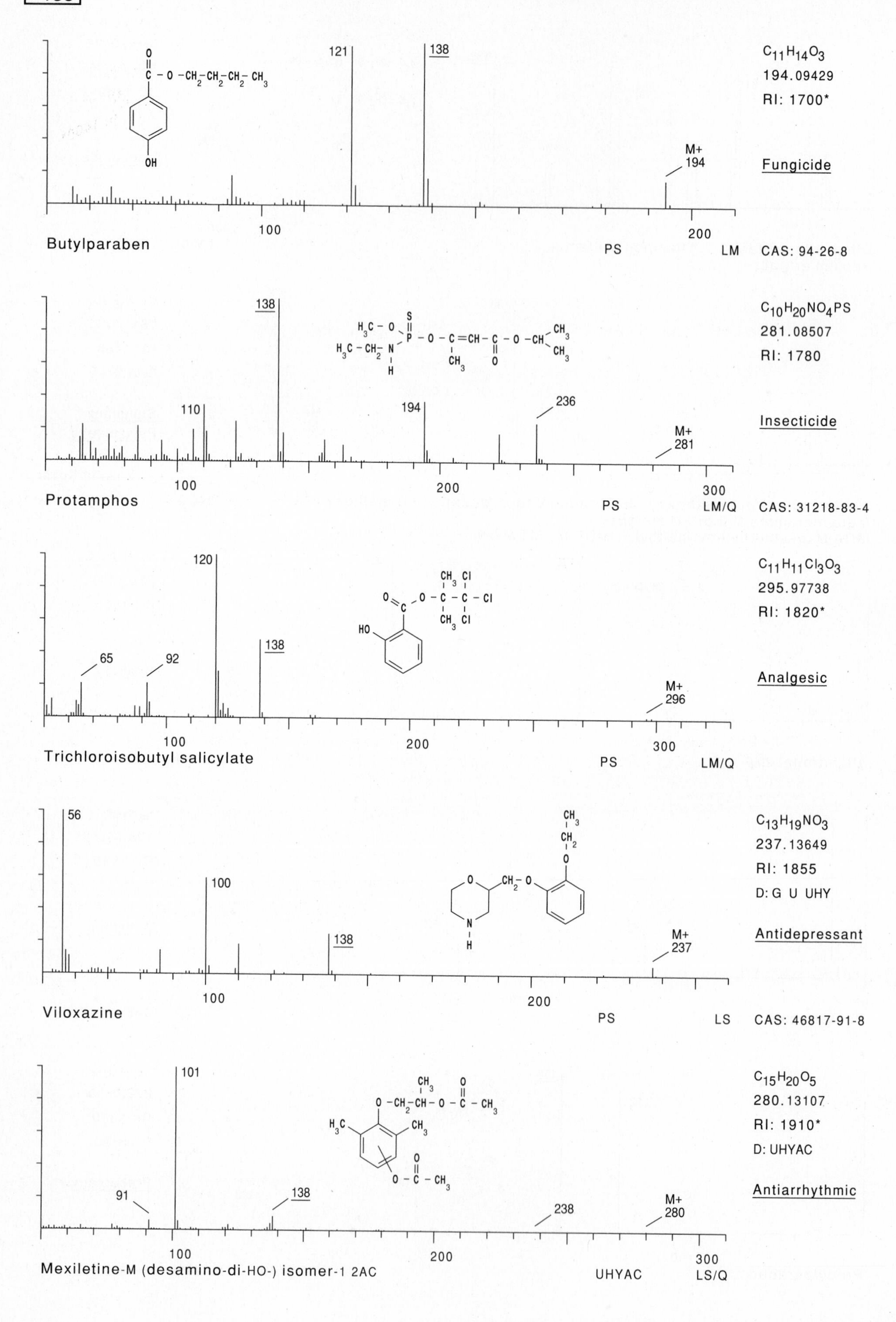

121
138
M+
194
100
200
Butylparaben
PS
LM
C11H14O3
194.09429
RI: 1700*
Fungicide
CAS: 94-26-8
OH
138
110
194
236
M+
281
100
200
300
Protamphos
PS
LM/Q
C10H20NO4PS
281.08507
RI: 1780
Insecticide
CAS: 31218-83-4
120
138
65
92
M+
296
100
200
300
Trichloroisobutyl salicylate
PS
LM/Q
C11H11Cl3O3
295.97738
RI: 1820*
Analgesic
56
100
138
M+
237
100
200
Viloxazine
PS
LS
C13H19NO3
237.13649
RI: 1855
D: G U UHY
Antidepressant
CAS: 46817-91-8
101
91
138
238
M+
280
100
200
300
Mexiletine-M (desamino-di-HO-) isomer-1 2AC
UHYAC
LS/Q
C15H20O5
280.13107
RI: 1910*
D: UHYAC
Antiarrhythmic

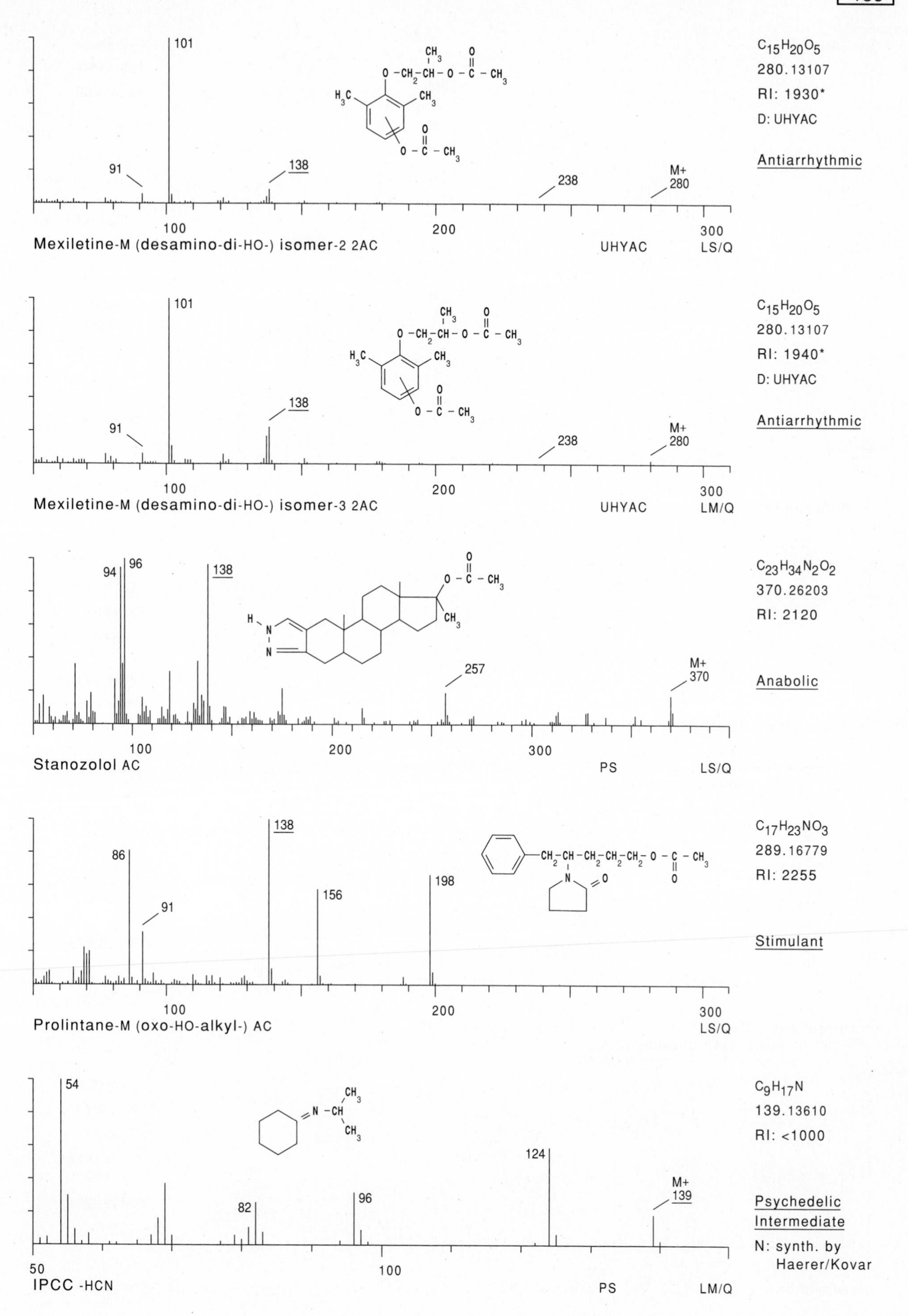

$C_{15}H_{20}O_5$
280.13107
RI: 1930*
D: UHYAC

Antiarrhythmic

$C_{15}H_{20}O_5$
280.13107
RI: 1940*
D: UHYAC

Antiarrhythmic

$C_{23}H_{34}N_2O_2$
370.26203
RI: 2120

Anabolic

$C_{17}H_{23}NO_3$
289.16779
RI: 2255

Stimulant

$C_9H_{17}N$
139.13610
RI: <1000

Psychedelic
Intermediate

N: synth. by Haerer/Kovar

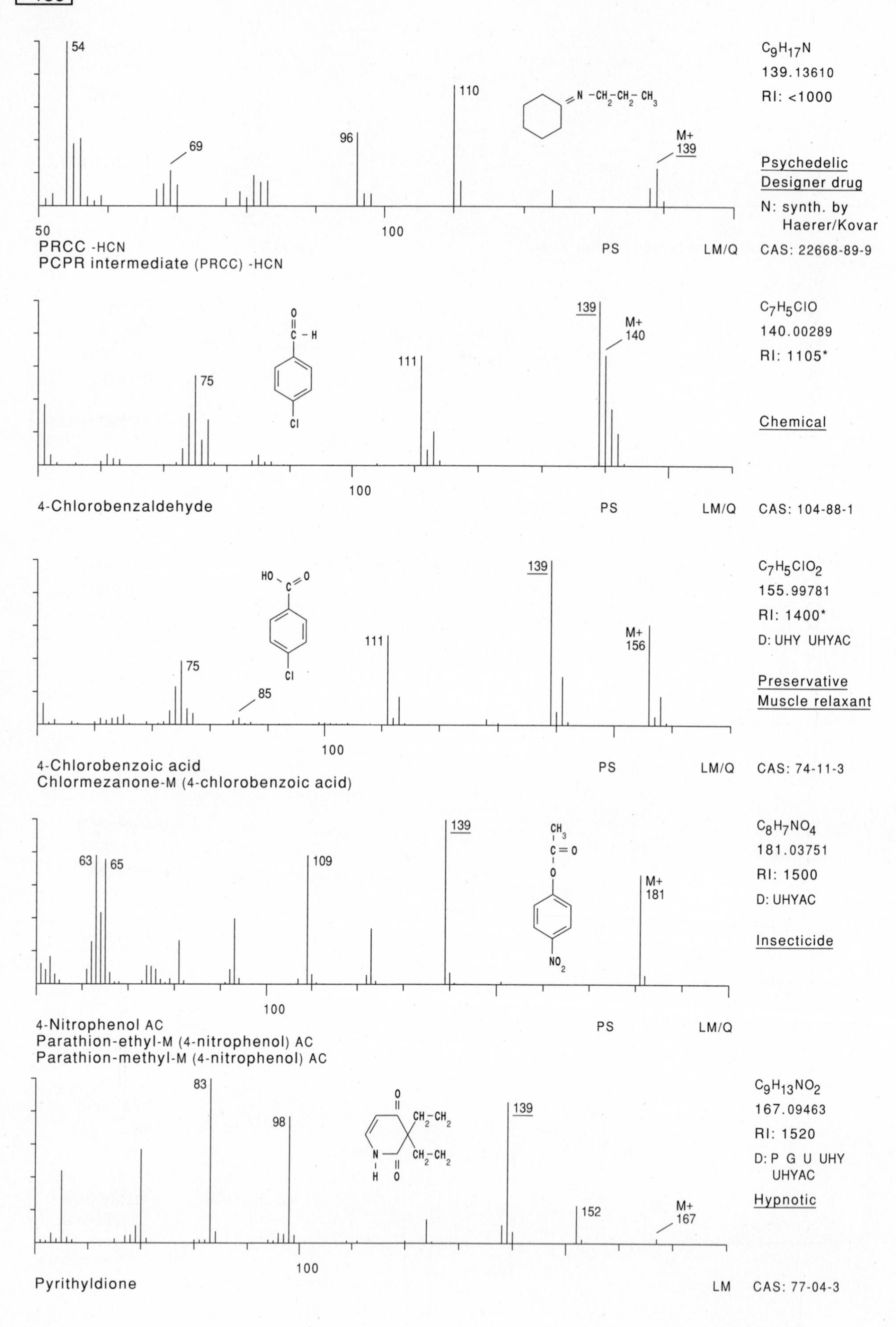

54
69
96
110
M+
139
50
100
PRCC -HCN
PCPR intermediate (PRCC) -HCN
PS
LM/Q
C9H17N
139.13610
RI: <1000
Psychedelic
Designer drug
N: synth. by Haerer/Kovar
CAS: 22668-89-9
75
111
139
M+
140
100
4-Chlorobenzaldehyde
PS
LM/Q
C7H5ClO
140.00289
RI: 1105*
Chemical
CAS: 104-88-1
75
85
111
139
M+
156
100
4-Chlorobenzoic acid
Chlormezanone-M (4-chlorobenzoic acid)
PS
LM/Q
C7H5ClO2
155.99781
RI: 1400*
D: UHY UHYAC
Preservative
Muscle relaxant
CAS: 74-11-3
63
65
109
139
M+
181
100
4-Nitrophenol AC
Parathion-ethyl-M (4-nitrophenol) AC
Parathion-methyl-M (4-nitrophenol) AC
PS
LM/Q
C8H7NO4
181.03751
RI: 1500
D: UHYAC
Insecticide
83
98
139
152
M+
167
100
Pyrithyldione
LM
C9H13NO2
167.09463
RI: 1520
D: P G U UHY UHYAC
Hypnotic
CAS: 77-04-3

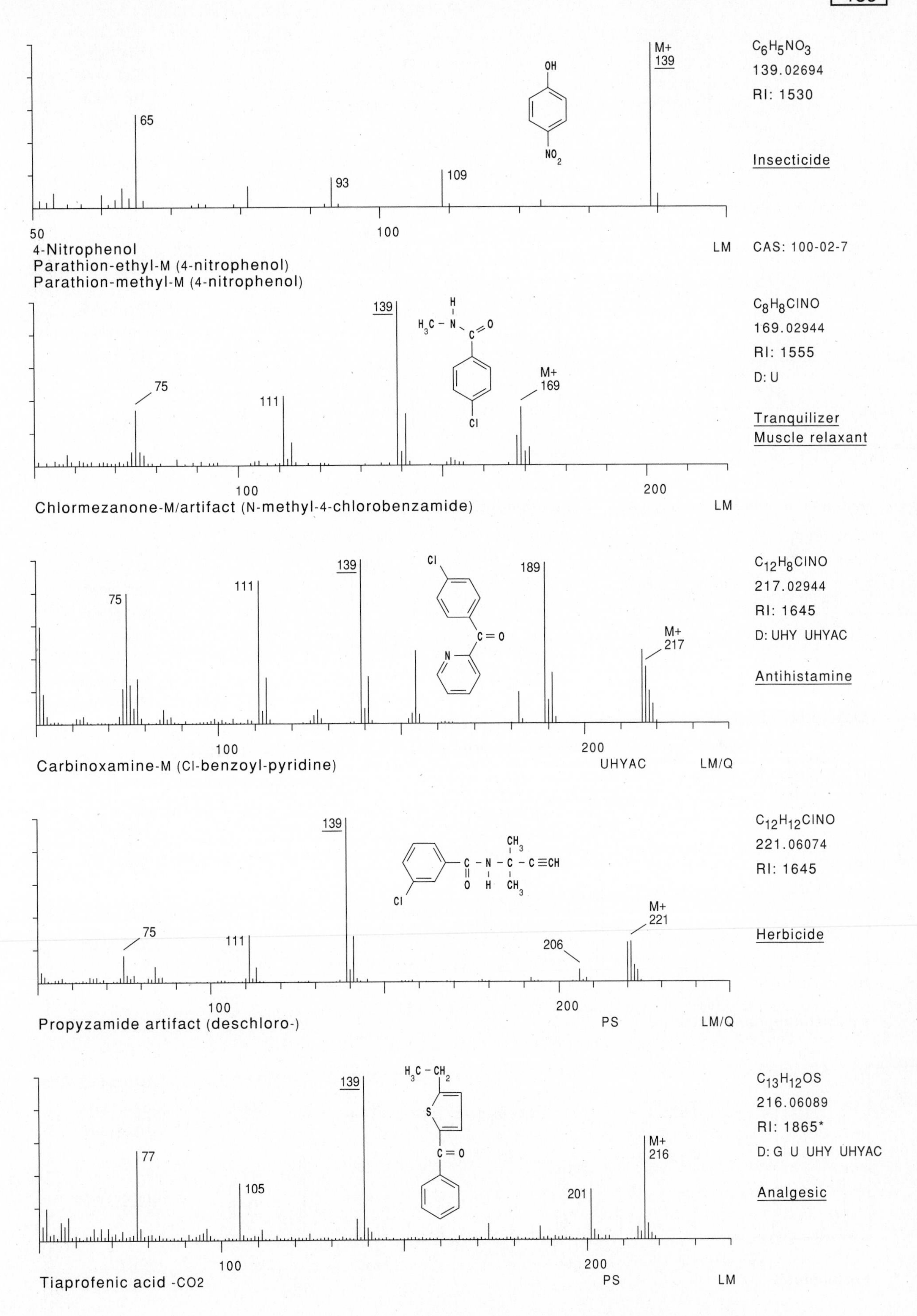

4-Nitrophenol
Parathion-ethyl-M (4-nitrophenol)
Parathion-methyl-M (4-nitrophenol)

$C_6H_5NO_3$
139.02694
RI: 1530
Insecticide
CAS: 100-02-7

Chlormezanone-M/artifact (N-methyl-4-chlorobenzamide)

C_8H_8ClNO
169.02944
RI: 1555
D: U
Tranquilizer
Muscle relaxant

Carbinoxamine-M (Cl-benzoyl-pyridine)

$C_{12}H_8ClNO$
217.02944
RI: 1645
D: UHY UHYAC
Antihistamine

Propyzamide artifact (deschloro-)

$C_{12}H_{12}ClNO$
221.06074
RI: 1645
Herbicide

Tiaprofenic acid -CO2

$C_{13}H_{12}OS$
216.06089
RI: 1865*
D: G U UHY UHYAC
Analgesic

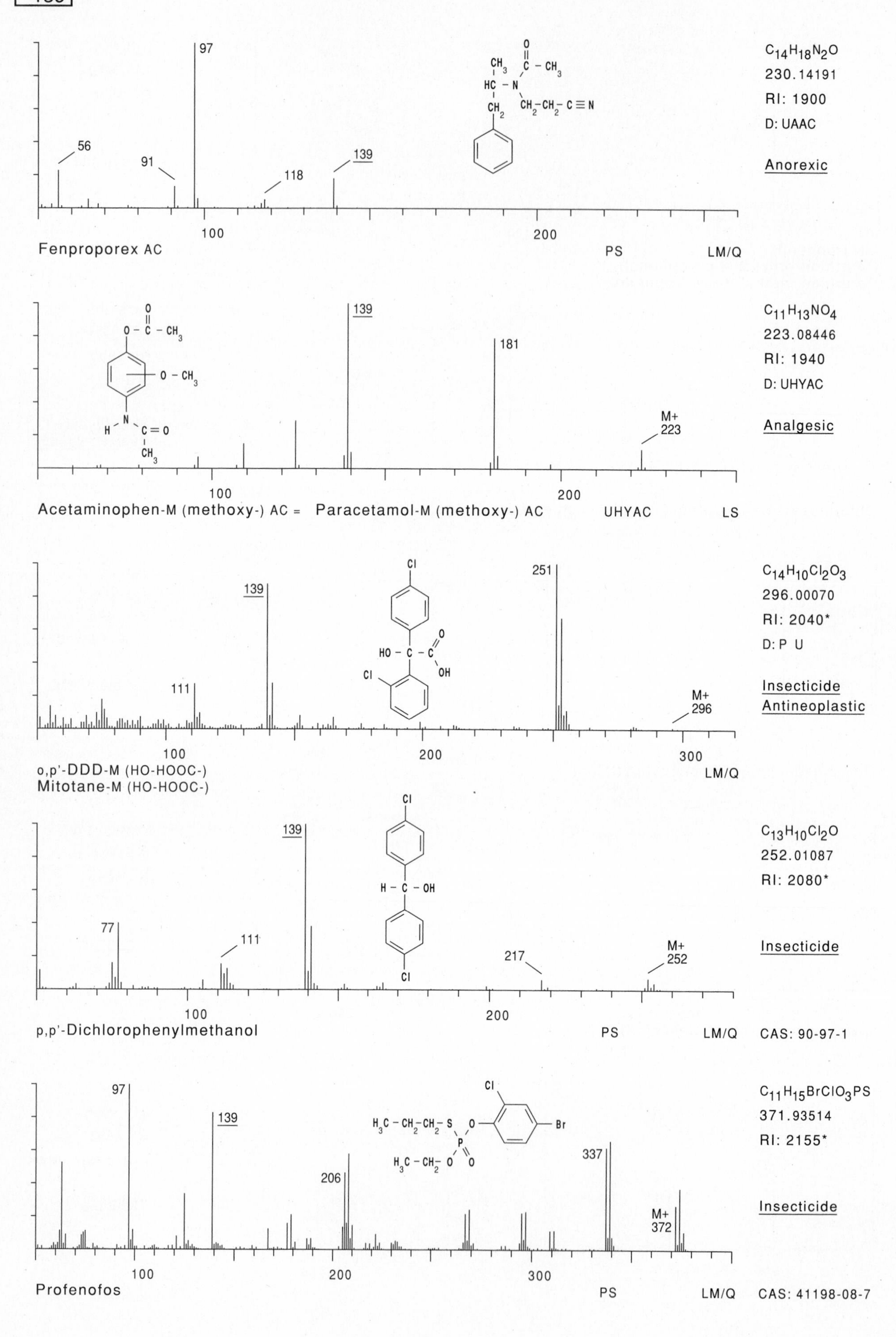

97
56
91
139
118
100
200
Fenproporex AC
PS
LM/Q
$C_{14}H_{18}N_2O$
230.14191
RI: 1900
D: UAAC
Anorexic
139
181
M+
223
100
200
Acetaminophen-M (methoxy-) AC = Paracetamol-M (methoxy-) AC
UHYAC
LS
$C_{11}H_{13}NO_4$
223.08446
RI: 1940
D: UHYAC
Analgesic
251
139
111
M+
296
100
200
300
o,p'-DDD-M (HO-HOOC-)
Mitotane-M (HO-HOOC-)
LM/Q
$C_{14}H_{10}Cl_2O_3$
296.00070
RI: 2040*
D: P U
Insecticide
Antineoplastic
139
77
111
217
M+
252
100
200
p,p'-Dichlorophenylmethanol
PS
LM/Q
CAS: 90-97-1
$C_{13}H_{10}Cl_2O$
252.01087
RI: 2080*
Insecticide
97
139
206
337
M+
372
100
200
300
Profenofos
PS
LM/Q
CAS: 41198-08-7
$C_{11}H_{15}BrClO_3PS$
371.93514
RI: 2155*
Insecticide

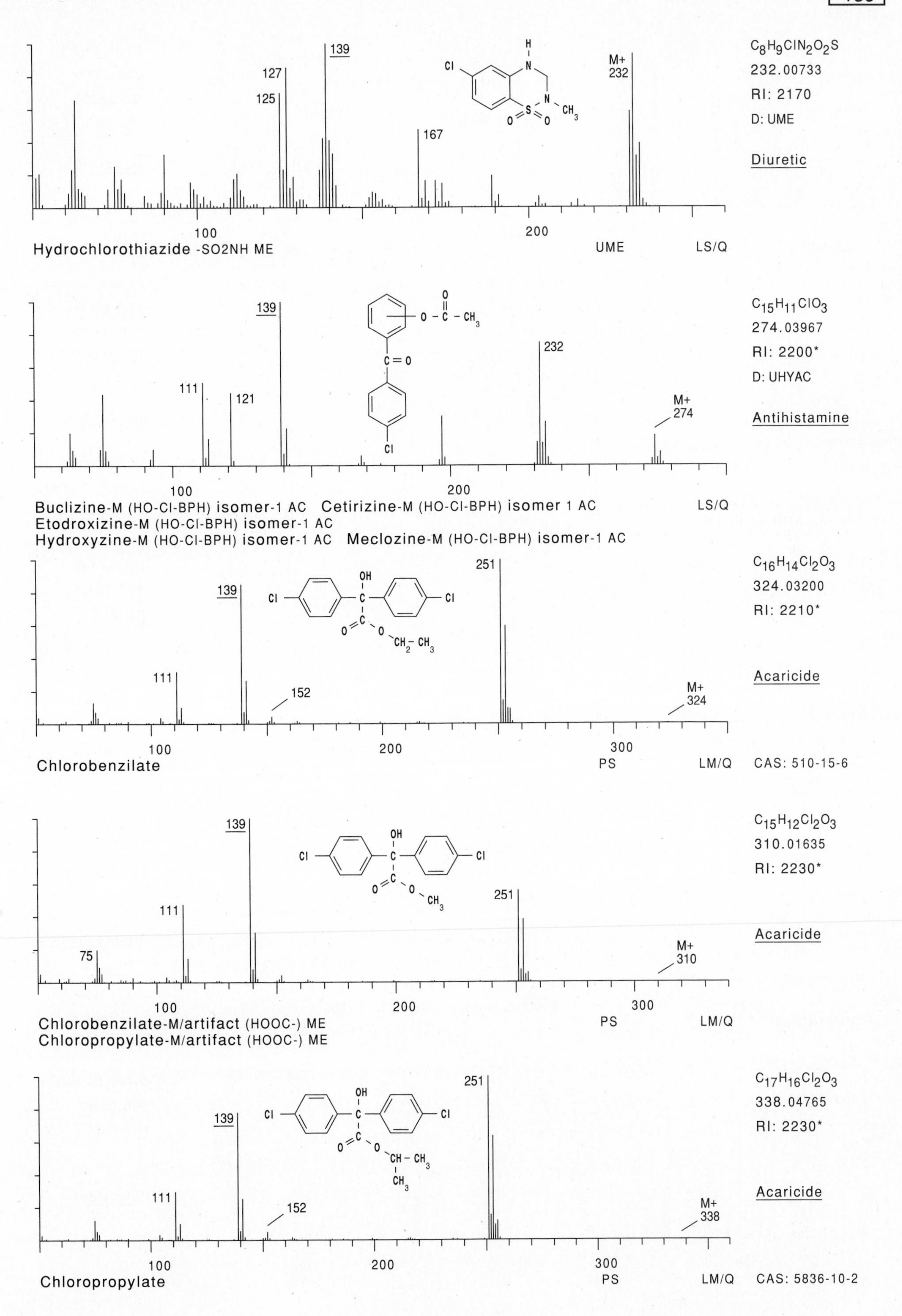

$C_8H_9ClN_2O_2S$
232.00733
RI: 2170
D: UME

Diuretic

$C_{15}H_{11}ClO_3$
274.03967
RI: 2200*
D: UHYAC

Antihistamine

$C_{16}H_{14}Cl_2O_3$
324.03200
RI: 2210*

Acaricide

CAS: 510-15-6

$C_{15}H_{12}Cl_2O_3$
310.01635
RI: 2230*

Acaricide

$C_{17}H_{16}Cl_2O_3$
338.04765
RI: 2230*

Acaricide

CAS: 5836-10-2

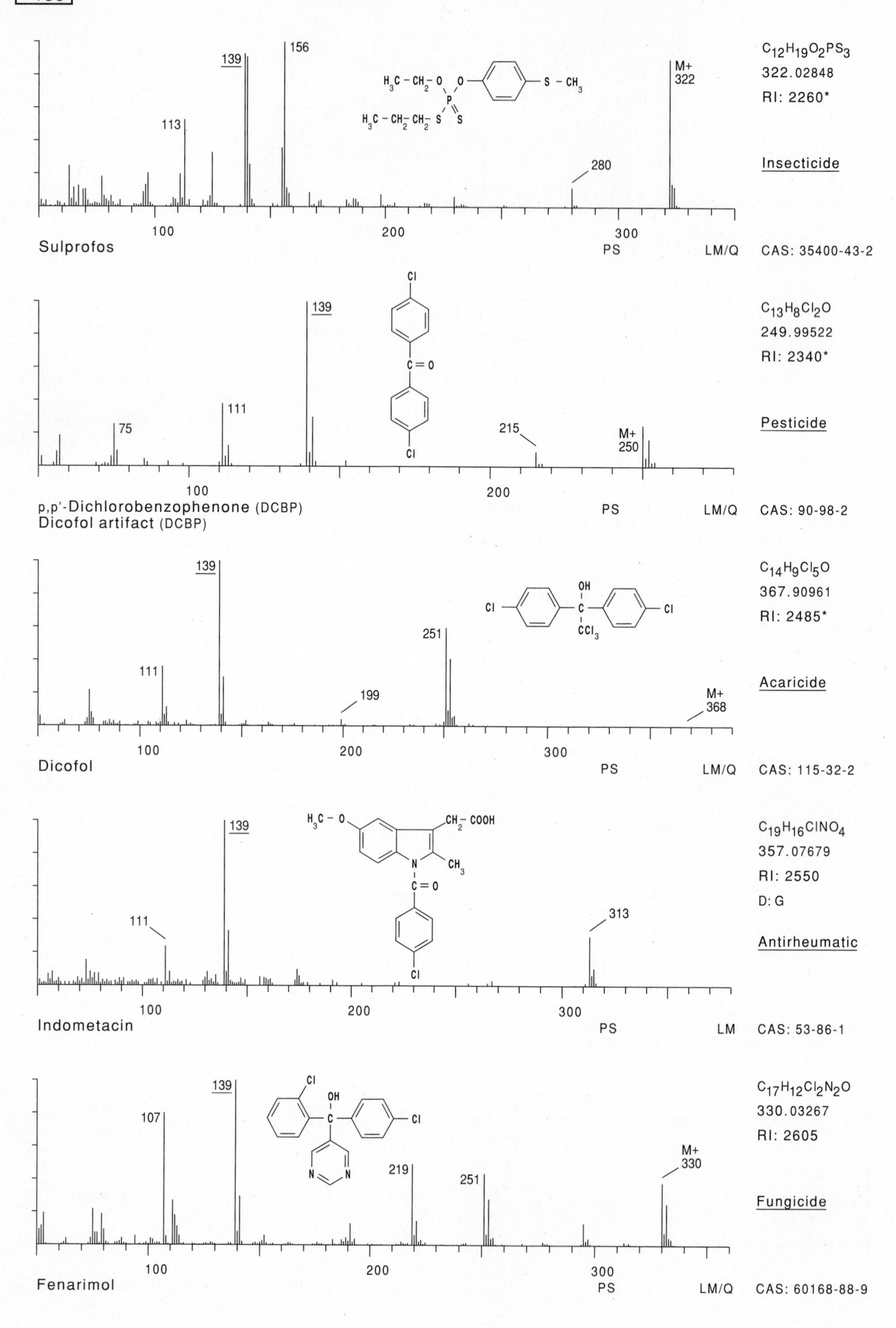
139
156
113
M+
322
280
100
200
300
Sulprofos
PS
LM/Q
$C_{12}H_{19}O_2PS_3$
322.02848
RI: 2260*
Insecticide
CAS: 35400-43-2
139
111
75
215
M+
250
100
200
p,p'-Dichlorobenzophenone (DCBP)
Dicofol artifact (DCBP)
PS
LM/Q
$C_{13}H_8Cl_2O$
249.99522
RI: 2340*
Pesticide
CAS: 90-98-2
139
251
111
199
M+
368
100
200
300
Dicofol
PS
LM/Q
$C_{14}H_9Cl_5O$
367.90961
RI: 2485*
Acaricide
CAS: 115-32-2
139
111
313
100
200
300
Indometacin
PS
LM
$C_{19}H_{16}ClNO_4$
357.07679
RI: 2550
D: G
Antirheumatic
CAS: 53-86-1
139
107
219
251
M+
330
100
200
300
Fenarimol
PS
LM/Q
$C_{17}H_{12}Cl_2N_2O$
330.03267
RI: 2605
Fungicide
CAS: 60168-88-9

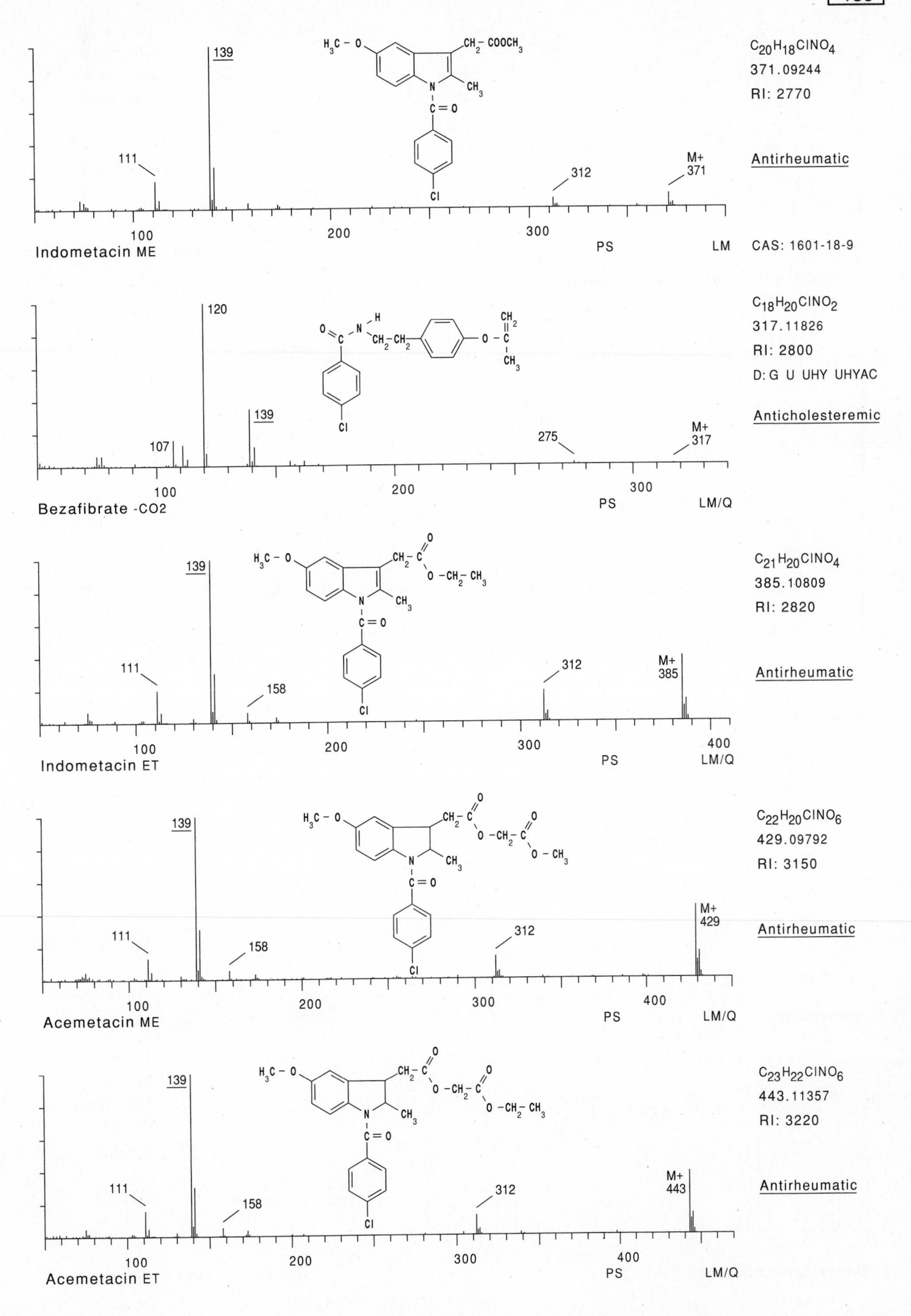

$C_{20}H_{18}ClNO_4$
371.09244
RI: 2770

Antirheumatic

CAS: 1601-18-9

$C_{18}H_{20}ClNO_2$
317.11826
RI: 2800
D: G U UHY UHYAC

Anticholesteremic

$C_{21}H_{20}ClNO_4$
385.10809
RI: 2820

Antirheumatic

$C_{22}H_{20}ClNO_6$
429.09792
RI: 3150

Antirheumatic

$C_{23}H_{22}ClNO_6$
443.11357
RI: 3220

Antirheumatic

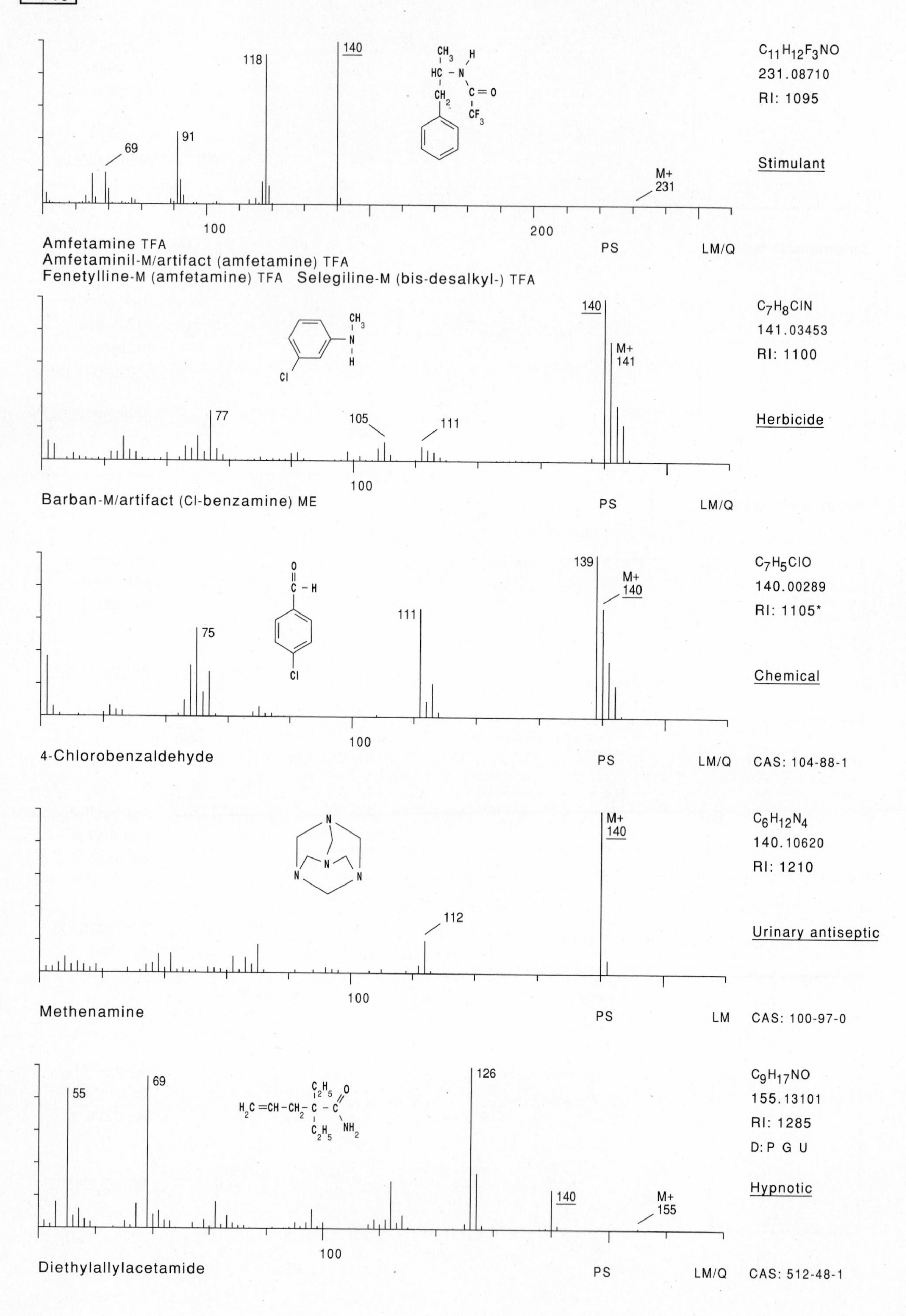

140
118
140
91
69
M+
231
100
200
PS
LM/Q
$C_{11}H_{12}F_3NO$
231.08710
RI: 1095
Stimulant
Amfetamine TFA
Amfetaminil-M/artifact (amfetamine) TFA
Fenetylline-M (amfetamine) TFA Selegiline-M (bis-desalkyl-) TFA
140
M+
141
77
105
111
100
PS
LM/Q
C_7H_8ClN
141.03453
RI: 1100
Herbicide
Barban-M/artifact (Cl-benzamine) ME
139
M+
140
111
75
100
PS
LM/Q
C_7H_5ClO
140.00289
RI: 1105*
Chemical
4-Chlorobenzaldehyde
CAS: 104-88-1
M+
140
112
100
PS
LM
$C_6H_{12}N_4$
140.10620
RI: 1210
Urinary antiseptic
Methenamine
CAS: 100-97-0
55
69
126
140
M+
155
100
PS
LM/Q
$C_9H_{17}NO$
155.13101
RI: 1285
D: P G U
Hypnotic
Diethylallylacetamide
CAS: 512-48-1

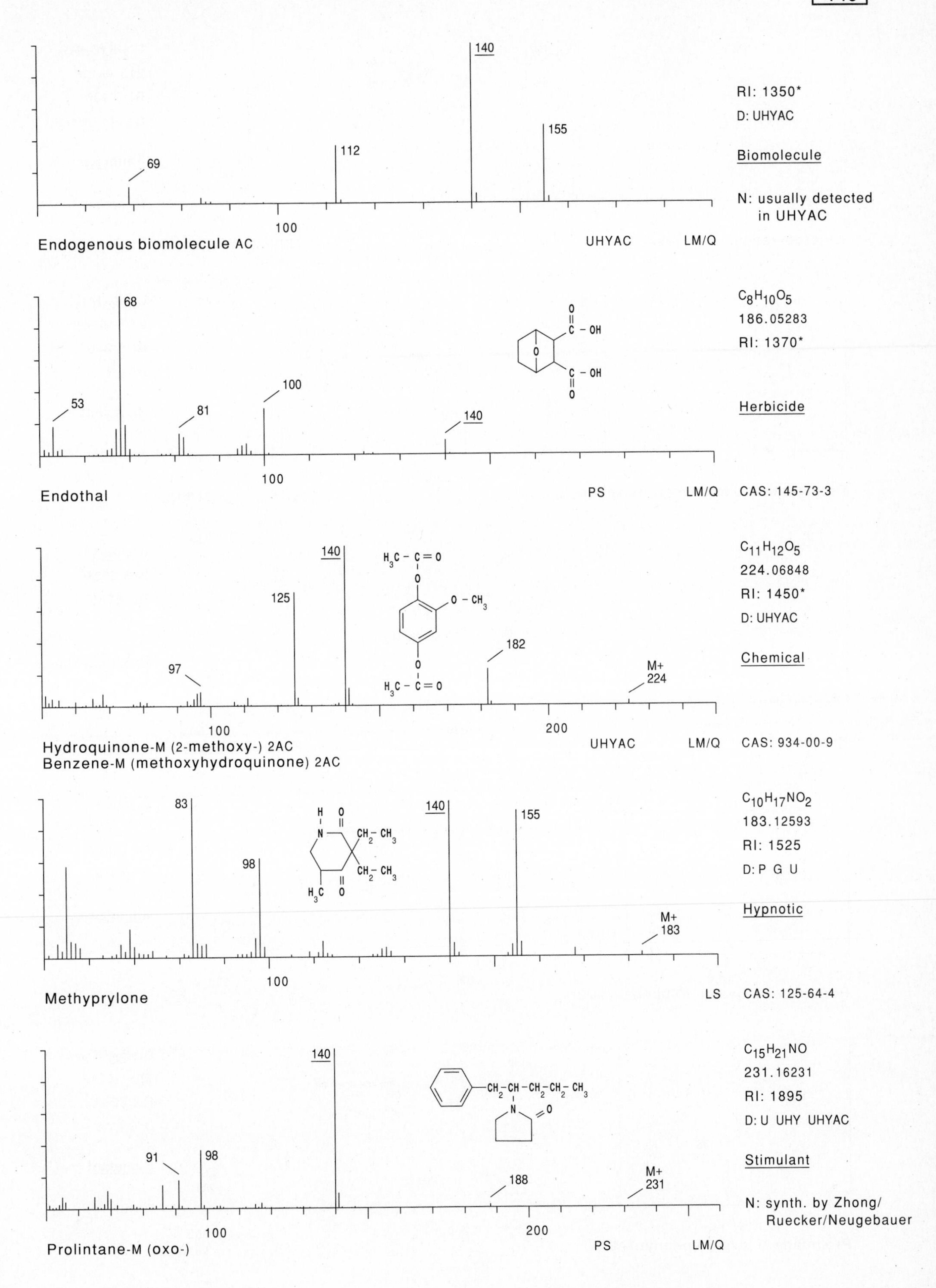
140
155
112
69
100
Endogenous biomolecule AC
UHYAC
LM/Q
RI: 1350*
D: UHYAC
Biomolecule
N: usually detected in UHYAC
68
53
81
100
140
100
Endothal
PS
LM/Q
C8H10O5
186.05283
RI: 1370*
Herbicide
CAS: 145-73-3
140
125
97
182
M+ 224
100
200
Hydroquinone-M (2-methoxy-) 2AC
Benzene-M (methoxyhydroquinone) 2AC
UHYAC
LM/Q
C11H12O5
224.06848
RI: 1450*
D: UHYAC
Chemical
CAS: 934-00-9
83
98
140
155
M+ 183
100
Methyprylone
LS
C10H17NO2
183.12593
RI: 1525
D: P G U
Hypnotic
CAS: 125-64-4
140
91
98
188
M+ 231
100
200
Prolintane-M (oxo-)
PS
LM/Q
C15H21NO
231.16231
RI: 1895
D: U UHY UHYAC
Stimulant
N: synth. by Zhong/Ruecker/Neugebauer

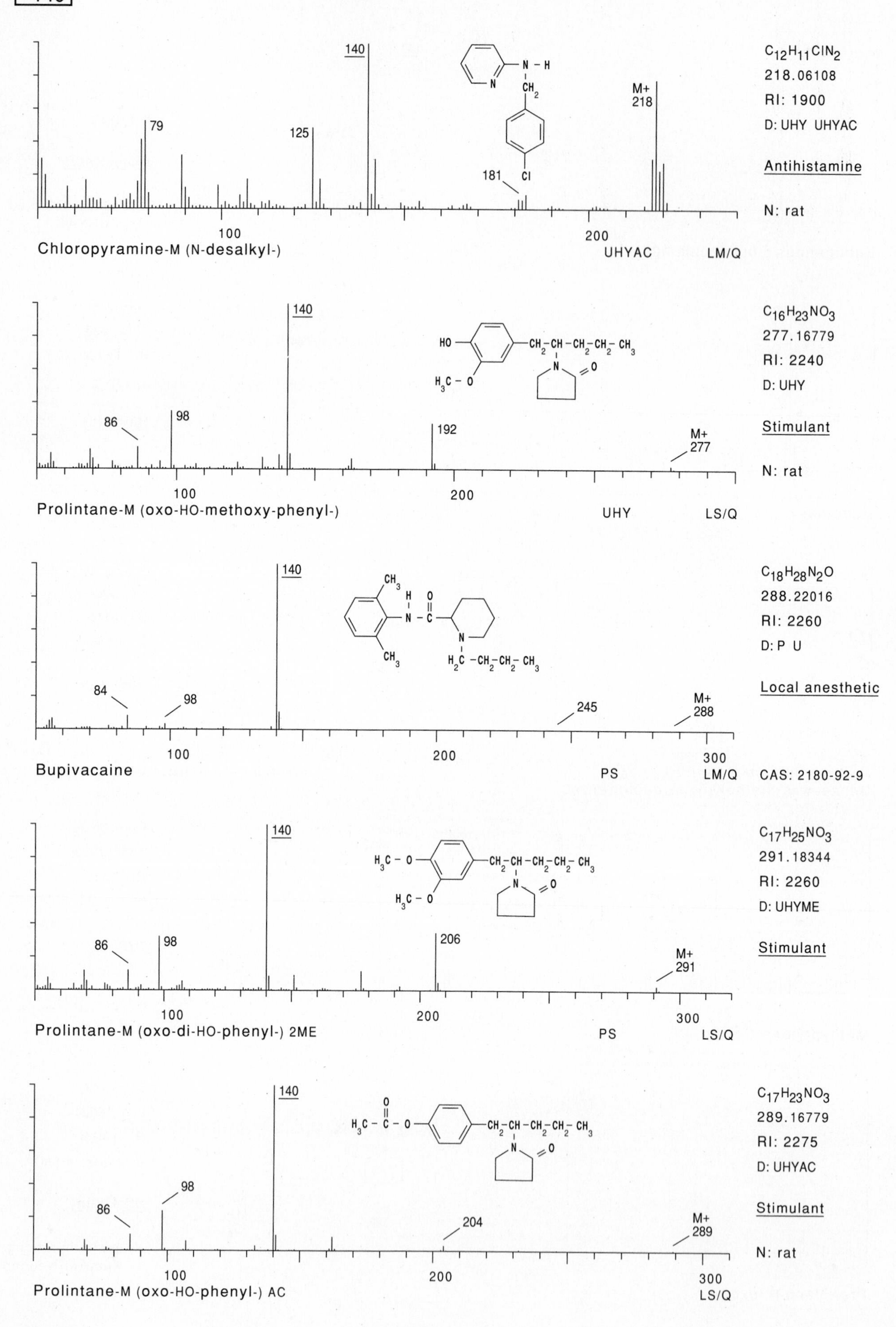
140
79
125
181
M+
218
100
200
Chloropyramine-M (N-desalkyl-)
UHYAC
LM/Q
$C_{12}H_{11}ClN_2$
218.06108
RI: 1900
D: UHY UHYAC
Antihistamine
N: rat
140
86
98
192
M+
277
100
200
Prolintane-M (oxo-HO-methoxy-phenyl-)
UHY
LS/Q
$C_{16}H_{23}NO_3$
277.16779
RI: 2240
D: UHY
Stimulant
N: rat
140
84
98
245
M+
288
100
200
300
Bupivacaine
PS
LM/Q
$C_{18}H_{28}N_2O$
288.22016
RI: 2260
D: P U
Local anesthetic
CAS: 2180-92-9
140
86
98
206
M+
291
100
200
300
Prolintane-M (oxo-di-HO-phenyl-) 2ME
PS
LS/Q
$C_{17}H_{25}NO_3$
291.18344
RI: 2260
D: UHYME
Stimulant
140
86
98
204
M+
289
100
200
300
Prolintane-M (oxo-HO-phenyl-) AC
LS/Q
$C_{17}H_{23}NO_3$
289.16779
RI: 2275
D: UHYAC
Stimulant
N: rat

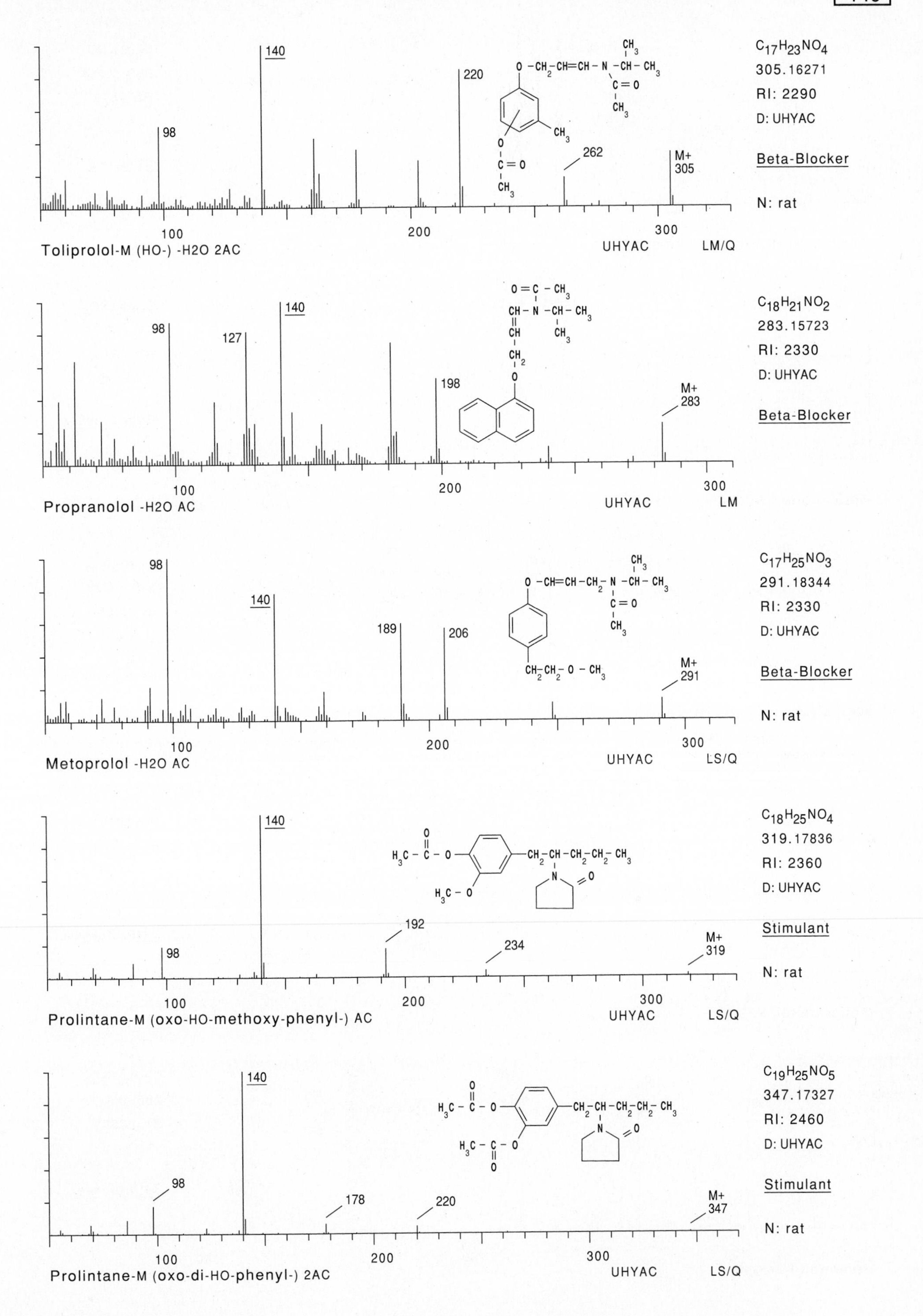
Toliprolol-M (HO-) -H2O 2AC
UHYAC LM/Q
C17H23NO4
305.16271
RI: 2290
D: UHYAC
Beta-Blocker
N: rat
Propranolol -H2O AC
UHYAC LM
C18H21NO2
283.15723
RI: 2330
D: UHYAC
Beta-Blocker
Metoprolol -H2O AC
UHYAC LS/Q
C17H25NO3
291.18344
RI: 2330
D: UHYAC
Beta-Blocker
N: rat
Prolintane-M (oxo-HO-methoxy-phenyl-) AC
UHYAC LS/Q
C18H25NO4
319.17836
RI: 2360
D: UHYAC
Stimulant
N: rat
Prolintane-M (oxo-di-HO-phenyl-) 2AC
UHYAC LS/Q
C19H25NO5
347.17327
RI: 2460
D: UHYAC
Stimulant
N: rat

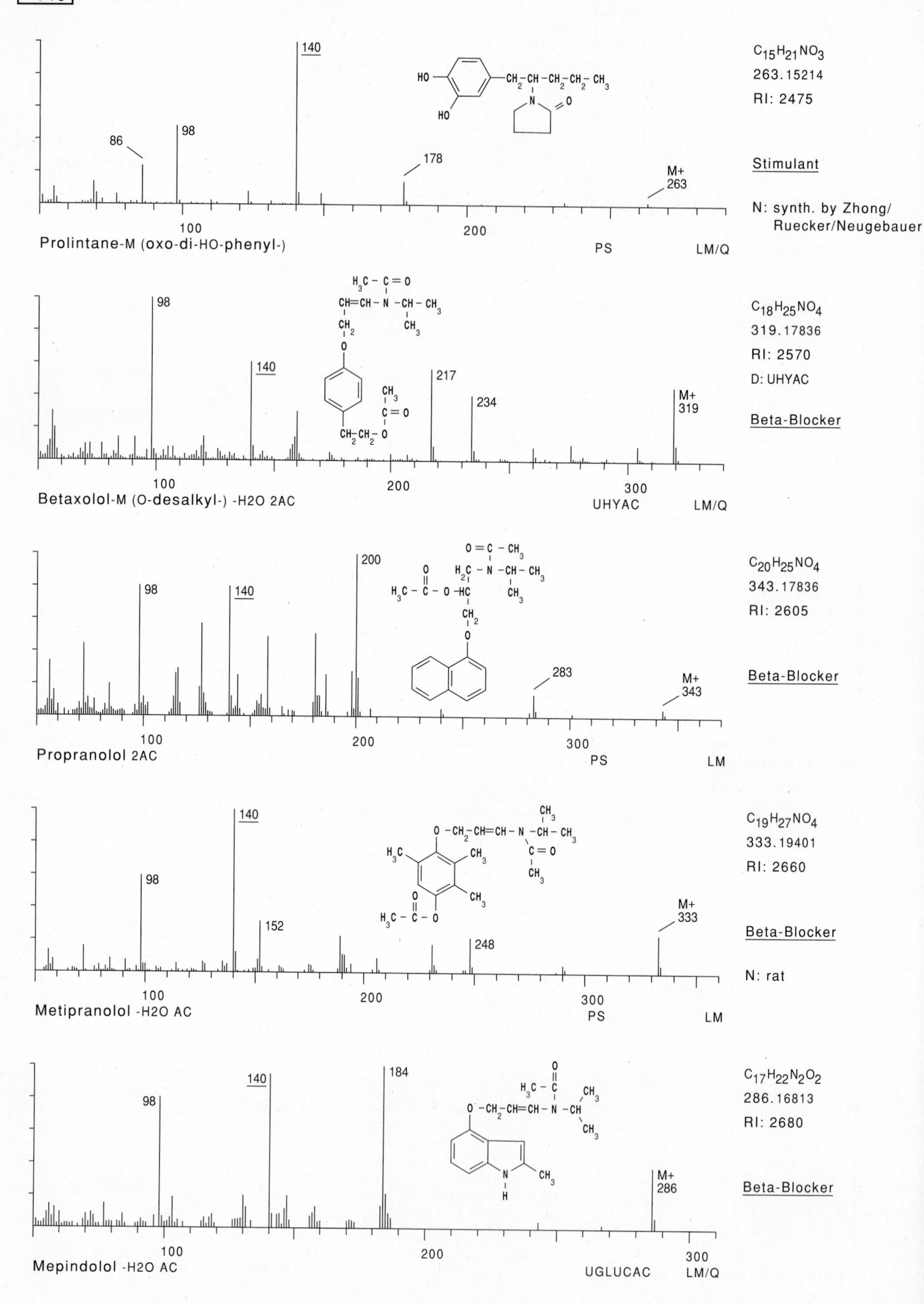

Prolintane-M (oxo-di-HO-phenyl-)
C15H21NO3
263.15214
RI: 2475
Stimulant
N: synth. by Zhong/Ruecker/Neugebauer
PS
LM/Q
Betaxolol-M (O-desalkyl-) -H2O 2AC
C18H25NO4
319.17836
RI: 2570
D: UHYAC
Beta-Blocker
UHYAC
LM/Q
Propranolol 2AC
C20H25NO4
343.17836
RI: 2605
Beta-Blocker
PS
LM
Metipranolol -H2O AC
C19H27NO4
333.19401
RI: 2660
Beta-Blocker
N: rat
PS
LM
Mepindolol -H2O AC
C17H22N2O2
286.16813
RI: 2680
Beta-Blocker
UGLUCAC
LM/Q

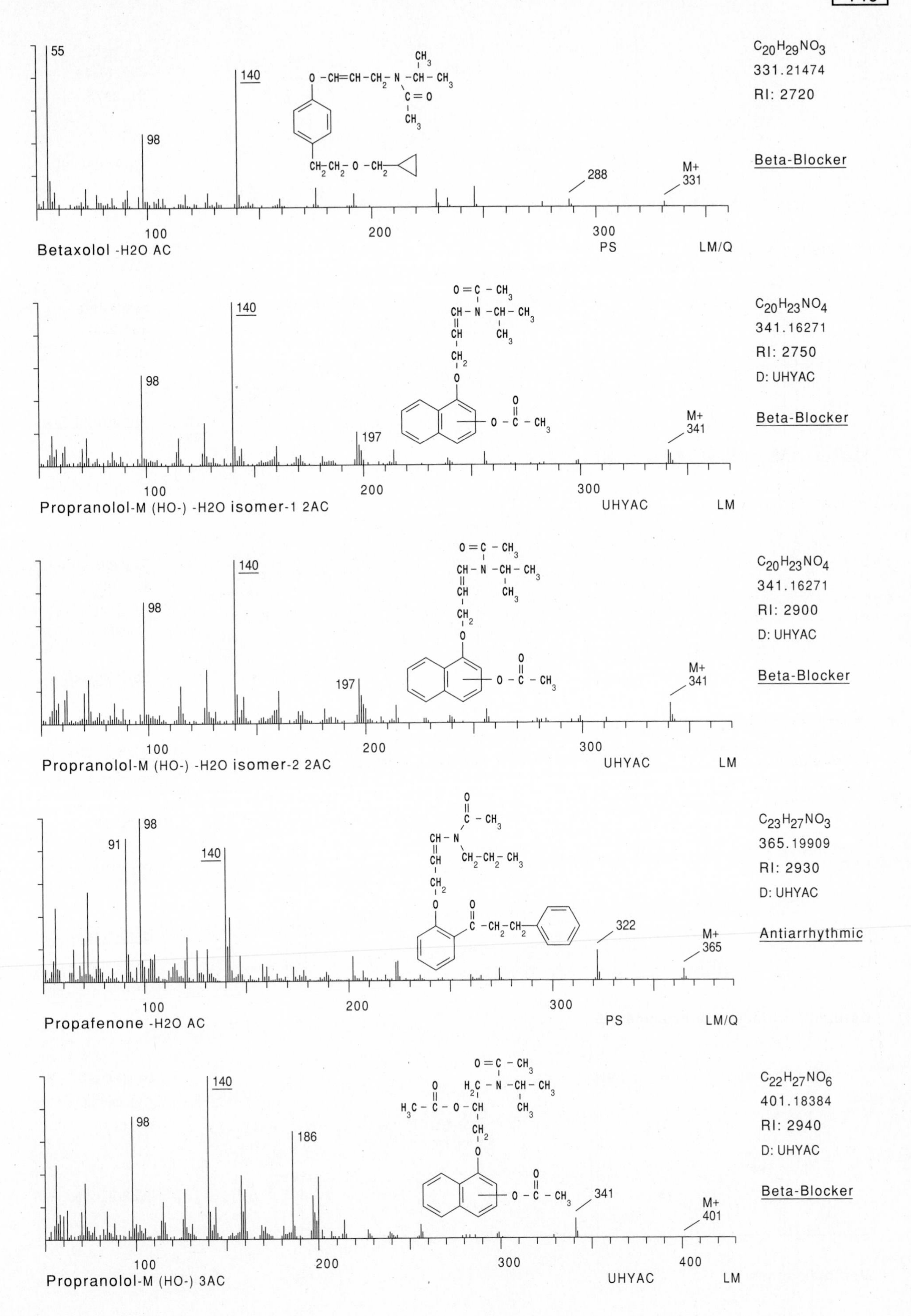

55
140
98
288
M+
331
100
200
300
PS
LM/Q
Betaxolol -H2O AC
C20H29NO3
331.21474
RI: 2720
Beta-Blocker
140
98
197
M+
341
100
200
300
UHYAC
LM
Propranolol-M (HO-) -H2O isomer-1 2AC
C20H23NO4
341.16271
RI: 2750
D: UHYAC
Beta-Blocker
140
98
197
M+
341
100
200
300
UHYAC
LM
Propranolol-M (HO-) -H2O isomer-2 2AC
C20H23NO4
341.16271
RI: 2900
D: UHYAC
Beta-Blocker
98
91
140
322
M+
365
100
200
300
PS
LM/Q
Propafenone -H2O AC
C23H27NO3
365.19909
RI: 2930
D: UHYAC
Antiarrhythmic
140
98
186
341
M+
401
100
200
300
400
UHYAC
LM
Propranolol-M (HO-) 3AC
C22H27NO6
401.18384
RI: 2940
D: UHYAC
Beta-Blocker

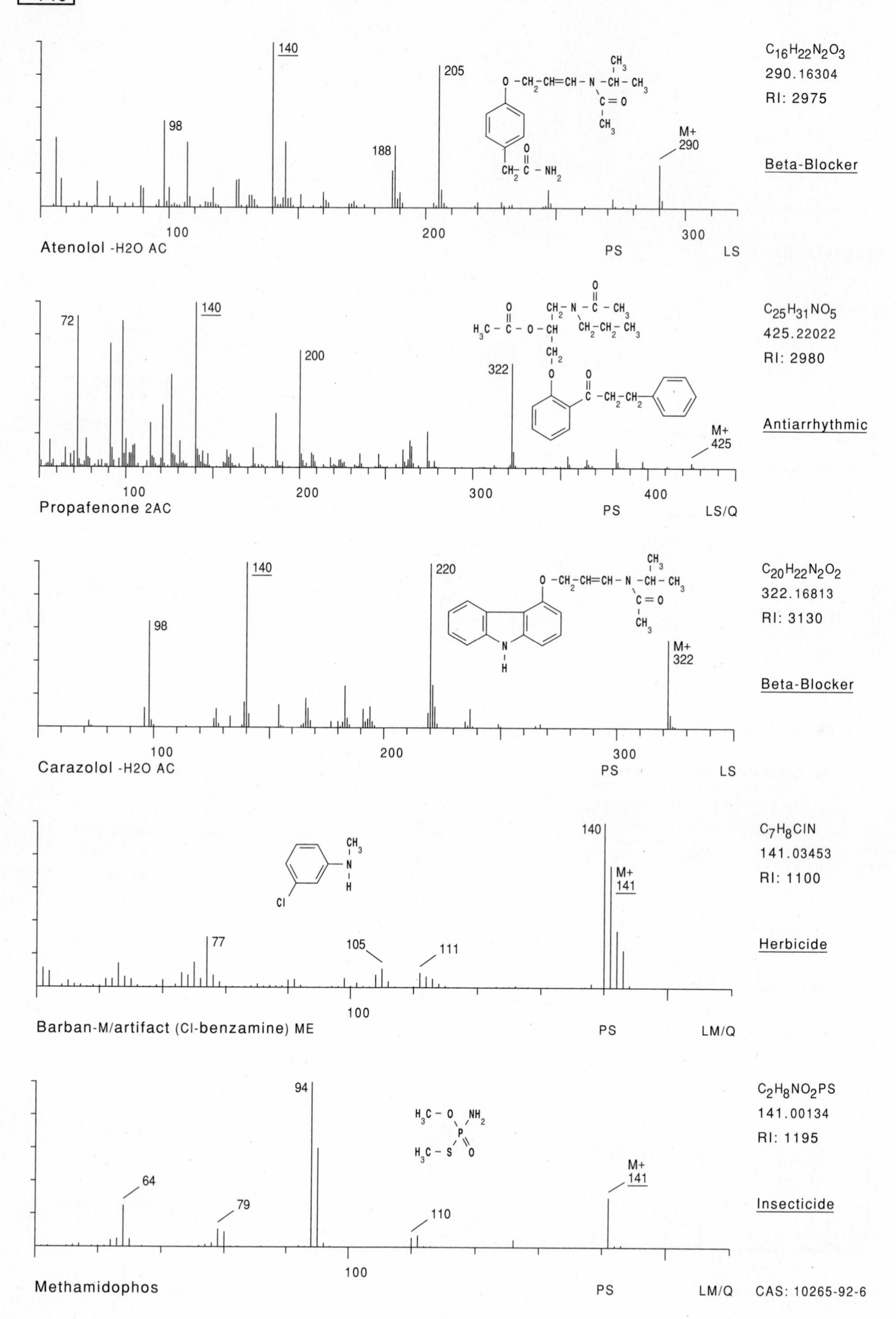

140
205
98
188
M+
290
100
200
300
Atenolol -H2O AC
PS
LS
$C_{16}H_{22}N_2O_3$
290.16304
RI: 2975
Beta-Blocker
72
140
200
322
M+
425
100
200
300
400
Propafenone 2AC
PS
LS/Q
$C_{25}H_{31}NO_5$
425.22022
RI: 2980
Antiarrhythmic
140
220
98
M+
322
100
200
300
Carazolol -H2O AC
PS
LS
$C_{20}H_{22}N_2O_2$
322.16813
RI: 3130
Beta-Blocker
140
M+
141
77
105
111
100
Barban-M/artifact (Cl-benzamine) ME
PS
LM/Q
C_7H_8ClN
141.03453
RI: 1100
Herbicide
94
64
79
110
M+
141
100
Methamidophos
PS
LM/Q
$C_2H_8NO_2PS$
141.00134
RI: 1195
Insecticide
CAS: 10265-92-6

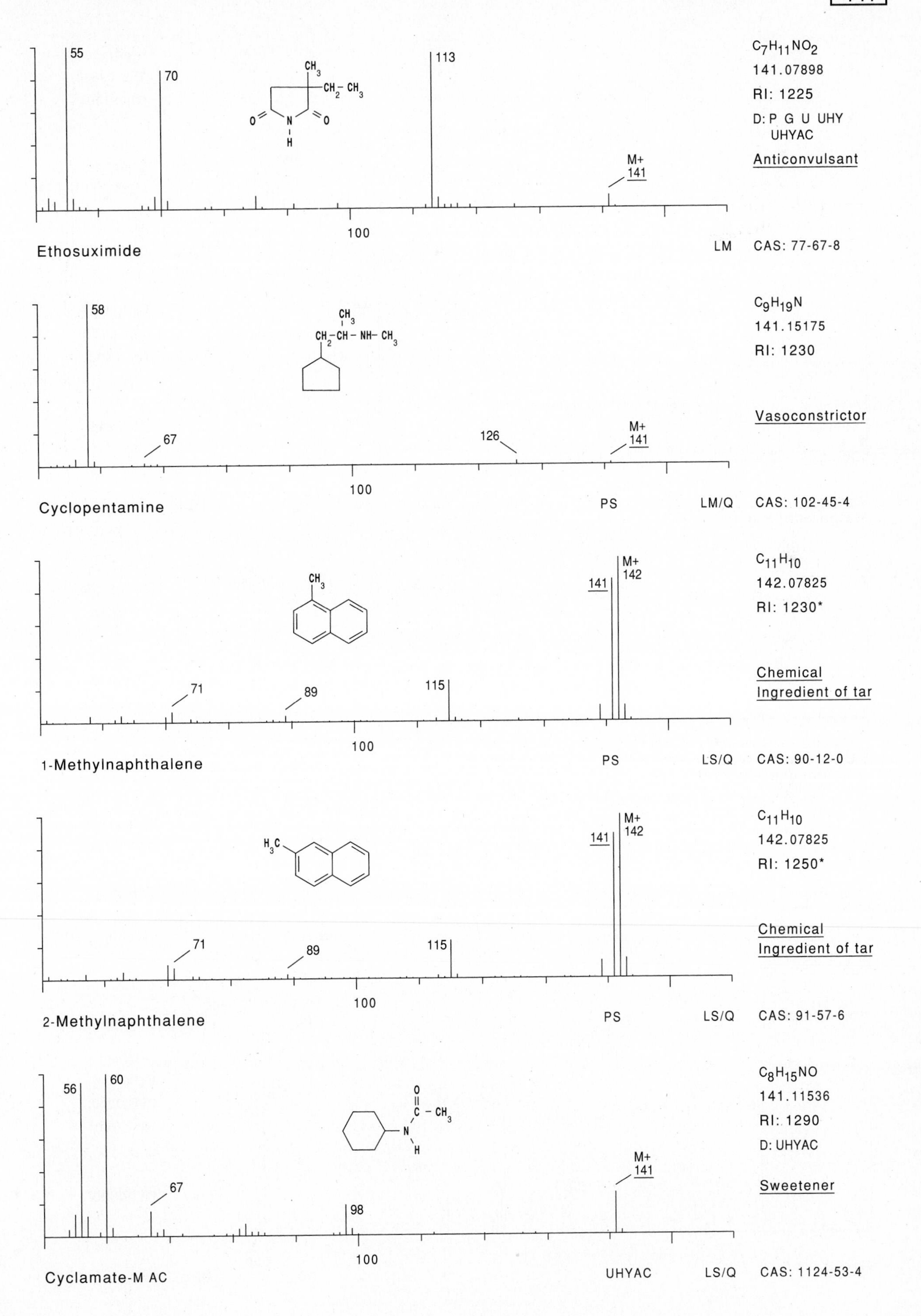

55
70
113
M+ 141
100
Ethosuximide
LM
$C_7H_{11}NO_2$
141.07898
RI: 1225
D: P G U UHY UHYAC
Anticonvulsant
CAS: 77-67-8
58
67
126
M+ 141
100
Cyclopentamine
PS
LM/Q
$C_9H_{19}N$
141.15175
RI: 1230
Vasoconstrictor
CAS: 102-45-4
71
89
115
141
M+ 142
100
1-Methylnaphthalene
PS
LS/Q
$C_{11}H_{10}$
142.07825
RI: 1230*
Chemical Ingredient of tar
CAS: 90-12-0
71
89
115
141
M+ 142
100
2-Methylnaphthalene
PS
LS/Q
$C_{11}H_{10}$
142.07825
RI: 1250*
Chemical Ingredient of tar
CAS: 91-57-6
56
60
67
98
M+ 141
100
Cyclamate-M AC
UHYAC
LS/Q
$C_8H_{15}NO$
141.11536
RI: 1290
D: UHYAC
Sweetener
CAS: 1124-53-4

141

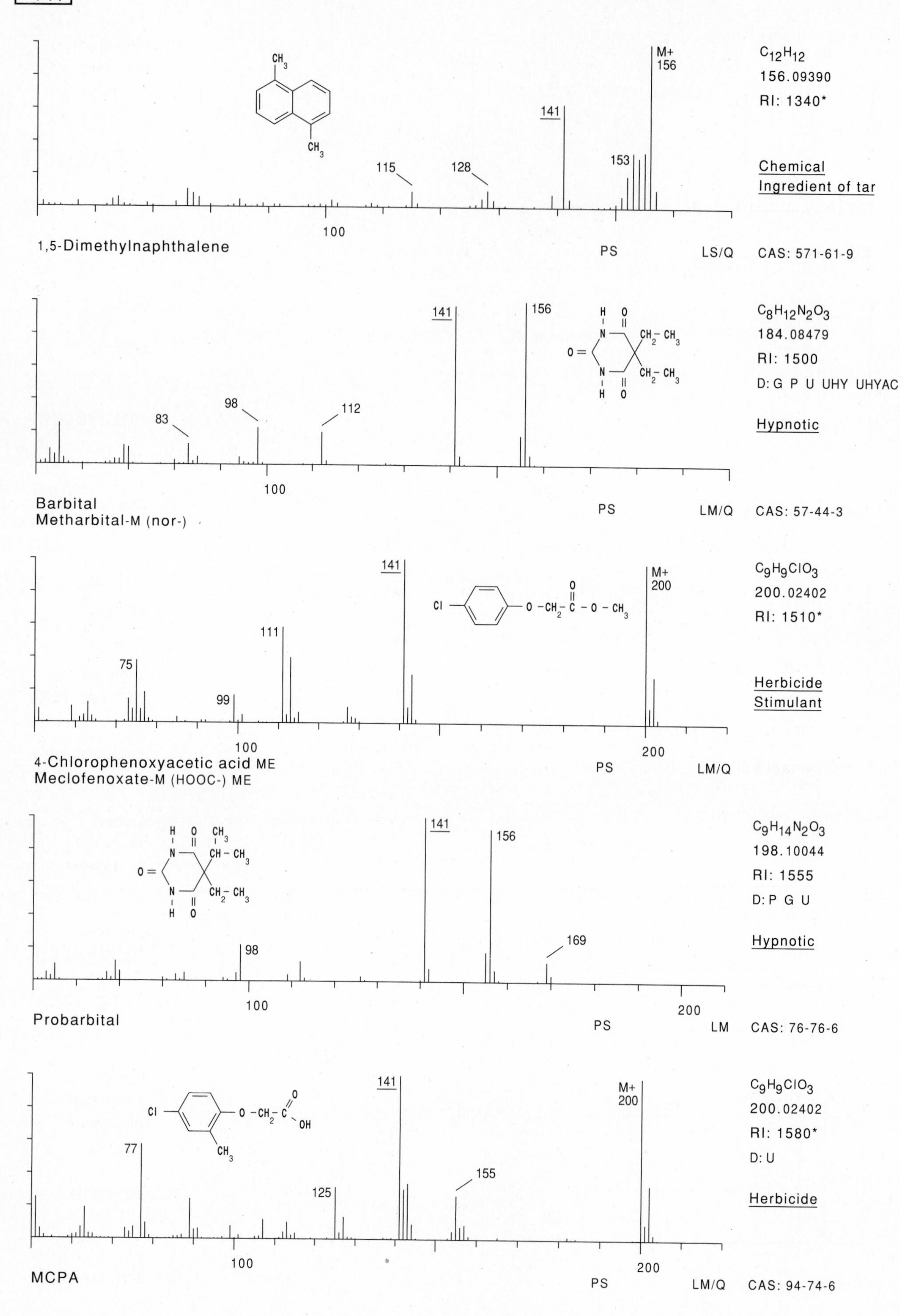

M+ 156
141
115
128
153
100
1,5-Dimethylnaphthalene
PS
LS/Q
C12H12
156.09390
RI: 1340*
Chemical
Ingredient of tar
CAS: 571-61-9
141
156
83
98
112
100
Barbital
Metharbital-M (nor-)
PS
LM/Q
C8H12N2O3
184.08479
RI: 1500
D: G P U UHY UHYAC
Hypnotic
CAS: 57-44-3
141
M+ 200
111
75
99
100
200
4-Chlorophenoxyacetic acid ME
Meclofenoxate-M (HOOC-) ME
PS
LM/Q
C9H9ClO3
200.02402
RI: 1510*
Herbicide
Stimulant
141
156
98
169
100
200
Probarbital
PS
LM
C9H14N2O3
198.10044
RI: 1555
D: P G U
Hypnotic
CAS: 76-76-6
141
M+ 200
77
125
155
100
200
MCPA
PS
LM/Q
C9H9ClO3
200.02402
RI: 1580*
D: U
Herbicide
CAS: 94-74-6

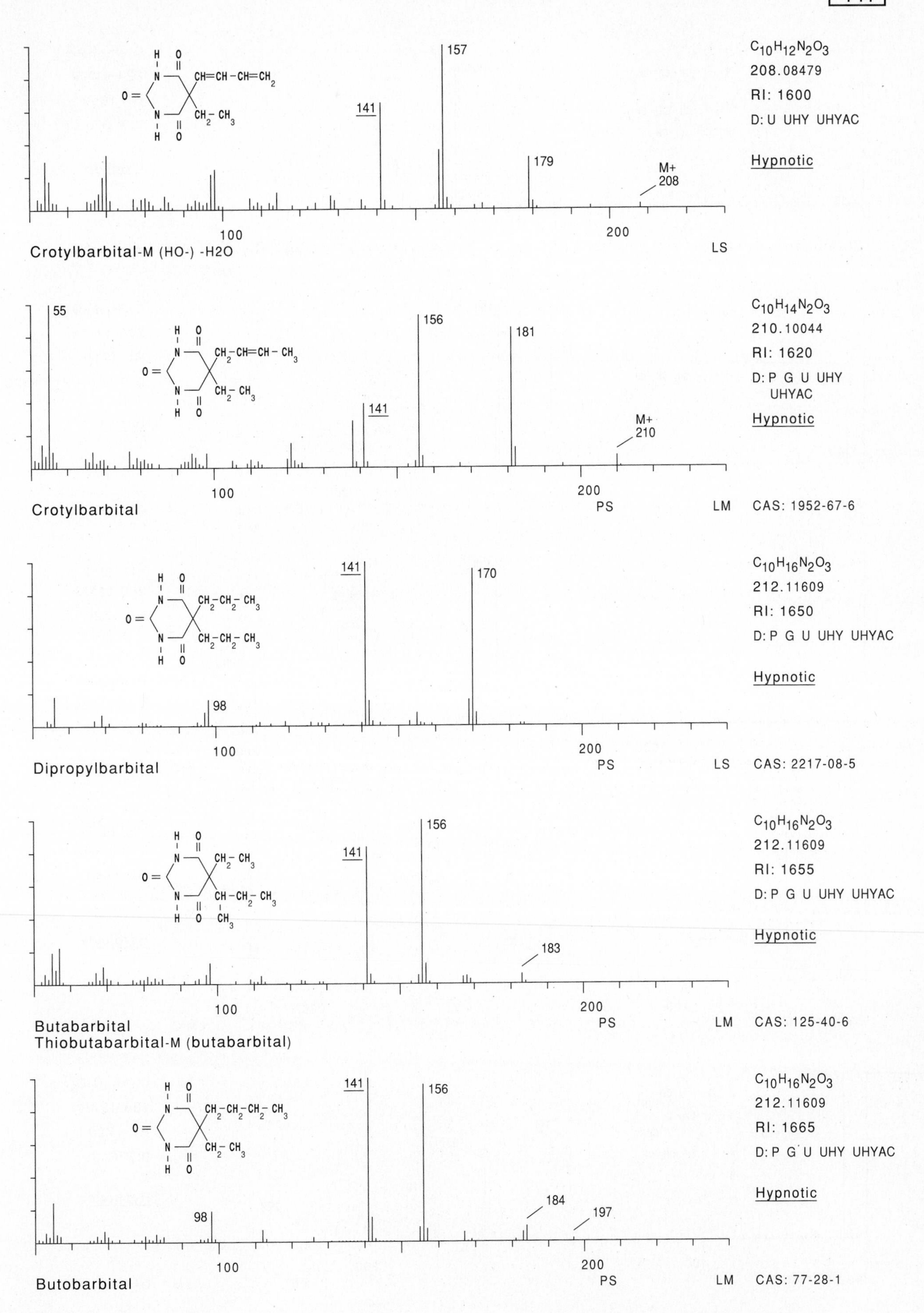

C10H12N2O3
208.08479
RI: 1600
D: U UHY UHYAC
Hypnotic
157
141
179
M+
208
100
200
LS
Crotylbarbital-M (HO-) -H2O
C10H14N2O3
210.10044
RI: 1620
D: P G U UHY
UHYAC
Hypnotic
55
156
181
141
M+
210
100
200
PS
LM
Crotylbarbital
CAS: 1952-67-6
C10H16N2O3
212.11609
RI: 1650
D: P G U UHY UHYAC
Hypnotic
141
170
98
100
200
PS
LS
Dipropylbarbital
CAS: 2217-08-5
C10H16N2O3
212.11609
RI: 1655
D: P G U UHY UHYAC
Hypnotic
156
141
183
100
200
PS
LM
Butabarbital
Thiobutabarbital-M (butabarbital)
CAS: 125-40-6
C10H16N2O3
212.11609
RI: 1665
D: P G U UHY UHYAC
Hypnotic
141
156
98
184
197
100
200
PS
LM
Butobarbital
CAS: 77-28-1

141

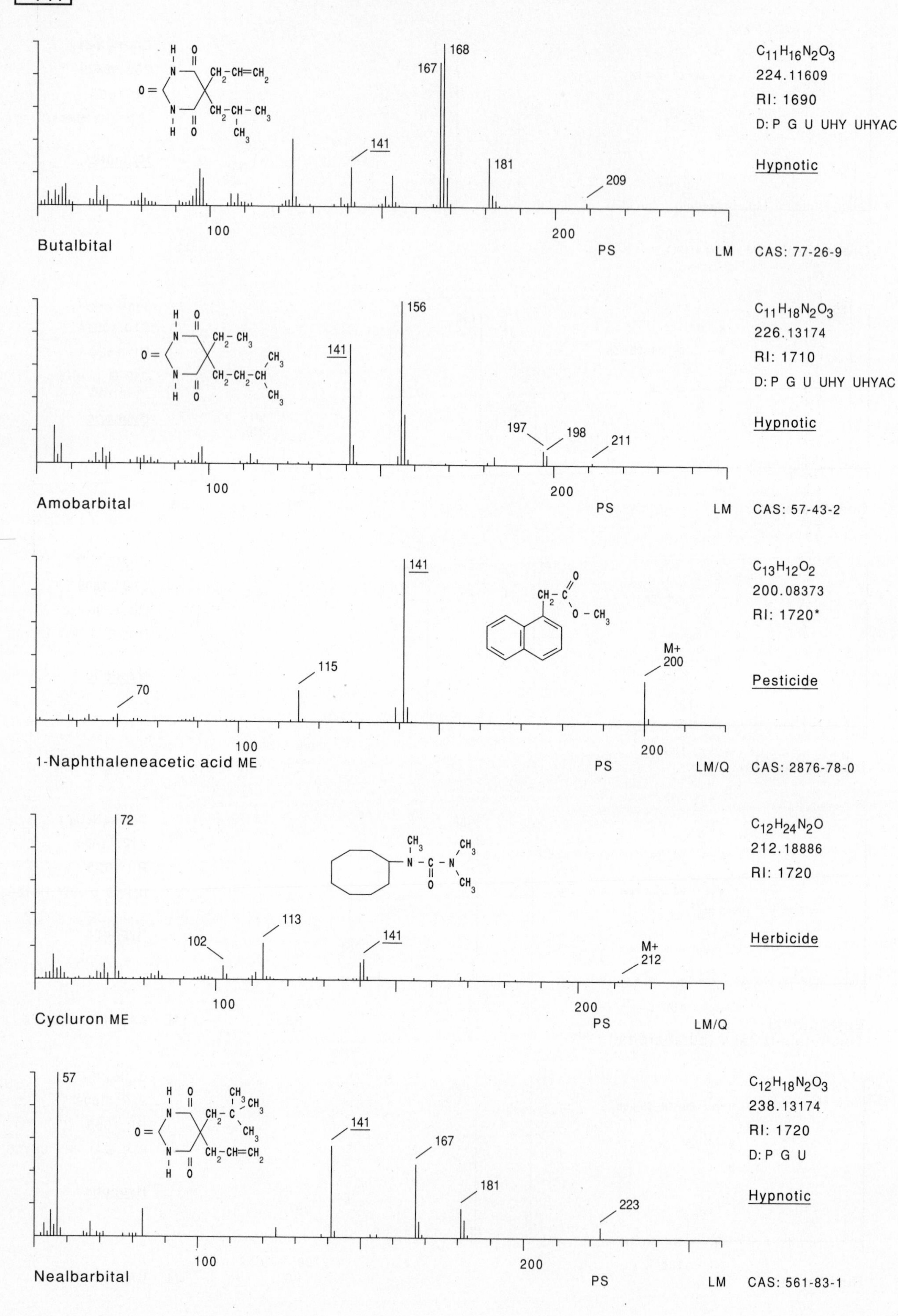
168
167
141
181
209
100
200
Butalbital
PS
LM
C11H16N2O3
224.11609
RI: 1690
D: P G U UHY UHYAC
Hypnotic
CAS: 77-26-9
156
141
197
198
211
100
200
Amobarbital
PS
LM
C11H18N2O3
226.13174
RI: 1710
D: P G U UHY UHYAC
Hypnotic
CAS: 57-43-2
141
115
70
M+
200
100
200
1-Naphthaleneacetic acid ME
PS
LM/Q
C13H12O2
200.08373
RI: 1720*
Pesticide
CAS: 2876-78-0
72
113
102
141
M+
212
100
200
Cycluron ME
PS
LM/Q
C12H24N2O
212.18886
RI: 1720
Herbicide
57
141
167
181
223
100
200
Nealbarbital
PS
LM
C12H18N2O3
238.13174
RI: 1720
D: P G U
Hypnotic
CAS: 561-83-1

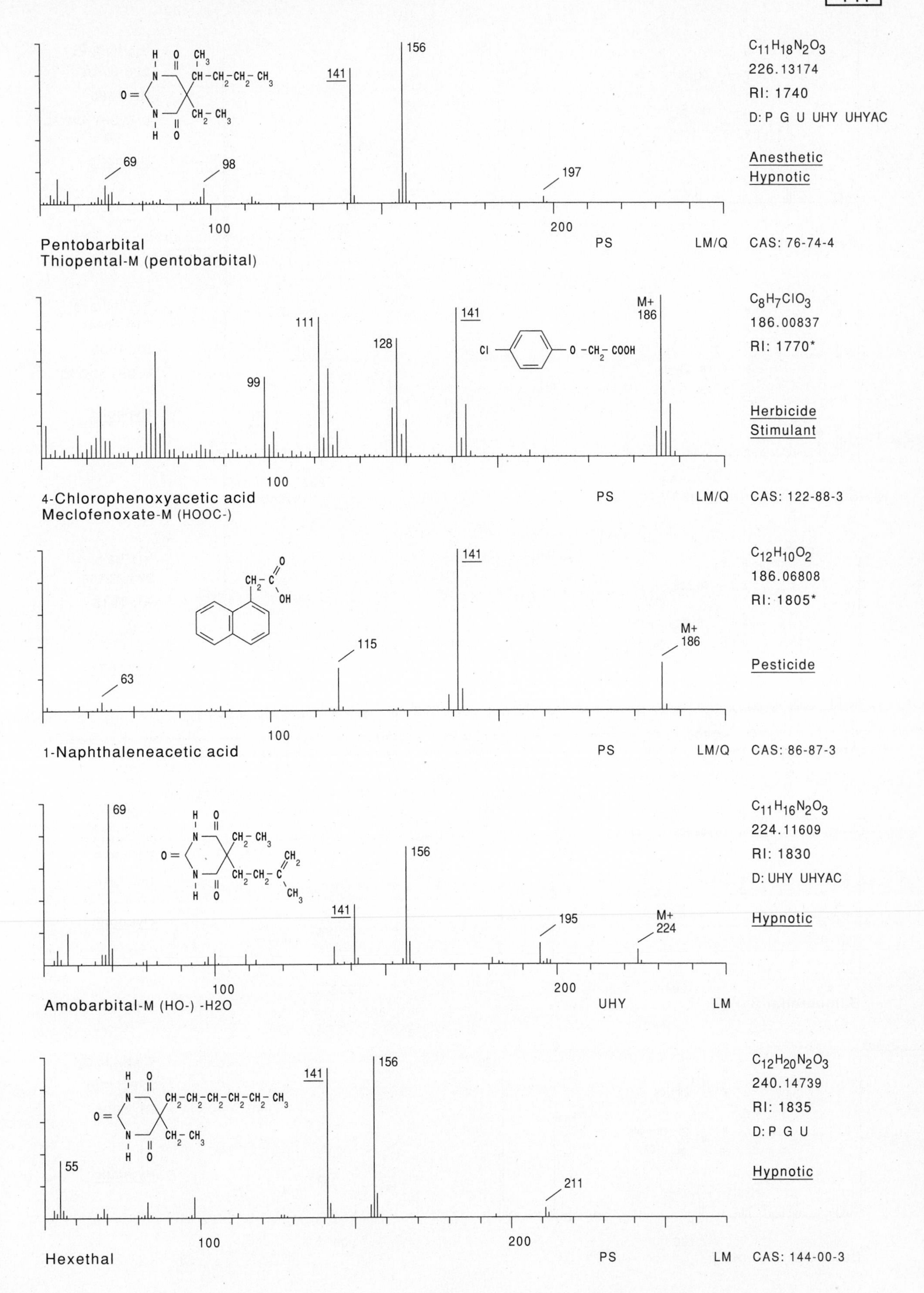

$C_{11}H_{18}N_2O_3$
226.13174
RI: 1740
D: P G U UHY UHYAC

Anesthetic
Hypnotic

Pentobarbital
Thiopental-M (pentobarbital)

CAS: 76-74-4

$C_8H_7ClO_3$
186.00837
RI: 1770*

Herbicide
Stimulant

4-Chlorophenoxyacetic acid
Meclofenoxate-M (HOOC-)

CAS: 122-88-3

$C_{12}H_{10}O_2$
186.06808
RI: 1805*

Pesticide

1-Naphthaleneacetic acid

CAS: 86-87-3

$C_{11}H_{16}N_2O_3$
224.11609
RI: 1830
D: UHY UHYAC

Hypnotic

Amobarbital-M (HO-) -H2O

$C_{12}H_{20}N_2O_3$
240.14739
RI: 1835
D: P G U

Hypnotic

Hexethal

CAS: 144-00-3

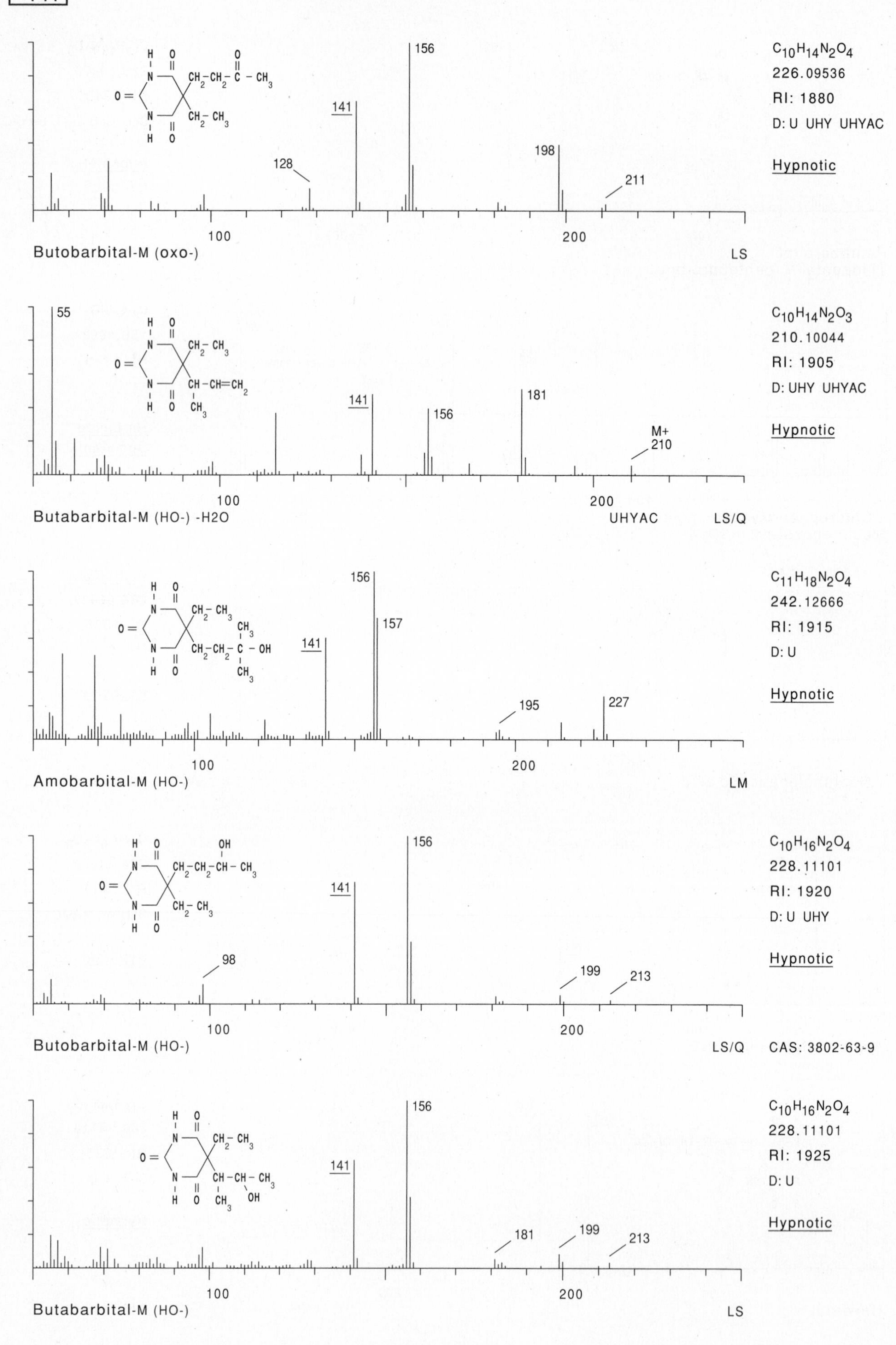
156
141
128
198
211
100
200
Butobarbital-M (oxo-)
LS
C10H14N2O4
226.09536
RI: 1880
D: U UHY UHYAC
Hypnotic
55
141
156
181
M+
210
100
200
Butabarbital-M (HO-) -H2O
UHYAC
LS/Q
C10H14N2O3
210.10044
RI: 1905
D: UHY UHYAC
Hypnotic
156
157
141
195
227
100
200
Amobarbital-M (HO-)
LM
C11H18N2O4
242.12666
RI: 1915
D: U
Hypnotic
156
141
98
199
213
100
200
Butobarbital-M (HO-)
LS/Q
C10H16N2O4
228.11101
RI: 1920
D: U UHY
Hypnotic
CAS: 3802-63-9
156
141
181
199
213
100
200
Butabarbital-M (HO-)
LS
C10H16N2O4
228.11101
RI: 1925
D: U
Hypnotic

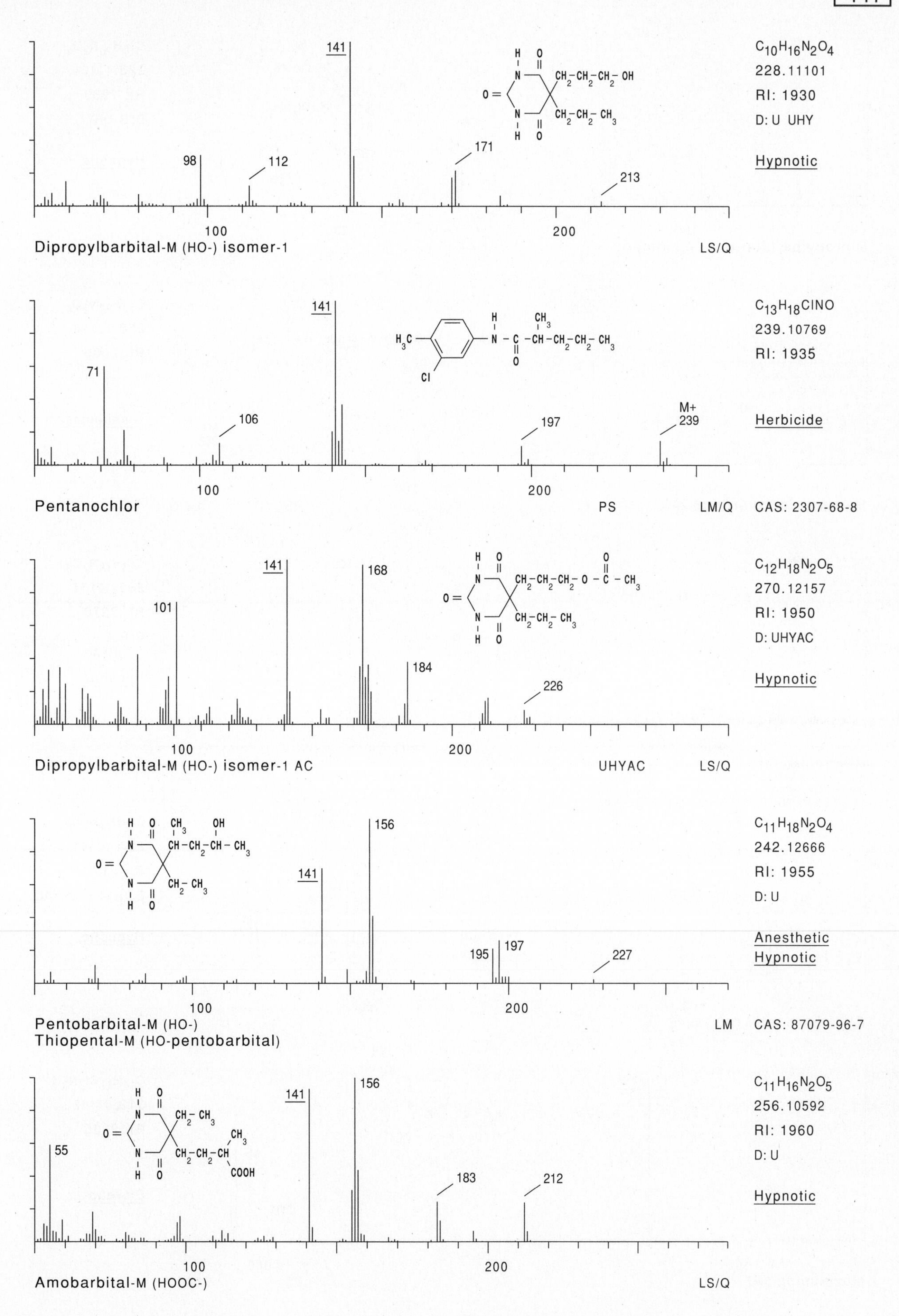

141
98
112
171
213
100
200
Dipropylbarbital-M (HO-) isomer-1
LS/Q
C10H16N2O4
228.11101
RI: 1930
D: U UHY
Hypnotic
141
71
106
197
M+
239
100
200
Pentanochlor
PS
LM/Q
C13H18ClNO
239.10769
RI: 1935
Herbicide
CAS: 2307-68-8
141
168
101
184
226
100
200
Dipropylbarbital-M (HO-) isomer-1 AC
UHYAC
LS/Q
C12H18N2O5
270.12157
RI: 1950
D: UHYAC
Hypnotic
156
141
195
197
227
100
200
Pentobarbital-M (HO-)
Thiopental-M (HO-pentobarbital)
LM
C11H18N2O4
242.12666
RI: 1955
D: U
Anesthetic
Hypnotic
CAS: 87079-96-7
141
156
55
183
212
100
200
Amobarbital-M (HOOC-)
LS/Q
C11H16N2O5
256.10592
RI: 1960
D: U
Hypnotic

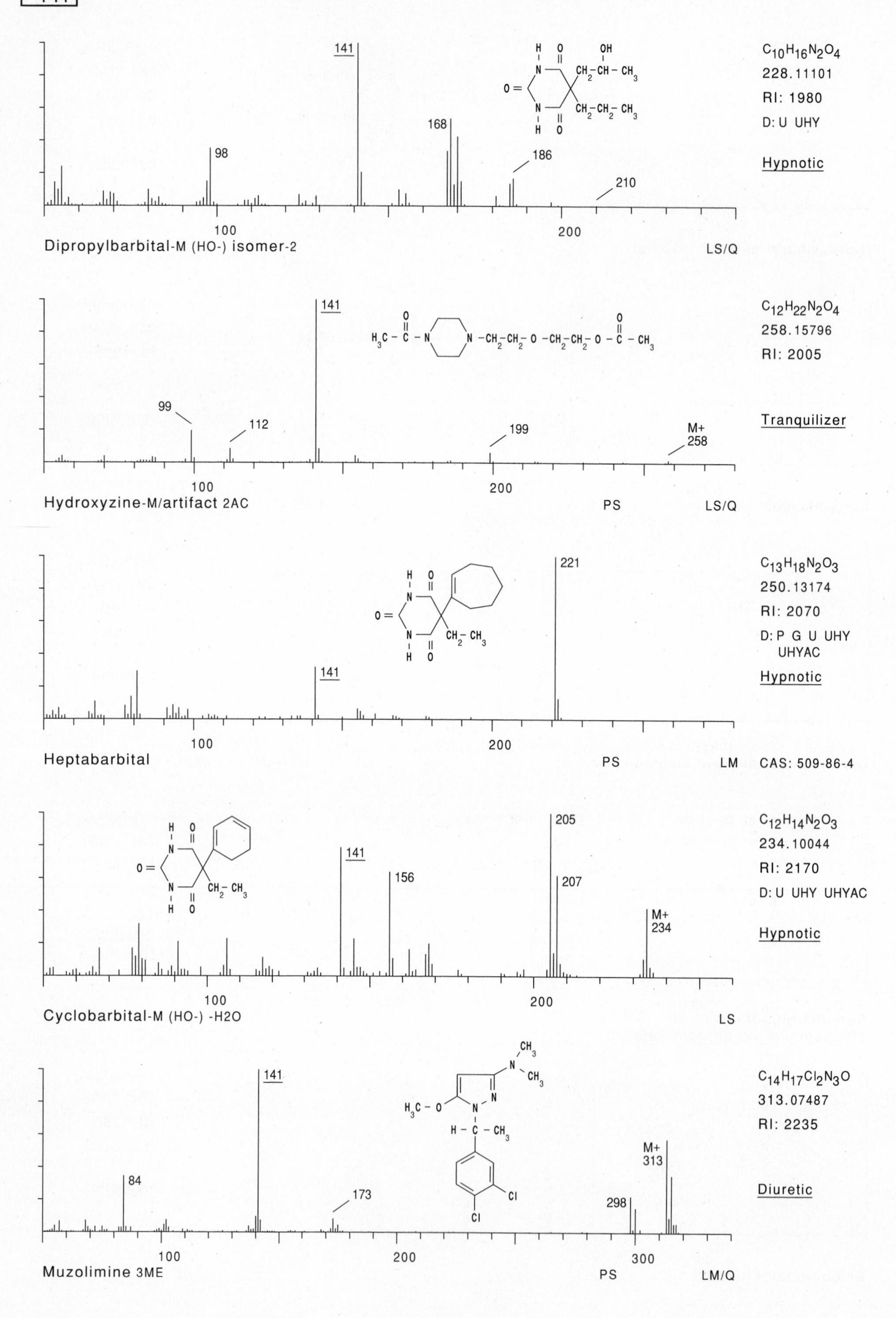
Dipropylbarbital-M (HO-) isomer-2
LS/Q
$C_{10}H_{16}N_2O_4$
228.11101
RI: 1980
D: U UHY
Hypnotic
Hydroxyzine-M/artifact 2AC
PS
LS/Q
$C_{12}H_{22}N_2O_4$
258.15796
RI: 2005
Tranquilizer
Heptabarbital
PS
LM
$C_{13}H_{18}N_2O_3$
250.13174
RI: 2070
D: P G U UHY UHYAC
Hypnotic
CAS: 509-86-4
Cyclobarbital-M (HO-) -H2O
LS
$C_{12}H_{14}N_2O_3$
234.10044
RI: 2170
D: U UHY UHYAC
Hypnotic
Muzolimine 3ME
PS
LM/Q
$C_{14}H_{17}Cl_2N_3O$
313.07487
RI: 2235
Diuretic

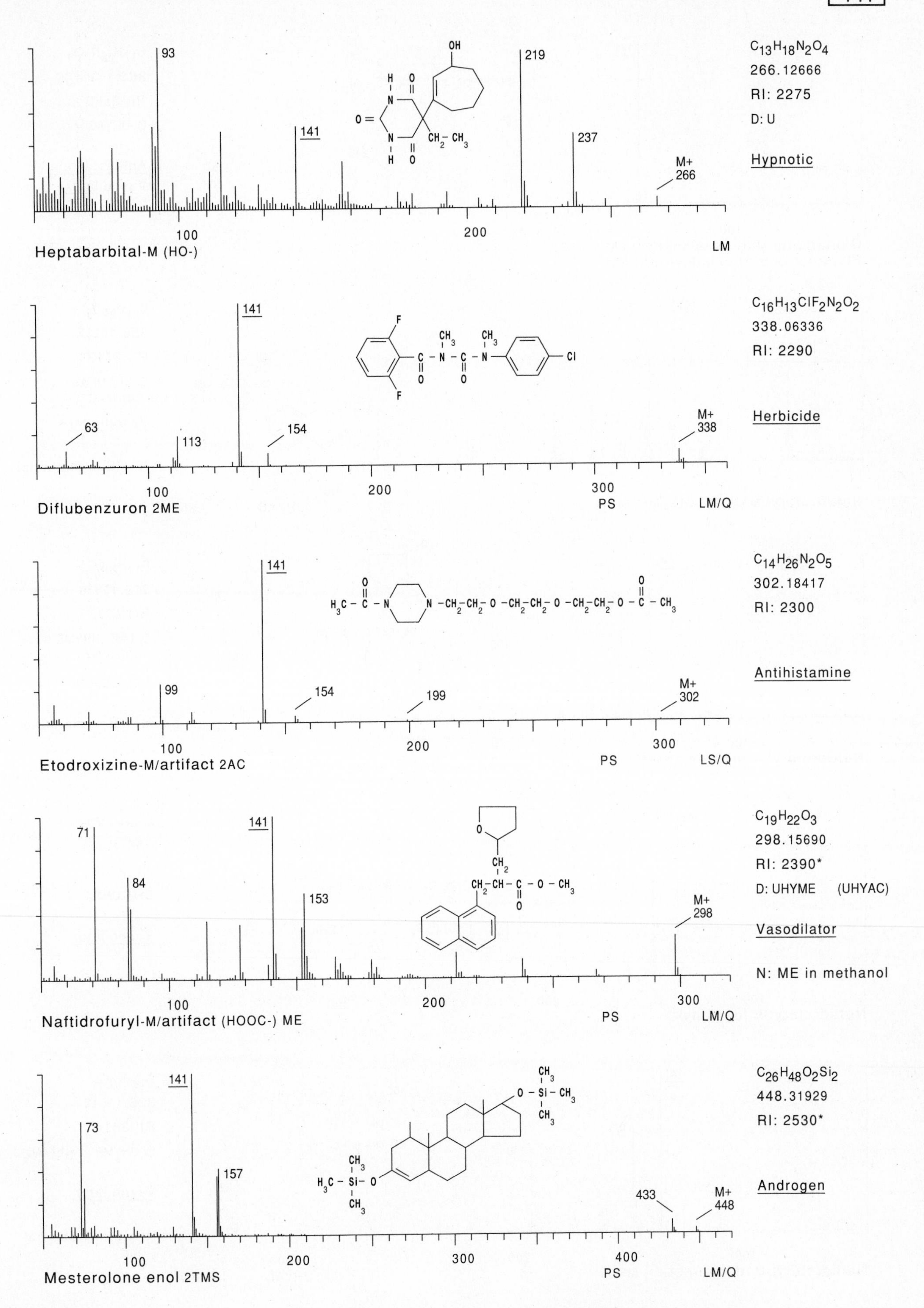

$C_{13}H_{18}N_2O_4$
266.12666
RI: 2275
D: U
Hypnotic

$C_{16}H_{13}ClF_2N_2O_2$
338.06336
RI: 2290
Herbicide

$C_{14}H_{26}N_2O_5$
302.18417
RI: 2300
Antihistamine

$C_{19}H_{22}O_3$
298.15690
RI: 2390*
D: UHYME (UHYAC)
Vasodilator
N: ME in methanol

$C_{26}H_{48}O_2Si_2$
448.31929
RI: 2530*
Androgen

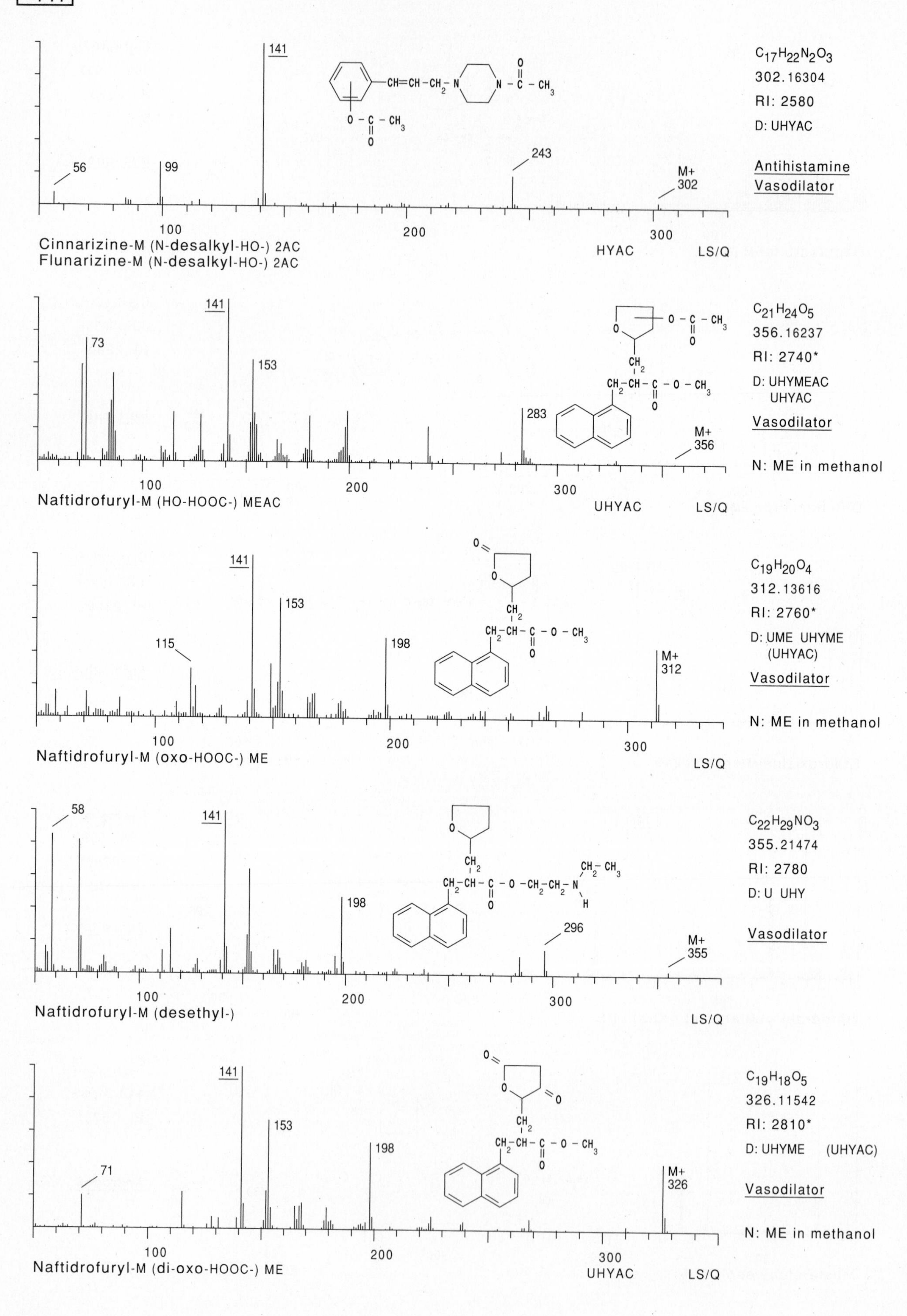
141
56
99
243
M+
302
100
200
300
Cinnarizine-M (N-desalkyl-HO-) 2AC
Flunarizine-M (N-desalkyl-HO-) 2AC
HYAC
LS/Q
$C_{17}H_{22}N_2O_3$
302.16304
RI: 2580
D: UHYAC
Antihistamine
Vasodilator
141
73
153
283
M+
356
100
200
300
Naftidrofuryl-M (HO-HOOC-) MEAC
UHYAC
LS/Q
$C_{21}H_{24}O_5$
356.16237
RI: 2740*
D: UHYMEAC
UHYAC
Vasodilator
N: ME in methanol
141
153
115
198
M+
312
100
200
300
Naftidrofuryl-M (oxo-HOOC-) ME
LS/Q
$C_{19}H_{20}O_4$
312.13616
RI: 2760*
D: UME UHYME
(UHYAC)
Vasodilator
N: ME in methanol
58
141
198
296
M+
355
100
200
300
Naftidrofuryl-M (desethyl-)
LS/Q
$C_{22}H_{29}NO_3$
355.21474
RI: 2780
D: U UHY
Vasodilator
141
153
198
71
M+
326
100
200
300
Naftidrofuryl-M (di-oxo-HOOC-) ME
UHYAC
LS/Q
$C_{19}H_{18}O_5$
326.11542
RI: 2810*
D: UHYME (UHYAC)
Vasodilator
N: ME in methanol

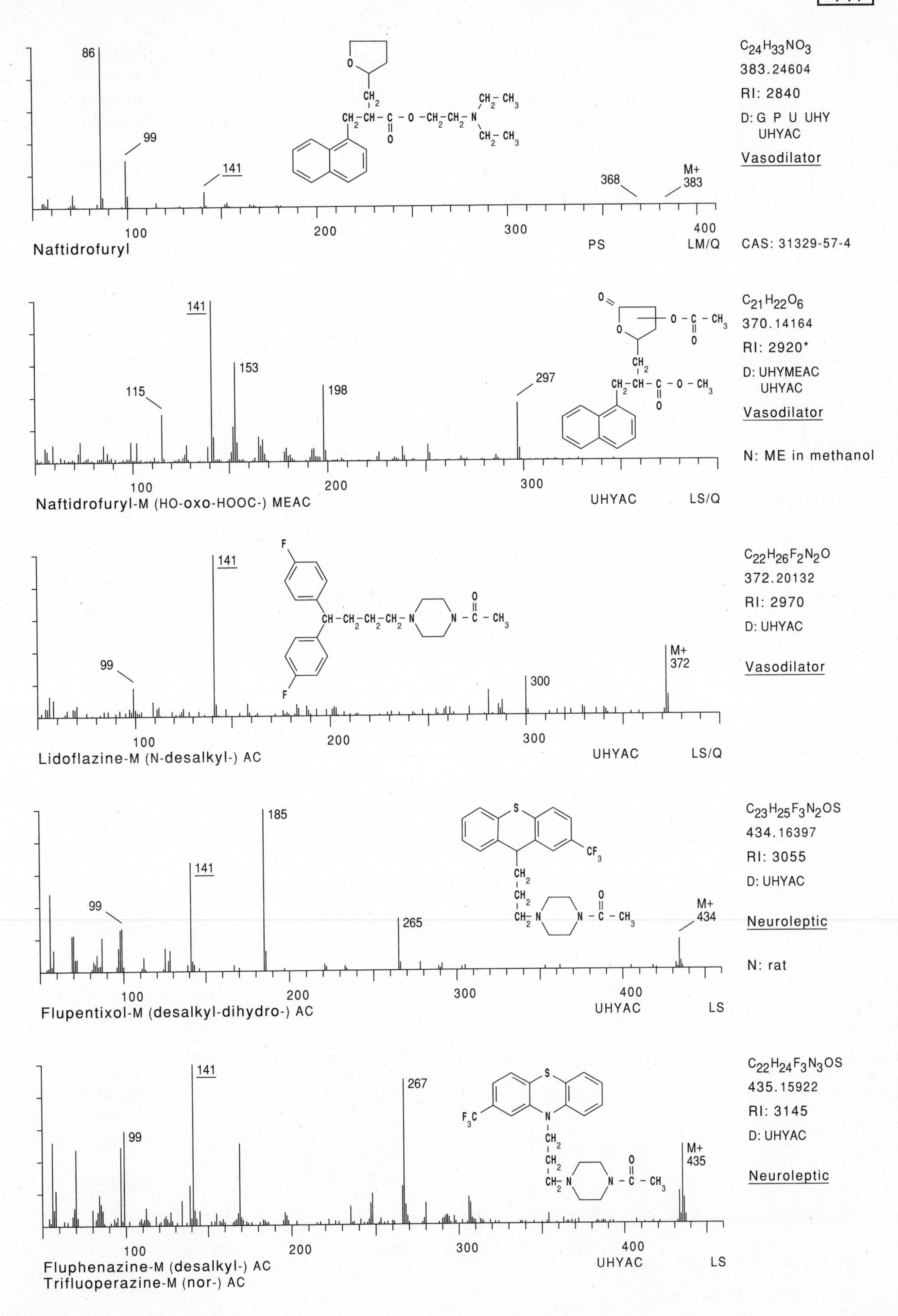

Naftidrofuryl
$C_{24}H_{33}NO_3$
383.24604
RI: 2840
D: G P U UHY UHYAC
Vasodilator
CAS: 31329-57-4
PS LM/Q
Naftidrofuryl-M (HO-oxo-HOOC-) MEAC
$C_{21}H_{22}O_6$
370.14164
RI: 2920*
D: UHYMEAC UHYAC
Vasodilator
N: ME in methanol
UHYAC LS/Q
Lidoflazine-M (N-desalkyl-) AC
$C_{22}H_{26}F_2N_2O$
372.20132
RI: 2970
D: UHYAC
Vasodilator
UHYAC LS/Q
Flupentixol-M (desalkyl-dihydro-) AC
$C_{23}H_{25}F_3N_2OS$
434.16397
RI: 3055
D: UHYAC
Neuroleptic
N: rat
UHYAC LS
Fluphenazine-M (desalkyl-) AC
Trifluoperazine-M (nor-) AC
$C_{22}H_{24}F_3N_3OS$
435.15922
RI: 3145
D: UHYAC
Neuroleptic
UHYAC LS

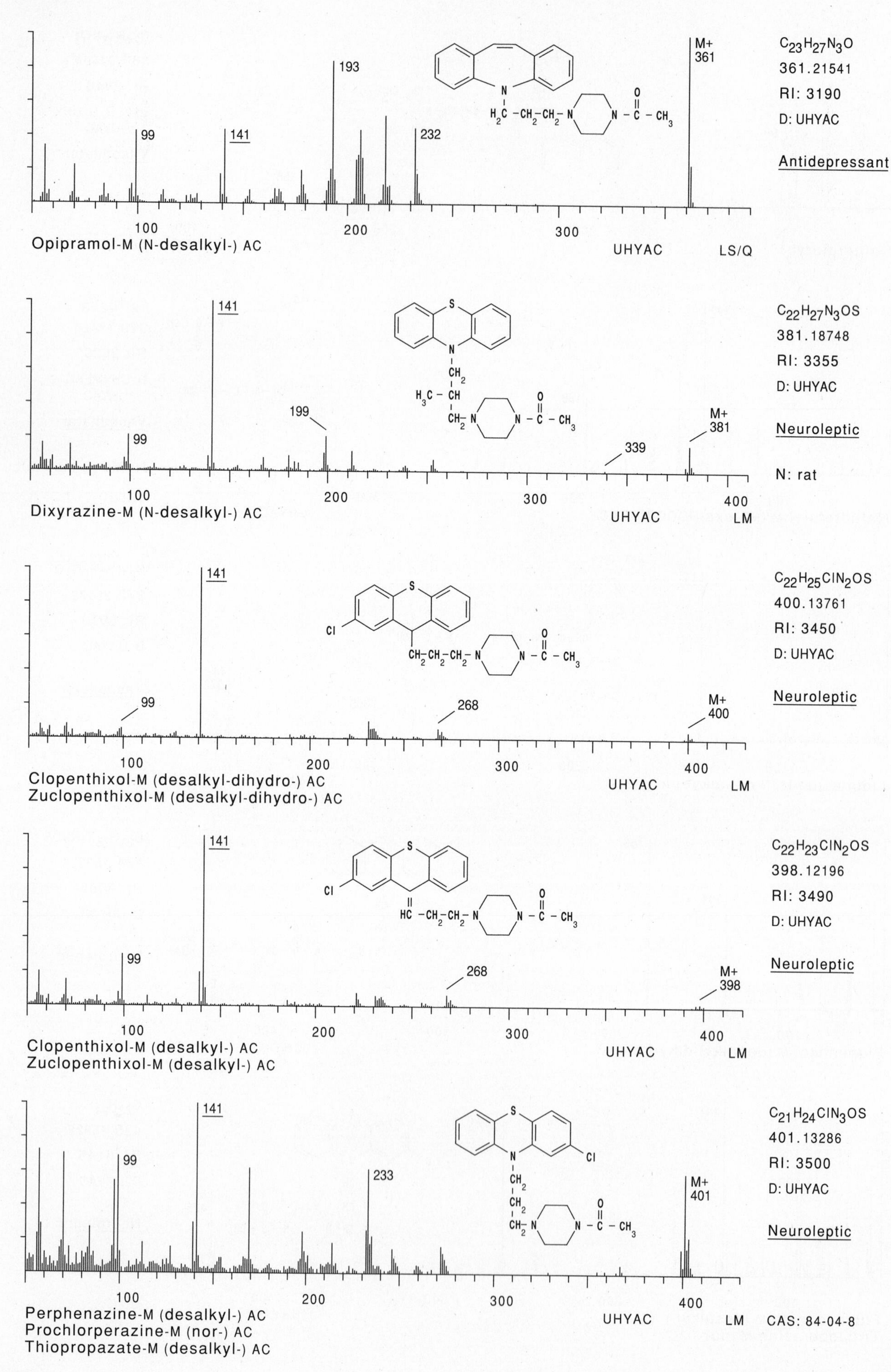

M+
361
193
99
141
232
100
200
300
Opipramol-M (N-desalkyl-) AC
UHYAC
LS/Q
$C_{23}H_{27}N_3O$
361.21541
RI: 3190
D: UHYAC
Antidepressant
141
199
99
339
M+
381
100
200
300
400
Dixyrazine-M (N-desalkyl-) AC
UHYAC
LM
$C_{22}H_{27}N_3OS$
381.18748
RI: 3355
D: UHYAC
Neuroleptic
N: rat
141
99
268
M+
400
100
200
300
400
Clopenthixol-M (desalkyl-dihydro-) AC
Zuclopenthixol-M (desalkyl-dihydro-) AC
UHYAC
LM
$C_{22}H_{25}ClN_2OS$
400.13761
RI: 3450
D: UHYAC
Neuroleptic
141
99
268
M+
398
100
200
300
400
Clopenthixol-M (desalkyl-) AC
Zuclopenthixol-M (desalkyl-) AC
UHYAC
LM
$C_{22}H_{23}ClN_2OS$
398.12196
RI: 3490
D: UHYAC
Neuroleptic
141
99
233
M+
401
100
200
300
400
Perphenazine-M (desalkyl-) AC
Prochlorperazine-M (nor-) AC
Thiopropazate-M (desalkyl-) AC
UHYAC
LM
CAS: 84-04-8
$C_{21}H_{24}ClN_3OS$
401.13286
RI: 3500
D: UHYAC
Neuroleptic

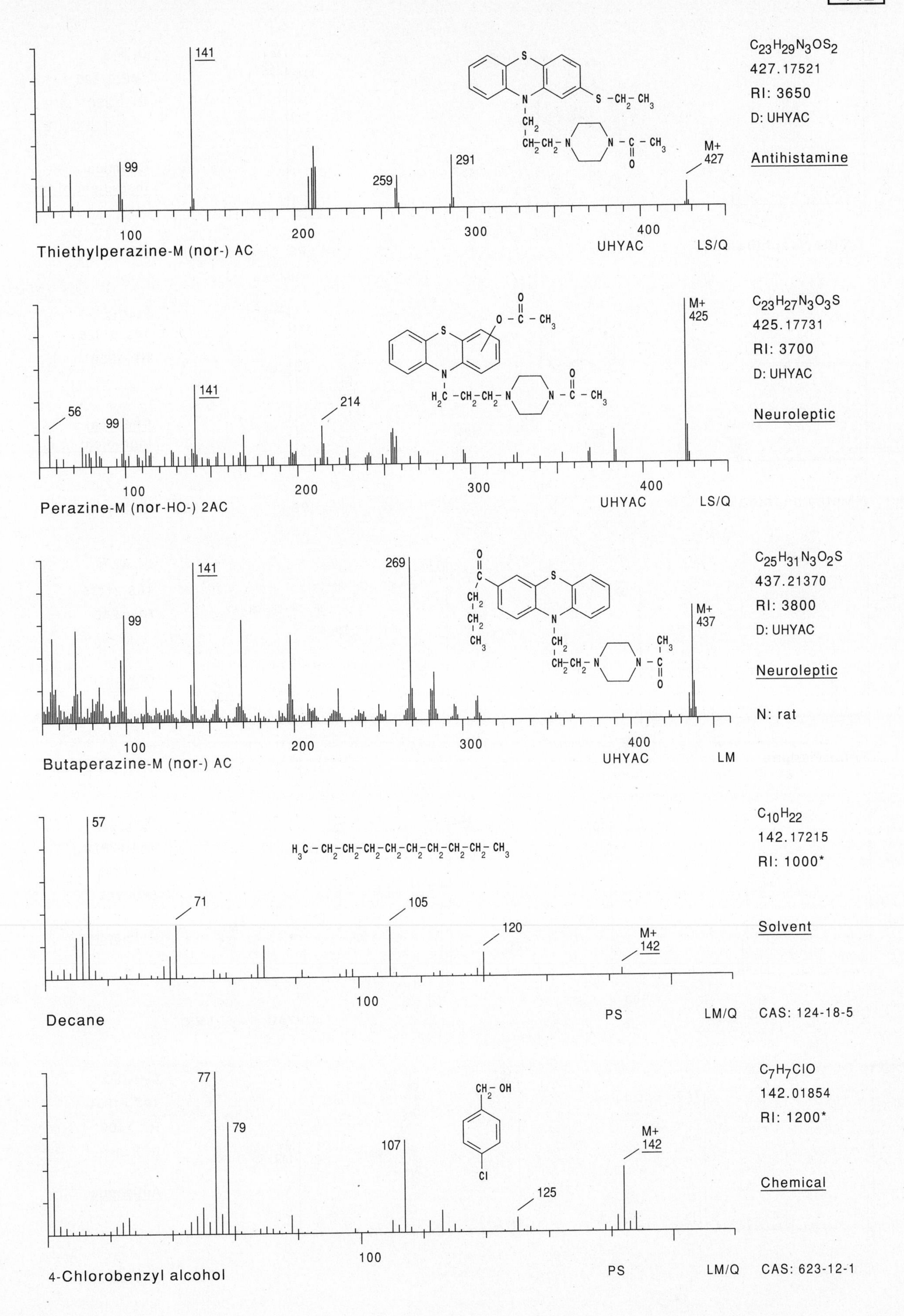

$C_{23}H_{29}N_3OS_2$
427.17521
RI: 3650
D: UHYAC
Antihistamine

$C_{23}H_{27}N_3O_3S$
425.17731
RI: 3700
D: UHYAC
Neuroleptic

$C_{25}H_{31}N_3O_2S$
437.21370
RI: 3800
D: UHYAC
Neuroleptic
N: rat

$C_{10}H_{22}$
142.17215
RI: 1000*
Solvent
CAS: 124-18-5

C_7H_7ClO
142.01854
RI: 1200*
Chemical
CAS: 623-12-1

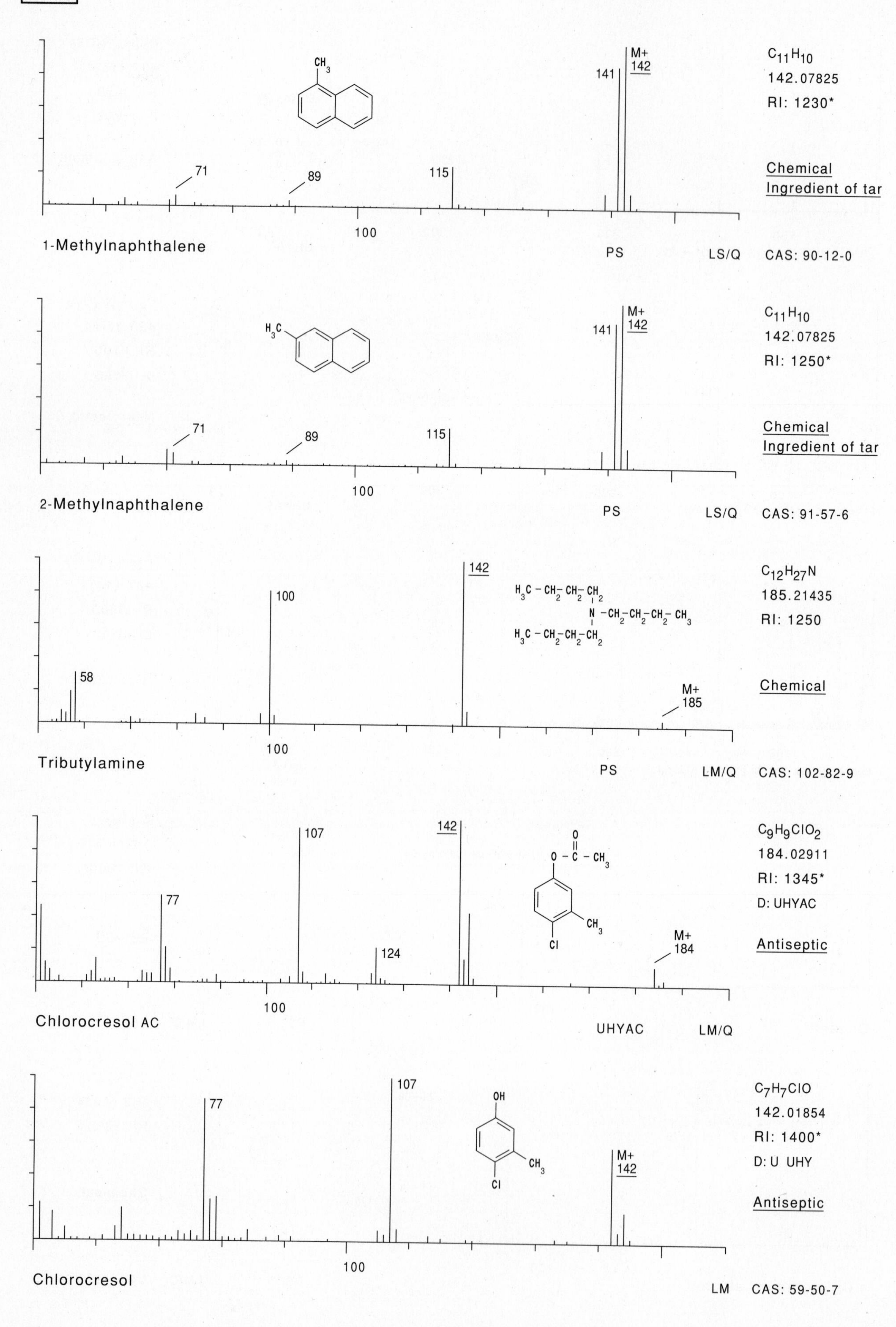

CH3
M+
142
141
71
89
115
100
$C_{11}H_{10}$
142.07825
RI: 1230*
Chemical
Ingredient of tar
1-Methylnaphthalene
PS
LS/Q
CAS: 90-12-0
H3C
M+
142
141
71
89
115
100
$C_{11}H_{10}$
142.07825
RI: 1250*
Chemical
Ingredient of tar
2-Methylnaphthalene
PS
LS/Q
CAS: 91-57-6
142
100
58
M+
185
$C_{12}H_{27}N$
185.21435
RI: 1250
Chemical
Tributylamine
PS
LM/Q
CAS: 102-82-9
107
142
77
124
M+
184
$C_9H_9ClO_2$
184.02911
RI: 1345*
D: UHYAC
Antiseptic
Chlorocresol AC
UHYAC
LM/Q
107
77
OH
CH3
Cl
M+
142
C_7H_7ClO
142.01854
RI: 1400*
D: U UHY
Antiseptic
Chlorocresol
LM
CAS: 59-50-7

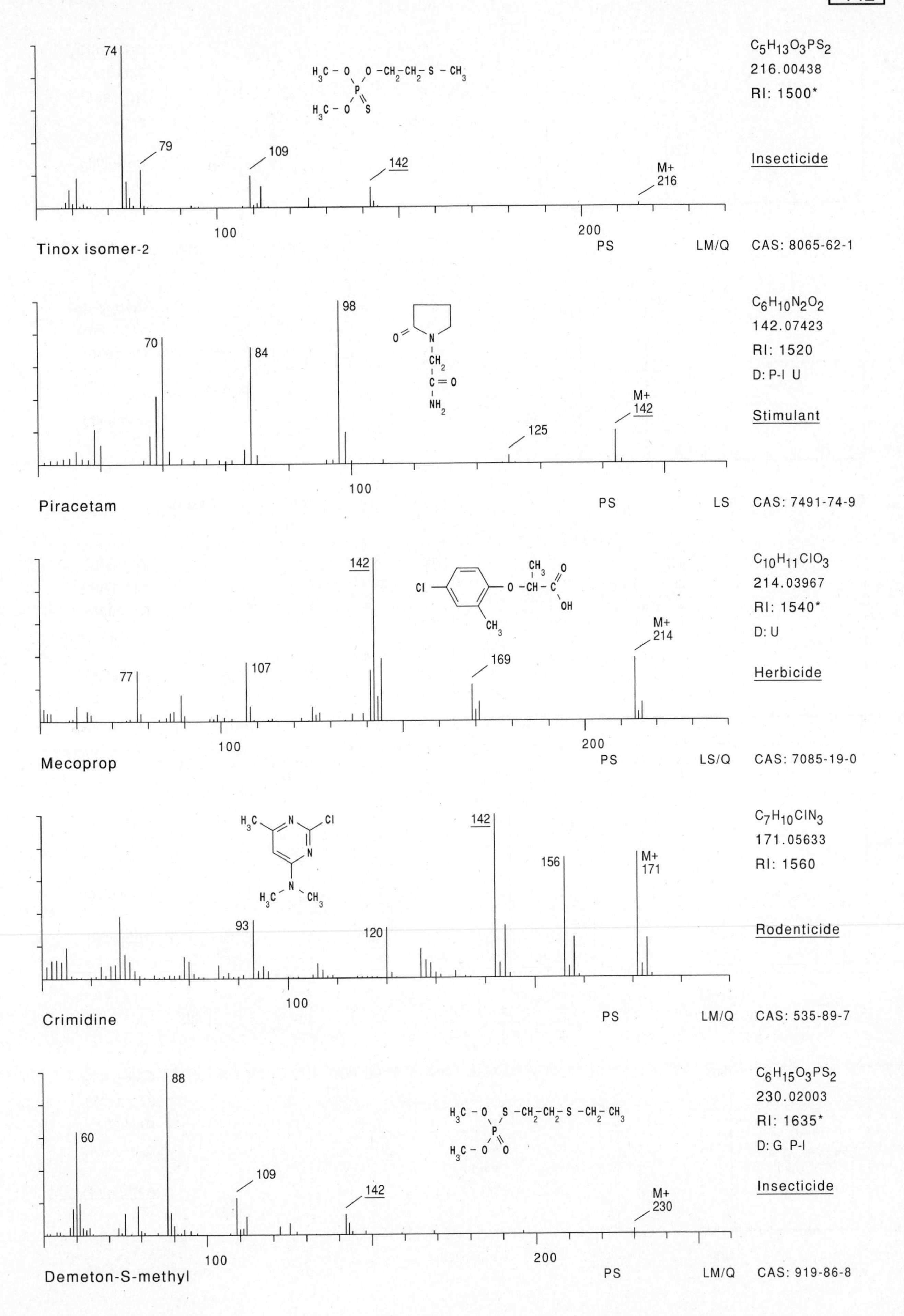

$C_5H_{13}O_3PS_2$
216.00438
RI: 1500*

Insecticide

CAS: 8065-62-1

$C_6H_{10}N_2O_2$
142.07423
RI: 1520
D: P-I U

Stimulant

CAS: 7491-74-9

$C_{10}H_{11}ClO_3$
214.03967
RI: 1540*
D: U

Herbicide

CAS: 7085-19-0

$C_7H_{10}ClN_3$
171.05633
RI: 1560

Rodenticide

CAS: 535-89-7

$C_6H_{15}O_3PS_2$
230.02003
RI: 1635*
D: G P-I

Insecticide

CAS: 919-86-8

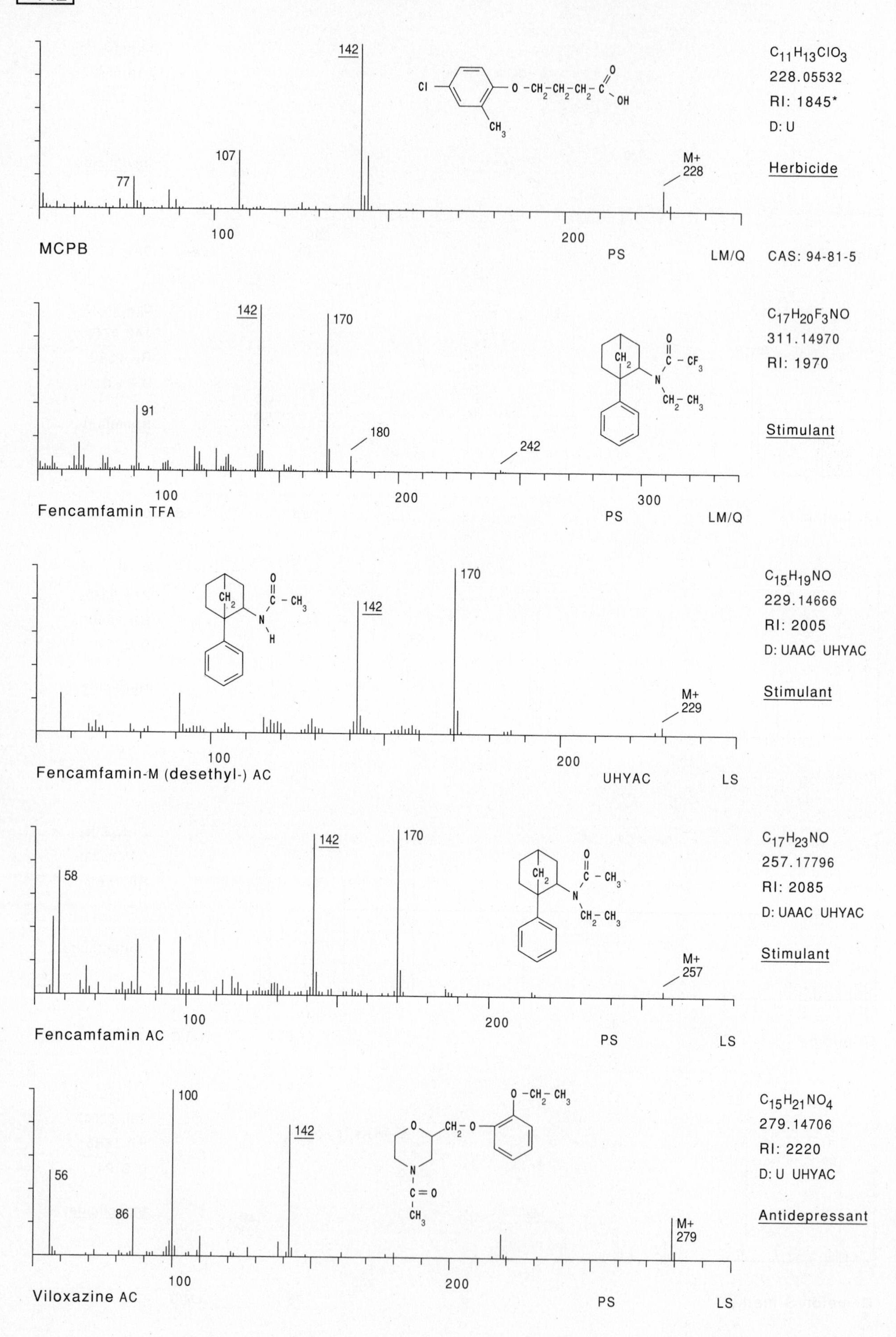
142
107
77
M+
228
100
200
PS
LM/Q
MCPB
$C_{11}H_{13}ClO_3$
228.05532
RI: 1845*
D: U
Herbicide
CAS: 94-81-5
142
170
91
180
242
100
200
300
PS
LM/Q
Fencamfamin TFA
$C_{17}H_{20}F_3NO$
311.14970
RI: 1970
Stimulant
170
142
M+
229
100
200
UHYAC
LS
Fencamfamin-M (desethyl-) AC
$C_{15}H_{19}NO$
229.14666
RI: 2005
D: UAAC UHYAC
Stimulant
142
170
58
M+
257
100
200
PS
LS
Fencamfamin AC
$C_{17}H_{23}NO$
257.17796
RI: 2085
D: UAAC UHYAC
Stimulant
100
142
56
86
M+
279
100
200
PS
LS
Viloxazine AC
$C_{15}H_{21}NO_4$
279.14706
RI: 2220
D: U UHYAC
Antidepressant

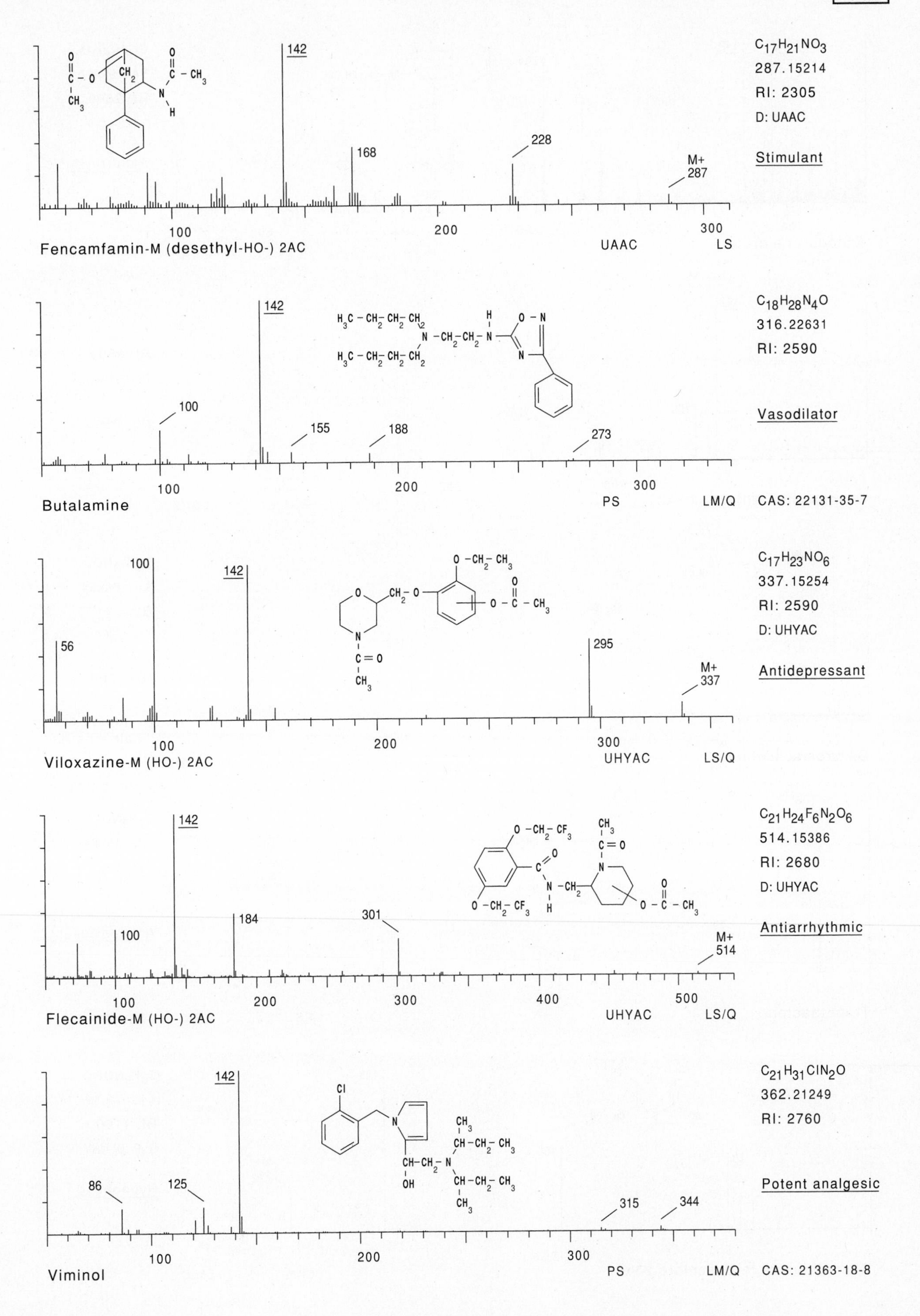

$C_{17}H_{21}NO_3$
287.15214
RI: 2305
D: UAAC

Stimulant

$C_{18}H_{28}N_4O$
316.22631
RI: 2590

Vasodilator

CAS: 22131-35-7

$C_{17}H_{23}NO_6$
337.15254
RI: 2590
D: UHYAC

Antidepressant

$C_{21}H_{24}F_6N_2O_6$
514.15386
RI: 2680
D: UHYAC

Antiarrhythmic

$C_{21}H_{31}ClN_2O$
362.21249
RI: 2760

Potent analgesic

CAS: 21363-18-8

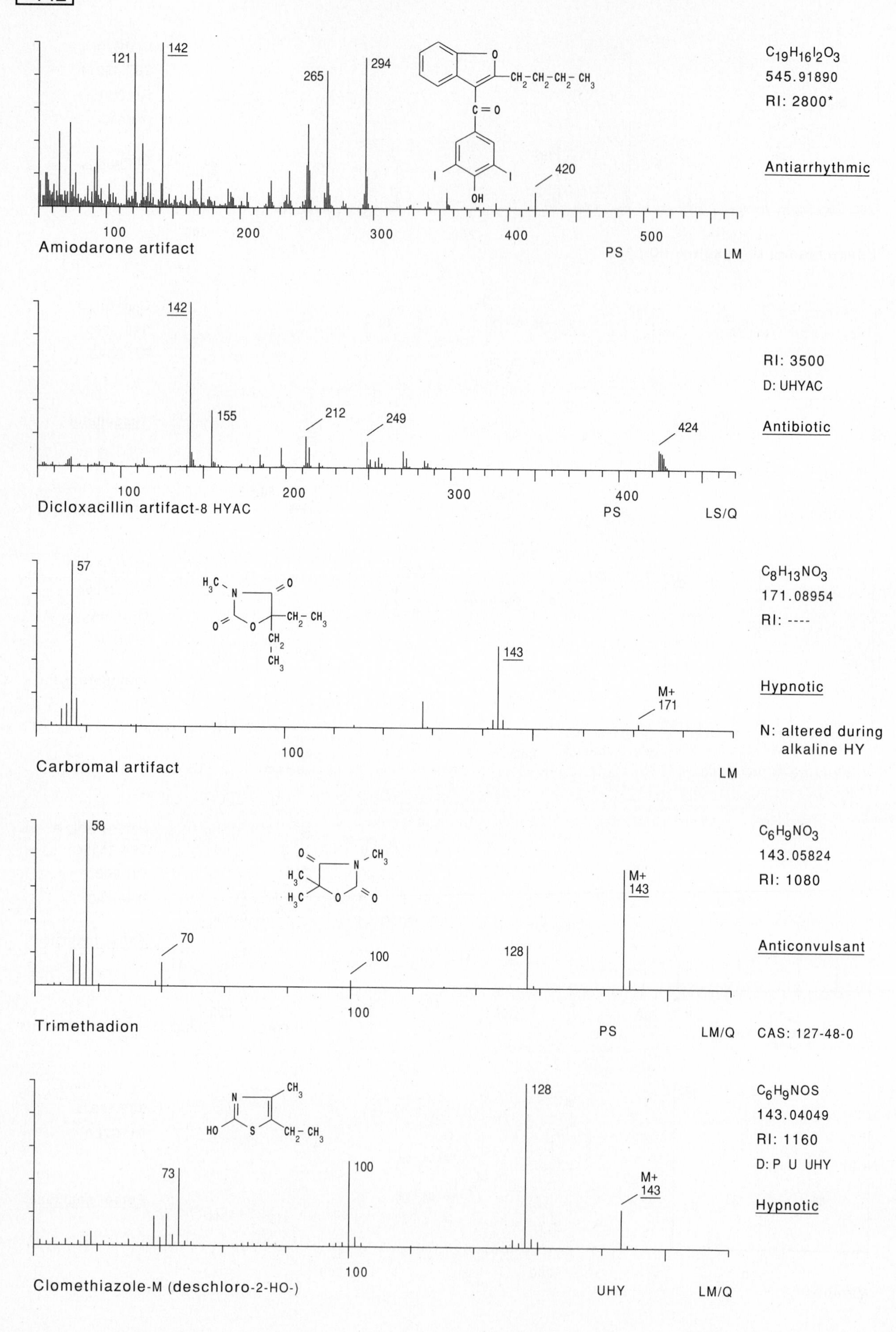

Amiodarone artifact
PS
LM
$C_{19}H_{16}I_2O_3$
545.91890
RI: 2800*
Antiarrhythmic
Dicloxacillin artifact-8 HYAC
PS
LS/Q
RI: 3500
D: UHYAC
Antibiotic
Carbromal artifact
LM
$C_8H_{13}NO_3$
171.08954
RI: ----
Hypnotic
N: altered during alkaline HY
Trimethadion
PS
LM/Q
$C_6H_9NO_3$
143.05824
RI: 1080
Anticonvulsant
CAS: 127-48-0
Clomethiazole-M (deschloro-2-HO-)
UHY
LM/Q
C_6H_9NOS
143.04049
RI: 1160
D: P U UHY
Hypnotic

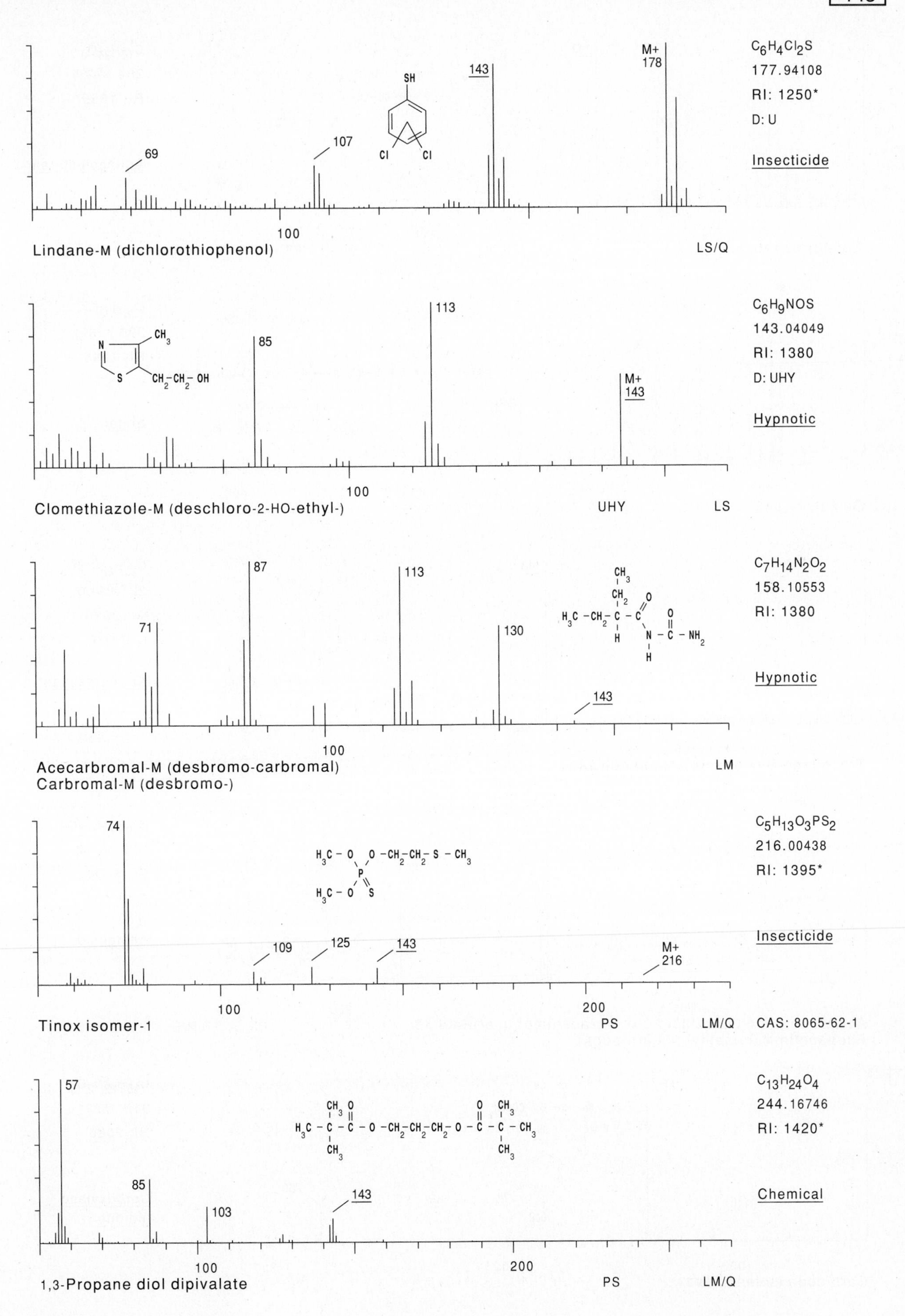

$C_6H_4Cl_2S$
177.94108
RI: 1250*
D: U
Insecticide

C_6H_9NOS
143.04049
RI: 1380
D: UHY
Hypnotic

$C_7H_{14}N_2O_2$
158.10553
RI: 1380
Hypnotic

$C_5H_{13}O_3PS_2$
216.00438
RI: 1395*
Insecticide
CAS: 8065-62-1

$C_{13}H_{24}O_4$
244.16746
RI: 1420*
Chemical

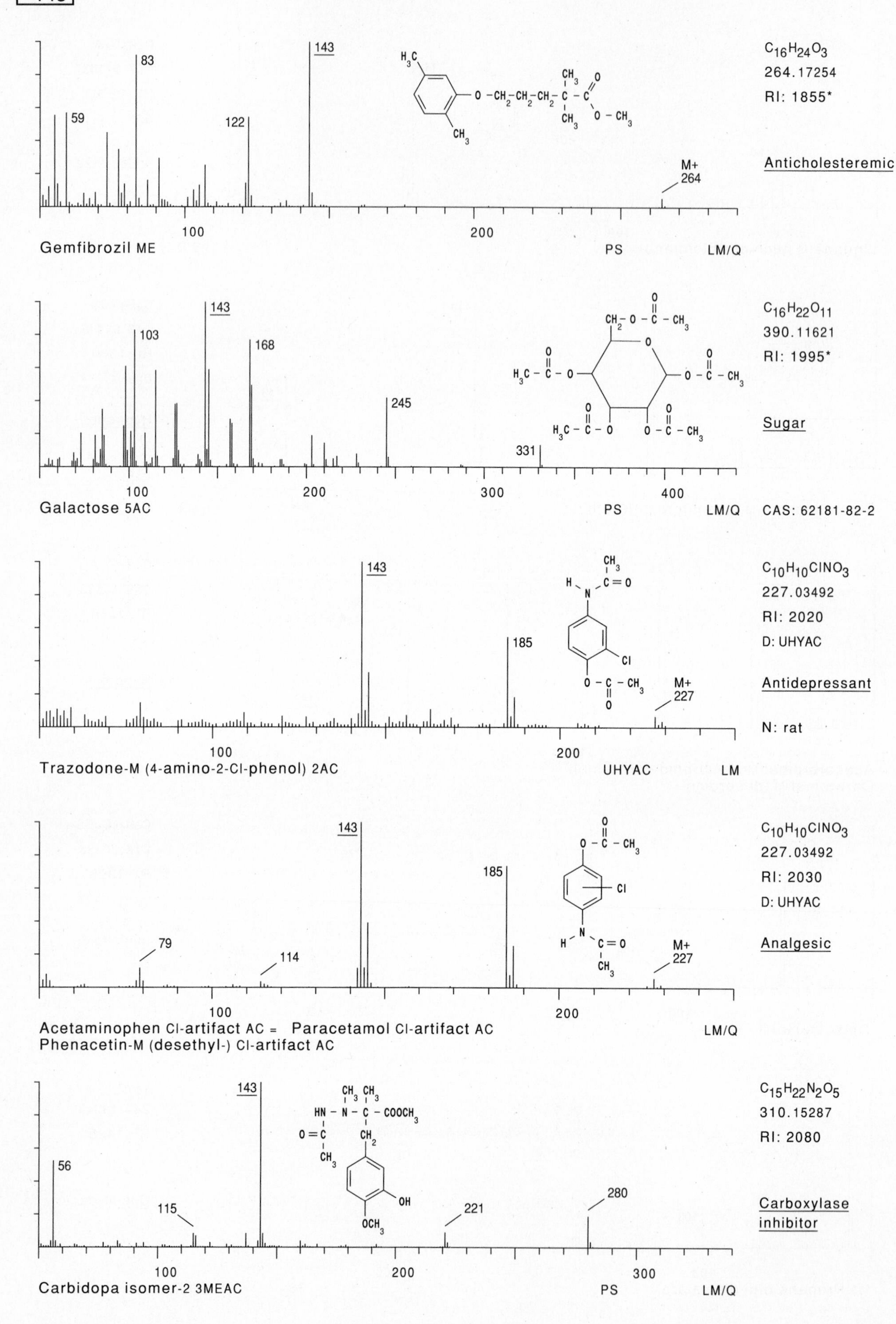
143
83
59
122
M+
264
100
200
Gemfibrozil ME
PS
LM/Q
$C_{16}H_{24}O_3$
264.17254
RI: 1855*
Anticholesteremic
143
103
168
245
331
100
200
300
400
Galactose 5AC
PS
LM/Q
$C_{16}H_{22}O_{11}$
390.11621
RI: 1995*
Sugar
CAS: 62181-82-2
143
185
M+
227
100
200
Trazodone-M (4-amino-2-Cl-phenol) 2AC
UHYAC
LM
$C_{10}H_{10}ClNO_3$
227.03492
RI: 2020
D: UHYAC
Antidepressant
N: rat
143
185
79
114
M+
227
100
200
Acetaminophen Cl-artifact AC = Paracetamol Cl-artifact AC
Phenacetin-M (desethyl-) Cl-artifact AC
LM/Q
$C_{10}H_{10}ClNO_3$
227.03492
RI: 2030
D: UHYAC
Analgesic
143
56
115
221
280
100
200
300
Carbidopa isomer-2 3MEAC
PS
LM/Q
$C_{15}H_{22}N_2O_5$
310.15287
RI: 2080
Carboxylase inhibitor

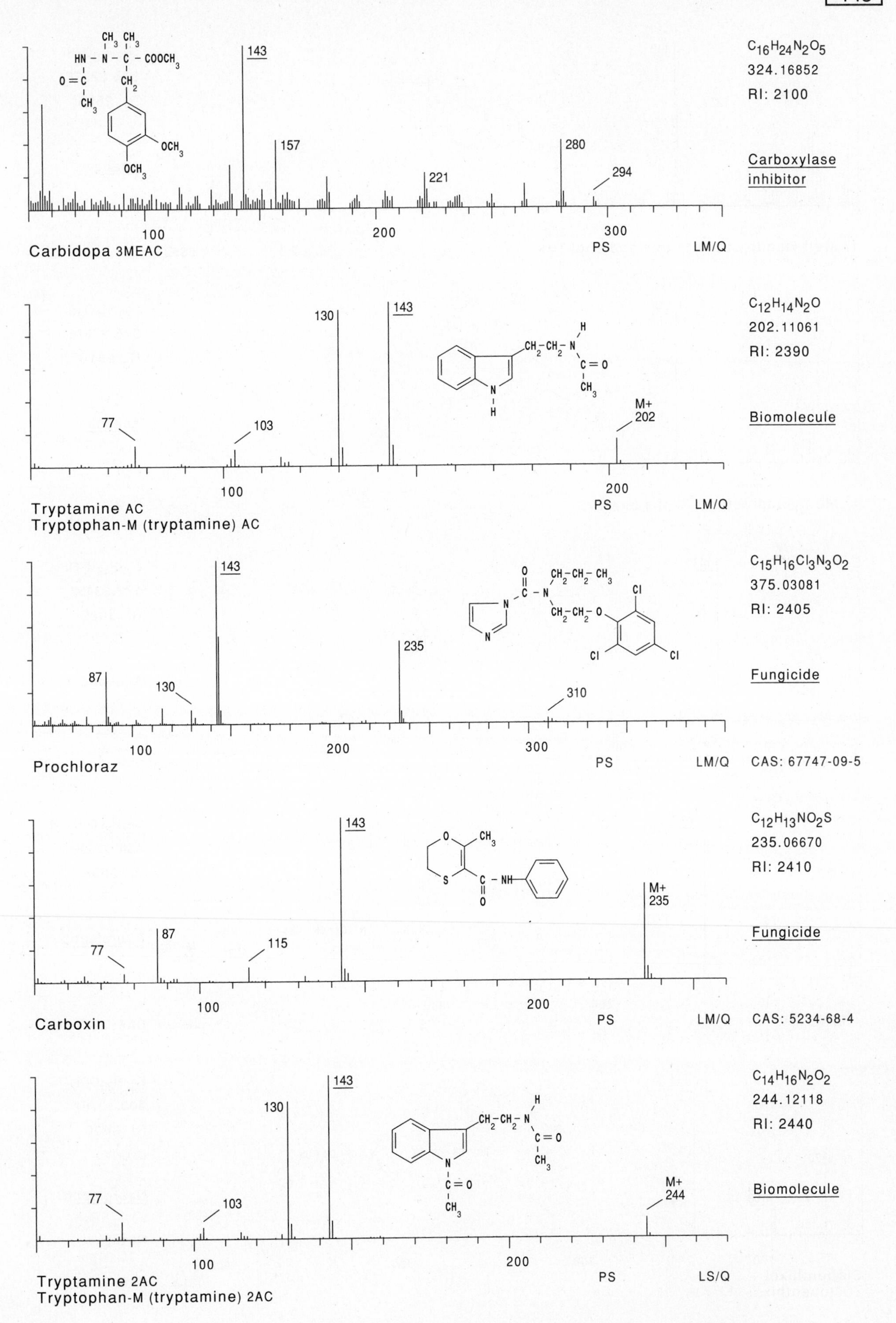

Carbidopa 3MEAC

$C_{16}H_{24}N_2O_5$
324.16852
RI: 2100

Carboxylase inhibitor

Tryptamine AC
Tryptophan-M (tryptamine) AC

$C_{12}H_{14}N_2O$
202.11061
RI: 2390

Biomolecule

Prochloraz

$C_{15}H_{16}Cl_3N_3O_2$
375.03081
RI: 2405

Fungicide

CAS: 67747-09-5

Carboxin

$C_{12}H_{13}NO_2S$
235.06670
RI: 2410

Fungicide

CAS: 5234-68-4

Tryptamine 2AC
Tryptophan-M (tryptamine) 2AC

$C_{14}H_{16}N_2O_2$
244.12118
RI: 2440

Biomolecule

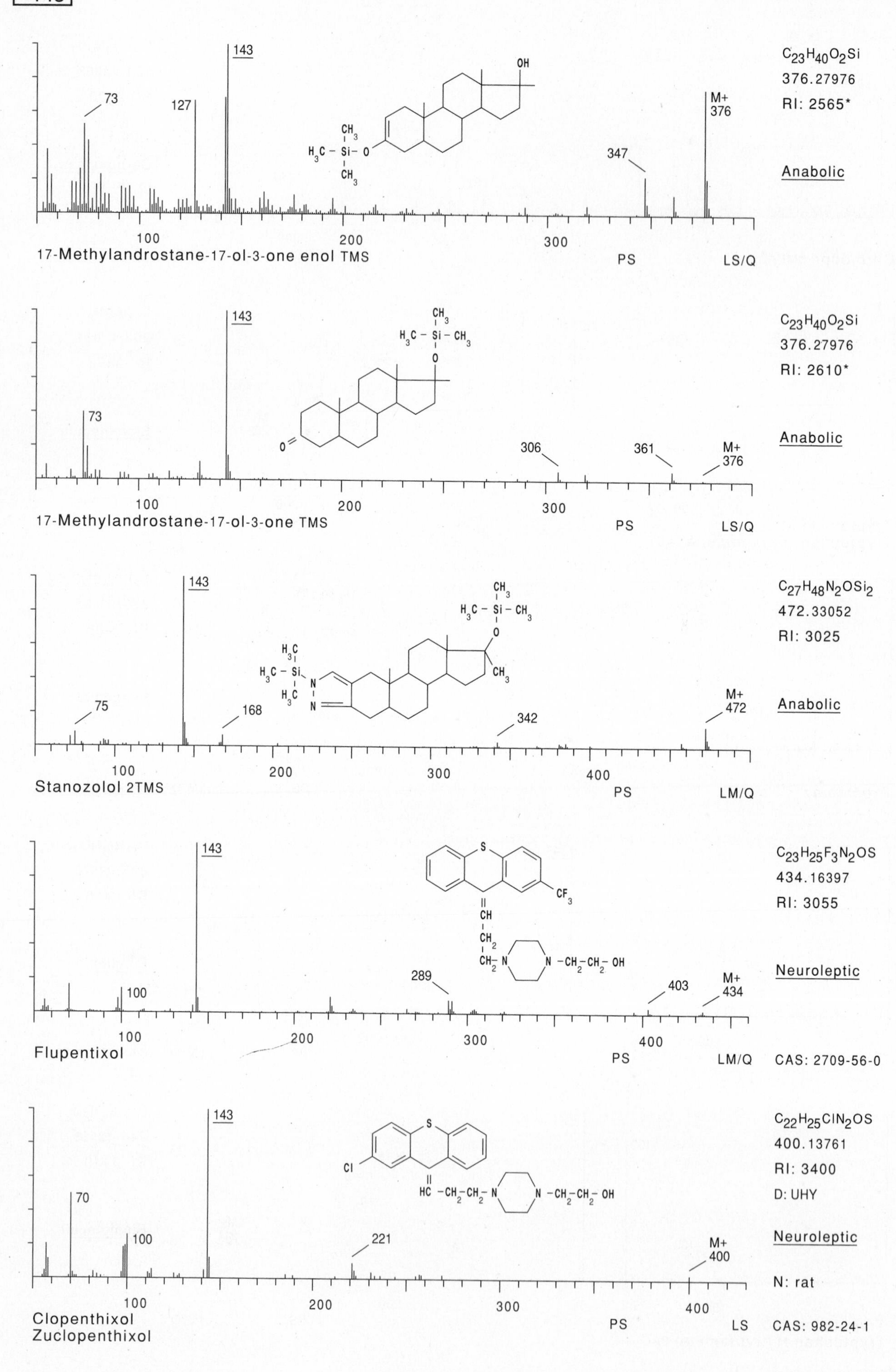

143
73
127
M+
376
347
OH
100
200
300
17-Methylandrostane-17-ol-3-one enol TMS
PS
LS/Q
$C_{23}H_{40}O_2Si$
376.27976
RI: 2565*
Anabolic
143
73
306
361
M+
376
17-Methylandrostane-17-ol-3-one TMS
PS
LS/Q
$C_{23}H_{40}O_2Si$
376.27976
RI: 2610*
Anabolic
143
75
168
342
M+
472
400
Stanozolol 2TMS
PS
LM/Q
$C_{27}H_{48}N_2OSi_2$
472.33052
RI: 3025
Anabolic
143
100
289
403
M+
434
Flupentixol
PS
LM/Q
$C_{23}H_{25}F_3N_2OS$
434.16397
RI: 3055
Neuroleptic
CAS: 2709-56-0
143
70
100
221
M+
400
Clopenthixol
Zuclopenthixol
PS
LS
$C_{22}H_{25}ClN_2OS$
400.13761
RI: 3400
D: UHY
Neuroleptic
N: rat
CAS: 982-24-1

144

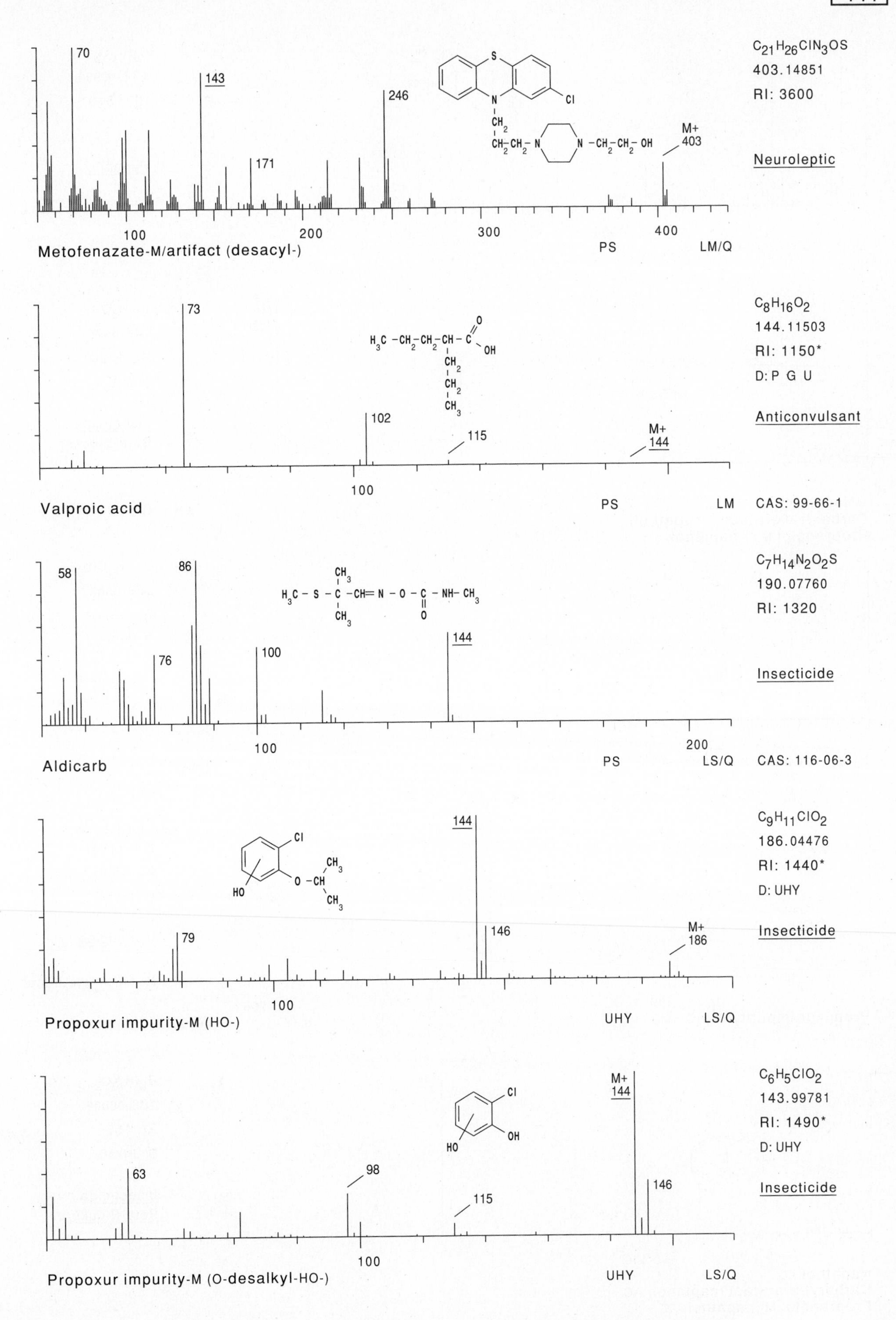
70
143
246
171
M+
403
100
200
300
400
Metofenazate-M/artifact (desacyl-)
PS
LM/Q
C21H26ClN3OS
403.14851
RI: 3600
Neuroleptic
73
102
115
M+
144
100
Valproic acid
PS
LM
C8H16O2
144.11503
RI: 1150*
D: P G U
Anticonvulsant
CAS: 99-66-1
58
86
76
100
144
100
200
Aldicarb
PS
LS/Q
C7H14N2O2S
190.07760
RI: 1320
Insecticide
CAS: 116-06-3
144
79
146
M+
186
100
Propoxur impurity-M (HO-)
UHY
LS/Q
C9H11ClO2
186.04476
RI: 1440*
D: UHY
Insecticide
M+
144
63
98
115
146
100
Propoxur impurity-M (O-desalkyl-HO-)
UHY
LS/Q
C6H5ClO2
143.99781
RI: 1490*
D: UHY
Insecticide

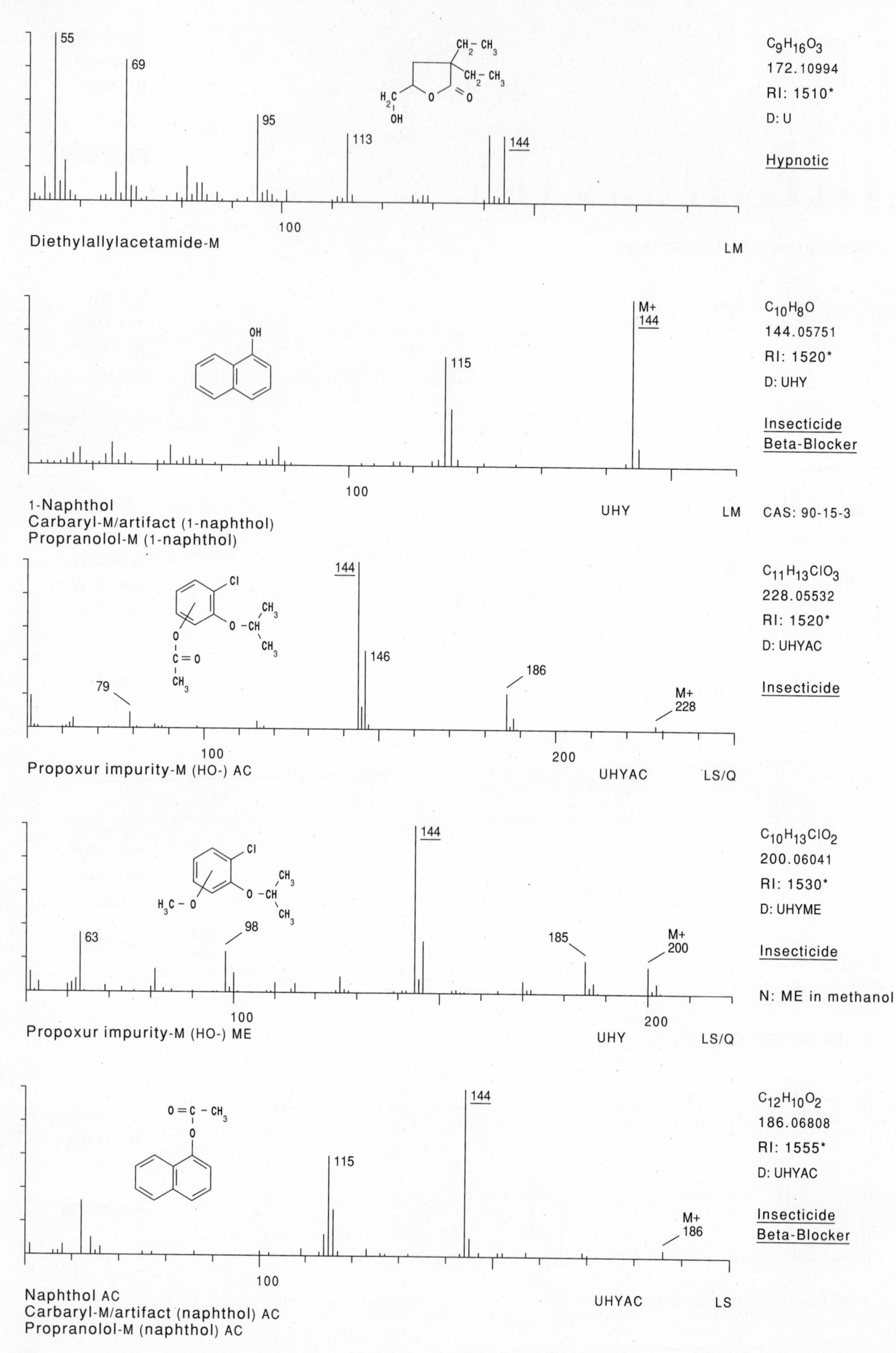

55
69
95
113
144
100
Diethylallylacetamide-M
LM
$C_9H_{16}O_3$
172.10994
RI: 1510*
D: U
Hypnotic
M+
144
115
100
1-Naphthol
Carbaryl-M/artifact (1-naphthol)
Propranolol-M (1-naphthol)
UHY
LM
$C_{10}H_8O$
144.05751
RI: 1520*
D: UHY
Insecticide
Beta-Blocker
CAS: 90-15-3
144
146
79
186
M+
228
100
200
Propoxur impurity-M (HO-) AC
UHYAC
LS/Q
$C_{11}H_{13}ClO_3$
228.05532
RI: 1520*
D: UHYAC
Insecticide
144
63
98
185
M+
200
100
200
Propoxur impurity-M (HO-) ME
UHY
LS/Q
$C_{10}H_{13}ClO_2$
200.06041
RI: 1530*
D: UHYME
Insecticide
N: ME in methanol
144
115
M+
186
100
Naphthol AC
Carbaryl-M/artifact (naphthol) AC
Propranolol-M (naphthol) AC
UHYAC
LS
$C_{12}H_{10}O_2$
186.06808
RI: 1555*
D: UHYAC
Insecticide
Beta-Blocker

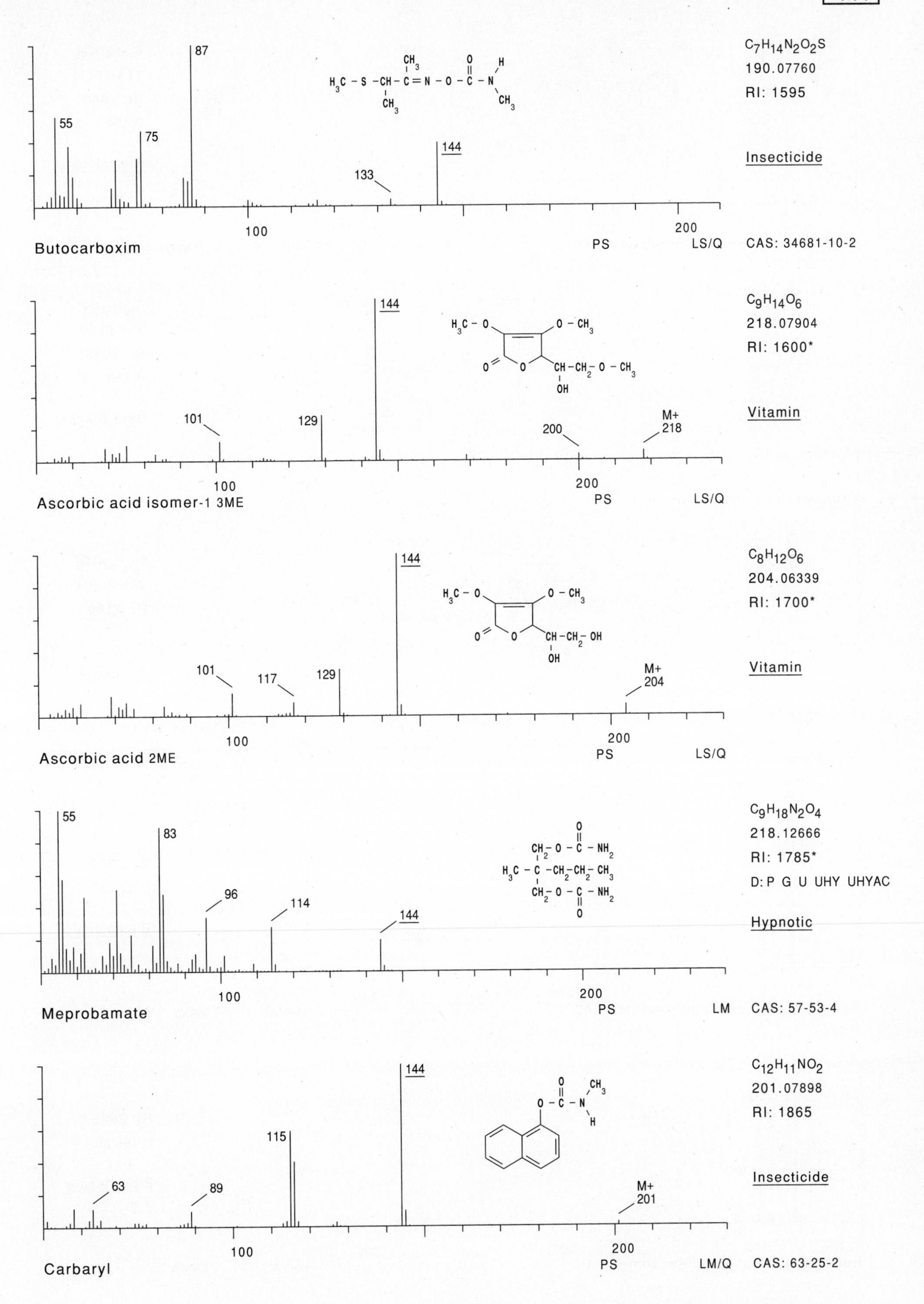
87
55
75
144
133
100
200
Butocarboxim
PS
LS/Q
$C_7H_{14}N_2O_2S$
190.07760
RI: 1595
Insecticide
CAS: 34681-10-2
144
101
129
200
M+
218
100
200
Ascorbic acid isomer-1 3ME
PS
LS/Q
$C_9H_{14}O_6$
218.07904
RI: 1600*
Vitamin
144
101
117
129
M+
204
100
200
Ascorbic acid 2ME
PS
LS/Q
$C_8H_{12}O_6$
204.06339
RI: 1700*
Vitamin
55
83
96
114
144
100
200
Meprobamate
PS
LM
$C_9H_{18}N_2O_4$
218.12666
RI: 1785*
D: P G U UHY UHYAC
Hypnotic
CAS: 57-53-4
144
115
63
89
M+
201
100
200
Carbaryl
PS
LM/Q
$C_{12}H_{11}NO_2$
201.07898
RI: 1865
Insecticide
CAS: 63-25-2

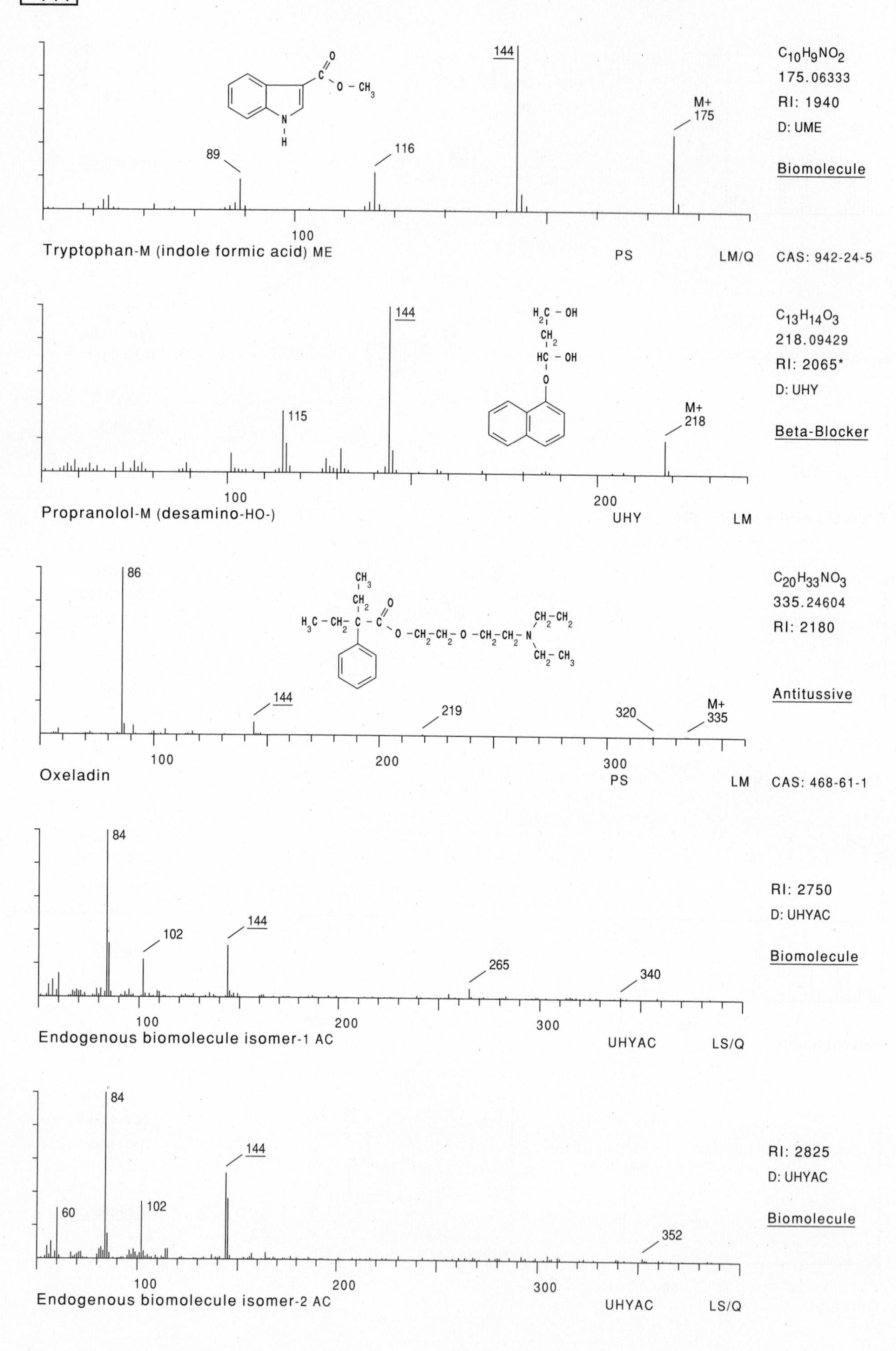

144
M+
175
89
116
100
Tryptophan-M (indole formic acid) ME
PS
LM/Q
$C_{10}H_9NO_2$
175.06333
RI: 1940
D: UME
Biomolecule
CAS: 942-24-5
144
115
M+
218
100
200
Propranolol-M (desamino-HO-)
UHY
LM
$C_{13}H_{14}O_3$
218.09429
RI: 2065*
D: UHY
Beta-Blocker
86
144
219
320
M+
335
100
200
300
Oxeladin
PS
LM
$C_{20}H_{33}NO_3$
335.24604
RI: 2180
Antitussive
CAS: 468-61-1
84
102
144
265
340
100
200
300
Endogenous biomolecule isomer-1 AC
UHYAC
LS/Q
RI: 2750
D: UHYAC
Biomolecule
84
60
102
144
352
100
200
300
Endogenous biomolecule isomer-2 AC
UHYAC
LS/Q
RI: 2825
D: UHYAC
Biomolecule

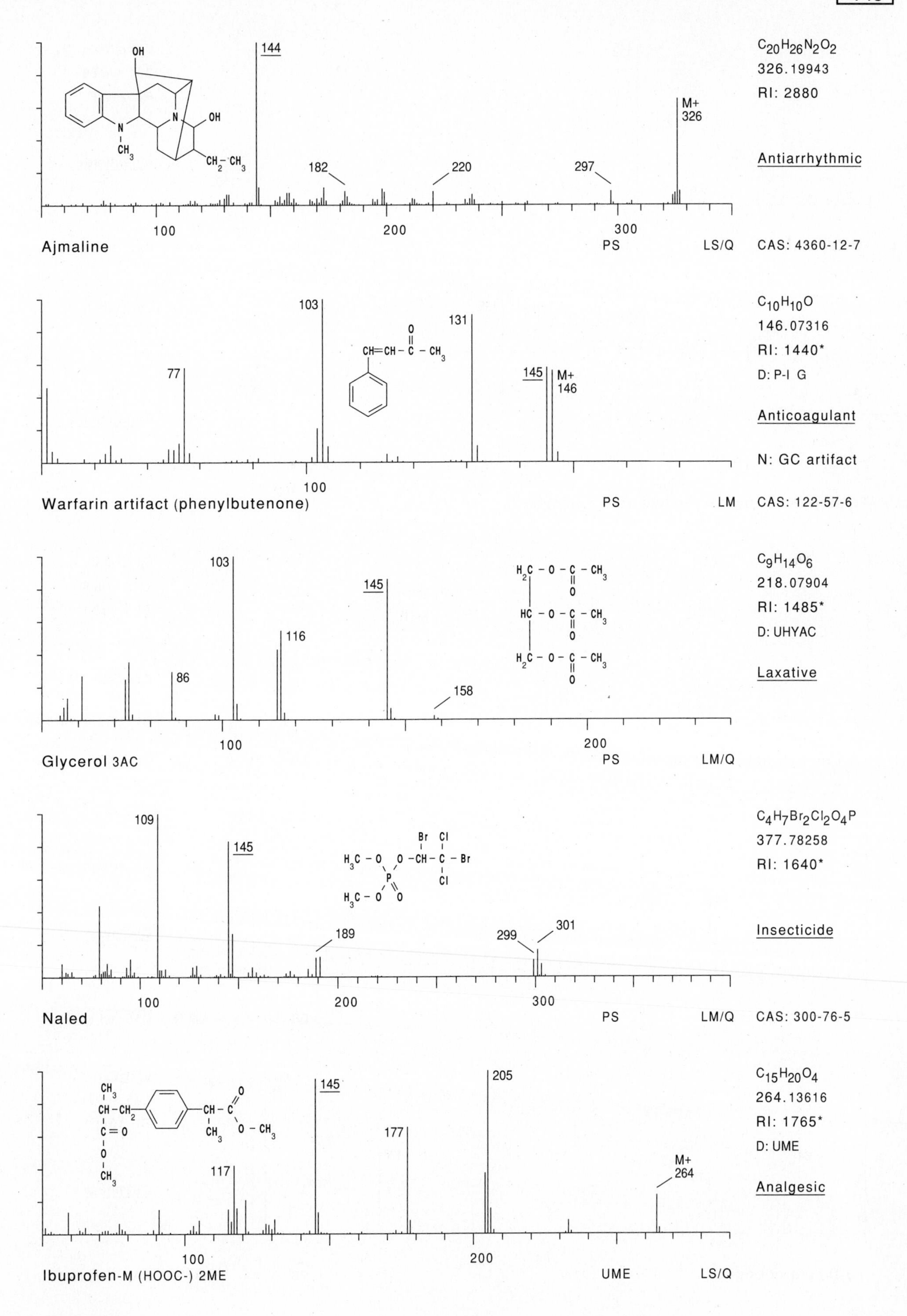

Ajmaline
C20H26N2O2
326.19943
RI: 2880
Antiarrhythmic
CAS: 4360-12-7
PS
LS/Q
144
182
220
297
M+
326
OH
N
CH3
CH2-CH3
100
200
300
Warfarin artifact (phenylbutenone)
C10H10O
146.07316
RI: 1440*
D: P-I G
Anticoagulant
N: GC artifact
CAS: 122-57-6
PS
LM
77
103
131
145
M+
146
CH=CH-C-CH3
Glycerol 3AC
C9H14O6
218.07904
RI: 1485*
D: UHYAC
Laxative
PS
LM/Q
86
103
116
145
158
H2C-O-C-CH3
HC-O-C-CH3
Naled
C4H7Br2Cl2O4P
377.78258
RI: 1640*
Insecticide
CAS: 300-76-5
PS
LM/Q
109
145
189
299
301
Br
Cl
H3C-O
O-CH-C-Br
P
H3C-O
Ibuprofen-M (HOOC-) 2ME
C15H20O4
264.13616
RI: 1765*
D: UME
Analgesic
UME
LS/Q
117
145
177
205
M+
264
CH3
CH-CH2
C=O
O
CH-C
CH3
O-CH3

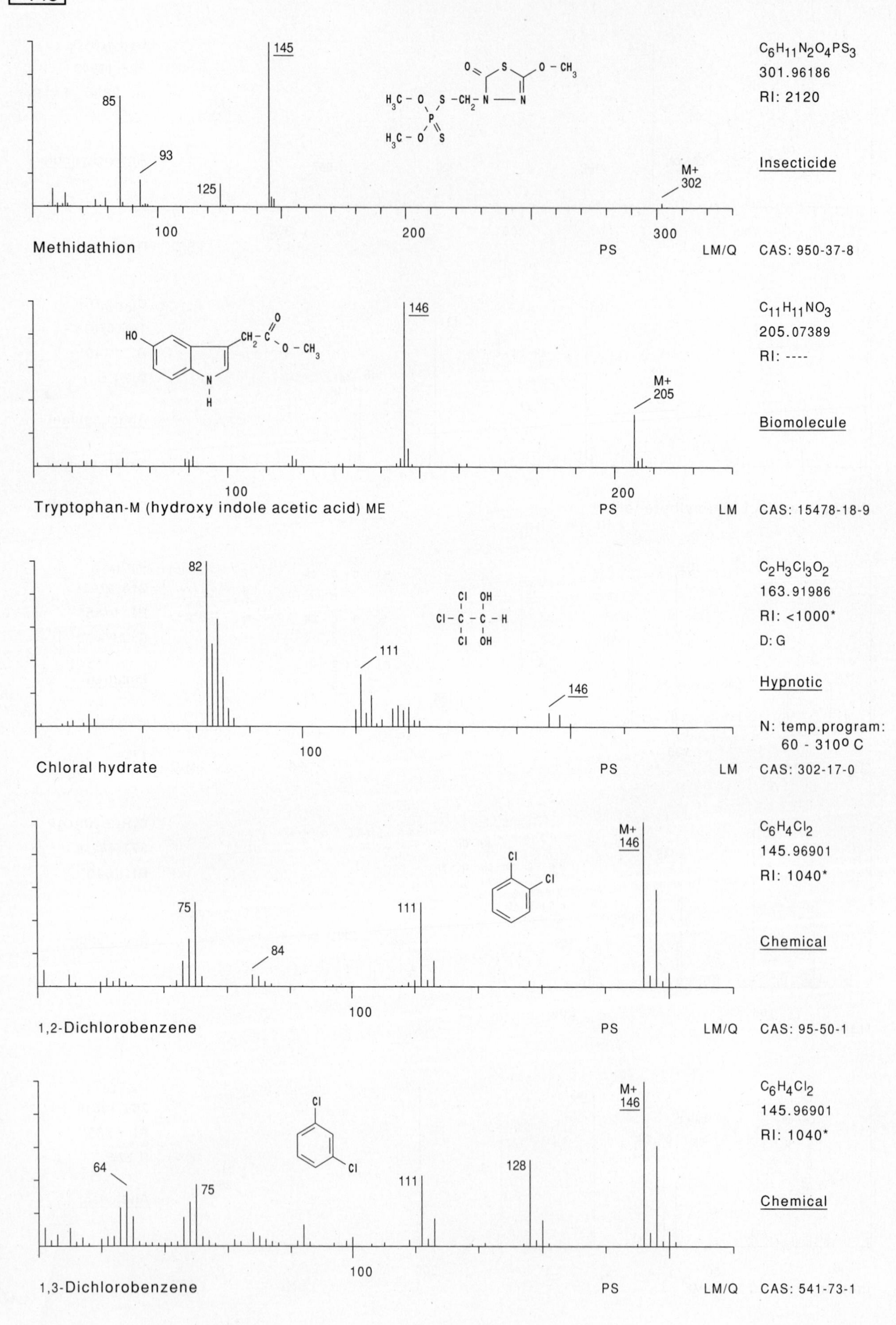

$C_6H_{11}N_2O_4PS_3$
301.96186
RI: 2120

Insecticide

CAS: 950-37-8

$C_{11}H_{11}NO_3$
205.07389
RI: ----

Biomolecule

CAS: 15478-18-9

$C_2H_3Cl_3O_2$
163.91986
RI: <1000*
D: G

Hypnotic

N: temp.program: 60 - 310° C

CAS: 302-17-0

$C_6H_4Cl_2$
145.96901
RI: 1040*

Chemical

CAS: 95-50-1

$C_6H_4Cl_2$
145.96901
RI: 1040*

Chemical

CAS: 541-73-1

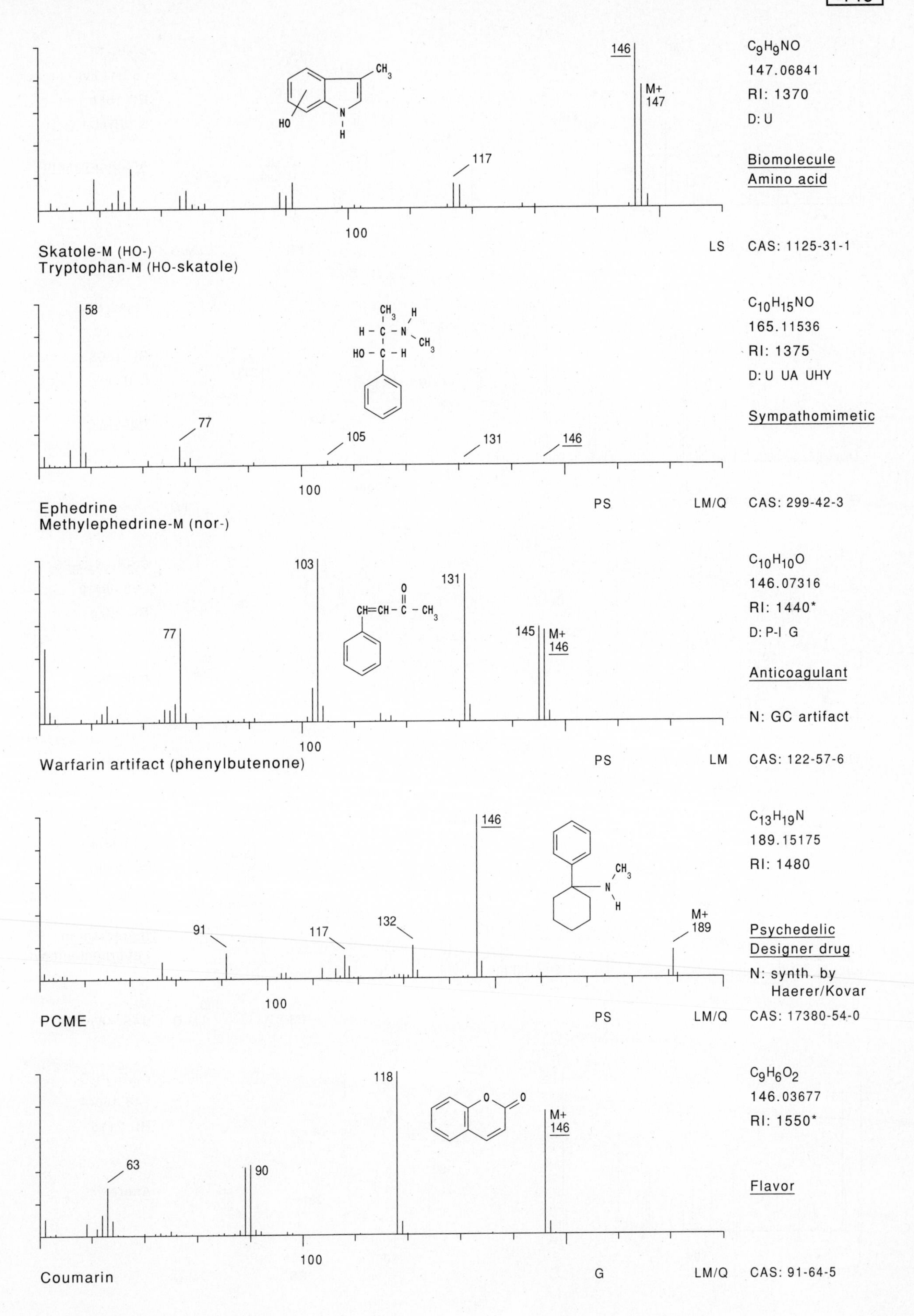

C_9H_9NO
147.06841
RI: 1370
D: U

Biomolecule
Amino acid

CAS: 1125-31-1

$C_{10}H_{15}NO$
165.11536
RI: 1375
D: U UA UHY

Sympathomimetic

CAS: 299-42-3

$C_{10}H_{10}O$
146.07316
RI: 1440*
D: P-I G

Anticoagulant

N: GC artifact

CAS: 122-57-6

$C_{13}H_{19}N$
189.15175
RI: 1480

Psychedelic
Designer drug
N: synth. by
Haerer/Kovar
CAS: 17380-54-0

$C_9H_6O_2$
146.03677
RI: 1550*

Flavor

CAS: 91-64-5

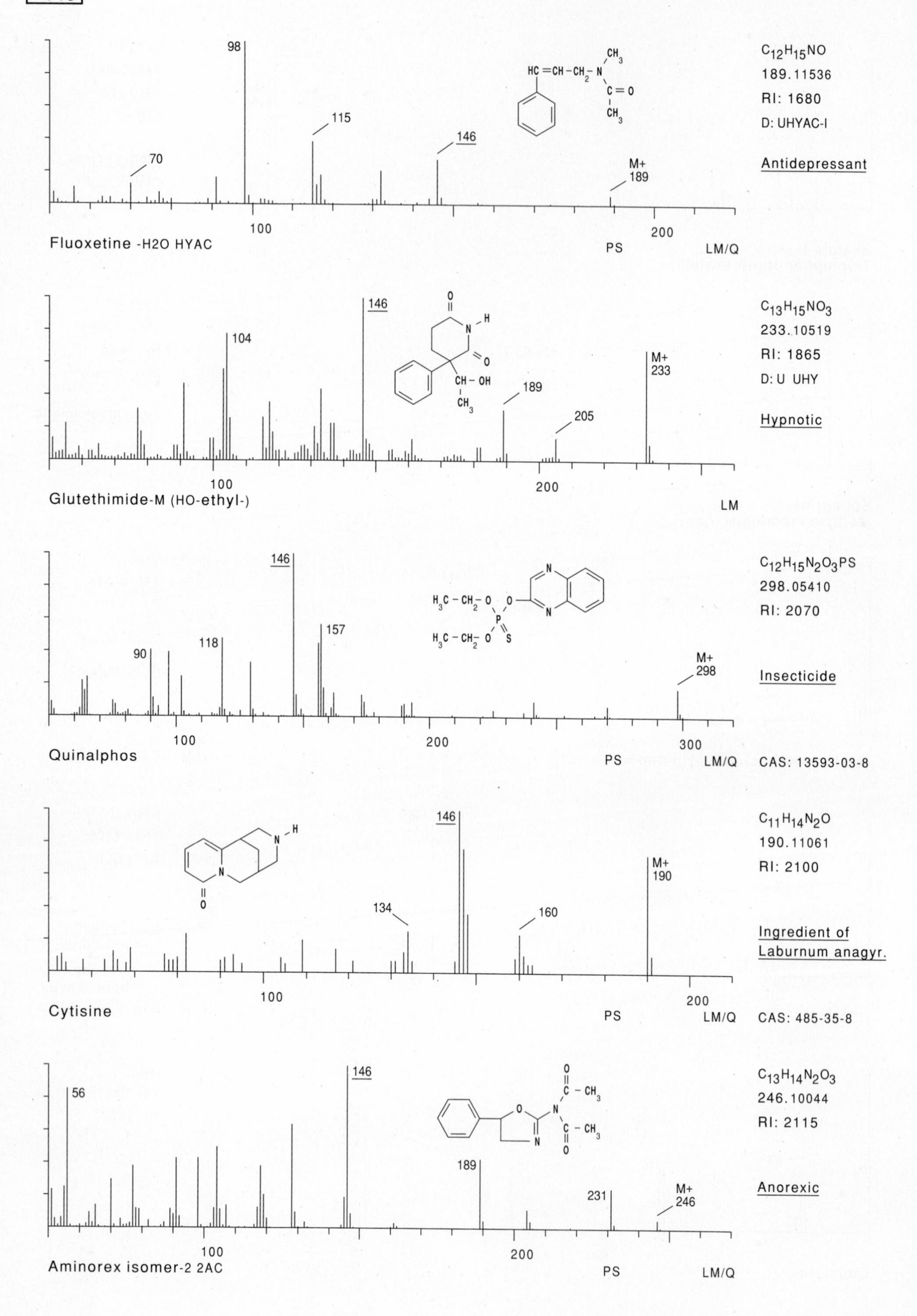

98
115
146
70
M+
189
100
200
PS
LM/Q
Fluoxetine -H2O HYAC
$C_{12}H_{15}NO$
189.11536
RI: 1680
D: UHYAC-I
Antidepressant
146
104
M+
233
189
205
100
200
LM
Glutethimide-M (HO-ethyl-)
$C_{13}H_{15}NO_3$
233.10519
RI: 1865
D: U UHY
Hypnotic
146
157
118
90
M+
298
100
200
300
PS
LM/Q
Quinalphos
$C_{12}H_{15}N_2O_3PS$
298.05410
RI: 2070
Insecticide
CAS: 13593-03-8
146
M+
190
134
160
100
200
PS
LM/Q
Cytisine
$C_{11}H_{14}N_2O$
190.11061
RI: 2100
Ingredient of
Laburnum anagyr.
CAS: 485-35-8
146
56
189
231
M+
246
100
200
PS
LM/Q
Aminorex isomer-2 2AC
$C_{13}H_{14}N_2O_3$
246.10044
RI: 2115
Anorexic

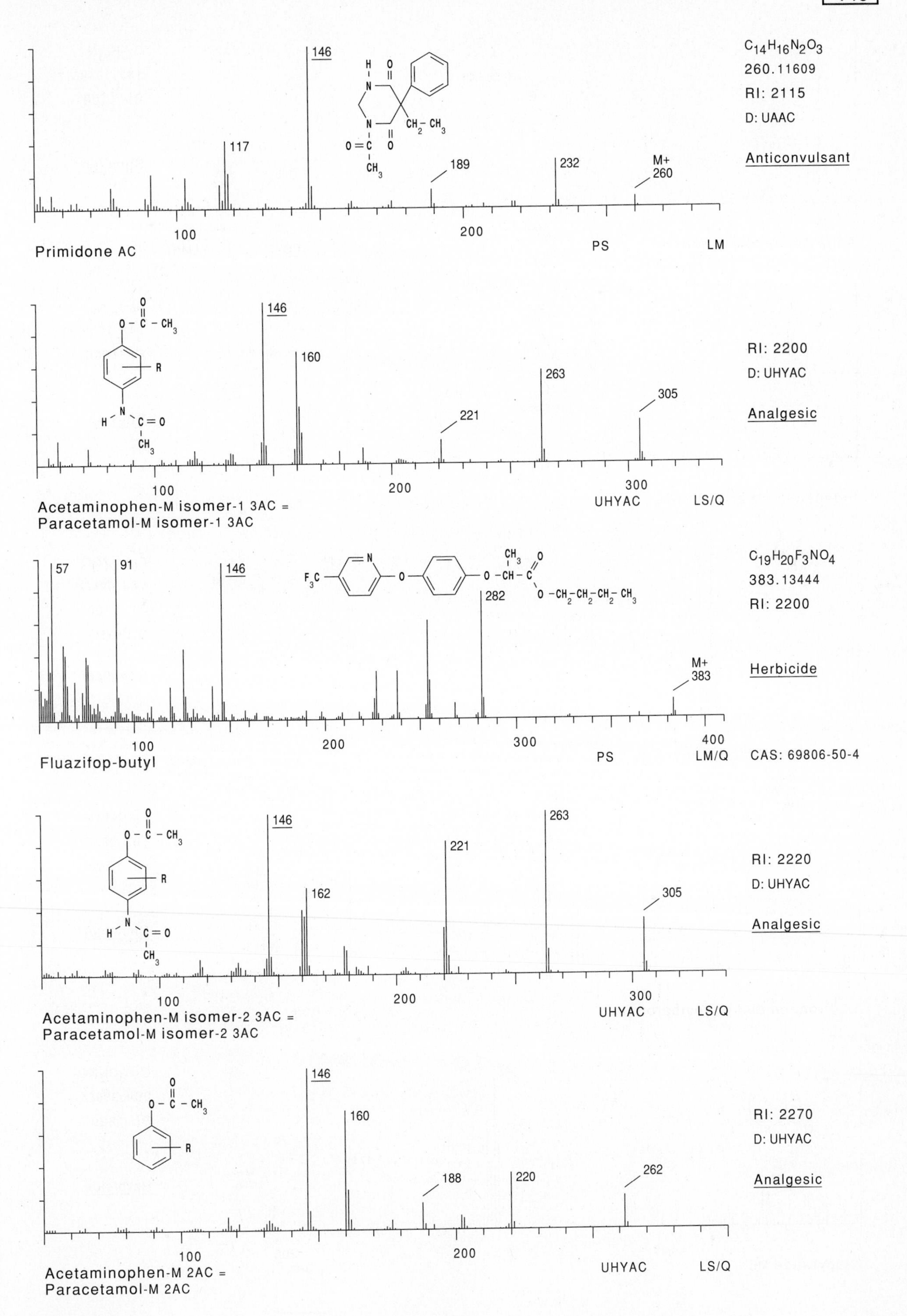
146
117
189
232
M+
260
100
200
Primidone AC
PS
LM
$C_{14}H_{16}N_2O_3$
260.11609
RI: 2115
D: UAAC
Anticonvulsant
146
160
221
263
305
100
200
300
Acetaminophen-M isomer-1 3AC =
Paracetamol-M isomer-1 3AC
UHYAC
LS/Q
RI: 2200
D: UHYAC
Analgesic
57
91
146
282
M+
383
100
200
300
400
Fluazifop-butyl
PS
LM/Q
$C_{19}H_{20}F_3NO_4$
383.13444
RI: 2200
Herbicide
CAS: 69806-50-4
146
263
221
162
305
100
200
300
Acetaminophen-M isomer-2 3AC =
Paracetamol-M isomer-2 3AC
UHYAC
LS/Q
RI: 2220
D: UHYAC
Analgesic
146
160
188
220
262
100
200
Acetaminophen-M 2AC =
Paracetamol-M 2AC
UHYAC
LS/Q
RI: 2270
D: UHYAC
Analgesic

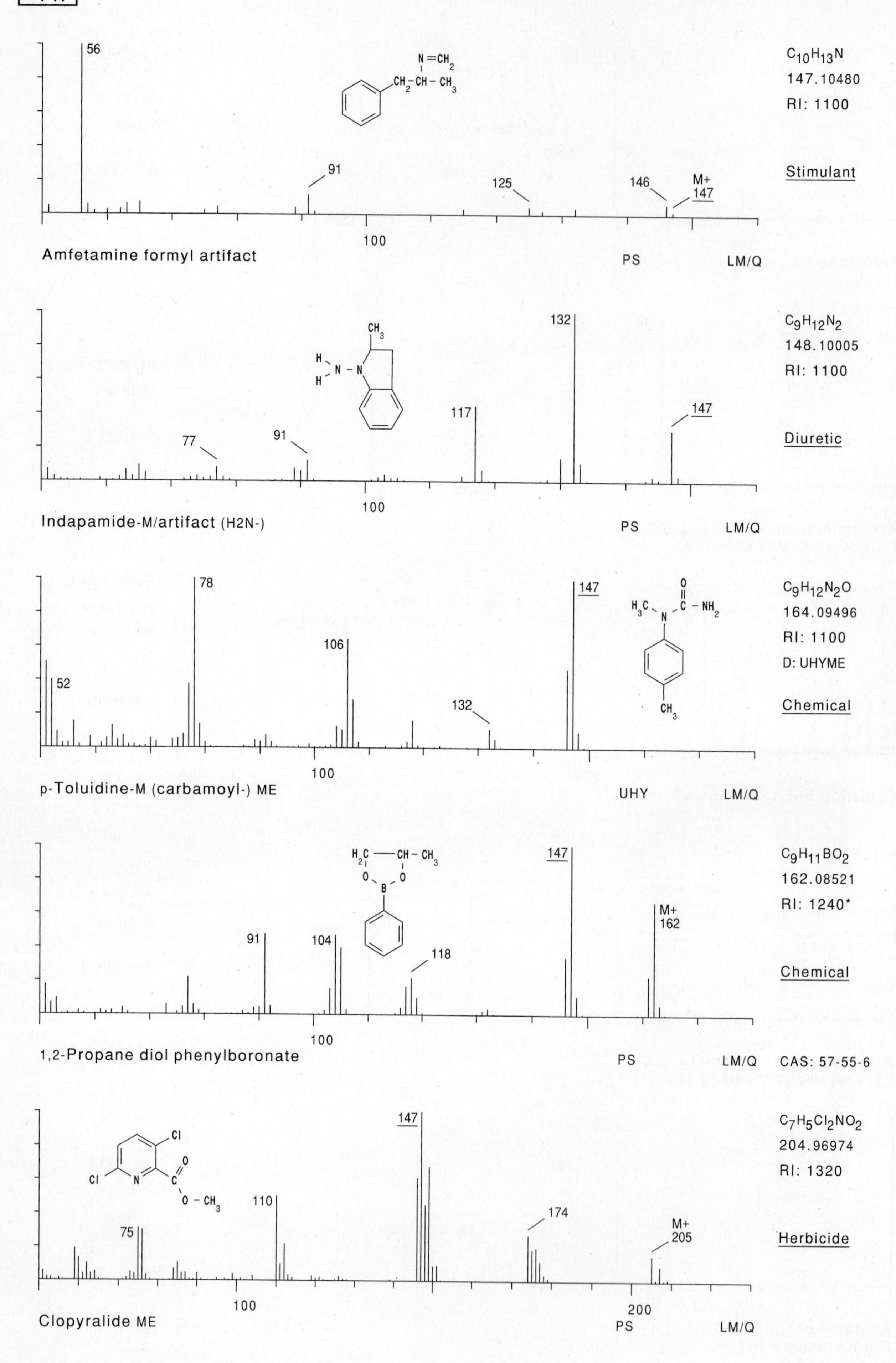
56
N=CH2
CH2-CH-CH3
91
125
146
M+
147
100
$C_{10}H_{13}N$
147.10480
RI: 1100
Stimulant
Amfetamine formyl artifact
PS
LM/Q
CH3
H
N-N
H
132
117
147
77
91
100
$C_9H_{12}N_2$
148.10005
RI: 1100
Diuretic
Indapamide-M/artifact (H2N-)
PS
LM/Q
78
147
O
H3C
N
C-NH2
106
52
132
CH3
100
$C_9H_{12}N_2O$
164.09496
RI: 1100
D: UHYME
Chemical
p-Toluidine-M (carbamoyl-) ME
UHY
LM/Q
H2C
CH-CH3
O
B
O
147
M+
162
91
104
118
100
$C_9H_{11}BO_2$
162.08521
RI: 1240*
Chemical
1,2-Propane diol phenylboronate
PS
LM/Q
CAS: 57-55-6
Cl
O
Cl
N
C
O-CH3
147
110
174
75
M+
205
100
200
$C_7H_5Cl_2NO_2$
204.96974
RI: 1320
Herbicide
Clopyralide ME
PS
LM/Q

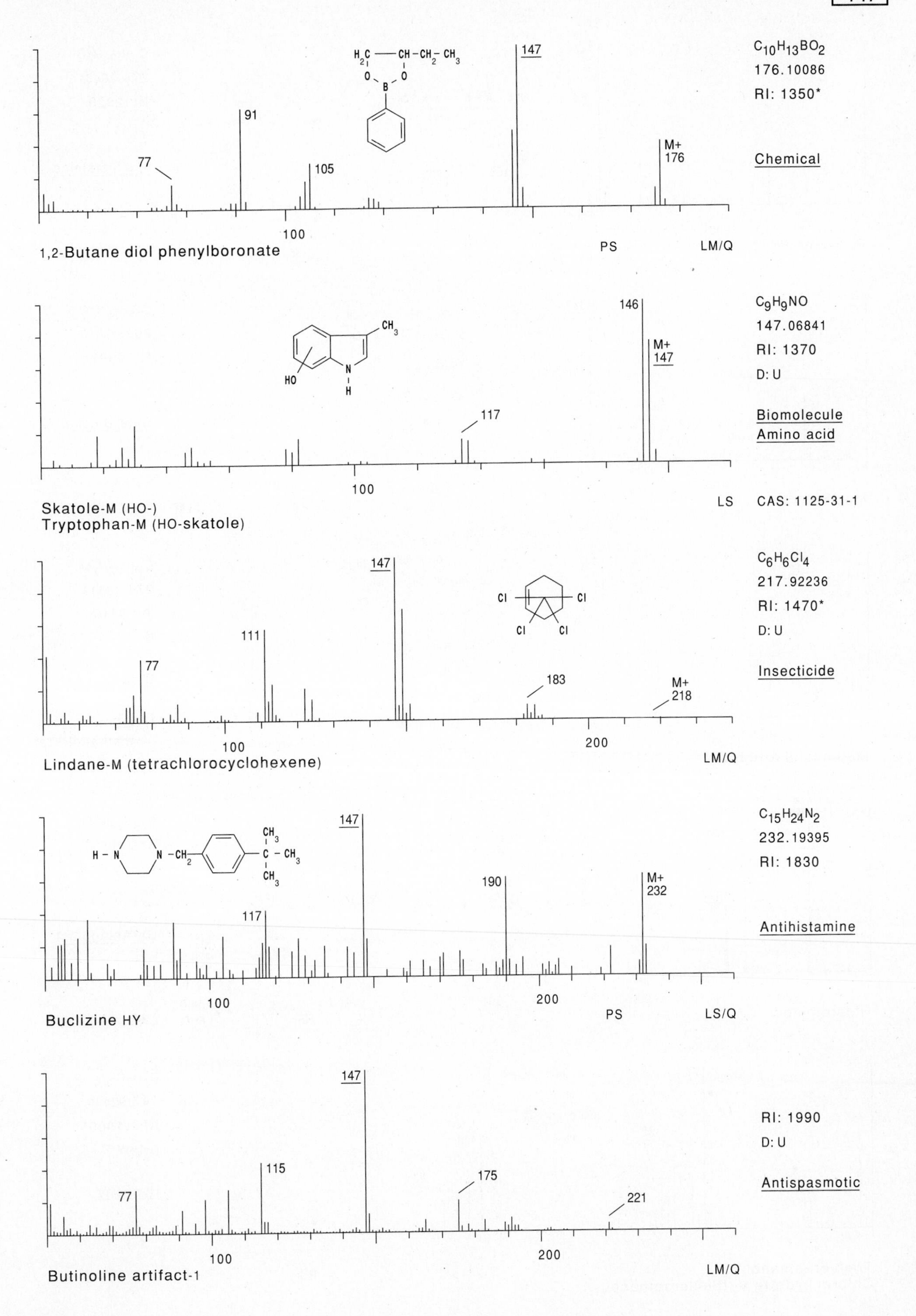

147
91
77
105
M+
176
100
1,2-Butane diol phenylboronate
PS
LM/Q
$C_{10}H_{13}BO_2$
176.10086
RI: 1350*
Chemical
146
M+
147
117
100
Skatole-M (HO-)
Tryptophan-M (HO-skatole)
LS
C_9H_9NO
147.06841
RI: 1370
D: U
Biomolecule
Amino acid
CAS: 1125-31-1
147
111
77
183
M+
218
100
200
Lindane-M (tetrachlorocyclohexene)
LM/Q
$C_6H_6Cl_4$
217.92236
RI: 1470*
D: U
Insecticide
147
190
117
M+
232
100
200
Buclizine HY
PS
LS/Q
$C_{15}H_{24}N_2$
232.19395
RI: 1830
Antihistamine
147
115
175
77
221
100
200
Butinoline artifact-1
LM/Q
RI: 1990
D: U
Antispasmotic

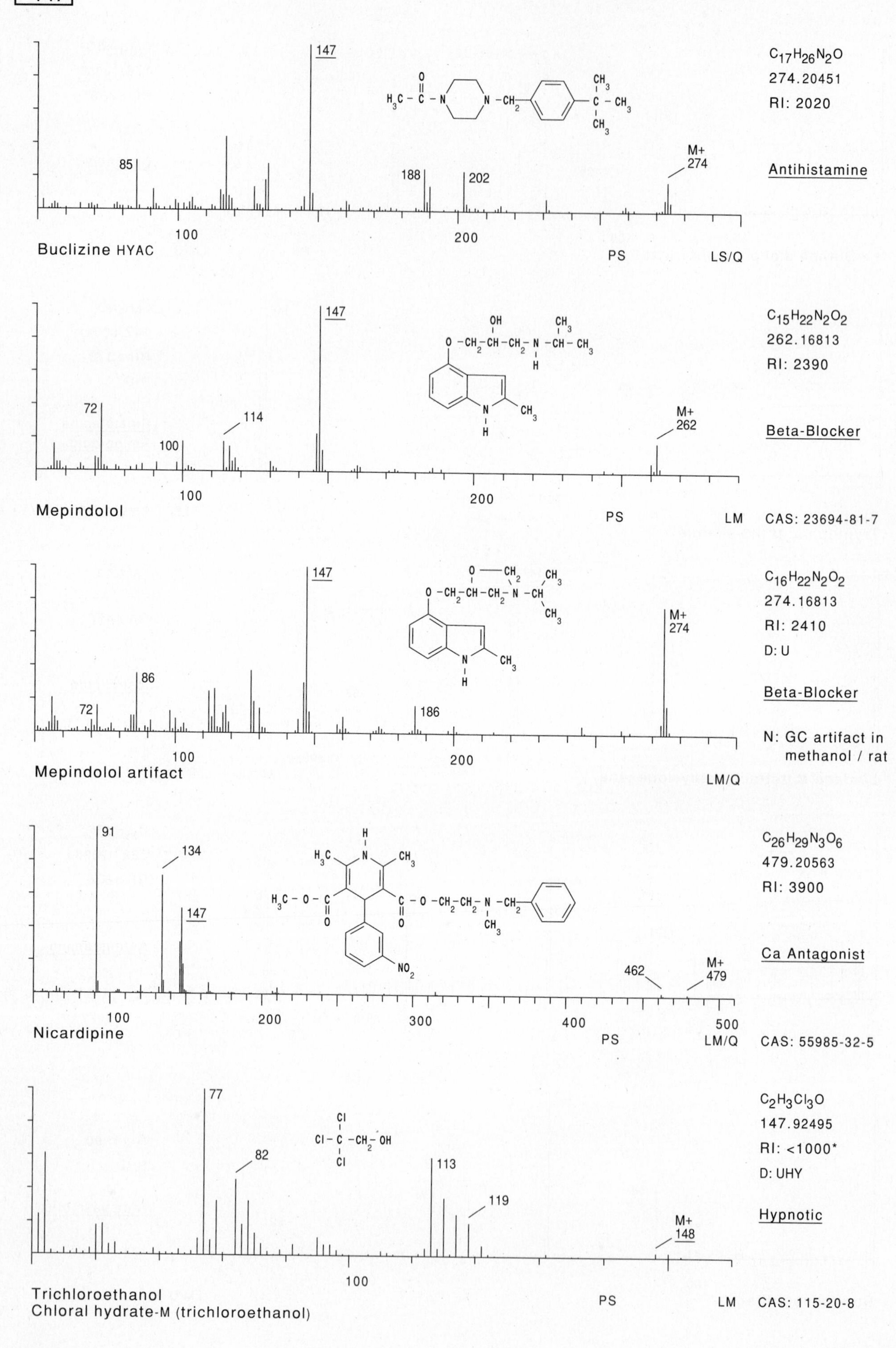
147
85
188
202
M+
274
100
200
Buclizine HYAC
PS
LS/Q
$C_{17}H_{26}N_2O$
274.20451
RI: 2020
Antihistamine
147
72
114
100
M+
262
Mepindolol
PS
LM
$C_{15}H_{22}N_2O_2$
262.16813
RI: 2390
Beta-Blocker
CAS: 23694-81-7
147
86
72
186
M+
274
Mepindolol artifact
LM/Q
$C_{16}H_{22}N_2O_2$
274.16813
RI: 2410
D: U
Beta-Blocker
N: GC artifact in
methanol / rat
91
134
147
462
M+
479
100
200
300
400
500
Nicardipine
PS
LM/Q
$C_{26}H_{29}N_3O_6$
479.20563
RI: 3900
Ca Antagonist
CAS: 55985-32-5
77
82
113
119
M+
148
Trichloroethanol
Chloral hydrate-M (trichloroethanol)
PS
LM
$C_2H_3Cl_3O$
147.92495
RI: <1000*
D: UHY
Hypnotic
CAS: 115-20-8

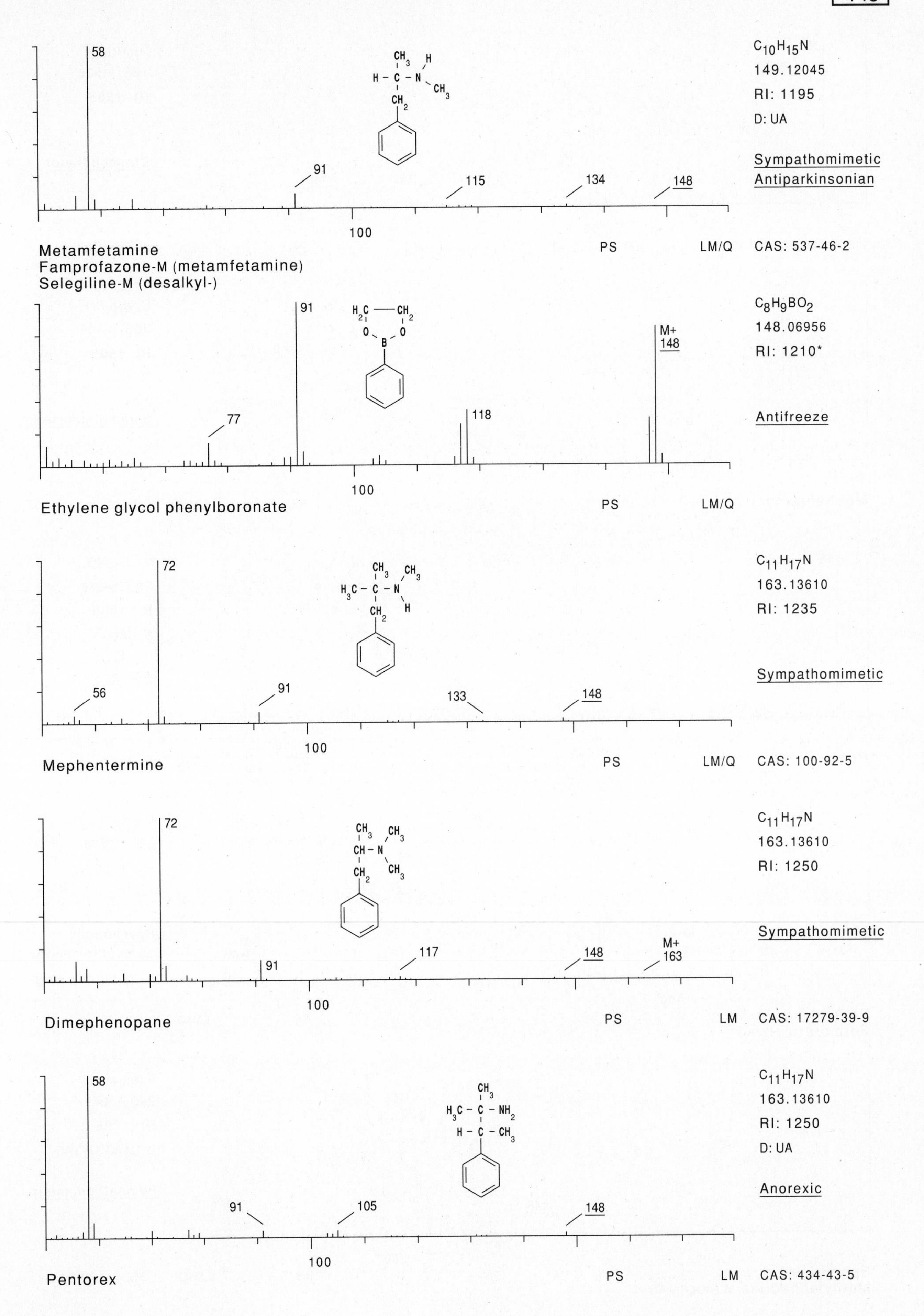

58
91
115
134
148
100
$C_{10}H_{15}N$
149.12045
RI: 1195
D: UA
Sympathomimetic
Antiparkinsonian
Metamfetamine
Famprofazone-M (metamfetamine)
Selegiline-M (desalkyl-)
PS
LM/Q
CAS: 537-46-2
91
M+
148
77
118
100
$C_8H_9BO_2$
148.06956
RI: 1210*
Antifreeze
Ethylene glycol phenylboronate
PS
LM/Q
72
56
91
133
148
100
$C_{11}H_{17}N$
163.13610
RI: 1235
Sympathomimetic
Mephentermine
PS
LM/Q
CAS: 100-92-5
72
91
117
148
M+
163
100
$C_{11}H_{17}N$
163.13610
RI: 1250
Sympathomimetic
Dimephenopane
PS
LM
CAS: 17279-39-9
58
91
105
148
100
$C_{11}H_{17}N$
163.13610
RI: 1250
D: UA
Anorexic
Pentorex
PS
LM
CAS: 434-43-5

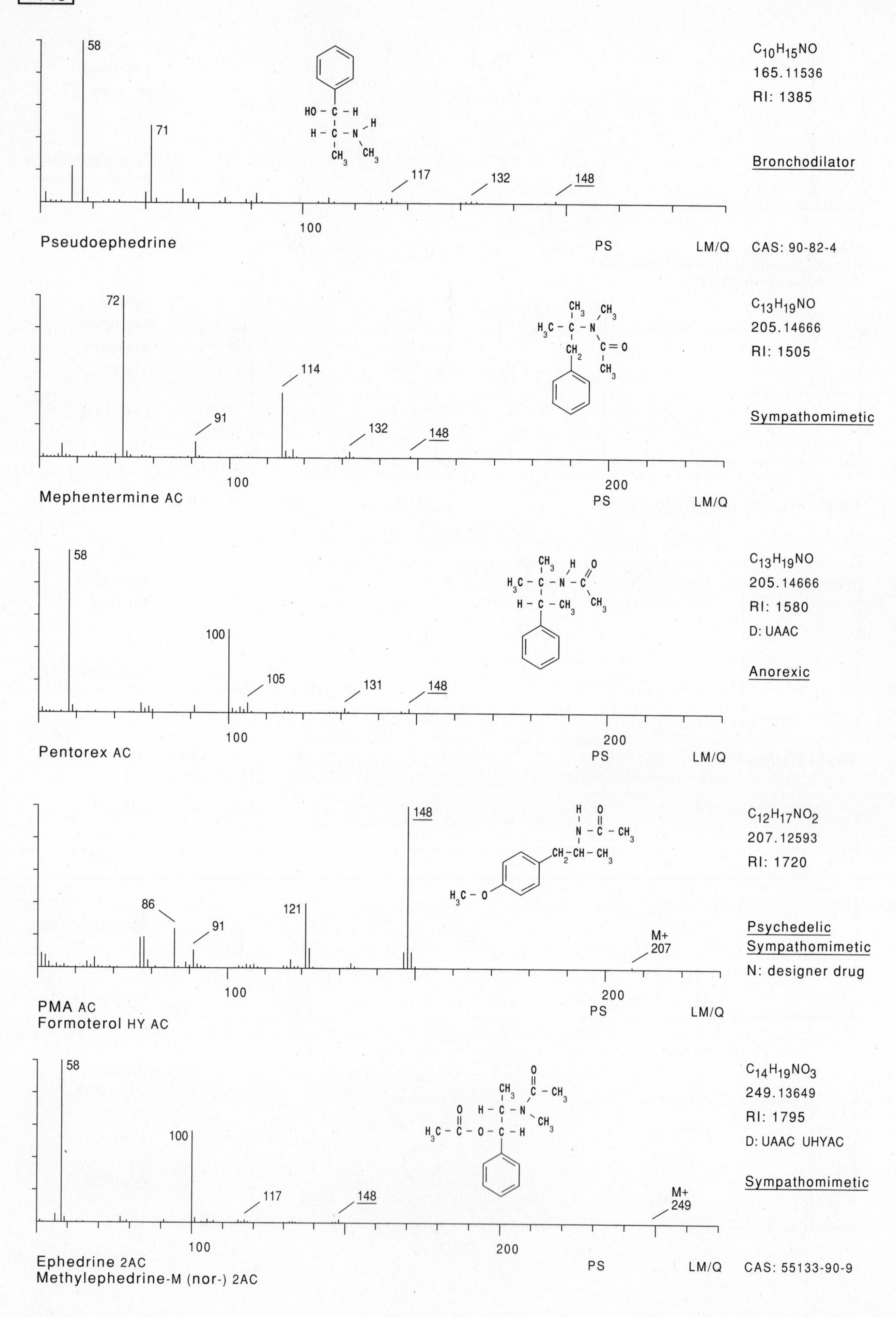
58
71
117
132
148
100
Pseudoephedrine
PS
LM/Q
$C_{10}H_{15}NO$
165.11536
RI: 1385
Bronchodilator
CAS: 90-82-4
72
114
91
132
148
100
200
Mephentermine AC
PS
LM/Q
$C_{13}H_{19}NO$
205.14666
RI: 1505
Sympathomimetic
58
100
105
131
148
100
200
Pentorex AC
PS
LM/Q
$C_{13}H_{19}NO$
205.14666
RI: 1580
D: UAAC
Anorexic
148
86
91
121
M+
207
100
200
PMA AC
Formoterol HY AC
PS
LM/Q
$C_{12}H_{17}NO_2$
207.12593
RI: 1720
Psychedelic
Sympathomimetic
N: designer drug
58
100
117
148
M+
249
100
200
Ephedrine 2AC
Methylephedrine-M (nor-) 2AC
PS
LM/Q
$C_{14}H_{19}NO_3$
249.13649
RI: 1795
D: UAAC UHYAC
Sympathomimetic
CAS: 55133-90-9

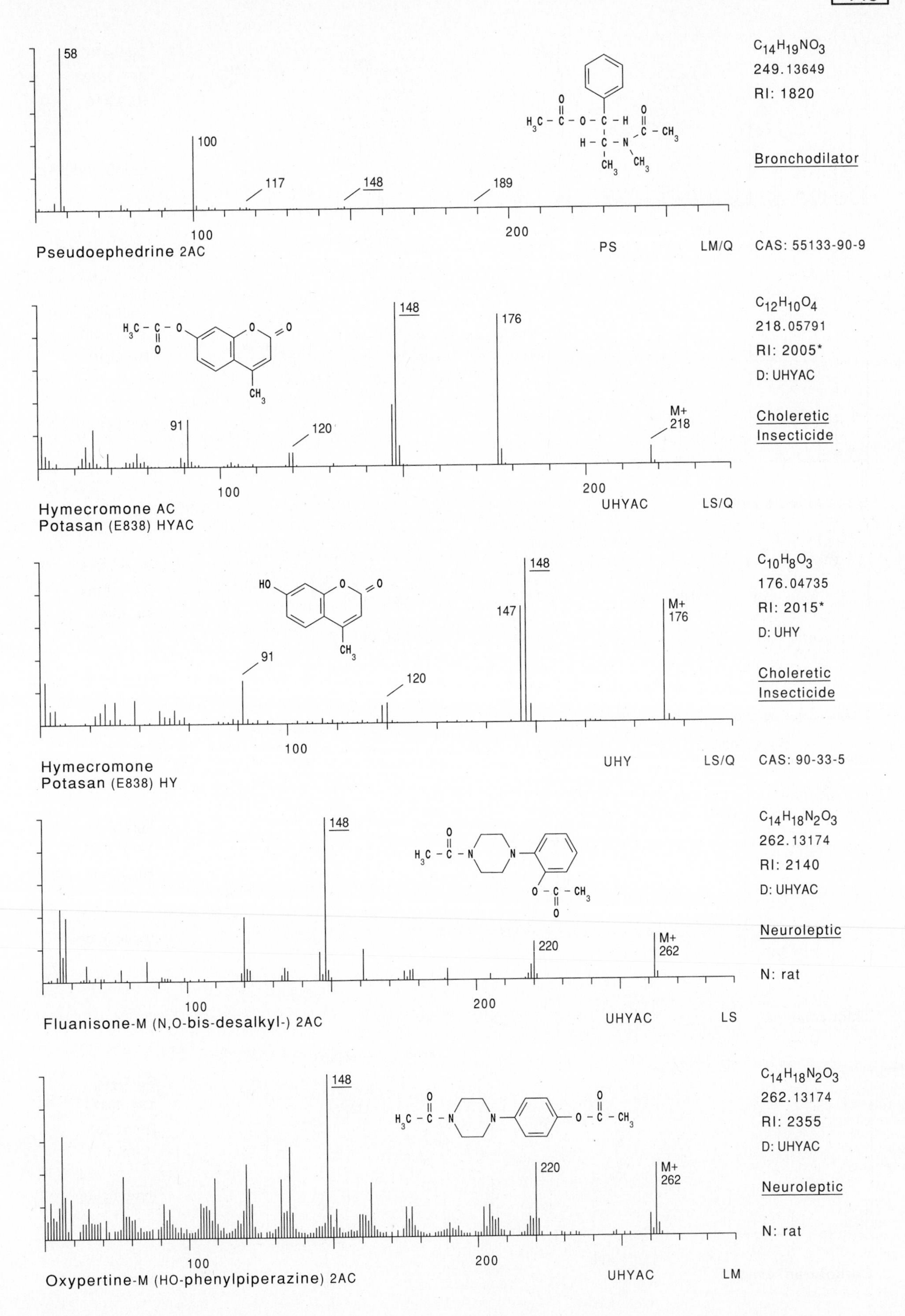

$C_{14}H_{19}NO_3$
249.13649
RI: 1820

Bronchodilator

CAS: 55133-90-9

$C_{12}H_{10}O_4$
218.05791
RI: 2005*
D: UHYAC

Choleretic
Insecticide

$C_{10}H_8O_3$
176.04735
RI: 2015*
D: UHY

Choleretic
Insecticide

CAS: 90-33-5

$C_{14}H_{18}N_2O_3$
262.13174
RI: 2140
D: UHYAC

Neuroleptic

N: rat

$C_{14}H_{18}N_2O_3$
262.13174
RI: 2355
D: UHYAC

Neuroleptic

N: rat

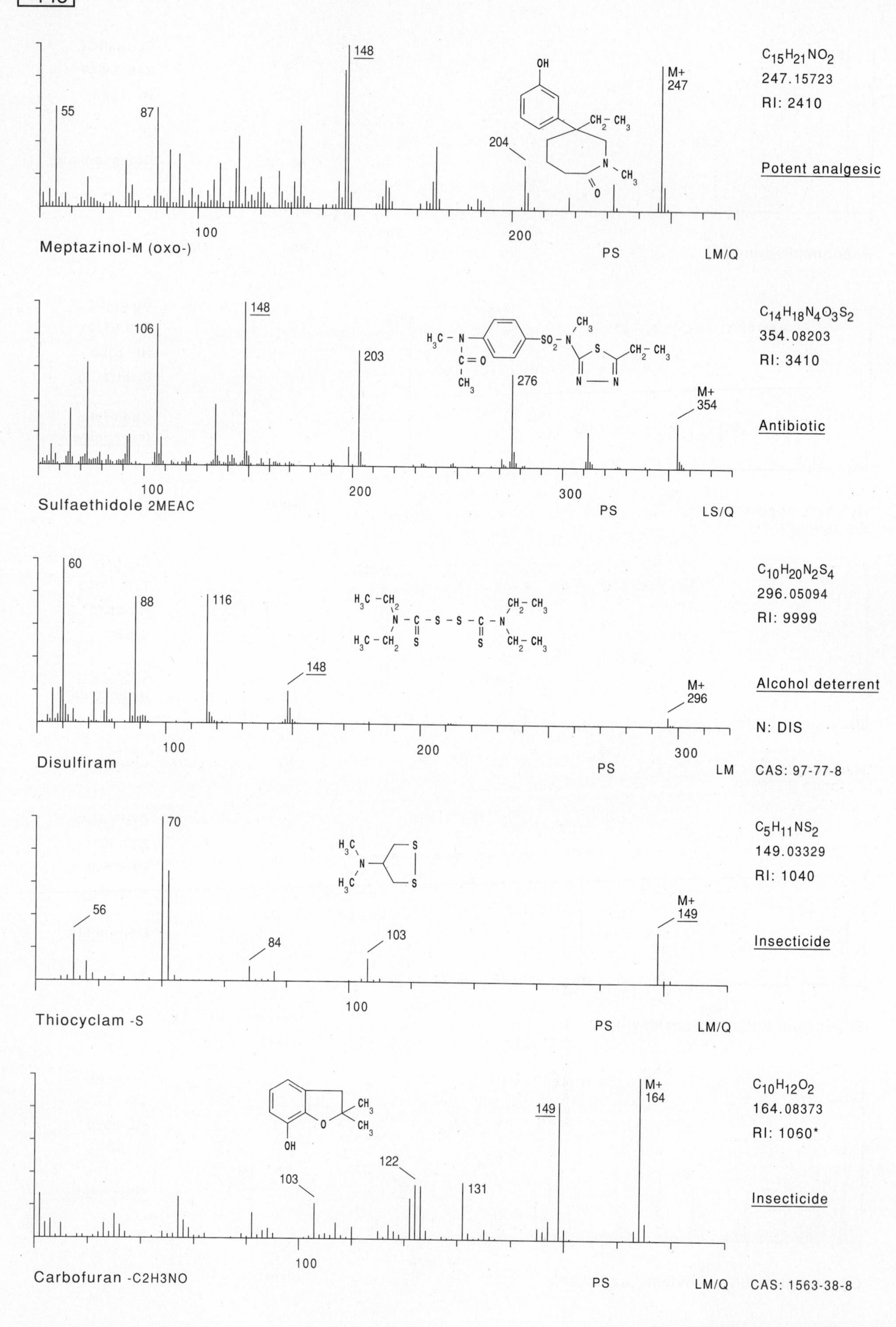
148
55
87
204
M+
247
100
200
Meptazinol-M (oxo-)
PS
LM/Q
$C_{15}H_{21}NO_2$
247.15723
RI: 2410
Potent analgesic
148
106
203
276
M+
354
100
200
300
Sulfaethidole 2MEAC
PS
LS/Q
$C_{14}H_{18}N_4O_3S_2$
354.08203
RI: 3410
Antibiotic
60
88
116
148
M+
296
100
200
300
Disulfiram
PS
LM
$C_{10}H_{20}N_2S_4$
296.05094
RI: 9999
Alcohol deterrent
N: DIS
CAS: 97-77-8
70
56
84
103
M+
149
100
Thiocyclam -S
PS
LM/Q
$C_5H_{11}NS_2$
149.03329
RI: 1040
Insecticide
M+
164
149
122
103
131
100
Carbofuran -C2H3NO
PS
LM/Q
CAS: 1563-38-8
$C_{10}H_{12}O_2$
164.08373
RI: 1060*
Insecticide

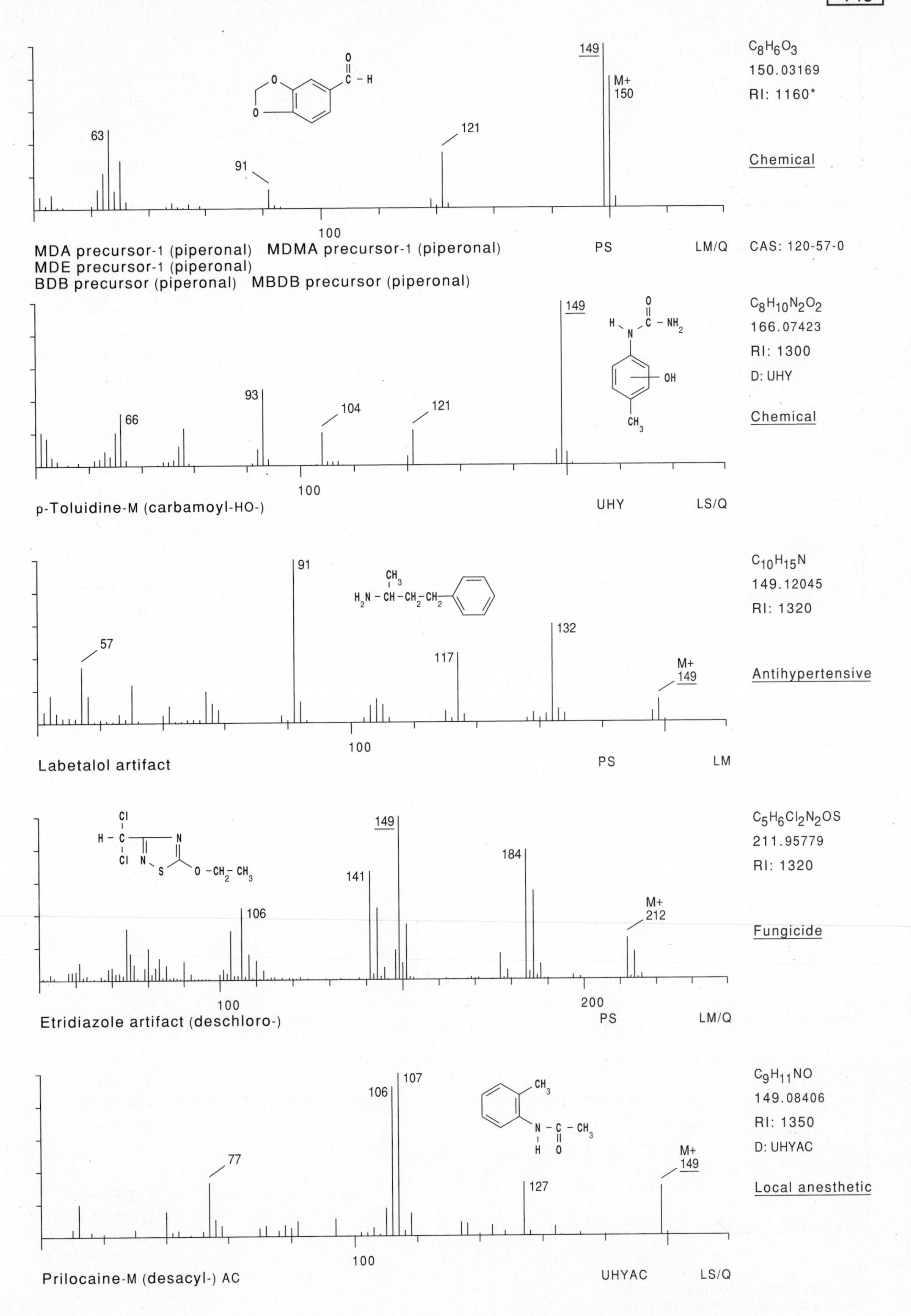

$C_8H_6O_3$
150.03169
RI: 1160*
Chemical
PS
LM/Q
CAS: 120-57-0
MDA precursor-1 (piperonal) MDMA precursor-1 (piperonal)
MDE precursor-1 (piperonal)
BDB precursor (piperonal) MBDB precursor (piperonal)
$C_8H_{10}N_2O_2$
166.07423
RI: 1300
D: UHY
Chemical
UHY
LS/Q
p-Toluidine-M (carbamoyl-HO-)
$C_{10}H_{15}N$
149.12045
RI: 1320
Antihypertensive
PS
LM
Labetalol artifact
$C_5H_6Cl_2N_2OS$
211.95779
RI: 1320
Fungicide
PS
LM/Q
Etridiazole artifact (deschloro-)
$C_9H_{11}NO$
149.08406
RI: 1350
D: UHYAC
Local anesthetic
UHYAC
LS/Q
Prilocaine-M (desacyl-) AC

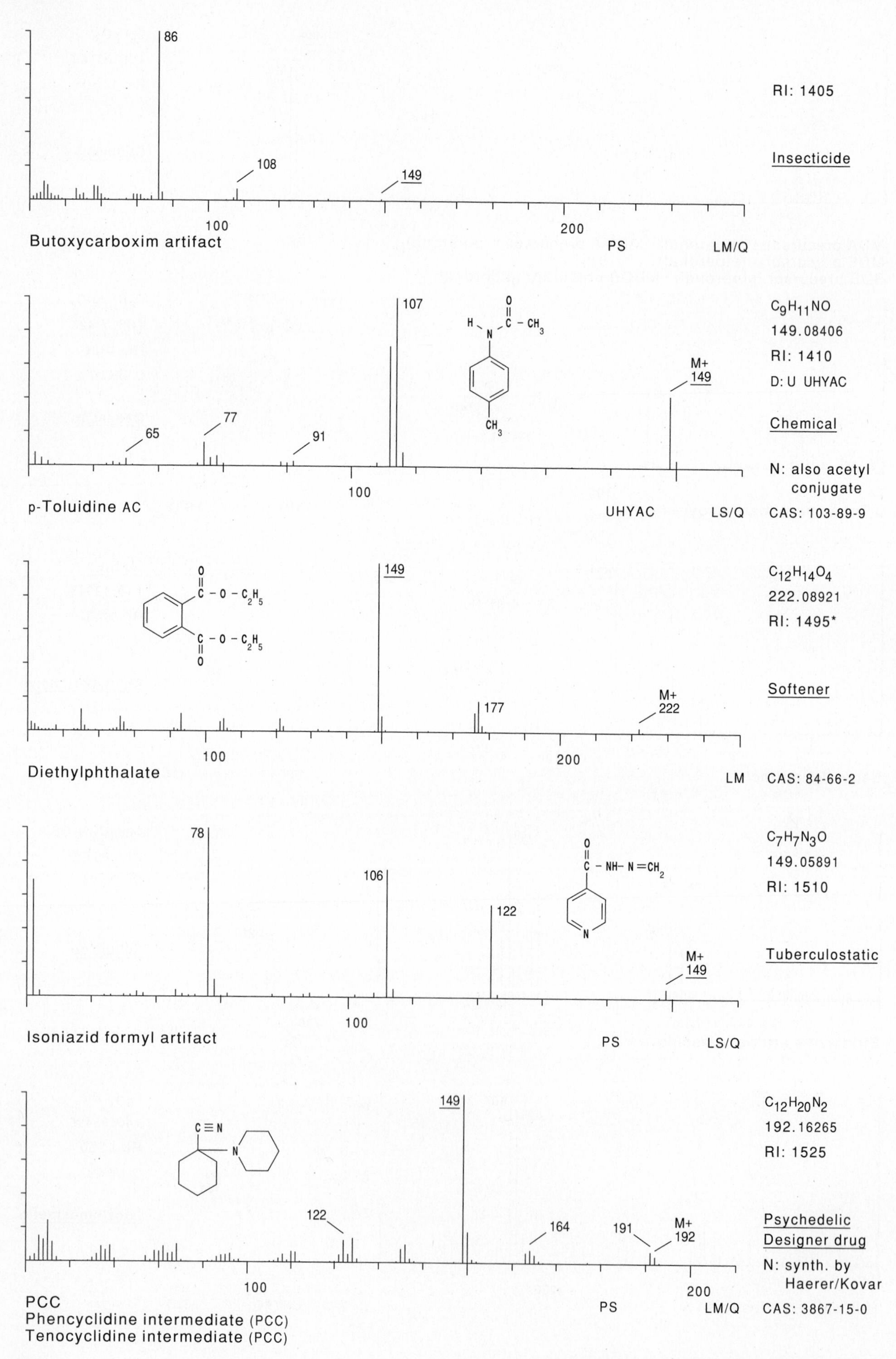

86
108
149
100
200
RI: 1405
Insecticide
Butoxycarboxim artifact
PS
LM/Q
107
M+
149
65
77
91
100
$C_9H_{11}NO$
149.08406
RI: 1410
D: U UHYAC
Chemical
N: also acetyl conjugate
p-Toluidine AC
UHYAC
LS/Q
CAS: 103-89-9
149
177
M+
222
100
200
$C_{12}H_{14}O_4$
222.08921
RI: 1495*
Softener
Diethylphthalate
LM
CAS: 84-66-2
78
106
122
M+
149
100
$C_7H_7N_3O$
149.05891
RI: 1510
Tuberculostatic
Isoniazid formyl artifact
PS
LS/Q
149
122
164
191
M+
192
100
200
$C_{12}H_{20}N_2$
192.16265
RI: 1525
Psychedelic
Designer drug
N: synth. by Haerer/Kovar
PCC
Phencyclidine intermediate (PCC)
Tenocyclidine intermediate (PCC)
PS
LM/Q
CAS: 3867-15-0

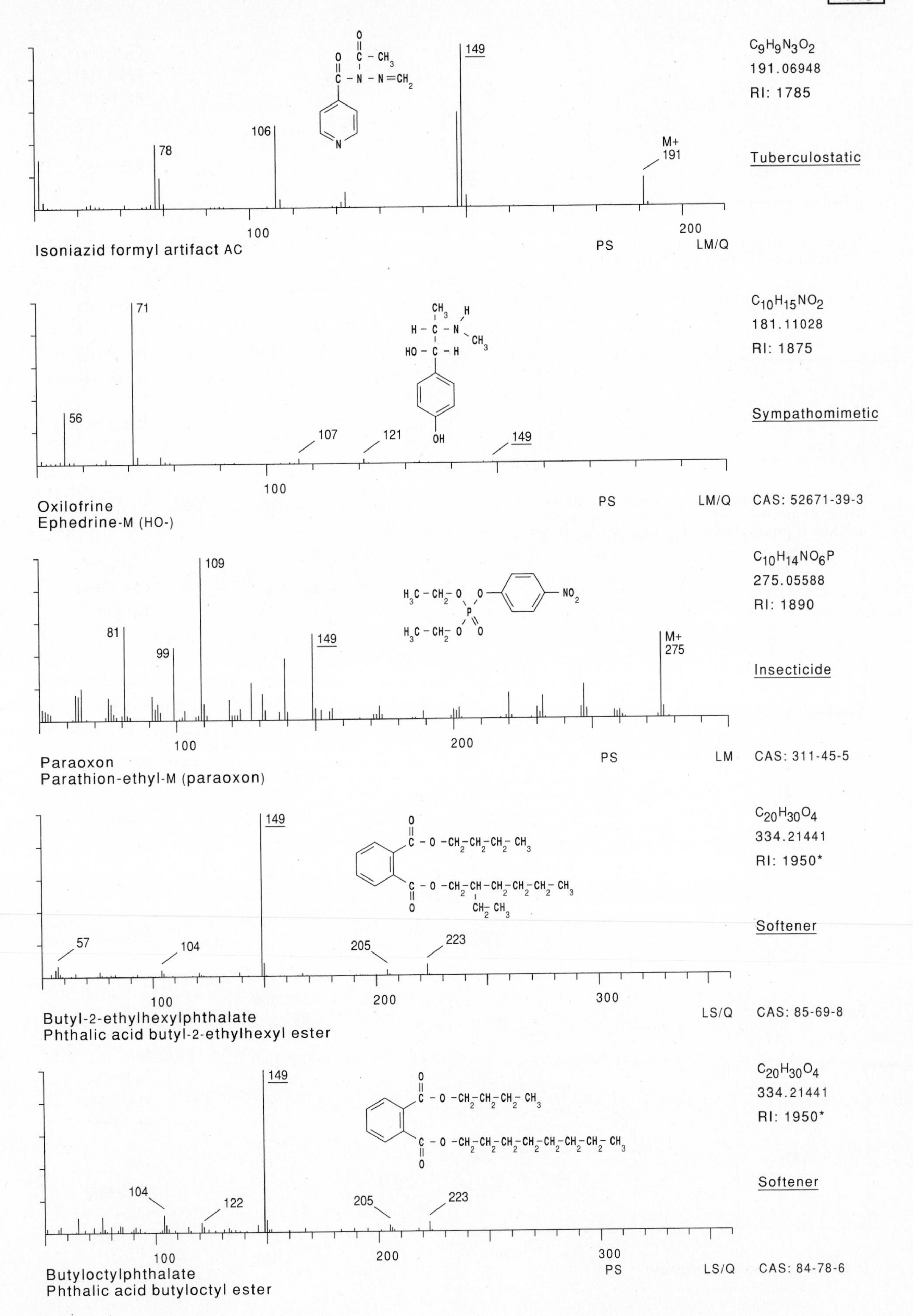
149
106
78
M+
191
100
200
Isoniazid formyl artifact AC
PS
LM/Q
C9H9N3O2
191.06948
RI: 1785
Tuberculostatic
71
56
107
121
149
OH
100
Oxilofrine
Ephedrine-M (HO-)
PS
LM/Q
C10H15NO2
181.11028
RI: 1875
Sympathomimetic
CAS: 52671-39-3
109
81
99
149
M+
275
NO2
100
200
Paraoxon
Parathion-ethyl-M (paraoxon)
PS
LM
C10H14NO6P
275.05588
RI: 1890
Insecticide
CAS: 311-45-5
149
57
104
205
223
100
200
300
Butyl-2-ethylhexylphthalate
Phthalic acid butyl-2-ethylhexyl ester
LS/Q
C20H30O4
334.21441
RI: 1950*
Softener
CAS: 85-69-8
149
104
122
205
223
100
200
300
Butyloctylphthalate
Phthalic acid butyloctyl ester
PS
LS/Q
C20H30O4
334.21441
RI: 1950*
Softener
CAS: 84-78-6

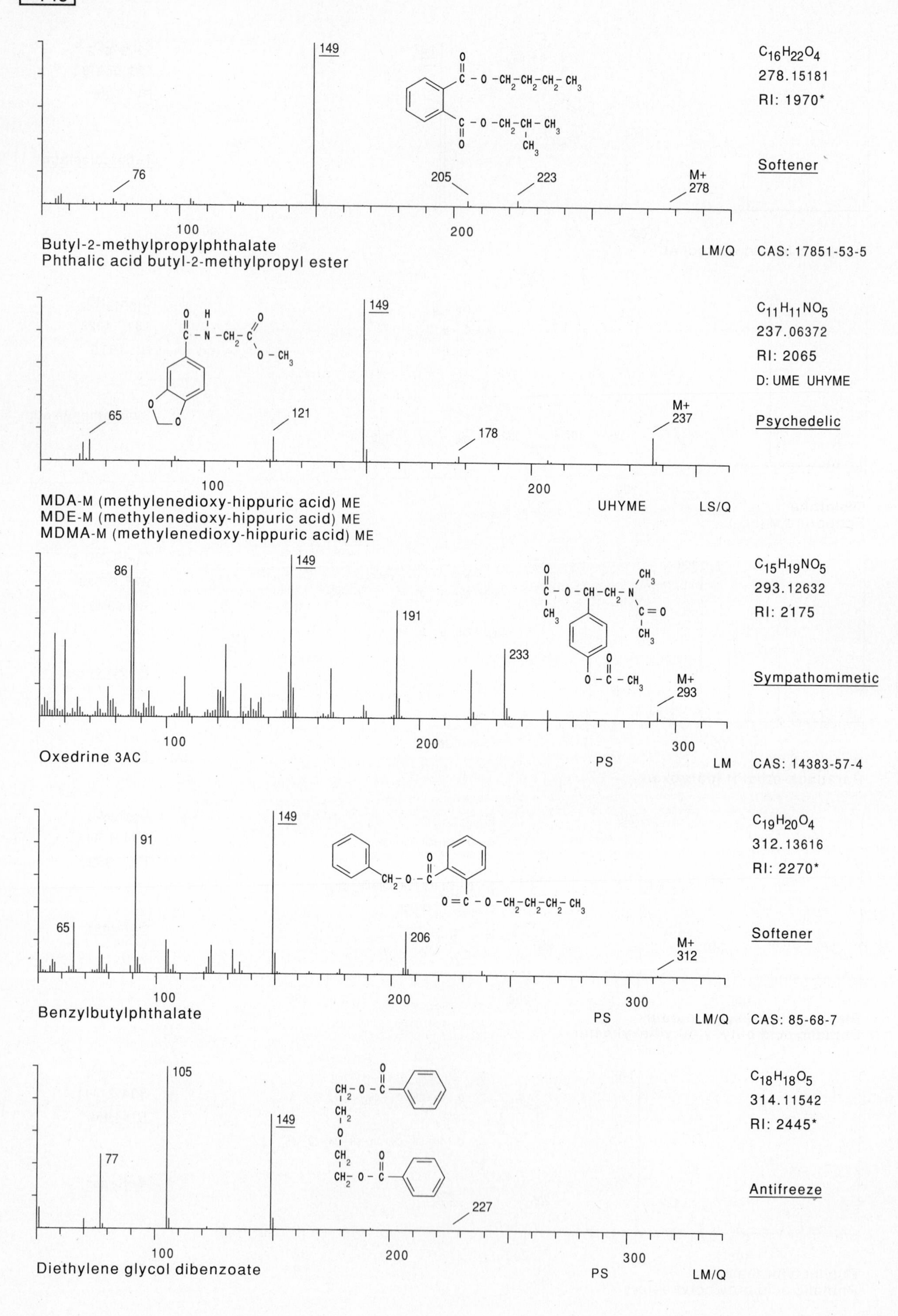
149
76
205
223
M+
278
100
200
Butyl-2-methylpropylphthalate
Phthalic acid butyl-2-methylpropyl ester
LM/Q
C16H22O4
278.15181
RI: 1970*
Softener
CAS: 17851-53-5
149
65
121
178
M+
237
100
200
UHYME
LS/Q
C11H11NO5
237.06372
RI: 2065
D: UME UHYME
Psychedelic
MDA-M (methylenedioxy-hippuric acid) ME
MDE-M (methylenedioxy-hippuric acid) ME
MDMA-M (methylenedioxy-hippuric acid) ME
86
149
191
233
M+
293
100
200
300
Oxedrine 3AC
PS
LM
C15H19NO5
293.12632
RI: 2175
Sympathomimetic
CAS: 14383-57-4
149
91
65
206
M+
312
100
200
300
Benzylbutylphthalate
PS
LM/Q
C19H20O4
312.13616
RI: 2270*
Softener
CAS: 85-68-7
105
149
77
227
100
200
300
Diethylene glycol dibenzoate
PS
LM/Q
C18H18O5
314.11542
RI: 2445*
Antifreeze

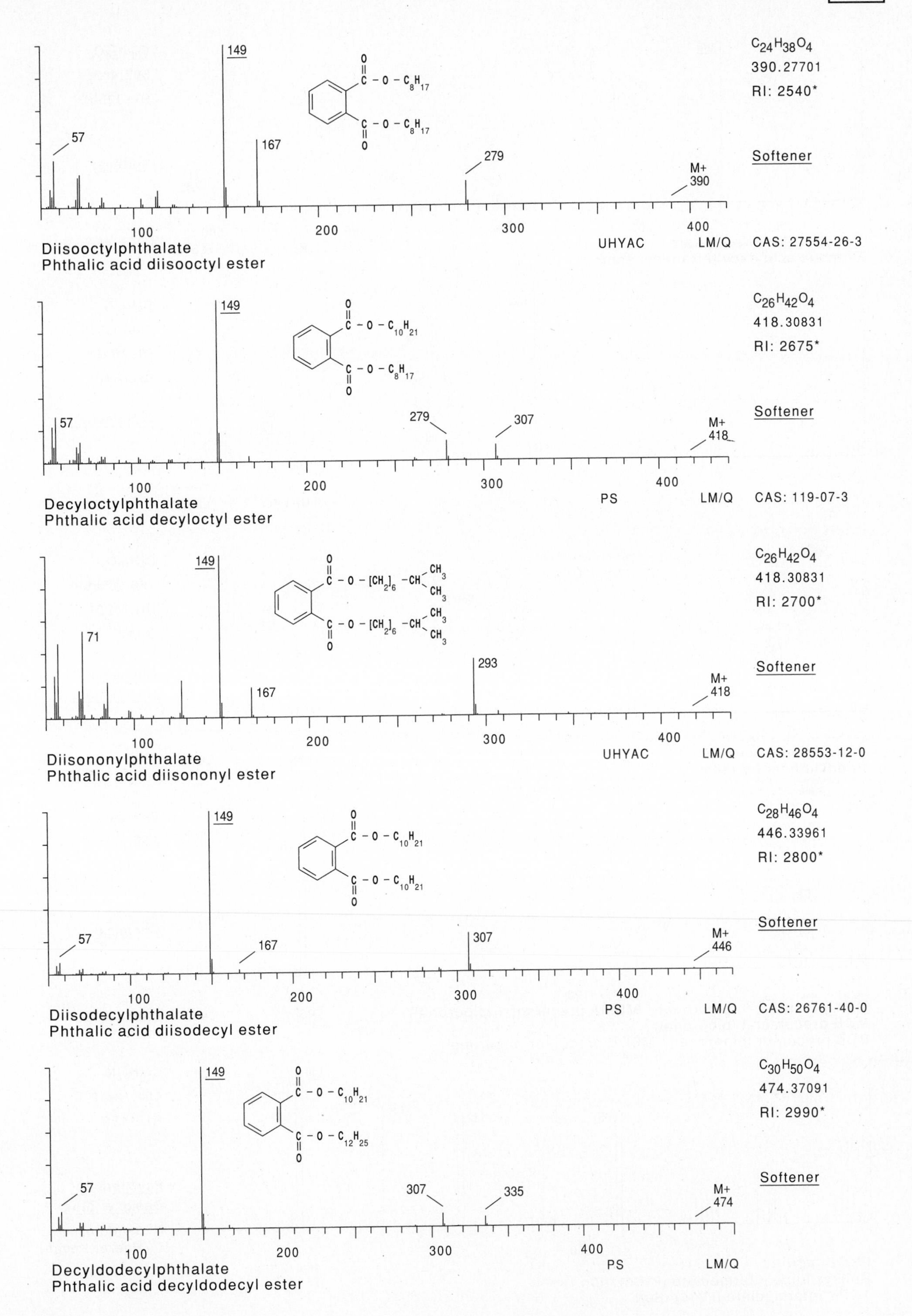

$C_{24}H_{38}O_4$
390.27701
RI: 2540*
Softener
CAS: 27554-26-3

$C_{26}H_{42}O_4$
418.30831
RI: 2675*
Softener
CAS: 119-07-3

$C_{26}H_{42}O_4$
418.30831
RI: 2700*
Softener
CAS: 28553-12-0

$C_{28}H_{46}O_4$
446.33961
RI: 2800*
Softener
CAS: 26761-40-0

$C_{30}H_{50}O_4$
474.37091
RI: 2990*
Softener

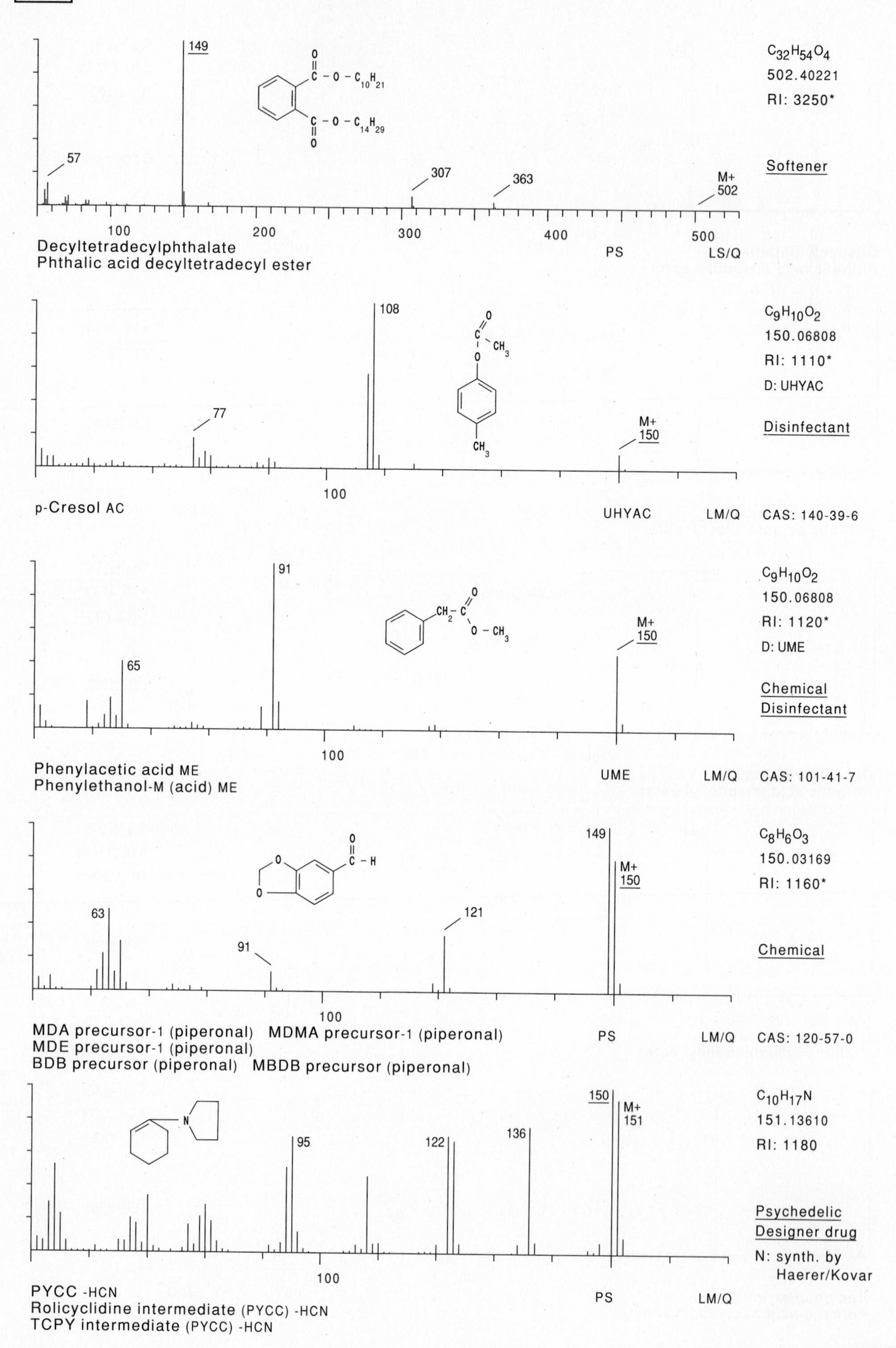

149
57
307
363
M+
502
100
200
300
400
500
PS
LS/Q
$C_{32}H_{54}O_4$
502.40221
RI: 3250*
Softener
Decyltetradecylphthalate
Phthalic acid decyltetradecyl ester
108
77
M+
150
100
UHYAC
LM/Q
$C_9H_{10}O_2$
150.06808
RI: 1110*
D: UHYAC
Disinfectant
p-Cresol AC
CAS: 140-39-6
91
65
M+
150
100
UME
LM/Q
$C_9H_{10}O_2$
150.06808
RI: 1120*
D: UME
Chemical
Disinfectant
Phenylacetic acid ME
Phenylethanol-M (acid) ME
CAS: 101-41-7
149
M+
150
63
121
91
100
PS
LM/Q
$C_8H_6O_3$
150.03169
RI: 1160*
Chemical
MDA precursor-1 (piperonal) MDMA precursor-1 (piperonal)
MDE precursor-1 (piperonal)
BDB precursor (piperonal) MBDB precursor (piperonal)
CAS: 120-57-0
150
M+
151
95
122
136
100
PS
LM/Q
$C_{10}H_{17}N$
151.13610
RI: 1180
Psychedelic
Designer drug
N: synth. by
Haerer/Kovar
PYCC -HCN
Rolicyclidine intermediate (PYCC) -HCN
TCPY intermediate (PYCC) -HCN

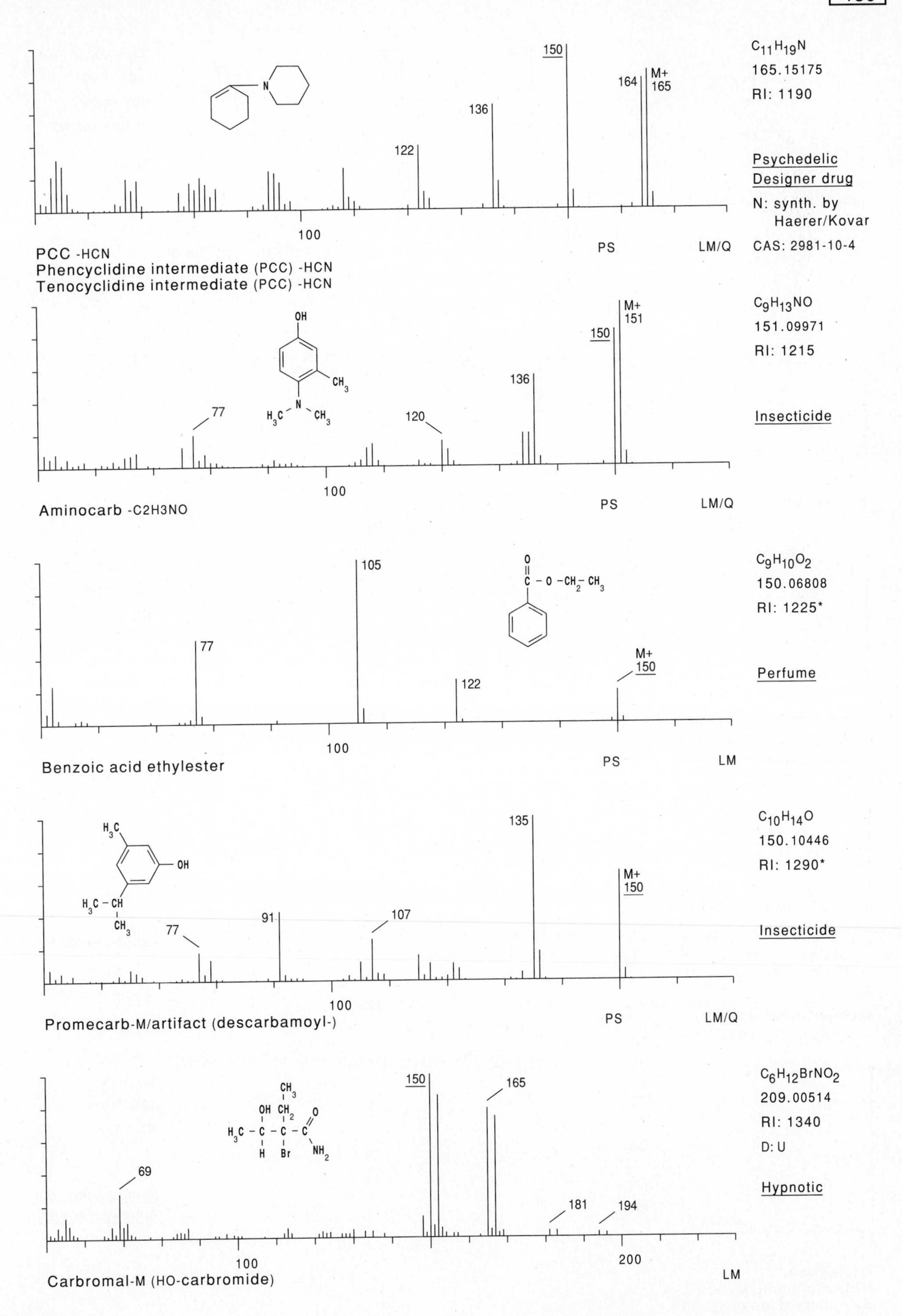
150
164
M+
165
136
122
100
PS
LM/Q
$C_{11}H_{19}N$
165.15175
RI: 1190
Psychedelic
Designer drug
N: synth. by
Haerer/Kovar
CAS: 2981-10-4
PCC -HCN
Phencyclidine intermediate (PCC) -HCN
Tenocyclidine intermediate (PCC) -HCN
OH
CH3
H3C N CH3
M+
151
150
136
120
77
100
PS
LM/Q
$C_9H_{13}NO$
151.09971
RI: 1215
Insecticide
Aminocarb -C2H3NO
105
77
122
M+
150
C - O - CH2 - CH3
100
PS
LM
$C_9H_{10}O_2$
150.06808
RI: 1225*
Perfume
Benzoic acid ethylester
135
M+
150
91
107
77
H3C
OH
H3C - CH
CH3
100
PS
LM/Q
$C_{10}H_{14}O$
150.10446
RI: 1290*
Insecticide
Promecarb-M/artifact (descarbamoyl-)
150
165
69
181
194
CH3
OH CH2
H3C - C - C - C
H Br NH2
100
200
LM
$C_6H_{12}BrNO_2$
209.00514
RI: 1340
D: U
Hypnotic
Carbromal-M (HO-carbromide)

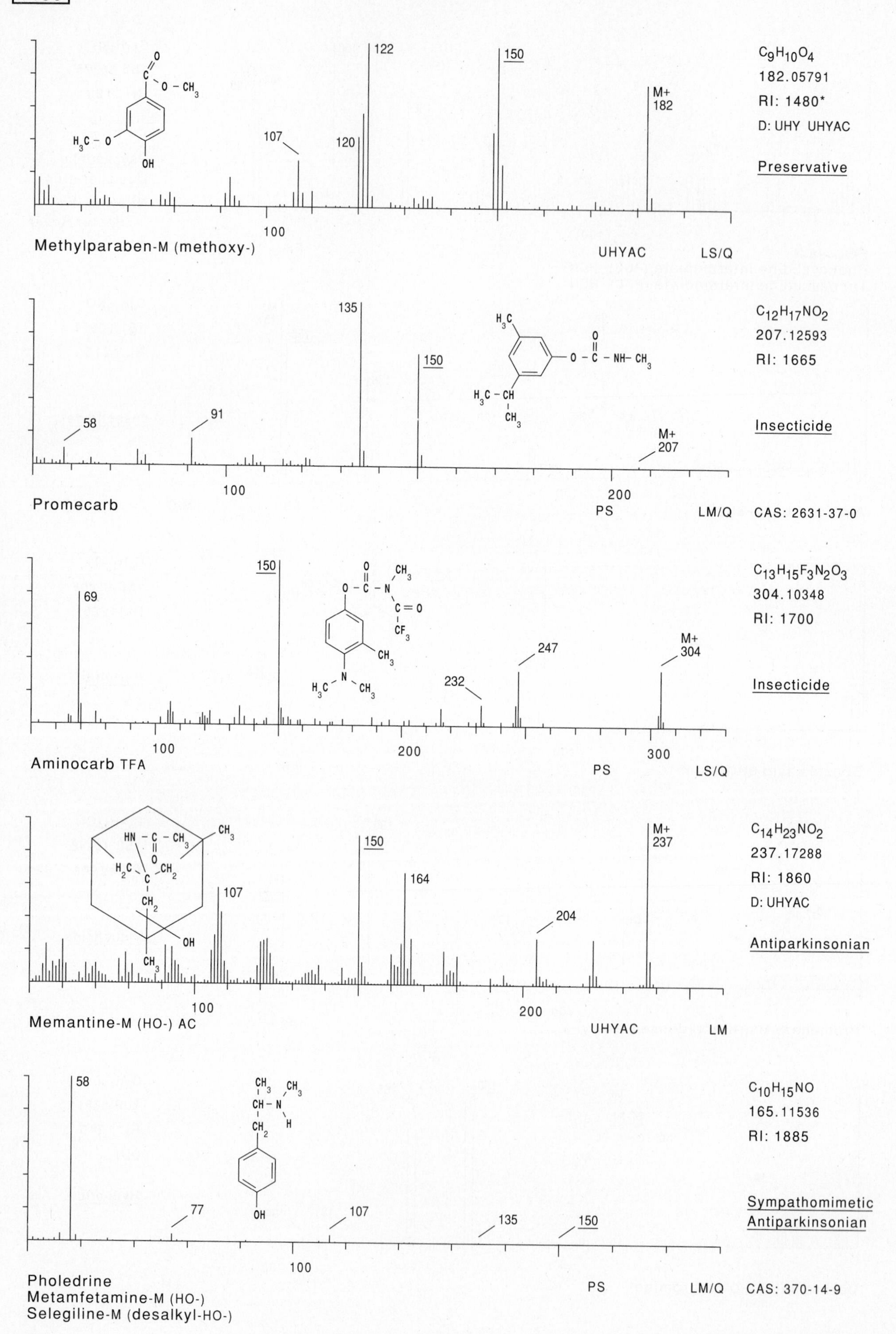

122
150
M+
182
107
120
100
Methylparaben-M (methoxy-)
UHYAC
LS/Q
C9H10O4
182.05791
RI: 1480*
D: UHY UHYAC
Preservative
135
150
58
91
M+
207
100
200
Promecarb
PS
LM/Q
C12H17NO2
207.12593
RI: 1665
Insecticide
CAS: 2631-37-0
150
69
247
M+
304
232
100
200
300
Aminocarb TFA
PS
LS/Q
C13H15F3N2O3
304.10348
RI: 1700
Insecticide
M+
237
150
164
107
204
100
200
Memantine-M (HO-) AC
UHYAC
LM
C14H23NO2
237.17288
RI: 1860
D: UHYAC
Antiparkinsonian
58
77
107
135
150
100
Pholedrine
Metamfetamine-M (HO-)
Selegiline-M (desalkyl-HO-)
PS
LM/Q
C10H15NO
165.11536
RI: 1885
Sympathomimetic
Antiparkinsonian
CAS: 370-14-9

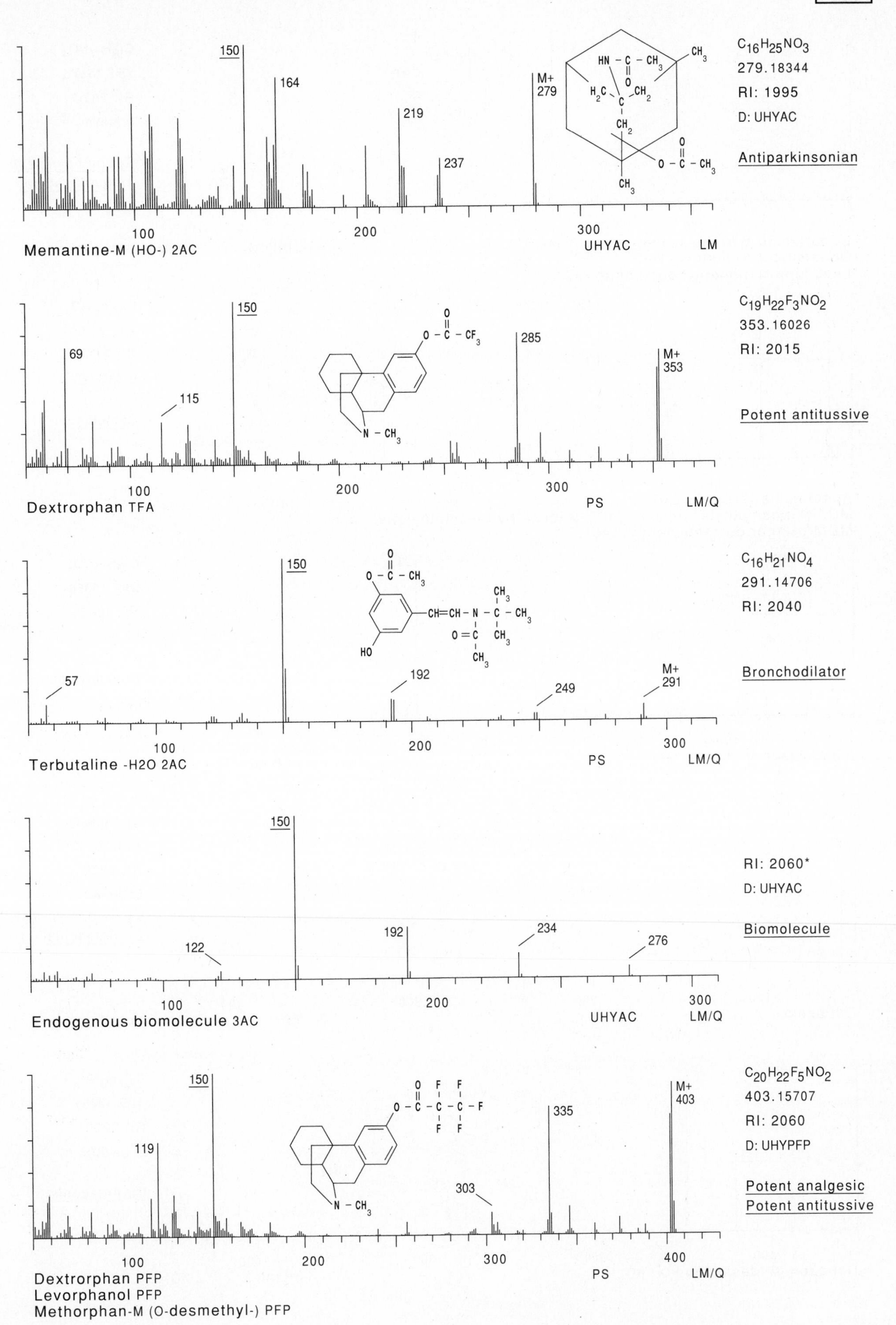
150
164
219
237
M+
279
100
200
300
Memantine-M (HO-) 2AC
UHYAC
LM
C16H25NO3
279.18344
RI: 1995
D: UHYAC
Antiparkinsonian
150
69
115
285
M+
353
Dextrorphan TFA
PS
LM/Q
C19H22F3NO2
353.16026
RI: 2015
Potent antitussive
150
57
192
249
M+
291
Terbutaline -H2O 2AC
PS
LM/Q
C16H21NO4
291.14706
RI: 2040
Bronchodilator
150
122
192
234
276
Endogenous biomolecule 3AC
UHYAC
LM/Q
RI: 2060*
D: UHYAC
Biomolecule
150
119
303
335
M+
403
400
Dextrorphan PFP
Levorphanol PFP
Methorphan-M (O-desmethyl-) PFP
PS
LM/Q
C20H22F5NO2
403.15707
RI: 2060
D: UHYPFP
Potent analgesic
Potent antitussive

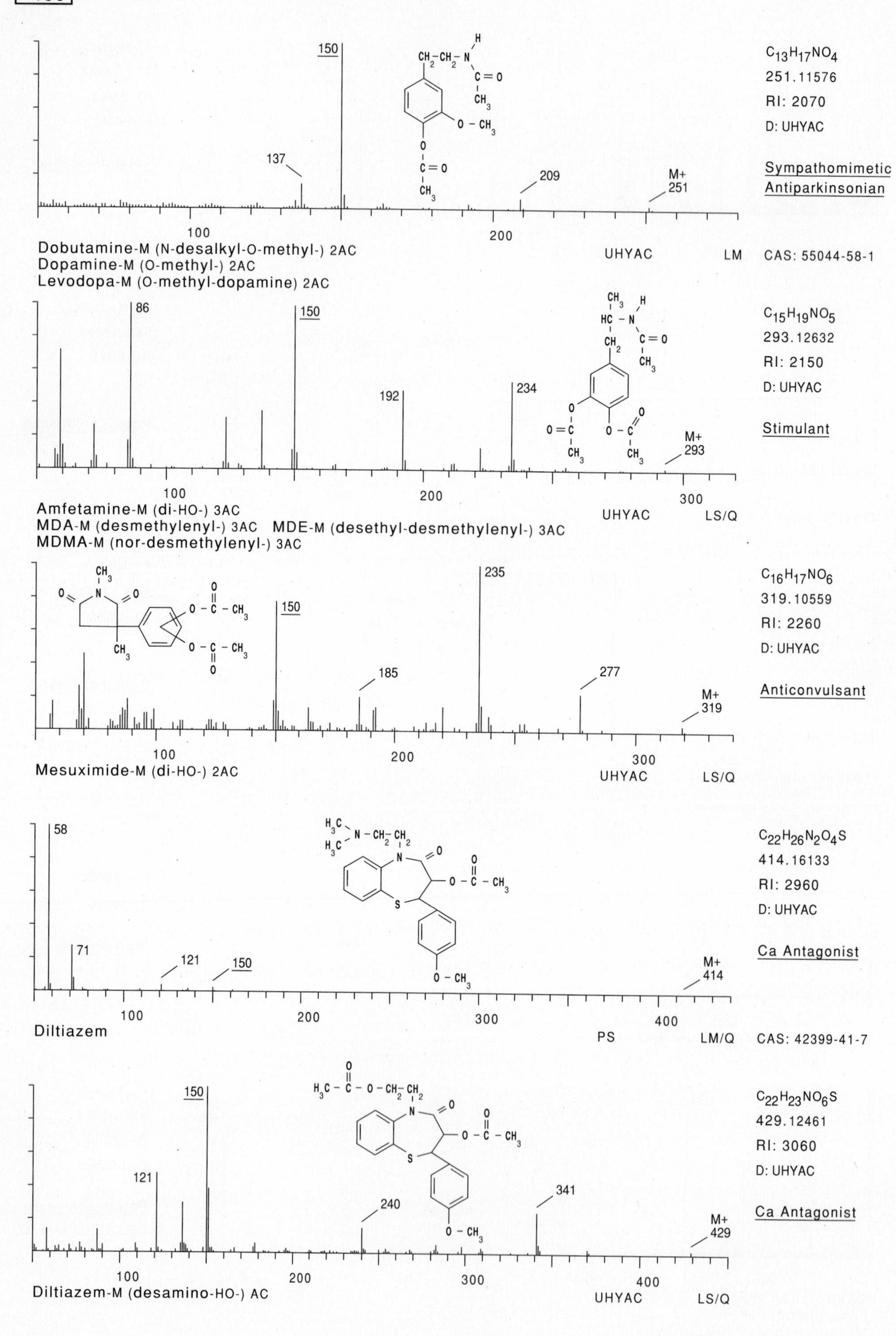
150
$C_{13}H_{17}NO_4$
251.11576
RI: 2070
D: UHYAC
Sympathomimetic
Antiparkinsonian
Dobutamine-M (N-desalkyl-O-methyl-) 2AC
Dopamine-M (O-methyl-) 2AC
Levodopa-M (O-methyl-dopamine) 2AC
UHYAC
LM
CAS: 55044-58-1
$C_{15}H_{19}NO_5$
293.12632
RI: 2150
D: UHYAC
Stimulant
Amfetamine-M (di-HO-) 3AC
MDA-M (desmethylenyl-) 3AC MDE-M (desethyl-desmethylenyl-) 3AC
MDMA-M (nor-desmethylenyl-) 3AC
UHYAC
LS/Q
$C_{16}H_{17}NO_6$
319.10559
RI: 2260
D: UHYAC
Anticonvulsant
Mesuximide-M (di-HO-) 2AC
UHYAC
LS/Q
$C_{22}H_{26}N_2O_4S$
414.16133
RI: 2960
D: UHYAC
Ca Antagonist
Diltiazem
PS
LM/Q
CAS: 42399-41-7
$C_{22}H_{23}NO_6S$
429.12461
RI: 3060
D: UHYAC
Ca Antagonist
Diltiazem-M (desamino-HO-) AC
UHYAC
LS/Q

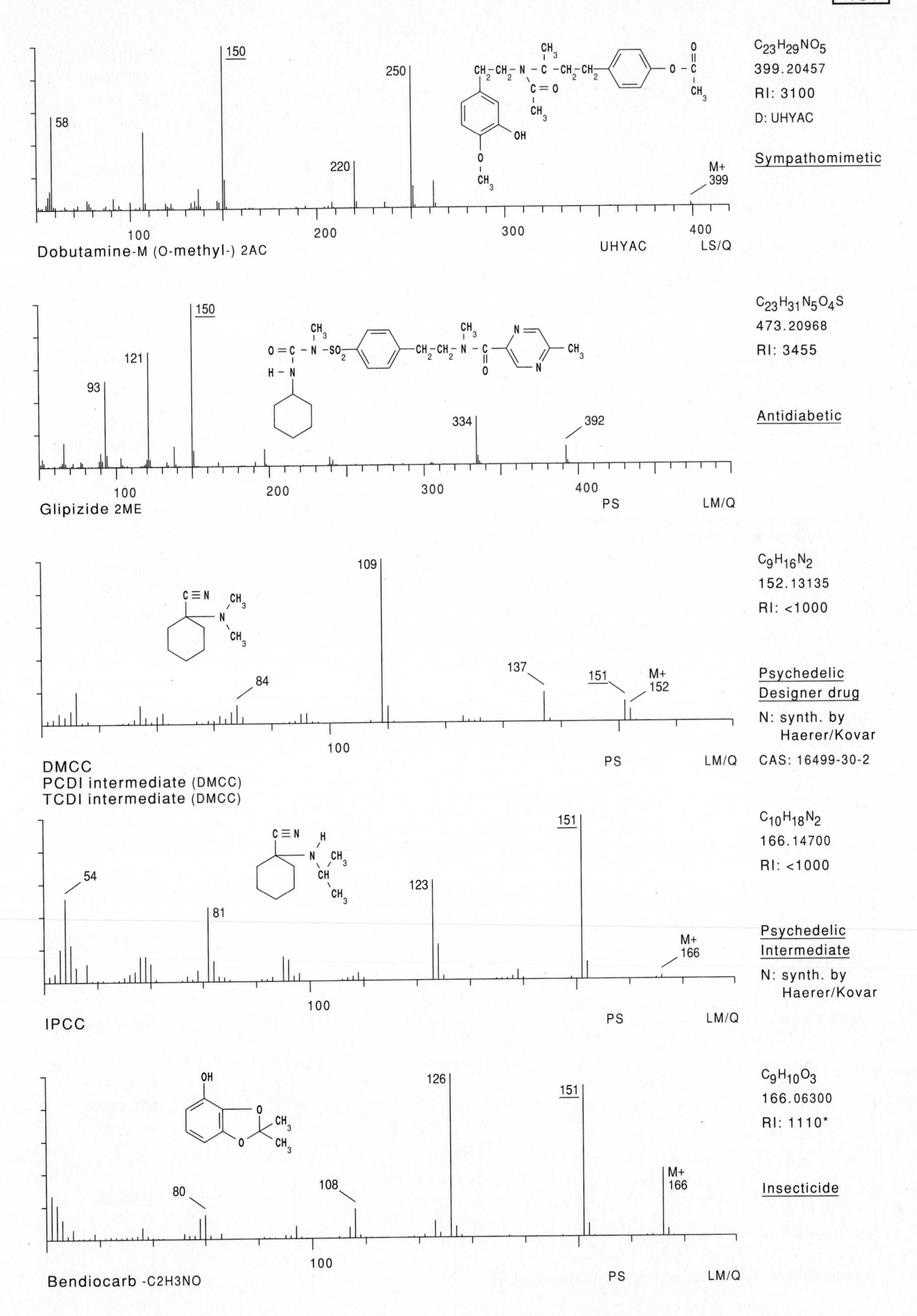
C23H29NO5
399.20457
RI: 3100
D: UHYAC
Sympathomimetic
M+ 399
UHYAC
LS/Q
Dobutamine-M (O-methyl-) 2AC
C23H31N5O4S
473.20968
RI: 3455
Antidiabetic
PS
LM/Q
Glipizide 2ME
C9H16N2
152.13135
RI: <1000
Psychedelic
Designer drug
N: synth. by Haerer/Kovar
CAS: 16499-30-2
M+ 152
PS
LM/Q
DMCC
PCDI intermediate (DMCC)
TCDI intermediate (DMCC)
C10H18N2
166.14700
RI: <1000
Psychedelic
Intermediate
N: synth. by Haerer/Kovar
M+ 166
PS
LM/Q
IPCC
C9H10O3
166.06300
RI: 1110*
Insecticide
M+ 166
PS
LM/Q
Bendiocarb -C2H3NO

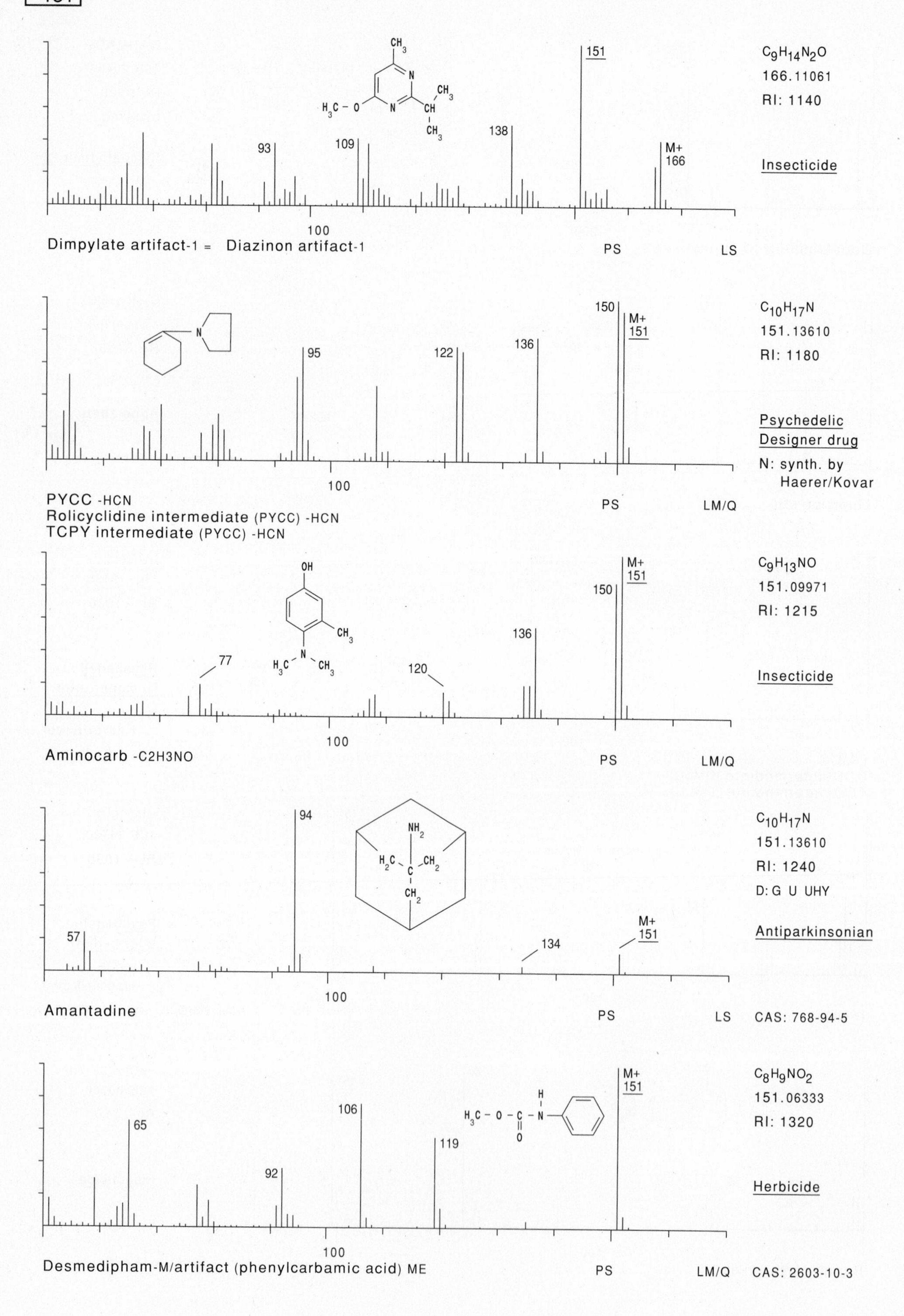
151
M+
166
93
109
138
100
PS
LS
$C_9H_{14}N_2O$
166.11061
RI: 1140
Insecticide
Dimpylate artifact-1 = Diazinon artifact-1
150
M+
151
95
122
136
100
PS
LM/Q
$C_{10}H_{17}N$
151.13610
RI: 1180
Psychedelic
Designer drug
N: synth. by
Haerer/Kovar
PYCC -HCN
Rolicyclidine intermediate (PYCC) -HCN
TCPY intermediate (PYCC) -HCN
M+
151
150
136
77
120
100
PS
LM/Q
$C_9H_{13}NO$
151.09971
RI: 1215
Insecticide
Aminocarb -C2H3NO
94
57
134
M+
151
100
PS
LS
$C_{10}H_{17}N$
151.13610
RI: 1240
D: G U UHY
Antiparkinsonian
Amantadine
CAS: 768-94-5
M+
151
106
65
119
92
100
PS
LM/Q
$C_8H_9NO_2$
151.06333
RI: 1320
Herbicide
Desmedipham-M/artifact (phenylcarbamic acid) ME
CAS: 2603-10-3

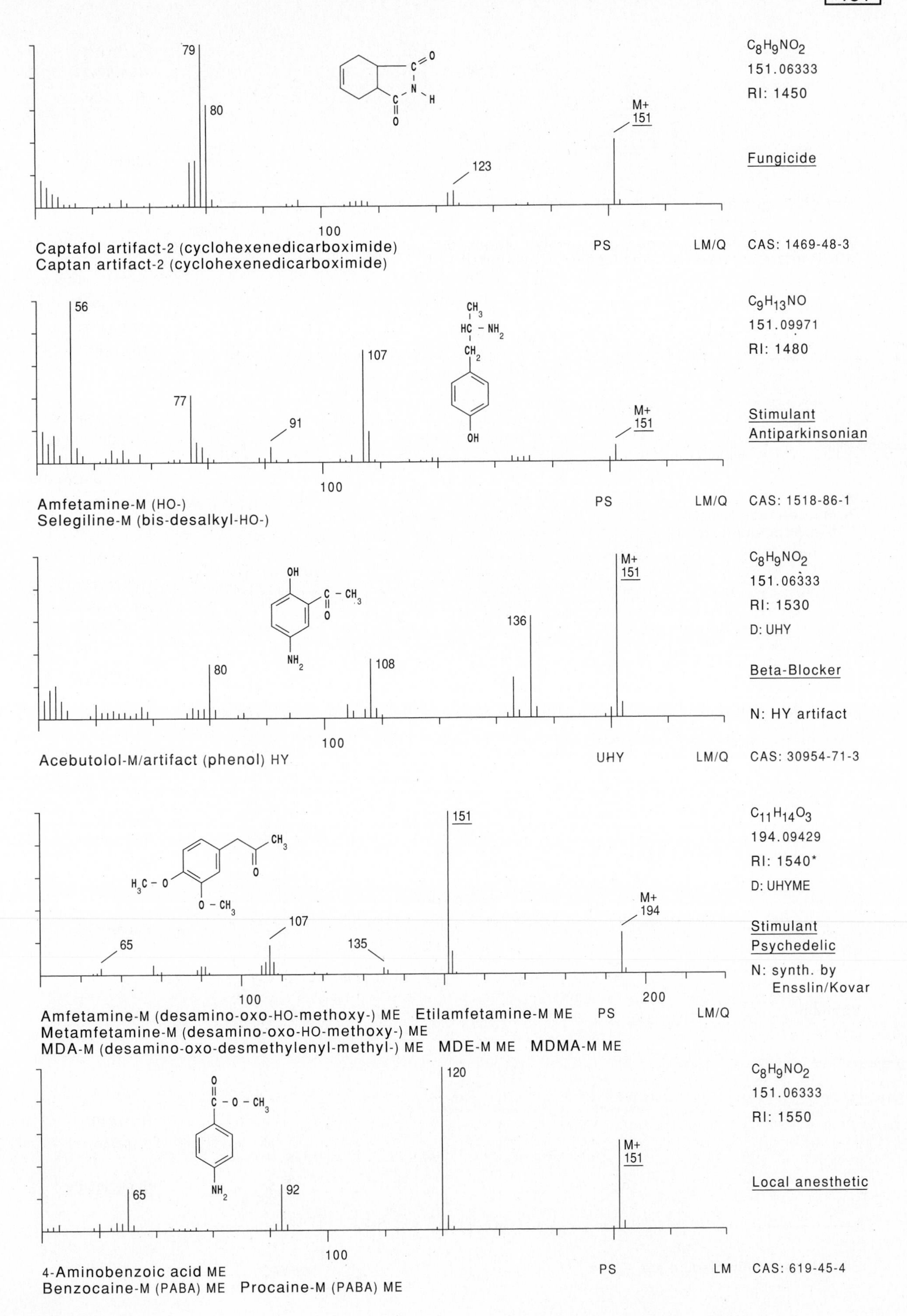
79
80
123
M+
151
100
$C_8H_9NO_2$
151.06333
RI: 1450
Fungicide
Captafol artifact-2 (cyclohexenedicarboximide)
Captan artifact-2 (cyclohexenedicarboximide)
PS
LM/Q
CAS: 1469-48-3
56
77
91
107
M+
151
100
$C_9H_{13}NO$
151.09971
RI: 1480
Stimulant
Antiparkinsonian
Amfetamine-M (HO-)
Selegiline-M (bis-desalkyl-HO-)
PS
LM/Q
CAS: 1518-86-1
80
108
136
M+
151
100
$C_8H_9NO_2$
151.06333
RI: 1530
D: UHY
Beta-Blocker
N: HY artifact
Acebutolol-M/artifact (phenol) HY
UHY
LM/Q
CAS: 30954-71-3
65
107
135
151
M+
194
100
200
$C_{11}H_{14}O_3$
194.09429
RI: 1540*
D: UHYME
Stimulant
Psychedelic
N: synth. by Ensslin/Kovar
Amfetamine-M (desamino-oxo-HO-methoxy-) ME
Etilamfetamine-M ME
PS
LM/Q
Metamfetamine-M (desamino-oxo-HO-methoxy-) ME
MDA-M (desamino-oxo-desmethylenyl-methyl-) ME
MDE-M ME
MDMA-M ME
65
92
120
M+
151
100
$C_8H_9NO_2$
151.06333
RI: 1550
Local anesthetic
4-Aminobenzoic acid ME
Benzocaine-M (PABA) ME
Procaine-M (PABA) ME
PS
LM
CAS: 619-45-4

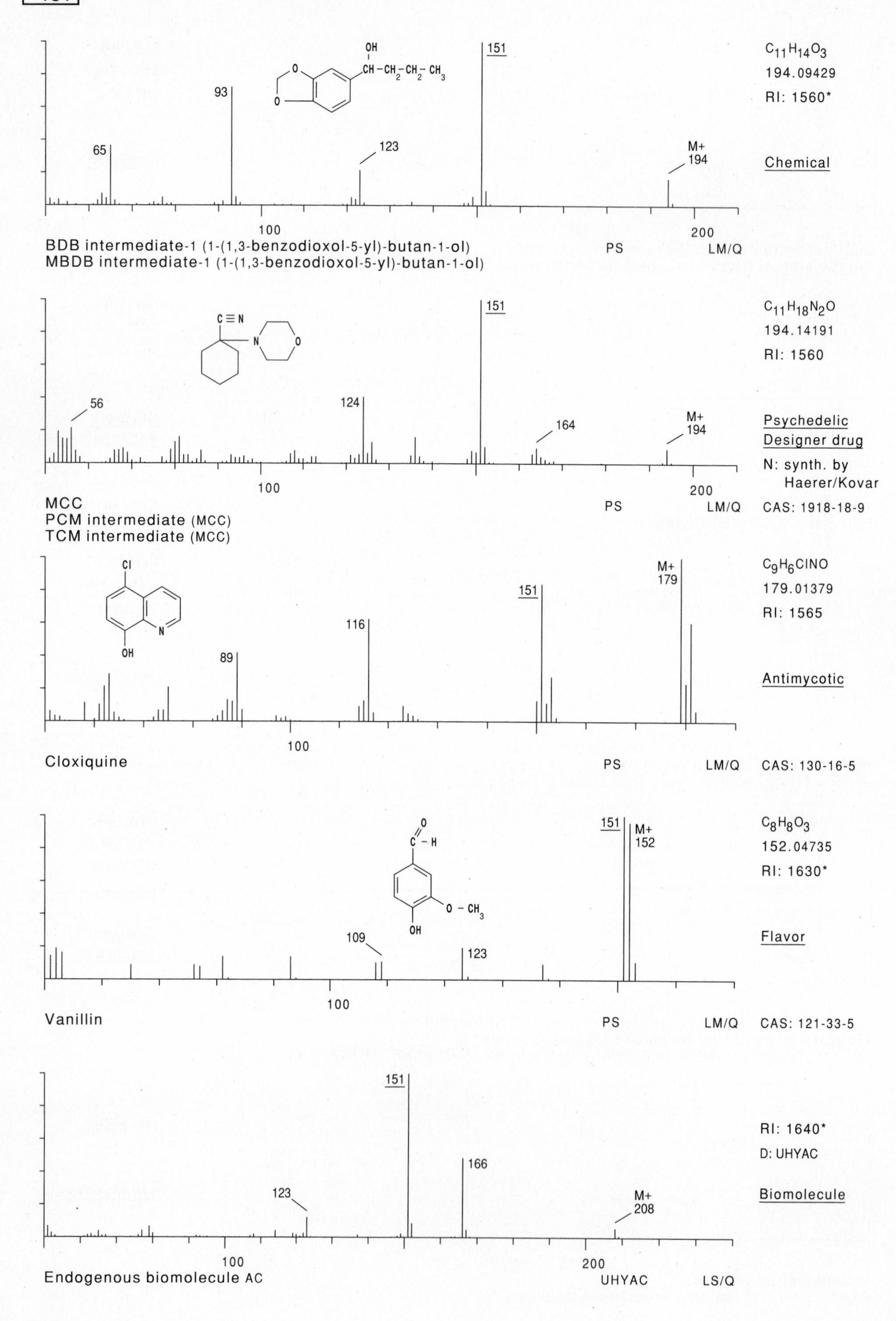
151
93
65
123
M+
194
100
200
$C_{11}H_{14}O_3$
194.09429
RI: 1560*
Chemical
BDB intermediate-1 (1-(1,3-benzodioxol-5-yl)-butan-1-ol)
MBDB intermediate-1 (1-(1,3-benzodioxol-5-yl)-butan-1-ol)
PS
LM/Q
151
56
124
164
M+
194
100
200
$C_{11}H_{18}N_2O$
194.14191
RI: 1560
Psychedelic
Designer drug
N: synth. by
Haerer/Kovar
MCC
PCM intermediate (MCC)
TCM intermediate (MCC)
PS
LM/Q
CAS: 1918-18-9
M+
179
151
116
89
100
C_9H_6ClNO
179.01379
RI: 1565
Antimycotic
Cloxiquine
PS
LM/Q
CAS: 130-16-5
151
M+
152
109
123
100
$C_8H_8O_3$
152.04735
RI: 1630*
Flavor
Vanillin
PS
LM/Q
CAS: 121-33-5
151
166
123
M+
208
100
200
RI: 1640*
D: UHYAC
Biomolecule
Endogenous biomolecule AC
UHYAC
LS/Q

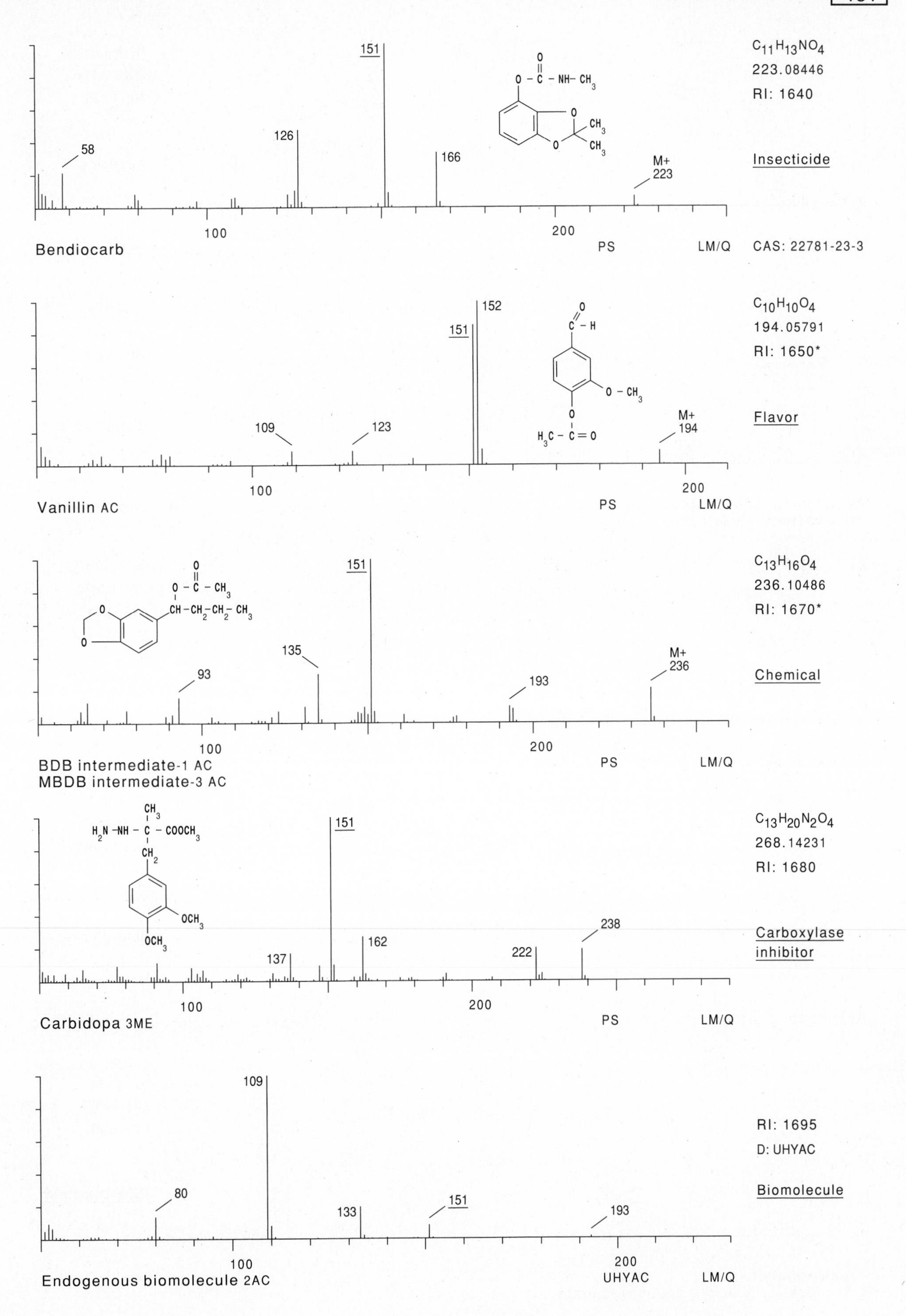

$C_{11}H_{13}NO_4$
223.08446
RI: 1640
Insecticide
CAS: 22781-23-3

$C_{10}H_{10}O_4$
194.05791
RI: 1650*
Flavor

$C_{13}H_{16}O_4$
236.10486
RI: 1670*
Chemical

$C_{13}H_{20}N_2O_4$
268.14231
RI: 1680
Carboxylase inhibitor

RI: 1695
D: UHYAC
Biomolecule

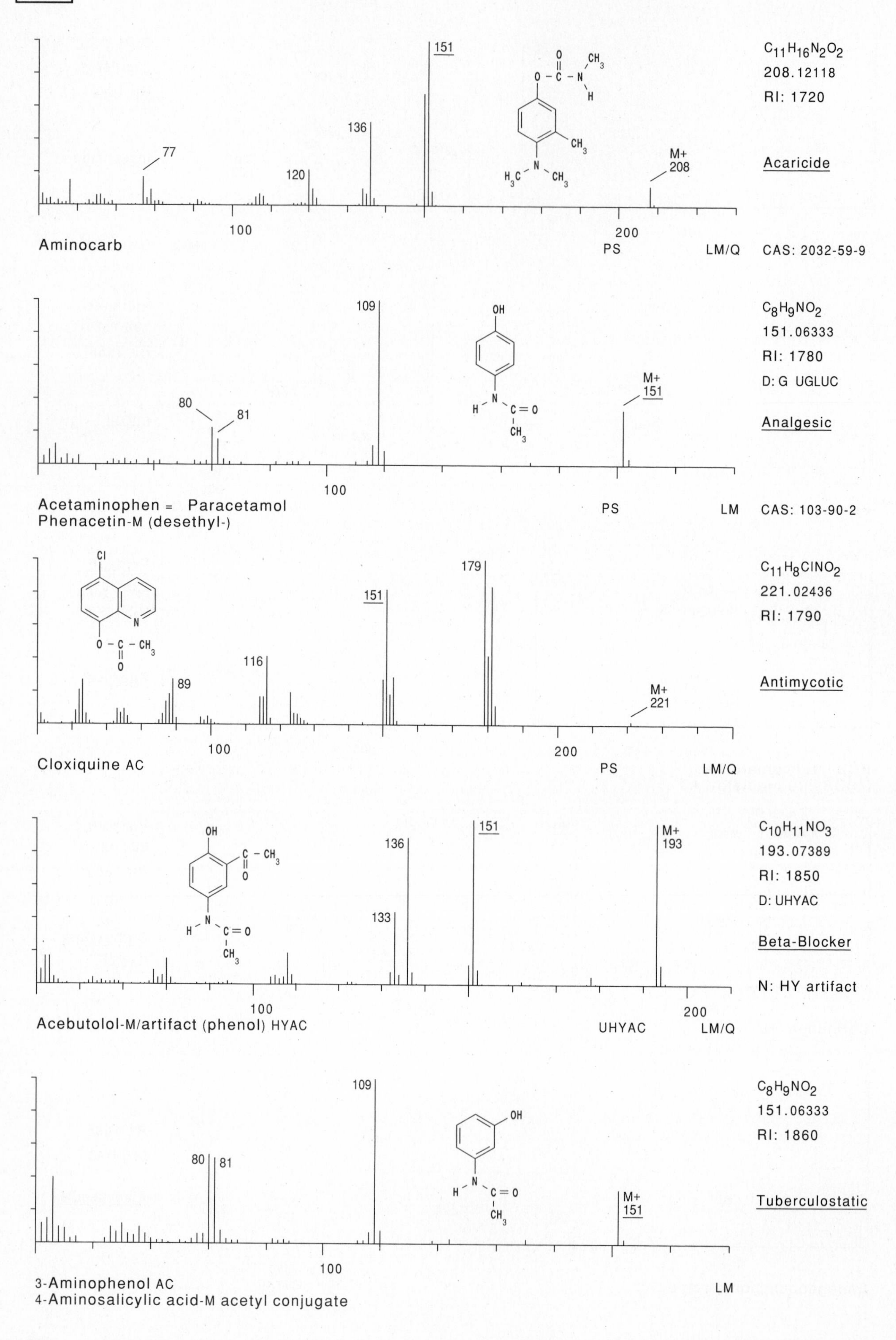

151
136
120
77
M+
208
100
200
Aminocarb
PS
LM/Q
$C_{11}H_{16}N_2O_2$
208.12118
RI: 1720
Acaricide
CAS: 2032-59-9
109
80
81
M+
151
100
Acetaminophen = Paracetamol
Phenacetin-M (desethyl-)
PS
LM
$C_8H_9NO_2$
151.06333
RI: 1780
D: G UGLUC
Analgesic
CAS: 103-90-2
179
151
116
89
M+
221
100
200
Cloxiquine AC
PS
LM/Q
$C_{11}H_8ClNO_2$
221.02436
RI: 1790
Antimycotic
151
136
133
M+
193
100
200
Acebutolol-M/artifact (phenol) HYAC
UHYAC
LM/Q
$C_{10}H_{11}NO_3$
193.07389
RI: 1850
D: UHYAC
Beta-Blocker
N: HY artifact
109
80
81
M+
151
100
3-Aminophenol AC
4-Aminosalicylic acid-M acetyl conjugate
LM
$C_8H_9NO_2$
151.06333
RI: 1860
Tuberculostatic

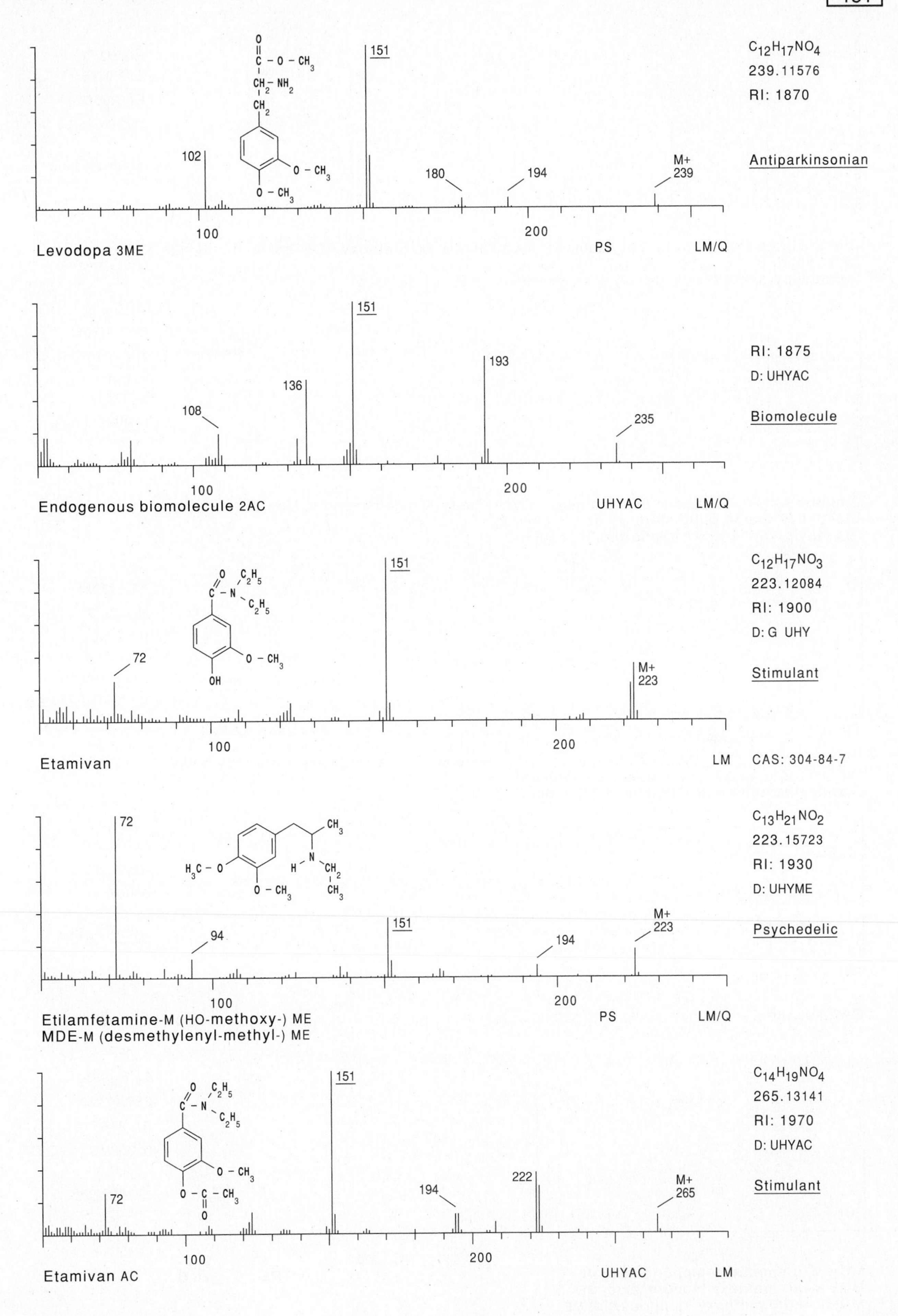

151
102
180
194
M+
239
100
200
Levodopa 3ME
PS
LM/Q
$C_{12}H_{17}NO_4$
239.11576
RI: 1870
Antiparkinsonian
151
193
136
108
235
100
200
Endogenous biomolecule 2AC
UHYAC
LM/Q
RI: 1875
D: UHYAC
Biomolecule
151
72
M+
223
100
200
Etamivan
LM
$C_{12}H_{17}NO_3$
223.12084
RI: 1900
D: G UHY
Stimulant
CAS: 304-84-7
72
151
94
194
M+
223
100
200
Etilamfetamine-M (HO-methoxy-) ME
MDE-M (desmethylenyl-methyl-) ME
PS
LM/Q
$C_{13}H_{21}NO_2$
223.15723
RI: 1930
D: UHYME
Psychedelic
151
72
194
222
M+
265
100
200
Etamivan AC
UHYAC
LM
$C_{14}H_{19}NO_4$
265.13141
RI: 1970
D: UHYAC
Stimulant

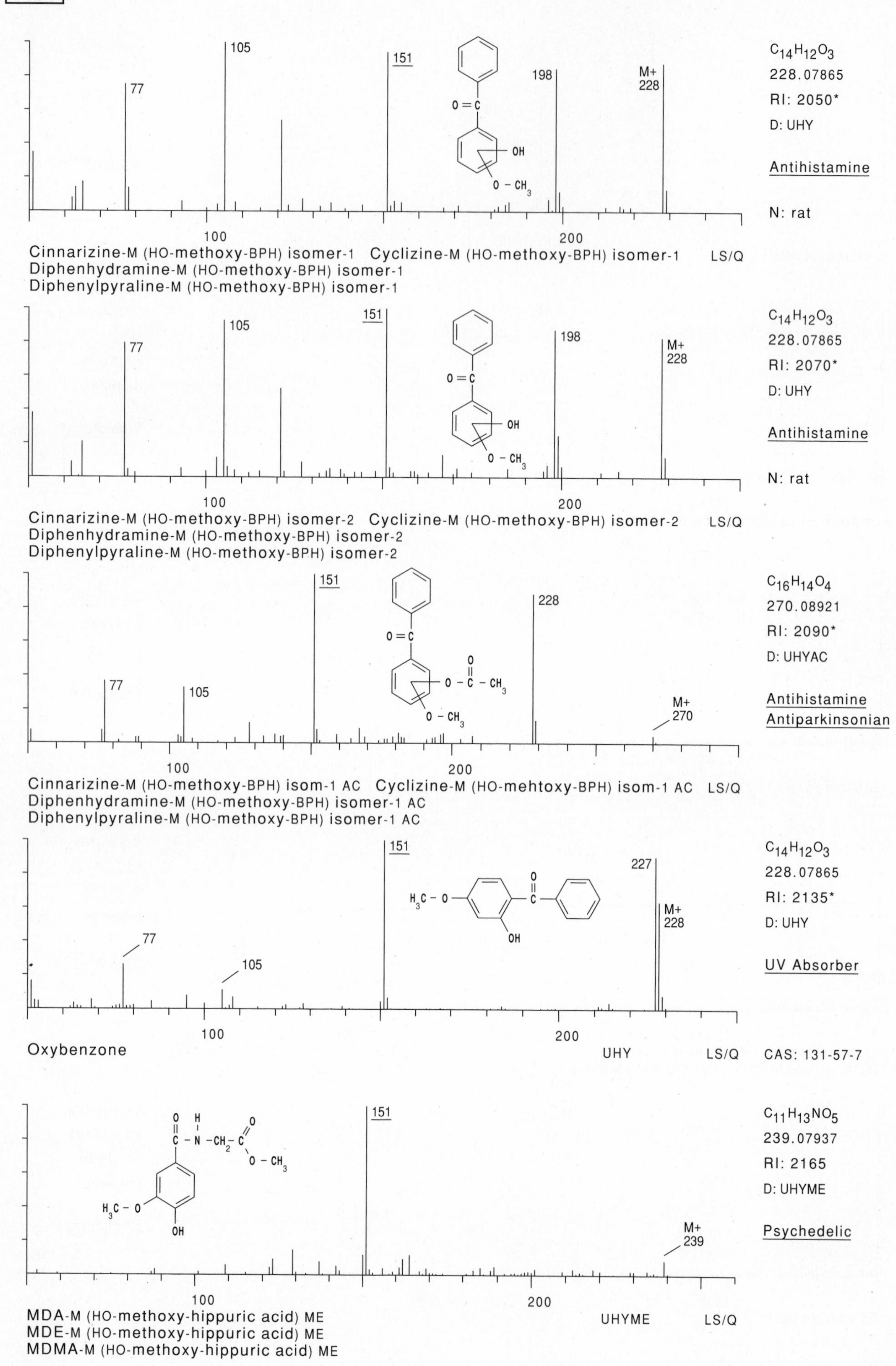

105
151
77
198
M+
228
O = C
OH
O - CH3
100
200
Cinnarizine-M (HO-methoxy-BPH) isomer-1 Cyclizine-M (HO-methoxy-BPH) isomer-1 LS/Q
Diphenhydramine-M (HO-methoxy-BPH) isomer-1
Diphenylpyraline-M (HO-methoxy-BPH) isomer-1
$C_{14}H_{12}O_3$
228.07865
RI: 2050*
D: UHY
Antihistamine
N: rat
151
105
77
198
M+
228
100
200
Cinnarizine-M (HO-methoxy-BPH) isomer-2 Cyclizine-M (HO-methoxy-BPH) isomer-2 LS/Q
Diphenhydramine-M (HO-methoxy-BPH) isomer-2
Diphenylpyraline-M (HO-methoxy-BPH) isomer-2
$C_{14}H_{12}O_3$
228.07865
RI: 2070*
D: UHY
Antihistamine
N: rat
151
228
77
105
O - C - CH3
M+
270
100
200
Cinnarizine-M (HO-methoxy-BPH) isom-1 AC Cyclizine-M (HO-mehtoxy-BPH) isom-1 AC LS/Q
Diphenhydramine-M (HO-methoxy-BPH) isomer-1 AC
Diphenylpyraline-M (HO-methoxy-BPH) isomer-1 AC
$C_{16}H_{14}O_4$
270.08921
RI: 2090*
D: UHYAC
Antihistamine
Antiparkinsonian
151
227
M+
228
77
105
H3C - O
OH
100
200
Oxybenzone
UHY
LS/Q
$C_{14}H_{12}O_3$
228.07865
RI: 2135*
D: UHY
UV Absorber
CAS: 131-57-7
151
C - N - CH2 - C
O - CH3
H3C - O
OH
M+
239
100
200
MDA-M (HO-methoxy-hippuric acid) ME
MDE-M (HO-methoxy-hippuric acid) ME
MDMA-M (HO-methoxy-hippuric acid) ME
UHYME
LS/Q
$C_{11}H_{13}NO_5$
239.07937
RI: 2165
D: UHYME
Psychedelic

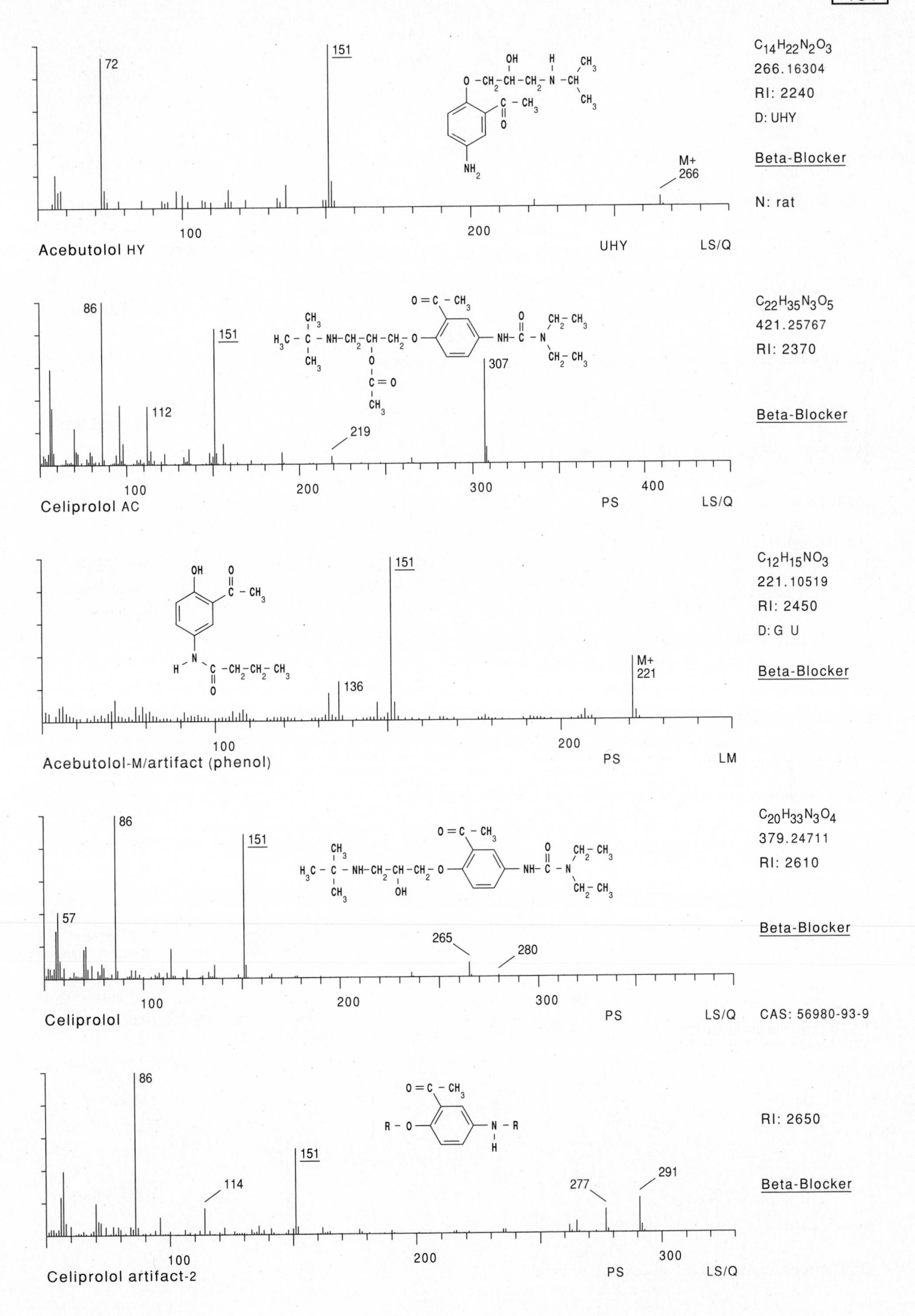

Acebutolol HY
UHY LS/Q
$C_{14}H_{22}N_2O_3$
266.16304
RI: 2240
D: UHY
Beta-Blocker
N: rat
Celiprolol AC
PS LS/Q
$C_{22}H_{35}N_3O_5$
421.25767
RI: 2370
Beta-Blocker
Acebutolol-M/artifact (phenol)
PS LM
$C_{12}H_{15}NO_3$
221.10519
RI: 2450
D: G U
Beta-Blocker
Celiprolol
PS LS/Q
$C_{20}H_{33}N_3O_4$
379.24711
RI: 2610
Beta-Blocker
CAS: 56980-93-9
Celiprolol artifact-2
PS LS/Q
RI: 2650
Beta-Blocker

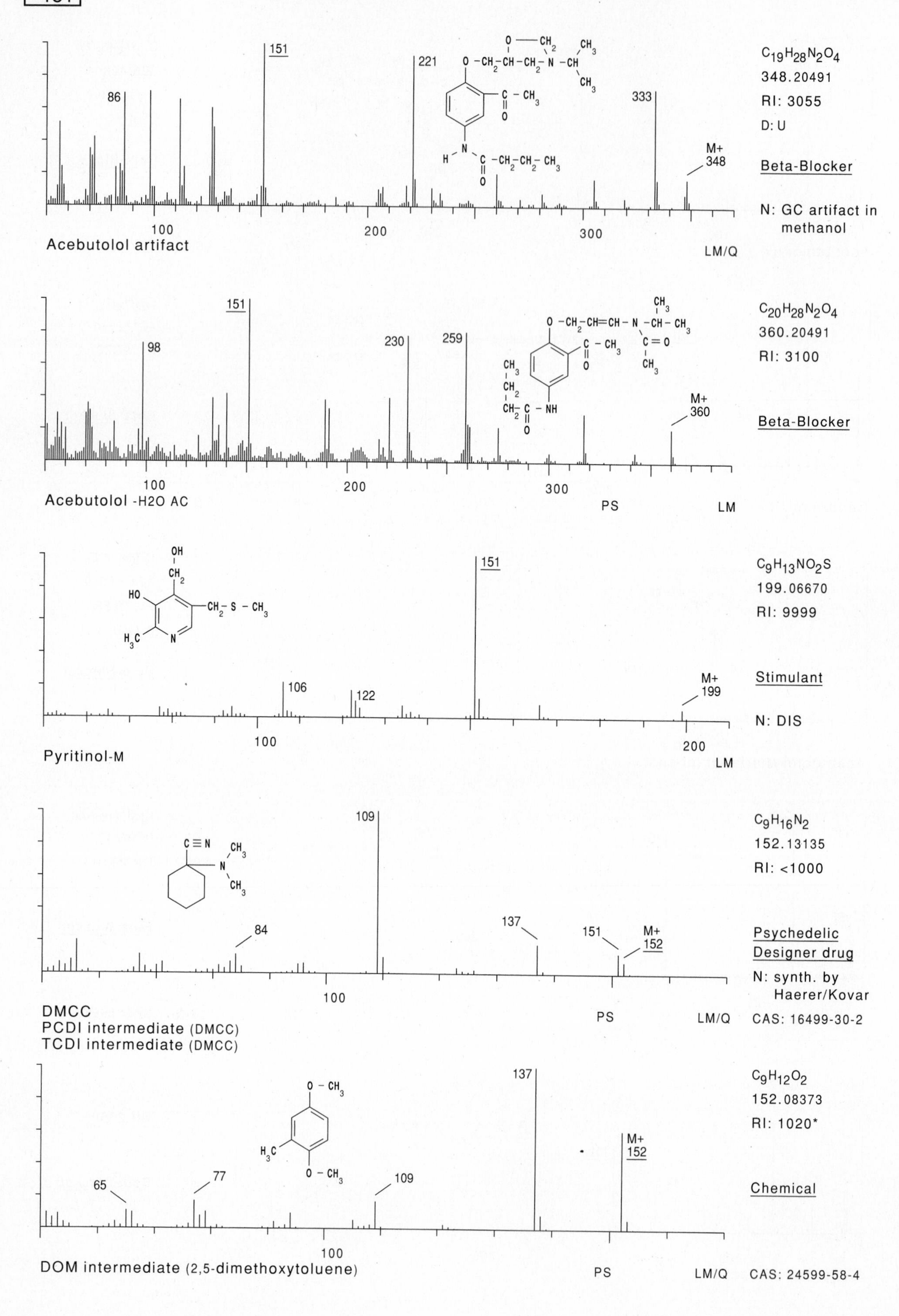
151
221
86
333
M+
348
100
200
300
Acebutolol artifact
LM/Q
C19H28N2O4
348.20491
RI: 3055
D: U
Beta-Blocker
N: GC artifact in
methanol
151
98
230
259
M+
360
100
200
300
Acebutolol -H2O AC
PS
LM
C20H28N2O4
360.20491
RI: 3100
Beta-Blocker
151
106
122
M+
199
100
200
Pyritinol-M
LM
C9H13NO2S
199.06670
RI: 9999
Stimulant
N: DIS
109
84
137
151
M+
152
100
DMCC
PCDI intermediate (DMCC)
TCDI intermediate (DMCC)
PS
LM/Q
C9H16N2
152.13135
RI: <1000
Psychedelic
Designer drug
N: synth. by
Haerer/Kovar
CAS: 16499-30-2
137
M+
152
65
77
109
100
DOM intermediate (2,5-dimethoxytoluene)
PS
LM/Q
C9H12O2
152.08373
RI: 1020*
Chemical
CAS: 24599-58-4

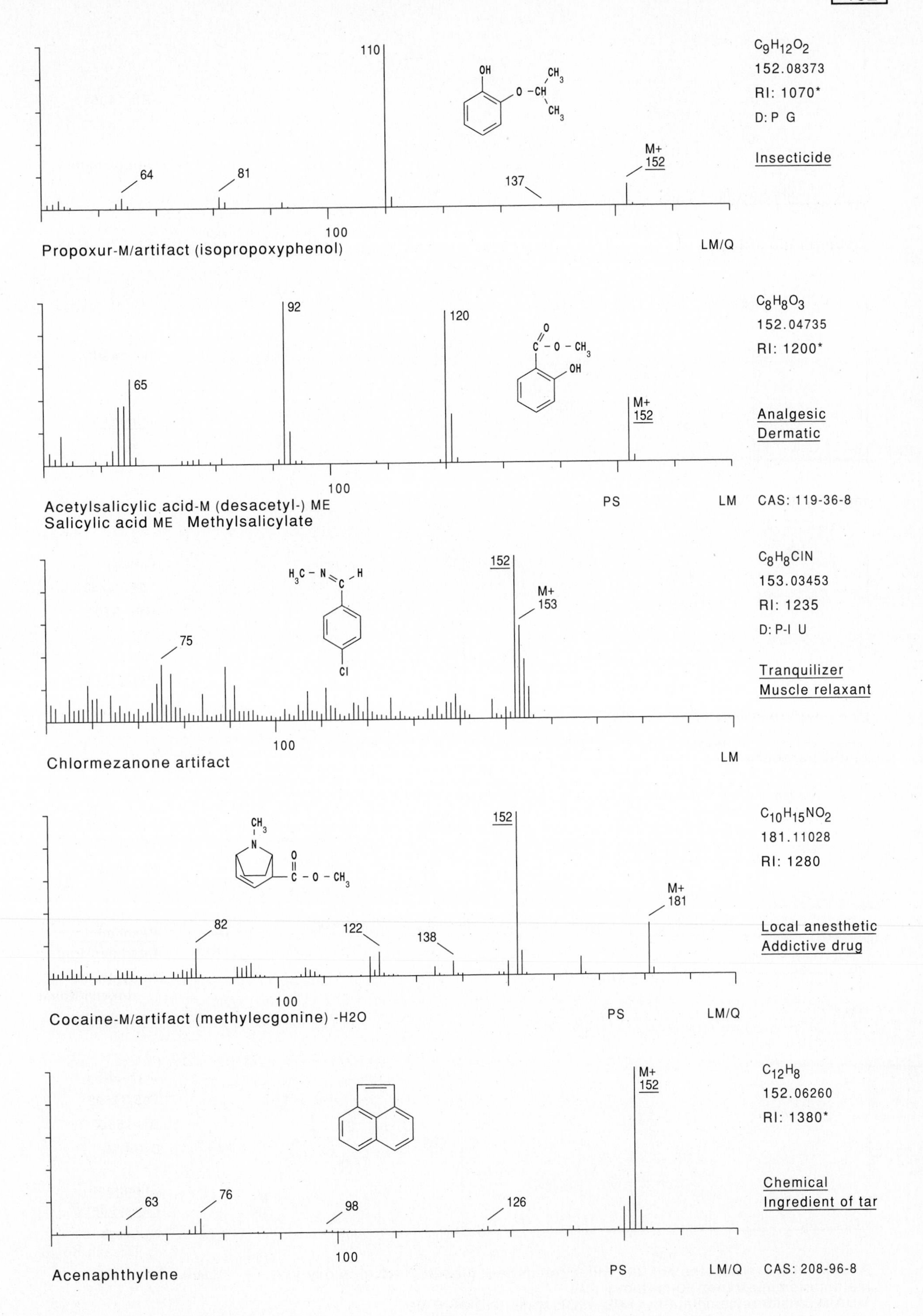
110
64
81
137
M+
152
100
LM/Q
$C_9H_{12}O_2$
152.08373
RI: 1070*
D: P G
Insecticide
Propoxur-M/artifact (isopropoxyphenol)
92
120
65
M+
152
100
PS
LM
$C_8H_8O_3$
152.04735
RI: 1200*
Analgesic
Dermatic
CAS: 119-36-8
Acetylsalicylic acid-M (desacetyl-) ME
Salicylic acid ME Methylsalicylate
152
M+
153
75
Cl
100
LM
C_8H_8ClN
153.03453
RI: 1235
D: P-I U
Tranquilizer
Muscle relaxant
Chlormezanone artifact
152
82
122
138
M+
181
100
PS
LM/Q
$C_{10}H_{15}NO_2$
181.11028
RI: 1280
Local anesthetic
Addictive drug
Cocaine-M/artifact (methylecgonine) -H2O
M+
152
63
76
98
126
100
PS
LM/Q
$C_{12}H_8$
152.06260
RI: 1380*
Chemical
Ingredient of tar
CAS: 208-96-8
Acenaphthylene

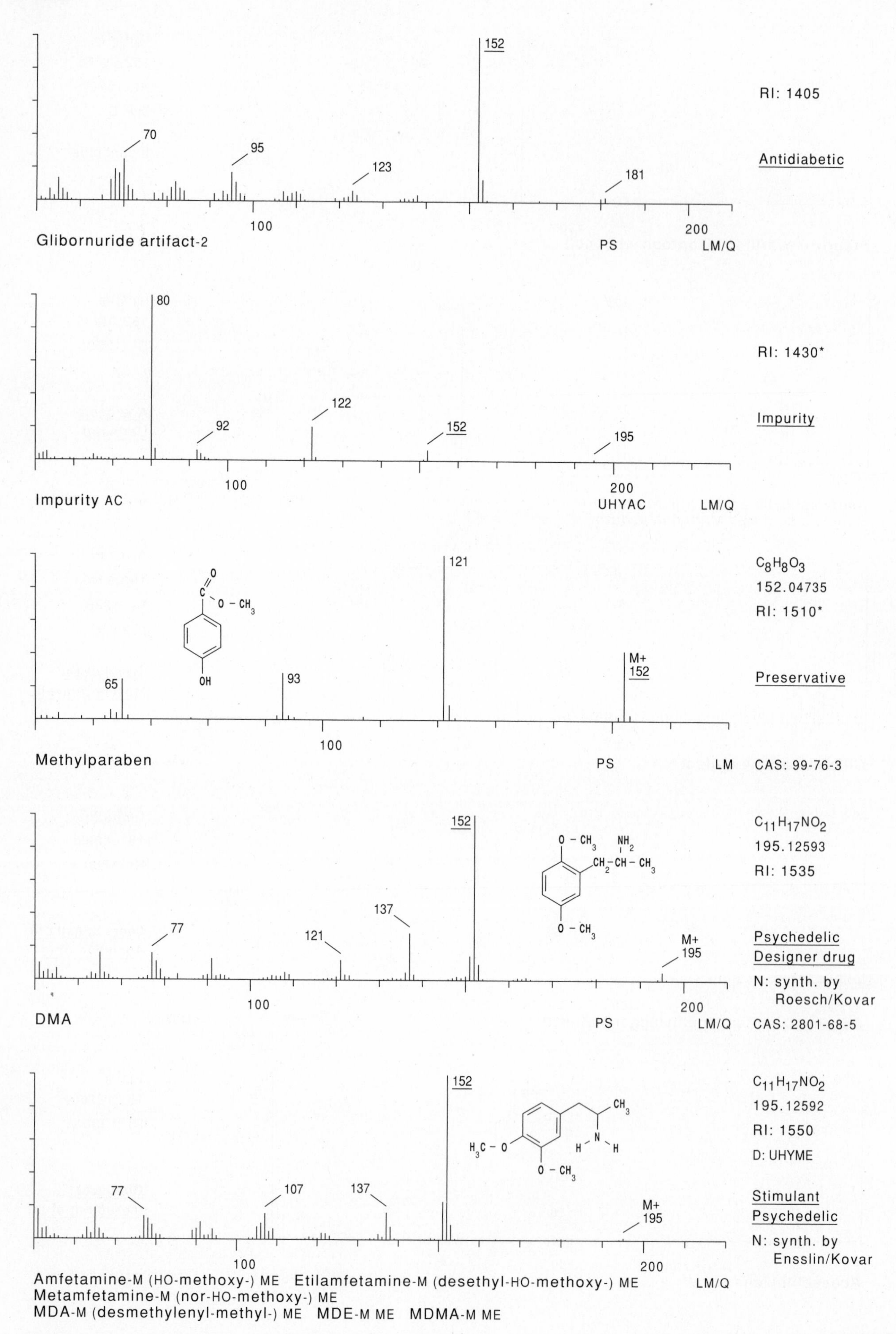
152
70
95
123
181
100
200
RI: 1405
Antidiabetic
Glibornuride artifact-2
PS
LM/Q
80
122
92
152
195
100
200
RI: 1430*
Impurity
Impurity AC
UHYAC
LM/Q
121
C8H8O3
152.04735
RI: 1510*
65
93
M+
152
Preservative
100
Methylparaben
PS
LM
CAS: 99-76-3
152
C11H17NO2
195.12593
RI: 1535
137
77
121
M+
195
Psychedelic
Designer drug
N: synth. by
Roesch/Kovar
100
200
DMA
PS
LM/Q
CAS: 2801-68-5
152
C11H17NO2
195.12592
RI: 1550
D: UHYME
77
107
137
M+
195
Stimulant
Psychedelic
N: synth. by
Ensslin/Kovar
100
200
LM/Q
Amfetamine-M (HO-methoxy-) ME Etilamfetamine-M (desethyl-HO-methoxy-) ME
Metamfetamine-M (nor-HO-methoxy-) ME
MDA-M (desmethylenyl-methyl-) ME MDE-M ME MDMA-M ME

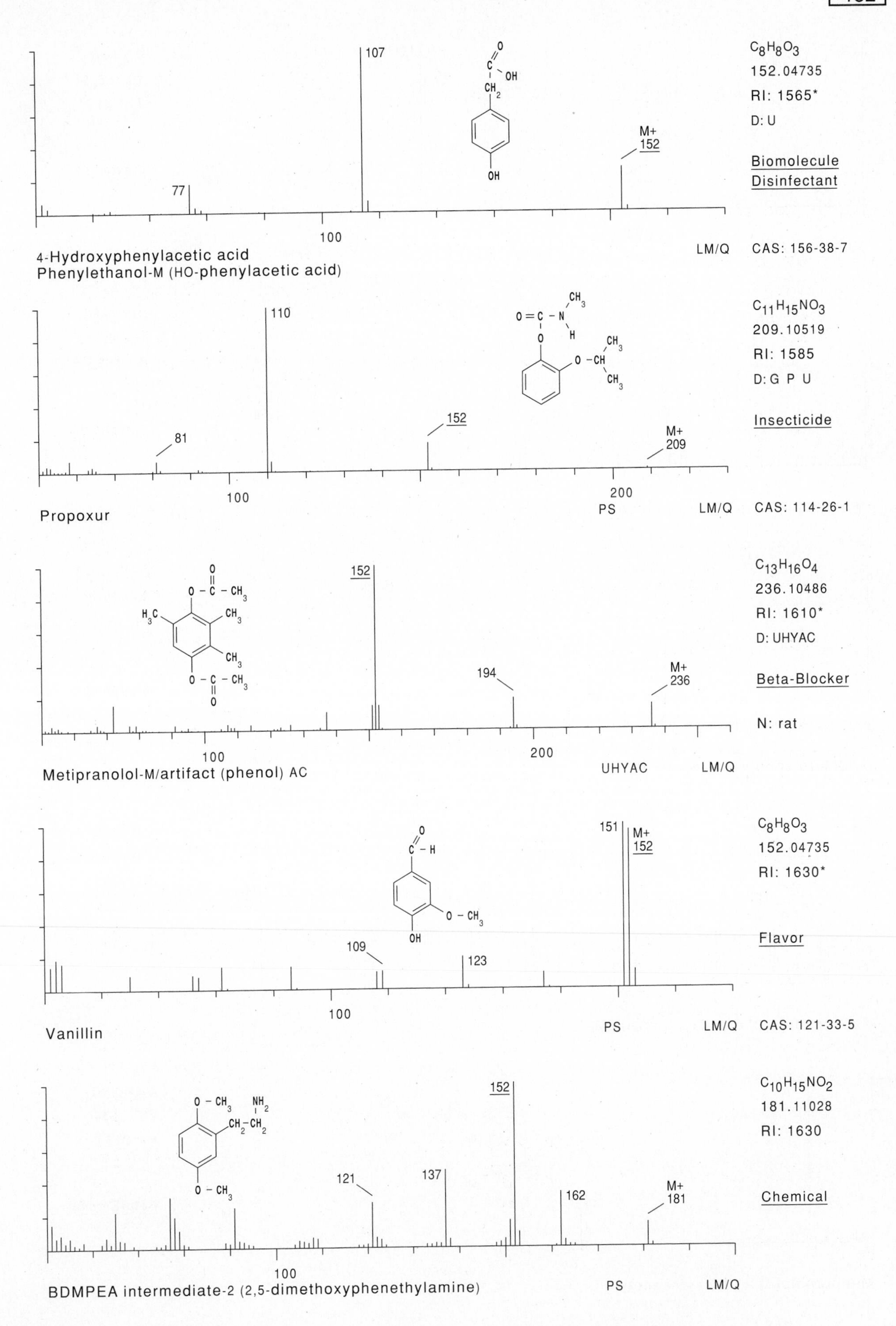

4-Hydroxyphenylacetic acid
Phenylethanol-M (HO-phenylacetic acid)

$C_8H_8O_3$
152.04735
RI: 1565*
D: U
Biomolecule
Disinfectant
CAS: 156-38-7

Propoxur

$C_{11}H_{15}NO_3$
209.10519
RI: 1585
D: G P U
Insecticide
CAS: 114-26-1

Metipranolol-M/artifact (phenol) AC

$C_{13}H_{16}O_4$
236.10486
RI: 1610*
D: UHYAC
Beta-Blocker
N: rat

Vanillin

$C_8H_8O_3$
152.04735
RI: 1630*
Flavor
CAS: 121-33-5

BDMPEA intermediate-2 (2,5-dimethoxyphenethylamine)

$C_{10}H_{15}NO_2$
181.11028
RI: 1630
Chemical

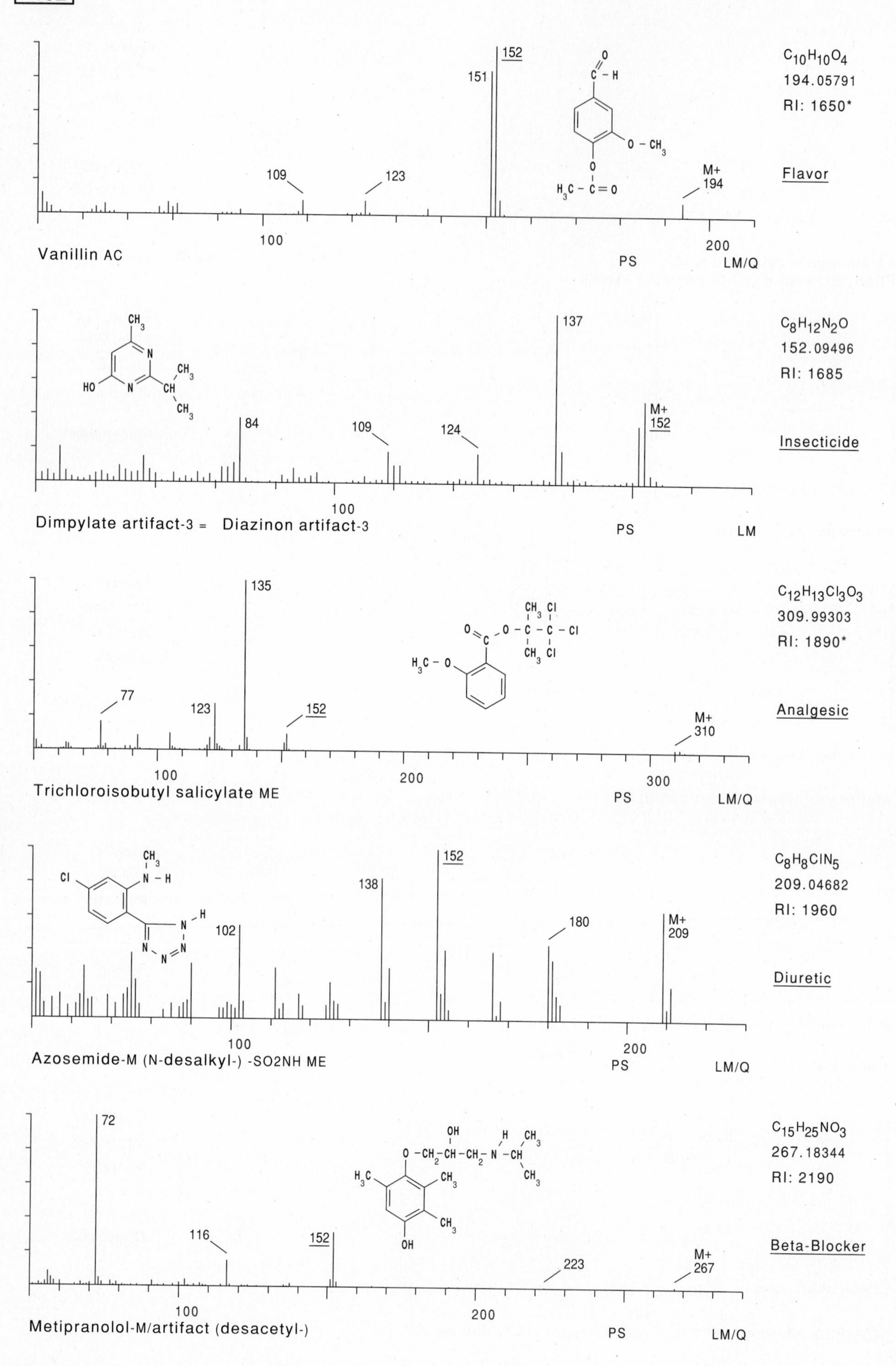

152
151
109
123
M+
194
100
200
Vanillin AC
PS
LM/Q
C10H10O4
194.05791
RI: 1650*
Flavor
137
84
109
124
M+
152
Dimpylate artifact-3 = Diazinon artifact-3
LM
C8H12N2O
152.09496
RI: 1685
Insecticide
135
77
123
152
M+
310
300
Trichloroisobutyl salicylate ME
C12H13Cl3O3
309.99303
RI: 1890*
Analgesic
152
138
102
180
M+
209
Azosemide-M (N-desalkyl-) -SO2NH ME
C8H8ClN5
209.04682
RI: 1960
Diuretic
72
116
223
M+
267
Metipranolol-M/artifact (desacetyl-)
C15H25NO3
267.18344
RI: 2190
Beta-Blocker

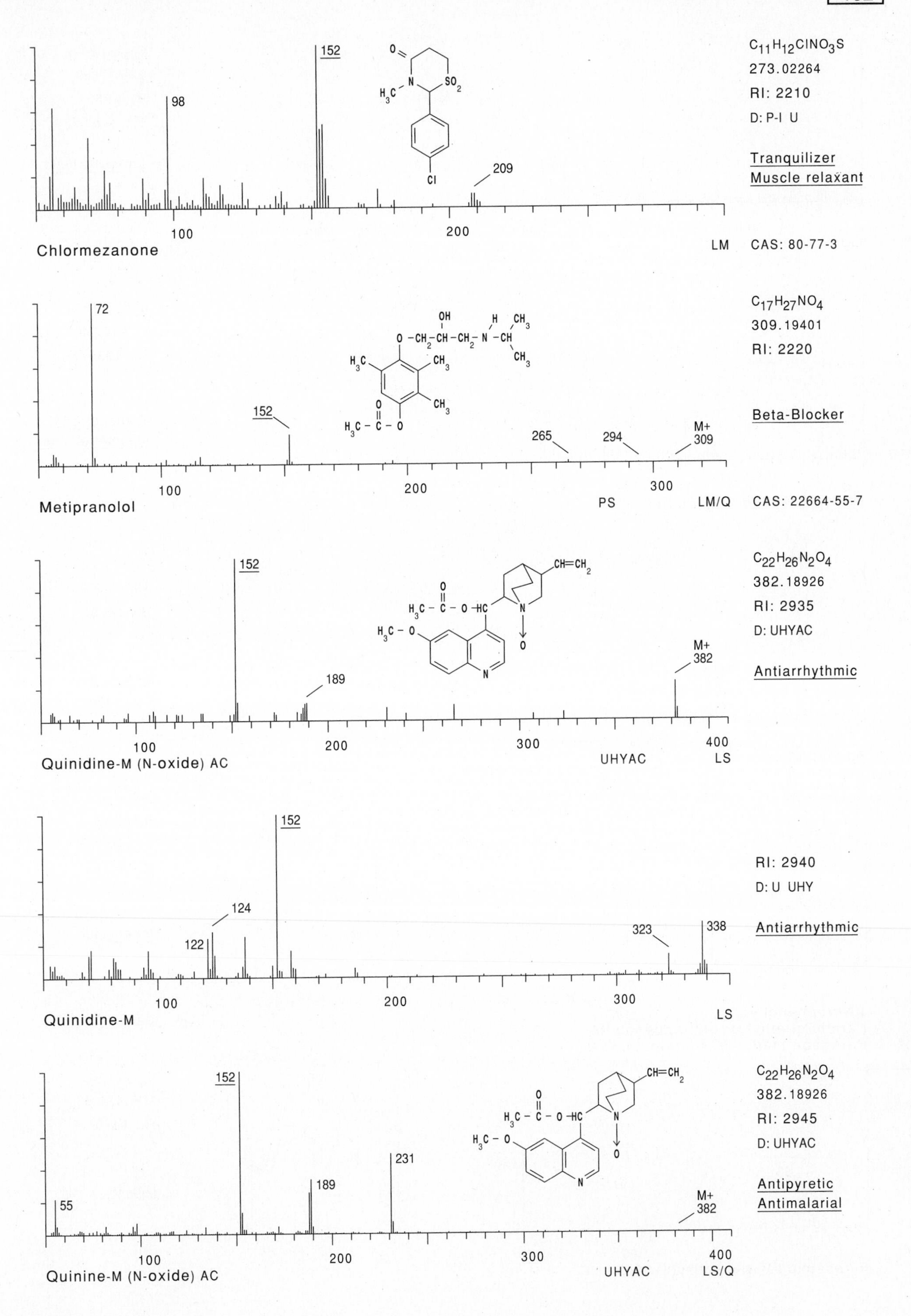

$C_{11}H_{12}ClNO_3S$
273.02264
RI: 2210
D: P-I U
Tranquilizer
Muscle relaxant
Chlormezanone
LM
CAS: 80-77-3
152
98
209
100
200
$C_{17}H_{27}NO_4$
309.19401
RI: 2220
Beta-Blocker
72
152
265
294
M+
309
100
200
300
Metipranolol
PS
LM/Q
CAS: 22664-55-7
$C_{22}H_{26}N_2O_4$
382.18926
RI: 2935
D: UHYAC
Antiarrhythmic
152
189
M+
382
100
200
300
400
Quinidine-M (N-oxide) AC
UHYAC
LS
RI: 2940
D: U UHY
Antiarrhythmic
152
124
122
323
338
100
200
300
Quinidine-M
LS
$C_{22}H_{26}N_2O_4$
382.18926
RI: 2945
D: UHYAC
Antipyretic
Antimalarial
152
231
189
55
M+
382
100
200
300
400
Quinine-M (N-oxide) AC
UHYAC
LS/Q

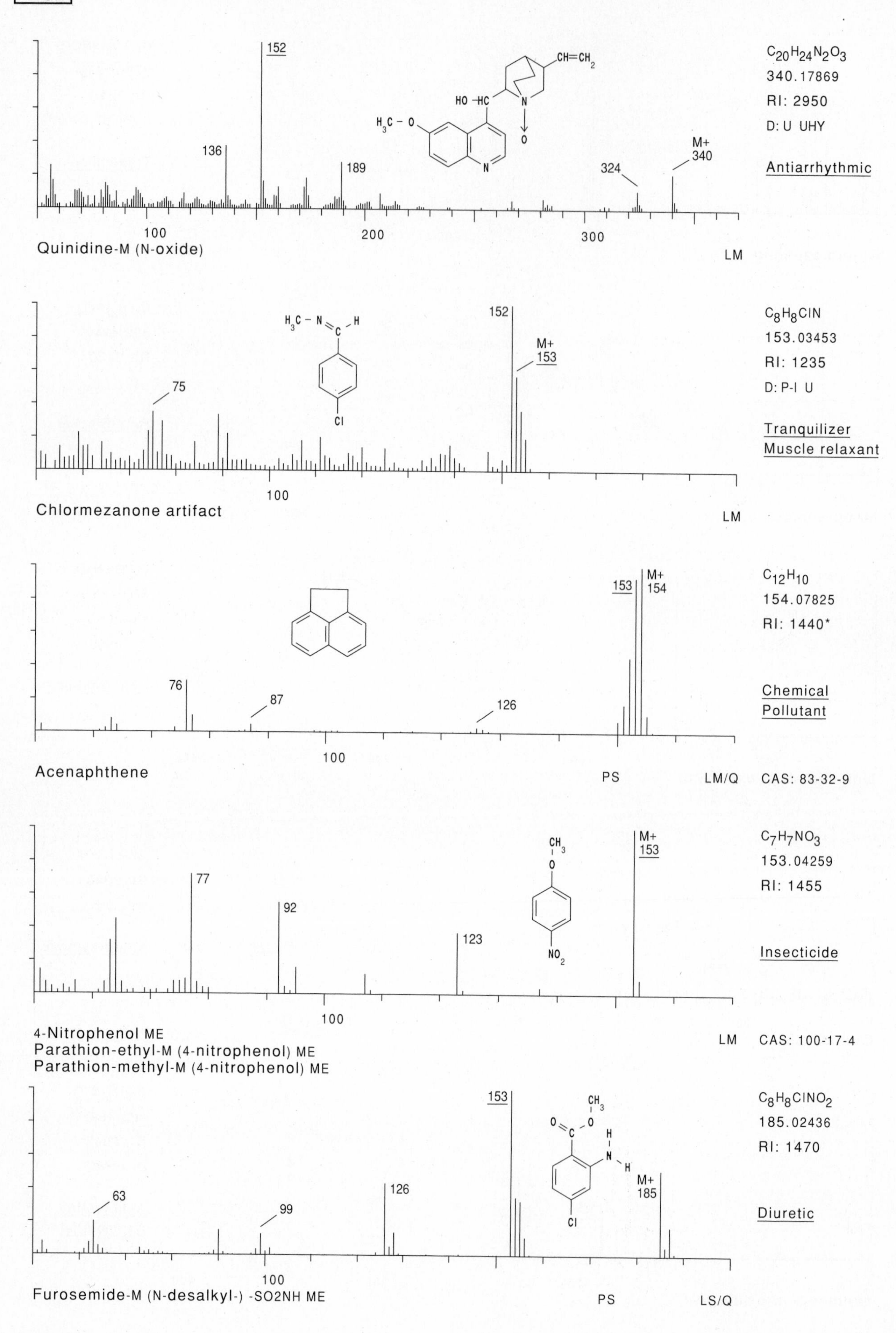

152
136
189
324
M+
340
100
200
300
Quinidine-M (N-oxide)
LM
$C_{20}H_{24}N_2O_3$
340.17869
RI: 2950
D: U UHY
Antiarrhythmic
152
M+
153
75
100
Chlormezanone artifact
LM
C_8H_8ClN
153.03453
RI: 1235
D: P-I U
Tranquilizer
Muscle relaxant
153
M+
154
76
87
126
100
Acenaphthene
PS
LM/Q
$C_{12}H_{10}$
154.07825
RI: 1440*
Chemical
Pollutant
CAS: 83-32-9
M+
153
77
92
123
100
4-Nitrophenol ME
Parathion-ethyl-M (4-nitrophenol) ME
Parathion-methyl-M (4-nitrophenol) ME
LM
$C_7H_7NO_3$
153.04259
RI: 1455
Insecticide
CAS: 100-17-4
153
63
99
126
M+
185
100
Furosemide-M (N-desalkyl-) -SO2NH ME
PS
LS/Q
$C_8H_8ClNO_2$
185.02436
RI: 1470
Diuretic

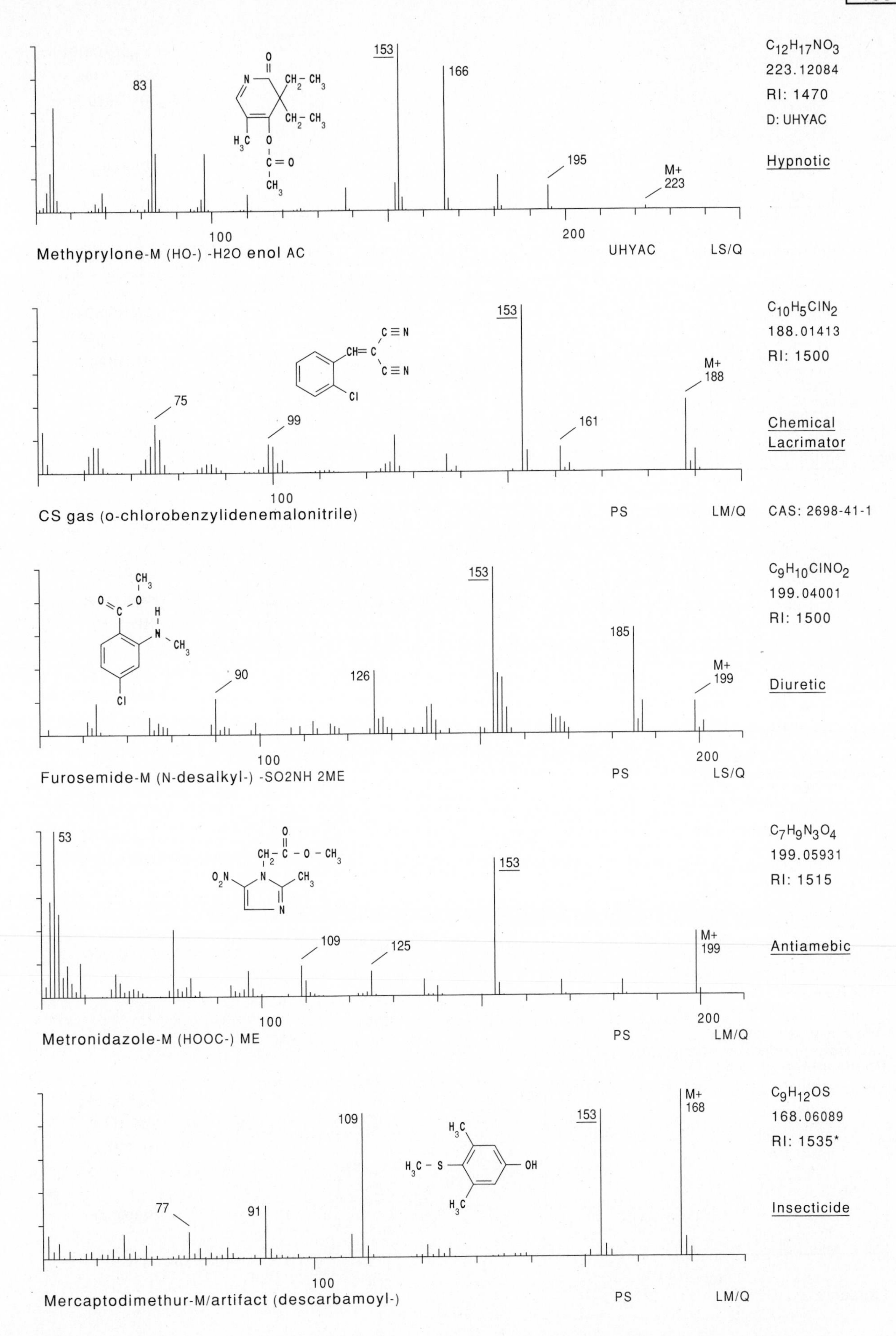
153
166
83
195
M+
223
100
200
Methyprylone-M (HO-) -H2O enol AC
UHYAC
LS/Q
C12H17NO3
223.12084
RI: 1470
D: UHYAC
Hypnotic
153
75
99
161
M+
188
100
CS gas (o-chlorobenzylidenemalonitrile)
PS
LM/Q
C10H5ClN2
188.01413
RI: 1500
Chemical
Lacrimator
CAS: 2698-41-1
153
185
90
126
M+
199
100
200
Furosemide-M (N-desalkyl-) -SO2NH 2ME
PS
LS/Q
C9H10ClNO2
199.04001
RI: 1500
Diuretic
53
153
109
125
M+
199
100
200
Metronidazole-M (HOOC-) ME
PS
LM/Q
C7H9N3O4
199.05931
RI: 1515
Antiamebic
M+
168
109
153
77
91
100
Mercaptodimethur-M/artifact (descarbamoyl-)
PS
LM/Q
C9H12OS
168.06089
RI: 1535*
Insecticide

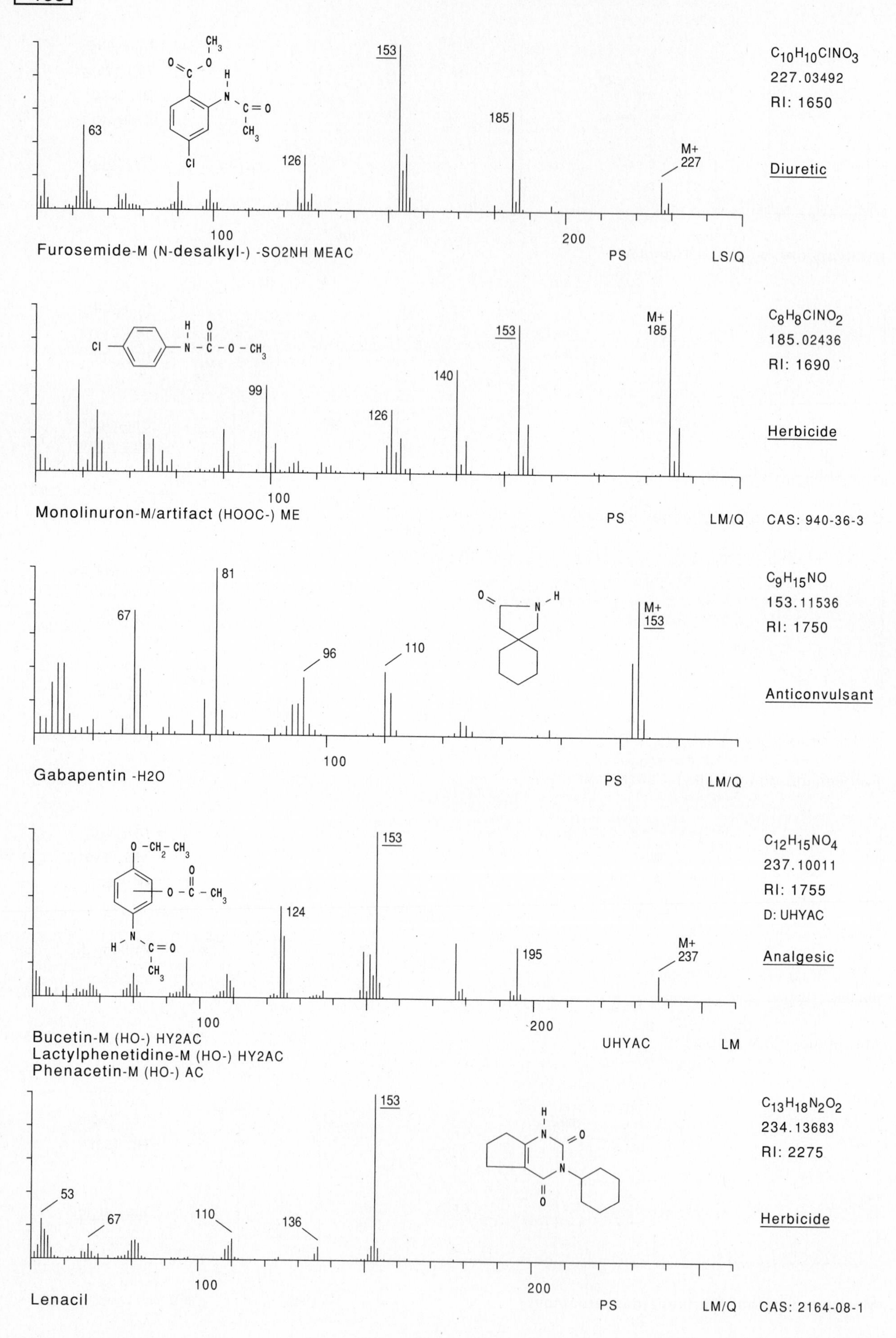
153
63
126
185
M+
227
100
200
PS
LS/Q
$C_{10}H_{10}ClNO_3$
227.03492
RI: 1650
Diuretic
Furosemide-M (N-desalkyl-) -SO2NH MEAC
M+
185
153
140
99
126
100
PS
LM/Q
$C_8H_8ClNO_2$
185.02436
RI: 1690
Herbicide
Monolinuron-M/artifact (HOOC-) ME
CAS: 940-36-3
81
67
96
110
M+
153
100
PS
LM/Q
$C_9H_{15}NO$
153.11536
RI: 1750
Anticonvulsant
Gabapentin -H2O
153
124
195
M+
237
100
200
UHYAC
LM
$C_{12}H_{15}NO_4$
237.10011
RI: 1755
D: UHYAC
Analgesic
Bucetin-M (HO-) HY2AC
Lactylphenetidine-M (HO-) HY2AC
Phenacetin-M (HO-) AC
153
53
67
110
136
100
200
PS
LM/Q
$C_{13}H_{18}N_2O_2$
234.13683
RI: 2275
Herbicide
Lenacil
CAS: 2164-08-1

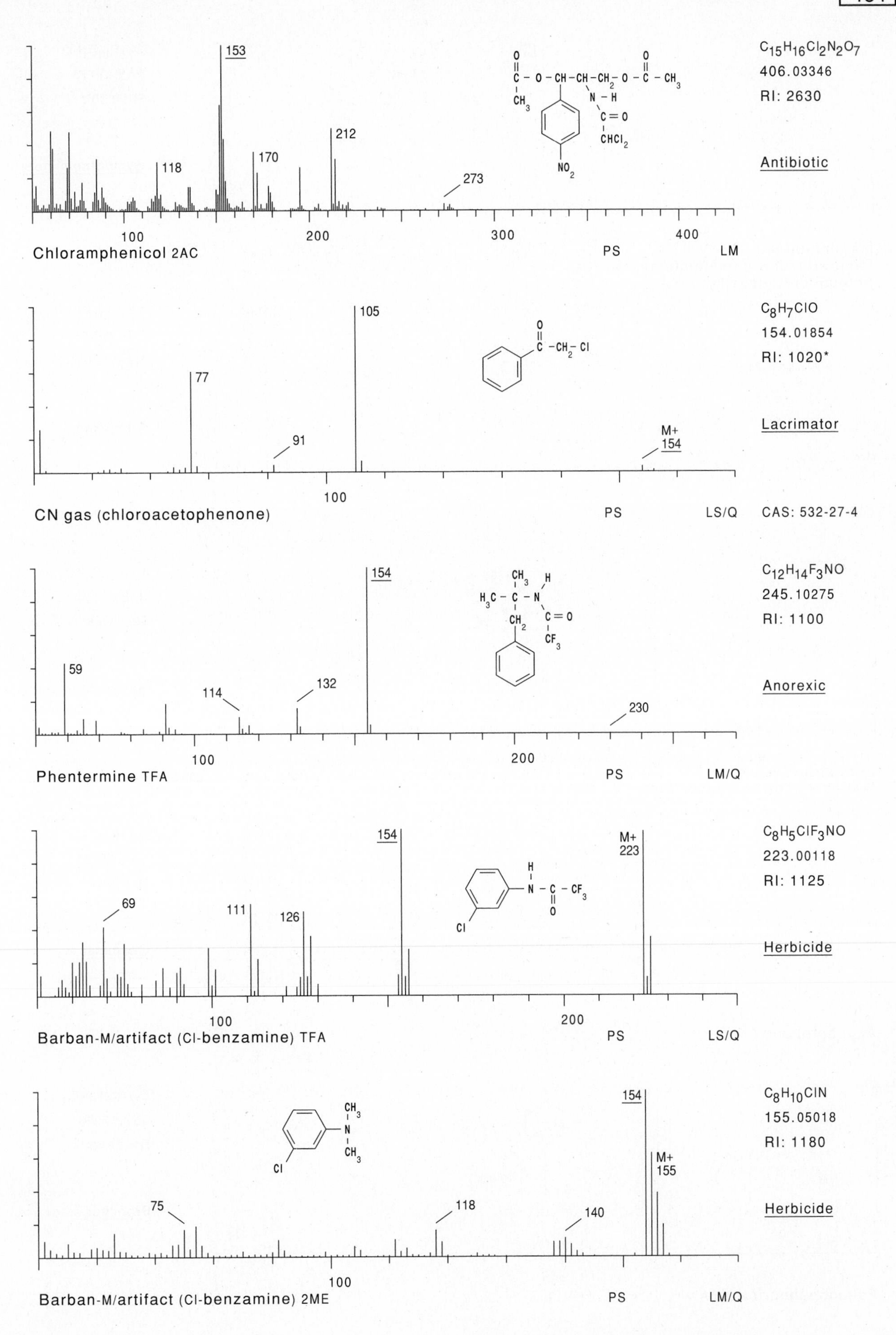

Chloramphenicol 2AC

$C_{15}H_{16}Cl_2N_2O_7$
406.03346
RI: 2630

Antibiotic

CN gas (chloroacetophenone)

C_8H_7ClO
154.01854
RI: 1020*

Lacrimator

CAS: 532-27-4

Phentermine TFA

$C_{12}H_{14}F_3NO$
245.10275
RI: 1100

Anorexic

Barban-M/artifact (Cl-benzamine) TFA

$C_8H_5ClF_3NO$
223.00118
RI: 1125

Herbicide

Barban-M/artifact (Cl-benzamine) 2ME

$C_8H_{10}ClN$
155.05018
RI: 1180

Herbicide

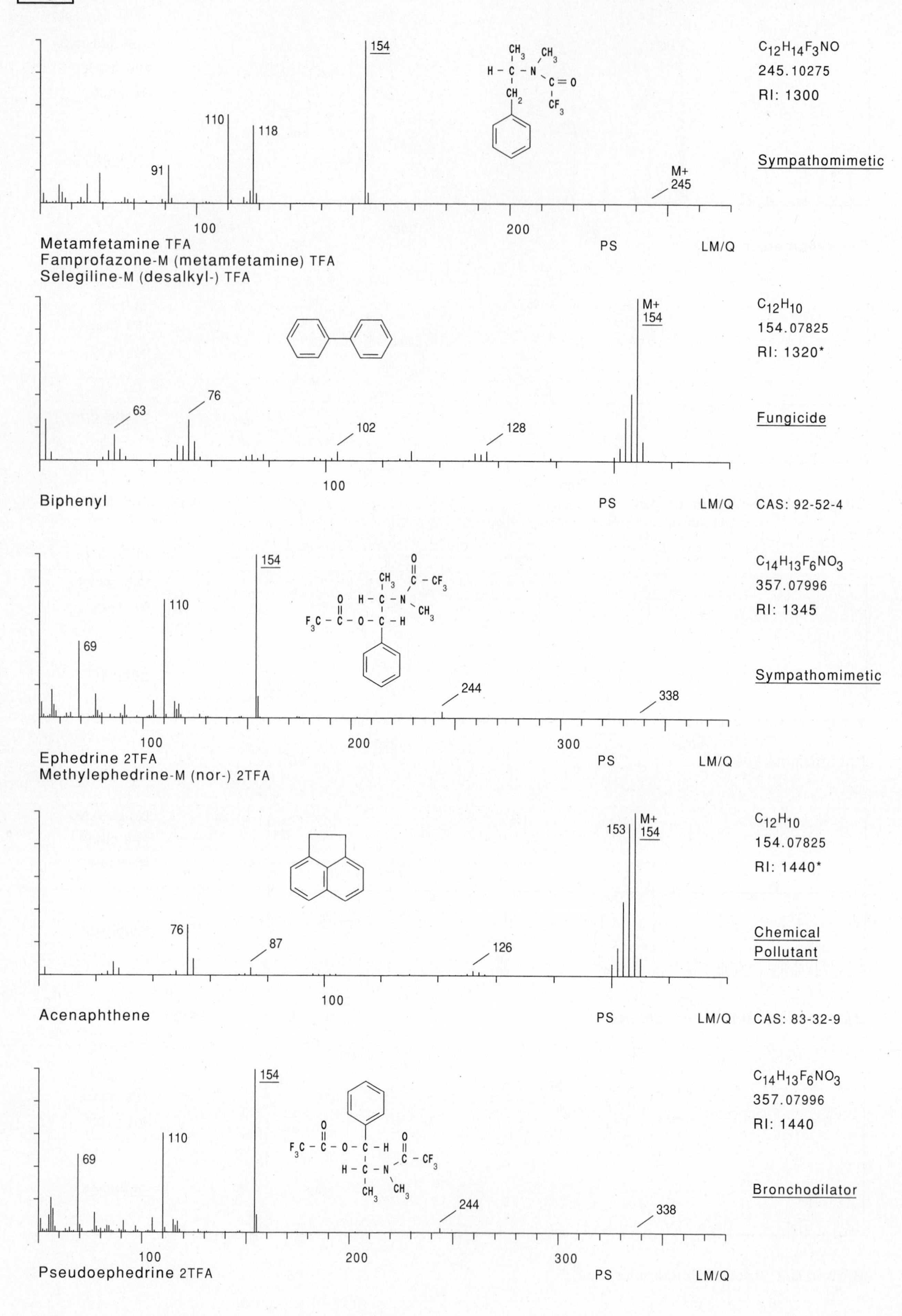
154
110
118
91
M+
245
100
200
PS
LM/Q
$C_{12}H_{14}F_3NO$
245.10275
RI: 1300
Sympathomimetic
Metamfetamine TFA
Famprofazone-M (metamfetamine) TFA
Selegiline-M (desalkyl-) TFA
M+
154
76
63
102
128
100
PS
LM/Q
$C_{12}H_{10}$
154.07825
RI: 1320*
Fungicide
Biphenyl
CAS: 92-52-4
154
110
69
244
338
100
200
300
PS
LM/Q
$C_{14}H_{13}F_6NO_3$
357.07996
RI: 1345
Sympathomimetic
Ephedrine 2TFA
Methylephedrine-M (nor-) 2TFA
M+
153
154
76
87
126
100
PS
LM/Q
$C_{12}H_{10}$
154.07825
RI: 1440*
Chemical
Pollutant
Acenaphthene
CAS: 83-32-9
154
110
69
244
338
100
200
300
PS
LM/Q
$C_{14}H_{13}F_6NO_3$
357.07996
RI: 1440
Bronchodilator
Pseudoephedrine 2TFA

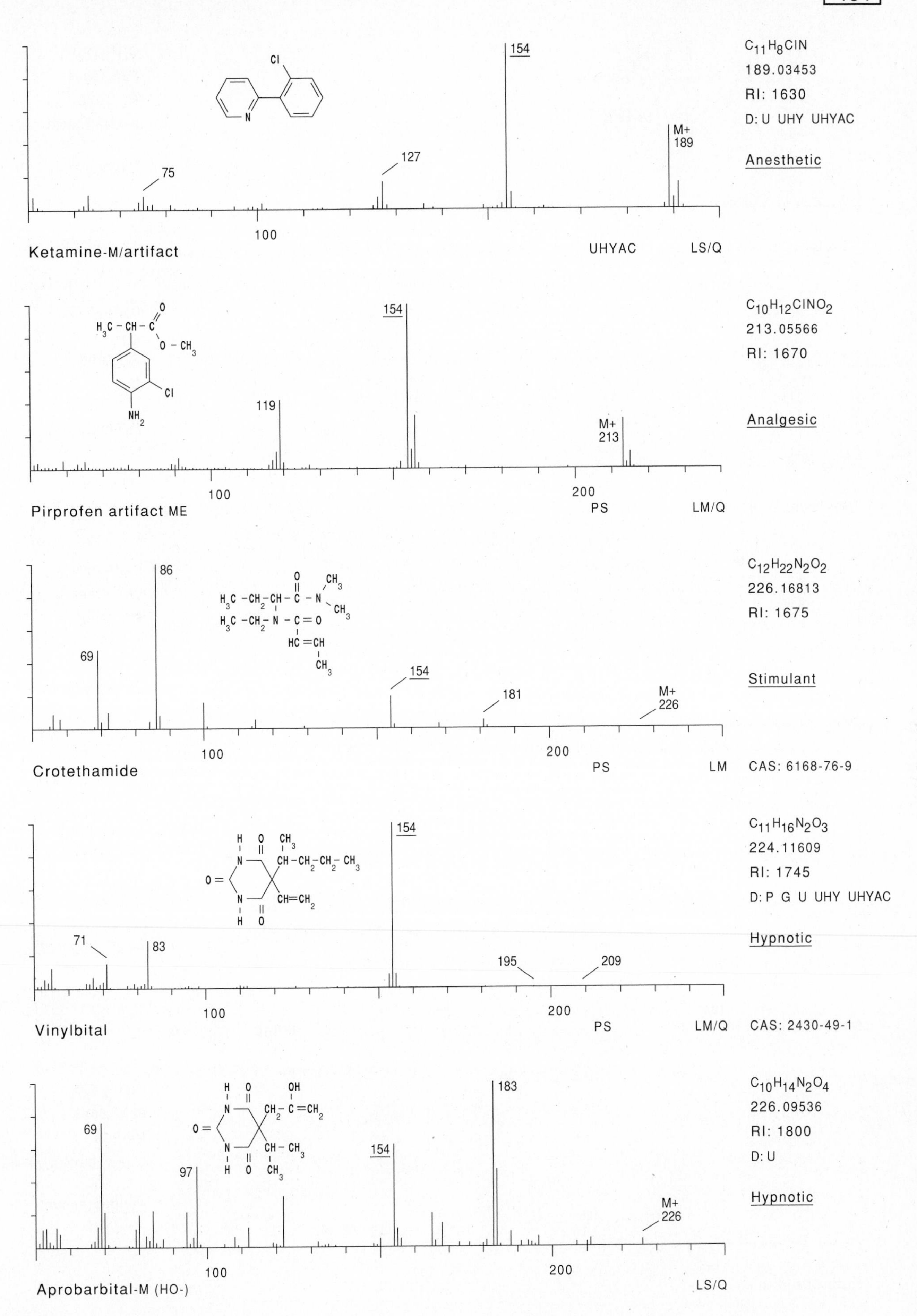

$C_{11}H_8ClN$
189.03453
RI: 1630
D: U UHY UHYAC
Anesthetic

$C_{10}H_{12}ClNO_2$
213.05566
RI: 1670
Analgesic

$C_{12}H_{22}N_2O_2$
226.16813
RI: 1675
Stimulant
CAS: 6168-76-9

$C_{11}H_{16}N_2O_3$
224.11609
RI: 1745
D: P G U UHY UHYAC
Hypnotic
CAS: 2430-49-1

$C_{10}H_{14}N_2O_4$
226.09536
RI: 1800
D: U
Hypnotic

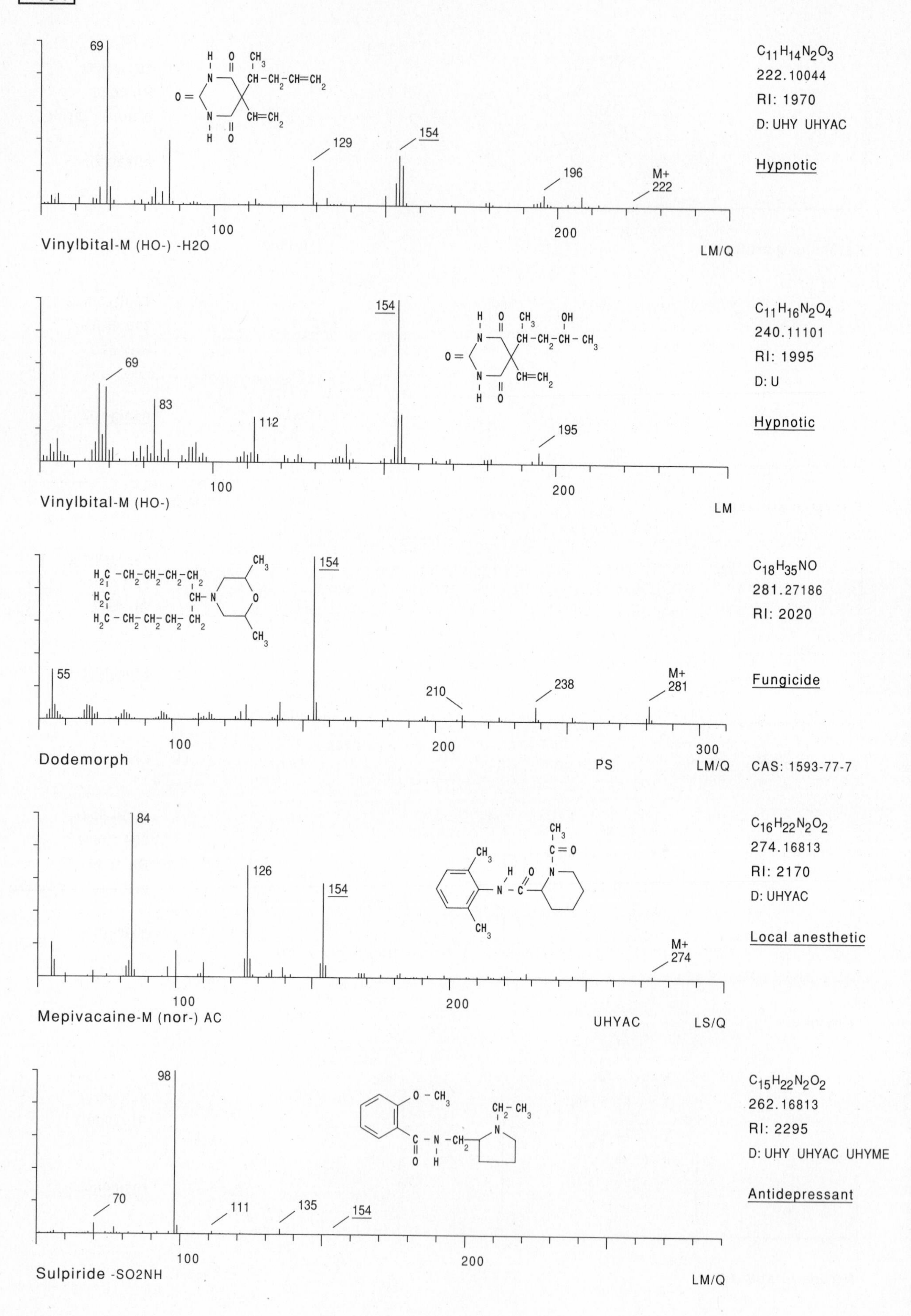
69
129
154
196
M+
222
100
200
Vinylbital-M (HO-) -H2O
LM/Q
$C_{11}H_{14}N_2O_3$
222.10044
RI: 1970
D: UHY UHYAC
Hypnotic
154
69
83
112
195
100
200
Vinylbital-M (HO-)
LM
$C_{11}H_{16}N_2O_4$
240.11101
RI: 1995
D: U
Hypnotic
154
55
210
238
M+
281
100
200
300
Dodemorph
PS
LM/Q
$C_{18}H_{35}NO$
281.27186
RI: 2020
Fungicide
CAS: 1593-77-7
84
126
154
M+
274
100
200
Mepivacaine-M (nor-) AC
UHYAC
LS/Q
$C_{16}H_{22}N_2O_2$
274.16813
RI: 2170
D: UHYAC
Local anesthetic
98
70
111
135
154
100
200
Sulpiride -SO2NH
LM/Q
$C_{15}H_{22}N_2O_2$
262.16813
RI: 2295
D: UHY UHYAC UHYME
Antidepressant

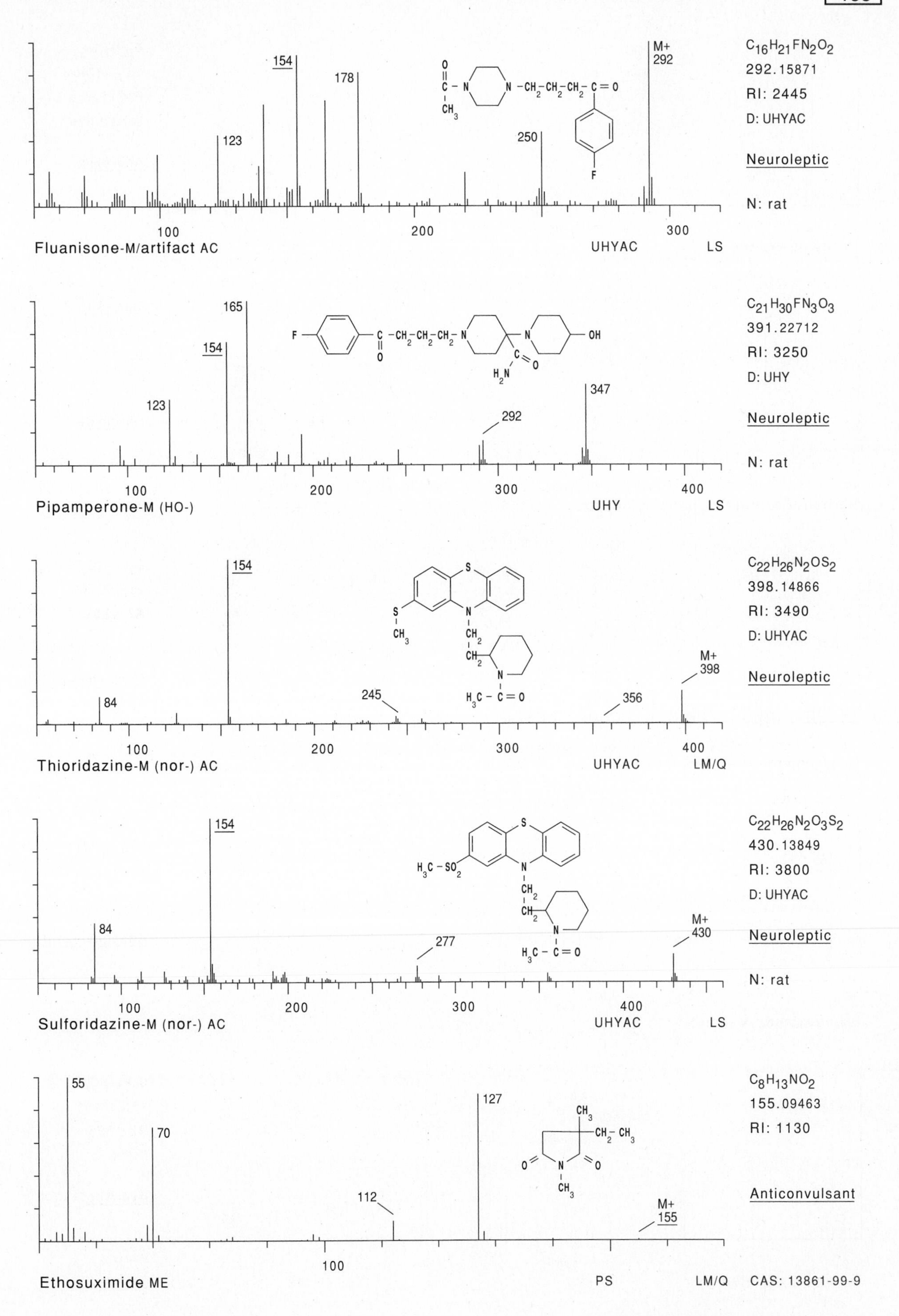

$C_{16}H_{21}FN_2O_2$
292.15871
RI: 2445
D: UHYAC

Neuroleptic

N: rat

$C_{21}H_{30}FN_3O_3$
391.22712
RI: 3250
D: UHY

Neuroleptic

N: rat

$C_{22}H_{26}N_2OS_2$
398.14866
RI: 3490
D: UHYAC

Neuroleptic

$C_{22}H_{26}N_2O_3S_2$
430.13849
RI: 3800
D: UHYAC

Neuroleptic

N: rat

$C_8H_{13}NO_2$
155.09463
RI: 1130

Anticonvulsant

CAS: 13861-99-9

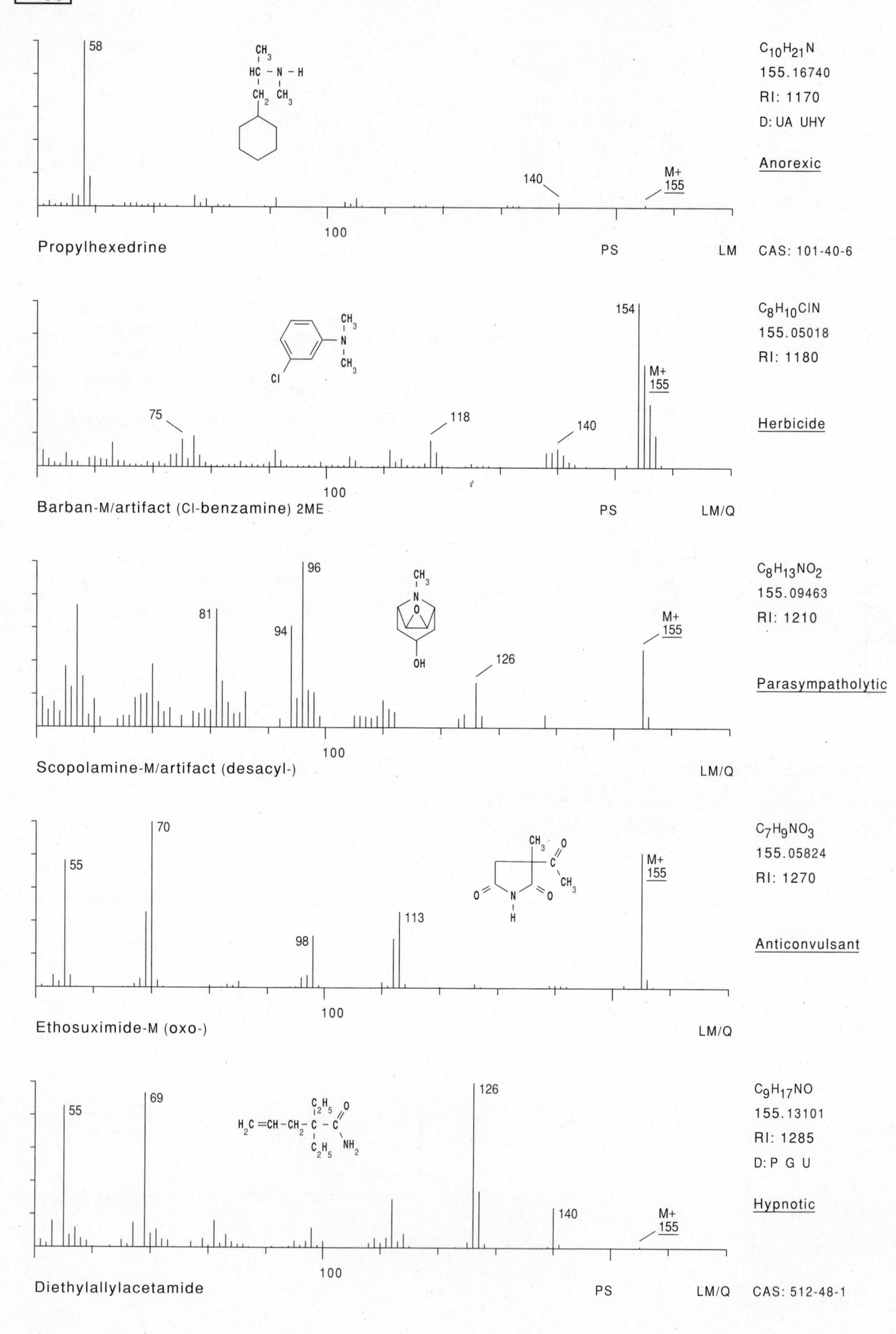

Propylhexedrine

$C_{10}H_{21}N$
155.16740
RI: 1170
D: UA UHY
Anorexic
CAS: 101-40-6

Barban-M/artifact (Cl-benzamine) 2ME

$C_8H_{10}ClN$
155.05018
RI: 1180
Herbicide

Scopolamine-M/artifact (desacyl-)

$C_8H_{13}NO_2$
155.09463
RI: 1210
Parasympatholytic

Ethosuximide-M (oxo-)

$C_7H_9NO_3$
155.05824
RI: 1270
Anticonvulsant

Diethylallylacetamide

$C_9H_{17}NO$
155.13101
RI: 1285
D: P G U
Hypnotic
CAS: 512-48-1

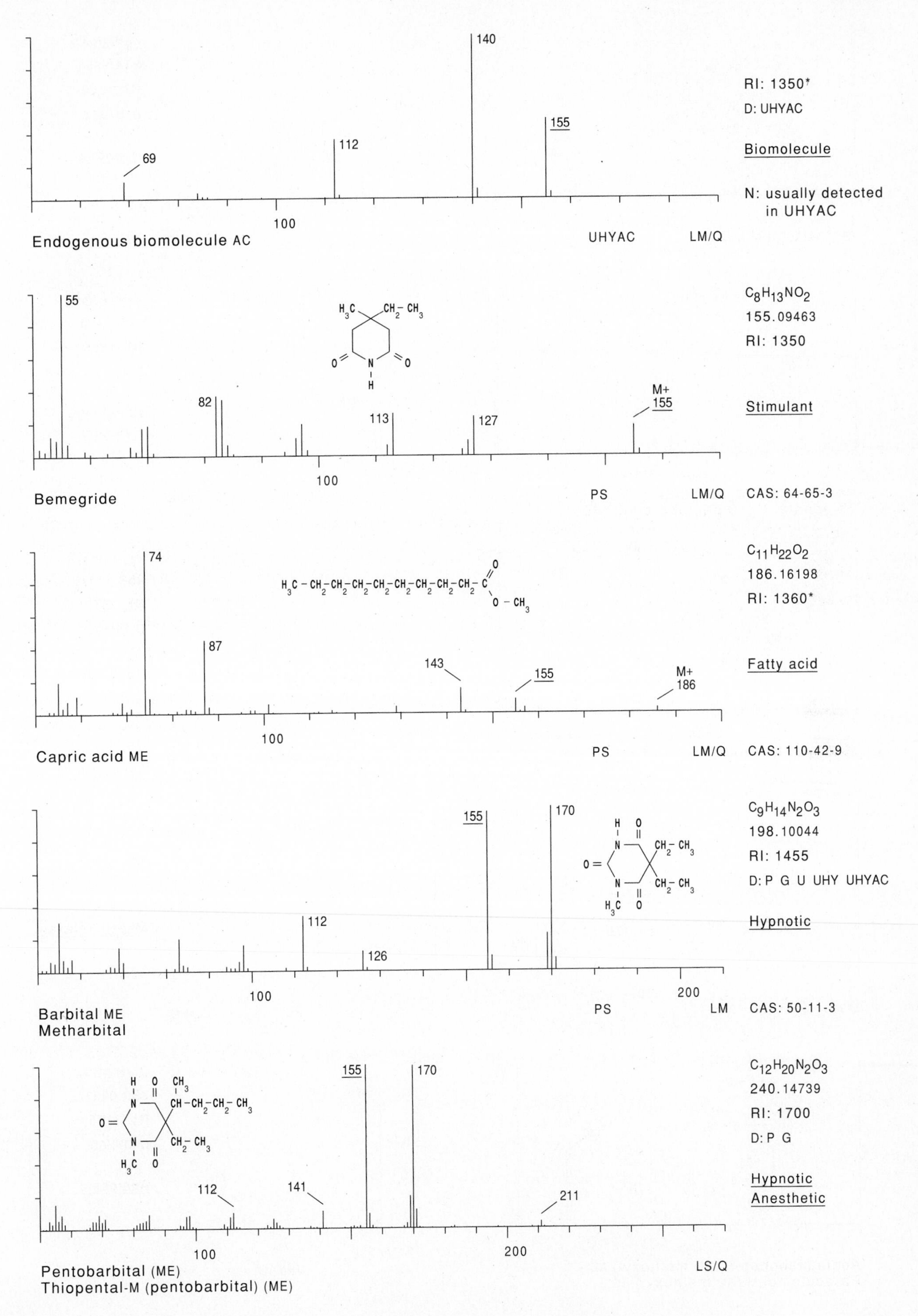
140
155
112
69
100
RI: 1350*
D: UHYAC
Biomolecule
N: usually detected in UHYAC
Endogenous biomolecule AC
UHYAC
LM/Q
55
82
113
127
M+
155
100
$C_8H_{13}NO_2$
155.09463
RI: 1350
Stimulant
Bemegride
PS
LM/Q
CAS: 64-65-3
74
87
143
155
M+
186
100
$C_{11}H_{22}O_2$
186.16198
RI: 1360*
Fatty acid
Capric acid ME
PS
LM/Q
CAS: 110-42-9
155
170
112
126
100
200
$C_9H_{14}N_2O_3$
198.10044
RI: 1455
D: P G U UHY UHYAC
Hypnotic
Barbital ME
Metharbital
PS
LM
CAS: 50-11-3
155
170
112
141
211
100
200
$C_{12}H_{20}N_2O_3$
240.14739
RI: 1700
D: P G
Hypnotic
Anesthetic
Pentobarbital (ME)
Thiopental-M (pentobarbital) (ME)
LS/Q

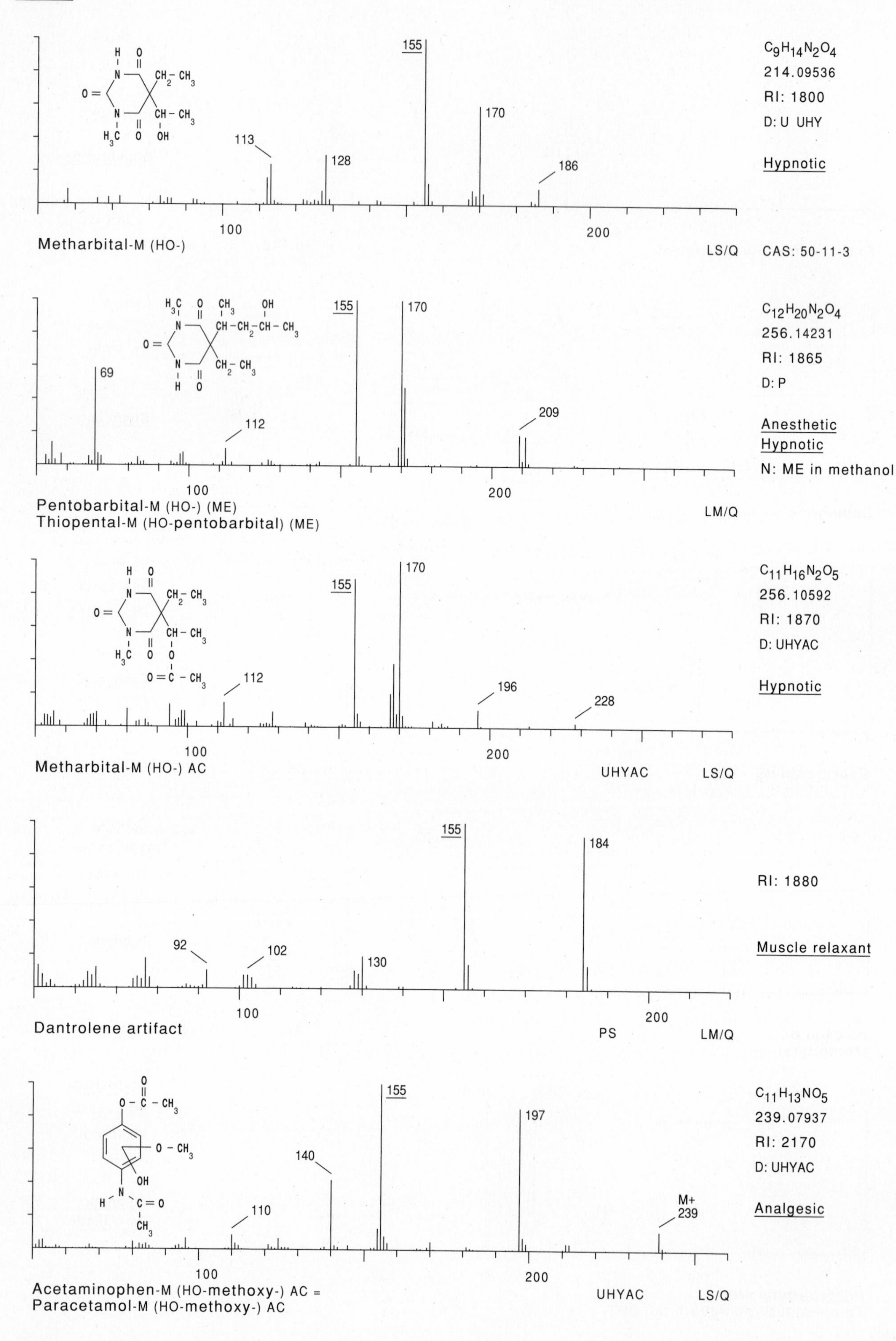
155
170
113
128
186
100
200
Metharbital-M (HO-)
LS/Q
$C_9H_{14}N_2O_4$
214.09536
RI: 1800
D: U UHY
Hypnotic
CAS: 50-11-3
155
170
69
112
209
100
200
Pentobarbital-M (HO-) (ME)
Thiopental-M (HO-pentobarbital) (ME)
LM/Q
$C_{12}H_{20}N_2O_4$
256.14231
RI: 1865
D: P
Anesthetic
Hypnotic
N: ME in methanol
170
155
112
196
228
100
200
Metharbital-M (HO-) AC
UHYAC
LS/Q
$C_{11}H_{16}N_2O_5$
256.10592
RI: 1870
D: UHYAC
Hypnotic
155
184
92
102
130
100
200
Dantrolene artifact
PS
LM/Q
RI: 1880
Muscle relaxant
155
197
140
110
M+
239
100
200
Acetaminophen-M (HO-methoxy-) AC =
Paracetamol-M (HO-methoxy-) AC
UHYAC
LS/Q
$C_{11}H_{13}NO_5$
239.07937
RI: 2170
D: UHYAC
Analgesic

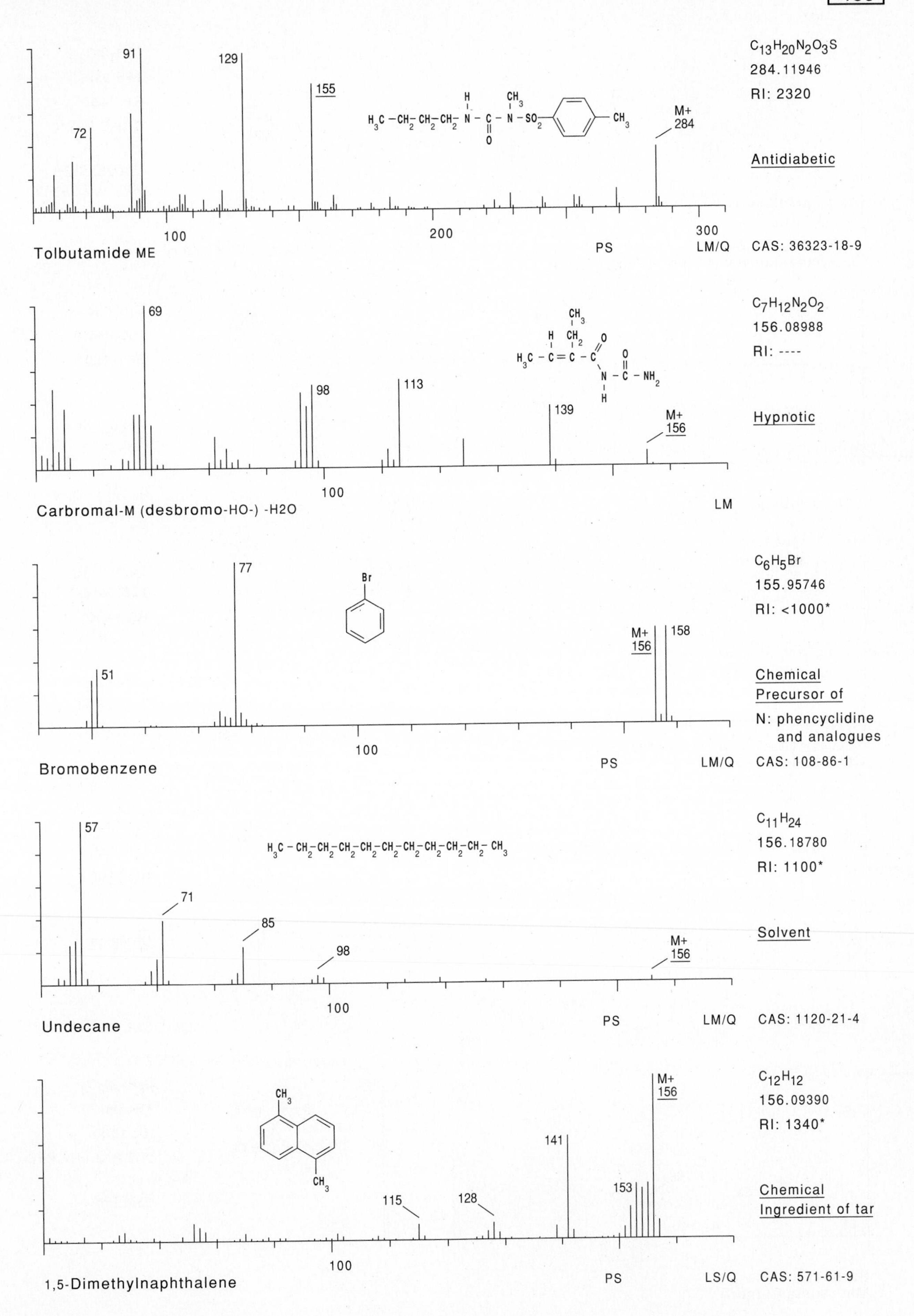

Tolbutamide ME

$C_{13}H_{20}N_2O_3S$
284.11946
RI: 2320

Antidiabetic

CAS: 36323-18-9

Carbromal-M (desbromo-HO-) -H2O

$C_7H_{12}N_2O_2$
156.08988
RI: ----

Hypnotic

Bromobenzene

C_6H_5Br
155.95746
RI: <1000*

Chemical
Precursor of
N: phencyclidine and analogues

CAS: 108-86-1

Undecane

$C_{11}H_{24}$
156.18780
RI: 1100*

Solvent

CAS: 1120-21-4

1,5-Dimethylnaphthalene

$C_{12}H_{12}$
156.09390
RI: 1340*

Chemical
Ingredient of tar

CAS: 571-61-9

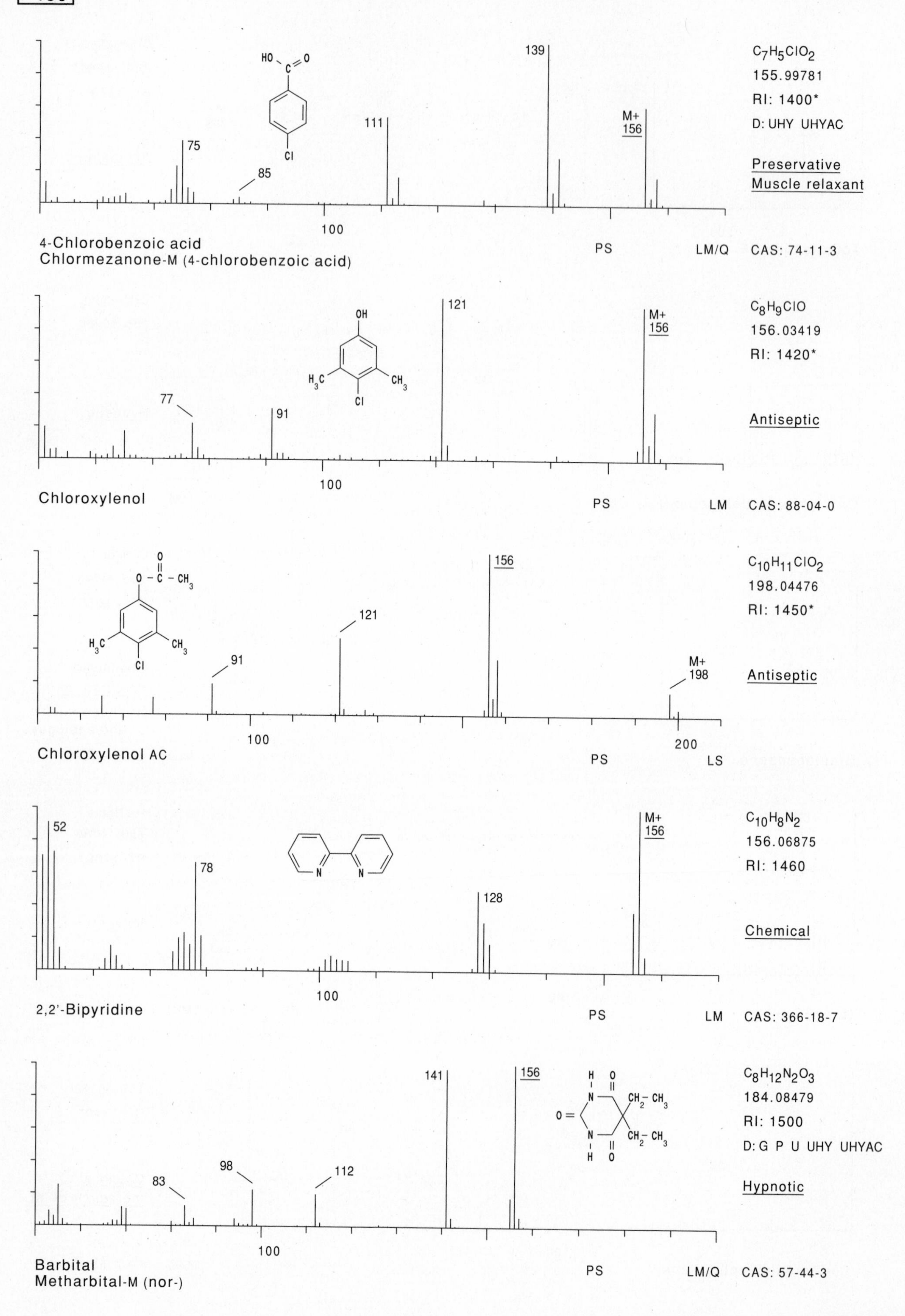

139
111
75
85
M+
156
100
4-Chlorobenzoic acid
Chlormezanone-M (4-chlorobenzoic acid)
PS
LM/Q
C7H5ClO2
155.99781
RI: 1400*
D: UHY UHYAC
Preservative
Muscle relaxant
CAS: 74-11-3
121
77
91
M+
156
100
Chloroxylenol
PS
LM
C8H9ClO
156.03419
RI: 1420*
Antiseptic
CAS: 88-04-0
156
121
91
M+
198
100
200
Chloroxylenol AC
PS
LS
C10H11ClO2
198.04476
RI: 1450*
Antiseptic
52
78
128
M+
156
100
2,2'-Bipyridine
PS
LM
C10H8N2
156.06875
RI: 1460
Chemical
CAS: 366-18-7
141
156
83
98
112
100
Barbital
Metharbital-M (nor-)
PS
LM/Q
C8H12N2O3
184.08479
RI: 1500
D: G P U UHY UHYAC
Hypnotic
CAS: 57-44-3

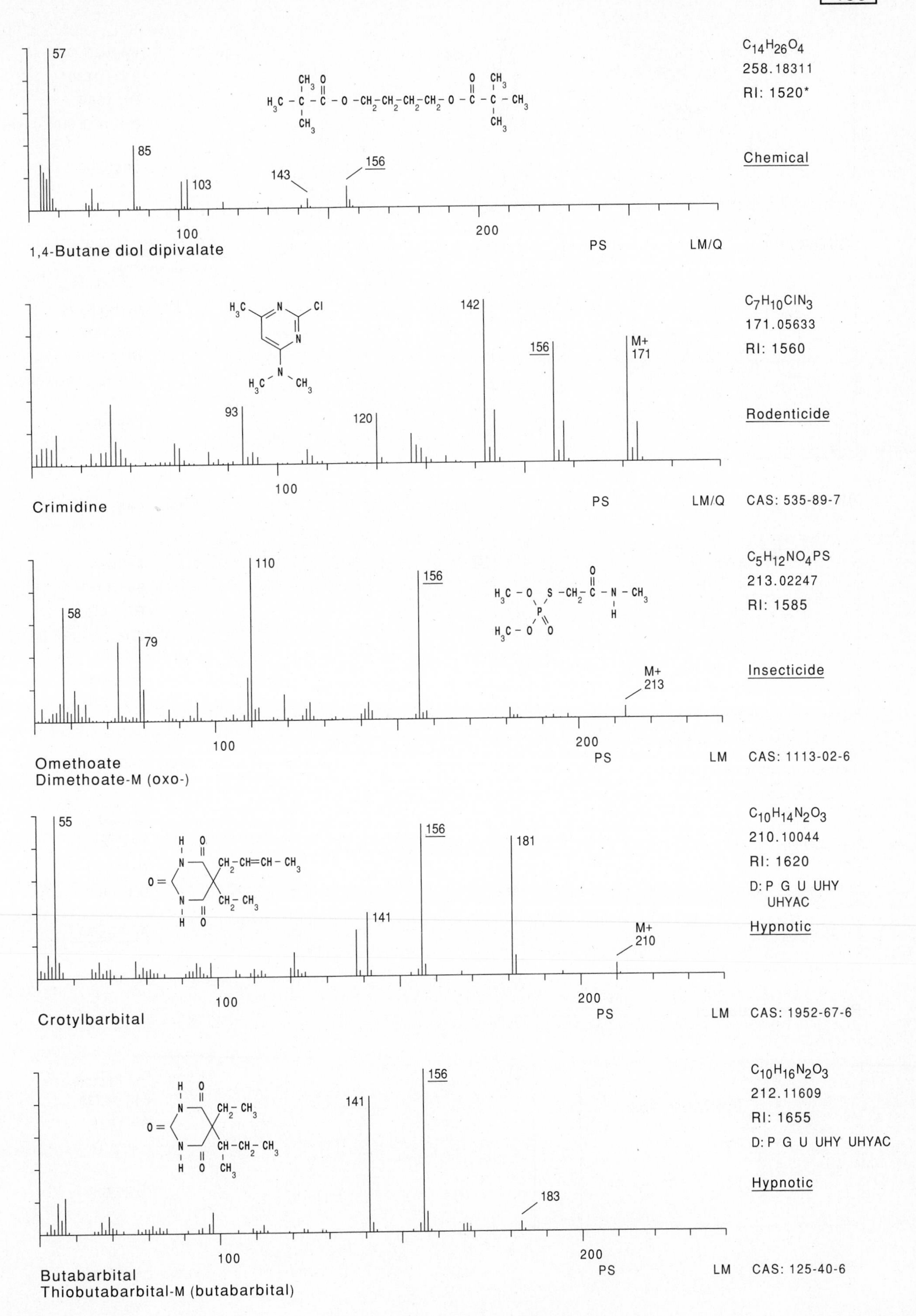
57
85
103
143
156
100
200
1,4-Butane diol dipivalate
PS
LM/Q
C14H26O4
258.18311
RI: 1520*
Chemical
142
156
M+
171
93
120
100
Crimidine
PS
LM/Q
C7H10ClN3
171.05633
RI: 1560
Rodenticide
CAS: 535-89-7
110
156
58
79
M+
213
100
200
Omethoate
Dimethoate-M (oxo-)
PS
LM
C5H12NO4PS
213.02247
RI: 1585
Insecticide
CAS: 1113-02-6
55
156
181
141
M+
210
100
200
Crotylbarbital
PS
LM
C10H14N2O3
210.10044
RI: 1620
D: P G U UHY
UHYAC
Hypnotic
CAS: 1952-67-6
156
141
183
100
200
Butabarbital
Thiobutabarbital-M (butabarbital)
PS
LM
C10H16N2O3
212.11609
RI: 1655
D: P G U UHY UHYAC
Hypnotic
CAS: 125-40-6

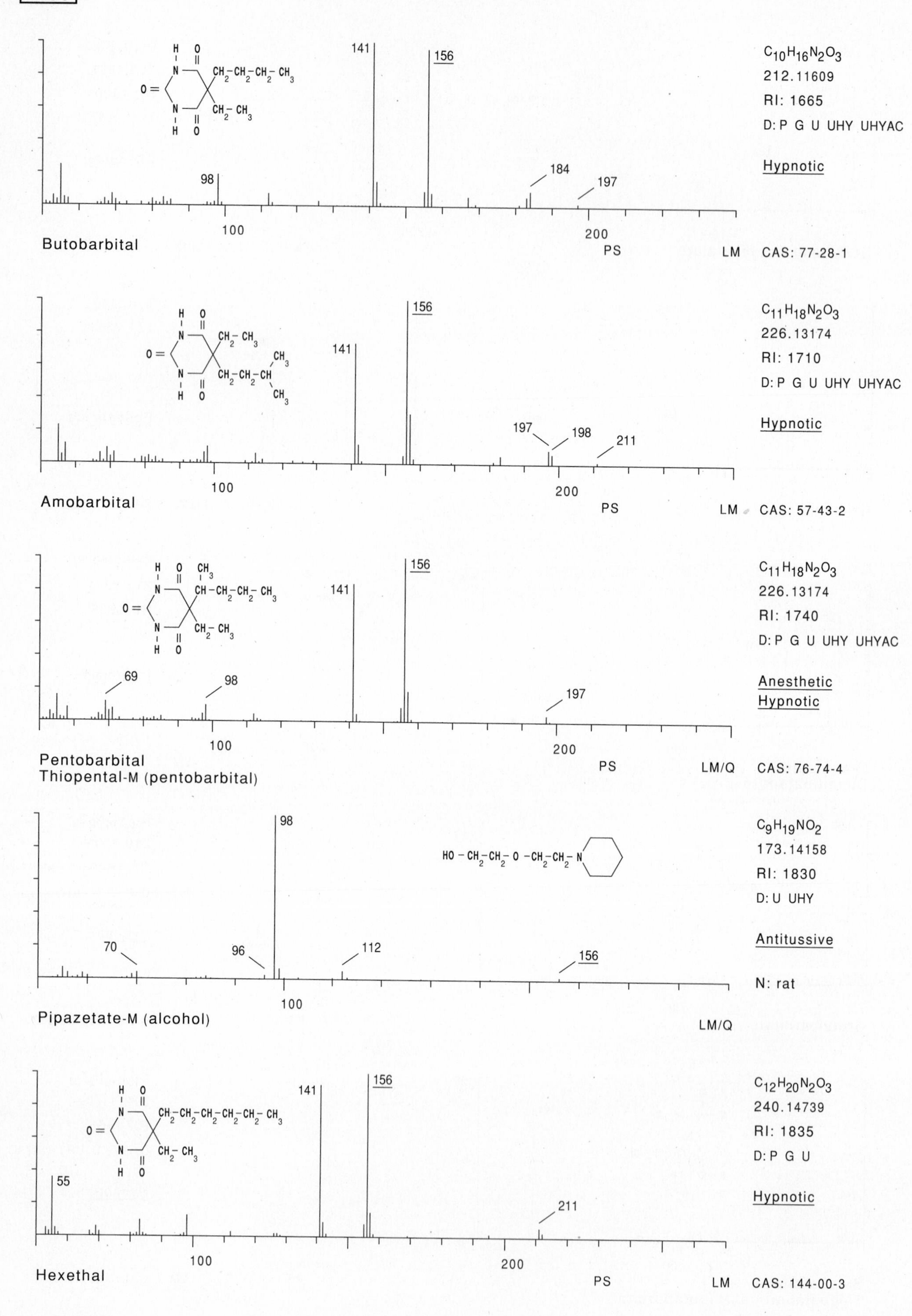

141
156
98
184
197
100
200
Butobarbital
PS
LM
C10H16N2O3
212.11609
RI: 1665
D: P G U UHY UHYAC
Hypnotic
CAS: 77-28-1
156
141
197
198
211
100
200
Amobarbital
PS
LM
C11H18N2O3
226.13174
RI: 1710
D: P G U UHY UHYAC
Hypnotic
CAS: 57-43-2
156
141
69
98
197
100
200
Pentobarbital
Thiopental-M (pentobarbital)
PS
LM/Q
C11H18N2O3
226.13174
RI: 1740
D: P G U UHY UHYAC
Anesthetic
Hypnotic
CAS: 76-74-4
98
70
96
112
156
100
Pipazetate-M (alcohol)
LM/Q
C9H19NO2
173.14158
RI: 1830
D: U UHY
Antitussive
N: rat
156
141
55
211
100
200
Hexethal
PS
LM
C12H20N2O3
240.14739
RI: 1835
D: P G U
Hypnotic
CAS: 144-00-3

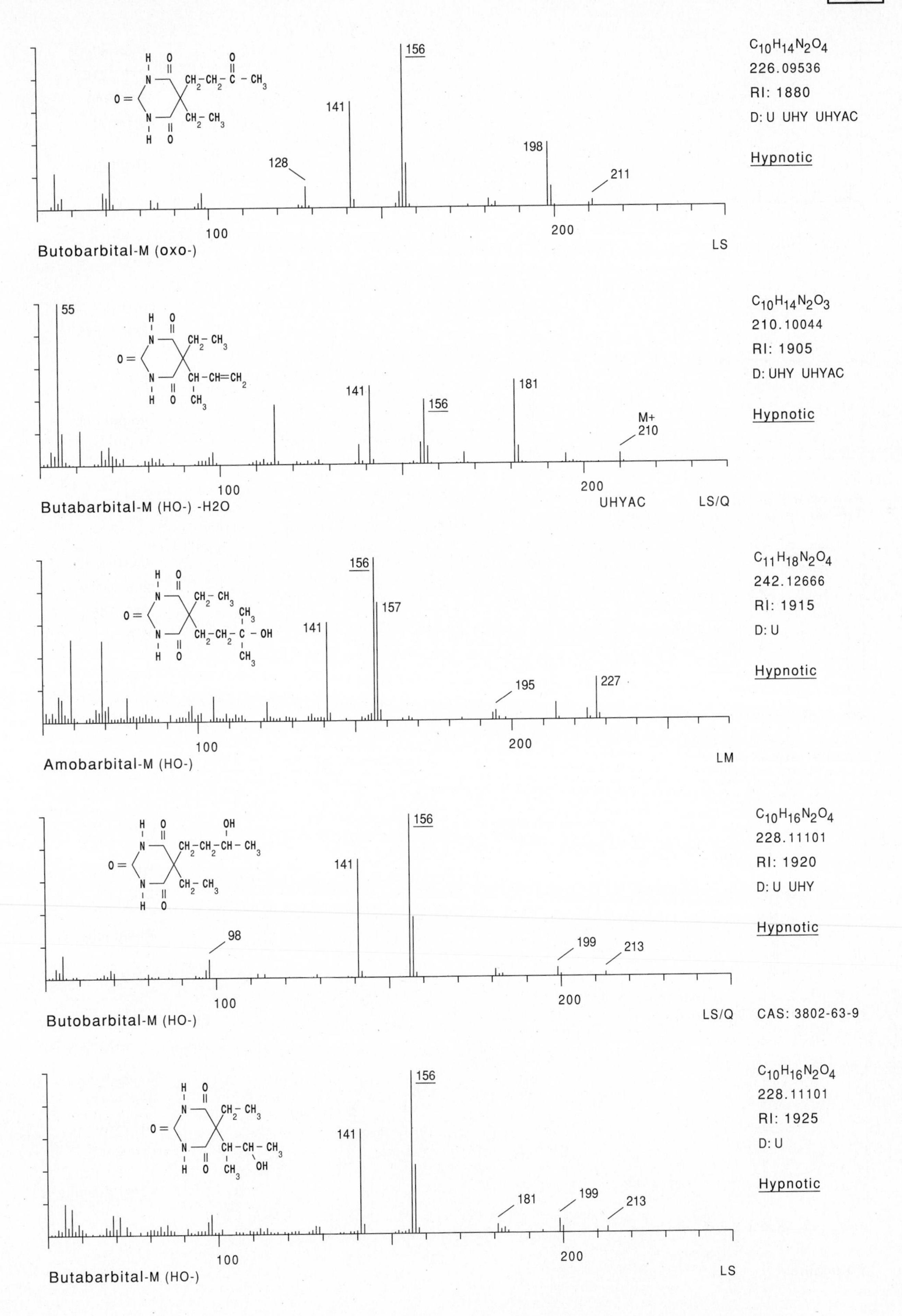

Butobarbital-M (oxo-)

$C_{10}H_{14}N_2O_4$
226.09536
RI: 1880
D: U UHY UHYAC
Hypnotic

Butabarbital-M (HO-) -H2O

$C_{10}H_{14}N_2O_3$
210.10044
RI: 1905
D: UHY UHYAC
Hypnotic

Amobarbital-M (HO-)

$C_{11}H_{18}N_2O_4$
242.12666
RI: 1915
D: U
Hypnotic

Butobarbital-M (HO-)

$C_{10}H_{16}N_2O_4$
228.11101
RI: 1920
D: U UHY
Hypnotic
CAS: 3802-63-9

Butabarbital-M (HO-)

$C_{10}H_{16}N_2O_4$
228.11101
RI: 1925
D: U
Hypnotic

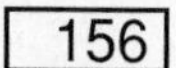

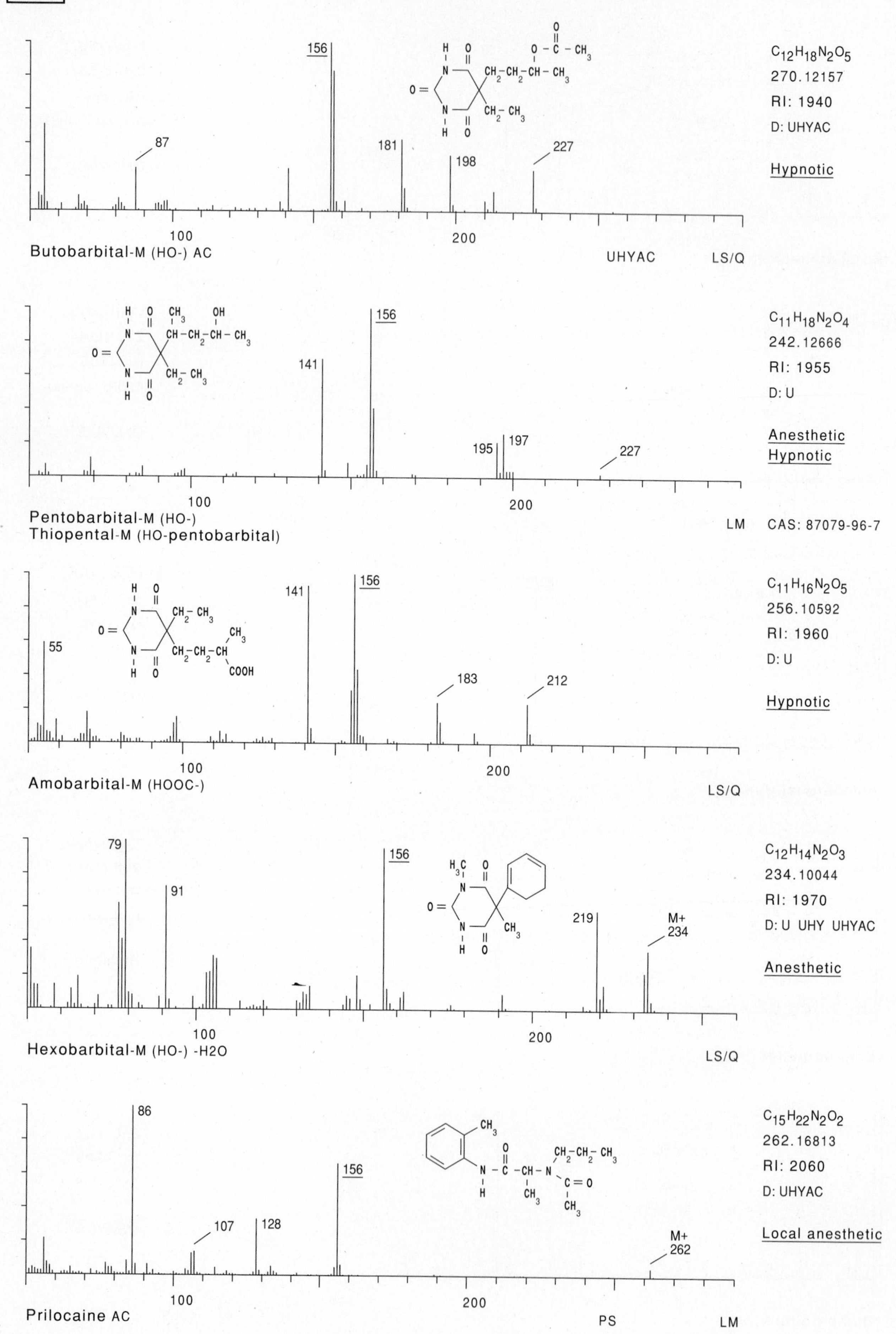

156
87
181
198
227
100
200
Butobarbital-M (HO-) AC
UHYAC
LS/Q
$C_{12}H_{18}N_2O_5$
270.12157
RI: 1940
D: UHYAC
Hypnotic
156
141
195
197
227
100
200
Pentobarbital-M (HO-)
Thiopental-M (HO-pentobarbital)
LM
$C_{11}H_{18}N_2O_4$
242.12666
RI: 1955
D: U
Anesthetic
Hypnotic
CAS: 87079-96-7
141
156
55
183
212
100
200
Amobarbital-M (HOOC-)
LS/Q
$C_{11}H_{16}N_2O_5$
256.10592
RI: 1960
D: U
Hypnotic
79
91
156
219
M+
234
100
200
Hexobarbital-M (HO-) -H2O
LS/Q
$C_{12}H_{14}N_2O_3$
234.10044
RI: 1970
D: U UHY UHYAC
Anesthetic
86
156
107
128
M+
262
100
200
Prilocaine AC
PS
LM
$C_{15}H_{22}N_2O_2$
262.16813
RI: 2060
D: UHYAC
Local anesthetic

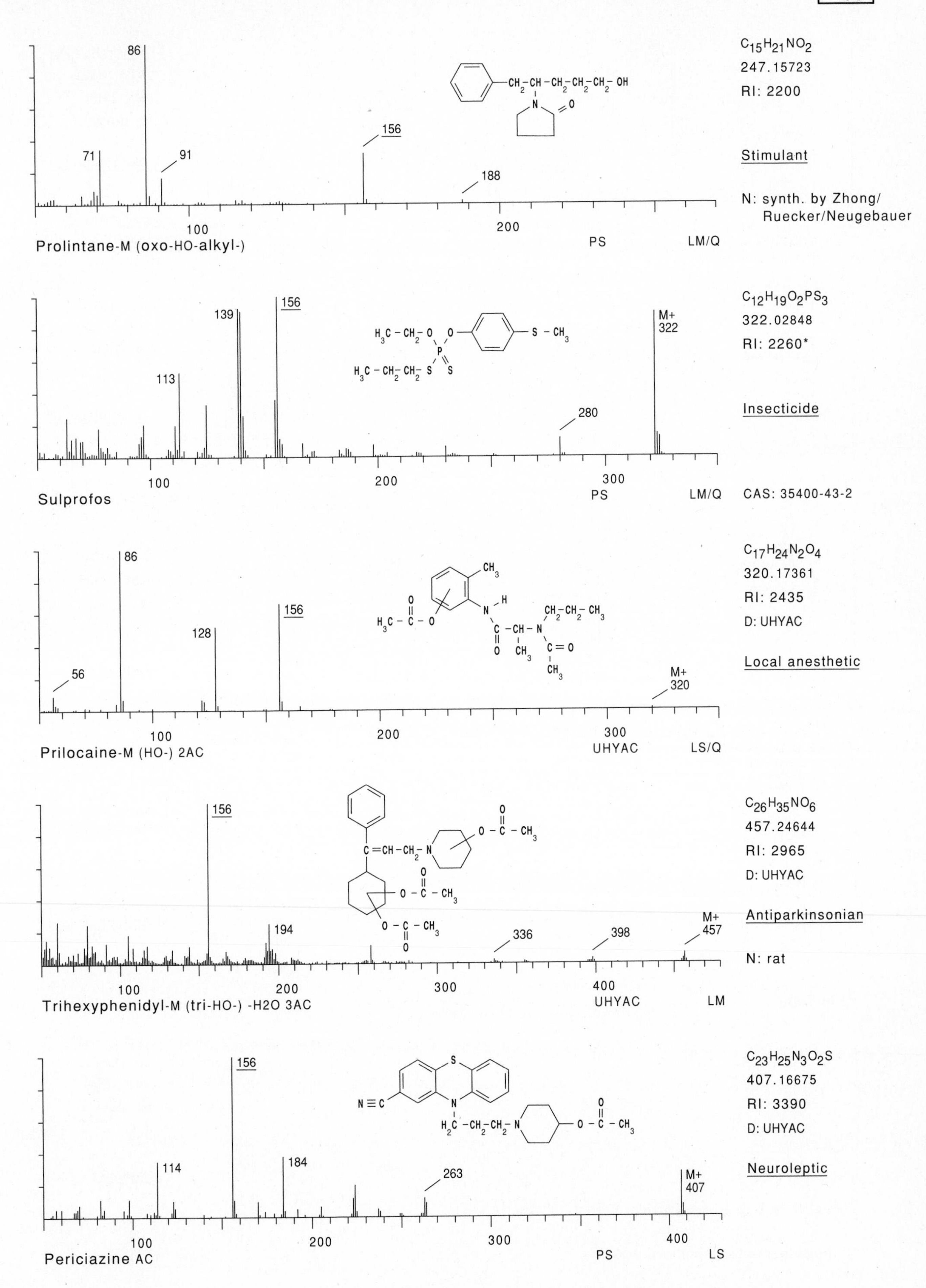

$C_{15}H_{21}NO_2$
247.15723
RI: 2200

Stimulant

N: synth. by Zhong/ Ruecker/Neugebauer

$C_{12}H_{19}O_2PS_3$
322.02848
RI: 2260*

Insecticide

CAS: 35400-43-2

$C_{17}H_{24}N_2O_4$
320.17361
RI: 2435
D: UHYAC

Local anesthetic

$C_{26}H_{35}NO_6$
457.24644
RI: 2965
D: UHYAC

Antiparkinsonian

N: rat

$C_{23}H_{25}N_3O_2S$
407.16675
RI: 3390
D: UHYAC

Neuroleptic

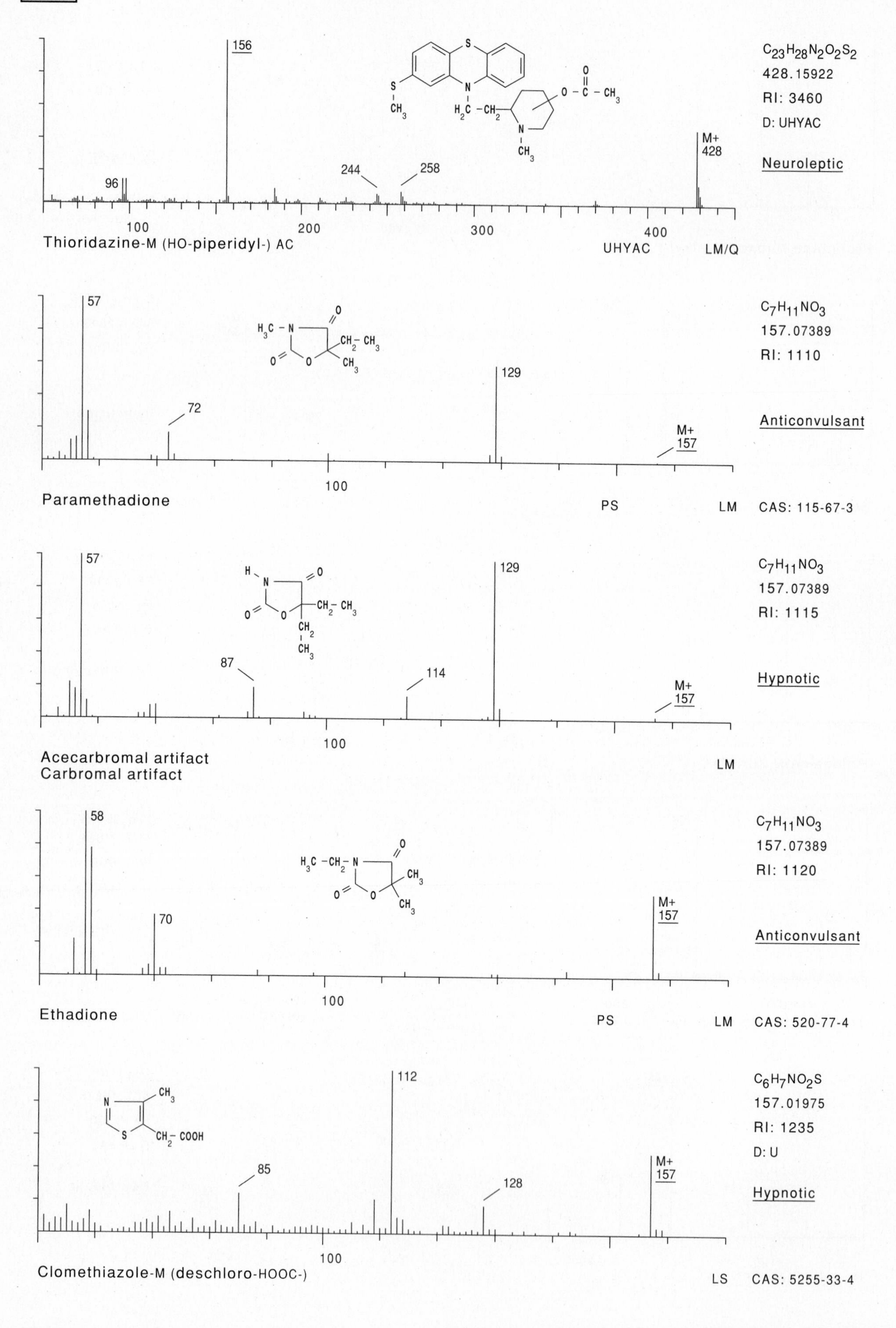
156
96
244
258
M+
428
100
200
300
400
Thioridazine-M (HO-piperidyl-) AC
UHYAC
LM/Q
C23H28N2O2S2
428.15922
RI: 3460
D: UHYAC
Neuroleptic
57
72
129
M+
157
100
Paramethadione
PS
LM
C7H11NO3
157.07389
RI: 1110
Anticonvulsant
CAS: 115-67-3
57
87
114
129
M+
157
100
Acecarbromal artifact
Carbromal artifact
LM
C7H11NO3
157.07389
RI: 1115
Hypnotic
58
70
M+
157
100
Ethadione
PS
LM
C7H11NO3
157.07389
RI: 1120
Anticonvulsant
CAS: 520-77-4
112
85
128
M+
157
100
Clomethiazole-M (deschloro-HOOC-)
LS
C6H7NO2S
157.01975
RI: 1235
D: U
Hypnotic
CAS: 5255-33-4

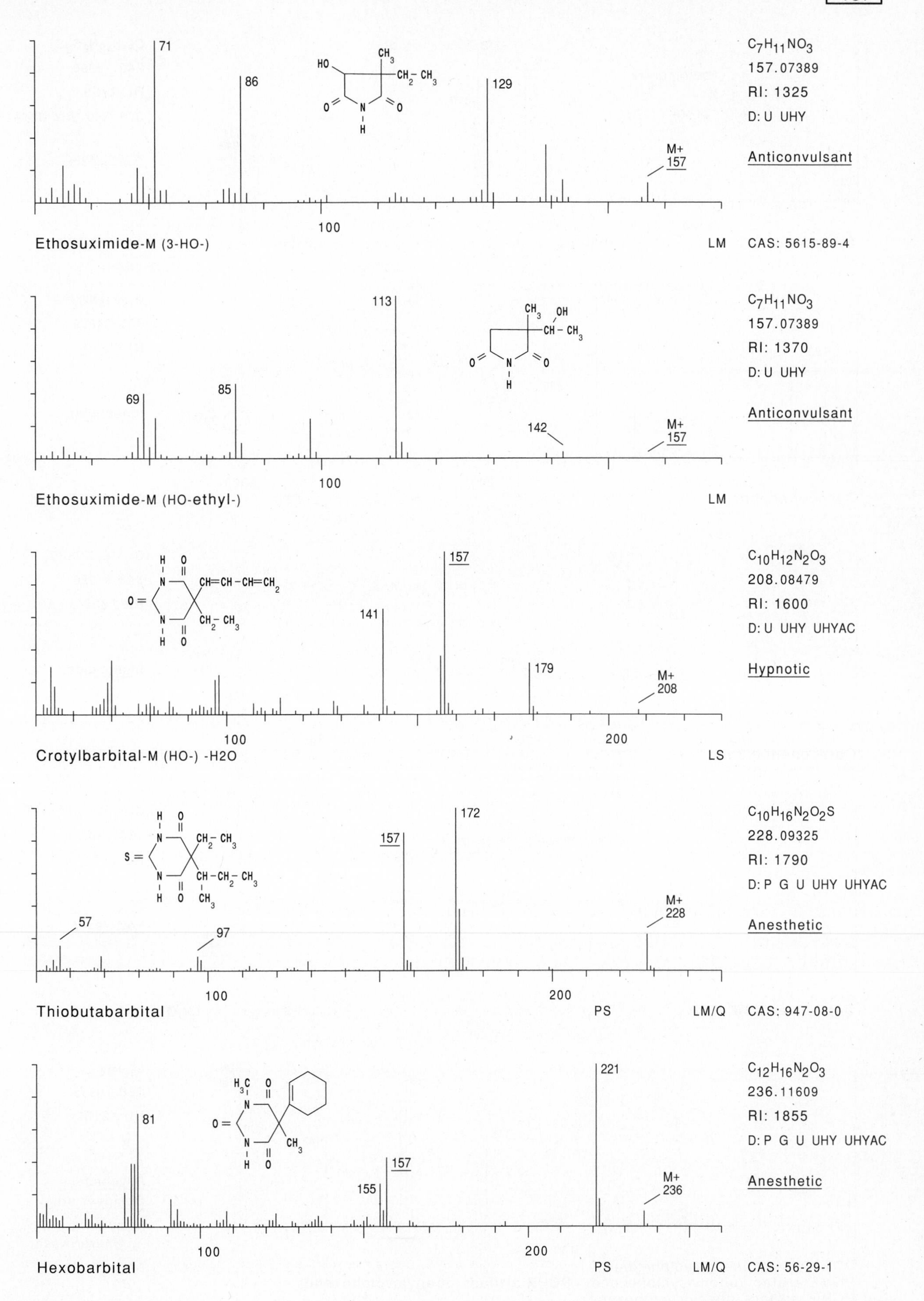

C7H11NO3
157.07389
RI: 1325
D: U UHY
Anticonvulsant
71
86
129
M+
157
100
Ethosuximide-M (3-HO-)
LM
CAS: 5615-89-4
C7H11NO3
157.07389
RI: 1370
D: U UHY
Anticonvulsant
113
69
85
142
M+
157
100
Ethosuximide-M (HO-ethyl-)
LM
C10H12N2O3
208.08479
RI: 1600
D: U UHY UHYAC
Hypnotic
157
141
179
M+
208
100
200
Crotylbarbital-M (HO-) -H2O
LS
C10H16N2O2S
228.09325
RI: 1790
D: P G U UHY UHYAC
Anesthetic
172
157
57
97
M+
228
100
200
Thiobutabarbital
PS
LM/Q
CAS: 947-08-0
C12H16N2O3
236.11609
RI: 1855
D: P G U UHY UHYAC
Anesthetic
221
81
157
155
M+
236
100
200
Hexobarbital
PS
LM/Q
CAS: 56-29-1

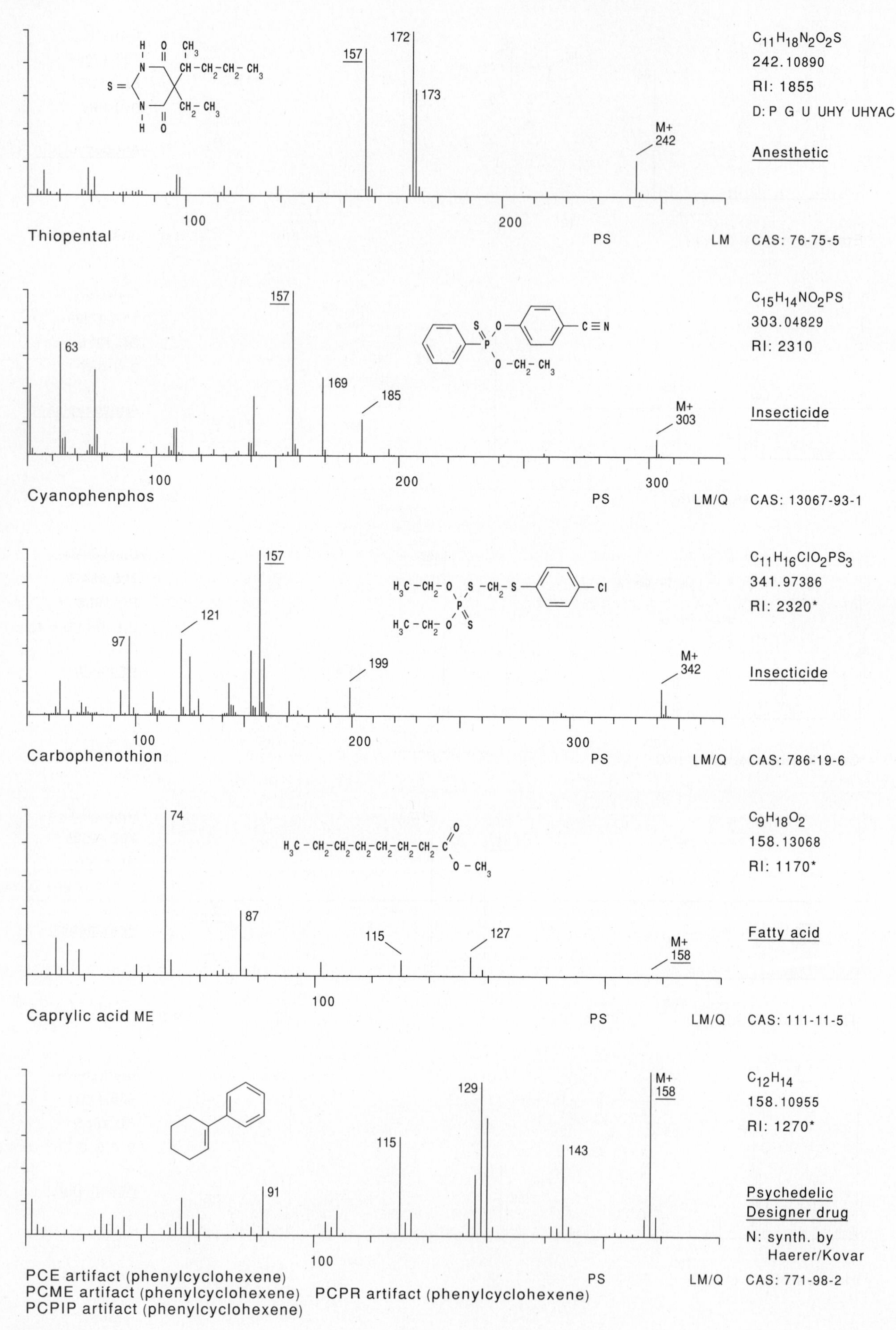

172
157
173
M+
242
100
200
Thiopental
PS
LM
$C_{11}H_{18}N_2O_2S$
242.10890
RI: 1855
D: P G U UHY UHYAC
Anesthetic
CAS: 76-75-5
157
63
169
185
M+
303
100
200
300
Cyanophenphos
PS
LM/Q
$C_{15}H_{14}NO_2PS$
303.04829
RI: 2310
Insecticide
CAS: 13067-93-1
157
121
97
199
M+
342
100
200
300
Carbophenothion
PS
LM/Q
$C_{11}H_{16}ClO_2PS_3$
341.97386
RI: 2320*
Insecticide
CAS: 786-19-6
74
87
115
127
M+
158
100
Caprylic acid ME
PS
LM/Q
$C_9H_{18}O_2$
158.13068
RI: 1170*
Fatty acid
CAS: 111-11-5
129
M+
158
115
143
91
100
PCE artifact (phenylcyclohexene)
PCME artifact (phenylcyclohexene) PCPR artifact (phenylcyclohexene)
PCPIP artifact (phenylcyclohexene)
PS
LM/Q
$C_{12}H_{14}$
158.10955
RI: 1270*
Psychedelic
Designer drug
N: synth. by Haerer/Kovar
CAS: 771-98-2

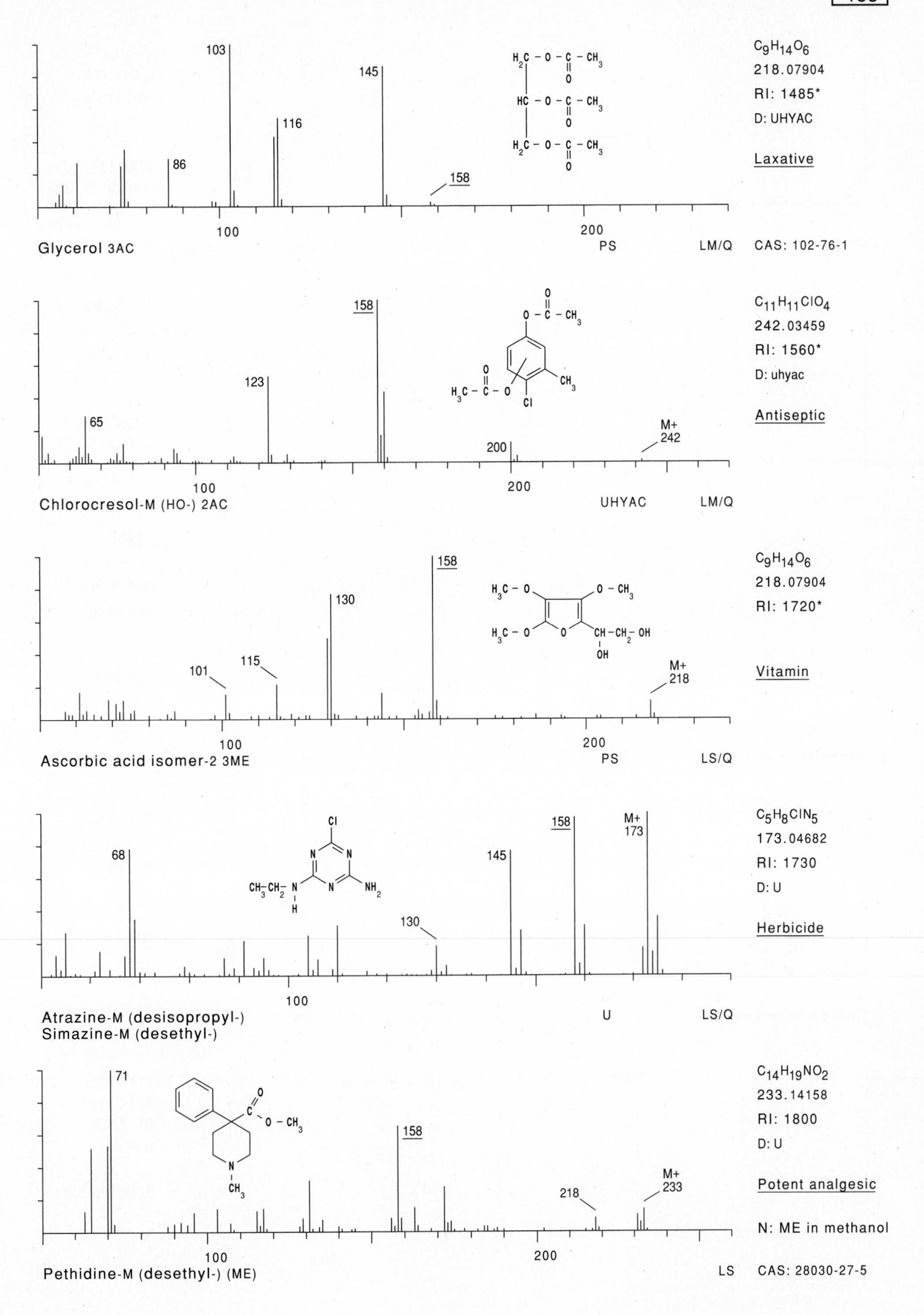

Glycerol 3AC
PS
LM/Q
C9H14O6
218.07904
RI: 1485*
D: UHYAC
Laxative
CAS: 102-76-1
Chlorocresol-M (HO-) 2AC
UHYAC
LM/Q
C11H11ClO4
242.03459
RI: 1560*
D: uhyac
Antiseptic
Ascorbic acid isomer-2 3ME
PS
LS/Q
C9H14O6
218.07904
RI: 1720*
Vitamin
Atrazine-M (desisopropyl-)
Simazine-M (desethyl-)
U
LS/Q
C5H8ClN5
173.04682
RI: 1730
D: U
Herbicide
Pethidine-M (desethyl-) (ME)
LS
C14H19NO2
233.14158
RI: 1800
D: U
Potent analgesic
N: ME in methanol
CAS: 28030-27-5

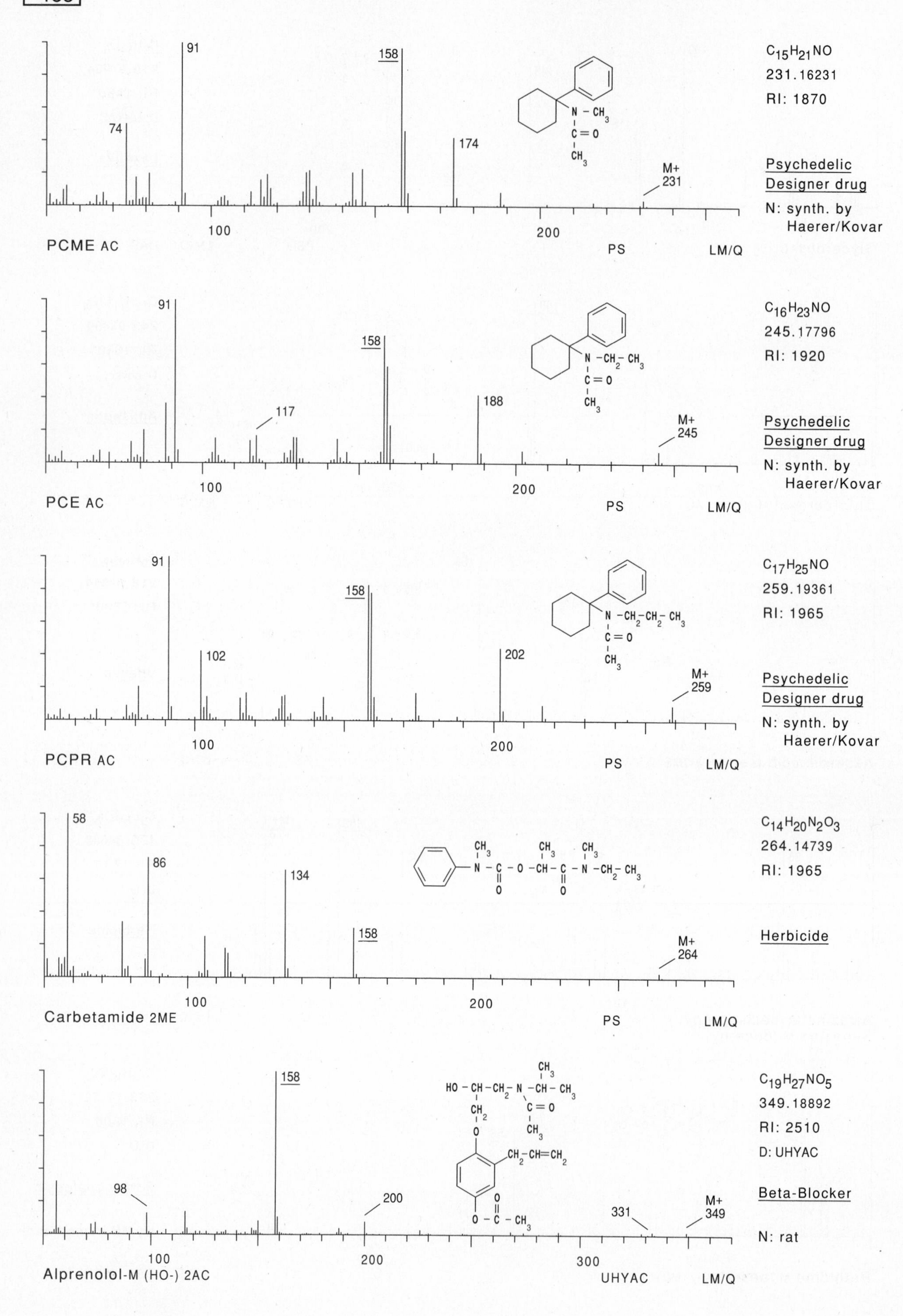
PCME AC
$C_{15}H_{21}NO$
231.16231
RI: 1870
Psychedelic
Designer drug
N: synth. by Haerer/Kovar
PS
LM/Q
PCE AC
$C_{16}H_{23}NO$
245.17796
RI: 1920
Psychedelic
Designer drug
N: synth. by Haerer/Kovar
PS
LM/Q
PCPR AC
$C_{17}H_{25}NO$
259.19361
RI: 1965
Psychedelic
Designer drug
N: synth. by Haerer/Kovar
PS
LM/Q
Carbetamide 2ME
$C_{14}H_{20}N_2O_3$
264.14739
RI: 1965
Herbicide
PS
LM/Q
Alprenolol-M (HO-) 2AC
$C_{19}H_{27}NO_5$
349.18892
RI: 2510
D: UHYAC
Beta-Blocker
N: rat
UHYAC
LM/Q

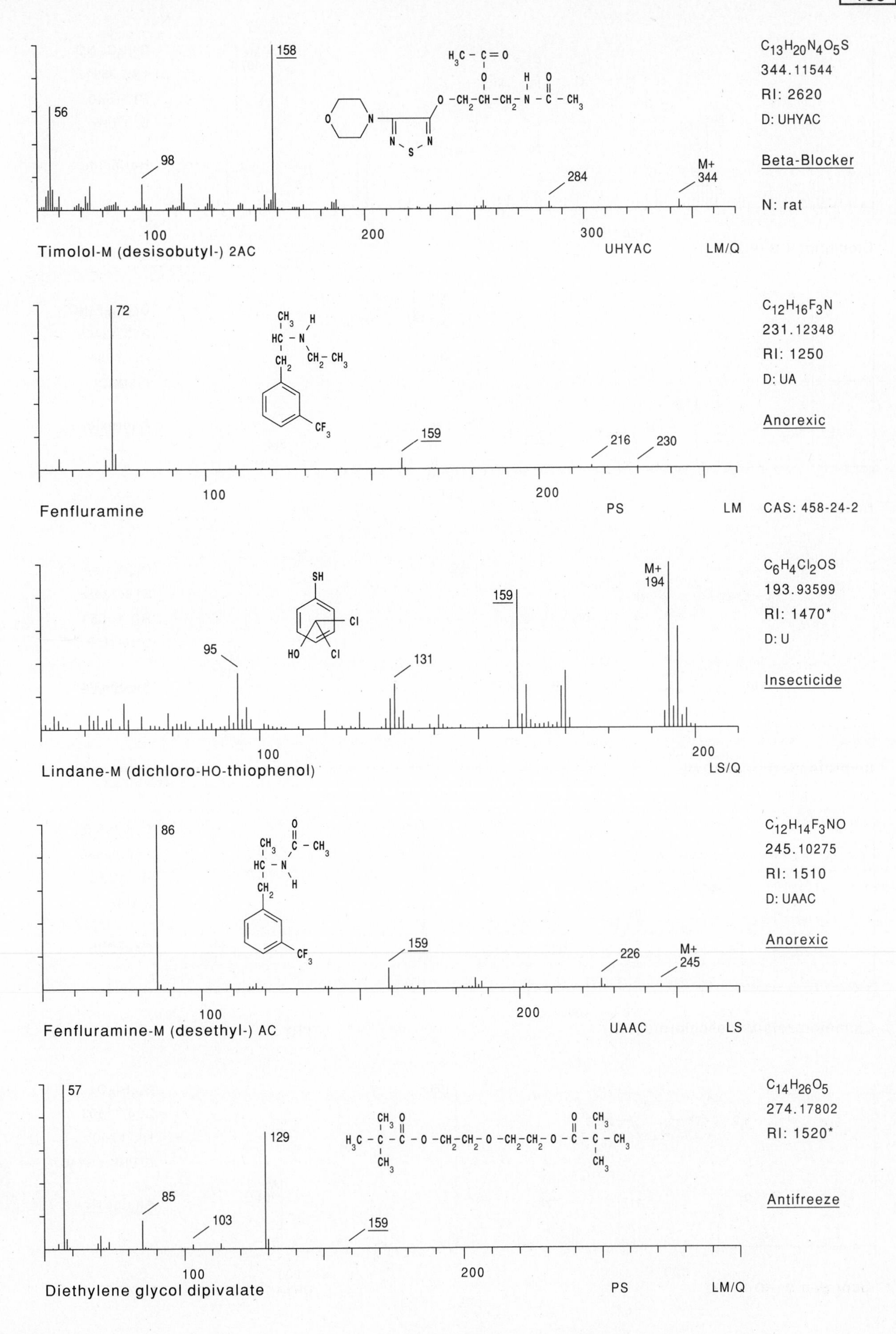
158
56
98
284
M+
344
100
200
300
Timolol-M (desisobutyl-) 2AC
UHYAC
LM/Q
C13H20N4O5S
344.11544
RI: 2620
D: UHYAC
Beta-Blocker
N: rat
72
159
216
230
100
200
Fenfluramine
PS
LM
C12H16F3N
231.12348
RI: 1250
D: UA
Anorexic
CAS: 458-24-2
M+
194
159
95
131
100
200
Lindane-M (dichloro-HO-thiophenol)
LS/Q
C6H4Cl2OS
193.93599
RI: 1470*
D: U
Insecticide
86
159
226
M+
245
100
200
Fenfluramine-M (desethyl-) AC
UAAC
LS
C12H14F3NO
245.10275
RI: 1510
D: UAAC
Anorexic
57
129
85
103
159
100
200
Diethylene glycol dipivalate
PS
LM/Q
C14H26O5
274.17802
RI: 1520*
Antifreeze

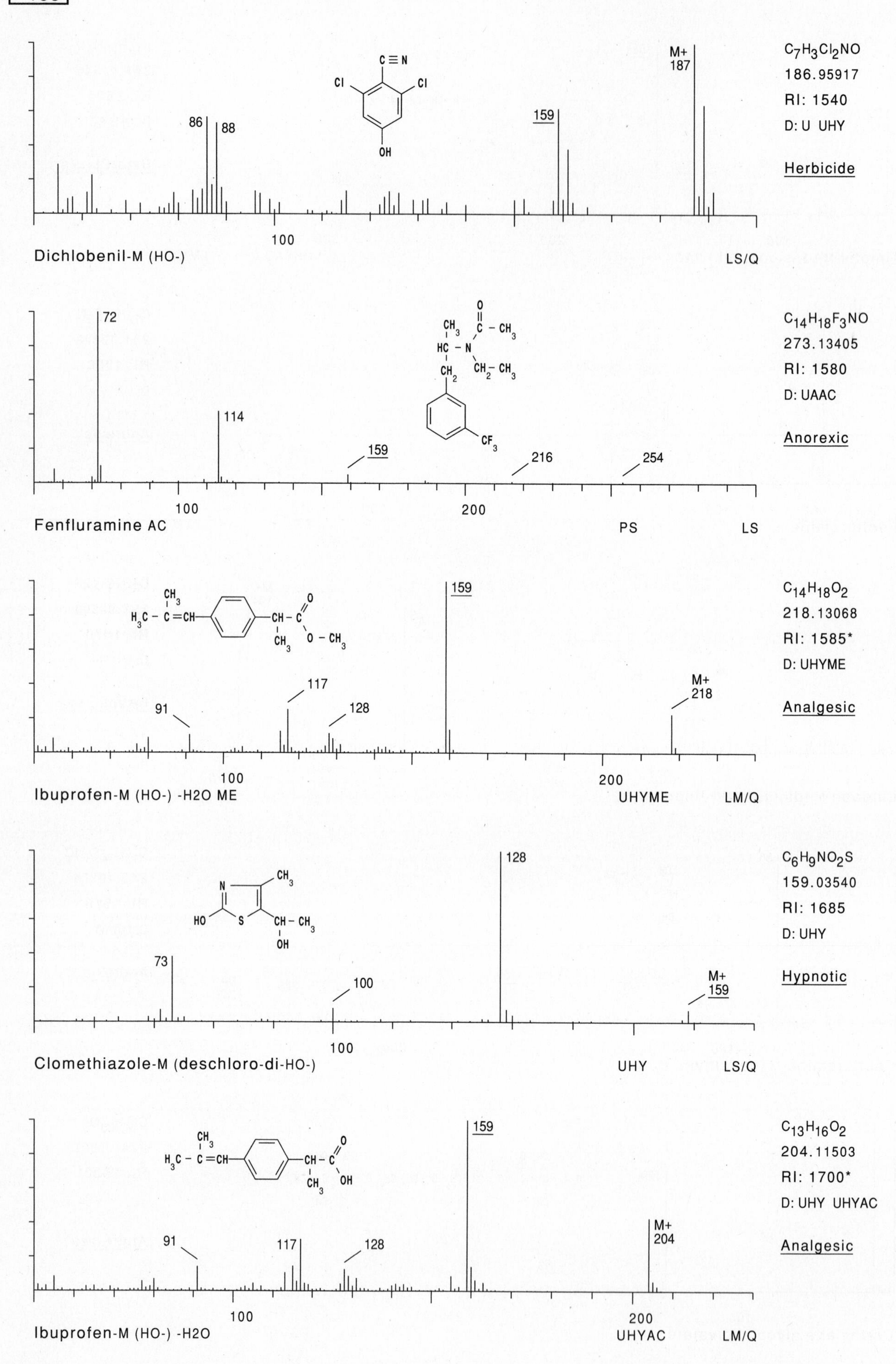

M+
187
86
88
159
100
Dichlobenil-M (HO-)
LS/Q
$C_7H_3Cl_2NO$
186.95917
RI: 1540
D: U UHY
Herbicide
72
114
159
216
254
100
200
Fenfluramine AC
PS
LS
$C_{14}H_{18}F_3NO$
273.13405
RI: 1580
D: UAAC
Anorexic
159
M+
218
117
91
128
100
200
Ibuprofen-M (HO-) -H2O ME
UHYME
LM/Q
$C_{14}H_{18}O_2$
218.13068
RI: 1585*
D: UHYME
Analgesic
128
73
100
M+
159
100
Clomethiazole-M (deschloro-di-HO-)
UHY
LS/Q
$C_6H_9NO_2S$
159.03540
RI: 1685
D: UHY
Hypnotic
159
M+
204
91
117
128
100
200
UHYAC
Ibuprofen-M (HO-) -H2O
LM/Q
$C_{13}H_{16}O_2$
204.11503
RI: 1700*
D: UHY UHYAC
Analgesic

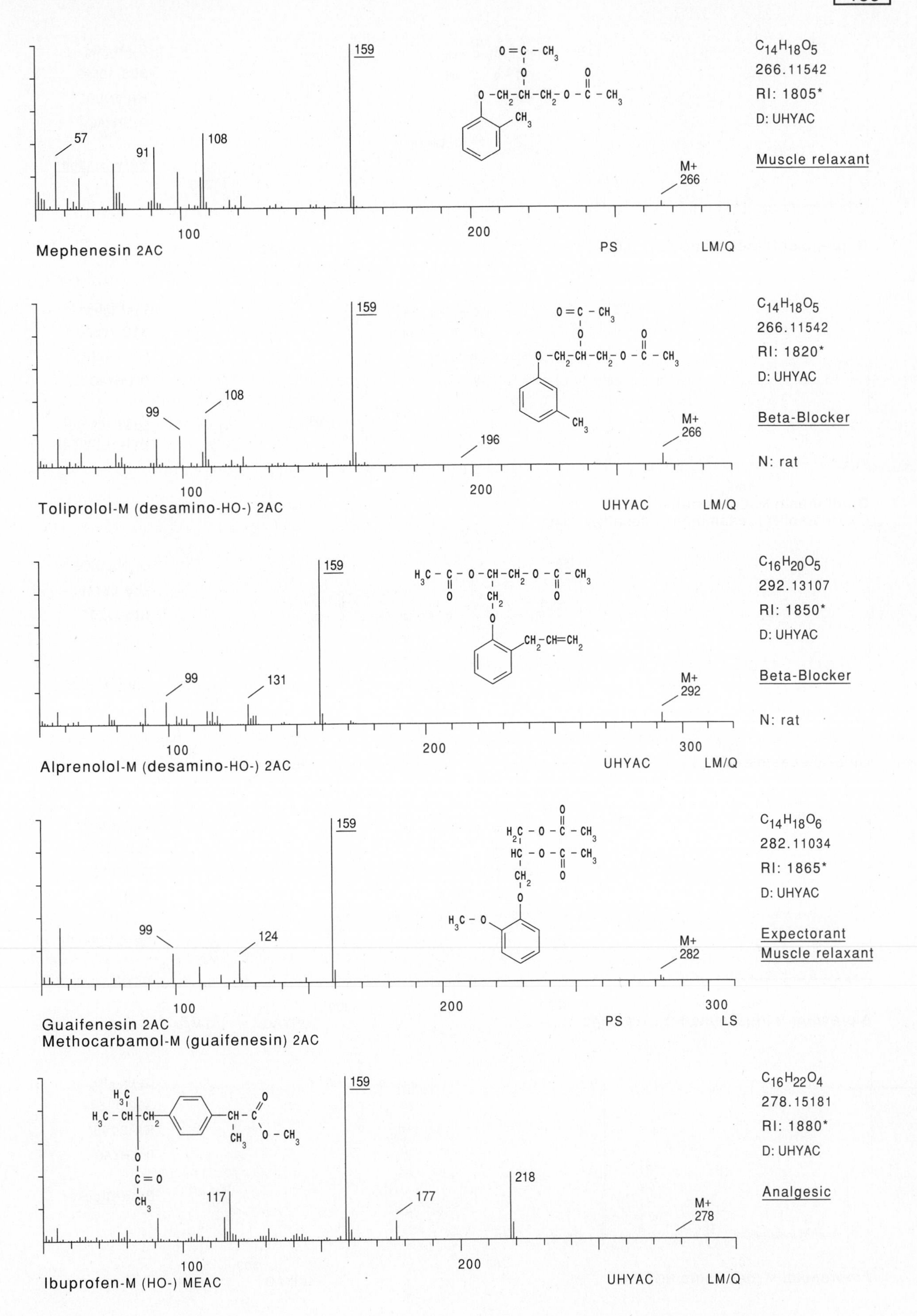

159
57
91
108
M+
266
100
200
Mephenesin 2AC
PS
LM/Q
C14H18O5
266.11542
RI: 1805*
D: UHYAC
Muscle relaxant
159
99
108
196
M+
266
100
200
Toliprolol-M (desamino-HO-) 2AC
UHYAC
LM/Q
C14H18O5
266.11542
RI: 1820*
D: UHYAC
Beta-Blocker
N: rat
159
99
131
M+
292
100
200
300
Alprenolol-M (desamino-HO-) 2AC
UHYAC
LM/Q
C16H20O5
292.13107
RI: 1850*
D: UHYAC
Beta-Blocker
N: rat
159
99
124
M+
282
100
200
300
Guaifenesin 2AC
Methocarbamol-M (guaifenesin) 2AC
PS
LS
C14H18O6
282.11034
RI: 1865*
D: UHYAC
Expectorant
Muscle relaxant
159
117
177
218
M+
278
100
200
Ibuprofen-M (HO-) MEAC
UHYAC
LM/Q
C16H22O4
278.15181
RI: 1880*
D: UHYAC
Analgesic

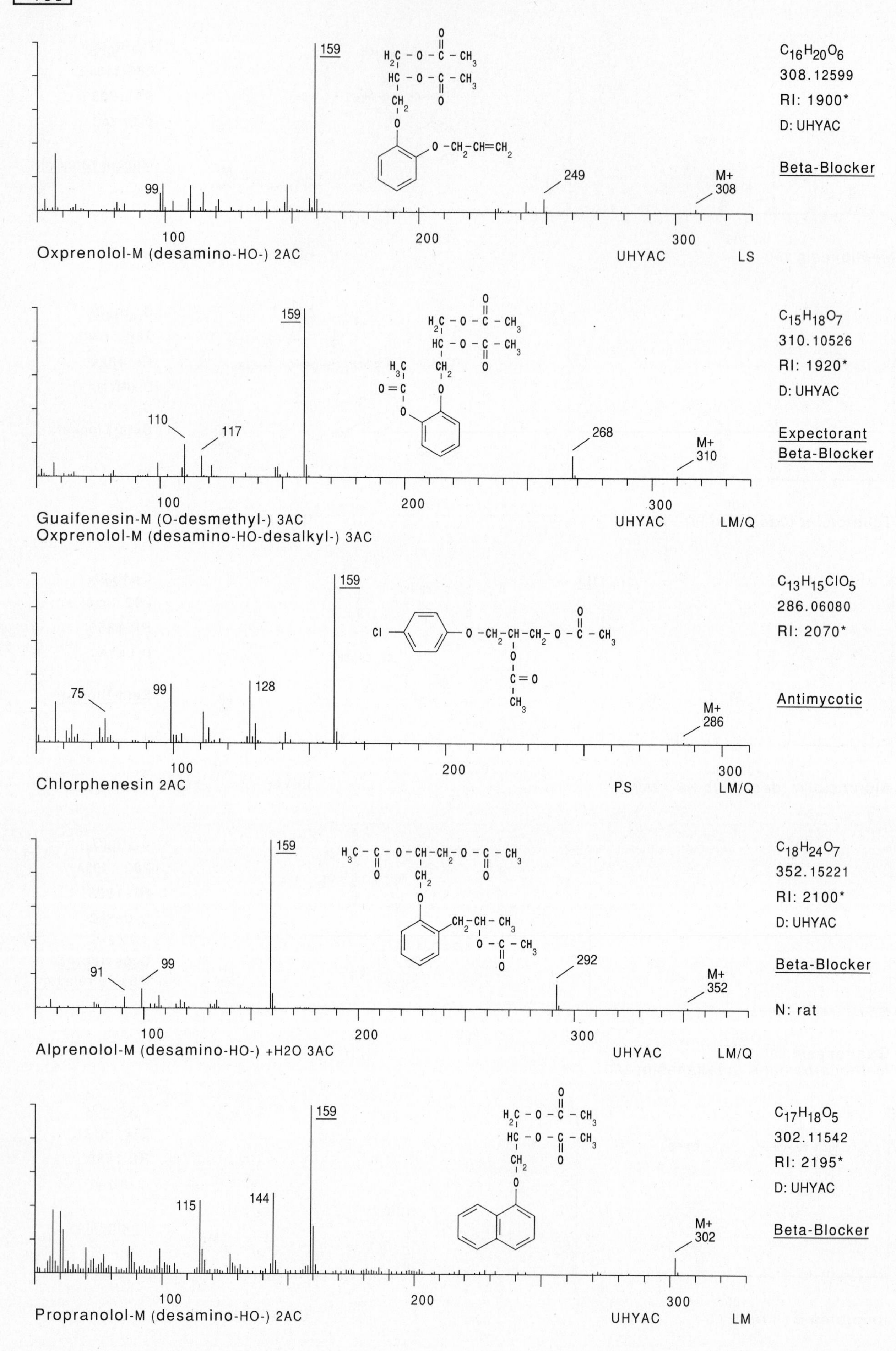

159
99
249
M+
308
100
200
300
Oxprenolol-M (desamino-HO-) 2AC
UHYAC
LS
C16H20O6
308.12599
RI: 1900*
D: UHYAC
Beta-Blocker
159
110
117
268
M+
310
100
200
300
Guaifenesin-M (O-desmethyl-) 3AC
Oxprenolol-M (desamino-HO-desalkyl-) 3AC
UHYAC
LM/Q
C15H18O7
310.10526
RI: 1920*
D: UHYAC
Expectorant
Beta-Blocker
159
75
99
128
M+
286
100
200
300
Chlorphenesin 2AC
PS
LM/Q
C13H15ClO5
286.06080
RI: 2070*
Antimycotic
159
91
99
292
M+
352
100
200
300
Alprenolol-M (desamino-HO-) +H2O 3AC
UHYAC
LM/Q
C18H24O7
352.15221
RI: 2100*
D: UHYAC
Beta-Blocker
N: rat
159
115
144
M+
302
100
200
300
Propranolol-M (desamino-HO-) 2AC
UHYAC
LM
C17H18O5
302.11542
RI: 2195*
D: UHYAC
Beta-Blocker

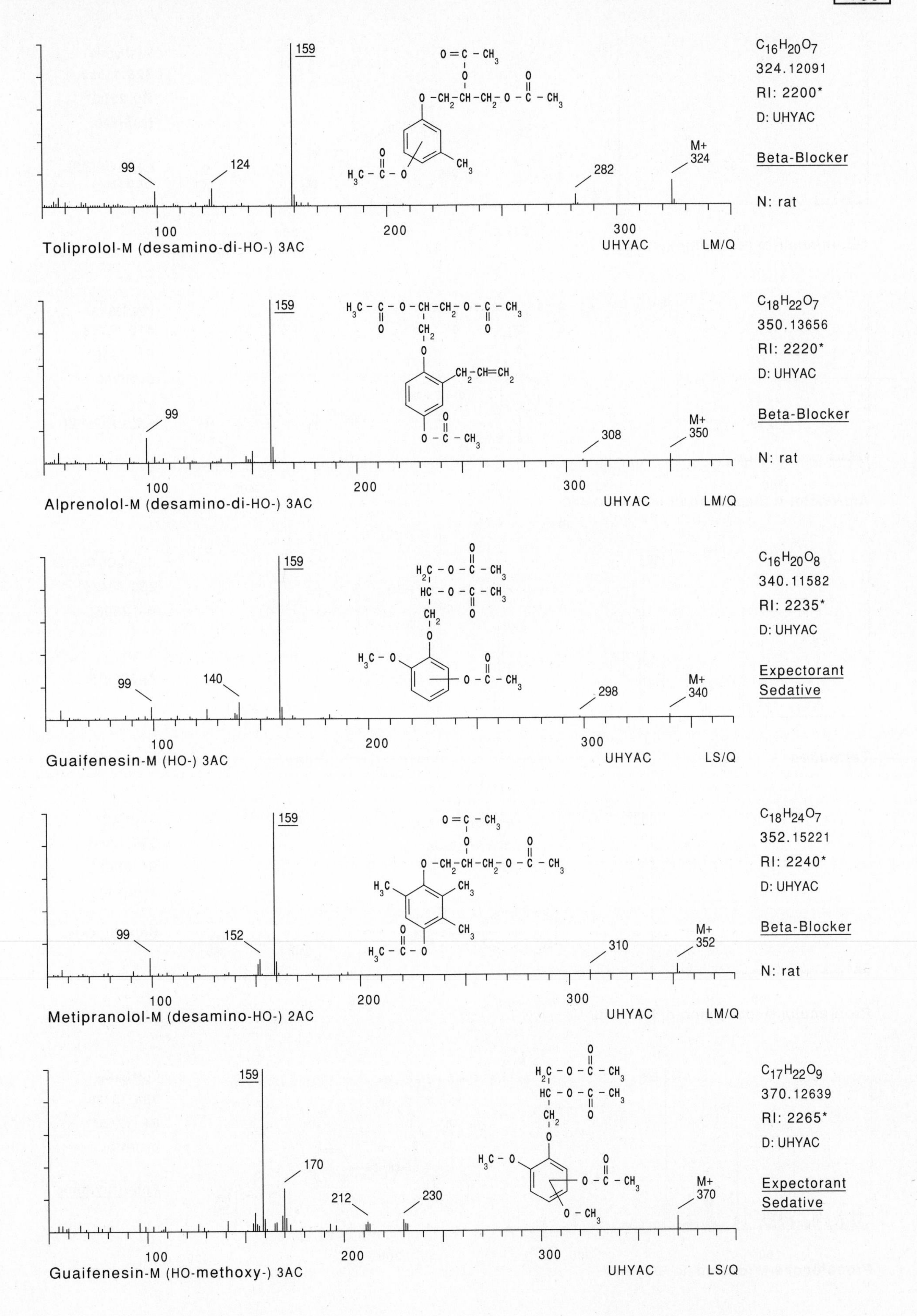

$C_{16}H_{20}O_7$
324.12091
RI: 2200*
D: UHYAC
Beta-Blocker
N: rat

$C_{18}H_{22}O_7$
350.13656
RI: 2220*
D: UHYAC
Beta-Blocker
N: rat

$C_{16}H_{20}O_8$
340.11582
RI: 2235*
D: UHYAC
Expectorant
Sedative

$C_{18}H_{24}O_7$
352.15221
RI: 2240*
D: UHYAC
Beta-Blocker
N: rat

$C_{17}H_{22}O_9$
370.12639
RI: 2265*
D: UHYAC
Expectorant
Sedative

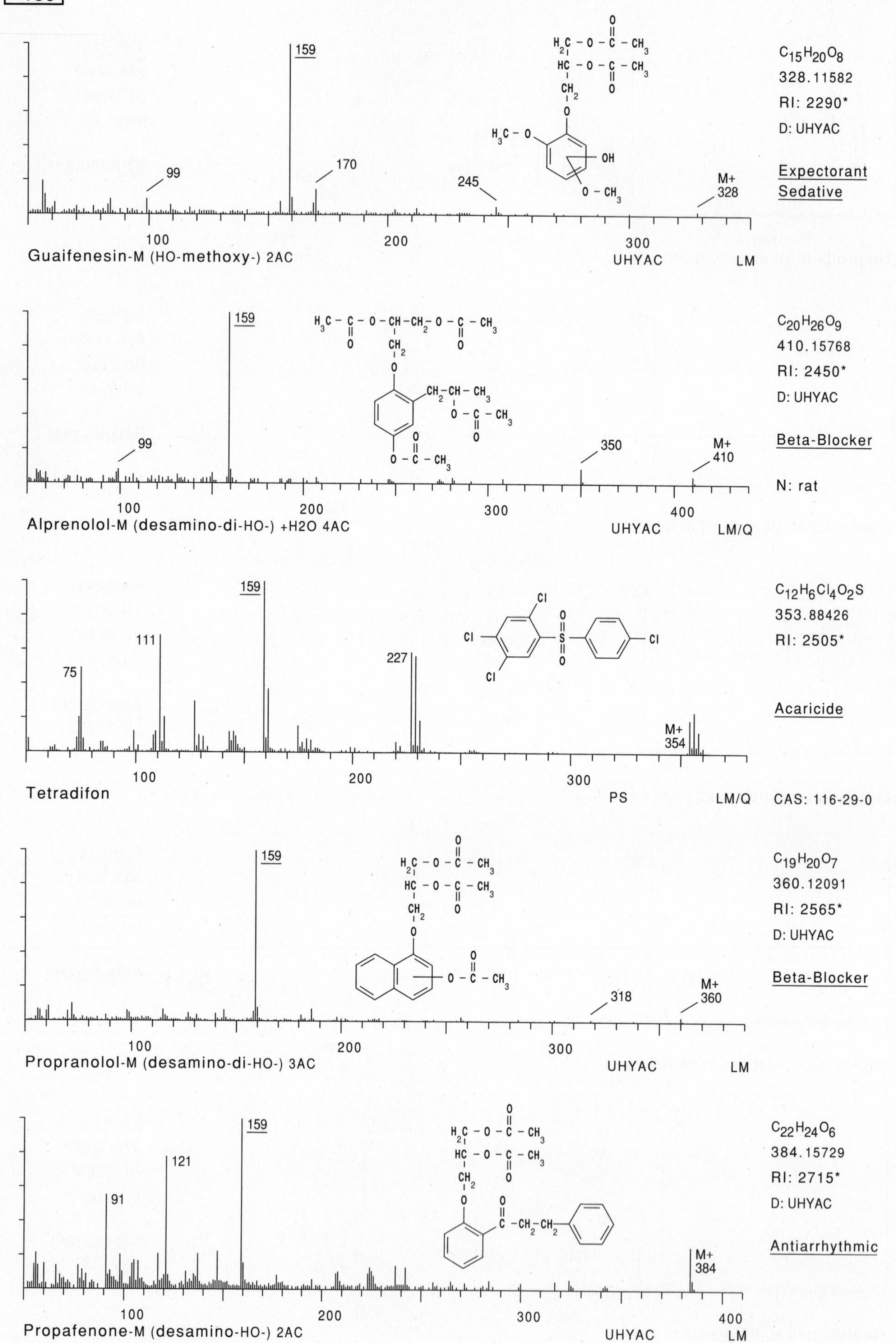
159
99
170
245
M+
328
100
200
300
Guaifenesin-M (HO-methoxy-) 2AC
UHYAC
LM
$C_{15}H_{20}O_8$
328.11582
RI: 2290*
D: UHYAC
Expectorant
Sedative
159
99
350
M+
410
100
200
300
400
Alprenolol-M (desamino-di-HO-) +H2O 4AC
UHYAC
LM/Q
$C_{20}H_{26}O_9$
410.15768
RI: 2450*
D: UHYAC
Beta-Blocker
N: rat
159
111
75
227
M+
354
100
200
300
Tetradifon
PS
LM/Q
$C_{12}H_6Cl_4O_2S$
353.88426
RI: 2505*
Acaricide
CAS: 116-29-0
159
318
M+
360
100
200
300
Propranolol-M (desamino-di-HO-) 3AC
UHYAC
LM
$C_{19}H_{20}O_7$
360.12091
RI: 2565*
D: UHYAC
Beta-Blocker
159
121
91
M+
384
100
200
300
400
Propafenone-M (desamino-HO-) 2AC
UHYAC
LM
$C_{22}H_{24}O_6$
384.15729
RI: 2715*
D: UHYAC
Antiarrhythmic

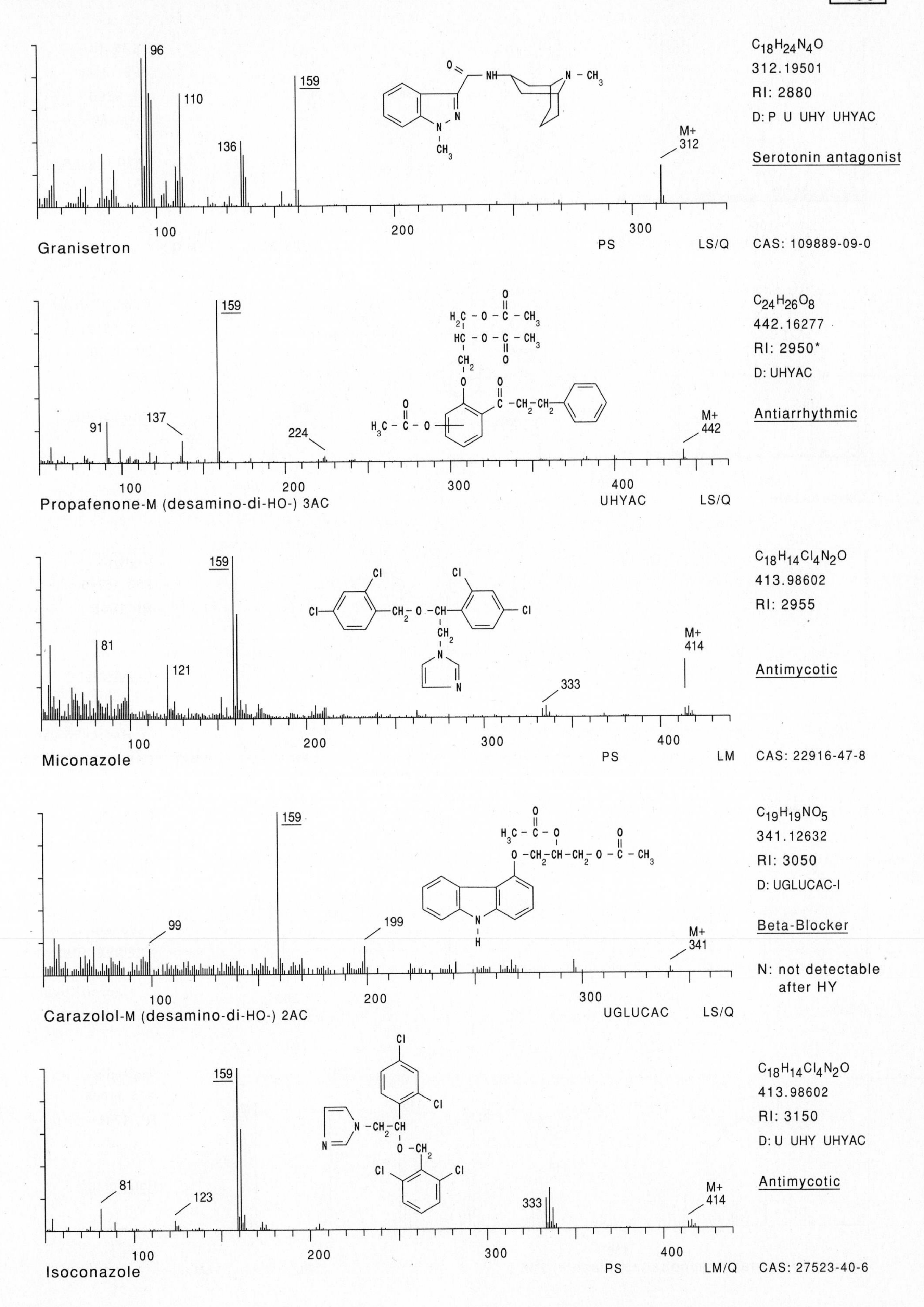

96
110
136
159
M+
312
100
200
300
PS
LS/Q
Granisetron
$C_{18}H_{24}N_4O$
312.19501
RI: 2880
D: P U UHY UHYAC
Serotonin antagonist
CAS: 109889-09-0
91
137
159
224
M+
442
100
200
300
400
UHYAC
LS/Q
Propafenone-M (desamino-di-HO-) 3AC
$C_{24}H_{26}O_8$
442.16277
RI: 2950*
D: UHYAC
Antiarrhythmic
81
121
159
333
M+
414
100
200
300
400
PS
LM
Miconazole
$C_{18}H_{14}Cl_4N_2O$
413.98602
RI: 2955
Antimycotic
CAS: 22916-47-8
99
159
199
M+
341
100
200
300
UGLUCAC
LS/Q
Carazolol-M (desamino-di-HO-) 2AC
$C_{19}H_{19}NO_5$
341.12632
RI: 3050
D: UGLUCAC-I
Beta-Blocker
N: not detectable after HY
81
123
159
333
M+
414
100
200
300
400
PS
LM/Q
Isoconazole
$C_{18}H_{14}Cl_4N_2O$
413.98602
RI: 3150
D: U UHY UHYAC
Antimycotic
CAS: 27523-40-6

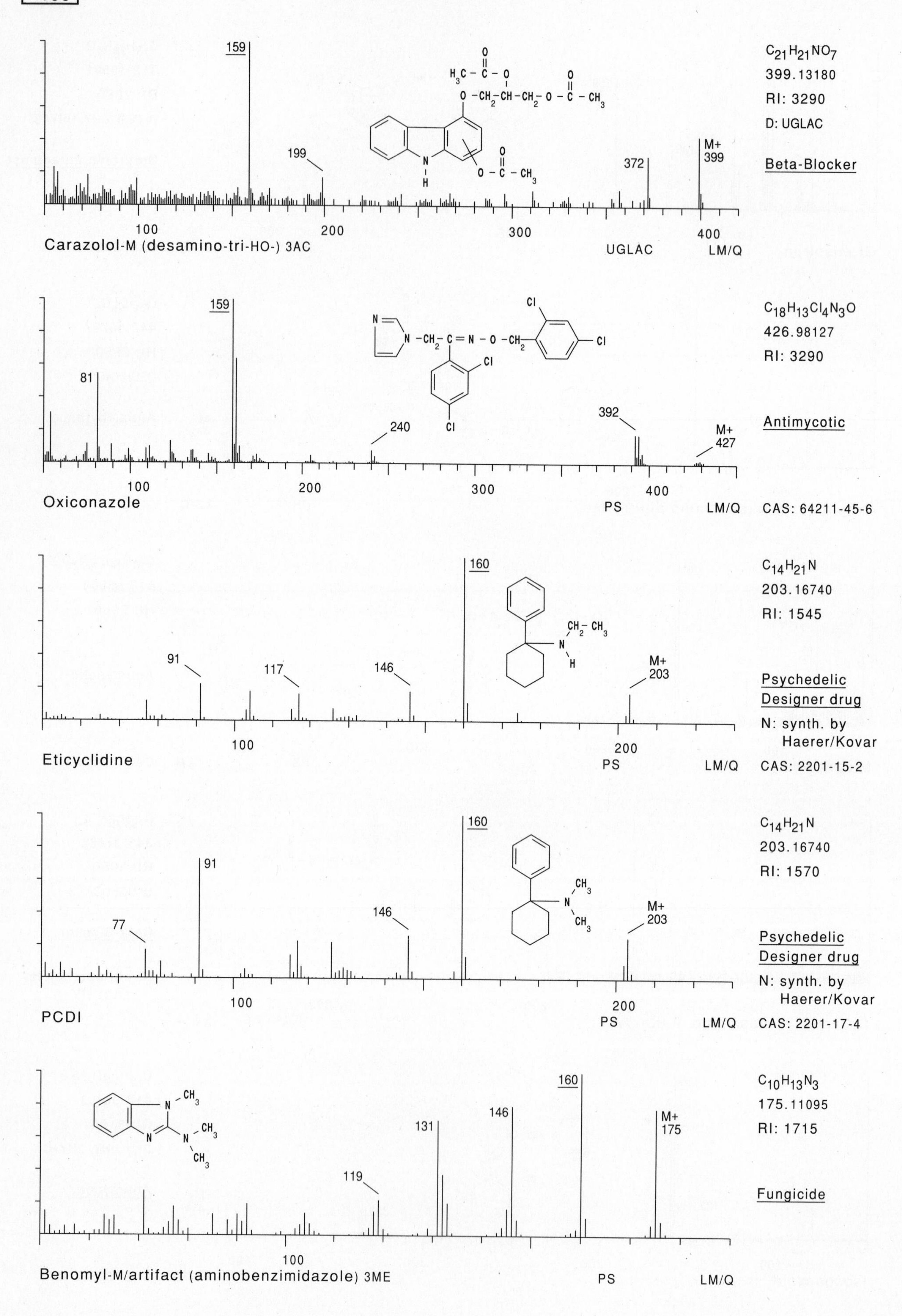
159
199
372
M+
399
100
200
300
400
UGLAC
LM/Q
Carazolol-M (desamino-tri-HO-) 3AC
$C_{21}H_{21}NO_7$
399.13180
RI: 3290
D: UGLAC
Beta-Blocker
159
81
240
392
M+
427
100
200
300
400
PS
LM/Q
Oxiconazole
$C_{18}H_{13}Cl_4N_3O$
426.98127
RI: 3290
Antimycotic
CAS: 64211-45-6
160
91
117
146
M+
203
100
200
PS
LM/Q
Eticyclidine
$C_{14}H_{21}N$
203.16740
RI: 1545
Psychedelic
Designer drug
N: synth. by
Haerer/Kovar
CAS: 2201-15-2
160
91
77
146
M+
203
100
200
PS
LM/Q
PCDI
$C_{14}H_{21}N$
203.16740
RI: 1570
Psychedelic
Designer drug
N: synth. by
Haerer/Kovar
CAS: 2201-17-4
160
146
131
119
M+
175
100
PS
LM/Q
Benomyl-M/artifact (aminobenzimidazole) 3ME
$C_{10}H_{13}N_3$
175.11095
RI: 1715
Fungicide

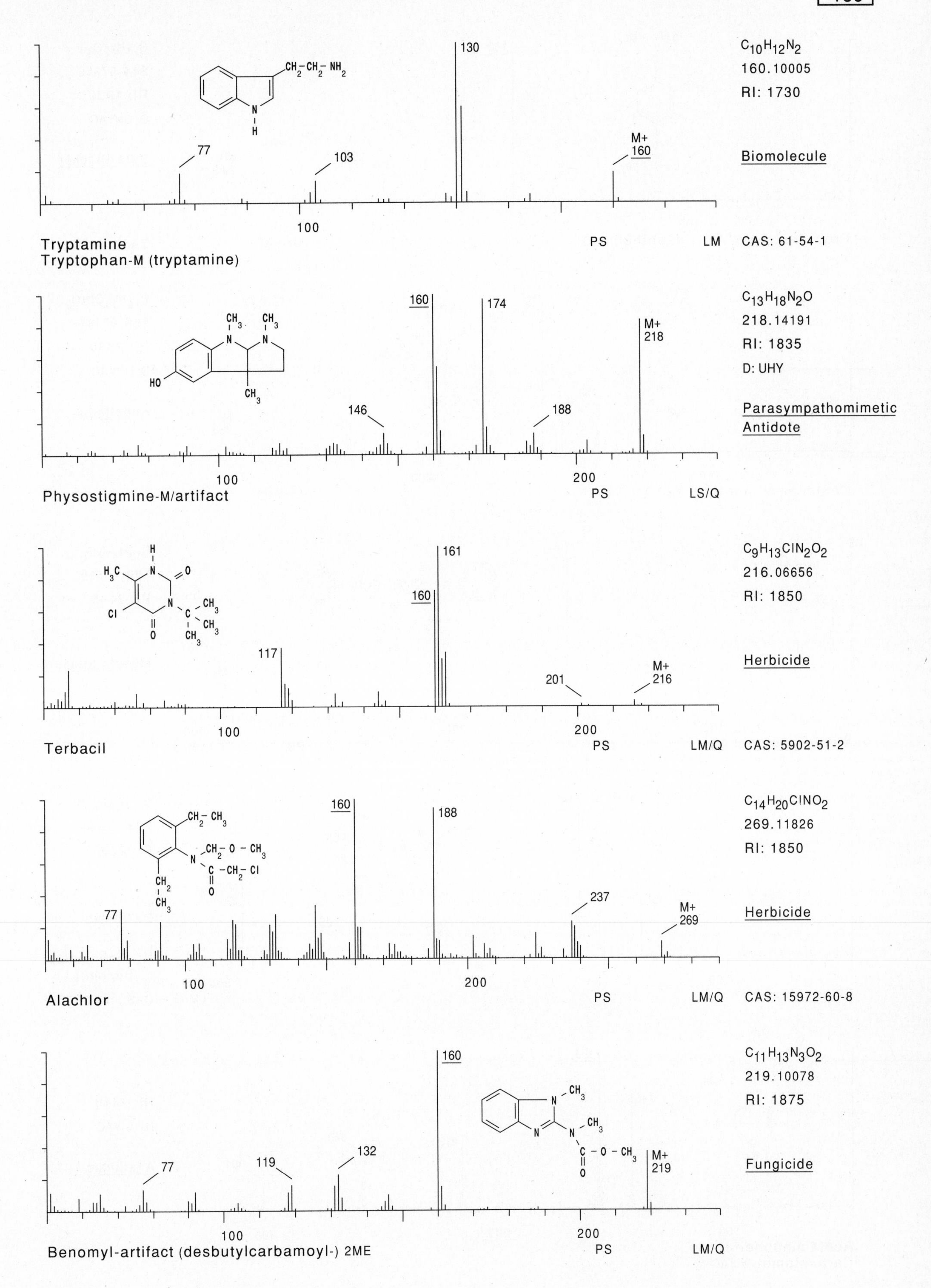

Tryptamine
Tryptophan-M (tryptamine)

$C_{10}H_{12}N_2$
160.10005
RI: 1730

Biomolecule

CAS: 61-54-1

Physostigmine-M/artifact

$C_{13}H_{18}N_2O$
218.14191
RI: 1835
D: UHY

Parasympathomimetic
Antidote

Terbacil

$C_9H_{13}ClN_2O_2$
216.06656
RI: 1850

Herbicide

CAS: 5902-51-2

Alachlor

$C_{14}H_{20}ClNO_2$
269.11826
RI: 1850

Herbicide

CAS: 15972-60-8

Benomyl-artifact (desbutylcarbamoyl-) 2ME

$C_{11}H_{13}N_3O_2$
219.10078
RI: 1875

Fungicide

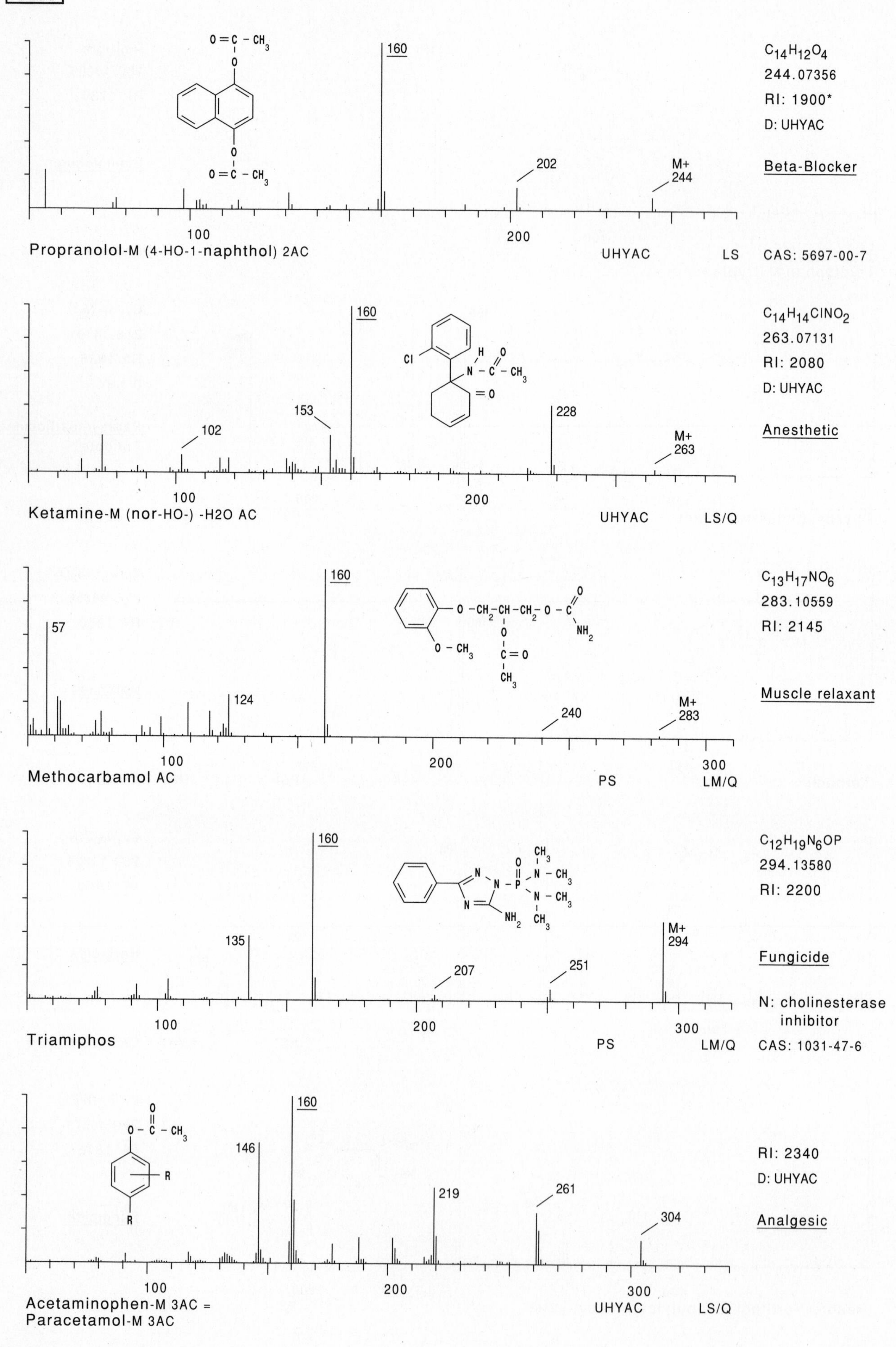
160
202
M+
244
100
200
Propranolol-M (4-HO-1-naphthol) 2AC
UHYAC
LS
$C_{14}H_{12}O_4$
244.07356
RI: 1900*
D: UHYAC
Beta-Blocker
CAS: 5697-00-7
160
153
102
228
M+
263
100
200
Ketamine-M (nor-HO-) -H2O AC
UHYAC
LS/Q
$C_{14}H_{14}ClNO_2$
263.07131
RI: 2080
D: UHYAC
Anesthetic
160
57
124
240
M+
283
100
200
300
Methocarbamol AC
PS
LM/Q
$C_{13}H_{17}NO_6$
283.10559
RI: 2145
Muscle relaxant
160
135
M+
294
207
251
100
200
300
Triamiphos
PS
LM/Q
$C_{12}H_{19}N_6OP$
294.13580
RI: 2200
Fungicide
N: cholinesterase inhibitor
CAS: 1031-47-6
160
146
219
261
304
100
200
300
Acetaminophen-M 3AC =
Paracetamol-M 3AC
UHYAC
LS/Q
RI: 2340
D: UHYAC
Analgesic

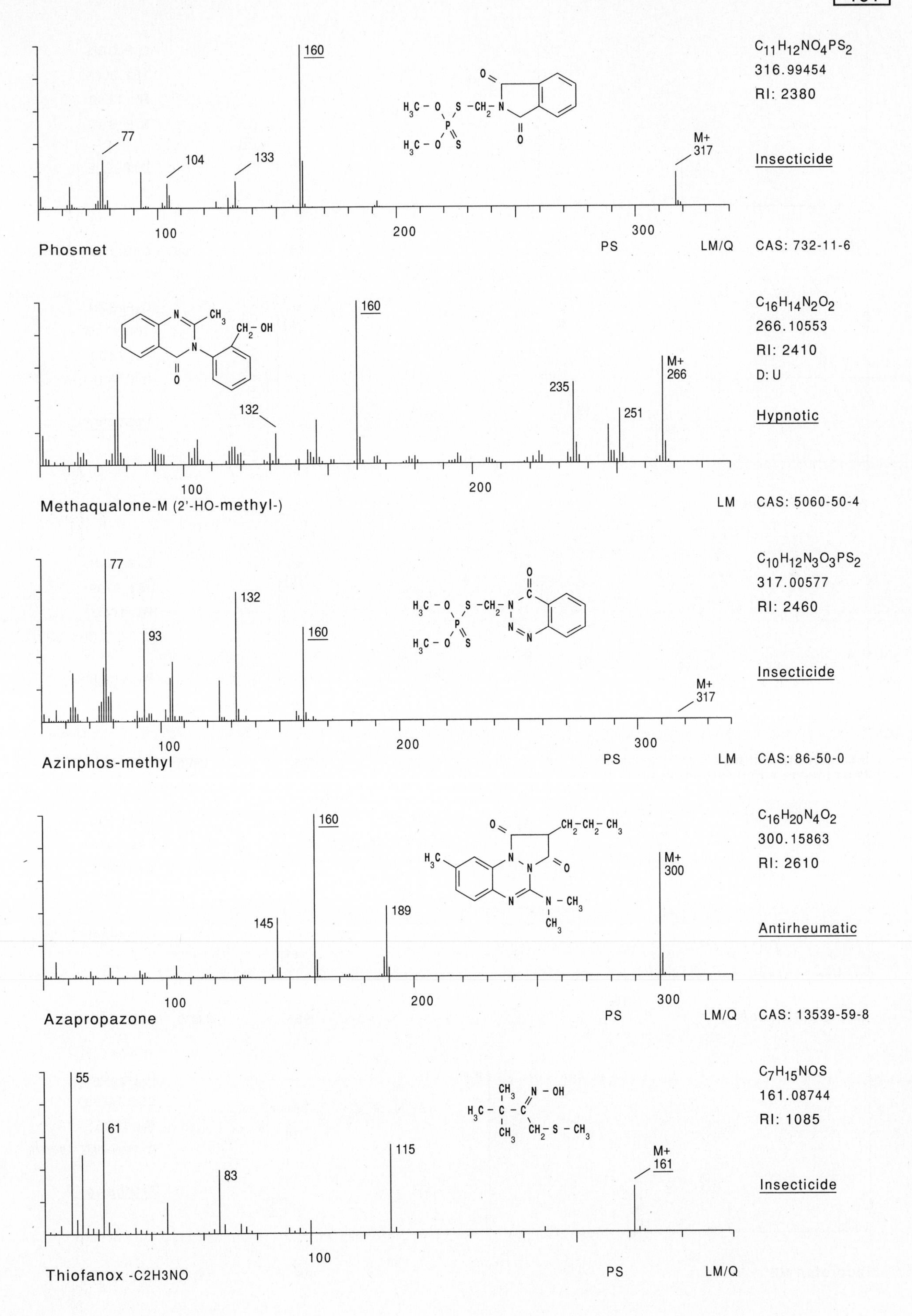

160
77
104
133
M+
317
100
200
300
PS
LM/Q
Phosmet
$C_{11}H_{12}NO_4PS_2$
316.99454
RI: 2380
Insecticide
CAS: 732-11-6
160
132
235
251
M+
266
100
200
LM
Methaqualone-M (2'-HO-methyl-)
$C_{16}H_{14}N_2O_2$
266.10553
RI: 2410
D: U
Hypnotic
CAS: 5060-50-4
77
132
93
160
M+
317
100
200
300
PS
LM
Azinphos-methyl
$C_{10}H_{12}N_3O_3PS_2$
317.00577
RI: 2460
Insecticide
CAS: 86-50-0
160
145
189
M+
300
100
200
300
PS
LM/Q
Azapropazone
$C_{16}H_{20}N_4O_2$
300.15863
RI: 2610
Antirheumatic
CAS: 13539-59-8
55
61
83
115
M+
161
100
PS
LM/Q
Thiofanox -C2H3NO
$C_7H_{15}NOS$
161.08744
RI: 1085
Insecticide

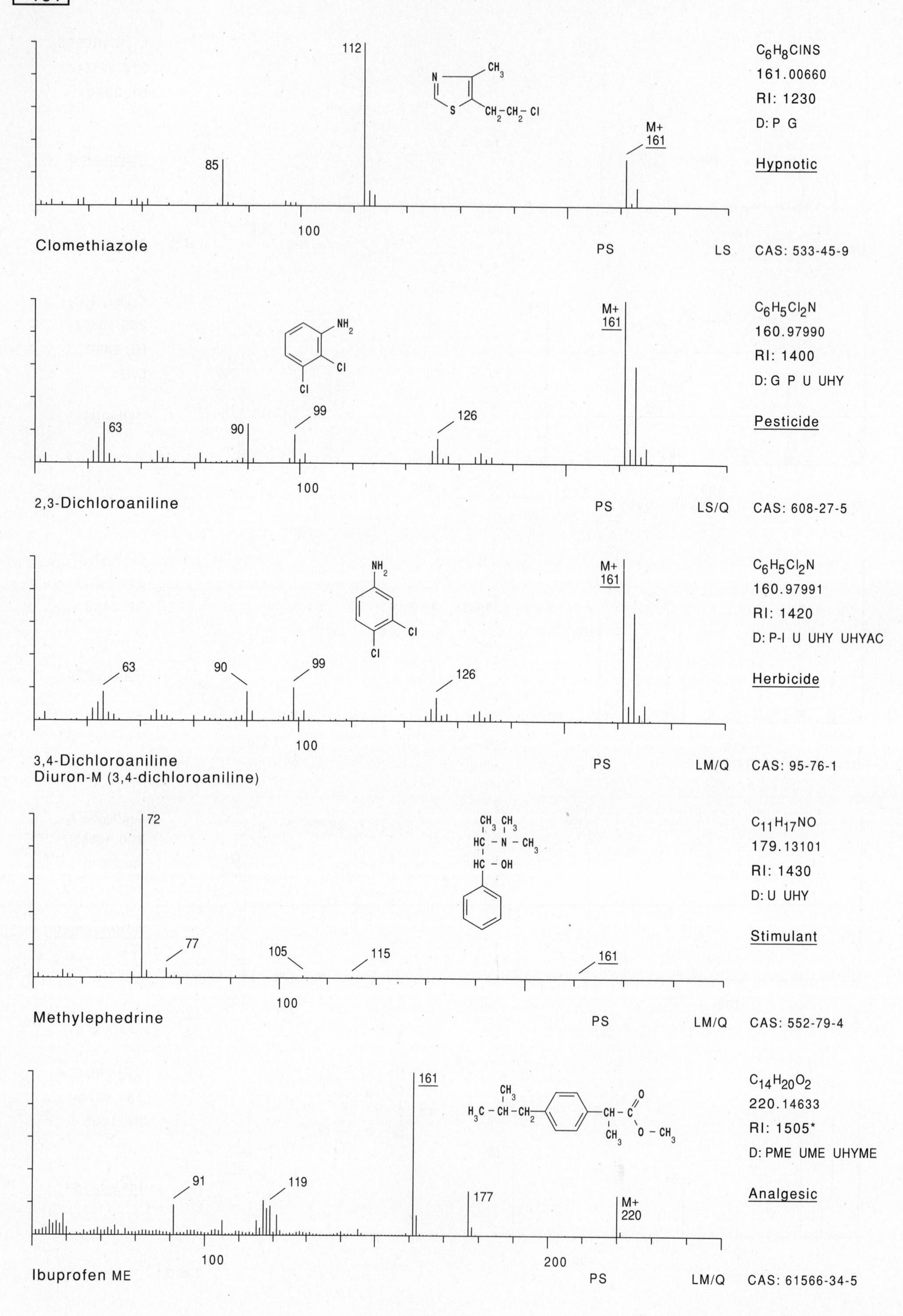
112
85
M+
161
100
Clomethiazole
PS
LS
C6H8ClNS
161.00660
RI: 1230
D: P G
Hypnotic
CAS: 533-45-9
M+
161
63
90
99
126
2,3-Dichloroaniline
LS/Q
C6H5Cl2N
160.97990
RI: 1400
D: G P U UHY
Pesticide
CAS: 608-27-5
3,4-Dichloroaniline
Diuron-M (3,4-dichloroaniline)
LM/Q
C6H5Cl2N
160.97991
RI: 1420
D: P-I U UHY UHYAC
Herbicide
CAS: 95-76-1
72
77
105
115
161
Methylephedrine
C11H17NO
179.13101
RI: 1430
D: U UHY
Stimulant
CAS: 552-79-4
91
119
177
M+
220
200
Ibuprofen ME
C14H20O2
220.14633
RI: 1505*
D: PME UME UHYME
Analgesic
CAS: 61566-34-5

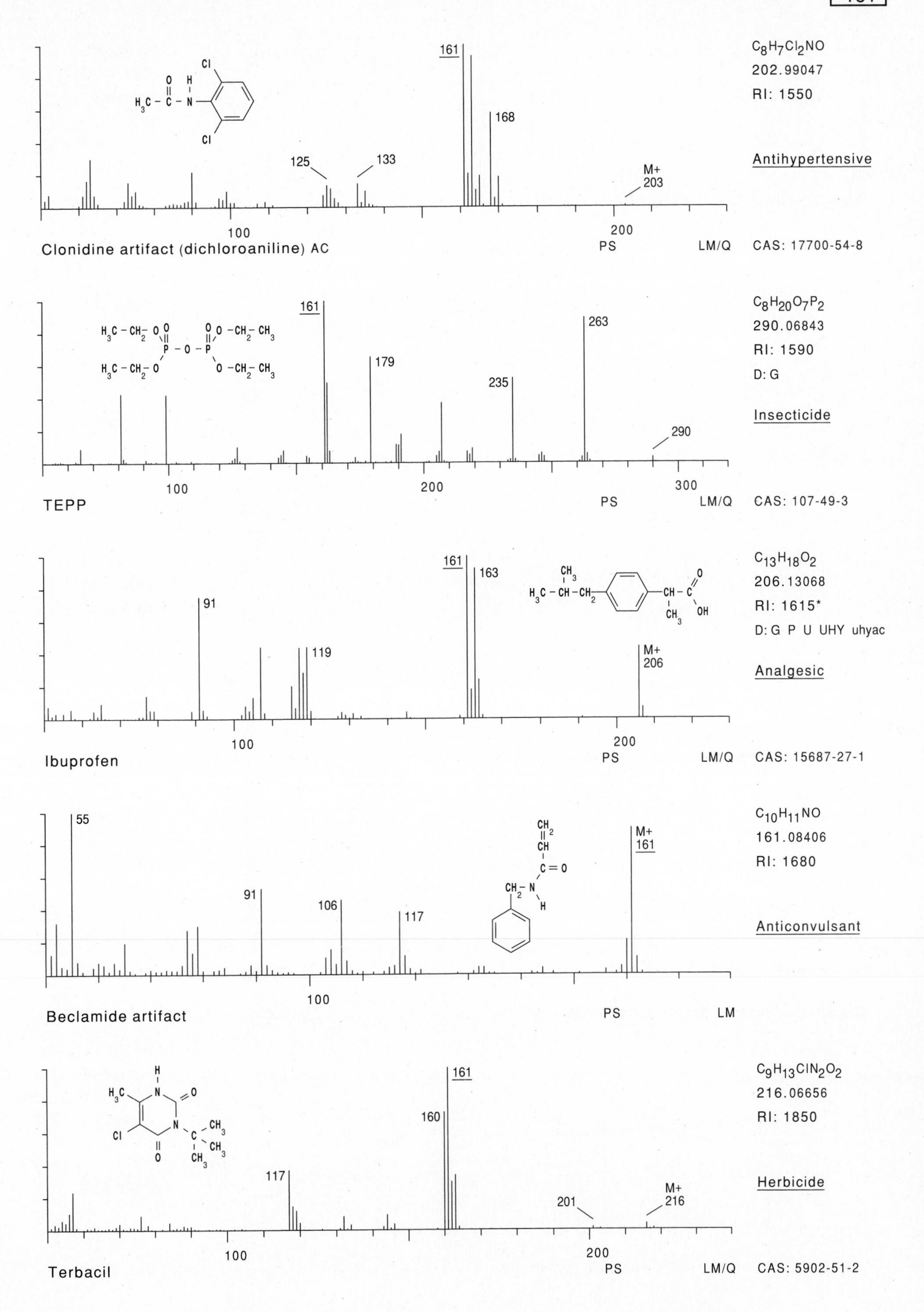

$C_8H_7Cl_2NO$
202.99047
RI: 1550
Antihypertensive
CAS: 17700-54-8

$C_8H_{20}O_7P_2$
290.06843
RI: 1590
D: G
Insecticide
CAS: 107-49-3

$C_{13}H_{18}O_2$
206.13068
RI: 1615*
D: G P U UHY uhyac
Analgesic
CAS: 15687-27-1

$C_{10}H_{11}NO$
161.08406
RI: 1680
Anticonvulsant

$C_9H_{13}ClN_2O_2$
216.06656
RI: 1850
Herbicide
CAS: 5902-51-2

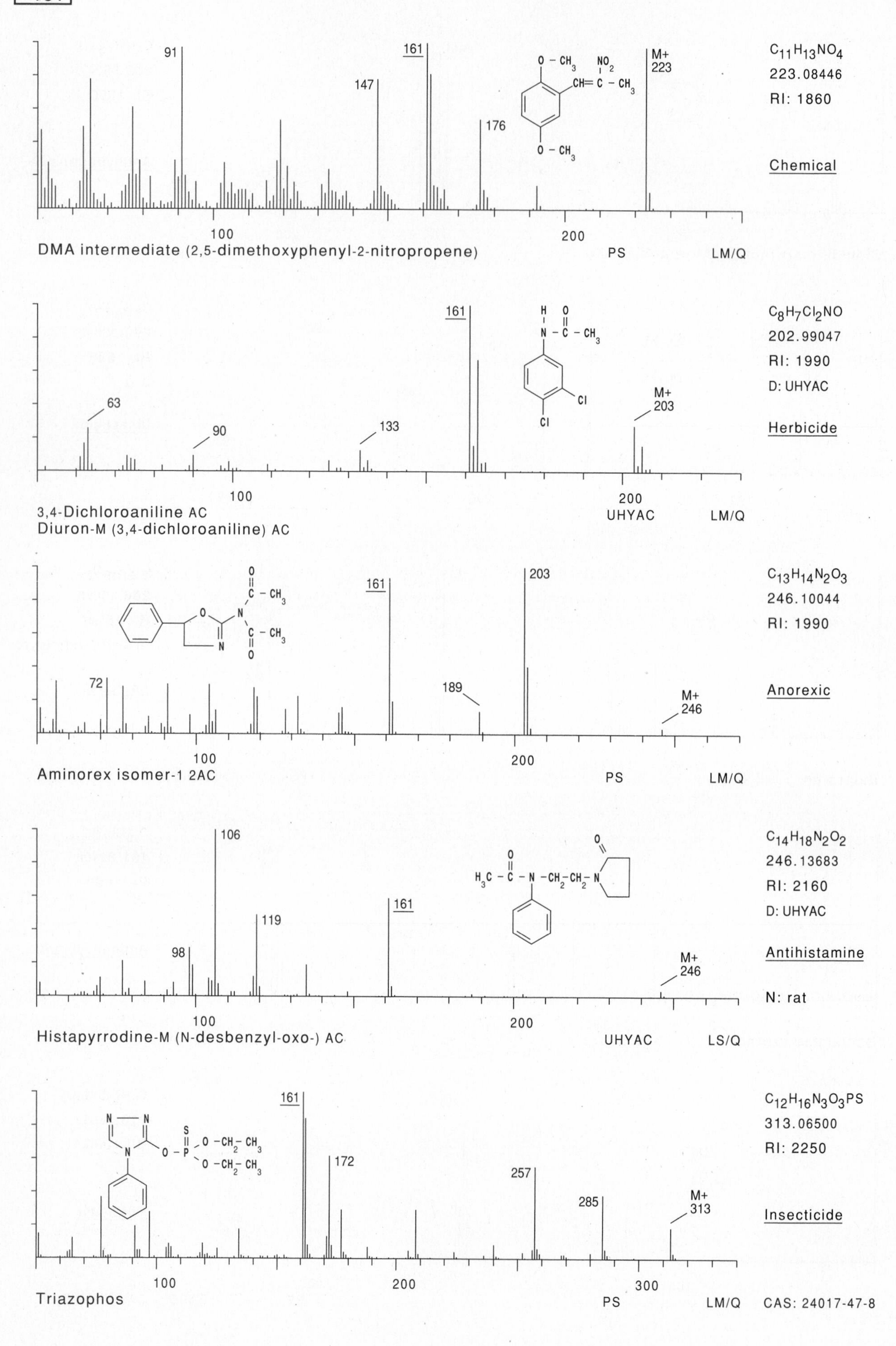

91
161
147
176
M+
223
100
200
DMA intermediate (2,5-dimethoxyphenyl-2-nitropropene)
PS
LM/Q
$C_{11}H_{13}NO_4$
223.08446
RI: 1860
Chemical
161
63
90
133
M+
203
3,4-Dichloroaniline AC
Diuron-M (3,4-dichloroaniline) AC
UHYAC
LM/Q
$C_8H_7Cl_2NO$
202.99047
RI: 1990
D: UHYAC
Herbicide
161
203
72
189
M+
246
Aminorex isomer-1 2AC
PS
LM/Q
$C_{13}H_{14}N_2O_3$
246.10044
RI: 1990
Anorexic
106
161
119
98
M+
246
Histapyrrodine-M (N-desbenzyl-oxo-) AC
UHYAC
LS/Q
$C_{14}H_{18}N_2O_2$
246.13683
RI: 2160
D: UHYAC
Antihistamine
N: rat
161
172
257
285
M+
313
300
Triazophos
PS
LM/Q
$C_{12}H_{16}N_3O_3PS$
313.06500
RI: 2250
Insecticide
CAS: 24017-47-8

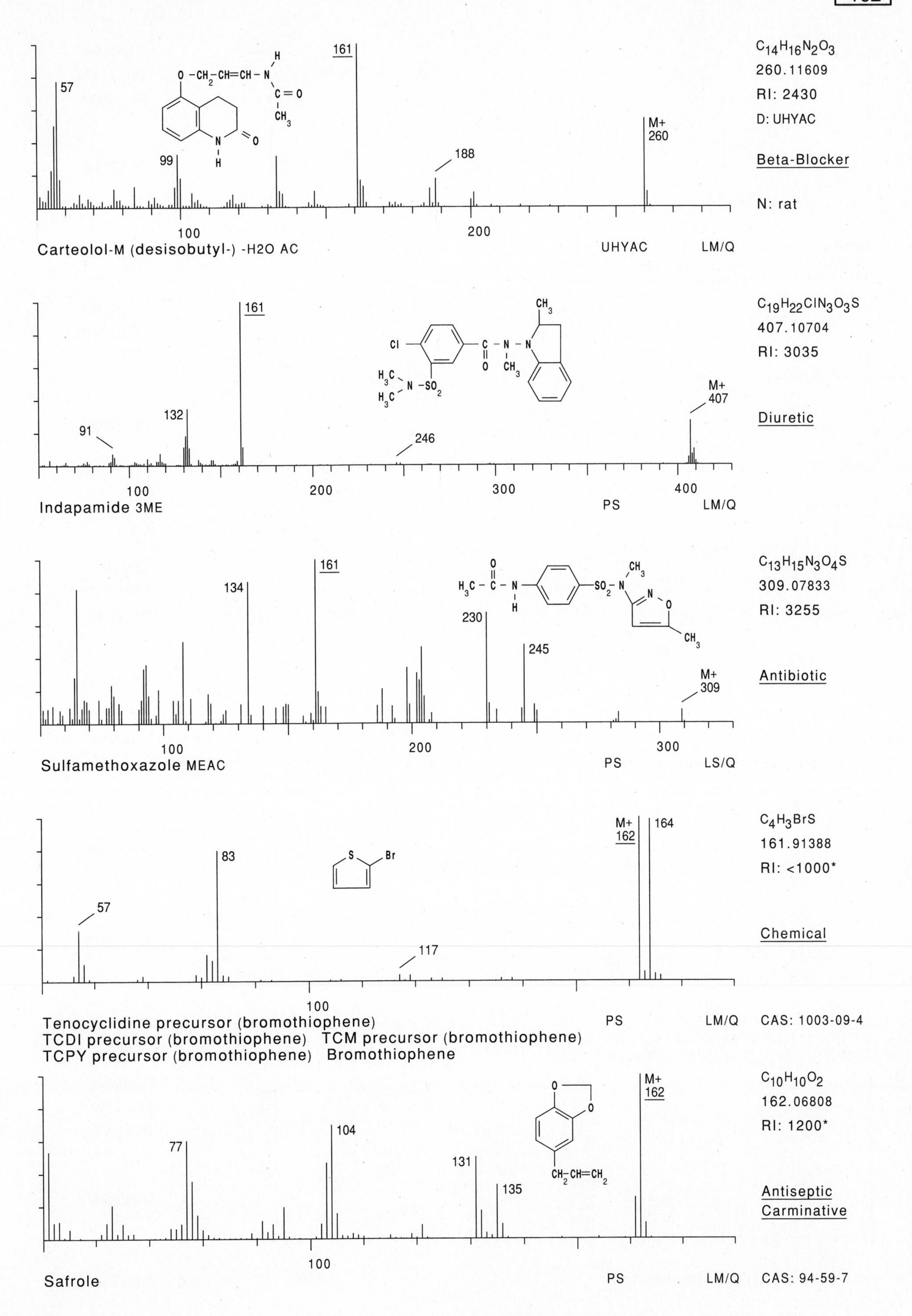

$C_{14}H_{16}N_2O_3$
260.11609
RI: 2430
D: UHYAC

Beta-Blocker

N: rat

$C_{19}H_{22}ClN_3O_3S$
407.10704
RI: 3035

Diuretic

$C_{13}H_{15}N_3O_4S$
309.07833
RI: 3255

Antibiotic

C_4H_3BrS
161.91388
RI: <1000*

Chemical

CAS: 1003-09-4

$C_{10}H_{10}O_2$
162.06808
RI: 1200*

Antiseptic
Carminative

CAS: 94-59-7

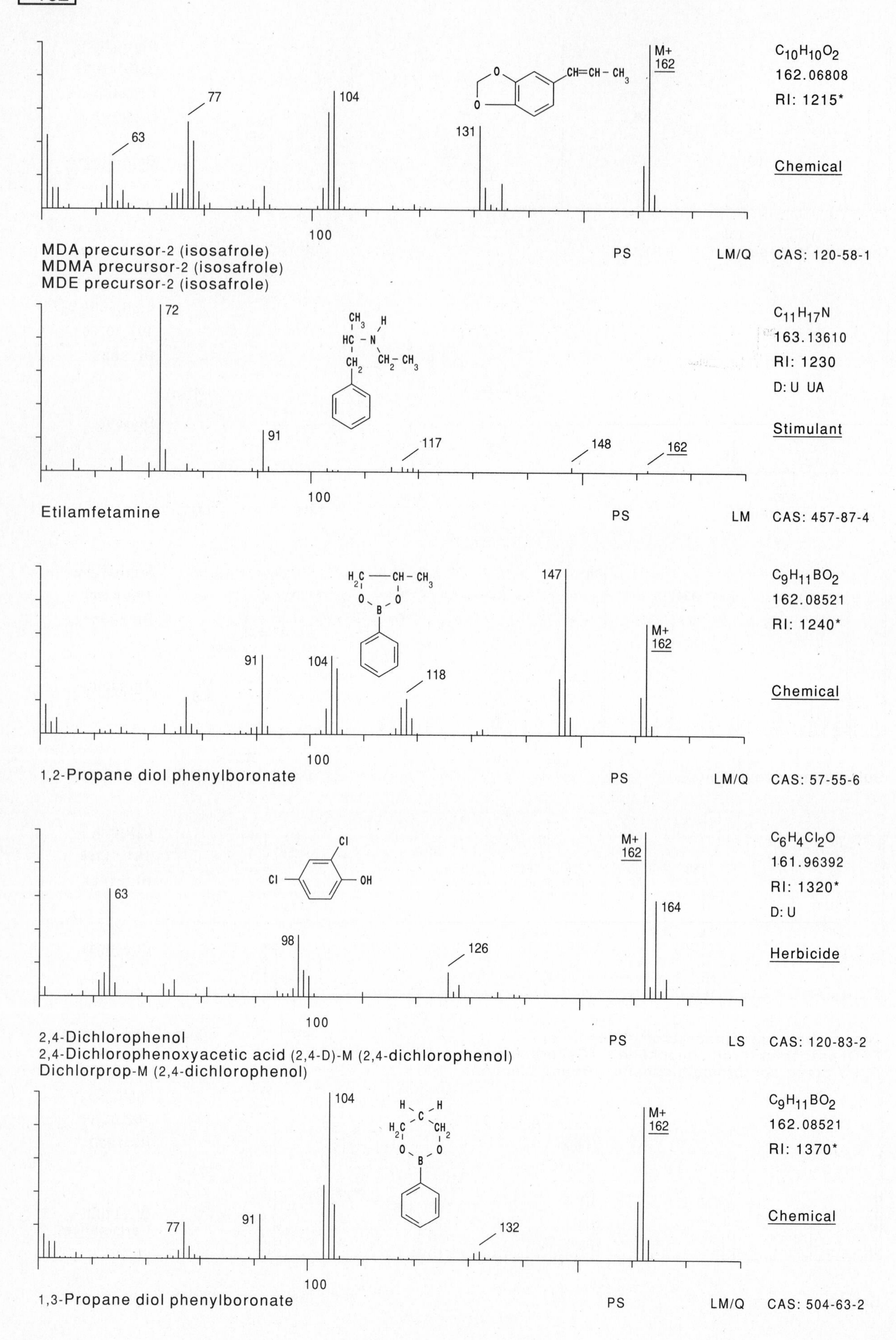
M+
162
77
104
63
131
CH=CH-CH3
100
$C_{10}H_{10}O_2$
162.06808
RI: 1215*
Chemical
MDA precursor-2 (isosafrole)
MDMA precursor-2 (isosafrole)
MDE precursor-2 (isosafrole)
PS
LM/Q
CAS: 120-58-1
72
91
117
148
162
100
$C_{11}H_{17}N$
163.13610
RI: 1230
D: U UA
Stimulant
Etilamfetamine
PS
LM
CAS: 457-87-4
147
M+
162
91
104
118
100
$C_9H_{11}BO_2$
162.08521
RI: 1240*
Chemical
1,2-Propane diol phenylboronate
PS
LM/Q
CAS: 57-55-6
M+
162
164
63
98
126
Cl
OH
100
$C_6H_4Cl_2O$
161.96392
RI: 1320*
D: U
Herbicide
2,4-Dichlorophenol
2,4-Dichlorophenoxyacetic acid (2,4-D)-M (2,4-dichlorophenol)
Dichlorprop-M (2,4-dichlorophenol)
PS
LS
CAS: 120-83-2
104
M+
162
77
91
132
100
$C_9H_{11}BO_2$
162.08521
RI: 1370*
Chemical
1,3-Propane diol phenylboronate
PS
LM/Q
CAS: 504-63-2

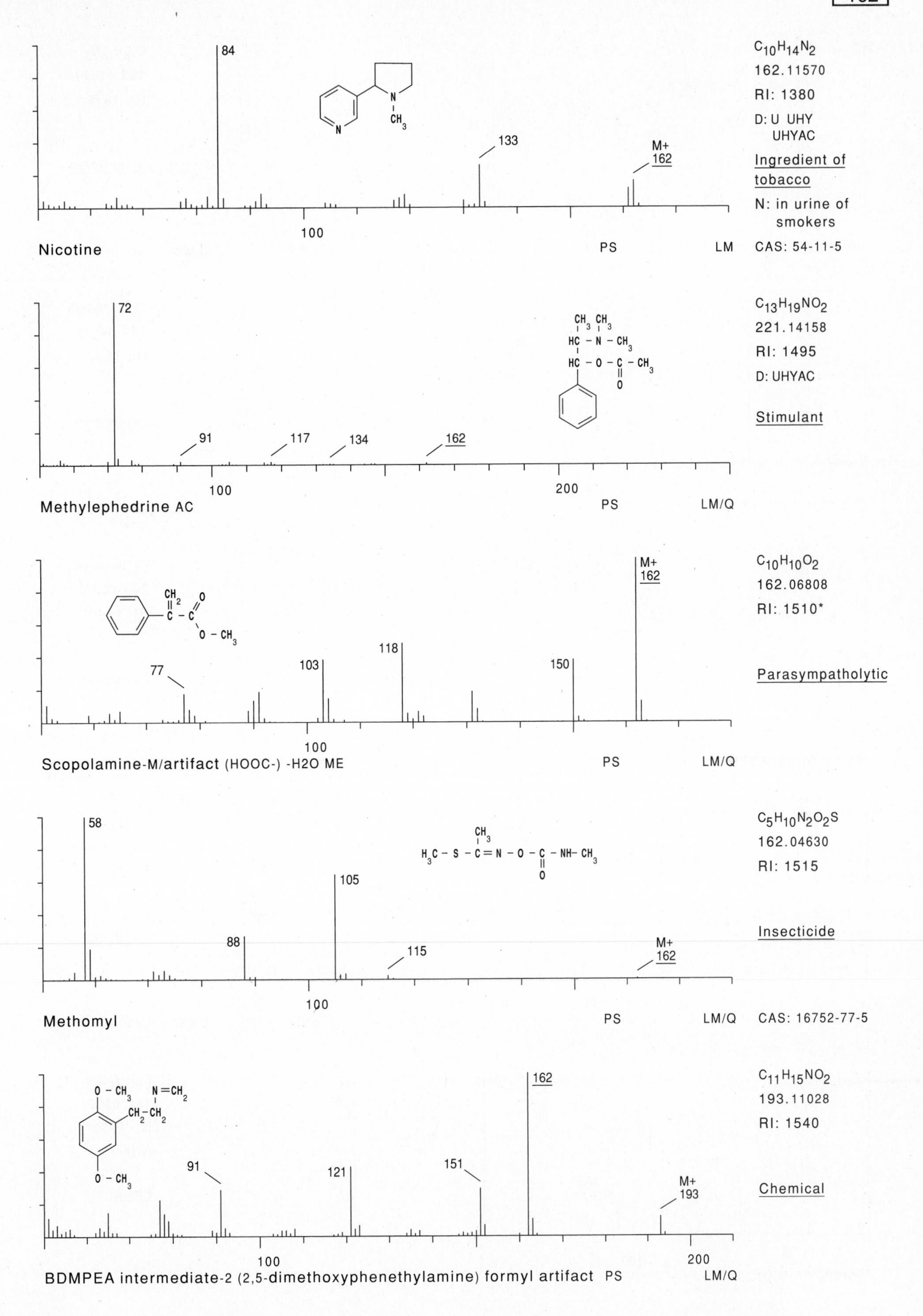
84
133
M+
162
100
Nicotine
PS
LM
$C_{10}H_{14}N_2$
162.11570
RI: 1380
D: U UHY
UHYAC
Ingredient of tobacco
N: in urine of smokers
CAS: 54-11-5
72
91
117
134
162
100
200
Methylephedrine AC
PS
LM/Q
$C_{13}H_{19}NO_2$
221.14158
RI: 1495
D: UHYAC
Stimulant
M+
162
77
103
118
150
100
Scopolamine-M/artifact (HOOC-) -H2O ME
PS
LM/Q
$C_{10}H_{10}O_2$
162.06808
RI: 1510*
Parasympatholytic
58
105
88
115
M+
162
100
Methomyl
PS
LM/Q
$C_5H_{10}N_2O_2S$
162.04630
RI: 1515
Insecticide
CAS: 16752-77-5
162
91
121
151
M+
193
100
200
BDMPEA intermediate-2 (2,5-dimethoxyphenethylamine) formyl artifact
PS
LM/Q
$C_{11}H_{15}NO_2$
193.11028
RI: 1540
Chemical

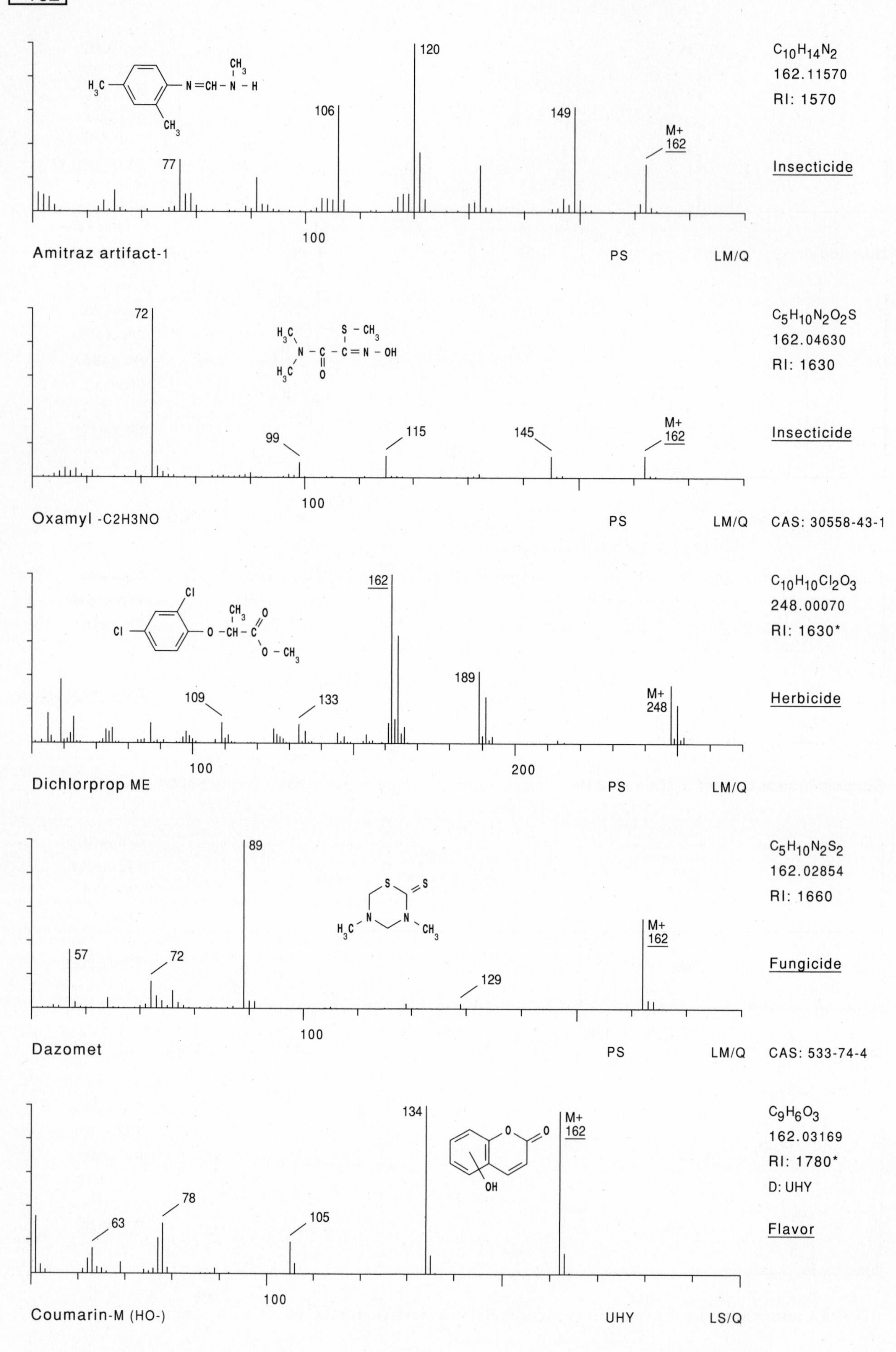

120
106
149
77
M+
162
100
Amitraz artifact-1
PS
LM/Q
$C_{10}H_{14}N_2$
162.11570
RI: 1570
Insecticide
72
99
115
145
M+
162
100
Oxamyl -C2H3NO
PS
LM/Q
$C_5H_{10}N_2O_2S$
162.04630
RI: 1630
Insecticide
CAS: 30558-43-1
162
189
109
133
M+
248
100
200
Dichlorprop ME
PS
LM/Q
$C_{10}H_{10}Cl_2O_3$
248.00070
RI: 1630*
Herbicide
89
57
72
129
M+
162
100
Dazomet
PS
LM/Q
$C_5H_{10}N_2S_2$
162.02854
RI: 1660
Fungicide
CAS: 533-74-4
134
M+
162
63
78
105
100
Coumarin-M (HO-)
UHY
LS/Q
$C_9H_6O_3$
162.03169
RI: 1780*
D: UHY
Flavor

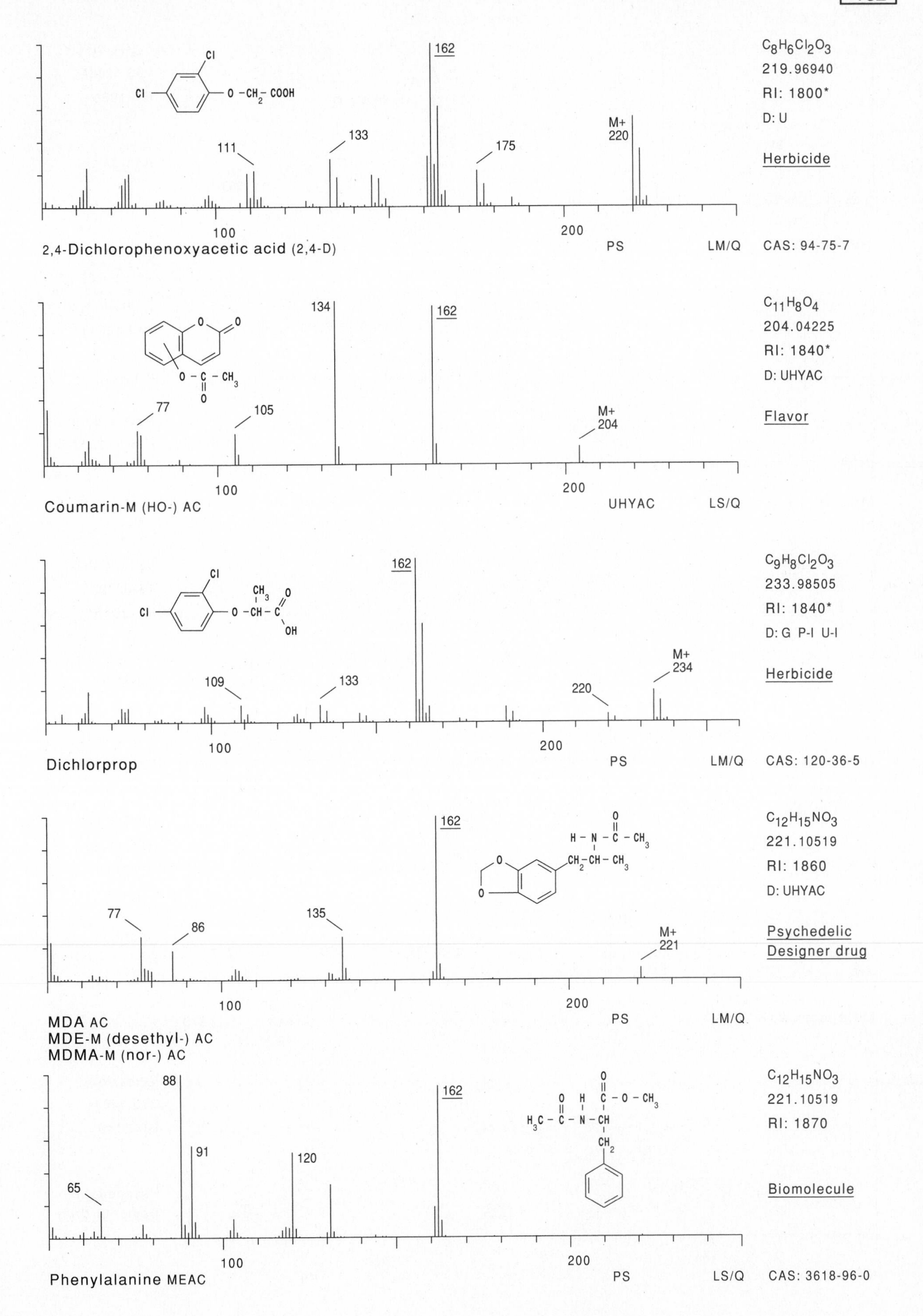

2,4-Dichlorophenoxyacetic acid (2,4-D)
C8H6Cl2O3
219.96940
RI: 1800*
D: U
Herbicide
PS
LM/Q
CAS: 94-75-7
Cl
O -CH2- COOH
111
133
162
175
M+
220
100
200
Coumarin-M (HO-) AC
C11H8O4
204.04225
RI: 1840*
D: UHYAC
Flavor
UHYAC
LS/Q
77
105
134
162
M+
204
Dichlorprop
C9H8Cl2O3
233.98505
RI: 1840*
D: G P-I U-I
Herbicide
PS
LM/Q
CAS: 120-36-5
109
133
162
220
M+
234
MDA AC
MDE-M (desethyl-) AC
MDMA-M (nor-) AC
C12H15NO3
221.10519
RI: 1860
D: UHYAC
Psychedelic
Designer drug
PS
LM/Q
77
86
135
162
M+
221
Phenylalanine MEAC
C12H15NO3
221.10519
RI: 1870
Biomolecule
PS
LS/Q
CAS: 3618-96-0
65
88
91
120
162

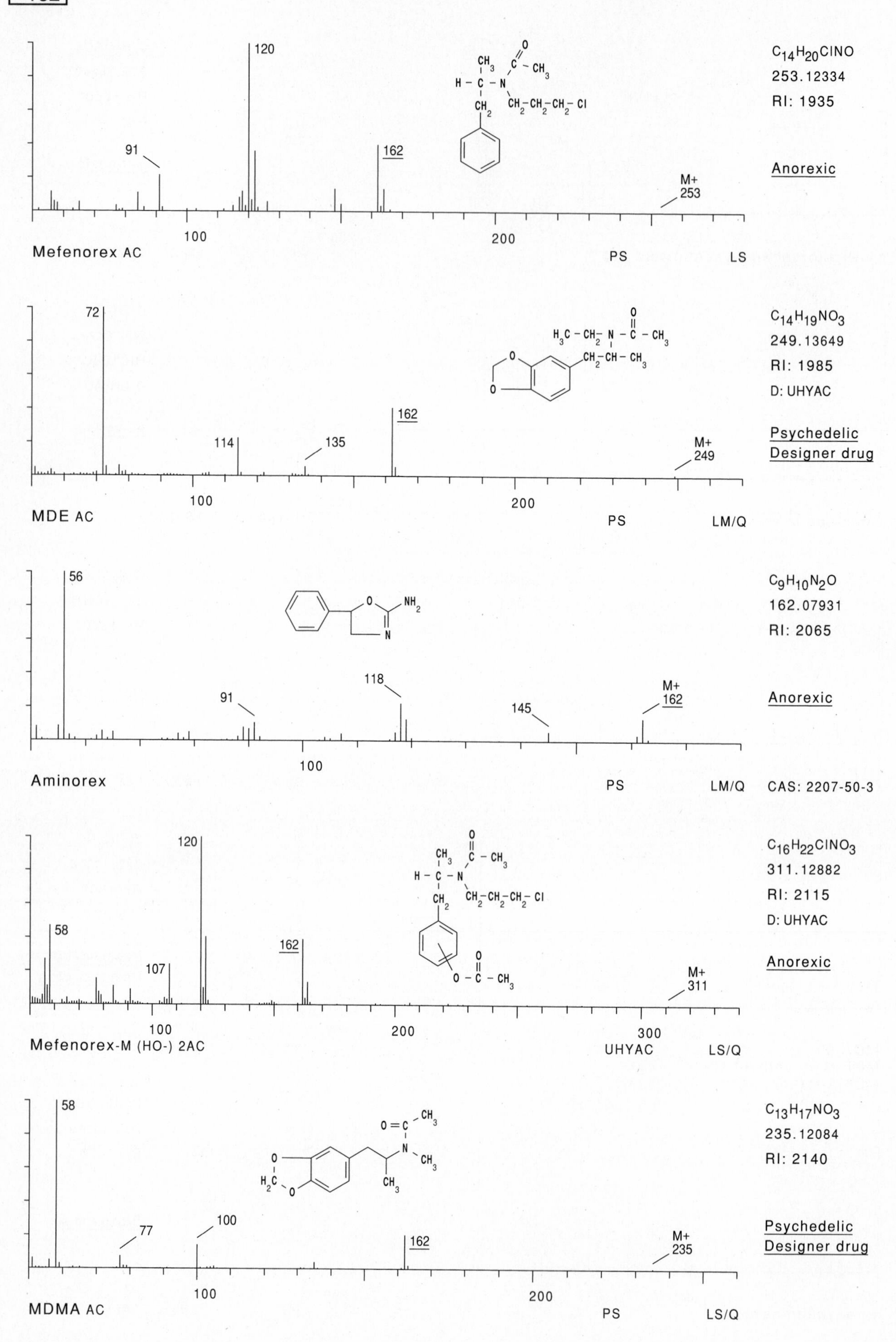

120
91
162
M+
253
100
200
Mefenorex AC
PS
LS
$C_{14}H_{20}ClNO$
253.12334
RI: 1935
Anorexic
72
114
135
162
M+
249
100
200
MDE AC
PS
LM/Q
$C_{14}H_{19}NO_3$
249.13649
RI: 1985
D: UHYAC
Psychedelic
Designer drug
56
91
118
145
M+
162
100
Aminorex
PS
LM/Q
$C_9H_{10}N_2O$
162.07931
RI: 2065
Anorexic
CAS: 2207-50-3
120
58
107
162
M+
311
100
200
300
Mefenorex-M (HO-) 2AC
UHYAC
LS/Q
$C_{16}H_{22}ClNO_3$
311.12882
RI: 2115
D: UHYAC
Anorexic
58
77
100
162
M+
235
100
200
MDMA AC
PS
LS/Q
$C_{13}H_{17}NO_3$
235.12084
RI: 2140
Psychedelic
Designer drug

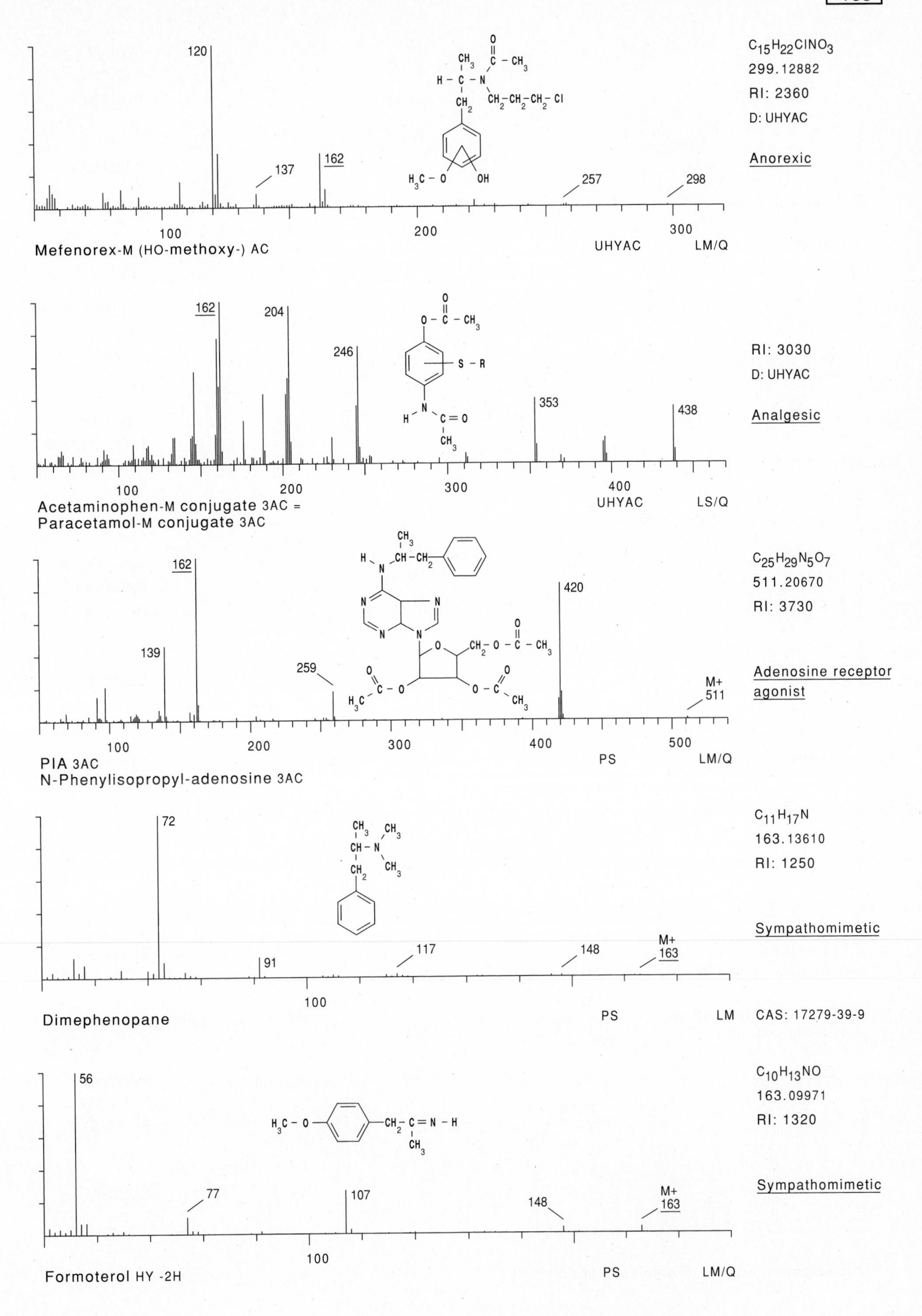
163
Mefenorex-M (HO-methoxy-) AC
UHYAC
LM/Q
C15H22ClNO3
299.12882
RI: 2360
D: UHYAC
Anorexic
Acetaminophen-M conjugate 3AC =
Paracetamol-M conjugate 3AC
UHYAC
LS/Q
RI: 3030
D: UHYAC
Analgesic
PIA 3AC
N-Phenylisopropyl-adenosine 3AC
PS
LM/Q
C25H29N5O7
511.20670
RI: 3730
Adenosine receptor agonist
Dimephenopane
PS
LM
C11H17N
163.13610
RI: 1250
Sympathomimetic
CAS: 17279-39-9
Formoterol HY -2H
PS
LM/Q
C10H13NO
163.09971
RI: 1320
Sympathomimetic

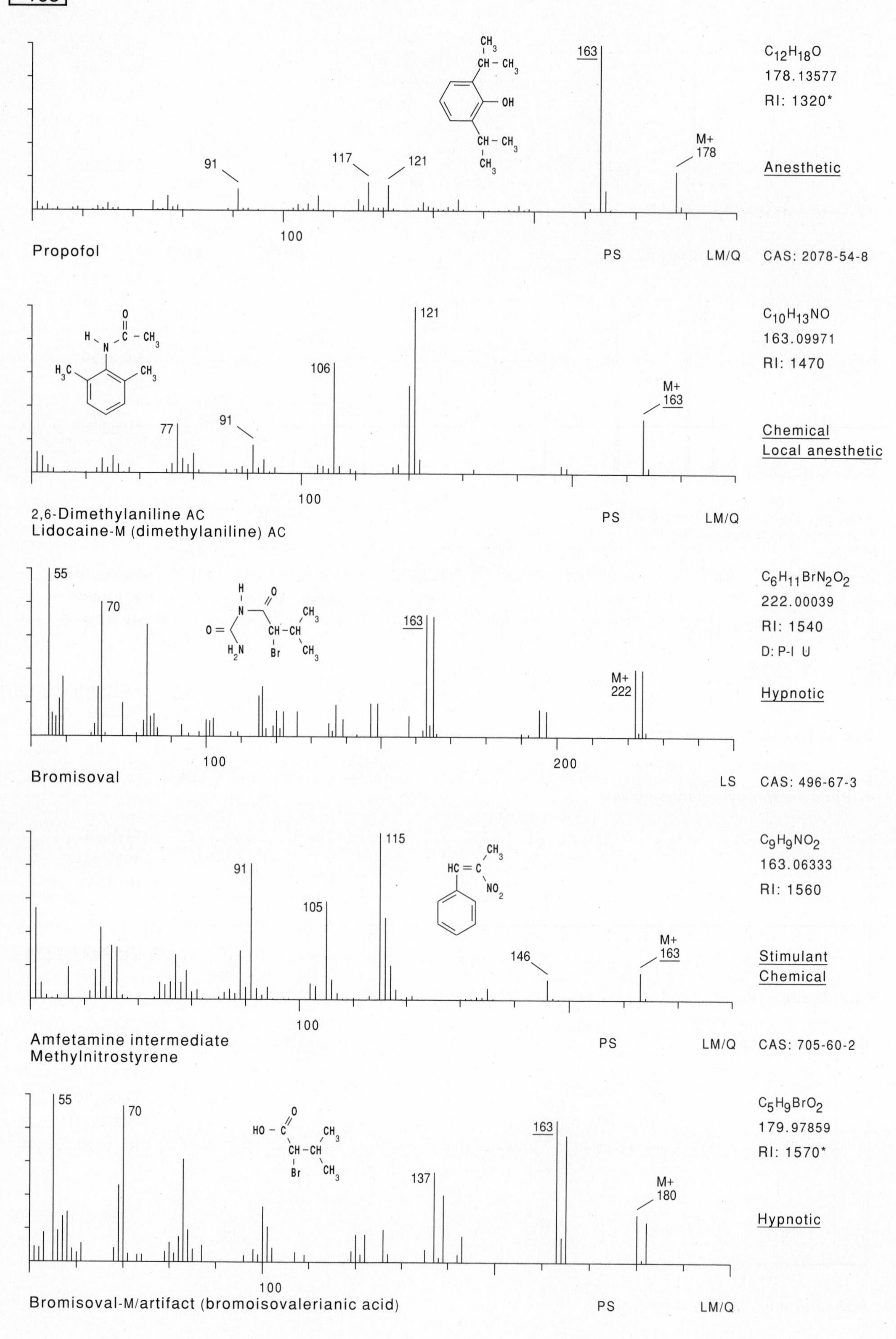
163
91
117
121
M+
178
100
CH3
CH - CH3
OH
CH - CH3
CH3
C12H18O
178.13577
RI: 1320*
Anesthetic
Propofol
PS
LM/Q
CAS: 2078-54-8
121
106
91
77
M+
163
100
C10H13NO
163.09971
RI: 1470
Chemical
Local anesthetic
2,6-Dimethylaniline AC
Lidocaine-M (dimethylaniline) AC
PS
LM/Q
55
70
163
M+
222
100
200
C6H11BrN2O2
222.00039
RI: 1540
D: P-I U
Hypnotic
Bromisoval
LS
CAS: 496-67-3
115
91
105
146
M+
163
100
C9H9NO2
163.06333
RI: 1560
Stimulant
Chemical
Amfetamine intermediate
Methylnitrostyrene
PS
LM/Q
CAS: 705-60-2
55
70
163
137
M+
180
100
C5H9BrO2
179.97859
RI: 1570*
Hypnotic
Bromisoval-M/artifact (bromoisovalerianic acid)
PS
LM/Q

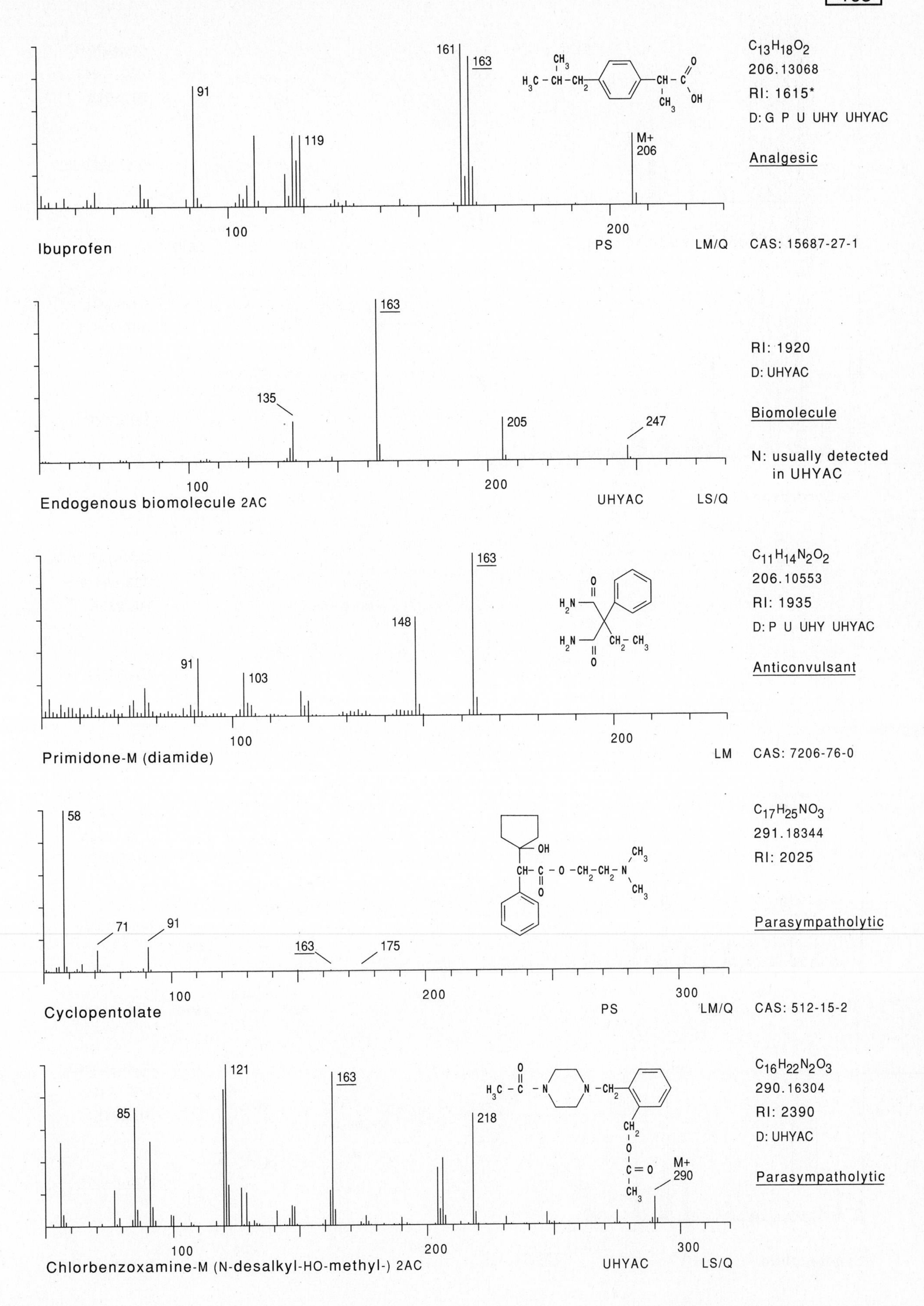

161
163
91
119
M+
206
100
200
PS
LM/Q
Ibuprofen
$C_{13}H_{18}O_2$
206.13068
RI: 1615*
D: G P U UHY UHYAC
Analgesic
CAS: 15687-27-1
163
135
205
247
100
200
UHYAC
LS/Q
Endogenous biomolecule 2AC
RI: 1920
D: UHYAC
Biomolecule
N: usually detected in UHYAC
163
148
91
103
100
200
LM
Primidone-M (diamide)
$C_{11}H_{14}N_2O_2$
206.10553
RI: 1935
D: P U UHY UHYAC
Anticonvulsant
CAS: 7206-76-0
58
71
91
163
175
100
200
300
PS
LM/Q
Cyclopentolate
$C_{17}H_{25}NO_3$
291.18344
RI: 2025
Parasympatholytic
CAS: 512-15-2
121
163
85
218
M+
290
100
200
300
UHYAC
LS/Q
Chlorbenzoxamine-M (N-desalkyl-HO-methyl-) 2AC
$C_{16}H_{22}N_2O_3$
290.16304
RI: 2390
D: UHYAC
Parasympatholytic

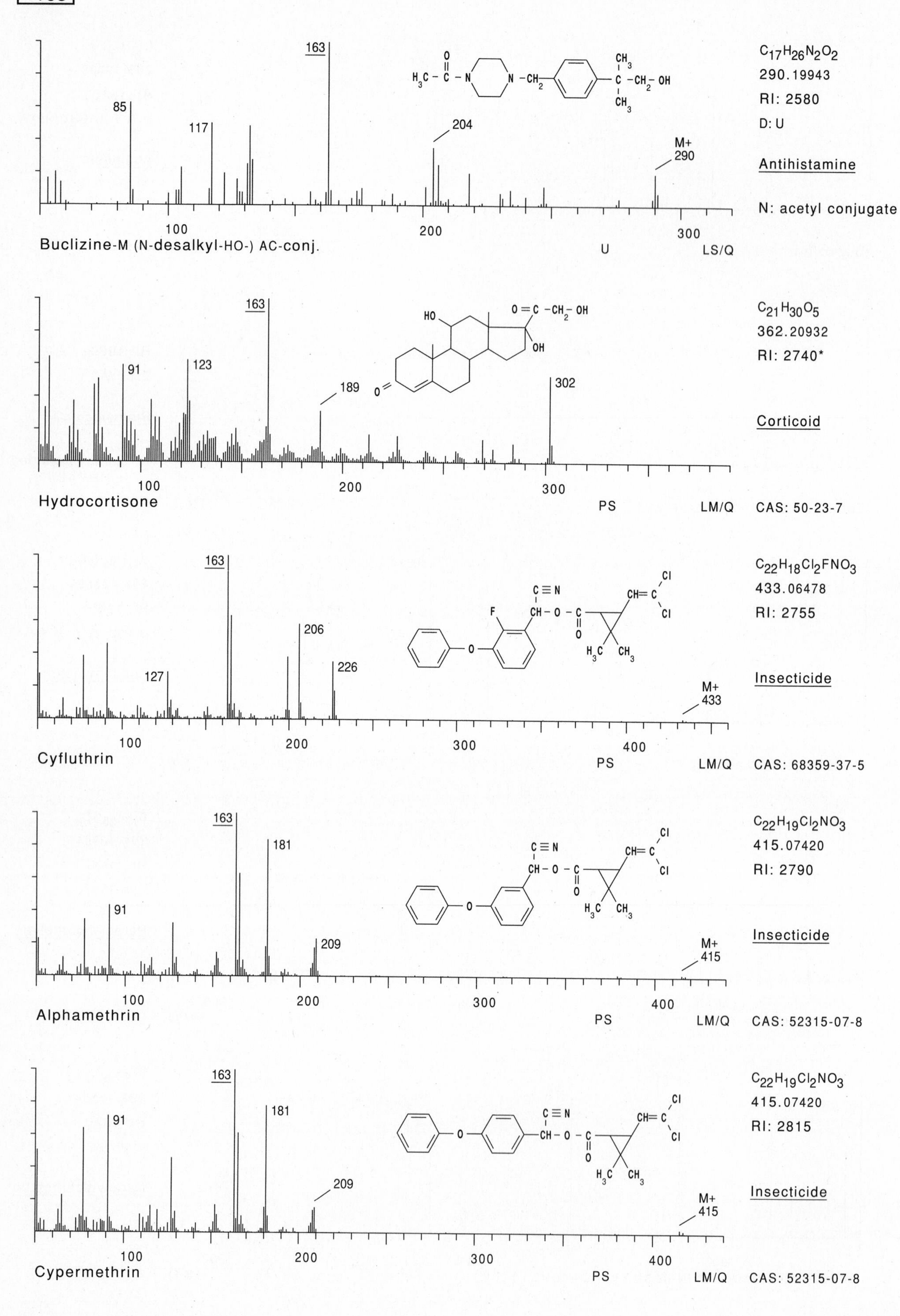
163
85
117
204
M+
290
100
200
300
Buclizine-M (N-desalkyl-HO-) AC-conj.
U
LS/Q
$C_{17}H_{26}N_2O_2$
290.19943
RI: 2580
D: U
Antihistamine
N: acetyl conjugate
163
91
123
189
302
100
200
300
Hydrocortisone
PS
LM/Q
$C_{21}H_{30}O_5$
362.20932
RI: 2740*
Corticoid
CAS: 50-23-7
163
206
127
226
M+
433
100
200
300
400
Cyfluthrin
PS
LM/Q
$C_{22}H_{18}Cl_2FNO_3$
433.06478
RI: 2755
Insecticide
CAS: 68359-37-5
163
181
91
209
M+
415
100
200
300
400
Alphamethrin
PS
LM/Q
$C_{22}H_{19}Cl_2NO_3$
415.07420
RI: 2790
Insecticide
CAS: 52315-07-8
163
91
181
209
M+
415
100
200
300
400
Cypermethrin
PS
LM/Q
$C_{22}H_{19}Cl_2NO_3$
415.07420
RI: 2815
Insecticide
CAS: 52315-07-8

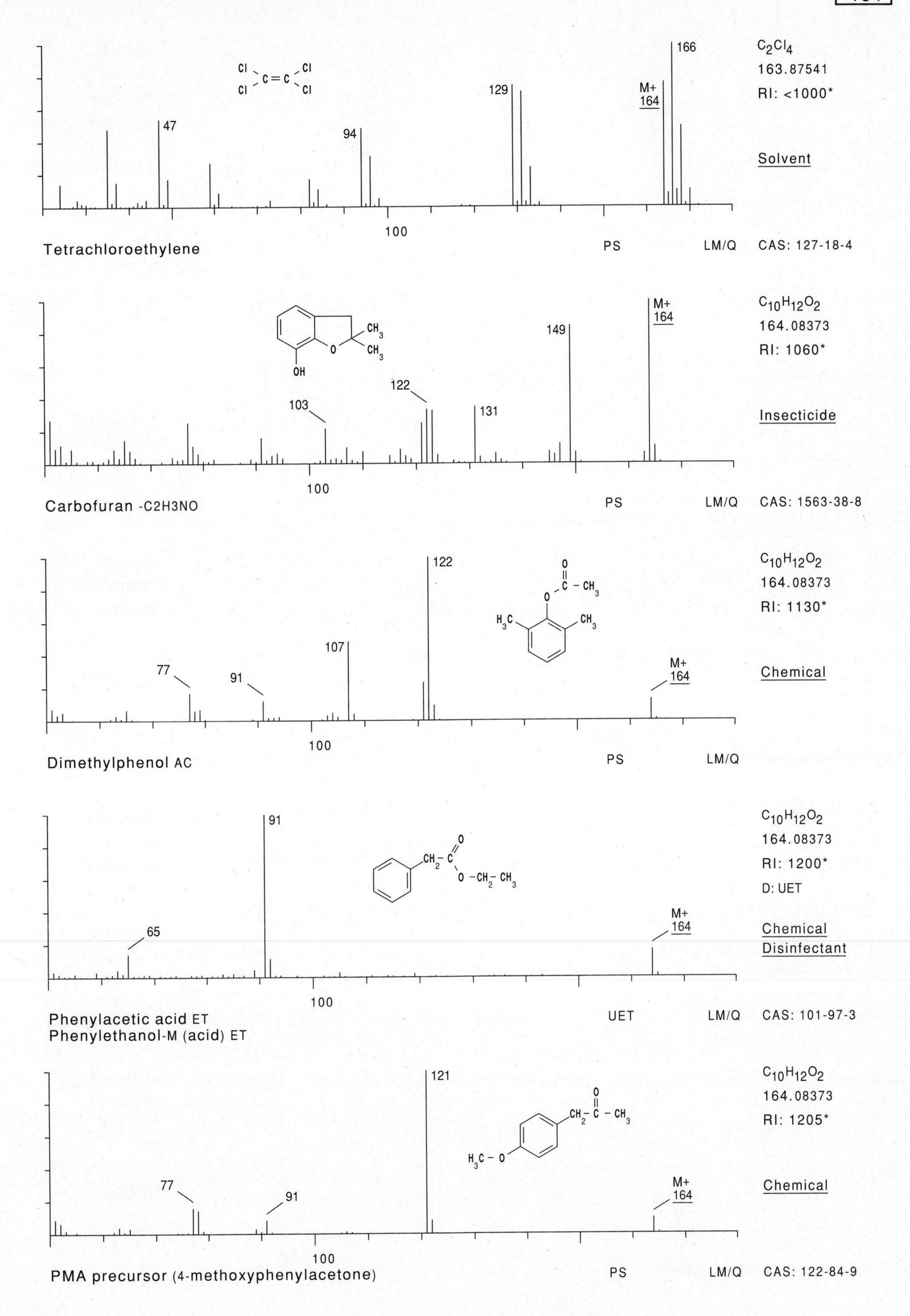

166
129
M+
164
47
94
100
Tetrachloroethylene
PS
LM/Q
C_2Cl_4
163.87541
RI: <1000*
Solvent
CAS: 127-18-4
M+
164
149
122
103
131
CH3
CH3
OH
O
100
Carbofuran -C2H3NO
PS
LM/Q
$C_{10}H_{12}O_2$
164.08373
RI: 1060*
Insecticide
CAS: 1563-38-8
122
107
77
91
M+
164
100
Dimethylphenol AC
PS
LM/Q
$C_{10}H_{12}O_2$
164.08373
RI: 1130*
Chemical
91
65
M+
164
100
Phenylacetic acid ET
Phenylethanol-M (acid) ET
UET
LM/Q
$C_{10}H_{12}O_2$
164.08373
RI: 1200*
D: UET
Chemical
Disinfectant
CAS: 101-97-3
121
77
91
M+
164
100
PMA precursor (4-methoxyphenylacetone)
PS
LM/Q
$C_{10}H_{12}O_2$
164.08373
RI: 1205*
Chemical
CAS: 122-84-9

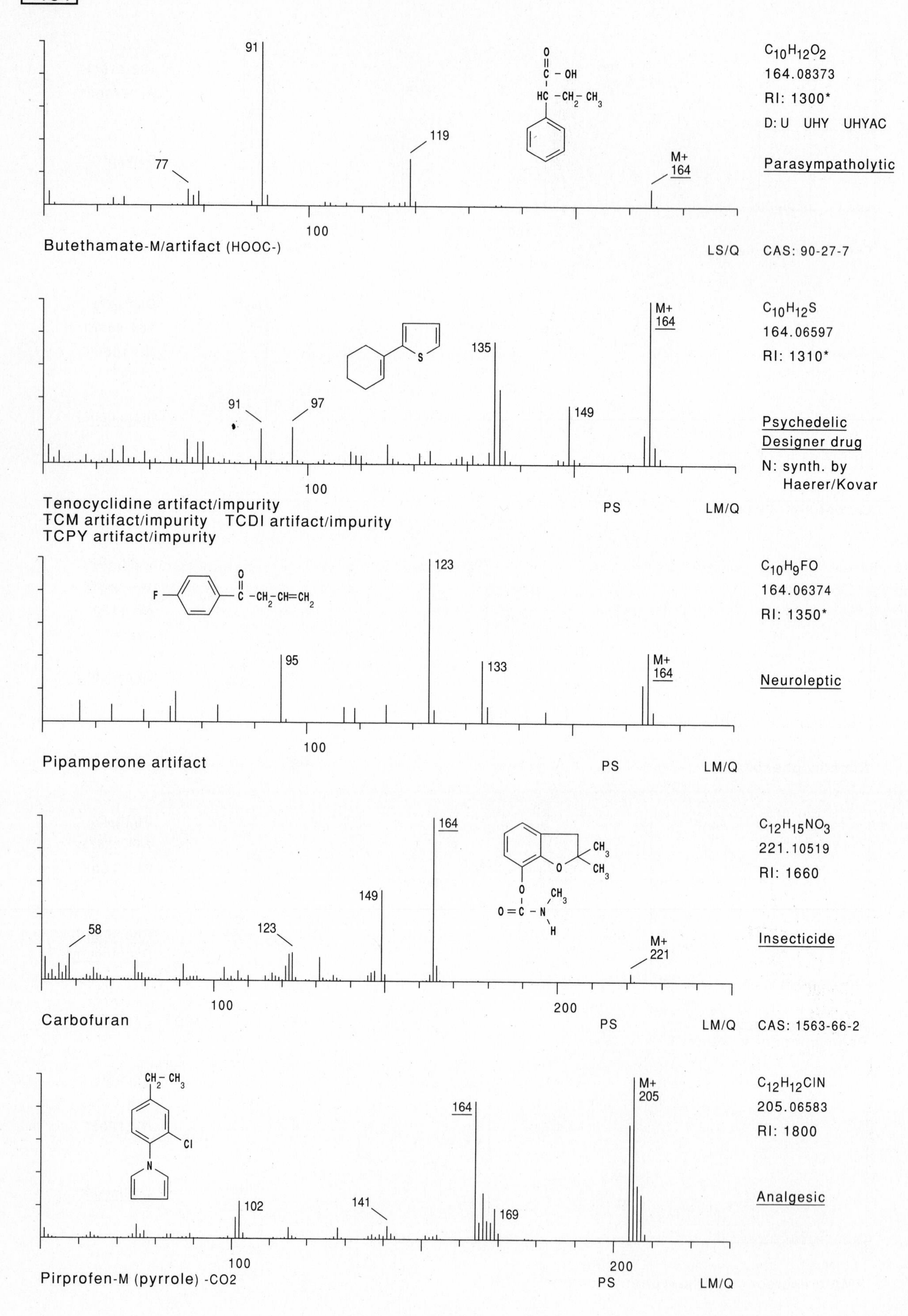
91
77
119
M+
164
100
Butethamate-M/artifact (HOOC-)
LS/Q
$C_{10}H_{12}O_2$
164.08373
RI: 1300*
D: U UHY UHYAC
Parasympatholytic
CAS: 90-27-7
M+
164
135
91
97
149
100
Tenocyclidine artifact/impurity
TCM artifact/impurity TCDI artifact/impurity
TCPY artifact/impurity
PS
LM/Q
$C_{10}H_{12}S$
164.06597
RI: 1310*
Psychedelic
Designer drug
N: synth. by
Haerer/Kovar
123
95
133
M+
164
100
Pipamperone artifact
PS
LM/Q
$C_{10}H_9FO$
164.06374
RI: 1350*
Neuroleptic
164
149
58
123
M+
221
100
200
Carbofuran
PS
LM/Q
$C_{12}H_{15}NO_3$
221.10519
RI: 1660
Insecticide
CAS: 1563-66-2
M+
205
164
102
141
169
100
200
Pirprofen-M (pyrrole) -CO2
PS
LM/Q
$C_{12}H_{12}ClN$
205.06583
RI: 1800
Analgesic

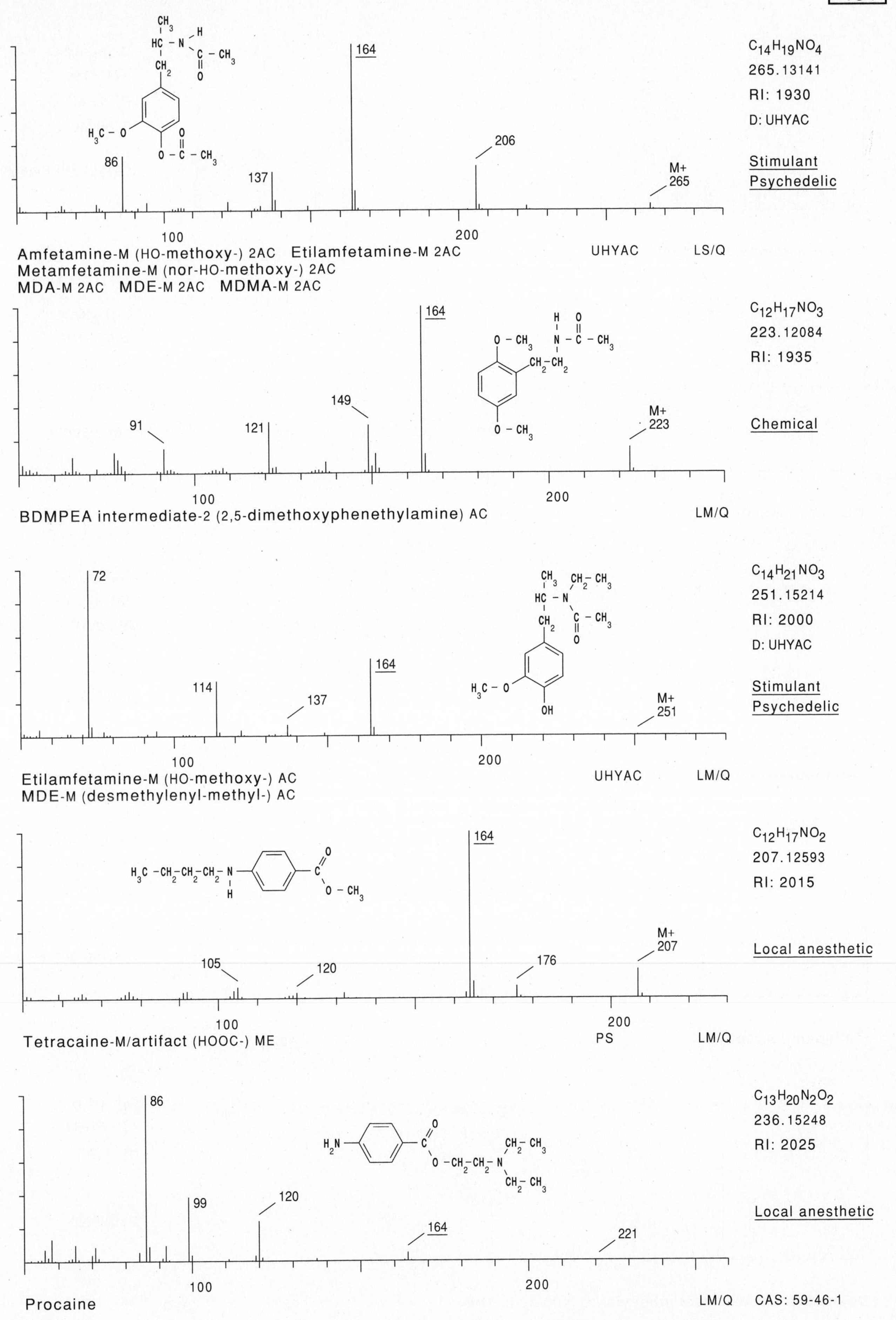

$C_{14}H_{19}NO_4$
265.13141
RI: 1930
D: UHYAC
Stimulant
Psychedelic

Amfetamine-M (HO-methoxy-) 2AC Etilamfetamine-M 2AC
Metamfetamine-M (nor-HO-methoxy-) 2AC
MDA-M 2AC MDE-M 2AC MDMA-M 2AC

$C_{12}H_{17}NO_3$
223.12084
RI: 1935
Chemical

BDMPEA intermediate-2 (2,5-dimethoxyphenethylamine) AC

$C_{14}H_{21}NO_3$
251.15214
RI: 2000
D: UHYAC
Stimulant
Psychedelic

Etilamfetamine-M (HO-methoxy-) AC
MDE-M (desmethylenyl-methyl-) AC

$C_{12}H_{17}NO_2$
207.12593
RI: 2015
Local anesthetic

Tetracaine-M/artifact (HOOC-) ME

$C_{13}H_{20}N_2O_2$
236.15248
RI: 2025
Local anesthetic

Procaine

CAS: 59-46-1

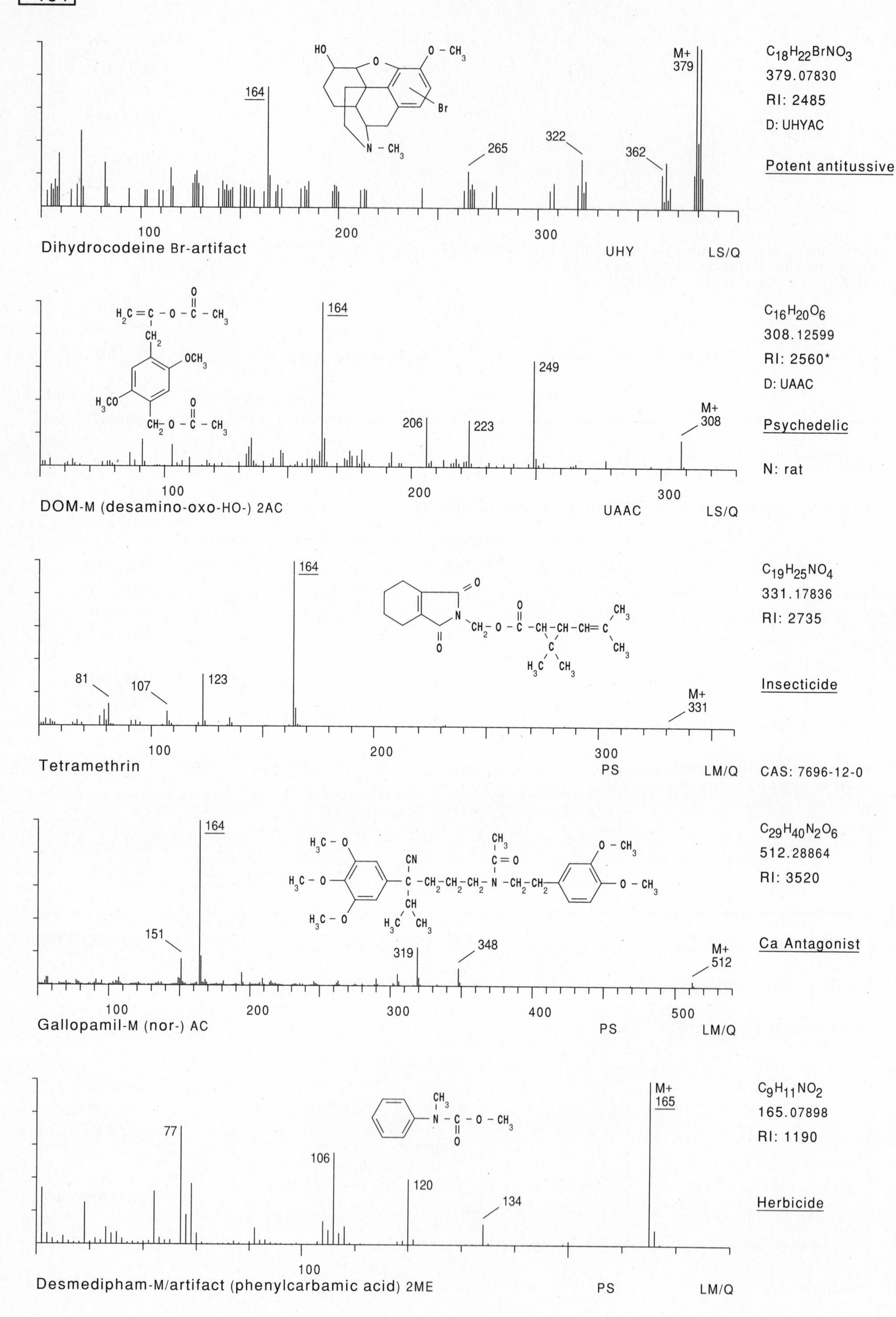

M+ 379
164
265
322
362
100
200
300
Dihydrocodeine Br-artifact
UHY
LS/Q
$C_{18}H_{22}BrNO_3$
379.07830
RI: 2485
D: UHYAC
Potent antitussive
164
249
206
223
M+ 308
100
200
300
DOM-M (desamino-oxo-HO-) 2AC
UAAC
LS/Q
$C_{16}H_{20}O_6$
308.12599
RI: 2560*
D: UAAC
Psychedelic
N: rat
164
81
107
123
M+ 331
100
200
300
Tetramethrin
PS
LM/Q
$C_{19}H_{25}NO_4$
331.17836
RI: 2735
Insecticide
CAS: 7696-12-0
164
151
319
348
M+ 512
100
200
300
400
500
Gallopamil-M (nor-) AC
PS
LM/Q
$C_{29}H_{40}N_2O_6$
512.28864
RI: 3520
Ca Antagonist
M+ 165
77
106
120
134
100
Desmedipham-M/artifact (phenylcarbamic acid) 2ME
PS
LM/Q
$C_9H_{11}NO_2$
165.07898
RI: 1190
Herbicide

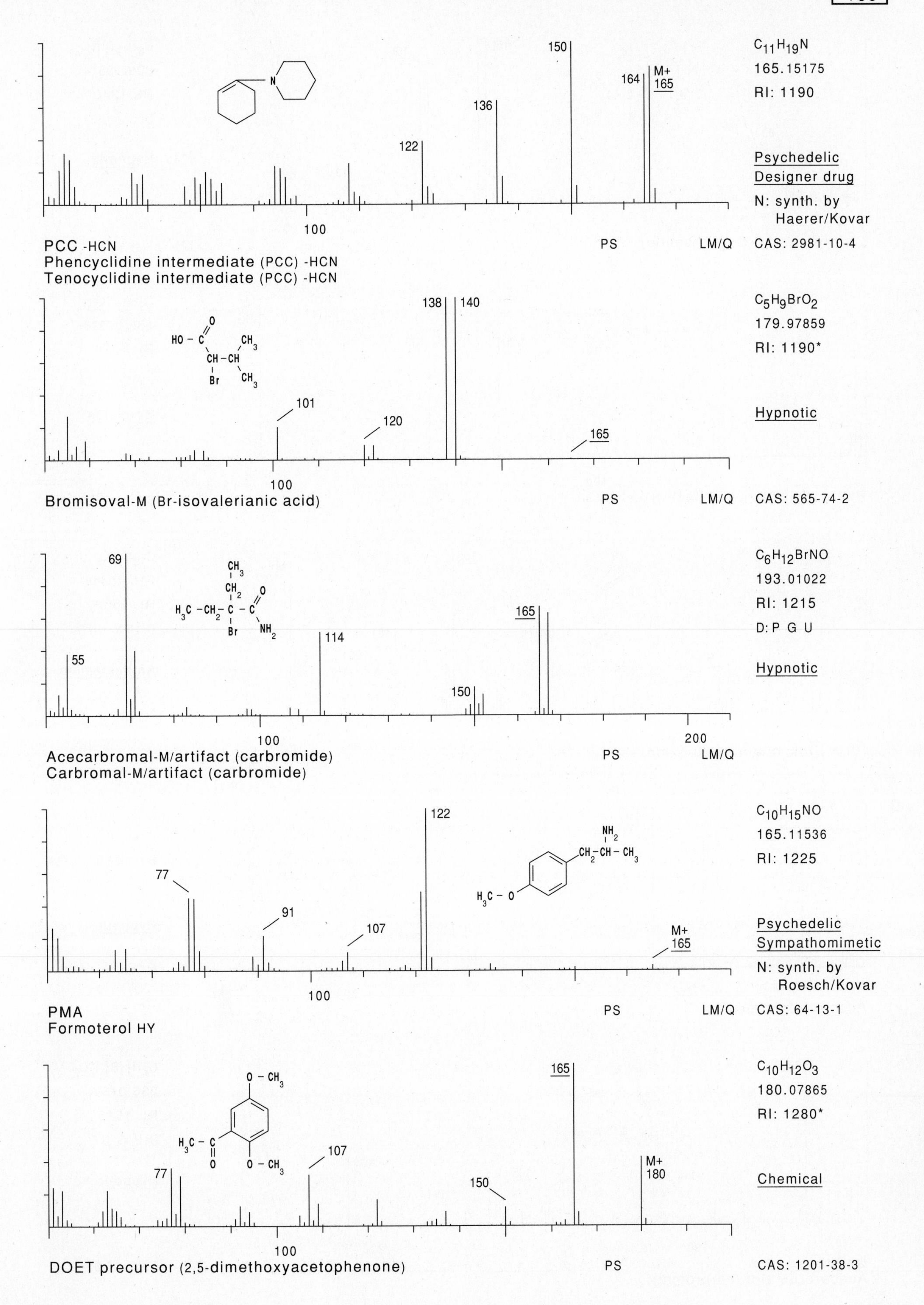

$C_{11}H_{19}N$
165.15175
RI: 1190

Psychedelic
Designer drug
N: synth. by Haerer/Kovar

PCC -HCN
Phencyclidine intermediate (PCC) -HCN
Tenocyclidine intermediate (PCC) -HCN

CAS: 2981-10-4

$C_5H_9BrO_2$
179.97859
RI: 1190*

Hypnotic

Bromisoval-M (Br-isovalerianic acid)

CAS: 565-74-2

$C_6H_{12}BrNO$
193.01022
RI: 1215
D:P G U

Hypnotic

Acecarbromal-M/artifact (carbromide)
Carbromal-M/artifact (carbromide)

$C_{10}H_{15}NO$
165.11536
RI: 1225

Psychedelic
Sympathomimetic
N: synth. by Roesch/Kovar

PMA
Formoterol HY

CAS: 64-13-1

$C_{10}H_{12}O_3$
180.07865
RI: 1280*

Chemical

DOET precursor (2,5-dimethoxyacetophenone)

CAS: 1201-38-3

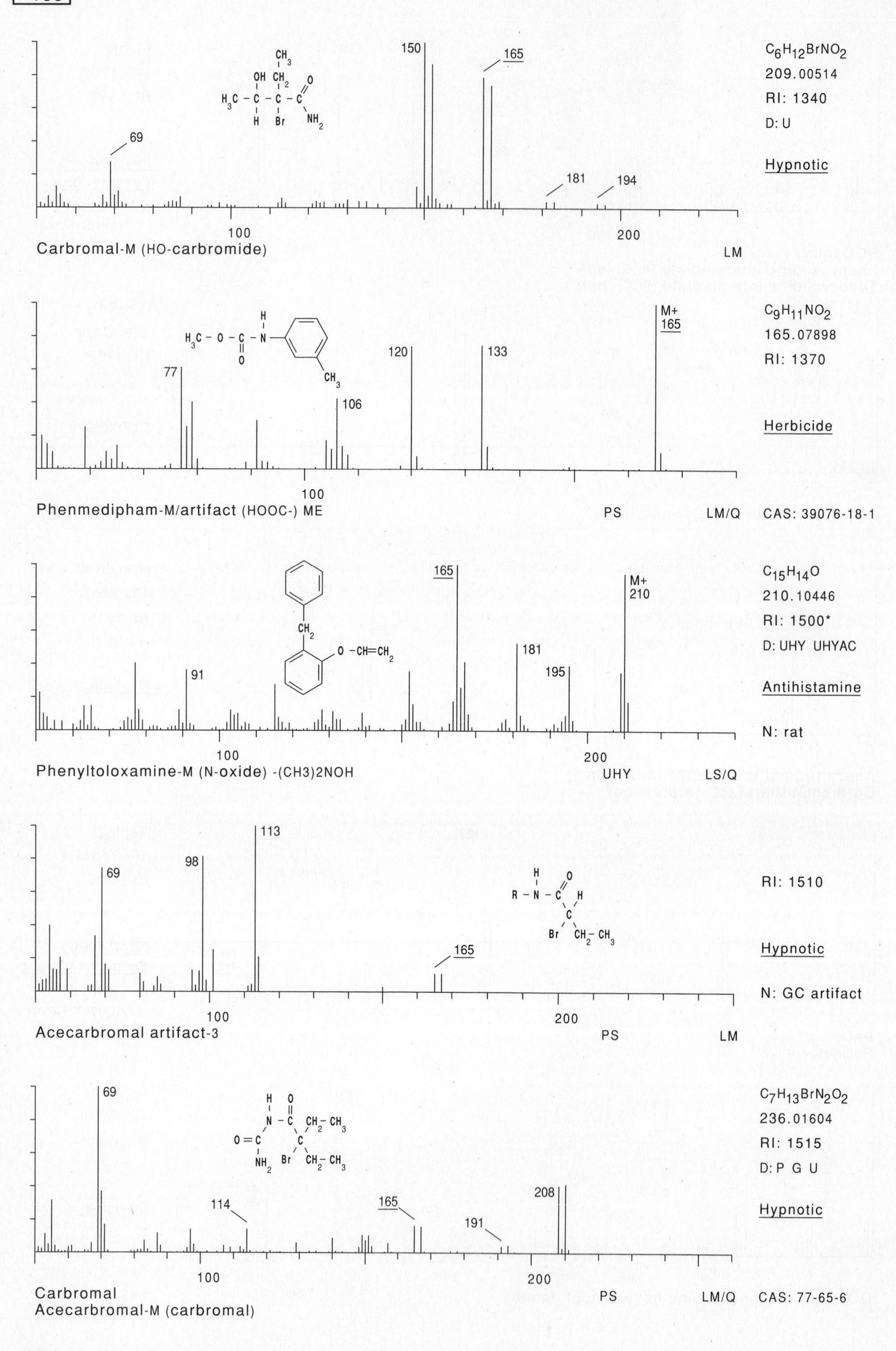
150
165
69
181
194
100
200
Carbromal-M (HO-carbromide)
LM
$C_6H_{12}BrNO_2$
209.00514
RI: 1340
D: U
Hypnotic
M+ 165
120
133
77
106
Phenmedipham-M/artifact (HOOC-) ME
PS
LM/Q
$C_9H_{11}NO_2$
165.07898
RI: 1370
Herbicide
CAS: 39076-18-1
165
M+ 210
181
195
91
Phenyltoloxamine-M (N-oxide) -(CH3)2NOH
UHY
LS/Q
$C_{15}H_{14}O$
210.10446
RI: 1500*
D: UHY UHYAC
Antihistamine
N: rat
113
98
69
165
Acecarbromal artifact-3
PS
LM
RI: 1510
Hypnotic
N: GC artifact
69
114
165
191
208
Carbromal
Acecarbromal-M (carbromal)
PS
LM/Q
$C_7H_{13}BrN_2O_2$
236.01604
RI: 1515
D: P G U
Hypnotic
CAS: 77-65-6

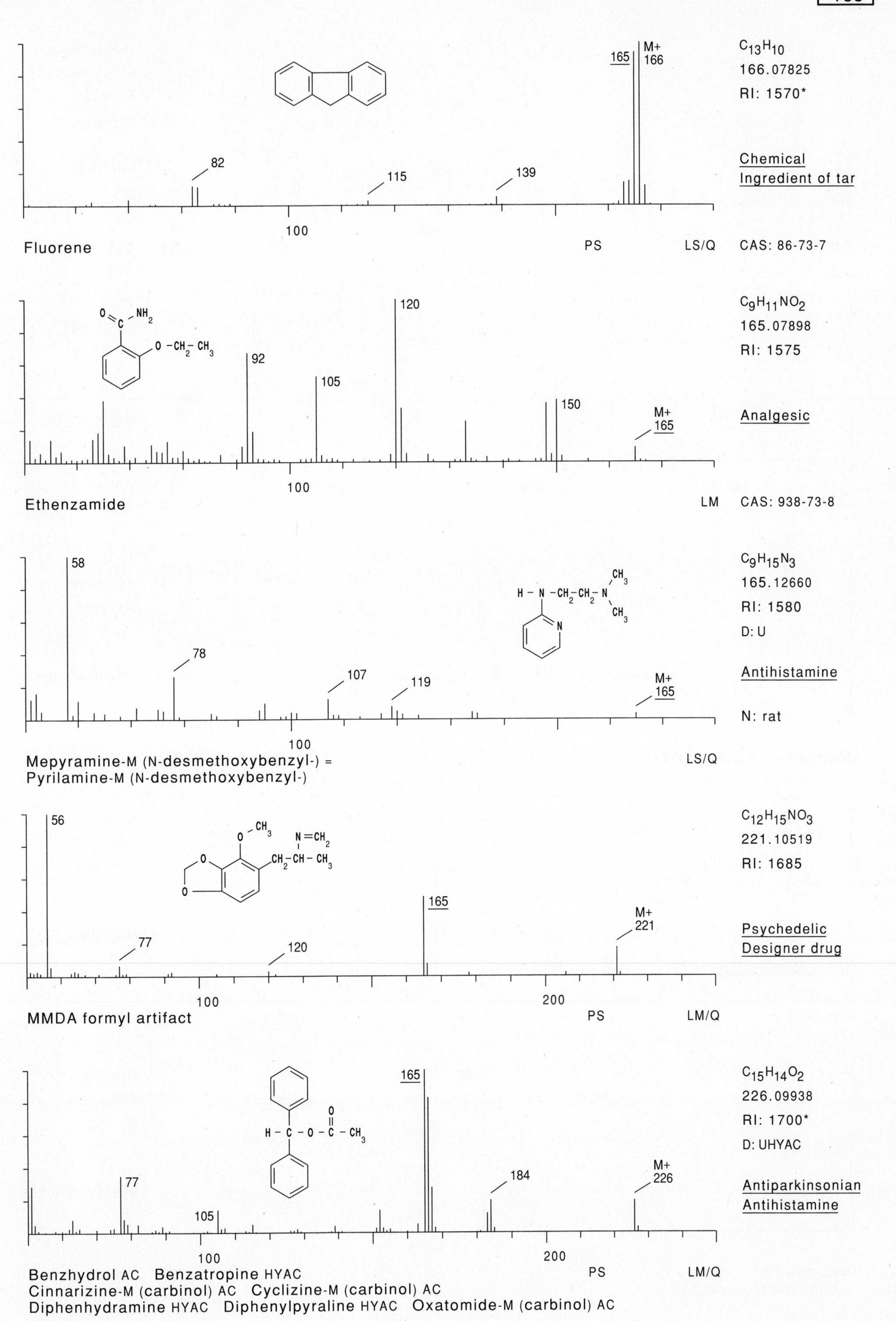

Benzhydrol AC Benzatropine HYAC
Cinnarizine-M (carbinol) AC Cyclizine-M (carbinol) AC
Diphenhydramine HYAC Diphenylpyraline HYAC Oxatomide-M (carbinol) AC

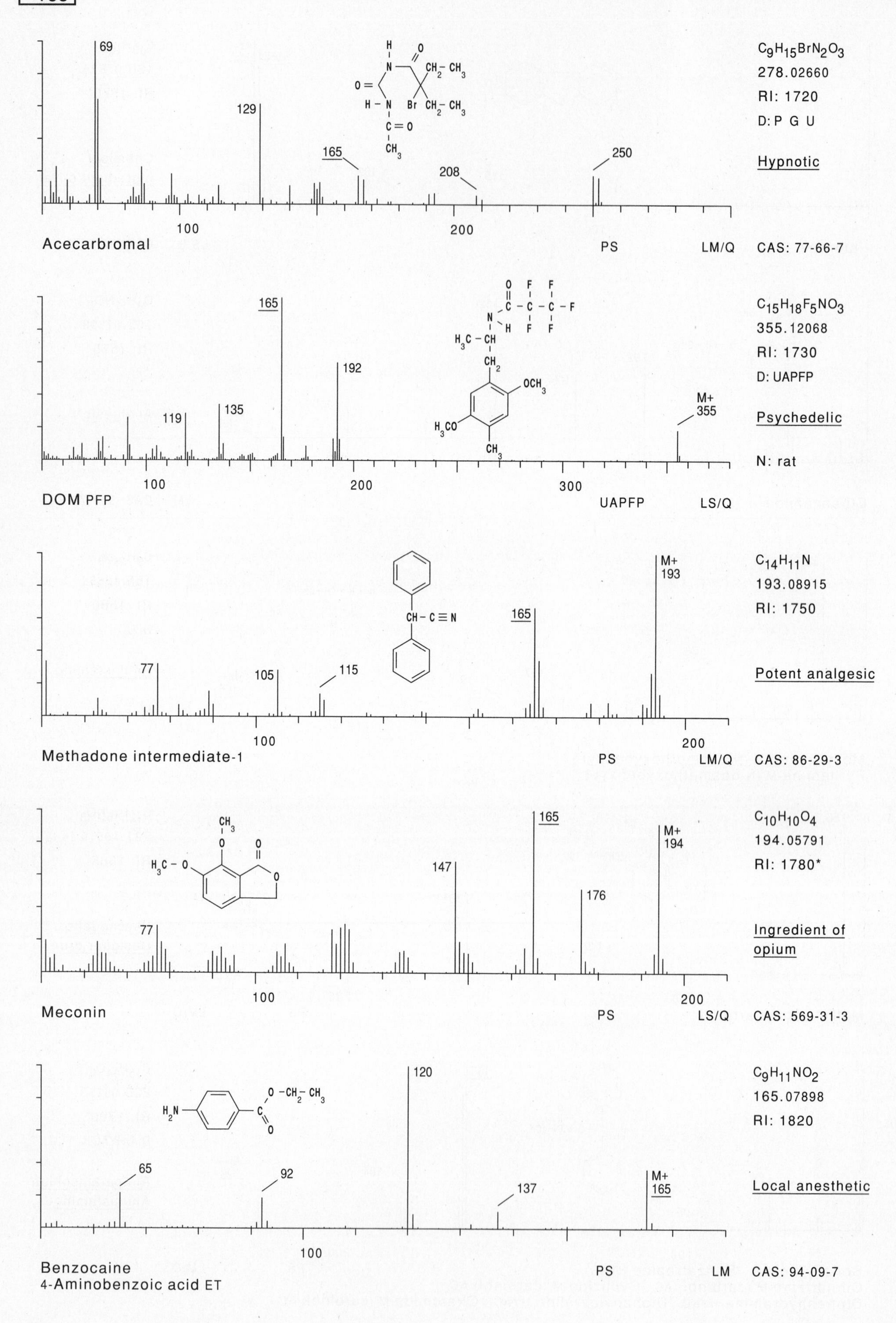
69
129
165
208
250
100
200
Acecarbromal
PS
LM/Q
$C_9H_{15}BrN_2O_3$
278.02660
RI: 1720
D: P G U
Hypnotic
CAS: 77-66-7
165
192
119
135
M+
355
100
200
300
DOM PFP
UAPFP
LS/Q
$C_{15}H_{18}F_5NO_3$
355.12068
RI: 1730
D: UAPFP
Psychedelic
N: rat
M+
193
165
77
105
115
100
200
Methadone intermediate-1
PS
LM/Q
$C_{14}H_{11}N$
193.08915
RI: 1750
Potent analgesic
CAS: 86-29-3
165
M+
194
147
176
77
100
200
Meconin
PS
LS/Q
$C_{10}H_{10}O_4$
194.05791
RI: 1780*
Ingredient of opium
CAS: 569-31-3
120
65
92
137
M+
165
100
Benzocaine
4-Aminobenzoic acid ET
PS
LM
$C_9H_{11}NO_2$
165.07898
RI: 1820
Local anesthetic
CAS: 94-09-7

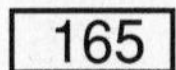

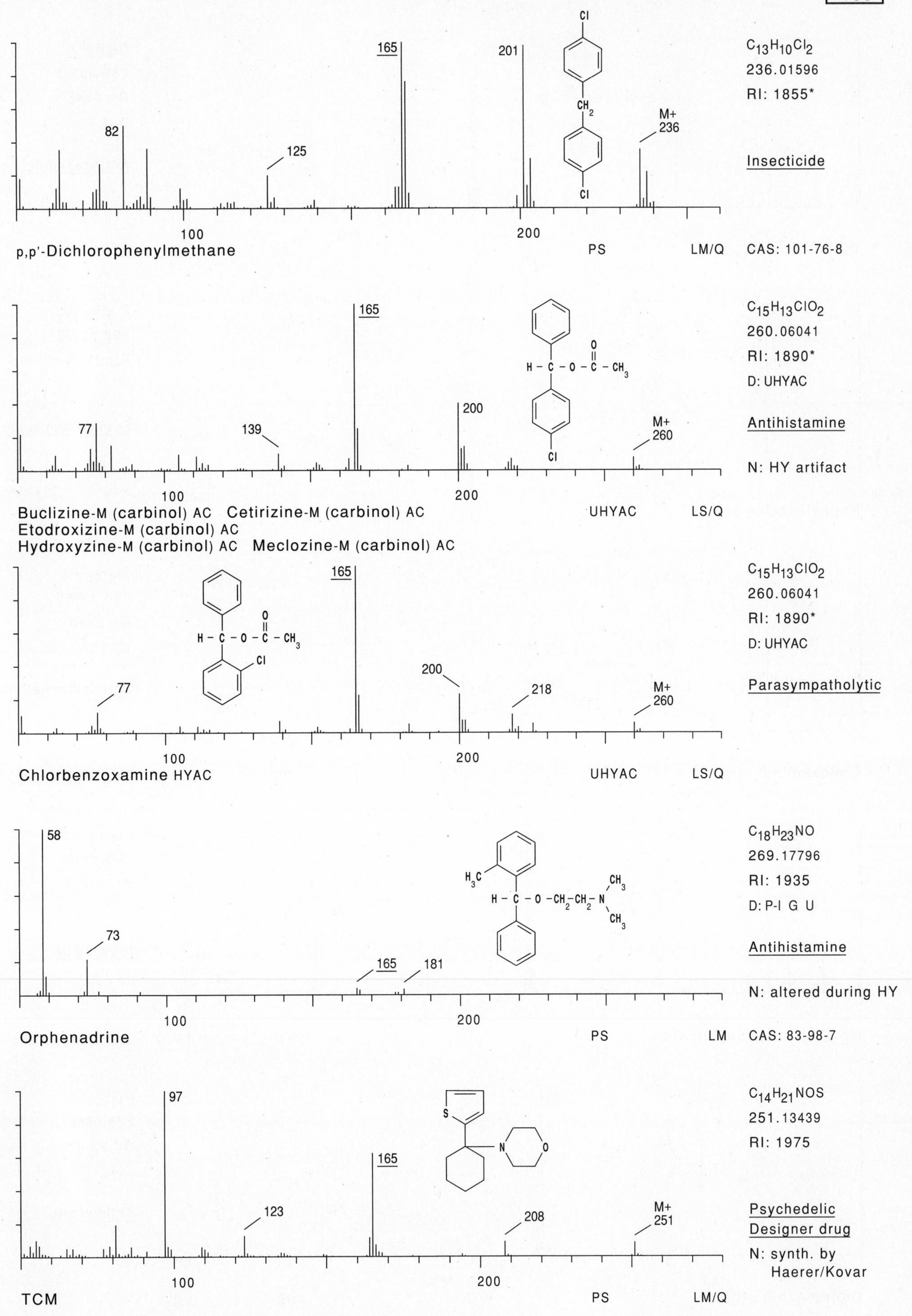

$C_{13}H_{10}Cl_2$
236.01596
RI: 1855*

Insecticide

CAS: 101-76-8

$C_{15}H_{13}ClO_2$
260.06041
RI: 1890*
D: UHYAC

Antihistamine

N: HY artifact

$C_{15}H_{13}ClO_2$
260.06041
RI: 1890*
D: UHYAC

Parasympatholytic

$C_{18}H_{23}NO$
269.17796
RI: 1935
D: P-I G U

Antihistamine

N: altered during HY

CAS: 83-98-7

$C_{14}H_{21}NOS$
251.13439
RI: 1975

Psychedelic
Designer drug

N: synth. by
Haerer/Kovar

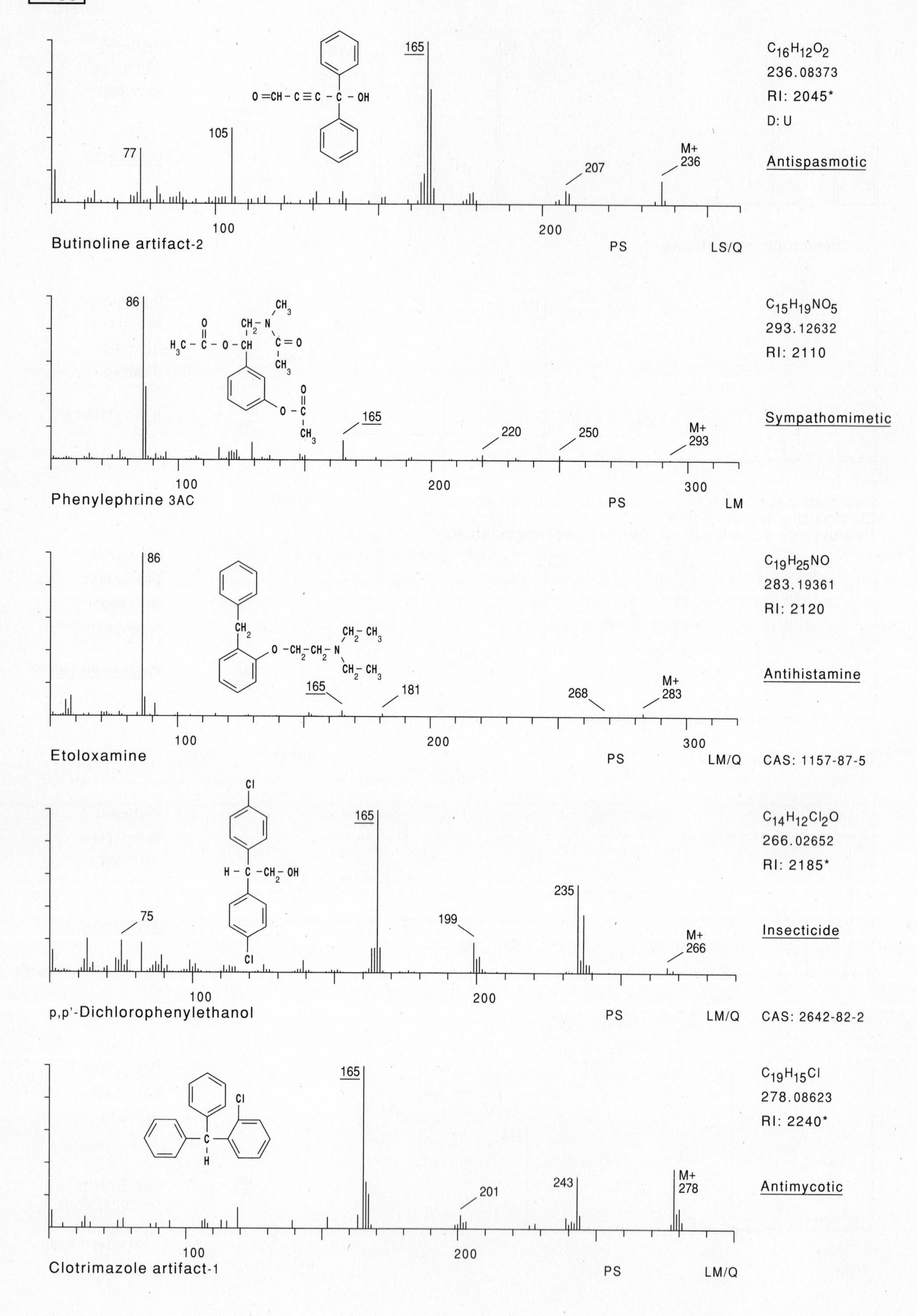

O =CH – C ≡ C – C – OH
165
105
77
207
M+
236
100
200
PS
LS/Q
Butinoline artifact-2
$C_{16}H_{12}O_2$
236.08373
RI: 2045*
D: U
Antispasmotic
86
165
220
250
M+
293
100
200
300
PS
LM
Phenylephrine 3AC
$C_{15}H_{19}NO_5$
293.12632
RI: 2110
Sympathomimetic
86
165
181
268
M+
283
100
200
300
PS
LM/Q
Etoloxamine
$C_{19}H_{25}NO$
283.19361
RI: 2120
Antihistamine
CAS: 1157-87-5
H – C – CH2– OH
165
75
199
235
M+
266
100
200
PS
LM/Q
p,p'-Dichlorophenylethanol
$C_{14}H_{12}Cl_2O$
266.02652
RI: 2185*
Insecticide
CAS: 2642-82-2
165
201
243
M+
278
100
200
PS
LM/Q
Clotrimazole artifact-1
$C_{19}H_{15}Cl$
278.08623
RI: 2240*
Antimycotic

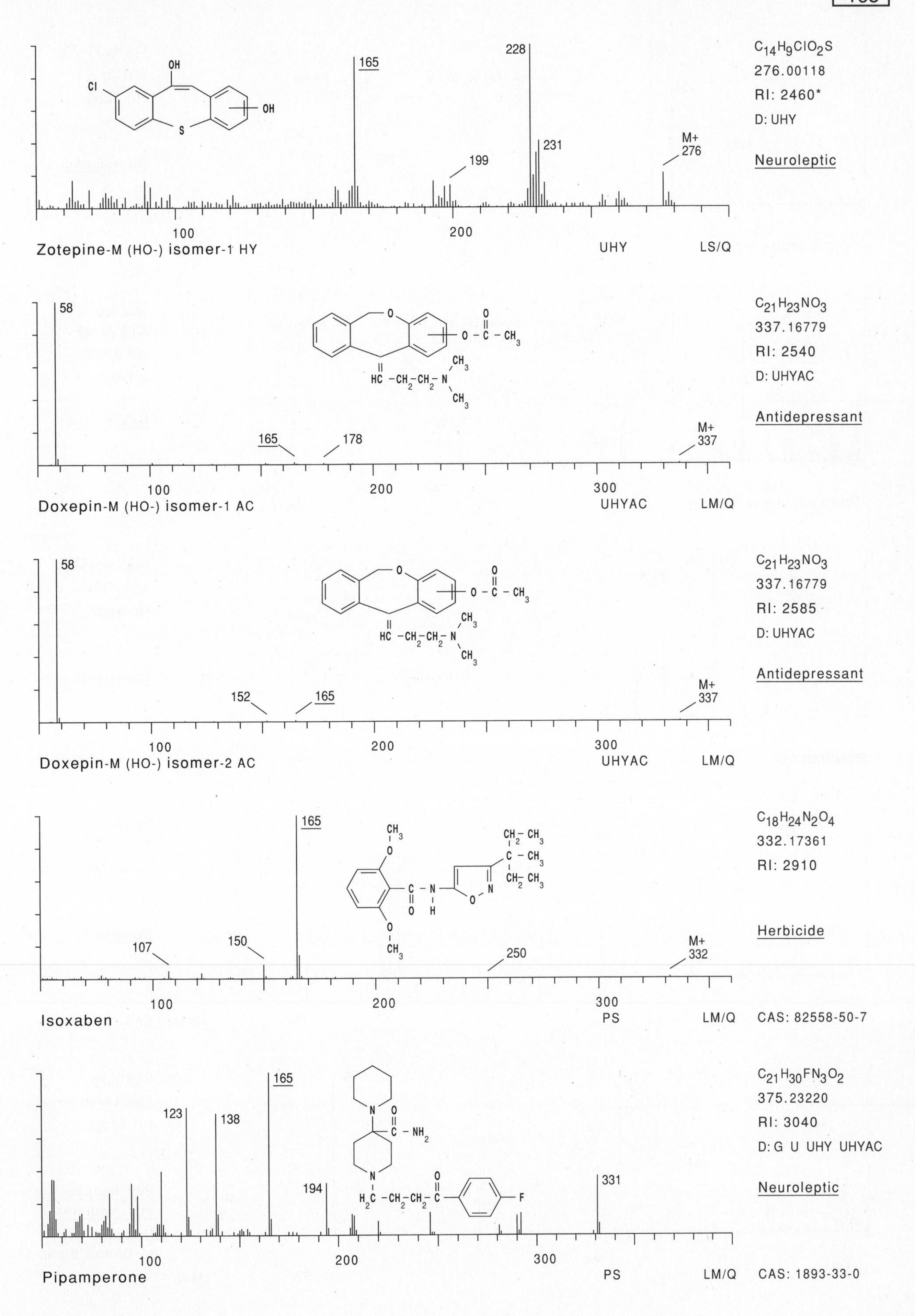

Zotepine-M (HO-) isomer-1 HY

$C_{14}H_{9}ClO_{2}S$
276.00118
RI: 2460*
D: UHY

Neuroleptic

Doxepin-M (HO-) isomer-1 AC

$C_{21}H_{23}NO_{3}$
337.16779
RI: 2540
D: UHYAC

Antidepressant

Doxepin-M (HO-) isomer-2 AC

$C_{21}H_{23}NO_{3}$
337.16779
RI: 2585
D: UHYAC

Antidepressant

Isoxaben

$C_{18}H_{24}N_{2}O_{4}$
332.17361
RI: 2910

Herbicide

CAS: 82558-50-7

Pipamperone

$C_{21}H_{30}FN_{3}O_{2}$
375.23220
RI: 3040
D: G U UHY UHYAC

Neuroleptic

CAS: 1893-33-0

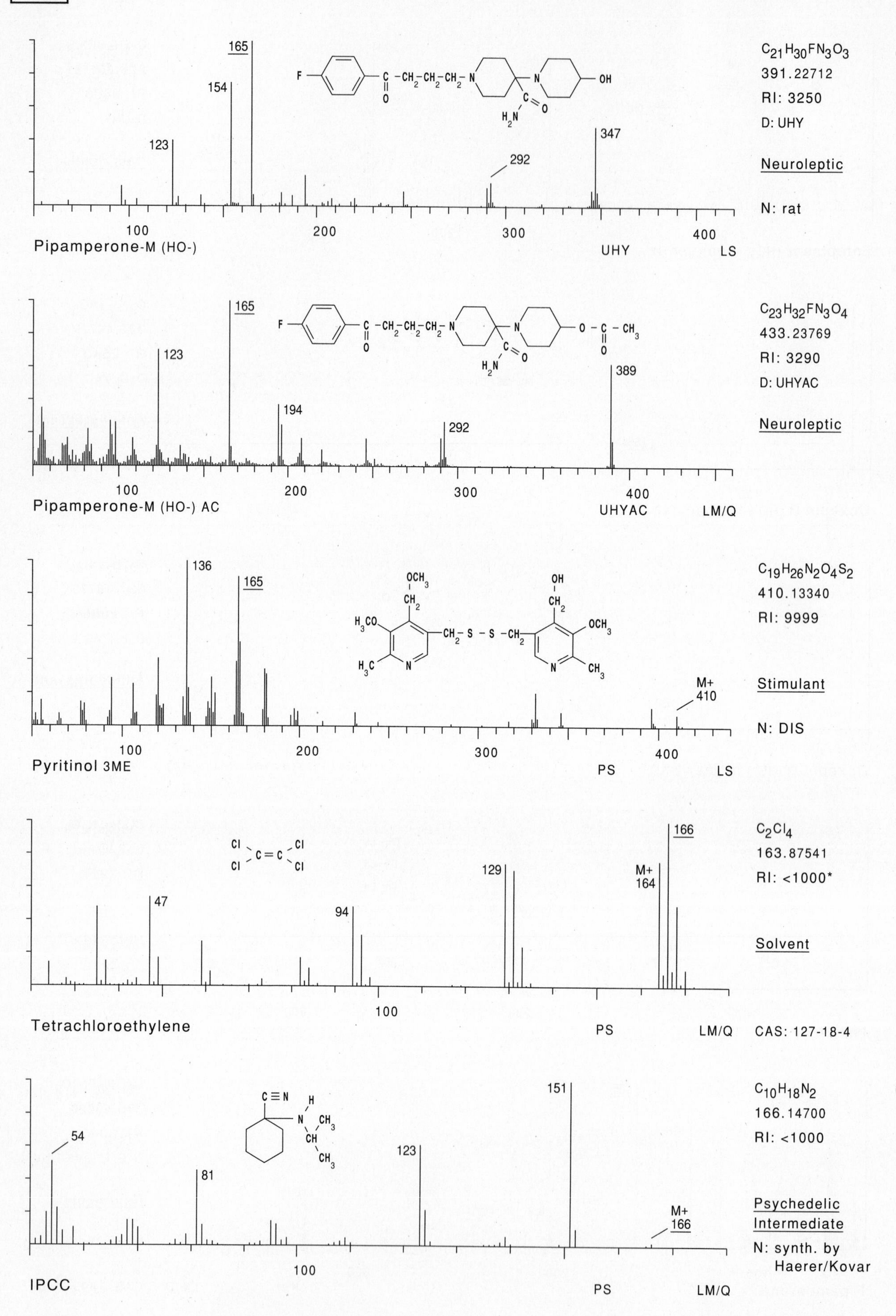

165
154
123
347
292
100
200
300
400
Pipamperone-M (HO-)
UHY
LS
$C_{21}H_{30}FN_3O_3$
391.22712
RI: 3250
D: UHY
Neuroleptic
N: rat
165
123
389
194
292
Pipamperone-M (HO-) AC
UHYAC
LM/Q
$C_{23}H_{32}FN_3O_4$
433.23769
RI: 3290
D: UHYAC
Neuroleptic
136
165
M+
410
Pyritinol 3ME
PS
LS
$C_{19}H_{26}N_2O_4S_2$
410.13340
RI: 9999
Stimulant
N: DIS
166
129
M+
164
47
94
Tetrachloroethylene
PS
LM/Q
C_2Cl_4
163.87541
RI: <1000*
Solvent
CAS: 127-18-4
151
54
123
81
M+
166
IPCC
PS
LM/Q
$C_{10}H_{18}N_2$
166.14700
RI: <1000
Psychedelic
Intermediate
N: synth. by
Haerer/Kovar

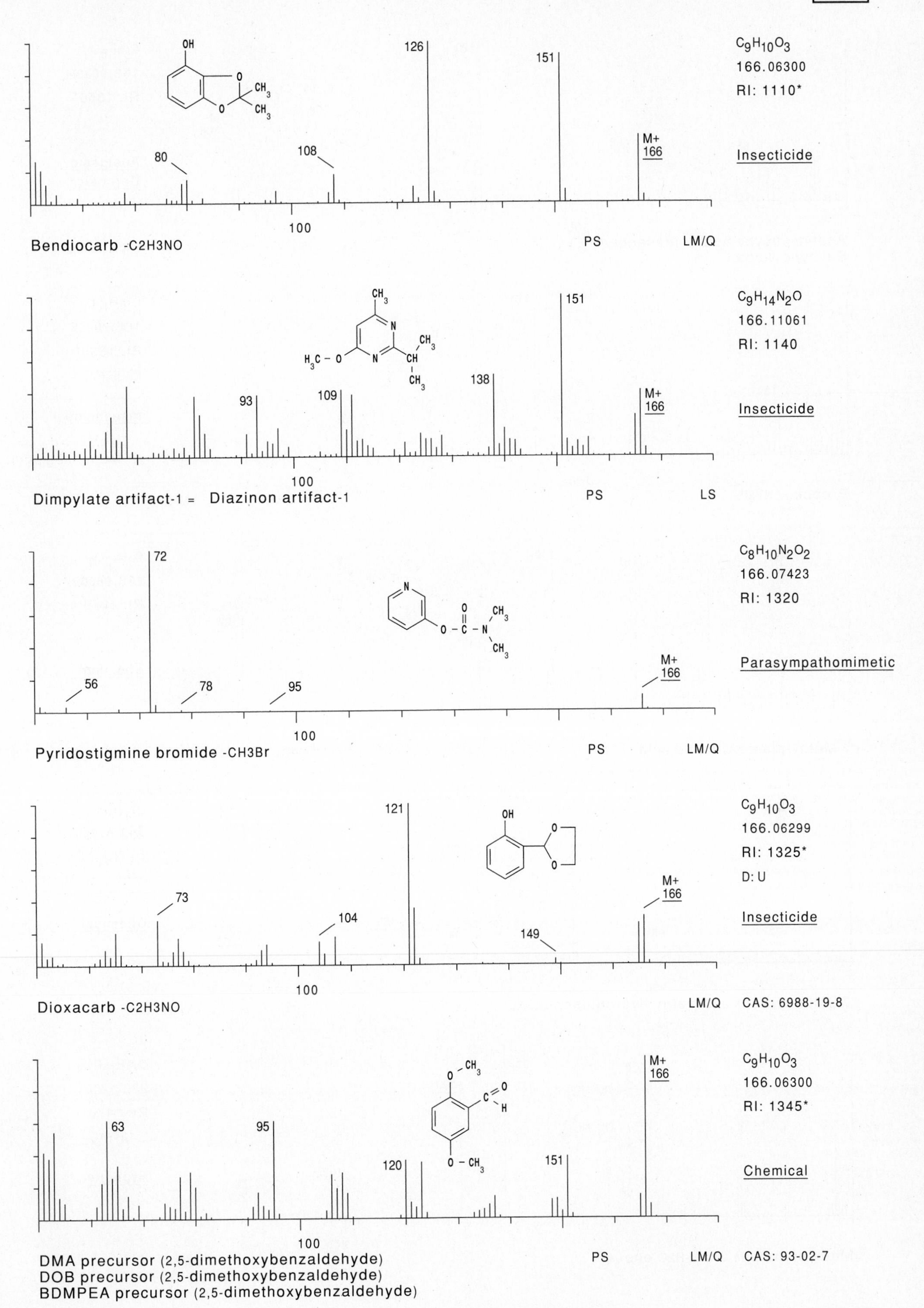
166
$C_9H_{10}O_3$
166.06300
RI: 1110*
Insecticide
OH
O
O
CH_3
CH_3
80
108
126
151
M+
166
100
Bendiocarb -C2H3NO
PS
LM/Q
$C_9H_{14}N_2O$
166.11061
RI: 1140
Insecticide
CH_3
N
N
$H_3C - O$
CH
CH_3
CH_3
93
109
138
151
M+
166
100
Dimpylate artifact-1 = Diazinon artifact-1
PS
LS
$C_8H_{10}N_2O_2$
166.07423
RI: 1320
Parasympathomimetic
N
O
O - C - N
CH_3
CH_3
56
72
78
95
M+
166
100
Pyridostigmine bromide -CH3Br
PS
LM/Q
$C_9H_{10}O_3$
166.06299
RI: 1325*
D: U
Insecticide
OH
O
O
73
104
121
149
M+
166
100
Dioxacarb -C2H3NO
LM/Q
CAS: 6988-19-8
$C_9H_{10}O_3$
166.06300
RI: 1345*
Chemical
O - CH_3
C = O
H
O - CH_3
63
95
120
151
M+
166
100
DMA precursor (2,5-dimethoxybenzaldehyde)
DOB precursor (2,5-dimethoxybenzaldehyde)
BDMPEA precursor (2,5-dimethoxybenzaldehyde)
PS
LM/Q
CAS: 93-02-7

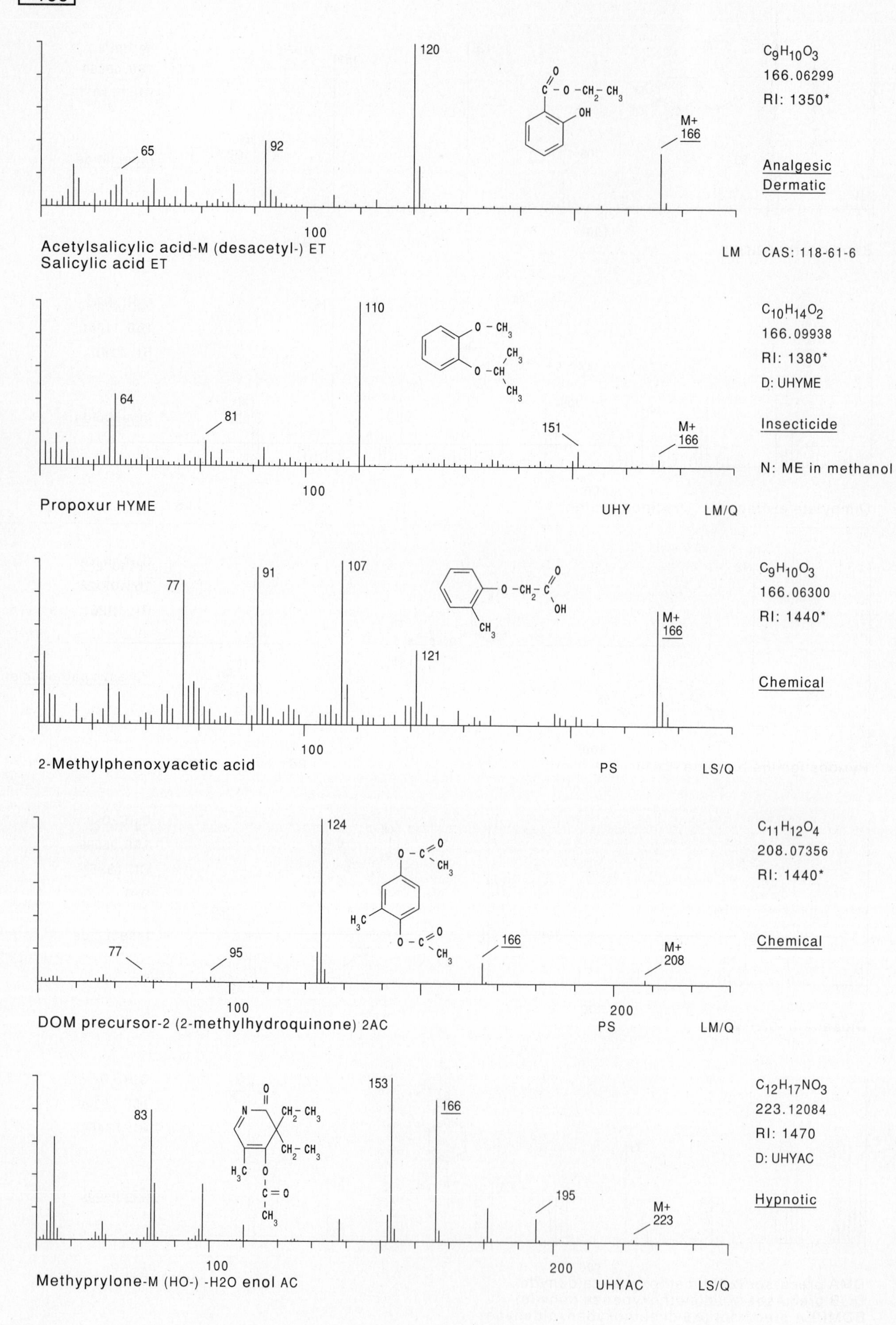

120
92
65
M+
166
$C_9H_{10}O_3$
166.06299
RI: 1350*
Analgesic
Dermatic
100
Acetylsalicylic acid-M (desacetyl-) ET
Salicylic acid ET
LM
CAS: 118-61-6
110
64
81
151
M+
166
$C_{10}H_{14}O_2$
166.09938
RI: 1380*
D: UHYME
Insecticide
N: ME in methanol
100
Propoxur HYME
UHY
LM/Q
107
91
77
121
M+
166
$C_9H_{10}O_3$
166.06300
RI: 1440*
Chemical
100
2-Methylphenoxyacetic acid
PS
LS/Q
124
77
95
166
M+
208
$C_{11}H_{12}O_4$
208.07356
RI: 1440*
Chemical
100
200
DOM precursor-2 (2-methylhydroquinone) 2AC
PS
LM/Q
153
166
83
195
M+
223
$C_{12}H_{17}NO_3$
223.12084
RI: 1470
D: UHYAC
Hypnotic
100
200
Methyprylone-M (HO-) -H2O enol AC
UHYAC
LS/Q

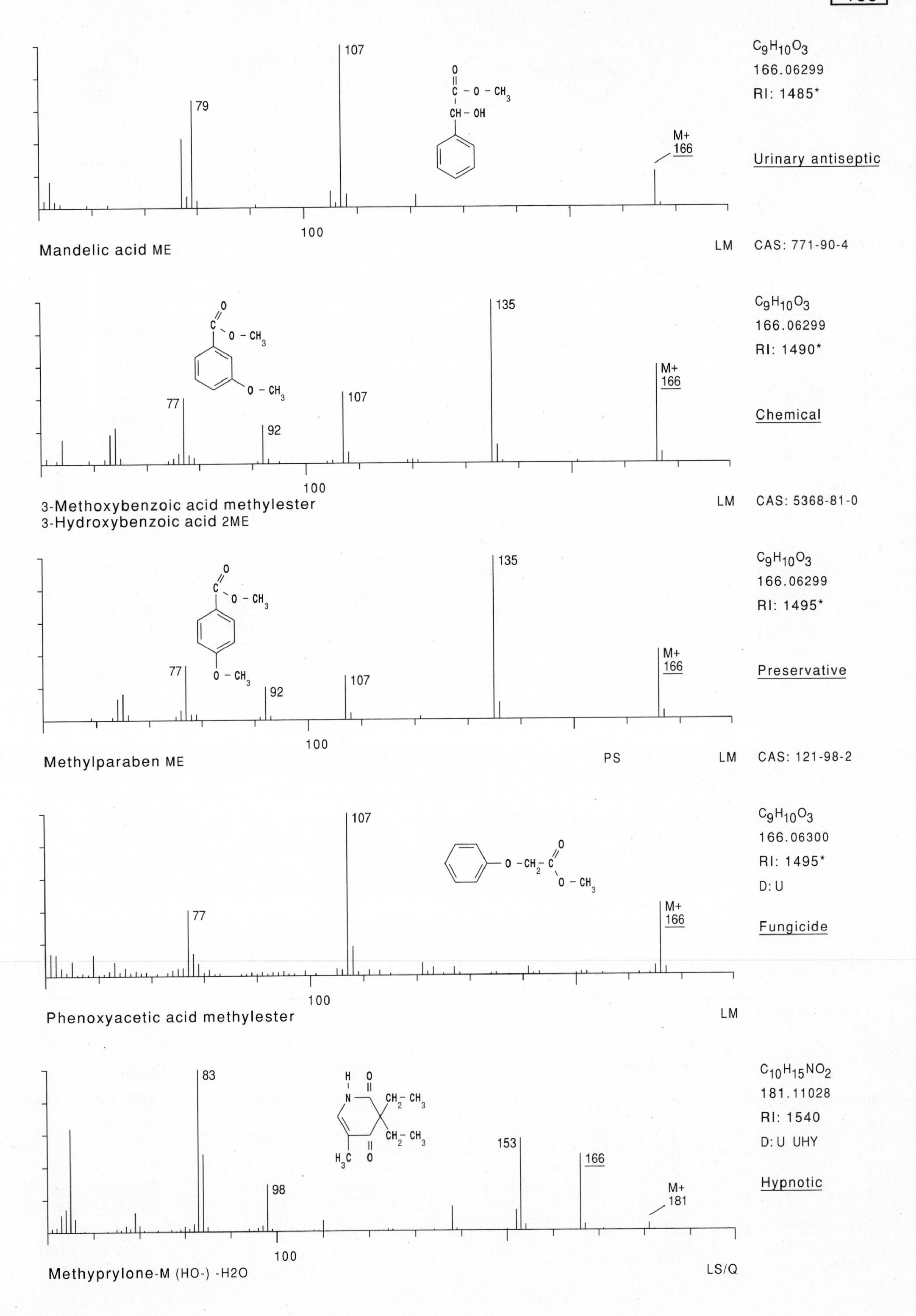

Mandelic acid ME

LM

$C_9H_{10}O_3$
166.06299
RI: 1485*
Urinary antiseptic
CAS: 771-90-4

3-Methoxybenzoic acid methylester
3-Hydroxybenzoic acid 2ME

LM

$C_9H_{10}O_3$
166.06299
RI: 1490*
Chemical
CAS: 5368-81-0

Methylparaben ME

PS LM

$C_9H_{10}O_3$
166.06299
RI: 1495*
Preservative
CAS: 121-98-2

Phenoxyacetic acid methylester

LM

$C_9H_{10}O_3$
166.06300
RI: 1495*
D: U
Fungicide

Methyprylone-M (HO-) -H2O

LS/Q

$C_{10}H_{15}NO_2$
181.11028
RI: 1540
D: U UHY
Hypnotic

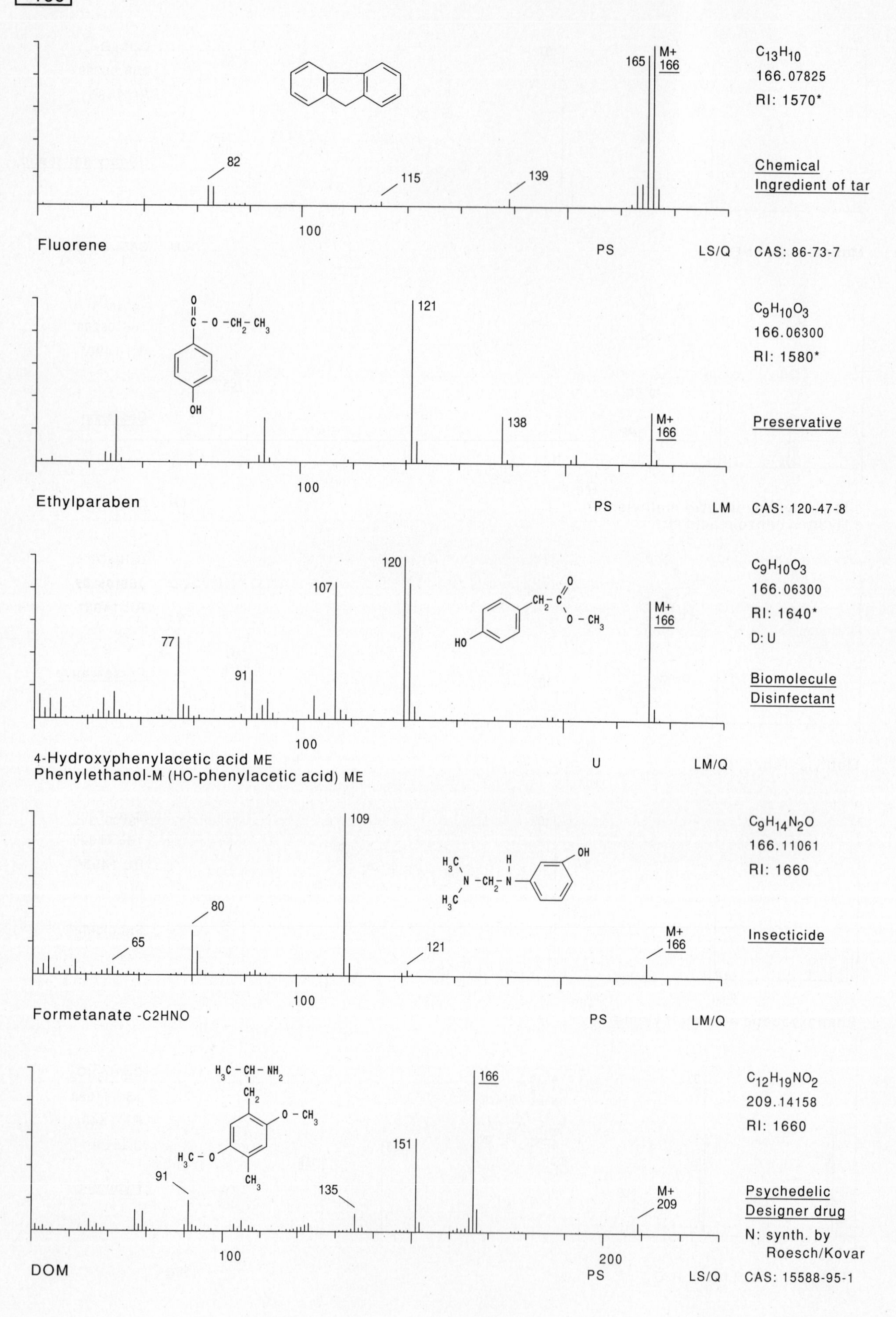

165
M+
166
82
115
139
100
Fluorene
PS
LS/Q
$C_{13}H_{10}$
166.07825
RI: 1570*
Chemical
Ingredient of tar
CAS: 86-73-7
121
138
M+
166
100
Ethylparaben
PS
LM
$C_9H_{10}O_3$
166.06300
RI: 1580*
Preservative
CAS: 120-47-8
120
107
77
91
M+
166
100
4-Hydroxyphenylacetic acid ME
Phenylethanol-M (HO-phenylacetic acid) ME
U
LM/Q
$C_9H_{10}O_3$
166.06300
RI: 1640*
D: U
Biomolecule
Disinfectant
109
80
65
121
M+
166
100
Formetanate -C2HNO
PS
LM/Q
$C_9H_{14}N_2O$
166.11061
RI: 1660
Insecticide
166
151
91
135
M+
209
100
200
DOM
PS
LS/Q
$C_{12}H_{19}NO_2$
209.14158
RI: 1660
Psychedelic
Designer drug
N: synth. by
Roesch/Kovar
CAS: 15588-95-1

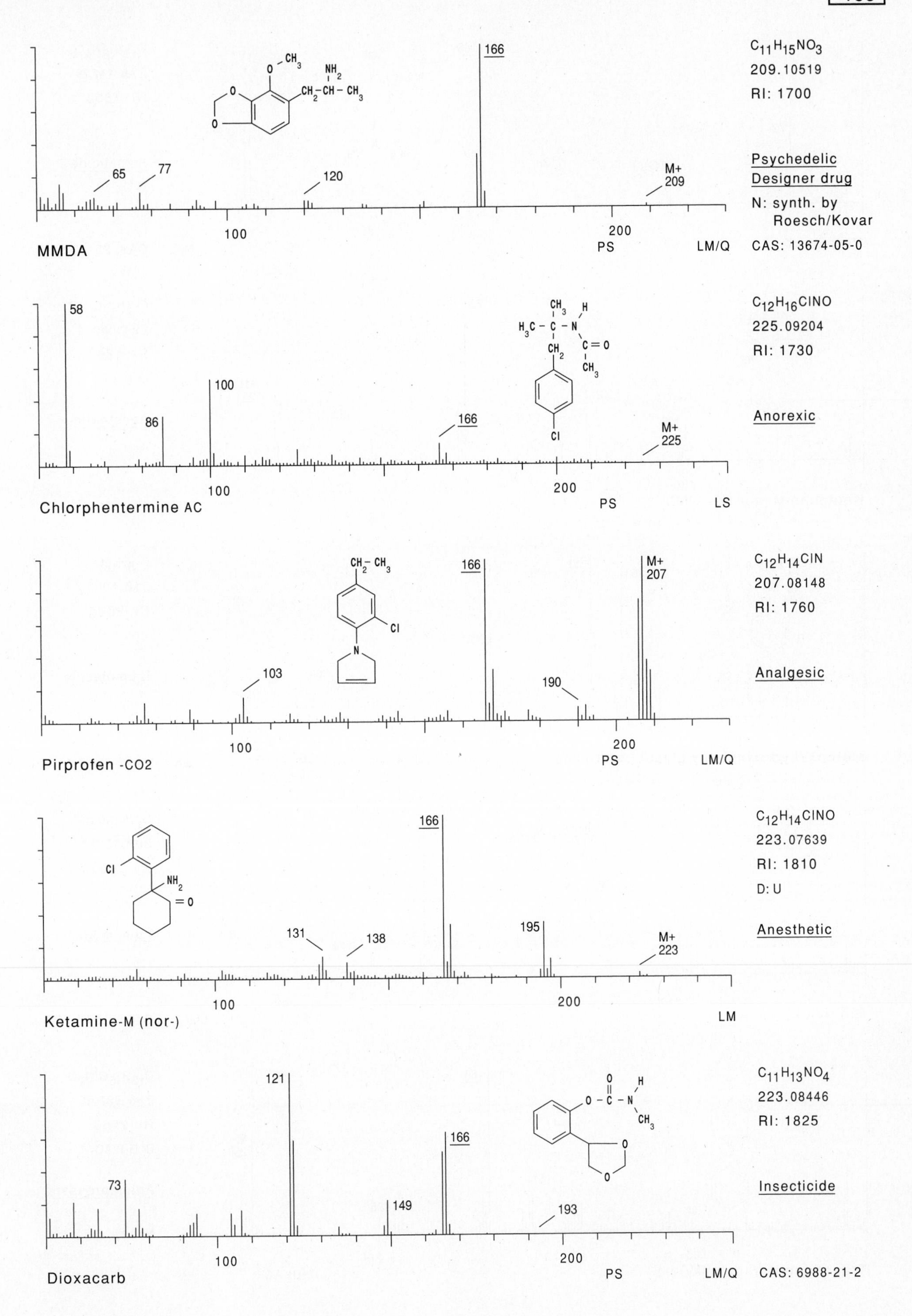
166
MMDA
$C_{11}H_{15}NO_3$
209.10519
RI: 1700
Psychedelic
Designer drug
N: synth. by
Roesch/Kovar
CAS: 13674-05-0
65
77
120
166
M+
209
100
200
PS
LM/Q
Chlorphentermine AC
$C_{12}H_{16}ClNO$
225.09204
RI: 1730
Anorexic
58
86
100
166
M+
225
PS
LS
Pirprofen -CO2
$C_{12}H_{14}ClN$
207.08148
RI: 1760
Analgesic
103
166
190
M+
207
PS
LM/Q
Ketamine-M (nor-)
$C_{12}H_{14}ClNO$
223.07639
RI: 1810
D: U
Anesthetic
131
138
166
195
M+
223
LM
Dioxacarb
$C_{11}H_{13}NO_4$
223.08446
RI: 1825
Insecticide
CAS: 6988-21-2
73
121
149
166
193
PS
LM/Q

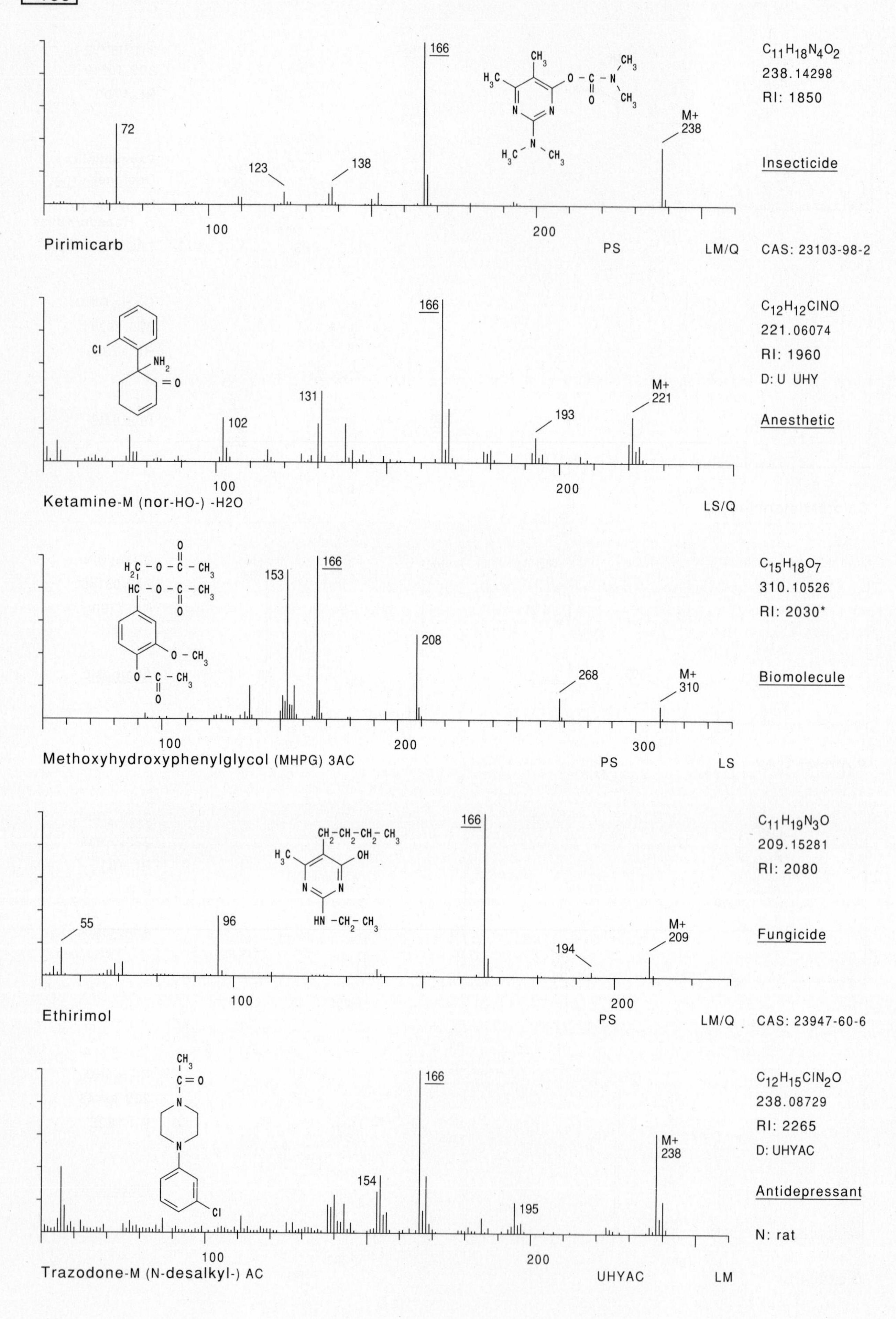

166
72
123
138
M+
238
100
200
PS
LM/Q
C11H18N4O2
238.14298
RI: 1850
Insecticide
CAS: 23103-98-2
Pirimicarb
166
131
102
193
M+
221
LS/Q
C12H12ClNO
221.06074
RI: 1960
D: U UHY
Anesthetic
Ketamine-M (nor-HO-) -H2O
153
166
208
268
M+
310
300
PS
LS
C15H18O7
310.10526
RI: 2030*
Biomolecule
Methoxyhydroxyphenylglycol (MHPG) 3AC
166
55
96
194
M+
209
PS
LM/Q
C11H19N3O
209.15281
RI: 2080
Fungicide
CAS: 23947-60-6
Ethirimol
166
154
195
M+
238
UHYAC
LM
C12H15ClN2O
238.08729
RI: 2265
D: UHYAC
Antidepressant
N: rat
Trazodone-M (N-desalkyl-) AC

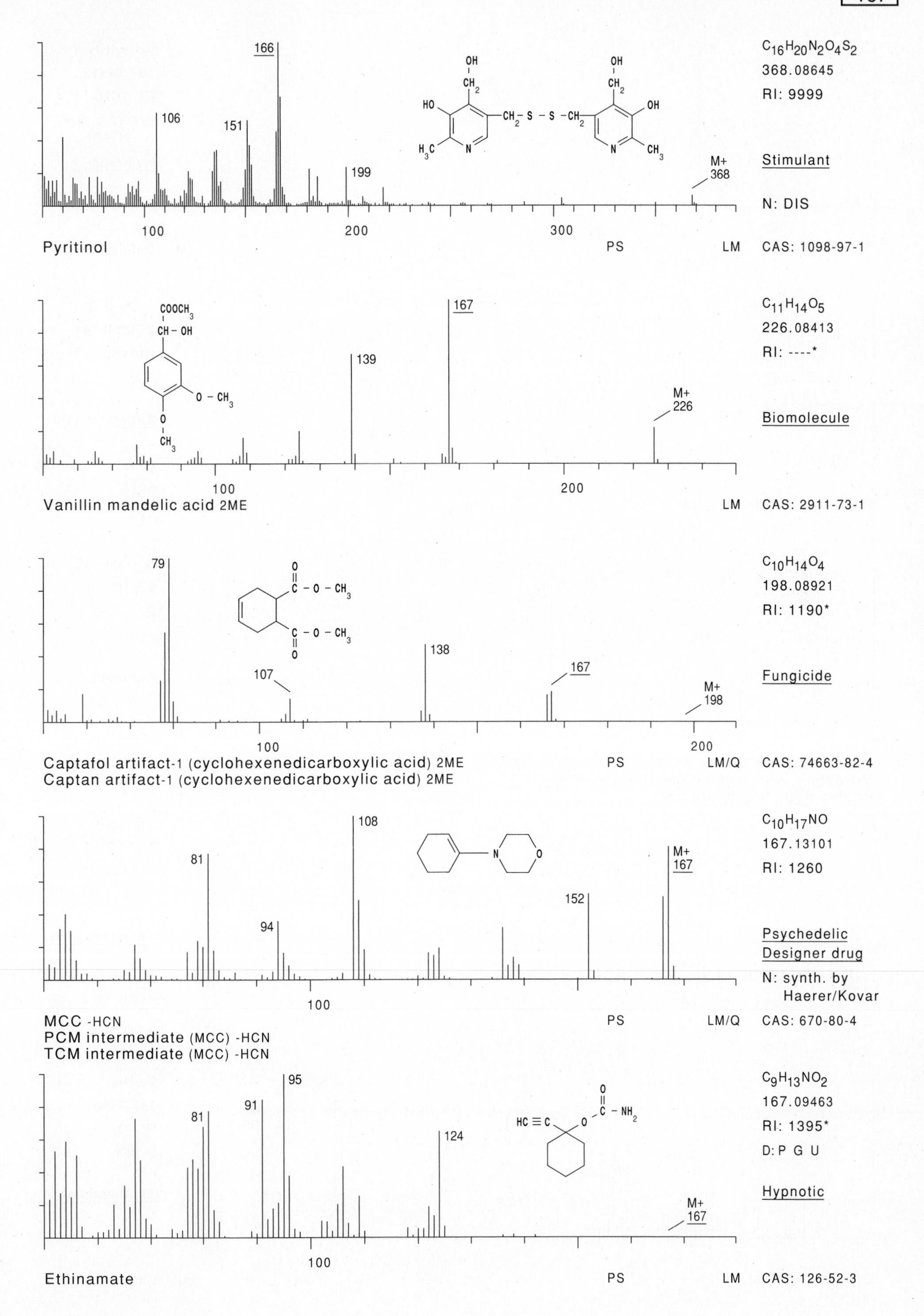

$C_{16}H_{20}N_2O_4S_2$
368.08645
RI: 9999
Stimulant
N: DIS
Pyritinol
PS
LM
CAS: 1098-97-1
$C_{11}H_{14}O_5$
226.08413
RI: ----*
Biomolecule
Vanillin mandelic acid 2ME
LM
CAS: 2911-73-1
$C_{10}H_{14}O_4$
198.08921
RI: 1190*
Fungicide
Captafol artifact-1 (cyclohexenedicarboxylic acid) 2ME
Captan artifact-1 (cyclohexenedicarboxylic acid) 2ME
PS
LM/Q
CAS: 74663-82-4
$C_{10}H_{17}NO$
167.13101
RI: 1260
Psychedelic
Designer drug
N: synth. by Haerer/Kovar
MCC -HCN
PCM intermediate (MCC) -HCN
TCM intermediate (MCC) -HCN
PS
LM/Q
CAS: 670-80-4
$C_9H_{13}NO_2$
167.09463
RI: 1395*
D: P G U
Hypnotic
Ethinamate
PS
LM
CAS: 126-52-3

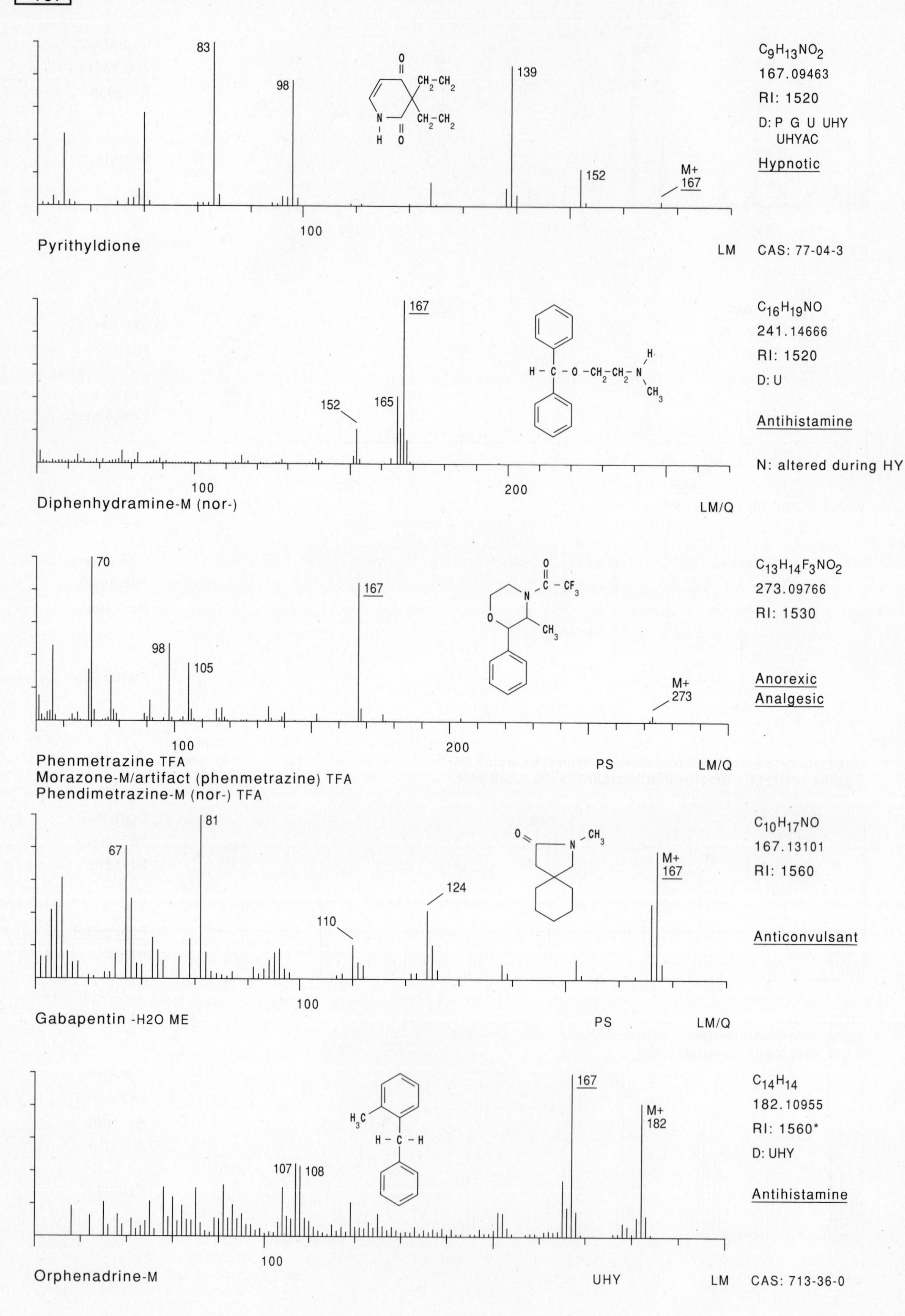

83
98
139
152
M+
167
100
Pyrithyldione
LM
C9H13NO2
167.09463
RI: 1520
D: P G U UHY
UHYAC
Hypnotic
CAS: 77-04-3
167
152
165
100
200
Diphenhydramine-M (nor-)
LM/Q
C16H19NO
241.14666
RI: 1520
D: U
Antihistamine
N: altered during HY
70
167
98
105
M+
273
100
200
Phenmetrazine TFA
Morazone-M/artifact (phenmetrazine) TFA
Phendimetrazine-M (nor-) TFA
PS
LM/Q
C13H14F3NO2
273.09766
RI: 1530
Anorexic
Analgesic
81
67
M+
167
124
110
100
Gabapentin -H2O ME
PS
LM/Q
C10H17NO
167.13101
RI: 1560
Anticonvulsant
167
M+
182
107
108
100
Orphenadrine-M
UHY
LM
C14H14
182.10955
RI: 1560*
D: UHY
Antihistamine
CAS: 713-36-0

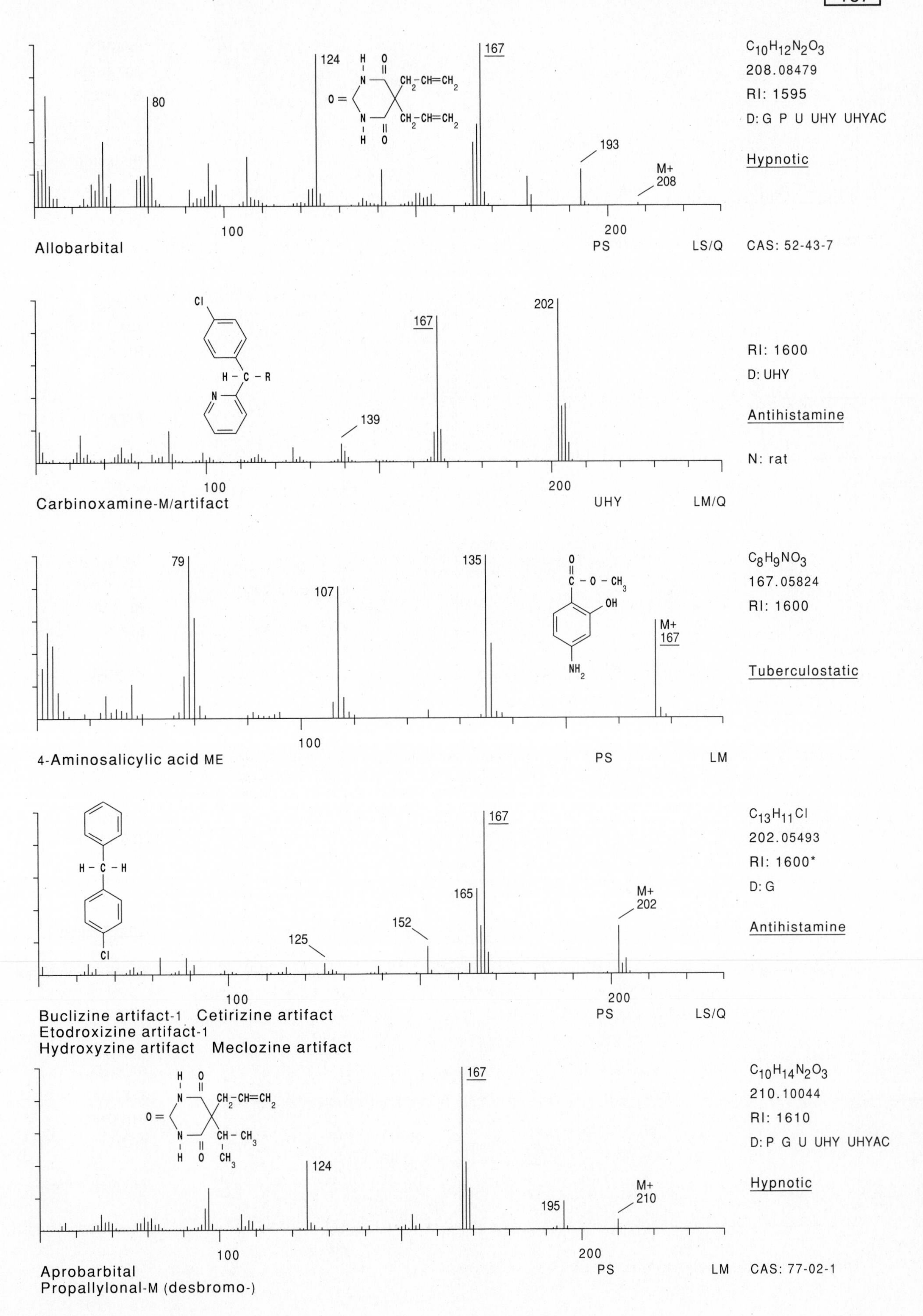

124
167
80
193
M+
208
100
200
PS
LS/Q
Allobarbital
$C_{10}H_{12}N_2O_3$
208.08479
RI: 1595
D: G P U UHY UHYAC
Hypnotic
CAS: 52-43-7
202
167
139
100
200
UHY
LM/Q
Carbinoxamine-M/artifact
RI: 1600
D: UHY
Antihistamine
N: rat
79
135
107
M+
167
100
PS
LM
4-Aminosalicylic acid ME
$C_8H_9NO_3$
167.05824
RI: 1600
Tuberculostatic
167
165
M+
202
152
125
100
200
PS
LS/Q
Buclizine artifact-1 Cetirizine artifact
Etodroxizine artifact-1
Hydroxyzine artifact Meclozine artifact
$C_{13}H_{11}Cl$
202.05493
RI: 1600*
D: G
Antihistamine
167
124
195
M+
210
100
200
PS
LM
Aprobarbital
Propallylonal-M (desbromo-)
$C_{10}H_{14}N_2O_3$
210.10044
RI: 1610
D: P G U UHY UHYAC
Hypnotic
CAS: 77-02-1

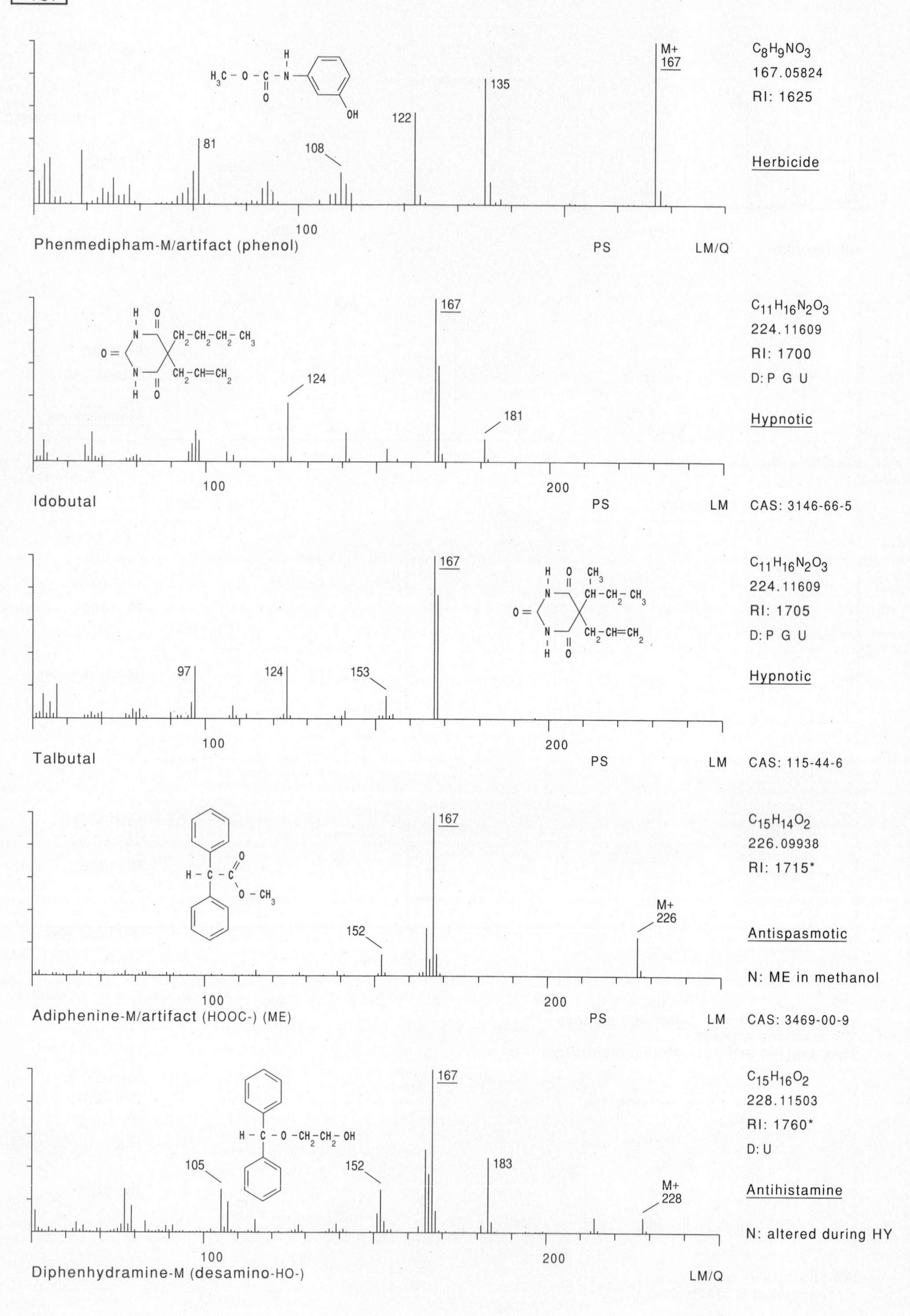

M+
167
135
122
81
108
100
Phenmedipham-M/artifact (phenol)
PS
LM/Q
$C_8H_9NO_3$
167.05824
RI: 1625
Herbicide
167
124
181
100
200
Idobutal
PS
LM
$C_{11}H_{16}N_2O_3$
224.11609
RI: 1700
D: P G U
Hypnotic
CAS: 3146-66-5
167
97
124
153
100
200
Talbutal
PS
LM
$C_{11}H_{16}N_2O_3$
224.11609
RI: 1705
D: P G U
Hypnotic
CAS: 115-44-6
167
152
M+
226
100
200
Adiphenine-M/artifact (HOOC-) (ME)
PS
LM
$C_{15}H_{14}O_2$
226.09938
RI: 1715*
Antispasmotic
N: ME in methanol
CAS: 3469-00-9
167
105
152
183
M+
228
100
200
Diphenhydramine-M (desamino-HO-)
LM/Q
$C_{15}H_{16}O_2$
228.11503
RI: 1760*
D: U
Antihistamine
N: altered during HY

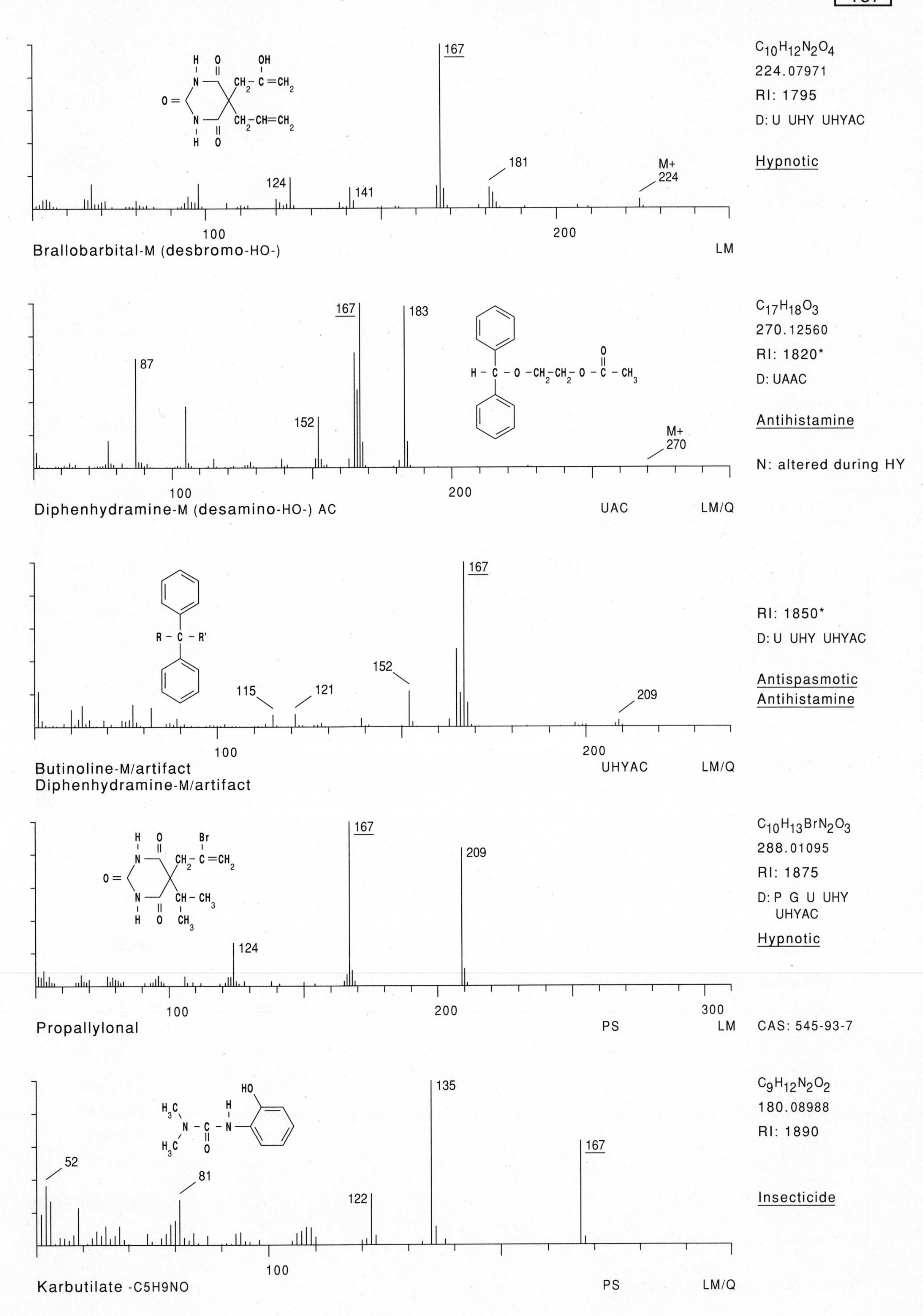

Brallobarbital-M (desbromo-HO-)
LM
$C_{10}H_{12}N_2O_4$
224.07971
RI: 1795
D: U UHY UHYAC
Hypnotic
Diphenhydramine-M (desamino-HO-) AC
UAC
LM/Q
$C_{17}H_{18}O_3$
270.12560
RI: 1820*
D: UAAC
Antihistamine
N: altered during HY
Butinoline-M/artifact
Diphenhydramine-M/artifact
UHYAC
LM/Q
RI: 1850*
D: U UHY UHYAC
Antispasmotic
Antihistamine
Propallylonal
PS
LM
$C_{10}H_{13}BrN_2O_3$
288.01095
RI: 1875
D: P G U UHY UHYAC
Hypnotic
CAS: 545-93-7
Karbutilate -C5H9NO
PS
LM/Q
$C_9H_{12}N_2O_2$
180.08988
RI: 1890
Insecticide

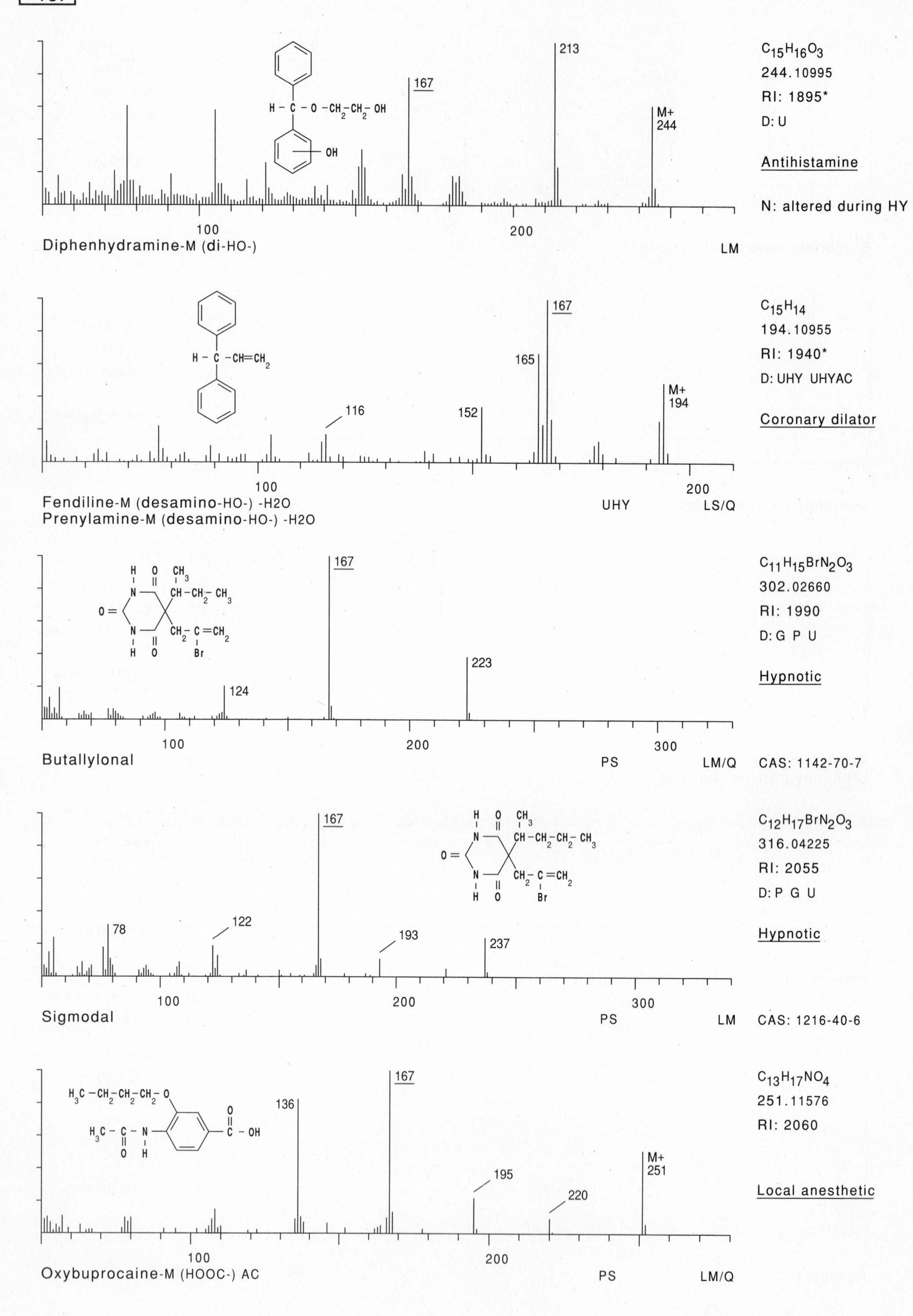

213
167
M+
244
H − C − O −CH2−CH2− OH
OH
100
200
Diphenhydramine-M (di-HO-)
LM
C15H16O3
244.10995
RI: 1895*
D: U
Antihistamine
N: altered during HY
167
165
152
116
M+
194
H − C −CH=CH2
100
200
Fendiline-M (desamino-HO-) -H2O
Prenylamine-M (desamino-HO-) -H2O
UHY
LS/Q
C15H14
194.10955
RI: 1940*
D: UHY UHYAC
Coronary dilator
167
223
124
100
200
300
Butallylonal
PS
LM/Q
C11H15BrN2O3
302.02660
RI: 1990
D: G P U
Hypnotic
CAS: 1142-70-7
167
78
122
193
237
100
200
300
Sigmodal
PS
LM
C12H17BrN2O3
316.04225
RI: 2055
D: P G U
Hypnotic
CAS: 1216-40-6
167
136
195
220
M+
251
H3C −CH2−CH2−CH2− O
H3C − C − N
C − OH
100
200
Oxybuprocaine-M (HOOC-) AC
PS
LM/Q
C13H17NO4
251.11576
RI: 2060
Local anesthetic

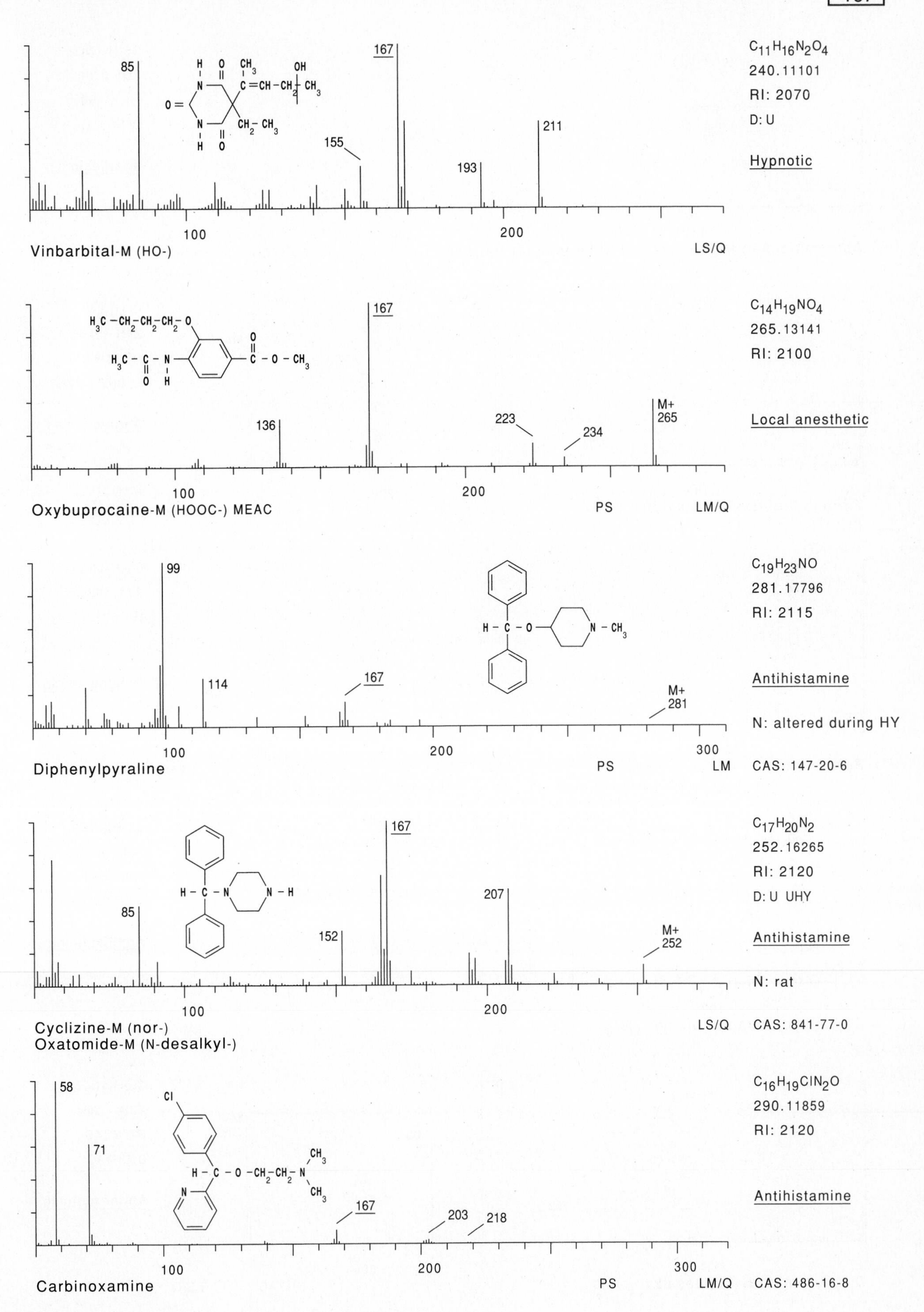

85
167
155
193
211
100
200
Vinbarbital-M (HO-)
LS/Q
$C_{11}H_{16}N_2O_4$
240.11101
RI: 2070
D: U
Hypnotic
167
136
223
234
M+
265
100
200
Oxybuprocaine-M (HOOC-) MEAC
PS
LM/Q
$C_{14}H_{19}NO_4$
265.13141
RI: 2100
Local anesthetic
99
114
167
M+
281
100
200
300
Diphenylpyraline
PS
LM
$C_{19}H_{23}NO$
281.17796
RI: 2115
Antihistamine
N: altered during HY
CAS: 147-20-6
167
85
207
152
M+
252
100
200
Cyclizine-M (nor-)
Oxatomide-M (N-desalkyl-)
LS/Q
$C_{17}H_{20}N_2$
252.16265
RI: 2120
D: U UHY
Antihistamine
N: rat
CAS: 841-77-0
58
71
167
203
218
100
200
300
Carbinoxamine
PS
LM/Q
$C_{16}H_{19}ClN_2O$
290.11859
RI: 2120
Antihistamine
CAS: 486-16-8

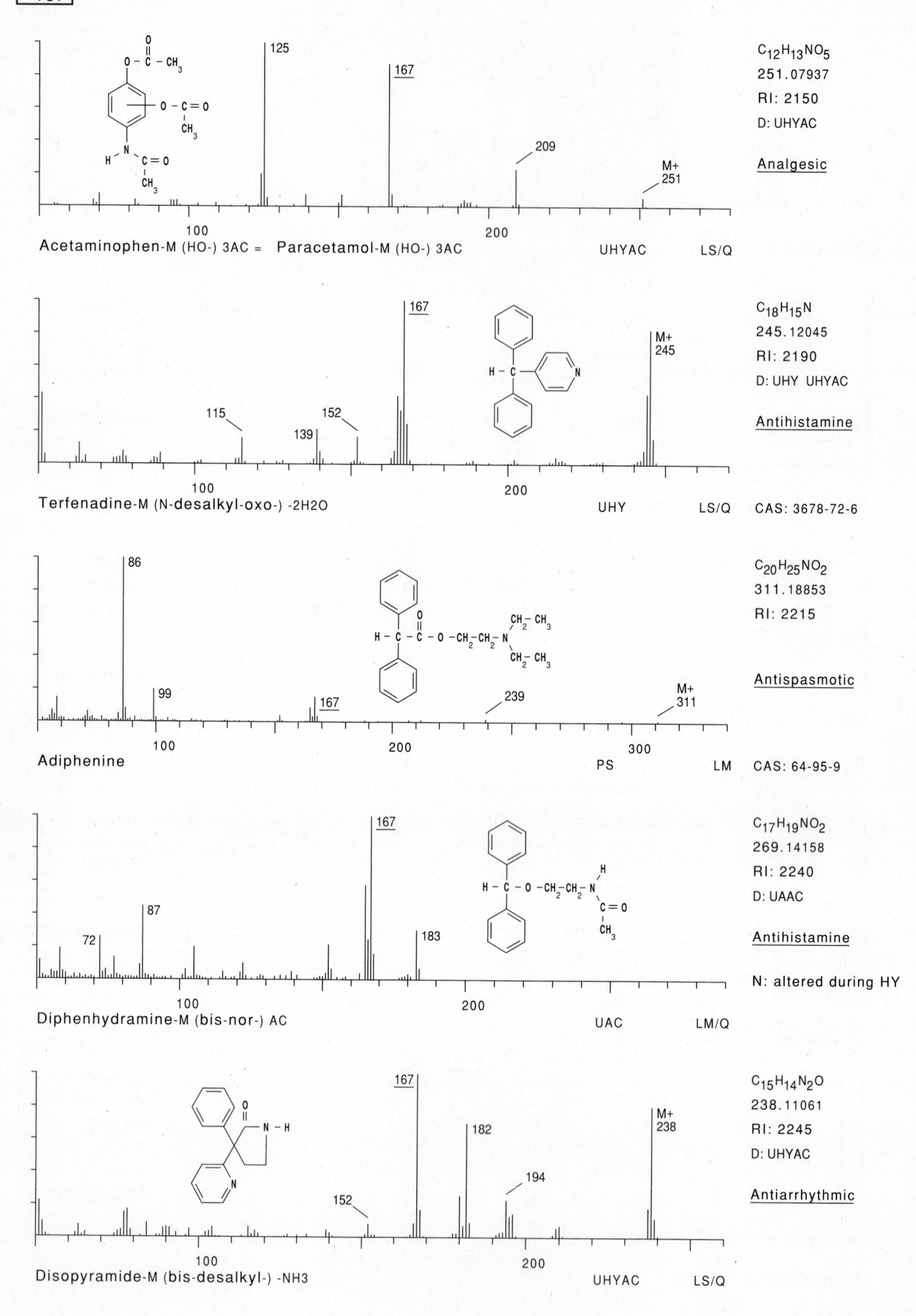
125
167
209
M+
251
100
200
Acetaminophen-M (HO-) 3AC = Paracetamol-M (HO-) 3AC
UHYAC
LS/Q
C12H13NO5
251.07937
RI: 2150
D: UHYAC
Analgesic
167
M+
245
115
152
139
100
200
Terfenadine-M (N-desalkyl-oxo-) -2H2O
UHY
LS/Q
C18H15N
245.12045
RI: 2190
D: UHY UHYAC
Antihistamine
CAS: 3678-72-6
86
99
167
239
M+
311
100
200
300
Adiphenine
PS
LM
C20H25NO2
311.18853
RI: 2215
Antispasmotic
CAS: 64-95-9
167
87
72
183
100
200
Diphenhydramine-M (bis-nor-) AC
UAC
LM/Q
C17H19NO2
269.14158
RI: 2240
D: UAAC
Antihistamine
N: altered during HY
167
182
M+
238
194
152
100
200
Disopyramide-M (bis-desalkyl-) -NH3
UHYAC
LS/Q
C15H14N2O
238.11061
RI: 2245
D: UHYAC
Antiarrhythmic

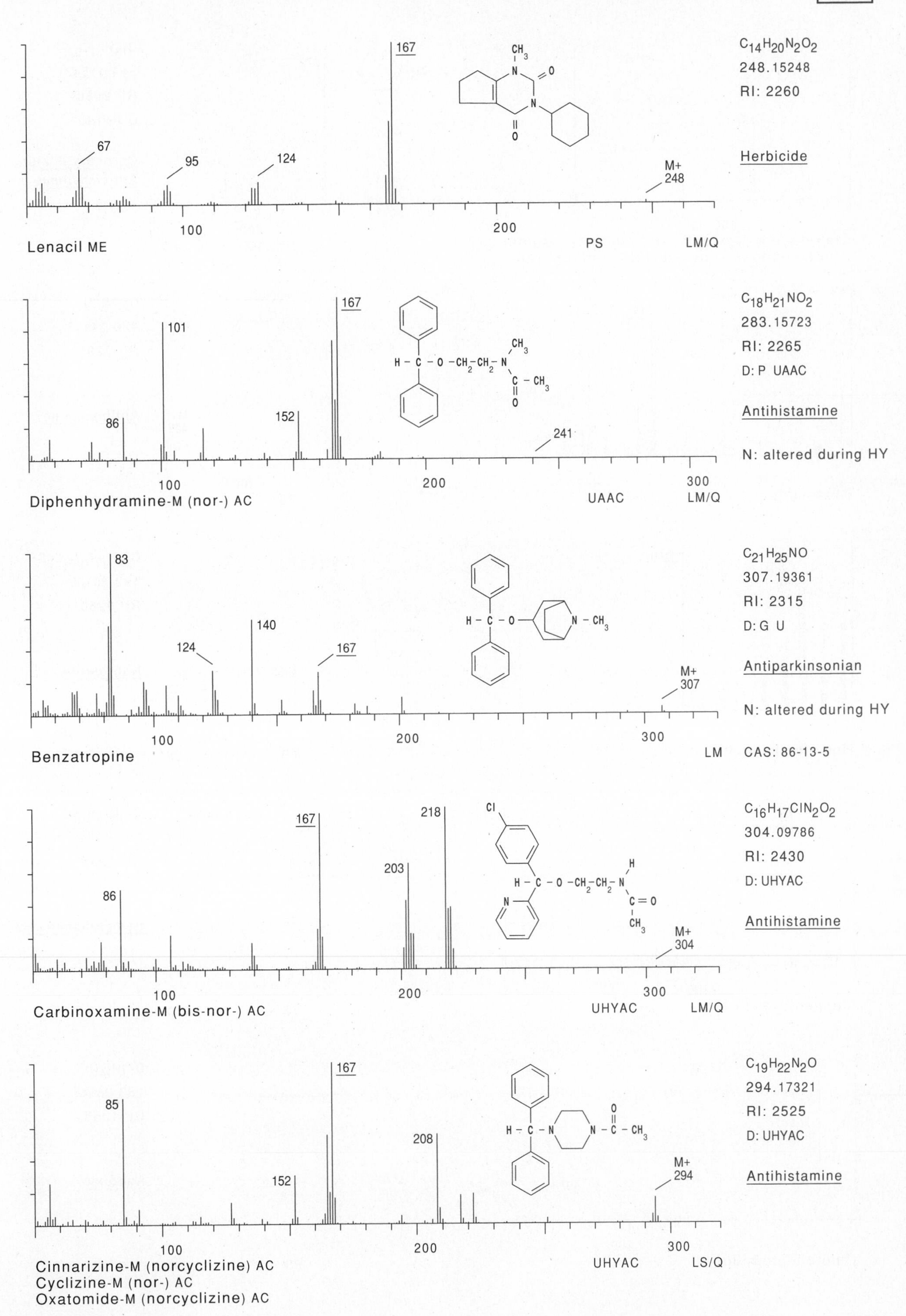

Lenacil ME
PS LM/Q
C14H20N2O2
248.15248
RI: 2260
Herbicide
M+ 248
Diphenhydramine-M (nor-) AC
UAAC LM/Q
C18H21NO2
283.15723
RI: 2265
D: P UAAC
Antihistamine
N: altered during HY
Benzatropine
LM
C21H25NO
307.19361
RI: 2315
D: G U
Antiparkinsonian
N: altered during HY
CAS: 86-13-5
M+ 307
Carbinoxamine-M (bis-nor-) AC
UHYAC LM/Q
C16H17ClN2O2
304.09786
RI: 2430
D: UHYAC
Antihistamine
M+ 304
Cinnarizine-M (norcyclizine) AC
Cyclizine-M (nor-) AC
Oxatomide-M (norcyclizine) AC
UHYAC LS/Q
C19H22N2O
294.17321
RI: 2525
D: UHYAC
Antihistamine
M+ 294

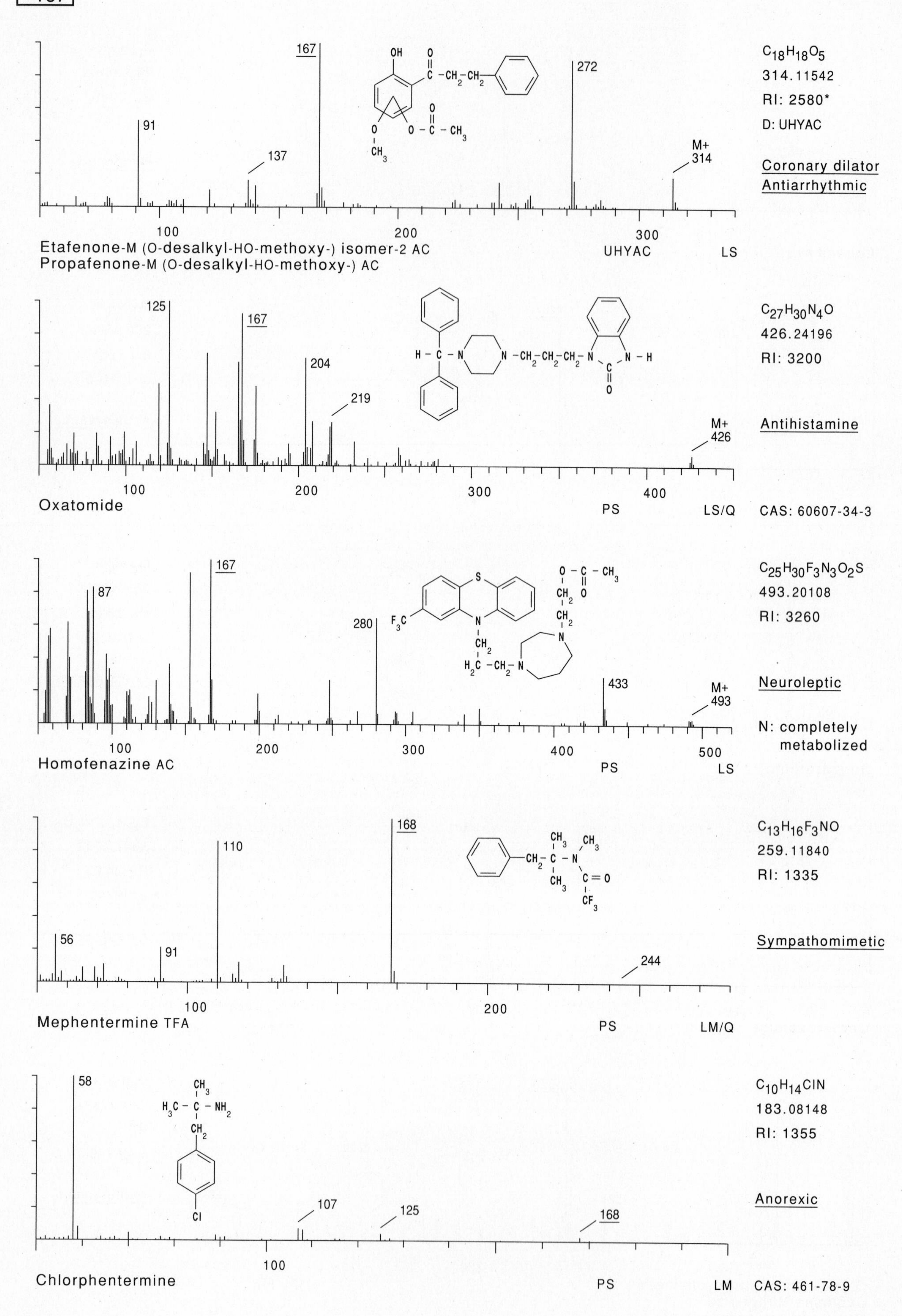
167
91
137
272
M+
314
100
200
300
Etafenone-M (O-desalkyl-HO-methoxy-) isomer-2 AC
Propafenone-M (O-desalkyl-HO-methoxy-) AC
UHYAC
LS
C18H18O5
314.11542
RI: 2580*
D: UHYAC
Coronary dilator
Antiarrhythmic
125
167
204
219
M+
426
100
200
300
400
Oxatomide
PS
LS/Q
C27H30N4O
426.24196
RI: 3200
Antihistamine
CAS: 60607-34-3
167
87
280
433
M+
493
100
200
300
400
500
Homofenazine AC
PS
LS
C25H30F3N3O2S
493.20108
RI: 3260
Neuroleptic
N: completely metabolized
168
110
56
91
244
100
200
Mephentermine TFA
PS
LM/Q
C13H16F3NO
259.11840
RI: 1335
Sympathomimetic
58
107
125
168
100
Chlorphentermine
PS
LM
C10H14ClN
183.08148
RI: 1355
Anorexic
CAS: 461-78-9

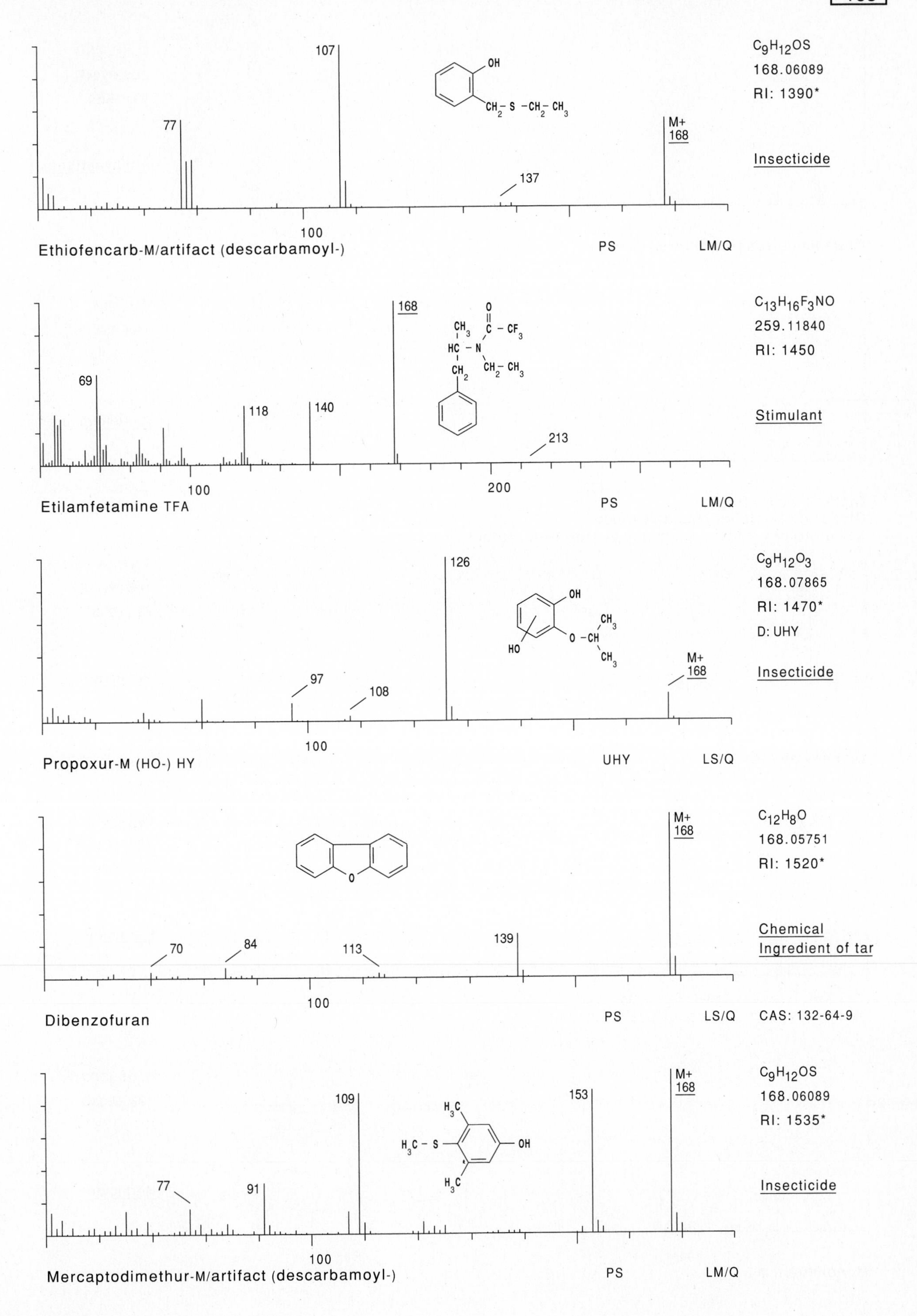

C9H12OS
168.06089
RI: 1390*
Insecticide
Ethiofencarb-M/artifact (descarbamoyl-)
PS
LM/Q
C13H16F3NO
259.11840
RI: 1450
Stimulant
Etilamfetamine TFA
PS
LM/Q
C9H12O3
168.07865
RI: 1470*
D: UHY
Insecticide
Propoxur-M (HO-) HY
UHY
LS/Q
C12H8O
168.05751
RI: 1520*
Chemical
Ingredient of tar
Dibenzofuran
PS
LS/Q
CAS: 132-64-9
C9H12OS
168.06089
RI: 1535*
Insecticide
Mercaptodimethur-M/artifact (descarbamoyl-)
PS
LM/Q

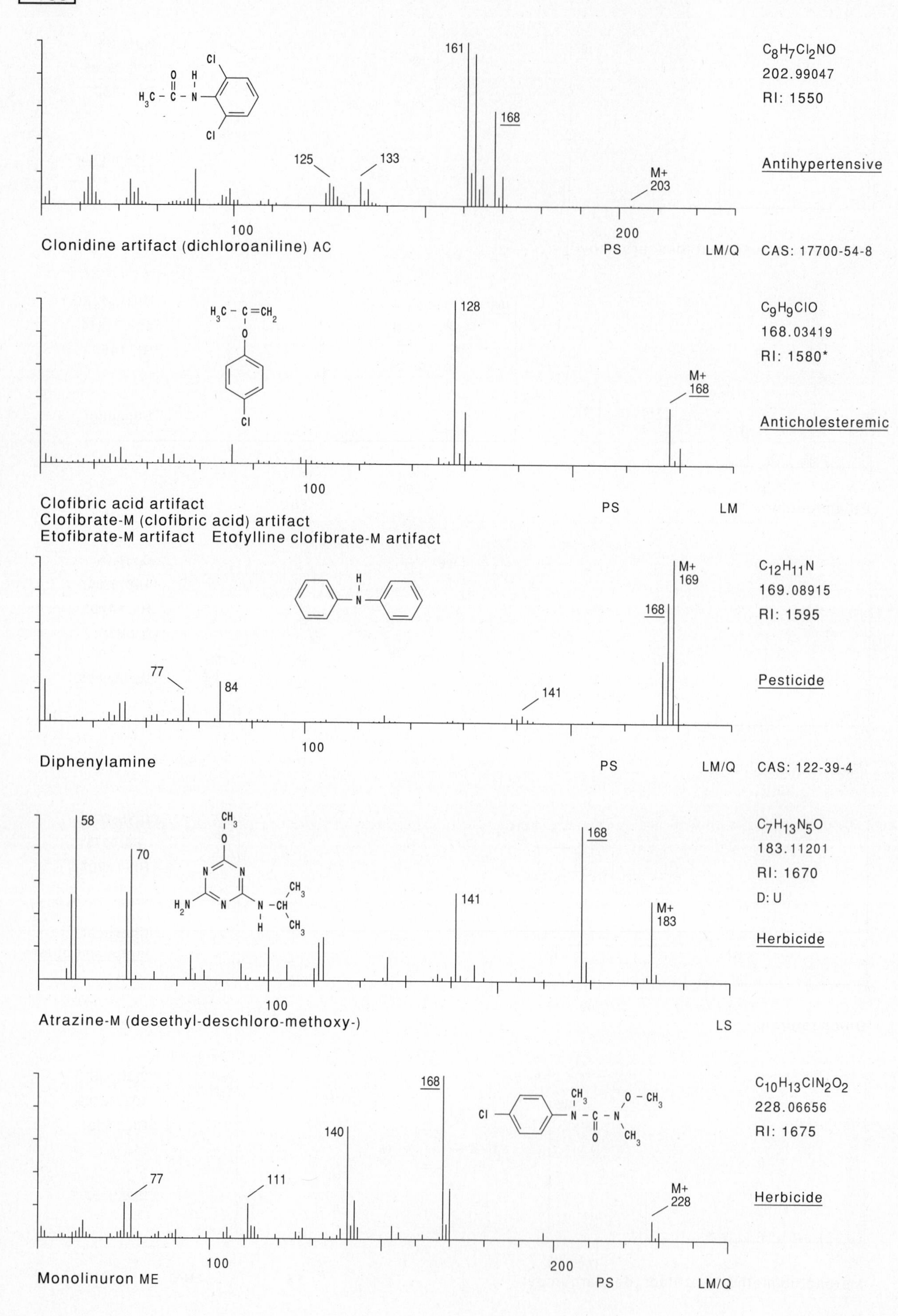

161
168
125
133
M+
203
100
200
Clonidine artifact (dichloroaniline) AC
PS
LM/Q
$C_8H_7Cl_2NO$
202.99047
RI: 1550
Antihypertensive
CAS: 17700-54-8
128
M+
168
100
Clofibric acid artifact
Clofibrate-M (clofibric acid) artifact
Etofibrate-M artifact Etofylline clofibrate-M artifact
PS
LM
C_9H_9ClO
168.03419
RI: 1580*
Anticholesteremic
M+
169
168
77
84
141
100
Diphenylamine
PS
LM/Q
$C_{12}H_{11}N$
169.08915
RI: 1595
Pesticide
CAS: 122-39-4
58
70
168
141
M+
183
100
Atrazine-M (desethyl-deschloro-methoxy-)
LS
$C_7H_{13}N_5O$
183.11201
RI: 1670
D: U
Herbicide
168
140
77
111
M+
228
100
200
Monolinuron ME
PS
LM/Q
$C_{10}H_{13}ClN_2O_2$
228.06656
RI: 1675
Herbicide

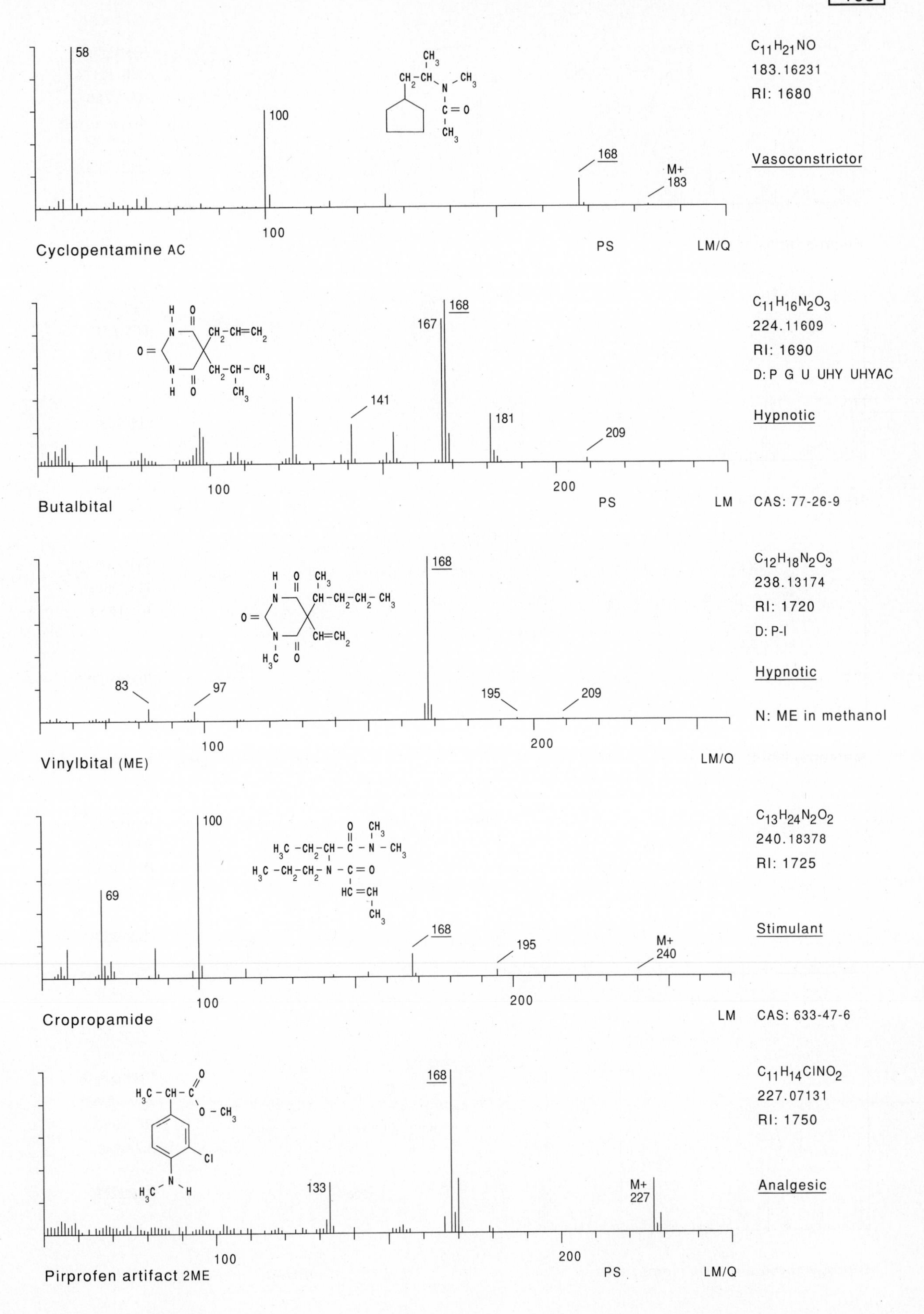
58
100
168
M+
183
100
Cyclopentamine AC
PS
LM/Q
C11H21NO
183.16231
RI: 1680
Vasoconstrictor
168
167
141
181
209
100
200
Butalbital
PS
LM
C11H16N2O3
224.11609
RI: 1690
D: P G U UHY UHYAC
Hypnotic
CAS: 77-26-9
168
83
97
195
209
100
200
Vinylbital (ME)
LM/Q
C12H18N2O3
238.13174
RI: 1720
D: P-I
Hypnotic
N: ME in methanol
100
69
168
195
M+
240
100
200
Cropropamide
LM
C13H24N2O2
240.18378
RI: 1725
Stimulant
CAS: 633-47-6
168
133
M+
227
100
200
Pirprofen artifact 2ME
PS
LM/Q
C11H14ClNO2
227.07131
RI: 1750
Analgesic

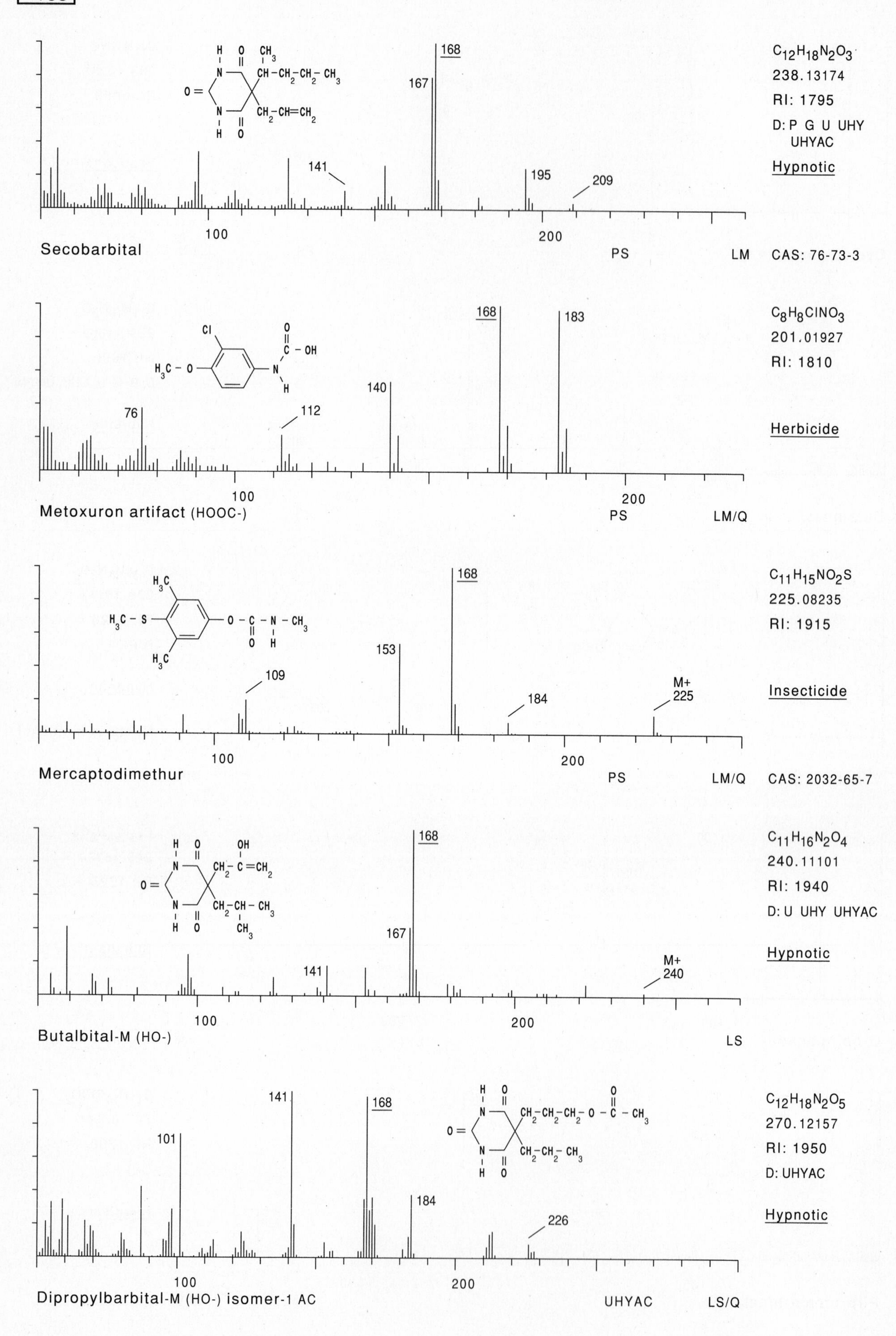
168
167
141
195
209
100
200
Secobarbital
PS
LM
$C_{12}H_{18}N_2O_3$
238.13174
RI: 1795
D: P G U UHY UHYAC
Hypnotic
CAS: 76-73-3
168
183
140
112
76
Metoxuron artifact (HOOC-)
PS
LM/Q
$C_8H_8ClNO_3$
201.01927
RI: 1810
Herbicide
168
153
109
184
M+ 225
Mercaptodimethur
PS
LM/Q
$C_{11}H_{15}NO_2S$
225.08235
RI: 1915
Insecticide
CAS: 2032-65-7
168
167
141
M+ 240
Butalbital-M (HO-)
LS
$C_{11}H_{16}N_2O_4$
240.11101
RI: 1940
D: U UHY UHYAC
Hypnotic
141
168
101
184
226
Dipropylbarbital-M (HO-) isomer-1 AC
UHYAC
LS/Q
$C_{12}H_{18}N_2O_5$
270.12157
RI: 1950
D: UHYAC
Hypnotic

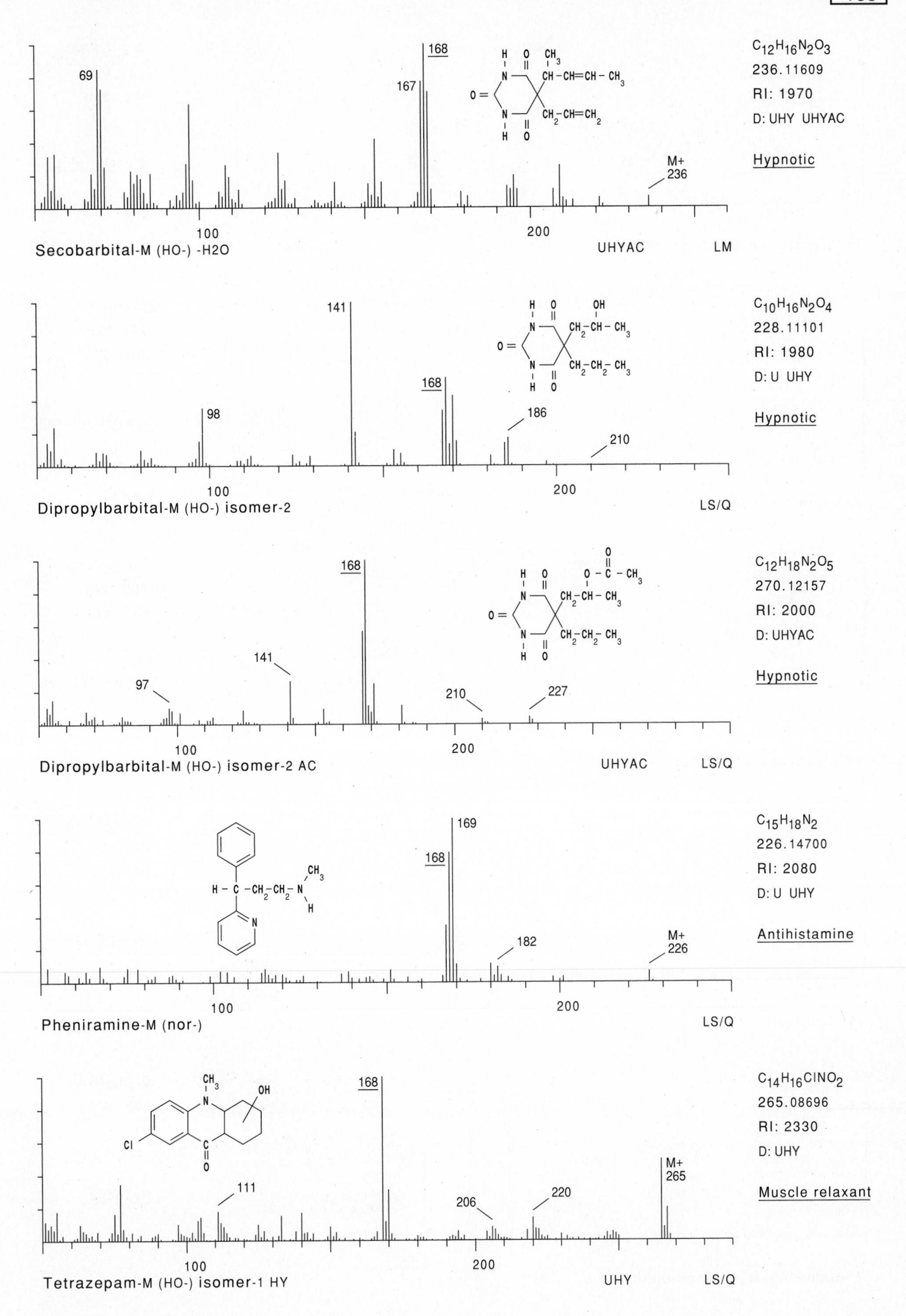

Secobarbital-M (HO-) -H2O
UHYAC
LM
$C_{12}H_{16}N_2O_3$
236.11609
RI: 1970
D: UHY UHYAC
Hypnotic
Dipropylbarbital-M (HO-) isomer-2
LS/Q
$C_{10}H_{16}N_2O_4$
228.11101
RI: 1980
D: U UHY
Hypnotic
Dipropylbarbital-M (HO-) isomer-2 AC
UHYAC
LS/Q
$C_{12}H_{18}N_2O_5$
270.12157
RI: 2000
D: UHYAC
Hypnotic
Pheniramine-M (nor-)
LS/Q
$C_{15}H_{18}N_2$
226.14700
RI: 2080
D: U UHY
Antihistamine
Tetrazepam-M (HO-) isomer-1 HY
UHY
LS/Q
$C_{14}H_{16}ClNO_2$
265.08696
RI: 2330
D: UHY
Muscle relaxant

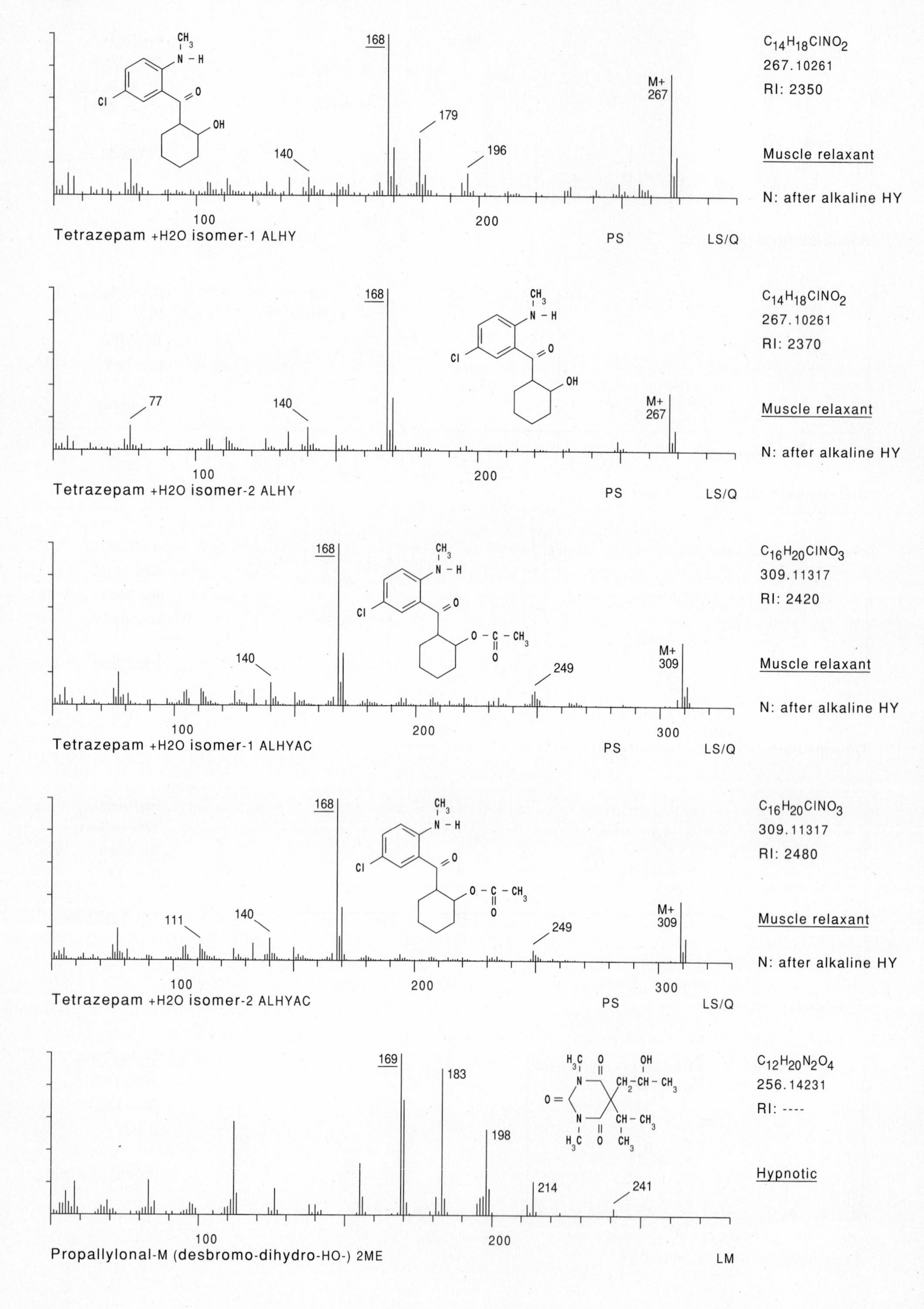

168
179
140
196
M+
267
100
200
Tetrazepam +H2O isomer-1 ALHY
PS
LS/Q
C14H18ClNO2
267.10261
RI: 2350
Muscle relaxant
N: after alkaline HY
168
77
140
M+
267
100
200
Tetrazepam +H2O isomer-2 ALHY
PS
LS/Q
C14H18ClNO2
267.10261
RI: 2370
Muscle relaxant
N: after alkaline HY
168
140
249
M+
309
100
200
300
Tetrazepam +H2O isomer-1 ALHYAC
PS
LS/Q
C16H20ClNO3
309.11317
RI: 2420
Muscle relaxant
N: after alkaline HY
168
111
140
249
M+
309
100
200
300
Tetrazepam +H2O isomer-2 ALHYAC
PS
LS/Q
C16H20ClNO3
309.11317
RI: 2480
Muscle relaxant
N: after alkaline HY
169
183
198
214
241
100
200
Propallylonal-M (desbromo-dihydro-HO-) 2ME
LM
C12H20N2O4
256.14231
RI: ----
Hypnotic

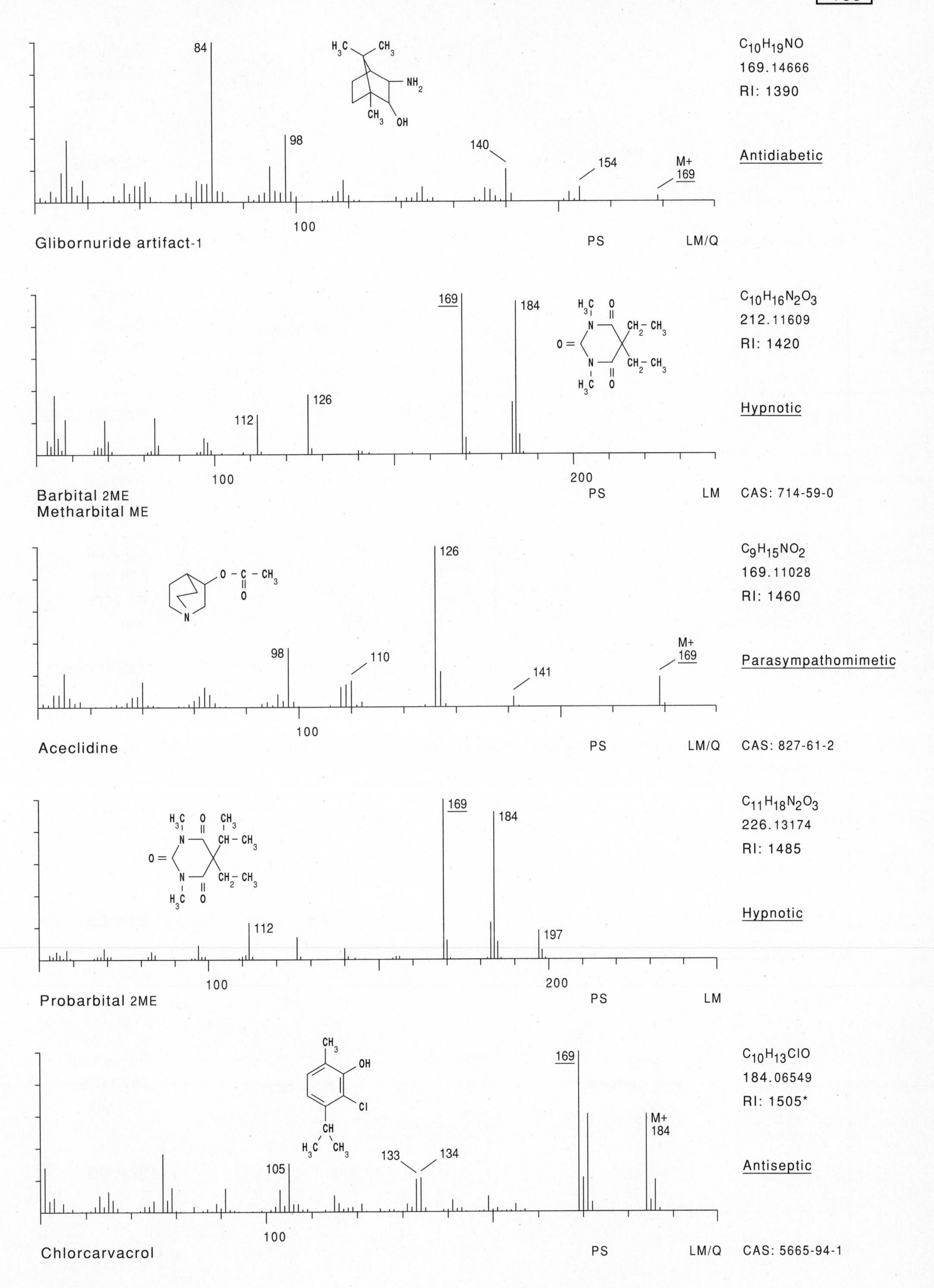

84
98
140
154
M+
169
100
Glibornuride artifact-1
PS
LM/Q
$C_{10}H_{19}NO$
169.14666
RI: 1390
Antidiabetic
169
184
126
112
100
200
Barbital 2ME
Metharbital ME
PS
LM
$C_{10}H_{16}N_2O_3$
212.11609
RI: 1420
Hypnotic
CAS: 714-59-0
126
98
110
141
M+
169
100
Aceclidine
PS
LM/Q
$C_9H_{15}NO_2$
169.11028
RI: 1460
Parasympathomimetic
CAS: 827-61-2
169
184
112
197
100
200
Probarbital 2ME
PS
LM
$C_{11}H_{18}N_2O_3$
226.13174
RI: 1485
Hypnotic
169
M+
184
105
133
134
100
Chlorcarvacrol
PS
LM/Q
$C_{10}H_{13}ClO$
184.06549
RI: 1505*
Antiseptic
CAS: 5665-94-1

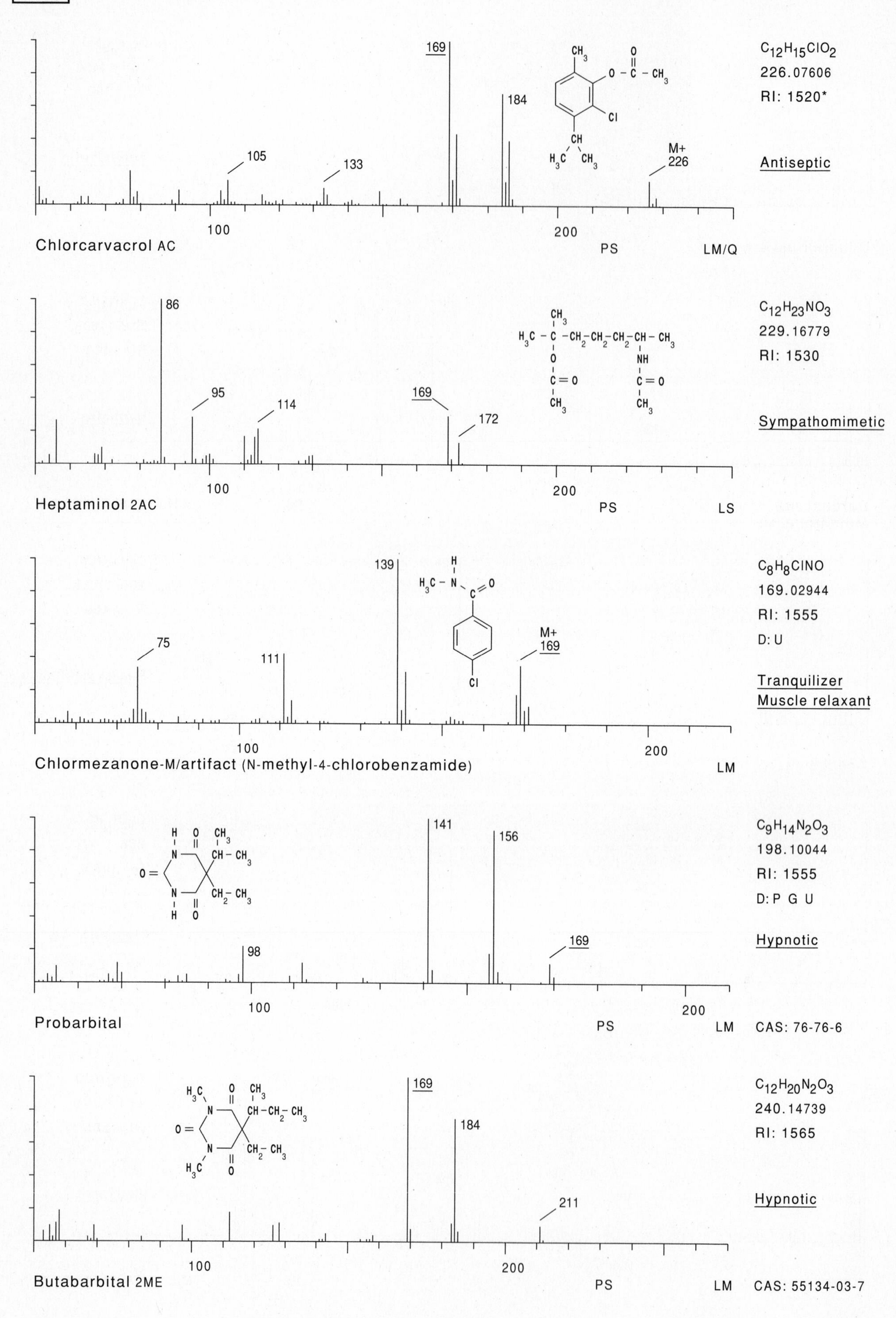

Chlorcarvacrol AC
$C_{12}H_{15}ClO_2$
226.07606
RI: 1520*
Antiseptic
169
184
105
133
M+ 226
100
200
PS
LM/Q
Heptaminol 2AC
$C_{12}H_{23}NO_3$
229.16779
RI: 1530
Sympathomimetic
86
95
114
169
172
100
200
PS
LS
Chlormezanone-M/artifact (N-methyl-4-chlorobenzamide)
C_8H_8ClNO
169.02944
RI: 1555
D: U
Tranquilizer
Muscle relaxant
139
75
111
M+ 169
100
200
LM
Probarbital
$C_9H_{14}N_2O_3$
198.10044
RI: 1555
D: P G U
Hypnotic
CAS: 76-76-6
141
156
98
169
100
200
PS
LM
Butabarbital 2ME
$C_{12}H_{20}N_2O_3$
240.14739
RI: 1565
Hypnotic
CAS: 55134-03-7
169
184
211
100
200
PS
LM

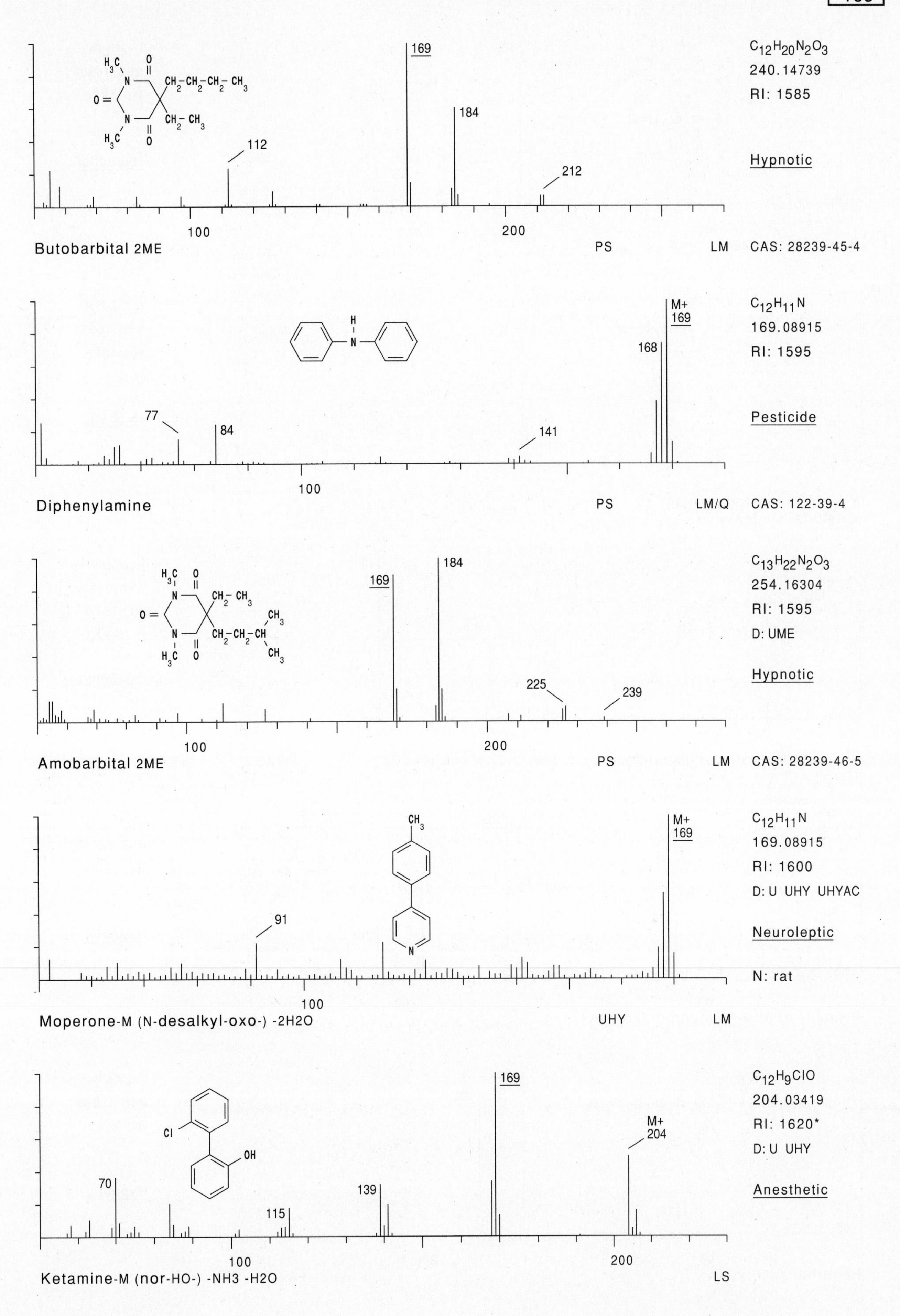
169
184
112
212
100
200
Butobarbital 2ME
PS
LM
$C_{12}H_{20}N_2O_3$
240.14739
RI: 1585
Hypnotic
CAS: 28239-45-4
M+
169
168
77
84
141
Diphenylamine
LM/Q
$C_{12}H_{11}N$
169.08915
RI: 1595
Pesticide
CAS: 122-39-4
184
169
225
239
Amobarbital 2ME
$C_{13}H_{22}N_2O_3$
254.16304
RI: 1595
D: UME
Hypnotic
CAS: 28239-46-5
91
Moperone-M (N-desalkyl-oxo-) -2H2O
UHY
RI: 1600
D: U UHY UHYAC
Neuroleptic
N: rat
M+
204
70
139
115
Ketamine-M (nor-HO-) -NH3 -H2O
LS
$C_{12}H_9ClO$
204.03419
RI: 1620*
D: U UHY
Anesthetic

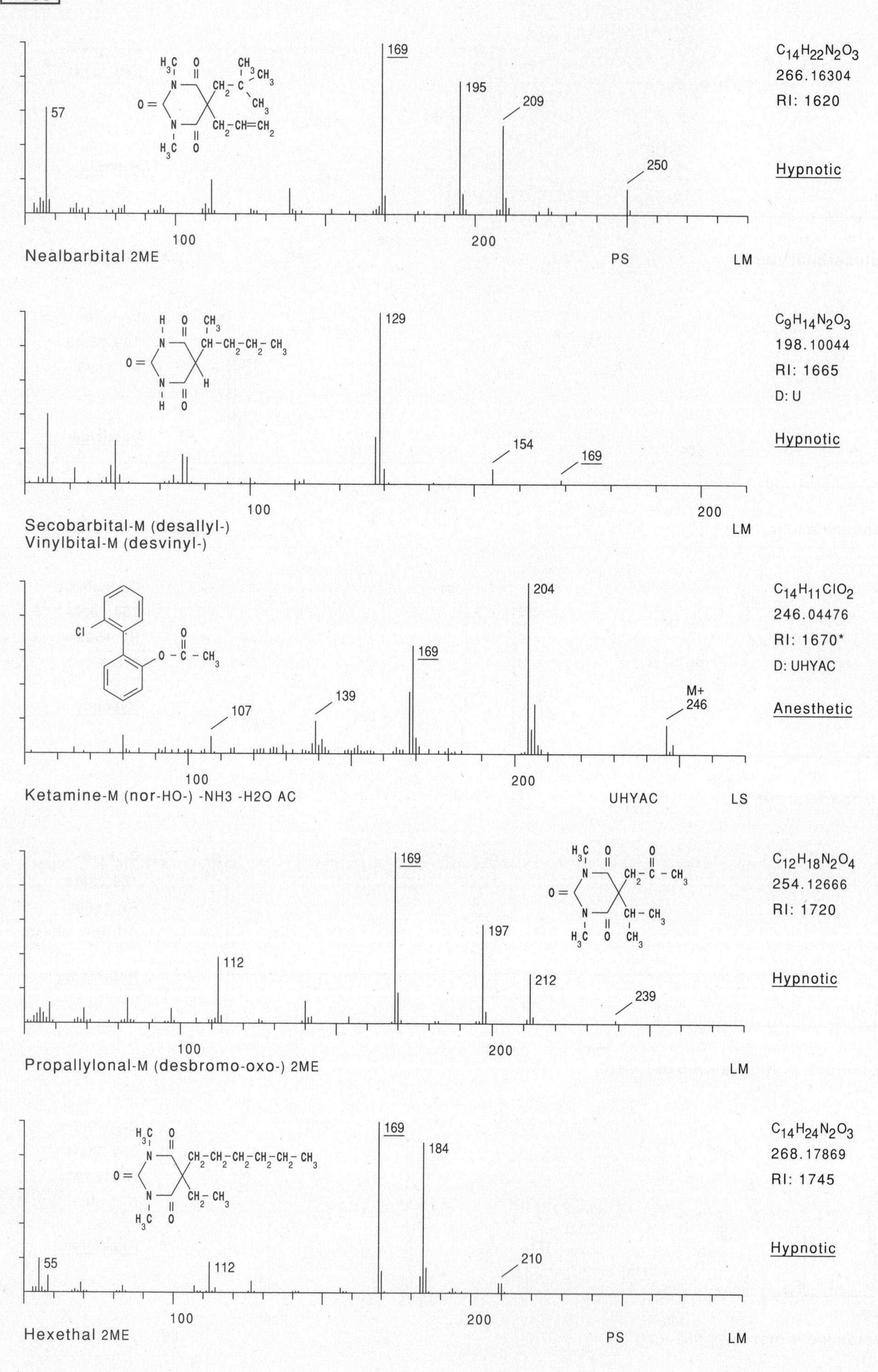

C14H22N2O3
266.16304
RI: 1620
Hypnotic
Nealbarbital 2ME
PS
LM
169
195
209
57
250
100
200
C9H14N2O3
198.10044
RI: 1665
D: U
Hypnotic
Secobarbital-M (desallyl-)
Vinylbital-M (desvinyl-)
LM
129
154
169
100
200
C14H11ClO2
246.04476
RI: 1670*
D: UHYAC
Anesthetic
Ketamine-M (nor-HO-) -NH3 -H2O AC
UHYAC
LS
204
169
139
107
M+
246
100
200
C12H18N2O4
254.12666
RI: 1720
Hypnotic
Propallylonal-M (desbromo-oxo-) 2ME
LM
169
197
112
212
239
100
200
C14H24N2O3
268.17869
RI: 1745
Hypnotic
Hexethal 2ME
PS
LM
169
184
55
112
210
100
200

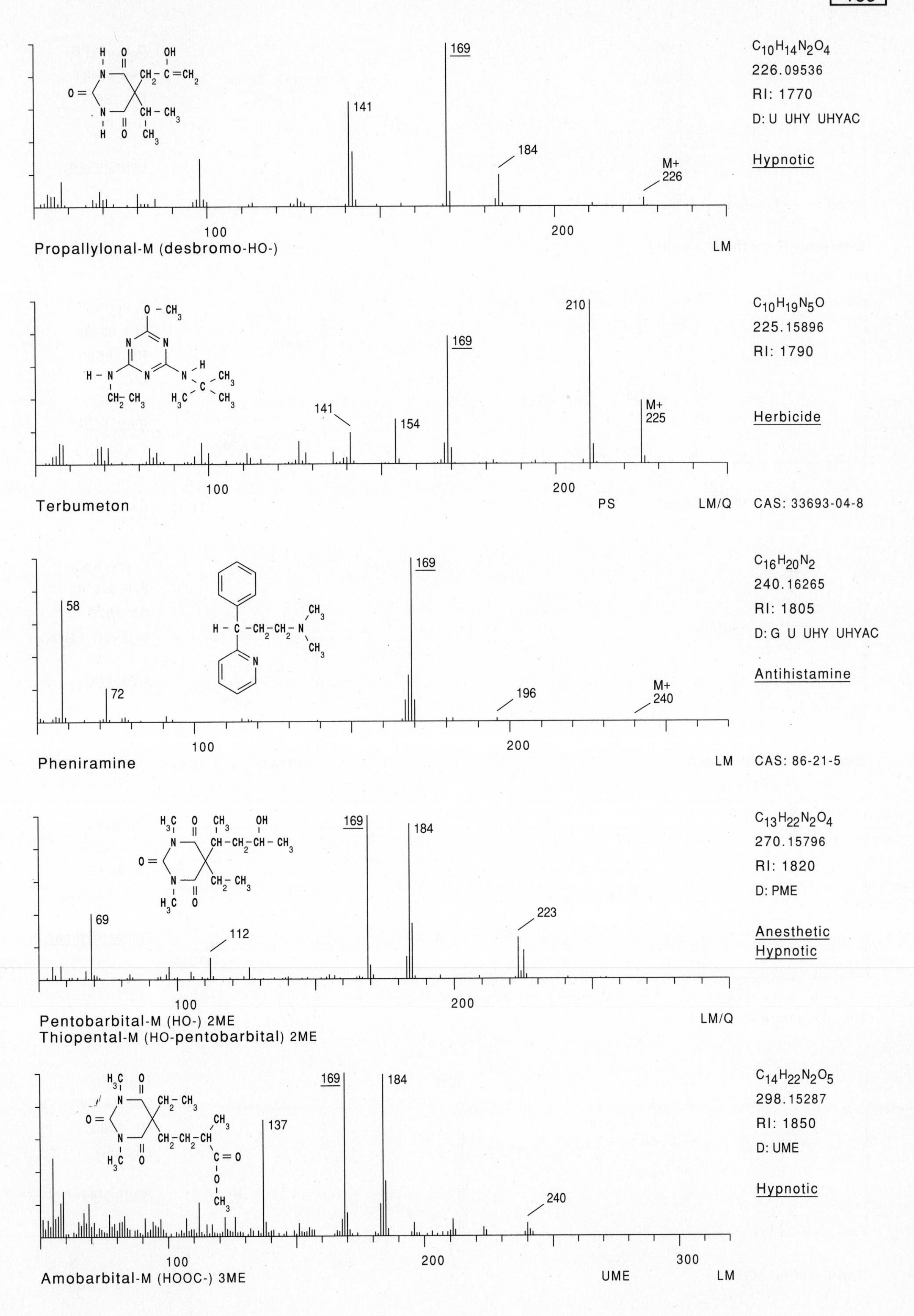
169
141
184
M+
226
100
200
Propallylonal-M (desbromo-HO-)
LM
C10H14N2O4
226.09536
RI: 1770
D: U UHY UHYAC
Hypnotic
210
169
141
154
M+
225
100
200
Terbumeton
PS
LM/Q
C10H19N5O
225.15896
RI: 1790
Herbicide
CAS: 33693-04-8
169
58
72
196
M+
240
100
200
Pheniramine
LM
C16H20N2
240.16265
RI: 1805
D: G U UHY UHYAC
Antihistamine
CAS: 86-21-5
169
184
69
112
223
100
200
Pentobarbital-M (HO-) 2ME
Thiopental-M (HO-pentobarbital) 2ME
LM/Q
C13H22N2O4
270.15796
RI: 1820
D: PME
Anesthetic
Hypnotic
169
184
137
240
100
200
300
Amobarbital-M (HOOC-) 3ME
UME
LM
C14H22N2O5
298.15287
RI: 1850
D: UME
Hypnotic

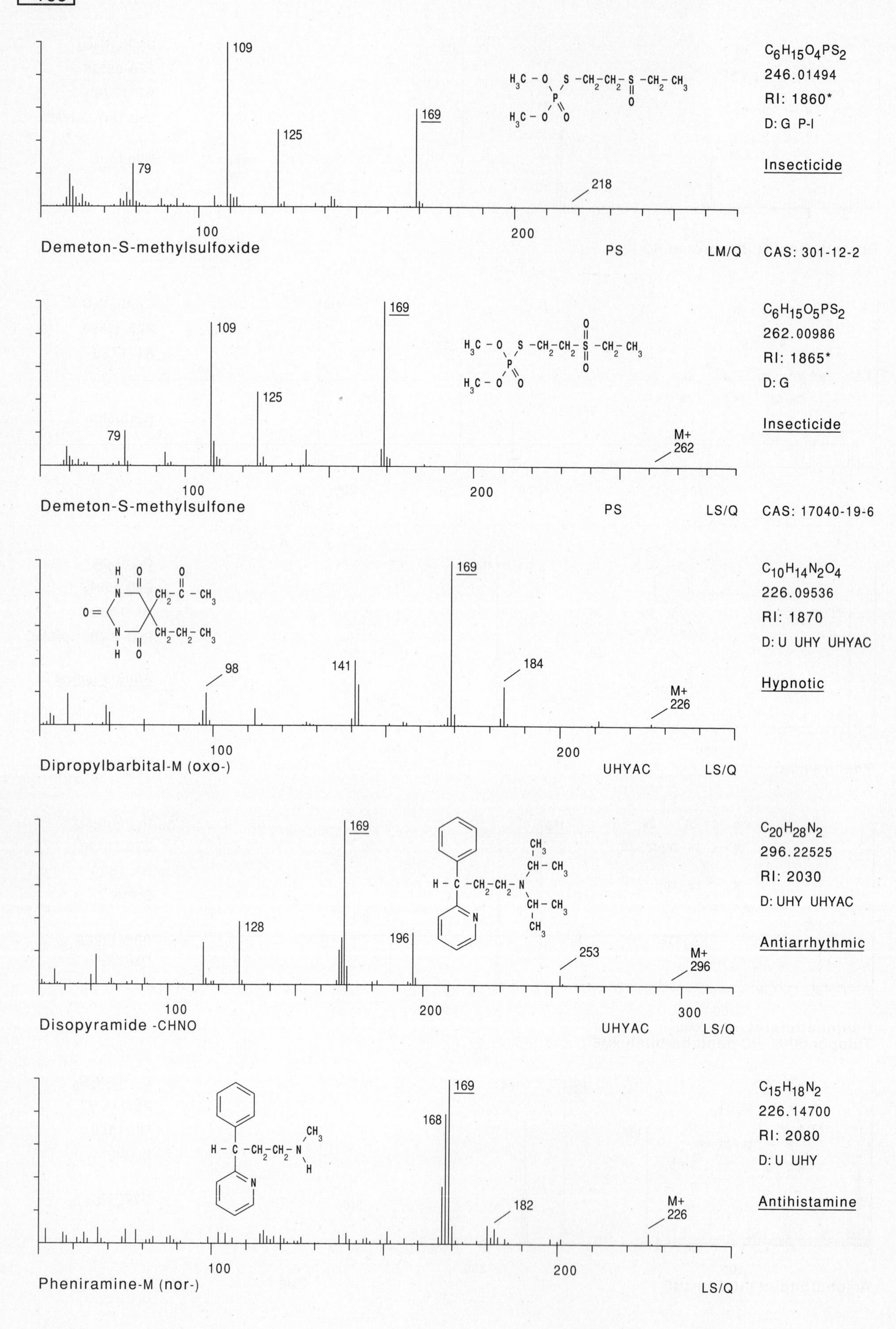
169
Demeton-S-methylsulfoxide
PS
LM/Q
C6H15O4PS2
246.01494
RI: 1860*
D: G P-I
Insecticide
CAS: 301-12-2
79
109
125
169
218
100
200
Demeton-S-methylsulfone
PS
LS/Q
C6H15O5PS2
262.00986
RI: 1865*
D: G
Insecticide
CAS: 17040-19-6
79
109
125
169
M+
262
100
200
Dipropylbarbital-M (oxo-)
UHYAC
LS/Q
C10H14N2O4
226.09536
RI: 1870
D: U UHY UHYAC
Hypnotic
98
141
169
184
M+
226
100
200
Disopyramide -CHNO
UHYAC
LS/Q
C20H28N2
296.22525
RI: 2030
D: UHY UHYAC
Antiarrhythmic
128
169
196
253
M+
296
100
200
300
Pheniramine-M (nor-)
LS/Q
C15H18N2
226.14700
RI: 2080
D: U UHY
Antihistamine
168
169
182
M+
226
100
200

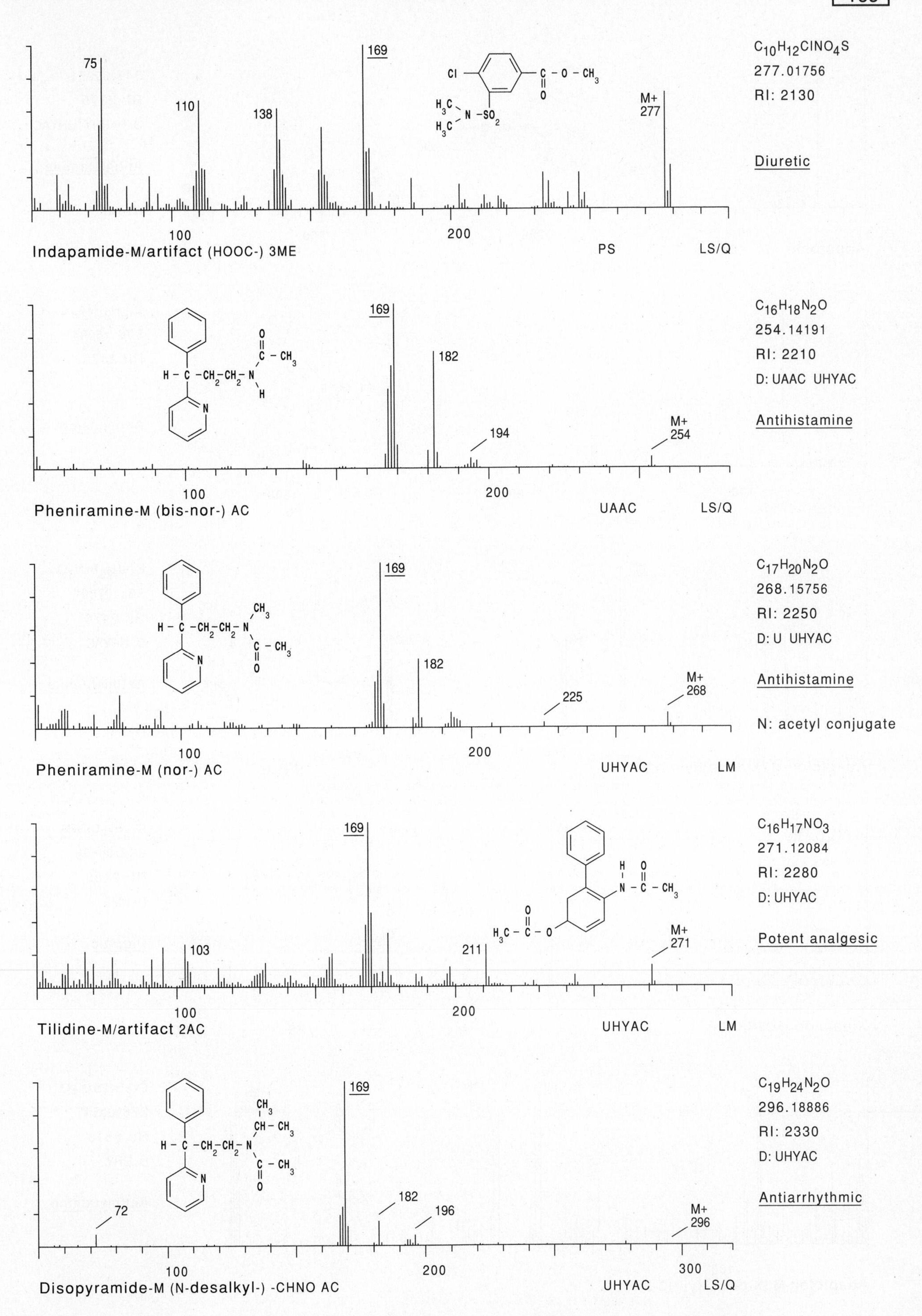

$C_{10}H_{12}ClNO_4S$
277.01756
RI: 2130

Diuretic

$C_{16}H_{18}N_2O$
254.14191
RI: 2210
D: UAAC UHYAC

Antihistamine

$C_{17}H_{20}N_2O$
268.15756
RI: 2250
D: U UHYAC

Antihistamine

N: acetyl conjugate

$C_{16}H_{17}NO_3$
271.12084
RI: 2280
D: UHYAC

Potent analgesic

$C_{19}H_{24}N_2O$
296.18886
RI: 2330
D: UHYAC

Antiarrhythmic

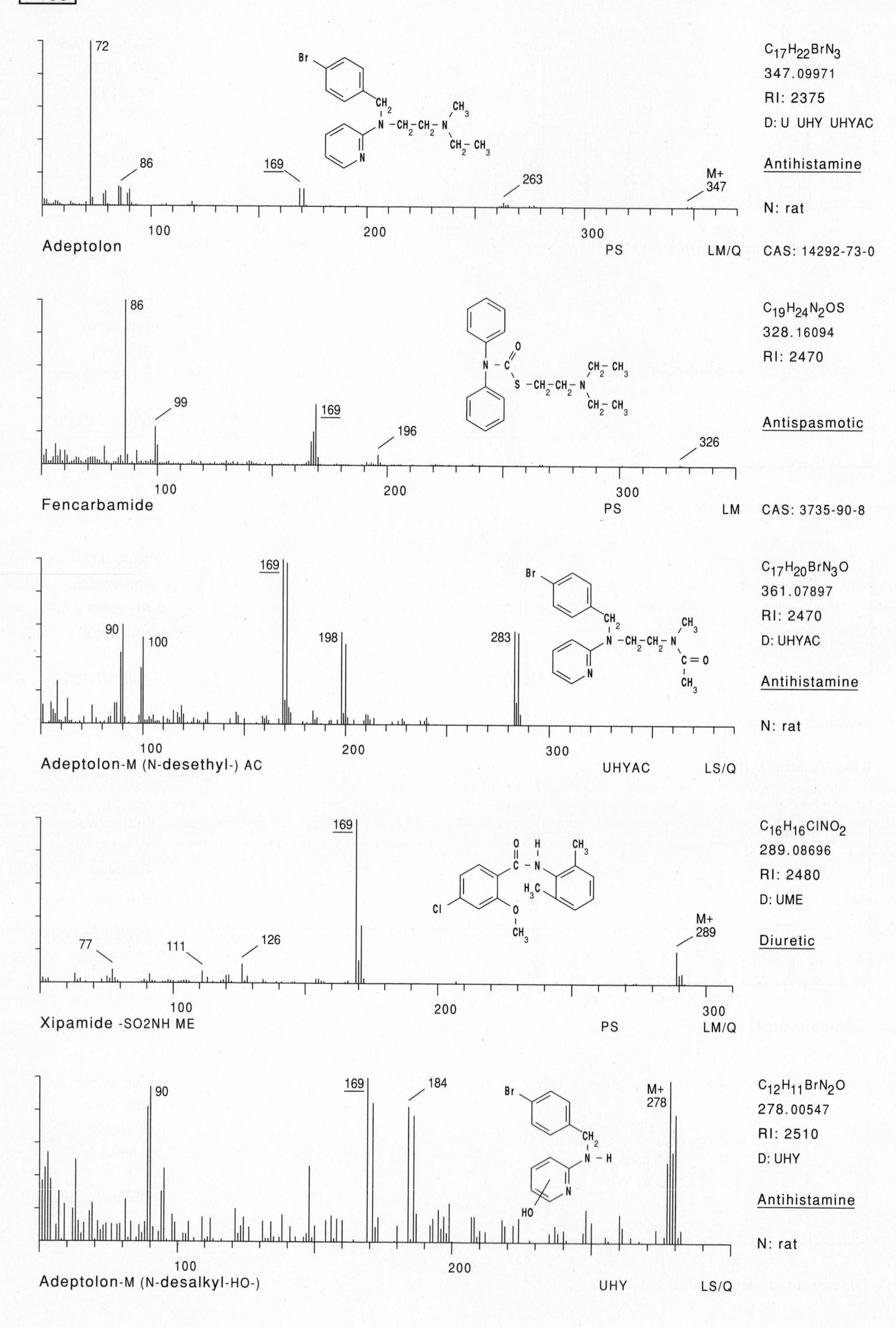
72
86
169
263
M+
347
100
200
300
Adeptolon
PS
LM/Q
$C_{17}H_{22}BrN_3$
347.09971
RI: 2375
D: U UHY UHYAC
Antihistamine
N: rat
CAS: 14292-73-0
86
99
169
196
326
100
200
300
Fencarbamide
PS
LM
$C_{19}H_{24}N_2OS$
328.16094
RI: 2470
Antispasmotic
CAS: 3735-90-8
169
90
100
198
283
100
200
300
Adeptolon-M (N-desethyl-) AC
UHYAC
LS/Q
$C_{17}H_{20}BrN_3O$
361.07897
RI: 2470
D: UHYAC
Antihistamine
N: rat
169
77
111
126
M+
289
100
200
300
Xipamide -SO2NH ME
PS
LM/Q
$C_{16}H_{16}ClNO_2$
289.08696
RI: 2480
D: UME
Diuretic
90
169
184
M+
278
100
200
Adeptolon-M (N-desalkyl-HO-)
UHY
LS/Q
$C_{12}H_{11}BrN_2O$
278.00547
RI: 2510
D: UHY
Antihistamine
N: rat

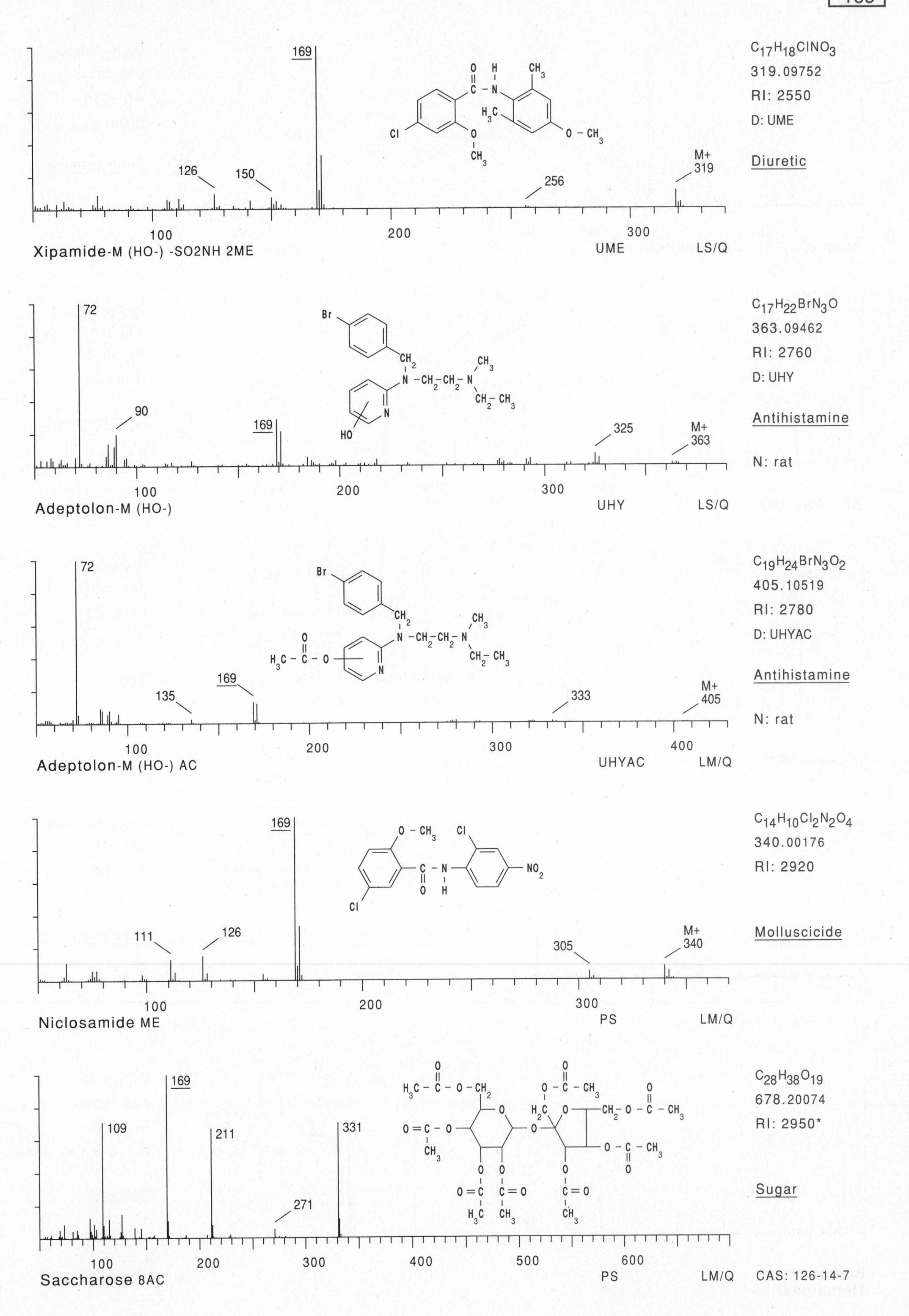

$C_{17}H_{18}ClNO_3$
319.09752
RI: 2550
D: UME
Diuretic

$C_{17}H_{22}BrN_3O$
363.09462
RI: 2760
D: UHY
Antihistamine
N: rat

$C_{19}H_{24}BrN_3O_2$
405.10519
RI: 2780
D: UHYAC
Antihistamine
N: rat

$C_{14}H_{10}Cl_2N_2O_4$
340.00176
RI: 2920
Molluscicide

$C_{28}H_{38}O_{19}$
678.20074
RI: 2950*
Sugar
CAS: 126-14-7

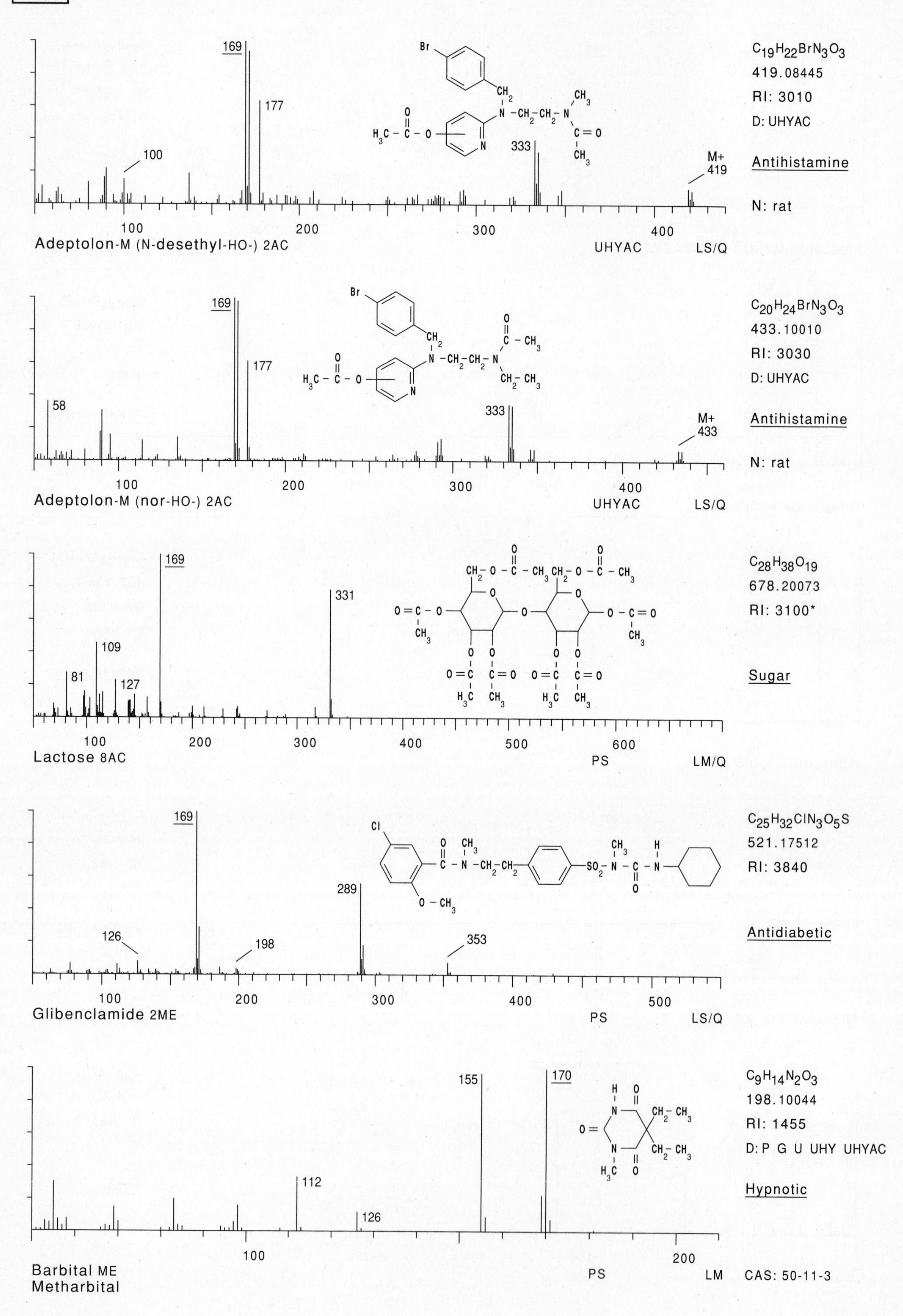

169
177
100
333
M+
419
100
200
300
400
Adeptolon-M (N-desethyl-HO-) 2AC
UHYAC
LS/Q
$C_{19}H_{22}BrN_3O_3$
419.08445
RI: 3010
D: UHYAC
Antihistamine
N: rat
169
177
58
333
M+
433
Adeptolon-M (nor-HO-) 2AC
UHYAC
LS/Q
$C_{20}H_{24}BrN_3O_3$
433.10010
RI: 3030
D: UHYAC
Antihistamine
N: rat
169
331
109
81
127
Lactose 8AC
PS
LM/Q
$C_{28}H_{38}O_{19}$
678.20073
RI: 3100*
Sugar
169
289
126
198
353
Glibenclamide 2ME
PS
LS/Q
$C_{25}H_{32}ClN_3O_5S$
521.17512
RI: 3840
Antidiabetic
155
170
112
126
Barbital ME
Metharbital
PS
LM
$C_9H_{14}N_2O_3$
198.10044
RI: 1455
D: P G U UHY UHYAC
Hypnotic
CAS: 50-11-3

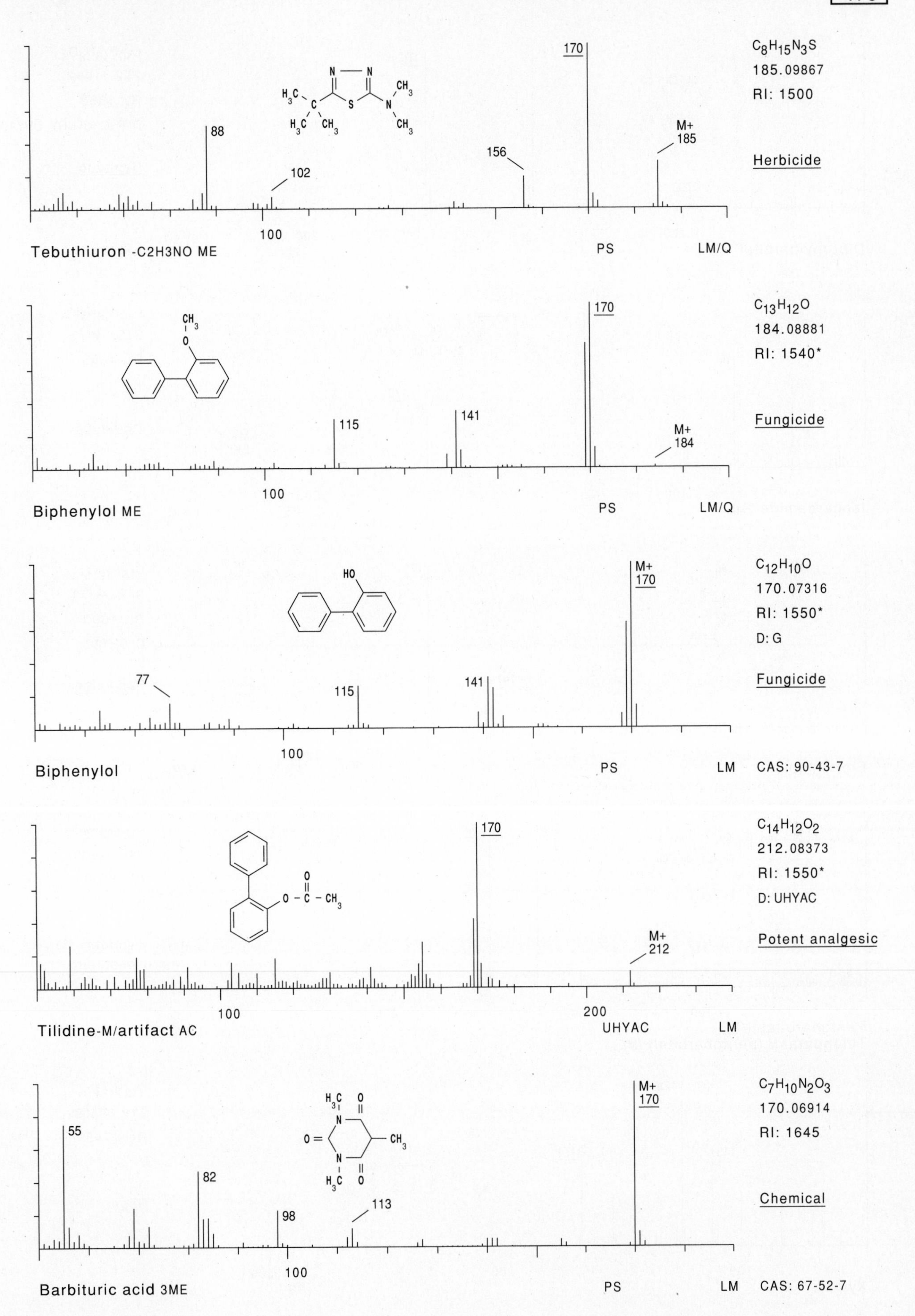

170
88
102
156
M+
185
100
Tebuthiuron -C2H3NO ME
PS
LM/Q
C8H15N3S
185.09867
RI: 1500
Herbicide
170
115
141
M+
184
100
Biphenylol ME
PS
LM/Q
C13H12O
184.08881
RI: 1540*
Fungicide
M+
170
77
115
141
100
Biphenylol
PS
LM
C12H10O
170.07316
RI: 1550*
D: G
Fungicide
CAS: 90-43-7
170
M+
212
100
200
Tilidine-M/artifact AC
UHYAC
LM
C14H12O2
212.08373
RI: 1550*
D: UHYAC
Potent analgesic
M+
170
55
82
98
113
100
Barbituric acid 3ME
PS
LM
C7H10N2O3
170.06914
RI: 1645
Chemical
CAS: 67-52-7

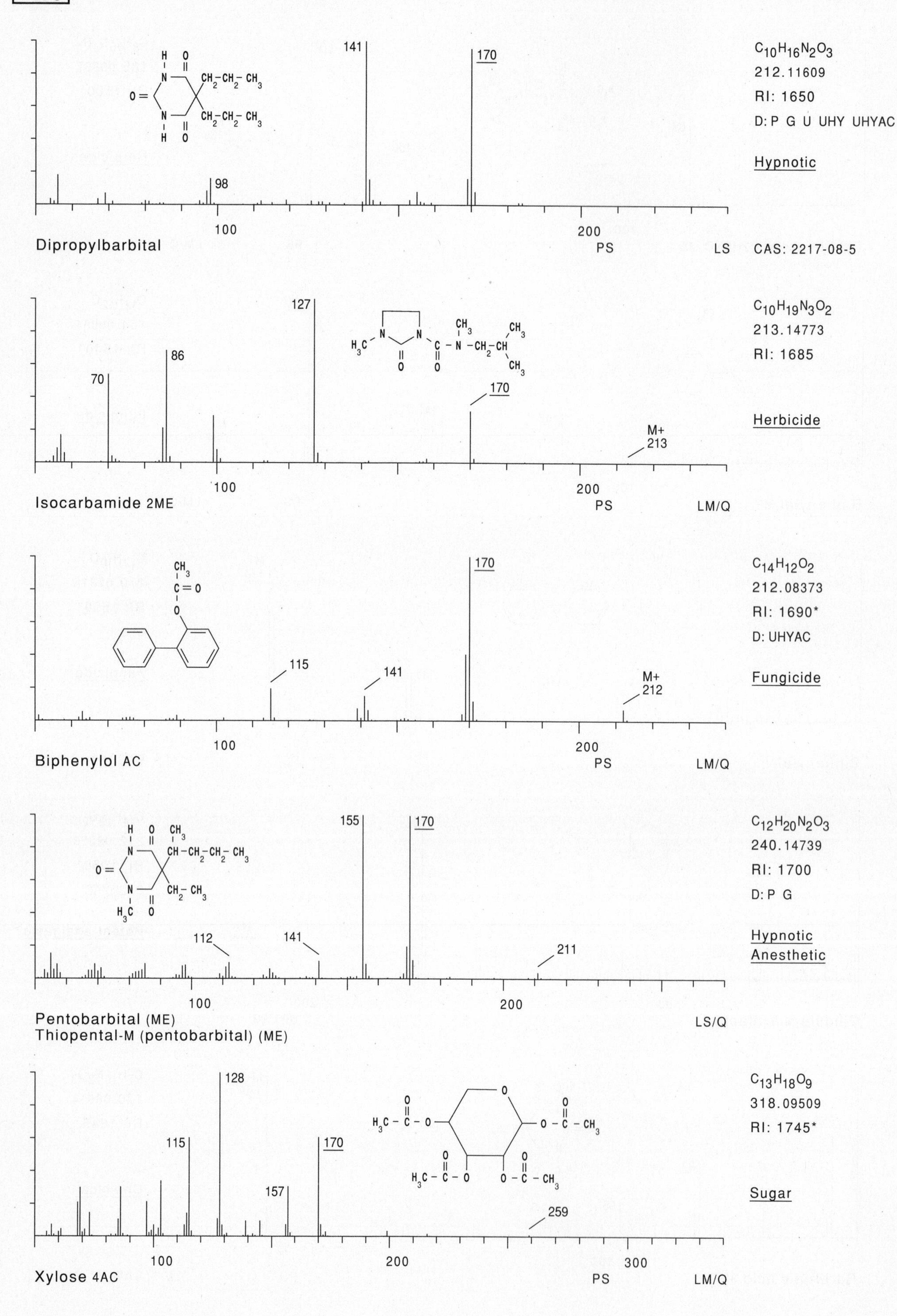
141
170
98
100
200
Dipropylbarbital
PS
LS
C10H16N2O3
212.11609
RI: 1650
D: P G U UHY UHYAC
Hypnotic
CAS: 2217-08-5
127
86
70
170
M+
213
100
200
Isocarbamide 2ME
PS
LM/Q
C10H19N3O2
213.14773
RI: 1685
Herbicide
170
115
141
M+
212
100
200
Biphenylol AC
PS
LM/Q
C14H12O2
212.08373
RI: 1690*
D: UHYAC
Fungicide
155
170
112
141
211
100
200
Pentobarbital (ME)
Thiopental-M (pentobarbital) (ME)
LS/Q
C12H20N2O3
240.14739
RI: 1700
D: P G
Hypnotic
Anesthetic
128
115
170
157
259
100
200
300
Xylose 4AC
PS
LM/Q
C13H18O9
318.09509
RI: 1745*
Sugar

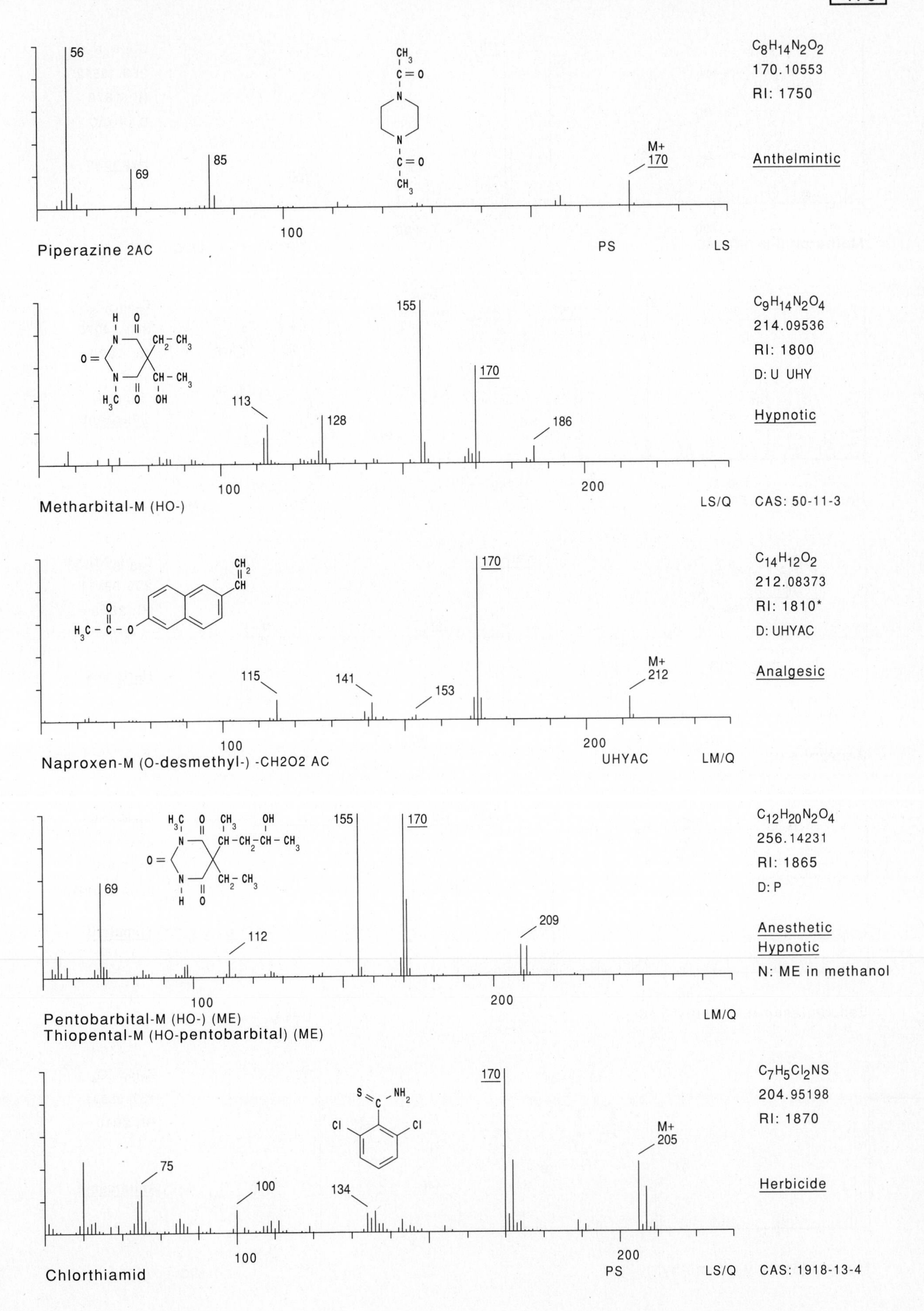

56
69
85
M+
170
100
Piperazine 2AC
PS
LS
C8H14N2O2
170.10553
RI: 1750
Anthelmintic
155
170
113
128
186
100
200
Metharbital-M (HO-)
LS/Q
C9H14N2O4
214.09536
RI: 1800
D: U UHY
Hypnotic
CAS: 50-11-3
170
115
141
153
M+
212
100
200
Naproxen-M (O-desmethyl-) -CH2O2 AC
UHYAC
LM/Q
C14H12O2
212.08373
RI: 1810*
D: UHYAC
Analgesic
155
170
69
112
209
100
200
Pentobarbital-M (HO-) (ME)
Thiopental-M (HO-pentobarbital) (ME)
LM/Q
C12H20N2O4
256.14231
RI: 1865
D: P
Anesthetic
Hypnotic
N: ME in methanol
170
M+
205
75
100
134
100
200
Chlorthiamid
PS
LS/Q
C7H5Cl2NS
204.95198
RI: 1870
Herbicide
CAS: 1918-13-4

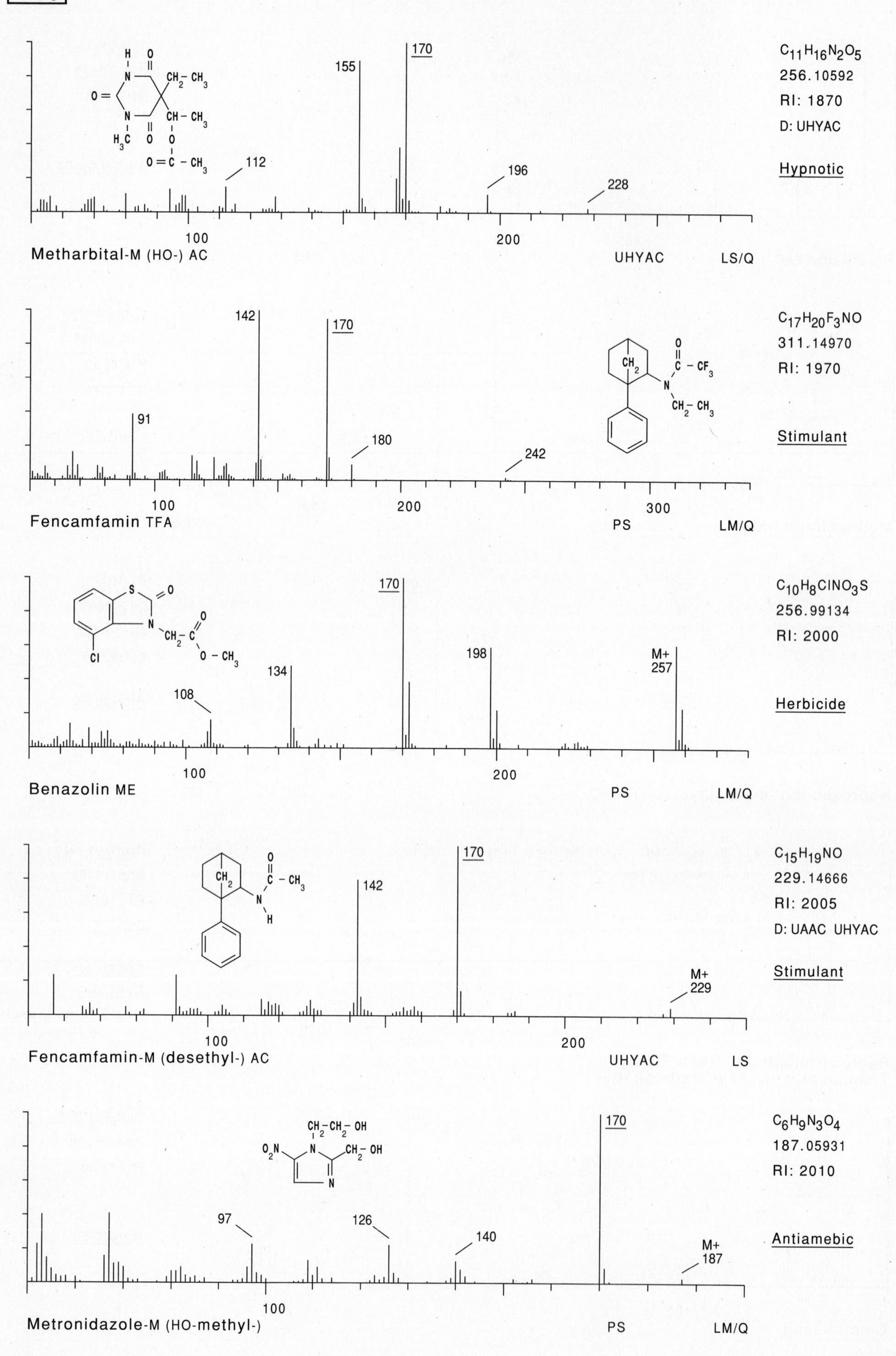
170
155
112
196
228
100
200
Metharbital-M (HO-) AC
UHYAC
LS/Q
C11H16N2O5
256.10592
RI: 1870
D: UHYAC
Hypnotic
142
170
91
180
242
100
200
300
Fencamfamin TFA
PS
LM/Q
C17H20F3NO
311.14970
RI: 1970
Stimulant
170
198
134
108
M+
257
100
200
Benazolin ME
PS
LM/Q
C10H8ClNO3S
256.99134
RI: 2000
Herbicide
170
142
M+
229
100
200
Fencamfamin-M (desethyl-) AC
UHYAC
LS
C15H19NO
229.14666
RI: 2005
D: UAAC UHYAC
Stimulant
170
97
126
140
M+
187
100
Metronidazole-M (HO-methyl-)
PS
LM/Q
C6H9N3O4
187.05931
RI: 2010
Antiamebic

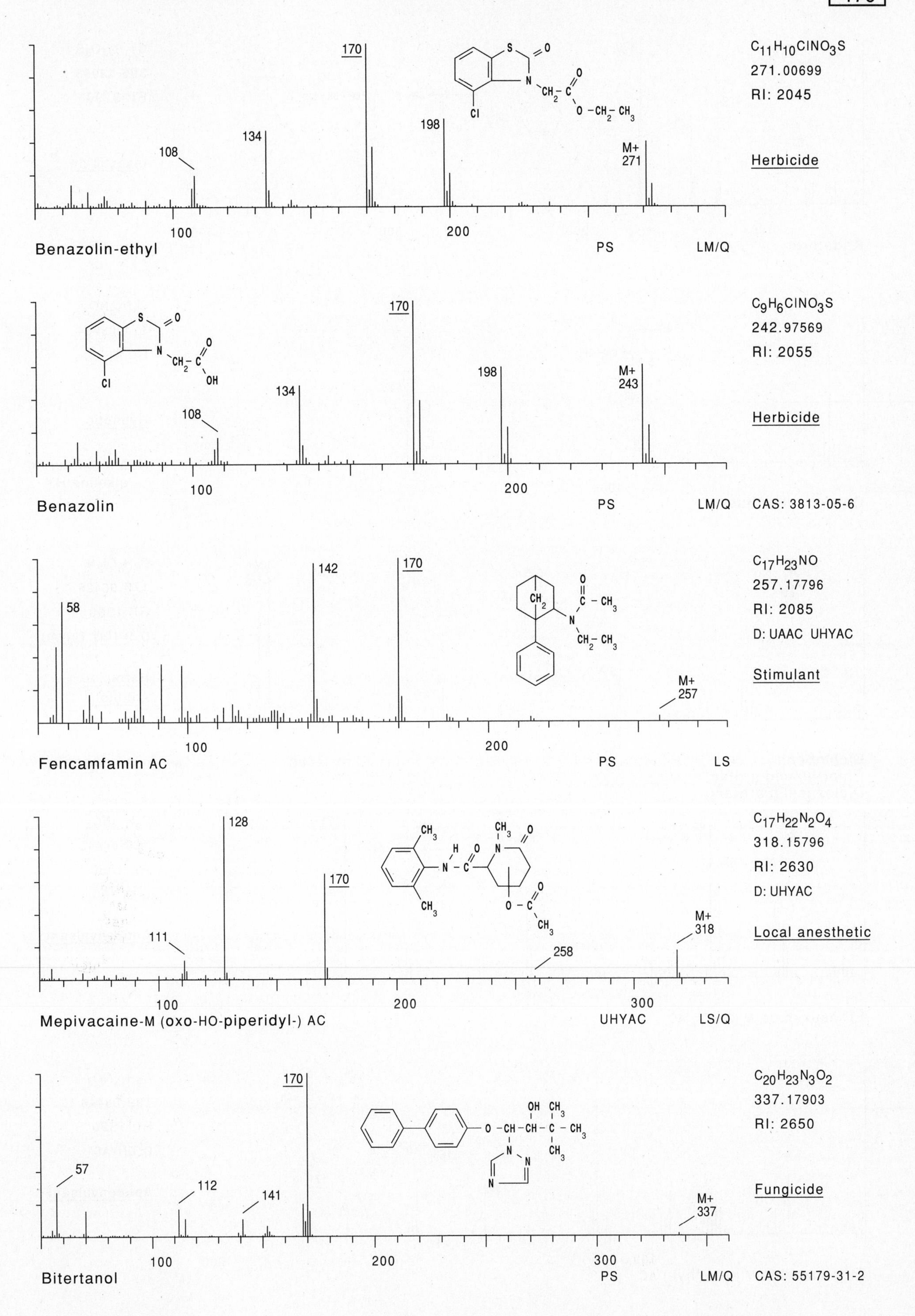

Benazolin-ethyl
C11H10ClNO3S
271.00699
RI: 2045
Herbicide
108
134
170
198
M+
271
100
200
PS
LM/Q
Benazolin
C9H6ClNO3S
242.97569
RI: 2055
Herbicide
108
134
170
198
M+
243
100
200
PS
LM/Q
CAS: 3813-05-6
Fencamfamin AC
C17H23NO
257.17796
RI: 2085
D: UAAC UHYAC
Stimulant
58
142
170
M+
257
100
200
PS
LS
Mepivacaine-M (oxo-HO-piperidyl-) AC
C17H22N2O4
318.15796
RI: 2630
D: UHYAC
Local anesthetic
111
128
170
258
M+
318
100
200
300
UHYAC
LS/Q
Bitertanol
C20H23N3O2
337.17903
RI: 2650
Fungicide
57
112
141
170
M+
337
100
200
300
PS
LM/Q
CAS: 55179-31-2

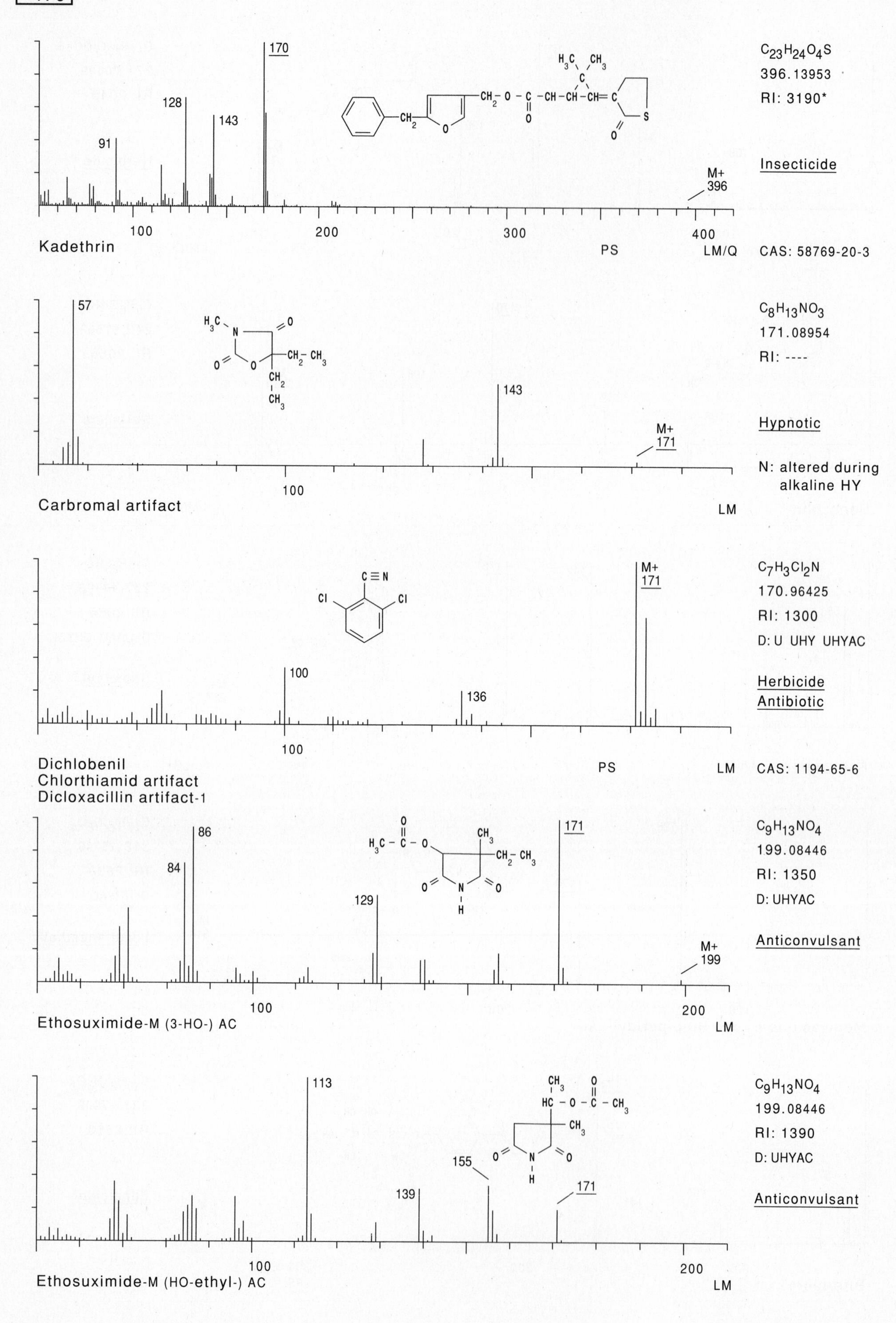

Kadethrin
$C_{23}H_{24}O_4S$
396.13953
RI: 3190*
Insecticide
PS
LM/Q
CAS: 58769-20-3
Carbromal artifact
$C_8H_{13}NO_3$
171.08954
RI: ----
Hypnotic
N: altered during alkaline HY
LM
Dichlobenil
Chlorthiamid artifact
Dicloxacillin artifact-1
$C_7H_3Cl_2N$
170.96425
RI: 1300
D: U UHY UHYAC
Herbicide
Antibiotic
PS
LM
CAS: 1194-65-6
Ethosuximide-M (3-HO-) AC
$C_9H_{13}NO_4$
199.08446
RI: 1350
D: UHYAC
Anticonvulsant
LM
Ethosuximide-M (HO-ethyl-) AC
$C_9H_{13}NO_4$
199.08446
RI: 1390
D: UHYAC
Anticonvulsant
LM

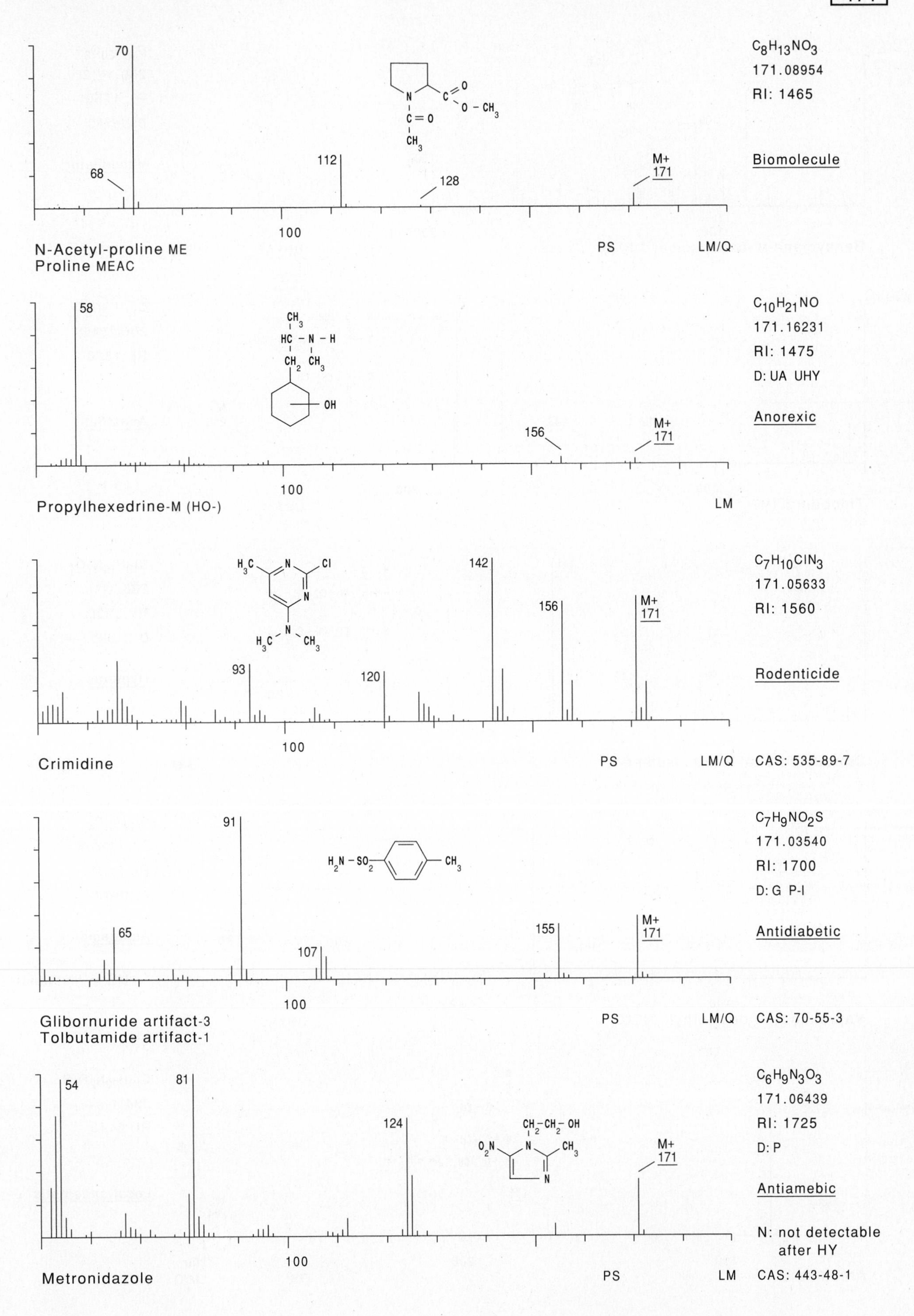
70
68
112
128
M+
171
100
N-Acetyl-proline ME
Proline MEAC
PS
LM/Q
C8H13NO3
171.08954
RI: 1465
Biomolecule
58
156
M+
171
100
Propylhexedrine-M (HO-)
LM
C10H21NO
171.16231
RI: 1475
D: UA UHY
Anorexic
142
156
M+
171
93
120
100
Crimidine
PS
LM/Q
C7H10ClN3
171.05633
RI: 1560
Rodenticide
CAS: 535-89-7
91
65
107
155
M+
171
100
Glibornuride artifact-3
Tolbutamide artifact-1
PS
LM/Q
C7H9NO2S
171.03540
RI: 1700
D: G P-I
Antidiabetic
CAS: 70-55-3
54
81
124
M+
171
100
Metronidazole
PS
LM
C6H9N3O3
171.06439
RI: 1725
D: P
Antiamebic
N: not detectable
after HY
CAS: 443-48-1

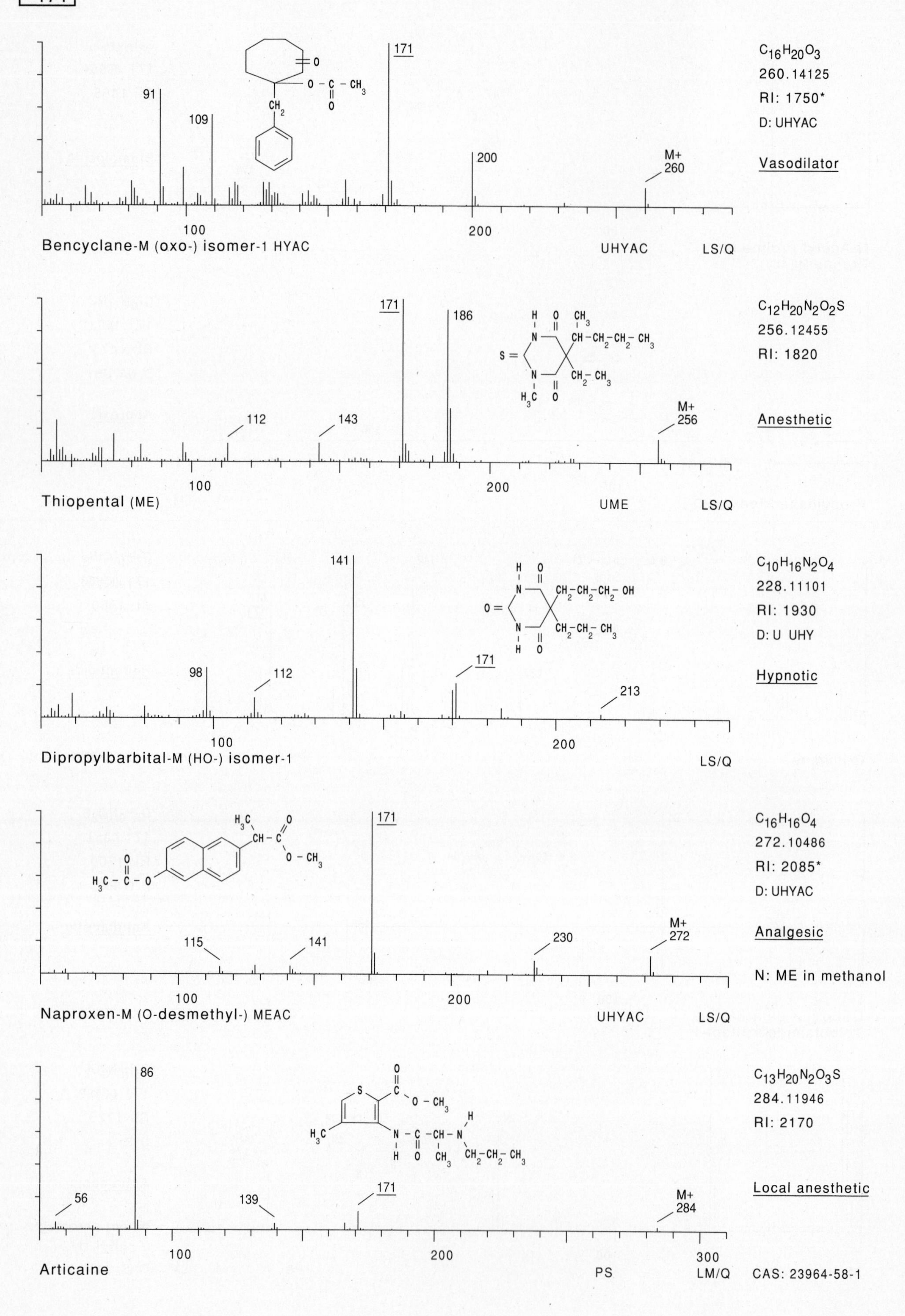

171
91
109
200
M+
260
100
200
Bencyclane-M (oxo-) isomer-1 HYAC
UHYAC
LS/Q
C16H20O3
260.14125
RI: 1750*
D: UHYAC
Vasodilator
171
186
112
143
M+
256
100
200
Thiopental (ME)
UME
LS/Q
C12H20N2O2S
256.12455
RI: 1820
Anesthetic
141
98
112
171
213
100
200
Dipropylbarbital-M (HO-) isomer-1
LS/Q
C10H16N2O4
228.11101
RI: 1930
D: U UHY
Hypnotic
171
115
141
230
M+
272
100
200
Naproxen-M (O-desmethyl-) MEAC
UHYAC
LS/Q
C16H16O4
272.10486
RI: 2085*
D: UHYAC
Analgesic
N: ME in methanol
86
56
139
171
M+
284
100
200
300
Articaine
PS
LM/Q
C13H20N2O3S
284.11946
RI: 2170
Local anesthetic
CAS: 23964-58-1

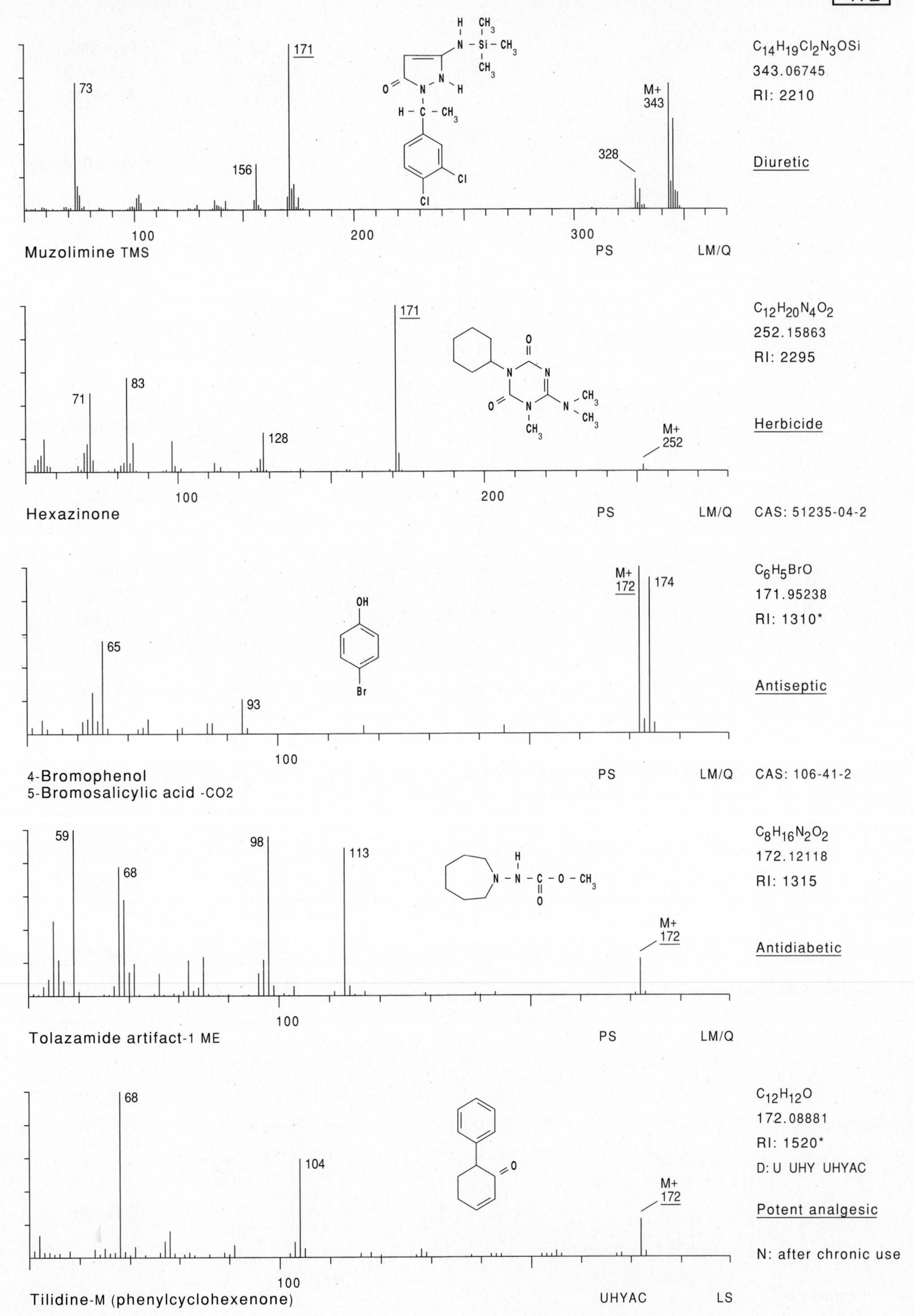

172
73
171
156
328
M+
343
100
200
300
Muzolimine TMS
PS
LM/Q
C14H19Cl2N3OSi
343.06745
RI: 2210
Diuretic
71
83
128
171
M+
252
100
200
Hexazinone
PS
LM/Q
C12H20N4O2
252.15863
RI: 2295
Herbicide
CAS: 51235-04-2
65
93
M+
172
174
100
4-Bromophenol
5-Bromosalicylic acid -CO2
PS
LM/Q
C6H5BrO
171.95238
RI: 1310*
Antiseptic
CAS: 106-41-2
59
68
98
113
M+
172
100
Tolazamide artifact-1 ME
PS
LM/Q
C8H16N2O2
172.12118
RI: 1315
Antidiabetic
68
104
M+
172
100
Tilidine-M (phenylcyclohexenone)
UHYAC
LS
C12H12O
172.08881
RI: 1520*
D: U UHY UHYAC
Potent analgesic
N: after chronic use

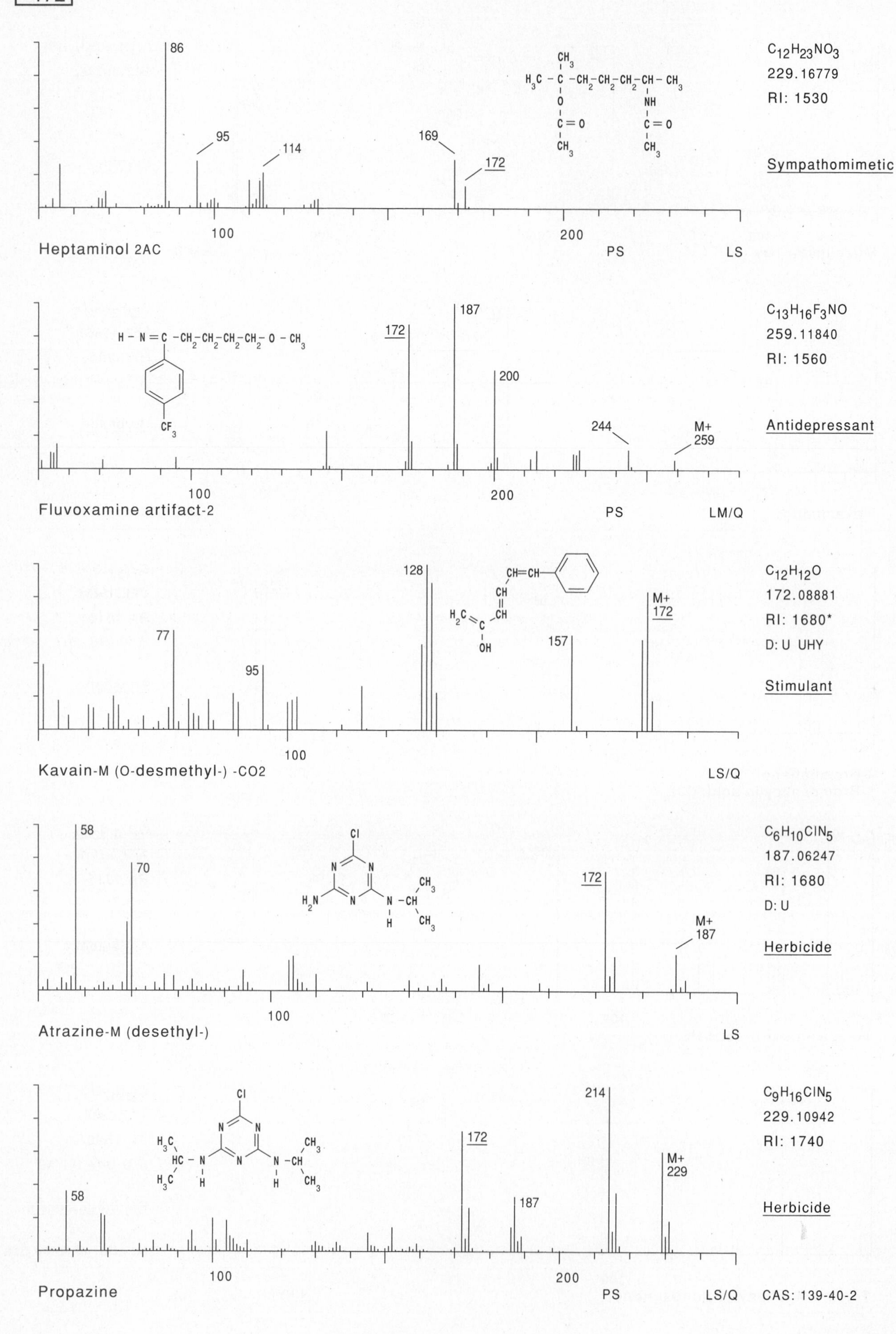
86
95
114
169
172
100
200
$C_{12}H_{23}NO_3$
229.16779
RI: 1530
Sympathomimetic
Heptaminol 2AC
PS
LS
187
172
200
244
M+
259
$C_{13}H_{16}F_3NO$
259.11840
RI: 1560
Antidepressant
Fluvoxamine artifact-2
PS
LM/Q
128
77
95
157
M+
172
$C_{12}H_{12}O$
172.08881
RI: 1680*
D: U UHY
Stimulant
Kavain-M (O-desmethyl-) -CO2
LS/Q
58
70
172
M+
187
$C_6H_{10}ClN_5$
187.06247
RI: 1680
D: U
Herbicide
Atrazine-M (desethyl-)
LS
214
172
187
58
M+
229
$C_9H_{16}ClN_5$
229.10942
RI: 1740
Herbicide
Propazine
PS
LS/Q
CAS: 139-40-2

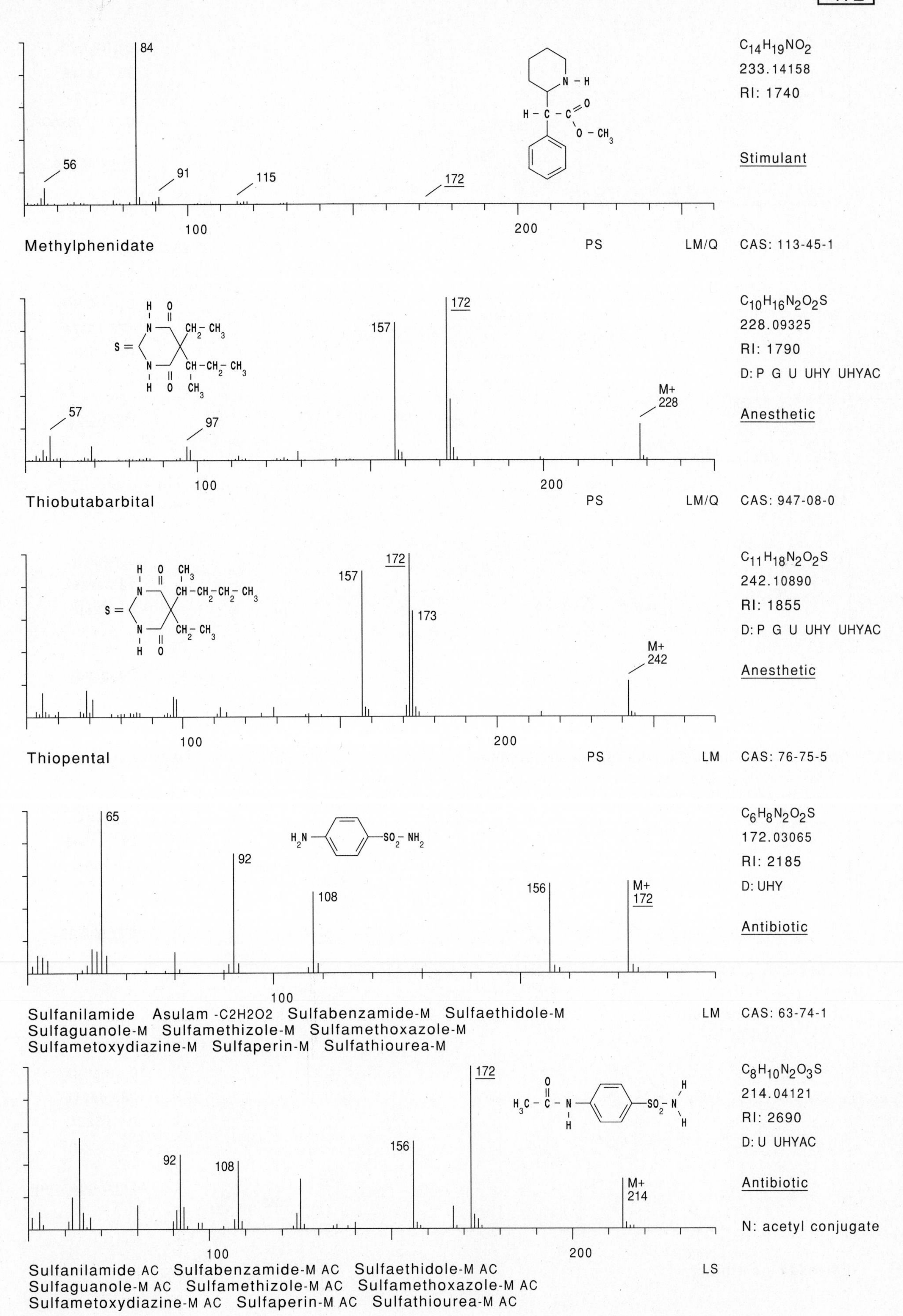
172
84
56
91
115
172
100
200
Methylphenidate
PS
LM/Q
C14H19NO2
233.14158
RI: 1740
Stimulant
CAS: 113-45-1
172
157
57
97
M+
228
100
200
Thiobutabarbital
PS
LM/Q
C10H16N2O2S
228.09325
RI: 1790
D: P G U UHY UHYAC
Anesthetic
CAS: 947-08-0
172
157
173
M+
242
100
200
Thiopental
PS
LM
C11H18N2O2S
242.10890
RI: 1855
D: P G U UHY UHYAC
Anesthetic
CAS: 76-75-5
65
92
108
156
M+
172
100
Sulfanilamide Asulam -C2H2O2 Sulfabenzamide-M Sulfaethidole-M
Sulfaguanole-M Sulfamethizole-M Sulfamethoxazole-M
Sulfametoxydiazine-M Sulfaperin-M Sulfathiourea-M
LM
C6H8N2O2S
172.03065
RI: 2185
D: UHY
Antibiotic
CAS: 63-74-1
172
92
108
156
M+
214
100
200
Sulfanilamide AC Sulfabenzamide-M AC Sulfaethidole-M AC
Sulfaguanole-M AC Sulfamethizole-M AC Sulfamethoxazole-M AC
Sulfametoxydiazine-M AC Sulfaperin-M AC Sulfathiourea-M AC
LS
C8H10N2O3S
214.04121
RI: 2690
D: U UHYAC
Antibiotic
N: acetyl conjugate

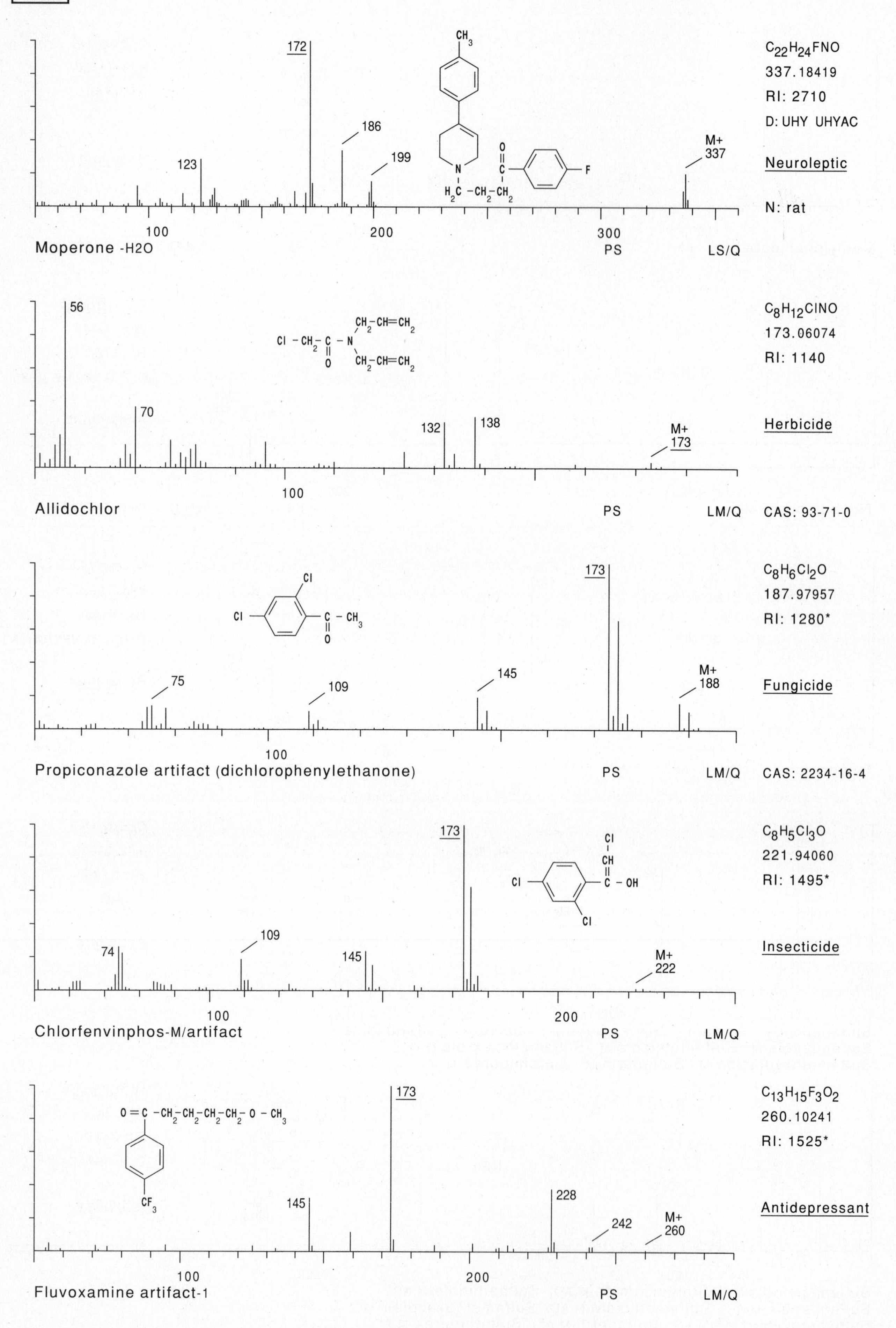

$C_{22}H_{24}FNO$
337.18419
RI: 2710
D: UHY UHYAC
Neuroleptic
N: rat

$C_8H_{12}ClNO$
173.06074
RI: 1140
Herbicide
CAS: 93-71-0

$C_8H_6Cl_2O$
187.97957
RI: 1280*
Fungicide
CAS: 2234-16-4

$C_8H_5Cl_3O$
221.94060
RI: 1495*
Insecticide

$C_{13}H_{15}F_3O_2$
260.10241
RI: 1525*
Antidepressant

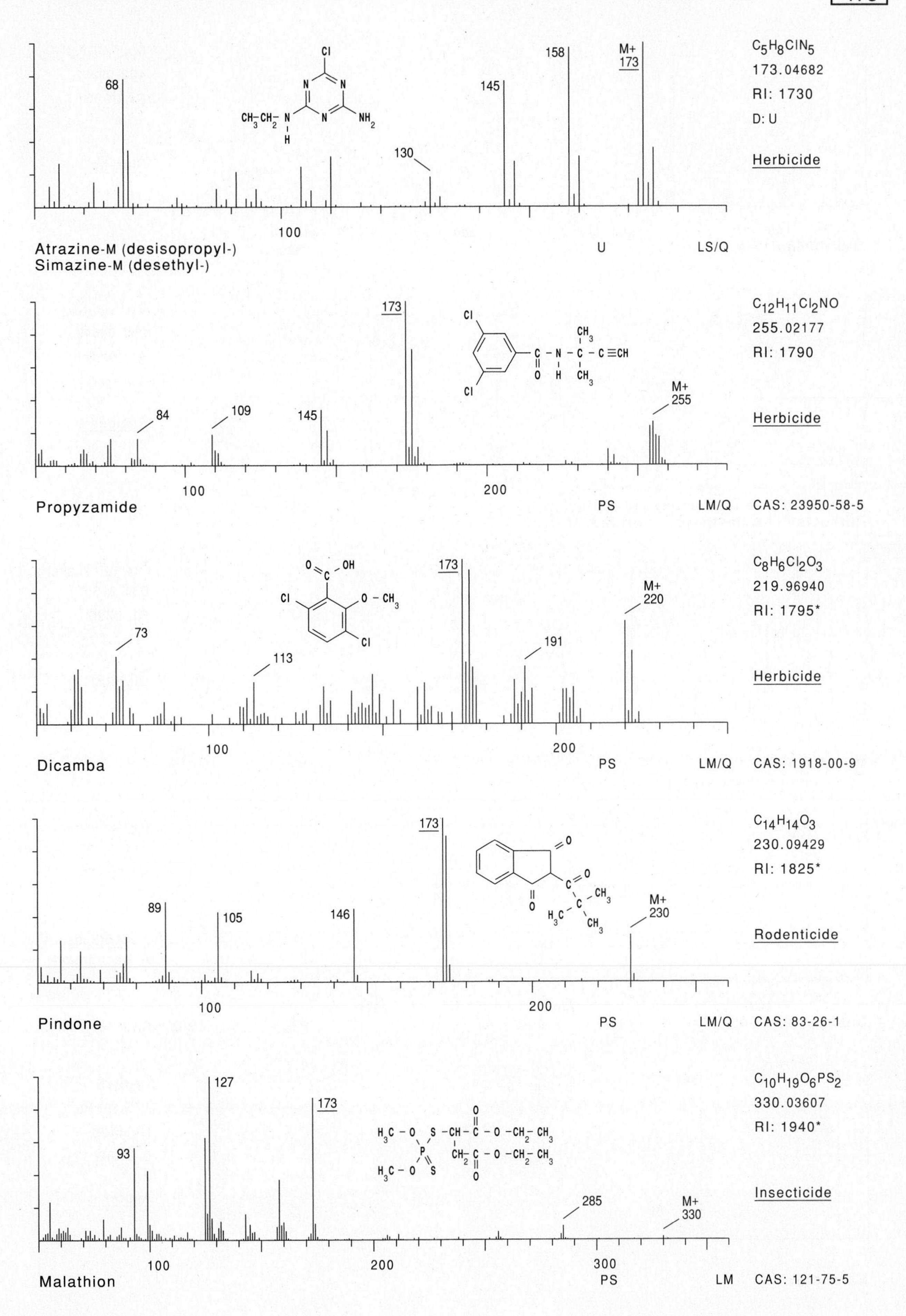

68
158
M+
173
145
130
100
Atrazine-M (desisopropyl-)
Simazine-M (desethyl-)
U
LS/Q
C5H8ClN5
173.04682
RI: 1730
D: U
Herbicide
173
84
109
145
M+
255
100
200
Propyzamide
PS
LM/Q
C12H11Cl2NO
255.02177
RI: 1790
Herbicide
CAS: 23950-58-5
73
113
173
191
M+
220
100
200
Dicamba
PS
LM/Q
C8H6Cl2O3
219.96940
RI: 1795*
Herbicide
CAS: 1918-00-9
89
105
146
173
M+
230
100
200
Pindone
PS
LM/Q
C14H14O3
230.09429
RI: 1825*
Rodenticide
CAS: 83-26-1
93
127
173
285
M+
330
100
200
300
Malathion
PS
LM
C10H19O6PS2
330.03607
RI: 1940*
Insecticide
CAS: 121-75-5

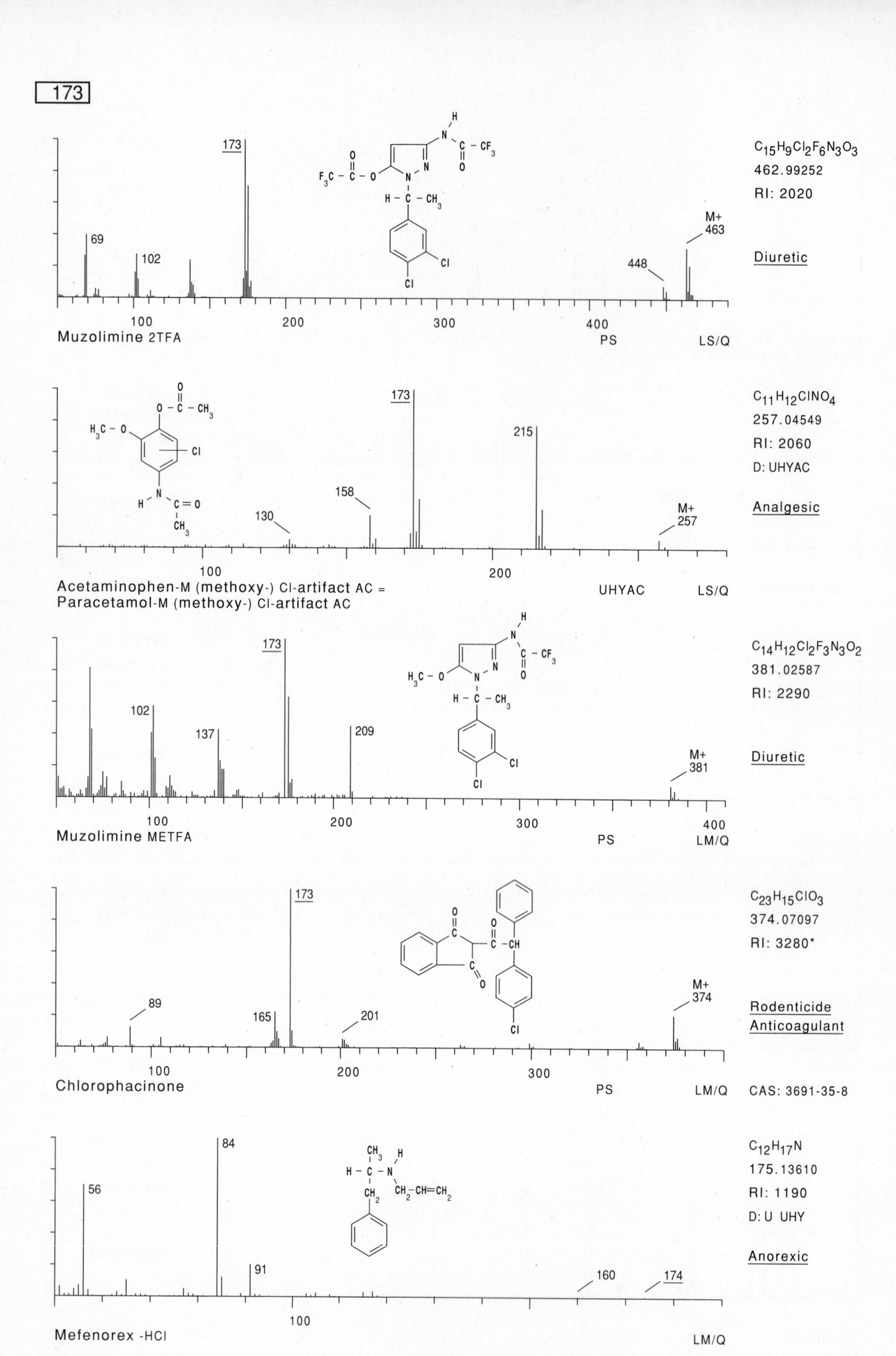

173
69
102
448
M+
463
100
200
300
400
Muzolimine 2TFA
PS
LS/Q
C15H9Cl2F6N3O3
462.99252
RI: 2020
Diuretic
173
215
158
130
M+
257
100
200
Acetaminophen-M (methoxy-) Cl-artifact AC =
Paracetamol-M (methoxy-) Cl-artifact AC
UHYAC
LS/Q
C11H12ClNO4
257.04549
RI: 2060
D: UHYAC
Analgesic
173
102
137
209
M+
381
100
200
300
400
Muzolimine METFA
PS
LM/Q
C14H12Cl2F3N3O2
381.02587
RI: 2290
Diuretic
173
89
165
201
M+
374
100
200
300
Chlorophacinone
PS
LM/Q
C23H15ClO3
374.07097
RI: 3280*
Rodenticide
Anticoagulant
CAS: 3691-35-8
84
56
91
160
174
100
Mefenorex -HCl
LM/Q
C12H17N
175.13610
RI: 1190
D: U UHY
Anorexic

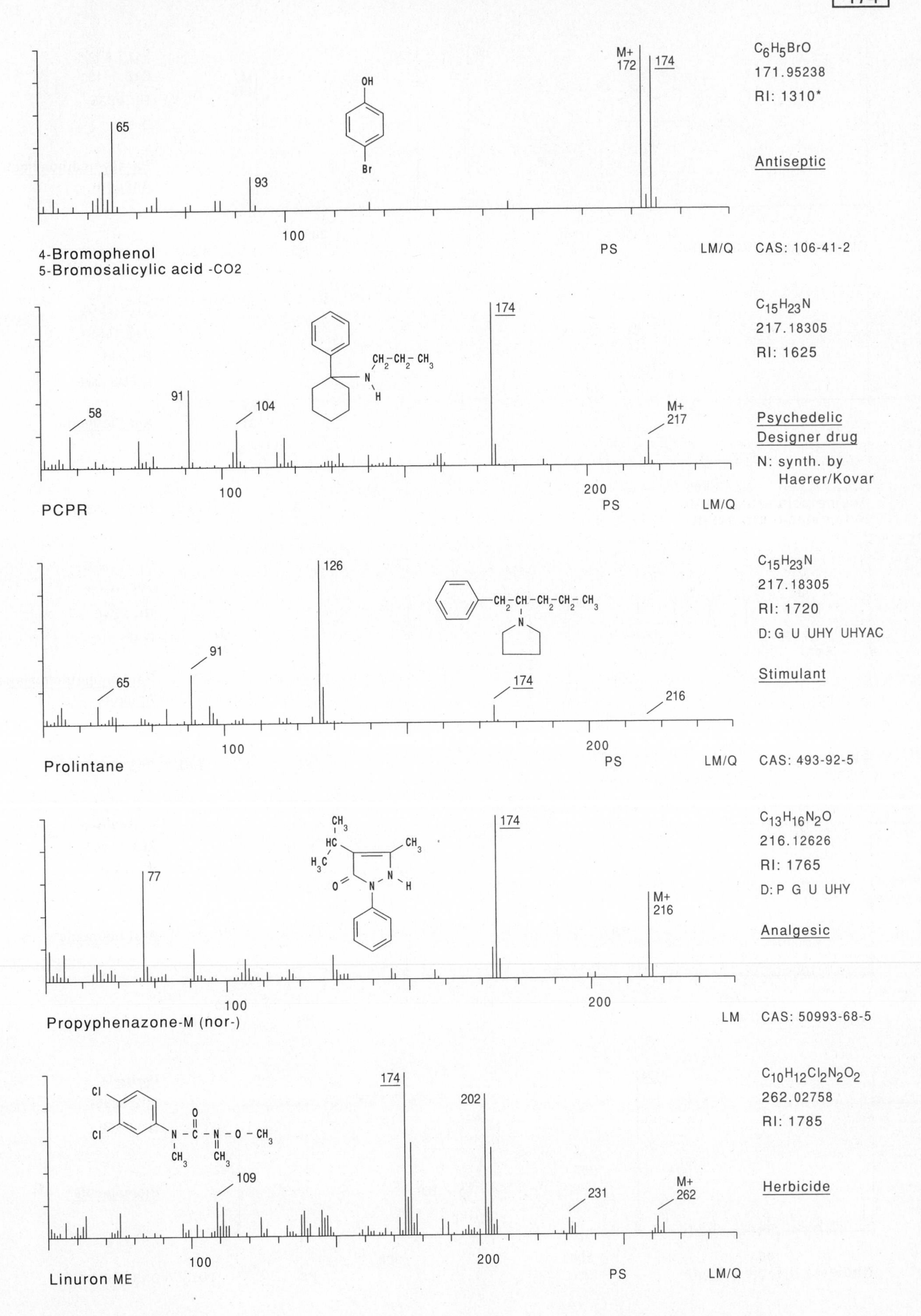

C_6H_5BrO
171.95238
RI: 1310*

Antiseptic

4-Bromophenol
5-Bromosalicylic acid -CO2

CAS: 106-41-2

$C_{15}H_{23}N$
217.18305
RI: 1625

Psychedelic
Designer drug
N: synth. by Haerer/Kovar

PCPR

$C_{15}H_{23}N$
217.18305
RI: 1720
D: G U UHY UHYAC

Stimulant

Prolintane

CAS: 493-92-5

$C_{13}H_{16}N_2O$
216.12626
RI: 1765
D: P G U UHY

Analgesic

Propyphenazone-M (nor-)

CAS: 50993-68-5

$C_{10}H_{12}Cl_2N_2O_2$
262.02758
RI: 1785

Herbicide

Linuron ME

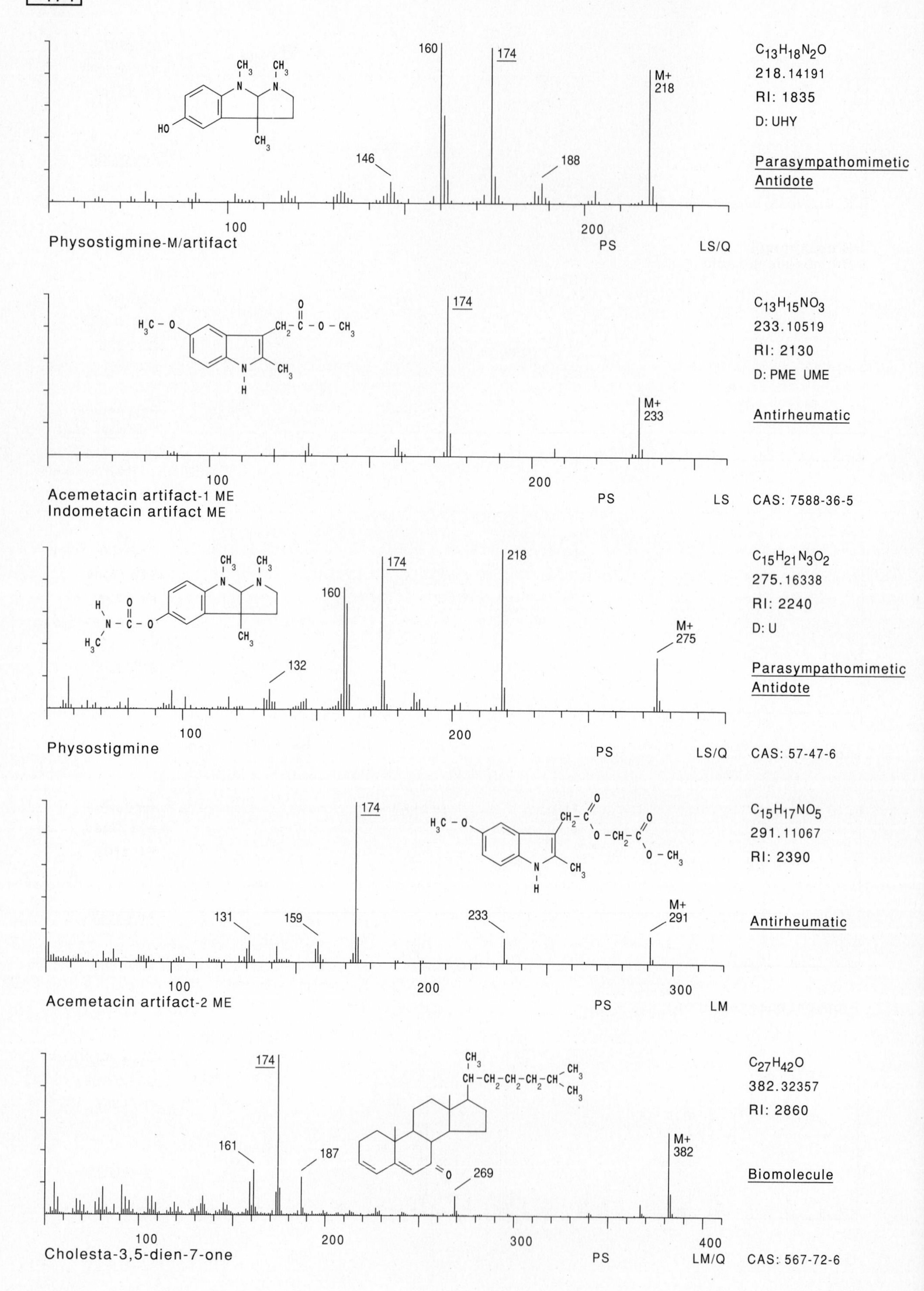

160
174
M+
218
146
188
100
200
Physostigmine-M/artifact
PS
LS/Q
C13H18N2O
218.14191
RI: 1835
D: UHY
Parasympathomimetic
Antidote
174
M+
233
100
200
Acemetacin artifact-1 ME
Indometacin artifact ME
PS
LS
C13H15NO3
233.10519
RI: 2130
D: PME UME
Antirheumatic
CAS: 7588-36-5
174
218
160
M+
275
132
100
200
Physostigmine
PS
LS/Q
C15H21N3O2
275.16338
RI: 2240
D: U
Parasympathomimetic
Antidote
CAS: 57-47-6
174
131
159
233
M+
291
100
200
300
Acemetacin artifact-2 ME
PS
LM
C15H17NO5
291.11067
RI: 2390
Antirheumatic
174
161
187
269
M+
382
100
200
300
400
Cholesta-3,5-dien-7-one
PS
LM/Q
C27H42O
382.32357
RI: 2860
Biomolecule
CAS: 567-72-6

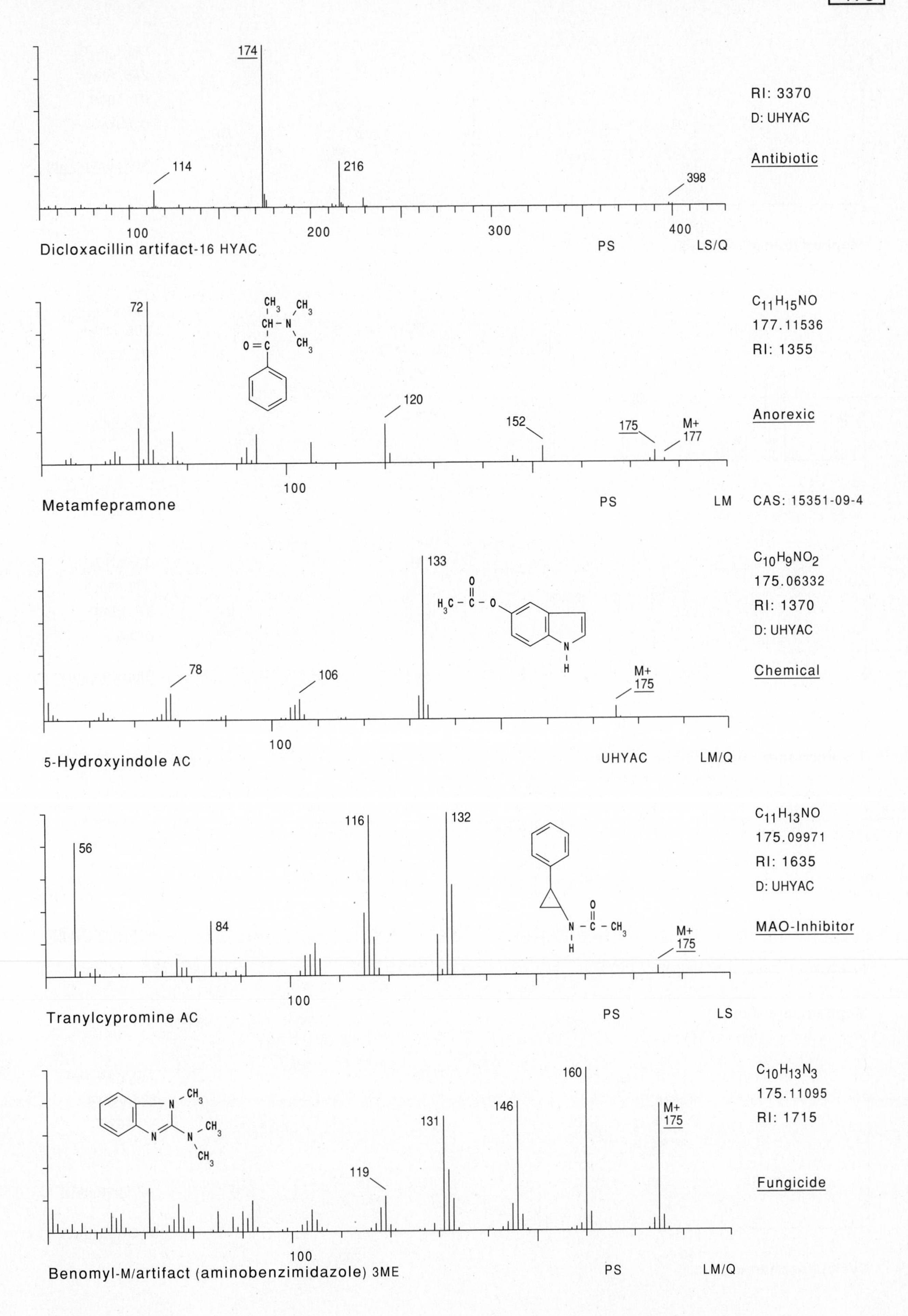

174
114
216
398
100
200
300
400
Dicloxacillin artifact-16 HYAC
PS
LS/Q
RI: 3370
D: UHYAC
Antibiotic
72
CH3 CH3
CH - N
O = C
CH3
120
152
175
M+
177
100
Metamfepramone
PS
LM
C11H15NO
177.11536
RI: 1355
Anorexic
CAS: 15351-09-4
133
O
H3C - C - O
N
H
78
106
M+
175
100
5-Hydroxyindole AC
UHYAC
LM/Q
C10H9NO2
175.06332
RI: 1370
D: UHYAC
Chemical
116
132
56
84
O
N - C - CH3
H
M+
175
100
Tranylcypromine AC
PS
LS
C11H13NO
175.09971
RI: 1635
D: UHYAC
MAO-Inhibitor
N - CH3
N
N - CH3
CH3
160
146
131
119
M+
175
100
Benomyl-M/artifact (aminobenzimidazole) 3ME
PS
LM/Q
C10H13N3
175.11095
RI: 1715
Fungicide

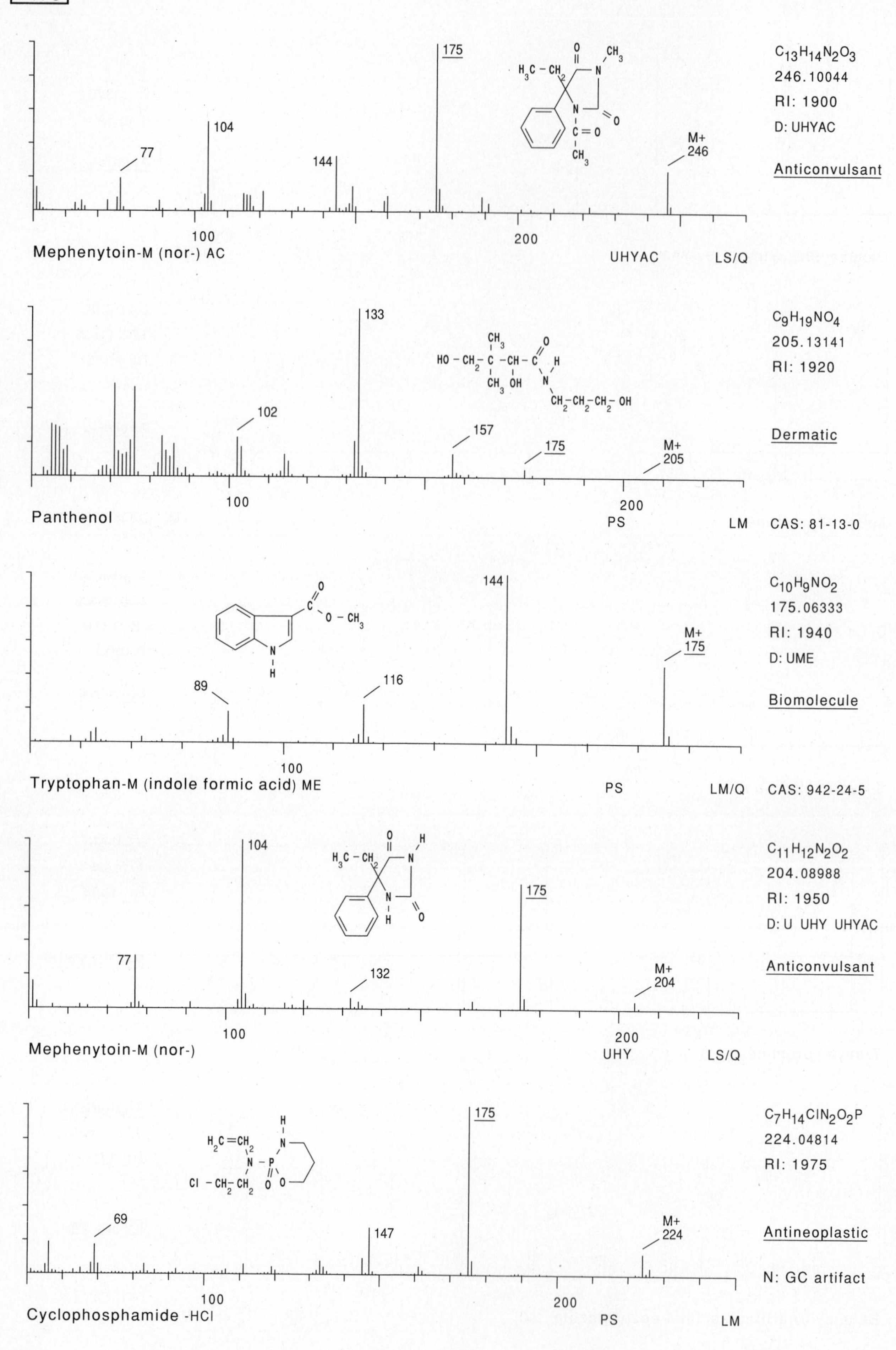

Mephenytoin-M (nor-) AC
UHYAC
LS/Q
C13H14N2O3
246.10044
RI: 1900
D: UHYAC
Anticonvulsant
Panthenol
PS
LM
CAS: 81-13-0
C9H19NO4
205.13141
RI: 1920
Dermatic
Tryptophan-M (indole formic acid) ME
PS
LM/Q
CAS: 942-24-5
C10H9NO2
175.06333
RI: 1940
D: UME
Biomolecule
Mephenytoin-M (nor-)
UHY
LS/Q
C11H12N2O2
204.08988
RI: 1950
D: U UHY UHYAC
Anticonvulsant
Cyclophosphamide -HCl
PS
LM
C7H14ClN2O2P
224.04814
RI: 1975
Antineoplastic
N: GC artifact

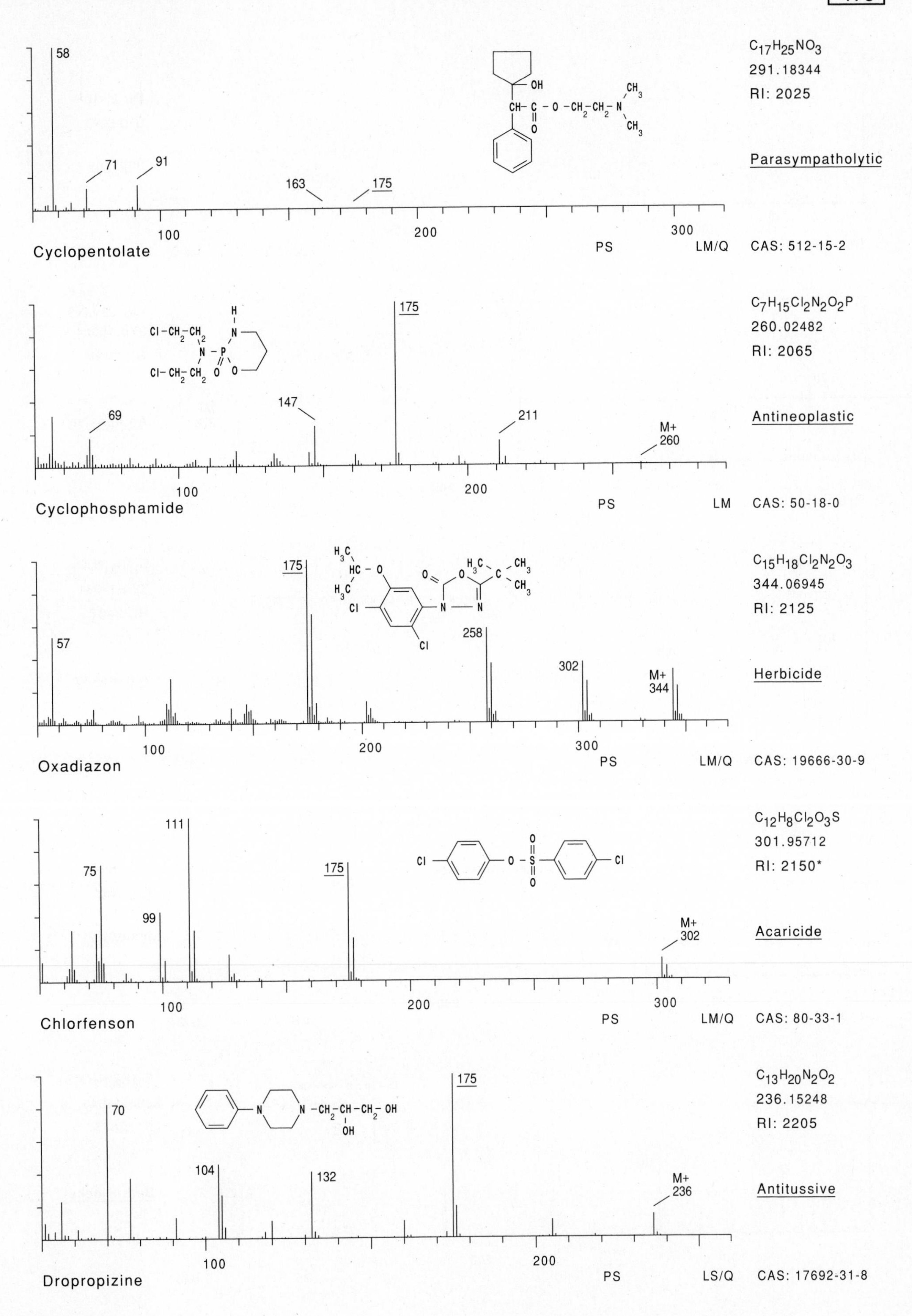
Cyclopentolate
$C_{17}H_{25}NO_3$
291.18344
RI: 2025
Parasympatholytic
CAS: 512-15-2
58
71
91
163
175
100
200
300
PS
LM/Q
Cyclophosphamide
$C_7H_{15}Cl_2N_2O_2P$
260.02482
RI: 2065
Antineoplastic
CAS: 50-18-0
69
147
175
211
M+
260
PS
LM
Oxadiazon
$C_{15}H_{18}Cl_2N_2O_3$
344.06945
RI: 2125
Herbicide
CAS: 19666-30-9
57
175
258
302
M+
344
PS
LM/Q
Chlorfenson
$C_{12}H_8Cl_2O_3S$
301.95712
RI: 2150*
Acaricide
CAS: 80-33-1
75
99
111
175
M+
302
PS
LM/Q
Dropropizine
$C_{13}H_{20}N_2O_2$
236.15248
RI: 2205
Antitussive
CAS: 17692-31-8
70
104
132
175
M+
236
PS
LS/Q

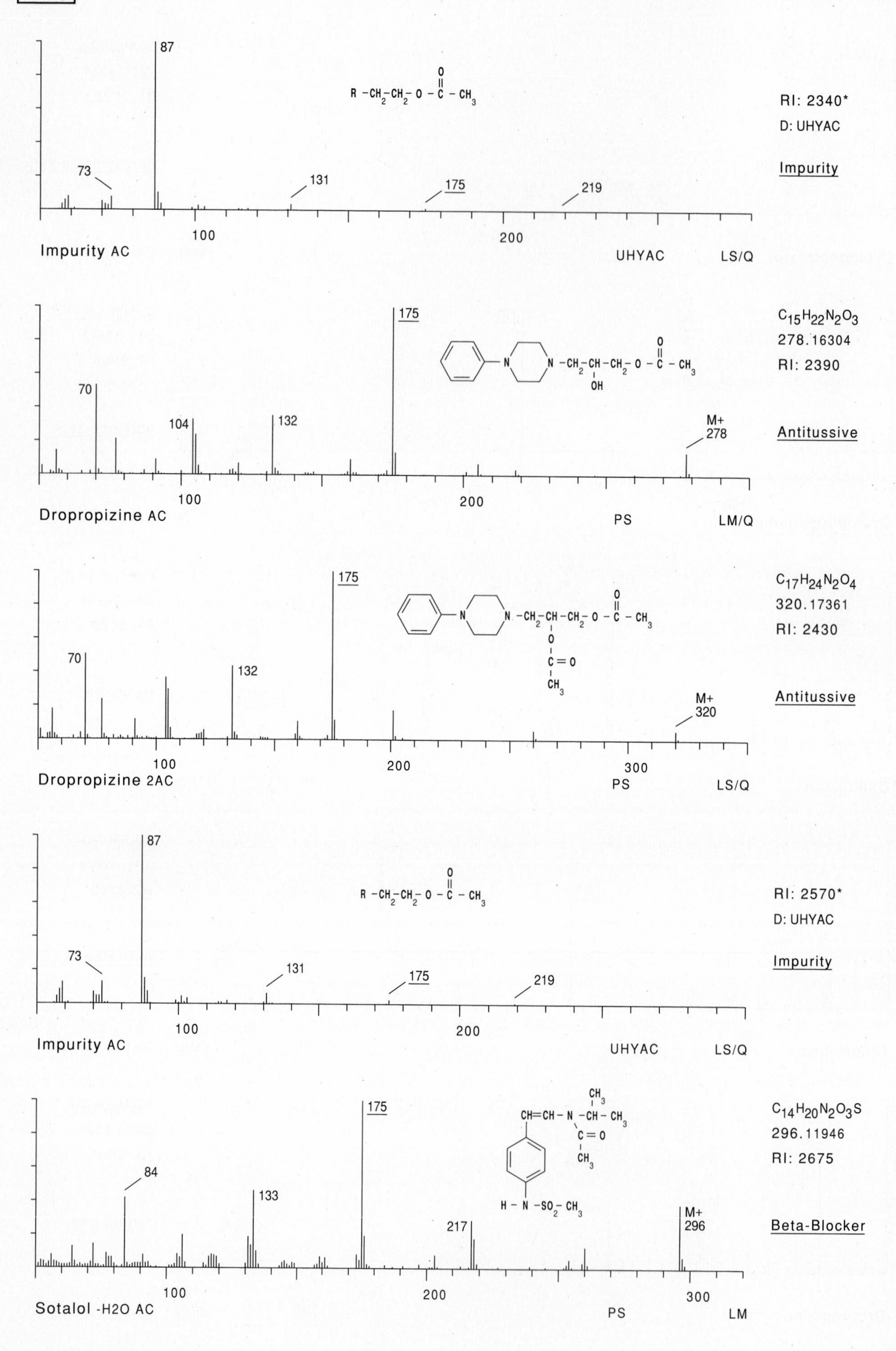
87
R –CH2–CH2–O–C(=O)–CH3
RI: 2340*
D: UHYAC
73
131
175
219
Impurity
100
200
Impurity AC
UHYAC
LS/Q
175
C15H22N2O3
278.16304
RI: 2390
70
104
132
M+
278
Antitussive
Dropropizine AC
PS
LM/Q
175
C17H24N2O4
320.17361
RI: 2430
70
132
M+
320
Antitussive
300
Dropropizine 2AC
PS
LS/Q
87
RI: 2570*
D: UHYAC
73
131
175
219
Impurity
Impurity AC
UHYAC
LS/Q
175
C14H20N2O3S
296.11946
RI: 2675
84
133
217
M+
296
Beta-Blocker
Sotalol -H2O AC
PS
LM

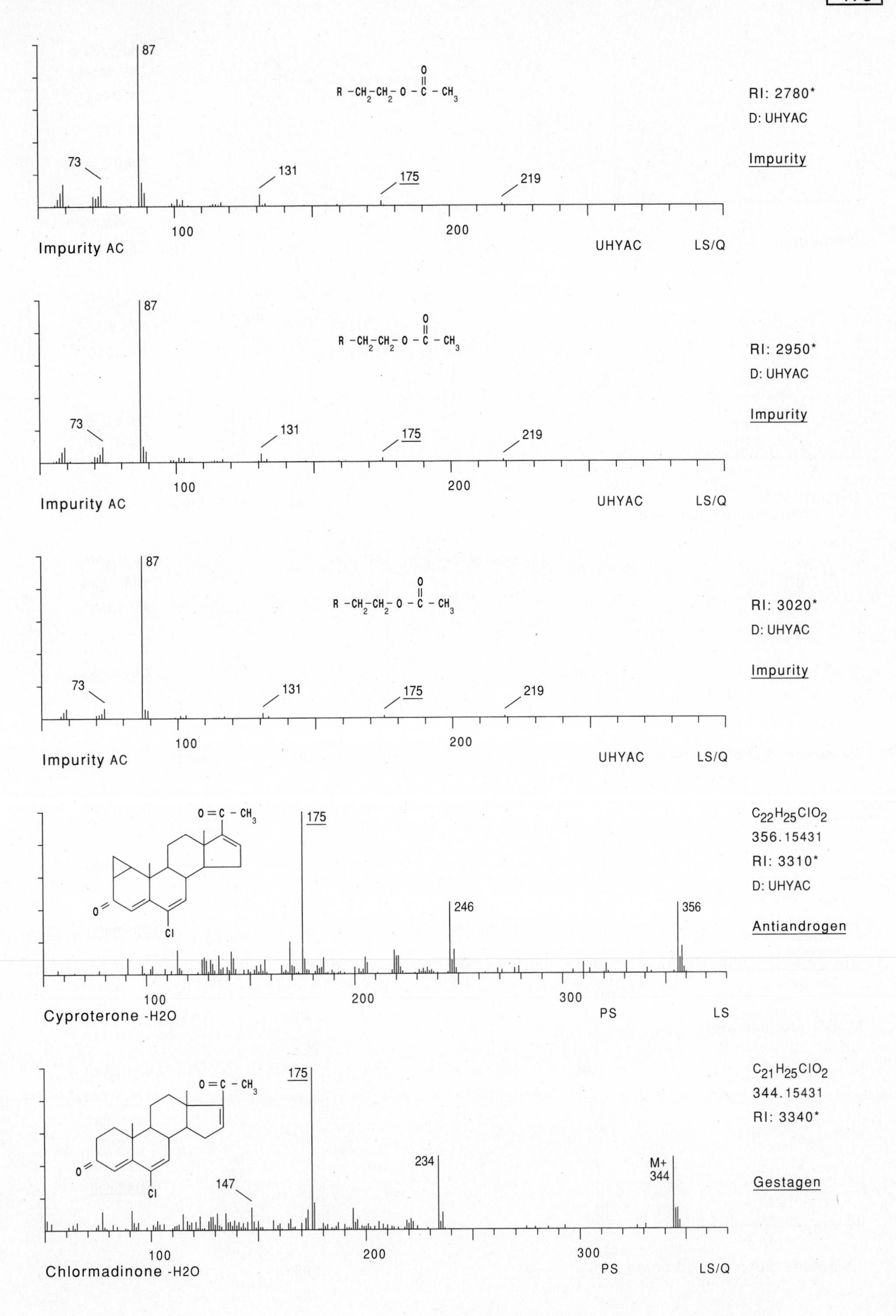
87
73
131
175
219
100
200
Impurity AC
UHYAC
LS/Q
RI: 2780*
D: UHYAC
Impurity
RI: 2950*
D: UHYAC
Impurity
RI: 3020*
D: UHYAC
Impurity
175
246
356
300
Cyproterone -H2O
PS
LS
$C_{22}H_{25}ClO_2$
356.15431
RI: 3310*
D: UHYAC
Antiandrogen
175
147
234
M+
344
Chlormadinone -H2O
PS
LS/Q
$C_{21}H_{25}ClO_2$
344.15431
RI: 3340*
Gestagen

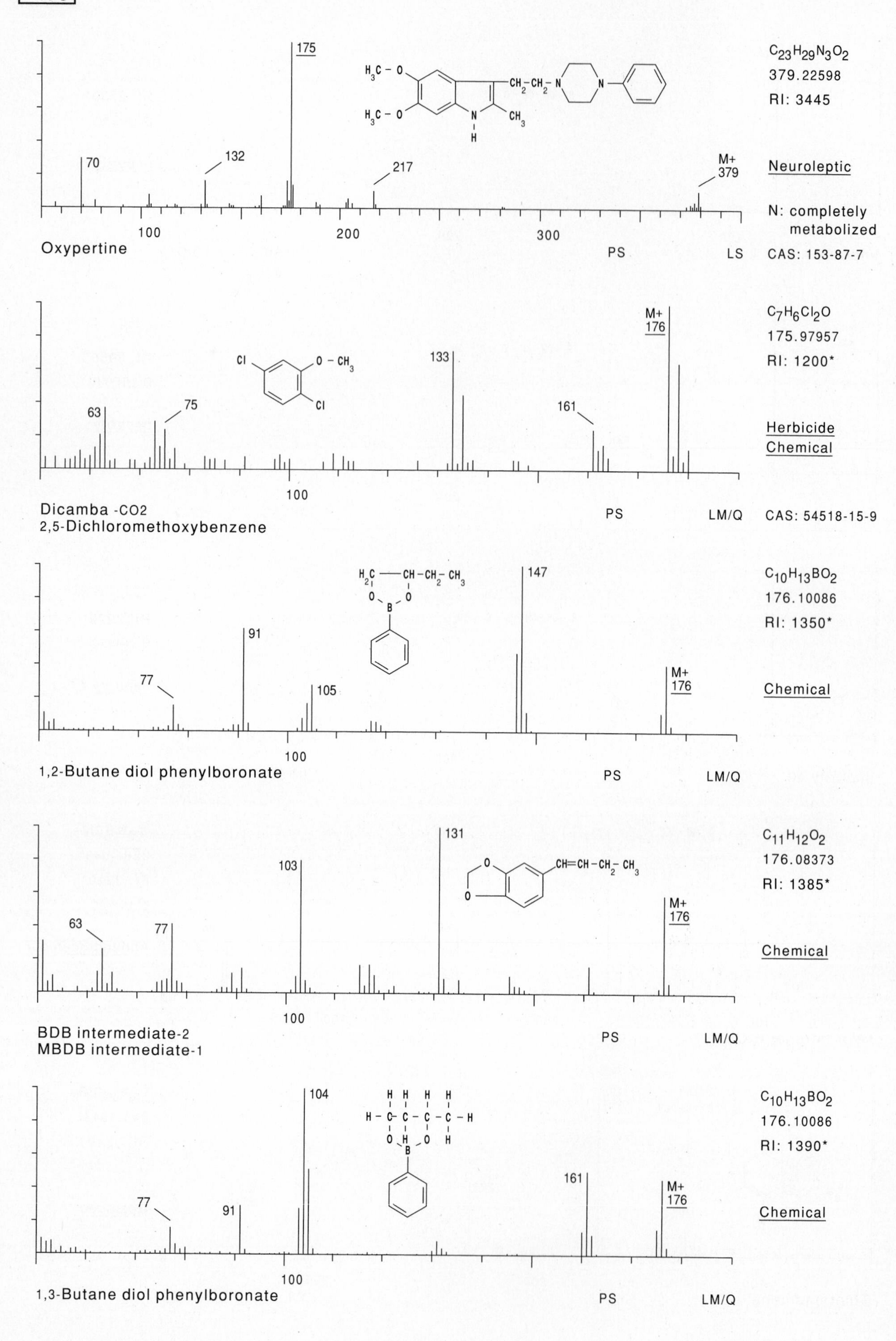
175
70
132
217
M+
379
100
200
300
Oxypertine
PS
LS
$C_{23}H_{29}N_3O_2$
379.22598
RI: 3445
Neuroleptic
N: completely metabolized
CAS: 153-87-7
M+
176
133
63
75
161
100
Dicamba -CO2
2,5-Dichloromethoxybenzene
PS
LM/Q
$C_7H_6Cl_2O$
175.97957
RI: 1200*
Herbicide
Chemical
CAS: 54518-15-9
147
91
77
105
M+
176
100
1,2-Butane diol phenylboronate
PS
LM/Q
$C_{10}H_{13}BO_2$
176.10086
RI: 1350*
Chemical
131
103
63
77
M+
176
100
BDB intermediate-2
MBDB intermediate-1
PS
LM/Q
$C_{11}H_{12}O_2$
176.08373
RI: 1385*
Chemical
104
77
91
161
M+
176
100
1,3-Butane diol phenylboronate
PS
LM/Q
$C_{10}H_{13}BO_2$
176.10086
RI: 1390*
Chemical

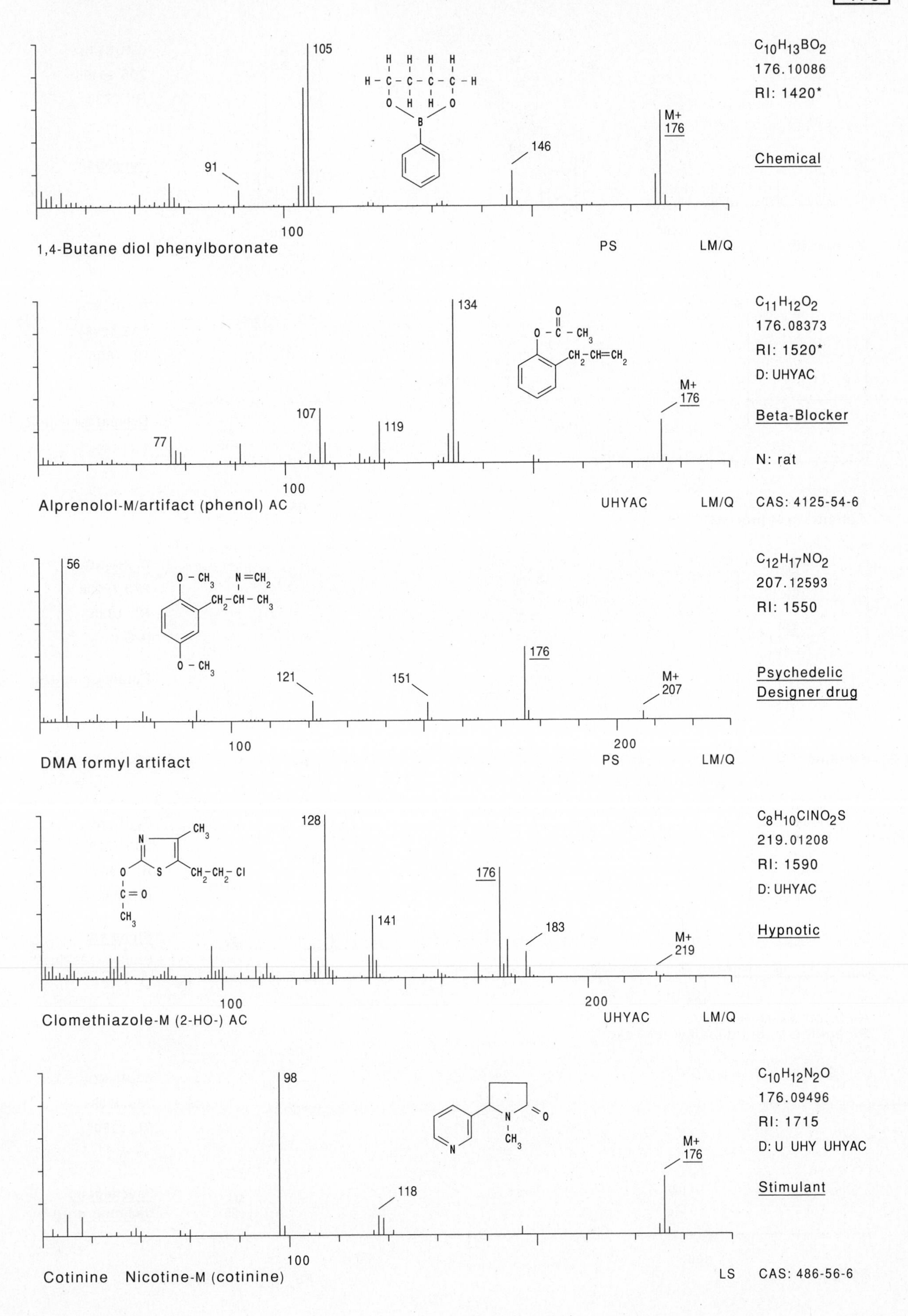

105
91
146
M+ 176
100
1,4-Butane diol phenylboronate
PS
LM/Q
$C_{10}H_{13}BO_2$
176.10086
RI: 1420*
Chemical
134
107
119
77
M+ 176
100
Alprenolol-M/artifact (phenol) AC
UHYAC
LM/Q
$C_{11}H_{12}O_2$
176.08373
RI: 1520*
D: UHYAC
Beta-Blocker
N: rat
CAS: 4125-54-6
56
176
121
151
M+ 207
100
200
DMA formyl artifact
PS
LM/Q
$C_{12}H_{17}NO_2$
207.12593
RI: 1550
Psychedelic
Designer drug
128
176
141
183
M+ 219
100
200
Clomethiazole-M (2-HO-) AC
UHYAC
LM/Q
$C_8H_{10}ClNO_2S$
219.01208
RI: 1590
D: UHYAC
Hypnotic
98
118
M+ 176
100
Cotinine Nicotine-M (cotinine)
LS
$C_{10}H_{12}N_2O$
176.09496
RI: 1715
D: U UHY UHYAC
Stimulant
CAS: 486-56-6

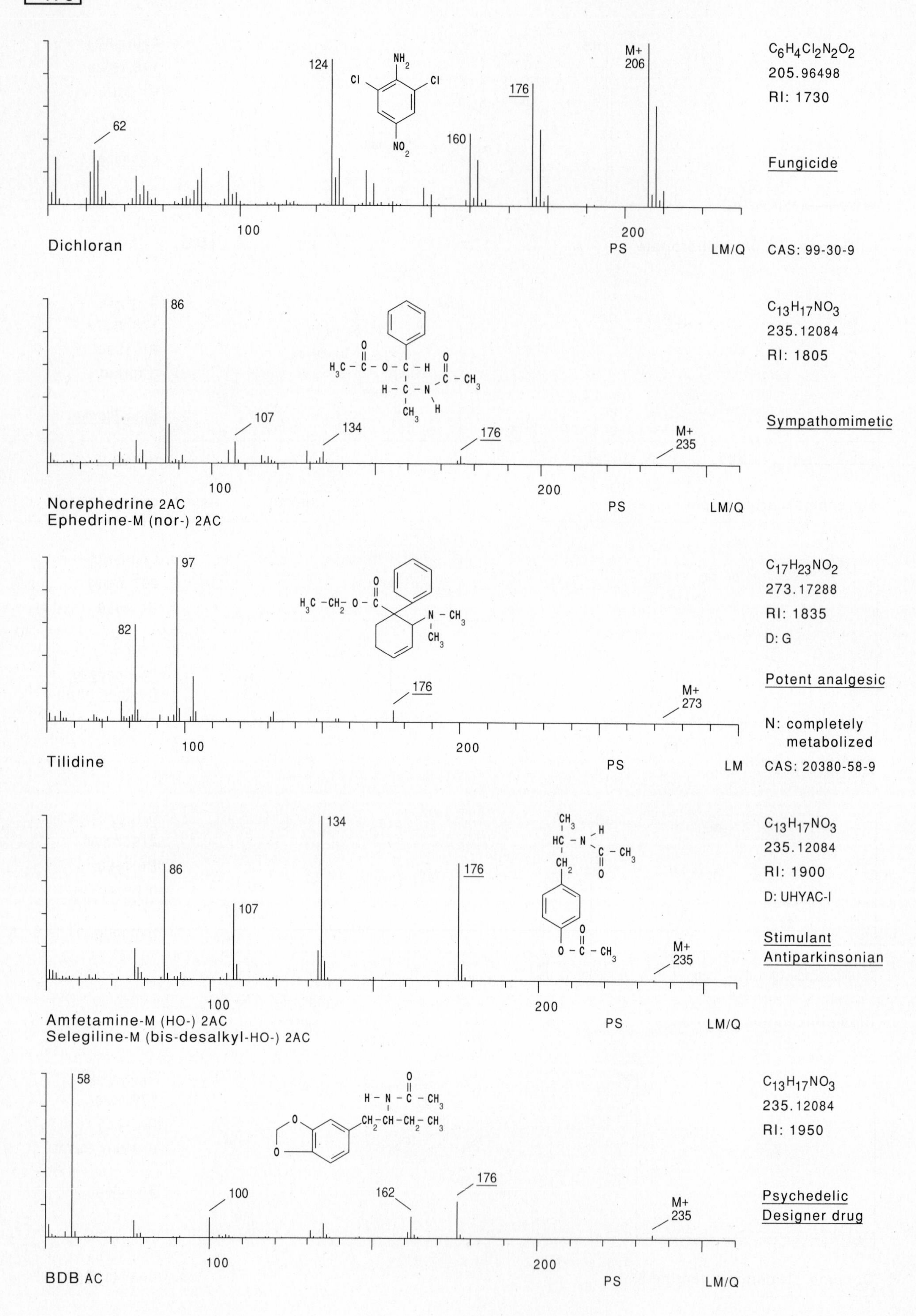

124
NH2
Cl
Cl
NO2
M+
206
176
160
62
100
200
Dichloran
PS
LM/Q
C6H4Cl2N2O2
205.96498
RI: 1730
Fungicide
CAS: 99-30-9
86
107
134
176
M+
235
100
200
Norephedrine 2AC
Ephedrine-M (nor-) 2AC
PS
LM/Q
C13H17NO3
235.12084
RI: 1805
Sympathomimetic
97
82
176
M+
273
100
200
Tilidine
PS
LM
C17H23NO2
273.17288
RI: 1835
D: G
Potent analgesic
N: completely metabolized
CAS: 20380-58-9
134
86
107
176
M+
235
100
200
Amfetamine-M (HO-) 2AC
Selegiline-M (bis-desalkyl-HO-) 2AC
PS
LM/Q
C13H17NO3
235.12084
RI: 1900
D: UHYAC-I
Stimulant
Antiparkinsonian
58
100
162
176
M+
235
100
200
BDB AC
PS
LM/Q
C13H17NO3
235.12084
RI: 1950
Psychedelic
Designer drug

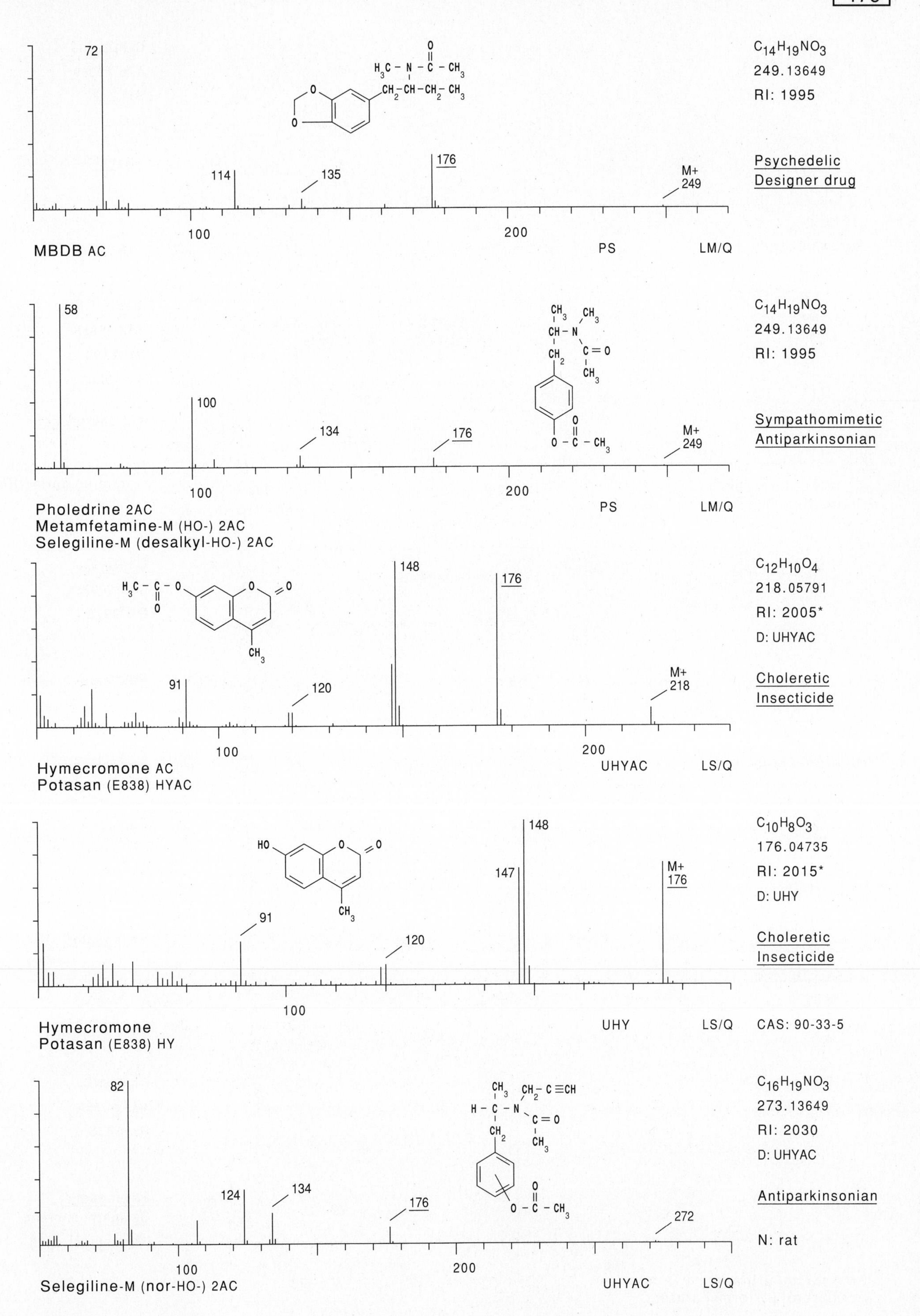

MBDB AC — PS — LM/Q

$C_{14}H_{19}NO_3$
249.13649
RI: 1995

Psychedelic
Designer drug

Pholedrine 2AC
Metamfetamine-M (HO-) 2AC
Selegiline-M (desalkyl-HO-) 2AC — PS — LM/Q

$C_{14}H_{19}NO_3$
249.13649
RI: 1995

Sympathomimetic
Antiparkinsonian

Hymecromone AC
Potasan (E838) HYAC — UHYAC — LS/Q

$C_{12}H_{10}O_4$
218.05791
RI: 2005*
D: UHYAC

Choleretic
Insecticide

Hymecromone
Potasan (E838) HY — UHY — LS/Q

$C_{10}H_8O_3$
176.04735
RI: 2015*
D: UHY

Choleretic
Insecticide

CAS: 90-33-5

Selegiline-M (nor-HO-) 2AC — UHYAC — LS/Q

$C_{16}H_{19}NO_3$
273.13649
RI: 2030
D: UHYAC

Antiparkinsonian

N: rat

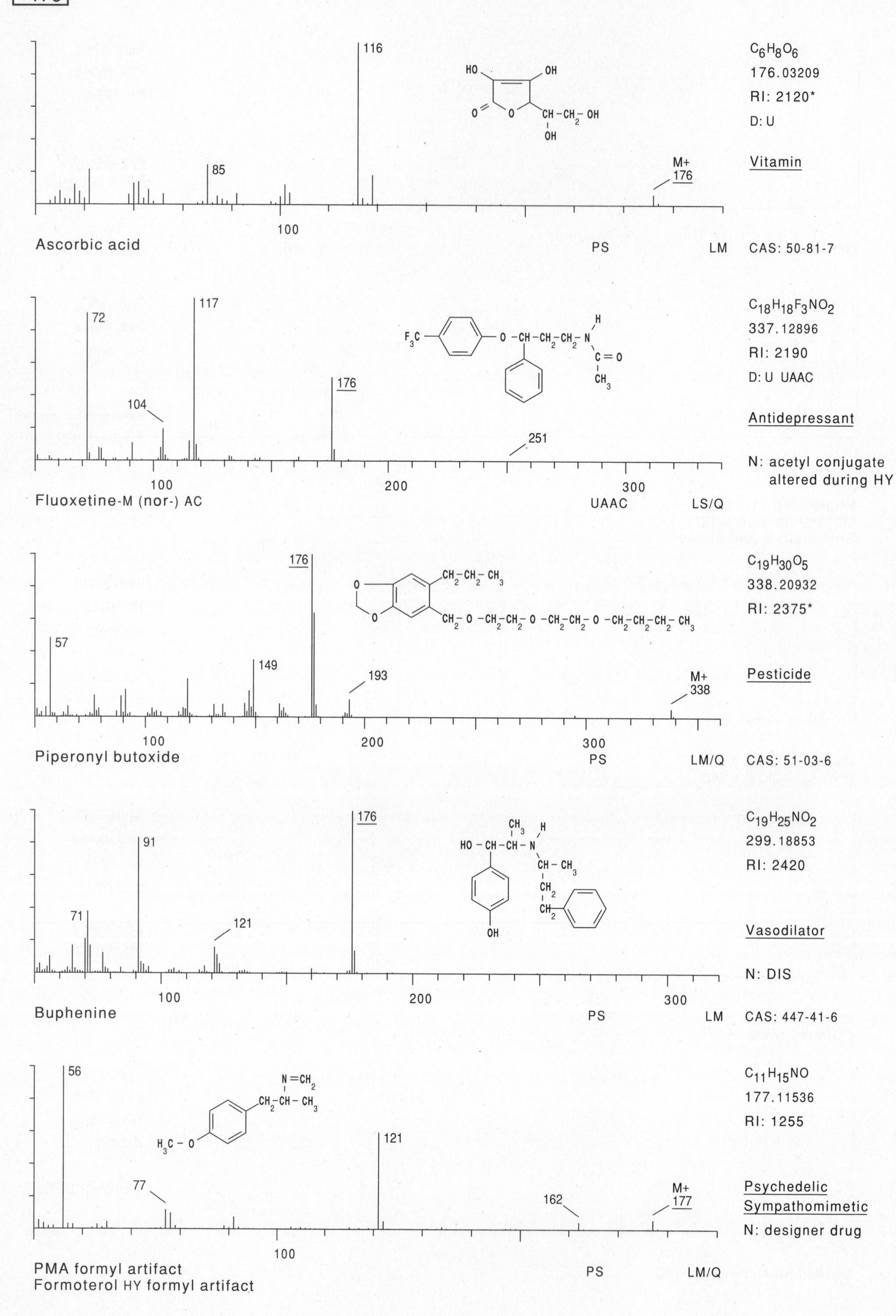
176
116
85
M+
176
100
Ascorbic acid
PS
LM
$C_6H_8O_6$
176.03209
RI: 2120*
D: U
Vitamin
CAS: 50-81-7
72
117
104
176
251
100
200
300
Fluoxetine-M (nor-) AC
UAAC
LS/Q
$C_{18}H_{18}F_3NO_2$
337.12896
RI: 2190
D: U UAAC
Antidepressant
N: acetyl conjugate altered during HY
176
57
149
193
M+
338
100
200
300
Piperonyl butoxide
PS
LM/Q
$C_{19}H_{30}O_5$
338.20932
RI: 2375*
Pesticide
CAS: 51-03-6
176
91
71
121
100
200
300
Buphenine
PS
LM
$C_{19}H_{25}NO_2$
299.18853
RI: 2420
Vasodilator
N: DIS
CAS: 447-41-6
56
121
77
162
M+
177
100
PMA formyl artifact
Formoterol HY formyl artifact
PS
LM/Q
$C_{11}H_{15}NO$
177.11536
RI: 1255
Psychedelic
Sympathomimetic
N: designer drug

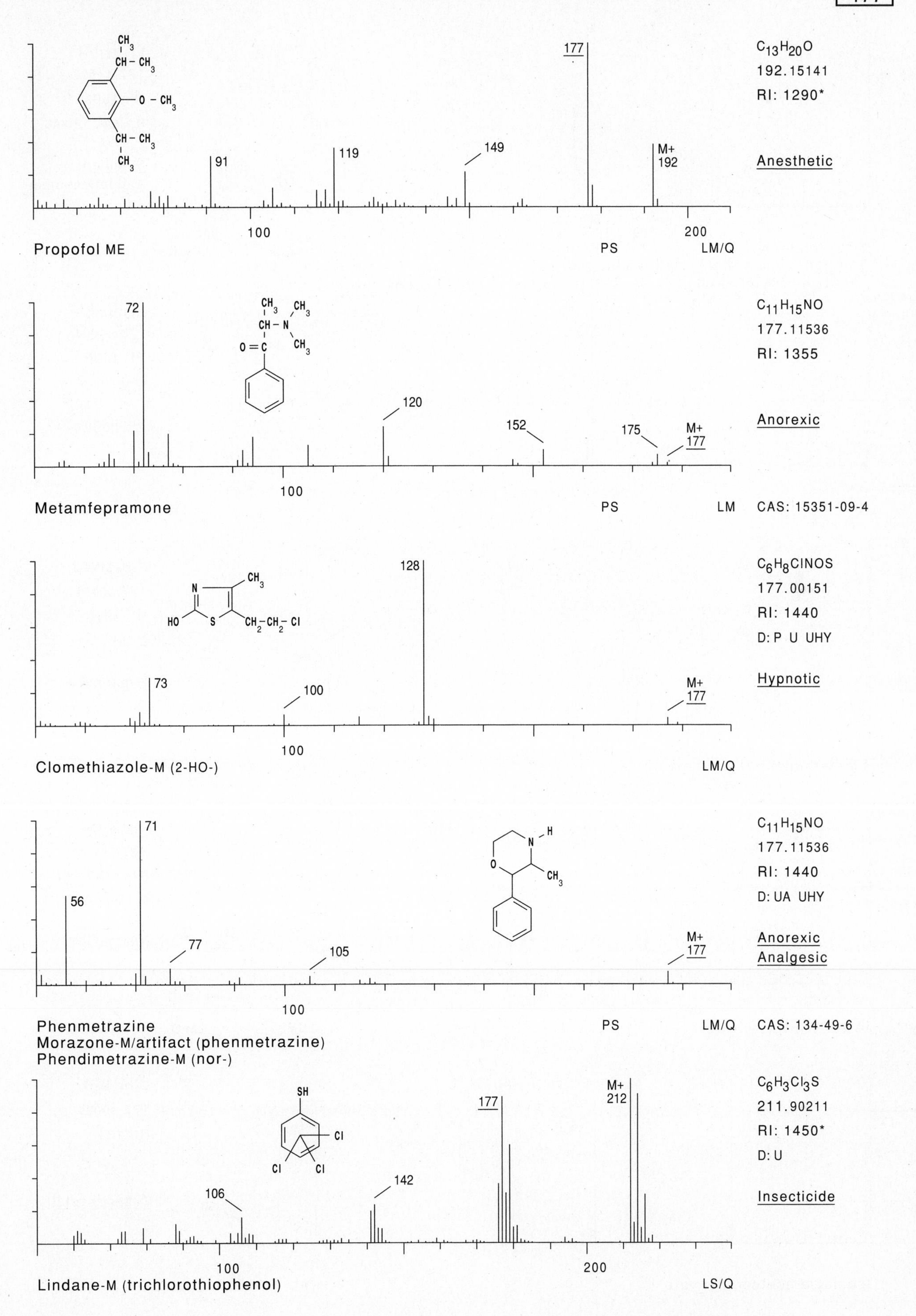

177
91
119
149
M+
192
100
200
Propofol ME
PS
LM/Q
C13H20O
192.15141
RI: 1290*
Anesthetic
72
120
152
175
M+
177
100
Metamfepramone
PS
LM
C11H15NO
177.11536
RI: 1355
Anorexic
CAS: 15351-09-4
128
73
100
M+
177
100
Clomethiazole-M (2-HO-)
LM/Q
C6H8ClNOS
177.00151
RI: 1440
D: P U UHY
Hypnotic
71
56
77
105
M+
177
100
Phenmetrazine
Morazone-M/artifact (phenmetrazine)
Phendimetrazine-M (nor-)
PS
LM/Q
C11H15NO
177.11536
RI: 1440
D: UA UHY
Anorexic
Analgesic
CAS: 134-49-6
177
M+
212
106
142
100
200
Lindane-M (trichlorothiophenol)
LS/Q
C6H3Cl3S
211.90211
RI: 1450*
D: U
Insecticide

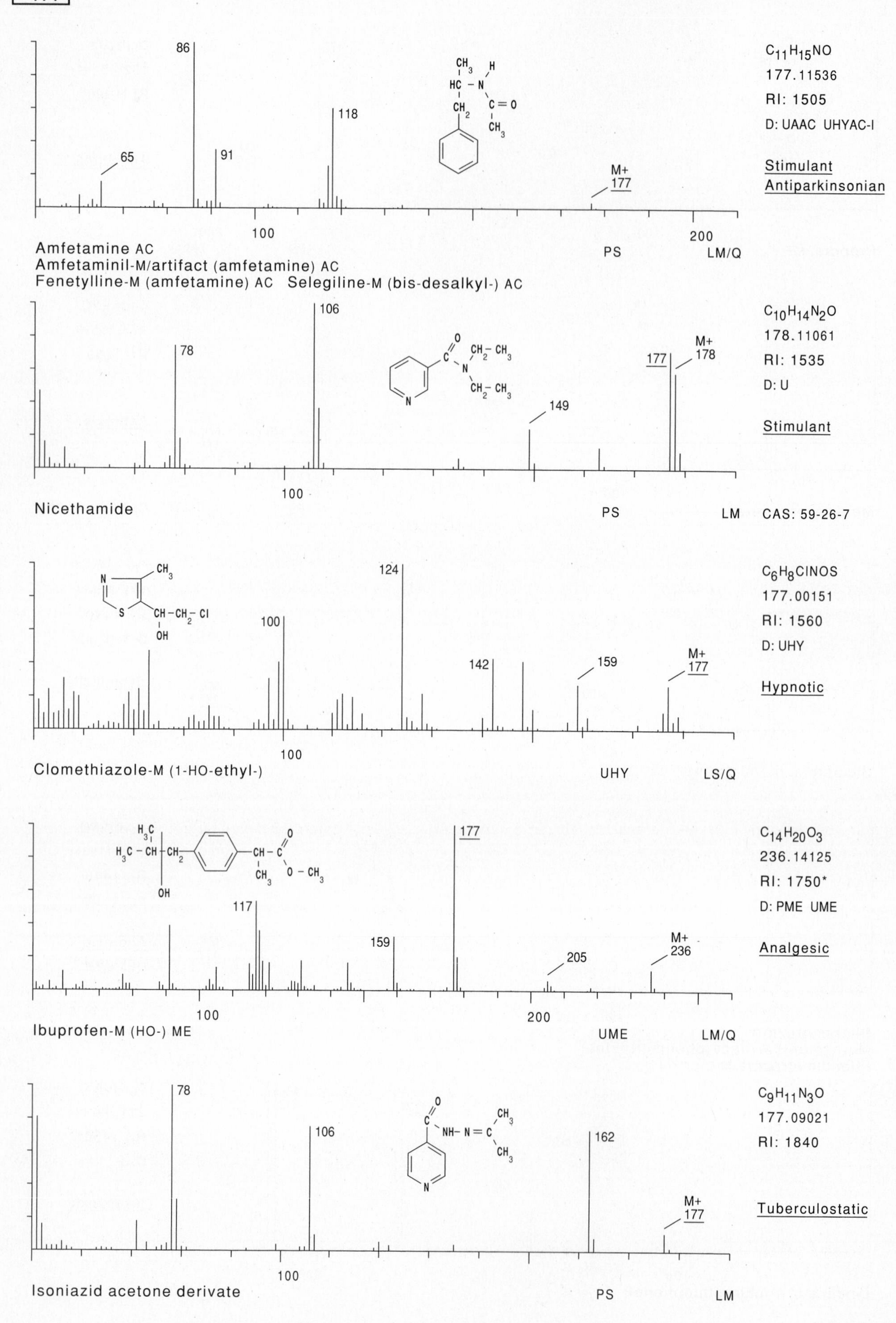
86
118
91
65
M+
177
100
200
Amfetamine AC
Amfetaminil-M/artifact (amfetamine) AC
Fenetylline-M (amfetamine) AC Selegiline-M (bis-desalkyl-) AC
PS
LM/Q
C11H15NO
177.11536
RI: 1505
D: UAAC UHYAC-I
Stimulant
Antiparkinsonian
106
78
177
M+
178
149
100
Nicethamide
PS
LM
C10H14N2O
178.11061
RI: 1535
D: U
Stimulant
CAS: 59-26-7
124
100
142
159
M+
177
100
Clomethiazole-M (1-HO-ethyl-)
UHY
LS/Q
C6H8ClNOS
177.00151
RI: 1560
D: UHY
Hypnotic
177
117
159
205
M+
236
100
200
Ibuprofen-M (HO-) ME
UME
LM/Q
C14H20O3
236.14125
RI: 1750*
D: PME UME
Analgesic
78
106
162
M+
177
100
Isoniazid acetone derivate
PS
LM
C9H11N3O
177.09021
RI: 1840
Tuberculostatic

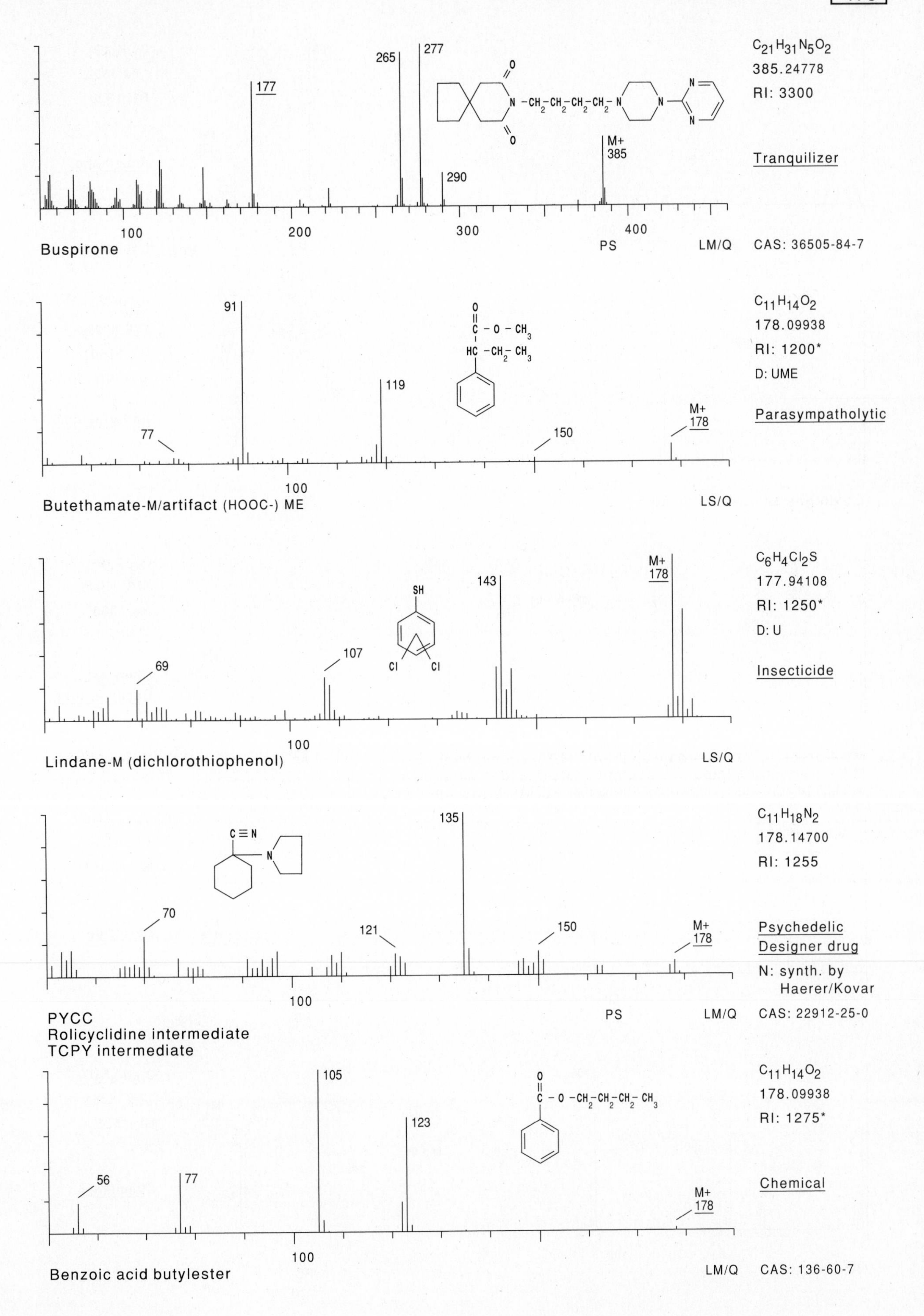

$C_{21}H_{31}N_5O_2$
385.24778
RI: 3300

Tranquilizer

CAS: 36505-84-7

$C_{11}H_{14}O_2$
178.09938
RI: 1200*
D: UME

Parasympatholytic

$C_6H_4Cl_2S$
177.94108
RI: 1250*
D: U

Insecticide

$C_{11}H_{18}N_2$
178.14700
RI: 1255

Psychedelic
Designer drug

N: synth. by
Haerer/Kovar

CAS: 22912-25-0

$C_{11}H_{14}O_2$
178.09938
RI: 1275*

Chemical

CAS: 136-60-7

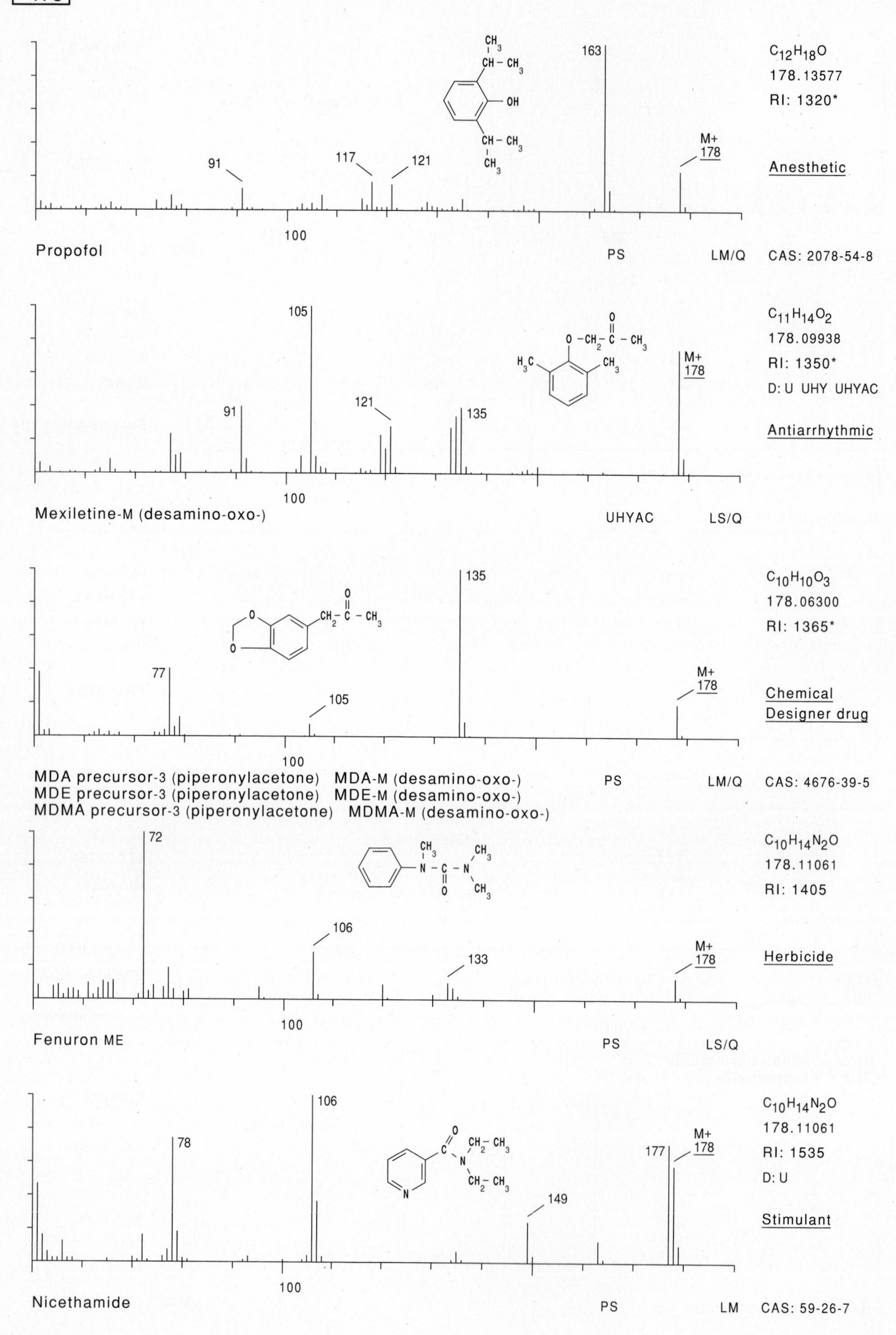

163
91
117
121
M+
178
100
$C_{12}H_{18}O$
178.13577
RI: 1320*
Anesthetic
Propofol
PS
LM/Q
CAS: 2078-54-8
105
91
121
135
M+
178
100
$C_{11}H_{14}O_2$
178.09938
RI: 1350*
D: U UHY UHYAC
Antiarrhythmic
Mexiletine-M (desamino-oxo-)
UHYAC
LS/Q
135
77
105
M+
178
100
$C_{10}H_{10}O_3$
178.06300
RI: 1365*
Chemical
Designer drug
MDA precursor-3 (piperonylacetone) MDA-M (desamino-oxo-)
MDE precursor-3 (piperonylacetone) MDE-M (desamino-oxo-)
MDMA precursor-3 (piperonylacetone) MDMA-M (desamino-oxo-)
PS
LM/Q
CAS: 4676-39-5
72
106
133
M+
178
100
$C_{10}H_{14}N_2O$
178.11061
RI: 1405
Herbicide
Fenuron ME
PS
LS/Q
106
78
177
M+
178
149
100
$C_{10}H_{14}N_2O$
178.11061
RI: 1535
D: U
Stimulant
Nicethamide
PS
LM
CAS: 59-26-7

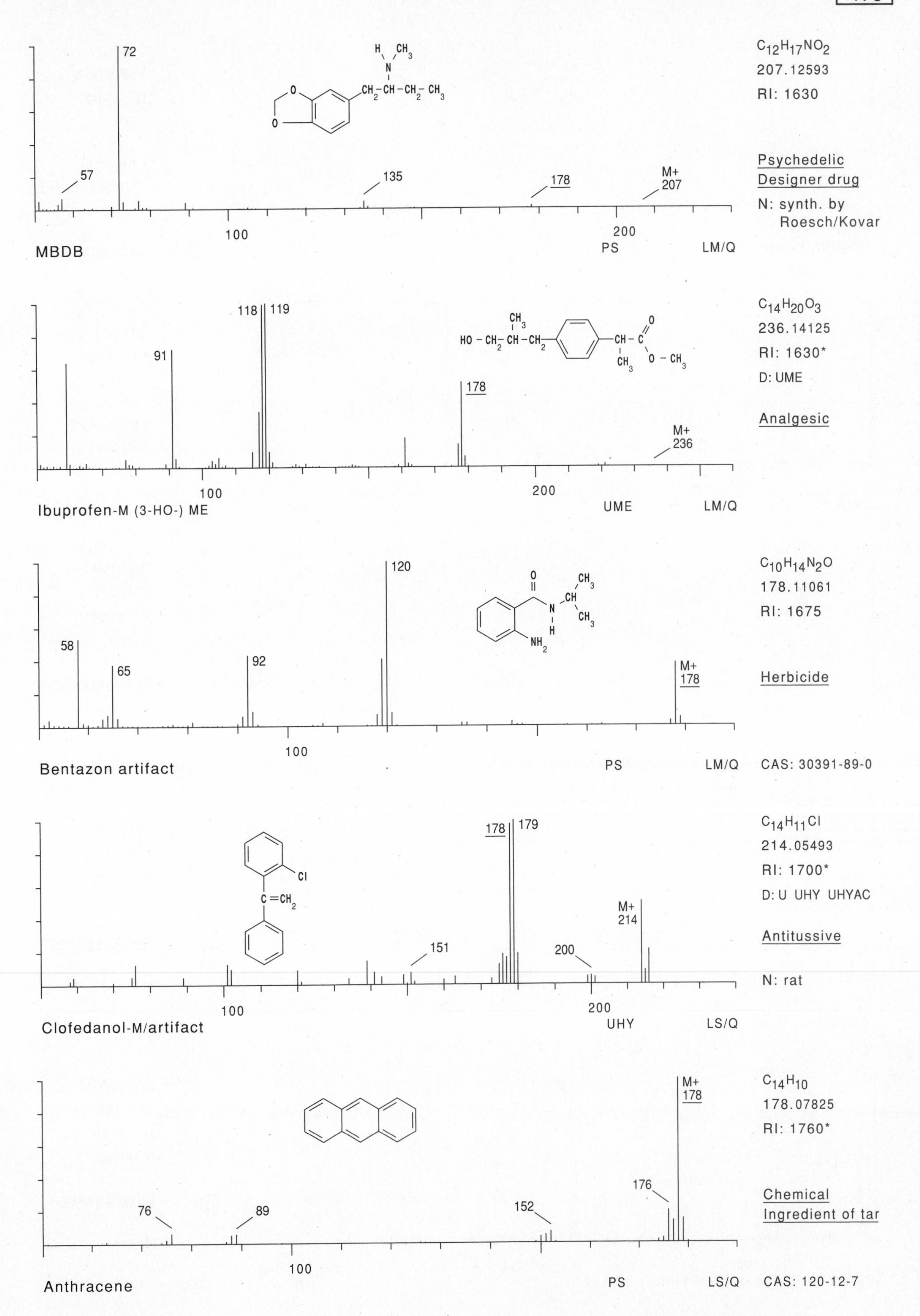

$C_{12}H_{17}NO_2$
207.12593
RI: 1630

Psychedelic
Designer drug

N: synth. by Roesch/Kovar

$C_{14}H_{20}O_3$
236.14125
RI: 1630*
D: UME

Analgesic

$C_{10}H_{14}N_2O$
178.11061
RI: 1675

Herbicide

CAS: 30391-89-0

$C_{14}H_{11}Cl$
214.05493
RI: 1700*
D: U UHY UHYAC

Antitussive

N: rat

$C_{14}H_{10}$
178.07825
RI: 1760*

Chemical
Ingredient of tar

CAS: 120-12-7

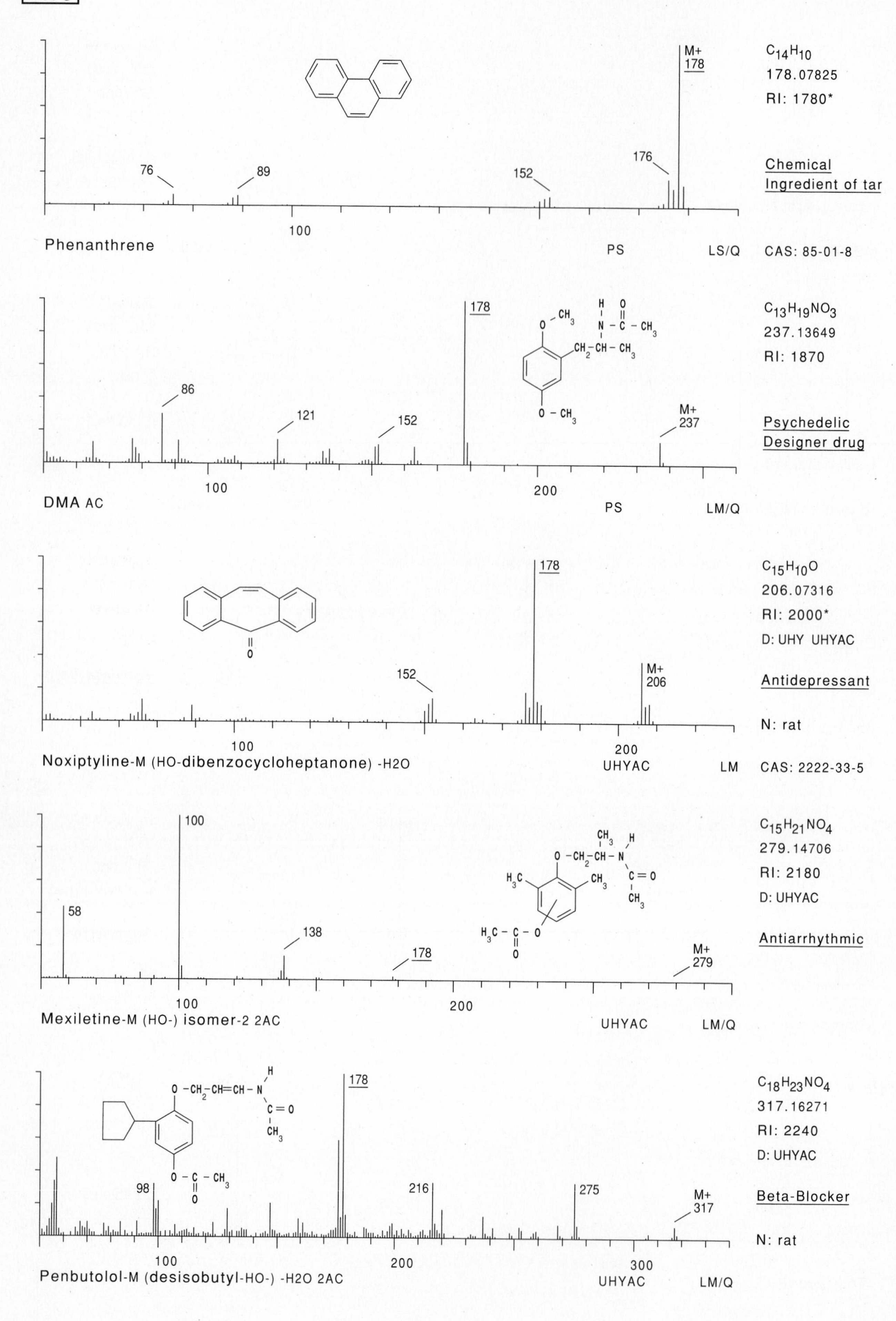
M+ 178
76
89
152
176
100
Phenanthrene
PS
LS/Q
$C_{14}H_{10}$
178.07825
RI: 1780*
Chemical Ingredient of tar
CAS: 85-01-8
178
86
121
152
M+ 237
100
200
DMA AC
PS
LM/Q
$C_{13}H_{19}NO_3$
237.13649
RI: 1870
Psychedelic Designer drug
178
152
M+ 206
100
200
Noxiptyline-M (HO-dibenzocycloheptanone) -H2O
UHYAC
LM
$C_{15}H_{10}O$
206.07316
RI: 2000*
D: UHY UHYAC
Antidepressant
N: rat
CAS: 2222-33-5
100
58
138
178
M+ 279
100
200
Mexiletine-M (HO-) isomer-2 2AC
UHYAC
LM/Q
$C_{15}H_{21}NO_4$
279.14706
RI: 2180
D: UHYAC
Antiarrhythmic
178
98
216
275
M+ 317
100
200
300
Penbutolol-M (desisobutyl-HO-) -H2O 2AC
UHYAC
LM/Q
$C_{18}H_{23}NO_4$
317.16271
RI: 2240
D: UHYAC
Beta-Blocker
N: rat

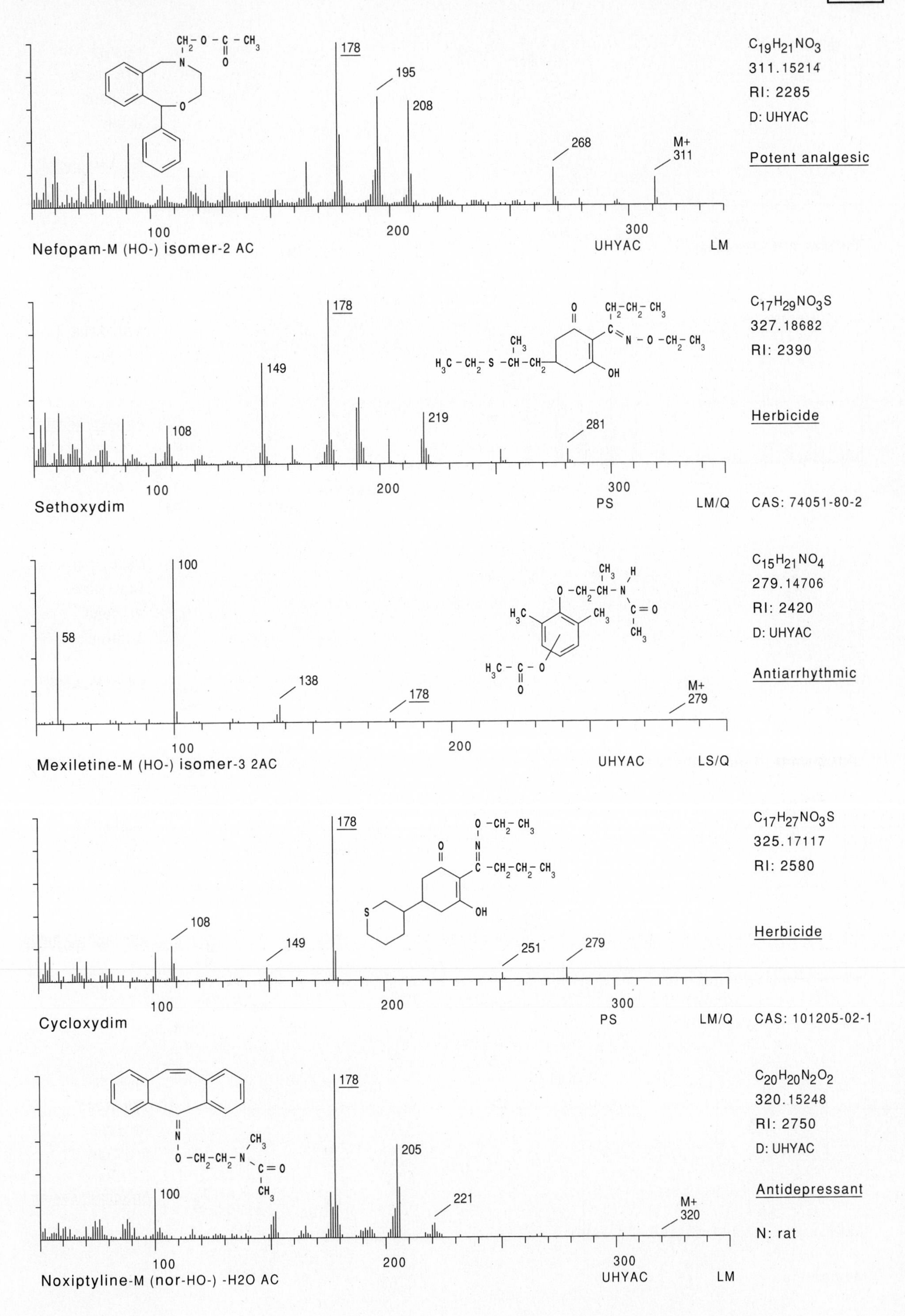

Nefopam-M (HO-) isomer-2 AC
UHYAC
LM
178
195
208
268
M+
311
C19H21NO3
311.15214
RI: 2285
D: UHYAC
Potent analgesic
Sethoxydim
PS
LM/Q
178
149
108
219
281
C17H29NO3S
327.18682
RI: 2390
Herbicide
CAS: 74051-80-2
Mexiletine-M (HO-) isomer-3 2AC
UHYAC
LS/Q
100
58
138
178
M+
279
C15H21NO4
279.14706
RI: 2420
D: UHYAC
Antiarrhythmic
Cycloxydim
PS
LM/Q
178
108
149
251
279
C17H27NO3S
325.17117
RI: 2580
Herbicide
CAS: 101205-02-1
Noxiptyline-M (nor-HO-) -H2O AC
UHYAC
LM
178
205
100
221
M+
320
C20H20N2O2
320.15248
RI: 2750
D: UHYAC
Antidepressant
N: rat

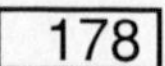
178

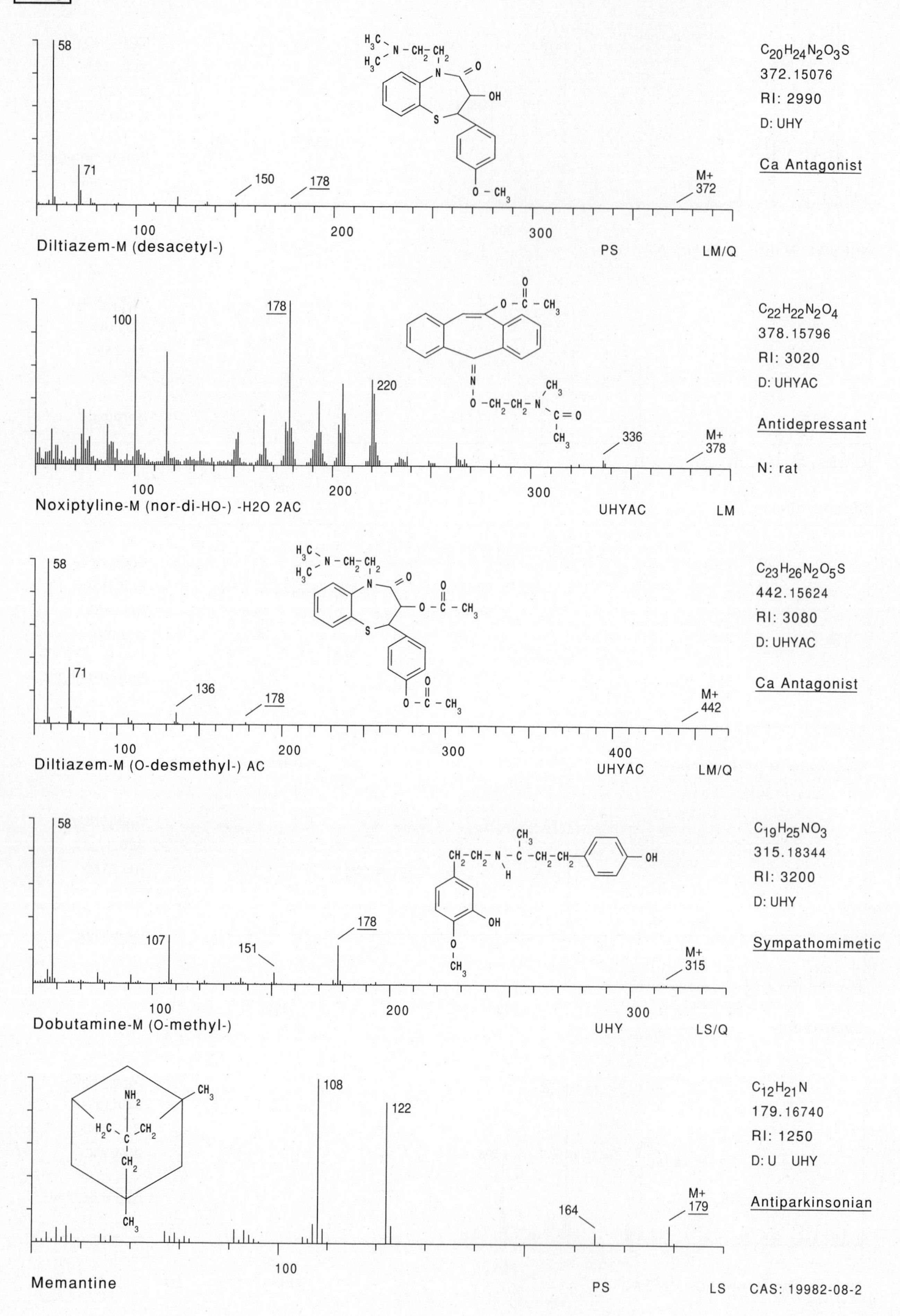

58
71
150
178
M+
372
100
200
300
Diltiazem-M (desacetyl-)
PS
LM/Q
C20H24N2O3S
372.15076
RI: 2990
D: UHY
Ca Antagonist
100
178
220
336
M+
378
Noxiptyline-M (nor-di-HO-) -H2O 2AC
UHYAC
LM
C22H22N2O4
378.15796
RI: 3020
D: UHYAC
Antidepressant
N: rat
58
71
136
178
M+
442
400
Diltiazem-M (O-desmethyl-) AC
UHYAC
LM/Q
C23H26N2O5S
442.15624
RI: 3080
D: UHYAC
Ca Antagonist
58
107
151
178
M+
315
Dobutamine-M (O-methyl-)
UHY
LS/Q
C19H25NO3
315.18344
RI: 3200
D: UHY
Sympathomimetic
108
122
164
M+
179
Memantine
PS
LS
C12H21N
179.16740
RI: 1250
D: U UHY
Antiparkinsonian
CAS: 19982-08-2

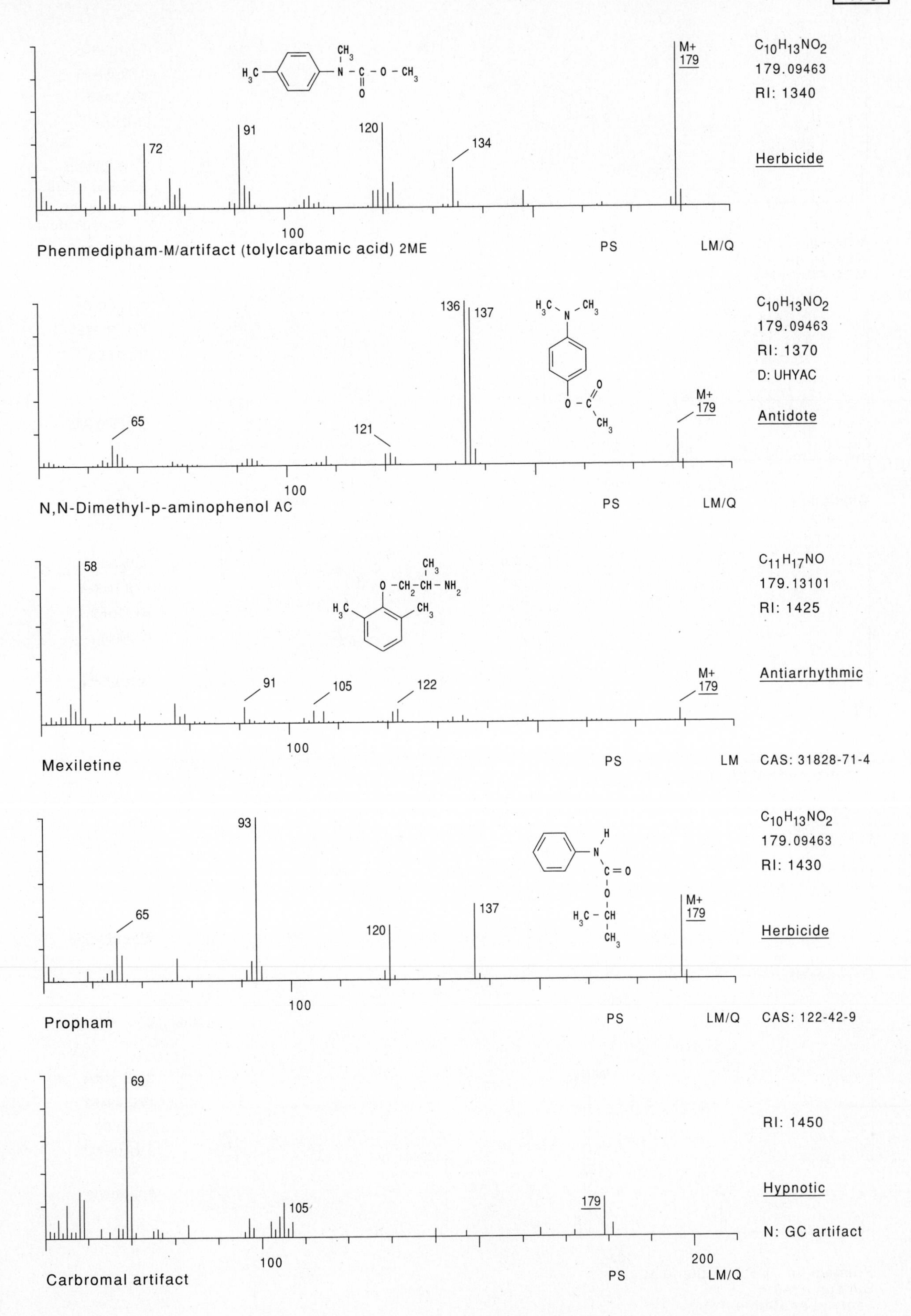

Phenmedipham-M/artifact (tolylcarbamic acid) 2ME

$C_{10}H_{13}NO_2$
179.09463
RI: 1340
Herbicide

N,N-Dimethyl-p-aminophenol AC

$C_{10}H_{13}NO_2$
179.09463
RI: 1370
D: UHYAC
Antidote

Mexiletine

$C_{11}H_{17}NO$
179.13101
RI: 1425
Antiarrhythmic
CAS: 31828-71-4

Propham

$C_{10}H_{13}NO_2$
179.09463
RI: 1430
Herbicide
CAS: 122-42-9

Carbromal artifact

RI: 1450
Hypnotic
N: GC artifact

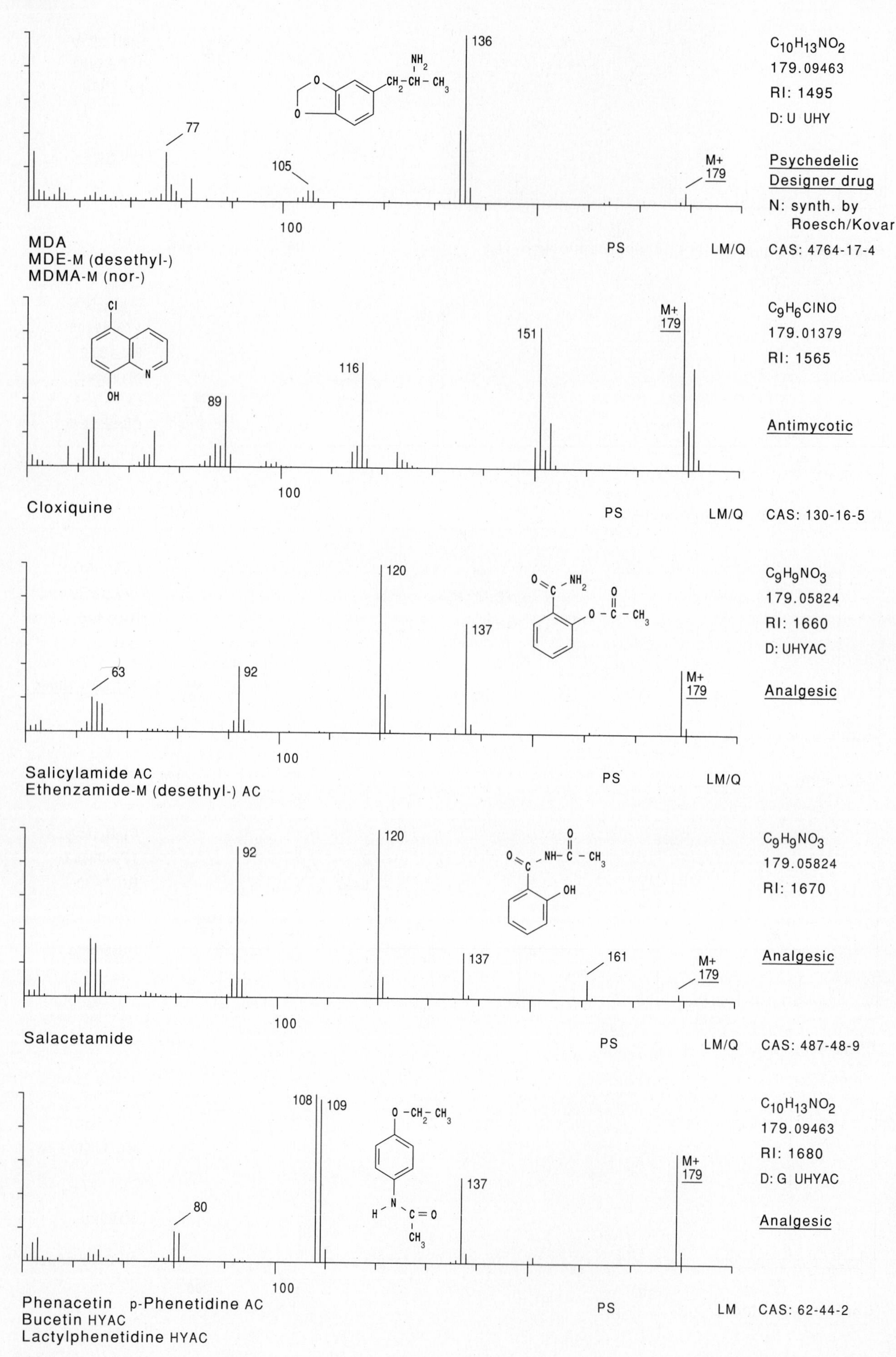

136
77
105
M+
179
100
MDA
MDE-M (desethyl-)
MDMA-M (nor-)
PS
LM/Q
C10H13NO2
179.09463
RI: 1495
D: U UHY
Psychedelic
Designer drug
N: synth. by
Roesch/Kovar
CAS: 4764-17-4
Cl
OH
N
M+
179
151
116
89
100
Cloxiquine
PS
LM/Q
C9H6ClNO
179.01379
RI: 1565
Antimycotic
CAS: 130-16-5
120
137
63
92
M+
179
100
Salicylamide AC
Ethenzamide-M (desethyl-) AC
PS
LM/Q
C9H9NO3
179.05824
RI: 1660
D: UHYAC
Analgesic
120
92
137
161
M+
179
OH
100
Salacetamide
PS
LM/Q
C9H9NO3
179.05824
RI: 1670
Analgesic
CAS: 487-48-9
108
109
80
137
M+
179
100
Phenacetin p-Phenetidine AC
Bucetin HYAC
Lactylphenetidine HYAC
PS
LM
C10H13NO2
179.09463
RI: 1680
D: G UHYAC
Analgesic
CAS: 62-44-2

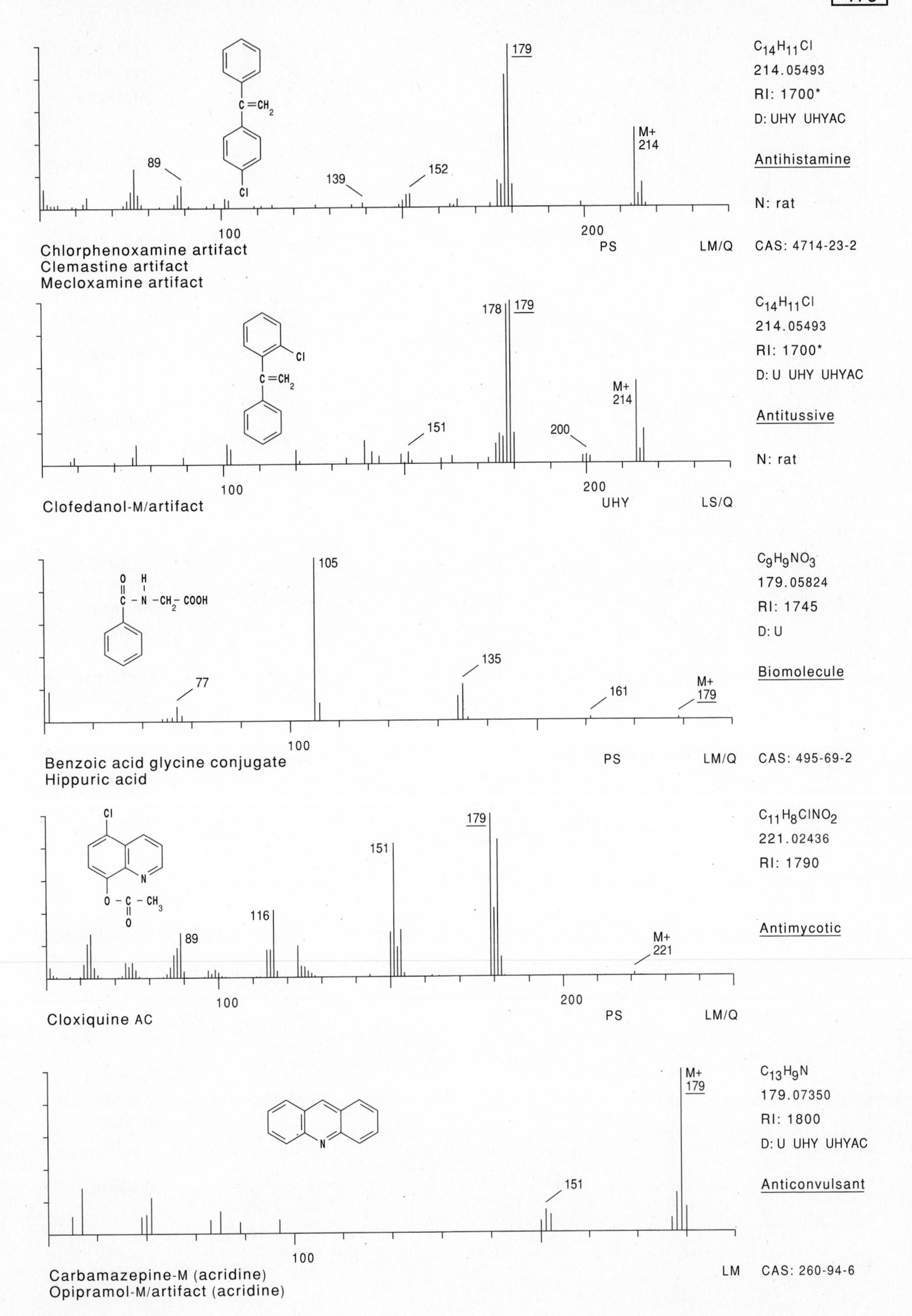
179
C14H11Cl
214.05493
RI: 1700*
D: UHY UHYAC
Antihistamine
N: rat
CAS: 4714-23-2
Chlorphenoxamine artifact
Clemastine artifact
Mecloxamine artifact
PS
LM/Q
C14H11Cl
214.05493
RI: 1700*
D: U UHY UHYAC
Antitussive
N: rat
Clofedanol-M/artifact
UHY
LS/Q
C9H9NO3
179.05824
RI: 1745
D: U
Biomolecule
CAS: 495-69-2
Benzoic acid glycine conjugate
Hippuric acid
PS
LM/Q
C11H8ClNO2
221.02436
RI: 1790
Antimycotic
Cloxiquine AC
PS
LM/Q
C13H9N
179.07350
RI: 1800
D: U UHY UHYAC
Anticonvulsant
CAS: 260-94-6
Carbamazepine-M (acridine)
Opipramol-M/artifact (acridine)
LM

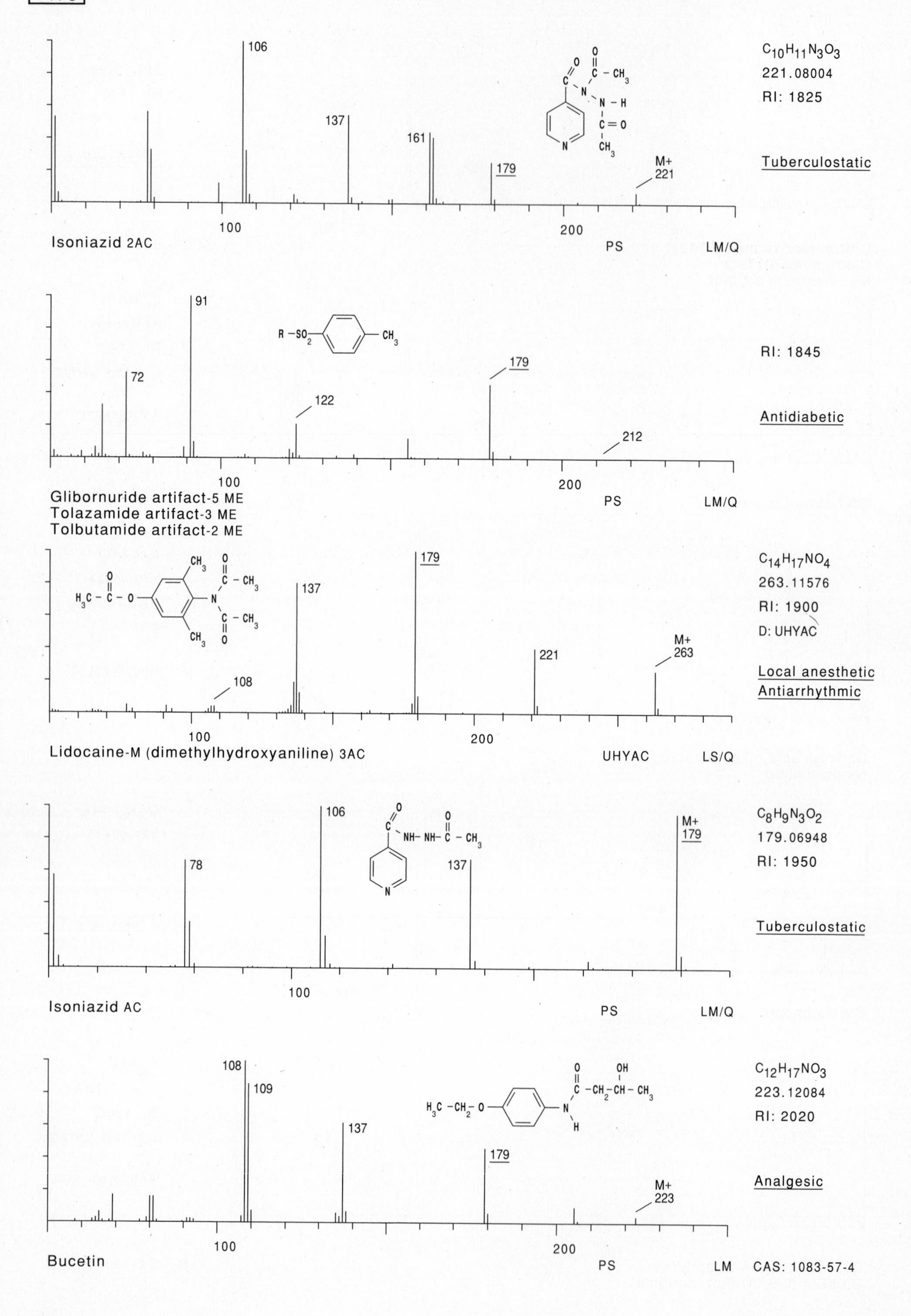
106
137
161
179
M+
221
100
200
Isoniazid 2AC
PS
LM/Q
$C_{10}H_{11}N_3O_3$
221.08004
RI: 1825
Tuberculostatic
91
72
122
179
212
100
200
RI: 1845
Antidiabetic
Glibornuride artifact-5 ME
Tolazamide artifact-3 ME
Tolbutamide artifact-2 ME
PS
LM/Q
179
137
108
221
M+
263
100
200
$C_{14}H_{17}NO_4$
263.11576
RI: 1900
D: UHYAC
Local anesthetic
Antiarrhythmic
Lidocaine-M (dimethylhydroxyaniline) 3AC
UHYAC
LS/Q
106
78
137
M+
179
100
$C_8H_9N_3O_2$
179.06948
RI: 1950
Tuberculostatic
Isoniazid AC
PS
LM/Q
108
109
137
179
M+
223
100
200
$C_{12}H_{17}NO_3$
223.12084
RI: 2020
Analgesic
Bucetin
PS
LM
CAS: 1083-57-4

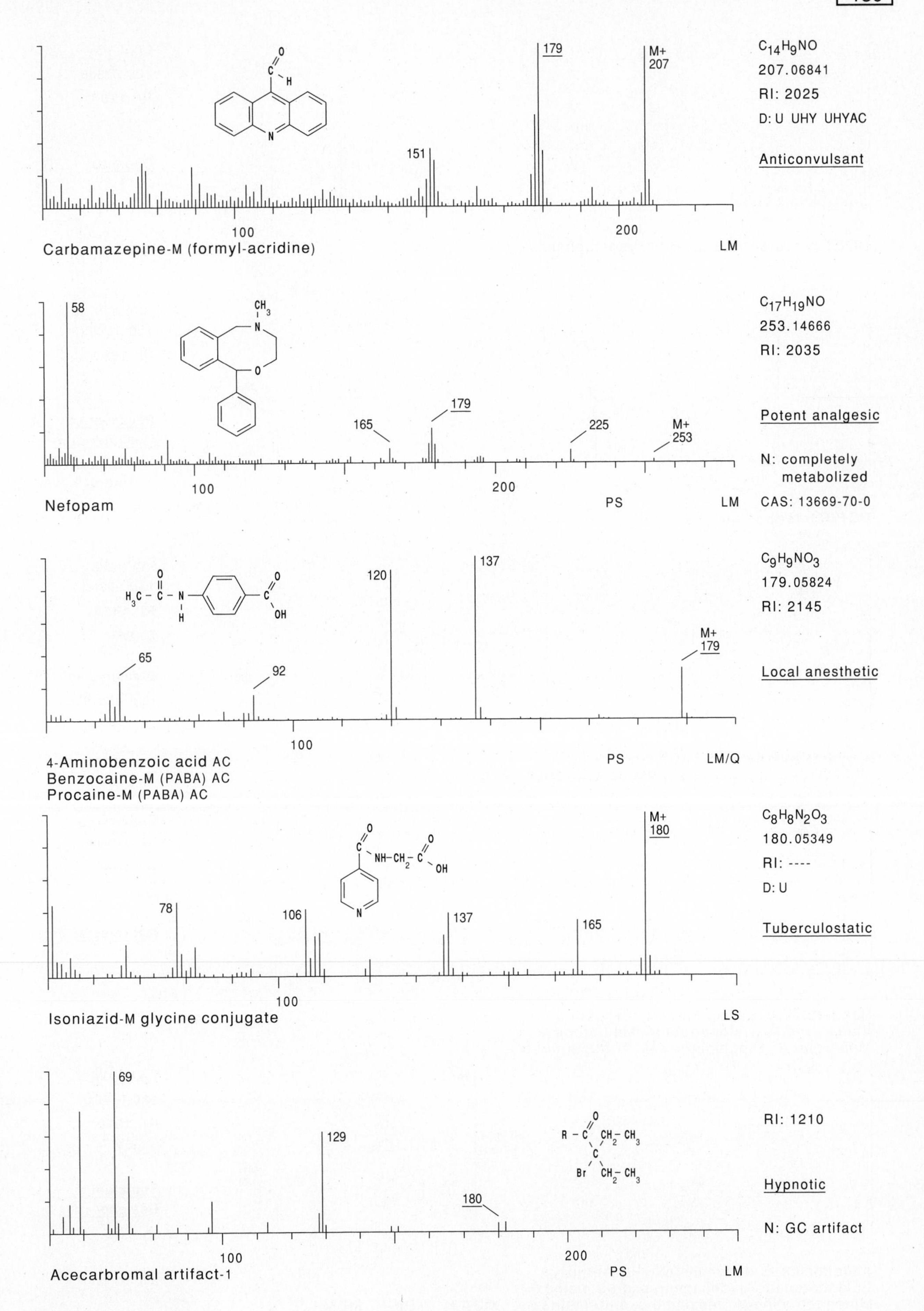

$C_{14}H_9NO$
207.06841
RI: 2025
D: U UHY UHYAC

Anticonvulsant

$C_{17}H_{19}NO$
253.14666
RI: 2035

Potent analgesic

N: completely metabolized
CAS: 13669-70-0

$C_9H_9NO_3$
179.05824
RI: 2145

Local anesthetic

$C_8H_8N_2O_3$
180.05349
RI: ----
D: U

Tuberculostatic

RI: 1210

Hypnotic

N: GC artifact

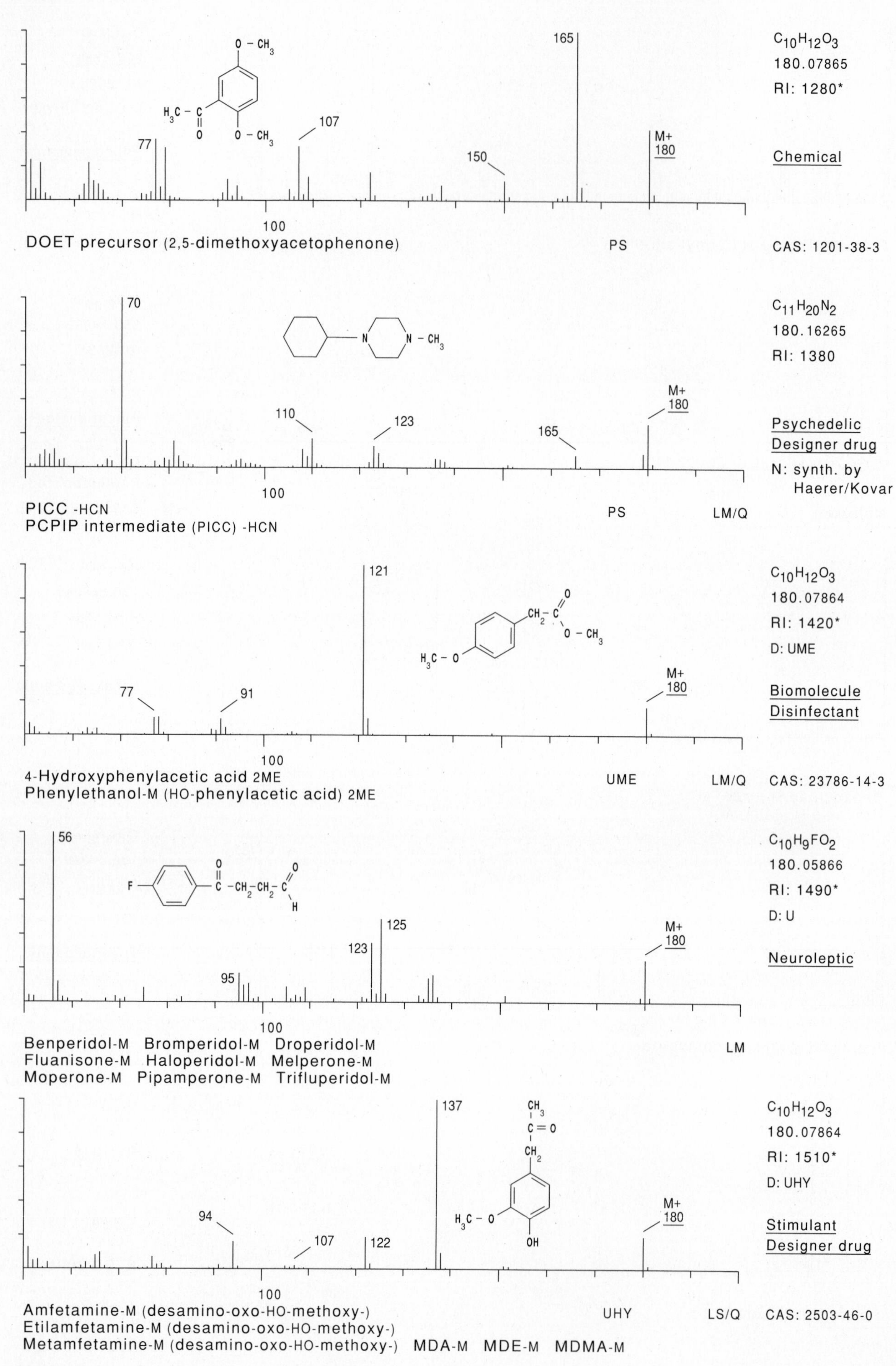
C10H12O3
180.07865
RI: 1280*
Chemical
165
107
77
150
M+ 180
100
DOET precursor (2,5-dimethoxyacetophenone)
PS
CAS: 1201-38-3
C11H20N2
180.16265
RI: 1380
Psychedelic
Designer drug
N: synth. by Haerer/Kovar
70
110
123
165
M+ 180
100
PICC -HCN
PCPIP intermediate (PICC) -HCN
PS
LM/Q
C10H12O3
180.07864
RI: 1420*
D: UME
Biomolecule
Disinfectant
121
77
91
M+ 180
100
4-Hydroxyphenylacetic acid 2ME
Phenylethanol-M (HO-phenylacetic acid) 2ME
UME
LM/Q
CAS: 23786-14-3
C10H9FO2
180.05866
RI: 1490*
D: U
Neuroleptic
56
125
123
95
M+ 180
100
Benperidol-M Bromperidol-M Droperidol-M
Fluanisone-M Haloperidol-M Melperone-M
Moperone-M Pipamperone-M Trifluperidol-M
LM
C10H12O3
180.07864
RI: 1510*
D: UHY
Stimulant
Designer drug
137
94
107
122
M+ 180
100
Amfetamine-M (desamino-oxo-HO-methoxy-)
Etilamfetamine-M (desamino-oxo-HO-methoxy-)
Metamfetamine-M (desamino-oxo-HO-methoxy-) MDA-M MDE-M MDMA-M
UHY
LS/Q
CAS: 2503-46-0

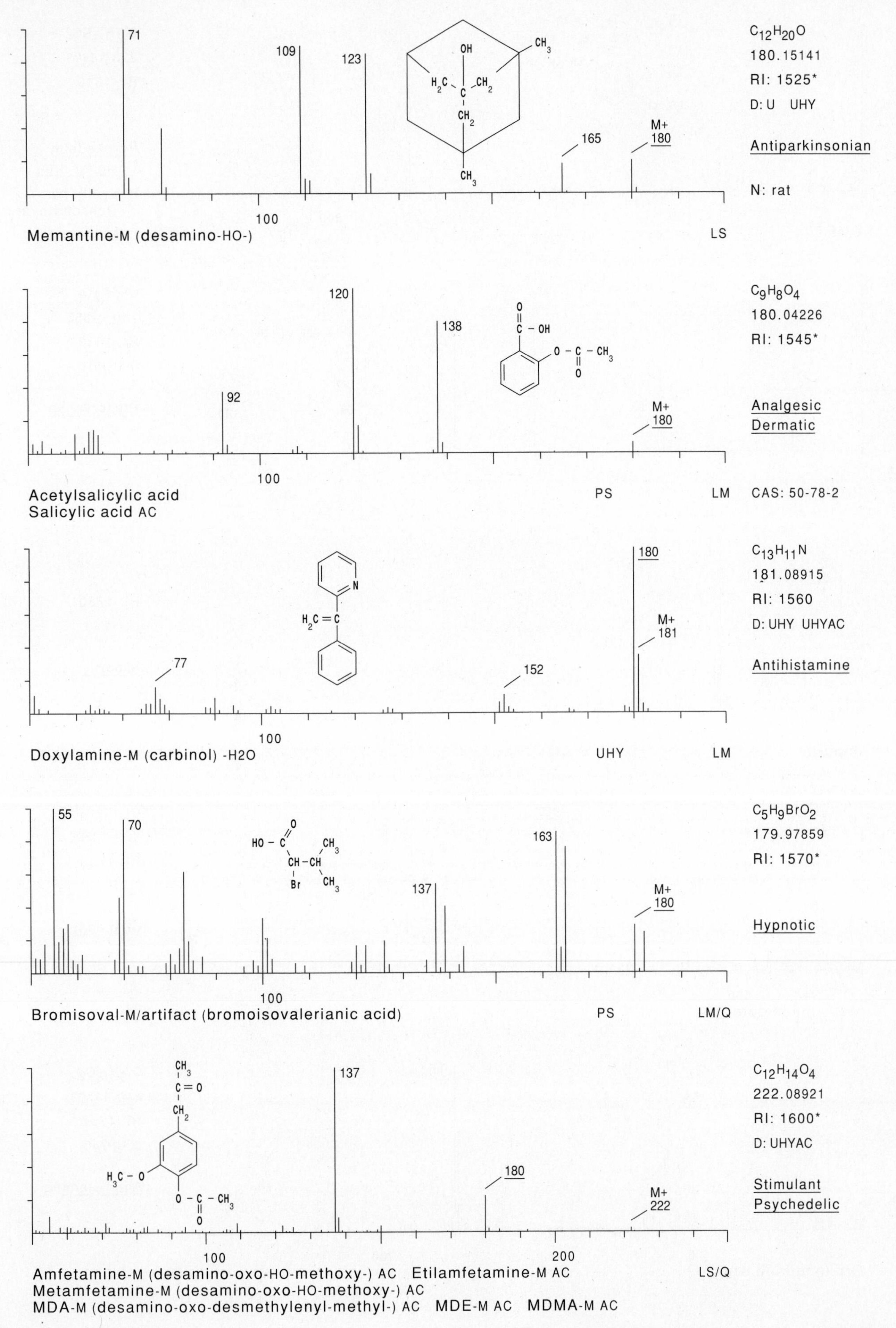
71
109
123
OH
CH3
H2C C CH2
CH2
CH3
165
M+
180
100
Memantine-M (desamino-HO-)
LS
C12H20O
180.15141
RI: 1525*
D: U UHY
Antiparkinsonian
N: rat
120
138
92
O
C - OH
O - C - CH3
M+
180
100
Acetylsalicylic acid
Salicylic acid AC
PS
LM
C9H8O4
180.04226
RI: 1545*
Analgesic
Dermatic
CAS: 50-78-2
180
N
H2C = C
M+
181
77
152
100
Doxylamine-M (carbinol) -H2O
UHY
LM
C13H11N
181.08915
RI: 1560
D: UHY UHYAC
Antihistamine
55
70
HO - C
CH3
CH - CH
Br
163
137
M+
180
100
Bromisoval-M/artifact (bromoisovalerianic acid)
PS
LM/Q
C5H9BrO2
179.97859
RI: 1570*
Hypnotic
CH3
C = O
CH2
H3C - O
O - C - CH3
137
180
M+
222
100
200
Amfetamine-M (desamino-oxo-HO-methoxy-) AC Etilamfetamine-M AC
Metamfetamine-M (desamino-oxo-HO-methoxy-) AC
MDA-M (desamino-oxo-desmethylenyl-methyl-) AC MDE-M AC MDMA-M AC
LS/Q
C12H14O4
222.08921
RI: 1600*
D: UHYAC
Stimulant
Psychedelic

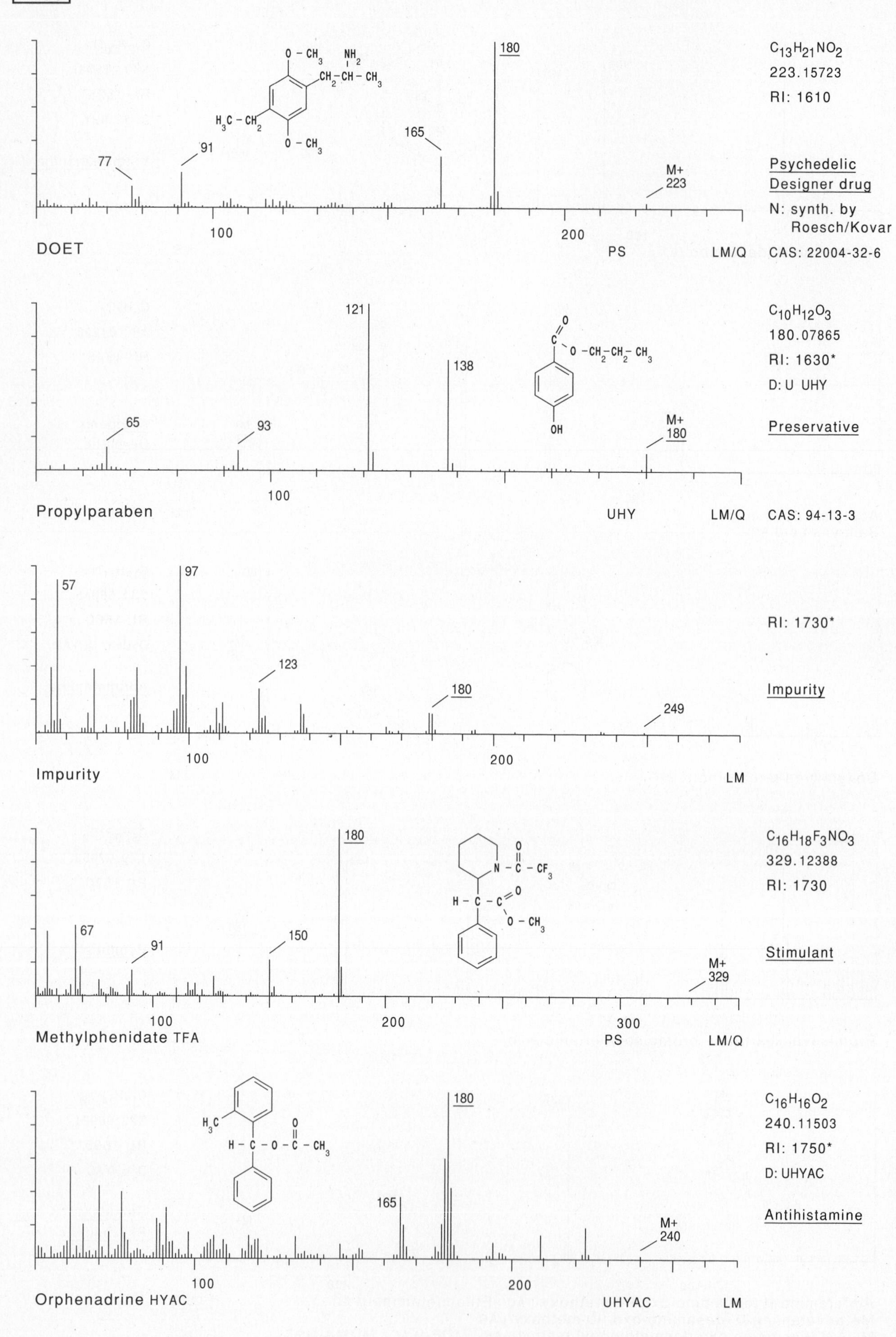

180
165
91
77
M+
223
100
200
DOET
PS
LM/Q
C13H21NO2
223.15723
RI: 1610
Psychedelic
Designer drug
N: synth. by
Roesch/Kovar
CAS: 22004-32-6
121
138
65
93
M+
180
100
Propylparaben
UHY
LM/Q
C10H12O3
180.07865
RI: 1630*
D: U UHY
Preservative
CAS: 94-13-3
97
57
123
180
249
100
200
Impurity
LM
RI: 1730*
Impurity
180
67
91
150
M+
329
100
200
300
Methylphenidate TFA
PS
LM/Q
C16H18F3NO3
329.12388
RI: 1730
Stimulant
180
165
M+
240
100
200
Orphenadrine HYAC
UHYAC
LM
C16H16O2
240.11503
RI: 1750*
D: UHYAC
Antihistamine

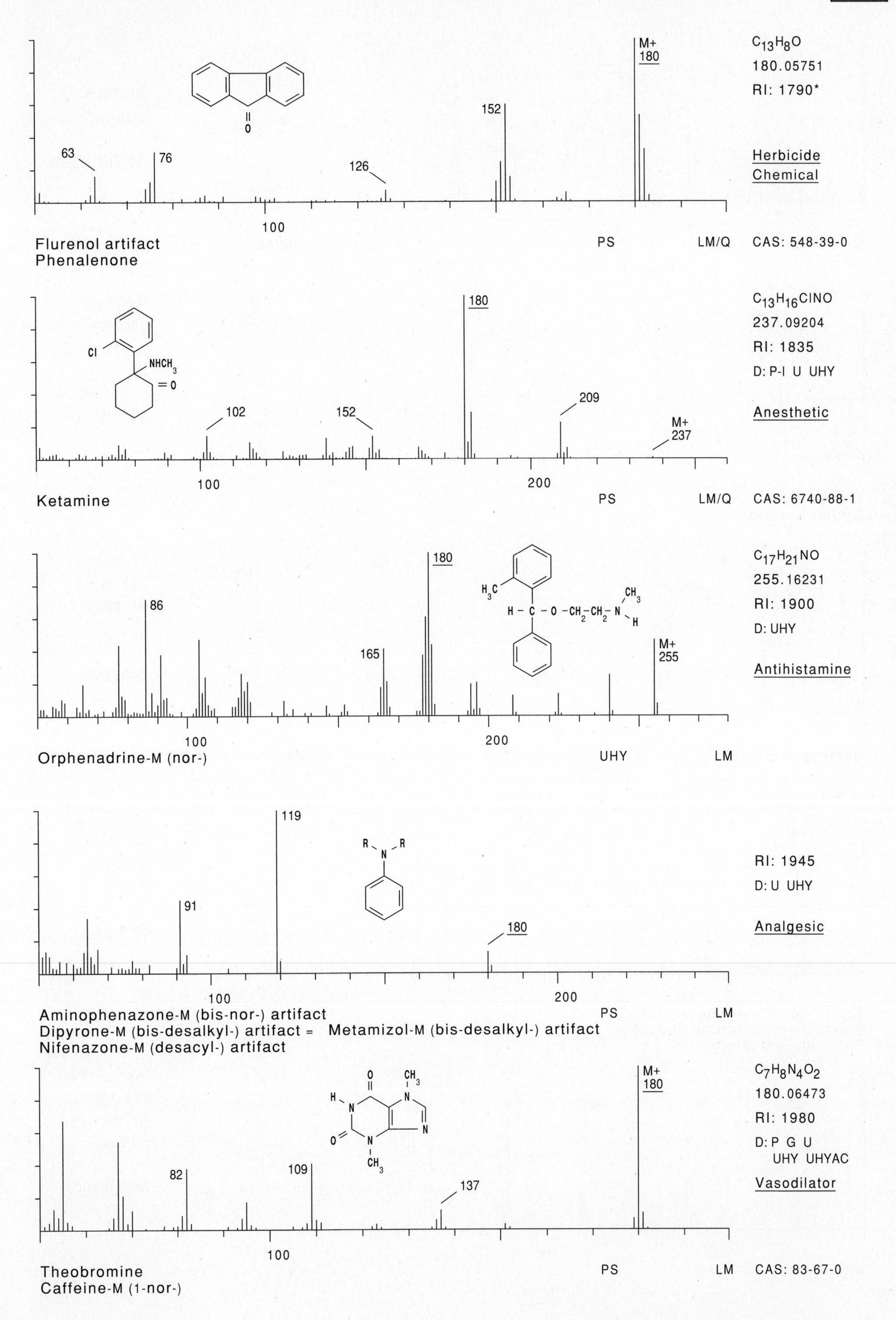

$C_{13}H_8O$
180.05751
RI: 1790*

Herbicide
Chemical

Flurenol artifact
Phenalenone

CAS: 548-39-0

$C_{13}H_{16}ClNO$
237.09204
RI: 1835
D: P-I U UHY

Anesthetic

Ketamine

CAS: 6740-88-1

$C_{17}H_{21}NO$
255.16231
RI: 1900
D: UHY

Antihistamine

Orphenadrine-M (nor-)

RI: 1945
D: U UHY

Analgesic

Aminophenazone-M (bis-nor-) artifact
Dipyrone-M (bis-desalkyl-) artifact = Metamizol-M (bis-desalkyl-) artifact
Nifenazone-M (desacyl-) artifact

$C_7H_8N_4O_2$
180.06473
RI: 1980
D: P G U
UHY UHYAC

Vasodilator

Theobromine
Caffeine-M (1-nor-)

CAS: 83-67-0

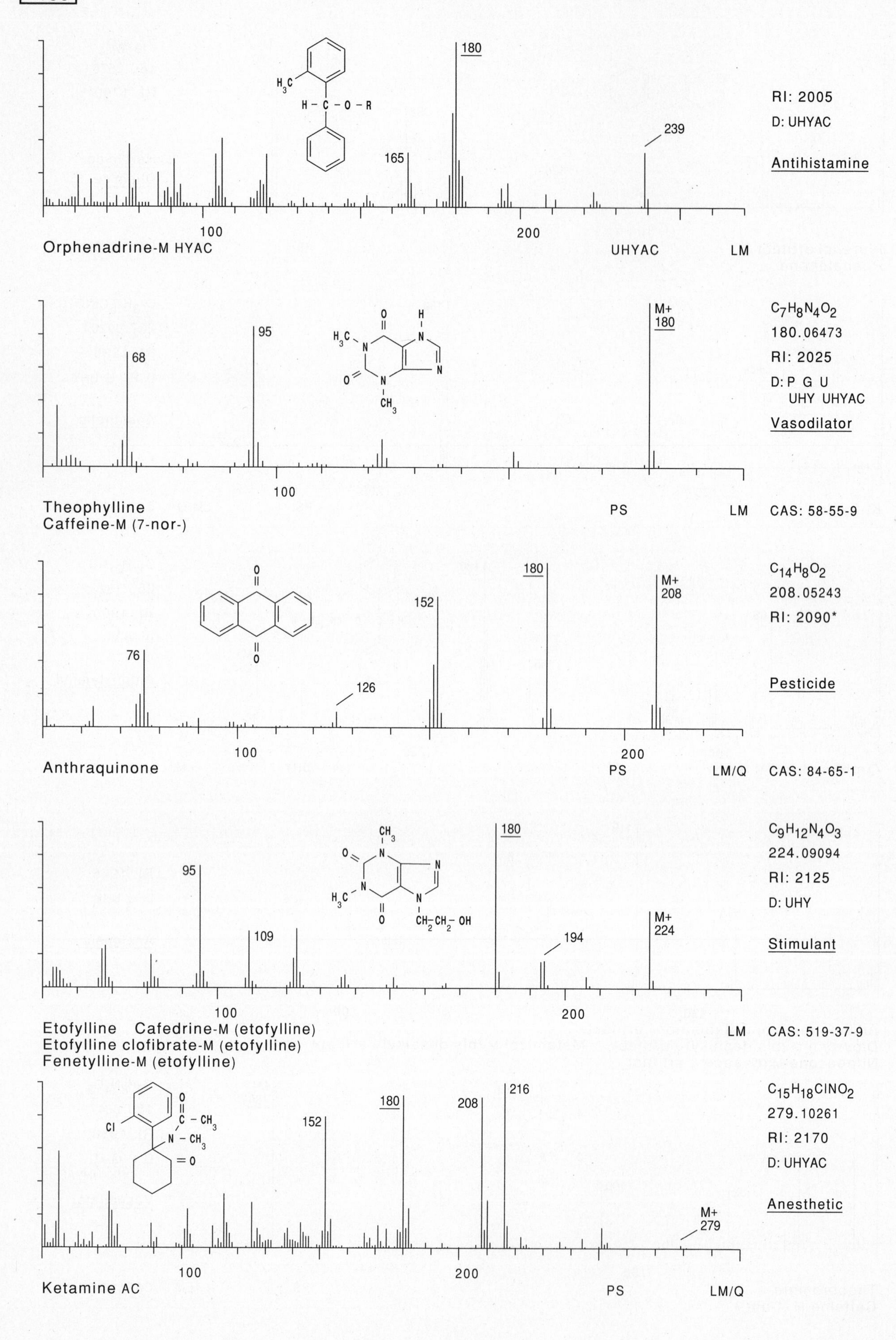
180
180
H3C
H – C – O – R
165
239
100
200
RI: 2005
D: UHYAC
Antihistamine
Orphenadrine-M HYAC
UHYAC
LM
O
H
H3C
N
N
N
O
N
CH3
68
95
M+
180
100
C7H8N4O2
180.06473
RI: 2025
D: P G U
UHY UHYAC
Vasodilator
Theophylline
Caffeine-M (7-nor-)
PS
LM
CAS: 58-55-9
O
O
76
126
152
180
M+
208
100
200
C14H8O2
208.05243
RI: 2090*
Pesticide
Anthraquinone
PS
LM/Q
CAS: 84-65-1
CH3
O
N
N
N
N
H3C
O
CH2–CH2–OH
95
109
180
194
M+
224
100
200
C9H12N4O3
224.09094
RI: 2125
D: UHY
Stimulant
Etofylline Cafedrine-M (etofylline)
Etofylline clofibrate-M (etofylline)
Fenetylline-M (etofylline)
LM
CAS: 519-37-9
Cl
O
C – CH3
N – CH3
=O
152
180
208
216
M+
279
100
200
C15H18ClNO2
279.10261
RI: 2170
D: UHYAC
Anesthetic
Ketamine AC
PS
LM/Q

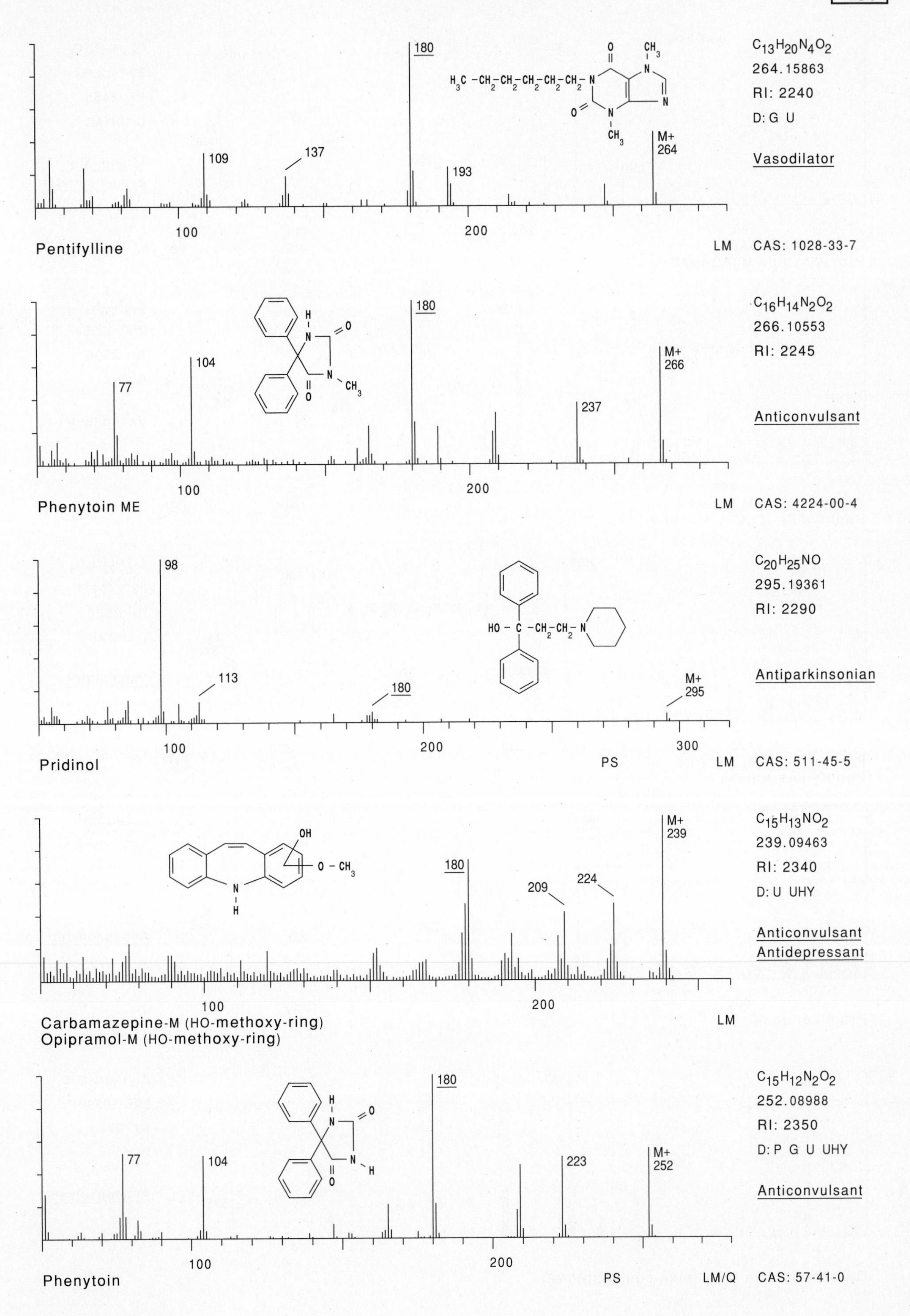

Pentifylline
$C_{13}H_{20}N_4O_2$
264.15863
RI: 2240
D: G U
Vasodilator
LM CAS: 1028-33-7

Phenytoin ME
$C_{16}H_{14}N_2O_2$
266.10553
RI: 2245
Anticonvulsant
LM CAS: 4224-00-4

Pridinol
$C_{20}H_{25}NO$
295.19361
RI: 2290
Antiparkinsonian
PS LM CAS: 511-45-5

Carbamazepine-M (HO-methoxy-ring)
Opipramol-M (HO-methoxy-ring)
$C_{15}H_{13}NO_2$
239.09463
RI: 2340
D: U UHY
Anticonvulsant
Antidepressant
LM

Phenytoin
$C_{15}H_{12}N_2O_2$
252.08988
RI: 2350
D: P G U UHY
Anticonvulsant
PS LM/Q CAS: 57-41-0

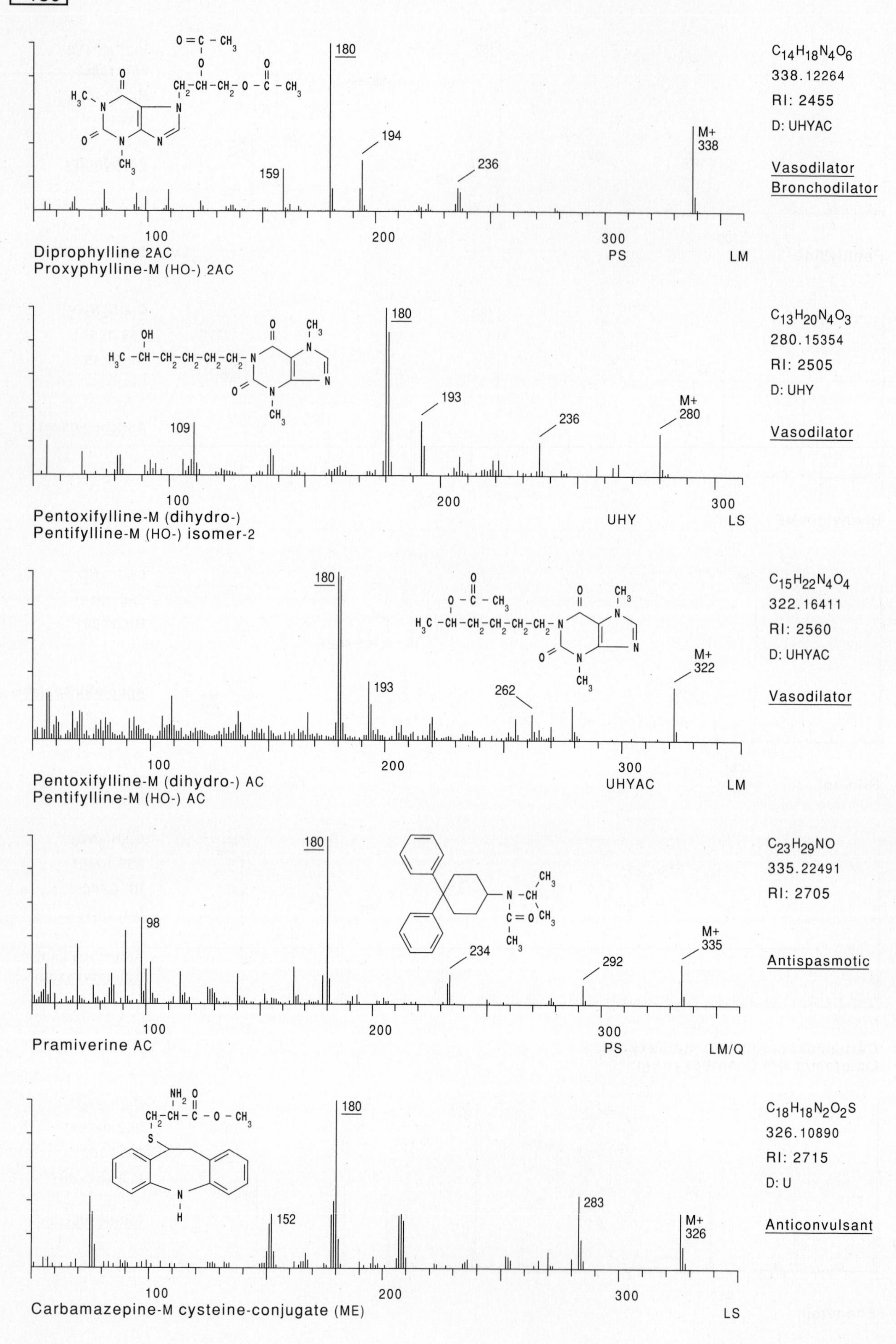
180
194
159
236
M+
338
100
200
300
PS
LM
$C_{14}H_{18}N_4O_6$
338.12264
RI: 2455
D: UHYAC
Vasodilator
Bronchodilator
Diprophylline 2AC
Proxyphylline-M (HO-) 2AC
180
193
109
236
M+
280
100
200
300
UHY
LS
$C_{13}H_{20}N_4O_3$
280.15354
RI: 2505
D: UHY
Vasodilator
Pentoxifylline-M (dihydro-)
Pentifylline-M (HO-) isomer-2
180
193
262
M+
322
100
200
300
UHYAC
LM
$C_{15}H_{22}N_4O_4$
322.16411
RI: 2560
D: UHYAC
Vasodilator
Pentoxifylline-M (dihydro-) AC
Pentifylline-M (HO-) AC
180
98
234
292
M+
335
100
200
300
PS
LM/Q
$C_{23}H_{29}NO$
335.22491
RI: 2705
Antispasmotic
Pramiverine AC
180
152
283
M+
326
100
200
300
LS
$C_{18}H_{18}N_2O_2S$
326.10890
RI: 2715
D: U
Anticonvulsant
Carbamazepine-M cysteine-conjugate (ME)

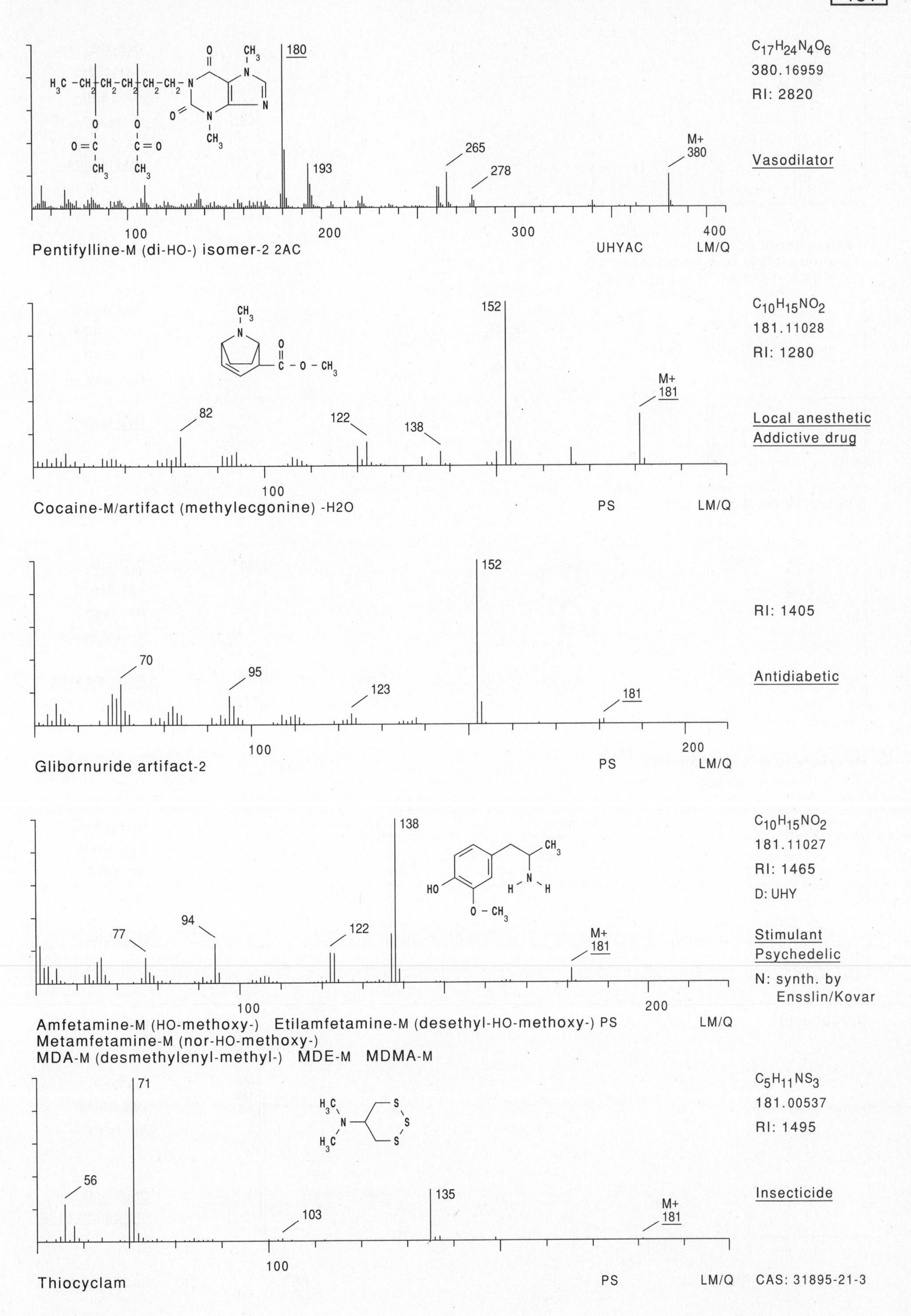
$C_{17}H_{24}N_4O_6$
380.16959
RI: 2820
Vasodilator
Pentifylline-M (di-HO-) isomer-2 2AC
UHYAC
LM/Q
$C_{10}H_{15}NO_2$
181.11028
RI: 1280
Local anesthetic
Addictive drug
Cocaine-M/artifact (methylecgonine) -H2O
PS
LM/Q
RI: 1405
Antidiabetic
Glibornuride artifact-2
PS
LM/Q
$C_{10}H_{15}NO_2$
181.11027
RI: 1465
D: UHY
Stimulant
Psychedelic
N: synth. by Ensslin/Kovar
Amfetamine-M (HO-methoxy-) Etilamfetamine-M (desethyl-HO-methoxy-) PS
LM/Q
Metamfetamine-M (nor-HO-methoxy-)
MDA-M (desmethylenyl-methyl-) MDE-M MDMA-M
$C_5H_{11}NS_3$
181.00537
RI: 1495
Insecticide
Thiocyclam
PS
LM/Q
CAS: 31895-21-3

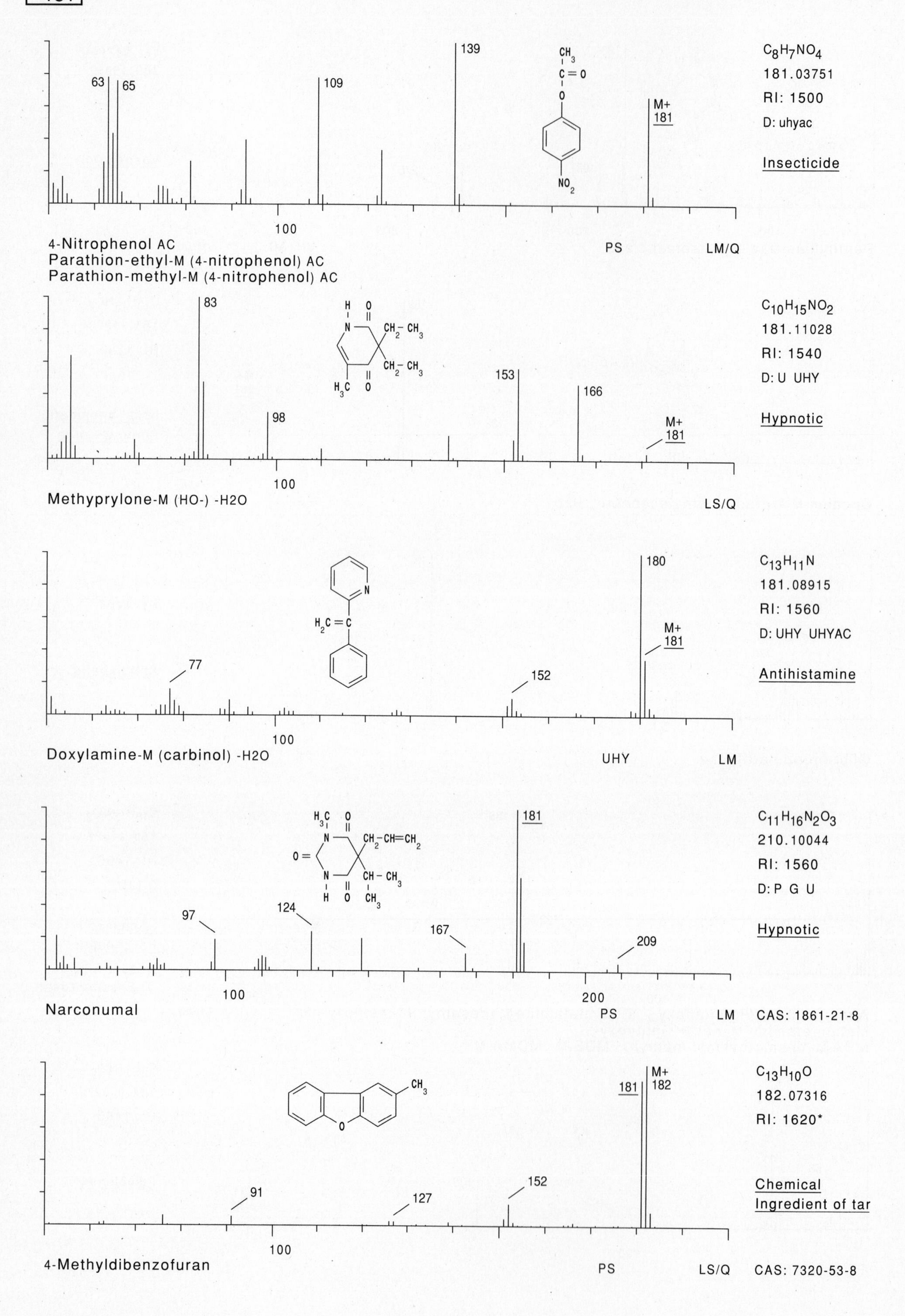

$C_8H_7NO_4$
181.03751
RI: 1500
D: uhyac
Insecticide

4-Nitrophenol AC
Parathion-ethyl-M (4-nitrophenol) AC
Parathion-methyl-M (4-nitrophenol) AC

$C_{10}H_{15}NO_2$
181.11028
RI: 1540
D: U UHY
Hypnotic

Methyprylone-M (HO-) -H2O

$C_{13}H_{11}N$
181.08915
RI: 1560
D: UHY UHYAC
Antihistamine

Doxylamine-M (carbinol) -H2O

$C_{11}H_{16}N_2O_3$
210.10044
RI: 1560
D: P G U
Hypnotic

Narconumal
CAS: 1861-21-8

$C_{13}H_{10}O$
182.07316
RI: 1620*
Chemical
Ingredient of tar

4-Methyldibenzofuran
CAS: 7320-53-8

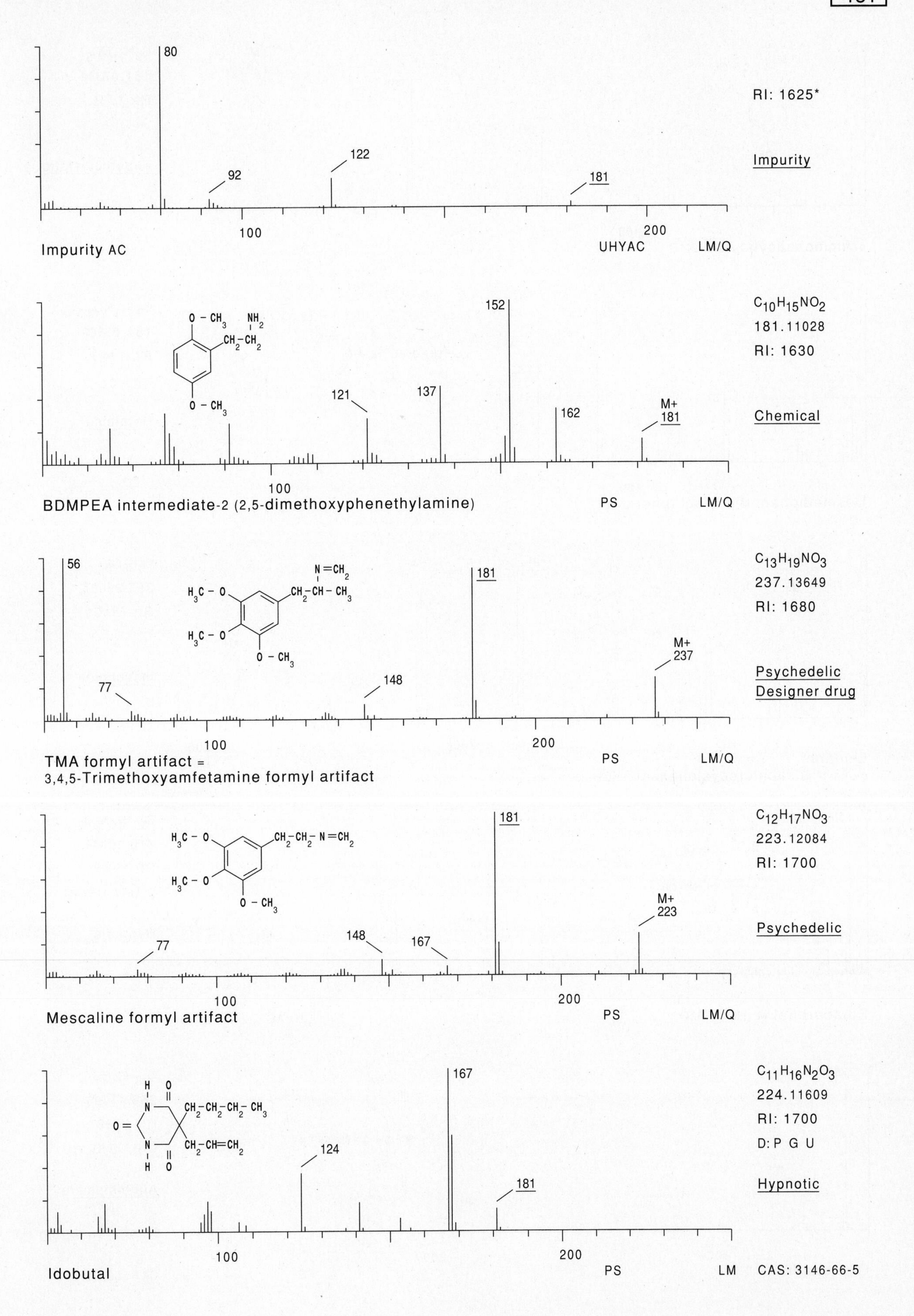
80
122
92
181
100
200
Impurity AC
UHYAC
LM/Q
RI: 1625*
Impurity
152
O - CH3
NH2
CH2-CH2
O - CH3
121
137
162
M+
181
100
BDMPEA intermediate-2 (2,5-dimethoxyphenethylamine)
PS
LM/Q
C10H15NO2
181.11028
RI: 1630
Chemical
56
N=CH2
H3C - O
CH2-CH-CH3
H3C - O
O - CH3
181
M+
237
148
77
100
200
TMA formyl artifact =
3,4,5-Trimethoxyamfetamine formyl artifact
PS
LM/Q
C13H19NO3
237.13649
RI: 1680
Psychedelic
Designer drug
181
H3C - O
CH2-CH2-N=CH2
H3C - O
O - CH3
M+
223
148
167
77
100
200
Mescaline formyl artifact
PS
LM/Q
C12H17NO3
223.12084
RI: 1700
Psychedelic
167
H O
N
CH2-CH2-CH2-CH3
O =
CH2-CH=CH2
N
H O
124
181
100
200
Idobutal
PS
LM
C11H16N2O3
224.11609
RI: 1700
D: P G U
Hypnotic
CAS: 3146-66-5

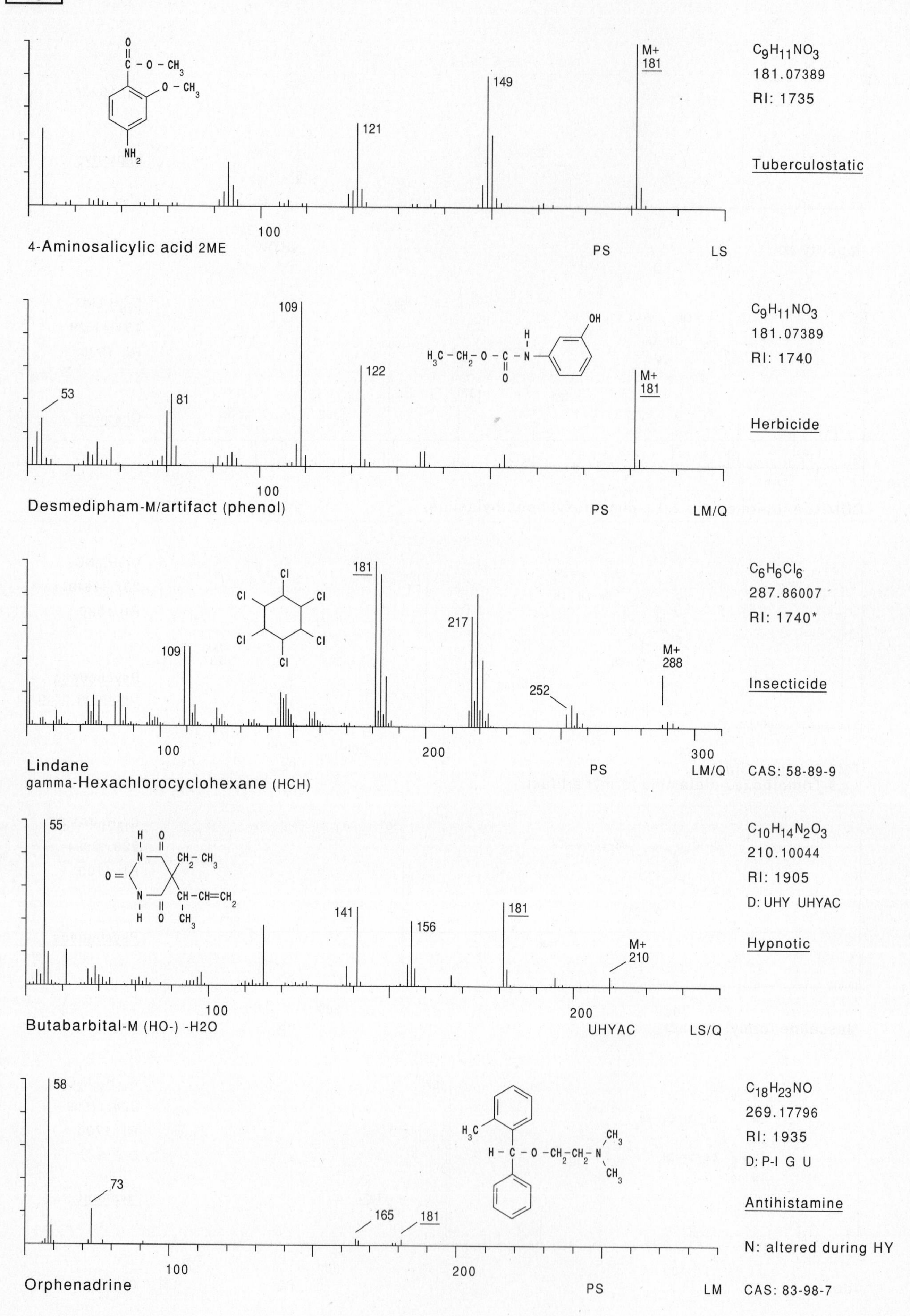
M+
181
149
121
100
4-Aminosalicylic acid 2ME
PS
LS
$C_9H_{11}NO_3$
181.07389
RI: 1735
Tuberculostatic
109
122
53
81
M+
181
100
Desmedipham-M/artifact (phenol)
PS
LM/Q
$C_9H_{11}NO_3$
181.07389
RI: 1740
Herbicide
181
217
109
252
M+
288
100
200
300
Lindane
gamma-Hexachlorocyclohexane (HCH)
PS
LM/Q
$C_6H_6Cl_6$
287.86007
RI: 1740*
Insecticide
CAS: 58-89-9
55
141
156
181
M+
210
100
200
Butabarbital-M (HO-) -H2O
UHYAC
LS/Q
$C_{10}H_{14}N_2O_3$
210.10044
RI: 1905
D: UHY UHYAC
Hypnotic
58
73
165
181
100
200
Orphenadrine
PS
LM
$C_{18}H_{23}NO$
269.17796
RI: 1935
D: P-I G U
Antihistamine
N: altered during HY
CAS: 83-98-7

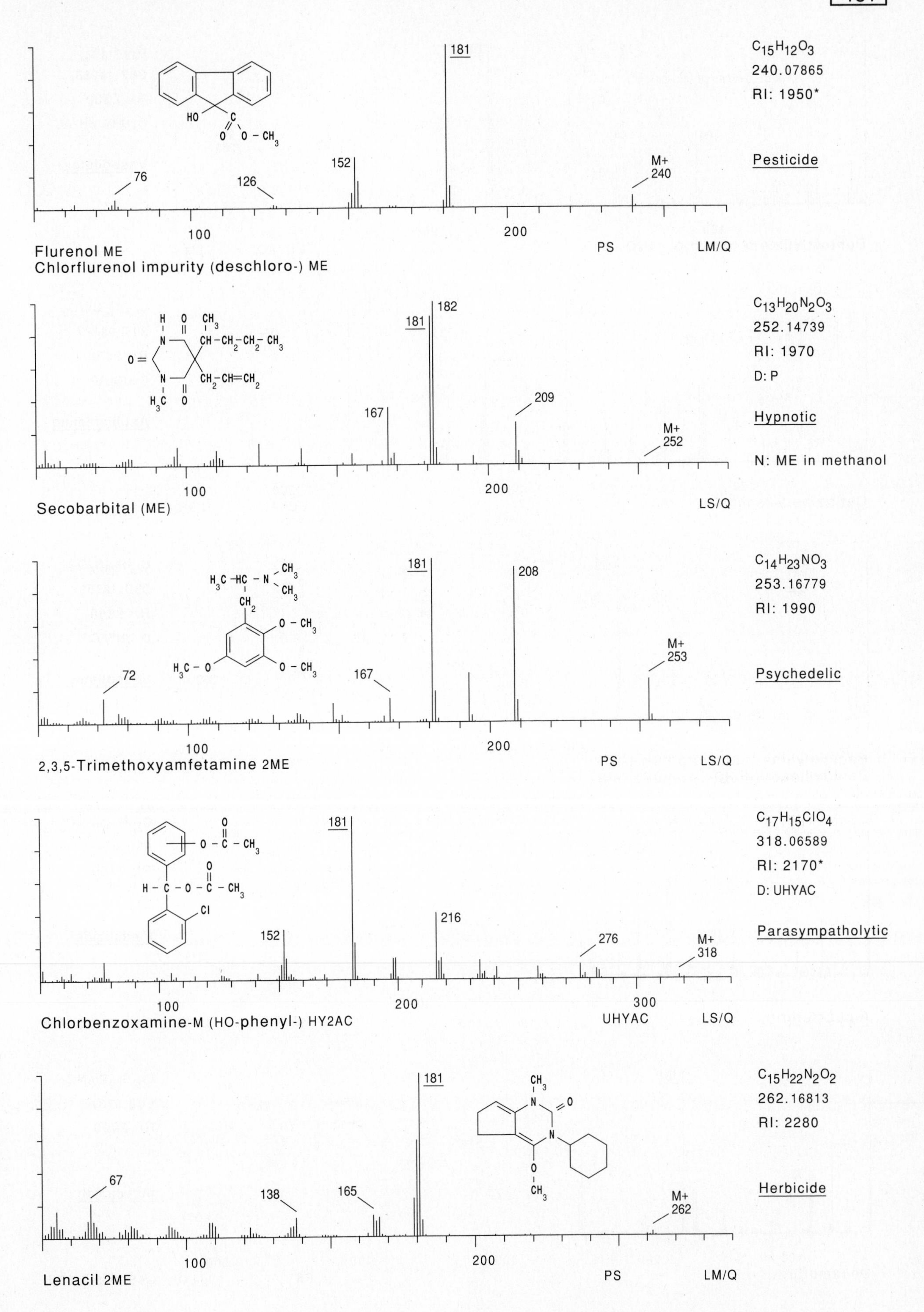

181
152
126
76
M+
240
100
200
PS
LM/Q
C15H12O3
240.07865
RI: 1950*
Pesticide
Flurenol ME
Chlorflurenol impurity (deschloro-) ME
182
181
167
209
M+
252
100
200
LS/Q
C13H20N2O3
252.14739
RI: 1970
D: P
Hypnotic
N: ME in methanol
Secobarbital (ME)
181
208
72
167
M+
253
100
200
PS
LS/Q
C14H23NO3
253.16779
RI: 1990
Psychedelic
2,3,5-Trimethoxyamfetamine 2ME
181
152
216
276
M+
318
100
200
300
UHYAC
LS/Q
C17H15ClO4
318.06589
RI: 2170*
D: UHYAC
Parasympatholytic
Chlorbenzoxamine-M (HO-phenyl-) HY2AC
181
67
138
165
M+
262
100
200
PS
LM/Q
C15H22N2O2
262.16813
RI: 2280
Herbicide
Lenacil 2ME

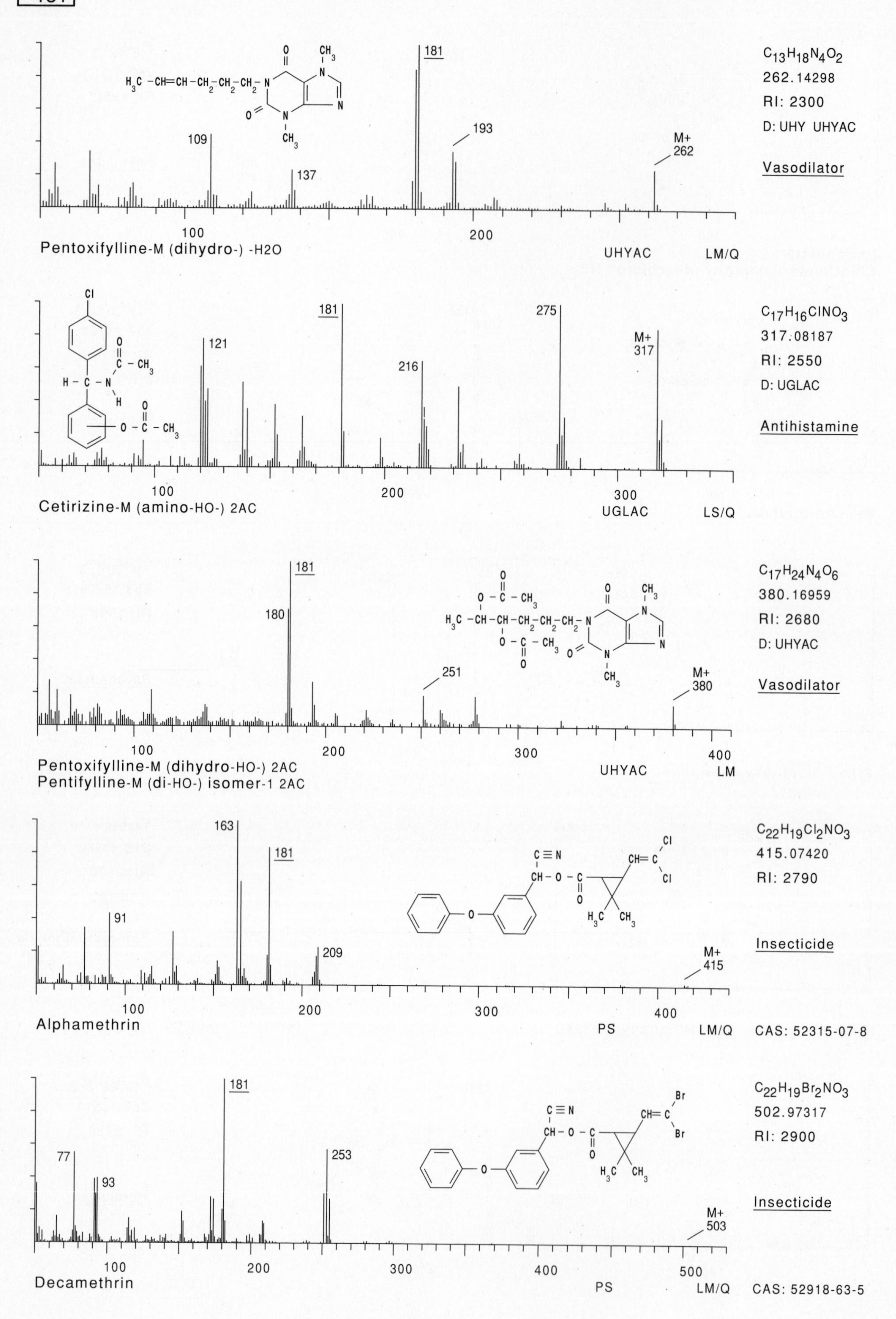
181
109
137
193
M+
262
100
200
Pentoxifylline-M (dihydro-) -H2O
UHYAC
LM/Q
$C_{13}H_{18}N_4O_2$
262.14298
RI: 2300
D: UHY UHYAC
Vasodilator
121
181
216
275
M+
317
100
200
300
Cetirizine-M (amino-HO-) 2AC
UGLAC
LS/Q
$C_{17}H_{16}ClNO_3$
317.08187
RI: 2550
D: UGLAC
Antihistamine
181
180
251
M+
380
100
200
300
400
Pentoxifylline-M (dihydro-HO-) 2AC
Pentifylline-M (di-HO-) isomer-1 2AC
UHYAC
LM
$C_{17}H_{24}N_4O_6$
380.16959
RI: 2680
D: UHYAC
Vasodilator
163
181
91
209
M+
415
100
200
300
400
Alphamethrin
PS
LM/Q
$C_{22}H_{19}Cl_2NO_3$
415.07420
RI: 2790
Insecticide
CAS: 52315-07-8
181
77
93
253
M+
503
100
200
300
400
500
Decamethrin
PS
LM/Q
$C_{22}H_{19}Br_2NO_3$
502.97317
RI: 2900
Insecticide
CAS: 52918-63-5

182

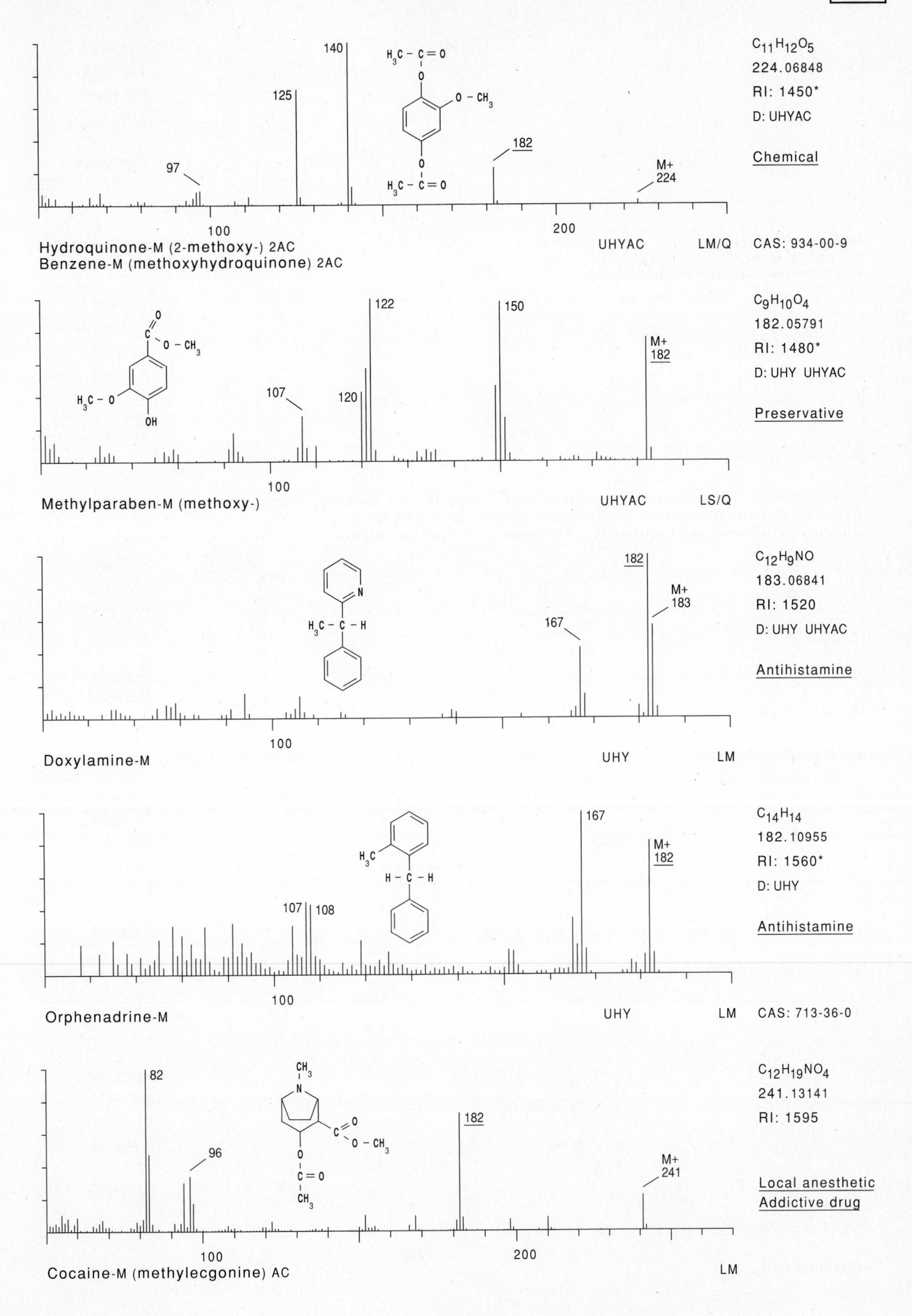

140
125
97
182
M+
224
100
200
C11H12O5
224.06848
RI: 1450*
D: UHYAC
Chemical
UHYAC
LM/Q
CAS: 934-00-9
Hydroquinone-M (2-methoxy-) 2AC
Benzene-M (methoxyhydroquinone) 2AC
122
150
M+
182
107
120
100
C9H10O4
182.05791
RI: 1480*
D: UHY UHYAC
Preservative
Methylparaben-M (methoxy-)
UHYAC
LS/Q
182
M+
183
167
100
C12H9NO
183.06841
RI: 1520
D: UHY UHYAC
Antihistamine
Doxylamine-M
UHY
LM
167
M+
182
107
108
100
C14H14
182.10955
RI: 1560*
D: UHY
Antihistamine
Orphenadrine-M
UHY
LM
CAS: 713-36-0
82
96
182
M+
241
100
200
C12H19NO4
241.13141
RI: 1595
Local anesthetic
Addictive drug
Cocaine-M (methylecgonine) AC
LM

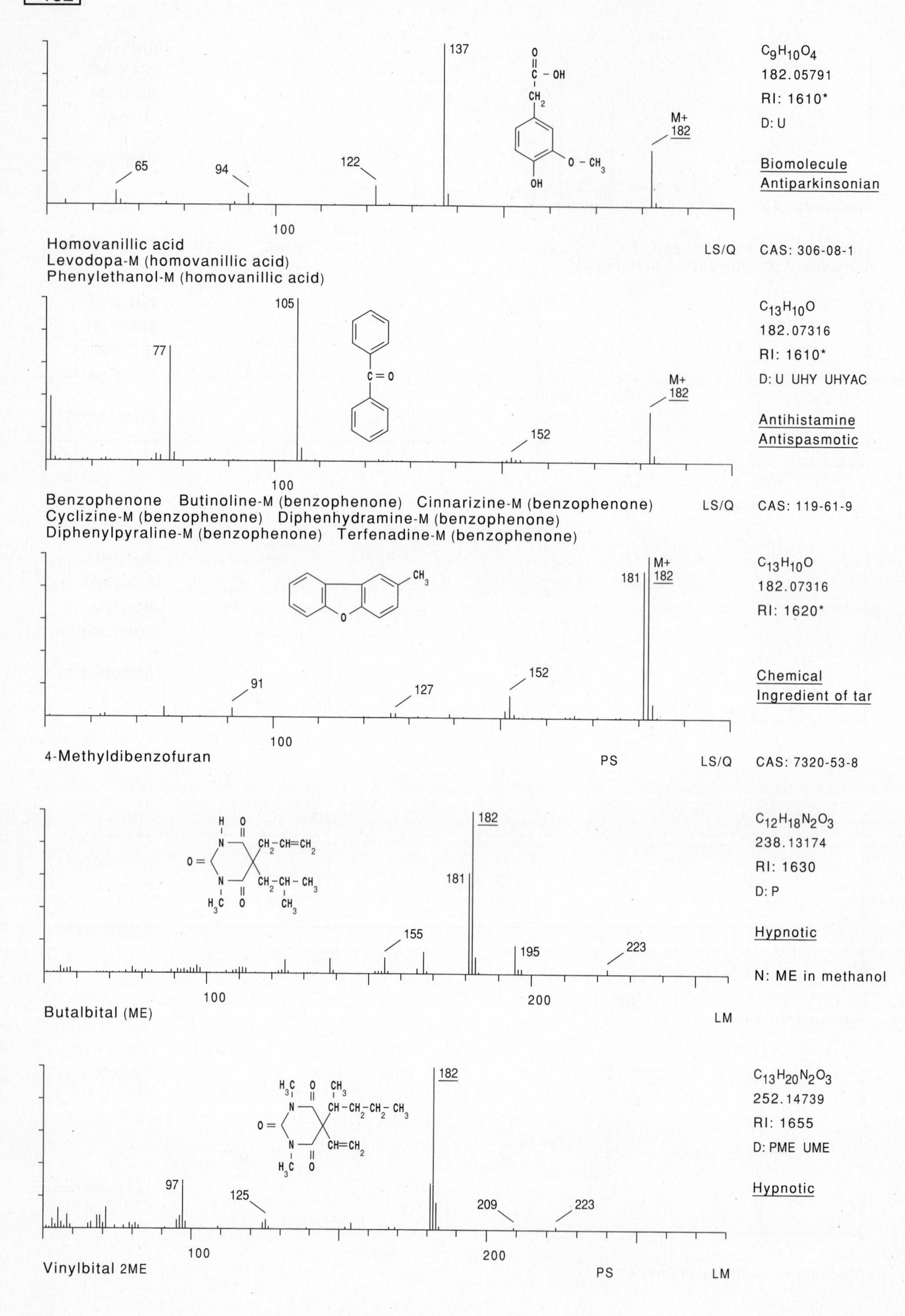

137
65
94
122
M+
182
100
C9H10O4
182.05791
RI: 1610*
D: U
Biomolecule
Antiparkinsonian
Homovanillic acid
Levodopa-M (homovanillic acid)
Phenylethanol-M (homovanillic acid)
LS/Q
CAS: 306-08-1
105
77
152
C13H10O
182.07316
RI: 1610*
D: U UHY UHYAC
Antihistamine
Antispasmotic
Benzophenone Butinoline-M (benzophenone) Cinnarizine-M (benzophenone)
Cyclizine-M (benzophenone) Diphenhydramine-M (benzophenone)
Diphenylpyraline-M (benzophenone) Terfenadine-M (benzophenone)
CAS: 119-61-9
181
91
127
RI: 1620*
Chemical
Ingredient of tar
4-Methyldibenzofuran
PS
CAS: 7320-53-8
155
195
223
200
C12H18N2O3
238.13174
RI: 1630
D: P
Hypnotic
N: ME in methanol
Butalbital (ME)
LM
97
125
209
C13H20N2O3
252.14739
RI: 1655
D: PME UME
Vinylbital 2ME

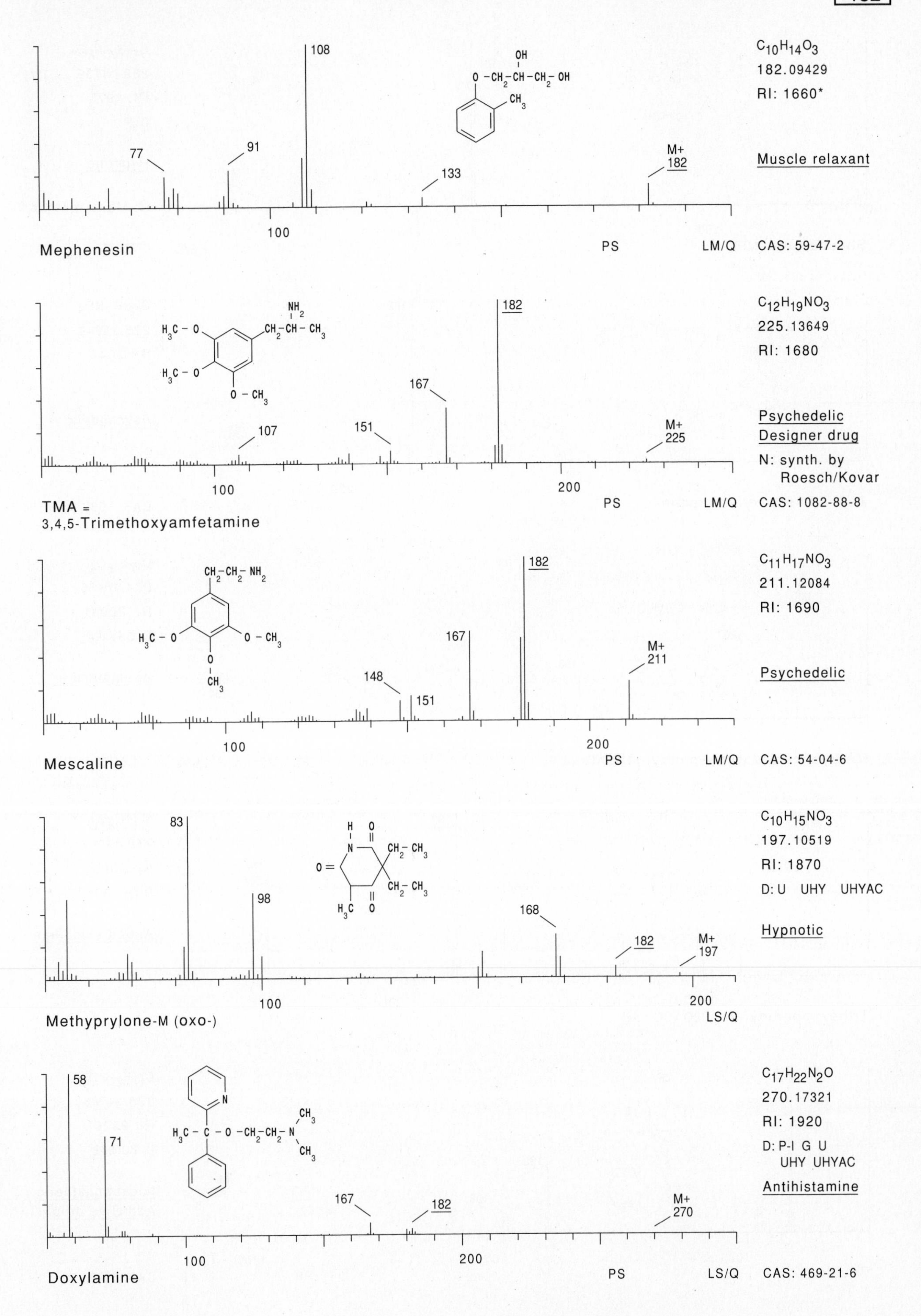

108
91
77
133
M+
182
100
Mephenesin
PS
LM/Q
$C_{10}H_{14}O_3$
182.09429
RI: 1660*
Muscle relaxant
CAS: 59-47-2
182
167
151
107
M+
225
100
200
TMA =
3,4,5-Trimethoxyamfetamine
PS
LM/Q
$C_{12}H_{19}NO_3$
225.13649
RI: 1680
Psychedelic
Designer drug
N: synth. by
Roesch/Kovar
CAS: 1082-88-8
182
167
148
151
M+
211
100
200
Mescaline
PS
LM/Q
$C_{11}H_{17}NO_3$
211.12084
RI: 1690
Psychedelic
CAS: 54-04-6
83
98
168
182
M+
197
100
200
Methyprylone-M (oxo-)
LS/Q
$C_{10}H_{15}NO_3$
197.10519
RI: 1870
D: U UHY UHYAC
Hypnotic
58
71
167
182
M+
270
100
200
Doxylamine
PS
LS/Q
$C_{17}H_{22}N_2O$
270.17321
RI: 1920
D: P-I G U
UHY UHYAC
Antihistamine
CAS: 469-21-6

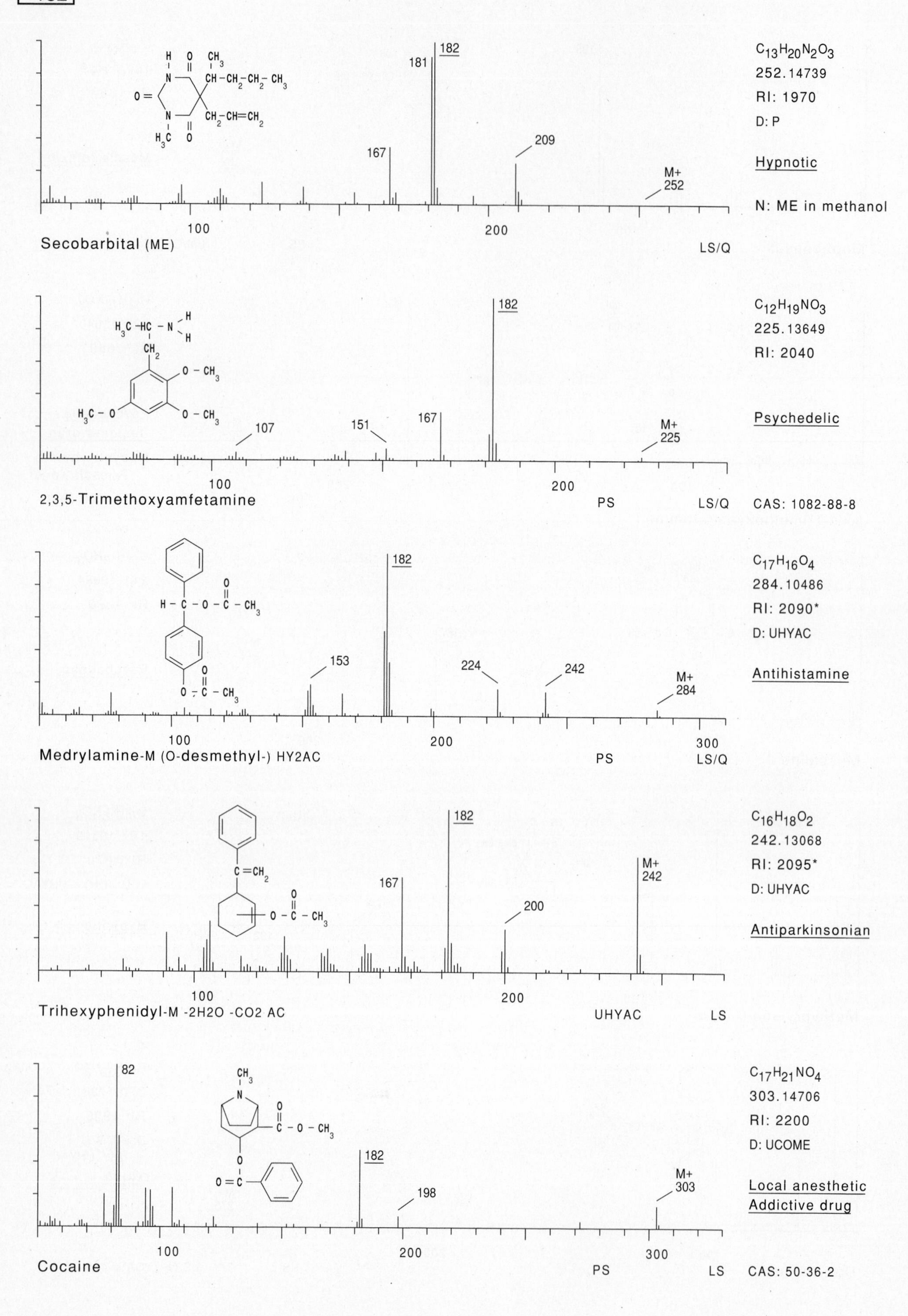

182
181
167
209
M+
252
100
200
Secobarbital (ME)
LS/Q
$C_{13}H_{20}N_2O_3$
252.14739
RI: 1970
D: P
Hypnotic
N: ME in methanol
182
107
151
167
M+
225
100
200
2,3,5-Trimethoxyamfetamine
PS
LS/Q
$C_{12}H_{19}NO_3$
225.13649
RI: 2040
Psychedelic
CAS: 1082-88-8
182
153
224
242
M+
284
100
200
300
Medrylamine-M (O-desmethyl-) HY2AC
PS
LS/Q
$C_{17}H_{16}O_4$
284.10486
RI: 2090*
D: UHYAC
Antihistamine
182
167
200
M+
242
100
200
Trihexyphenidyl-M -2H2O -CO2 AC
UHYAC
LS
$C_{16}H_{18}O_2$
242.13068
RI: 2095*
D: UHYAC
Antiparkinsonian
82
182
198
M+
303
100
200
300
Cocaine
PS
LS
$C_{17}H_{21}NO_4$
303.14706
RI: 2200
D: UCOME
Local anesthetic
Addictive drug
CAS: 50-36-2

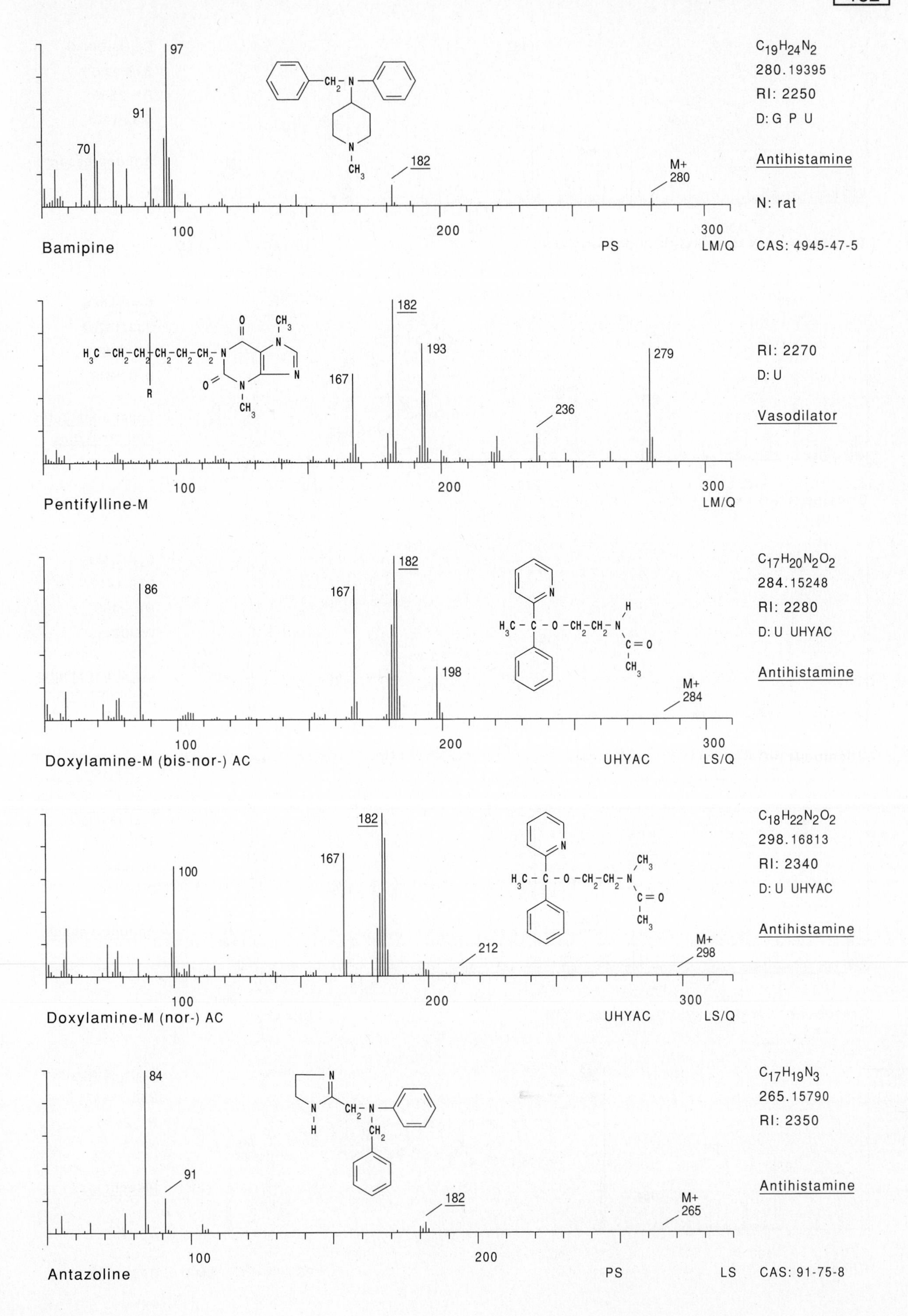

Bamipine
C19H24N2
280.19395
RI: 2250
D: G P U
Antihistamine
N: rat
CAS: 4945-47-5
PS
LM/Q
Pentifylline-M
RI: 2270
D: U
Vasodilator
LM/Q
Doxylamine-M (bis-nor-) AC
C17H20N2O2
284.15248
RI: 2280
D: U UHYAC
Antihistamine
UHYAC
LS/Q
Doxylamine-M (nor-) AC
C18H22N2O2
298.16813
RI: 2340
D: U UHYAC
Antihistamine
UHYAC
LS/Q
Antazoline
C17H19N3
265.15790
RI: 2350
Antihistamine
PS
LS
CAS: 91-75-8

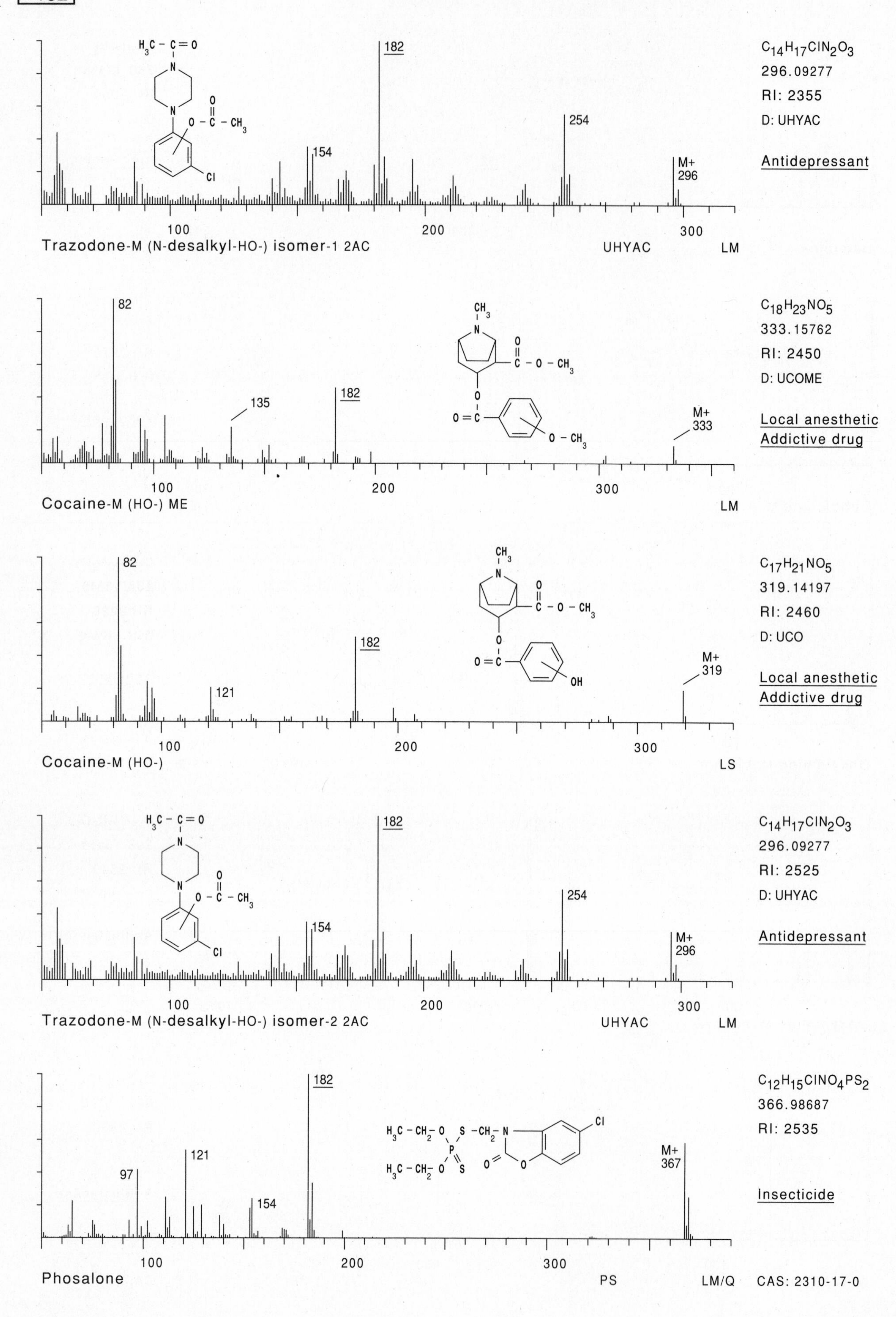

182
154
254
M+
296
100
200
300
Trazodone-M (N-desalkyl-HO-) isomer-1 2AC
UHYAC
LM
$C_{14}H_{17}ClN_2O_3$
296.09277
RI: 2355
D: UHYAC
Antidepressant
82
135
182
M+
333
Cocaine-M (HO-) ME
LM
$C_{18}H_{23}NO_5$
333.15762
RI: 2450
D: UCOME
Local anesthetic
Addictive drug
82
121
182
M+
319
Cocaine-M (HO-)
LS
$C_{17}H_{21}NO_5$
319.14197
RI: 2460
D: UCO
Local anesthetic
Addictive drug
182
154
254
M+
296
Trazodone-M (N-desalkyl-HO-) isomer-2 2AC
UHYAC
LM
$C_{14}H_{17}ClN_2O_3$
296.09277
RI: 2525
D: UHYAC
Antidepressant
182
121
97
154
M+
367
Phosalone
PS
LM/Q
$C_{12}H_{15}ClNO_4PS_2$
366.98687
RI: 2535
Insecticide
CAS: 2310-17-0

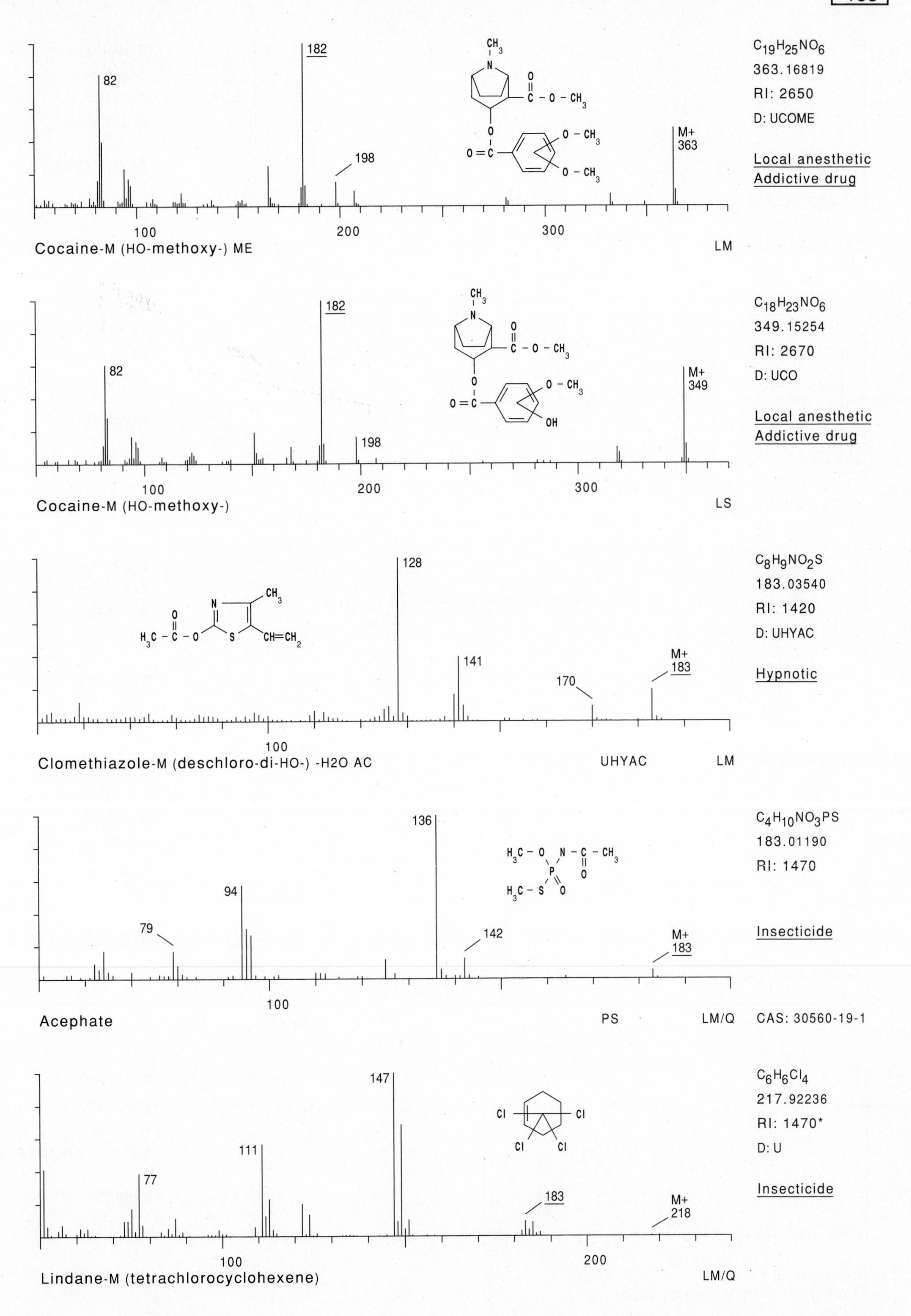

82
182
198
M+ 363
100
200
300
Cocaine-M (HO-methoxy-) ME
LM
$C_{19}H_{25}NO_6$
363.16819
RI: 2650
D: UCOME
Local anesthetic
Addictive drug
82
182
198
M+ 349
100
200
300
Cocaine-M (HO-methoxy-)
LS
$C_{18}H_{23}NO_6$
349.15254
RI: 2670
D: UCO
Local anesthetic
Addictive drug
128
141
170
M+ 183
100
Clomethiazole-M (deschloro-di-HO-) -H2O AC
UHYAC
LM
$C_8H_9NO_2S$
183.03540
RI: 1420
D: UHYAC
Hypnotic
136
94
79
142
M+ 183
100
Acephate
PS
LM/Q
$C_4H_{10}NO_3PS$
183.01190
RI: 1470
Insecticide
CAS: 30560-19-1
147
111
77
183
M+ 218
100
200
Lindane-M (tetrachlorocyclohexene)
LM/Q
$C_6H_6Cl_4$
217.92236
RI: 1470*
D: U
Insecticide

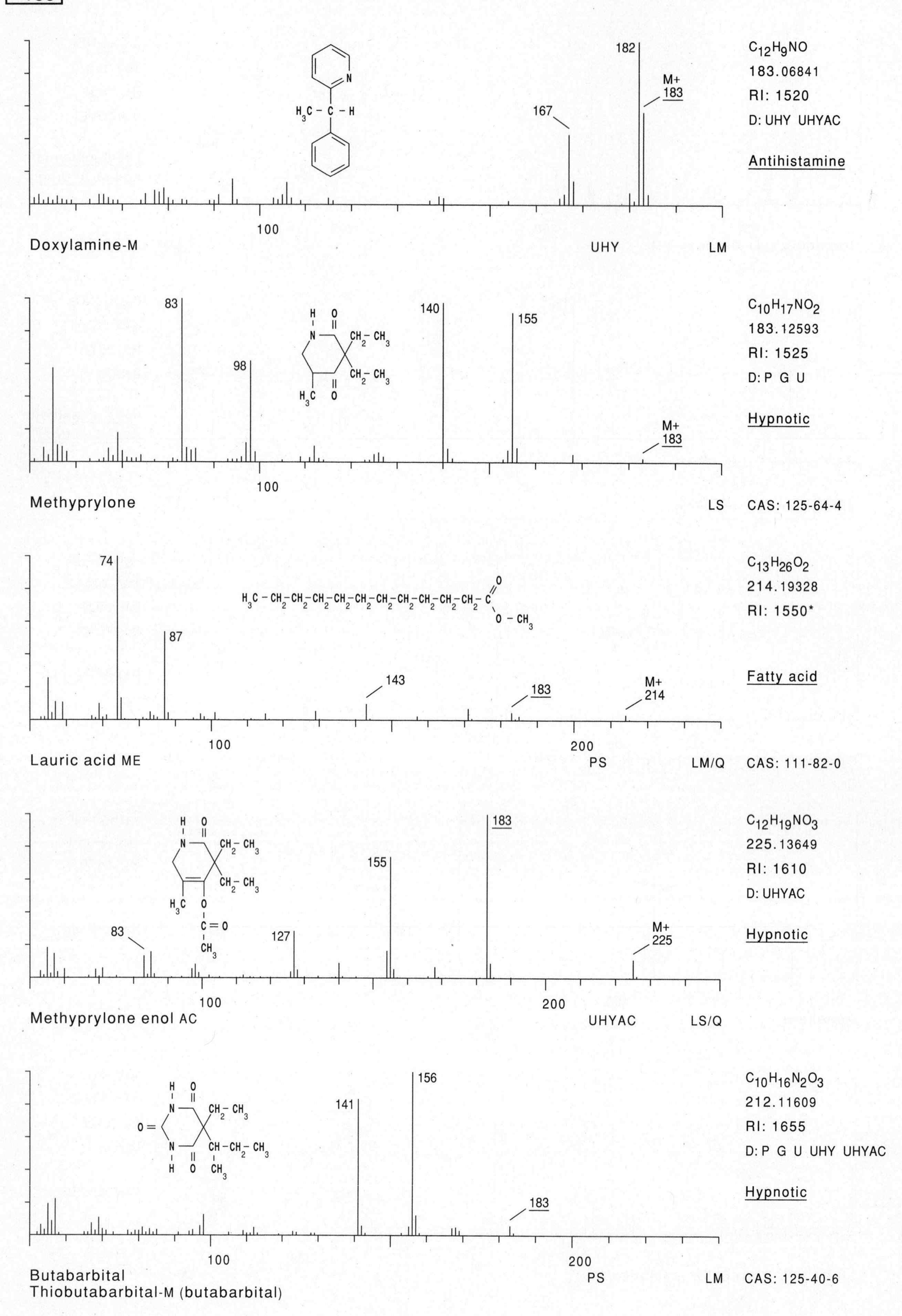

182
M+
183
167
100
Doxylamine-M
UHY
LM
$C_{12}H_9NO$
183.06841
RI: 1520
D: UHY UHYAC
Antihistamine
83
140
155
98
M+
183
100
Methyprylone
LS
CAS: 125-64-4
$C_{10}H_{17}NO_2$
183.12593
RI: 1525
D: P G U
Hypnotic
74
87
143
183
M+
214
100
200
Lauric acid ME
PS
LM/Q
CAS: 111-82-0
$C_{13}H_{26}O_2$
214.19328
RI: 1550*
Fatty acid
183
155
127
83
M+
225
100
200
Methyprylone enol AC
UHYAC
LS/Q
$C_{12}H_{19}NO_3$
225.13649
RI: 1610
D: UHYAC
Hypnotic
156
141
183
100
200
Butabarbital
Thiobutabarbital-M (butabarbital)
PS
LM
CAS: 125-40-6
$C_{10}H_{16}N_2O_3$
212.11609
RI: 1655
D: P G U UHY UHYAC
Hypnotic

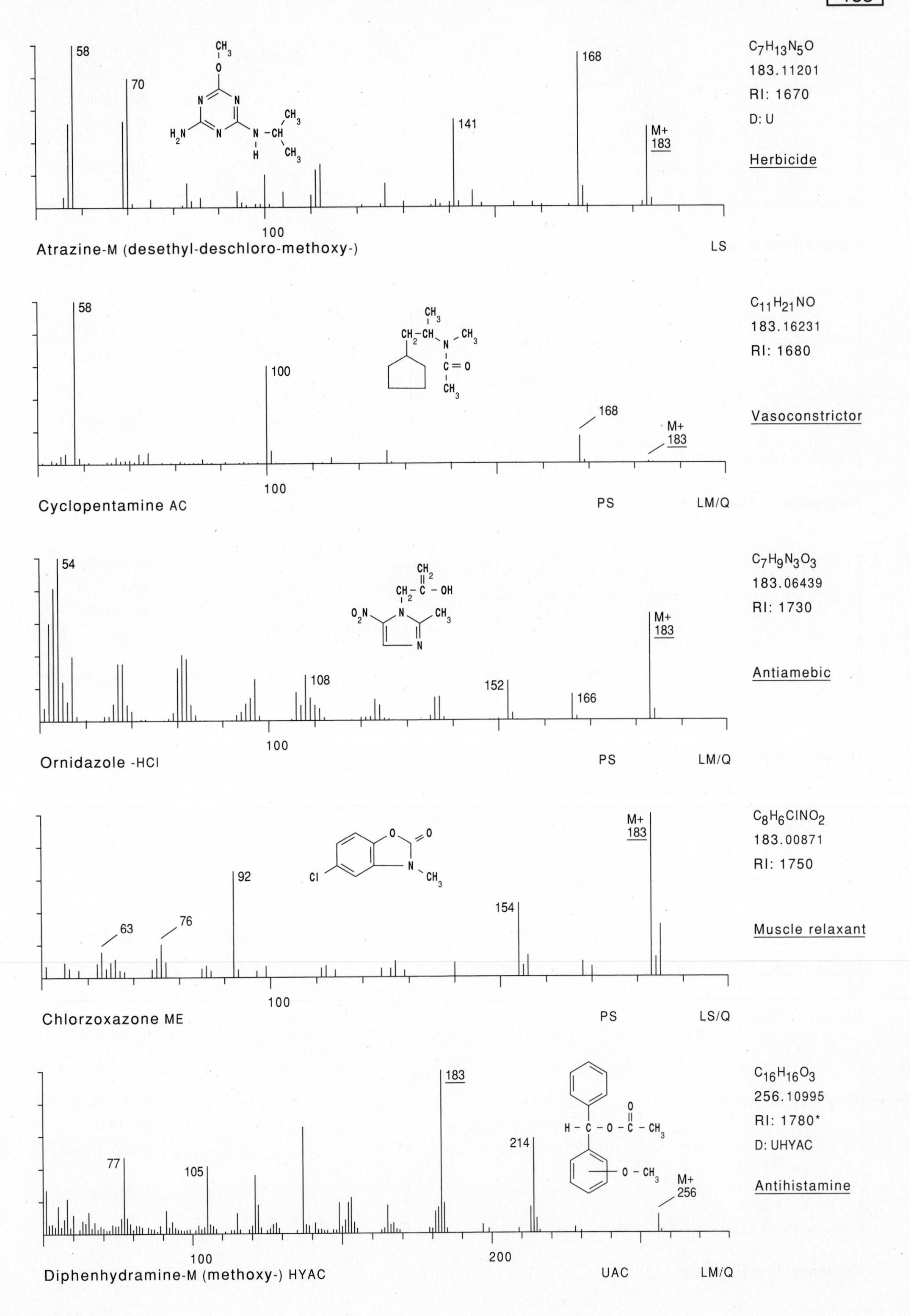

Atrazine-M (desethyl-deschloro-methoxy-)
LS
C7H13N5O
183.11201
RI: 1670
D: U
Herbicide
58
70
141
168
M+
183
100
Cyclopentamine AC
PS
LM/Q
C11H21NO
183.16231
RI: 1680
Vasoconstrictor
58
100
168
M+
183
Ornidazole -HCl
PS
LM/Q
C7H9N3O3
183.06439
RI: 1730
Antiamebic
54
108
152
166
M+
183
Chlorzoxazone ME
PS
LS/Q
C8H6ClNO2
183.00871
RI: 1750
Muscle relaxant
63
76
92
154
M+
183
Diphenhydramine-M (methoxy-) HYAC
UAC
LM/Q
C16H16O3
256.10995
RI: 1780*
D: UHYAC
Antihistamine
77
105
183
214
M+
256
200

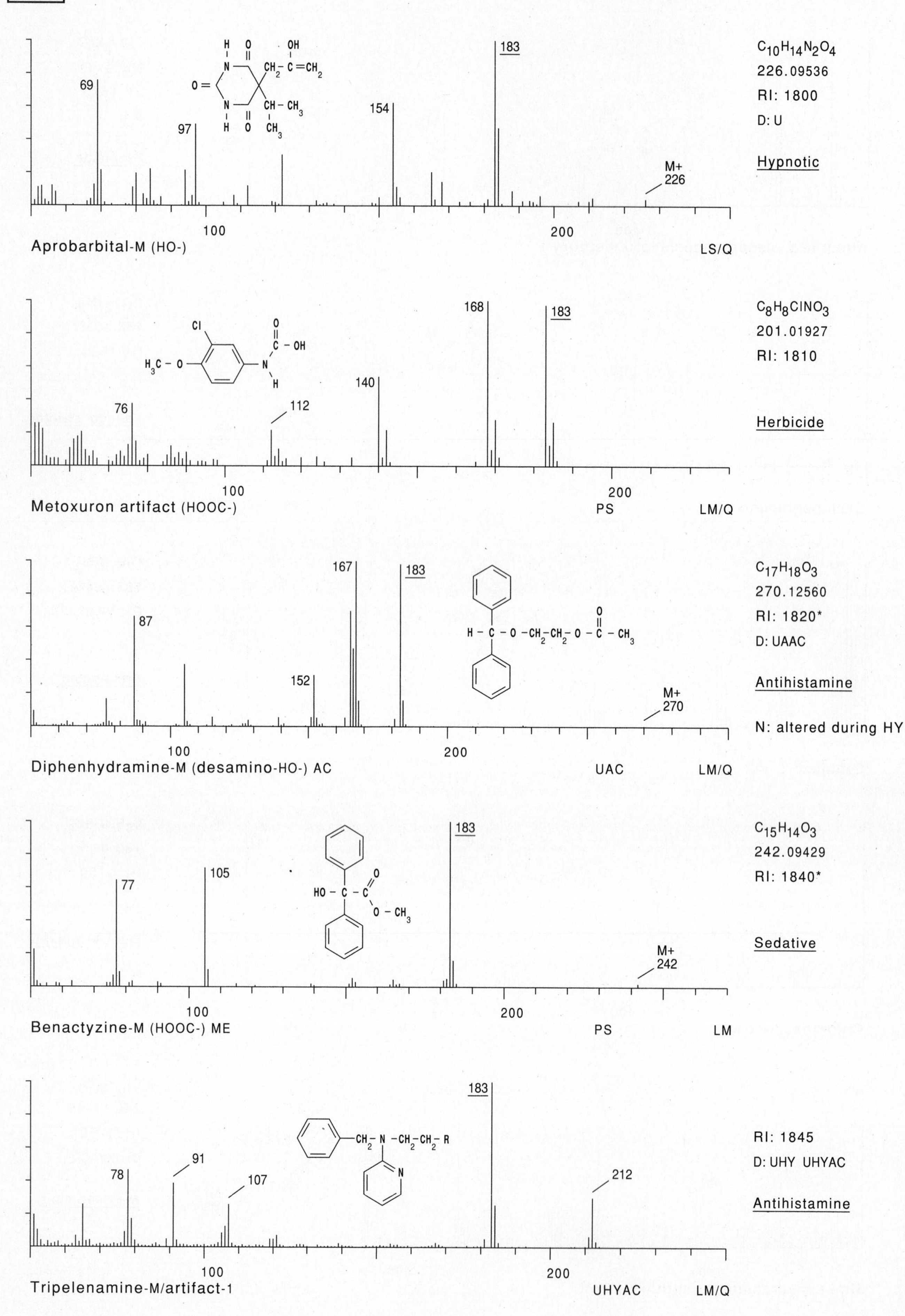
183
69
97
154
M+
226
100
200
Aprobarbital-M (HO-)
LS/Q
$C_{10}H_{14}N_2O_4$
226.09536
RI: 1800
D: U
Hypnotic
168
183
76
112
140
100
200
Metoxuron artifact (HOOC-)
PS
LM/Q
$C_8H_8ClNO_3$
201.01927
RI: 1810
Herbicide
167
183
87
152
M+
270
100
200
Diphenhydramine-M (desamino-HO-) AC
UAC
LM/Q
$C_{17}H_{18}O_3$
270.12560
RI: 1820*
D: UAAC
Antihistamine
N: altered during HY
183
77
105
M+
242
100
200
Benactyzine-M (HOOC-) ME
PS
LM
$C_{15}H_{14}O_3$
242.09429
RI: 1840*
Sedative
183
78
91
107
212
100
200
Tripelenamine-M/artifact-1
UHYAC
LM/Q
RI: 1845
D: UHY UHYAC
Antihistamine

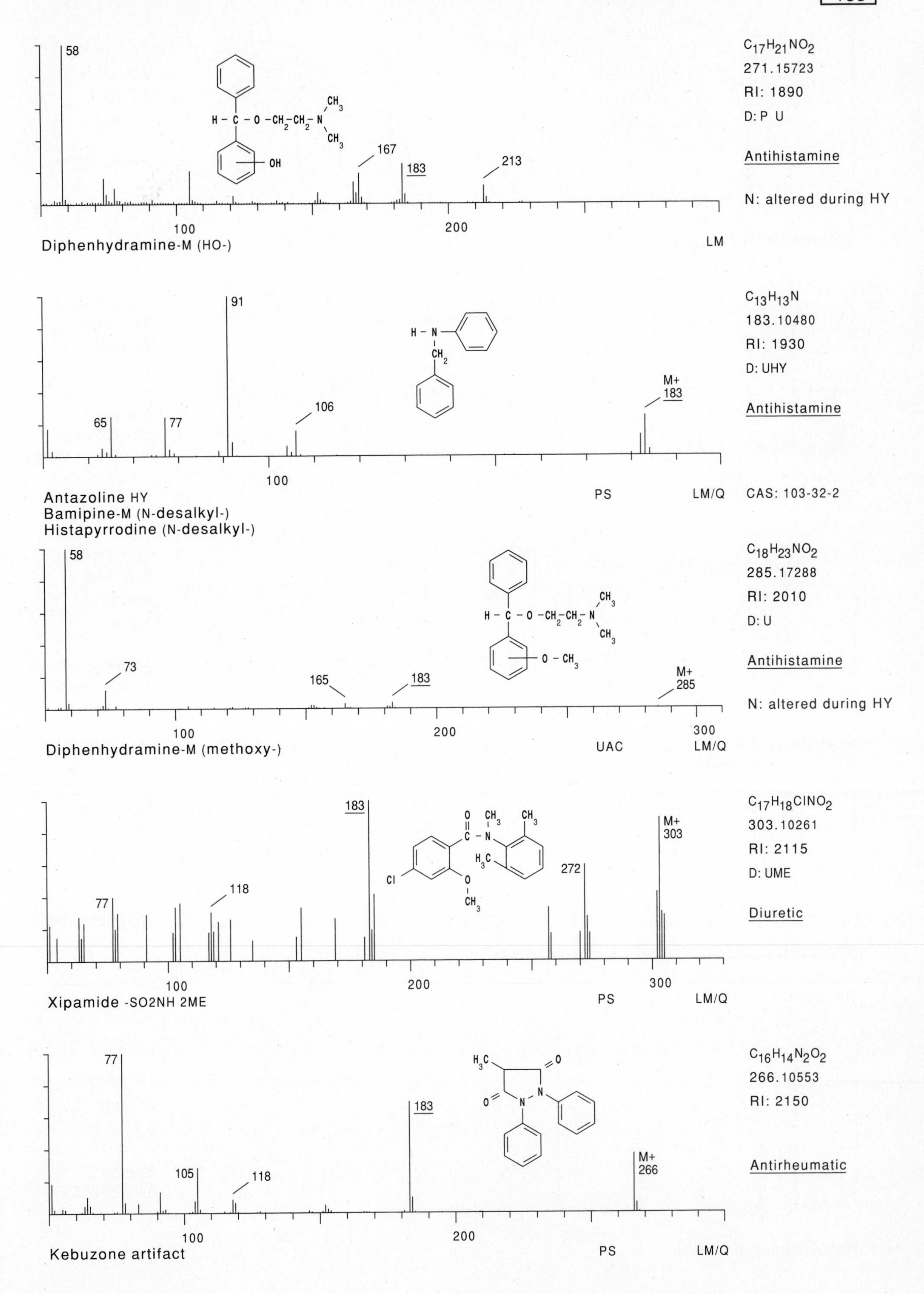

$C_{17}H_{21}NO_2$
271.15723
RI: 1890
D: P U
Antihistamine
N: altered during HY
Diphenhydramine-M (HO-)
LM
$C_{13}H_{13}N$
183.10480
RI: 1930
D: UHY
Antihistamine
Antazoline HY
Bamipine-M (N-desalkyl-)
Histapyrrodine (N-desalkyl-)
PS
LM/Q
CAS: 103-32-2
$C_{18}H_{23}NO_2$
285.17288
RI: 2010
D: U
Antihistamine
N: altered during HY
Diphenhydramine-M (methoxy-)
UAC
LM/Q
$C_{17}H_{18}ClNO_2$
303.10261
RI: 2115
D: UME
Diuretic
Xipamide -SO2NH 2ME
PS
LM/Q
$C_{16}H_{14}N_2O_2$
266.10553
RI: 2150
Antirheumatic
Kebuzone artifact
PS
LM/Q

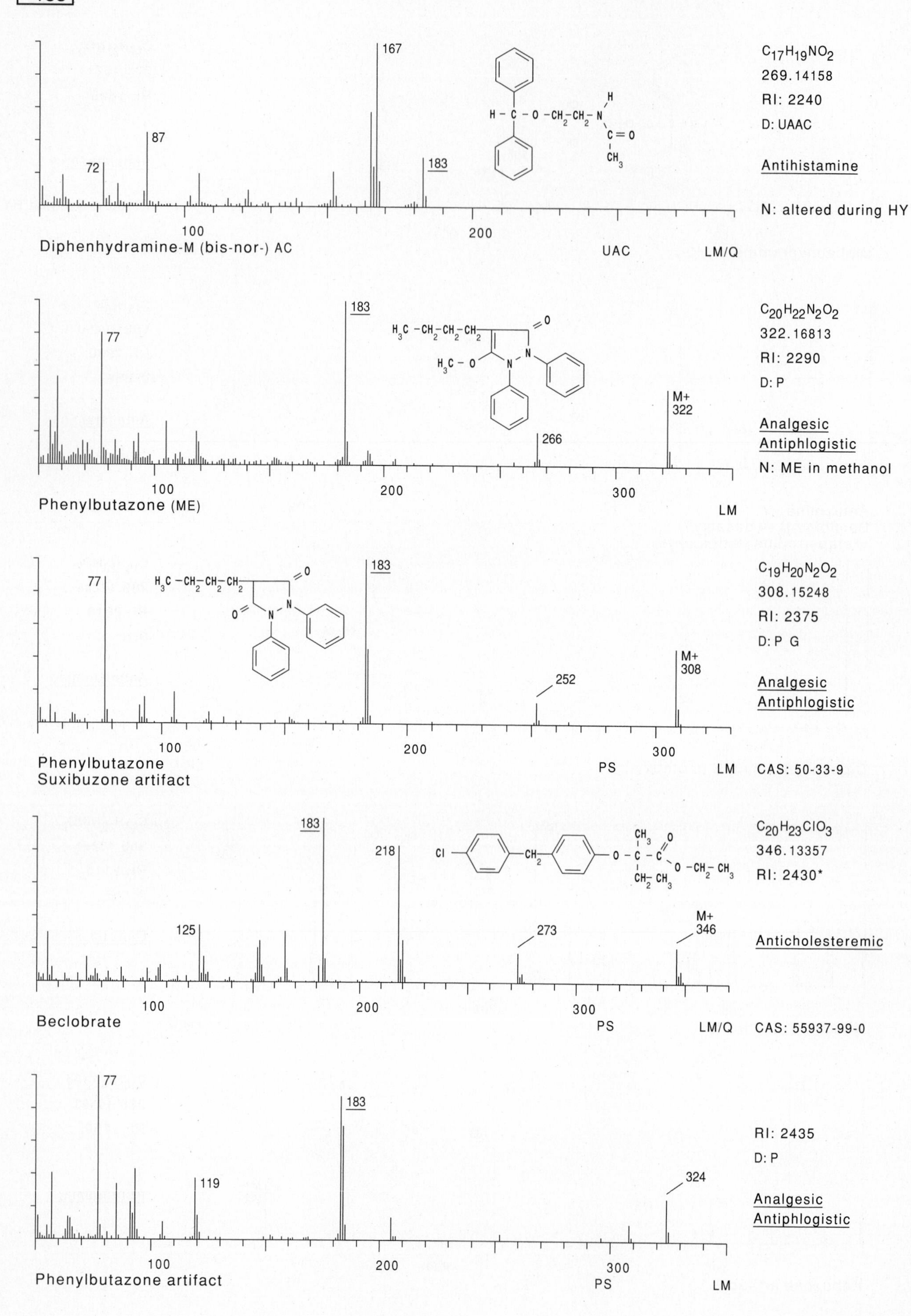
167
87
72
183
H
H - C - O - CH2 - CH2 - N
C = O
CH3
C17H19NO2
269.14158
RI: 2240
D: UAAC
Antihistamine
N: altered during HY
100
200
Diphenhydramine-M (bis-nor-) AC
UAC
LM/Q
183
77
H3C - CH2 - CH2 - CH2
O
N
N
H3C - O
M+
322
266
C20H22N2O2
322.16813
RI: 2290
D: P
Analgesic
Antiphlogistic
N: ME in methanol
100
200
300
Phenylbutazone (ME)
LM
183
77
H3C - CH2 - CH2 - CH2
O
O
N
N
M+
308
252
C19H20N2O2
308.15248
RI: 2375
D: P G
Analgesic
Antiphlogistic
100
200
300
Phenylbutazone
Suxibuzone artifact
PS
LM
CAS: 50-33-9
183
218
Cl
CH2
O - C - C
CH3
O
O - CH2 - CH3
CH2 CH3
125
273
M+
346
C20H23ClO3
346.13357
RI: 2430*
Anticholesteremic
100
200
300
Beclobrate
PS
LM/Q
CAS: 55937-99-0
77
183
119
324
RI: 2435
D: P
Analgesic
Antiphlogistic
100
200
300
Phenylbutazone artifact
PS
LM

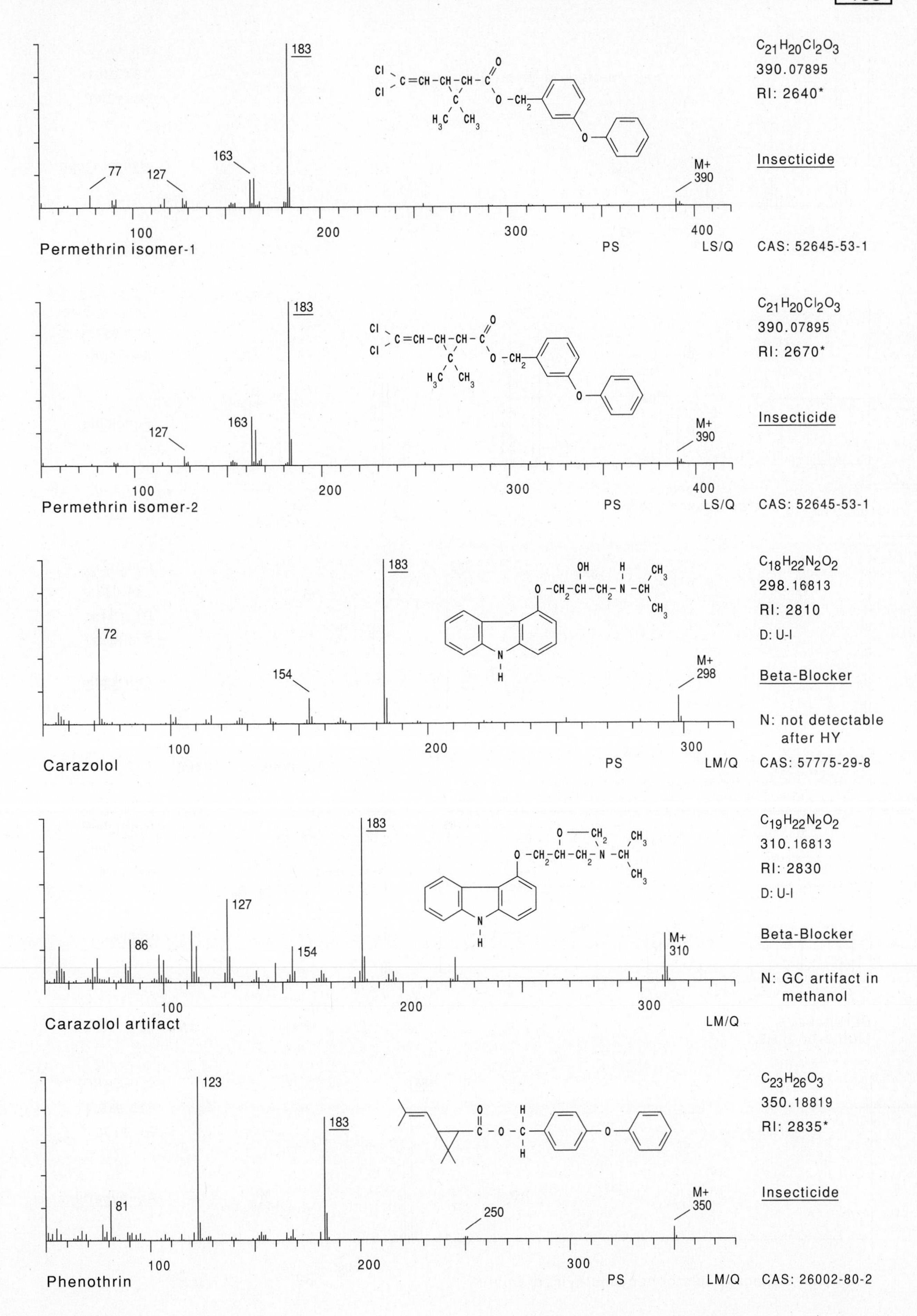
183
77
127
163
M+
390
100
200
300
400
Permethrin isomer-1
PS
LS/Q
$C_{21}H_{20}Cl_2O_3$
390.07895
RI: 2640*
Insecticide
CAS: 52645-53-1
183
127
163
M+
390
Permethrin isomer-2
PS
LS/Q
$C_{21}H_{20}Cl_2O_3$
390.07895
RI: 2670*
Insecticide
CAS: 52645-53-1
183
72
154
M+
298
Carazolol
PS
LM/Q
$C_{18}H_{22}N_2O_2$
298.16813
RI: 2810
D: U-I
Beta-Blocker
N: not detectable after HY
CAS: 57775-29-8
183
127
86
154
M+
310
Carazolol artifact
LM/Q
$C_{19}H_{22}N_2O_2$
310.16813
RI: 2830
D: U-I
Beta-Blocker
N: GC artifact in methanol
123
183
81
250
M+
350
Phenothrin
PS
LM/Q
$C_{23}H_{26}O_3$
350.18819
RI: 2835*
Insecticide
CAS: 26002-80-2

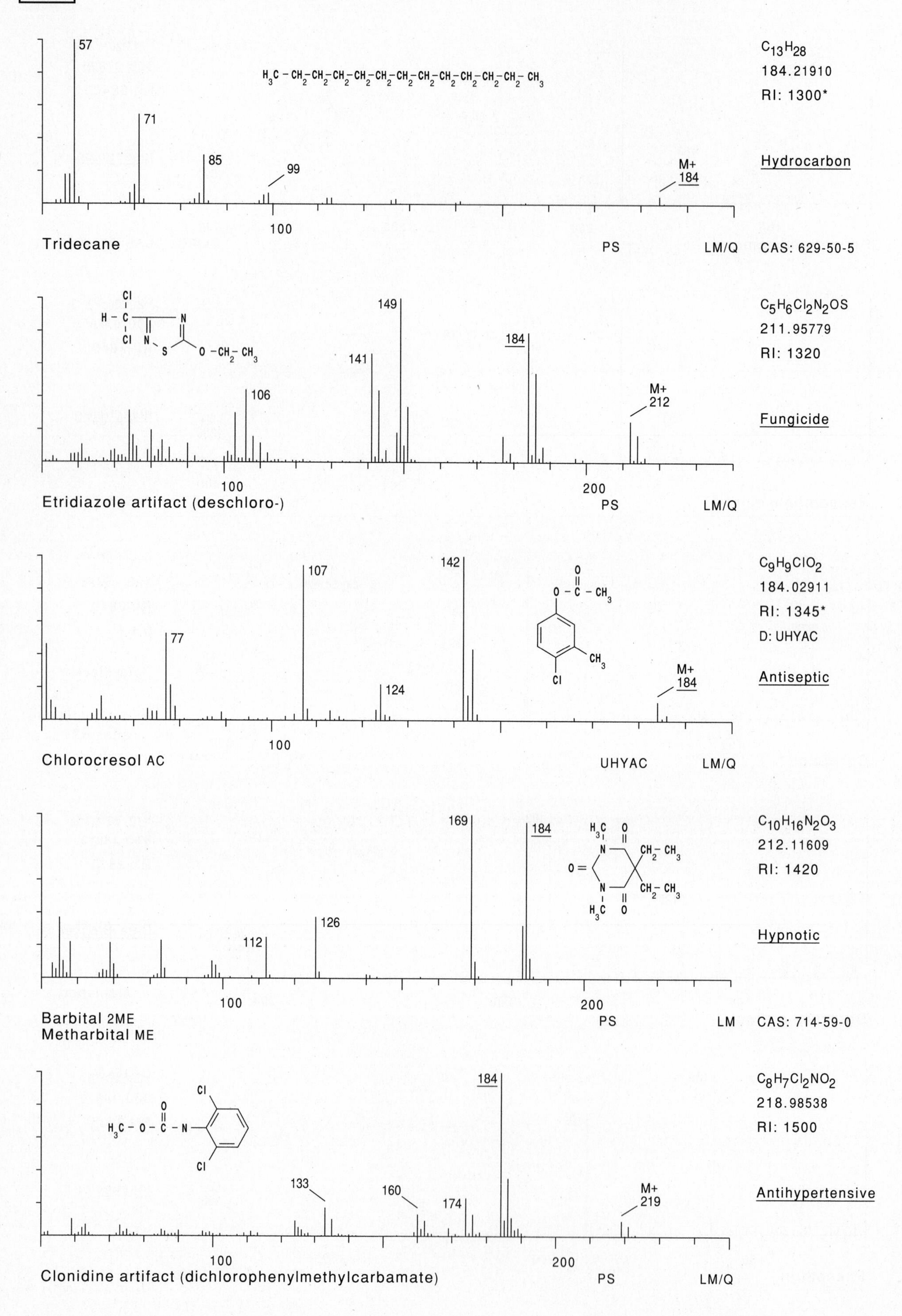
57
71
85
99
M+
184
100
$H_3C-CH_2-CH_2-CH_2-CH_2-CH_2-CH_2-CH_2-CH_2-CH_2-CH_2-CH_2-CH_3$
Tridecane
PS
LM/Q
$C_{13}H_{28}$
184.21910
RI: 1300*
Hydrocarbon
CAS: 629-50-5
149
184
141
106
M+
212
100
200
Etridiazole artifact (deschloro-)
PS
LM/Q
$C_5H_6Cl_2N_2OS$
211.95779
RI: 1320
Fungicide
142
107
77
124
M+
184
100
Chlorocresol AC
UHYAC
LM/Q
$C_9H_9ClO_2$
184.02911
RI: 1345*
D: UHYAC
Antiseptic
169
184
126
112
100
200
Barbital 2ME
Metharbital ME
PS
LM
$C_{10}H_{16}N_2O_3$
212.11609
RI: 1420
Hypnotic
CAS: 714-59-0
184
133
160
174
M+
219
100
200
Clonidine artifact (dichlorophenylmethylcarbamate)
PS
LM/Q
$C_8H_7Cl_2NO_2$
218.98538
RI: 1500
Antihypertensive

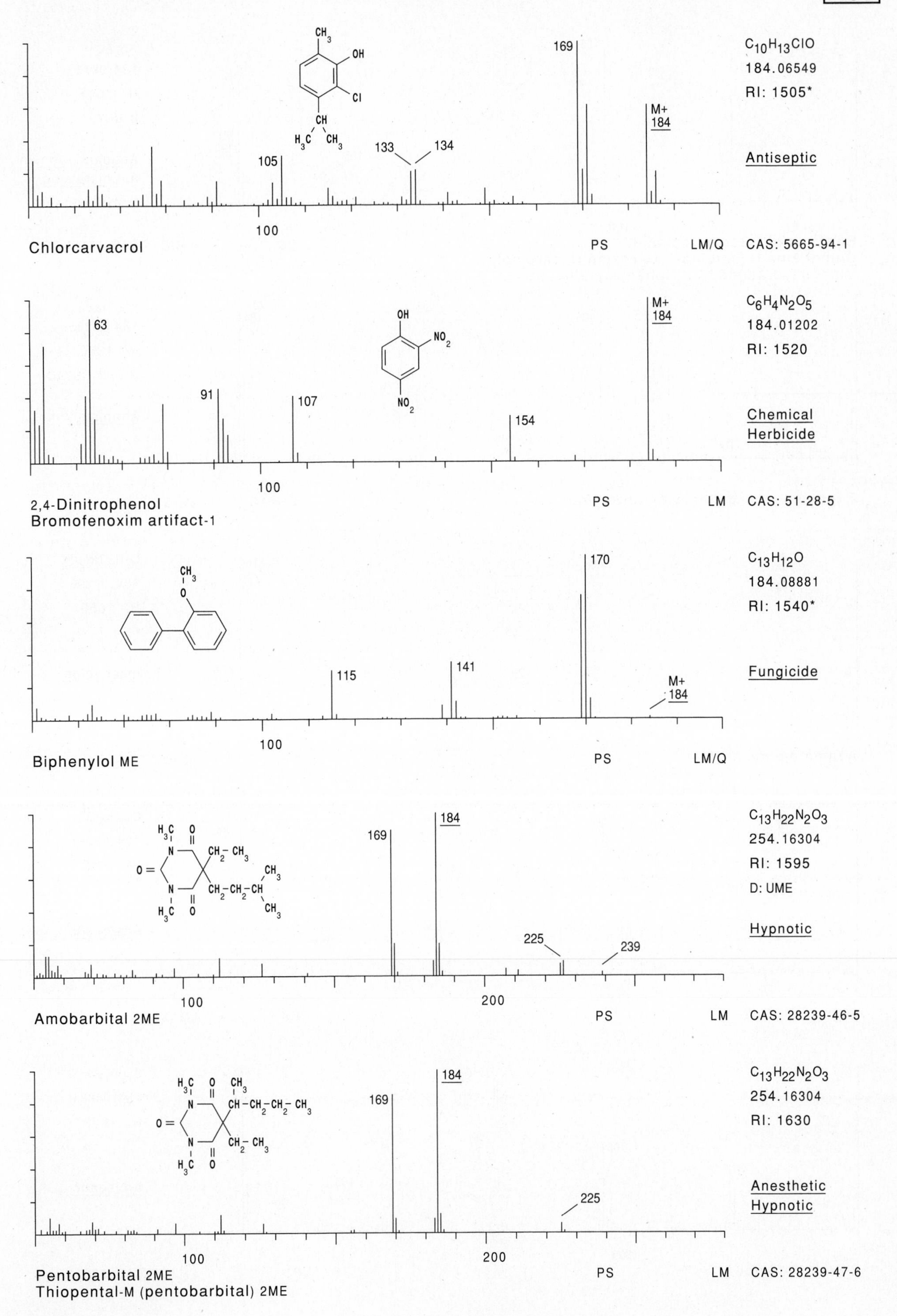

Chlorcarvacrol
$C_{10}H_{13}ClO$
184.06549
RI: 1505*
Antiseptic
CAS: 5665-94-1
PS
LM/Q
M+
184
169
133
134
105
100
2,4-Dinitrophenol
Bromofenoxim artifact-1
$C_6H_4N_2O_5$
184.01202
RI: 1520
Chemical
Herbicide
CAS: 51-28-5
PS
LM
M+
184
63
91
107
154
100
Biphenylol ME
$C_{13}H_{12}O$
184.08881
RI: 1540*
Fungicide
PS
LM/Q
170
115
141
M+
184
100
Amobarbital 2ME
$C_{13}H_{22}N_2O_3$
254.16304
RI: 1595
D: UME
Hypnotic
CAS: 28239-46-5
PS
LM
184
169
225
239
100
200
Pentobarbital 2ME
Thiopental-M (pentobarbital) 2ME
$C_{13}H_{22}N_2O_3$
254.16304
RI: 1630
Anesthetic
Hypnotic
CAS: 28239-47-6
PS
LM
184
169
225
100
200

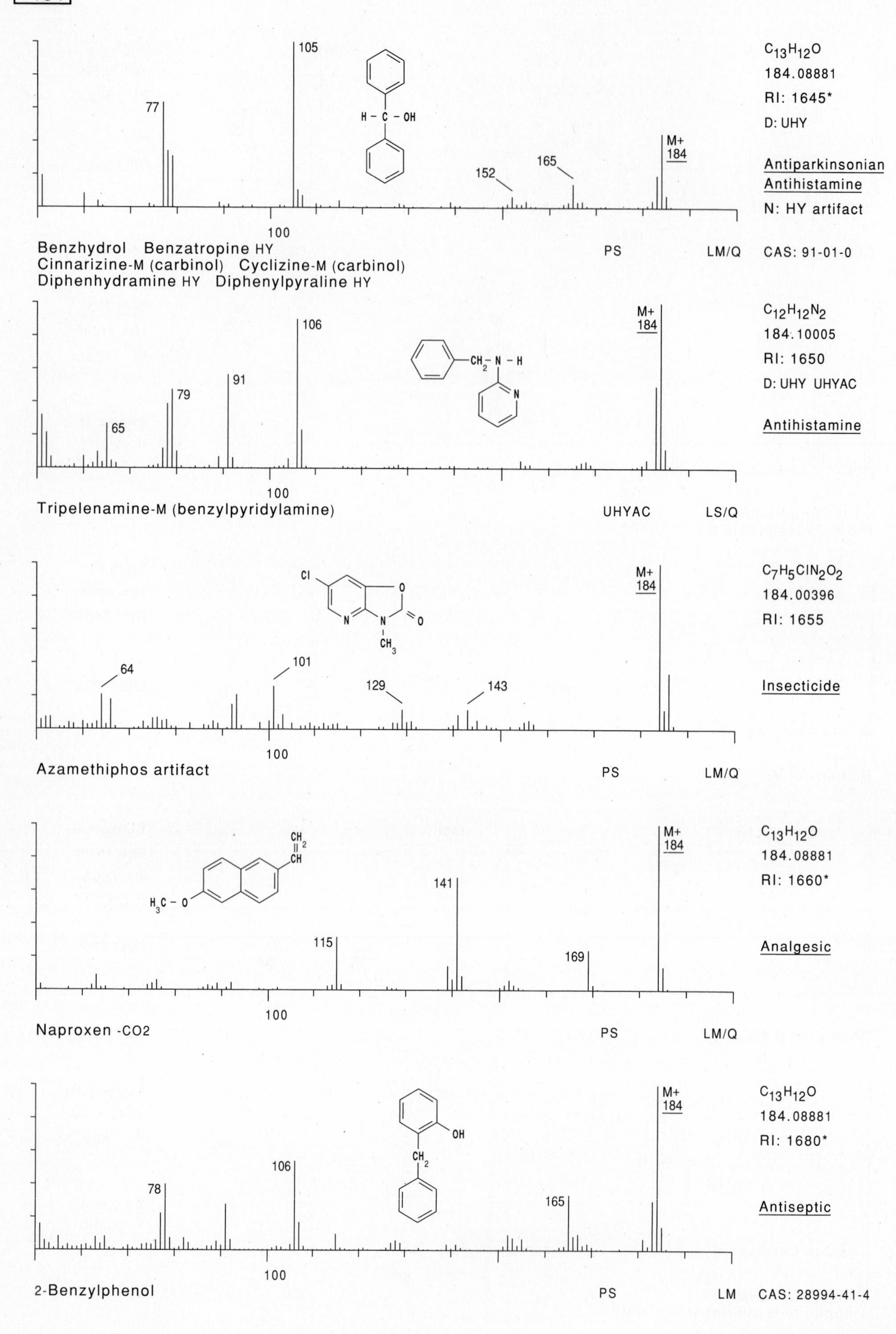
105
77
H - C - OH
152
165
M+
184
100
Benzhydrol Benzatropine HY
Cinnarizine-M (carbinol) Cyclizine-M (carbinol)
Diphenhydramine HY Diphenylpyraline HY
PS
LM/Q
$C_{13}H_{12}O$
184.08881
RI: 1645*
D: UHY
Antiparkinsonian
Antihistamine
N: HY artifact
CAS: 91-01-0
M+
184
106
CH2 - N - H
N
91
79
65
100
Tripelenamine-M (benzylpyridylamine)
UHYAC
LS/Q
$C_{12}H_{12}N_2$
184.10005
RI: 1650
D: UHY UHYAC
Antihistamine
M+
184
Cl
O
N
N
O
CH3
64
101
129
143
100
Azamethiphos artifact
PS
LM/Q
$C_7H_5ClN_2O_2$
184.00396
RI: 1655
Insecticide
M+
184
CH2
CH
H3C - O
141
115
169
100
Naproxen -CO2
PS
LM/Q
$C_{13}H_{12}O$
184.08881
RI: 1660*
Analgesic
M+
184
OH
CH2
106
78
165
100
2-Benzylphenol
PS
LM
$C_{13}H_{12}O$
184.08881
RI: 1680*
Antiseptic
CAS: 28994-41-4

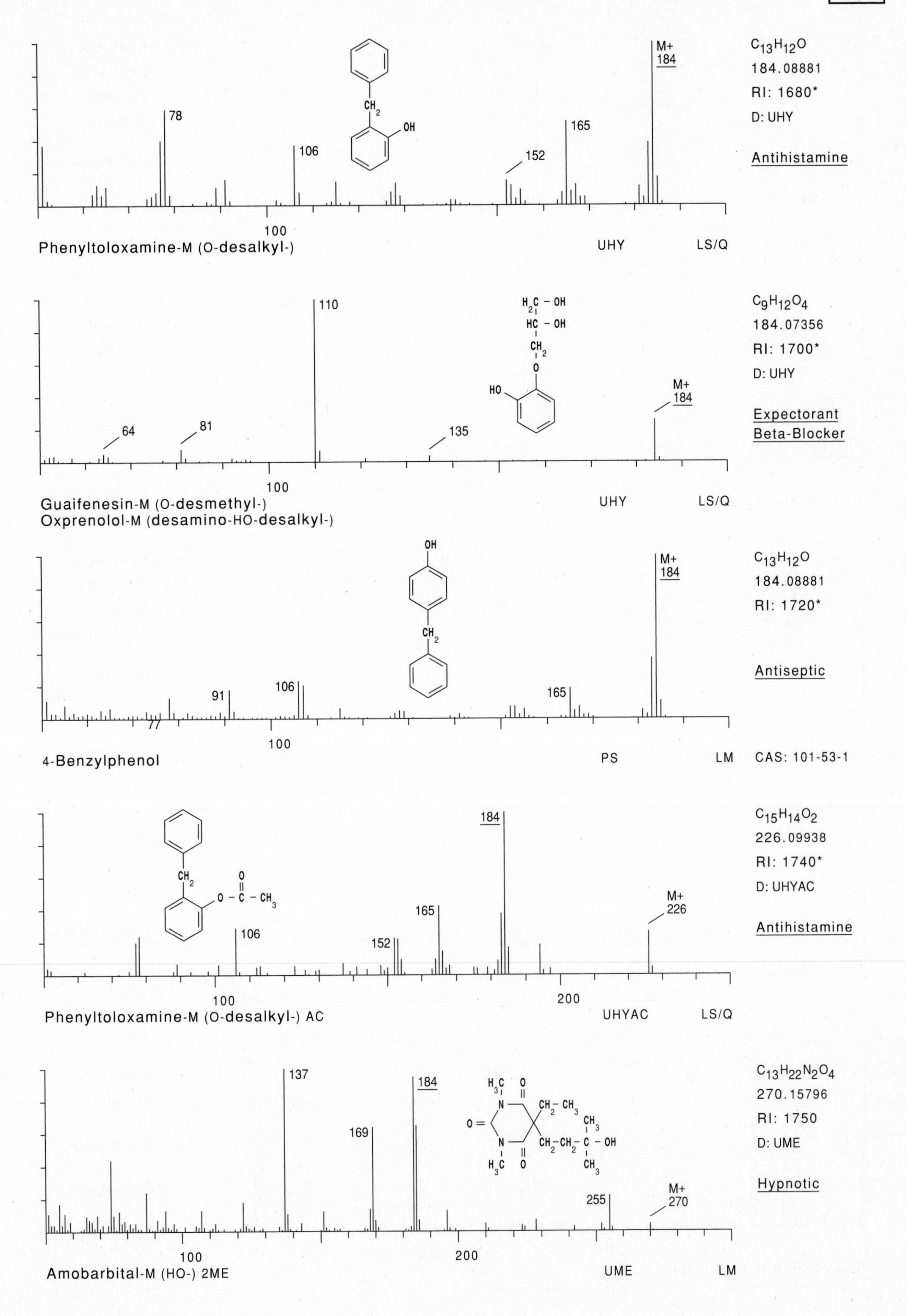
M+
184
78
165
106
152
OH
100
Phenyltoloxamine-M (O-desalkyl-)
UHY
LS/Q
C13H12O
184.08881
RI: 1680*
D: UHY
Antihistamine
110
H2C - OH
HC - OH
CH2
O
HO
M+
184
64
81
135
100
Guaifenesin-M (O-desmethyl-)
Oxprenolol-M (desamino-HO-desalkyl-)
UHY
LS/Q
C9H12O4
184.07356
RI: 1700*
D: UHY
Expectorant
Beta-Blocker
OH
M+
184
106
91
165
100
4-Benzylphenol
PS
LM
C13H12O
184.08881
RI: 1720*
Antiseptic
CAS: 101-53-1
184
M+
226
165
106
152
100
200
Phenyltoloxamine-M (O-desalkyl-) AC
UHYAC
LS/Q
C15H14O2
226.09938
RI: 1740*
D: UHYAC
Antihistamine
137
184
169
255
M+
270
100
200
Amobarbital-M (HO-) 2ME
UME
LM
C13H22N2O4
270.15796
RI: 1750
D: UME
Hypnotic

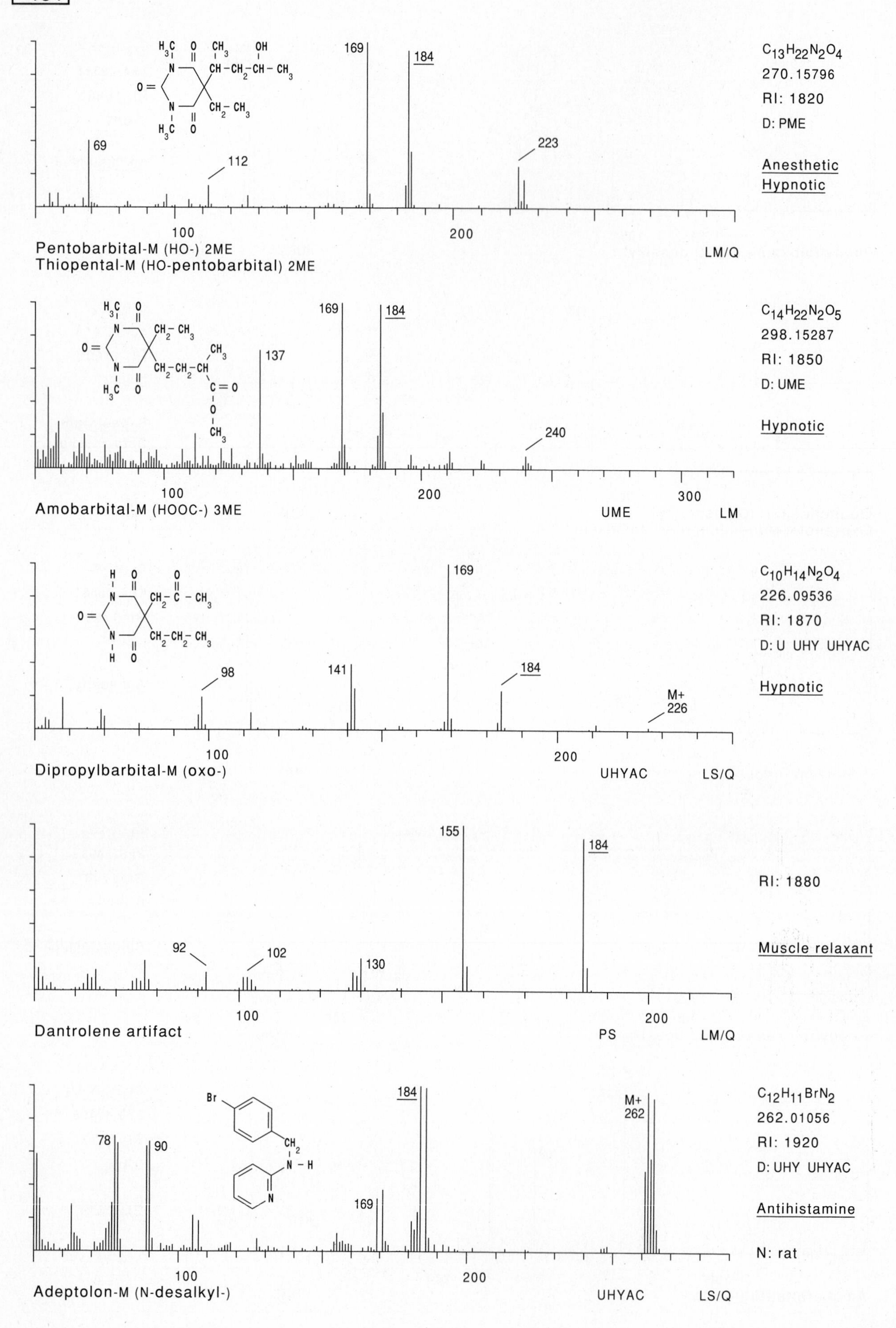

169
184
69
112
223
100
200
LM/Q
Pentobarbital-M (HO-) 2ME
Thiopental-M (HO-pentobarbital) 2ME
C13H22N2O4
270.15796
RI: 1820
D: PME
Anesthetic
Hypnotic
169
184
137
240
100
200
300
UME
LM
Amobarbital-M (HOOC-) 3ME
C14H22N2O5
298.15287
RI: 1850
D: UME
Hypnotic
169
98
141
184
M+
226
100
200
UHYAC
LS/Q
Dipropylbarbital-M (oxo-)
C10H14N2O4
226.09536
RI: 1870
D: U UHY UHYAC
Hypnotic
155
184
92
102
130
100
200
PS
LM/Q
Dantrolene artifact
RI: 1880
Muscle relaxant
184
M+
262
78
90
169
100
200
UHYAC
LS/Q
Adeptolon-M (N-desalkyl-)
C12H11BrN2
262.01056
RI: 1920
D: UHY UHYAC
Antihistamine
N: rat

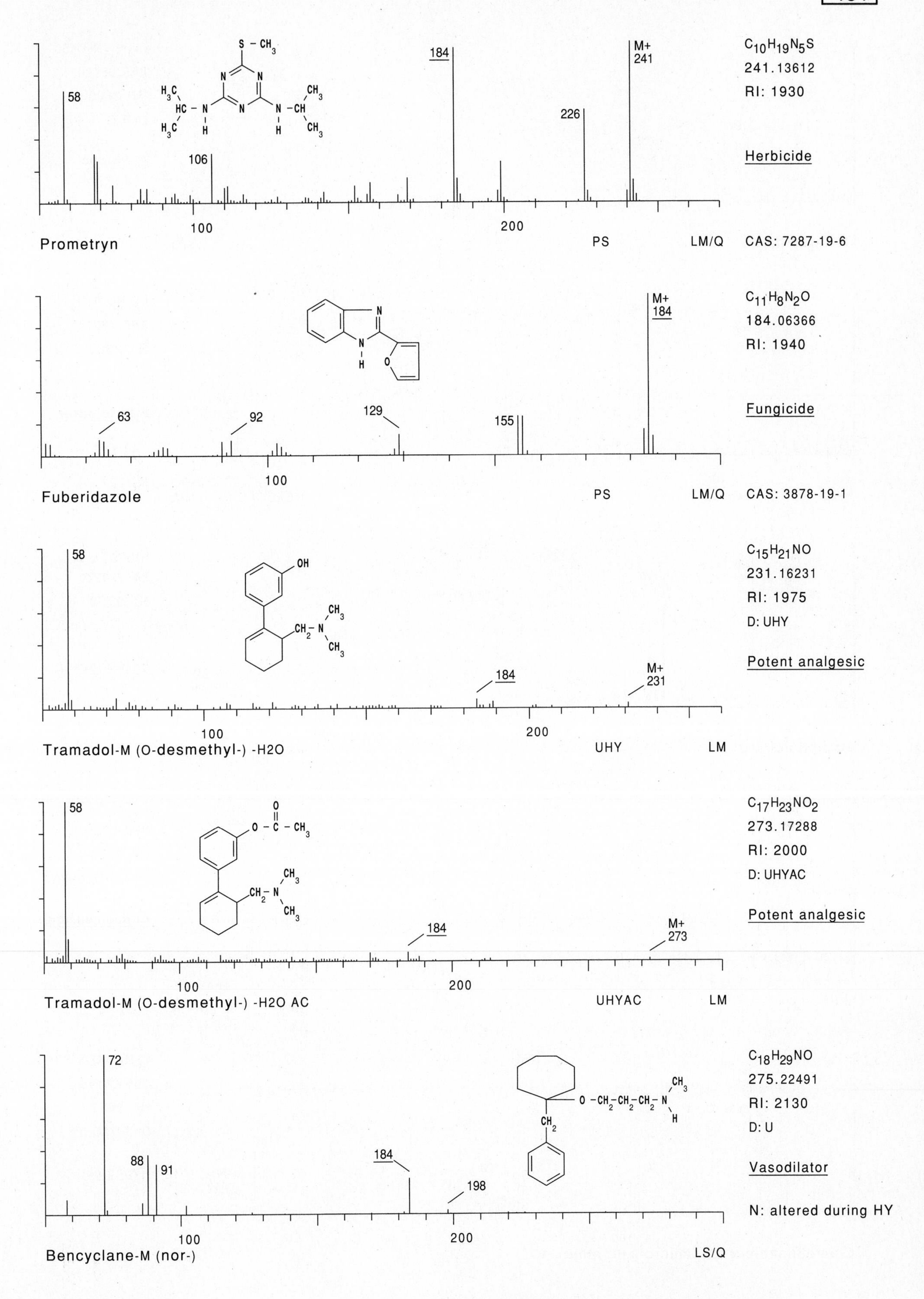

184
58
106
184
226
M+
241
100
200
Prometryn
PS
LM/Q
C10H19N5S
241.13612
RI: 1930
Herbicide
CAS: 7287-19-6
63
92
129
155
M+
184
100
Fuberidazole
PS
LM/Q
C11H8N2O
184.06366
RI: 1940
Fungicide
CAS: 3878-19-1
58
184
M+
231
100
200
Tramadol-M (O-desmethyl-) -H2O
UHY
LM
C15H21NO
231.16231
RI: 1975
D: UHY
Potent analgesic
58
184
M+
273
100
200
Tramadol-M (O-desmethyl-) -H2O AC
UHYAC
LM
C17H23NO2
273.17288
RI: 2000
D: UHYAC
Potent analgesic
72
88
91
184
198
100
200
Bencyclane-M (nor-)
LS/Q
C18H29NO
275.22491
RI: 2130
D: U
Vasodilator
N: altered during HY

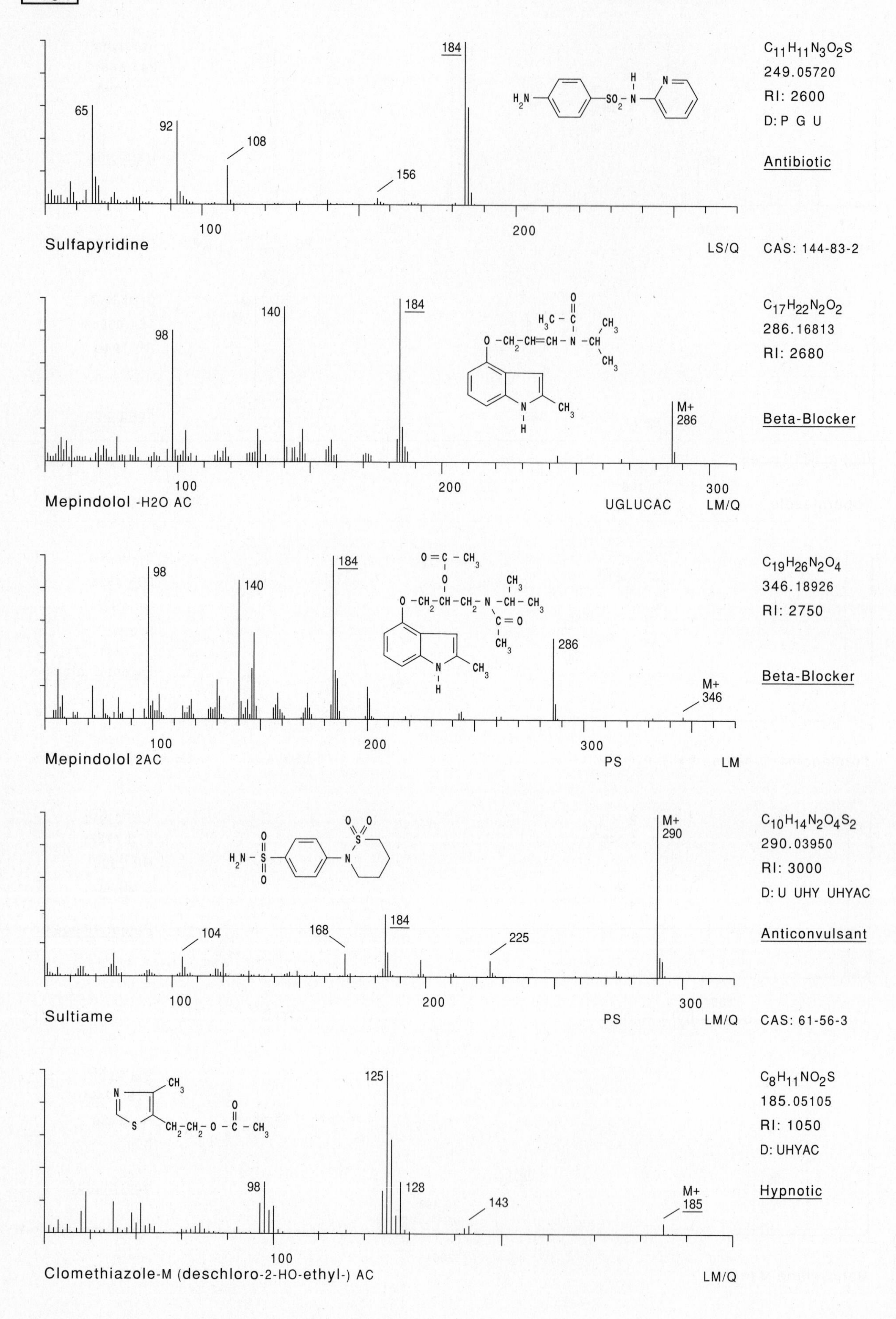
184
65
92
108
156
100
200
Sulfapyridine
LS/Q
$C_{11}H_{11}N_3O_2S$
249.05720
RI: 2600
D: P G U
Antibiotic
CAS: 144-83-2
98
140
184
M+
286
100
200
300
Mepindolol -H2O AC
UGLUCAC
LM/Q
$C_{17}H_{22}N_2O_2$
286.16813
RI: 2680
Beta-Blocker
98
140
184
286
M+
346
100
200
300
Mepindolol 2AC
PS
LM
$C_{19}H_{26}N_2O_4$
346.18926
RI: 2750
Beta-Blocker
M+
290
104
168
184
225
100
200
300
Sultiame
PS
LM/Q
$C_{10}H_{14}N_2O_4S_2$
290.03950
RI: 3000
D: U UHY UHYAC
Anticonvulsant
CAS: 61-56-3
125
98
128
143
M+
185
100
Clomethiazole-M (deschloro-2-HO-ethyl-) AC
LM/Q
$C_8H_{11}NO_2S$
185.05105
RI: 1050
D: UHYAC
Hypnotic

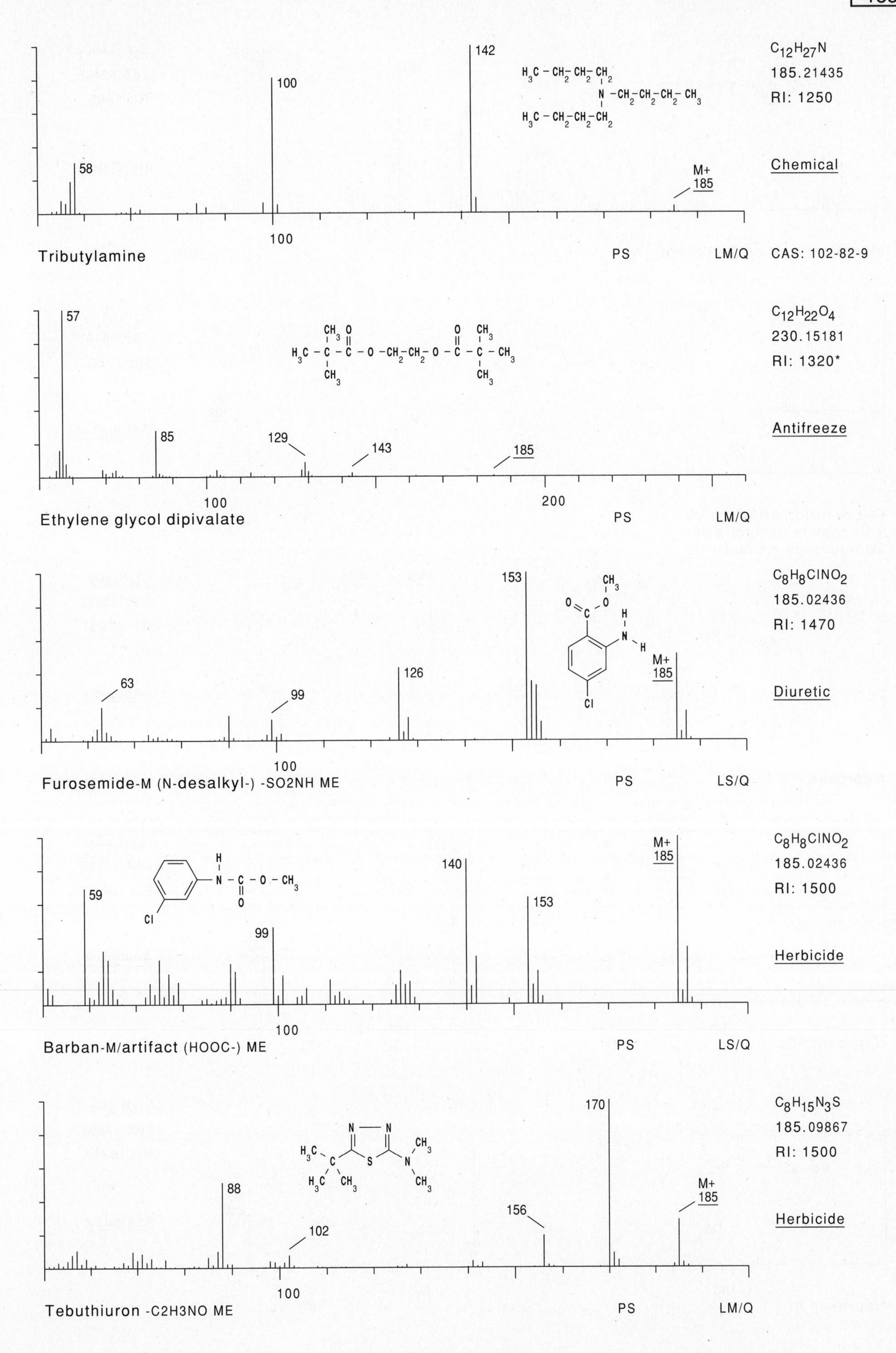

$C_{12}H_{27}N$
185.21435
RI: 1250

Chemical

CAS: 102-82-9

$C_{12}H_{22}O_4$
230.15181
RI: 1320*

Antifreeze

$C_8H_8ClNO_2$
185.02436
RI: 1470

Diuretic

$C_8H_8ClNO_2$
185.02436
RI: 1500

Herbicide

$C_8H_{15}N_3S$
185.09867
RI: 1500

Herbicide

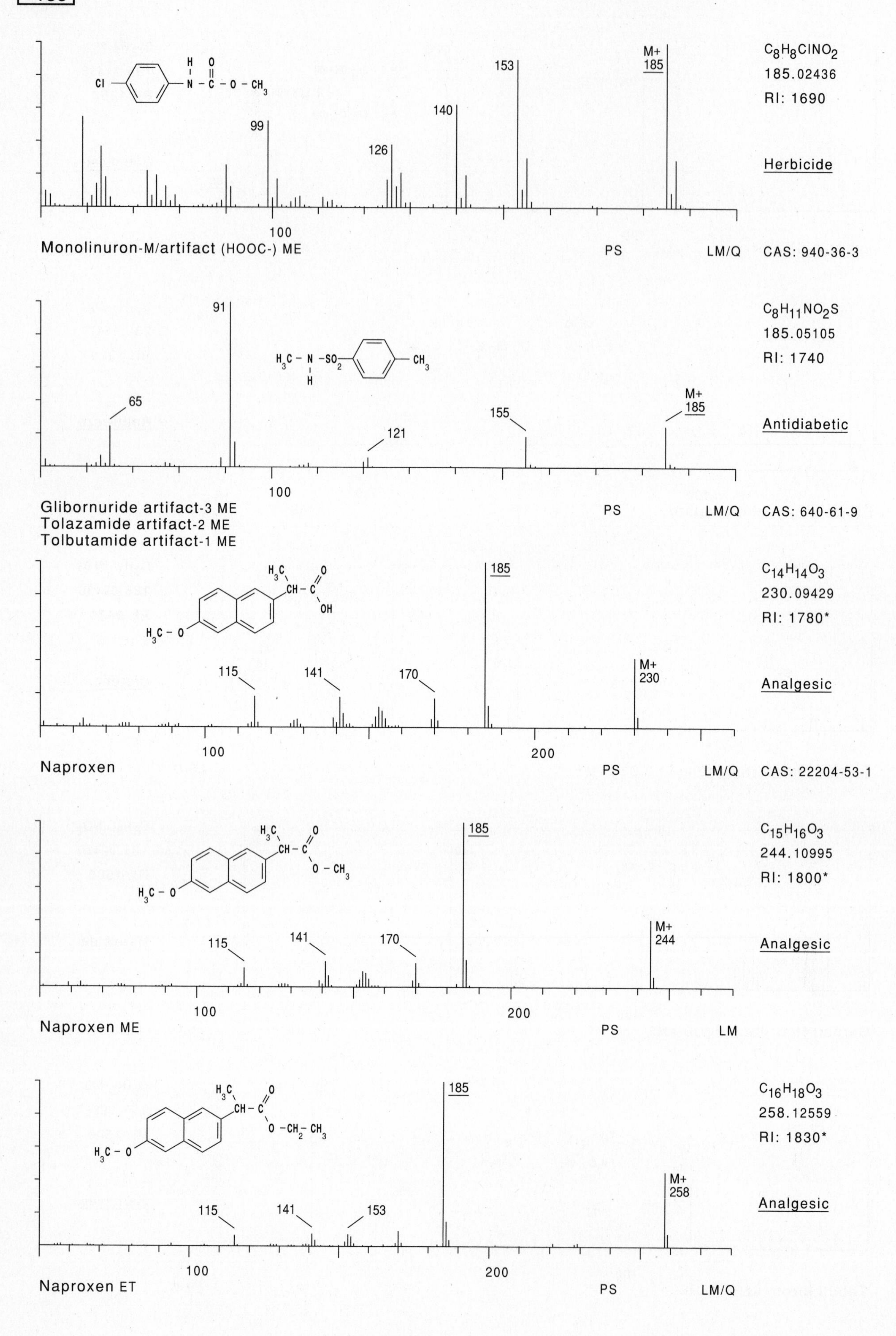

Cl
H
O
N - C - O - CH3
M+
185
153
140
99
126
100
C8H8ClNO2
185.02436
RI: 1690
Herbicide
Monolinuron-M/artifact (HOOC-) ME
PS
LM/Q
CAS: 940-36-3
91
H3C - N - SO2
H
CH3
65
M+
185
155
121
100
C8H11NO2S
185.05105
RI: 1740
Antidiabetic
Glibornuride artifact-3 ME
Tolazamide artifact-2 ME
Tolbutamide artifact-1 ME
PS
LM/Q
CAS: 640-61-9
H3C
O
CH - C
OH
H3C - O
185
115
141
170
M+
230
100
200
C14H14O3
230.09429
RI: 1780*
Analgesic
Naproxen
PS
LM/Q
CAS: 22204-53-1
H3C
O
CH - C
O - CH3
H3C - O
185
115
141
170
M+
244
100
200
C15H16O3
244.10995
RI: 1800*
Analgesic
Naproxen ME
PS
LM
H3C
O
CH - C
O - CH2 - CH3
H3C - O
185
115
141
153
M+
258
100
200
C16H18O3
258.12559
RI: 1830*
Analgesic
Naproxen ET
PS
LM/Q

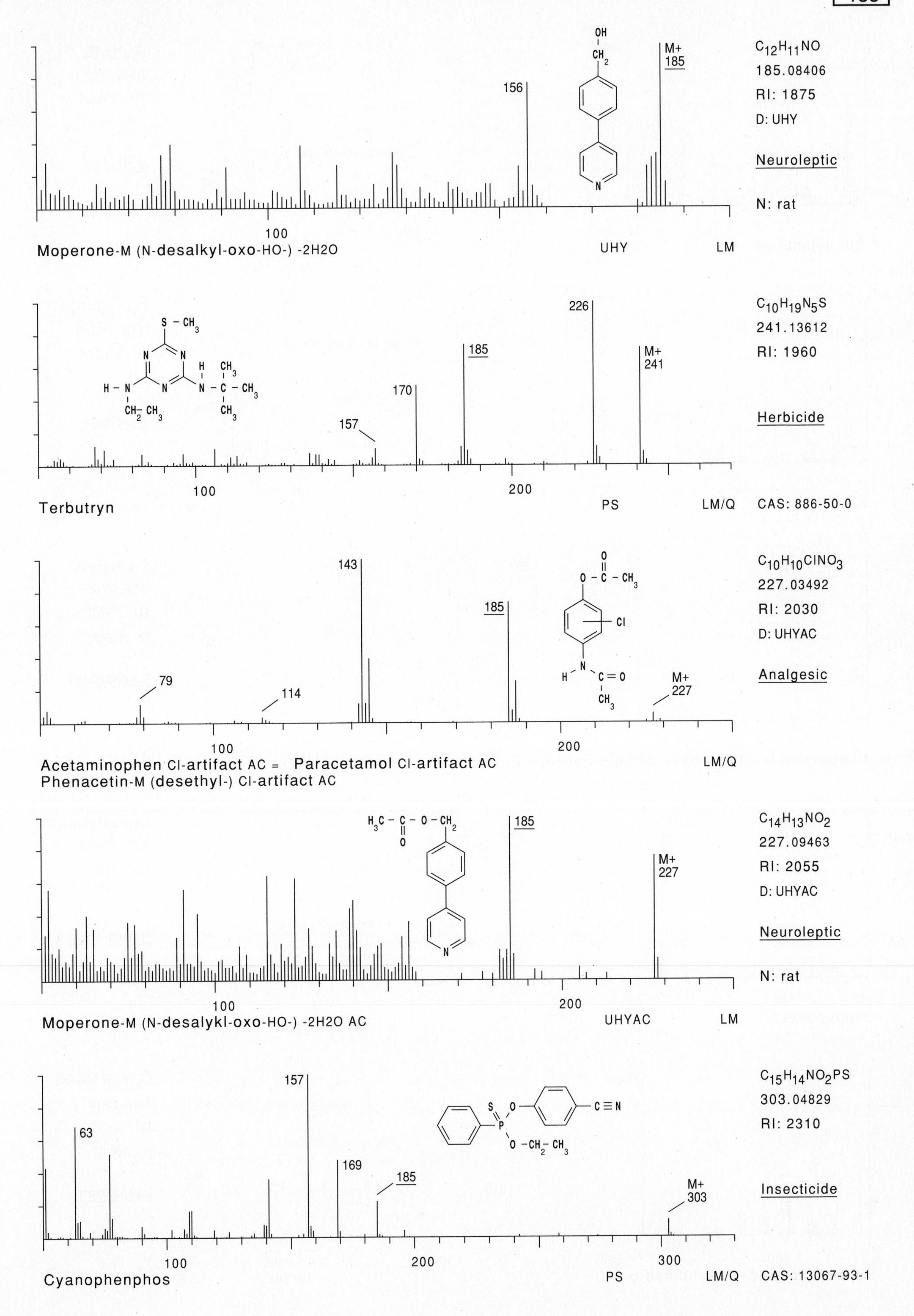

M+
185
156
100
Moperone-M (N-desalkyl-oxo-HO-) -2H2O
UHY
LM
$C_{12}H_{11}NO$
185.08406
RI: 1875
D: UHY
Neuroleptic
N: rat
226
185
M+
241
170
157
100
200
Terbutryn
PS
LM/Q
$C_{10}H_{19}N_5S$
241.13612
RI: 1960
Herbicide
CAS: 886-50-0
143
185
79
114
M+
227
100
200
Acetaminophen Cl-artifact AC = Paracetamol Cl-artifact AC
Phenacetin-M (desethyl-) Cl-artifact AC
LM/Q
$C_{10}H_{10}ClNO_3$
227.03492
RI: 2030
D: UHYAC
Analgesic
185
M+
227
100
200
Moperone-M (N-desalykl-oxo-HO-) -2H2O AC
UHYAC
LM
$C_{14}H_{13}NO_2$
227.09463
RI: 2055
D: UHYAC
Neuroleptic
N: rat
157
63
169
185
M+
303
100
200
300
Cyanophenphos
PS
LM/Q
$C_{15}H_{14}NO_2PS$
303.04829
RI: 2310
Insecticide
CAS: 13067-93-1

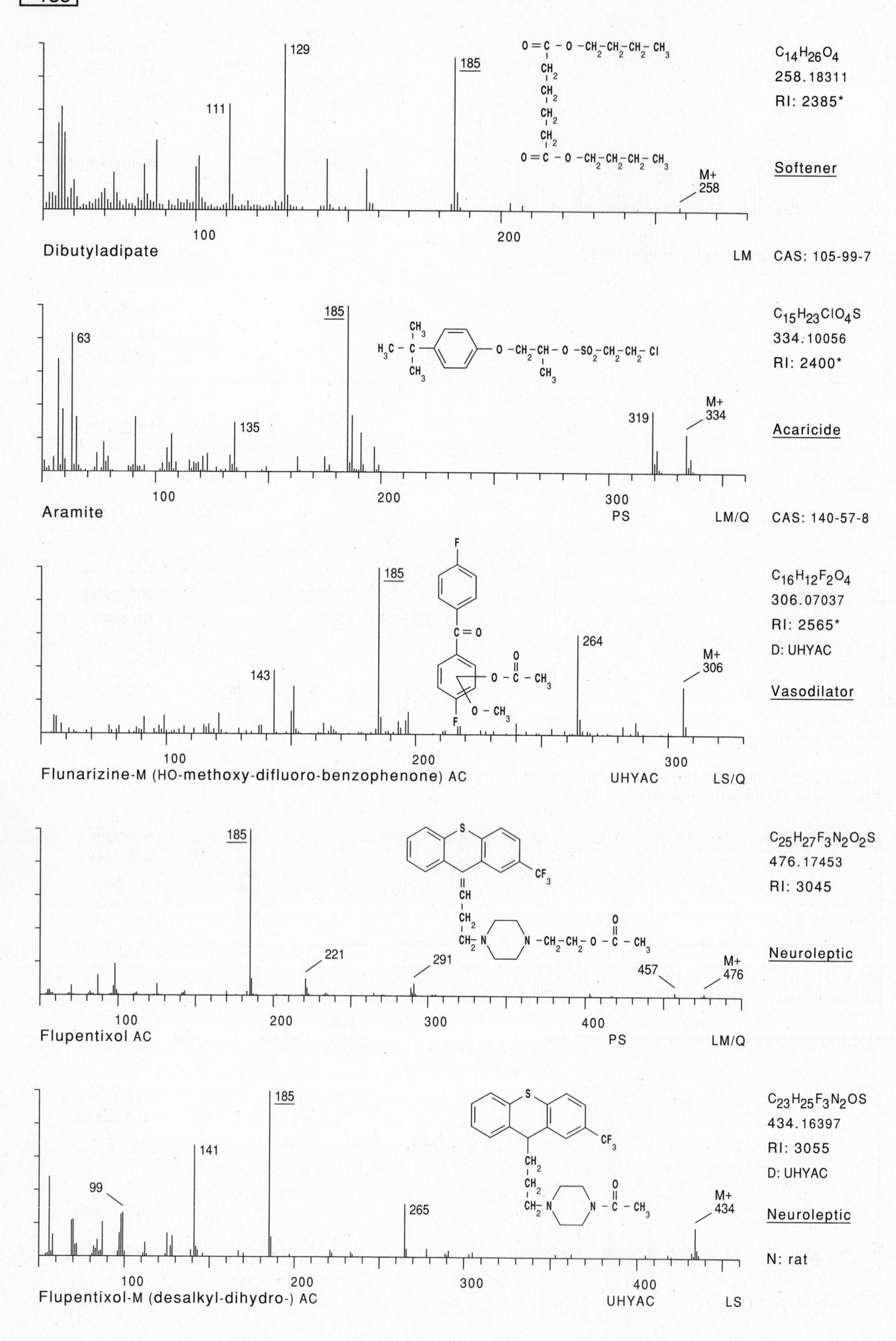
129
185
111
M+
258
100
200
Dibutyladipate
LM
$C_{14}H_{26}O_4$
258.18311
RI: 2385*
Softener
CAS: 105-99-7
185
63
135
319
M+
334
Aramite
PS
LM/Q
$C_{15}H_{23}ClO_4S$
334.10056
RI: 2400*
Acaricide
CAS: 140-57-8
185
264
143
M+
306
Flunarizine-M (HO-methoxy-difluoro-benzophenone) AC
UHYAC
LS/Q
$C_{16}H_{12}F_2O_4$
306.07037
RI: 2565*
D: UHYAC
Vasodilator
185
221
291
457
M+
476
Flupentixol AC
PS
LM/Q
$C_{25}H_{27}F_3N_2O_2S$
476.17453
RI: 3045
Neuroleptic
185
141
99
265
M+
434
Flupentixol-M (desalkyl-dihydro-) AC
UHYAC
LS
$C_{23}H_{25}F_3N_2OS$
434.16397
RI: 3055
D: UHYAC
Neuroleptic
N: rat

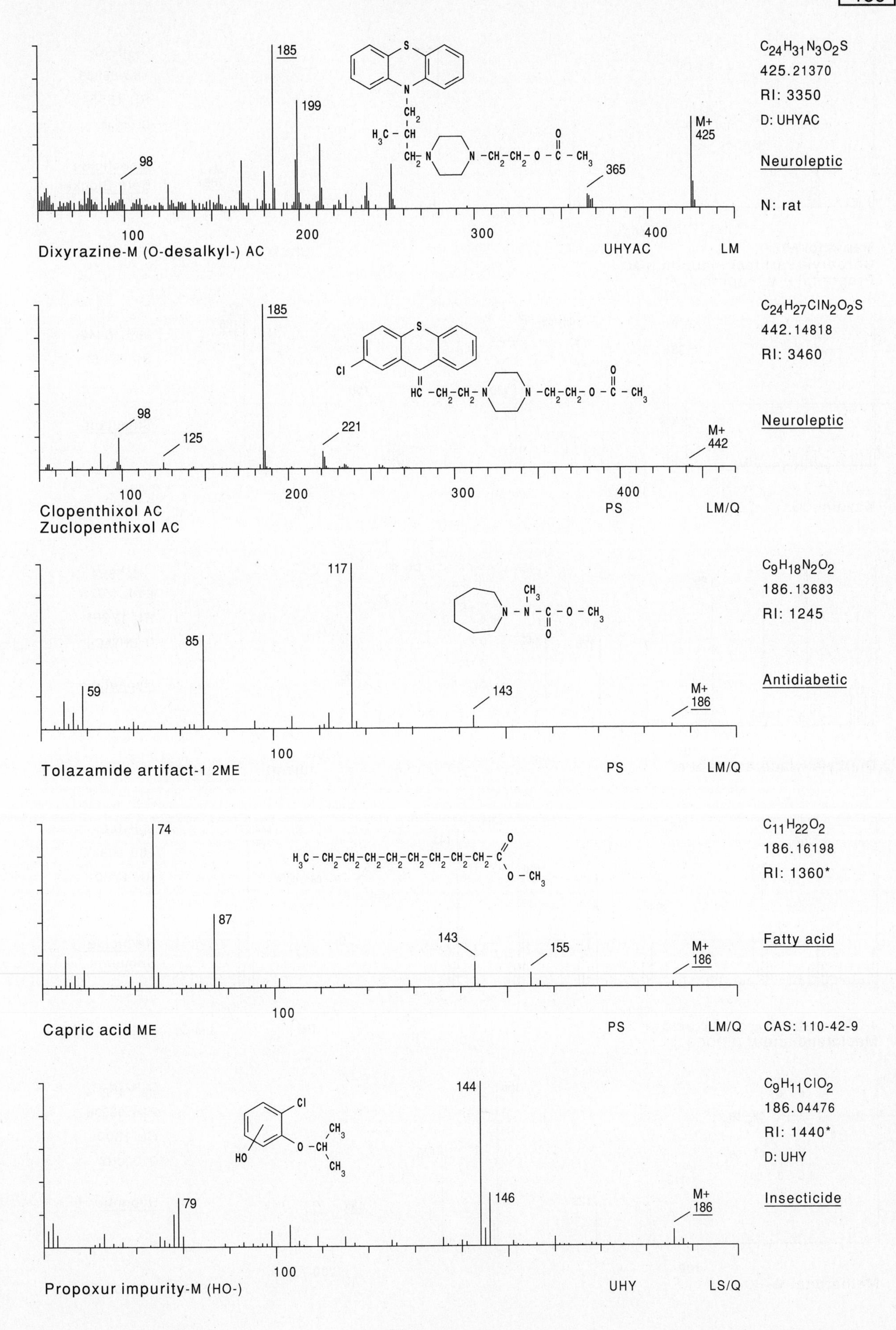

Dixyrazine-M (O-desalkyl-) AC
UHYAC
LM
C24H31N3O2S
425.21370
RI: 3350
D: UHYAC
Neuroleptic
N: rat
Clopenthixol AC
Zuclopenthixol AC
PS
LM/Q
C24H27ClN2O2S
442.14818
RI: 3460
Neuroleptic
Tolazamide artifact-1 2ME
PS
LM/Q
C9H18N2O2
186.13683
RI: 1245
Antidiabetic
Capric acid ME
PS
LM/Q
C11H22O2
186.16198
RI: 1360*
Fatty acid
CAS: 110-42-9
Propoxur impurity-M (HO-)
UHY
LS/Q
C9H11ClO2
186.04476
RI: 1440*
D: UHY
Insecticide

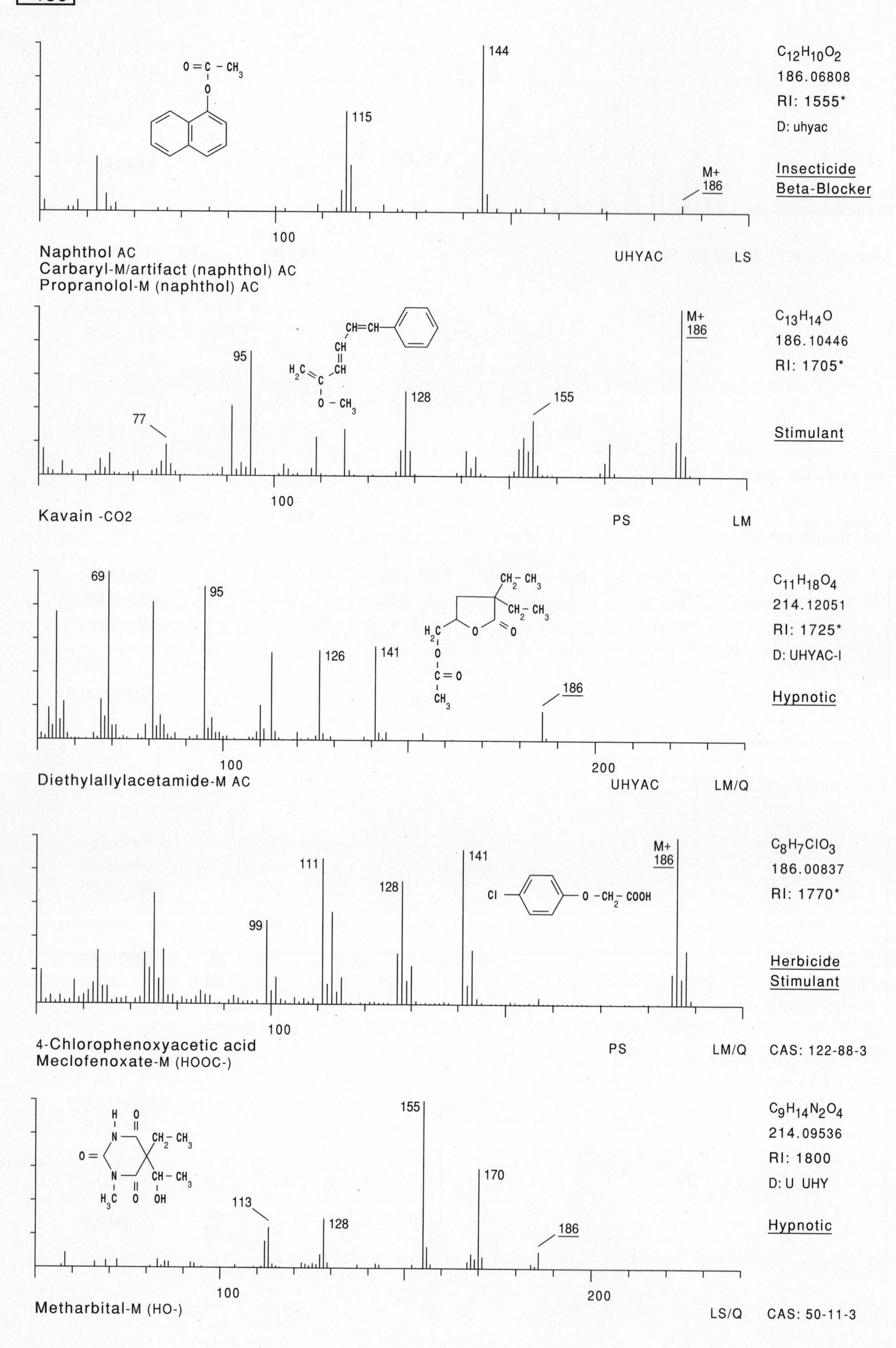
144
115
M+
186
100
UHYAC
LS
$C_{12}H_{10}O_2$
186.06808
RI: 1555*
D: uhyac
Insecticide
Beta-Blocker
Naphthol AC
Carbaryl-M/artifact (naphthol) AC
Propranolol-M (naphthol) AC
M+
186
95
128
155
77
100
PS
LM
$C_{13}H_{14}O$
186.10446
RI: 1705*
Stimulant
Kavain -CO2
69
95
126
141
186
100
200
UHYAC
LM/Q
$C_{11}H_{18}O_4$
214.12051
RI: 1725*
D: UHYAC-I
Hypnotic
Diethylallylacetamide-M AC
M+
186
111
141
128
99
100
PS
LM/Q
$C_8H_7ClO_3$
186.00837
RI: 1770*
Herbicide
Stimulant
4-Chlorophenoxyacetic acid
Meclofenoxate-M (HOOC-)
CAS: 122-88-3
155
170
113
128
186
100
200
LS/Q
$C_9H_{14}N_2O_4$
214.09536
RI: 1800
D: U UHY
Hypnotic
Metharbital-M (HO-)
CAS: 50-11-3

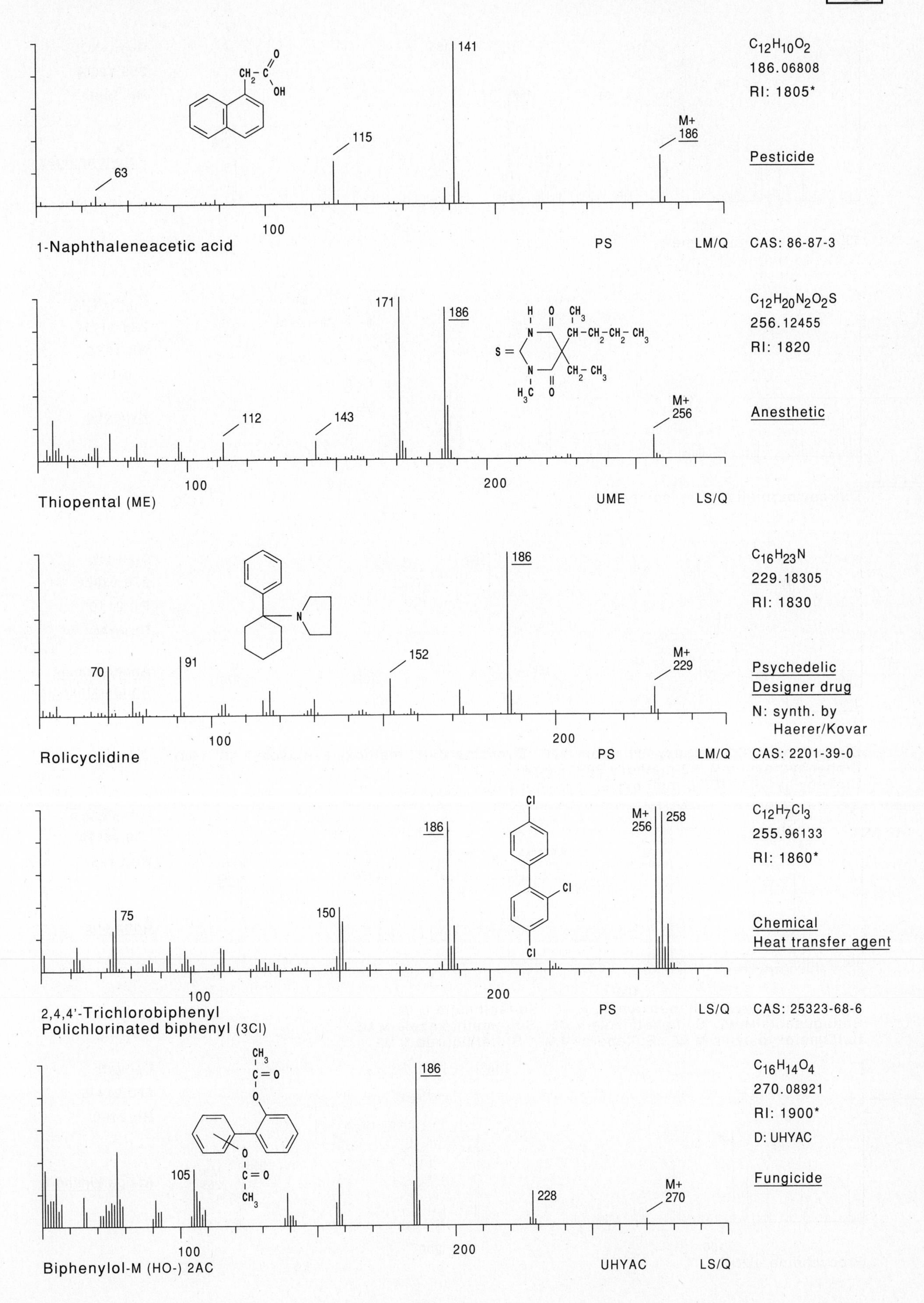
141
115
63
M+
186
100
1-Naphthaleneacetic acid
PS
LM/Q
C12H10O2
186.06808
RI: 1805*
Pesticide
CAS: 86-87-3
171
186
112
143
M+
256
100
200
Thiopental (ME)
UME
LS/Q
C12H20N2O2S
256.12455
RI: 1820
Anesthetic
186
70
91
152
M+
229
100
200
Rolicyclidine
PS
LM/Q
C16H23N
229.18305
RI: 1830
Psychedelic
Designer drug
N: synth. by
Haerer/Kovar
CAS: 2201-39-0
186
M+
256
258
75
150
Cl
100
200
2,4,4'-Trichlorobiphenyl
Polichlorinated biphenyl (3Cl)
PS
LS/Q
C12H7Cl3
255.96133
RI: 1860*
Chemical
Heat transfer agent
CAS: 25323-68-6
186
105
228
M+
270
100
200
Biphenylol-M (HO-) 2AC
UHYAC
LS/Q
C16H14O4
270.08921
RI: 1900*
D: UHYAC
Fungicide

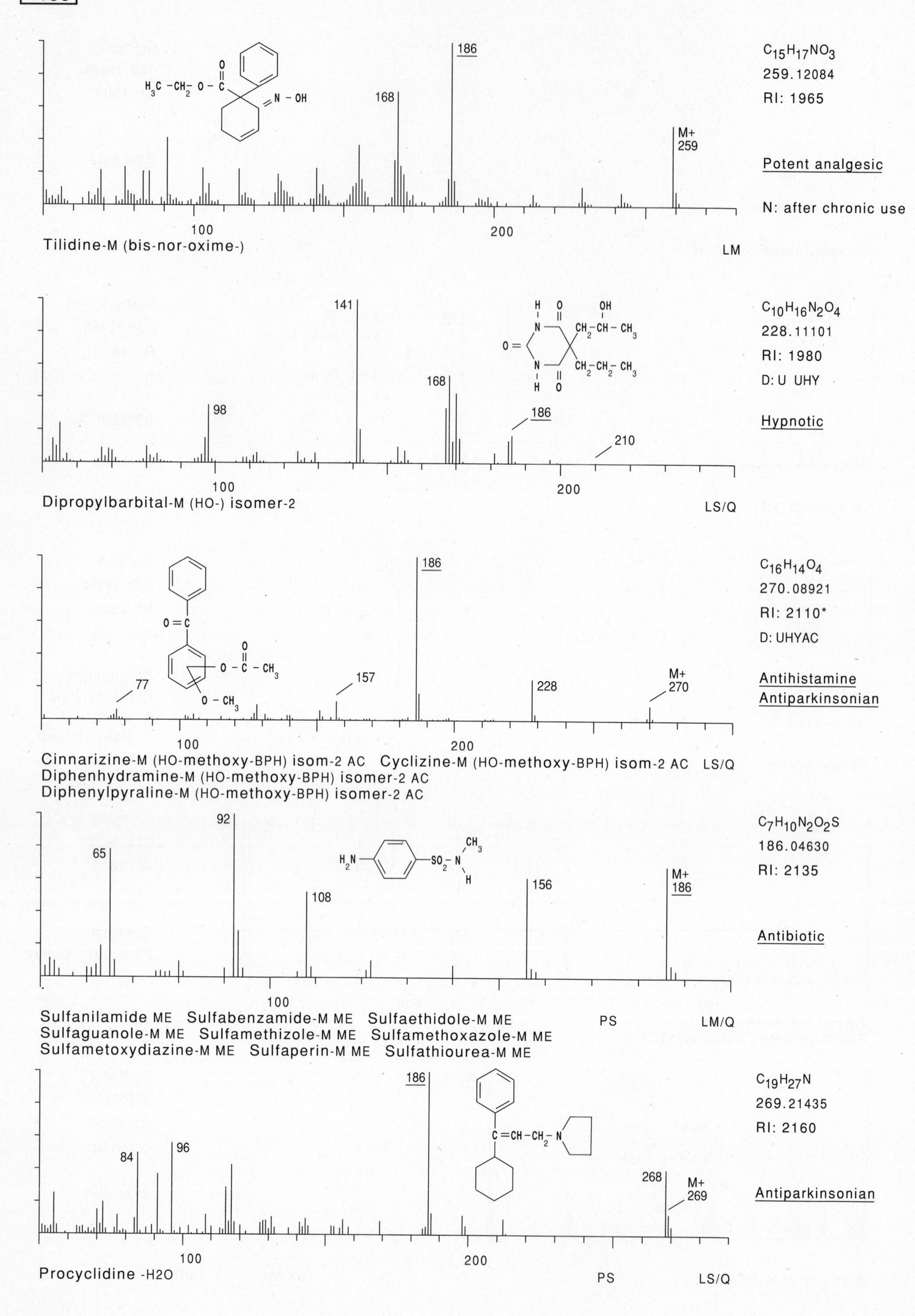

186
168
M+
259
100
200
Tilidine-M (bis-nor-oxime-)
LM
$C_{15}H_{17}NO_3$
259.12084
RI: 1965
Potent analgesic
N: after chronic use
141
98
168
186
210
100
200
Dipropylbarbital-M (HO-) isomer-2
LS/Q
$C_{10}H_{16}N_2O_4$
228.11101
RI: 1980
D: U UHY
Hypnotic
186
77
157
228
M+
270
100
200
Cinnarizine-M (HO-methoxy-BPH) isom-2 AC Cyclizine-M (HO-methoxy-BPH) isom-2 AC
LS/Q
Diphenhydramine-M (HO-methoxy-BPH) isomer-2 AC
Diphenylpyraline-M (HO-methoxy-BPH) isomer-2 AC
$C_{16}H_{14}O_4$
270.08921
RI: 2110*
D: UHYAC
Antihistamine
Antiparkinsonian
92
65
108
156
M+
186
100
Sulfanilamide ME Sulfabenzamide-M ME Sulfaethidole-M ME
PS
LM/Q
Sulfaguanole-M ME Sulfamethizole-M ME Sulfamethoxazole-M ME
Sulfametoxydiazine-M ME Sulfaperin-M ME Sulfathiourea-M ME
$C_7H_{10}N_2O_2S$
186.04630
RI: 2135
Antibiotic
186
84
96
268
M+
269
100
200
Procyclidine -H2O
PS
LS/Q
$C_{19}H_{27}N$
269.21435
RI: 2160
Antiparkinsonian

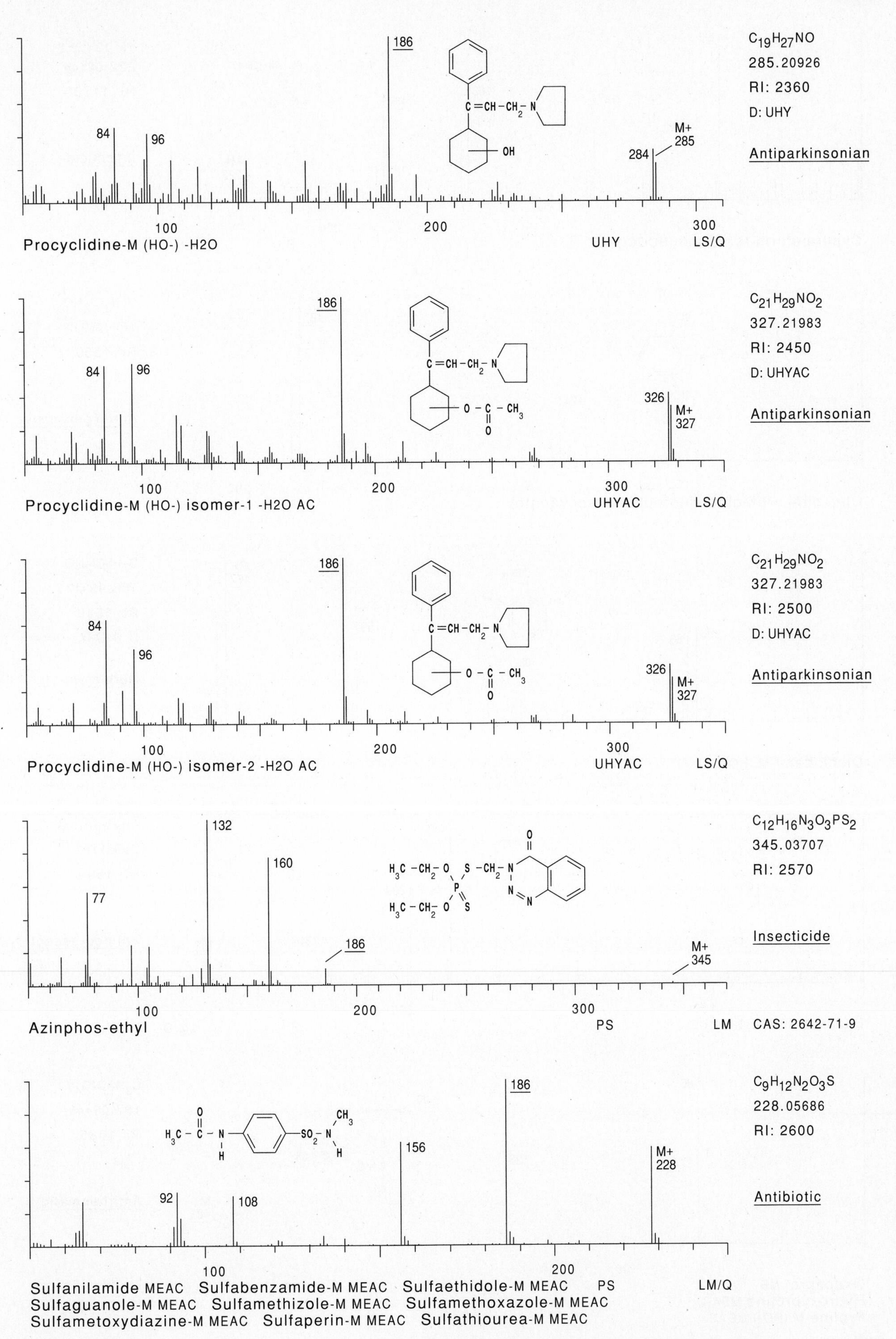

$C_{19}H_{27}NO$
285.20926
RI: 2360
D: UHY
Antiparkinsonian

$C_{21}H_{29}NO_2$
327.21983
RI: 2450
D: UHYAC
Antiparkinsonian

$C_{21}H_{29}NO_2$
327.21983
RI: 2500
D: UHYAC
Antiparkinsonian

$C_{12}H_{16}N_3O_3PS_2$
345.03707
RI: 2570
Insecticide
CAS: 2642-71-9

$C_9H_{12}N_2O_3S$
228.05686
RI: 2600
Antibiotic

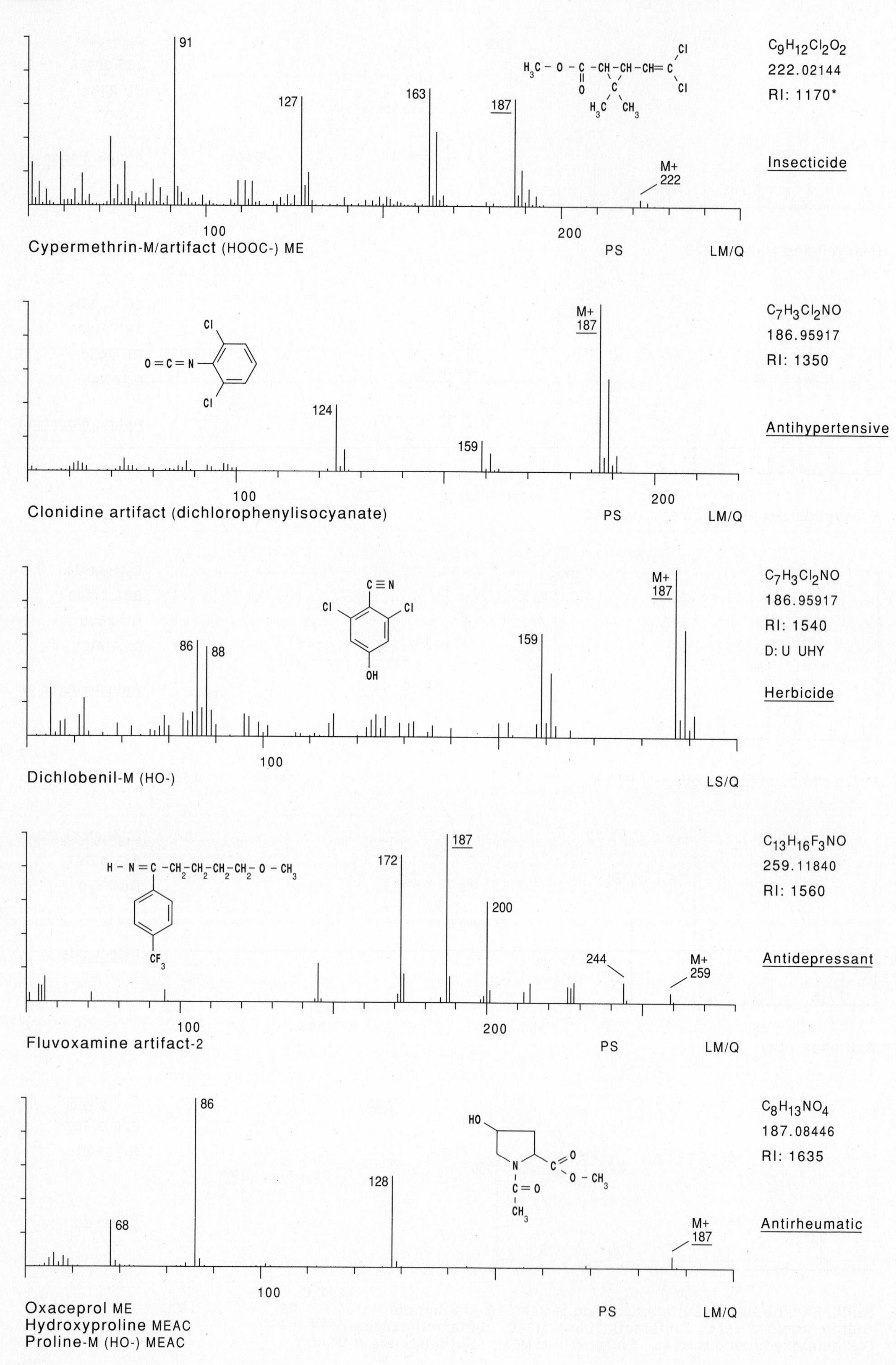
91
127
163
187
M+
222
100
200
Cypermethrin-M/artifact (HOOC-) ME
PS
LM/Q
$C_9H_{12}Cl_2O_2$
222.02144
RI: 1170*
Insecticide
M+
187
124
159
100
200
Clonidine artifact (dichlorophenylisocyanate)
PS
LM/Q
$C_7H_3Cl_2NO$
186.95917
RI: 1350
Antihypertensive
M+
187
86
88
159
100
Dichlobenil-M (HO-)
LS/Q
$C_7H_3Cl_2NO$
186.95917
RI: 1540
D: U UHY
Herbicide
187
172
200
244
M+
259
100
200
Fluvoxamine artifact-2
PS
LM/Q
$C_{13}H_{16}F_3NO$
259.11840
RI: 1560
Antidepressant
86
128
68
M+
187
100
Oxaceprol ME
Hydroxyproline MEAC
Proline-M (HO-) MEAC
PS
LM/Q
$C_8H_{13}NO_4$
187.08446
RI: 1635
Antirheumatic

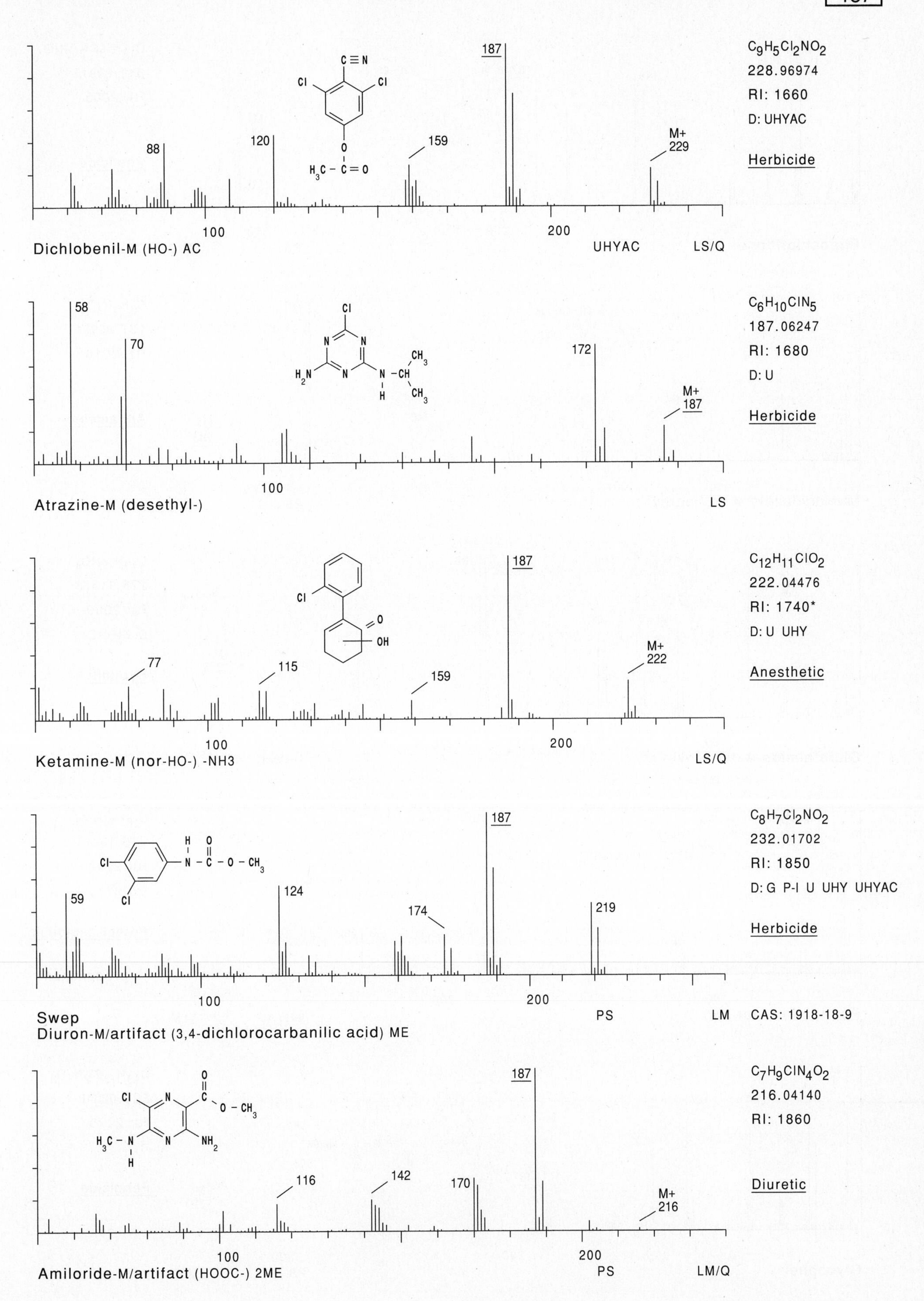

$C_9H_5Cl_2NO_2$
228.96974
RI: 1660
D: UHYAC
Herbicide
Dichlobenil-M (HO-) AC
UHYAC
LS/Q
$C_6H_{10}ClN_5$
187.06247
RI: 1680
D: U
Herbicide
Atrazine-M (desethyl-)
LS
$C_{12}H_{11}ClO_2$
222.04476
RI: 1740*
D: U UHY
Anesthetic
Ketamine-M (nor-HO-) -NH3
LS/Q
$C_8H_7Cl_2NO_2$
232.01702
RI: 1850
D: G P-I U UHY UHYAC
Herbicide
Swep
Diuron-M/artifact (3,4-dichlorocarbanilic acid) ME
PS
LM
CAS: 1918-18-9
$C_7H_9ClN_4O_2$
216.04140
RI: 1860
Diuretic
Amiloride-M/artifact (HOOC-) 2ME
PS
LM/Q

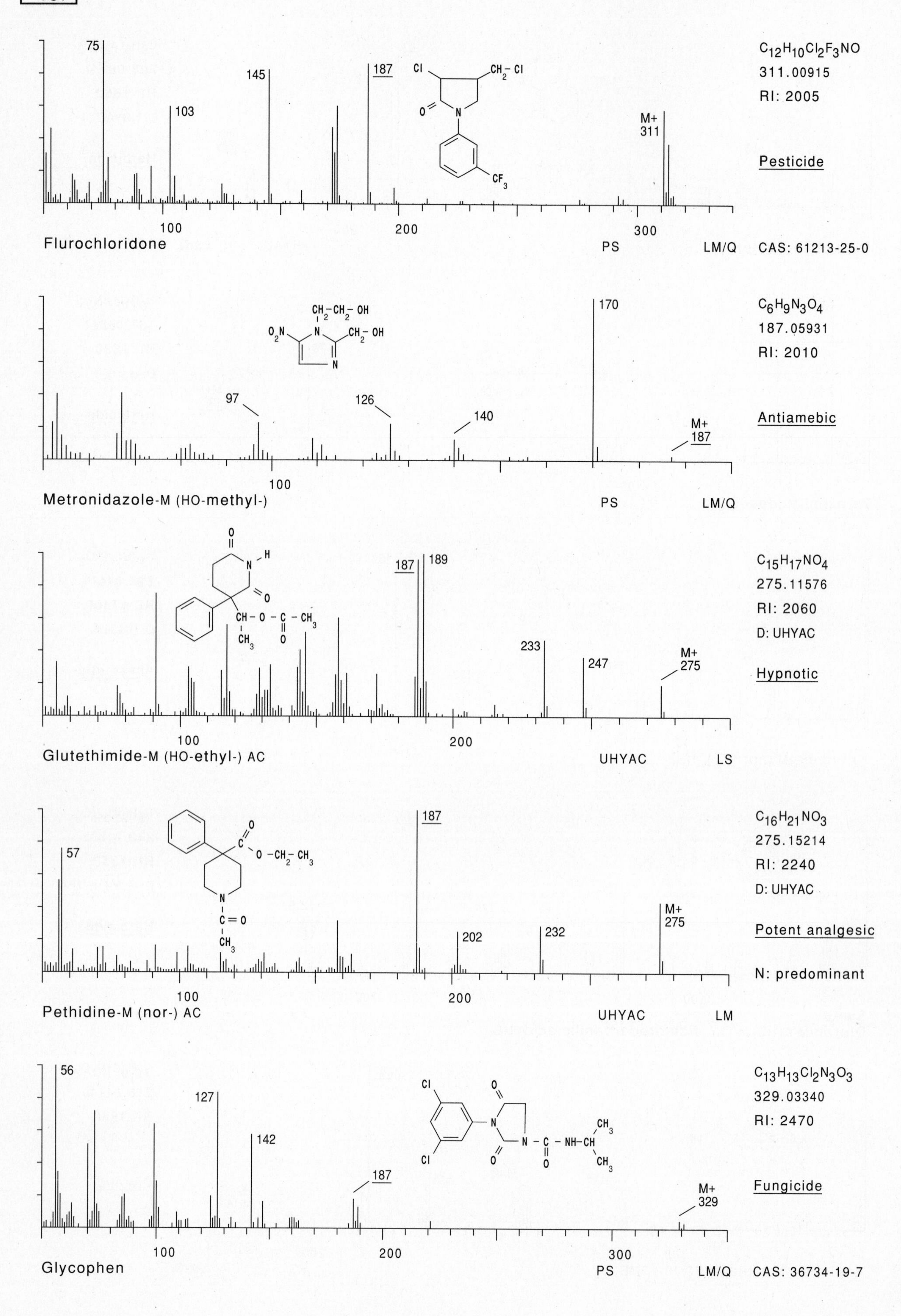

Flurochloridone
$C_{12}H_{10}Cl_2F_3NO$
311.00915
RI: 2005
Pesticide
CAS: 61213-25-0
PS
LM/Q
Metronidazole-M (HO-methyl-)
$C_6H_9N_3O_4$
187.05931
RI: 2010
Antiamebic
PS
LM/Q
Glutethimide-M (HO-ethyl-) AC
$C_{15}H_{17}NO_4$
275.11576
RI: 2060
D: UHYAC
Hypnotic
UHYAC
LS
Pethidine-M (nor-) AC
$C_{16}H_{21}NO_3$
275.15214
RI: 2240
D: UHYAC
Potent analgesic
N: predominant
UHYAC
LM
Glycophen
$C_{13}H_{13}Cl_2N_3O_3$
329.03340
RI: 2470
Fungicide
CAS: 36734-19-7
PS
LM/Q

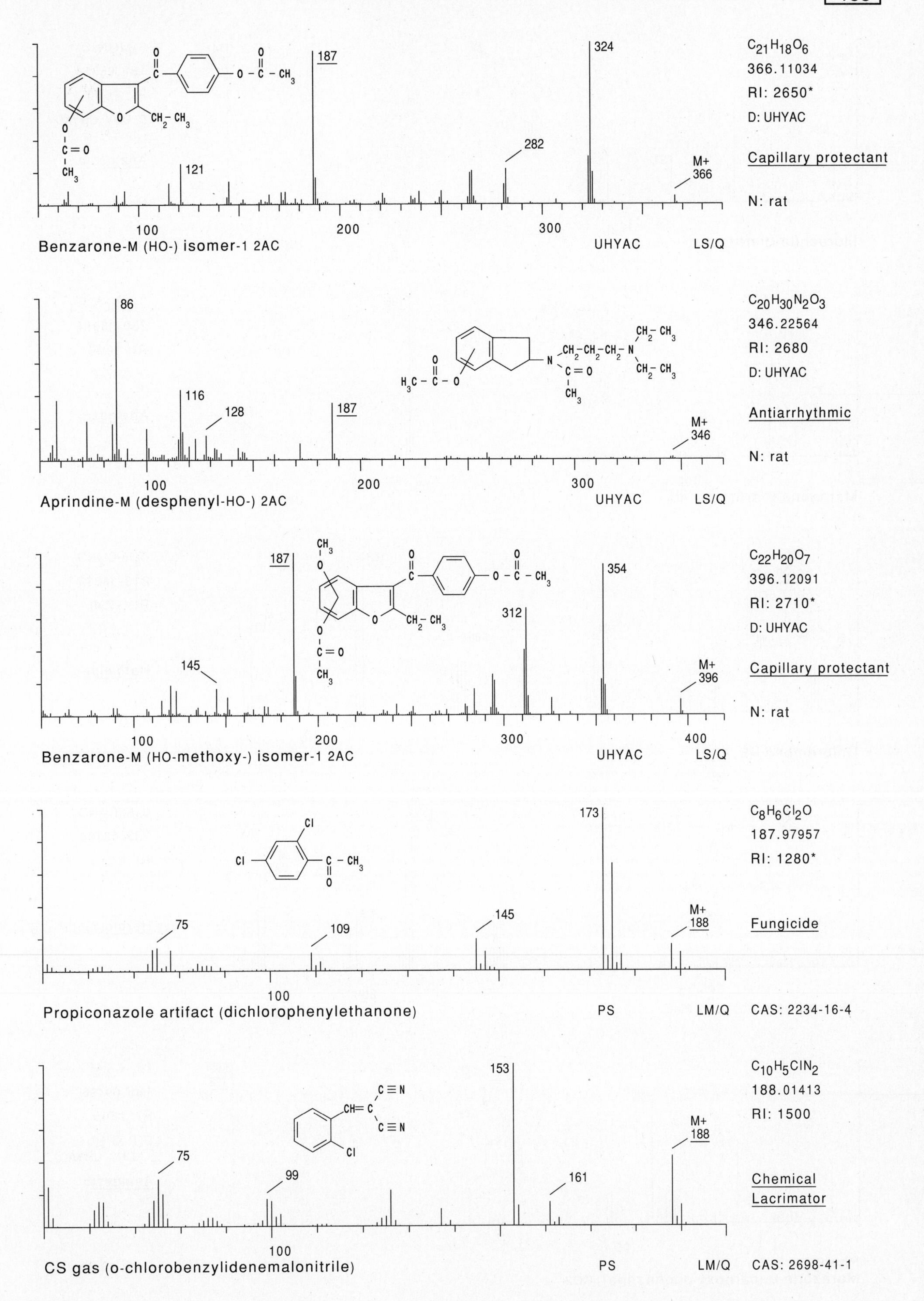

187
324
121
282
M+
366
100
200
300
Benzarone-M (HO-) isomer-1 2AC
UHYAC
LS/Q
$C_{21}H_{18}O_6$
366.11034
RI: 2650*
D: UHYAC
Capillary protectant
N: rat
86
116
128
187
M+
346
100
200
300
Aprindine-M (desphenyl-HO-) 2AC
UHYAC
LS/Q
$C_{20}H_{30}N_2O_3$
346.22564
RI: 2680
D: UHYAC
Antiarrhythmic
N: rat
187
354
312
145
M+
396
100
200
300
400
Benzarone-M (HO-methoxy-) isomer-1 2AC
UHYAC
LS/Q
$C_{22}H_{20}O_7$
396.12091
RI: 2710*
D: UHYAC
Capillary protectant
N: rat
173
75
109
145
M+
188
100
Propiconazole artifact (dichlorophenylethanone)
PS
LM/Q
$C_8H_6Cl_2O$
187.97957
RI: 1280*
Fungicide
CAS: 2234-16-4
153
75
99
161
M+
188
100
CS gas (o-chlorobenzylidenemalonitrile)
PS
LM/Q
$C_{10}H_5ClN_2$
188.01413
RI: 1500
Chemical
Lacrimator
CAS: 2698-41-1

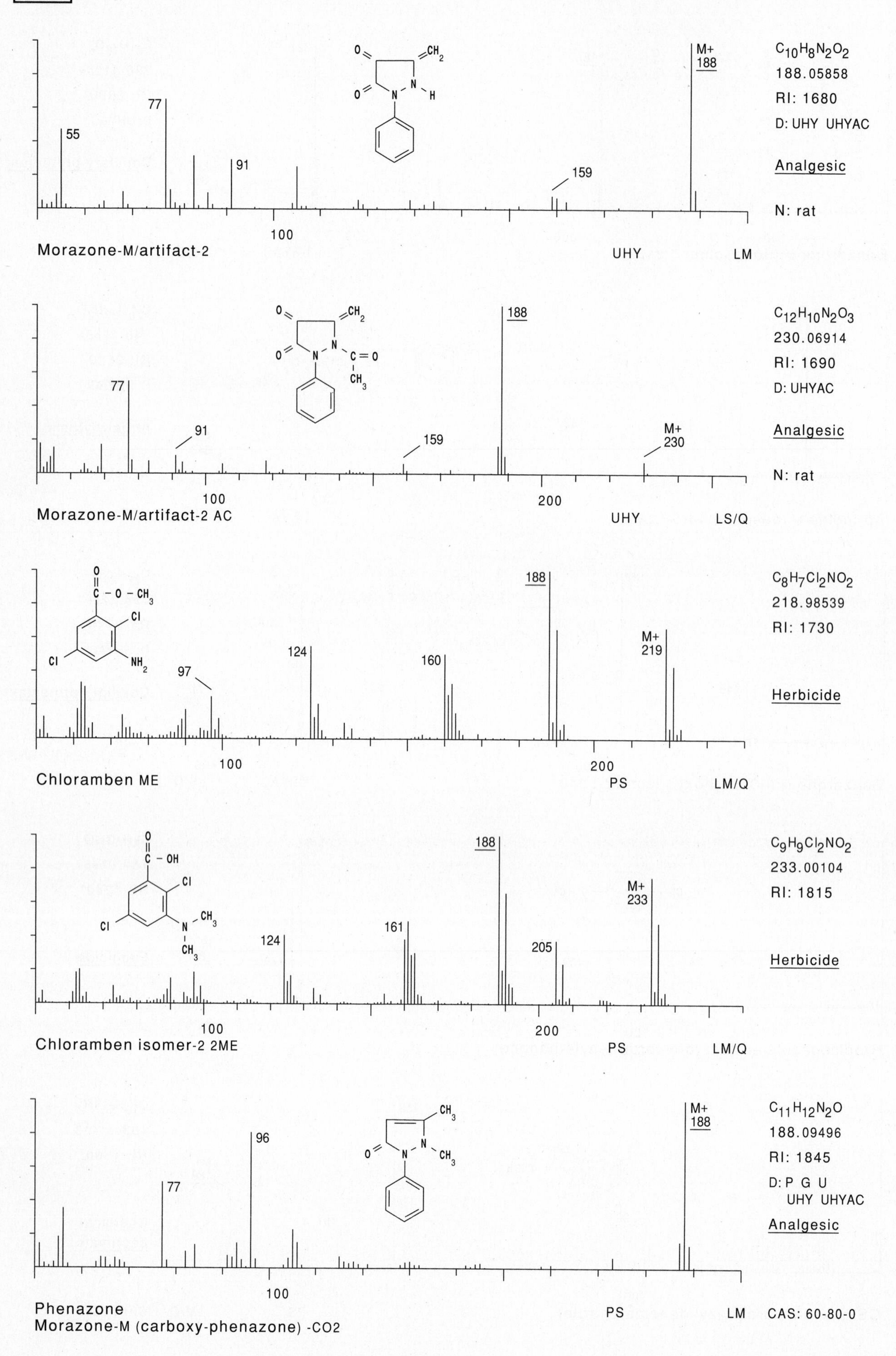
M+
188
55
77
91
159
100
Morazone-M/artifact-2
UHY
LM
$C_{10}H_8N_2O_2$
188.05858
RI: 1680
D: UHY UHYAC
Analgesic
N: rat
188
77
91
159
M+
230
100
200
Morazone-M/artifact-2 AC
UHY
LS/Q
$C_{12}H_{10}N_2O_3$
230.06914
RI: 1690
D: UHYAC
Analgesic
N: rat
188
124
160
97
M+
219
100
200
Chloramben ME
PS
LM/Q
$C_8H_7Cl_2NO_2$
218.98539
RI: 1730
Herbicide
188
M+
233
124
161
205
100
200
Chloramben isomer-2 2ME
PS
LM/Q
$C_9H_9Cl_2NO_2$
233.00104
RI: 1815
Herbicide
M+
188
96
77
100
Phenazone
Morazone-M (carboxy-phenazone) -CO2
PS
LM
CAS: 60-80-0
$C_{11}H_{12}N_2O$
188.09496
RI: 1845
D: P G U
UHY UHYAC
Analgesic

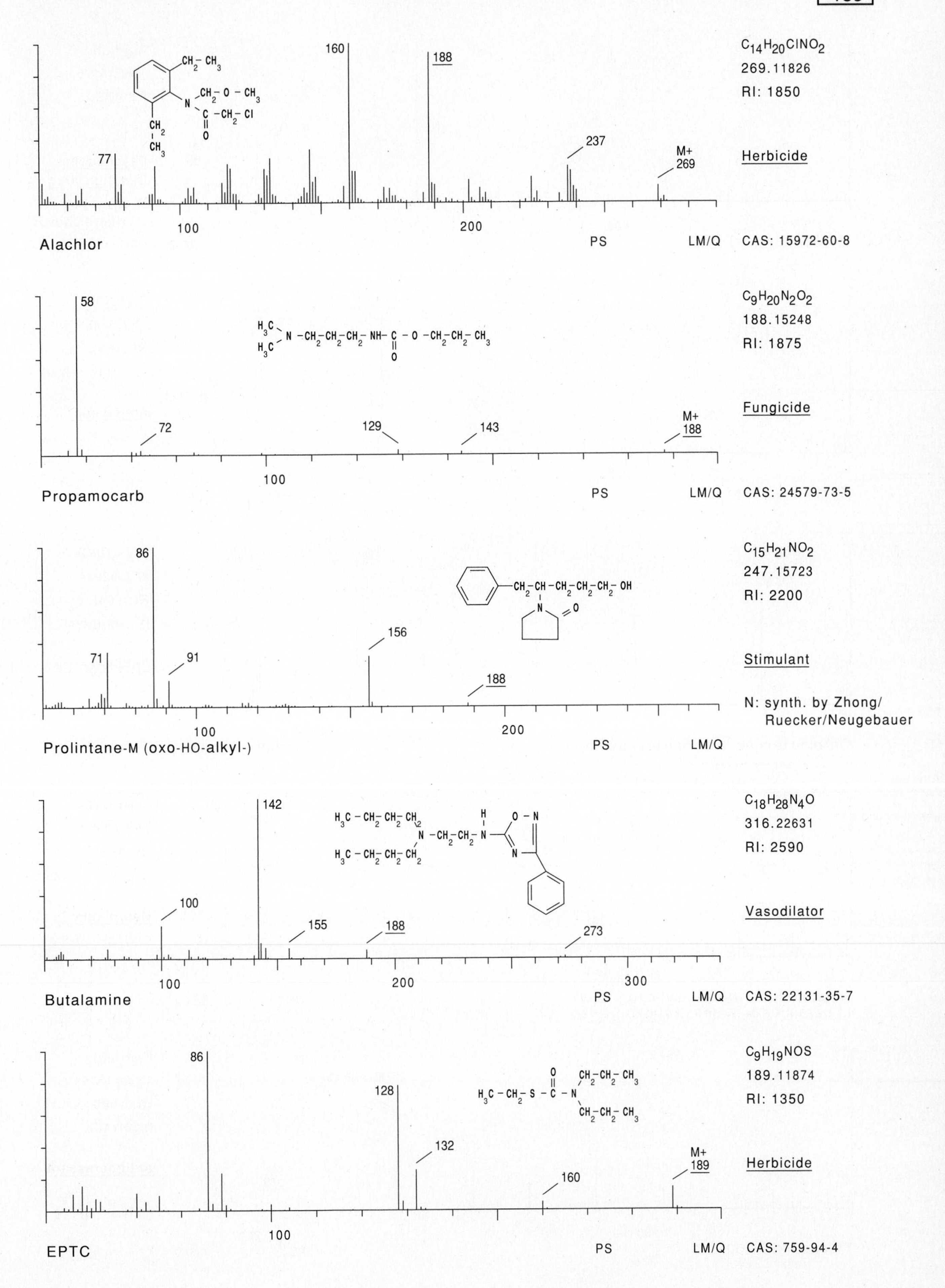

160
188
77
237
M+
269
100
200
Alachlor
PS
LM/Q
C14H20ClNO2
269.11826
RI: 1850
Herbicide
CAS: 15972-60-8
58
72
129
143
M+
188
100
Propamocarb
PS
LM/Q
C9H20N2O2
188.15248
RI: 1875
Fungicide
CAS: 24579-73-5
86
71
91
156
188
100
200
Prolintane-M (oxo-HO-alkyl-)
PS
LM/Q
C15H21NO2
247.15723
RI: 2200
Stimulant
N: synth. by Zhong/
Ruecker/Neugebauer
142
100
155
188
273
100
200
300
Butalamine
PS
LM/Q
C18H28N4O
316.22631
RI: 2590
Vasodilator
CAS: 22131-35-7
86
128
132
160
M+
189
100
EPTC
PS
LM/Q
C9H19NOS
189.11874
RI: 1350
Herbicide
CAS: 759-94-4

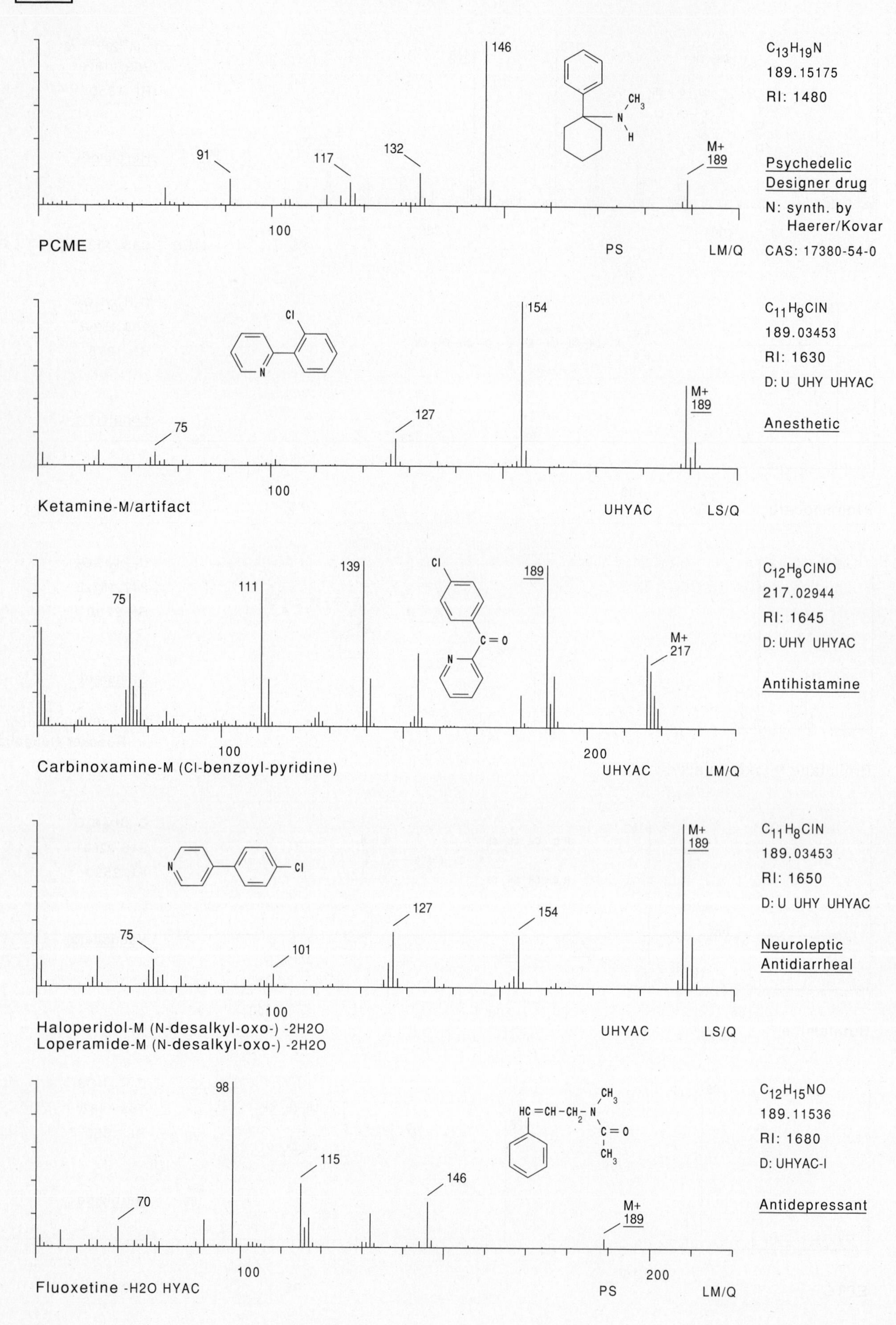

146
91
117
132
M+
189
100
CH3
N
H
PCME
PS
LM/Q
$C_{13}H_{19}N$
189.15175
RI: 1480
Psychedelic
Designer drug
N: synth. by Haerer/Kovar
CAS: 17380-54-0
154
Cl
N
127
75
M+
189
100
Ketamine-M/artifact
UHYAC
LS/Q
$C_{11}H_8ClN$
189.03453
RI: 1630
D: U UHY UHYAC
Anesthetic
139
111
75
189
Cl
C=O
N
M+
217
100
200
Carbinoxamine-M (Cl-benzoyl-pyridine)
UHYAC
LM/Q
$C_{12}H_8ClNO$
217.02944
RI: 1645
D: UHY UHYAC
Antihistamine
M+
189
N
Cl
127
154
75
101
100
Haloperidol-M (N-desalkyl-oxo-) -2H2O
Loperamide-M (N-desalkyl-oxo-) -2H2O
UHYAC
LS/Q
$C_{11}H_8ClN$
189.03453
RI: 1650
D: U UHY UHYAC
Neuroleptic
Antidiarrheal
98
HC=CH-CH2-N
CH3
C=O
CH3
115
146
70
M+
189
100
200
Fluoxetine -H2O HYAC
PS
LM/Q
$C_{12}H_{15}NO$
189.11536
RI: 1680
D: UHYAC-I
Antidepressant

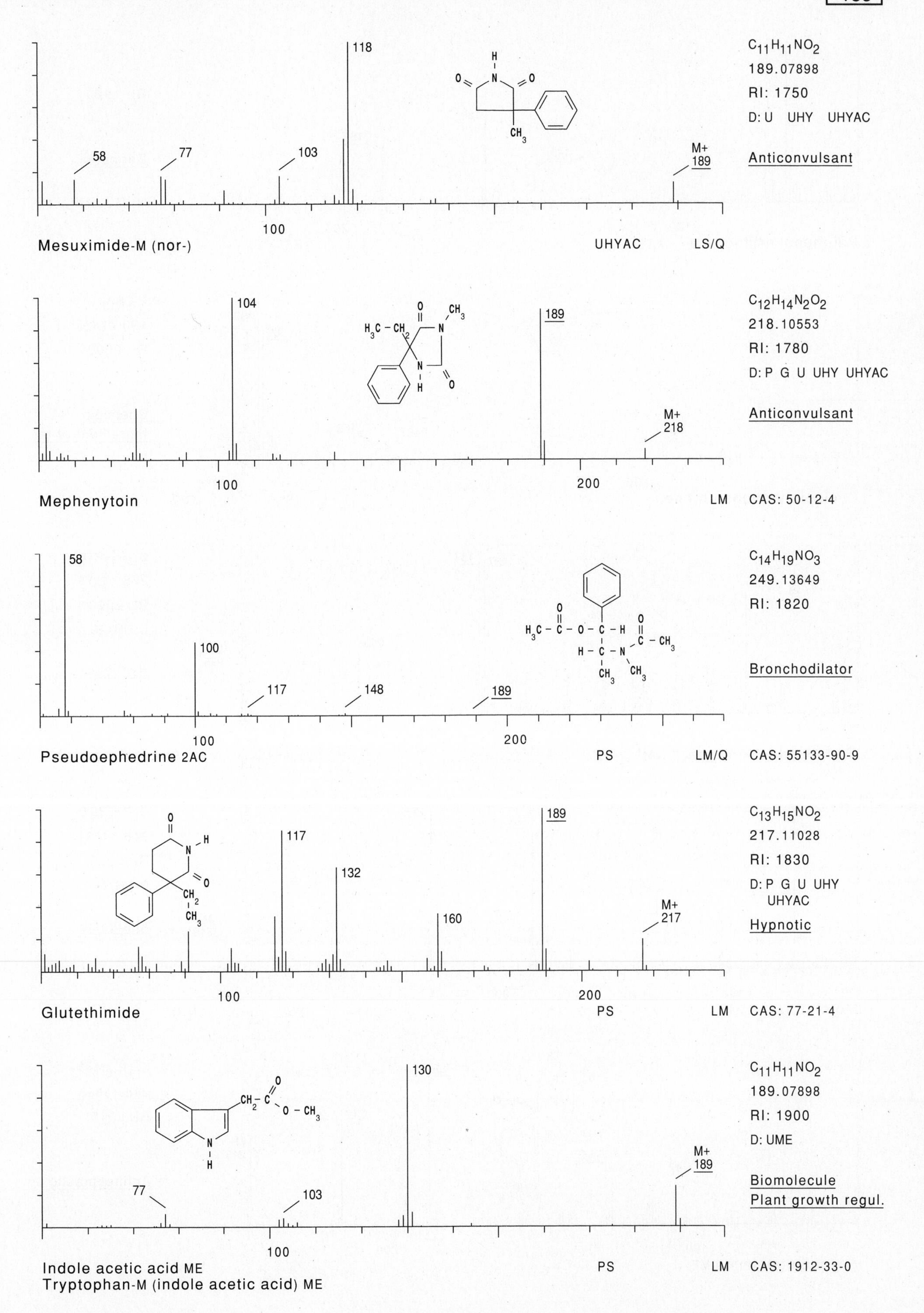
118
58
77
103
M+
189
100
Mesuximide-M (nor-)
UHYAC
LS/Q
$C_{11}H_{11}NO_2$
189.07898
RI: 1750
D: U UHY UHYAC
Anticonvulsant
104
189
M+
218
100
200
Mephenytoin
LM
$C_{12}H_{14}N_2O_2$
218.10553
RI: 1780
D: P G U UHY UHYAC
Anticonvulsant
CAS: 50-12-4
58
100
117
148
189
100
200
Pseudoephedrine 2AC
PS
LM/Q
$C_{14}H_{19}NO_3$
249.13649
RI: 1820
Bronchodilator
CAS: 55133-90-9
117
132
160
189
M+
217
100
200
Glutethimide
PS
LM
$C_{13}H_{15}NO_2$
217.11028
RI: 1830
D: P G U UHY
UHYAC
Hypnotic
CAS: 77-21-4
130
77
103
M+
189
100
Indole acetic acid ME
Tryptophan-M (indole acetic acid) ME
PS
LM
$C_{11}H_{11}NO_2$
189.07898
RI: 1900
D: UME
Biomolecule
Plant growth regul.
CAS: 1912-33-0

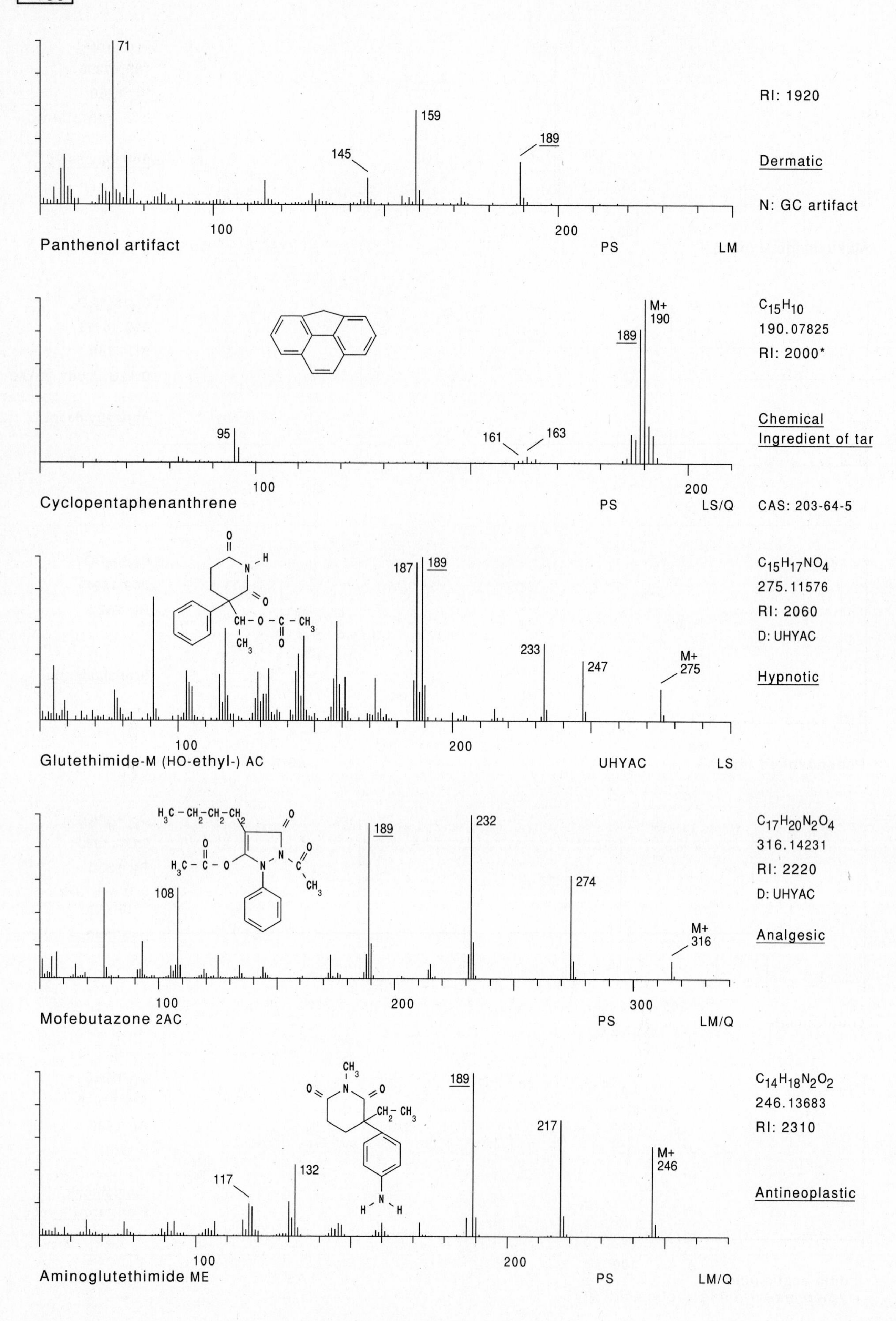
71
159
145
189
100
200
Panthenol artifact
PS
LM
RI: 1920
Dermatic
N: GC artifact
M+
190
189
95
161
163
100
200
Cyclopentaphenanthrene
PS
LS/Q
C15H10
190.07825
RI: 2000*
Chemical
Ingredient of tar
CAS: 203-64-5
187
189
233
247
M+
275
100
200
Glutethimide-M (HO-ethyl-) AC
UHYAC
LS
C15H17NO4
275.11576
RI: 2060
D: UHYAC
Hypnotic
189
232
274
108
M+
316
100
200
300
Mofebutazone 2AC
PS
LM/Q
C17H20N2O4
316.14231
RI: 2220
D: UHYAC
Analgesic
189
217
M+
246
132
117
100
200
Aminoglutethimide ME
PS
LM/Q
C14H18N2O2
246.13683
RI: 2310
Antineoplastic

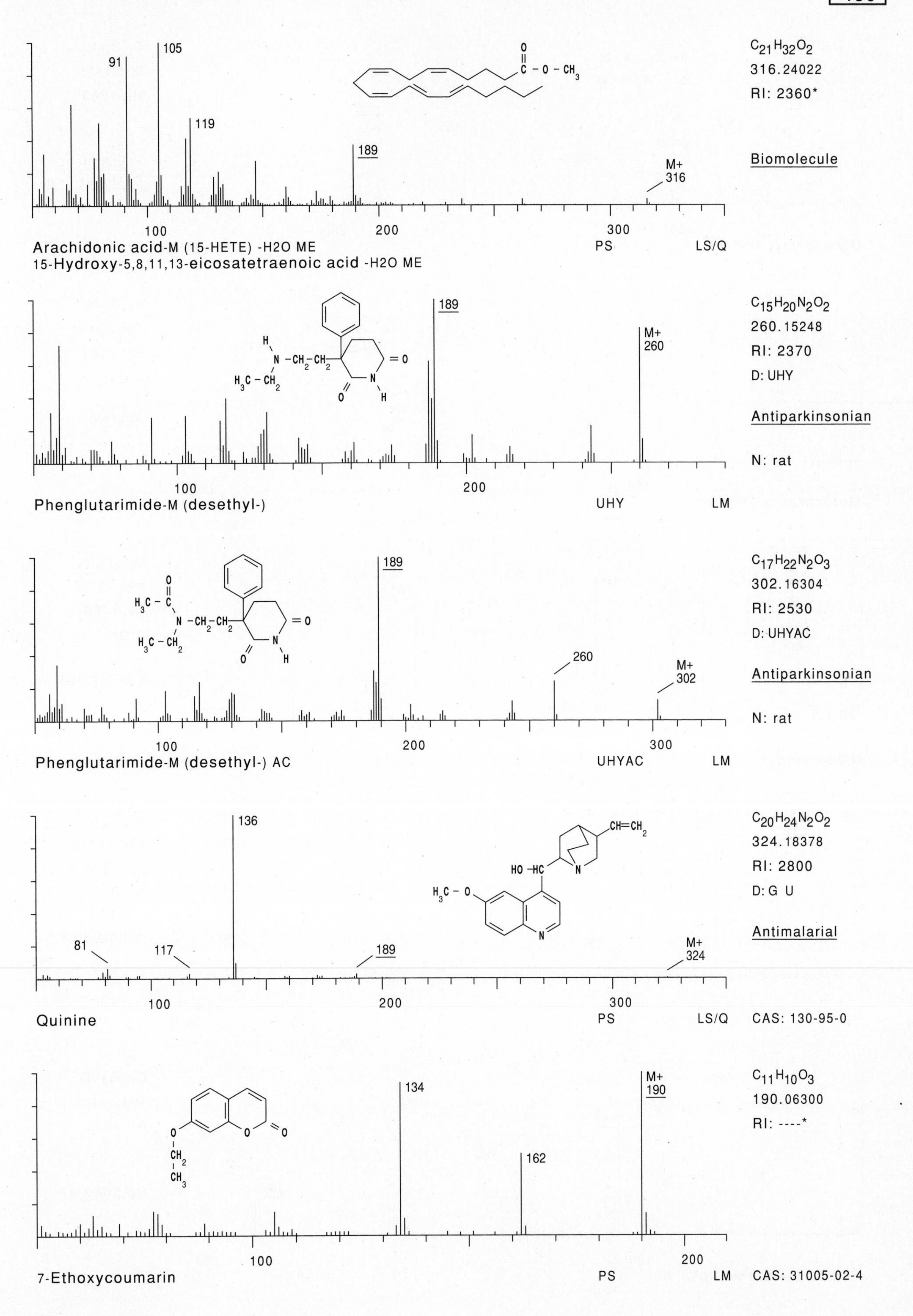

$C_{21}H_{32}O_2$
316.24022
RI: 2360*

Biomolecule

$C_{15}H_{20}N_2O_2$
260.15248
RI: 2370
D: UHY

Antiparkinsonian

N: rat

$C_{17}H_{22}N_2O_3$
302.16304
RI: 2530
D: UHYAC

Antiparkinsonian

N: rat

$C_{20}H_{24}N_2O_2$
324.18378
RI: 2800
D: G U

Antimalarial

CAS: 130-95-0

$C_{11}H_{10}O_3$
190.06300
RI: ----*

CAS: 31005-02-4

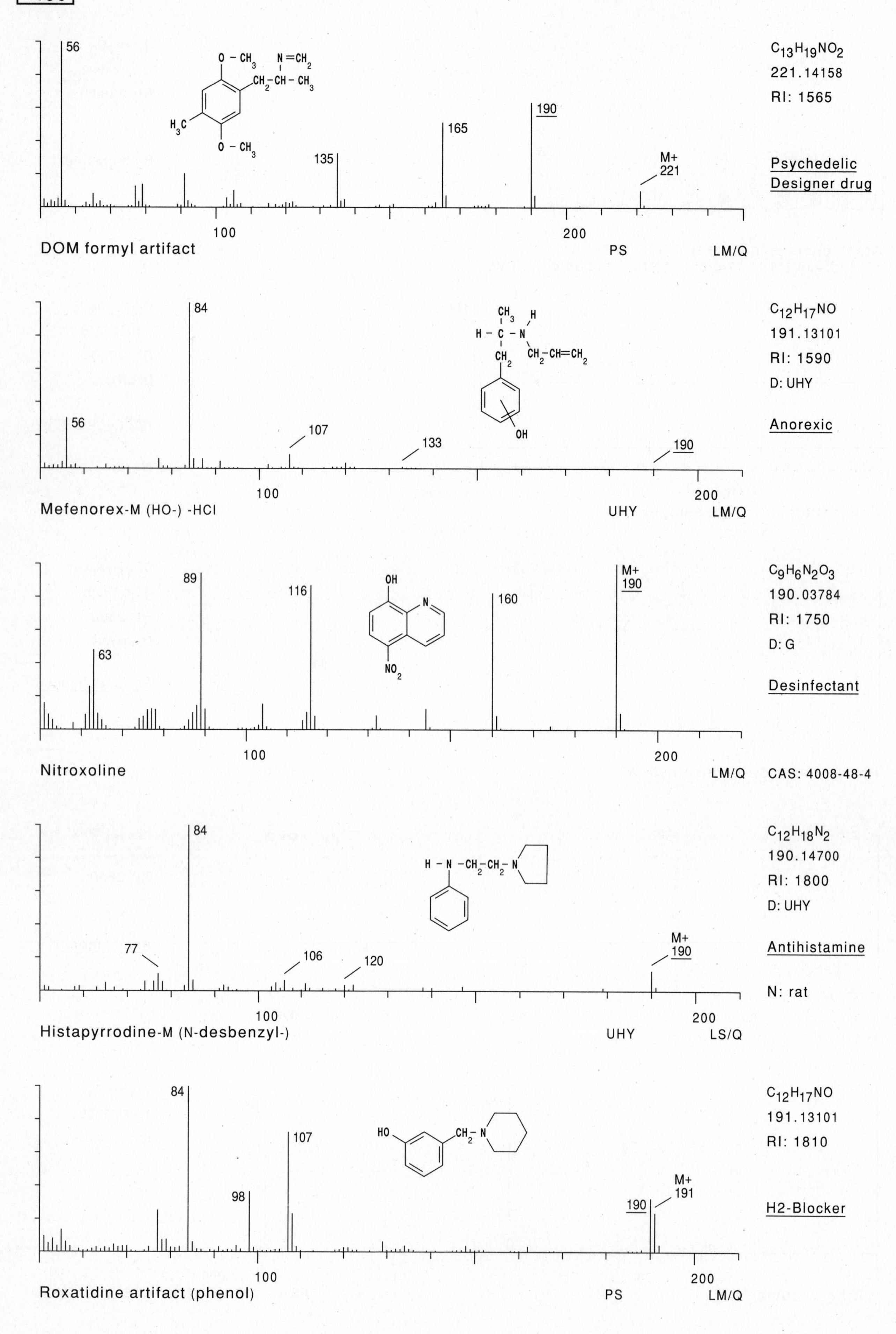

56
190
165
135
M+
221
100
200
C13H19NO2
221.14158
RI: 1565
Psychedelic
Designer drug
DOM formyl artifact
PS
LM/Q
84
56
107
133
190
100
200
C12H17NO
191.13101
RI: 1590
D: UHY
Anorexic
Mefenorex-M (HO-) -HCl
UHY
LM/Q
89
116
160
M+
190
63
100
200
C9H6N2O3
190.03784
RI: 1750
D: G
Desinfectant
Nitroxoline
LM/Q
CAS: 4008-48-4
84
77
106
120
M+
190
100
200
C12H18N2
190.14700
RI: 1800
D: UHY
Antihistamine
N: rat
Histapyrrodine-M (N-desbenzyl-)
UHY
LS/Q
84
107
98
190
M+
191
100
200
C12H17NO
191.13101
RI: 1810
H2-Blocker
Roxatidine artifact (phenol)
PS
LM/Q

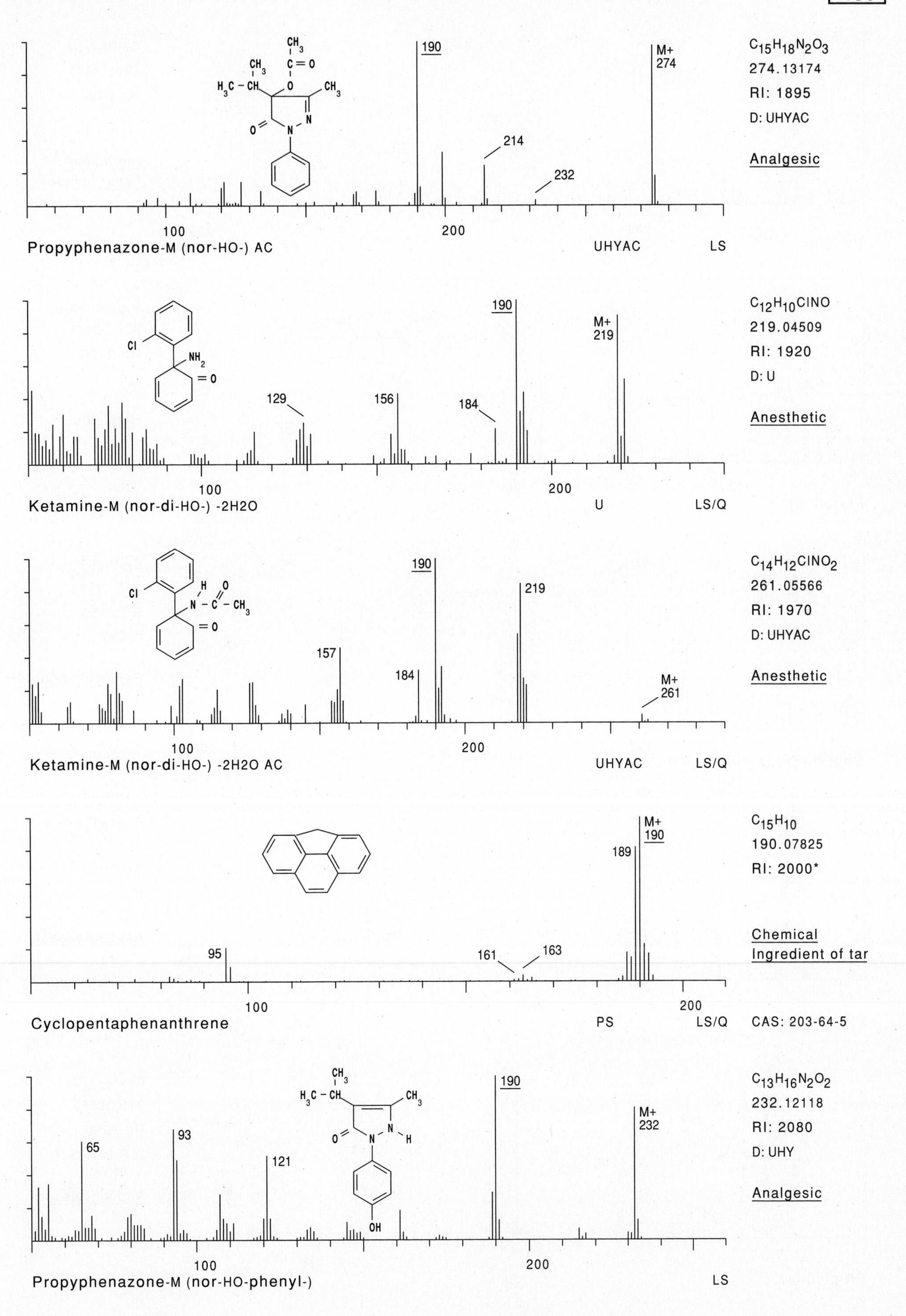

190
M+
274
214
232
100
200
Propyphenazone-M (nor-HO-) AC
UHYAC
LS
C15H18N2O3
274.13174
RI: 1895
D: UHYAC
Analgesic
190
M+
219
129
156
184
100
200
Ketamine-M (nor-di-HO-) -2H2O
U
LS/Q
C12H10ClNO
219.04509
RI: 1920
D: U
Anesthetic
190
219
157
184
M+
261
100
200
Ketamine-M (nor-di-HO-) -2H2O AC
UHYAC
LS/Q
C14H12ClNO2
261.05566
RI: 1970
D: UHYAC
Anesthetic
M+
190
189
95
161
163
100
200
Cyclopentaphenanthrene
PS
LS/Q
C15H10
190.07825
RI: 2000*
Chemical
Ingredient of tar
CAS: 203-64-5
190
M+
232
65
93
121
100
200
Propyphenazone-M (nor-HO-phenyl-)
LS
C13H16N2O2
232.12118
RI: 2080
D: UHY
Analgesic

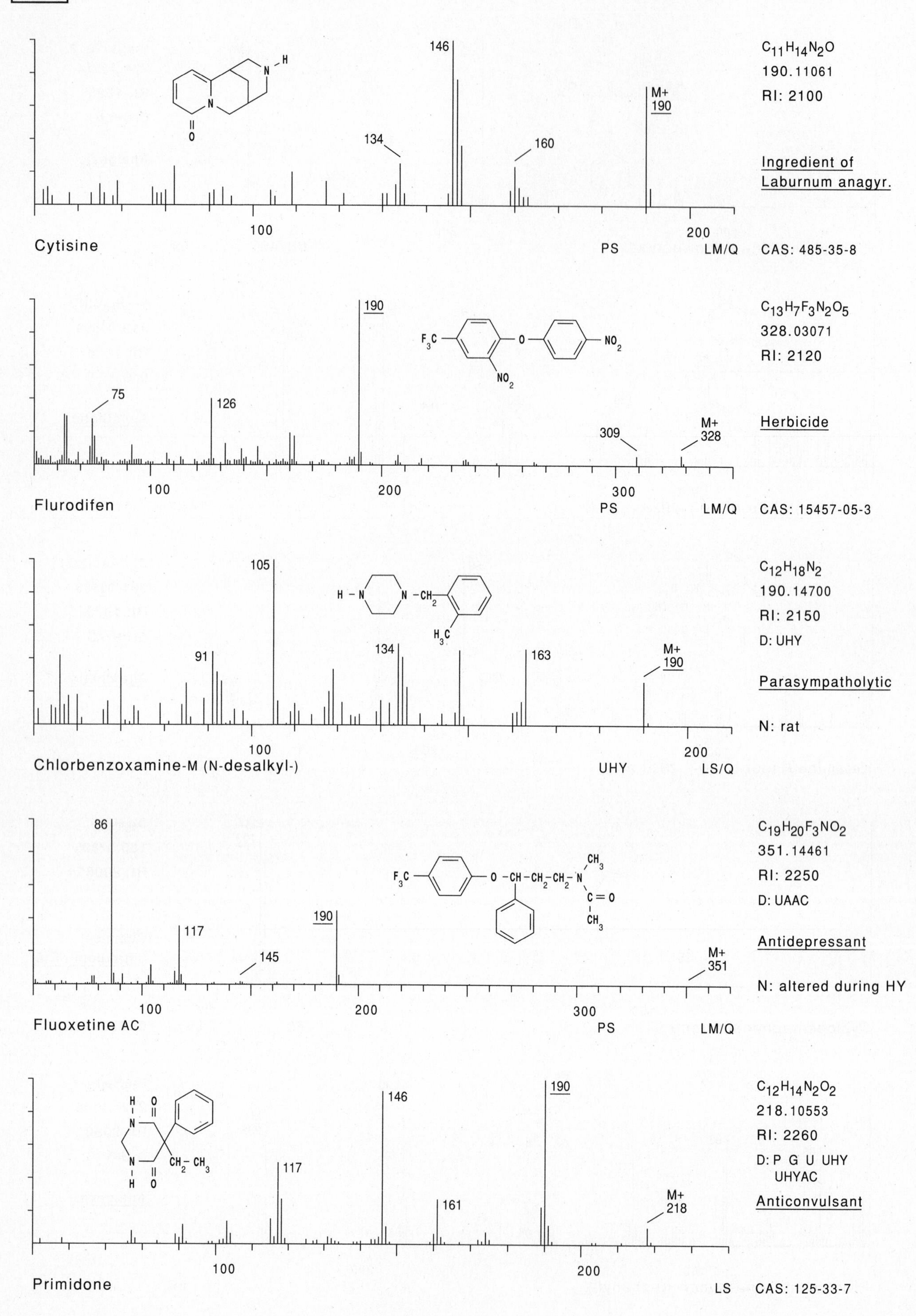

Cytisine
$C_{11}H_{14}N_2O$
190.11061
RI: 2100
Ingredient of Laburnum anagyr.
CAS: 485-35-8
PS LM/Q
M+ 190
146
134
160
Flurodifen
$C_{13}H_7F_3N_2O_5$
328.03071
RI: 2120
Herbicide
CAS: 15457-05-3
PS LM/Q
190
75
126
309
M+ 328
Chlorbenzoxamine-M (N-desalkyl-)
$C_{12}H_{18}N_2$
190.14700
RI: 2150
D: UHY
Parasympatholytic
N: rat
UHY LS/Q
105
91
134
163
M+ 190
Fluoxetine AC
$C_{19}H_{20}F_3NO_2$
351.14461
RI: 2250
D: UAAC
Antidepressant
N: altered during HY
PS LM/Q
86
117
190
145
M+ 351
Primidone
$C_{12}H_{14}N_2O_2$
218.10553
RI: 2260
D: P G U UHY UHYAC
Anticonvulsant
CAS: 125-33-7
LS
190
146
117
161
M+ 218

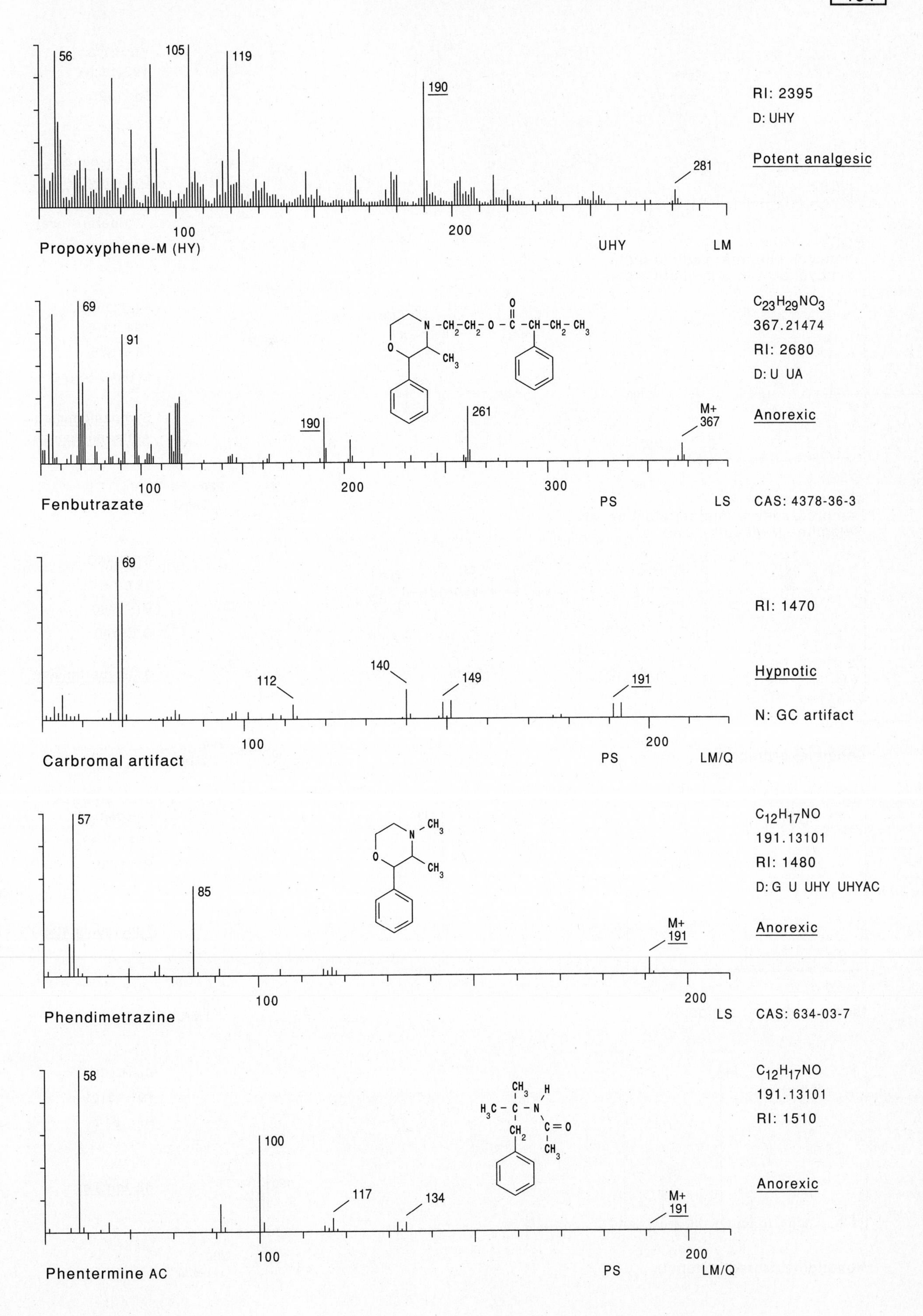

56
105
119
190
281
100
200
Propoxyphene-M (HY)
UHY
LM
RI: 2395
D: UHY
Potent analgesic
69
91
190
261
M+
367
100
200
300
Fenbutrazate
PS
LS
$C_{23}H_{29}NO_3$
367.21474
RI: 2680
D: U UA
Anorexic
CAS: 4378-36-3
69
112
140
149
191
100
200
Carbromal artifact
PS
LM/Q
RI: 1470
Hypnotic
N: GC artifact
57
85
M+
191
100
200
Phendimetrazine
LS
$C_{12}H_{17}NO$
191.13101
RI: 1480
D: G U UHY UHYAC
Anorexic
CAS: 634-03-7
58
100
117
134
M+
191
100
200
Phentermine AC
PS
LM/Q
$C_{12}H_{17}NO$
191.13101
RI: 1510
Anorexic

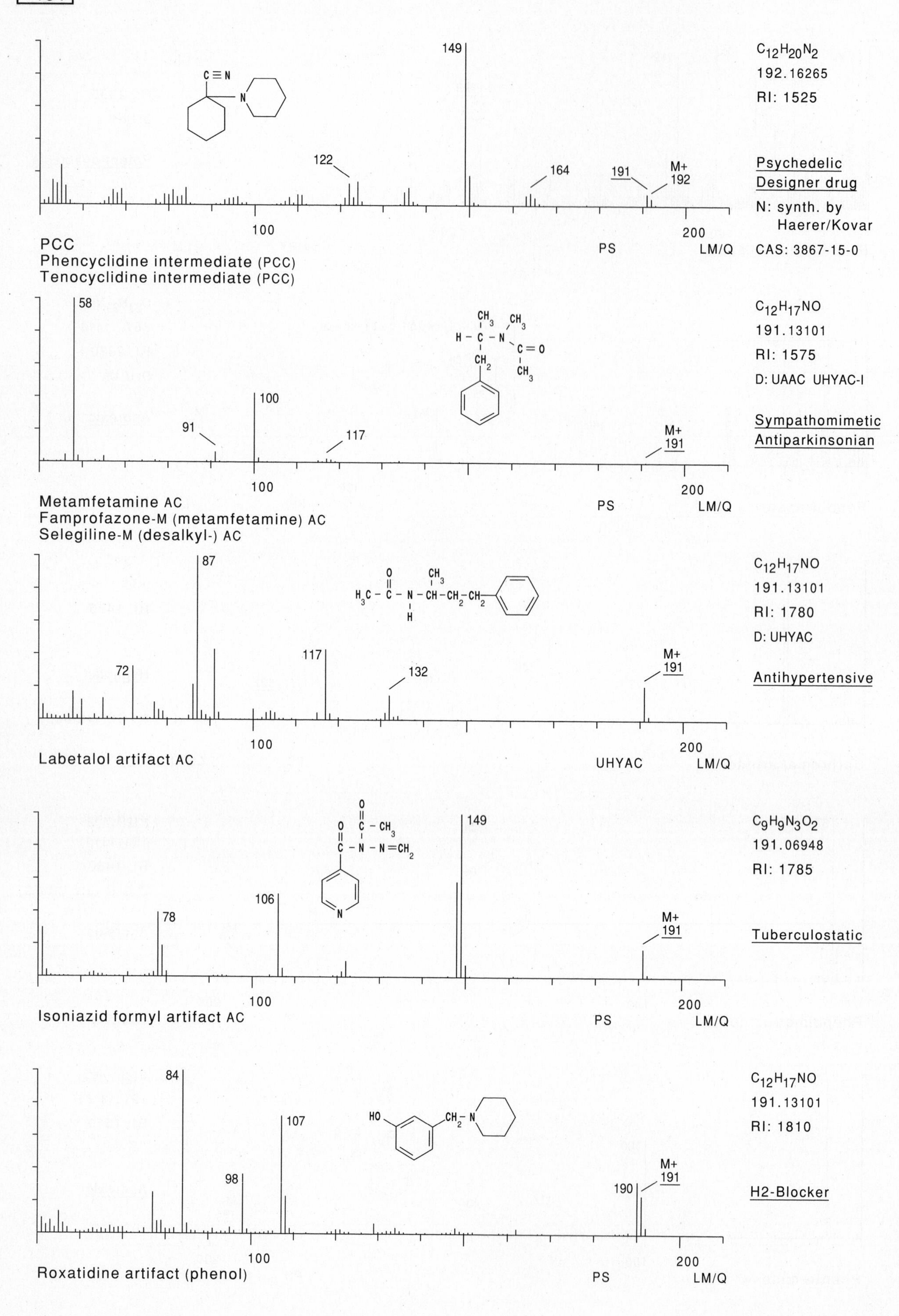

PCC
Phencyclidine intermediate (PCC)
Tenocyclidine intermediate (PCC)

$C_{12}H_{20}N_2$
192.16265
RI: 1525

Psychedelic
Designer drug

N: synth. by Haerer/Kovar
CAS: 3867-15-0

Metamfetamine AC
Famprofazone-M (metamfetamine) AC
Selegiline-M (desalkyl-) AC

$C_{12}H_{17}NO$
191.13101
RI: 1575
D: UAAC UHYAC-I

Sympathomimetic
Antiparkinsonian

Labetalol artifact AC

$C_{12}H_{17}NO$
191.13101
RI: 1780
D: UHYAC

Antihypertensive

Isoniazid formyl artifact AC

$C_9H_9N_3O_2$
191.06948
RI: 1785

Tuberculostatic

Roxatidine artifact (phenol)

$C_{12}H_{17}NO$
191.13101
RI: 1810

H2-Blocker

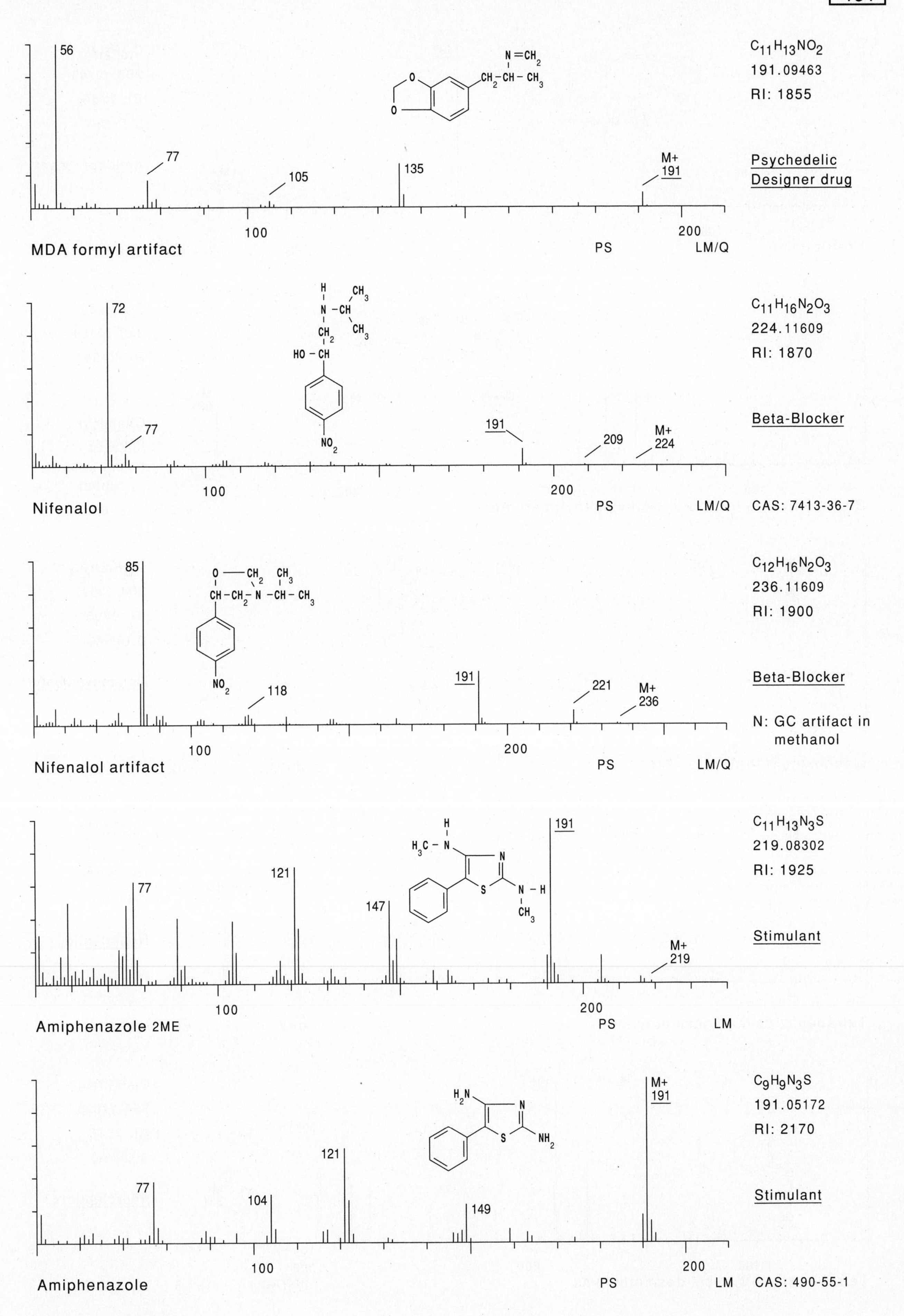

$C_{11}H_{13}NO_2$
191.09463
RI: 1855

Psychedelic Designer drug

$C_{11}H_{16}N_2O_3$
224.11609
RI: 1870

Beta-Blocker

CAS: 7413-36-7

$C_{12}H_{16}N_2O_3$
236.11609
RI: 1900

Beta-Blocker

N: GC artifact in methanol

$C_{11}H_{13}N_3S$
219.08302
RI: 1925

Stimulant

$C_9H_9N_3S$
191.05172
RI: 2170

Stimulant

CAS: 490-55-1

191

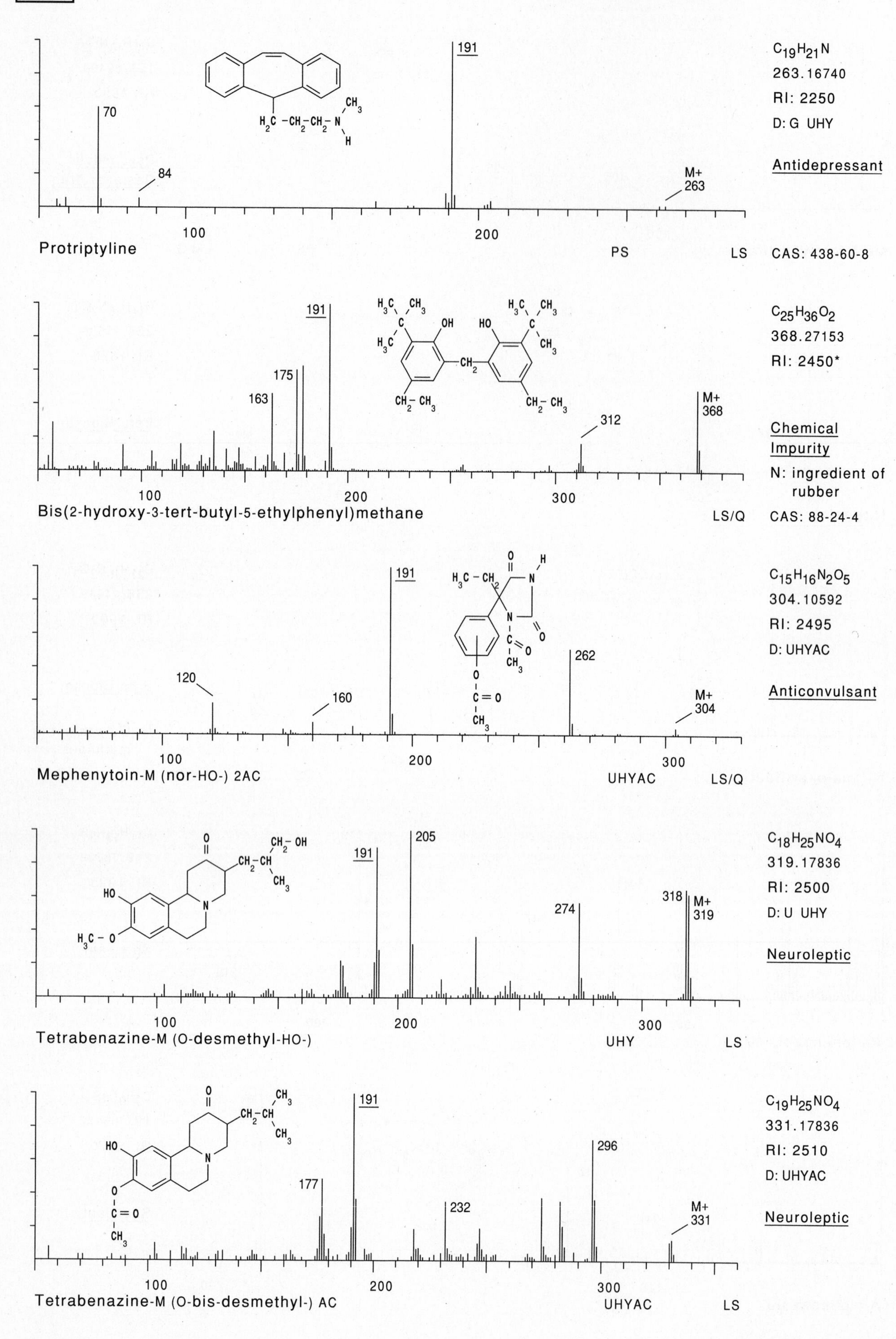
191
70
84
M+
263
100
200
Protriptyline
PS
LS
C19H21N
263.16740
RI: 2250
D: G UHY
Antidepressant
CAS: 438-60-8
191
175
163
312
M+
368
100
200
300
Bis(2-hydroxy-3-tert-butyl-5-ethylphenyl)methane
LS/Q
C25H36O2
368.27153
RI: 2450*
Chemical Impurity
N: ingredient of rubber
CAS: 88-24-4
191
120
160
262
M+
304
100
200
300
Mephenytoin-M (nor-HO-) 2AC
UHYAC
LS/Q
C15H16N2O5
304.10592
RI: 2495
D: UHYAC
Anticonvulsant
205
191
274
318
M+
319
100
200
300
Tetrabenazine-M (O-desmethyl-HO-)
UHY
LS
C18H25NO4
319.17836
RI: 2500
D: U UHY
Neuroleptic
191
296
177
232
M+
331
100
200
300
Tetrabenazine-M (O-bis-desmethyl-) AC
UHYAC
LS
C19H25NO4
331.17836
RI: 2510
D: UHYAC
Neuroleptic

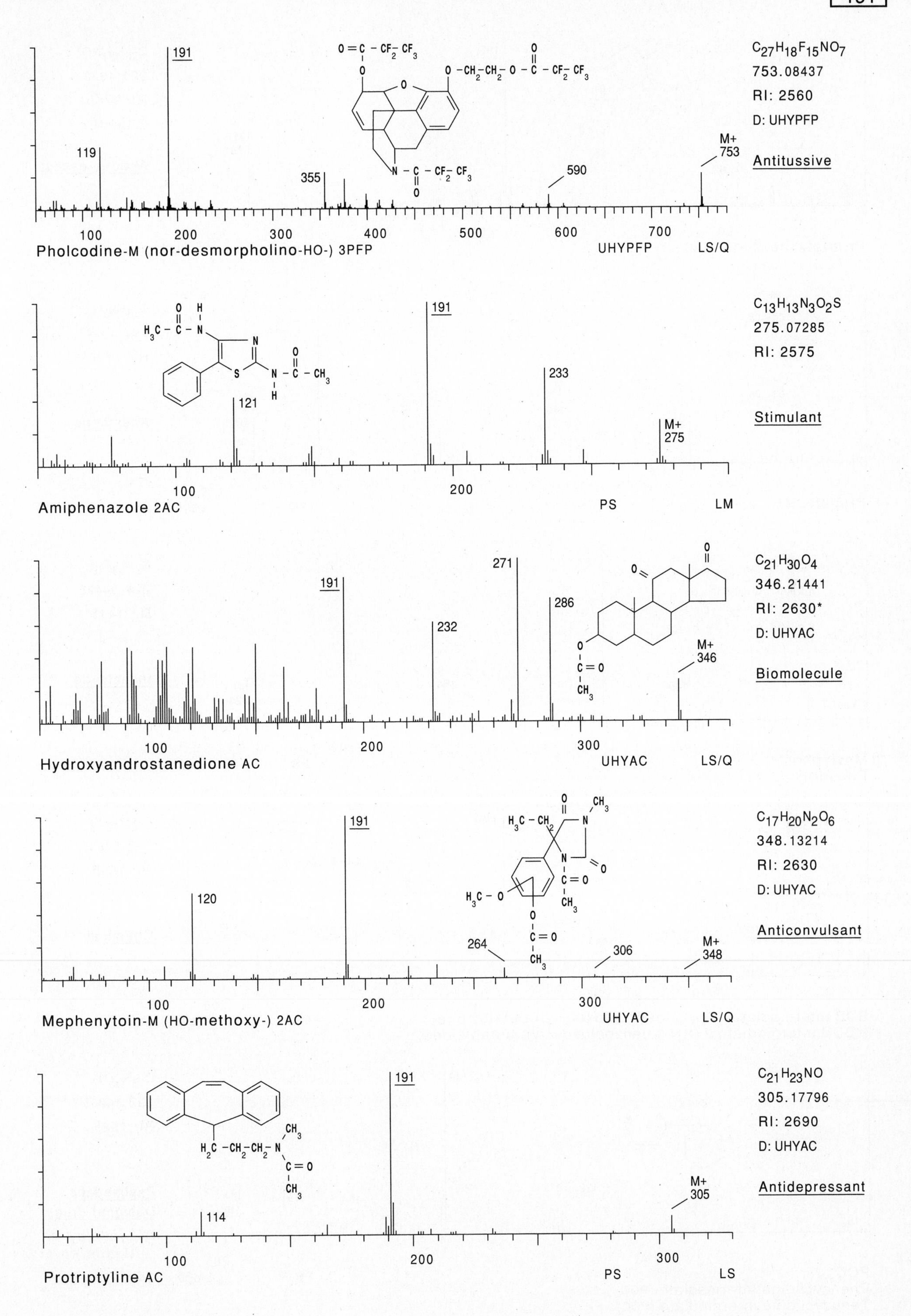

$C_{27}H_{18}F_{15}NO_7$
753.08437
RI: 2560
D: UHYPFP
Antitussive

$C_{13}H_{13}N_3O_2S$
275.07285
RI: 2575
Stimulant

$C_{21}H_{30}O_4$
346.21441
RI: 2630*
D: UHYAC
Biomolecule

$C_{17}H_{20}N_2O_6$
348.13214
RI: 2630
D: UHYAC
Anticonvulsant

$C_{21}H_{23}NO$
305.17796
RI: 2690
D: UHYAC
Antidepressant

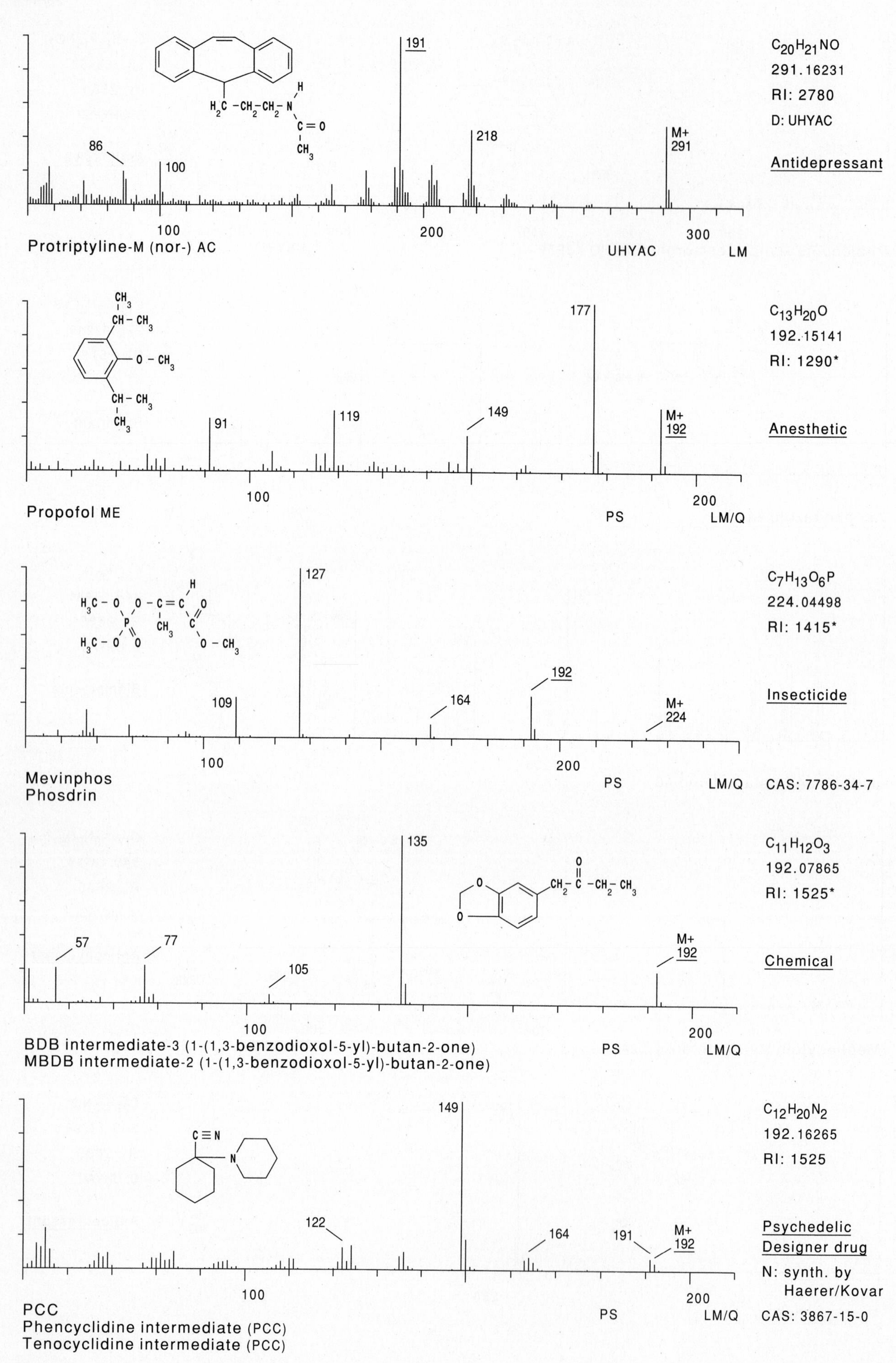

191
218
86
100
M+
291
100
200
300
Protriptyline-M (nor-) AC
UHYAC
LM
$C_{20}H_{21}NO$
291.16231
RI: 2780
D: UHYAC
Antidepressant
177
91
119
149
M+
192
100
200
Propofol ME
PS
LM/Q
$C_{13}H_{20}O$
192.15141
RI: 1290*
Anesthetic
127
109
164
192
M+
224
100
200
Mevinphos
Phosdrin
PS
LM/Q
$C_7H_{13}O_6P$
224.04498
RI: 1415*
Insecticide
CAS: 7786-34-7
135
57
77
105
M+
192
100
200
BDB intermediate-3 (1-(1,3-benzodioxol-5-yl)-butan-2-one)
MBDB intermediate-2 (1-(1,3-benzodioxol-5-yl)-butan-2-one)
PS
LM/Q
$C_{11}H_{12}O_3$
192.07865
RI: 1525*
Chemical
149
122
164
191
M+
192
100
200
PCC
Phencyclidine intermediate (PCC)
Tenocyclidine intermediate (PCC)
PS
LM/Q
$C_{12}H_{20}N_2$
192.16265
RI: 1525
Psychedelic
Designer drug
N: synth. by
Haerer/Kovar
CAS: 3867-15-0

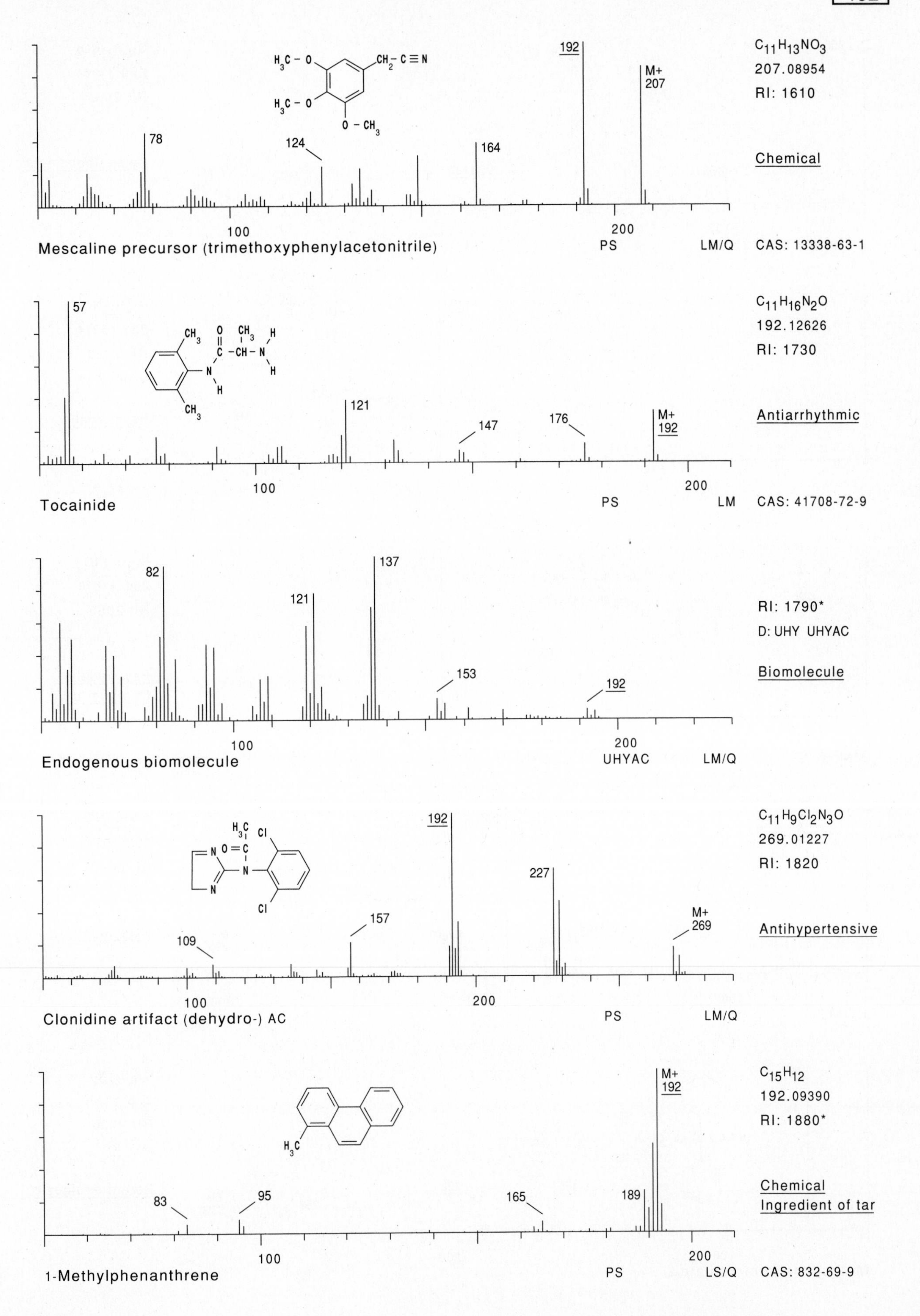

$C_{11}H_{13}NO_3$
207.08954
RI: 1610
Chemical
Mescaline precursor (trimethoxyphenylacetonitrile)
PS
LM/Q
CAS: 13338-63-1
$C_{11}H_{16}N_2O$
192.12626
RI: 1730
Antiarrhythmic
Tocainide
PS
LM
CAS: 41708-72-9
RI: 1790*
D: UHY UHYAC
Biomolecule
Endogenous biomolecule
UHYAC
LM/Q
$C_{11}H_9Cl_2N_3O$
269.01227
RI: 1820
Antihypertensive
Clonidine artifact (dehydro-) AC
PS
LM/Q
$C_{15}H_{12}$
192.09390
RI: 1880*
Chemical
Ingredient of tar
1-Methylphenanthrene
PS
LS/Q
CAS: 832-69-9

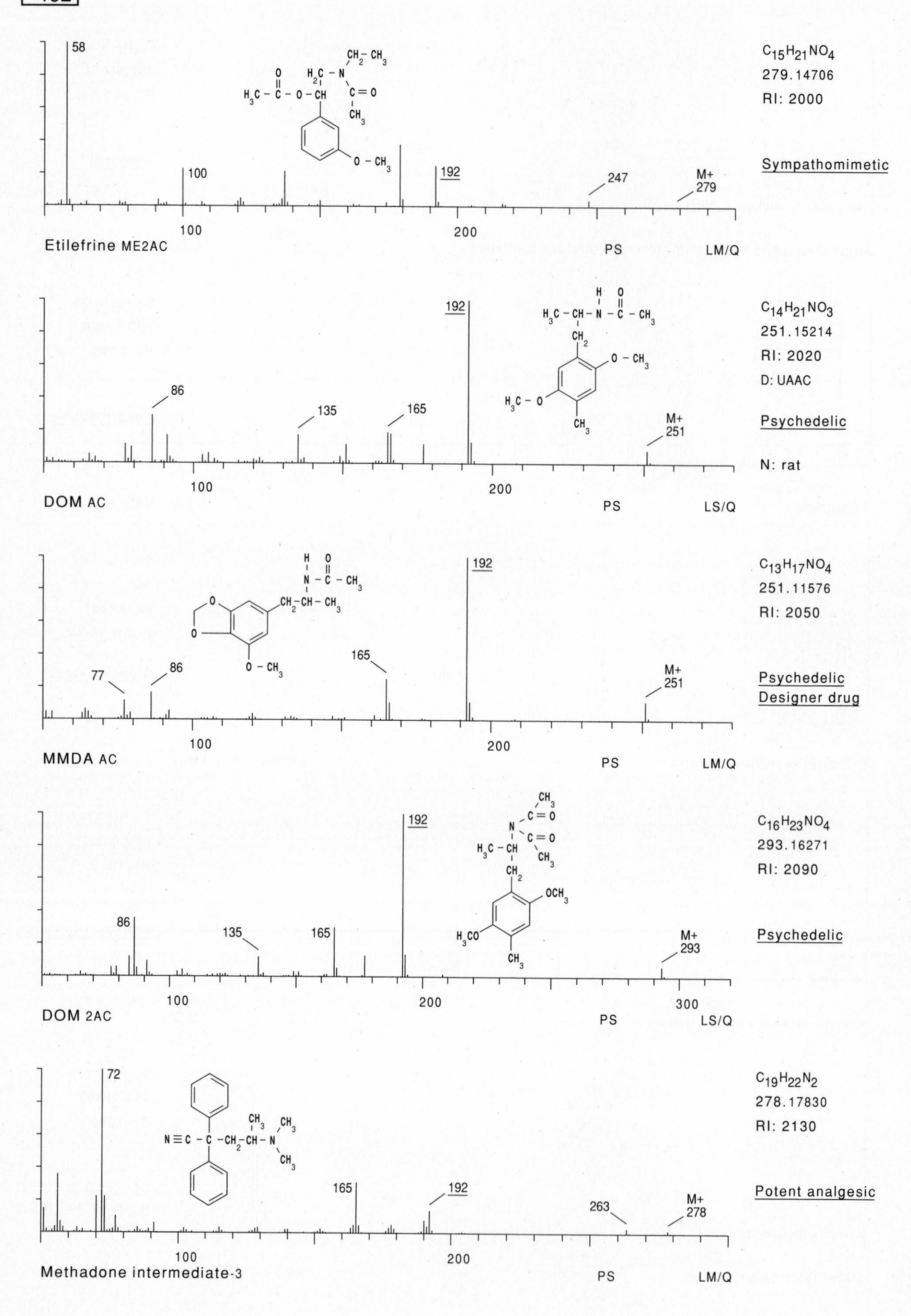

58
100
192
247
M+
279
100
200
Etilefrine ME2AC
PS
LM/Q
C15H21NO4
279.14706
RI: 2000
Sympathomimetic
192
86
135
165
M+
251
100
200
DOM AC
PS
LS/Q
C14H21NO3
251.15214
RI: 2020
D: UAAC
Psychedelic
N: rat
192
77
86
165
M+
251
100
200
MMDA AC
PS
LM/Q
C13H17NO4
251.11576
RI: 2050
Psychedelic
Designer drug
192
86
135
165
M+
293
100
200
300
DOM 2AC
PS
LS/Q
C16H23NO4
293.16271
RI: 2090
Psychedelic
72
165
192
263
M+
278
100
200
Methadone intermediate-3
PS
LM/Q
C19H22N2
278.17830
RI: 2130
Potent analgesic

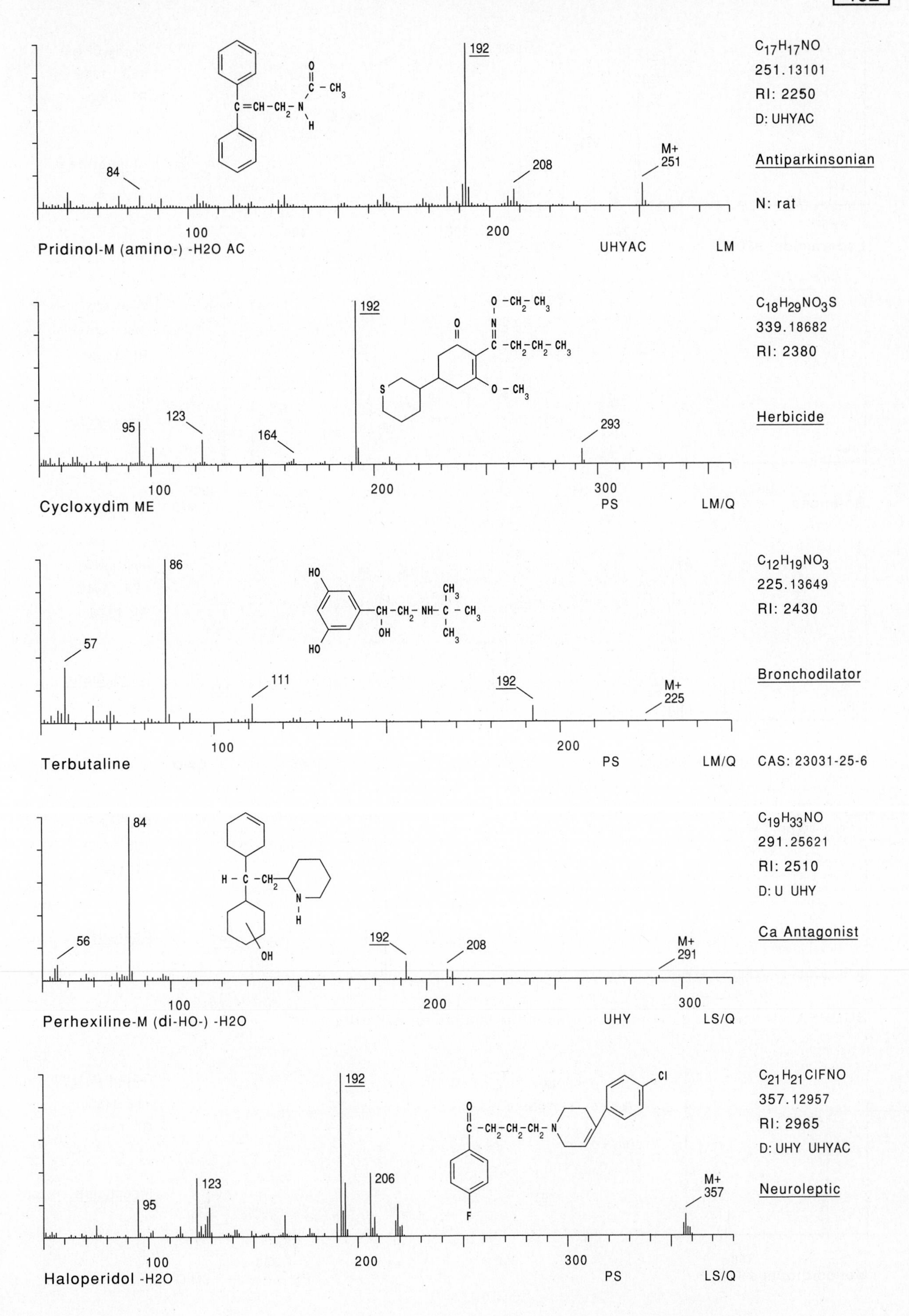

Pridinol-M (amino-) -H2O AC
UHYAC LM
$C_{17}H_{17}NO$
251.13101
RI: 2250
D: UHYAC
Antiparkinsonian
N: rat
84
192
208
M+ 251
Cycloxydim ME
PS LM/Q
$C_{18}H_{29}NO_3S$
339.18682
RI: 2380
Herbicide
95
123
164
192
293
Terbutaline
PS LM/Q
$C_{12}H_{19}NO_3$
225.13649
RI: 2430
Bronchodilator
CAS: 23031-25-6
57
86
111
192
M+ 225
Perhexiline-M (di-HO-) -H2O
UHY LS/Q
$C_{19}H_{33}NO$
291.25621
RI: 2510
D: U UHY
Ca Antagonist
56
84
192
208
M+ 291
Haloperidol -H2O
PS LS/Q
$C_{21}H_{21}ClFNO$
357.12957
RI: 2965
D: UHY UHYAC
Neuroleptic
95
123
192
206
M+ 357

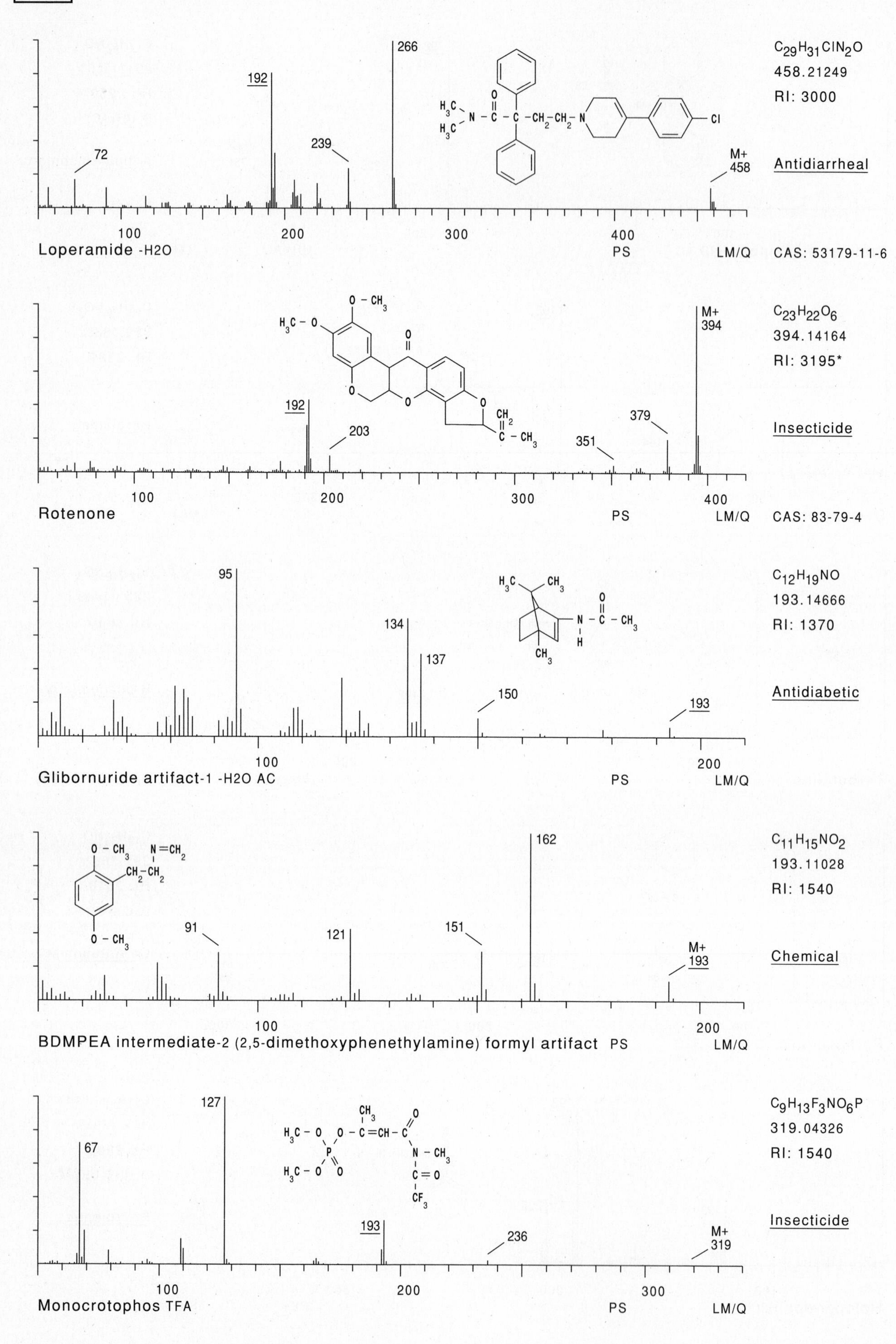

266
192
239
72
M+
458
100
200
300
400
PS
LM/Q
Loperamide -H2O
$C_{29}H_{31}ClN_2O$
458.21249
RI: 3000
Antidiarrheal
CAS: 53179-11-6
M+
394
192
203
379
351
Rotenone
$C_{23}H_{22}O_6$
394.14164
RI: 3195*
Insecticide
CAS: 83-79-4
95
134
137
150
193
Glibornuride artifact-1 -H2O AC
$C_{12}H_{19}NO$
193.14666
RI: 1370
Antidiabetic
162
91
121
151
M+
193
BDMPEA intermediate-2 (2,5-dimethoxyphenethylamine) formyl artifact
$C_{11}H_{15}NO_2$
193.11028
RI: 1540
Chemical
127
67
193
236
M+
319
Monocrotophos TFA
$C_9H_{13}F_3NO_6P$
319.04326
RI: 1540
Insecticide

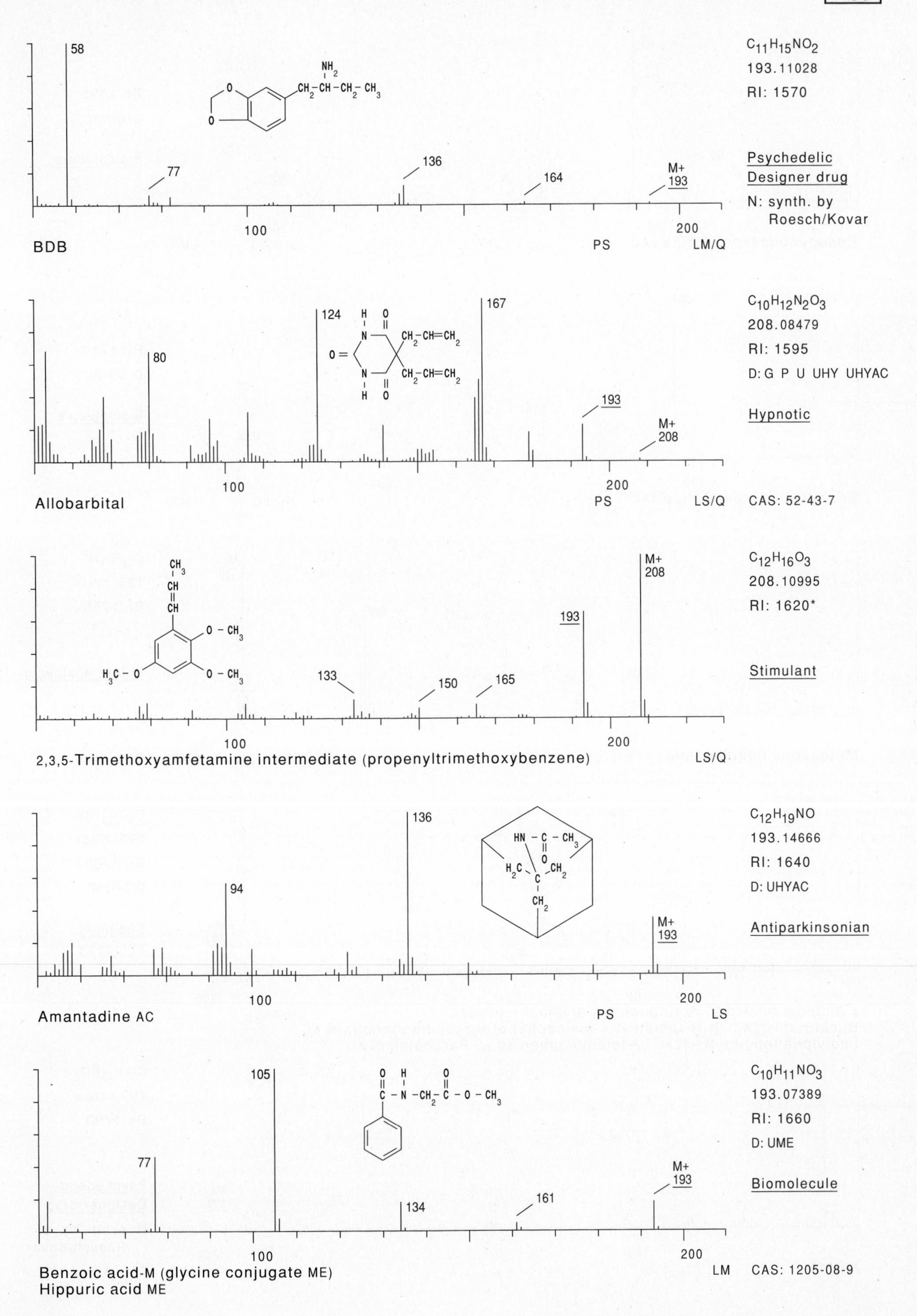
58
77
136
164
M+
193
100
200
PS
LM/Q
BDB
C11H15NO2
193.11028
RI: 1570
Psychedelic
Designer drug
N: synth. by
Roesch/Kovar
80
124
167
193
M+
208
100
200
PS
LS/Q
Allobarbital
C10H12N2O3
208.08479
RI: 1595
D: G P U UHY UHYAC
Hypnotic
CAS: 52-43-7
133
150
165
193
M+
208
100
200
LS/Q
2,3,5-Trimethoxyamfetamine intermediate (propenyltrimethoxybenzene)
C12H16O3
208.10995
RI: 1620*
Stimulant
94
136
M+
193
100
200
PS
LS
Amantadine AC
C12H19NO
193.14666
RI: 1640
D: UHYAC
Antiparkinsonian
77
105
134
161
M+
193
100
200
LM
Benzoic acid-M (glycine conjugate ME)
Hippuric acid ME
C10H11NO3
193.07389
RI: 1660
D: UME
Biomolecule
CAS: 1205-08-9

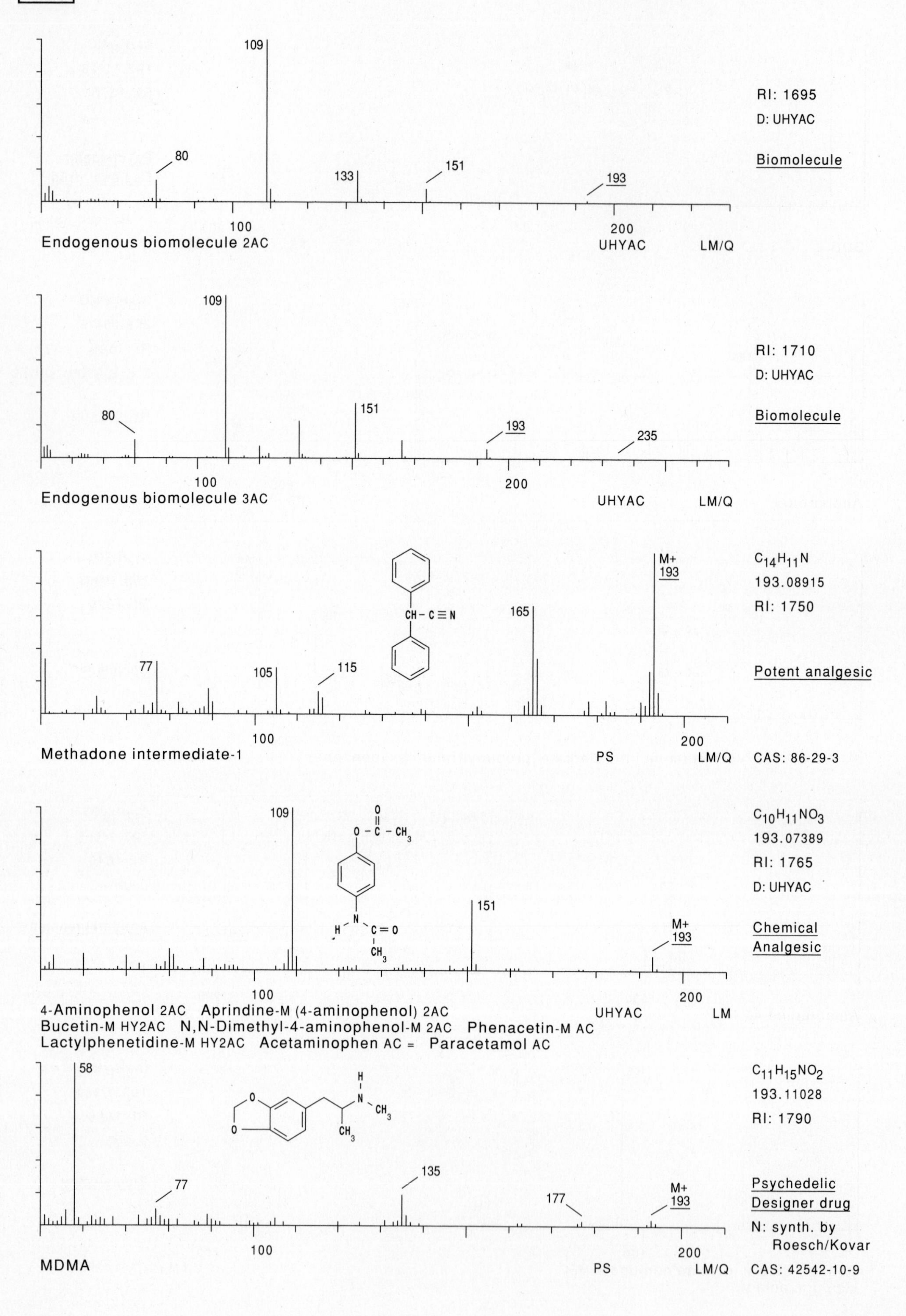

109
80
133
151
193
100
200
Endogenous biomolecule 2AC
UHYAC
LM/Q
RI: 1695
D: UHYAC
Biomolecule
109
80
151
193
235
100
200
Endogenous biomolecule 3AC
UHYAC
LM/Q
RI: 1710
D: UHYAC
Biomolecule
CH - C ≡ N
M+
193
165
77
105
115
100
200
Methadone intermediate-1
PS
LM/Q
C14H11N
193.08915
RI: 1750
Potent analgesic
CAS: 86-29-3
109
O - C - CH3
151
M+
193
100
200
4-Aminophenol 2AC Aprindine-M (4-aminophenol) 2AC
Bucetin-M HY2AC N,N-Dimethyl-4-aminophenol-M 2AC Phenacetin-M AC
Lactylphenetidine-M HY2AC Acetaminophen AC = Paracetamol AC
UHYAC
LM
C10H11NO3
193.07389
RI: 1765
D: UHYAC
Chemical
Analgesic
58
135
77
177
M+
193
100
200
MDMA
PS
LM/Q
C11H15NO2
193.11028
RI: 1790
Psychedelic
Designer drug
N: synth. by
Roesch/Kovar
CAS: 42542-10-9

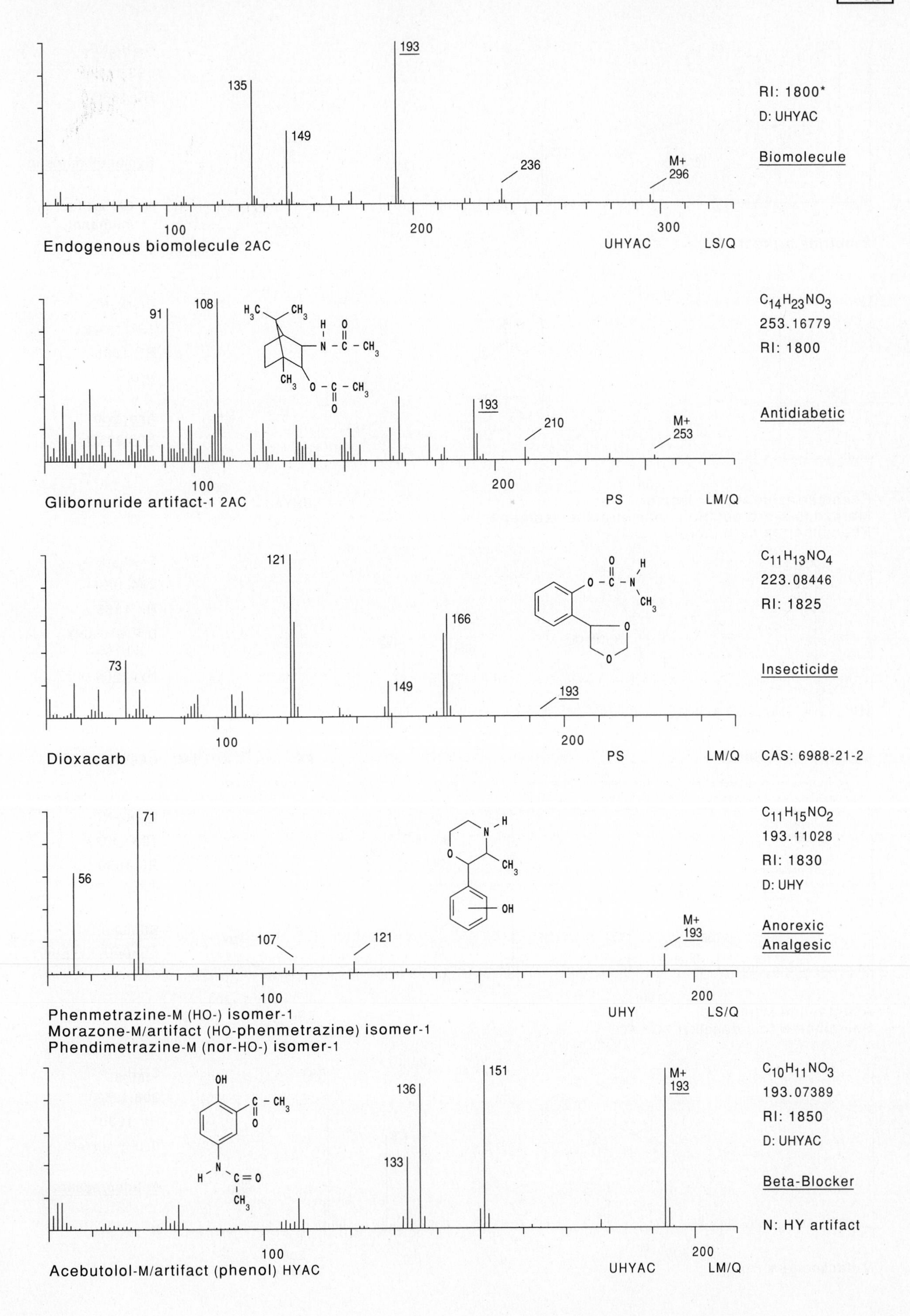

193
135
149
236
M+
296
100
200
300
Endogenous biomolecule 2AC
UHYAC
LS/Q
RI: 1800*
D: UHYAC
Biomolecule
91
108
193
210
M+
253
100
200
Glibornuride artifact-1 2AC
PS
LM/Q
$C_{14}H_{23}NO_3$
253.16779
RI: 1800
Antidiabetic
121
166
73
149
193
100
200
Dioxacarb
PS
LM/Q
$C_{11}H_{13}NO_4$
223.08446
RI: 1825
Insecticide
CAS: 6988-21-2
71
56
107
121
M+
193
100
200
Phenmetrazine-M (HO-) isomer-1
Morazone-M/artifact (HO-phenmetrazine) isomer-1
Phendimetrazine-M (nor-HO-) isomer-1
UHY
LS/Q
$C_{11}H_{15}NO_2$
193.11028
RI: 1830
D: UHY
Anorexic
Analgesic
151
136
133
M+
193
100
200
Acebutolol-M/artifact (phenol) HYAC
UHYAC
LM/Q
$C_{10}H_{11}NO_3$
193.07389
RI: 1850
D: UHYAC
Beta-Blocker
N: HY artifact

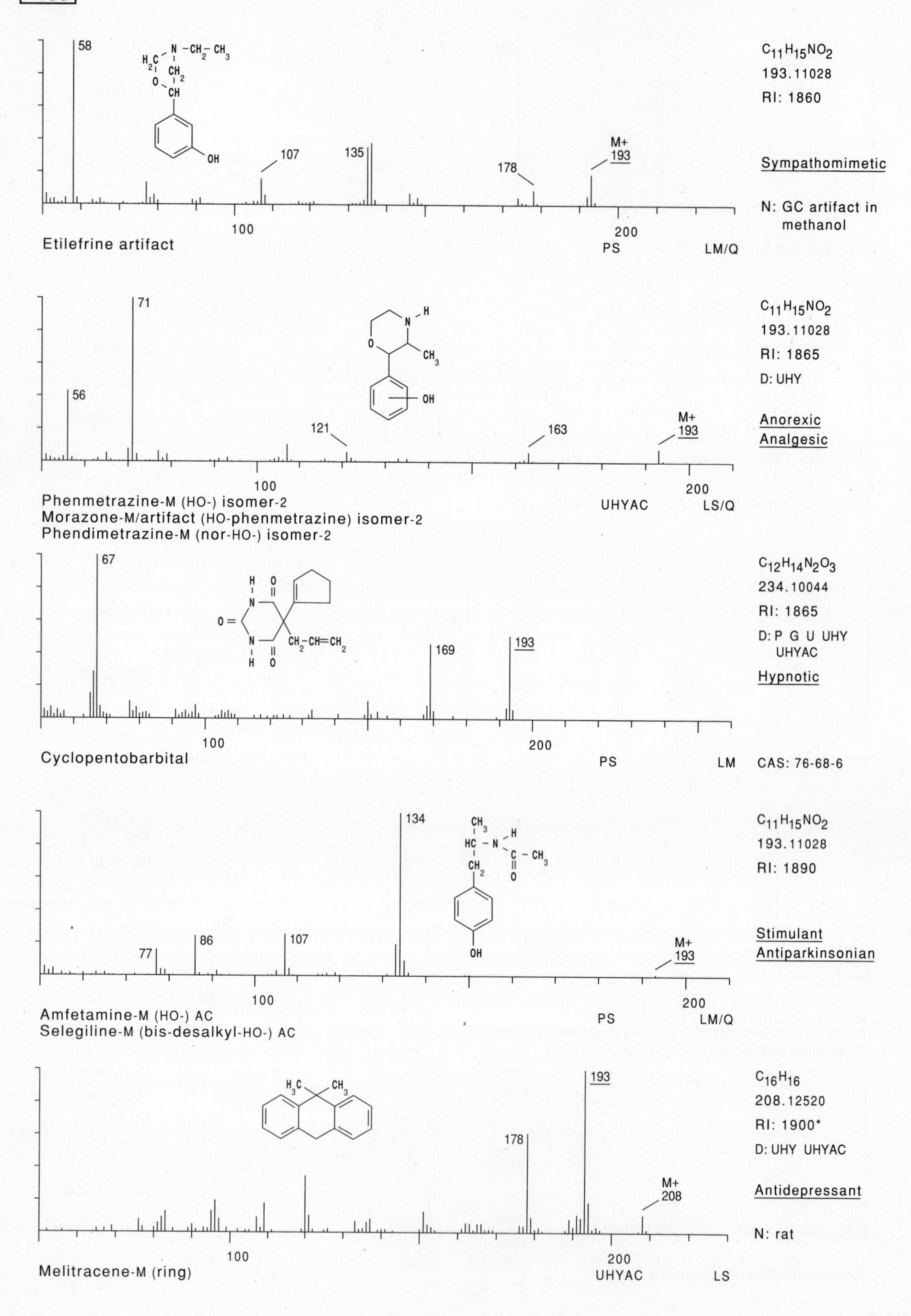
193
58
107
135
178
M+
193
100
200
Etilefrine artifact
PS
LM/Q
$C_{11}H_{15}NO_2$
193.11028
RI: 1860
Sympathomimetic
N: GC artifact in methanol
71
56
121
163
M+
193
100
200
Phenmetrazine-M (HO-) isomer-2
Morazone-M/artifact (HO-phenmetrazine) isomer-2
Phendimetrazine-M (nor-HO-) isomer-2
UHYAC
LS/Q
$C_{11}H_{15}NO_2$
193.11028
RI: 1865
D: UHY
Anorexic
Analgesic
67
169
193
100
200
Cyclopentobarbital
PS
LM
$C_{12}H_{14}N_2O_3$
234.10044
RI: 1865
D: P G U UHY UHYAC
Hypnotic
CAS: 76-68-6
134
77
86
107
M+
193
100
200
Amfetamine-M (HO-) AC
Selegiline-M (bis-desalkyl-HO-) AC
PS
LM/Q
$C_{11}H_{15}NO_2$
193.11028
RI: 1890
Stimulant
Antiparkinsonian
193
178
M+
208
100
200
Melitracene-M (ring)
UHYAC
LS
$C_{16}H_{16}$
208.12520
RI: 1900*
D: UHY UHYAC
Antidepressant
N: rat

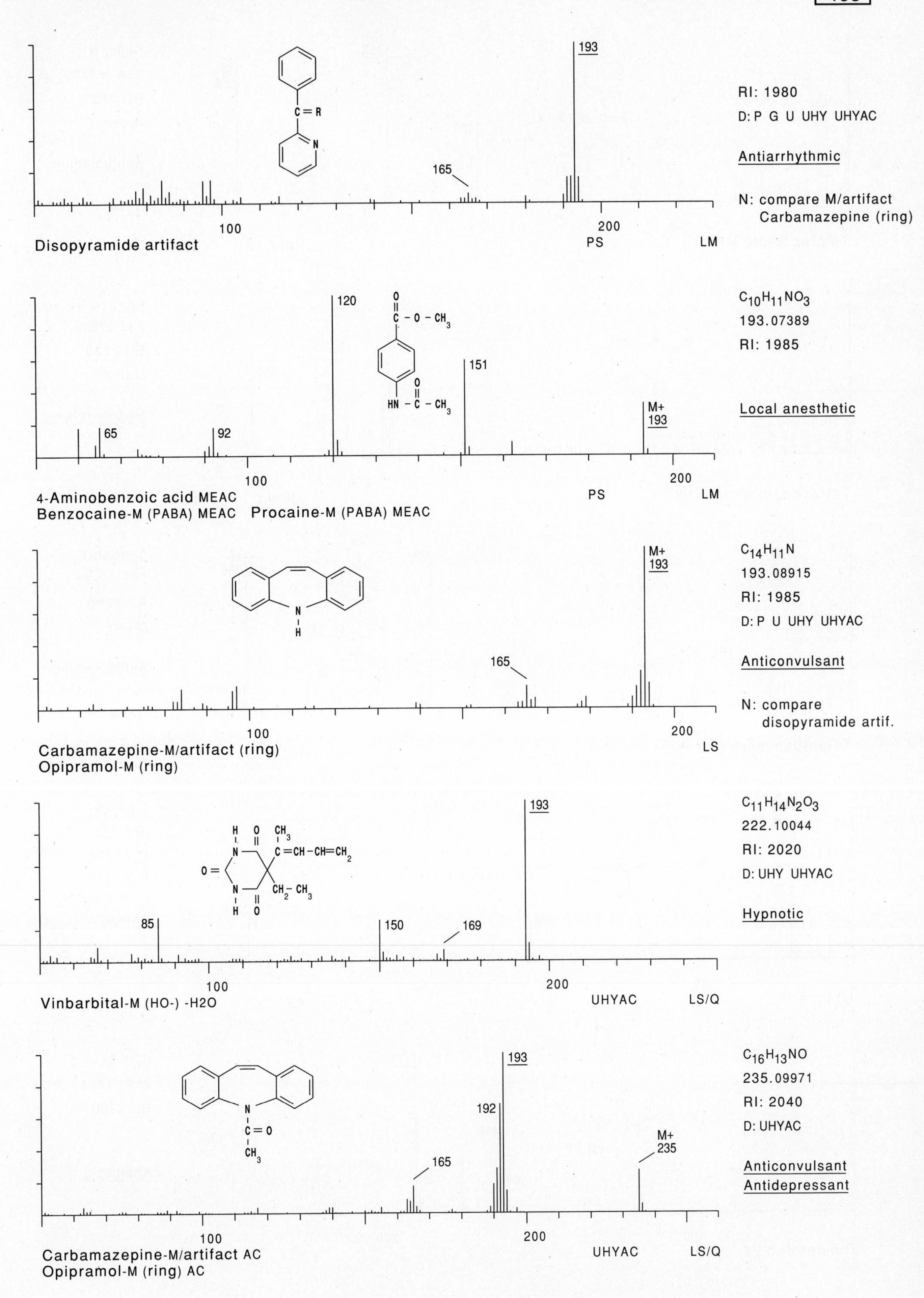
193
C = R
N
165
100
200
PS
LM
Disopyramide artifact
RI: 1980
D: P G U UHY UHYAC
Antiarrhythmic
N: compare M/artifact
Carbamazepine (ring)
120
O
C - O - CH3
151
O
HN - C - CH3
M+
193
65
92
100
200
PS
LM
4-Aminobenzoic acid MEAC
Benzocaine-M (PABA) MEAC
Procaine-M (PABA) MEAC
C10H11NO3
193.07389
RI: 1985
Local anesthetic
M+
193
N
H
165
100
200
LS
Carbamazepine-M/artifact (ring)
Opipramol-M (ring)
C14H11N
193.08915
RI: 1985
D: P U UHY UHYAC
Anticonvulsant
N: compare
disopyramide artif.
193
H O CH3
N C=CH-CH=CH2
O=
N CH2-CH3
H O
85
150
169
100
200
UHYAC
LS/Q
Vinbarbital-M (HO-) -H2O
C11H14N2O3
222.10044
RI: 2020
D: UHY UHYAC
Hypnotic
193
192
N
C = O
CH3
M+
235
165
100
200
UHYAC
LS/Q
Carbamazepine-M/artifact AC
Opipramol-M (ring) AC
C16H13NO
235.09971
RI: 2040
D: UHYAC
Anticonvulsant
Antidepressant

193

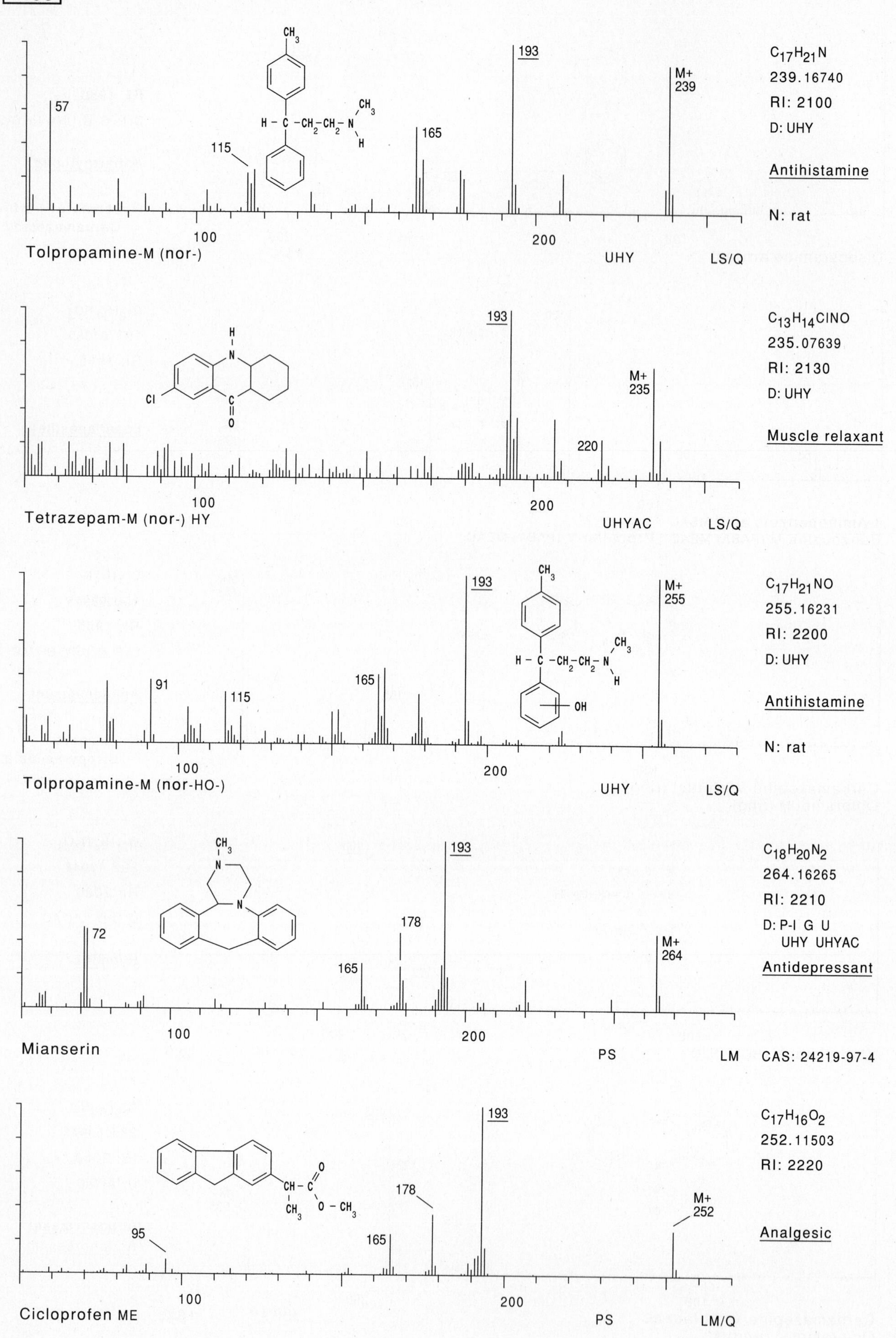

Tolpropamine-M (nor-)
UHY
LS/Q
C17H21N
239.16740
RI: 2100
D: UHY
Antihistamine
N: rat
Tetrazepam-M (nor-) HY
UHYAC
LS/Q
C13H14ClNO
235.07639
RI: 2130
D: UHY
Muscle relaxant
Tolpropamine-M (nor-HO-)
UHY
LS/Q
C17H21NO
255.16231
RI: 2200
D: UHY
Antihistamine
N: rat
Mianserin
PS
LM
C18H20N2
264.16265
RI: 2210
D: P-I G U
UHY UHYAC
Antidepressant
CAS: 24219-97-4
Cicloprofen ME
PS
LM/Q
C17H16O2
252.11503
RI: 2220
Analgesic

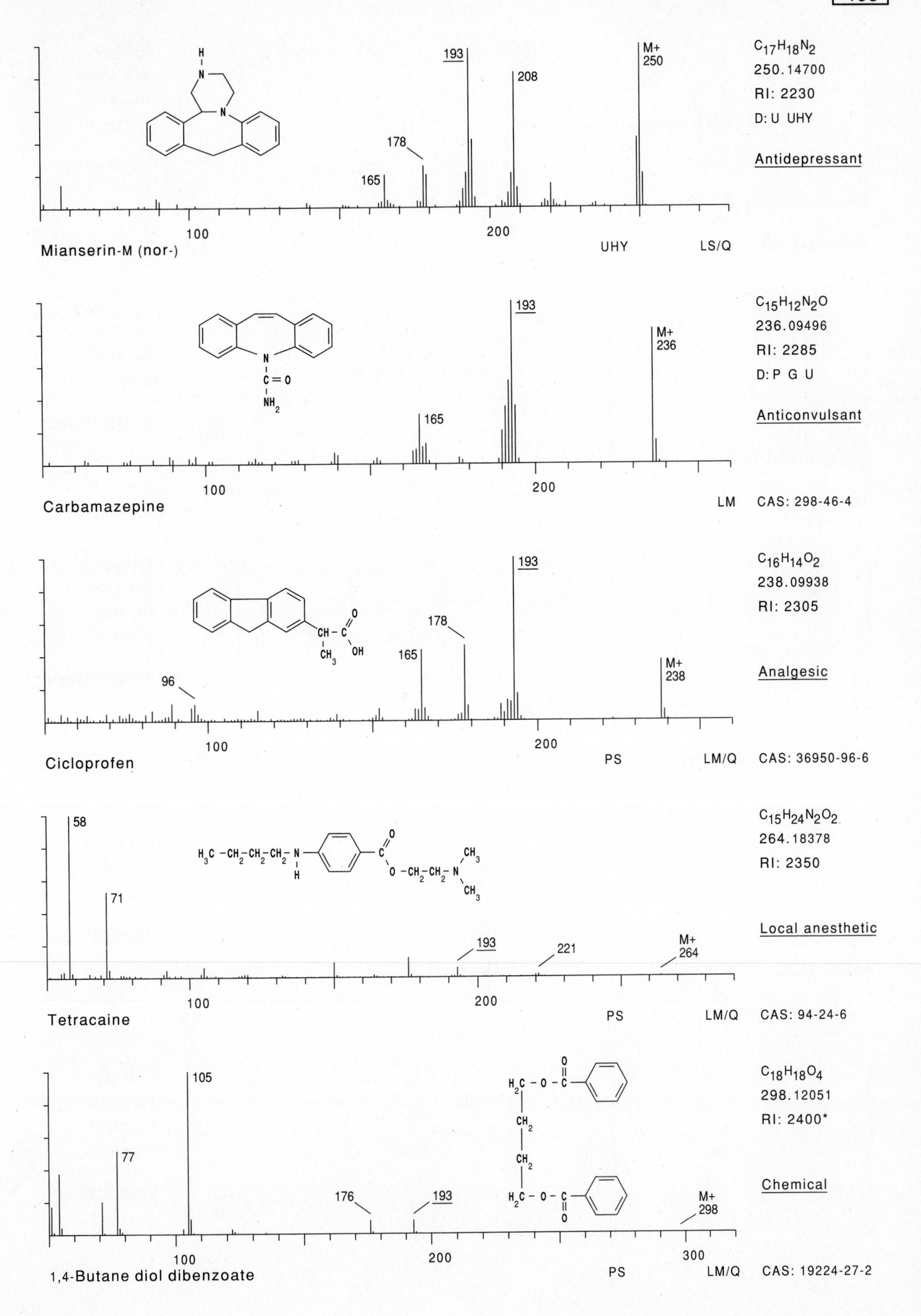
193
208
M+
250
178
165
100
200
Mianserin-M (nor-)
UHY
LS/Q
C17H18N2
250.14700
RI: 2230
D: U UHY
Antidepressant
193
M+
236
165
100
200
Carbamazepine
LM
C15H12N2O
236.09496
RI: 2285
D: P G U
Anticonvulsant
CAS: 298-46-4
193
178
165
96
M+
238
100
200
Cicloprofen
PS
LM/Q
C16H14O2
238.09938
RI: 2305
Analgesic
CAS: 36950-96-6
58
71
193
221
M+
264
100
200
Tetracaine
PS
LM/Q
C15H24N2O2
264.18378
RI: 2350
Local anesthetic
CAS: 94-24-6
105
77
176
193
M+
298
100
200
300
1,4-Butane diol dibenzoate
PS
LM/Q
C18H18O4
298.12051
RI: 2400*
Chemical
CAS: 19224-27-2

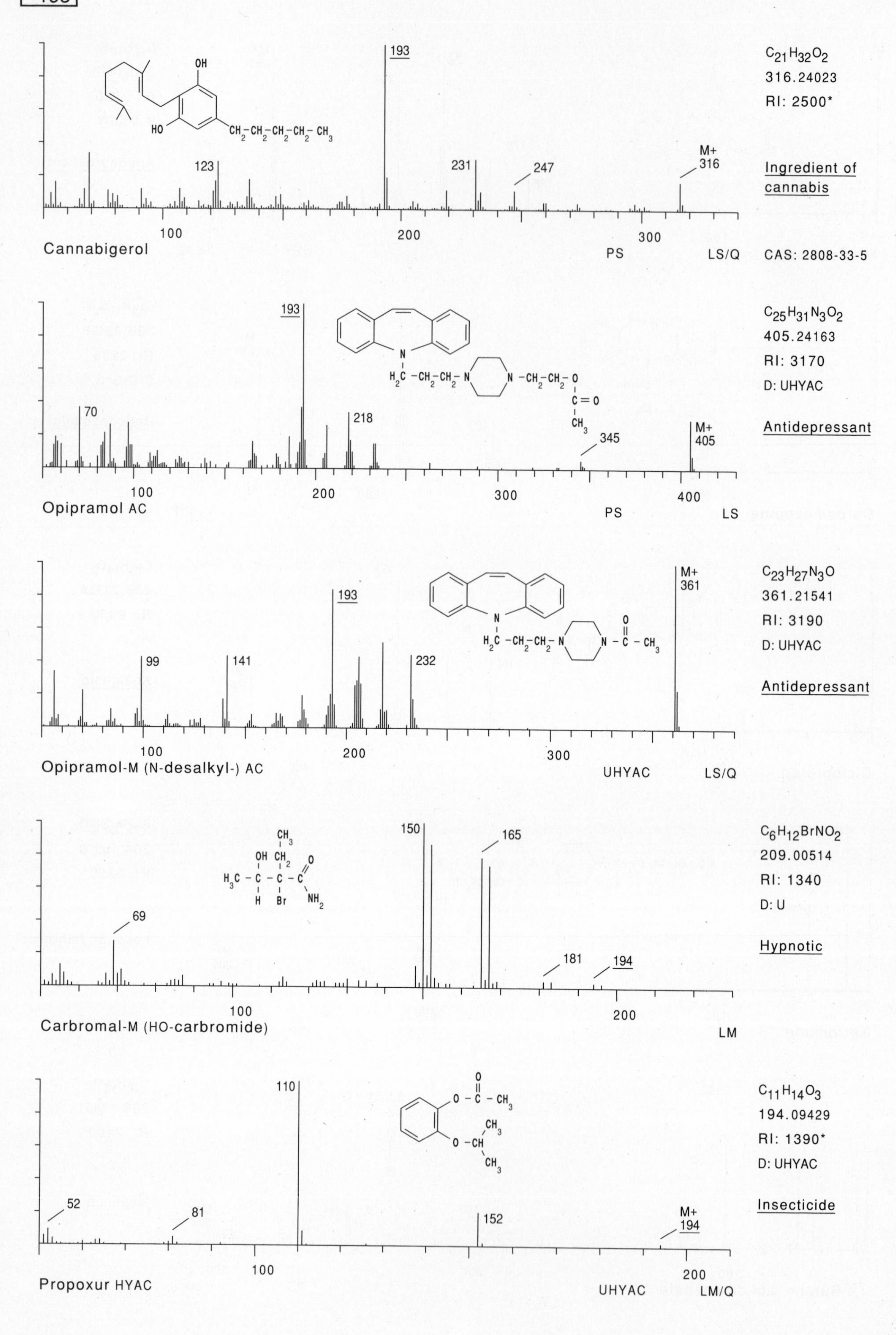
193
Cannabigerol
C21H32O2
316.24023
RI: 2500*
Ingredient of cannabis
CAS: 2808-33-5
PS
LS/Q
Opipramol AC
C25H31N3O2
405.24163
RI: 3170
D: UHYAC
Antidepressant
PS
LS
Opipramol-M (N-desalkyl-) AC
C23H27N3O
361.21541
RI: 3190
D: UHYAC
Antidepressant
UHYAC
LS/Q
Carbromal-M (HO-carbromide)
C6H12BrNO2
209.00514
RI: 1340
D: U
Hypnotic
LM
Propoxur HYAC
C11H14O3
194.09429
RI: 1390*
D: UHYAC
Insecticide
UHYAC
LM/Q

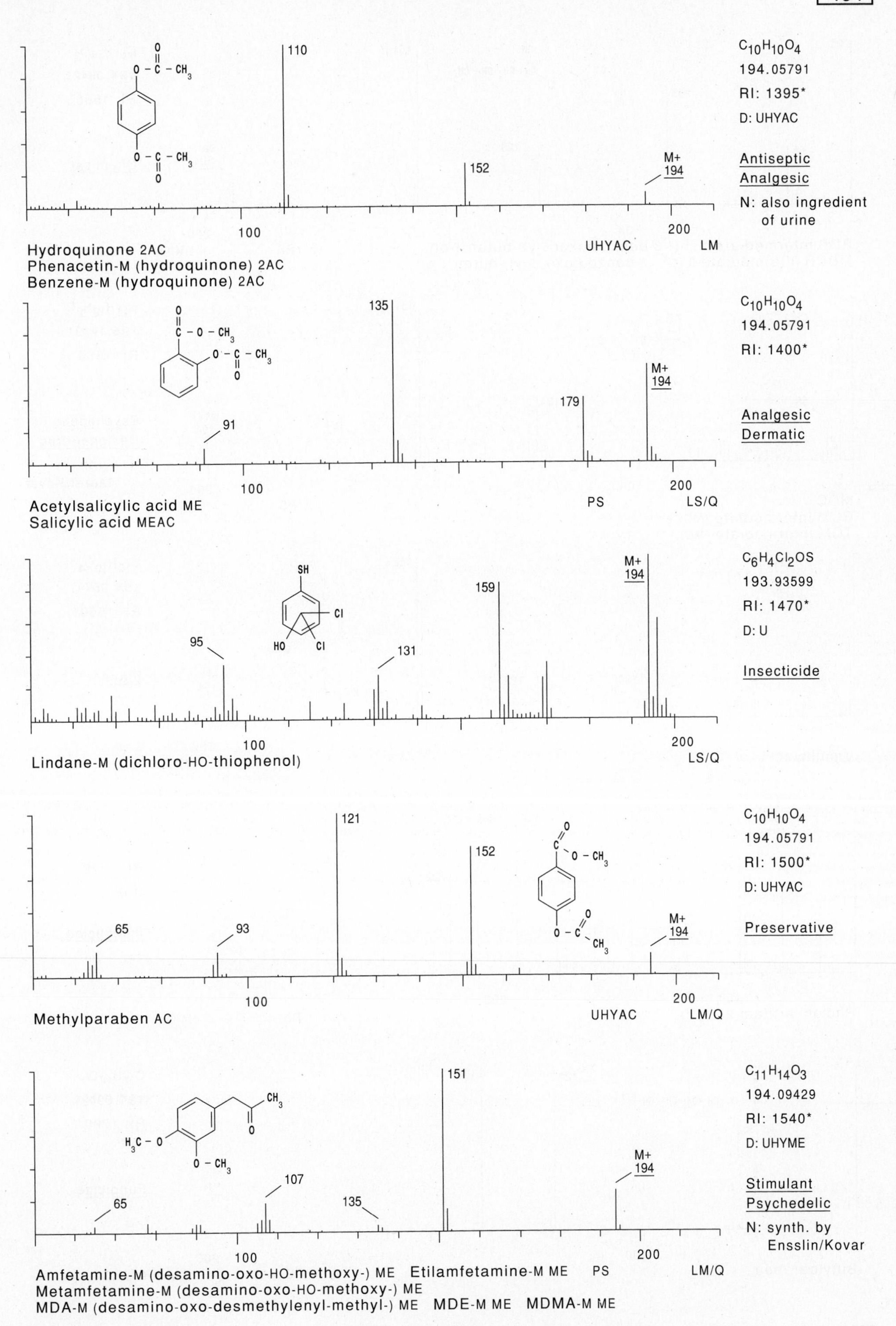

110
152
M+
194
100
200
UHYAC
LM
$C_{10}H_{10}O_4$
194.05791
RI: 1395*
D: UHYAC
Antiseptic
Analgesic
N: also ingredient of urine
Hydroquinone 2AC
Phenacetin-M (hydroquinone) 2AC
Benzene-M (hydroquinone) 2AC
135
179
91
PS
LS/Q
$C_{10}H_{10}O_4$
194.05791
RI: 1400*
Analgesic
Dermatic
Acetylsalicylic acid ME
Salicylic acid MEAC
159
95
131
$C_6H_4Cl_2OS$
193.93599
RI: 1470*
D: U
Insecticide
Lindane-M (dichloro-HO-thiophenol)
121
65
93
$C_{10}H_{10}O_4$
194.05791
RI: 1500*
D: UHYAC
Preservative
Methylparaben AC
151
107
135
$C_{11}H_{14}O_3$
194.09429
RI: 1540*
D: UHYME
Stimulant
Psychedelic
N: synth. by Ensslin/Kovar
Amfetamine-M (desamino-oxo-HO-methoxy-) ME Etilamfetamine-M ME
Metamfetamine-M (desamino-oxo-HO-methoxy-) ME
MDA-M (desamino-oxo-desmethylenyl-methyl-) ME MDE-M ME MDMA-M ME

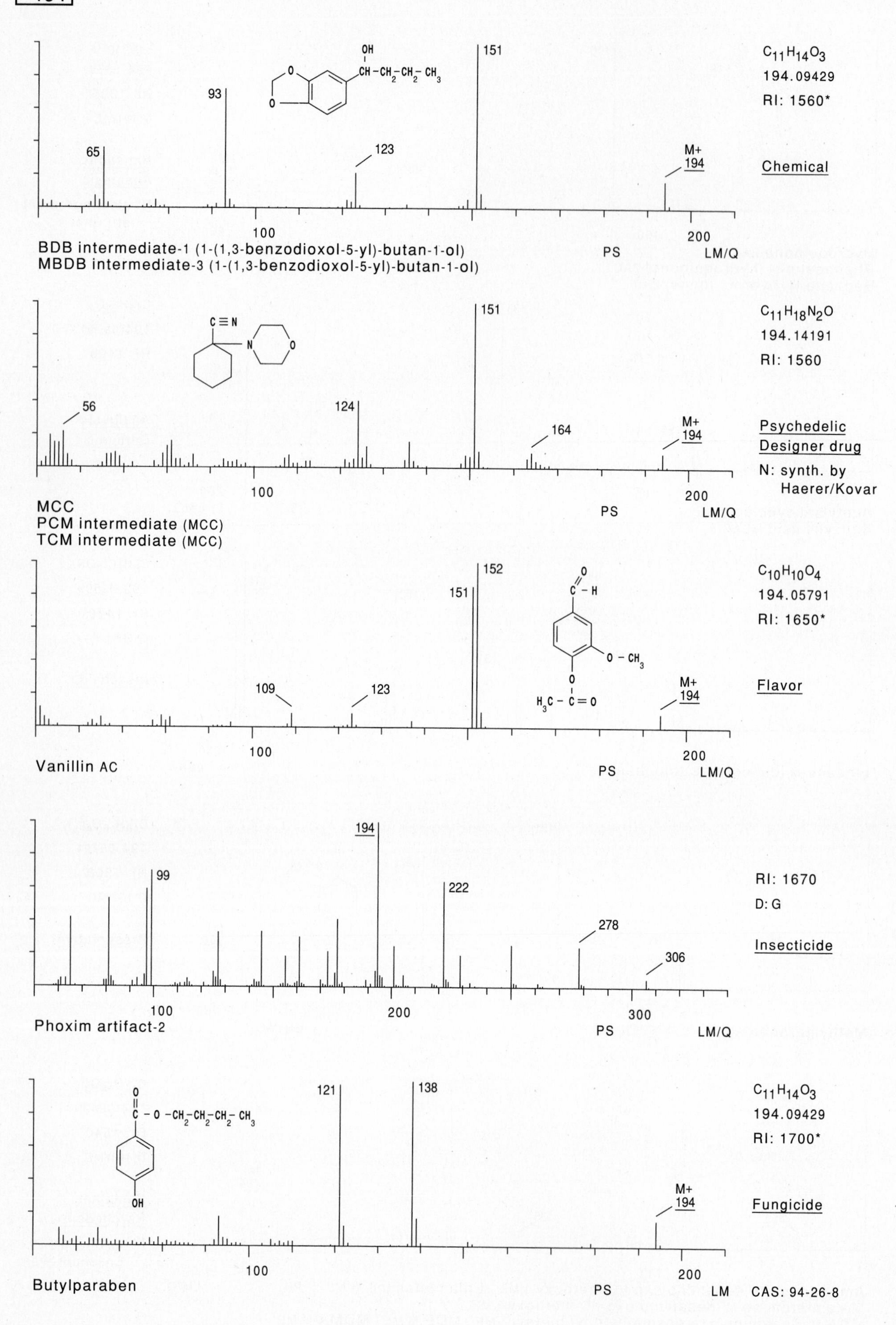
OH
CH-CH2-CH2-CH3
151
93
65
123
M+
194
100
200
PS
LM/Q
$C_{11}H_{14}O_3$
194.09429
RI: 1560*
Chemical
BDB intermediate-1 (1-(1,3-benzodioxol-5-yl)-butan-1-ol)
MBDB intermediate-3 (1-(1,3-benzodioxol-5-yl)-butan-1-ol)
C≡N
N
O
151
56
124
164
M+
194
100
200
PS
LM/Q
$C_{11}H_{18}N_2O$
194.14191
RI: 1560
Psychedelic
Designer drug
N: synth. by
Haerer/Kovar
MCC
PCM intermediate (MCC)
TCM intermediate (MCC)
152
151
O
C-H
O-CH3
H3C-C=O
109
123
M+
194
100
200
PS
LM/Q
$C_{10}H_{10}O_4$
194.05791
RI: 1650*
Flavor
Vanillin AC
194
99
222
278
306
100
200
300
PS
LM/Q
RI: 1670
D: G
Insecticide
Phoxim artifact-2
O
C-O-CH2-CH2-CH2-CH3
OH
121
138
M+
194
100
200
PS
LM
$C_{11}H_{14}O_3$
194.09429
RI: 1700*
Fungicide
Butylparaben
CAS: 94-26-8

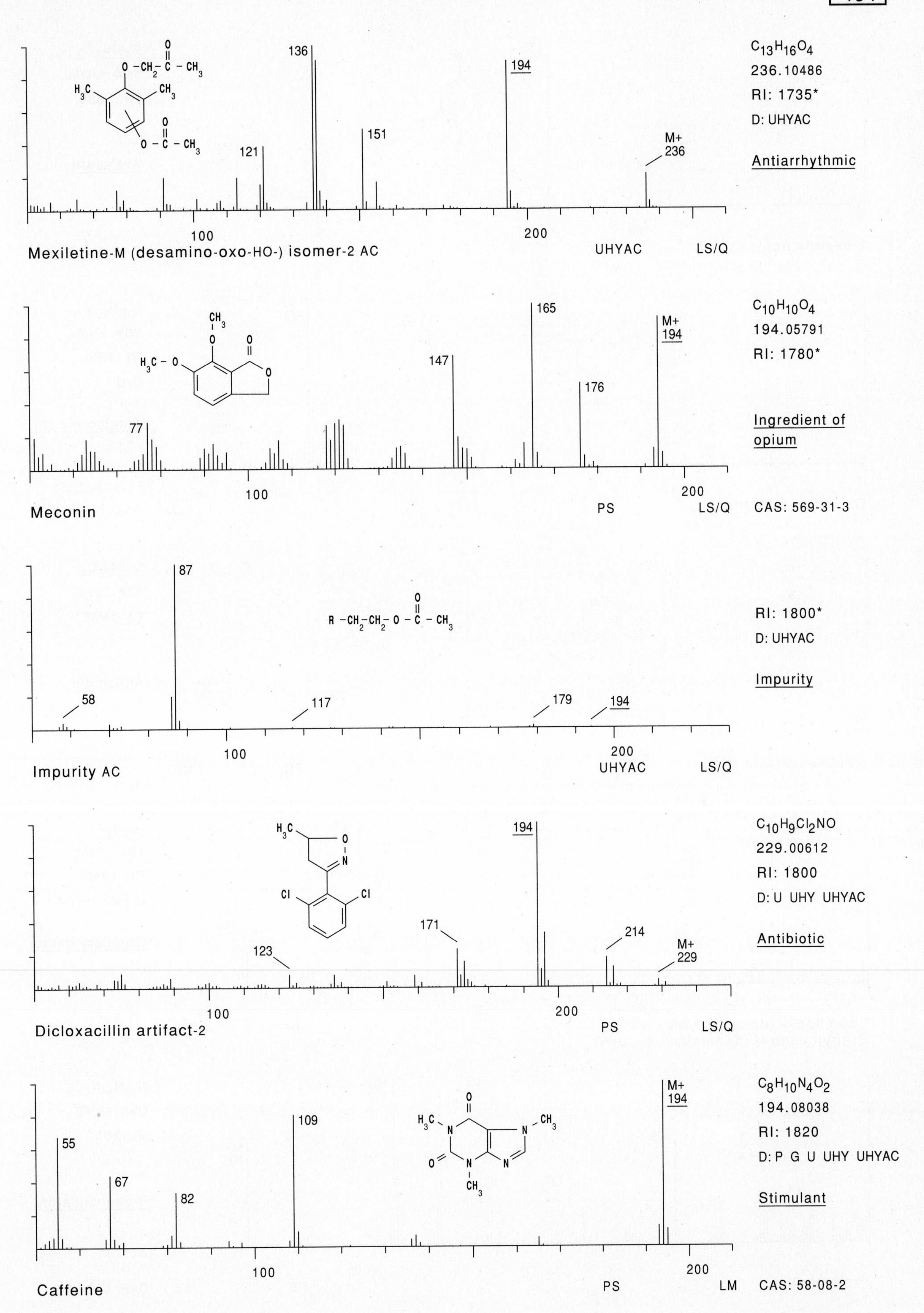

136
194
151
121
M+
236
100
200
Mexiletine-M (desamino-oxo-HO-) isomer-2 AC
UHYAC
LS/Q
C13H16O4
236.10486
RI: 1735*
D: UHYAC
Antiarrhythmic
165
M+
194
147
176
77
100
200
Meconin
PS
LS/Q
C10H10O4
194.05791
RI: 1780*
Ingredient of opium
CAS: 569-31-3
87
58
117
179
194
100
200
Impurity AC
UHYAC
LS/Q
RI: 1800*
D: UHYAC
Impurity
194
171
214
123
M+
229
100
200
Dicloxacillin artifact-2
PS
LS/Q
C10H9Cl2NO
229.00612
RI: 1800
D: U UHY UHYAC
Antibiotic
M+
194
109
55
67
82
100
200
Caffeine
PS
LM
C8H10N4O2
194.08038
RI: 1820
D: P G U UHY UHYAC
Stimulant
CAS: 58-08-2

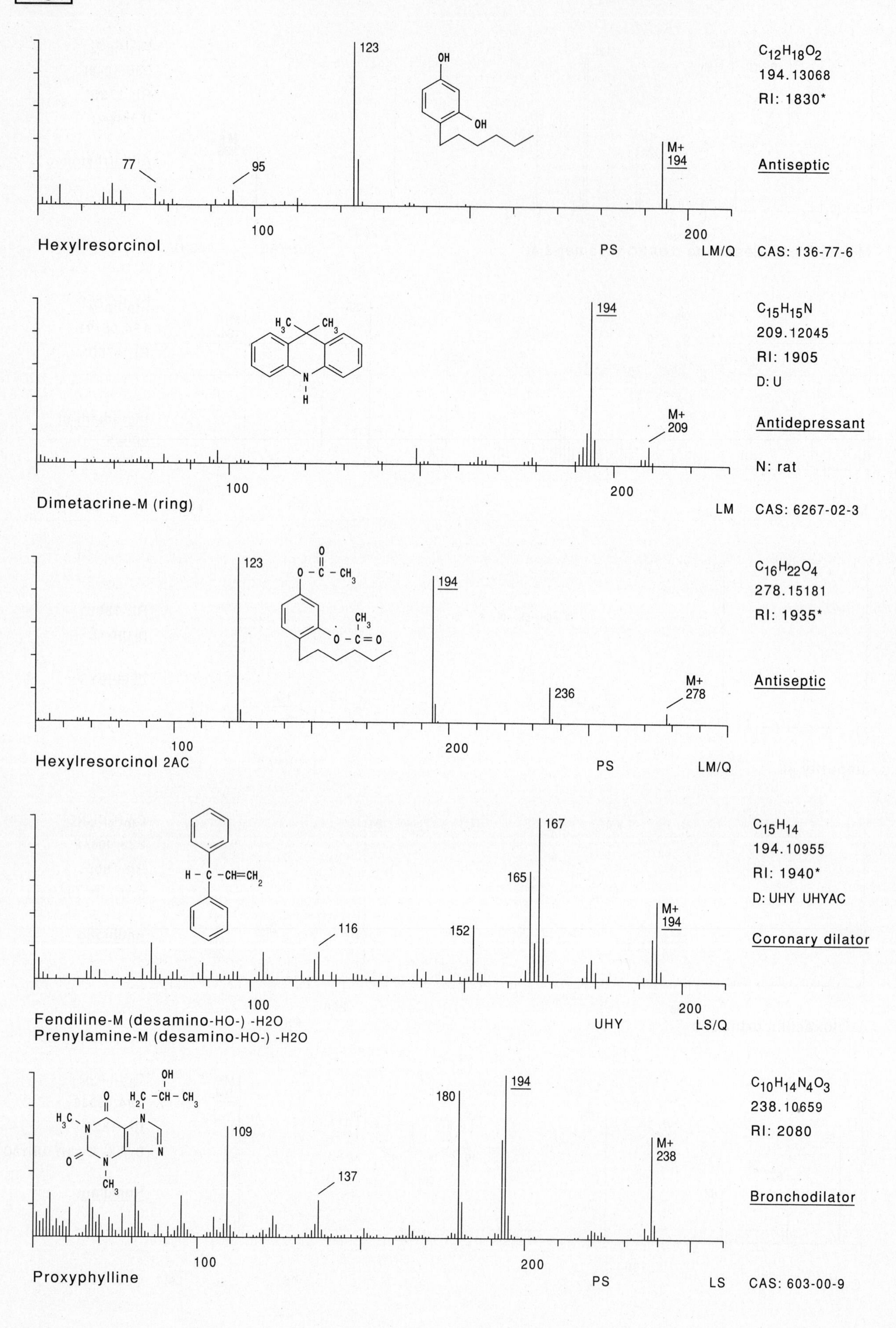

123
OH
OH
77
95
M+
194
100
200
Hexylresorcinol
PS
LM/Q
C12H18O2
194.13068
RI: 1830*
Antiseptic
CAS: 136-77-6
194
H3C CH3
N
H
M+
209
100
200
Dimetacrine-M (ring)
LM
C15H15N
209.12045
RI: 1905
D: U
Antidepressant
N: rat
CAS: 6267-02-3
123
O
O - C - CH3
194
CH3
O - C = O
236
M+
278
100
200
Hexylresorcinol 2AC
PS
LM/Q
C16H22O4
278.15181
RI: 1935*
Antiseptic
H - C - CH=CH2
167
165
152
116
M+
194
100
200
Fendiline-M (desamino-HO-) -H2O
Prenylamine-M (desamino-HO-) -H2O
UHY
LS/Q
C15H14
194.10955
RI: 1940*
D: UHY UHYAC
Coronary dilator
OH
O
H2C - CH - CH3
H3C
N
N
N
O
N
CH3
194
180
109
137
M+
238
100
200
Proxyphylline
PS
LS
C10H14N4O3
238.10659
RI: 2080
Bronchodilator
CAS: 603-00-9

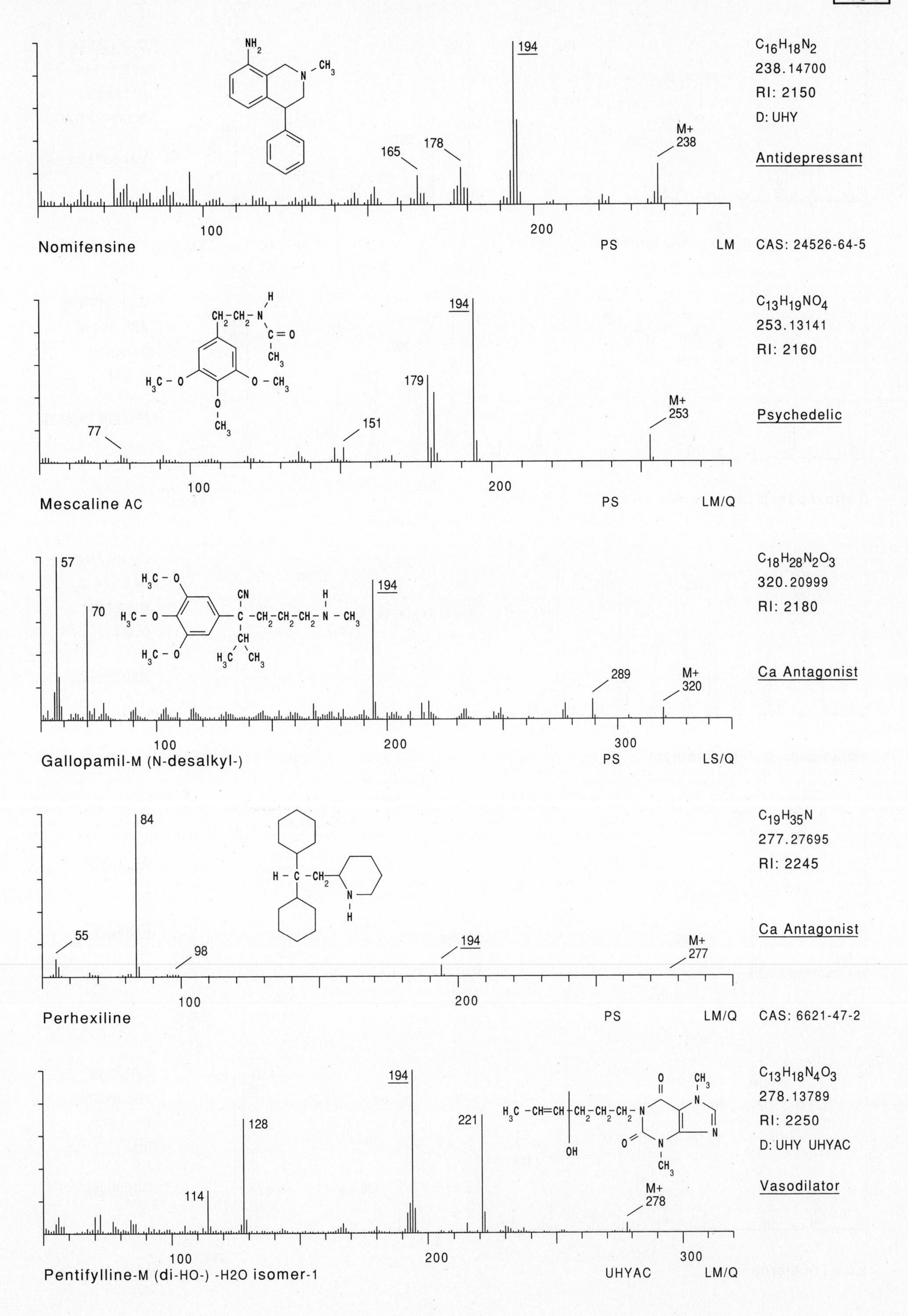

$C_{16}H_{18}N_2$
238.14700
RI: 2150
D: UHY
Antidepressant
CAS: 24526-64-5

$C_{13}H_{19}NO_4$
253.13141
RI: 2160
Psychedelic

$C_{18}H_{28}N_2O_3$
320.20999
RI: 2180
Ca Antagonist

$C_{19}H_{35}N$
277.27695
RI: 2245
Ca Antagonist
CAS: 6621-47-2

$C_{13}H_{18}N_4O_3$
278.13789
RI: 2250
D: UHY UHYAC
Vasodilator

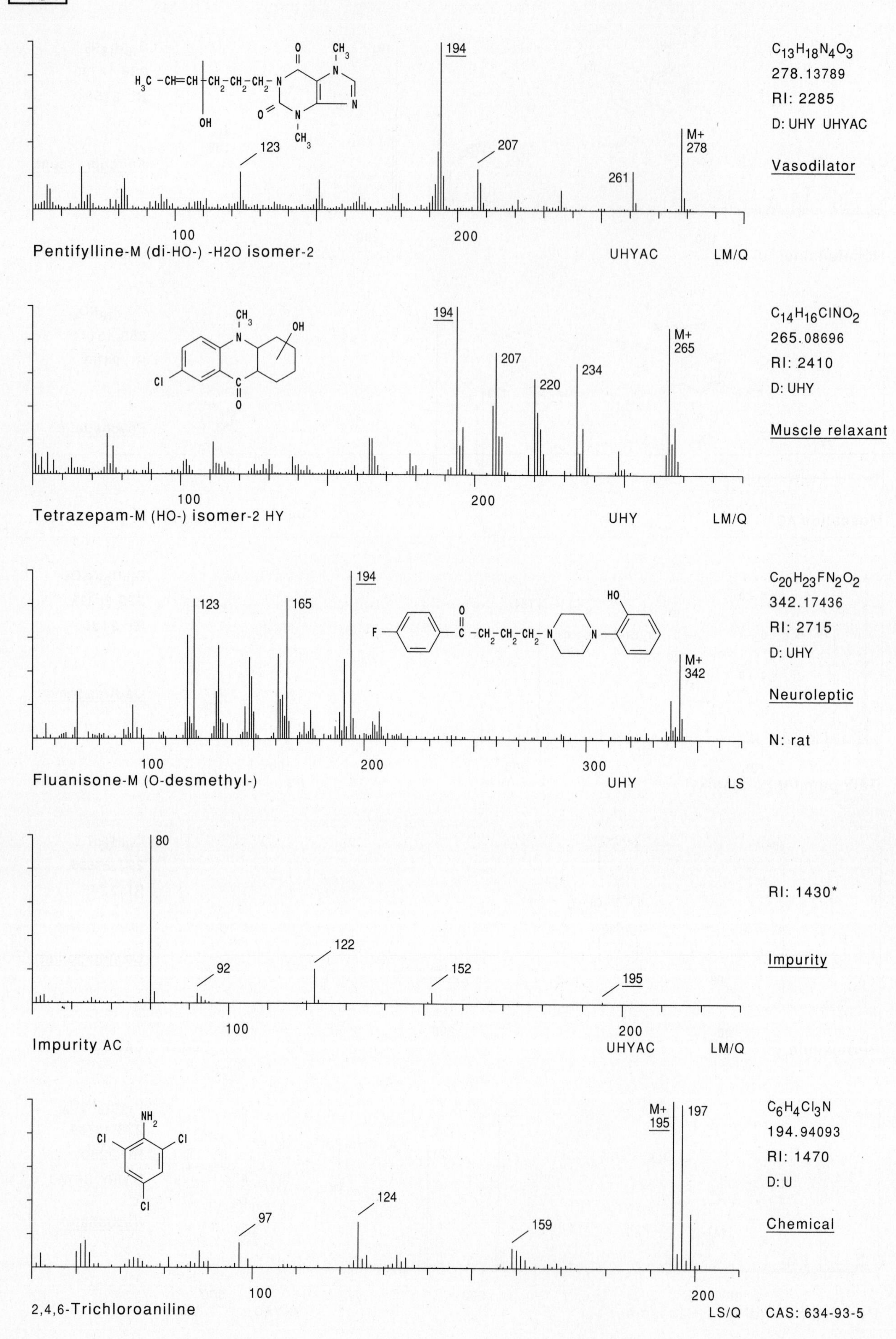

194
123
207
261
M+ 278
100
200
Pentifylline-M (di-HO-) -H2O isomer-2
UHYAC
LM/Q
$C_{13}H_{18}N_4O_3$
278.13789
RI: 2285
D: UHY UHYAC
Vasodilator
194
207
220
234
M+ 265
100
200
Tetrazepam-M (HO-) isomer-2 HY
UHY
LM/Q
$C_{14}H_{16}ClNO_2$
265.08696
RI: 2410
D: UHY
Muscle relaxant
123
165
194
M+ 342
100
200
300
Fluanisone-M (O-desmethyl-)
UHY
LS
$C_{20}H_{23}FN_2O_2$
342.17436
RI: 2715
D: UHY
Neuroleptic
N: rat
80
92
122
152
195
100
200
Impurity AC
UHYAC
LM/Q
RI: 1430*
Impurity
M+ 195
197
97
124
159
100
200
2,4,6-Trichloroaniline
LS/Q
$C_6H_4Cl_3N$
194.94093
RI: 1470
D: U
Chemical
CAS: 634-93-5

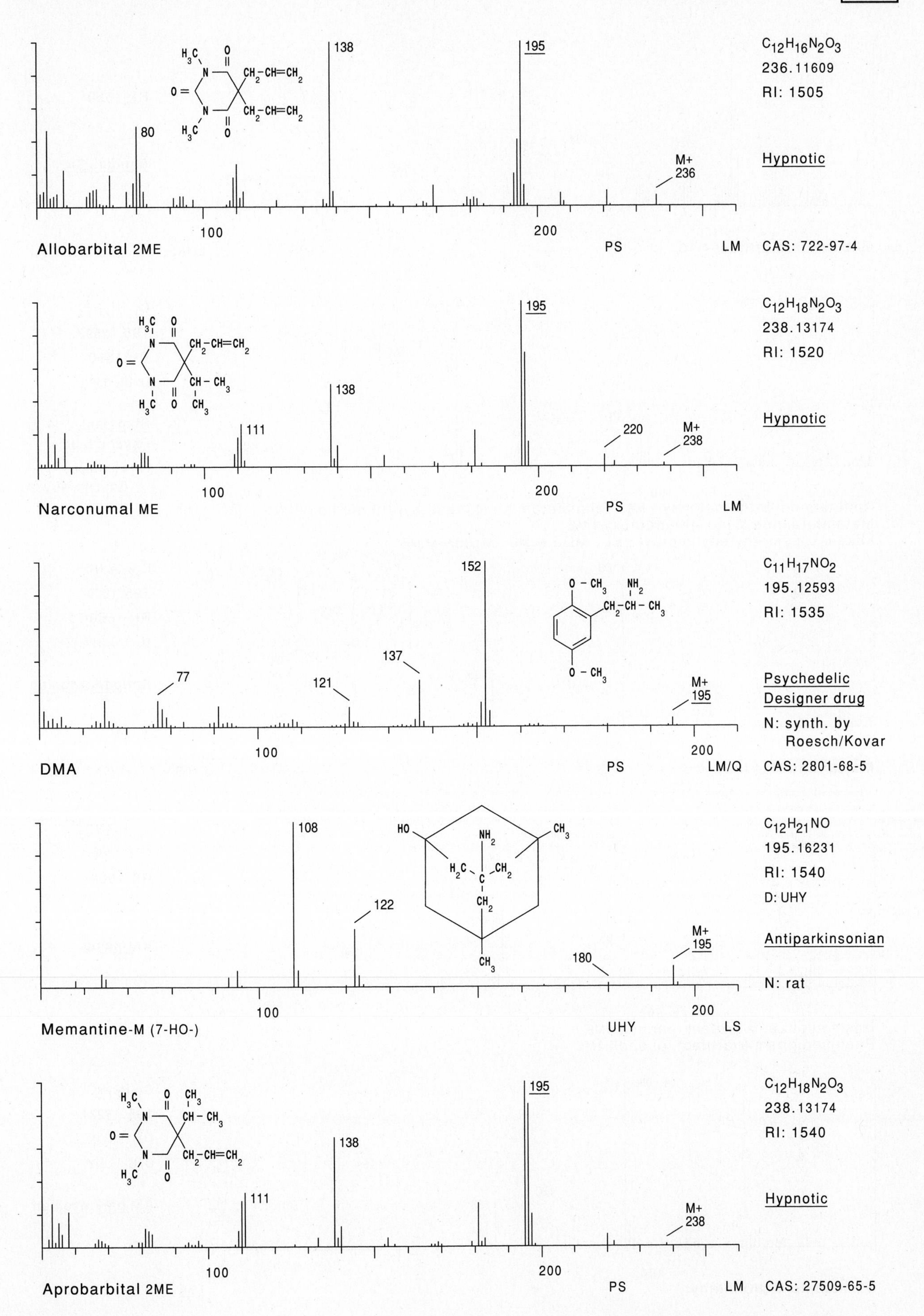

138
195
80
M+
236
100
200
Allobarbital 2ME
PS
LM
C12H16N2O3
236.11609
RI: 1505
Hypnotic
CAS: 722-97-4
195
138
111
220
M+
238
100
200
Narconumal ME
PS
LM
C12H18N2O3
238.13174
RI: 1520
Hypnotic
152
137
121
77
M+
195
100
200
DMA
PS
LM/Q
C11H17NO2
195.12593
RI: 1535
Psychedelic
Designer drug
N: synth. by
Roesch/Kovar
CAS: 2801-68-5
108
122
180
M+
195
100
200
Memantine-M (7-HO-)
UHY
LS
C12H21NO
195.16231
RI: 1540
D: UHY
Antiparkinsonian
N: rat
195
138
111
M+
238
100
200
Aprobarbital 2ME
PS
LM
C12H18N2O3
238.13174
RI: 1540
Hypnotic
CAS: 27509-65-5

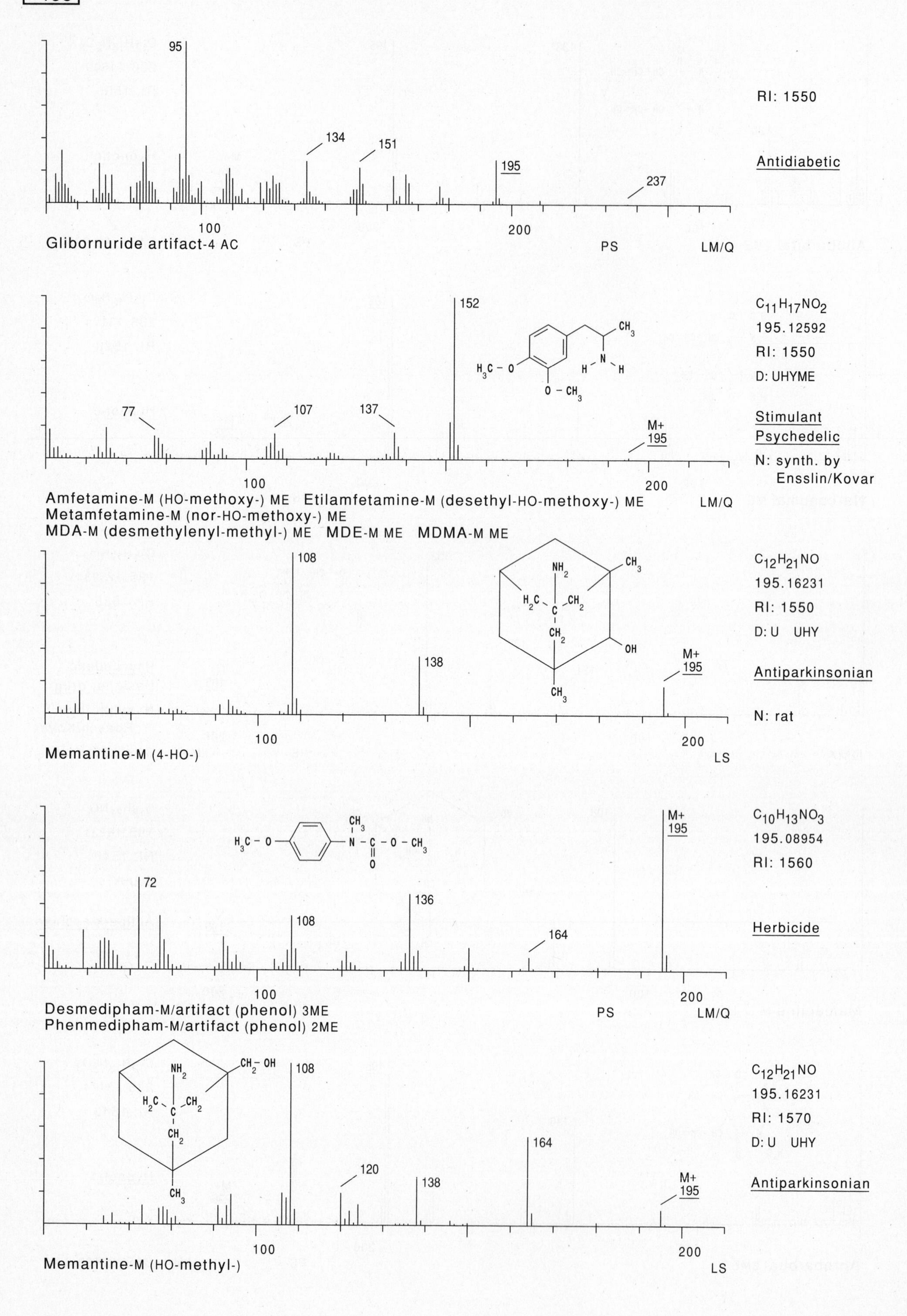

RI: 1550
Antidiabetic
Glibornuride artifact-4 AC
PS
LM/Q
$C_{11}H_{17}NO_2$
195.12592
RI: 1550
D: UHYME
Stimulant
Psychedelic
N: synth. by Ensslin/Kovar
Amfetamine-M (HO-methoxy-) ME Etilamfetamine-M (desethyl-HO-methoxy-) ME
Metamfetamine-M (nor-HO-methoxy-) ME
MDA-M (desmethylenyl-methyl-) ME MDE-M ME MDMA-M ME
LM/Q
$C_{12}H_{21}NO$
195.16231
RI: 1550
D: U UHY
Antiparkinsonian
N: rat
Memantine-M (4-HO-)
LS
$C_{10}H_{13}NO_3$
195.08954
RI: 1560
Herbicide
Desmedipham-M/artifact (phenol) 3ME
Phenmedipham-M/artifact (phenol) 2ME
PS
LM/Q
$C_{12}H_{21}NO$
195.16231
RI: 1570
D: U UHY
Antiparkinsonian
Memantine-M (HO-methyl-)
LS

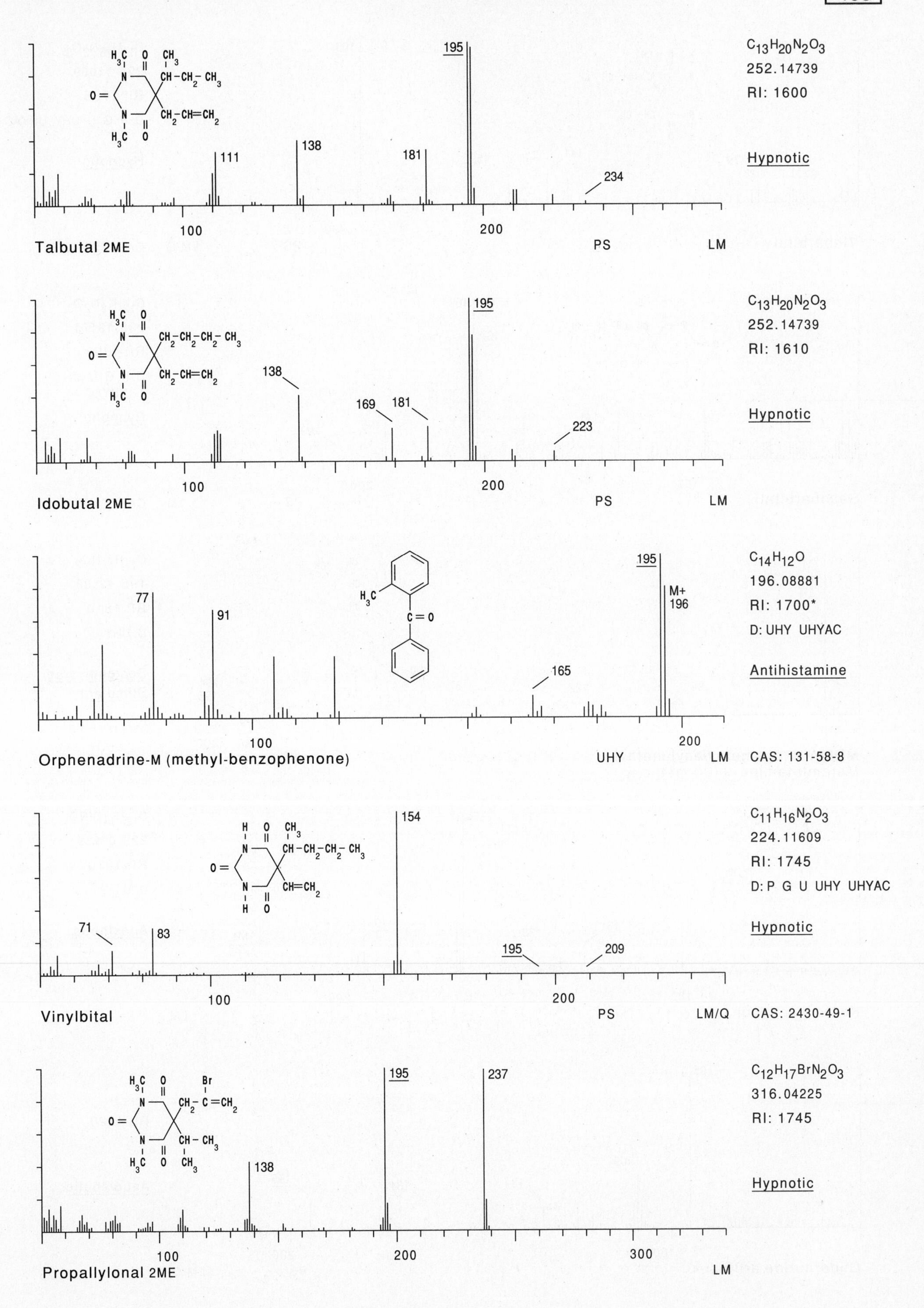
C13H20N2O3
252.14739
RI: 1600
Hypnotic
Talbutal 2ME
PS
LM
C13H20N2O3
252.14739
RI: 1610
Hypnotic
Idobutal 2ME
PS
LM
C14H12O
196.08881
RI: 1700*
D: UHY UHYAC
Antihistamine
Orphenadrine-M (methyl-benzophenone)
UHY
LM
CAS: 131-58-8
C11H16N2O3
224.11609
RI: 1745
D: P G U UHY UHYAC
Hypnotic
Vinylbital
PS
LM/Q
CAS: 2430-49-1
C12H17BrN2O3
316.04225
RI: 1745
Hypnotic
Propallylonal 2ME
LM

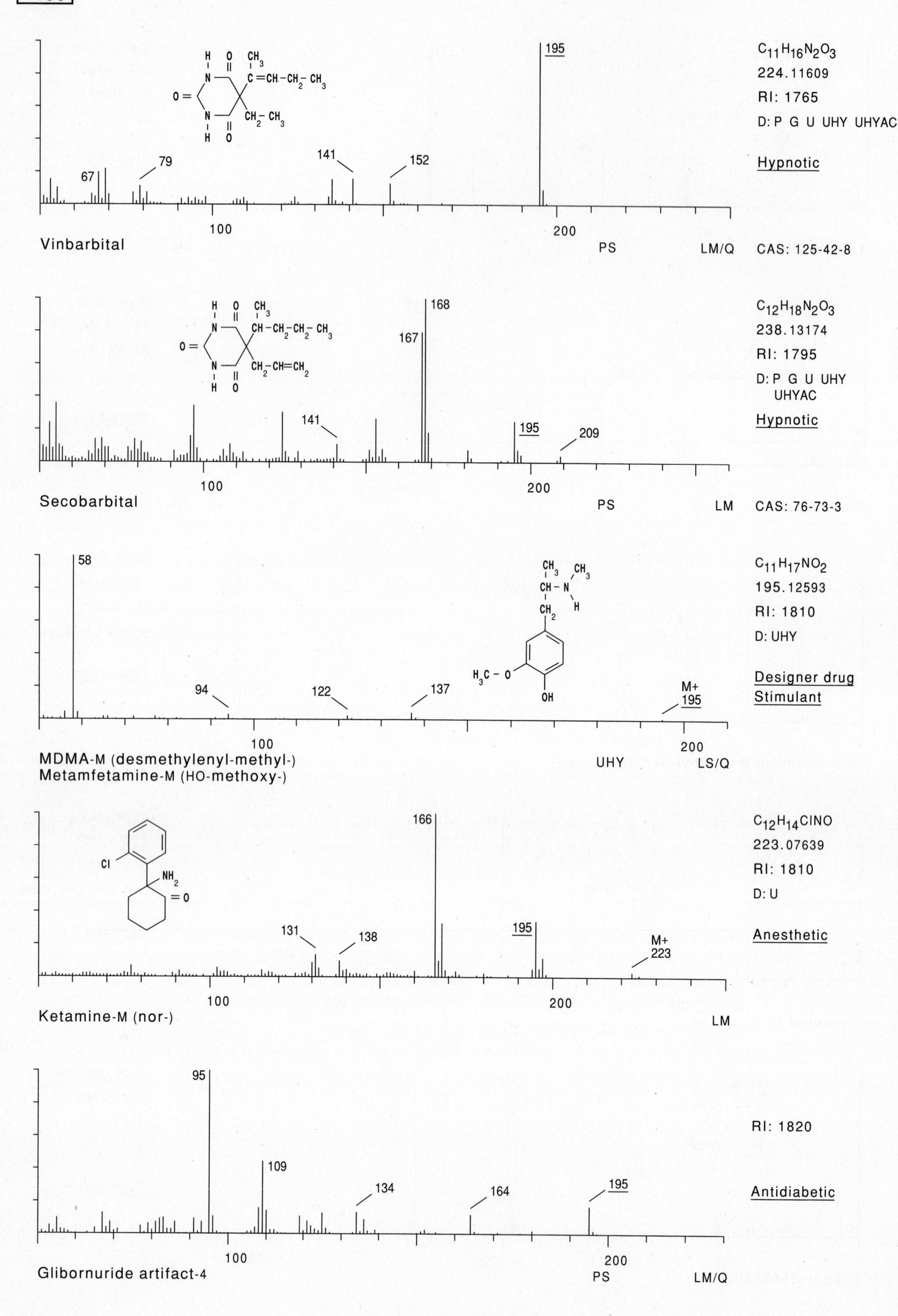
195
141
152
79
67
100
200
PS
LM/Q
Vinbarbital
$C_{11}H_{16}N_2O_3$
224.11609
RI: 1765
D: P G U UHY UHYAC
Hypnotic
CAS: 125-42-8
168
167
141
195
209
100
200
PS
LM
Secobarbital
$C_{12}H_{18}N_2O_3$
238.13174
RI: 1795
D: P G U UHY UHYAC
Hypnotic
CAS: 76-73-3
58
94
122
137
M+ 195
100
200
UHY
LS/Q
MDMA-M (desmethylenyl-methyl-)
Metamfetamine-M (HO-methoxy-)
$C_{11}H_{17}NO_2$
195.12593
RI: 1810
D: UHY
Designer drug
Stimulant
166
131
138
195
M+ 223
100
200
LM
Ketamine-M (nor-)
$C_{12}H_{14}ClNO$
223.07639
RI: 1810
D: U
Anesthetic
95
109
134
164
195
100
200
PS
LM/Q
Glibornuride artifact-4
RI: 1820
Antidiabetic

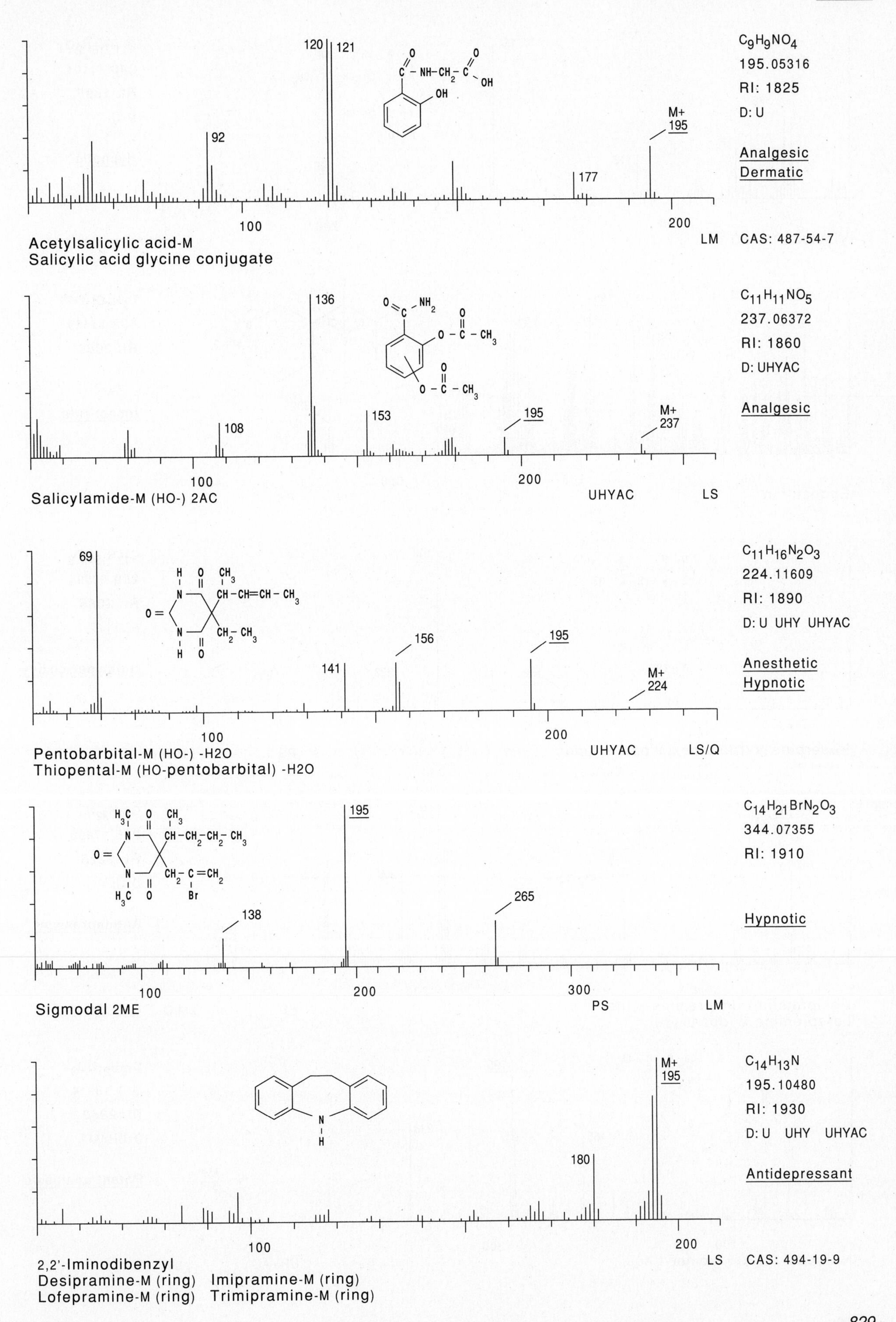
120
121
92
177
M+
195
100
200
LM
Acetylsalicylic acid-M
Salicylic acid glycine conjugate
C9H9NO4
195.05316
RI: 1825
D: U
Analgesic
Dermatic
CAS: 487-54-7
136
108
153
195
M+
237
100
200
UHYAC
LS
Salicylamide-M (HO-) 2AC
C11H11NO5
237.06372
RI: 1860
D: UHYAC
Analgesic
69
141
156
195
M+
224
100
200
UHYAC
LS/Q
Pentobarbital-M (HO-) -H2O
Thiopental-M (HO-pentobarbital) -H2O
C11H16N2O3
224.11609
RI: 1890
D: U UHY UHYAC
Anesthetic
Hypnotic
195
138
265
100
200
300
PS
LM
Sigmodal 2ME
C14H21BrN2O3
344.07355
RI: 1910
Hypnotic
M+
195
180
100
200
LS
2,2'-Iminodibenzyl
Desipramine-M (ring)
Imipramine-M (ring)
Lofepramine-M (ring)
Trimipramine-M (ring)
C14H13N
195.10480
RI: 1930
D: U UHY UHYAC
Antidepressant
CAS: 494-19-9

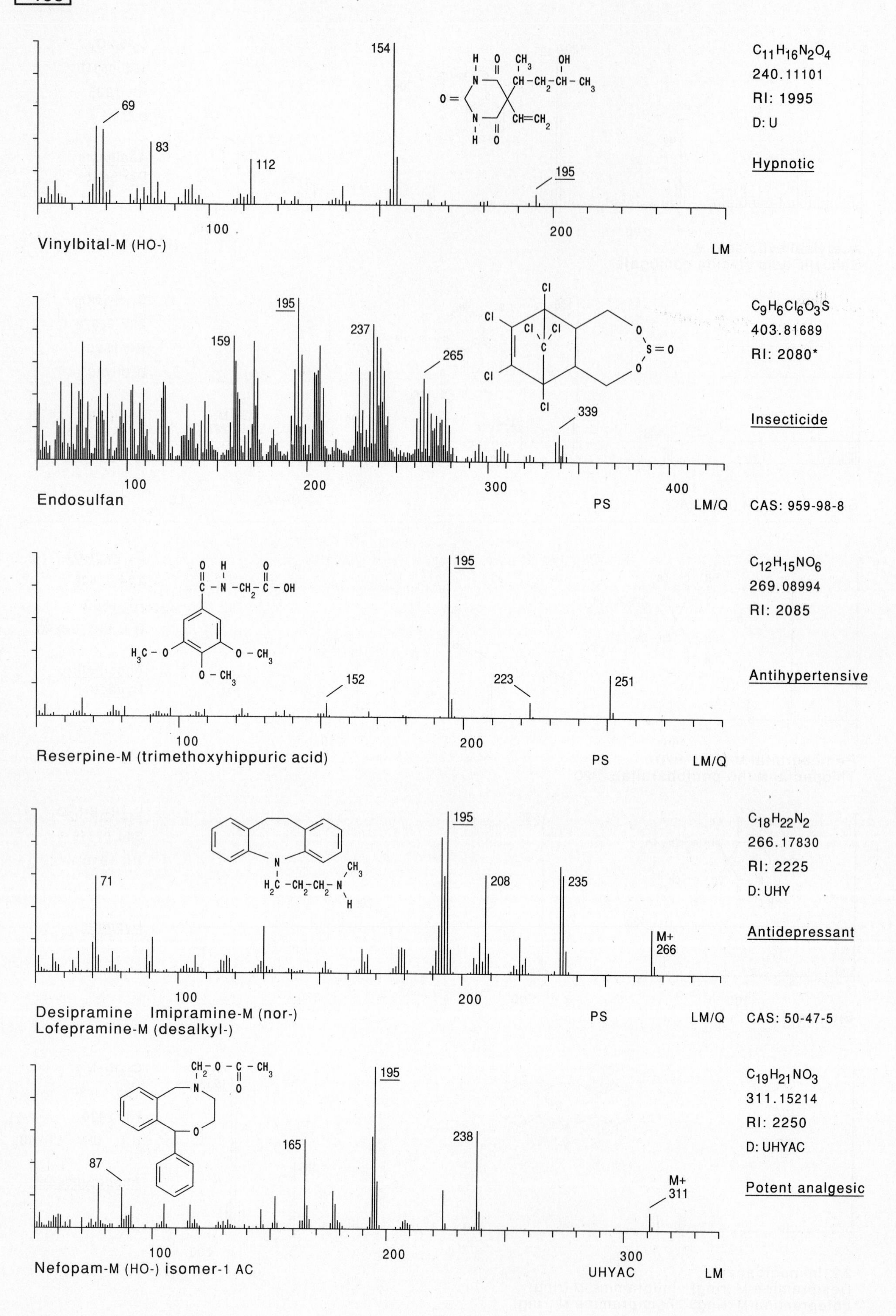
154
69
83
112
195
100
200
Vinylbital-M (HO-)
LM
$C_{11}H_{16}N_2O_4$
240.11101
RI: 1995
D: U
Hypnotic
195
159
237
265
339
100
200
300
400
Endosulfan
PS
LM/Q
$C_9H_6Cl_6O_3S$
403.81689
RI: 2080*
Insecticide
CAS: 959-98-8
195
152
223
251
100
200
Reserpine-M (trimethoxyhippuric acid)
PS
LM/Q
$C_{12}H_{15}NO_6$
269.08994
RI: 2085
Antihypertensive
195
71
208
235
M+
266
100
200
Desipramine Imipramine-M (nor-)
Lofepramine-M (desalkyl-)
PS
LM/Q
$C_{18}H_{22}N_2$
266.17830
RI: 2225
D: UHY
Antidepressant
CAS: 50-47-5
195
165
238
87
M+
311
100
200
300
Nefopam-M (HO-) isomer-1 AC
UHYAC
LM
$C_{19}H_{21}NO_3$
311.15214
RI: 2250
D: UHYAC
Potent analgesic

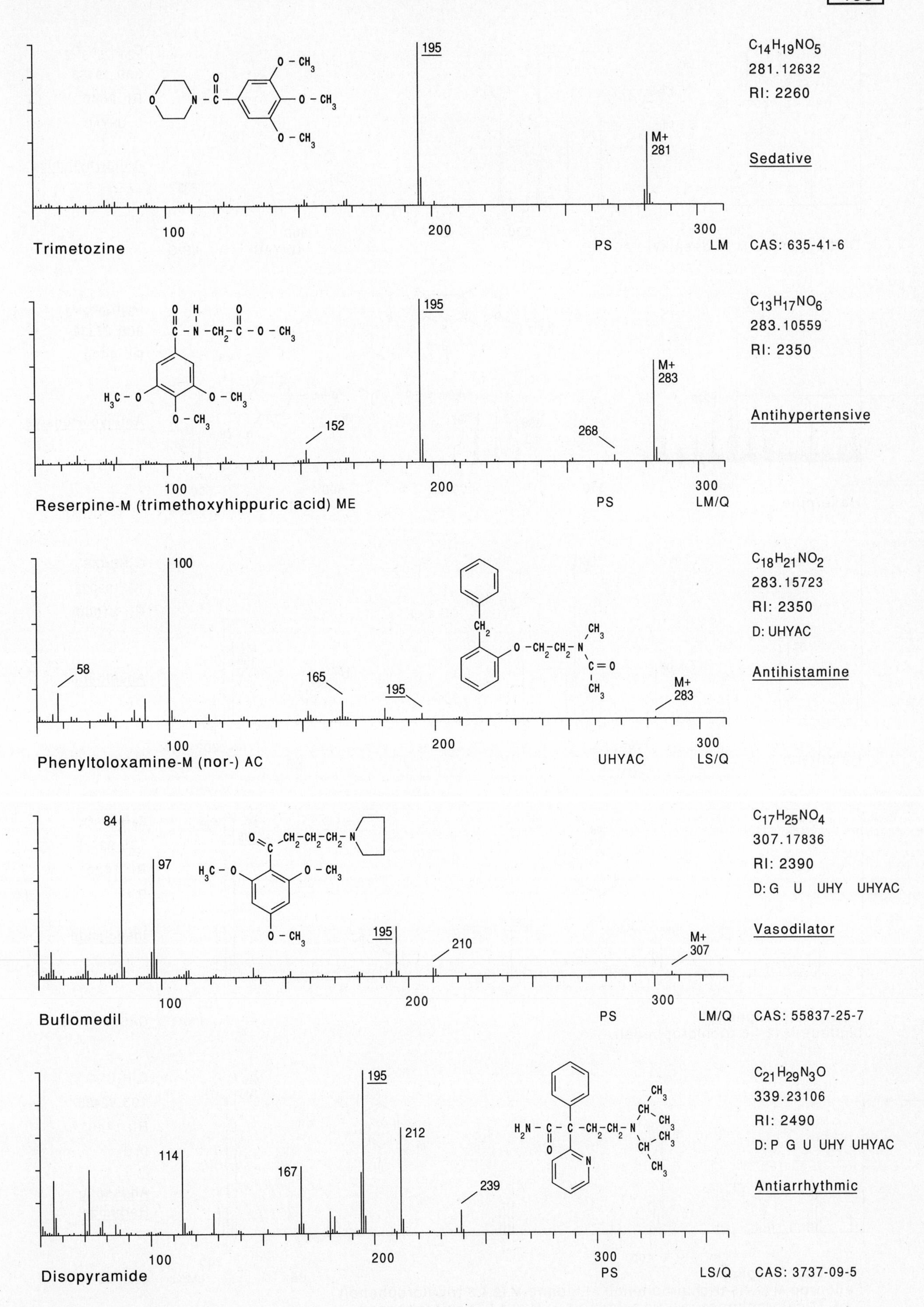

Trimetozine
C14H19NO5
281.12632
RI: 2260
Sedative
195
M+ 281
PS LM
CAS: 635-41-6
Reserpine-M (trimethoxyhippuric acid) ME
C13H17NO6
283.10559
RI: 2350
Antihypertensive
195
152
268
M+ 283
PS LM/Q
Phenyltoloxamine-M (nor-) AC
C18H21NO2
283.15723
RI: 2350
D: UHYAC
Antihistamine
100
58
165
195
M+ 283
UHYAC LS/Q
Buflomedil
C17H25NO4
307.17836
RI: 2390
D: G U UHY UHYAC
Vasodilator
84
97
195
210
M+ 307
PS LM/Q
CAS: 55837-25-7
Disopyramide
C21H29N3O
339.23106
RI: 2490
D: P G U UHY UHYAC
Antiarrhythmic
195
212
114
167
239
PS LS/Q
CAS: 3737-09-5

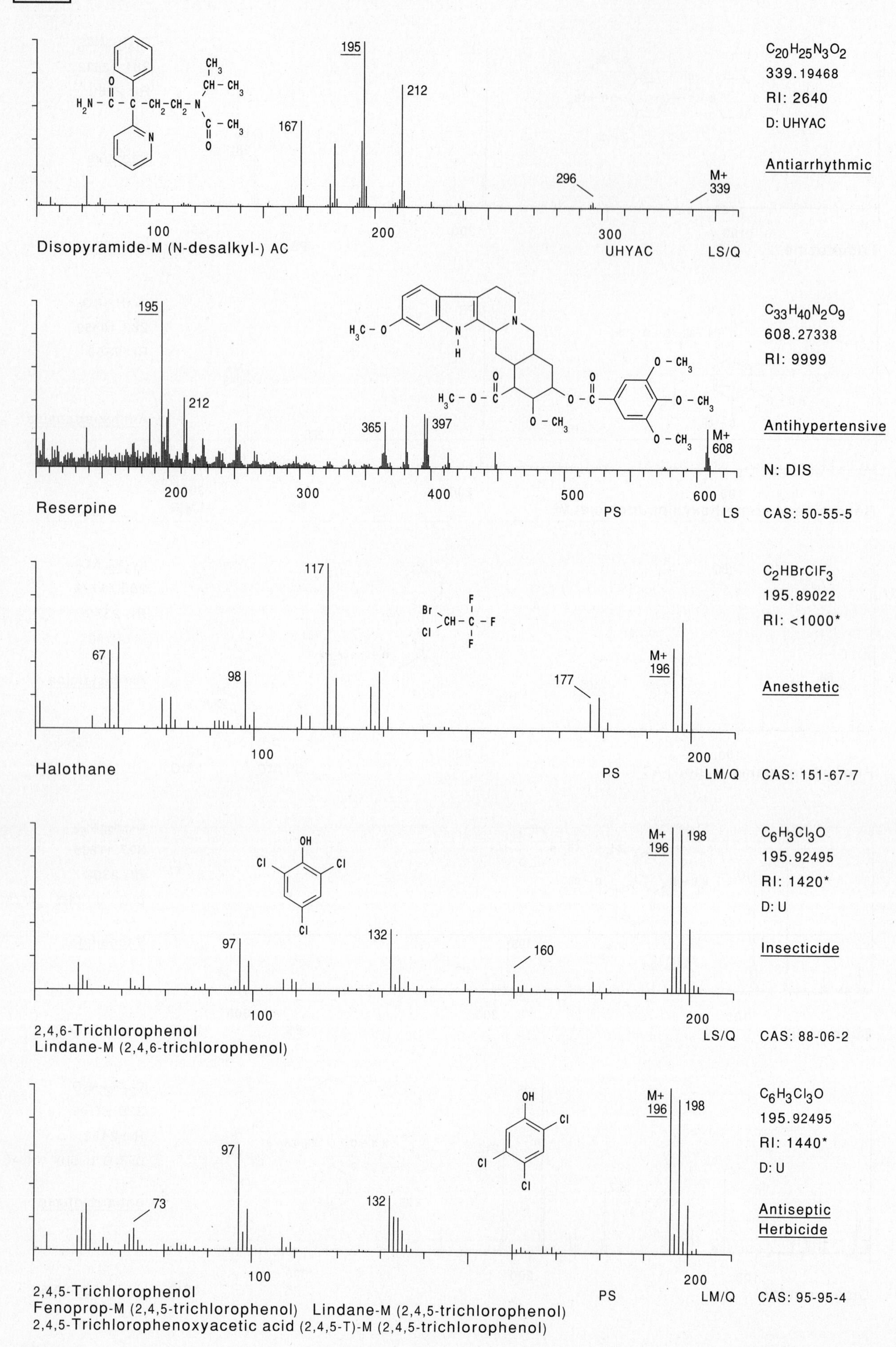
195
212
167
296
M+
339
100
200
300
Disopyramide-M (N-desalkyl-) AC
UHYAC
LS/Q
$C_{20}H_{25}N_3O_2$
339.19468
RI: 2640
D: UHYAC
Antiarrhythmic
195
212
365
397
M+
608
200
300
400
500
600
Reserpine
PS
LS
$C_{33}H_{40}N_2O_9$
608.27338
RI: 9999
Antihypertensive
N: DIS
CAS: 50-55-5
117
67
98
177
M+
196
100
200
Halothane
PS
LM/Q
$C_2HBrClF_3$
195.89022
RI: <1000*
Anesthetic
CAS: 151-67-7
M+
196
198
97
132
160
100
200
2,4,6-Trichlorophenol
Lindane-M (2,4,6-trichlorophenol)
LS/Q
$C_6H_3Cl_3O$
195.92495
RI: 1420*
D: U
Insecticide
CAS: 88-06-2
M+
196
198
97
73
132
100
200
2,4,5-Trichlorophenol
Fenoprop-M (2,4,5-trichlorophenol) Lindane-M (2,4,5-trichlorophenol)
2,4,5-Trichlorophenoxyacetic acid (2,4,5-T)-M (2,4,5-trichlorophenol)
PS
LM/Q
$C_6H_3Cl_3O$
195.92495
RI: 1440*
D: U
Antiseptic
Herbicide
CAS: 95-95-4

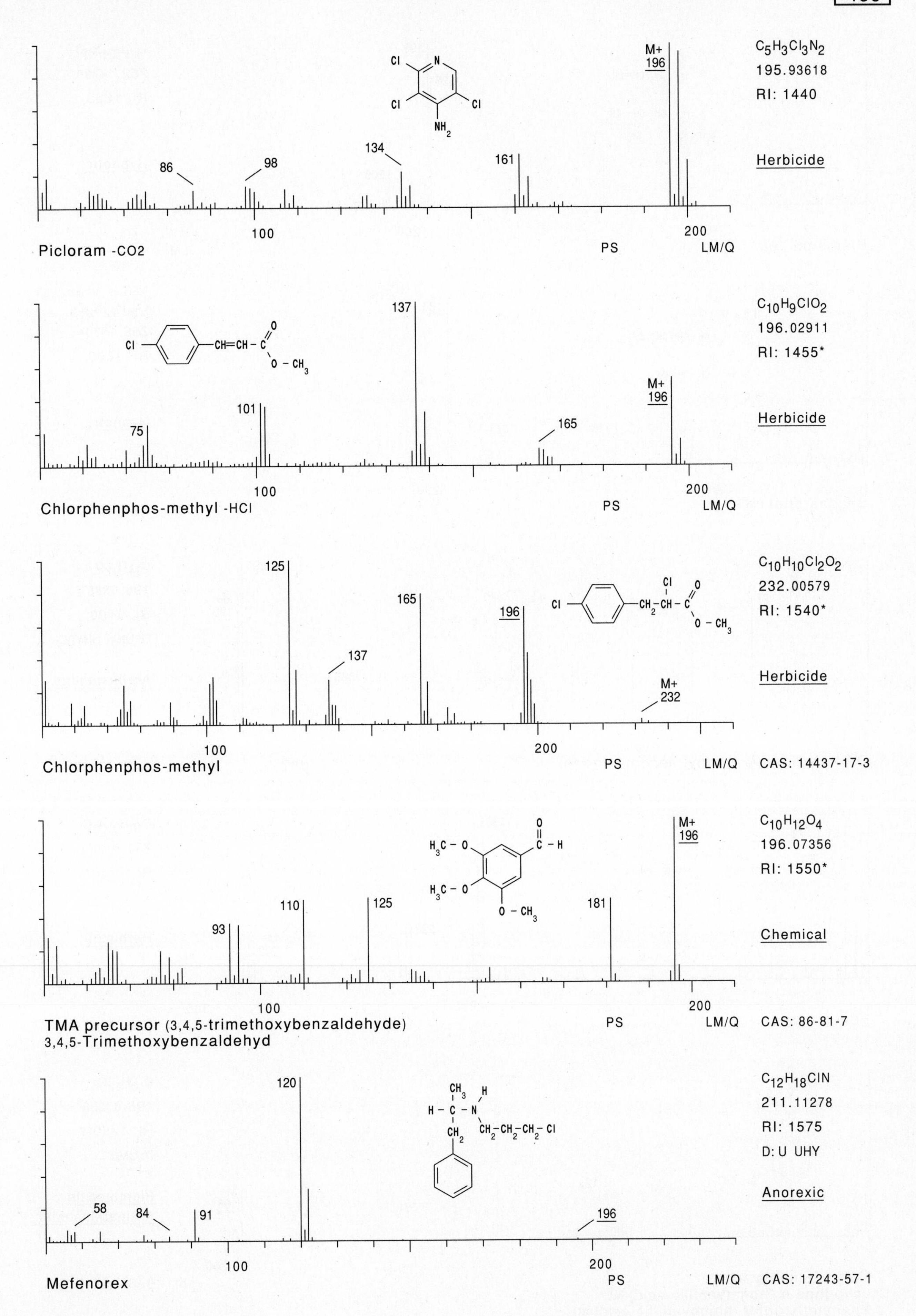

$C_5H_3Cl_3N_2$
195.93618
RI: 1440
Herbicide

$C_{10}H_9ClO_2$
196.02911
RI: 1455*
Herbicide

$C_{10}H_{10}Cl_2O_2$
232.00579
RI: 1540*
Herbicide
CAS: 14437-17-3

$C_{10}H_{12}O_4$
196.07356
RI: 1550*
Chemical
CAS: 86-81-7

$C_{12}H_{18}ClN$
211.11278
RI: 1575
D: U UHY
Anorexic
CAS: 17243-57-1

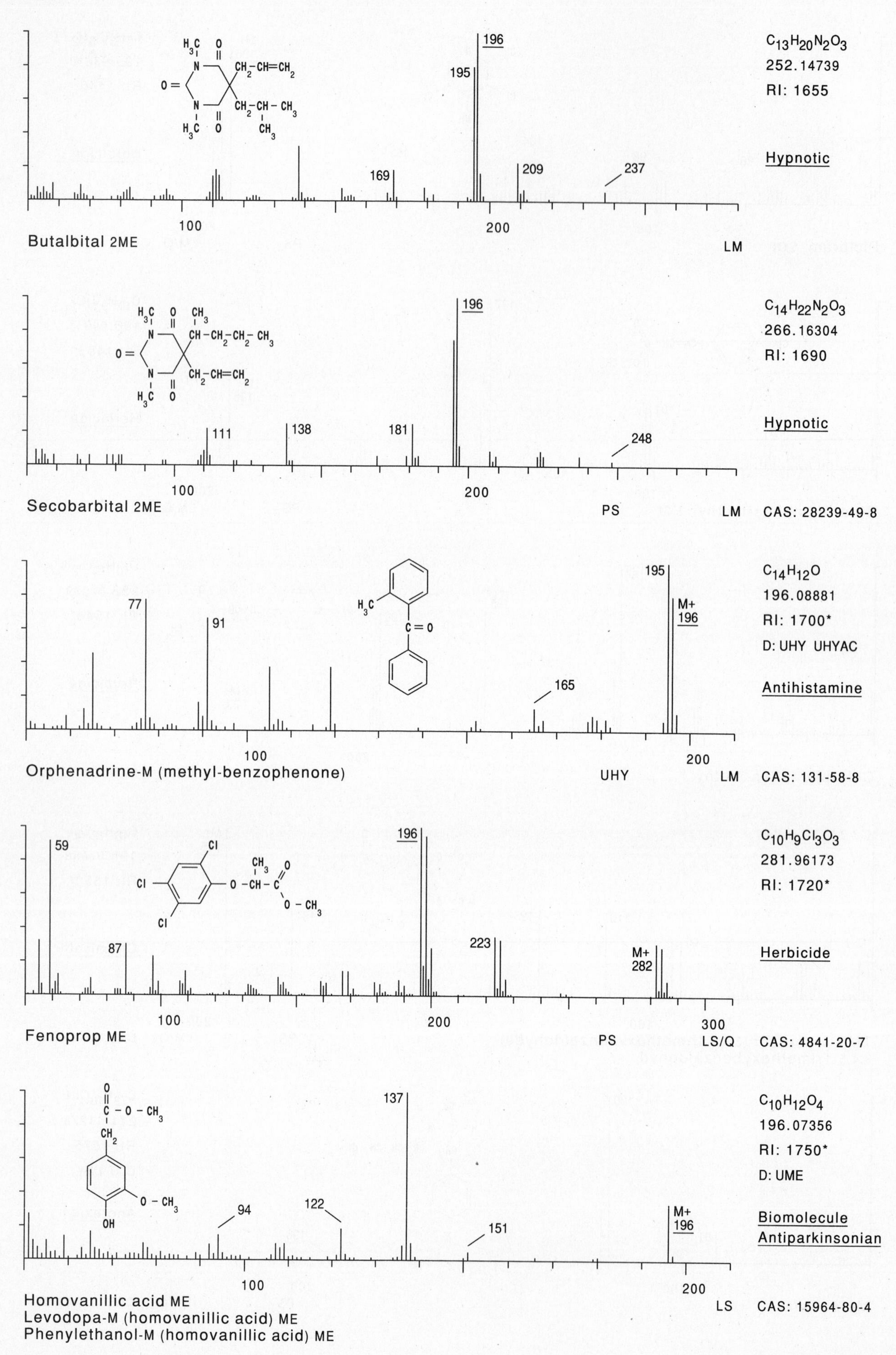
196
195
169
209
237
100
200
$C_{13}H_{20}N_2O_3$
252.14739
RI: 1655
Hypnotic
Butalbital 2ME
LM
196
111
138
181
248
100
200
$C_{14}H_{22}N_2O_3$
266.16304
RI: 1690
Hypnotic
Secobarbital 2ME
PS
LM
CAS: 28239-49-8
195
M+
196
77
91
165
100
200
$C_{14}H_{12}O$
196.08881
RI: 1700*
D: UHY UHYAC
Antihistamine
Orphenadrine-M (methyl-benzophenone)
UHY
LM
CAS: 131-58-8
196
59
87
223
M+
282
100
200
300
$C_{10}H_9Cl_3O_3$
281.96173
RI: 1720*
Herbicide
Fenoprop ME
PS
LS/Q
CAS: 4841-20-7
137
94
122
151
M+
196
100
200
$C_{10}H_{12}O_4$
196.07356
RI: 1750*
D: UME
Biomolecule
Antiparkinsonian
Homovanillic acid ME
Levodopa-M (homovanillic acid) ME
Phenylethanol-M (homovanillic acid) ME
LS
CAS: 15964-80-4

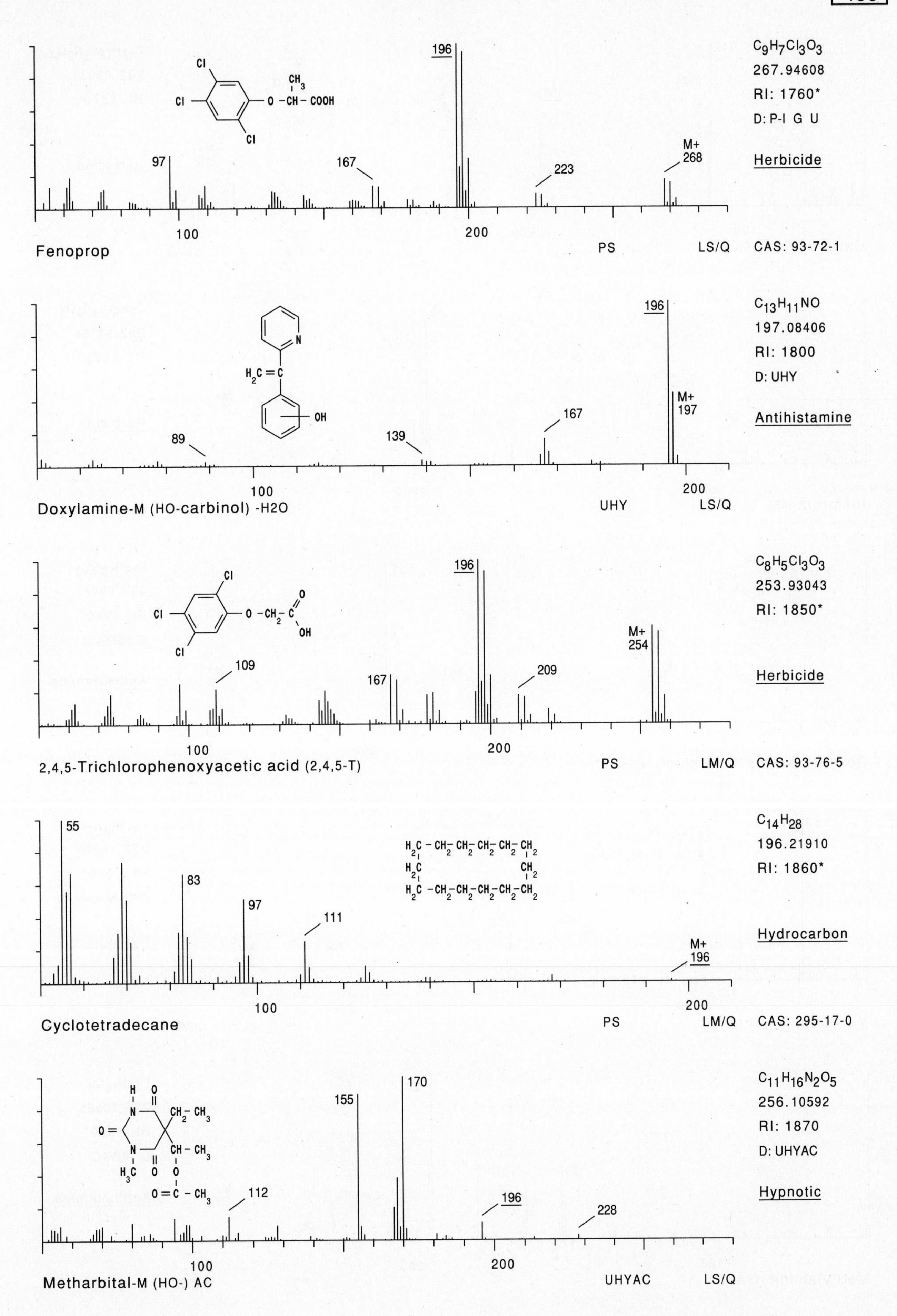
Fenoprop
PS LS/Q
C9H7Cl3O3
267.94608
RI: 1760*
D: P-I G U
Herbicide
CAS: 93-72-1
97
167
196
223
M+
268
100
200
Doxylamine-M (HO-carbinol) -H2O
UHY LS/Q
C13H11NO
197.08406
RI: 1800
D: UHY
Antihistamine
89
139
167
196
M+
197
2,4,5-Trichlorophenoxyacetic acid (2,4,5-T)
PS LM/Q
C8H5Cl3O3
253.93043
RI: 1850*
Herbicide
CAS: 93-76-5
109
167
196
209
M+
254
Cyclotetradecane
PS LM/Q
C14H28
196.21910
RI: 1860*
Hydrocarbon
CAS: 295-17-0
55
83
97
111
M+
196
Metharbital-M (HO-) AC
UHYAC LS/Q
C11H16N2O5
256.10592
RI: 1870
D: UHYAC
Hypnotic
112
155
170
196
228

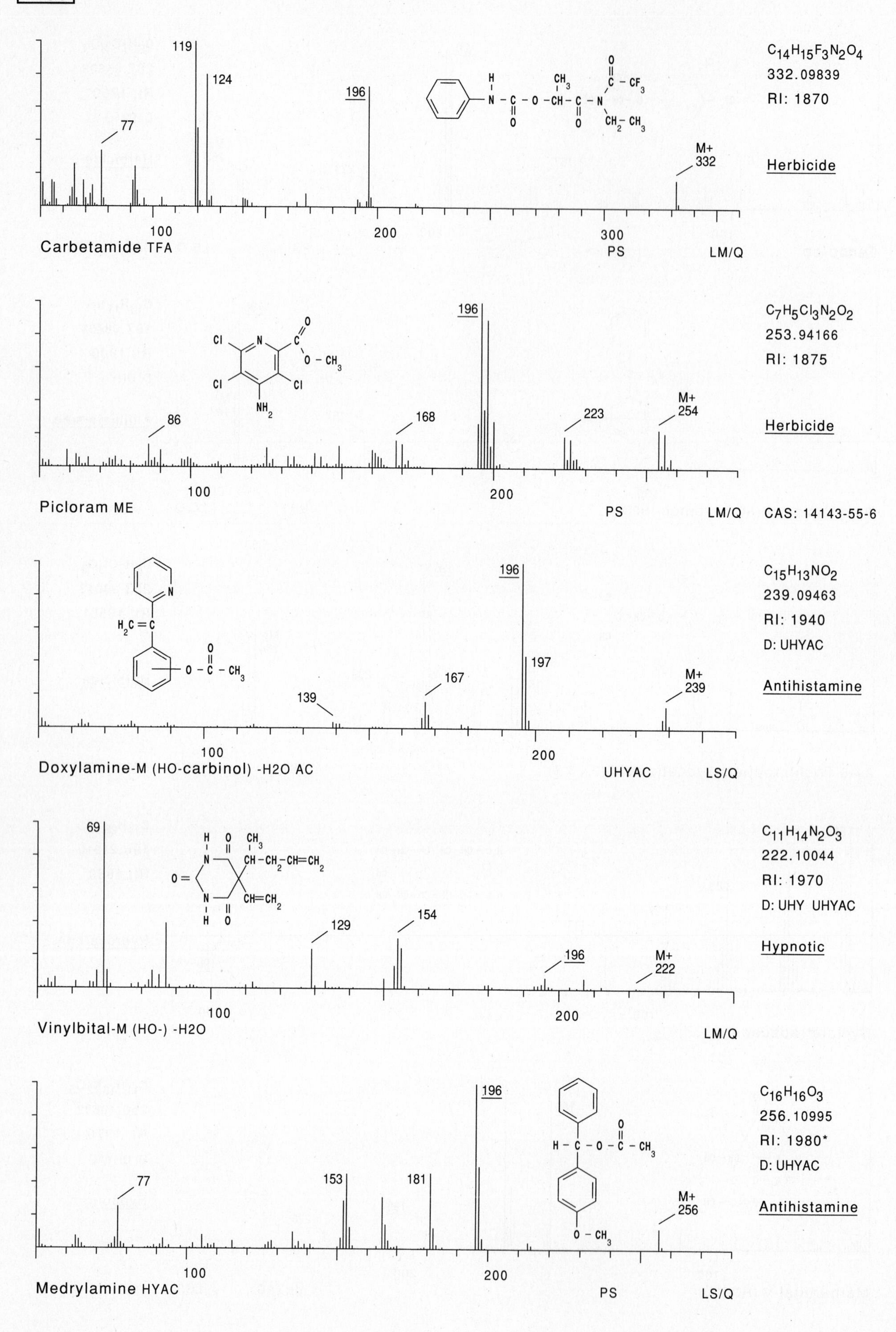
119
124
196
77
M+
332
100
200
300
Carbetamide TFA
PS
LM/Q
C14H15F3N2O4
332.09839
RI: 1870
Herbicide
196
86
168
223
M+
254
Picloram ME
C7H5Cl3N2O2
253.94166
RI: 1875
Herbicide
CAS: 14143-55-6
196
197
167
139
M+
239
Doxylamine-M (HO-carbinol) -H2O AC
UHYAC
LS/Q
C15H13NO2
239.09463
RI: 1940
D: UHYAC
Antihistamine
69
129
154
196
M+
222
Vinylbital-M (HO-) -H2O
C11H14N2O3
222.10044
RI: 1970
D: UHY UHYAC
Hypnotic
196
77
153
181
M+
256
Medrylamine HYAC
C16H16O3
256.10995
RI: 1980*
D: UHYAC
Antihistamine

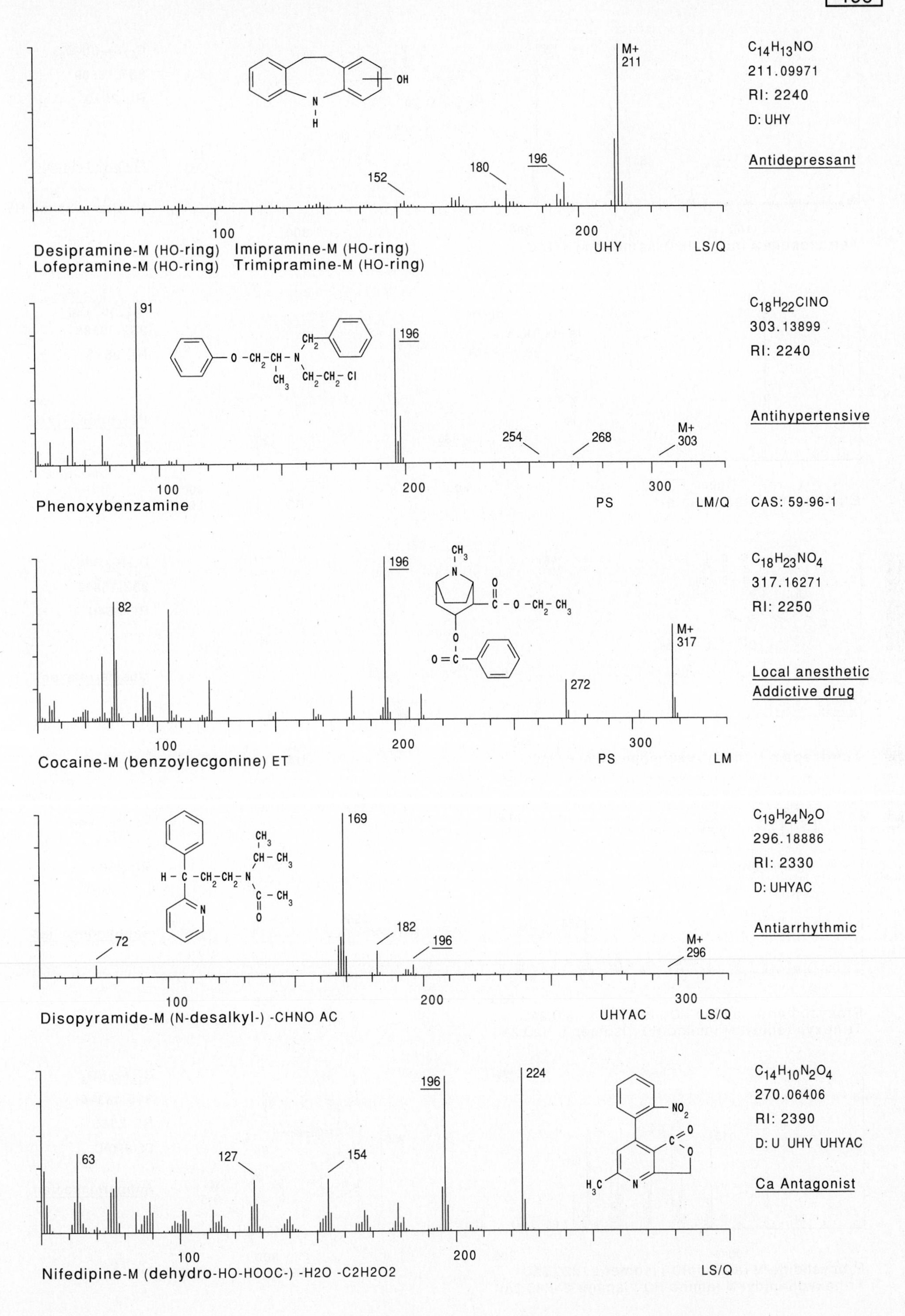
196
M+
211
152
180
196
100
200
UHY
LS/Q
Desipramine-M (HO-ring) Imipramine-M (HO-ring)
Lofepramine-M (HO-ring) Trimipramine-M (HO-ring)
C14H13NO
211.09971
RI: 2240
D: UHY
Antidepressant
91
196
254
268
M+
303
100
200
300
PS
LM/Q
Phenoxybenzamine
C18H22ClNO
303.13899
RI: 2240
Antihypertensive
CAS: 59-96-1
196
82
272
M+
317
100
200
300
PS
LM
Cocaine-M (benzoylecgonine) ET
C18H23NO4
317.16271
RI: 2250
Local anesthetic
Addictive drug
169
72
182
196
M+
296
100
200
300
UHYAC
LS/Q
Disopyramide-M (N-desalkyl-) -CHNO AC
C19H24N2O
296.18886
RI: 2330
D: UHYAC
Antiarrhythmic
196
224
63
127
154
100
200
LS/Q
Nifedipine-M (dehydro-HO-HOOC-) -H2O -C2H2O2
C14H10N2O4
270.06406
RI: 2390
D: U UHY UHYAC
Ca Antagonist

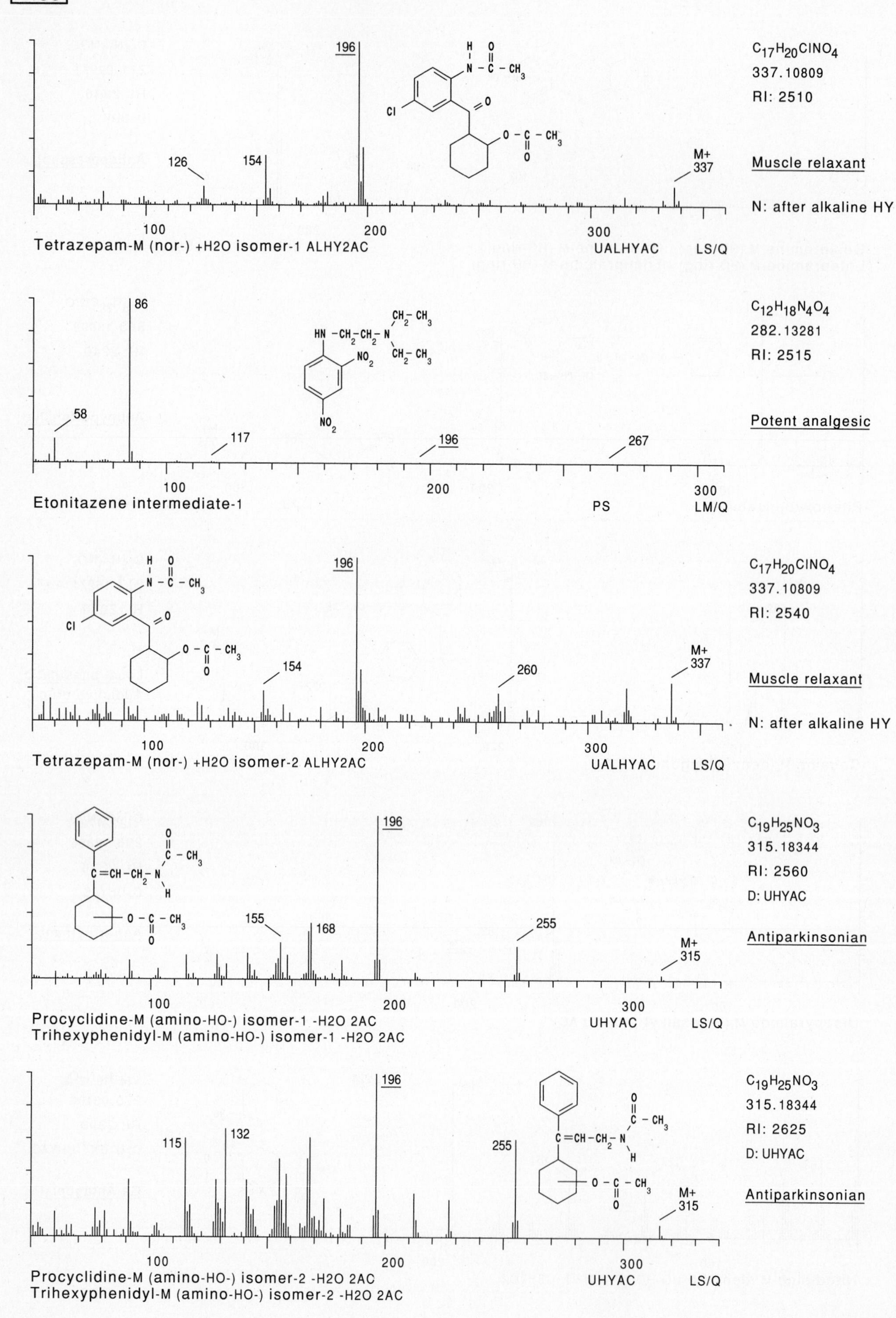

196
126
154
M+
337
100
200
300
Tetrazepam-M (nor-) +H2O isomer-1 ALHY2AC
UALHYAC
LS/Q
$C_{17}H_{20}ClNO_4$
337.10809
RI: 2510
Muscle relaxant
N: after alkaline HY
86
58
117
196
267
100
200
300
Etonitazene intermediate-1
PS
LM/Q
$C_{12}H_{18}N_4O_4$
282.13281
RI: 2515
Potent analgesic
196
154
260
M+
337
Tetrazepam-M (nor-) +H2O isomer-2 ALHY2AC
UALHYAC
LS/Q
$C_{17}H_{20}ClNO_4$
337.10809
RI: 2540
Muscle relaxant
N: after alkaline HY
196
155
168
255
M+
315
Procyclidine-M (amino-HO-) isomer-1 -H2O 2AC
Trihexyphenidyl-M (amino-HO-) isomer-1 -H2O 2AC
UHYAC
LS/Q
$C_{19}H_{25}NO_3$
315.18344
RI: 2560
D: UHYAC
Antiparkinsonian
196
115
132
255
M+
315
Procyclidine-M (amino-HO-) isomer-2 -H2O 2AC
Trihexyphenidyl-M (amino-HO-) isomer-2 -H2O 2AC
UHYAC
LS/Q
$C_{19}H_{25}NO_3$
315.18344
RI: 2625
D: UHYAC
Antiparkinsonian

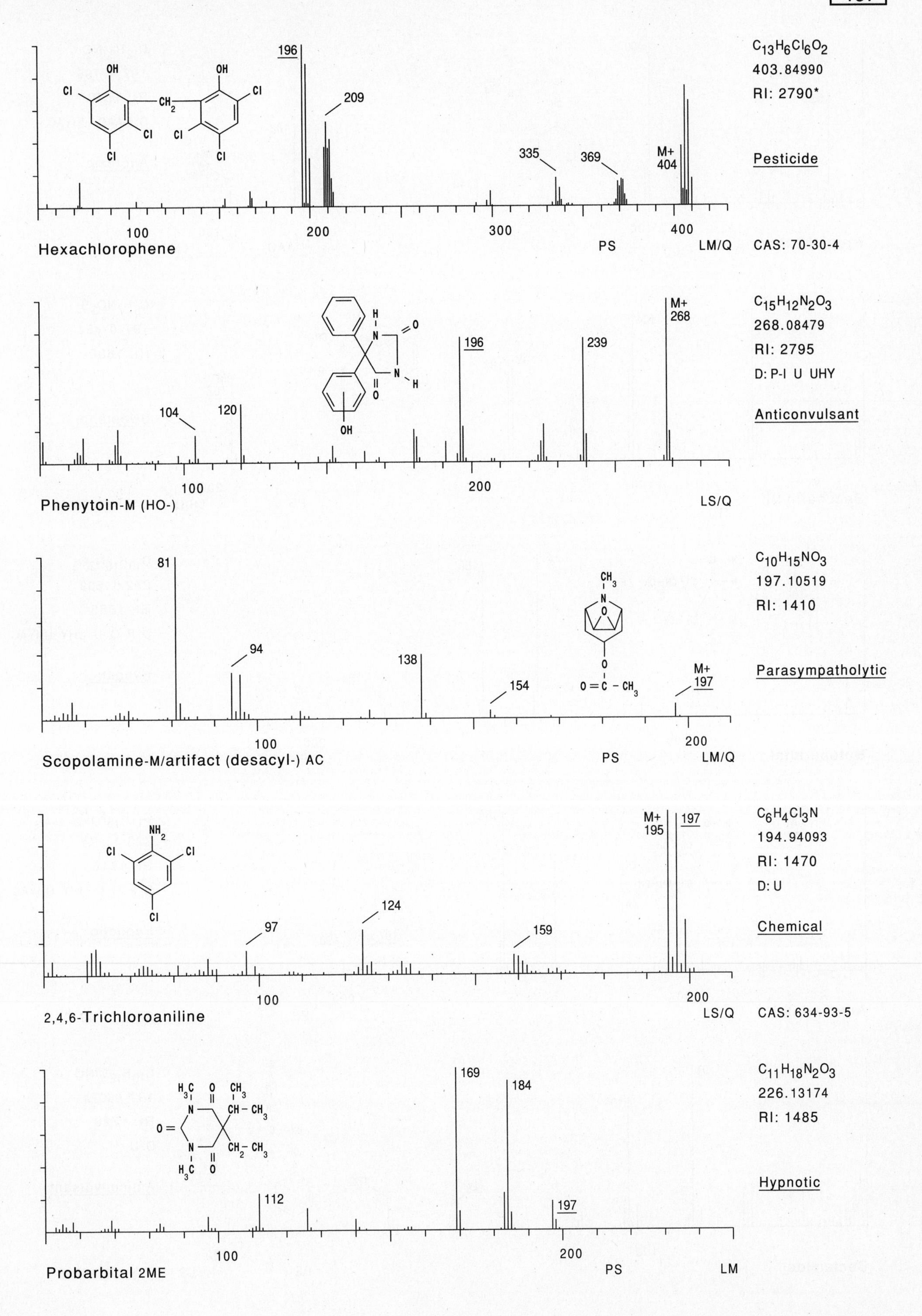

Hexachlorophene

$C_{13}H_6Cl_6O_2$
403.84990
RI: 2790*
Pesticide
CAS: 70-30-4

Phenytoin-M (HO-)

$C_{15}H_{12}N_2O_3$
268.08479
RI: 2795
D: P-I U UHY
Anticonvulsant

Scopolamine-M/artifact (desacyl-) AC

$C_{10}H_{15}NO_3$
197.10519
RI: 1410
Parasympatholytic

2,4,6-Trichloroaniline

$C_6H_4Cl_3N$
194.94093
RI: 1470
D: U
Chemical
CAS: 634-93-5

Probarbital 2ME

$C_{11}H_{18}N_2O_3$
226.13174
RI: 1485
Hypnotic

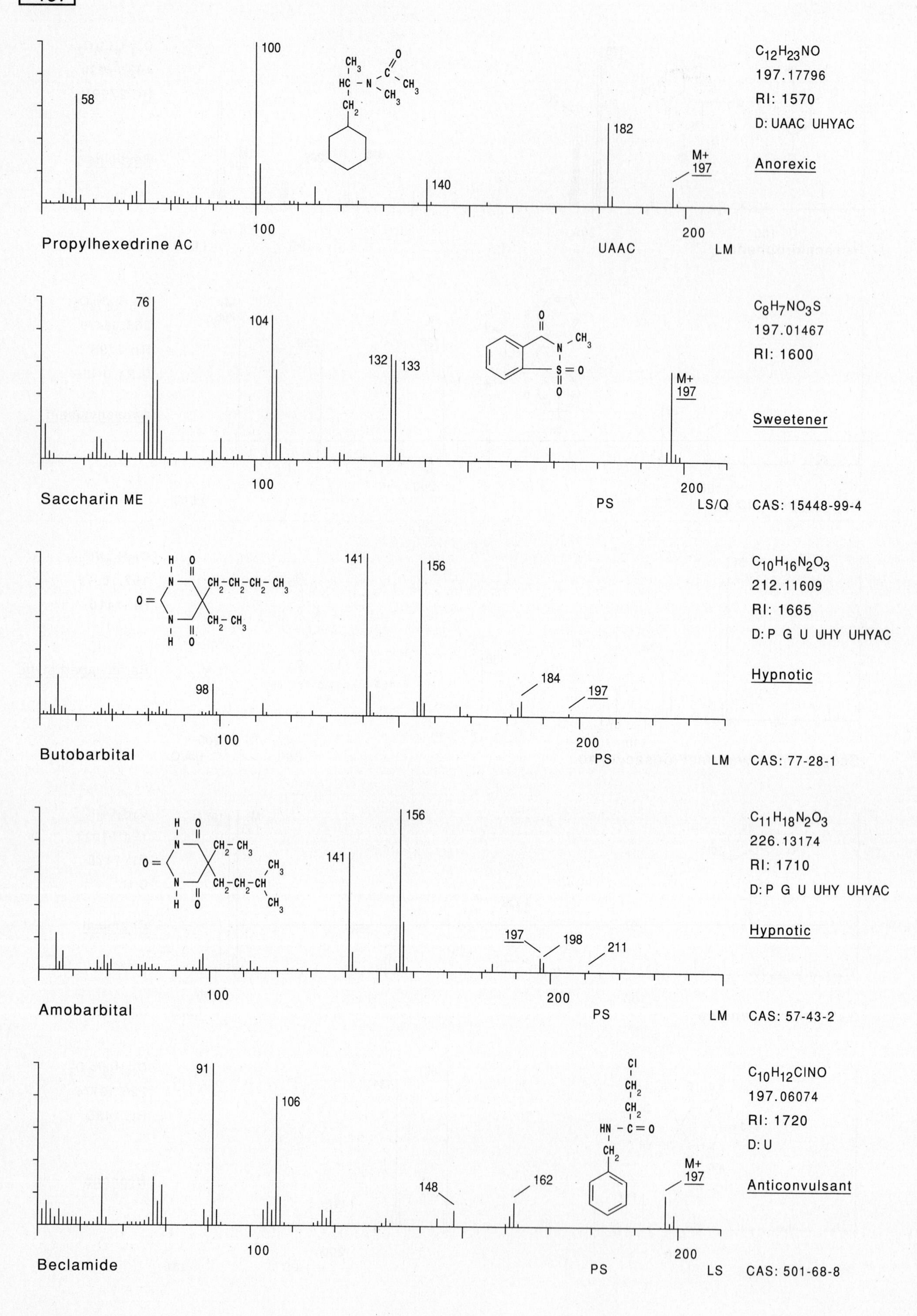

197
58
100
182
M+
197
140
100
200
Propylhexedrine AC
UAAC
LM
C12H23NO
197.17796
RI: 1570
D: UAAC UHYAC
Anorexic
76
104
132
133
M+
197
100
200
Saccharin ME
PS
LS/Q
C8H7NO3S
197.01467
RI: 1600
Sweetener
CAS: 15448-99-4
141
156
98
184
197
100
200
Butobarbital
PS
LM
C10H16N2O3
212.11609
RI: 1665
D: P G U UHY UHYAC
Hypnotic
CAS: 77-28-1
156
141
197
198
211
100
200
Amobarbital
PS
LM
C11H18N2O3
226.13174
RI: 1710
D: P G U UHY UHYAC
Hypnotic
CAS: 57-43-2
91
106
148
162
M+
197
100
200
Beclamide
PS
LS
C10H12ClNO
197.06074
RI: 1720
D: U
Anticonvulsant
CAS: 501-68-8

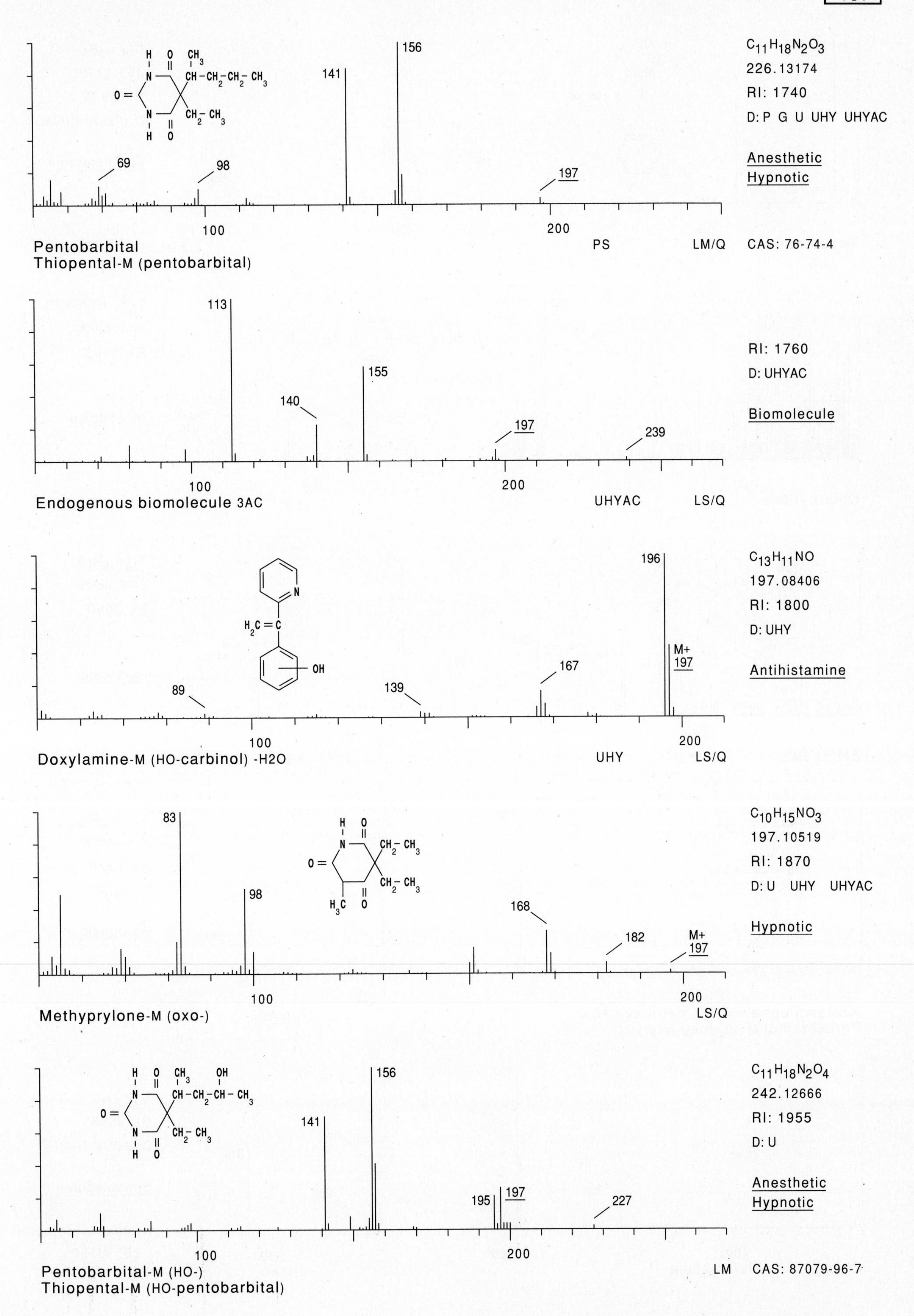
156
141
69
98
197
100
200
Pentobarbital
Thiopental-M (pentobarbital)
PS
LM/Q
$C_{11}H_{18}N_2O_3$
226.13174
RI: 1740
D: P G U UHY UHYAC
Anesthetic
Hypnotic
CAS: 76-74-4
113
155
140
197
239
100
200
Endogenous biomolecule 3AC
UHYAC
LS/Q
RI: 1760
D: UHYAC
Biomolecule
196
M+
197
167
139
89
OH
100
200
Doxylamine-M (HO-carbinol) -H2O
UHY
LS/Q
$C_{13}H_{11}NO$
197.08406
RI: 1800
D: UHY
Antihistamine
83
98
168
182
M+
197
100
200
Methyprylone-M (oxo-)
LS/Q
$C_{10}H_{15}NO_3$
197.10519
RI: 1870
D: U UHY UHYAC
Hypnotic
156
141
195
197
227
100
200
Pentobarbital-M (HO-)
Thiopental-M (HO-pentobarbital)
LM
$C_{11}H_{18}N_2O_4$
242.12666
RI: 1955
D: U
Anesthetic
Hypnotic
CAS: 87079-96-7

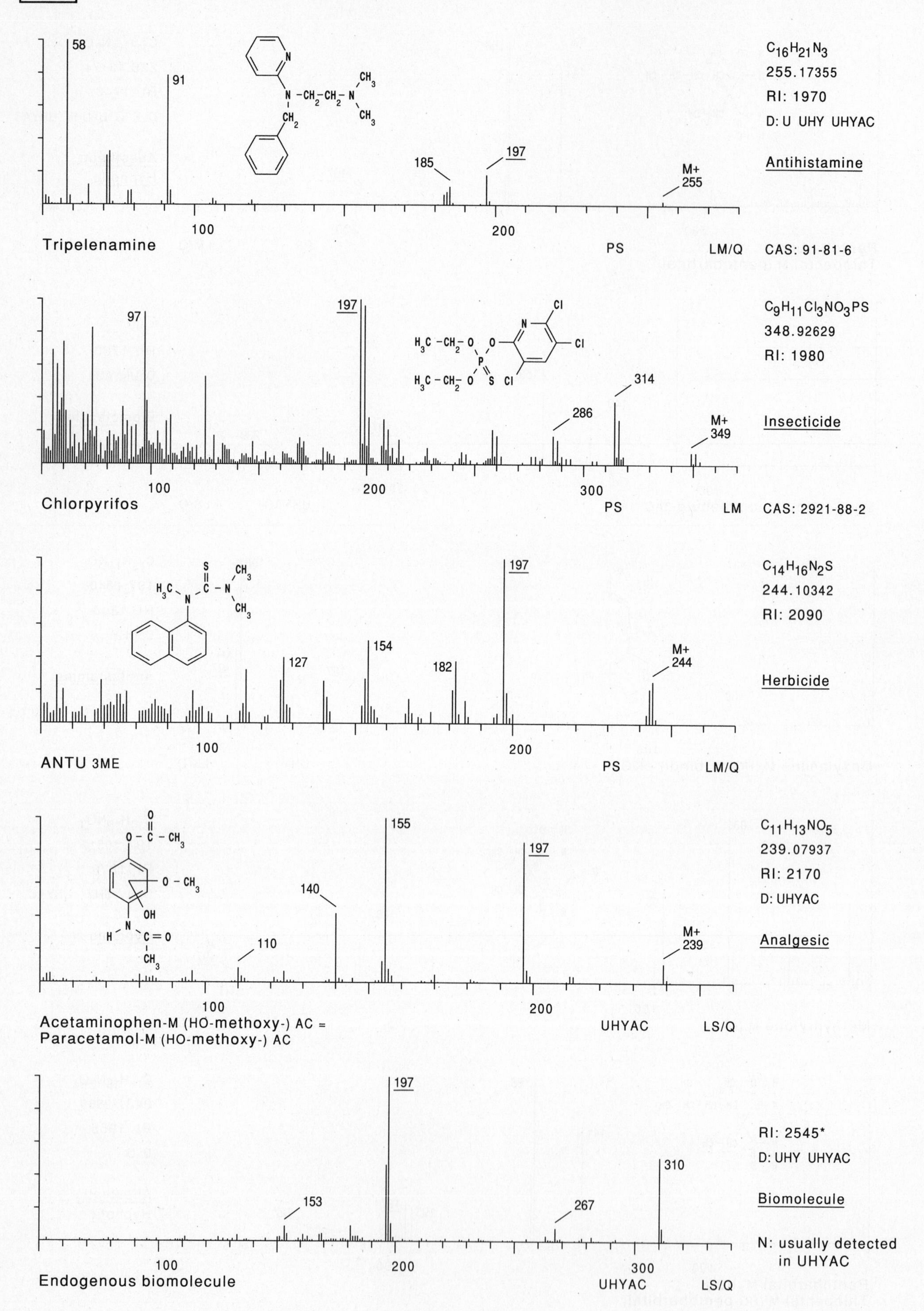

Tripelenamine
$C_{16}H_{21}N_3$
255.17355
RI: 1970
D: U UHY UHYAC
Antihistamine
PS
LM/Q
CAS: 91-81-6
Chlorpyrifos
$C_9H_{11}Cl_3NO_3PS$
348.92629
RI: 1980
Insecticide
PS
LM
CAS: 2921-88-2
ANTU 3ME
$C_{14}H_{16}N_2S$
244.10342
RI: 2090
Herbicide
PS
LM/Q
Acetaminophen-M (HO-methoxy-) AC =
Paracetamol-M (HO-methoxy-) AC
$C_{11}H_{13}NO_5$
239.07937
RI: 2170
D: UHYAC
Analgesic
UHYAC
LS/Q
Endogenous biomolecule
RI: 2545*
D: UHY UHYAC
Biomolecule
N: usually detected in UHYAC
UHYAC
LS/Q

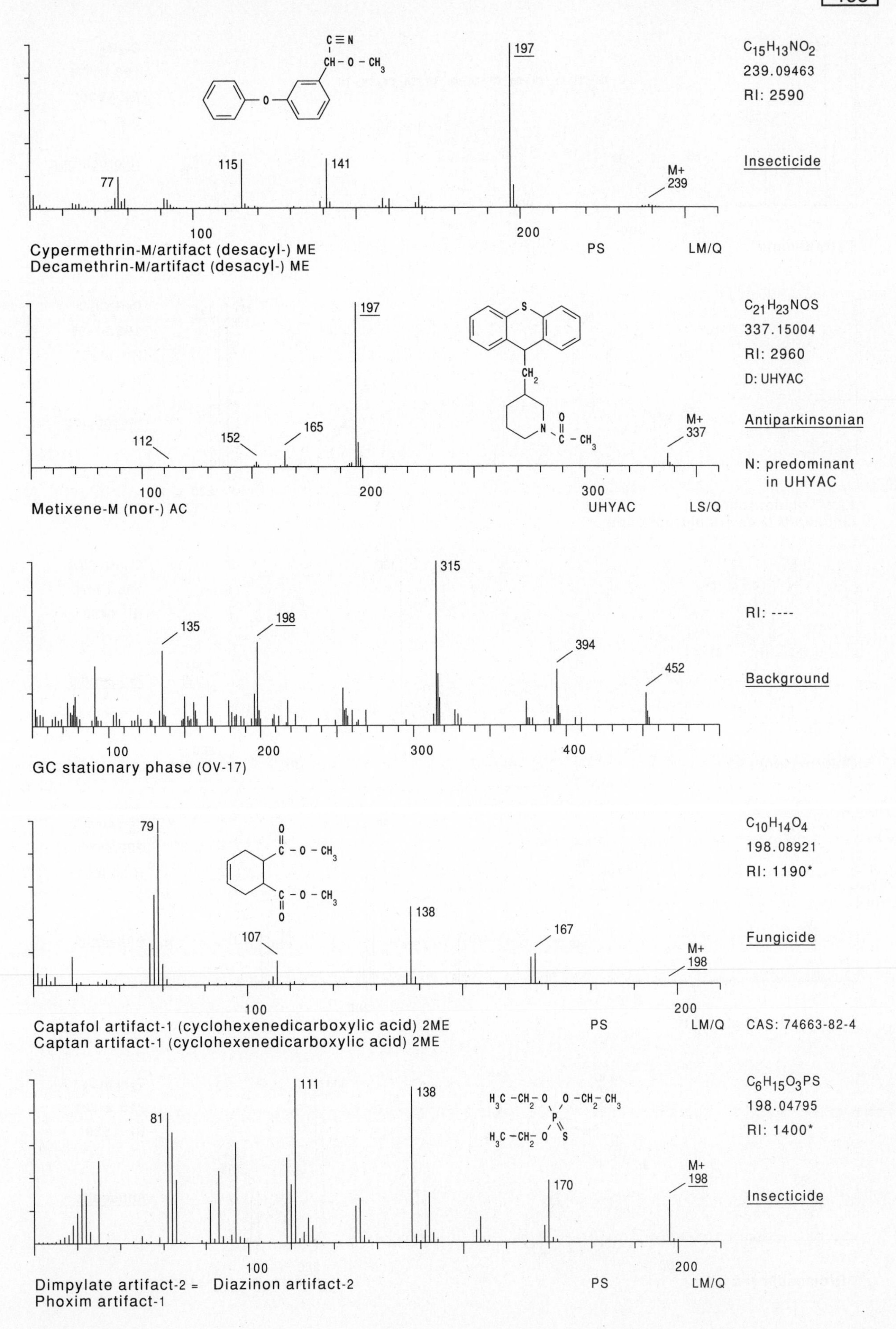

197
141
115
77
M+
239
100
200
PS
LM/Q
$C_{15}H_{13}NO_2$
239.09463
RI: 2590
Insecticide
Cypermethrin-M/artifact (desacyl-) ME
Decamethrin-M/artifact (desacyl-) ME
197
165
152
112
M+
337
100
200
300
UHYAC
LS/Q
$C_{21}H_{23}NOS$
337.15004
RI: 2960
D: UHYAC
Antiparkinsonian
N: predominant in UHYAC
Metixene-M (nor-) AC
315
198
135
394
452
100
200
300
400
RI: ----
Background
GC stationary phase (OV-17)
79
138
107
167
M+
198
100
200
PS
LM/Q
$C_{10}H_{14}O_4$
198.08921
RI: 1190*
Fungicide
Captafol artifact-1 (cyclohexenedicarboxylic acid) 2ME
Captan artifact-1 (cyclohexenedicarboxylic acid) 2ME
CAS: 74663-82-4
111
138
81
170
M+
198
100
200
PS
LM/Q
$C_6H_{15}O_3PS$
198.04795
RI: 1400*
Insecticide
Dimpylate artifact-2 = Diazinon artifact-2
Phoxim artifact-1

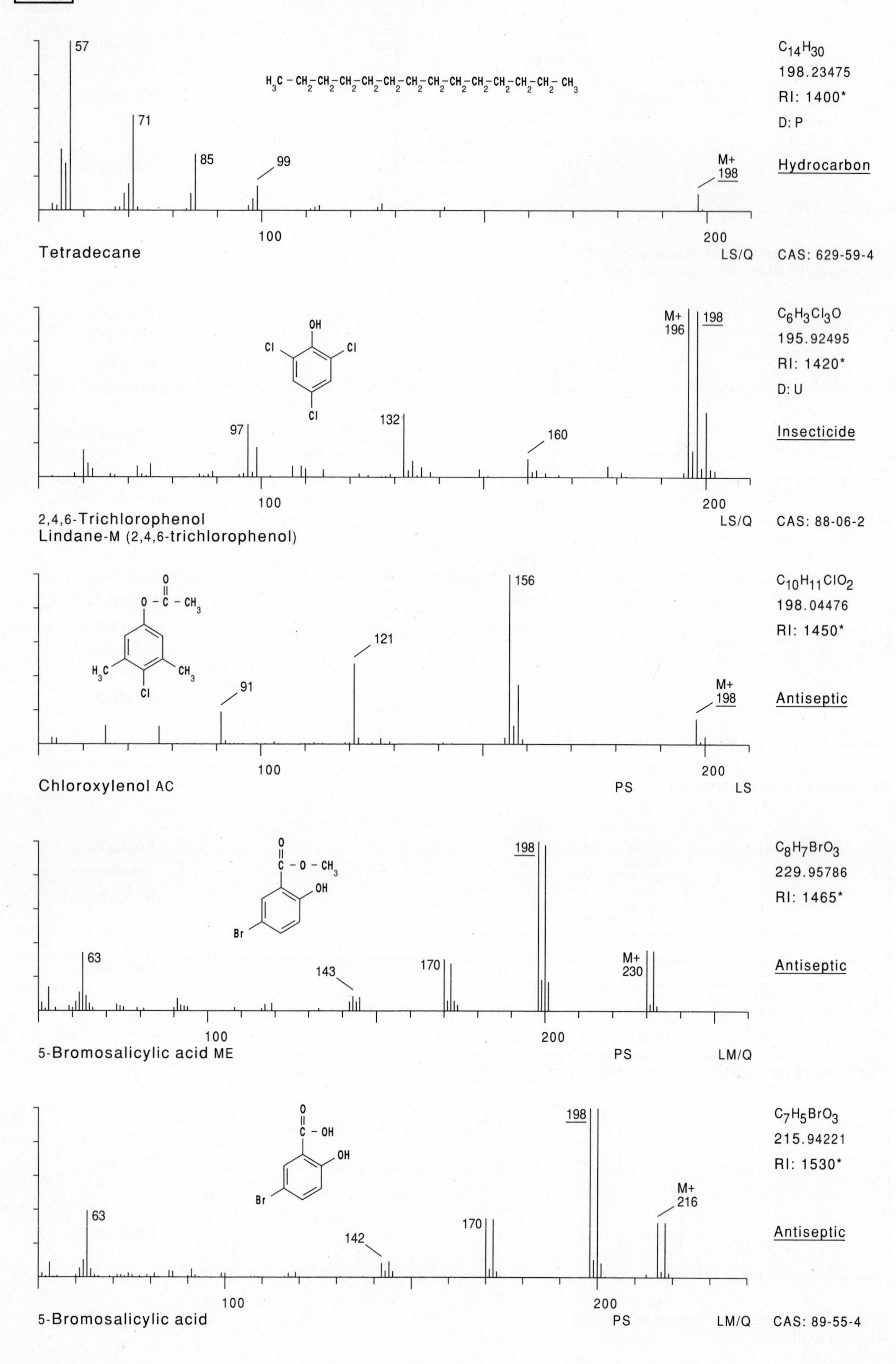
57
71
85
99
M+
198
100
200
Tetradecane
LS/Q
C14H30
198.23475
RI: 1400*
D: P
Hydrocarbon
CAS: 629-59-4
M+
196
198
97
132
160
OH
Cl
Cl
Cl
2,4,6-Trichlorophenol
Lindane-M (2,4,6-trichlorophenol)
C6H3Cl3O
195.92495
RI: 1420*
D: U
Insecticide
CAS: 88-06-2
156
121
91
M+
198
Chloroxylenol AC
PS
LS
C10H11ClO2
198.04476
RI: 1450*
Antiseptic
198
63
143
170
M+
230
OH
Br
5-Bromosalicylic acid ME
PS
LM/Q
C8H7BrO3
229.95786
RI: 1465*
Antiseptic
198
63
142
170
M+
216
C - OH
OH
Br
5-Bromosalicylic acid
PS
LM/Q
C7H5BrO3
215.94221
RI: 1530*
Antiseptic
CAS: 89-55-4

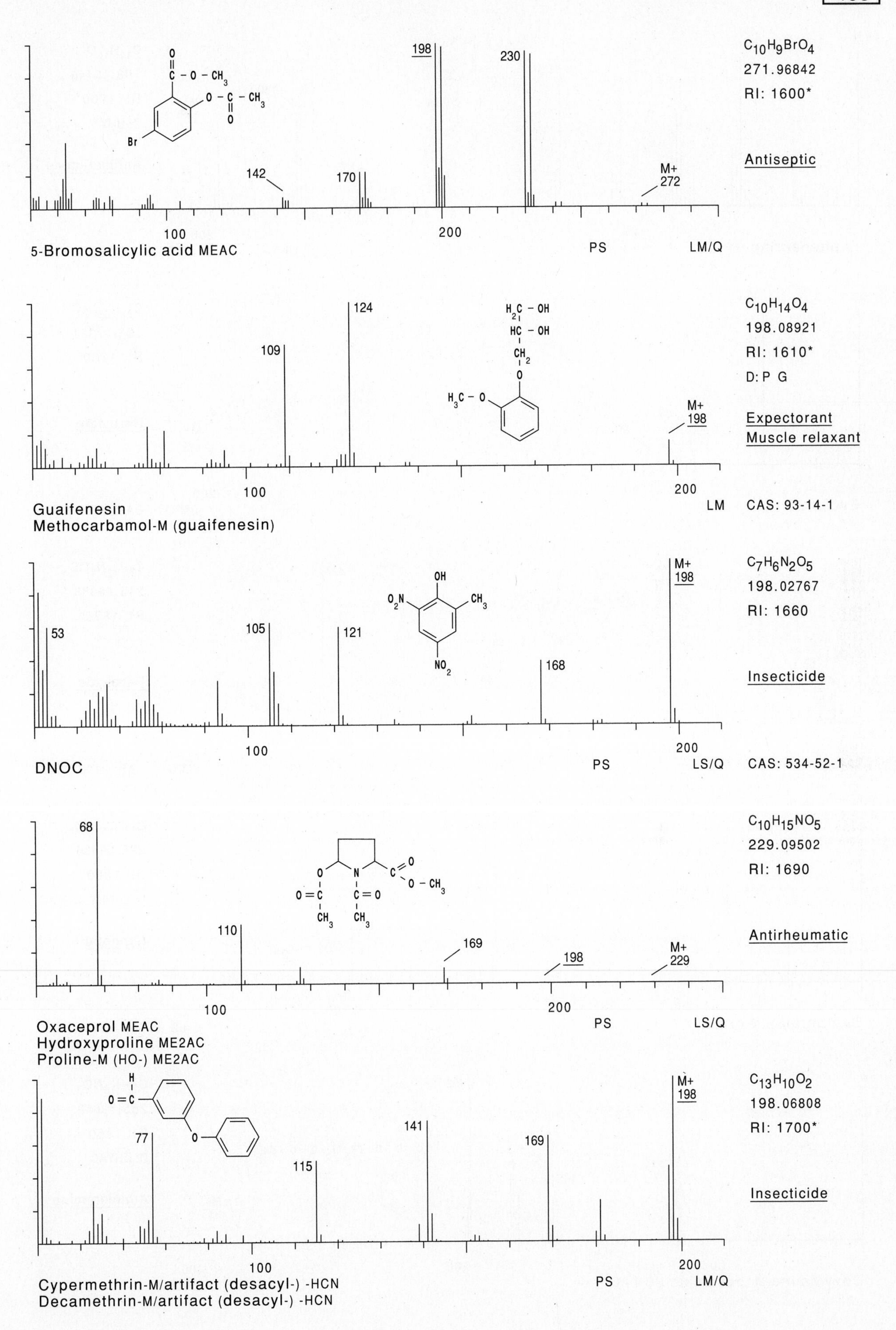

198
230
142
170
M+
272
100
200
5-Bromosalicylic acid MEAC
PS
LM/Q
$C_{10}H_9BrO_4$
271.96842
RI: 1600*
Antiseptic
124
109
M+
198
100
200
Guaifenesin
Methocarbamol-M (guaifenesin)
LM
$C_{10}H_{14}O_4$
198.08921
RI: 1610*
D:P G
Expectorant
Muscle relaxant
CAS: 93-14-1
M+
198
53
105
121
168
100
200
DNOC
PS
LS/Q
$C_7H_6N_2O_5$
198.02767
RI: 1660
Insecticide
CAS: 534-52-1
68
110
169
198
M+
229
100
200
Oxaceprol MEAC
Hydroxyproline ME2AC
Proline-M (HO-) ME2AC
PS
LS/Q
$C_{10}H_{15}NO_5$
229.09502
RI: 1690
Antirheumatic
M+
198
77
115
141
169
100
200
Cypermethrin-M/artifact (desacyl-) -HCN
Decamethrin-M/artifact (desacyl-) -HCN
PS
LM/Q
$C_{13}H_{10}O_2$
198.06808
RI: 1700*
Insecticide

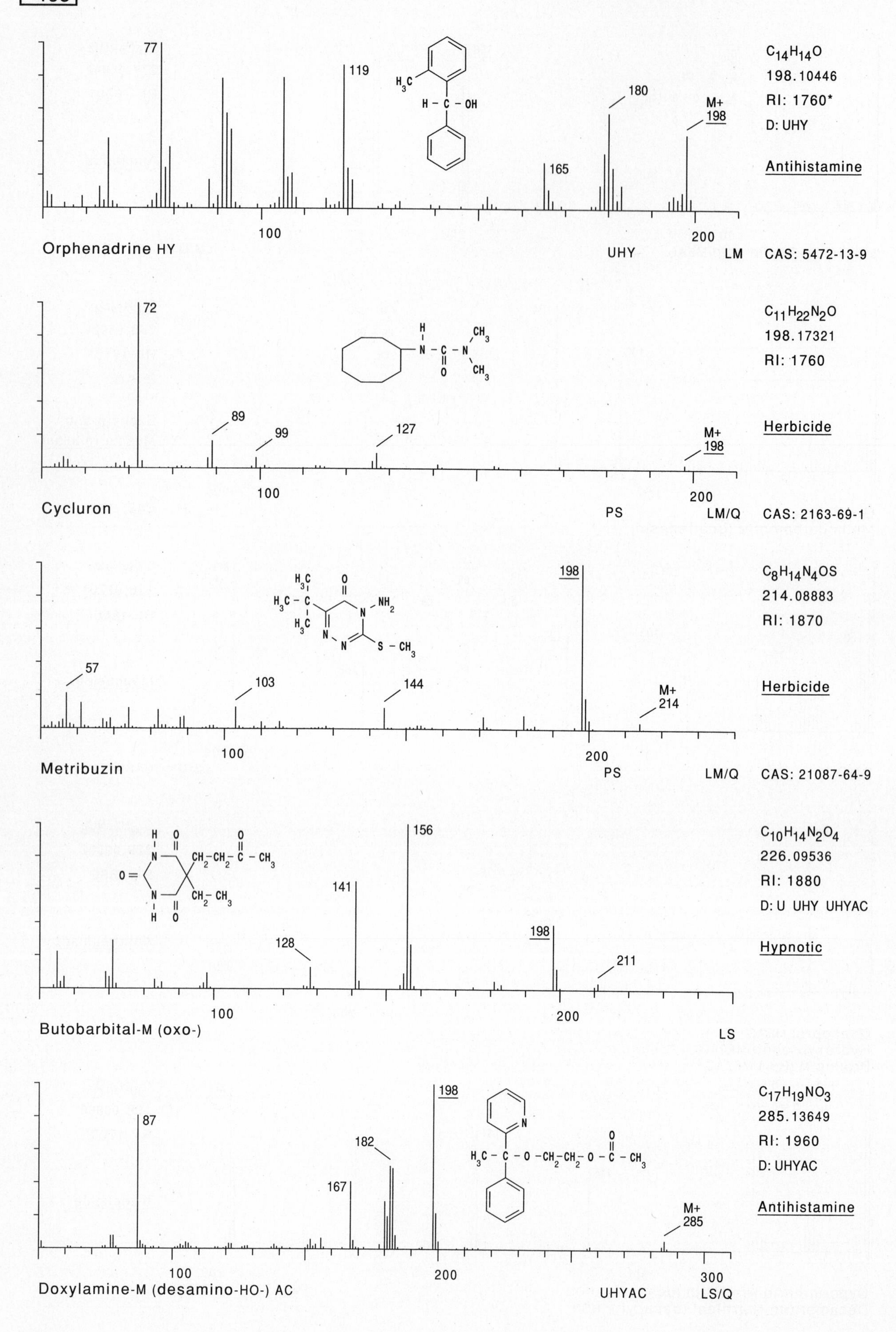

77
119
180
M+
198
165
100
200
Orphenadrine HY
UHY
LM
$C_{14}H_{14}O$
198.10446
RI: 1760*
D: UHY
Antihistamine
CAS: 5472-13-9
72
89
99
127
M+
198
100
200
Cycluron
PS
LM/Q
$C_{11}H_{22}N_2O$
198.17321
RI: 1760
Herbicide
CAS: 2163-69-1
198
57
103
144
M+
214
100
200
Metribuzin
PS
LM/Q
$C_8H_{14}N_4OS$
214.08883
RI: 1870
Herbicide
CAS: 21087-64-9
156
141
128
198
211
100
200
Butobarbital-M (oxo-)
LS
$C_{10}H_{14}N_2O_4$
226.09536
RI: 1880
D: U UHY UHYAC
Hypnotic
198
87
182
167
M+
285
100
200
300
Doxylamine-M (desamino-HO-) AC
UHYAC
LS/Q
$C_{17}H_{19}NO_3$
285.13649
RI: 1960
D: UHYAC
Antihistamine

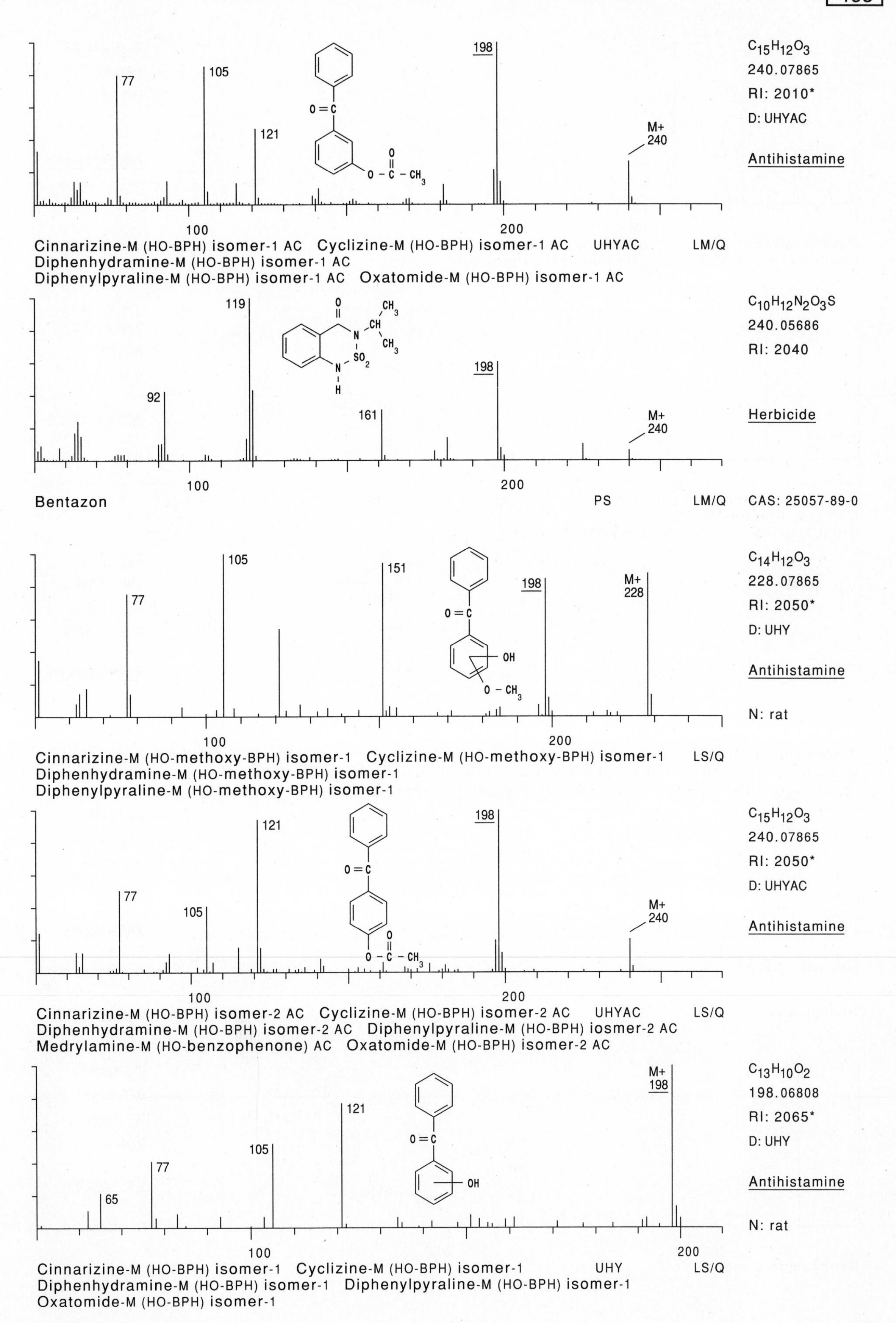

Cinnarizine-M (HO-BPH) isomer-1 AC Cyclizine-M (HO-BPH) isomer-1 AC UHYAC LM/Q
Diphenhydramine-M (HO-BPH) isomer-1 AC
Diphenylpyraline-M (HO-BPH) isomer-1 AC Oxatomide-M (HO-BPH) isomer-1 AC

$C_{15}H_{12}O_3$
240.07865
RI: 2010*
D: UHYAC

Antihistamine

Bentazon PS LM/Q

$C_{10}H_{12}N_2O_3S$
240.05686
RI: 2040

Herbicide

CAS: 25057-89-0

Cinnarizine-M (HO-methoxy-BPH) isomer-1 Cyclizine-M (HO-methoxy-BPH) isomer-1 LS/Q
Diphenhydramine-M (HO-methoxy-BPH) isomer-1
Diphenylpyraline-M (HO-methoxy-BPH) isomer-1

$C_{14}H_{12}O_3$
228.07865
RI: 2050*
D: UHY

Antihistamine

N: rat

Cinnarizine-M (HO-BPH) isomer-2 AC Cyclizine-M (HO-BPH) isomer-2 AC UHYAC LS/Q
Diphenhydramine-M (HO-BPH) isomer-2 AC Diphenylpyraline-M (HO-BPH) iosmer-2 AC
Medrylamine-M (HO-benzophenone) AC Oxatomide-M (HO-BPH) isomer-2 AC

$C_{15}H_{12}O_3$
240.07865
RI: 2050*
D: UHYAC

Antihistamine

Cinnarizine-M (HO-BPH) isomer-1 Cyclizine-M (HO-BPH) isomer-1 UHY LS/Q
Diphenhydramine-M (HO-BPH) isomer-1 Diphenylpyraline-M (HO-BPH) isomer-1
Oxatomide-M (HO-BPH) isomer-1

$C_{13}H_{10}O_2$
198.06808
RI: 2065*
D: UHY

Antihistamine

N: rat

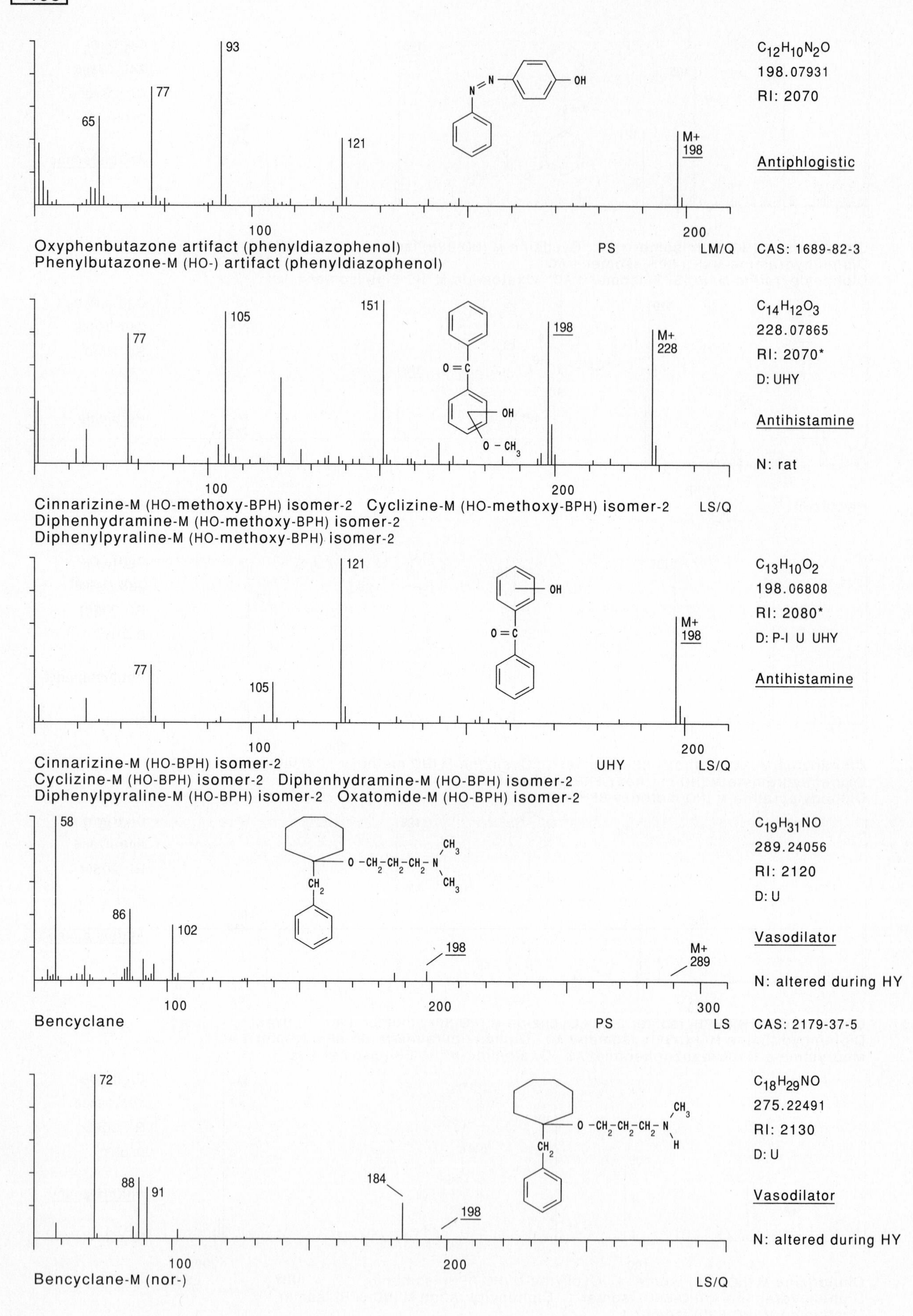
198
93
77
65
121
M+
198
OH
$C_{12}H_{10}N_2O$
198.07931
RI: 2070
Antiphlogistic
100
200
Oxyphenbutazone artifact (phenyldiazophenol)
Phenylbutazone-M (HO-) artifact (phenyldiazophenol)
PS
LM/Q
CAS: 1689-82-3
151
105
77
198
M+
228
O=C
OH
O - CH3
$C_{14}H_{12}O_3$
228.07865
RI: 2070*
D: UHY
Antihistamine
N: rat
100
200
Cinnarizine-M (HO-methoxy-BPH) isomer-2 Cyclizine-M (HO-methoxy-BPH) isomer-2
LS/Q
Diphenhydramine-M (HO-methoxy-BPH) isomer-2
Diphenylpyraline-M (HO-methoxy-BPH) isomer-2
121
OH
M+
198
O=C
77
105
$C_{13}H_{10}O_2$
198.06808
RI: 2080*
D: P-I U UHY
Antihistamine
100
200
Cinnarizine-M (HO-BPH) isomer-2
UHY
LS/Q
Cyclizine-M (HO-BPH) isomer-2 Diphenhydramine-M (HO-BPH) isomer-2
Diphenylpyraline-M (HO-BPH) isomer-2 Oxatomide-M (HO-BPH) isomer-2
58
86
102
198
M+
289
$C_{19}H_{31}NO$
289.24056
RI: 2120
D: U
Vasodilator
N: altered during HY
100
200
300
Bencyclane
PS
LS
CAS: 2179-37-5
72
88
91
184
198
$C_{18}H_{29}NO$
275.22491
RI: 2130
D: U
Vasodilator
N: altered during HY
100
200
Bencyclane-M (nor-)
LS/Q

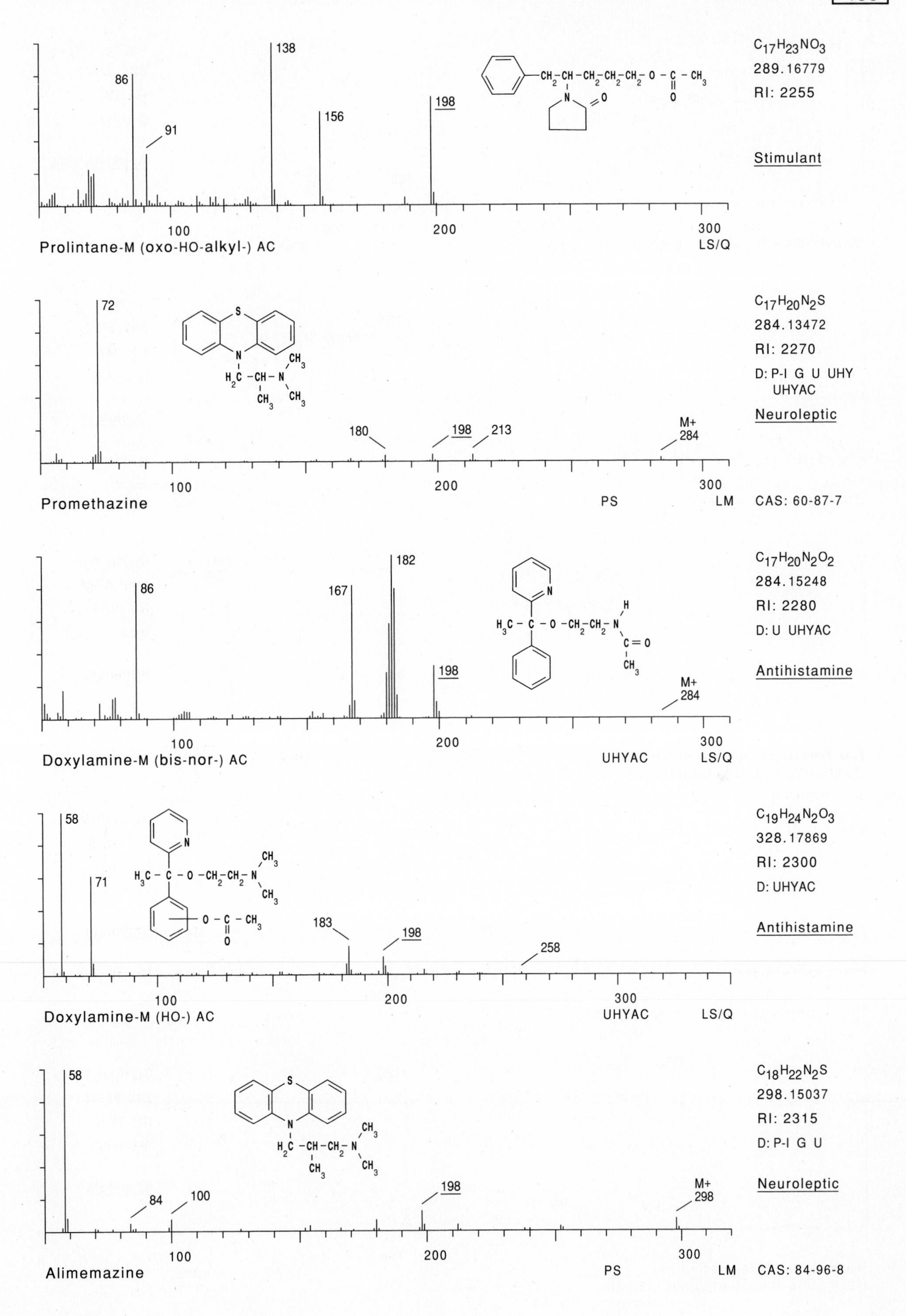
Prolintane-M (oxo-HO-alkyl-) AC
LS/Q
C17H23NO3
289.16779
RI: 2255
Stimulant
Promethazine
PS
LM
C17H20N2S
284.13472
RI: 2270
D: P-I G U UHY UHYAC
Neuroleptic
CAS: 60-87-7
Doxylamine-M (bis-nor-) AC
UHYAC
LS/Q
C17H20N2O2
284.15248
RI: 2280
D: U UHYAC
Antihistamine
Doxylamine-M (HO-) AC
UHYAC
LS/Q
C19H24N2O3
328.17869
RI: 2300
D: UHYAC
Antihistamine
Alimemazine
PS
LM
C18H22N2S
298.15037
RI: 2315
D: P-I G U
Neuroleptic
CAS: 84-96-8

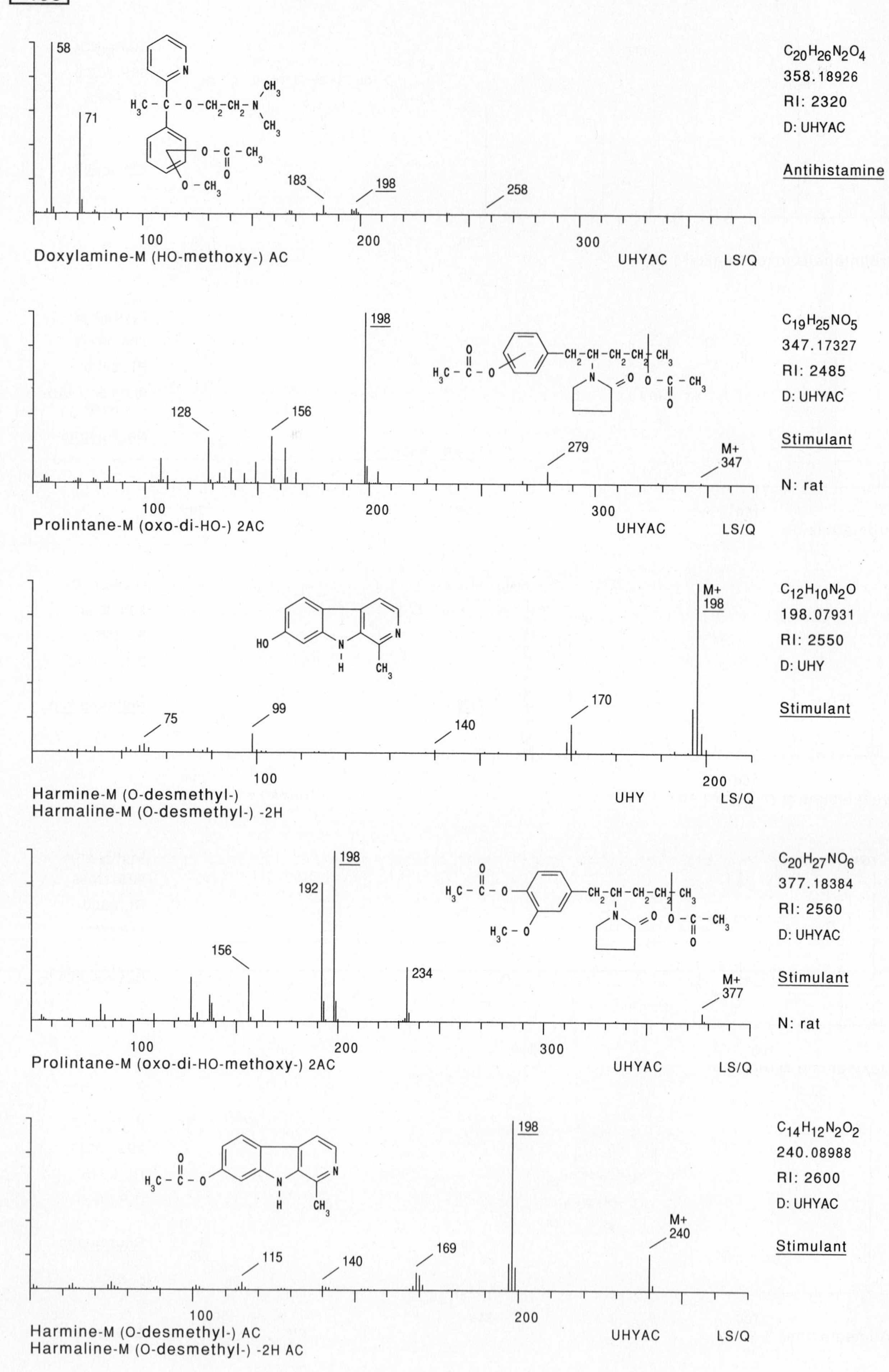
58
71
183
198
258
100
200
300
Doxylamine-M (HO-methoxy-) AC
UHYAC
LS/Q
$C_{20}H_{26}N_2O_4$
358.18926
RI: 2320
D: UHYAC
Antihistamine
198
128
156
279
M+
347
100
200
300
Prolintane-M (oxo-di-HO-) 2AC
UHYAC
LS/Q
$C_{19}H_{25}NO_5$
347.17327
RI: 2485
D: UHYAC
Stimulant
N: rat
M+
198
75
99
140
170
100
200
Harmine-M (O-desmethyl-)
Harmaline-M (O-desmethyl-) -2H
UHY
LS/Q
$C_{12}H_{10}N_2O$
198.07931
RI: 2550
D: UHY
Stimulant
198
192
156
234
M+
377
100
200
300
Prolintane-M (oxo-di-HO-methoxy-) 2AC
UHYAC
LS/Q
$C_{20}H_{27}NO_6$
377.18384
RI: 2560
D: UHYAC
Stimulant
N: rat
198
M+
240
115
140
169
100
200
Harmine-M (O-desmethyl-) AC
Harmaline-M (O-desmethyl-) -2H AC
UHYAC
LS/Q
$C_{14}H_{12}N_2O_2$
240.08988
RI: 2600
D: UHYAC
Stimulant

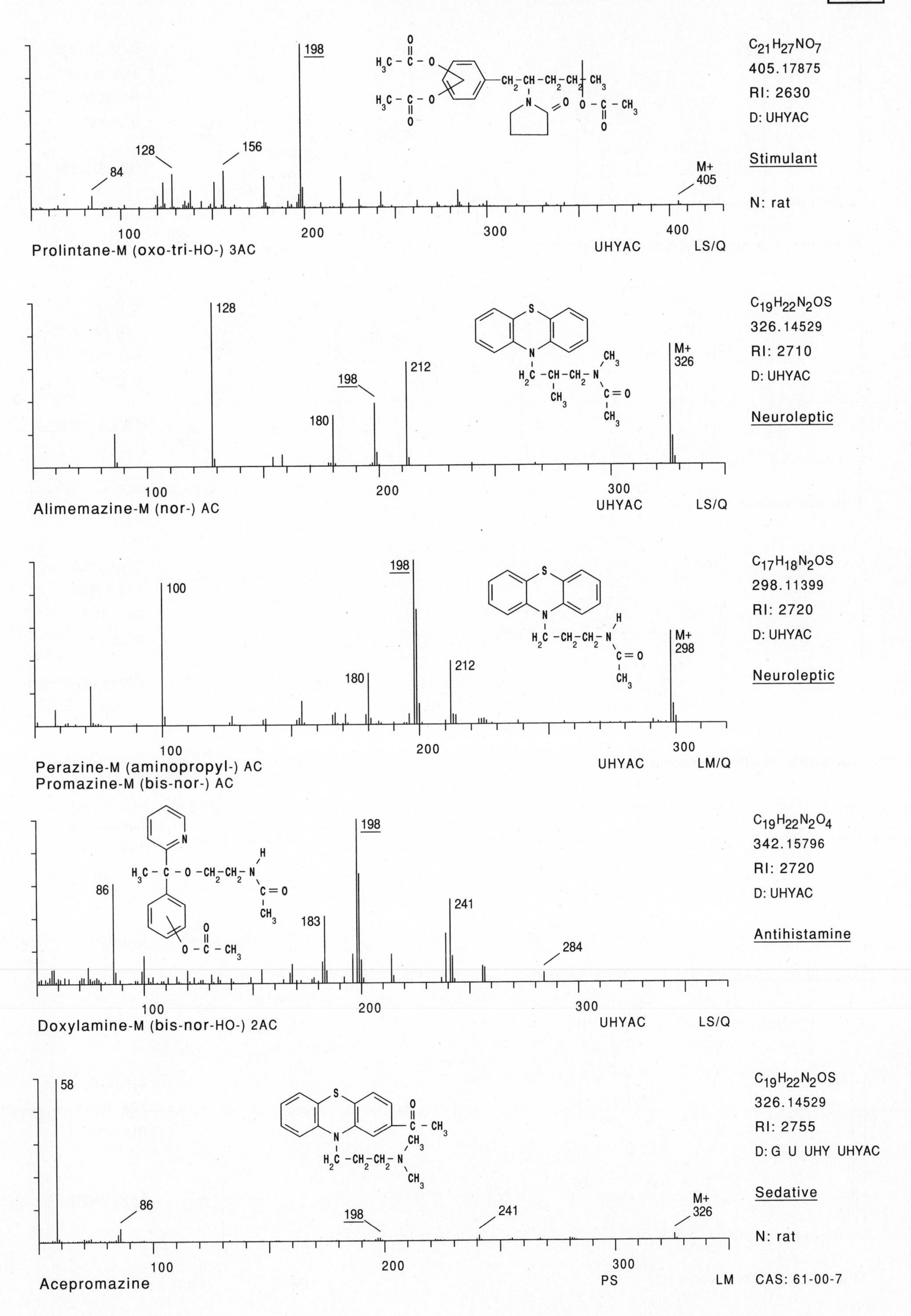
198
128
156
84
M+
405
100
200
300
400
Prolintane-M (oxo-tri-HO-) 3AC
UHYAC
LS/Q
$C_{21}H_{27}NO_7$
405.17875
RI: 2630
D: UHYAC
Stimulant
N: rat
128
212
198
180
M+
326
100
200
300
Alimemazine-M (nor-) AC
UHYAC
LS/Q
$C_{19}H_{22}N_2OS$
326.14529
RI: 2710
D: UHYAC
Neuroleptic
198
100
M+
298
212
180
100
200
300
Perazine-M (aminopropyl-) AC
Promazine-M (bis-nor-) AC
UHYAC
LM/Q
$C_{17}H_{18}N_2OS$
298.11399
RI: 2720
D: UHYAC
Neuroleptic
198
86
241
183
284
100
200
300
Doxylamine-M (bis-nor-HO-) 2AC
UHYAC
LS/Q
$C_{19}H_{22}N_2O_4$
342.15796
RI: 2720
D: UHYAC
Antihistamine
58
86
198
241
M+
326
100
200
300
Acepromazine
PS
LM
$C_{19}H_{22}N_2OS$
326.14529
RI: 2755
D: G U UHY UHYAC
Sedative
N: rat
CAS: 61-00-7

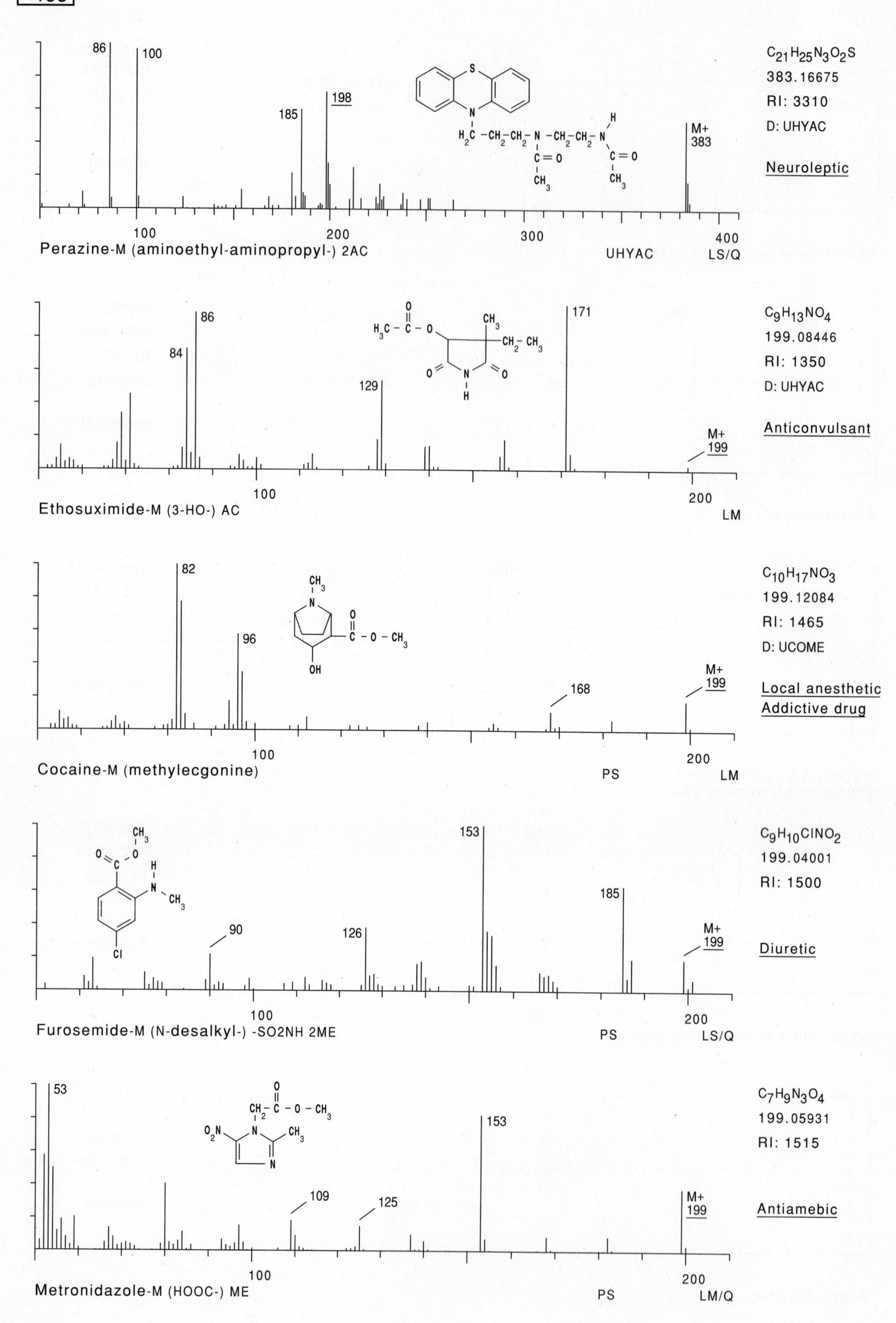
86
100
185
198
M+
383
100
200
300
400
Perazine-M (aminoethyl-aminopropyl-) 2AC
UHYAC
LS/Q
C21H25N3O2S
383.16675
RI: 3310
D: UHYAC
Neuroleptic
86
84
129
171
M+
199
100
200
LM
Ethosuximide-M (3-HO-) AC
C9H13NO4
199.08446
RI: 1350
D: UHYAC
Anticonvulsant
82
96
168
M+
199
100
200
PS
LM
Cocaine-M (methylecgonine)
C10H17NO3
199.12084
RI: 1465
D: UCOME
Local anesthetic
Addictive drug
153
185
90
126
M+
199
100
200
PS
LS/Q
Furosemide-M (N-desalkyl-) -SO2NH 2ME
C9H10ClNO2
199.04001
RI: 1500
Diuretic
53
153
109
125
M+
199
100
200
PS
LM/Q
Metronidazole-M (HOOC-) ME
C7H9N3O4
199.05931
RI: 1515
Antiamebic

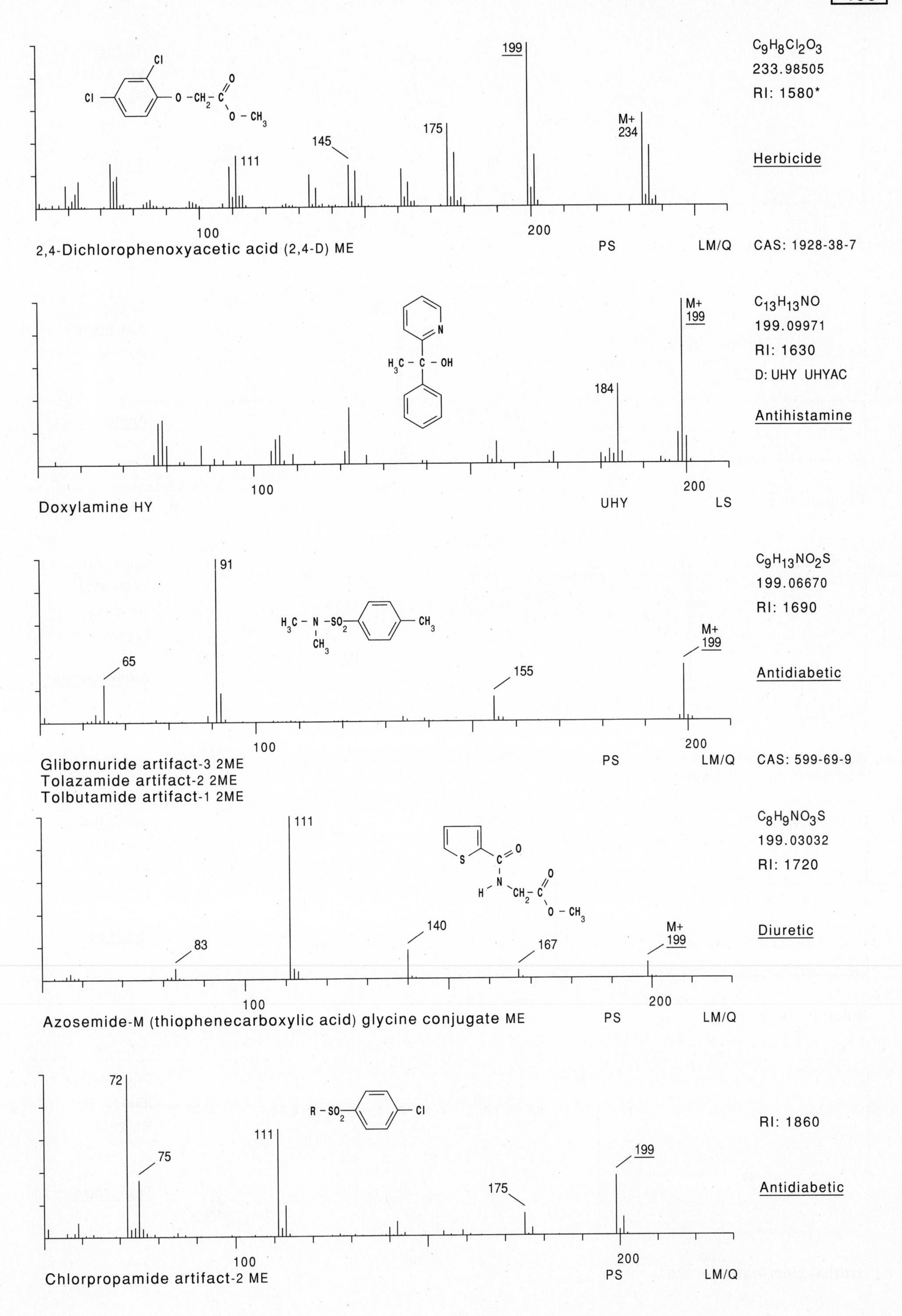
$C_9H_8Cl_2O_3$
233.98505
RI: 1580*
Herbicide
199
175
145
111
M+
234
100
200
2,4-Dichlorophenoxyacetic acid (2,4-D) ME
PS
LM/Q
CAS: 1928-38-7
$C_{13}H_{13}NO$
199.09971
RI: 1630
D: UHY UHYAC
Antihistamine
M+
199
184
100
200
Doxylamine HY
UHY
LS
$C_9H_{13}NO_2S$
199.06670
RI: 1690
Antidiabetic
91
65
155
M+
199
100
200
Glibornuride artifact-3 2ME
Tolazamide artifact-2 2ME
Tolbutamide artifact-1 2ME
PS
LM/Q
CAS: 599-69-9
$C_8H_9NO_3S$
199.03032
RI: 1720
Diuretic
111
83
140
167
M+
199
100
200
Azosemide-M (thiophenecarboxylic acid) glycine conjugate ME
PS
LM/Q
RI: 1860
Antidiabetic
72
75
111
175
199
100
200
Chlorpropamide artifact-2 ME
PS
LM/Q

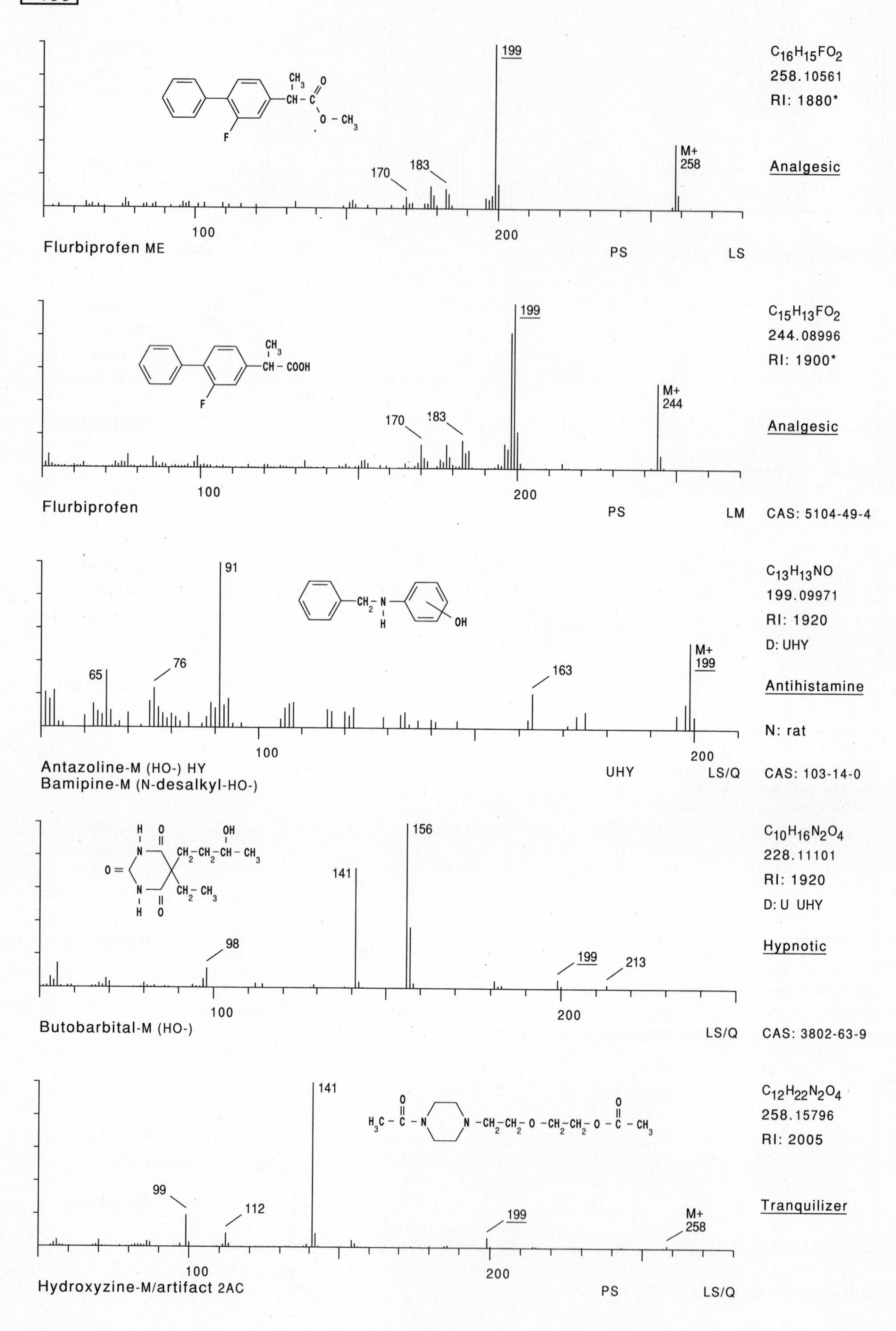

199
170
183
M+
258
100
200
Flurbiprofen ME
PS
LS
$C_{16}H_{15}FO_2$
258.10561
RI: 1880*
Analgesic
199
170
183
M+
244
100
200
Flurbiprofen
PS
LM
$C_{15}H_{13}FO_2$
244.08996
RI: 1900*
Analgesic
CAS: 5104-49-4
91
65
76
163
M+
199
100
200
Antazoline-M (HO-) HY
Bamipine-M (N-desalkyl-HO-)
UHY
LS/Q
$C_{13}H_{13}NO$
199.09971
RI: 1920
D: UHY
Antihistamine
N: rat
CAS: 103-14-0
156
141
98
199
213
100
200
Butobarbital-M (HO-)
LS/Q
$C_{10}H_{16}N_2O_4$
228.11101
RI: 1920
D: U UHY
Hypnotic
CAS: 3802-63-9
141
99
112
199
M+
258
100
200
Hydroxyzine-M/artifact 2AC
PS
LS/Q
$C_{12}H_{22}N_2O_4$
258.15796
RI: 2005
Tranquilizer

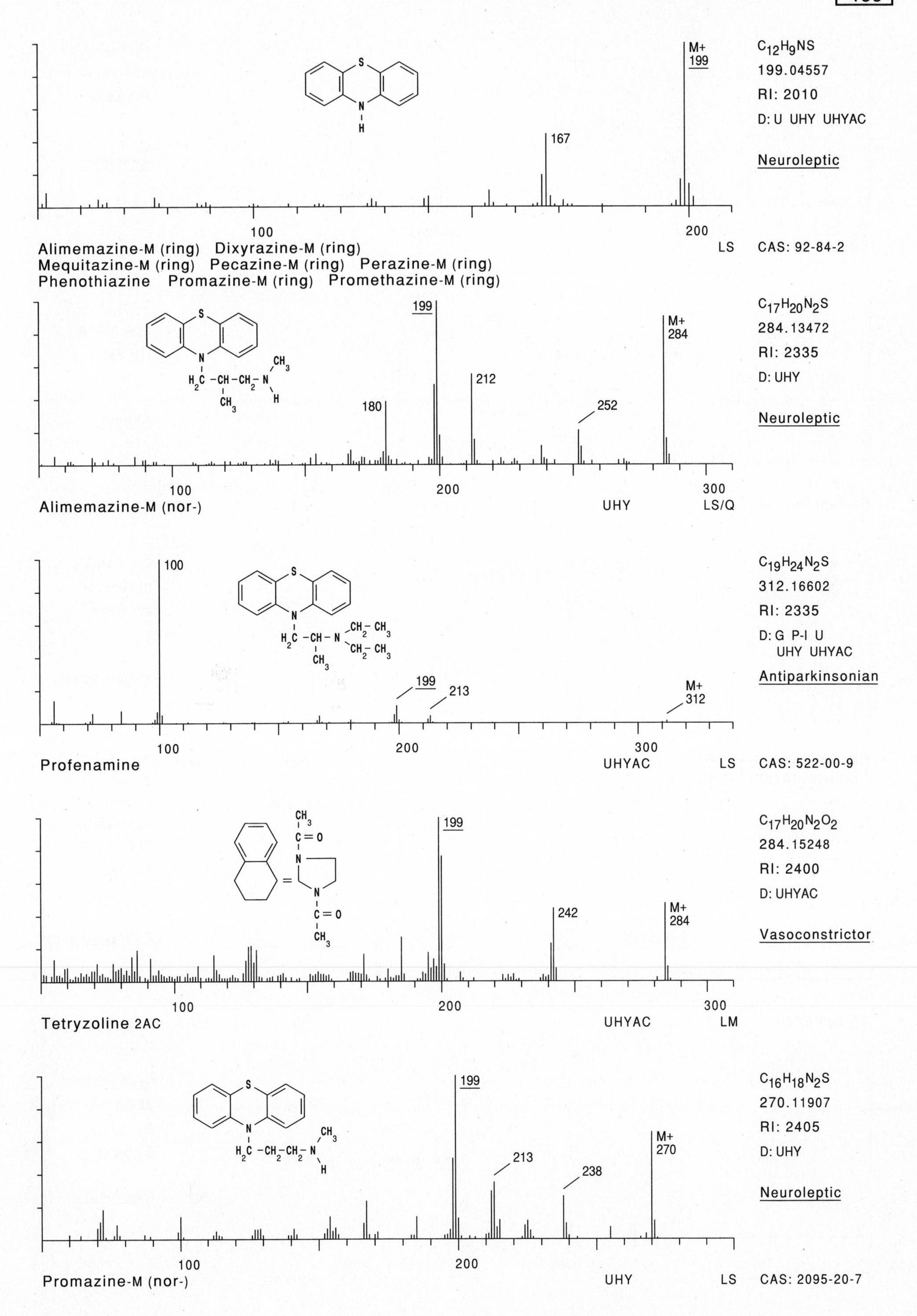
M+
199
167
100
200
LS
C12H9NS
199.04557
RI: 2010
D: U UHY UHYAC
Neuroleptic
CAS: 92-84-2
Alimemazine-M (ring) Dixyrazine-M (ring)
Mequitazine-M (ring) Pecazine-M (ring) Perazine-M (ring)
Phenothiazine Promazine-M (ring) Promethazine-M (ring)
199
M+
284
212
180
252
100
200
300
UHY
LS/Q
C17H20N2S
284.13472
RI: 2335
D: UHY
Neuroleptic
Alimemazine-M (nor-)
100
199
213
M+
312
100
200
300
UHYAC
LS
C19H24N2S
312.16602
RI: 2335
D: G P-I U
UHY UHYAC
Antiparkinsonian
CAS: 522-00-9
Profenamine
199
242
M+
284
100
200
300
UHYAC
LM
C17H20N2O2
284.15248
RI: 2400
D: UHYAC
Vasoconstrictor
Tetryzoline 2AC
199
M+
270
213
238
100
200
UHY
LS
C16H18N2S
270.11907
RI: 2405
D: UHY
Neuroleptic
CAS: 2095-20-7
Promazine-M (nor-)

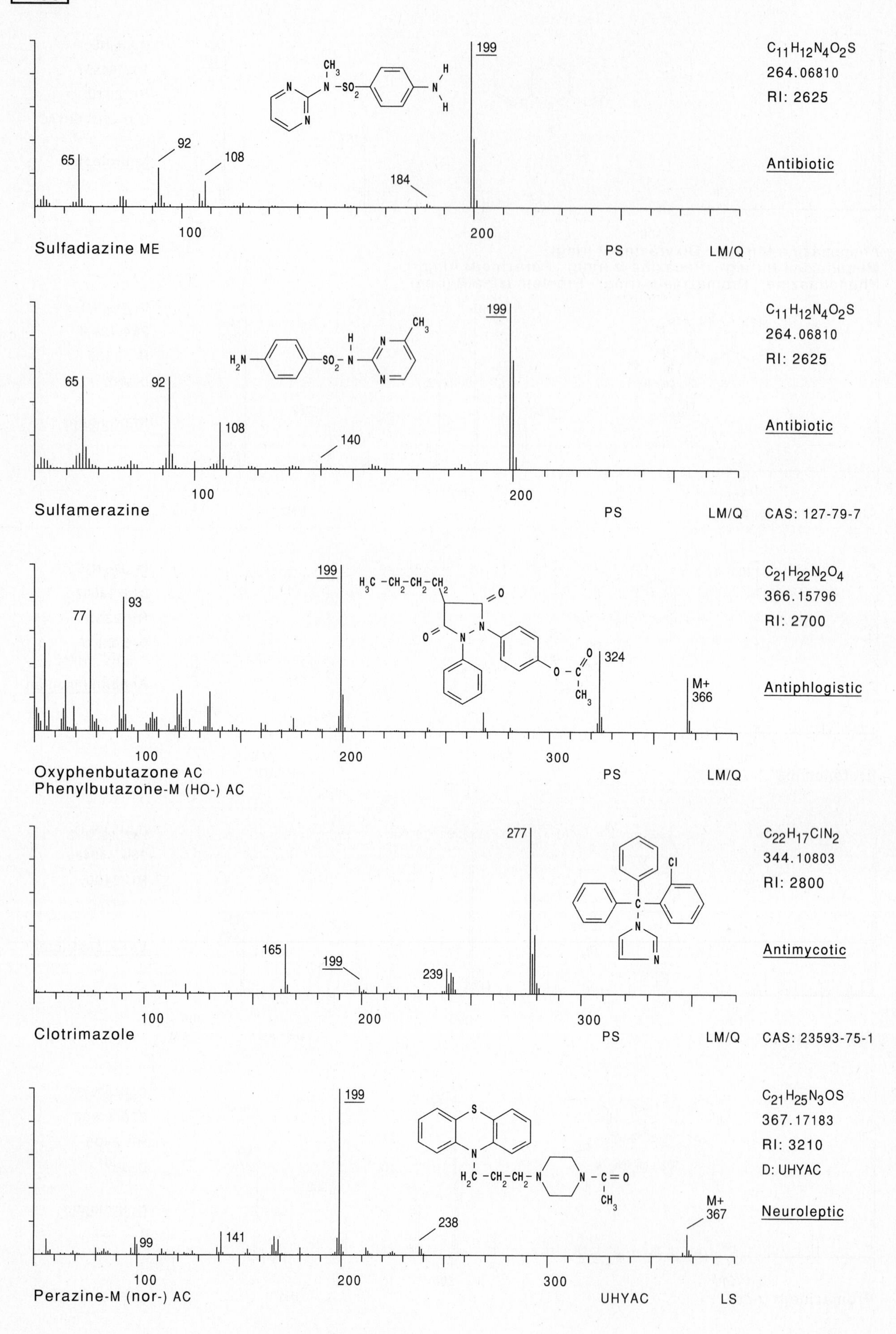

199
92
108
65
184
100
200
Sulfadiazine ME
PS
LM/Q
$C_{11}H_{12}N_4O_2S$
264.06810
RI: 2625
Antibiotic
199
65
92
108
140
100
200
Sulfamerazine
PS
LM/Q
$C_{11}H_{12}N_4O_2S$
264.06810
RI: 2625
Antibiotic
CAS: 127-79-7
199
93
77
324
M+
366
100
200
300
Oxyphenbutazone AC
Phenylbutazone-M (HO-) AC
PS
LM/Q
$C_{21}H_{22}N_2O_4$
366.15796
RI: 2700
Antiphlogistic
277
165
199
239
100
200
300
Clotrimazole
PS
LM/Q
$C_{22}H_{17}ClN_2$
344.10803
RI: 2800
Antimycotic
CAS: 23593-75-1
199
99
141
238
M+
367
100
200
300
Perazine-M (nor-) AC
UHYAC
LS
$C_{21}H_{25}N_3OS$
367.17183
RI: 3210
D: UHYAC
Neuroleptic

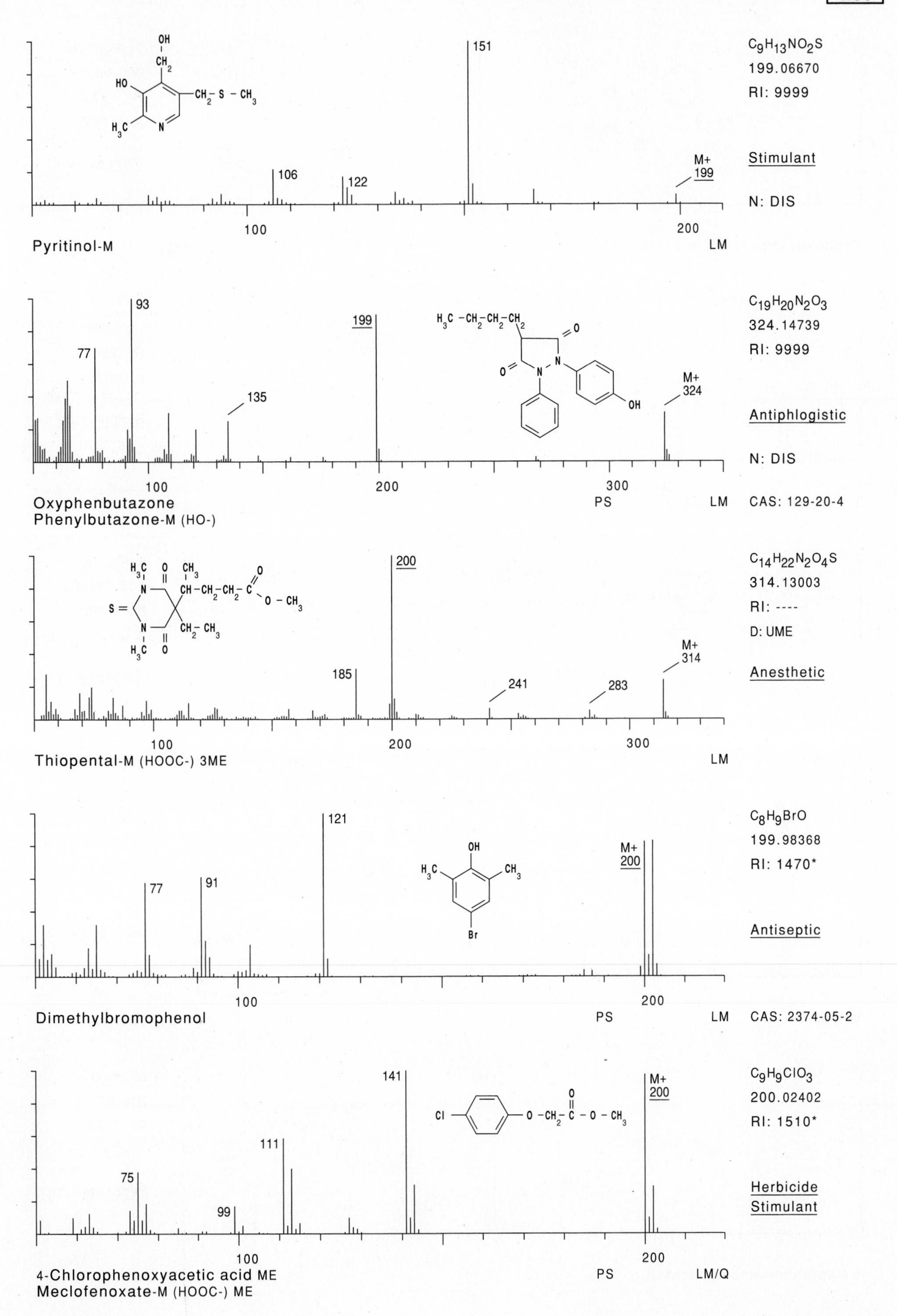

151
106
122
M+
199
100
200
LM
Pyritinol-M
$C_9H_{13}NO_2S$
199.06670
RI: 9999
Stimulant
N: DIS
93
77
199
135
M+
324
100
200
300
PS
LM
Oxyphenbutazone
Phenylbutazone-M (HO-)
$C_{19}H_{20}N_2O_3$
324.14739
RI: 9999
Antiphlogistic
N: DIS
CAS: 129-20-4
200
185
241
283
M+
314
100
200
300
LM
Thiopental-M (HOOC-) 3ME
$C_{14}H_{22}N_2O_4S$
314.13003
RI: ----
D: UME
Anesthetic
121
77
91
M+
200
100
200
PS
LM
Dimethylbromophenol
C_8H_9BrO
199.98368
RI: 1470*
Antiseptic
CAS: 2374-05-2
141
111
75
99
M+
200
100
200
PS
LM/Q
4-Chlorophenoxyacetic acid ME
Meclofenoxate-M (HOOC-) ME
$C_9H_9ClO_3$
200.02402
RI: 1510*
Herbicide
Stimulant

200

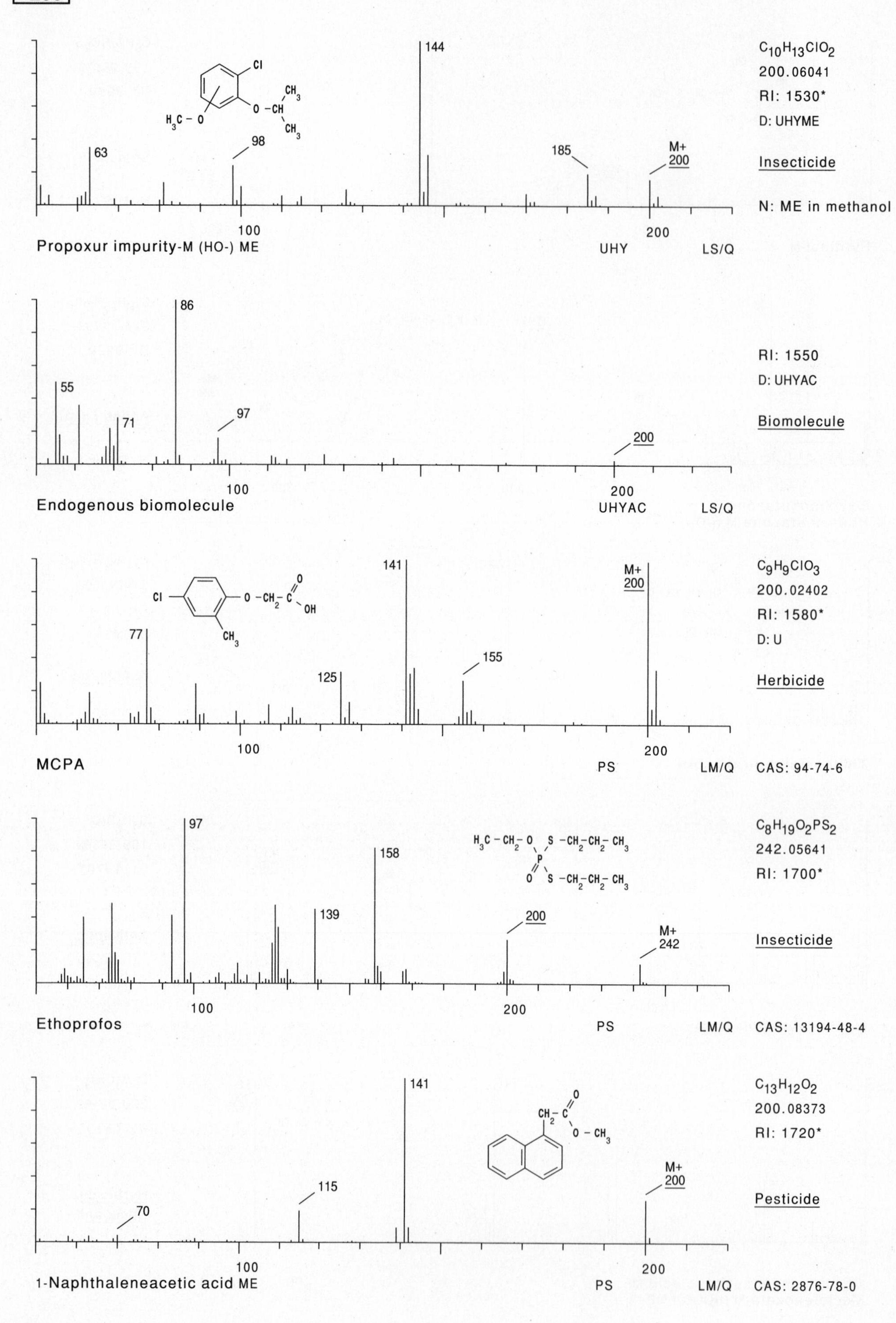

144
Cl
CH3
O - CH
H3C - O
CH3
63
98
185
M+
200
100
200
C10H13ClO2
200.06041
RI: 1530*
D: UHYME
Insecticide
N: ME in methanol
Propoxur impurity-M (HO-) ME
UHY
LS/Q
86
55
71
97
200
100
200
RI: 1550
D: UHYAC
Biomolecule
Endogenous biomolecule
UHYAC
LS/Q
141
M+
200
Cl
O - CH2 - C
O
OH
CH3
77
155
125
100
200
C9H9ClO3
200.02402
RI: 1580*
D: U
Herbicide
MCPA
PS
LM/Q
CAS: 94-74-6
97
158
H3C - CH2 - O
S - CH2 - CH2 - CH3
P
O
S - CH2 - CH2 - CH3
139
200
M+
242
100
200
C8H19O2PS2
242.05641
RI: 1700*
Insecticide
Ethoprofos
PS
LM/Q
CAS: 13194-48-4
141
CH2 - C
O
O - CH3
115
70
M+
200
100
200
C13H12O2
200.08373
RI: 1720*
Pesticide
1-Naphthaleneacetic acid ME
PS
LM/Q
CAS: 2876-78-0

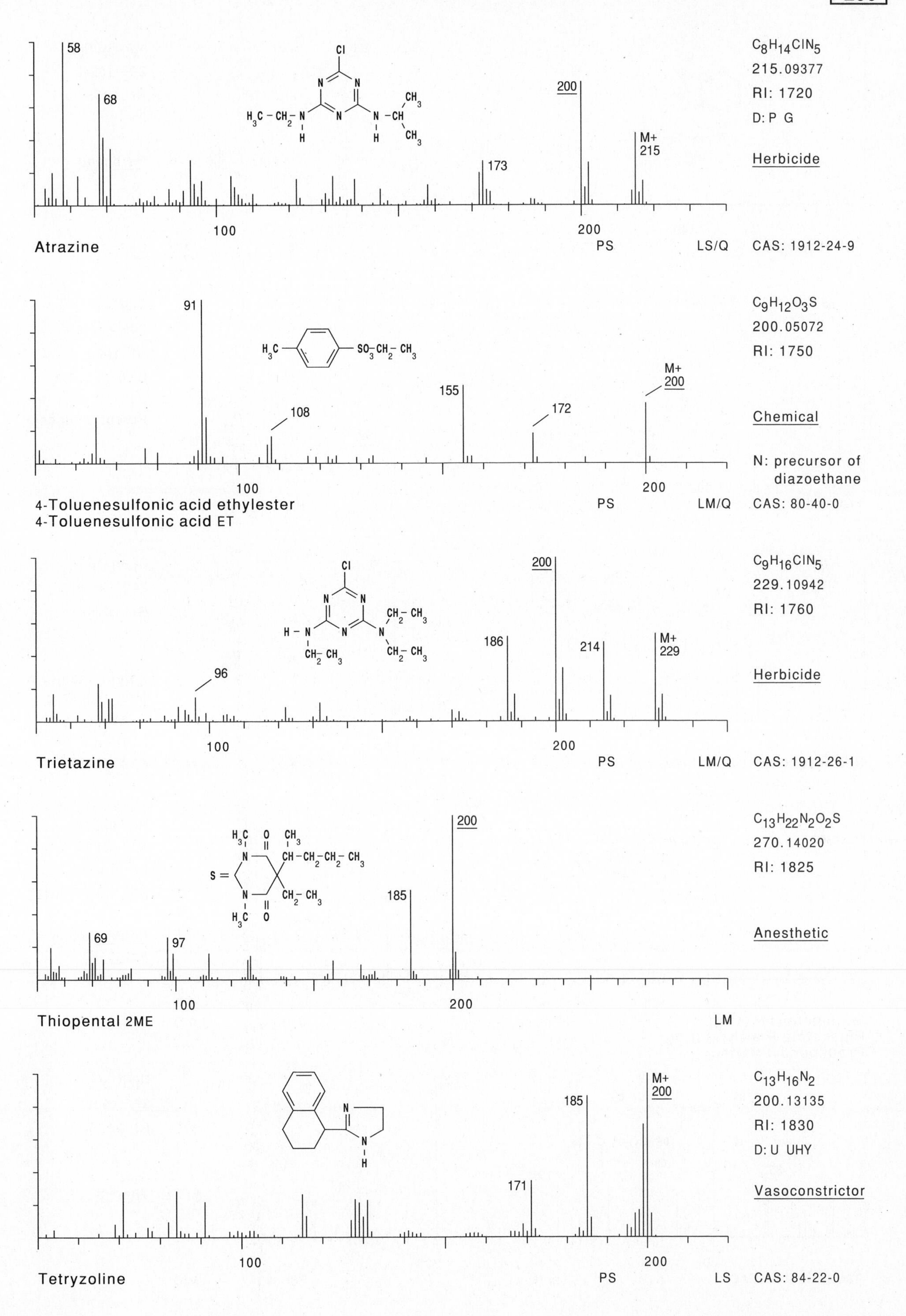

Atrazine

$C_8H_{14}ClN_5$
215.09377
RI: 1720
D: P G
Herbicide
CAS: 1912-24-9

4-Toluenesulfonic acid ethylester
4-Toluenesulfonic acid ET

$C_9H_{12}O_3S$
200.05072
RI: 1750
Chemical
N: precursor of diazoethane
CAS: 80-40-0

Trietazine

$C_9H_{16}ClN_5$
229.10942
RI: 1760
Herbicide
CAS: 1912-26-1

Thiopental 2ME

$C_{13}H_{22}N_2O_2S$
270.14020
RI: 1825
Anesthetic

Tetryzoline

$C_{13}H_{16}N_2$
200.13135
RI: 1830
D: U UHY
Vasoconstrictor
CAS: 84-22-0

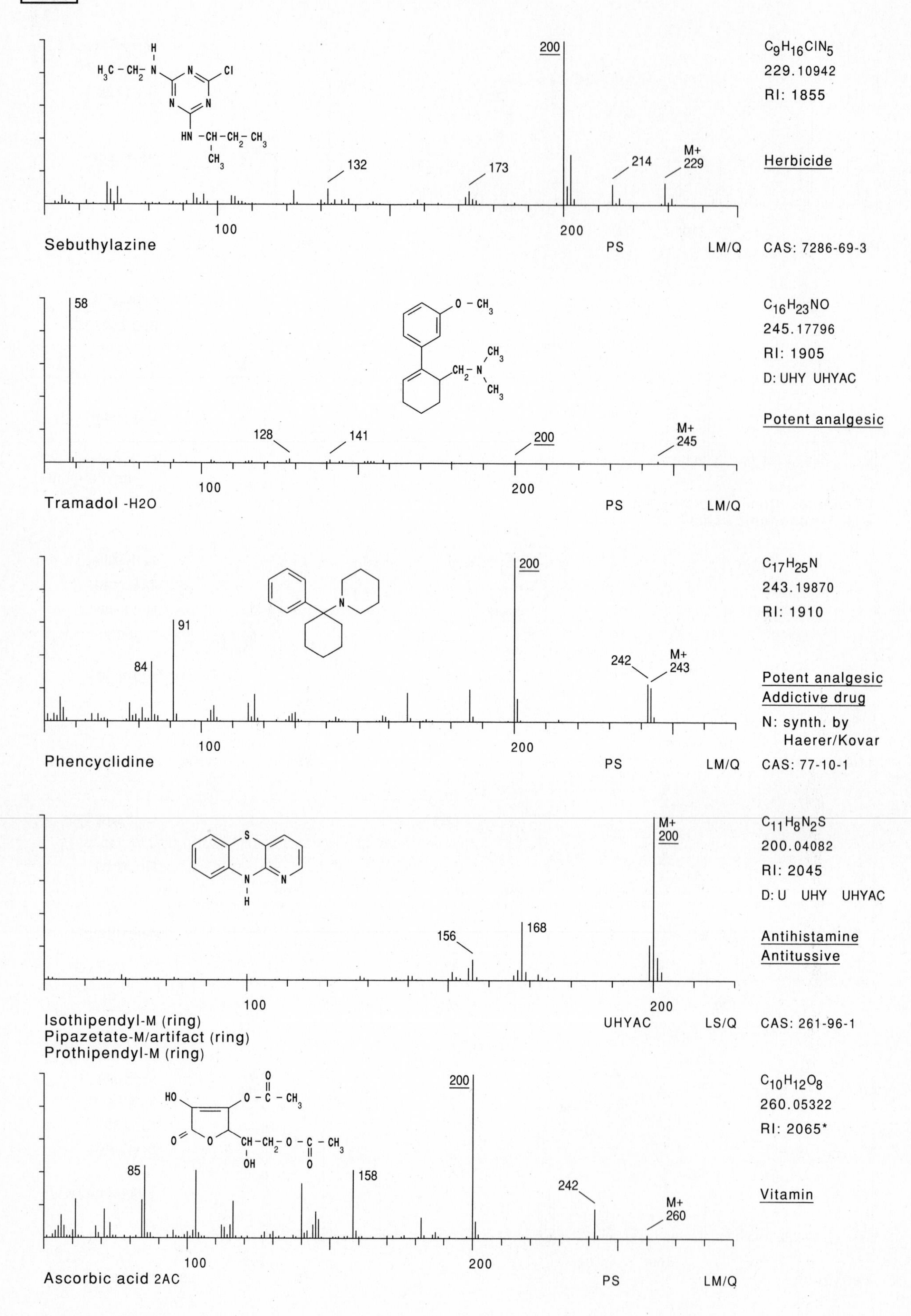

$C_9H_{16}ClN_5$
229.10942
RI: 1855
Herbicide
Sebuthylazine
PS
LM/Q
CAS: 7286-69-3
132
173
200
214
M+
229
100
200
$C_{16}H_{23}NO$
245.17796
RI: 1905
D: UHY UHYAC
Potent analgesic
Tramadol -H2O
PS
LM/Q
58
128
141
200
M+
245
$C_{17}H_{25}N$
243.19870
RI: 1910
Potent analgesic
Addictive drug
N: synth. by
Haerer/Kovar
Phencyclidine
PS
LM/Q
CAS: 77-10-1
84
91
200
242
M+
243
$C_{11}H_8N_2S$
200.04082
RI: 2045
D: U UHY UHYAC
Antihistamine
Antitussive
Isothipendyl-M (ring)
Pipazetate-M/artifact (ring)
Prothipendyl-M (ring)
UHYAC
LS/Q
CAS: 261-96-1
156
168
M+
200
$C_{10}H_{12}O_8$
260.05322
RI: 2065*
Vitamin
Ascorbic acid 2AC
PS
LM/Q
85
158
200
242
M+
260

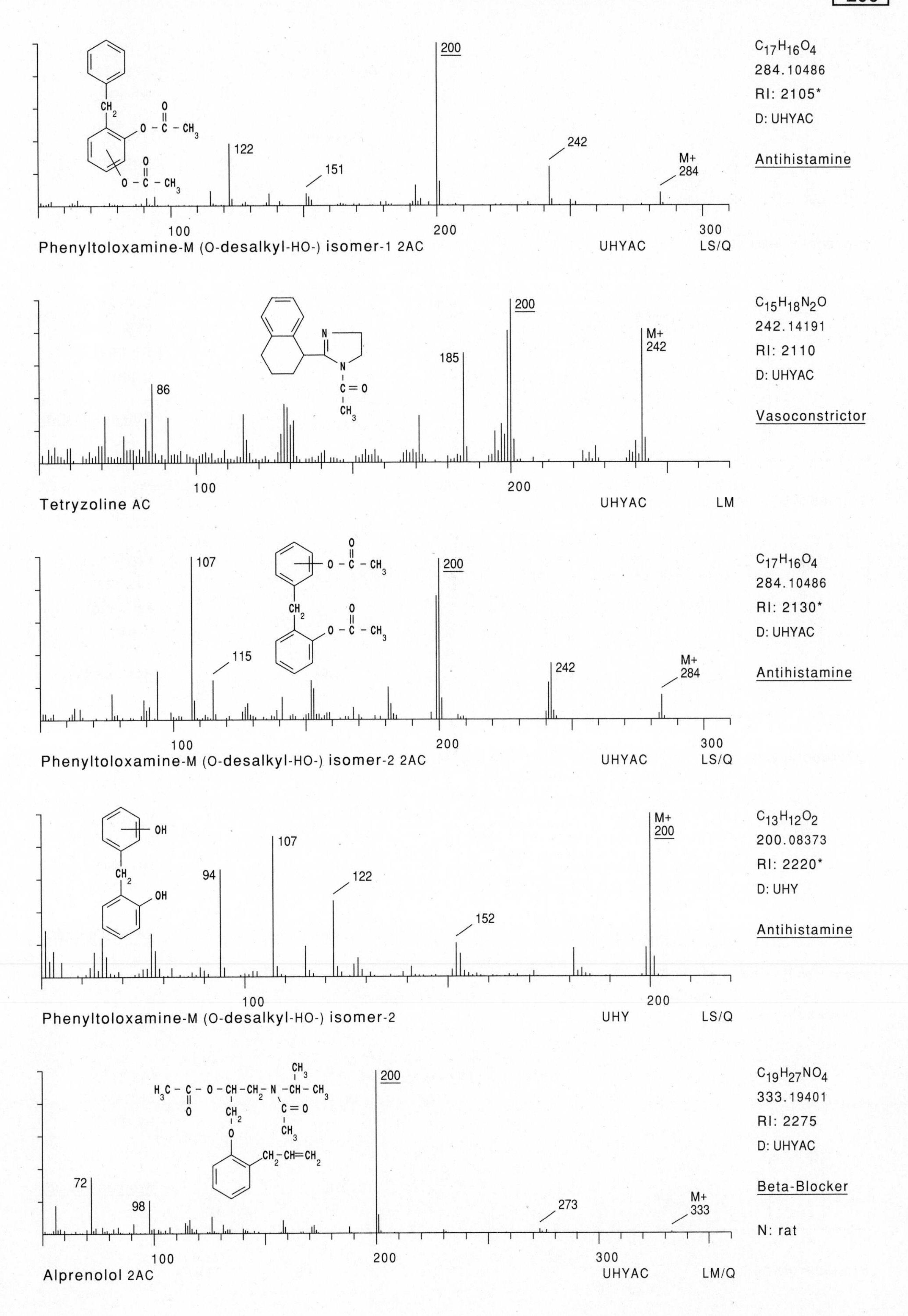

C17H16O4
284.10486
RI: 2105*
D: UHYAC
Antihistamine
Phenyltoloxamine-M (O-desalkyl-HO-) isomer-1 2AC
UHYAC
LS/Q
C15H18N2O
242.14191
RI: 2110
D: UHYAC
Vasoconstrictor
Tetryzoline AC
UHYAC
LM
C17H16O4
284.10486
RI: 2130*
D: UHYAC
Antihistamine
Phenyltoloxamine-M (O-desalkyl-HO-) isomer-2 2AC
UHYAC
LS/Q
C13H12O2
200.08373
RI: 2220*
D: UHY
Antihistamine
Phenyltoloxamine-M (O-desalkyl-HO-) isomer-2
UHY
LS/Q
C19H27NO4
333.19401
RI: 2275
D: UHYAC
Beta-Blocker
N: rat
Alprenolol 2AC
UHYAC
LM/Q

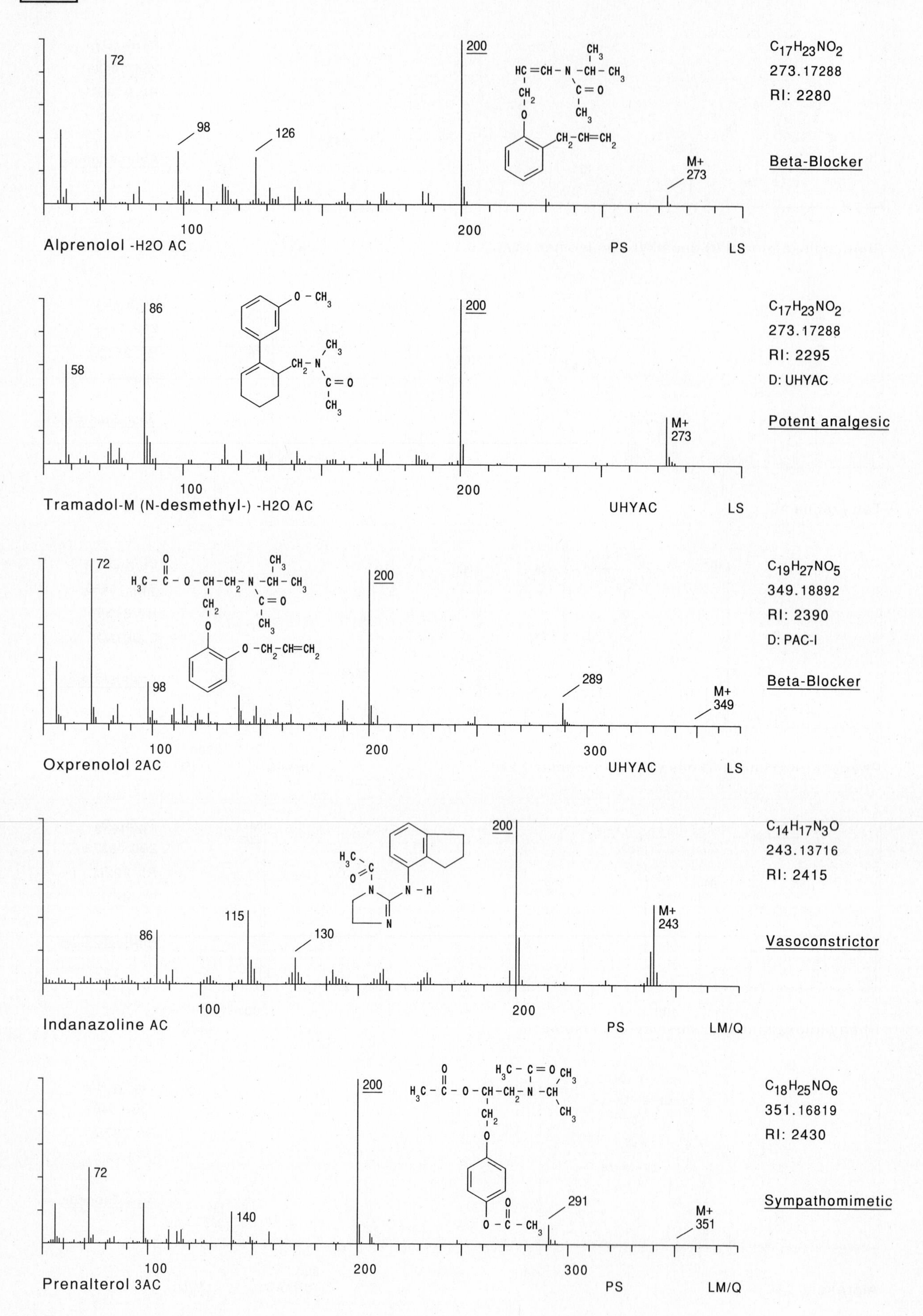

72
98
126
200
M+
273
100
200
Alprenolol -H2O AC
PS
LS
$C_{17}H_{23}NO_2$
273.17288
RI: 2280
Beta-Blocker
58
86
200
M+
273
100
200
Tramadol-M (N-desmethyl-) -H2O AC
UHYAC
LS
$C_{17}H_{23}NO_2$
273.17288
RI: 2295
D: UHYAC
Potent analgesic
72
98
200
289
M+
349
100
200
300
Oxprenolol 2AC
UHYAC
LS
$C_{19}H_{27}NO_5$
349.18892
RI: 2390
D: PAC-I
Beta-Blocker
86
115
130
200
M+
243
100
200
Indanazoline AC
PS
LM/Q
$C_{14}H_{17}N_3O$
243.13716
RI: 2415
Vasoconstrictor
72
140
200
291
M+
351
100
200
300
Prenalterol 3AC
PS
LM/Q
$C_{18}H_{25}NO_6$
351.16819
RI: 2430
Sympathomimetic

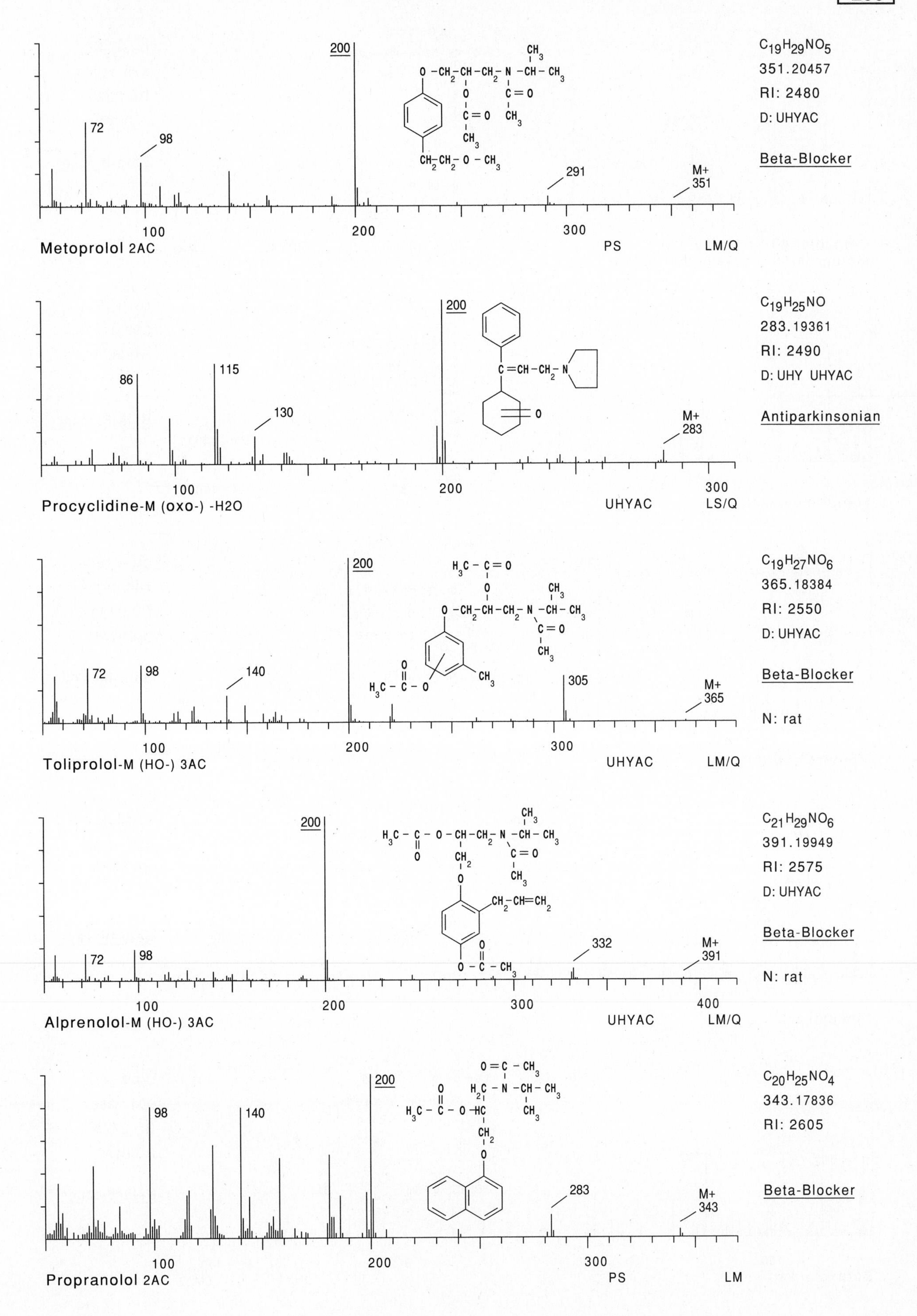

C19H29NO5
351.20457
RI: 2480
D: UHYAC
Beta-Blocker
M+ 351
Metoprolol 2AC
PS
LM/Q
C19H25NO
283.19361
RI: 2490
D: UHY UHYAC
Antiparkinsonian
M+ 283
Procyclidine-M (oxo-) -H2O
UHYAC
LS/Q
C19H27NO6
365.18384
RI: 2550
D: UHYAC
Beta-Blocker
N: rat
M+ 365
Toliprolol-M (HO-) 3AC
UHYAC
LM/Q
C21H29NO6
391.19949
RI: 2575
D: UHYAC
Beta-Blocker
N: rat
M+ 391
Alprenolol-M (HO-) 3AC
UHYAC
LM/Q
C20H25NO4
343.17836
RI: 2605
Beta-Blocker
M+ 343
Propranolol 2AC
PS
LM

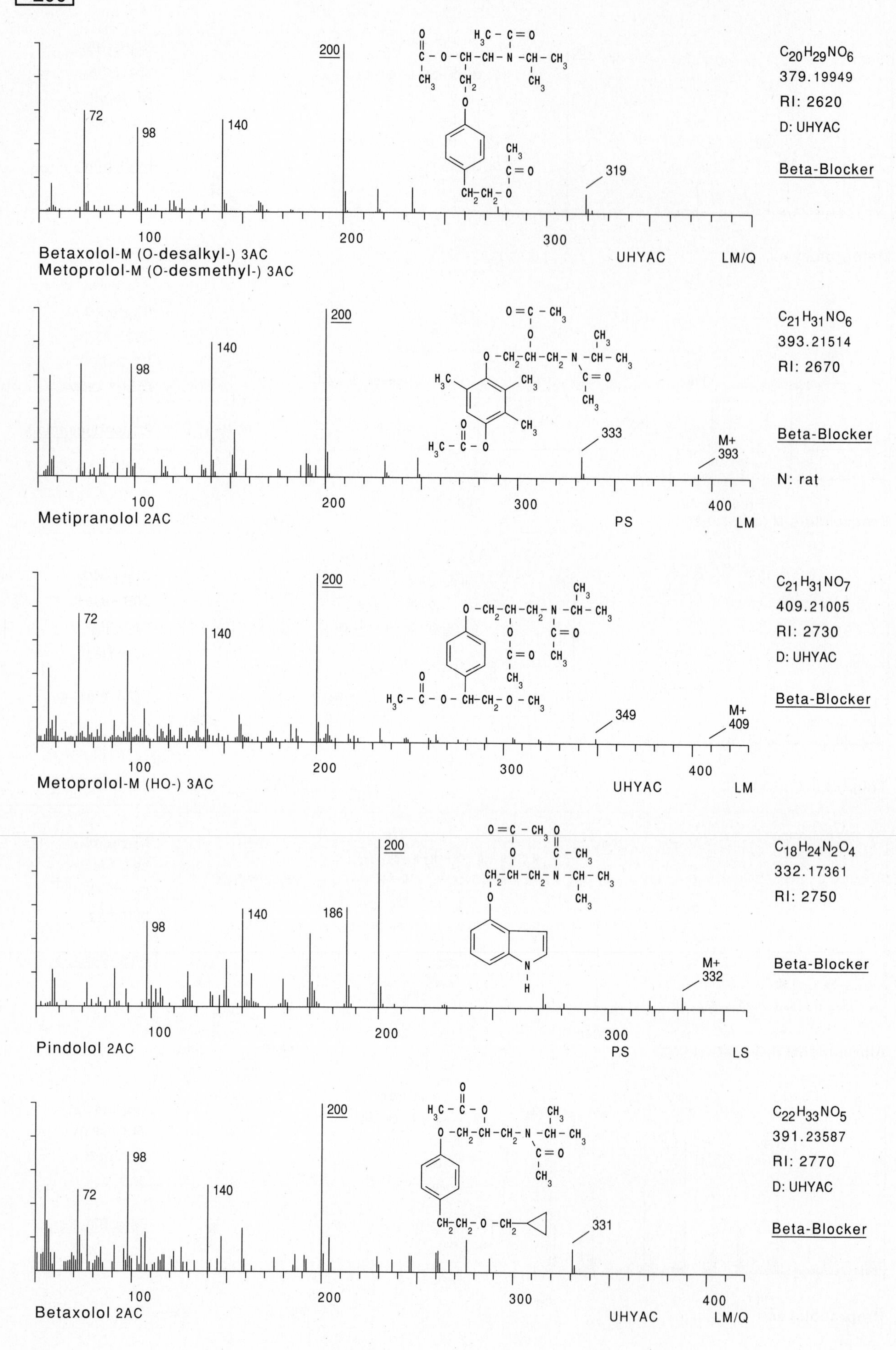
200
72
98
140
319
100
200
300
UHYAC
LM/Q
Betaxolol-M (O-desalkyl-) 3AC
Metoprolol-M (O-desmethyl-) 3AC
C20H29NO6
379.19949
RI: 2620
D: UHYAC
Beta-Blocker
200
140
98
333
M+
393
Metipranolol 2AC
PS
LM
C21H31NO6
393.21514
RI: 2670
Beta-Blocker
N: rat
200
72
140
349
M+
409
Metoprolol-M (HO-) 3AC
UHYAC
LM
C21H31NO7
409.21005
RI: 2730
D: UHYAC
Beta-Blocker
200
98
140
186
M+
332
Pindolol 2AC
PS
LS
C18H24N2O4
332.17361
RI: 2750
Beta-Blocker
200
98
72
140
331
Betaxolol 2AC
UHYAC
LM/Q
C22H33NO5
391.23587
RI: 2770
D: UHYAC
Beta-Blocker

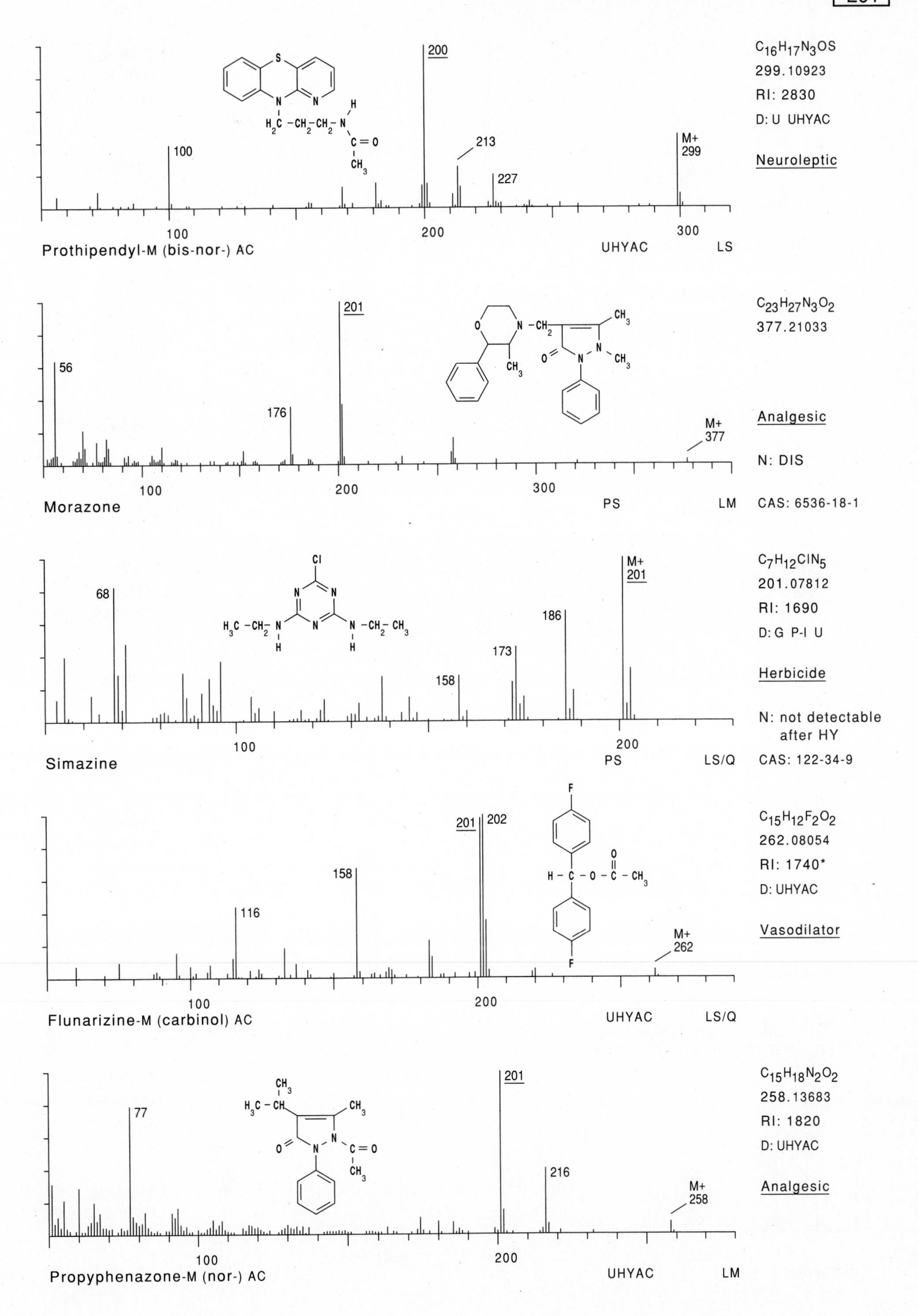

$C_{16}H_{17}N_3OS$
299.10923
RI: 2830
D: U UHYAC

Neuroleptic

$C_{23}H_{27}N_3O_2$
377.21033

Analgesic

N: DIS

CAS: 6536-18-1

$C_7H_{12}ClN_5$
201.07812
RI: 1690
D: G P-I U

Herbicide

N: not detectable after HY

CAS: 122-34-9

$C_{15}H_{12}F_2O_2$
262.08054
RI: 1740*
D: UHYAC

Vasodilator

$C_{15}H_{18}N_2O_2$
258.13683
RI: 1820
D: UHYAC

Analgesic

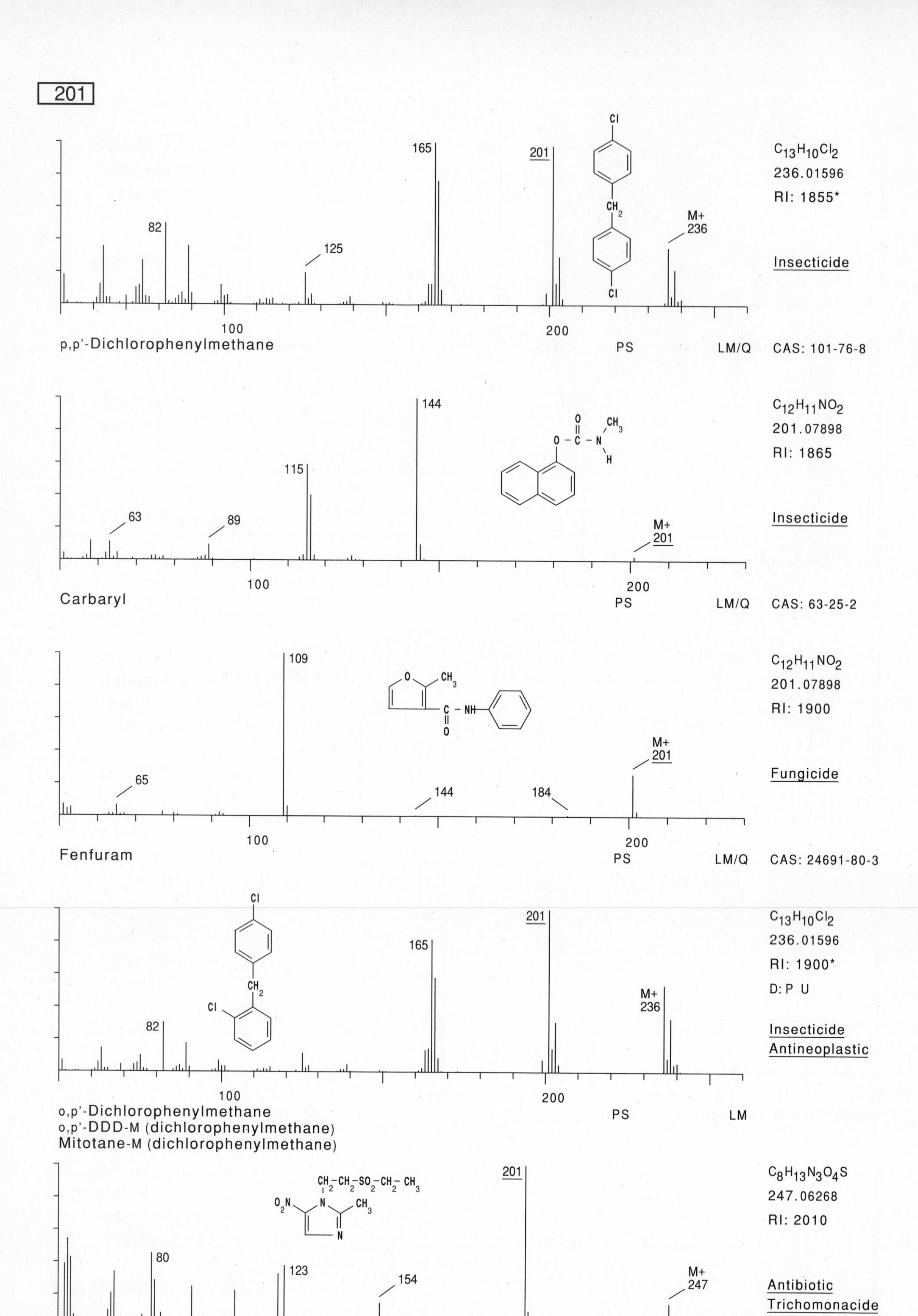
165
201
Cl
CH2
Cl
82
125
M+
236
100
200
$C_{13}H_{10}Cl_2$
236.01596
RI: 1855*
Insecticide
p,p'-Dichlorophenylmethane
PS
LM/Q
CAS: 101-76-8
144
115
63
89
M+
201
O
CH3
O - C - N
H
$C_{12}H_{11}NO_2$
201.07898
RI: 1865
Insecticide
Carbaryl
CAS: 63-25-2
109
O
CH3
C - NH
O
65
144
184
M+
201
$C_{12}H_{11}NO_2$
201.07898
RI: 1900
Fungicide
Fenfuram
CAS: 24691-80-3
Cl
CH2
Cl
201
165
M+
236
82
$C_{13}H_{10}Cl_2$
236.01596
RI: 1900*
D: P U
Insecticide
Antineoplastic
o,p'-Dichlorophenylmethane
o,p'-DDD-M (dichlorophenylmethane)
Mitotane-M (dichlorophenylmethane)
LM
CH2-CH2-SO2-CH2-CH3
O2N
N
CH3
N
201
80
123
154
M+
247
$C_8H_{13}N_3O_4S$
247.06268
RI: 2010
Antibiotic
Trichomonacide
Tinidazole
CAS: 19387-91-8

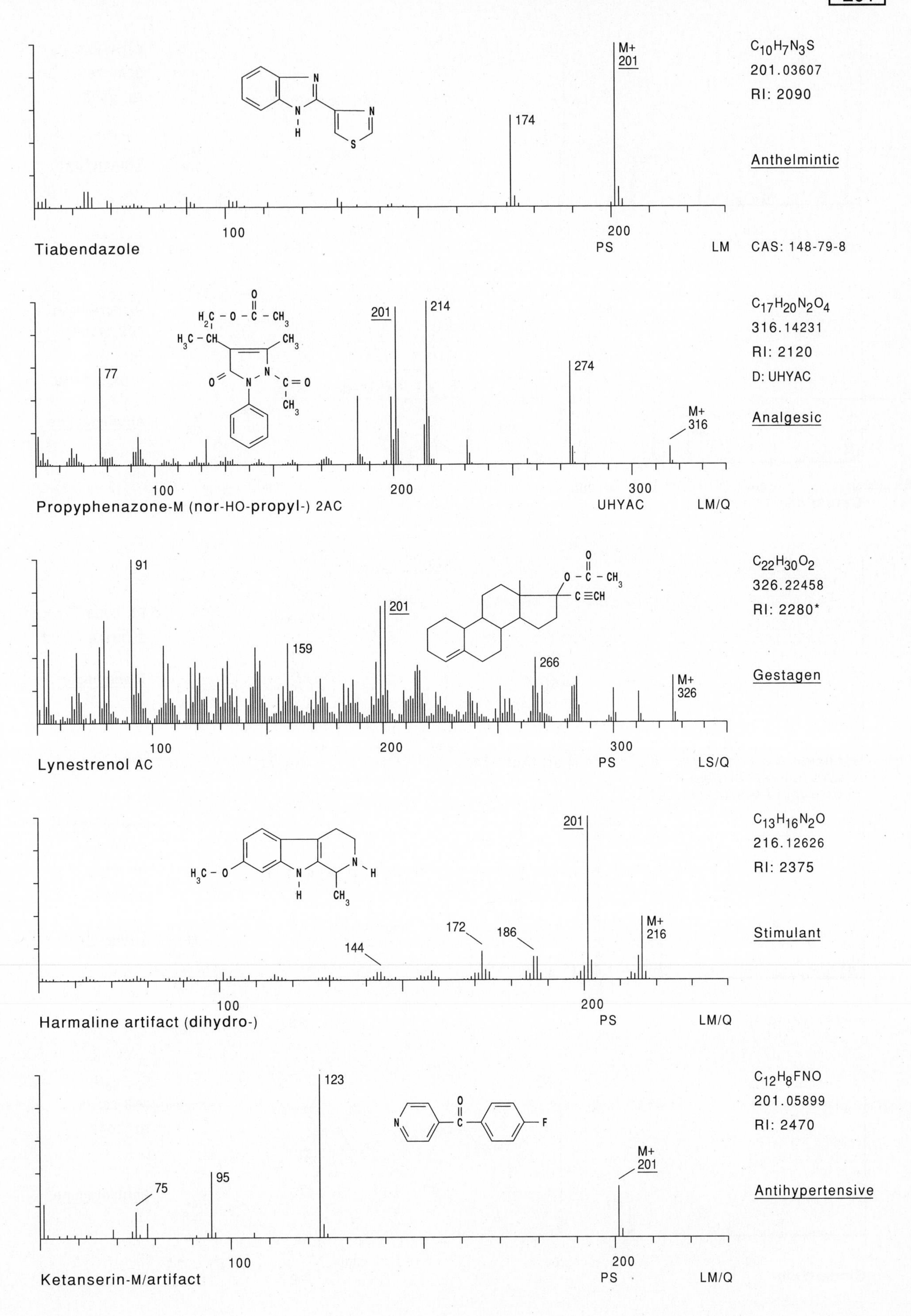
M+
201
174
100
200
Tiabendazole
PS
LM
$C_{10}H_7N_3S$
201.03607
RI: 2090
Anthelmintic
CAS: 148-79-8
201
214
77
274
M+
316
100
200
300
Propyphenazone-M (nor-HO-propyl-) 2AC
UHYAC
LM/Q
$C_{17}H_{20}N_2O_4$
316.14231
RI: 2120
D: UHYAC
Analgesic
91
201
159
266
M+
326
100
200
300
Lynestrenol AC
PS
LS/Q
$C_{22}H_{30}O_2$
326.22458
RI: 2280*
Gestagen
201
172
186
144
M+
216
100
200
Harmaline artifact (dihydro-)
PS
LM/Q
$C_{13}H_{16}N_2O$
216.12626
RI: 2375
Stimulant
123
95
75
M+
201
100
200
Ketanserin-M/artifact
PS
LM/Q
$C_{12}H_8FNO$
201.05899
RI: 2470
Antihypertensive

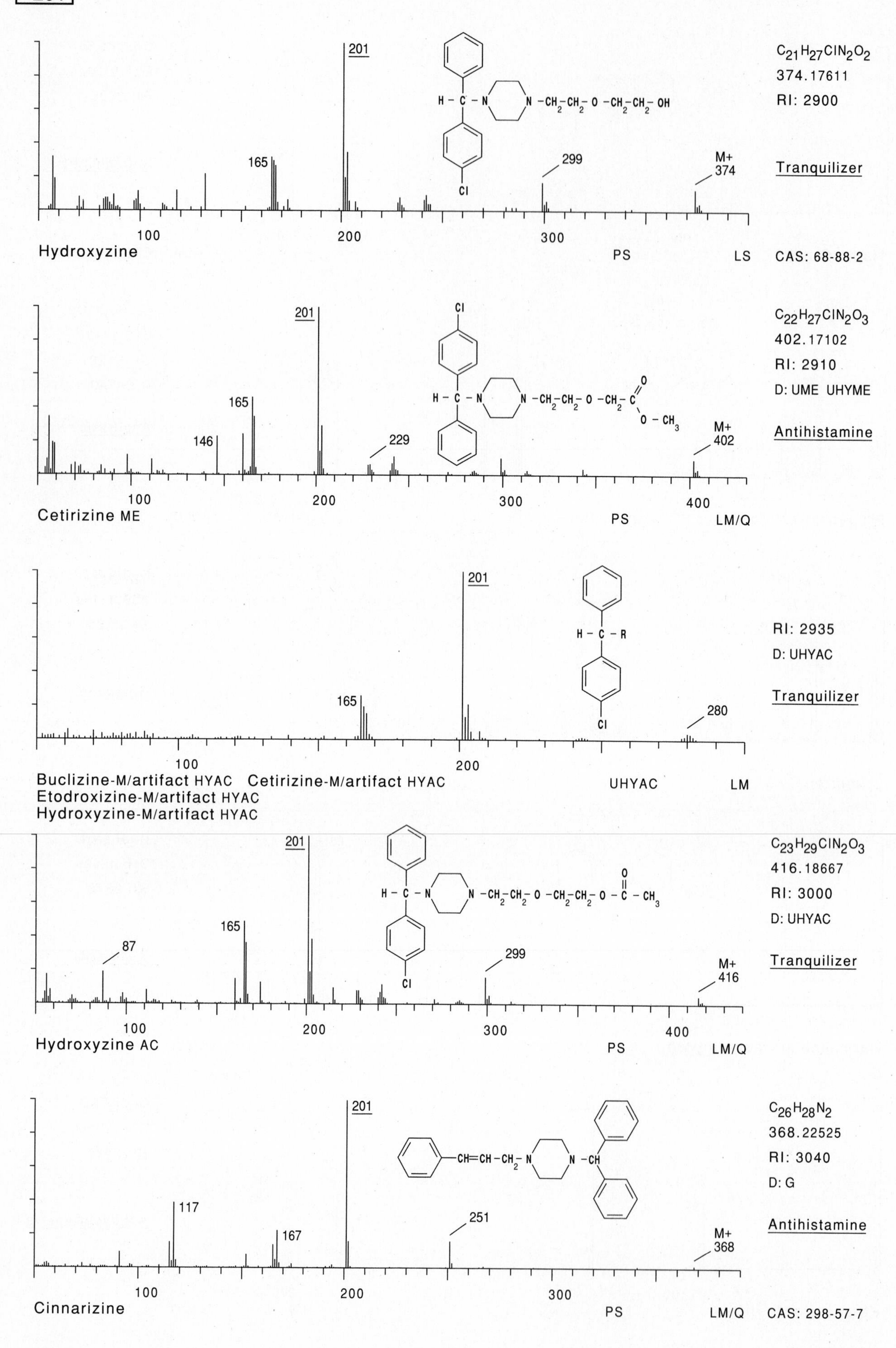

201
165
299
M+
374
100
200
300
Hydroxyzine
PS
LS
$C_{21}H_{27}ClN_2O_2$
374.17611
RI: 2900
Tranquilizer
CAS: 68-88-2
201
165
146
229
M+
402
100
200
300
400
Cetirizine ME
PS
LM/Q
$C_{22}H_{27}ClN_2O_3$
402.17102
RI: 2910
D: UME UHYME
Antihistamine
201
165
280
100
200
Buclizine-M/artifact HYAC Cetirizine-M/artifact HYAC
Etodroxizine-M/artifact HYAC
Hydroxyzine-M/artifact HYAC
UHYAC
LM
RI: 2935
D: UHYAC
Tranquilizer
201
165
87
299
M+
416
100
200
300
400
Hydroxyzine AC
PS
LM/Q
$C_{23}H_{29}ClN_2O_3$
416.18667
RI: 3000
D: UHYAC
Tranquilizer
201
117
167
251
M+
368
100
200
300
Cinnarizine
PS
LM/Q
$C_{26}H_{28}N_2$
368.22525
RI: 3040
D: G
Antihistamine
CAS: 298-57-7

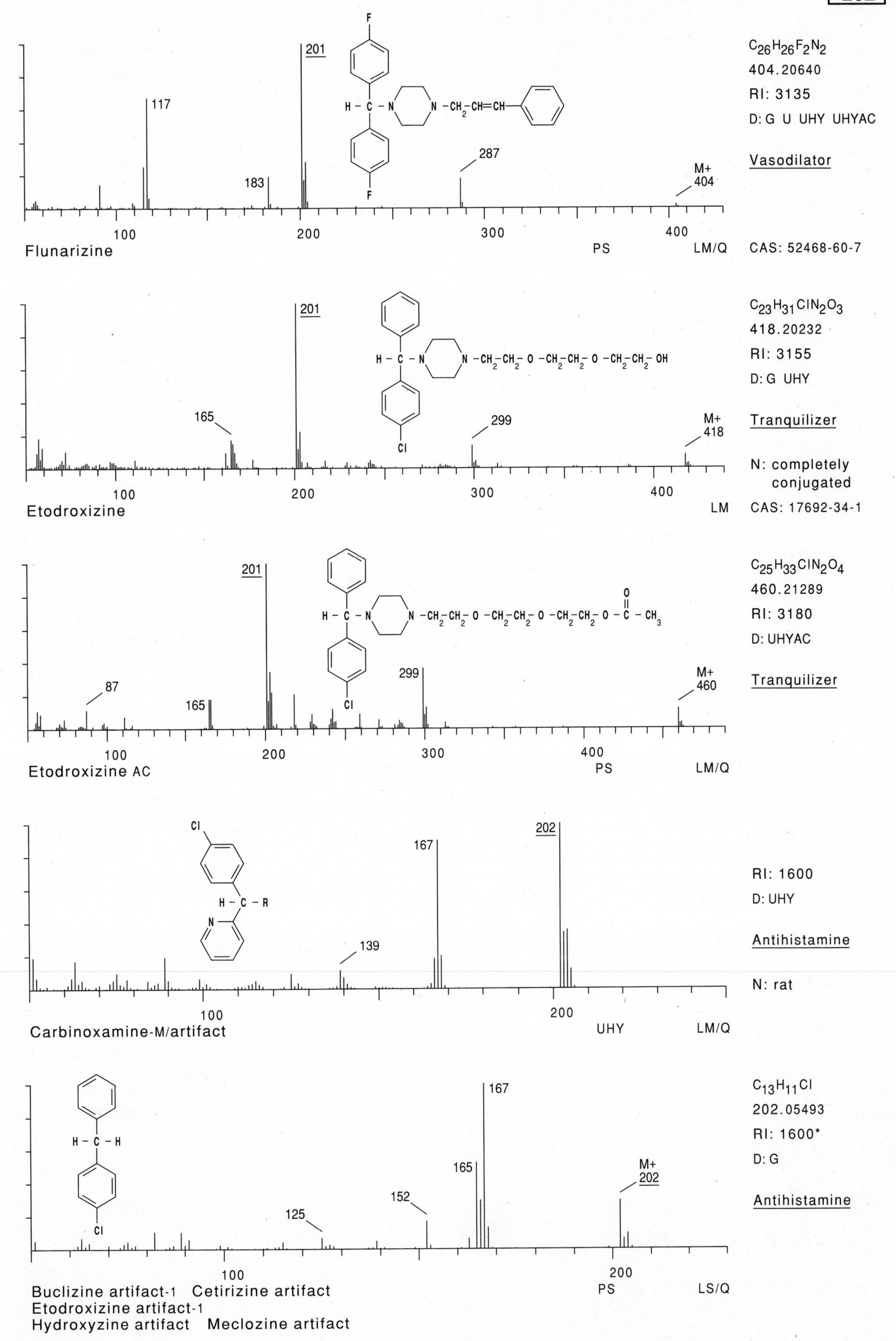

201
117
183
287
M+ 404
100
200
300
400
Flunarizine
PS
LM/Q
$C_{26}H_{26}F_2N_2$
404.20640
RI: 3135
D: G U UHY UHYAC
Vasodilator
CAS: 52468-60-7
201
165
299
M+ 418
Etodroxizine
LM
$C_{23}H_{31}ClN_2O_3$
418.20232
RI: 3155
D: G UHY
Tranquilizer
N: completely conjugated
CAS: 17692-34-1
201
87
165
299
M+ 460
Etodroxizine AC
PS
LM/Q
$C_{25}H_{33}ClN_2O_4$
460.21289
RI: 3180
D: UHYAC
Tranquilizer
202
167
139
Carbinoxamine-M/artifact
UHY
LM/Q
RI: 1600
D: UHY
Antihistamine
N: rat
167
165
152
125
M+ 202
Buclizine artifact-1 Cetirizine artifact
Etodroxizine artifact-1
Hydroxyzine artifact Meclozine artifact
PS
LS/Q
$C_{13}H_{11}Cl$
202.05493
RI: 1600*
D: G
Antihistamine

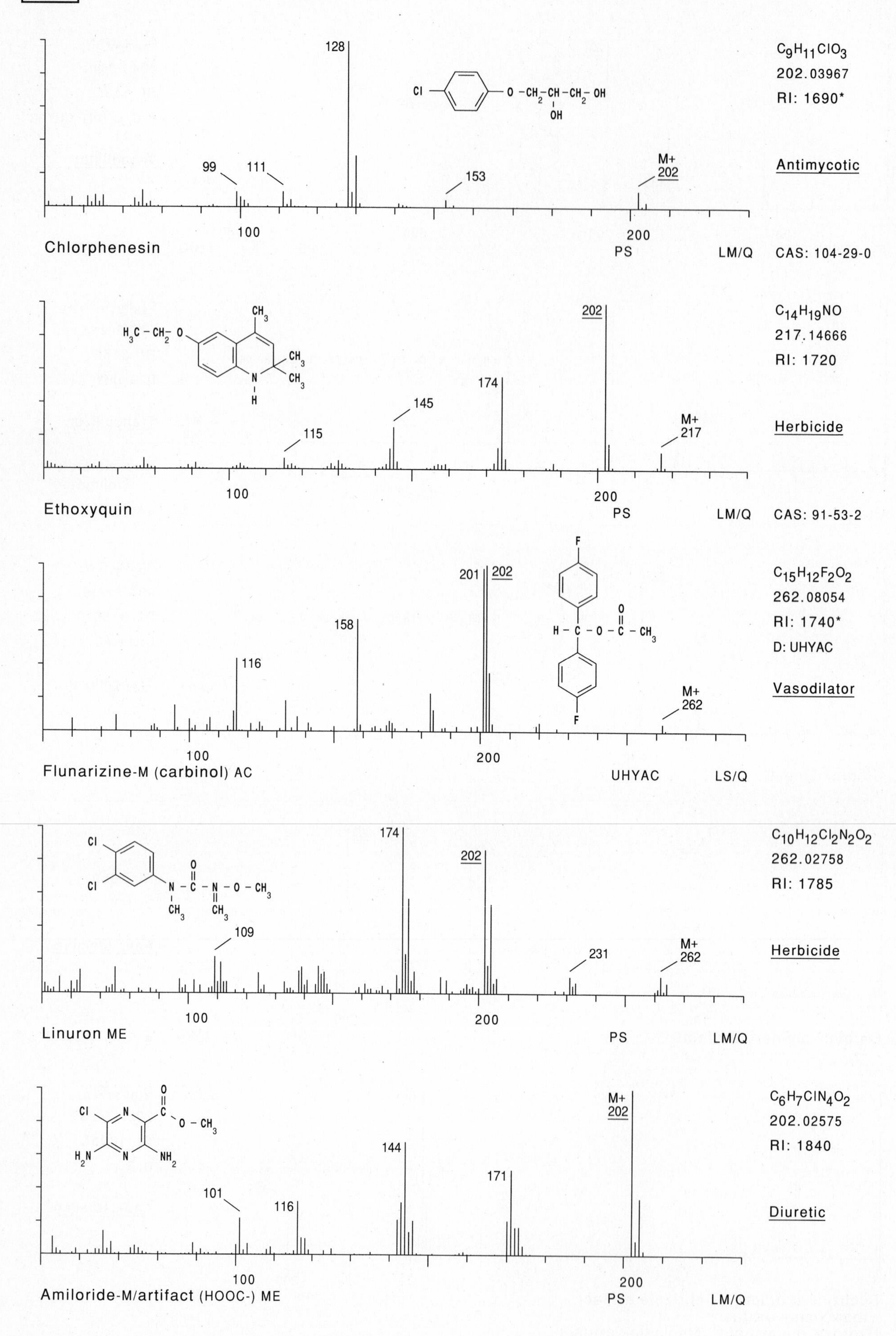

128
Cl O - CH2 - CH - CH2 - OH
OH
99
111
153
M+
202
100
200
Chlorphenesin
PS
LM/Q
C9H11ClO3
202.03967
RI: 1690*
Antimycotic
CAS: 104-29-0
CH3
H3C - CH2 - O
CH3
CH3
N
H
202
174
145
115
M+
217
100
200
Ethoxyquin
PS
LM/Q
C14H19NO
217.14666
RI: 1720
Herbicide
CAS: 91-53-2
F
201
202
158
116
H - C - O - C - CH3
O
F
M+
262
100
200
Flunarizine-M (carbinol) AC
UHYAC
LS/Q
C15H12F2O2
262.08054
RI: 1740*
D: UHYAC
Vasodilator
Cl
Cl
N - C - N - O - CH3
O
CH3
CH3
174
202
109
231
M+
262
100
200
Linuron ME
PS
LM/Q
C10H12Cl2N2O2
262.02758
RI: 1785
Herbicide
Cl
N
O
C
O - CH3
H2N
N
NH2
M+
202
144
171
101
116
100
200
Amiloride-M/artifact (HOOC-) ME
PS
LM/Q
C6H7ClN4O2
202.02575
RI: 1840
Diuretic

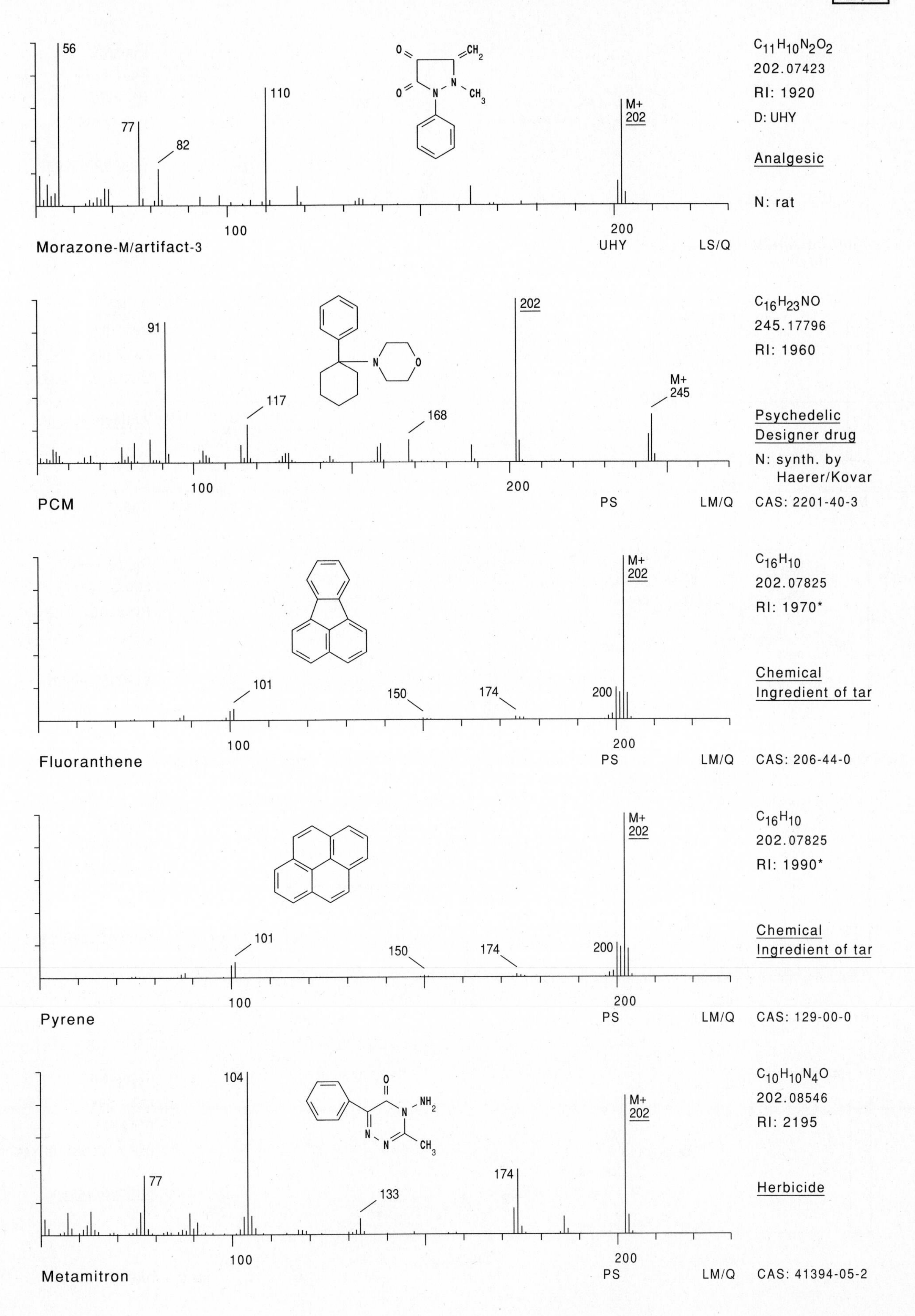

56
77
82
110
M+
202
100
200
Morazone-M/artifact-3
UHY
LS/Q
C11H10N2O2
202.07423
RI: 1920
D: UHY
Analgesic
N: rat
91
117
168
202
M+
245
100
200
PCM
PS
LM/Q
C16H23NO
245.17796
RI: 1960
Psychedelic
Designer drug
N: synth. by
Haerer/Kovar
CAS: 2201-40-3
M+
202
101
150
174
200
100
200
Fluoranthene
PS
LM/Q
C16H10
202.07825
RI: 1970*
Chemical
Ingredient of tar
CAS: 206-44-0
M+
202
101
150
174
200
100
200
Pyrene
PS
LM/Q
C16H10
202.07825
RI: 1990*
Chemical
Ingredient of tar
CAS: 129-00-0
104
77
133
174
M+
202
100
200
Metamitron
PS
LM/Q
C10H10N4O
202.08546
RI: 2195
Herbicide
CAS: 41394-05-2

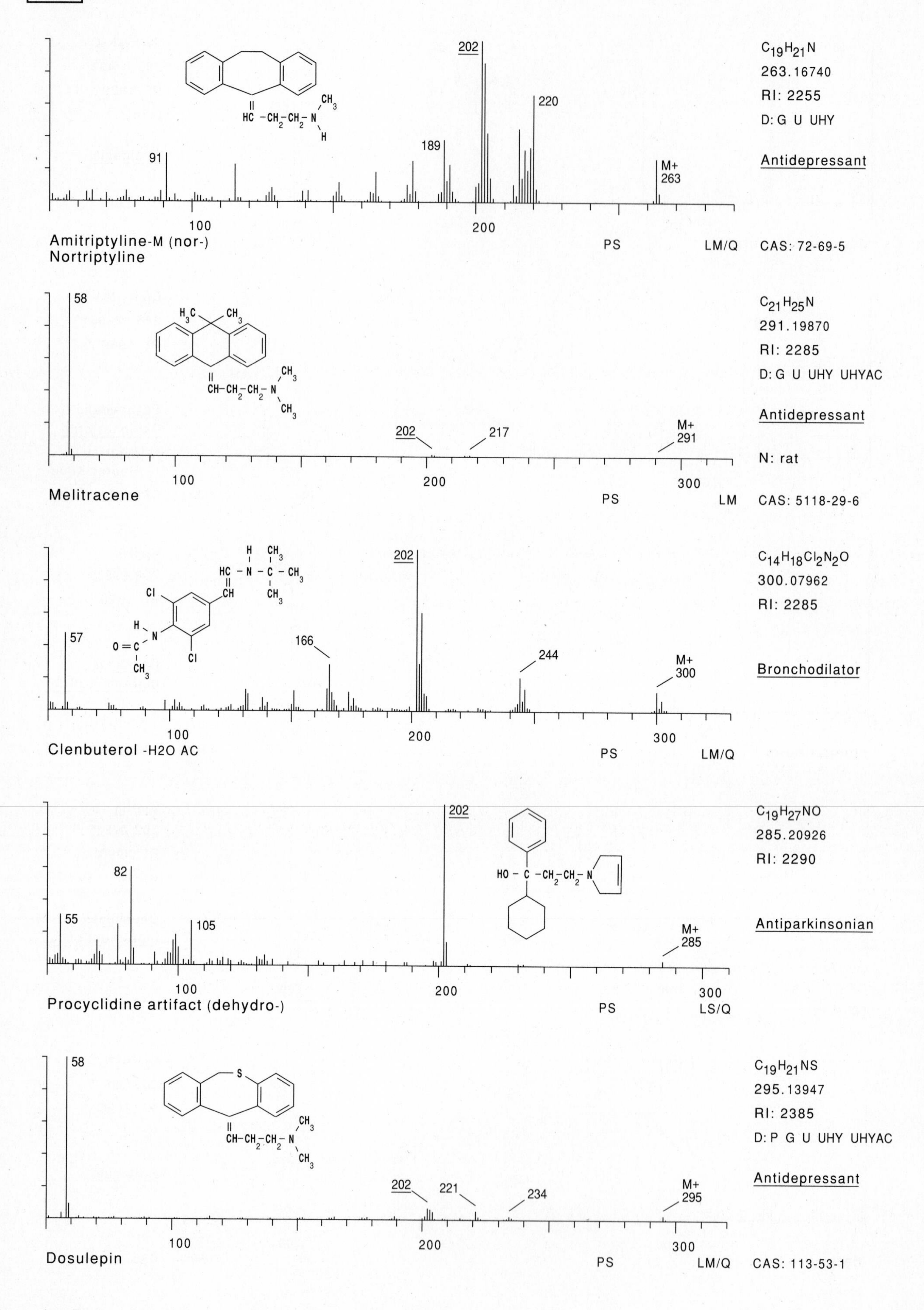

202
220
189
91
M+
263
100
200
Amitriptyline-M (nor-)
Nortriptyline
PS
LM/Q
C19H21N
263.16740
RI: 2255
D: G U UHY
Antidepressant
CAS: 72-69-5
58
202
217
M+
291
100
200
300
Melitracene
PS
LM
C21H25N
291.19870
RI: 2285
D: G U UHY UHYAC
Antidepressant
N: rat
CAS: 5118-29-6
202
57
166
244
M+
300
100
200
300
Clenbuterol -H2O AC
PS
LM/Q
C14H18Cl2N2O
300.07962
RI: 2285
Bronchodilator
202
82
55
105
M+
285
100
200
300
Procyclidine artifact (dehydro-)
PS
LS/Q
C19H27NO
285.20926
RI: 2290
Antiparkinsonian
58
202
221
234
M+
295
100
200
300
Dosulepin
PS
LM/Q
C19H21NS
295.13947
RI: 2385
D: P G U UHY UHYAC
Antidepressant
CAS: 113-53-1

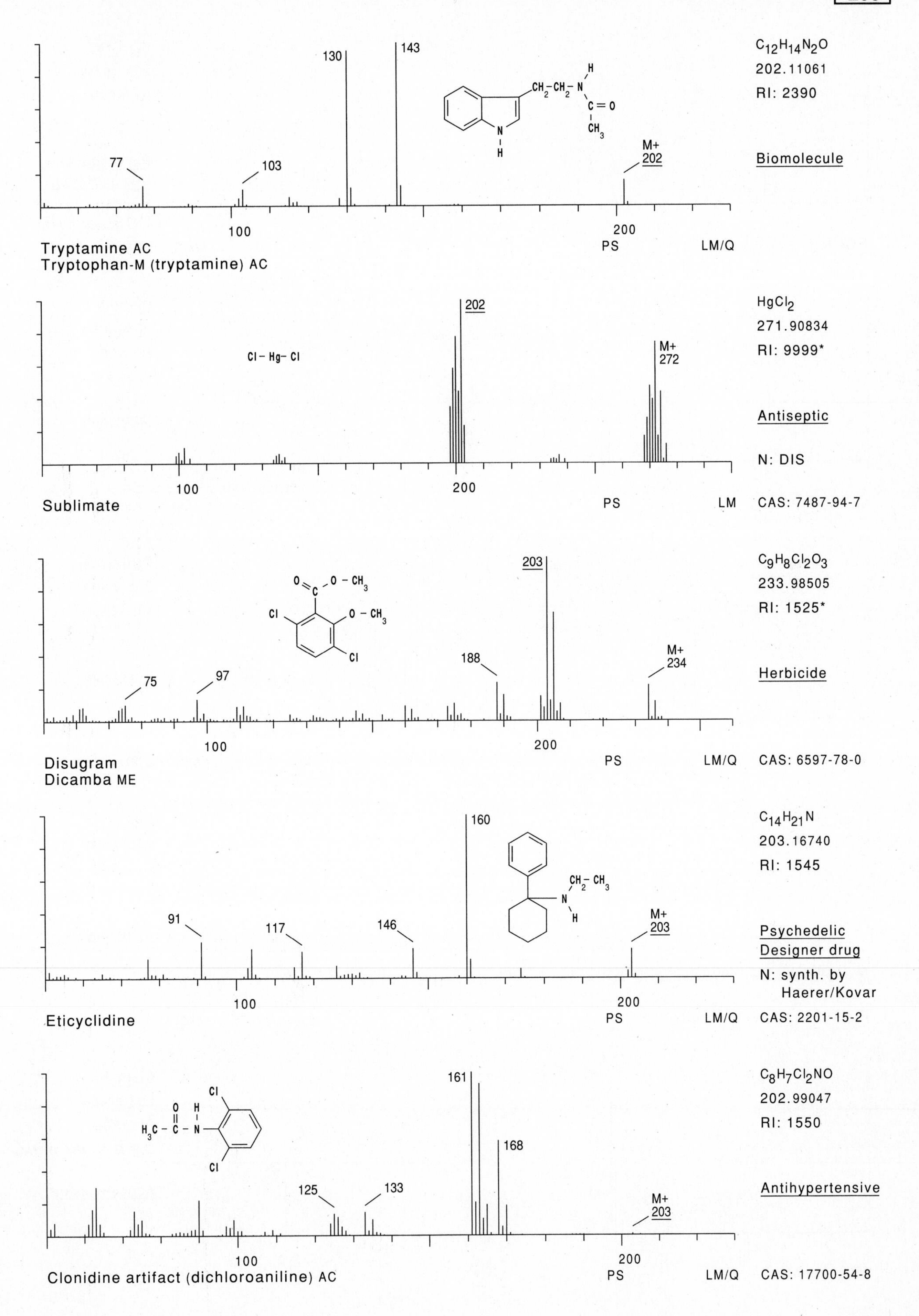

$C_{12}H_{14}N_2O$
202.11061
RI: 2390

Biomolecule

$HgCl_2$
271.90834
RI: 9999*

Antiseptic

N: DIS

CAS: 7487-94-7

$C_9H_8Cl_2O_3$
233.98505
RI: 1525*

Herbicide

CAS: 6597-78-0

$C_{14}H_{21}N$
203.16740
RI: 1545

Psychedelic
Designer drug

N: synth. by Haerer/Kovar

CAS: 2201-15-2

$C_8H_7Cl_2NO$
202.99047
RI: 1550

Antihypertensive

CAS: 17700-54-8

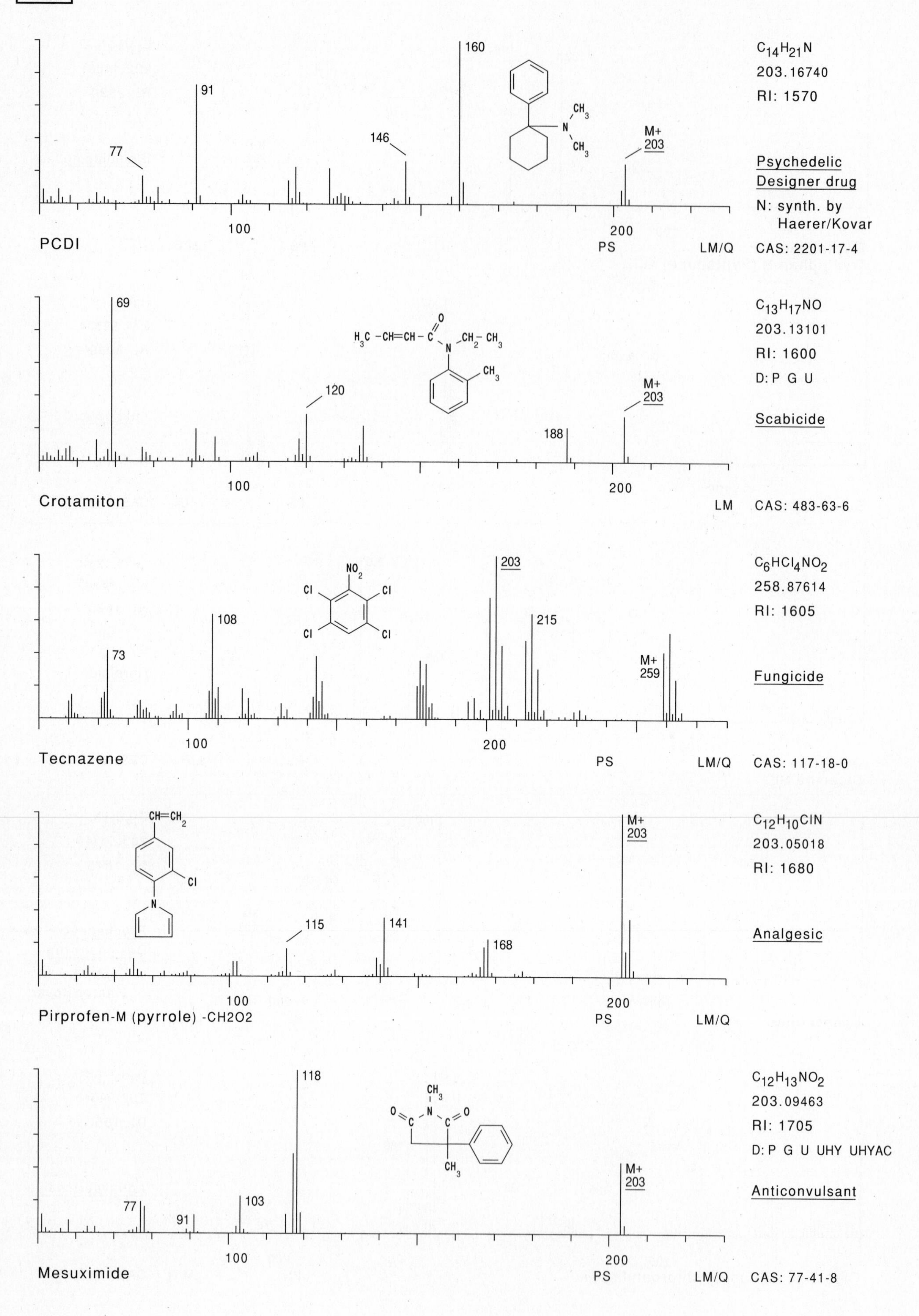

PCDI

$C_{14}H_{21}N$
203.16740
RI: 1570

Psychedelic
Designer drug
N: synth. by Haerer/Kovar
CAS: 2201-17-4

Crotamiton

$C_{13}H_{17}NO$
203.13101
RI: 1600
D: P G U

Scabicide
CAS: 483-63-6

Tecnazene

$C_6HCl_4NO_2$
258.87614
RI: 1605

Fungicide
CAS: 117-18-0

Pirprofen-M (pyrrole) -CH2O2

$C_{12}H_{10}ClN$
203.05018
RI: 1680

Analgesic

Mesuximide

$C_{12}H_{13}NO_2$
203.09463
RI: 1705
D: P G U UHY UHYAC

Anticonvulsant
CAS: 77-41-8

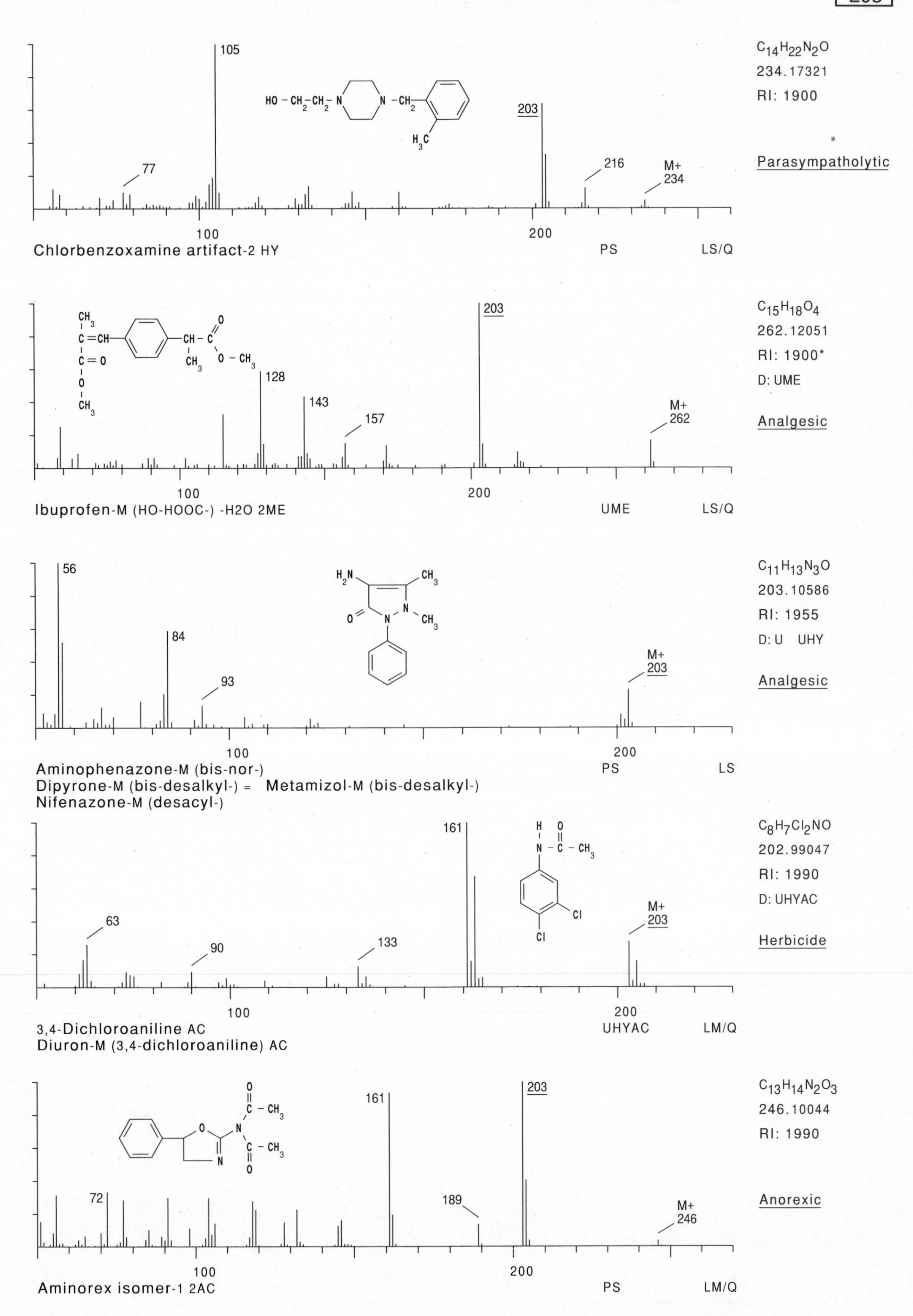
105
HO - CH2 - CH2 - N N - CH2
H3C
203
77
216
M+
234
100
200
Chlorbenzoxamine artifact-2 HY
PS
LS/Q
C14H22N2O
234.17321
RI: 1900
Parasympatholytic
CH3
C = CH
C = O
O
CH3
CH - C
CH3
O
O - CH3
203
128
143
157
M+
262
100
200
Ibuprofen-M (HO-HOOC-) -H2O 2ME
UME
LS/Q
C15H18O4
262.12051
RI: 1900*
D: UME
Analgesic
56
H2N
CH3
O
N
N
CH3
84
93
M+
203
100
200
Aminophenazone-M (bis-nor-)
Dipyrone-M (bis-desalkyl-) = Metamizol-M (bis-desalkyl-)
Nifenazone-M (desacyl-)
PS
LS
C11H13N3O
203.10586
RI: 1955
D: U UHY
Analgesic
161
H O
N - C - CH3
Cl
Cl
63
90
133
M+
203
100
200
3,4-Dichloroaniline AC
Diuron-M (3,4-dichloroaniline) AC
UHYAC
LM/Q
C8H7Cl2NO
202.99047
RI: 1990
D: UHYAC
Herbicide
O
C - CH3
N
C - CH3
O
N
O
161
203
72
189
M+
246
100
200
Aminorex isomer-1 2AC
PS
LM/Q
C13H14N2O3
246.10044
RI: 1990
Anorexic

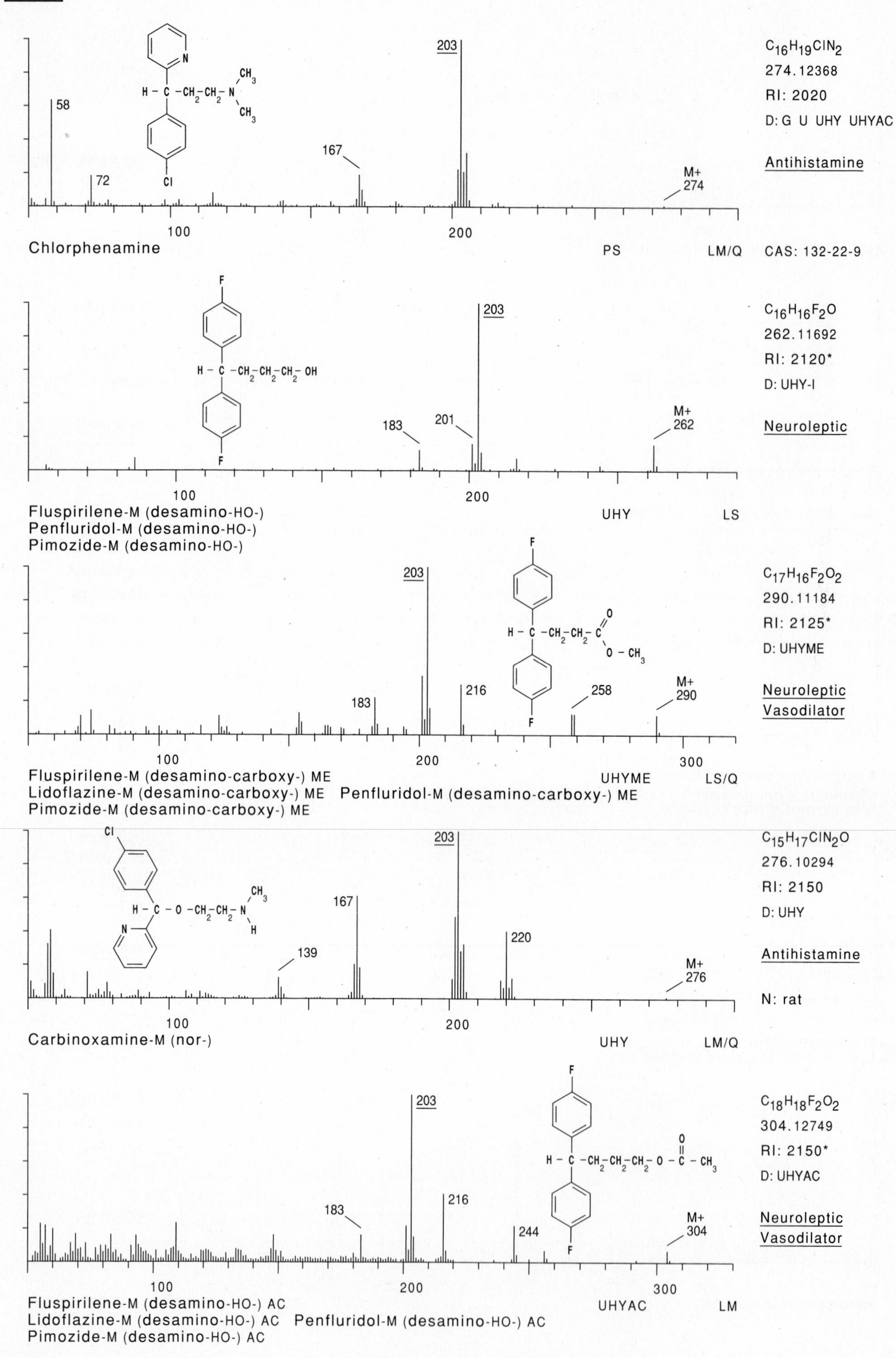

203
58
72
167
M+
274
100
200
H - C - CH2 - CH2 - N
CH3
CH3
N
Cl
Chlorphenamine
PS
LM/Q
C16H19ClN2
274.12368
RI: 2020
D: G U UHY UHYAC
Antihistamine
CAS: 132-22-9
203
183
201
M+
262
H - C - CH2 - CH2 - CH2 - OH
F
F
100
200
Fluspirilene-M (desamino-HO-)
Penfluridol-M (desamino-HO-)
Pimozide-M (desamino-HO-)
UHY
LS
C16H16F2O
262.11692
RI: 2120*
D: UHY-I
Neuroleptic
203
216
183
258
M+
290
H - C - CH2 - CH2 - C
O
O - CH3
F
F
100
200
300
Fluspirilene-M (desamino-carboxy-) ME
Lidoflazine-M (desamino-carboxy-) ME
Penfluridol-M (desamino-carboxy-) ME
Pimozide-M (desamino-carboxy-) ME
UHYME
LS/Q
C17H16F2O2
290.11184
RI: 2125*
D: UHYME
Neuroleptic
Vasodilator
203
167
220
139
M+
276
Cl
H - C - O - CH2 - CH2 - N
CH3
H
N
100
200
Carbinoxamine-M (nor-)
UHY
LM/Q
C15H17ClN2O
276.10294
RI: 2150
D: UHY
Antihistamine
N: rat
203
216
183
244
M+
304
H - C - CH2 - CH2 - CH2 - O - C - CH3
O
F
F
100
200
300
Fluspirilene-M (desamino-HO-) AC
Lidoflazine-M (desamino-HO-) AC
Penfluridol-M (desamino-HO-) AC
Pimozide-M (desamino-HO-) AC
UHYAC
LM
C18H18F2O2
304.12749
RI: 2150*
D: UHYAC
Neuroleptic
Vasodilator

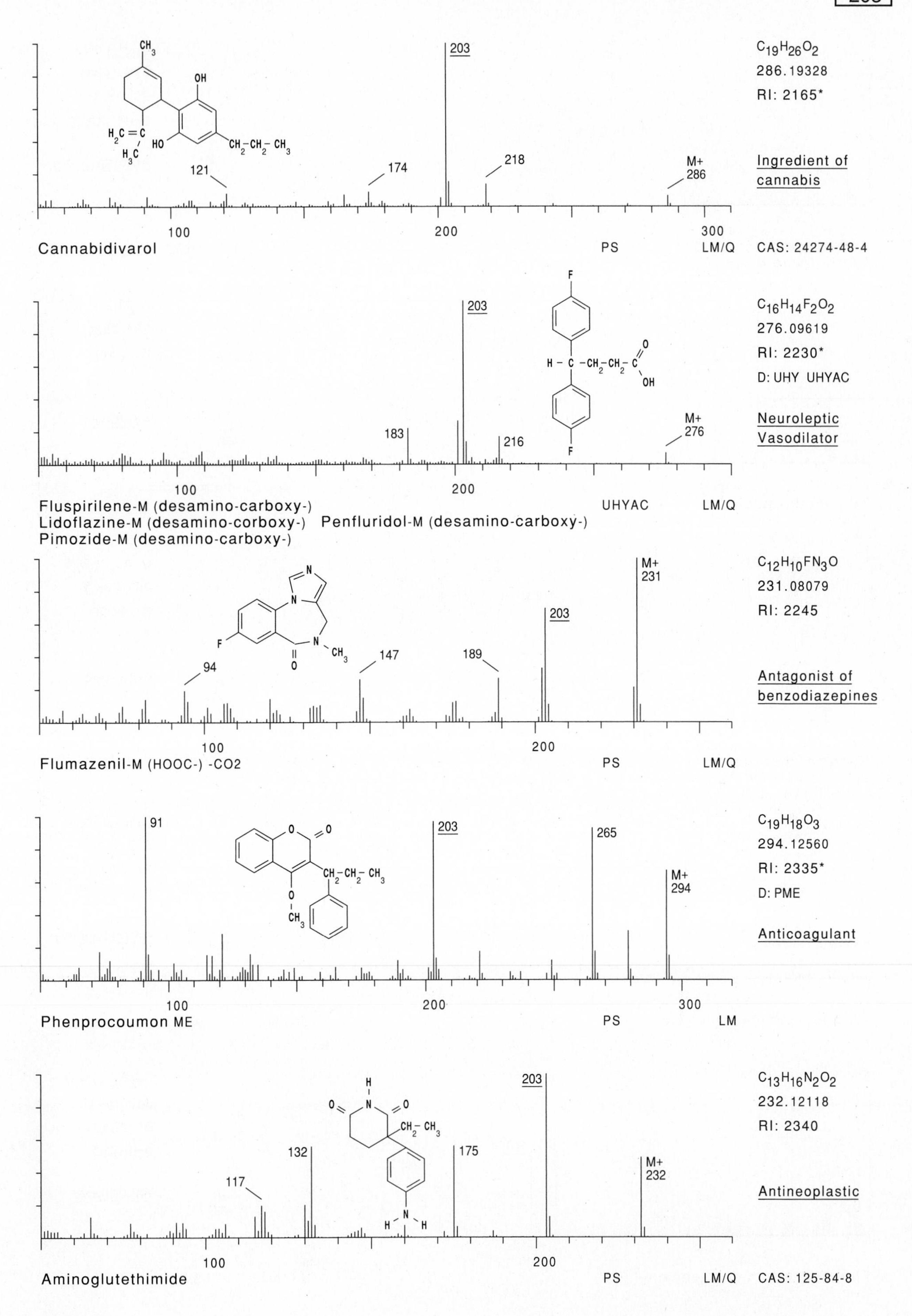

Cannabidivarol
$C_{19}H_{26}O_2$
286.19328
RI: 2165*
Ingredient of cannabis
PS LM/Q
CAS: 24274-48-4
Fluspirilene-M (desamino-carboxy-)
Lidoflazine-M (desamino-corboxy-) Penfluridol-M (desamino-carboxy-)
Pimozide-M (desamino-carboxy-)
$C_{16}H_{14}F_2O_2$
276.09619
RI: 2230*
D: UHY UHYAC
Neuroleptic
Vasodilator
UHYAC LM/Q
Flumazenil-M (HOOC-) -CO2
$C_{12}H_{10}FN_3O$
231.08079
RI: 2245
Antagonist of benzodiazepines
PS LM/Q
Phenprocoumon ME
$C_{19}H_{18}O_3$
294.12560
RI: 2335*
D: PME
Anticoagulant
PS LM
Aminoglutethimide
$C_{13}H_{16}N_2O_2$
232.12118
RI: 2340
Antineoplastic
PS LM/Q
CAS: 125-84-8

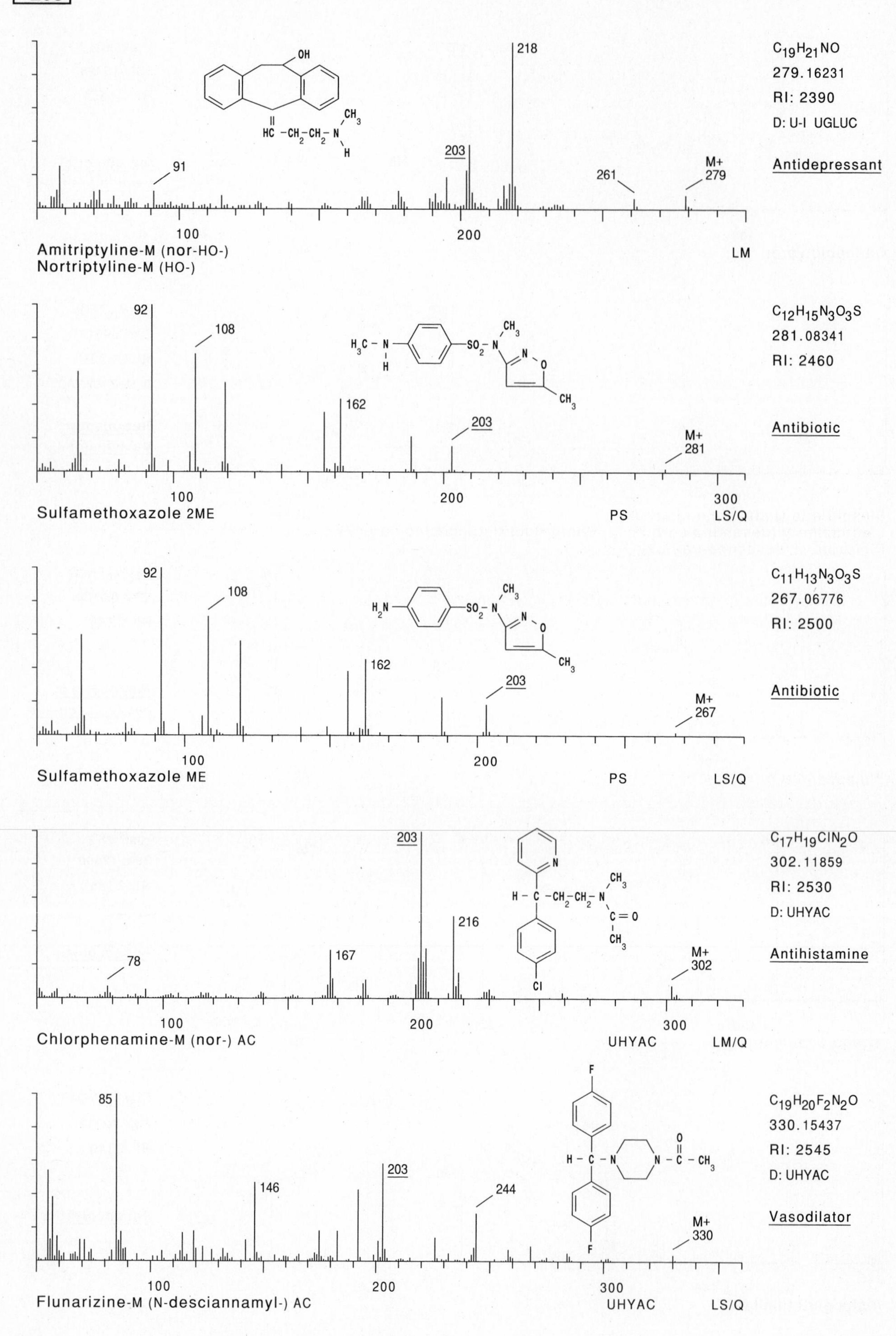

218
203
91
261
M+ 279
100
200
LM
Amitriptyline-M (nor-HO-)
Nortriptyline-M (HO-)
$C_{19}H_{21}NO$
279.16231
RI: 2390
D: U-I UGLUC
Antidepressant
92
108
162
203
M+ 281
100
200
300
PS
LS/Q
Sulfamethoxazole 2ME
$C_{12}H_{15}N_3O_3S$
281.08341
RI: 2460
Antibiotic
92
108
162
203
M+ 267
100
200
PS
LS/Q
Sulfamethoxazole ME
$C_{11}H_{13}N_3O_3S$
267.06776
RI: 2500
Antibiotic
203
216
167
78
M+ 302
100
200
300
UHYAC
LM/Q
Chlorphenamine-M (nor-) AC
$C_{17}H_{19}ClN_2O$
302.11859
RI: 2530
D: UHYAC
Antihistamine
85
203
146
244
M+ 330
100
200
300
UHYAC
LS/Q
Flunarizine-M (N-desciannamyl-) AC
$C_{19}H_{20}F_2N_2O$
330.15437
RI: 2545
D: UHYAC
Vasodilator

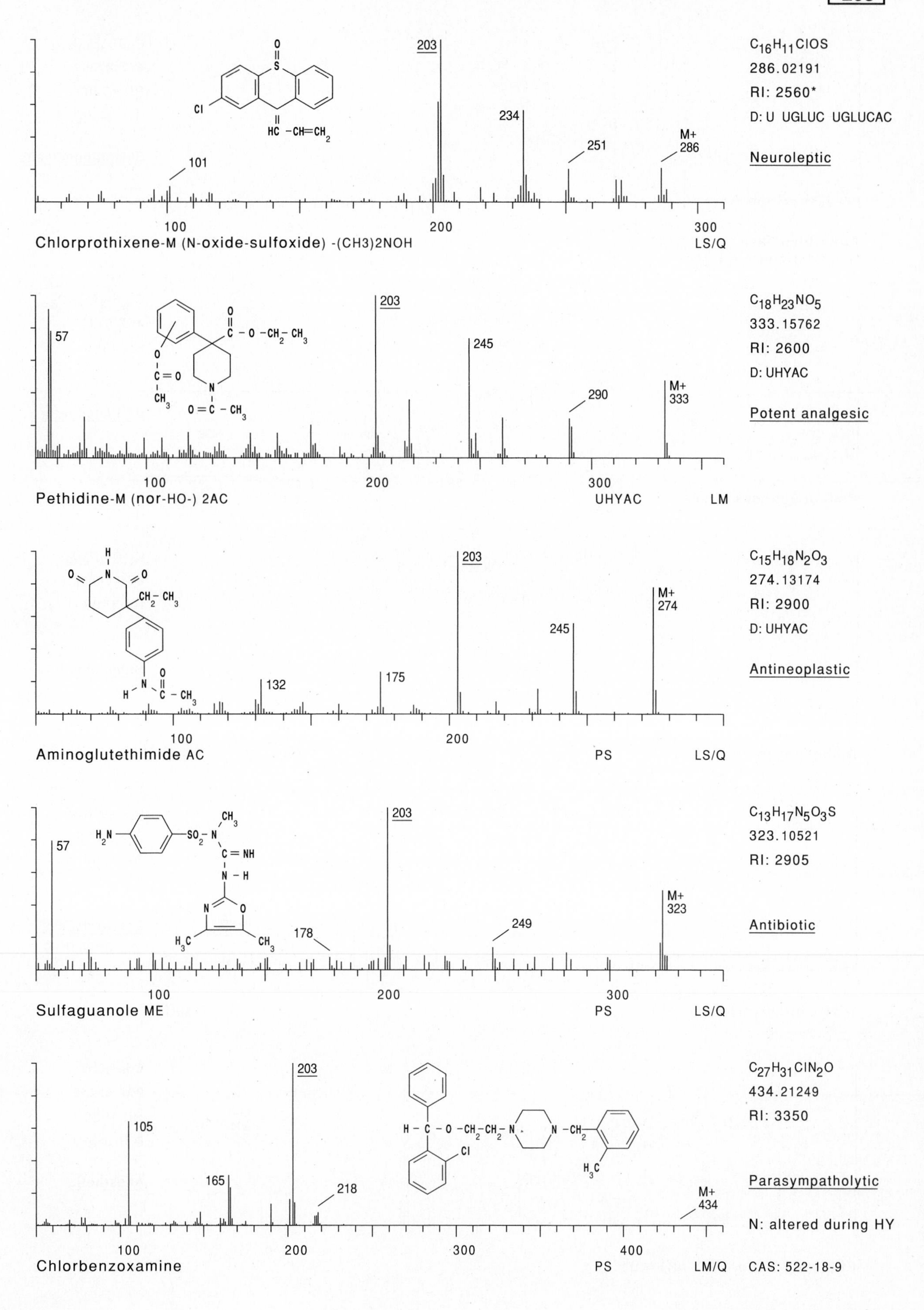

Chlorprothixene-M (N-oxide-sulfoxide) -(CH3)2NOH
LS/Q
$C_{16}H_{11}ClOS$
286.02191
RI: 2560*
D: U UGLUC UGLUCAC
Neuroleptic
203
234
251
M+ 286
101
Pethidine-M (nor-HO-) 2AC
UHYAC
LM
$C_{18}H_{23}NO_5$
333.15762
RI: 2600
D: UHYAC
Potent analgesic
57
203
245
290
M+ 333
Aminoglutethimide AC
PS
LS/Q
$C_{15}H_{18}N_2O_3$
274.13174
RI: 2900
D: UHYAC
Antineoplastic
132
175
203
245
M+ 274
Sulfaguanole ME
PS
LS/Q
$C_{13}H_{17}N_5O_3S$
323.10521
RI: 2905
Antibiotic
57
178
203
249
M+ 323
Chlorbenzoxamine
PS
LM/Q
$C_{27}H_{31}ClN_2O$
434.21249
RI: 3350
Parasympatholytic
N: altered during HY
CAS: 522-18-9
105
165
203
218
M+ 434

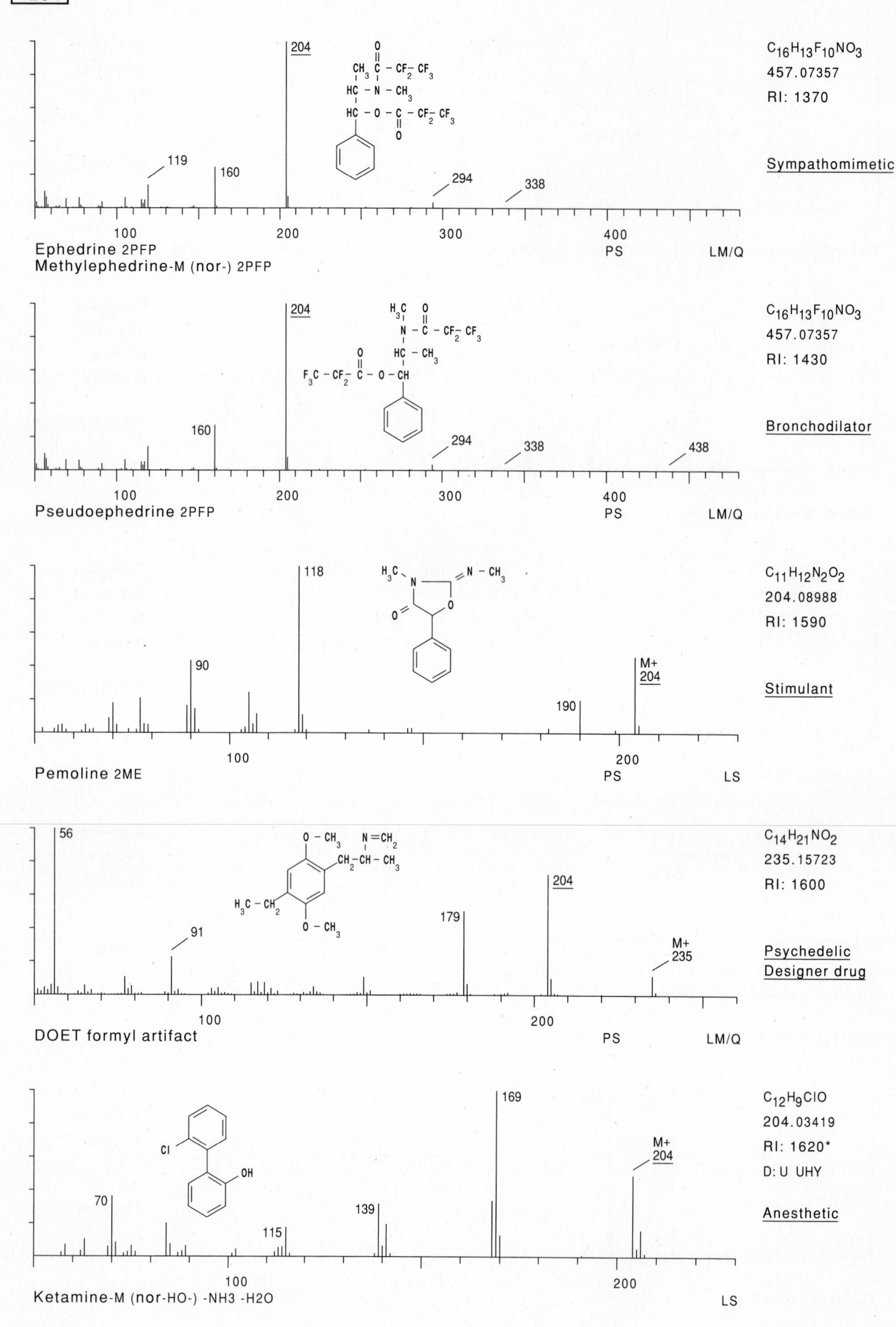
204
119
160
294
338
100
200
300
400
PS
LM/Q
Ephedrine 2PFP
Methylephedrine-M (nor-) 2PFP
$C_{16}H_{13}F_{10}NO_3$
457.07357
RI: 1370
Sympathomimetic
204
160
294
338
438
Pseudoephedrine 2PFP
$C_{16}H_{13}F_{10}NO_3$
457.07357
RI: 1430
Bronchodilator
118
90
M+
204
190
Pemoline 2ME
LS
$C_{11}H_{12}N_2O_2$
204.08988
RI: 1590
Stimulant
56
204
179
91
M+
235
DOET formyl artifact
$C_{14}H_{21}NO_2$
235.15723
RI: 1600
Psychedelic
Designer drug
169
M+
204
70
139
115
Ketamine-M (nor-HO-) -NH3 -H2O
$C_{12}H_9ClO$
204.03419
RI: 1620*
D: U UHY
Anesthetic

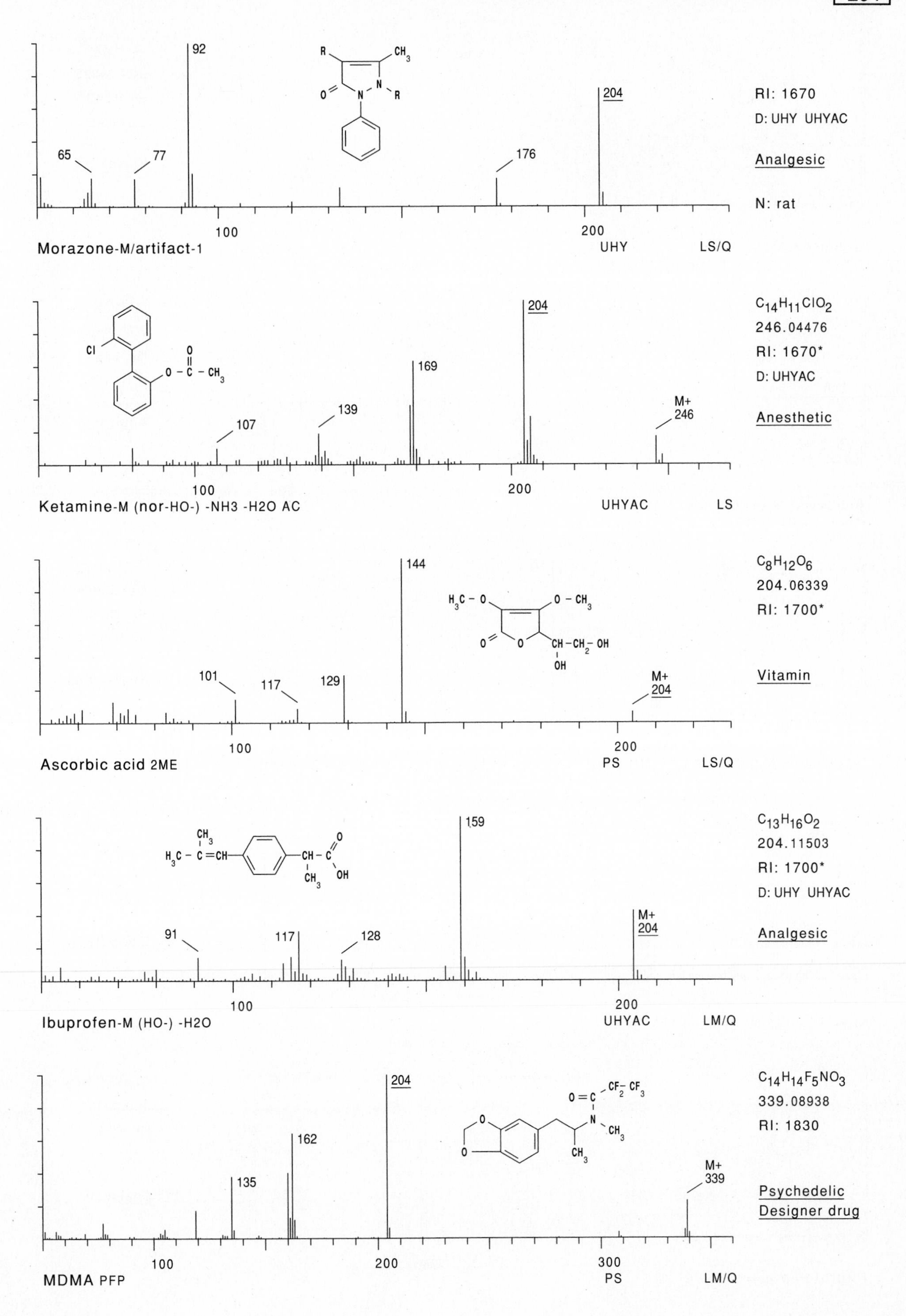
92
204
65
77
176
100
200
Morazone-M/artifact-1
UHY
LS/Q
RI: 1670
D: UHY UHYAC
Analgesic
N: rat
204
169
139
107
M+
246
100
200
Ketamine-M (nor-HO-) -NH3 -H2O AC
UHYAC
LS
C14H11ClO2
246.04476
RI: 1670*
D: UHYAC
Anesthetic
144
101
117
129
M+
204
100
200
Ascorbic acid 2ME
PS
LS/Q
C8H12O6
204.06339
RI: 1700*
Vitamin
159
91
117
128
M+
204
100
200
Ibuprofen-M (HO-) -H2O
UHYAC
LM/Q
C13H16O2
204.11503
RI: 1700*
D: UHY UHYAC
Analgesic
204
162
135
M+
339
100
200
300
MDMA PFP
PS
LM/Q
C14H14F5NO3
339.08938
RI: 1830
Psychedelic
Designer drug

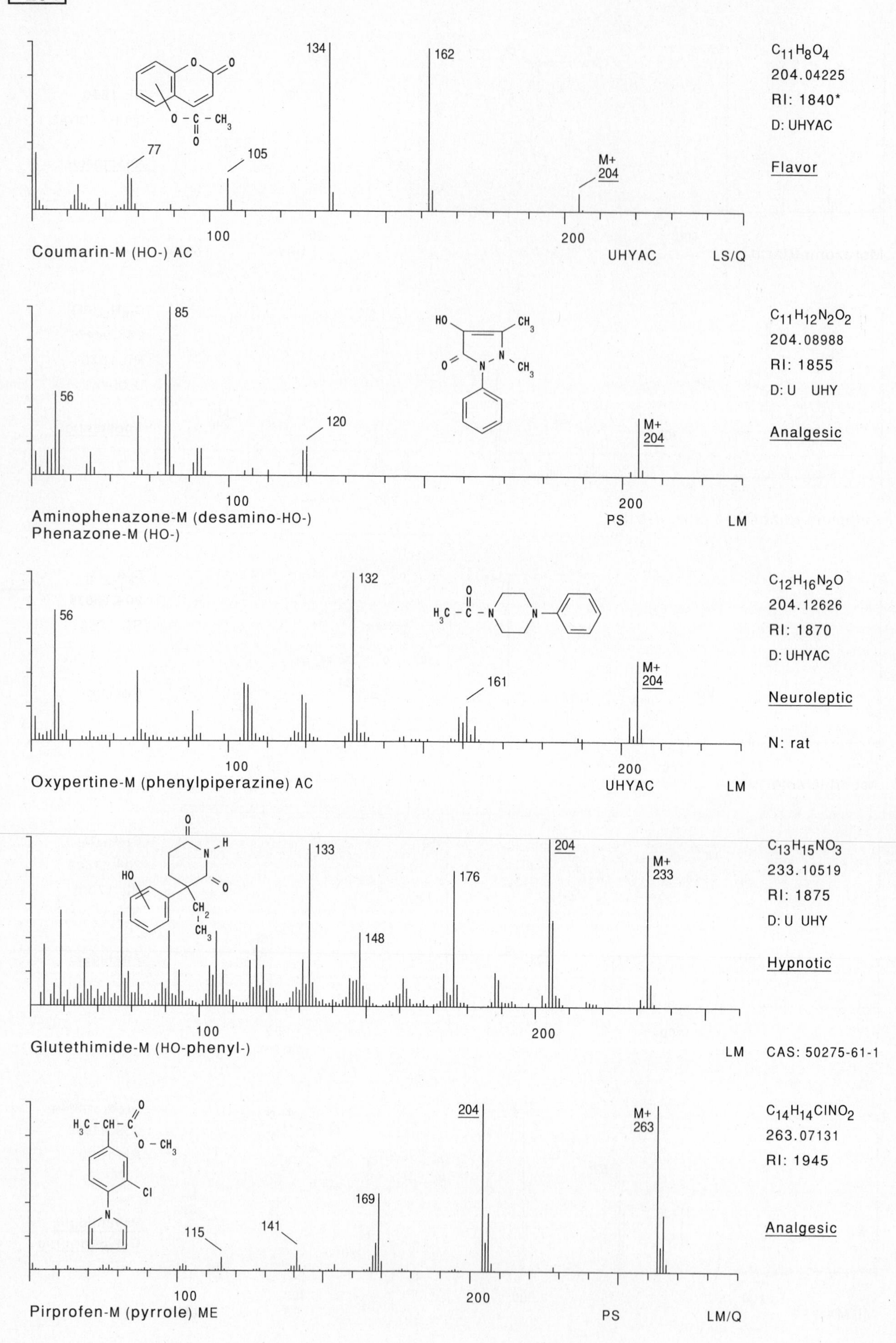
134
162
77
105
M+
204
100
200
Coumarin-M (HO-) AC
UHYAC
LS/Q
$C_{11}H_8O_4$
204.04225
RI: 1840*
D: UHYAC
Flavor
85
56
120
M+
204
Aminophenazone-M (desamino-HO-)
Phenazone-M (HO-)
PS
LM
$C_{11}H_{12}N_2O_2$
204.08988
RI: 1855
D: U UHY
Analgesic
132
56
161
Oxypertine-M (phenylpiperazine) AC
UHYAC
$C_{12}H_{16}N_2O$
204.12626
RI: 1870
D: UHYAC
Neuroleptic
N: rat
133
176
148
233
Glutethimide-M (HO-phenyl-)
$C_{13}H_{15}NO_3$
233.10519
RI: 1875
D: U UHY
Hypnotic
CAS: 50275-61-1
169
115
141
263
Pirprofen-M (pyrrole) ME
LM/Q
$C_{14}H_{14}ClNO_2$
263.07131
RI: 1945
Analgesic

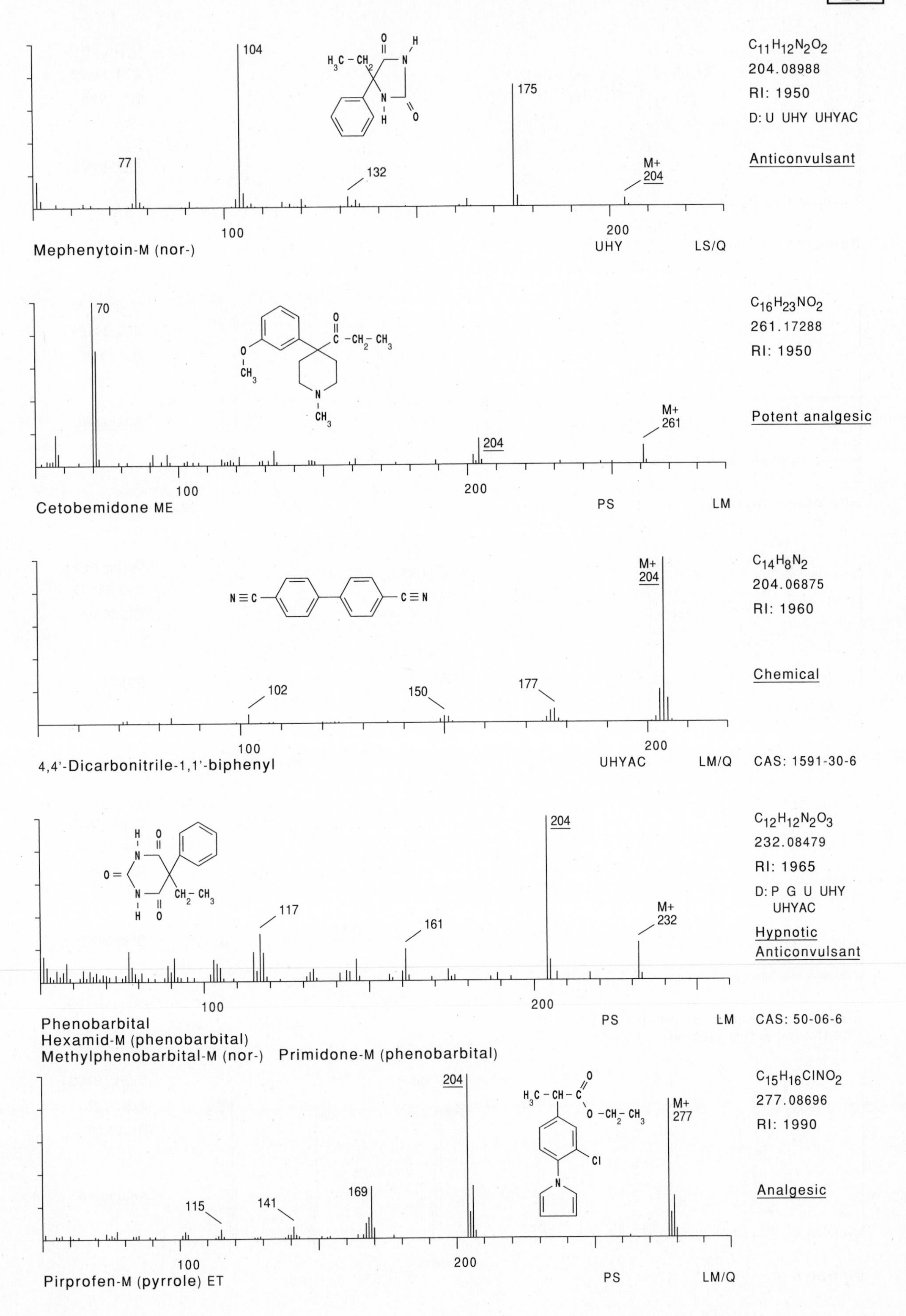

Mephenytoin-M (nor-)
UHY
LS/Q
$C_{11}H_{12}N_2O_2$
204.08988
RI: 1950
D: U UHY UHYAC
Anticonvulsant
Cetobemidone ME
PS
LM
$C_{16}H_{23}NO_2$
261.17288
RI: 1950
Potent analgesic
4,4'-Dicarbonitrile-1,1'-biphenyl
UHYAC
LM/Q
$C_{14}H_8N_2$
204.06875
RI: 1960
Chemical
CAS: 1591-30-6
Phenobarbital
Hexamid-M (phenobarbital)
Methylphenobarbital-M (nor-)
Primidone-M (phenobarbital)
PS
LM
$C_{12}H_{12}N_2O_3$
232.08479
RI: 1965
D: P G U UHY UHYAC
Hypnotic
Anticonvulsant
CAS: 50-06-6
Pirprofen-M (pyrrole) ET
PS
LM/Q
$C_{15}H_{16}ClNO_2$
277.08696
RI: 1990
Analgesic

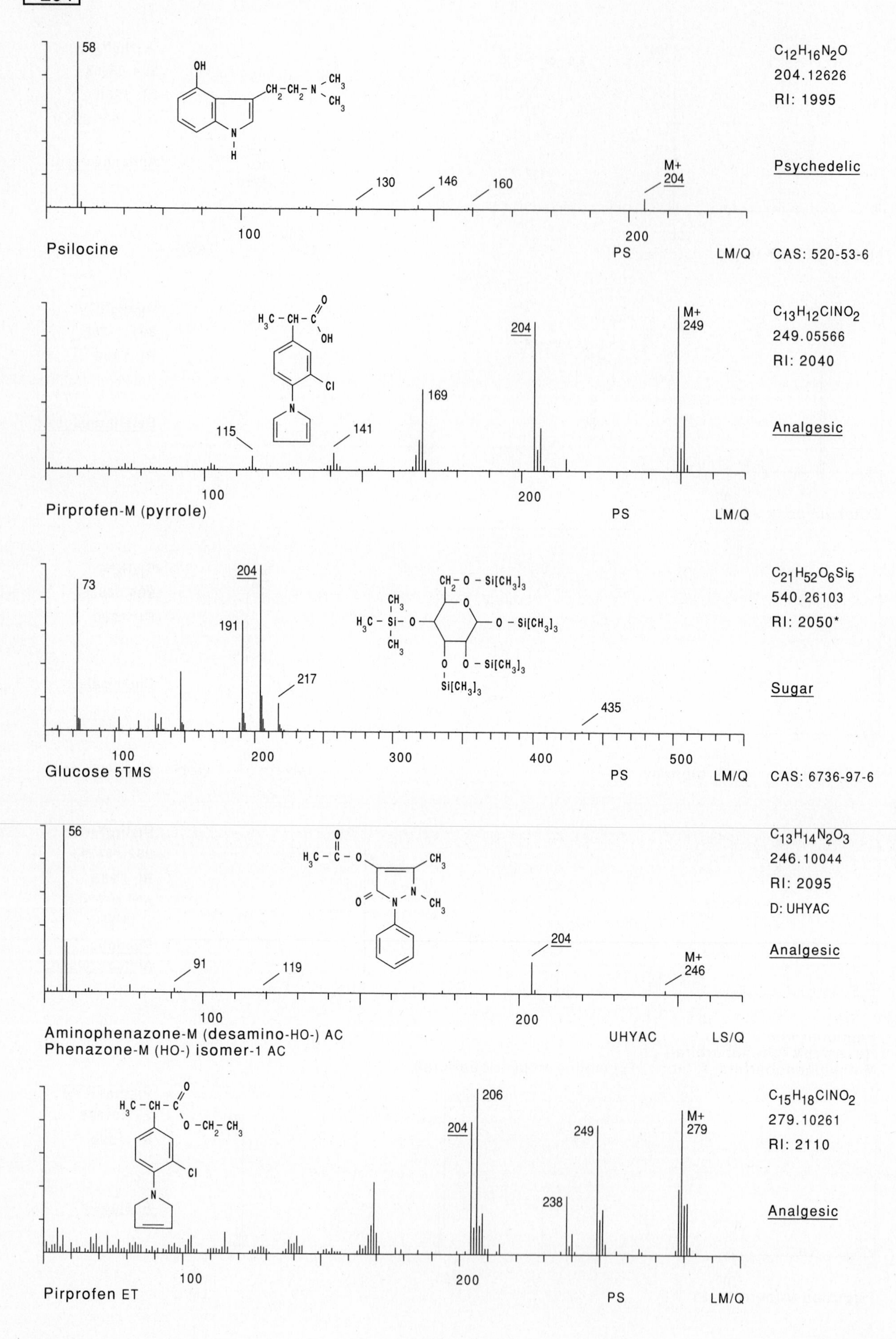

C12H16N2O
204.12626
RI: 1995
Psychedelic
Psilocine
PS
LM/Q
CAS: 520-53-6
C13H12ClNO2
249.05566
RI: 2040
Analgesic
Pirprofen-M (pyrrole)
PS
LM/Q
C21H52O6Si5
540.26103
RI: 2050*
Sugar
Glucose 5TMS
PS
LM/Q
CAS: 6736-97-6
C13H14N2O3
246.10044
RI: 2095
D: UHYAC
Analgesic
Aminophenazone-M (desamino-HO-) AC
Phenazone-M (HO-) isomer-1 AC
UHYAC
LS/Q
C15H18ClNO2
279.10261
RI: 2110
Analgesic
Pirprofen ET
PS
LM/Q

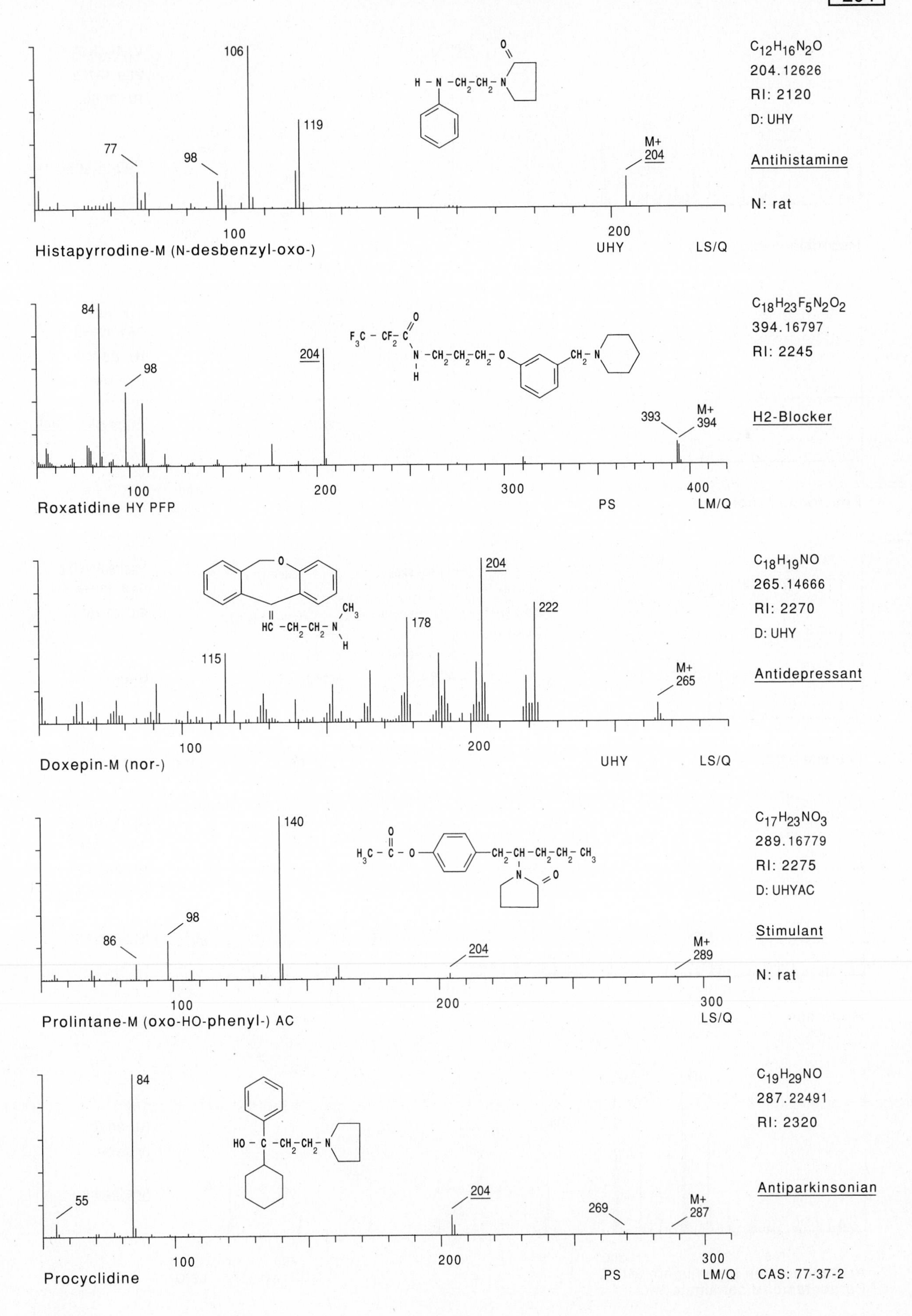

106
119
77
98
M+
204
100
200
Histapyrrodine-M (N-desbenzyl-oxo-)
UHY
LS/Q
C12H16N2O
204.12626
RI: 2120
D: UHY
Antihistamine
N: rat
84
98
204
393
M+
394
100
200
300
400
Roxatidine HY PFP
PS
LM/Q
C18H23F5N2O2
394.16797
RI: 2245
H2-Blocker
204
222
178
115
M+
265
100
200
Doxepin-M (nor-)
UHY
LS/Q
C18H19NO
265.14666
RI: 2270
D: UHY
Antidepressant
140
98
86
204
M+
289
100
200
300
Prolintane-M (oxo-HO-phenyl-) AC
LS/Q
C17H23NO3
289.16779
RI: 2275
D: UHYAC
Stimulant
N: rat
84
55
204
269
M+
287
100
200
300
Procyclidine
PS
LM/Q
C19H29NO
287.22491
RI: 2320
Antiparkinsonian
CAS: 77-37-2

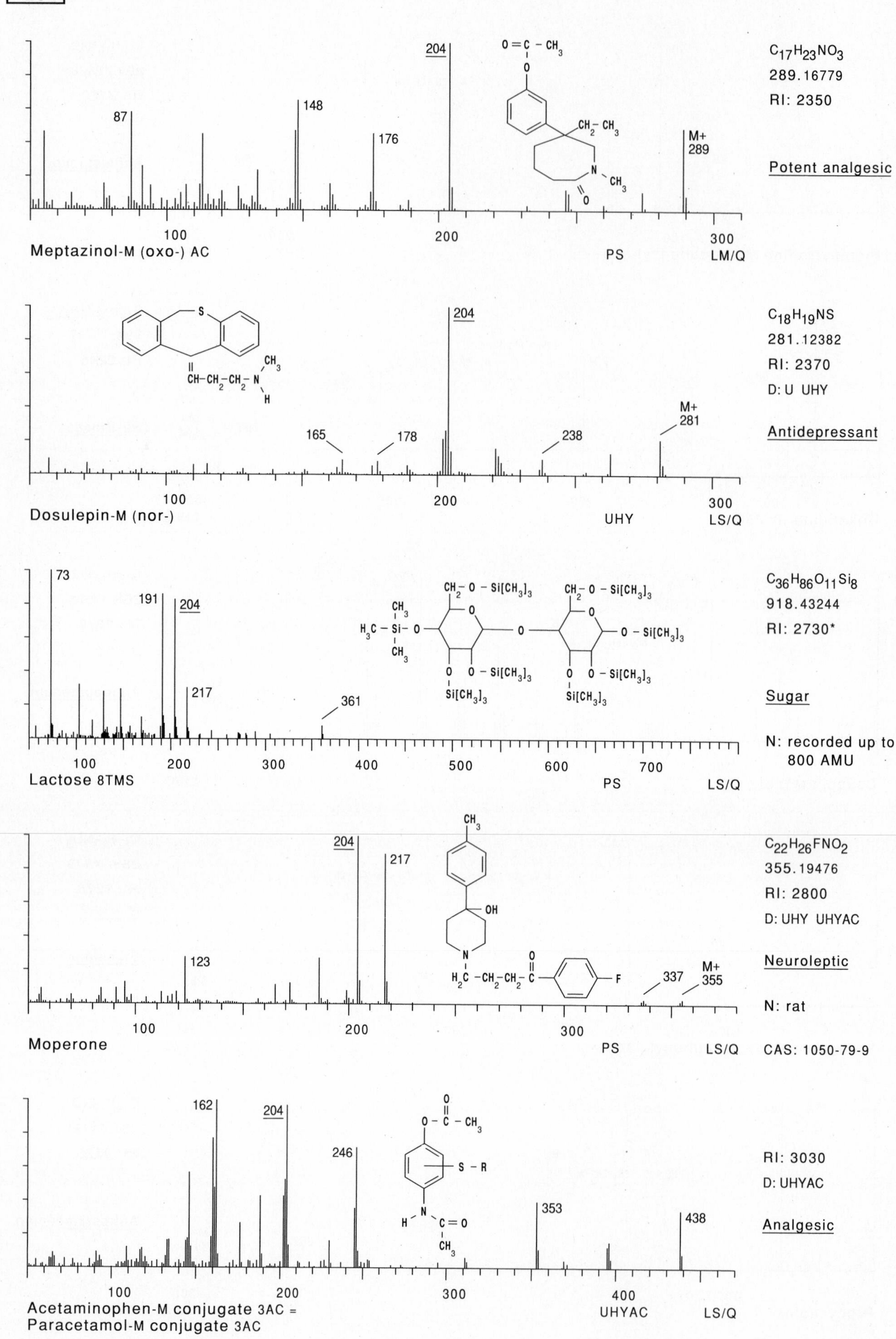
Meptazinol-M (oxo-) AC
PS
LM/Q
$C_{17}H_{23}NO_3$
289.16779
RI: 2350
Potent analgesic
Dosulepin-M (nor-)
UHY
LS/Q
$C_{18}H_{19}NS$
281.12382
RI: 2370
D: U UHY
Antidepressant
Lactose 8TMS
PS
LS/Q
$C_{36}H_{86}O_{11}Si_8$
918.43244
RI: 2730*
Sugar
N: recorded up to 800 AMU
Moperone
PS
LS/Q
$C_{22}H_{26}FNO_2$
355.19476
RI: 2800
D: UHY UHYAC
Neuroleptic
N: rat
CAS: 1050-79-9
Acetaminophen-M conjugate 3AC =
Paracetamol-M conjugate 3AC
UHYAC
LS/Q
RI: 3030
D: UHYAC
Analgesic

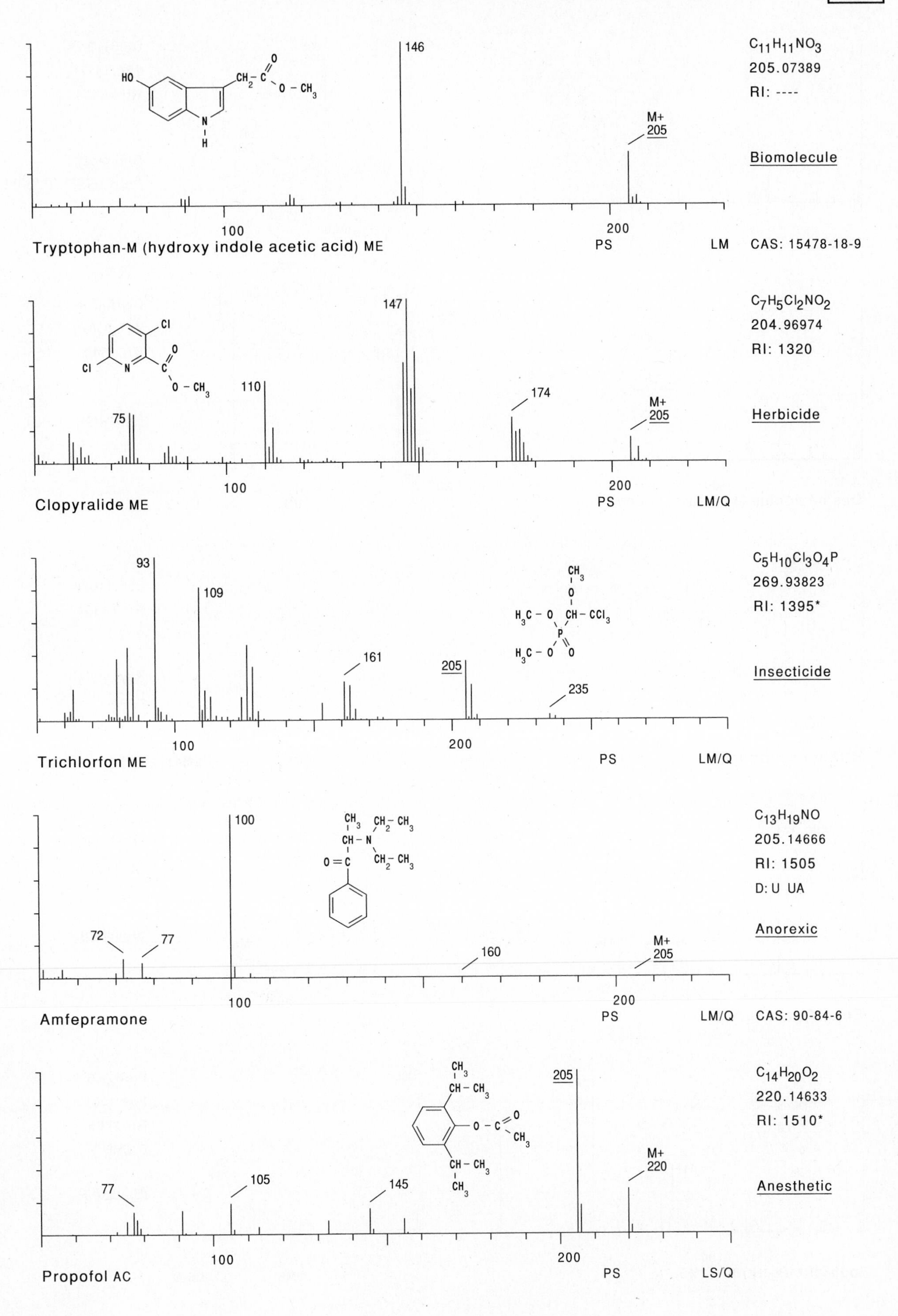

146
M+
205
100
200
Tryptophan-M (hydroxy indole acetic acid) ME
PS
LM
C11H11NO3
205.07389
RI: ----
Biomolecule
CAS: 15478-18-9
147
110
174
75
M+
205
100
200
Clopyralide ME
PS
LM/Q
C7H5Cl2NO2
204.96974
RI: 1320
Herbicide
93
109
161
205
235
100
200
Trichlorfon ME
PS
LM/Q
C5H10Cl3O4P
269.93823
RI: 1395*
Insecticide
100
72
77
160
M+
205
100
200
Amfepramone
PS
LM/Q
C13H19NO
205.14666
RI: 1505
D: U UA
Anorexic
CAS: 90-84-6
205
M+
220
77
105
145
100
200
Propofol AC
PS
LS/Q
C14H20O2
220.14633
RI: 1510*
Anesthetic

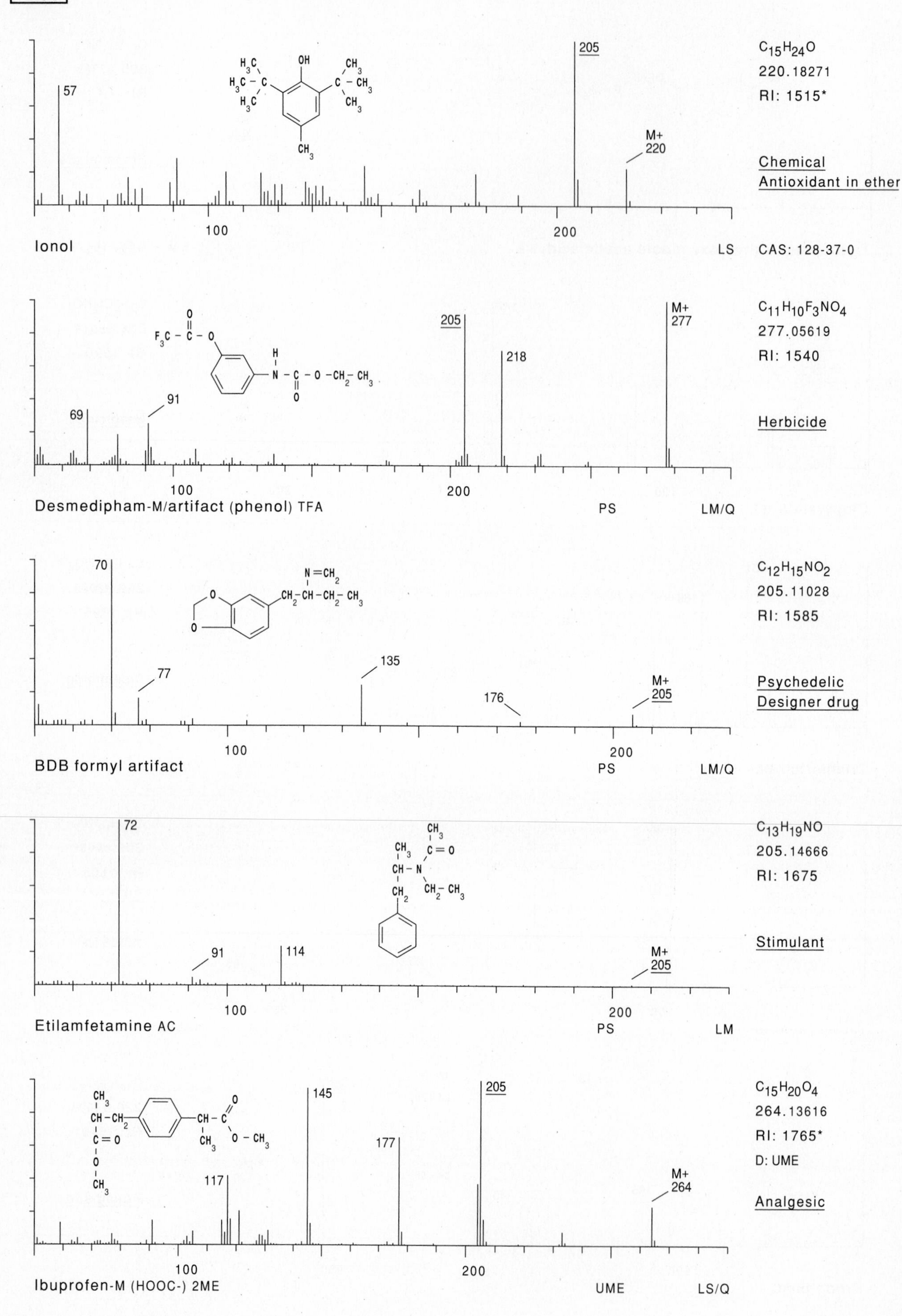

57
205
M+
220
100
200
Ionol
LS
C15H24O
220.18271
RI: 1515*
Chemical
Antioxidant in ether
CAS: 128-37-0
69
91
205
218
M+
277
100
200
Desmedipham-M/artifact (phenol) TFA
PS
LM/Q
C11H10F3NO4
277.05619
RI: 1540
Herbicide
70
77
135
176
M+
205
100
200
BDB formyl artifact
PS
LM/Q
C12H15NO2
205.11028
RI: 1585
Psychedelic
Designer drug
72
91
114
M+
205
100
200
Etilamfetamine AC
PS
LM
C13H19NO
205.14666
RI: 1675
Stimulant
117
145
177
205
M+
264
100
200
Ibuprofen-M (HOOC-) 2ME
UME
LS/Q
C15H20O4
264.13616
RI: 1765*
D: UME
Analgesic

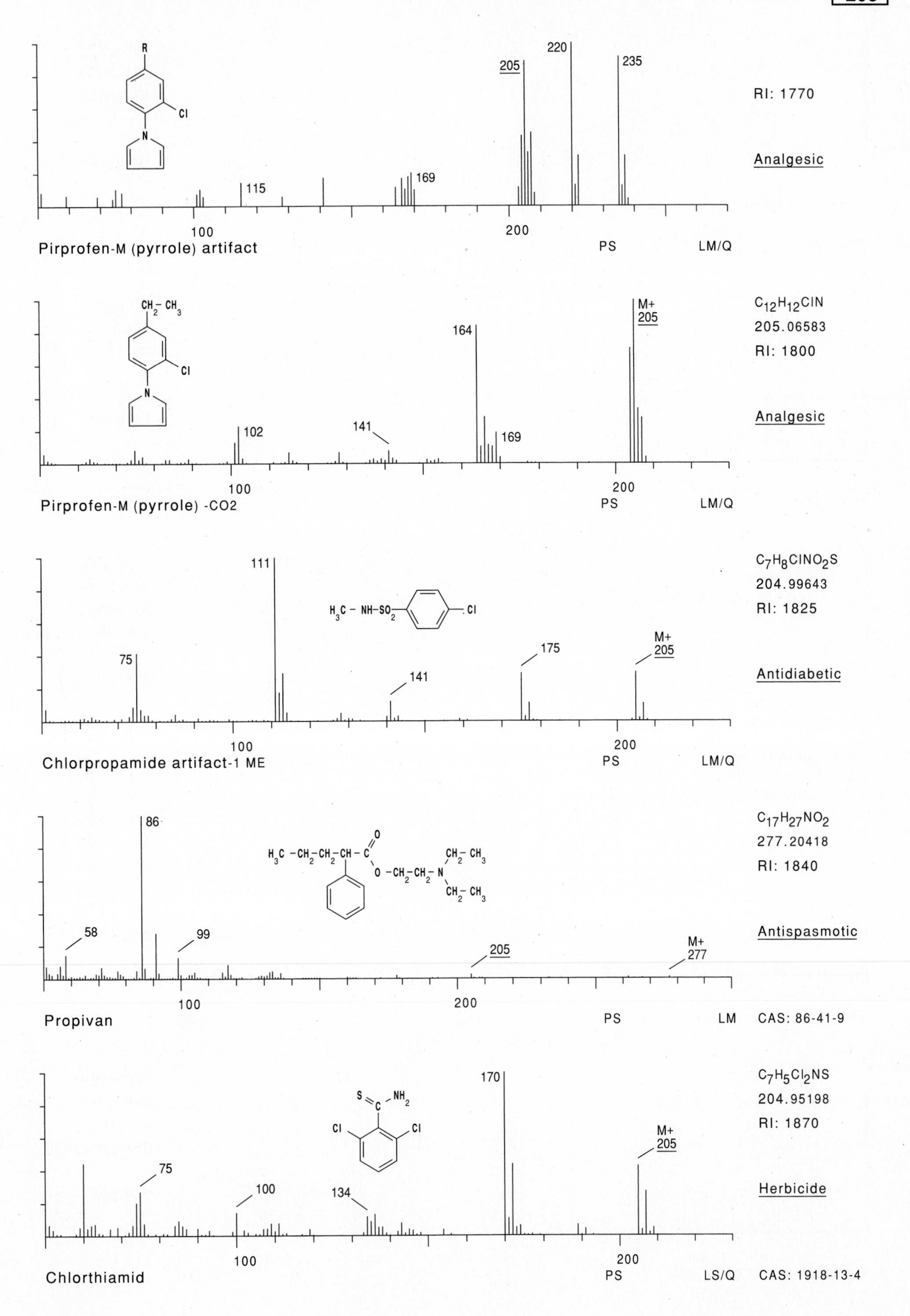

Pirprofen-M (pyrrole) artifact
RI: 1770
Analgesic
PS
LM/Q
Pirprofen-M (pyrrole) -CO2
$C_{12}H_{12}ClN$
205.06583
RI: 1800
Analgesic
PS
LM/Q
Chlorpropamide artifact-1 ME
$C_7H_8ClNO_2S$
204.99643
RI: 1825
Antidiabetic
PS
LM/Q
Propivan
$C_{17}H_{27}NO_2$
277.20418
RI: 1840
Antispasmotic
PS
LM
CAS: 86-41-9
Chlorthiamid
$C_7H_5Cl_2NS$
204.95198
RI: 1870
Herbicide
PS
LS/Q
CAS: 1918-13-4

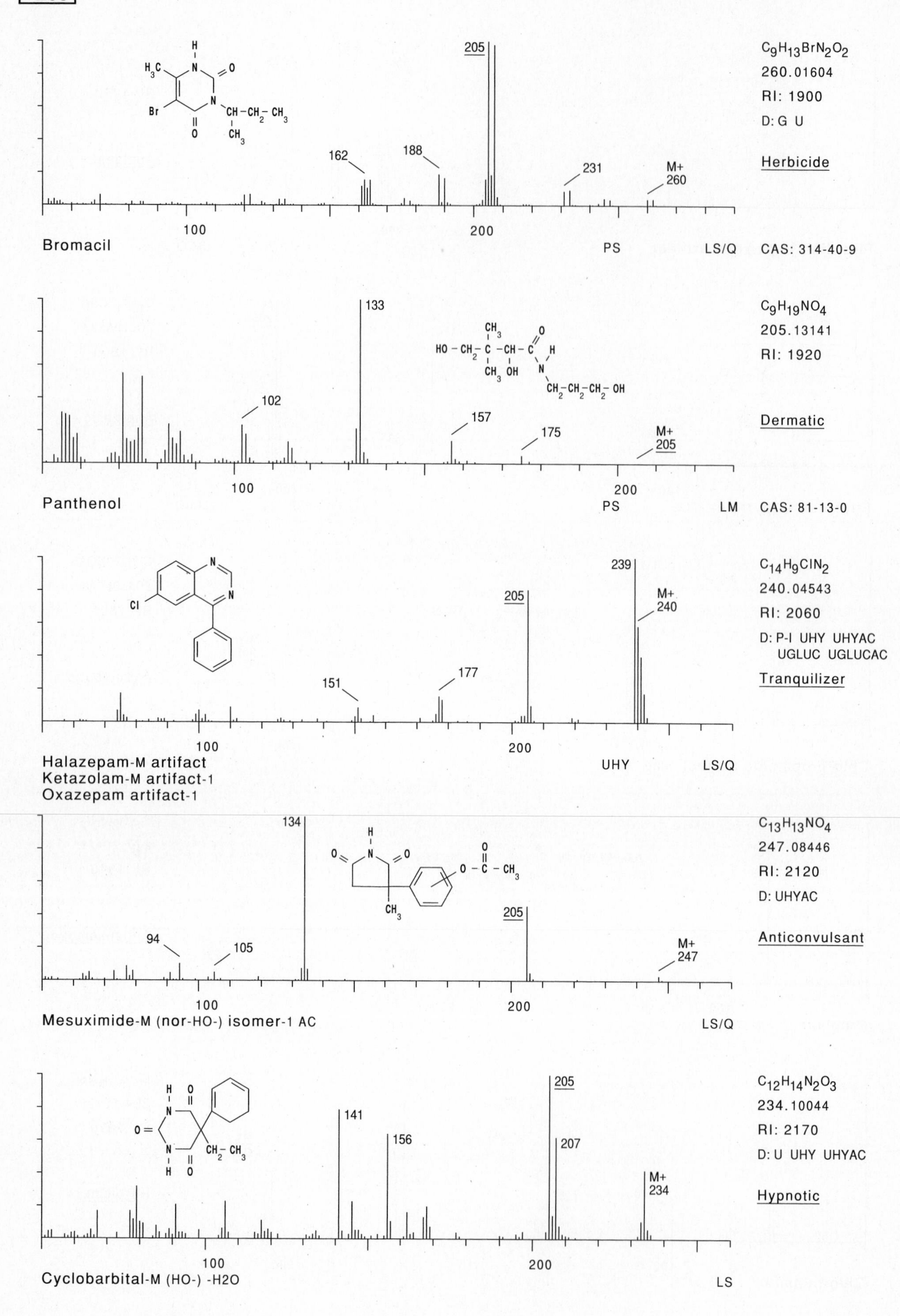
205
162
188
231
M+
260
100
200
Bromacil
PS
LS/Q
$C_9H_{13}BrN_2O_2$
260.01604
RI: 1900
D: G U
Herbicide
CAS: 314-40-9
133
102
157
175
M+
205
100
200
Panthenol
PS
LM
$C_9H_{19}NO_4$
205.13141
RI: 1920
Dermatic
CAS: 81-13-0
239
M+
240
205
151
177
100
200
Halazepam-M artifact
Ketazolam-M artifact-1
Oxazepam artifact-1
UHY
LS/Q
$C_{14}H_9ClN_2$
240.04543
RI: 2060
D: P-I UHY UHYAC
UGLUC UGLUCAC
Tranquilizer
134
205
94
105
M+
247
100
200
Mesuximide-M (nor-HO-) isomer-1 AC
LS/Q
$C_{13}H_{13}NO_4$
247.08446
RI: 2120
D: UHYAC
Anticonvulsant
205
141
156
207
M+
234
100
200
Cyclobarbital-M (HO-) -H2O
LS
$C_{12}H_{14}N_2O_3$
234.10044
RI: 2170
D: U UHY UHYAC
Hypnotic

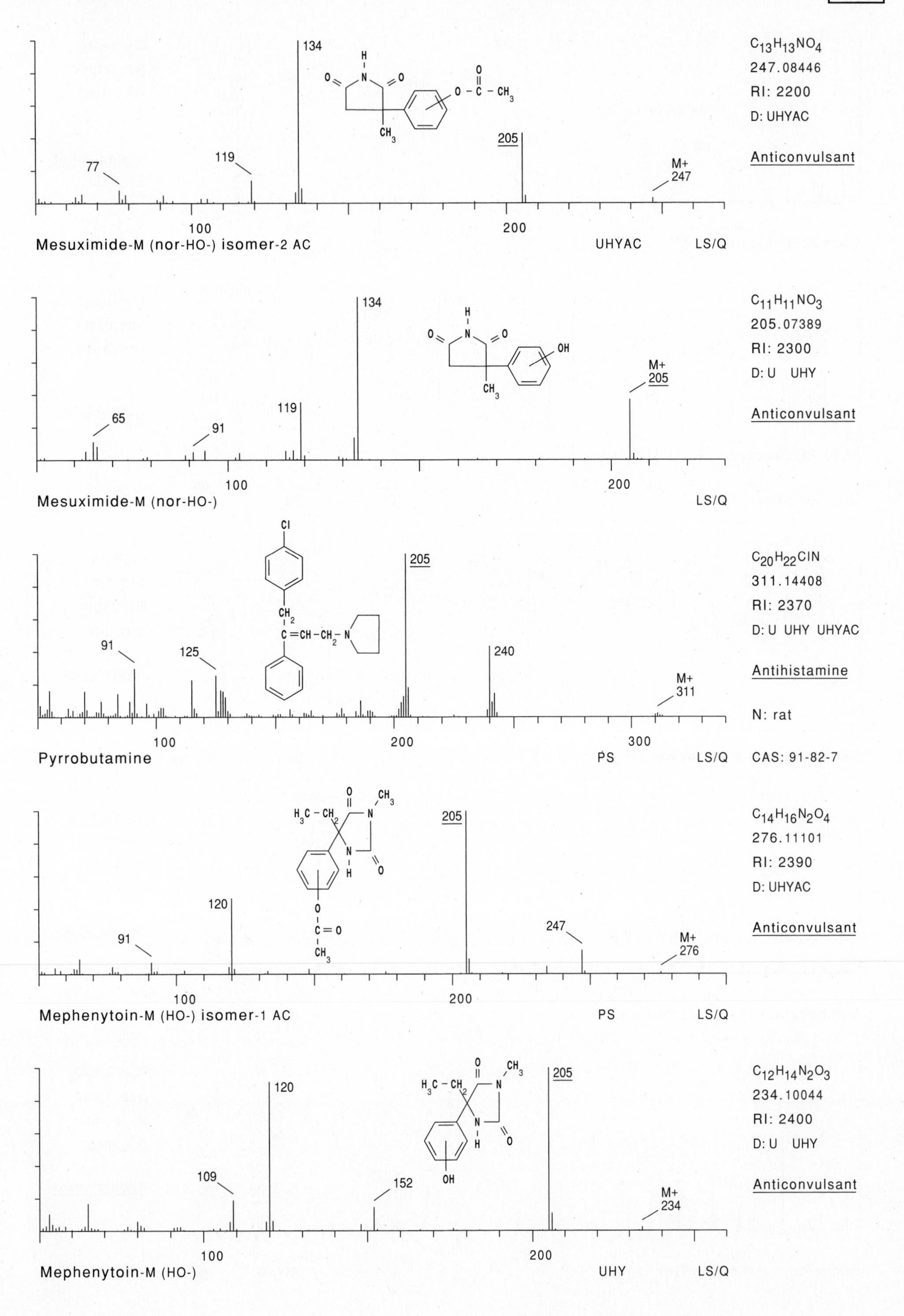

134
119
77
205
M+
247
100
200
Mesuximide-M (nor-HO-) isomer-2 AC
UHYAC
LS/Q
$C_{13}H_{13}NO_4$
247.08446
RI: 2200
D: UHYAC
Anticonvulsant
134
119
65
91
M+
205
100
200
Mesuximide-M (nor-HO-)
LS/Q
$C_{11}H_{11}NO_3$
205.07389
RI: 2300
D: U UHY
Anticonvulsant
Cl
205
240
91
125
M+
311
100
200
300
Pyrrobutamine
PS
LS/Q
$C_{20}H_{22}ClN$
311.14408
RI: 2370
D: U UHY UHYAC
Antihistamine
N: rat
CAS: 91-82-7
205
120
91
247
M+
276
100
200
Mephenytoin-M (HO-) isomer-1 AC
PS
LS/Q
$C_{14}H_{16}N_2O_4$
276.11101
RI: 2390
D: UHYAC
Anticonvulsant
120
205
109
152
M+
234
100
200
Mephenytoin-M (HO-)
UHY
LS/Q
$C_{12}H_{14}N_2O_3$
234.10044
RI: 2400
D: U UHY
Anticonvulsant

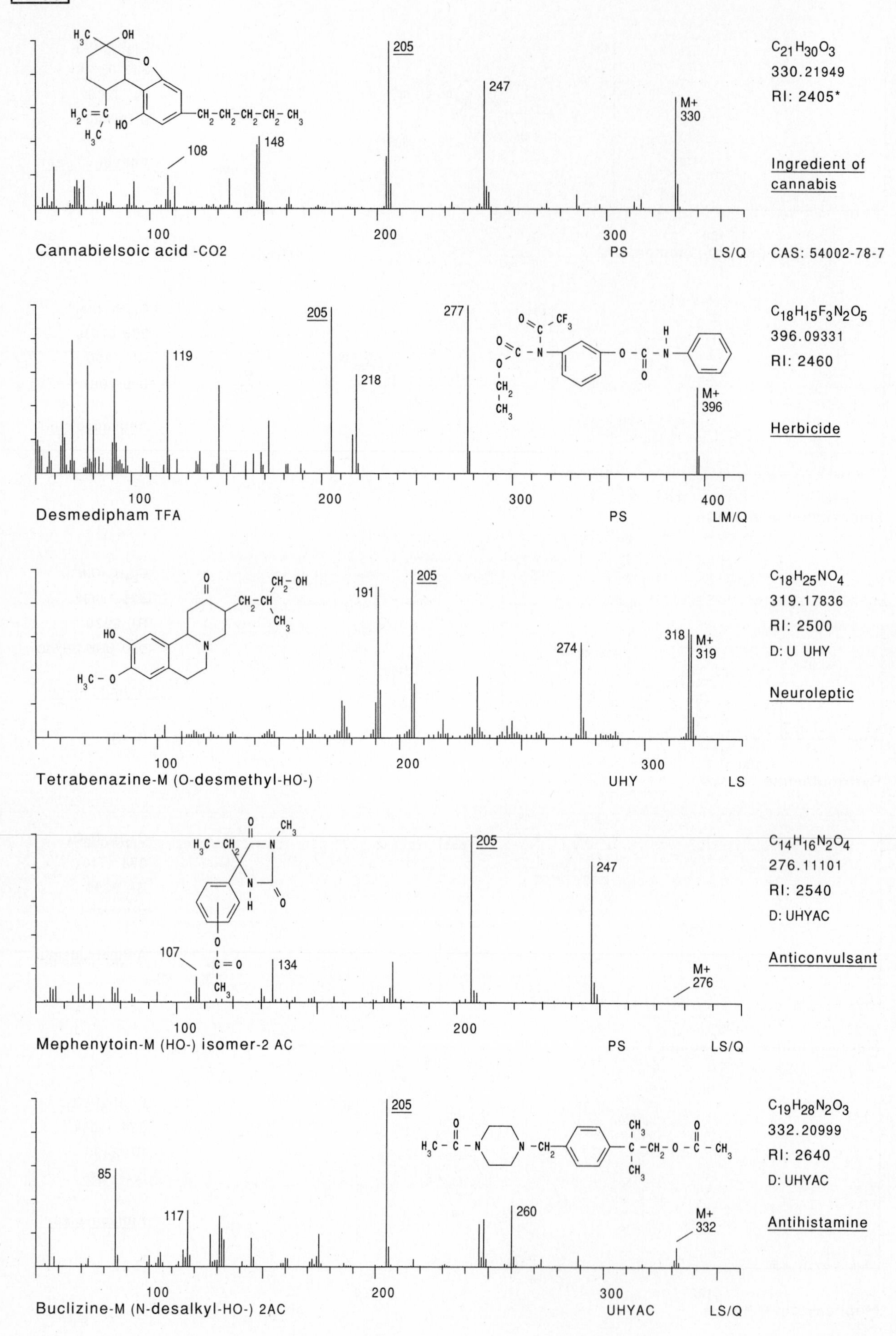

$C_{21}H_{30}O_3$
330.21949
RI: 2405*

Ingredient of cannabis

CAS: 54002-78-7

$C_{18}H_{15}F_3N_2O_5$
396.09331
RI: 2460

Herbicide

$C_{18}H_{25}NO_4$
319.17836
RI: 2500
D: U UHY

Neuroleptic

$C_{14}H_{16}N_2O_4$
276.11101
RI: 2540
D: UHYAC

Anticonvulsant

$C_{19}H_{28}N_2O_3$
332.20999
RI: 2640
D: UHYAC

Antihistamine

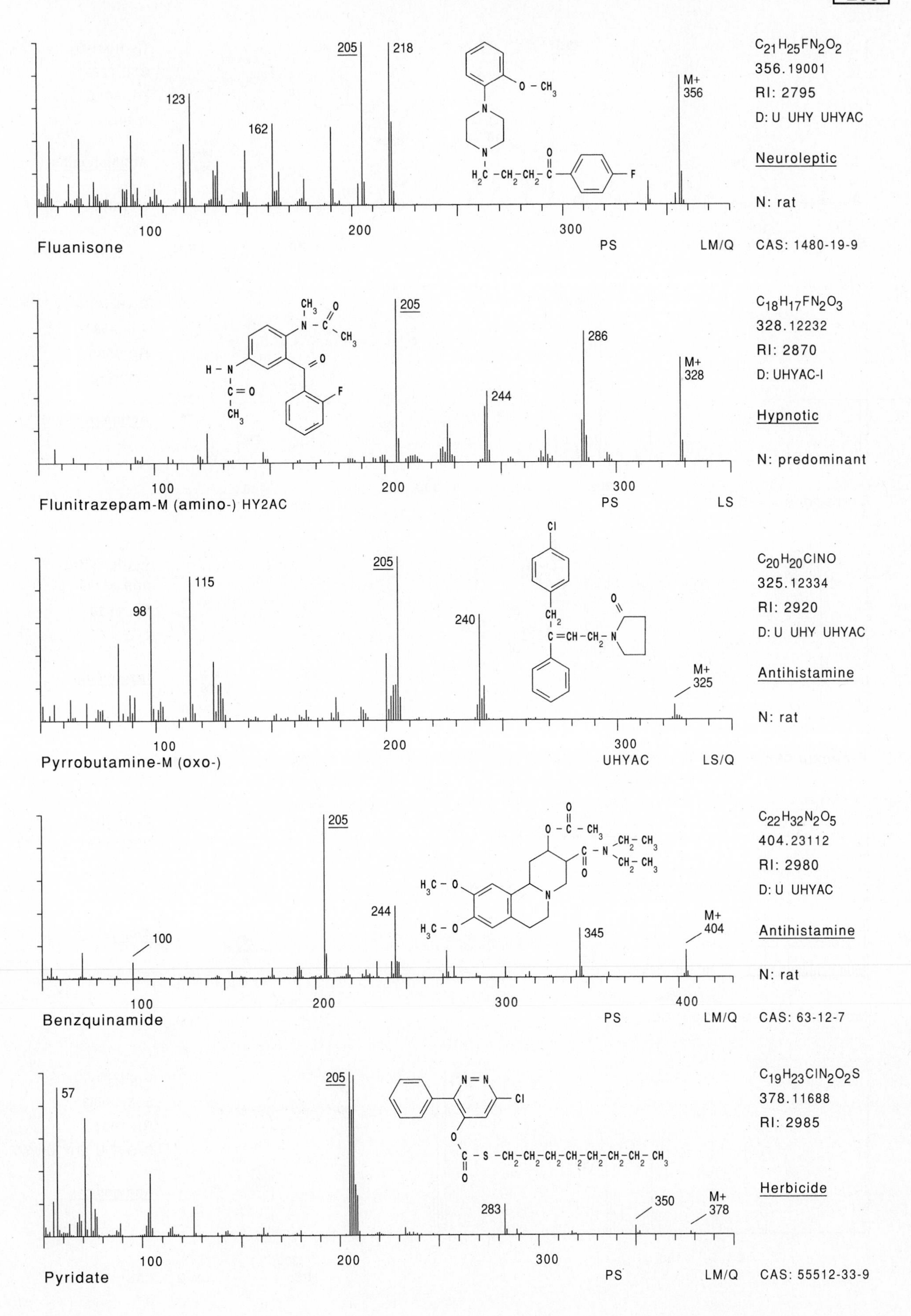
205
218
123
162
M+
356
100
200
300
Fluanisone
PS
LM/Q
C21H25FN2O2
356.19001
RI: 2795
D: U UHY UHYAC
Neuroleptic
N: rat
CAS: 1480-19-9
205
286
M+
328
244
100
200
300
Flunitrazepam-M (amino-) HY2AC
PS
LS
C18H17FN2O3
328.12232
RI: 2870
D: UHYAC-I
Hypnotic
N: predominant
205
115
98
240
M+
325
100
200
300
Pyrrobutamine-M (oxo-)
UHYAC
LS/Q
C20H20ClNO
325.12334
RI: 2920
D: U UHY UHYAC
Antihistamine
N: rat
205
244
100
345
M+
404
100
200
300
400
Benzquinamide
PS
LM/Q
C22H32N2O5
404.23112
RI: 2980
D: U UHYAC
Antihistamine
N: rat
CAS: 63-12-7
205
57
283
350
M+
378
100
200
300
Pyridate
PS
LM/Q
C19H23ClN2O2S
378.11688
RI: 2985
Herbicide
CAS: 55512-33-9

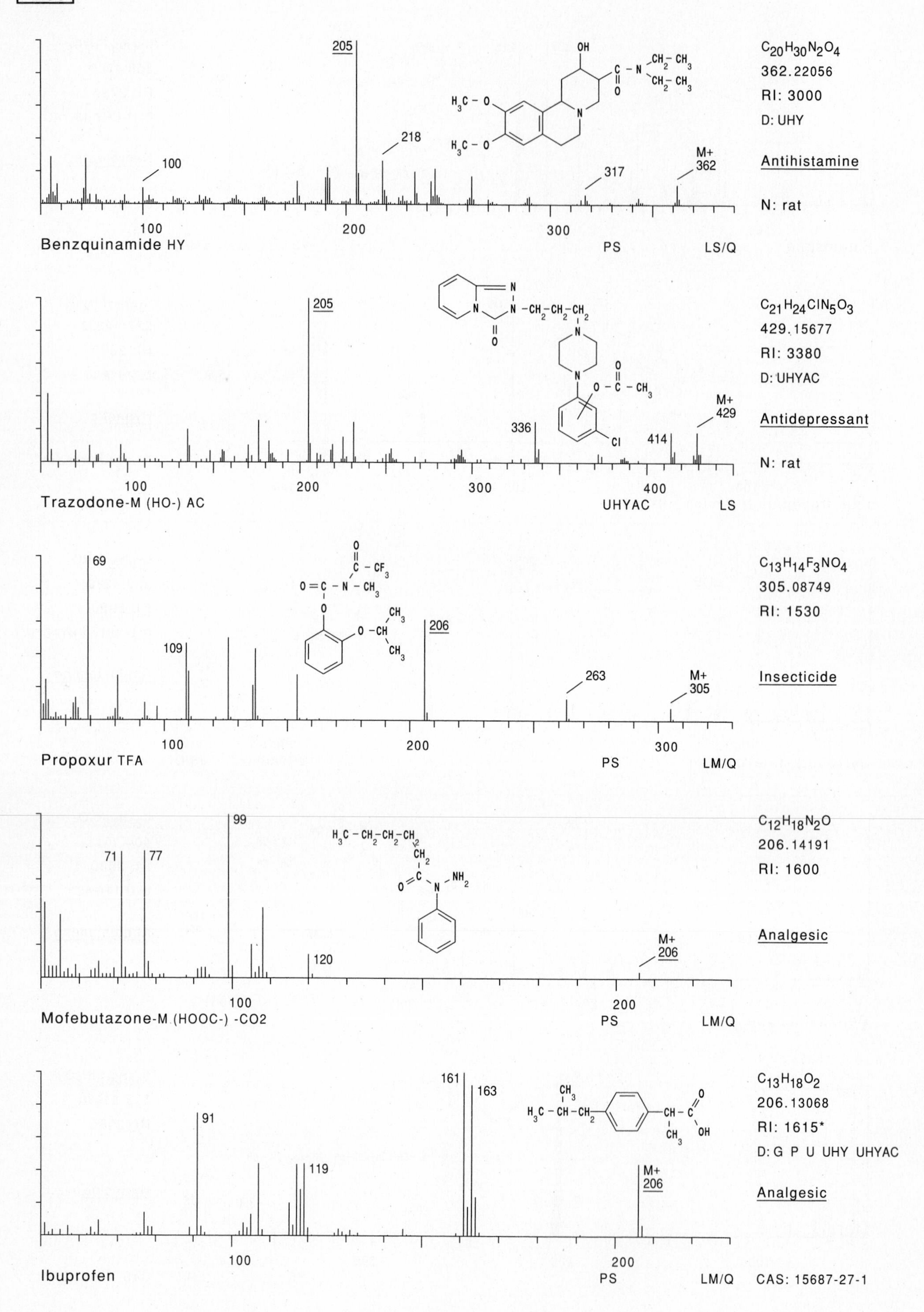
205
218
100
317
M+
362
100
200
300
Benzquinamide HY
PS
LS/Q
C20H30N2O4
362.22056
RI: 3000
D: UHY
Antihistamine
N: rat
205
336
414
M+
429
100
200
300
400
Trazodone-M (HO-) AC
UHYAC
LS
C21H24ClN5O3
429.15677
RI: 3380
D: UHYAC
Antidepressant
N: rat
69
109
206
263
M+
305
100
200
300
Propoxur TFA
PS
LM/Q
C13H14F3NO4
305.08749
RI: 1530
Insecticide
99
71
77
120
M+
206
100
200
Mofebutazone-M (HOOC-) -CO2
PS
LM/Q
C12H18N2O
206.14191
RI: 1600
Analgesic
161
163
91
119
M+
206
100
200
Ibuprofen
PS
LM/Q
C13H18O2
206.13068
RI: 1615*
D: G P U UHY UHYAC
Analgesic
CAS: 15687-27-1

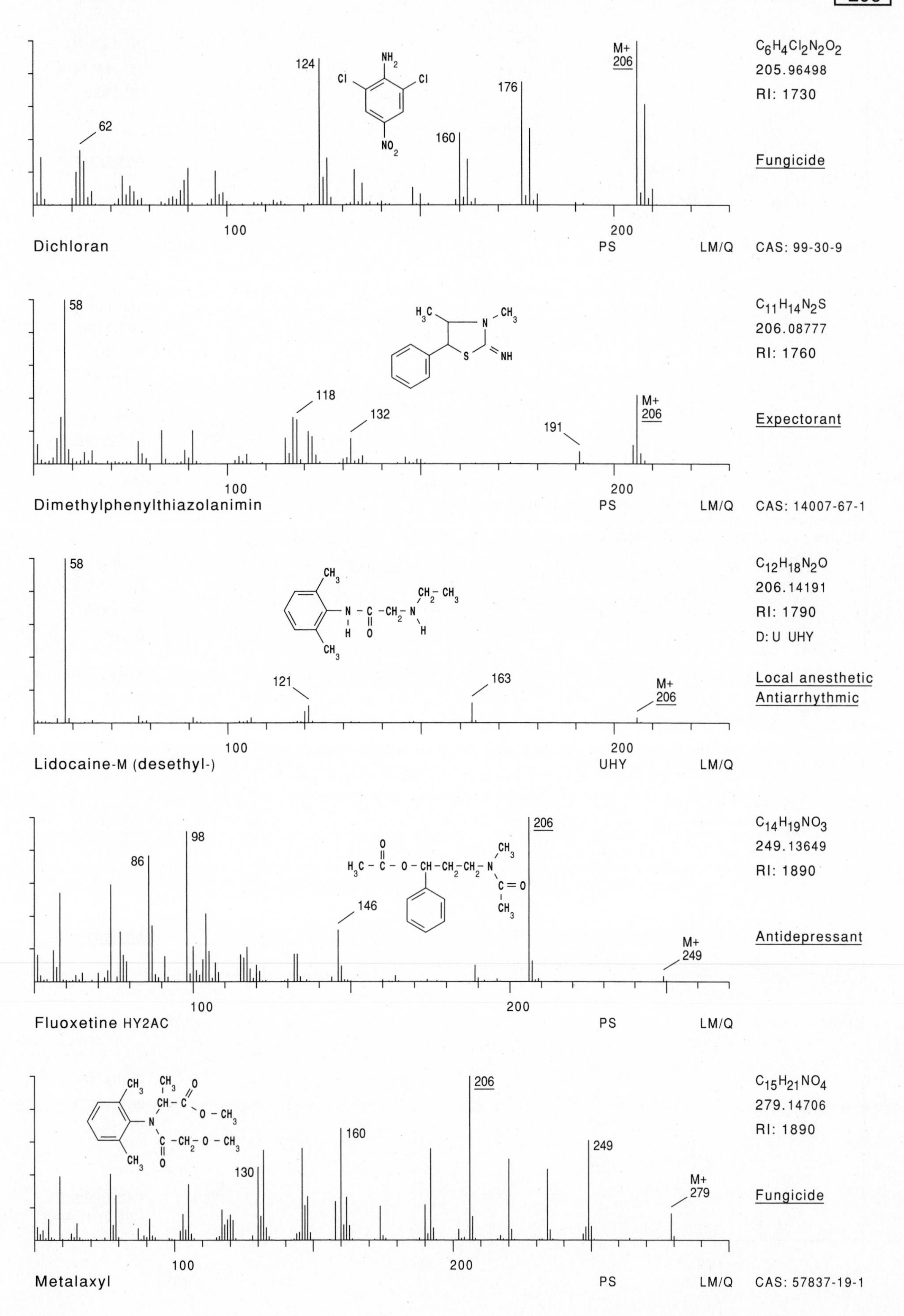
124
62
160
176
M+
206
100
200
Dichloran
PS
LM/Q
C6H4Cl2N2O2
205.96498
RI: 1730
Fungicide
CAS: 99-30-9
58
118
132
191
M+
206
100
200
Dimethylphenylthiazolanimin
PS
LM/Q
C11H14N2S
206.08777
RI: 1760
Expectorant
CAS: 14007-67-1
58
121
163
M+
206
100
200
Lidocaine-M (desethyl-)
UHY
LM/Q
C12H18N2O
206.14191
RI: 1790
D: U UHY
Local anesthetic
Antiarrhythmic
98
86
206
146
M+
249
100
200
Fluoxetine HY2AC
PS
LM/Q
C14H19NO3
249.13649
RI: 1890
Antidepressant
206
160
130
249
M+
279
100
200
Metalaxyl
PS
LM/Q
C15H21NO4
279.14706
RI: 1890
Fungicide
CAS: 57837-19-1

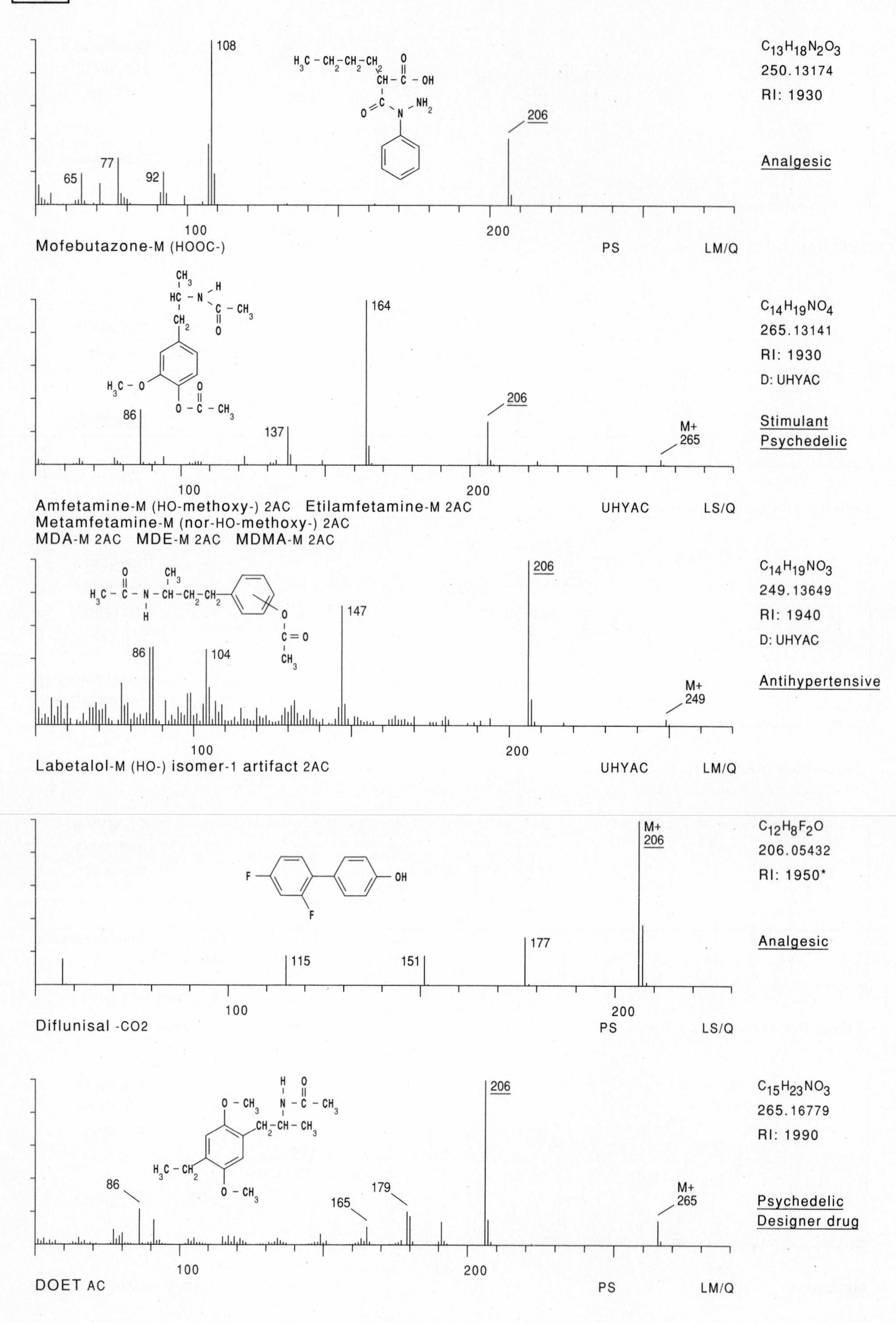
108
65
77
92
206
100
200
$C_{13}H_{18}N_2O_3$
250.13174
RI: 1930
Analgesic
Mofebutazone-M (HOOC-)
PS
LM/Q
164
86
137
206
M+
265
$C_{14}H_{19}NO_4$
265.13141
RI: 1930
D: UHYAC
Stimulant
Psychedelic
Amfetamine-M (HO-methoxy-) 2AC Etilamfetamine-M 2AC
Metamfetamine-M (nor-HO-methoxy-) 2AC
MDA-M 2AC MDE-M 2AC MDMA-M 2AC
UHYAC
LS/Q
206
147
86
104
M+
249
$C_{14}H_{19}NO_3$
249.13649
RI: 1940
D: UHYAC
Antihypertensive
Labetalol-M (HO-) isomer-1 artifact 2AC
UHYAC
LM/Q
M+
206
115
151
177
$C_{12}H_8F_2O$
206.05432
RI: 1950*
Analgesic
Diflunisal -CO2
PS
LS/Q
206
86
165
179
M+
265
$C_{15}H_{23}NO_3$
265.16779
RI: 1990
Psychedelic
Designer drug
DOET AC
PS
LM/Q

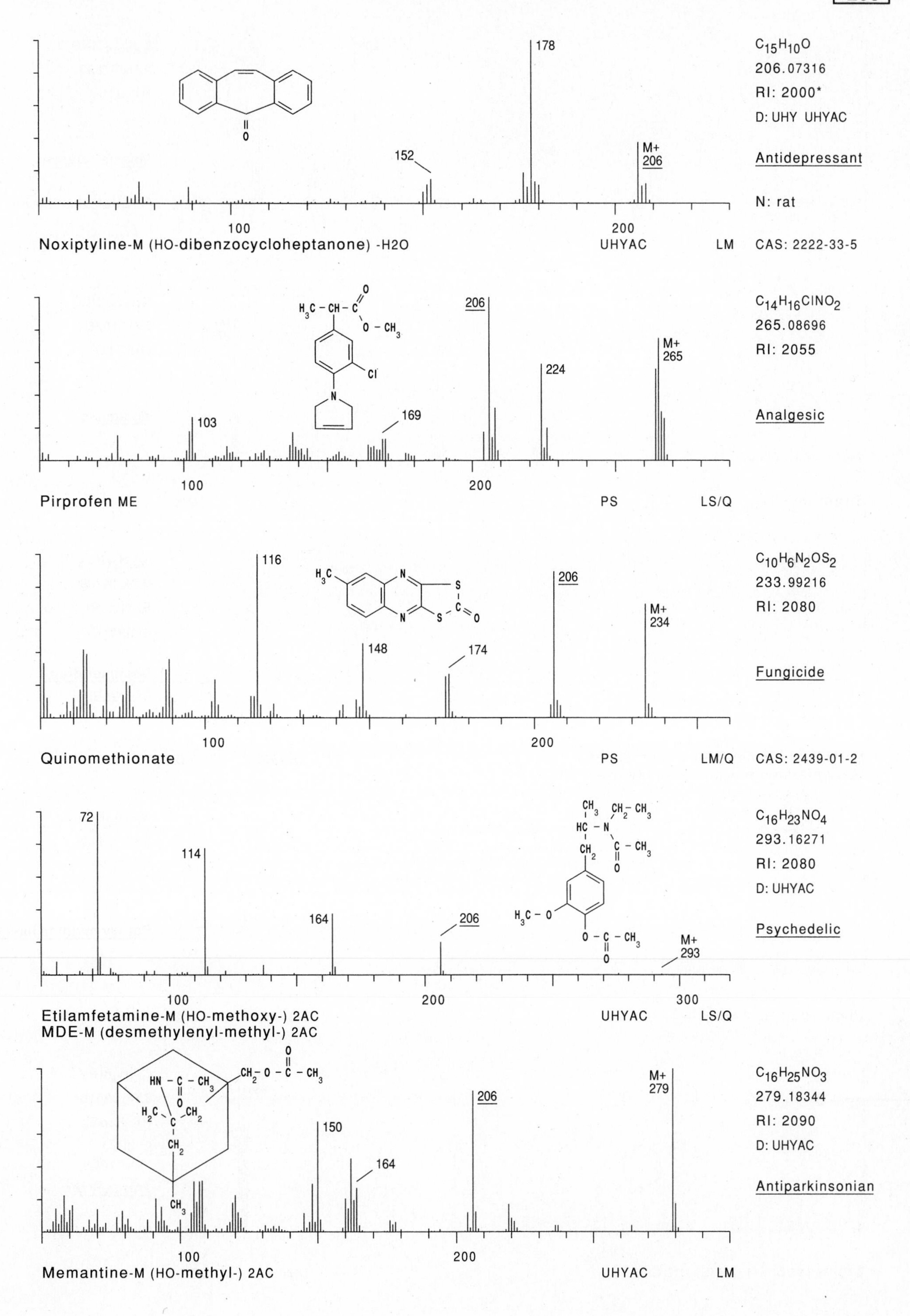
178
152
M+
206
100
200
Noxiptyline-M (HO-dibenzocycloheptanone) -H2O
UHYAC
LM
$C_{15}H_{10}O$
206.07316
RI: 2000*
D: UHY UHYAC
Antidepressant
N: rat
CAS: 2222-33-5
206
M+
265
224
103
169
Cl
Pirprofen ME
PS
LS/Q
$C_{14}H_{16}ClNO_2$
265.08696
RI: 2055
Analgesic
116
206
M+
234
148
174
Quinomethionate
PS
LM/Q
$C_{10}H_6N_2OS_2$
233.99216
RI: 2080
Fungicide
CAS: 2439-01-2
72
114
164
206
M+
293
300
Etilamfetamine-M (HO-methoxy-) 2AC
MDE-M (desmethylenyl-methyl-) 2AC
UHYAC
LS/Q
$C_{16}H_{23}NO_4$
293.16271
RI: 2080
D: UHYAC
Psychedelic
M+
279
206
150
164
Memantine-M (HO-methyl-) 2AC
UHYAC
LM
$C_{16}H_{25}NO_3$
279.18344
RI: 2090
D: UHYAC
Antiparkinsonian

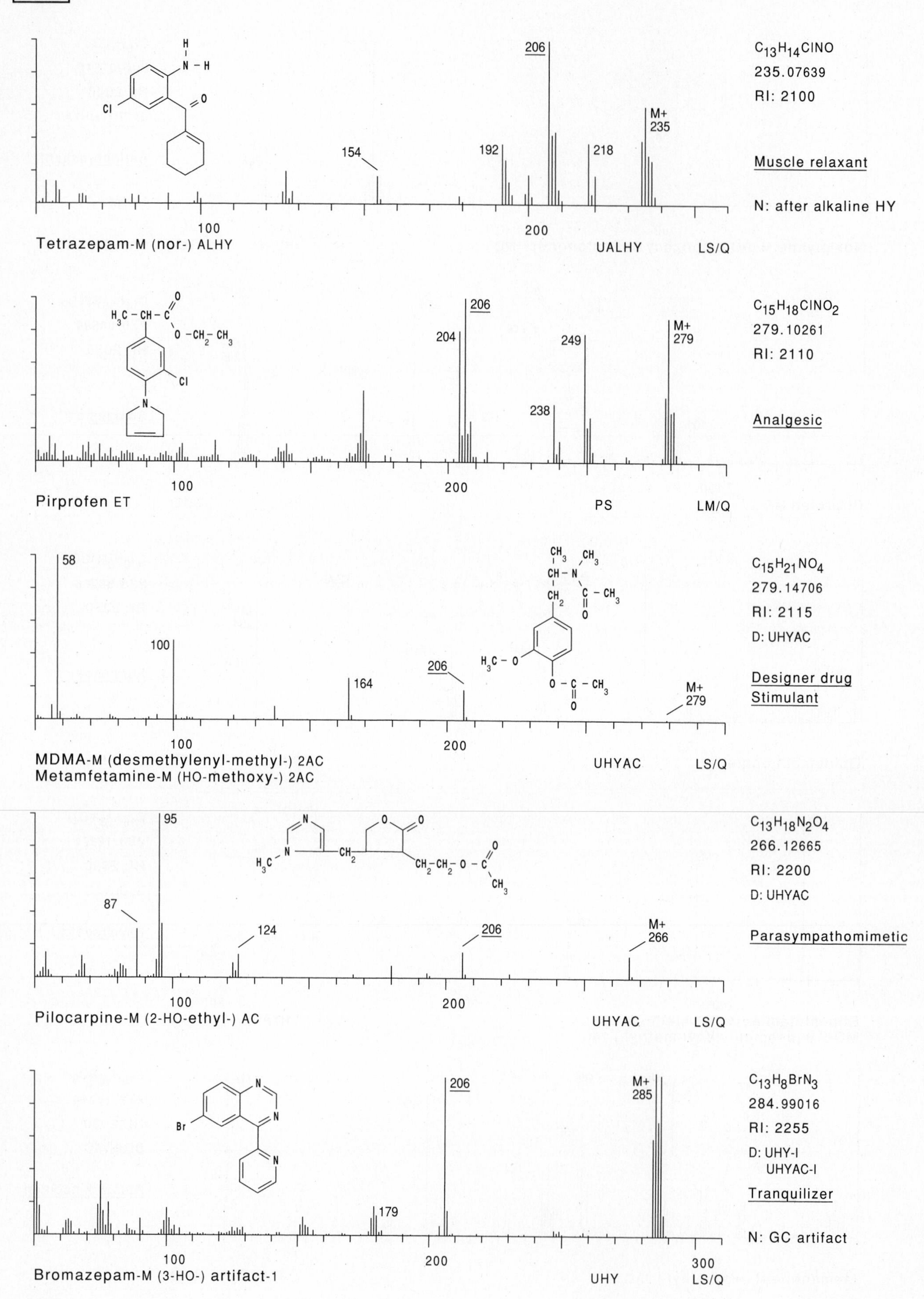

206
154
192
218
M+
235
100
200
Tetrazepam-M (nor-) ALHY
UALHY
LS/Q
C13H14ClNO
235.07639
RI: 2100
Muscle relaxant
N: after alkaline HY
206
204
249
M+
279
238
100
200
Pirprofen ET
PS
LM/Q
C15H18ClNO2
279.10261
RI: 2110
Analgesic
58
100
164
206
M+
279
100
200
MDMA-M (desmethylenyl-methyl-) 2AC
Metamfetamine-M (HO-methoxy-) 2AC
UHYAC
LS/Q
C15H21NO4
279.14706
RI: 2115
D: UHYAC
Designer drug
Stimulant
95
87
124
206
M+
266
100
200
Pilocarpine-M (2-HO-ethyl-) AC
UHYAC
LS/Q
C13H18N2O4
266.12665
RI: 2200
D: UHYAC
Parasympathomimetic
206
M+
285
179
100
200
300
Bromazepam-M (3-HO-) artifact-1
UHY
LS/Q
C13H8BrN3
284.99016
RI: 2255
D: UHY-I
UHYAC-I
Tranquilizer
N: GC artifact

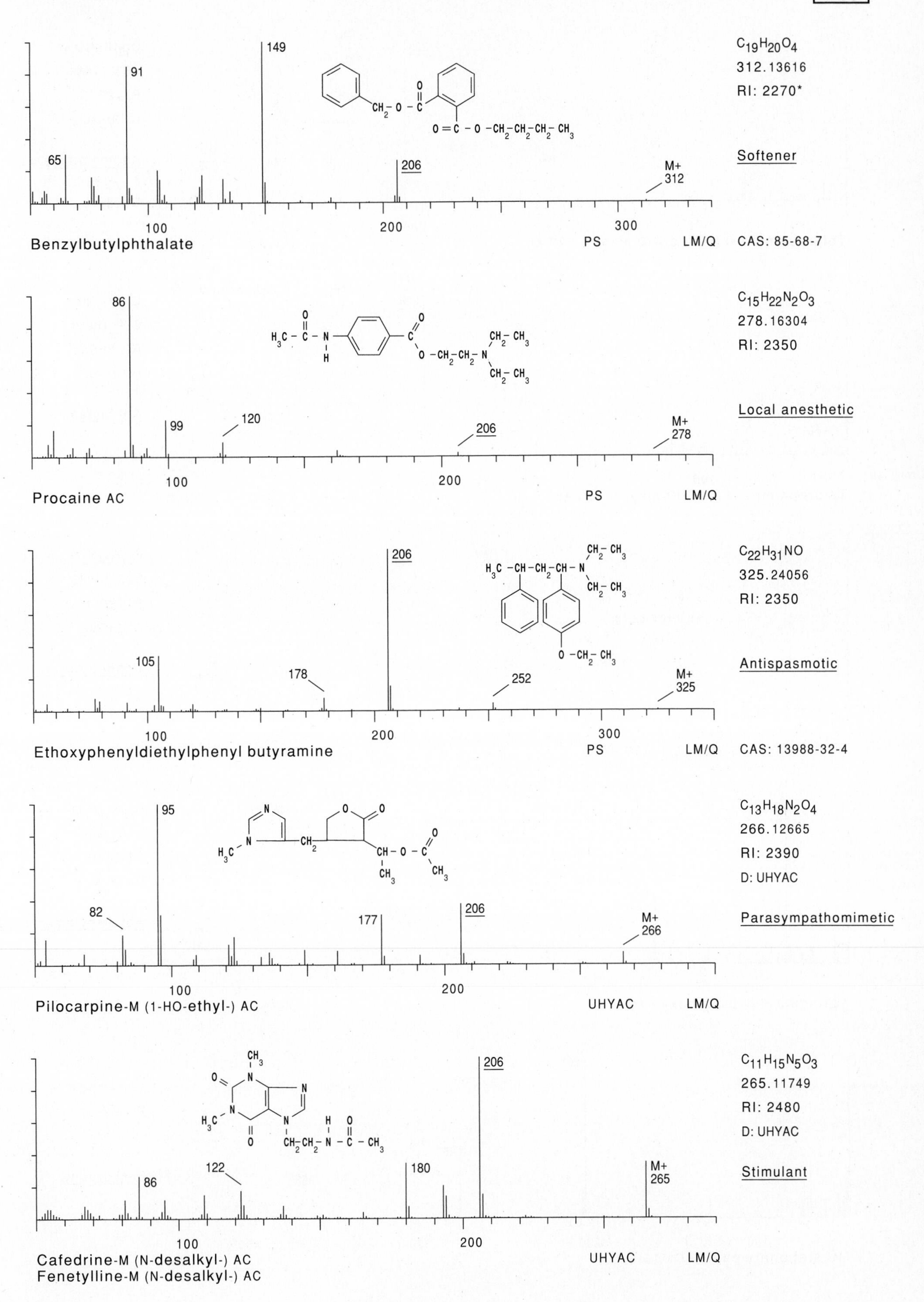

Benzylbutylphthalate
$C_{19}H_{20}O_4$
312.13616
RI: 2270*
Softener
CAS: 85-68-7
PS LM/Q
Procaine AC
$C_{15}H_{22}N_2O_3$
278.16304
RI: 2350
Local anesthetic
PS LM/Q
Ethoxyphenyldiethylphenyl butyramine
$C_{22}H_{31}NO$
325.24056
RI: 2350
Antispasmodic
CAS: 13988-32-4
PS LM/Q
Pilocarpine-M (1-HO-ethyl-) AC
$C_{13}H_{18}N_2O_4$
266.12665
RI: 2390
D: UHYAC
Parasympathomimetic
UHYAC LM/Q
Cafedrine-M (N-desalkyl-) AC
Fenetylline-M (N-desalkyl-) AC
$C_{11}H_{15}N_5O_3$
265.11749
RI: 2480
D: UHYAC
Stimulant
UHYAC LM/Q

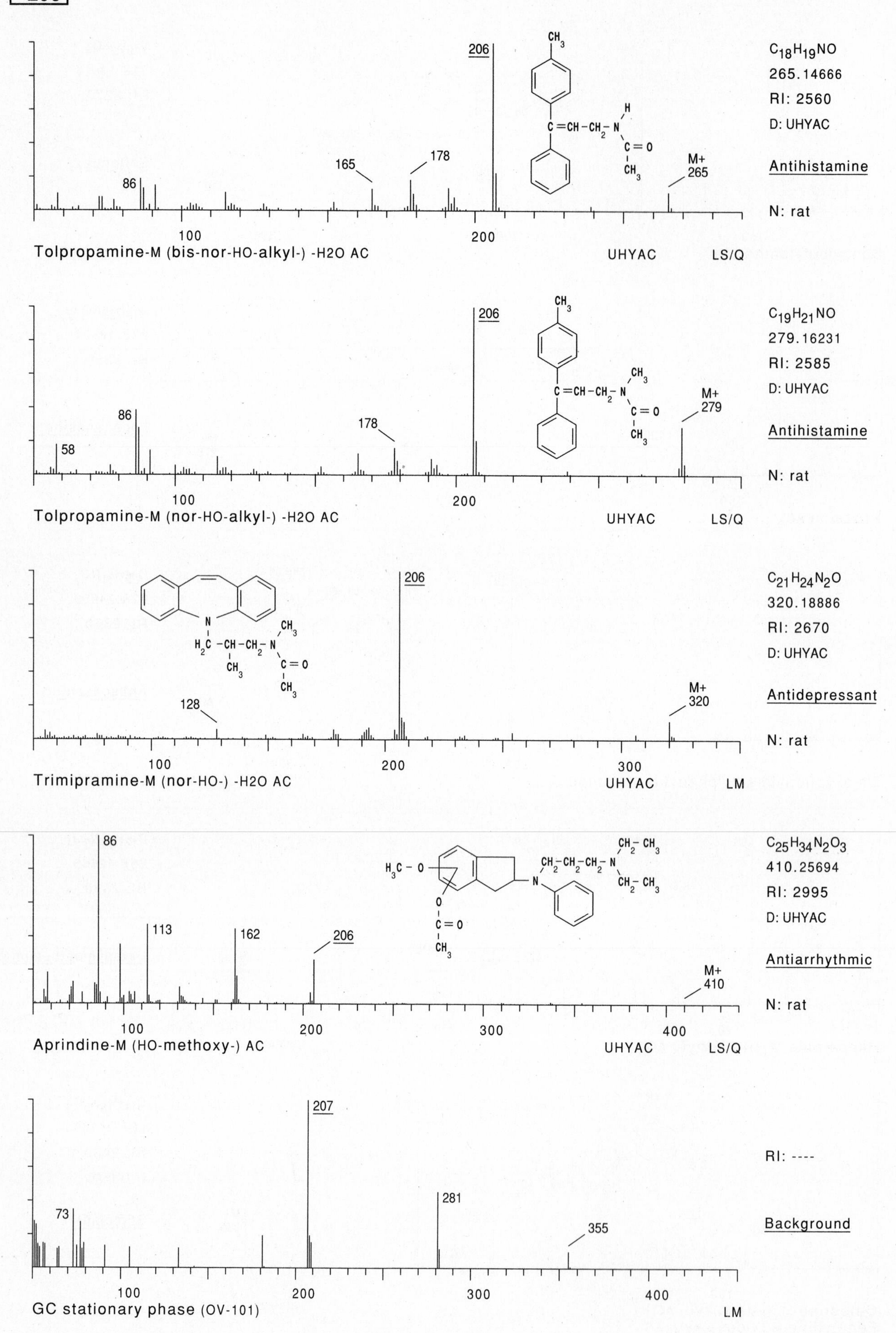
206
CH3
H
C=CH-CH2-N
C=O
CH3
165
178
86
M+
265
100
200
Tolpropamine-M (bis-nor-HO-alkyl-) -H2O AC
UHYAC
LS/Q
C18H19NO
265.14666
RI: 2560
D: UHYAC
Antihistamine
N: rat
206
CH3
CH3
C=CH-CH2-N
C=O
CH3
M+
279
86
178
58
100
200
Tolpropamine-M (nor-HO-alkyl-) -H2O AC
UHYAC
LS/Q
C19H21NO
279.16231
RI: 2585
D: UHYAC
Antihistamine
N: rat
206
N
CH3
H2C -CH-CH2-N
CH3
C=O
CH3
128
M+
320
100
200
300
Trimipramine-M (nor-HO-) -H2O AC
UHYAC
LM
C21H24N2O
320.18886
RI: 2670
D: UHYAC
Antidepressant
N: rat
86
CH2-CH3
H3C-O
CH2-CH2-CH2-N
N
CH2-CH3
O
C=O
CH3
113
162
206
M+
410
100
200
300
400
Aprindine-M (HO-methoxy-) AC
UHYAC
LS/Q
C25H34N2O3
410.25694
RI: 2995
D: UHYAC
Antiarrhythmic
N: rat
207
281
73
355
100
200
300
400
GC stationary phase (OV-101)
LM
RI: ----
Background

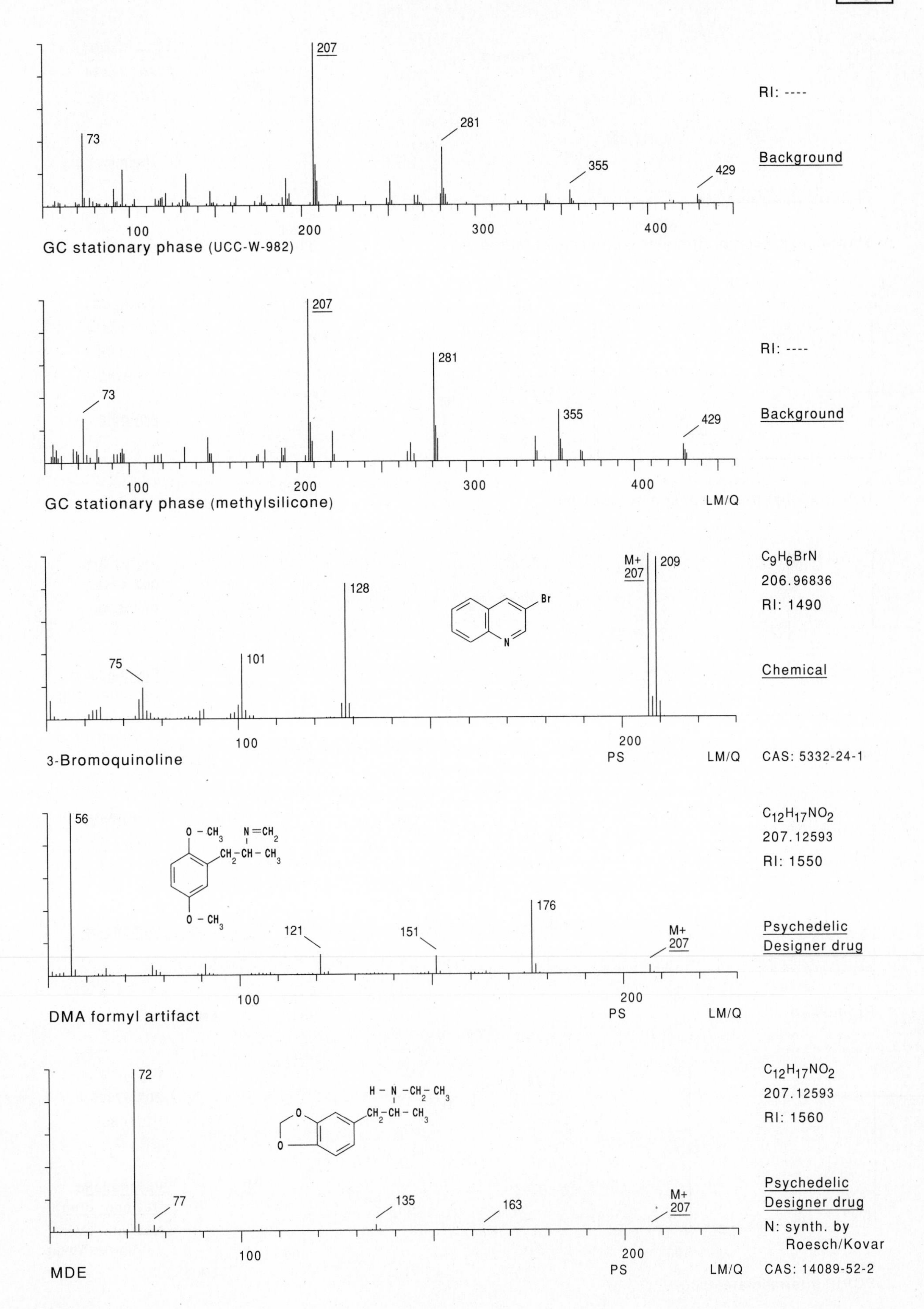
207
73
281
355
429
100
200
300
400
GC stationary phase (UCC-W-982)
RI: ----
Background
207
281
73
355
429
100
200
300
400
LM/Q
GC stationary phase (methylsilicone)
RI: ----
Background
M+
207
209
128
Br
N
75
101
100
200
PS
LM/Q
3-Bromoquinoline
C9H6BrN
206.96836
RI: 1490
Chemical
CAS: 5332-24-1
56
O - CH3
N = CH2
CH2 - CH - CH3
O - CH3
176
121
151
M+
207
100
200
PS
LM/Q
DMA formyl artifact
C12H17NO2
207.12593
RI: 1550
Psychedelic
Designer drug
72
H - N - CH2 - CH3
CH2 - CH - CH3
O
O
77
135
163
M+
207
100
200
PS
LM/Q
MDE
C12H17NO2
207.12593
RI: 1560
Psychedelic
Designer drug
N: synth. by
Roesch/Kovar
CAS: 14089-52-2

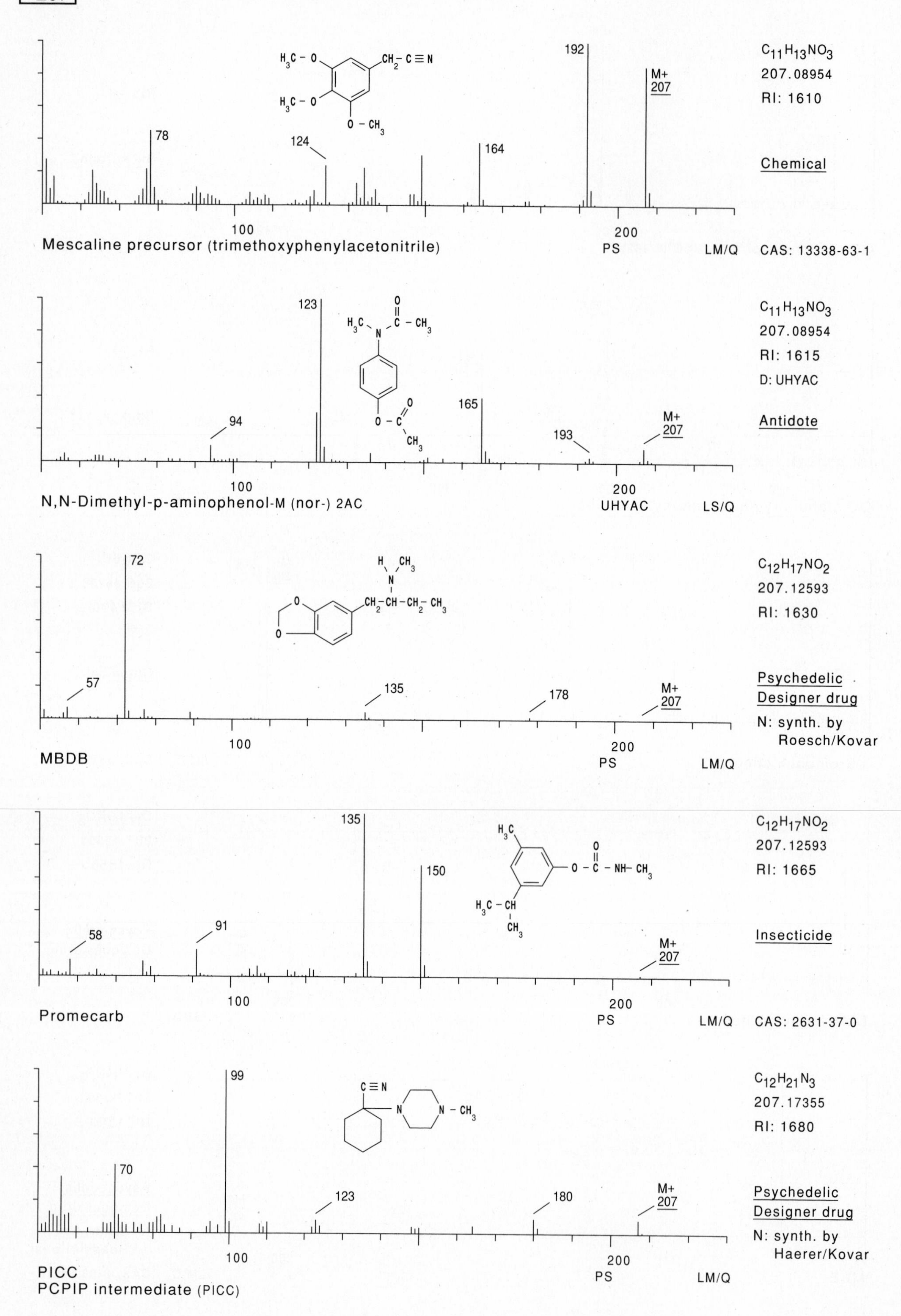
78
124
164
192
M+
207
100
200
Mescaline precursor (trimethoxyphenylacetonitrile)
PS
LM/Q
C11H13NO3
207.08954
RI: 1610
Chemical
CAS: 13338-63-1
123
94
165
193
M+
207
100
200
N,N-Dimethyl-p-aminophenol-M (nor-) 2AC
UHYAC
LS/Q
C11H13NO3
207.08954
RI: 1615
D: UHYAC
Antidote
72
57
135
178
M+
207
100
200
MBDB
PS
LM/Q
C12H17NO2
207.12593
RI: 1630
Psychedelic
Designer drug
N: synth. by
Roesch/Kovar
135
150
58
91
M+
207
100
200
Promecarb
PS
LM/Q
C12H17NO2
207.12593
RI: 1665
Insecticide
CAS: 2631-37-0
99
70
123
180
M+
207
100
200
PICC
PCPIP intermediate (PICC)
PS
LM/Q
C12H21N3
207.17355
RI: 1680
Psychedelic
Designer drug
N: synth. by
Haerer/Kovar

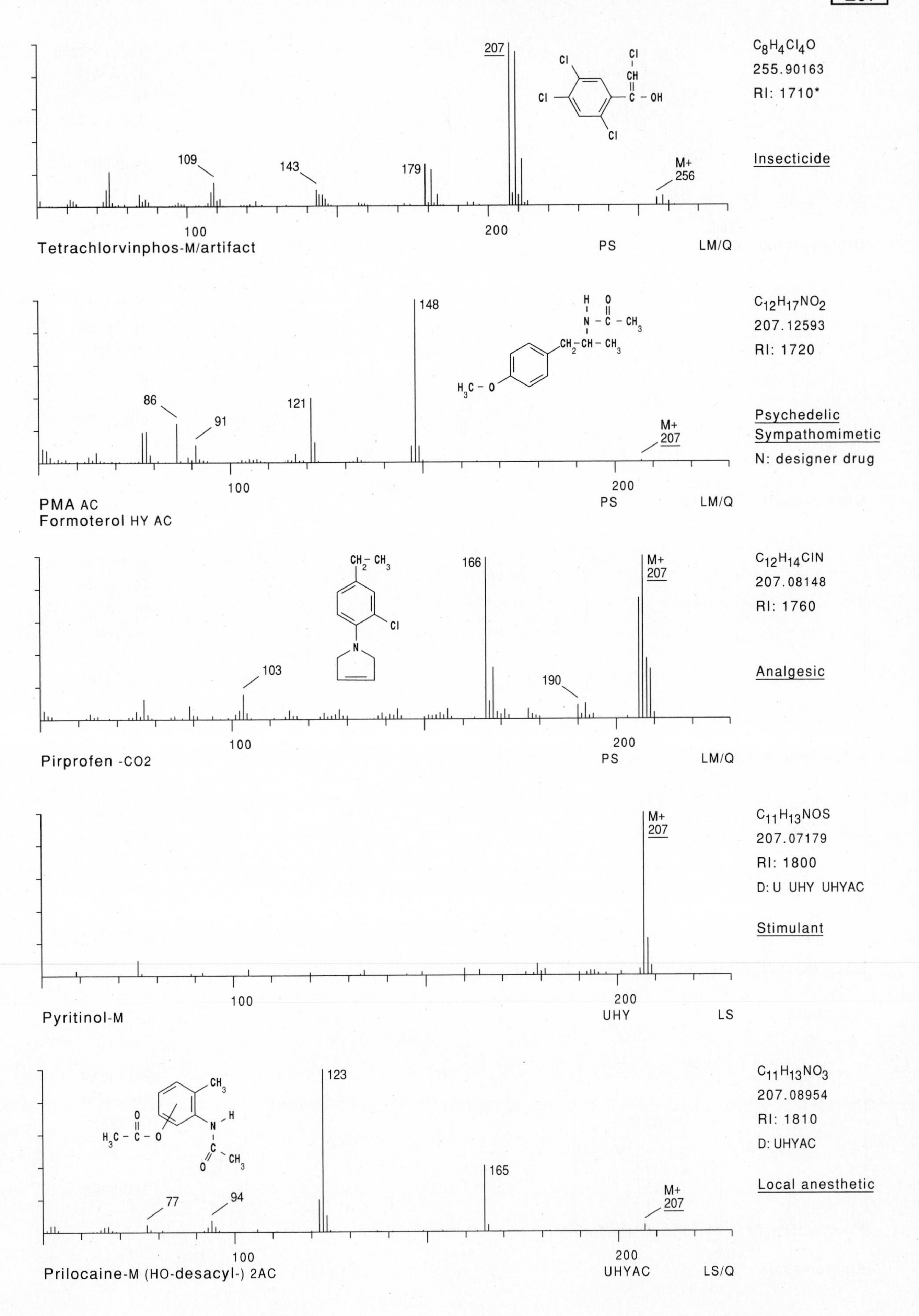

Tetrachlorvinphos-M/artifact
PS LM/Q
$C_8H_4Cl_4O$
255.90163
RI: 1710*
Insecticide
M+ 256
PMA AC
Formoterol HY AC
PS LM/Q
$C_{12}H_{17}NO_2$
207.12593
RI: 1720
Psychedelic
Sympathomimetic
N: designer drug
M+ 207
Pirprofen -CO2
PS LM/Q
$C_{12}H_{14}ClN$
207.08148
RI: 1760
Analgesic
M+ 207
Pyritinol-M
UHY LS
$C_{11}H_{13}NOS$
207.07179
RI: 1800
D: U UHY UHYAC
Stimulant
M+ 207
Prilocaine-M (HO-desacyl-) 2AC
UHYAC LS/Q
$C_{11}H_{13}NO_3$
207.08954
RI: 1810
D: UHYAC
Local anesthetic
M+ 207

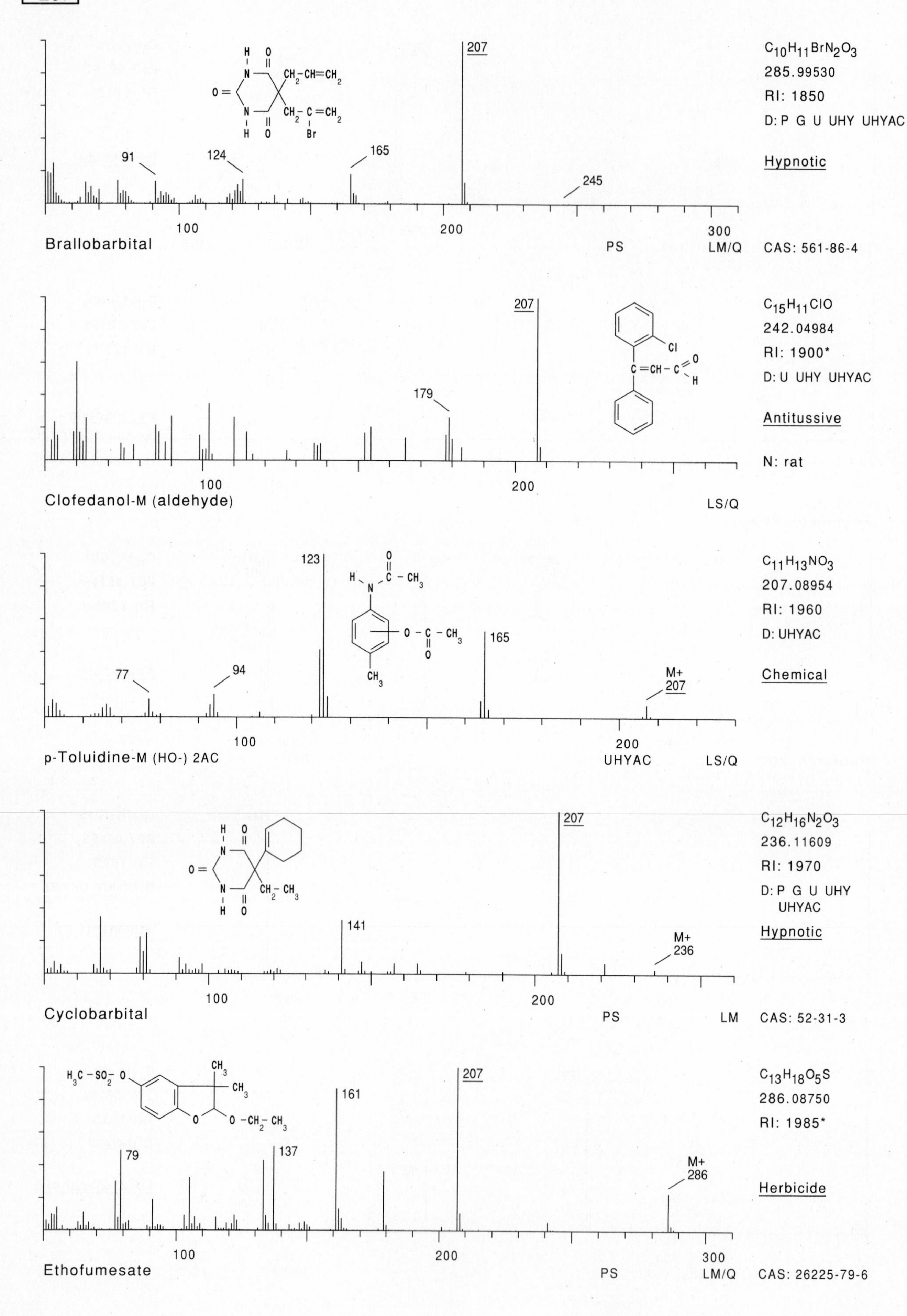

Brallobarbital
$C_{10}H_{11}BrN_2O_3$
285.99530
RI: 1850
D: P G U UHY UHYAC
Hypnotic
CAS: 561-86-4
PS
LM/Q
207
165
124
91
245
100
200
300
Clofedanol-M (aldehyde)
$C_{15}H_{11}ClO$
242.04984
RI: 1900*
D: U UHY UHYAC
Antitussive
N: rat
LS/Q
207
179
100
200
p-Toluidine-M (HO-) 2AC
$C_{11}H_{13}NO_3$
207.08954
RI: 1960
D: UHYAC
Chemical
UHYAC
LS/Q
123
165
77
94
M+
207
100
200
Cyclobarbital
$C_{12}H_{16}N_2O_3$
236.11609
RI: 1970
D: P G U UHY UHYAC
Hypnotic
CAS: 52-31-3
PS
LM
207
141
M+
236
100
200
Ethofumesate
$C_{13}H_{18}O_5S$
286.08750
RI: 1985*
Herbicide
CAS: 26225-79-6
PS
LM/Q
207
161
137
79
M+
286
100
200
300

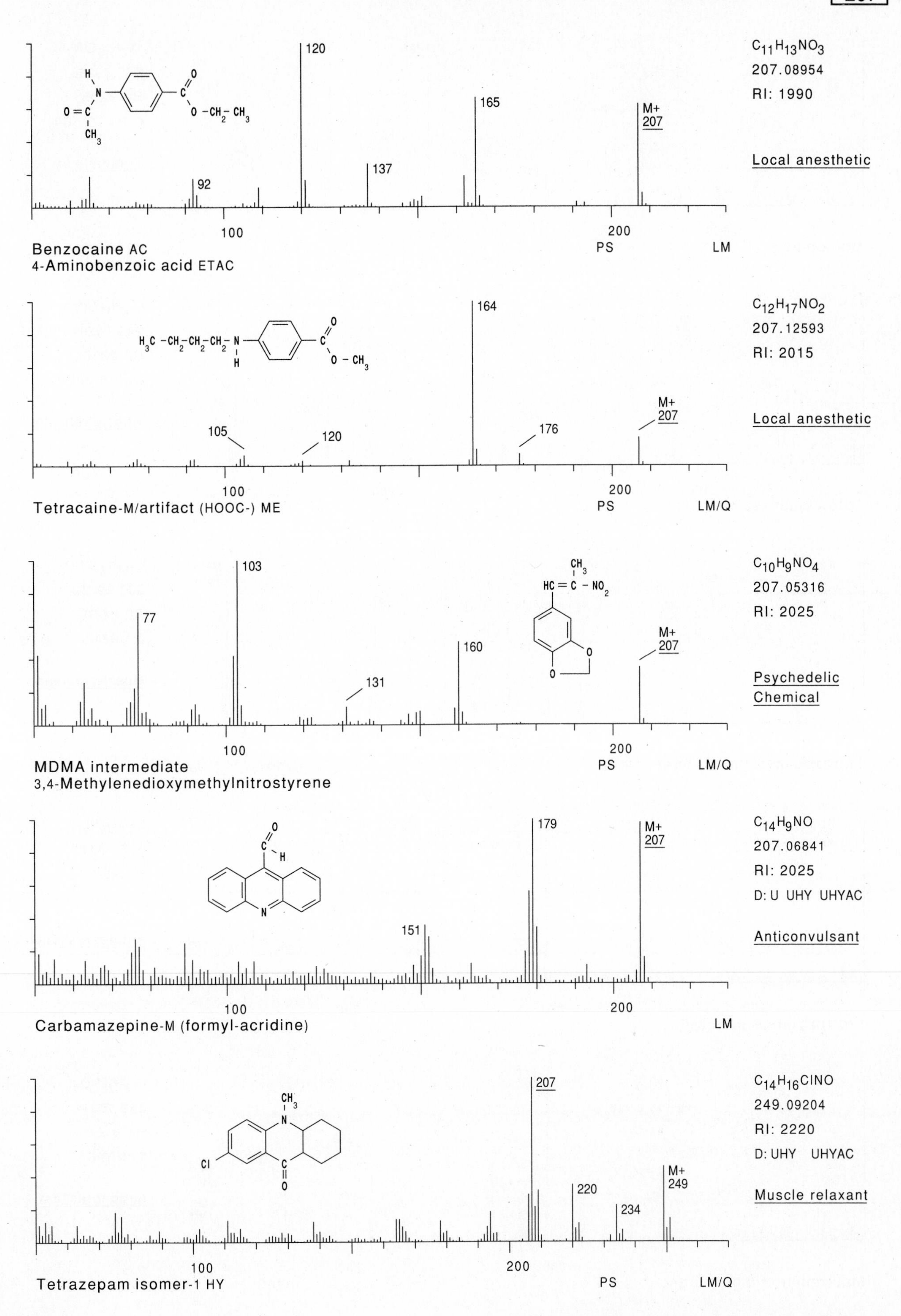
120
165
M+
207
137
92
100
200
Benzocaine AC
4-Aminobenzoic acid ETAC
PS
LM
C11H13NO3
207.08954
RI: 1990
Local anesthetic
164
M+
207
176
105
120
100
200
Tetracaine-M/artifact (HOOC-) ME
PS
LM/Q
C12H17NO2
207.12593
RI: 2015
Local anesthetic
103
77
160
131
M+
207
100
200
MDMA intermediate
3,4-Methylenedioxymethylnitrostyrene
PS
LM/Q
C10H9NO4
207.05316
RI: 2025
Psychedelic
Chemical
179
M+
207
151
100
200
Carbamazepine-M (formyl-acridine)
LM
C14H9NO
207.06841
RI: 2025
D: U UHY UHYAC
Anticonvulsant
207
220
234
M+
249
100
200
Tetrazepam isomer-1 HY
PS
LM/Q
C14H16ClNO
249.09204
RI: 2220
D: UHY UHYAC
Muscle relaxant

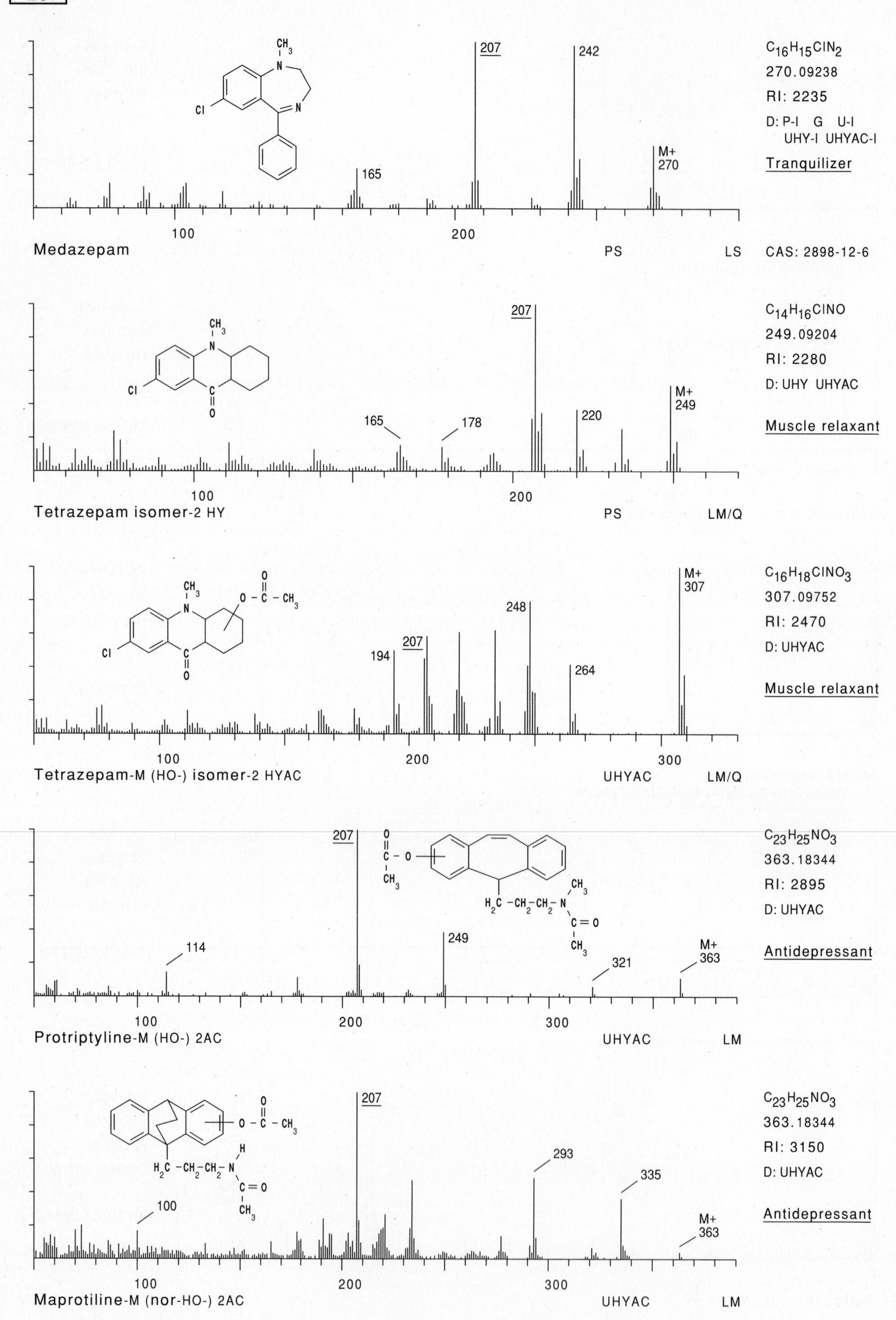

$C_{16}H_{15}ClN_2$
270.09238
RI: 2235
D: P-I G U-I UHY-I UHYAC-I
Tranquilizer
CAS: 2898-12-6

$C_{14}H_{16}ClNO$
249.09204
RI: 2280
D: UHY UHYAC
Muscle relaxant

$C_{16}H_{18}ClNO_3$
307.09752
RI: 2470
D: UHYAC
Muscle relaxant

$C_{23}H_{25}NO_3$
363.18344
RI: 2895
D: UHYAC
Antidepressant

$C_{23}H_{25}NO_3$
363.18344
RI: 3150
D: UHYAC
Antidepressant

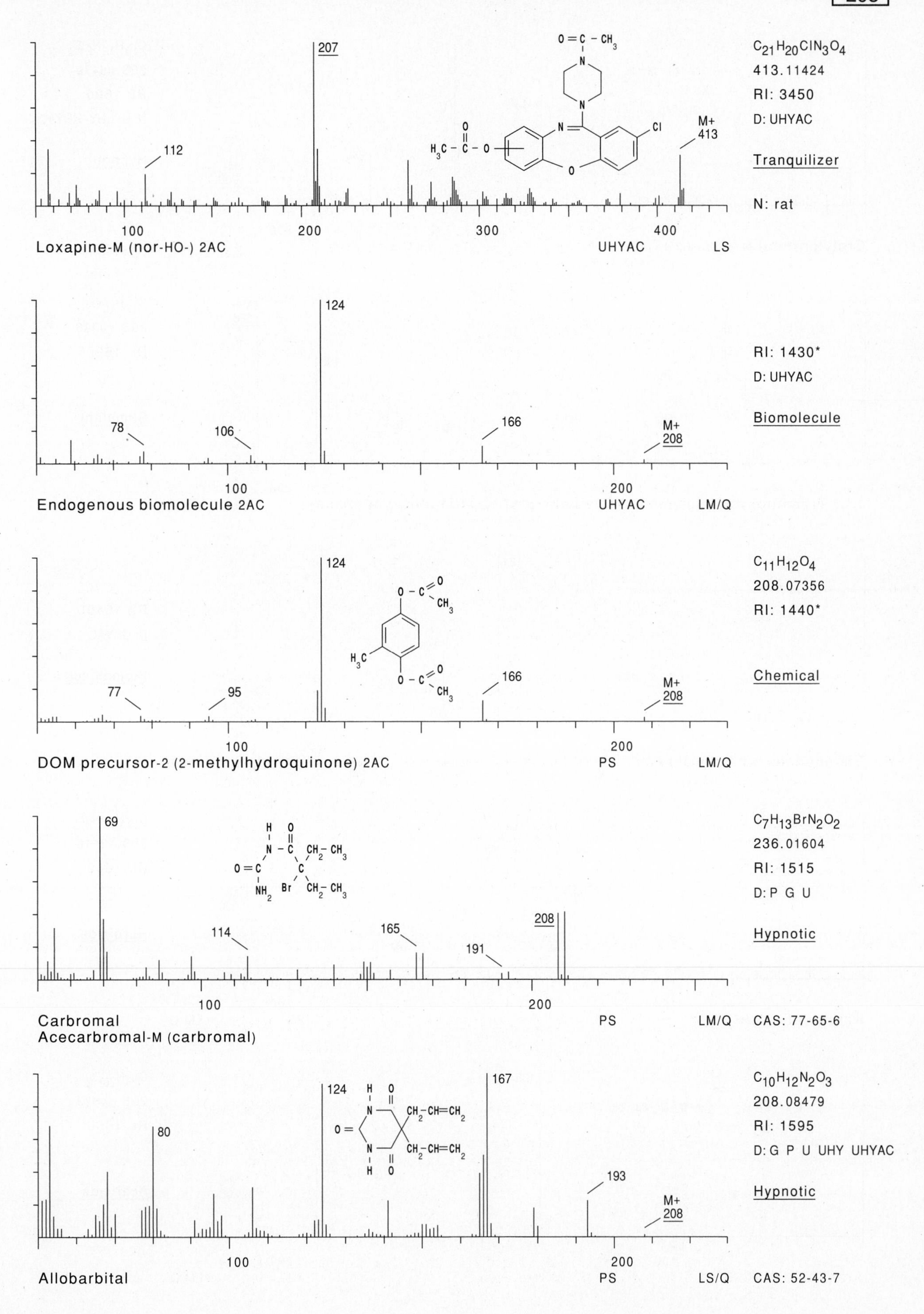

207
112
M+
413
100
200
300
400
Loxapine-M (nor-HO-) 2AC
UHYAC
LS
C21H20ClN3O4
413.11424
RI: 3450
D: UHYAC
Tranquilizer
N: rat
124
78
106
166
M+
208
100
200
Endogenous biomolecule 2AC
UHYAC
LM/Q
RI: 1430*
D: UHYAC
Biomolecule
124
77
95
166
M+
208
100
200
DOM precursor-2 (2-methylhydroquinone) 2AC
PS
LM/Q
C11H12O4
208.07356
RI: 1440*
Chemical
69
114
165
191
208
100
200
Carbromal
Acecarbromal-M (carbromal)
PS
LM/Q
C7H13BrN2O2
236.01604
RI: 1515
D: P G U
Hypnotic
CAS: 77-65-6
167
124
80
193
M+
208
100
200
Allobarbital
PS
LS/Q
C10H12N2O3
208.08479
RI: 1595
D: G P U UHY UHYAC
Hypnotic
CAS: 52-43-7

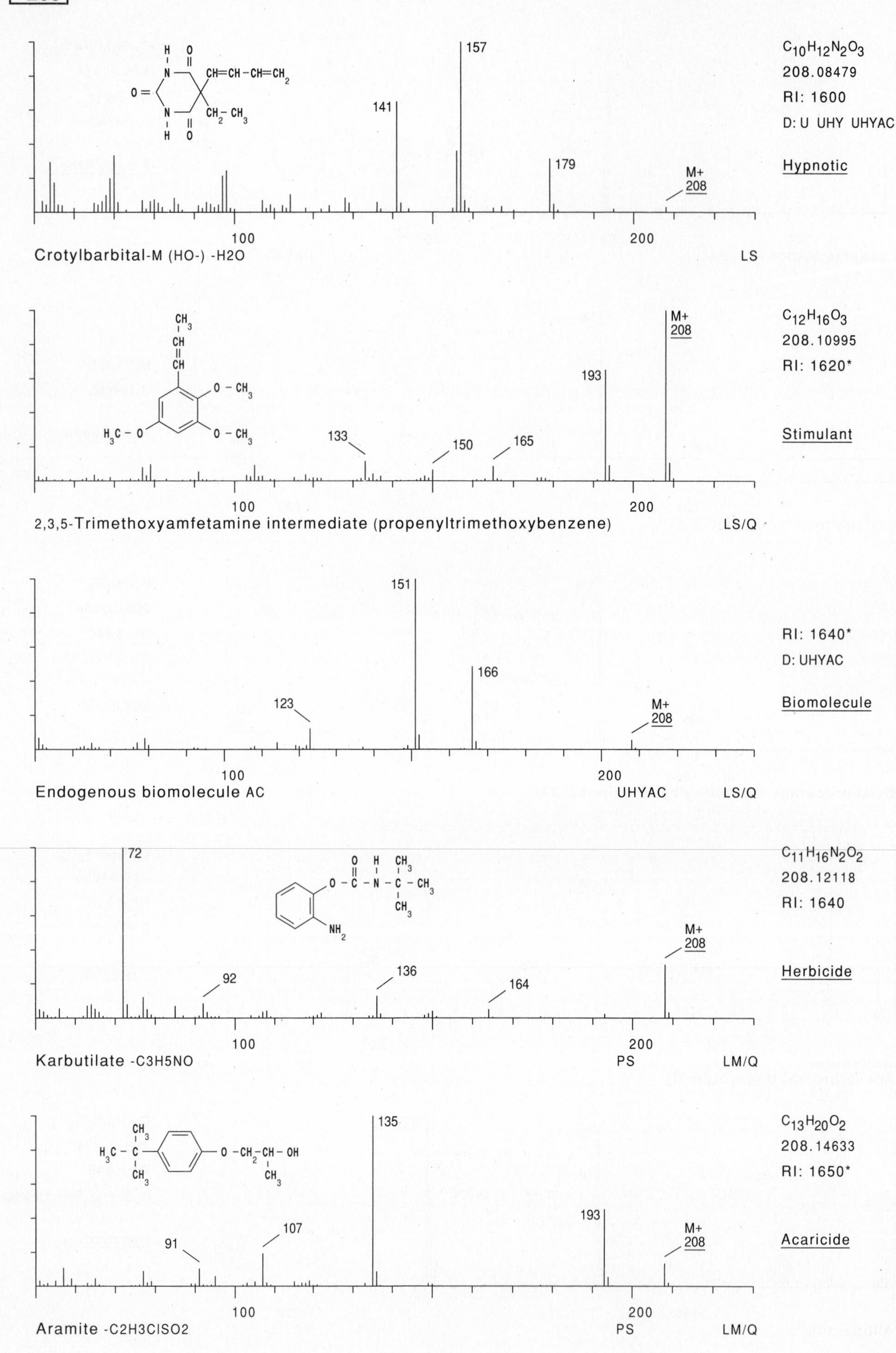
157
141
179
M+
208
100
200
Crotylbarbital-M (HO-) -H2O
LS
$C_{10}H_{12}N_2O_3$
208.08479
RI: 1600
D: U UHY UHYAC
Hypnotic
M+
208
193
133
150
165
100
200
2,3,5-Trimethoxyamfetamine intermediate (propenyltrimethoxybenzene)
LS/Q
$C_{12}H_{16}O_3$
208.10995
RI: 1620*
Stimulant
151
166
123
M+
208
100
200
Endogenous biomolecule AC
UHYAC
LS/Q
RI: 1640*
D: UHYAC
Biomolecule
72
M+
208
92
136
164
100
200
Karbutilate -C3H5NO
PS
LM/Q
$C_{11}H_{16}N_2O_2$
208.12118
RI: 1640
Herbicide
135
193
107
91
M+
208
100
200
Aramite -C2H3ClSO2
PS
LM/Q
$C_{13}H_{20}O_2$
208.14633
RI: 1650*
Acaricide

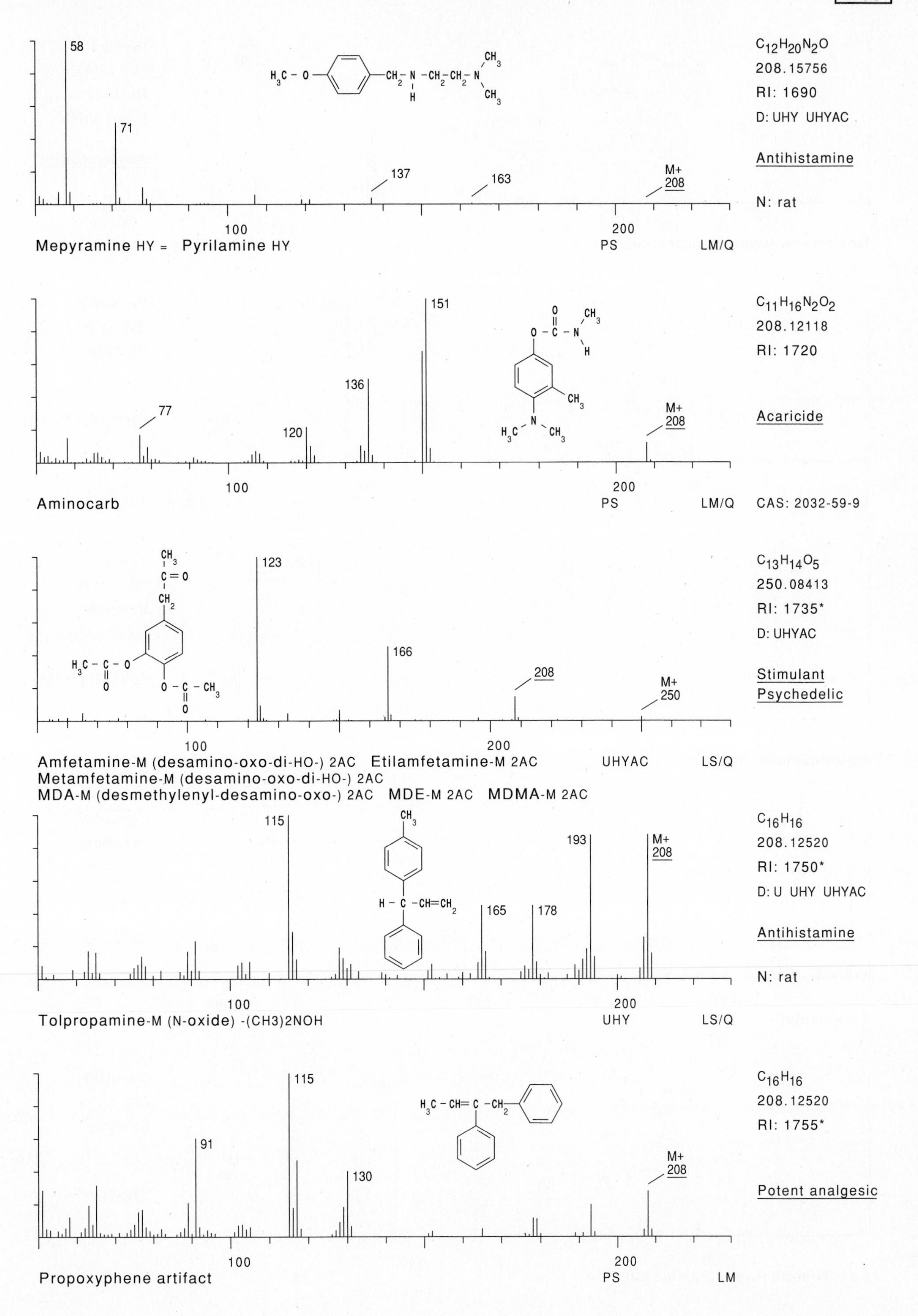

$C_{12}H_{20}N_2O$
208.15756
RI: 1690
D: UHY UHYAC

Antihistamine

N: rat

$C_{11}H_{16}N_2O_2$
208.12118
RI: 1720

Acaricide

CAS: 2032-59-9

$C_{13}H_{14}O_5$
250.08413
RI: 1735*
D: UHYAC

Stimulant
Psychedelic

$C_{16}H_{16}$
208.12520
RI: 1750*
D: U UHY UHYAC

Antihistamine

N: rat

$C_{16}H_{16}$
208.12520
RI: 1755*

Potent analgesic

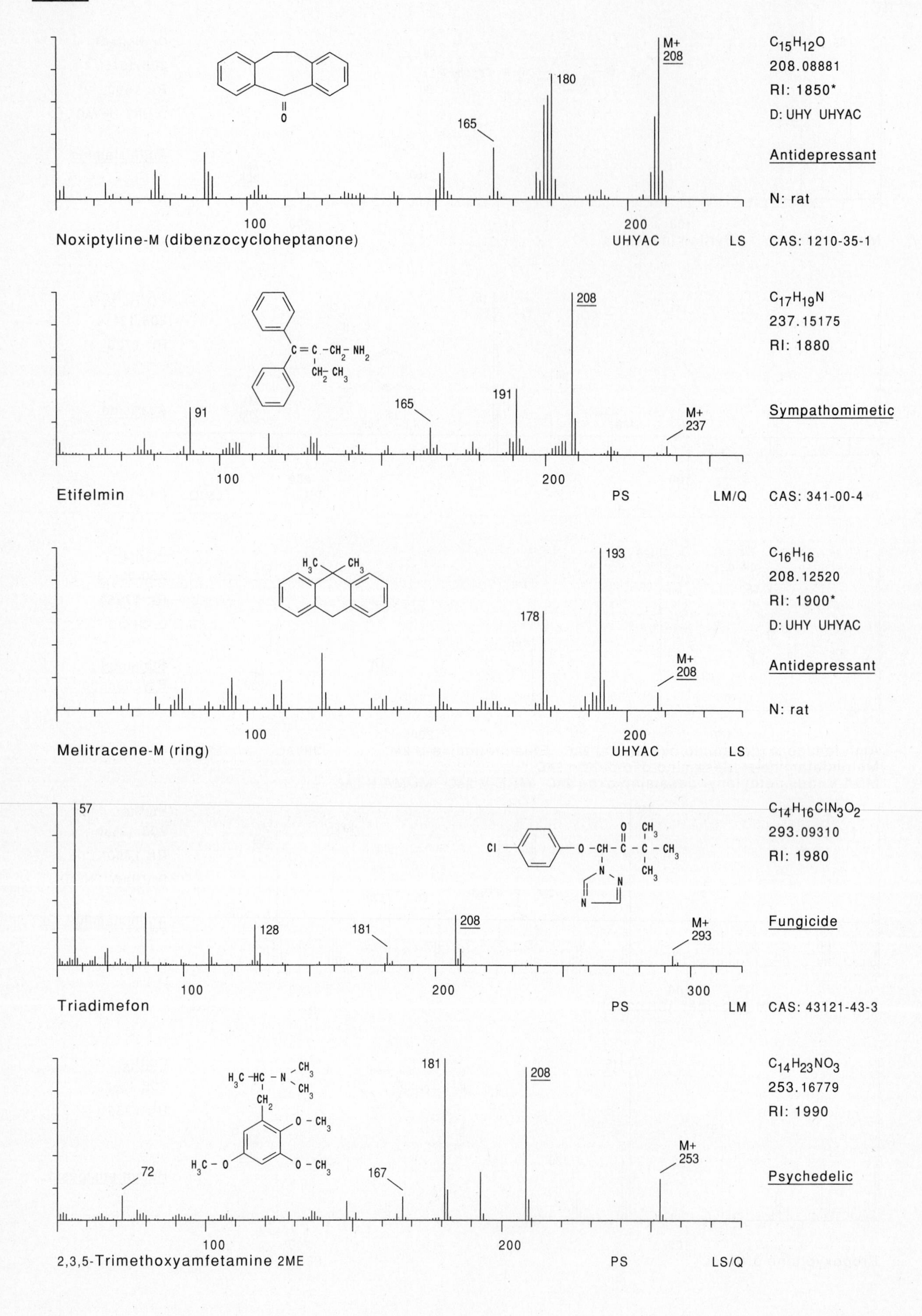
M+
208
180
165
100
200
Noxiptyline-M (dibenzocycloheptanone)
UHYAC
LS
$C_{15}H_{12}O$
208.08881
RI: 1850*
D: UHY UHYAC
Antidepressant
N: rat
CAS: 1210-35-1
208
191
165
91
M+
237
100
200
Etifelmin
PS
LM/Q
$C_{17}H_{19}N$
237.15175
RI: 1880
Sympathomimetic
CAS: 341-00-4
193
178
M+
208
100
200
Melitracene-M (ring)
UHYAC
LS
$C_{16}H_{16}$
208.12520
RI: 1900*
D: UHY UHYAC
Antidepressant
N: rat
57
128
181
208
M+
293
100
200
300
Triadimefon
PS
LM
$C_{14}H_{16}ClN_3O_2$
293.09310
RI: 1980
Fungicide
CAS: 43121-43-3
181
208
72
167
M+
253
100
200
2,3,5-Trimethoxyamfetamine 2ME
PS
LS/Q
$C_{14}H_{23}NO_3$
253.16779
RI: 1990
Psychedelic

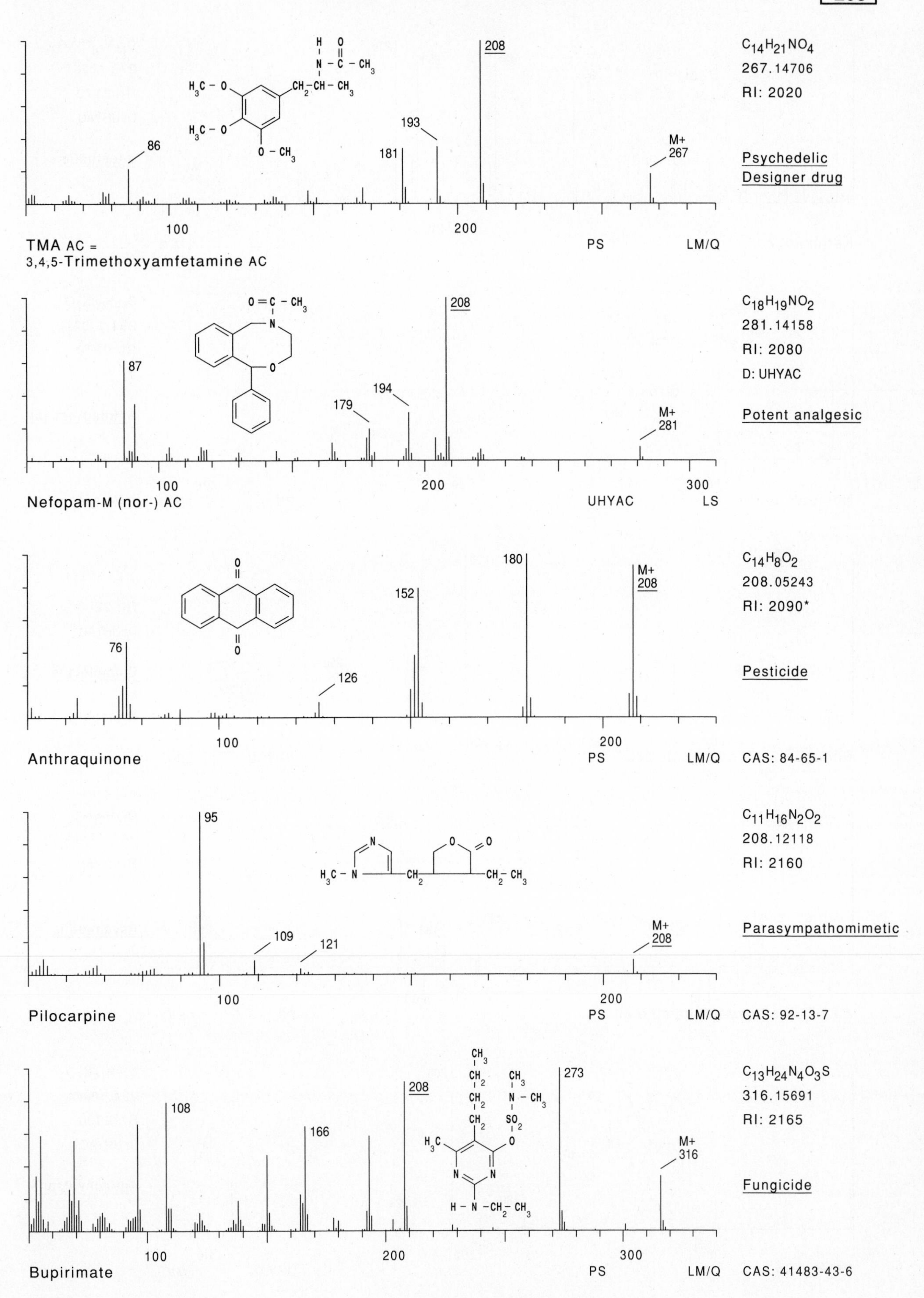
C14H21NO4
267.14706
RI: 2020
Psychedelic
Designer drug
TMA AC =
3,4,5-Trimethoxyamfetamine AC
PS
LM/Q
C18H19NO2
281.14158
RI: 2080
D: UHYAC
Potent analgesic
Nefopam-M (nor-) AC
UHYAC
LS
C14H8O2
208.05243
RI: 2090*
Pesticide
Anthraquinone
PS
LM/Q
CAS: 84-65-1
C11H16N2O2
208.12118
RI: 2160
Parasympathomimetic
Pilocarpine
PS
LM/Q
CAS: 92-13-7
C13H24N4O3S
316.15691
RI: 2165
Fungicide
Bupirimate
PS
LM/Q
CAS: 41483-43-6

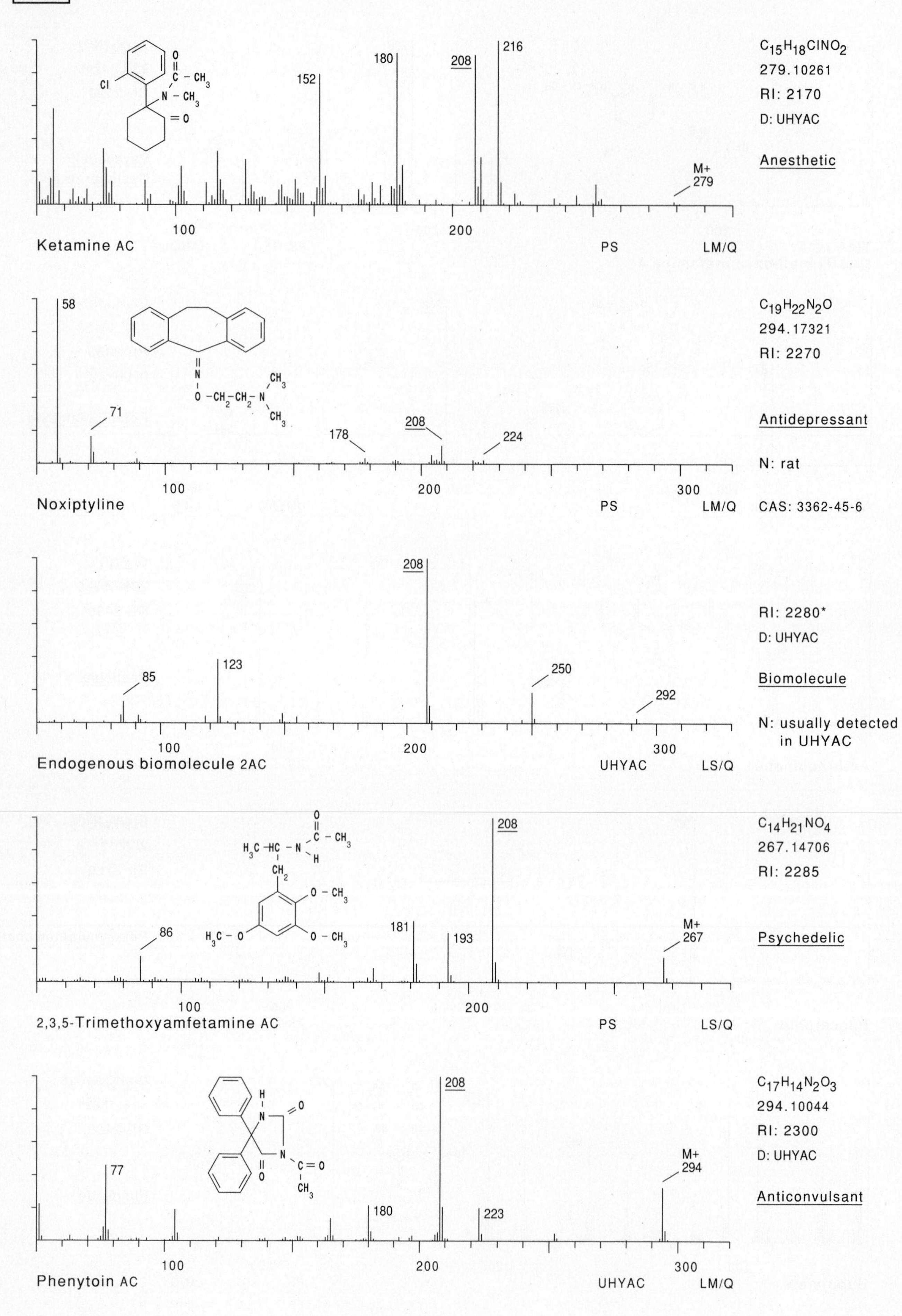

216
180
208
152
Cl
O
C - CH3
N - CH3
=O
M+
279
100
200
Ketamine AC
PS
LM/Q
C15H18ClNO2
279.10261
RI: 2170
D: UHYAC
Anesthetic
58
71
178
208
224
N
O - CH2 - CH2 - N
CH3
CH3
100
200
300
Noxiptyline
PS
LM/Q
C19H22N2O
294.17321
RI: 2270
Antidepressant
N: rat
CAS: 3362-45-6
208
123
85
250
292
100
200
300
Endogenous biomolecule 2AC
UHYAC
LS/Q
RI: 2280*
D: UHYAC
Biomolecule
N: usually detected
in UHYAC
208
O
C - CH3
H3C - HC - N
H
CH2
O - CH3
H3C - O
O - CH3
86
181
193
M+
267
100
200
2,3,5-Trimethoxyamfetamine AC
PS
LS/Q
C14H21NO4
267.14706
RI: 2285
Psychedelic
208
H
N
=O
N
O
C = O
CH3
77
180
223
M+
294
100
200
300
Phenytoin AC
UHYAC
LM/Q
C17H14N2O3
294.10044
RI: 2300
D: UHYAC
Anticonvulsant

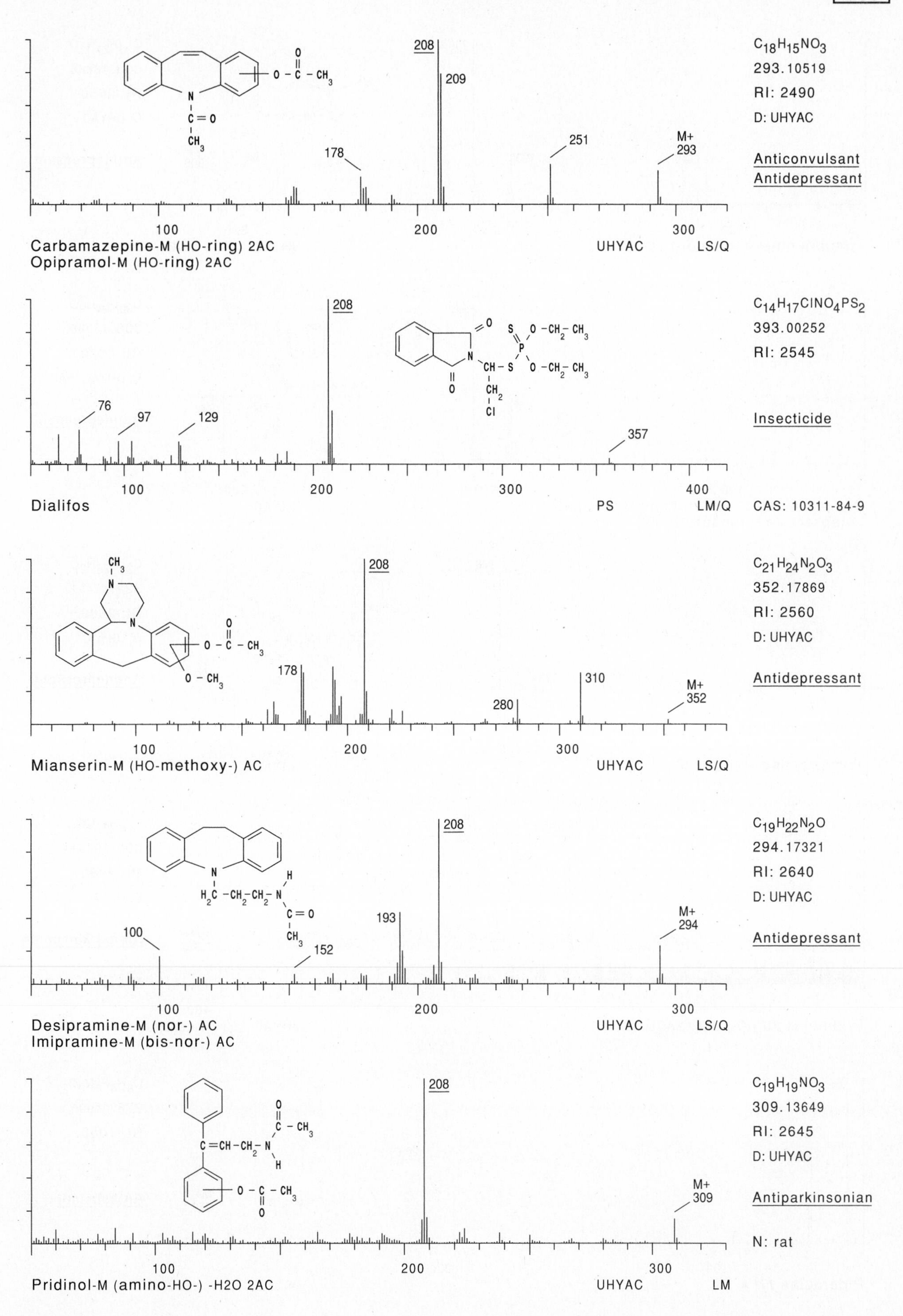

208
209
178
251
M+
293
100
200
300
Carbamazepine-M (HO-ring) 2AC
Opipramol-M (HO-ring) 2AC
UHYAC
LS/Q
$C_{18}H_{15}NO_3$
293.10519
RI: 2490
D: UHYAC
Anticonvulsant
Antidepressant
208
76
97
129
357
100
200
300
400
Dialifos
PS
LM/Q
$C_{14}H_{17}ClNO_4PS_2$
393.00252
RI: 2545
Insecticide
CAS: 10311-84-9
208
178
310
280
M+
352
100
200
300
Mianserin-M (HO-methoxy-) AC
UHYAC
LS/Q
$C_{21}H_{24}N_2O_3$
352.17869
RI: 2560
D: UHYAC
Antidepressant
208
193
100
152
M+
294
100
200
300
Desipramine-M (nor-) AC
Imipramine-M (bis-nor-) AC
UHYAC
LS/Q
$C_{19}H_{22}N_2O$
294.17321
RI: 2640
D: UHYAC
Antidepressant
208
M+
309
100
200
300
Pridinol-M (amino-HO-) -H2O 2AC
UHYAC
LM
$C_{19}H_{19}NO_3$
309.13649
RI: 2645
D: UHYAC
Antiparkinsonian
N: rat

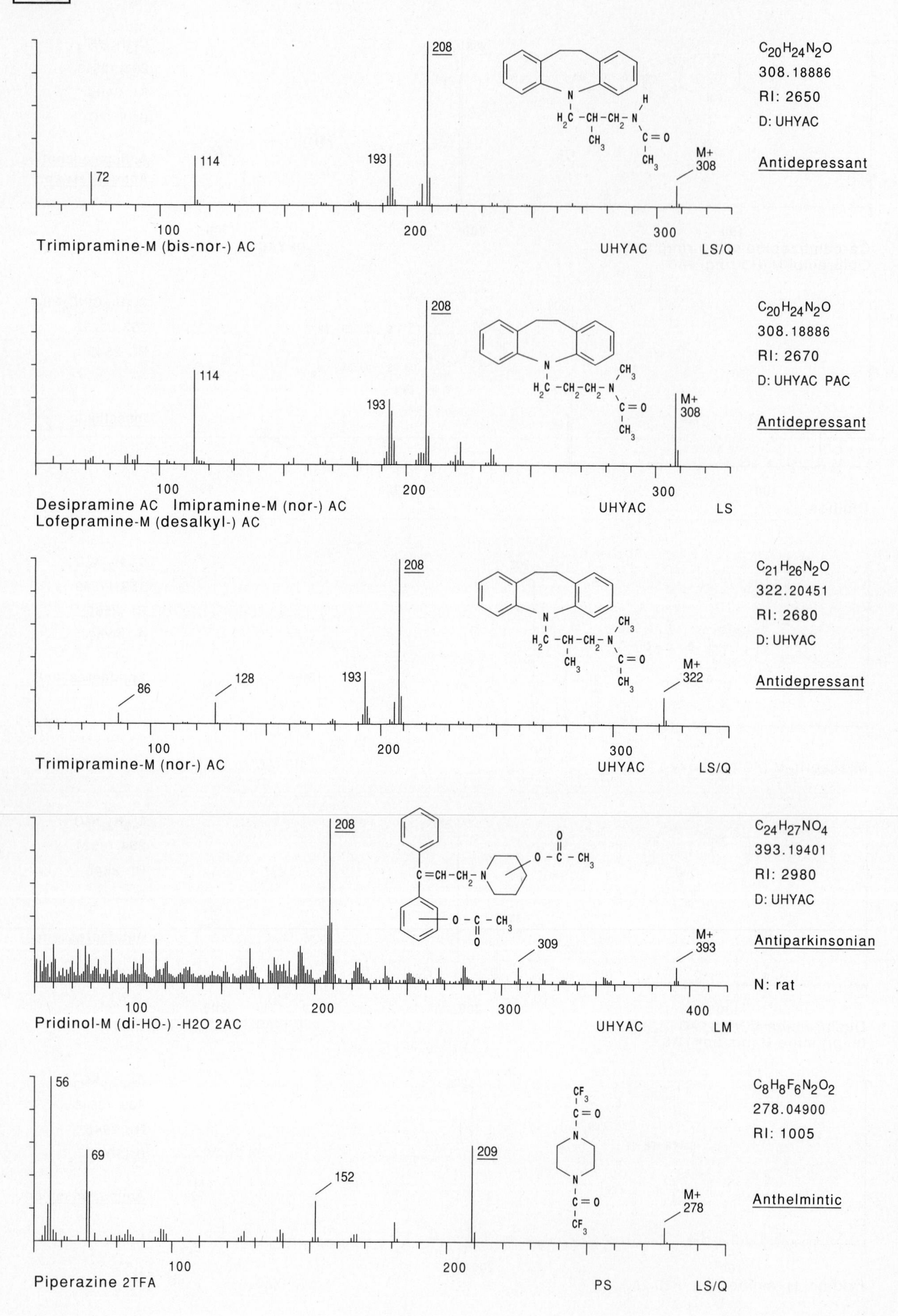
208
114
72
193
M+
308
$C_{20}H_{24}N_2O$
308.18886
RI: 2650
D: UHYAC
Antidepressant
100
200
300
Trimipramine-M (bis-nor-) AC
UHYAC
LS/Q
208
114
193
M+
308
$C_{20}H_{24}N_2O$
308.18886
RI: 2670
D: UHYAC PAC
Antidepressant
Desipramine AC Imipramine-M (nor-) AC
Lofepramine-M (desalkyl-) AC
UHYAC
LS
208
86
128
193
M+
322
$C_{21}H_{26}N_2O$
322.20451
RI: 2680
D: UHYAC
Antidepressant
Trimipramine-M (nor-) AC
UHYAC
LS/Q
208
309
M+
393
$C_{24}H_{27}NO_4$
393.19401
RI: 2980
D: UHYAC
Antiparkinsonian
N: rat
400
Pridinol-M (di-HO-) -H2O 2AC
UHYAC
LM
56
69
152
209
M+
278
$C_8H_8F_6N_2O_2$
278.04900
RI: 1005
Anthelmintic
Piperazine 2TFA
PS
LS/Q

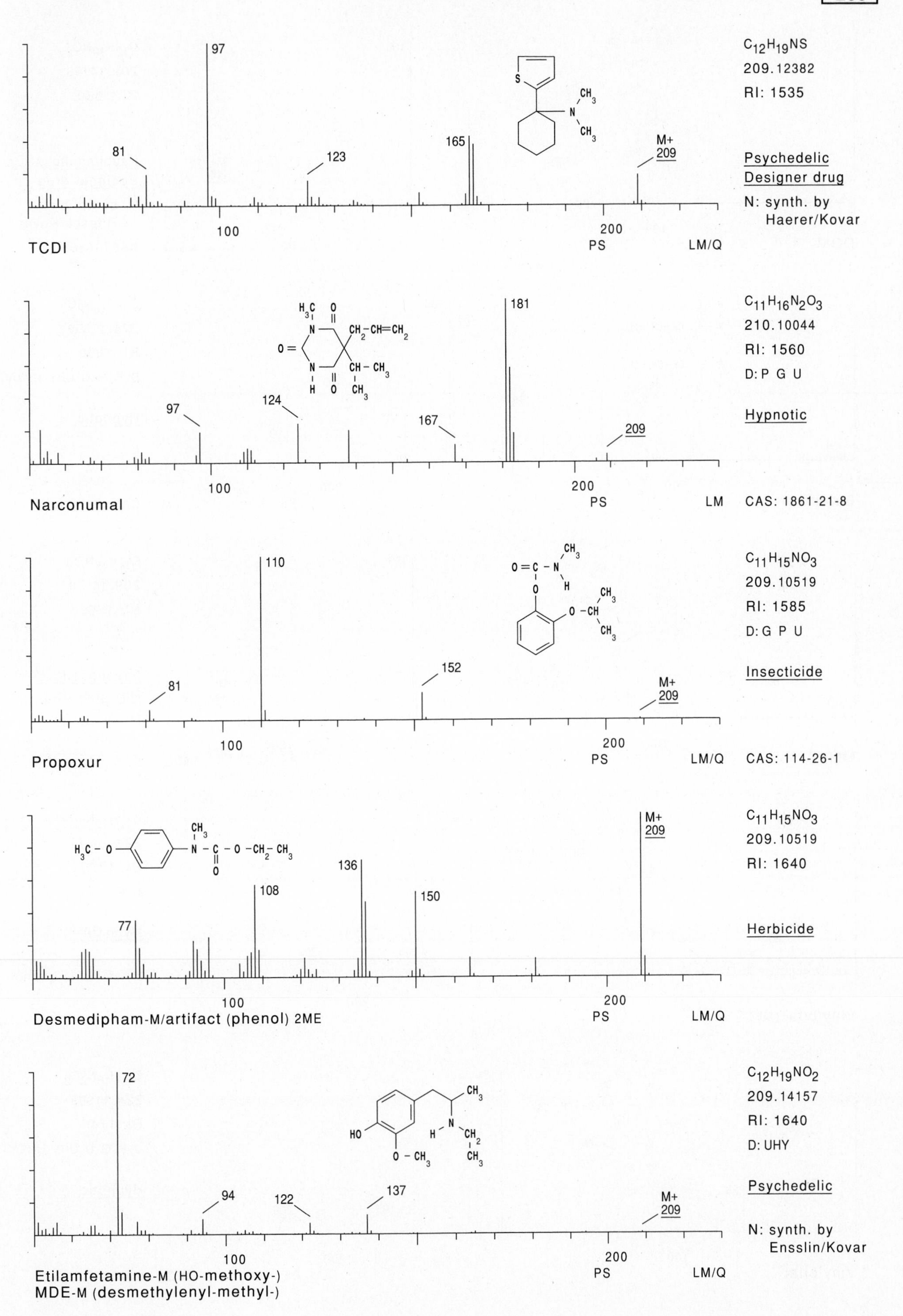
97
81
123
165
M+
209
100
200
PS
LM/Q
TCDI
$C_{12}H_{19}NS$
209.12382
RI: 1535
Psychedelic
Designer drug
N: synth. by
Haerer/Kovar
181
97
124
167
209
100
200
PS
LM
Narconumal
$C_{11}H_{16}N_2O_3$
210.10044
RI: 1560
D: P G U
Hypnotic
CAS: 1861-21-8
110
81
152
M+
209
100
200
PS
LM/Q
Propoxur
$C_{11}H_{15}NO_3$
209.10519
RI: 1585
D: G P U
Insecticide
CAS: 114-26-1
M+
209
136
108
150
77
100
200
PS
LM/Q
Desmedipham-M/artifact (phenol) 2ME
$C_{11}H_{15}NO_3$
209.10519
RI: 1640
Herbicide
72
94
122
137
M+
209
100
200
PS
LM/Q
Etilamfetamine-M (HO-methoxy-)
MDE-M (desmethylenyl-methyl-)
$C_{12}H_{19}NO_2$
209.14157
RI: 1640
D: UHY
Psychedelic
N: synth. by
Ensslin/Kovar

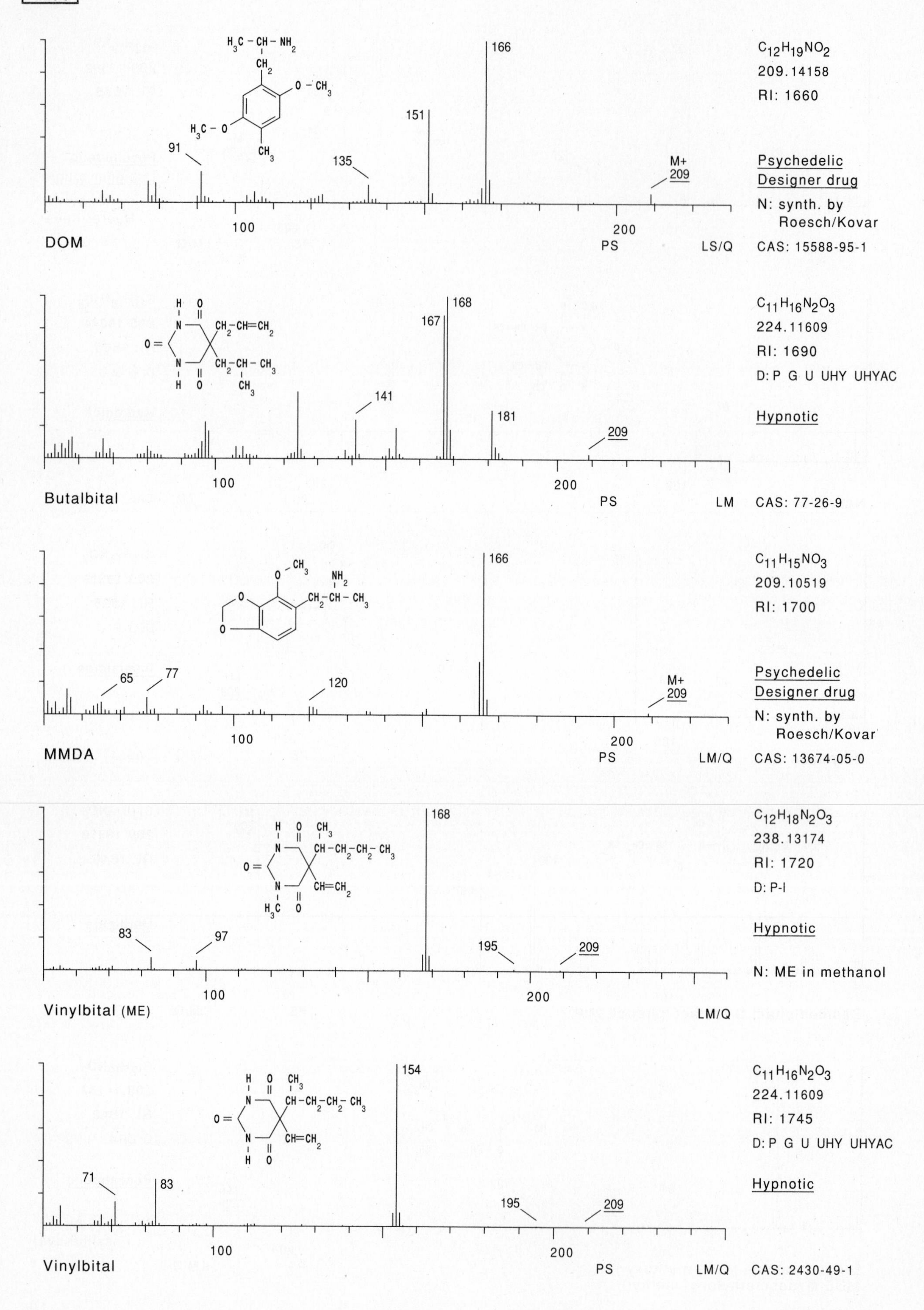

DOM
91
135
151
166
M+
209
100
200
PS
LS/Q
C12H19NO2
209.14158
RI: 1660
Psychedelic
Designer drug
N: synth. by Roesch/Kovar
CAS: 15588-95-1
Butalbital
141
167
168
181
209
100
200
PS
LM
C11H16N2O3
224.11609
RI: 1690
D: P G U UHY UHYAC
Hypnotic
CAS: 77-26-9
MMDA
65
77
120
166
M+
209
100
200
PS
LM/Q
C11H15NO3
209.10519
RI: 1700
Psychedelic
Designer drug
N: synth. by Roesch/Kovar
CAS: 13674-05-0
Vinylbital (ME)
83
97
168
195
209
100
200
LM/Q
C12H18N2O3
238.13174
RI: 1720
D: P-I
Hypnotic
N: ME in methanol
Vinylbital
71
83
154
195
209
100
200
PS
LM/Q
C11H16N2O3
224.11609
RI: 1745
D: P G U UHY UHYAC
Hypnotic
CAS: 2430-49-1

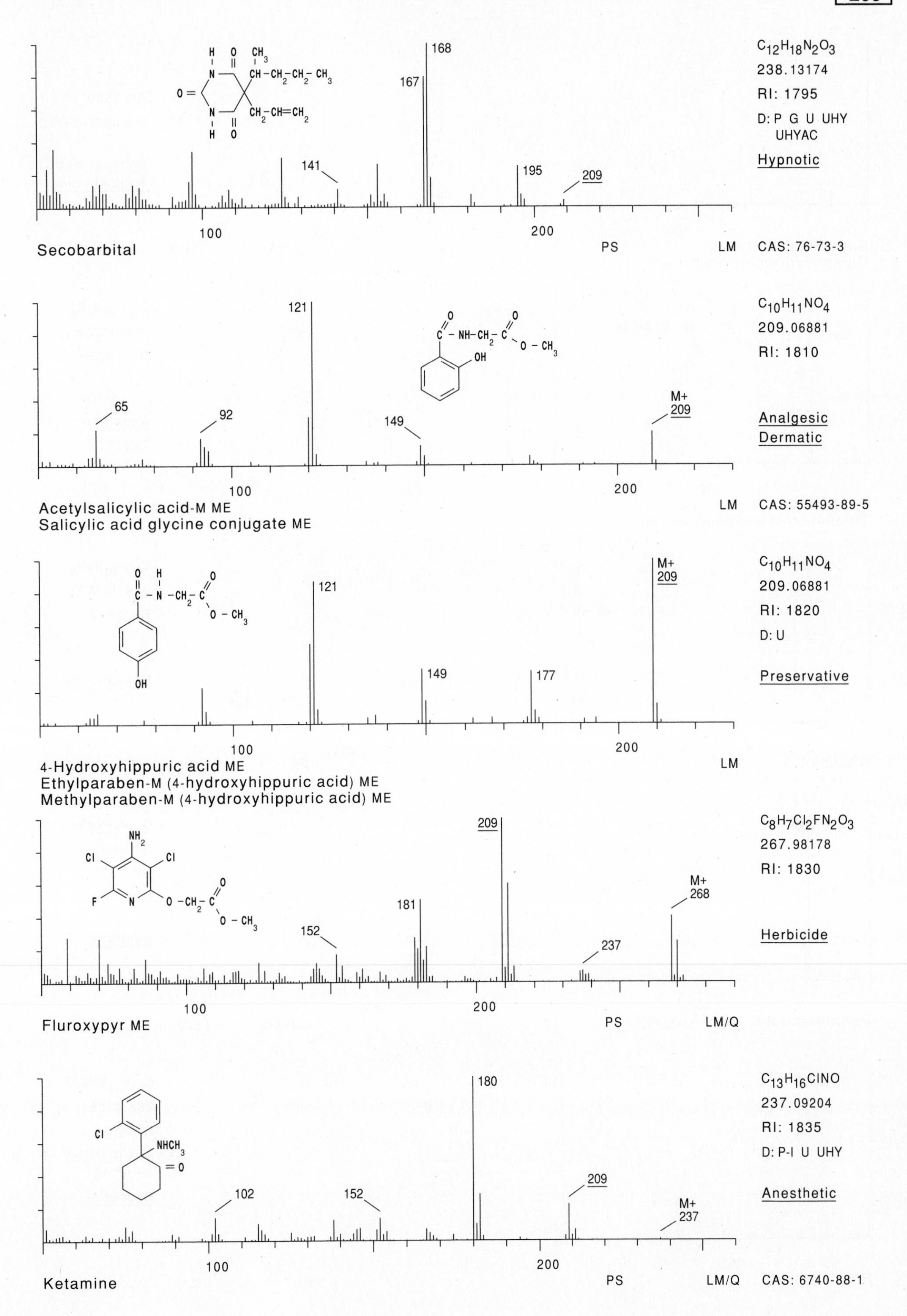

168
167
141
195
209
100
200
Secobarbital
PS
LM
C12H18N2O3
238.13174
RI: 1795
D: P G U UHY
UHYAC
Hypnotic
CAS: 76-73-3
121
65
92
149
M+
209
100
200
Acetylsalicylic acid-M ME
Salicylic acid glycine conjugate ME
LM
C10H11NO4
209.06881
RI: 1810
Analgesic
Dermatic
CAS: 55493-89-5
M+
209
121
149
177
100
200
4-Hydroxyhippuric acid ME
Ethylparaben-M (4-hydroxyhippuric acid) ME
Methylparaben-M (4-hydroxyhippuric acid) ME
LM
C10H11NO4
209.06881
RI: 1820
D: U
Preservative
209
181
152
237
M+
268
100
200
Fluroxypyr ME
PS
LM/Q
C8H7Cl2FN2O3
267.98178
RI: 1830
Herbicide
180
102
152
209
M+
237
100
200
Ketamine
PS
LM/Q
C13H16ClNO
237.09204
RI: 1835
D: P-I U UHY
Anesthetic
CAS: 6740-88-1

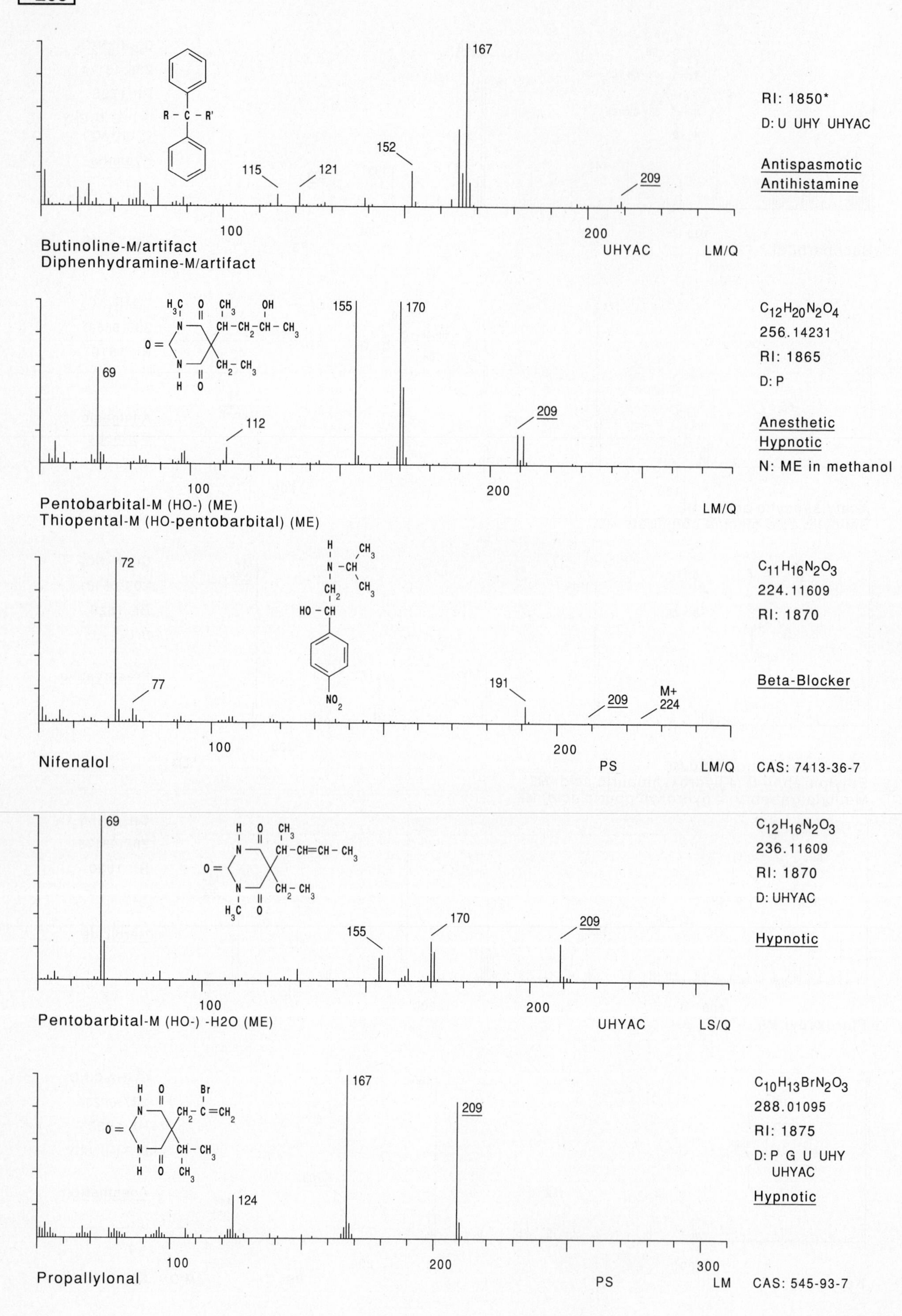

167
152
115
121
209
100
200
UHYAC
LM/Q
RI: 1850*
D: U UHY UHYAC
Antispasmotic
Antihistamine
Butinoline-M/artifact
Diphenhydramine-M/artifact
155
170
69
112
209
100
200
LM/Q
C12H20N2O4
256.14231
RI: 1865
D: P
Anesthetic
Hypnotic
N: ME in methanol
Pentobarbital-M (HO-) (ME)
Thiopental-M (HO-pentobarbital) (ME)
72
77
191
209
M+
224
100
200
PS
LM/Q
C11H16N2O3
224.11609
RI: 1870
Beta-Blocker
Nifenalol
CAS: 7413-36-7
69
155
170
209
100
200
UHYAC
LS/Q
C12H16N2O3
236.11609
RI: 1870
D: UHYAC
Hypnotic
Pentobarbital-M (HO-) -H2O (ME)
167
209
124
100
200
300
PS
LM
C10H13BrN2O3
288.01095
RI: 1875
D: P G U UHY
UHYAC
Hypnotic
Propallylonal
CAS: 545-93-7

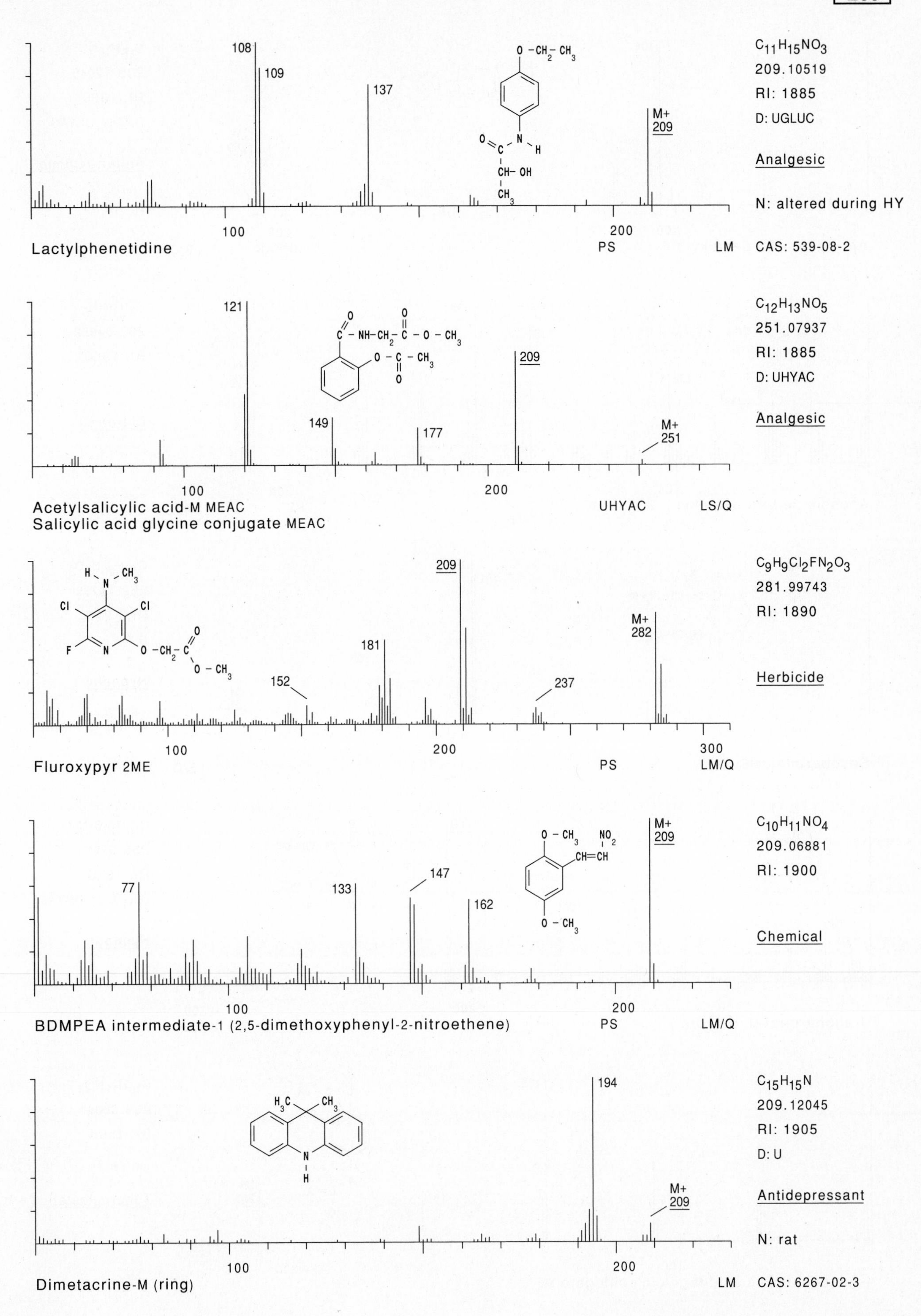

108
109
137
M+
209
100
200
Lactylphenetidine
PS
LM
$C_{11}H_{15}NO_3$
209.10519
RI: 1885
D: UGLUC
Analgesic
N: altered during HY
CAS: 539-08-2
121
149
177
209
M+
251
100
200
Acetylsalicylic acid-M MEAC
Salicylic acid glycine conjugate MEAC
UHYAC
LS/Q
$C_{12}H_{13}NO_5$
251.07937
RI: 1885
D: UHYAC
Analgesic
209
181
152
237
M+
282
100
200
300
Fluroxypyr 2ME
PS
LM/Q
$C_9H_9Cl_2FN_2O_3$
281.99743
RI: 1890
Herbicide
M+
209
77
133
147
162
100
200
BDMPEA intermediate-1 (2,5-dimethoxyphenyl-2-nitroethene)
PS
LM/Q
$C_{10}H_{11}NO_4$
209.06881
RI: 1900
Chemical
194
M+
209
100
200
Dimetacrine-M (ring)
LM
$C_{15}H_{15}N$
209.12045
RI: 1905
D: U
Antidepressant
N: rat
CAS: 6267-02-3

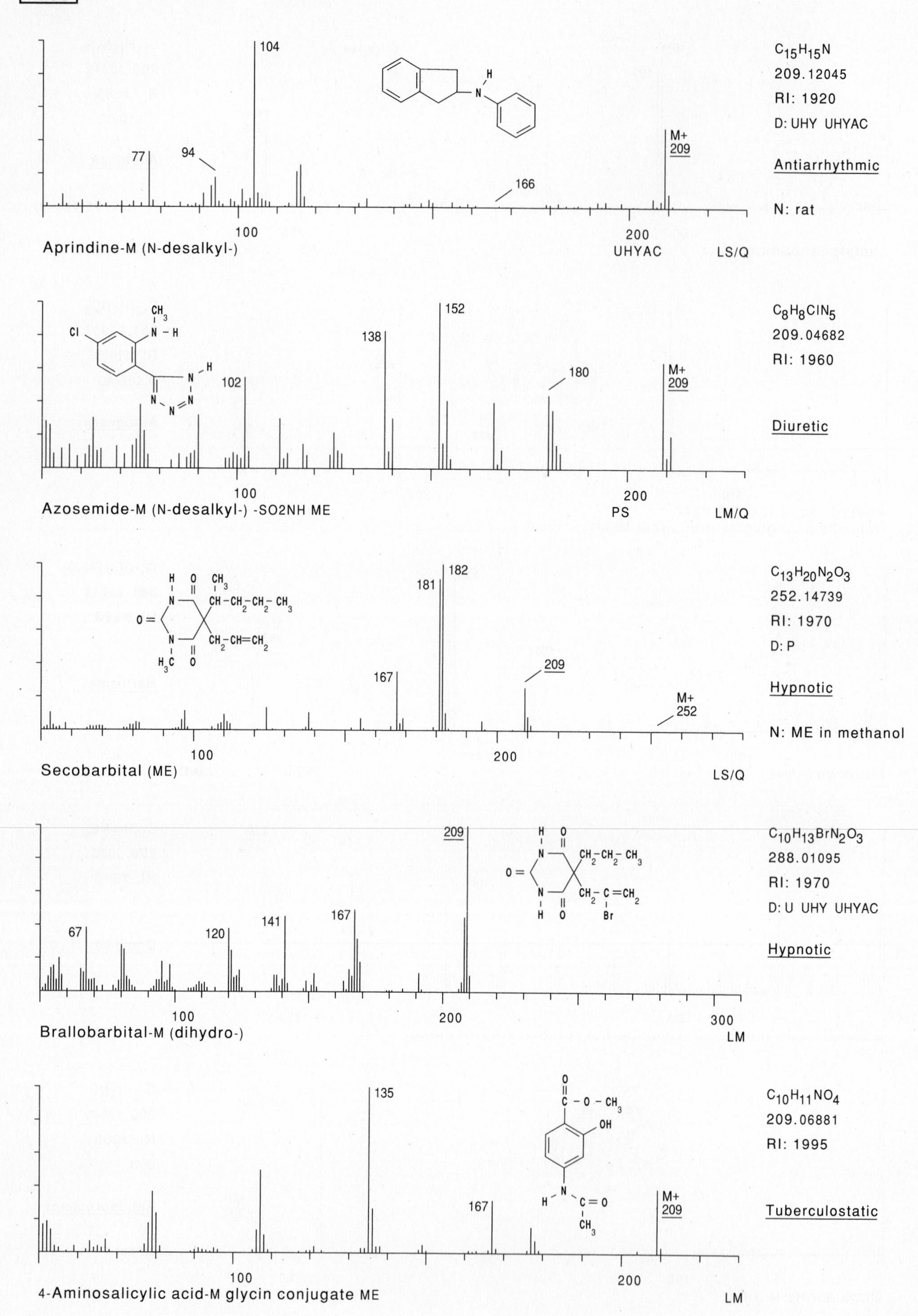

104
77
94
166
M+
209
100
200
Aprindine-M (N-desalkyl-)
UHYAC
LS/Q
$C_{15}H_{15}N$
209.12045
RI: 1920
D: UHY UHYAC
Antiarrhythmic
N: rat
152
138
102
180
M+
209
100
200
Azosemide-M (N-desalkyl-) -SO2NH ME
PS
LM/Q
$C_8H_8ClN_5$
209.04682
RI: 1960
Diuretic
182
181
167
209
M+
252
100
200
Secobarbital (ME)
LS/Q
$C_{13}H_{20}N_2O_3$
252.14739
RI: 1970
D: P
Hypnotic
N: ME in methanol
209
67
120
141
167
100
200
300
Brallobarbital-M (dihydro-)
LM
$C_{10}H_{13}BrN_2O_3$
288.01095
RI: 1970
D: U UHY UHYAC
Hypnotic
135
167
M+
209
100
200
4-Aminosalicylic acid-M glycin conjugate ME
LM
$C_{10}H_{11}NO_4$
209.06881
RI: 1995
Tuberculostatic

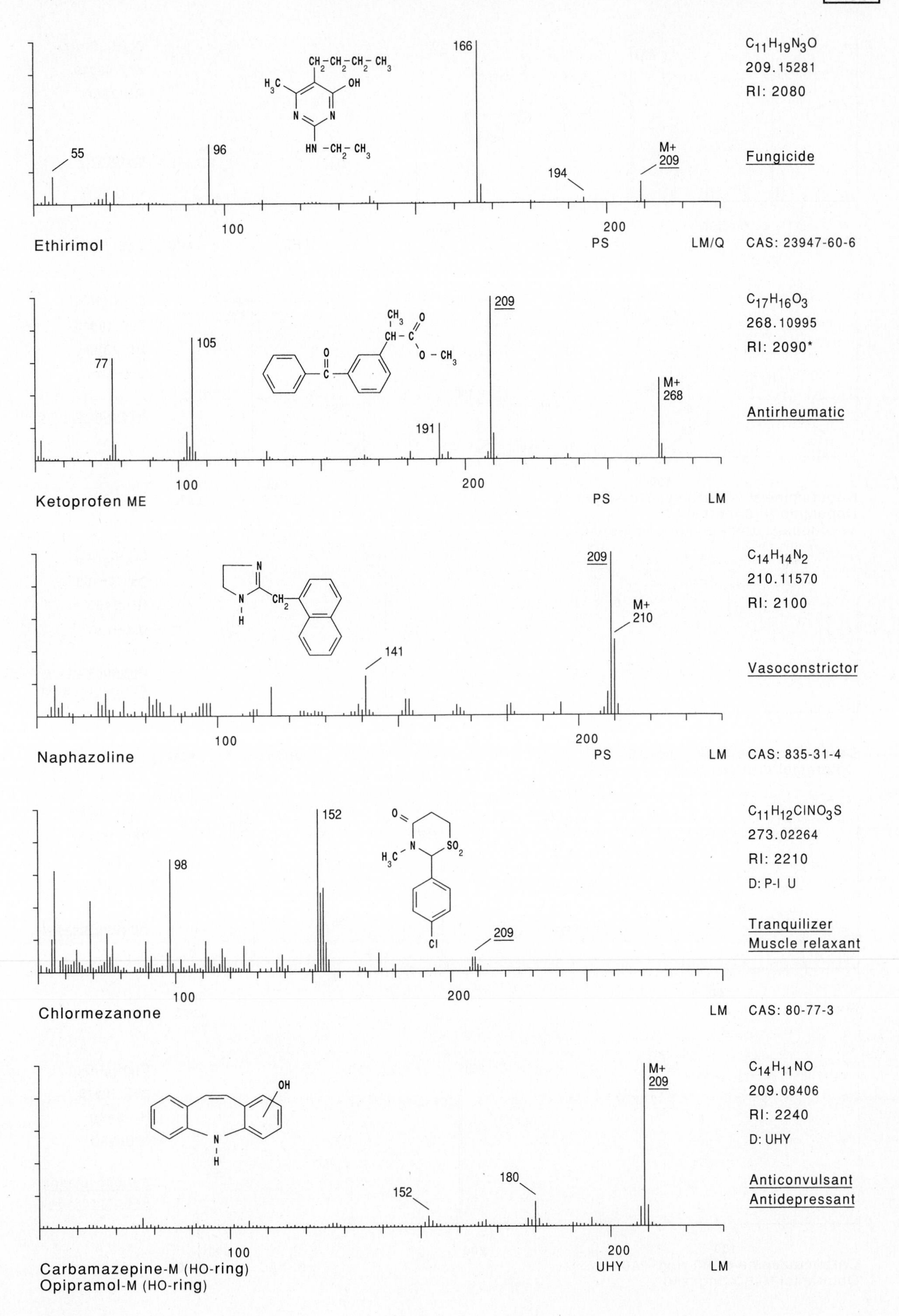

166
55
96
194
M+
209
100
200
PS
LM/Q
Ethirimol
$C_{11}H_{19}N_3O$
209.15281
RI: 2080
Fungicide
CAS: 23947-60-6
209
77
105
191
M+
268
100
200
PS
LM
Ketoprofen ME
$C_{17}H_{16}O_3$
268.10995
RI: 2090*
Antirheumatic
209
M+
210
141
100
200
PS
LM
Naphazoline
$C_{14}H_{14}N_2$
210.11570
RI: 2100
Vasoconstrictor
CAS: 835-31-4
152
98
209
100
200
LM
Chlormezanone
$C_{11}H_{12}ClNO_3S$
273.02264
RI: 2210
D: P-I U
Tranquilizer
Muscle relaxant
CAS: 80-77-3
M+
209
180
152
100
200
UHY
LM
Carbamazepine-M (HO-ring)
Opipramol-M (HO-ring)
$C_{14}H_{11}NO$
209.08406
RI: 2240
D: UHY
Anticonvulsant
Antidepressant

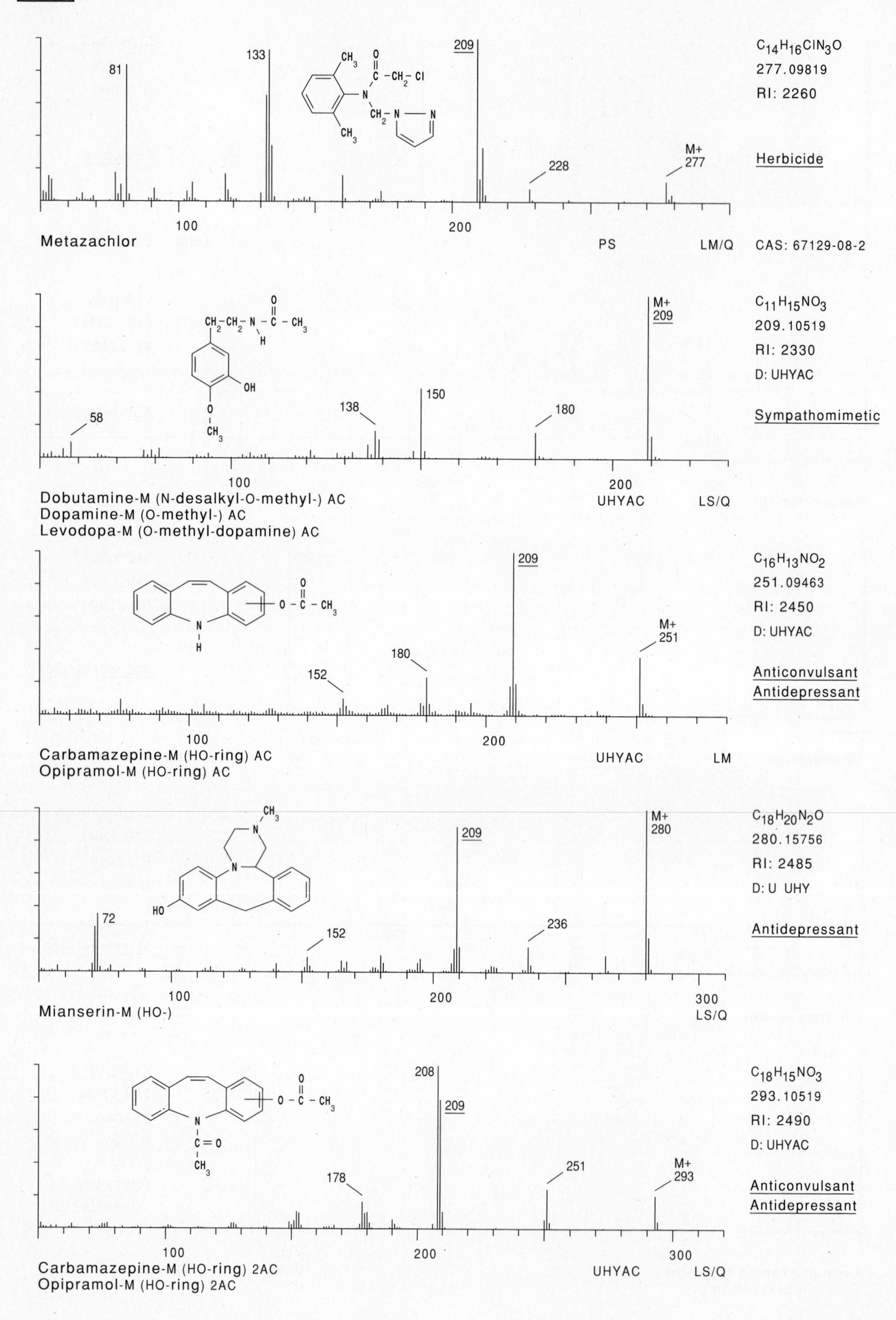

Metazachlor
$C_{14}H_{16}ClN_3O$
277.09819
RI: 2260
Herbicide
PS
LM/Q
CAS: 67129-08-2
81
133
209
228
M+
277
Dobutamine-M (N-desalkyl-O-methyl-) AC
Dopamine-M (O-methyl-) AC
Levodopa-M (O-methyl-dopamine) AC
$C_{11}H_{15}NO_3$
209.10519
RI: 2330
D: UHYAC
Sympathomimetic
UHYAC
LS/Q
58
138
150
180
M+
209
Carbamazepine-M (HO-ring) AC
Opipramol-M (HO-ring) AC
$C_{16}H_{13}NO_2$
251.09463
RI: 2450
D: UHYAC
Anticonvulsant
Antidepressant
UHYAC
LM
152
180
209
M+
251
Mianserin-M (HO-)
$C_{18}H_{20}N_2O$
280.15756
RI: 2485
D: U UHY
Antidepressant
LS/Q
72
152
209
236
M+
280
Carbamazepine-M (HO-ring) 2AC
Opipramol-M (HO-ring) 2AC
$C_{18}H_{15}NO_3$
293.10519
RI: 2490
D: UHYAC
Anticonvulsant
Antidepressant
UHYAC
LS/Q
178
208
209
251
M+
293
100
200
300

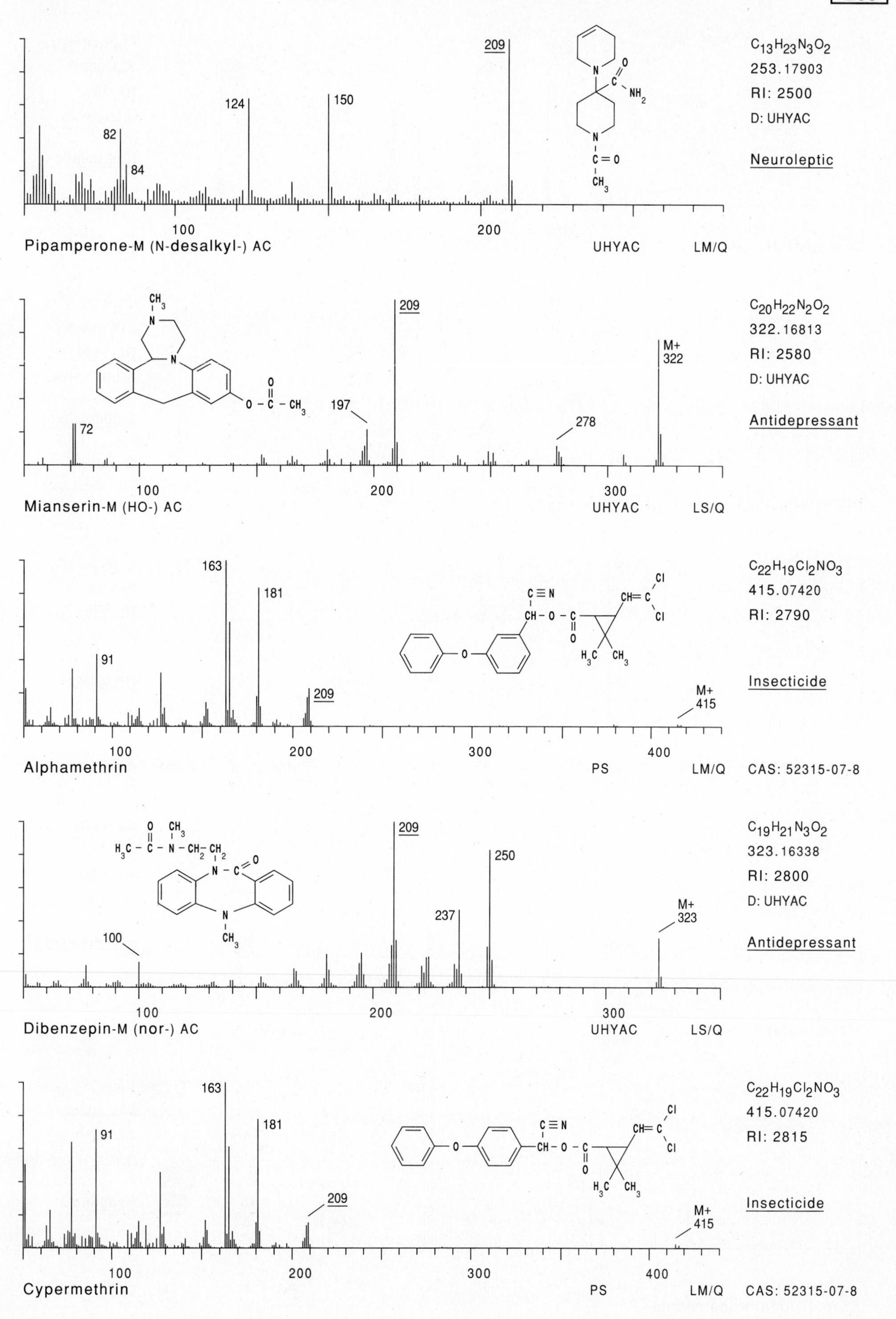

209
150
124
82
84
100
200
Pipamperone-M (N-desalkyl-) AC
UHYAC
LM/Q
C13H23N3O2
253.17903
RI: 2500
D: UHYAC
Neuroleptic
209
M+
322
197
278
72
100
200
300
Mianserin-M (HO-) AC
UHYAC
LS/Q
C20H22N2O2
322.16813
RI: 2580
D: UHYAC
Antidepressant
163
181
91
209
M+
415
100
200
300
400
Alphamethrin
PS
LM/Q
C22H19Cl2NO3
415.07420
RI: 2790
Insecticide
CAS: 52315-07-8
209
250
237
M+
323
100
100
200
300
Dibenzepin-M (nor-) AC
UHYAC
LS/Q
C19H21N3O2
323.16338
RI: 2800
D: UHYAC
Antidepressant
163
181
91
209
M+
415
100
200
300
400
Cypermethrin
PS
LM/Q
C22H19Cl2NO3
415.07420
RI: 2815
Insecticide
CAS: 52315-07-8

209

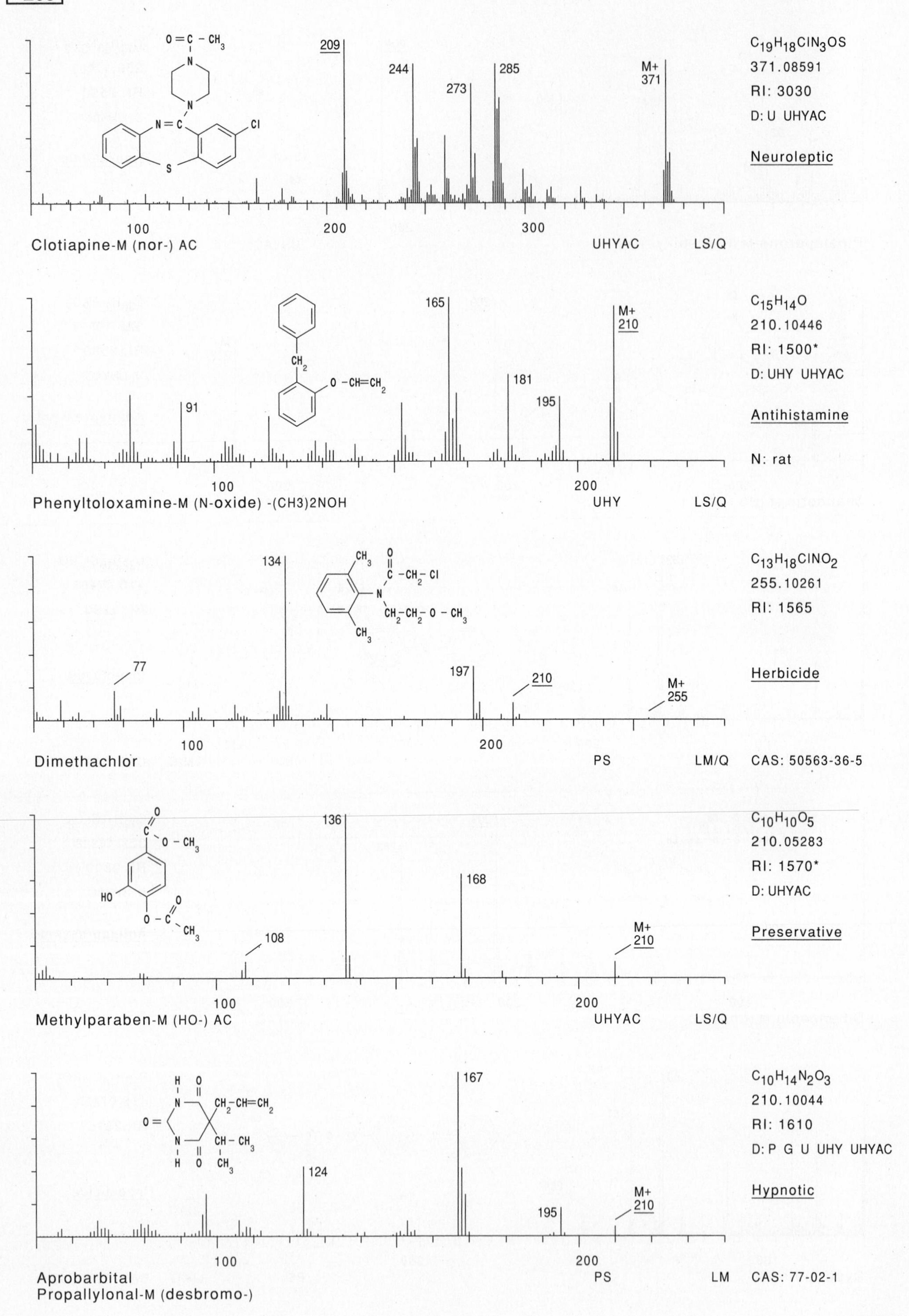

Clotiapine-M (nor-) AC
UHYAC
LS/Q
$C_{19}H_{18}ClN_3OS$
371.08591
RI: 3030
D: U UHYAC
Neuroleptic
209
244
273
285
M+
371
Phenyltoloxamine-M (N-oxide) -(CH3)2NOH
UHY
LS/Q
$C_{15}H_{14}O$
210.10446
RI: 1500*
D: UHY UHYAC
Antihistamine
N: rat
91
165
181
195
M+
210
Dimethachlor
PS
LM/Q
$C_{13}H_{18}ClNO_2$
255.10261
RI: 1565
Herbicide
CAS: 50563-36-5
77
134
197
210
M+
255
Methylparaben-M (HO-) AC
UHYAC
LS/Q
$C_{10}H_{10}O_5$
210.05283
RI: 1570*
D: UHYAC
Preservative
108
136
168
M+
210
Aprobarbital
Propallylonal-M (desbromo-)
PS
LM
$C_{10}H_{14}N_2O_3$
210.10044
RI: 1610
D: P G U UHY UHYAC
Hypnotic
CAS: 77-02-1
124
167
195
M+
210

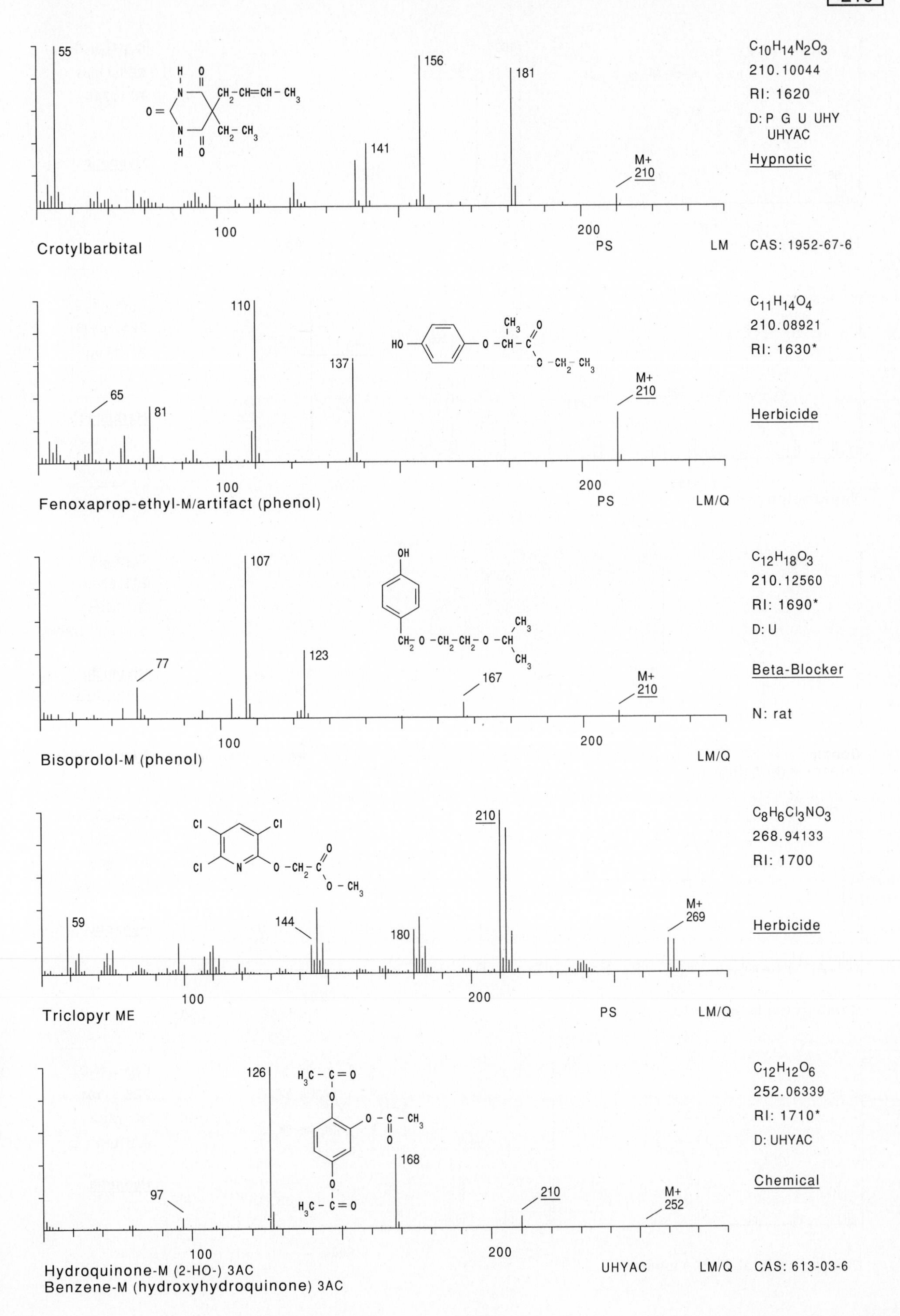

$C_{10}H_{14}N_2O_3$
210.10044
RI: 1620
D: P G U UHY UHYAC
Hypnotic
CAS: 1952-67-6

$C_{11}H_{14}O_4$
210.08921
RI: 1630*
Herbicide

$C_{12}H_{18}O_3$
210.12560
RI: 1690*
D: U
Beta-Blocker
N: rat

$C_8H_6Cl_3NO_3$
268.94133
RI: 1700
Herbicide

$C_{12}H_{12}O_6$
252.06339
RI: 1710*
D: UHYAC
Chemical
CAS: 613-03-6

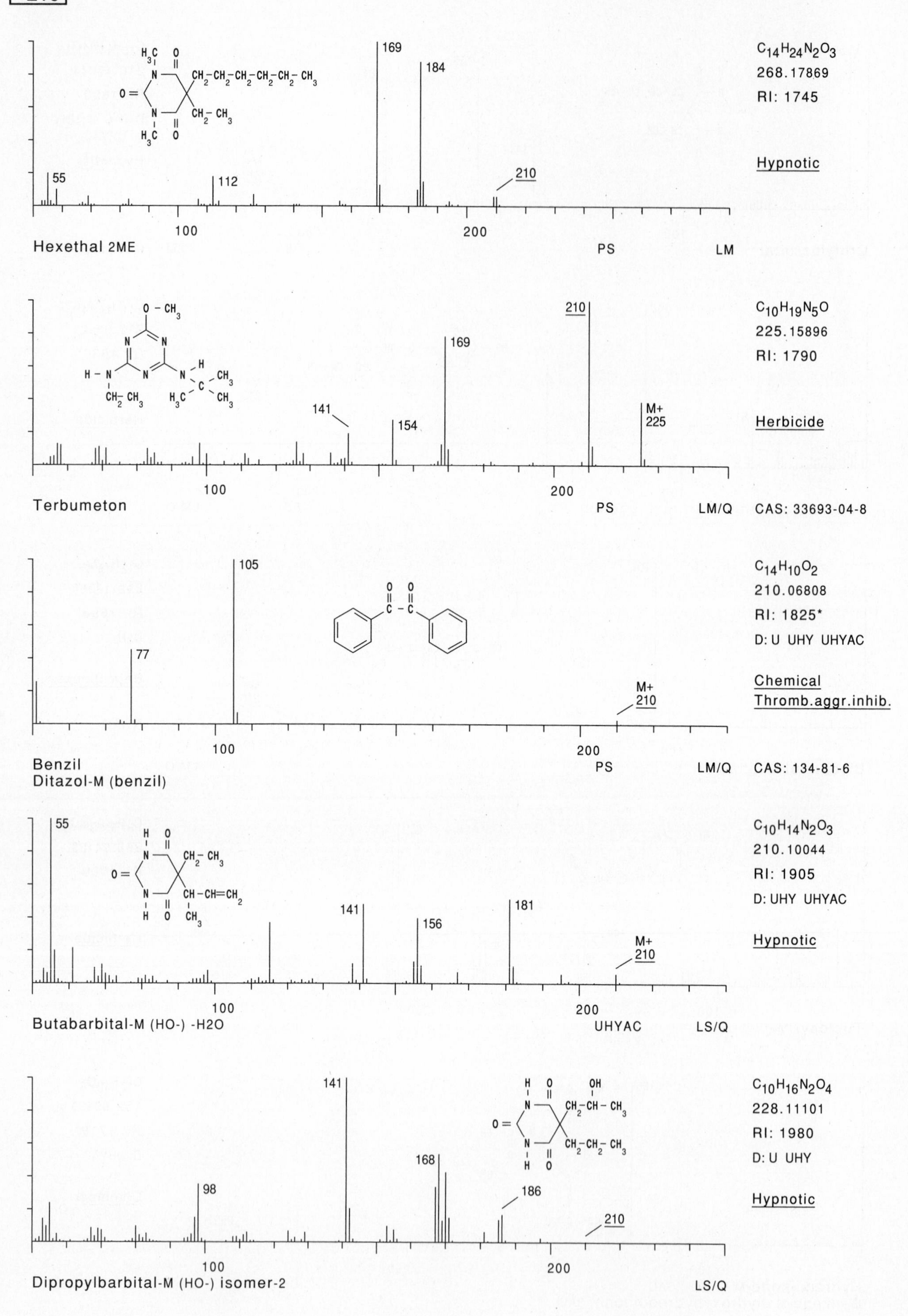
210
169
184
55
112
210
100
200
Hexethal 2ME
PS
LM
$C_{14}H_{24}N_2O_3$
268.17869
RI: 1745
Hypnotic
210
169
141
154
M+
225
100
200
Terbumeton
PS
LM/Q
$C_{10}H_{19}N_5O$
225.15896
RI: 1790
Herbicide
CAS: 33693-04-8
105
77
M+
210
100
200
Benzil
Ditazol-M (benzil)
PS
LM/Q
$C_{14}H_{10}O_2$
210.06808
RI: 1825*
D: U UHY UHYAC
Chemical
Thromb.aggr.inhib.
CAS: 134-81-6
55
141
156
181
M+
210
100
200
Butabarbital-M (HO-) -H2O
UHYAC
LS/Q
$C_{10}H_{14}N_2O_3$
210.10044
RI: 1905
D: UHY UHYAC
Hypnotic
141
98
168
186
210
100
200
Dipropylbarbital-M (HO-) isomer-2
LS/Q
$C_{10}H_{16}N_2O_4$
228.11101
RI: 1980
D: U UHY
Hypnotic

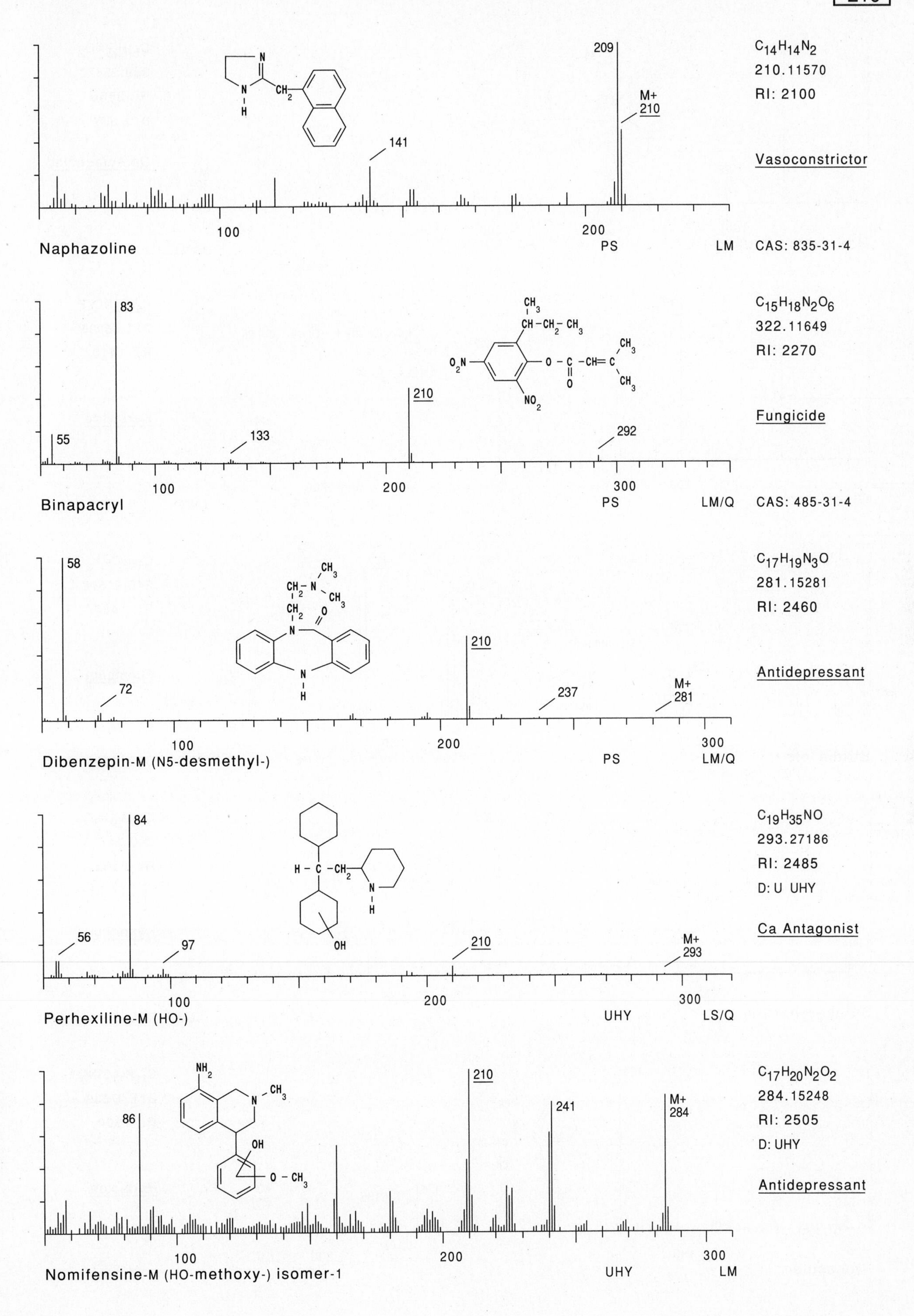

Naphazoline
$C_{14}H_{14}N_2$
210.11570
RI: 2100
Vasoconstrictor
CAS: 835-31-4
PS
LM
Binapacryl
$C_{15}H_{18}N_2O_6$
322.11649
RI: 2270
Fungicide
CAS: 485-31-4
PS
LM/Q
Dibenzepin-M (N5-desmethyl-)
$C_{17}H_{19}N_3O$
281.15281
RI: 2460
Antidepressant
PS
LM/Q
Perhexiline-M (HO-)
$C_{19}H_{35}NO$
293.27186
RI: 2485
D: U UHY
Ca Antagonist
UHY
LS/Q
Nomifensine-M (HO-methoxy-) isomer-1
$C_{17}H_{20}N_2O_2$
284.15248
RI: 2505
D: UHY
Antidepressant
UHY
LM

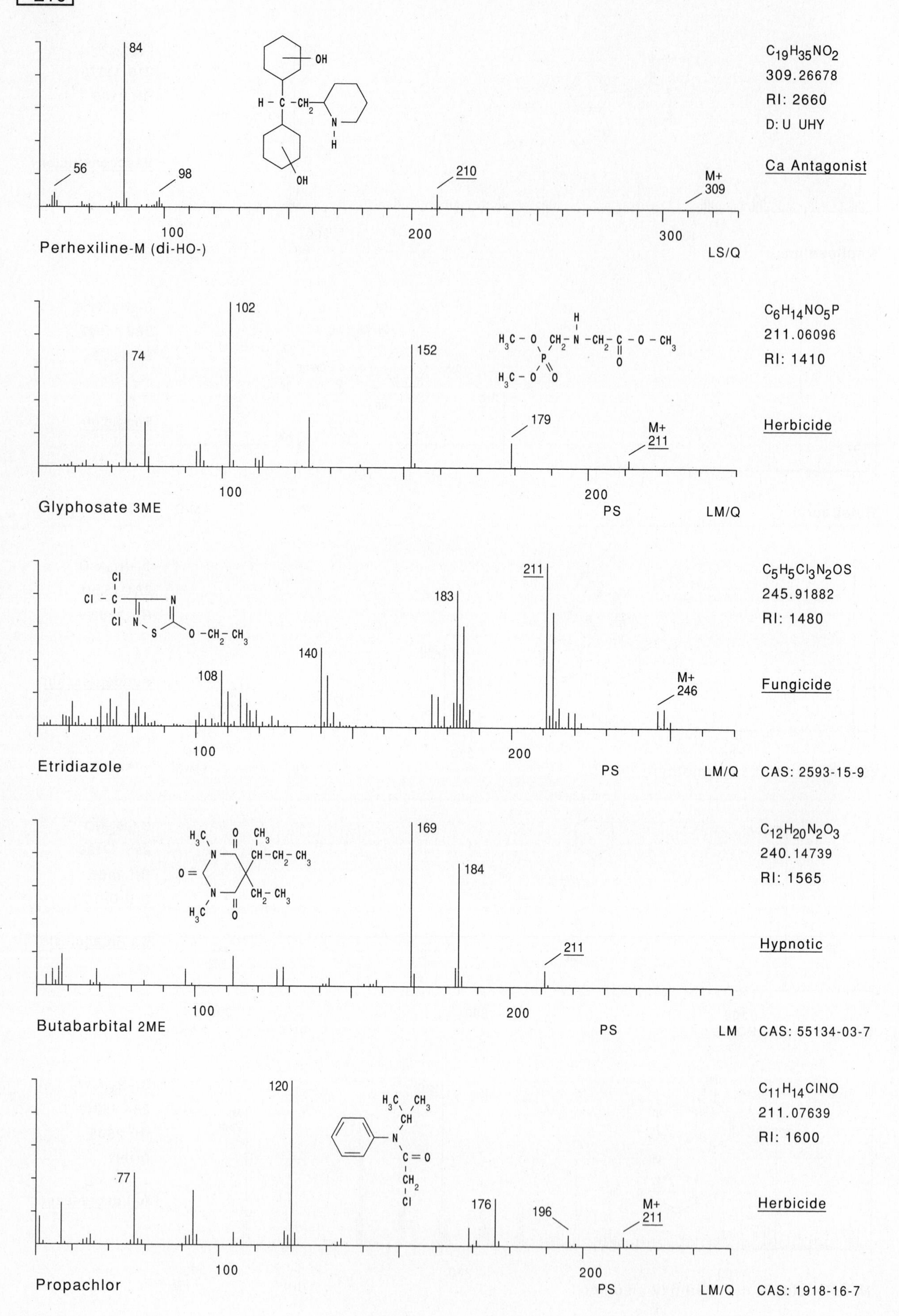

84
56
98
210
M+
309
OH
H - C - CH2
N
H
100
200
300
Perhexiline-M (di-HO-)
LS/Q
C19H35NO2
309.26678
RI: 2660
D: U UHY
Ca Antagonist
102
74
152
179
M+
211
Glyphosate 3ME
PS
LM/Q
C6H14NO5P
211.06096
RI: 1410
Herbicide
211
183
140
108
M+
246
Etridiazole
C5H5Cl3N2OS
245.91882
RI: 1480
Fungicide
CAS: 2593-15-9
169
184
211
Butabarbital 2ME
LM
C12H20N2O3
240.14739
RI: 1565
Hypnotic
CAS: 55134-03-7
120
77
176
196
M+
211
Propachlor
C11H14ClNO
211.07639
RI: 1600
Herbicide
CAS: 1918-16-7

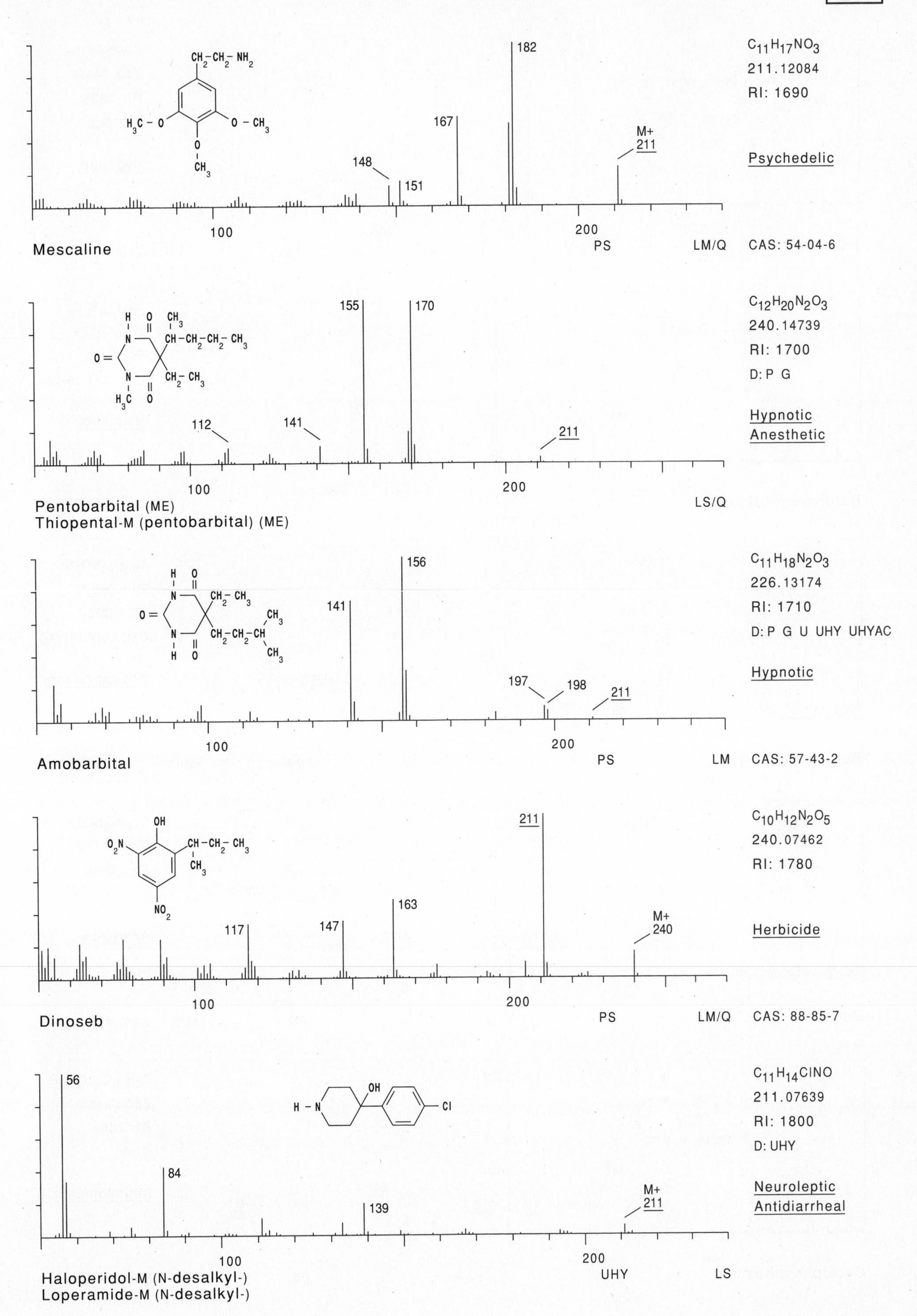

Mescaline
C11H17NO3
211.12084
RI: 1690
Psychedelic
CAS: 54-04-6
PS
LM/Q
182
167
148
151
M+
211
100
200
Pentobarbital (ME)
Thiopental-M (pentobarbital) (ME)
C12H20N2O3
240.14739
RI: 1700
D: P G
Hypnotic
Anesthetic
LS/Q
155
170
112
141
211
Amobarbital
C11H18N2O3
226.13174
RI: 1710
D: P G U UHY UHYAC
Hypnotic
CAS: 57-43-2
PS
LM
156
141
197
198
211
Dinoseb
C10H12N2O5
240.07462
RI: 1780
Herbicide
CAS: 88-85-7
PS
LM/Q
211
163
147
117
M+
240
Haloperidol-M (N-desalkyl-)
Loperamide-M (N-desalkyl-)
C11H14ClNO
211.07639
RI: 1800
D: UHY
Neuroleptic
Antidiarrheal
UHY
LS
56
84
139
M+
211

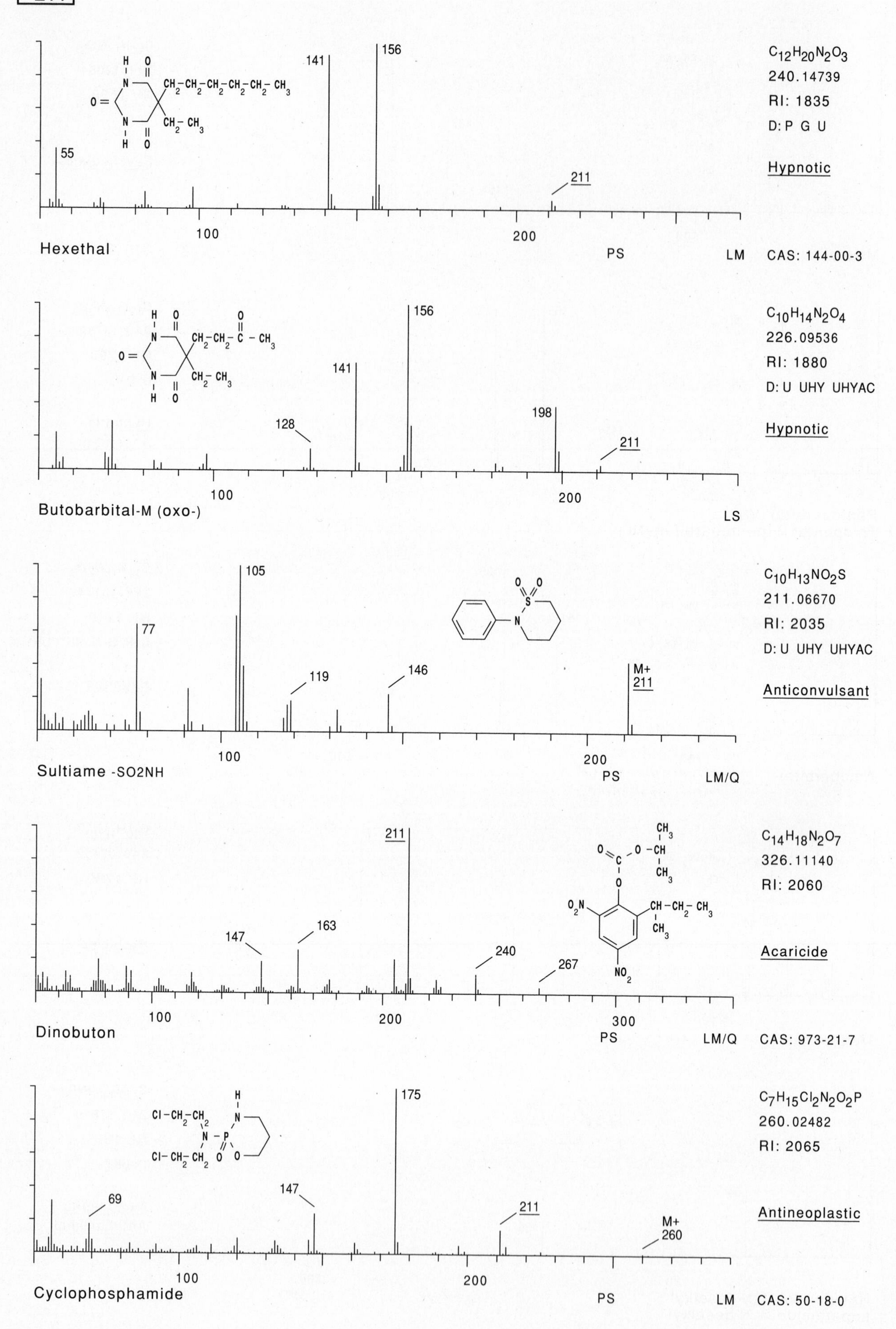

Hexethal
$C_{12}H_{20}N_2O_3$
240.14739
RI: 1835
D:P G U
Hypnotic
PS
LM
CAS: 144-00-3
Butobarbital-M (oxo-)
$C_{10}H_{14}N_2O_4$
226.09536
RI: 1880
D: U UHY UHYAC
Hypnotic
LS
Sultiame -SO2NH
$C_{10}H_{13}NO_2S$
211.06670
RI: 2035
D: U UHY UHYAC
Anticonvulsant
PS
LM/Q
Dinobuton
$C_{14}H_{18}N_2O_7$
326.11140
RI: 2060
Acaricide
PS
LM/Q
CAS: 973-21-7
Cyclophosphamide
$C_7H_{15}Cl_2N_2O_2P$
260.02482
RI: 2065
Antineoplastic
PS
LM
CAS: 50-18-0

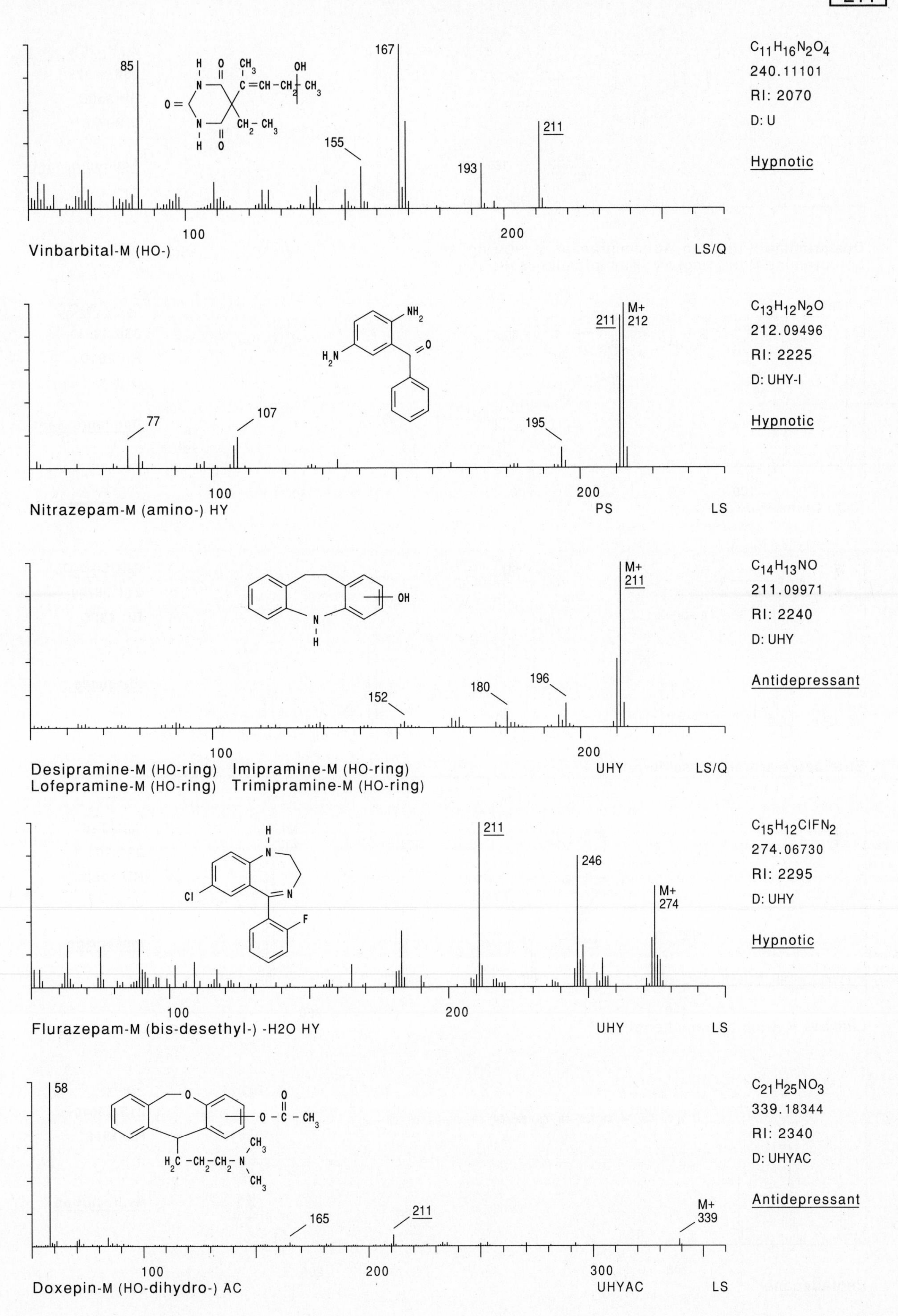

85
167
155
193
211
100
200
Vinbarbital-M (HO-)
LS/Q
$C_{11}H_{16}N_2O_4$
240.11101
RI: 2070
D: U
Hypnotic
77
107
195
211
M+
212
100
200
Nitrazepam-M (amino-) HY
PS
LS
$C_{13}H_{12}N_2O$
212.09496
RI: 2225
D: UHY-I
Hypnotic
152
180
196
M+
211
100
200
Desipramine-M (HO-ring) Imipramine-M (HO-ring)
Lofepramine-M (HO-ring) Trimipramine-M (HO-ring)
UHY
LS/Q
$C_{14}H_{13}NO$
211.09971
RI: 2240
D: UHY
Antidepressant
211
246
M+
274
100
200
Flurazepam-M (bis-desethyl-) -H2O HY
UHY
LS
$C_{15}H_{12}ClFN_2$
274.06730
RI: 2295
D: UHY
Hypnotic
58
165
211
M+
339
100
200
300
Doxepin-M (HO-dihydro-) AC
UHYAC
LS
$C_{21}H_{25}NO_3$
339.18344
RI: 2340
D: UHYAC
Antidepressant

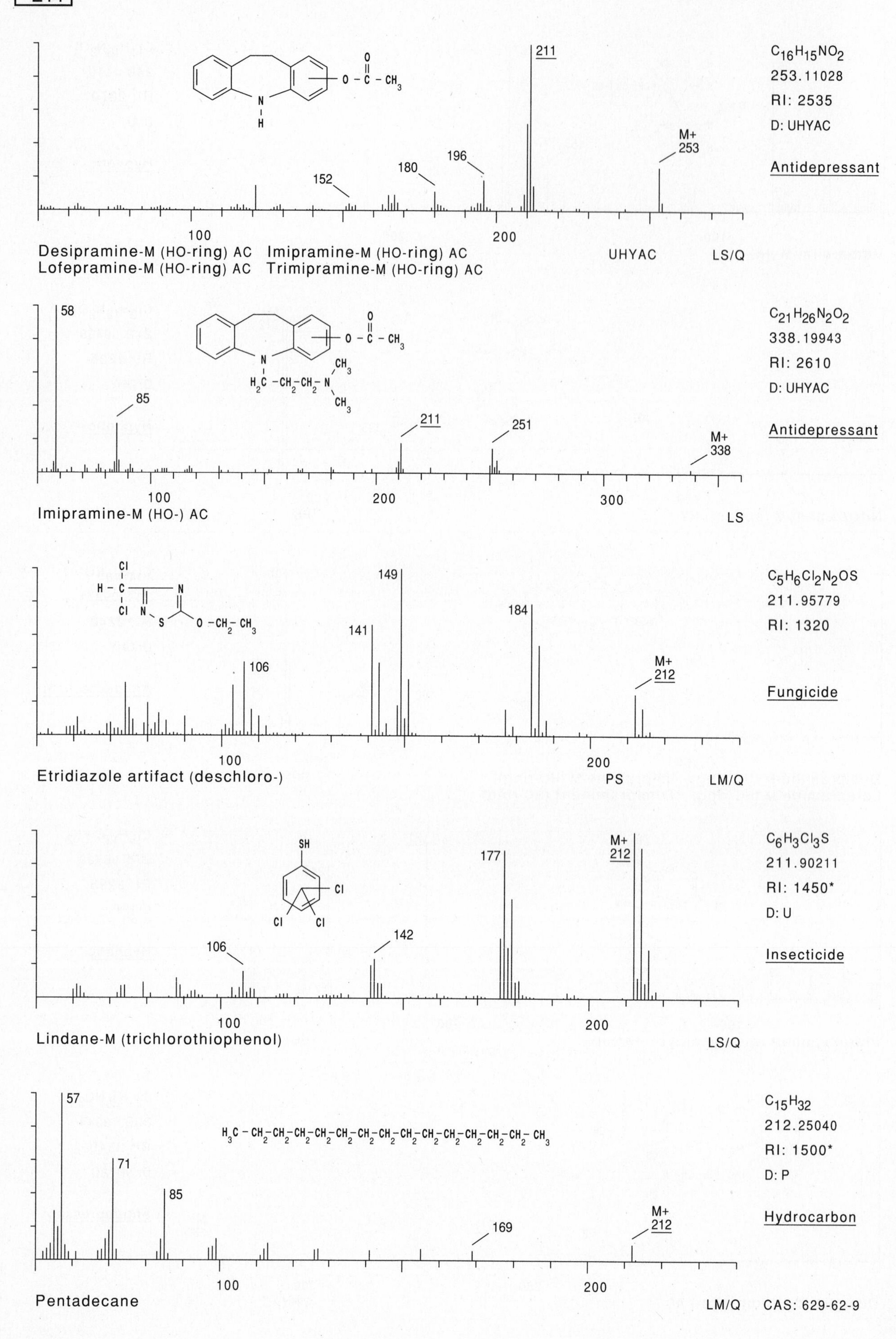
211
M+
253
196
180
152
100
200
Desipramine-M (HO-ring) AC Imipramine-M (HO-ring) AC
Lofepramine-M (HO-ring) AC Trimipramine-M (HO-ring) AC
UHYAC
LS/Q
C16H15NO2
253.11028
RI: 2535
D: UHYAC
Antidepressant
58
85
211
251
M+
338
100
200
300
Imipramine-M (HO-) AC
LS
C21H26N2O2
338.19943
RI: 2610
D: UHYAC
Antidepressant
149
184
141
106
M+
212
100
200
Etridiazole artifact (deschloro-)
PS
LM/Q
C5H6Cl2N2OS
211.95779
RI: 1320
Fungicide
M+
212
177
142
106
100
200
Lindane-M (trichlorothiophenol)
LS/Q
C6H3Cl3S
211.90211
RI: 1450*
D: U
Insecticide
57
71
85
169
M+
212
100
200
Pentadecane
LM/Q
C15H32
212.25040
RI: 1500*
D: P
Hydrocarbon
CAS: 629-62-9

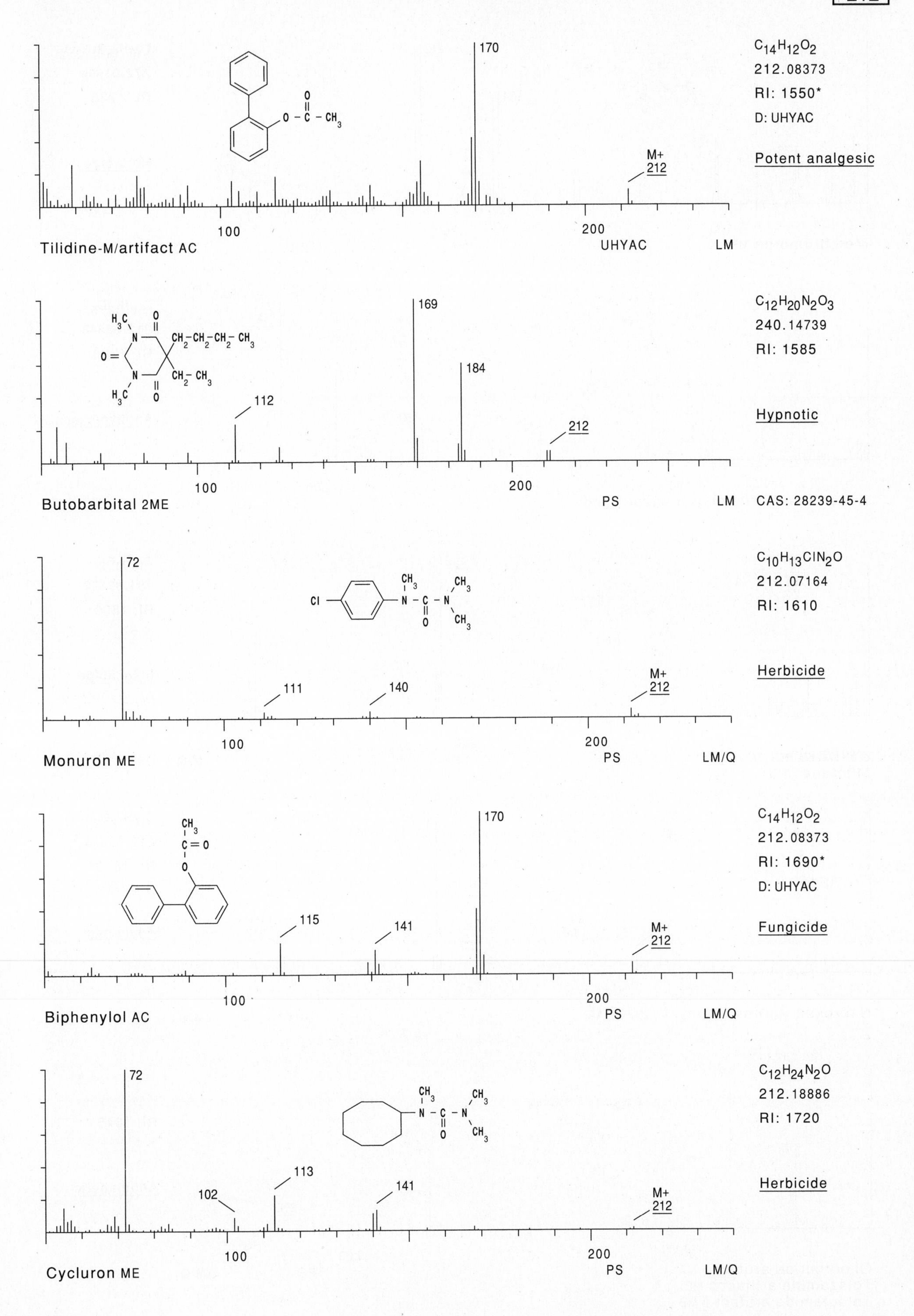
170
M+
212
100
200
Tilidine-M/artifact AC
UHYAC
LM
$C_{14}H_{12}O_2$
212.08373
RI: 1550*
D: UHYAC
Potent analgesic
169
184
112
212
100
200
Butobarbital 2ME
PS
LM
$C_{12}H_{20}N_2O_3$
240.14739
RI: 1585
Hypnotic
CAS: 28239-45-4
72
111
140
M+
212
100
200
Monuron ME
PS
LM/Q
$C_{10}H_{13}ClN_2O$
212.07164
RI: 1610
Herbicide
170
115
141
M+
212
100
200
Biphenylol AC
PS
LM/Q
$C_{14}H_{12}O_2$
212.08373
RI: 1690*
D: UHYAC
Fungicide
72
102
113
141
M+
212
100
200
Cycluron ME
PS
LM/Q
$C_{12}H_{24}N_2O$
212.18886
RI: 1720
Herbicide

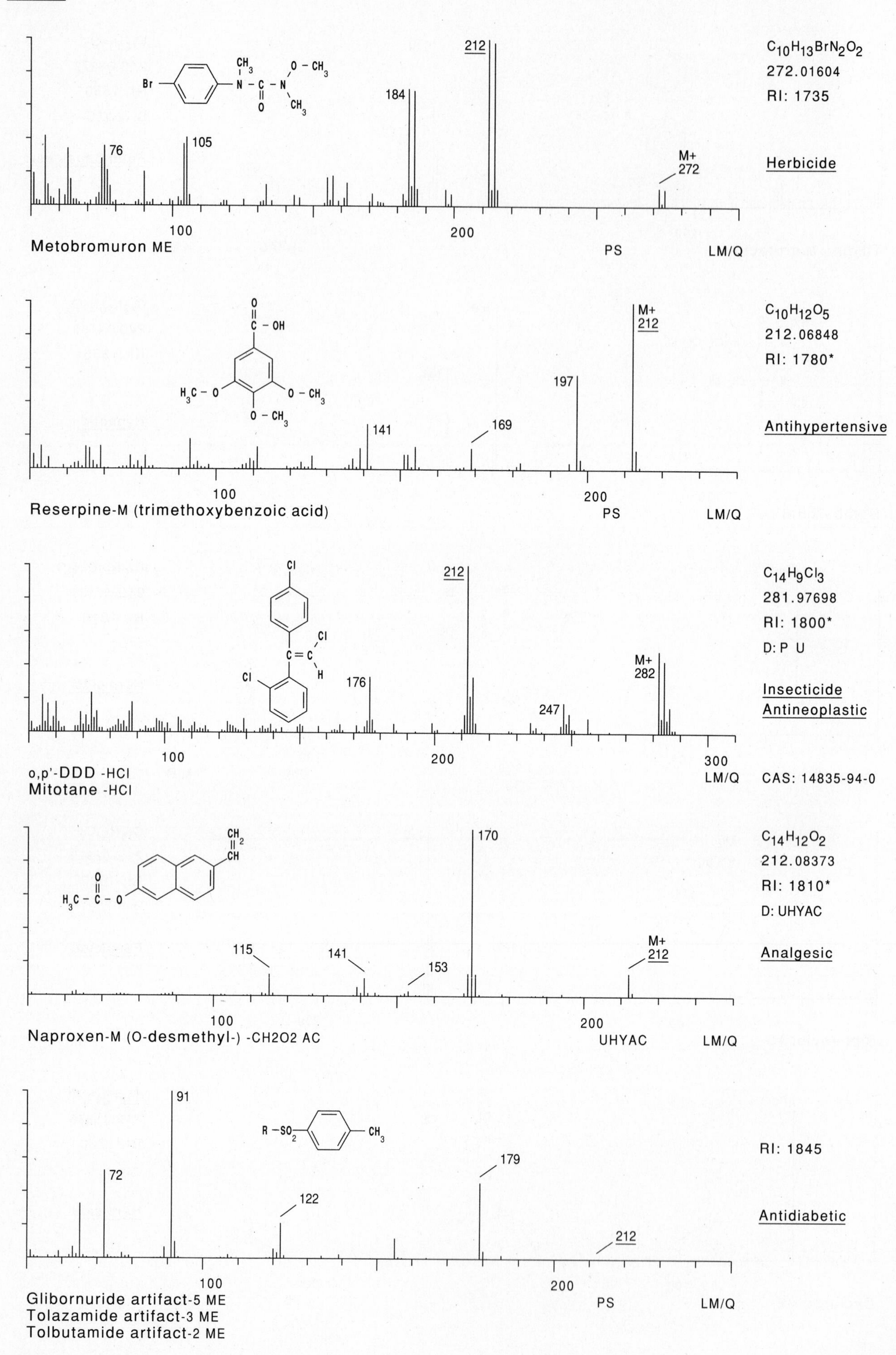

212
184
105
76
M+
272
100
200
PS
LM/Q
Metobromuron ME
$C_{10}H_{13}BrN_2O_2$
272.01604
RI: 1735
Herbicide
M+
212
197
141
169
100
200
PS
LM/Q
Reserpine-M (trimethoxybenzoic acid)
$C_{10}H_{12}O_5$
212.06848
RI: 1780*
Antihypertensive
212
176
M+
282
247
100
200
300
LM/Q
o,p'-DDD -HCl
Mitotane -HCl
$C_{14}H_9Cl_3$
281.97698
RI: 1800*
D: P U
Insecticide
Antineoplastic
CAS: 14835-94-0
170
115
141
153
M+
212
100
200
UHYAC
LM/Q
Naproxen-M (O-desmethyl-) -CH2O2 AC
$C_{14}H_{12}O_2$
212.08373
RI: 1810*
D: UHYAC
Analgesic
91
72
179
122
212
100
200
PS
LM/Q
Glibornuride artifact-5 ME
Tolazamide artifact-3 ME
Tolbutamide artifact-2 ME
RI: 1845
Antidiabetic

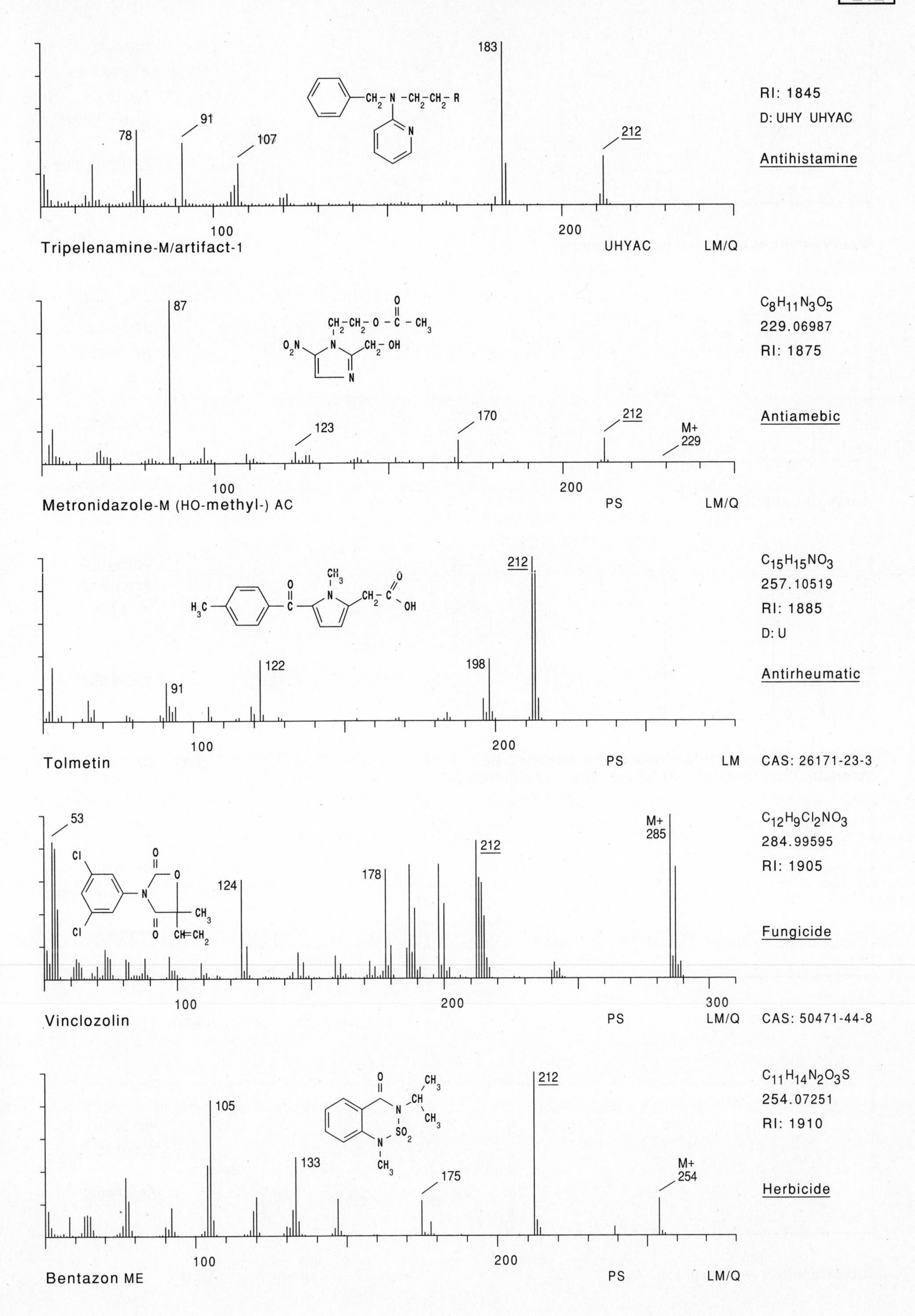

Tripelenamine-M/artifact-1
UHYAC
LM/Q
RI: 1845
D: UHY UHYAC
Antihistamine
Metronidazole-M (HO-methyl-) AC
PS
LM/Q
$C_8H_{11}N_3O_5$
229.06987
RI: 1875
Antiamebic
Tolmetin
PS
LM
CAS: 26171-23-3
$C_{15}H_{15}NO_3$
257.10519
RI: 1885
D: U
Antirheumatic
Vinclozolin
PS
LM/Q
CAS: 50471-44-8
$C_{12}H_9Cl_2NO_3$
284.99595
RI: 1905
Fungicide
Bentazon ME
PS
LM/Q
$C_{11}H_{14}N_2O_3S$
254.07251
RI: 1910
Herbicide

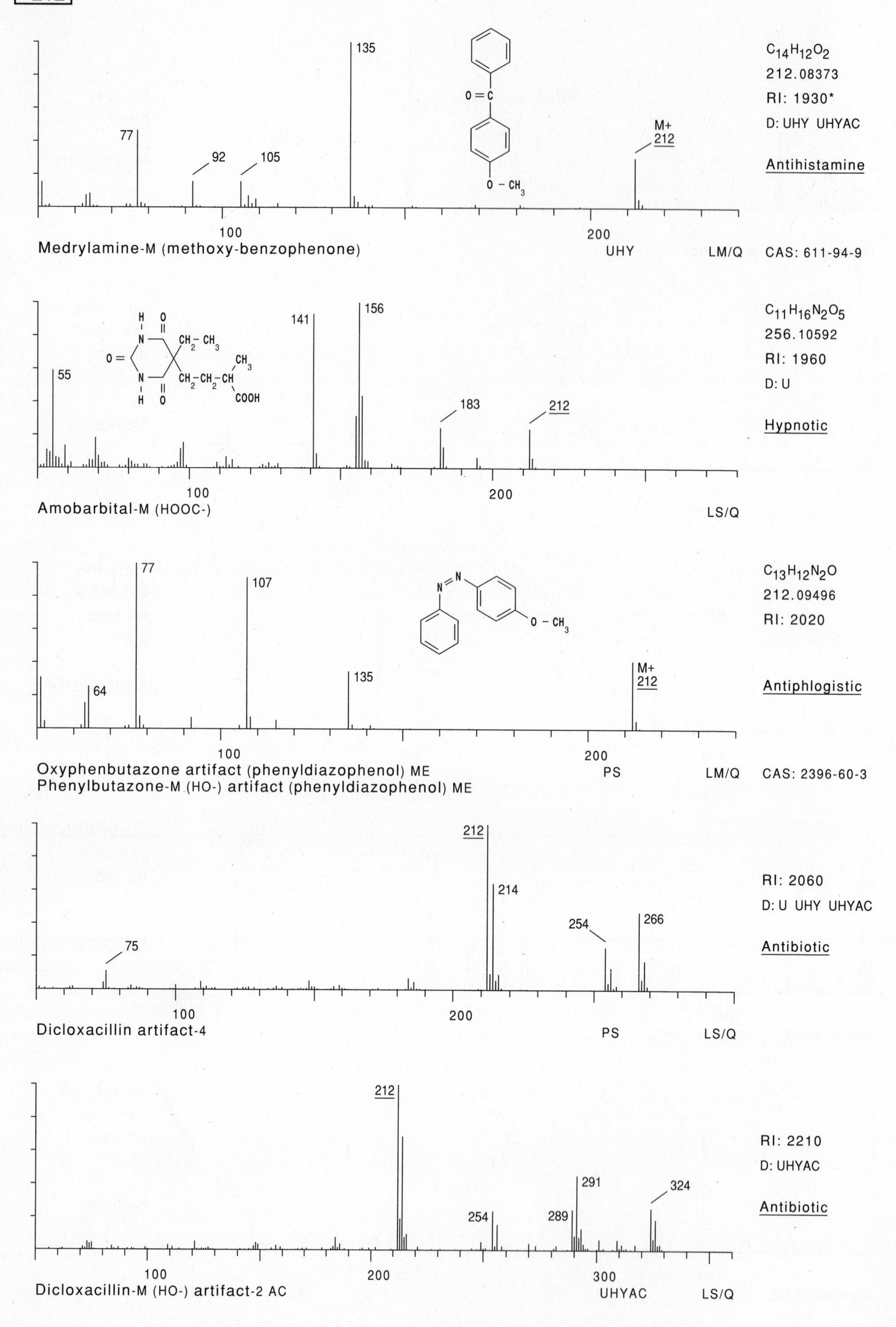

135
77
92
105
M+
212
O – CH3
100
200
Medrylamine-M (methoxy-benzophenone)
UHY
LM/Q
$C_{14}H_{12}O_2$
212.08373
RI: 1930*
D: UHY UHYAC
Antihistamine
CAS: 611-94-9
141
156
55
183
212
COOH
100
200
Amobarbital-M (HOOC-)
LS/Q
$C_{11}H_{16}N_2O_5$
256.10592
RI: 1960
D: U
Hypnotic
77
107
64
135
M+
212
100
200
Oxyphenbutazone artifact (phenyldiazophenol) ME
Phenylbutazone-M (HO-) artifact (phenyldiazophenol) ME
PS
LM/Q
$C_{13}H_{12}N_2O$
212.09496
RI: 2020
Antiphlogistic
CAS: 2396-60-3
212
214
254
266
75
100
200
Dicloxacillin artifact-4
PS
LS/Q
RI: 2060
D: U UHY UHYAC
Antibiotic
212
254
289
291
324
100
200
300
Dicloxacillin-M (HO-) artifact-2 AC
UHYAC
LS/Q
RI: 2210
D: UHYAC
Antibiotic

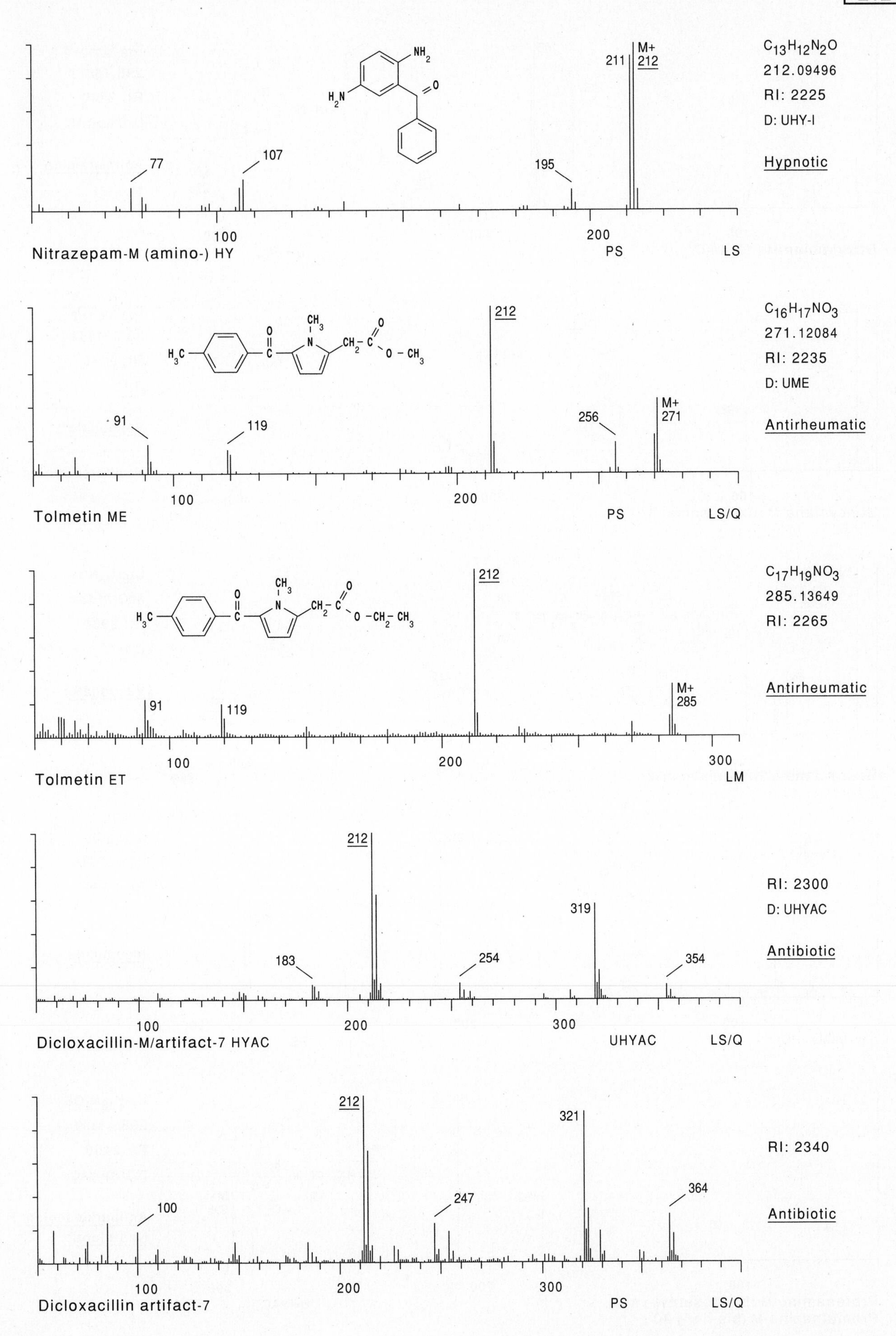

211
M+
212
77
107
195
100
200
Nitrazepam-M (amino-) HY
PS
LS
$C_{13}H_{12}N_2O$
212.09496
RI: 2225
D: UHY-I
Hypnotic
212
91
119
256
M+
271
100
200
Tolmetin ME
PS
LS/Q
$C_{16}H_{17}NO_3$
271.12084
RI: 2235
D: UME
Antirheumatic
212
91
119
M+
285
100
200
300
Tolmetin ET
LM
$C_{17}H_{19}NO_3$
285.13649
RI: 2265
Antirheumatic
212
183
254
319
354
100
200
300
Dicloxacillin-M/artifact-7 HYAC
UHYAC
LS/Q
RI: 2300
D: UHYAC
Antibiotic
212
321
100
247
364
100
200
300
Dicloxacillin artifact-7
PS
LS/Q
RI: 2340
Antibiotic

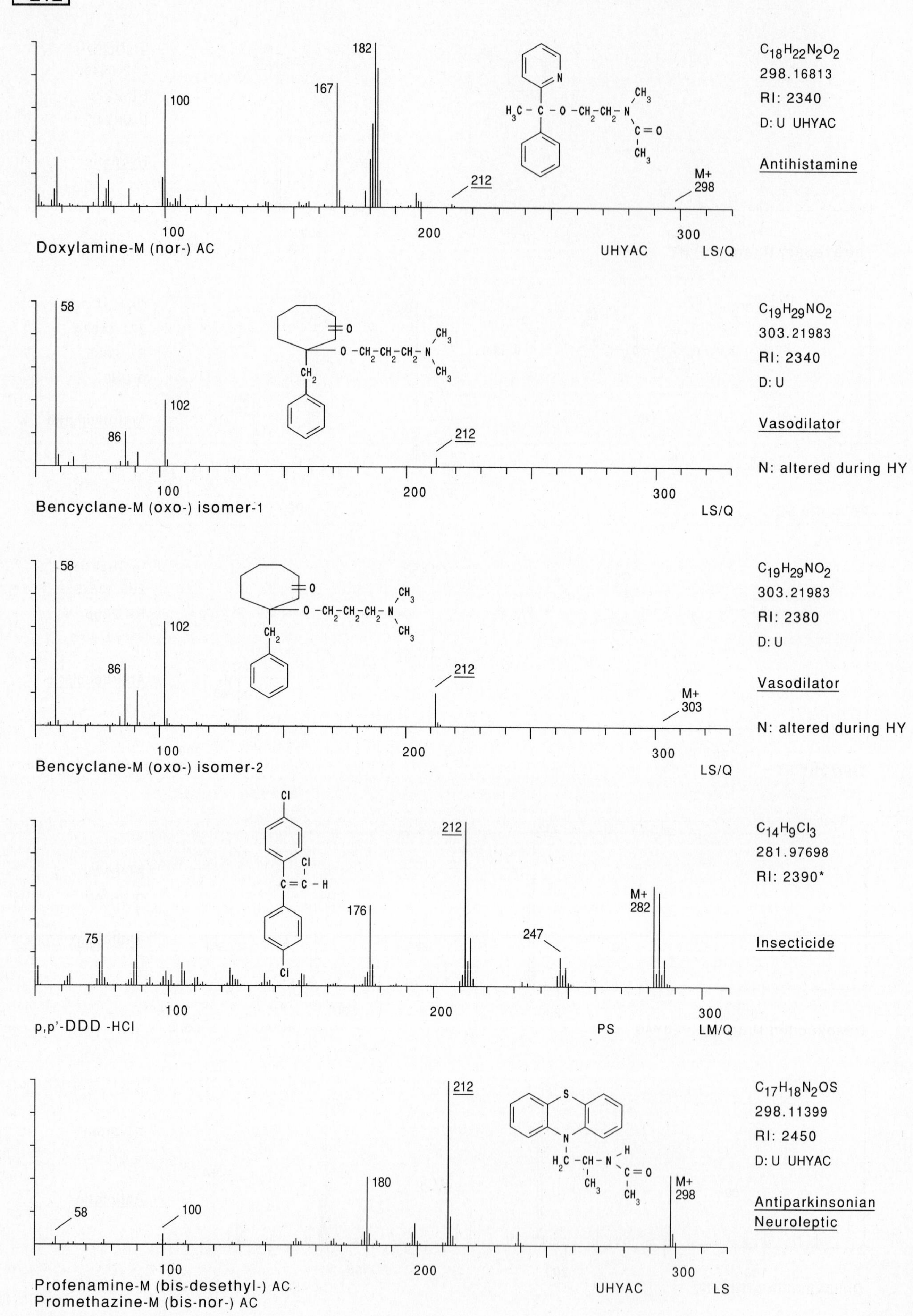
182
167
100
212
M+
298
100
200
300
Doxylamine-M (nor-) AC
UHYAC
LS/Q
$C_{18}H_{22}N_2O_2$
298.16813
RI: 2340
D: U UHYAC
Antihistamine
58
102
86
212
Bencyclane-M (oxo-) isomer-1
LS/Q
$C_{19}H_{29}NO_2$
303.21983
RI: 2340
D: U
Vasodilator
N: altered during HY
58
102
86
212
M+
303
Bencyclane-M (oxo-) isomer-2
LS/Q
$C_{19}H_{29}NO_2$
303.21983
RI: 2380
D: U
Vasodilator
N: altered during HY
212
176
75
247
M+
282
p,p'-DDD -HCl
PS
LM/Q
$C_{14}H_9Cl_3$
281.97698
RI: 2390*
Insecticide
212
180
58
100
M+
298
Profenamine-M (bis-desethyl-) AC
Promethazine-M (bis-nor-) AC
UHYAC
LS
$C_{17}H_{18}N_2OS$
298.11399
RI: 2450
D: U UHYAC
Antiparkinsonian
Neuroleptic

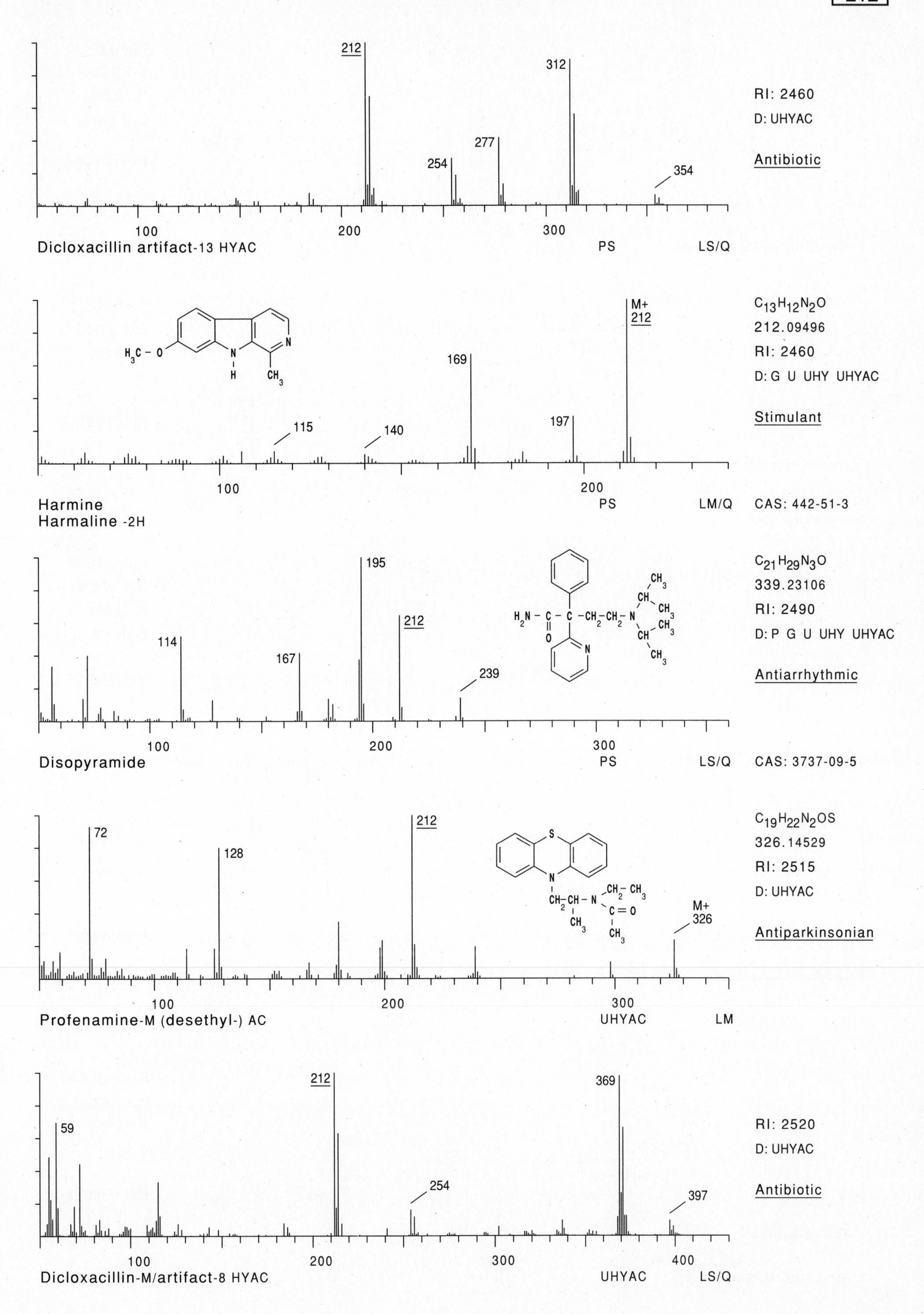
212
312
277
254
354
100
200
300
Dicloxacillin artifact-13 HYAC
PS
LS/Q
RI: 2460
D: UHYAC
Antibiotic
M+
212
169
197
115
140
100
200
Harmine
Harmaline -2H
PS
LM/Q
$C_{13}H_{12}N_2O$
212.09496
RI: 2460
D: G U UHY UHYAC
Stimulant
CAS: 442-51-3
195
212
114
167
239
100
200
300
Disopyramide
PS
LS/Q
$C_{21}H_{29}N_3O$
339.23106
RI: 2490
D: P G U UHY UHYAC
Antiarrhythmic
CAS: 3737-09-5
72
212
128
M+
326
100
200
300
Profenamine-M (desethyl-) AC
UHYAC
LM
$C_{19}H_{22}N_2OS$
326.14529
RI: 2515
D: UHYAC
Antiparkinsonian
212
369
59
254
397
100
200
300
400
Dicloxacillin-M/artifact-8 HYAC
UHYAC
LS/Q
RI: 2520
D: UHYAC
Antibiotic

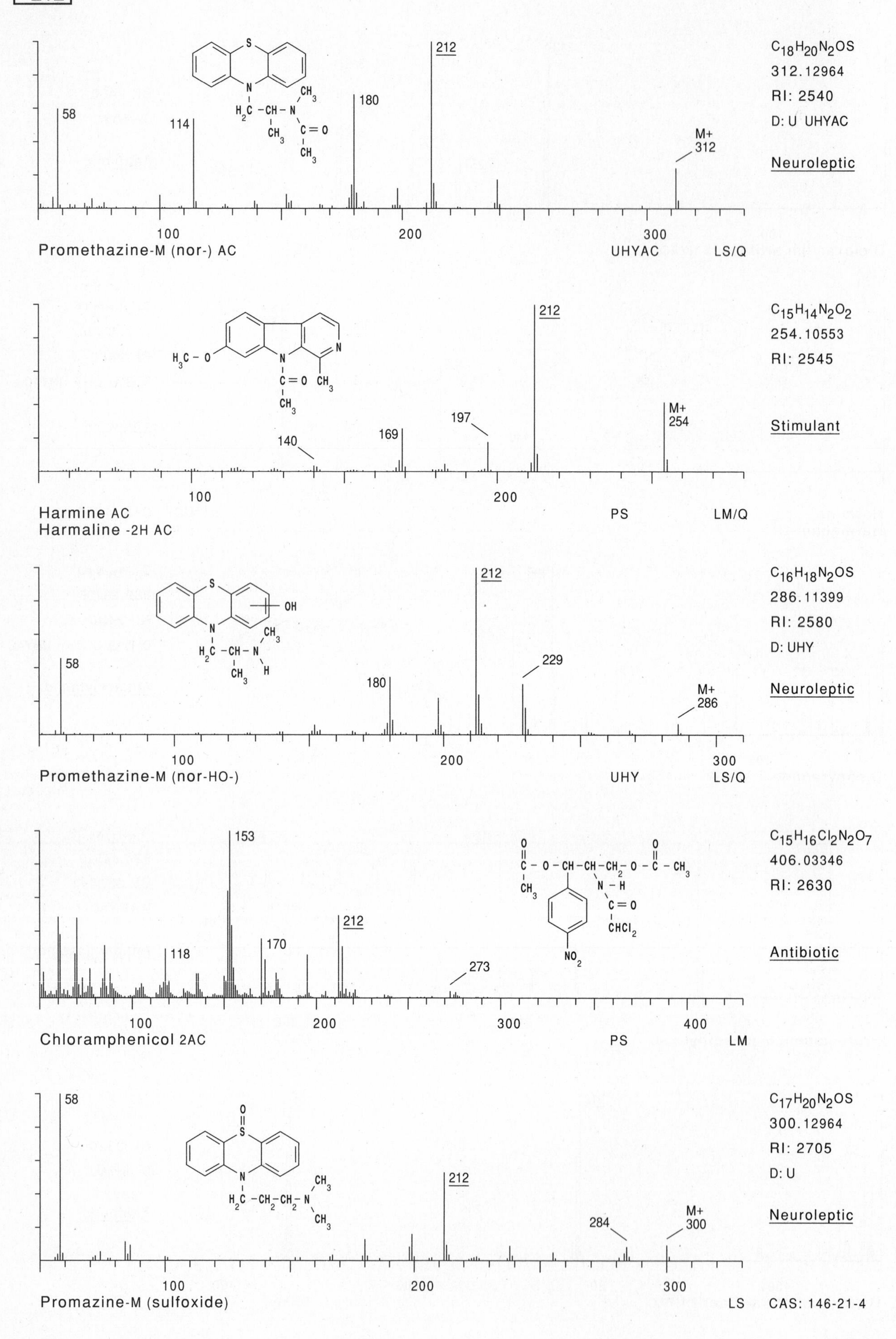

212
180
58
114
M+
312
100
200
300
Promethazine-M (nor-) AC
UHYAC
LS/Q
$C_{18}H_{20}N_2OS$
312.12964
RI: 2540
D: U UHYAC
Neuroleptic
212
M+
254
197
169
140
100
200
Harmine AC
Harmaline -2H AC
PS
LM/Q
$C_{15}H_{14}N_2O_2$
254.10553
RI: 2545
Stimulant
212
58
229
180
M+
286
100
200
300
Promethazine-M (nor-HO-)
UHY
LS/Q
$C_{16}H_{18}N_2OS$
286.11399
RI: 2580
D: UHY
Neuroleptic
153
212
170
118
273
100
200
300
400
Chloramphenicol 2AC
PS
LM
$C_{15}H_{16}Cl_2N_2O_7$
406.03346
RI: 2630
Antibiotic
58
212
284
M+
300
100
200
300
Promazine-M (sulfoxide)
LS
CAS: 146-21-4
$C_{17}H_{20}N_2OS$
300.12964
RI: 2705
D: U
Neuroleptic

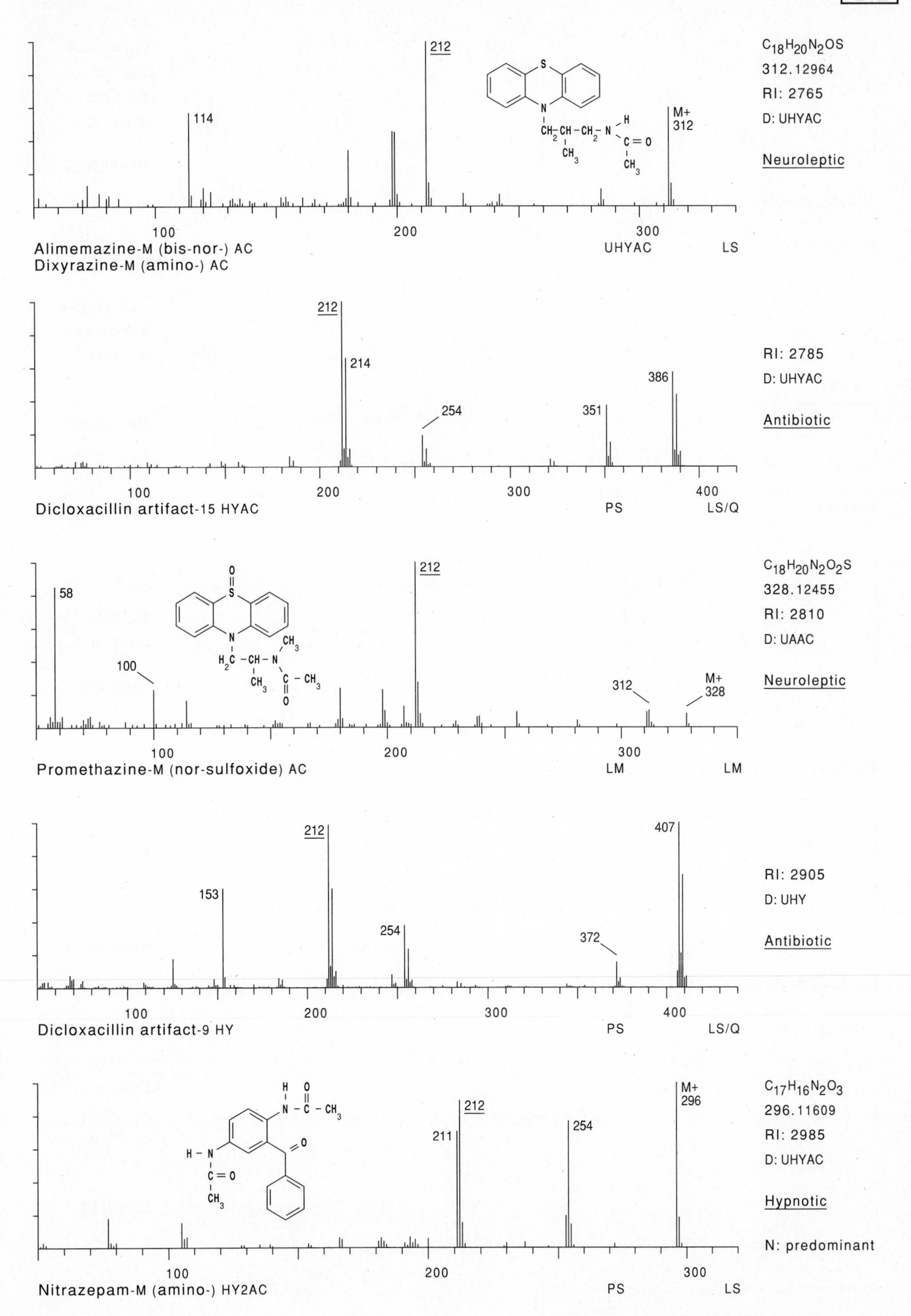

212
114
M+
312
100
200
300
Alimemazine-M (bis-nor-) AC
Dixyrazine-M (amino-) AC
UHYAC
LS
C18H20N2OS
312.12964
RI: 2765
D: UHYAC
Neuroleptic
212
214
254
351
386
100
200
300
400
Dicloxacillin artifact-15 HYAC
PS
LS/Q
RI: 2785
D: UHYAC
Antibiotic
58
100
212
312
M+
328
100
200
300
Promethazine-M (nor-sulfoxide) AC
LM
LM
C18H20N2O2S
328.12455
RI: 2810
D: UAAC
Neuroleptic
153
212
254
372
407
100
200
300
400
Dicloxacillin artifact-9 HY
PS
LS/Q
RI: 2905
D: UHY
Antibiotic
211
212
254
M+
296
100
200
300
Nitrazepam-M (amino-) HY2AC
PS
LS
C17H16N2O3
296.11609
RI: 2985
D: UHYAC
Hypnotic
N: predominant

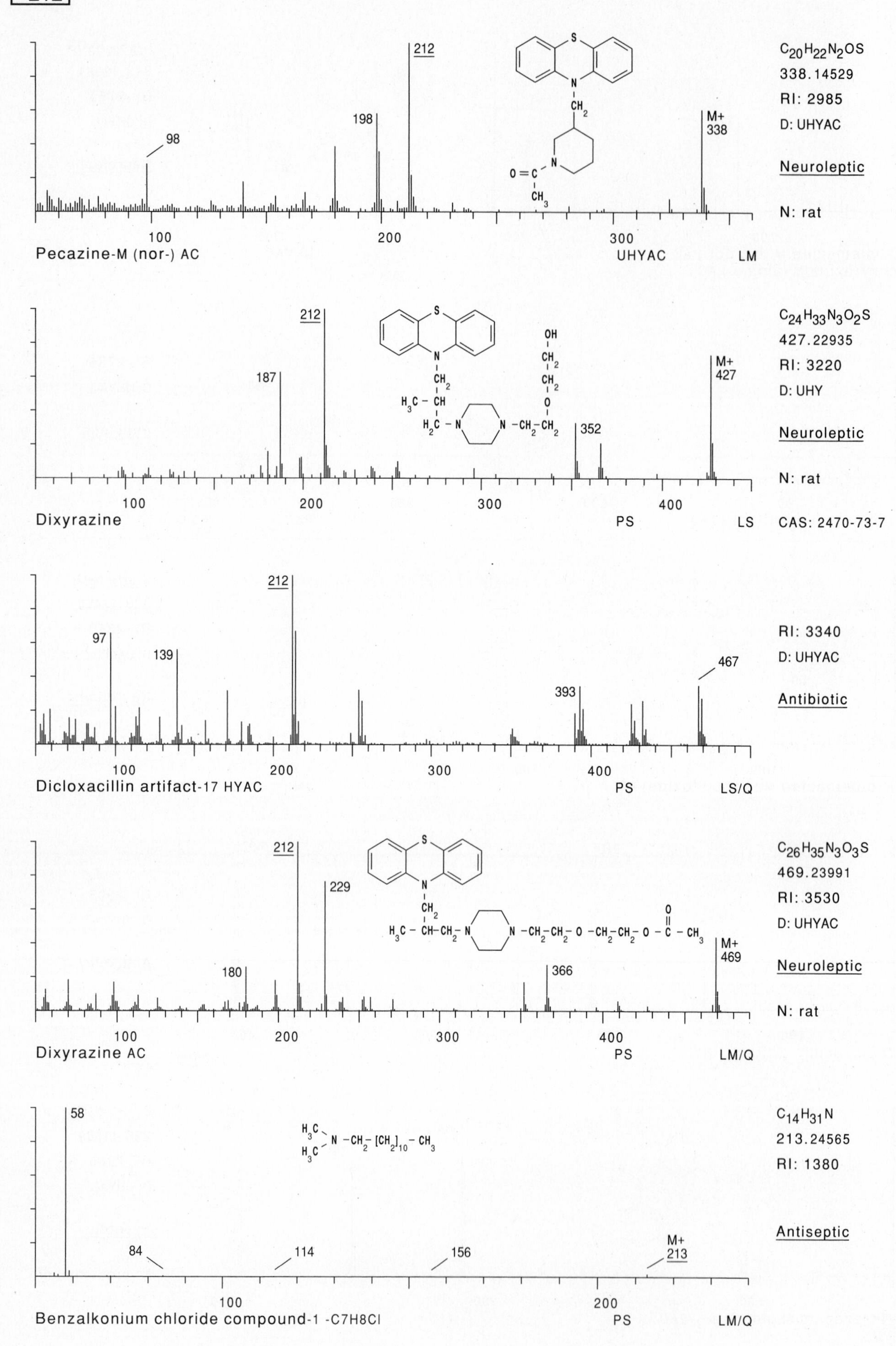
Pecazine-M (nor-) AC
UHYAC
LM
$C_{20}H_{22}N_2OS$
338.14529
RI: 2985
D: UHYAC
Neuroleptic
N: rat
98
198
212
M+ 338
Dixyrazine
PS
LS
$C_{24}H_{33}N_3O_2S$
427.22935
RI: 3220
D: UHY
Neuroleptic
N: rat
CAS: 2470-73-7
187
212
352
M+ 427
Dicloxacillin artifact-17 HYAC
PS
LS/Q
RI: 3340
D: UHYAC
Antibiotic
97
139
212
393
467
Dixyrazine AC
PS
LM/Q
$C_{26}H_{35}N_3O_3S$
469.23991
RI: 3530
D: UHYAC
Neuroleptic
N: rat
180
212
229
366
M+ 469
Benzalkonium chloride compound-1 -C7H8Cl
PS
LM/Q
$C_{14}H_{31}N$
213.24565
RI: 1380
Antiseptic
58
84
114
156
M+ 213

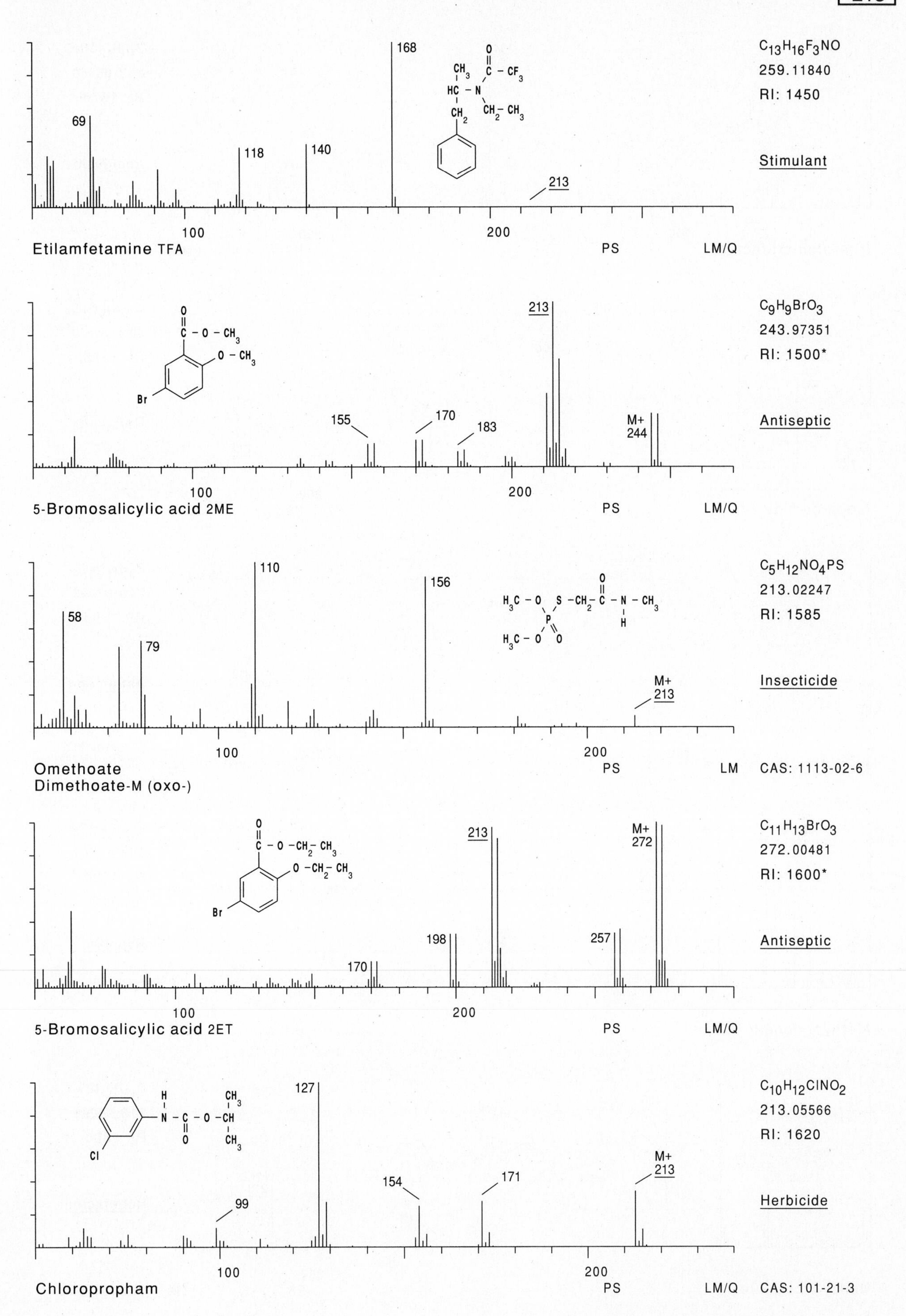

$C_{13}H_{16}F_3NO$
259.11840
RI: 1450
Stimulant

$C_9H_9BrO_3$
243.97351
RI: 1500*
Antiseptic

$C_5H_{12}NO_4PS$
213.02247
RI: 1585
Insecticide
CAS: 1113-02-6

$C_{11}H_{13}BrO_3$
272.00481
RI: 1600*
Antiseptic

$C_{10}H_{12}ClNO_2$
213.05566
RI: 1620
Herbicide
CAS: 101-21-3

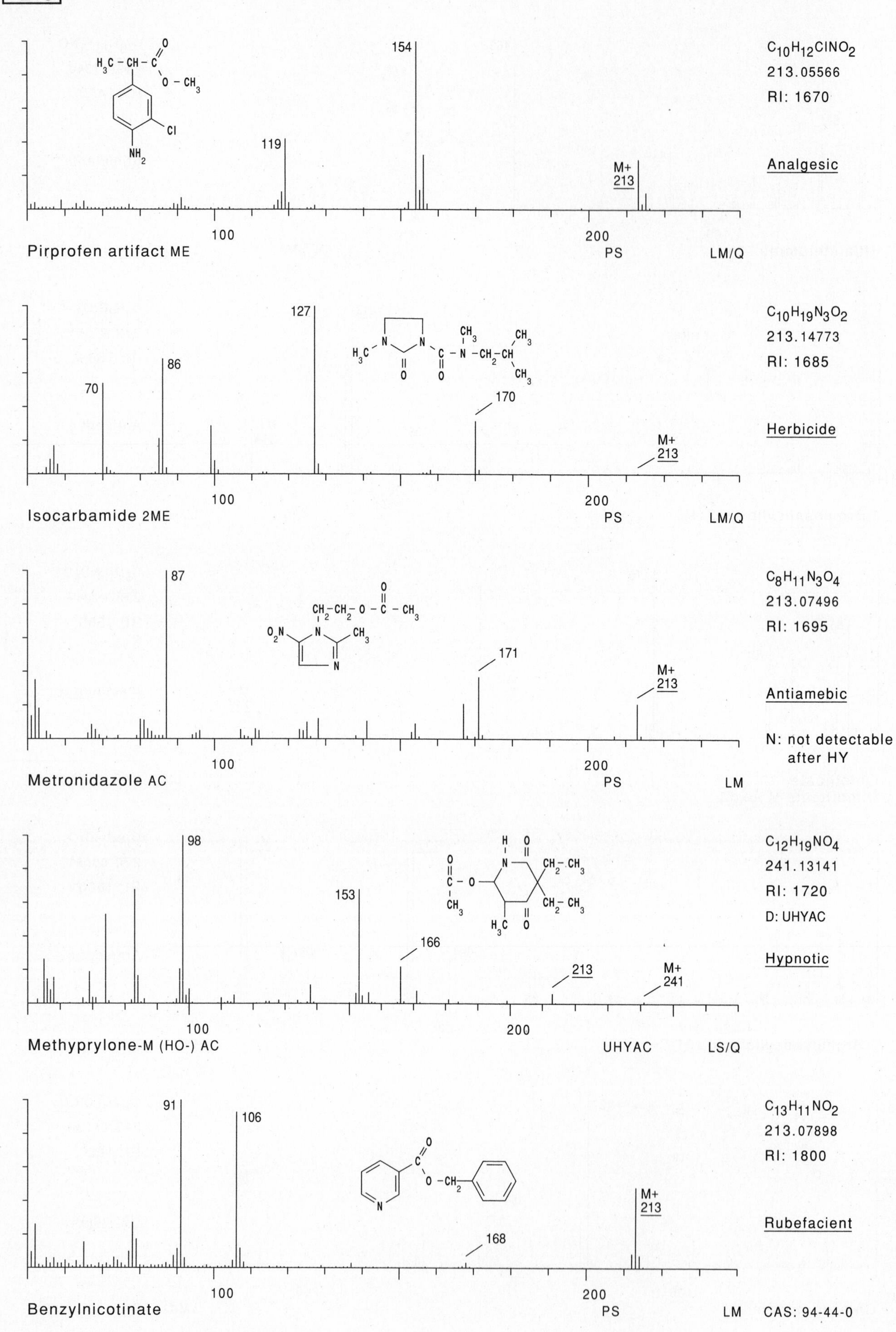
154
119
M+
213
100
200
Pirprofen artifact ME
PS
LM/Q
$C_{10}H_{12}ClNO_2$
213.05566
RI: 1670
Analgesic
127
86
70
170
M+
213
Isocarbamide 2ME
LM/Q
$C_{10}H_{19}N_3O_2$
213.14773
RI: 1685
Herbicide
87
171
M+
213
Metronidazole AC
LM
$C_8H_{11}N_3O_4$
213.07496
RI: 1695
Antiamebic
N: not detectable after HY
98
153
166
213
M+
241
Methyprylone-M (HO-) AC
UHYAC
LS/Q
$C_{12}H_{19}NO_4$
241.13141
RI: 1720
D: UHYAC
Hypnotic
91
106
168
M+
213
Benzylnicotinate
$C_{13}H_{11}NO_2$
213.07898
RI: 1800
Rubefacient
CAS: 94-44-0

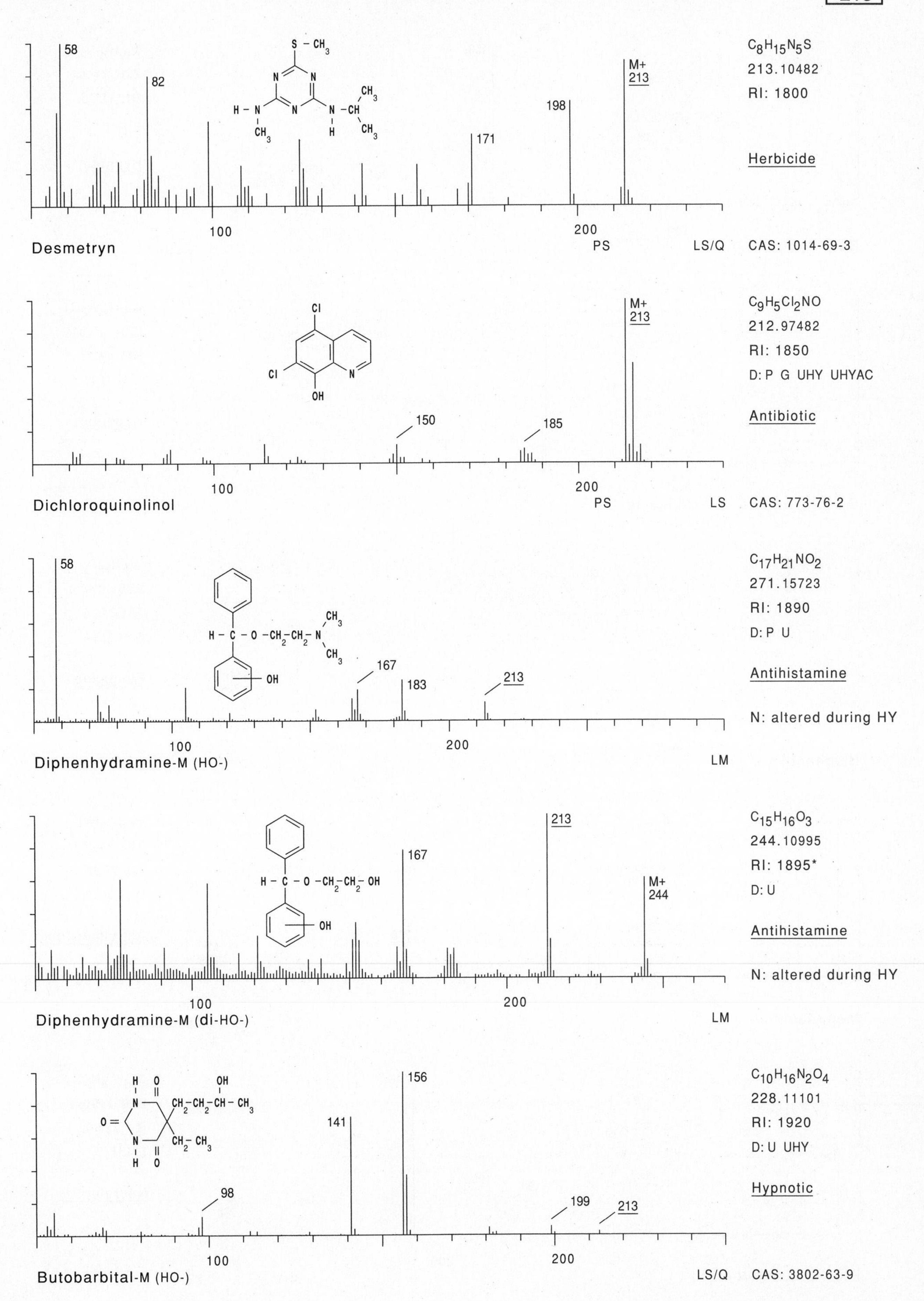

$C_8H_{15}N_5S$
213.10482
RI: 1800

Herbicide

CAS: 1014-69-3

$C_9H_5Cl_2NO$
212.97482
RI: 1850
D: P G UHY UHYAC

Antibiotic

CAS: 773-76-2

$C_{17}H_{21}NO_2$
271.15723
RI: 1890
D: P U

Antihistamine

N: altered during HY

$C_{15}H_{16}O_3$
244.10995
RI: 1895*
D: U

Antihistamine

N: altered during HY

$C_{10}H_{16}N_2O_4$
228.11101
RI: 1920
D: U UHY

Hypnotic

CAS: 3802-63-9

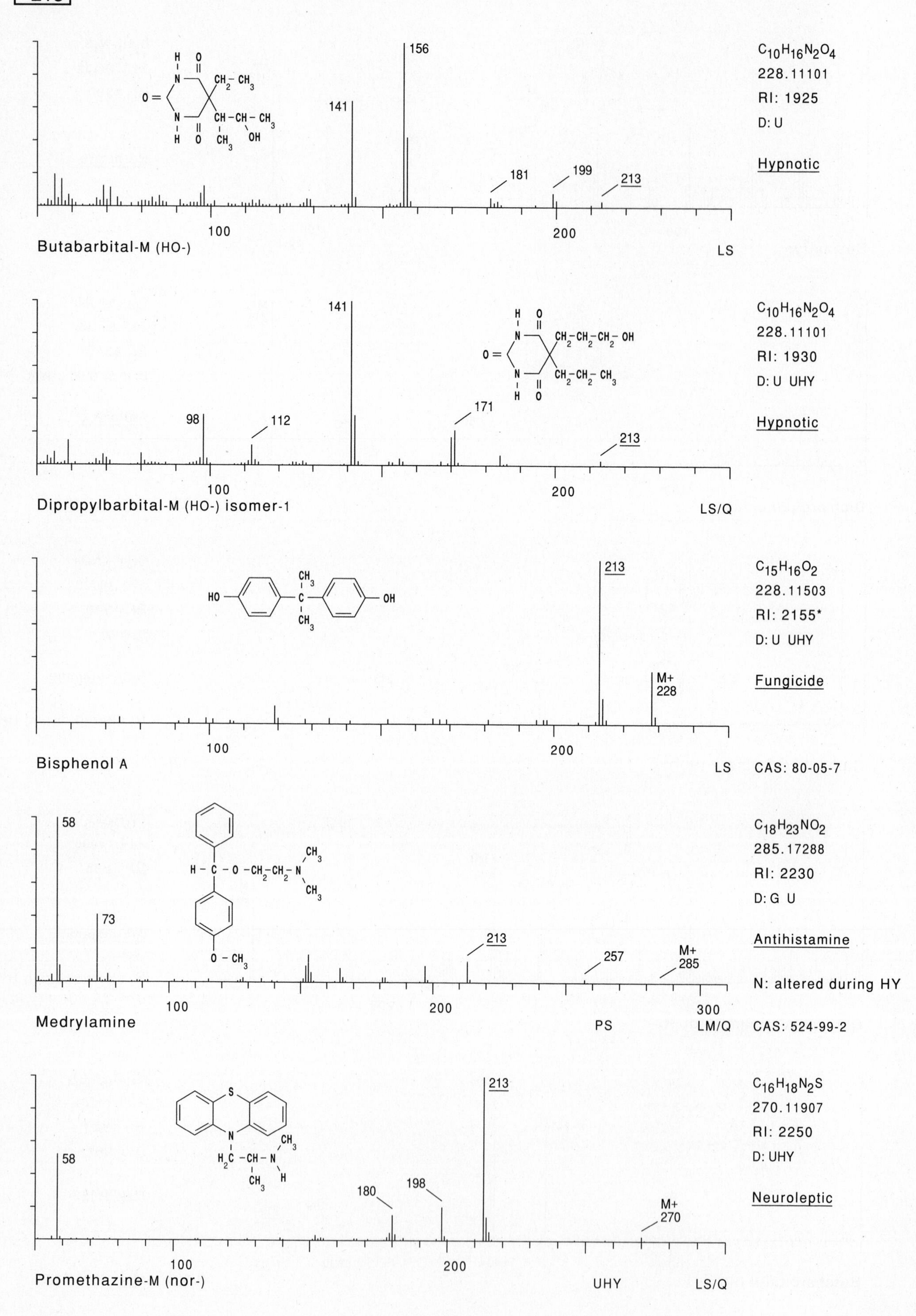
156
141
181
199
213
100
200
Butabarbital-M (HO-)
LS
$C_{10}H_{16}N_2O_4$
228.11101
RI: 1925
D: U
Hypnotic
141
98
112
171
213
Dipropylbarbital-M (HO-) isomer-1
LS/Q
$C_{10}H_{16}N_2O_4$
228.11101
RI: 1930
D: U UHY
Hypnotic
213
M+ 228
Bisphenol A
LS
$C_{15}H_{16}O_2$
228.11503
RI: 2155*
D: U UHY
Fungicide
CAS: 80-05-7
58
73
213
257
M+ 285
300
Medrylamine
PS
LM/Q
$C_{18}H_{23}NO_2$
285.17288
RI: 2230
D: G U
Antihistamine
N: altered during HY
CAS: 524-99-2
213
58
180
198
M+ 270
Promethazine-M (nor-)
UHY
LS/Q
$C_{16}H_{18}N_2S$
270.11907
RI: 2250
D: UHY
Neuroleptic

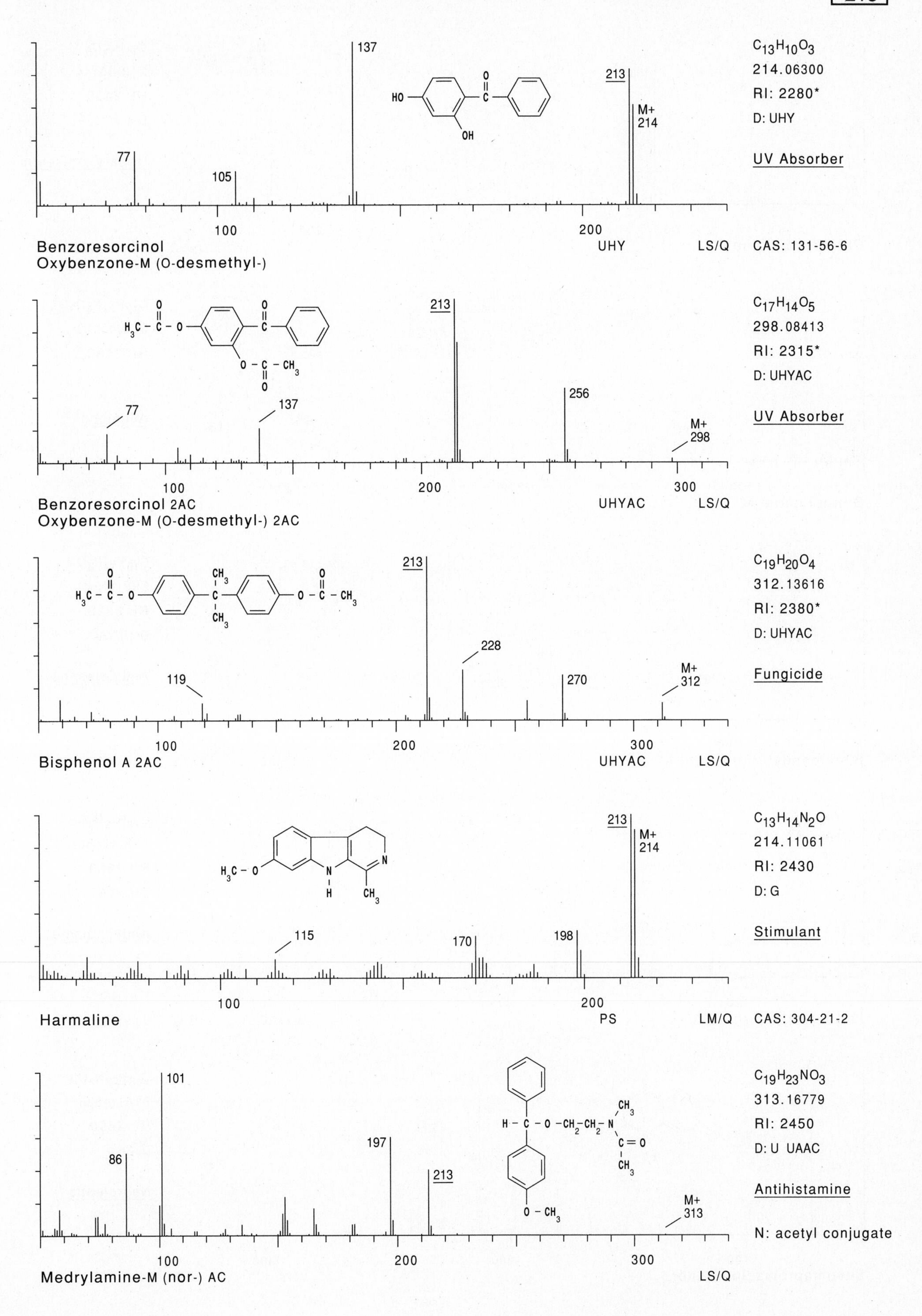
137
213
M+
214
77
105
100
200
HO
OH
Benzoresorcinol
Oxybenzone-M (O-desmethyl-)
UHY
LS/Q
C13H10O3
214.06300
RI: 2280*
D: UHY
UV Absorber
CAS: 131-56-6
213
256
77
137
M+
298
100
200
300
Benzoresorcinol 2AC
Oxybenzone-M (O-desmethyl-) 2AC
UHYAC
LS/Q
C17H14O5
298.08413
RI: 2315*
D: UHYAC
UV Absorber
213
228
119
270
M+
312
100
200
300
Bisphenol A 2AC
UHYAC
LS/Q
C19H20O4
312.13616
RI: 2380*
D: UHYAC
Fungicide
213
M+
214
115
170
198
100
200
Harmaline
PS
LM/Q
C13H14N2O
214.11061
RI: 2430
D: G
Stimulant
CAS: 304-21-2
101
86
197
213
M+
313
100
200
300
Medrylamine-M (nor-) AC
LS/Q
C19H23NO3
313.16779
RI: 2450
D: U UAAC
Antihistamine
N: acetyl conjugate

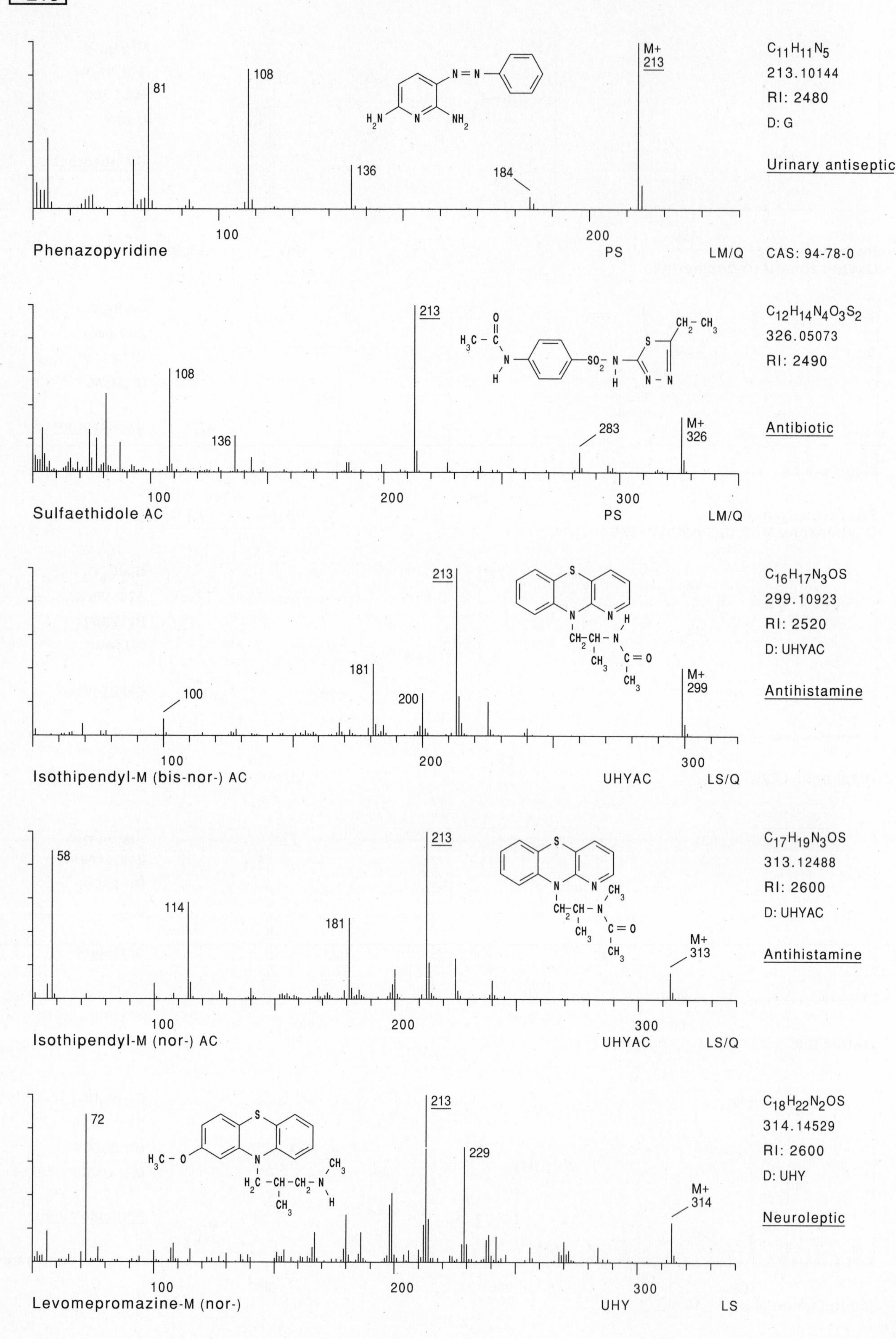

M+ 213
108
81
136
184
100
200
Phenazopyridine
PS
LM/Q
$C_{11}H_{11}N_5$
213.10144
RI: 2480
D: G
Urinary antiseptic
CAS: 94-78-0
213
108
136
283
M+ 326
100
200
300
Sulfaethidole AC
PS
LM/Q
$C_{12}H_{14}N_4O_3S_2$
326.05073
RI: 2490
Antibiotic
213
181
200
100
M+ 299
Isothipendyl-M (bis-nor-) AC
UHYAC
LS/Q
$C_{16}H_{17}N_3OS$
299.10923
RI: 2520
D: UHYAC
Antihistamine
213
58
114
181
M+ 313
Isothipendyl-M (nor-) AC
UHYAC
LS/Q
$C_{17}H_{19}N_3OS$
313.12488
RI: 2600
D: UHYAC
Antihistamine
213
72
229
M+ 314
Levomepromazine-M (nor-)
UHY
LS
$C_{18}H_{22}N_2OS$
314.14529
RI: 2600
D: UHY
Neuroleptic

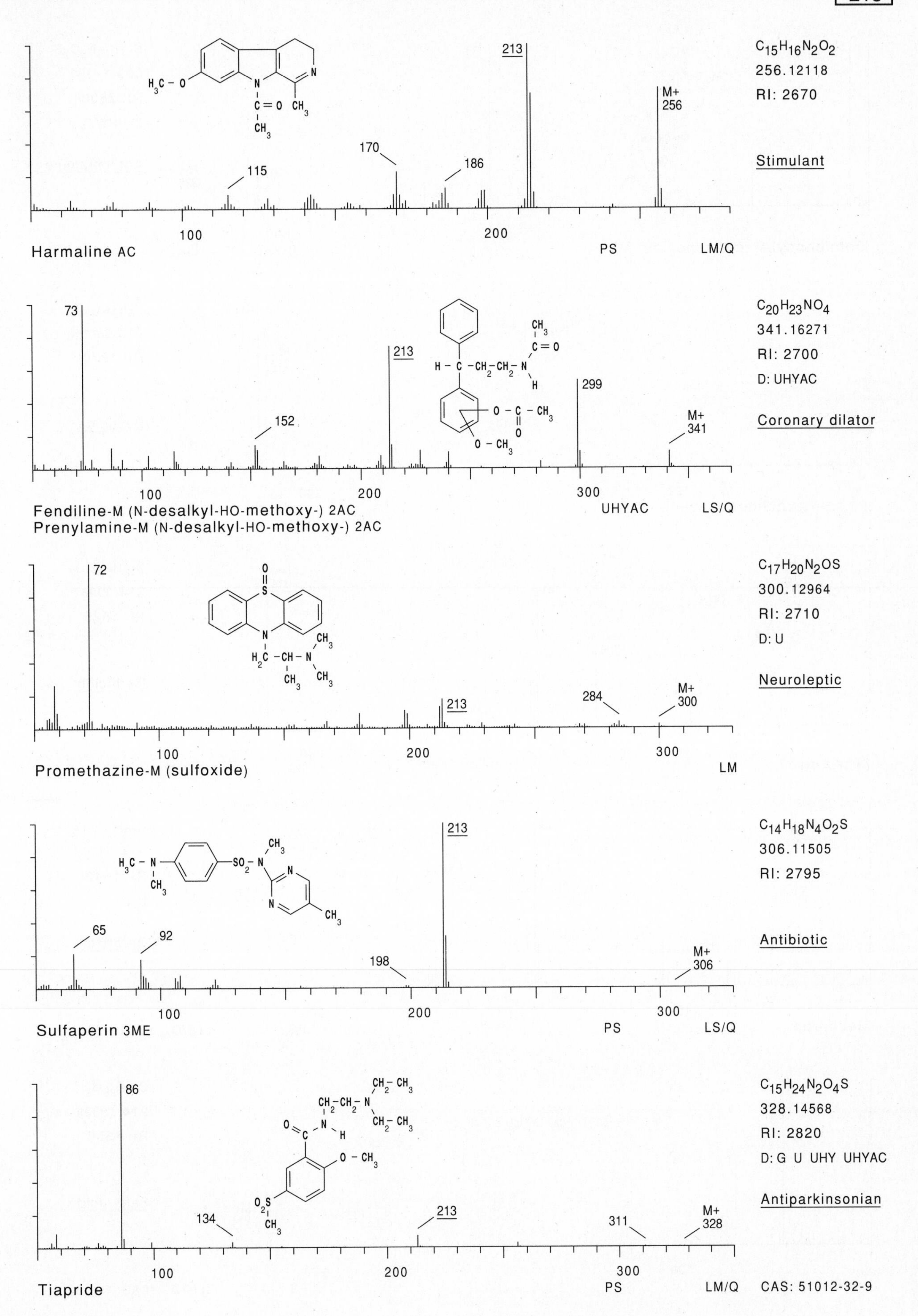

Harmaline AC

$C_{15}H_{16}N_2O_2$
256.12118
RI: 2670
Stimulant

Fendiline-M (N-desalkyl-HO-methoxy-) 2AC
Prenylamine-M (N-desalkyl-HO-methoxy-) 2AC

$C_{20}H_{23}NO_4$
341.16271
RI: 2700
D: UHYAC
Coronary dilator

Promethazine-M (sulfoxide)

$C_{17}H_{20}N_2OS$
300.12964
RI: 2710
D: U
Neuroleptic

Sulfaperin 3ME

$C_{14}H_{18}N_4O_2S$
306.11505
RI: 2795
Antibiotic

Tiapride

$C_{15}H_{24}N_2O_4S$
328.14568
RI: 2820
D: G U UHY UHYAC
Antiparkinsonian
CAS: 51012-32-9

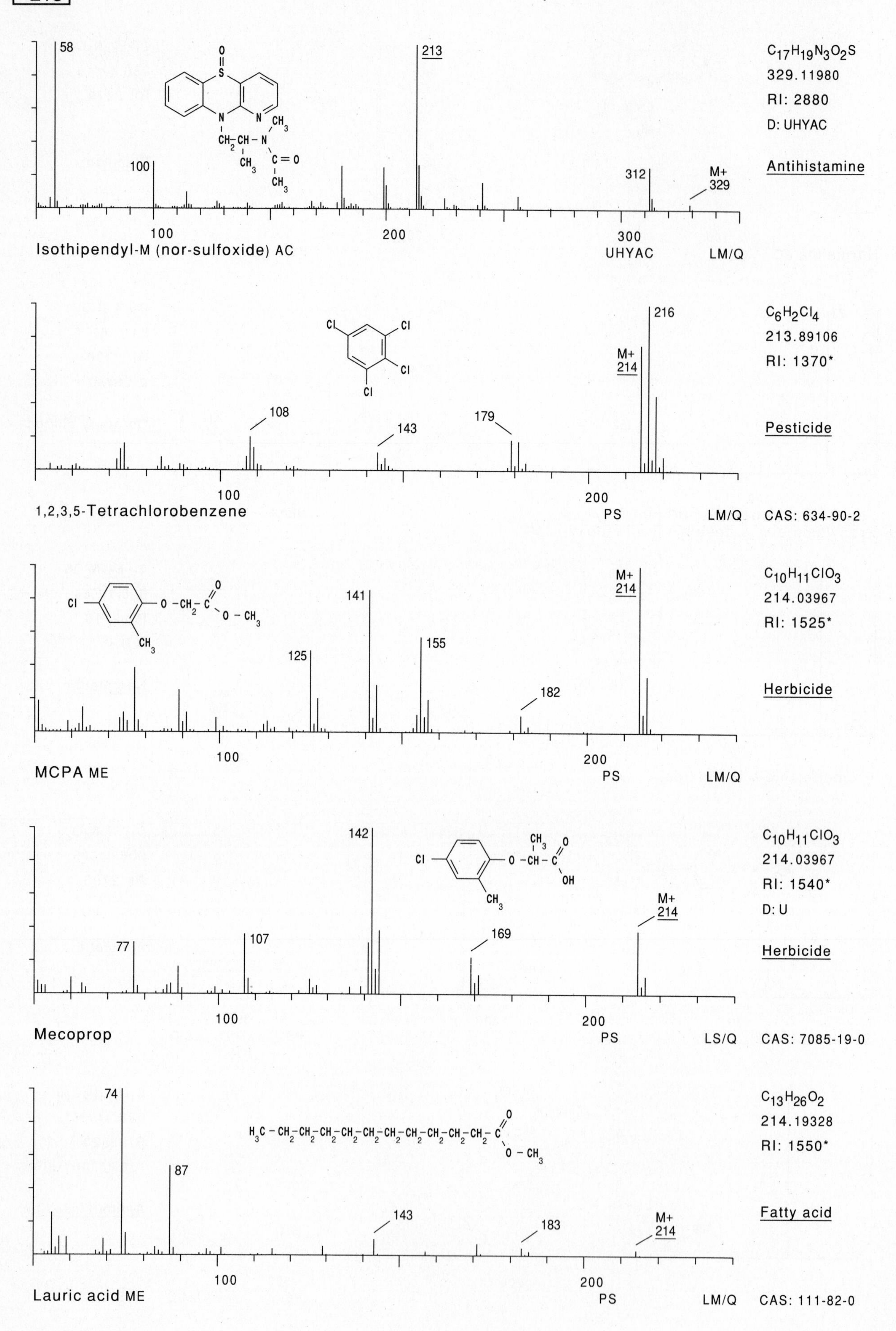
58
213
100
312
M+
329
100
200
300
Isothipendyl-M (nor-sulfoxide) AC
UHYAC
LM/Q
C17H19N3O2S
329.11980
RI: 2880
D: UHYAC
Antihistamine
216
M+
214
108
143
179
100
200
1,2,3,5-Tetrachlorobenzene
PS
LM/Q
C6H2Cl4
213.89106
RI: 1370*
Pesticide
CAS: 634-90-2
M+
214
141
125
155
182
100
200
MCPA ME
PS
LM/Q
C10H11ClO3
214.03967
RI: 1525*
Herbicide
142
M+
214
77
107
169
100
200
Mecoprop
PS
LS/Q
C10H11ClO3
214.03967
RI: 1540*
D: U
Herbicide
CAS: 7085-19-0
74
87
143
183
M+
214
100
200
Lauric acid ME
PS
LM/Q
C13H26O2
214.19328
RI: 1550*
Fatty acid
CAS: 111-82-0

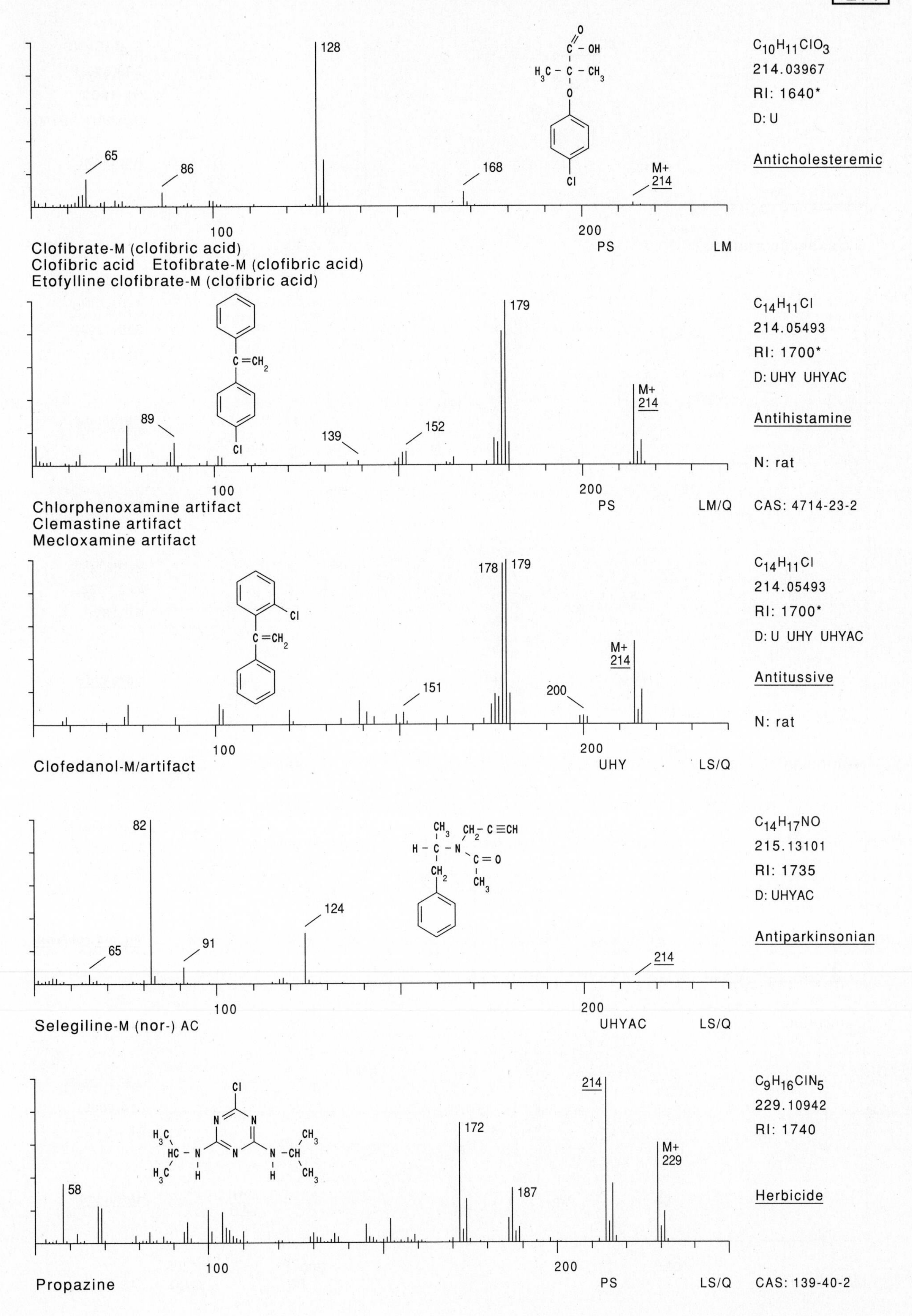

$C_{10}H_{11}ClO_3$
214.03967
RI: 1640*
D: U
Anticholesteremic

$C_{14}H_{11}Cl$
214.05493
RI: 1700*
D: UHY UHYAC
Antihistamine
N: rat
CAS: 4714-23-2

$C_{14}H_{11}Cl$
214.05493
RI: 1700*
D: U UHY UHYAC
Antitussive
N: rat

$C_{14}H_{17}NO$
215.13101
RI: 1735
D: UHYAC
Antiparkinsonian

$C_9H_{16}ClN_5$
229.10942
RI: 1740
Herbicide
CAS: 139-40-2

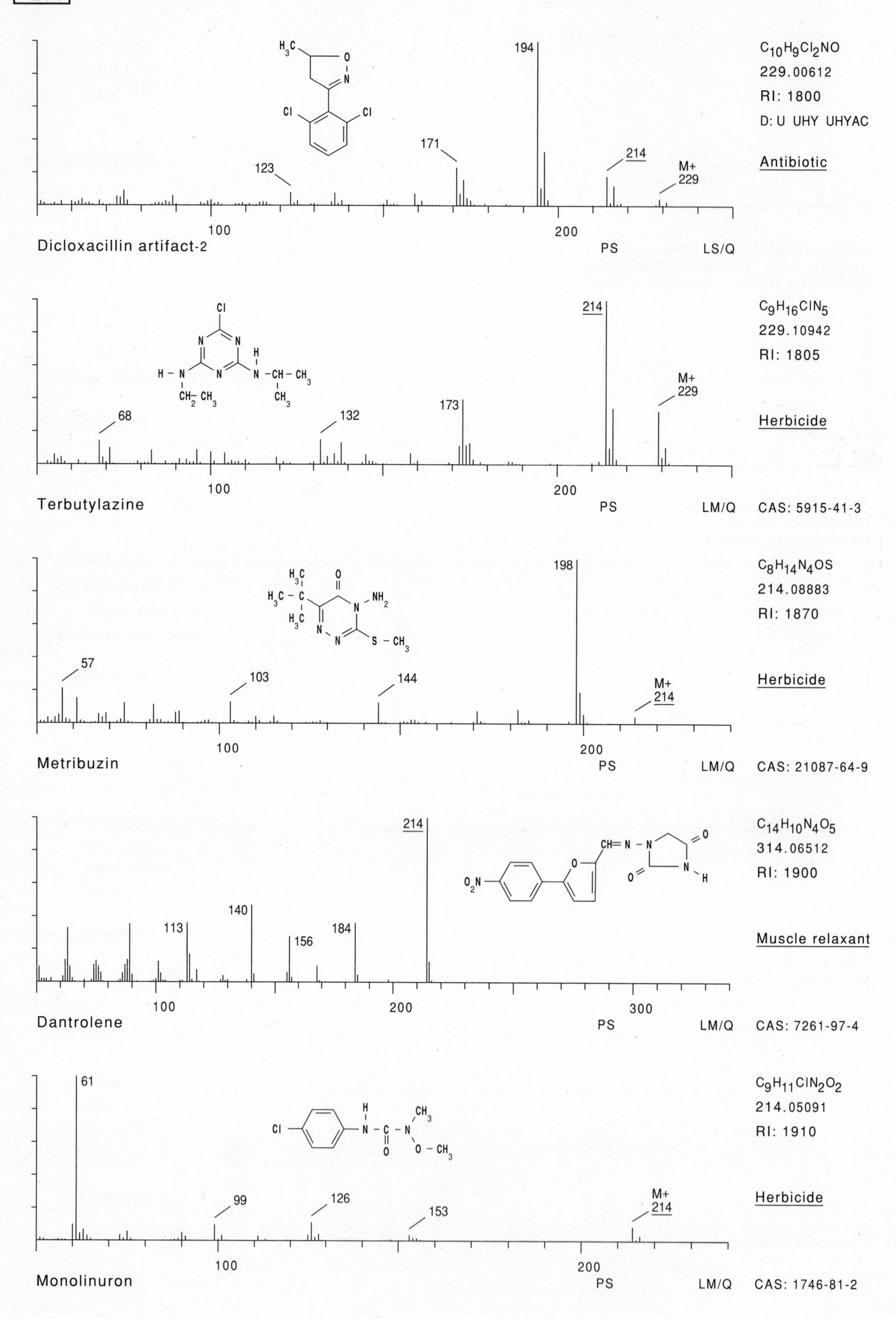

194
171
123
214
M+
229
100
200
$C_{10}H_9Cl_2NO$
229.00612
RI: 1800
D: U UHY UHYAC
Antibiotic
Dicloxacillin artifact-2
PS
LS/Q
214
173
132
68
M+
229
$C_9H_{16}ClN_5$
229.10942
RI: 1805
Herbicide
Terbutylazine
LM/Q
CAS: 5915-41-3
198
57
103
144
M+
214
$C_8H_{14}N_4OS$
214.08883
RI: 1870
Herbicide
Metribuzin
CAS: 21087-64-9
214
140
113
184
156
300
$C_{14}H_{10}N_4O_5$
314.06512
RI: 1900
Muscle relaxant
Dantrolene
CAS: 7261-97-4
61
99
126
153
M+
214
$C_9H_{11}ClN_2O_2$
214.05091
RI: 1910
Herbicide
Monolinuron
CAS: 1746-81-2

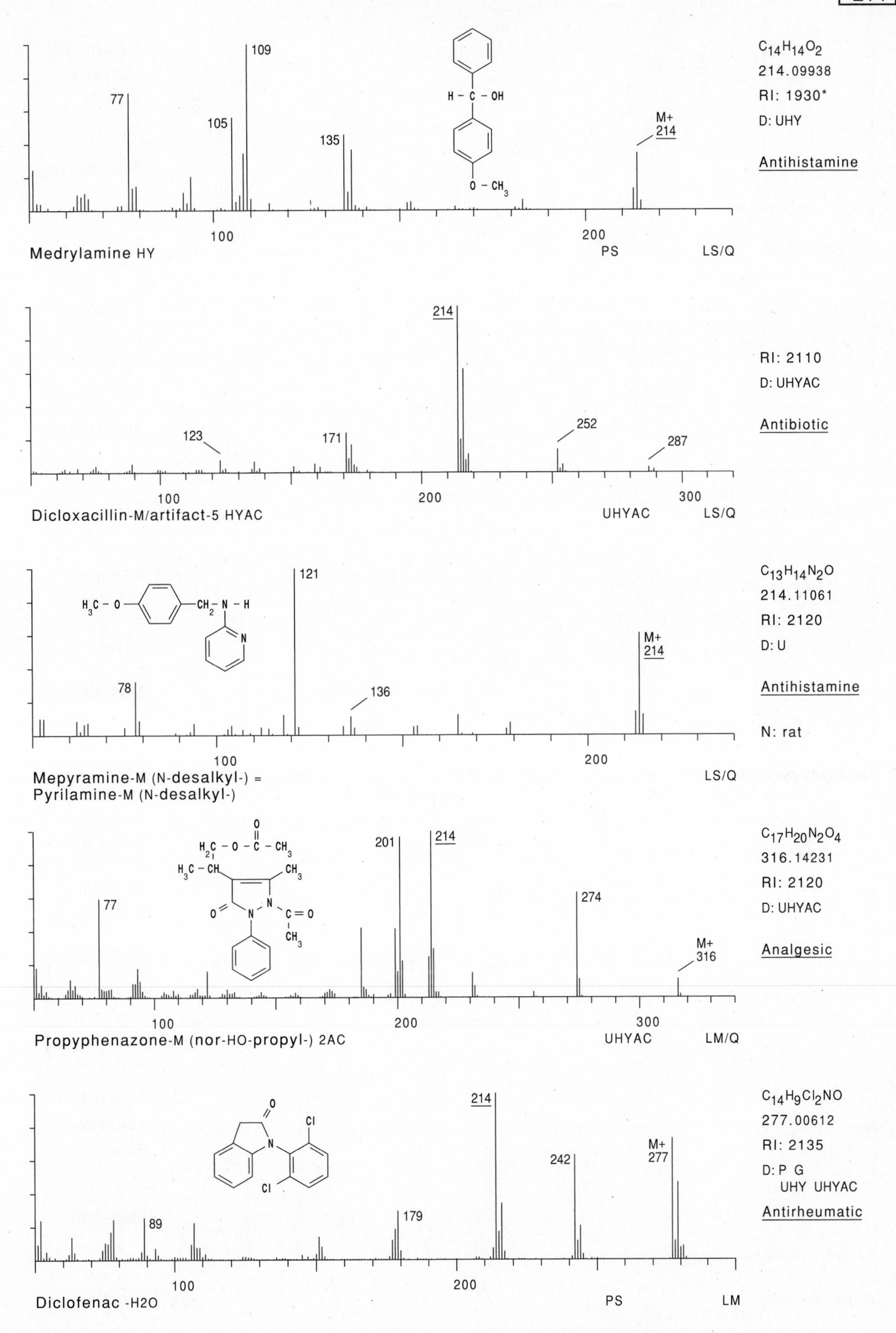

Medrylamine HY
PS LS/Q
C14H14O2
214.09938
RI: 1930*
D: UHY
Antihistamine
Dicloxacillin-M/artifact-5 HYAC
UHYAC LS/Q
RI: 2110
D: UHYAC
Antibiotic
Mepyramine-M (N-desalkyl-) =
Pyrilamine-M (N-desalkyl-)
LS/Q
C13H14N2O
214.11061
RI: 2120
D: U
Antihistamine
N: rat
Propyphenazone-M (nor-HO-propyl-) 2AC
UHYAC LM/Q
C17H20N2O4
316.14231
RI: 2120
D: UHYAC
Analgesic
Diclofenac -H2O
PS LM
C14H9Cl2NO
277.00612
RI: 2135
D: P G
UHY UHYAC
Antirheumatic

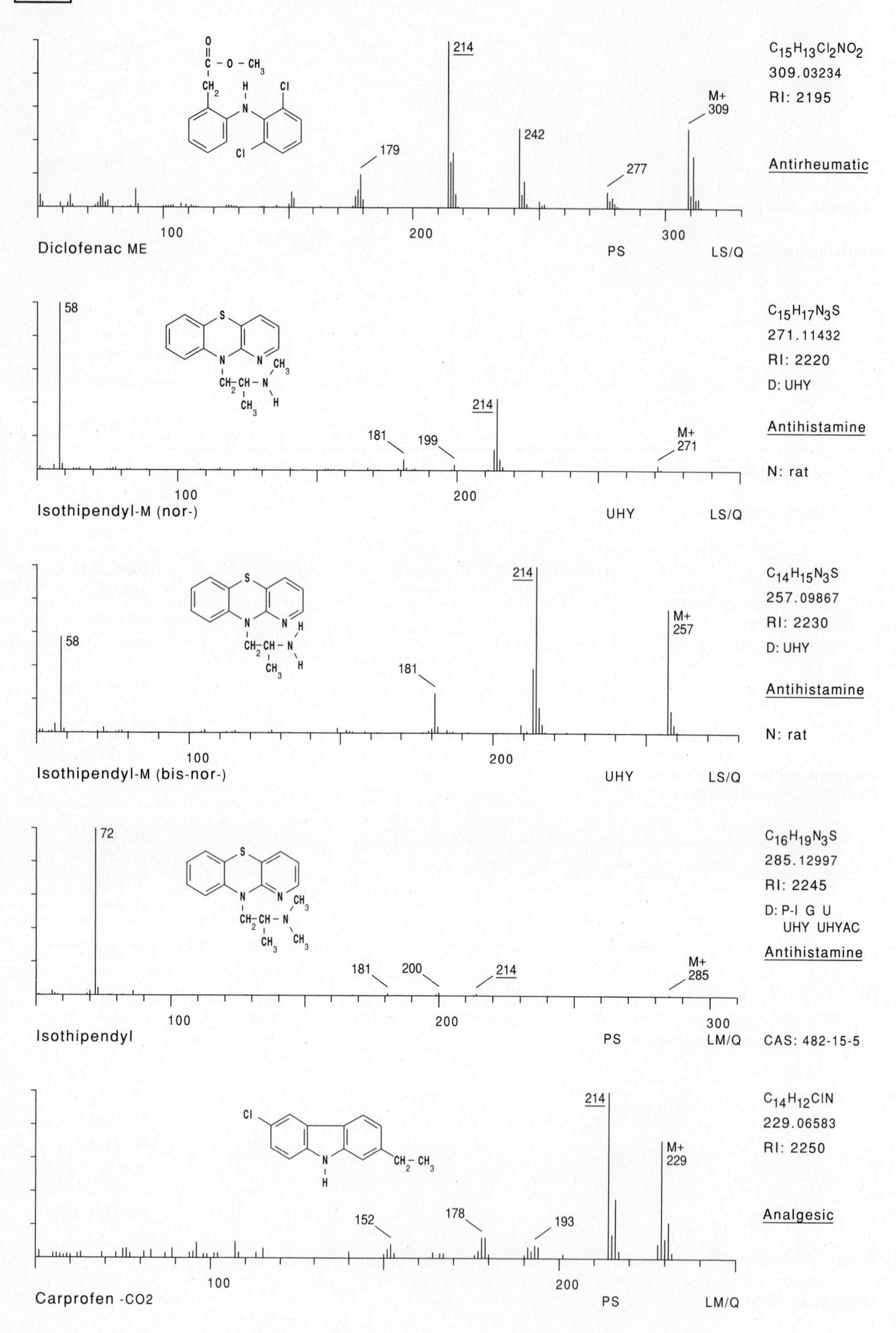

214
C15H13Cl2NO2
309.03234
RI: 2195
M+ 309
242
179
277
Antirheumatic
100
200
300
Diclofenac ME
PS
LS/Q
58
C15H17N3S
271.11432
RI: 2220
D: UHY
214
181
199
M+ 271
Antihistamine
N: rat
Isothipendyl-M (nor-)
UHY
LS/Q
214
C14H15N3S
257.09867
RI: 2230
D: UHY
M+ 257
58
181
Antihistamine
N: rat
Isothipendyl-M (bis-nor-)
UHY
LS/Q
72
C16H19N3S
285.12997
RI: 2245
D: P-I G U
UHY UHYAC
Antihistamine
181
200
214
M+ 285
Isothipendyl
PS
LM/Q
CAS: 482-15-5
214
C14H12ClN
229.06583
RI: 2250
M+ 229
152
178
193
Analgesic
Carprofen -CO2
PS
LM/Q

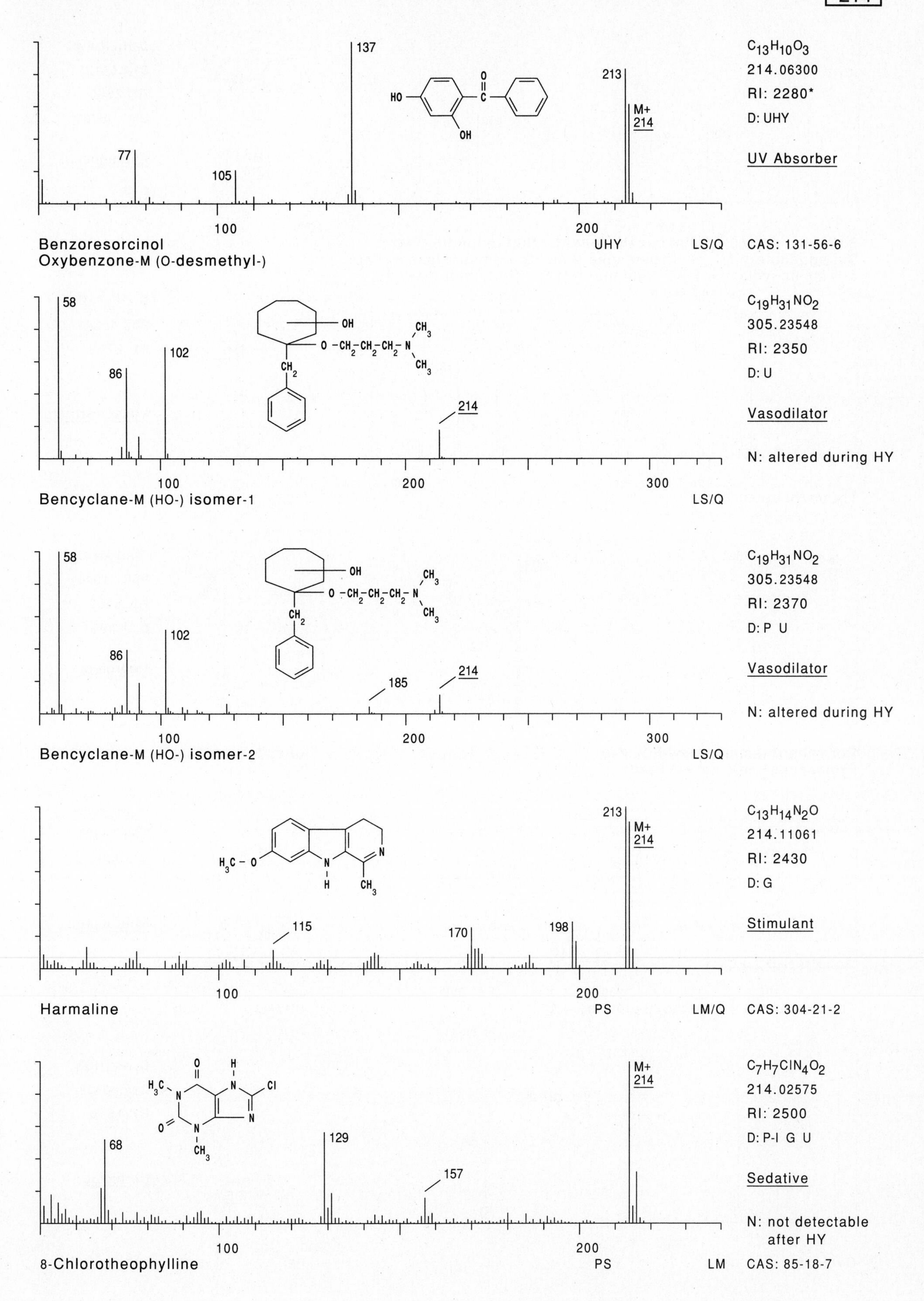
214
137
213
M+ 214
77
105
100
200
HO
OH
$C_{13}H_{10}O_3$
214.06300
RI: 2280*
D: UHY
UV Absorber
Benzoresorcinol
Oxybenzone-M (O-desmethyl-)
UHY
LS/Q
CAS: 131-56-6
58
86
102
214
OH
O -CH2-CH2-CH2- N
CH3
CH2
100
200
300
$C_{19}H_{31}NO_2$
305.23548
RI: 2350
D: U
Vasodilator
N: altered during HY
Bencyclane-M (HO-) isomer-1
LS/Q
58
86
102
185
214
$C_{19}H_{31}NO_2$
305.23548
RI: 2370
D: P U
Vasodilator
N: altered during HY
Bencyclane-M (HO-) isomer-2
LS/Q
213
M+ 214
115
170
198
H3C - O
N
H
CH3
$C_{13}H_{14}N_2O$
214.11061
RI: 2430
D: G
Stimulant
Harmaline
PS
LM/Q
CAS: 304-21-2
M+ 214
68
129
157
H3C
Cl
$C_7H_7ClN_4O_2$
214.02575
RI: 2500
D: P-I G U
Sedative
N: not detectable after HY
8-Chlorotheophylline
PS
LM
CAS: 85-18-7

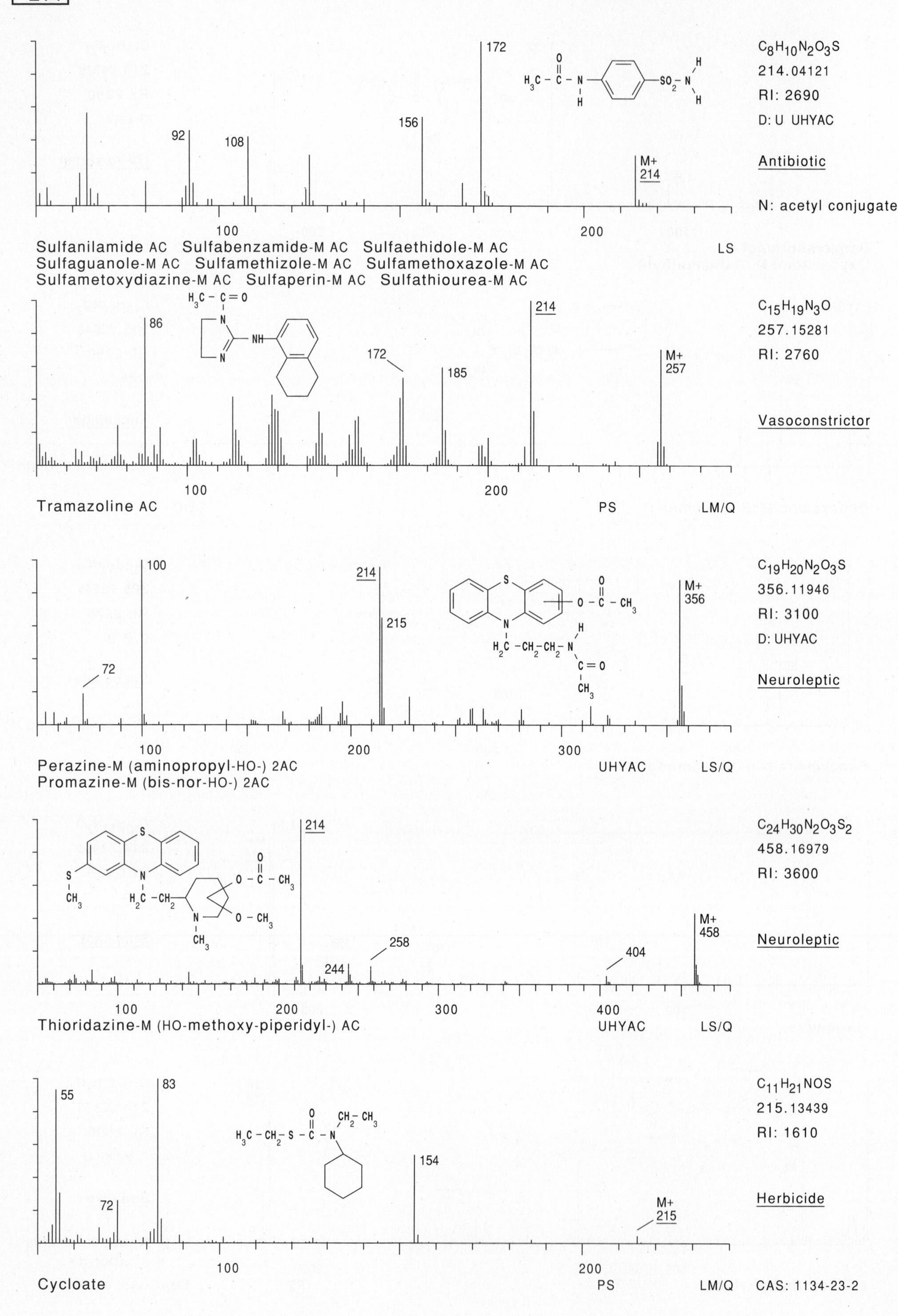

C8H10N2O3S
214.04121
RI: 2690
D: U UHYAC
Antibiotic
N: acetyl conjugate
M+ 214
Sulfanilamide AC Sulfabenzamide-M AC Sulfaethidole-M AC
Sulfaguanole-M AC Sulfamethizole-M AC Sulfamethoxazole-M AC
Sulfametoxydiazine-M AC Sulfaperin-M AC Sulfathiourea-M AC
LS
C15H19N3O
257.15281
RI: 2760
Vasoconstrictor
M+ 257
Tramazoline AC
PS
LM/Q
C19H20N2O3S
356.11946
RI: 3100
D: UHYAC
Neuroleptic
M+ 356
Perazine-M (aminopropyl-HO-) 2AC
Promazine-M (bis-nor-HO-) 2AC
UHYAC
LS/Q
C24H30N2O3S2
458.16979
RI: 3600
Neuroleptic
M+ 458
Thioridazine-M (HO-methoxy-piperidyl-) AC
UHYAC
LS/Q
C11H21NOS
215.13439
RI: 1610
Herbicide
M+ 215
Cycloate
PS
LM/Q
CAS: 1134-23-2

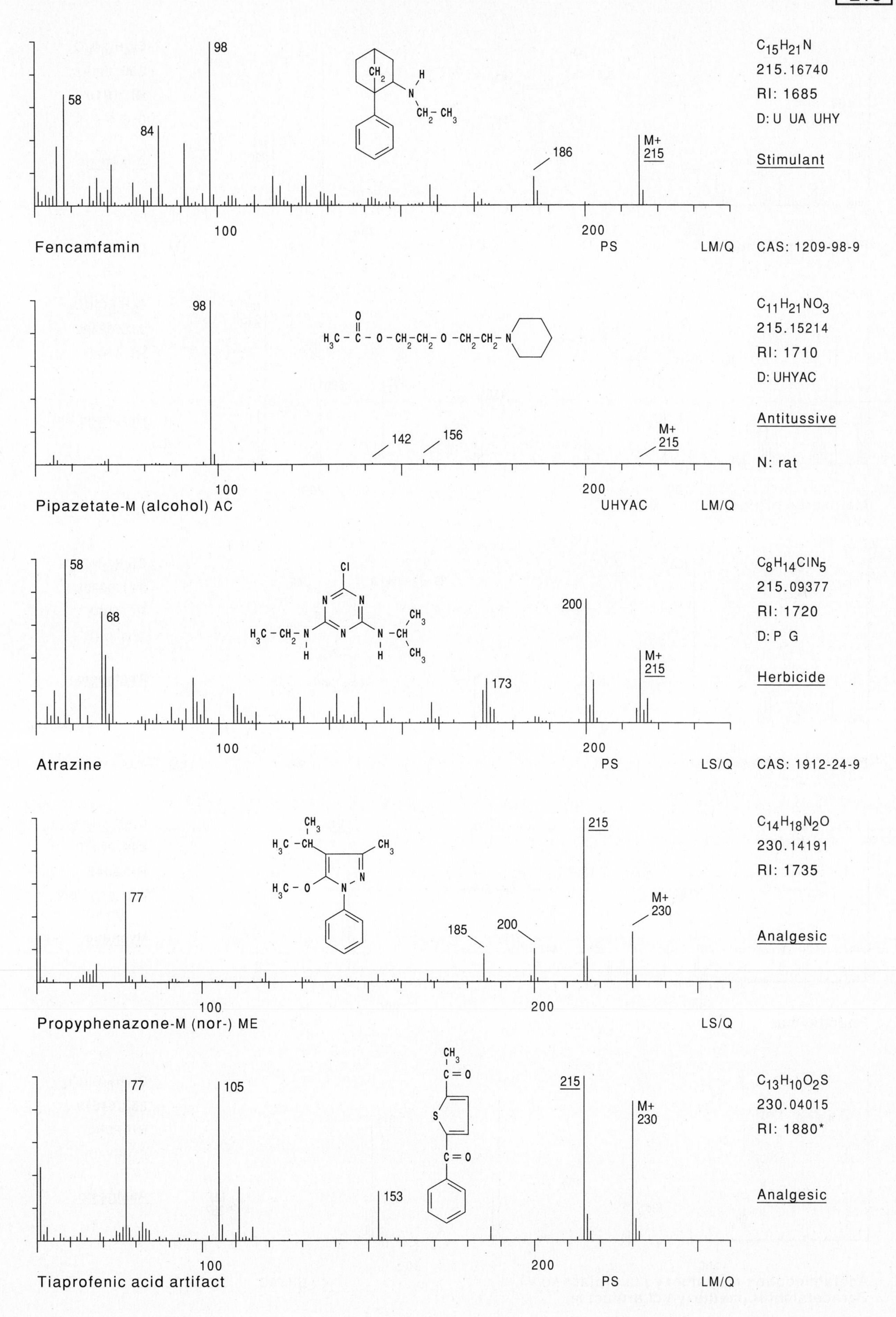

215
Fencamfamin
$C_{15}H_{21}N$
215.16740
RI: 1685
D: U UA UHY
Stimulant
CAS: 1209-98-9
PS
LM/Q
Pipazetate-M (alcohol) AC
$C_{11}H_{21}NO_3$
215.15214
RI: 1710
D: UHYAC
Antitussive
N: rat
UHYAC
LM/Q
Atrazine
$C_8H_{14}ClN_5$
215.09377
RI: 1720
D: P G
Herbicide
CAS: 1912-24-9
PS
LS/Q
Propyphenazone-M (nor-) ME
$C_{14}H_{18}N_2O$
230.14191
RI: 1735
Analgesic
LS/Q
Tiaprofenic acid artifact
$C_{13}H_{10}O_2S$
230.04015
RI: 1880*
Analgesic
PS
LM/Q

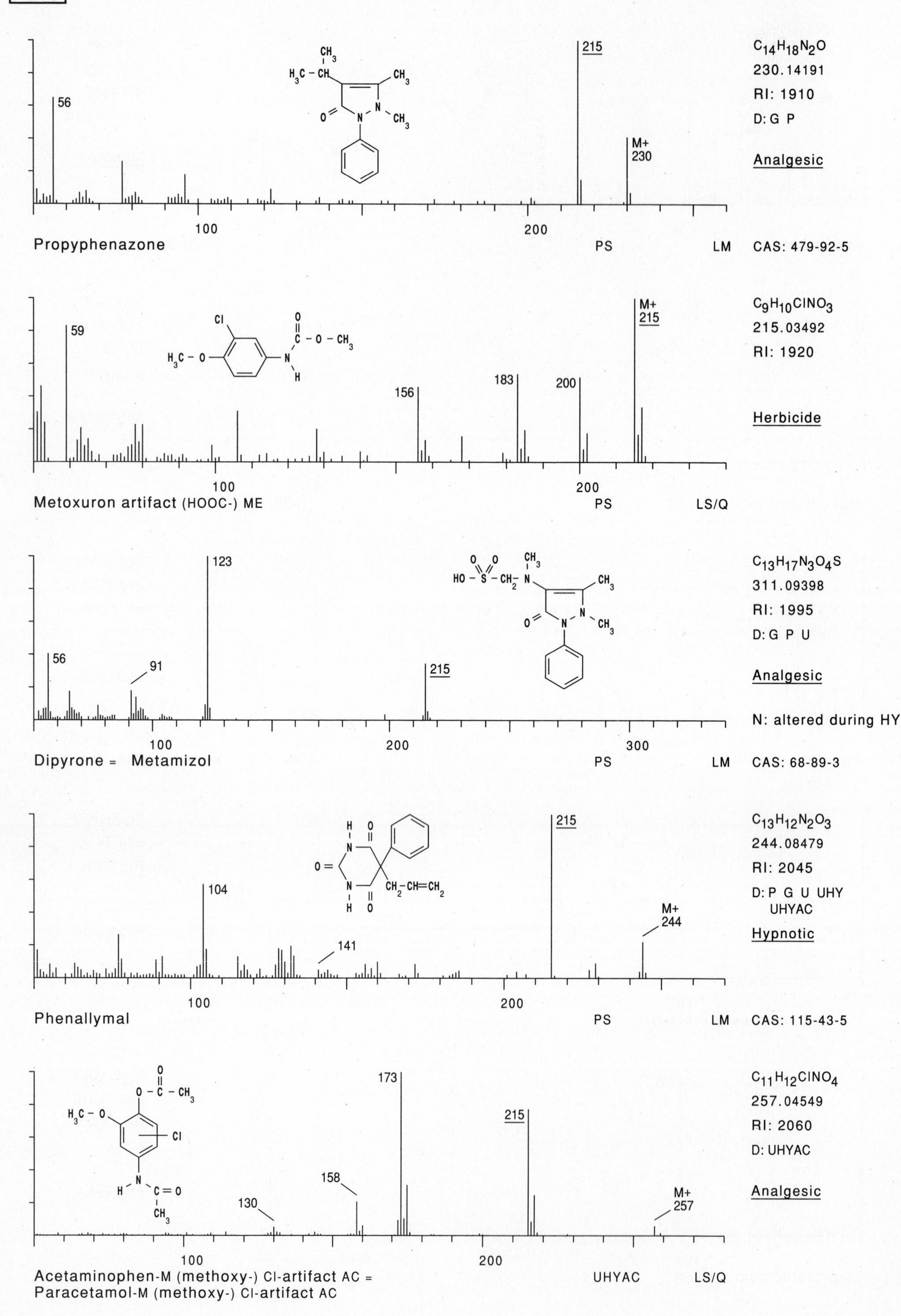
56
215
M+
230
100
200
Propyphenazone
PS
LM
$C_{14}H_{18}N_2O$
230.14191
RI: 1910
D: G P
Analgesic
CAS: 479-92-5
59
156
183
200
M+
215
Metoxuron artifact (HOOC-) ME
LS/Q
$C_9H_{10}ClNO_3$
215.03492
RI: 1920
Herbicide
123
56
91
300
Dipyrone = Metamizol
$C_{13}H_{17}N_3O_4S$
311.09398
RI: 1995
D: G P U
N: altered during HY
CAS: 68-89-3
104
141
M+
244
Phenallymal
$C_{13}H_{12}N_2O_3$
244.08479
RI: 2045
D: P G U UHY UHYAC
Hypnotic
CAS: 115-43-5
173
158
130
M+
257
Acetaminophen-M (methoxy-) Cl-artifact AC =
Paracetamol-M (methoxy-) Cl-artifact AC
UHYAC
$C_{11}H_{12}ClNO_4$
257.04549
RI: 2060
D: UHYAC

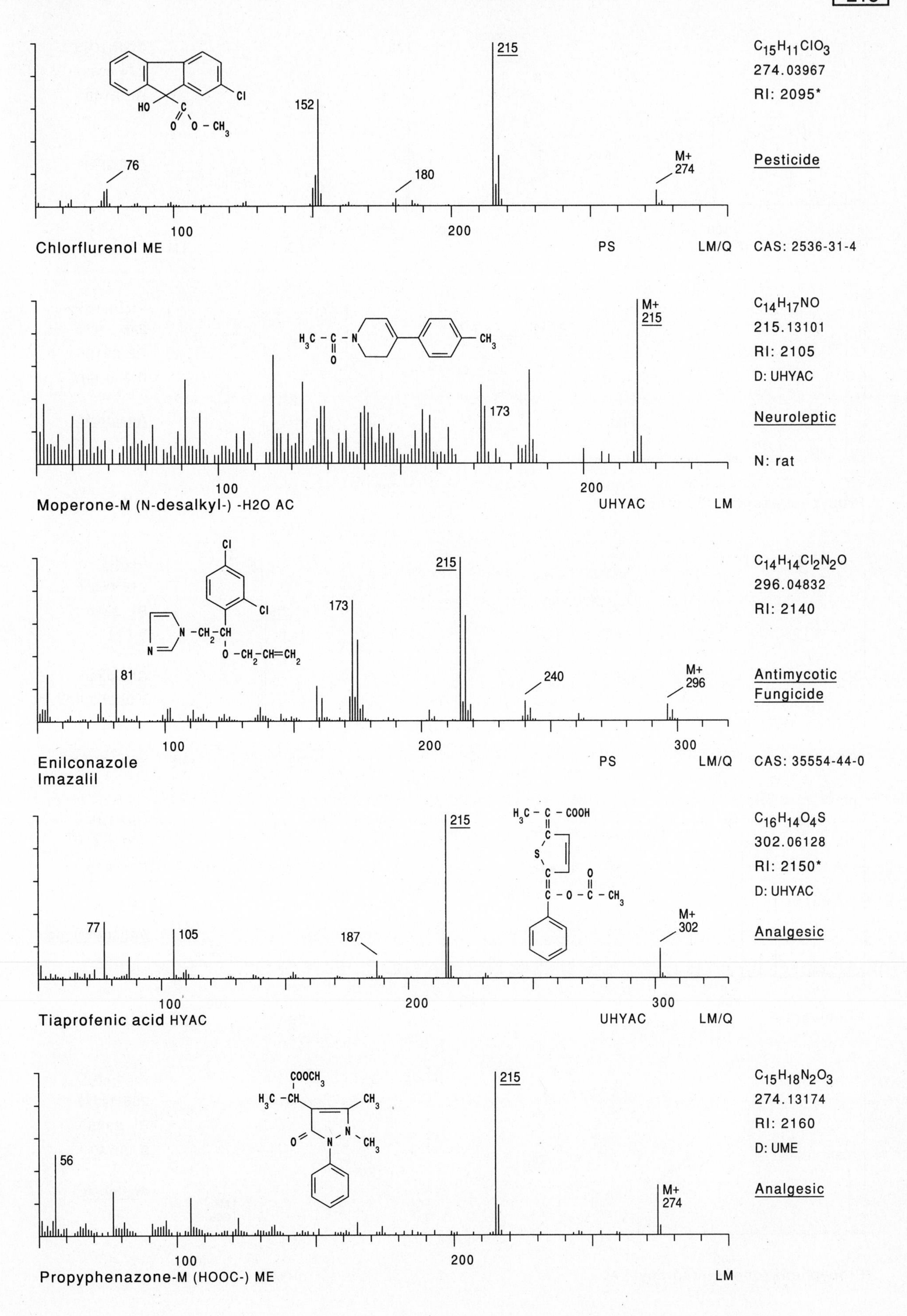

215
C15H11ClO3
274.03967
RI: 2095*
Pesticide
Chlorflurenol ME
PS
LM/Q
CAS: 2536-31-4
C14H17NO
215.13101
RI: 2105
D: UHYAC
Neuroleptic
N: rat
Moperone-M (N-desalkyl-) -H2O AC
UHYAC
LM
C14H14Cl2N2O
296.04832
RI: 2140
Antimycotic
Fungicide
Enilconazole
Imazalil
PS
LM/Q
CAS: 35554-44-0
C16H14O4S
302.06128
RI: 2150*
D: UHYAC
Analgesic
Tiaprofenic acid HYAC
UHYAC
LM/Q
C15H18N2O3
274.13174
RI: 2160
D: UME
Analgesic
Propyphenazone-M (HOOC-) ME
LM

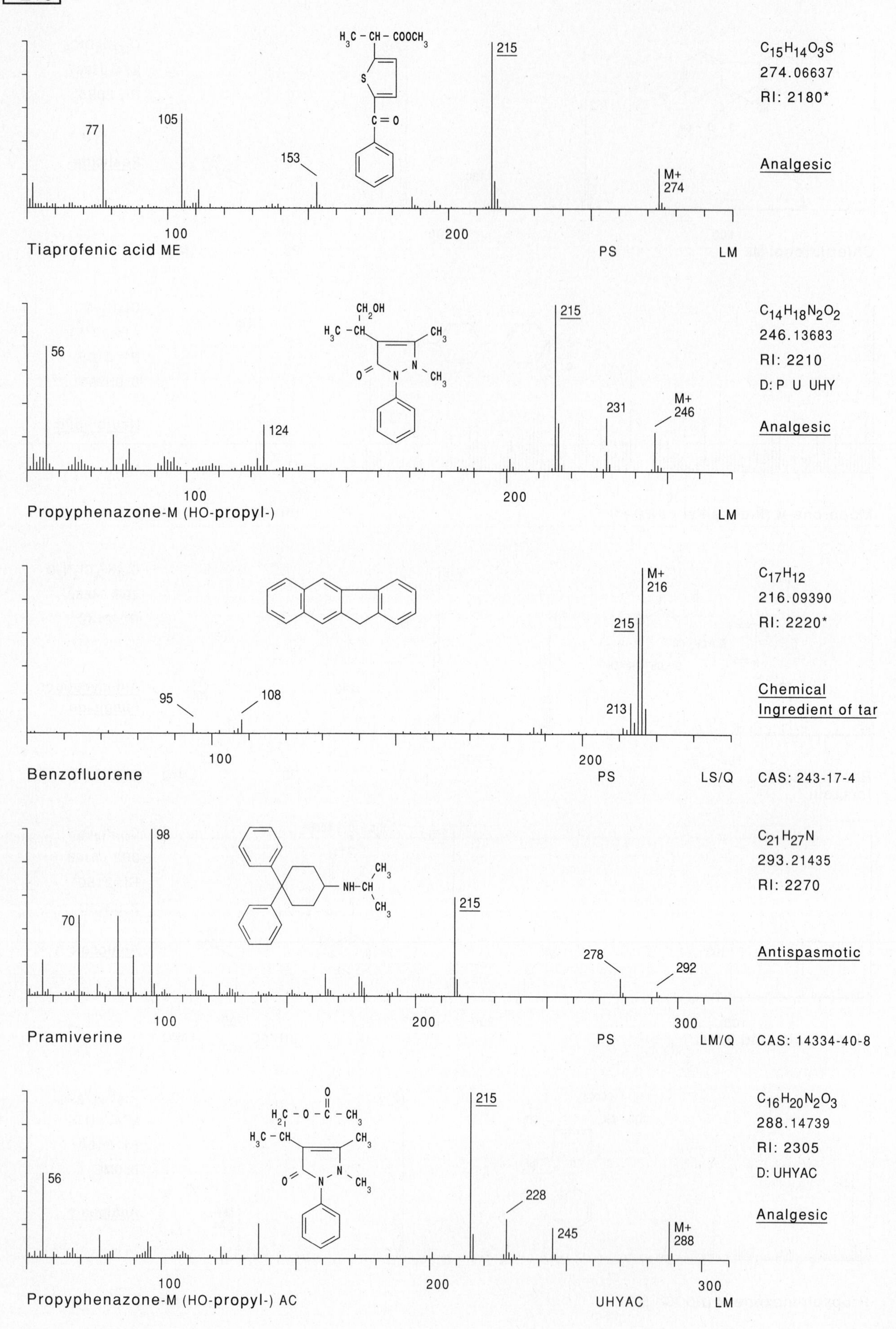
Tiaprofenic acid ME
PS LM
$C_{15}H_{14}O_3S$
274.06637
RI: 2180*
Analgesic
Propyphenazone-M (HO-propyl-)
LM
$C_{14}H_{18}N_2O_2$
246.13683
RI: 2210
D: P U UHY
Analgesic
Benzofluorene
PS LS/Q
$C_{17}H_{12}$
216.09390
RI: 2220*
Chemical
Ingredient of tar
CAS: 243-17-4
Pramiverine
PS LM/Q
$C_{21}H_{27}N$
293.21435
RI: 2270
Antispasmotic
CAS: 14334-40-8
Propyphenazone-M (HO-propyl-) AC
UHYAC LM
$C_{16}H_{20}N_2O_3$
288.14739
RI: 2305
D: UHYAC
Analgesic

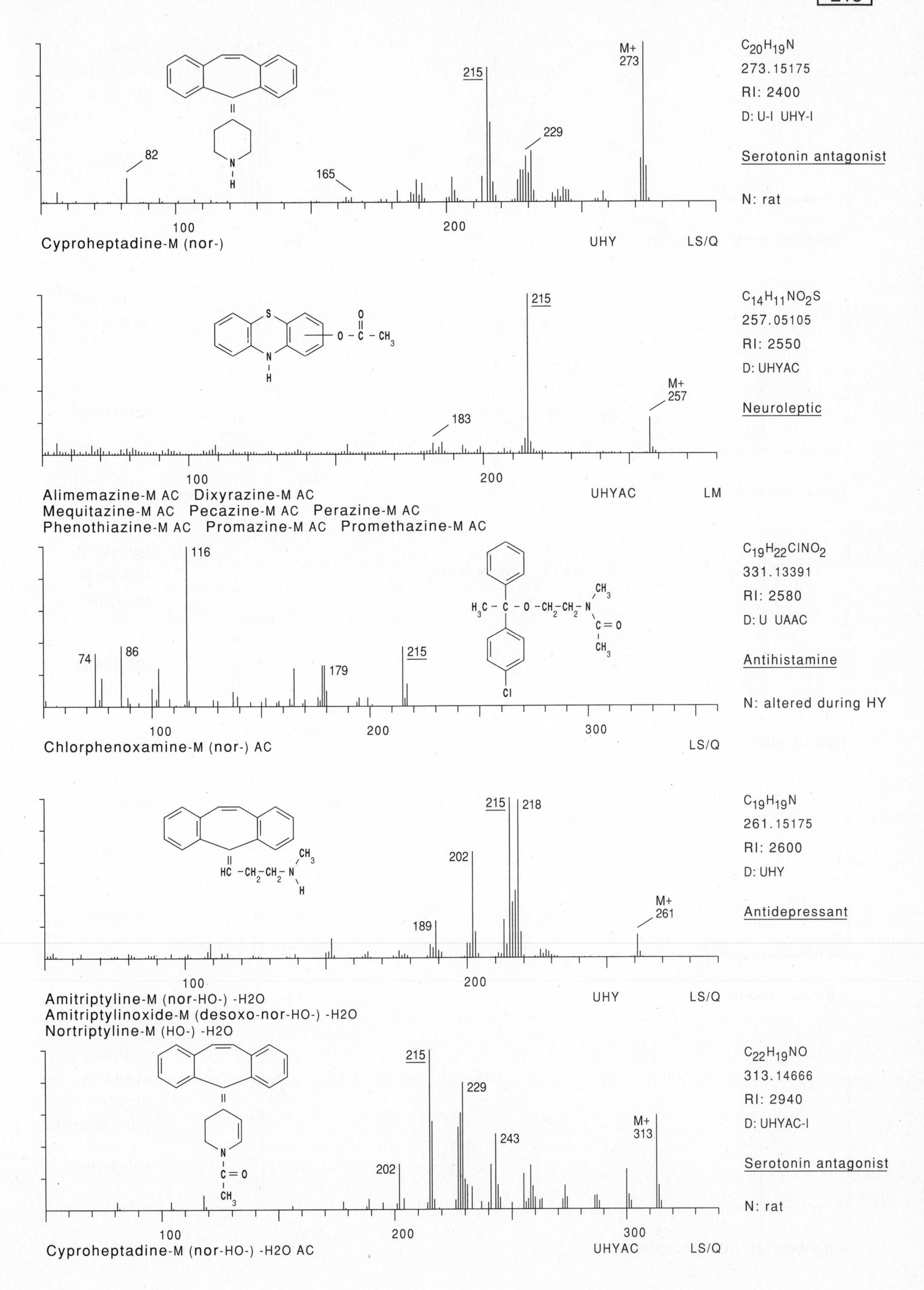

M+
273
215
229
82
165
100
200
Cyproheptadine-M (nor-)
UHY
LS/Q
$C_{20}H_{19}N$
273.15175
RI: 2400
D: U-I UHY-I
Serotonin antagonist
N: rat
215
M+
257
183
100
200
Alimemazine-M AC Dixyrazine-M AC
Mequitazine-M AC Pecazine-M AC Perazine-M AC
Phenothiazine-M AC Promazine-M AC Promethazine-M AC
UHYAC
LM
$C_{14}H_{11}NO_2S$
257.05105
RI: 2550
D: UHYAC
Neuroleptic
116
74
86
179
215
100
200
300
Chlorphenoxamine-M (nor-) AC
LS/Q
$C_{19}H_{22}ClNO_2$
331.13391
RI: 2580
D: U UAAC
Antihistamine
N: altered during HY
215
218
202
189
M+
261
100
200
Amitriptyline-M (nor-HO-) -H2O
Amitriptylinoxide-M (desoxo-nor-HO-) -H2O
Nortriptyline-M (HO-) -H2O
UHY
LS/Q
$C_{19}H_{19}N$
261.15175
RI: 2600
D: UHY
Antidepressant
215
229
243
202
M+
313
100
200
300
Cyproheptadine-M (nor-HO-) -H2O AC
UHYAC
LS/Q
$C_{22}H_{19}NO$
313.14666
RI: 2940
D: UHYAC-I
Serotonin antagonist
N: rat

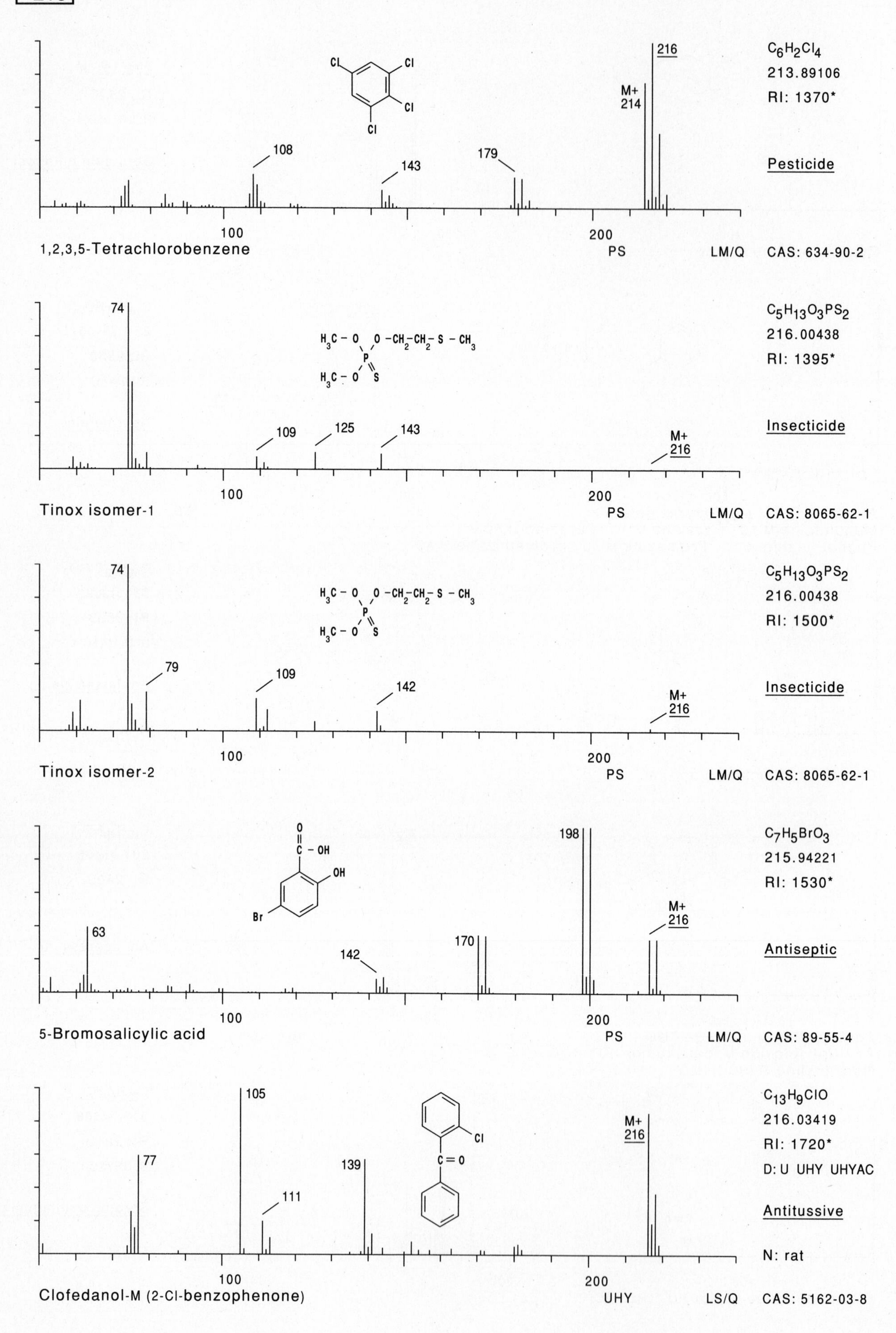

216
M+
214
108
143
179
100
200
1,2,3,5-Tetrachlorobenzene
PS
LM/Q
$C_6H_2Cl_4$
213.89106
RI: 1370*
Pesticide
CAS: 634-90-2
74
109
125
143
M+
216
Tinox isomer-1
LM/Q
$C_5H_{13}O_3PS_2$
216.00438
RI: 1395*
Insecticide
CAS: 8065-62-1
74
79
109
142
M+
216
Tinox isomer-2
$C_5H_{13}O_3PS_2$
216.00438
RI: 1500*
Insecticide
CAS: 8065-62-1
198
63
170
142
M+
216
5-Bromosalicylic acid
$C_7H_5BrO_3$
215.94221
RI: 1530*
Antiseptic
CAS: 89-55-4
105
77
139
111
M+
216
Clofedanol-M (2-Cl-benzophenone)
UHY
LS/Q
$C_{13}H_9ClO$
216.03419
RI: 1720*
D: U UHY UHYAC
Antitussive
N: rat
CAS: 5162-03-8

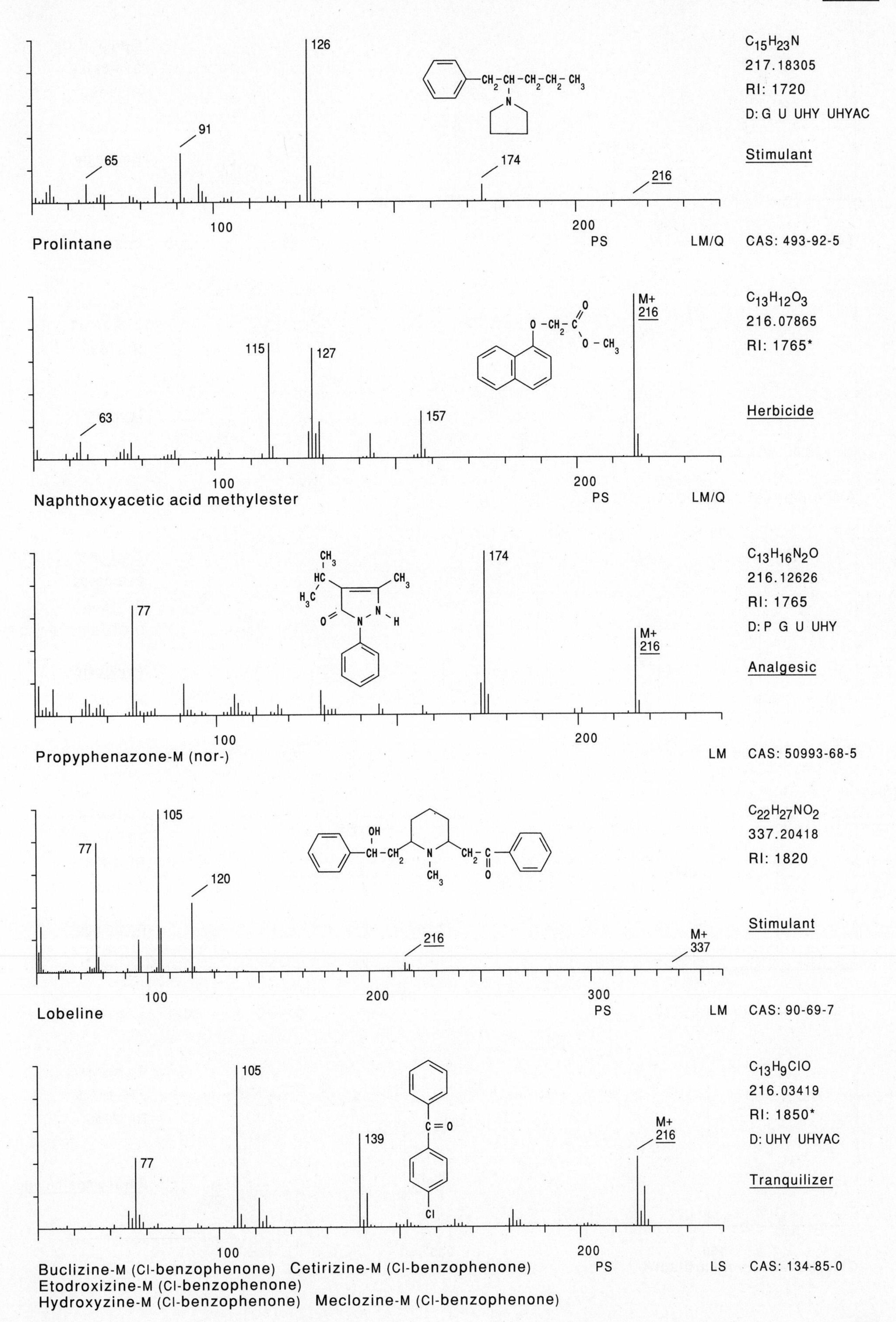

126
91
65
174
216
100
200
Prolintane
PS
LM/Q
$C_{15}H_{23}N$
217.18305
RI: 1720
D: G U UHY UHYAC
Stimulant
CAS: 493-92-5
M+ 216
115
127
63
157
Naphthoxyacetic acid methylester
$C_{13}H_{12}O_3$
216.07865
RI: 1765*
Herbicide
174
77
Propyphenazone-M (nor-)
LM
$C_{13}H_{16}N_2O$
216.12626
RI: 1765
D: P G U UHY
Analgesic
CAS: 50993-68-5
105
77
120
M+ 337
300
Lobeline
$C_{22}H_{27}NO_2$
337.20418
RI: 1820
Stimulant
CAS: 90-69-7
139
Buclizine-M (Cl-benzophenone) Cetirizine-M (Cl-benzophenone)
Etodroxizine-M (Cl-benzophenone)
Hydroxyzine-M (Cl-benzophenone) Meclozine-M (Cl-benzophenone)
LS
$C_{13}H_9ClO$
216.03419
RI: 1850*
D: UHY UHYAC
Tranquilizer
CAS: 134-85-0

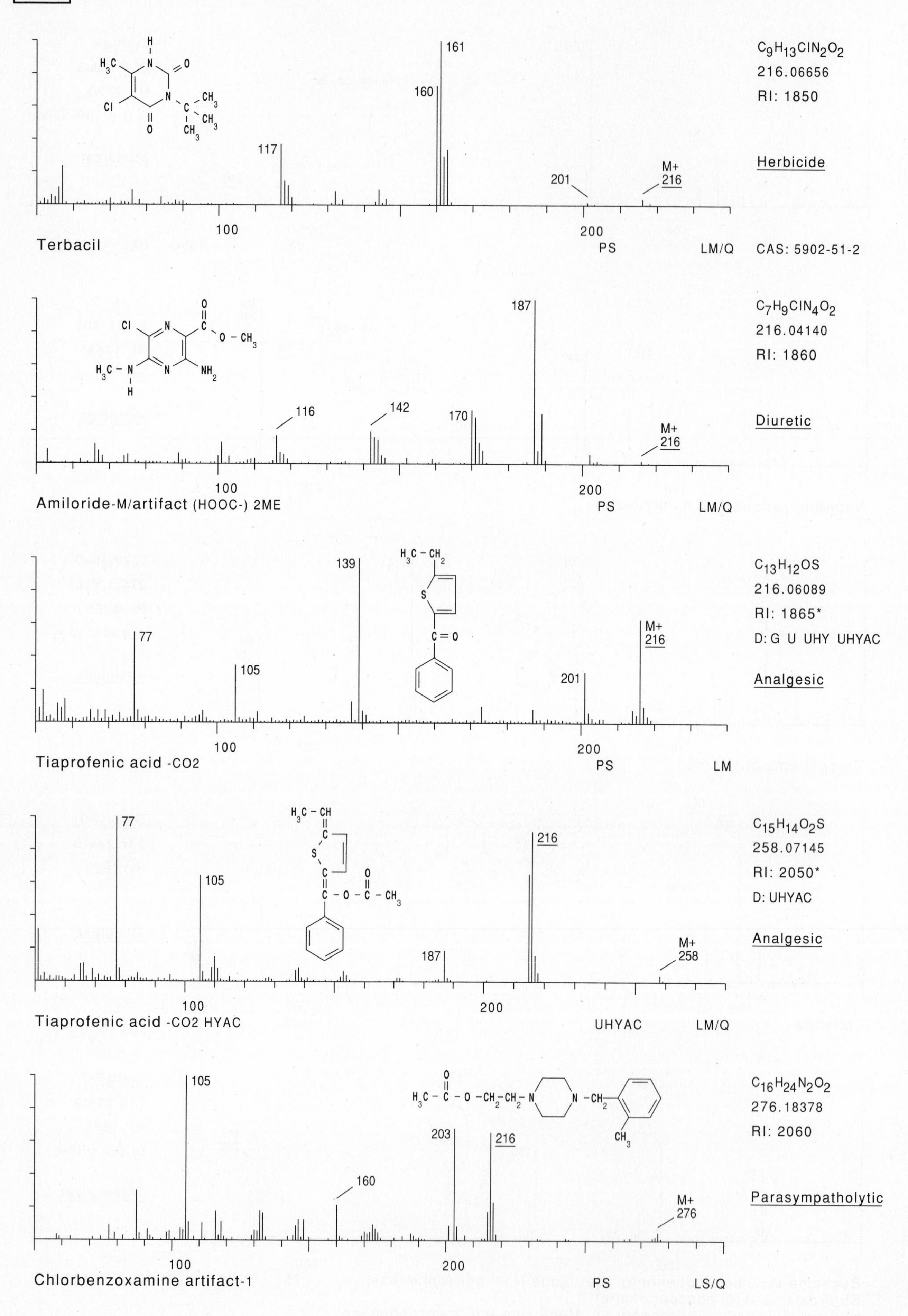

161
160
117
201
M+
216
100
200
PS
LM/Q
Terbacil
$C_9H_{13}ClN_2O_2$
216.06656
RI: 1850
Herbicide
CAS: 5902-51-2
187
116
142
170
M+
216
100
200
PS
LM/Q
Amiloride-M/artifact (HOOC-) 2ME
$C_7H_9ClN_4O_2$
216.04140
RI: 1860
Diuretic
139
77
105
201
M+
216
100
200
PS
LM
Tiaprofenic acid -CO2
$C_{13}H_{12}OS$
216.06089
RI: 1865*
D: G U UHY UHYAC
Analgesic
77
105
187
216
M+
258
100
200
UHYAC
LM/Q
Tiaprofenic acid -CO2 HYAC
$C_{15}H_{14}O_2S$
258.07145
RI: 2050*
D: UHYAC
Analgesic
105
203
216
160
M+
276
100
200
PS
LS/Q
Chlorbenzoxamine artifact-1
$C_{16}H_{24}N_2O_2$
276.18378
RI: 2060
Parasympatholytic

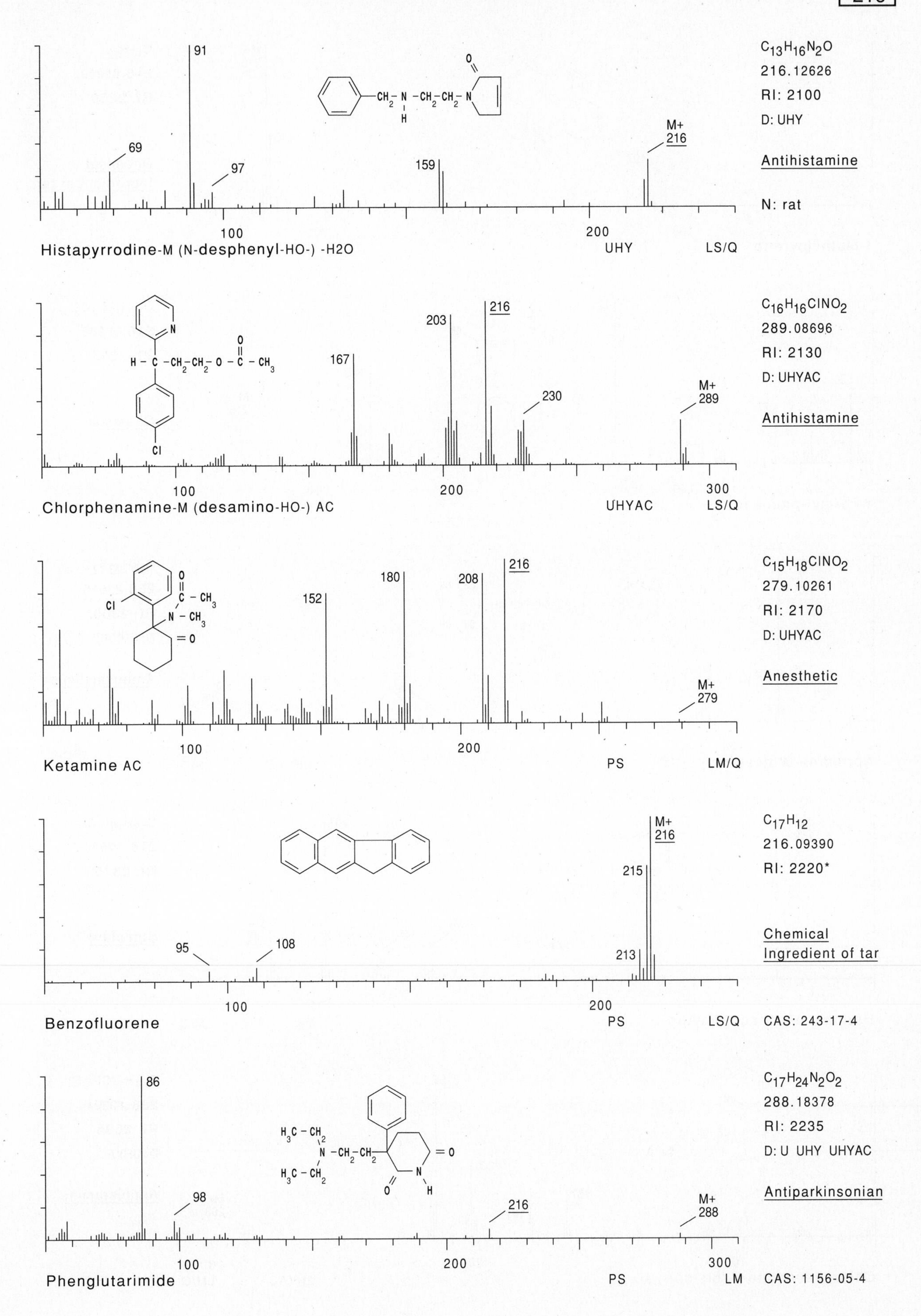

91
69
97
159
M+
216
Histapyrrodine-M (N-desphenyl-HO-) -H2O
UHY
LS/Q
C13H16N2O
216.12626
RI: 2100
D: UHY
Antihistamine
N: rat
216
203
167
230
M+
289
Chlorphenamine-M (desamino-HO-) AC
UHYAC
LS/Q
C16H16ClNO2
289.08696
RI: 2130
D: UHYAC
Antihistamine
216
180
208
152
M+
279
Ketamine AC
PS
LM/Q
C15H18ClNO2
279.10261
RI: 2170
D: UHYAC
Anesthetic
M+
216
215
213
95
108
Benzofluorene
PS
LS/Q
C17H12
216.09390
RI: 2220*
Chemical
Ingredient of tar
CAS: 243-17-4
86
98
216
M+
288
Phenglutarimide
PS
LM
C17H24N2O2
288.18378
RI: 2235
D: U UHY UHYAC
Antiparkinsonian
CAS: 1156-05-4

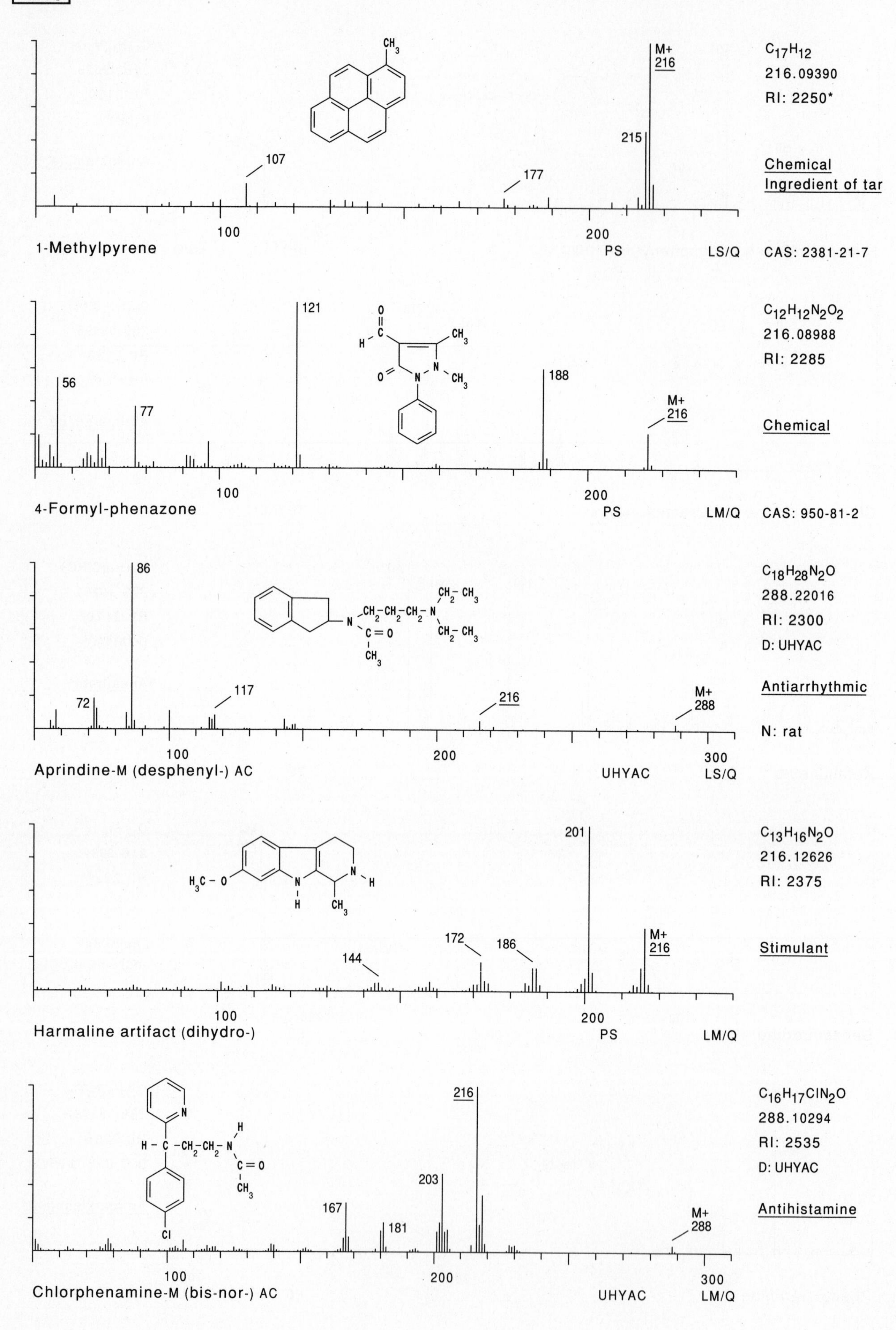

CH3
M+
216
215
107
177
100
200
1-Methylpyrene
PS
LS/Q
C17H12
216.09390
RI: 2250*
Chemical
Ingredient of tar
CAS: 2381-21-7
121
56
77
188
M+
216
100
200
4-Formyl-phenazone
PS
LM/Q
C12H12N2O2
216.08988
RI: 2285
Chemical
CAS: 950-81-2
86
72
117
216
M+
288
100
200
300
Aprindine-M (desphenyl-) AC
UHYAC
LS/Q
C18H28N2O
288.22016
RI: 2300
D: UHYAC
Antiarrhythmic
N: rat
201
144
172
186
M+
216
100
200
Harmaline artifact (dihydro-)
PS
LM/Q
C13H16N2O
216.12626
RI: 2375
Stimulant
216
203
167
181
M+
288
100
200
300
Chlorphenamine-M (bis-nor-) AC
UHYAC
LM/Q
C16H17ClN2O
288.10294
RI: 2535
D: UHYAC
Antihistamine

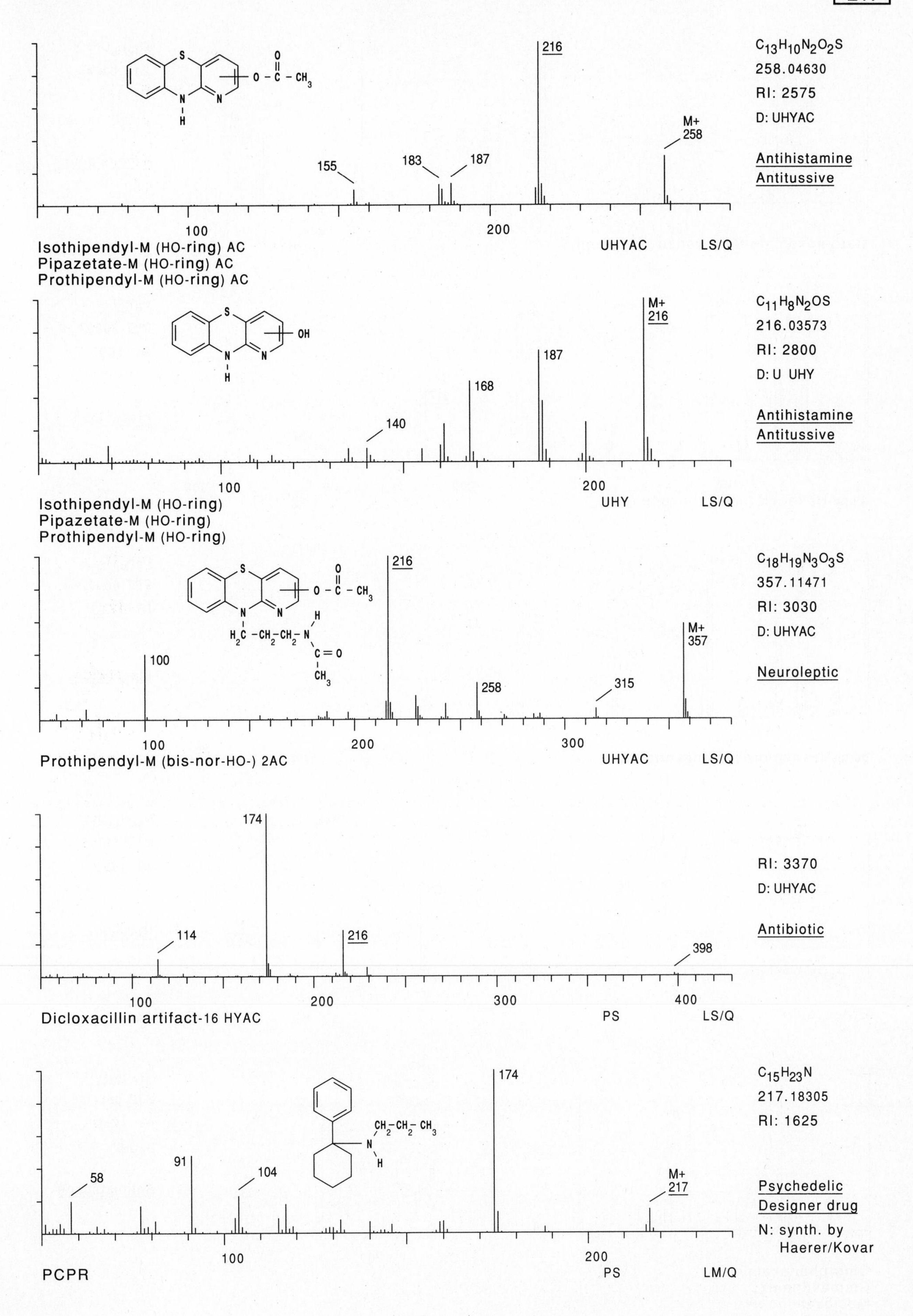

Isothipendyl-M (HO-ring) AC
Pipazetate-M (HO-ring) AC
Prothipendyl-M (HO-ring) AC

$C_{13}H_{10}N_2O_2S$
258.04630
RI: 2575
D: UHYAC

Antihistamine
Antitussive

Isothipendyl-M (HO-ring)
Pipazetate-M (HO-ring)
Prothipendyl-M (HO-ring)

$C_{11}H_8N_2OS$
216.03573
RI: 2800
D: U UHY

Antihistamine
Antitussive

Prothipendyl-M (bis-nor-HO-) 2AC

$C_{18}H_{19}N_3O_3S$
357.11471
RI: 3030
D: UHYAC

Neuroleptic

Dicloxacillin artifact-16 HYAC

RI: 3370
D: UHYAC

Antibiotic

PCPR

$C_{15}H_{23}N$
217.18305
RI: 1625

Psychedelic
Designer drug

N: synth. by
Haerer/Kovar

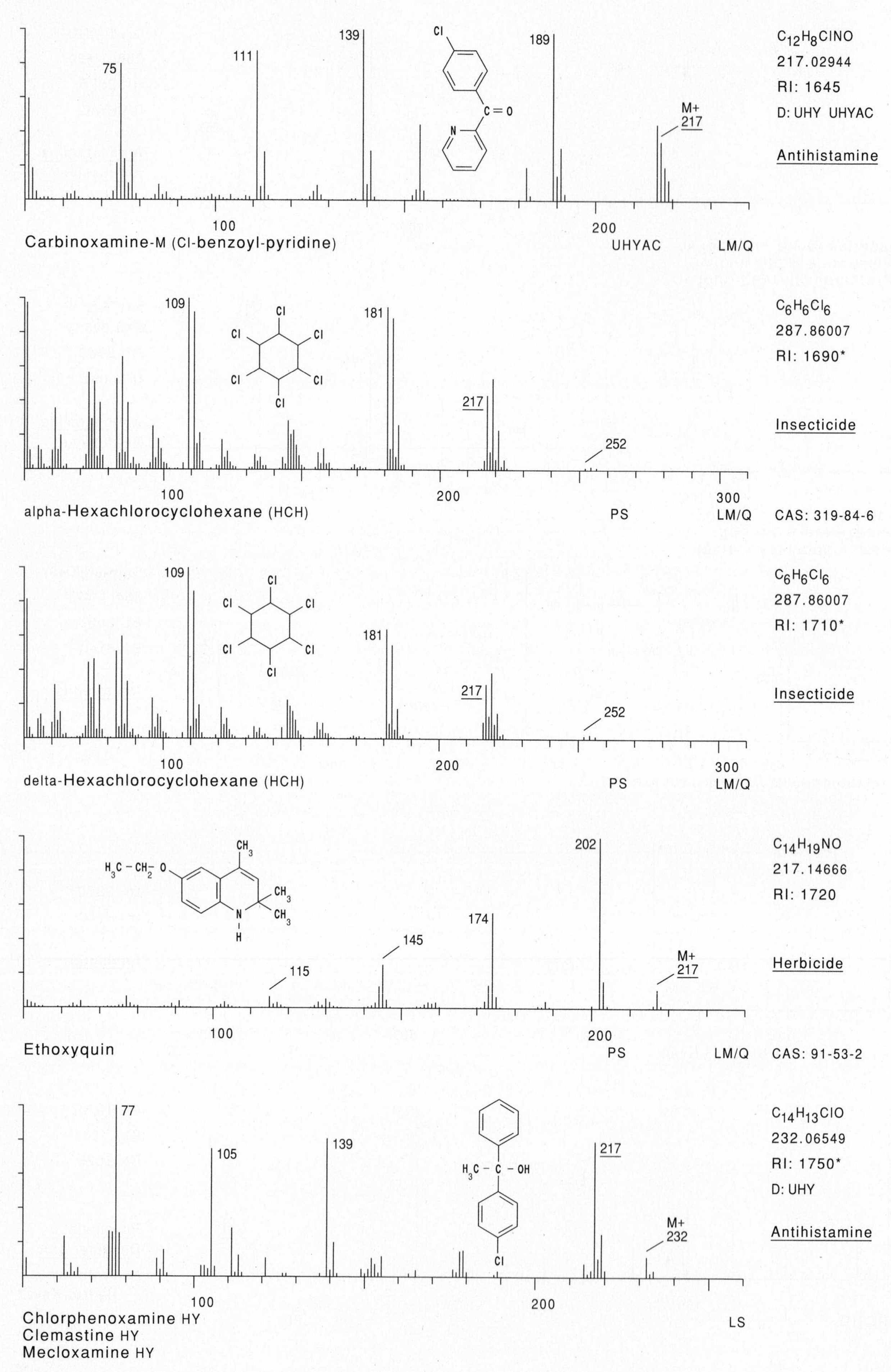

75
111
139
189
Cl
C=O
N
M+
217
100
200
$C_{12}H_8ClNO$
217.02944
RI: 1645
D: UHY UHYAC
Antihistamine
Carbinoxamine-M (Cl-benzoyl-pyridine)
UHYAC
LM/Q
109
181
217
252
Cl
300
$C_6H_6Cl_6$
287.86007
RI: 1690*
Insecticide
alpha-Hexachlorocyclohexane (HCH)
PS
LM/Q
CAS: 319-84-6
109
181
217
252
$C_6H_6Cl_6$
287.86007
RI: 1710*
Insecticide
delta-Hexachlorocyclohexane (HCH)
PS
LM/Q
H_3C-CH_2-O
CH_3
N
H
202
174
145
115
M+
217
$C_{14}H_{19}NO$
217.14666
RI: 1720
Herbicide
Ethoxyquin
PS
LM/Q
CAS: 91-53-2
77
105
139
217
$H_3C-C-OH$
Cl
M+
232
$C_{14}H_{13}ClO$
232.06549
RI: 1750*
D: UHY
Antihistamine
Chlorphenoxamine HY
Clemastine HY
Mecloxamine HY
LS

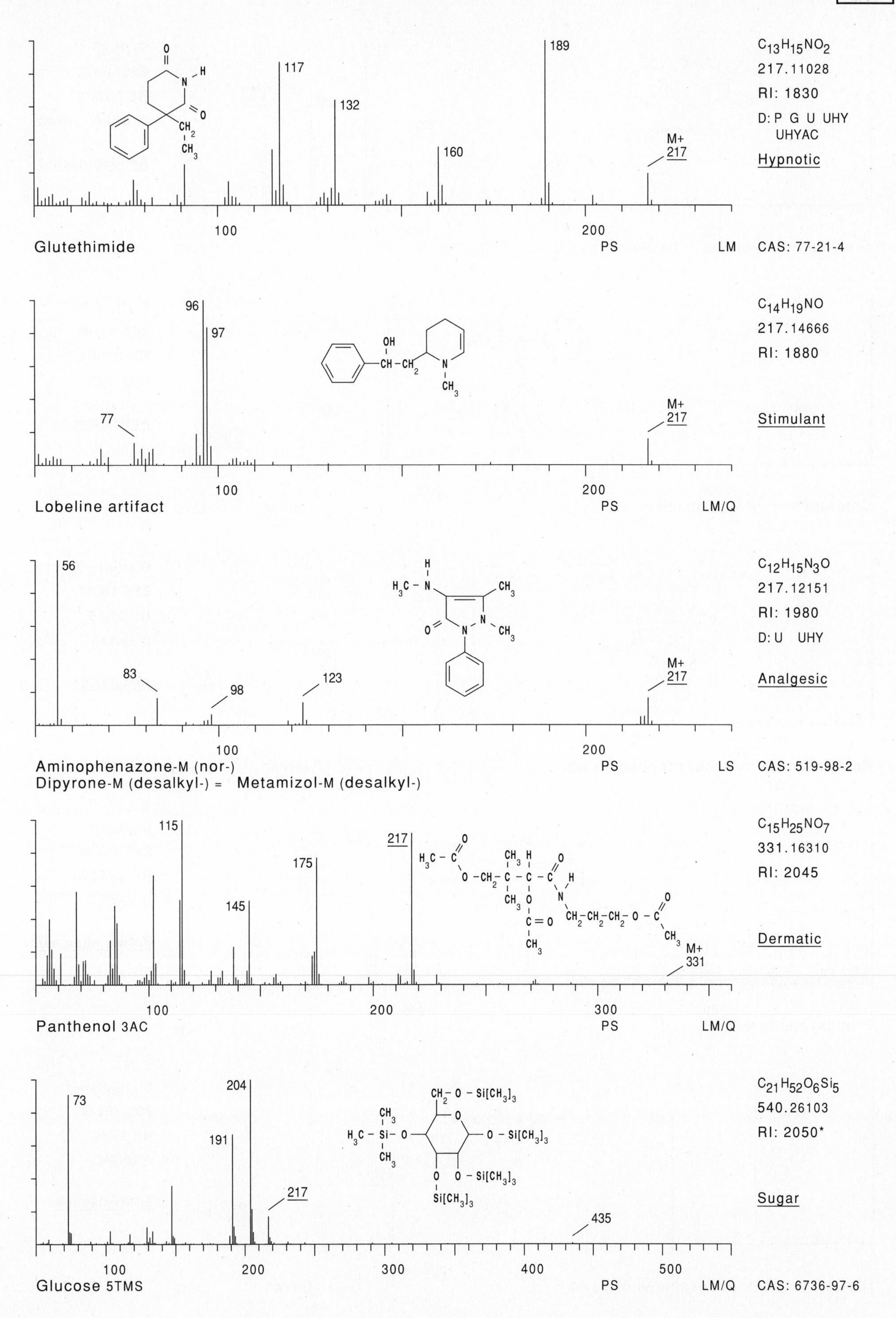

$C_{13}H_{15}NO_2$
217.11028
RI: 1830
D: P G U UHY UHYAC
Hypnotic
Glutethimide
PS LM
CAS: 77-21-4
$C_{14}H_{19}NO$
217.14666
RI: 1880
Stimulant
Lobeline artifact
PS LM/Q
$C_{12}H_{15}N_3O$
217.12151
RI: 1980
D: U UHY
Analgesic
Aminophenazone-M (nor-)
Dipyrone-M (desalkyl-) = Metamizol-M (desalkyl-)
PS LS
CAS: 519-98-2
$C_{15}H_{25}NO_7$
331.16310
RI: 2045
Dermatic
Panthenol 3AC
PS LM/Q
$C_{21}H_{52}O_6Si_5$
540.26103
RI: 2050*
Sugar
Glucose 5TMS
PS LM/Q
CAS: 6736-97-6

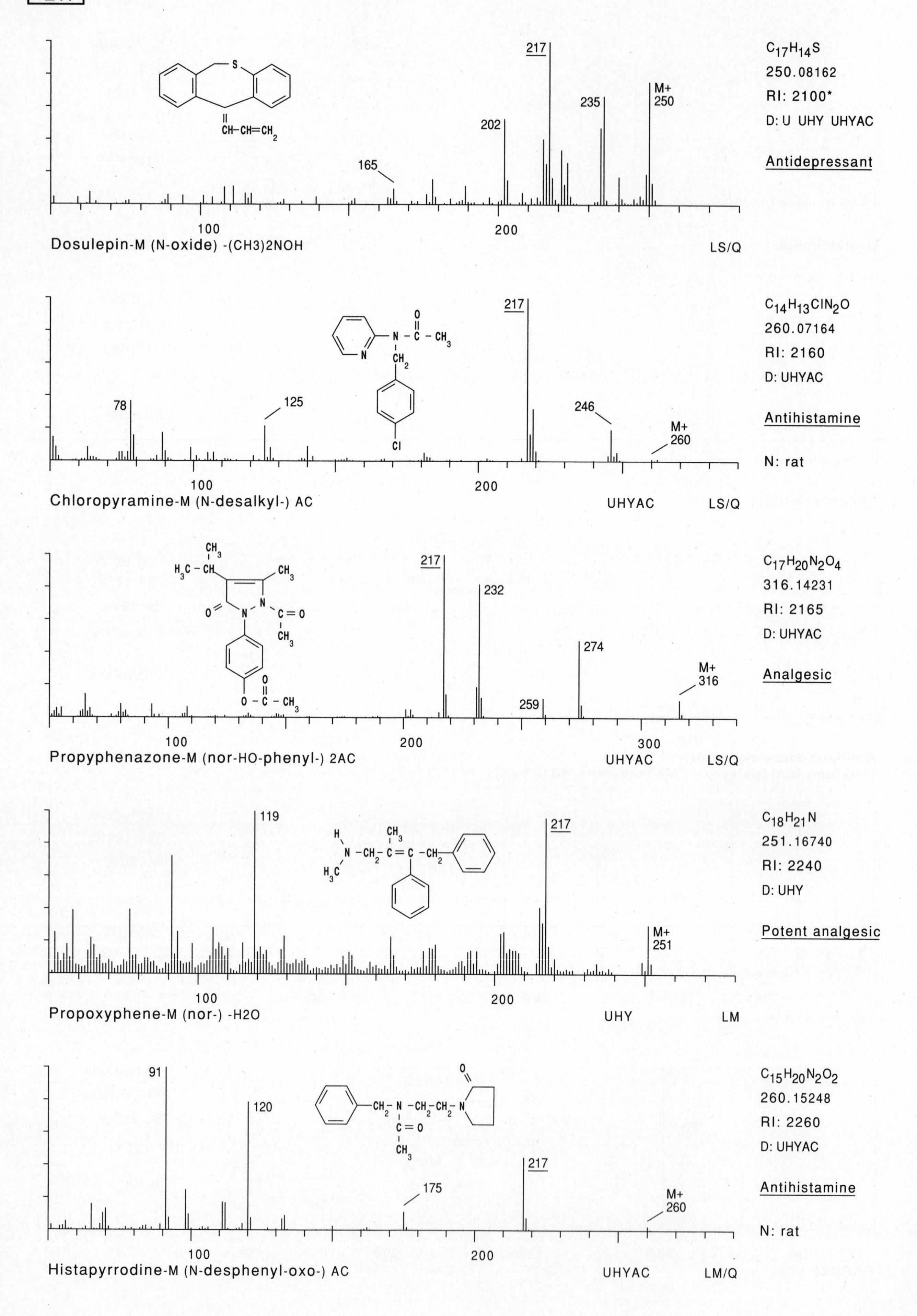

217
202
235
M+
250
165
100
200
Dosulepin-M (N-oxide) -(CH3)2NOH
LS/Q
C17H14S
250.08162
RI: 2100*
D: U UHY UHYAC
Antidepressant
217
78
125
246
M+
260
100
200
Chloropyramine-M (N-desalkyl-) AC
UHYAC
LS/Q
C14H13ClN2O
260.07164
RI: 2160
D: UHYAC
Antihistamine
N: rat
217
232
274
259
M+
316
100
200
300
Propyphenazone-M (nor-HO-phenyl-) 2AC
UHYAC
LS/Q
C17H20N2O4
316.14231
RI: 2165
D: UHYAC
Analgesic
119
217
M+
251
100
200
Propoxyphene-M (nor-) -H2O
UHY
LM
C18H21N
251.16740
RI: 2240
D: UHY
Potent analgesic
91
120
217
175
M+
260
100
200
Histapyrrodine-M (N-desphenyl-oxo-) AC
UHYAC
LM/Q
C15H20N2O2
260.15248
RI: 2260
D: UHYAC
Antihistamine
N: rat

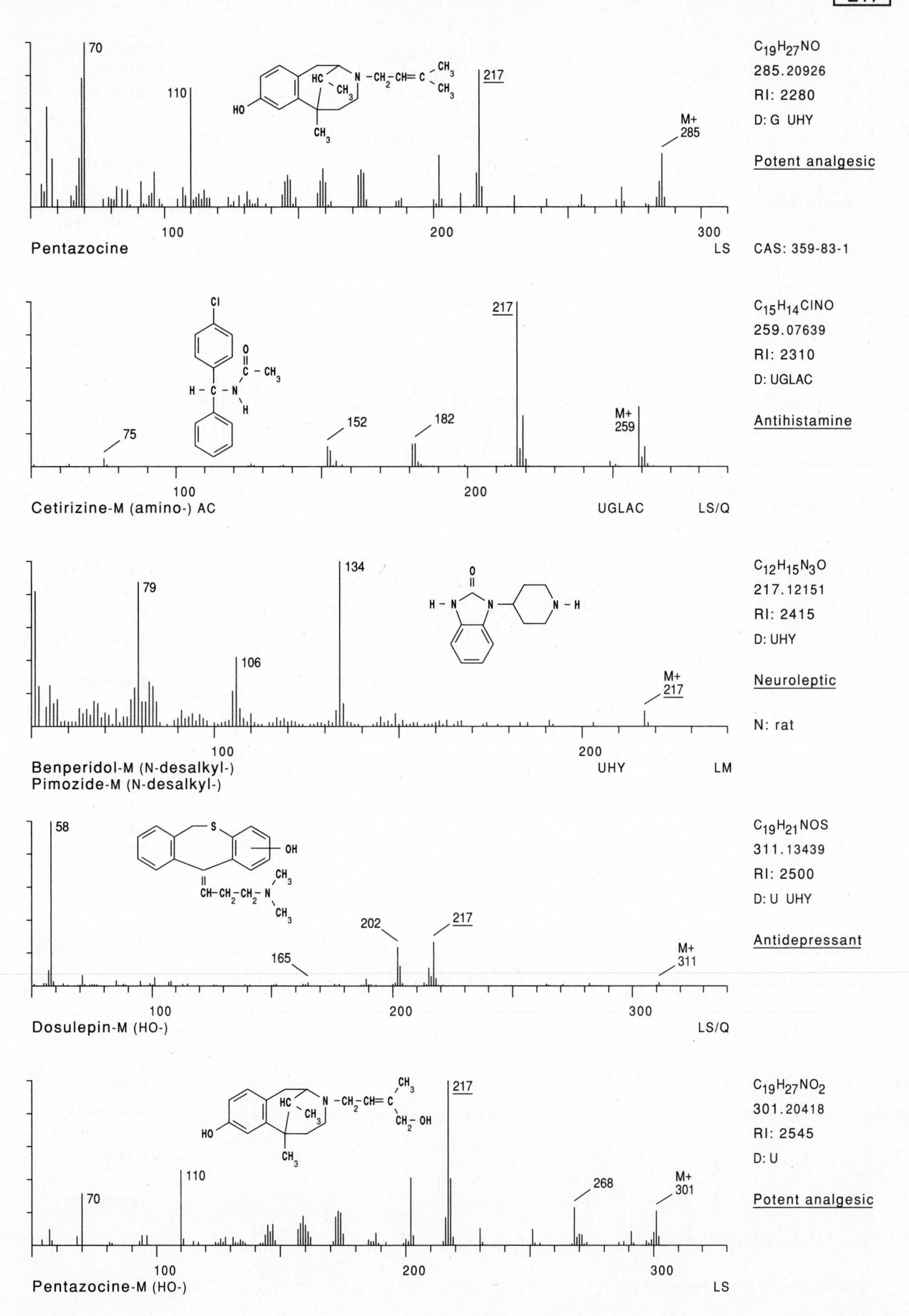

Pentazocine
LS
C19H27NO
285.20926
RI: 2280
D: G UHY
Potent analgesic
CAS: 359-83-1
Cetirizine-M (amino-) AC
UGLAC
LS/Q
C15H14ClNO
259.07639
RI: 2310
D: UGLAC
Antihistamine
Benperidol-M (N-desalkyl-)
Pimozide-M (N-desalkyl-)
UHY
LM
C12H15N3O
217.12151
RI: 2415
D: UHY
Neuroleptic
N: rat
Dosulepin-M (HO-)
LS/Q
C19H21NOS
311.13439
RI: 2500
D: U UHY
Antidepressant
Pentazocine-M (HO-)
LS
C19H27NO2
301.20418
RI: 2545
D: U
Potent analgesic

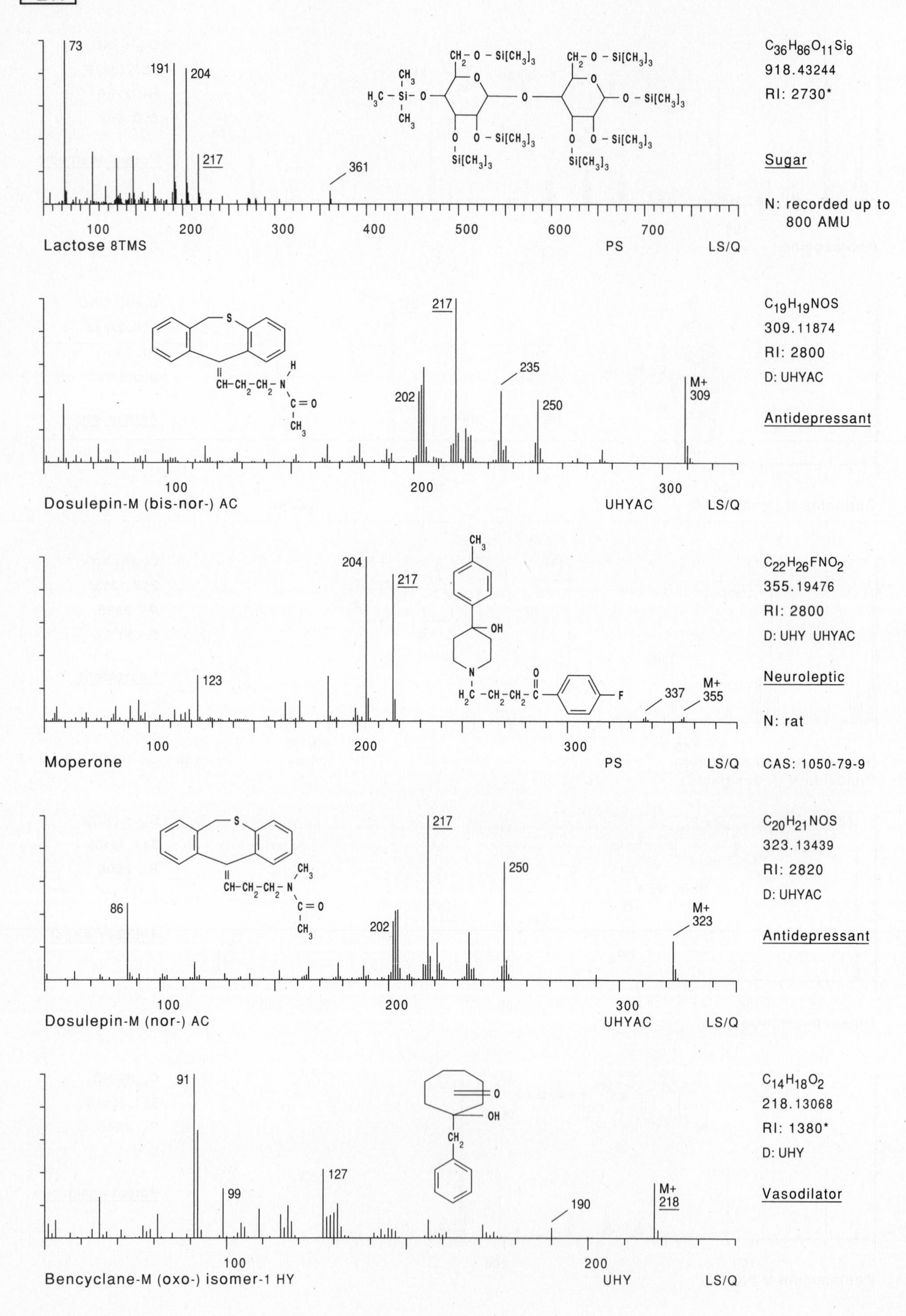

Lactose 8TMS
PS
LS/Q
C36H86O11Si8
918.43244
RI: 2730*
Sugar
N: recorded up to 800 AMU
Dosulepin-M (bis-nor-) AC
UHYAC
LS/Q
C19H19NOS
309.11874
RI: 2800
D: UHYAC
Antidepressant
Moperone
PS
LS/Q
C22H26FNO2
355.19476
RI: 2800
D: UHY UHYAC
Neuroleptic
N: rat
CAS: 1050-79-9
Dosulepin-M (nor-) AC
UHYAC
LS/Q
C20H21NOS
323.13439
RI: 2820
D: UHYAC
Antidepressant
Bencyclane-M (oxo-) isomer-1 HY
UHY
LS/Q
C14H18O2
218.13068
RI: 1380*
D: UHY
Vasodilator

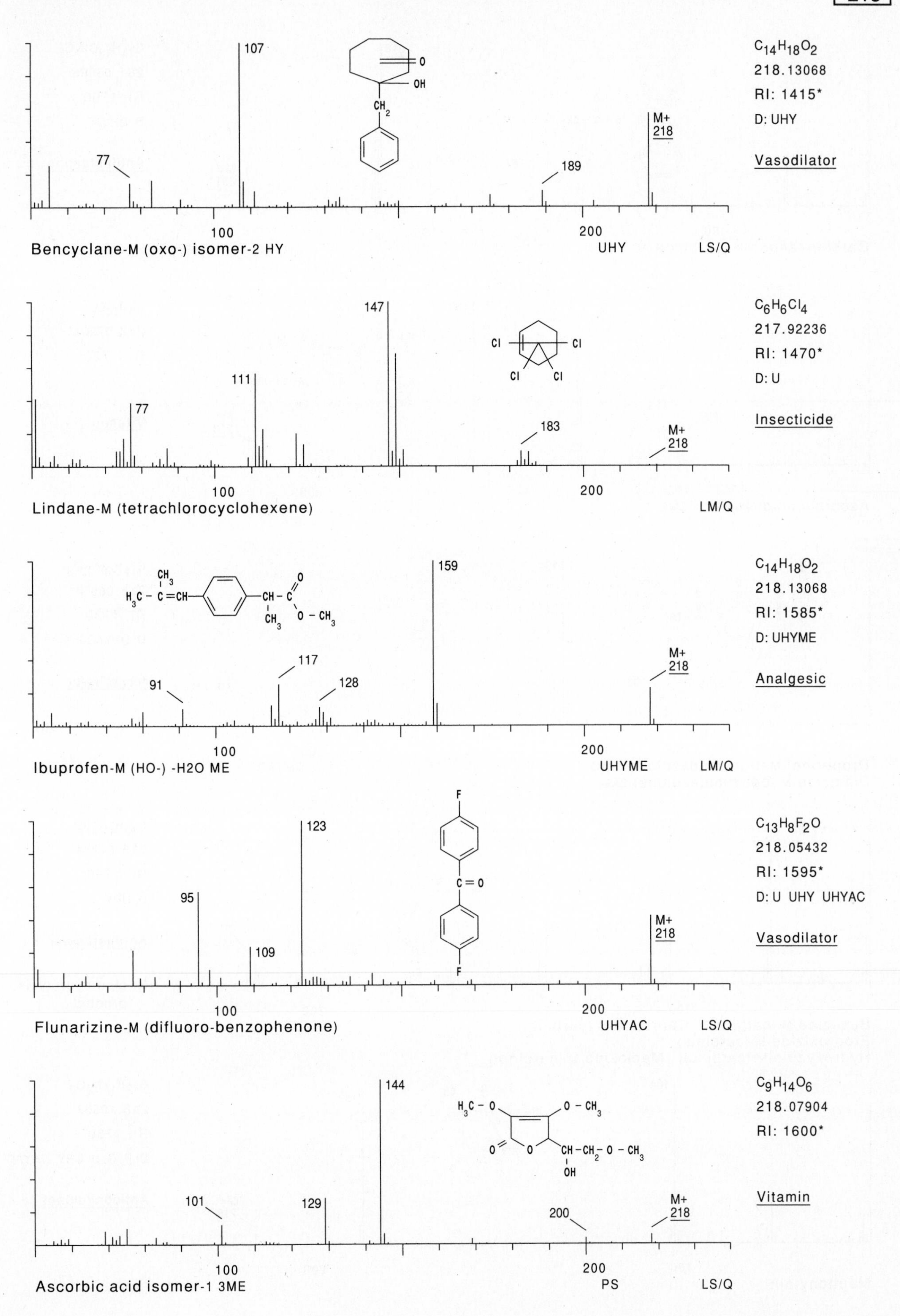

$C_{14}H_{18}O_2$
218.13068
RI: 1415*
D: UHY

Vasodilator

$C_6H_6Cl_4$
217.92236
RI: 1470*
D: U

Insecticide

$C_{14}H_{18}O_2$
218.13068
RI: 1585*
D: UHYME

Analgesic

$C_{13}H_8F_2O$
218.05432
RI: 1595*
D: U UHY UHYAC

Vasodilator

$C_9H_{14}O_6$
218.07904
RI: 1600*

Vitamin

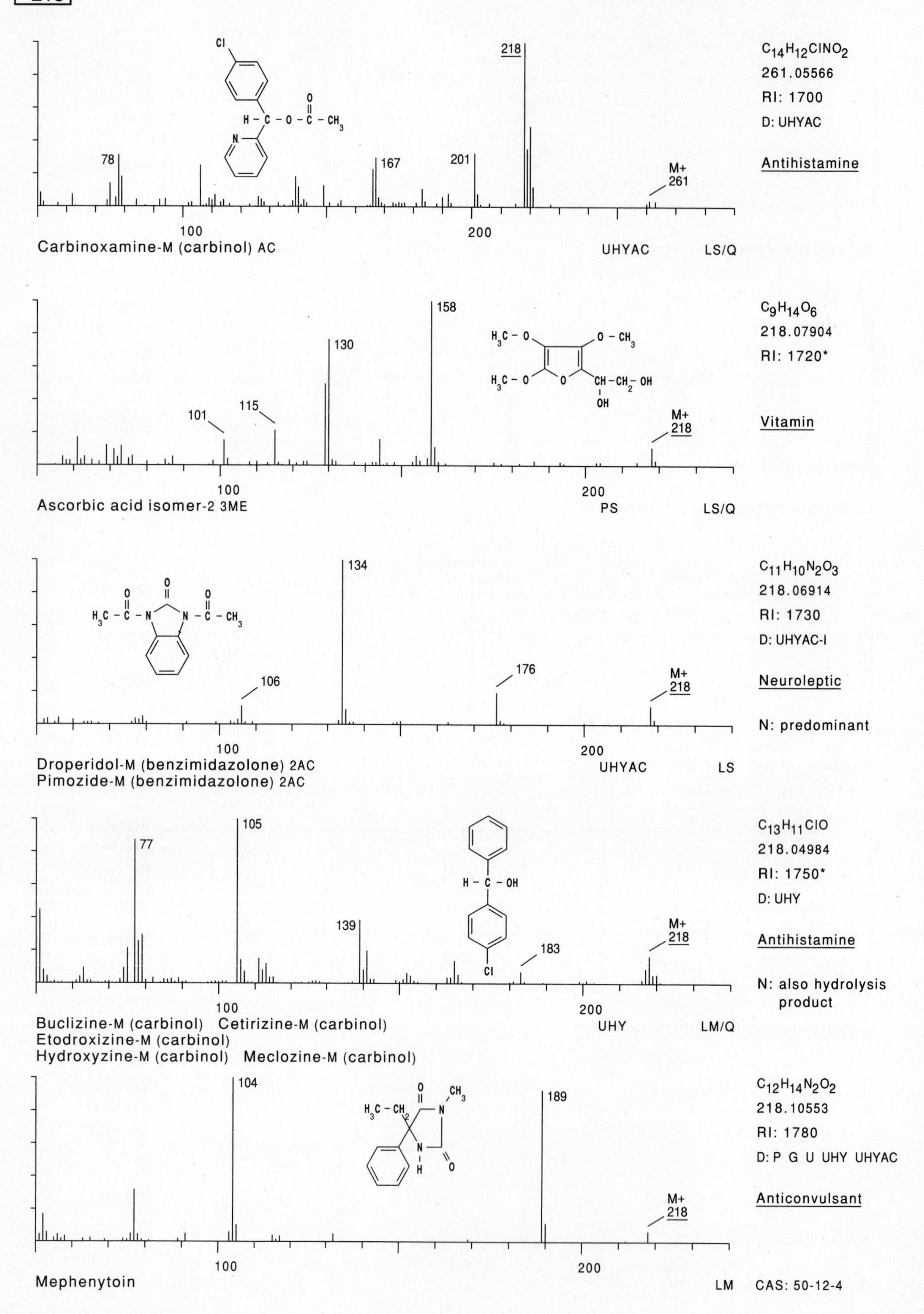

218
78
167
201
M+ 261
100
200
Carbinoxamine-M (carbinol) AC
UHYAC
LS/Q
$C_{14}H_{12}ClNO_2$
261.05566
RI: 1700
D: UHYAC
Antihistamine
158
130
115
101
M+ 218
Ascorbic acid isomer-2 3ME
PS
LS/Q
$C_9H_{14}O_6$
218.07904
RI: 1720*
Vitamin
134
106
176
M+ 218
Droperidol-M (benzimidazolone) 2AC
Pimozide-M (benzimidazolone) 2AC
UHYAC
LS
$C_{11}H_{10}N_2O_3$
218.06914
RI: 1730
D: UHYAC-I
Neuroleptic
N: predominant
105
77
139
183
M+ 218
Buclizine-M (carbinol) Cetirizine-M (carbinol)
Etodroxizine-M (carbinol)
Hydroxyzine-M (carbinol) Meclozine-M (carbinol)
UHY
LM/Q
$C_{13}H_{11}ClO$
218.04984
RI: 1750*
D: UHY
Antihistamine
N: also hydrolysis product
104
189
M+ 218
Mephenytoin
LM
CAS: 50-12-4
$C_{12}H_{14}N_2O_2$
218.10553
RI: 1780
D: P G U UHY UHYAC
Anticonvulsant

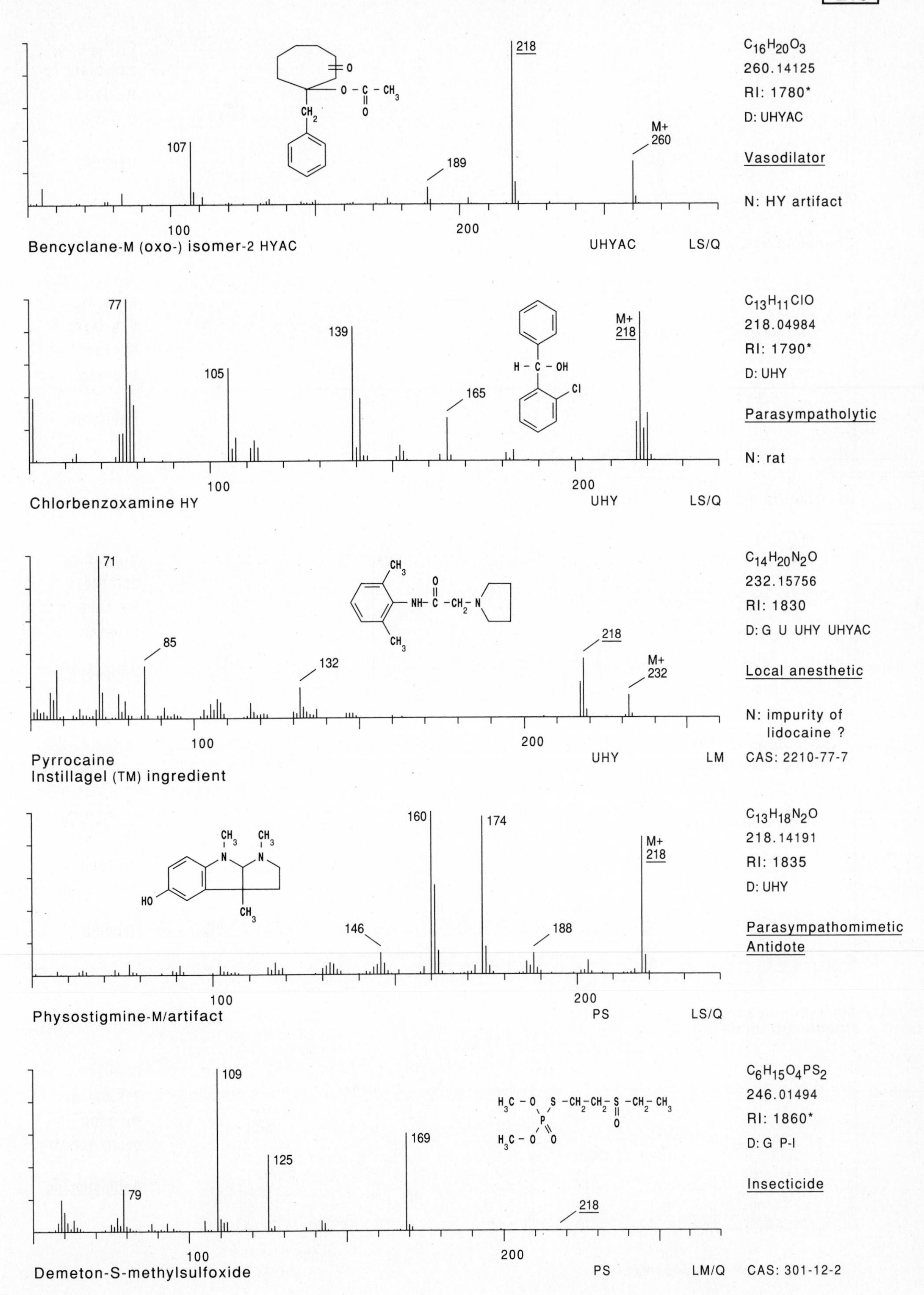

218
M+
260
107
189
100
200
Bencyclane-M (oxo-) isomer-2 HYAC
UHYAC
LS/Q
$C_{16}H_{20}O_3$
260.14125
RI: 1780*
D: UHYAC
Vasodilator
N: HY artifact
77
139
105
165
M+
218
H - C - OH
Cl
Chlorbenzoxamine HY
UHY
LS/Q
$C_{13}H_{11}ClO$
218.04984
RI: 1790*
D: UHY
Parasympatholytic
N: rat
71
85
132
218
M+
232
Pyrrocaine
Instillagel (TM) ingredient
UHY
LM
$C_{14}H_{20}N_2O$
232.15756
RI: 1830
D: G U UHY UHYAC
Local anesthetic
N: impurity of lidocaine ?
CAS: 2210-77-7
160
174
M+
218
146
188
HO
Physostigmine-M/artifact
PS
LS/Q
$C_{13}H_{18}N_2O$
218.14191
RI: 1835
D: UHY
Parasympathomimetic
Antidote
109
169
125
79
218
Demeton-S-methylsulfoxide
PS
LM/Q
$C_6H_{15}O_4PS_2$
246.01494
RI: 1860*
D: G P-I
Insecticide
CAS: 301-12-2

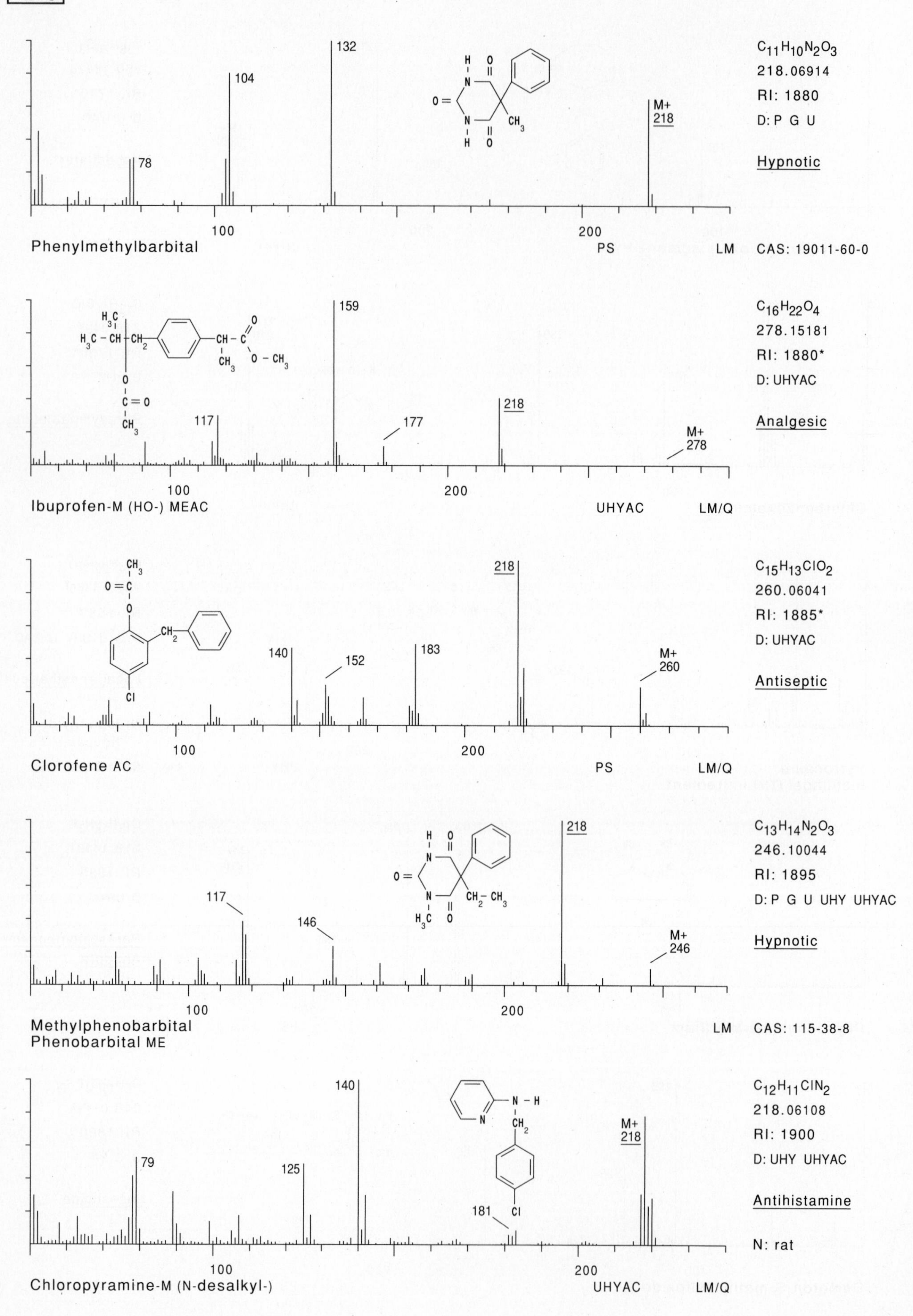

132
104
78
M+
218
100
200
Phenylmethylbarbital
PS
LM
C11H10N2O3
218.06914
RI: 1880
D: P G U
Hypnotic
CAS: 19011-60-0
159
218
117
177
M+
278
Ibuprofen-M (HO-) MEAC
UHYAC
LM/Q
C16H22O4
278.15181
RI: 1880*
D: UHYAC
Analgesic
218
140
152
183
M+
260
Clorofene AC
PS
LM/Q
C15H13ClO2
260.06041
RI: 1885*
D: UHYAC
Antiseptic
218
117
146
M+
246
Methylphenobarbital
Phenobarbital ME
LM
C13H14N2O3
246.10044
RI: 1895
D: P G U UHY UHYAC
Hypnotic
CAS: 115-38-8
140
79
125
181
M+
218
Chloropyramine-M (N-desalkyl-)
UHYAC
LM/Q
C12H11ClN2
218.06108
RI: 1900
D: UHY UHYAC
Antihistamine
N: rat

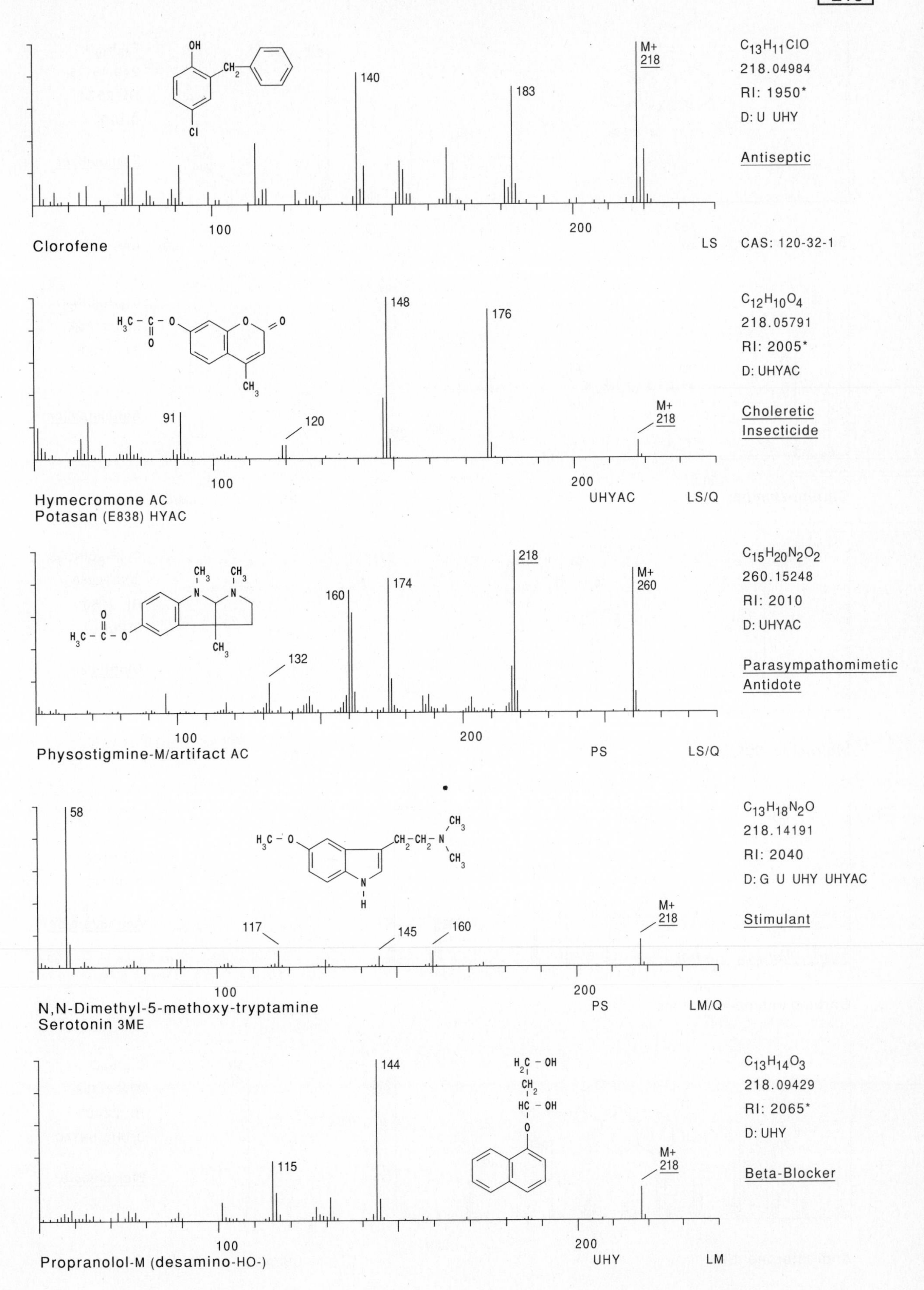

$C_{13}H_{11}ClO$
218.04984
RI: 1950*
D: U UHY
Antiseptic
CAS: 120-32-1

$C_{12}H_{10}O_4$
218.05791
RI: 2005*
D: UHYAC
Choleretic
Insecticide

$C_{15}H_{20}N_2O_2$
260.15248
RI: 2010
D: UHYAC
Parasympathomimetic
Antidote

$C_{13}H_{18}N_2O$
218.14191
RI: 2040
D: G U UHY UHYAC
Stimulant

$C_{13}H_{14}O_3$
218.09429
RI: 2065*
D: UHY
Beta-Blocker

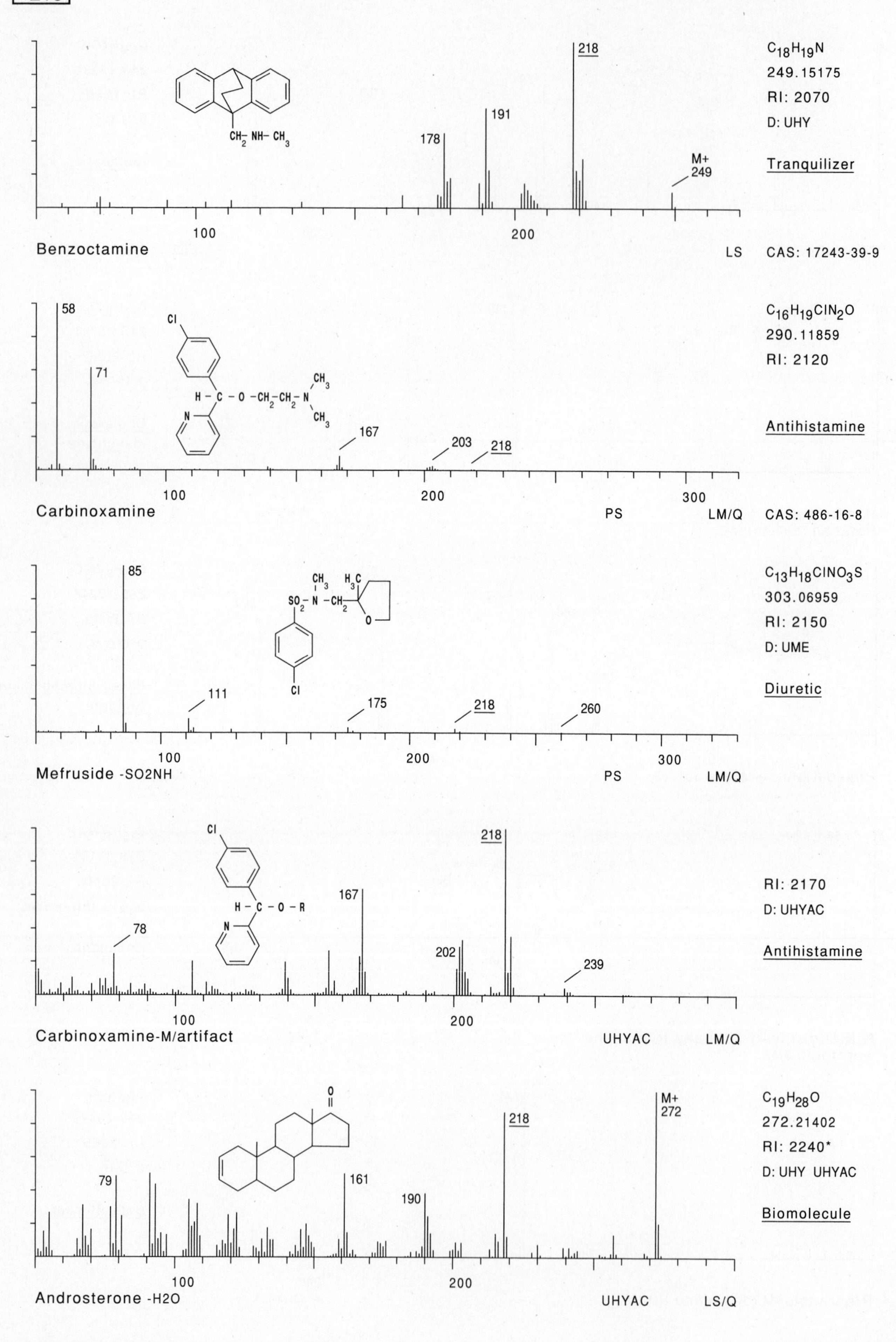

Benzoctamine
$C_{18}H_{19}N$
249.15175
RI: 2070
D: UHY
Tranquilizer
LS
CAS: 17243-39-9
Carbinoxamine
$C_{16}H_{19}ClN_2O$
290.11859
RI: 2120
Antihistamine
PS
LM/Q
CAS: 486-16-8
Mefruside -SO2NH
$C_{13}H_{18}ClNO_3S$
303.06959
RI: 2150
D: UME
Diuretic
PS
LM/Q
Carbinoxamine-M/artifact
RI: 2170
D: UHYAC
Antihistamine
UHYAC
LM/Q
Androsterone -H2O
$C_{19}H_{28}O$
272.21402
RI: 2240*
D: UHY UHYAC
Biomolecule
UHYAC
LS/Q

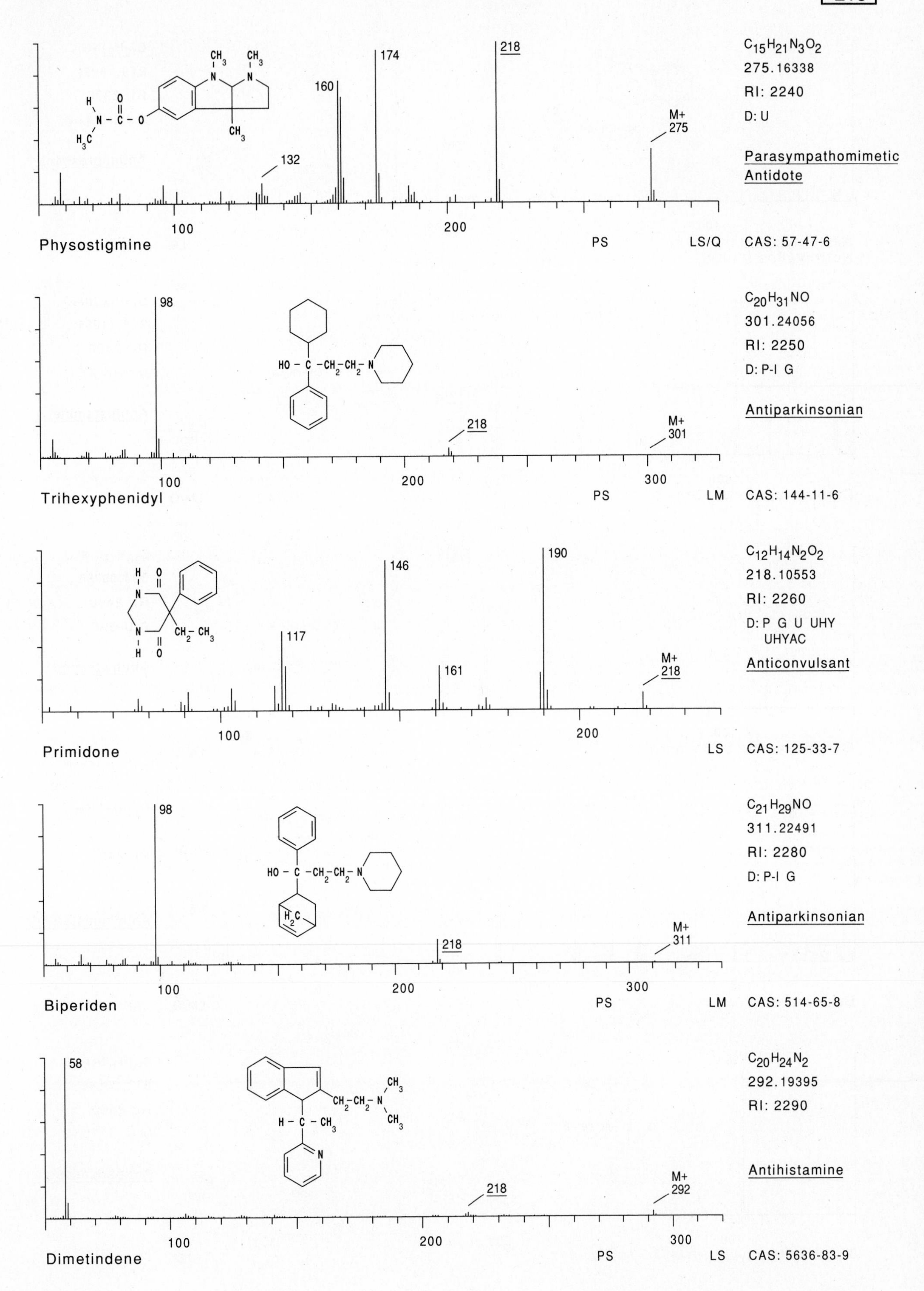

Physostigmine
$C_{15}H_{21}N_3O_2$
275.16338
RI: 2240
D: U
Parasympathomimetic
Antidote
PS LS/Q
CAS: 57-47-6
Trihexyphenidyl
$C_{20}H_{31}NO$
301.24056
RI: 2250
D: P-I G
Antiparkinsonian
PS LM
CAS: 144-11-6
Primidone
$C_{12}H_{14}N_2O_2$
218.10553
RI: 2260
D: P G U UHY
UHYAC
Anticonvulsant
LS
CAS: 125-33-7
Biperiden
$C_{21}H_{29}NO$
311.22491
RI: 2280
D: P-I G
Antiparkinsonian
PS LM
CAS: 514-65-8
Dimetindene
$C_{20}H_{24}N_2$
292.19395
RI: 2290
Antihistamine
PS LS
CAS: 5636-83-9

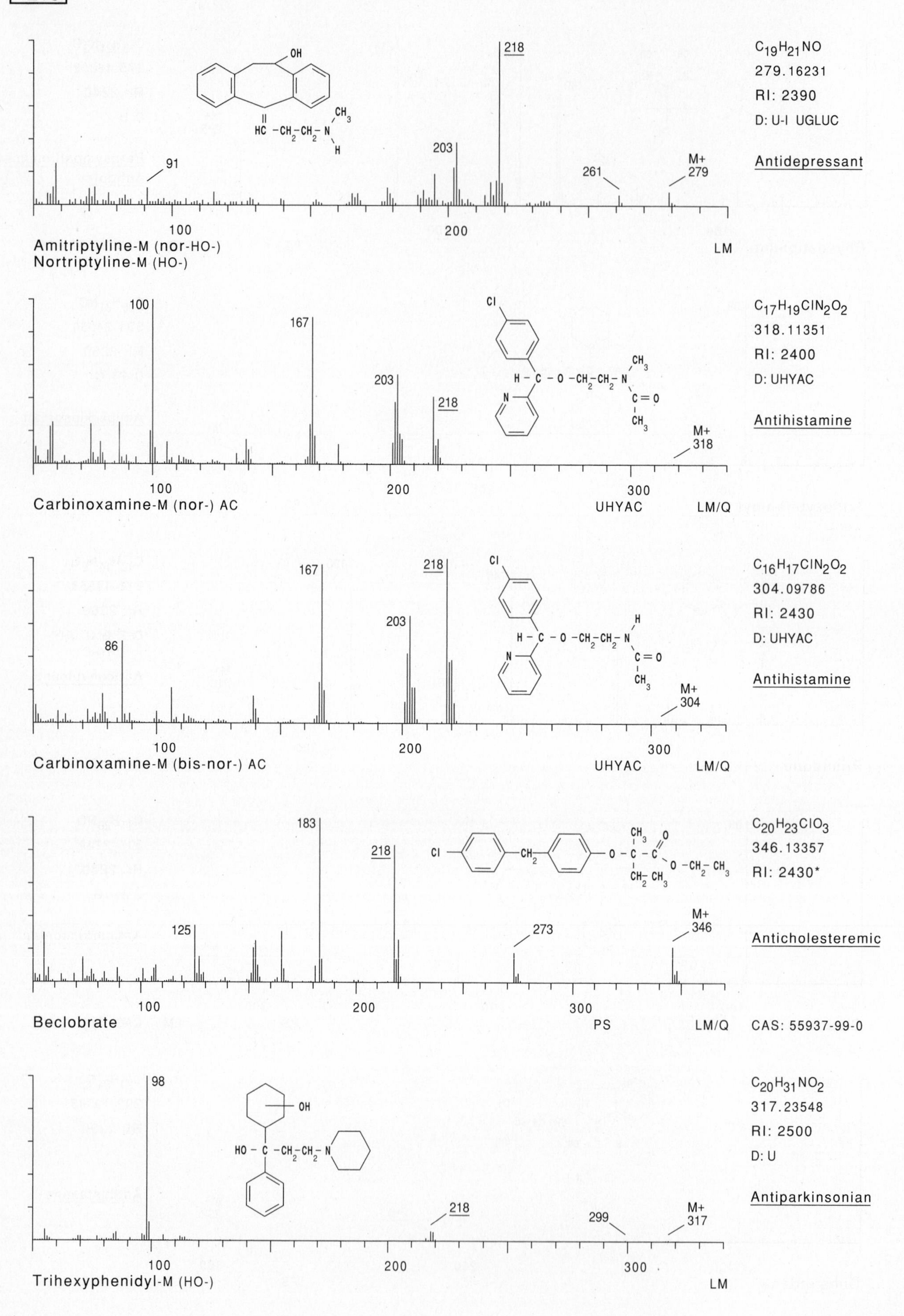

Amitriptyline-M (nor-HO-)
Nortriptyline-M (HO-)

$C_{19}H_{21}NO$
279.16231
RI: 2390
D: U-I UGLUC

Antidepressant

Carbinoxamine-M (nor-) AC

$C_{17}H_{19}ClN_2O_2$
318.11351
RI: 2400
D: UHYAC

Antihistamine

Carbinoxamine-M (bis-nor-) AC

$C_{16}H_{17}ClN_2O_2$
304.09786
RI: 2430
D: UHYAC

Antihistamine

Beclobrate

$C_{20}H_{23}ClO_3$
346.13357
RI: 2430*

Anticholesteremic

CAS: 55937-99-0

Trihexyphenidyl-M (HO-)

$C_{20}H_{31}NO_2$
317.23548
RI: 2500
D: U

Antiparkinsonian

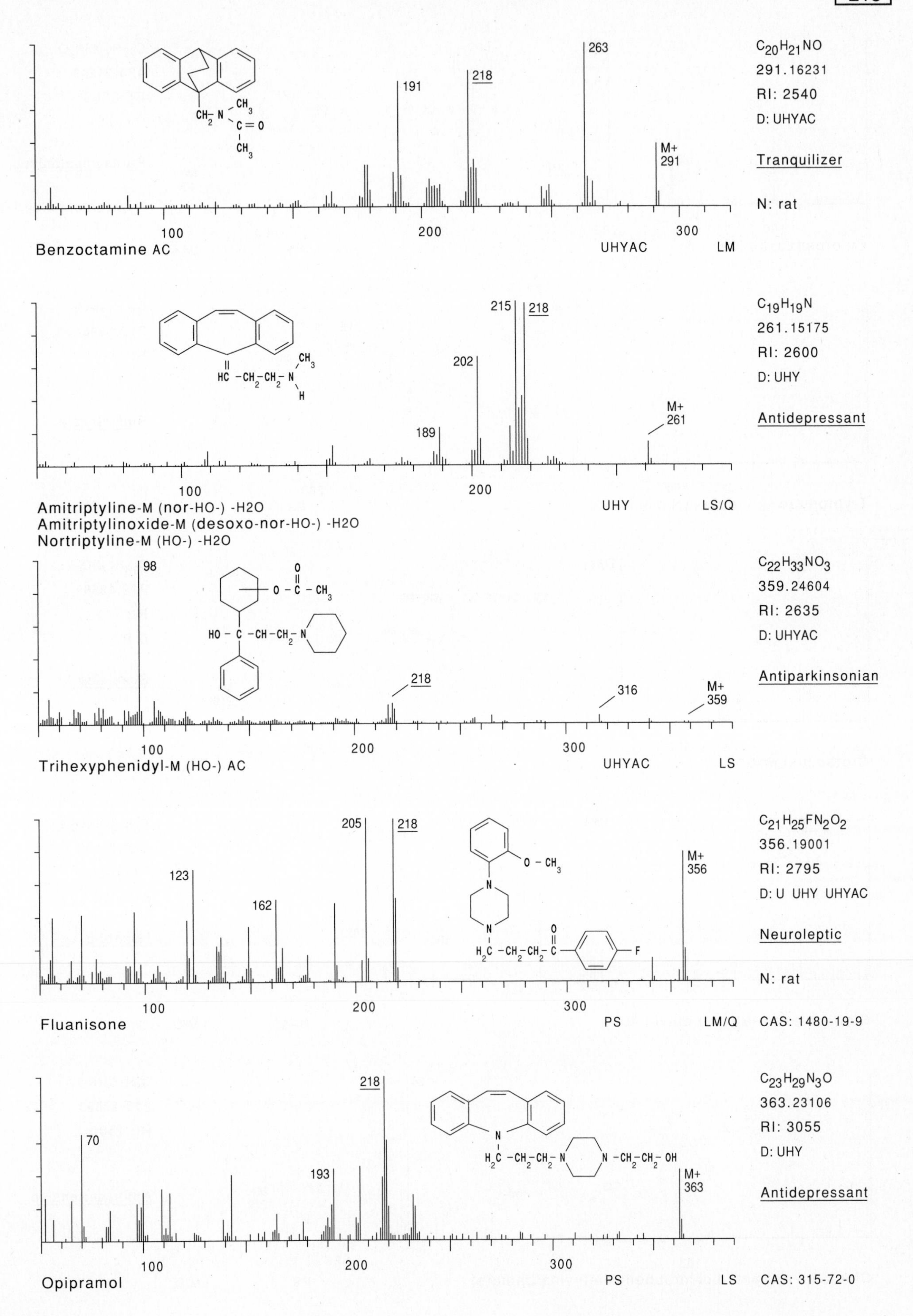

$C_{20}H_{21}NO$
291.16231
RI: 2540
D: UHYAC
Tranquilizer
N: rat

$C_{19}H_{19}N$
261.15175
RI: 2600
D: UHY
Antidepressant

$C_{22}H_{33}NO_3$
359.24604
RI: 2635
D: UHYAC
Antiparkinsonian

$C_{21}H_{25}FN_2O_2$
356.19001
RI: 2795
D: U UHY UHYAC
Neuroleptic
N: rat
CAS: 1480-19-9

$C_{23}H_{29}N_3O$
363.23106
RI: 3055
D: UHY
Antidepressant
CAS: 315-72-0

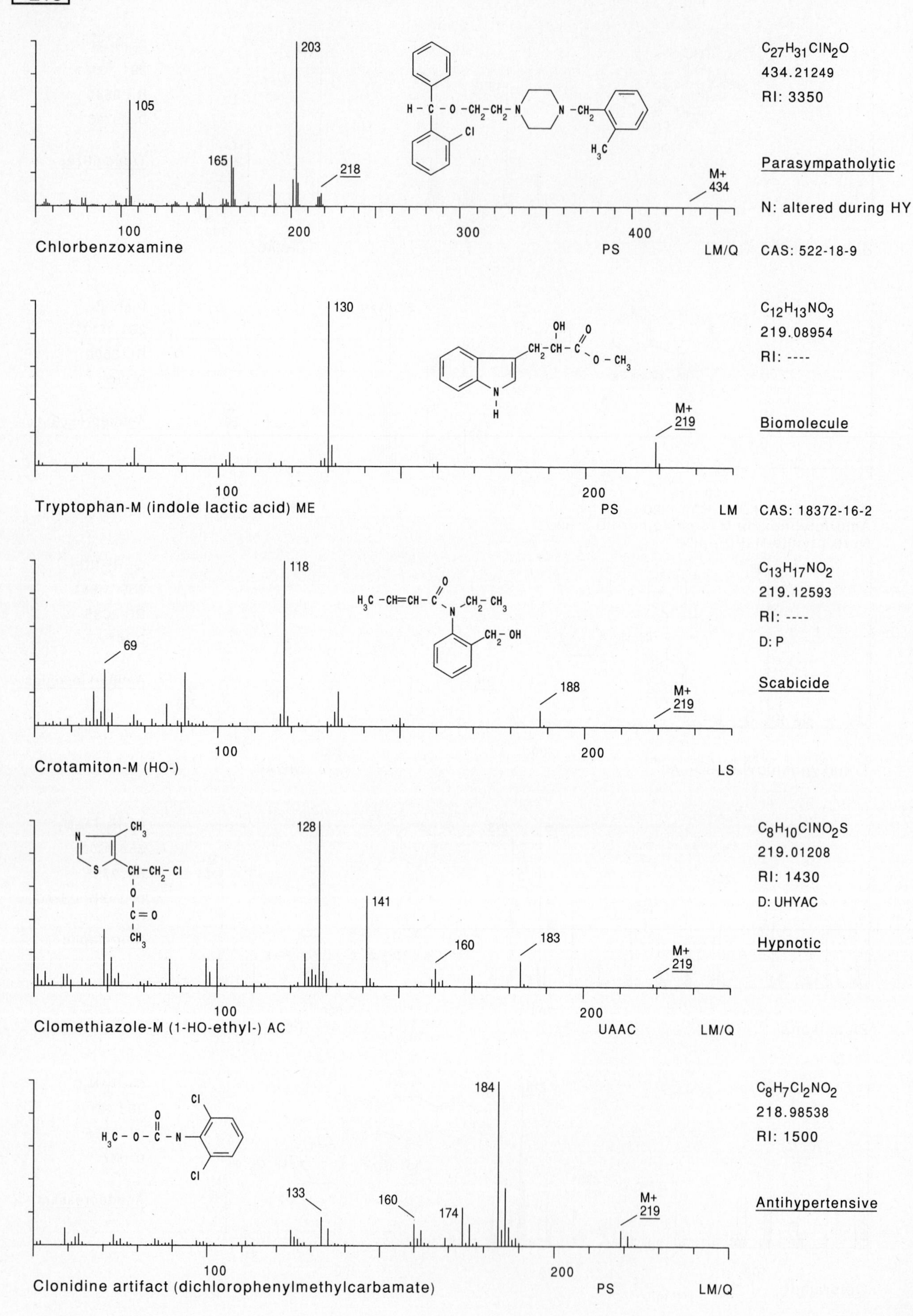

203
105
165
218
M+
434
100
200
300
400
Chlorbenzoxamine
PS
LM/Q
C27H31ClN2O
434.21249
RI: 3350
Parasympatholytic
N: altered during HY
CAS: 522-18-9
130
M+
219
100
200
Tryptophan-M (indole lactic acid) ME
PS
LM
C12H13NO3
219.08954
RI: ----
Biomolecule
CAS: 18372-16-2
118
69
188
M+
219
100
200
Crotamiton-M (HO-)
LS
C13H17NO2
219.12593
RI: ----
D: P
Scabicide
128
141
160
183
M+
219
100
200
Clomethiazole-M (1-HO-ethyl-) AC
UAAC
LM/Q
C8H10ClNO2S
219.01208
RI: 1430
D: UHYAC
Hypnotic
184
133
160
174
M+
219
100
200
Clonidine artifact (dichlorophenylmethylcarbamate)
PS
LM/Q
C8H7Cl2NO2
218.98538
RI: 1500
Antihypertensive

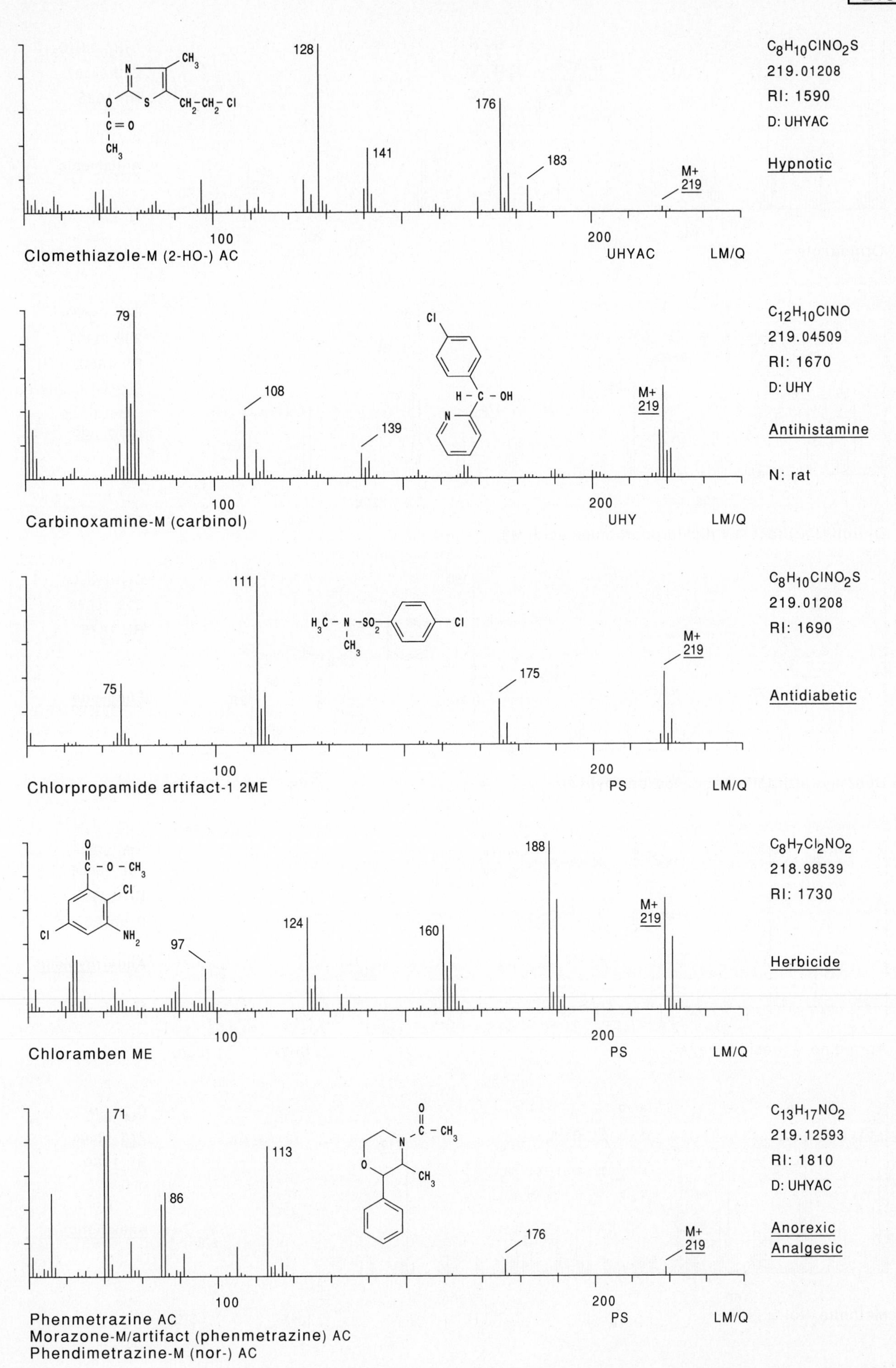

$C_8H_{10}ClNO_2S$
219.01208
RI: 1590
D: UHYAC
Hypnotic

$C_{12}H_{10}ClNO$
219.04509
RI: 1670
D: UHY
Antihistamine
N: rat

$C_8H_{10}ClNO_2S$
219.01208
RI: 1690
Antidiabetic

$C_8H_7Cl_2NO_2$
218.98539
RI: 1730
Herbicide

$C_{13}H_{17}NO_2$
219.12593
RI: 1810
D: UHYAC
Anorexic
Analgesic

219

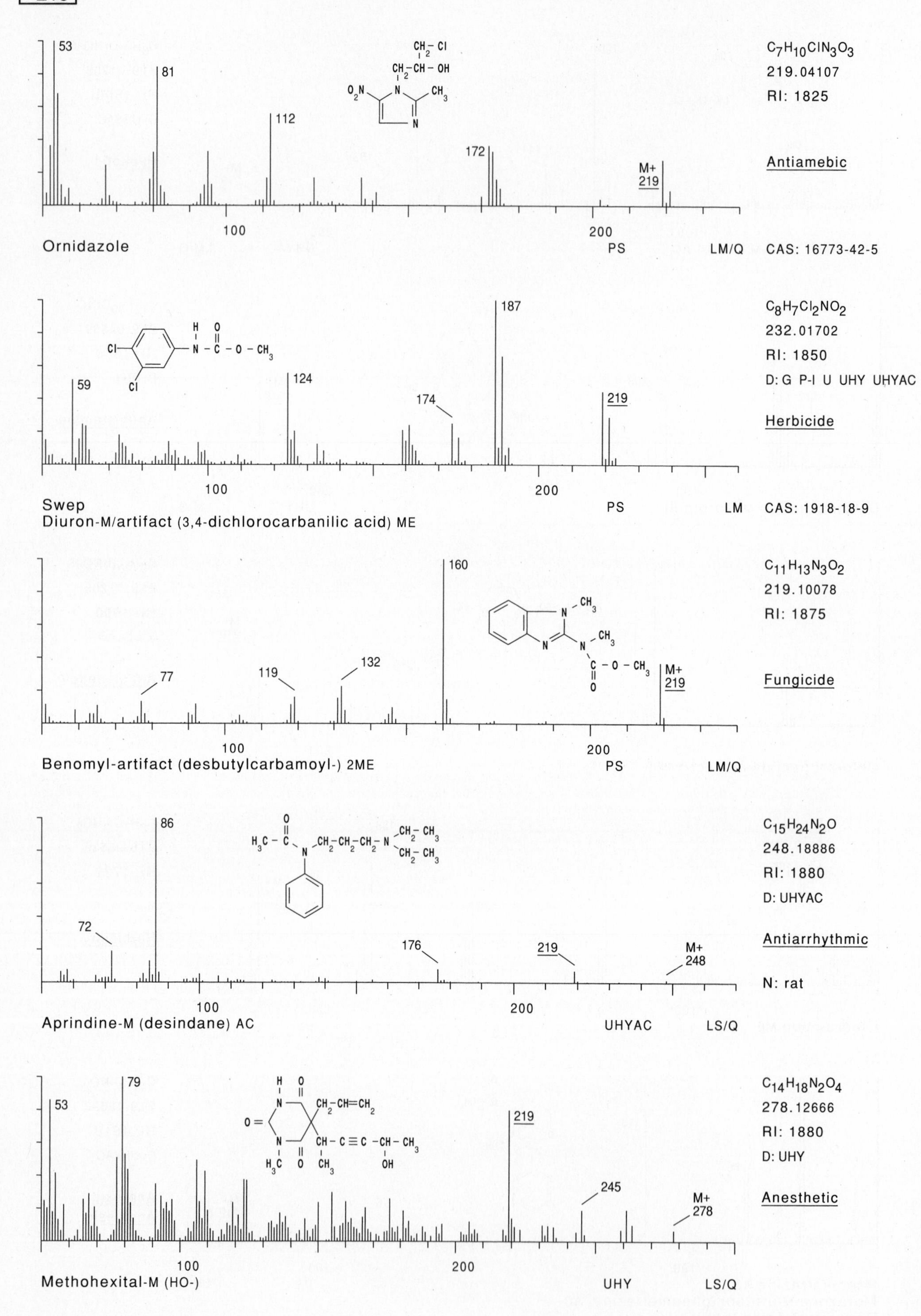
Ornidazole
C7H10ClN3O3
219.04107
RI: 1825
Antiamebic
CAS: 16773-42-5
PS
LM/Q
M+
219
53
81
112
172
Swep
Diuron-M/artifact (3,4-dichlorocarbanilic acid) ME
C8H7Cl2NO2
232.01702
RI: 1850
D: G P-I U UHY UHYAC
Herbicide
CAS: 1918-18-9
PS
LM
59
124
174
187
219
Benomyl-artifact (desbutylcarbamoyl-) 2ME
C11H13N3O2
219.10078
RI: 1875
Fungicide
PS
LM/Q
77
119
132
160
M+
219
Aprindine-M (desindane) AC
C15H24N2O
248.18886
RI: 1880
D: UHYAC
Antiarrhythmic
N: rat
UHYAC
LS/Q
72
86
176
219
M+
248
Methohexital-M (HO-)
C14H18N2O4
278.12666
RI: 1880
D: UHY
Anesthetic
UHY
LS/Q
53
79
219
245
M+
278
100
200

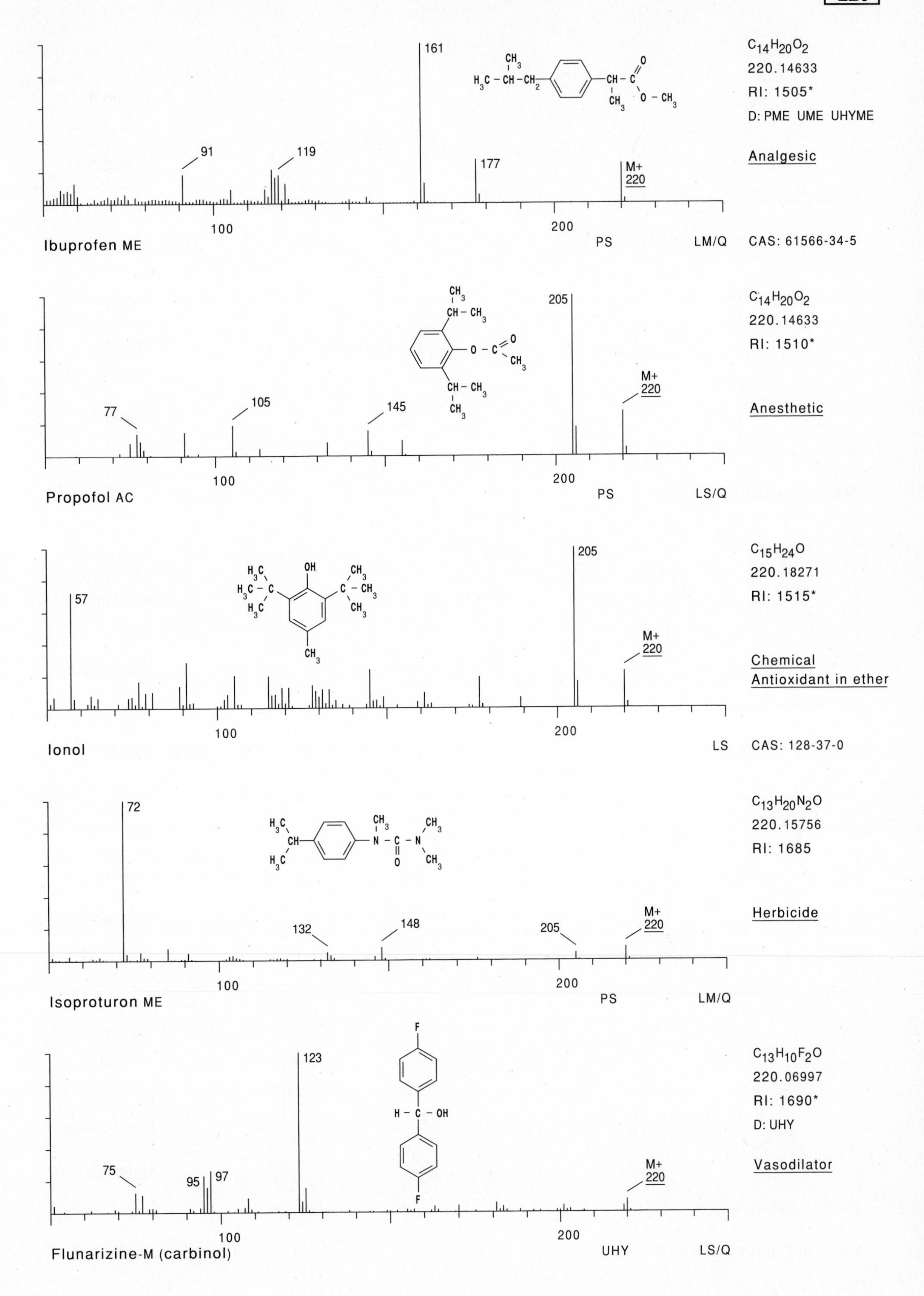
161
91
119
177
M+
220
100
200
PS
LM/Q
Ibuprofen ME
$C_{14}H_{20}O_2$
220.14633
RI: 1505*
D: PME UME UHYME
Analgesic
CAS: 61566-34-5
205
77
105
145
M+
220
100
200
PS
LS/Q
Propofol AC
$C_{14}H_{20}O_2$
220.14633
RI: 1510*
Anesthetic
205
57
M+
220
100
200
LS
Ionol
$C_{15}H_{24}O$
220.18271
RI: 1515*
Chemical
Antioxidant in ether
CAS: 128-37-0
72
132
148
205
M+
220
100
200
PS
LM/Q
Isoproturon ME
$C_{13}H_{20}N_2O$
220.15756
RI: 1685
Herbicide
123
75
95
97
M+
220
100
200
UHY
LS/Q
Flunarizine-M (carbinol)
$C_{13}H_{10}F_2O$
220.06997
RI: 1690*
D: UHY
Vasodilator

220

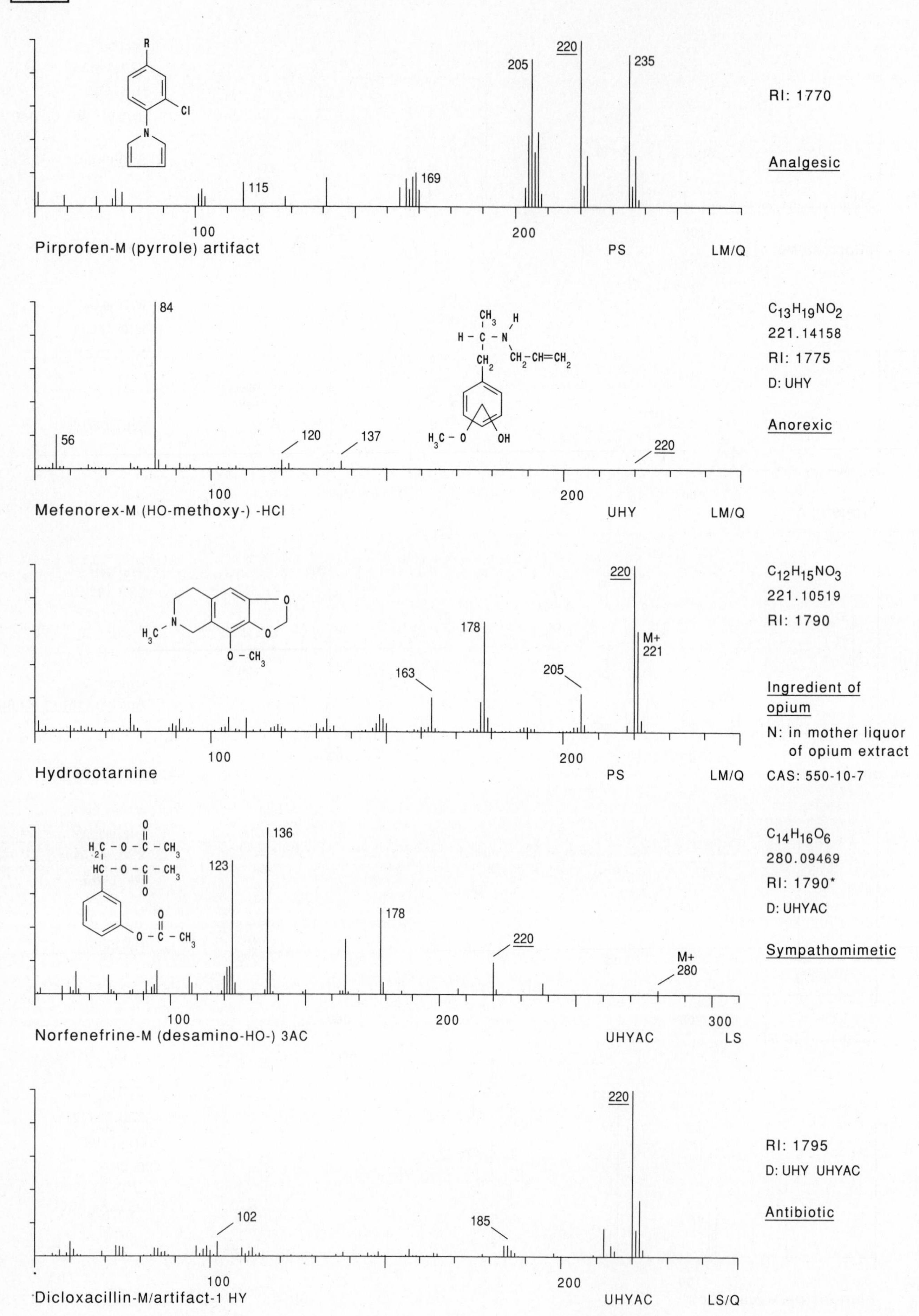
Pirprofen-M (pyrrole) artifact
PS LM/Q
RI: 1770
Analgesic
205
220
235
115
169
Mefenorex-M (HO-methoxy-) -HCl
UHY LM/Q
C13H19NO2
221.14158
RI: 1775
D: UHY
Anorexic
84
56
120
137
220
Hydrocotarnine
PS LM/Q
C12H15NO3
221.10519
RI: 1790
Ingredient of opium
N: in mother liquor of opium extract
CAS: 550-10-7
220
178
M+ 221
163
205
Norfenefrine-M (desamino-HO-) 3AC
UHYAC LS
C14H16O6
280.09469
RI: 1790*
D: UHYAC
Sympathomimetic
136
123
178
220
M+ 280
Dicloxacillin-M/artifact-1 HY
UHYAC LS/Q
RI: 1795
D: UHY UHYAC
Antibiotic
220
102
185

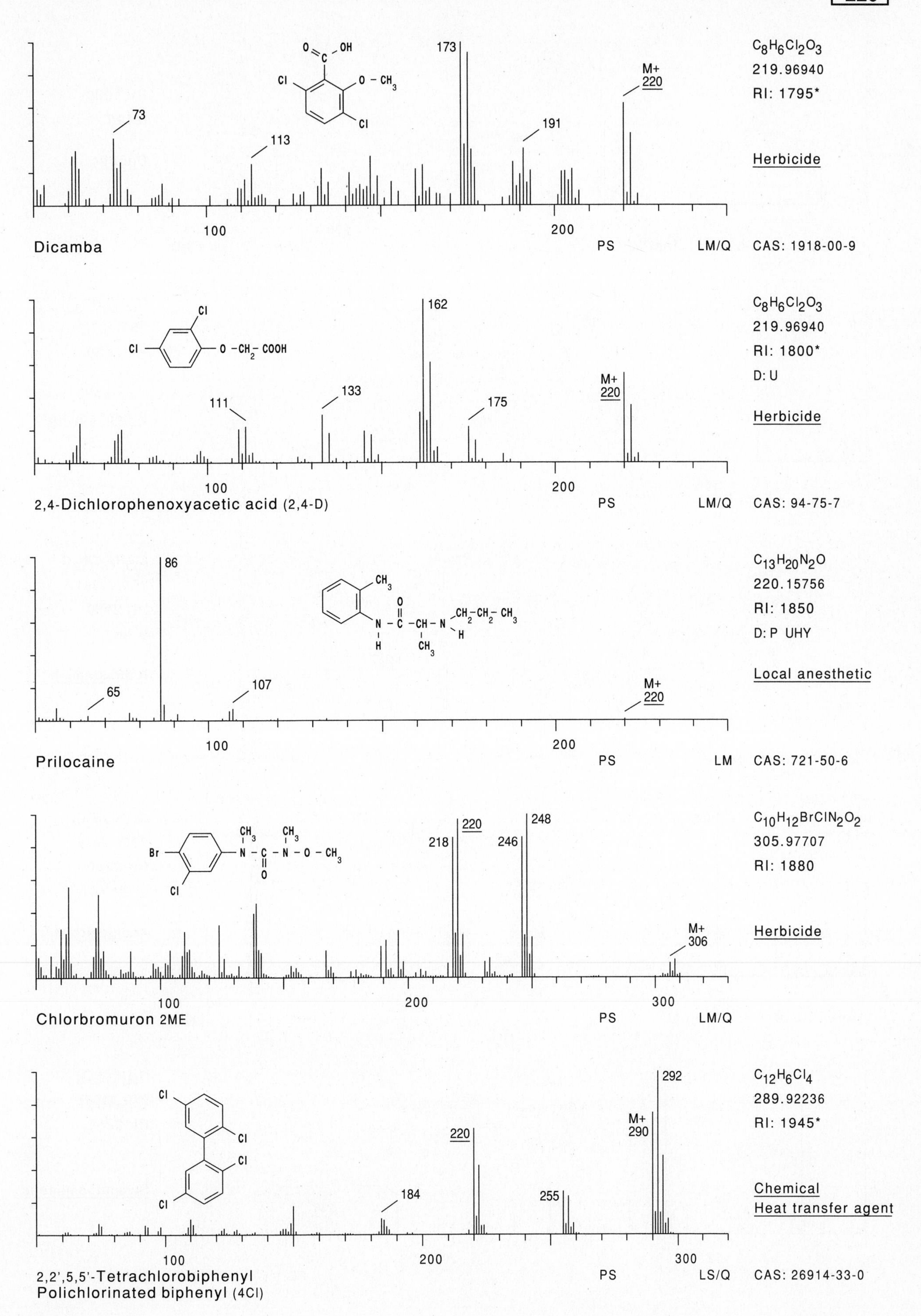

Dicamba PS LM/Q

$C_8H_6Cl_2O_3$
219.96940
RI: 1795*

Herbicide

CAS: 1918-00-9

2,4-Dichlorophenoxyacetic acid (2,4-D) PS LM/Q

$C_8H_6Cl_2O_3$
219.96940
RI: 1800*
D: U

Herbicide

CAS: 94-75-7

Prilocaine PS LM

$C_{13}H_{20}N_2O$
220.15756
RI: 1850
D: P UHY

Local anesthetic

CAS: 721-50-6

Chlorbromuron 2ME PS LM/Q

$C_{10}H_{12}BrClN_2O_2$
305.97707
RI: 1880

Herbicide

2,2',5,5'-Tetrachlorobiphenyl
Polichlorinated biphenyl (4Cl) PS LS/Q

$C_{12}H_6Cl_4$
289.92236
RI: 1945*

Chemical
Heat transfer agent

CAS: 26914-33-0

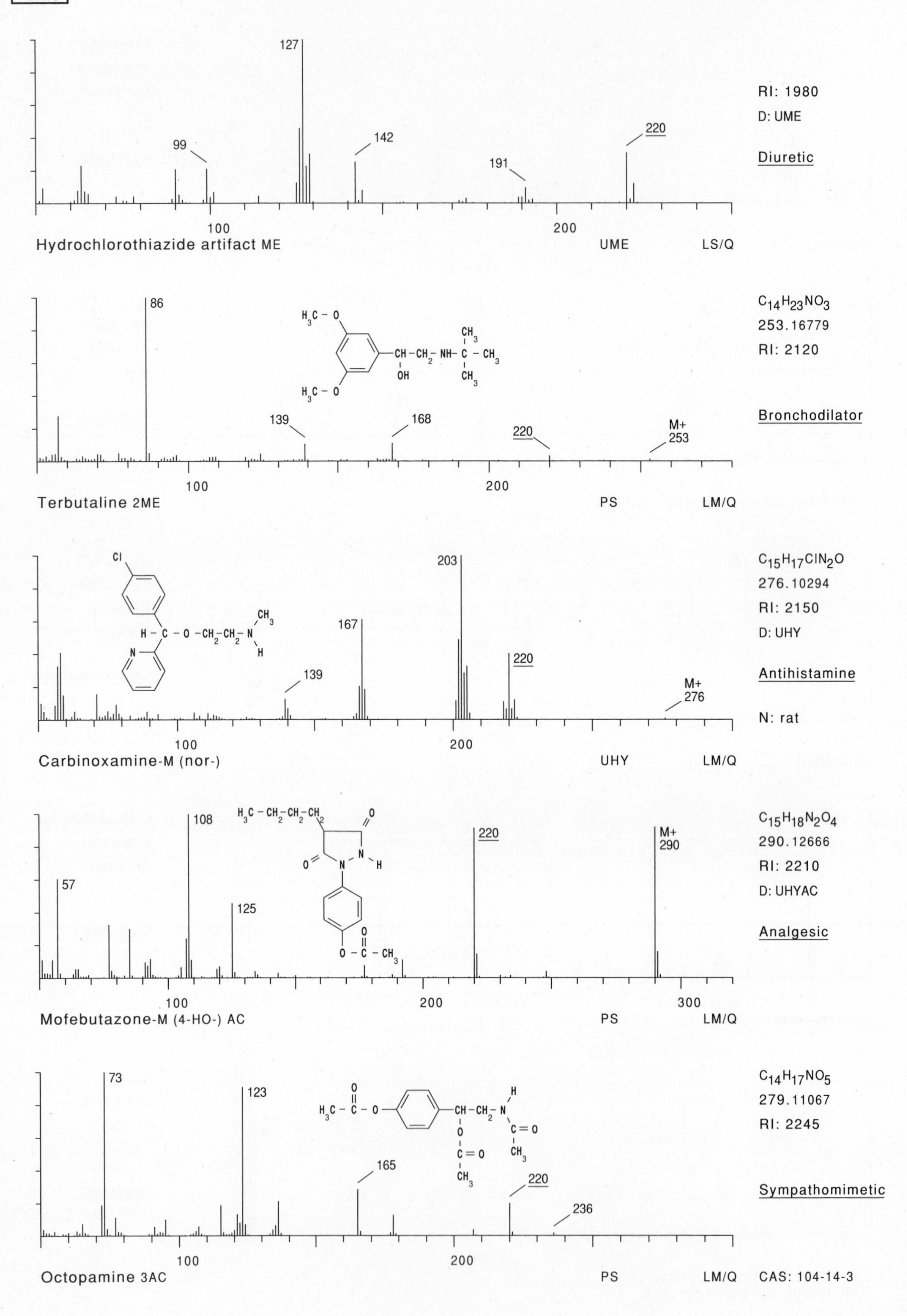

127
99
142
191
220
100
200
Hydrochlorothiazide artifact ME
UME
LS/Q
RI: 1980
D: UME
Diuretic
86
139
168
220
M+
253
100
200
Terbutaline 2ME
PS
LM/Q
C14H23NO3
253.16779
RI: 2120
Bronchodilator
203
167
139
220
M+
276
100
200
Carbinoxamine-M (nor-)
UHY
LM/Q
C15H17ClN2O
276.10294
RI: 2150
D: UHY
Antihistamine
N: rat
108
57
125
220
M+
290
100
200
300
Mofebutazone-M (4-HO-) AC
PS
LM/Q
C15H18N2O4
290.12666
RI: 2210
D: UHYAC
Analgesic
73
123
165
220
236
100
200
Octopamine 3AC
PS
LM/Q
C14H17NO5
279.11067
RI: 2245
Sympathomimetic
CAS: 104-14-3

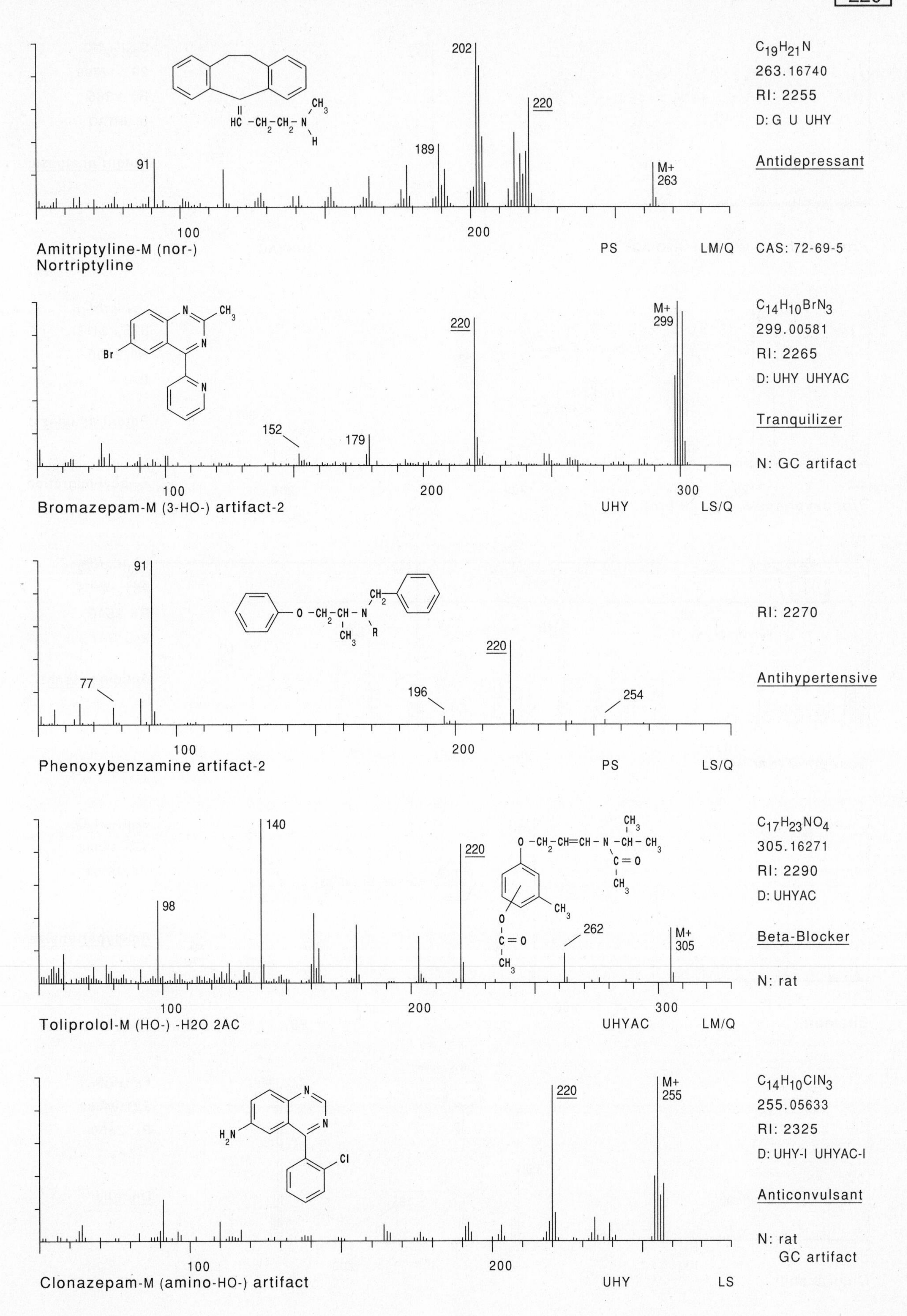
202
220
189
91
M+
263
100
200
Amitriptyline-M (nor-)
Nortriptyline
PS
LM/Q
C19H21N
263.16740
RI: 2255
D: G U UHY
Antidepressant
CAS: 72-69-5
M+
299
220
152
179
100
200
300
Bromazepam-M (3-HO-) artifact-2
UHY
LS/Q
C14H10BrN3
299.00581
RI: 2265
D: UHY UHYAC
Tranquilizer
N: GC artifact
91
220
77
196
254
100
200
Phenoxybenzamine artifact-2
PS
LS/Q
RI: 2270
Antihypertensive
140
220
98
262
M+
305
100
200
300
Toliprolol-M (HO-) -H2O 2AC
UHYAC
LM/Q
C17H23NO4
305.16271
RI: 2290
D: UHYAC
Beta-Blocker
N: rat
220
M+
255
100
200
Clonazepam-M (amino-HO-) artifact
UHY
LS
C14H10ClN3
255.05633
RI: 2325
D: UHY-I UHYAC-I
Anticonvulsant
N: rat
GC artifact

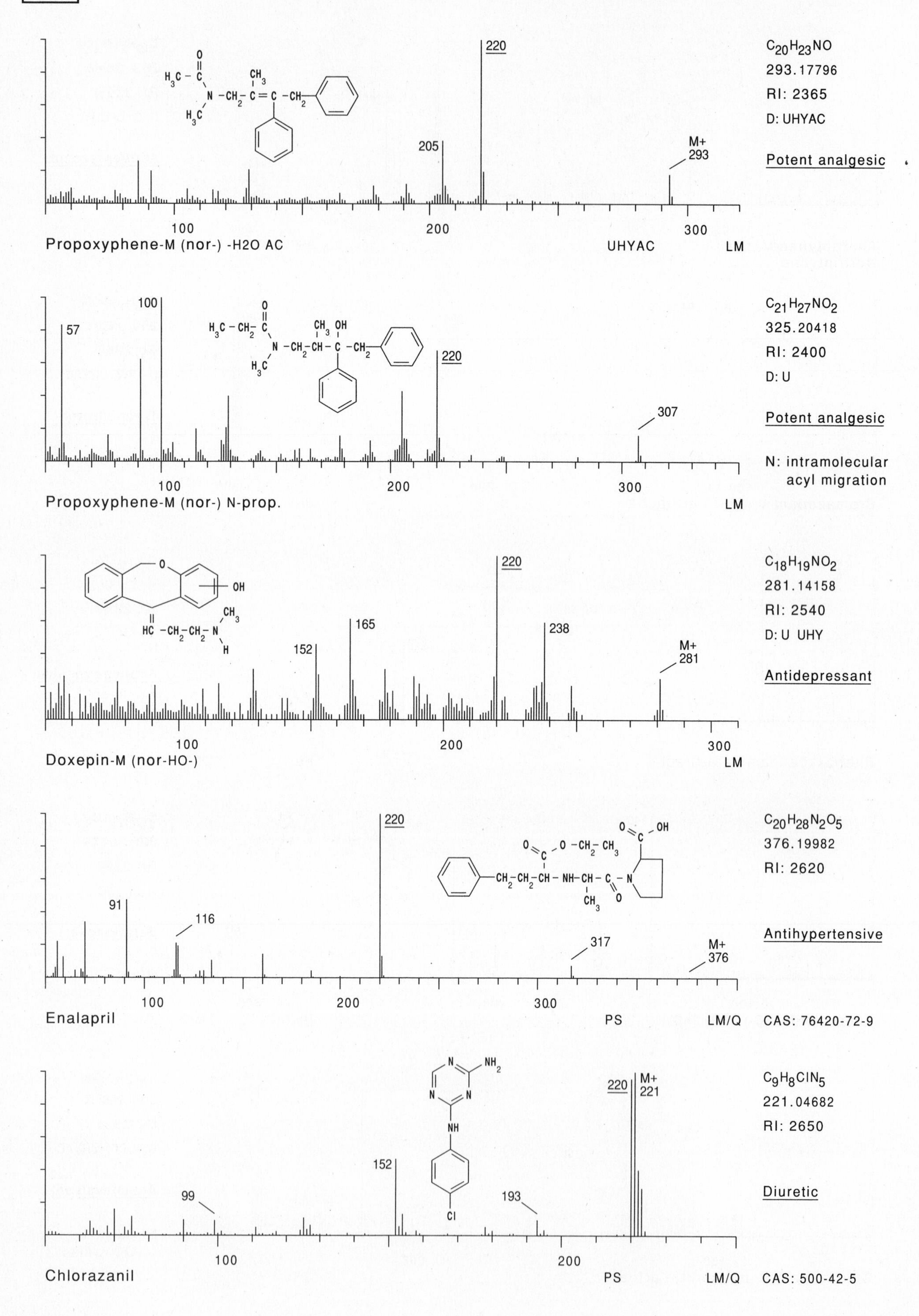

220
205
M+
293
100
200
300
Propoxyphene-M (nor-) -H2O AC
UHYAC
LM
$C_{20}H_{23}NO$
293.17796
RI: 2365
D: UHYAC
Potent analgesic
100
57
220
307
Propoxyphene-M (nor-) N-prop.
LM
$C_{21}H_{27}NO_2$
325.20418
RI: 2400
D: U
Potent analgesic
N: intramolecular acyl migration
220
165
238
152
M+
281
Doxepin-M (nor-HO-)
LM
$C_{18}H_{19}NO_2$
281.14158
RI: 2540
D: U UHY
Antidepressant
220
91
116
317
M+
376
Enalapril
PS
LM/Q
$C_{20}H_{28}N_2O_5$
376.19982
RI: 2620
Antihypertensive
CAS: 76420-72-9
220
M+
221
152
99
193
Chlorazanil
PS
LM/Q
$C_9H_8ClN_5$
221.04682
RI: 2650
Diuretic
CAS: 500-42-5

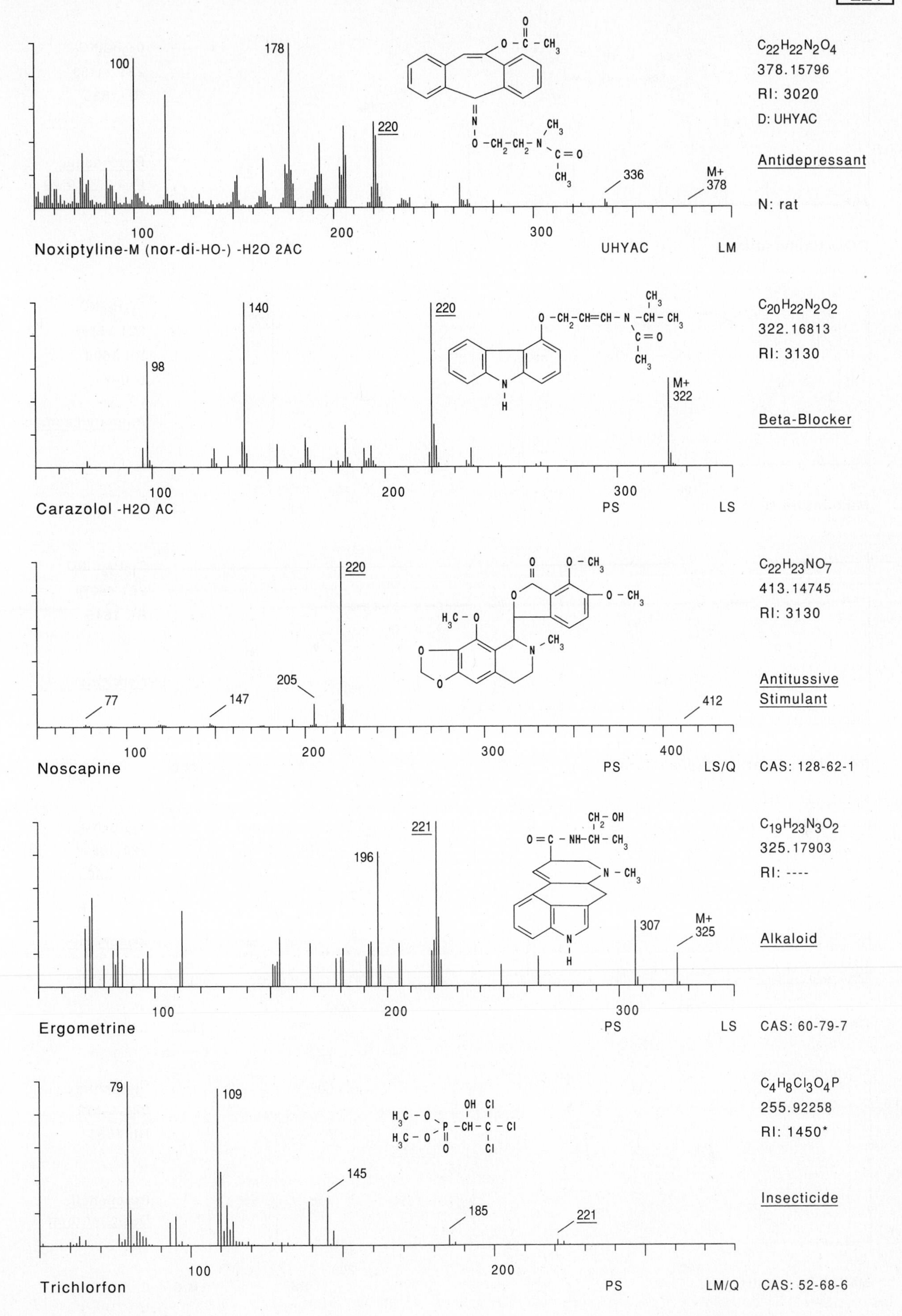
Noxiptyline-M (nor-di-HO-) -H2O 2AC
UHYAC LM
$C_{22}H_{22}N_2O_4$
378.15796
RI: 3020
D: UHYAC
Antidepressant
N: rat
Carazolol -H2O AC
PS LS
$C_{20}H_{22}N_2O_2$
322.16813
RI: 3130
Beta-Blocker
Noscapine
PS LS/Q
$C_{22}H_{23}NO_7$
413.14745
RI: 3130
Antitussive
Stimulant
CAS: 128-62-1
Ergometrine
PS LS
$C_{19}H_{23}N_3O_2$
325.17903
RI: ----
Alkaloid
CAS: 60-79-7
Trichlorfon
PS LM/Q
$C_4H_8Cl_3O_4P$
255.92258
RI: 1450*
Insecticide
CAS: 52-68-6

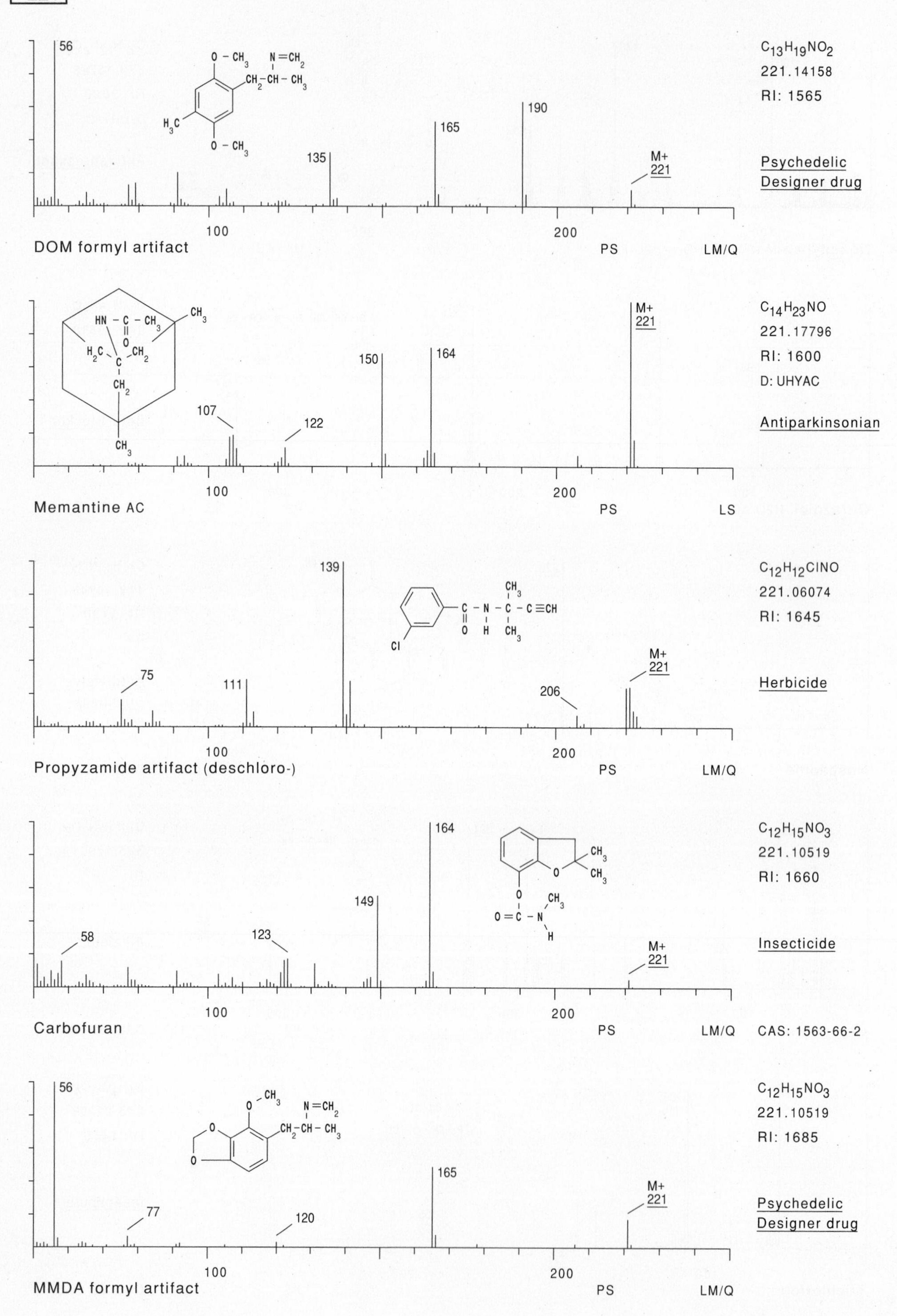
56
190
165
135
M+
221
100
200
DOM formyl artifact
PS
LM/Q
C13H19NO2
221.14158
RI: 1565
Psychedelic
Designer drug
M+
221
150
164
107
122
Memantine AC
PS
LS
C14H23NO
221.17796
RI: 1600
D: UHYAC
Antiparkinsonian
139
75
111
206
M+
221
Propyzamide artifact (deschloro-)
PS
LM/Q
C12H12ClNO
221.06074
RI: 1645
Herbicide
164
149
58
123
M+
221
Carbofuran
PS
LM/Q
C12H15NO3
221.10519
RI: 1660
Insecticide
CAS: 1563-66-2
56
165
77
120
M+
221
MMDA formyl artifact
PS
LM/Q
C12H15NO3
221.10519
RI: 1685
Psychedelic
Designer drug

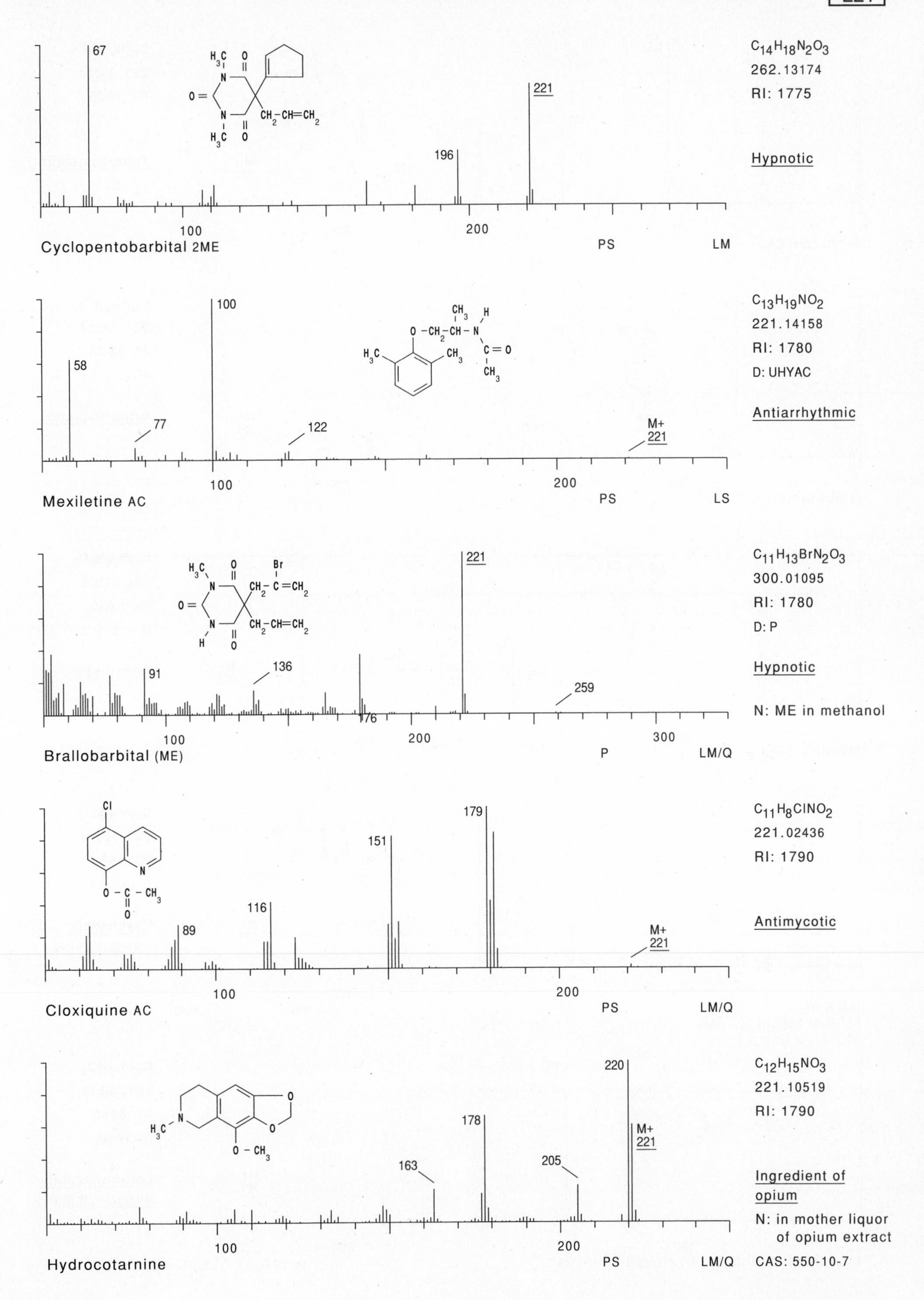

67
221
196
100
200
Cyclopentobarbital 2ME
PS
LM
$C_{14}H_{18}N_2O_3$
262.13174
RI: 1775
Hypnotic
100
58
77
122
M+
221
100
200
Mexiletine AC
PS
LS
$C_{13}H_{19}NO_2$
221.14158
RI: 1780
D: UHYAC
Antiarrhythmic
221
91
136
259
176
100
200
300
Brallobarbital (ME)
P
LM/Q
$C_{11}H_{13}BrN_2O_3$
300.01095
RI: 1780
D: P
Hypnotic
N: ME in methanol
179
151
116
89
M+
221
100
200
Cloxiquine AC
PS
LM/Q
$C_{11}H_8ClNO_2$
221.02436
RI: 1790
Antimycotic
220
178
M+
221
163
205
100
200
Hydrocotarnine
PS
LM/Q
$C_{12}H_{15}NO_3$
221.10519
RI: 1790
Ingredient of opium
N: in mother liquor of opium extract
CAS: 550-10-7

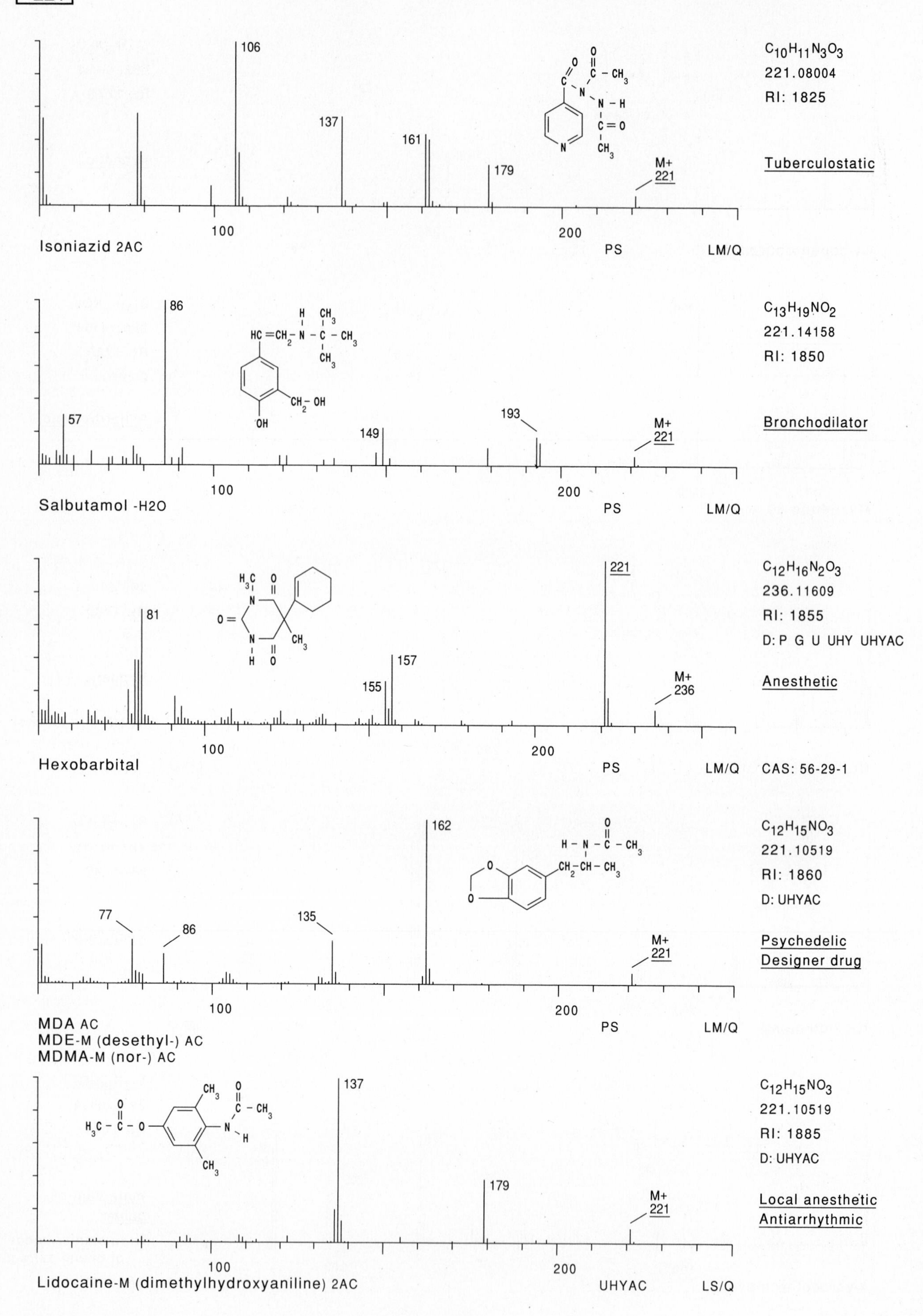

106
137
161
179
M+
221
100
200
Isoniazid 2AC
PS
LM/Q
$C_{10}H_{11}N_3O_3$
221.08004
RI: 1825
Tuberculostatic
86
57
149
193
M+
221
100
200
Salbutamol -H2O
PS
LM/Q
$C_{13}H_{19}NO_2$
221.14158
RI: 1850
Bronchodilator
221
81
157
155
M+
236
100
200
Hexobarbital
PS
LM/Q
$C_{12}H_{16}N_2O_3$
236.11609
RI: 1855
D: P G U UHY UHYAC
Anesthetic
CAS: 56-29-1
162
77
86
135
M+
221
100
200
MDA AC
MDE-M (desethyl-) AC
MDMA-M (nor-) AC
PS
LM/Q
$C_{12}H_{15}NO_3$
221.10519
RI: 1860
D: UHYAC
Psychedelic
Designer drug
137
179
M+
221
100
200
Lidocaine-M (dimethylhydroxyaniline) 2AC
UHYAC
LS/Q
$C_{12}H_{15}NO_3$
221.10519
RI: 1885
D: UHYAC
Local anesthetic
Antiarrhythmic

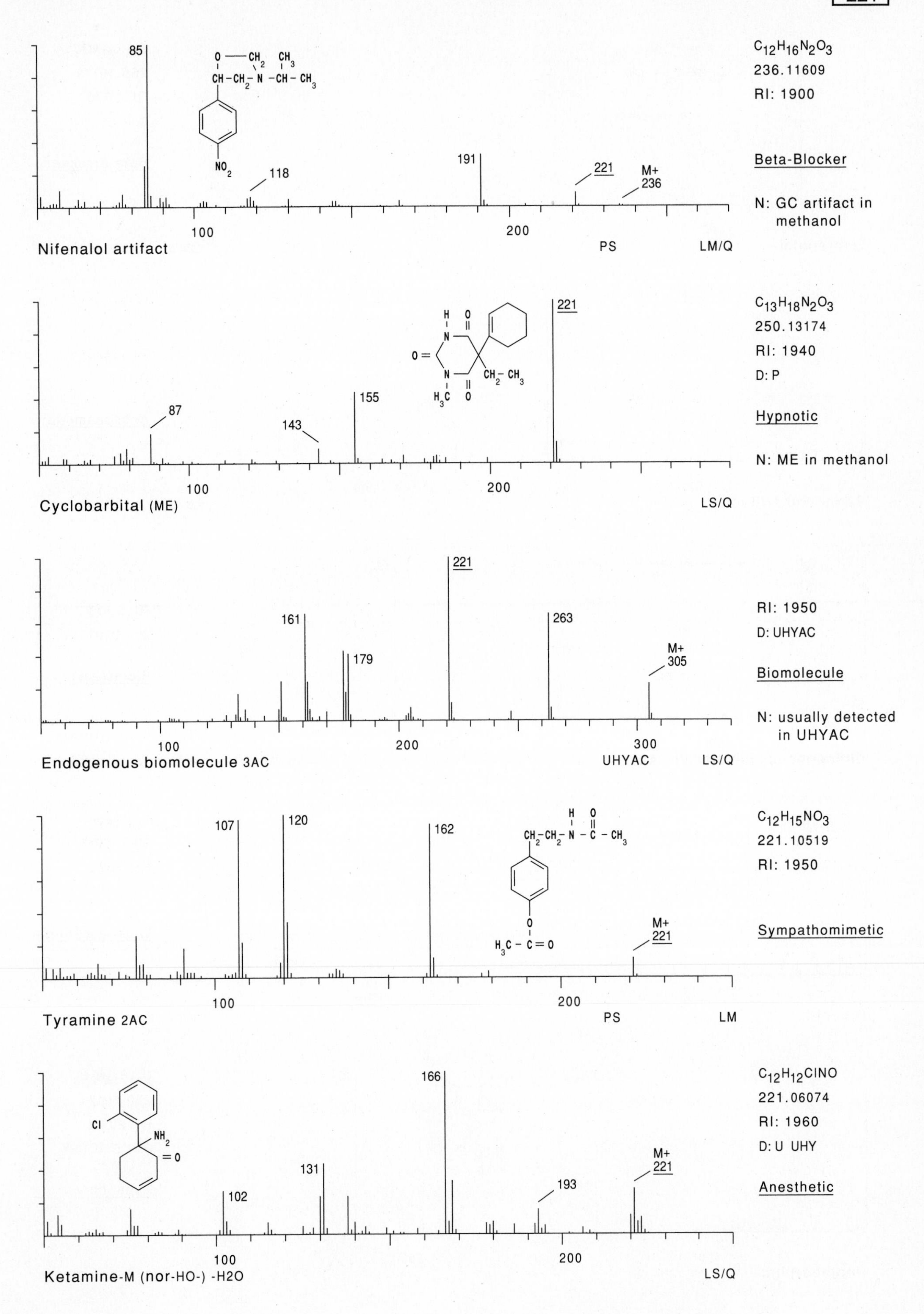

85
118
191
221
M+
236
100
200
Nifenalol artifact
PS
LM/Q
C12H16N2O3
236.11609
RI: 1900
Beta-Blocker
N: GC artifact in methanol
221
155
143
87
100
200
Cyclobarbital (ME)
LS/Q
C13H18N2O3
250.13174
RI: 1940
D: P
Hypnotic
N: ME in methanol
221
161
179
263
M+
305
100
200
300
Endogenous biomolecule 3AC
UHYAC
LS/Q
RI: 1950
D: UHYAC
Biomolecule
N: usually detected in UHYAC
107
120
162
M+
221
100
200
Tyramine 2AC
PS
LM
C12H15NO3
221.10519
RI: 1950
Sympathomimetic
166
131
102
193
M+
221
100
200
Ketamine-M (nor-HO-) -H2O
LS/Q
C12H12ClNO
221.06074
RI: 1960
D: U UHY
Anesthetic

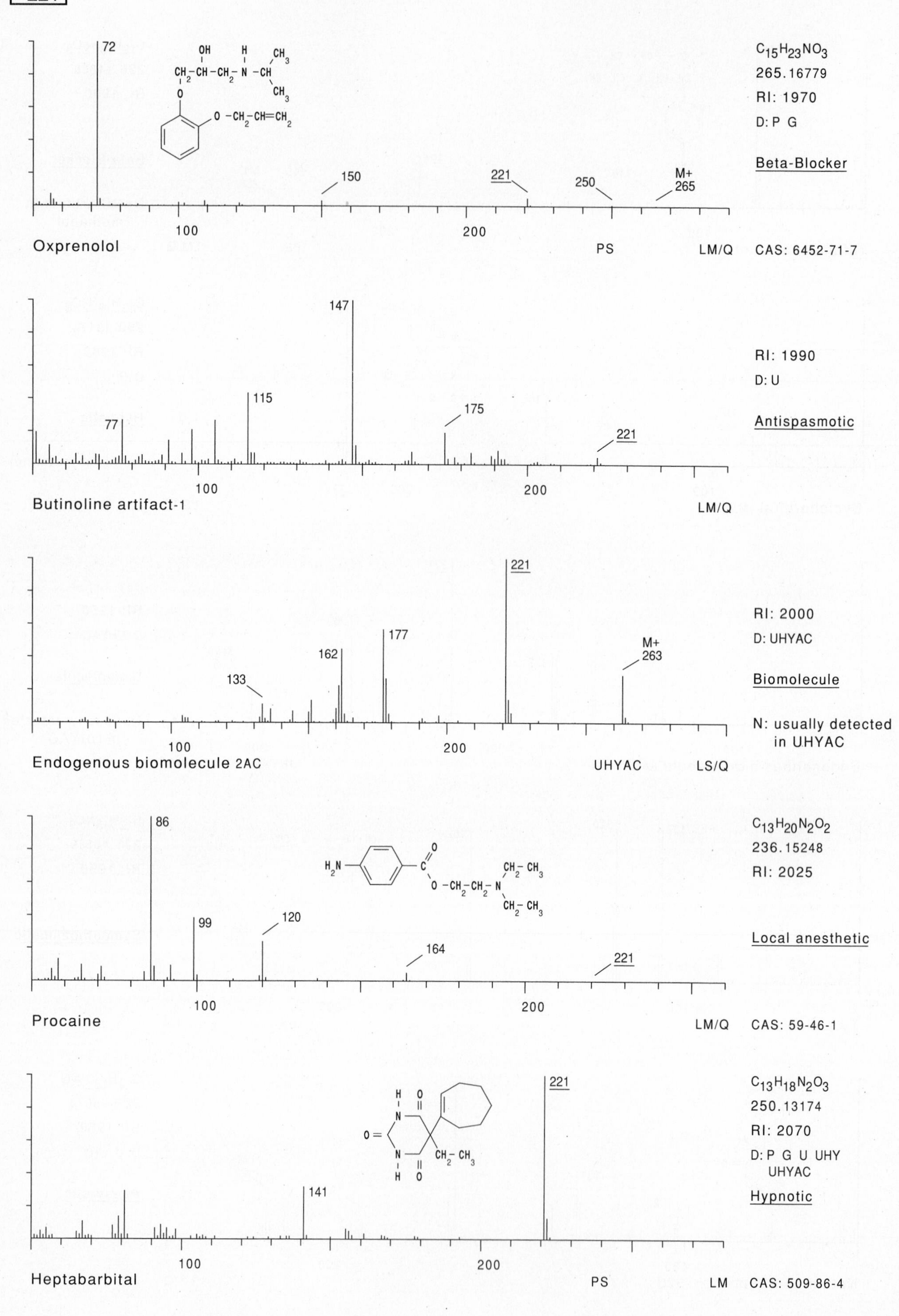

72
150
221
250
M+
265
100
200
Oxprenolol
PS
LM/Q
$C_{15}H_{23}NO_3$
265.16779
RI: 1970
D: P G
Beta-Blocker
CAS: 6452-71-7
147
115
77
175
221
Butinoline artifact-1
LM/Q
RI: 1990
D: U
Antispasmotic
221
177
162
133
M+
263
Endogenous biomolecule 2AC
UHYAC
LS/Q
RI: 2000
D: UHYAC
Biomolecule
N: usually detected in UHYAC
86
99
120
164
221
Procaine
LM/Q
$C_{13}H_{20}N_2O_2$
236.15248
RI: 2025
Local anesthetic
CAS: 59-46-1
221
141
Heptabarbital
PS
LM
$C_{13}H_{18}N_2O_3$
250.13174
RI: 2070
D: P G U UHY UHYAC
Hypnotic
CAS: 509-86-4

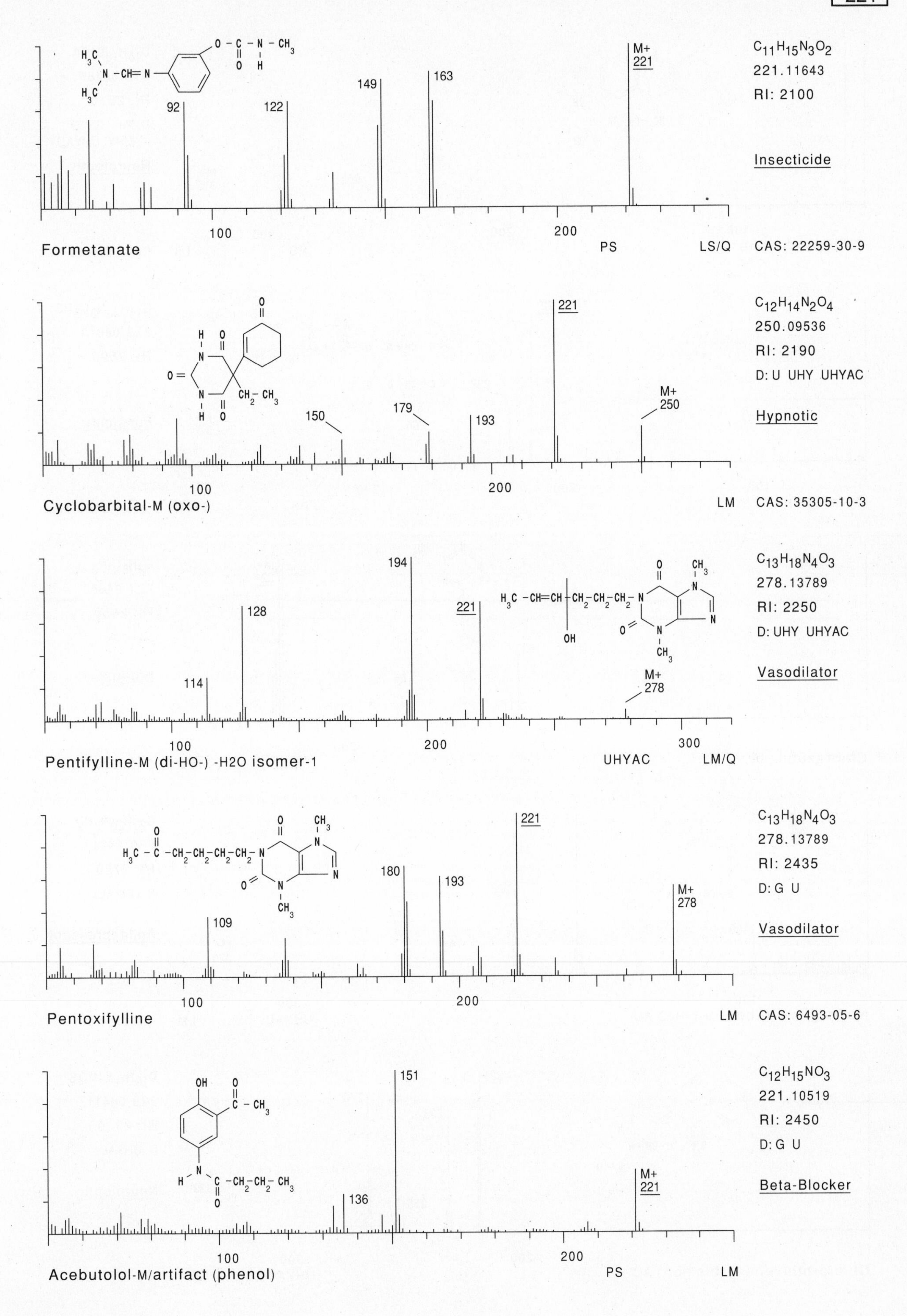

$C_{11}H_{15}N_3O_2$
221.11643
RI: 2100
Insecticide
CAS: 22259-30-9

$C_{12}H_{14}N_2O_4$
250.09536
RI: 2190
D: U UHY UHYAC
Hypnotic
CAS: 35305-10-3

$C_{13}H_{18}N_4O_3$
278.13789
RI: 2250
D: UHY UHYAC
Vasodilator

$C_{13}H_{18}N_4O_3$
278.13789
RI: 2435
D: G U
Vasodilator
CAS: 6493-05-6

$C_{12}H_{15}NO_3$
221.10519
RI: 2450
D: G U
Beta-Blocker

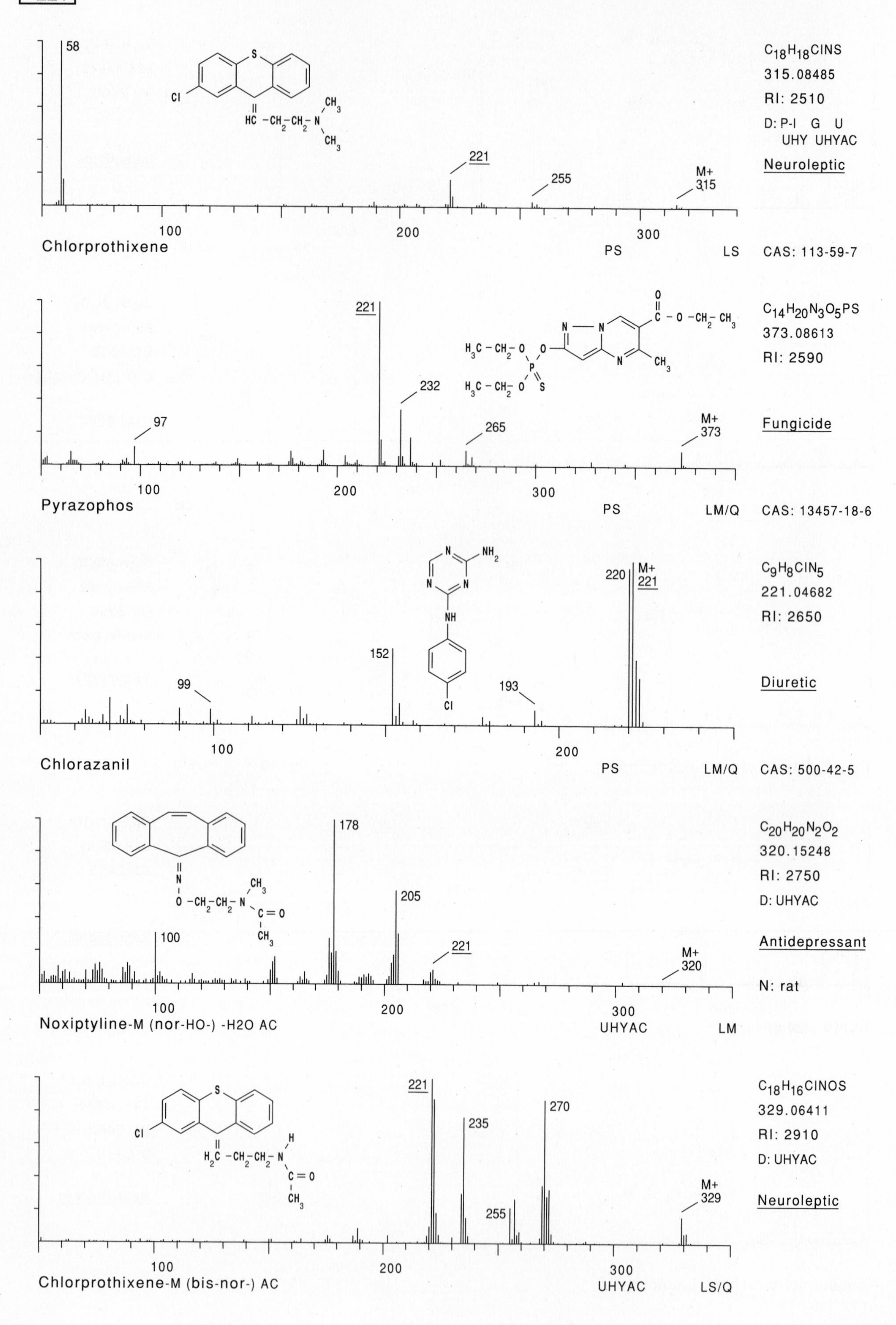
58
221
255
M+
315
100
200
300
Chlorprothixene
PS
LS
$C_{18}H_{18}ClNS$
315.08485
RI: 2510
D: P-I G U
UHY UHYAC
Neuroleptic
CAS: 113-59-7
221
232
97
265
M+
373
Pyrazophos
LM/Q
$C_{14}H_{20}N_3O_5PS$
373.08613
RI: 2590
Fungicide
CAS: 13457-18-6
220
M+
221
152
99
193
Chlorazanil
$C_9H_8ClN_5$
221.04682
RI: 2650
Diuretic
CAS: 500-42-5
178
205
100
221
M+
320
Noxiptyline-M (nor-HO-) -H2O AC
UHYAC
LM
$C_{20}H_{20}N_2O_2$
320.15248
RI: 2750
D: UHYAC
Antidepressant
N: rat
221
235
270
255
M+
329
Chlorprothixene-M (bis-nor-) AC
LS/Q
$C_{18}H_{16}ClNOS$
329.06411
RI: 2910
D: UHYAC
Neuroleptic

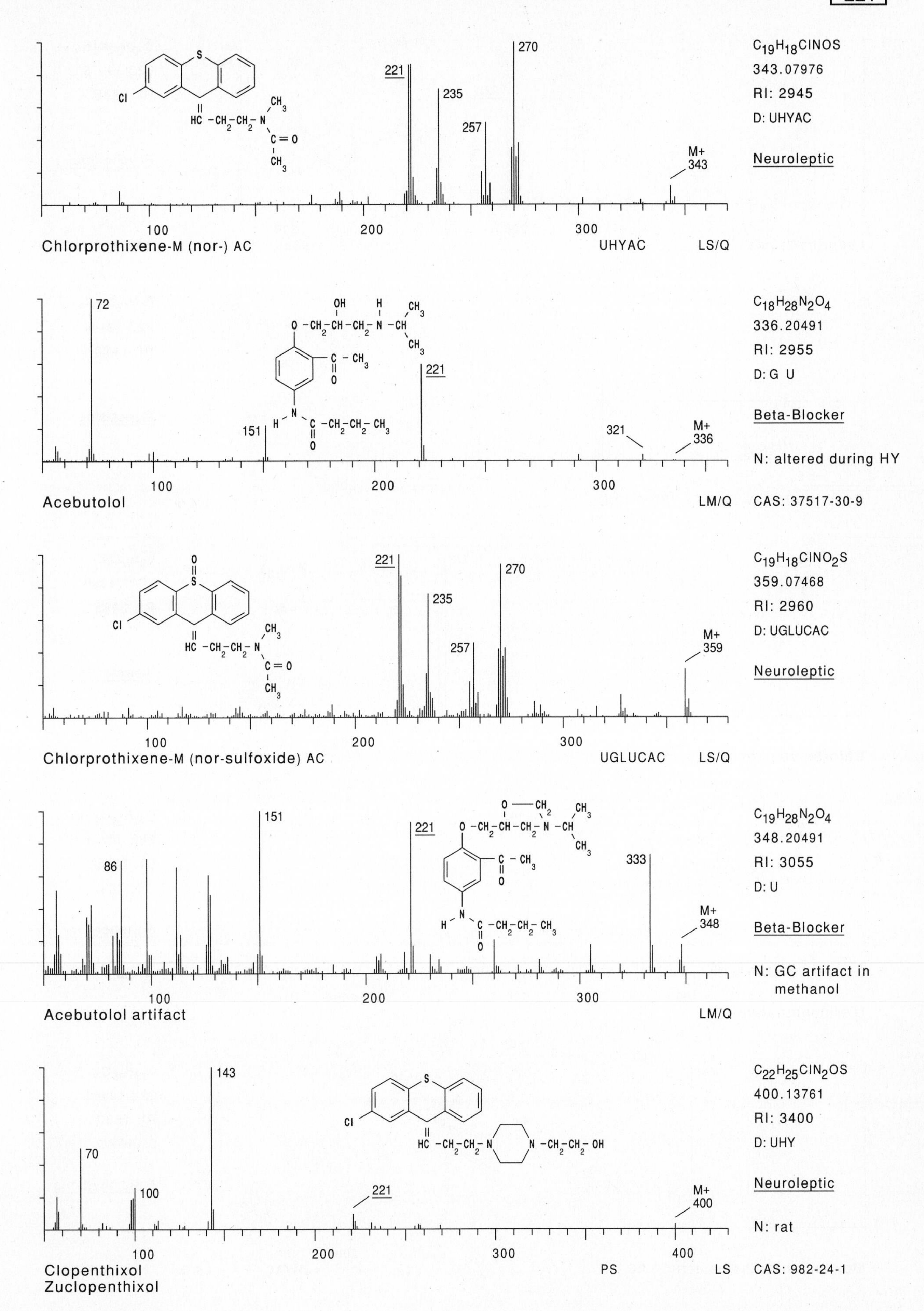

270
221
235
257
M+
343
100
200
300
Chlorprothixene-M (nor-) AC
UHYAC
LS/Q
$C_{19}H_{18}ClNOS$
343.07976
RI: 2945
D: UHYAC
Neuroleptic
72
221
151
321
M+
336
Acebutolol
LM/Q
$C_{18}H_{28}N_2O_4$
336.20491
RI: 2955
D: G U
Beta-Blocker
N: altered during HY
CAS: 37517-30-9
221
270
235
257
M+
359
Chlorprothixene-M (nor-sulfoxide) AC
UGLUCAC
LS/Q
$C_{19}H_{18}ClNO_2S$
359.07468
RI: 2960
D: UGLUCAC
Neuroleptic
151
221
86
333
M+
348
Acebutolol artifact
LM/Q
$C_{19}H_{28}N_2O_4$
348.20491
RI: 3055
D: U
Beta-Blocker
N: GC artifact in methanol
143
70
100
221
M+
400
400
Clopenthixol
Zuclopenthixol
PS
LS
$C_{22}H_{25}ClN_2OS$
400.13761
RI: 3400
D: UHY
Neuroleptic
N: rat
CAS: 982-24-1

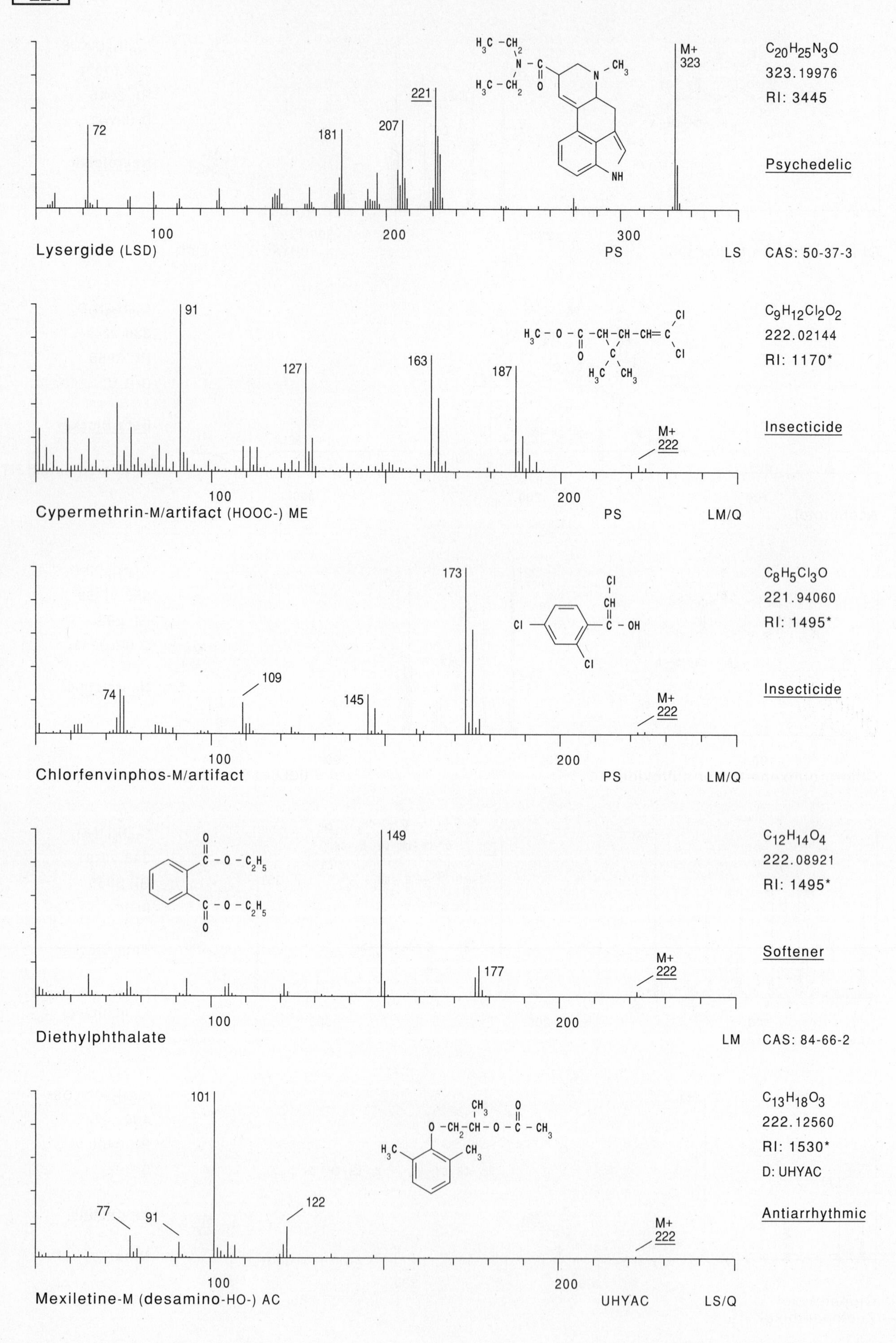

72
181
207
221
M+
323
100
200
300
Lysergide (LSD)
PS
LS
$C_{20}H_{25}N_3O$
323.19976
RI: 3445
Psychedelic
CAS: 50-37-3
91
127
163
187
M+
222
Cypermethrin-M/artifact (HOOC-) ME
LM/Q
$C_9H_{12}Cl_2O_2$
222.02144
RI: 1170*
Insecticide
173
74
109
145
Chlorfenvinphos-M/artifact
$C_8H_5Cl_3O$
221.94060
RI: 1495*
Insecticide
149
177
Diethylphthalate
LM
$C_{12}H_{14}O_4$
222.08921
RI: 1495*
Softener
CAS: 84-66-2
101
77
91
122
Mexiletine-M (desamino-HO-) AC
UHYAC
LS/Q
$C_{13}H_{18}O_3$
222.12560
RI: 1530*
D: UHYAC
Antiarrhythmic